Comparative Biology of the Normal Lung

Second Edition

Comparative Biology of the Normal Lung

Second Edition

Edited by

Richard A. Parent, PhD, DABT, FATS, RAC, ERT
Consultox Ltd, Damariscotta, Maine

Section editors

Section I

Kent E. Pinkerton, PhD
Professor, Center for Health and the Environment, University of California, Davis

Laura S. Van Winkle, PhD
Adjunct Professor, Center for Health and the Environment, University of California, Davis

Charles G. Plopper, PhD
Professor Emeritus, California Primate Research Center, University of California, Davis;
Professor Emeritus, Anatomy, Physiology and Cell Biology, University of California, Davis

Section II

Daniel L. Costa, MS, ScD
National Program Director for Air, Climate & Energy, Office of Research and Development, USEPA, North Carolina;
Adjunct Professor, Environmental Science & Engineering, UNC Gillings School of Global Public Health, Chapel Hill, North Carolina

Jeffrey S. Tepper, PhD, DABT
Consultant and Principal, Tepper Nonclinical Consulting, San Carlos, California

Section III

Debra L. Laskin, PhD
Distinguished Professor and Chair, Department of Pharmacology and Toxicology, Ernest Mario School of Pharmacy
Rutgers University, Piscataway, New Jersey

Andrew J. Gow, PhD
Associate Professor, Department of Pharmacology and Toxicology, Ernest Mario School of Pharmacy, Rutgers University, Piscataway, New Jersey

Section IV

Richard B. Schlesinger, PhD
Associate Dean and Professor, Dyson College of Arts and Sciences, Pace University, New York

Gary R. Burleson, PhD
President and CEO, Burleson Research Technologies, Inc. (BRT), Morrisville, North Carolina

ELSEVIER

AMSTERDAM • BOSTON • HEIDELBERG • LONDON • NEW YORK • OXFORD • PARIS
SAN DIEGO • SAN FRANCISCO • SINGAPORE • SYDNEY • TOKYO

Academic Press is an imprint of Elsevier

Academic Press is an imprint of Elsevier
125 London Wall, London EC2Y 5AS, UK
525 B Street, Suite 1800, San Diego, CA 92101-4495, USA
225 Wyman Street, Waltham, MA 02451, USA
The Boulevard, Langford Lane, Kidlington, Oxford OX5 1GB, UK

Second Edition 2015

First Edition 1992, Published by CRC Press, Inc.

Library of Congress Cataloging-in-Publication Data
A catalog record for this book is available from the Library of Congress

British Library Cataloguing-in-Publication Data
A catalogue record for this book is available from the British Library

ISBN: 978-0-12-404577-4

For information on all Academic Press publications
visit our website at http://store.elsevier.com/

Publisher: Christine Minihane
Acquisition Editor: Kristine Jones
Editorial Project Manager: Molly McLaughlin
Production Project Manager: Caroline Johnson
Designer: Alan Studholme

Contents

Section II
Comparative Respiratory Physiology of the Normal Lung
Daniel L. Costa and Jeffrey S. Tepper

Section III
Comparative Biology of the Normal Lung
Debra L. Laskin and Andrew J. Gow

List of Contributors

Saurabh Aggarwal MD, PhD, Vascular Biology Center, Georgia Regents University, Augusta, GA, USA; Department of Veterans Affairs Medical Center, Northport, NY, USA

Jason H.T. Bates PhD, DSc, Department of Medicine, University of Vermont, Burlington, VT, USA

Stephen M. Black PhD, Vascular Biology Center, Georgia Regents University, Augusta, GA, USA; Department of Veterans Affairs Medical Center, Northport, NY, USA

Peter F. Bove PhD, Department of Medicine, Cystic Fibrosis/Pulmonary Research and Treatment Center, The University of North Carolina, Chapel Hill, Chapel Hill, NC, USA

James S. Brown MSPH, PhD, National Center for Environmental Assessment, US Environmental Protection Agency, Research Triangle Park, Durham, NC, USA

Jared M. Brown PhD, Department of Pharmaceutical Sciences, Skaggs School of Pharmacy, University of Colorado Anschutz Medical Campus, Aurora, CO, USA

Gary R. Burleson PhD, BRT-Burleson Research Technologies, Inc. Morrisville, NC, USA

Stefanie C.M. Burleson PhD, BRT-Burleson Research Technologies, Inc. Morrisville, NC, USA; East Carolina University, Greenville, NC, USA

Florence G. Burleson PhD, BRT-Burleson Research Technologies, Inc. Morrisville, NC, USA

Alejandro Castañeda MS, Center for Health and the Environment, University of California, Davis, Davis, CA, USA

James F. Collawn PhD, Department of Cell, Developmental and Integrative Biology, School of Medicine, University of Alabama at Birmingham, Birmingham, AL, USA; Pulmonary Injury and Repair Center, School of Medicine, University of Alabama at Birmingham, Birmingham, AL, USA

Donald N. Cook PhD, Laboratory of Respiratory Biology, Division of Intramural Research, National Institute of Environmental Health Sciences, NIH, Research Triangle Park, NC, USA

Daniel L. Costa MS, ScD, US Environmental Protection Agency, Research Triangle Park, Durham, NC, USA

Elvira C. Covarrubias, University of California, Davis, Davis, CA, USA; University of Rochester, Rochester, NY, USA

James D. Crapo MD, National Jewish Health, Denver, CO, USA; Department of Medicine, National Jewish Health Center, Denver, CO, USA

Albert van der Vliet PhD, Department of Pathology, College of Medicine, University of Vermont, Burlington, VT, USA

Aimen K. Farraj PhD, DABT, National Health and Environmental Effects Research Laboratory, US Environmental Protection Agency, Research Triangle Park, NC, USA

Jane E. Farrington MS, Laboratory of Cellular & Molecular Biophysics, National Institute of Child Health and Human Development, National Institutes of Health, Bethesda, MD, USA

Robert B. Fick MD, Merck Research Laboratories, Palo Alto, CA, USA

Aron B. Fisher MD, Institute for Environmental Medicine and Department of Physiology, Perelman School of Medicine, University of Pennsylvania, Philadelphia, PA, USA

Henry J. Forman PhD, Department of Environmental Health sciences, School of Public Health, University of Alabama at Birmingham, Birmingham, Alabama, USA

W. Michael Foster PhD, Division of Pulmonary and Critical Care Medicine, Duke University Medical Center, Durham, NC, USA

Peter Gehr PhD, Institute of Anatomy, University of Berne, Berne, Switzerland

Andrew J. Gow PhD, Department of Medicine, Duke University Medical Center, Durham, NC, USA

Christine M. Gross BS, Vascular Biology Center, Georgia Regents University, Augusta, GA, USA; Department of Veterans Affairs Medical Center, Northport, NY, USA

Ryan Hadley MD, Division of Pulmonary and Critical Care Medicine, Department of Internal Medicine, University of Michigan Medical School, Ann Arbor, MI, USA

Jack R. Harkema DVM, PhD, DACVP, Department of Pathobiology and Diagnostic investigation, Center for Integrative Toxicology, College of Veterinary Medicine, Michigan State University, East Lansing, MI, USA

Mehdi Saeed Hazari MSc, PhD, National Health and Environmental Effects Research Laboratory, US Environmental Protection Agency, Research Triangle Park, NC, USA

Daniel Hirsch MD, Division of Neonatology, Department of Pediatrics, Rutgers Robert Wood Johnson Medical School, New Brunswick, NJ, USA

Dallas M. Hyde PhD, Department of Anatomy, Physiology and Cell Biology, School of Veterinary Medicine, University of California, Davis, Davis, CA, USA

Victor J. Johnson PhD, BRT-Burleson Research Technologies, Morrisville, NC, USA

James H. Jones PhD, Department of Physiological Sciences, School of Veterinary Medicine, University of California, Davis, Davis, CA, USA; Cardiovascular Research Institute, University of California, San Francisco, San Francisco, CA, USA

Pranita Katwa PhD, Department of Pathology and Laboratory Medicine, Alpert Medical School, Brown University, Providence, RI, USA; Department of Pharmacology & Toxicology, Brody School of Medicine, East Carolina University, Greenville, NC, USA

Murat Kaynar MD, PhD, Critical Care Medicine, University of Pittsburgh, Pittsburgh, PA, USA; University of Pittsburgh School of Medicine, Pittsburgh, PA, USA; University of Pittsburgh Graduate School Public Health, Pittsburgh, PA, USA

Debra L. Laskin PhD, Department of Pharmacology and Toxicology, Ernest Mario School of Pharmacy, Rutgers University, Piscataway, NJ, USA

Jeffrey D. Laskin PhD, Environmental & Occupational Health Sciences Institute, Robert Wood Johnson Medical School, Rutgers University, Piscataway, NJ, USA

Hui-Hua Li MD, Department of Environmental and Occupational Health, University of Pittsburgh, Pittsburgh, PA, USA; University of Pittsburgh School of Medicine, Pittsburgh, PA, USA; University of Pittsburgh Graduate School Public Health, Pittsburgh, PA, USA

James D. Londino PhD, Department of Anesthesiology, School of Medicine, University of Alabama at Birmingham, Birmingham, AL, USA; Department of Cell, Developmental and Integrative Biology, School of Medicine, University of Alabama at Birmingham, Birmingham, AL, USA; Pulmonary Injury and Repair Center, School of Medicine, University of Alabama at Birmingham, Birmingham, AL, USA

Kim E. Longworth PhD, Department of Physiological Sciences, School of Veterinary Medicine, University of California, Davis, Davis, CA, USA; Cardiovascular Research Institute, University of California, San Francisco, San Francisco, CA, USA

Michael I. Luster PhD, Department of Occupational Medicine, West Virginia University, WV, USA

Rama Malaviya PhD, Department of Pharmacology and Toxicology, Ernest Mario School of Pharmacy, Rutgers University, Piscataway, NJ, USA

Mark D. Mannie PhD, East Carolina University, Greenville, NC, USA

Andrew T. Mariassy PhD, Department of Research, Mt Sinai Medical Center, University of Miami, Miami, FL, USA

Sadis Matalon PhD, MSC, DrSc (Hon), Department of Anesthesiology, School of Medicine, University of Alabama at Birmingham, Birmingham, AL, USA; Department of Cell, Developmental and Integrative Biology, School of Medicine, University of Alabama at Birmingham, Birmingham, AL, USA; Pulmonary Injury and Repair Center, School of Medicine, University of Alabama at Birmingham, Birmingham, AL, USA

Stephanie A. Matthes PhD, Division of Pulmonary and Critical Care Medicine, Department of Internal Medicine, University of Michigan Medical School, Ann Arbor, MI, USA

John T. McBride MD, University of California, Davis, Davis, CA, USA; University of Rochester, Rochester, NY, USA

Robert R. Mercer PhD, National Institute of Occupational Safety and Health, Morgantown, WV, USA

John B. Morris PhD, Department of Pharmaceutical Sciences, University of Connecticut, Storrs, CT, USA

Hideki Nakano PhD, Laboratory of Respiratory Biology, Division of Intramural Research, National Institute of Environmental Health Sciences, NIH, Research Triangle Park, NC, USA

Stephen G. Olmstead PhD, East Carolina University, Greenville, NC, USA

Janice L. Peake BS, Department of Anatomy, Physiology and Cell Biology, School of Veterinary Medicine, University of California, Davis, Davis, CA, USA

Kent E. Pinkerton PhD, Department of Anatomy, Physiology and Cell Biology, School of Veterinary Medicine, University of California, Davis, Davis, CA, USA; University of Rochester, Rochester, NY, USA; Center for Health and the Environment, University of California, Davis, Davis, CA, USA

Bruce R. Pitt PhD, Department of Bioengineering, University of Pittsburgh, Pittsburgh, PA, USA; Department of Environmental and Occupational Health, University of Pittsburgh, Pittsburgh, PA, USA; University of Pittsburgh School of Medicine, Pittsburgh, PA, USA; University of Pittsburgh Graduate School Public Health, Pittsburgh, PA, USA

Charles G. Plopper PhD, Department of Anatomy, Physiology and Cell Biology, School of Veterinary Medicine, University of California, Davis, Davis, CA, USA; University of Rochester, Rochester, NY, USA

Robert J. Porcelli PhD, Department of Veterans Affairs Medical Center, Northport, NY, USA

Scott H. Randell PhD, Department of Cell Biology and Physiology, The University of North Carolina, Chapel Hill, Chapel Hill, NC, USA

Susan D. Reynolds PhD, Department of Pediatrics, National Jewish Health, Denver, CO, USA

Jesse Roman MD, Department of Internal Medicine, University of Louisville School of Medicine, Louisville, KY, USA

Philip L. Sannes PhD, Department of Molecular Biomedical Sciences, Center for Comparative Medicine and Translational Research, College of Veterinary Medicine, North Carolina State University, Raleigh, NC, USA

Rashmin C. Savani MBChB, Division of Neonatal-Perinatal Medicine and The Center for Pulmonary & Vascular Biology, University of Texas Southwestern Medical Center, Dallas, TX, USA

Richard B. Schlesinger PhD, Dyson College of Arts and Sciences, Pace University, New York, NY, USA

Madhuri Singal PhD, RRT, Research Institute for Fragrance Materials Inc, Woodcliff Lake, NJ, USA

Suzette Smiley-Jewell PhD, University of California, Davis, Davis, CA, USA; University of Rochester, Rochester, NY, USA

Bernd W. Spur PhD, Department of Cell Biology, Rowan University School of Osteopathic Medicine, Stratford, NJ, USA

Claudette M. St. Croix PhD, Department of Environmental and Occupational Health, University of Pittsburgh, Pittsburgh, PA, USA; University of Pittsburgh School of Medicine, Pittsburgh, PA, USA; University of Pittsburgh Graduate School Public Health, Pittsburgh, PA, USA

Judith A. St. George PhD, California National Primate Research Center, University of California, Davis, Davis, CA, USA

Jeffrey S. Tepper PhD, Tepper Nonclinical Consulting, San Carlos, CA, USA

Kalidasan Thambiayya PhD, Department of Bioengineering, University of Pittsburgh, Pittsburgh, PA, USA; University of Pittsburgh School of Medicine, Pittsburgh, PA, USA; University of Pittsburgh Graduate School Public Health, Pittsburgh, PA, USA

Michael R. Van Scott PhD, East Carolina University, Greenville, NC, USA

Laura S. Van Winkle BS, PhD, Department of Anatomy, Physiology and Cell Biology, School of Veterinary Medicine, University of California, Davis, Davis, CA, USA; University of Rochester, Rochester, NY, USA

Peter D. Wagner MD, Department of Medicine, University of California, San Diego, La Jolla, CA, USA

Karla A. Wasserloos MS, Department of Environmental and Occupational Health, University of Pittsburgh, Pittsburgh. PA, USA; University of Pittsburgh School of Medicine, Pittsburgh, PA, USA; University of Pittsburgh Graduate School Public Health, Pittsburgh, PA, USA

Barry Weinberger MD, Division of Neonatology, Department of Pediatrics, Rutgers Robert Wood Johnson Medical School, New Brunswick, NJ, USA

Eric S. White MD, Division of Pulmonary and Critical Care Medicine, Department of Internal Medicine, University of Michigan Medical School, Ann Arbor, MI, USA

Kingsley Yin PhD, Department of Cell Biology, Rowan University School of Osteopathic Medicine, Stratford, NJ, USA

Berran Yucesoy MS, PhD, Toxicology and Molecular Biology Branch, Health Effects Laboratory Division, NIOSH/CDC, Morgantown, WV, USA

Graeme R. Zosky PhD, MBiostat, School of Medicine, University of Tasmania, Hobart, Australia

Section I

Structural and Cellular Diversity of the Mammalian Respiratory System

Kent E. Pinkerton, Laura S. Van Winkle and Charles G. Plopper

Chapter 1

Overview of Diversity in the Respiratory System of Mammals

Kent E. Pinkerton, Laura S. Van Winkle and Charles G. Plopper

One of the principal functions of the respiratory system in mammals is gas exchange. To accomplish this task requires the presence of a membrane with an extremely large surface area but with minimal thickness. The surface area of this membrane is approximately 25 times larger than that of the integumentary surface for any given mammalian species. The tissue that separates the air surface of this membrane from the large vascular bed associated with it is limited in thickness to facilitate the requirements for rapid and efficient gas diffusion. Since gas exchange (oxygen and carbon dioxide) between the vascular system and external air is vital to the overall metabolic function of the organism, the capillary bed adjacent to the gas-exchange air surface has a similar surface area. The existence of such a large surface area creates a considerable number of challenges in terms of packaging and architectural organization to accommodate such a large gas flow and volume change.

Another set of challenges encompasses those parameters that need to be developed to protect this gas-exchange membrane from injury. In mammals, a series of respiratory structures exists that provide for the filtration of incoming air to remove particulates and reactive gases, as well as the means for the clearance of trapped materials. Both clearance and filtration tasks require the production of secretory products to facilitate trapping and removal. A significant amount of energy is expended in the transport or ciliary movement to clear these secretions from the system. These protective secretions found in the airways also play a major role in regulating the temperature and the humidity of the respired air. Airway secretions facilitate the warming and humidification of inspired air to body temperature, saturation, and the recovery of a portion of the heat and fluid on expiration to reduce heat loss and to prevent desiccation of the membrane. The majority of the protective functions are carried out by the upper conducting airways. In mammals, these include the nasal cavity, the larynx, and a series of branched tubes, i.e., the tracheobronchial airway tree. The geometry of the tracheobronchial airway tree is also part of the strategy for directing air to such an immense membrane surface to facilitate gas exchange. The gas-exchange area is organized into small air pockets or alveoli, which open into cylindrical channels or alveolar ducts. In mammals, the millions of alveoli are organized into groups that receive air from the most distal branches of the airway tree. The presence of a large capillary bed, which receives the entire cardiac output from the right side of the heart with each contraction, results in a significant role for the pulmonary vascular bed and associated cellular structures in a variety of metabolic and regulatory functions, other than those directly related to respiration. Even though the physiological requirements and accompanying biological problems to be solved to ensure adequate gas exchange for the organism are the same for all species of mammals, it is becoming readily apparent that there are major differences in the structural organization and cellular composition of the system in different species of mammals.

The structures that provide the initial interface between the organism and the external air, the nasal cavity and the larynx, exhibit a significant amount of interspecies variability in both architectural organization and cellular composition. For the nasal cavity, a number of general architectural features are common to all mammals (Chapter 2). The cavity is bilateral and is separated by a central septum. The primary filtration network in the nasal cavity consists of a series of turbinates, which are outgrowths of the lateral and dorsal walls. The olfactory system is restricted to the most posterior portion of the cavity. It is housed in chambers that are outside the direct path of airflow. The differences between species are numerous. One of the most highly variable areas is the organization of the turbinates. There is significant variability in the number of turbinates, their extent within the nasal cavity, the number of branches, and the overall complexity. In some species, such as primates, the turbinate structure is minimal, while in others, such as the dog, it is extremely complex. The same variability holds for the overall surface area of the nasal cavity and the volumes of specific chambers and airspaces. There are also substantial species differences in many of the critical architectural features, which could be expected to have a major impact on airflow properties and

Comparative Biology of the Normal Lung. http://dx.doi.org/10.1016/B978-0-12-404577-4.00001-1

on air filtration capabilities. The cellular populations that line the nasal cavities are also extremely heterogeneous and species specific in their composition. Quantitative studies have not been performed on a large number of species (Chapter 2). For the features that have been evaluated, the differences in epithelial cells lining the passages are significant. These include most parameters that could be considered relevant functional characteristics of epithelial populations, including cell density, epithelial composition, distribution of specific differentiated cell types, metabolic potential of specific subpopulations, and the number and extent of epithelial derivatives and their distribution (Chapter 2).

The organization and basic structural units of the lung, as well as the composition of the tracheobronchial airway tree, exhibit a significant amount of interspecies diversity in mammals (Chapters 3, 4, 5, and 6). One common feature is a relatively dichotomous branching pattern in which the daughter branches are slightly smaller in cross-sectional area than the parent (Chapter 4). Another feature of the branching pattern is that eventually the smallest and most distal daughter branches join the most proximal aspects of the gas-exchange area, or parenchyma (Chapters 3–8). This airway tree has a significant extracellular connective tissue component, as well as a continuous epithelial lining, which is joined to that of the more proximal conducting airways and to the gas-exchange epithelium. If the architectural organization and distribution of the tracheobronchial airway tree is evaluated quantitatively, very few features appear to be consistent from species to species (Chapter 4). The average distance through this tree from a proximal reference point, generally the larynx, to the gas-exchange area, is highly variable. This is also true for the shortest pathway to the gas-exchange area and the longest pathway. The size of daughter branches and the angle of branching is also a species-specific characteristic. In some species both daughter branches are approximately half the cross-sectional area of the parent and divide at nearly the same branch angle. In other species, there can be as much as one order of magnitude difference in cross-sectional area between two daughter branches and between a daughter branch and the parent branch. There are significant differences in branch angle and in the actual volume and surface area of individual daughter branches. The organization of the branching pattern itself is also highly species specific. In many species, primarily ungulates, there is an extra bronchus, which arises from the trachea proximal to the major bifurcations of the trachea (Chapters 3 and 4). The cellular composition of the tracheobronchial airway tree also has a large number of species-specific characteristics (Chapters 5 and 6). There are differences in the epithelial populations of the tracheobronchial tree if any of the following are compared: (1) the same airway level in different species; (2) different airway levels in the same species; (3) the types of cells present in different airway generations (Chapters 5 and 6); (4) the overall abundance of secretory product and the number of cells in different airways; (5) the carbohydrate content of secretory granules (Chapter 5); and (6) the ultrastructural composition of the same cell type in different species and different airway levels (Chapters 5 and 7). The organization of the transition zone between the tracheobronchial tree and the gas-exchange area is similar between mammalian species only in that there is continuity between the epithelium and extracellular components of the conducting airways and the gas-exchange area. There is significant variation in the overall organization and distribution of this system and the extent of the interaction that occurs between gas-exchange tissue and conducting airway tissue (Chapters 3, 7, and 8). The epithelial cells lining the bronchiolar, or more distal, portion of the tracheobronchial tree have a significant amount of variability in terms of abundance, types of cells present, and ultrastructural features of the cells in the adult animal and in the secretory product that these cells produce (Chapters 6 and 7).

The organization of the lungs themselves also has a significant amount of species-specific variability. Even when lungs and associated structures are evaluated at a subgross level (Chapter 3), virtually none of the features appears to be consistent for all species. While all species have a pleural surface, the thickness, vascularity, and organization of the mesothelium appear to vary. The same is true for the extent and distribution of secondary septation within the lung and the distribution of connective tissue elements. While the lungs of all species have a blood vascular supply that includes both a systemic circulation (bronchial artery) and the system that distributes the blood from the right heart (pulmonary circulation), the size, organization, and supply zones of this system are highly species specific. When the functional unit of gas exchange within the lung, the pulmonary acinus, is evaluated in terms of its architectural and cellular composition, a significant degree of species variability, even in this basic component, is evident (Chapter 8). The size of individual units and the organization of the units is species specific. The size of the entire membrane surface and its thickness, while roughly approximating metabolic requirements, can vary by as much as a factor of two in different species with a reasonably similar metabolic body size. The cellular composition and organization of the blood/air barrier has a number of common features, including two epithelial cell types, one secretory (the type II cell) and the other a barrier-maintaining squamous cell (the type I cell) (Chapter 9). Both of these cell types are juxtaposed to the capillary bed lined by endothelial cells. The cellular organization of the gas-exchange area shows variability in abundance, size, and organization of individual cells (Chapter 9). Even the secretory proteins of the epithelial cells vary from species to species (Chapters 5, 6 and 11). The extracellular matrix and basement membrane components also show many species-specific features (Chapter 10).

The purpose of this revised volume is to update and address these issues in a quantitative fashion and to emphasize the species-specific differences that may be relevant to understanding the pathobiological and toxicological susceptibility of the respiratory system of different species. While the basic organization of the lung in mammals seems to be a class-wide feature, the strategies that different species have developed in order to confront the various problems of providing an adequate surface membrane for gas exchange and for protecting that membrane are highly diverse.

Chapter 2

Comparative Anatomy and Epithelial Cell Biology of the Nose

Jack R. Harkema

Chapter Outline

1. INTRODUCTION

The nose is the portal of entry for inspired air. Besides serving as the sensory organ for smell (olfaction), the nose also functions as a guardian of the distal respiratory tract by filtering, humidifying, and heating the inhaled air. In addition, it protects the delicate gas-exchange regions of the lung by effectively absorbing water-soluble and reactive gases and vapors, trapping inhaled particles, metabolizing airborne xenobiotics, and triggering protective physiologic responses through its cellular chemosensors. Human nasal airways and associated paranasal sinuses may be afflicted by many diseases caused by inhalation exposures to chemical toxicants, infectious agents, and allergens. Like other organs, the overall health of the nose is also dependent on other factors, such as age, gender, and nutrition. Unlike the lower respiratory tract and lung, the nasal mucous membranes are populated by numerous microorganisms, known as the nasal microbiome (Faust et al., 2012; Frank et al., 2010).

This previously underinvestigated part of the respiratory tract has gained more respect due to research findings about this enigmatic organ in a variety of scientific disciplines, including neuroscience, immunology, molecular biology, and toxicology. The discoveries made in recent years in the area of olfactory science were highlighted worldwide when Drs Linda Buck and Richard Axel were awarded the 2004 Nobel Prize in Physiology or Medicine for their pioneering work on the cellular and molecular mechanisms of olfaction (Mombaerts, 2004).

This chapter provides a comparative overview of nasal anatomy, histology, and cell biology among laboratory rodents (mice and rats), nonhuman primates (macaque monkeys), and where, possible, humans. Special emphasis is placed on the surface epithelial cells that line the nasal airways and are often the first cellular contacts with inhaled air.

2. GROSS ANATOMY

The nasal cavity of humans and other mammals is divided into two main air passages by the nasal septum. Each nasal passage extends from the nostrils (nares) to the nasopharynx (Figure 1). The nasopharynx is defined as the airway posterior to the termination of the nasal septum and proximal to the termination of the soft palate. Inhaled air flows through the nostril openings, or nares, into the vestibule, which is a slight dilatation just inside the nares, or nostrils, and before the main chamber of the nose. Unlike the more distal main chamber that is surrounded by bone, the nasal vestibule is surrounded by flexible cartilage. The luminal surface is lined by a squamous epithelium similar to that of external skin. In humans, unlike laboratory mice, the nasal vestibule also contains varying numbers of hair follicles near the nostrils.

After passing through the nasal vestibule, inhaled air courses through the narrowest part of the entire respiratory tract, the nasal valve (ostium internum), before entering the main nasal chamber. A lateral wall, septal wall, roof, and floor define each nasal passage of the main chamber. The lumen of the main chamber is lined by well-vascularized

Comparative Biology of the Normal Lung. http://dx.doi.org/10.1016/B978-0-12-404577-4.00002-3

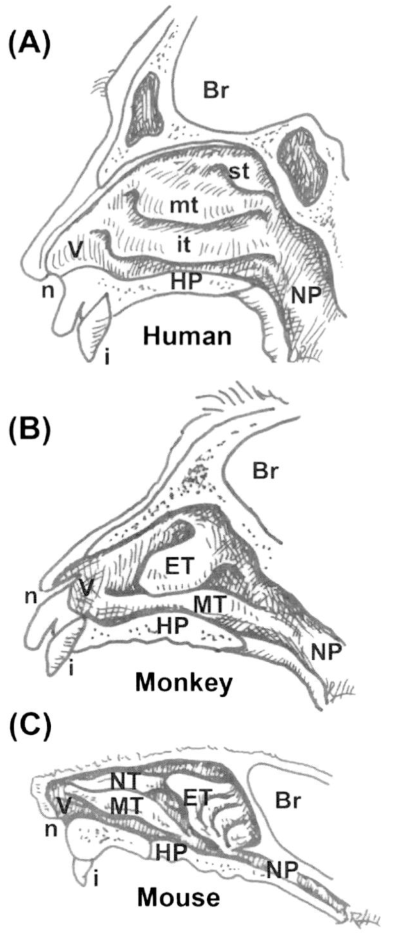

FIGURE 1 Schematic illustrations of the right nasal passages from a human (A), monkey (B), and mouse (C). The septum has been removed, exposing the lateral wall. ET, ethmoid turbinate(s); HP, hard palate; i, incisor tooth; it, inferior turbinate; mt, middle turbinate; MT, maxilloturbinate; n, nostril (naris); NP, nasopharynx; NT, nasoturbinate; st, superior turbinate; V, nasal vestibule; Br, brain.

and innervated mucous membranes (nasal mucosa) that are covered by a continuous layer of mucus. The nasal mucous layer is moved distally, by the synchronous beating of motile cilia, to the oropharynx where it is swallowed into the esophagus.

Turbinates—bony structures lined by the well-vascularized mucosal tissue—project into the airway lumen from the lateral walls into the main chamber of the nose in laboratory animals and humans (Figure 1). Nasal turbinates increase the inner surface area of the nose, which is important in the filtering, humidifying, and warming of the inspired air. The turbinated main chamber of the human nose is approximately 5–8 cm long, 6–10 cm^3 in volume (one passage), but has a total surface area (both nasal passages) of 150–200 cm^2 (Guilmette et al., 1989). In contrast, the nasal airway of 16-week-old male B65C3F1 mice is 5 mm in length, 0.03 cm^3 (30 μl) in total volume (both passages), and has a total surface area of 2.90 cm^2 (Gross et al., 1982). Relative surface area (i.e., surface area/volume) of the nasal cavity, however, is approximately five times greater in the laboratory mouse than in the human, due principally to the greater complexity of murine nasal turbinates.

Although there are some general similarities between the nasal passages of rodents, nonhuman primates, and humans, there are also striking species differences in nasal architecture (Figure 1). From a comparative viewpoint, humans have relatively simple noses with breathing as the primary function (microsmatic mammals), while mice and rats have more complex noses with olfaction as the primary function (macrosmatic mammals). In addition, the nasal and oral cavities of humans (and most nonhuman primates) are arranged in a manner to allow for both nasal and oral breathing. Laboratory rats and mice are obligate nose breathers, due to the close apposition of the epiglottis to the soft palate. Interspecies variability in nasal gross anatomy has been emphasized in previous reviews (Gross and Morgan, 1992; Harkema et al., 2006; Herbert and Leininger, 1999; Negus, 1958).

Marked differences in nasal airflow patterns between humans and laboratory rodents are primarily due to the species-specific variation in number and shape of the nasal turbinates. The human nose has three turbinates: the superior, middle, and inferior (Figure 1). These structures are relatively simple in shape compared to turbinates of rodents, which have complex folding and branching patterns. In laboratory rats and mice, evolutionary pressures concerned chiefly with olfactory function and dentition have defined the shape of the turbinates and the type and distribution of the cells lining these structures. The complex naso- and maxilloturbinates of obligate nose breathing rodents are thought to provide better protection of the lower respiratory tract due to enhanced filtration, absorption, and disposal of airborne particles and gases, as compared to the simple structure of the middle and inferior turbinates in the human nose. The highly complex shape of the ethmoid turbinates, lined predominantly by olfactory neuroepithelium, in the distal half of the rodent nasal cavity is suitably designed for acute olfaction. Differences in the complexity of the gross turbinate structure at different levels throughout the nasal airways of the laboratory mouse and macaque monkey are illustrated in Figures 2 and 3, respectively.

Luminal surfaces of the nasal mucosa (with the exception of the most proximal regions of the nasal vestibule) are lined by airway lining fluid composed primarily of mucus, a watery and sticky glycoprotein. It is secreted from mucous (goblet) cells in surface epithelium and from underlying glands in the lamina propria. Its physical and chemical properties are well suited for its role in upper airway defense by trapping inhaled particles and scrubbing out certain reactive gases. The synchronized beating of airway surface cilia propels the nasal mucus at different speeds and directions depending on the intranasal location. Mucus covering the olfactory mucosa moves very slowly, with a turnover time of probably several days. In contrast,

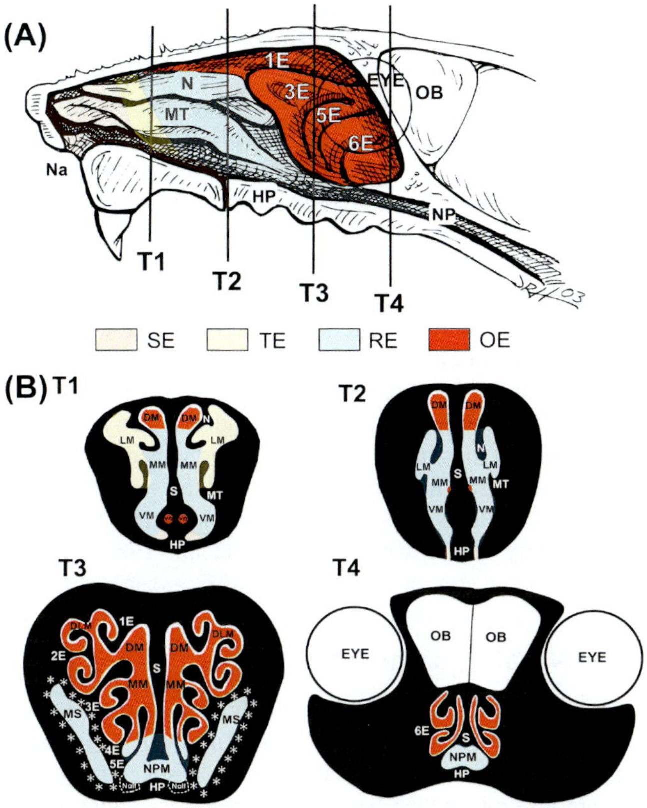

FIGURE 2 (A) Diagrammatic representation of the right nasal passage of a mouse. The nasal septum has been removed to expose the turbinates (N, nasoturbinate; MT, maxilloturbinate; 1E, 3E, 5E, and 6E, ethmoid turbinates). Vertical lines represent the rostral surfaces of transverse sections (T1−T4) taken at different levels throughout the nasal passage. Intranasal areas represent the regions lined by different surface epithelia: squamous epithelium (SE); transitional epithelium (TE); respiratory epithelium (RE); red, olfactory epithelium (OE). (B) Transverse sections at different levels of the nasal cavity (T1−T4). DM, dorsal medial meatus; HP, hard palate; LM, lateral meatus; MM, medial meatus; MS, maxillary sinus; MT, maxilloturbinate; N, nasoturbinate; NALT, nasal-associated lymphoid tissue; NPM, nasopharyngeal meatus; OB, olfactory bulb of brain; S, septum; VM, ventral meatus; VO, vomeronasal organ; 1E−6E, ethmoid turbinates. Asterisks in T3 around MS, lateral nasal gland.

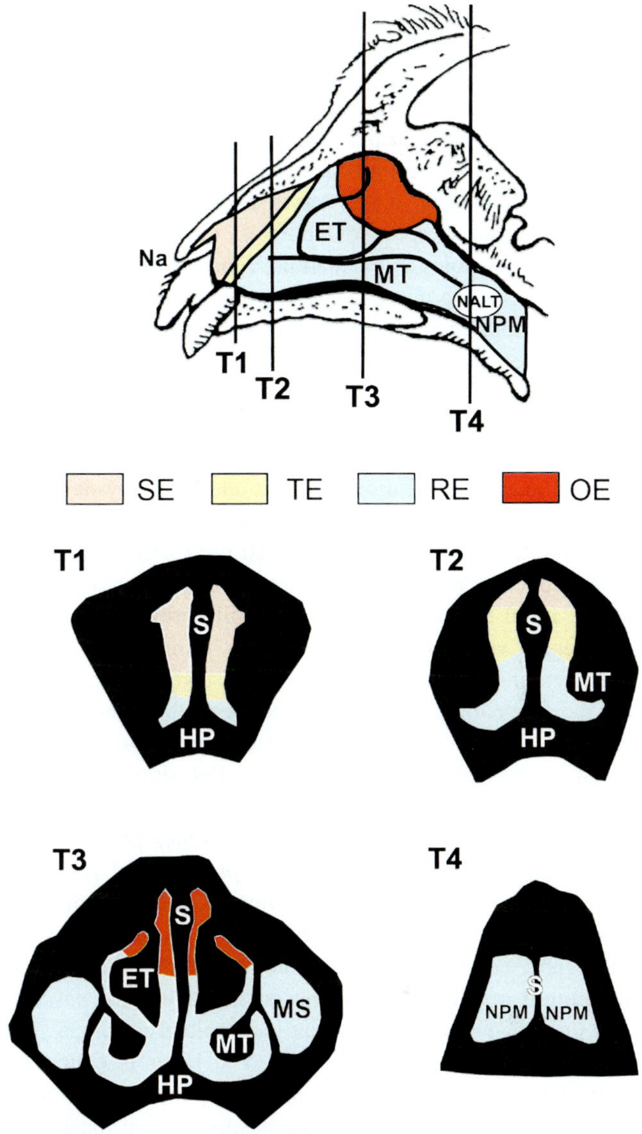

FIGURE 3 Schematic representation of the lateral wall of the right nasal passage from the nose of a *Rhesus macaque* monkey. The nasal septum has been removed to expose the turbinates (maxilloturbinate and ethmoid turbinate). Vertical lines represent the rostral surfaces of transverse sections T1−T4 taken at different levels throughout the nasal cavity. Intranasal areas represent the regions lined by different surface epithelia: squamous epithelium (SE); transitional epithelium (TE); respiratory epithelium (RE); olfactory epithelium (OE). Cross-sections at different levels of the nasal cavity (T1−T4). ET, ethmoid turbinate; HP, hard palate; MS, maxillary sinus; MT, maxilloturbinate; NALT, nasal-associated lymphoid tissue; NPM, nasopharyngeal meatus; S, septum.

the mucus covering the respiratory mucosa is driven along rapidly (1−30 mm/min) by synchronized beating of cilia with an estimated turnover time of about 10 min in the laboratory rat (Morgan et al., 1984).

Nasal mucus containing entrapped materials ultimately is propelled by beating cilia to the naso- and oropharynx, and then swallowed into the esophagus and cleared through the digestive tract. The nasal mucocilary apparatus (i.e., mucus and cilia) exhibits a range of responses to inhaled xenobiotic agents and can be a sensitive indicator of toxicity. Because this upper airway apparatus is one of the first lines of defense against inhaled microbial pathogens and chemical toxicants, compromises in nasal defense capabilities could lead to infections in both the upper or lower respiratory tract, including the lung.

3. PREPARATION OF NASAL TISSUES FOR MICROSCOPIC EXAMINATION

Excellent detailed descriptions of how to prepare the nasal cavities of laboratory rodents for light microscopic examination can be found in the published literature (Young, 1981; Uraih et al., 1990; Mery et al., 1994; Harkema et al., 2006).

Special attention to fixation, decalcification, and uniform sampling is required due to the anatomical complexity of this organ and its encasement within bone. The first systematic approach for sectioning the nasal passages of laboratory rodents for toxicological studies was described by Young (1981; Figure 2). Although others have made minor variations to this methodology, it still serves as the best approach for consistent tissue sampling of laboratory rodent nasal passages for routine histologic evaluation. Similar tissue selection approaches for the microscopic examination of the nasal airways of nonhuman primates have been used by others (Harkema et al., 1987a, 1987b; Carey et al., 2007; Figure 3). More sections may be needed under certain research or diagnostic circumstances.

The four transverse nasal tissue sections are taken at specific anatomic locations that are identified by dental or palatine landmarks. In rodents (mice and rats), the most proximal tissue section, T1, is taken immediately posterior to the upper incisor teeth and contains the proximal aspects of the nasal septum, naso- and maxilloturbinates, and the intraseptal vomeronasal organ as its prominent morphologic features. Tissue section 2 (T2) is taken at the level of the incisive papilla and duct that connects the nasal and oral cavities. The principal morphologic features in T2 are the distal aspects of the naso- and maxilloturbinates and the midseptum. Sometimes, a cross-section of the very proximal tip of the third ethmoid turbinate is also present in the dorsomedial aspect of one or both of the nasal passages, depending on the cut of this section. The third transverse tissue section (T3) is taken at the level of the second palatal ridge and contains the elaborate ethmoid turbinates, an intramural maxillary sinus adjacent to each nasal passage, and the distal nasal septum that incompletely separates the two nasal passages at this level. The latter is called the "septal window," or the opening that joins the two nasal airways with the entrance to the nasopharygeal meatus.

The most distal nasal tissue section (T4) is taken at the level of the first molar teeth and contains the cross-sectional profiles of the distal aspects of the nasal septum and ethmoid turbinates, and the proximal nasopharyngeal meatus. Figure 2 also provides a general map of the intranasal distribution of the principal nasal epithelia that line the luminal surfaces of the nasal airways of the laboratory mouse. Descriptions of the histologic makeup of these nasal epithelia are provided below.

4. HISTOLOGY AND CELL BIOLOGY OF THE NASAL EPITHELIUM

Besides differences in the gross architecture of the nose in rodents, nonhuman primates, and humans, there are also species differences in the surface epithelial cell populations lining the nasal passages. These differences among species are found in the distribution of nasal epithelial populations and in the types of nasal cells within these populations. There are, however, four distinct nasal epithelial populations in humans and laboratory animals. These include the squamous epithelium (SE), which is primarily restricted to the nasal vestibule; ciliated, pseudostratified, cuboidal/columnar epithelium, or respiratory epithelium (RE), in the main chamber and nasopharynx; poorly ciliated cuboidal/columnar epithelium, or often termed transitional epithelium (TE), lying between SE and RE in the proximal aspect of the main chamber; and olfactory epithelium (OE), located in the dorsal or dorsoposterior aspect of the nasal cavity. The distribution of nasal airway epithelia in the nasal cavity of the mouse and monkey is also illustrated in Figures 2 and 3, respectively. The reader is also referred to other reports for a more thorough and detailed description of the intranasal distribution of airway surface epithelia in laboratory rodents and nonhuman primates (Kepler et al., 1995; Mery et al., 1994; Harkema et al., 1987a, 1987b, 2006, 2012; Carey et al., 2007).

4.1 Nasal Squamous Epithelium

The nasal vestibule of laboratory animals and humans is completely lined by a stratified SE (Figure 4). It is composed of basal cells along the basal lamina and several layers of squamous cells, which become progressively flatter toward the luminal surface of the airway. This region of the nasal mucosa probably functions similar to the epidermis of skin, to protect the underlying tissues from potentially harmful environmental agents. The T1 section of mice and rats contains a small amount of squamous epithelium lining the floor of the ventral meatus in each nasal passage. Although the cellular composition is similar among rodent and primate species, SE is much thicker in primates (Figure 4).

4.2 Nasal Transitional Epithelium

Distal to the stratified squamous epithelium and proximal to the ciliated RE is a narrow zone of poorly or nonciliated, microvilli-covered cuboidal/columnar surface epithelium, which has been referred to as nasal nonciliated or scantly ciliated RE or nasal TE (Figure 5). Common, distinctive features of TE in mice, monkeys, and humans include the following: (1) an anatomical location in the proximal aspect of the nasal cavity between the SE and the RE; (2) the presence of numerous nonciliated, cuboidal/columnar surface cells and basal cells; (3) few ciliated cells and a scarcity of mucous (goblet) cells; and (4) an abrupt morphological border with SE but a less abrupt border with RE. In T1 of mice, the TE lines the luminal surfaces of the lateral wall, maxilloturbinate, and medial aspect of the nasoturbinate. These epithelial surfaces border the lateral meatus in the proximal nasal airways. The TE of rats and mice is thin (1–2 cells thick), pseudostratified, and composed of three

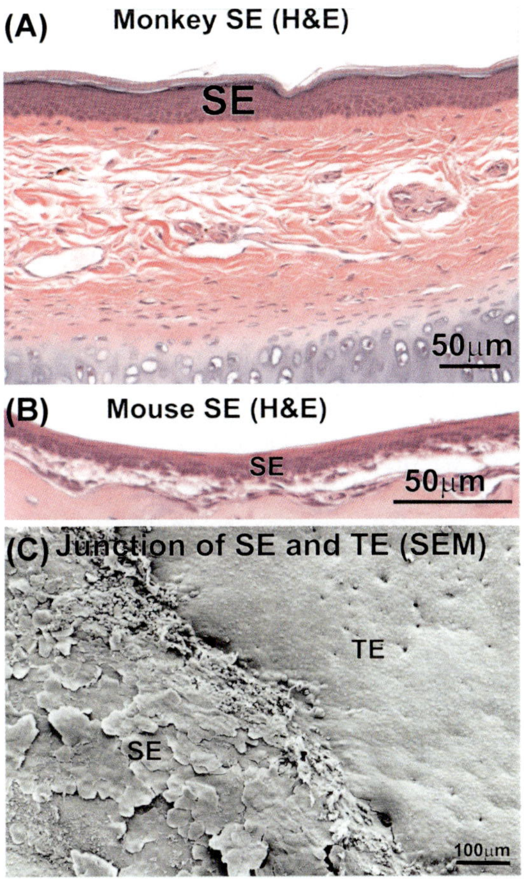

FIGURE 4 Light photomicrographs of squamous epithelium (SE) lining the proximal nasal airway of a *Rhesus macaque* monkey and C57BL/6 mouse, (A) and (B) respectively, stained with hematoxylin and eosin (H&E). (C) Scanning electron micrograph (SEM) of the junction between squamous and transitional epithelium, SE and TE respectively, in the proximal nasal cavity of a macaque monkey, between the nasal vestibule and main chamber.

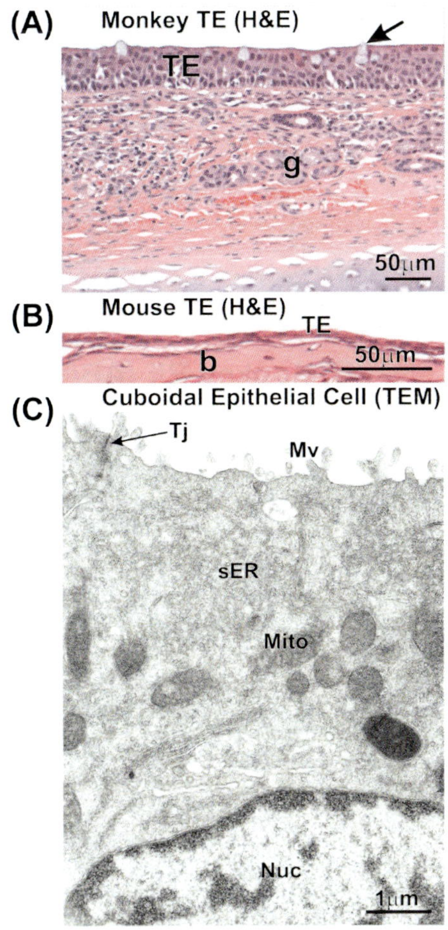

FIGURE 5 Light photomicrographs of the nasal transitional epithelium (TE) from the macaque monkey (A) and C57BL/6 mouse (B), stained with hematoxylin and eosin (HE). Arrow, mucous cell; g, gland in the lamina propria underlying the surface epithelium; b, turbinate bone. (C) Transmission electron micrograph (TEM) of the apical region of a nonciliated cuboidal epithelial cell in the nasal transitional epithelium of a macaque monkey. Tj, tight junction between epithelial cells; Mv, microvilli on luminal surface; sER, abundant smooth endoplasmic reticulum in cytoplasm; Mito, mitochondria; Nuc, cellular nucleus.

nonciliated cell types (basal, columnar, and cuboidal) and scattered ciliated cells. In contrast, the TE of primates is thick (4–5 cells), stratified, and composed of at least five different epithelial cell types, including mucous cells (Figure 5).

Luminal, nonciliated cells in the TE of rodents (mice and rats) rarely contain secretory granules but do have abundant smooth endoplasmic reticulum (sER) in their apices (Figure 5). sER is an important intracellular site for xenobiotic-metabolizing enzymes, including cytochrome p450 proteins. The prominent presence of this organelle in these cells, as well as in the sustentacular cells of OE, suggests metabolizing capability for certain inhaled and noninhaled xenobiotics.

In both mice and humans, the lamina propria beneath the TE and RE, described below, is a highly vascularized and innervated, loose fibroelastic connective tissue containing serous and seromucous secretory glands. Cavernous venous plexuses, referred venous plexuses, sometimes referred to as "swell bodies," are also present in the lamina propria of the nasal mucosa and are located in distinct regions along the nasal airway. Dilation of these vessels is thought to alter nasal airflow by thickening the mucosa, narrowing air passages, and diverting inhaled air. Prominent swell bodies in the mouse and rat are found in the maxilloturbinates and lateral wall in the proximal nasal airways.

4.3 Nasal Respiratory Epithelium

The majority of the nonolfactory nasal epithelium of mice, monkeys, and humans is RE (Figures 2 and 3). RE is a pseudostratified, cuboidal/columnar, ciliated epithelium (Figure 6). Most of the nasal cavity in humans and nonhuman primates is lined by RE (Harkema et al., 1987a; Figure 3). In laboratory mice, however, less than half of the nasal cavity

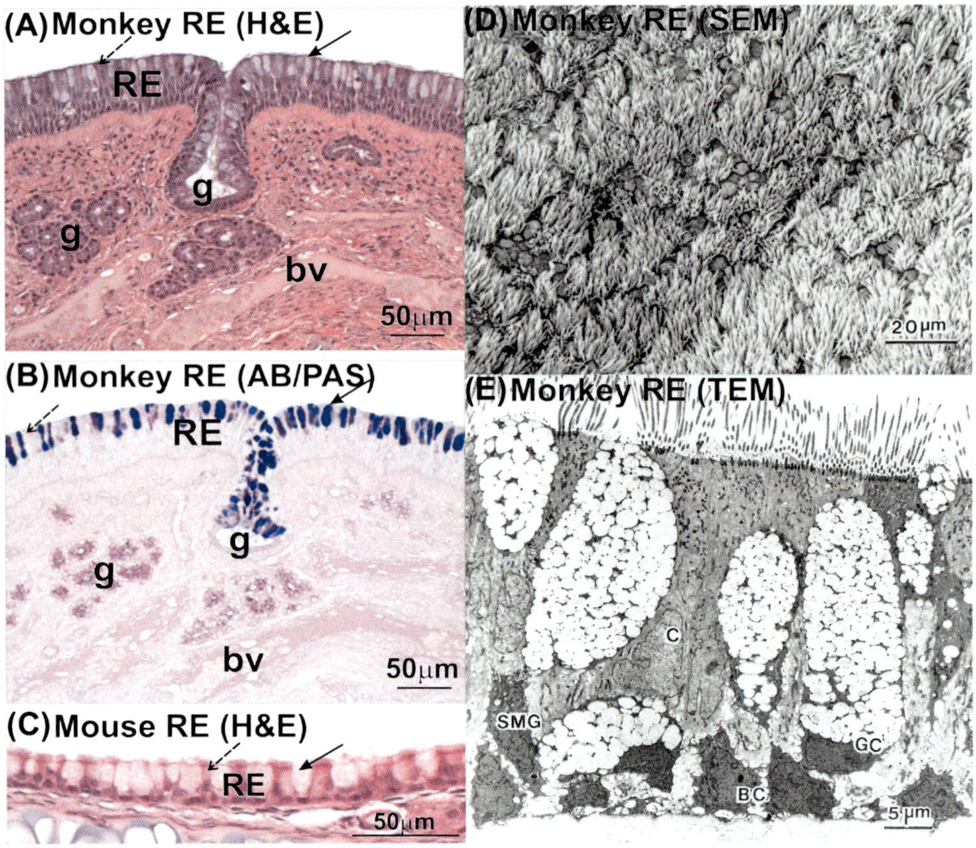

FIGURE 6 Light photomicrographs of nasal respiratory epithelium (RE) from a macaque monkey (A, B) and C57BL/6 mouse (C) stained with hematoxylin and eosin (H&E; A and C) and Alcian blue (pH 2.5)/periodic acid Schiff (AB/PAS; B). g, secretory gland in lamina propria; bv, blood vessel in lamina propria; solid arrow, mucous goblet cell; stippled arrow, ciliated cell. (D) Scanning electron micrograph (SEM) of the ciliated luminal surface of the nasal respiratory epithelium from a macaque monkey. (E) Transmission electron micrograph (TEM) of the ciliated nasal respiratory epithelium from a macaque monkey. C, ciliated cell; GC, mucous goblet cell; SMG; small mucous granule cell; BC, basal cell.

is covered by RE (Gross et al., 1982). In both rodent and primate species, the maxillary sinus is also lined by RE. Like SE, RE in primates is much thicker than that of mice and other rodents (Figure 6).

RE in the nasal cavity of the mouse is composed of morphologically distinct cell types that include numerous ciliated (most abundant), mucous, and basal cells, and lesser numbers of nonciliated columnar/cuboidal cells, including solitary chemoreceptor and brush cells that have distinctive apical microvilli. In both primate and murine RE, the thickness and number of ciliated and mucous cells varies according to intranasal location. In mice, for example, mucous goblet cells are conspicuous in the RE lining the proximal, mid-, and ventral septum (T1; Figure 2) and the nasopharyngeal meatus (T4; Figure 2). Mucous cells, however, are scarce in RE lining of the more distal midseptum, lateral wall, nasoturbinate, and maxilloturbinate in T2 and the maxillary sinus in T3 (Figure 2). In the RE of human and nonhuman primates, there is a gradual anterior to posterior increase in mucous goblet cell density along both the nasal septal and lateral walls (Harkema et al., 1987b; Morgensen and Tos, 1977).

Numerous secretory glands, containing serous and/or mucous cells, are located in the lamina propria beneath RE predominantly in the proximal septum and lateral walls of mice. The largest nasal secretory gland in the mouse and rat is the bilateral lateral nasal gland (LNG), also referred to as Steno's gland, which surrounds the maxillary sinus located in the lateral wall adjacent to each nasal passage (T3; Figure 2; Bojsen-Moller, 1964). Similar nasal glands are also found in the primate nose, including that of humans (Bojsen-Moller, 1967).

Secretory contents of the LNG drain into the nasal vestibule. This nasal gland is a major site for the synthesis and secretion of odorant-binding proteins that are odorant carriers in nasal mucus. It also synthesizes large amounts of immunoglobulin A, which is important for immune defense of the upper respiratory tract, and testosterone and salivary androgen-binding proteins, which are likely important in olfaction and reproductive behavior (Zhou et al., 2009). Because of its high metabolic capacity, it is also a potential target site for chemical toxicity caused by secondary metabolites of inhaled or ingested xenonobiotic compounds.

4.4 Olfactory Epithelium

The most prominent difference in nasal epithelium between mice and humans is the amount of the main nasal airway chamber that is lined by OE (Figures 2 and 3). A greater percentage of the nasal cavity of mice, which have an acute sense of smell, is lined by OE as compared to that of monkeys or humans, whose sense of smell is not as well developed. Gross et al. (1982) morphometrically determined that close to 50% of the murine nasal cavity is lined by this unique sensory neuroepithelium. The OE of humans is limited to a small area of 500 mm^2 in the mid-dorsal nasal passages, which is only 3% of the total surface area of the nasal cavity (Sorokin, 1988). In contrast, OE lines most of the ethmoid turbinates, roof, lateral wall, and septum in the distal nasal airways of mice (T2, T3, and T4; Figure 2).

In mice, monkeys, and humans, OE is a pseudostratified, columnar neuroepithelium containing three principal epithelial cell types. These are the olfactory sensory neurons (OSNs), sustentacular cells, and basal cells (Figure 7). OSNs are bipolar neuronal cells interposed among the sustentacular cells. Mature OSNs immunohistochemically express olfactory marker protein (OMP), which is used for light microscopic detection of these unique neuroepithelial cells (Figure 7) found in both rodents and primates, including humans. The cellular composition and epithelial thickness are remarkably similar among primate and rodent species (Figure 7).

The dendritic portions of OSNs extend above the epithelial surface and terminate into a bulbous olfactory knob, from which protrude 10–15 immotile cilia (Figure 7). These long and thin cilia, about 50 μm in length and 0.1–0.3 μm in diameter, are enmeshed with each other and with microvilli in the surface fluid, and provide an extensive surface area for reception of odorants. The ciliary membranes of the OSNs contain the odorant receptors (ORs) responsible for the chemical interaction with and initial detection of inhaled odors. ORs are G protein-coupled, seven transmembrane proteins that are encoded by the largest gene families known to exist in a given animal genome (Mombaerts, 2001). Odorant genes were discovered by Linda Buck and Richard Axel, who were awarded the 2004 Nobel Prize in Physiology or Medicine for their landmark work in the cellular and molecular biology of olfaction (Buck and Axel, 1991). Their elegant and novel work was one of the first applications of degenerate polymerase chain reaction. It is now estimated that there are 500–1000 OR genes in the mouse (Buck, 1992), and ~1000 sequences in humans, residing in multiple clusters spread throughout the genome, with more than half being pseudogenes (Mombaerts, 2001).

A single OR gene is expressed in a minute subset of OSNs, with the current belief that each OSN expresses only a single OR (one receptor–one neuron rule). Interestingly, mouse OR genes are expressed in OSNs within one of four even-sized, distinct topographical zones in the OE lining the nasal cavity (Ressler et al., 1994). OSNs expressing a given OR are distributed in a random, punctate manner within a zone. There are approximately 2 million OSNs within the nasal cavity of a mouse.

The axon of the OSN originates from the base of the cell and passes through the basal lamina to join axons from other OSNs forming nonmyelinated nerve fascicles, or bundles, in the lamina propria. These olfactory nerves perforate the bony cribiform plate, which separates the nasal cavity from the brain, and form the outer olfactory nerve layer of the olfactory bulb. Axons of OSNs that express the same OR gene converge with extreme precision on ~2000 signal-processing modules called "glomeruli" that reside in distinct locations within the olfactory bulb. Glomeruli are relatively large spherical neuropils (100–200 μm in diameter) in which the axons of OSNs form synaptic connections on the dendrites of mitral and tufted cells, the output neurons of the olfactory bulb. Transmission of olfactory information is further sent through the axons of the mitral and tufted cells to the olfactory cortex.

The OE also contains two types of basal cells—horizontal (HBC) and globose (GBC). HBCs are thin cells located along the basal lamina, which share many of the same morphologic and histochemical features as the basal cells of nasal RE (e.g., contain keratins). In contrast, GBCs are morphologically more round or oval and are located above the HBCs. These cells have a more electron-lucent cytoplasm than the HBCs and are not immunohistochemically reactive for keratin. Some of the GBCs are the progenitor cells for OSNs, while the HBCs give rise to GBCs (Suzuki and Takeda, 1991; Jang et al., 2003). Multipotent basal cells within the OE or in Bowman's gland (BG) ducts are the likely progenitors for sustentacular cells, which are described next.

Sustentacular (supporting) cells are columnar epithelial cells that span the entire thickness of the OE from the airway surface to the basal lamina. The distinct oval nuclei of the sustentacular cells are aligned in a single row along the apical aspect of the OE and are the most apically located epithelial nuclei within the mammalian OE (Figure 7). The supranuclear portion of the cell is broad, while the portion of the cell below the nucleus tapers to a thin foot-like process that attaches to the basal lamina. These supporting cells surround the OSNs, making multiple contacts with OSNs through fine cellular extensions. The apical surfaces of sustentacular cells are lined by numerous long microvilli that intermingle with the thin cilia of the OSNs along the surface of the airway lumen. The supranuclear cytoplasm of sustentacular cells has abundant sER and xenobiotic-metabolizing enzymes (e.g., cytochrome P-450, flavin-containing monooxygenases, *N*-acetyltransferases). The metabolism in these cells may be important in detoxification of inhaled xenobiotics and in the function of smell (Ding and Dahl, 2003; Ling et al., 2004). Sustentacular cells are also thought to contribute to the regulation of the ionic composition of the overlying mucous layer, which undoubtedly

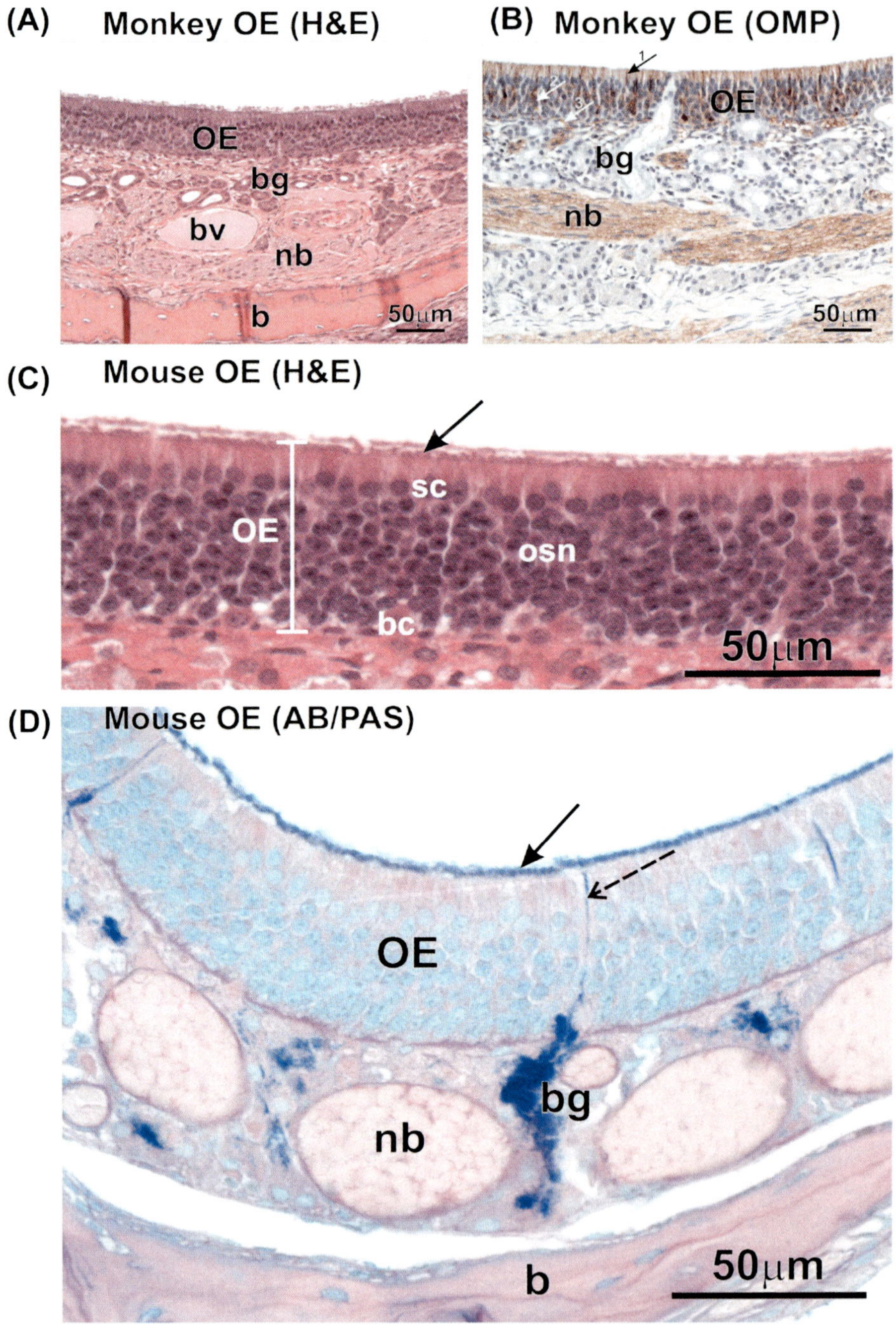

FIGURE 7 Light photomicrographs of the nasal olfactory epithelium (OE) in the macaque monkey (A, B) and C57BL/6 mouse (C, D). Nasal tissues in (A) and (B) are stained with hematoxylin and eosin (H&E), in (B) immunohistochemically stained for olfactory marker protein (OMP; dark stain in OE and nerve bundles (nb) in lamina propria) and (D) histochemically stained with Alcian blue (pH 2.5)/periodic acid Schiff (AB/PAS) for acidic and neutral mucosubstances (dark stain). bg; Bowman's gland; solid arrow, immotile olfactory cilia that are part of the dendritic portion (OMP positive) of olfactory sensory neurons in the OE. In (B) arrows 1, 2, and 3 are OMP-positive olfactory cilia, cell body of olfactory sensory neuron, and axon of olfactory sensory neuron, respectively. In (C) sc, sustentacular cells; OSN; olfactory sensory neurons; BC, basal cells. The stippled arrow in (D) is the AB/PAS-stained mucosubstances in the intraepithelial duct of Bowman's gland (bg).

affects the chemical interactions between odors and their ORs. Mammalian sustentacular cells do not contain mucin glycoproteins, which are characteristic of the columnar mucus-secreting epithelial cells in the nasal RE (e.g., mucous goblet cells).

In addition to these principal epithelial cells in the OE of mice and humans, there are substantial numbers of different types of apically located microvillous cells whose function(s) are not fully understood. One subgroup of microvillous cells in OE has a morphology like microvillous sensory neurons of the vomeronasal organ (described below), with an axonal component projecting to the olfactory bulb (Mombaerts, 2008). These cells also express the transient receptor potential channel C2 (TRP C6). The TRP ion channel superfamily functions as diverse cellular sensors. Another group of microvillous cells are nonneuronal, lacking a basal/axonal segment. This subgroup of microvillous cells in mice express TRP C6, phospholipase C beta-2 (PLCβ2), and the type 3 inositol trixphosphate receptor (IP3R3), suggesting that they are secondary chemosensory cells (Elsaesser et al., 2005; Hegg et al., 2010). Some of these cells also express transient receptor channel M5 (TRP M5), and others do not (Hansen and Finger, 2008). It has been suggested that these cells may interface with peptidergic nerve fibers immunoreactive with neuropeptides (e.g., substance P), suggesting that these solitary chemosensory cells may transmit sensation of chemical irritants via the trigeminal nerve.

The production and secretion of mucus covering the luminal surface of OE is restricted to the subepithelial BG (Figure 7). These distinct mucus-producing secretory glands are located in the underlying lamina propria and interspersed among the olfactory nerve bundles. They are simple tubular-type glands composed of small compact acini. Ducts from these glands transverse the basal lamina at regular intervals and extend through the OE to the luminal surface. BG contain copious amounts of neutral and acidic mucosubstances that contribute to the mucous layer covering the luminal surface of the OE. Like the sustentacular cells, both the acinar and duct cells of BG also contain many xenobiotic-metabolizing enzymes.

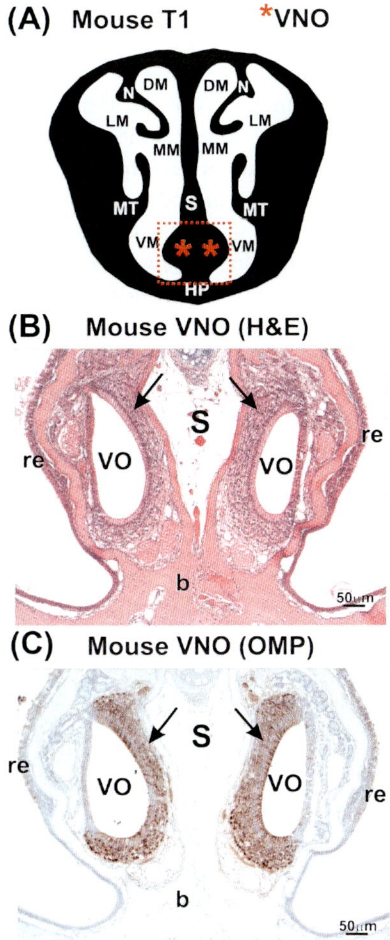

FIGURE 8 (T1) Schematic representation of the cross-sectional profile of the proximal aspect of the murine nose (T1 level). Mice have several accessory olfactory organs that are missing or poorly developed in humans and nonhuman primates. DM, dorsomeatus; HP, hard palate; LM, lateral meatus; MM, middle meatus; MT; maxilloturbinate; N, nasoturbinate; S, septum; VM, ventral meatus; asterisk, vomeronasal organ VNO. The rectangle is the intranasal location for the light photomicrographs of the ventral septum containing the bilateral vomeronasal organ (*) and histochemically stained with hematoxylin and eosin (H&E; B) and immunohistochemically for olfactory marker protein (OMP; C). S, septum; re, respiratory epithelium; VO, vomeronasal organ; b, bone. In (C) OMP-stained olfactory sensory neurons in the medial aspect of the vomeronasal organ are lined by olfactory epithelium (arrow).

Accessory Olfactory Organs in Mice

Besides the OE in the main nasal chamber described above, there are three accessory olfactory organs found in the nasal airways of rats and mice, but they are not well developed or functional in the human nose. These include the vomeronasal organ (VNO), the septal organ of Masera (SOM), and the septal organ of Grüneberg (SOG). VNO, also referred to as the organ of Jacobson, is a paired tubular diverticulum located in the ventral portion of the proximal septum (T1) of the mouse (Figure 8). The lateral wall of each diverticulum is lined by a thick nonsensory, columnar epithelium, while the medial wall is lined by a neuroepithelium. The sensory nerves in the VNO have long microvilli but lack the cilia found in OSNs of the OE (i.e., neuronal microvillous cells described above). Like OSNs in the main nasal chamber of the mouse, these olfactory neurons contain OMP (Figure 8).

The axonal portions of OE in VNO synapse with glomeruli located in the accessory olfactory bulb rather than those in the olfactory bulb. A small duct extends from each diverticulum to the proximal nasal airway lumen near the nasal opening of the nasopalatine canal that connects the nasal cavity with the oral cavity. A primary function of the VNO is to detect pheromones, airborne chemical messages or signals, sent between individual mice, which are essential for normal reproductive physiology and behavior. Initial pheromone detection in VNO is triggered by its binding to G protein-coupled receptors followed by the activation of TRP C2 (Zufall, 2005; Kiselyov et al., 2010).

Another small, but distinct, chemosensory structure found in the noses of rodents, including mice, is the SOM (Ma et al., 2003). It is a small patch of olfactory neuroepithelium with OMP-positive OSNs surrounded by RE and lying bilaterally in the ventral nasal septum near the entrances of the nasopalatine ducts (T2; Figure 9). It constitutes approximately 1% of the OE in the main chamber of the mouse nose. Neuronal projections from the SOM connect to distinct regions in the main olfactory bulb. SOM, however, has not been identified in the human nose. For the mouse, it has been suggested that the SOM may have an alerting function by sensing airborne orders during quiet breathing when intranasal airflows do not reach the main OE in the more dorsal regions of the nasal cavity. In light of the anatomical location of the SOM near the intranasal opening of the nasopalatine duct connecting the oral and nasal cavities, it may also detect odors from the mouth and act as a chemosensory surveyor of foods.

A third accessory population of OMP-expressing sensory neurons is the SOG (Grüneburg, 1973; Storan and Key, 2006). It is located bilaterally in a discrete area of the dorsorostral tip of the nasal septum in the mouse and rat. Because it is located rostral to the routine T1 nasal section, this accessory olfactory organ would not be examined with the routine tissue sectioning of the mouse nasal cavity for light microscopy.

Like the SOM, the axons of the SOG project to a distinct subset of glomeruli in the main olfactory bulb. Unlike OSNs of the MOE, VNO, and SOM, which are all part of a luminal surface epithelium, the OSNs of the SOG are located in clusters in the lamina propria beneath OMP-negative surface epithelium. Although the function of SOG is unknown, its rostral location suggests that it may have a special function for early detection of distinct odorants.

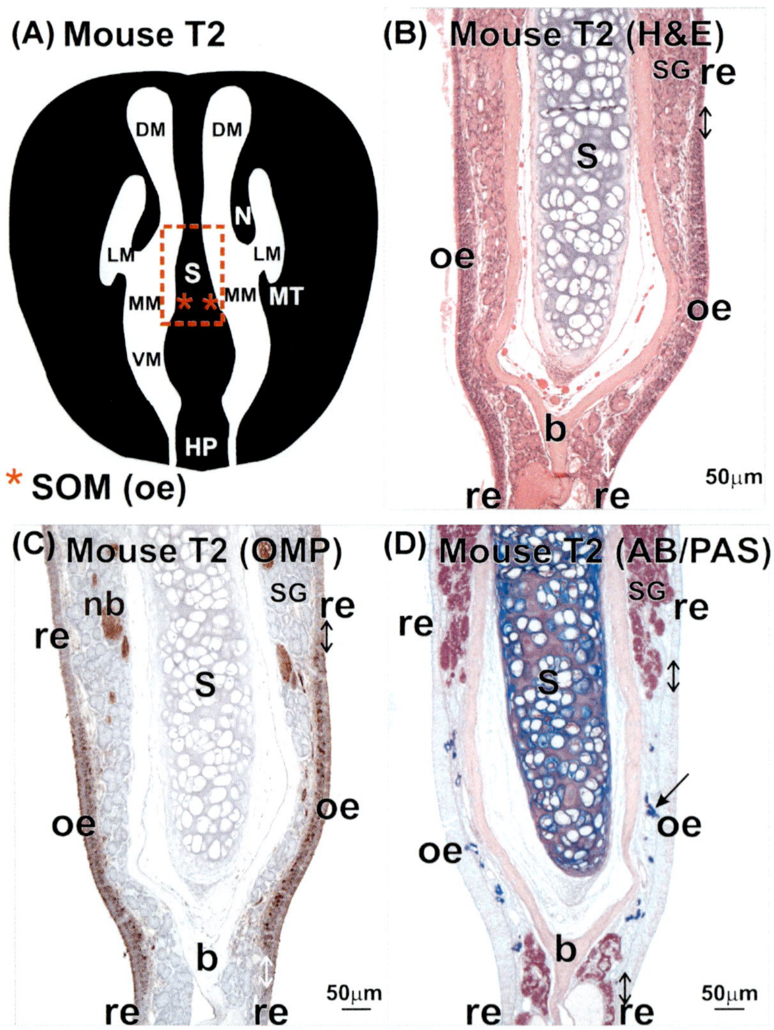

FIGURE 9 (T2) Schematic representation of the cross-sectional profile of the murine nose at level T2. DM, dorsomeatus; HP, hard palate; LM, lateral meatus; MM, middle meatus; MT, maxilloturbinate; N, nasoturbinate; S, septum; VM, ventral meatus. The rectangle is the intranasal location for the light photomicrographs of the midseptum stained with hematoxylin and eosin (H&E; B), immunohistochemically for olfactory marker protein (OMP; C) and histochemically with Alcian blue (pH 2.5)/periodic acid Schiff (AB/PAS; stains for mucosubstances in mucous cells; D). b, bone; oe, olfactory epithelium (septal organ of Massera); re, respiratory epithelium; S, septal cartilage; double arrows, transition between epithelial types; arrow, AB/PAS-stained mucosubstances in Bowman's glands.

4.5 Lymphoepithelium and Nasal-Associated Lymphoid Tissue

In addition to the four principal nasal epithelia already described, there is another specialized epithelium, lymphoepithelium (LE), in the nasal airways of rats and mice; this covers discrete focal aggregates of nasal-associated lymphoid tissue (NALT) in the underlying lamina propria (Figure 10). In mice, NALT with associated LE is restricted to the ventral aspects of the lateral walls at the opening of the nasopharyngeal duct (T3) (Kuper, 2006). The overlying LE is composed of cuboidal ciliated cells, a few mucous cells, and numerous nonciliated, cuboidal cells with luminal microvilli (so-called membranous or M cells) similar to those in the gut- and bronchus-associated lymphoid tissues in the intestinal and lower respiratory tracts, respectively. M cells are thought to be involved in the uptake and translocation of inhaled antigen from the nasal lumen to the underlying lymphoid structures.

NALT, with its specialized LE, has also been described in the nasopharyngeal airways of nonhuman primates, but these lymphoid structures with LE are more numerous and are located on both the lateral and septal walls of the proximal nasopharynx of the monkey (Harkema et al., 1987a; Carey et al., 2007; Figure 10). The correlate of NALT in humans is Waldeyer's ring, the oropharyngeal lymphoid tissues composed of the adenoid, and the bilateral tubule, palatine, and lingual tonsils (Brandtzaeg, 1984).

The location of NALT at the entrance of the nasopharyngeal duct is a very strategic position because most of the nasal secretions and inhaled air, both presumably laden with antigenic material, flow over this area. Although the function of NALT and its place in the general mucosal-associated lymphoid system are not fully understood, these mucosal lymphoid tissues may have an important function in regional immune defense of the upper airways.

Nasal Trigeminal Nerves

Inhaled chemicals may neurogenically stimulate not only the OE in the main chamber and accessory olfactory organs but also chemoreceptors of nasal trigeminal nerves, which are part of the somatic sensory system for the eyes, nose, and mouth. Trigeminal chemoreceptors are present throughout the nasal epithelium lining the murine and human nasal cavity. Simulation of trigeminal nerve fibers produces sensations described as irritating, painful, burning, cooling, tingling, stinging, or pungent. Chemical stimulation of these sensory nerves may also result in protective reflexes causing increased secretions (e.g., nasal mucus), decreased respiration rate (e.g., apnea), and reduction in the nasal airways due to vascular congestion and mucosal swelling. Nasal trigeminal nerve fibers that respond to irritants release neuropeptides, such as substance P and calcitonin gene-related peptides, and are presumably polymodal nociceptors.

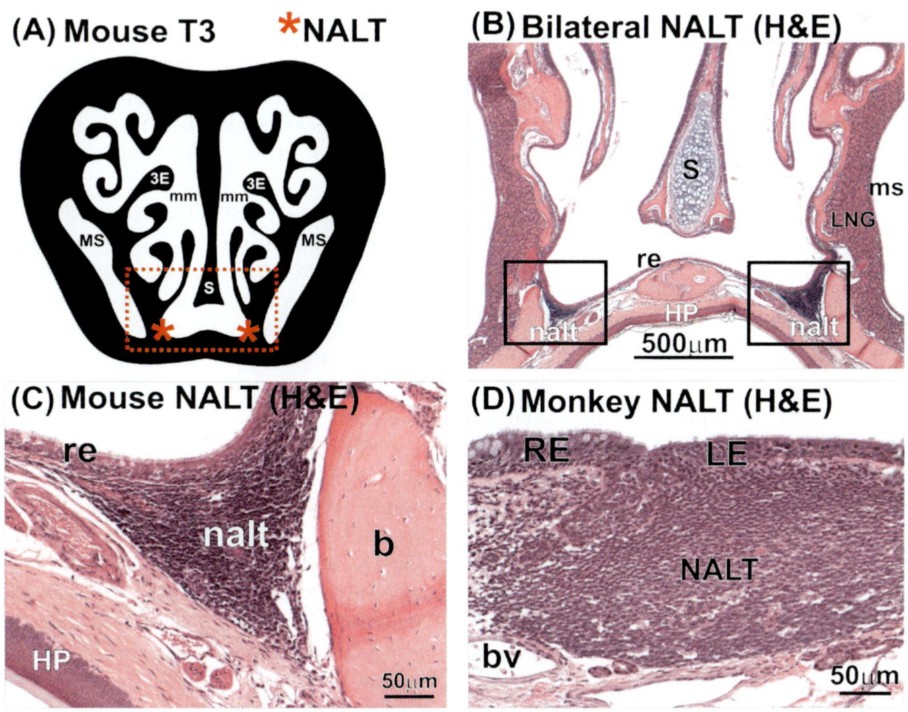

FIGURE 10 (A) Schematic representation of the cross-sectional profile of the murine nose at level T3. 3E, third ethmoid turbinate; mm; middle meatus; MS, maxillary sinus; S, septum. The rectangle is the intranasal location for the light photomicrographs of the nasopharyngeal meatus containing the bilateral nasal-associated lymphoid tissue (*NALT; B, C). (D) Light photomicrograph of NALT in the macaque monkey. Nasal tissues are stained with hematoxylin and eosin (H&E). b, bone; HP, hard palate; LNG, lateral nasal gland; NALT, nasal-associated lymphoid tissue; RE, respiratory epithelium; LE, lymphoepithelium.

Nasal trigeminal nerve fibers ramify repeatedly in the nasal mucosa, and their intraepithelial nerve endings extend close to the airway surface, just below the apical tight junctions connecting epithelial cells. Interestingly, some of these trigeminal nerve fibers in the murine nasal epithelium, predominantly in the RE lining the proximal nasal airways, are intimately associated with and synapse with solitary chemoreceptive cells (SCCs), distinct non-ciliated epithelial cells with loose apical microvilli. These epithelial cells are morphologically distinct from brush cells, also found in the RE, which have a stiff tuft of apical microvilli. SCCs express elements of the taste transduction cascade, including Tas1R and Tas2 receptor molecules, α-gustducin, PLCβ2, and TrpM5. They detect inhaled irritants and other xenobiotic substances, which trigger trigeminally mediated airway reflexes. SCCs in mice are located in nasal, laryngeal, and tracheobronchial airways, but only in the nasal epithelium due to SCC synapse with trigeminal nerve fibers (Tizzano et al., 2011). In contrast to the nasal SCCs that rely on taste receptors, the free trigeminal nerve endings in the nasal epithelium mediate irritant responses through transient receptor potential (TRP) cationic channels (e.g., TrpV1, TrpA1), acid sensing ion channels, and other ion channels. Therefore, free trigeminal nerve fibers and SCCs respond to different chemical agents.

Nasal Blood Vessels

As mentioned previously, the lamina propria in both murine and human nasal mucosa has a rich and complex network of blood vessels, with each of the epithelial regions receiving blood from a separate arterial supply (Sorokin, 1988). The vascular system in the nose is composed of resistance and capacitance vessels. Resistance vessels are small arteries, arterioles, and arteriovenous anatomoses. Blood flow to the mucosa is regulated by constriction and dilatation of these vessels.

A unique feature of the nasal vasculature is the large venous sinusoids (i.e., capacitance vessels, venous erectile tissue, or swell bodies) that are distributed throughout the mucosa. In humans and mice, these blood vessels are well developed in specific sites of the anterior or proximal aspects of the nasal passages. Capacitance vessels have dense adrenergic innervations, and the congestion and constriction of these vascular structures are regulated by the sympathetic nerve supply to the nose (Olsson and Bende, 1986). Congestion of blood in these vessels can change mucosal thickness, resulting in an altering of nasal airflow patterns and nasal resistance.

5. CONCLUSION

In this chapter, some of the major similarities and differences in gross and microscopic anatomy among laboratory animals and humans have been highlighted, with special emphasis on the surface epithelial cells that line the nasal airways. This comparative overview has emphasized the complexity and diversity of the nasal airways. In general, mice and other rodents have much more complex turbinate structures than do primates, including humans. In addition, the nasal cavity of primates is lined by much less OE than that of rodents. The cellular composition of the nasal epithelium is dependent on both the intranasal location and the animal species.

REFERENCES

Bojsen-Moller, F., 1964. Anat. Rec. 150, 11–24.
Bojsen-Moller, F., 1967. Rev. Laryngol. Otol. Rhinol. (Bord) 88 (Suppl.), 71–75.
Brandtzaeg, P., 1984. In: Bienenstock, J. (Ed.), Immunology of the Lung and Upper Respiratory Tract. McGraw-Hill, New York, pp. 28–95.
Buck, L., Axel, R., 1991. Cell 65, 175–187.
Buck, L.B., 1992. Curr. Opin. Neurobiol. 2, 282–288.
Carey, S.A., et al., 2007. Toxicol. Pathol. 35, 27–40.
Ding, X., Dahl, A.R., 2003. In: Doty, R.L. (Ed.), Handbook of Olfaction and Gustation. Marcel Dekker, Inc., New York, pp. 51–73.
Elsaesser, R., et al., 2005. Eur. J. Neurosci. 21, 2692–2700.
Faust, K., et al., 2012. PLoS Comput. Biol. 8, 7.
Frank, D.N., et al., 2010. PLoS One 17, 5.
Gross, E.A., Morgan, K.T., 1992. In: Parent, R.A. (Ed.), Comparative Biology of the Normal Lung. CRC Press, Inc., Boca Raton, pp. 7–25.
Gross, E.A., et al., 1982. J. Anat. 135, 83–88.
Grüneburg, H., 1973. Z. Anat. Entwicklungsgesch 140, 39–52.
Guilmette, R.A., Wicks, J.D., Wolff, R.K., 1989. J. Aerosol. Med. 2, 365–377.
Hansen, A., Finger, T.E., 2008. BMC Neurosci. 9, 115.
Harkema, J.R., et al., 1987a. Am. J. Anat. 180, 266–279.
Harkema, J.R., et al., 1987b. J. Histochem. Cytochem. 35, 279–286.
Harkema, J.R., Carey, S.A., Wagner, J.G., 2006. Toxicol. Pathol. 34, 252–269.
Hegg, C.C., et al., 2010. Eur. J. Neurosci. 32, 1632–1645.
Herbert, R.A., Leininger, J.L., 1999. In: Maronpot, R., Boorman, G.A., Gaul, B.W. (Eds.), Pathology of the Mouse: Reference and Atlas. Cache River Press, Vienna, pp. 259–292.
Jang, W., Youngentob, S.L., Schwob, J.E., 2003. J. Comp. Neurol. 460, 123–140.
Kepler, G.M., et al., 1995. Inhal. Toxicol. 7, 1207–1224.
Kiselyov, K., van Rossum, D.B., Patterson, R.L., 2010. Vitam. Horm. 83, 197–213.
Kuper, C.F., 2006. Toxicol. Pathol. 34, 609–615.
Ling, G., et al., 2004. Chem. Biol. Interact. 147, 247–258.
Ma, M., et al., 2003. J. Neurosci. 23, 317–324.
Mery, S., et al., 1994. Toxicol. Pathol. 22, 353–372.
Mombaerts, P., 2001. Annu. Rev. Genomics Hum. Genet. 2, 493–510.
Mombaerts, P., 2004. N. Engl. J. Med. 351, 2579–2580.
Mombaerts, P., 2008. Chem. Senses 33. Abstr. S28.
Morgan, K.T., et al., 1984. Am. Rev. Respir. Dis. 130, 275–281.
Morgensen, C., Tos, M., 1977. Anat. Anaz. 142, 322–330.
Negus, V.E., 1958. In: Comparative Anatomy and Physiology of the Nose and Paranasal Sinuses. E&S Livingstone, Edinburgh.
Olsson, P., Bende, M., 1986. Acta Otolaryngol. 102, 482–487.

Ressler, K.J., Sullivan, S.L., Buck, L.B., 1994. Curr. Opin. Neurobiol. 4, 588–596.

Sorokin, S.P., 1988. In: Weiss, L. (Ed.), Cell and Tissue Biology: A Textbook of Histology. Urban & Schwarzenberg, pp. 751–814.

Storan, M.J., Key, B., 2006. J. Comp. Neurol. 494, 834–844.

Suzuki, Y., Takeda, M., 1991. Cell. Tissue Res. 266, 239–245.

Tizzano, M., et al., 2011. BMC Pulm. Med. 11, 3.

Uraih, L., Morgan, K.T., Maronpot, R.R., 1990. Environ. Health Perspect. 85, 161–352.

Young, J.T., 1981. Fundam. Appl. Toxicol. 1, 309–312.

Zhou, X., et al., 2009. J. Steroid Biochem. Mol. Biol. 117, 81–86.

Zufall, F., 2005. Naunyn. Schmiedeb. Arch. Pharmacol. 371, 245–250.

Chapter 3

Gross and Subgross Anatomy of Lungs, Pleura, Connective Tissue Septa, Distal Airways, and Structural Units

Janice L. Peake and Kent E. Pinkerton*

Chapter Outline

1. INTRODUCTION

In the lung, there is much anatomical diversity between species, which may affect how the animal responds to disease. To correlate animal studies to human disease, we need to understand what these differences are. These differences can include the branching geometry of the airways, variations in breathing rate and depth, as well as the subgross architecture of the lungs. The purpose of this chapter is to examine the basic anatomical units of the lungs in a variety of mammalian species and discuss their functional significance.

McLaughlin et al. (1961, 1966) correlated existing information on the comparative subgross morphology of lungs and reported observations made primarily with a dissecting light microscope on serial subgross sections of lungs whose vessels had been injected with colored latex. Species differences in morphology of lungs have been published by Phalen et al. (1975), including an extensive bibliography on lung morphology. The Division of Lung Diseases of the National Institutes of Health's Heart, Lung and Blood Institute sponsored a workshop entitled *Comparative Biology of the Lung*, which published findings in a special edition of the *American Review of Respiratory Disease* (American Thoracic Society, 1983). Some of this information concerning the lung morphology of common domestic animals is now found in standard textbooks on veterinary anatomy (Dyce et al., 1987, 2009). Similar information concerning common laboratory mammals was more difficult to find until the first publication of this book, *Comparative Biology of the Normal Lung*, in 1992.

This chapter reviews the comparative morphology of the pulmonary pleura, segmental and interlobular septa, terminal airways, and structural units for a number of mammalian species. Some methods for anatomical identification are reviewed, and the advantages and disadvantages for each are discussed. Examples of the biological significance of gross and subgross species differences in lung morphology are discussed.

2. OBSERVATIONS

2.1 Pulmonary Pleura

The pleura is recognized as more than just a mechanical envelope, and its significance to both normal function and reactions to pulmonary and systemic diseases has been reviewed (*The Pleura in Health and Disease*, 1985; Sahn, 1988). The basic structure of the mesothelium (Wang, 1985) and the connective tissue components (Rennard et al., 1985) have been well described. The pulmonary (visceral) pleura of human and large animal lungs tends to

* With acknowledgment to Walter S. Tyler, Professor Emeritus, for providing the basis for this updated chapter from the 1st edition of *Comparative Biology of the Normal Lung*.

Comparative Biology of the Normal Lung. http://dx.doi.org/10.1016/B978-0-12-404577-4.00003-5

be thick, whereas the pulmonary pleura of lungs from smaller animals, including common laboratory species, tends to be thin (Table 1). Although thick and thin are relative terms that do not precisely describe the thickness of the pleura, species differences are obvious (Figure 1). The variable thickness of the pulmonary pleura is due to the submesothelial layer containing the connective tissue components, blood vessels, and lymphatics. In mammals, pulmonary pleural thickness varies from 20 to 80 μm (Albertine et al., 1982; Mariassy and Wheeldon, 1983; Negrini and Moriondo, 2013). The pulmonary pleura consists predominantly of a monolayer of squamous mesothelial cells with microvilli resting on a thin basement membrane firmly attached to a dense connective tissue consisting of elastic and collagenous fibers (Figure 2). It is also important to note that the thickness of the pulmonary pleura in an individual lung varies with the anatomical location examined (Mariassy and Wheeldon, 1983). To date, there are no precise data available on the thickness of the pleura in specific regions of the lungs from most species.

The complexity of the blood supply to the pulmonary or visceral pleura of the human lungs has been reviewed by Staub et al. (1985), Bemaudin and Fleury (1985), and Sahn (1988). They concluded that the pulmonary pleura of human lungs is supplied by the bronchial artery, as described by Miller (1950), McLaughlin et al. (1961), and Naigaishi (1972), and not by the pulmonary artery, as described by von Hayek (1960). Because the pulmonary pleura of human lungs is thick, this conclusion supports the generalizations of McLaughlin et al. (1961, 1966) that thick pleura is supplied by branches of the bronchial artery and thin pleura is supplied by the pulmonary artery. Differences in the pressure of these two systems of vessels are likely to influence the rates of pleural fluid formation.

Although the lymphatic drainage of the human lung has been well described and reviewed (Bernaudin and Fleury, 1985), there have been only a few studies of other species (Leak and Jamuar, 1983). Animals with thick pulmonary pleura tend to have the most extensive network of lymphatics (Leak and Jamuar, 1983; Bernaudin and Fleury, 1985; Negrini and Moriondo, 2013). Of the animals with thin pulmonary pleura, the dog appears to have the most extensive pleural lymphatics (Naigaishi, 1972; Leak and Jamuar, 1983; Bernaudin and Fleury, 1985). Other animals with thin pulmonary pleura, such as the mouse, rat, and rabbit, have few pleural lymphatic vessels (Leak and Jamuar, 1983).

2.2 Segmental and Interlobular Septa

Several orders of connective tissue septa branch from the pleura and penetrate into the pulmonary parenchyma. Although all lungs have similar structural units, the visibility of bronchopulmonary segments and lobules either from the pleural surface or from cut surfaces depends upon the amount of connective tissue that separates these bronchopulmonary segments and lobules. Connective tissue between bronchopulmonary segments commonly represents only a moderate thickening of the interlobular connective tissue. In lungs from macaque monkeys, dogs, cats, and common laboratory species, the separation between either adjacent bronchopulmonary segments or lobules is not obvious either from the pleural surface or from cut surfaces. Lobules are easily seen from the pleural surface of human lungs and the lungs from cattle, sheep, swine, and horses, due to the larger amount of interlobular connective tissue in these species (Table 1). Although it appears that bronchopulmonary segments are also well separated by connective tissue in these species, the precise boundaries are often difficult to determine.

It is the completeness of interlobular septa, rather than the amount of interlobular connective tissue, that determines a number of functions, in particular the extent of collateral ventilation between adjacent lobules (Van Allen et al., 1931; Woolcock and Macklem, 1971; Robinson, 1982). Morphologically, completeness of connective tissue septa must be evaluated using three-dimensional approaches, such as dissections of individual lobules (von Hayek, 1960), examination of serial slices as used by McLaughlin et al. (1961, 1966), or by computed tomography (CT) as used by Webb et al. (1988) and Chen et al. (1992). Observations of serial gross slices using a dissecting light microscope revealed that interlobular connective tissue completely separated lobules in lungs from cattle, pigs, and sheep; however, in human lungs and lungs from horses, the separation of lobules was incomplete (Table 1). Interlobular septa can be seen in isolated normal human and animal lungs using high-resolution computed tomography (HRCT) (Webb et al., 1988; Chen et al., 1992). When lungs from pigs, dogs, rabbits, and sheep were examined using the same HRCT equipment and techniques, interlobular connective tissues were readily seen in the lungs from pigs but not in those from dogs, rabbits, or sheep (Chen et al., 1992).

The 1982 review of species differences in pulmonary anatomy and physiologic measurements of collateral ventilation by Robinson (1982) confirmed the morphologic observations of McLaughlin et al. (1961, 1966), with the exception of sheep, which have been shown by Kuriyama and Wagner (1981) to have collateral ventilation. Because the morphology of sheep lungs, as reported by McLaughlin et al. (1961), did not fit either the functional data or HRCT studies, we have examined the extent of connective tissue septa in sheep lungs. Lobules appear to be completely separated when sheep lungs are viewed either from the pleural surface or from cut surfaces of fixed sheep lungs (Figure 3). When blocks from gross slices of sheep lung are

TABLE 1 Comparative Lung Biology: Morphologic Features of Pleura, Interlobular and Segmental Septa, and Distal Airways

	Human	Macaque Monkey	Dog, Cat	Ferret	Mouse, Rat, Gerbil, Hamster, Guinea Pig, Rabbit	Horse, Sheep	Ox, Pig
Pleura	Thick	Thin	Thin	Thin	Thin	Thick	Thick
Interlobular and segmental connective tissue	Extensive, interlobular partially surrounds many lobules	Little	Little, if any	Little	Little, if any	Extensive,[a] interlobular partially surrounds many lobules	Extensive[a] interlobular surrounds many lobules completely
Nonrespiratory bronchiole commonly only one (nonalveolarized)	Several generations TB ends in respiratory bronchioles	Fewer generations, commonly only one TB ends in respiratory bronchioles	Fewer generations TB ends in respiratory bronchioles	Several generations TB ends in respiratory bronchioles	Several generations TB ends in alveolar ducts or very short respiratory bronchioles	Several generations TB ends in alveolar ducts or very short respiratory bronchioles	Several generations TB ends in alveolar ducts or very short respiratory bronchioles
Respiratory bronchiole (alveolarized)	1–3 generations	Several generations	Several generations	Several generations	Absent or a single short generation	Absent or a single short generation	Absent or a single short generation

TB, terminal nonrespiratory bronchiole.
[a]*The interlobular connective tissue of the sheep appears extensive and lobules appear completely separated in gross preparations, but not in light microscopy, scanning electron microscopy, or high-resolution computed tomography.*

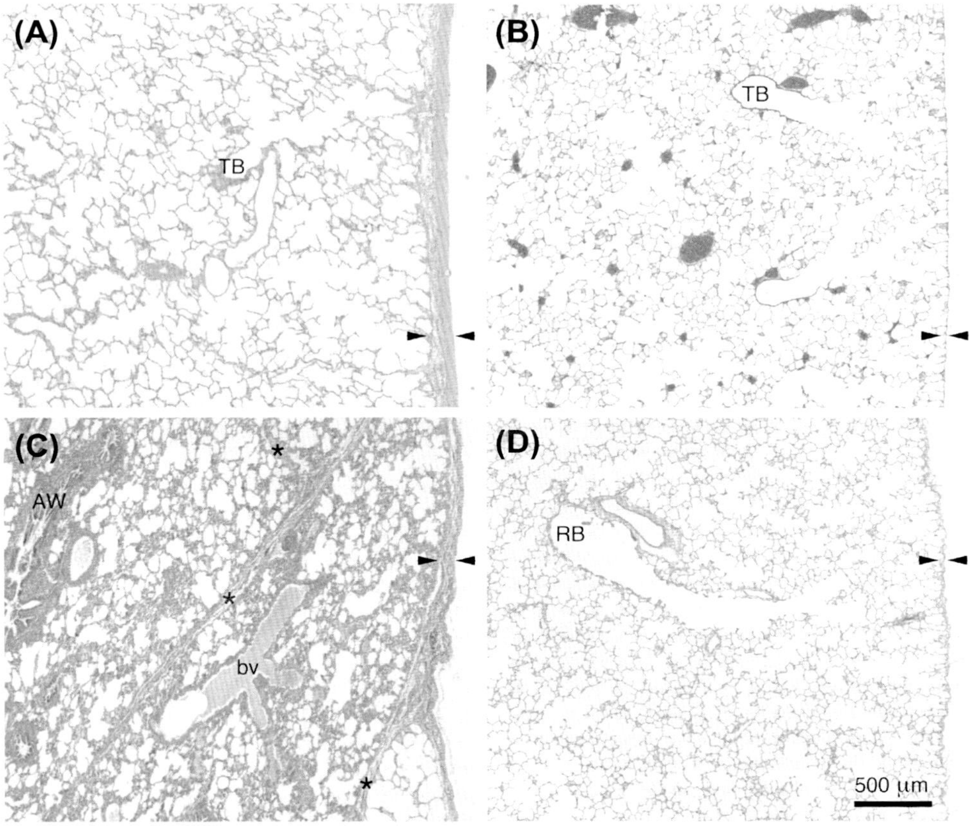

FIGURE 1 Pulmonary pleura and interlobular connective tissue from several species. Arrowheads designate the thick or thin pleura. Light micrographs of inflation-fixed, hematoxylin- and eosin-stained paraffin sections of lung from (A) horse, (B) rat, (C) pig, and (D) monkey. The horse and rat have terminal bronchioles (TB) and the monkey has respiratory bronchioles (RB) ending in alveolar ducts. A trichrome-stained, immersion-fixed paraffin section of a (C) pig lung shows extensive interlobular connective tissue (*). An airway (AW) and blood vessel (bv) are labeled.

embedded in paraffin and 6-μm sections are cut and stained in the usual manner, connective tissue septa are less common than would be expected from the gross appearance. Most of the few septa found are incomplete. Much of this connective tissue appears to be minor extensions of the perivascular tissues around bronchial arteries going to the pleura (Figure 3). These septa are distributed differently and are less complete than those seen in histological sections of pig lungs (Figure 1(C)). Complete interlobular septa are rare, but we did find a few. The spacing and interconnections of these complete septa suggest that they may be between pulmonary segments rather than between lobules. Interlobular septa appear incomplete in critical point-dried sheep lungs prepared for scanning electron microscopy (SEM). This is in contrast to observations of identically prepared lungs from cattle (Mariassy et al., 1975). Our gross observations of fixed sheep lungs are in accord with those made by McLaughlin et al. (1961), but those investigators did not follow their gross and subgross observations either with histological sections (due to the difficulty of cutting sections that contain latex in the blood vessels; O'Callaghan et al., 1987), with SEM (which was not commonly used to study the lung until the 1970s; Nowell and Tyler, 1971), or with HRCT (which was not available for such studies until the late 1980s; Webb et al., 1988; Chen et al., 1992). Thus, studies using additional techniques indicate that interlobular septa in sheep lungs are incomplete, rather than complete, as reported by McLaughlin et al. in 1961 (Table 1).

2.3 Distal Airways

The distal airways of the human lung consist of several generations of nonrespiratory bronchioles, the last of which is called the *terminal bronchiole*, followed by about one to three generations of respiratory bronchioles and about three to five generations of alveolar ducts (Miller, 1950; von Hayek, 1960; Weibel, 1963; Pinkerton et al., 2000). There have been many reports that the distal airways of animals differ from those of humans and from each other. The information, however, is at times inconsistent because of the differences in methods and definitions. The definitions used in this review are listed in Table 2.

Definitions are a particularly difficult problem in an area of special interest: the distinctions between respiratory bronchioles and alveolar ducts. Studies that use negative

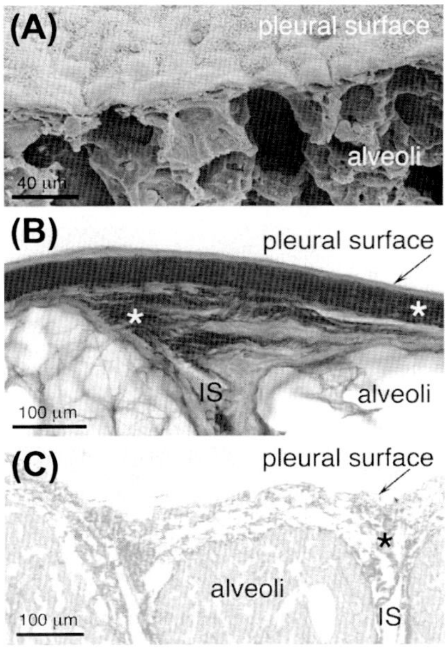

FIGURE 2 Pulmonary pleura, interlobular septa (IS), and underlying alveoli. The pleural surface is covered by a simple squamous epithelium with microvilli projecting from the surface as seen in the scanning electron micrograph (SEM) from a Sprague-Dawley rat in panel (A). Panel (B) is an elastin-stained paraffin section showing the thick, black elastic layer (white asterisk) in the pig pleura extending into the IS. Panel (C) is a Picro Sirius Red-stained paraffin section from a fetal lamb lung showing collagenous fibers (black asterisk) found in the pleura and IS.

casts or replicas of airways must rely upon the relative degree of "alveolarization" of these distal airways. When alveoli completely cover an airway, the airway is considered an alveolar duct, whereas when the airway has fewer alveoli, it is considered a respiratory bronchiole (Schreider and Raabe, 1981; Phalen and Oldham, 1983). Studies of subgross slices using dissecting microscopes commonly apply the same criterion, but from the luminal surface. In such preparations, additional criteria from Table 2 can also be applied, as the epithelial and subepithelial tissues can be seen and photographed (McLaughlin et al., 1961), but not at high magnification or resolution. Examination of thin slices by conventional transmission light microscopy (LM) has the advantage of better resolution of epithelium and other tissues, but it has the potential disadvantage of misinterpretation of poorly alveolarized respiratory bronchioles as nonrespiratory bronchioles if the alveoli are not in the plane of section. That potential problem becomes greater as the tissue block size and section thickness decrease to that required for transmission electron microscopy (TEM). Most TEM studies now use LM or SEM to determine the precise sites sampled for TEM examination. Although SEM is limited to the examination of surfaces, it is very useful for studies of the distal airways, as large blocks can be examined at very low or high magnification at high resolution, and the surface view provides much of the needed information in a readily comprehended form. Both cartilage and glands can be readily identified using SEM. Because airway epithelial cells have a typical surface appearance in each species, local differences in the epithelium caused by vessels on the abluminal airway surface can also be determined (Tyler and Plopper, 1985).

Sampling for routine paraffin embedding results in randomly oriented airway profiles. In contrast, airways can be longitudinally bisected by dissection of either wet (Plopper et al., 1983; Bastacky et al., 1983) or dried tissue (Tyler et al., 1985). Bisection of wet tissue has the advantage that both halves of each airway can be examined using either the same or other techniques. In addition, this dissection method allows for the identification of precise airway generations (Pinkerton et al., 1993; Peake et al.,

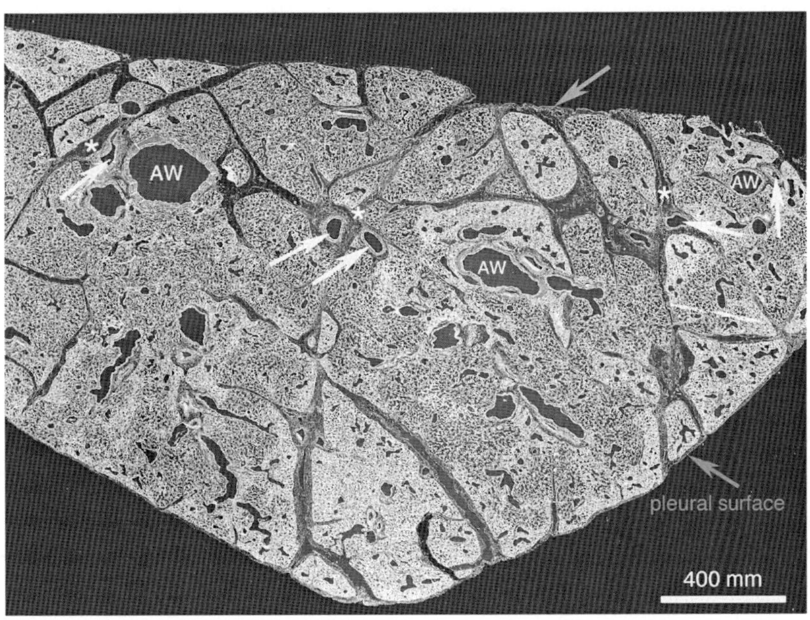

FIGURE 3 Paraffin section stained with hematoxylin and eosin from a fetal lamb lung 121 days of gestational age. Note the interlobular connective tissue septa (white asterisks) leading from the pleura (gray arrows) to the blood vessels (white arrows). A few airways (AWs) are labeled.

TABLE 2 Airway Definitions*

Bronchus
Complete walls formed by irregular plates or plaques of cartilage in loose connective tissue with elastic and collagenous fibers and smooth muscle. Mucous or mixed glands in the lamina propria or submucosa. Ciliated pseudostratified epithelium with mucous (goblet) cells. Cells as described by Jeffery (1983), Spicer et al. (1983), and Mariassy (in Chapter 5 of this volume).
Nonrespiratory Bronchiole
Walls of smooth muscle and loose connective tissue. No cartilage, no glands. Interalveolar septa attached to their abluminal surface. Ciliated columnar to squamous epithelium. Cells types as described by Plopper (1983) and in Chapter 8 of this volume. No mucous (goblet) cells. Terminal bronchioles are the most distal, highest generation, bronchioles which are not alveolarized. They are the last of the conducting airways.
Respiratory Bronchiole
Transition between conducting airways and respiratory air spaces. Structure similar to nonrespiratory bronchioles, except for the presence of openings in the walls for alveoli and shorter epithelium. Low generation respiratory bronchiole may be "poorly alveolarized" (i.e., have few alveoli). Higher generation respiratory bronchioles are typically "well alveolarized" (i.e., have numerous alveoli).
Alveolar Ducts
Alveoli open into the lumen all around the circumference. Spiral smooth muscle. Type I or II alveolar epithelium.

*Airways are defined and named by their characteristic structure. No single characteristic is diagnostic.

2000; Coppens et al., 2007). Tissues examined in the SEM remain suitable for further study using LM or TEM (Bastacky et al., 1983). Three-dimensional reconstructions of distal airways made from serial sections can provide both morphologic and morphometric information and appear to be close to the optimal approach (Mercer and Crapo, 1987, 1988, 1989 and Chapter 10 of this volume). Computer-assisted three-dimensional reconstructions are less time-consuming and provide more potential for morphometry than physical models. Computer programs for three-dimensional reconstruction are available for both small and large computers. Data input may be by manual digitization of images or by direct electronic digitization of images. The types of slices and the mode of imaging can vary from physical thick or thin sections, to optical sections using confocal microscopy, to X-rays using CT (Webb et al., 1988). Technological improvements will continue to increase the amount of information available using these methods and will also make them less time-consuming. Recently, micro-CT was used to quantitate volumetric and morphologic changes in soft tissue, respiratory tracts, and vascularization in fibrotic, emphysematous, and non-diseased human lung specimens (Kampschulte et al., 2013).

The comparative morphologic aspects of the distal airways from various species are presented in Table 1. Although the total number of airway generations was not counted in most of the studies from which Table 1 was developed, the number of generations between the last bronchus, as identified by the presence of glands or cartilage in its walls, and the first alveolarized bronchiole can be readily determined in subgross slices and SEMs. By these methods, several species, such as cats, dogs (McLaughlin et al., 1961), and macaque monkeys (Castleman et al., 1975), have fewer generations of nonrespiratory bronchioles than other animals or humans (Table 1). There are two patterns of structures that directly follow the last nonrespiratory or terminal bronchiole (Figure 4).

The terminal bronchiole is followed by one to three generations of respiratory bronchioles in humans (Weibel, 1963; Pinkerton et al., 2000) and macaque monkeys (Castleman et al., 1975), and several generations of respiratory bronchioles in dogs (Hyde et al., 1978), cats (Hyde et al., 1985), and ferrets (Hyde et al., 1979) (Table 1). The SEM in Figure 5(A) illustrates this type of terminal bronchiolar ending. In mice (Zitnik et al., 1978), rats (Schwartz et al., 1976), and gerbils respiratory bronchioles are absent and terminal bronchioles open directly into alveolar ducts (Figure 5(B)).

Respiratory bronchioles are poorly developed in hamsters (Kennedy et al., 1978), rabbits, and guinea pigs, which

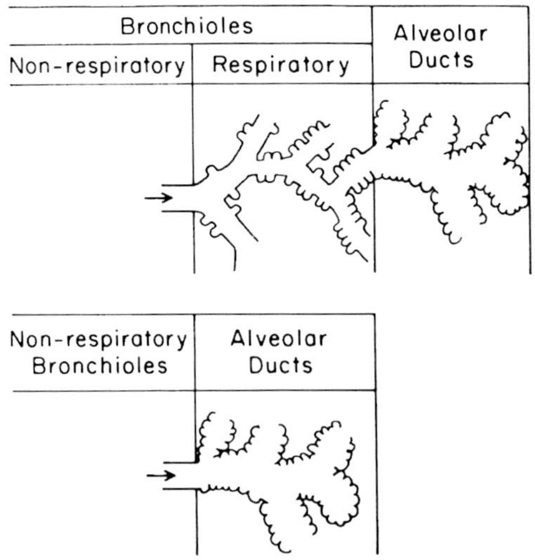

FIGURE 4 Diagram showing a terminal bronchiole followed by a respiratory bronchiole and terminating in alveolar ducts versus a terminal bronchiole that opens directly into alveolar ducts.

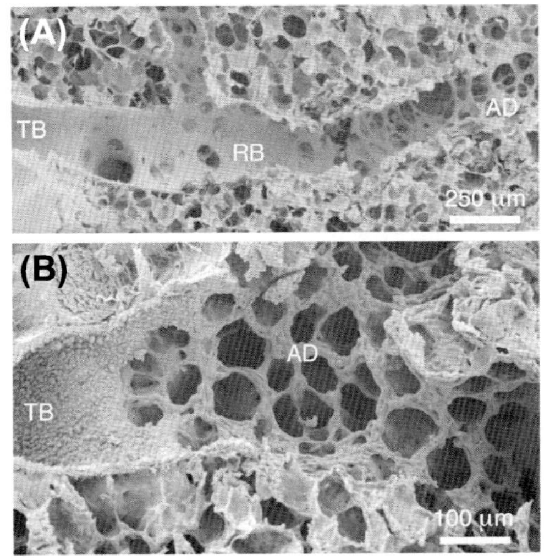

FIGURE 5 A scanning electron micrograph (SEM) from a dog lung (panel A) shows a terminal bronchiole (TB) that is followed by a respiratory bronchiole (RB) before opening into alveolar ducts (AD). The SEM from a rat lung (panel B) shows a terminal bronchiole that opens directly into alveolar ducts.

may have a single short generation or none. In horses (Tyler et al., 1971), ox (Mariassy et al., 1975), sheep, and pigs, respiratory bronchioles are also absent, or at best, only a single short generation is present.

2.4 Structural Units

Individual lobes of lungs are convenient morphologic reference points for descriptive studies and samples for morphologic studies and for in vitro physiologic and biochemical studies. The lungs from humans and domestic animals are divided into different lobes (Table 3). The human left lung is divided into two lobes, the superior and inferior, and the right lung is divided into three lobes by the addition of a middle lobe (International Anatomical Nomenclature Committee, 1983). Because the terms *cranial* and *caudal* are more appropriate terms of direction in quadrupeds than are *superior* and *inferior,* the two lobes of the left lungs from all of the common domestic animals are called cranial and caudal (International Committee Veterinary Gross Anatomical Nomenclature, 1983). In all common domestic animals except the horse, the left cranial lobe is further divided into cranial and caudal parts by an intralobar fissure. The right lungs from the domestic animals have four lobes, cranial, middle, caudal, and accessory, with the exception of the horse. In ruminants and pigs, the right cranial lobe is served by a tracheal bronchus. The right cranial lobe of ruminants is divided into cranial and caudal parts because of the further division of the tracheal bronchus. The accessory lobe is located between the heart and diaphragm, bounded laterally by the fold of the caudal vena cava and medially by the mediastinum. In horses, with the exception of the accessory lobe of the right lung, the division of lungs into lobes is very incomplete, but, based on distribution of bronchi, both lungs are described as having cranial and caudal lobes. Of the common domestic animals, the lobes of the lungs are most completely separated in dogs and least separated in horses. The right lungs

TABLE 3 Terminology of Lung Lobes

Humans	Animals	
NA	NAV	Old
Left Lung		
Superior	Cranial[a]	
	Cranial part	Apical
	Caudal part	Cardiac
Inferior	Caudal[a]	Diaphragmatic
Right Lung		
Superior	Cranial[b]	Apical
Middle	Middle	Cardiac
Inferior	Caudal	Diaphragmatic
	Accessory	Intermediate

Abbreviations: NA, *Nomina Anatomica* terms (1983); NAV, *Nomina Anatomica Veterinaria* terms (1983); Old, old terminology found in many references.
[a]*The left lungs of small laboratory mammals (mouse, rat, hamster, gerbil) are not divided, but those of larger ones (guinea pig, rabbit) are.*
[b]*The right cranial lobe of ruminates is divided into cranial and caudal parts.*

from laboratory mammals, including nonhuman primates (Lineback, 1965), are divided into the same four lobes as those of domestic animals. The left lungs from the small laboratory mammals (e.g., mice, rats, hamsters, gerbils) are not divided, but those from larger mammals (e.g., guinea pigs, rabbits) are divided.

Bronchopulmonary segments are the next smaller unit of lungs. Most often, bronchopulmonary segments are supplied by the first generation of bronchi distal to the lobar bronchi. They are roughly pyramidal in shape and have their bases at the pleural surface. Animals with extensive interlobular connective tissue tend to also have extensive connective tissue between segments. In those species, segments are convenient sample sites, as they can be surgically resected with minimal air leakage and hemorrhage. Segments are not discernable from the pleural surface in species that have little or no interlobular and intersegmental connective tissue. In all species, bronchopulmonary segments are named according to the bronchus that supplies them (International Committee on Veterinary Gross Anatomical Nomenclature, 1983) and can be identified by injection of colored material into that bronchus.

As previously stated, lobules are easily seen through the pleura in lungs from humans and from animals with extensive interlobular connective tissue, such as cattle and pigs. Lobules may also be readily dissected in human lungs (von Hayek, 1960). These lobules correspond to the secondary lobules according to Miller (1950). The *Nomina Histologica,* approved in 1980 by both the International Congress of Anatomists and the World Association of Veterinary Anatomists, follows the nomenclature of Miller (1950). In both sources, secondary lobules are defined by the connective tissue that surrounds them, rather than the airway that supplies them. *Lobule* is a useful, descriptive term in lungs with large amounts of interlobular connective tissue, such as cattle and pigs, even though within the same lung, lobules vary greatly in size and presumably contents. In a footnote, the *Nomina Histologica* states that a secondary lobule consists of a group of "about 50 primary lobules."

A secondary lobule contains several acini, each supplied by a terminal bronchiole (Nagaishi, 1972; Weibel, 1983). Because the acinus is a functional unit as well as a structural unit, the acinus is generally considered to include not only the terminal bronchiole and all of the respiratory bronchioles, alveolar ducts, and alveoli supplied by that terminal bronchiole, but also the vessels and nerves that supply that portion of parenchyma. The differences in distal airways described above result in large species differences in acinar size. Species with several generations of respiratory bronchioles will have much larger acini than those in lungs from species that have either no respiratory bronchioles or only a single short generation of respiratory bronchioles.

Both Miller (1950) and the *Nomina Histologica* define primary lobules as consisting of a respiratory bronchiole together with its associated alveolar ducts, alveolar sacs, and alveoli. This definition of a primary lobule corresponds to that for the "ventilatory unit" described by Storey and Staub (1962) and adopted by Mercer and Crapo (1988). Please refer to Chapter 10 for additional information. This unit is obviously appropriate and useful in animals with well-developed respiratory bronchioles, such as humans and the cats studied by Storey and Staub, but appears less useful in animals whose respiratory bronchioles are either poorly developed or absent (Table 1).

3. DISCUSSION

The biological significance of thick or thin pulmonary pleuras and the extensive interlobular connective tissue that accompanies thick pleura appears to relate to the completeness of the interlobular septa formed by that connective tissue. Thick pleura and the extensive interlobular connective tissue that forms complete interlobular septa, as seen in the ox and pig (Table 1), prevent collateral ventilation between adjacent lobules (Van Allen et al., 1931; Woolcock and Macklen, 1971). Dogs, which have thin pleuras and lack interlobular septa, have excellent collateral ventilation (Woolcock and Macklen, 1971), which occurs through respiratory bronchioles (Martin, 1966). Horse lungs, like human lungs (Table 1), have a thick pleura and extensive interlobular connective tissue, which forms incomplete interlobular septa. Horses have greater collateral flow resistance than do dogs (Robinson and Sorenson, 1978). The report of collateral ventilation in sheep (Kuriyama and Wagner, 1981) and the lack of complete interlobular septa reported by Chen et al. (1992) using HRCT, and as reported here by LM and SEM, eliminate the remaining exception (Robinson, 1982).

Most significant effects of the species variation in pleura and interlobular septa result from these differences in collateral ventilation (Robinson, 1982). After distal airway obstruction, animals without collateral ventilation tend to develop atelectasis, whereas those with good collateral ventilation do not (Robinson, 1982). Complete interlobular septa apparently also reduce the spread of infections between adjacent lobules, as domestic animals without collateral ventilation, such as the ox and pig, tend to develop bronchopneumonia with consolidated and healthy lobules frequently adjacent to each other in the same lobe, whereas in animals with excellent collateral ventilation, such as the dog, pneumonia is more diffuse within a lobe (Dungworth, 1985).

Robinson (1982) also reviewed the literature relative to species differences in vascular reactivity to hypoxia. He noted that some species with poor collateral ventilation had strong hypoxic vasoconstrictor response, and that some with good collateral ventilation had a weaker hypoxic vasoconstrictor response. This was confirmed by Elliott et al. (1991), who reported that the hyperoxic pulmonary vasoconstrictor response was strongest in pigs, weakest in dogs, and intermediate in horses. Robinson (1982) speculated that in matching ventilation and perfusion, some species rely primarily on hypoxic vasoconstriction and others on collateral ventilation. Additional species need to be studied to determine whether these observations are limited to the species studied or are generally true for animals with similar lung anatomy. Thus, comparatively simple species differences in morphologic aspects may be correlated with basic mechanisms of lung function.

Much of the biological significance of the species differences in distal airways relates to the resulting differences in the acinus, the functional unit of the lung, and the critical sensitivity of the junction of the conducting and respiratory portions of the lung to inhaled particles and gases. The acinus consists of the terminal bronchiole and all of the air spaces supplied by it (Nagaishi, 1972; Schreider and Raabe, 1981; Weibel, 1983). In animals without respiratory bronchioles, the terminal bronchiole ends by forming several alveolar ducts, resulting in a comparatively simple acinus that most likely has a relatively small volume. In humans and animals that have respiratory bronchioles, the acinus is more complex and probably larger, as it includes the respiratory bronchioles, in addition to all of the structures of several simple acini. These differences in acinar size are readily apparent in Weibel (1983, see Figure 16). The complex acini of species with respiratory bronchioles have airway cell types associated with respiratory bronchioles, in addition to the alveolar cell types that are common to both simple and complex acini. These species differences are of critical importance, as the proximal portion of the acinus is the most vulnerable part of the lung to some air pollutants (Schwartz et al., 1976; Hyde et al., 1978, 1985, 1988; Zitnik et al., 1978; Castleman et al., 1980; Boorman et al., 1980; Tyler et al., 1988) and to cigarette smoke in young humans (Niewoehner et al., 1974), and is the site for centriacinar emphysema. The larger volume of acini with respiratory bronchioles should result in a larger dose of inhaled toxicants to the sensitive cells of the centriacinar region in humans and in animals with respiratory bronchioles.

4. SUMMARY

Many animals are highly suitable for studies to determine the general reactions of cells or tissues to injury and disease. Using a two-species approach, animals possessing variations in structural tissue units, such as terminal and respiratory bronchioles, interlobular connective tissue, and pleural thickness, can be used for experimental studies that will later be extrapolated to humans. It is increasingly obvious that an understanding of biological processes at both the molecular and submolecular levels and at the whole-animal level will be needed to provide answers to many questions concerning diseases of humans.

REFERENCES

Albertine, K.H., Wiener-Kronish, J.P., Roos, P.J., Staub, N.C., 1982. Structure, blood supply, and lymphatic vessels of the sheep's visceral pleura. Am. J. Anat. 165, 277–294.

Bastacky, J., Hayes, T.L., von Schmidt, B., 1983. Lung structure as revealed by microdissection: positional morphology of human lung. Am. Rev. Respir. Dis. 128, S7.

Bernaudin, J.-F., Fleury, J., 1985. Anatomy of the blood and lymphatic circulation of the pleural serosa. In: Chretien, J., Bignon, J., Hirsch, A. (Eds.), The Pleura in Health and Disease. Marcel Dekker, New York (Chapter 6).

Boorman, G.A., Schwartz, L.W., Dungworth, D.L., 1980. Pulmonary effects of prolonged ozone insult in rats. Morphometric evaluation of the central acinus. Lab. Invest. 43, 108.

Castleman, W.L., Dungworth, D.L., Tyler, W.S., 1975. Intrapulmonary airway morphology in three species of monkeys: a correlated scanning and transmission electron microscopic study. Am. J. Anat. 142, 107.

Castleman, W.L., Dungworth, D.L., Schwartz, L.W., Tyler, W.S., 1980. Acute respiratory bronchiolitis. An ultrastructural and autoradiographic study of epithelial cell injury and renewal in rhesus monkeys exposed to ozone. Am. J. Pathol. 98, 811.

Chen, Q., Kelin, J.S., Gamsu, G., Webb, W.R., Lynch, D., 1992. High-resolution computed tomography of the mammalian lung. Am. J. Vet. Res. 53, 1218–1224.

Chretien, J., Bignon, J., Hirsch, A. (Eds.), 1985. The Pleura in Health and Disease. Marcel Dekker, New York.

Coppens, J.T., Van Winkle, L.S., Pinkerton, K.E., Plopper, C.G., 2007. Distribution of Clara cell secretory protein expression in the tracheobronchial airways of rhesus monkeys. Am. J. Physiol. Lung Cell. Mol. Physiol. 292 (5), L1155–L1162.

Dungworth, D.L., 1985. The respiratory system. In: Jubb, K.V.F., Kennedy, P.C., Palmer, N. (Eds.), Pathology of Domestic Animals, third ed., vol. 2. Academic Press, New York, pp. 413–556.

Dyce, K.M., Sack, W.O., Wensing, C.J.G., 1987. Textbook of Veterinary Anatomy, first ed. W.B. Saunders, Philadelphia.

Dyce, K.M., Sack, W.O., Wensing, C.J.G., 2009. Textbook of Veterinary Anatomy, fourth ed. W.B. Saunders, Philadelphia.

Elliott, A.R., Steffey, E.P., Jarvis, K., Marshall, B.E., 1991. Unilateral hypoxic pulmonary vasoconstriction in three species: the dog, pony, and miniature swine. Respir. Physiol. 85 (3), 355–369.

Hyde, D., Orthoefer, J., Dungworth, D., Tyler, W., Carter, R., Lum, H., 1978. Morphometric and morphologic evaluation of pulmonary lesions in beagle dogs chronically exposed to high ambient levels of air pollutants. Lab. Invest. 38, 455.

Hyde, D.M., Samuelson, D.A., Blakeney, W.H., Kosch, P.C., 1979. A correlative light microscopy, transmission and scanning electron microscopy study of the ferret lung. In: Johari, O., Becker, R.P. (Eds.), Scanning Electron Microscopy1197911H. SEM, Inc., AMF O'Hare, IL, pp. 891–898.

Hyde, D.M., Plopper, C.G., Weir, A.J., Murnane, R.D., Warren, D.L., Last, J.A., Pepelko, W.E., 1985. Peribronchiolar fibrosis in lungs of cats chronically exposed to diesel exhaust. Lab. Invest. 52 (2), 195.

Hyde, D.M., Plopper, C.G., Harkema, J.R., St George, J.A., Tyler, W.S., Dungworth, D.L., 1988. Ozone-induced structural changes in monkey respiratory system. In: Atmospheric Ozone Research and Its Policy Implications. Elsevier Science Publishers, B.V, pp. 523–532.

International Anatomical Nomenclature Committee, 1983. Nomina Anatomica, fifth ed. Williams & Wilkins, Baltimore/London.

International Anatomical Nomenclature Committee (Subcommittee on Histology), 1983. Nomina Histologica, 2nd ed. Included in Nomina Anatomica, fifth ed. Williams & Wilkins, Baltimore/London.

Jeffery, P.K., 1983. Morphologic features of airway surface epithelial cells and glands. Am. Rev. Respir. Dis. 128, S14.

Kampschulte, M., Schneider, C.R., Litzlbauer, H.D., Tscholl, D., Schneider, C., Zeiner, C., Krombach, G.A., Ritman, E.L., Bohle, R.M., Langheinrich, A.C., 2013. Quantitative 3D micro-CT imaging of human lung tissue. Fortschr Röntgenstr 185, 869–876.

Kennedy, A.R., Desrosiers, A., Terzaghi, M., Little, J.B., 1978. Morphometric and histological analysis of the lungs of syrian golden hamsters. J. Anat. 125, 527.

Kuriyama, T., Wagner Jr., W.W., 1981. Collateral ventilation may protect against high-altitude pulmonary hypertension. J. Appl. Physiol. Respir. Environ. Exerc. Physiol. 51, 1251.

Leak, L.V., Jamuar, M.P., 1983. Ultrastructure of pulmonary lymphatic vessels. Am. Rev. Respir. Dis. 128, S59.

Lineback, P., 1965. The respiratory, digestive and urinary systems. In: Hartman, C.G., Straus Jr., W.L. (Eds.), The Anatomy of the Rhesus Monkey. Hafner Publishing, New York, pp. 210–230.

Mariassy, A.T., Plopper, C.G., Dungworth, D.L., 1975. Characteristics of bovine lung as observed by scanning electron microscopy. Anat. Rec. 183, 13.

Mariassy, A.T., Wheeldon, E.B., 1983. The pleura: a combined light microscopic, scanning and transmission electron microscopic study in the sheep. I. Normal pleura. Exp. Lung Res. 4, 293.

Martin, H.B., 1966. Respiratory bronchioles as the pathway for collateral ventilation. J. Appl. Physiol. 21, 1443.

McLaughlin, R.F., Tyler, W.S., Canada, R.O., 1961. A study of the subgross pulmonary anatomy in various mammals. Am. J. Anat. 108, 149.

McLaughlin Jr., R.F., Tyler, W.S., Canada, R.O., 1966. Subgross pulmonary anatomy of the rabbit, rat, and guinea pig, with additional notes on the human lung. Am. Rev. Respir. Dis. 94, 380.

Mercer, R.R., Crapo, J.D., 1987. Three-dimensional reconstruction of the rat acinus. J. Appl. Physiol. 63, 785.

Mercer, R.R., Crapo, J.D., 1988. Structure of the gas exchange region of the lungs determined by three-dimensional reconstructions. In: Gardner, D.E., Crapo, J.D., Massaro, E.J. (Eds.), Toxicology of the Lung. Raven Press, New York, pp. 43–70.

Mercer, R.R., Crapo, J.D., 1989. Anatomical modeling of microdosimetry of inhaled particles and gases in the lung. In: Extrapolation of Dosimetric Relationships for Inhaled Particles and Gases. Academic Press, New York, pp. 69–78.

Miller, W.S., 1950. The Lung, second ed. Charles C. Thomas, Springfield, IL.

Naigaishi, C., 1972. Functional Anatomy and Histology of the Lung. University Park Press, Baltimore, pp. 79–179.

Negrini, D., Moriondo, A., 2013. Pleural function and lymphatics: review. Acta Physiol. 207, 244–259.

Niewoehner, D.E., Kleinennan, J., Rice, D.B., 1974. Pathologic changes in the peripheral airways of young cigarette smokers. N. Engl. J. Med. 291, 755.

Nowell, J.A., Tyler, W.S., 1971. Scanning electron microscopy of the surface morphology of mammalian lungs. Am. Rev. Respir. Dis. 103, 313.

O'Callaghan, M.W., Pascoe, J.R., Tyler, W.S., Mason, D.K., 1987. Exercise-induced pulmonary haemorrhage in the horse: results of a detailed clinical, post mortem and imaging study. V. Microscopic observations. Equine Vet. J. 19 (5), 411.

Peake, J.L., Reynolds, S.D., Stripp, B.R., Stephens, K.E., Pinkerton, K.E., 2000. Alteration of pulmonary neuroendocrine cells during epithelial repair of naphthalene-induced airway injury. Am. J. Pathol. 156 (1), 279–286.

Phalen, R.F., Yeh, H.C., Raabe, O.G., Velasquez, D.J., 1975. Selected References on Lung Airway Structure and the Deposition of Inhaled Particles (A Bibliography). Lovelace Foundation, Albuquerque, NM.

Phalen, R.F., Oldham, M.J., 1983. Tracheobronchial airway structure as revealed by casting techniques. Am. Rev. Respir. Dis. 128, SI.

Pinkerton, K.E., Gallen, J.T., Mercer, R.R., Wong, V.C., Plopper, C.G., Tarkington, B.K., 1993. Aerosolized fluorescent microspheres detected in the lung using confocal scanning laser microscopy. Microsc. Res. Technol. 26 (5), 437–443.

Pinkerton, K.E., Green, F.H., Saiki, C., Vallyathan, V., Plopper, C.G., Gopal, V., Hung, D., Bahne, E.B., Lin, S.S., Menache, M.G., Schenker, M.B., 2000. Distribution of particulate matter and tissue remodeling in the human lung. Environ. Health Perspect. 108 (11), 1063–1069.

Plopper, C.G., 1983. Comparative morphologic features of bronchiolar epithelial cells: the clara cell. Am. Rev. Respir. Dis. 128, S37.

Plopper, C.G., Mariassy, A.T., Lollini, L.O., 1983. Structure as revealed by airway dissection: a comparison of mammalian lungs. Am. Rev. Respir. Dis. 128, S4.

Rennard, S.I., Jaurand, M.-C., Bignon, J., Ferrans, V.J., Crystal, R.G., 1985. Connective tissue matrix of the pleura. In: Chretien, J., Bignon, J., Hirsch, A. (Eds.), The Pleura in Health and Disease. Marcel Dekker, New York (Chapter 4).

Robinson, N.E., 1982. Some functional consequences of species differences in lung anatomy. Adv. Vet. Sci. Comp. Med. 26, 1.

Robinson, N.E., Sorenson, P.R., 1978. Collateral flow resistance and time constants in dog and horse lungs. J. Appl. Physiol. Respir. Environ. Exerc. Physiol. 44, 63.

Sahn, S.A., 1988. The pleura. Am. Rev. Respir. Dis. 138, 184.

Schreider, J.P., Raabe, O.G., 1981. Structure of the human acinus. Am. J. Anat. 162, 221.

Schwartz, L.W., Dungworth, D.L., Mustafa, M.G., Tarkington, B.K., Tyler, W.S., 1976. Pulmonary responses of rats to ambient levels of ozone. Lab. Invest. 34, 656.

Spicer, S.S., Schulte, B.A., Thomopoulos, G.N., 1983. Histochemical properties of the respiratory tract epithelium in different species. Am. Rev. Resp. Dis. 128, S20.

Staub, N.C., Wiener-Kronish, J.P., Albertine, K.H., 1985. Transport through the pleura. In: Chretien, J., Bignon, J., Hirsch, A. (Eds.), The Pleura in Health and Disease. Marcel Dekker, New York (Chapter 10).

Storey, W.F., Staub, N.C., 1962. Ventilation of terminal air units. J. Appl. Physiol. 17 (3), 391.

Tyler, N.K., Plopper, C.G., 1985. Morphology of the distal conducting airways in rhesus monkey lungs. Anat. Rec. 211, 295.

Tyler, W.S., Gillespie, J.R., Nowell, J.A., 1971. Modern functional morphology of the equine lung. Equine Vet. J. 3, 84.

Tyler, W.S., Coalson, J.J., Stripp, B., Guest (Eds.), 1983. American Thoracic Society: comparative biology of the lung. Am. Rev. Respir. Dis. 128 (Suppl. 2).

Tyler, W.S., Dungworth, D.L., Plopper, C.G., Hyde, D.M., Tyler, N.K., 1985. Structural evaluation of the respiratory system. Fundam. Appl. Toxicol 5, 405.

Tyler, W.S., Tyler, N.K., Last, J.A., Gillespie, M.J., Barstow, T.J., 1988. Comparison of daily and seasonal exposures of young monkeys to ozone. Toxicology 50, 131.

Van Allen, C.M., Lindskog, G.E., Richter, H.G., 1931. Collateral respiration. Transfer of air collaterally between pulmonary lobules. J. Clin. Invest. 10, 559.

von Hayek, H., 1960. The Human Lung (V.E. Krahl, Trans.). Hafner Publishing, New York.

Wang, N.-S., 1985. Mesothelial cells in situ. In: Chretien, J., Bignon, J., Hirsch, A. (Eds.), The Pleura in Health and Disease. Marcel Dekker, New York (Chapter 2).

Webb, W.R., Stein, M.G., Finkbeiner, W.E., Im, J.-G., Lynch, D., Gamsu, G., 1988. Normal and diseased isolated lungs: high-resolution CT. Radiology 166, 81.

Weibel, E.R., 1963. Morphometry of the Human Lung. Academic Press, New York.

Weibel, E.R., 1983. Is the lung built reasonably? Am. Rev. Respir. Dis. 128, 752.

Woolcock, A.J., Macklem, P.T., 1971. Mechanical factors influencing collateral ventilation in human, dog, and pig lungs. J. Appl. Physiol. 30, 99.

Zitnik, L.A., Schwartz, L.W., McQuillen, N.K., Zee, Y.C., Osebold, J.W., 1978. Pulmonary changes induced by low-level ozone: morphological observations. J. Environ. Pathol. Toxicol. 1, 365.

Chapter 4

Architecture of the Tracheobronchial Tree

Kent E. Pinkerton, Laura S. Van Winkle, Charles G. Plopper, Suzette Smiley-Jewell, Elvira C. Covarrubias and John T. McBride

Chapter Outline

1. INTRODUCTION

The structure of the mammalian tracheobronchial tree fulfills a number of important and occasionally contradictory requirements. The geometry of the complex system of branching airways allows for efficient transport of gas to and from the lung periphery, while occupying a minimum of volume within the neck and thorax (Figure 1). Individual airways are adequately stiff to sustain the high airflows necessary for extremes of exertion but adequately compliant to narrow during coughing and to permit the removal of accumulated mucus and debris. In addition, bronchial architecture is a major determinant of the pattern of deposition of inhaled particles, a function with important implications for health and disease.

Parameters of airway geometry that determine the mechanical characteristics and function of the tracheobronchial tree include not only the diameter and length of the individual airway branches but also the overall branching pattern, the angles of branching, the orientation of the system to gravity, and airway stiffness. Airway stiffness depends on the structure of the airway wall, particularly the orientation and size of incomplete cartilage rings, cartilage plates, and attachments to surrounding tissue.

Attempts to quantitatively describe airway geometry have been undertaken for several purposes. The most frequent have been to understand how the structure of the airways determines lung function or how differences in airway geometry between species or between individuals of the same species influence function or susceptibility to disease. This issue has been of particular interest to inhalation toxicologists concerned with how differences in airway geometry between experimental animals and humans might influence aerosol deposition and the outcome of exposure studies. A second purpose has been to quantitate airway size or geometry in particular individuals so that the effects on airway structure of an experimental manipulation, such as exposure to a suspected toxin, can be investigated.

Although tracheal and central airway morphometry are relatively straightforward, morphometric analyses of the airways peripheral to the segmental bronchi are often confusing, primarily because of the difficulty of quantitatively describing a system that includes thousands of individual branches. Because collection of the data necessary to describe the airways of even a single individual is a formidable task, such data have been generated for no more than a few individuals of each of a few species. However, more recent approaches have taken advantage of state-of-the art imaging modalities to more efficiently view the three-dimensional nature of the bronchial airways, which will be discussed in this chapter, including corrosion casting, magnetic resonance imaging (MRI), and computed tomography (CT).

2. TRACHEA AND AIRWAYS

The geometry and stiffness of the trachea and the lobar and segmental bronchi are of interest because these airways contribute a major portion of the overall resistance to

Comparative Biology of the Normal Lung. http://dx.doi.org/10.1016/B978-0-12-404577-4.00004-7

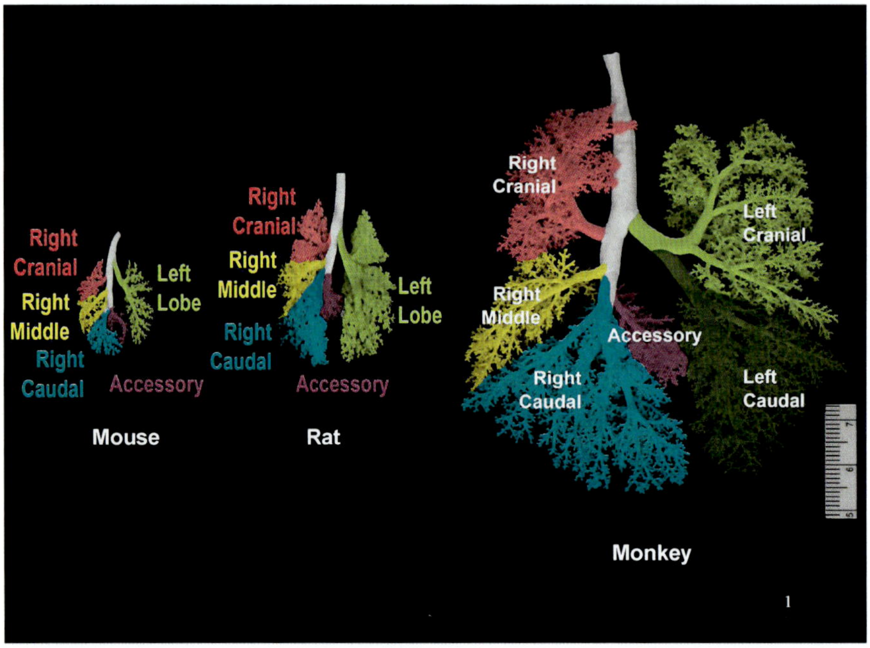

FIGURE 1 Comparison of mouse, rat, and monkey airways. For the mouse and the rat there are a total of five lobes, one on the left and four on the right. For the nonhuman primate lung there are a total of six lobes, two on the left and four on the right. A millimeter ruler is shown to identify the relative size of each lung airway corrosion cast.

airflow below the larynx (Pedley et al., 1970) and because maximal expiratory flow at most lung volumes in humans and many other species is limited by the development of critical flow conditions at one or more points within these airways as a consequence of dynamic airway narrowing (Dawson and Elliott, 1977) (Wilson et al., 1980). Both of these phenomena result from the fact that, in all species studied, the cross-sectional area of the airway system is smallest and flow velocities are greatest in the trachea and central airways.

2.1 Tracheal Dimensions

Tracheal length and diameter have been measured in a number of species. In general, tracheal dimensions can be predicted from body size, as illustrated by the linear relationship between the logarithm of tracheal dimensions and that of body mass (Figure 1). Leith has argued (Leith, 1982) that the fact that tracheal size scales with the 0.67 power of body weight suggests that the necessity to maintain maximal flows during extremes of exercise is a limiting constraint of tracheal design. The ratio of tracheal diameter to tracheal length is relatively constant among species, approximating 1:8. For particular species, unique constraints, such as an unusual shape or mobility of the neck and thorax, influence airway structure; for example, the giraffe's trachea is considerably longer than expected for body mass, but a little wider (Warren, 1974). Animals that dissipate body heat by panting may have relatively large central airways in order to minimize the impedance of rapid, shallow respirations. In general, however, the functional consequences of differences between species in central airway geometry have not been documented.

2.2 Central Airway Branching

Branching of the central airways is less regular than that of the intralobar airways, perhaps because the need to reach the various lobes while avoiding the heart and mediastinal structures outweighs other considerations. The overall pattern of central airway branching is relatively consistent across species, a major exception being the takeoff of a right upper lobe bronchus from the mid-trachea in a number of species, including pigs, sheep, and horses. Central airway branching has been described quantitatively in a variety of species, including the human, dog, rat, hamster, rabbit, and donkey (Nikiforov and Schlesinger, 1985; Raabe et al., 1976; Woldehiwot and Horsfield, 1978; Schlesinger and McFadden, 1981).

2.3 Tracheal Compliance

The stiffness of the trachea is largely determined in humans by 19–20 incomplete cartilage rings, and a similar pattern is found in most other mammals. Overall tracheal stiffness also depends, however, on tone in the muscular membrane that makes up the posterior tracheal wall, even in excised lungs (Hahn et al., 1980), longitudinal tension (Hyatt et al., 1981), and the relationship of the trachea and central airways to the surrounding mediastinal structures. Therefore, compliance measured in excised tissue may differ from that present in vivo, and tracheal compliance in vivo may itself vary considerably from time to time. The area or pressure/volume relationships for central airways of humans and dogs have been measured under a variety of circumstances (Croteau and Cook, 1961; Nakamura et al., 1981; Prakash and Hyatt, 1978; Martin and Proctor, 1958; Pardaens et al., 1972).

3. INTRALOBAR AIRWAYS

The intralobar conducting airways include those distal to the segmental bronchi and proximal to the gas-exchanging portion of the lung. In many species (rodents, cows, pigs, sheep, horses), airways completely lined by airway epithelium (conducting airways) abruptly branch into alveolar ducts (airways with no airway epithelium), which continue to ramify within in the acinus. In other species (human, monkey, ferret, dog, cat), conducting airways and alveolar ducts are separated by one or more generations of airways whose airway epithelium is interrupted by areas of gas-exchanging epithelium (respiratory bronchioles). Respiratory bronchioles are considered, by convention, to comprise part of the gas-exchanging lung parenchyma, and the conducting airways to end with the "terminal" bronchioles—the most peripheral airways along each path, which are completely lined by airway epithelium.

Chapter 7 discusses the respiratory bronchiole in great depth. Some data on the morphometry of the respiratory airways are available (Parker et al., 1971); this discussion will be limited to that of the conducting airways.

Morphometric analysis of the gas-exchanging lung parenchyma is usually based on stereological analysis of two-dimensional histologic sections. The theoretical basis of such techniques, however, depends on the assumption that the structures being measured are randomly oriented. The airways are highly ordered, so that although measurements of airway profiles on histologic sections have been useful for some purposes, the potential of this method is limited. Rigorous morphometric analysis of such a highly ordered system requires information about the position of measured airways within the system, information that usually depends on the consideration of a large sample, if not all, of the airways within a lobe or lung. Such measurements can theoretically be made from radiographic images (Yeh et al., 1975; Nadel et al., 1968), but most studies have been based on the analysis of corrosion casts, which, if carefully prepared, closely preserve the organization and dimensions of the airways (Horsfield and Cumming 1968; Phalen, 1978). Detailed measurements of a single airway cast have required up to 1 year for a skilled technician (Raabe et al., 1976). Limited measurements sufficient to quantitate airway dimensions in a single lobe, on the other hand, may be completed in as little as one day (McBride and Chuang, 1985). Differences in techniques of cast preparation undoubtedly explain differences in details of morphometric measurements between various studies, but the extent of such differences, and even the reproducibility of particular techniques, has seldom been studied (McBride and Chuang, 1985; Nikiforov and Schlesinger, 1985).

Although some primary measurements of airways are available (Raabe et al., 1976), most investigators have fit raw data into a scheme of airway geometry that allows important aspects of airway structure (such as the asymmetry of airway branching or the dimensions of airways at various levels) to be described with one or more calculated parameters. Some of these parameters are conceptually simple, such as the mean diameter of airways at a given number of bifurcations from the trachea. Others are less straightforward, such as the branching ratio of Horsfield and Cumming or the inverse power law index of West et al. (1986). Measurements from one or more individuals of a particular species fitted to such schemes have been used as models of airway geometry suitable for theoretical studies or comparisons between species. No particular scheme or model is correct or incorrect; each has potential advantages and disadvantages for describing or allowing comparisons of one or another aspect of airway architecture. Recent methods to visualize and analyze corrosion casts of rat airways with CT suggest we may be able to use nominal geometric parameters to eventually describe the architecture of the airways as a means of constructing more realistic airway models for particle transport, as well as normal and abnormal respiratory physiology (Lee et al., 2008a,b).

The following sections briefly summarize the major methods that have been used to analyze intralobar airway geometry, models of airway structure that have been developed using these methods, and an overview of several major aspects of airway morphometry.

3.1 Early Studies

A number of investigators prior to 1960 measured airway casts in order to analyze bronchial architecture (Ewart, 1889; Rohrer, 1915; Findeisen, 1935; Barnett, 1957; Rahn and Ross, 1957; Ross, 1957). These studies provided a basis of information that recent investigators have solidified and expanded.

3.2 Weibel

Weibel (1963) published the first comprehensive analysis of airway morphometry. He measured diameters and lengths of individual airway branches of a human lung cast and calculated the mean dimensions of airways in each "generation," defined as airways separated from the trachea by the same number of bifurcations. The trachea was considered generation 0, the two mainstem bronchi generation 1, and so on to the periphery. He estimated a mean of 13 (range 6–23) generations of airways between the trachea and the terminal bronchioles and 67,000 terminal bronchioles in the human lung, and used these data to develop a simple symmetrical model of human airway architecture (Table 1), which has been used frequently by subsequent investigators. The strength of this analysis and model is its simplicity—the lung is assumed to be a symmetrically bifurcating system, with all airways of a particular generation having the length and diameter, found to be the mean value for airways of that generation encountered in the lung cast. It has the important deficiency, however, of ignoring the asymmetry of airway branching; for instance, the mean cross-sectional area of airways in the sixth generation is 20 mm^2, but this generation includes large subsegmental airways in the lower lobe with areas of 85 mm^2 and peripheral bronchioles in the upper lobes with areas of 7 mm^2, clearly airways at different functional levels in the bronchial tree. Weibel recognized this problem and developed a mathematical description of the asymmetry of branching (Weibel, 1963), which, however, has not been used by subsequent investigators.

3.3 Horsfield and Cumming

Horsfield and Cumming applied to airway morphometry several variations of a mathematical system developed for the analysis of river systems and other natural branching trees (Horsfield, 1987). The main strength of this analysis is its ability to describe the asymmetry of branching and to identify airways at similar functional levels. Rather than classifying individual airways by their relationship to the trachea (generation), each airway is identified by its relationship to the periphery (order). A peripheral unit, such as the terminal bronchiole or the first airway in any pathway narrower than a given diameter is defined, and all airways in the cast meeting that criterion are considered as members of the first order. The analysis then moves centrally, with the parent airway at each bifurcation assigned an order based on the order of the two daughter airways. Although several systems of ordering have been used and described in detail (Horsfield, 1986), this discussion will deal primarily with "Horsfield" ordering, in which the parent airway is assigned an order one greater than that of the major daughter. With this technique, airways of the same order are more uniform in size than airways of the same

TABLE 1 Weibel Symmetrical Model of Human Airways[a]

Generation (Trachea = 1)	Number of Airways	Diameter (mm)	Length (mm)	S^b (cm^b)
1	1	18	120	2.54
2	2	12.2	47.6	2.33
3	4	8.3	19	2.13
4	8	5.6	7.6	2
5	16	4.5	12.7	2.48
6	32	3.5	10.7	3.11
7	64	2.8	9	3.96
8	128	2.3	7.6	5.1
9	256	1.86	6.4	6.95
10	512	1.54	5.4	9.56
11	1024	1.3	4.6	13.4
12	2048	1.09	3.9	19.6
13	4096	0.95	3.3	28.8
14	8192	0.82	2.7	44.5
15	16,384	0.74	2.3	69.4
16	32,768	0.66	2	113
17	65,536	0.6	1.65	180

[a]*Adapted from Weibel (1963).*
[b]*S = summed cross-sectional area of airways in each generation.*

generation using Weibel's method, so airway dimensions can be characterized with more precision.

Horsfield has used two methods to describe the degree of branching asymmetry. The first involves comparing the orders of the two daughter airways at each bifurcation. At symmetric bifurcations of an airway of order x, both daughters will be order $x - 1$. At asymmetric branchings, the major (larger) daughter will have order $x - 1$ and the minor daughter order $x - n$, where n is greater than 1 and increases with greater degrees of asymmetry. The difference between the orders of the two daughters $(n - 1)$ is called *delta*. The mean value of delta for all bifurcations of airways of a particular order describes the average asymmetry of branching at that level. A second method for describing the average asymmetry of branching is based on the number of branches of each order. If branching is symmetric, the number of airways of order $x - 1$ will be twice the number of order x airways. With increasing asymmetry, the number of order $x - 1$ airways will be progressively fewer. The ratio of order $x - 1$ to order x airways, called the *branching ratio*, is 2 for symmetric branching and approaches 1 for asymmetric branching. The branching ratio can be calculated for a particular order or estimated for the whole lung from a semilogarithmic plot of airway number as a function of order (Figure 2). Various values of the branching ratio correspond to particular values of delta; for instance, a branching ratio of 1.38 corresponds to a delta value of 3, a situation in which, on average, each order x airway divides into one airway of order $x - 1$ and one of order $x - 4$.

Horsfield models of airway geometry for the human and the dog (Tables 2 and 3) include mean values for airway dimensions and delta for each of 28 (human) and 43 (dog) airway orders between the trachea and terminal bronchioles. In both cases data are available for individual lobes, as well as for the lung as a whole. While these models allow calculations that take into account branching asymmetry, they also represent a simplification of reality by assuming that all bifurcations of airways of a particular order have the same degree of asymmetry. These models do not include data on branching angles.

3.4 Raabe, Phalen, Yeh et al.

In an effort to develop data suitable for calculations of aerosol deposition, Raabe et al. (1976) made extensive measurements of casts of individuals of four species (human, dog, rat, hamster). These data are unique in including measurements from more than one species obtained with a single technique, in being comprehensive (including organization, dimensions, orientation to gravity, and branching angle), and in being published in a form that makes primary measurements easily available to other investigators. Data for the rat and human have been used to develop models of airway architecture for theoretical deposition studies (Yeh et al., 1979; Yeh and Schum, 1980). These models describe completely for each species a single "typical" pathway from trachea to terminal bronchiole. To develop the rat lung model, the number of bifurcations along each of the 2500 pathways was counted, and the median number was

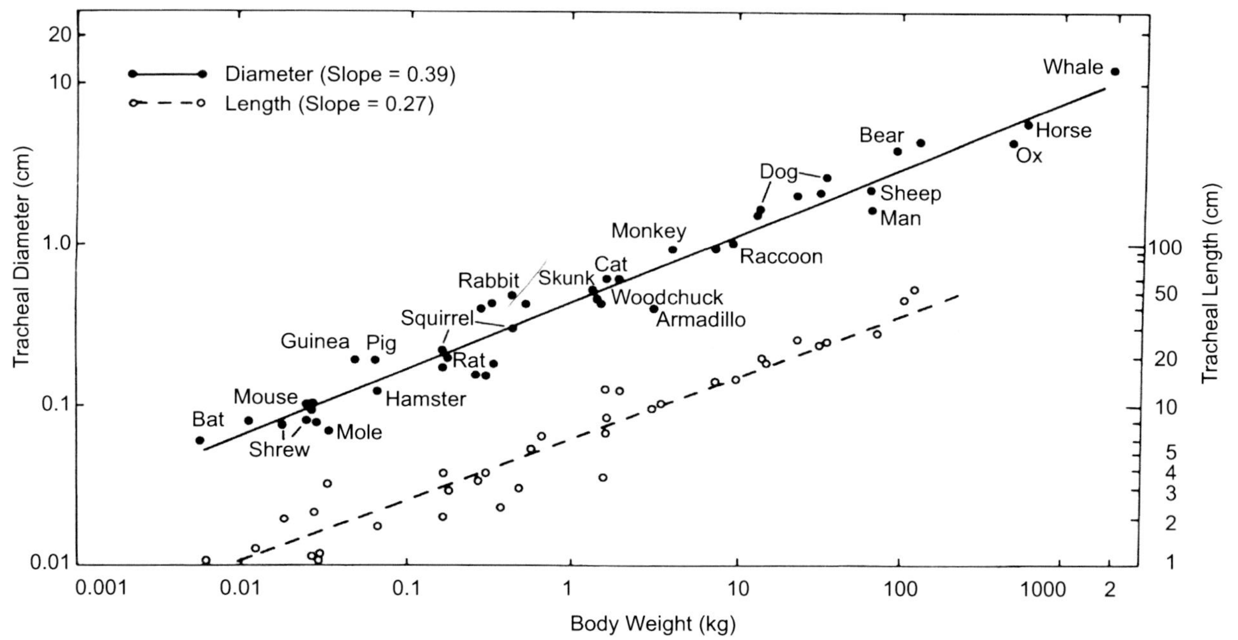

FIGURE 2 Relationship between body mass and the logarithms of tracheal diameter and length for a variety of species. *From Tenney and Bartlett (1967). Used by permission.*

TABLE 2 Horsfield/Fredberg Model of Human Airways[a]

Order "down" (Trachea = 1)	Horsfield order (TB = 1)	Delta	Diameter (mm)	Length (mm)
1	28	1	16	100
2	27	2	12	50
3	26	3	11	22
4	25	3	8	11
5	24	3	7.3	10.5
6	23	3	5.9	11.3
7	22	3	5.9	11.3
8	21	3	5.4	9.7
9	20	3	4.3	10.8
10	19	3	3.5	9.5
11	18	3	3.5	8.6
12	17	3	3.1	9.9
13	16	3	2.9	8
14	15	3	2.8	9.2
15	14	3	2.7	8.2
16	13	3	2.5	8.1
17	12	3	2.4	7.7
18	11	3	2.18	6.4
19	10	3	2	6.3
20	9	3	1.8	5.17
21	8	3	1.6	4.8
22	7	2	1.4	4.2
23	6	2	1.1	3.6
24	5	1	0.95	3.1
25	4	0	0.76	2.5
26	3	0	0.63	1.1
27	2	0	0.53	1.31
28	1	0	0.48	1.05

[a]*Adapted from Horsfield (1987) and Fredberg and Hoenig (1978).*

determined to be 16. All pathways within the lung with that number of bifurcations were then identified, and mean values were calculated for dimensions, branching angles, and the inclination to gravity of each generation of airway along this typical pathway (Table 4). Overall branching asymmetry was estimated by calculating a branching ratio

TABLE 3 Horsfield Model of Canine Airways[a]

Order "down" (Trachea = 1)	Horsfield order TB = 1)	Delta	Diameter (mm)	Length (mm)
1	43	2	21	200
2	42	2	21	7.5
3	41	2	20.3	18
4	40	10	11.5	10.1
5	39	10	10.8	9.61
6	38	10	10.1	9.11
7	37	10	9.39	8.63
8	36	10	8.76	8.18
9	35	10	8.18	7.75
10	34	10	7.64	7.34
11	33	10	7.14	6.96
12	32	10	6.66	6.6
13	31	10	6.22	6.25
14	30	10	5.81	5.92
15	29	10	5.43	5.61
16	28	10	5.07	5.32
17	27	10	4.73	5.04
18	26	10	4.42	4.78
19	25	10	4.13	4.53
20	24	10	3.85	4.29
21	23	10	3.6	4.07
22	22	10	3.36	3.85
23	21	4	3.14	3.65
24	20	4	2.93	3.46
25	19	4	2.74	3.28
26	18	4	2.55	3.11
27	17	4	2.38	2.94
28	16	4	2.23	2.79
29	15	4	2.08	2.64
30	14	4	1.94	2.51
31	13	4	1.81	2.37
32	12	4	1.69	2.25
33	11	4	1.58	2.13
34	10	4	1.48	2.02
35	9	4	1.39	1.91

Continued

TABLE 3 Horsfield Model of Canine Airways[a]—cont'd

Order "down" (Trachea = 1)	Horsfield order TB = 1)	Delta	Diameter (mm)	Length (mm)
36	8	4	1.24	1.81
37	7	4	1.1	1.72
38	6	4	0.983	1.63
39	5	4	0.876	1.54
40	4	4	0.781	1.46
41	3	4	0.697	1.39
42	2	0	0.621	1.31
43	1	0	0.554	1.25

[a]*Adapted from Horsfield et al. (1978).*

based on the total number of terminal bronchioles and the number of bifurcations along the typical pathway, with the assumption that asymmetry was similar at all levels in the system.

A similar process was used to develop the human lung model (Table 5). The human data suggested that airway branching was relatively symmetrical because the mean number of bifurcations was 16 and the total number of terminal bronchioles was 26,000; therefore this model is a symmetrical one, similar to that of Weibel. These models also include values for all pertinent aspects of airway geometry that are important in aerosol deposition calculations. On the other hand, the "typical pathway" model also represents a dramatic simplification of the great variability in airway architecture among the various pathways of the normal lung.

3.5 West, Goldberger et al.

West, Goldberger et al. (West et al., 1986) have used renormalization group theory, a field of mathematics related to fractal geometry, to develop a set of parameters

TABLE 4 Yeh Model of Airways of the Rat[a]

Generation (Trachea = 1)	Number of Airways	Diameter (mm)	Length (mm)	θ^b (°)	ψ^c (°)	S^d (cm^b)
1	1	3.4	26.8	0	86	0.091
2	2	2.9	7.15	15	90	0.132
3	3	2.63	4	43	86	0.163
4	5	2.03	1.76	36	71	0.162
5	8	1.63	2.08	32	59	0.17
6	14	1.34	1.17	22	58	0.197
7	23	1.23	1.14	16	61	0.273
8	38	1.12	1.3	17	58	0.374
9	65	0.95	0.99	20	55	0.461
10	109	0.87	0.91	15	58	0.648
11	184	0.78	0.96	16	61	0.879
12	309	0.7	0.73	17	56	1.189
13	521	0.58	0.75	17	58	1.377
14	877	0.49	0.6	22	58	1.654
15	1477	0.36	0.55	24	57	1.503
16	2487	0.2	0.35	44	58	0.781

[a]*Adapted from Yeh and Schum (1980).*
[b]*θ = branching angle between path of daughter and that of parent.*
[c]*ψ = orientation to gravity with 90° corresponding to horizontal.*
[d]*S = summed cross-sectional area of airways in each generation.*

TABLE 5 Yeh Model of Human Airways[a]

Generation (Trachea = 1)	Number of Airways	Diameter (mm)	Length (mm)	θ^b (°)	ψ^c (°)	S^d (cm^2)
1	1	20.1	100	0	0	3.17
2	2	15.6	43.6	33	20	3.82
3	4	11.3	17.8	34	31	4.01
4	8	8.27	9.65	22	43	4.3
5	16	6.51	9.95	20	39	5.33
6	32	5.74	10.1	18	39	8.28
7	64	4.35	8.9	19	40	9.51
8	128	3.73	9.62	22	36	13.99
9	256	3.22	8.67	28	39	20.85
10	512	2.57	6.67	22	45	26.56
11	1024	1.98	5.56	33	43	31.53
12	2048	1.56	4.46	34	45	39.14
13	4096	1.18	3.59	37	45	44.79
14	8192	0.92	2.75	39	60	54.46
15	16,384	0.73	2.12	39	60	68.57
16	32,768	0.6	1.68	51	60	92.65

[a]*Adapted from Yeh and Schum (1980).*
[b]*θ = branching angle between path of daughter and that of parent.*
[c]*ψ = orientation to gravity with 90° corresponding to horizontal.*
[d]*S = summed cross-sectional area of airways in each generation.*

for describing the relationship between airway generation and the mean dimensions of airways in each generation. As previously mentioned, the variance in diameters and lengths of airways in each generation resulting from the asymmetry of branching usually represents a major limitation to this type of analysis. Renormalization group theory, however, does not assume symmetric branching but rather assumes that with each bifurcation there is a given distribution of possible relationships between the diameters of parent and daughter airways. When the actual data of Weibel and of Raabe et al. are fitted to West's equation, parameters can be derived that describe the relationship between mean diameters or lengths and generation with remarkable precision (Figure 3). These parameters, which include the inverse power law index, or slope factor, and the period of modulation, may prove useful in cross-species or intraspecies comparisons but have not been used by other investigators to date. This analysis is particularly interesting, because it captures the early development of the airways when mechanisms that control airway branching presumably result in the recapitulation of a similar range of branching patterns at progressively peripheral levels during fetal lung development.

3.6 Stereological Measures of Bronchiolar Diameters

Several investigators have measured the diameters of airway profiles on histologic sections in order to document the size of peripheral airways in normal lungs and in response to experimental manipulations (Matsuba and Thurlbeck, 1971; Castleman, 1984; Castleman et al., 1988). Lung sections are examined, and airways that meet particular criteria, such as those with circular or elliptical profiles and those without cartilage within the airway wall, are identified and the smallest diameter is measured. The mean diameter of measured airways is then calculated and used to compare airway dimensions between experimental groups. Because the airways are not randomly oriented, it is impossible to be certain that differences identified with this technique reflect differences in airway diameters; differences in sampling and differences in the lengths of particular airways might also be responsible.

3.7 McBride and Chuang

While comprehensive methods of airway morphometry have described the overall scheme of airway branching and

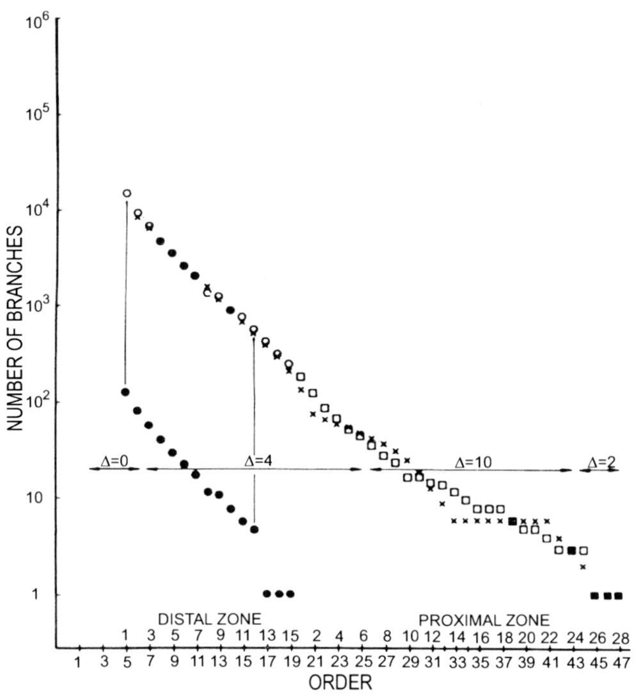

FIGURE 3 Relationship between the logarithm of the number of airways in each Horsfield order and airway order for the dog. The branching ratio, approximated by the slope of the regression line, is different for proximal and distal airway orders. Corresponding values for delta are indicated. Order 1 airways in this illustration correspond to respiratory airways several bifurcations distal to terminal bronchioles. Closed circles and open squares represent actual measurements. *From Horsfield et al. (1978). Used by permission.*

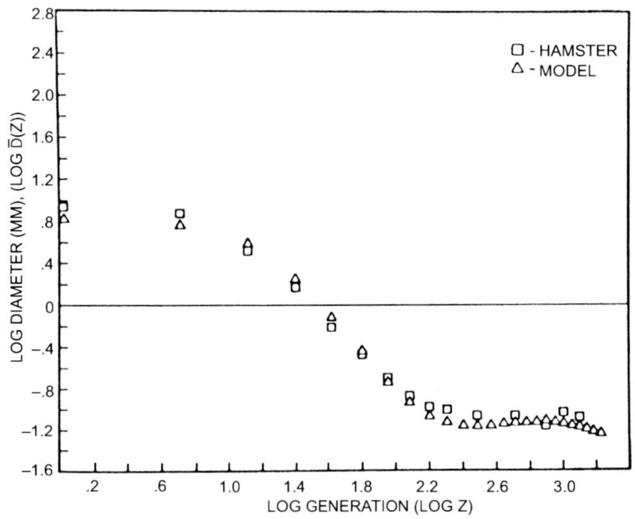

FIGURE 4 Logarithm of mean diameters for airways of each generation from the hamster data of Raabe et al., plotted as a function of the logarithm of generation (squares). When the data are fitted to the equation of West et al., parameters can be identified that give values (triangles) that agree closely with the primary measurements. *From West et al. (1986). Used by permission.*

dimensions, they have not provided an efficient method for comparing the dimensions of airways at given levels in the bronchial tree among individuals of the same species. For instance, although the mean dimensions of airways of a given Horsfield order within an individual can be calculated relatively precisely, the variation among individuals is too great to allow sensitive studies of difference in airway dimensions between experimental and control groups (McBride and Chuang, 1985). This variability results primarily from the inability to select precisely analogous terminal units in different animals. McBride and Chuang have described a technique for comparing the dimensions of airways at various levels among individuals that is not influenced by differences in the number of terminal units. This method is based on the relationship between the cross-sectional area of an airway and the number and size of terminal bronchioles subtended by that airway, a relationship that has been directly or indirectly recognized by many previous investigators (Ross, 1957; Hooper, 1977) and one that reflects the fact that airways of a given diameter tend to subtend a similar volume of lung or number of terminal units. In studies of casts from ferrets, the number and size of terminal bronchioles served by each left-lower lobe airway was documented by a relatively rapid technique. The ratio of the summed cross-sectional area of the terminal bronchioles served by a given airway and the summed cross-sectional areas of all terminal bronchioles in the lobe was considered to be an estimate of the fraction of the lobe subtended by that airway. The relationship between the actual cross-sectional area of 120 airways and the fractions of the lobe those airways were estimated to serve (Figure 4) describes a slightly curved line on log/log coordinates for airways from subsegmental bronchi that serve 10–15% of the lobe to distal airways that serve less than 0.5% of the lobe. This relationship can then be used to estimate the characteristic size of airways at three arbitrarily selected levels corresponding to subsegmental, intermediate, and peripheral airways. This method is relatively powerful because airways at identical levels in the various individual animals can be identified with precision, regardless of the number or size of terminal units selected for the analysis in various individual animals While this technique is not particularly useful in considering the overall geometry or function of the lung, it does allow practical studies of differences in airway dimensions among groups of individuals of the same species (McBride, 1985).

3.8 Magnetic Resonance Imaging and Computed Tomography

MRI allows for soft tissue differentiation of an entire organ. MRI has been successfully used to measure in situ airway length, diameter, and gravitational, bifurcation, and rotational angles for multiple airway generations in rats (Oakes et al., 2012). The measurements were found to agree with previous casting studies, such as by Raabe et al. (1976), except that major bifurcation angles were found larger and airway diameters were found smaller.

High-resolution computed tomography (HRCT) has been used to measure airway size and function. HRCT allows resolution of structures at hundreds of micrometers. It has been especially helpful in measuring airway wall thickness and airway lumen area in patients with asthma and chronic pulmonary obstructive disease (Nakano et al., 2002). A rapidly emerging field is optical coherence tomography, which can be used to study lung structure at a scale of tens of micrometers, including individual alveoli (McLaughlin et al., 2014). Similar to MRI and HRCT, it can be used in vivo, which is a boon to respiratory physiologists and physicians.

3.9 Specific Aspects of Intralobar Airway Morphometry

It is difficult to make firm generalizations about airway morphometry in any species or to compare airway structure among species because data are available for only a few individuals and knowledge about interindividual variation within species is relatively primitive.

Overall Number of Conducting Airways

In a bifurcating system, whether symmetric or asymmetric, the number of terminal units (terminal bronchioles) is directly related to the number of individual airways (or the number of bifurcations) within the system. Various investigators suggest a value of around 25,000 for the number of terminal bronchioles in the human lung (Table 6). The variability of these estimates is not surprising; if the level of airway identified as a terminal bronchiole varied because of biologic variation or technical considerations by just one generation of airways either way, the number counted would differ by a factor of two. There is only a rough correlation between body size and the number and average dimensions of terminal bronchioles reported for various species. This should not be overinterpreted, however, because the number of generations of respiratory bronchioles and alveolar ducts, and the number and size of alveoli, also may vary among species.

TABLE 6 **Number and Dimensions of Mammalian Terminal Bronchioles**

Species	Study	Number	Diameter (mm)	Length (mm)
Human	Weibel (1963)	6500	0.6	1.65
	Horsfield et al. (1971)	33,000	0.6	1.68
	Yeh and Schum (1980)	27,00	0.5–0.66	1.18–1.25
Dog	Horsfield and Cumming (1976)	18,000–36,000	0.2	0.5
	Ross (1957)	21,000	0.2–0.27	
Ferret	McBride (1985)	1700		
Rat		2500	0.2	0.35
Hamster	Raabe et al. (1976)	1200	0.2–0.3	0.4

Pattern of Intralobar Airway Branching

The basic pattern of airway branching in all species studied is dichotomous, presumably reflecting the dichotomous process of airway branching during morphogenesis. Rarely, an airway appears to branch into three daughters, but for purposes of analysis this can be considered to represent sequential bifurcations separated by an airway of zero length. The airways of all species branch asymmetrically. The degree of asymmetry, however, appears to differ among species, and to a lesser extent among lobes in individuals and among airways at different levels within the same lobe. The human lung has the least asymmetry of branching among species studied to date. In contrast, many mammals have dramatic asymmetry with airway systems in which major daughters make up longitudinal pathways that comprise nearly straight trunks giving off smaller minor daughters at relatively large angles. Each side branch itself gives rise to a relatively straight longitudinal pathway of major daughters giving off smaller side branches, and so on to the periphery (Figure 5). This pattern resembles the monopodial branching of pine trees where sprouts appear laterally along preexisting branches. The basis of asymmetric branching is unclear, but the shape of the lung or lobe may contribute; relatively spherical lobes, such as the lower lobes of the human, are relatively symmetric, while long, narrow lobes in small mammals are relatively asymmetric.

The asymmetry of airway branching can be assessed by the number of airways in successive orders (the branching ratio), by the relationship between the order of two daughter airways at a bifurcation (delta), or by the distribution of the number of bifurcations between the trachea

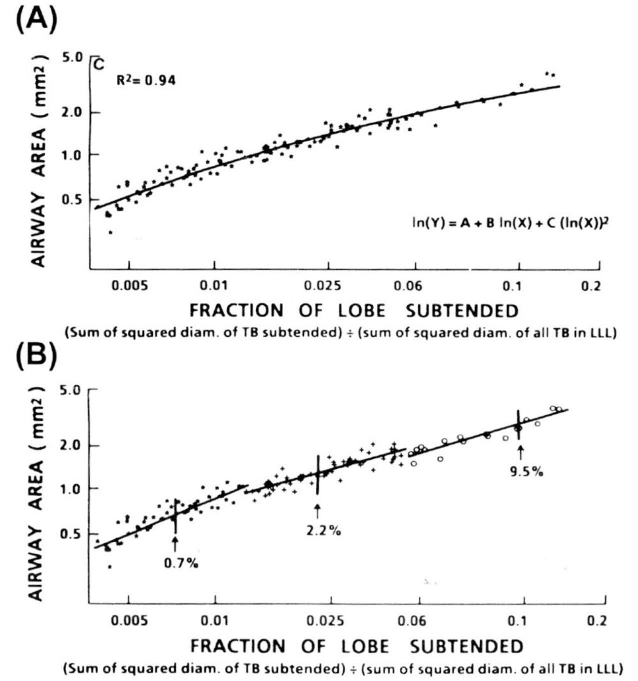

FIGURE 5 (A) Measured cross-sectional areas of 120 airways from a ferret cast are plotted as a function of the lobe each airway was estimated to subtend, based on the size and number of its terminal bronchioles. The data describe a quadratic relationship on logarithmic axes. (B) The same data are divided into three grouped airways (peripheral, intermediate, and subsegmental), and the characteristic airway cross-sectional area at three arbitrary levels determined from regressions are calculated in the same way for each animal in control and experimental groups to identify changes in airway dimensions related to a particular manipulation. *From McBride and Chuang (1985). Used by permission.*

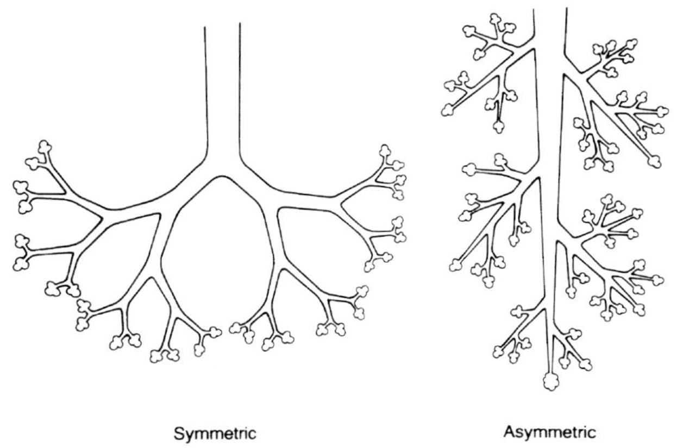

FIGURE 6 Schematic representation of a relatively symmetrically branching airway system and one that branches asymmetrically. The asymmetrical system resembles a pine tree, with nearly straight major pathways giving off progressively smaller side branches at irregular intervals.

and respiratory bronchioles down the various paths within a lung (Yeh et al., 1979). With symmetric branching there is the same number of bifurcations along all pathways, and with greater degrees of asymmetry, the distribution widens. A commonly used method for assessing branching asymmetry has been the comparison of the diameters or lengths of the two daughter airways at a bifurcation. With this technique, the ratio of the diameter or length of the minor daughter to that of the major daughter at bifurcations of parent airways of the same generation or of similar sizes can be used to estimate differences in asymmetry at various levels in the bronchial tree without the need to analyze the entire airway system (Figure 6).

There is considerable variation in overall branching patterns among species. Horsfield et al. (Horsfield, 1986) have calculated an overall branching ratio of approximately 1.38 for humans and 1.33 for dogs, corresponding to models in which an airway of order x divides into one daughter of order $x + 1$ and one daughter of order $x + 3$ (in the human) or $x + 4$ (in the dog). This greater asymmetry in the dog is reflected also in the fact that while the dog and human have a similar number of terminal bronchioles (and therefore the same number of individual airways within the bronchial tree), the dog has 43 Horsfield orders of airways, while the human has only 28. The analysis of Phalen et al. (Figure 6) also suggests that the human has relatively less asymmetry than the dog, rat, or hamster. These data do not justify a conclusion that larger lungs are more symmetric; even larger animals, such as the donkey (Schlesinger and McFadden, 1981), have relatively asymmetric airways. Lung shape, rather than size, may be an important determinant of branching asymmetry (Figures 7 and 8).

The asymmetry of branching appears to vary with the level within the lung (Figure 9). Many models of intralobar airway geometry have assumed that the pattern of branching at successive generations or orders is uniform; that is, that the pattern of branching among all airways of a particular order or generation repeats the pattern of the previous order or generation (but on a smaller scale). This assumption holds relatively well as a first approximation but not on closer inspection. In general, the branching of the central airways is markedly asymmetric; in several species, asymmetry has been shown to decrease at more peripheral levels. Branching in the human is asymmetric through most of the intralobar airways with a delta value of 3 but becomes nearly symmetric in the last two orders before the terminal bronchioles. In the dog (Figure 2) asymmetry is more pronounced centrally; the value for delta ranges from 10 in the major intralobar airways, to 4 in the intermediate airways, to 0 (symmetrical branching) in the last two orders of conducting airways. Ratios of daughter airway diameters for parent airways of various size classes in four species were calculated and, in general, branching becomes more symmetrical in more peripheral airways (Figure 6).

The degree of branching asymmetry also appears to vary between lobes. In the human (Horsfield et al., 1971)

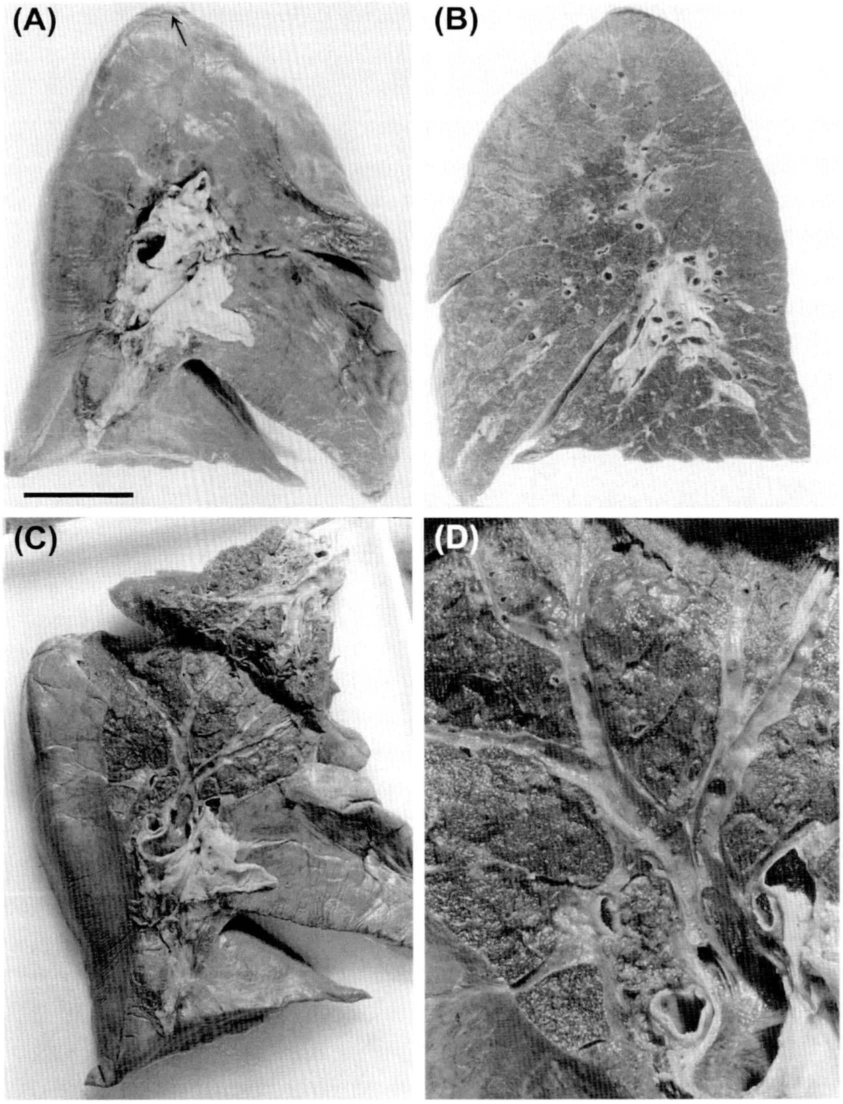

FIGURE 7 Medial surface and shape of the left human lung (A) and the cut, sagittal surface of the same lung (B). Symmetrically distinct airway paths beginning at the left mainstem bronchus are followed to the apico-posterior and apico-anterior regions of the left upper lobe (C). A magnified view of the microdissected airways (D). Scale bar in (A) equals 5 cm.

the value of delta for individual segments ranged from 2 to 4. In the rat (Yeh et al., 1979), the branching ratio ranged from 1.56 to 1.72 in various lobes, leading Yeh to suggest that such differences in airway branching might explain intralobar differences in aerosol deposition that have been reported in this species.

Branching Angles

The angle formed between a daughter airway and the path of its parent airway varies with the degree of asymmetry of the bifurcation and the level of the bifurcation within the lung. The fact that the minor daughter deviates from the path of the parent to a greater extent than the major daughter has been recognized for many years (Ewart, 1889; Barnett, 1957). When this is most apparent, the airway system resembles a "pine tree" (Figure 5)—the major daughter deviates little if at all from the parent, while the minor daughter leaves nearly perpendicularly. The relationship between the daughter airway size and the branching angle has been further confirmed by analyses of Thurlbeck and Horsfield and of Raabe et al. (Figures 6–8), as well as more recent analysis by Lee et al. (2008a). Mean branching angles have been said to increase toward the periphery of the lung, but this tendency is not consistent, as can be seen in the rat lung model of Yeh et al. (Table 4) and the work of Lee et al. (2008b). Thurlbeck and Horsfield (1980) could find no systematic relationship between order and branching angle in data for the rat, hamster, and human.

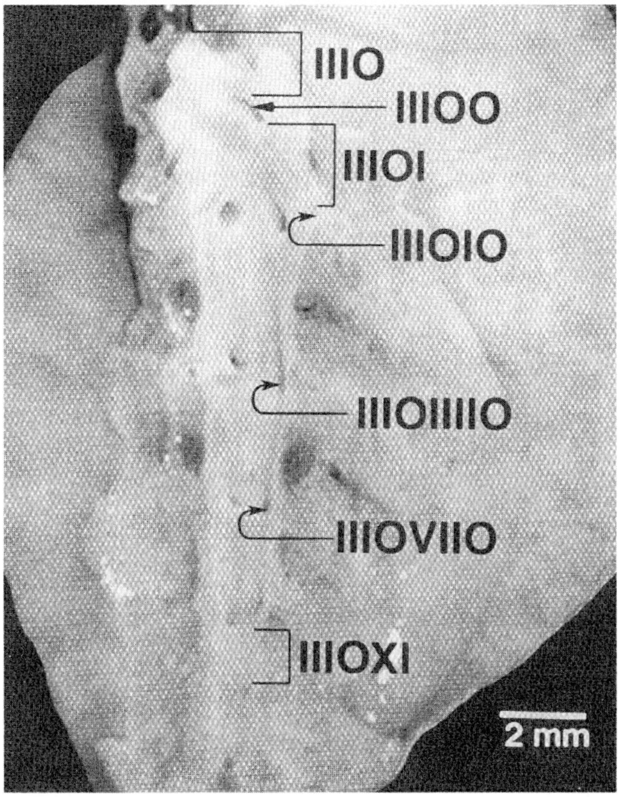

FIGURE 8 Microdissected asymmetric axial pathway of the infracardiac lung lobe of a 50-day-old rat. The branching pattern of the axial pathway is designated by the binary classification system. Selected major daughter airway segments are indicated by brackets and minor daughter airways branches by arrows. Accordingly, the first major daughter airway segment of the right infracardiac lobe is designated as IIOI and the first minor daughter as IIIOO. Scale bar, 2 mm.

Airway Dimensions

With successive branching to the periphery, the airways become narrower and shorter along each longitudinal pathway. Recently reviewed arguments show that optimal design for the bronchial tree would involve a geometry minimizing one or more factors, such as power expenditure, dead space volume, resistance, or entropy production, and that these considerations would be satisfied by a consistent decrease in the dimensions of successive generations or orders approximately in proportion to the cubed root of the branching ratio. Therefore, if the branching ratio is 2, as in the symmetric Weibel model, the diameter or length of airways in a particular generation or order would be smaller than those of parent airways by the cubed root of 2 (or 1.26). Using Horsfield orders, the factor would be the cubed root of 1.3−1.4 for the human or dog. The human data of Weibel and the human and dog data of Horsfield approximate these predictions. When this is the case, the summed cross-sectional area of airways in successively peripheral orders or generations increases dramatically toward the periphery, an observation also supported by human and dog measurements. This prediction may be less valid in the rat, however, where Yeh (Table 4) found that there was not a consistent increase in airway summed cross-sectional area with successively peripheral generations. The extent of the decrease in the mean size of airways in each generation or order of airways is reflected in the various airway models illustrated previously. The comparability of the relationships between airway diameter and order across species is illustrated in Figure 8.

There is only a weak correlation between body size and the diameters of terminal bronchioles (Table 6), but this may be explained by the differences in the number and complexity of respiratory airways and alveolar ducts.

Attempts to quantitate airway length have been less successful than attempts to quantitate airway diameter. In general, the relationship between level within the lobe and mean airway length is similar to that between level and diameter, but the variability of length measurements is greater. This reflects the fact that although the diameter of individual airways is not influenced by local branching patterns, but is tightly coupled to the relative flow through the airway, the length of individual airways, as usually defined, is determined primarily by the distance between successive bifurcations. This length often approaches zero, particularly in relatively asymmetric systems. Other methods of characterizing the longitudinal dimension of airways have included the analysis of path length, or the distance between the trachea or bronchus and the terminal bronchiole (Ross, 1957; Horsfield and Cumming, 1968), and measurements of the distance between airways serving particular fractions of a lobe and the terminal bronchioles subtending the major pathway to the periphery.

3.10 Compliance of Intralobar Airways

The compliance of individual intrapulmonary airways is important in considerations of expiratory flow limitation but has been studied extensively only in the dog and in humans (Hughes et al., 1972; Martin and Proctor, 1958; Nakamura et al., 1981). Isolated peripheral airways that have less intrinsic airway wall support are more compliant than central airways but derive support in situ from attachments to the surrounding parenchyma. Peripheral airways are, however, more compliant than central airways, particularly at low lung volumes.

3.11 Overall Concepts of Intralobar Airway Morphometry

Despite the complexity of intralobar airway morphometry, several basic concepts can be defended without confusion: (1) All mammalian airway systems studied to date can be

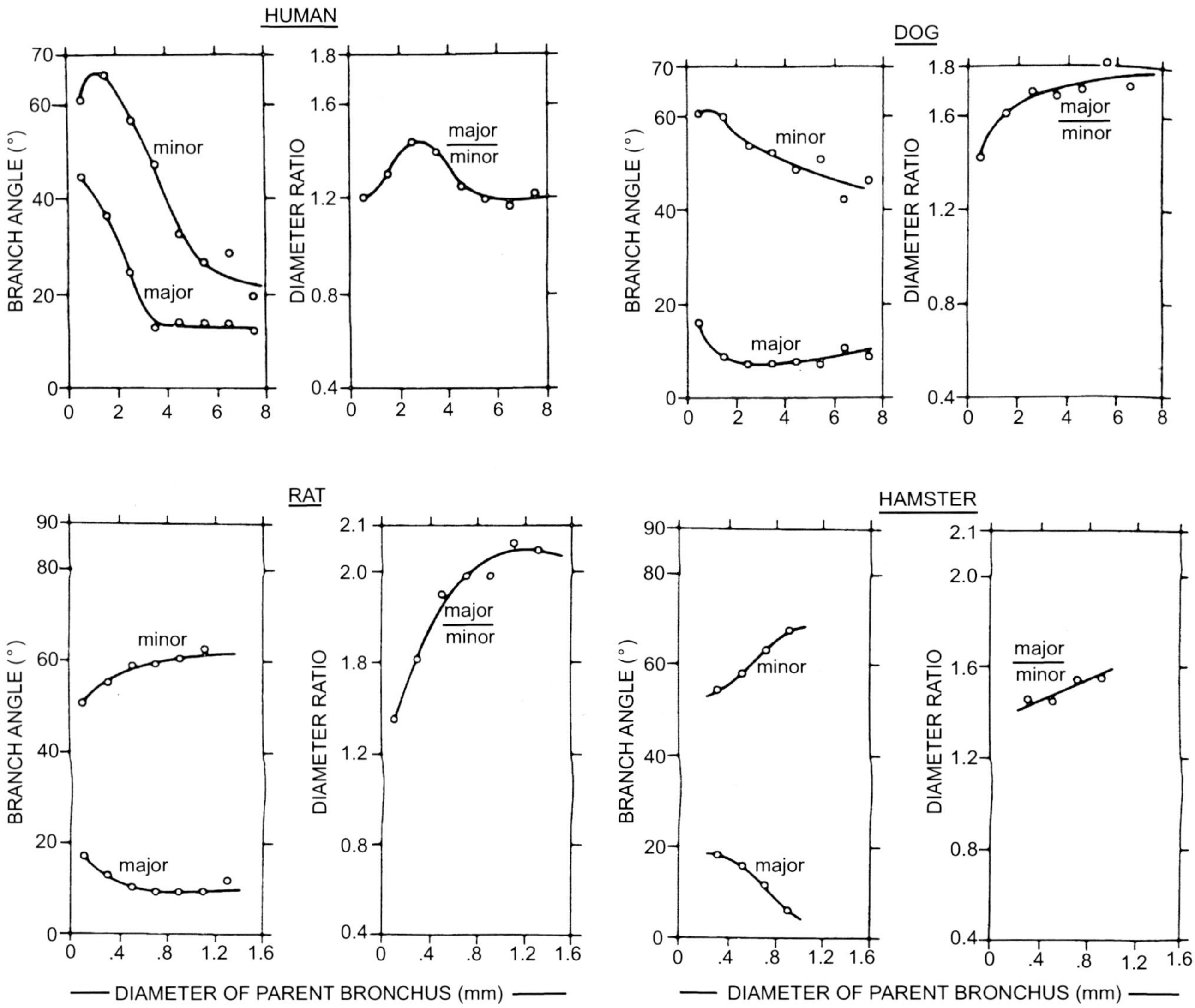

FIGURE 9 The asymmetry of branching of parent airways of different sizes (larger airways correspond to central and smaller to relatively peripheral levels) for data from four different species. Asymmetry is measured by (1) the ratio of the diameter of the larger daughter to that of the smaller daughter and (2) the branching angles of each daughter form the pathway of the parent. Asymmetry tends to be greater in central airways for all species except the human. *From Phalen et al. (1978). Used with permission.*

assumed to branch dichotomously. (2) The majority of bifurcations in all species are asymmetric, but the degree of asymmetry varies among species and between central and peripheral location within each species. (3) The angle of branching varies as a function of the relationship between daughter airway dimension at each bifurcation. (4) The summed cross-sectional area of airways at successively peripheral levels generally increases, although the extent of this increase may depend on species. (5) The overall geometry of the mammalian bronchial tree seems to allow efficient gas transport, to require a minimum of tissue and space within the thorax, and to allow for the maximal flow necessary for extremes of exertion. (6) While differences in airway geometry among species clearly exist, the functional implications of such differences have, at present, not been clearly established.

4. NOVEL APPROACHES FOR THE STUDY OF THE TRACHEOBRONCHIAL TREE

As described throughout this chapter, the tracheobronchial tree represents a simple, yet highly complex, structure for examination and understanding. Of the more than 40 different cell types identified within the respiratory system, at least 12 are found in the epithelial population that line the tracheobronchial tree (Breeze and Wheeldon, 1977; Jeffrey, 1983; Plopper et al., 1983; Plopper et al., 1987;

Widdicombe and Pack, 1982). This section provides a brief introduction of methods to study cells and tissues of airway generation–specific regions within different locations of the bronchial tree for a wide variety of species. These approaches typically are dependent upon the use of inflation-fixed lung for airway microdissection or agarose-infused lungs for isolating airways to perform biochemical, metabolic, or mRNA analyses.

4.1 Airway Microdissection

Microdissection of airways can be performed on any lung tissues that have been adequately inflation-fixed. Figure 7 shows a left human lung in which the steps of airway microdissection have been illustrated. The lung or lobe is cannulated and fixed by inflation with glutaraldehyde-paraformaldehyde (1–10%) at 30 cm H_2O pressure. After 4 h of fixation under pressure, the lung volume can be measured by fluid displacement. Any axial pathway (from the mainstem or segmental bronchus to terminal and/or respiratory bronchioles) and its distal side branches can be exposed by microdissection into a costal and a mediastinal half (Pinkerton et al., 1986; Pinkerton et al., 2000; Avadhanam et al., 1997). The number of branches is determined by a direct count of all airway branches beginning at the lobar bronchus and continuing along the axial path to the first proximal respiratory bronchiole (Figures 7 and 12). A binary system can also be used for unique identification of asymmetric airway branching using "1" to designate a major daughter airway and "0" for a minor daughter airway (Figure 8). The precise location of any airway in the bronchial tree can be identified by the airway branching history designated by the sequence of "1" and "0." Airway generation–specific tissues can be processed for embedment in paraffin, glycomethacrylate, or plastic resin (Spur's or epon) for further microscopic evaluation. In all cases of airway microdissection, axial pathways, and distal side branches (daughter airways) can be studied with the unaided eye, a dissecting microscope, or with a confocal laser scanning microscope, especially if fluorescent staining of tissues is employed (Tran et al., 2004) using long working distance water-immersion objectives (Figures 10 and 11).

4.2 Isolated, Fixed Whole-Mount Airway Preparations

Whole-mount airway preparations isolated from the lungs of mice are ideal due to the lack of cartilage within airway walls, although other species such as rats, hamsters, and monkeys can also be prepared using this method. Typically, the right middle lobe is used for airway isolation, due to its single dominant, fairly straight axial pathway. To isolate this entire axial airway path

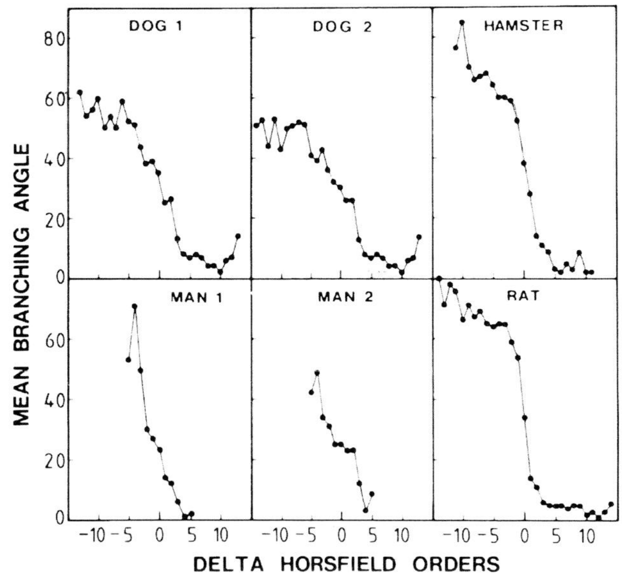

FIGURE 10 Branching angles for airways from four species are plotted as a function of the asymmetry of branching estimated by values of delta (see text). Smaller daughters have negative delta values, which become more negative for increasing degrees of asymmetry, and larger daughters have positive values of delta, which increase with increasing asymmetry. Branching angles in all species are greater for smaller daughters than for larger daughters, a finding that is more marked with greater degrees of asymmetry. *From Thurlbeck and Horsfield (1980). Used with permission.*

from the surrounding parenchyma, the lobe is rapidly infused with 95% ethanol:glacial acetic acid (99:1) to visually bring the lung lobe to full inflation. This fixative allows the main axial airway path to be easily separated from the surrounding parenchyma and vasculature by microdissection with the aid of a dissecting microscope (Figure 13). All daughter airway branches are trimmed off from the axial pathway using microsurgical scissors and the aid of a dissecting microscope (Plopper, 1990; Avadhanam et al., 1997). The axial airway is cut open longitudinally to facilitate complete viewing of the entire airway. Flattening of the isolated whole-mount preparation can be done on a glass slide using a coverslip to allow for easy microscopic observation using bright field or confocal laser scanning microscopy. This method allows the entire airway path to be seen *en face* (Figure 13). Further immunostaining of the whole-mount preparation can be performed for visualization of specific cell types (such as neuroendocrine cells) along the entire airway path. Portions of the whole-mount can also be embedded in glycomethacrylate for further confirmation of cell identification (Figure 14). If immunostaining is not done immediately, the microdissected whole-mount airway can be stored in 70% ethanol.

Whole-mount airways can be categorized into two distinct anatomical features: (1) airway bifurcations

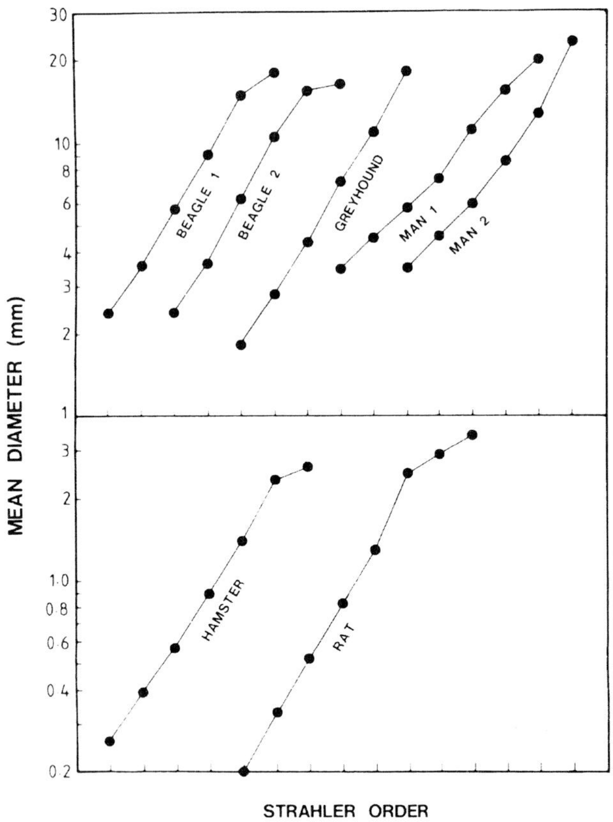

FIGURE 11 The relationship between the logarithm of the mean airway diameter and Strahler order for casts from four different species. The slope of this relationship appears to be relatively constant across species for intralobar airways. *From Thurlbeck and Horsfield (1980). Used with permission.*

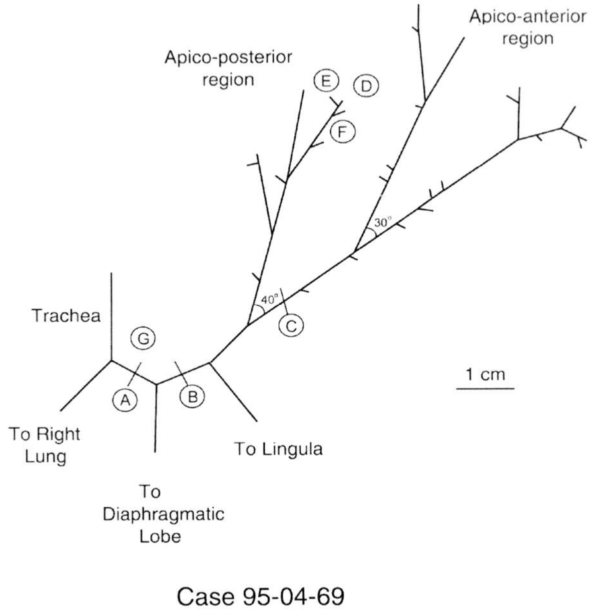

FIGURE 12 Stick diagram of the airway branching pattern from the left upper lobe of a 45-year-old farm laborer (employed for 15 years) who had lived for 17 years in Fresno County. He was a nonsmoker. Beginning at the left mainstem bronchus, airway paths leading to the apico-posterior and apico-anterior regions of the lungs were followed by microdissection. Letters in circles indicate sites taken for embedment and histologic examination. A, second generation airway (left mainstem bronchus); B, third-generation airway; C, fifth-generation airway; D,E,F, first-generation respiratory bronchioles beyond the tenth-generation airway; G, hilar lymph node.

consisting of the branch point ridge and (2) airway segments or generations forming the region between airway branch points (Avadhanam et al., 1997; Peake et al., 2000). For ease in identification, airway bifurcations can be categorized numerically from a proximal to distal direction beginning at the first intralobar bifurcation of the right middle lobe. This numbering approach can be used in place of the more traditional binary classification system (Avadhanam et al., 1997; Figure 8).

4.3 Metabolic/Biochemical/Gene Expression of Isolated Unfixed Tracheobronchial Airways

Methods have been described for isolation of the tracheobronchial tree of mice, rats, and monkeys for metabolic, protein, and gene expression analysis (Plopper et al., 1991). This procedure involves cannulation of either the trachea or lobar bronchus, followed by full visual inflation of the lungs and/or lobe with 1% low-melting-point agarose (FMC Bioproducts, Rockland, ME). The inflated lung/lobe is quickly immersed in ice-cold Waymouth's medium (Gibco Laboratories, Greenland, NY). Under a dissecting microscope and with the lungs placed on a cold plate, the airways can be separated from the blood vessels and the parenchyma. Different segments of the airway subcompartments, including the major daughter, minor daughter, and distal bronchioles, can be obtained. Each of these isolated airways can be used for microsomal, protein, RNA, or DNA analysis. All steps for

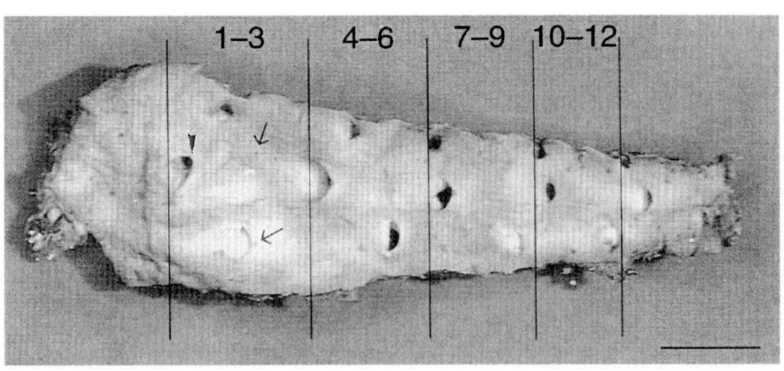

FIGURE 13 Whole-mount airway preparation of the main pathway of the mouse right middle lobe. The numerical designation of airway bifurcation ridges begins with the first intralobar bifurcation. The numerical designation of airway generations of the axial pathway also begins immediately after the first intralobar airway branch point (arrowhead). Airway generations 1–3, 4–6, 7–9, and 10–12 are identified by the vertical lines superimposed over the image. NEBs (arrows) can be visualized sporadically dotting the surfaces of the airway. Scale bar, 1 mm.

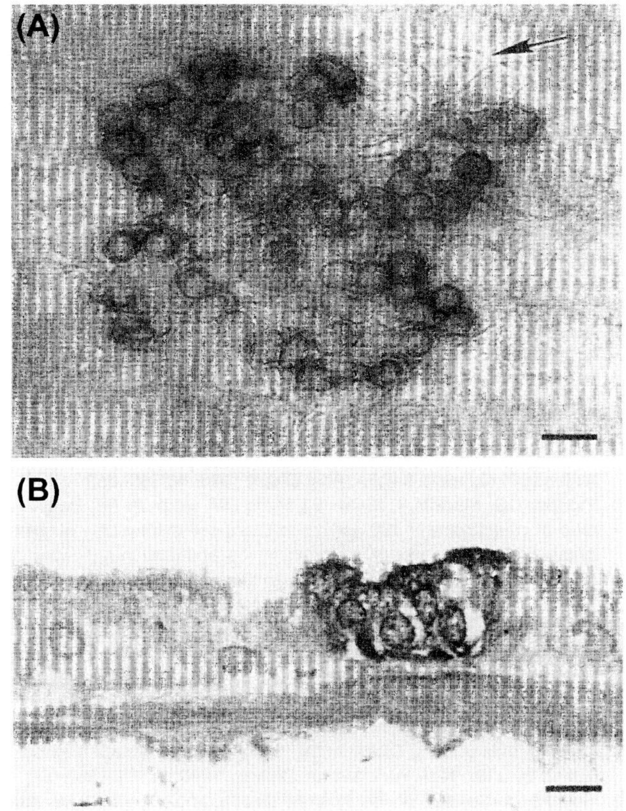

FIGURE 14 NEB with small CGRP-IR cells intermixed with CGRP-negative bronchiolar epithelial cells of the airway from a naphthalene-treated mouse whole-mount airway preparation. These unstained cells may be PNECs, not expressing this marker, or they could be PNEC precursors. A: The NEB appears to be surrounded by larger epithelial cells (arrow). In the 1.5-μm-thick plastic section from a whole-mount airway preparation that had been immunostained before embedding and sectioning (B), the nuclear density of the NEB is greater than that of the surrounding CGRP-negative epithelial cells. Scale bar, 10 μm.

microsomal or protein preparations should be carried out at 0–4 °C. Microsomes are obtained by differential centrifugation using standard techniques. Microsomal preparations are resuspended in 0.1 M phosphate buffer (pH 7.4), and protein concentrations were determined with the micro Bio-Rad assay. A variety of assays can be performed, in particular a number of the cytochrome P450 isoforms (CYP1A1, CYP2B, and CYP2E1) found in the lungs. Enzyme activities can be measured for CYP1A1 and CYP2B by *O*-dealkylation of ethoxyresorufin and pentoxyresorufin (Sigma), respectively (Lee et al., 1998).

Although the preparation of RNA from the tracheobronchial tree can be done using lungs inflated with low-temperature agarose, more recent studies have demonstrated RNA-later solutions are better to inflate the lungs and to preserve RNA integrity. An advantage of RNA-later is the ability to store the inflated lungs in RNA-later solution at room temperature or in the refrigerator until ready for performing airway microdissection (Baker et al., 2004).

REFERENCES

Avadhanam, K.P., Plopper, C.G., Pinkerton, K.E., 1997. Mapping the distribution of neuroepithelial bodies of the rat lung. A whole-mount immunohistochemical approach. Am. J. Pathol. 150 (3), 851–859.

Baker, G.L., Shultz, M.A., Fanucchi, M.V., Morin, D.M., Buckpitt, A.R., Plopper, C.G., 2004. Assessing gene expression in lung subcompartments utilizing in situ RNA preservation. Toxicol. Sci. 77, 135–141.

Barnett, C.H., 1957. A note on the dimensions of the bronchial tree. Thorax 12, 175–176.

Breeze, R.G., Wheeldon, E.B., 1977. The cells of the pulmonary airways. Am. Rev. Respir. Dis. 116, 705–777.

Castleman, W.L., 1984. Alterations in pulmonary ultrastructure and morphometric parameters induced by parainfluenza (Sendai) virus in rats during postnatal growth. Am. J. Pathol. 114, 322–335.

Castleman, W.L., Sorkness, R.L., Lemanske, R.F., Grasee, F., Suyemoto, M.M., 1988. Neonatal bronchiolitis and pneumonia induces bronchiolar hypoplasia and alveolar dysplasia in rats. Lab. Invest. 59, 387–396.

Croteau, J.R., Cook, C.D., 1961. Volume-pressure and length-tension measurements in human tracheal and bronchial segments. J. Appl. Physiol. 16, 170–173.

Dawson, S.V., Elliott, E.A., 1977. Wave-speed limitation on expiratory flow—a unifying concept. J. Appl. Physiol. 43, 498–515.

Ewart, W., 1889. The Bronchi and Pulmonary Blood-Vessels. Churchill, London.

Findeisen, W., 1935. Uber das absetzen Kleiner, in der Luft Suspendierter Teilchen in der Menschlichen Lunge bei der Atmung. Arch. Ges. Physiol. 236, 367–379.

Fredberg, J.J., Hoenig, A., 1978. Mechanical response of the lungs at high frequencies. J. Biomech. Eng. 100, 57–66.

Hahn, H.L., Watson, A., Wilson, A.G., Pride, N.B., 1980. Influence of bronchomotor tone on airway dimensions and resistance in excised dog lungs. J. Appl. Physiol. 49, 270–278.

Hooper, G., 1977. Diameters of bronchi at asymmetrical divisions. Respir. Physiol. 31, 291–294.

Horsfield, K., 1987. Morphometry of airways. In: Fishman, A.P. (Ed.), Handbook of Physiology. Respiration. Sect. 3, vol. III. American Physiological Society, Bethesda, pp. 75–81 pt. 1.

Horsfield, K., Cumming, G., 1976. Morphology of the bronchial tree in the dog. Respir. Physiol. 26, 173–182.

Horsfield, K., Dart, G., Olson, D.E., Filley, F.G., Cumming, G., 1971. Models of the human bronchial tree. J. Appl. Physiol. 31, 207–217.

Horsfield, K., Kemp, W., Phillips, S., 1978. An asymmetrical model of the airways of the dog lung. J. Appl. Physiol. 52, 21–26.

Hughes, J.M.B., Hoppin, F.G., Wilson, A.G., 1972. Use of stereoscopic X-ray pairs for measurements of airway length and diameter in situ. Br. J. Radiol. 45, 477–485.

Hyatt, R.E., Offord, K.P., Lai-Fook, S.J., 1981. Effect of length changes on bronchial diameters. J. Appl. Physiol. 50, 1168–1172.

Jeffrey, P.K., 1983. Morphologic features of airway surface epithelial cells and glands. Am. Rev. Respir. Dis. 128, S14–S20.

Lee, C., Watt, K.C., Chang, A.M., Plopper, C.G., Buckpitt, A.R., Pinkerton, K.E., 1998. Site-selective differences in cytochrome P450 isoform activities. Comparison of expression in rat and rhesus monkey lung and induction in rats. Drug Metab. Dispos. 26, 396–400.

Lee, D.Y., Park, S.S., Ban-Weiss, G.A., Fanucchi, M.V., Plopper, C.G., Wexler, A.S., 2008a. Bifurcation model for characterization of pulmonary architecture. Anat. Rec. 291, 379–389.

Lee, D.Y., Fanucchi, M.V., Plopper, C.G., Fung, J., Wexler, A.S., 2008b. Pulmonary architecture in the conducting regions of six rats. Anat. Rec. 291, 916–926.

Leith, D.E., 1982. Mammalian tracheal dimensions: scaling and physiology. J. Appl. Physiol. 55, 196–200.

Martin, H.B., Proctor, D.F., 1958. Pressure-volume measurements on dog bronchi. J. Appl. Physiol. 13, 337–343.

Matsuba, K., Thurlbeck, W.M., 1971. The number and dimensions of small airways in non-emphysematous lungs. Am. Rev. Respir. Dis. 104, 516–524.

McBride, J., 1985. Postpneumonectomy airway growth in the ferret. J. Appl. Physiol. 58, 1010–1014.

McBride, J.T., Chuang, C., 1985. A technique for quantitating airway size from bronchial casts. J. Appl. Physiol. 58, 1015–1022.

McLaughlin, R.A., Noble, P.B., Sampson, D.D., 2014. Optical coherence tomography in respiratory science and medicine: from airways to alveoli. Physiology 29, 369–380.

Nadel, J.A., Wolfe, W.G., Graf, P.D., 1968. Powdered tantalum as a medium for bronchography in dog and human lungs. Invest. Radiol. 3, 229–238.

Nakano, Y., Muller, N.L., King, G.G., Niimi, A., Kalloger, S.E., Mishima, M., Pare, P.D., 2002. Quantitative assessment of airway remodeling using high-resolution CT. Chest 122, 271S–275S.

Nakamura, M., Sasaki, H., Sekizawa, K., Ishii, M., Takishima, T., Hoppin, F.G., 1981. Series distribution of airway collapsibility in dogs. J. Appl. Physiol. 50, 325–333.

Nikiforov, A., Schlesinger, R.B., 1985. Morphometric variability of the human upper bronchial tree. Respir. Physiol. 59, 289–299.

Oakes, J.M., Scadeng, M., Breen, E.C., Marsden, A.L., Darquenne, C., 2012. Rat airway morphometry measured from in situ MRI-based geometric models. J. Appl. Physiol. 112, 1921–1931.

Pardaens, J., Van De Woestihne, K.P., Clement, J., 1972. A physical model of expiration. J. Appl. Physiol. 33, 479–490.

Parker, H., Horsfield, K., Cumming, G., 1971. Morphology of distal airways in the human lung. J. Appl. Physiol. 31, 386–391.

Peake, J.L., Reynolds, S.D., Stripp, B.R., Stephens, K.E., Pinkerton, K.E., 2000. Alteration of pulmonary neuroendocrine cells during epithelial repair of naphthalene-induced airway injury. Am. J. Pathol. 156 (1), 279–286.

Pedley, T.J., Schroter, R.C., Sudlow, M.F., 1970. The prediction of pressure drop and variation of resistance within the human bronchial airways. Respir. Physiol. 9, 387–405.

Phalen, R.F., Yeh, H.C., Schum, F.M., Raabe, O.G., 1978. Application of an idealized model to morphometry of the mammalian tracheobronchial tree. Anat. Rec. 190, 167–176.

Pinkerton, K.E., Plopper, C.G., Mercer, R.R., Roggli, V.L., Patra, A.L., Brody, A.R., Crapo, J.D., 1986. Airway branching patterns influence asbestos fiber location and the extent of tissue injury in the pulmonary parenchyma. Lab. Invest. 55 (6), 688–695.

Pinkerton, K.E., Green, F.H., Saiki, C., Vallyathan, V., Plopper, C.G., Gopal, V., Hung, D., Bahne, E.B., Lin, S.S., Menache, M.G., Schenker, M.B., 2000. Distribution of particulate matter and tissue remodeling in the human lung. Environ. Health Perspect. 108 (11), 1063–1069.

Prakash, U.B.S., Hyatt, R.E., 1978. Static mechanical properties of bronchi in normal excised human lungs. J. Appl. Physiol. 45, 45–50.

Plopper, C.G., 1990. Structural methods for studying bronchiolar epithelial cells. In: Gil, J. (Ed.), Models of Lung Disease: Microscopy and Structural Methods. Marcel Dekker, New York, pp. 537–559.

Plopper, C.G., Halsebo, J.E., Berger, W.J., Sonstegard, K.S., Nettesheim, P., 1983. Distribution of nonciliated bronchiolar epithelium (Clara) cells in intra- and extrapulmonary airways of the rabbit. Exp. Lung Res. 5, 79–98.

Plopper, C.G., Cranz, D.L., Kemp, L., Serabjit-Singh, C.J., Philpot, R.M., 1987. Immunohistochemical demonstration of cytochrome P-450 monooxygenase in Clara cells throughout the tracheobronchial airways of the rabbit. Exp. Lung Res. 13, 59–68.

Plopper, C.G., Chang, A., Pang, A., Buckpitt, A., 1991. Use of microdissected airways to better define metabolism and cytotoxicity in murine bronchiolar epithelium. Exp. Lung Res. 17, 181–196.

Raabe, O.G., Yeh, H.C., Schum, G.M., Phalen, R.M., 1976. Tracheobronchial Geometry: Human, Dog, Rat, Hamster. Lovelace Foundation, Albuquerque, NM.

Rahn, H., Ross, B.B., 1957. Bronchial tree casts, lobe weights and anatomical dead space measurements in the dog's lung. J. Appl. Physiol. 10, 154–157.

Rohrer, F., 1915. Der Stromungswiderstand in der Menschlichen Atemwegen and der Einfluss der Unregelmassigen Verzweigung des Bronchialsystems auf der Atmungsverlauf in Verschiedenen Lungenbezinken. Arch. Ges. Physiol. 162, 225–259.

Ross, B.B., 1957. Influence of bronchial tree structure on ventilation in the dog's lung as inferred from measurements of plastic cast. J. Appl. Physiol. 10, 1–14.

Schlesinger, R.B., McFadden, L.A., 1981. Comparative morphometry of the upper bronchial tree in six mammalian species. Anat. Rec. 199, 99–108.

Thurlbeck, A., Horsfield, K., 1980. Branching angles in the bronchial tree related to order of branching. Respir. Physiol. 41, 173–181.

Tran, M.U., Weir, A.J., Fanucchi, M.V., Murphy, A.E., Van Winkle, L.S., Evans, M.J., Smiley-Jewell, S.M., Miller, L., Schelegle, E.S., Gershwin, L.J., Hyde, D.M., Plopper, C.G., 2004. Smooth muscle development during postnatal growth of distal bronchioles in infant rhesus monkeys. J. Appl. Physiol. 97, 2364–2371.

Warren, J.V., 1974. The physiology of the giraffe. Sci. Am. 231, 96–105.

Weibel, E.R., 1963. Morphometry of the Human Lung. Academic Press, New York.

West, B.J., Bhargava, V., Goldberger, A.L., 1986. Beyond the principle of similitude: renormalization in the bronchial tree. J. Appl. Physiol. 60, 1089–1097.

Widdicombe, J.G., Pack, R.J., 1982. The Clara cell. Eur. J. Respir. Dis. 63, 202–220.

Wilson, T.A., Hyatt, R.E., Rodarte, J.R., 1980. The mechanisms that limit expiratory flow. Lung 158, 193–200.

Woldehiwot, Z., Horsfield, K., 1978. Diameter, length and branching angles of the upper airways in the dog lung. Respir. Physiol. 33, 213–218.

Yeh, H.C., Schum, G.M., 1980. Models of human lung airways and their application to inhaled particle deposition. Bull. Math. Biol. 42, 461–480.

Yeh, H.C., Hulvert, A.J., Phalen, R.F., Velasquez, D.J., Harris, T.D., 1975. A stereoradiographic technique and its application to the evaluation of lung casts. Invest. Radiol. 10, 351–357.

Yeh, H.C., Schum, F.M., Duggan, M.T., 1979. Anatomic models of the tracheobronchial and pulmonary regions of the rat. Anat. Rec. 195, 482–492.

Chapter 5

Secretory Glycoconjugates of the Trachea and Bronchi

Judith A. St George

Chapter Outline

1. INTRODUCTION

The secretory cells of the tracheobronchial epithelium and submucosal glands collectively produce the mucous blanket of the conducting airways. There is, however, wide variation from species to species in the type of secretory cells at a given airway level, the numbers of secretory cells present, as well as the presence and abundance of submucosal glands. Some species, including the rat, mouse, rabbit, hamster, and guinea pig, have few if any glands; other species, such as the human, monkey, dog, cat, sheep, pig, and cow, have well-developed glands (Jeffery, 1983; Choi et al., 2000). These differences may affect the amount and properties of mucus available on the epithelial surface for mucociliary clearance. Mucociliary clearance is important in the innate defense of the lungs, where contaminants of inhaled air are trapped or dissolved in the mucous layer, then removed by ciliary transport. Optimal clearance is dependent on a complex interaction between the mucous layer, the underlying periciliary fluid, and ciliary action (Button et al., 2012). Although ciliated cells are present in relatively constant levels in most airway levels of most species, the content and sources of the mucous blanket vary considerably. The sources include mucous and serous cells of both the epithelial surface and submucosal glands. In some species, the Clara cell or noncilated bronchiolar epithelial cell may also contribute to the mucous layer (Plopper et al., 1984). This variability in sources of mucus suggests variation in the composition and properties of mucus. Several methods have been used to define mucous composition, including carbohydrate histochemistry and cytochemistry (Spicer et al., 1983), autoradiography (Lamb and Reid, 1969), immunohistochemistry (St George et al., 1985), and biochemistry (Rose et al., 1979; Gendler and Spicer, 1995; Thornton et al., 2008; Rose and Voynow, 2006). All but the last of these methods have the advantage of localizing the constituents of mucus in situ.

The goals of this chapter are to introduce the reader to the composition and sources of the mucous blanket, as well as to highlight the differences between species that may influence how experimental results in animal models can be extrapolated to humans.

2. GEL-FORMING MUCIN

Mucous clearance is central to the innate defense of the lungs (Knowles and Boucher, 2002). However, overproduction and/or compromised clearance are characteristic of all airway inflammatory respiratory diseases, including chronic obstructive pulmonary disease, cystic fibrosis (CF), and asthma. In addition, the secretions from patients with these conditions are different due to differences in proportion of MUC gene products, glycosylation (Kirkham et al., 2002), the ratio of solids to liquid (Martens et al., 2011), or combinations of these (for a review, see Knowles and Boucher, 2002; Thornton et al., 2008). The gel-forming mucins are the key to proper function. The apical secretions covering the airway surfaces are not homogenous but rather

Comparative Biology of the Normal Lung. http://dx.doi.org/10.1016/B978-0-12-404577-4.00005-9

a complex combination of polymeric mucins, proteins, lipids, water, ions, and more. The complexity of this fluid and the nature of mucins have made characterization difficult, in large part due to the challenges of isolating and purifying mucins. Since publication of the first edition of *Comparative Biology of the Normal Lung*, significant advances have been made in the understanding of mucins as molecular tools have become available. The most important of these has been the identification of the mucin genes (MUC genes in humans).

More than 20 human and murine mucin genes have been described, but not all MUC gene products have similar protein backbones (Dekker et al., 2002; Rose and Voynow, 2006). The mucin family is broadly defined as proteins containing O-linked oligosaccharides as greater than 50% of the mass. Although several MUC genes are expressed in the lung, the two major gel-forming mucins are MUC5AC and MUC5B. In general, MUC5AC has been localized to the goblet cell of the epithelial surface, while MUC5B has been identified in the submucosal glands (Hovenberg et al., 1996). The protein backbones from these two genes are characterized by variable numbers of tandem repeats (TRs) that contain proline and are rich in serine and theronine, the sites for O-glycosylation (for a review, see Rose and Voynow, 2006). There are cysteine-rich domains that alternate with the TR regions. In addition, the N and C terminal ends contain cysteine-rich domains similar to domains contained within von Willebrand factor. TR domains are typically not conserved between species, but non-TRs appear highly conserved (Rose and Voynow, 2006). An individual mucin gene cannot identify a specific cell type as several MUC genes are expressed within the same or different cell types of the epithelia or glands (Audie et al., 1993).

In summary, from this very brief overview, it is clear that mucin is a complex mixture of mucin gene products and other constituents that influence mucous properties. Infection and inflammation affect expression levels, changes in glycosylation, and the ratio of gene products. Knowledge of the animal model with regard to mucin gene expression will be important to relate data back to humans, for example, MUC5AC is a major component of normal mucus in humans; however, in normal rat airways, MUC5AC expression is very low to nonexistent, but it can be induced significantly by exposure to irritants (Borchers et al., 1998)

3. HISTOCHEMISTRY AND CYTOCHEMISTRY

Given the large contribution of carbohydrates to the composition and properties of mucus, it is important to characterize this portion of mucin molecule. Carbohydrate histochemistry and cytochemistry have been used to localize carbohydrates containing vicinal hydroxyl groups, carboxyl groups, and sulfate esters. Tables 1 and 2 summarize the histochemically defined content of the secretory cells of the epithelial surface (Table 1) and submucosal glands (Table 2). Human respiratory mucous cells of both the epithelial surface and glands were shown to contain acidic glycoconjugates due to the presence of sulfate and/or sialic acid, whereas neutral glycoconjugates were localized in serous cells of submucosal glands (Lamb and Reid, 1969). With the examination of additional species, it was found that although mucous cells generally contain acidic mucins, the predominance of either sulfo- or sialomucin varies with the species. For example, sulfomucins predominate in the mucous cells of surface epithelium in dogs (Spicer et al., 1971), cats (Jeffery, 1977), rabbits (Plopper et al., 1984), and macaque monkeys (St George et al., 1984a). In the epithelial surface of sheep (Mariassy et al., 1988a) and humans (Lamb and Reid, 1969 and Spicer et al., 1971), either sulfo- or sialomucin predominates, but this varies according to airway level. In sheep, mucous cells of airway generations greater than 14 in the left cranial lobe or 22 in the left caudal lobe contain sulfomucin (Mariassy et al., 1988a). More proximal generations are lined by mucous cells with either sialo- or sulfomucins. The nasal cavity and distal bronchioles of humans (Thaete et al., 1981) and the nasal cavity of macaque monkeys (Harkema et al., 1987) contain predominantly sulfomucins. By contrast, in both rats and mice (McCarthy and Reid, 1964), sialomucins predominate in epithelial mucous cells of airway surfaces when mucous cells are found. The predominant secretory cell of rat tracheobronchial epithelium, however, is a serous cell containing a neutral glycoconjugate (Spicer et al., 1980).

Acidic mucins are found in the mucous cells of submucosal glands as well. Sulfomucins are contained in glandular mucous cells of most species, including the rat, mouse (McCarthy and Reid, 1964), dog (Spicer et al., 1971), sheep (Mariassy et al., 1988a), pig (Jones et al., 1975), and rhesus monkey (St George et al., 1986a). In rats, the distribution of acidic mucins varies with the location within the gland, with mucous tubules containing sulfomucins and mucous ducts containing sialomucins (Mochizuki et al., 1982). In submucosal glands of human respiratory airways, sialo- and sulfomucins are present in roughly equal proportions (Lamb and Reid, 1969).

Ultrastructural studies using carbohydrate cytochemical methods have been used to examine the intracellular distribution of respiratory glycoconjugates. On the whole, these have confirmed and extended the observations made at the light microscopic level. Differences in carbohydrate content have been demonstrated within individual granules of serous and mucous cells. Both cell types may contain monophasic, biphasic, or triphasic granules. Mucous granules of rabbit trachea are biphasic, with an outer cortex

TABLE 1 Carbohydrate Content of Tracheal Epithelium

			Carbohydrate Content		
Species	Cell Type	Abundance	PAS	AB	HID
Hamster[a]	Clara	+ + +	+	–	–
	Mucous	+	+	+	–
Rat[b]	Serous	+ + +	+	–	–
	Mucous	+	+	+	–
Mouse[b]	Mucous	+	+	+	–
Rabbit[c]	Mucous	+	+	+	+
	Clara	+ + +	+/–	–	–
Canine[d]	Mucous	+ +	+	+	+
Cat[e]	Mucous	+ +	+	+	
	Serous	+	ND	ND	ND
Pig[f]	Mucous	+	+	+ and –	+ and –
Sheep[g]	Mucous	+ +	+	+	+ and –
Rhesus[h]	Mucous	+ +	+	+	+
Human[i]	Mucous	+ + +	+	+	+

Note: PAS = periodic acid Schiff, reacts with vicinal hydroxyl groups; AB = alcian blue, at pH 2.6 reacts with acidic glycoconjugates; HID = high-iron diamine, reacts with acidic glycoconjugates containing sulfate esters; ND = not determined.
[a]*Emura and Mohr (1975).*
[b]*McCarthy and Reid (1964).*
[c]*Plopper et al. (1984).*
[d]*Spicer et al. (1971).*
[e]*Jeffery (1977).*
[f]*Jones et al. (1975).*
[g]*Mariassy et al. (1988a).*
[h]*St George et al. (1984a).*
[i]*Lamb and Reid (1969).*

where the sulfated glycoconjugate is concentrated (Plopper et al., 1984). By contrast, in rhesus tracheal mucous cells, the sulfated material is concentrated within the inner one or two core regions (St George et al., 1984a). In human epithelial mucous cells, the cores have been described as invariably negative for carbohydrate (Thaete et al., 1981).

The content of the nonciliated bronchiolar-cell or Clara-cell secretory granules has also been characterized using histochemical methods for glycoproteins and lipids (for a review, see Plopper, 1983). These studies have shown that Clara-cell granules do not contain acidic glycoconjugates in any species, including the human (Cutz and Conen, 1971), mouse (Luke and Spicer, 1966 and Pack et al., 1981), rat (Spicer et al., 1980), hamster and guinea pig (Luke and Spicer, 1966), rabbit (Plopper et al., 1984), or cat (Klika and Petrik, 1969). Carbohydrates containing vicinal hydroxyl groups have been demonstrated consistently in some species and variably in the mouse and rat (Plopper, 1983). There is considerable histochemical evidence suggesting that Clara cells synthesize lipids or phospholipids (Cutz and Conen, 1971; Azzopardi and Thurlbeck, 1969), but whether these components are incorporated into secretory granules has yet to be convincingly documented. Chapter 7 in this volume discusses the immunochemical characterization of the secretory product of nonciliated bronchiolar epithelial cells.

4. LECTIN HISTOCHEMISTRY AND CYTOCHEMISTRY

More specific information on the composition of the glycoconjugates contained within the mucous and serous cells of the respiratory system was provided with the use of lectins. Lectins, with specificity for sugars primarily in the terminal position, were used to localize these sugars in complex carbohydrates. A summary of lectin reactivity is presented in Table 3. The majority of the work in this area was done by Spicer and colleagues, and was reviewed for the human, rat, and mouse in 1983 (Spicer et al., 1983). The results from the lectin studies provided additional

TABLE 2 Carbohydrate Content of Tracheal Submucosal Glands

Species	Abundance	Secretory Cell	Carbohydrate Content		
			PAS	AB	HID
Hamster[a]	+/−	Mucous	+	+	−
Rat[b]	+	Serous	+	−	−
		Mucous	+	+	+ and −
Mouse[b]	+/−	Serous	+		−
		Mucous	+	+	+ and −
Rabbit[c]	+/−	Mucous	+	+	+
Canine[d]	++	Serous	+	−	−
		Mucous	+	+	+ and −
Cat[e]	++++	Serous	+	+	+ and −
		Mucous	+	+	+
Pig[f]	++	Serous	+	−	−
		Mucous	+	+	+ and −
Sheep[g]	++	Serous	+	−	−
		Mucous	+	+	+
Rhesus[h]	++	Serous	+	−	−
		Mucous	+	+	+
Human[i]	+++	Serous	+	−	−
		Mucous	+	+	+ and −

Note: PAS = periodic acid Schiff, reacts with vicinal hydroxyl groups; AB = alcian blue, at pH 2.6 reacts with acidic glycoconjugates; HID = high-iron diamine, reacts with acidic glycoconjugates containing sulfate esters; ND = not determined.
[a]*Emura and Mohr (1975).*
[b]*McCarthy and Reid (1964).*
[c]*Plopper et al. (1984).*
[d]*Spicer et al. (1971).*
[e]*Jeffery (1977).*
[f]*Jones et al. (1975).*
[g]*Mariassy et al. (1988a).*
[h]*St George et al. (1984a).*
[i]*Lamb and Reid (1969).*

information and further complicated the interpretation of secretory cell content. For example, histochemical analysis had not indicated the heterogeneity of content of both serous and mucous cells that became apparent with lectin application. An important result not reflected in Table 3 is that although some cells of a specific type contain a sugar, not all will, nor will they contain it in the same concentration. Additionally, in human airways, the sugar content of mucous, but not serous, cells varies with blood group antigen and presumably with the secretory status of a given individual (Spicer et al., 1983). Reactivity with lectins was used to define a specific cell type or was used as a marker for a secretory product (Wasano et al., 1988a). With the characterization of lectin reactivity established in the airway cells of healthy animals, investigators employed lectins to detect shifts in secretory products with injury or disease (Inai et al., 1987; Mariassy et al., 1989).

Lectins applied at the level of the electron microscope further elucidated the nature of the airway secretory product(s). Wasano et al. (1988b) examined mucous cells in the hamster trachea and demonstrated that glycoconjugates with discrete terminal sugars are localized within distinct regions of the Golgi apparatus. Sugars appeared in a sequential fashion from *cis*-to *trans*-cistemae. *N*-acetyl galactosamine was the only sugar detected in the *cis*-cistemae; then *N*-acetyl glucosamine and galactose, and lastly fucose and sialic acid, were detected in the *trans*-cistemae. This sequence of sugars coincides with the order

TABLE 3 Lectin Reactivity in Airway Surface Epithelium

Lectin									
Species/Cell Type	**Sugar Specificity**	**LCA Man Glc Glc NAc**	**WGA NANA Glc NAc**	**BSAI Gal**	**DBA Gal NAc**	**SBA Gal-Gal Nac Gal Gal Nac**	**PNA Gal Gal NAc**	**RCA Gal Gal NAc**	**UEA Fuc**
Human bronchi[a]	Mucous	–	++++	+	+	+++	++++	++++	++++
Sheep trachea-bronchiole[b]	Mucous M1			–	+++	++	+++		++++
	Mucous M2			–	+++	++	+++		++++
	Mucous M3			+++	+	–	–		–
Rat trachea[c]	Mucous	–		–	+	+	–(+++)	+	+++
Rhesus trachea	Mucous	++++	++++	++++	–	–(+++)[c]	–(+++)		+
Lectin Reactivity of Airway Submucosal Glands									
		LCA	WGA	BSAI	DBA	SBA	PNA	RCA	UEA
Human bronchi	Mucous	–	+	+	+	+++	++++	+++	++++
	Serous	+++	+++	–	–	++	+	++	–
Sheep	Mucous 4			–	++++	++	–	–	++
	Serous			–	+++	++	–		++
Mouse trachea[a]	Mucous	–		–	–	+	–(+++)	–	+++
	Serous			–	++++	++++	++++	++++	–
Rat trachea	Mucous	++		-	++	++	–(++–)	–	+++
	Serous			–	–	–	–(++–+)	–	–
Rhesus trachea	Mucous	++++	++++	+++	–	–(++++)	–(++–+)		++++
	Serous	++	++	+	–	–(++)	–(++)		+

Note: LCA = *Lotus tetragonolobus*; WGA = wheat germ agglutinin; BSA I = *Bandeirea simplicifolia* I; DBA = *Dolichos biflorus*; SBA = Glycine max; PNA = *Arachis hypogea*; RCA = *Ricinus communis*; UEA = *Ulex europeus*; Man = mannose; Glc = glucose; GlcNAc = *N*-acetyl glucosamine; NANA = *N*-acetyl neuraminic acid (sialic acid); Gal = galactose; Ga1NAc = *N*-acetylgalactosamine; Fuc = fucose.

[a] *Spicer et al. (1983).*

[b] *Mariassy et al. (1988b).*

[c] *Reaction in parentheses is after neuraminidase treatment.*

of glycosylation of *O*-linked respiratory mucins demonstrated using biochemical methods (Rana et al., 1986).

5. IMMUNOHISTOCHEMISTRY

Immunohistochemical methods can be used to define more precisely the variations in content within cells or granules. For example, the enzymes lysozyme (Bowes et al., 1981) and lactoferrin (Bowes and Corrin, 1977) have been localized within secretory cells of the respiratory tract. Lysozyme, which does not contain carbohydrates, has been localized within serous cells of submucosal glands, leading Spicer and colleagues to suggest that the carbohydrate-negative portion of serous granules is lysozyme (Spicer et al., 1980).

Immunohistochemical studies using monoclonal antibodies against respiratory secretory products have expanded the previously discussed heterogeneity (St George et al., 1984b, 1985; Basbaum et al., 1984). We have used monoclonal antibodies against rhesus or rabbit tracheal secretions to show several antigenically distinct subpopulations of mucous cells (St George et al., 1985). When serial sections of histochemically and immunohistochemically stained samples are compared, secretory cells are found that do not vary histochemically but vary immunohistochemically (St George et al., 1984b). Application of the monoclonal antibodies to four airway levels distal to the trachea in rhesus monkey indicate that the more distal the airway, the less the secretory content resembles that in trachea (St George et al., 1986b). This variation in content suggests variations in the biophysical properties of secretion at different airway levels.

We have also used the panel of monoclonal antibodies to examine the appearance of secretory products in the airways of developing fetal monkey lungs. From that study, we have concluded that the antigens of adult trachea are present in neonatal trachea at birth but that the antigens appear sequentially rather than simultaneously. The results of these studies using monoclonal antibodies are probably detecting differences in the glycosylation of mucous glycoconjugates, because characterization of the epitopes of our antibodies (Lin et al., 1989) and those of others (Basbaum et al., 1986) have indicated a carbohydrate specificity. Antibodies to the protein portion of mucus have been used to localize and quantify mucins within normal human airways and compare the results to those from a patient with CF (Burgel et al., 2007).

Although the secretory cells are thought to be the primary source of the mucous blanket, other cell types within the epithelial surface and glands may modify or supplement the fluid lining. Modification to the fluid lining includes the secretion of water and ions. These changes influence the properties of the mucous blanket and are therefore important to the function of mucociliary clearance. One other cell type that can participate in this process is the ciliated cell. Several studies have demonstrated the presence of glycoprotein on the surface of ciliated cells (Spicer et al., 1971; Jeffery, 1977). We have observed vesicles or granules containing mucous antigens in ciliated cells (St George et al., 1984b), suggesting either endocytosis or exocytosis of components of the mucous blanket.

6. QUANTITATION

With certain toxicological insults or respiratory diseases, profound changes occur in the volume of secretion produced and stored in the airways. It has therefore been important to develop accurate methods to quantify the volume of stored secretory products under normal as well as abnormal conditions. Previous methods have included determining the mean gland-to-wall ratio (Reid Index, Reid, 1960) or point counting to quantify submucosal glands (Bedrossian et al., 1971). We have used a computerized morphometric method to assess the volume of the secretory product per unit surface area and have demonstrated that there is considerable variation at different airway levels in the amount of glycoconjugate stored in airway epithelium of the rhesus monkey (Heidsiek et al., 1987; Plopper et al., 1988). The product of the tracheal surface was predominantly acidic, while that in submucosal glands was neutral. The trachea stored at least twice as much per unit of surface area, as compared to distal airways (generation 11). Additionally, a change in secretory product from sulfomucin to sialomucin was detected distally. One surprising feature of these studies was the contribution to the total secretory product by submucosal glands versus surface epithelium, as suggested by the amount of stored product in either surface epithelium or submucosal glands. Several investigators have indicated that the overwhelming majority of secretion is from submucosal glands (Reid, 1960). However, morphometric analysis of stored product in rhesus trachea demonstrated that the contribution by the submucosal glands was less than 50% of the total product (Heidsiek et al., 1987). These same quantitative methods were used to immunochemically compare the secretory cells from normal airways with those from an asthma patient (Ordonez et al., 2001).

7. CONCLUSIONS

Several methods, including biochemical, histochemical, and immunohistochemical techniques, have been used to define the content of secretory cells of airway epithelium. Application of these methods has demonstrated the extensive heterogeneity in content when comparing secretory cells from the epithelial surface to those in submucosal glands, in the same cell type at the same airway level, and in secretory cells from different airway levels, as well as from one species to another. Many tools are now in place to more fully understand the secretory response and the airway secretory products; however, the investigator should consider the variations that occur at different airway levels in the same species, as well as differences between species. The influence of secretion by different sites (surface vs glands and proximal vs distal) on the properties of the mucous blanket and the importance in differences in control in these regions remains to be elucidated. Importantly, extrapolation of results from one species to another should be done, bearing in mind that the aforementioned differences may dictate a very different secretory response to a given toxicological insult.

REFERENCES

Audie, J.P., et al., 1993. Expression of human mucin genes in respiratory, digestive, and reproductive tracts ascertained by in situ hybridization. J. Histochem. Cytochem. 41, 1479–1485.

Azzopardi, A., Thurlbeck, W.M., 1969. The histochemistry of the nonciliated bronchiolar epithelial cells. Am. Rev. Respir. Dis. 99, 516.

Basbaum, C.B., et al., 1984. Monoclonal antibodies as probes for unique antigens in secretory cells of mixed exocrine organs. Proc. Natl. Acad. Sci. USA 81, 4419.

Basbaum, C.B., et al., 1986. Tracheal carbohydrate antigens identified by monoclonal antibodies. Arch. Biochem. Biophys. 249, 363.

Bedrossian, C.W.M., Anderson, A.E., Foraker, A.G., 1971. Comparison of methods for quantitating bronchial morphology. Thorax 26, 406.

Borchers, M.T., Wert, S.E., Leikauf, G.D., 1998. Acrolein-induced MUC5ac expression in rat airways. Am. J. Physiol. 274, L573–L581.

Bowes, D., Corrin, B., 1977. Ultrastructural immunocytochemical localization of lysozyme in human bronchial glands. Thorax 32, 163.

Burgel, P.R, Montani, D., Danel, C., Dusser, D.J., Nadel, J.A., 2007. A morphometric study of mucins and small airway plugging in cystic fibrosis. Thorax 62, 153–161.

Bowes, D., Clark, A.E., Corrin, B., 1981. Ultrastructural localization of lactoferrin and glycoprotein in human bronchial glands. Thorax 36, 108.

Button, B., et al., 2012. A periciliary brush promotes the lung health by separating the mucus layer from airway epithelia. Science 337, 937–994.

Choi, H.K., Finkbeiner, W.E., Widdicombe, J.H., 2000. A comparative study of mammalian tracheal mucous glands. J. Anat. 197 (Pt 3), 361–372.

Cutz, E., Conen, P.E., 1971. Ultrastructure and cytochemistry of Clara cells. Am. J. Pathol. 62 (1), 127.

Dekker, J., et al., 2002. The MUC family: an obituary. Trends Biochem. Sci. 27, 126–131.

Emura, M., Mohr, U., 1975. Morphological studies on the development of tracheal epithelium in the Syrian golden hamster. Z. Versuchstierk. 17 (1), 14–26.

Gendler, S.J., Spicer, A.P., 1995. Epithelial mucin genes. Annu. Rev. Physiol. 57, 607–634.

Harkema, J.R., et al., 1987. Regional differences in quantities of histochemically detectable mucosubstances in nasal, paranasal, and nasopharyngeal epithelium of the bonnet monkey. J. Histochem. Cytochem. 35, 279.

Heidsiek, J.G., et al., 1987. Quantitative histochemistry of mucosubstance in tracheal epithelium of the macaque monkey. J. Histochem. Cytochem. 35, 435.

Hovenberg, H.W., Davies, J.R., Carlstedt, I., 1996. Different mucins are produced by the surface epithelium and the submucosa in human trachea: identification of MUC5AC as a major mucin from the goblet cells. Biochem. J. 318 (Pt 1), 319–324.

Inai, K., et al., 1987. An altered lectin binding to mucus glycoprotein in goblet cells of human tracheobronchial epithelium among former mustard-gas workers. Acta Pathol. Jpn. 37, 537.

Jeffery, P.K., 1977. Structure and function of mucus-secreting cells of cat and goose airway epithelium. In: Respiratory Tract Mucus. Elsevier-North Holland, New York, pp. 5–19.

Jeffery, P.K., 1983. Morphologic features of airway surface epithelial cells and glands. Am. Rev. Resp. Dis. 128, S14.

Jones, R.A., Baskerville, A., Reid, L.M., 1975. Histochemical identification of glycoproteins in pig bronchial epithelium: (a) normal and (b) hypertrophied from enzootic pneumonia. J. Pathol. 116, 1.

Kirkham, S., et al., 2002. Heterogeneity of airways mucus: variations in the amounts and glycoforms of the major oligomeric mucins MUC5AC and MUC5B. Biochem. J. 361, 537–546.

Klika, E., Petrik, P., 1969. A study of the structure of the lung alveolus and bronchiolar epithelium. Acta Histochem. (Jena) 20, 331.

Knowles, M.R., Boucher, R.C., 2002. Mucus clearance as a primary innate defense mechanism for mammalian airways. J. Clin. Invest. 109, 571–577.

Lamb, D., Reid, L.M., 1969. Histochemical types of acidic glycoprotein produced by mucous cells of the tracheobronchial glands in man. J. Pathol. 98, 213.

Lin, H., et al., 1989. An ELISA method for the quantitation of tracheal mucins from human and nonhuman primates. Am. J. Respir. Cell. Mol. Biol. 1, 41.

Luke, J.L., Spicer, S.S., 1966. Histochemistry of surface epithelium and plural mucins in mammalian lung. Lab. Invest. 14, 2101.

Mariassy, A.T., et al., 1988a. Tracheobronchial epithelium of the sheep. III. Carbohydrate histochemical and cytochemical characterization of secretory epithelial cells. Anat. Rec. 221, 540.

Mariassy, A.T., et al., 1988b. Tracheobronchial epithelium of the sheep. IV. Lectin histochemical characterization of secretory epithelial cells. Anat. Rec. 222, 49.

Mariassy, A.T., et al., 1989. Lectin-detectable effects of localized pneumonia on airway mucous cell populations: role of cyclooxygenase metabolites. Exp. Lung Res. 15, 113.

Martens, C.J., et al., 2011. Mucous solids and liquid secretion by airways: studies with normal pig, cystic fibrosis human, and non-cystic fibrosis human bronchi. Am. J. Physiol. Lung Cell. Mol. Physiol. 301, L236–L246.

McCarthy, C., Reid, L.M., 1964. Acid mucopolysaccharides in the bronchial tree in the mouse and rat (sialomucins and sulphate). Q. J. Exper. Physiol. 49, 81.

Mochizuki, I., et al., 1982. Carbohydrate histochemistry of rat respiratory glands. Anat. Rec. 202, 45.

Ordonez, C.L., et al., 2001. Mild and moderate asthma is associated with airway goblet cell hyperplasia and abnormalities in mucin gene expression. Am. J. Respir. Crit. Care Med. 163, 517–523.

Pack, R.J., Al-Ugaily, L.H., Morris, G., 1981. The cells of the tracheobronchial epithelium of the mouse: a quantitative light and electron microscopic study. J. Anat. 132, 71.

Plopper, C.G., 1983. Comparative morphologic features of bronchiolar epithelial cells: the Clara cell. Am. Rev. Respir. Dis. 128, S37.

Plopper, C.G., et al., 1984. Carbohydrate cytochemistry of tracheobronchial airway epithelium of the rabbit. J. Histochem. Cytochem. 32, 209.

Plopper, C.G., et al., 1988. Tracheobronchial epithelium in the adult rhesus monkey: a quantitative histochemical and ultrastructural study. Am. J. Anat. 184, 31.

Rana, S.S., Chandrasekaran, E.V., Mendicino, J., 1986. Structures of the sialylated oligosaccharide chains in swine trachea mucin glycoproteins. J. Biol. Chem. 262, 3654.

Reid, L.M., 1960. Measurement of the bronchial mucous gland layer: a diagnostic yardstick in chronic bronchitis. Thorax 15, 132.

Rose, M.C., Lynn, W.S., Kaufman, B., 1979. Resolution of the major components of human lung mucosal gel and their capabilities for reaggregation and gel formation. Biochemistry 18, 4030.

Rose, M.C., Voynow, J.A., 2006. Respiratory tract mucin genes and mucin glycoproteins in health and disease. Physiol. Rev. 86, 245–278.

Spicer, S.S., et al., 1971. Histochemistry of mucosubstances in the canine and human respiratory tract. Lab. Invest. 25, 483.

Spicer, S.S., et al., 1980. Complex carbohydrates of rat tracheobronchial surface epithelium visualized ultrastructurally. Am. J. Anat. 158, 93.

Spicer, S.S., Schulte, B.A., Thomopoulos, G.N., 1983. Histochemical properties of the respiratory tract in different species. Am. Rev. Respir. Dis. 128, S20.

St George, J.A., Nishio, S.J., Plopper, C.G., 1984a. Carbohydrate cytochemistry of the rhesus monkey tracheal epithelium. Anat. Rec. 210, 293.

St George, J.A., et al., 1984b. An immunocytochemical/histochemical approach to tracheobronchial mucus characterization in the rabbit. Am. Rev. Respir. Dis. 130, 124.

St George, J.A., et al., 1985. An immunohistochemical characterization of rhesus monkey respiratory secretions using monoclonal antibodies. Am. Rev. Respir. Dis. 132, 556.

St George, J.A., et al., 1986a. Carbohydrate cytochemistry of rhesus monkey tracheal submucosal glands. Anat. Rec. 216, 60.

St George, J.A., et al., 1986b. Variation of respiratory mucins with airway level in rhesus monkey. Am. Rev. Respir. Dis. 133, A294.

Thaete, L.C., Spicer, S.S., Spock, A., 1981. Histology, ultrastructure, and carbohydrate cytochemistry of surface and glandular epithelium of human nasal mucosa. Am. J. Anat. 162, 243.

Thornton, D.J., Rousseau, K., McGuckin, M.A., 2008. Structure and function of the polymeric mucins in airways mucus. Annu. Rev. Physiol. 70, 459–486.

Wasano, K., et al., 1988a. Membrane differentiation markers of airway epithelial secretory cells. J. Histochem. Cytochem 36 (2), 167.

Wasano, K., Nakamura, K., Yamamoto, T., 1988b. Lectin-gold cytochemistry of mucin oligosaccharide biosynthesis in golgi apparatus of airway secretory cells of the hamster. Anat. Rec. 221, 635.

Chapter 6

Epithelial Cells of Trachea and Bronchi

Susan D. Reynolds, Kent E. Pinkerton and Andrew T. Mariassy

Chapter Outline

1. INTRODUCTION

The trachea, bronchi, and bronchioles constitute the branched conduit for transport of gases to and from the lung parenchyma. The airways regulate the flow and condition the inhaled air. To accomplish these tasks, mammalian airways have acquired a number of adaptive mechanisms and functions, which are reflected in their structure.

The airway wall consists of the following:

(1) A well-vascularized mucous membrane, composed of an epithelium and lamina propria, which protects and cleanses the lung and conditions the air

(2) A submucosa with glands that elaborate the airway secretions

(3) Coils of smooth muscle and a webbing of elastic connective tissue inserted into the lung scaffolding, which modulate airway caliber

(4) A sleeve of coarse connective tissue and cartilaginous plates that maintain the airway patency

(5) Autonomic and local innervation of smooth muscle, glands, vasculature, and mucosa, all of which participate in the homeostasis of the lung environment.

The focus of this chapter is on the comparative morphology of the tracheobronchial airway tree. Special emphasis is given to the morphology and morphometry of epithelial cell subtypes that comprise lining the mucosal surface, the epithelium.

2. DEFINITIONS AND NOMENCLATURE

The number of cartilaginous airways and branching patterns varies significantly among humans, macaque monkeys, pigs, sheep, rats, and diving mammals (Hislop et al., 1984; Mariassy and Plopper, 1984). Furthermore, a number of the structural features that are used to define airway divisions are highly variable among species. As a consequence, the definitions and description of airway divisions presented here should be regarded as generalizations. Species-specific information is provided in later sections.

2.1 Positional Landmarks

Overview: Division of the airways into anatomical entities is based on their position in the airway tree and on a combination of structural features, including airway wall composition, mucosal structure, and epithelial cell composition. Thus, a description of the airways and their anatomic position within the tracheobronchial tree requires a definition of the general airway structure.

Axial determinates: Two axes are used to describe the position of anatomical entities within the airway tree. The proximal (rostral) to distal axis runs parallel to the spine with proximal defined as the region that is closest to the head. The dorsal-to-ventral axis runs perpendicular to the spine, with dorsal being closest to the spine. These terms are most useful in positional coordinates within the trachea and bronchi.

Airway generation: As the airway branches, axial designations become less useful. Airways that are not surrounded by parenchymal tissue and contain few branches are termed extrapulmonary airways. The remainder of airways are termed intrapulmonary. An alternative system uses airway caliber to distinguish position. Larger airways are termed proximal, whereas smaller airways are termed distal.

Comparative Biology of the Normal Lung. http://dx.doi.org/10.1016/B978-0-12-404577-4.00006-0

Airway generation and position within the tracheobronchial tree are more precisely defined using a binary numbering system (Mariassy and Plopper, 1983; Plopper et al., 1983a, 1989; Peake et al., 2000; Hare, 1955). This approach allows precise localization and correlation of airways within an individual lung and facilitates the comparison of airways between control and test animals and among different species.

2.2 Structural Landmarks

Cartilaginous and noncartilaginous airways: The airways can also be stratified based on the presence of cartilage rings. Although this metric was used to define comparable regions of the human and mouse lung (Rock et al., 2010), cartilage location and size exhibits significant variability. Importantly, the number of cartilaginous airways and pattern varies among humans, monkeys, pigs, sheep, rats, and diving mammals (Hislop et al., 1984; Mariassy and Plopper, 1984; Ninomiya et al., 2005). Small rodents (e.g., hamster) lack cartilage in segmental bronchi (Kennedy et al., 1978). In contrast, in large animals (especially marine mammals), the cartilage is massive with respect to the fraction of the total airway wall it occupies, but it also extends into the distal tracheobronchial tree. In these species, cartilage surrounds respiratory bronchioles (Drabek and Kooyman, 1983).

Epithelial subtypes: The airway epithelium also varies with location. A *stratified* epithelium consists of multiple rows of cells; the only cells that are in direct contact with the basement membrane are those that are adjacent to the basement membrane. A stratified epithelium is typical of the epidermis but has been described in normal and diseased airways (Ghosh et al., 2013; Araya et al., 2007). In a *pseudostratified* epithelium, the nuclei form two or more rows but all epithelial cells have direct contact with the basement membrane. In the human airway, this epithelial subtype is identified throughout the airways of normal subjects as well as in the epithelial lining of the honeycombing cyst in idiopathic pulmonary fibrosis (Ghosh et al., 2013; Seibold et al., 2013). Finally, a *simple* epithelium contains a single row of cells, with all cells being in contact with the basement membrane. This epithelial subtype is typical in the distal airways of mice.

In stratified and pseudostratified epithelia, cells that are adjacent to the basement membrane are termed *abluminal.* In contrast, cells whose apical surface contacts the airway lumen are termed *luminal.* In pseudostratified epithelia, abluminal cells typically exhibit a greater projection onto the basement membrane than do luminal cells.

Epithelial cell shape is described as *pyramidal* (triangular), *bottle or flask* (cell width at the center is greater than at the luminal or abluminal ends), *columnar* (height greater than width), *cuboidal* (height equals width), or *squamous* (height less than width). A pseudostratified epithelium contains pyramidal abluminal cells and columnar luminal cells. In contrast, simple epithelia are described as columnar, cuboidal, or squamous.

3. AIRWAY REGIONS

3.1 Overview

A more generally applicable description of the tracheobronchial tree uses airway position along the proximal–distal axis and airway caliber to specify regions. According to this nomenclature, the airway is divided into the following: (1) trachea; (2) primary (principal), secondary (lobar), tertiary (segmental), and subsegmental bronchi; and (3) bronchioles, terminal bronchioles, and respiratory bronchioles. The respiratory bronchioles open into alveolar ducts and alveoli. However, alveolar ducts and alveoli are not considered to be part of the conducting airway.

3.2 Trachea (Tracheal Airway)

The trachea extends from the cricoid cartilage of the larynx to the bifurcation, or carina, where it divides into right and left main-stem (primary) bronchi. The characteristic feature of the trachea is the relatively regular sequence of C-shaped plates of hyaline cartilage interconnected by annular ligaments. The gaps in the cartilages are bridged by trachealis muscle, which is composed of smooth muscle fibers. This region of the trachea is referred to as the membranous trachea and is positioned dorsally. In general, the trachea is lined with ciliated pseudostratified columnar epithelium. However, the cellular composition and thickness of this lining varies with position along the proximal–distal and dorsal–ventral axes, and among species.

3.3 Bronchi (Bronchial Airway)

The term *bronchus* generally refers to large airways distal to the trachea. The second division forms the secondary or lobar bronchi. These branches supply individual lung lobes. The third division is into segmental or tertiary bronchi, which supply individual lung segments. The bronchus walls contain cartilaginous plates, smooth muscle, and submucosal glands. These airways are lined with ciliated pseudostratified columnar epithelium. The cellular composition and thickness of this lining varies with airway generation and between species.

The subsegmental bronchi include the remainder of the bronchial portion of the tree. Division of the subsegmental airways is species-specific and variability in the branching pattern is discussed in Chapter 5. These airways are lined with ciliated, columnar epithelium. The organization, cellular composition, and thickness of this lining varies with airway generation and between species.

3.4 Bronchiole (Bronchiolar Airway)

Bronchioles are defined as distal airways that are the continuation of bronchi but are contained within the lung lobe. Consequently, bronchioles are considered to be intrapulmonary airways. The caliber of bronchioles is less than that of the bronchi. Bronchiolar airways lack cartilage and submucosal glands but have a prominent smooth muscle component. The bronchiolar epithelial lining varies from columnar to cuboidal.

3.5 Terminal Bronchiole (Terminal Bronchiolar Airway)

The terminal bronchiole is the most distal generation of nonalveolarized conducting airways and has the smallest caliber of all airway regions. Terminal bronchioles are supported by perobronchiolar attachments to alveolar structures (Linhartova et al., 1971, 1982). Their epithelial lining consists of cuboidal cells.

3.6 Respiratory Bronchiole (Respiratory Bronchiolar Airway)

Respiratory bronchioles form the transition between conducting and respiratory regions of the lungs. The walls of these airways contain alveolar out-pockets, which are lined with squamous (type I) and cuboidal (type II) epithelial cells. In the rhesus monkey, the respiratory bronchiole consists of two epithelial subregions: a pseudostratified and a simple cuboidal epithelium (Tyler et al., 1988). In contrast with humans and nonhuman primates, the mouse airway does not contain respiratory bronchioles.

4. COMPARATIVE STRUCTURE OF THE TRACHEOBRONCHIAL TREE

4.1 Structural Similarities

Overview: The tracheobronchial wall exhibits several structural similarities from the trachea through the respiratory bronchioles. These include a *mucosa,* a continuous epithelial lining with subjacent fine connective tissue, and a *submucosa* with smooth muscle, connective tissue and glands. The airways are loosely attached to their surrounding tissues by the *adventitia*.

Mucosa: The innermost continuous lining of the tracheobronchial tree is provided by an epithelium that rests on a basement membrane. This epithelial lining, together with a layer of fine connective tissue (the lamina propria), is referred to as the mucosa. This layer is delineated from the submucosa is by a fine smooth muscle layer referred to as the muscularis mucosae. In the tracheobronchial tree, the division of smooth muscle into layers, as in the gut, is indistinct, and most of the smooth muscle of muscularis mucosae is replaced by a rich network of elastic tissue.

Lamina propria: An ill-defined connective tissue layer, 50 to 100 Rm in thickness, attaches the basement membrane to the airway wall. The lamina propria consists of reticular, elastic, and coarser collagenous fibers and contains an extensive capillary bed not readily apparent in immersion-fixed specimens. The extent of this vascular bed is best appreciated when lungs are fixed by vascular perfusion (Mariassy et al., 1991). A network of capillary and venous channels becomes apparent. The submucosal vascular channels are especially extensive in rabbits (Hughes, 1965) and sheep (Hill et al., 1989).

Submucosa: This is a layer of connective tissue that contains the submucosal glands, tracheobronchial vasculature and innervation. The smooth muscle here is not arranged into discernible layers. Instead, it forms an irregular spiral, spanning the airway wall partially, as in the trachea, or completely surrounds the airways, as in the rest of the airway tree. In the majority of tertiary and smaller bronchi, a relatively distinct pattern of inner, quasi-longitudinal and outer circular smooth muscle spirals are discernible. The submucosa is surrounded with cartilage and coarse connective tissue of the airway wall. In bronchioles, the muscle layers become a single band of tightly spiraled smooth muscle; submucosal glands and cartilages are absent.

Adventitia: The loose connective tissue of the airway extends to the accompanying vessels and forms a sheet that delineates the common peribronchial and perivascular space. This connective tissue extends into the lung septal scaffolding from the tracheobronchial airways and vasculature to the pleura, where it blends into the pleural connective tissue.

4.2 Structural Distinctions

Trachea: Tracheal cross-sectional shape, area, and location of the bronchi varies significantly among domestic animals (Schummer et al., 1986). For instance, in artiodactyla (cloven-hoofed animals), the right cranial lobe receives a lobar bronchus directly from the trachea 5−10 cm above the carina.

The tracheal lamina propria contains an abundance of elastic fibers that run longitudinally and allow recoil of the tracheal structures after extensive neck movement. In carnivores, ruminants, and humans, a considerable area of the tracheal wall is spanned by the trachealis muscle and is referred to as the membranous trachea (pars membranacea). A very loose connective tissue layer is located between the cartilage and the external surface of the trachealis muscle. In most mammals, the trachealis muscle is inserted either into the inside perimeter or the ends of the cartilage. However, in carnivores, the muscle inserts on the cartilage periphery, allowing for extreme changes in the tracheal cross-sectional area (Schummer et al., 1986).

Bronchi: There is considerable variation of bronchial structure among animals and humans. This variation includes differences in length and size of the airways, airway wall components (e.g., epithelial lining, submucosal glands, cartilage), and smooth muscle distribution, vascularization, and innervation.

The reasons for such structural variation are not entirely clear. Some of these logically stem from adaptation to a specific environment, such as the extensive cartilage supporting the airways of diving mammals. Others, such as the extent and abundance of submucosal glands or distinct cell populations, are more difficult to interpret. An analysis of mouse bronchi, indicate that a lack submucosal glands is compensated for by expression of submucosal gland protein in the surface epithelium (Di, 2011). Interspecies differences in the number of cartilaginous airways and the pattern of branching between humans, monkeys, pigs, rats, and diving mammals have been reported (Hislop et al., 1984).

5. COMPARATIVE MORPHOLOGY OF THE AIRWAY EPITHELIUM

5.1 Morphometry

Fixation: A number of considerations must be addressed when evaluating the quantitative data characterizing airway epithelial cells. The lung, especially the airway wall, contains smooth muscle and an extensive connective tissue mass. Both of these tissues are likely to contract during fixation and tissue processing, especially in airways fixed by fluid immersion.

Instillation of fixative into the airways under a head of pressure opposes the reduction in airway diameter and cross-sectional area. Similarly, less shrinkage of the airways occurs when airways are fixed in situ through the relevant vasculature. Comparison of airway cross-sectional diameters in fixed lungs and airway casts of the rabbit indicate that there is little shrinkage with airway infusion fixation (Plopper et al., 1983b). Because changes in airway cross-sectional area directly affect epithelial height, cell population density, and epithelial cell morphology, results obtained using distinct fixation methods should not be compared.

Quantification: Variation in epithelial cell morphology can be overcome by use of defined sampling methodology (Mariassy and Plopper, 1984; Plopper et al., 1989; Hsia et al., 2010; Hyde et al., 2007; Plopper, 1983). Volume densities (V_V) are determined by point (P) counting using a grid of evenly spaced points. V_V is calculated by the formula, $V_V = P_P = P_N/P_T$. P_P is the point fraction of P_N and is determined by counting the number of test points intersecting the structure of interest divided by the total points intersecting the reference space (P_T). The reference space is frequently the epithelium. V_V is presented in $\mu m^3/\mu m^3$. The surface area of the basement membrane per reference volume (S_V) is determined by counting intercepts (I) between the surface and grid of evenly spaced cycloid lines. S_V was calculated using the formula, $S_V = 2I_o/L_T$. In this formula, I_0 is the number of intersections between the cycloid grid and a surface, the basement membrane. L_T is the total length of the test line within the reference volume, the epithelium. The thickness of a structure or the volume per unit area of basement (Vs) membrane is calculated by determining the arithmetic mean thickness: $\tau = V_V/S_V$. This parameter is presented in $\mu m^3/\mu m^2$ (Hyde et al., 2007).

5.2 Epithelial Structure

The tracheobronchial epithelium varies from tall columnar pseudostratified in the trachea, through simple low cuboidal in the bronchioles, to the simple squamous epithelium of the alveolar out-pockets in respiratory bronchioles. Surface epithelial cell abundance (Figure 1 and Tables 2–5) varies among species and airway generations. Similarly, airway epithelial height (Table 1) varies among animal species and airway generations. In general, a positive correlation exists between the animal size and epithelial height and a negative correlation is found between animal size and airway generation number. Additional variations of the epithelial thickness are known to occur in the trachea when cartilaginous and membranous regions are compared (Wilson et al., 1984). The histochemistry and cytochemistry of the tracheobronchial tree is reviewed in greater detail in Chapter 5.

5.3 Nonsecretory Epithelial Cell Types

Introduction: The conducting airway epithelium is typically characterized as a *mucociliated* epithelium. Consequently, epithelial cells are subtyped first as *ciliated* or nonciliated cells. The latter cell population can be further subdivided into *nonsecretory* and *secretory* cells. Finally, the nonsecretory cell population contains several subpopulations including the *basal* and *myoepithelial* cell types which are discussed in this section.

Ciliated Cell: The morphological features of this airway epithelial cell type have been reviewed (Breeze and Wheeldon, 1977; Sturgess and Turner, 1984; Kuhn, 1976) (Figures 2–4). This section is focused on the quantitative aspects of the ciliated cells, their relationship to neighboring epithelial cells, and their structure–function correlates.

Ciliated cells extend through the entire thickness of the epithelium, and their height corresponds to the epithelial height (Table 1). Although classically defined as columnar,

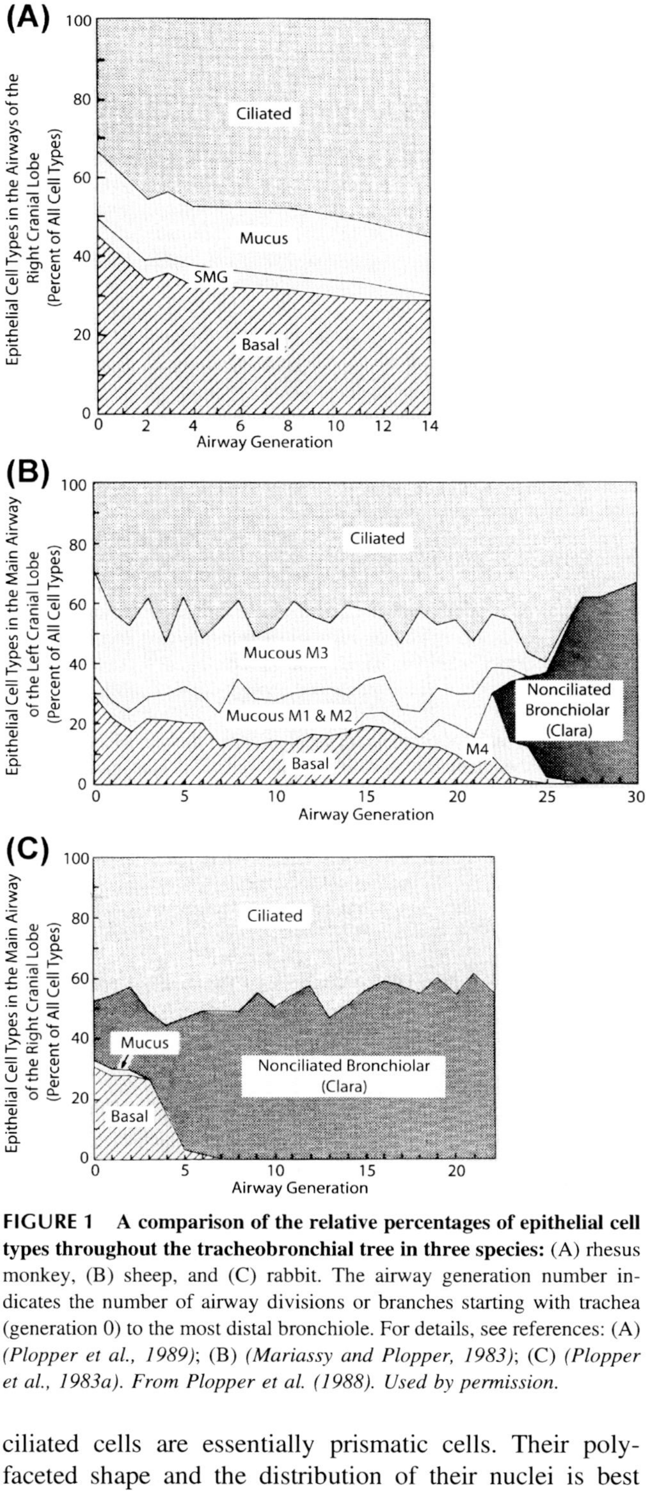

FIGURE 1 A comparison of the relative percentages of epithelial cell types throughout the tracheobronchial tree in three species: (A) rhesus monkey, (B) sheep, and (C) rabbit. The airway generation number indicates the number of airway divisions or branches starting with trachea (generation 0) to the most distal bronchiole. For details, see references: (A) *(Plopper et al., 1989)*; (B) *(Mariassy and Plopper, 1983)*; (C) *(Plopper et al., 1983a). From Plopper et al. (1988). Used by permission.*

ciliated cells are essentially prismatic cells. Their polyfaceted shape and the distribution of their nuclei is best appreciated in sections of epithelium parallel to the basement membrane. From their shape, it is evident that ciliated cells are extremely malleable and conform their shape to the shape of neighboring cells, with the exception of their large vesicular nuclei.

The apical surface of ciliated cells contains cilia and microvilli, both of which project into the airway lumen. The number of the cilia per cell varies from 50−100 (Fawcet, 1986) to 300 (Jeffery, 1987). The most commonly quoted ciliary dimensions are 6 pm in length and 0.3 pm in diameter; these values very likely originate from measurements of human tracheal ciliated cells (Rhodin, 1966). During human fetal development, ciliary length varies from 5.5 to 10.5 ptin (Gaillard et al., 1989). In the trachea of domestic animals, the length (in pm) of cilia are as follows: horse, 6.7 ± 0.6; cattle, 4.2 ± 0.5; pig, 5.0 ± 0.0; cat, 5.1 ± 0.1; rabbit, 4.6 ± 0.6; guinea pig, 5.0 ± 0.0; rat, 5.0 ± 0.0; mouse, 4.6 ± 1.1; hamster, 4.2 ± 0.6 (Pavelka et al., 1976).

Microvilli are 2−3 pm in length and 0.1 pm in diameter. They contain parallel filaments and their surface is lined by the glycocalyx (Breeze and Wheeldon, 1977). In humans, microvillar length was reported to range from 0.7 to 3.01 m, the diameter from 0.12 to 0.13 pm, and microvillus density from 10 to 15 per sectioned ciliated cell profile (Gilljam et al., 1987; Sturgess, 1984).

Ciliated cells are firmly attached at their luminal end to neighboring cells by junctional complexes. The lateral attachment is variable and consists of desmosomes. Ciliated cells, with some exceptions (e.g., dog; Kawanami et al., 1979), do not have hemidesmosomes on the basal surface.

The cytoplasm of ciliated cells generally is less electron dense than that of nonciliated cells. There is little granular endoplasmic reticulum (GER). A variable amount of glycogen is usually present. The Golgi apparatus varies from a few cistemae to multiple complexes. The upper one third of the ciliated cell, just below the terminal web, contains a high density of mitochondria. Agranular endoplasmic reticulum (AER) and vesicles are present in ciliated cells (Rhodin, 1966). Another organelle frequently found in abundance in ciliated cells is the lysosome. The glycoconjugates in these organelles makes them conspicuous, even at the light microscopic level, with routine carbohydrate histochemistry. The co-location of an extremely large surface area and the presence of transport vesicles, AER, mitochondria, microtubules, and lysosomes suggest that ciliated cells are involved in transepithelial transport.

One of the most consistent features of the surface epithelial population of different species is the abundance and distribution of ciliated cells. This cell type is between 30% and 50% of the total epithelial population in the trachea (Figure 1; Table 2). In those species that have been studied, the proportion of ciliated cells increases in the primary bronchi (Table 3) of some species and decreases in others (Table 2). Ciliated cells are between 40% and 50% of the population in more distal bronchi (Figure 1; Tables 4 and 5). Despite this relative similarity, some heterogeneity in ciliated cell distribution has been reported. For instance, in the cat, the tracheal pseudostratified epithelium varies from a smooth contour to longitudinally oriented ciliated cell lined clefts (Tandler et al., 1983a,b). Similar ciliated

TABLE 1 Height of the Airway Epithelium in Adult Mammalians

	Trachea		Bronchi					
	n	(Generation 0)	*n*	(Generations 1–2)	*n*	(Generations 4–6)	*n*	(Generations 7–11)
Primates								
Human (Pavelka et al., 1976; Plopper et al., 1983c; Korhonen et al., 1969)	3	50–100 (Korhonen et al., 1969); 43 ± 2	5	50-; 39- (Plopper et al., 1983c; Korhonen et al., 1969)'	–	–	–	–
Bonnet monkey (*M. radiata*) (Wilson et al., 1984)	3	20 ± 7	3	21 ± 6	–	–	–	–
Rhesus monkey (*M. mullata*) (Plopper et al., 1989)	5	28 ± 5	3	19 ± 1	5	17 ± 1	5	13 ± 1
Carnivores								
Dog (Widdicombe et al., 1986)	5	26 ± 4	–	20–50	–	–	–	–
Cat (Pavelka et al., 1976)	5	20–24 (Evans et al., 1989); 24±	–	–	–	–	–	–
Ferret (Leigh et al., 1986; Evans et al., 1989)	4	–	–	–	–	–	–	–
Perissodactyla								
Horse (Pavelka et al., 1976)	3	48 ± 3	–	–	–	–	–	–
Artiodactyla								
Cow (Pavelka et al., 1976)	5	60 ± 3	–	–	–	–	–	–
Sheep (Mariassy and Plopper, 1983; Evans et al., 1989)	5	41–57 (Evans et al., 1989); 59 ± 7[a]	6	40 ± 3	8	32 ± 4[a]	6	30 ± 5[a]
Pig (Pavelka et al., 1976)	5	46 ± 5	3	35–50 (Pavelka et al., 1976)	–	–	–	–
Rodentia								
Guinea pig (Pavelka et al., 1976)	5	30 ± 3	–	15–30 (Pavelka et al., 1976)	–	–	–	–
Hamster (Plopper et al., 1983c; Kennedy et al., 1978; Evans et al., 1989)	3	20 ± 0 (Kennedy et al., 1978); 8–14 (Evans et al., 1989)	–	151 ± 11	5	20–	–	–
Rat (Plopper et al., 1983c; Nikula et al., 1988; Pack et al., 1980)	3	24–1 (Nikula et al., 1988)	20	13–	6	13–	–	–
Mouse (Pack et al., 1980; Evans et al., 1989)	6	11–14 (Evans et al., 1989)	–	–	3	8–16	–	–
Lagomorpha								
Rabbit (Plopper et al., 1983a)	–	21 ± 1	4	22 ± 3	6	9 ± 2	5	9 ± 1

– Data not available.
± SD.
Jeffery and Reid (1975); Widdicombe et al. (1986).
[a]*insufflation fixed, epon embedded trachea (personal observation, Mariassy, A.T.).*

TABLE 2 Population Densities (% Cells) of the Tracheal Surface Epithelium in Adult Mammals and Juvenile Birds

	Cells/mm		Epithelial Cells							
			Mucous							
	n	Basal Lamina	Basal	Ciliated	(Goblet)	Serous	Other[a]	Clara	Brush	Unidentified
Primates										
Human (Pavelka et al., 1976; Korhonen et al., 1969)	3	303 ± 20	33	49	9	0	9	0	–	–
Bonnet monkey (*M. radiata*) (Wilson et al., 1984)	3	266 ± 12	31	41	0	0	16 SMGC	0	<1	4
Rhesus monkey (*M. mullata*) (Plopper et al., 1989)	5	181 ± 51	42	33	17	0	4 SMGC	0	<1	4
Carnivores										
Dog (Widdicombe et al., 1986)	5	–	–	–	–	–	–	–	–	–
Cat (Pavelka et al., 1976)	5	273 ± 15	37	36	20	0	5	0	<1	1
Ferret (Leigh et al., 1986)	4	183–	25	54	22	0	–	0	Rare	–
Perissodactyla										
Horse (Pavelka et al., 1976)	3	307 ± 23	31	46	5	0	18	0	–	–
Artiodactyla										
Cow (Pavelka et al., 1976)	5	323 ± 24	31	42	4	0	23	0	<1	–
Sheep (Mariassy and Plopper, 1983)	5	414 ± 33	29	31	5	0	36 M3 (Mariassy and Plopper, 1983)	0	<1	0
Pig (Pavelka et al., 1976)	5	303 ± 17	31	43	3	0	23	0	<1	–
Rodentia										
Guinea pig (Pavelka et al., 1976)	5	307 ± 5	34	32	5	0	29	0	<0	–
Hamster (Plopper et al., 1983c; Kennedy et al., 1978)	3	151 ± 11	6	48	0	0	0	41	<1	5
Rat (Plopper et al., 1983c; Evans et al., 1989)	3	148 ± 3	13	41	1	39	0	0	1	6
Rat (Nikula et al., 1988)	5	168 ± 12	21	32	2	42	1	0	>1	1
Rat (Jeffery and Reid, 1975)	–	142–	27	33	<1	27	13	0	–	–
Mouse (Pack et al., 1980)	6	215–	10	39	<1	0	1	49	<1	–
Lagomorpha										
Rabbit (Plopper et al., 1983a)	–	211 ± 30	28	43	1	0	0	18	Rare	9
Avia										
Goose (Phipps and Richardson, 1976)	–	213–	33	50	12	<0.01	0	0	0	5

– Data not available.
± SD.
SMGC, small mucous granule cell.
[a]*Nonciliated cells, mostly eccrine (confirm in reference).*

TABLE 3 Population Densities (% Cells) of the Bronchial Surface Epithelium in Adult Mammals (Mainstem, Primary Bronchi)

		Cells/mm	Epithelial Cells							
					Mucous					
	n	Basal Lamina	Basal	Ciliated	(Goblet)	Serous	Other[a]	Clara	Brush	Unidentified
Primates										
Human (Gilljam et al., 1987)	5	–	6 ± 1	56 ± 10	26 ± 5	Scarce	31 ± 2	0	–	–
Bonnet monkey (*M. radiata*) (Wilson et al., 1984)	3	210 ± 16	35	33	10	0	15 SMGC	0	–	6
Rhesus monkey (*M. mullata*) (Plopper et al., 1989)	5	175 ± 46	32	44	15	0	5 SMGC	0	–	3
Carnivores										
Dog	–	–	–	–	–	–	–	–	–	–
Cat	–	–	–	–	–	–	–	–	–	–
Ferret (Robinson et al., 1986)	3	–	–	–	50/mm	–	20/mm	0	–	–
Perissodactyla										
Horse	–	–	–	–	–	–	–	–	–	–
Artiodactyla										
Cow	–	–	–	–	–	–	–	–	–	–
Sheep (Mariassy and Plopper, 1983)	6	285 ± 24	18	48	4	0	30 M3 (Christensen et al., 1987)	0	<1	0
Pig	–	–	–	–	–	–	–	–	–	–
Rodentia										
Guinea pig	–	–	–	–	–	–	–	–	–	–
Hamster (Christensen et al., 1987)	6	–	22	49	4	0.2	8	15	Rare	1
Rat (Jeffery and Reid, 1975)	20	126–	27	35	<1	21	16	0	–	–
Mouse (Pack et al., 1980)	3	215–	4	47	1	0	2	46	<1	0
Lagomorpha										
Rabbit (Plopper et al., 1983a)	4	194 ± 17	27	43	1	0	7	22	<1	–

– Data not available.
± SD.
SMGC, small mucous granule cell.
[a]*Nonciliated cells, mostly eccrine (confirm in reference).*

TABLE 4 Population Densities (% Cells) of the Bronchial Surface Epithelium in Adult Mammals (Lobar Bronchi, Generations 2–6)

		Cells/mm	Epithelial Cells							
					Mucous					
	n	Basal Lamina	Basal	Ciliated	(Goblet)	Serous	Other[a]	Clara	Brush	Unidentified
Primates										
Human	–	–	–	–	–	–	–	–	–	–
Bonnet monkey (*M. radiata*)	–	–	–	–	–	–	–	–	–	–
Rhesus monkey (*M. mullata*) (Plopper et al., 1989)	5	184 ± 49	32	47	15	0	5 SMGC	0	–	2
Carnivores										
Dog	–	–	–	–	–	–	–	–	–	–
Cat	–	–	–	–	–	–	–	–	–	–
Ferret	–	–	–	–	–	–	–	–	–	–
Perissodactyla										
Horse	–	–	–	–	–	–	–	–	–	–
Artiodactyla										
Cow	–	–	–	–	–	–	–	–	–	–
Sheep (Mariassy and Plopper, 1983)	8	248 ± 23	19	39	12	0	30 M3 (Plopper et al., 1983a)	0	<1	0
Pig	–	–	–	–	–	–	–	–	–	–
Rodentia										
Guinea pig	–	–	–	–	–	–	–	–	–	–
Hamster (Kennedy et al., 1978)	5	179–	18	35	27	0	12	7	<1	1
Rat (Jeffery and Reid, 1975)	6	116–	14	53	<1	20	12	0	Infrequent	
Mouse (Pack et al., 1980)	3	199–	1	36	0	0	2	61	<1	<1
Lagomorpha										
Rabbit (Plopper et al., 1983a)	6	114 ± 12	2	49	0	0	7	41	<1	–

– Data not available.
± SD.
SMGC, small mucous granule cell.
[a]*Nonciliated cells, mostly eccrine (confirm in reference).*

TABLE 5 Population Densities (% Cells) of the Bronchial Surface Epithelium in Adult Mammals Segmental Bronchi, Generations 7–11

		Cells/mm	Epithelial Cells							
					Mucous					
	n	Basal Lamina	Basal	Ciliated	Goblet	Serous	Other[a]	Clara	Brush	Unidentified
Primates										
Human	–	–	–	–	–	–	–	–	–	–
Bonnet monkey (*M. radiata*)	–	–	–	–	–	–	–	–	–	–
Rhesus monkey (*M. mullata*) (Plopper et al., 1989)	5	158 ± 15	29	49	14	0	3 SMGC	0	–	2
Carnivores										
Dog	–	–	–	–	–	–	–	–	–	–
Cat	–	–	–	–	–	–	–	–	–	–
Ferret	–	–	–	–	–	–	–	–	–	–
Perissodactyla										
Horse	–	–	–	–	–	–	–	–	–	–
Artiodactyla										
Cow	–	–	–	–	–	–	–	–	–	–
Sheep (Mariassy and Plopper, 1983)	8	223 ± 17	18	43	8	0	31 M3 (Mariassy and Plopper, 1983)	0	<1	0
Pig	–	–	–	–	–	–	–	–	–	–
Rodentia										
Guinea pig	–	–	–	–	–	–	–	–	–	–
Hamster	–	–	–	–	–	–	–	–	–	–
Rat	–	–	–	–	–	–	–	–	–	–
Mouse	–	–	–	–	–	–	–	–	–	–
Lagomorpha										
Rabbit (Plopper et al., 1983a)	5	147 ± 18	0	49	0	0	4	47	–	0

– Data not available.
± SD.
SMGC, small mucous granule cells.
[a]*Nonciliated cells, mostly eccrine (confirm in reference).*

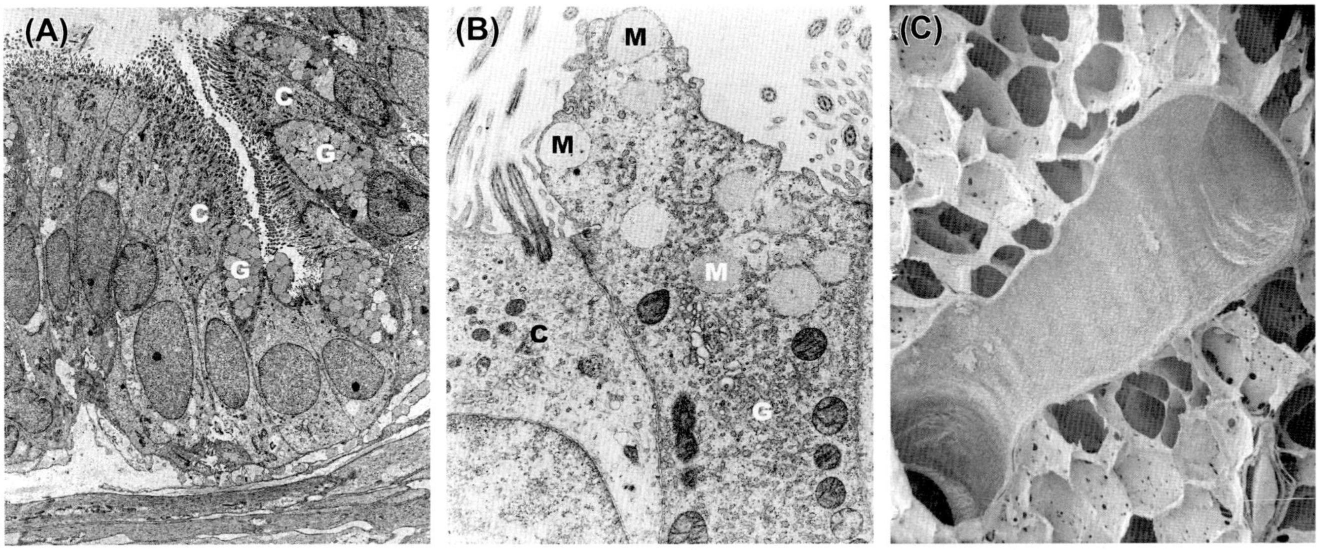

FIGURE 2 Ultrastructure of the pseudostratified columnar epithelium. (A, B) Transmission electron micrograph of epithelial cells from sheep bronchi, generations 8–15. (A) Transmission electron micrograph of goblet (G) cells that are sub-classified as M1 mucous cells. M1 cells contain coarse electron-dense secretory granules and are surrounded by basal cells and ciliated cells (C). (B) Transmission electron micrograph of a goblet cell that is sub-classified as a M2 mucus cell. M2 cells contain coarse electron-lucent secretory granules and are flanked by ciliated cells (C) and by profiles of two M3 mucous cells that contain small nucleated granules and abundant AER. (C) Scanning electron micrograph of the rhesus monkey bronchi, generation 4. Small mucous granule cell (SMGC) and M1 mucous cells are shown.

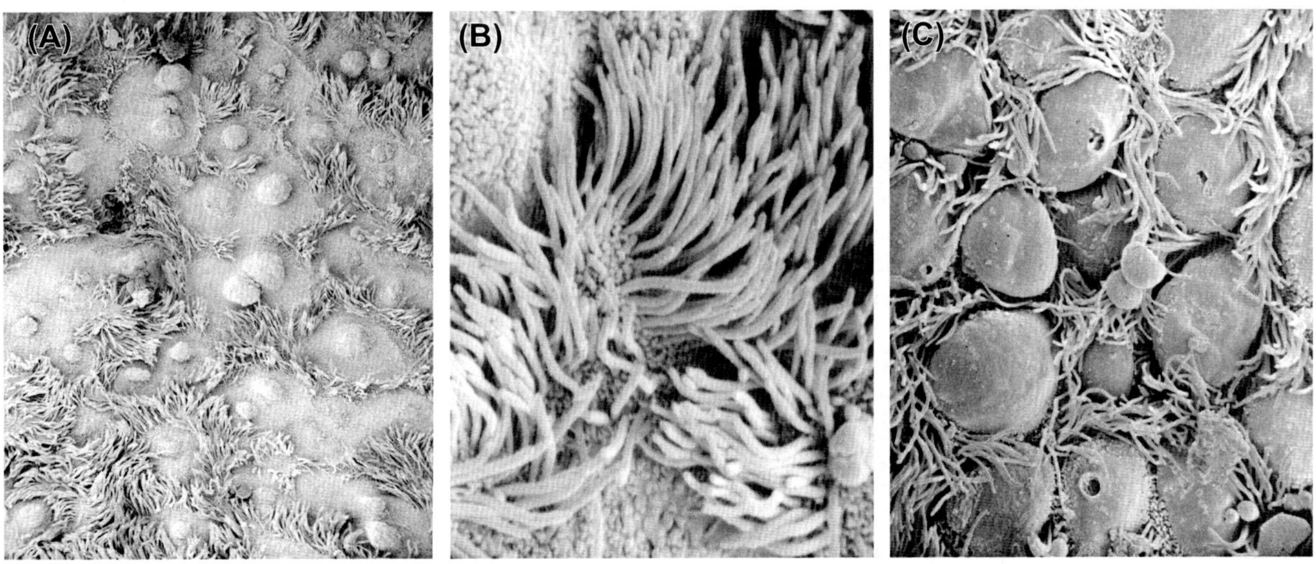

FIGURE 3 Ultrastructure of the tracheobronchial glands. (A) Scanning electron-microscopic image of the rhesus monkey tracheal submucosal gland. Ciliated containing apical cilia and domed serous cells are interspersed within the epithelium. (B) Scanning electron-microscopic image of the sheep bronchus, generation 22. Two M4 mucus cells are located among ciliated cells. M4 cells contain large secretory granules having an electron-dense core and a lucent halo. These cells appear to be very similar to the sero-mucous cells in the submucosal glands. (C) Scanning electron-microscopic image of the sheep submucosal gland, generation 7. Serous cells form the majority of the gland epithelium.

cell population density variation was reported for the posterior trachea of the bonnet monkey (Wilson et al., 1984).

Basal cell: The basal cell is a relatively small (5–15 m), dense, pleomorphic, low polygonal cell. The shape of basal cells is profoundly affected by the constriction of the airways. Thus, basal cells can be triangular or tall/spindle-shaped. Their cytoplasm characteristically contains bundles of intermediate or tonofilaments, extending from hemidesmosomes facing the basement membrane to numerous desmosomes attaching the basal cell to its neighboring epithelial cells (Rhodin, 1985). In hamsters, tonofilaments are absent (Plopper

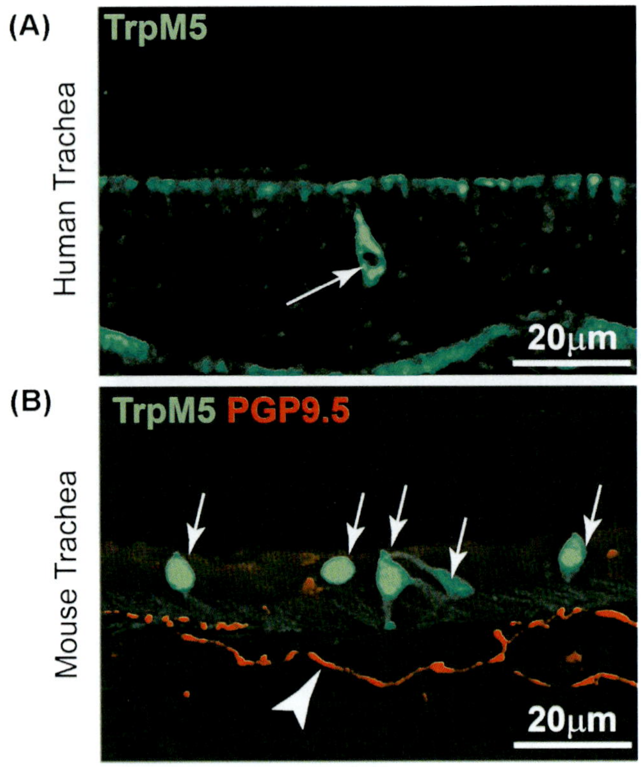

FIGURE 4 Brush cells located in the pseudostratified epithelium of the trachea. (A,B) Immunofluorescence detection of TrpM5 (green) in human (A) and TrpM5 (green) and PGP9.5 (red) in mouse (B). Arrows, brush cells; arrowhead, nerve.

et al., 1983c). The electron-dense cytoplasm contains sparse organelles, which include ribosomes and mitochondria. The basal cell nucleus is large relative to cell size and a high nuclear-to-cytoplasmic ratio is a characteristic feature. Nuclear chromatin is condensed and peripherally marginated and is most often considerably denser than in the nuclei of other neighboring epithelial cells.

Basal cells contact the basal lamina. They are the only cells, at least in the trachea, having hemidesmosomal attachments to the basal lamina (Evans et al., 1989). An exception to this has been reported in the dog (Kawanami et al., 1979). Basal cells play an important role in the attachment of columnar cells, especially in airways with high pseudo-stratified epithelium (Evans et al., 1989). In such areas, basal cells occupy as much as 95% of the contact surface of the basement membrane, even though they represent only 7–10% of the cell population (Evans et al., 1989). This, and direct morphologic and quantitative data, show that the columnar cells contact the basement membrane through a disproportionally small area, and that these attachments lack hemidesmosomes.

The majority of columnar cell adhesion is mediated through numerous desmosomes that connect columnar cells to basal cells. Direct contacts between columnar cells and the basement membrane account for only 3% of attachments. Evans and colleagues reported a close correlation (r = 0.97) between the number of basal epithelial cells and the height of the epithelium across examined species (Evans et al., 1989). The same group also indicated a positive correlation between basal cell number per unit length of the basement membrane and animal size. Population densities of basal cells in the trachea of 15 mammalian species are shown in Table 2. The percentage of basal cells ranges from 6 in the hamster through 33 in humans and 42 in rhesus monkeys.

Basal cells are part of the epithelial population in the trachea and in all generations of bronchi proximal to generation 3 in all species studied (Tables 2–4). However, the numerical density of basal cells has not been reported for intrapulmonary bronchi in many species. Basal cells have been reported in bronchioles of primates (see Chapter 8), but not in bronchi distal to generation 6 in other species (Table 5).

Basal cell frequency is altered by aging, injury, and disease. Wansleeben reported that epithelial cell density and basal cell frequency decreases in male and female C57Bl/6 mice by 22 months of age (Wansleeben et al., 2014). In mice, epithelial cell death in response to acid or naphthalene exposure results in transient basal cell hyperplasia (Cole et al., 2010; Rock et al., 2009). These hyperplastic basal cells increase expression of the intermediate filament protein keratin 14, exhibit an increased mitotic index, and have been termed reparative basal cells (Cole et al., 2010). Similar cells were identified in human tracheobronchial tissue and may be indicative of an injury/repair process (Ghosh et al., 2013; Hegab et al., 2011).

Myoepithelial cell: Myoepithelial cells have been extensively described in the salivary gland. In this tissue, they are fusiform or stellate cells that are adjacent to the basement membrane of glandular acini and smaller ducts (Pinkstaff, 1980; Shear, 1966). In the lung, myoepithelial cells are limited to the submucosal gland.

Myoepithelial cells exhibit histological and morphological similarity to smooth muscle and are generally considered to be contractile. Rha reported that cat tracheal submucosal gland myoepithelial, serous, and mucus cells were in close contact with substance P immunoreactive nerve fibers (Rha et al., 1994). However, no contraction was detected when porcine submucosal gland secretion was induced by substance P or other agonists that increased glandular secretion rate (Khansaheb et al., 2011). Shimura and colleagues suggested that myoepithelial cell contraction in cats was primarily dependent on cholinergic nerve stimulation and muscarinic receptors. Adrenergic nerves regulated contraction to a smaller extent (Shimura et al., 1987). In canine submucosal glands, the adrenergic terminals were associated with myoepithelial and gland cells (Hisa et al., 1982).

Several reports of epithelial–myoepithelial carcinoma suggest that myoepithelial cells can proliferate (Doganay et al., 2003; Wilson and Moran, 1997). This function is supported by lineage tracing analysis in human (Engelhardt et al., 1995) and mouse injury/repair (Borthwick et al., 2001) and basal cell purification studies (Hegab et al., 2011, 2012).

In humans, myoepithelal cell hypertrophy was associated with fatal asthma, and an increase in intracellular myofilaments suggested increased contractility (Green et al., 2010). Xie et al. demonstrated that calcitonin gene-related peptide (CGRP; see below) activates CFTR-dependent submucosal gland secretion and that the putative CGRP-to-myoepithelial cell signaling pathway is hyperactivated in cystic fibrosis patients and in the pig, ferret, and mouse models of cystic fibrosis (Xie et al., 2011).

5.4 Chemosensory Epithelial Cells

Introduction: Organs and tissues interact with the environment via physical and chemical sensors. Light, sound, and pressure are detected by specialized receptor cells, including photoreceptors for vision (Zeiss and Johnson, 2004), auditory hair cells for hearing (Lopez-Schier, 2004), and Merkel cells for touch (Vaigot et al., 1987). In contrast, the chemical senses utilize four chemosensory systems: olfaction, taste, solitary chemoreceptor cells (SCC), and free nerve endings (Tizzano and Finger, 2013).

Differences in structure, innervation, and utilization distinguish the chemosensory systems. Furthermore, chemosensory systems are subcategorized as internal or external. Internal sensory systems, such as the carotid body, monitor oxygen concentration in the tissues that are not directly exposed to the environment (Milsom and Burleson, 2007). In contrast, external sensory systems, including the pulmonary neuroepithelial body (NEB) (see below), evaluate oxygen levels in inspired air (Cutz et al., 2013).

The cells that comprise many external chemosensory systems (including olfaction, taste buds, and nasal solitary chemosensory cells) are replaced at a rate that is similar to that of surrounding epithelium (Moulton, 1974; Beidler and Smallman, 1965; Gulbransen et al., 2008). Similarly, solitary chemosensory cells, taste receptor cells (Farbman, 1980), and both main and accessory olfactory receptor neurons (Moulton, 1974; Beidler and Smallman, 1965) are renewed through an organism's life.

The conducting airway epithelium senses the luminal environment via two, if not more, chemosensory systems. Solitary chemosensory cells, which were shown to be synonymous with the mysterious *brush cell* (Reid et al., 2005), detect bitter tastants, including bacterial quorum sensing molecules (Krasteva et al., 2012). In contrast, NEBs sense hypoxic conditions during the postnatal period (Cutz et al., 2013). Many additional NEB functions have been discovered over the past three decades, suggesting that additional brush cell functions also await discovery.

Brush cell: The brush cell has many names, including tuft cell, fibrillovesicular cell, multivesicular cell, and caveolated cell. The brush cell is characterized by an apical cluster of rigid, densely packed, large (2–3 pm) and long microvilli. Each microvillus contains core filaments that typically extend in bundles through the terminal web and end in the supranuclear cytoplasm.

The brush cell has been identified in a majority of examined species, including mammals and aquatic vertebrates (Tables 2–5; Breeze and Wheeldon, 1977; Tizzano and Finger, 2013). This cell type has been described at all levels of the respiratory tract, including the trachea and bronchi of many species (Tizzano et al., 2006; Tizzano et al., 2011; Merigo et al., 2005; Barham et al., 2013) and in the alveolus of rats (Meyrick and Reid, 1968). Brush cells are relatively rare and are always less than 1% of the cell epithelial cell population in all species studied, including the mouse (Saunders et al., 2013).

Renewal of chemosensory cells provides a mechanism by which cells damaged by occasional exposure to toxic chemicals can be replaced (Gulbransen et al., 2008). In contrast with most epithelial chemosensory cells, brush cells in the trachea of adult mice are a relatively static population. In vitro studies suggest that brush cells are replaced through proliferation and differentiation of a non-brush cell type, potentially the basal cell (Saunders et al., 2013).

Brush cell functions were previously thought to include monitoring the depth of the periciliary fluid. However, studies have identified a critical role in chemoreception, particularly in sampling the chemical composition of the airway lining fluid (Krasteva and Kummer, 2012; Kinnamon and Reynolds, 2009; Kinnamon, 2012; Finger et al., 2003). Consequently, brush cells are now categorized as lung epithelial chemosensory cells.

Brush cells express components of the canonical sweet, umami (glutamate), and bitter taste transduction pathways (Kinnamon and Reynolds, 2009; Kinnamon, 2012). Receptors for sweet and umami tastants are G-protein coupled receptors termed T1R, whereas bitter compounds are detected by a distinct G-protein coupled receptor, the T2R (Kinnamon, 2012). Both receptor types signal through second messenger signaling mechanisms, activate phospholipase Cβ2, and isositol trisphosphate, and mediate Ca^{2+} release from internal stores by TrpM5 (Krasteva et al., 2012; Tizzano et al., 2011; Merigo et al., 2005; Finger et al., 2003). These signals result in brush cell plasma membrane depolarization, acetylcholine release, and activation of gustatory afferents that are in close apposition to the brush cell body. Vagal nerve activation results in protective respiratory reflexes (Krasteva et al., 2011).

Studies have demonstrated that brush cells sense bitter tastants, such as bacterial quorum sensing molecules, and regulate respiration rate (Krasteva et al., 2011; Krasteva et al., 2012). Analysis of ciliary beat frequency indicates that T2R activation on ciliated epithelial cells increases ciliary beat frequency (Shah et al., 2009) and implies a similar function for T2R activation on brush cells. Finally, brush cells activate mast cells and neurogenic inflammation via stimulation of cholinergic nerves (Saunders et al., 2014).

Pulmonary neuroendocrine cells: Pulmonary neuroendocrine cells (PNECs) have been referred to by numerous names, including helle zelle (clear cells), Feyrter's cells, Kultschitzky cells, AUPD (amine precursor uptake and decarboxylation), endocrine-like cells, enterochromaffin-like cells, and small granule cells (Pearsall, 1989). Johnson and Georgieff coined the term PNEC and suggested a role for these cells in the pathobiology of lung disease (Johnson and Georgieff, 1989).

The morphology of PNECs has been extensively reviewed (Breeze and Wheeldon, 1977; Lauweryns et al., 1972; Sorokin et al., 1983). PNECs are typically bottle-shaped and their cytoplasm contains numerous dense core vesicles. These vesicles contain neurotransmitters including amines (serotonin, 5-HT) and peptides (calcitonin gene related peptide (CGRP), substance P, protein gene product 9.5, and chromogranin A). A full list of NEC products and markers was assembled by Cutz and coworkers (Cutz et al., 2013).

PNECs exist as single cells (solitary PNECs) or in clusters of 2–25 cells termed neuroepithelial bodies (NEBs). Both solitary NECs and/or NEBs are innervated in rabbit (Lauweryns and Cokelaere, 1973; Lauweryns et al., 1977, 1978), although there may be a noninnervated population in rat (Larson et al., 2003). In innervated NEB of rat, mouse, and rabbit, vagal afferents that are derived from the nodose ganglion form arborizations within and surrounding NEBs (Brouns et al., 2006, 2009; Adriaensen and Timmermans, 2004). These afferents are mylenated in the rat and mouse (Lauweryns et al., 1985). However, very few NEBs are mylenated in the neonatal rat (Cutz et al., 2013). Furthermore, the dorsal root ganglion forms a nerve plexus at NEBs within the rat airway (Brouns et al., 2006, 2009).

NEBs exhibit age-related variations in ultrastructure. In prenatal mice and humans, NEB access to the airway lumen is limited by nonciliated cells, termed Clara cells (see below) (Haller, 1994; Khoor et al., 1996). The PNEC surface that is exposed to the lumen is partially covered with small protrusions and microvilli. Small bare areas that lack these structures are associated with vesicles. As mice age, the protrusions become more regular and the bare areas become more common.

PNEC and NEB function varies with developmental age. PNECs are the first specialized epithelial cell type to be detected during lung development (Sorokin et al., 1983; Linnoila, 2006). Early PNECs express serotonin but have few dense core vesicles. Pulse/chase studies in hamsters demonstrated that NEB were foci for mitotic epithelial cells and suggested that PNEC/NEB elaborated epithelial growth factors (Hoyt et al., 1990). In guinea pigs and rats, CGRP functions as a mitogen for tracheal epithelial cells (White et al., 1993; Hoyt et al., 1993). In humans and mice, gastrin-releasing peptide (the major neuropeptide generated by NEB in these humans) is elaborated by NEB and stimulates epithelial growth (Sunday et al., 1990, 1993).

During the transition to air breathing and during the early postnatal period, NEBs function as oxygen sensors. Lauweryns and colleagues demonstrated that PNECs degranulate in response to hypoxia but not hypoxemia (Lauweryns and Cokelaere, 1973; Lauweryns et al., 1977, 1978). Degranulation results in amine release and may promote the transition from the fetal to the postnatal pulmonary circulation and/or balance perfusion (Cutz et al., 2013). Cutz and colleagues demonstrated that PNECs detect hypoxia using signaling systems found in other O_2-sensing cells. Hypoxia inhibits an O_2-sensitive K-channel, causes membrane depolarization, activates voltage-gated Ca^{2+} channels, and releases serotonin (Weir et al., 2005; Lopez-Barneo, 1996; Archer and Michelakis, 2002). PNECs also use an alternative O_2 sensor, NADPH oxidase (Wang et al., 1996). Gene knockout studies in mice indicate that this system is unique to NEBs and that it is not utilized by other O_2-sensing cell types (Kazemian et al., 2001; Fu et al., 2000).

Rabbit PNECs also sense the concentration of CO_2 and acid in the apical lining fluid. The molecular mechanism regulating PNEC detection of CO_2 and acid has not been determined but is thought to be similar to that found in other CO_2/acid sensory cells. It is not known whether PNEC/NEB detect O_2, CO_2, or acid in the adult lung (reviewed in Cutz et al., 2013).

NEBs were thought to be preferentially located at airway branch points (Lauweryns et al., 1972). However, analysis of adult mouse airways demonstrated that NEBs were common at branch points and in the airway walls opposite to the branch point (Peake et al., 2000). Following injury by the Clara cell toxicant naphthalene, PNECs proliferate and NEBs become hypertrophic (increase number of cells per NEB) and hyperplastic (increased numbers of NEB per airway) (Reynolds et al., 2000; Stevens et al., 1997). Interestingly, the NEB area increases after naphthalene injury (Peake et al., 2000). Clara cells that overlay the NEB in the steady state become interspersed with PNEC and the Clara cells increase their projection on the basement membrane. The mechanism(s) regulating this morphological change and its impact on epithelial repair are unknown.

5.5 Secretory Cells

Overview: Secretory cells were originally distinguished on the basis of ultrastructural characteristics. *Mucous cells* contain relatively homogeneous granules. *Seromucous cells* contain dense, osmiophilic, proteinaceous granule core, and a surrounding halo. *Serous cells* are distinguished from other secretory cell types by position and are abundant in many glands in a variety of animal species and humans (Basbaum et al., 1990). *Clara cells* are nonciliated exocrine cells that contain spherical to ovoid-shaped, membrane-bound secretory granules. *Degranulated exocrine cells* (intermediate) are morphologically similar to secretory cells. However, they contain a dense cytoplasm, a small nucleus, and lack substantial secretory granules. *Small mucous granule cells* (SMGCs) contain a few relatively small secretory granules. Their morphological appearance was interpreted as representing mucus-secreting cells in the early stages of the secretory cycle.

Mucous cell (Goblet): Morphological characteristics of the mucous cell have been examined in detail in a number of species, including humans (Rhodin, 1966) (Sturgess, 1984; Basbaum and Finkbeiner, 1989). Additional information may be found in comparative studies (Pavelka et al., 1976; Plopper et al., 1983c; Jeffery et al., 1982; Jeffery, 1983).

Because most routine fixatives do not preserve the carbohydrate component of the mucous secretory granules and tissue embedding further elutes their content, mucous cells in such preparations appear empty. However, in well-preserved histological specimens, carbohydrates can be divided into neutral, acidic, and/or acidic sulfated subsets using histochemical methods (e.g., periodic acid Schiff [PAS], Alcian blue-PAS, and high-iron diamine [HID]). Further histochemical characterization of the carbohydrate mucus composition can be based on lectin (specific carbohydrate binding molecule) histochemistry (Thaete et al., 1981; Mazzuca et al., 1982; Mariassy et al., 1988a).

In the trachea, most mucous cells are columnar and only rare cells have the classical goblet shape. When sectioned parallel to the basement membrane, mucus cells have a rounded profile. Apical attachment of mucous cells is made by junctional complexes. Desmosomal attachments anchor the mucous cell to its neighbors on the lateral surfaces. Contacts with the basal lamina are often lacking in longitudinal sections due to the attenuated cytoplasm forming the stem.

Mucous cell shape and the organelles detected in each cell profile depend upon the plane of section. Up to 60% of the mucous cell profile is occupied by membrane-bound secretory granules (Mariassy and Plopper, 1984; Wilson et al., 1984; Rhodin, 1966; Plopper et al., 1983c). These granules vary in their morphologic and histochemical properties. Three-dimensional reconstruction of mucous goblet cells showed that this variation reflects different stages of mucous cell secretion (Adler et al., 1982). Similarly, the apical surface of mucus cells varies with the stage of secretion.

The electron microscopic image of the mucous secretory granules in well-preserved specimens varies from homogeneous, electron-lucent, through flocculent heterogeneous granules to coarsely granular with variable electron density (Mariassy and Plopper, 1984; Krause and Leeson, 1973; Mariassy et al., 1988b; Hyde et al., 1979). The internal pattern of granules also varies, presumably due to the variation in glycoconjugate moieties.

The degree of granule osmiophilia is thought to correspond to the concentration of the proteinaceous component. The electron density of mucous cell granules may vary, but they lack a dense, osmiophilic, proteinaceous granule core or a surrounding halo. Cytochemical differentiation of mucus cell secretory granules has been evaluated ultrastructurally (Spicer and Schulte, 1982; Spicer et al., 1983) using a periodic acid-thiocarbohydrazide method that localized hexose-rich neutral glycoconjugates (GCs), dialyzed iron (DI) and HID localized carboxylated and sulfated GCs, respectively (Mariassy et al., 1988b; Plopper et al., 1989).

The organelles of mucous cells include one or more Golgi bodies with associated vesicles and GER profiles between the secretory granules. The percentage of the cellular space occupied by organelles and the nucleus vary inversely with the percentage of stored granules. Quantitative estimates of cellular organelles in the secretory cells of sheep defined morphologic differences among the six secretory cell varieties examined (Mariassy and Plopper, 1984). In mucous cells that are distended with granules, the nucleus is dense and compressed into the base of the cell. Conversely, in the actively metabolizing mucous cells, the nucleus is large and round to oval in shape. Nuclear chromatin is heterogeneous and considerably less electron dense. The cytosol of mucous cells is more electron-dense than that of ciliated and indeterminate cells. This higher electron density persists after the extrusion of secretory granules.

Serous cell: Serous cells are found in abundance in many glands in a variety of animal species and humans (Basbaum et al., 1990). However, significant numbers surface airway epithelial serous cells are found only in pathogen-free rats (Jeffery and Reid, 1975; Tables 2–5). Depending on the source, the percentage of serous cells varies from 27% to 42% (Jeffery and Reid, 1975; Plopper et al., 1983c; Nikula et al., 1988). Jeffery reported serous cells in pathogen-free cats, juvenile hamsters, and human fetal tissue (Jeffery, 1983). However, it is not clear how frequent these cells are. In the adult hamster trachea, one study reported no serous cells (Plopper et al., 1983c), whereas another reported that 1.3% of secretory cells in the bronchus have morphologic features of serous cells (Christensen et al., 1987). These include cells containing electron-dense homogeneous secretory granules, extensive GER profiles, a large vesicular nucleus, and no AER. In the snake (*Natrix maura*), secretory cells are similar to those found in lizards and are similar to serous cells in mammals (Zepeda et al., 1995).

Functionally, serous cells have been shown to elaborate a number of proteins, including lysozyme, lactoferrin, peroxidase, antileukeoprotease, proline-rich proteins (PRPs), albumin, and specific high-sulfate-containing carbohydrates. Serous cells of the rat lung proliferate in response to nitrogen dioxide exposure (Evans et al., 1986).

A cell type with generally serous cell characteristics that contained disc-shaped (0.13 urn) and rod-shaped (0.4 ilin) granules was described in the dog (Frasca et al., 1968), human (Plopper et al., 1983c), and cat (Jeffery, 1987). Frasca and colleagues referred to these cells as *special type cells* (Frasca et al., 1968). However, the function of this cell is not known and there are no quantitative data available on this cell.

Nonciliated bronchiolar (Clara) cell: This type of cell is discussed in greater detail in Chapter 8. The cell phenotype, described in the trachea and bronchi of mice (Pack et al., 1980), hamsters (Plopper et al., 1983c; Christensen et al., 1987), and rabbits (Plopper et al., 1983a), is very similar to the nonciliated bronchiolar (or Clara) cell, observed only in the distal airways of other species (Plopper et al., 1983c). Clara (Clara, 1937) described the histology of this cell type in humans and rabbits. Plopper reviewed Clara cell morphology and function, and defined the criteria for the identification of this cell type (Plopper et al., 1983a; Plopper et al., 1980a). Although the Clara cell was renamed the Club cell (Winkelmann and Noack, 2010), the term Clara cell will be used in this chapter to avoid confusion.

Clara cells are nonciliated exocrine cells that contain spherical to ovoid-shaped, membrane-bound secretory granules. In many species, their cell apex contains extensive arrays of AER and protrudes into the airway lumen. GER is found in the basolateral perinuclear regions. The glycogen content of these cells is variable and species dependent. Glycogen is absent from cells of species that have Clara cells in the upper airways (e.g., mouse, hamster, and rabbit).

Bronchiolar Clara cells have been examined in a number of species (Plopper, 1983). Based on their ultrastructural characteristics, the animals were divided into three major groups. In the first group, which included mouse, hamster, rabbit, rat, guinea pig, sheep, horse (Plopper et al., 1980b, 1980c), and llama (Heath et al., 1980), Clara cells have extensive AER, secretory granules, and no glycogen. Clara cells in the mouse and rat lung proliferate in response to naphthalene or oxidant-gas induced injury (Evans et al., 1978; Van Winkle et al., 1995; Stripp et al., 1995). In the second group, which included humans and three species of primates (Plopper et al., 1980a; Castleman et al., 1975), Clara cells contain abundant GER, small numbers of secretory granules, and sparse AER and glycogen. The final group includes cat, dog, ox (Plopper et al., 1980a, 1980c), and ferret (Hyde et al., 1979). In these species, bronchiolar Clara cells contain substantial stores of glycogen, few ovoid granules, and lack any significant AER. Clara cells in adult animals more closely resemble the undifferentiated cells in fetal animals than cells in adults of other species. The abundance of Clara cells in proximal airways is highly species-specific (Tables 2–5).

Sero-mucous cell: There are numerous references to cell types that contain nucleated secretory granules composed of an electron-lucent rim and an electron-dense core (Mariassy and Plopper, 1983; Krause and Leeson, 1973; Jeffery and Reid, 1975). Because morphologic criteria for their definition overlap the characteristics of serous, mucous, and even nonciliated bronchiolar cells, cells with this type of granule will be referred to as *sero-mucous cells.* The main reason for adopting this term is that changes in cell populations occur in response to various forms of airway irritation and injury. These range from the presence of mucous cells in non-pathogen-free rats and rabbits to frank mucous metaplasia in response to chronic irritation. Sero-mucous cells are found in rats exposed to tobacco smoke and in unaffected human airways from lobes resected for carcinoma (Jeffery, 1987). Exposure to sulfur dioxide and isoprenaline sulfite appears to stimulate the formation of mucous granules in nonciliated bronchiolar (Clara) cells (Jeffery, 1987).

In sheep, mucous cells with nucleated granules (Mariassy and Plopper, 1983) were reported in distal airways and submucosal glands (Figure 3(B) and (C)). These mucous cells were colocalized with nonciliated bronchiolar cells. Their distribution in the airways was also inversely proportional to the distribution of submucosal glands. Many of these cells have, in addition to nucleated granules, discrete homogeneous granules, as are found in serous and in nonciliated bronchiolar cells, indicating a transition between the two cell types. These cells are included in the "other" category in Tables 2–5

Unidentified cells: The description of this group of cells is rather difficult because there is little established agreement on either their morphologic characteristics or their role in airway epithelium. Furthermore, these cells have been given various descriptive names, including degranulated exocrine cell (intermediate), preciliated cell, and indeterminate.

Degranulated exocrine cells, as their name implies, appear similar to secretory cells. They contain a dense cytoplasm and small nucleus, but lack substantial secretory granules. The organelles are often abundant. Indeterminate or unidentified cells include all other epithelial cells that lack distinct morphologic characteristic for their categorization. As suggested by Jeffery (1978), this is a classification cell category rather than a specific cell type.

Intermediate cells also include preciliated and other cells that lack the morphologic characteristics of other cell types (Breeze and Wheeldon, 1977). The preciliated cell has fibrogranular condensations, basal bodies, numerous microvilli, multiple pairs of centrioles and electron-lucent cytoplasm. Analysis of this cell type was facilitated by discovery that FoxJ1 deletion in mice stabilized the phenotype (You et al., 2004). Brody and colleagues demonstrated a cause and consequence relationship between cells exhibiting a primary cilium and those that go on to become multiciliated cells in vitro (Jain et al., 2010).

Small mucous granule cells: SMGCs were described in humans (McDowell et al., 1978) and in bonnet and rhesus monkeys (Plopper et al., 1989; Wilson et al., 1984) as mucous cells containing few small (0.05–0.02 pm) secretory granules. Their morphological appearance was interpreted as representing mucus-secreting cells in the early stages of the secretory cycle. A similar cell type was reported in the hamster (Becci et al., 1978); however, later studies showed that these cells contain extensive AER and were very similar to nonciliated bronchiolar (Clara) cells (Plopper et al., 1983c; Christensen et al., 1987).

In sheep, a secretory cell designated as M3 mucous cell has similar features, including extensive AER (35%) of cell volume, small secretory granules, and dense cytoplasm (Mariassy and Plopper, 1984, 1983; Mariassy et al., 1988a; Mariassy et al., 1988b). This cell constituted 36% of the epithelial cells in the trachea and approximately 25% in the rest of the airway tree (Table 4; Mariassy and Plopper, 1983). The exact function of SMGCs in the hamster and sheep is not known. It seems likely from the content of AER and the widespread distribution in the airways that they play a role in xenobiotic metabolism.

REFERENCES

Araya, J., Cambier, S., Markovics, J.A., Wolters, P., Jablons, D., Hill, A., Finkbeiner, W., Jones, K., Broaddus, V.C., Sheppard, D., et al., 2007. Squamous metaplasia amplifies pathologic epithelial-mesenchymal interactions in copd patients. J. Clin. Invest. 117 (11), 3551–3562, 17965775.

Adriaensen, D., Timmermans, J.P., 2004. Purinergic signalling in the lung: important in asthma and copd? Curr. Opin. Pharmacol. 4 (3), 207–214, 15140410.

Archer, S., Michelakis, E., 2002. The mechanism(s) of hypoxic pulmonary vasoconstriction: potassium channels, redox O(2) sensors, and controversies. News Physiol. Sci. 17, 131–137, 12136039.

Adler, K.B., Hardwick, D.H., Craighead, J.E., 1982. Porcine tracheal goblet cell ultrastructure: a three-dimensional reconstruction. Exp. Lung Res. 3 (1), 69–80, 7060544.

Breeze, R.G., Wheeldon, E.B., 1977. The cells of the pulmonary airways. Am. Rev. Respir. Dis. 116 (4), 705–777, 921054.

Borthwick, D.W., Shahbazian, M., Krantz, Q.T., Dorin, J.R., Randell, S.H., 2001. Evidence for stem-cell niches in the tracheal epithelium. Am. J. Respir. Cell. Mol. Biol. 24 (6), 662–670, 11415930.

Beidler, L.M., Smallman, R.L., 1965. Renewal of cells within taste buds. J. Cell. Biol. 27 (2), 263–272, 5884625.

Barham, H.P., Cooper, S.E., Anderson, C.B., Tizzano, M., Kingdom, T.T., Finger, T.E., Kinnamon, S.C., Ramakrishnan, V.R., 2013. Solitary chemosensory cells and bitter taste receptor signaling in human sinonasal mucosa. Int. Forum Allergy Rhinol. (6), 450–457, 23404938.

Brouns, I., Pintelon, I., De Proost, I., Alewaters, R., Timmermans, J.P., Adriaensen, D., 2006. Neurochemical characterisation of sensory receptors in airway smooth muscle: comparison with pulmonary neuroepithelial bodies. Histochem. Cell. Biol. 125 (4), 351–367, 16283357.

Brouns, I., Oztay, F., Pintelon, I., De Proost, I., Lembrechts, R., Timmermans, J.P., Adriaensen, D., 2009. Neurochemical pattern of the complex innervation of neuroepithelial bodies in mouse lungs. Histochem. Cell. Biol. 131 (1), 55–74, 18762965.

Basbaum, C.B., Finkbeiner, W.E., 1989. Mucus Producing Cells of the Airway. Marcel Dekker, New York.

Basbaum, C.B., Jany, B., Finkbeiner, W.E., 1990. The serous cell. Annu. Rev. Physiol. 52, 97–113, 2158772.

Becci, P.J., McDowell, E.M., Trump, B.F., 1978. The respiratory epithelium. II. Hamster trachea, bronchus, and bronchioles. J. Natl. Cancer Inst. 61 (2), 551–561, 355650.

Cole, B.B., Smith, R.W., Jenkins, K.M., Graham, B.B., Reynolds, P.R., Reynolds, S.D., 2010. Tracheal basal cells: a facultative progenitor cell pool. Am. J. Pathol. 177 (1), 362–376, 20522644.

Cutz, E., Pan, J., Yeger, H., Domnik, N.J., Fisher, J.T., 2013. Recent advances and contraversies on the role of pulmonary neuroepithelial bodies as airway sensors. Semin. Cell. Dev. Biol. 24 (1), 40–50, 23022441.

Christensen, T.G., Breuer, R., Hornstra, L.J., Lucey, E.C., Snider, G.L., 1987. The ultrastructure of hamster bronchial epithelium. Exp. Lung Res. 13 (3), 253–277, 3691409.

Clara, M., 1937. Zur histobiologie des bronchal epithels. Z Mikrosk Anat Forsch 41, 321.

Castleman, W.L., Dungworth, D.L., Tyler, W.S., 1975. Intrapulmonary airway morphology in three species of monkeys: a correlated scanning and transmission electron microscopic study. Am. J. Anat. 142 (1), 107–121, 804810.

Di, Y.P., 2011. Functional roles of splunc1 in the innate immune response against gram-negative bacteria. Biochem. Soc. Trans. 39 (4), 1051–1055, 21787346.

Drabek, C.M., Kooyman, G.L., 1983. Terminal airway embryology of the delphinid porpoises, *Stenella attenuata* and *S. longirostris*. J. Morphol. 175 (1), 65–72, 6842612.

Doganay, L., Bilgi, S., Ozdil, A., Yoruk, Y., Altaner, S., Kutlu, K., 2003. Epithelial-myoepithelial carcinoma of the lung. A case report and review of the literature. Arch. Pathol. Lab. Med. 127 (4), e177–180, 12683896.

Evans, M.J., Cox, R.A., Shami, S.G., Wilson, B., Plopper, C.G., 1989. The role of basal cells in attachment of columnar cells to the basal lamina of the trachea. Am. J. Respir. Cell. Mol. Biol. 1 (6), 463–469, 2637758.

Engelhardt, J.F., Schlossberg, H., Yankaskas, J.R., Dudus, L., 1995. Progenitor cells of the adult human airway involved in submucosal gland development. Development 121 (7), 2031–2046, 7635050.

Evans, M.J., Shami, S.G., Cabral-Anderson, L.J., Dekker, N.P., 1986. Role of nonciliated cells in renewal of the bronchial epithelium of rats exposed to NO_2. Am. J. Pathol. 123 (1), 126–133, 3963147.

Evans, M.J., Cabral-Anderson, L.J., Freeman, G., 1978. Role of the Clara cell in renewal of the bronchiolar epithelium. Lab. Invest. 38 (6), 648–653, 661220.

Fawcet D.W., 1986. Philadelphia: W.B. Saunders Company.

Farbman, A.I., 1980. Renewal of taste bud cells in rat circumvallate papillae. Cell. Tissue Kinet. 13 (4), 349–357, 7428010.

Finger, T.E., Bottger, B., Hansen, A., Anderson, K.T., Alimohammadi, H., Silver, W.L., 2003. Solitary chemoreceptor cells in the nasal cavity serve as sentinels of respiration. Proc. Natl. Acad. Sci. U. S. A. 100 (15), 8981–8986, 12857948.

Fu, X.W., Wang, D., Nurse, C.A., Dinauer, M.C., Cutz, E., 2000. NADPH oxidase is an O_2 sensor in airway chemoreceptors: evidence from K^+ current modulation in wild-type and oxidase-deficient mice. Proc. Natl. Acad. Sci. U. S. A. 97 (8), 4374–4379, 10760304.

Frasca, J.M., Auerbach, O., Parks, V.R., Jamieson, J.D., 1968. Electron microscopic observations of the bronchial epithelium of dogs. I. Control dogs. Exp. Mol. Pathol. 9 (3), 363–379, 4952083.

Ghosh, M., Ahmad, S., Jian, A., Li, B., Smith, R.W., Helm, K.M., Seibold, M.A., Groshong, S.D., White, C.W., Reynolds, S.D., 2013. Human tracheobronchial basal cells: normal versus remodeling/repairing phenotypes in vivo and in vitro. Am. J. Respir. Cell. Mol. Biol. (49), 1127–1134, 23927678.

Gaillard, D.A., Lallement, A.V., Petit, A.F., Puchelle, E.S., 1989. In vivo ciliogenesis in human fetal tracheal epithelium. Am. J. Anat. 185 (4), 415–428, 2782282.

Gilljam, H., Motakefi, A.M., Robertson, B., Strandvik, B., 1987. Ultrastructure of the bronchial epithelium in adult patients with cystic fibrosis. Eur. J. Respir. Dis. 71 (3), 187–194, 3678420.

Green, F.H., Williams, D.J., James, A., McPhee, L.J., Mitchell, I., Mauad, T., 2010. Increased myoepithelial cells of bronchial submucosal glands in fatal asthma. Thorax 65 (1), 32–38, 19996346.

Gulbransen, B.D., Clapp, T.R., Finger, T.E., Kinnamon, S.C., 2008. Nasal solitary chemoreceptor cell responses to bitter and trigeminal stimulants in vitro. J. Neurophysiol. 99 (6), 2929–2937, 18417634.

Hislop, A., Howard, S., Fairweather, D.V., 1984. Morphometric studies on the structural development of the lung in *Macaca fascicularis* during fetal and postnatal life. J. Anat. 138 (Pt 1), 95–112, 6706842.

Hare, W.C., 1955. The broncho-pulmonary segments in the sheep. J. Anat. 89 (3), 387–402, 13251969.

Hughes, T., 1965. Microcirculation of the tracheobronchial tree. Nature 206 (982), 425–426, 4953699.

Hill, P., Goulding, D., Webber, S.E., Widdicombe, J.G., 1989. Blood sinuses in the submucosa of the large airways of the sheep. J. Anat. 162, 235–247, 2808119.

Hsia, C.C., Hyde, D.M., Ochs, M., Weibel, E.R., 2010. An official research policy statement of the American thoracic society/European respiratory society: standards for quantitative assessment of lung structure. Am. J. Respir. Crit. Care Med. 181 (4), 394–418, 20130146.

Hyde, D.M., Tyler, N.K., Plopper, C.G., 2007. Morphometry of the respiratory tract: avoiding the sampling, size, orientation, and reference traps. Toxicol. Pathol. 35 (1), 41–48, 17325971.

Hegab, A.E., Ha, V.L., Gilbert, J.L., Zhang, K.X., Malkoski, S.P., Chon, A.T., Darmawan, D.O., Bisht, B., Ooi, A.T., Pellegrini, M., et al., 2011. A novel stem/progenitor cell population from murine tracheal submucosal gland ducts with multipotent regenerative potential. Stem Cells (29), 1283–1293, 21710468.

Hisa, Y., Matsui, T., Fukui, K., Ibata, Y., Mizukoshi, O., 1982. Ultrastructural and fluorescence histochemical studies on the sympathetic innervation of the canine laryngeal glands. Acta Otolaryngol. 93 (1–2), 119–122, 7064687.

Hegab, A.E., Ha, V.L., Darmawan, D.O., Gilbert, J.L., Ooi, A.T., Attiga, Y.S., Bisht, B., Nickerson, D.W., Gomperts, B.N., 2012. Isolation and in vitro characterization of basal and submucosal gland duct stem/progenitor cells from human proximal airways. Stem Cells Transl. Med. 1 (10), 719–724, 23197663.

Haller, C.J., 1994. A scanning and transmission electron microscopic study of the development of the surface structure of neuroepithelial bodies in the mouse lung. Micron 25 (6), 527–538, 7881894.

Hoyt Jr, R.F., McNelly, N.A., Sorokin, S.P., 1990. Dynamics of neuroepithelial body (neb) formation in developing hamster lung: light microscopic autoradiography after 3H-thymidine labeling in vivo. Anat. Rec. 227 (3), 340–350, 2372138.

Hoyt, R., McNeely, N.A., Sorokin, S.P. 1993. Calcitonin gene-related peptice (CGRP) as regional mitogen for tracheobronchial epithelium of organ cultured fetal rat lungs. Am. Rev. Resp. Dis. 147: A498.

Hyde, D.M., Samuelson, D.A., Blakeney, W.H., Kosch, P.C., 1979. A correlative light microscopy, transmission and scanning electron microscopy study of the ferret lung. Scan Electron Microsc. (3), 891–898, 524059.

Heath, D., Smith, P., Biggar, R., 1980. Clara cells in llamas born and living at high and low altitudes. Br. J. Dis. Chest 74 (1), 75–80, 7356916.

Jeffery, P.K., 1987. Structure and Function of Adult Tracheobronchial Epitheluim. Churchill Livingstone, London.

Johnson, D.E., Georgieff, M.K., 1989. Pulmonary neuroendocrine cells. Their secretory products and their potential roles in health and chronic lung disease in infancy. Am. Rev. Respir. Dis. 140 (6), 1807–1812, 2574971.

Jeffery, P.K., Ayers, M., Rogers, D., 1982. The mechanisms and control of bronchial mucous cell hyperplasia. Adv. Exp. Med. Biol. 144, 399–409, 7044067.

Jeffery, P.K., 1983. Morphologic features of airway surface epithelial cells and glands. Am. Rev. Respir. Dis. 128 (2 Pt 2), S14–S20, 6881701.

Jeffery, P.K., Reid, L., 1975. New observations of rat airway epithelium: a quantitative and electron microscopic study. J. Anat. 120 (Pt 2), 295–320, 1201964.

Jeffery, P.K., 1978. Structure and function of mucus-secreting cells of cat and goose airway epithelium.

Jain, R., Pan, J., Driscoll, J.A., Wisner, J.W., Huang, T., Gunsten, S.P., You, Y., Brody, S.L., 2010. Temporal relationship between primary and motile ciliogenesis in airway epithelial cells. Am. J. Respir. Cell. Mol. Biol. 43 (6), 731–739, 20118219.

Kennedy, A.R., Desrosiers, A., Terzaghi, M., Little, J.B., 1978. Morphometric and histological analysis of the lungs of syrian golden hamsters. J. Anat. 125 (Pt 3), 527–553, 640957.

Kuhn, C.I., 1976. The Cells of the Lung and Their Organelles. Marcel Dekker, New York.

Kawanami, O., Ferrans, V.J., Crystal, R.G., 1979. Anchoring fibrils in the normal canine respiratory system. Am. Rev. Respir. Dis. 120 (3), 595–611, 484934.

Khansaheb, M., Choi, J.Y., Joo, N.S., Yang, Y.M., Krouse, M., Wine, J.J., 2011. Properties of substance p-stimulated mucus secretion from porcine tracheal submucosal glands. Am. J. Physiol. Lung Cell. Mol. Physiol. 300 (3), L370–L379, 21131402.

Krasteva, G., Canning, B.J., Papadakis, T., Kummer, W., 2012. Cholinergic brush cells in the trachea mediate respiratory responses to quorum sensing molecules. Life Sci. 91 (21–22), 992–996, 22749863.

Krasteva, G., Kummer, W., 2012. "Tasting" the airway lining fluid. Histochem Cell. Biol. 138 (3), 365–383, 22777347.

Kinnamon, S.C., Reynolds, S.D., 2009. Cell biology. using taste to clear the air(ways). Science 325 (5944), 1081–1082, 19713515.

Kinnamon, S.C., 2012. Taste receptor signalling - from tongues to lungs. Acta Physiol. (Oxf) 204 (2), 158–168, 21481196.

Krasteva, G., Canning, B.J., Hartmann, P., Veres, T.Z., Papadakis, T., Muhlfeld, C., Schliecker, K., Tallini, Y.N., Braun, A., Hackstein, H., et al., 2011. Cholinergic chemosensory cells in the trachea regulate breathing. Proc. Natl. Acad. Sci. U. S. A. 108 (23), 9478–9483, 21606356.

Khoor, A., Gray, M.E., Singh, G., Stahlman, M.T., 1996. Ontogeny of clara cell-specific protein and its mRNA: their association with neuroepithelial bodies in human fetal lung and in bronchopulmonary dysplasia. J. Histochem. Cytochem. 44 (12), 1429–1438, 8985135.

Kazemian, P., Stephenson, R., Yeger, H., Cutz, E., 2001. Respiratory control in neonatal mice with NADPH oxidase deficiency. Respir. Physiol. 126 (2), 89–101, 11348637.

Krause, W.J., Leeson, C.R., 1973. The postnatal development of the respiratory system of the opossum. I. Light and scanning electron microscopy. Am. J. Anat. 137 (3), 337–355, 4716357.

Korhonen, L.K., Holopainen, E., Paavolainen, M., 1969. Some histochemical characteristics of tracheobronchial tree and pulmonary neoplasms. Acta Histochem. 32 (1), 57–73, 4188703.

Linhartova, A., Anderson Jr., A., Foraker, A.G., 1971. Radial traction and bronchiolar obstruction in pulmonary emphysema. Observed and theoretical aspects. Arch. Pathol. 92 (5), 384–391, 5110576.

Linhartova, A., Anderson Jr., A.E., Foraker, A.G., 1982. Affixment arrangements of peribronchiolar alveoli in normal and emphysematous lungs. Arch. Pathol. Lab. Med. 106 (10), 499–502, 6896979.

Lopez-Schier, H., 2004. Regeneration: did you hear the news? Curr. Biol. 14 (3), R127–R128, 14986686.

Lauweryns, J.M., Cokelaere, M., Theunynck, P., 1972. Neuro-epithelial bodies in the respiratory mucosa of various mammals. A light optical, histochemical and ultrastructural investigation. Z Zellforsch Mikrosk Anat 135 (4), 569–592, 4346123.

Lauweryns, J.M., Cokelaere, M., 1973. Hypoxia-sensitive neuro-epithelial bodies. Intrapulmonary secretory neuroreceptors, modulated by the CNS. Z Zellforsch Mikrosk Anat 145 (4), 521–540, 4774984.

Lauweryns, J.M., Cokelaere, M., Deleersynder, M., Liebens, M., 1977. Intrapulmonary neuro-epithelial bodies in newborn rabbits. Influence of hypoxia, hyperoxia, hypercapnia, nicotine, reserpine, l-DOPA and 5-HTP. Cell. Tissue Res. 182 (4), 425–440, 922815.

Lauweryns, J.M., Cokelaere, M., Lerut, T., Theunynck, P., 1978. Cross-circulation studies on the influence of hypoxia and hypoxaemia on neuro-epithelial bodies in young rabbits. Cell. Tissue Res. 193 (3), 373–386, 728949.

Larson, S.D., Schelegle, E.S., Hyde, D.M., Plopper, C.G., 2003. The three-dimensional distribution of nerves along the entire intrapulmonary airway tree of the adult rat and the anatomical relationship between nerves and neuroepithelial bodies. Am. J. Respir. Cell. Mol. Biol. 28 (5), 592–599, 12707015.

Lauweryns, J.M., Van Lommel, A.T., Dom, R.J., 1985. Innervation of rabbit intrapulmonary neuroepithelial bodies. Quantitative and qualitative ultrastructural study after vagotomy. J. Neurol. Sci. 67 (1), 81–92, 3981214.

Linnoila, R.I., 2006. Functional facets of the pulmonary neuroendocrine system. Lab. Invest 86 (5), 425–444, 16568108.

Lopez-Barneo, J., 1996. Oxygen-sensing by ion channels and the regulation of cellular functions. Trends Neurosci. 19 (10), 435–440, 8888521.

Leigh, M.W., Gambling, T.M., Carson, J.L., Collier, A.M., Wood, R.E., Boat, T.F., 1986. Postnatal development of tracheal surface epithelium and submucosal glands in the ferret. Exp. Lung Res. 10 (2), 153–169, 2420581.

Mariassy, A.T., Plopper, C.G., 1984. Tracheobronchial epithelium of the sheep: II. Ultrastructural and morphometric analysis of the epithelial secretory cell types. Anat. Rec. 209 (4), 523–534, 6476418.

Mariassy, A.T., Plopper, C.G., 1983. Tracheobronchial epithelium of the sheep: I. Quantitative light-microscopic study of epithelial cell abundance, and distribution. Anat. Rec. 205 (3), 263–275, 6837941.

Mariassy, A.T., Gazeroglu, H., Wanner, A., 1991. Morphometry of the subepithelial circulation in sheep airways. Effect of vascular congestion. Am. Rev. Respir. Dis. 143 (1), 162–166, 1986673.

Milsom, W.K., Burleson, M.L., 2007. Peripheral arterial chemoreceptors and the evolution of the carotid body. Respir. Physiol. Neurobiol. 157 (1), 4–11, 17353155.

Moulton, D.G., 1974. Dynamics of cell populations in the olfactory epithelium. Ann. N. Y. Acad. Sci. 237 (0), 52–61, 4214101.

Merigo, F., Benati, D., Tizzano, M., Osculati, F., Sbarbati, A., 2005. Alpha-gustducin immunoreactivity in the airways. Cell. Tissue Res. 319 (2), 211–219, 15654652.

Meyrick, B., Reid, L., 1968. The alveolar brush cell in rat lung—a third pneumonocyte. J. Ultrastruct. Res. 23 (1), 71–80, 5670842.

Mazzuca, M., Lhermitte, M., Lafitte, J.J., Roussel, P., 1982. Use of lectins for detection of glycoconjugates in the glandular cells of the human bronchial mucosa. J. Histochem. Cytochem. 30 (9), 956–966, 7130674.

Mariassy, A.T., Plopper, C.G., St George, J.A., Wilson, D.W., 1988a. Tracheobronchial epithelium of the sheep: IV. Lectin histochemical characterization of secretory epithelial cells. Anat. Rec. 222 (1), 49–59, 3189886.

Mariassy, A.T., St George, J.A., Nishio, S.J., Plopper, C.G., 1988b. Tracheobronchial epithelium of the sheep: III. Carbohydrate histochemical and cytochemical characterization of secretory epithelial cells. Anat. Rec. 221 (1), 540–549, 3389536.

McDowell, E.M., Barrett, L.A., Glavin, F., Harris, C.C., Trump, B.F., 1978. The respiratory epithelium. I. Human bronchus. J. Natl. Cancer Inst. 61 (2), 539–549, 355649.

Ninomiya, H., Inomata, T., Shirouzu, H., Katsumata, E., 2005. Microanatomy of the terminal air spaces of baird's beaked whale (*Berardius bairdii*) lungs. J. Vet. Med. Sci. 67 (5), 473–479, 15942131.

Nikula, K.J., Wilson, D.W., Giri, S.N., Plopper, C.G., Dungworth, D.L., 1988. The response of the rat tracheal epithelium to ozone exposure. Injury, adaptation, and repair. Am. J. Pathol. 131 (2), 373–384, 3358460.

Plopper, C.G., Halsebo, J.E., Berger, W.J., Sonstegard, K.S., Nettesheim, P., 1983a. Distribution of nonciliated bronchiolar epithelial (clara) cells in intra- and extrapulmonary airways of the rabbit. Exp. Lung Res. 5 (2), 79–98, 6628348.

Plopper, C.G., Heidsiek, J.G., Weir, A.J., George, J.A., Hyde, D.M., 1989. Tracheobronchial epithelium in the adult rhesus monkey: a quantitative histochemical and ultrastructural study. Am. J. Anat. 184 (1), 31–40, 2916437.

Peake, J.L., Reynolds, S.D., Stripp, B.R., Stephens, K.E., Pinkerton, K.E., 2000. Alteration of pulmonary neuroendocrine cells during epithelial repair of naphthalene-induced airway injury. Am. J. Pathol. 156 (1), 279–286, 10623676.

Plopper, C.G., Mariassy, A.T., Lollini, L.O., 1983b. Structure as revealed by airway dissection. A comparison of mammalian lungs. Am. Rev. Respir. Dis. 128 (2 Pt 2), S4–S7, 6881706.

Plopper, C.G., 1983. Comparative morphologic features of bronchiolar epithelial cells. The Clara cell. Am. Rev. Respir. Dis. 128 (2 Pt 2), S37–S41, 6881705.

Pavelka, M., Ronge, H.R., Stockinger, G., 1976. Comparative study of tracheal epithelium of different mammals. Acta Anat. (Basel) 94 (2), 262–282, 961348.

Plopper, C.G., Mariassy, A.T., Wilson, D.W., Alley, J.L., Nishio, S.J., Nettesheim, P., 1983c. Comparison of nonciliated tracheal epithelial cells in six mammalian species: ultrastructure and population densities. Exp. Lung Res. 5 (4), 281–294, 6662075.

Pinkstaff, C.A., 1980. The cytology of salivary glands. Int. Rev. Cytol. 63, 141–261, 395130.

Pearsall, A.D., 1989. Ultrastructural architecture of pulmonary small-granule cell clusters in adult syrian golden hamster. Am. J. Anat. 186 (2), 173–185, 2573267.

Pack, R.J., Al-Ugaily, L.H., Morris, G., Widdicombe, J.G., 1980. The distribution and structure of cells in the tracheal epithelium of the mouse. Cell. Tissue Res. 208 (1), 65–84, 6248229.

Plopper, C.G., Hill, L.H., Mariassy, A.T., 1980a. Ultrastructure of the nonciliated bronchiolar epithelial (Clara) cell of mammalian lung. Iii. A study of man with comparison of 15 mammalian species. Exp. Lung Res. 1 (2), 171–180, 7227344.

Plopper, C.G., Mariassy, A.T., Hill, L.H., 1980b. Ultrastructure of the nonciliated bronchiolar epithelial (Clara) cell of mammalian lung: I. A comparison of rabbit, guinea pig, rat, hamster, and mouse. Exp. Lung Res. 1 (2), 139–154, 7227343.

Plopper, C.G., Mariassy, A.T., Hill, L.H., 1980c. Ultrastructure of the nonciliated bronchiolar epithelial (Clara) cell of mammalian lung: II. A comparison of horse, steer, sheep, dog, and cat. Exp. Lung Res. 1 (2), 155–169, 7194780.

Plopper, C.G., et al., 1988. Inhalation Toxicology. Springer-Verlag, Berlin, 25–40.

Phipps, R.J., Richardson, P.S., Jun 1976. Proceedings: The nervous and pharmacological control of tracheal mucus secretion in the goose. J. Physiol. 258 (2), 116P–117P, 957149.

Rock, J.R., Randell, S.H., Hogan, B.L., 2010. Airway basal stem cells: a perspective on their roles in epithelial homeostasis and remodeling. Dis. Model Mech. 9–10, 545–556, 20699479.

Rhodin, J.A., 1966. The ciliated cell. Ultrastructure and function of the human tracheal mucosa. Am. Rev. Respir. Dis. 93 (Suppl. 3), 1–15, 5954680.

Rhodin, J.A.G., 1985. Respiratory System. Oxford University Press, New York.

Rock, J.R., Onaitis, M.W., Rawlins, E.L., Lu, Y., Clark, C.P., Xue, Y., Randell, S.H., Hogan, B.L., 2009. Basal cells as stem cells of the mouse trachea and human airway epithelium. Proc. Natl. Acad. Sci. U. S. A. 106 (31), 12771–12775, 19625615.

Rha, K.S., Majima, Y., Sakakura, Y., Yasui, Y., Nakano, K., Ishihara, A., 1994. Distribution of substance p immunoreactive nerve fibers in the tracheal submucosal gland of cats. Ann. Otol. Rhinol. Laryngol. 103 (3), 222–226, 7510096.

Reid, L., Meyrick, B., Antony, V.B., Chang, L.Y., Crapo, J.D., Reynolds, H.Y., 2005. The mysterious pulmonary brush cell: a cell in search of a function. Am. J. Respir. Crit. Care Med. 172 (1), 136–139, 15817800.

Reynolds, S.D., Giangreco, A., Power, J.H., Stripp, B.R., 2000. Neuroepithelial bodies of pulmonary airways serve as a reservoir of progenitor cells capable of epithelial regeneration. Am. J. Pathol. 156 (1), 269–278, 10623675.

Robinson, N.P., Venning, L., Kyle, H., Widdicombe, J.G., 1986. Quantitation of the secretory cells of the ferret tracheobronchial tree. J. Anat. 145, 173–188, 3429303.

Seibold, M.A., Smith, R.W., Urbanek, C., Groshong, S.D., Cosgrove, G.P., Brown, K.K., Schwarz, M.I., Schwartz, D.A., Reynolds, S.D., 2013. The idiopathic pulmonary fibrosis honeycomb cyst contains a mucocilary pseudostratified epithelium. PLoS One 8 (3), e58658, 23527003.

Schummer, A., Nickel, R., Sack, W.O., 1986. Respiratory System. Verlag Paul Parey, Berlin.

Sturgess, J.M., Turner, J.A., 1984. Ultrastructural pathology of cilia in the immotile cilia syndrome. Perspect. Pediatr. Pathol. 8 (2), 133–161, 6377227.

Sturgess, J.M., 1984. Structure of the human airway mucosa. Semin. Resp. Med. 5, 301.

Shear, M., 1966. The structure and function of myoepithelial cells in salivary glands. Arch. Oral Biol. 11 (8), 769–780, 5226772.

Shimura, S., Sasaki, T., Okayama, H., Sasaki, H., Takishima, T., 1987. Neural control of contraction in isolated submucosal gland from feline trachea. J. Appl. Physiol. (1985) 62 (6), 2404–2409, 3301788.

Saunders, C.J., Reynolds, S.D., Finger, T.E., 2013. Chemosensory brush cells of the trachea: a stable population in a dynamic epithelium. Am. J. Respir. Cell. Mol. Biol. 49, 190–196, 23526223.

Shah, A.S., Ben-Shahar, Y., Moninger, T.O., Kline, J.N., Welsh, M.J., 2009. Motile cilia of human airway epithelia are chemosensory. Science 325, 1131–1134, 19628819.

Saunders, C.J., Christensen, M., Finger, T.E., Tizzano, M., 2014. Cholinergic neurotransmission links solitary chemosensory cells to nasal inflammation. Proc. Natl. Acad. Sci. U. S. A. 111 (16), 6075–6080, 24711432.

Sorokin, S.P., Hoyt Jr., R.F., Pearsall, A.D., 1983. Comparative biology of small granule cells and neuroepithelial bodies in the respiratory system. Short review. Am. Rev. Respir. Dis. 128 (2 Pt 2), S26–S31, 6881703.

Sunday, M.E., Hua, J., Reyes, B., Masui, H., Torday, J.S., 1993. Anti-bombesin monoclonal antibodies modulate fetal mouse lung growth and maturation in utero and in organ cultures. Anat. Rec. 236 (1), 25–32; discussion 33–24, 8507013.

Sunday, M.E., Hua, J., Dai, H.B., Nusrat, A., Torday, J.S., 1990. Bombesin increases fetal lung growth and maturation in utero and in organ culture. Am. J. Respir. Cell. Mol. Biol. 3 (3), 199–205, 2390263.

Stevens, T.P., McBride, J.T., Peake, J.L., Pinkerton, K.E., Stripp, B.R., 1997. Cell proliferation contributes to PNEC hyperplasia after acute airway injury. Am. J. Physiol. 272 (3 Pt 1), L486–L493, 9124606.

Spicer, S.S., Schulte, B.A., 1982. Ultrastructural methods for localizing complex carbohydrates. Hum. Pathol. 13 (4), 343–354, 6176522.

Spicer, S.S., Schulte, B.A., Thomopoulos, G.N., 1983. Histochemical properties of the respiratory tract epithelium in different species. Am. Rev. Respir. Dis. 128 (2 Pt 2), S20–S26, 6881702.

Stripp, B.R., Maxson, K., Mera, R., Singh, G., 1995. Plasticity of airway cell proliferation and gene expression after acute naphthalene injury. Am. J. Physiol. 269 (6 Pt 1), L791–L799, 8572241.

Tyler, N.K., Hyde, D.M., Hendrickx, A.G., Plopper, C.G., 1988. Morphogenesis of the respiratory bronchiole in rhesus monkey lungs. Am. J. Anat. 182 (3), 215–223, 3213820.

Tandler, B., Sherman, J.M., Boat, T.F., Wood, R.E., 1983a. Surface architecture of the mucosal epithelium of the cat trachea: II. Structure and dynamics of the membranous portion. Am. J. Anat. 168 (2), 133–144, 6650431.

Tandler, B., Sherman, J.M., Boat, T.F., 1983b. Surface architecture of the mucosal epithelium of the cat trachea: I. Cartilaginous portion. Am. J. Anat. 168 (2), 119–131, 6650430.

Tizzano, M., Finger, T.E., 2013. Chemosensors in the nose: guardians of the airways. Physiol. (Bethesda) 28 (1), 51–60, 23280357.

Tizzano, M., Merigo, F., Sbarbati, A., 2006. Evidence of solitary chemosensory cells in a large mammal: the diffuse chemosensory system in *Bos taurus* airways. J. Anat. 209 (3), 333–337, 16928202.

Tizzano, M., Cristofoletti, M., Sbarbati, A., Finger, T.E., 2011. Expression of taste receptors in solitary chemosensory cells of rodent airways. BMC Pulm. Med. 11, 3, 21232137.

Thaete, L.G., Spicer, S.S., Spock, A., 1981. Histology, ultrastructure, and carbohydrate cytochemistry of surface and glandular epithelium of human nasal mucosa. Am. J. Anat. 162 (3), 243–263, 7315752.

Vaigot, P., Pisani, A., Darmon, Y.M., Ortonne, J.P., 1987. The majority of epidermal merkel cells are non-proliferative: a quantitative immunofluorescence analysis. Acta Derm. Venereol. 67 (6), 517–520, 2451378.

Van Winkle, L.S., Buckpitt, A.R., Nishio, S.J., Isaac, J.M., Plopper, C.G., 1995. Cellular response in naphthalene-induced Clara cell injury and bronchiolar epithelial repair in mice. Am. J. Physiol. 269 (6 Pt 1), L800–L818, 8572242.

Wilson, D.W., Plopper, C.G., Hyde, D.M., 1984. The tracheobronchial epithelium of the bonnet monkey (*Macaca radiata*): a quantitative ultrastructural study. Am. J. Anat. 171 (1), 25–40, 6486068.

Wansleeben, C., Bowie, E., Hotten, D.F., Yu, Y.R., Hogan, B.L., 2014. Age-related changes in the cellular composition and epithelial organization of the mouse trachea. PLoS One 9 (3), e93496, 24675804.

Wilson, R.W., Moran, C.A., 1997. Epithelial-myoepithelial carcinoma of the lung: immunohistochemical and ultrastructural observations and review of the literature. Hum. Pathol. 28 (5), 631–635, 9158714.

White, S.R., Hershenson, M.B., Sigrist, K.S., Zimmermann, A., Solway, J., 1993. Proliferation of guinea pig tracheal epithelial cells induced by calcitonin gene-related peptide. Am. J. Respir. Cell. Mol. Biol. 8 (6), 592–596, 8323744.

Weir, E.K., Lopez-Barneo, J., Buckler, K.J., Archer, S.L., 2005. Acute oxygen-sensing mechanisms. N. Engl. J. Med. 353 (19), 2042–2055, 16282179.

Wang, D., Youngson, C., Wong, V., Yeger, H., Dinauer, M.C., Vega-Saenz Miera, E., Rudy, B., Cutz, E., 1996. NADPH-oxidase and a hydrogen peroxide-sensitive K+ channel may function as an oxygen sensor complex in airway chemoreceptors and small cell lung carcinoma cell lines. Proc. Natl. Acad. Sci. U. S. A. 93 (23), 13182–13187, 8917565.

Winkelmann, A., Noack, T., 2010. The Clara cell: a "Third reich eponym"? Eur. Respir. J. 36 (4), 722–727, 20223917.

Widdicombe, J.H., Gashi, A.A., Basbaum, C.B., Nathanson, I.T., 1986. Structural changes associated with fluid absorption by dog tracheal epithelium. Exp. Lung Res. 10 (1), 57–69, 3948811.

Xie, W., Fisher, J.T., Lynch, T.J., Luo, M., Evans, T.I., Neff, T.L., Zhou, W., Zhang, Y., Ou, Y., Bunnett, N.W., et al., 2011. CGRP induction in cystic fibrosis airways alters the submucosal gland progenitor cell niche in mice. J. Clin. Invest 121 (8), 3144–3158, 21765217.

You, Y., Huang, T., Richer, E.J., Schmidt, J.E., Zabner, J., Borok, Z., Brody, S.L., 2004. Role of f-box factor foxj1 in differentiation of ciliated airway epithelial cells. Am. J. Physiol. Lung Cell. Mol. Physiol. 286 (4), L650–L657, 12818891.

Zeiss, C.J., Johnson, E.A., 2004. Proliferation of microglia, but not photoreceptors, in the outer nuclear layer of the rd-1 mouse. Invest Ophthalmol. Vis. Sci. 45 (3), 971–976, 14985319.

Zepeda, M.L., Chinoy, M.R., Wilson, J.M., 1995. Characterization of stem cells in human airway capable of reconstituting a fully differentiated bronchial epithelium. Somat. Cell. Mol. Genet. 21 (1), 61–73, 7541563.

Chapter 7

Epithelial Cells of the Bronchiole

Charles G. Plopper and Dallas M. Hyde

Chapter Outline

1. INTRODUCTION

This chapter focuses on the epithelium of the distal conducting airways, or bronchioles, with emphasis on the area of transition between conducting airways and the gas-exchange area. The bronchiolar region has a substantial variation from one species of mammal to the next. This variability includes the microenvironment in which epithelial cells are found, particularly the zone of transition between bronchiolar, non-gas exchange regions and the alveolar gas-exchange regions. There is also considerable interspecies variability in the general composition of the bronchiolar epithelium. The principal alterations involve the distribution, number, and percentage of the population that is nonciliated cells. In addition, there is substantial variability in the composition and differentiated expression of the nonciliated epithelial population in this region. Chapter 9 deals with the secretory products produced by this nonciliated cell population in different species of mammals. Many of these aspects have also been reviewed elsewhere (Plopper et al., 1983; Widdicombe and Pack, 1982; Plopper and Dungworth, 1987; Massaro, 1989; Devereux et al., 1989; Plopper, 1990).

2. MICROENVIRONMENT OF BRONCHIOLAR EPITHELIUM

The most consistent location of bronchiolar epithelium is the proximal, or central, portion of the pulmonary acinus (Figure 1). This is the zone of transition between conducting airways and the gas-exchange area. The most distal conducting airways proximal to the pulmonary acinus are termed *bronchioles*. The microenvironment in which the bronchiolar epithelial cell population finds itself varies greatly from species to species in a number of factors, including the following: (1) the relative position of the most distal bronchiole within the airway tree, (2) the extent of the transitional zone, and (3) the interrelationship between connective tissue, vasculature, and epithelial elements in the gas-exchange area with those of the conducting airway adjacent to it. The term *transitional bronchiole* has been used to describe the extension of the conducting airway (bronchiole) that is in closest proximity to the most proximal gas-exchange unit, or alveolus (Haefeli-Bleuer and

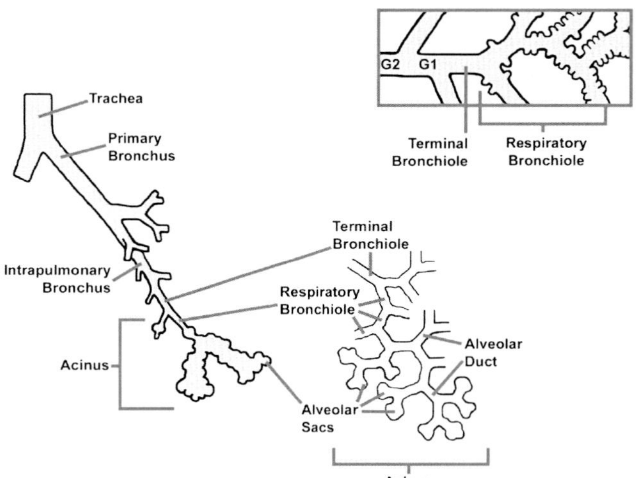

FIGURE 1 Diagrammatic representation of the transition zone between the tracheobronchial airway tree and the fundamental unit of the gas exchange area, the acinus.

Comparative Biology of the Normal Lung. http://dx.doi.org/10.1016/B978-0-12-404577-4.00007-2

Weibel, 1988; Rodriguez et al., 1987). The number of generations of branching in which this transitional (termed terminal or respiratory) bronchiole is situated in relation to the trachea ranges from 6 to 25 (mean = 14) generations in human lungs (Florens et al., 2011; Rodriguez et al., 1987), from 9 to 22 generations in Macaque monkeys (Tyler and Plopper, 1985), from 6 to 21 generations (mean = 12) in the mouse (Onuma et al., 2001), from 6 to 25 generations (mean = 15) in the rat (Lee et al., 2008; Haefeli-Bleuer and Weibel, 1988), from 12 to 26 (mean = 18) in the rabbit (Haefeli-Bleuer and Weibel, 1988), from and 30 to 35 in the sheep (Mariassy and Plopper, 1983).

Table 1 compares the centriacinar conducting airway organization in a number of mammalian species. As illustrated in Figure 2, this transitional zone is very short in a large number of species, including most of the smaller laboratory species and many of the larger domestic animals. The transition zone in these species is composed primarily of a short or long nonalveolarized (or terminal) bronchiole, which joins one or more airways in which the bronchiolar epithelium is interdigitated with alveolar outpockets. This interdigitation extends less than one generation of branching and generally involves an area of less than two to three alveolar outpocketings in extent. As a consequence, the extension of the bronchiolar epithelium into matrix zones where alveolar epithelium differentiates is minimal in these species. By contrast (Table 1), most of the carnivores that have been studied, as well as primates, appear to have an extensive transitional, or respiratory, bronchiolar zone. This refers to a transitional area where numerous generations (usually three or more) of branching of airways contain both a population characteristic of more proximal conducting airways and an epithelial population associated with alveolar outpocketings of the gas-exchange area.

The organization of the peribronchiolar matrix associated with bronchiolar epithelium has a number of common characteristics between species. These include the presence of smooth muscle adjacent to the basal lamina, extensive collagen interspersed with elastic fibers, and the absence of large numbers of capillaries, as would be observed in a

TABLE 1 Comparison of Species Differences in Microenvironment of Bronchiolar Cells: Centriacinar Organization and Tracheobronchial Distribution

	Transitional or Respiratory Bronchiole[a]		Nonciliated Bronchiolar Cells Found in[f]	
Species	**Extensive**	**Minimal**	**Trachea**	**Lobar Bronchus**
Mouse		+	>50%	>60%[g]
Hamster		+	>40%	>15%[h]
Rat		+	0	0
Guinea pig		+	0	0
Rabbit		+	>15%	>25%[i]
Dog	+		0	0
Cat	+		0	0
Ferret	+[b*]		0	0
Macaque monkey	+[c]		0	0
Sheep	+[e]	+	0	0
Pig		+	0	0
Horse		+	0	0
Cow		+[d]	0	0
Human	+		0	0

[a]*Based on Tyler (1983).*
[b]*From Hyde et al. (1979).*
[c]*From Tyler and Plopper (1985).*
[d]*From Mariassy et al. (1975).*
[e]*Based on Haefeli-Bleuer and Weibel (1988).*
[f]*From Plopper (1983).*
[g]*From Pack et al. (1980, 1981).*
[h]*From Christensen et al. (1987a,b).*
[i]*From Plopper et al. (1983).*
**Sterner-Kock et al. (2000).*

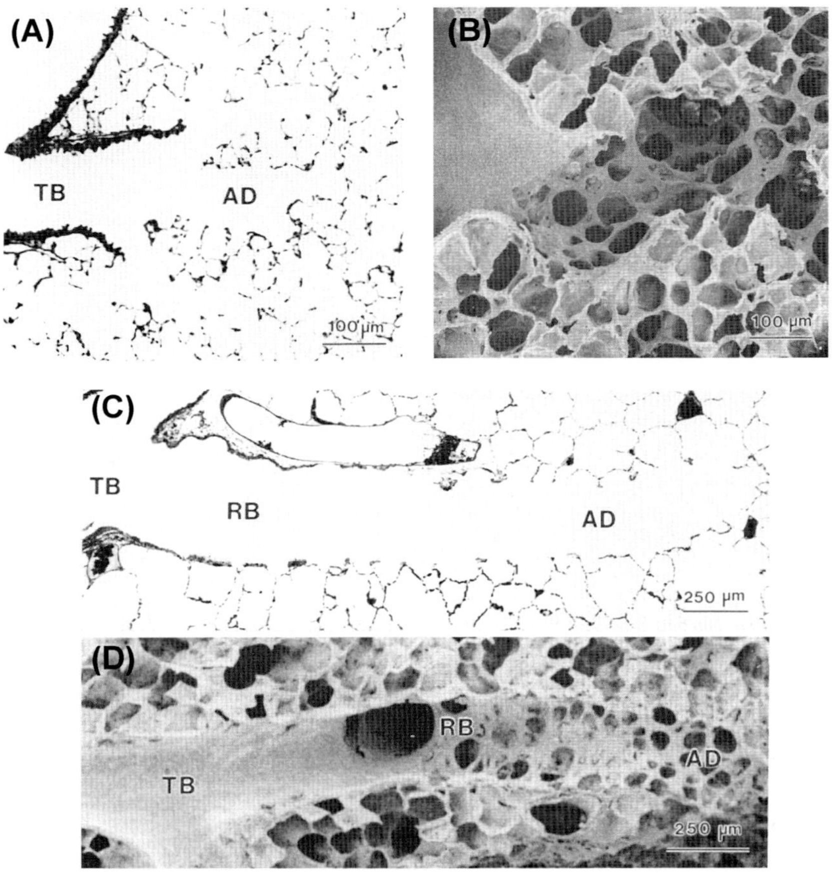

FIGURE 2 A comparison of the microanatomical differences in the transitional zone between species (mouse, A; rat, B) where the change from terminal bronchiole (TB) to alveolar duct (AD) is abrupt and species (cat, C, D) where the change is gradual and bronchiolar epithelium and alveolar epithelium are found in an extensive respiratory bronchiole (RB).

gas-exchange area. Unlike the gas-exchange epithelium, the capillaries that are present are not closely associated with the basal lamina of the epithelium. Vessels associated with the bronchiolar epithelial population include the pulmonary arteriole. An aspect of variability between species is the amount of cartilage that can be found in the transitional zone. In most species, cartilage is not observed in any of the distal conducting airways. However, this does not appear to be the case with the respiratory bronchiole in nonhuman primates (Tyler and Plopper, 1985), in which cartilage plates have been observed at the bifurcation points in the proximal respiratory bronchioles. Although the matrix composition of the alveolar gas-exchange portions of the lung have been studied in some detail, this is not the case for the matrix composition associated with bronchiolar epithelium.

Further complicating our understanding of the role that matrix may play in the regulation of differentiated expression in bronchiolar epithelium is the fact that the two principal cell types observed in the bronchiolar epithelial population of different species vary considerably in the extent to which they are found throughout the rest of the conducting airway tree. Ciliated cells are found throughout the airway tree, but nonciliated bronchiolar cells are not (Table 1). In only three species have nonciliated cells, which closely resemble the nonciliated populations in the bronchioles, been identified in the trachea and bronchi. These three species are the mouse, hamster, and rabbit.

3. COMPOSITION OF BRONCHIOLAR EPITHELIUM

Table 2 and Figures 3 and 4 compare species differences in the cellular composition of the most distal bronchiolar regions. They compare the relative proportions of cell types observed in terminal bronchioles and in respiratory bronchioles. For most species, the nonciliated population represents the majority of the cells in the population (Figure 3(A)). The relative proportions of the nonciliated bronchiolar epithelial cells (new proposed nomenclature: bronchiolar exocrine club, or "Club," cells; Winkelmann and Noack, 2010) and ciliated cells in these species are approximately equal, with the nonciliated cells being the majority. This appears to be the case for all of the species

TABLE 2 Comparison of Species Differences in Cellular Composition of Centriacinar Bronchiolar Epithelium

	Cell Types in Terminal Bronchioles				Cell Types in Respiratory Bronchioles				
Species	**Club**	**Ciliated**	**Goblet**	**Basal**	**Club**	**Ciliated**	**Goblet**	**Basal**	**References**
Mouse	>50%	<50%	–	–	N/A	N/A	N/A	N/A	Pack et al. (1980, 1981)
Hamster	>50%	<50%	–	–	N/A	N/A	N/A	N/A	Plopper et al. (1980a), Plopper (1983)
Rat	>50%	<50%	–	–	N/A	N/A	N/A	N/A	Chang et al. (1988), Young et al. (1986), Plopper et al. (1994)
Guinea pig	>50%	<50%	–	–	N/A	N/A	N/A	N/A	Plopper et al. (1980a)
Rabbit	>50%	<50%	–	–	N/A	N/A	N/A	N/A	Plopper et al. (1980a), Hyde et al. (1983)
Dog	>95%	<5%	–	–	>95%	<5%	–	–	Plopper et al. (1980b)
Cat	>95%	<5%	–	–	>95%	<5%	–	–	Hyde et al. (1985, Plopper et al. (1980b)
Macaque monkey	–	~50%	~20%	~10%	>90%	<10%	+	+	Plopper et al. (1989), Tyler and Plopper (1985)
Sheep	>60%	<40%	–	–	N/A	N/A	N/A	N/A	Mariassy and Plopper (1983), Plopper et al. (1980b)
Pig	+	+	–	–	N/A	N/A	N/A	N/A	Baskerville and Olson (1969)
Horse	>50%	<50%	–	–	N/A	N/A	N/A	N/A	Plopper et al. (1980b)
Cow	>50%	<50%	–	–	N/A	N/A	N/A	N/A	Mariassy et al. (1975), Plopper et al. (1980b)
Human	11 –41%	+	2%	+	22%	+	0%	+	Shijubo et al. (1997), Boers et al. (1999), Shijubo et al. (1999), Cutz and Conen (1971), Jarkovska (1970), Plopper et al. (1980a)

Note: N/A, not applicable; –, not present; +, present in variable amounts.

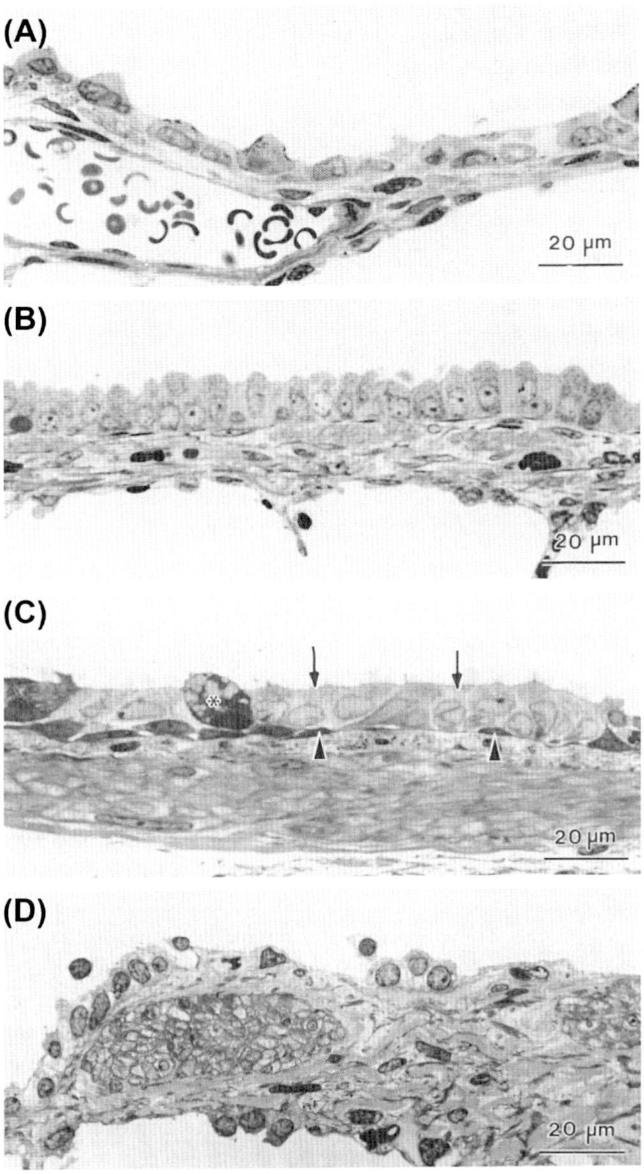

FIGURE 3 A comparison of the microanatomical appearance of the bronchiolar epithelium in the distal bronchiole of a species (rabbit, A) with an abundance of both nonciliated and ciliated cells, with a species (cat, B) where numerous generations of respiratory bronchioles are lined by nonciliated bronchiolar cells, and with a species (rhesus monkey, C and D) in which two bronchiolar epithelial populations (C: one with goblet (*), ciliated (arrows), and basal (arrowheads) cells; D: the other with Club cells) share the airway lumen with alveolar epithelium.

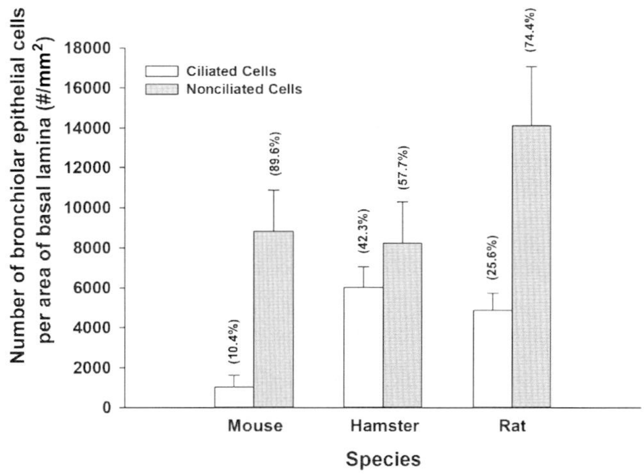

FIGURE 4 A morphometric comparison of the percentages of ciliated and Club cells in the terminal bronchioles of mouse, hamster, and rat. Data were calculated from measurements of nuclear volume and morphometric determinations of volume density of nuclei for both cell types (mean ± 1 SD).

without extensive respiratory bronchioles. In carnivores, which have a substantial respiratory, as well as a terminal, bronchiolar region, the epithelial populations are predominantly nonciliated in both the terminal and the respiratory bronchioles (Figure 3(B)). In contrast, the terminal and respiratory bronchiolar epithelium of primates appears to be much more heterogeneous than that for carnivores. The epithelial composition in the terminal bronchioles includes ciliated cells, which appear to be the majority; a secretory cell closely resembling the goblet cell with PAS and alcian blue positive granules; and a basal cell population (Figure 3(C)). The latter is defined based on its extensive attachment to the basal lamina and the lack of a cytoplasmic portion that extends all the way to the airway lumen. The respiratory bronchiole appears to contain the same epithelial population as in the terminal bronchiole on one portion of respiratory bronchiolar lumen and a second portion that is primarily nonciliated cells that do not resemble goblet cells, with a small proportion of ciliated cells and no basal cells (Figure 3(D)). Studies focusing on humans using CC10 expression suggest that this may also be the case for humans. Figure 4 compares the percentages of ciliated and nonciliated cells in terminal bronchioles of the mouse, hamster, and rat, based on morphometric estimations of nuclear size and volume per unit surface area. A small proportion of the nonciliated population may have the characteristics of stem cells (Garcia et al., 2012).

As illustrated in Table 3, the actual number (numerical density) of epithelial cells and nonciliated cells within the epithelial population of bronchioles has a considerable amount of interspecies variability. This table summarizes the current information on the five species in which quantitative data are available. These particular numbers are based on morphometric estimations of volume per unit surface area and determinations of nuclear volume by three-dimensional reconstruction. One thing that appears to be a significant variable factor is the density of the total epithelial population lining the bronchioles. This compares bronchiolar density for terminal bronchioles in the rat, the rabbit, and humans and respiratory bronchioles in the cat and the bonnet monkey. The human has the most and the

TABLE 3 Comparison of Numerical Density of Bronchiolar Epithelium and Density and Percentage of Club Cells in the Bronchiolar Epithelial Population of Adults

		Club Cells	
Species	Bronchiolar Epithelium Density[a] (#/mm^2)	Density[a] (#/,mm^2)	% Population
Rat[b]	17,070 ± 791	4336 ± 201	25.4
Rabbit[c]	15,073 ± 706	9261 ± 434	61.4
Cat[d]	19,532 ± 383	19,532 ± 383	100
Bonnet monkey[e]	9565 ± 304	8800 ± 280	92
Human[f]	30,300 ± 5600	10,600 ± 800	34

[a]*Mean + SD.*
[b]*From Young et al. (1986), Chang et al. (1988).*
[c]*From Hyde et al. (1983).*
[d]*From Hyde et al. (1985).*
[e]*From Moffatt et al. (1987).*
[f]*From Mercer et al. (1994).*

bonnet monkey the least bronchiolar density. Even more striking is the organization of the population. Nonciliated cells are the majority in three of these species, and in one they are almost 100%. In the rat and human, ciliated cells appear to be the predominant cell type.

4. ULTRASTRUCTURE OF NONCILIATED BRONCHIOLAR EPITHELIAL CELLS

Table 4 compares, in relative terms, the cytoplasmic organization of Club cells in a variety of species. The major conclusion from these data are that there are very few ultrastructural features of the nonciliated cell population found in the most distal conducting airways that are consistent between species. Smooth endoplasmic reticulum is the predominant cytoplasmic constituent in a large number of species (Figure 5(A)). It is the dominant feature of the apical portion of the cell in these species. This particular organelle does not appear to be a significant portion of the cytoplasm in carnivores or primates. The other major differing feature is the abundance of cytoplasmic glycogen. In most species, this composes less than 10% of the cytoplasmic volume. In carnivores, particularly the dog and cat, it is the predominant cytoplasmic constituent (Figure 5(B)).

Both glycogen and endoplasmic reticulum appear to be minor constituents of the cytoplasm in nonciliated bronchiolar cells in primates. Most of the species exhibit secretory granules. The content of these granules is not completely defined for all species. Enlarged mitochondria with a dense matrix and few cristae are observed in the Club cells of a number of species, including the mouse, rat, guinea pig, rabbit, cat, and pig (Table 4). The enlarged mitochondria contribute to the large portion of the cytoplasmic volume committed to mitochondria by mice and guinea pig, which is twice or more than that for any other species. The two most consistent features of Club cells in all species are the nucleus and lateral extensions of the cytoplasm. These occupy the lateral portions of the cells between the basal lamina and the apical junctional complex. They have been observed in all species and are one feature that clearly separates them from the other nonciliated cuboidal epithelial cells found in the central acinus, the type II alveolar cell.

The differences between the ultrastructure of Club cells are further magnified when the absolute volumes of constituents are determined (Table 5). The overall cell volume is relatively similar. The nucleus and every other constituent vary from a minimum of 50% (nucleus) to as much as 200-fold (glycogen).

5. SUMMARY

The bronchiolar epithelium of the distal conducting airways has one of the highest degrees of diversity of any region of the lung. Almost any aspect relevant to the biology of this cell population is highly variable. All the parameters of their environment, including their position in relation to the tracheobronchial tree, their interaction with gas-exchange epithelium in the transitional zone, and the extracellular matrix composition, are very species specific. The makeup of the population varies in terms of which cell types are present, the relative proportion of

TABLE 4 Comparison of Relative Proportions (Percentages)[a] of Cellular Components in Club Cells

Species	Nucleus	Agranular Endoplasmic Reticulum	Secretory Granules	Cytoplasmic Glycogen	Mitochondria	Large Mitochondria	Lateral Cytoplasmic Extensions
Mouse	21.8 ± 06.5	54.8 ± 07.5	+	0	34.7 ± 6.4	+	+
Hamster	25.2 ± 06.1	79.3 ± 07.6	+	0	10.7 ± 4.4	–	+
Rat	28.5 ± 10.4	66.2 ± 09.4	+	00.1 ± 00.4	16.3 ± 6.0	+	+
Guinea pig	28.6 ± 08.9	58.3 ± 09.0	+	00.0 ± 08.5	25.1 ± 0.0	+	+
Rabbit	23.8 ± 08.8	61.6 ± 05.4	+	07.0 ± 05.4	19.1 ± 7.6	+	+
Dog	23.4 ± 11.4	24.7 ± 08.6	+	57.1 ± 13.7	08.0 ± 6.7	–	+
Cat	26.7 ± 09.2	10.7 ± 06.7	–	61.3 ± 10.1	19.5 ± 9.6	+	+
Macaque monkey	28.6 ± 04.4	05.2 ± 03.3	+	0	14.1 ± 2.8	–	+
Sheep	26.6 ± 10.3	64.6 ± 18.3	+	06.8 ± 12.1	13.8 ± 8.8	–	+
Pig	?	?	+	?	?	+	+
Horse	08.6 ± 04.1	70.6 ± 04.5	+	0	10.6 ± 4.4	–	+
Cow	27.7 ± 10.1	21.9 ± 10.1	+	62.3 ± 11.5	12.0 ± 7.5	–	+
Human	41.9 ± 10.0	03.1 ± 03.5	+	04.6 ± 05.8	15.3 ± 6.6	–	+

[a] *As percent of cytoplasmic volume; mean + SD; + = as present; – = not present.*

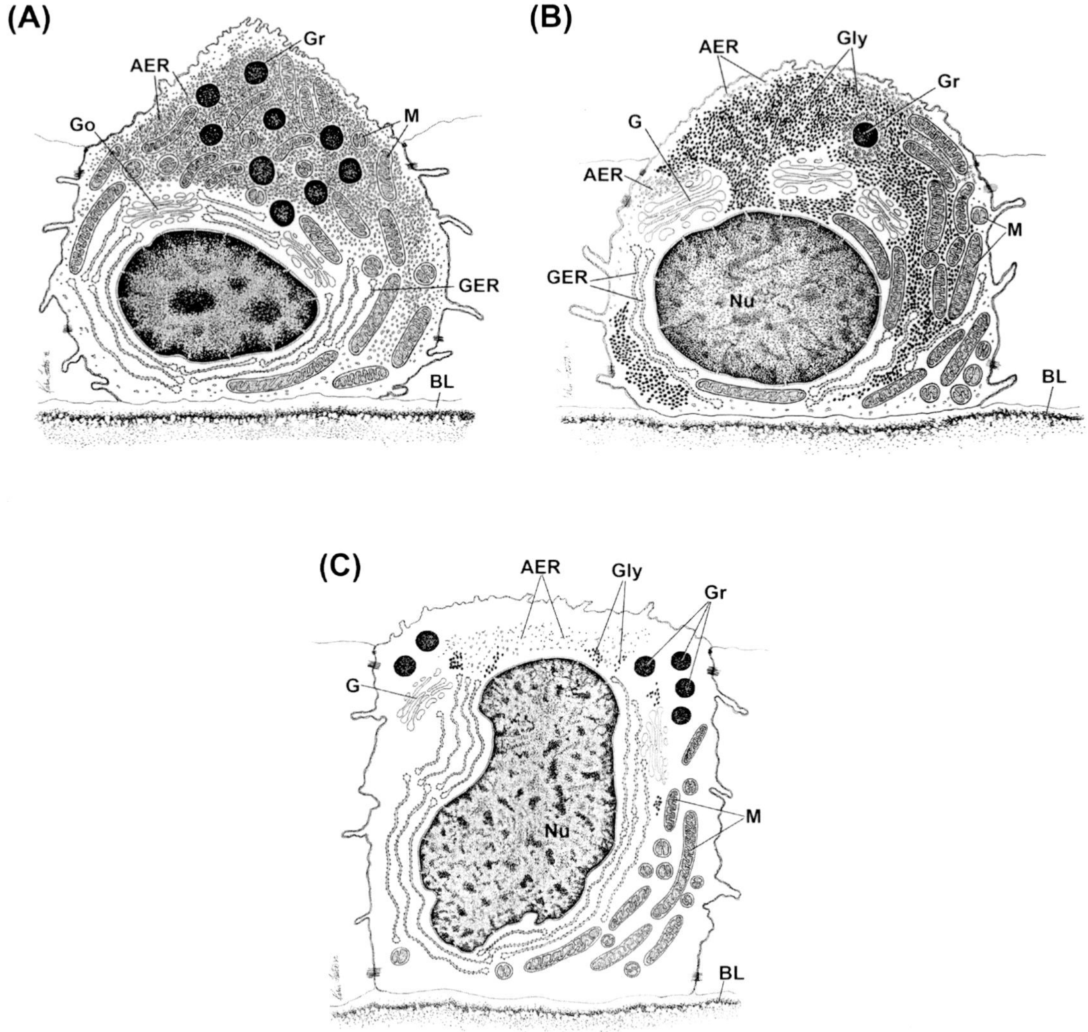

FIGURE 5 A comparison of the ultrastructural appearance of Club cells in species with abundant agranular endoplasmic reticulum (AER) (A), such as most laboratory mammals; with species (B) where cytoplasmic glycogen (Gly) predominates, such as carnivores; and with primates (C) where neither component is abundant. Compare with data in Tables 4 and 5. Drawings based on data in Table 5. Golgi apparatus (Go); secretory granules (Gr); mitochondria (M); basal lamina (BL); granular endoplasmic reticulum (GER); and nucleus (N).

TABLE 5 Comparison of Volumes (μm^3) of Cellular Components in Club Cells

Component	Rabbit[a]	Cat[b]	Bonnet Monkey[c]
Cell	471.8 ± 51.8	496.1 + 40.8	418.9 ± 36.1
Nucleus	083.0 ± 07.0	132.5 + 45.6	119.9 ± 18.4
SER	213.4 ± 51.0	038.9 + 24.4	015.6 ± 09.9
Glycogen	037.5 ± 13.1	222.9 + 36.4	0
RER	025.8 ± 19.4	001.56 + 03.38	038.9 ± 01.2
Mitochondria	075.4 ± 17.7	070.9 + 34.9	042.0 ± 08.3
Granules	012.5 ± 06.7	0	006.3 ± 01.4
Golgi apparatus	005.7 ± 05.7	014.6 + 13.1	024.1 ± 02.8

Mean ± 1 SD; SER, smooth endoplasmic reticulum; RER, rough endoplasmic reticulum.
[a]*From Hyde et al. (1983).*
[b]*From Hyde et al. (1985).*
[c]*From Moffatt et al. (1987).*

different cell types within the population, and the absolute density of cells: all are species-specific characteristics. The ultrastructural features of the nonciliated cell (Club cell) are also variable by species, both in terms of which components are present within the cell and the relative abundance of each structure.

REFERENCES

Baskerville, A., Olson, I.A., 1969. The ultrastructure of the bronchiolar epithelium of the pig (Abstr). Proc. Anat. Soc. G. B. & I 15–16.

Boers, J.E., Ambergen, A.W., Thunnissen, F.B., 1999. Number and proliferation of clara cells in normal human airway epithelium. Am. J. Respir. Crit. Care Med. 159, 1585–1591.

Chang, L.Y., Mercer, R.R., Stockstill, B.L., et al., 1988. Effects of low levels of NO_2 on terminal bronchiolar cells and its relative toxicity compared to O_3. Toxicol. Appl. Pharmacol. 96, 451–464.

Christensen, T.G., Breuer, R., Hornstra, L.J., Lucey, E.C., Snider, G.L., 1987a. The ultrastructure of hamster bronchial epithelium. Exp. Lung Res. 13, 253–277.

Christensen, T.G., Breuer, R., Hornstra, L.J., Lucey, E.C., Stone, P.J., Snider, G.L., 1987b. An ultrastructural study of the response of hamster bronchial epithelium to human neutrophil elastase. Exp. Lung Res. 13, 279–297.

Cutz, E., Conen, P.E., 1971. Ultrastructure and cytochemistry of Clara cells. Am. J. Pathol. 62, 127–141.

Devereux, T.R., Domin, B.A., Philpot, R.M., 1989. Xenobiotic metabolism by isolated pulmonary cells. Pharmacol. Ther. 41, 243–256.

Florens, M., Sapoval, B., Filoche, M., 2011. An anatomical and functional model of the human tracheobronchial tree. J. Appl. Physiol. 110, 756–763.

Garcia, O., Carraro, G., Navarro, S., et al., 2012. Cell-based therapies for lung disease. Br. Med. Bull. 101, 147–161.

Haefeli-Bleuer, B., Weibel, E.R., 1988. Morphometry of the human pulmonary acinus. Anat. Rec. 220, 401–414.

Hyde, D.M., Samuelson, D.A., Blakeney, W.H., Kosch, P.C., 1979. A correlative light microscopy, transmission and scanning electron microscopy study of the ferret lung. Scan Electron Microsc. 891–898.

Hyde, D.M., Plopper, C.G., Kass, P.H., Alley, J.L., 1983. Estimation of cell numbers and volumes of bronchiolar epithelium during rabbit lung maturation. Am. J. Anat. 167, 359–370.

Hyde, D.M., Plopper, C.G., Weir, A.J., et al., 1985. Peribronchiolar fibrosis in lungs of cats chronically exposed to diesel exhaust. Lab. Invest. 52, 195–206.

Jarkovska, D., 1970. Ultrastructure of the epithelium of the respiratory bronchioles in man. Folia Morphol. 18, 352–358 passim.

Lee, D., Fanucchi, M.V., Plopper, C.G., Fung, J., Wexler, A.S., 2008. Pulmonary architecture in the conducting regions of six rats. Anat. Rec. (Hoboken) 291, 916–926.

Mariassy, A.T., Plopper, C.G., 1983. Tracheobronchial epithelium of the sheep: I. Quantitative light-microscopic study of epithelial cell abundance, and distribution. Anat. Rec. 205, 263–275.

Mariassy, A.T., Plopper, C.G., Dungworth, D.L., 1975. Characteristics of bovine lung as observed by scanning electron microscopy. Anat. Rec. 183, 13–26.

Massaro, G.D., 1989. Nonciliated bronchiolar epithelial (clara) cells. In: Massaro, D. (Ed.), Lung Cell Biology. Marcel Dekker, New York, pp. 81–114.

Mercer, R.R., Russell, M.L., Roggli, V.L., Crapo, J.D., 1994. Cell number and distribution in human and rat airways. Am. J. Respir. Cell Mol. Biol. 10, 613–624.

Moffatt, R.K., Hyde, D.M., Plopper, C.G., Tyler, W.S., Putney, L.F., 1987. Ozone-induced adaptive and reactive cellular changes in respiratory bronchioles of bonnet monkeys. Exp. Lung Res. 12, 57–74.

Onuma, K., Ebina, M., Takahashi, T., Nukiwa, T., 2001. Irregularity of airway branching in a mouse bronchial tree: a 3-D morphometric study. Tohoku J. Exp. Med. 194, 157–164.

Pack, R.J., Al-Ugaily, L.H., Morris, G., Widdicombe, J.G., 1980. The distribution and structure of cells in the tracheal epithelium of the mouse. Cell. Tissue Res. 208, 65–84.

Pack, R.J., Al-Ugaily, L.H., Morris, G., 1981. The cells of the tracheobronchial epithelium of the mouse: a quantitative light and electron microscope study. J. Anat. 132, 71–84.

Plopper, C.G., 1983. Comparative morphologic features of bronchiolar epithelial cells. The Clara cell. Am. Rev. Respir. Dis. 128, S37–S41.

Plopper, C.G., 1990. Clara cells. In: Crystal, R.G., West, J.B. (Eds.), The Lung: Scientific Foundations. Raven Press, New York.

Plopper, C.G., Dungworth, D.L., 1987. Structure, function, cell injury and cell renewal of bronchiolar and alveolar epithelium. In: McDowell, E.M. (Ed.), Lung Carcinomas. Churchill Livingstone, London, pp. 94–128.

Plopper, C.G., Mariassy, A.T., Hill, L.H., 1980a. Ultrastructure of the nonciliated bronchiolar epithelial (Clara) cell of mammalian lung: I. A comparison of rabbit, guinea pig, rat, hamster, and mouse. Exp. Lung Res. 1, 139–154.

Plopper, C.G., Mariassy, A.T., Hill, L.H., 1980b. Ultrastructure of the nonciliated bronchiolar epithelial (Clara) cell of mammalian lung: II. A comparison of horse, steer, sheep, dog, and cat. Exp. Lung Res. 1, 155–169.

Plopper, C.G., Halsebo, J.E., Berger, W.J., Sonstegard, K.S., Nettesheim, P., 1983. Distribution of nonciliated bronchiolar epithelial (Clara) cells in intra- and extrapulmonary airways of the rabbit. Exp. Lung Res. 5, 79–98.

Plopper, C.G., Heidsiek, J.G., Weir, A.J., George, J.A., Hyde, D.M., 1989. Tracheobronchial epithelium in the adult rhesus monkey: a quantitative histochemical and ultrastructural study. Am. J. Anat. 184, 31–40.

Plopper, C.G., Chu, F.P., Haselton, C.J., Peake, J., Wu, J., Pinkerton, K.E., 1994. Dose-dependent tolerance to ozone. I. Tracheobronchial epithelial reorganization in rats after 20 months' exposure. Am. J. Pathol. 144, 404–420.

Rodriguez, M., Bur, S., Favre, A., Weibel, E.R., 1987. Pulmonary acinus: geometry and morphometry of the peripheral airway system in rat and rabbit. Am. J. Anat. 180, 143–155.

Shijubo, N., Itoh, Y., Yamaguchi, T., et al., 1997. Serum and BAL Clara cell 10 kDa protein (CC10) levels and CC10-positive bronchiolar cells are decreased in smokers. Eur. Respir. J. 10, 1108–1114.

Shijubo, N., Itoh, Y., Yamaguchi, T., et al., 1999. Clara cell protein-positive epithelial cells are reduced in small airways of asthmatics. Am. J. Respir. Crit. Care Med. 160, 930–933.

Sterner-Kock, A., Kock, M., Braun, R., Hyde, D.M., 2000. Ozone-induced epithelial injury in the ferret is similar to nonhuman primates. Am. J. Respir. Crit. Care Med. 162, 1152–1156.

Tyler, W.S., 1983. Comparative subgross anatomy of lungs. Pleuras, interlobular septa, and distal airways. Am. Rev. Respir. Dis. 128, S32–S36.

Tyler, N.K., Plopper, C.G., 1985. Morphology of the distal conducting airways in rhesus monkey lungs. Anat. Rec. 211, 295–303.

Widdicombe, J.G., Pack, R.J., 1982. The clara cell. Eur. J. Respir. Dis. 63, 202–220.

Winkelmann, A., Noack, T., 2010. The clara cell: a "Third Reich eponym"? Eur. Respir. J. 36, 722–727.

Young, S.L., Fram, E.K., Randell, S.H., 1986. Quantitative three-dimensional reconstruction and carbohydrate cytochemistry of rat nonciliated bronchiolar (Clara) cells. Am. Rev. Respir. Dis. 133, 899–907.

Chapter 8

Architecture of the Gas Exchange Region of the Lungs

Robert R. Mercer and James D. Crapo

Chapter Outline

1. INTRODUCTION

The architectural arrangement of the small airways and distal gas-exchange region in the lungs is obviously critical in determining the sites of deposition of gases and particles, as well as in subsequent tissue responses due to injury. In this chapter, we will cover the results of quantitative determination of the structure of the small airways and gas-exchange regions of the lungs. Because extrapolation of experimental studies in laboratory animals to the adverse health effect in humans is a central interest, we will be particularly interested in comparing the structure of the gas-exchange region of different species and the implications that these differences in structure have on dosimetry. The complex architecture of the pulmonary airway system has made the calculation of dose of inhaled pollutants delivered to specific sites in different species a difficult process. This difficulty is particularly apparent in the lack of structural data describing the gas-exchange regions of the lungs where low level, chronic exposures to reactive gases and aerosols have their most significant effects. The focus on injury in the gas-exchange region has been demonstrated for inhaled oxidant gases (Crapo et al., 1984), asbestos (Pinkerton et al., 1986), silica (Scabilloni et al., 2005), and newly developed nanomaterials such as titanium dioxide (McKinney et al., 2012) and carbon nanotubes (Porter et al., 2012; Mercer et al., 2013).

Detailed and quantitative assessments of the structure of the gas-exchange region in the lungs is a relatively recent task undertaken by a number of laboratories. For example, Weibel's (1963) model of the respiring region of the human lung was not a direct measurement and was based on the assumptions that the airways branch dichotomously up to the bronchiole-alveolar duct junction, with three generations of alveolar ducts proceeding out to the alveolar sac. The model of Yeh et al. (1979), based on the study of a single corrosion cast of a rat lung, also included an extrapolation of the acinar region of the rat using Weibel's model of the human lung scaled down to the alveolar dimensions in the rat lung. The advent of new stereological methods for estimating size and number, the development of quantitative methods for analyzing structure using serial section reconstructions, and the analysis of casts made of the gas-exchange region in the lungs have provided a more rigorous and detailed knowledge of the structural arrangement of the gas-exchange region in the lungs. Data from these techniques are being used to determine the extent to which differences in acinar structure between different species influence pulmonary dosimetry in laboratory animals and humans, and the significance that variations in acinar size, branching pattern, and other structural features may have on the homogeneity of lesions following inhalation of airborne pollutants.

2. DEFINITIONS

There is a wide range of terms used in describing the terminal airways and gas-exchange region of the lungs.

Comparative Biology of the Normal Lung. http://dx.doi.org/10.1016/B978-0-12-404577-4.00008-4

The pulmonary acinus is perhaps the most commonly used definition, which we can use as a basis for quantitation of lung structure. The word acinus, Latin for grape-like, is a general term used in anatomical nomenclature to indicate a sac-like structure. In the case of the lungs, this nomenclature arises quite naturally from examination of casts of the lung where one finds an airway branch that connects a distal unit containing the gas-exchange surfaces of the lung in the form of respiratory bronchioles, alveoli, and alveolar ducts. Although this definition is relatively simple in concept, there is considerable difficulty in applying it to quantitative studies of lung structure in different species.

The difficulty is particularly apparent when one examines smaller laboratory animals that either lack or have only rudimentary respiratory bronchioles. In these cases, Yeh et al. (1979) have defined the acinus to include the alveoli and alveolar ducts distal to the final airway epithelial termination, with the terminal bronchiole being the airway proximal to the epithelial termination. Rodriguez et al. (1987) have taken it to be the largest parenchymal lung unit where all airways are alveolated. When applying the definitions used by Yeh et al. (1979) or Rodriguez et al. (1987), there is considerable difficulty in comparing gas-exchange structures across different species. Mice do not have respiratory bronchioles. Respiratory bronchioles in the rat are limited to very short segments with only occasional alveolar outpockets, while larger species, such as the human, have one or more generations of respiratory bronchioles (Figure 1). Applying the definition of Yeh et al. (1979) or that of Rodriguez et al. (1987), it is a common occurrence to make the decision concerning the size of the collection of alveoli and alveolar ducts that constitutes an acinus based on the presence or absence of one alveolar outpocket in the short, final airway segment of the rat.

Fortunately there is a definition, corresponding to a collection of alveoli and alveolar ducts, which has a sac-like outer boundary, can easily be identified and counted, and eliminates difficulties in comparison of species with widely differing branching patterns in their small airways and absent or extensive respiratory bronchioles. Storey and Staub (1962) proposed a basic unit of ventilation or, as we will refer to it, a ventilatory unit, which consists of the alveoli and alveolar ducts distal to the transition from one bronchiole to an alveolar duct system. Based on serial section reconstructions, we have demonstrated (Mercer and Crapo, 1987b) that the alveoli and alveolar ducts of ventilatory units occur in a highly interconnected pattern and do not protrude into adjacent units. Thus, this definition leads to subdivisions of the lung consisting of a highly interconnected collection of alveoli and alveolar ducts, as shown in Figure 1. This definition may be used to compare species differences in gas-exchange structure because the bronchiole-to-alveolar duct junction is unambiguously present in all species, whether the immediately proximal airway a respiratory or terminal bronchiole.

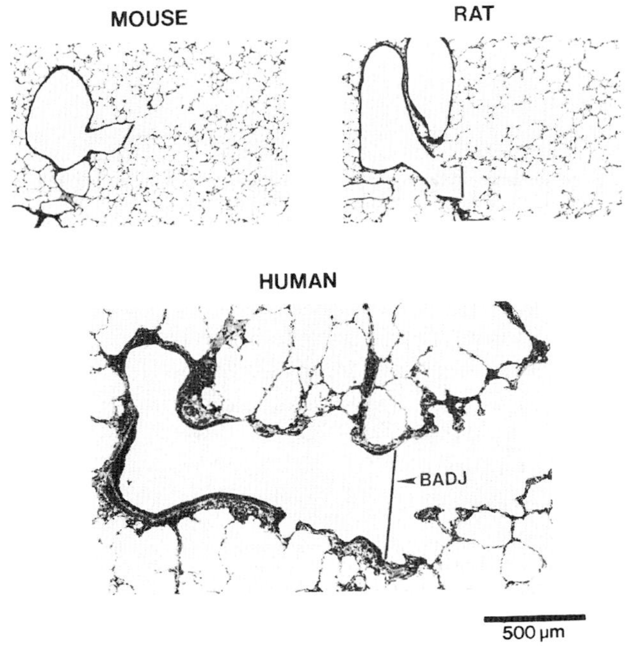

FIGURE 1 Light micrographs of bronchiole–alveolar duct junctions of mouse, rat, and human lungs. These micrographs demonstrate the bronchiole–alveolar duct junctions of various species. There is significant variation in the geometry of the airway immediately proximal to the bronchiole–alveolar duct junction. For instance, there are no alveoli in the walls of the very short final airway of the mouse, while the airway in the human lung is 1–2 mm in length and contains numerous alveolar outpockets.

The ventilatory unit is functionally important because it is the smallest common denominator in determining the distribution of inspired gas to the gas-exchange surfaces of the lungs. Thus, the ventilatory unit has a central role when considering what effect variations in size and architecture have in determining the dose of inhaled pollutants delivered to the gas-exchange region.

Figure 2 provides a schematic view of a lung illustrating the various anatomical descriptions that are used in subsequent descriptions. These include the subdivision of the lung that consists of the final terminal bronchiole dividing the distal alveoli-alveolar ducts into two or more units, distal respiratory bronchioles and ventilatory units. We note that this subdivision is also sometimes referred to as an acinus. However, it is clear from serial section analysis of the gas-exchange region of different species, such as illustrated in Figure 2, that more than one sac-like cluster of alveoli and alveolar ducts is distal to a single terminal bronchiole. The gas-exchange structure within each ventilatory unit is composed of alveolar ducts and the encircling alveoli.

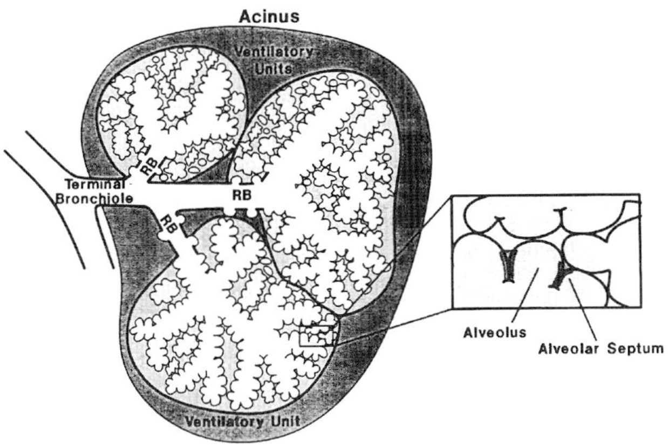

FIGURE 2 Components of the acinus. This figure illustrates the components of the acinus. These include the terminal bronchiole dividing the distal alveoli-alveolar ducts into two or more ventilatory units and the respiratory bronchioles (RB). Each ventilatory unit consists of the alveoli and alveolar ducts distal to a single bronchiole–alveolar duct junction. Three ventilator units are illustrated in this example.

3. METHODS OF STUDY

The basic methods used to study the gas-exchange region of the lungs consist of: (1) study of the small airways by casting techniques; (2) determination of the number of gas-exchange structures, either acini, ventilatory units, or alveoli, by the use of stereological techniques; and (3) determination of the structural relationships between different components by quantitative measurements of number, size, and diffusional characteristics using serial section analysis techniques. Each of these methods has inherent advantages and limitations. For instance, casting of the lungs is relatively simple to carry out, and the subsequent manual measurements do not require elaborate instruments. However, casting techniques do not preserve the delicate arrangement of the surfactant-lined surfaces of alveoli and alveolar ducts, because the installation of the fluid casting material destroys the critical air-to-tissue barrier and reexpands the lung in a nonhomogeneous manner (Gil et al., 1979). On the other hand, stereological methods for estimating number are efficient and accurate, and can be applied to lungs fixed in any manner (Mercer et al., 1987; Randell et al., 1989). However, the quantitative descriptions provided by stereological techniques are limited to determining the number of units and the average size of the unit, and do not allow the measurement of parameters critical to the characterization of the gas-exchange process, such as alveolar duct diameter or length.

Quantitative serial section analysis technique may also be used to analyze lungs preserved in any of a number of ways. In particular, the most appropriate of these is to use vascular perfusion-fixed lung preserved at levels of lung volume comparable to functional residual capacity. In this way, the structure of the lungs may be preserved in a state most nearly approximating that present during normal breathing. Since serial section analysis techniques are based on entry of the three-dimensional structure of the gas-exchange region of the lungs into a computer, the types of analysis that may be performed are virtually unlimited. Indeed, quantitative serial section analysis techniques have been used to determine ventilatory unit size, number of alveoli per ventilatory unit, alveolar duct branching pattern, and alveolar duct length, and to provide an estimate of the diffusional distance present in the gas-exchange region of different species. The major limitation of this technique is the difficulty and expense of entry of the structure into digital form for numerical and visual analysis. In particular, previous studies have usually included hundreds or even thousands of sections to be entered. Recent studies have shown that the number of sections required to achieve a high accuracy in numerical analysis based on serial sections is significantly smaller than that used in previous studies (Mercer et al., 1990a,b). In addition, advances based on systematic sampling (Gundersen and Jensen, 1987) in a series of sections have been developed that do not require computer entry (Mercer and Crapo, 1989).

4. COMPARISONS OF STRUCTURE ACROSS SPECIES

Over the size of mammalian lungs from shrew to whale, the mechanical characteristics of the lungs determined from pressure versus volume behavior and other measures of pulmonary function are remarkably similar (Leith, 1976; Schroter, 1980). As we shall demonstrate, this similarity in mechanical function is paralleled by similarity in the structural determinates of gas exchange.

Table 1 displays the quantitative results describing the acinar structure reported for various species by a number of investigators. Although there are some discrepancies and differences in methodology, the results demonstrate a remarkably smooth transition in number, volume, and dimensions of the acinus as one progresses from mouse to human lungs. Over the range from mouse to human lung, there is an approximately 50-fold increase in acinar number, a 170-fold increase in acinar volume, and a 7-fold increase in acinar diameter.

In spite of these large changes in volume, surface area, and other absolute measures, the pattern or structural arrangement of the gas-exchange unit is remarkably similar. For instance, the average number of generations of alveolar ducts in a single unit is approximately the same across species. This is somewhat unexpected, since different investigators approximate the closely spaced and irregular-shaped branches of the alveolar ducts in different manners. For instance, Parker et al. (1971) stated that essentially all trifurcations could be subdivided into

TABLE 1 Acinar Morphometry

Species	References	TBD (mm)	#/Lung	V (mm^3)	D or L (mm)	# Alveoli	Alveolar Duct Generations
Human	Pump (1964)	0.58		1.33–30.9		15,000	
	Horsfield and Cuming (1968), Parker et al. (1971)	0.60	27,992			10,714	6
	Hansen and Ampaga (1975), Hansen et al. (1975)	0.58	23,000	160.8	7.04 (L)	14,000–20,000	9
	Boyden (1971)		80,000	15.6			2–5
	Schreider and Raabe (1981)	0.43			5.1 (L)	7100	8–12
	Haefeli-Bleuer and Weibel (1988)	0.50	26,000–32,000	187.0	8.8 (L)	10,344	9
	Mercer, personal communication	0.5	43,000	200.0	6.0 (L)	8000	9
Baboon	Mercer and Crapo (1988b)	0.4	32,000	15.0	2.9 (D)	5094	8–2
Rabbit	Kliment (1973)		17,900	2.54			
	Rodriguez et al. (1987)			3.46	1.95 (L)		6
Guinea pig	Kliment (1973)		5100	1.25			
	Mercer, personal communication		4097	1.09	1.56 (D)	6890	9–12
Rat	Kliment (1973)		2500	1.3			
	Yeh et al. (1979)		2487	5.06			
	Mercer and Crapo (1987b)	0.21	2020	1.9	1.5 (D)	5243	10–12
	Rodriguez et al. (1987)			1.46	1.46 (D)		6
Mouse	Mercer and Crapo (1988b)		1025	0.3	0.9 (D)	4208	8–11

TBD, terminal bronchiole diameter; #/Lung, # units per lung; V, unit gas exchange volume; D or L, unit diameter or length.
Alveoli, # alveoli per unit; Alveolar duct generations, average # of alveolar generations from first alveolar duct to alveolar duct of a terminal sac.

successive bifurcations separated by a very narrow alveolar duct segment. In large part, this technical issue reflects the fact that the convenient, right circular cylinder model is used to approximate the actual structure of the alveolar ducts.

In our studies of alveolar ducts, we have found that the length of the alveolar duct segments between bifurcations was approximately equal to the alveolar duct diameter (Mercer and Crapo, 1987b). It was also found that the distance for each unique path of the alveolar duct network from the bronchiole-alveolar duct junction to the most distal region in the acinus was approximately equal to the ventilatory unit diameter. These results are not consistent with the traditional descriptions of the alveolar duct network as a linear, fanlike network of long, narrow tubes. It is thus more appropriate to consider the matrix of alveoli and alveolar ducts distal to the bronchiole-alveolar duct junctions as a three-dimensional lattice or honeycomb that fills a semispherical space.

In this respect, it is of particular interest to note that the number of alveoli used to fill the space occupied by the acinus is relatively constant between species. We have found in our studies that the ventilatory unit size and alveolar size increase in direct proportion in various species, as indicated in Figure 3 (Mercer and Crapo, 1988a; Mercer et al., 1991, 1994). The results from Figure 3, combined with the relatively constant number of alveoli per acinus, indicate that alveoli and alveolar duct structures in larger species are a scaled-up version of that present in smaller species, with no additional complexity in the arrangement or number of alveolar ducts and alveoli.

Not all aspects of acinar structure are similar across species. Most obvious of these is the difference in the structure and arrangement of the bronchioles supplying the gas-exchange region of different species. In addition to the differences in airways proximal to the bronchiole-alveolar duct junction, one finds significant differences

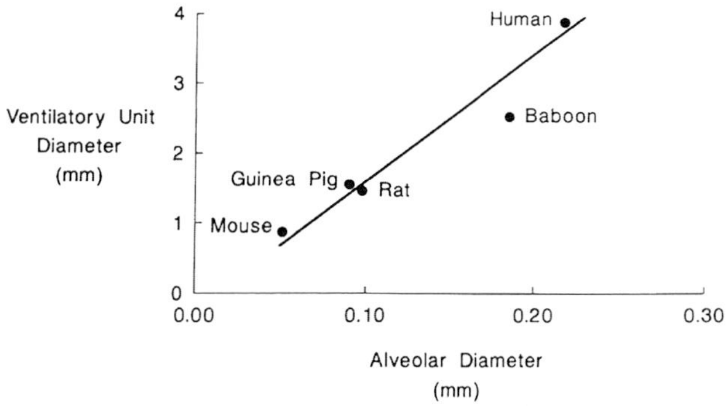

FIGURE 3 Alveolar versus ventilatory unit diameter. The alveolar diameter versus ventilatory unit diameter from lungs of different species preserved by vascular perfusion fixation at lung volumes comparable for functional residual capacity. A remarkable similarity in structure is demonstrated by the fact that the ratio of ventilator unit diameter to alveolar diameter is constant over the large range of lung size examined, with the ventilatory unit being 17.5 alveolar diameters in size.

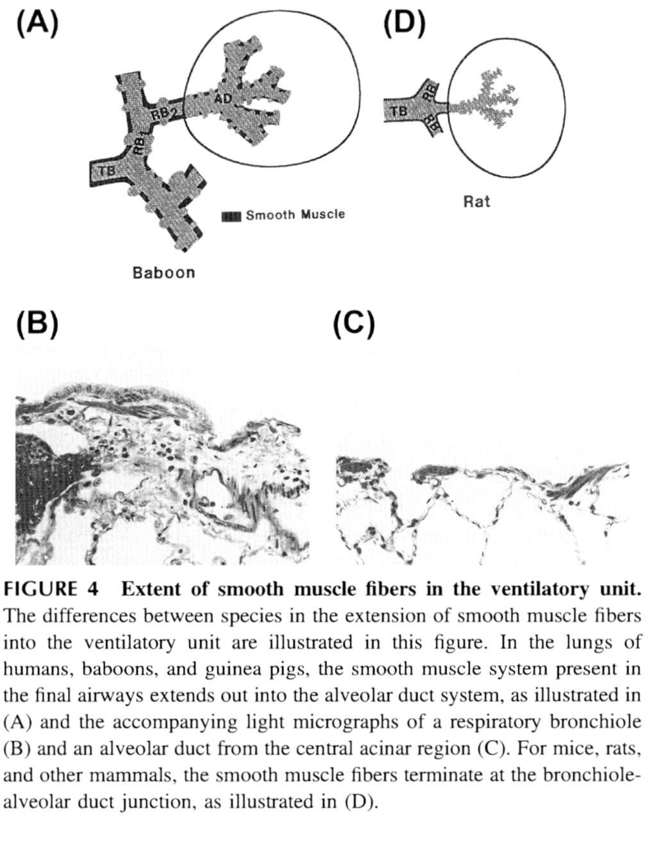

FIGURE 4 Extent of smooth muscle fibers in the ventilatory unit. The differences between species in the extension of smooth muscle fibers into the ventilatory unit are illustrated in this figure. In the lungs of humans, baboons, and guinea pigs, the smooth muscle system present in the final airways extends out into the alveolar duct system, as illustrated in (A) and the accompanying light micrographs of a respiratory bronchiole (B) and an alveolar duct from the central acinar region (C). For mice, rats, and other mammals, the smooth muscle fibers terminate at the bronchiole-alveolar duct junction, as illustrated in (D).

with regard to the extension of smooth muscle from the bronchiole–alveolar duct junction out into the alveolar ducts. For guinea pig (Nadel, 1964), baboon, and human lungs (Macklin, 1929), the fibers of smooth muscle derived from the airways extends out into the first few generations of alveolar ducts, as is illustrated in Figure 4. In the lungs of rats and mice, smooth-muscle fibers do not extend past the bronchiole–alveolar duct junction. This network of smooth muscle can profoundly affect gas exchange, as originally demonstrated by Nadel et al. (1964) in the guinea pig.

Because this network of smooth muscle is just beneath the thin alveolar epithelium, it is obviously more susceptible to injury by reactive airborne gases that can penetrate several microns into alveolar tissue (Miller et al., 1982; Overton et al., 1989) than smooth muscle fibers beneath the much thicker airway epithelium. Thus, the smooth muscle fibers in the alveolar region are susceptible to attack by inhaled pollutants and may play a role in the rapid changes in respiratory mechanics following very short exposures to ozone in these larger species (Fouke et al., 1988).

5. VARIATION IN SIZE

Analysis of the branching patterns of the airways using casts (Horsfield and Cumming, 1968; Yeh et al., 1979; Ross, 1957) and by serial section analysis (Mercer and Crapo, 1989; Mercer et al., 1991) has demonstrated that there are significant variations in the volume of anatomic dead space from trachea out to the acinus. For instance, Horsfield and Cumming (1968) calculated that a greater than four-fold difference in transit time exists from the canna to distal respiratory bronchioles based on analysis of a single human lung cast. Such variations in airway dead space when applied to problems of either gas exchange or dosimetry of inhaled pollutants can lead to significant regional variation between different lobes (Overton et al., 1989).

A major assumption made in all models of gas exchange or dosimetry based on the analysis of airway dead space variations has been that the ventilation to each acinus and its ventilatory unit subdivisions is constant throughout the lungs (Miller et al., 1982; Overton et al., 1989). Thus, current models of gas exchange and dosimetry of the lungs have been based on a homogeneously ventilating lung, where acinar size and ventilation are constant between different units and regions. However, as early as the studies of Boyden (1971), it has been known that there is a considerable variation in the size of different acini due to different growth rates during development. For instance, Boyden (1971) states that "one leaves this study feeling that so great is the fight for space in the post natal years of growth and differentiation that the end-result is a highly variable and individualist pattern." Boyden made these conclusions based on physical models derived from serial sections of lungs fixed by intratracheal instillation of fixative. Rodriguez et al. (1987) and Haefeli-Bleuer and Weibel (1988), using casting techniques in lungs, preserved near-total lung capacity and concluded that there were variations in size of acini over the entire lung. However, neither of these studies looked at lungs perfused by vascular perfusion fixation and therefore do not describe the distributions of acinar size that would be present at normal levels

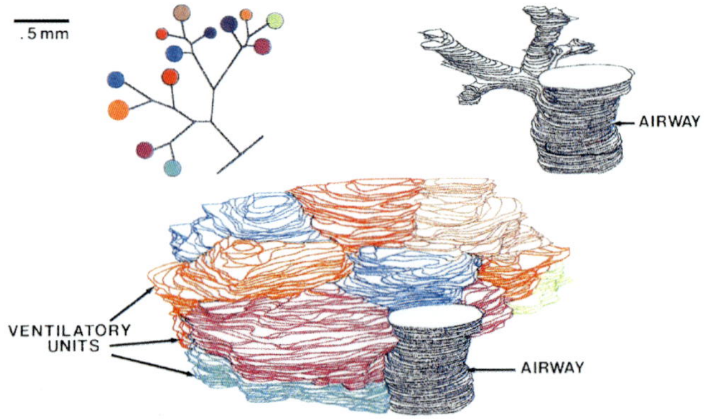

FIGURE 5 Hidden line view of a three-dimensional reconstruction of the airways and distal ventilatory units from a rat lung. A stick-like figure of the airway branching pattern with color-coded circles representing the ventilator units is given in the upper left. The diameter of each circle is proportional to the alveolar air volume of each ventilator unit. Three-dimensional reconstructions of the airways distal to the main stem bronchus and the outer boundary of each ventilator unit are given in the upper right and bottom middle of the figure, respectively. This particular example was chosen as it is relatively simple, containing only 13 ventilatory units and 23 airway segments. Typical reconstructions of the airway branching pattern off of the main bronchus contained 40–80 ventilatory units and 80–120 airway segments.

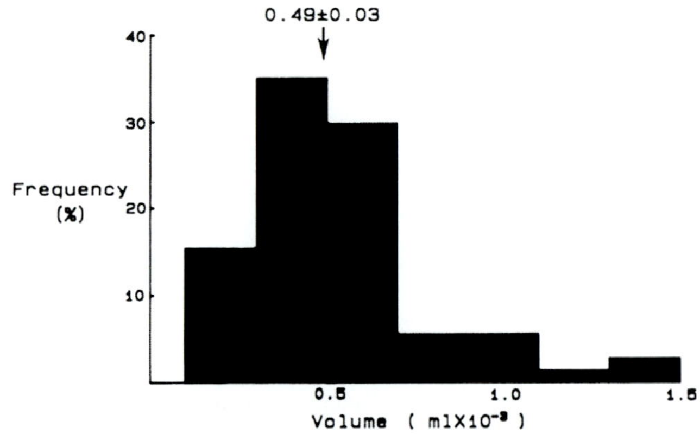

FIGURE 6 Ventilatory unit size distribution of the rat. The airspace volume of each ventilatory unit distal to a single airway branch off of the main stem bronchus was determined using serial sections, such as illustrated in Figure 5. This figure gives a histogram of the volume distribution of the 43 ventilatory units distal to the main stem bronchus. The histogram demonstrates a typical 10-fold variation in ventilatory unit size. The ventilatory units ventilated by these airways accounted for approximately 2% of the total lung volume alveolar volume.

of ventilation. It might be concluded from the casting studies, for instance, that differences in the degree of filling or in the compliance of different units resulted in the large variations in unit volume.

We have examined the size distribution of ventilatory units within the lungs by carrying out serial section analysis on in situ perfusion-fixed lungs at levels of lung inflation comparable to normal ventilation (Mercer and Crapo, 1989; Mercer et al., 1991). The branching patterns of airways distal to main stem bronchi of the left lungs and the outer boundaries of the ventilatory units ventilated by these airways in the lungs of rats were examined by serial-section analysis techniques. Figure 5 shows an example reconstruction from a relatively small airways off of the main stem bronchus. A reconstruction of the airways is given in the upper right of the figure and a reconstruction of the outer boundaries is shown in the lower half of the figure. From these serial sections it is possible to determine the gas volume of each ventilator unit as well as the branching pattern of alveolar ducts and alveolar sacs (terminal alveolar ducts) and the dead space/branching pattern of proximal airways.

Histograms of the volume of the ventilatory units connected to the airways in these serial section analyses have been used to determine the variation in gas-exchange unit volume between ventilatory units at functional residual capacity. An example of the histogram from one such reconstruction containing 43 ventilatory units of the rat is shown in Figure 6. The results demonstrate that ventilatory units have a considerable variation in size within local clusters of units ventilated by common airways. Indeed, it is typical to find a 15-fold difference in the size of ventilatory units distal to a single bronchus. In addition, these studies demonstrate that the size of individual alveoli within the large units is comparable to the alveolar size within the small units, and the alveolar epithelial surface to volume ratio is approximately the same for large and small ventilatory units (Mercer et al., 1991). These results would appear to exclude the potential for compliance effects causing the difference in size.

One of the major influences of variation in airway dead space and ventilatory unit size will be to vary the fraction of fresh inspired air received by each unit (V_A). The distribution of the ratio of ventilatory unit ventilation to ventilatory unit volume (V_A/V) is given in Figure 7. The results demonstrate a significant variation in the fraction of fresh inspired gas received by different ventilatory units. When comparable calculations are used to determine the variation in dose due to an inhaled, reactive, gaseous pollutant (e.g., ozone), these differences in ventilation are a significant factor in determining the variation in dose to the different ventilatory units. Approximately 5% of the ventilatory units are located on low dead space paths and have a larger than average ventilation (Mercer and Crapo, 1988b). These units have been predicted to receive a threefold or greater relative exposure to an inhaled gas due to this local variation (Mercer and Crapo, 1988b; Mercer et al., 1991). Because a comparable variation in the volume of human acini has been demonstrated by Haefeli-Bleuer and Weibel (1988), similar variations in ventilation and airborne pollutant dosimetry would also be expected. The results suggest that variation in the volume of ventilatory units is important in determining the distribution of inspired

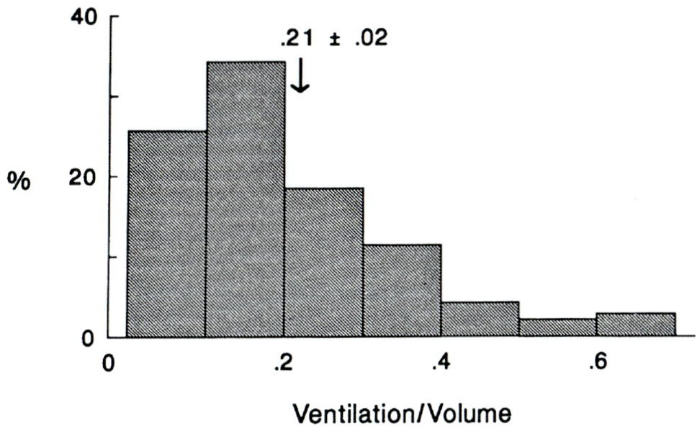

FIGURE 7 Distribution of the ratio of inspired gas ventilating each ventilatory unit to the ventilatory unit volume. This figure gives the histogram for the ratio of ventilatory unit ventilation to ventilatory unit volume determined from ventilation of each ventilatory unit supplied by the airways analyzed in Figure 6. The histogram demonstrates the significant variation in specific ventilation of the ventilatory units due to variations in airway dead space and ventilatory unit size. For instance, approximately 10% of the ventilatory units have a specific ventilation that is twice the mean value of 0.21, while 25% of the ventilatory units have a specific ventilation that is half the mean.

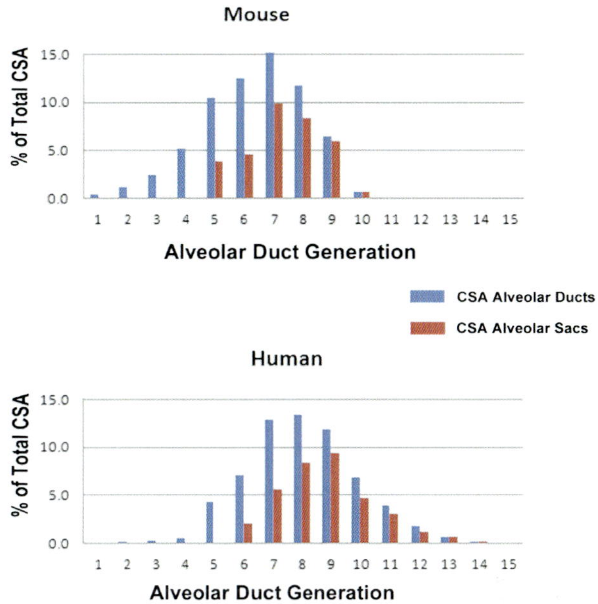

FIGURE 8 Comparison of the branching pattern of alveolar ducts and alveolar sacs (terminal alveolar ducts) in the ventilator units of mice versus human lungs. Figure gives plots of how the cross-sectional area of the alveolar ducts (with connected alveoli) and alveolar sacs (terminal alveolar ducts) change with generation number with the first generation beginning at the bronchiole to alveolar duct junction in the ventilator units of mouse and human lungs. The principal difference between the alveolar duct/alveolar sac branching pattern between the much smaller mouse lung and larger human lung appears to be the later occurrence of terminating alveolar sacs and a great number of subsequent generations in the larger human lung.

gas and may be particularly important in understanding the magnitude and heterogeneity of damage caused by inhalation of toxic substances. Models predicting the injury pattern due to these local variations in branching pattern and ventilatory unit size have predicted that there should be such a heterogeneity in damage (Mercer et al., 1991; Overton et al., 1996). Conformation of this pattern of injury with differences in local branching pattern and ventilator unit size has been demonstrated in the case of ozone inhalation injury (Pinkerton et al., 1992).

The serial section analysis of ventilator unit structure can also be used to investigate how the alveolar duct branching pattern varies across species. Figure 8 give example plots of how the cross-sectional area of the alveolar ducts (with connected alveoli) and alveolar sections (terminal alveolar ducts) changes with alveolar duct generation in the ventilator units of mouse and human lungs. As indicated by the lack of terminal alveolar sacs in generations 1 through 4, the first generations do not contain terminating alveolar ducts (alveolar sacs) but instead are associated with rapid dissemination of gas to more distal gas exchange regions. Across species, approximately 65 percent of the volume is present in the alveolar ducts and associated alveoli, which provide communications with more distal gas exchange regions, and 35 percent of the volume in a ventilator unit is contained in the alveolar sacs. A general bell-shaped pattern for the distribution of gas volume in the ventilatory is evident in both the mouse and human lungs. The principal difference between the alveolar duct/alveolar sac branching pattern between smaller and larger mammalian lungs appears to be the delay in occurrence of terminating alveolar sacs and a great number of subsequent generations. Data for these graphs of mouse, and human, ventilator unit branching patterns as well as those for rat and baboon are given in Table 2.

6. VASCULAR PERFUSION OF THE ACINUS

Not only is ventilation critical in determining dose–response relationships within the gas-exchange region of the lung but ventilation–perfusion relationships are also important, particularly for systemic delivery of inhaled gases. Of particular importance is the role variations in pulmonary capillary transit time have in determining the degree of oxygen and carbon dioxide exchange as this determines the time that transiting red blood cells have to equilibrate with the oxygen partial pressure in the lungs and to release carbon dioxide. Previous investigators have used analysis of single sections to estimate pulmonary capillary length (Staub and Schultz, 1968; Sobin et al., 1980). For instance, Sobin et al. (1980) used vascular tracers that subdivide each planar section into arteriole, venous, and capillary perfusion regions. By appropriate morphometric analysis, these investigators were able to estimate the

TABLE 2 Species Comparison of Alveolar Duct and Alveolar Sac Branching Structure[a]

Species	Generation	1	2	3	4	5	6	7	8	9	10	11	12	13	14	15
Mouse Cd-1																
	Alveolar duct	0.4	1.2	2.4	5.3	10.6	12.5	15.2	11.8	6.5	0.7					
	Alveolar sac	0.0	0.0	0.0	0.0	3.8	4.6	9.9	8.4	6.0	0.7					
Rat SD																
	Alveolar duct	0.4	1.0	2.2	5.2	8.8	15.3	11.9	6.6	6.4	4.7	3.3	1.4			
	Alveolar sac	0.0	0.0	0.0	0.7	1.9	7.9	9.3	2.1	4.2	3.0	2.2	1.5			
Baboon																
	Alveolar duct	0.1	0.1	0.3	0.5	4.2	8.5	12.8	14.6	11.5	6.7	4.0	1.6	0.7	0.1	
	Alveolar sac	0.0	0.0	0.0	0.0	0.0	2.0	5.7	8.6	8.0	4.7	3.4	1.1	0.5	0.1	
Human																
	Alveolar duct	0.1	0.2	0.3	0.6	4.3	7.1	12.9	13.4	12.0	6.8	3.9	1.8	0.7	0.3	0.1
	Alveolar sac	0.0	0.0	0.0	0.0	0.1	2.1	5.6	8.4	9.5	4.7	3.2	1.2	0.7	0.2	0.1

[a]*Table 2 gives the volume of alveolar air in alveolar ducts (including surrounding alveoli) and in alveolar sacs (terminal alveolar ducts with surrounding alveoli) versus generation from the bronchiole–alveolar duct junction expressed as a percentage of total ventilatory unit gas volume.*

TABLE 3 Pulmonary Capillary Length

Species	Capillary Length (μm)	References
Dog	600–800	Staub and Schultz (1968)
Cat	556 ± 286	Sobin et al. (1980)
Cat	600–800	Staub and Schultz (1968)
Rabbit	550–650	Staub and Schultz (1968)
Rat	205 ± 8	Mercer and Crapo (1987b)

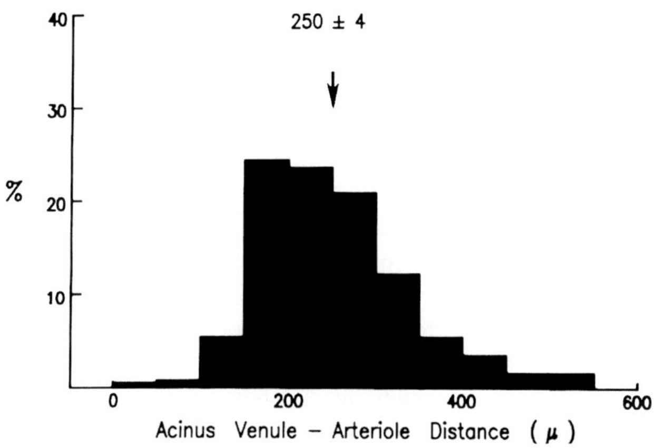

FIGURE 9 The distribution of distances from pulmonary venules to the nearest pulmonary arteriole for blood vessels perfusing an acinus of the rat lung.

average capillary length. The results obtained for a number of species are given in Table 3.

In addition to determining the average pulmonary capillary length, it is important to determine how the pulmonary capillary length varies within the acinus. Histograms of the acinar venule—arteriole distance, such as that given for the rat in Figure 9, suggests a significant variation in pulmonary capillary length. Based on the distances from venule to nearest arteriole and the assumption that red blood cell transit time is proportional to the square of length, a 25-fold variation in transit time could be expected in the acinus. Given the high blood flow rates present during exercise, the variation in transit time indicates that there is insufficient time for equilibration of a significant fraction of red blood cells during passage through the ventilatory unit.

7. SIZE LIMITATIONS IN GAS EXCHANGE

Given the quantitative descriptions of acinar structure and variations in acinar size, it is obviously of interest to consider the implications in terms of gas-exchange capacity. Illustrated in Figure 10 is a spherical model for gas

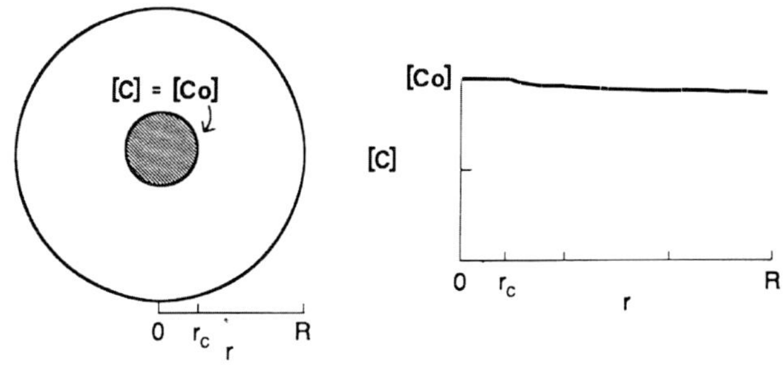

FIGURE 10 Model for determination of gas concentration in ventilatory units. This figure illustrates the results of a spherical model for the determination of the concentration gradients in the ventilatory unit. On the left side, the central core of the sphere is illustrated. The central core is assumed to be maintained at a constant concentration by ventilation with diffusion occurring from r_c, to the outer boundary at R. In both the convective and diffuse regions, it is assumed that the uptake of oxygen is occurring according to a first-order reaction (K). On the right side, the concentration profile is given for the human lung under resting conditions (K = 0.8%/s).

exchange in the ventilatory unit based on a central region maintained at a constant concentration by ventilation and an outer region where diffusion occurs. For both regions, the uptake of oxygen is proportional to the concentration.

Aside from the details of solving the equations describing this model (Appendix A), the salient features reside in three species-dependent parameters: (1) the oxygen uptake coefficient (K), (2) the size of the ventilatory unit (R), and (3) the fraction of the ventilatory unit maintained at a constant concentration core (r_c) by ventilation. The reaction or uptake coefficient K, which is the instantaneous rate of change in oxygen concentration, can be determined from measurements of oxygen uptake at rest and exercise (Lechner, 1978). The size of the ventilatory unit (R) is obtained from serial section reconstructions of lung preserved by vascular perfusion of fixative at functional residual volume. The size of the constant concentration region, r_c, is determined from the ratio of alveolar ventilation to ventilatory unit volume. The values used in the determination of oxygen uptake for rat and human lungs at rest and during exercise are given in Table 4, which also gives whole lung oxygen uptake (VO_2), dead space volume (V_D), parenchymal gas volume (V_P), and functional residual capacity (FRC), as well as the relevant model parameters.

Because the rate of oxygen uptake is low at rest, there is no significant gradient in oxygen concentration for small or large species, even though the constant concentration region is quite small, as illustrated in Figure 11. Of more interest is the absence of a significant concentration gradient under model conditions corresponding to exercise. This is predominately due to the dramatic increase in tidal volume, which approaches vital capacity at high levels of exercise (Murray, 1986). These results indicate that diffusional limitations in the transport of oxygen do not occur in

TABLE 4 Respiratory Parameters in Gas Exchange

Species	Condition	VO_2 (ml/min)	V_D (ml)	V_p (ml)	FRC (ml)	K %/s	r_c (mm)	R (mm)	% Efficiency
Rat[a]	Rest	4.7	0.8	1.9	3.9	7	0.49	0.8	99
	Exercise	24.5	0.8	5.7	3.9	29	0.81	0.9	97
Human	Rest	250	190	600	2200	0.8	3.3	6.0	99
	Exercise	4000	190	2400	2200	9.3	5.7	6.6	95

[a]*Data from various sources: Mercer and Crapo (1988b), Lechner (1978), and Stahl (1967).*

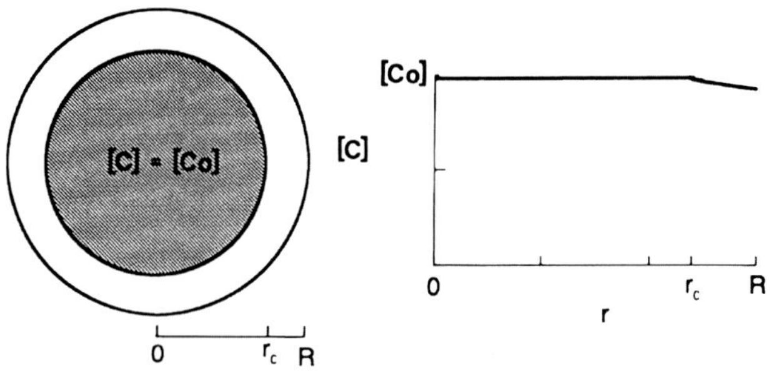

FIGURE 11 Concentration gradients during high rates of uptake. This figure demonstrates the conditions present during exercise where tidal volume exceeds functional residual capacity ($r_c = 0.8R$) and the uptake of oxygen is high ($K = 14\%/s$). Under these conditions, no significant concentration gradients are predicted because of the high alveolar ventilation.

larger lungs. On the other hand, the preceding predicted variations in red blood cell transit times at high rates of exercise would appear to limit equilibration of red blood cells within the acinus and suggests that inhomogeneity in perfusion may limit maximal oxygen uptake. This influence on the inhomogeneity of perfusion may be a factor in the increased dispersion of perfusion and evidence that red blood cells failed to equilibrate on exit from the capillaries with exercise at simulated altitude (Gale et al., 1985; Wagner et al., 1986).

Weibel (1983), in a comparison of morphometric estimates of maximal oxygen uptake with physiologic determinations, concluded that there were significant diffusional limitations in larger species. This was based on modeling of diffusional uptake of oxygen in a thick slab in the absence of convective transport (DuBois and Rogers, 1968). Under resting conditions, convective transport within the acinus is only a minor component of oxygen transport, as illustrated by the size of the convective region in Figure 11. However, at moderate to high levels of exercise, as illustrated by the large increase in convective region in Figure 11, alveolar ventilation provided by increases in respiratory frequency and tidal volume is a significant contributor to convective transport in the acinus and cannot be neglected in an analysis of oxygen transport.

8. APPENDIX A

To determine the extent to which species variations in acinar size and intralung variations in acinar size influence gas exchange, a number of simplifying assumptions concerning the structure of the ventilatory unit must be made. The first of these will be to consider the ventilatory unit as a sphere that has a central core maintained at a constant concentration C_o by ventilation. This assumption obviously differs from the traditional models, which assume a linear pattern of alveolar ducts that branch in a dichotomous pattern. However, as we have previously demonstrated, the branching pattern of the alveolar ducts is equivalent to a three-dimensional lattice or open framework.

Thus, we will consider a spherical model for determination of the concentration gradients in the ventilatory unit, as illustrated in Figure 11. This model consists of a central core that is maintained at a constant concentration by ventilation with diffusion occurring from rc to the outer boundary at R. In both the convective and diffusive regions, it will be assumed that the uptake of oxygen is linearly proportional to the oxygen concentration according to a first-order reaction coefficient k. Under these assumptions the differential equation for transport in the diffusion region is given by Eqn (1). The solution to this equation (Hershey, 1974) is given in Eqn (4), where the constants a and a_2 are chosen to solve the corresponding boundary conditions given in Eqns (2) and (3).

$$\frac{D}{r^2}\frac{\partial}{\partial r}\left[r^2\frac{\partial C}{\partial r}\right] - KC = 0 \tag{1}$$

$$C = C_o \text{ at } r = r_c \tag{2}$$

$$\frac{\partial C}{\partial r} = 0 \text{ at } r = R \tag{3}$$

$$C = \frac{C_o}{r}\left(a_1 \cosh\sqrt{\frac{Kr^2}{D}} + a_2 \sinh\sqrt{\frac{Kr^2}{D}}\right) \tag{4}$$

DISCLAIMER

The findings and conclusions in this report are those of the authors and do not necessarily represent the views of the National Institute for Occupational Safety and Health.

ACKNOWLEDGMENTS

This work was supported by US EPA Cooperative Agreement CR813113. Although the research described in this article has been funded by the US EPA, it has not been subjected to the Agency's required policy review and therefore does not necessarily reflect the views of the Agency and no official endorsement should be inferred.

REFERENCES

Boyden, E.A., 1971. The structure of the pulmonary acinus in a child of six years and eight months. Am. J. Anat. 132 (3), 275–299.

Crapo, J.D., Barry, B.E., Change, L., Mercer, R.R., 1984. Alterations in lung structure caused by inhalation of oxidants. J. Toxicol. Environ. Health 13, 301–321.

DuBois, A.B., Rogers, R.M., 1968. Respiratory factors determining the tissue concentrations of inhaled toxic substances. Respir. Physiol. 5, 34–52.

Fouke, J.M., Delemos, R.A., McFadden Jr., E.R., 1988. Airway response to ultra short-term exposure to ozone. Am. Rev. Respir. Dis. 137, 326–330.

Gale, G.E., Torre-Bueno, J.R., Moon, R.E., Saltzman, H.A., Wagner, P.D., 1985. Ventilation-perfusion inequality in normal humans during exercise at sea level and simulated altitude. J. Appl. Physiol. 58, 978–988.

Gil, J., Bachofen, H., Gehr, P., Weibel, E.R., 1979. Alveolar volume-surface area relation in air- and saline-filled lungs fixed by vascular perfusion. J. Appl. Physiol. 47, 990–1001.

Gundersen, H.J.G., Jensen, E.B., 1987. The efficiency of systematic sampling in stereology and its prediction. J. Microsc. 147, 229–263.

Hansen, J.E., Ampaya, E.P., 1975. Human air space shapes, sizes, areas, and volumes. J. Appl. Physiol. 38, 990–995.

Hansen, J.E., Ampaya, E.P., Bryant, G.H., Navin, J.J., 1975. Branching pattern of airways and air spaces of a single human terminal bronchiole. J. Appl. Physiol. 38 (6), 983–989.

Haefeli-Bleuer, B., Weibel, E.R., 1988. Morphometry of the human pulmonary acinus. Anat. Rec. 220, 401–414.

Hershey, D., 1974. Transport Analysis. Plenum Publishing Corp, New York, N.Y.

Horsfield, K., Cumming, G., 1968. Morphology of the bronchial tree in man. J. Appl. Physiol. 24 (3), 373–383.

Kliment, V., 1973. Similarity and dimensional analysis, evaluation of aerosol deposition in the lungs of laboratory animals and man. Folia Morphol. (Praha) 21 (1), 59–64.

Lechner, A.J., 1978. The scaling of maximal oxygen consumption and pulmonary dimensions in small mammals. Respir. Physiol. 34, 29–44.

Leith, D.E., 1976. Comparative mammalian respiratory mechanics. Physiologist 19 (4), 485–510.

Macklin, C.C., 1929. The musculature of the bronchi and lungs. Physiol. Rev. 9 (1), 1–60.

Mercer, R.R., Crapo, J.D., 1987a. Determination of the pulmonary capillary length within the acinus of the rat. Fed. Proc. 46, 520.

Mercer, R.R., Crapo, J.D., 1987b. Three-dimensional reconstruction of the acinus of the rat. J. Appl. Physiol. 63, 785–794.

Mercer, R.R., Laco, J.M., Crapo, J.D., 1987. Three-dimensional reconstruction of alveoli in the rat lung for determination of pressure-volume relationships. J. Appl. Physiol. 62, 1480–1487.

Mercer, R.R., Crapo, J.D., 1988a. Structure of the gas exchange region of the lungs determined by three-dimensional reconstruction. In: Toxicology of the Lung. Raven Press, New York, pp. 43–70.

Mercer, R.R., Crapo, J.D., 1988b. Comparison of gas exchange structure across species. Am. Rev. Respir. Dis. 137, 409.

Mercer, R.R., Crapo, J.D., 1989. Anatomical Modeling of Microdosimetry of Inhaled Particles and Gases in the Lung. Extrapolation of Dosimetric Relationships for Inhaled Particles and Gases. Academic Press, New York, pp. 69–78.

Mercer, R.R., Randell, S.H., Young, S.L., 1990a. Measurement of boundaries using a digitizer tablet. J. Microsc. 160, 97–105.

Mercer, R.R., McCauley, G.M., Anjilvel, S., 1990b. Approximation of surfaces in quantitative 3-D reconstructions. IEEE Trans. BME 37, 1136–1146.

Mercer, R.R., Anjilvel, S., Miller, F.J., Crapo, J.D., 1991. Inhomogeneity of ventilatory unit volume and its effects on reactive gas uptake. J. Appl. Physiol. 70, 2193–2205.

Mercer, R.R., Russell, M.L., Crapo, J.D., 1994. Alveolar septal structure in different species. J. Appl. Physiol. 77, 1060–1066.

Mercer, R.R., Scabilloni, J.F., Hubbs, A.F., Battelli, L.A., McKinney, W., Friend, S., Wolfarth, M.G., Andrew, M., Castanova, V., Porter, D.W., 2013. Distribution and fibrotic response following inhalation exposure to multi-walled carbon nanotubes. Part Fibre Toxicol. 10, 13.

McKinney, W., Jackson, M., Sager, T.M., Reynolds, J.S., Chen, B.T., Afshari, A., Krajnak, K., Waugh, S., Johnson, C., Mercer, R.R., Frazer, D.G., Thomas, T.A., Castranova, V., 2012. Pulmonary and cardiovascular responses of rats after inhalation of a commercial antimicrobial spray containing titanium dioxide nanoparticles. Inhal. Toxicol. 24, 447–457.

Miller, F.J., Overton, J.H., Myers, E.T., Graham, J.A., 1982. Pulmonary Dosimetry of Nitrogen Dioxide in Animals and Man. Elsevier Scientific Publishing Company, Amsterdam, The Netherlands, pp. 377–386.

Murray, J.F., 1986. The Normal Lung. W.B. Saunders Co., Philadelphia.

Nadel, J.A., Colebatch, H.J.H., Olsen, C.R., 1964. Location and mechanism of airway constriction after barium sulfate microembolism. J. Appl. Physiol. 19 (3), 387–394.

Overton, J.H., Barnett, A.E., Graham, R.C., 1989. Significances of the variability of tracheobronchial airway paths and their air flow rates to dosimetry model predictions of the absorption of gases. In: Extrapolation of Dosimetric Relationships for Inhaled Particles and Gases. Academic Press, New York, pp. 273–291.

Overton, J.H., Graham, R.C., Menache, M.G., Mercer, R.R., Miller, F.J., 1996. Influence of tracheobronchial region expansion and volume on reactive gas uptake and interspecies dose extrapolations. Inhal. Toxicol. 8, 723–745.

Parker, H., Horsfield, K., Cumming, G., 1971. Morphology of distal airways in the human lung. J. Appl. Physiol. 31, 386–391.

Pinkerton, K.E., Plopper, C.G., Mercer, R.R., Roggli, V.L., Patra, A.L., Brody, A.R., Crapo, J.D., 1986. Airway branching patterns influence asbestos fiber location and the extent of tissue injury in the pulmonary parenchyma. Lab. Invest. 55, 688–695.

Pinkerton, K.E., Mercer, R.R., Plopper, C.G., Crapo, J.D., 1992. Distribution of injury and microdosimetry of ozone in the ventilatory unit of the rat. J. Appl. Physiol. 73, 817–824.

Porter, D.W., Hubbs, A.F., Chen, T.H.B., McKinney, W., Mercer, R.R., Wolfarth, M.G., Battelli, L., Wu, N., Sriran, K., Leonard, S., Andrew, M.E., Willard, P., Tsuruoka, T., Endo, M., Tsukada, T., Munekane, F., Frazer, D.G., Castranova, V., 2012. Acute pulmonary dose-responses to inhaled multi-walled carbon nanotubes. Nanotoxicology. Epub.

Pump, K.K., 1964. The morphology of the finer branches of the bronchial tree of the human lung. Dis. Chest. 46, 379–398.

Randell, S.H., Mercer, R.R., Young, S.Y., 1989. Postnatal growth of pulmonary acini and alveoli in normal and oxygen-exposed rats studied by serial section reconstructions. Am. J. Anat. 186, 55–68.

Rodriguez, M., Bur, S., Favre, A., Weibel, E.R., 1987. Pulmonary acinus: geometry and morphometry of the peripheral airway system in rat and rabbit. Am. J. Anat. 180, 143–155.

Ross, B.B., 1957. Influence of bronchial tree structure on ventilation in the dog's lung as inferred from measurements of a plastic cast. J. Appl. Physiol. 10 (1), 1–14.

Scabilloni, J.F., Wang, L., Antonini, J.M., Roberts, J.R., Castranova, V., Mercer, R.R., 2005. Matrix metalloproteinase induction in fibrosis and fibrotic nodule formation due to silica inhalation. Am. J. Physiol. Lung Cell Mol. Physiol. 288 (4), L709–L717.

Schreider, J.P., Raabe, O.G., 1981. Structure of the human respiratory acinus. Am. J. Anat. 162, 221–232.

Schroter, R.C., 1980. Quantitative comparisons of mammalian lung pressure volume curves. Respir. Physiol. 42, 101–107.

Sobin, S.S., Fung, Y.C., Lindal, R.G., Tremer, H.M., Clark, L., 1980. Topology of pulmonary arterioles, capillaries, and venules in the car. Microvasc. Res. 19, 217–233.

Stahl, W.R., 1967. Scaling of respiratory variables in mammals. J. Appl. Physiol. 22 (3), 453–460.

Staub, N.C., Schultz, E.L., 1968. Pulmonary capillary length in dog, cat and rabbit. Respir. Physiol. 5, 371–378.

Storey, W.F., Staub, N.C., 1962. Ventilation of terminal air units. J. Appl. Physiol. 17, 391–397.

Wagner, P.D., Gale, G.E., Moon, R.E., Torre-Dueno, J.R., Stolp, B.W., Saltzman, H.A., 1986. Pulmonary gas exchange in humans exercising at sea level and simulated altitude. J. Appl. Physiol. 61, 260–270.

Weibel, E.R., 1963. Morphometry of the Human Lung. Springer-Verlag, Berlin, and Academic Press, New York.

Weibel, E.R., 1983. Is the lung built reasonably? Am. Rev. Respir. Dis. 128, 752–760.

Yeh, H.C., Schum, G.M., Duggan, M.T., 1979. Anatomic models of the tracheobronchial and pulmonary regions of the rat. Anat. Rec. 195, 483–492.

Chapter 9

Architecture and Cellular Composition of the Air–Blood Tissue Barrier

Kent E. Pinkerton, Peter Gehr, Alejandro Castañeda and James D. Crapo

Chapter Outline

1. INTRODUCTION

There are over 4000 different varieties of mammals. Among the terrestrial mammals, body size ranges from the 2-g Etruscan shrew to the 6600-kg African elephant. Lung tissue mass involved in gas exchange in this class of animals also covers a wide spectrum. Differences in the dimensions of the mammalian respiratory system are intriguing, since those pulmonary tissues involved in gas exchange are strikingly similar in organization and cellular composition, yet serve vastly different oxygen consumption needs in a highly efficient manner for animals spanning more than a 3-million-fold difference in total body mass (Gehr et al., 1981). The purpose of this chapter is to review and to compare the anatomy and cellular composition of the alveolar region in mammalian lungs.

2. ORGANIZATION OF THE GAS EXCHANGE REGION OF THE LUNGS

The lung parenchyma is that portion of the lungs involved in gas exchange. The most prominent structure in this region is the alveolus (Figure 1). Each alveolus in the lung parenchyma opens directly into an alveolar duct or occasionally, in a limited number of species, into a respiratory bronchiole. Alveolar ducts can be as few as two or as many as six generations. An alveolar sac represents the last and most peripheral generation of an alveolar duct. Alveoli and alveolar ducts arising from a single conducting airway constitute a pulmonary acinus. The size of the pulmonary acinus is dependent in part on the anatomical makeup of the transitional zone between the last conducting airways (called terminal bronchioles) and the alveolar ducts. In some species, this transition is extremely abrupt, going directly from the terminal bronchiole to the alveolar duct. Examples of this type of transition are found in the lungs of rodents, sheep, pigs, cows, and horses. In other species, airways beyond the terminal bronchiole contain alveolar outpocketings within their walls. These airways, called respiratory bronchioles, may consist of one to three generations before reaching an alveolar duct. Respiratory bronchioles are found in ferrets, cats, dogs, and primates. Since the proximal portions of the pulmonary acinus formed by respiratory bronchioles have been described in detail in Chapters 3, 6, and 7, only those more distal lung regions composed entirely of alveoli and alveolar ducts will be discussed in this chapter. A very recent publication on assessing the morphology of pulmonary acini (Vasilescu et al., 2012) presents methods for three-dimensional reconstruction of the acinus architecture that allow quantitative comparison of the acini of animals with different physiological needs over the whole size spectrum of the mammalian lung. For additional information on the structure of the pulmonary acinus, see Chapter 8.

The parenchyma composes 80–90% of the total lung volume when intratracheal instillation of glutaraldehyde-based fixatives is done under physiologic conditions (70–80% of total lung capacity) (Hayatdavoudi et al., 1980). In the rat, the parenchymal lung fraction is 81% (Burri and Weibel, 1971; Crapo et al., 1980; Pinkerton et al., 1982), in dogs 85% (Crapo et al., 1983), in pigs and horses 86% (Pinkerton et al., 1989; Gehr and Erni, 1980), and in humans 85–90% (Weibel, 1963; Gehr et al., 1978). These values reflect a high degree of organizational homogeneity of the

Comparative Biology of the Normal Lung. http://dx.doi.org/10.1016/B978-0-12-404577-4.00009-6

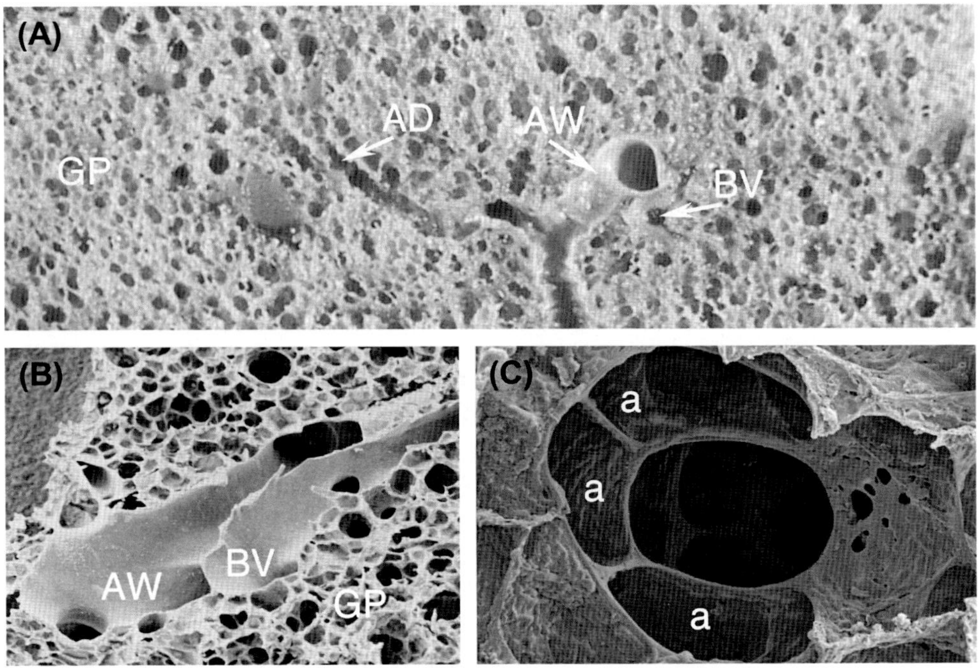

FIGURE 1 Gas exchange parenchyma. (A) Gross anatomical view of human lung, showing gas exchange parenchyma (GP), airways (AW), and blood vessels (BV). *Courtesy Institute of Anatomy, University of Berne, Switzerland.* Magnification 5×. (B) Scanning electron micrograph of hamster lung, showing gas exchange parenchyma (GP) and paired airway (AW) and pulmonary artery (BV) (Burri, 1974; Burri et al., 1974; Pinkerton et al., 1982; Pinkerton et al., 2015). Magnification 40×. (C) Scanning electron micrograph of human lung, showing an alveolar duct with alveoli (a) concentrically arranged around it. *From Gehr et al. (1978).* Magnification 170×.

parenchymal and nonparenchymal lung fractions over a large range of mammalian sizes.

Body weight, lung volume, alveolar and capillary surface areas, capillary volume, and harmonic mean thickness of the air–blood tissue barrier for a variety of mammalian species are given in Table 1. These species include commonly used laboratory research animals, domestic and nondomestic animals, and humans. Rats and dogs are further divided according to strain, age, sex, and/or weight.

The organization and packaging of two large surfaces for gas exchange into the limited volume of the lungs is remarkable. This arrangement, in the form of alveoli, interfaces an air surface to a blood surface of similar proportions separated by a thin tissue barrier (Figure 2) A review by Pinkerton and colleagues (Pinkerton et al., 2015) provides further details. With large differences in lung size among mammalian species (Table 1), it would follow that alveolar size may vary from species to species. However, for a number of animals, alveolar size and surface density are independent of body mass. An example is the suni, which has a body mass of 3.5 kg with alveoli of similar size and surface density as the 100 kg wildebeest (Gehr et al., 1981).

The ratio of total capillary surface area to alveolar surface area in the majority of mammals is slightly less than 1. This ratio ranges from 0.75 to 0.95 for most species. Capillaries are organized within the alveolar septa as a single sheet separated from the air space by a thin barrier of tissue formed by epithelial, interstitial, and endothelial compartments. The harmonic mean tissue thickness of the air–blood barrier (T_{ht}, Table 1) ranges from 0.27 μm in the shrew to 0.67 μm in the baboon. A thin tissue barrier provides an extremely efficient means of gas transfer over a large surface area, which juxtaposes the air spaces of the lungs with the pulmonary capillary bed. The integrity of these delicate alveolar septa is maintained in large measure by a network of collagen and elastin fibers that alternate on different sides of the capillary bed, coursing through the septa to form "thick" and "thin" portions of the alveolar wall to allow for a highly efficient structural makeup to facilitate gas diffusion (Weibel, 1979). In addition to collagen and elastin, the thick portion of the alveolar wall contains interstitial cells and extracellular matrix sandwiched between epithelial and endothelial layers, while the thin portion is formed by a single fused basement membrane wedged between epithelial and endothelial cell layers (Figure 2). The thin portion of the alveolar septal wall is a highly efficient region of the air–blood barrier of the lungs for the transfer of gases to and from the capillary blood. It is this portion of the air–blood barrier that is maximally exposed to the alveolar air space when the lungs are fixed by vascular perfusion under conditions approaching functional residual capacity (Gil et al., 1979).

The allometric relationship of body mass to a number of parenchymal lung values listed in Table 1 are shown as plots in Figures 3–6. Open circles designate African mammals, while closed circles designate laboratory animals, domesticated species, and humans. In most cases, African species fit well into allometric regressions with other mammalian species. This observation is of interest since a number of African mammals, such as gazelles and wildebeests, with greater physical activity and higher

TABLE 1 Comparative Anatomy of the Lung Parenchyma and Air–Blood Tissue Barrier[a]

Species	N	Body Weight (g)	Lung Volume (ml)	Aveolar Surface Area (Both Lungs) cm^2	Capillary Surface Area (Both Lungs) cm^2	Capillary Volume (Both Lungs) cm^2	Th Tissue (I-Lm)	References
Shrew (*Suncus etrascus*)	4	2.6 ± 0.2	0.10 ± 0.01	170 ± 10	130 ± 15	0.0118 ± 0.032	0.27 ± 0.02	Gehr et al. (1980)
White mouse (*Mus musculus*)	5	23 ± 2	0.74 ± 0.07	680 ± 85	590 ± 60	0.084 ± 0.009	0.32 ± 0.01	Geelhaar and Weibel (1971)
Waltzing mouse (*Mus wagneri*)	5	13 ± 1	0.58 ± 0.06	630 ± 40	540 ± 30	0.065 ± 0.008	0.26 ± 0.002	Geelhaar and Weibel (1971)
Syrian golden hamster (*Mesocricetus auratus*)	4	118 ± 7	2.81 ± 0.24	2760 ± 250	2410 ± 190	0.294 ± 0.011	0.39 ± 0.10	Gehr (unpublished data)
White rat (*Rattus rattus*)	8	140 ± 7	6.34 ± 0.25	3880 ± 190	4070 ± 200	0.480 ± 0.022	0.37 ± 0.02	Burri and Weibel (1971)
White rat (Sprague–Dawley)	6	360 ± 4	10.82 ± 0.38	4865 ± 380	4270 ± 385	0.631 ± 0.07	0.40 ± 0.02	Crapo et al. (1980)
White rat (Fischer 344)								
Male: 5 months	4	289 ± 13	8.60 ± 0.31	3915 ± 390	3830 ± 395	0.65 ± 0.06	0.38 ± 0.03	Pinkerton et al. (1982)
Female: 5 months	4	182 ± 5	7.48 ± 0.10	3420 ± 125	3260 ± 185	0.46 ± 0.10	0.34 ± 0.01	Pinkerton et al. (1982)
Male: 26 months	4	391 ± 11	12.67 ± 0.74	4630 ± 440	4490 ± 485	0.67 ± 0.10	0.37 ± 0.01	Pinkerton et al. (1982)
Female: 26 months	4	298 ± 7	9.39 ± 0.40	4020 ± 25	3570 ± 165	0.34 ± 0.05	0.37 ± 0.01	Pinkerton et al. (1982)
Guinea pig (*Cavia porcellus*)	15	429 ± 11	13.04 ± 3.03	9100 ± 280	7400 ± 230	1.50 ± 0.08	0.42 ± 0.01	Forrest and Weibel (1975)
Rabbit (*Oryctolagus cuniculus*)	6	3560	79.2	58,600 ± 12,400	47,000 ± 8800	7.15 ± 1.88	0.50 ± 0.04	Gehr et al. (1981)
Dwarf mongoose (*Helogale pervula*)	3	52,800 ± 9800	30.6 ± 5.6	16,100 ± 2600	14,600 ± 3400	2.06 ± 0.52	0.39 ± 0.02	Gehr et al. (1981)
Genet cat (*Genetta tigrina*)	2	137,200 ± 4300	99.0 ± 12.2	56,300 ± 6400	42,300 ± 1600	5.04 ± 0.63	0.51 ± 0.02	Gehr et al. (1981)
Dog (*Canis familiaris*)	3	5400	284.2	182,000 ± 135,000	141,000 ± 111,000	26.0 ± 24.9	0.43 ± 0.02	Gehr et al. (1981)
Dog (*C. familiaris*)	8	11,200 ± 400	736 ± 25	407,000 ± 39,000	329,000 ± 16,000	50.2 ± 5.0	0.46 ± 0.01	Siegwart et al. (1971)
Dog (*C. familiaris*)	4	16,000 ± 3000	1322 ± 64	510,000 ± 10,000	570,000 ± 20,000	92 ± 5	0.45 ± 0.01	Crapo et al. (1983)
Dog (*C. familiaris*)	6	22,800 ± 600	1501 ± 74	897,000 ± 69,000	718,000 ± 69,000	71.8 ± 4.5	0.48 ± 0.01	

Continued

TABLE 1 Comparative Anatomy of the Lung Parenchyma and Air–Blood Tissue Barrier[a]—cont'd

Species	N	Body Weight (g)	Lung Volume (ml)	Aveolar Surface Area (Both Lungs) cm^2	Capillary Surface Area (Both Lungs) cm^2	Capillary Volume (Both Lungs) cm^2	Th Tissue (I-Lm)	References
								Siegwart et al. (1971)
Dog (*C. familiaris*)	5	46,100	2888	1,769,000 ± 456,000	1,319,000 ± 375,000	234 ± 69	0.53 ± 0.08	Gehr et al. (1981)
Camel (*Camelus dromedarus*)	2	231,700 ± 2,700	15,900 ± 1400	4,305,000 ± 584,000	2,726,000 ± 292,000	378 ± 100	0.60 ± 0.06	Gehr et al. (1981)
Giraffe (*Giraffa camelopardalis*)	1	3,83,000	21,000	63,61,000	55,16,000	965	0.6	Gehr et al. (1981)
Suni (*Nesotragus moschatus*)	2	3300 ± 300	209.4 ± 0.6	96,900 ± 5500	81,300 ± 13,000	12.4 ± 0.7	0.56 ± 0.09	Gehr et al. (1981)
Dik–dik (*Madoqua kirkii*)	2	4200 ± 100	313.4 ± 1.2	146,000 ± 700	130,000 ± 6550	22.6 ± 3.3	0.43 ± 0.02	Gehr et al. (1981)
Wildebeest (*Connochaetes tauriras*)	1	1,02,000	7678	39,08,000	28,13,000	472	0.37	Gehr et al. (1981)
Waterbuck (*Kobus defassa*)	2	109,800 ± 16,300	7835 ± 1550	3,829,000 ± 950,000	3,378,000 ± 460,000	584 ± 98	0.46 ± 0.04	Gehr et al. (1981)
African goat (*Capra hircus*)	2	20,900 ± 1000	1370 ± 15	449,000 ± 12,000	439,000 ± 12,000	101 ± 8	0.54 ± 0.03	Gehr et al. (1981)
African sheep (*Ovis aries*)	2	21,800 ± 200	17,055 ± 435	671,000 ± 71,000	645,000 ± 139,000	146 ± 35	0.53 ± 0.05	Gehr et al. (1981)
Zebu cattle (*Bos indicus*)	4	192,500 ± 24,000	10,145 ± 1960	3,850,000 ± 420,000	3,795,000 ± 392,000	700 ± 124	0.50 ± 0.04	Gehr et al. (1981)
Swiss cow (*Bos taurus*)	1	7,00,000	22,450	1,28,30,000	1,13,80,000	2770	0.51	Gehr et al. (1981)
Horse (*Equis caballus*)	2	510,000 ± 0	37,650 ± 1,050	24,560,000 ± 124,000	16,630,000 ± 1,080,000	2800 ± 300	0.60 ± 0.02	Gehr and Emi (1980)
Monkey (*Macaca irus*)	6	3710	184.2	133,000 ± 12,700	116,000 ± 15,400	15.5 ± 2.7	0.50 ± 0.03	Gehr et al. (1981)
Baboon (*Papio papio*)	5	29,000 ± 3000	2393 ± 100	496,000 ± 77,000	386,000 ± 95,000	44 ± 17	0.67 ± 0.06	Crapo et al. (1980)
Man (*Homo sapiens*)	8	74,000 ± 4000	4341 ± 285	1,430,000 ± 120,000	1,260,000 ± 120,000	213 ± 31	0.62 ± 0.04	Gehr et al. (1978)

[a]*All values are mean ± SEM.*

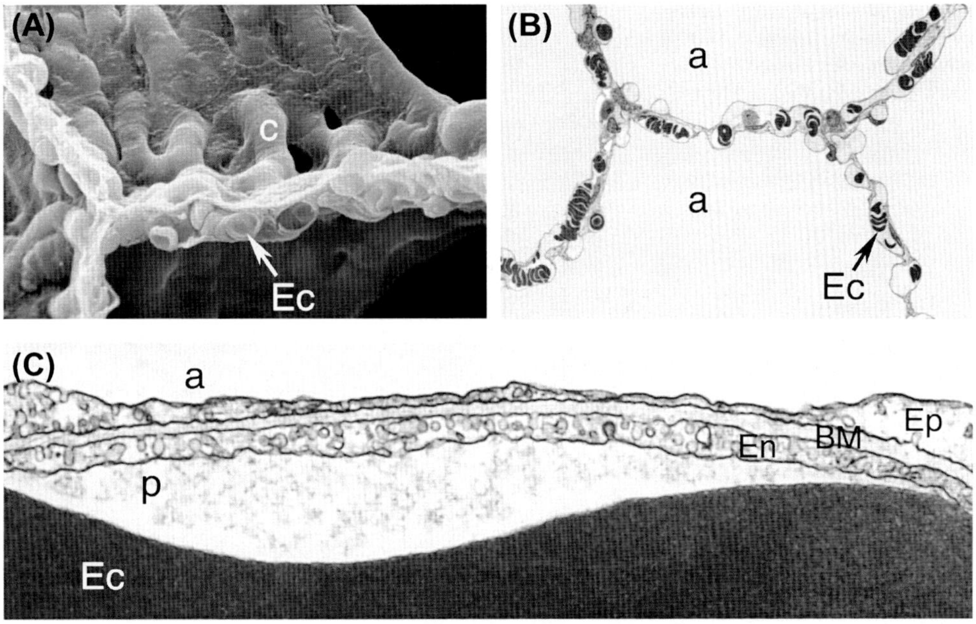

FIGURE 2 **Gas exchange apparatus of human lung.** (A) Scanning electron micrograph of the cut surface of the interalveolar septa (IS) showing capillary (C) with erythrocytes (Ec) in it. Also shown is a surface view of the alveolar wall with the capillaries (c) bulging into the alveolar space. Alveoli (a). *From Gehr et al. (1978)*. Magnification 500×. (B) Transmission electron micrograph of interalveolar septa (IS), showing the capillaries meandering around the connective tissue frame from one alveolus (a) to its neighbor, where they are bulging into the alveolar space. Erythrocytes (Ec). *From Gehr et al. (1978)*. Magnification 400×. (C) Transmission electron micrograph at higher magnification, showing the air–blood tissue barrier consisting of very thin squamous cellular layers: the type I epithelial layer (Ep) and the endothelial layer (En) with fused basement membranes (BM) of the two cellular layers. Alveoli (a), capillary blood plasma (p), erythrocyte (Ec). Magnification 30,000×.

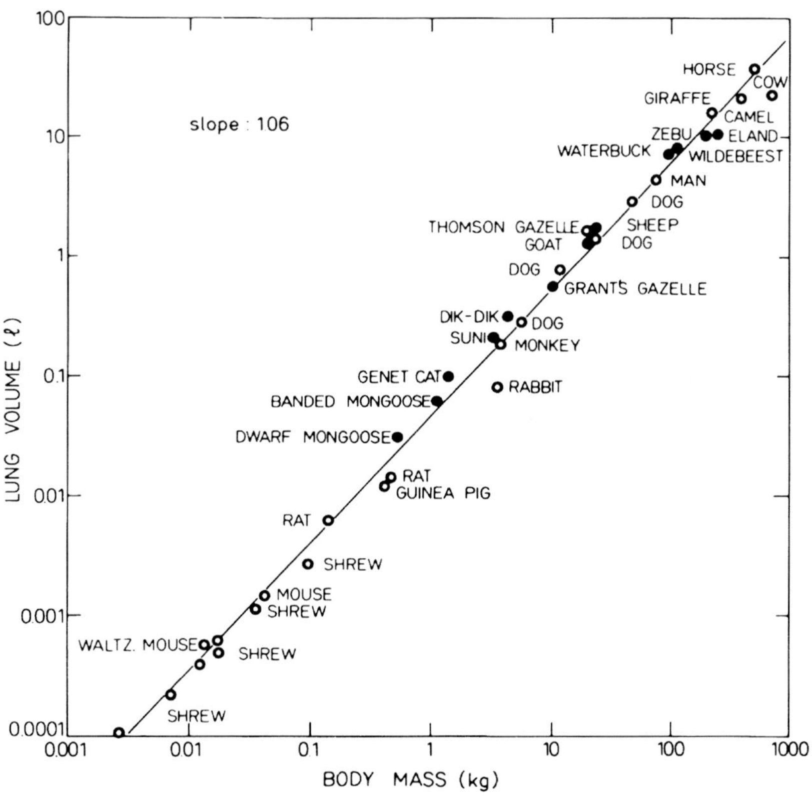

FIGURE 3 **Allometric plot of mean lung volume to mean body mass for mammalian species. Closed circles are African species and open circles are other species.** *From Gehr et al., 1981. J. Respir. Physiol. 44, 61–86. With permission.*

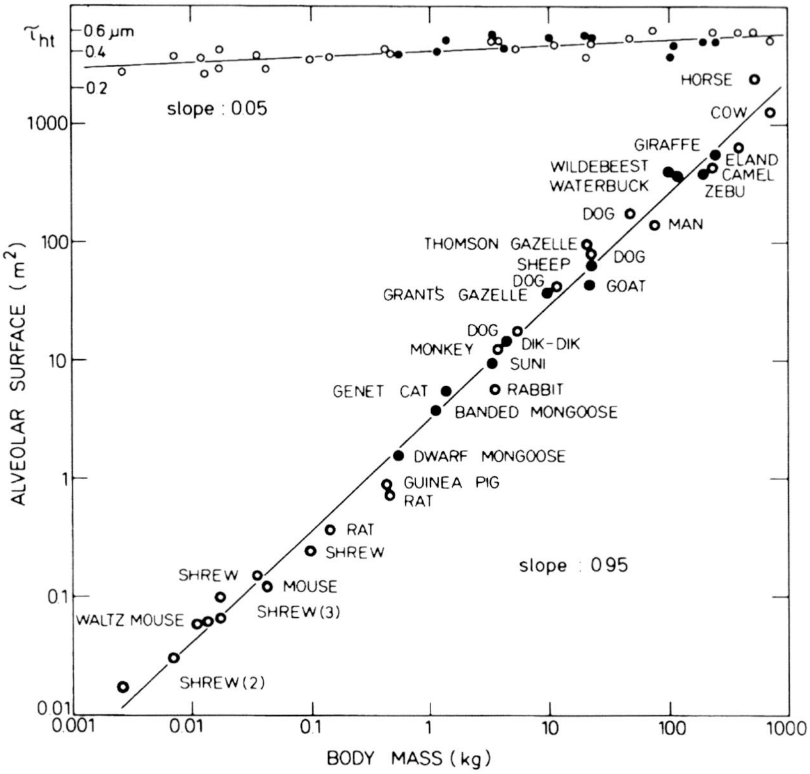

FIGURE 4 Allometric plot of alveolar surface area and harmonic mean tissue thickness (T_5) to body mass for mammalian species. *From Gehr et al., 1981. J. Respir. Physiol. 44, 61–86. With permission.*

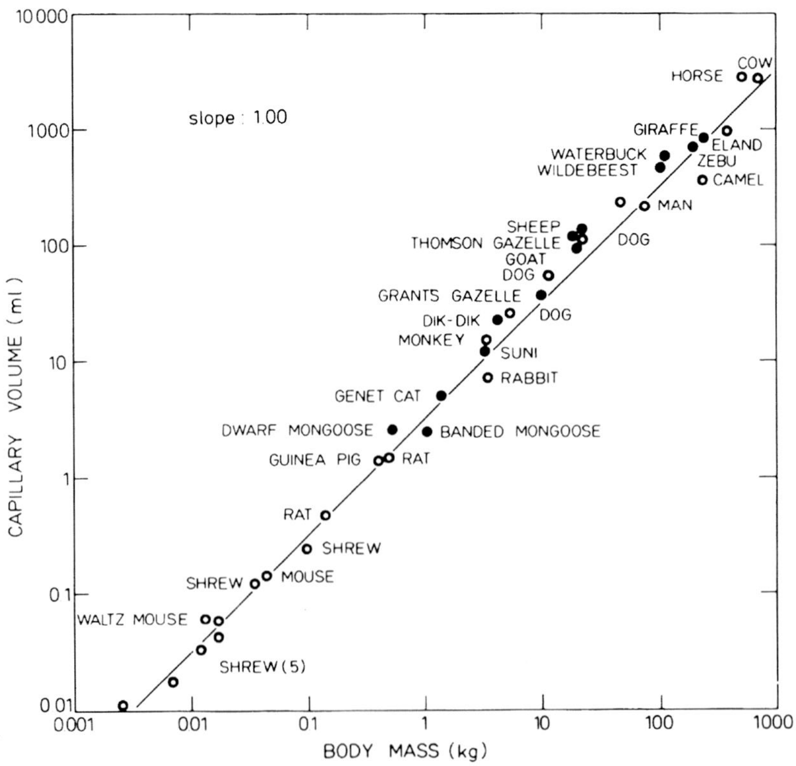

FIGURE 5 Allometric plot of capillary volume to body mass for mammalian species. *From Gehr et al., 1981. J Respir. Physiol. 44, 61–86. With permission.*

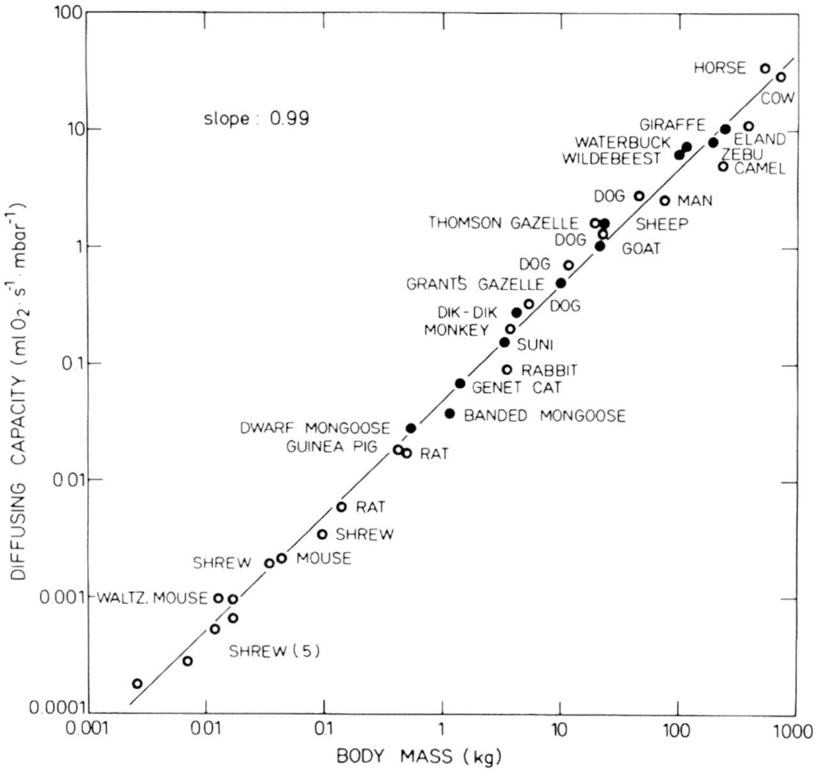

FIGURE 6 Allometric plot of pulmonary diffusing capacity to body mass for mammalian species. *From Gehr et al., 1981. J. Respir. Physiol. 44, 61–86. With permission.*

relative oxygen consumption do not have a greater lung volume to body mass ratio than do less active domesticated species. The waltzing mouse, which spends most of its waking hours in constant motion, also fits well into allometric regressions with other less active species. These observations suggest that the organization and architecture of the lungs provide an adequate reserve for a wide range of activity. The relationship of lung volume to body mass shown in Figure 3 scales linearly with an allometric slope of 1.06. In a similar manner, alveolar surface area and harmonic mean tissue thickness to body mass (Figure 4) have allometric slopes of 0.95 and 0.05, respectively, while capillary volume to body mass (Figure 5) has an allometric slope of 1.00. Capillary volume to alveolar surface area (capillary loading) increases slightly with increasing body mass. The increase in T_h with increasing body mass is statistically significant. The allometric slope of harmonic mean tissue thickness (T_{ht}) shown in Figure 3 has a value of 0.05.

Pulmonary diffusing capacity has traditionally been measured using physiologic techniques (Roughton and Forster, 1957). Weibel described a method for estimating oxygen-diffusing capacity using stereological techniques (Weibel, 1971). Diffusing capacity based on alveolar tissue densities, surface areas, and harmonic mean thicknesses of the air–blood barrier for species shown in Figures 3–5 is given in Figure 6 with an allometric slope of 0.99.

3. CELLULAR COMPOSITION OF THE AIR–BLOOD TISSUE BARRIER

Total tissue volumes, surface areas, and mean tissue thicknesses of the air–blood barrier in the alveolar lung region of the rat, dog, baboon, and human are shown in Table 2. Fischer 344 and Sprague–Dawley rats, two strains of rats commonly used for toxicological research, are compared. Volume and surface area differences for each of the alveolar tissue compartments are due primarily to differences in lung size relative to total body mass for each species studied. When tissue compartment volumes are expressed as a fraction of the total alveolar tissue volume, a marked similarity between species is apparent (Figure 7). This interspecies similarity is further observed when the epithelial, interstitial, and endothelial compartments are expressed as an arithmetic mean tissue thickness or average tissue thickness of the air-to-blood tissue barrier (Weibel, 1979). Most species show similar characteristics in the thickness of these compartments, with the exception of baboons and shews having "thinner" type II cells and humans having a thicker interstitium (Gehr et al., 1980).

The five major cell types in the alveolar region of the lungs in mammals are listed in Table 3. These include alveolar type I and type II epithelial cells, interstitial cells, endothelial cells, and alveolar macrophages. A sixth cell type, dendritic cells, has also been found in the alveolar

TABLE 2 Total Tissue Volumes, Surface Areas and Mean Tissue Thickness in the Alveolar Region of Normal Mammalian Lungs[a]

	Fischer 344 Rat[b] (n = 4)	Sprague–Dawley Rat (n = 8)	Dog (n = 4)	Baboon (n = 5)	Human[c] (n = 8)
Body weight, kg	0.29 ± 0.01a	0.36 ± 0.01	16 ± 3	29 ± 3	74 ± 4
Lung volume, ml	8.6 ± 0.31a	10.55 ± 0.37a	1322 ± 64	2393 ± 100	4341 ± 284
Total volumes, cm^3/ both lungs	5.978 ± 0.197	7.216 ± 0.278a	914 ± 52	1851 ± 24	3422 ± 223
Air	0.649 ± 0.057a	0.659 ± 0.055a	92 ± 5b	44 ± 17b	169 ± 24
Capillary lumen	0.428 ± 0.046a	0.671 ± 0.041a	78 ± 4b	68 ± 9b	314 ± 41
Tissue	0.082 ± 0.006a	0.144 ± 0.010a	16.5 ± 1.9b	14.4 ± 2.3b	32.5 ± 3.9
Type I epithelium	0.037 ± 0.009a	0.053 ± 0.009a	5.6 ± 0.5b	3.5 ± 0.8b	32.1 ± 5.0
Type II epithelium	0.068 ± 0.005a	0.079 ± 0.015a	12.9 ± 0.7b	9.4 ± 1.6b	54.0 ± 7.0
Cellular interstitium	0.128 ± 0.016a	0.214 ± 0.016a	22.8 ± 0.7b	24.6 ± 3.7b	98.3 ± 12.4
Noncellular interstitium	0.094 ± 0.009a	0.156 ± 0.007a	17.6 ± 1.6	13.4 ± 2.6	42.6 ± 5.4
Endothelium	0.019 ± 0.007a	0.025 ± 0.006a	2.5 ± 1.1b	2.4 ± 1.1b	54.7 ± 15.7
Macrophages					
Surface area, m^2/both lungs[d]					
Alveolar epithelium	0.391 ± 0.39a	0.387 ± 0.025a	51.0 ± 1.0b	47.7 ± 7.7b	89.0 ± 8.0
Type I	0.015 ± 0.005a	0.015 ± 0.002a	1.0 ± 0.2	1.9 ± 0.3	7.0 ± 1.0
Type II	0.383 ± 0.39a	0.452 ± 0.035a	57.0 ± 2.0	38.6 ± 9.5	91.0 ± 9.0
Capillary endothelium					
Arithmetic mean, tissue thickness, μm					
Epithelium					
Type I	0.212 ± 0.008b	0.384 ± 0.038a	0.327 ± 0.043ab	0.308 ± 0.021ab	0.361 ± 0.024a
Type II	2.758 ± 0.424ab	3.653 ± 0.266a	4.138 ± 0.340ac	1.839 ± 0.141b	5.019 ± 0.551c
Interstitium	0.500 ± 0.028a	0.693 ± 0.058a	0.658 ± 0.033a	0.847 ± 0.140a	1.634 ± 0.164
Endothelium	0.246 ± 0.011a	0.358 ± 0.031a	0.308 ± 0.019a	0.361 ± 0.038ab	0.474 ± 0.052b

[a]*All values are mean ± SEM. This table is modified from Crapo et al., 1983. For comparisons between species all data connected by the same letter subscript are not statistically different from each other.*
[b]*From Pinkerton et al. (1982).*
[c]*From Crapo et al. (1982).*
[d]*Type I surface area (SA) is the SA of basement membrane under Type I cells; Type II SA is the air surface of Type II cells excluding the extra SA contributed by microvilli; endothelial SA is the luminal surface of the endothelial cells.*

wall associated with immune defense in the lung parenchyma, as well as in the lung airways and pleura (Sertl et al., 1986). All six cell types will be discussed for different mammalian species.

Although lung size varies greatly between species, there is a high correlation in the distribution for the five major cell types in the lungs and in the average size of each cell type between species. The alveolar epithelium, composed of type I and type II cells, represents one-fourth of the total alveolar cell number (Figure 8). Alveolar type I cells have thin, attenuated cytoplasm and are two to four times larger in volume and cover 20 to 40 times greater surface area per cell than type II cells. Alveolar type III cells, also called brush cells, are cuboidal with abundant perinuclear cytoplasm. Type III cells are rarely seen in the alveolar lung parenchyma but have been found in highest concentrations in rats

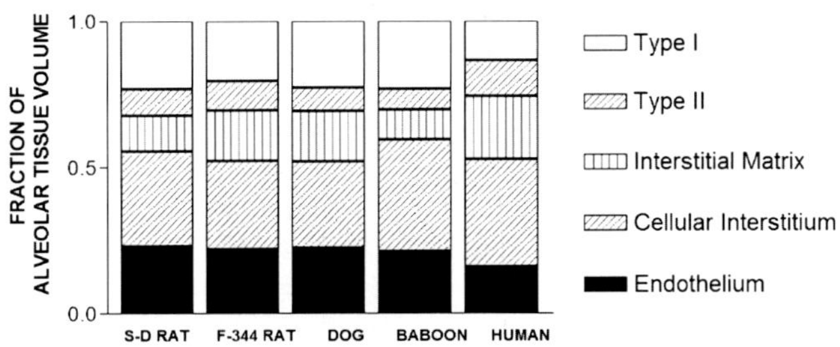

FIGURE 7 Composition of the alveolar septum in mammalian lungs. Each tissue compartment is expressed as a fraction of the total septal wall volume. Alveolar macrophages and capillary plasma and blood are excluded from the measurement.

in alveolar regions close to bronchiolar–alveolar duct junctions (Chang et al., 1986). The uniformity of epithelial cell size between species (Table 3) for alveolar type I and type II cells is striking, since large differences in the surface area of a single alveolus can exist between species (Figure 9). For example, the rat alveolus has a surface area of 14,000 μm^2, which is covered by an average of two type I cells and three type II cells. In contrast, the human alveolus has a surface area of 200,000–300,000 μm^2 and is covered by 32 type I cells and 51 type II cells (Crapo et al., 1989). These findings suggest that type I and type II cell shape and size are independent of alveolar size. It also follows that these cellular characteristics are likely to provide significant functional integrity and homeostatic balance for the epithelium in most mammalian species. The interaction of oxidant gases and particulate material with the epithelial lining cells of the air–blood tissue barrier may also be highly influential in the homeostatic balance of the epithelium in virtually every species (Sherman et al., 2005; Madl et al., 2010).

The subcellular composition of the alveolar type II cell in the lungs of a variety of mammalian species is shown in Table 4. Alveolar type II cells are of particular importance in the synthesis and secretion of pulmonary surfactant to

TABLE 3 Characteristics of Cells From the Alveolar Region of Normal Mammalian Lungs[a]

	Fischer 344 Rat[b]	Sprague–Dawley Rat	Dog	Baboon	Human[c]
Total number of cells/lung, 10^9	0.67 ± 0.02a	0.89 ± 0.04a	114 ± 13b	99 ± 9b	230 ± 25
Total lung cells, %					
Alveolar type I	8.1 ± 0.3a	8.9 ± 0.9a	12.5 ± 1.7b	11.8 ± 0.6b	8.3 ± 0.06a
Alveolar type II	12.1 ± 0.7a	14.2 ± 0.7ab	11.8 ± 0.6a	7.7 ± 1.0	15.9 ± 0.8b
Endothelial	51.1 ± 1.7a	42.2 ± 1.1a	45.7 ± 0.8a	36.3 ± 2.4	30.2 ± 2.4
Interstitial	24.4 ± 0.7a	27.7 ± 1.8a	26.6 ± 0.7a	41.8 ± 2.7	36.1 ± 1.0
Macrophage	4.3 ± 1.0a	3.0 ± 0.3a	3.4 ± 0.6a	2.3 ± 0.7a	9.4 ± 2.2
Alveolar surface covered, %					
Alveolar type I	96.4 ± 0.5a	96.2 ± 0.5a	97.3 ± 0.4a	96.0 ± 0.6a	92.9 ± 1.0
Alveolar type II	3.6 ± 0.3a	3.8 ± 0.5a	2.7 ± 1.0a	4.0 ± 0.6a	7.1 ± 1.0a
Alveolar cell volume, μm^3					
Alveolar type I	1530 ± 121a	2042 ± 374a	1196 ± 88a	1224 ± 136a	1764 ± 155
Alveolar type II	455 ± 108a	443 ± 80a	428 ± 37a	539 ± 184a	889 ± 101
Endothelial	275 ± 25a	387 ± 30a	343 ± 19a	365 ± 61a	632 ± 64
Interstitial	427 ± 55a	331 ± 67ab	440 ± 47a	227 ± 30b	637 ± 26
Macrophage	639 ± 131a	1058 ± 257a	654 ± 116a	1059 ± 287a	2492 ± 167
Alveolar cell surface area, μm^2					
Alveolar type I	7287 ± 755b	5320 ± 694ab	3794 ± 487a	4004 ± 383a	5098 ± 659a
Alveolar type II	185 ± 56ab	123 ± 20a	107 ± 15a	285 ± 85b	183 ± 14ab
Endothelial	1121 ± 95a	1105 ± 72a	1137 ± 127a	1040 ± 209a	1353 ± 67a

[a]*All values are mean ± SEM. This table is modified from Crapo et al. (1983). For comparisons between species, letter subscripts indicate those values that are not statistically different from other values having the same letter subscript.*
[b]*From Pinkerton et al. (1982).*
[c]*From Crapo et al. (1983).*

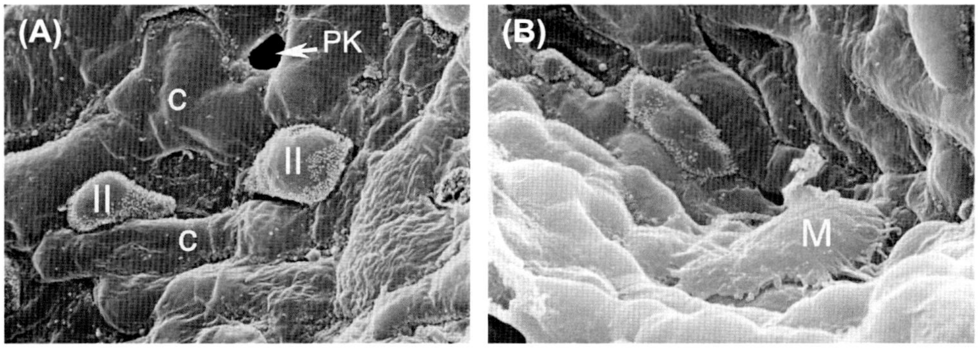

FIGURE 8 Alveolar surface. (A) Scanning electron micrograph of human lung, showing capillaries (c) bulging into the alveolar space and two epithelial type II cells (II) in between. Note interalveolar pore, pore of Kohn (PK). Magnification 800×. (B) Scanning electron micrograph of human lung, showing the same structure as in (A) and, in addition, an alveolar macrophage (M) with many cellular processes reaching out for foreign material to be taken up. From *Gehr et al., 1978*. Magnification 900×.

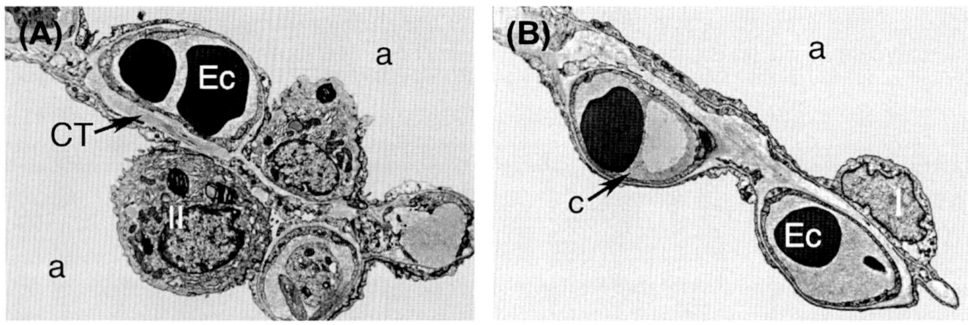

FIGURE 9 Interalveolar septa at entrance ring, human lung. (A) Transmission electron micrograph showing an alveolar entrance ring enforced with connective tissue (CT) fibers. Note the cuboidal type II cells (type II) producing surfactant and the squamous type I epithelial cells (type I). Capillaries (C), erythrocytes (Ec), alveoli (a). Magnification 2500×. (B) Transmission electron micrograph showing another view of an alveolar entrance ring with the thin squamous epithelial type I cell layers and a nucleus of a type I cell (type I). Capillaries (c), erythrocytes (Ec), alveoli (a). Magnification 240×.

line alveolar air spaces as a means of reducing surface tension and preventing alveolar collapse in the lungs during respiration (Pattle, 1965; Bachofen and Schürch, 2001). Surfactant function is also critical in defense and particle displacement. Alveolar type II cells also serve as the stem cells for the generation of daughter type II cells and can undergo changes in cell shape and organelle content to form alveolar type I cells (Evans et al., 1972). Although the generation and secretion of lamellar bodies in type II cells can occur over a short period of time (see the review of Wright and Clements, 1987), lamellar body volume fraction in type II cells is remarkably constant under normal conditions (Young et al., 1985). Lamellar body volume fraction is similar for a wide variety of mammalian species, with ferrets, rabbits, and humans having the highest values, and baboons and neonatal rats the lowest. Mitochondrial volume fraction in alveolar type II cells also demonstrates a high degree of homogeneity between species, with values ranging between 5% and 9% (Table 4). Miller et al. (1982, 1986) have described lipid droplets and cisternal bodies in type II cells of ferrets, guinea pigs, and dogs. Cisternal bodies have not been described in other species, while lipid droplets have been observed occasionally in baboons and rats. The volume fractions of type II cell organelles in rabbits and humans shown in Table 4 were derived from the morphometric analysis of multiple type II cell profiles sequentially sampled from three to six sites per case. More than 2000 points falling on type II cell profiles from each case were counted to yield a relative sampling error of 2–5% for each organelle. Please refer to *Stereological Methods* by E.R. Weibel (1979) and *Unbiased Stereology* by Howard and Reed (2005) for a more complete description of this sampling scheme.

The most populous cells in the lung parenchyma are interstitial and endothelial cells. Mesenchymal interstitial cells form a heterogeneous population of cells composed of fibroblasts and pericytes. Interstitial macrophages, monocytes, dendritic cells, and lymphocytes are also present but in fewer numbers. The cell differential of the interstitium for rats and humans is shown in Table 5. Although the total interstitial cell number per lung is several hundred-fold greater in humans than in rats, the proportion of specific interstitial cell types is virtually identical for these two species. Endothelial cells, like type I cells, have thin attenuated cytoplasm but are smaller in volume and size, covering about one-fourth the area of a type I cell.

Alveolar macrophages represent the most variable lung cell type in mammalian species (Migliaccio and

TABLE 4 Alveolar Type II Cell Composition (%)

Species	N	Nucleus	Cytoplasm	Mitochondria	Lamellar Bodies	References
Rat[a]						
Neonate	4	–	–	5 ± 0.1	8 ± 0.3	Vidic and Burri (1981)
Sprague Dawley	5	24 ± 2	55 ± 1	8 ± 0.5	12 ± 0.5	Young et al. (1985)
Fischer 344	4	21 ± 1	54 ± 1	9 ± 0.3	15 ± 0.4	Pinkerton et al. (1990)
Ferret	3	–	–	–	20 ± 1	Miller et al. (1982)
Rabbit	7	21 ± 1	51 ± 2	8 ± 0.3	21 ± 1	Pinkerton (unpublished data)
Baboon	8	29 ± 3	53 ± 2	7 ± 1.0	9 ± 1	Young, S.L. (personal communication)
Human	7	22 ± 2	56 ± 2	5 ± 0.2	17 ± 2	Pinkerton (unpublished data)

Values are mean ± SEM; N = sample size.
[a]*Neonatal rats are 21 days of age; Sprague Dawley and Fischer 344 rats are male adults.*

TABLE 5 Differential Count of Parenchymal Insterstitial Cells of the Lungs

N	Fischer 344 Rat[a] 12	Human[b] 8
Total interstitial	$142 \pm 12 \times 10^6$	$84 \pm 10 \times 10^9$
Cell number/lung[c]		
Percent distribution of cell types		
Fibroblast	58%	51%
Pericyte	21%	17%
Monocyte–macrophage	16%	12%
Lymphocyte	5%	7%
Unknown	3%	7%

[a]*Values are mean ± SEM; N = sample size.*
[b]*From Pinkerton et al. (1982).*
[c]*From Pinkerton et al. (1988).*

Holian, 2010). Alveolar macrophage number and size within the air spaces of the lungs are influenced by a variety of conditions, which may result in large fluctuations over a short period of time. These fluctuations in cell number may lead to lung injury or reflect a diverse number of disease processes occurring in the lungs. Alveolar macrophages contribute to the sterility of the lung by removing foreign respired matter through phagocytic uptake into cytoplasmic vacuoles that fuse with primary lysosomes to form phagolysosomes. These particles are subsequently degraded by enzymatic digestion or removed from the lungs by the migration of the alveolar macrophages out of the lungs via the mucociliary escalator of the tracheobronchial tree or into the lung interstitium and lymphatics. Alveolar macrophage number may increase in the lungs by in situ cell division of resident alveolar macrophages or by the egress of interstitial macrophages or blood-borne monocytes (which have matured into macrophages in the interstitium) into alveolar air spaces.

Stone and colleagues have morphometrically estimated the number of macrophages per alveolus (Stone et al. 1992). It is essential to keep in mind that the number of macrophages found in the lungs can be a consequence of multiple factors, including age, injury, and exposure to a wide variety of environmental air pollutants. For additional details regarding these three potential sources for the recruitment of alveolar macrophages, please refer to an excellent review by Migliaccio and Holian (2010). Comparative studies indicate that alveolar macrophage number is significantly higher in "normal" human lungs compared to other species. This increase in cell number may be due to smoking exposures and/or other environmental pollutants from which laboratory species would be shielded. Confirmation of the influence of smoking on the macrophage population in the human lung is implied in the work of Crapo et al. (1982), who found nonsmokers to have significantly lower fractions of alveolar macrophages present in the lungs than smokers.

Dendritic cells have been identified in the interstitium of humans, rats, and mice between type I pneumocytes of the alveolar wall, from which the dendritic processes are extended into the alveolar lumen (Sertl et al., 1986). Dendritic cells have also been identified in the visceral pleura, where their processes are extended parallel to the mesothelial lining. In rats, the number of dendritic cells decreases with the size of the airway but increases in the lung parenchyma (Wikstrom and Stumbles, 2007). Compared to their airway counterparts, parenchymal dendritic cells phagocytose greater amounts of antigens, have diminished ability to activate lymphocytes, and slower turnover rates (72 h half-life) (Holt et al., 1994). The faster turnover rate (12 h half-life) of airway dendritic cells is attributed to a quicker migration rate to regional lymph nodes. Thus, parenchymal dendritic cells may help salvage debris and induce tolerance in the lower airways and parenchyma. In contrast, airway dendritic cells may be important for sampling antigens and initiating an adaptive immune response in the bronchial tree, where initiation of infection is more common. Both classes of dendritic cells, conventional and plasmacytoid, are found in the lung parenchyma.

Conventional dendritic cells continuously sample the lung environment by ingesting foreign pathogens. Like alveolar macrophages, conventional dendritic cells degrade pathogens; however, their primary role is not pathogen clearance. Instead, conventional dendritic cells display pathogenic antigens on their surface and migrate to the regional lymph nodes, where they activate lymphocyte cells to induce long-term memory responses against specific pathogens should that pathogen be encountered in the future. Thus, conventional dendritic cells bridge the gap between early innate immune responses and adaptive immune responses that confer immunological memory. Plasmacytoid dendritic cells, on the other hand, are not involved in migration to the lymph nodes and activation of lymphocyte cells. Instead, they initiate viral immune responses through the uptake of viral particles and secretion of very high levels of interferons or cytokines that interfere with viral replication and activate other innate immune cells. In BALB/c mice, conventional dendritic cells compromise approximately 2% of total cells in the main conducting airways and 1% of total cells in the lung parenchyma (von Garnier et al., 2005), while plasmacytoid dendritic cells make up about 0.5% of total cells of the main conducting airways and 2% total cells of the lung parenchyma (Wikstrom and Stumbles, 2007).

4. SUMMARY

Comparative studies of mammals demonstrate great diversity in the volume and surface area of alveolar lung tissues, but a striking similarity in the relative proportion, composition, and size of cell types present in this region of the lungs. Although exceptions exist, alveolar cells in mammalian lungs conform to similar patterns of structure and function that are unique to each parenchymal cell type. However, it is important to keep in mind that species differences in the architectural design and size of pulmonary acini forming the lung parenchyma, along with differences in the upper respiratory tract (i.e., nasal passages, larynx, and tracheobronchial tree), may significantly affect the relative dose of inhaled pollutants and chemicals delivered to individual cells of the lung parenchyma. Adsorption of inhaled substances along the upper respiratory tract will be different for different species. Acinar size along with differences in alveolar wall thickness and number of cells per alveoli among species may also alter the volume of gases, and thus the biological response of inhaled substances in the lower respiratory tract. The combination of any of these factors may contribute to marked variations in alveolar cell response of the air–blood tissue barrier to toxic substances in the mammalian lung.

REFERENCES

Bachofen, H., Schürch, S., 2001. Alveolar surface forces and lung architecture. Comp. Biochem. Physiol. A Mol. Integr. Physiol. 129, 183–193. Review.

Burri, P.H., 1974. The postnatal growth of the rat lung, III: Morphology. Anat. Rec. 180, 77–98.

Burri, P.H., Weibel, E.R., 1971. Morphometric estimation of pulmonary diffusion capacity. II. Effect of PO, on the growing rat lung to hypoxia and hyperoxia. Respir. Physiol. 11, 247–264.

Burri, P.H., Dbaly, J., Weibel, E.R., 1974. The postnatal growth of the rat lung: I. Morphometry. Anat. Rec. 178, 711–730.

Crapo, J.D., Barry, B.E., Foscue, H.A., Shelburne, J., 1980. Structural and biochemical changes in rat lungs occurring during exposures to lethal and adaptive doses of oxygen. Am. Rev. Respir. Dis. 122, 123–145.

Crapo, J.D., Barry, B.E., Gehr, P., Bachofen, M., Weibel, E.R., 1982. Cell numbers and cell characteristics of the normal human lung. Am. Rev. Respir. Dis. 126, 332–337.

Crapo, J.D., Young, S.L., Fram, E.K., Pinkerton, K.E., Barry, B.E., Crapo, R.O., 1983. Morphometric characteristics of cells in alveolar region of mammalian lungs. Am. Rev. Respir. Dis. 128, S42–S46.

Crapo, K.A., Mercer, R.R., Gehr, P., Stockstill, B., Crapo, J.D., 1989. Allometric relationships of cell numbers and size in the mammalian lung. Am. Rev. Respir. Dis. A289.

Chang, L.Y., Mercer, R.R., Crapo, J.D., 1986. Differential distribution of brush cells in the rat lung. Anat. Rec. 216, 49–54.

Evans, M.J., Stephens, J., Cabral, L.J., Freeman, G.T., 1972. Cell renewal in the lungs of rats exposed to low levels of NO. Arch. Environ. Health 24, 180–188.

Forrest, J.B., Weibel, E.R., 1975. Morphometric estimation of pulmonary diffusion capacity. VII. The normal Guinea pig lung. Respir. Physiol. 24, 191–202.

Geelhaar, A., Weibel, E.R., 1971. Morphometric estimation of pulmonary diffusion capacity. III. The effect of increased oxygen consumption in Japanese waltzing mice. Respir. Physiol. 11, 354–366.

Gehr, P., Erni, H., 1980. Morphometric estimation of pulmonary diffusion capacity in two horse lungs. Respir. Physiol. 41, 199–210.

Gehr, P., Bachofen, M., Weibel, E.R., 1978. The normal human lung: ultrastructure and morphometric estimation of diffusion capacity. Respir. Physiol. 32, 121–140.

Gehr, P., Sehovic, S., Burfi, P.H., Claassen, H., Weibel, E.R., 1980. The lung of shrews: morphometric estimation of diffusion capacity. Respir. Physiol. 40, 33–47.

Gehr, P., Mwangi, D.K., Ammann, A., Maloiy, G.M.O., Taylor, C.R., Weibel, E.R., 1981. Design of the mammalian respiratory system. V. Scaling morphometric pulmonary diffusing capacity to body mass: wild and domestic mammals. Respir. Physiol. 44, 61–86.

Gil, J., Bachofen, H., Gehr, P., Weibel, E.R., 1979. Alveolar volume-surface area relation in air and saline-filled lungs fixed by vascular perfusion. J. Appl. Physiol. 47, 990–1001.

Hayatdavoudi, G., Crapo, J.D., Miller, F.J., O'Neil, J.J., 1980. Factors determining degree of inflation in intratracheally fixed rat lungs. J. Appl. Physiol. 48, 389–393.

Holt, P.G., Haining, S., Nelson, D.J., Sedgwick, J.D., 1994. Origin and steady-state turnover of class II MHC-bearing dendritic cells in the epithelium of the conducting airways. J. Immunol. 153, 256–261.

Howard, C.V., Reed, M.G., 2005. Unbiased Stereology: Three Dimensional Measurement in Microscopy, second ed. Taylor & Francis, Inc., Abingdon, UK.

Madl, A.K., Carosino, C., Pinkerton, K.E., 2010. Airborne particles and structural Remodeling of the lung. In: Gehr, P., Blank, F., Muhlfeld, C., Rothen-Rutishauser, B. (Eds.), Particle-Lung Interactions, second ed. Informa Healthcare, New York.

Migliaccio, C.T., Holian, A., 2010. Inflammatory cells of the lung: macrophages. In: McQueen, C.A. (Editor-in-Chief), Comprehensive Toxicology. In: Yost, G.S. (Volume Editor), Respiratory Toxicology, second ed., vol. 8. Elsevier Ltd., pp. 93–114.

Miller, M.L., Andringa, A., Vinegar, A., 1982. Ultrastructure and morphometry of the alveolar type II cell of the ferret. J. Ultrastruct. Res. 79, 85–91.

Miller, M.L., Andringa, A., Adams, W., Radike, M.J., 1986. Intracistemal protein in the type H pneumocyte of the ferret, guinea pig and mongrel dog. J. Ultrastruct. Res. 95, 131–141.

Pinkerton, K.E., Herring, M.J., Hyde, D.M., Green, F.H.Y., 2015. Normal aging of the lung. In: Harding, R., Pinkerton, K.E. (Eds.), The Lung: Development, Aging and the Environment, Second Edition. Academic Press. ISBN: 9780127999418.

Pattle, R.E., 1965. Surface lining of lung alveoli. Physiol. Rev. 45, 48–79.

Pinkerton, K.E., Barry, B.E., O'Neil, J.J., Raub, J.A., Pratt, P.C., Crapo, J.D., 1982. Morphologic changes in the lung during the life-span of Fischer 344 rats. Am. J. Anat. 164, 155–174.

Pinkerton, K.E., Mercer, R.R., Crapo, J.D., 1988. Interstitial characteristics of the normal human lung. Am. Rev. Respir. Dis. 137, 344.

Pinkerton, K.E., Kendall, J.K., Randall, G.C., Chechowitz, M.A., Hyde, D.M., Plopper, C.G., 1989. Hypophysectomy and porcine fetal lung development. Am. J. Respir. Cell. Mol. Biol. 1, 319–328.

Pinkerton, K.E., Young, S.L., Fram, E.K., Crapo, J.D., 1990. Alveolar type II cell responses to chronic inhalation of chrysotile asbestos. Am. J. Respir. Cell. Mol. Biol. 3 (6), 543–552.

Roughton, J.F.W., Forster, R.E., 1957. Relative importance of diffusion and chemical reaction rates in determining rate of exchange of gases in the human lung, with special reference to true diffusing capacity of pulmonary membrane and volume of blood in the lung capillaries. J. Appl. Physiol. 11, 290–302.

Sertl, K., Takemura, T., Tschachler, E., Ferrans, V.J., Kaliner, M.A., Shevach, E.M., 1986. Dendritic cells with antigen-presenting capability reside in airway epithelium, lung parenchyma, and visceral pleura. J. Exp. Med 163, 436–451.

Sherman, J.G., Peake, J.L., Wong, V.J., Pinkerton, K.E., 2005. Air pollutants, epithelial surfaces, and lung disease. In: Foster, W.M., Costa, D.L. (Eds.), Air Pollutants and the Respiratory Tract, vol. 204. Taylor & Francis Group, Boca Raton, Florida, pp. 157–193.

Siegwart, B., Gehr, P., Gil, J., Wiebel, E.R., 1971. Morphometric estimation of pulmonary diffusion capacity. IV. The normal dog. Respir. Physiol. 13, 141–159.

Stone, K.C., Mercer, R.R., Freeman, B.A., Chang, L.Y., Crapo, J.D., 1992. Distribution of lung cell numbers and volumes between alveolar and nonalveolar tissue. Am. Rev. Respir. Dis. 146, 454–456. PMID: 1489139.

Vasilescu, D.M., Gao, Z., Saha, P.K., Yin, L., Wang, G., Haefeli-Bleuer, B., Ochs, M., Weibel, E.R., Hoffman, E.A., 2012. Assessment of morphometry of pulmonary acini in mouse lungs by nondestructive imaging using multiscale microcomputed tomography. Proc. Natl. Acad. Sci. U. S. A. 109, 17105–17110.

Vidic, B., Burri, P.H., 1981. Quantitative cellular and subcellular changes in the rat type II pneumocyte during early postnatal development. Am. Rev. Respir. Dis. 124, 174–178.

von Garnier, C., Filgueira, L., Wikstrom, M., Smith, M., Thomas, J.A., Strickland, D.H., Holt, P.G., Stumbles, P.A., 2005. Anatomical location determines the distribution and function of dendritic cells and other APCs in the respiratory tract. J. Immunol. 175, 1609–1618.

Weibel, E.R., 1963. Morphometry of the Human Lung. Springer Verlag, New York.

Weibel, E.R., 1971. Morphometric estimation of pulmonary diffusing capacity. I. Model and method. Respir. Physiol. 11, 54–75.

Weibel, E.R., 1979. Stereological Methods, vol. I. Academic Press, London.

Wikstrom, M.E., Stumbles, P.A., 2007. Mouse respiratory tract dendritic cell subsets and the immunological fate of inhaled antigens. Immunol. Cell Biol. 85, 182–188.

Wright, J.R., Clements, J.A., 1987. Metabolism and turnover of lung surfactant. Am. Rev. Respir. Dis. 135, 426–444.

Young, S.L., Fram, E.K., Craig, B.L., 1985. Three-dimensional reconstruction and quantitative analysis of rat lung type II cells: a computer-based study. Am. J. Anat. 174, 1–14.

Chapter 10

Basement Membranes and the Extracellular Matrix

Jane E. Farrington and Philip L. Sannes

Chapter Outline

1. INTRODUCTION

Cytochemistry is the visualization of the microanatomical location of specific chemical constituents within cells and their environments. In the strictest terms, this visualization applies to the level of the electron microscope; whereas histochemistry is used when referring to the level of the light microscope. For purposes of this discussion, they will often be used somewhat interchangeably. The "histochemist/cytochemist" attempts to use the unique chemical nature of a complex biologic structure or one or more of its specific components to design a chemical reaction that results in a product that can be detected (visualized) by light or electron microscopy. This may entail the exploitation of the hydrolytic capacity of an enzyme to cleave a synthetic substrate, whose products can be detected with a secondary reaction to reveal the subcellular location of the enzyme (e.g., acid phosphatase within lysosomes). Certain chemical constituents can be cleverly detected by often simple oxidation or reduction reactions whose products can be made visible through secondary reactants, such as heavy metals or certain dyes. When the antigenic reactivity of specific cellular constituents is exploited for localization, the visualization is referred to as immunocytochemistry. Both approaches have proven to be most valuable in appreciating the distribution of the various known components of extracellular matrices (ECMs) in the lung—both fibrous and nonfibrous elements. Traditionally, cytochemists have attempted to bridge the naturally occurring gaps between biochemists and morphologists, whose scientific questions may differ by what is a biological system's composition versus its structure. The cytochemist sees the question as one of *structural composition*, a perspective that has drawn increasing interest over the last 10 years. This growth is evidenced by a progressive move toward more hybridization of the "traditional disciplines," so that cytochemical techniques are used more and more frequently by both biochemists and morphologists.

In classic histological terms, ECMs include fibrous connective tissue elements (collagen and elastic fibers) and ground substance. A more modern perspective has evolved to coincide with a greater knowledge of ECM biochemistry and composition, which divides it into four molecular categories: the collagens, elastic fibers, structural glycoproteins, and proteoglycans (Hay, 1981). Elastic fibers can legitimately be grouped with the structural glycoproteins (Amenta et al., 1988). Figure 1 represents what might be the considered "optimal" preservation of morphological detail of ECM found in lung tissue. Collagen fibers, seen in cross-section, have little electron density or morphological detail. Similarly, elastic fibers have minimal electron density, although distinct regions of the microfibrillar and elastin components are distinguishable. The protein cores of proteoglycans are preserved in this preparation as very thin, electron-dense filaments, and appear to fill the interstitial matrix regions. Basement membranes, which lie directly subjacent to epithelial and a few other select

Comparative Biology of the Normal Lung. http://dx.doi.org/10.1016/B978-0-12-404577-4.00010-2

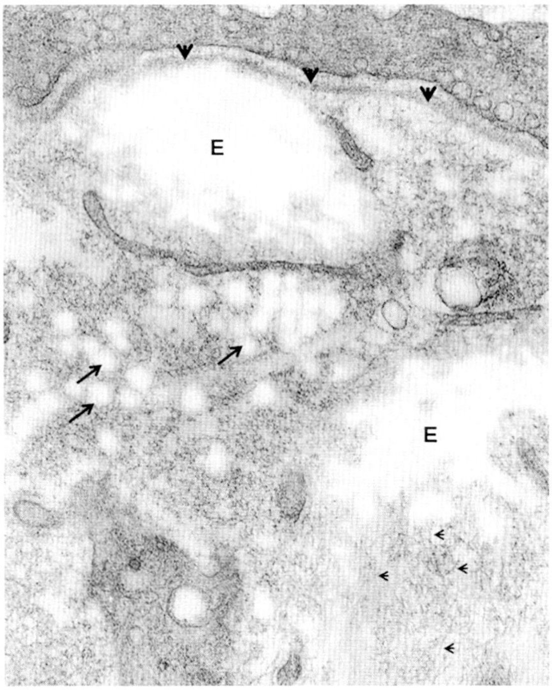

FIGURE 1 Electron micrograph of a rat trachea fixed 2 h with 1% glutaraldehyde—1% paraformaldehyde in 0.1 M cacodylate—HCl buffer, pH 7.4, and postfixed 2 h with 1% osmium tetroxide. Collagen fibers (large arrows) and elastic fibers (E) are observed in cross-section. The lamina densa of the basement membrane of a capillary is seen in profile (large arrowheads). The protein cores of proteoglycans are visible as electron-dense filaments (small arrowheads). Counterstained with uranyl acetate and lead citrate. Magnification ×75,000.

specialized cell populations, are seen as three distinct regions morphologically: (1) an electron-dense region referred to as the lamina densa (sometimes called the basal lamina); (2) an electron-lucent region between the lamina densa and epithelial cell, called the lamina lucida; and (3) a region connecting the lamina densa, with the underlying ECM referred to as the lamina reticularis. In regions where basement membranes of adjacent tissues *fuse*, such as the alveolar region of the lung, the terminology for the segments of basement membrane change. The lamina lucida (alveolar side) is called the lamina rara externa, and the lamina reticularis (endothelial side) is referred to as the lamina rara interna. This appropriately reflects the fact that a very specialized basement membrane is formed when it is sandwiched between adjacent epithelia, thus forming the very thinnest of barriers.

All of these morphological components and their attendant terminologies are critical for understanding the anatomical relationships of most of the basic constituents of pulmonary connective tissue matrices. Not too many years ago, this may have been more than sufficient information, but our knowledge of the importance of ECM has progressed very rapidly, and questions of the specific composition of these constituents is of enormous value. The purpose of this section is to digest and summarize the existing body of knowledge on the cytochemical localization of specific components found in ECM in the lung. In many cases, the knowledge base is sparse or incomplete, which in part reflects the lack of attention this specialized area has received from pulmonary researchers. However, interest has grown considerably in recent years, and it is hoped that what is offered in the text and tables will be as useful to those who have dedicated their research careers to this area as it is to those whose interest may be just commencing. It is the intent of this work to focus strictly on the cytochemical reactivity of ECM, which will include all the major molecular components mentioned above. A special discussion is reserved for basement membranes because they represent a particularly important expression of the disposition of the various ECM components within the lung. A more biochemical perspective of pulmonary ECM is covered in considerable detail in Chapter 4. The functional significance of established cytochemical findings is discussed, when appropriate, in the context of our present understanding of structure/function relationships.

2. METHODOLOGICAL APPROACHES

The use of specific techniques should always be driven by the scientific question being posed. However, the large range of cytochemical methods from which to choose (Table 1) often leaves the researcher with a knowledge gap that may limit his or her ability to attempt some of the more sophisticated approaches. This is sometimes more of a psychological than truly practical question, but it is nonetheless real. Most of the established methods for localization of ECM-related molecules are relatively straightforward, whereas others are limited by the source and/or supply of antibody.

The purpose of this section is to present established methods that have been successfully applied to or specifically developed for the localization of ECM components in the lung. It should be noted that more general reviews on the subject of pulmonary cytochemistry have been published (Spicer and Schulte, 1989; Schulte et al., 1989). These deal with many technical/procedural details that will not be presented here. This discussion will be limited to methodologic applications relevant to ECM only.

2.1 General Methods for Staining ECM

A wide variety of methods are available that are highly useful to the pulmonary researcher for staining ECM for evaluation by the light microscope. These methods are found in abundance in any technique book on histology, histochemistry, or histopathology. For collagen fibers, methods typically include acid analine dyes, ammoniacal silver solutions, phosphotungstic, and phosphmolybdic

TABLE 1 Cytochemical Reactivity of Pulmonary Connective Tissues

Stain Technique	Specificity	Reactive Site/Structure	Region Studied		Species	References
			Airway	Alveolus		
Ruthenium red	Anionic end groups	Collagen fiber surfaces	+	+	R	Vaccaro and Brody, 1979
					R	Vaccaro et al., 1985
		Basement membrane	NR	+	R	Vaccaro and Brody, 1981
					R	Grant et al., 1983
					M	Penney and Rosenkrans, 1984
					R	Vaccaro et al., 1985
					M	Rosenkrans and Penney, 1985
					M	Rosenkrans and Penney, 1986
					R	Brody et al., 1982
Dialyzed (colloidal) iron	Anionic end groups	Collagen fibers	+	+	R	Katsuyama and Spicer, 1977
		Elastic fibers	+	+	R	Sannes, 1986
		Basement membrane	±	±	R	Sannes, 1986
		Cartilage	+		R	Sannes, 1986
High-iron diamine	Sulfated end groups	Collagen fibers	+	+	R	Katsuyama and Spicer, 1977
		Elastic fibers	+	+	R	Sannes, 1986
		Basement membrane	+	+	R	Sannes, 1984; Khosla et al., 1994
		Cartilage	+	+	R	Sannes, 1986
Cuprolinic blue	Anionic end groups	Collagen fibers	NR	+	M	Van Kuppevelt et al., 1984a
		Elastic fibers	NR	NR	M	Van Kuppevelt et al., 1984b
		Basement membrane	NR	+	H	Van Kuppevelt et al., 1985

Key: NR = not reported, R = rat, M = mouse, H = human.

acid hematoxylin methods (which include the classic trichrome stains), and periodic acid-leuchofuchsin methods. Choosing which is most useful depends upon the contrast or color combination desired or the manner in which the tissue is prepared (manner and length of fixation). For example, the acid aniline dyes have little use when staining 1- to 2-pm-thick glycol methacrylate sections, and new or adaptations of established methods must be employed to fit a specific need (Sannes, 1987). In other cases, one method is as useful as another. This is also the case with elastic fibers, for which a number of excellent methods exist; some of the more popular methods include Verhoef's iodine-iron hematoxylin, orcein, resorcin fuchsin, and aldehyde fuchsin.

All of these methods enable relatively accurate resolution of these ECM components, principally because they are of sufficient size and contain an adequate concentration of molecules that react with sufficient intensity with the respective dye solution to be detected. Their true specificity may in some cases be questionable, because the reactive molecules may in fact be complex glycosaminoglycans, glycoproteins, and proteoglycans, now known to be associated with these larger fibers. Nevertheless, such methods are highly useful in the evaluation of fibrous ECM components. The nonfibrous ones present more of a problem because they are either extracted during tissue processing or are not present in sufficient quantities within microscopically visible foci to be detected with these less specific methods. It therefore becomes necessary to use immunologic means of detection, which afford a reasonable degree of enhancement of the component of interest. These more specific methods are discussed below.

2.2 Detection by Charge/Charge Density

ECMs, by their very nature, are highly negatively charged (anionic) macromolecules. In large part, this is due to the presence of carboxyl and sulfate groups, which in biologic tissues can express a net negative charge. These characteristics have been exploited by cytochemists to demonstrate their location. Relevant studies involving pulmonary tissues are summarized in Table 1.

Ruthenium Red

Ruthenium red (RR) is a polycationic dye that has been used to sustain negatively charged molecular species in tissue sections (Luft, 1971). In the electron microscope, stained areas are electron dense and thus are easily distinguished from those that are unstained. RR has been used in variety of different protocols, which vary depending upon the results desired, but most generally it is used alone or in combination with osmium for postfixation preservation of anionic sites within tissues. Because of its highly charged nature, it does not penetrate cells or dense connective tissues very well, so that it often is necessary to "permeabilize" the tissue either before or simultaneously with RR treatment. This was done in the first RR study performed on pulmonary tissue (Vaccaro and Brody, 1979), where 1% RR was used as part of a 0.1% Triton X-100 solution containing 0.15 M cacodylate buffer (pH 7.4). The reaction product they observed was 10- to 40-nm thick and granular in nature, typical of RR. They reported it to be concentrated in a rather regular pattern at the alveolar side of the alveolar basement membrane (ABM), and with the major doublet band of collagen fibers. Enzymatic digestions (done prior to fixation treatments) with collagenases or glycosidases revealed that the former were likely to be heparin/heparan sulfate in nature, while the latter were probably composed of chondroitin/dermatan sulfate. RR can also be used in combination with mordants (e.g., tannic acid) or other dyes (e.g., alcian blue, acridine orange, safranin O), which help preserve the integrity of complex carbohydrates, of which as much as 40% are lost in routine fixation/dehydration protocols for tissue preparation (Chen and Wight, 1984). Using a combination of tannic acid and RR, Vaccaro and Brody (1979) further demonstrated a fine filamentous network in which collagen fibers are enmeshed. This material was believed to be protein or glycoprotein in composition.

A number of other studies have used modifications of RR solutions to obtain data on anionic sites in pulmonary tissues. Using a similar approach to their earlier work, Vaccaro and Brody (1979, 1981) reported that the anionic sites observed in the ABM of the adult rat (heparan sulfate) were different from the RR-reactive sites in the unfused capillary basement membrane, which represented heparan sulfate plus several other proteoglycans. Shortly thereafter, Brody et al. (1982) demonstrated that during early postnatal development of the rat, there was twice as much RR-reactive material (heparan sulfate) in the ABM at birth than at 8 days of life, and 5 times that found in the adult. They also showed that the asymmetric distribution of anionic sites seen in the ABM of the adult did not appear until 8 days of life. At earlier times, they were distributed equally on both sides of the ABM.

These important studies and attendant technical approaches have provided critical insights into the anatomical distribution of specific ECM components in the lung and have been the basis for a number of other studies that have focused on certain pathologic outcomes (Grant et al., 1983; Penney and Rosenkrans, 1984; Vaccaro et al., 1985; Rosenkrans and Penney, 1985, 1986). The areas of increased density fill the regions of ABM when seen in cross-sectional profile, but are more discrete and filamentous in nature when viewed in horizontal (on-face) sections. The resulting visual composition appears to reflect the proposed microarchitectural arrangement of type IV collagen within the lamina densa of basement membranes (Yurchenko and Furthmayr, 1984).

Dialyzed Iron

Dialyzed iron (DI) is similar to RR, as it too is a polycationic dye that reacts with anionic sights in tissue sections (Wetzel et al., 1966). Generally, it has not been as popular as RR because DI's penetration characteristics are even less effective. Because the incubation solution is iron based, the reaction product is a very fine, electron-dense particulate. Several studies have demonstrated its usefulness in the lung. Katsuyama and Spicer (1977) were the first to employ the method on pulmonary tissues and described reactivity on epithelial cell surfaces and basement membrane in the alveolus of the rat. Sannes (1986) extended these findings to show DI reactivity in ECM of the alveolus and bronchial tree. Notable observations included reactive sites surrounding collagen fibers and fiber bundles, and within the microfibrillar component of elastic fibers. These sites were believed to be indicative of glycosaminoglycans and glycoproteins known to be associated with these structural components of ECM.

High-Iron Diamine

High-iron diamine (HID) is a mixture of the *meta* and *para* isomers of *N,N*-dimethyl-phenylenediamine hydrochloride and ferric chloride (Spicer, 1965). Unlike RR and DI, which are nonspecific cationic dyes that will react with any site where there is sufficient anionic charge to produce a visible reaction product, HID is uniquely reactive for sulfated end groups (esters) of glycosaminoglycans and glycoproteins (Gad and Sylven, 1969), which make the method a powerful tool for examining the distribution of sulfated components of ECM. It also has the distinct advantage of penetrating cells and tissues very effectively. Katsuyama and Spicer (1977) first used the technique on pulmonary tissue and described reactivity on the apical surface of type II pneumocytes and on the alveolar side of the ABM (the lamina rara externa). In a later work, Sannes et al. (1979) developed a method that increased the sensitivity of detection of HID reaction product using a silver intensification procedure, which enabled the quantification of reactive sites in ECM (Sannes, 1984). This latter investigation demonstrated that the basement membrane microdomains associated with type I pneumocytes (Figure 2) were more reactive (more sulfated) than those of the type II pneumocytes (Figure 3). In a separate study, Sannes (1986) described highly organized HID reactivity surrounding collagen fibers, within the elastin component of elastic fibers and filling the interstitial spaces between fibers and cells (Figure 4). The discrete silver grains, indicative of HID-reactive sulfate groups, are readily identifiable and can easily demonstrate any regional or

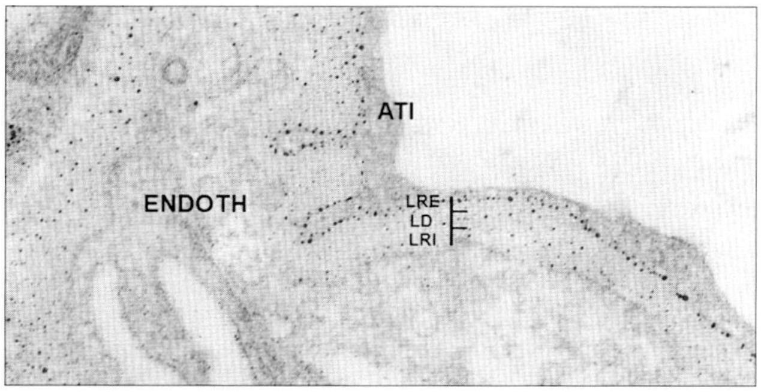

FIGURE 2 Electron micrograph of rat lung processed by the high-iron diamine (HID) procedure followed by a thiocarbohydrazide-silver proteinate intensification procedure. HID-reactive sites, representing sulfated end groups of glycosaminoglycans and glycoproteins, were observed to be concentrated in the lamina rara externa (LRE) of the alveolar basement membrane beneath a type I pneumocyte (ATI), shared with an adjacent endothelial cell (ENDOTH). The lamina densa (LD) and lamina rara intema (LRI) have very little HID reactivity. Magnification ×80,000.

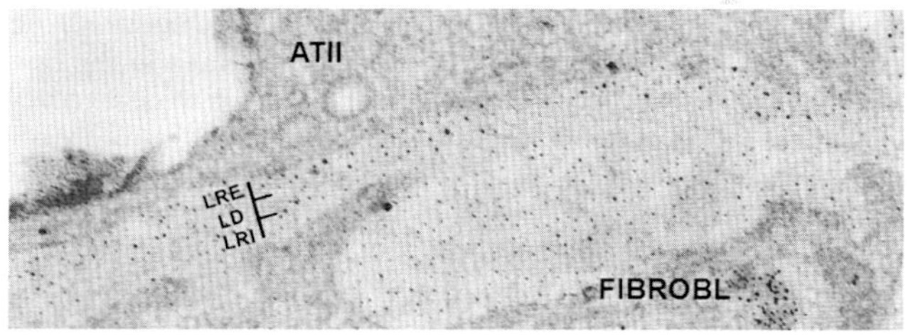

FIGURE 3 Electron micrograph of rat alveolar region prepared in the same fashion as Figure 2. The basement membrane associated with the alveolar type II cell (ATII) has a lamina rara externa (LRE) that is much less reactive with high-iron diamine (HID) than the LRE associated with the type I cell in Figure 2. The lamina densa (LD) and lamina rara interna (LRI) of all regions appear comparable in HID reactivity. An adjacent fibroblast (FIBROBL) is also depicted. Magnification ×80,000.

site-specific concentration of sulfates in a remarkably sensitive fashion. Figure 4 shows the intense reactivity of the elastin component of elastic fibers, while Figure 5 demonstrates the pattern of reactivity associated with the outer surface of collagen fibers.

Cuprolinic Blue

One of the more novel cytochemical methods to be employed on pulmonary tissues to demonstrate proteoglycans is cuprolinic blue. This phthalocyanin dye, when used according to strict electrolyte concentrations in the incubation medium, imparts a unique contrast to tissue proteoglycans when viewed with the electron microscope (Scott, 1980). Reactive sites are filamentous, rather than granular or punctate in nature, and would appear to demonstrate sulfated anionic glycosaminoglycan side chains collapsed upon the protein core of the proteoglycan. Van Kuppevelt et al. (1984a, 1984b, 1985) have effectively applied this technique to the lung and have presented

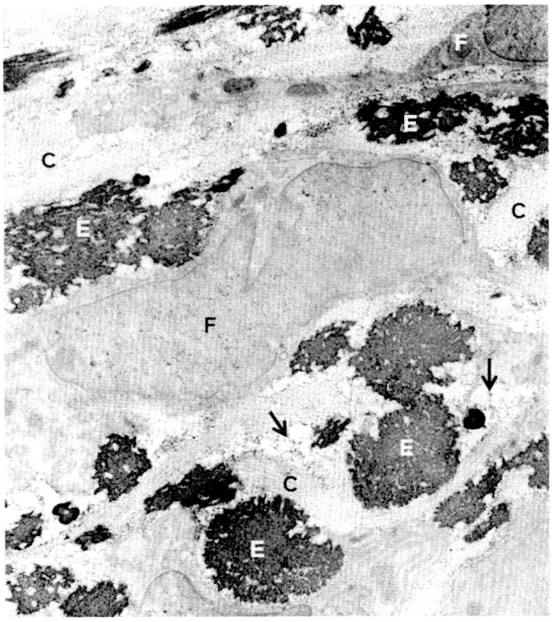

FIGURE 4 Electron micrograph of interstitial connective tissue matrix from the trachea of a 26-day-old rat prepared in the same manner as described in Figures 3–6, with high-iron diamine (HID). Fibroblasts (F) are surrounded by the heavily HID-reactive elastin component of elastic fibers (E), and bundles of collagen fibers (C), which are HID-reactive on their surfaces. Interstitial spaces contain HID-reactive filaments (arrows). Magnification ×20,000.

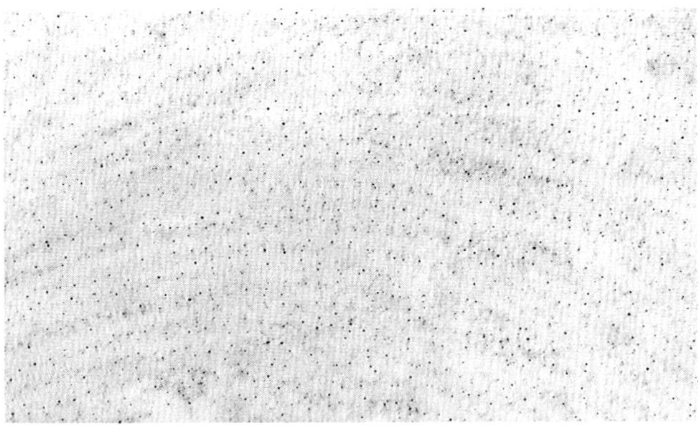

FIGURE 5 Electron micrograph representing a higher power magnification of a region similar to that depicted in Figure 4, showing a bundle of collagen fibers in cross-section. The silver grains indicative of HID-reaction product appear at periodic intervals (65–75 nm) along the outer surface of the fibers. Magnification ×80,000.

remarkable findings. They showed that the length of filaments at reactive sites was longer in regions associated with type I cells compared with type II cells in mouse (Van Kuppevelt et al., 1984a) and humans (Van Kuppevelt et al., 1985). These sites were demonstrated to be heparan sulfate in nature. They also reported reactive sites in close contact with collagen fibers located at regular (40 nm) intervals along their length, which were indicative of the presence of

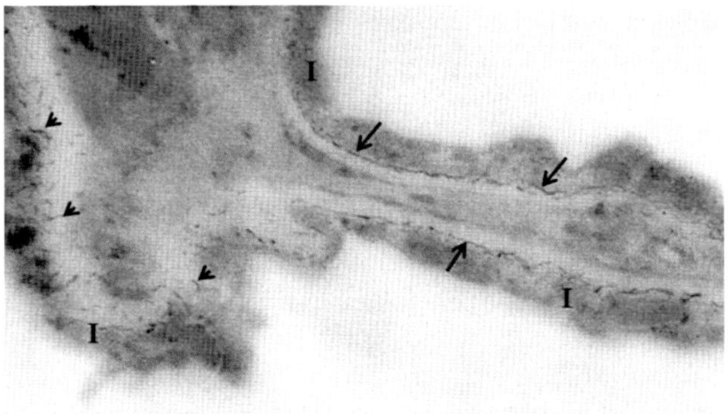

FIGURE 6 Electron micrograph of rat pulmonary alveolus stained with cuprolinic blue according to the procedure of Van Kuppevelt et al. (1984a). Electron-dense filaments indicative of cuprolinic blue reactivity are detectable in region of the lamina rara externa (LRE) in cross-section of the alveolar basement membrane beneath type I cells (I). The arrangement of reactive filaments is slightly different when seen in a horizontal section (arrowheads). A = alveolus. Magnification ×60,000.

dermatan sulfate. The reactivity of ABM with cuprolinic blue is demonstrated in Figure 6.

Immunostaining technology has proven extremely useful in the detection and localization of a wide variety of components of ECM in the lung at both the light and electron microscope levels. The obvious advantage of this approach is the specificity of detection, depending upon the purity and cross-reactivity (if any) of the antibody preparation. Of course, this also presents a possible obstacle due to the availability and supply of antibodies of interest. Commercial selection of such products is expanding rapidly, however, and as a consequence, so is the breadth of investigative potential in this area. Relevant studies are summarized in Table 2.

A number of different collagen types have been localized in pulmonary tissues by immunostaining. The first immunostaining study on lung collagens was by Madri and Furthmayr (1980), who demonstrated types I, III, IV, and V in normal and fibrotic human lungs by light microscopy. Types I and II were found in the interstitium of the alveolar septae, while type III was also found around pulmonary vessels. Types IV and V collagen both appeared in basement membranes, while type V showed some reactivity in interstitium. These results were confirmed by immunocytochemical results at the level of the electron microscope (Sano et al., 1981; Gil and Martinez-Hernandez, 1984; Amenta et al., 1988). Type VI collagen has been visualized in the alveolar septum, airway interstitium, and surrounding blood vessels (Amenta et al., 1988), similar to that shown for type III. Type X has been demonstrated to be confined to the smooth muscle cells of vessel walls and alveolar septae (Gibson et al., 1986). Types I, III, and V have been localized within consecutive layers of the subepithelial basement membrane zone (Evans et al., 2002).

TABLE 2 Immunocytochemical Reactivity of Pulmonary Connective Tissue Matrix Components

Antigen	Reactive Site/Structure	Species	LM	EM	References
Collagen					
Type I	Interstitium (alveolar septae)	H	+	−	Madri and Furthmayr, 1980
	Major airways; vascular structures	R	+	+	Gil and Martinez-Hernandez, 1984
	Interstitium	P	+		Evans et al., 2002
Type II	Bronchial cartilage	H	+		Madri and Furthmayr, 1980
Type III	Alveolar septae; vascular structures	H	+		Madri and Furthmayr, 1980
	Interstitium	R	+	+	Amenta et al., 1988
	Interstitium of airways	P	+		Evans et al., 2002
Type IV	Basement membrane	H	+		Madri and Furthmayr, 1980
		R	+	+	Amenta et al., 1988
		B	+		Sano et al., 1981
Type V	Interstitium	H	+		Madri and Furthmayr, 1980
	Cell surfaces	B	+		Sano et al., 1981
	Interstitium of airways	P	+		Evans et al., 2002
Type VI	Alveolar septae; vascular structures	R	+	+	Amenta et al., 1988
	Interstitium				
Type X	Smooth muscle cell surfaces	A	+		Gibson et al., 1986
Elastin	Elastin component of elastic fibers	H	+		Fukuda et al., 1988
Fibronectin	Interstitial spaces	R	+	+	Rosenkrans et al., 1983
	Cell surfaces	R	+	+	Gil and Martinez-Hernandez, 1984
		H	+	+	Torikata et al., 1985
		H	+		Grigioni et al., 1987
Laminin	Basement membrane	M	+		Bender et al., 1981
		R	+		Laurie et al., 1983
		R	+	+	Gil and Martinez-Hernandez, 1984
		H	+		Grigioni et al., 1987
Entactin	Basement membrane	M	+		Bender et al., 1981
		R	+		Sannes et al., 1993
Heparan sulfate proteoglycan	Basement membrane (trachea only)	R	+		Laurie et al., 1983
	Basement membrane (alveoli)	R	+		Sannes et al., 1993
Chondroitin sulfate proteoglcyan	Basement membrane	R	+		Sannes et al., 1993
Decorin	Basement membrane; interstitium	H	+		Bensadoun et al., 1996
		R	+		Wang et al., 2002
Thrombospondin	Basement membrane (trachealglands)	H	+		Wight et al., 1985
	Alveolar and ductal epithelium				
Osteonectin (BM-40; SPARC (and endothelium))		M*	+		Sage et al., 1980

Keys: LM = light microscopy, EM = electron microscopy, H = human, R = rat, B = bovine, A = chicken, P = primate, M = mouse, M* = mouse fetal tissues only.

Other ECM components have been immunolocalized with highly specific antibodies. Damian et al. (1979) and Fukuda et al. (1984) have immunolocalized the elastin component of elastic fibers by electron microscopy. These reagents did not cross-react with the microfibrillar component of elastic fibers or any other ECM element. Fibronectin has been demonstrated by immunostaining methods on epithelial cell surfaces and interstitial spaces (Rosenkrans et al., 1983; Gil and Martinez-Hernandez, 1984; Torikata et al., 1985; Grigioni et al., 1987). It is occasionally found in basement membranes in some preparations, but it is believed that its "spotty" distribution results from circulating fibronectin being "trapped" there, and not a real structural presence. Laminin is present in all basement membrane regions so far investigated, and the lung is no exception (Bender et al., 1981; Laurie et al., 1983; Gil and Martinez-Hernandez, 1984; Grigioni et al., 1987 Sannes et al., 1993). Entactin has not been studied in great detail in the lung, but not surprisingly, it was seen confined to basement membranes (Bender et al., 1981; Sannes et al., 1993). Heparan sulfate proteoglycan was found in large quantities in the lung (Laurie et al., 1983; Sannes et al., 1993), as well as chondroitin sulfate, chondroitin sulfate proteoglycan (Sannes et al., 1993; Wang et al., 2002), decorin (Bensadoun et al., 1996; Wang et al., 2002), syndecans (Frevert and Sannes, 2005), and versican (Bensadoun et al., 1996). Thrombospondin is an endogenous component of ECM as well, and it has been immunolocalized in the basement membranes of glandular areas of developing lung (Wight et al., 1985). Osteonectin (BM-40, SPARC) is a secreted glycoprotein that appears to be incorporated into ECM, and it has also been reported to be associated with alveolar and ductal epithelium and endothelium in the fetal mouse lung (Sage et al., 1980). It is not found in the adult lung. Interestingly, growth factors, such as fibroblast growth factor-2 (FGF-2), have also been immunohistochemically localized to basement membranes throughout the lung (Sannes et al., 1992).

3. DISCUSSION AND CONCLUSIONS

The cytochemical studies described above have served two very important roles in understanding the ECM of the lung: (1) they provide data essential for accurately interpreting biochemical and physiological data and (2) they enable the microdissection of anatomical sites that cannot be analyzed by any other means. It is well established that there can be major shifts in the composition of ECM during normal development of the lung and pathologic changes (Horwitz and Crystal, 1975; Madri and Furthmayr, 1980; Radhakrishnamurthy et al., 1980; Konno et al., 1982; Karlinsky, 1982). In the Madri and Furthmayr study (1980), an excellent correlation of histochemical with biochemical data was demonstrated for the changes in collagen types with the development of pulmonary fibrosis. They showed that type I collagen was markedly increased in thickened septal regions and that type III was significantly reduced and confined to perivascular regions. Type V was markedly increased in the interstitium and was found in areas of smooth muscle proliferation. These results are particularly interesting because they demonstrate the importance of correlating histochemical and biochemical findings. In this case, the regionally specific changes they observed virtually required histochemistry to enable any interpretation of data. It is known that the amount of elastic fibers changes little during the development of emphysema, but their functional characteristics are dramatically altered. Fukuda et al. (1988) showed with immunocytochemical methods that four different types of abnormal elastic fibers could be identified. Such discrimination would not be possible biochemically or with routine morphology, or even less-specific cytochemistry. The key advantage of this approach, again, is the location of the component of interest combined with its composition. Horwitz and Crystal (1975) showed that hyaluronic acid and chondroitin sulfate are the predominating glycosaminoglycans in early lung development, while dermatan sulfate, heparin, and heparan sulfate increase with maturation and are present in larger amounts in adult life. Although no study has been done on whole lung correlating these findings with specific cytochemistry with each of these components, Brody et al. (1982) described decreases in basement membrane heparan sulfate from birth to adult, reflecting in part the fusing of capillary and ABM and the development of asymmetry of heparan sulfate within its structure. This is important because it again demonstrates that regional changes may not reflect those seen in whole lung and therefore the need to consider the location of a quantitative change, which is often best achieved cytochemically.

Cytochemical observations by themselves have been enormously useful in helping to determine functional correlates to sometimes complex biologic problems. An excellent example of this is the work that has been done on pulmonary basement membranes. Katsuyama and Spicer (1977) were the first to embark on a cytochemical investigation of the ABM. They described what they called a "cation-retaining layer" within the ABM, composed of negatively charged macromolecules that prevent the movement of cations from the basement membrane to the alveolar space. To demonstrate this layer, they used DI, HID, and pyroantimonate osmium (PAO). The results indicated for the first time that the cytochemical reactivity of the ABM was asymmetrical; that is, the alveolar side or the lamina rara externa was more reactive than the other two basement membrane regions. Subsequently, Vaccaro and Brody (1979) described and quantitated RR-reactive anionic sites within the ABM. Like the reaction product seen with DI, HID, and PAO, the RR reaction product was found principally in the lamina rara externa, and probably reflected the presence of heparan sulfate. However, the visualized product generated by RR was more granular in nature, and therefore lent itself to quantitation. Using this approach, the authors later showed that the composition of the ABM was different from that which surrounds capillary endothelium (CBM), which was composed of several other glycosaminoglycans, in addition to heparan sulfate (Vaccaro and Brody, 1981).

Brody et al. (1982) then demonstrated that, in developing lungs, fusion of ABM and CBM was accompanied by a significant redistribution of basement membrane components. At birth, there is no asymmetric arrangement of RR-reactive sites in the ABM, representing heparan sulfate. Rather, they are equally distributed on the alveolar and interstitial sides. The reactivity on the capillary side decreased steadily with age. This seemed to correlate with a fivefold decrease in the accumulation of heparan sulfate from birth to the adult (Brody et al., 1982). Grant et al. (1983) were the first to describe a difference in distribution of RR-reactive anionic sites within the ABM between type I and type II pneumocytes. Sannes (1984) and Van Kuppevelt et al. (1984a, 1985) showed similar findings using very different cytochemical methods. In the former work, a silver-intensification modification of the HID method (Sannes et al., 1979) was employed to obtain a more sensitive and quantitative evaluation of the ABM than originally described by Katsuyama and Spicer (1977). It was shown that the HID-reactive anionic sites found in the lamina rara externa were found in significantly greater numbers beneath type I cells than beneath type II cells (Sannes, 1984). In fact, it clearly showed that the asymmetry originally seen in the ABM by Katsuyama and Spicer (1977), and later by Vaccaro and Brody (1979), was only found to be associated with type I cells, not type II cells. Van Kuppevelt et al. (1984a) used cuprolinic blue to demonstrate similar differences, but the nature of the reaction product was markedly different. Cuprolinic blue apparently demonstrates proteoglycans by reacting with the charged end groups of glycosaminoglycans in the presence of a critical concentration of electrolytes (Scott, 1973). The reaction product "collapses" onto the protein core of the proteoglycan, forming an electron-dense filament, whose length can be varied within the basement membrane, while the protein core is not. Alternatively, the peroxidase label may be insufficiently discrete and therefore unreliable to reveal the real microarchitectural asymmetry of heparan sulfate. This question could be better addressed with an immunolocalization performed using a more sensitive immunogold technique.

All of these cytochemical observations have provided great insight into understanding the structural composition of pulmonary basement membranes. More importantly, however, the compositional differences in ECM documented in these studies may prove to be critical determinants of the processes of proliferation, migration, and differentiation (Kainulainen et al., 1998). In vitro studies have documented the sensitivity of isolated type II pneumocytes to different ECM substrata (Clark et al., 1986; Rannels et al., 1987a, 1987b; Shannon et al., 1987; Sannes et al., 1996, 1998; Isakson et al., 2001; Isakson et al., 2002; Li et al., 2002; Newman et al., 2004; Leiner et al., 2006). It has been demonstrated that specific components of the ABM seen cytochemically, such as heparan sulfate proteoglycans and laminin (Rannels et al., 1987a, 1987b; Sannes et al., 1993), have very specific effects on the differentiation of type II cells into type I cells (Sannes et al., 1998), a critical event in maintaining stable cell populations and achieving effective repair following injury/inflammation in the pulmonary alveolus.

ACKNOWLEDGMENTS

This work was supported in part from Public Health Service Grants HL-44497 and HL-95411, and grants from the North Carolina State University College of Veterinary Medicine.

REFERENCES

Amenta, P.S., Gil, J., Martinez-Hernandez, A., 1988. J. Histochem. Cytochem. 36, 167–175.

Bender, B.L., et al., 1981. Am. J. Pathol. 103, 419–425.

Bensadoun, E.S., et al., 1996. Am. J. Respir. Crit. Care Med. 154 (6 Pt 1), 1819–1828.

Brody, J.S., et al., 1982. J. Cell. Biol. 95, 394–401.

Chen, K., Wight, T.N., 1984. J. Histochem. Cytochem. 32, 347–352.

Clark, R.A., et al., 1986. J. Clin. Invest. 77, 1831–1838.

Damian, V.V., et al., 1979. Am. J. Pathol. 96, 439–449.

Evans, M.J., et al., 2002. Am. J. Physiol. Lung Cell. Mol. Physiol. 283 (6), L1263–L1270.

Frevert, C.W., Sannes, P.L., 2005. Eur. Respir. Rev. 14 (97), 137–144.

Fukuda, Y., Ferrans, V.J., Crystal, R.G., 1984. Am. J. Anat. 170, 597.

Fukuda, Y., et al., 1988. Histochem. Cytochem. 36, 866–876.

Gad, A., Sylven, B., 1969. J. Histochem. Cytochem. 17, 156–169.

Gibson, G.J., Bearman, C.H., Flint, M.H., 1986. Coll. Rel. Res. 6, 163–175.

Gil, J., Martinez-Hernandez, A., 1984. J. Histochem. Cytochem. 32, 230–243.

Grant, M.M., Cutts, N.R., Brody, J.S., 1983. Dev. Biol. 97, 173–183.

Grigioni, W.F., et al., 1987. Am. J. Pathol. 128, 217–228.

Hay, E.D., 1981. In: Hay, E.D. (Ed.), Cell Biology of Extracellular Matrix. Plenum Press, New York, pp. 1–4.

Horwitz, A., Crystal, R.J., 1975. J. Clin. Invest. 56, 1312–1328.

Isakson, B.E., et al., 2001. Am. J. Physiol. Cell. Physiol. 281 (4), C1291–C1299.

Isakson, B.E., et al., 2002. Vitro Cell. Dev. Biol. Anim. 38 (8), 443–449.

Kainulainen, V., et al., 1998. J. Biol. Chem. 8:273 (19), 11563–11569.

Karlinsky, J., 1982. Am. Rev. Resp. Dis. 125, 85–101.

Katsuyama, T., Spicer, S.S., 1977. Lab. Invest. 36, 428–436.

Khosla, J., Correa, M.T., Sannes, P.L., 1994. Am. J. Respir. Cell. Mol. Biol. 10 (4), 462–469.

Konno, K., et al., 1982. Am. Rev. Resp. Dis. 126, 797–806.
Laurie, G.W., Leblond, C.P., Martin, G.R., 1983. Am. J. Anat. 167, 71–89.
Leiner, K.A., et al., 2006. Am. J. Respir. Cell. Mol. Biol. 35 (5), 611–618.
Li, C.M., et al., 2002. Am. J. Respir. Cell. Mol. Biol. 27 (3), 345–352.
Luft, J.H., 1971. Anat. Rec. 171, 347–366.
Madri, J.A., Furthmayr, H., 1980. Hum. Pathol. 11, 353–362.
Newman, D.R., et al., 2004. Am. J. Physiol. Lung Cell. Mol. Physiol. 287 (1), L191–L200.
Penney, D.P., Rosenkrans, W.A., 1984. Rad. Res. 99, 410–419.
Radhakrishnamurthy, B., Williams, J., Berenson, G.S., 1980. Proc. Soc. Exp. Biol. Med. 164, 287–301.
Rannels, S.R., et al., 1987a (Cell Physiol. 22). Am. J. Physiol. 253, C759–C768.
Rannels, S.R., et al., 1987b (Cell Physiol. 22). Am. J. Physiol. 253, C835–C844.
Rosenkrans, W.A., Penney, D.P., 1986. Anat. Rec. 215, 127–139.
Rosenkrans, W.A., et al., 1983. Cell. Tissue Res. 233, 113–125.
Rosenkrans, W.A., Penney, D.P., 1985. Int. J. Rad. Oncol. Biol. Phys. 11, 1629–1631.
Sage, H., et al., 1980. J. Histochem. Cytochem 37, 819–828.
Sannes, P.L., 1984. J. Histochem. Cytochem. 32, 827–832.
Sannes, P.L., 1986. Histochemistry 84, 49–57.
Sannes, P.L., 1987. Stain Technol. 62, 280–286.
Sannes, P.L., Spicer, S.S., Katsuyama, T., 1979. J. Histochem. Cytochem. 27, 1108–1113.
Sannes, P.L., Burch, K.K., Khosla, J., 1992. Am. J. Respir. Cell. Mol. Biol. 7 (2), 230–237.
Sannes, P.L., et al., 1993. Am. J. Respir. Cell. Mol. Biol. 8 (3), 245–251.
Sannes, P.L., Khosla, J., Cheng, P.W., 1996. Am. J. Physiol. 271 (5 Pt 1), L688–L697.
Sannes, P.L., et al., 1998. Am. J. Physiol. 275 (4 Pt 1), L701–L708.
Sano, J., et al., 1981. Biomed. Res. 2, 20–34.
Schulte, B.A., Harley, R.A., Spicer, S.S., 1989. In: Gil, J. (Ed.), Structural Methods in Experimental Lung Research. Marcel Dekker, New York.
Scott, J.E., 1973. Biochem. Soc. Trans. 1, 787.
Scott, J.E., 1980. Biochem. J. 187, 887–892.
Shannon, J.M., Mason, R.J., Jennings, S.D., 1987. Biochim. Biophys. Acta 931, 143–155.
Spicer, S.S., 1965. J. Histochem. Cytochem. 13, 211–219.
Spicer, S.S., Schulte, B.A., 1989. In: Gil, J. (Ed.), Structural Methods in Experimental Lung Research. Marcel Dekker, New York.
Torikata, C., et al., 1985. Lab. Invest 52, 399–408.
Vaccaro, C.A., Brody, J.S., 1979. Am. Rev. Resp. Dis. 120, 901–912.
Vaccaro, C.A., Brody, J.S., 1981. J. Cell. Biol. 91, 427–435.
Vaccaro, C.A., Brody, J.S., Snider, G.L., 1985. Am. Rev. Respir. Dis. 132, 905–912.
Van Kuppevelt, T.H.M.S.M., Cremers, F.P.M., Kuyper, C.M.A., 1984a. Histochem. J. 16, 657–666.
Van Kuppevelt, T.H.M.S.M., et al., 1984b. Histochem. J. 16, 671–682.
Van Kuppevelt, T.H.M.S.M., et al., 1985. Eur. J. Cell. Biol. 36, 74.
Wang, Y., et al., 2002. Am. J. Physiol. Lung Cell. Mol. Physiol. 282 (3), L484–L490.
Wetzel, M.G., Wetzel, B.K., Spicer, S.S., 1966. J. Cell. Biol. 30, 299–313.
Wight, T.N., et al., 1985. J. Histochem. Cytochem. 33, 295–306.
Yurchenko, P.D., Furthmayr, H., 1984. Biochemistry 23, 1839–1851.

Chapter 11

Comparative Mammalian Lung Primary Surface Epithelial Cell Culture

Peter F. Bove and Scott H. Randell

Chapter Outline

1. INTRODUCTION

Normal adult human lungs contain approximately 1500 miles (2400 km) of airways and more than 300 million alveoli, with a total surface area of about 750 square feet (70 m^2), equal to half of a tennis court. Epithelial cells lining the airways and alveoli provide critical physiologic functions and are central to the development of disease. Lung epithelial cell culture is a vital part of the research and development enterprise, enabling mechanistic approaches to scientific questions that complement in vivo studies. Lung epithelial cell culture paralleled the development of cell culture in general and has incrementally advanced, allowing more faithful reproductions of key molecular and physiologic processes relevant to lung biology and disease. The purpose of this chapter is to review the literature, focusing on the history, development, and current state of the art of lung epithelial cell culture from mammalian species. We examine methods and results including milestones, the overall approach, and key points, but we do not include detailed technical protocols, for which the reader is referred to the original publications. We explore the use of lung organ cultures and in vivo grafting for developmental biology, infectious disease, cancer, transplantation, and lung regenerative medicine research, although not comprehensively. We focus instead on the current status of primary airway and alveolar surface epithelial cell cultures from dissociated cells.

2. BRIEF HISTORY OF LUNG EPITHELIAL CELL CULTURE

In 1882–1883, Sydney Ringer published observations that ex vivo beating of the frog heart required solutions with specific ionic compositions (reviewed in Miller, 2004). In 1885, Wilhelm Roux explanted fragments of chick embryos into warm saline solution (discussed in Oppenheimer, 1966). In 1907, Ross G. Harrison documented nerve cell fiber growth from a transected frog embryo cultivated in clotted lymph (Harrison, 1907). Driven by vaccine development in the twentieth century, mammalian cell culture incrementally advanced by use of trypsin (Rous and Jones, 1916), an aseptic technique supported by specialized flasks and roller bottles (Carrel, 1923; Gey, 1933), antibiotics (e.g., Williams et al., 1959), chemically defined media (e.g., Evans et al., 1953), and serum-free media (e.g., Ham, 1965). Culture of epithelial cells was reported in 1922 (Ebeling and Fischer, 1922) and a continuous human epithelial cell line, HeLa, was reported in 1953 (Scherer et al., 1953). A lung cancer epithelial cell line was reported in 1955 (Frisch et al., 1955) and cell growth from normal human bronchus explants was reported in 1958 (Hoch-Ligeti and Hobbs, 1958).

3. LUNG ORGAN CULTURES AND IN VIVO GRAFTING

Cell and organ culture is rooted in experimental embryology, which was historically performed using sea urchin,

Comparative Biology of the Normal Lung. http://dx.doi.org/10.1016/B978-0-12-404577-4.00011-4

amphibian, or chick embryos. The lung is particularly compelling for developmental biology studies due to extensive branching morphogenesis, the intimate determinative relationship between developing endoderm and mesenchyme, and the need for lung developmental maturity at birth. The explanted embryonic lung continues to grow and develop ex vivo. In 1961, Sorokin summarized prior work and reported the effects of various growth conditions on cultured rat and guinea pig embryonic lungs or lung segments (Sorokin, 1961). Budding of the endodermally derived mouse lung epithelium depends on the surrounding mesenchyme (Alescio and Colombo, 1967) and recombination studies in various species illustrate reciprocal organ- and region-specific communication between epithelium and mesenchyme (Shannon, 1994; Shannon et al., 1998); (and references within). Since that time, literally hundreds of studies exploring spatial (e.g., Lazarus et al., 2011), cellular (e.g., Schnatwinkel and Niswander, 2013), and molecular (e.g., Miller et al., 2012) regulation of lung development have been performed using organ cultures from various mammals, with many studies using genetically manipulated mice.

Respiratory tract organ cultures have long been used as an infection model (e.g., Bang and Niven, 1958; Hoorn, 1964). It was noted that some viruses could not be grown in trypsin dissociated undifferentiated cells in flasks, but they could be grown in lung organ cultures (Tyrrell and Bynoe, 1965). Virus infection of lung organ cultures enabled the study of both virus growth and cytopathology in differentiated cells (Craighead, 1966). Similar to developmental biology studies, there have been many infectious disease investigations using lung organ cultures, including a study of influenza that incorporated precision-cut human lung slices with increased standardization and throughput (Wu et al., 2012). The observation that infectious agent tropism and the biologic response are likely to be different in poorly versus well-differentiated cells has driven improvements in respiratory tract epithelial cell culture models.

Prior to the development of sophisticated in vitro culture methods for dissociated cells, transplantation of organs or organ fragments into visible tissue compartments, such as the frog lymph sac or eye, enabled observation of tissue growth and development. In pioneering studies, chick lung rudiments were grafted into the chicken chorioallantoic membrane (Rudnick, 1933). Some lung organ culture studies incorporated ex vivo manipulation followed by transplantation into a compatible host. For example, normal mouse lung organ fragments were treated with a carcinogen in culture and then subcutaneously transplanted to observe the long-term development of adenomas (Laws and Flaks, 1966). The mouse is currently the preeminent model for lung cancer development (reviewed in Kwon and Berns, 2013) and xenotransplantation of human lung cancer cells into immune-deficient mice is widely used for phenotypic characterization and other purposes (e.g., Harris et al., 2011). Prior to the development of in vitro conditions supporting lung epithelial cell differentiation, a donor rat trachea was decellularized, inoculated with rat tracheal epithelial cells cultured on plastic, and then transplanted subcutaneously in vivo into isogenic rats, enabling organotypic epithelial cell differentiation (Terzaghi et al., 1978). In a variation, rabbit tracheal versus small airway cells were inoculated into a denuded rat trachea and transplanted subcutaneously into nude mice to observe cell-type differentiation potential (Hook et al., 1987). This model was subsequently used to examine the growth of basal versus columnar rat tracheal epithelial cells (Liu et al., 1994) and to study human cystic fibrosis (CF) versus non-CF epithelial cells (Zhang et al., 1996). Heterotopic subcutaneous transplantation of nondenuded mouse tracheas into isogenic versus allogenic mouse hosts is a commonly used model for transplant rejection (reviewed in Jungraithmayr et al., 2013). Lung regenerative medicine is a rapidly evolving field rationally predicated on a better understanding of stem cell biology, and transplantation is a key method for testing lung somatic stem and progenitor cells and the progeny of embryonic and induced pluripotent stem cells for studies of lung cell therapy and creation of bioengineered lung tissue (reviewed in Patel et al., 2012; Garcia et al., 2012; Rock and Konigshoff, 2012; Kotton, 2012; Brouwer et al., 2013).

4. MODERN AIRWAY EPITHELIAL CELL CULTURES[a]

Primary airway epithelial cell cultures were historically created from the rat (Heckman et al., 1978) and human (Lechner et al., 1982) by explanting minced conducting airway tissue fragments and harvesting epithelial cell outgrowths. Subsequent studies used proteases for primary epithelial cell dissociation (Gruenert et al., 1995). Epithelial cells originating from airways with a pseudostratified in vivo morphology grown as monolayers on conventional plastic culture dishes proliferated and could be passaged, but the cells became squamous, losing differentiated features. When the cells were allowed to form three-dimensional spheroid aggregates (Jorissen et al., 1989) or were grown on thick collagen gels (Wu et al., 1985; Benali et al., 1993), they polarized and began to recapitulate the characteristic pseudostratified mucociliary morphology found in vivo. A fundamental practical breakthrough was the culture of airway epithelial cells on porous supports with media in the basolateral compartment and exposure of the apical surface to air (Whitcutt et al., 1988). Human airway epithelial cells

[a] Portions of this section are paraphrased from our prior detailed methods chapter on human airway epithelial cell cultures (Fulcher et al., 2005).

TABLE 1 Key References for Airway Epithelial Cell Culture from Various Mammalian Species

Genus and Species	Common Name	Notes on Types of Cultures and Parameters Studied	References
Homo sapiens	Human	Clonal growth of human bronchial epithelial cells	(Lechner et al., 1981)
		Detailed description of a media for proliferation of human bronchial epithelial cells on plastic	(Lechner and LaVeck, 1985)
		Differentiation of serially passaged cells at an ALI	(Gray et al., 1996)
		Current detailed methods for ALI cultures using nonproprietary reagents	(Fulcher and Randell, 2013)
		Current detailed methods for ALI cultures using proprietary reagents	(Neuberger et al., 2011)
Rattus norvegicus	Norway rat	Epithelial outgrowths from explants	(Heckman et al., 1978)
		Monolayer cultures, studies of transformation	(Nettesheim and Barrett, 1984)
		ALI cultures, effects of substratum, retinoids, and submersion	(Kaartinen et al., 1993)
Mus musculus	Laboratory mouse	Dissociated cells grown in serum-free media	(Nedrud and Wu, 1982)
		Polarized cultures, ion transport	(Clarke et al., 1992)
		ALI cultures	(You et al., 2002)
		Current detailed methods for ALI cultures	(You and Brody, 2013)
Macaca mulatta	Rhesus monkey	Extracellular matrix effects on monkey and human cell mucin production	(Robinson and Wu, 1993)
Mesocricetus auratus	Syrian golden hamster	Tracheal epithelial cell proliferation	(Goldman and Baseman, 1980b)
		Mucin secretion	(Goldman and Baseman, 1980a)
		Mucin secretion	(Kim et al., 1989)
		ALI cultures	(Rowe et al., 2004)
Oryctolagus cuniculus	European rabbit	Proliferation in defined serum-free media	(Wu and Smith, 1982)
		Matrix effects on differentiation	(Baeza-Squiban et al., 1994)
		ALI cultures	(Tournier et al., 1998)
Cavia porcellus	Guinea pig	ALI cultures	(Adler et al., 1990)

Continued

TABLE 1 Key References for Airway Epithelial Cell Culture from Various Mammalian Species—cont'd

Genus and Species	Common Name	Notes on Types of Cultures and Parameters Studied	References
Canis lupus familiaris	Domestic dog	Electrical properties on membrane supports	(Coleman et al., 1984)
		Surface mucin release by proteases on plastic	(Varsano et al., 1987)
Felis catus	Domestic cat	Epithelial sheet removal	(Tandler et al., 1981)
		Dissociated cell culture, collagen gel substrate, mucin secretion	(Rieves et al., 1992)
Equus ferus caballus	Domestic horse	Adult tracheal cell growth on collagen-coated plastic	(Sime et al., 1997)
		ALI cultures	(Schwab et al., 2010)
		ALI cultures	(Abraham et al., 2011)
Sus domesticus	Domestic pig	ALI cultures	(Young et al., 2000)
		ALI cultures	(Khoufache et al., 2010)
Bos primigenius	Domestic cattle	Fetal tracheal epithelial cell isolation	(Schumann et al., 1988)
		Adult bronchial epithelial cell growth on collagen-coated plastic	(Beckmann et al., 1992)
		ALI cultures	(Kondo et al., 1993)
Ovis aries	Domestic sheep	ALI cultures	(Radi and Ackermann, 2004)
Mustela putorius furo	Ferret	ALI cultures, influenza virus tropism	(Zeng et al., 2013)
Multiple species	Human, pig, ferret, mouse	Comparative electrophysiology	(Liu et al., 2007)

ALI, air-liquid interface.

TABLE 2 Chronological Advancement of Alveolar Type II (ATII) Cell Culture Methods from Mammalian Species

Species/Age	Notes on Protocols and Results	References
Adult rabbit	First report of ATII cell isolation.	(Kikkawa and Yoneda, 1974)
Adult rabbit	ATII cell isolation followed by high-density cell plating on plastic ±10% fetal bovine serum, lamellar bodies present for 3–5 days followed by loss of ATII cell type-specific phenotype.	(Diglio and Kikkawa, 1977)
Adult rat	Trypsin digestion and cell separation on a discontinuous albumin density gradient via unit gravity velocity sedimentation.	(Mason et al., 1977)
Fetal rat	Enzymatic dissociation/digestion (trypsin, collagenase, and chick serum), crude extract inoculated into a Gelfoam® sponge.	(Douglas and Teel, 1976)
Adult rat	Clonal culture technique, develops "L-2" ATII cell type.	(Douglas and Farrell, 1976)
Adult rat	Elastase digestion, discontinuous density centrifugation, and plastic adherence.	(Dobbs and Mason, 1979)
Adult rat	Centrifugal elutriation (counterflow centrifugation) improves ATII cell purity.	(Greenleaf et al., 1979)
Adult rat	Culture of ATII cells on floating collagen membranes.	(Geppert et al., 1980)
Adult rat	Transepithelial transport of ATII cells in primary culture on plastic, plastic coated with ECM from bovine corneal endothelial cells, or Millipore filters.	(Mason et al., 1982)
Adult human	Dissociation with porcine pancreatic elastase, then discontinuous metrizamide density gradient and differential adherence, culture on bovine corneal endothelial cell ECM.	(Robinson et al., 1984)
Adult rat	Improved isolation with more representative recovery, retention of cell polarity, and increased purity via macrophage and lymphocyte depletion.	(Weller and Karnovsky, 1986)
Adult rat	Isolation up to 95% purity by flow cytometry.	(Wilson et al., 1986)
Fetal human	Isolation and characterization of ATII cells from 18 to 24 weeks' gestation human fetal lung.	(Ballard et al., 1986)
Adult rat	Functional differentiation of ATII cells in vitro, effects of cell shape, cell–ECM interactions, and cell–cell interactions.	(Shannon et al., 1987)
Adult mouse	Isolation and identification of murine ATII cells, ~65% purity.	(Massey et al., 1987)
Adult rat	Improved ATII cell differentiation in vitro, effect of hydrocortisone and cyclic AMP.	(Kawada et al., 1988)
Adult rat	High electrical resistance ATII cell monolayers, studied bioelectric properties including active sodium transport.	(Cheek et al., 1989)
Adult hamster	Isolation of ATII and club (Clara) cells, biotransformation activity.	(Myles et al., 1989)
Adult rat, adult human	Effects of two ECMs on morphologic and biochemical properties of rat and human ATII cells, Matrigel® versus bovine corneal endothelial cell ECM.	(Edelson et al., 1989)
Adult human	Stimulation of DNA synthesis by insulin and endothelial cell growth supplement	(Mason et al., 1990)
Adult rat	Improved maintenance of adult rat ATII cell differentiation in vitro; effects of serum-free, hormonally defined medium; and a reconstituted basement membrane.	(Kawada et al., 1990)

Continued

TABLE 2 Chronological Advancement of Alveolar Type II (ATII) Cell Culture Methods from Mammalian Species—cont'd

Species/Age	Notes on Protocols and Results	References
Adult rat	Modulation of ATII cell differentiation and function via culture on attached or floating collagen gels with embedded fibroblasts.	(Shannon et al., 1992)
Adult rat	Proliferation of ATII cells in low-density primary cultures in specialized media.	(Leslie et al., 1993)
Adult rat	Developed defined serum-free medium for primary ATII culture, allowed formation of monolayers with characteristic bioelectric properties and phenotype.	(Borok et al., 1994)
Adult rat	Enhanced RGD peptide-dependent adhesion on ProNectin-F-coated dishes in the presence of 0.5 mM Mn^{++}.	(Lwebuga-Mukasa, 1994)
Adult rat	KGF increases SPA and B mRNA.	(Sugahara et al., 1995)
Adult mouse	Isolation via instilled dispase and agarose, mechanical digestion, and negative cell selection, culture on fibronectin-coated dishes in DMEM 10% fetal bovine serum.	(Corti et al., 1996)
Adult rat	Effects of epidermal growth factor on alveolar epithelial junctional permeability and active sodium transport	(Borok et al., 1996)
Adult rat	KGF increases SPA and SPD mRNA levels and secretion, effects of ECM, KGF, combination.	(Xu et al., 1998)
Adult rat	KGF modulates ATII cell phenotype in vitro, expression of aquaporin 5.	(Borok et al., 1998)
Adult rat	KGF regulates lung epithelial cell proliferation and surfactant protein gene expression in adult rat lung in situ.	(Yano et al., 2000)
Adult rat	Co-culture with lung fibroblasts in culture improves SPA and SPC mRNA expression and differentiation via KGF.	(Shannon et al., 2001)
Guinea pig	Culture using an acellular human amniotic membrane.	(Sakamoto et al., 2001)
Fetal rat	ATII cells isolated from explant cultures of hormone-treated fetal rat lungs.	(Gonzales et al., 2001)
Adult rat	Growth factors, corticosteroids, rat serum, and a fibroblast feeder layer produce a polarized epithelium that secretes SPA and SPD.	(Mason et al., 2002)
Adult rat	Transforming growth factor-beta antagonizes ATII cell proliferation induced by KGF.	(Zhang et al., 2004)
Adult porcine	Porcine ATII cells in primary culture, morphological, bioelectrical, and immunocytochemical characterization.	(Steimer et al., 2006)
Adult human	Differentiated human ATII cells and reversibility of their phenotype in vitro.	(Wang et al., 2007)
Adult mouse	Effects of ECM on transport properties and transdifferentiation into an alveolar epithelial type I (AT1) cell-like phenotype.	(Demaio et al., 2009)
Adult human	Bioelectric properties of cultured ATII cells in response to local nucleotide signaling.	(Bove et al., 2010)
Adult porcine	CFTR-mediated transepithelial liquid transport.	(Li et al., 2012)

ECM, extracellular matrix; KGF, keratinocyte growth factor; SPA, SPB, SPD, surfactant proteins A, B, and D.

are typically procured using the original monolayer techniques (Lechner and LaVeck, 1985) and are plated on porous supports at an air-liquid interface (ALI) in modified media (Gray et al., 1996). The cells grow and differentiate at an ALI and exhibit mucociliary transport (Matsui et al., 1998a,b), resistance to gene therapy vectors (Pickles et al., 1998), and cell type-specific virus tropism (Zhang et al., 2002). Gene expression of human bronchial epithelial cells in these cultures appears similar to freshly harvested in vivo specimens (Pezzulo et al., 2011). These cultures are widely used for basic and applied studies of airway biology and disease, including proof of concept and milestone preclinical testing for pharmaceutical development and for toxicology. There are multiple commercial sources for human primary airway epithelial cells and specialized media. Monolayer and ALI airway epithelial cell cultures have been created from many species, and some require species-specific modifications of the method involving a choice of porous support, extracellular matrix coating, and media composition. A noncomprehensive list of airway epithelial cultures from various mammalian species and their key features are given in Table 1.

5. MODERN ALVEOLAR EPITHELIAL CELL CULTURES

Alveoli comprise more than 99% of the internal surface area of the lung. Maintaining their normal structure and milieu is critical for their vital gas exchange function. Two cell types line the alveolar surface, the squamous alveolar type I (ATI) cell and the cuboidal type II (ATII) cell. ATI cells are large (50- to 100-μm diameter), flat (~25–100 nm) squamous cells making up approximately 8% of the cells in the lungs and covering more than 95% of the alveolar surface. ATII cells are smaller (~10 μm in diameter), cuboidal in shape, comprise almost 15% of all lung cells, and cover the remaining 3–5% of the alveolar surface. ATII cells synthesize, secrete, and recycle pulmonary surfactant and actively transport ions. Moreover, ATII cells in vivo serve as progenitor cells and proliferate in response to lung injury. Due to extreme cellular heterogeneity of the lungs, it is difficult to attribute specific functions to ATII cells from studies of mixed cell populations or whole lungs. To date, no immortal cell lines exhibit the full range of primary ATII cell functions. For these reasons, methods have been developed to isolate and culture primary ATII cells to study their biochemistry and cell and molecular biology and function. Here, we focus on the evolution of mammalian ATII cell isolation and culture techniques from different mammals, as well as the challenge of maintaining their phenotype in vitro.

Kikkawa and Yoneda first isolated ATII cells from rabbits (Kikkawa and Yoneda, 1974; Diglio and Kikkawa, 1977), while Mason et al. and Dobbs et al. isolated ATII cells from adult rats (Mason et al., 1977; Dobbs and Mason, 1979). Douglas et al. used fetal (Douglas and Teel, 1976) and adult (Douglas and Farrell, 1976) rats. Trypsin and/or elastase instilled down the airway was used to selectively target ATII cell release from the basal lamina. Protocols typically yielded crude initial cell mixtures; subsequent steps were used to optimize ATII purity, including separation based on size, immunoglobulin G panning, discontinuous density centrifugation, adherence in primary culture, or fluorescence staining and cell sorting. The evolution of ATII cell culture methods from various mammalian species is given in Table 2. Compared to ATII cells, there are fewer studies of ATI cells, but these cells can be selectively isolated and cultured, as described in detail for rats by Gonzalez and Dobbs (2013).

REFERENCES

Abraham, G., Zizzadoro, C., Kacza, J., Ellenberger, C., Abs, V., Franke, J., Schoon, H.A., Seeger, J., Tesfaigzi, Y., Ungemach, F.R., 2011. Growth and differentiation of primary and passaged equine bronchial epithelial cells under conventional and air-liquid-interface culture conditions. BMC Vet. Res. 7, 26.

Adler, K.B., Cheng, P.W., Kim, K.C., 1990. Characterization of guinea pig tracheal epithelial cells maintained in biphasic organotypic culture: cellular composition and biochemical analysis of released glycoconjugates. Am. J. Respir. Cell. Mol. Biol. 2, 145–154.

Alescio, T., Colombo, P.E., 1967. A quantitative assessment of mesenchymal contribution to epithelial growth rate in mouse embryonic lung developing in vitro. J. Embryol. Exp. Morphol. 17, 213–227.

Baeza-Squiban, A., Boisvieux-Ulrich, E., Guilianelli, C., Houcine, O., Geraud, G., Guennou, C., Marano, F., 1994. Extracellular matrix-dependent differentiation of rabbit tracheal epithelial cells in primary culture. In Vitro Cell Dev. Biol. Anim. 30A, 56–67.

Ballard, P.L., Ertsey, R., Gonzales, L.K., Liley, H.G., Williams, M.C., 1986. Isolation and characterization of differentiated alveolar type II cells from fetal human lung. Biochim. Biophys. Acta. 883, 335–344.

Bang, F.B., Niven, J.S., 1958. A study of infection in organised tissue cultures. Br. J. Exp. Pathol. 39, 317–322.

Beckmann, J.D., Takizawa, H., Romberger, D., Illig, M., Claassen, L., Rickard, K., Rennard, S.I., 1992. Serum-free culture of fractionated bovine bronchial epithelial cells. In Vitro Cell Dev. Biol. 28A, 39–46.

Benali, R., Tournier, J.M., Chevillard, M., Zahm, J.M., Klossek, J.M., Hinnrasky, J., Gaillard, D., Maquart, F.X., Puchelle, E., 1993. Tubule formation by human surface respiratory epithelial cells cultured in a three-dimensional collagen lattice. Am. J. Physiol. 264, L183–L192.

Borok, Z., Danto, S.I., Zabski, S.M., Crandall, E.D., 1994. Defined medium for primary culture de novo of adult rat alveolar epithelial cells. In Vitro Cell Dev. Biol. Anim. 30A, 99–104.

Borok, Z., Hami, A., Danto, S.I., Lubman, R.L., Kim, K.J., Crandall, E.D., 1996. Effects of EGF on alveolar epithelial junctional permeability and active sodium transport. Am. J. Physiol. 270, L559–L565.

Borok, Z., Lubman, R.L., Danto, S.I., Zhang, X.L., Zabski, S.M., King, L.S., Lee, D.M., Agre, P., Crandall, E.D., 1998. Keratinocyte growth factor modulates alveolar epithelial cell phenotype *in vitro*: expression of aquaporin 5. Am. J. Respir. Cell. Mol. Biol. 18, 554–561.

Bove, P.F., Grubb, B.R., Okada, S.F., Ribeiro, C.M., Rogers, T.D., Randell, S.H., O'Neal, W.K., Boucher, R.C., 2010. Human alveolar type II cells secrete and absorb liquid in response to local nucleotide signaling. J. Biol. Chem. 285, 34939–34949.

Brouwer, K.M., Hoogenkamp, H.R., Daamen, W.F., van Kuppevelt, T.H., 2013. Regenerative medicine for the respiratory system: distant future or tomorrow's treatment? Am. J. Respir. Crit. Care Med. 187, 468–475.

Carrel, A., 1923. A method for the physiological study OF tissues in vitro. J. Exp. Med. 38, 407–418.

Cheek, J.M., Kim, K.J., Crandall, E.D., 1989. Tight monolayers of rat alveolar epithelial cells: bioelectric properties and active sodium transport. Am. J. Physiol. 256, C688–C693.

Clarke, L.L., Burns, K.A., Bayle, J.-Y., Boucher, R.C., Van Scott, M.R., 1992. Sodium and chloride conductive pathways in cultured mouse tracheal epithelium. Am. J. Physiol. 263, L519–L525.

Coleman, D.L., Tuet, I.K., Widdicombe, J.H., 1984. Electrical properties of dog tracheal epithelial cells grown in monolayer culture. Am. J. Physiol. 246, C355–C359.

Corti, M., Brody, A.R., Harrison, J.H., 1996. Isolation and primary culture of murine alveolar type II cells. Am. J. Respir. Cell. Mol. Biol. 14, 309–315.

Craighead, J.E., 1966. Growth of Myxovirus parainfluenza type 3 in organ cultures of guinea pig tissue. J. Bacteriol. 92, 751–761.

Demaio, L., Tseng, W., Balverde, Z., Alvarez, J.R., Kim, K.J., Kelley, D.G., Senior, R.M., Crandall, E.D., Borok, Z., 2009. Characterization of mouse alveolar epithelial cell monolayers. Am. J. Physiol. Lung Cell. Mol. Physiol. 296, L1051–L1058.

Diglio, C.A., Kikkawa, Y., 1977. The type II epithelial cells of the lung. IV. Adaption and behavior of isolated type II cells in culture. Lab. Invest. 37, 622–631.

Dobbs, L.G., Mason, R.J., 1979. Pulmonary alveolar type II cells isolated from rats. Release of phosphatidylcholine in response to beta-adrenergic stimulation. J. Clin. Invest. 63, 378–387.

Douglas, W.H., Farrell, P.M., 1976. Isolation of cells that retain differentiated functions in vitro: properties of clonally isolated type II alveolar pneymonocytes. Environ. Health Perspect. 16, 83–88.

Douglas, W.H., Teel, R.W., 1976. An organotypic in vitro model system for studying pulmonary surfactant production by type II alveolar pneumonocytes. Am. Rev. Respir. Dis. 113, 17–23.

Ebeling, A.H., Fischer, A., 1922. Mixed cultures of pure strains of fibroblasts and epithelial cells. J. Exp. Med. 36, 285–289.

Edelson, J.D., Shannon, J.M., Mason, R.J., 1989. Effects of two extracellular matrices on morphologic and biochemical properties of human type II cells in vitro. Am. Rev. Respir. Dis. 140, 1398–1404.

Evans, V.J., Shannon Jr., J.E., Bryant, J.C., Waltz, H.K., Earle, W.R., Sanford, K.K., 1953. A quantitative study of the effect of certain chemically defined media on the proliferation in vitro of strain L cells from the mouse. J. Natl. Cancer Inst. 13, 773–784.

Frisch, A.W., Jentoft, V., BAarger, R., Losli, E.J., 1955. A human epithelium-like cell (maben) derived from an adenocarcinoma of lung; isolation, continuous propagation, and effects of selected viruses. Am. J. Clin. Pathol. 25, 1107–1112.

Fulcher, M.L., Gabriel, S., Burns, K.A., Yankaskas, J.R., Randell, S.H., 2005. Well-differentiated human airway epithelial cell cultures. Methods Mol. Med. 107, 183–206.

Fulcher, M.L., Randell, S.H., 2013. Human nasal and tracheo-bronchial respiratory epithelial cell culture. Methods Mol. Biol. 945, 109–121.

Garcia, O., Carraro, G., Navarro, S., Bertoncello, I., McQualter, J., Driscoll, B., Jesudason, E., Warburton, D., 2012. Cell-based therapies for lung disease. Br. Med. Bull. 101, 147–161.

Geppert, E.F., Williams, M.C., Mason, R.J., 1980. Primary culture of rat alveolar type II Cells on floating collagen membranes. Morphological and biochemical observations. Exp. Cell. Res. 128, 363–374.

Gey, G.O., 1933. An improved technic for massive cultures. Am. J. Cancer 17, 752–756.

Goldman, W.E., Baseman, J.B., 1980a. Glycoprotein secretion by cultured hamster trachea epithelial cells: a model system for in vitro studies of mucus synthesis. In Vitro 16, 320–329.

Goldman, W.E., Baseman, J.B., 1980b. Selective isolation and culture of a proliferating epithelial cell population from the hamster trachea. In Vitro 16, 313–319.

Gonzales, L.W., Angampalli, S., Guttentag, S.H., Beers, M.F., Feinstein, S.I., Matlapudi, A., Ballard, P.L., 2001. Maintenance of differentiated function of the surfactant system in human fetal lung type II epithelial cells cultured on plastic. Pediatr. Pathol. Mol. Med. 20, 387–412.

Gonzalez, R.F., Dobbs, L.G., 2013. Isolation and culture of alveolar epithelial Type I and Type II cells from rat lungs. Methods Mol. Biol. 945, 145–159.

Gray, T.E., Guzman, K., Davis, C.W., Abdullah, L.H., Nettesheim, P., 1996. Mucociliary differentiation of serially passaged normal human tracheobronchial epithelial cells. Am. J. Respir. Cell. Mol. Biol. 14, 104–112.

Greenleaf, R.D., Mason, R.J., Williams, M.C., 1979. Isolation of alveolar type II cells by centrifugal elutriation. In Vitro 15, 673–684.

Gruenert, D.C., Finkbeiner, W.E., Widdicombe, J.H., 1995. Culture and transformation of human airway epithelial cells. Am. J. Physiol. 268, L347–L360.

Ham, R.G., 1965. Clonal growth of mammalian cells in a chemically defined, synthetic medium. Proc. Natl. Acad. Sci. U. S. A. 53, 288–293.

Harris Jr, J.E., Shin, J., Lee, B., Pelosky, K., Hooker, C.M., Harbom, K., Hulbert, A., Zahnow, C., Yang, S.C., Baylin, S., Brayton, C., Brock, M.V., 2011. A murine xenograft model of spontaneous metastases of human lung adenocarcinoma. J. Surg. Res. 171, e75–e79.

Harrison, R.G., 1907. Observations on the living developing nerve fiber. Anat. Rec. 1, 116–118.

Heckman, C.A., Marchok, A.C., Nettesheim, P., 1978. Respiratory tract epithelium in primary culture: concurrent growth and differentiation during establishment. J. Cell. Sci. 32, 269–291.

Hoch-Ligeti, C., Hobbs, J.P., 1958. Behavior of explants from human adult bronchial epithelium in vitro. Proc. Soc. Exp. Biol. Med. 97, 59–62.

Hook, G.E., Brody, A.R., Cameron, G.S., Jetten, A.M., Gilmore, L.B., Nettesheim, P., 1987. Repopulation of denuded tracheas by Clara cells isolated from the lungs of rabbits. Exp. Lung Res. 12, 311–329.

Hoorn, B., 1964. Respiratory viruses in model experiments. Acta Otolaryngol. Suppl. 188 (Suppl.).

Jorissen, M., Van Der Schueren, B., van den Berghe, H., Cassiman, J.J., 1989. The preservation and regeneration of cilia on human nasal epithelial cells cultured in vitro. Arch. Otorhinolaryngol. 246, 308–314.

Jungraithmayr, W., Jang, J.H., Schrepfer, S., Inci, I., Weder, W., 2013. Small animal models of experimental obliterative bronchiolitis. Am. J. Respir. Cell. Mol. Biol. Jun. 48 (6), 675–684.

Kaartinen, L., Nettesheim, P., Adler, K.B., Randell, S.H., 1993. Rat tracheal epithelial cell differentiation in vitro. In Vitro Cell Dev. Biol. Anim. 29A, 481—492.

Kawada, H., Shannon, J.M., Mason, R.J., 1988. Improved maintenance of adult rat alveolar type II cell differentiation in vitro: effect of hydrocortisone and cyclic AMP. Biochim. Biophys. Acta 972, 152—166.

Kawada, H., Shannon, J.M., Mason, R.J., 1990. Improved maintenance of adult rat alveolar type II cell differentiation in vitro: effect of serum-free, hormonally defined medium and a reconstituted basement membrane. Am. J. Respir. Cell. Mol. Biol. 3, 33—43.

Khoufache, K., Cabaret, O., Farrugia, C., Rivollet, D., Alliot, A., Allaire, E., Cordonnier, C., Bretagne, S., Botterel, F., 2010. Primary in vitro culture of porcine tracheal epithelial cells in an air-liquid interface as a model to study airway epithelium and *Aspergillus fumigatus* interactions. Med. Mycol. 48, 1049—1055.

Kikkawa, Y., Yoneda, K., 1974. The type II epithelial cell of the lung. I. Method of isolation. Lab. Invest. 30, 76—84.

Kim, K.C., Opaskar-Hincman, H., Bhaskar, K.R., 1989. Secretions from primary hamster tracheal surface epithelial cells in culture: mucin-like glycoproteins, proteoglycans, and lipids. Exp. Lung Res. 15, 299—314.

Kondo, M., Finkbeiner, W.E., Widdicombe, J.H., 1993. Cultures of bovine tracheal epithelium with differentiated ultrastructure and ion transport. In Vitro Cell Dev. Biol. 29A, 19—24.

Kotton, D.N., 2012. Next-generation regeneration: the hope and hype of lung stem cell research. Am. J. Respir. Crit. Care Med. 185, 1255—1260.

Kwon, M.C., Berns, A., 2013. Mouse models for lung cancer. Mol. Oncol. 7, 165—177.

Laws, J.O., Flaks, A., 1966. Pulmonary adenomata induced by carcinogen treatment in organ culture. Br. J. Cancer 20, 550—554.

Lazarus, A., Del-Moral, P.M., Ilovich, O., Mishani, E., Warburton, D., Keshet, E., 2011. A perfusion-independent role of blood vessels in determining branching stereotypy of lung airways. Development 138, 2359—2368.

Lechner, J.F., LaVeck, M.A., 1985. A serum-free method for culturing normal human bronchial epithelial cells at clonal density. J. Tiss. Cult. Meth. 9, 43—48.

Lechner, J.F., Haugen, A., Autrup, H., McClendon, I.A., Trump, B.F., Harris, C.C., 1981. Clonal growth of epithelial cells from normal adult human bronchus. Cancer Res. 41, 2294—2304.

Lechner, J.F., Haugen, A., McLendon, I.A., Pettis, E.W., 1982. Clonal growth of normal adult human bronchial epithelial cells in a serum-free medium. In Vitro 18, 633—642.

Leslie, C.C., McCormick-Shannon, K., Mason, R.J., Shannon, J.M., 1993. Proliferation of rat alveolar epithelial cells in low density primary culture. Am. J. Respir. Cell. Mol. Biol. 9, 64—72.

Li, X., Comellas, A.P., Karp, P.H., Ernst, S.E., Moninger, T.O., Gansemer, N.D., Taft, P.J., Pezzulo, A.A., Rector, M.V., Rossen, N., Stoltz, D.A., McCray Jr, P.B., Welsh, M.J., Zabner, J., 2012. CFTR is required for maximal transepithelial liquid transport in pig alveolar epithelia. Am. J. Physiol. Lung Cell. Mol. Physiol. 303, L152—L160.

Liu, J.Y., Nettesheim, P., Randell, S.H., 1994. Growth and differentiation of tracheal epithelial progenitor cells. Am. J. Physiol. Lung Cell. Mol. Physiol. 266, L296—L307.

Liu, X., Luo, M., Zhang, L., Ding, W., Yan, Z., Engelhardt, J.F., 2007. Bioelectric properties of chloride channels in human, pig, ferret, and mouse airway epithelia. Am. J. Respir. Cell. Mol. Biol. 36, 313—323.

Lwebuga-Mukasa, J.S., 1994. A Mn(2+)-enhanced, RGD-dependent adhesion technique for isolation of adult rat type II alveolar epithelial cells for immediate functional studies. Am. J. Respir. Cell. Mol. Biol. 10, 347—354.

Mason, R.J., Apostolou, S., Power, J., Robinson, P., 1990. Human alveolar type II cells: stimulation of DNA synthesis by insulin and endothelial cell growth supplement. Am. J. Respir. Cell. Mol. Biol. 3, 571—577.

Mason, R.J., Lewis, M.C., Edeen, K.E., McCormick-Shannon, K., Nielsen, L.D., Shannon, J.M., 2002. Maintenance of surfactant protein A and D secretion by rat alveolar type II cells in vitro. Am. J. Physiol. Lung Cell. Mol. Physiol. 282, L249—L258.

Mason, R.J., Williams, M.C., Greenleaf, R.D., Clements, J.A., 1977. Isolation and properties of type II alveolar cells from rat lung. Am. Rev. Respir. Dis. 115, 1015—1026.

Mason, R.J., Williams, M.C., Widdicombe, J.H., Sanders, M.J., Misfeldt, D.S., Berry, L.C., 1982. Transepithelial transport by pulmonary alveolar type II cells in primary culture. Proc. Natl. Acad. Sci. U. S. A. 79, 6033—6037.

Massey, T.E., Geddes, B.A., Forkert, P.G., 1987. Isolation of nonciliated bronchiolar epithelial (Clara) cells and alveolar type II cells from mouse lungs. Can. J. Physiol. Pharmacol. 65, 2368—2372.

Matsui, H., Grubb, B.R., Tarran, R., Randell, S.H., Gatzy, J.T., Davis, C.W., Boucher, R.C., 1998a. Evidence for periciliary liquid layer depletion, not abnormal ion composition, in the pathogenesis of cystic fibrosis airways disease. Cell 95, 1005—1015.

Matsui, H., Randell, S.H., Peretti, S.W., Davis, C.W., Boucher, R.C., 1998b. Coordinated clearance of periciliary liquid and mucus from airway surfaces. J. Clin. Invest. 102, 1125—1131.

Miller, D.J., 2004. Sydney Ringer; physiological saline, calcium and the contraction of the heart. J. Physiol. 555, 585—587.

Miller, M.F., Cohen, E.D., Baggs, J.E., Lu, M.M., Hogenesch, J.B., Morrisey, E.E., 2012. Wnt ligands signal in a cooperative manner to promote foregut organogenesis. Proc. Natl. Acad. Sci. U. S. A. 109, 15348—15353.

Myles, J.E., Geddes, B.A., Massey, T.E., 1989. Biotransformation activities in Clara and alveolar type II cells isolated from hamster lungs. Res. Commun. Chem. Pathol. Pharmacol. 66, 297—309.

Nedrud, J.G., Wu, R., 1982. Growth of murine tracheal epithelial cells in hormone supplemented medium on collagen substrates. Fed. Proc. 41.

Nettesheim, P., Barrett, J.C., 1984. Tracheal epithelial cell transformation: a model system for studies on neoplastic progression. Crit. Rev. Toxicol. 12, 215—239.

Neuberger, T., Burton, B., Clark, H., Van, G.F., 2011. Use of primary cultures of human bronchial epithelial cells isolated from cystic fibrosis patients for the pre-clinical testing of CFTR modulators. Methods Mol. Biol. 741, 39—54.

Oppenheimer, J.M., 1966. Ross Harrison's contributions to experimental embryology. Bull. Hist. Med. 40, 525—543.

Patel, B., Gauvin, R., Absar, S., Gupta, V., Gupta, N., Nahar, K., Khademhosseini, A., Ahsan, F., 2012. Computational and bioengineered lungs as alternatives to whole animal, isolated organ, and cell-based lung models. Am. J. Physiol. Lung Cell. Mol. Physiol. 303, L733—L747.

Pezzulo, A.A., Starner, T.D., Scheetz, T.E., Traver, G.L., Tilley, A.E., Harvey, B.G., Crystal, R.G., McCray Jr, P.B., Zabner, J., 2011. The air-liquid interface and use of primary cell cultures are important to recapitulate the transcriptional profile of in vivo airway epithelia. Am. J. Physiol. Lung Cell. Mol. Physiol. 300, L25—L31.

Pickles, R.J., McCarty, D., Matsui, H., Hart, P.J., Randell, S.H., Boucher, R.C., 1998. Limited entry of adenoviral vectors into well differentiated airway epithelium is responsible for inefficient gene transfer. J. Virol. 72, 6014–6023.

Radi, Z.A., Ackermann, M.R., 2004. Growth of differentiated ovine tracheal epithelial cells in vitro. J. Vet. Med. A Physiol. Pathol. Clin. Med. 51, 167–170.

Rieves, R.D., Goff, J., Wu, T., Larivee, P., Logun, C., Shelhamer, J.H., 1992. Airway epithelial cell mucin release: immunologic quantitation and response to platelet-activating factor. Am. J. Respir. Cell. Mol. Biol. 6, 158–167.

Robinson, C.B., Wu, R., 1993. Mucin synthesis and secretion by cultured tracheal cells: effects of collagen gel substratum thickness. In Vitro Cell Dev. Biol. Anim. 29A, 469–477.

Robinson, P.C., Voelker, D.R., Mason, R.J., 1984. Isolation and culture of human alveolar type II epithelial cells. Characterization of their phospholipid secretion. Am. Rev. Respir. Dis. 130, 1156–1160.

Rock, J., Konigshoff, M., 2012. Endogenous lung regeneration: potential and limitations. Am. J. Respir. Crit. Care Med. 186, 1213–1219.

Rous, P., Jones, F.S., 1916. A method for obtaining suspensions of living cells from the fixed tissues, and for the plating out of individual cells. J. Exp. Med. 23, 549–555.

Rowe, R.K., Brody, S.L., Pekosz, A., 2004. Differentiated cultures of primary hamster tracheal airway epithelial cells. In Vitro Cell Dev. Biol. Anim. 40, 303–311.

Rudnick, D., 1933. Developmental capacities of the chick lung in chorioallantoic grafts. J. Exp. Zool. 66, 125–154.

Sakamoto, T., Hirano, K., Morishima, Y., Masuyama, K., Ishii, Y., Nomura, A., Uchida, Y., Ohtsuka, M., Sekizawa, K., 2001. Maintenance of the differentiated type II cell characteristics by culture on an acellular human amnion membrane. In Vitro Cell Dev. Biol. Anim. 37, 471–479.

Scherer, W.F., Syverton, J.T., Gey, G.O., 1953. Studies on the propagation in vitro of poliomyelitis viruses. IV. Viral multiplication in a stable strain of human malignant epithelial cells (strain HeLa) derived from an epidermoid carcinoma of the cervix. J. Exp. Med. 97, 695–710.

Schnatwinkel, C., Niswander, L., 2013. Multiparametric image analysis of lung-branching morphogenesis. Dev. Dyn. 242, 622–637.

Schumann, B.L., Cody, T.E., Miller, M.L., Leikauf, G.D., 1988. Isolation, characterization, and long-term culture of fetal bovine tracheal epithelial cells. In Vitro Cell. Dev. Biol. 24, 211–216.

Schwab, U.E., Fulcher, M.L., Randell, S.H., Flaminio, M.J., Russell, D.G., 2010. Equine bronchial epithelial cells differentiate into ciliated and mucus producing cells in vitro. In Vitro Cell Dev. Biol. Anim. 46, 102–106.

Shannon, J.M., 1994. Induction of alveolar type II cell differentiation in fetal tracheal epithelium by grafted distal lung mesenchyme. Dev. Biol. 166, 600–614.

Shannon, J.M., Jennings, S.D., Nielsen, L.D., 1992. Modulation of alveolar type II cell differentiated function in vitro. Am. J. Physiol. 262, L427–L436.

Shannon, J.M., Mason, R.J., Jennings, S.D., 1987. Functional differentiation of alveolar type II epithelial cells in vitro: effects of cell shape, cell-matrix interactions and cell-cell interactions. Biochim. Biophys. Acta 931, 143–156.

Shannon, J.M., Nielsen, L.D., Gebb, S.A., Randell, S.H., 1998. Mesenchyme specifies epithelial differentiation in reciprocal recombinants of embryonic lung and trachea. Dev. Dyn. 212, 482–494.

Shannon, J.M., Pan, T., Nielsen, L.D., Edeen, K.E., Mason, R.J., 2001. Lung fibroblasts improve differentiation of rat type II cells in primary culture. Am. J. Respir. Cell. Mol. Biol. 24, 235–244.

Sime, A., McKellar, Q., Nolan, A., 1997. Method for the growth of equine airway epithelial cells in culture. Res. Vet. Sci. 62, 30–33.

Sorokin, S., 1961. A study of development in organ cultures of mammalian lungs. Dev. Biol. 3, 60–83.

Steimer, A., Laue, M., Franke, H., Haltner-Ukomado, E., Lehr, C.M., 2006. Porcine alveolar epithelial cells in primary culture: morphological, bioelectrical and immunocytochemical characterization. Pharm. Res. 23, 2078–2093.

Sugahara, K., Rubin, J.S., Mason, R.J., Aronsen, E.L., Shannon, J.M., 1995. Keratinocyte growth factor increases mRNAs for SP-A and SP-B in adult rat alveolar type II cells in culture. Am. J. Physiol. 269, L344–L350.

Tandler, B., Sherman, J., Boat, T.F., 1981. EDTA-mediated separation of cat tracheal lining epithelium. Am. Rev. Respir. Dis. 124, 469–475.

Terzaghi, M., Nettesheim, P., Williams, M.L., 1978. Repopulation of denuded tracheal grafts with normal, preneoplastic, and neoplastic epithelial cell populations. Cancer Res. 38, 4546–4553.

Tournier, F., Laoukili, J., Giuliani, I., Gendron, M.C., Guennou, C., Marano, F., 1998. Ciliated differentiation of rabbit tracheal epithelial cells in vitro. Eur. J. Cell. Biol. 77, 205–213.

Tyrrell, D.A., Bynoe, M.L., 1965. Cultivation of a novel type of common-cold virus in organ cultures. Br. Med. J. 1, 1467–1470.

Varsano, S., Basbaum, C.B., Forsberg, L.S., Borson, D.B., Caughey, G., Nadel, J.A., 1987. Dog tracheal epithelial cells in culture synthesize sulfated macromolecular glycoconjugates and release them from the cell surface upon exposure to extracellular proteinases. Exp. Lung Res. 13, 157–184.

Wang, J., Edeen, K., Manzer, R., Chang, Y., Wang, S., Chen, X., Funk, C.J., Cosgrove, G.P., Fang, X., Mason, R.J., 2007. Differentiated human alveolar epithelial cells and reversibility of their phenotype in vitro. Am. J. Respir. Cell. Mol. Biol. 36, 661–668.

Weller, N.K., Karnovsky, M.J., 1986. Improved isolation of rat lung alveolar type II cells. More representative recovery and retention of cell polarity. Am. J. Pathol. 122, 92–100.

Whitcutt, M.J., Adler, K., Wu, R., 1988. A biphasic chamber system for maintaining polarity of differentiation of cultured respiratory tract epithelial cells. In Vitro Cell Dev. Biol. 24, 420–428.

Williams, W.J., Larson, E., Phillips, T.L., 1959. A neoplastic connective tissue mast cell capable of continuous growth in tissue culture. J. Biophys. Biochem. Cytol. 6, 361–368.

Wilson, J.S., Steinkamp, J.A., Lehnert, B.E., 1986. Isolation of viable type II alveolar epithelial cells by flow cytometry. Cytometry 7, 157–162.

Wu, R., Smith, D., 1982. Continuous multiplication of rabbit tracheal epithelial cells in a defined, hormone-supplemented medium. In Vitro 18, 800–812.

Wu, R., Yankaskas, J., Cheng, E., Knowles, M.R., Boucher, R., 1985. Growth and differentiation of human nasal epithelial cells in culture. Serum-free, hormone-supplemented medium and proteoglycan synthesis. Am. Rev. Respir. Dis. 132, 311–320.

Wu, W., Zhang, W., Booth, J.L., Metcalf, J.P., 2012. Influenza A(H1N1) pdm09 virus suppresses RIG-I initiated innate antiviral responses in the human lung. PLoS ONE 7, e49856.

Xu, X., McCormick-Shannon, K., Voelker, D.R., Mason, R.J., 1998. KGF increases SP-A and SP-D mRNA levels and secretion in cultured rat alveolar type II cells. Am. J. Respir. Cell. Mol. Biol. 18, 168–178.

Yano, T., Mason, R.J., Pan, T., Deterding, R.R., Nielsen, L.D., Shannon, J.M., 2000. KGF regulates pulmonary epithelial proliferation and surfactant protein gene expression in adult rat lung. Am. J. Physiol. Lung Cell Mol. Physiol. 279, L1146–L1158.

You, Y., Brody, S.L., 2013. Culture and differentiation of mouse tracheal epithelial cells. Methods Mol. Biol. 945, 123–143.

You, Y., Richer, E.J., Huang, T., Brody, S.L., 2002. Growth and differentiation of mouse tracheal epithelial cells: selection of a proliferative population. Am. J. Physiol. 283, L1315–L1321.

Young, T.F., Thacker, E.L., Erickson, B.Z., Ross, R.F., 2000. A tissue culture system to study respiratory ciliary epithelial adherence of selected swine mycoplasmas. Vet. Microbiol. 71, 269–279.

Zeng, H., Goldsmith, C.S., Maines, T.R., Belser, J.A., Gustin, K.M., Pekosz, A., Zaki, S.R., Katz, J.M., Tumpey, T.M., 2013. Tropism and infectivity of influenza virus, including highly pathogenic avian H5N1 virus, in ferret tracheal differentiated primary epithelial cell cultures. J. Virol. 87, 2597–2607.

Zhang, F., Nielsen, L.D., Lucas, J.J., Mason, R.J., 2004. Transforming growth factor-beta antagonizes alveolar type II cell proliferation induced by keratinocyte growth factor. Am. J. Respir. Cell. Mol. Biol. 31, 679–686.

Zhang, L., Peeples, M.E., Boucher, R.C., Collins, P.L., Pickles, R.J., 2002. Respiratory syncytial virus infection of human airway epithelial cells is polarized, specific to ciliated cells, and without obvious cytopathology. J. Virol. 76, 5654–5666.

Zhang, Y., Yankaskas, J., Wilson, J., Engelhardt, J.F., 1996. In vivo analysis of fluid transport in cystic fibrosis airway epithelia of bronchial xenografts. Am. J. Physiol. 270, C1326–C1335.

Section II

Comparative Respiratory Physiology of the Normal Lung

Daniel L. Costa and Jeffrey S. Tepper

Chapter 12

Gas Exchange at Rest and during Exercise in Mammals

Peter D. Wagner, James H. Jones and Kim E. Longworth

Chapter Outline

1. INTRODUCTION

In this second edition, the scope of this chapter has been broadened to include consideration of gas exchange from air to mitochondria as an integrated transport system. As with pulmonary gas exchange per se, the principles are similar across the mammalian species, permitting a general, species-independent description. This new section will address how the lungs/chest wall, heart, blood, and muscles work together as a system to dictate the flow of O_2 from air to mitochondria. The changes in this edition to the pre-existing sections in the first edition are to illustrate principles with concrete examples and to provide some figures, of which there were none in the first edition. However, the basic architecture and content have not been disturbed as they depict well the basic gas exchange principles common to all mammals as well as the comparative abilities for respiration across the range of mammals, from tiny to enormous.

2. ABSOLUTE VERSUS ALLOMETRIC COMPARISONS IN PHYSIOLOGY

If comparisons between respiratory structures and functions in different species of mammals are to be meaningful, it is necessary to distinguish between comparisons that are made in absolute versus relative terms. *Absolute* comparisons directly relate the value of a structural or functional variable in one species to those in other species, such as comparing tidal volumes (ml vs. ml). The potential problem with making absolute comparisons among different species (or even individuals within a species) is that differences in body sizes inherently dictate that there will be large differences in the absolute values of many respiratory variables (e.g., volumes), whereas other variables are independent of body size (e.g., binding capacity of hemoglobin molecules with O_2). For some variables, direct comparison of their absolute values may be appropriate; for other variables, the effects of

Comparative Biology of the Normal Lung. http://dx.doi.org/10.1016/B978-0-12-404577-4.00012-6

differences in body size must be taken into account by expressing the variables in *relative*, usually *specific*, terms (i.e., *mass specific*, with the value divided by body mass, such as ml/kg; Royal Society, 1975).

Attempts to make meaningful comparisons among respiratory variables are further complicated by the fact that many respiratory functions, including the fundamental determinant of gas exchange, the whole body rate of oxygen consumption($\dot{V}O_2$),[1] vary in a regular manner with body size both at rest (Kleiber, 1932; Brody et al., 1934; Schmidt–Nielsen, 1984) and at maximal rates of oxygen consumption ($\dot{V}O_2max$) (Seeherman et al., 1981; Taylor and Jones, 1987; Taylor et al., 1988). These *allometric* variations in physiological function are typically described by a simple power function of the form

$$Y = a \cdot M^b \tag{1}$$

where Y is the variable being measured, a is a mass coefficient, M is body mass, and b is a mass exponent. For ease in making comparisons, allometric data are frequently logarithmically transformed, such as

$$\log Y = \log a + (b \log M), \tag{2}$$

allowing allometric relationships to be compared using linearized functions (Schmidt–Nielsen, 1984). When making comparisons among respiratory variables in different species of mammals, in some instances it may be more meaningful to compare absolute values, whereas in others it may be more useful to compare relative values, with the latter expressed as specific variables (divided by M) or as values that have been standardized in terms of the allometric effect (divided by M^b) due to differences in body size of the animals being compared. A number of useful allometric equations related to respiratory function have been compiled by Peters (1983), Calder (1984), Schmidt–Nielsen (1984) and others are presented in tables in chapters by Tepper and Costa (Chapter ##) and Hazari (Chapter ##).

3. DEMAND FOR OXYGEN AND FLUX THROUGH THE RESPIRATORY SYSTEM

Under steady-state conditions, oxidative metabolism supplies the mammalian body with essentially all of the metabolic energy needed to maintain the supply of high-energy phosphate compounds required to maintain homeostasis. For oxidative metabolism to continue at steady state, O_2 molecules must enter the body and pass through each of the serial transport steps of the respiratory system (ventilatory convection, pulmonary diffusion, circulatory convection, and peripheral tissue diffusion) at a rate that is exactly matched to the rate at which they are being consumed by oxidative phosphorylation on the inner mitochondrial membranes.[2] Transient changes in demand, such as what occurs from rest to exercise, may temporarily alter the equilibrium of the system and tap labile O_2 stores (e.g., O_2 bound to hemoglobin or myoglobin). However, at steady state, measurements of whole body O_2 consumption and CO_2 production (($\dot{V}O_2$) and ($\dot{V}CO_2$)) are accurate estimates of the rates at which O_2 and CO_2 molecules are moving across the pulmonary gas exchanger between ventilated gas and circulating blood. Whether the body's global demand for O_2 results from a high rate of oxidative metabolism in a small mass of tissue or a lower rate of metabolism in a larger mass, the functional demand on the lung is the same. Measurement of the rate of whole body aerobic metabolism at steady state, or ($\dot{V}O_2$), is therefore a direct measure of the rate at which the pulmonary gas exchanger is functioning to meet the body's demand.

When the demand for O_2 that is set in the peripheral tissues increases, gas exchange rates at each of the respiratory system's transport steps must adjust proportionally in order to maintain the supply of O_2 to the tissues. Dejours (1975) proposed that the O_2 transport system functions as a cascade delivery system, in which the difference in the concentration of O_2 at the top of the cascade (ambient air) and the bottom (mitochondrial inner membranes) drives the flux of O_2 through the system (and the reverse for CO_2). Because the partial pressure of O_2 at the mitochondrion is thought to be always near 0 torr (Weibel, 1984b),[3] the difference between ambient and mitochondrial O_2 concentrations is maintained at an essentially constant value, even as flux through the system changes dramatically. This change in flux with a constant absolute driving potential is accomplished by adjustments in the conductances of diffusion barriers within the O_2 transport system, as well as localized increases in O_2 and CO_2 partial pressure differences driving diffusion across individual barriers.

This chapter describes the quantitative mechanisms by which changes in O_2 and CO_2 transport across the pulmonary gas exchanger occur, whether the changes in specific ($\dot{V}O_2$) result from differences in body size or from

1. Rates of O_2 consumption and CO_2 production are properly expressed in terms of quantity of gas per unit time, i.e., M_{O2}, where M is the quantity of gas in moles (Piiper et al., 1971). However, because of the traditional expression of gas flux in energy metabolism in terms of the volume of gas, corrected to STPD conditions, the traditional symbol, VO_2 is used in this chapter. The two units are interconvertible using the ideal gas law, such that 22.39 ml [STPD] = 1 mmol (although CO_2 deviates slightly from being an ideal gas).

2. In oxidative metabolism, consumption of 1 ml [STPD] O_2 liberates approximately 20.083 J energy (Schmidt–Nielsen, 1990); 1 ml [STPD] O_2 equals 4.4663×10^{-2} mmol O_2; thus, an O_2 consumption rate of 1 ml [STPD] O_2/s = 20.1 W, and a VO_2 of 1 mmol O_2/s ≈ 450 W.

3. The proper SI unit for pressure is the kPa, which is equivalent to 7. 50 torr (Mechtly, 1973). Because of the traditional expression of gas partial pressures in torr, it is used in this chapter for clarity.

activity during exercise. Although the discussion will be limited to pulmonary gas exchange, it should be noted at the outset that the function of the lung as a gas exchanger is intimately influenced by changes in the function of the peripheral gas exchanger, where the demand for O_2 is set. Increased extraction of O_2 from capillary blood and increased specific cardiac output, resulting in decreased time for gas exchange in the pulmonary capillaries, increase the integrated partial pressure difference for O_2 (and CO_2) across the alveolar-capillary membrane and, consequently, the rates of pulmonary gas transfer. Therefore, adjustments within the pulmonary gas exchanger to meet the body's demands for higher gas flux must be recognized as only one component of an integrated O_2 delivery system, which is influenced by every component in the system (Jones and Karas, 1988).

4. MEASURING RATES OF WHOLE BODY OXYGEN CONSUMPTION

Indirect calorimetry techniques for measuring whole body ($\dot{V}O_2$) fall into two general categories: *closed-circuit* and *open-circuit* systems. Closed-circuit and open-circuit indirect calorimetry systems and techniques have been extensively reviewed by Kleiber (1975), McLean and Tobin (1987), and Kaufmann et al. (1989).

Closed-circuit systems require that an animal be placed in a gas-tight chamber filled with air or a gas mixture containing O_2, and the CO_2 produced by the animal is absorbed. The volume of O_2 consumed per unit time is measured by the decrease in pressure in the chamber as the CO_2 is absorbed, or by the volume of O_2 that must be added to the chamber to maintain gas pressure. The earliest application of this approach was Lavoisier's (1780) bell-jar respirometer, which was greatly improved by Regnault and Reiset (1849), who passed the gas in the chamber over a CO_2 absorber and measured the volume of O_2 added to maintain the pressure in the chamber. Further modifications led to devices, such as the well-known Krogh (1913) respirometer for measuring the metabolic rate in humans, in which the volume of gas respired is measured on a kymograph on a breath-to-breath basis, and the rate at which the volume of gas in the system decreases (following CO_2 absorption) is equal to the subject's rate of O_2 consumption.

Open-circuit systems do not require that the animal be kept in a closed chamber. Instead, the animal is placed in a box or wears a loose-fitting mask through which ambient air, or a gas of controlled composition, flows past the animal's face and expired gas is collected into a vacuum line in which it mixes with the gas (usually air) that constitutes the bias flow. The ($\dot{V}O_2$) of the animal is calculated either from the flow rate through the system and the decrease in O_2 concentration in the expired gas line, or else by introducing a calibrated flow of diluent gas into the system to achieve the same decrease in O_2 concentration as was caused by the animal. When an open-circuit system is used in exercise studies, it is particularly important that the resistance at the mask inlet be minimized in order to keep the pressure drop at the animal's face as low as possible (ideally <1 cm H_2O) so that the animal does not have to inspire against a significant subatmospheric pressure.

One of the earliest open-circuit systems was designed by Pettenkofer (1862), who pumped air out of a chamber that housed a subject, measured the flow rate, and captured and measured the amount of CO_2 leaving the chamber, allowing ($\dot{V}CO_2$) to be calculated. This procedure was modified by Tigerstedt (1906) to allow continuous gas analysis of the air flowing out of the chamber, permitting both ($\dot{V}O_2$ and $\dot{V}CO_2$) to be measured. Haldane (1892) used a gravimetric apparatus consisting of a series of traps for H_2O and CO_2 through which expired gas from the animal flowed, allowing him to measure pre- and post-experimental mass changes in the animal and the absorbents, demonstrating the mass of CO_2 and H_2O produced by the animal and allowing the mass of O_2 consumed to be calculated.

A partial open-circuit system was developed by Douglas (1911), in which a subject wearing a mouthpiece (or mask for animals) breathed through a two-way valve, inhaling ambient air and expiring into a diffusion-resistant bag. After a measured interval of gas collection, the contents of the bag could be emptied through a volume meter and the composition of the gas analyzed, allowing $\dot{V}O_2$ and $\dot{V}CO_2$ to be calculated. A potential problem with this technique for use in exercise studies is that the subject must breathe against the increased impedance of the valve(s) in the system, and this can bias the ($\dot{V}O_2$) measured in some subjects. Shephard (1955) has critically reviewed this problem.

Subsequent developments in open-circuit techniques include monitoring gas compositions in real time with rapidly responding gas analyzers (Depocas and Hart, 1957; Tucker, 1968; Withers, 1977) while measuring H_2O vapor concentration and the flow rate of gas through the line. Water vapor can be frozen out of gas samples or collected on an absorbent and weighed, or measured directly with a dew-point hygrometer. Application of these techniques requires the use of calibrated gas analyzers and a flow transducer calibrated for flow rates in the bias line (approximately 10 times the animal's minute ventilation), which can be quite high for large mammals, particularly during exercise. These values are then used to calculate the volumes of CO_2 and H_2O added by the animal to the flow, after which the quantity of O_2 removed can be determined.

The *N_2-dilution technique* (Fedak et al., 1981) simplifies the open-circuit approach to measuring $\dot{V}O_2$ and $\dot{V}CO_2$. In this technique, measurements of flow rate through the box or mask and the concentration of O_2 in the expired gas line (ΔC) are made with the animal connected to the system. Then, with the animal removed, a calibrated flow of N_2 ($\dot{V}^*_{N_2}$) is introduced at the box or mask to cause an identical deflection in the expired O_2 concentration during calibration (ΔC^*) at the same expired gas flow rate as when the animal was connected:

$$\dot{V}O_2 = 0.02094 \cdot \frac{\dot{V}^*_{N_2}}{0.8} \cdot \frac{\Delta C^*}{\Delta C} \qquad (3)$$

This equation applies to the situation in which H_2O and CO_2 are absorbed from the sample line prior to entering the O_2 analyzer. Unlike the other open-circuit techniques, the N_2-dilution technique does not require the use of a calibrated O_2 analyzer, as long as it responds linearly and has identical output when the gas compositions are the same during measurement and calibration. Similarly, the flow transducer for the expired gas line need not be calibrated, as long as its output permits the flow through the system during calibration to be matched to that when the animal was connected. The latter point is particularly helpful when measuring ($\dot{V}O_2$) in large exercising animals, as the bias flow in the system must be approximately 5–10 times the animal's minute ventilation to prevent rebreathing, necessitating flows as high as 10^5 l/min for horses (Jones et al., 1989a), which can be difficult to accurately calibrate. Using the N_2-dilution technique, it is only necessary to calibrate the flow rate of N_2 diluent into the system, which is only 2–4% of the flow in the expired line and is within the range of many accurate and widely available flow transducers. The accuracy of the N_2-dilution technique for ($\dot{V}O_2$) measurements is $\pm 3\%$ and avoids the potential for multiplicative errors that can occur with methods in which a number of calibrated analyzers and transducers are required. Furthermore, the technique is not biased by leaks in the flow system, as long as the leaks occur at a constant rate, are of constant gas composition, and are of the same magnitude during calibration as when the animal is attached to the system.

An additional technique that has been used to measure ($\dot{V}O_2$) in mammals that are anesthetized, at rest, or exercising, is the application of the Fick principle (Fick, 1870). In this method, no attempt is made to measure ventilation or concentrations of gases ventilated. Instead, cardiac output ($\dot{Q}_T$) is measured directly with dye or thermal dilution, or with an electromagnetic or Doppler flowmeter placed on the aorta, and the concentrations of O_2 in the arterial (CaO_2) and mixed venous ($C\bar{v}O_2$) blood are measured. From these values, $\dot{V}O_2$ is calculated:

$$\dot{V}O_2 = \dot{Q}_T \cdot (CaO_2 - C\bar{v}O_2) \qquad (4)$$

This approach has been successfully used to obtain $\dot{V}O_2$ measurements from chronically instrumented animals during exercise (Ordway et al., 1984; Musch et al., 1985). Caution must be taken in applying this technique to ensure that the method used to measure cardiac output and blood O_2 concentrations are accurate. This is a problem when analyzing blood from different mammalian species, as some instruments designed for human blood (e.g., CO oximeters or hemoximeters) may not give accurate readings with nonhuman blood. Such instruments must be validated by tonometry against reference techniques, such as the $Fe(CN)_6$ method (Tucker, 1967), a galvanic cell (Grubb and Mills, 1981), or ideally, a direct manometric technique (Van Slyke and Neill, 1924; Natelson, 1951). For a review of these methods, see Wells and Weber (1989).

4.1 Resting Rates of Oxygen Consumption and Body Size

Nineteenth century physiologists recognized that the basal, or *standard*, rates of energy metabolism[4] in mammals are correlated with body size. Sarrus et Rameaux (1839) proposed that metabolic rate varies in direct proportion to the surface area of an animal ($\propto M^{\frac{2}{3}}$, its surface for heat loss), thus stating the so-called Surface Law. Regnault and Reiset's (1849) data suggested that smaller animals have higher specific rates of metabolism $\dot{V}O_2/M$, as Bergmann and Leuckart (1855) concluded. Rubner (1883) and Richet (1889) empirically demonstrated that specific metabolism is directly related to body surface area (i.e., proportional to $M^{\frac{2}{3}}$) intraspecifically in dogs and rabbits, respectively. Voit (1901) extended this argument to suggest that the same relationship exists interspecifically as well.

Kleiber (1932) analyzed the relationship between body size and standard metabolism with measurements made in 10 mammals and 3 birds across a greater size range, from rat (0.15 kg) to steer (679 kg). He found the allometric relationship between metabolic rate and M to have a mass exponent of 0.74, which was significantly different from the 0.67 exponent predicted by the Surface Law. A short time later, Brody et al. (1934) and Benedict (1938) published measurements of metabolic rates and body sizes of larger numbers of species spanning even greater size ranges, and confirmed that the interspecific mass exponent for standard metabolism (and $\dot{V}O_2$) is approximately 0.73–0.75. Kleiber (1961) rounded the exponent to 0.75 (which is not statistically distinguishable from the exponents derived from any of these empirical data) in order to facilitate calculation of the value as proportional to $M^{\frac{3}{4}}$ versus $M^{\frac{2}{3}}$ predicted by the Surface Law.

Measurements of standard metabolism and O_2 consumption for numerous species of mammals are tabulated in the above references, and values for a number of species are

4. The standard rate of energy metabolism is defined as the rate of energy turnover ($\propto \dot{V}O_2$) when the animal is in its thermoneutral zone, post-absorptive, and doing no external work on the environment.

in Table 1. Table 2 shows maximal oxygen consumption rates, while allometric equations summarizing the findings of several major comparative studies of standard and maximal rates of O_2 consumption are presented in Table 3.

It should be noted that not all investigators agree with the interpretation that interspecific comparisons yield a

TABLE 1 Values of Standard (Basal) Oxygen Consumption for Representative Mammalian Species of Different Body Mass

Species	M (kg)	$\dot{V}O_2$ std (ml [STPD] O_2/(s kg))	Reference
Horse *Equus caballus*	703	0.040	Benedict, 1938
Ox *Bos taurus*	700	0.031	Brody, 1945
Pig *Sus scrofa*	122	0.048	Benedict, 1938
Human *Homo sapiens*	71.5	0.058	Altman and Dittmer, 1971
Sheep *Ovis aries*	51.5	0.071	Altman and Dittmer, 1971
Goat *Capra hircus*	36	0.054	Benedict, 1938
Dog *Canis familiaris*	13.2	0.026	Altman and Dittmer, 1971
Rabbit *Oryctolagus cuniculus*	4	0.183	Harkness and Wagner, 1983
Rhesus monkey *Macaca mulatta*	3.2	0.034	Altman and Dittmer, 1971
Cat *Felis catus*	3	0.122	Benedict, 1938
Guinea pig *Cavia porcellus*	0.900	0.230	Harkness and Wagner, 1983
Rat *Rattus norvegicus*	0.375	0.247	Harkness and Wagner, 1983
Hamster *Mesocricetus auratus*	0.115	0.278	Harkness and Wagner, 1983
Gerbil *Meriones unguiculatus*	0.090	0.389	Harkness and Wagner, 1983
Mouse *Mus musculus*	0.030	0.528	Harkness and Wagner, 1983

TABLE 2 Maximal Oxygen Consumption for Mammals

Species	*M* (kg)	$\dot{V}O_2$ max (ml [STPD] O_2/(s kg))	Reference
Horse, thoroughbred *Equus caballus*	466	2.76	Birks et al., 1991
Horse, standardbred *E. caballus*	453	2.23	Jones et al., 1989a
Ox, steer *Bos taurus*	449	0.85	Jones et al., 1989a
Eland *Taurotragus oryx*	240	0.60	Taylor et al., 1981
Zebu ox *B. indicus*	232	0.49	Taylor et al., 1981
Pony *E. caballus*	171	1.48	Taylor et al., 1987b
Hereford calf *B. taurus*	141	0.61	Taylor et al., 1987b
Waterbuck *Kobus deffassa*	109.8	0.79	Taylor et al., 1981
Pony *E. caballus*	105.0	1.61	Seeherman et al., 1981
Wildebeest *Connochaetes taurinus*	102.0	0.73	Taylor et al., 1981
Human, untrained *Homo sapiens*	69.8	0.81	Williams et al., 1986
Human, trained *H. sapiens*	63.9	1.18	Williams et al., 1986
Goat *Capra hircus*	30.0	0.95	Taylor and Jones, 1987
Lion *Panthera leo*	30.0	1.00	Seeherman et al., 1981
Dog *Canis familiaris*	28.0	2.29	Taylor and Jones, 1987
Wolf *C. lupus*	27.6	2.60	Weibel et al., 1983
Dog *C. familiaris*	25.3	2.67	Weibel et al., 1983
African sheep *Ovis aries*	21.8	0.78	Taylor et al., 1981

Continued

TABLE 2 Maximal Oxygen Consumption for Mammals—cont'd

Species	*M* (kg)	$\dot{V}O_2$ max (ml [STPD] O_2/(s kg))	Reference
Dog *C. familiaris*	21.0	2.64	Seeherman et al., 1981
African goat *Capra hircus*	20.9	0.87	Taylor et al., 1981
Pig *Sus scrofa*	18.5	1.56	Seeherman et al., 1981
Coyote *Canis latrans*	12.4	3.10	Weibel et al., 1983
Grant's gazelle *Gazella granti*	10.1	0.89	Taylor et al., 1981
Red fox *Vulpes vulpes*	4.61	2.89	Weibel et al., 1983
Blue fox *Alopex lagopus*	4.40	3.62	Longworth et al., 1989
Dik–dik *Madoqua kirkii*	4.2	0.91	Taylor et al., 1981
Spring hares *Pedetes capensis*	3.00	1.62	Seeherman et al., 1981
Genet cat *Genetta tirina*	1.4	1.8	Taylor et al., 1981
Banded mongoose *Mungos mungo*	1.14	1.9	Taylor et al., 1981
Rat kangaroo *Bettongia penicillata*	1.10	2.95	Seeherman et al., 1981
Guinea pig *Cavia porcellus*	0.841	1.12	Pasquis et al., 1970
Dwarf mongoose *Helogale pervula*	0.58	2.1	Taylor et al., 1981
White rat *Rattus norvegicus*	0.376	1.51	Pasquis et al., 1970
White rat, untrained *R. norvegicus*	—	1.21	Conley et al., 1985
White rat, trained *R. norvegicus*	—	1.56	Conley et al., 1985
White rat *R. norvegicus*	0.205	1.61	Seeherman et al., 1981
Hamster *Mesocricetus auratus*	0.100	1.97	Pasquis et al., 1970
Chipmunk *Tamias striatus*	0.0902	3.97	Seeherman et al., 1981
Merriam's chipmunk *Eutamias merriami*	0.075	1.96	Lechner, 1978
Lemming *Dicrostonyx groenlandicus*	0.061	2.05	Lechner, 1978
Mouse *Mus musculus*	0.033	2.89	Pasquis et al., 1970
Deer mouse *Peromyscus maniculatus*	—	2.95	Chappell, 1984
European wood-mouse *Apodemus sylvaticus*	0.020	4.4	Hoppeler et al., 1984
Pygmy mouse *Baiomys taylori*	0.0072	4.36	Seeherman et al., 1981

Note: Values of maximal oxygen consumption for mammals of different body mass measured with animals running on a motorized treadmill.

Continued

mass exponent of 3/4. In particular, Heusner (1982a,b) has argued that the intraspecific mass exponent of 2/3 is maintained within species across the size range and that interspecific data, therefore, represent comparisons of groups of data with different mass coefficients (a in Eqns (1) and (2)). Feldman and McMahon (1983) have responded that—regardless of whether the relationship is viewed as a series of intraspecific regressions of smaller size range, each with a slope of 2/3 and with a increasing as $M^{0.008}$ between species, or as a relationship with constant a and a slope of 3/4—the functional result is the same and is *not* a simple Surface Law relationship between species. A further suggestion that the "true" value of the mass exponent for interspecific standard metabolism is approximately 3/4 is the repeated observation that physiological frequencies that support the transport of O_2 through the respiratory system at rest, such as ventilatory and cardiac frequencies, minute ventilation, and cardiac output, all vary as $M^{\frac{3}{4}}$ (Stahl, 1967; Schmidt–Nielsen, 1984).

The practical implication of the above discussion is that, in making any sort of comparison or extrapolation of physiological variables that are associated with oxidative metabolism and the O_2 transport system, it is imperative to consider the inherent differences in *physiological time*

TABLE 3 Allometry of O_2 Consumption and Flux

Taxon	Status	Intercept	Slope	N	*M* Range	References
Mammals +7 birds	Basal	10.2	0.734	19	0.016-3833	Brody, 1945
Mammals	Rest	10.7	0.739	10	0.173-679	Kleiber, 1932
Mammals	Rest	11.6	0.76			Stahl, 1967
Mammals	Std	9.8	0.756	12	0.021-600	Kleiber, 1961
Small mammals	Std	9.13	0.727	56	0.003-3.71	Lechner, 1978
Fossorial mammals	Basal	5.62	0.557	16	0.021-0.63	McNab, 1979
Monotremes	Rest	4.72	0.79	6	0.73-4.9	Konoplev et al., 1978
Dasyurid marsupials	Basal	6.75	0.74	12	0.008-5.05	MacMillen and Nelson, 1969
Marsupials	Std	6.94	0.737	8	0.014-32.49	Dawson and Hulbert, 1970
Marsupials	Rest	14.4	0.74	29	0.01-5.1	Konoplev et al., 1978
Insectivores (>25 g)	Rest	21.4	0.73	12	0.03-0.89	Konoplev et al., 1978
Chiropterans	Rest	18.8	0.73	17	0.005-0.59	Konoplev et al., 1978
Edentates	Rest	14.6	0.77	12	1.6-4.7	Konoplev et al., 1978
Lagomorphs	Rest	23.7	0.72	11	0.07-5.9	Konoplev et al., 1978
Rodents	Rest	24.9	0.69	133	0.007-0.96	Konoplev et al., 1978
Higher primates	Rest	37.6	0.73	10	0.23-71	Konoplev et al., 1978
Carnivores	Rest	30.0	0.73	24	0.17-26	Konoplev et al., 1978
Artiodactyls and perissodactyls	Rest	34.6	0.73	19	13-665	Konoplev et al., 1978
Mammals	Max	100	0.845	45	0.007-677	Taylor et al., 1981
Mammals	Max	107	0.854	19	0.020-469	Taylor et al., 1988
Mammals	Max	128	0.856	33	0.007-469	Calculated*
Mammals	Max	115	0.809	22	0.007-263	Taylor et al., 1981
Canids	Max	219	0.905	4	4.6-27.6	Weibel et al., 1983
Mammals	Max	54.0	0.678	14	0.003-2.54	Lechner, 1978
Mammals	Max	67.5	0.73	4	0.033-0.841	Pasquis et al., 1970

Note: Allometry of oxygen consumption ($\dot{V}O_2$, ml [STPD] O_2/min)) under resting (rest), basal (basal), standard (std), and maximal (max) conditions. $\dot{V}O_2$ is related to body mass (*M*, in kg) by the function: $\dot{V}O_2 = a \cdot M^b$ where a is the intercept of the log-transformed linear regression, and b is the slope of the log-transformed linear regression. N is the number of species. In some cases, regression equations were calculated from data given in the reference. Units have been converted for equivalence, where necessary.
*Values of these regressions derived from data in studies by Taylor et al. (1981); Weibel et al. (1983); Hoppeler et al. (1984); Taylor et al. (1987b); Jones et al. (1989); and Longworth et al. (1989).

($\propto M^{\frac{3}{4}}$) that exist among mammals simply due to differences in their body sizes (Lindstedt and Calder, 1981). Allometric effects on standard metabolism will ultimately affect nearly all variables associated with gas exchange (e.g., ventilatory frequency and minute ventilation), thus dictating that physiological responses on an *absolute time* scale (h, min, s, independent of *M*) are not physiologically equivalent for mammals of different size. These effects may severely bias the interpretation of data collected on an absolute time scale, such as exposure to a toxicant for 1 h (measured on an absolute time scale), when the animals being studied are of different *M* and, therefore, ventilating and receiving very different doses of the toxicant in relation to their physiological time scales.

4.2 Maximum Rates of Oxygen Consumption and Body Size

When a mammal exercises, it increases the turnover rate of high-energy phosphate compounds in its working muscle and increases the demand for O_2 in the mitochondria. This demand is passed through the serial steps of the respiratory system as an increase in the pressure differences driving diffusion of O_2 and CO_2, resulting in higher fluxes. With increasing exercise intensity, there is a linear increase in the rate of O_2 uptake (Margaria et al., 1933; Seeherman et al., 1981) to a certain point, beyond which $\dot{V}O_2$ remains constant at higher speeds. At intensities above that at which the rate of O_2 uptake plateaus, the turnover rate of high-energy phosphate continues to increase, but the source of metabolic energy that enables this increased turnover to occur is anaerobic glycolysis, with subsequent increases in the concentration of lactate in the muscle and blood. The maximum rate of O_2 consumption ($\dot{V}O_2max$) that a mammal can attain is empirically defined by the following:

1. A plateau in $\dot{V}O_2$ with increasing running speed
2. A respiratory quotient ($R = \dot{V}CO_2/\dot{V}O_2$) greater than 1.0
3. A net rate of lactate accumulation in the plasma sufficient to account for the increased metabolic energy demand beyond that supplied aerobically, with an accumulation rate of approximately 11 mmol/(kg min) being stoichiometrically equivalent (in terms of phosphorylation potential) to a $\dot{V}O_2$ of 1 ml [STPD] O_2/(s kg) (Seeherman et al., 1981).

If oxidative capacities of animals are to be compared, it is essential that true *maximal* rates of aerobic metabolism be elicited. Comparing submaximal rates of $\dot{V}O_2$ is not particularly meaningful, as the values chosen may represent different fractions of the animals' true maximum aerobic capacities. Because the maximum demand for O_2 that an animal's muscles can generate is, to some degree, influenced by how those muscles are recruited, the activity in which the animal is engaged may influence the magnitude of ($\dot{V}O_2max$) measured (Astrand and Rodahl, 1986). For instance, an animal trained for running may not recruit as large a muscle mass when swimming and would, therefore, have a lower ($\dot{V}O_2max$) while swimming than when engaged in the activity for which its muscles and respiratory system were trained. Similarly, cold exposure, frequently used to elicit high rates of aerobic metabolism in small mammals, yields significantly lower maximal rates of O_2 consumption than the same animals achieve when ($\dot{V}O_2max$) is measured during exercise (Seeherman et al., 1981; Hoppeler et al., 1984; Conley et al., 1985). Although many studies report maximum measurable specific muscle $\dot{V}O_2$ elicited during electrical stimulation of isolated perfused muscles, these values are at best less than one-half the calculated specific muscle $\dot{V}O_2$ in whole animals during conscious maximal aerobic exercise. Therefore, if the lung is to be studied at its maximal functional capacity, it is necessary to do so during exercise in an awake animal.

To date, the most comprehensive data set collected to compare maximum rates of O_2 consumption in mammals has been accumulated by Taylor and collaborators (Taylor et al., 1981, 1987a, 1988; Weibel et al., 1983; Hoppeler et al., 1984; Conley et al., 1985; Taylor and Jones, 1987; Jones et al., 1989a; Longworth et al., 1989). These data are summarized in Table 2 and allometric equations derived from them are in Table 3. One of the great strengths of this data set is that the measurements were all made by the same investigators using the same criteria for validating that ($\dot{V}O_2max$) had been obtained, thus reducing the variability inherent in comparing data gathered from different laboratories, each with its own biases. The original allometric analysis of these data for 22 cursorial species suggested that $\dot{V}O_2$ varied as $M^{0.81}$ (Taylor et al., 1981), a mass exponent not significantly different from Kleiber's 0.75 for standard $\dot{V}O_2$, indicating that all mammals have the capacity to increase $\dot{V}O_2$ approximately 10-fold above their resting rates. The addition of data from more species has increased the mass exponent for ($\dot{V}O_2max$) to 0.86 (Taylor et al., 1988), which is significantly different from 0.75, indicating that as M increases, mammals have greater *aerobic scopes*, or capacities to increase $\dot{V}O_2$ above resting rates. Because the mass exponents for both standard and maximum rates of O_2 consumption are less than 1, the data also indicate that smaller mammals, simply by virtue of their size, have higher specific rates of O_2 consumption $\dot{V}O_2/M$ at rest and maximal exercise than do larger mammals.

5. DIFFUSION OF GASES

Movement of O_2 through the respiratory system from ambient air to the site of oxidative phosphorylation on the inner mitochondrial membrane occurs by two transport processes: convection and diffusion. Convection, or bulk transport, occurs during ventilation of gas in and out of the thorax and circulation of blood between the lungs and peripheral tissues. It is defined as either the product of pump frequency (*f*) and pump volume (V), or else as the quotient of potential (ΔP) and resistance to flow (R):

$$\dot{V} = f \cdot V \frac{\Delta P}{R} \tag{5}$$

where $\dot{V}$ is the flow rate (volume/time) of the medium being pumped. The amount of a substance (e.g., O_2) delivered by convection is, therefore, the flow rate *x* ([concentration in] — [concentration out]), equivalent to Eqn (4).

Whether the transport of O_2 and CO_2 between alveolar gas and pulmonary capillary blood occurs due to active secretion or to passive diffusion alone was a question initially raised in disputes between Ludwig (1865) and Pfluger (1872, 1875). Bohr (1891, 1909) rekindled the argument and was joined by Haldane (1892) in arguing that O_2 and CO_2 were actively secreted, and Krogh (1919) and Barcroft et al. (1920) rebutted that passive diffusion alone could explain the transfer of respiratory gases. Although the current thinking is that alveolar-capillary O_2 transport occurs exclusively by molecular diffusion, some experiments have again raised the possibility that CO_2 may be actively transported to some degree (Jones et al., 1969; Gurtner et al., 1969; Robertson and Hlastala, 1977), and the controversy has yet to be fully resolved (Forster, 1977;

Gurtner, 1977; Scheid and Piiper, 1980; Scotto et al., 1984; Anthonisen and Fleetham, 1987). This topic has not been further pursued and most in the field continue to feel that molecular diffusion and not active transport is responsible for CO_2 exchange.

Molecular diffusion of O_2 and CO_2 occurs across the pulmonary gas exchanger that separates alveolar gas from pulmonary capillary blood, and the peripheral gas exchanger that separates capillary blood from mitochondria. The fundamental principles of simple binary diffusion were first delineated by Fick (1855) and have been reviewed by Forster (1964) and Chang (1987). Fick postulated that the mass flux of a substance in direction x (J_x) is directly related to its concentration gradient in the x direction (dC/dx) and a diffusion coefficient (D) that is a property of the medium through which diffusion is occurring:

$$J_x = -D \cdot \frac{\mathrm{d}C}{\mathrm{d}x} \tag{6}$$

For diffusion across a planar membrane, Fick expanded this relationship to the familiar form of his Second Law:

$$\frac{\mathrm{dQ}}{\mathrm{d}t} = -AK\frac{\partial P}{\partial x} \tag{7}$$

where dQ/dt is the instantaneous rate of gas flux (ml [STPD]/s) perpendicularly across a membrane of area A (cm^2), D is in cm^2/s, and flux is driven by a concentration gradient $\partial C/\partial x$, where C is in ml [STPD]/cm^3 and x is in cm. In order to make this relationship more useful in physiology, Krogh (1919) incorporated gas solubility (α) into the diffusion coefficient (D), creating a diffusion constant, K, that relates the flux of gas across a membrane to the difference in partial pressure of the gas (∂P) across that membrane:

$$\frac{\mathrm{dQ}}{\mathrm{d}t} = -A \cdot K \cdot \frac{\partial P}{\partial x} \tag{8}$$

This differential equation establishes the principle that net movement of a gas occurs from a region of higher partial pressure to one of lower partial pressure, and further, that the rate of such net movement at any instant is proportional to the partial pressure difference between the regions at that same point in time. However, the equation needs to be solved (i.e., integrated) if one is to know whether, within available red cell capillary transit time, diffusion equilibration has occurred—or not.

Two approaches to solving this equation for the respiratory gases have been applied; one is to approximate their dissociation curves by a straight line, the slope of which has been designated as "β" (Piiper and Scheid, 1981). This has the great advantage of allowing the equation to be solved algebraically, which in turn allows general principles to be stated. However, the obvious inaccuracy that this approximation requires means that for gases with distinctly nonlinear dissociation curves, O_2 in particular, the quantitative outcomes from using this approach are not accurate and must be interpreted accordingly. The second approach has been to use actual dissociation curves (Wagner and West, 1972) but employ forward numerical integration procedures starting with entry conditions (i.e., composition of mixed venous blood entering the pulmonary capillary) and numerically execute Eqn (8) in successive, tiny, time increments until the total time has reached the available transit time, at which point the calculations are stopped. This approach, using actual dissociation curves, allows realistic quantitative results, but the outcome is specific to the conditions and values of input variables and thus not generalizable. As a result, the two approaches are complementary and both offer value.

Equation (8) can be integrated over the time available for gas exchange. Using the first approach (i.e., assuming a linear O_2Hb dissociation curve), the result is as follows:

$$(\mathrm{P_A} - \mathrm{P_c}(t)) = (\mathrm{P_A} - \mathrm{P_v}) \cdot \exp\left(\frac{-D_\mathrm{L} \cdot t}{(\beta \cdot \mathrm{V_c})}\right) \tag{8a}$$

In this equation, P_A and P_v are partial pressures of the gas in the alveolar gas and mixed venous blood (considered constant in time), and $P_c(t)$ is partial pressure in the capillary blood at time t in the red cell's transit through the capillary starting at P_v when $t = 0$ and ending at the end-capillary (or arterial, usually taken to be the same) partial pressure at the end of the capillary, when $t = V_c/\dot{Q}_T$ (see Section 11). D_L is the total pulmonary diffusing capacity for the gas, β is, as stated above, the (constant) slope of the gas dissociation curve, Vc is the total volume of blood contained within the capillaries, and $\dot{Q}_T$ is total pulmonary blood flow rate.

If we wish to determine the end-capillary partial pressure, it is found by substituting $t = V_c/\dot{Q}_T$ (the total time spent by a red cell in the capillaries) into Eqn (8a), yielding Eqn (8b):

$$(\mathrm{P_A} - \mathrm{P_a}) = (\mathrm{P_A} - \mathrm{P_v}) \cdot \exp\left(\frac{-D_\mathrm{L}}{(\beta \cdot \dot{\mathrm{Q}}_\mathrm{T})}\right) \tag{8b}$$

where Pa is the end-capillary partial pressure, usually equated to arterial (hence the designation "a"). When using these equations to evaluate $P_c(t)$ (or P_a) for given values of P_A, P_v, D_L, β, and $\dot{Q}_T$, it is critical to use correct units. In particular, D_L must be in the same units as $\beta \cdot \dot{Q}_T$. D_L is usually expressed as ml gas/(min torr). If $\dot{Q}_T$ is expressed in liters blood/min, β must be expressed in ml gas per liter blood per torr partial pressure. Similarly, P_A, $P_c(t)$, P_a and P_v must all be expressed in the same units as one another—all torr or all kPa, for example.

Using this approach, capillary partial pressure can be calculated from beginning to end of the capillary and its time course profile plotted (Figure 1). In this particular calculation, normal resting human values for necessary

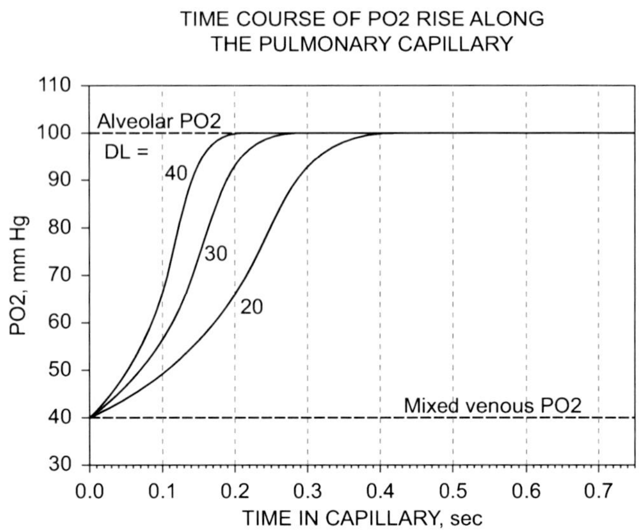

FIGURE 1 Time course of increase in capillary P_{O_2} from mixed venous (40 mmHg) at $t = 0$ to alveolar (100 mmHg) at $t = 0.75$ sec, which is the normal capillary transit time, determined from the ratio of capillary blood volume (75 ml) to cardiac output (6 l/min) in resting humans. Even if D_L is as low as 20 ml/(min mmHg), diffusive equilibration is essentially complete halfway down the capillary. Calculations are for resting, normoxic conditions, and use the actual O_2Hb dissociation curve.

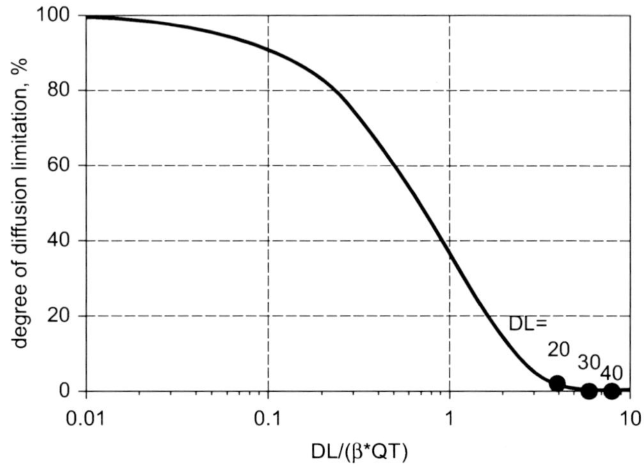

FIGURE 2 **Diffusion limitation as a function of the ratio ($D_L/\beta*Q_T$).** 100% means that there is no O_2 diffusive transport at all; 0% means complete equilibration of capillary with alveolar P_{O_2}. Note that in normal resting humans, over the range of D_L from 20 to 40 ml/min/mm Hg, this calculation, based on the linear approximation to the O_2Hb dissociation curve, predicts almost no diffusion limitation.

variables were used, and the second approach—i.e., using the actual O_2Hb dissociation curve and a numerical forward integration procedure—was applied, to avoid the inaccuracies of assuming a linear dissociation curve. It shows that normally there is complete capillary-alveolar diffusion equilibration by the first one-third to one-half of the capillary transit of the red cell.

If P_a is calculated and found to essentially equal P_A, as in Figure 1, we say that diffusion equilibration has occurred, and thus there is no limitation to gas exchange imposed by the diffusive process. On the other hand, if P_a is less than P_A, there is diffusion limitation of gas exchange, and gas flux across the alveolar membrane has been accordingly reduced.

The exponent $D_L/(\beta\cdot\dot{Q}_T)$ in Eqn (8b) was developed and explored in detail by Piiper and Scheid in the 1980s (Piiper and Scheid, 1981) and illustrates a critical, unifying principle shown in Figure 2. This figure shows the degree of diffusion limitation associated with different values of $D_L/(\beta\cdot\dot{Q}_T)$. Its beauty lies in its generality—whether diffusion limitation occurs or not is a function of this exponent—simply, the ratio of diffusive conductance across the alveolar membrane (D_L, ml/min/mm Hg) to perfusive conductance (β (ml/(l torr)))$\cdot\dot{Q}_T$ (l/min)) for the gas in blood. Absolute values of D_L, β, and $\dot{Q}_T$ are not per se important; it is the net value of the ratio. This makes for simple comparisons not just within and between species, but between different gases.

The three points shown on this curve (Figure 2) reflect the same values as shown in Figure 1. Comparing Figures 1 and 2 reveals an aspect important to O_2 diffusion in the lungs. In Figure 1, at $D_L = 20$ ml/(min torr), equilibration is complete well before the end of the capillary. However, Figure 2 indicates that for the same D_L, β, and $\dot{Q}_T$, there is about 2% failure of diffusive equilibration. This difference illustrates the approximation error inherent in assuming a linear O_2Hb dissociation curve, and further shows that the nonlinear nature of the curve actually enhances the rate of diffusive equilibration in the lungs. This happens because early in the capillary transit, gas exchange is occurring on the steep portion of the O_2Hb curve, allowing large increases in O_2 concentration in blood without much rise in P_{O_2}. The lower the capillary P_{O_2}, the larger the driving gradient for O_2, explaining the advantage of the O_2Hb curve shape.

When $D_L/\beta\cdot\dot{Q}_T$ is compared for O_2 and CO_2 at rest in normoxia, the following is found: D_L is about 20 times higher for CO_2 due to its 20-fold higher solubility in the alveolar membrane; β is about 10 times higher for CO_2, reflecting the steeper CO_2 than O_2 dissociation curve, and so $D_L/\beta\cdot\dot{Q}_T$ is about twice as high for CO_2 as for O_2, making it less vulnerable to diffusion limitation than O_2.

During exercise, the value of β for O_2 increases substantially because venous P_{O_2} falls onto the steep part of the O_2 dissociation curve, while arterial P_{O_2} moves little if at all. But for CO_2, its β changes little as the CO_2 curve is quasi-linear over a wide range. Both D_L (for O_2 and CO_2) and $\dot{Q}_T$ increase with exercise. Taken together, these changes cause the ratio $D_L/\beta\cdot\dot{Q}_T$ to fall for O_2 but change little for CO_2. Thus, O_2 is even more subject to diffusion limitation than is CO_2 during exercise. Exercise in hypoxia further increases β for O_2, but has little effect on CO_2, so that $D_L/\beta\cdot\dot{Q}_T$ falls still further for O_2 without real change

TABLE 4 Diffusion Limitation of O_2 Uptake, Predicted From the $D_L/(\beta \cdot \dot{Q}_T)$ Relationship, using Typical Normal Human Values of DL, β and $\dot{Q}_T$

Condition	DL	β (Figure 3)	$\dot{Q}_T$	$D_L/(\beta \cdot \dot{Q}_T)$	O_2 Exchange Reduction, %
Normoxia Rest	30	0.8	6	6.3	0.2
Normoxia Exercise	60	1.8	24	1.4	25
Hypoxia Rest	40	1.9	7	3.0	5
Hypoxia exercise	80	4.1	24	0.8	44

for CO_2, and O_2 diffusion limitation may be substantial (West and Wagner, 1980; Torre-Bueno et al., 1985; Wagner et al., 1987).

Table 4, paired with Figure 3, exemplifies these concepts for O_2 by using typical human values for D_L, β and $\dot{Q}_T$ at rest and during exercise in normoxia and hypoxia. It is clear that hypoxia and exercise individually, and especially together, are challenging in terms of reaching diffusion equilibration—even for normal healthy individuals. There is

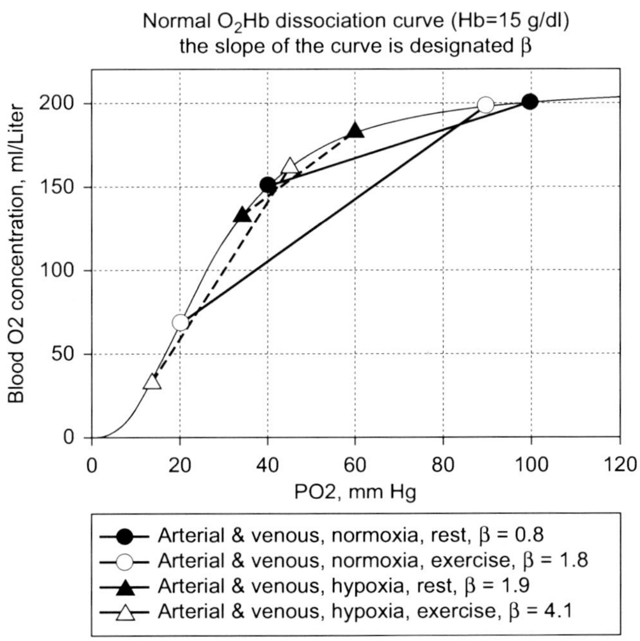

FIGURE 3 The normal O_2Hb dissociation curve showing typical values for both P_{O_2} and O_2 concentration in arterial and venous blood for rest and exercise in normoxia and hypoxia. The straight lines between arterial and venous pairs show the average slope of the curve (β) for each condition. Hypoxia and exercise each steepen the average slope; combined they act synergistically. Increases in β reduce the rate of diffusive equilibration.

abundant experimental evidence of rapid desaturation of Hb from rest to exercise, especially in hypoxia, but even in normoxia (Powers et al., 1989; Wagner et al., 1986).

As $D_L/(\beta \cdot \dot{Q}_T)$ would suggest, exceptional athletes with very high maximal cardiac output ($\dot{Q}_T$), but normal Hb (and thus normal β) and lungs that are usually not exceptional in alveolar surface area for diffusion (and thus normal or near-normal D_L) are especially at risk of diffusion limitation, and this has been observed experimentally (Powers et al., 1989).

Applying $D_L/(\beta \cdot \dot{Q}_T)$ concepts to carbon monoxide (CO) is instructive. Compared to O_2, D_L for CO is similar to that for O_2 because they have similar water solubilities and similar molecular weights, and of course in a given lung, ($\dot{Q}_T$) is the same when computing $D_L/(\beta \cdot \dot{Q}_T)$ for each gas. However, the slope of the COHb dissociation curve in its steep portion is some 240 times greater than that for O_2—the P_{50} is only about 0.1 torr. We must use the steep portion in this discussion, because if Hb were fully saturated with CO, life would not be possible. As a result, β is about 240 times greater for CO than for O_2, and $D_L/(\beta \cdot \dot{Q}_T)$ is thus about 240 times smaller for CO than for O_2. This, in turn, means that CO exchange across the alveolar membrane is always diffusion-limited, even at rest. Indeed, this is precisely why CO is used as the gas to measure diffusing capacity.

Finally, from $D_L/(\beta \cdot \dot{Q}_T)$, we can predict that inert gases, that are carried in blood only in physical solution, would never experience diffusion limitation. This is because for any one such gas, its solubility in blood (i.e., β) and in the alveolar membrane (which determines D_L) are fairly similar. For O_2 on the other hand (in normoxia, at rest), its "solubility" in blood (β as defined previously) is some thirty-fold higher than its solubility in water (determining D_L). Thus, $D_L/(\beta \cdot \dot{Q}_T)$ is about 30 times higher for inert gases than for O_2, and this means that by the time observable diffusion limitation occurred for an inert gas, O_2 diffusive exchange would have been reduced to levels incapable of supporting life. This difference only increases during exercise because β increases for O_2 (Figure 3), but stays essentially constant for inert gases.

Values for diffusion constants and coefficients of pertinent substances and tissues are in Table 5. For practical purposes in assessing the diffusion of gases across the alveolar-capillary membrane in the lung, the key principles expressed by this relationship are as

1. The flux of a gas across a membrane is directly proportional to the surface area of the membrane.
2. The flux of a gas across a membrane is inversely proportional to the thickness of the membrane.
3. The flux of a gas across a membrane is directly proportional to the partial pressure difference of the gas across the membrane.

TABLE 5 Physical Constants and Coefficients Related to Diffusion in Respiratory Physiology

Medium	T (°C)	D × 10^4 (cm^2/min)	α (ml/ml atm)	*K* × 10^5 (cm^2/min atm)	Reference
Oxygen					
Water	20	12.4	0.031	3.84	Altman and Dittmer (1971)
	37	18.2	0.024	4.36	Altman and Dittmer (1971)
Muscle tissue	37	7.0	0.0235	1.65	Altman and Dittmer (1971)
Connective tissue	37	5.8	0.0235	1.35	Altman and Dittmer (1971)
Lung tissues	37	13.8	0.018	2.50	Altman and Dittmer (1971)
Serum	37	11.2	0.0214	2.39	Altman and Dittmer (1971)
Blood	37		0.0227		Altman and Dittmer (1971)
Carbon Dioxide					
Water	37	15.3	0.567	86.7	Altman and Dittmer (1971)
Muscle tissue	22	6.0	0.78	47.0	Altman and Dittmer (1971)
Blood	38		0.47		Altman and Dittmer (1971)
Carbon Monoxide					
Lung tissues	37	12.0[a]	0.0179[b]	2.15	Altman and Dittmer (1971)[a]
					Power (1968)[b]
Plasma	37	14.8[a]	0.0189[b]	2.28	Altman and Dittmer (1971)[a]
					Power (1968)[b]

Note: Diffusion coefficient (D); Bunsen solubility coefficient (α); and Krogh's (1919) diffusion constant (*K*) for oxygen, carbon dioxide, and carbon monoxide in various media. *K* is the product of D and α.

4. The flux of a gas across a membrane is directly proportional to the solubility of the gas in the material comprising the membrane.
5. The flux of a gas across a membrane is inversely proportional to the square root of the molecular weight of the gas, Graham's (1833) Law.

5.1 Facilitated Diffusion

Diffusion of O_2 through solutions of hemoglobin and myoglobin solutions maintained at low partial pressures of O_2 (P_{O_2}) occurs at rates higher than can be explained by simple molecular diffusion (Klug et al., 1956; Wittenberg, 1959; Scholander, 1960). This enhancement of diffusion is termed *facilitated diffusion* and appears to be due to simple molecular diffusion of O_2 combined with the translational diffusion of hemoglobin or myoglobin molecules that have O_2 molecules bound to them. Because of the great number of binding sites for O_2 in a physiologic hemoglobin or myoglobin solution, the facilitation of O_2 delivery is significant, despite the fact that the hemoglobin and myoglobin molecules are large and diffuse very slowly (Kreuzer and Hoofd, 1987). Such facilitation may have marked effects on O_2 transport within the erythrocytes during pulmonary gas exchange.

Transport of CO_2 is also facilitated, although not by a carrier, as with O_2 but rather, by its conversion to HCO_3^- by carbonic anhydrase and its movement along a concentration gradient in parallel with CO_2 diffusion. Experiments in which pulmonary carbonic anhydrase activity was eliminated with acetazolamide resulted in large decreases in the capacity of the lungs to transport CO_2 (Hill and Enns, 1981). Nevertheless, the physiological role of facilitated CO_2 diffusion has not been well documented (Kreuzer and Hoofd, 1987), and remains unclear today.

6. ALVEOLAR-CAPILLARY CONDUCTANCE: PULMONARY DIFFUSING CAPACITY (TRANSFER FACTOR)

The ease with which a gas moves by diffusion across a barrier can be expressed as a conductance term ($G = 1$/resistance), denoting the quantity of gas transferred per unit time per unit driving pressure (partial pressure difference across the barrier). This conductance term incorporates within it considerations of the surface area and thickness of the barrier, as well as properties of the gas and material comprising the barrier, i.e., molecular size and solubility (*K* in Eqn (8)).

The conductance with which the lungs exchange any gas between alveolar gas and pulmonary capillary blood is referred to as the *pulmonary diffusing capacity* (D_L), or *pulmonary transfer factor* for that gas. The latter term is more appropriate because the transfer is not a capacity, and for O_2, CO_2 and CO, chemical reaction, as well as diffusion, is involved in the transfer. Nevertheless, the term *diffusing capacity* is widely used in the United States and for clarity will be used in this chapter. Estimates of the value of pulmonary diffusing capacity can be made in living animals using in vivo physiologic techniques or by analyzing the structural components of the lung with morphometric techniques.

For those gases that chemically combine with hemoglobin (e.g., O_2 and CO), the Roughton and Forster (1957) model for diffusing capacity (D_L) divides the transfer of gas into two parts:

$$\frac{1}{D_L} = \frac{1}{D_M} + \frac{1}{\theta \cdot V(c)} \tag{9}$$

where the reciprocal of the diffusing capacity of the whole lung (D_L, ml gas [STPD]/s·torr) is the sum of two serial resistances. D_M is the diffusing capacity of the alveolar-capillary membrane alone, and $\theta \cdot V(c)$, the diffusing capacity of the erythrocytes (D_e) in the pulmonary capillary, is the product of the rate of chemical reaction of the gas with whole blood (θ, in ml gas [STPD] ml/blood·s·torr) and pulmonary capillary blood volume (V(c), in ml). For gases, e.g., O_2, that chemically react with hemoglobin in the blood in a nonlinear manner, θ changes with O_2 saturation of the hemoglobin, and D_L for O_2 changes as the saturation of hemoglobin changes.

Functionally, the transfer of oxygen ($\dot{V}O_2$) in the lung is described as the product of the diffusing capacity for oxygen ($D_{L_{O_2}}$) and the mean O_2 partial pressure difference between alveolar gas and capillary blood. Thus, $D_{L_{O_2}}$ could potentially be calculated by the following equation:

$$D_{L_{O_2}} = \frac{\dot{V}O_2}{\overline{P}A_{O_2} - \overline{P}c_{O_2}} \tag{10}$$

where $\overline{P}A_{O_2}$ is the average O_2 partial pressure in the alveoli and $\overline{P}c_{O_2}$, is the mean O_2 tension in the pulmonary capillary. Practically, alveolar P_{O_2} can be estimated in either of two ways: from an end-tidal sample of gas, or by using the ideal alveolar gas equation (Riley et al., 1946), which requires an analysis of CO_2 tension in an arterial blood sample and assumes that arterial P_{CO_2} is equal to alveolar P_{CO_2} (Enghoff, 1938).

Mean pulmonary capillary P_{O_2} is more difficult to determine, because the P_{O_2} of the blood as it passes through the pulmonary capillary changes nonlinearly as O_2 concentration increases. In addition, end-capillary P_{O_2}, cannot be directly measured or realistically assumed to be equal to arterial P_{O_2}, as there may be variable amounts of venous admixture in arterial blood. Heterogeneity in alveolar ventilation-perfusion and diffusion-perfusion ratios can also have substantial effects on arterial P_{O_2}.

6.1 Use of CO to Estimate $D_{L_{O_2}}$

The necessity of determining a value for $\overline{P}_{CO_2}$, is obviated if D_L for carbon monoxide ($D_{L_{CO}}$) is estimated instead of $D_{L_{O_2}}$. Krogh (1915) first developed this technique, capitalizing on the fact that the affinity of hemoglobin for CO is 210 times higher than that for O_2. Thus, as long as alveolar CO concentration is low (inspired fraction <0.01) and exposure time is brief, all CO is rapidly bound to hemoglobin, and P_{CO} in the capillary blood is effectively 0 torr. To calculate $D_{L_{CO}}$, only the rate of removal of CO needs to be measured and alveolar P_{CO} needs to be estimated. $D_{L_{CO}}$ is useful as a comparative index of pulmonary conductance, but it cannot be used as a direct estimate of $D_{L_{O_2}}$ because of differences in the chemical reaction rates of CO and O_2 with hemoglobin. The alveolar membrane component of pulmonary diffusing capacity for O_2 ($D_{M_{O_2}}$) can be calculated from $D_{M_{CO}}$ and the ratios of the square roots of molecular weight (MW) and solubilities (α) for O_2 and CO (Forster, 1987):

$$D_{M_{O_2}} = D_{M_{CO}} \cdot \frac{\alpha_{O_2}}{\alpha_{CO}} \cdot \frac{\sqrt{MW_{CO}}}{\sqrt{MW_{O_2}}} \tag{11}$$

The ratio $D_{M_{O_2}}/D_{M_{CO}}$ is approximately 1.23.

Although measurement of the uptake rate of CO across the lung yields an estimate of DM_{CO} that can be used to calculate a value for $D_{M_{O_2}}$, estimating D_e is more problematic. Because D_e is the product of capillary volume, V(c), and θ, the rate of gas uptake by the blood, factors determining the value of D_e include diffusion of the gas into the erythrocyte and chemical reaction of the gas with hemoglobin, thus the value of θ must be determined empirically for each gas. If the major resistance to gas transport in the pulmonary capillaries ($1/D_e$) were diffusion rather than chemical reaction, $D_{L_{O_2}}$ would be determined almost exclusively by diffusive properties of the gas (α_{O_2} and $MW^{-1/2}$), and would have a value close to $1.23 \times D_{L_{CO}}$. Experimental work and theoretical modeling by Yamaguchi et al. (1985) and Scheid et al. (1986) suggest that diffusion of O_2 is more important in limiting the kinetics of O_2 binding in the pulmonary capillary than the reaction rate of O_2 with hemoglobin.

The relative contributions of the serial resistances $1/D_M$ and $1/D_e$ to the overall resistance to diffusion in the lung ($1/D_L$) has not been resolved. Most physiologic estimates produce results consistent with the Roughton and Forster (1957) model that divides the resistance to diffusion approximately equally between membrane and erythrocyte components (Sackner et al., 1975; Crapo et al., 1982). Morphometric estimates place the majority of the resistance to diffusion in the erythrocyte and estimate the ratio D_M/D_e as high as 6 (Weibel, 1972). Crapo and Crapo (1983) and Crapo et al. (1986) compared physiologic and morphometric estimates of $D_{L_{CO}}$ in the same dogs and confirmed this discrepancy. Forster (1987) suggests that the individual components of the Roughton-Forster equation have a different functional significance as estimated by physiologic and morphometric techniques; the two estimates may, therefore, not be comparable. More recent measurements of nitric oxide (NO) diffusing capacity ($D_{L_{NO}}$) have been made to estimate D_M alone (Tamhane et al., 2001). $D_{L_{NO}}$ is about fourfold higher than $D_{L_{CO}}$ (Tamhane et al., 2001), suggesting that the membrane component of diffusion resistance is actually very low. However, because $D_{L_{NO}}$ is so high (30% higher than predicted from simultaneous $D_{L_{CO}}$ measurements), just what it measures is still in need of clarification.

6.2 Measurement of $D_{L_{CO}}$

All techniques for measuring $D_{L_{CO}}$ are based on estimating a solution to Eqn (10), with CO replacing O_2. In every technique, alveolar CO fraction is maintained at less than 0.01 to ensure that CO bound to hemoglobin in the blood is not at a sufficiently high concentration to create a significant arteriovenous concentration difference. If the concentrations of CO are kept low and the times of exposure are short, the P_{CO} in pulmonary capillary blood ($\bar{P}c_{CO}$ in Eqn (10)) can be assumed to be near 0 torr and ignored in the calculations. Gas samples are analyzed for CO concentration with an infrared analyzer, gas chromatograph, or mass spectrometer if ^{13}CO or $C^{18}O$ is used (MW 29 or 30, allowing separation from the ubiquitous N_2 signal with an MW of 28) (Forster, 1987). The methods described below were originally developed to measure $D_{L_{CO}}$ in awake humans; however, as indicated, they have been adapted for use in awake and anesthetized nonhuman mammals. Thus, the volumes of gas mixtures used and times allowed for equilibration of gas samples must be adjusted when species other than humans are studied.

6.3 Steady-State $D_{L_{CO}}$

In the steady-state method for $D_{L_{CO}}$ the subject breathes a gas mixture containing 0.1–0.3% CO until alveolar P_{CO} reaches a plateau (approximately 12 breaths in humans). The transfer rate of CO, $\dot{V}CO$ is calculated by subtracting the amount of CO expired (concentration × volume) from the amount inspired (concentration × volume) during the measuring interval. Alveolar P_{CO} is calculated using a modified form of the alveolar air equation by measuring arterial P_{CO_2} as an estimate of alveolar P_{CO_2} ($\bar{P}A_{CO_2}$) measuring inspired P_{CO} (PI_{CO}) and CO fractions (FI_{CO}), and mixed expired CO and CO_2 fractions ($F\bar{E}_{CO}$ and $F\bar{E}_{CO_2}$) (Filley et al., 1954):

$$\bar{P}A_{CO} = PI_{CO} - \bar{P}A_{CO_2} \cdot \left(\frac{FI_{CO} - F\bar{E}_{CO}}{F\bar{E}_{CO_2}} \right) \quad (12)$$

or by sampling end-tidal gas (Bates et al., 1955). As long as the inspired CO concentration is low and the breathing period is kept short, mean capillary P_{CO} is low enough to be considered 0 torr. This technique has been used in animal studies (Mauderly, 1972), but has not been as widely utilized as the other methods.

6.4 Single-Breath $D_{L_{CO}}$

In this method the subject inspires to total lung capacity a gas mixture containing approximately 0.3% CO; 10% of an insoluble inert gas such as He, Ne, or Ar, and 20–21% O_2. After breath holding for 10 s (in humans), the subject exhales and an expiratory gas sample is analyzed for CO and the inert gas. $\dot{V}CO$ is calculated from the inspired and expired concentrations of CO and the alveolar gas volume (functional residual capacity plus inspired volume), which is calculated from the dilution of the inspired inert gas. Mean alveolar P_{CO} can also be calculated from the concentrations of CO at the beginning and end of the measurement (Forster et al., 1954, 1955; Ogilvie et al., 1957; Bates et al., 1971; Cotes, 1975). The advantage of this method over the measurement of steady-state $D_{L_{CO}}$ is that no arterial blood sample is required. Because the single-breath technique requires a cooperative ventilatory maneuver by the subject, it can only practically be used in nonhuman species that are anesthetized so as to allow the investigator to inflate and deflate the lungs with the mixture of marker gases. When applying this technique, it is important to realize that the volume to which the lungs are inflated directly affects the results (Cassidy et al., 1980). If the lungs are to be inflated to total lung capacity, the investigator must monitor transpulmonary pressure during inflation to ensure that 30 cm H_2O is achieved. This method has been widely applied to anesthetized nonhuman mammals (Robinson et al., 1972; Takezawa et al., 1980; Weibel et al., 1983). Representative values of single-breath $D_{L_{CO}}$ measured in a number of species are in Table 6.

6.5 Rebreathing $D_{L_{CO}}$

This method for determining $D_{L_{CO}}$ is fundamentally similar to the single-breath method but extends the total dilution volume (total alveolar volume in the single-breath method) to include the volume of a bag filled with air and low concentrations of CO and an inert gas. The subject breathes rapidly for a short interval until the gas in the bag is well mixed with the gas in the lung. $\dot{V}CO$ is calculated similarly as in the single-breath method, except that the alveolar volume is replaced by the total volume of the lung-bag system, again determined by dilution of the inert gas, and the alveolar P_{CO} can be calculated as follows:

$$PA_{CO} = PI_{CO} \cdot \frac{FA_{He}}{FI_{He}} \quad (13)$$

where FA_{He} and FI_{He} represent the alveolar and inspired fractions, respectively, of He (or other inert gas). The advantage of this method over the single-breath $D_{L_{CO}}$ is that it tends to compensate for uneven distribution of ventilation and perfusion in the lung. Techniques have been devised to apply this method to anesthetized

TABLE 6 Physiologic Measurements of $D_{L_{CO}}$

Species	M (kg)	Condition	$D_{L_{CO}}$ (ml · min^{-1} · torr^{-1})	Technique	Reference
Human *Homo sapiens*	—	Rest Exer	21 (34 refs)40 (27 refs)	ss	Piiper and Scheid, 1980
Human *H. sapiens*	—	Rest Exer	46 55	Rbsi	Meyer et al., 1981
Human *H. sapiens*	—	Rest Exer	29 (8 refs) 46 (4 refs)	rb	Piiper and Scheid, 1980
Human *H. sapiens*	—	Rest Exer	34 (45 refs) 47 (12 refs)	sb	Piiper and Scheid, 1980
Dog *Canis familiaris*	16	Anes	52	rb	Crapo and Crapo, 1983
Coyote *C. latrans*	12.4	Anes	16.3	sb	Weibel et al., 1983
Red fox *Vulpes vulpes*	4.6	Anes	4.0	sb	Weibel et al., 1983
Cat *Felis domesticus*	3.3	Anes	1.31	sb	Watanabe and Frank, 1975
Cat *F. domesticus*	2.9	Anes	1.2	sb	Vreim and Staub, 1973
Rat *Rattus norvegicus*	0.338 0.493	Anes	0.338 0.493	rb	Johanson et al., 1973
Rat *R. norvegicus*	0.203	Anes	0.084	ss	Turek and Pierce, 1973
Rat *R. norvegicus*	0.128 0.186	♂Anes ♀Anes	0.146 0.151	sb	Takezawa et al., 1980
Hamster *Mesocricetus auratus*	0.126	Anes	0.136	rb	Snider et al., 1978

Note: Measured values of physiologic pulmonary diffusing capacity for carbon monoxide $D_{L_{CO}}$ in various species of mammals. Condition denotes the activity of the subject during the measurement: rest; anes, anesthetized; exer, exercise; HC, hypercapnic; NC, normocapnic; HO, hyperoxic; NO, normoxic. Technique indicates the method used to measure $D_{L_{CO}}$: sb, single breath; rb, rebreathing; ss, steady state; rbsi, rebreathing with stable isotopes.

(Crapo and Crapo, 1983; Crapo et al., 1988), resting (Amis et al., 1984a), and exercising (Carlin et al., 1988) animals. Representative values are in Table 6.

6.6 Relationship between Different Methods

In humans, single-breath $D_{L_{CO}}$ is about 60% greater than steady-state $D_{L_{CO}}$ at rest, mainly because single-breath $D_{L_{CO}}$ is measured at total lung capacity and steady-state $D_{L_{CO}}$ is measured at small ventilatory volumes around functional residual capacity. However, the discrepancy may also exist due to there being a more homogenous distribution of intrapulmonary gases in the steady-state method. Rebreathing $D_{L_{CO}}$ is about 30% greater than single-breath $D_{L_{CO}}$ when they are measured at the same lung volumes, presumably because of a more homogenous gas distribution in the rebreathing method (Piiper and Scheid, 1980).

6.7 Measurement of $D_{L_{O_2}}$: Steady-State $D_{L_{O_2}}$

To determine $D_{L_{O_2}}$ from Eqn (10), measurements of the transfer rate of O_2 ($\dot{V}O_2$) and the mean alveolar to pulmonary capillary P_{O_2} difference are required. $\dot{V}O_2$ can easily be measured, and although the alveolar P_{O_2} is neither uniform nor constant throughout the lung, a value for PA_{O_2}, can be obtained using the ideal alveolar gas equation. As discussed above, the problems with measuring $D_{L_{O_2}}$ lie in estimating mean capillary P_{O_2}. Although the P_{O_2} of blood entering the pulmonary

TABLE 7 Measured Values of Physiologic Pulmonary Diffusing Capacity for Oxygen ($D_{L_{O_2}}$)

Species	M (kg)	Condition	$D_{L_{O_2}}$ (ml · min^{-1} · torr^{-1})	Technique	Reference
Human *Homo sapiens*	80	Exer HO	77.1	eoBi	Rosenhamer et al., 1971
Human *H. sapiens*	—	Rest Exer	54 65	Rbsi	Meyer et al., 1981
Human *H. sapiens*	—	Rest Exer	25-33 (4 refs) 40-43 (3 refs)	sb	Piiper and Scheid, 1980
Human *H. sapiens*	—	Rest Exer	22-53 (6 refs) 43-107 (3 refs)	rb	Piiper and Scheid, 1980
Human *H. sapiens*	—	Rest Exer	26-47 (3 refs) 57-92 (4 refs)	ss	Piiper and Scheid, 1980
Human *H. sapiens*	—	Exer NO Exer HO	54.1 88.3	ss	Hempleman and Hammond, 1988
Dog *C. familiaris*	28.3	Rest HO NC HC Exer HO NC HC	29 39 29 39 74 76	ss	Scotto et al., 1987

Note: Condition denotes the activity of the subject during the measurement: rest; exer, exercise; HC, hypercapnic; NC, normocapnic; HO, hypoxic; NO, normoxic. Technique indicates the method used to measure $D_{L_{O_2}}$: sb, single-breath; rb, rebreathing; ss, steady state; rbsi, rebreathing with stable isotopes; eoBi, ear oximeter and Bohr integration.

capillary can be obtained by sampling mixed venous blood, the P_{O_2} of end-capillary blood is not the same as that in arterial blood. However, Lilienthal et al. (1946) and Riley and Cournand (1949) developed a method for estimating the alveolar to end-capillary P_{O_2} difference by dividing the alveolar to arterial P_{O_2} difference (PA_{O_2} — Pa_{O_2}) into two components: that due to incomplete equilibration (diffusion limitation) and that due to physiologic shunt (areas with a low ventilation/perfusion ratio and venous admixture). The latter is almost eliminated by giving the subject 12–14% O_2 to breathe.

When a normal subject breathes room air, the arterial P_{O_2} and O_2 saturation (Sa_{O_2}) are on the flattened upper portion of the oxy–hemoglobin equilibrium curve (OEC). Thus, a small decrease in Sa_{O_2} due to physiologic shunt results in a large decrease in Pa_{O_2}. However, when a subject breathes 12–14% O_2, both Sa_{O_2}, and Pa_{O_2} fall to the steep portion of the OEC, and a decrease in Sa_{O_2} of the same magnitude due to physiologic shunt results in a much smaller decrease in Pa_{O_2}. Therefore, most of the PA_{O_2}–Pa_{O_2} in hypoxia is due to diffusion limitation, and Pa_{O_2} is similar to end-capillary P_{O_2}. In practice, the experimental data in normoxia and hypoxia are used in a trial-end-error procedure to calculate the alveolar to mean capillary P_{O_2} and $D_{L_{O_2}}$ (Forster, 1987). Development of this method represented an important advance in estimating $D_{L_{O_2}}$, however, its assumptions about the effects of uneven ventilation/perfusion ratios are not valid and lead to an underestimate of $D_{L_{O_2}}$ (Riley and Permutt, 1973; Piiper and Scheid, 1980). Reported values are in Table 7.

6.8 Single-Breath and Rebreathing $D_{L_{O_2}}$

The single-breath and rebreathing methods for estimating $D_{L_{CO}}$ can be modified to estimate $D_{L_{O_2}}$ by using isotopic O_2. Both require measurement of pulmonary blood flow using an inert gas and simultaneous measurement of $\dot{V}O_2$. The single-breath method uses the stable isotope $^{18}O_2$, and the rebreathing method uses the natural isotope $^{16}O_2$. Both methods are described by Forster et al. (1986) and Forster (1987).

6.9 Morphometric Measurement of D_L

A different approach to estimating the diffusive conductance of the pulmonary gas exchanger is to directly measure the alveolar tissue surface area and thickness available for diffusion and to calculate the conductance, given the diffusion coefficients for respiratory gases (Table 5). A morphometric technique for estimating $D_{L_{O_2}}$ by stereologic methods was developed by Weibel (1970/1971), and the outline below is based on his refinements of the approach (Weibel, 1984a,b). In this technique, a random sampling procedure is applied as sections of lung are systematically analyzed at the gross macroscopic level, through low- and high-power light microscopic levels, and finally at the level of electron microscopy.

In the morphometric method of estimating D_L, quantities of structural parameters are estimated with respect to reference spaces of equal or higher dimension (e.g., surface areas are measured with respect to the volumes in which the surfaces are contained). Because the areas and volumes measured are functions of the degree to which the airways and vessels are inflated and distended during fixation, care must be taken to maintain instillation/perfusion pressures for fixatives at physiologically meaningful values. Test grids are utilized to superimpose patterns of points, lines, and areas on projected images of sections of tissue samples, and the systems to which the measurements are referenced become the areas of the microscope fields, the lengths of the test lines on the grids, or the numbers of points on the grids that fall on the structures being measured. If a test grid is applied to the sample at random, the probability that a test point will fall on a given component (e.g., a capillary profile or alveolar septum) is proportional to the volume of that component in the reference space. Thus, the volume density (V_v(comp, ref) = volume of component/volume of reference space) of a component in the reference space is the ratio of points falling on each:

$$V_v\,(\text{comp}, \text{ref}) = P_p\,(\text{comp}, \text{ref}) \tag{14}$$

The reference spaces are arranged in hierarchical order, such that (1) total lung volume (L; the reference space) is divided into the volumes of its constituent lobes (to weight the contribution of each lobe according to its contribution to total lung volume); (2) lobar volume (reference) is divided into *coarse nonparenchyma* (structures >1 mm in diameter or thickness, too large to contribute significantly to gas exchange) and *coarse parenchyma* (cp); (3) *coarse parenchyma* (reference) is divided into *fine nonparenchyma* (conducting airways, epithelium, connective tissue, and vessels >20 μm in diameter, again too large to contribute substantially to gas exchange) and *fine parenchyma* (fp); and (4) *fine parenchyma* (reference) is divided into the volume of gas in the alveoli and the volume of gas-exchange *septum* (s), to which the volumes of capillaries and alveolar epithelium and connective tissue are referenced when the point counts are made on electron micrographs. A series of V_v ratios, each using the next higher level as its reference space, is accumulated and ultimately refers to the total lung volume as the reference space to determine the absolute volume of each component of interest in the entire lung:

$$V(\text{comp}) = \frac{V(\text{comp})}{V(S)}\cdot\frac{V(S)}{V(\text{fp})}\cdot\frac{V(\text{fp})}{V(\text{cp})}\cdot\frac{V(\text{cp})}{V(L)}\cdot V(L) \tag{15}$$

$$= V_v(\text{comp}, S)\cdot V_v(s, \text{fp})\cdot V_v(\text{fp}, \text{cp})\cdot V_v(\text{cp}, L)\cdot V(L) \tag{16}$$

where the modifiers of volumes (V) and volume densities (V_v) refer to the abbreviations of reference spaces listed above, and volume densities are calculated as in Eqn (14).

Surface densities (S_v(comp, ref) = surface area of component/volume of reference space) can be estimated with a related counting method. The surface boundary of a capillary or alveolus appears on a section as a linear trace of the profile. The probability that a test line of given length on the test grid will intersect the boundary trace of the component of interest is directly proportional to the surface density of the object in the reference space, and the volume of the reference space is defined by the number of points falling on it:

$$S_V = 2\cdot\frac{I}{L_T} \tag{17}$$

where I is the number of intersections of the test line with the boundary trace of the component and L_T is the length of the test line.

The thicknesses of the alveolar-capillary and plasma barriers cannot be estimated by point-counting methods. Instead, a test grid with continuous horizontal and vertical lines is superimposed on the sample, and a ruler is used to measure the length of any test line that intercepts both the alveolar and capillary surfaces, or both the capillary and erythrocyte surfaces, depending on whether the thickness of alveolar septal tissues or the pulmonary capillary plasma layer is being estimated. Because the diffusion pathway of a gas molecule from erythrocyte to alveolar gas space can be along any vector, these two-dimensional intercept lengths are tabulated according to their class lengths on a logarithmically scaled ruler and are then summated to calculate a *harmonic mean thickness* (Weibel, 1980) for the barrier of interest, alveolar septal tissue (τ_{ht}, or capillary plasma (τ_{hp})):

$$\tau_h = \frac{2}{3}\cdot\ell_h \tag{18}$$

$$\ell_h = \frac{\sum ni}{\sum n_i\cdot\frac{1}{\ell_h}} \tag{19}$$

where ℓ_h is the harmonic mean of the intercept lengths and n_i is the number of intercepts of length ℓ_i.

Estimates of alveolar and capillary volumes, surface areas, and harmonic mean barrier thicknesses are used to calculate morphometric DL_{O_2}. The membrane diffusing capacity (D_M in Eqn (9)) is composed of two resistances in series:

$$\frac{1}{D_M} = \frac{1}{D_t} + \frac{1}{D_p} \tag{20}$$

where $1/D_t$ is the reciprocal of the tissue component of diffusing capacity, i.e., the resistance to diffusion across the tissue barrier composed of alveolar epithelium, interstitium, and capillary endothelium, and $1/D_p$ is the reciprocal of the plasma diffusing capacity, the resistance to diffusion across the plasma barrier. The full equation describing the morphometric D_L is, therefore:

$$\frac{1}{D_L} = \frac{1}{D_t} + \frac{1}{D_p} + \frac{1}{D_e} \quad (21)$$

where $1/D_e$ is the reciprocal of the erythrocyte diffusing capacity, defined as $\theta \cdot V(c)$ [Eqn (9)].

The tissue and plasma diffusing capacities are conductance terms and are described by Fick's Second Law as being proportional to the ratios of their surface areas to thicknesses, multiplied by a permeation coefficient. However, the equations by which they are calculated are more complex because neither the tissue nor the plasma barrier is a simple sheet of constant thickness. The tissue diffusing capacity is calculated as:

$$D_t = K_t \cdot \frac{S(A) + S(C)}{2} \cdot \frac{1}{\tau_{ht}} \quad (22)$$

where K_t is a permeation coefficient (Table 5), the product of the solubility and diffusion coefficients for a given gas in tissue; S(A) and S(c) are the surface areas of the alveoli and capillaries, respectively; and τ_{ht} is the harmonic mean thickness of the tissue barrier (Weibel, 1980). The plasma diffusing capacity is calculated as

$$D_P = K_P \cdot \frac{S(c)}{\tau_{hp}} \quad (23)$$

where K_p is the permeation coefficient for a given gas in plasma (Table 5) and τ_{hp} is the harmonic mean thickness of the tissue barrier. Reported values for morphometric $D_{L_{O_2}}$ are in Table 8.

7. O_2 AND CO_2 UPTAKE KINETICS OF RED BLOOD CELLS

The coefficient θ is the rate of uptake of O_2 or CO by whole blood. This variable is determined by both a diffusive step across the erythrocyte membrane and through the cell, and a chemical reactive step in which O_2 or CO combines with hemoglobin. In the case of θ_{CO} this step includes displacement by CO of O_2 bound to hemoglobin and the subsequent formation of carboxyhemoglobin.

θ_{O_2} for whole blood is dependent on hemoglobin concentration, O_2 saturation of the blood, and temperature. θ_{O_2} can be calculated with a modified form of Staub et al. (1961, 1962) equation:

$$\theta_{O_2} = k'_c \cdot (\alpha_{O_2} \cdot 0.0587) \cdot (1 - S_{O_2}) \cdot 0.0133 \cdot [Hb] \quad (24)$$

where k'_c is the reaction velocity of O_2 with hemoglobin in 1/(mM s), determined empirically (Holland et al., 1977), α_{O_2} is the Bunsen solubility coefficient of O_2 in ml [STPD] O_2/(ml atm), S_{O_2} is the fractional O_2 saturation of the blood, [Hb] is hemoglobin concentration in g/d1, 0.0133 is the O_2 carrying capacity of hemoglobin [Hüfner's (1877) constant], and 0.0587 is a conversion factor to obtain θ_{O_2} in ml [STPD] O_2/(ml blood s torr).

In 1923, a technique was developed for measuring rapid chemical reactions and measurements were begun of k'_c and θ_{O_2} in hemoglobin solutions (Hartridge and Roughton, 1923). In the 1950s, Roughton and co-workers (Roughton and Forster, 1957; Roughton et al., 1959) extensively studied θ_{CO} and θ_{O_2} on whole red cells spectrophotometrically, using a continuous-flow apparatus. In the early 1960s Staub et al. (1961, 1962) made measurements of θ_{O_2} at different hemoglobin S_{O_2} using an O_2 electrode. They showed that although increased with increasing S_{O_2}, θ_{O_2} decreased when S_{O_2} exceeded approximately 0.70 because of the decrease in available hemoglobin binding sites. The dependence of θ_{O_2} on S_{O_2} was confirmed by others (Holland et al., 1977) and is widely accepted.

Holland and Forster (1966), using a stopped-flow technique, found k'_c and, therefore θ_{O_2} to be dependent on red cell size, such that k'_c was higher in blood from animals with smaller erythrocytes, presumably because diffusion distances within the cell are shorter (Forster, 1987). However, this result was questioned by later workers, who suggested it was due to the relative differences in an unstirred layer of buffer around red cells of different sizes in the stopped-flow apparatus (Huxley and Kutchai, 1983; Forster, 1987). Further work focused on determining the effects of this unstirred boundary layer around the red cells during in vitro measurements. This unstirred layer was shown to be about 1–2 μm at the start of the reaction, but increased as the reaction proceeded (Coin and Olson, 1979; Huxley and Kutchai, 1983). The unstirred layer is responsible for most of the resistance to O_2 uptake by the red cell in the apparatus, the erythrocyte membrane offering little resistance by comparison. Reported values of k'_c and θ are given in Table 9.

Holland et al. (1985) used Na-dithionite in the reaction buffer to scavenge O_2 as it passed out of the cell. Thus, O_2 release experiments could be performed without an unstirred layer. By using this technique, Holland et al. (1985), and later McKay and Holland et al. (1988), determined that the unstirred layer slowed O_2 uptake by the cell by a factor of 1.3–2.0. Newer values for θ_{O_2} take this unstirred layer into account and are, consequently, higher than earlier measurements (Yamaguchi et al., 1985; Hallam et al., 1989).

8. CHANGES IN PULMONARY DIFFUSING CAPACITY (D_L)

8.1 Changes in D_L with Exercise: Physiologic

With the development of the CO technique for measuring D_L, Krogh and Krogh (1909/1910) demonstrated that $D_{L_{CO}}$ increases with exercise. This increase is apparently due to a combination of the changes that occur in pulmonary blood flow ($\dot{Q}$), capillary surface area, alveolar ventilation ($\dot{V}A$), and

TABLE 8 Pulmonary Morphometric Variables and Diffusing Capacity

Species	M (kg)	V_L (ml)	S(A) (m^2)	S(c) (m^2)	V(c) (ml)	τ_{ht} μm	τ_{hp} μm	θ_{O_2} ml · ml^{-1} · min^{-1} · $torr^{-1}$	($ml \cdot min^{-1} \cdot torr^{-1}$)					Reference
									$D_{t_{O_2}}$	$D_{P_{O_2}}$	$D_{e_{O_2}}$	$D_{M_{O_2}}$	$D_{L_{O_2}}$	
Ox														
Bos indicus	193	10,145	385	380	700	0.50	0.215	1.5					641	Gehr et al., 1981a
B. taurus	700	22,450	1283	1138	2770	0.51		1.5					2347	
Horse														
Equus caballus	510	37,650	2456	1663	2800	0.60	0.21	1.5					2726	Gehr et al., 1981a
Ox														
B. taurus	474	25,500	735	654	1289	0.650	0.210	2.84	3555	10,551	3271	2673	1536	Constantinopol et al., 1989
Horse														
E. caballus	447	46,500	1171	1140	1940	0.643	0.165	3.90	5954	23,494	7563	4747	2897	Constantinopol et al., 1989
Giraffe														
Giraffa camelopardalis	383	21,000	636	552	965	0.60	0.220	1.5					894	Gehr et al., 1981a
Eland														
Taurotragus oryx	240	10,668	557	461	833	0.50	0.142	1.5					805	Gehr et al., 1981a
Camel														
Camelus dromedarus	232	15,900	431	273	378	0.60	0.143	1.5					395	Gehr et al., 1981a
Pony														
E. caballus	176	9777	325	309	548	0.511	0.164	3.26	2059	6220	1764	1542	827	Weibel et al., 1987
Ox calf														
B. taurus	149	5397	213	206	356	0.422	0.190	2.10	1636	3522	751	1118	446	Weibel et al., 1987
Waterbuck														
Kobus defassa	110	7836	383	338	584	0.46	0.170	1.5					579	Gehr et al., 1981a
Wildebeest														
Connochaetes taurinus	102	7678	391	281	473	0.37	0.219	1.5					504	Gehr et al., 1981a
Human														
Homo sapiens	74.0	4341	143	126	213	0.62		1.5					198	Gehr et al., 1981a
Dog														
Canis familiaris	28.2	1587	77.6	61.2	118	0.509	0.117	3.84	450	1733	453	355	199	Weibel et al., 1987
Goat														
Capra hircus	27.7	1742	67.3	59.0	121	0.521	0.227	2.19	396	924	266	271	132	Weibel et al., 1987
Wolf														
Canis lupus	27.6	2112	105	74.4	112	0.493	0.136	1.23					105.2	Weibel et al., 1983

Continued

TABLE 8 Pulmonary Morphometric Variables and Diffusing Capacity—cont'd

Species	*M* (kg)	V_L (ml)	S(A) (m^2)	S(c) (m^2)	V(c) (ml)	τ_{ht} μm	τ_{hp} μm	θ_{O_2} $ml \cdot ml^{-1} \cdot min^{-1} \cdot torr^{-1}$	($ml \cdot min^{-1} \cdot torr^{-1}$) $D_{t_{O_2}}$	$D_{p_{O_2}}$	$D_{e_{O_2}}$	$D_{M_{O_2}}$	$D_{L_{O_2}}$	Reference
Dog *C. familiaris*	25.3	2561	90.8	54.6	100	0.430	0.138	1.23					89.9	Weibel et al., 1983
Dog *C. familiaris*	5.4 11.6 22.3 46.1	284 764 1406 2888	18.2 43.2 82.7 177	14.1 35.5 65.5 132	26.0 55.8 100 234	0.43 0.47 0.50 0.53		1.5 1.5 1.5 1.5					26.0 57.3 110 227	Gehr et al., 1981b
Sheep *Ovis aries*	21.8	1706	67.1	64.5	146	0.53	0.324	1.5					124	Gehr et al., 1981a
Goat *Capra hircus*	20.9	1370	44.9	43.9	101	0.54	0.321	1.5					85.0	Gehr et al., 1981a
Gazelle *Gazella granti* *G. thomsoni*	10.1 19.5	562 1670	39.9 95.0	30.3 78.8	37.0 119	0.54 0.37	0.115 0.220	1.5 1.5					40.1 129	Gehr et al., 1981a
Coyote *Canis latrans*	12.4	1160	47.5	33.1	51.9	0.494	0.166	1.23					49.6	Weibel et al., 1983
Red fox *Vulpes vulpes*	4.61	290	7.75	6.69	11.7	0.470	0.157	1.23					10.4	Weibel et al., 1983
Blue fox *Alopex lagopus*	4.27	206	13.4	9.14	13.2	0.497	0.103	3.31	75.0	289	43.6	59.4	25.1	Longworth et al., 1989
Dik–dik *Madoqua kirkii*	4.2	313	14.6	13.0	22.6	0.43	0.164	1.5					22.7	Gehr et al., 1981a
Monkey *Macaca irus*	3.71	184	13.3	11.6	15.5	0.50		1.5					16.4	Gehr et al., 1981a
Rabbit *Oryctolagus cuniculus*	3.56	79.2	5.86	4.70	7.15	0.50		1.5					7.34	Gehr et al., 1981a
Suni *Nesotragus moschatus*	3.3	209	9.69	8.13	12.5	0.56	0.182	1.5					12.3	Gehr et al., 1981a
Genet cat *Genetta tigrina*	1.37	99.0	5.63	4.23	5.04	0.51	0.210	1.5					5.52	Gehr et al., 1981a
Mongoose *Helogale pervula* *Mungos mungo*	0.53 1.14	30.6 63.3	1.61 3.94	1.46 2.60	2.06 2.53	0.39 0.41	0.167 0.181	1.5 1.5					2.24 3.04	Gehr et al., 1981a
Guinea pig *Cavia procellus*	0.429	13.0	0.91	0.74	1.46	0.42		1.5					1.43	Gehr et al., 1981a

Rat														
Rattus rattus	0.140	6.34	0.388	0.407	0.480	0.37		1.5					0.0543	Gehr et al., 1981a
	0.457	13.4	0.786	0.816	1.48	0.40		1.5					1.46	
Marsupial rat														
Dasyuroides bymei	0.120	6.7	0.37	0.29	0.28	0.38	0.103	4.2	3.0	13	1.2		0.840	Hallam et al., 1989
Mouse														
Mus musculus	0.042	1.45	0.125	0.113	0.147	0.29		1.5					0.171	Gehr et al., 1981a
M. wagneri	0.013	0.571	0.063	0.054	0.065	0.26		1.5					0.0776	
Bat														
Pipistrellus pipistrellus	0.0051	0.31	0.118	0.033	0.03	0.206	0.213	2.3	0.52	1.35	0.07	0.37	0.06	Maina and King, 1984
Miniopterus minor	0.0091	0.65	0.054	0.048	0.04	0.216	0.165	2.3	0.68	1.04	0.09	0.41	0.07	
Tadarida mops	0.024	2.3	0.172	0.146	0.14	0.221	0.170	2.3	2.06	3.48	0.31	1.28	0.24	
Cynopterus brachyotus	0.0367	2.9	0.162	0.115	0.08	0.282	0.097	2.3	1.31	5.75	0.18	1.01	0.15	
Cheiromeles torquatus	0.1727	10.1	0.70	0.576	0.45	0.202	0.190	2.3	9.26	14.46	1.01	5.55	0.72	
Shrew														
Sorex minutus	0.0029	0.11	0.024	0.017	0.011	0.26	0.09	1.5				0.196	0.0152	Gehr et al., 1981a
Neomys fodiens	0.017	0.61	0.104	0.077	0.060	0.29	0.09	1.5				0.804	0.0768	
Suncus etruscus	0.0026	0.10	0.017	0.013	0.012	0.27	0.14	1.5				0.128	0.0144	
Crocidura juvenata	0.007	0.22	0.031	0.023	0.018	0.37	0.10	1.5				0.381	0.0224	
C. russula	0.0122	0.39	0.062	0.042	0.034	0.36	0.11	1.5				0.372	0.0424	
C. poensis	0.017	0.49	0.067	0.050	0.043	0.42	0.14	1.5				0.937	0.0520	
C. flavescens	0.035	1.13	0.156	0.123	0.135	0.38	0.13	1.5				1.64	0.158	
C. giffardi	0.096	2.7	0.25	0.202	0.245	0.35		1.5					0.283	

Note: Morphometrically measured values of structures related to pulmonary gas exchange and morphometric estimates of pulmonary diffusing capacity for oxygen (D_{LO_2}). Body mass is *M*, lung volume, V_L; alveolar surface area, S(A); pulmonary capillary surface area, S(c); pulmonary capillary volume, V(c); harmonic mean thickness of the alveolar tissue barrier, τ_{ht}; harmonic mean thickness of the capillary plasma barrier, τ_{hp}; chemical reaction rate of O_2 with hemoglobin, θ_{O2}; tissue component of diffusing capacity, D_{tO_2}; plasma component of diffusing capacity, D_{pO_2}; erthryocyte component of diffusing capacity, D_{eO_2}; membrane component of diffusing capacity, D_{MO_2}; and pulmonary diffusing capacity, D_{LO_2}.

TABLE 9 Reaction Rates of O_2 with Blood

Species	S_{O_2} (%)	k'_c ($mM^{-1} \cdot s^{-1}$)	θ_{O_2} $ml \cdot ml^{-1} \cdot min^{-1} \cdot torr^{-1}$	Technique	Reference
Horse, at $\dot{V}O_2$ MAX	60	440	3.90	Calc USL	Constantinopol et al., 1989
Equus caballus					
Steer, at $\dot{V}O_2$ MAX	60	440	2.84	Calc USL	Constantinopol et al., 1989
Bos taurus					
Pony, at $\dot{V}O_2$ MAX	60	440	3.26	Calc USL*	Weibel et al., 1987
E. caballus	60	100	2.10		
Calf, at $\dot{V}O_2$ MAX	60	440	3.29	Calc USL*	Weibel et al., 1987
B. taurus	60	85	1.29		
Human	0	164	2.5	rr O_2e	Staub, 1962
Homo sapiens	38 60 91 98	280 450 1420 1630	2.7 2.8 2.0 0.5		
H. sapiens					
Human, adult	0	99	1.58	sf sp	Holland et al., 1977
H. sapiens	47	169	1.44		
	73	319	1.38		
Human newborn	0	89	1.42	sf sp	Holland et al., 1977
H. sapiens	26 30 63	112 139 207	1.33 1.55 1.23		
Human	70	99	—	sf sp di	Holland et al., 1985
H. sapiens					
Dog, at $\dot{V}O_2$ MAX	60	440	3.84	Calc USL*	Weibel et al., 1987
Canis familiaris	60	68	2.72		
Goat, at $\dot{V}O_2$ MAX	60	440	2.19	Calc USL*	Weibel et al., 1987
Capra hircus	60	145	3.13		
Goat	70	145	—	sf sp di	Holland et al., 1985
C. hircus					
Sheep	70	115	—	sf sp di	Holland et al., 1985
Ovis aries					
Fox, at $\dot{V}O_2$ MAX	60	440	3.31	Calc USL	Longworth et al., 1989
Alopex lagopus					
Marsupial rat	—	217	4.20	sf sp USL	Hallam et al., 1989
Dasyuroides byrnei					

Note: Reported values for chemical reaction rates of oxygen with blood (θ_{O_2}). SO_2 is the percent saturation of hemoglobin with O_2 for which the measurement was made; k'_c, the reaction velocity of O_2 with hemoglobin; θ, the measured or calculated value in ml [STPD] O_2/(ml blood min torr); and Technique, the method used to obtain the value; rr, rapid reaction chamber; sf, stopped flow apparatus; O_2e, O_2 electrode; sp, spectrophotometer; di, with dithionite; calc, calculated; and USL, corrected for unstirred layer.
*θ_{O_2} *was calculated two ways for each species: using a species-independent value of* k'_c *(top line), and using a species-specific value of* k'_c *(bottom line) obtained from Holland et al. (1966).*

pulmonary volume between rest and exercise, rather than the increase in ($\dot{V}O_2$), per se. Kotter et al. (1969) examined the effects of raising O_2 consumption with 2,4-dinitrophenol on $D_{L_{CO}}$ and $D_{L_{O_2}}$. They found that a 600% increase in $\dot{V}O_2$ resulted in only a 15% increase in $D_{L_{CO}}$, suggesting that O_2 flux itself has little effect on diffusing capacity.

In general, the increase in $\dot{Q}$ and alveolar volume that occurs during exercise affects D_L by increasing the effective gas-blood contact area. The relative contribution of an increase in each variable is not completely understood. Although one study (Gurtner and Fowler, 1971) attributed all of the increase in $D_{L_{CO}}$ with exercise to the increase in alveolar volume, numerous other workers have measured increases in $D_{L_{CO}}$ with increases in $\dot{Q}$ (Forster and Crandall, 1976; Forster, 1987; Hsia et al., 2007, 2008; Tamhane et al., 2001). Pulmonary capillary blood volume doubles from rest to exercise in humans (Johnson et al., 1960), and presumably, similar changes occur in other mammals. An increase in capillary blood volume can increase the capillary surface area available for diffusion, either by increasing the capillary diameter of already patent vessels or by recruitment of capillaries not perfused at rest.

Although the increase with exercise in $\dot{Q}$ and alveolar volume can increase D_L, the magnitude of the increase may be underestimated if the mismatching of alveolar ventilation/perfusion ratios ($\dot{V}A/Q$) in the lung increases during exercise. At rest, the unequal vertical distribution of $\dot{Q}$ due to gravity (West, 1965) is not matched by $\dot{V}A$, and the inhomogeneous distribution of $\dot{V}A/Q$ leads to an underestimate of D_L (Chinet et al., 1971). During exercise, $\dot{V}A/Q$ ratios change little in dogs (Sylvester et al., 1981) or horses (Wagner et al., 1989), but become less homogenous in humans (Hammond et al., 1986; Wagner et al., 1986), leading to further underestimation of D_L by standard techniques. To overcome this problem, Hammond and Hempelmann (1987) modeled pulmonary gas exchange during heavy exercise in humans (>42 ml · [STPD] · O_2/s (no *M* given to permit mass-specific calculation)). They measured $\dot{V}A/Q$ distributions with the multiple inert gas elimination technique (see below) and compared $D_{L_{O_2}}$ estimates made assuming either homogenous distribution of $D_{L_{O_2}}$ throughout the lung or the distribution of $D_{L_{O_2}}$ weighted to the distribution of $\dot{Q}$ or $\dot{V}A$ from the $\dot{V}A/Q$ measurements. They found significant effects on the estimates of $D_{L_{O_2}}$ of both the measured $\dot{V}A/Q$ inhomogeneities and the calculated $D_L/\dot{Q}$ in-homogeneities; estimates that accounted for in-homogeneities were 20–30% higher than the uniform model at both rest and exercise.

8.2 Changes in D_L with Exercise: Morphometric

The morphometric model of D_L was presented by Weibel (1970/71) as a "truly maximal" estimate of diffusing capacity. One of the assumptions in the model is that all of the alveolar and capillary tissues participate in gas exchange. As such, morphometric D_L would be expected to be greater than any physiologic measurement of D_L made at rest, and morphometric D_L values should be closer to physiologic values measured during exercise (Forster, 1964, 1987; Crapo and Crapo, 1983).

Siegwart et al. (1971) measured a morphometric $D_{L_{O_2}}$ for dog lungs that was more than three times higher than that measured physiologically during exercise or elevated $\dot{V}O_2$ It is likely, however, that the exercise intensities attained were submaximal and may not have elicited a maximal increase in physiologic $D_{L_{O_2}}$. In a study on dogs, Scotto et al. (1987) measured a physiologic $D_{L_{O_2}}$ of 0.06 ml [STPD] O_2/(s torr kg)), less than one-half that measured morphometrically (0.12 ml [STPD] O_2/(s torr kg)) by Weibel et al. (1987). However, Scotto's dogs were running at speeds less than 40% of those used by Taylor et al. (1987) to elicit $\dot{V}O_2$max on a similar incline. Similarly, the cardiac outputs of Scotto's dogs were 5 ml/(s kg), less than 40% of those measured in Taylor and Weibel's dogs (14.4 ml/(s kg)). Clearly, the two measurements of $D_{L_{O_2}}$ were not made under similar maximal conditions.

Karas et al. (1987) attempted to evaluate the reliability of morphometric $D_{L_{O_2}}$ estimates by running goats on a treadmill in normoxia and progressive hypoxia until $\dot{V}O_2$max decreased, then analyzed the magnitude of PA_{O_2}–Pa_{O_2} observed versus that predicted by their morphometric $D_{L_{O_2}}$ and a simple Bohr (1909) integration model of pulmonary capillary P_{O_2} equilibration. The results indicated that PA_{O_2}–Pa_{O_2} increased as predicted when the inspired O_2 fraction decreased to the value predicted by the model, between 0.14 and 0.15, arguing for the validity of the morphometric estimate.

Hammond and Hemplemann (1987) evaluated the effects of functional inhomogeneities on estimates of $D_{L_{O_2}}$ by determining a physiologic DL_{O_2} in humans during severe exercise of 1.8 ml [STPD] O_2/(s torr)), while accounting for $\dot{V}A/Q$ and $D_L/\dot{Q}$ inhomogeneities (see above). This value is only about 20% lower than the morphometric estimate for humans of 2.3 ml [STPD] O_2/(s torr)) (Weibel, 1984b), and suggests the morphometric $D_{L_{O_2}}$ may represent a meaningful estimate of maximal diffusing capacity.

8.3 Changes in D_L with Body Size

There are few comprehensive studies of changes in inter-specific physiologic D_L in mammals. In general, techniques and conditions vary substantially between measurements made in different laboratories, making comparisons of data to obtain a general allometric relationship somewhat unreliable. However, Stahl (1967) compiled such data from the literature for CO and O_2 diffusing capacities. He obtained logarithmic regressions of D_L as functions of

TABLE 10 Allometry of Pulmonary Diffusing Capacity

Taxon	Method	Intercept	Slope	N	*M* Range	Reference
$D_{L_{CO}}$ ml [STPD] CO · min^{-1} · $torr^{-1}$; pulmonary diffusing capacity for CO (physiologic)						
Mammals		0.22	1.14			Stahl, 1967
Mammals		0.614	0.985		0.025-500	O'Neil and Leith, 1980
Canids	sb	0.592	1.275	4	4.6-27.6	Weibel et al., 1983
Laboratory mammals	sb	0.4680	0.74	4	0.040-3.50	Takezawa et al., 1980
Dog *Canis familiaris*	sb	1.67	0.79	1	7.3-63	Robinson et al., 1972
Rabbit *Oryctolagus cuniculus*	sb	0.4074	0.92	1	1.30-3.50	Takezawa et al., 1980
Guinea pig *Cavia porcellus*	sb	0.4571	0.74	1	0.250-1.000	Takezawa et al., 1980
Rat *Rattus norvegicus*	sb	0.4467	0.75	1	0.050-0.400	Takezawa et al., 1980
Hamster *Mesocricetus auratus*	sb	0.7943	0.91	1	0.040-0.120	Takezawa et al., 1980
Mouse *Mus musculus*	rb	0.00106	0.971	1	0.020-0.039	Depledge, 1985
$D_{L_{O_2}}$ ml [STPD] O_2 · min^{-1} · $torr^{-1}$; pulmonary diffusing capacity for O_2 (morphometric)						
Mammals		0.16	1.18			Stahl, 1967
Mammals		3.92	0.991	33	0.003-700	Gehr et al., 1981a
Mammals		6.49	0.962	47	0.003-700	Calculated*
Canids		1.92	1.25	4	4.6-27.6	Weibel et al., 1983
Dog *C. familiaris*		6.56	1.00	1	2.65-57	Gehr et al., 1981b
Bats		6.25	0.932	5	0.005-0.173	Maina and King, 1984

Note: Allometry of pulmonary diffusing capacity for carbon monoxide ($D_{L_{CO}}$, physiologic) and oxygen ($D_{L_{O_2}}$, morphometric); both are in ml [STPD]/(min torr). D_L is related to body mass (*M*, in kg) by the function: $D_L = a \cdot M^b$ where a is the intercept of the log-transformed linear regression, and b is the slope of the log-transformed linear regression. Method indicates if $D_{L_{CO}}$ was determined by the single-breath (sb) or rebreathing (rb) technique. N is the number of species. In some cases, regression equations were calculated from data given in the reference. Units have been converted for equivalence, where necessary.
* Values for these regressions derived from data in studies by Gehr et al. (1981a); Weibel et al. (1983, 1987); Constantinopol et al. (1989); and Longworth et al. (1989).

body mass (i.e., mass exponents, b, in Eqn (1)) with slopes of 1.14 for $D_{L_{CO}}$ and 1.18 for $D_{L_{O_2}}$ (Table 10). These slopes suggest that specific D_L (D_L/M) increases as body size increases. In contrast, Takezawa et al. (1980) measured single-breath $D_{L_{CO}}$ in hamsters, rats, guinea pigs, and rabbits spanning a size range from 0.04 to 4 kg and obtained a mass exponent for $D_{L_{CO}}$ as a function of M of 0.74 (Table 10), indicating that $D_{L_{CO}}$ increases almost directly proportionally to standard $\dot{V}O_2$ rather than to body mass.

More comprehensive analyses of the relationship between body size and pulmonary diffusing capacity have been made for morphometric $D_{L_{O_2}}$. The largest body of work in this area has been by Weibel and colleagues (Gehr et al., 1981a; Weibel et al., 1983, 1987; Constantinopol et al., 1989; Longworth et al., 1989). They have measured morphometric $D_{L_{O_2}}$ for nearly 40 species of mammals, ranging in size from 0.003 to 474 kg. The advantage of comparing data from a single laboratory is that any potential bias in estimating morphometric $D_{L_{O_2}}$ (discussed previously) is unlikely to be systematically different for different-sized mammals. Thus, even if the absolute values of morphometric $D_{L_{O_2}}$ are not accurate, changes with body size should still be valid.

Gehr et al. (1981a) found that morphometric $D_{L_{O_2}}$ was related to body mass with a mass exponent of 0.99 (Table 10), a value not significantly different from direct proportionality between $D_{L_{O_2}}$ and body mass. With respect to the pulmonary structural components that determine $D_{L_{O_2}}$, they found that lung volume, alveolar surface area, capillary surface area, and capillary volume all changed directly proportionally (b = 1.0) with body mass (Table 11). The alveolar-capillary

TABLE 11 Allometry of Pulmonary Structural Variables

Taxon	Intercept	Slope	N	*M* Range	Reference
V_L, ml; lung volume					
Mammals	56.7	1.02	21	0.01-2000	Tenney and Remmers, 1963
Mammals	40.0	1.021	13	0.003-3.71	Lechner, 1978
Mammals	46.0	1.059	33	0.003-700	Gehr et al., 1981a
Mammals	47.5	1.060	47	0.003-700	Calculated*
Canids	54.9	1.15	4	4.6-27.6	Weibel et al., 1973
Dog *Canis familiaris*	52.8	1.07	1	2.65-57	Gehr et al., 1981b
Bats	96.4	1.07	5	0.005-0.173	Maina and King, 1984
S(A), m^2; alveolar surface area					
Mammals	1.87	0.888	13	0.003-3.71	L,echner,, 1978
Mammals	3.34	0.949	33	0.003-700	Gehr et al., 1981a
Mammals	3.36	0.935	47	0.003-700	Calculated*
Canids	1.12	1.38	4	4.6-27.6	Weibel et al., 1983
Dog *C. familiaris*	3.12	1.05	1	2.65-57	Gehr et al., 1981b
Bats[†]	5.18	1.01	5	0.005-0.173	Maina and King, 1984
S(c), m^2; pulmonary capillary surface area					
Mammals	2.73	0.952	33	0.003-700	Gehr et al., 1981a
Mammals	2.72	0.941	47	0.003-700	Calculate*
Canids	1.14	1.25	4	4.6-27.6	Weibel et al., 1983
Dog *C. familiaris*	2.53	1.05	1	2.65-57	Gehr et al., 1981b
V(c), ml; pulmonary capillary volume					
Mammals	3.20	1.000	33	0.003-700	Gehr et al., 1981a
Mammals	3.63	1.009	47	0.003-700	Calculated*
Canids	1.92	1.23	4	4.6-27.6	Weibel et al., 1983
Dog *C. familiaris*	3.28	1.12	1	2.65-57	Gehr et al., 1981b
Bats	3.73	0.954	5	0.005-0.173	Maina and King, 1984
τ_{ht}, μm; harmonic mean thickness of alveolar membrane					
Mammals	0.416	0.050	33	0.003-700	Gehr et al., 1981a
Mammals	0.413	0.053	47	0.003-700	Calculated*
Canids	0.476	0.000	4	4.6-27.6	Weibel et al., 1983
Dog *C. familiaris*	0.39	0.08	1	2.65-57	Gehr et al., 1981b
Bats	0.249	0.021	5	0.005-0.173	Maina and King, 1984

Note: Allometry of pulmonary structural variables that contribute to gas conductance of the lung and morphometric estimates of pulmonary diffusing capacity ($D_{L_{O_2}}$). Variables are related to body mass (*M*, in kg) by the function: $Y = a \cdot M^b$ where Y is the variable, a is the intercept of the log-transformed linear regression, and b is the slope of the log-transformed linear regression. N is the number of species. In some cases, regression equations were calculated from data given in the reference. Units have been converted for equivalence, where necessary.
*Values of these regressions derived from data in studies by Gehr et al. (1981a), Weibel et al. (1983, 1987), Constantinopol et al. (1989), and Longworth et al. (1989).
[†]These authors combined S(A) and S(c) to derive a functional "tissue" surface area, S(t).

barrier thickness did increase slightly (b = 0.05) as body size increased, but this change was not great enough to alter the apparent direct proportionality of $D_{L_{O_2}}$.

Direct proportionality between lung volume and body mass has also been found by other workers (Tenney and Remmers, 1963; Stahl, 1967; Lechner, 1978) and is consistent with the direct proportionality reported between physiological lung capacities and body mass (Leith, 1983). However, there are differences in alveolar surface area measured by different workers (Table 11). Tenney and Remmers (1963) reported measurements on 21 species spanning five orders of magnitude in body mass. They found that alveolar surface area changed in direct proportion to whole body $\dot{V}O_2$ ($\propto M^{\frac{3}{4}}$) rather than body mass. This result suggested that specific alveolar surface area is greater in smaller mammals. However, Gehr et al. (1981a) pointed out that four of the five largest species in this study were marine mammals that have a strikingly different pulmonary morphology than terrestrial mammals, making it inappropriate to include them in the comparison. This reasoning is particularly relevant because the marine species are all at one end of the size range and would, therefore, exert extreme leverage on the slope of the regression line. Gehr et al. recalculated the Tenney and Remmers data without the marine species and obtained a relationship indicating direct proportionality between alveolar surface area and body mass. Lechner (1978) obtained a mass exponent of 0.89 for the regression of alveolar surface area on body mass for 13 species ranging from 2 to 3700 g. However, the surface areas reported were cited from seven different studies and might not, therefore, be comparable. In view of the thickening of the alveolar-capillary barrier with body size (Gehr et al., 1981a), alveolar surface area alone is probably not, by itself, an appropriate index of $D_{L_{O_2}}$.

The relationship between size and $D_{L_{O_2}}$ has functional significance when related to rates of O_2 transfer in the lung. Taylor, Weibel, and co-workers (Weibel and Taylor, 1981; Taylor et al., 1987) compared the relationships of $D_{L_{O_2}}$ and $\dot{V}O_2max$ to M (Tables 3 and 10), and, in turn, the relationship between $D_{L_{O_2}}$ and $\dot{V}O_2max$ as M changes. They felt that comparison of $D_{L_{O_2}}$ to $\dot{V}O_2max$, rather than standard (basal) $\dot{V}O_2$ was most appropriate because $\dot{V}O_2max$ represents the functional limit for which the structural components of the pulmonary gas exchanger are presumably designed (Weibel and Taylor, 1981). If $D_{L_{O_2}}$ and $\dot{V}O_2max$ changed differently with size, then, according to Eqn (10), the alveolar to capillary P_{O_2} difference must change with size. Weibel et al. (1981) concluded that the disparate scaling of $D_{L_{O_2}}$ and $\dot{V}O_2max$ with M meant that the P_{O_2} difference between alveolar gas and capillary blood in the lung decreases with increasing body mass.

In order to critically examine the relationship between $D_{L_{O_2}}$ and $\dot{V}O_2max$, it is most appropriate to make the comparison in species in which both variables have been measured. Lechner (1978) found that in mammals spanning over three orders of magnitude in M, both alveolar surface area and $\dot{V}O_2max$ increased more slowly than did M ($b < 1.0$), and he concluded that $D_{L_{O_2}}$ may set a constraint on $\dot{V}O_2max$ as M increases. However, of the 13 species for which he reported $\dot{V}O_2max$, only in five were alveolar surface areas determined, adding significant variance to the comparison.

Taylor et al. (1988) compared $D_{L_{O_2}}$ and $\dot{V}O_2max$ in the same individuals of 18 mammalian species spanning three orders of magnitude in M. The allometric relationship they obtained agreed with their previous studies: $D_{L_{O_2}}$ varied as M with a mass exponent of 1.08 and $\dot{V}O_2max$ with a mass exponent of 0.86. In these 18 species it appears that the P_{O_2} difference across the lung decreases as body mass increases, or restated, smaller mammals must have a larger P_{O_2} difference for diffusion in the lung at $\dot{V}O_2max$ than do larger mammals, because they have relatively less D_L available to facilitate maximal O_2 transfer. Longworth et al. (1989) found that in one small mammal, the 4 kg fox, this larger pressure head could be achieved by a higher specific cardiac output and specific alveolar ventilation during maximal exercise.

9. CARBON DIOXIDE TRANSPORT

Inert gases, by definition, are carried in the blood only in dissolved form at concentrations proportional to their partial pressures, as described by Henry's Law. Oxygen and CO are both carried by the blood in physical solution and chemically bound to hemoglobin, and their chemical reaction rates have been discussed (Eqn (24), Table 12). Unlike these gases, CO_2 is transported in the blood both in solution and in multiple chemically bound forms, although only CO_2 that is physically dissolved actually diffuses across the alveolar-capillary membrane. Despite its 21-fold greater solubility than O_2 in blood, only approximately 5% of the CO_2 transported in blood is carried in solution (Klocke, 1987). Carbon dioxide reversibly reacts with two major components of interest in the blood: proteins, with which it forms carbamates (plasma proteins in plasma, hemoglobin in erythrocytes), and water, with which it hydrates to form carbonic acid, which then dissociates into bicarbonate ions and protons. The time course of the reactions with these components is potentially important for the kinetics of CO_2 exchange, as it has been estimated that dissolved CO_2 only constitutes approximately 8% of the CO_2 exchanged in the lung at rest, while that bound to proteins accounts for 13%, and the HCO^{-3} portion for 79%.

The hydration/dehydration reaction of CO_2 and H_2O is slow, requiring many seconds to proceed to completion if uncatalyzed; carbonic anhydrase (CA) increases the rate of

TABLE 12 Blood Respiratory Variables

Species	*M* (kg)	[Hb] ($g \cdot dl^{-1}$)	Hct (%)	O_2 capacity (ml [STPD] $O_2 \cdot dl^{-1}$)	P_{50} (torr)	Bohr Effect ($\Delta \log P_{50}/\Delta pH$)	Haldane Effect (ml [STPD] $CO_2 \cdot dl^{-1}$)	Temperature Coefficient ($\Delta \log P_{50}/\Delta T$)	Reference
African elephant *Loxodonta africana a.*	2000	15.3	42.7	20.5	23.2	−0.351	5.5		Dhindsa et al., 1972
Asian elephant *Elephas maximus*	1400	14.9	42	20.0	25.2	−0.351	5.5		Dhindsa et al., 1972
White rhinoceros *Cerathotherium sinum*					20	−0.62			Baumann et al., 1984
Ox *Bos taurus*		11.5	40		31.5	−0.49		0.023	Altman and Dittmer, 1971, Ross-Fanelli et al., 1964, Willford and Hill, 1986
Horse *Equus caballus*		11.1	33.4		25.1	−0.45		0.023	Altman and Dittmer, 1971, Ross-Fanelli et al., 1964, Willford and Hill, 1986
Camel *Camelus dromedarus*		9.41	22.1	12.6	25.9				Riegel et al., 1967
Tapir *Tapirus terrestris*						−0.58			Baumann et al., 1984
Man *Homo sapiens*	67.0		44.8		26.5	−0.47		0.024	Astrup et al., 1965, Lahiri, 1975
Bladdemose seal *Cystophora cristata*	50	26.4	63	36	24	−0.66	8.4		Clausen and Ersland, 1969
Pronghom antelope *Antilocapra americana*	38.4	16.8	42.7	22.5	28.3				Dhindsa et al., 1974
Orangutan, juvenile *Pongo pygmaeus*	34.6	12.9	40.9	17.4	24.4	−0.55			Riegel et al., 1966
Sheep *Ovis aries*	33.0		35.0		32.0	−0.49			Lahiri, 1975
Chimpanzee *Pan troglodytes*	30.4	13.5	41.2	17.3	26.7	−0.462	4.7		Lenfant and Aucutt, 1969
Pig *Sus scrofa*	30				35.7	−0.441		0.016	Willford and Hill, 1986
Goat *Capra hircus*	25.0		34.0		29.3	−0.48			Lahiri, 1975
Dog *Canis familiaris*	20.0		40.0		29.0	−0.48		0.022	Lahiri, 1975, Reeves et al., 1982

Continued

TABLE 12 Blood Respiratory Variables—cont'd

Species	*M* (kg)	[Hb] (g · dl^{-1})	Hct (%)	O_2 capacity (ml [STPD] $O_2 \cdot dl^{-1}$)	P_{50} (torr)	Bohr Effect ($\Delta \log P_{50}/\Delta pH$)	Haldane Effect (ml [STPD] $CO_2 \cdot dl^{-1}$)	Temperature Coefficient ($\Delta \log P_{50}/\Delta T$)	Reference
Gorilla, juvenile *Gorilla gorilla*	16.5	12.3	42.0	16.4	25.4	−0.48			Riegel et al., 1966
Kangaroo *Macropus giganteus*	15	18.6	53.0	24.9	27.5	−0.54			Bartels et al., 1966
Wallaby *M. eugenii*					32				Holland et al., 1988
Baboons *Papio anubis*	10.36	11.7	37.1	15.4	37.2	−0.550	4.7		Lenfant and Aucutt, 1969
Pigtail monkey *Macaca nemestrina*	9.50	12.6	42.0	16.6	36.7	−0.520	4.5		Lenfant and Aucutt, 1969
Tasmanian devil *Sarcophilus harrisii*	8	20.1	47.0	26.9	41.2	−0.47			Bartels et al., 1966
Woodchuck *Marmot monax*	4.45	13.5	39.4		27.8	−0.72			Boggs et al., 1984
Rhesus monkey *Macaca mulatta*	3.94	12.9	41.5	16.9	35.2	−0.518	4.4		Lenfant and Aucutt, 1969
Armadillo *Dasypus novemcinctus*	3.9	11.1	31	14.8	23.3	−0.55			Dhindsa et al., 1971
Pangolin *Manis pentadactyla*	3.6		36.5		25	−0.51			Weber et al., 1986
Opossum *Didelphis virginianis*	3.40		31.5		38.7	−0.49			Lahiri, 1975
Rabbit *Oryctolagus cuniculus*	3.10		35.0		30.0	−0.43			Lahiri, 1975
Cat *Felts domesticus*	2.64		42		35.0	−0.50			Lahiri, 1975
Echidna *Tachyglossus setosus*	2.1	17.6	48	21.9	21.3	−0.49		0.014	Lahiri, 1975
Platypus *Ornithorynchus anatinus*	1.8	18.3	52	22.7	27.2	−0.56		0.014	Parer and Metcalfe, 1967
Prairie dog *Cynomys ludoriciana*	1.28	15.1	47.3		22				Boggs et al., 1984
Muskrat *Ondatra zibethica*	0.97	13.3	36.2		26.1	−0.66			Rothstein et al., 1984

Squirrel monkey *Saimiri sciureus*	0.95	14.3	42.0	17.7	35.5	−0.542	4.5		Lenfant and Aucutt, 1969
Hedgehog *Erinaceus europaeus*	0.820		47.0	19.0	36	−0.49		0.017	Bartels et al., 1969, Clausen and Ersland, 1968
Guinea pig *Cavia porcellus*	0.669		41.1		26.7	−0.48			Lahiri, 1975
Ground squirrel *Spermophilus beecheyi*	0.598	15.1	48.7		26				Boggs et al., 1984
White *rat* *Rattus norvegicus*	0.555		41.9		36.0	−0.52			Lahiri, 1975
Guinea pig *Cavia porcellus*	0.291	12.6			25.3				Soulard et al., 1983
Mole rat *Spaiax ehrenbergi*	0.196	15.0	45.6		29.5	−0.53			Boggs et al., 1984
Bat *Rousettus aegyptiacus*	0.146	20.0	55		30.8	−0.55			Jiirgens et al., 1981
Golden hamster *Mesocricetus auratus*	0.135		42.4		26.0	−0.44			Lahiri, 1975
Pocket gopher *Thomomys bottae*	0.135	17.1	46		33.3	−0.61			Boggs et al., 1984
Naked mole rat *Heterocephalus glaber*	0.106	13.6	46		23.3	−0.43			Boggs et al., 1984
Gerbil *Meriones unguiculatus*	0.078		31.2		28.5	−0.51			Lahiri, 1975
Mouse *Mus musculus*	0.029		36.4		34.7	−0.50			Lahiri, 1975
Bat *Pipistrellus pipistrellus*	0.005	24.4	61.5		36.6	−0.47			Jiirgens et al., 1981
Mole *Talpa europa*			47.2	22.4	24	−0.47			Bartels et al., 1969
Shrew *Crocidura russula*			35.5	21.7	37	−0.63			Bartels et al., 1969

Note: Body mass is *M*; hemoglobin concentration, [Hb]; hematocrit, Hct; O_2 carrying capacity of the blood, O_2 capacity; oxygen half-saturation pressure, P_{50}; and oxygen equilibrium curve, temperature coefficient.

this reaction by 10,000 to 16,000× (Kemohan et al., 1963). Although CA is not present in plasma (Maren, 1967), it is highly concentrated in erythrocytes (Roughton, 1964), and its activity varies inversely with body mass, and thus, proportionally to specific rates of CO_2 production ($\dot{V}CO_2/M$) and elimination (Larimer and Schmidt–Nielsen, 1960; Magid, 1967, Table 13). Carbonic anhydrase has been localized on the membranes of pulmonary capillary endothelial cells (Effros et al., 1980; Lonnerholm, 1982; Ryan et al., 1982), and experimental evidence demonstrates that plasma HCO^{-3} has direct access to CA (Crandall and O'Brasky, 1978; Effros et al., 1978; Klocke, 1978), thus, the potentially slow plasma dehydration reaction may be directly catalyzed, although most of the HCO^{-3} probably diffuses into the erythrocyte for dehydration to be catalyzed by the much higher concentration of CA there (Klocke, 1987).

Carbon dioxide bound as plasma carbamate probably contributes insignificantly to CO_2 exchange, however, that bound to hemoglobin, the oxylabile portion, may play a significant role. This is because the binding affinity of hemoglobin for CO_2 changes with the oxygenation of hemoglobin and the concomitant release of protons, commonly referred to as the Haldane effect (Christiansen et al., 1914). For a given P_{CO_2}, deoxygenated hemoglobin binds more CO_2 than does oxygenated hemoglobin, resulting in maintenance of a higher pressure head for diffusion of CO_2 as hemoglobin is oxygenated and CO_2 is liberated during the erythrocyte's transit of the pulmonary capillary, functionally increasing the capacitance of the blood for CO_2. The contribution of the Haldane effect to CO_2 excretion in the lung has been estimated as 35% (Klocke, 1987) to 46% (Hill et al., 1973). The magnitude of the Haldane effect is generally expressed as the change in CO_2 concentration between deoxygenated and fully oxygenated hemoglobin at a P_{CO_2} of 40 torr and is a function of hemoglobin concentration. Representative values are given in Table 12 and show no body-size dependence.

10. OXYGEN TRANSPORT

Oxygen is carried in the blood both in solution and chemically bound to hemoglobin. The vast majority of O_2 transported and delivered to the tissues is that bound to hemoglobin, with each hemoglobin molecule having the potential to carry four O_2 molecules bound to it. Hemoglobin molecules exhibit cooperative ligand bonding, thus the relationship between the P_{O_2} in the blood and the quantity of O_2 bound to the hemoglobin is described by a sigmoidal oxygen equilibrium curve that is subject to allosteric modification. The total quantity of O_2 carried in the blood is directly proportional to the concentration of hemoglobin in the blood (Table 12), which in turn, is proportional to the hematocrit, or packed cell volume (Table 12). Despite the higher $\dot{V}O_2/M$ in smaller mammals, no systematic relationship exists between hemoglobin concentration or hematocrit and body size (Table 13; Schmidt–Nielsen, 1984).

TABLE 13 Allometry of Blood Variables

Taxon	Intercept	Slope	N	*M* Range	Reference
Blood volume ($cm^3 \cdot kg^{-1}$)					
Eutherians	65.6	1.02	97	$0.01–10^5$	Stahl, 1967
Mammals	76	1.0			Prothero, 1980
Hematocrit (%)					
Mammals	45.8	−0.01	123		Prothero, 1980
Oxygen half-saturation pressure (P_{50}, torr)					
Eutherians and marsupials	39.4	−0.03	89	0.021-6.35	Prothero, 1980
Marsupials	32.1	−0.074	7		Bland and Holland, 1977
Small mammals	50.3	−0.054	17		Schmidt-Nielsen and Larimer, 1958
Bohr effect ($\Delta \log P_{50}/\Delta pH$)					
Mammals (hemoglobin solution)	0.76	−0.0596	10	0.03-3140	Riggs, 1960
Erythrocyte carbonic anhydrase (U/μl erythrocyte* or U/100 μg hemoglobin)**					
Mammals*	1.146	−0.107	15	0.006-601	Larimer and Schmidt-Nielsen, 1960
Mammals**	0.695	−0.107	13	0.025-4000	Magid, 1967

Note: Variables are related to body mass (M, in kg) by the function: $Y = a \cdot M^b$, where Y is the variable, a is the intercept of the log-transformed linear regression, and b is the slope of the log-transformed linear regression. N is the number of species. In some cases, regression equations were calculated from data given in the reference. Units have been converted for equivalence, where necessary.

The ease with which hemoglobin binds O_2, or its O_2 affinity, is described by the P_{O_2} required to saturate half of the hemoglobin binding sites (P_{50}, or partial pressure resulting in 50% saturation). The P_{50} of the blood has great significance for gas exchange in the lungs and peripheral tissues, as the quantity of O_2 bound to the hemoglobin (proportional to its saturation) depends on the P_{O_2}, as does the pressure head for diffusion. Blood with a higher P_{50} will have a higher P_{O_2} at any given saturation than blood with a lower P_{50}, yielding a higher pressure head for diffusion into peripheral tissues (higher P_{O_2} in the blood at any saturation), but a lower pressure head for diffusion in the lung from alveolar gas to capillary blood.

Values of P_{50} measured in a number of mammalian species are in Table 12. It is well established that smaller mammals, with higher $\dot{V}O_2/M$, have higher P_{50} values than do larger mammals (Table 12; Schmidt–Nielsen and Larimer, 1958; Bland and Holland, 1977; Prothero, 1980), although specific environmental demands (e.g., adaptation to a fossorial niche) may affect individual species.

Bohr et al. (1904) noted that the affinity of hemoglobin for O_2 decreases with increasing P_{CO_2}, a phenomenon called the Bohr effect. The alteration of O_2 affinity is caused primarily by binding of protons to the hemoglobin molecule with the addition of CO_2, although CO_2 does have a small direct effect, and the Bohr effect is typically expressed as the change in O_2 affinity for a unit change in pH ($\Delta \log P_{50}/\Delta pH$); the value is negative because P_{50} decreases as pH increases. The overall significance of the Bohr effect for O_2 delivery is twofold: first, the release of protons by the hemoglobin when O_2 is bound buffers the pH change in the blood as CO_2 diffuses out of the pulmonary capillary (with the opposite effect in peripheral tissues); second, the increase in O_2 affinity as CO_2 is removed from the blood maintains a lower P_{O_2} in the pulmonary capillary as O_2 diffuses in, thus maintaining a larger $PA_{O_2}-\bar{P}_{CO_2}$, enhancing O_2 transport and resulting in a larger change in O_2 concentration in the blood for the same change in P_{O_2} between mixed venous and arterial blood. Analogous to the Haldane effect for CO_2, the Bohr effect creates a larger functional capacitance for O_2 in the blood.

Riggs (1960) observed that the magnitude of the Bohr effect (its absolute value) in hemoglobin solutions increases regularly with decreasing body size (Table 11), suggesting that it may facilitate O_2 exchange more for smaller mammals with higher $\dot{V}O_2/M$ (Schmidt–Nielsen, 1984). However, Bartels (1964) has noted that the change in Bohr effect with body size is much smaller in whole blood than in isolated hemoglobin solution (Table 12), thus minimizing this effect.

11. CAPILLARY TRANSIT (CONTACT) TIME

The mean contact time ($\bar{t}_c$) between the erythrocytes and the alveolar-capillary membrane is given by the ratio of pulmonary capillary volume [V(c)] and cardiac output ($\dot{Q}T$):

$$\bar{t}_c = \frac{V(c)}{\dot{Q}_T} \tag{25}$$

Lindstedt (1984) has demonstrated that because V(c) allometrically increases more ($\propto M^{1.0}$, Table 13) with increases in body size than does $\dot{Q}T$ ($\propto M^{0.8}$, Stahl, 1967), small mammals would appear to have less time available for blood in their pulmonary capillaries to equilibrate with alveolar gas than do large mammals. During exercise at $\dot{V}O_2$max, the pulmonary $\bar{t}_c$ has been estimated to vary from 0.13 s for 4.4-kg foxes (Longworth et al., 1989) to 0.4–0.5 s for 460-kg horses and steers (Constantinopol et al., 1989). The shorter contact times for diffusion would appear to be a potential problem for allowing equilibration between alveolar gas and arterial blood for smaller mammals; however, Longworth et al. (1989) data demonstrated almost no $PA_{O_2}-Pa_{O_2}$ difference in the 4-kg foxes at $\dot{V}_{O_2}$max, despite their high $\dot{V}_{O_2}\max/M$ and small $\bar{t}_c$. This suggests that even this short contact time in the fox does not result in diffusion limitation.

12. PARTIAL PRESSURES OF ALVEOLAR GAS

Fick's second law of diffusion (Eqn (7)) indicates that the flux of a gas will be driven by the difference in partial pressures of the gas across a barrier, e.g., the alveolar-capillary membrane, and Eqn (10) shows that the rates at which respiratory gases are transferred across the lung will be functions of both the mean alveolar pressures of the gases and their mean pressures in the pulmonary capillary blood. The values of these gas pressures are affected by different factors for different gases. The effects of hemoglobin's concentration and allosteric binding properties on the partial pressures of O_2 and CO_2 in the pulmonary capillary blood have been discussed previously.

The partial pressure of CO_2 in alveolar gas (PA_{CO_2}, torr) is a function of the rates at which CO_2 is being delivered to the alveolus for removal ($\dot{V}CO_2$, ml [STPD] CO_2/(min kg)) and the rate at which it is being removed from the alveolus, or alveolar ventilation ($\dot{V}A$, ml [BTPS]/(min kg)), as stated in the classic alveolar ventilation equation (Fenn et al., 1946):

$$PA_{CO_2} = K \cdot \frac{\dot{V}_{CO_2}}{\dot{V}_A} \tag{26}$$

where K is a factor that converts gas fractions to partial pressures and BTPS volumes to STPD. Minute ventilation and standard $\dot{V}O_2$ vary in mammals according to allometric equations with indistinguishable mass exponents (Stahl, 1967; Calder, 1984; Schmidt–Nielsen, 1984). If the respiratory exchange ratio ($R = \dot{V}CO_2/\dot{V}O_2$) were similar and dead space were approximately a constant fraction of tidal volume in all mammals (no data conflict with these assumptions), then it would appear that PA_{CO_2} in all mammals at rest must be nearly identical, approximately 40 torr.

Lahiri (1975) suggested that very small mammals might have lower blood and alveolar P_{CO_2} than larger mammals, however, measurements of P_{CO_2} from tissue gas pockets found no difference between values in shrews, bats, and pocket mice, and those in larger mammals (Tenney and Morrison, 1967). Alveolar P_{CO_2} is generally regarded as being approximately 40 torr and independent of body size in all mammals, except unusually adapted species, e.g., fossorial rodents (Dejours, 1981; Schmidt–Nielsen, 1984). Because of the high diffusivity of CO_2, (Table 5), Pa_{CO_2} is expected to usually be nearly identical to PA_{CO_2}, with little diffusion limitation.

Whereas PA_{CO_2} is determined by the ratio of $\dot{V}CO_2$ to $\dot{V}A$, the calculation of a value for PA_{O_2}, is complicated by the fact that it is affected by differences in the rates at which O_2 is being removed from the lung ($\dot{V}O_2$) and the rate at which CO_2 is entering it ($\dot{V}CO_2$), expressed as the respiratory exchange ratio ($R = \dot{V}CO_2/\dot{V}O_2$). Ideal alveolar P_{O_2} can be calculated using modifications of the concepts derived by Fenn et al. (1946):

$$PA_{O_2} = PI_{O_2} - PA_{CO_2} \cdot \left[FI_{O_2} + \frac{(1 - FI_{O_2})}{R} \right] \quad (27)$$

where PI_{O_2} is inspired P_{O_2} and FI_{O_2} is the inspired fraction of O_2. The PA_{O_2} is a function, therefore, of PA_{CO_2} (determined by $\dot{V}CO_2$ and $\dot{V}A$) and $\dot{V}O_2$ (determines R). The assessment of PA_{O_2}–Pa_{O_2} indicates whether the lung is functioning well as a gas exchanger, or whether oxygenation of the arterial blood is hampered by diffusion limitation, right to left vascular shunts, or mismatching of alveolar ventilation and perfusion. If resting PA_{CO_2} and R are similar for different-sized mammals, there is no reason to expect PA_{O_2} or Pa_{O_2} to vary systematically with body size.

13. BLOOD GAS VALUES

Although no systematic differences in resting blood gas values among mammals of different size are reported, during maximal exercise such differences are marked. Longworth et al. (1989) compared arterial blood gas data from 4-kg foxes exercising at $\dot{V}O_2max$ at sea level breathing air with those from larger mammals up to 460 kg (Taylor et al., 1987b; Jones et al., 1989a) and demonstrated that Pa_{CO_2} at $\dot{V}O_2max$ varied proportionally with body mass, and Pa_{O_2}, varied inversely, with values ranging from 19 to 47 torr (Pa_{CO_2}) and 120 to 87 torr (Pa_{O_2}). More severe hypoxemia (to 69 torr) has been reported in 460-kg thoroughbred horses (Jones, 1989) with higher $\dot{V}O_2max$ than the standardbreds used by Jones et al. (1989a). The reasons why small mammals should hyperventilate at $\dot{V}O_2max$ while large mammals hypoventilate, are not understood and are the subject of current research. However, data from the galloping horse show not only hypercapnia but also considerable diffusion limitation of O_2 exchange (Wagner et al., 1989) as well as severe pulmonary arterial hypertension (Erickson et al., 1992). Taken together, these findings clearly show that the equine lung is substantially underdeveloped for the O_2 and CO_2 transport requirements necessary for heavy exercise.

A critical factor in measuring blood gases for evaluation of pulmonary gas exchange function is that the values measured by the electrodes in a blood-gas analyzer, typically at a temperature of 37 °C, should be corrected to the animal's body temperature. The reason for this is that the solubilities of the gases, as well as the binding affinities with hemoglobin, decrease with increasing temperature. The temperature coefficient (Δ log $P_{50}/\Delta T$) for temperature changes in O_2 is on the order of 0.02 (Astrup et al., 1965; Reeves et al., 1982) for many species of mammals, however, significant variation exists (Table 13). A variety of algorithms are available for correcting P_{O_2}, P_{CO_2}, and pH for temperature changes (Ashwood et al., 1983). For resting, steady-state measurements, correction of blood gases to rectal temperature is usually adequate, as mammals are relatively isothermic. However, during heavy exercise, redistribution of blood flow from viscera to working muscle may result in differences between rectal and pulmonary arterial temperatures of several °C, yielding errors of up to 20 torr in calculated Pa_{O_2}, if the measured values were corrected to rectal temperature rather than to pulmonary arterial temperature (Jones et al., 1989b). Intramuscular temperatures may be even higher than pulmonary arterial temperature, thus evaluation of gas exchange during exercise requires direct measurement of temperature at the site of gas exchange.

14. ARTERIAL HYPOXEMIA

Because increased Pa_{CO_2} is a stronger stimulant of ventilation than decreased Pa_{O_2}, impairment of pulmonary gas exchange usually causes alterations in alveolar ventilation to maintain normal Pa_{CO_2} while Pa_{O_2} is not as closely regulated. Furthermore, the high diffusivity of CO_2 relative to O_2, means that CO_2 rarely fails to equilibrate between alveolar gas and capillary blood. As a result, Pa_{O_2} is the variable most affected by impairment of the pulmonary gas exchanger.

Arterial hypoxemia, defined as a significant reduction in O_2, concentration in the blood, is a term frequently used to indicate a reduction in Ca_{O_2} due to reduced Pa_{O_2}, as opposed to alterations in the blood's capacitance for O_2, e.g., anemia or carboxemia. Five primary causes of arterial hypoxemia are recognized: (1) reduced inspired P_{O_2}, either due to decreased FI_{O_2} or decreased total gas pressure, e.g., at altitude; (2) hypoventilation, in which case increased PA_{CO_2} reduces PA_{O_2} (Eqn (27)); (3) diffusion limitation, discussed previously; (4) shunt, in which some mixed venous blood mixes with blood that has already undergone alveolar gas exchange, thereby lowering the O_2 concentration of the arterial mixture and the Pa_{O_2}; and (5) ventilation-perfusion inhomogeneities, in which the ratios of alveolar ventilation and blood flow in different groups of alveoli vary, reducing Ca_{O_2}, and Pa_{O_2}. The first two mechanisms cause arterial hypoxemia by directly reducing PA_{O_2}. Diffusion limitation results in hypoxemia when the pulmonary capillary P_{O_2} does not increase sufficiently to be close to PA_{O_2} before exiting the capillary. Shunt directly lowers Pa_{O_2} by reducing the concentration of O_2 in the arterial blood after venous admixture occurs.

Ventilation-perfusion mismatching causes a decrease in Pa_{O_2} by two mechanisms. First of all, alveolar units with a high $\dot{V}A/\dot{Q}$ ratio (i.e., hyperventilated alveoli) yield capillary blood with a high Pa_{O_2}, in contrast to low $\dot{V}A/\dot{Q}$ units (i.e., hypoventilated alveoli) that yield capillary blood with a low Pa_{O_2}. The outputs of these two groups do not counterbalance each other, however, as the sigmoid shape of the O_2-hemoglobin equilibrium curve results in high Pa_{O_2} blood carrying only a slightly greater concentration of O_2 than blood from alveoli with a normal $\dot{V}A/\dot{Q}$ ratio, in which the hemoglobin is already 95–97% saturated with O_2. In contrast, blood emanating from the low $\dot{V}A/\dot{Q}$ units with a low Pa_{O_2} has a markedly decreased O_2 concentration, thus, blood from the low $\dot{V}A/\dot{Q}$ units has an O_2 concentration that is reduced more than the O_2 concentration in the blood from the high $\dot{V}A/\dot{Q}$ units is increased. A second mechanism that compounds this effect in lowering Pa_{O_2}, is that the inherently greater blood flow to low $\dot{V}A/\dot{Q}$ units means that there will inevitably be *more* blood contributed to the arterial pool from the low $\dot{V}A/\dot{Q}$ units than from the high $\dot{V}A/\dot{Q}$ regions, amplifying the reduction in Ca_{O_2} and Pa_{O_2}, due to the hypoxemic blood from the low $\dot{V}A/\dot{Q}$ regions.

15. VENTILATION-PERFUSION DISTRIBUTIONS

It has long been recognized that inhomogeneity in the distribution of ventilation and perfusion to alveoli must lead to impaired gas exchange (Haldane, 1922). Methods to assess the distribution of ventilation and perfusion have included the use of inhaled or infused radioactive marker gases, which have demonstrated vertical gradients of pulmonary ventilation and perfusion in conscious, upright humans (Knipping et al., 1955; West and Dollery, 1960; Ball et al., 1962; West, 1964) and horses (Amis et al., 1984b), although not in greyhounds (Amis et al., 1982). Although the topographic distributions of ventilation and perfusion measured by these techniques are useful in detecting gross differences in $\dot{V}A/\dot{Q}$, they are incapable of resolving local inhomogeneities that may have great functional significance.

Physiologic methods to assess continuous distributions of $\dot{V}A/\dot{Q}$ have progressed ever since the experimental evidence of Krogh and Lindhard (1914, 1917) demonstrated that the distribution of alveolar ventilation is not uniform. Major breakthroughs in this area came with the development of the *O_2-CO_2 diagram* by Fenn et al. (1946) and its utilization in quantifying ventilation-perfusion relationships with a log normal distribution (Rahn, 1949; Rahn and Fenn, 1955). Simultaneously, Lilienthal et al. (1946) were developing concepts of venous admixture that led to Riley and Cournand (1949, 1951) developing the *three-compartment model*, in which the lung is modeled as containing an ideal, evenly ventilated, and perfused compartment, a nonventilated (shunt) compartment, and a nonperfused (dead space) compartment, and the *four-quadrant diagram*, relating partial pressures and concentrations of O_2 and CO_2 in various ideal and mixed gas and blood components [e.g., mixed expired vs. (ideal) alveolar component of expired gas, mixed arterial vs. (ideal) alveolar component of arterial blood] for different respiratory exchange ratios (Riley and Permutt, 1964). Other approaches have used wash-in or washout of insoluble gases (e.g., Briscoe, 1959; Briscoe et al., 1960) or alteration of increased O_2 pressure to compare its effects on alveolar-arterial P_{O_2} differences with those expected from different $\dot{V}A/\dot{Q}$ distributions (Lenfant and Okubo, 1968); these methods assume that the distributions being measured are unaffected by the experimental perturbations.

In 1967, the use of inert gas washout to analyze $\dot{V}A/\dot{Q}$ inequality in dog lungs was pioneered (Farhi and Yokoyama, 1967). Information regarding $\dot{V}A/\dot{Q}$ distribution is obtained from inert gas elimination rates, because for any given $\dot{V}A/\dot{Q}$ ratio, a less soluble gas will be excreted from the blood to alveolar gas more completely than will a more soluble gas. As a result, if multiple inert gases are utilized, it is possible to estimate the distribution of ventilation and perfusion ratios in the lung based on the ratios of inert gases of different solubilities that are excreted into alveolar gas or retained in arterial blood as gases are exchanged in the lung (Yokoyama and Farhi, 1967; Wagner et al., 1974b; Wagner and West, 1980).

The theoretical basis for the analysis of continuous distributions of $\dot{V}A/\dot{Q}$ ratios was developed by Kety (1951), Noehren (1962), and Farhi (1967). Consideration of mass balance indicates that for each inert gas, its retention (R), defined as the ratio of partial pressure of the gas in the capillary (P_C) to that in the mixed venous blood ($P_{\bar{v}}$) and its excretion (E), defined as the ratio of alveolar (P_A) and mixed venous partial pressures, is as follows.

$$\frac{P_C}{P_{\bar{v}}} = \frac{P_A}{P_{\bar{v}}} = \frac{\lambda}{\lambda + \frac{\dot{V}A}{\dot{Q}}} \quad (28)$$

where λ is the blood/gas partition coefficient for that gas, i.e., its solubility, and $\dot{V}A/\dot{Q}$ is the ventilation-perfusion ratio of that alveolus. As each inert gas with its individual λ is retained by each alveolus with its individual $\dot{V}A/\dot{Q}$ ratio according to Eqn (28), then the total lung retention (yielding the mixed arterial concentration) will be a blood flow-weighted mean of these compartmental values:

$$R = \frac{P_a}{P_{\bar{v}}} = \frac{1}{\dot{Q}} \sum_{i=1}^{n} \frac{\dot{Q}_i \cdot \lambda}{\lambda + \frac{\dot{V}_{A_i}}{\dot{Q}_i}}, \tag{29}$$

where $\dot{Q}$ is total pulmonary blood flow, $\dot{Q}_i$ is blood flow to the ith alveolus, and $\dot{V}_{A_i}$ is ventilation to the ith alveolus. Similarly, the excretion of the gas from each alveolus will yield a mixed alveolar concentration according to a ventilation-weighted mean:

$$E = \frac{P_A}{P_{\bar{v}}} = \frac{1}{\dot{V}_A} \sum_{i=1}^{n} \frac{\dot{V}_{A_i} \cdot \lambda}{\lambda + \frac{\dot{V}_{A_i}}{\dot{Q}_i}}, \tag{30}$$

where $\dot{V}_A$ is total alveolar ventilation. If retentions and excretions of inert gases of different λ are measured during steady-state conditions, it is possible to calculate a distribution of $\dot{V}A/\dot{Q}$ ratios in the lung that would produce the observed retention/excretion pattern.

In practical terms, the multiple-inert gas elimination technique (MIGET) has been developed to yield considerable information about distributions of ventilation and perfusion (Wagner et al., 1974; West and Wagner, 1977; Wagner and West, 1980). The technique typically uses six inert gases (e.g., acetone, diethylether, enflurane or halothane, cyclopropane, ethane, and sulfur hexafluoride) that have considerably different blood/gas partition coefficients spanning a range of nearly 10^5. The gases are infused into a peripheral vein, and mixed expired, arterial, and/or mixed venous blood samples are drawn simultaneously in order to determine the retentions and excretions of the gases. Analysis of the gases, after extraction from the blood samples, is done with a flame ionization or electron capture (SF_6 only) detector on a gas chromatograph (Wagner et al., 1974; Wagner and Lopez, 1984). Blood/gas partition coefficients are then measured for the gases, and the resulting retention/excretion vs. λ relationships are utilized to calculate $\dot{V}A/\dot{Q}$ distributions in the lung (Olszowka and Wagner, 1980).

The retention/excretion data are used to simultaneously solve the six linear equations that describe the retention and excretion of each of the inert gases with different λ in order to find the distribution of $\dot{V}A/\dot{Q}$ ratios in the lung that best fits the measured values. Typically, the data are fit to a model with 50 $\dot{V}A/\dot{Q}$ values logarithmically scaled from 0.001 to 100, plus shunt ($\dot{V}A/\dot{Q} = 0$) and dead space ($\dot{V}A/\dot{Q} = \infty$) (Olszowka and Wagner, 1980). The measured $\dot{V}A/\dot{Q}$ distribution allows estimates to be made of the degree to which Pa_{O_2} is lowered due to shunt and $\dot{V}A/\dot{Q}$ mismatch, thereby allowing the contribution of diffusion limitation to $PA_{O_2}-Pa_{O_2}$ to be assessed. Alteration of a number of different parameters related to the dispersion of the $\dot{V}A/\dot{Q}$ distribution (e.g., log SD of the $\dot{V}A$ or $\dot{Q}$ data) can affect gas exchange (Hlastala and Robertson, 1978; Neufeld et al., 1978; Gale et al., 1985).

Studies of $\dot{V}A/\dot{Q}$ distributions in resting or anesthetized humans, dogs, and horses have found relatively good matching of ventilation and perfusion, occasionally with some ventilation of high $\dot{V}A/\dot{Q}$ regions that are poorly perfused, but in general with log SD of the $\dot{V}A/\dot{Q}$ values <0.6 (Dueck et al., 1977; Hedenstierna et al., 1987; Wagner et al., 1987). During exercise, Wagner et al. (1986) have shown that in humans at sea level $\dot{V}A/\dot{Q}$ distributions become less homogenous, contributing to an increase in $PA_{O_2}Pa_{O_2}$ with intense exercise. Diffusion limitation is also often observed. Why these changes occur remains uncertain still today but the prevailing hypothesis is transient interstitial edema. In contrast, horses demonstrate no decrease in $\dot{V}A/\dot{Q}$ matching with heavy exercise, despite their development of marked arterial hypoxemia (Wagner et al., 1989). This demonstrates the substantial degree of diffusion limitation horses show during heavy exercise. Insufficient data are available from other species of mammals to determine a pattern of change in $\dot{V}A/\dot{Q}$ distributions with exercise.

16. INTEGRATION OF TRANSPORT PROCESSES FOR O_2 BETWEEN THE AIR AND THE MUSCLE MITOCHONDRIA

The entire preceding discussion has examined O_2 transport functions individually, but to understand the limits to mitochondrial O_2 availability and utilization, one must consider the entire O_2 transport pathway as a single integrated system. Figure 4 shows a simple cartoon of the mammalian O_2 transport system. It shows that the transport system includes the lungs, heart, circulation (vasculature and the blood within it) and the muscles. It also shows the principal transport functions at each step. This figure clearly implies that O_2 transport consists of an in series system of conductances, and they are listed on the right side of the figure: ventilation (convective), alveolar-capillary (AC) diffusion, circulation (convective), and capillary to

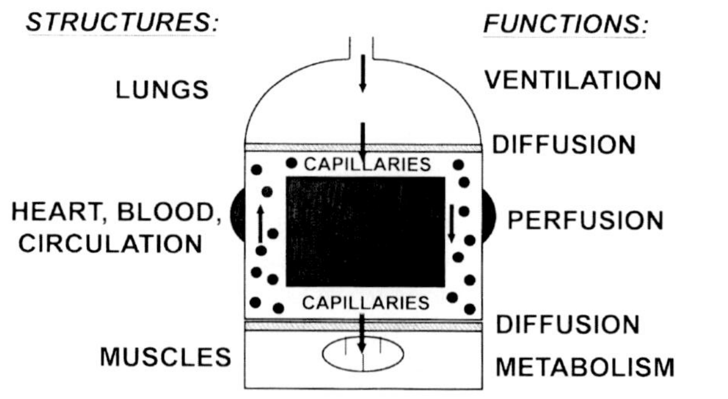

FIGURE 4 Diagram of the major steps in the O_2 transport pathway from the air to the mitochondria. See text for further details.

mitochondrion (CM) diffusion within muscle. While two of the processes are based on convection, and two on diffusion, there is a unifying principle that at each step, mass of O_2 is conserved—O_2 is neither gained nor lost in each of the transport processes at each step.

This mass conservation principle can be expressed quantitatively with a transport equation describing each step. Here the linear approximation to the O_2Hb dissociation curves has been used to simplify the system and allow the principles to be laid out:

$$\text{Ventilation:}\quad \dot{V}O_2 = \dot{V}I \cdot k \cdot PI_{O_2} - \dot{V}A \cdot k \cdot PA_{O_2} \quad (31)$$

$$\text{Alveolar-capillary Diffusion:}\quad PA_{O_2} - Pa_{O_2} = (PA_{O_2} - PV_{O_2}) \cdot \exp\left(\frac{-D_L}{(\beta \cdot \dot{Q}_T)}\right) \quad (8b)$$

$$\text{Circulation:}\quad \dot{V}O_2 = \dot{Q}_T \cdot (Ca_{O_2} - Cv_{O_2}) \quad (32)$$

$$\text{Capillary-mitochondrial Diffusion:}\quad (Pa_{O_2} - Pv_{O_2}) = (Pa_{O_2} - Pm_{O_2}) \cdot \exp\left(\frac{-D_M}{(\beta \cdot \dot{Q}_T)}\right) \quad (33)$$

In the above equations, PI_{O_2}, PA_{O_2}, Pa_{O_2}, Pv_{O_2} and Pm_{O_2} are the O_2 partial pressures in inspired air, alveolar gas, arterial blood, venous blood and mitochondria respectively, while k is a constant relating partial pressure to fractional concentration of O_2 in the gas phase, equal to (1/total dry gas pressure). β remains as the linear approximation to the slope of the O_2Hb dissociation curve.$\dot{V}_{O_2}$ is the actual O_2 consumption rate, $\dot{V}I$ and $\dot{V}A$ are inspired and expired alveolar ventilation (usually approximated to be equal), and $\dot{Q}_T$ is cardiac output. D_L and D_M are lung and muscle O_2 diffusing capacities respectively. Note that Eqns (8b) and (32) embody the linear approximation to the O_2Hb curve, allowing the diffusion processes in lung and muscle to be described by the simple equations indicated, but the other three equations are general and work for both the linear and actual dissociation curves.

Finally, since mitochondrial use of O_2 does depend on available O_2 (via the law of mass action), $\dot{V}O_2$ is a quantifiable function of Pm_{O_2}. This function is found to be hyperbolic (Gnaiger et al., 1998) (Scandurra and Gnaiger, 2010) as predicted by Michaelis–Menten theory. The determinants of the relationship are two constants: maximal mitochondrial $\dot{V}O_2$ defined as the $\dot{V}O_2$ attainable when O_2 is in excess (Vmax), and the P_{50} of the hyperbolic relationship (i.e., the mitochondrial P_{O_2} when $\dot{V}O_2$ is 50% of Vmax). Thus:

$$\dot{V}O_2 = \text{Vmax} \cdot \left(\frac{Pm_{O_2}}{Pm_{O_2} + P_{50}}\right) \quad (34)$$

These equations conserve O_2 at every step by ensuring that, as each of the P_{O_2}'s appear in different steps, they are numerically the same whichever step equation they are part of, and similarly, that $\dot{V}O_2$ is the same at every step. If one wished to solve these equations using actual dissociation curves, Eqns (8b) and (32) would have to be replaced by the numerical integrative procedures mentioned above, both in the lungs Eqn (8b) and muscles Eqn (32), where, because the principles are identical, the equations are also identical except for different boundary conditions and diffusing capacities.

The most important messages from Eqns (31)–(34) are as follows.

First, for given input values of inspired P_{O_2}, ventilation, cardiac output, [Hb] and β, lung and muscle diffusing capacities, and mitochondrial O_2 utilization kinetics (Vmax and P_{50}), the five equations, contain exactly five unknowns (PA_{O_2}, Pa_{O_2}, Pv_{O_2}, Pm_{O_2} and $\dot{V}O_2$). This means that they can be solved uniquely for these five variables. Thus, the given input variables uniquely combine to determine $\dot{V}O_2$ and the O_2 associated partial pressures at each step of the transport pathway.

This analysis describes the general case, but an important specific case is seen when mitochondrial P_{O_2}, Pm_{O_2}, is high enough that $\dot{V}O_2$ essentially equals Vmax. In that case, and at all higher values of Pm_{O_2}, $\dot{V}O_2$ loses its dependence on O_2 transport and its determining conductances defined by the above equations, and sits at Vmax, which cannot of course be exceeded no matter how high the Pm_{O_2}.

The second important message is that all organs and functional steps in the transport pathway (see Figure 4)—not just one—play a quantitative role in determining actual $\dot{V}O_2$. This is illustrated in Figure 5, taken from (Wagner, 1996) where changes in the value of any single variable affect $\dot{V}O_2$ rather similarly across a wide range, both when above and when below normal. The calculations of Figure 5 reflect the solution to the five Eqns (31)–(34).

The third important message is that the determining variables act (1) directly in the step of their principal involvement (e.g., $\dot{Q}_T$ in circulating blood between lung and muscles) and also (2) indirectly in other organs/functions (e.g., $\dot{Q}_T$ in affecting D_L $(\beta \cdot \dot{Q}_T)$ and $D_M/(\beta \cdot \dot{Q}_T)$). But note the directionality: an increase in $\dot{Q}_T$ *enhances* circulatory O_2 transport (see Eqn (32)) but *impedes* diffusive exchange in both lungs and muscles (see Eqns (8b) and (33)).

The fourth important message is that, as shown in Figure 5, increases in the value of any single conductance do little to improve $\dot{V}O_2$, but decreases in any single conductance have a major negative impact on $\dot{V}O_2$. This behavior embodies that of a traditional "bucket brigade", where with several strong people and one weak person, the rate of passing buckets along the line is no more than that which the weakest person can manage.

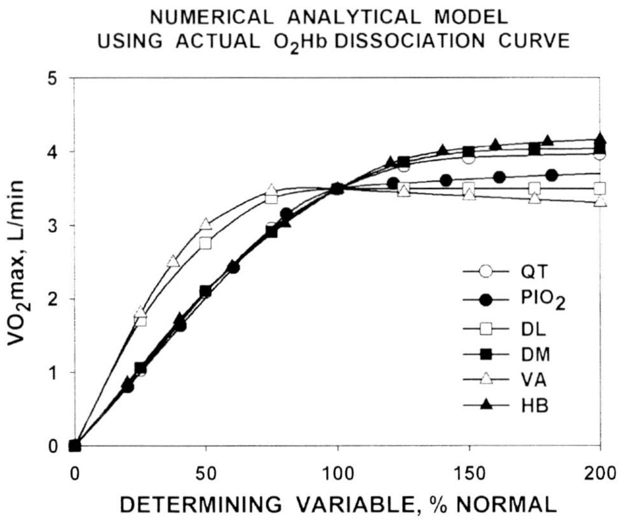

FIGURE 5 Dependence of maximal $\dot{V}O_2$($\dot{V}O_2$max) on the values of the major variables involved in O_2 transport. Each curve shows the effects of changing the single indicated variable while keeping all others constant. This figure shows that each step of the pathway has roughly similar influence on $\dot{V}O_2$max, and that the effects are asymmetrical. Thus, decreases in any variable greatly impair O_2 transport while increases provide little benefit, much like the behavior of a "bucket brigade".

Finally, as a set of general principles based on the most basic of physical laws—conservation of mass—this systems analysis of O_2 transport can be applied across the mammalian world, using values of the above-mentioned transport variables appropriate to each species, and transport equations that account for anatomical and functional differences.

17. SUMMARY

Gas exchange is a fundamentally important function of the lung. The same basic mechanisms of diffusion and convection govern gas exchange and gas transport in all mammals. The demand for O_2 set by the body varies allometrically with changes in body size, so that smaller mammals have a higher mass-specific demand than do larger mammals. Structural variables associated with the lung and O_2 transport system (surface area, membrane thickness, capillary volume, blood O_2 capacitance) do not change allometrically on a mass-specific basis, therefore, the enhanced gas exchange that smaller mammals exhibit results mainly from changes in functional parameters in the system, primarily ventilatory and cardiac frequencies. These size-dependent differences in structure–function relationships change regularly with body size and result in greater variation in aerobic demand and gas-exchange rates than exists within mammals of any given size. Extrapolation of the basic principles and mechanisms of gas exchange in mammals requires careful consideration of allometric relationships associated with respiratory function, particularly during exercise, when the O_2 transport system may be utilized to its maximal capacity.

REFERENCES

Altman, P.L., Dittmer, D.S., 1971. Biological Handbooks: Respiration and Circulation, vol. 930. Federation of American Societies for Experimental Biology, Bethesda, MD.

Amis, T.C., Jones, H.A., Rhodes, C.G., Heather, J.D., Hughes, J.M.B., 1982. Regional distribution of pulmonary ventilation and perfusion in the conscious dog. Am. J. Vet. Res. 43, 1972–1977.

Amis, T.C., Jones, H.A., 1984a. Measurement of functional residual capacity and pulmonary carbon monoxide uptake in conscious Greyhounds. Am. J. Vet. Res. 45, 1447–1450.

Amis, T.C., Pascoe, J.R., Hornof, W., 1984b. Topographic distribution of pulmonary ventilation and perfusion in the horse. Am. J. Vet. Res. 45, 1597–1601.

Anthonisen, N.R., Fleetham, J.A., 1987. Ventilation: total, alveolar, and dead space. In: Farhi, L.E., Tenney, S.M. (Eds.), Handbook of Physiology. Sect. 3: The Respiratory System. American Physiological Society, Bethesda, MD, pp. 113–129.

Ashwood, E.G., Kost, G., Kenny, M., 1983. Temperature correction of blood-gas and pH measurements. Clin. Chem. 29, 1877–1885.

Astrand, P., Rodahl, K., 1986. Textbook of Work Physiology, third ed., vol. 681. McGraw Hill, New York.

Astrup, P., Engel, K., Severinghaus, J.W., Munson, E., 1965. Scand. J. Clin. Lab. Invest. 17, 515–523.

Ball, W.C., Stewart, P.B., Newsham, G.S., Bates, D.V., 1962. Regional pulmonary function studied with xenon^{133}. J. Clin. Invest. 41, 519–531.

Barcroft, J., Cook, A., Hartridge, H., Parsons, T.R., 1920. The flow of oxygen through the pulmonary epithelium. J. Physiol. Lond. 53, 450–472.

Bartels, H., 1964. Comparative physiology of oxygen transport in mammals. Lancet 1964, 599–604.

Bartels, H., Riegel, K., Kleihauer, E., 1966. Comparative studies of the respiratory functions of mammalian blood. II. Marsupialia: great grey kangaroo and Tasmanian devil. Respir. Physiol. 1, 145–150.

Bartels, H., Schmelzle, R., Ulrich, S., 1969. Comparative studies of the respiratory functions of mammalian blood. V. Insectivora: shrew, mole and nonhibernating and hibernating hedgehog. Respir. Physiol. 7, 278–286.

Bates, D.V., Macklem, P.T., Christie, R.V., 1971. Respiratory Function in Disease, second ed., vol. 584. W.B. Saunders, Philadelphia.

Baumann, R., Mazur, G., Braunitzer, G., 1984. Oxygen binding properties of hemoglobin from the white rhinoceros (B2-GLU) and the tapir. Respir. Physiol. 56, 1–9.

Benedict, F.G., 1938. Vital Energetics: A Study in Comparative Basal Metabolism, vol. 215. Carnegie Institute of Washington, Washington, D.C.

Bergmann, C., Leuckart, R., 1855. Anatomisch-Physiologische Ubersicht Des Tierreichs, vol. 247. J.B. Muller, Stuttgart.

Birks, E.K., Jones, J.H., Vandervort, L.J., Priest, A.K., Berry, J.D., 1991. Plasma lactate kinetics during exercise. In: Persson, S.G.B., Lindholm, A., Jeffcott, L.B. (Eds.), Equine Exercise Physiology 3, Almqvist and Wiksell Tryckeri AB. Uppsala, Sweden, pp. 179–187.

Bland, D.K., Holland, R.A.B., 1977. Oxygen affinity and 2,3-diphosphoglycerate in blood of Australian marsupials of differing body size. Respir. Physiol. 31, 279–290.

Boggs, D.F., Kilgore, D.L., Birchard, G.F., 1984. Respiratory physiology of burrowing mammals and birds. Comp. Biochem. Physiol. 77A, 1–7.

Bohr, C., 1891. Ueber die Lungenatmung. Skand. Arch. Physiol. 2, 236–268.

Bohr, C., Hasselbach, K., Krogh, A., 1904. Ueber einen in biologischer Beziehung wichtigen Einfluss, den die Kohlensaurespannung des Blutes auf dessen Sauerstoffspannung ausbubt. Skand. Arch. Physiol. 16, 402–412.

Bohr, C., 1909. Ueber die spezifische Tatigkeit der Lungen bei der respiatorischen Gasaufnahme and ihr Verhalten zu der durch die Alveolarwand stattfindenden Gasdiffusion. Scand. Arch. Physiol. 22, 221–280.

Briscoe, W.A., 1959. A method for dealing with data concerning uneven ventilation of the lung and its effect on gas transfer. J. Appl. Physiol. 14, 291–298.

Briscoe, W.A., Cree, E.M., Filler, J., Houssay, H.E.J., Cournand, A., 1960. Lung volume, alveolar ventilation and perfusion interrelationships in chronic pulmonary emphysema. J. Appl. Physiol. 15, 785–795.

Brody, S., Procter, R.C., Ashworth, U.S., 1934. Basal metabolism, endogenous nitrogen, creatinine and neutral sulphur excretions as functions of body weight. Univ. Mo. Agric. Exp. Sta. Res. Bull. 220, 1–40.

Brody, S., 1945. Bioenergetics and Growth, vol. 915. Reinhold Publ. Corp., New York.

Calder, W.A., 1984. Size, Function, and Life History, vol. 431. Harvard University Press, Cambridge, MA.

Carlin, J.I., Cassidy, S.S., Rajagopal, U., Clifford, P.S., Johnson, R.L., 1988. Noninvasive diffusing capacity and cardiac output in exercising dogs. J. Appl. Physiol. 65, 669–674.

Cassidy, S.S., Ramanathan, M., Rose, G.L., Johnson, R.L., 1980. Hysteresis in the relation between diffusing capacity of lung and lung volume. J. Appl. Physiol. 49, 566–570.

Chang, H.K., 1987. Diffusion of gases. In: Farhi, L.E., Tenney, S.M. (Eds.), Handbook of Physiology. Sect. 3: The Respiratory System. American Physiological Society, Bethesda, MD, pp. 33–50.

Chappell, M.A., 1984. Maximum oxygen consumption during exercise and cold exposure in deer mice, *Peromyscus maniculatus*. Respir. Physiol. 55, 367–377.

Chinet, A., Micheli, J.L., Haab, P., 1971. Inhomogeneity effects on O_2 and CO pulmonary diffusing capacity estimates by steady-state methods. Theory Respir. Physiol. 13, 1–22.

Christiansen, J., Douglas, C.G., Haldane, I.S., 1914. The absorption and dissociation of carbon dioxide by human blood. J. Physiol. Lond. 48, 244–271.

Clausen, G., Ersland, A., 1968. The respiratory properties of the blood of the hibernating hedgehog, *Erinaceus europaeus* L. Respir. Physiol. 5, 221–233.

Clausen, G., Ersland, A., 1969. The respiratory properties of the blood of the bladdemose seal (*Cystophora cristata*). Respir. Physiol. 7, 1–6.

Coin, J.T., Olson, J.S., 1979. The rate of oxygen uptake by human red cells. J. Biol. Chem. 254, 1178–1190.

Conley, K.E., Weibel, E.R., Taylor, C.R., Hoppeler, H., 1985. Aerobic capacity estimated by exercise versus cold-exposure: endurance training effects in rats. Respir. Physiol. 62, 273–280.

Constantinopol, M., Jones, J.H., Weibel, E.R., Taylor, C.R., Lindholm, A., Karas, R.H., 1989. Oxygen transport during exercise in large mammals. II. Oxygen uptake by the pulmonary gas exchanger. J. Appl. Physiol. 67, 871–878.

Cotes, J.E., 1975. Lung Function, third ed., vol. 619. Blackwell, Oxford.

Crandall, E.D., O'Brasky, J.E., 1978. Direct evidence for participation of rat lung carbonic anhydrase in CO_2 reactions. J. Clin. Invest. 62, 618–622.

Crapo, J.D., Crapo, R.D., 1983. Comparison of total lung diffusion capacity and the membrane component of diffusion capacity as determined by physiologic and morphometric techniques. Respir. Physiol. 51, 183–194.

Crapo, J.D., Crapo, R.O., Jensen, R.L., Mercer, R.R., Weibel, E.R., 1988. Evaluation of lung diffusing capacity by physiological and morphometric techniques. J. Appl. Physiol. 64, 2083–2091.

Crapo, R.O., Crapo, J.D., Morris, R.H., 1982. Lung tissue and capillary volumes by rebreathing and morphometric techniques. Respir. Physiol. 49, 175–186.

Crapo, R.O., Crapo, J.D., Weibel, E.R., 1986. Comparison of physiologic and morphometric estimations of the components of CO diffusing capacity. Prog. Resp. Res. 21, 88–91.

Dawson, T.J., Hulbert, A.J., 1970. Standard metabolism, body temperature, and surface areas of Australian marsupials. Am. J. Physiol. 218, 1233–1238.

Dejours, P., 1975. Principles of Comparative Respiratory Physiology, vol. 253. Elsevier/North-Holland Biomedical Press, Amsterdam.

Dejours, P., 1981. Principles of Comparative Respiratory Physiology, second ed., vol. 265. Elsevier/North-Holland Biomedical Press, Amsterdam.

Depledge, M.H., 1985. Respiration and lung function in the mouse, *Mus musculus* (with a note on mass exponents and respiratory variables). Respir. Physiol. 60, 83–94.

Depocas, F., Hart, J.S., 1957. Use of the Pauling oxygen analyzer for measurement of oxygen consumption in animals in open-circuit systems and in a short-lag closed circuit apparatus. J. Appl. Physiol. 10, 388–392.

Dhindsa, D.S., Hoversland, A.S., Metcalfe, J., 1971. Comparative studies of the respiratory functions of mammalian blood. VII. Armadillo (*Dasypus noveminctus*). Respir. Physiol. 13, 198–208.

Dhindsa, D.S., Sedgwick, C.J., Metcalfe, J., 1972. Comparative studies of the respiratory functions of mammalian blood. VIII. Asian elephant (*Elephas maximus*) and African elephant (*Loxodonta africana*). Respir. Physiol. 14, 332–342.

Dhindsa, D.S., Metcalfe, J., McKean, T., Thorne, T., 1974. Comparative studies of the respiratory functions of mammalian blood. XI. Pronghorn antelope (*Antilocapra americana*). Respir. Physiol. 21, 297–306.

Douglas, C.G., 1911. A method for determining the total respiratory exchange in man. J. Physiol. Lond. 42, xvii–xviii.

Dueck, R., Wagner, P.D., West, J.B., 1977. Effects of positive end-expiratory pressure on gas exchange in dogs with normal and oedematous lungs. Anesthesiology 47, 359–366.

Effros, R.M., Chang, R.S.Y., Silverman, P., 1978. Acceleration of plasma bicarbonate conversion to carbon dioxide by pulmonary carbonic anhydrase. Science 199, 427–429.

Effros, R.M., Shapiro, L., Silverman, P., 1980. Carbonic anhydrase activity of rabbit lungs. J. Appl. Physiol. 49, 589–600.

Enghoff, H., 1938. Volumen inefficax. Uppsala Lak. For. Forh. 44, 191.

Erickson, B.K., Erickson, H.H., Coffman, J.R., 1992. Pulmonary artery and aortic pressure changes during high intensity treadmill exercise in the horse: effect of frusemide and phentolamine. Equine Vet. J. 24 (3), 215–219.

Farhi, L.E., 1967. Elimination of inert gas by the lung. Respir. Physiol. 3, 1–11.

Farhi, L.E., Yokoyama, T., 1967. Effects of ventilation-perfusion inequality on elimination of inert gases. Respir. Physiol. 3, 12–20.

Fedak, M.A., Rome, L., Seeherman, H.J., 1981. One-step N_2-dilution technique for calibrating open-circuit $\dot{V}O_2$ measuring systems. J. Appl. Physiol. 51, 772–776.

Feldman, H.A., McMahon, T.A., 1983. The 3/4 exponent for energy metabolism is not a statistical artifact. Respir. Physiol. 52, 149–163.

Fenn, W.O., Rahn, H., Otis, A.B., 1946. A theoretical study of the composition of alveolar air at altitude. Am. J. Physiol. 146, 637–653.

Fick, A., 1855. Ueber diffusion (Poggendorff's). Ann. Phys. Chem. 94, 59–86.

Fick, A., 1870. Ueber die Messung des Blutquantums in den Herzventrikeln. Sitzber. Phys. Med. Ges. Wurzbg. XVI.

Filley, G.F., Macintosh, D.J., Wright, G.W., 1954. Carbon monoxide uptake and pulmonary diffusing capacity in normal subjects at rest and during exercise. J. Clin. Invest. 33, 530–539.

Forster, R.E., Fowler, W.S., Bates, D.V., 1954. Considerations on the uptake of carbon monoxide by the lungs. J. Clin. Invest. 33, 1128–1134.

Forster, R.E., 1964. Diffusion of gases. In: Fenn, W.O., Rahn, H. (Eds.), Handbook of Physiology. Sect. 3: Respiration. American Physiological Society, Bethesda, MD, pp. 839–872.

Forster, R.E., Crandall, E.D., 1976. Pulmonary gas exchange. Annu. Rev. Physiol. 38, 69–93.

Forster, R.E., 1977. Can alveolar PCO_2 exceed pulmonary end-capillary CO_2? No. J. Appl. Physiol. 42, 323–328.

Forster, R.E., Dubois, A.B., Briscoe, W.A., Fisher, A.B., 1986. The Lung: Physiologic Basis of Pulmonary Function Tests, third ed., vol. 329. Year Book Publishers, Chicago.

Forster, R.E., 1987. Diffusion of gases across the alveolar membrane. In: Farhi, L.E., Tenney, S.M. (Eds.), Handbook of Physiology. Sect. 3: The Respiratory System. American Physiological Society, Bethesda, MD, pp. 71–88.

Gale, G.E., Torre-Bueno, J.R., Moon, R.E., Saltzman, H.A., Wagner, P.D., 1985. Ventilation-perfusion inequality in normal humans during exercise at sea level and simulated altitude. J. Appl. Physiol. 58, 978–988.

Gehr, P., Sehovic, S., Burri, P.H., Claassen, H., Weibel, E.R., 1980. The shrew lung: morphometric estimation of diffusion capacity. Respir. Physiol. 40, 33–47.

Gehr, P., Mwangi, D.K., Ammann, A., Maloiy, G.M.O., Taylor, C.R., Weibel, E.R., 1981a. Design of the mammalian respiratory system. V. Scaling morphometric diffusing capacity to body mass: wild and domestic mammals. Respir. Physiol. 44, 61–86.

Gehr, P., Siegwart, B., Weibel, E.R., 1981b. Allometric analysis of the morphometric pulmonary diffusing capacity in dogs. M. Morphol. 168, 5–15.

Gnaiger, E., Lassnig, B., Kuznetsov, A., Rieger, G., Margreiter, R., 1998. Mitochondrial oxygen affinity, respiratory flux control and excess capacity of cytochrome c oxidase. J. Exp. Biol. 201, 1129–1139.

Graham, T., 1833. On the law of the diffusion of gases, 269–276, 351–358 Philos. Mag. Ser. 3 (2), 175–190.

Grubb, B.R., Mills, C.D., 1981. Blood oxygen content in microliter samples using an easy-to-build galvanic oxygen cell. J. Appl. Physiol. 50, 456–464.

Gurtner, G.H., Song, S.H., Farhi, L.E., 1969. Alveolar to mixed venous P_{CO_2} difference under conditions of no gas exchange. Respir. Physiol. 7, 173–187.

Gurtner, G.H., Fowler, W.S., 1971. Interrelationship of factors affecting pulmonary diffusing capacity. J. Appl. Physiol. 30, 619–624.

Gurtner, G.H., 1977. Can alveolar PCO2 exceed pulmonary end-capillary CO_2? Yes. J. Appl. Physiol. 42, 323–328.

Haldane, J.S., 1892. A new form of apparatus for measuring the respiratory exchanges of animals. J. Physiol. Lond. 13, 419–430.

Haldane, J.S., Smith, J.L., 1897. The absorption of oxygen by the lungs. J. Physiol. Lond. 22, 231–258.

Haldane, I.S., 1922. Respiration, Yale. University Press, New Haven, CT.

Hallam, J., Dawson, T.J., Holland, R.A.B., 1989. Gas exchange in the lung of a dasyurid marsupial: morphometric estimation of diffusing capacity and blood oxygen uptake kinetics. Respir. Physiol. 77, 309–322.

Hammond, M.D., Gale, G.E., Kapitan, K.S., Ries, A., Wagner, P.D., 1986. Pulmonary gas exchange in humans during exercise at sea level. J. Appl. Physiol. 60, 1590–1598.

Hammond, M.D., Hempelman, S.C., 1987. Oxygen diffusing capacity estimates derived from measured $\dot{V}A/\dot{Q}$ distributions in man. Respir. Physiol. 69, 129–147.

Harkness, J.E., Wagner, J.E., 1983. The Biology and Medicine of Rabbits and Rodents, second ed., 210pp. Lea and Febiger, Philadelphia.

Hartridge, H., Roughton, F.J.W., 1923. A method of measuring the velocity of very rapid chemical reaction. Proc. R. Soc. Lond. Ser. A. 104, 376–394.

Hedenstiema, G., Nyman, G., Kvart, C., Funkquist, B., 1987. Ventilation-perfusion relationships in the standing horse: an inert gas elimination study. Equine Vet. J. 19, 514–519.

Heusner, A.A., 1982a. Energy metabolism and body size. I. Is the 0.75 mass exponent of Kleiber's equation a statistical artifact? Respir. Physiol. 48, 1–12.

Heusner, A.A., 1982b. Energy metabolism and body size. H. Dimensional analysis and energetic non-similarity. Respir. Physiol. 48, 13–25.

Hill, E.P., Power, G.G., Longo, L.D., 1973. Mathematical simulation of pulmonary O_2 and CO_2 exchange. Am. J. Physiol. 224, 904–917.

Hill, E.P., Enns, T., 1981. CO_2 diffusing capacity in isolated lungs. Prog. Respir. Res. 16, 172–176.

Hlastala, M.P., Robertson, H.T., 1978. Inert gas characteristics of the normal and abnormal lung. J. Appl. Physiol. 44, 258–266.

Holland, R.A.B., Forster, R.E., 1966. The effect of size of red cells on the kinetics of their oxygen uptake. J. Gen. Physiol. 49, 199–220.

Holland, R.A.B., Van Hezewijk, W., Zubzanda, J., 1977. Velocity of oxygen uptake by partially saturated adult and fetal human red cells. Respir. Physiol. 29, 303–314.

Holland, R.A.B., Shibata, H., Scheid, P., Piiper, J., 1985. Kinetics of O_2 uptake and release by red cells in stopped-flow apparatus: effects of unstirred layer. Respir. Physiol. 59, 71–91.

Holland, R.A.B., Rimes, A.F., Comis, A., Tyndale-Biscoe, C.H., 1988. Oxygen carriage and carbonic anhydrase activity in the blood of a marsupial, the Tammar Wallaby *(Macropus eugenii)*, during early development. Respir. Physiol. 73, 69–86.

Hoppeler, H., Lindstedt, S.L., Uhlmann, E., Niesel, A., Cruz-Orive, L.M., Weibel, E.R., 1984. Oxygen consumption and the composition of skeletal muscle tissue after training and inactivation in the European woodmouse (*Apodemus sylvaticus*). J. Comp. Physiol. B 155, 51–61.

Hilfner, C.G., 1877. Uber die Quantitat Sauerstoff welche 1 gramm Haemoglobin zu binden vermag. Z. Physiol. Chem. 1-2, 317.

Hsia, C.C.W., Johnson Jr., R.L., McDonough, P., Dane, D.M., Hurst, M.D., Fehmel, J.L., Wagner, H.E., Wagner, P.D., 2007. Residence at 3,800-m altitude for 5 mo in growing dogs enhances lung diffusing capacity for oxygen that persists at least 2.5 years. J. Appl. Physiol. 102, 1448–1455.

Hsia, C.C.W., Wagner, P.D., Dane, D.M., Wagner, H.E., Johnson Jr., R.L., 2008. Predicting diffusive alveolar oxygen transfer from carbon monoxide-diffusing capacity in exercising foxhounds. J. Appl. Physiol. 105, 1441–1447.

Huxley, V.H., Kutchai, H., 1983. Effect of diffusion boundary layers on the initial uptake of O_2, by red cells. Theory vs. experiment. Microvasc. Res. 26, 89–107.

Johanson, W.G., Pierce, A.K., 1973. Lung structure and function with age in normal rats and rats with papain emphysema. J. ClM. Invest. 52, 2921–2927.

Johnson, R.L., Spicer, W.S., Bishop, J.M., Forster, R.E., 1960. Pulmonary capillary blood volume, flow and diffusing capacity during exercise. J. Appl. Physiol. 15, 893–902.

Jones, J.H., Karas, R.H., 1988. Structural versus functional limitations to oxygen transport: is there a difference? Adv. Exp. Med. Biol. 227, 293–300.

Jones, J.H., 1989. Comparative exercise physiology: how do horses compare with other mammals? Assoc. Equine Sports Med. Qt. 4, 19–25.

Jones, J.H., Longworth, K.E., Lindholm, A., Conley, K.E., Karas, R.H., Kayar, S.R., Taylor, C.R., 1989a. Oxygen transport during exercise in large mammals. I. Adaptive variation in oxygen demand. J. Appl. Physiol. 67, 862–870.

Jones, J.H., Taylor, C.R., Lindholm, A., Straub, R., Longworth, K.E., Karas, R.H., 1989b. Blood gas measurements during exercise: errors due to temperature correction. J. Appl. Physiol. 67, 879–884.

Jones, N.L., Campbell, E.J.M., Edwards, R.H.T., Wikoff, W.G., 1969. Alveolar-to-blood PCO_2 difference during rebreathing in exercise. J. Appl. Physiol. 27, 356–360.

Jiirgens, K.D., Bartels, H., Bartels, R., 1981. Blood oxygen transport and organ weights of small bats and small non-flying mammals. Respir. Physiol. 45, 243–260.

Karas, R.H., Taylor, C.R., Jones, J.H., Lindstedt, S.L., Reeves, R.B., Weibel, E.R., 1987. Adaptive variation in the mammalian respiratory system in relation to energetic demand. VII. Flow of oxygen across the pulmonary gas exchanger. Respir. Physiol. 69, 101–115.

Kaufmann, R., Forstner, H., Wieser, W., 1989. Respirometry-methods and approaches. In: Bridges, C.R., Butler, P.J. (Eds.), Techniques in Comparative Respiratory Physiology. Cambridge University Press, Cambridge, pp. 51–76.

Kemohan, J.C., Forrest, W.W., Roughton, F.J.W., 1963. The activity of concentrated solutions of carbonic anhydrase. Biochim. Biophys. Acta 67, 31–41.

Kety, S., 1951. The theory and applications of the exchange of inert gas at the lungs and tissues. Pharmacol. Rev. 3, 1–41.

Kleiber, M., 1932. Body size and metabolism. Hilgardia 6, 315–353.

Kleiber, M., 1961. The Fire of Life: An Introduction to Animal Energetics, vol. 454. Wiley, New York.

Kleiber, M., 1975. The Fire of Life, second ed., vol. 453. Robert E. Krieger, Huntington, NY.

Klocke, R.A., 1978. Carbon dioxide transport. In: Robin, E.D. (Ed.), Extrapulmonary Manifestations of Respiratory Disease. Marcel Dekker, New York, pp. 315–343.

Klocke, R.A., 1987. Carbon dioxide transport. In: Farhi, L.E., Tenney, S.M. (Eds.), Handbook of Physiology. Sect. 3: The Respiratory System. American Physiological Society, Bethesda, MD, pp. 173–197.

Klug, A., Kreuzer, F., Roughton, F.J.W., 1956. The diffusion of oxygen in concentrated haemoglobin solutions. Hely. Physiol. Pharmacol. Acta 14, 121–128.

Knipping, H.W., Bolt, W., Venrath, H., Valentin, H., Ludes, H., Endler, P., 1955. Eine neue Methode zur Prufung der Herz- and Lungenfunktion. Die regionale Funktionanalyse in der Lungen-und Herzklinik mit Hilfe des radioaktiven Edelgases Xenon[133] (1 Isotopen Thoraiographie). Deut. Med. Wochschr. 80, 1146–1147.

Konoplev, V.A., Sokolov, V.E., Zotin, A.I., 1978. Criterion of orderliness and some problems of taxonomy. In: Lamprecht, I., Zotin, A.I. (Eds.), Thermodynamics of Biological Processes. Walter De Gruyter, Berlin, pp. 349–359.

Kotter, D., Huch, A., Stotz, H., Piiper, J., 1969. Single breath CO diffusing capacity in anesthetized dogs with increased oxygen consumption. Respir. Physiol. 6, 202–208.

Kreuzer, F., Hoofd, L., 1987. Facilitated diffusion of oxygen and carbon dioxide. In: Farhi, L.E., Tenney, S.M. (Eds.), Handbook of Physiology. Sect. 3: The Respiratory System. American Physiological Society, Bethesda, MD, pp. 89–111.

Krogh, A., Krogh, M., 1909/10. Rate of diffusion of CO into the lungs of man. Scand. Arch. Physiol. 23, 236–247.

Krogh, A., 1913. A bicycle ergometer and respiration apparatus for the experimental study of muscular work. Skand. Arch. Physiol. 30, 375–394.

Krogh, A., Lindhard, J., 1914. On the average composition of the alveolar air and its variations during the respiratory cycle. J. Physiol. Lond. 47, 431–445.

Krogh, A., Lindhard, J., 1917. The volume of dead space in breathing and the mixing of gases in the lungs of man. J. Physiol. Lond. 51, 59–90.

Krogh, A., 1919. The rate of diffusion of gases through animal tissues, with some remarks on the coefficient of invasion. J. Physiol. Lond. 52, 391–608.

Krogh, M., 1914–15. Diffusion of gases through the lungs of man. J. Physiol. Lond. 49, 271–300.

Lahiri, S., 1975. Blood oxygen affinity and alveolar ventilation in relation to body weight in mammals. Am. J. Physiol. 229, 529–536.

Larimer, J.L., Schmidt-Nielsen, K., 1960. A comparison of blood carbonic anhydrase of various mammals. Comp. Biochem. Physiol. 1, 19–23.

Lavoisier, A.-L., 1780. Experiences sur la respiration des Animaux, et sur les changemens qui arrivent a l'air en passant par leur poumon. In: Memoires De Mathematique Et De Physique... De l'Academie Royale Des Sciences. Annee 1777. Imprimerie Royale, Paris, pp. 185–194.

Lechner, A.J., 1978. The scaling of maximal oxygen consumption and pulmonary dimensions in small mammals. Respir. Physiol. 34, 29–44.

Leith, D.E., 1983. Comparative mammalian respiratory mechanics. Am. Rev. Respir. Dis. 128, S77–S82.

Lenfant, C., Okubo, T., 1968. Distribution function of pulmonary blood flow and ventilation-perfusion ratio in man. J. Appl. Physiol. 24, 668–677.

Lenfant, C., Aucutt, C., 1969. Respiratory properties of the blood of five species of monkeys. Respir. Physiol. 6, 284–291.

Lilienthal, J.L., Riley, R.L., Proemmel, D.D., Franke, R.E., 1946. An experimental analysis in man of the oxygen pressure gradient from alveolar air to arterial blood during rest and exercise at sea level and altitude. Am. J. Physiol. 147, 199–216.

Lindstedt, S.L., Calder, W.A., 1981. Body size, physiological time, and longevity of homeothermic animals. Q. Rev. Biol. 56, 1–16.

Lindstedt, S.L., 1984. Pulmonary transit time and diffusing capacity in mammals. Am. J. Physiol. 246, R384–R388.

Longworth, K.E., Jones, J.H., Bicudo, J.E.P.W., Taylor, C.R., Weibel, E.R., 1989. High rate of O_2 consumption in exercising foxes: large PO_2 difference drives diffusion across the lung. Respir. Physiol. 77, 263–276.

Lonnerholm, G., 1982. Pulmonary carbonic anhydrase in the human, monkey, and rat. J. Appl. Physiol. 52, 352–356.

Ludwig, C., 1865. Zusammenstellung der Untersuchungen ueber Blutgase. Z. Gesellsc. Arzte. Wien 1, 145.

MacMillen, R.E., Nelson, J.E., 1969. Bioenergetics and body size in dasyurid marsupials. Am. J. Physiol. 217, 1246–1251.

Magid, E., 1967. Activity of carbonic anhydrase in mammalian blood in relation to body size. Comp. Biochem. Physiol. 21, 357–360.

Maina, J.N., King, A.S., 1984. Correlations between structure and function in the design of the bat lung: a morphometric study. J. Exp. Biol. 111, 43–61.

Maren, T.H., 1967. Carbonic anhydrase: chemistry, physiology, and inhibition. Physiol. Rev. 47, 595–781.

Margaria, R., Edwards, H.T., Dill, D.B., 1933. The possible mechanism of contracting and paying the oxygen debt and the role of lactic acid in muscular contraction. Am. J. Physiol. 106, 689–715.

Mauderly, J.L., 1972. Steady-state carbon monoxide diffusing capacity of unanesthetized Beagle dogs. Am. J. Vet. Res. 33, 1485–1491.

McLean, J.A., Tobin, G., 1987. Animal and Human Calorimetry, vol. 338. Cambridge University Press, Cambridge.

McNab, B.K., 1979. The influence of body size on the energetics and distribution of fossorial and burrowing mammals. Ecology 60, 1010–1021.

Mechtly, E.A., 1973. The International System of Units. Physical Constants and Conversion Factors, 2nd Revision, NASA SP-7012, vol. 21. National Aeronautics and Space Administration, Washington, D.C.

Meyer, M., Scheid, P., Riepl, G., Wagner, H.-J., Piiper, J., 1981. Pulmonary diffusing capacities for O, and CO measured by a rebreathing technique. J. Appl. Physiol. 51, 1643–1650.

Musch, T.I., Haidet, G.C., Ordway, G.A., Longhurst, J.C., Mitchell, J.H., 1985. Dynamic exercise training in Foxhounds. I. Oxygen consumption and hemodynamic responses. J. Appl. Physiol. 59, 183–189.

Natelson, S., 1951. Routine use of ultra micro methods in the clinical laboratory. Am. J. Clin. Pathol. 21, 1153–1172.

Neufeld, G.R., Williams, J.J., Klineberg, P.L., Marshall, B.E., 1978. Inert gas a-A differences: a direct reflection of V/O distribution. J. Appl. Physiol. 44, 277–283.

Noehren, T.H., 1962. Pulmonary clearance of inert gases with particular reference to diethyl ether. J. Appl. Physiol. 17, 795–798.

O'Neil, J.J., Leith, D.E., 1980. Lung diffusing capacity scaled in mammals from 25 g to 500 kg. Fed. Proc. 39, 972.

Ogilvie, C.M., Forster, R.E., Blakemore, W.S., Morton, J.W., 1957. A standardized breath holding technique for the clinical measurement of diffusing capacity of the lung for carbon monoxide. J. Clin. Invest. 36, 1–17.

Olszowka, A.J., Wagner, P.D., 1980. Numerical analysis of gas exchange. In: West, J.B. (Ed.), Pulmonary Gas Exchange. Academic Press, New York, pp. 263–306.

Ordway, G.A., Floyd, D.L., Longhurst, J.C., Mitchell, J.H., 1984. Oxygen consumption and hemodynamic responses during graded treadmill exercise in the dog. J. Appl. Physiol. 57, 601–607.

Parer, J.T., Metcalfe, J., 1967a. Respiratory studies of monotremes. I. Blood of the platypus (*Ornithorynchus anatinus*). Respir. Physiol. 3, 136–142.

Pasquis, P., Lacaisse, A., Dejours, P., 1970. Maximal oxygen uptake in four species of small mammals. Respir. Physiol. 9, 298–309.

Peters, R.H., 1983. The Ecological Implications of Body Size, vol. 329. Cambridge University Press, New York.

Pettenkofer, M., 1862. Ueber die respiration. Ann. der Chemie Pharm. (Suppl. 2), 1–52.

Pfliiger, E.F.W., 1872. Ueber die diffusion des Sauerstoffs, den Ort and die Gesetze der Oxydationsprozesse im tierischen Organismus. Pfliigers Arch. Gesamte Physiol. Menschen Tiere. 6, 43.

Pfltiger, E.F.W., 1875. Beitrage zur Leine von der respiration. Pflugers Arch. Gesamte Physiol. Menschen Tiere. 10, 251.

Piiper, J., Scheid, P., 1980. Blood-gas equilibration in lungs. In: West, J.B. (Ed.), Pulmonary Gas Exchange. Academic Press, New York, pp. 131–171.

Piiper, J., Scheid, P., 1981. Model for capillary-alveolar equilibration with special reference to O_2 uptake in hypoxia. Respir. Physiol. 46, 193–208.

Piiper, J., Dejours, P., Haab, P., Rahn, H., 1971. Concepts and basic quantities in gas exchange physiology. Respir. Physiol. 13, 292–304.

Prothero, J., 1980. Scaling of blood parameters in mammals. Comp. Biochem. Physiol. 67A, 649–657.

Rahn, H., 1949. A concept of mean alveolar air and the ventilation-blood flow relationships during pulmonary gas exchange. Am. J. Physiol. 158, 21–30.

Rahn, H., Fenn, W.O., 1955. A Graphical Analysis of the Respiratory Gas Exchange. American Physiological Society, Washington, D.C.

Reeves, R.B., Park, J.S., Lapennas, G.N., Olszowka, A.J., 1982. Oxygen affinity and Bohr coefficients of dog blood. J. Appl. Physiol. 53, 87–95.

Regnault, V., Reiset, J., 1849. Recherches Chimiques sur la respiration des Animaux. Ann. De Chim De Phys. Ser. 3 (26), 299–519.

Richet, C., 1889. La Chaleur Animale. Felix Alcan, Paris.

Riegel, K., Bartels, H., Kleihauer, E., Lang, E.M., Metcalfe, J., 1966. Comparative studies of the respiratory functions of mammalian blood. I. Gorilla, chimpanzee and organgutan. Respir. Physiol. 1, 138–144.

Riegel, K., Bartels, H., El Yassin, D., Oufi, J., Kleihauer, E., Parer, J.T., Metcalfe, J., 1967. Comparative studies of the respiratory functions of mammalian blood. III. Fetal and adult dromedary camel blood. Respir. Physiol. 2, 173–181.

Riggs, A., 1960. The nature and significance of the Bohr effect in mammalian hemoglobins. J. Gen. Physiol. 43, 737–752.

Riley, R.L., Lilienthal, J.L., Proemmel, D.D., Franke, R.E., 1946. On the determination of the physiologically effective pressures of oxygen and carbon dioxide in alveolar air. Am. J. Physiol. 147, 191–198.

Riley, R.L., Cournand, A., 1949. "Ideal" alveolar air and the analysis of ventilation-perfusion relationships in the lungs. J. Appl. Physiol. 1, 825–847.

Riley, R.L., Cournand, A., 1951. Analysis of factors affecting the partial pressures of oxygen and carbon dioxide in gas and blood of lungs: theory. J. Appl. Physiol. 4, 77–101.

Riley, R.L., Permutt, S., 1964. The four quadrant diagram for analyzing the distribution of gas and blood in the lung. In: Fenn, W.O., Rahn, H. (Eds.), Handbook of Physiology. Section 3: Respiration. American Physiological Society, Bethesda, MD, pp. 1413–1423.

Riley, R.L., Permutt, S., 1973. Venous admixture component of the AaP_{O2} gradient. J. Appl. Physiol. 35, 430–431.

Robertson, H.T., Hlastala, M.P., 1977. Elevated alveolar Pc_t relative to predicted values during normal gas exchange. J. Appl. Physiol. 43, 357–364.

Robinson, N.E., Gillespie, J.R., Berry, J.D., Simpson, A., 1972. Lung compliance, lung volumes and single breath diffusing capacity in dogs. J. Appl. Physiol. 33, 808–812.

Rosenhamer, G.J., Frierson, W.O., McIlroy, M.B., 1971. A bloodless method for measurement of diffusing capacity of the lungs for oxygen. J. Appl. Physiol. 30, 603–610.

Rothstein, S., Jurgens, K.D., Bartels, H., 1984. Oxygen and carbon dioxide transport in the blood of the muskrat (*Ondatra zibethicus*). Respir. Physiol. 57, 15–22.

Roughton, F.J.W., Forster, R.E., 1957. Relative importance of diffusion and chemical reaction rates in determining rate of exchange of gases in the human lung, with special reference to true diffusing capacity of pulmonary membrane and volume of blood in the capillaries. J. Appl. Physiol. 11, 290–302.

Roughton, F.J.W., Forster, R.E., Cander, L., 1959. Rate at which carbon monoxide replaces oxygen from combination with human hemoglobin in solution and in the red cell. J. Appl. Physiol. 11, 269–276.

Roughton, F.J.W., 1964. Transport of oxygen and carbon dioxide. In: Fenn, W.O., Rahn, H. (Eds.), Handbook of Physiology, Sect. 3: Respiration. American Physiological Society, Bethesda, MD, pp. 767–825.

Royal Society Symbols Committee, 1975. Quantities, Units, and Symbols, second ed., vol. 55. The Royal Society, London.

Rubner, M., 1883. Ueber den Einfluss der Korpergrosse auf Stoffund Kraftwechsel. Z. Biol. 19, 535–562.

Ryan, U.S., Whitney, P.L., Ryan, J.W., 1982. Localization of carbonic anhydrase on pulmonary artery endothelial cells in culture. J. Appl. Physiol. 53, 914–919.

Sackner, M.A., Greeneltch, D., Heiman, M.S., Epstein, L.S., Atkins, N., 1975. Diffusing capacity, membrane diffusing capacity, capillary blood volume, pulmonary tissue and cardiac output measured by a rebreathing technique. Am. Rev. Respir. Dis. 111, 157–165.

Sarrus et Rameaux, 1839. Memoire Addresse a l'Academic Royale. Bull. Mem. Acad. R. Med. 3, 1094–1100.

Scandurra, F.M., Gnaiger, E., 2010. Cell respiration under hypoxia: facts and artefacts in mitochondrial oxygen kinetics. Adv. Exper. Med. Biol. 662, 7–25.

Scheid, P., Piiper, J., 1980. Blood/gas equilibration of carbon dioxide in lungs. A critical review. Respir. Physiol. 39, 1–31.

Scheid, P., Hook, C., Yamaguchi, H., Piiper, J., 1986. Factors limiting O_2 transfer of red blood cells: model analysis using results from stopped-flow experiments. Prog. Resp. Res. 21, 11–17.

Schmidt-Nielsen, K., Larimer, J.L., 1958. Oxygen dissociation curves of mammalian blood in relation to body size. Am. J. Physiol. 195, 424–428.

Schmidt-Nielsen, K., 1984. Scaling: Why Is Animal Size So Important?, vol. 241. Cambridge University Press, New York.

Schmidt-Nielsen, K., 1990. Animal Physiology: Adaptation and Environment, fourth ed., vol. 602. Cambridge University Press, New York.

Scholander, P.F., 1960. Oxygen transport through hemoglobin solutions. Science 131, 585–590.

Scotto, P., Rieke, H., Schmitt, H.J., Meyer, M., Piiper, J., 1984. Blood-gas equilibration of CO_2 and O_2 in lungs of awake dogs during prolonged rebreathing. J. Appl. Physiol. 57, 1354–1359.

Scotto, P., Ichinose, Y., Patane, L., Meyer, M., Piiper, J., 1987. Alveolar-capillary diffusion of oxygen in dogs exercising in hypoxia. Respir. Physiol. 68, 1–10.

Seeherman, H.J., Taylor, C.R., Maloiy, G.M.O., Armstrong, R.B., 1981. Design of the mammalian respiratory system. II. Measuring maximum aerobic capacity. Respir. Physiol. 44, 11–23.

Shephard, R.F., 1955. A critical examination of the Douglas bag technique. J. Physiol. Lond. 127, 515–524.

Siegwart, B., Gehr, P., Gil, J., Weibel, E.R., 1971. Morphometric estimation of pulmonary diffusing capacity. IV. The normal dog lung. Respir. Physiol. 13, 141–159.

Snider, G.L., Celli, B.R., Goldstein, R.H., O'Brien, J.J., Lucey, E.C., 1978. Chronic interstitial pulmonary fibrosis produced in hamsters by endotracheal bleomycin. Am. Rev. Respir. Dis. 117, 289–297.

Soulard, C.D., Teisseire, B.P., Teisseire, L.J., Herigault, R.A., 1983. Consequences of an acute increase in P50 in anaesthetized guinea pigs. Respir. Physiol. 51, 21–30.

Stahl, W.R., 1967. Scaling of respiratory variables in mammals. J. Appl. Physiol. 22, 453–460.

Staub, N.C., Bishop, J.M., Forster, R.E., 1961. Velocity of O_2 uptake by human red blood cells. J. Appl. Physiol. 16, 511–516.

Staub, N.C., Bishop, J.M., Forster, R.E., 1962. Importance of diffusion and chemical reaction rates in O_2 uptake in the lung. J. Appl. Physiol. 17, 21–27.

Sylvester, J.T., Cymerman, A., Gurtner, G., Hottenstein, O., Cote, M., Wolfe, D., 1981. Components of alveolar-arterial O_2 tension gradient during rest and exercise at sea level and high altitude. J. Appl. Physiol. 50, 1129–1139.

Takezawa, J., Miller, F.J., O'Neil, J.J., 1980. Single-breath diffusing capacity and lung volumes in small laboratory animals. J. Appl. Physiol. 48, 1052–1059.

Tamhane, R.M., Johnson Jr., R.L., Hsia, C.C.W., 2001. Pulmonary membrane diffusing capacity and capillary blood volume measured during exercise from nitric oxide uptake. Chest 120, 1850–1856.

Taylor, C.R., Maloiy, G.M.O., Weibel, E.R., Langman, V.A., Kamau, J.M.Z., Seeherman, H.J., Heglund, N.C., 1981. Design of the mammalian respiratory system. III. Scaling maximum aerobic capacity to body mass: wild and domestic mammals. Respir. Physiol. 44, 25–37.

Taylor, C.R., Jones, J.H., 1987. Maximal oxygen consumption in mammals. In: Dejours, P. (Ed.), Comparative Physiology of Environmental Adaptations, vol. 2. Karger, Basel, pp. 188–195.

Taylor, C.R., Karas, R.H., Weibel, E.R., Hoppeler, H., 1987a. Adaptive variation in the mammalian respiratory system in relation to energetic demand. Respir. Physiol. 69, 1–127.

Taylor, C.R., Karas, R.H., Weibel, E.R., Hoppeler, H., 1987b. Adaptive variation in the mammalian respiratory system in relation to energetic demand. H. Reaching the limits to oxygen flow. Respir. Physiol. 69, 7–26.

Taylor, C.R., Longworth, K.E., Hoppeler, H., 1988. Matching O_2 delivery to O_2 demand in muscle: II. Allometric variation in energy demand. Adv. Exp. Med. Biol. 227, 171–181.

Tenney, S.M., Remmers, J.E., 1963. Comparative quantitative morphology of the mammalian lung: diffusing area. Nature 197, 54–56.

Tenney, S.M., Morrison, D.H., 1967. Tissue gas tensions in small wild mammals. Respir. Physiol. 3, 160–165.

Tigerstedt, R., 1906. Der Respirationsapparat in Neuen Physiologischen Institut zu Helsingfors. Skand. Arch. Physiol. 18, 298–305.

Torre-Bueno, J., Wagner, P.D., Saltzman, H.A., Gale, G.E., Moon, R.E., 1985. Diffusion limitation in normal humans during exercise at sea level and simulated altitude. J. Appl. Physiol. 58 (3), 989–995.

Tucker, V.A., 1967. Method for oxygen content and dissociation curves on microliter blood samples. J. Appl. Physiol. 23, 410–414.

Tucker, V.A., 1968. Respiratory exchange and evaporative water loss in the flying budgerigar. J. Exp. Biol. 48, 67–87.

Turek, Z., Grandtner, M., Ringnalda, B.E.M., Kreuzer, F., 1973. Hypoxic pulmonary steady-state diffusing capacity of CO and cardiac output in rats born at a simulated altitude of 3500 m. Pflugers Arch. 340, 11–18.

Van Slyke, D.D., Neill, J.M., 1924. The determination of gases in blood and other solutions by vacuum extraction and manometric measurement. J. Biol. Chem. 61, 523–573.

Voit, E., 1901. Uber die Grosse des Energiebedarfs der Tiere in Hungerzustande. Z. Biol. 41, 113–154.

Vreim, C.E., Staub, N.C., 1973. Indirect and direct pulmonary capillary blood volume in anesthetized open-thorax cats. J. Appl. Physiol. 34, 452–459.

Wagner, P.D., Naumann, P.F., Laravuso, R.B., 1974a. Simultaneous measurement of eight foreign gases in blood by gas chromatography. J. Appl. Physiol. 36, 600–605.

Wagner, P.D., Saltzman, H.A., West, J.B., 1974b. Measurement of continuous distributions of ventilation-perfusion ratios: theory. J. Appl. Physiol. 36, 588–599.

Wagner, P.D., West, J.B., 1972. Effects of diffusion impairment on O_2 and CO_2 time courses in pulmonary capillaries. J. Appl. Physiol. 33, 62–71.

Wagner, P.D., West, J.B., 1980. Ventilation-perfusion relationships. In: West, J.B. (Ed.), Pulmonary Gas Exchange. Academic Press, New York, pp. 219–262.

Wagner, P.D., Lopez, F.A., 1984. Gas chromatography techniques in respiratory physiology. In: Techniques in the Life Sciences, Respiratory Physiology. Elsevier Scientific, Ireland, pp. 1–24.

Wagner, P.D., Gale, G.E., Moon, R.E., Torre-Bueno, J.R., Stolp, B.W., Saltzman, H.A., 1986. Pulmonary gas exchange in humans exercising at sea level and simulated altitude. J. Appl. Physiol. 61, 260–270.

Wagner, P.D., Hedenstierna, G., Bylin, G., 1987a. Ventilation-perfusion inequality in chronic asthma. Am. Rev. Respir. Dis. 136, 605–612.

Wagner, P.D., Sutton, J.R., Reeves, J.T., Cymerman, A., Groves, B.M., Malconian, M.K., 1987b. Operation Everest II: pulmonary gas exchange during a simulated ascent of Mt. Everest. J. Appl. Physiol. 63 (6), 2348–2359.

Wagner, P.D., Gillespie, J.R., Landgren, G.L., Fedde, M.R., Jones, B.W., DeBowes, R.M., Pieschl, R.L., Erickson, H.H., 1989. Mechanism of exercise-induced hypoxemia in horses. J. Appl. Physiol. 66, 1227–1233.

Wagner, P.D., 1996. A theoretical analysis of factors determining $\dot{V}O_2$ MAX at sea level and altitude. Resp. Physiol. 106 (3), 329–343.

Watanabe, S., Frank, R., 1975. Lung volumes, mechanics, and single-breath diffusing capacity in anesthetized cats. J. Appl. Physiol. 38, 1148–1152.

Weber, R.E., Heath, M.E., White, F.N., 1986. Oxygen binding functions of blood and hemoglobin from the Chinese pangolin, *Manis pentadactyla*: possible implications of burrowing and low body temperature. Respir. Physiol. 64, 103–112.

Weibel, E.R., 1970/71. Morphometric estimation of pulmonary diffusion capacity. I. Model and method. Respir. Physiol. 11, 54–75.

Weibel, E.R., 1972. Morphometric estimation of pulmonary diffusion capacity. V. Comparative morphometry of alveolar lungs. Respir. Physiol. 14, 26–43.

Weibel, E.R., 1980. Stereological Methods. Volume II. Theoretical Foundations, vol. 340. Academic Press, London.

Weibel, E.R., Taylor, C.R., 1981. Design of the mammalian respiratory system. Respir. Physiol. 44, 1–164.

Weibel, E.R., Gehr, P., Cruz-Orive, L.M., Muller, A.E., Mwangi, D.K., Haussener, V., 1981. Design of the mammalian respiratory system. IV. Morphometric estimation of pulmonary diffusing capacity; critical evaluation of a new sampling method. Respir. Physiol. 44, 39–59.

Weibel, E.R., Taylor, C.R., O'Neil, J.J., Leith, D.E., Gehr, P., Hoppeler, H., Langman, V., Baudinette, R.V., 1983. Maximal oxygen consumption and pulmonary diffusing capacity: a direct comparison of physiologic and morphometric measurements in canids. Respir. Physiol. 54, 173–188.

Weibel, E.R., 1984a. Morphometric and stereological methods in respiratory physiology, including fixation techniques. In: Techniques in the Life Sciences, Respiratory Physiology. Elsevier Scientific, Ireland, pp. 1–35.

Weibel, E.R., 1984b. The Pathway for Oxygen, 425pp. Harvard University Press, Cambridge, MA.

Weibel, E.R., Marques, L.B., Constantinopol, M., Doffey, F., Gehr, P., Taylor, C.R., 1987. Adaptive variation in the mammalian respiratory system in relation to energetic demand. VI. The pulmonary gas exchanger. Respir. Physiol. 69, 81–100.

Wells, R.M.G., Weber, R.E., 1989. The measurement of oxygen affinity in blood and haemoglobin solutions. In: Bridges, C.R., Butler, P.J. (Eds.), Techniques in Comparative Respiratory Physiology. Cambridge University Press, Cambridge, pp. 279–303.

West, J.B., Dollery, C.T., 1960. Distribution of blood flow and ventilation-perfusion ratio in the lung, measured with radioactive CO_2. J. Appl. Physiol. 15, 405–410.

West, J.B., 1964. Topographical distribution of blood flow in the lung. In: Fenn, W.O., Rahn, H. (Eds.), Handbook of Physiology. Sect. 3: Respiration. American Physiological Society, Bethesda, MD, pp. 1437–1451.

West, J.B., Dollery, C.T., Naimark, A., 1964. Distribution of blood flow in isolated lung: relation to vascular and alveolar pressures. J. Appl. Physiol. 19, 713–724.

West, J.B., Wagner, P.D., 1977. Pulmonary gas exchange. In: West, J.B. (Ed.), Bioengineering Aspects of the Lung. Marcel Dekker, New York, pp. 361–457.

West, J.B., Wagner, P.D., 1980. Predicted gas exchange on the summit of Mt. Everest. Respir. Physiol. 42, 1–16.

West, J.B., 1985. Respiratory Physiology—the Essentials, third ed. Williams and Wilkins, Baltimore, MD.

Willford, D.C., Hill, E.P., 1986. Modest effect of temperature on the porcine oxygen dissociation curve. Respir. Physiol. 64, 113–123.

Williams, J.H., Powers, S.K., Stuart, M.K., 1986. Hemoglobin denaturation in highly trained athletes during heavy exercise. Med. Sci. Sports Exerc. 18, 168–173.

Withers, P.C., 1977. Measurement of $\dot{V}O_2$, $\dot{V}CO_2$ and evaporative water loss with a flow-through mask. J. Appl. Physiol. 42, 120–123.

Wittenberg, J.B., 1959. Oxygen transport—a new function proposed for myoglobin. Biol. Bull. Woods Hole 117, 402–403.

Yamaguchi, K., Nguyen-Phu, D., Scheid, P., Piiper, J., 1985. Kinetics of O_2 uptake and release by human erythrocytes studied by a stop-flow technique. J. Appl. Physiol. 58, 1215–1224.

Yokoyama, T., Farhi, L.E., 1967. Study of ventilation-perfusion ratio distribution in the anesthetized dog by multiple inert gas washout. Respir. Physiol. 3, 166–176.

Chapter 13

Aging of the Normal Lung

Graeme R. Zosky

Chapter Outline

1. INTRODUCTION

Understanding how the lung grows and develops during the "normal" aging process is critical to our interpretation of changes in lung outcomes in both health and disease. Assessing lung development in humans is difficult as (1) we lack detailed structural information at key developmental time points, (2) postnatal longitudinal data on lung growth are usually inferred from measures of lung function rather than direct structural data, and (3) the available cross-sectional data are confounded by variability at the population level and limited by the ethical issues associated with probing the lung using quantitative imaging. As such, the bulk of our understanding of lung growth and development comes from data obtained from animal models. These studies have provided critical insight into the structural changes, and the key molecular cues that drive these changes, which occur during normal lung growth and development.

In any treatise on aging in the normal lung it is important to recognize the epochs that have been used to categorize the major structural events that occur during lung growth: (1) *in utero* lung formation, (2) the transition to air breathing and early postnatal lung development, (3) lung growth from early childhood into adulthood, and (4) changes in lung structure and function from its typical peak in early adulthood (postpuberty) into senescence. It should be pointed out here that there are several between species differences in mammals in the morphology of the primary lung structures and patterns of lung growth. In particular, there are key differences in the timing of formation of lung units between humans and the Muridae (mice and rats) with the latter being the group in which we have the most detailed information.

In this chapter our basic understanding of lung development, from birth through postnatal growth and into senescence, will be summarized using species-specific information. Reference will be made throughout to comparisons with our current understanding of normal human lung growth and development in order to provide a context for the discussion. This chapter will begin by focusing on changes in lung structure with age, followed by a discussion on how these structural changes are reflected in lung volume, lung mechanics, and gas exchange and will finish with a summary of our limited knowledge on changes in ventilation patterns with age. One of the challenges faced in uniting the disparate data that exists on changes in lung structure and function with age is the variability between species in lifespan and the age at which major developmental events occur (Table 1). In order to facilitate a cohesive discussion, relative age will be defined in terms of gross developmental stage (e.g., birth, postnatal, pre- and postpuberty) rather than in absolute terms for a given species.

One of the greatest limitations in generating a comparative summary of lung structure and aging is the variation between species in the environmental exposures that occur over a lifetime. The observed changes in lung structure in humans from birth, through adulthood, and into senescence are a result of a combination of the normal aging process (i.e., growth until adulthood followed by programmed decreases in physiological function) and the

Comparative Biology of the Normal Lung. http://dx.doi.org/10.1016/B978-0-12-404577-4.00013-8

TABLE 1 Representative Values for Age at Puberty and Lifespan of the Species Described in this Chapter

Generic Name	Reference Species[a]	Age at Puberty[b]	Lifespan[b]	Source
Human	*Homo sapiens*	11–15 years	80 years	Parent et al. (2003) Oeppen and Vaupel (2002)
Mouse	*Mus musculus*	4–6 weeks	2 years	Falconer (1984) Yuan et al. (2009)
Rat	*Rattus norvegicus*	5–7 weeks	2 years	Hagenauer et al. (2011) Yu et al. (1982)
Rabbit	*Oryctolagus cuniculus*	4–9 months	7.5 years	Kilborn et al. (2002)
Guinea pig	*Cavia porcellus*	6–11 weeks	8 years	Wolfer and Lipp (1995) Pamplona et al. (1998)
Hamster	*Mesocricetus auratus*	4–7 weeks	4 years	Schulz et al. (2004) Hart and Setlow (1974)
Ferret	*Mustela putorius furo*	8–12 months	7 years	Ryan (1984) Wolf et al. (2000)
Dog	*Canis lupus familiaris*	7–10 months	8–14 years	Kilborn et al. (2002) Speakman et al. (2003)
Sheep	*Ovis aries*	5.5 months	15 years	Suttie et al. (1991) Oberbauer et al. (1988)
Pig	*Sus scrofa domestica*	6 months	20 years	Sterning et al. (1998) Douglas (1972)
Nonhuman primate	*Macaca* spp.	2.5–3 years	35 years	Plant (2001) Shimura et al. (1986)

[a]*Most commonly used species.*
[b]*Ages are indicative.*

sum of a lifetime of (variable) environmental exposures, the cumulative effect of which will increase with age. As such, it becomes difficult to describe the aging of the "normal" lung even in healthy individuals. In contrast, most (but not all) laboratory animals are raised in controlled environments where the exposure, while not absent, remains relatively constant throughout life. So in assessing the comparative changes in lung structure and function that occur across ages between species, the context in which this aging process takes place needs to be considered, particularly when comparing laboratory animal data to human data.

2. LUNG STRUCTURE

2.1 Basic Patterns in Lung Growth and Development

Despite several decades of research, our understanding of normal lung growth and development is incomplete and continues to evolve as new data and technology for assessing lung structure emerge. The development of new imaging technologies and increases in computer processing power means that this area will continue to develop for many years to come as our capacity to probe the basic structures of the lung *in vivo* improves. The data that exist to date show that the basic patterns of lung growth are highly conserved within mammals, with most of the variation occurring in the relative timing of the formation of the primary lung structures (Thurlbeck, 1975; Zoetis and Hurtt, 2003), in particular, the developmental stage that the lung has reached at birth. The lung grows from the budding of the ventral side of the primitive foregut early in fetal development (Chinoy, 2003). This embryonic stage of development is followed by the pseudoglandular stage, which is characterized by the appearance of some of the major cell types present in the adult lung including airway epithelial cells, goblet cells, airway smooth muscle cells, and cartilage (Copland and Post, 2004; Perelman et al., 1981). The pseudoglandular phase gives rise to the canicular stage during which airway branching morphogenesis occurs, resulting in the formation of the tracheobronchial tree (Merkus et al., 1996). During the saccular stage, the lung parenchyma begins to form by the thinning of the connective tissue and the development of the surfactant producing system (Merkus et al., 1996). This is followed by the alveolar stage, which involves development of the primary gas exchange units primarily by a process of septation (Massaro and Massaro, 2002).

It is these latter stages in particular where the relative development of the lung at birth varies considerably between species. For example, while the alveoli have begun to form in humans at term, over 80% of alveolarization occurs during postnatal growth; whereas the Muridae are born without alveoli (Mauderly, 2000) and other species such as sheep are born with relatively well-developed lungs (Massaro and Massaro, 2002). The description of the processes and patterns of alveolarization that occur have received considerable attention in recent years, and it is becoming increasingly clear that new alveoli continue to form into early adulthood in most mammals including humans (Massaro et al., 1985; Mund et al., 2008; Narayanan et al., 2012).

As the body grows so does the lung in order to meet metabolic requirements, and this can be seen in changes in lung structure through childhood, puberty, and into adulthood. This postnatal lung growth into adulthood can occur by lung expansion and/or by the formation of new alveoli (Schittny et al., 2008), both of which appear to occur in humans (Narayanan et al., 2012; Zeltner et al., 1987). Like the developmental stage of the lung at birth, the changes that occur in both lung structure and function from early adulthood into senescence appear to vary considerably between species, although there are many gaps in our current data set. Our species-specific understanding of these arbitrary stages of lung development is still growing, and our current knowledge in this area is summarized in more detail below.

2.2 Changes in Lung Structure with Age

Mice

Through the seminal work by Metzger et al. (2008), we have an incredibly detailed understanding of *in utero* branching morphogenesis in mice. Such detail on lung branching is not available in any other species. Metzger et al. (2008) identified several discrete patterns of daughter branch formation during early development of the conducting airways. These patterns were linked to specific molecular cues, suggesting that these processes are tightly regulated by ordered encoding sequences. Mice, like rats, are born with lungs that are at the saccular stage of development. The process of alveolarization is rapid in the first weeks of postnatal growth such that the bulk of alveolar formation has occurred around weaning (prepuberty) (Massaro et al., 1985). Until recently, this was considered to be a discrete process whereby postpuberty lung growth into adulthood was largely a function of lung expansion. Recent detailed structural information using design-based stereological approaches to the quantification of lung structure have challenged this dogma and suggest that the formation of new alveoli by septation occurs (at least) into early adulthood (Mund et al., 2008).

There are a plethora of data on mouse lung structure assessed to varying levels of accuracy. However, very few studies have specifically examined changes in lung structure in relation to normal aging from adulthood into senescence. Unfortunately, there are conflicting data from many of the earlier studies such that several questions remain regarding the normal changes in lung structure with aging in mice. For example, Ranga and Kleinerman (1980) showed by electron microscopy that the number of interalveolar pores does not change with age in BALB/c mice, whereas Kawakami et al. (1984) demonstrated morphometrically that the number and total area of interalveolar pores increases rapidly between 38 days and 9 months of age in the same strain. This variation between studies highlights one of the difficulties in summarizing structural data from laboratory animal models *ex vivo*. Most studies have used a simplistic approach whereby the density of a given structure on histological sections is reported. Such density-based measurements can be difficult to interpret as they are heavily influenced by inflation state. This is particularly important when comparing lung structure between ages as we know that the pressure−volume (PV) relationship changes such that inflation levels will be vastly different between ages, even when a standardized pressure-based perfusion/instillation protocol is used. In order to overcome this problem, a reference volume is required (e.g., total lung volume), so that quantitative measures of structural elements of the lung can be calculated. This issue led to the development of standardized guidelines for the quantitative assessment of lung structure (Hsia et al., 2010).

Kawakami et al. (1984) used an approach whereby total lung volume postfixation was calculated by water immersion. This allowed the calculation of the total number of interalveolar pores, total volume of air in the alveoli, and alveolar ducts. Their data showed a period of rapid growth from 38 days of age to 9 months with increases in all of these structural measures and a plateau between 9 and 28 months of age in BALB/cNNia mice. These observations are consistent with substantial lung growth postpuberty, which stabilizes in late adulthood and into senescence. Interestingly, Kawakami et al. (1984) also showed an increase in alveolar wall thickness that was explained by increased collagen deposition with age based on previous data in rats (see below for discussion). More recent studies have compared lung structure between strains in mice from adulthood into senescence. Huang et al. (2007) showed that tissue density remained constant between 2 (adulthood) and 20 (aged) months of age in C57BL/6 mice, whereas tissue density decreased substantially between these ages in DBA/2J mice. Interestingly, the collagen content increased significantly in the latter strain with age but was stable in the C57BL/6 mice (Huang et al., 2007). Huang et al. also showed that the elastic fiber content of the lung decreased with age, which may be associated with early remodeling of the lung in late adulthood (Huang et al., 2007). Collectively, these studies suggest that mice display the full range from stable to "declining" lung structure (characterized by increased collagen deposition) into senescence depending on which strain is used. Interestingly, genomic analysis of the aging mouse lung suggests that these senescence-related aging events may be preprogrammed in a tissue-specific manner (Calvi et al., 2011), however, the relative importance of this "programmed" aging versus the impact of lifelong exposure to inhaled particles in the healthy lungs remains to be determined.

Rats

Prior to the transgenic "revolution," the most heavily studied laboratory animal, in terms of normal aging of the lung, was the rat. A number of studies have specifically examined structural changes in the lung with age in rats. These studies have consistently demonstrated significant increases in lung volume (alveolar size, total air volume, alveolar tissue volume, and capillary blood volume) with age with variations in the rate of change depending on the exact age and strain of rat studied (Johanson and Pierce, 1973; Pinkerton et al., 1982). These changes correlate strongly with body weight, suggesting that they are largely due to linear growth with body size.

The detailed work by Pinkerton and colleagues (Pinkerton et al., 1982), who examined Fisher-344 (F-344) rats from 1 week of age (preweaning) through puberty and up to 26 months of age, showed that the cellular components of the interstitium remained relatively unchanged after 5 months of age following a period of rapid growth to that point. In contrast, the noncellular (extracellular matrix) components of the interstitium increased beyond this point. This dysmorphic growth between the cellular and noncellular components of the lung parenchyma was followed up by ultrastructural studies showing increased collagen deposition with age. This was mostly observed in the alveolar septa. Nardell and Brody (1982) described in detail four key stages of alveolarization in the postnatal Sprague–Dawley lung: expansion, proliferation, elastin accumulation, and proportional growth. The earlier stages are characterized by rapid alveolar septation, while the latter two primarily consist of elastin deposition and alveolar expansion. Their data suggest that alveolar septation is largely complete by 3 weeks of age. Similarly, Bolle et al. (2008) have shown that alveolar septal thickness decreases rapidly (from 13.4 to 5.4 μm) between 1 and 3 weeks of age but changes little thereafter.

Systematic attempts have been made to quantify changes in lung structure from adulthood into senescence in rats. Johanson and Pierce (1973) demonstrated an increase in mean chord length (a measure of linear free space as a proxy for alveolar size) in the face of constant surface area in Sprague–Dawley rats, which is indicative of increased alveolar size without tissue destruction. A more recent study on Wistar rats showed that parameters associated with airspace size increased from adulthood and that this was associated with a decrease in tissue density (Escolar et al., 1994). However, when these data were converted into absolute values, there was no decrease in total tissue volume (Escolar et al., 1994). Importantly, Kerr et al. (1990) were able to demonstrate an increase in airspace size and collagen content without a corresponding increase in surface area or alveolar number in 2-year-old Sprague–Dawley rats, while no differences were observed in F-344 rats. Taken together, these observations demonstrate a lack of evidence for substantive emphysematous-type changes in the lung with age in rats in contrast to suggestions in other species (detailed below). Descriptive studies have suggested cellular changes and extension of the distal airways in senescent rats (Yamamoto et al., 2003); however, the value of these observations is again limited by the lack of quantitative data that can clearly summarize any apparent tissue destruction that would be indicative of age-related emphysema. Collectively, these studies in rats suggest little alveolarization postadulthood, an increase in alveolar size with age that is related to lung expansion rather than emphysema, and increased collagen deposition. These observations parallel those in mice suggesting consistency in lung structural changes with age in the Muridae.

Interestingly, rats are one of the few species where we have data on the postnatal growth of the airway (Lee et al., 2011). These data are limited to 2 (prepuberty) to 11 (adulthood) weeks of age and show that the growth of the airway tree is not linear, i.e., is not simply a function of age or body size. In particular, the (relative) growth of the length of the airway tree over this age range is faster than that of the airway diameter, and the association between airway size (length and diameter) and body size differs between airway generations (Lee et al., 2011). There are no data on changes in airway structure with senescence in rats.

Rabbits

Details on the changes in lung structure with age in rabbits are limited. Early studies suggest that emphysema is common among aging rabbits (Strawbridge, 1960), which was linked to collagen deposition in the capillaries, ischemia, and alveolar degradation (Boatman and Martin, 1960). A more recent study has shown that the structure of the airway changes between 1 and 6 months of age with older rabbits having proportionally less, normalized-to-ideal airway area, airway smooth muscle than younger animals (Tepper et al., 1995). In terms of alveolarization, the relative development of the lung at birth is similar to that observed in humans with some alveoli present and the bulk of them formed postnatally into early adulthood (Kovar et al., 2002). The formation of new alveoli corresponds to an increase in surface area of the lung parenchyma and thinning of the alveolar walls with age (Kovar et al., 2002). The morphometric data collected by Kovar et al., (2002) detailed alveolar development up to 9 months of age and showed that alveolar volume appeared to decrease between 4 and 9 months of age (Kovar et al., 2002), suggesting an increase in alveolar number without lung expansion. While some of the rabbits in the earlier study (Strawbridge, 1960) were 18 months old, these are clearly not in the range of senescence for rabbits. Likewise, caution needs to be applied when interpreting the data from this, and the more recent study, as neither of them used stereological techniques to probe lung morphology.

Guinea Pigs

The newborn guinea pig lung is at the other extreme of lung development compared to the Muridae in that the lungs have a "mature" phenotype (Ross and Thurlbeck, 1992; Collins et al., 1986; Lechner and Banchero, 1982). Data from fixed Hartley albino guinea pig lungs show that the number of alveoli (approximately 15 million) does not change from birth up to 3 weeks of age, whereas the mean interalveolar wall distance, mean alveolar chord length, mean alveolar duct cord length and parenchymal surface area all increase significantly during this period (Ross and Thurlbeck, 1992). This observation suggests that lung growth during this period of rapid somatic growth is largely due to alveolar expansion rather than an active process of septation (Ross and Thurlbeck, 1992). This is in stark contrast to the other species described in this chapter. Data beyond these ages suggest that body weight increases linearly, while there is a plateau in alveolar surface area and volume between 13 and 16 weeks of age (Lechner and Banchero, 1980). There are no data available on structural changes in the lung beyond early adulthood in healthy guinea pigs.

Hamsters

Very little data exist on the newborn structure of the hamster lung, although they appear to have lungs that resemble the morphology of the murine lung (Mauderly, 2000). Of all the species that have been discussed thus far, the hamster seems to have been studied with a focus on the senescent lung, although there are only a handful of studies in this species. This shift in focus seems to be due to the hamster's utility as a model for elastase-induced emphysema. Consequently, there is little information on changes in alveolar number, size, or architecture in early postnatal life. Based on the data that does exist, the morphology of the parenchyma remains very stable from prepuberty (5–6 weeks of age) through to late adulthood (58 weeks of age) in golden hamsters with little (physiologically significant) increases in the internal surface area of the lung or the mean linear intercept between these ages (Snider and Sherter, 1977). The stability of the mean linear intercept, a measure of the mean free distance between gas exchange surfaces (Hsia et al., 2010), with aging has been confirmed in other studies (Goldstein, 1982). However, this consistency in internal morphology is not reflected in the structural elements of the lung with 18-month-old hamsters having 85% more collagen than 4-month-old hamsters.

Ferrets

The developmental state of the ferret lung at birth, in comparison to the other species discussed, is unclear. There are few studies that have specifically compared lung structure or function in ferrets at different ages. With a view to describing the age-related changes in alveolar architecture, Wood and colleagues (Wood et al., 1998) examined morphometric patterns of alveolar septae in ferrets from prepuberty (8–10 weeks) up to adulthood (2 years). Their data suggest that between these ages there is an increase in parenchymal complexity; however, this is unlikely to be due to the formation of new alveoli by active septation.

Dogs

At term the dog lung is in a relatively immature state, has not yet entered the alveolar stage of development, and has a distinctly saccular morphology (Sipriani et al., 2009), in contrast to humans where alveolarization has clearly begun at birth. Interestingly, while the absolute number of studies on the normal aging of the canine lung is relatively low compared to other species, the breadth of ages where we have some data spans all ages from the *in utero* period and into senescence. This is probably due to the interest in dogs from a veterinary point of view. This context is important as it leads to a word of caution regarding the data that we do have on the dog lung compared to other species. Most of the data described for other mammals in this chapter, particularly in the smaller laboratory animals, comes from studies where the animals were housed in controlled laboratory conditions. In contrast, most of the dog studies use multiple breeds held or obtained from undetermined sources. As such, the environmental exposures that the dogs in these studies were exposed to are likely to be vastly different to those experienced by the other species detailed above. Consequently, the definition of the "normal" lung in the canine situation is shifted. This is highlighted by the appearance of anthracosis, suggesting long-term exposure to carbonaceous particles in older dogs in some studies (Hyde et al., 1977).

As previously mentioned, the canine lung is in a relatively immature state at birth with all of the alveoli being formed postnatally. Due to the size of the dog lung, it has also been possible to show that the maturation of the lung during *in utero* development occurs in a radial fashion with the lung units associated with the upper airway being more developed than those of the distal lung (Sipriani et al., 2009). There is a gap in our knowledge of normal lung morphology in the first months of life in dogs as most studies looking at the association between age and lung structure begin from approximately 6–12 months of age. Interestingly, dogs are one of the few species where an attempt has been made to examine postnatal airway growth. Horsfield (1977) showed that the airways grow in proportion to body weight ($BW^{1/3}$), which appears to be in contrast to the recent studies in rats detailed earlier (Lee et al., 2011). Hyde et al. (1977) conducted a detailed study of parenchymal morphology of beagles ranging from 9 months to 10 years of age. In this study, the authors described the changes in volume density of the alveoli and tissue, which decrease with age, and the alveolar ducts, which increase with age. The volume density is a measure of the proportion of the tissue represented by a particular structure on histology (e.g., proportion of tissue that is alveoli). The study by Hyde et al. (1977) used a stereological approach to the calculation of morphometric parameters, however, they did not use a "design-based" approach and were not able to provide (for example) absolute values for the number of alveoli due to the lack of a reference volume. In other words, they demonstrated that the proportion of tissue represented by alveoli in a given field of view decreased with age but could not estimate the *total* number of alveoli in the dogs. Unfortunately, this makes the data from this study difficult to interpret. As such, we still have a poor understanding of the normal structural changes in the canine lung with age. An observational study by Reif and Rhodes, (1966) suggested that the senescent dog lung is characterized by pleural thickening, bronchial thickening, interstitial fibrosis, and in some cases focal emphysema. However, this study was conducted on an uncontrolled cohort of dogs from general veterinary practice, and, as such, environmental exposures are likely to have played a large role in these observations.

Sheep

At birth, sheep have well-developed lungs and have progressed into the alveolar stage of lung development (Alcorn et al., 1981). Davies et al. (1988) conducted a detailed stereological assessment of the postnatal sheep lung between 2 and 171 days of age. Between these ages, there was an increase in the surface area of the lung and the volume of the alveolar septae (Davies et al., 1988). The number of alveoli increased in the first few weeks of life with a doubling in the number of alveoli between 2 and (approximately) 120 days of age (Davies et al., 1988). At birth, the number of alveoli is already relatively high, and the lack of a substantive increase in all of these indices reflects the relatively mature state of the ovine lung at birth (Davies et al., 1988).

Pigs

Like sheep, pigs are often used as a model of early lung development although there are key morphological differences, in particular, the relatively advanced development of the alveoli, which distinguishes them from humans (Winkler and Cheville, 1984). At birth, the porcine lung is clearly into the advanced alveolar stage of development (Winkler and Cheville, 1984). Postnatal formation of the alveoli appears to be rapid in the first 8 weeks of life (Winkler and Cheville, 1984), with the total number of alveoli increasing from 30×10^6 at 3 days of age to 102×10^6 at 8 weeks of age (Rendas et al., 1978). The number of alveoli continues to increase until around 16 weeks of age, by which time the number has doubled again (203×10^6) (Rendas et al., 1978). The alveolarization process and development of the pig lung in the first weeks of life is complex, resulting in an apparent increase in the thickness of the alveolar septae at 1–2 months of age compared to 2–5 days of age (Winkler and Cheville, 1984; Mansell et al., 1995), which is thought to be due to increased deposition of connective tissue, cellular activation, and/or the production of surfactant (Winkler and Cheville, 1985). As developmental studies on pigs are typically aimed at early life, there are very little data on lung structure or function beyond early adulthood.

Nonhuman Primates

Conventional wisdom suggested that while the early development of the nonhuman primate lung parallels that of humans, the alveolar structure at birth is relatively mature such that postnatal alveolar growth was originally thought to occur entirely by expansion rather than an increase in the number of alveoli (Hislop et al., 1984). However, the subsequent application of quantitative stereological techniques (Hsia et al., 2010) to this issue demonstrated that the alveolar number clearly increases postnatally into adulthood while alveolar size does not change (Hyde et al., 2007). The most rapid increase in tissue volume and airspace size in the parenchyma occurs in the rhesus monkey between birth and 2 years of age (Hyde et al., 2007), which is around the age of puberty (Plant, 2001). In one of the few studies to examine lung structure in nonhuman primates from early life to ages close to their maximum lifespan, Shimura and colleagues (Shimura et al., 1986) showed that the number of pores of Kohn per alveolar unit increases linearly between 1 month and 10 years of age but does not change thereafter (up to 31 years of age), suggesting that the lung does not grow beyond 10 years of age. Interestingly, Shimura and colleagues (Shimura et al., 1986) also suggest that there is evidence of panacinar emphysema in aged monkeys that was attributed to decreased surfactant production (quantified by the density of lamellar bodies). However, it should be noted that half of the animals studied showed evidence of lesions resulting from lung mite infection. This highlights the difficulty in separating the effect of the "normal" aging process from the cumulative effect of long-term environmental exposures on the lung.

Summary

Mammals display the full range of lung maturity at birth from the Muridae and dogs, where alveolarization is entirely postnatal, to the guinea pig where the full complement of alveoli appears to be present at birth. Likewise, it is clear that in some species (mice, rats, primates, pigs, rabbits) new alveoli form into early adulthood while in other species postnatal neoalveolarization is limited (guinea pigs, sheep) (Figure 1). There still remain significant gaps in our knowledge of species-specific changes in lung structure with age, and research is required in this field. In particular, studies on the "normal" changes in lung structure into senescence are warranted. In species that are kept under controlled conditions (e.g., mice, rats), it appears as though the aged lung is characterized by increased collagen deposition with no alveolar destruction, which is consistent with fibrotic pathology (depending on which strain is used). In contrast, in species where environmental exposures and infection are not controlled (e.g., dog, nonhuman primates), there is evidence of emphysematous changes in lung structure in the latter stages of life (Figure 1). Of course, in the case of the Muridae there will still be environmental stimuli, however, these will be vastly different to the exposures of the other species and this needs to be considered when investigating age-related changes in lung structure.

3. LUNG FUNCTION

3.1 Measuring Lung Function

In recent decades there have been substantial advances in the techniques available for assessing lung function in laboratory animals. Many of the issues related to size have been overcome with improvements in pressure transducer technology that ensure linear responses over the range of interest while keeping functional dead space to a minimum. The advent of commercial systems for the assessment of lung function has meant that many of the latest techniques are being used around the world. This has contributed to the

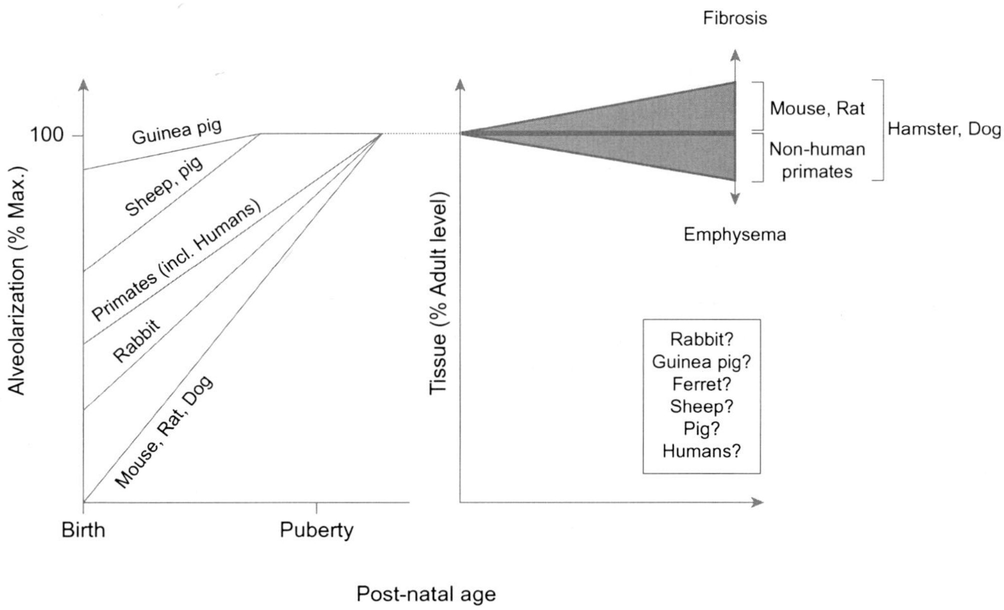

FIGURE 1 Schematic representation of our current understanding of alveolarization (left) from birth to early adulthood and senescence-related changes in tissue structure (right) for the species described in this chapter. The data available are currently not sufficient to draw true trajectories for alveolarization between birth and adulthood. As such, the lines drawn should only be considered as indicative of the pattern that may occur based on the data currently available.

recent increase in our knowledge of changes in lung function with age, particularly in the murine lung. However, there are two fundamental issues that need to be taken into account when interpreting the data available. The first is the compromise between accuracy and invasiveness. This problem has been termed the "phenotyping uncertainty principle" (Bates and Irvin, 2003).

At one end of the spectrum noninvasive techniques (e.g., whole body plethysmography in spontaneously breathing animals) are able to provide useful data on tidal volume, minute ventilation, and breathing frequency but have limited utility for assessing lung mechanics (Lundblad et al., 2007; Bates et al., 2004). At the other end of the spectrum are the invasive systems, which require the animal to be anaesthetized, surgically intubated, and mechanically ventilated but provide the most accurate data on lung mechanics (Bates and Irvin, 2003). The invasive approach is necessary due to our need to manually probe the system. In a perfect world we would be able to manipulate ventilation (e.g., breath holds) in conscious, spontaneously breathing laboratory animals, which is clearly not possible. This problem has complicated our ability to interpret even simple measures of lung function, in particular, total lung capacity (TLC). TLC in rodents is typically defined as the volume reached at some arbitrary transrespiratory pressure (P_{rs}). The problem is that the PV curve in a rodent never reaches a true plateau, and the lung can be artificially inflated to the point of rupture (Soutiere and Mitzner, 2004; Zosky et al., 2008). One could set this arbitrary P_{rs} as 20 cmH_2O as this is the point beyond which the PV curve appears to behave abnormally (Zosky et al., 2008); however, a lack of consensus regarding this issue precludes this standardization at present. These complications can make it difficult to accurately compare lung function data between studies. Our current understanding of the changes in lung function that occur with age are outlined below. For the purposes of this discussion, the data will be limited to that specifically related to static lung volume and lung mechanics. Details on tidal volumes, respiratory rates, and other ventilator parameters in conscious animals will be discussed in the subsequent section.

In order to provide a context for the following discussion, our current understanding of changes in lung volume and lung mechanics with age in humans should be considered. While the exact timing of changes in the magnitude of lung function differs depending on ethnicity and sex, the general pattern for most measures of lung function when plotted against age is the same. The absolute values for forced vital capacity, the volume of air expelled from TLC to residual volume (RV) during a maximum expiratory effort, and FEV_1, the volume of air expelled in the first second of a maximal expiratory effort from TLC, increase from birth to early adulthood (approx. 20–25 years) and steadily decline thereafter (Burrows et al., 1983).

3.2 Changes in Lung Function with Age

Mice

Lung volumes at various states of inflation have been measured *ex vivo* and *in vivo* in several strains of mice. Very few studies have examined the changes in static lung volumes that occur with age in mice. One recent study was able to show that thoracic gas volume (TGV) measured in anaesthetized, tracheostomized BALB/c mice at elastic-equilibrium lung volume (0 cmH_2O P_{rs}) increases rapidly between 2 (preweaning) and 6 (postpuberty) weeks of age. TGV at 0 cmH_2O P_{rs} continues to increase between 6 and 8 weeks of age (Bozanich et al., 2007). When lung volume was measured *ex vivo* in the same strain, following fixation a plateau appeared to occur around 5 weeks of age (Bozanich et al., 2005); however, it is difficult to compare these data sets as the latter will represent the lung volume approaching TLC (i.e., $P_{rs} >>> 0$) without the influence of the chest wall. A further study in C57BL/6 mice showed that lung volume at 0 and 30 cmH_2O P_{rs} continues to increase from 2 to 26 months of age (Huang et al., 2007), and the same data compared to another strain (DBA/2J) show that this pattern is conserved, although the magnitude of the increase was greater in the latter strain (Huang et al., 2007).

The changes in lung volume that have been described in these studies parallel the changes in lung function associated with normal postnatal aging into adulthood. Bozanich et al. (2007) have described a substantial decrease in airway resistance (R_{aw}), measured using a modification of the forced oscillation technique, which in BALB/c mice halves between 2 and 3 weeks of age (1.41 hPa/s/mL to 0.75 hPa/s/mL) and halves again between 3 and 8 weeks of age (0.36 hPa/s/mL). Both tissue damping (G), the frequency-dependent viscous resistance of the lung parenchyma, and tissue elastance (H), which indicates parenchymal stiffness, decrease steadily from 2 (G, 31 hPa/mL; H, 125 hPa/mL) to 8 (G, 9 hPa/mL; H, 39 hPa/mL) weeks of age with most of the change occurring between 2 and 6 weeks of age. Hysteresivity (G/H), which is the ratio of the dissipative and elastic properties of the lung parenchyma (Fredberg and Stamenovic, 1989), does not change significantly from 2 (preweaning) to 8 (adult) weeks of age suggesting that the fundamental mechanical properties of the lung tissue do not change over this period of growth. The data beyond this age in mice are more disjointed; however, a study by Huang and colleagues (Huang et al., 2007) compared the same parameters between 8 weeks and 20 months of age in C57BL/6 and DBA/2J mice and found that R_{aw}, G, and H continued to decrease with age. There were some deviations from this pattern with little change in R_{aw} between these ages in C57BL/6 mice and an increase in G/H in the DBA/2J mice, which was not observed in the other strain. These data suggest a continuing "improvement" in lung function throughout adulthood in mice.

Rats

There are several *ex vivo* studies that have measured lung volume in rats under varying conditions, and some studies have compared lung capacities *in vivo* for a given P_{rs}. The *in vivo* and *ex vivo* data that does exist suggest that total lung capacity, or some approximation of it at least, continues to increase with postnatal age into adulthood (Nardell and Brody, 1982; Gomes et al., 2001). For example, left lung volume measured by volume displacement in Sprague–Dawley rats increases from 0.43 mL at 10 days of age (preweaning) to 1.17 mL at 25 days of age (postweaning) up to 5.06 mL at 90 days of age (Gomes et al., 2001). *In vivo* the volume of the lung at 10 cmH_2O transpulmonary (P_{tp}) pressure increases linearly from 4 to 40 days postnatal age in Sprague–Dawley rats (Nardell and Brody, 1982). More recent studies that have compartmentalized lung volume measures (e.g., FRC, IC, RV, TLC) show that all of these parameters increase from early life into adulthood (90 days postnatal age) in an almost linear fashion (Bolle et al., 2008). Several earlier studies have examined similar measures in older rats (>18 months of age) and shown that this trend continues into late adulthood with functional residual capacity (FRC), vital capacity (VC), TLC, and RV continuing to increase with age well into senescence (Johanson and Pierce, 1973; Pinkerton et al., 1982; Mauderly, 1982; King, 1966). There is significant variation in the circumstances and techniques used to assess these measures; however, the consistency of the pattern observed across strains (Wistar, Fischer-344, Sprague–Dawley) suggests that this is the true pattern representing the association between lung volume and age in rats. This "continual" increase in lung volume with postnatal age parallels the linear increase in body weight seen in rats held under laboratory conditions.

Several attempts have been made to examine the ratio between TLC and body weight in rats and how this ratio changes with age. However, these analyses suffer from the same limitations outlined in mice whereby the P_{rs} used to assess TLC is arbitrary and is usually not associated with a true plateau in the PV relationship. Similarly, the use of body weight (BW) to normalize measures of lung volume is unlikely to be the best representative of body growth, particularly under the artificial living conditions created in the laboratory, and should instead be normalized to body length, an analysis that has yet to be conducted. Nevertheless, the ratio of TLC/BW tends to decrease with postnatal age (Bolle et al., 2008; Mauderly, 1982) suggesting dysmorphic somatic and lung growth, although, again, the caveat on using body weight as the standard for somatic growth should be acknowledged.

As you would predict with an increase in lung volume there is a leftward shift in the pressure–volume curve such that dynamic (C_{dyn}) and static (C_{st}) compliance increases with age in rats (Bolle et al., 2008) when assessed over a broad range of ages from 7 days to 90 days in Wistar–Kyoto rats. Studies at a finer level examining day-to-day variation in lung function during early postnatal growth (from 10 to 20 days of age) in Sprague–Dawley rats show that the increase in lung function may not be smooth, but the general trend is for an improvement in lung compliance with age (Broussard et al., 2006). Very little data exist on lung function *per se* in older or aged rats. One study suggests a pattern in lung function similar to that seen in humans with an improvement in quasi-static compliance in young adults compared to juveniles and a subsequent decrease in compliance into senescence (Mauderly, 1982). This observation is not consistent with the increases in lung volume described previously, however, it may reflect the increased collagen deposition that is thought to occur during the aging process in rats (Bellmunt et al., 1995).

Similar studies on invasive measures of lung mechanics like those described above in mice have been conducted in rats from early life into adulthood. As expected from the increase in lung volume with age there is a decrease in measures of lung mechanics including R_{aw}, G, H, respiratory system resistance (R_{rs}), and respiratory system elastance (E_{rs}) from birth up until adulthood in rats (Bolle et al., 2008; Gomes et al., 2001; Broussard et al., 2006). The decreases in these measures are most rapid over the first few weeks of life (Broussard et al., 2006) and tend to approach an asymptote at around 2 months of age (Bolle et al., 2008; Gomes et al., 2001). There are no data describing lung mechanics in senescent rats *in vivo*.

Rabbits

Lung volume at 20 cmH_2O P_{tp} increases substantially between 1 and 6 months of age in rabbits (23.3 vs 84.1 mL) (Shen et al., 2000). Unlike the Muridae, the lung volume of the rabbit, measured *ex vivo* by volume displacement, appears to increase with age up to a point (approx. 3 months) before reaching a plateau (Kovar et al., 2002). In contrast, *in vivo* studies suggest that other measures of lung volume (FRC) continue to increase into senescence (52.5 mL age 6–12 months vs 61.5 mL > 36 months) (Davidson et al., 1966). This observation may be due to lung growth or the onset of emphysematous changes in the lung. *In vivo* data on TLC into senescence in the rabbit are lacking, although one would predict that hyperinflation resulting from emphysematous alveolar expansion is likely.

In juvenile and adult rabbits there is no difference in V_{max} normalized pressure–volume curves (Shen et al., 2000). This is not the case into senescence with a significant leftward shift in the PV curve between 12 and 36 months of age (Davidson et al., 1966), which is consistent with a decrease in elastic recoil due to emphysematous changes in the lung (Martin et al., 1977). However, as discussed in the previous section, the structural data to support senescence-related emphysema in rabbits are lacking.

Studies on isolated rabbit airways show that the immature (1 month old) airway is more distensible than the mature airway (6 months old) (Shen et al., 2000), which is consistent with the structural differences observed between these ages (Tepper et al., 1995). There are very little data on invasive measures of lung mechanics in rabbits, and no study has compared lung mechanics between ages using contemporary measures of lung function. Nonelastic total respiratory system resistance calculated using pressure and flow measures in spontaneously breathing anaesthetized rabbits nearly doubles between 12 and 36 months of age (Davidson et al., 1966). This represents the in-phase component of resistance to flow and suggests some level of airway obstruction, although there is no morphological evidence to support this to date. These data are in contrast to the Muridae where lung function appears to continuously "improve" with age.

Guinea Pigs

As discussed earlier the guinea pig lung is at the extreme end of the "maturity" spectrum at birth with the gross morphology being akin to that of an adult lung in other mammalian species. Lung volume has been measured at different ages in guinea pigs both *in vivo* (Gaultier et al., 1984) and *ex vivo* (Ross and Thurlbeck, 1992; Lechner and Banchero, 1980; Forrest and Weibel, 1975). In the first 3 weeks of life, fixed lung volume ($P_{rs} = 25\ cmH_2O$) increases linearly in parallel with body weight (Ross and Thurlbeck, 1992). Between 4 and 9 weeks of age there continue to be large increases in fixed lung volume ($P_{rs} = 20-22\ cmH_2O$) (4–5 weeks, 11.0 mL; 8–9 weeks, 17.4 mL), which continues to increase into early adulthood (16–17 weeks, 19.5 mL) albeit at a slower rate of increase (Lechner and Banchero, 1980). Interestingly, the relationship between lung volume and body weight, perhaps reflecting the early structural maturation of the lung in this species, is very linear (Forrest and Weibel, 1975). There are no data on changes in lung volume associated with senescence in guinea pigs. There is only one study that has measured lung volume *in vivo* versus age in this species. Gaultier and colleagues (Gaultier et al., 1984) showed that FRC, measured by plethysmography, increases from 1.76 to 4.75 mL between 2 days of age and 4 weeks of age in guinea pigs. FRC plateaus between 4 and 10 weeks (4.60 mL) of age before increasing again between 10 and 16 weeks of age (6.03 mL). In contrast, TLC, measured at 30 cmH_2O transpulmonary pressure (P_{tp}), does not show a similar plateau, increasing from 5.9 mL at 2 days of age to 15.1, 19.2 and 23.9 mL at 4, 10, and 16 weeks of age, respectively (Gaultier et al., 1984).

There are very limited data on changes in lung mechanics with age in guinea pigs. The same study by Gaultier et al. (1984) showed that lung compliance (C_L) measured over the linear portion of the PV curve and from the shape constant (k) of the exponential function describing the PV curve increases significantly from 2 days of age (C_L, 1.53 mL/cmH_2O; $k = 0.146/cmH_2O$) to 4 weeks of age (C_L, 2.38 mL/cmH_2O; $k = 0.219/cmH_2O$) but changes very little after that up to 16 weeks of age. The dysmorphic increases in FRC, TLC, and lung compliance suggest that the normal aging of the guinea pig lung is not simply an increase in size of the newborn lung, despite the apparent level of lung maturity, although the nature of the changes in lung structure that are occurring are yet to be identified.

Hamsters

Hamsters have a rapid increase in body weight up until early adulthood, but beyond that, body size changes very little (Snider and Sherter, 1977). In parallel with these changes in body size there is a small, but significant, increase in TLC (25 cmH_2O P_{tp}) between 8 and 11 weeks of age with very little change thereafter. FRC does not change between 8 and 58 weeks of age (Snider and Sherter, 1977). These data match those of Goldstein (Goldstein, 1982) who found no difference in TLC or RV between 5 and 19 months of age in Syrian hamsters. In contrast, Mauderly (1979), who also found no difference in body weight with aging, demonstrated an increase in TLC (9.6 mL vs 11.1 mL) and FRC (3.5 mL vs 4.3 mL) in the same strain of hamster between 15 and 65 weeks of age. It is unclear why this discrepancy exists between these studies, however, there were methodological differences, and it should be noted that there were subtle differences in the exact ages measured. There are only limited data on changes in lung mechanics with age in hamsters. Snider and Sherter (1977) reported little change in static compliance between 9 and 58 weeks of age in hamsters, whereas Skornik and Brain (1990) reported significant decreases in lung resistance and dynamic compliance between adulthood and senescence in hamsters.

Ferrets

There are little to no data on changes in lung volume or lung mechanics with age in ferrets. Experiments focusing on intervention studies suggest that TLC increases between

6 and 10 weeks of age (Zhang et al., 1996); however, there are no data beyond this age.

Dogs

Data on the changes in lung volume with age in dogs are entirely from *in vivo* studies (Mauderly, 1974; Robinson and Gillespie, 1973; Quan et al., 1990). All of these studies describe an increase in lung volume with age. Quan et al. (1990) studied lung volume and function in beagles aged ~2–11 months of age. These authors found a linear association between age and FRC over this age range. In contrast, the study by Robinson and Gillespie (1973) did not find evidence for a change in FRC with age in beagles aged 289–3882 days. While Mauderly (1974), on the other hand, found an increase in FRC with age, this relationship was not strictly linear, with no difference in FRC between 12 months and 3 years of age but a significant increase in FRC between 3 (391 mL) and 10.5 (531 mL) years of age. The reasons for the discrepancy between this study and the previous study are unclear, although the regression approach used by Robinson and Gillespie (1973) was simplistic and may not have been sensitive to nonlinear associations. Only one study (Robinson and Gillespie, 1973) has examined TLC versus age in the canine lung and found no relationship between this parameter and age.

Changes in lung function with age in the above-mentioned studies paralleled the changes in lung volume that were described. Quan et al. (1990) showed a decrease in lung resistance in the first year of life, while dynamic compliance, static compliance, and specific conductance all increased with age. Robinson and Gillespie (1973) did not show any evidence for a change in lung compliance between dogs aged less than 500 days and dogs aged more than 3000 days, although their data were highly variable. In stark contrast to these data, Mauderly (1974) suggested that dynamic compliance decreases while pulmonary resistance increases into senescence in beagles. Taken together these studies suggest that changes in lung volume and function with age are (relatively) similar to that observed in humans, compared to the smaller laboratory species described above, with an increase in lung volume and improvement in lung function in the early years of life, a plateau in adulthood, and a decline in function into senescence.

Sheep

There are very little data on changes in lung volume with age in sheep. Davies and colleagues showed that lung volume, measured *ex vivo*, increases between 2 (206 mL) and 119 (912–948 mL) days of age and plateaus thereafter (up to 171 days of age). *In vivo* the only study that exists in this species shows a modest increase in FRC between 1 and 4 weeks of age from 96.6 to 160.7 mL (Johns et al., 1994). There are no published studies providing longitudinal data on lung function with age in sheep.

Pigs

Data on lung volume and function in pigs are limited. Rendas et al. (1978) describe a monotonic increase in *ex vivo* lung volume with age that parallels body weight. Lung volume doubles in the first 3 days of life (newborn, 56 mL; 3 days of age, 102 mL) and doubles again by 2–5 weeks of age (168–227 mL). Thereafter, lung volume increases exponentially such that an adult pig (195 kg) has a lung volume (4775 mL) that is two orders of magnitude larger than that of a newborn (1.3 kg) piglet (56 mL) (Rendas et al., 1978). Winkler and Cheville (1985) describe a less dramatic increase in lung volume, measured by volume displacement, over the first week of life (newborn, 87 mL; 7 days of age, 95 mL) with a tripling in lung volume by 30 days of age (272 mL). There is only one report of *in vivo* lung volume, which showed that, relative to body weight, there is no change in lung volume between 2 and 16 weeks of age (Rendas et al., 1978).

Changes in lung function with age have only been described in limited detail in the studies that examine the age-related changes in the porcine lung. Mansell et al. (1995) demonstrated that the bulk modulus (k), derived from the PV curve, increases between birth and 3–5 days of age ($k = 105\ cmH_2O$) before decreasing slightly at 25–30 days of age ($k = 71$) and remaining constant up to 80–85 days of age ($k = 76$). These data suggest that the newborn pig lung is relatively compliant and becomes relatively stiff over the first few days of life. Rendas et al. (1978) measured compliance relative to body weight and showed that this parameter remains constant between 2 and 16 weeks of age, indicating that absolute lung compliance increases with age postnatally in pigs.

Nonhuman Primates

There are very little data on changes in lung volume with age and none that has addressed age-related changes in lung function in nonhuman primates. The single study that has examined *in vivo* lung volume measured by nitrogen washout and CT showed that lung volume doubles between 13 weeks and 9 months of age in male baboons, which mirrors the doubling in body weight that is observed between these ages (Thomas et al., 1993). There are no data on changes in lung volume or lung mechanics with aging into adulthood or beyond in nonhuman primates.

Summary

There is considerable between-species variation in the pattern of change in lung volume and lung function

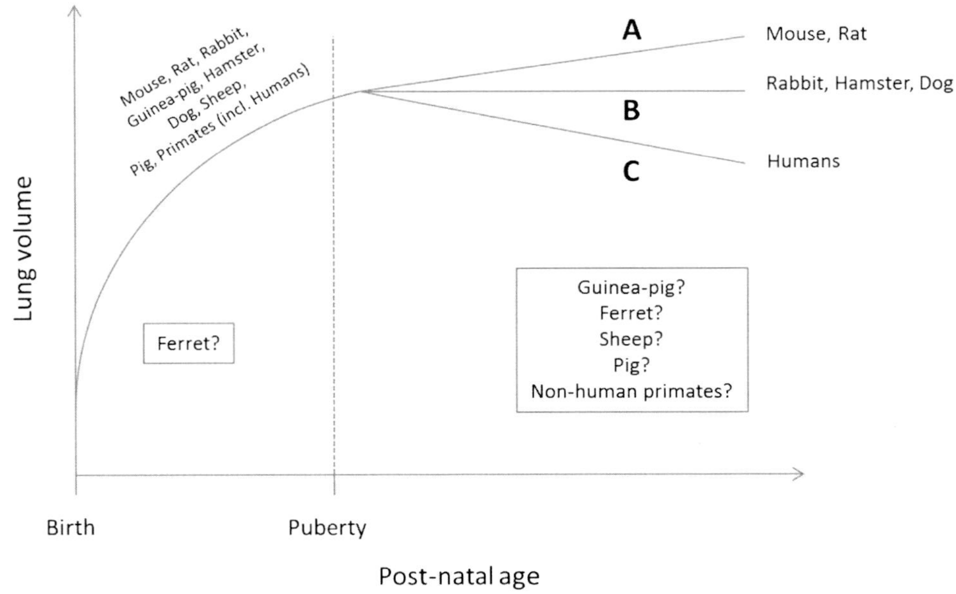

FIGURE 2 Schematic representation of the changes in lung volume (total capacity or equivalent) associated with age in the species described in this chapter. There are not sufficient data to produce meaningful trajectories between birth and adulthood, so for the sake of this schematic a single line has been used to represent the common increase in lung volume that has been observed in these species. There are three patterns from early adulthood into senescence: A, a continuing increase in lung volume; B, no change in lung volume; C, a decrease in lung volume (this has only been observed in humans to date as measured by FVC).

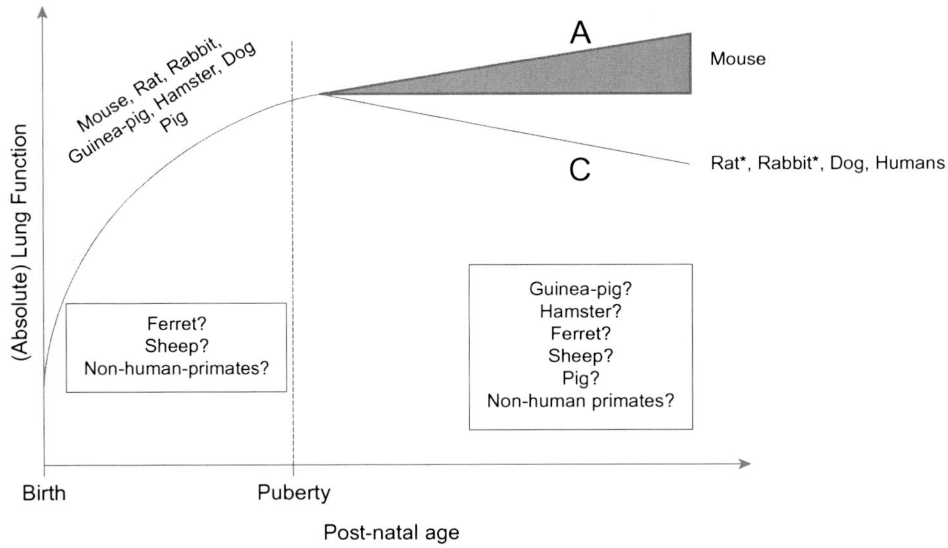

FIGURE 3 Schematic representation of the changes in lung function (absolute) associated with age in the species described in this chapter. There are not sufficient data to produce meaningful trajectories between birth and adulthood, so for the sake of this schematic a single line has been used to represent the common increase in lung function that has been observed in these species. There are two patterns from early adulthood into senescence: A, no change or an improvement in lung function (mice only with the pattern varying between strains); C, a decrease in lung function. *Data for the rat and rabbit are inferred from the data on lung function that are available, and these relationships are yet to be confirmed.

(particularly during adulthood) with age in mammals. There also remain several gaps in our understanding of, and little information regarding the processes (cellular and/or molecular) that control, postnatal changes in lung function with age. While the exact rate of change of absolute lung volume and function versus age between birth and their peak in adulthood is likely to differ between species, based on the data at hand it is not possible at this stage to accurately plot these relationships. However, we can say (with some level of confidence) that in all the species where data exist, both absolute lung volume (Figure 2) and function (Figure 3), whatever measure is used, increase from birth until adulthood in all the species that have been studied and there is no reason to suspect that this would differ in the species where data are lacking. In contrast, the changes in both lung volume and lung function beyond early adulthood appear to differ substantially between species. In both of these outcomes the traditional measures of these factors (i.e., FVC and FEV_1) decrease with age in humans. To date we do not have any data in other mammalian species to support a similar decrease in lung volume with increasing age into senescence. In contrast, lung volume in rabbits, hamsters, and dogs appears to remain relatively stable, while in the Muridae (both mice and rats) lung volume continues to increase with age (Figure 2). In some species (rat, rabbit, dog) there is some evidence to suggest that lung function decreases beyond adulthood; however, further studies are probably needed to confirm this pattern (Figure 3). In contrast, mouse lung function appears to remain stable or (in some strains at least) improve with aging (Figure 3).

4. GAS EXCHANGE

4.1 Recent Studies on Gas Exchange

While our knowledge in recent years on changes in lung structure and lung mechanics with age has increased exponentially as a result of growing interest in the utility of the Muridae as models for human disease, the same is not true for data on gas exchange. This is largely a function of the challenges associated with measuring gas concentrations in low volume samples on both the pulmonary and circulatory sides of the equation without compromising their physiological function. Many of the standard techniques for assessing gas exchange, such as carbon monoxide diffusing capacity (DL_{CO}), have been used for several decades; however, translation of these techniques has rarely been achieved in rats let alone mice. The advent of new imaging technologies using microparticle injection or contrast agents has partially rectified this issue, however, there are still very little data regarding the changes that occur in regard to gas exchange with age in mice. There are some data in rats describing age-related changes in gas exchange, however, most of these data, like that in other species, is several decades old. Few advances in age-related changes in gas exchange have been made in recent years. In humans, gas exchange, like lung structure and mechanics, "improves" in absolute terms with age until early adulthood before declining through late adulthood and into senescence. Interestingly, both absolute and volume-corrected DL_{CO} appear to decline from adulthood into senescence in humans (Stam et al., 1994). In contrast, absolute DL_{CO} increases while volume-corrected DL_{CO} remains constant in the first 2 years of life (Balinotti et al., 2009).

4.2 Changes in Gas Exchange with Age

Mice

As suggested above we lack detailed information regarding changes in gas exchange with age in mice; in particular, there are no functional data. In absolute terms, however, age-related changes in gas exchange can be inferred from structural data obtained by histology (Weibel, 1970). For example, by crudely assessing patterns in alveolar surface area, septal thickness, or tissue density, we may be able to infer how gas exchange may change in absolute terms; of course, this does not necessarily allow for volume-corrected estimates of gas exchange and ignores the circulatory side of the equation. Unfortunately, in the case of mice the structural changes, as outlined in Section 2.2.1, that occur with age vary considerably between studies and strain. As such it is likely that absolute gas exchange improves with age from birth until adulthood and from thereafter may stabilize, continue to improve, or decline, depending on which strain is examined.

Rats

While we are lacking direct functional and structural estimates of gas exchange capacity in mice, there are systematic age-related data for these parameters in rats. Johanson and Pierce (1973) demonstrated, functionally, that DL_{CO} increases between 4 and 18 months of age (4 months, $DL_{CO} = 0.15$ ml/min/mmHg; 18 months, $DL_{CO} = 0.24$ ml/min/mmHg), which can be largely accounted for by the associated increase in alveolar volume (V_A) between these ages (4 months, $DL_{CO}/V_A = 0.022$ mmHg/min; 18 months, $DL_{CO}/V_A = 0.028$ mmHg/min) in Sprague–Dawley rats. The observed increase in lung volume from early into late adulthood in this study paralleled the observed increase in internal surface area of the lung measured morphometrically (Johanson and Pierce, 1973). In contrast, Mauderly, (1982) demonstrated that DL_{CO}, in absolute terms and adjusted for lung volume, did not change between early adulthood (3 months) and senescence (27 months) in F-344 rats. Morphological estimates of pulmonary diffusion capacity in the same strain concur with this functional data whereby gas exchange capacity remains relatively unchanged from 5 to 26 months of age (Pinkerton et al., 1982).

Using a morphometric approach Pinkerton et al. (1982) clearly demonstrated a marked increase in pulmonary diffusion capacity just after birth (1 week of age) into adulthood (5 months) in F-344 rats. More recently, the potential mechanisms for increases in gas exchange in early life have been modeled in rats, suggesting alveolar expansion is the key means of increase in gas exchange capacity in early life (Blanco, 1995). These morphometrically established increases in diffusion capacity from early life into adulthood have been supported with functional measures of DL_{CO} showing a dramatic increase in this metric between 7 ($DL_{CO} = 9.1$ μmol/min/hPa) and 90 ($DL_{CO} = 214$ μmol/min/hPa) days of age in Wistar–Kyoto rats (Bolle et al., 2008). This increase in DL_{CO} corresponds to an increase in alveolar surface area (S_A), however, when specific diffusing capacity is calculated (DL_{CO}/S_A), the bulk of the increase in diffusing capacity occurs between 14 and 35 days of age and changes little thereafter (Bolle et al., 2008). Collectively, these data suggest that gas exchange capacity in rats increases from birth to early adulthood but tends to remain stable through late adulthood and into senescence.

Rabbits

To date there is only one study that has given any detail regarding gas exchange in rabbits. Davidson et al. (1966) showed no difference in arterial P_{O2} or P_{CO2} between young adult (6–12 months of age) and older (3–5 years) rabbits.

Guinea Pigs

There are very little data on gas exchange parameters in guinea pigs, particularly in relation to age. As detailed earlier the guinea pig is relatively mature at birth and postnatal changes in lung function differ from the Muridae in that there is no apparent period of rapid cellular proliferation in early postnatal life (Lechner and Banchero, 1982). In contrast, the lung grows steadily, primarily by expansion, from birth into adulthood, which is characterized by monotonic, exponential increases in alveolar and capillary surface area with age (Lechner and Banchero, 1982; Forrest and Weibel, 1975) and no change in barrier thickness (Lechner and Banchero, 1982). There are no functional measurements of gas exchange in guinea pigs that have examined a relationship with age. Morphometry-based estimates of diffusing capacity show a linear relationship between body weight and lung diffusing capacity in adult guinea pigs (Forrest and Weibel, 1975). On the basis of this observation, and the linear (exponential) association between age and other measures of lung morphometry, it could be inferred that there is a similar relationship between total lung diffusing capacity and age. There are no data on volume-corrected diffusing capacity or estimates of how gas exchange changes into late adulthood/senescence in guinea pigs.

Hamsters

There are little data on gas exchange in hamsters with age. There is a linear association between (log)DL_{CO} and (log) weight between 40 g and 120 g (Takezawa et al., 1980), which corresponds to 3–8 weeks of age. Interestingly, the slope and intercept of the relationship between DL_{CO} and body weight in "adult" hamsters, guinea pigs, rats, and rabbits is consistent (Takezawa et al., 1980).

Ferrets

There are no data on changes in gas exchange with age in ferrets.

Dogs

Of all the species described so far, the dog appears to be the most extensively studied in terms of functional measures of gas exchange in the lung. However, these data are mostly restricted to early and late adulthood. Mauderly (1974) found nonlinear associations between DL_{CO} and age in female beagles with a decrease between 1 (8.3 mL mmHg/min) and 3–4 (5.0 mL mmHg/min) years of age and a subsequent increase at 8–10.5 (7.2 mL mmHg/min) years of age. These data are inconsistent although may be explained by ongoing inflammatory responses to the environment and age-related changes in parenchymal structure. However, the apparent "improvement" in DL_{CO} between 3 and 8 years of age does not follow the fibrotic lesions and focal emphysema that have been reported in the aging canine lung (Reif and Rhodes, 1966). Interestingly, other measures of gas exchange capacity such as arterial P_{O2} and P_{CO2}, while statistically different between 1 and 10.5 years of age in beagles, do not change by a physiologically significant amount (Mauderly, 1974) and do not follow the large swings in DL_{CO} outlined above.

Robinson and Gillespie (1973) measured DL_{CO} in an assortment of male and female beagles ranging in age from 9 months to 10.5 years of age. Interestingly, they also found a complex association between DL_{CO} and age with a decrease between 9 months and 8 years of age and an apparent increase in older dogs (10.5 years of age). The decrease in DL_{CO} corresponded to the lowest capillary blood volume and was strongly associated with alveolar volume (Robinson and Gillespie, 1973). There have been no further advances in our understanding of the association between gas exchange parameters and age in dogs. The lack of a consistent pattern is likely to be due to the low numbers of animals at each age in the studies described and the confounding effect of the uncontrolled environment to which these animals are exposed.

Sheep

There is one study that has systematically examined gas exchange in sheep in the first 4 weeks of life. Johns et al. (1994) demonstrated that DL_{CO} doubles (1 week, 0.73 mL mmHg/min; 4 weeks 1.38 mL mmHg/min) between 1 and 4 weeks of age, however, this was entirely due to the increase in alveolar volume and did not change after correction for hemoglobin. This linear increase with body

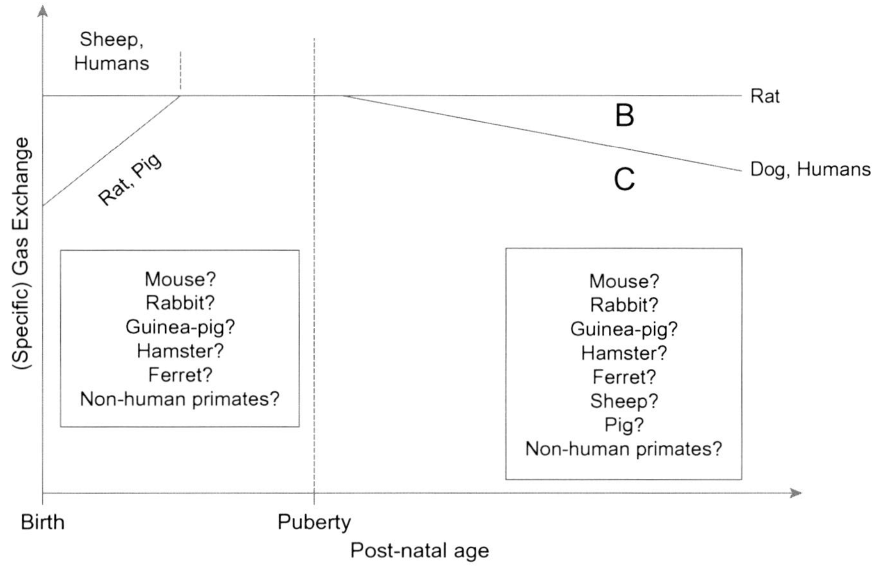

FIGURE 4 Schematic representation of the changes in specific (corrected for alveolar volume or a similar measure) gas exchange associated with age in the species described in this chapter. The data are very limited such that generalization regarding the comparative associations between species is difficult. Based on the data available it is apparent in some species that gas exchange is limited just after birth but rapidly improves in early postnatal life (rat and pig), while in other species (sheep) specific gas exchange changes vary little in early life. There are two patterns from early adulthood into senescence: B, no change in volume-corrected gas exchange (rats); C, a decrease in volume-corrected gas exchange (dogs).

size is consistent with the fact that the lung is well developed at birth in the sheep compared to most other species.

Pigs

In young pigs there are two studies, by the same group, that suggest oxygen diffusion is limited at birth and that this improves with age. Through a sequential series of experiments, Escourrou et al. (1988, 1993) were able to show that arterial P_{O_2} is lower in 2- and 4-week-old pigs compared to 8-week-old pigs. By ruling out an effect of shunting and showing that the distribution of perfusion decreased with age while ventilation remained constant (Escourrou et al., 1988), they were able to demonstrate that newborn pigs have limited pulmonary vascular reserve resulting in inadequate perfusion for the required diffusion of gas (Escourrou et al., 1993). There are no other studies on changes in gas exchange with age in pigs.

Nonhuman Primates

There are no studies examining gas exchange parameters against age in nonhuman primates.

Summary

Data within species on age-related changes in gas exchange are extremely limited. This is probably due to the technical difficulty of the measurements required to assess this parameter; particularly in smaller species. Based on the data we do have, absolute gas exchange (not surprisingly) tends to increase with postnatal growth into adulthood in all species, but it is unclear what happens thereafter. When gas exchange is expressed relative to alveolar volume or surface area, some between-species differences emerge. In both the rat and pig the available data suggest a functional limitation in gas exchange in very early postnatal life, whereas in the sheep this does not appear to occur (Figure 4). Beyond adulthood there are only two species where there are sufficient data to comment. In the rat there appears to be no senescence-related changes in specific gas exchange, while in the dog aging is associated with a decline in specific gas exchange (Figure 4). The latter is consistent with the pattern observed in humans (Stam et al., 1994).

5. PATTERNS OF VENTILATION

5.1 Basic Patterns in Ventilation

Patterns of ventilation in mammals change with age, particularly in early life, according to changes in metabolic demands. In humans, ventilation rate (f) decreases with age while tidal volume (V_t) increases (Mortola, 1987; Stahl, 1967; Rusconi et al., 1994). This pattern is most pronounced in the first 5–6 years of life (Aurora et al., 2005). The drivers of f and V_t are still poorly understood, but their product, minute volume (V_E), is thought to be related to metabolic rate (Gagliardi and Rusconi, 1997). There is a well-known relationship between metabolic rate and body size such that metabolic rate (energy expenditure per unit body size) decreases with age (Stahl, 1967). In order to meet higher metabolic demands, smaller animals (and individuals), relative to body size, have a higher minute volume. While this can be achieved by increasing f and/or V_t, it is generally thought that infant humans preferentially increase f as this is a more energy efficient strategy than increasing V_t (Gagliardi and Rusconi, 1997). There are little data, and certainly no systematic data, on changes in patterns of ventilation with senescence although the prevailing dogma is that f, V_t, and V_E change very little with aging beyond early adulthood. Given the simplicity in measuring these outcomes using standard plethysmography techniques, it is surprising how little is known about the association between ventilation rate, tidal volume, minute volume, and age. Our limited species-specific data are discussed in more detail below. In most instances the data have been extracted from control groups in studies examining ventilatory responses to specific perturbations, in particular, the ventilatory response to hypercapnia.

5.2 Changes in Ventilation Patterns with Age

Mice

There are very little data on ventilation parameters in relation to age in mice. Two studies by Bissonnette and Knopp (2001, 2004) detailed changes in ventilatory parameters in C57BL/6 mice from day 1 postnatal age to 30 days postnatal age (i.e., from birth until weaning). Ventilation rate increased slightly between 2 (144 breaths/min) and 4 (203 b/min) days of age but then changed very little across the ages studied (f = 220 breaths/min at 30 days of age) (Bissonnette and Knopp, 2001, 2004). Tidal volume increased exponentially with age (2 days of age, $V_t = 11.3\ \mu L$; 21.5 days of age, $V_t = 69.3\ \mu L$) (Bissonnette and Knopp, 2004) in parallel with body weight (1 day of age, wt = 2.2 g; 30 days of age wt = 14.5 g) (Bissonnette and Knopp, 2001). There was a corresponding increase in absolute minute volume, however, when expressed per gram of body weight there was no difference between ages. There are no data on ventilation parameters with senescence in mice.

Rats

There are several studies examining the development of the hypercapnic ventilatory response in rats, however, very few of them report baseline values of f, V_t, or V_E and instead describe their data on the basis of percentage changes in these parameters in response to CO_2 exposure. One longitudinal study by Abu-Shaweesh et al. (1999) in Sprague–Dawley rats ranging in ages from 5 to 42 days of age (10.1–161.3 g) measured f, V_t, and V_E but only reported baseline values for respiratory frequency. Their data showed an increase from 120 breaths/min to 155 breaths/min between 5 and 17 days of age then a decrease in f to 133 and 107 breaths/min at 23 and 42 days of age, respectively. There is only one other study examining ventilatory parameters with age in rats. Mauderly (1982) described an increase in f (48 breaths/min vs 54 breaths/min), V_t (1.1 mL vs 1.5 mL), and V_E (54 mL/min vs 82 mL/min) between the ages of 3 (early adulthood) and 18 months in Fischer-344 rats with no difference between 18 and 27 months of age and no difference across ages in V_E corrected for body weight.

Rabbits

There are two studies detailing the association between ventilatory parameters and age in rabbits. The first, by Grunstein and Tanaka (1985), described ventilatory changes from 1 to 26 days postnatal age. In this age range there was a linear decrease in f and linear increases in both Vt and V_E with age. When V_E was corrected for body mass there was no association with age, which is consistent with the descriptions in the other species listed above. Davidson et al. (1966) demonstrated that f continues to decrease and V_t continues to increase between 0.5 and 1.0 year of age and 3–5 years of age in rabbits. However, the increase in the V_t had begun to plateau by this stage such that V_E did not differ between these age classes (Davidson et al., 1966).

Guinea Pigs

There are very little data on ventilation and age in guinea pigs. Murphy et al. (1994) reported that tidal volume in anaesthetized guinea pigs does not change between 4 and 7 days of age ($V_t = 8.5$ mL) and adolescence ($V_t = 7.2$ mL). This is in stark contrast to the previous reports, particularly given the weight difference between these ages (86 g vs 442 g), and probably highlights the impact of anesthesia on ventilation patterns.

Hamsters

There are a few studies that have examined patterns in ventilation in relation to age in hamsters, however, no studies have examined these patterns in early life. An early study by Mauderly (1979) described no difference in f, V_t, or V_E between 15 and 65 weeks of age. However, these experiments were carried out under anesthesia, and subsequent studies in conscious hamsters did not support these data. Schlenker (1984) demonstrated that f decreases and V_t increases between 40 days and 140 days of age in hamsters. While these are in the lower age range compared to those used by Mauderly (1979), the latter time points (100 days of age and 140 days of age) used by Schlenker (1984) overlap Mauderly (1979) age range and clearly demonstrate an increase in V_t and a decrease in f. Schlenker, (1984) also showed that V_E increases between 40 and 140 days of age, although there appeared to be a plateau in V_E between 100 and 140 days of age. Skornik and Brain's study (Skornik and Brain, 1990), using hamsters between 18 and 98 weeks of age, clearly demonstrated a steady decrease in f

(18 weeks, f = 73 breaths/min; 98 weeks, f = 58.7 breaths/min), an increase in V_t (18 weeks, V_t = 1.10 mL; 98 weeks, V_t = 1.65 mL), and a corresponding increase in V_E (18 weeks, V_E = 80.7 mL/min; 96.5 mL/min). There was no change in V_E corrected for body weight between 18 and 98 weeks of age.

Ferrets

There are no systematic data on ventilation patterns in ferrets of different ages.

Dogs

Very little data exist on ventilation patterns in dogs. Mauderly (1974) described ventilation in restrained conscious purebred beagles. In contrast to reports in all of the previous species described above, respiratory rate increased between 1 and 10.5 years of age (1 year, f = 20 breaths/min; 10.5 years, f = 28 breaths/min), and V_t decreased (1 year, V_t = 181 mL; 10.5 years, V_t = 150 mL). There are, however, some discrepancies in the V_E data from this study, which suggests that there is no change in minute volume between ages despite the fact that the values provided how that V_E increases from 3.6 L/min−4.2 L/min.

Sheep

There are no data on changes in the pattern of ventilation with age in sheep.

Pigs

There is only one report describing changes in ventilation patterns with age in pigs. Rendas et al. (1978) showed that f decreases between 2 and 4 weeks of age (46 breaths/min) and 5−8 weeks of age (30 breaths/min) and remains constant thereafter over the ages measured (up to 16 weeks). Tidal volume increases between 2 and 16 weeks but does not change relative to body weight (Rendas et al., 1978).

Nonhuman Primates

There are no studies examining ventilation parameters against age in nonhuman primates.

Summary

As suggested earlier there are a severe lack of data on changes in the pattern of ventilation with age in mammals. This is particularly surprising given the ease with which respiratory frequency and tidal volume can be measured. What is perhaps even more surprising is the lack of consistency in the pattern of change in these parameters with age between species. This may be due to a lack of data and/or a reference point that allows for easy comparison between species. Almost every conceivable pattern was present in at least one of the species described. As such, it is difficult to make any generalizations regarding between-species differences as there appears to be almost no overlap in these patterns for the species described in this chapter. As a result, it was not possible to develop a schematic allowing simple between-species comparisons to be made like that detailed in the previous sections.

6. CONCLUDING REMARKS

In this chapter an attempt has been made to unify our current knowledge of the aging of the normal lung in mammals in a way that allows direct comparison between species. The breadth of this summary was limited to lung structure, lung function (volume and mechanics), gas exchange and patterns of ventilation. This summary has intentionally steered away from any discussion regarding the underlying molecular processes that drive aging processes, particularly into senescence, as the data are certainly not sufficient to produce a reasonable between-species comparison.

It was clear that there exists significant between species differences in the association between these parameters and age, particularly in the period between early adulthood and senescence. One of the difficulties in examining the primary driver(s) of these between-species differences is the lack of between-species consistency in the environmental exposures that were likely to be experienced by the animals in the studies that are described. While there are likely to be preprogrammed age-specific changes in the lung, any such patterns will be modified by environmental exposures—the cumulative effect of which will increase with age. As such, any between-species comparisons need to be considered in this context. It was also clear when compiling the studies that we lack systematic data in most species for many of the parameters that were described, which has limited many of the broad generalizations that were made throughout the chapter. Given the importance of age in determining the underlying structure and function of lung, and presumably the lungs' response to insults, it is important that future research efforts are directed toward understanding these processes.

REFERENCES

Abu-Shaweesh, J.M., Dreshaj, I.A., Thomas, A.J., Haxhiu, M.A., Strohl, K.P., Martin, R.J., 1999. Changes in respiratory timing induced by hypercapnia in maturing rats. J. Appl. Physiol. 87, 484−490.

Alcorn, D.G., Adamson, T.M., Maloney, J.E., Robinson, P.M., 1981. A morphologic and morphometric analysis of fetal lung development in the sheep. Anat. Rec. 201, 655−667.

Aurora, P., Kozlowska, W., Stocks, J., 2005. Gas mixing efficiency from birth to adulthood measured by multiple-breath washout. Respir. Physiol. Neurobiol. 148, 125−139.

Bolle, I., Eder, G., Takenaka, S., Ganguly, K., Karrasch, S., Zeller, C., Neuner, M., Kreyling, W.G., Tsuda, A., Schulz, H., 2008. Postnatal lung function in the developing rat. J. Appl. Physiol. 104, 1167–1176.

Boatman, E.S., Martin, H.B., 1960. Electron microscopy in pulmonary emphysema of rabbits. Am. J. Pathol. 37, 197–205.

Bates, J.H.T., Irvin, C.G., 2003. Measuring lung function in mice: the phenotyping uncertainty principle. J. Appl. Physiol. 94, 1297–1306.

Bates, J., Irvin, C., Brusasco, V., Drazen, J., Fredberg, J., Loring, S., Eidelman, D., Ludwig, M., Macklem, P., Martin, J., Hantos, Z., Hyatt, R., Lai-Fook, S., Leff, A., Solway, J., Lutchen, K., Suki, B., Mitzner, W., Pare, P., Pride, N., Sly, P., 2004. The use and misuse of Penh in animal models of lung disease. Am. J. Respir. Cell Mol. Biol. 31, 373–374.

Burrows, B., Cline, M.G., Knudson, R.J., Taussig, L.M., Lebowitz, M.D., 1983. A descriptive analysis of the growth and decline of the FVC and FEV. Chest 83, 717–724.

Bozanich, E.M., Janosi, T.Z., Collins, R.A., Thamrin, C., Turner, D.J., Hantos, Z., Sly, P.D., 2007. Methacholine responsiveness in mice from 2 to 8 wk of age. J. Appl. Physiol. 103, 542–546.

Bozanich, E.M., Collins, R.A., Thamrin, C., Hantos, Z., Sly, P.D., Turner, D.J., 2005. Developmental changes in airway and tissue mechanics in mice. J. Appl. Physiol. 99, 108–113.

Broussard, D., Larson, J.E., Craig Cohen, J., Lundblad, L.K.A., 2006. Developmental changes in respiratory mechanics in the neonatal rat. Exp. Lung Res. 32, 263–273.

Bellmunt, M.J., Portero, M., Pamplona, R., Muntaner, M., Prat, J., 1995. Age-related fluorescence in rat lung collagen. Lung 173, 177–185.

Balinotti, J.E., Tiller, C.J., Llapur, C.J., Jones, M.H., Kimmel, R.N., Coates, C.E., Katz, B.P., Nguyen, J.T., Tepper, R.S., 2009. Growth of the lung parenchyma in early life. Am. J. Respir. Crit. Care Med. 179, 134–137.

Blanco, L.N., 1995. Mechanisms for the generation of gas-exchange surface area in rat lung. Am. J. Physiol.-Lung Cell. Mol. Physiol. 13, L698–L708.

Bissonnette, J.M., Knopp, S.J., 2001. Developmental changes in the hypoxic ventilatory response in C57BL/6 mice. Respir. Physiol. 128, 179–186.

Bissonnette, J.M., Knopp, S.J., 2004. Hypercapnic ventilatory response in mice lacking the 65kDa isoform of glutamic acid decarboxylase (GAD65). Respir. Res. 5, 3.

Chinoy, M.R., 2003. Lung growth and development. Front. Biosci. 8, D392–D415.

Copland, I., Post, M., 2004. Lung development and fetal lung growth. Pediatr. Respir. Rev. 5, S259–S264.

Calvi, C.L., Podowski, M., D'Alessio, F.R., Metzger, S.L., Misono, K., Poonyagariyagorn, H., Lopez-Mercado, A., Ku, T., Lauer, T., Cheadle, C., Talbot Jr, C.C., Ji, C., McGrath-Morrow, S., King, L.S., Walston, J., Neptune, E.R., 2011. Critical transition in tissue homeostasis accompanies murine lung senescence. PLoS One 6, e20712.

Collins, M.H., Kleinerman, J., Moessinger, A.C., Collins, A.H., Stanley James, L., Blanc, W.A., 1986. Morphometric analysis of the growth of the normal fetal guinea pig lung. Anat. Rec. 216, 381–391.

Douglas, W.R., 1972. Of pigs and men and research. Origins Life Evol. Biospheres 3, 226–234.

Davies, P., Reid, L., Lister, G., Pitt, B., 1988. Postnatal growth of the sheep lung: a morphometric study. Anat. Rec. 220, 281–286.

Davidson, J.T., Wasserman, K., Lillington, G.A., William Schmidt, R., 1966. Effect of aging on respiratory mechanics and gas exchange in rabbits. J. Appl. Physiol. 21, 837–842.

Escolar, J.D., Gallego, B., Tejero, C., Escolar, M.A., 1994. Changes occurring with increasing age in the rat lung—morphometrical study. Anat. Rec. 239, 287–296.

Escourrou, P.J.L., Teisseire, B.P., Herigault, R.A., Vallez, M.O., Dupeyrat, A.J., Gaultier, C., 1988. Mechanism of improvement in pulmonary gas exchange during growth in awake piglets. J. Appl. Physiol. 65, 1055–1061.

Escourrou, P.J.L., Qi, X., Weiss, M., Mazmanian, G.M., Gaultier, C., Herve, P., 1993. Influence of pulmonary blood flow on gas exchange in piglets. J. Appl. Physiol. 75, 2478–2483.

Falconer, D.S., 1984. Weight and age at puberty in female and male mice of strains selected for large and small body size. Genet. Res. 44, 44–72.

Fredberg, J.J., Stamenovic, D., 1989. On the imperfect elasticity of lung tissue. J. Appl. Physiol. 67, 2408–2419.

Forrest, J.B., Weibel, E.R., 1975. Morphometric estimation of pulmonary diffusion capacity. Respir. Physiol. 24, 191–202.

Goldstein, R.H., 1982. Response of the aging lung to elastase injury. Am. Rev. Respir. Dis. 125, 295–298.

Gomes, R.F.M., Shardonofsky, F., Eidelman, D.H., Bates, J.H.T., 2001. Respiratory mechanics and lung development in the rat from early age to adulthood. J. Appl. Physiol. 90, 1631–1638.

Gaultier, C., Harf, A., Lorino, A.M., Atlan, G., 1984. Lung mechanics in growing guinea pigs. Respir. Physiol. 56, 217–228.

Gagliardi, L., Rusconi, F., 1997. Respiratory rate and body mass in the first three years of life. Archives Dis. Child. 76, 151–154.

Grunstein, M.M., Tanaka, D.T., 1985. Ontogeny of respiratory control and pulmonary mechanics in newborn rabbits. J. Appl. Physiol. 59, 1477–1486.

Hagenauer, M.H., King, A.F., Possidente, B., McGinnis, M.Y., Lumia, A.R., Peckham, E.M., Lee, T.M., 2011. Changes in circadian rhythms during puberty in *Rattus norvegicus*: developmental time course and gonadal dependency. Horm. Behav. 60, 46–57.

Hart, R.W., Setlow, R.B., 1974. Correlation between deoxyribonucleic acid excision-repair and life-span in a number of mammalian species. Proc. Natl. Acad. Sci. U.S.A 71, 2169–2173.

Hsia, C.C.W., Hyde, D.M., Ochs, M., Weibel, E.W., 2010. An official policy statement of the American Thoracic Society/European Respiratory Society: standards for quantitative assessment of lung structure. Am. J. Respir. Crit. Care Med. 181, 394–418.

Huang, K., Mitzner, W., Rabold, R., Schofield, B., Lee, H., Biswal, S., Tankersley, C.G., 2007. Variation in senescent-dependent lung changes in inbred mouse strains. J. Appl. Physiol. 102, 1632–1639.

Huang, K.W., Rabold, R., Schofield, B., Mitzner, W., Tankersley, C.G., 2007. Age-dependent changes of airway and lung parenchyma in C57BL/6J mice. J. Appl. Physiol. 102, 200–206.

Hyde, D.M., Robinson, N.E., Gillespie, J.R., Tyler, W.S., 1977. Morphometry of the distal air spaces in lungs of aging dogs. J. Appl. Physiol. 43, 86–91.

Horsfield, K., 1977. Postnatal growth of the dog's bronchial tree. Respir. Physiol. 29, 185–191.

Hislop, A., Howard, S., Fairweather, D.V.I., 1984. Morphometric studies on the structural development of the lung in *Macaca fascicularis* during fetal and postnatal life. J. Anat. 138, 95–112.

Hyde, D.M., Blozis, S.A., Avdalovic, M.V., Putney, L.F., Dettorre, R., Quesenberry, N.J., Singh, P., Tyler, N.K., 2007. Alveoli increase in number but not size from birth to adulthood in rhesus monkeys. Am. J. Physiol.-Lung Cell. Mol. Physiol. 293, L570–L579.

Johanson, W.G.J., Pierce, A.K., 1973. Lung structure and funcion with age in normal rats and rats with papain emphysema. J. Clin. Invest. 52, 2921.

Johns, D.P., Jakubowska, A.E., McCrabb, G.J., Harding, R., 1994. Measuring the lung diffusing capacity and functional residaul capacity in lambs during the first postnatal month. Pediatr. Pulmonol. 17, 347–353.

Kilborn, S.H., Trudel, G., Uhthoff, H., 2002. Review of growth plate closure compared with age at sexual maturity and lifespan in laboratory animals. J. Am. Assoc. Lab. Anim. Sci. 41, 21–26.

Kawakami, M., Paul, J.L., Thurlbeck, W.M., 1984. The effect of age on lung structure in male BALB/cNNia inbred mice. Am. J. Anat. 170, 1–21.

Kerr, J.S., Yu, S.Y., Riley, D.J., 1990. Strain specific respiratory air space enlargement in aged rats. Exp. Geront. 25, 563–574.

Kovar, J., Sly, P.D., Willet, K.E., 2002. Postnatal alveolar development of the rabbit. J. Appl. Physiol. 93, 629–635.

King, T.K.C., 1966. Measurement of functional residual capacity in the rat. J. Appl. Physiol. 21, 233–236.

Lee, D., Srirama, P.K., Wallis, C., Wexler, A.S., 2011. Postnatal growth of tracheobronchial airways of Sprague–Dawley rats. J. Anat. 218, 717–725.

Lechner, A.J., Banchero, N., 1982. Advanced pulmonary development in newborn guinea pigs (*Cavie porcellus*). Am. J. Anat. 163, 235–246.

Lechner, A.J., Banchero, N., 1980. Lung morphometry in guinea pigs acclimated to hypoxia during growth. Respir. Physiol. 42, 155–169.

Lundblad, L.K.A., Irvin, C.G., Hantos, Z., Sly, P., Mitzner, W., Bates, J.H.T., 2007. Penh is not a measure of airway resistance! Eur. Respir. J. 30, 805–816.

Merkus, P., ten Have-Opbroek, A.A.W., Quanjer, P.H., 1996. Human lung growth: a review. Pediatr. Pulmonol. 21, 383–397.

Massaro, D., Massaro, G.D., 2002. Pre- and postnatal lung development, maturation, and plasticity—invited review: pulmonary alveoli: formation, the "call for oxygen," and other regulators. Am. J. Physiol.Lung Cell. Mol. Physiol. 282, L345–L358.

Mauderly, J.L., 2000. Animal models for the effect of age on susceptibility to inhaled particulate matter. Inhal. Toxicol. 12, 863–900.

Massaro, D., Teich, N., Maxwell, S., Massaro, G.D., Whitney, P., 1985. Postnatal-development of alveoli - regulation and evidence for a critical period in rats. J. Clin. Invest. 76, 1297–1305.

Mund, S.I., Stampanoni, M., Schittny, J.C., August 2008. Developmental alveolarization of the mouse lung. Dev. Dyn. 237, 2108–2116.

Metzger, R.J., Klein, O.D., Martin, G.R., Krasnow, M.A., 2008. The branching programme of mouse lung development. Nature 453, 745–750.

Mansell, A.L., Collins, M.H., Johnson, E., Gil, J., 1995. Postnatal growth of lung parenchyma in the piglet: morphometry correlated mechanics. Anat. Rec. 241, 99–104.

Mauderly, J.L., 1982. The effect of age on respiratory function of Fischer-344 rats. Exp. Aging Res. 8, 31–36.

Martin, C.J., Chihara, S., Chang, D.B., 1977. A comparative study of the mechanical properties in aging alveolar wall. Am. Rev. Respir. Dis. 115, 981–988.

Mauderly, J.L., 1979. Ventilation, lung volumes and lung mechanics of young adult and old syrian hamsters. Exp. Aging Res. 5, 497–508.

Mauderly, J.L., 1974. Influence of sex and age on the pulmonary function of the unanesthetized beagle dogs. J. Gerontol. 29, 282–289.

Mortola, J.P., 1987. Dynamics of breathing in newborn animals. Physiol. Rev. 67, 187–243.

Murphy, T.M., Ray, D.W., Alger, L.E., Phillips, I.J., Roach, J.C., Leff, A.R., Solway, J., 1994. Ontogeny of dry gas hyperpnea-induced bronchoconstriction in guinea pigs. J. Appl. Physiol. 76, 1150–1155.

Narayanan, M., Owers-Bradley, J., Beardsmore, C.S., Mada, M., Ball, I., Garipov, R., Panesar, K.S., Kuehni, C.E., Spycher, B.D., Williams, S.E., Silverman, M., 2012. Alveolarization continues during childhood and adolescence. New evidence from helium-3 magnetic resonance. Am. J. Respir. Crit. Care Med. 185, 186–191.

Nardell, E.A., Brody, J.S., 1982. Determinants of mechanical properties of rat lung during postnatal development. J. Appl. Physiol. 53, 140–148.

Oeppen, J., Vaupel, J.W., 2002. Broken limits to life expectancy. Science 296, 1029–1031.

Oberbauer, A.M., Krook, L., Hogue, D.E., Currie, W.B., Thonney, M.L., 1988. Dietary calcium and metacarpal growth in ewes. J. Nutr. 118, 976–981.

Parent, A.-S., Teilmann, G., Juul, A., Skakkebaek, N.E., Toppari, J., Bourguignon, J.-P., 2003. The timing of normal puberty and the age limits of sexual precocity: variations around the world, secular trends and changes after migration. Endocr. Rev. 24, 668–693.

Plant, T.M., 2001. Neurobiological bases underlying the control of the onset of puberty in the rhesus monkey: a representative higher primate. Front. Neuroendocrin. 22, 107–139.

Pamplona, R., Portero-Otin, M., Riba, D., Ruiz, C., Prat, J., Bellmunt, M.J., Barja, G., 1998. Mitochondrial membrane peroxidizability index is inversely related to maximum life span in mammals. J. Lipid Res. 39, 1989–1994.

Perelman, R.H., Engle, M.J., Farrell, P.M., 1981. Perspectives on fetal lung development. Lung 159, 53–80.

Pinkerton, K.E., Barry, B.E., O'Neil, J.J., Raub, J.A., Pratt, P.C., Crapo, J.D., 1982. Morphologic changes in the lung during the lifespan of Fischer-344 rats. Am. J. Anat. 164, 155–174.

Quan, S.F., Lemen, R.J., Witten, M.L., Sherrill, D.L., Grad, R., Sobonya, R.E., Ray, C.G., 1990. Changes in lung mechanics and reactivity with age after viral bronchiolitis in beagle puppies. J. Appl. Physiol. 69, 2034–2042.

Ryan, K.L., 1984. Hormonal correlated of photoperiod-induced puberty in a reflex ovulator, the female ferret (*Mustela furo*). Biol. Reprod. 31, 925–935.

Ranga, V., Kleinerman, J., 1980. Interalveolar pores in mouse lungs—regional distribution and alterations with age. Am. Rev. Respir. Dis. 122, 477–481.

Ross, K.A., Thurlbeck, W.M., 1992. Lung growth in newborn guinea pigs: effects of endurance exercise. Respir. Physiol. 89, 353–364.

Reif, J.S., Rhodes, W.H., 1966. The lungs of aged dogs: a radiographic-morphologic correlation. Vet. Radiol. 7, 5–11.

Rendas, A., Branthwaite, M., Reid, L., 1978. Growth of pulmonary circulation in normal pig—structural analysis and cardiopulmonary function. J. Appl. Physiol. 45, 806–817.

Robinson, N.E., Gillespie, J.R., 1973. Lung volumes in aging beagle dogs. J. Appl. Physiol. 35, 317–321.

Rusconi, F., Castagneto, M., Porta, N., Gagliardi, L., Leo, G., Pellegatta, A., Razon, S., Braga, M., 1994. Reference values for respiratory rate in the first 3 years of life. Pediatrics 94, 350–355.

Schulz, K.M., Richardson, H.N., Zehr, J.L., Osetek, A.J., Menard, T.A., Sisk, C.L., 2004. Gonadal hormones masculinize and defeminize reproductive behaviors during puberty in the male Syrian hamster. Horm. Behav. 45, 242–249.

Suttie, J.M., Kostyo, J.L., Ebling, F.J.P., Wood, R.I., Bucholtz, D.C., Skottner, A., Adel, T.E., Towns, R.J., Foster, D.L., 1991. Metabolic interfaces between growth and reproduction. IV. Chronic pulsatile administration of growth hormone and the timing of puberty in the female sheep. Endocrinology 129, 2024–2032.

Sterning, M., Rydhmer, L., Eliasson-Selling, L., 1998. Relationships between age at puberty and interval from weaning to estrus and between estrus signs at puberty and after the first weaning in pigs. J. Anim. Sci. 76, 353–359.

Speakman, J.R., Van Acker, A., Harper, E.J., 2003. Age-related changes in the metabolism and body composition of three dog breeds and their relationship to life expectancy. Aging Cell 2, 265–275.

Shimura, S., Boatman, E.S., Martin, C.J., 1986. Effects of ageing on the alveolar pores of Kohn and on the cytoplasmic components of alveolar type II cells in monkey lungs. J. Pathol. 148, 1–11.

Schittny, J.C., Mund, S.I., Stampanoni, M., 2008. Evidence and structural mechanism for late lung alveolarization. Am. J. Physiol. Lung Cell. Mol. Physiol. 294, L246–L254.

Strawbridge, H.T.G., 1960. Chronic pulmonary epmhysema (an experimental study) II. Spontaneous pulmonary emphysema. Am. J. Pathol. 37, 309–331.

Snider, G.L., Sherter, C.B., 1977. A one-year study of the evolution of elastase-induced emphysema in hamsters. J. Appl. Physiol. 43, 721–729.

Sipriani, T.M., Grandi, F., da Silva, L.C.G., Maiorka, P.C., Vannucchi, C.I., 2009. Pulmonary maturation in canine foetuses from early pregnancy to parturition. Reprod. Domest. Anim. 44, 137–140.

Soutiere, S.E., Mitzner, W., 2004. On defining total lung capacity in the mouse. J. Appl. Physiol. 96, 1658–1664.

Shen, X., Ramchandani, R., Dunn, B., Lambert, R., Gunst, S.J., Tepper, R.S., 2000. Effect of transpulmonary pressure on airway diameter and responsiveness of immature and mature rabbits. J. Appl. Physiol. 89, 1584–1590.

Skornik, W.A., Brain, J.D., 1990. Breathing and lung mechanics in hamsters: effect of pentobarbital anaesthesia. J. Appl. Physiol. 68, 2536–2541.

Stam, H., Hrachovina, V., Stijnen, T., Versprille, A., 1994. Diffusing capacity dependent on lung volume and age in normal subjects. J. Appl. Physiol. 76, 2356–2363.

Stahl, W.R., 1967. Scaling of respiratory variables in mammals. J. Appl. Physiol. 22, 453–460.

Schlenker, E.H., 1984. An evaluation of ventilation in dystrophic Syrian hamsters. J. Appl. Physiol. 56, 914–921.

Thurlbeck, W.M., 1975. Postnatal-growth and development of lung. Am. Rev. Respir. Dis. 111, 803–844.

Tepper, R.S., Du, T., Styhler, A., Ludwig, M., Martin, J.G., 1995. Increased maximal pulmonary response to methacholine and airway smooth muscle in immature compare with mature rabbits. Am. J. Respir. Crit. Care Med. 151, 836–840.

Thomas, D.D., Standaert, T.A., Anton, W.R., Jones, D.R., Godwin, J.D., Raghu, G., Hodson, W.A., Allen, M.D., 1993. Growth potential of the transplanted lung in the infant primate. Ann. Thorac. Surg. 56, 1274–1278.

Takezawa, J., Miller, F.J., O'Neil, J.J., 1980. Single-breath diffusing capacity and lung volumes in small laboratory mammals. J. Appl. Physiol. 48, 1052–1059.

Wolfer, D.P., Lipp, H.-P., 1995. Evidence for physiological growth of hippocampal mossy fiber collaterals in the guinea pig during puberty and adulthood. Hippocampus 5, 329–340.

Wolf, K.N., Wildt, D.E., Vargas, A., Marinari, P.E., Kreeger, J.S., Ottinger, M.A., Howard, J.G., 2000. Age-dependent changes in sperm production, semen quality and testicular volume in the black-footed ferret (*Mustela nigripes*). Biol. Reprod. 63, 179–187.

Wood, J.P., Kolassa, J.E., McBride, J.T., 1998. Changes in alveolar septal border lengths with postnatal lung growth. Am. J. Physiol. Lung Cell. Mol. Physiol. 275, L1157–L1163.

Winkler, G.C., Cheville, N.F., 1984. The neonatal porcine lung: ultrastructural morphology and postnatal development of the terminal airways and alveolar region. Anat. Rec. 210, 303–313.

Winkler, G.C., Cheville, N.F., 1985. Morphometry of postnatal development in the porcine lung. Anat. Rec. 211, 427–433.

Weibel, E.R., 1970. Morphometric estimation of pulmonary diffusion capacity. Respir. Physiol. 11, 54–75.

Yuan, R., Tsaih, S.-W., Petkova, S.B., De Evsikova, C.M., Xing, S., Marion, M.A., Bogue, M.A., Mills, K.D., Peters, L.L., Bult, C.J., Rosen, C.J., Sundberg, J.P., Harrison, D.E., Churchill, G.A., Paigen, B., 2009. Aging in inbred strains of mice: study design and interim report on median lifespans and circulating IGF-1 levels. Aging Cell. 8, 277–287.

Yu, B.P., Masoro, E.J., Murata, I., Bertrand, H.A., Lynd, F.T., 1982. Life span study of SPF Fischer 344 male rats fed *ad libitum* or restricted diets: longevity, growth, lean body mass and disease. J. Gerontol. 37, 130–141.

Yamamoto, Y., Tanaka, A., Kanamaru, A., Tanaka, S., Tsubone, H., Atoji, Y., Suzuki, Y., 2003. Morphology of aging lung in F344/N rat: alveolar size, connective tissue and smooth muscle cell markers. Anat. Rec. 272A, 538–547.

Zoetis, T., Hurtt, M.E., 2003. Species comparison of lung development. Birth Defects Res. B Dev. Reprod. Toxicol. 68, 121–124.

Zeltner, T.B., Caduff, J.H., Gehr, P., Pfenninger, J., Burri, P.H., 1987. The postnatal development and growth of the human lung. I. Morphometry. Respir. Physiol. 67, 247–267.

Zosky, G.R., Janosi, T.Z., Adamicza, A., Bozanich, E.M., Cannizzaro, V., Larcombe, A.N., Turner, D.J., Sly, P.D., Hantos, Z., 2008. The bimodal quasi-static and dynamic elastance of the murine lung. J. Appl. Physiol. 105, 685–692.

Zhang, S., Garbutt, V., McBride, J.T., 1996. Strain-induced growth of the immature lung. J. Appl. Physiol. 81, 1471–1476.

Chapter 14

Pulmonary Hemodynamics

Saurabh Aggarwal, Christine M. Gross, Robert J. Porcelli and Stephen M. Black

Chapter Outline

1. PULMONARY HEMODYNAMICS: HISTORICAL PERSPECTIVE

During the past century, remarkable strides have been made in measuring and understanding pulmonary hemodynamics. In 1213, an Arab physician, Ibn al-Nafis, was the first person to accurately describe pulmonary circulation. Around 1546, Michael Servetus became the first European to describe the function of pulmonary circulation. In 1870, Adolph Fick proposed that the cardiac output could be calculated by dividing the uptake of oxygen by a time-specific difference in arteriovenous oxygen content. In 1910, Bornstein used nitrogen as the test gas to measure cardiac output and stated that the rate of absorption of an "inert" gas is proportional to volume of blood flowing in the pulmonary circulation. Therefore, the pulmonary blood flow may be determined by the difference between the partial pressures of the gas in alveolar air and in pulmonary arterial blood so long as its solubility coefficient in plasma is known. In 1912, August Krogh and Johannes Lindhard measured blood flow through human lung using nitrous oxide, a gas that is much more soluble than the inert gases. In 1955, G. de J. Lee and Arthur B. Dubois used a body plethysmograph (Lee and Dubois, 1955) to measure the instantaneous rate of absorption of nitrous oxide from the human lung, allowing the determination of pulmonary capillary blood flow at frequent intervals without the use of invasive techniques. Werner Frossman, along with Andre F. Cournand and Dickinson W. Richards, shared the Nobel Prize in 1956 for their roles in the introduction and standardization of cardiac catheterization. Karlman Wasserman and Julius H. Comroe used a modified Lee and DuBois method in 1961 to determine pulmonary capillary blood flow by measuring nitrous oxide absorption; however, the subjects' own thorax was used in the place of a body plethysmograph (Wasserman and Comroe, 1962). Finally, in 1970, William Ganz and Harold J. C. Swan introduced a multilumen, balloon-tipped catheter into the pulmonary artery and left atrium to simultaneously record their pressures (Swan et al., 1970). These historical events highlight both the discovery of the pulmonary circulation itself and the concepts that govern its normal physiology; historical perspectives of these landmark studies have been expertly detailed in other texts (Barer and Nusser, 1957a; Drougas et al., 1996; Holm et al., 1998). This chapter builds on that knowledge base by incorporating recent studies with historical works to take full advantage of the technologies defining or redefining many of the concepts of pulmonary circulatory physiology.

2. THE BASIC PRINCIPLES OF HEMODYNAMICS

Hemodynamics defines the relationship between blood pressure, blood flow, and resistance, which govern the

Comparative Biology of the Normal Lung. http://dx.doi.org/10.1016/B978-0-12-404577-4.00014-X

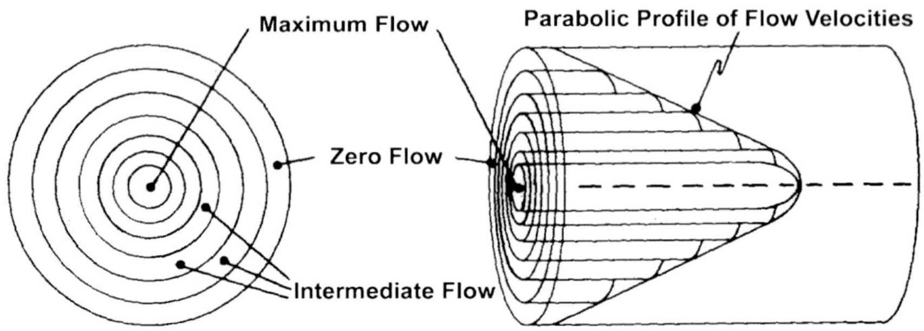

FIGURE 1 Diagrammatic depiction of flow laminae and the resulting parabolic profile of flow velocities. The flow down the axial portion of the tube or vessel is represented as maximum flow, whereas the surrounding laminae (or "rings" of flow) are each in turn, lesser flow rates. The flow nearest the wall is zero.

movement of blood through the circulatory system. The basic variables that are used in analyzing the circulation of blood are volume (V), flow (Q), pressure (P), resistance (R; or conductance), velocity (v), and vessel compliance (C). The relationship between blood flow, resistance, and pressure can be determined by using modified Ohm's law for the flow of electrons in an electrical circuit:

$$Q = \Delta P/R \qquad (1)$$

Blood flow is denoted in ml/min or l/min. In Eqn (1), ΔP (measured in mmHg) is the intravascular pressure gradient between the upstream (larger pressure) and downstream (smaller pressure) ends of the blood vessel and R is the hydraulic resistance to blood flow. Resistance to blood flow within a vessel is determined by three primary factors: vessel radius (r; or diameter), vessel length (L), and viscosity of the blood (η). The relationship between these three factors is known as Poiseuille's law (sometimes referred to as the Poiseuille–Hagen law) and is denoted by the following equation:

$$\text{Resistance}(R) = 8 \times \text{vessel length } (L) \times \text{blood viscosity } (\eta)/\pi \times \text{radius } (r)^4 \qquad (2)$$

We know that flow is pressure difference/resistance; therefore:

$$Q = \Delta P \times \pi \times (r)^4/8 \times \eta \times L \qquad (3)$$

The first parameter is the pressure difference or driving pressure for flow across the segment. This is presented in Eqn (3) as ΔP, or the measured intraluminal pressures at any two points a and b. These values in many vascular beds are simply the arteriovenous pressure differences. Although the driving pressures for flow in the lung are not always simple pressure differences between the pulmonary arteries and pulmonary veins, there are nevertheless pressure differentials that drive flow in this circulation as well. The most important parameter in this expression, however, is that of the radius of the tube or vessel lumen (r). In this instance, the flow will change as the fourth power of the change in the radius. For all other terms being equal, a 25% decrease in the radius of the tube will decrease flow by 68%. The significance of this relationship in blood vessels is magnified by the ability of these vessels to constrict or dilate due to the smooth muscle contained in the walls, which allows many vascular beds to have exquisite control of their own blood flow. The next parameter describes the contribution of the viscosity ($1/\eta$) of the perfusing fluid and the length of the vessel (1/l) to the overall flow. Equation (3) indicates that the viscosity of the fluid in the tube and the length of the vessel are inversely related to the flow. Thus, as the viscosity of the fluid or the length of the vessel increases, the flow will decrease. Under most circumstances, the vessel length does not change significantly and blood viscosity normally stays within a small range (except when hematocrit changes). Essentially, then, the most important elements governing blood flow are the driving pressures and the radii of the blood vessels, since the length of the vessels and the blood viscosities (hematocrit) either do not change or, if they do change, such changes are insignificant by comparison.

Another influential factor that determines the blood flow is the relationship between the cylindrical shape of the vessel to the parabolic profile of flow velocities (Figure 1). When one examines the flow through a tube or blood vessel in cross-section, there are "rings" or laminae of flow velocities. On the one hand, there is virtually no fluid flowing adjacent to the vessel wall due to the friction between the fluid and that wall. At the other extreme, the fluid in the center of the vessel is moving at some maximum flow. Between these two extremes are intermediate flow rates, which, if plotted with the flow rates at the wall and center of the tube, give a profile of flow that is parabolic. This flow condition has been termed *laminar flow*, and such flow will prevail as long as the tube is smooth and does not branch abruptly.

3. PULMONARY HEMODYNAMICS

Respiratory gas exchange is one of the main functions of the normal pulmonary circulation. To achieve this goal, the pulmonary circulation is a low-pressure, high-flow system with a great capability for the recruitment of unperfused vessels. To maintain this low transmural pressure, the walls of the pulmonary arteries are relatively thin. The anatomy of the pulmonary arteries changes from central conduit arteries to the peripheral resistance vessels. The distal-most segments of the precapillary arterioles comprise an endothelial layer underlined by a one to two layers of smooth muscle. This precapillary segment of the pulmonary vasculature is the site of the greatest decrease in pulmonary arterial pressure (PAP) and therefore contributes to the majority of the pulmonary vascular resistance (PVR) in the pulmonary circulation. Changes in tone or wall structure at this level can lead to large elevations in PAP.

The pulmonary circulation is the only system through which the entire cardiac output passes. Additionally, the pulmonary circulation has been shown to have minimal basal tone and minimal ability to change the forces that act upon it. Unlike many other organs, the lung does not, and cannot, control the blood flow delivered to it by the heart during normal steady states or during altered flow conditions. Despite this relative passivity, the lung is capable of significant redistributions of blood flow within itself, which, under normal circumstances, will accommodate any changes in blood flow imposed upon it, even when disease states compromise part of the vasculature itself. Thus, the physiology and pathophysiology of the pulmonary circulation is considerably more complicated than a first look might indicate.

The basic concepts of Poiseuille—Hagen law also apply to pulmonary hemodynamics and are centered around the relationship between the blood flow through the lung, the blood pressure differences, the pulmonary vessel radii, and the resistances across the pulmonary arteries, capillaries, and veins. PAP can be described by the following equation:

$$PAP = (CO \times PVR) + PVP \tag{4}$$

Here, CO is the cardiac output, which is approximately 5 l/min; PVR is the pulmonary vascular resistance, which is normally 1 mmHg/l/min; and PVP is the pulmonary venous pressure. The PVP is essentially the same as the left atrial pressure or pulmonary capillary wedge pressure (PCWP). Therefore, increases in CO, PVR, or PVP will lead to increases in PAP. Normally, pulmonary arterial systolic and diastolic pressures are approximately 25 and 10 mmHg, respectively; therefore, mean PAP is approximately 15 mmHg in humans. Because PVP is about 8 mmHg, the pressure gradient driving flow through the pulmonary circulation is rather small at approximately 7 mmHg (mean pulmonary arterial minus venous pressures). This is a much lower driving pressure for the blood flow than the systemic circulation, where the arterial-venous pressure gradient is approximately 90 mmHg in humans. Table 1 summarizes the pulmonary hemodynamic parameters for other species from different studies (Barer and Nusser, 1957a; Drougas et al., 1996; Holm et al., 1998; Morris et al., 2000; Grant and Canty, 1989; Baxter et al., 1952; Duke and Stedeford, 1960; Chen et al., 2005; Yuan et al., 2011; Liu et al., 2005; Emery et al., 2001; Fesler et al., 2006; Chick et al., 1988; Franceschi et al., 2013; Chaliki et al., 2002; Romand et al., 1994; Takeda et al., 1987; Johnson et al., 1992; Ravikumar et al., 2011; Reddy et al., 1995; Polglase et al., 2006; Newman et al., 1993; Nachar et al., 2001; Krebs et al., 2007; Hinder et al., 1998; Pennington et al., 1992; Patschova et al., 2010; Stickland et al., 2007; Li et al., 2013; Goetz et al., 1999; Tournoux et al., 2011; Ortiz et al., 2002; Ullrich et al., 1999). Although the blood flow through the pulmonary circulation is essentially the same as the blood flow through the systemic circulation, the pulmonary vascular resistance is 10- to 15-fold lower than the systemic vascular resistance. This explains why the mean pulmonary artery pressure is relatively low. In addition, the pulmonary blood vessels are capable of being passively distended, narrowed, or collapsed, as well as actively constricting or dilating. Thus, the luminal dimensions in a particular pulmonary blood vessel will depend, in part, on the level of vasoactivity (i.e., constriction), and, in part, on the transmural pressure, which is the sum of the intravascular and extravascular pressures to which that vessel wall is subjected. Other important factors influencing blood flow in the lung include the effects of gravity; the effects of inflation, pleural, and interstitial pressures on different segments of pulmonary circulation; and the baseline tone and reactivity of the blood vessels.

4. EFFECTS OF GRAVITY ON PULMONARY CIRCULATION

The effect of gravitational forces and the magnitude of that effect on pulmonary blood flow has been the subject of interest for some time. In human lung studies, West and Dollery (1960) first described gravitational differences in the uptake of radioactive carbon dioxide, a soluble perfusion-limited gas. In that study, the authors demonstrated that in an upright lung during breath-holding at 1 l above functional residual capacity (FRC), CO_2 uptake is higher in the dependent portions (base) of the lung than in the nondependent apices, consistent with relative blood flow in these regions (more in the base). In another study, it was shown that the alveolar pressure is uniform throughout the lung, whereas the pulmonary vascular pressures decrease about 1 cmH_2O/cm from the base to the apex of lung; therefore the blood flow to a given region will depend upon the vertical height of that region (Permutt et al., 1962). The role of gravity on lung perfusion has also been shown in humans in both erect or supine positions using ^{133}xenon (Anthonisen and Milic-Emili, 1966), magnetic resonance (Keilholz et al., 2001), and positron emission tomography (Brudin et al., 1994). In addition, using single-photon emission tomography (Kosuda et al., 2000), it was

TABLE 1 Pulmonary Hemodynamic Parameters

Species	Cardiac Output	Heart Rate	RAP	LAP/PCWP	PVR	MPAP	MPVP	PBF	References
Pig, 3–4 mo	7.19 ± 0.15 l/min	116 ± 4				21.4 ± 2.1 mmHg			Drougas et al. (1996)
Pig	210 ± 6 ml/min/kg	101 ± 2	0.8 ± 0.7	5.1 ± 0.8 mmHg	40.4 ± 2.6 mmHg kg min/l	13.8 ± 0.7 mmHg			Pennington et al. (1992)
Pig	3.4 ± 0.4 l/min	86 ± 16		8 ± 1 mmHg	2.9 ± 0.5 mmHg min/l	18 ± 3 mmHg			Holm et al. (1998)
Pig	2.65 ± 0.75 l/min	120.8 ± 13.1		8.4 ± 2.8 mmHg	4.5 ± 1.8 woods units	19.2 ± 3.3 mmHg			Morris et al. (2000)
Cat		227 ± 7				24.7 ± 1.4 cmH_2O			Grant and Canty, 1989
Cat				3–4 mmHg	0.09–0.12 mmHg/ml/min	19 ± 5.6 mmHg		280 ± 68.7 ml/min	Barer et al. (1957a)
Cat								121 ml/min/kg	Baxter et al. (1952)
Cat	170–334 ml/min			4–6.8 cmH_2O	0.034–0.145 cmH_2O	13–35 cmH_2O			Duke and Stedeford (1960)
Rat					0.05 ± 0.02 mmHg/ml/min	14.5 ± 0.8 mmHg			Chen et al. (2005)
Rat	126.8 ± 3.9 ml/min				0.13 ± 0.01 mmHg/min/ml	17.4 ± 1.0 mmHg			Yuan et al. (2011)
Rat		415 ± 15				14.2 ± 4.1 mmHg			Liu et al. (2005)
Rat						18.88 ± 1.60 mmHg	2.85 ± 0.47	53.91 ± 3.29 ml/min	Li et al. (2013)
Rat					0.55–0.65 mmHg/ml/min	16–20 mmHg			Emery et al. (2001)
Mouse	18 ml/min	445 ± 19							Tournoux et al. (2011)
Mouse			4.2 ± 0.4		0.9±0.8 mmHg min/min	13.2±1.9 mmHg			Ortiz et al., 2002
Mouse				6 ± 1 mmHg				80 ± 15 μl/min/g	Ullrich et al. (1999)
Dog	2.5 ± 0.2 l/min	81 ± 6	10 ± 1	12 ± 1 mmHg	124 ± 16 dynes/s/cm^5/m^2	16 ± 1 mmHg			Fesler et al. (2006)

Dog	5.4 ± 0.6 l/min	123 ± 6							Stickland et al. (2007)
Dog	108 ± 8 ml/min/kg	168 ± 3		9 ± 1 mmHg	0.14 ± 0.01 mmHg/min/kg/ml	24 ± 2 mmHg			Chick et al. (1988)
Dog		75 ± 19		11 ± 3 mmHg	366 ± 97 dynes/s/cm^{-5}/m^2	18 ± 3 mmHg			Franceschi et al. (2013)
Dog				13.4 ± 6.3 mmHg		20.2 ± 6.2 mmHg	17.1 ± 6.5		Chaliki et al. (2002)
Dog					437 ± 139 dynes/s/cm^5	19.8 ± 6.1 mmHg			Romand et al. (1994)
Dog								0.70 ± 0.04 l/min	Takeda et al. (1987)
Dog	3.29 ± 0.43 l/min							770 ± 533 ml/min	Johnson et al. (1992)
Dog		151 ± 21						179 ± 56 ml/min/kg	Ravikumar et al. (2011)
Horse	38 ± 7 l/min	32 ± 2				16 ± 3 mmHg			Patschova et al. (2010)
Horse		40 ± 3	7 ± 2	31 ± 2 mmHg		36 ± 2 mmHg	25 ± 2		Goetz et al. (1999)
Horse, galloping		225 ± 3	55 ± 5	81 ± 4 mmHg		98 ± 5 mmHg	64 ± 3		Goetz et al. (1999)
Sheep, 1 mo		141.8 ± 18.3	3.1 ± 0.6	6.2 ± 2.4 mmHg	20.6 ± 3.9 mmHg/l/min	16.2 ± 2.9 mmHg		515.9 ± 23.8 ml/min	Reddy et al. (1995)
Sheep, fetus		169 ± 15.5			2.3 ± 0.1 mmHg/ml/min	32.6 ± 0.7 mmHg		20.8 ± 4.2 ml/min	Polglase et al. (2006)
Sheep	4.4 ± 0.2 l/min			2.6 ± 0.9 cmH_2O	3.9 ± 0.2 cmH_2O/l/min	19.5 ± 1.7 cmH_2O			Newman et al. (1993)
Sheep	140 ± 3 ml/min/kg	128 ± 5			0.08 ± 0.001 mmHg/ml/min/kg	14 ± 0.7 mmHg			Nachar et al. (2001)
Sheep	3.7 ± 0.7 l/min	124 ± 18				21.7 ± 5 mmHg			Krebs et al. (2007)
Sheep	4.9 ± 0.2 l/min				161 ± 10 dynes/s/cm^5/m^2	21 ± 1 mmHg			Hinder et al. (1998)

Heart rate, beats per min; RAP, right atrial pressure (mmHg); LAP, left atrial pressure; PCWP, pulmonary capillary wedge pressure; PVR, pulmonary vascular resistance; MPAP, mean pulmonary artery pressure; MPVP, mean pulmonary venous pressure (mmHg); PBF, pulmonary blood flow.

illustrated that in an upright position, both blood flow and ventilation were greater in caudal regions than in horizontal postures. This redistribution was greater for blood flow than for ventilation, resulting in gradually decreasing ventilation-to-perfusion ratios from the apex to the base of the lung in an upright position (Liu et al., 2005). This ventilation–perfusion inequality is enough to lower the arterial partial pressure of oxygen by 5 mmHg less than the average alveolar air. Studies on humans experiencing weightlessness onboard a flying jet aircraft demonstrated that virtually all topographic inequality of ventilation, blood flow, and lung volume seen under gravitational conditions was abolished during short periods of zero gravity (Michels and West, 1978) and that the pulmonary blood flow became more uniform. Therefore, the major effect of gravity on the lung is to produce a vertical base-to-apex perfusion pressure gradient in the pulmonary circulation (Figure 2). In the upright position, the bases (or dependent portions) of the lung are fully perfused, while the apices are virtually unperfused (West et al., 1964; West and Dollery, 1965; West, 1977). Additionally, if the body position changes appreciably, this gradient reorients in the appropriate direction relative to the forces of gravity (Chaliki et al., 2002; Romand et al., 1994; Takeda et al., 1987). However, it is important to note that in quadrupeds such as horses (Hlastala et al., 1996), goats (Melsom et al., 1995), dogs (Parker et al., 1995), sheep (Walther et al., 1997), pigs (Glenny et al., 2000), and ponies (Jarvis et al., 1992), gravity may not be as important a determinant of pulmonary blood flow distribution because the posture of quadrupeds produces a relatively shorter vertical gradient compared with those of humans (Hughes, 1996). In contrast, baboons, like humans, spend most of their time in the upright posture and have remarkably similar pulmonary structures and physiology to humans, where gravity plays a significant role in pulmonary blood flow (Glenny et al., 1999).

4.1 Zones of the Lung

The effect of gravity on lung perfusion has traditionally been divided and functionally described as a four-zone system. The zonal model of lungs hypothesizes that gravity is the major determinant of regional perfusion, and therefore perfusion swells from nondependent to dependent regions of the lungs (Hughes et al., 1968a). According to the zonal framework, blood flow at any isogravitational plane is determined by the balance between alveolar pressure (P_{ALV}), pulmonary arterial blood pressure (P_{PA}), and pulmonary venous blood pressure (P_{PV}). The alveolar pressure is the force that acts on the adventitial aspect of the adjacent blood vessels, tending to push them towards collapse and thereby limiting blood flow. The PAP, which tends to distend these vessels, acts to oppose the effects of the alveolar pressure and thus to maintain blood flow. It is important to note that these perfusion differences between zones do not result from differences in vessel structure between the zones, because the vascular morphology within a given lung has been shown to be relatively constant (Wiebel, 1973).

4.2 Zone I

Briefly, zone I defines the condition in the apices or the top 4–5 cm of lung where alveolar pressures predominate, because they exceed both pulmonary arterial and pulmonary venous pressures ($P_{ALV} > P_{PA} > P_{PV}$). The net result of this situation is the closure of the alveolar and perialveolar vessels to flow (Permutt et al., 1962; Banister and Torrance, 1960) (Figure 2). The pressure required to close these vessels is termed the *critical closing pressure* (West and Dollery, 1965; Burton, 1951; Permutt and Riley, 1963), where the relative external pressure exceeds the internal (i.e., closure); the vessels subject to this closure seem to be capillaries and vessels up to 30 μm in dimension (West, 1977).

When defining zone I conditions, it is sometimes customary to view this portion of the lung as devoid of blood flow. However, this is not strictly true for several reasons. First, the bronchial circulation provides blood flow to this portion of the lung (Daly and Hebb, 1966). Second, the zonal pulmonary circulation provides some blood flow to this region through the perfusion of vessels that penetrate the region where the alveolar septa meet. These vessels have been defined as corner vessels, and they are subjected

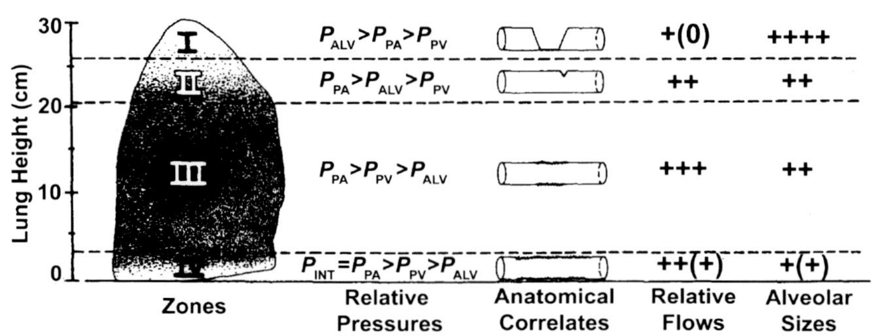

FIGURE 2 The diagrammatic composite of the pressure–flow relationships in the lung. In zone I, pulmonary alveolar pressure (P_{ALV}) is greater than the pulmonary arterial pressure (P_{PA}), while, P_{PA} is greater than the P_{ALV} in zone II. In zone III, P_{PA} is greater than pulmonary venous pressure (P_{PV}) and pulmonary interstitial pressure (P_{INT}) is equal to the P_{PA} in zone IV. In addition, see the text for a description and the referenced work.

to forces that maintain the blood flow in these vessels throughout the respiratory cycle. In fact, several investigators have observed open vessels in alveolar corners under zone I conditions (Goetz et al., 1999; West and Dollery, 1960) and have found that up to 15% of cardiac output goes through lungs in zone I (Koyama et al., 1989). This perfusion persists despite the fact that the vessels in the adjacent alveoli are collapsed as a result of the relatively high alveolar pressures (Anthonisen and Milic-Emili, 1966; Keilholz et al., 2001; Brudin et al., 1994; Kosuda et al., 2000). It has been suggested that these corner vessels in this zone contribute to gas exchange (Lamm et al., 1995) owing to their proximity to alveoli.

4.3 Zone II

The description of zone II conditions is somewhat more complex due to a number of different factors that influence blood flow in this region (West, 1977). In this zone, the PAP exceeds both the alveolar pressure and the pulmonary venous pressure, respectively ($P_{PA} > P_{ALV} > P_{PV}$). Thus, the pressure differential driving flow in this zone is the pulmonary arterial pressure minus the alveolar pressure (Figure 2).

An important factor influencing the blood flow in this zone is the differential narrowing of the vessels exposed to the alveolar pressures. The precapillary vessels of this zone are fully open because the pulmonary arterial pressure well exceeds the alveolar pressure ($P_{PA} > P_{ALV}$). However, as the intravascular pressure drops or dissipates down the alveolar vessels, the vessel segments downstream of the capillaries, or toward the venular end of the vasculature, will be compressed by the alveolar pressure (Grant and Canty, 1989; Pennington et al., 1992; Stickland et al., 2007; West et al., 1964). Therefore, as long as the blood pressure exceeds the alveolar pressure, complete collapse, as in Zone I, is averted and flow is maintained. This idea of a pliant tube being subjected to a collapsing pressure, which then offers resistance to flow, has been termed a Starling resistor (Permutt and Riley, 1963). This concept has uniformly been depicted as the pinching of the downstream end of the capillary segment of the vessels in zone II (Figure 2). This resistance is analogous to the compression of the blood vessels in the arm when blood pressure is measured with a sphygmomanometer (Permutt and Riley, 1963).

As long as the alveolar pressure exceeds the venous pressure ($P_{ALV} > P_{PV}$), the actual value of venous pressure is inconsequential to the dynamics of flow. To conceptualize this, start with zone III conditions, where blood flow is determined by the difference between the pulmonary arterial and venous pressures ($P_{PA} - P_{PV}$) and where pulmonary venous pressure exceeds alveolar pressure ($P_{PV} > P_{ALV}$). As one reduces the pulmonary venous pressure to a point just below the alveolar pressure, the driving pressure for flow becomes the difference between the pulmonary arterial and the alveolar pressures ($P_{PA} - P_{ALV}$) (Zone II). Much as the flow of water over a waterfall or sluice is unaffected by the distance the water ultimately falls, any value of venous pressure lower than the alveolar pressure will have no effect on blood flow. This is the reason why zone II conditions of flow have been often referred to as a vascular waterfall or sluice (Grant and Canty, 1989; Pennington et al., 1992; Stickland et al., 2007; Brudin et al., 1994).

4.4 Zone III

Descending farther from the apex towards the base, zone III includes the basal 20 cm of the lung where the pulmonary arterial pressures exceed the pulmonary venous and alveolar pressures, respectively ($P_{PA} > P_{PV} > P_{ALV}$). The pressure differential driving flow in this part of the lung is simply the pressure difference between the pulmonary arteries and the pulmonary veins ($P_{PA} - P_{PV}$). As a result, this zone is fully perfused (Glazier et al., 1969).

4.5 Zone IV

A fourth zone was added to the zonal model of the lungs when Hughes and coworkers (Hughes et al., 1968a) using intravenously injected radioactive xenon found a reduction in blood flow in the most dependent lung regions. Zone IV has been described as the very base of the lung in which the overlying column of lung increases the interstitial pressures (P_{INT}) on precapillary or extraalveolar vessels (Krebs et al., 2007; West and Dollery, 1965). In this instance, these extraalveolar vessels have been defined as vessels that are not subject to alveolar pressures; therefore, this definition is a physiological rather than a morphological one (Mead and Whittenberger, 1964). In this zone, the sum total of forces acting on these extraalveolar vessels functions to raise the vascular resistance, such that the flow in this zone is reduced toward zone II levels (West, 1977).

In zone IV, the pulmonary interstitial pressure (P_{INT}) becomes positive due to low lung volumes and/or the expansion of the pulmonary interstitial space by fluid and exceeds both P_{PV} and P_{ALV}. Consequently, flow in zone IV is determined by the arterio-interstitial pressure difference ($P_{PA} - P_{INT}$), which is less than the arteriovenous pressure difference ($P_{PA} - P_{PV}$); therefore, zone IV blood flow is less than zone III blood flow. The positive interstitial pressure causes extra-alveolar vessel compression, increased extra-alveolar vascular resistance, and decreased regional blood flow. An additional observation relative to this zone is that there is a smaller proportion of air space relative to the total zonal lung volume. Indeed, it has been shown that, under normal circumstances, the alveoli in this portion of the lung were relatively smaller than those in the apices of the same lung (West, 1977; Milic-Emili et al., 1966). This basal zone of reduced blood flow and smaller relative alveolar size could be normalized toward zone III conditions either by raising transpulmonary pressures (Hughes et al., 1968b) or with positive-pressure ventilation with pressures of up to 30 cm of water (Glazier et al., 1969).

It is important to note that the blood flow determinants in the lung are substantially different from those observed in the systemic circulation. Because the systemic vasculature does not have an equivalent to the alveolar pressure, it can be considered a fluid-filled system where the determinant of blood flow is simply the difference between the prevailing arterial and venous pressures. Interestingly, in studies where the lungs were filled with fluid to eliminate the effects of inflation and placed in fluid baths to minimize hydrostatic gradients, the pulmonary circulation became analogous to the systemic circulation, where the driving pressure for flow was simply the arterial minus the venous pressures (Lloyd and Wright, 1960). These data indicate the important role that alveolar pressures play in determining blood flow in zone II conditions.

Although this concept has been historically presented in the context of specific zones of blood flow through the lung, there are, in fact, no clear demarcating lines separating each of these zones. In reality, one flow condition blends into the next, as the shading of the lung in Figure 2 is meant to indicate (West and Dollery, 1965; Banister and Torrance, 1960). Therefore, it is possible that, with an increase in P_{PA}, zone I arteries may become zone II arteries and, as mean P_{PV} increases, zone II vessels may become zone III vessels. Zone III vessels may become highly distended such that they leak fluid and become converted to zone IV vessels. Figure 2 is also a summary list of each of these gradients in each of the zones described above.

One caveat to this broad scheme of hydrostatic gradients and their effects on pulmonary blood flow involves the size of the lung. Virtually all of the studies cited have been performed in the dog lung, which is a model large enough to have such gravity dependence. As such, the dog lung is a good correlative model of the human lung. However, when we examine smaller species, especially ones in which the gravitational gradient might be ventral to dorsal in orientation, the effects of gravity are much less notable. It has been reported that the hydrostatic pressure in the lung increases 1 cm H_2O for each centimeter of lung height above a given reference point (West and Dollery, 1965). It stands to reason that in the mouse or rat lung, which is only a few centimeters in its vertical dimension, these gravitational gradients may not be as significant as in larger models, such as the dog, sheep, cattle, or humans.

Interestingly, studies have indicated that pleural pressure gradients, which can also influence pulmonary blood flow, are also somewhat dependent upon the size of the animal (Agostoni and D'Angelo, 1970). This vertical gradient has been reported to be inversely related to species size and seemingly results from a more even distribution of pleural pressures at the gravitational top of the lung in larger species. However, such flattening of this vertical gradient seems to be limited, in that it ceases to be important in animals above a given body size. It is difficult to define whether these pleural pressure gradients result strictly from the forces of gravity or from some gravity-independent phenomenon that in some way limits the overall effects of gravity on pleural pressure.

4.6 Nongravitational Gradients

The zone model predicts uniform blood flow within isogravitational planes. However, many investigators have reported significant differences in the gradient of flow distribution between the supine and the prone postures, suggesting that factors other than gravity may also determine regional perfusion in the lungs (Glenny et al., 2000; Jarvis et al., 1992). Glenny et al. determined that under zone III conditions, gravity is not the principal determinant of regional perfusion distribution in supine and prone dogs (Glenny et al., 1991). In addition, using intravenous fluorescent microspheres, it was demonstrated that the gravitational flow gradients were absent in the prone position but present in the supine position in healthy sheep (Walther et al., 1997). Using single-photon emission computerized tomography of the lung in healthy humans, Hakim et al. demonstrated a concentric pattern of blood perfusion that decreased significantly from the center to the periphery, suggesting that blood flow rate was larger near the central region. They concluded that this gradient between the center and the periphery was gravity independent, as the subjects were laying in the supine position (Hakim et al., 1987). Therefore, all of these studies suggest that, in addition to the well-established gravitational gradient, there may be gravity-independent distributions of blood flow in the periphery of the lungs (Figure 3) (Glenny et al., 1999; Hughes et al., 1968a). However, many of the studies that have focused on gravity-independent distribution of pulmonary blood flow have shown either insignificant (Wiebel, 1973; Banister and Torrance, 1960; Burton, 1951) or highly significant effects on blood flow distribution (Permutt and Riley, 1963; Daly and Hebb, 1966; Koyama et al., 1989; Lamm et al., 1995; Glazier et al., 1969).

This gravity-independent peripheral redistribution may be an additional way in which the pulmonary vasculature adapts to a number of important gravity-independent stimuli, such as acute hypoxia. Indeed, a role for gravity-independent or lateral flow redistribution is suggested from studies in which the effect of positive end-expiratory pressures (PEEP) on pulmonary blood flow was examined (Herenstierna et al., 1979). These studies indicate that, along with the augmentation of the

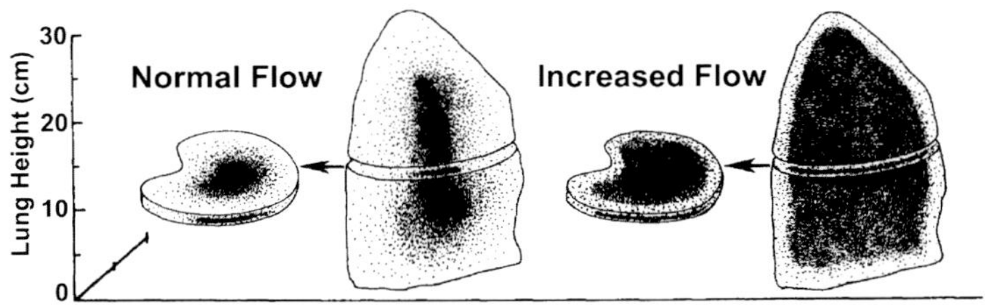

FIGURE 3 **The effects of increasing flow on the gravity-independent distribution of blood flow.** As flow to the lung increases, the blood is distributed up and out in the lung, while the core-to-peripherally oriented gradient in blood flow is maintained. See text for referenced work.

vertical gravitational gradient (>zone I), there is a peripheral distribution of blood flow away from the core of the lung as PEEP is applied. Other studies that have compared the effects of cardiac output on both the gravity-dependent and the gravity-independent distributions of blood flow have concluded that as cardiac output increases, blood flow everywhere in the lung also increases (Hakim et al., 1989). This results in the establishment of blood flow in zone I, as well as an increased flow to the peripheral regions of lung. Although the overall flow was increased in both the core and in the peripheral regions, it increased proportionally such that the core to peripherally oriented blood flow gradient was maintained (Hakim et al., 1989). Actual quantification of this gradient is difficult and depends upon a number of assumptions as to where these laterally oriented regions begin and end. However, studies have estimated the gradient to be between 3:1 to 4:1 (Greenleaf et al., 1974) and 10:1 (Glenny et al., 1999; Mead and Whittenberger, 1964), and it appears to be present both in dog and human lungs.

5. VENTILATORY EFFECTS ON PULMONARY HEMODYNAMICS

The discussion of the effects of inflation on the pulmonary circulation up to this point has focused on zonal blood flow through the gas-exchanging or alveolar vessels. These vessels compress and distort during lung inflation, and the degree to which these forces and blood pressures interact determines blood flow through that region of the lung (Franceschi et al., 2013; Keilholz et al., 2001; Milic-Emili et al., 1966). However, as the lung inflates, the inflation or pleural pressures have additional effects on other segments of the pulmonary vasculature. It has already been pointed out that inflation works to maintain blood flow through the corner vessels that course through the compartment where the alveolar septae meet (Anthonisen and Milic-Emili, 1966; Keilholz et al., 2001; Brudin et al., 1994; Kosuda et al., 2000). The precise mechanism by which these corner vessels maintain perfusion is not clear, but it may involve alveolar surface tensions and/or the forces that compress the alveolar septae, acting in some fashion to reduce their vascular resistance (West, 1977).

5.1 Positive Pressure Inflation

Positive pressure ventilation is mechanical ventilation in which air is transported into the airways and lungs under positive pressure, producing positive airway pressure during inspiration. Studies on the effects of positive inflation pressures on the other extra-alveolar vessels seem to demonstrate the expansion of the blood volume in these vessels while the alveolar vessels are being compressed (Macklin, 1946; Howell et al., 1961). The supposition is that as the positive pressures compresses the alveolar vessels, the excluded volume is accommodated by the extra-alveolar blood vessels. As a result, positive inflation pressures functionally reduce the pulmonary vasculature to a two-compartment system, consisting of a compressed compartment and an expanded compartment. Additionally, as positive pressure inflation increases the PVR in the compressible segments of zones II and III, zone I expands down the lung by the closure of additional vessels in the I—II interzonal region (West, 1977; Roos et al., 1961). The effect of PEEP, a type of positive pressure ventilation, on the regional distribution of blood flow in different postures, has recently been defined. PEEP caused an even redistribution of both ventilation and blood flow toward dependent regions (dorsal) of the lung, resulting in little change in the ventilation/perfusion (V/Q) correlation in humans lying in the supine position. However, in prone subjects, the redistribution toward dependent (ventral) regions was much greater for blood flow than for ventilation, causing increased V/Q mismatch (Glenny et al., 1991). Pulmonary perfusion is decreased to the nondependent lung and increased to the dependent lung in the lateral decubitus posture (LLD) (Mure et al., 1998). In addition, the use of unilateral PEEP to the dependent lung has been proposed to be superior to the conventional bilateral PEEP in producing a more even ventilation distribution and improving gas exchange (Hedenstierna et al., 1984).

5.2 Negative Pressure Inflation

Negative pressure ventilation is mechanical ventilation in which negative pressure is generated on the outside of the chest and transmitted to the interior to expand the lungs and allow air to flow in. Similar to positive pressure ventilation, negative inflation pressures also increases PVR in compressible segments at high inflation levels. At low lung volumes, either mode of inflation will, at first, reduce overall vascular resistance by presumably pulling open both corner and alveolar vessels as the lung begins to expand. However, the same endpoint of increased overall vascular resistance is reached at high inflation levels, irrespective of the mode by which the lung is inflated (Thomas et al., 1961; Culver and Butler, 1980). Additionally, there are also two vascular compartments that are affected when negative pressure is used to inflate the lung. The alveolar vessels are compressed as the lung inflates, while the extra-alveolar vessels, which are exposed to the negative pleural pressures, are distended. It is important to point out that these changes in vascular resistance that occur during inflation were only observed at the beginning and at the end of inspiration, while during most of the inspiratory effort notable vascular resistance changes did not occur. This implies that the increased resistance in the vessels directly exposed to the effects of inflation pressure is offset by the opposite effects of inflation on the extracapillary vessels during most of the respiratory cycle (Mead and Whittenberger, 1964).

Similar to PEEP, continuous negative external chest pressure (CNECP) applied around the chest and upper abdomen can increase FRC. However, CNECP has been shown to have an advantage in improving cardiac output compared to PEEP with identical increases in FRC. CNECP has also been shown to decrease pulmonary blood volume, pulmonary arterial pressure, and hydrostatic filtration pressures, thereby decreasing lung lymph flow (Krumpe and Gorin, 1981). However, the more negative pleural and interstitial pressures during CNECP might result in increased filtration of fluid into the lung interstitium, thereby causing pulmonary edema.

5.3 Deflation

The events that occur during deflation of the lung differ slightly from those observed during inspiration. As a result, a plot of the relationship between vascular pressure and lung volume exhibits a hysteresis between the inspiratory and the expiratory halves of the breath. It has been proposed that this results from the alveolar pressure having a greater effect on the vascular resistance during deflation (Pain and West, 1966). Studies have indicated that the net effect of the respiratory cycle on PVR results from both the prevailing alveolar pressures and alveolar surface tensions (Sun et al., 1987). As the lung inflates, the tendency of the alveolar pressure to compress the vasculature is opposed by the increasing surface tension, which acts to limit the inflation of the alveoli (Gil and Weibel, 1972). Thus, the ultimate effect of inspiration on the vasculature is somewhat less than would be expected from the alveolar pressures alone. During deflation, however, alveolar surface tension does not oppose the alveolar pressures to the same degree, and this allows the alveolar pressure to have a more direct effect on the PVR (Sun et al., 1987). PVR, when plotted against airway pressures, is found to be higher on deflation (Whittenberger et al., 1960).

One important caution concerning the results of these studies on the vascular consequences of lung inflation is that most studies have been performed under abnormal conditions where the parameter of interest was examined while the other important factors were carefully controlled. In truth, the net effects of inflation on the pulmonary circulation in a closed chest are extremely dynamic and depend upon the interaction of many forces. These forces include surface tension, elasticity of the tissues, smooth muscle tone, extravascular fluid balance, and the interdependence of the different tissue types, to name a few. Thus, studies of these dynamics are, at times, confounded, because some of these forces relate to each other in ways that are unpredictable and/or immeasurable.

6. SURFACE TENSION EFFECTS ON PULMONARY HEMODYNAMICS

The initial studies of the effects of alveolar surface tension on pulmonary vascular physiology were focused on the corner vessels and the alveolar capillaries. As we have seen, corner vessel flow is maintained even in areas of the lung where the alveolar capillaries are closed to flow (Ullrich et al., 1999; West and Dollery, 1960; Permutt et al., 1962; Anthonisen and Milic-Emili, 1966). It has been suggested that this observation results from the alveolar surface forces maintaining this patency (West, 1977). The pulmonary capillaries have also been proposed as a site of action for the alveolar surface tension. It has been observed that the blood flow in these capillaries occurs despite alveolar pressures being, at times, in excess of the capillary blood pressure (Lloyd and Wright, 1960; Thomas et al., 1961). Additional support for this suggestion derives from studies implying that capillary resistance increases when one removes the influence of the alveolar surface tension (Goetz et al., 1999; Hedenstierna et al., 1984), thus suggesting that the alveolar surface tension promotes capillary blood flow. Studies in which vascular resistance measurements were coupled with direct visualization of this microvasculature indicate that the pulmonary arterial and the venous vascular resistance increase with an increase in alveolar surface tension (by reducing surfactant), while the capillary resistance decreases with a subsequent increase in the capillary distension and recruitment (Nieman et al., 1988). Therefore, as the alveolar surface tension increases during the inspiration limb of the respiration, the pulmonary capillary resistance decreases, thereby facilitating ventilation-perfusion.

The importance of the alveolar surface tension in maintaining normal pulmonary vascular function extends beyond its simple effect on pulmonary vascular resistances. It has been suggested that a significant increase in alveolar surface tension can induce pulmonary edema, even when the capillary oncotic and hydrostatic pressures are normal (Culver and Butler, 1980; Krumpe and Gorin, 1981; Pain and West, 1966). The presence of such edema can, in turn, alter the relationship between the pulmonary arterial and alveolar pressures. Pulmonary edema has been shown to increase pulmonary vascular resistance at low lung volumes (Murray, 1978). However, whether the actual vessels involved are the alveolar (Muir et al., 1972, 1975) or extra-alveolar vessels (Krebs et al., 2007; Sun et al., 1987) is controversial. In addition, whether the observed resistance changes are related to the compression of the vasculature by the accumulating fluid in the interstitial spaces or to high alveolar pressures resulting from the trapping of air by peribronchial cuffs of edema fluid (Hughes and Rosenzweig, 1970) is unresolved (Ritchie et al., 1969). At higher lung volumes, on the other hand, overall vascular resistance increases to a greater degree in edematous lungs than in normal lungs. This has been, in part, attributed to the edematous lung being considerably less compliant and an increase in the inflation pressures, which are reflected as increases in vascular resistance (Murray, 1978). Thus, any interruption of alveolar surface forces that results in pulmonary edema may produce additional changes in pulmonary hemodynamics, which may exacerbate that same edematous condition, resulting in a vicious forward feedback loop.

7. PRESSURE EFFECTS ON PULMONARY BLOOD FLOW

Although the overall distribution of vascular pressures in the lung is significantly influenced by gravity, there are, in addition, other important factors that contribute to the net intravascular pressures and pressure pulses in the lung (Collins and Maccarro, 1979; Fishman, 1985; Milnor, 1989).

7.1 Pressure Pulses

Pulmonary arterial systolic and diastolic pressure pulse dampens and becomes out of phase as it progresses through the vasculature, including the pulmonary veins. The pressure pulses in the pulmonary veins have been shown to result from the transmission of these pulses from the pulmonary artery (Pinkerson, 1967; Szidon et al., 1968), although the pulses in the major pulmonary veins probably result, in part, from the action of the left atrium (Morgan et al., 1966a,b).

Muscular pulmonary arterioles are the primary site of vascular resistance. The greatest drop in pulmonary pressure and hence blood flow occurs during transition from the arterioles to the capillaries. A major factor that changes these pressure pulses is the mechanical distortion of the vessel wall. As the pressure pulse stretches the vessel wall, the transmission of that pulse to the adjacent downstream segment is dampened and temporally delayed. Although these pulses are generally transmitted orthograde down the pulmonary vasculature from artery to vein, some of the pulse energy is concomitantly reflected upstream, and this retrograde force tends to impede oncoming blood flow. Parenthetically, this phenomenon is one important element of the overall impedance of blood flow in the pulmonary circulation (Milnor, 1989).

This ability of the vessel wall to deform under pressure, or stress, is termed *strain*. The ability of the vessel wall to return to its normal state with the removal of that stress or strain is termed *elasticity*. Because these changes in the vessel wall are not instantaneous, it has been suggested that the vessel wall demonstrates a viscous property. Thus, these wall characteristics, taken together, have been defined as the viscoelasticity of the pulmonary blood vessels, and it appears that this viscoelasticity substantially contributes to the pressure—flow characteristics of the pulmonary circulation. Although viscoelasticity results from the structural elements in the vessel wall, it can be altered by changes in mean pulmonary arterial pressure, pressure pulse frequency, vessel tone, and level of lung inflation. Thus, as any of these influences change, the orthograde transmission of pressure and pressure pulses in the pulmonary circulation will also be altered.

7.2 Microvascular Pressures

The major reduction in the mean systemic arterial pressure has been historically reported to occur across the systemic capillary bed (Milnor, 1989). However, in recent years, the pre-capillaries segment, arterioles, have emerged as the major contributor to the pressure drop. This observation results from the marked increase in cross-sectional area of this vascular segment owing to thick smooth muscle layer and the resultant increase in the frictional loss of energy to the vessel walls. Quantification of these pressure differences in the pulmonary circulation is important both for the understanding of zonal blood flow and for the identification of vessels in the lung that respond during vasomotion. Early studies examining the longitudinal distribution of pulmonary vascular pressures, using the low-viscosity bolus method (Piiper, 1958), indicated that the overall pressure drop was evenly partitioned between the arterial and venous portions of the

TABLE 2 Longitudinal Pressure Distributions in the Lung Segmental Pressure Drop (% Total)

Species	Zone	Arterial	Capillary	Venous	References
Microvascular Pressures: Low Viscosity Bolus Technique					
Dog	III	46%	34%	20%	Brody et al. (1968)
Dog	II	−50%	50%?	–	Brody et al. (1968)
Microvascular Pressures: Outflow Occlusion Technique					
Dogs	III	43%	15%	42%	Hakim et al. (1982)
Dogs	III	39%	12%	49%	Michel et al. (1984)
Lambs	III	31%	42%	26%	Tod and Sylvester (1989)
Pig	III	20%	50%	30%	Rock et al. (1985)
Guinea pig	III	52%	0%	48%	Albert et al. (1989)
Rabbit	III	44%	0%	55%	Albert et al. (1989)
Microvascular Pressures: Micropuncture Technique					
Dog	III	40%	50%	10%	Bhattacharya et al. (1982)
Dog	III	30%	55%	20%	Shepard et al. (1988)
Dog	II	21%	38%	41%	Shepard et al. (1988)
Cat	III	27%	44%	29%	Nagasaka et al. (1984)
Rabbit	III	33%	60%	7%	Raj et al. (1986b)
Rabbit	II	28%	52%	20%	Bhattacharya et al. (1989)
Lamb	III	33%	29%	38%	Raj et al. (1986b)

vasculature (Table 2) (Szidon et al., 1968; Milnor, 1989; Pinkerson, 1967; Raj et al., 1986a,b; Bhattacharya et al., 1989, 1982; Nagasaka et al., 1984; Shepard et al., 1988; Albert et al., 1989; Rock et al., 1985; Tod and Sylvester, 1989; Michel et al., 1984; Hakim et al., 1982; Brody et al., 1968). Other studies, using the iso-gravimetric method, found that the pressure drop was either evenly distributed between the pre- and postcapillary midpoint (Piiper, 1958; Morgan et al., 1966a,b) or was slightly higher across the arterial compartment (Garr et al., 1967). One of the clear disadvantages of each of these techniques is their limitation in defining the pressure drop across the capillary segment as a separate entity.

A technique that is capable of defining the contribution of the middle, or capillary, segment to the overall pressure drop across the pulmonary circulation is the vascular occlusion technique (Hakim et al., 1979). The initial studies using this method demonstrated that approximately 58% of the total vascular resistance was upstream of the midpoint of the total pressure drop during control conditions (Hakim et al., 1979). Although this study seemingly confirmed the results of the other methods, it was not until subsequent studies used a double-occlusion technique that partitioning of the vasculature into three segments was successful. Some of these studies suggest that most of the pressure drop is evenly distributed between the pulmonary arteries and veins, with 15% occurring across the capillary segment (Table 2) (Michel et al., 1984; Hakim et al., 1982). Other studies using the same technique have similarly partitioned the vasculature and found the pressure drop across the capillary segment to be a more significant proportion of the total in dogs (46%) (Linehan et al., 1982), pigs (50%) (Rock et al., 1985), lambs (42%) (Tod and Sylvester, 1989), and rabbits (30%) (Albert et al., 1989). The only exception to these findings seems to be the results from the guinea pig lung, in which the total pressure drop was evenly distributed between the arterial and venous segments (52% and 48%, respectively), with virtually no pressure drop across the capillary segment (Albert et al., 1989).

Other assessments of these distributions have used direct measurements of segmental pressures with micropipettes (Bhattacharya and Staub, 1980); Table 2 summarizes the results from a number of studies using this technique. The most notable result is that the capillary segment is the location of the most significant pressure drop when the lung is in zone III conditions. The nearly unanimous definition of this vessel segment is the vessels between the 50-μm arterioles and venules. The upstream or arterial segment is the site of the next largest drop in microvascular pressures, followed by the venous segment. The most significant effect on these pressure distributions seems to be the zonal conditions in which the lung was studied. As the lung moves from zone III to zone II conditions, the major pressure drop across the capillary segment decreases, and the small pressure differential across the venous segment increases substantially.

The reasons for these differences may be related to the species studied, the experimental conditions under which the lung was studied, or the execution of the procedure in the hands of different investigators. This latter possibility seems the most likely because studies in which the different techniques were compared in the hands of the same investigator(s) have demonstrated that there is good correlation between them (Parker et al., 1983; Hakim and Kelly, 1989). For example, the comparison between the iso-gravimetric and venous occlusion methods in the same preparation indicated that the arterial and venous resistances measured by each technique were virtually identical (Parker et al., 1983). Furthermore, measurements of capillary pressures by each of these techniques also demonstrated a close correlation. Similar studies comparing the double-occlusion method with either micropipette measurements (Hakim and Kelly, 1989) or small catheters (Michel et al., 1984) have also demonstrated identical results between the methods.

The importance of these studies in defining the longitudinal distribution of the vascular pressures, or resistances, across the lung stands alone. However, many of these studies have also defined the segment that responds when the pulmonary vasculature is exposed to a number of vasoactive agents. Table 3 summarizes the location of resistance changes to a number of such vasoactive agents in the context of this three-compartment analysis of the pulmonary vasculature (Raj et al., 1986a,b; Nagasaka et al., 1984; Bhattacharya et al., 1982; Albert et al., 1989; Rock et al., 1985; Tod and Sylvester, 1989; Michel et al., 1984; Hakim et al., 1982, 1979; Brody et al., 1968; Linehan et al., 1982; Grimm et al., 1978; Siegel et al., 1989; Rippe et al., 1987; Tod et al., 1987; Townsley et al., 1985a; Kadowitz and Hyman, 1984; Noonan and Malik, 1986; Noonan et al., 1986; Younes et al., 1987).

Acute hypoxia primarily constricted the pulmonary arterial segment in all of the species studied. These species demonstrated a more even distribution of the effects of hypoxia across the vasculature. Although the absolute changes in vascular resistance differed appreciably from study to study, the finding of arterial hypoxic vasoconstriction was consistent, irrespective of the species studied or techniques used to discriminate the segment that was responding to the hypoxia stimulus.

The results with the biogenic amines were unambiguous. 5HT produced pulmonary arterial vasoconstriction, and this was observed with every drug concentration studied. Histamine, on the other hand, demonstrated predominant venoconstriction, with very little arterial vasoactivity by comparison. The only exception to this rule was the results in the rabbit, in which histamine is a potent arterial vasoconstrictor. Norepinephrine demonstrated predominant arterial vasoconstriction, with some studies indicating a venous component to this vasoactivity. This finding correlated well with the results with sympathetic stimulation, which seems to constrict these same vessel segments. Endothelial-derived peptide, endothelin (ET-1), has been shown to constrict arteries and veins with equal potency in guinea pigs (Ressmeyer et al., 2006) and rabbits, large veins in dogs, and small arteries in rats (Barnard et al., 1991). Another important group of agents studied in this compartmental fashion was arachidonic acid and a number of its metabolites. Arachidonic acid has been reported to constrict only pulmonary veins and capillaries. All of the products of arachidonic acid metabolism by the cyclooxygenase pathway induce vasoconstriction in the pulmonary circulation. Prostaglandin F2 alpha (PGF2α) constricts the arterial segment in the dog and all vascular segments in the lamb (Table 3). Prostaglandin F1 alpha (PGF1α) and E2 (PGE2), on the other hand, were shown to be potent venoconstrictors (Townsley et al., 1985b). The products of the lipoxygenase metabolic pathway of arachidonic acid, the leukotrienes, nearly all constricted the pulmonary veins to a greater extent than the arteries. The sole exception to this finding was again the rabbit, in which leukotriene C4 had no effect.

It seems clear from these studies that one can group vasoactive agents in a nontraditional classification scheme based on the segment of the pulmonary vasculature that responds to an agent rather than the biological classes of those agents. This would suggest that the pulmonary circulations of a number of species have developed unique distributions of receptors, membrane properties, or contractile mechanisms that produce markedly different vasoactive loci. Although this seemingly gives the pulmonary circulation the ability to segmentally respond to a number of stimuli, the true nature of the physiological consequences of this ability in vivo is unknown. Furthermore, although the data on the segmental vasoactivity in the lung are incomplete, one would hope that the additional data from ongoing studies will also be provided from this compartmental perspective.

7.3 Interstitial Pressures

The use of the micropuncture method has not been confined to the study of vascular pressures alone and has included studies on interstitial fluid pressures as well. Although these pressures are not a focus of this review, this information is included only as it applies to the pulmonary vascular pressures. The early studies of pulmonary interstitial

TABLE 3 Longitudinal Distribution of Vasoactivity

Agent	Species	% ↑ Rpv_T	Concentration	Segment Responding	Method	References
Hypoxia	Dog	~85	3% O_2	Arteries >> Veins	VO + LVB	Hakim et al. (1979)
	Dog	~100	3% O_2	Arteries >> Veins	VO + LVB	Hakim et al. (1979)
	Dog	~100	3% O_2	Arteries	LVB	Grimm et al. (1978)
	Cats	74	2% O_2	Arteries > Others	MP	Nagasaka et al. (1984)
	Lambs	60	8% O_2	All segments	MP	Raj et al. (1986a)
	Lambs (perinatal)	~53	4% O_2	Middle and arteries	VO	Tod and Sylvester (1989)
	Sheep	~30	0% O_2	Arteries	VO	Siegel et al. (1989)
	Pig	~37	7% O_2	Arteries and middle	VO	Rock et al. (1985)
5HT	Dog	~248 −380	0.5–2.5 μg	Arteries > Veins	VO	Rippe et al., (1987)
	Dog	~100	36–145 μg	Arteries	VO + LVB	Hakim et al. (1979)
	Dog	~70	90–200 μg	Arteries > Veins	VO	Hakim et al. (1982)
	Dog	~50	90–200 μg	Arteries > Veins	S CATH + VO	Michel et al. (1984)
	Dog	~300 −400	400 μg	Arteries	VO	Linehan et al. (1982)
	Dog	~100	2–10 μg	Arteries	LVB	Grimm et al. (1978)
	Dog	~100	100–150 μg	Arteries and veins	MP	Bhattacharya et al. (1982)
	Dog	51	1.5 μg	Arteries	LVB	Brody et al. (1968)
	Guinea pig	81	20, 200 μg	Arteries and veins	VO	Albert et al. (1989)
	Rabbit	352	20, 200 μg	Arteries	VO	Albert et al. (1989)
Histamine	Dog	~240 −430	5–10 μg	Veins	VO	Rippe et al. (1987)
	Dog	~200	2–10 μg	Veins	VO + LVB	Hakim et al. (1979)
	Dog	~100	100–400 μg	Veins >> Arteries	VO	Hakim et al. (1982)
	Dog	~50 −100	100–400 μg	Veins	S CATH + VO	Michel et al. (1984)
	Dog	~300 −380	400 μg	Veins	VO	Linehan et al. (1982)
	Dog	~100	36–145 μg	Veins	LVB	Grimm et al. (1978)
	Dog	23	3.0 μg	Veins	LVB	Brody et al. (1968)
	Guinea pig	189	15 μg	Veins > Arteries	VO	Albert et al. (1989)
	Rabbits	477	15 μg	Arteries	VO	Albert et al. (1989)
Norepinephrine	Dog	~186 −300	1.3–5 μg	Arteries and veins	VO	Rippe et al. (1987)
	Dog	~30 −100	10–100 μg	Arteriens > Veins	VO + LVB	Hakim et al. (1979)
	Dog	~200	10–100 μg	Arteries	LVB	Grimm et al. (1978)

TABLE 3 Longitudinal Distribution of Vasoactivity—cont'd

Agent	Species	% ↑ Rpv_T	Concentration	Segment Responding	Method	References
Sympathetic	Dog	~70 −100	20 v, 15 Hz, 2 ms	Arteries	VO + LVB	Hakim et al. (1979)
Stimulation	Dog	~270	20 v, 15 Hz, 2 ms	Arteries	LVB	Grimm et al. (1978)
Isoproterenol	Dog	−17	1.0 μg	Arteries	LVB	Brody et al. (1968)
Papaverine	Pig	~−58%	60 μg	No effect on distribution	VO	Tod et al. (1987)
Archadonic acid	Dog	126	0.45–14 (mg/100 g Lung)	Veins > Capillaries	VO	Townsley et al. (1985b)
$PGF_{2\alpha}$	Dog	~70 −100	0.5–3 μg	Arteries	VO + LVB	Hakim et al. (1979)
	Dog	~170	0.5–3 μg	Arteries	LVB	Grimm et al. (1978)
	Lambs	~50	~50 μg	All segments	VO	Tod and Sylvester (1989)
Leukoterines	Lambs	~160	1 μg	Arteries > Veins	Intact animal	Kadowitz and Hyman (1984)
	Sheep guinea pig	~125	2 μg	Veins > Arteries	VO	Noonan and Malik (1986)
	LTB_4	280	4 μg	Veins > Arteries	VO	Albert et al. (1989)
	LTC_4	350	4 μg	Veins > Arteries	VO	Albert et al. (1989)
	LTD_4	231	4 μg	Veins > Arteries	VO	Albert et al. (1989)
	LTC_4	~100	1 μg	Veins > Arteries	VO	Noonan et al. (1986)
	LTD_4	~150	1 μg	Veins > Arteries	VO	Noonan et al. (1986)
	Rabbit					
	LTC_4	No effect	4 μg		VO	Albert et al. (1989)
High flow (10x normal)	Dogs	~300	1000 ml/kg	Veins >> Arteries	GRAV	Younes et al. (1987)
Edema	Rabbit	0	Moderate	No change	MP	Raj et al. (1986b)
	Rabbit	~500	Severe	No effect on distribution	MP	Raj et al. (1986b)

Note: % ↑ Rpv_T = percent increase in total pulmonary vascular resistance; ~ = estimation based on authors' data; MP = micropipette; VO = vascular occlusion; LVB = low viscosity bolus; S CATH = small catheter; GRAV = gravimetric technique.

pressures used a number of different techniques and, as a result, have produced a number of different results (Lai-Fook, 1982; Bhattacharya et al., 1984). Early studies suggested that the effect of interstitial pressures on the pulmonary vasculature was the greatest at the bases of the lungs and the least at the apices (Krebs et al., 2007; Kosuda et al., 2000; Garr et al., 1967; Hakim et al., 1979). It was demonstrated in upright volunteers that the blood flow increased down the lung from zone I to zone III over most of the vertical height, but a small area of reduced blood flow was present at the base corresponding to zone IV. This reduction of pulmonary blood flow was thought to be the consequence of an increase in the interstitial pressure due to the reduced expansion of the lung parenchyma in the lower zones (Hughes et al., 1967). However, direct measurements of these pressures have revealed that such a gradient may not exist in the vertical plane but may, in fact, exist as a horizontal alveolar-hilar gradient (Bhattacharya et al., 1984; Hughes, 1971). As a result of these measurements, it has been suggested that the height-dependent gradient of interstitial pressures may be the indirect result of other phenomenon, such as the alveolar size gradient and the related variations in the elastic recoil of the lung in the closed-chested model (Bhattacharya et al., 1984). Although these direct measurement techniques have begun to provide data on the relationships between interstitial pressures, alveolar pressures, lung expansion, and pulmonary edema (Hakim et al., 1982; Hakim and Kelly, 1989; Bhattacharya and Staub, 1980; Parker et al., 1983), how all these forces influence the pulmonary vascular pressures, as measured by these methods, is still unknown.

7.4 Fetal and Neonatal Pressures

The vast majority of the studies that have defined segmental pulmonary vascular resistances have used the adult animal as the experimental model. However, a number of studies have examined the distribution of vascular resistances and blood flow in fetal and neonatal lungs. The pulmonary blood flow in the fetus is very low and reaches only 10–15% of the cardiac output near term. This blood flow is not related to respiratory function (Rudolph, 1979; Inselman and Mellins, 1981; Burri, 1985). The remaining blood flowing from the right ventricle passes through the patent ductus arteriosus and, under normal circumstances, never reaches the pulmonary circulation (Rudolph, 1979). In fetal lambs, it has been demonstrated that an important factor that maintains this minimal pulmonary blood flow is the high resident vascular resistance, which produces pulmonary arterial pressures that are equivalent to aortic pressures (Lewis et al., 1976). This high vascular resistance results from a number of features that are unique to the fetal pulmonary circulation. First, it was shown in the fetal lambs that the walls of vessels are thick with smooth muscle that impinges upon the vascular lumen, producing a fixed resistance to blood flow (Levin et al., 1976). Second, the fetal circulation has high levels of smooth muscle tone and exhibits hyperactivity, thereby adding to the overall vascular resistance (Lewis et al., 1976). These abnormally high resistances allow blood flow only during systole, while no flow occurs during diastole. In fact, studies have indicated that reverse flow actually occurs during late systole as a result of the high elastic recoil or viscoelasticity of the fetal pulmonary vasculature (Rudolph, 1977).

It has been generally accepted that the minimal blood flow in the fetal lung is more or less evenly distributed throughout the vasculature. However, in term fetal rabbits, using the plasma tracer fluorescein-isothiocyanate-labeled bovine serum albumin (FITC-BSA), Lipsett et al. demonstrated that fetal pulmonary blood flow is distributed only to a proportion of discrete lung lobules, while interspersed lobules receive no flow at all. The vascular resistance in these lobules alternates between high and low states with a periodicity of approximately 35 min. These lobules are therefore perfused for 13 min followed by a period of nonperfusion for 22 min. This circulatory pattern permits both the maintenance of high PVR and uniform lung development. Furthermore, by directing flow to only a portion of the vasculature, greater microvascular flow rates are achieved and hence the risk of blood stasis is reduced. Recruitment of these nonperfused regions at birth could thus produce a significant reduction in PVR (Lipsett et al., 1989; Lipsett and Gannon, 1991). The neonatal lung, on the other hand, has an even flow distribution to the entire vascular bed, suggesting that the birth-related changes in the pulmonary vasculature also result in flow distributions that are more typical of the adult.

These studies also confirm the well-known observation that there are substantial reductions in overall pulmonary vascular resistance that occur at birth. Other studies have shown that this birth-related change in vascular resistance in goats results from the expansion of (Gilbert et al., 1972), and the oxygen delivery to, the lung (Rudolph, 1977). For instance, it has been shown that in fetal lambs, an increase in oxygen tension dilates pulmonary vessels (Morin et al., 1988), while in human fetuses of late gestation, maternal hyperoxygenation significantly increases pulmonary blood flow (Rasanen et al., 1998). Furthermore, it has been indicated that these resistance changes occur across the middle segment of the vasculature (Tod and Sylvester, 1989), which correlates well with studies in lambs that have proposed this segment to be the site of the major vascular resistance in the fetus (Levin et al., 1976). The sharply reduced resistances in these vessels results from a number of changes that occur in the lung at, or shortly after, birth. First, it was observed in pigs that within minutes of birth there is the dilation of patent vessels and the recruitment of closed vessels into the now-functioning pulmonary circulation (Haworth and Hislop, 1981). These vascular changes may be similar to the effects of inflation of the lung at low lung volumes in which there is an opening of alveolar and corner vessels as ventilation is first initiated. A second major change is the significant loss of arterial smooth muscle during the first 24 h of extrautero life (Haworth and Hislop, 1981; Rendas et al., 1978). This change seems to be confined to vessels of $<200\ \mu m$ in dimension, whereas overall changes in the morphology of the human pulmonary vasculature occur more gradually (Hislop and Reid, 1973).

Other birth-related events that occur concurrently and sequentially to influence PVR include ventilation, increasing shear stress of blood flow, changes in the activities of a number of vasoactive agents and their signaling pathways, such as endothelium-derived nitric oxide (NO), PGI2, ET-1. In near-term fetal lambs, it has been shown that mechanical ventilation results in an increase in pulmonary blood flow by four- to five-fold within 1 h (Cornfield et al., 1992), and by the addition of nitro-L-arginine, this effect is reduced by 67%, suggesting a critical role for NO (Cornfield et al., 1992). Sudden surges in pulmonary blood flow induce shear stress at birth, which has been shown to further increase blood flow by activating endothelial nitric oxide synthase (eNOS) (Cornfield et al., 1992; Wedgwood et al., 2001) and upregulating its expression (Wedgwood et al., 2001). PGI2 and PGE2 are potent dilator prostanoids in the pulmonary vessels of the fetus and newborn. In intrapulmonary arteries of perinatal lambs from the third trimester to 4 weeks of age, the synthesis of PGI2 continuously increases. In lambs, PGE2 synthesis has been shown to increase during the third trimester but to decrease from late gestation to 1 week of age and remains unchanged postnatally (Shaul et al., 1999; Brannon et al., 1994). These factors contribute not only to the initial rapid decrease in PVR but also to the maintenance of low PVR in the newborn and adult. Indeed, although many of these changes occur within the first 2 weeks postpartum, the major remodeling of the pulmonary vasculature to an adult-like organ is not completed until 6 months (Haworth and Hislop, 1981) in pigs and is not fully matured until nearly 11 years of age in humans (Hislop and Reid, 1981).

The location of the major pressure drop, or resistance, in the neonatal lamb, goat, and rabbit is the arterial segment of the vasculature, which contributes in excess of 50% of the total resistance in each model (Tod and Sylvester, 1989; Ressmeyer et al., 2006; Hughes et al., 1967). Each of these species redistributes resistance differently with the aging of the vasculature. At 1 month, the major site of vascular resistance in the lamb is again the middle segment (Tod and Sylvester, 1989), whereas in the rabbit the major locus of resistance is still the arterial segment (Raj et al., 1986b). This again suggests that there are significant species differences that must be considered when one strives for a generalized understanding of pulmonary vascular function.

8. VISCOSITY EFFECTS ON PULMONARY HEMODYNAMICS

The effect of viscosity, as stated in the Poiseuille–Hagen formula, is inversely related to the fluid flow. Therefore, as the hematocrit increases, the frictional loss of energy to the vessel wall would also increase. This, in turn, would decrease flow. However, an examination of the flow characteristics of whole blood through blood vessels indicates that there are very few red cells near the vessel wall, while most of the red cells flow down the axial portion of the blood vessel (Bayliss, 1959,1965; Bugliarello and Hayden, 1962). It has been suggested that it is not the frictional loss of energy to the vessel wall that is significant, but the loss between the laminae of different flow velocities that is responsible for the observed effects of blood viscosity. Thus, as the hematocrit increases, so does the friction between the cells in the various flow laminae, and flow decreases appropriately.

It has also been shown that with water, the size of the tube through which the fluid is flowing has no real effect on the viscosity. This, however, is not the observation with blood flowing through a vasculature bed in which the vessels are of decreasing size (Milnor, 1989). Indeed, Fahraeus and Lindqvist demonstrated that, with blood, as the vessel diameter got smaller than 1 mm, the measured viscosity of the blood actually decreases (Fahraeus, 1921; Fahraeus and Lindqvist, 1931; Barbee and Cokelet, 1971). The reason(s) for this observation is unclear. One possibility is the aforementioned parabolic profile of the flow velocities, where the effect of these various laminae of flow velocities is to spin, or tumble, the cells as they flow through the vasculature. This, in turn, produces a pressure differential, or Bernoulli effect, that gathers the cells down the axis of the vessel, leaving predominately plasma at the vessel wall. As the vessels branch, the side vessels receive a higher proportion of plasma than red blood cells, a process termed plasma skimming, which, in turn, may be responsible for the decreased viscosity observed by Fahraeus and Lindqvist in the smaller vessels.

It is obvious that the overall effects of hematocrit on blood flow are clear, whereas the details are a bit more complex. For example, the axial accumulation of cells is a somewhat fixed effect in that it occurs at low flow rates and does not increase with increasing flows. Furthermore, the decreases in the apparent hematocrit occur in vessels of 1000–30 μm in dimension, while the opposite has been suggested to occur in vessels of <30 μm in dimension (Lewis et al., 1976; Levin et al., 1976; Rudolph, 1977; Lipsett et al., 1989).

The specific effects of hematocrit on the pulmonary circulation have been reported in a number of studies. First, it has been demonstrated that the small pulmonary vessels demonstrate lower hematocrit values than do the large pulmonary vessels (Levin et al., 1976; Rudolph, 1977; Lipsett and Gannon, 1991; Gilbert et al., 1972; Morin et al., 1988; Rasanen et al., 1998). Second, studies have demonstrated that as hematocrit increases there is a concomitant increase in pulmonary vascular resistance, which is magnified by increasing flow (Gehr et al., 1978; Vreim and Staub, 1973; Brudin et al., 1986). Although these studies were confined to the canine model, studies in humans suggest that hematocrit is indeed an important determinant of pulmonary blood flow and vascular resistance (Rosenthal and Fyler, 1974). Other studies have confirmed these findings (Wang et al., 1985) and have defined the segment in which hematocrit has most of its effect to be the capillary segment in the adult dog (Julien et al., 1985) and the arteries, microvessels, and small veins in the neonatal rabbit (Raj et al., 1989; Raj and Anderson, 1991).

It is clear that the information as to how hematocrit and the pulmonary vasculature relate to each other is incomplete. Yet, hematocrit is an important contributor to pulmonary hemodynamics, and the magnitude of its effect is probably dependent upon a number of other elements working in concert. These factors are likely to include the tone and distensibility of the vasculature, the pulmonary blood flow, and the overall vascular dimensions (Cornfield et al., 1992; Shaul et al., 1999; Brannon et al., 1994). Another important factor that interacts with hematocrit, and ultimately determines the actual pulmonary blood flow, is the presence or absence of laminar flow characteristics (Milnor, 1989). Any change in the laminar flow condition that results in turbulent flow is also likely to amplify the effect of hematocrit on the pulmonary blood flow. For example, any change in vascular morphology that results in altered branching patterns, changes in lumina dimensions, or smoothness of the vessel wall is likely to change not only the laminar flow state, but also the relation between hematocrit and pulmonary hemodynamics as well.

9. PULMONARY BLOOD VOLUME

In humans, the pulmonary circulation is an important reservoir with approximately 500–1000 ml or 10–20% of total blood volume. The pulmonary blood volume (Q_p) determines pulmonary blood flow by influencing PVR and PAP. An increase in Q_p has been shown to decrease PVR which ultimately increases pulmonary blood flow in isolated rabbit lungs (Scheeren et al., 1998). The pulmonary vasculature is less muscular and therefore is more compliant than the systemic circulation. The total volume in the lungs can readily change with alterations in intrathoracic pressures and by a number of abnormalities and vasoactive stimuli. For example, emphysema, pulmonary edema, adult respiratory distress syndrome (ARDS), pulmonary fibrosis, increases in intracranial pressures, and the use of PEEP decrease pulmonary blood volume (Hislop and Reid, 1981; Bayliss, 1959; Bayliss, 1965; Bugliarello and Hayden, 1962; Fahraeus, 1921; Fahraeus and Lindqvist, 1931). All of these entities produce their effect by changing pulmonary vascular compliance. On the other hand, congestive heart failure associated with cardiac abnormality or old age tends to increase pulmonary blood volume (Hislop and Reid, 1981; Barbee and Cokelet, 1971; Gehr et al., 1978). The effect of cardiac failure and aging on pulmonary blood volume is similar in that both lead to left ventricular dysfunction. Increased pulmonary blood flow associated with a decrease in cardiac output not only increases pulmonary blood volume but also distributes that volume in a more uniform way. This is distinctly different from increasing pulmonary arterial pressures, which produce similar uniform distributions of blood volume, but without an overall increase in that volume (Wollmer et al., 1984). In contrast, exercise generally has no effect on pulmonary blood volume unless the subject has preexisting heart disease, in which case exercise produces an increase in blood volume in the lung (Alexopoulos et al., 1988, 1989). Lastly, smoking and gender seem to have little impact on pulmonary blood volumes (Crapo et al., 1982; Brudin et al., 1987) until it is again associated with cardiac failure.

The effects of a number of vasoactive agents on pulmonary blood volume have also been examined in a number of species. It has been suggested that in dogs, cats, and ferrets, acute hypoxia and the biogenic amines (5HT, norepinephrine, and histamine) decrease pulmonary blood volume by constricting the pulmonary vasculature (Morgan et al., 1966a; Wang et al., 1985; Julien et al., 1985; Raj et al., 1989). Angiotensin II has been shown to increase pulmonary blood volume in dogs, presumably by having some effect on the left heart, despite the ability of angiotensin II to constrict the vasculature (Thorvaldson et al., 1984). Interestingly, isoproterenol was without influence on blood volume, despite its ability to dilate the pulmonary vasculature (Smiseth et al., 1988). However, nitric oxide has been shown to increase pulmonary blood volume by dilating pulmonary vasculature (Dahm et al., 2000). Furthermore, acute hypoxia seems to manifest its effect only at relatively low alveolar pressures, for when higher alveolar pressures are imposed on the lung, hypoxia ceases to produce any change in pulmonary blood volume (Thorvaldson et al., 1984).

The volume of blood in the lungs varies during different portions of both the cardiac and respiratory cycles. Thus, pulmonary blood volume is always in a dynamic state, and the measured volume is that momentary value that may not account for this dynamism. Despite these caveats, the examination of blood volume in the lung has been focused on four distinct areas of interest. These include the actual quantification of pulmonary blood volume, the distribution of that volume, the definition of conditions that alter this volume, and a description of the vessel segments that are involved in the accommodation of these changes in pulmonary blood volume.

9.1 Quantification of Pulmonary Blood Volumes

Quantifications of pulmonary blood volumes have used a number of techniques, and a review of the initial studies is presented in detail elsewhere (Harris and Heath, 1986). However, the blood volumes reported tend to reflect, in part, the methods used to determine these values and the species in which the blood volume is studied (Table 4). The pulmonary blood volume in humans, as determined by the Stewart–Hamilton dye-dilution technique, ranges between 211 and 310 cc (Stewart, 1921; Hamilton et al., 1928), whereas measurement by radiocardiographic methods indicate slightly higher values, ranging between 341 and 550 cc (Donato et al., 1962; Giuntini et al., 1963). In recent

TABLE 4 Pulmonary Blood Volumes

Species	Volume	Measurement Technique	References
Man	~211–310 cc	Dye dilution	Review, Harris and Heath (1986)
Man	~519 cc	Radiocardiographic	Giuntini et al. (1963)
Man	~341 cc	Radiocardiographic	Perasolo and Heisanken (1971)
Man	~550 cc	Radiocardiographic	Lewis et al. (1962)
Man	~508 cc	Direct measurements	Backmann and Hartung (1969)
Man	~365 cc/m^2	Dye dilution	Milnor et al. (1960)
Man	~322 cc/m^2	Dye dilution	Dock et al. (1961)
Man	~317 cc/m^2	Double indicator dilution	Varnauskas et al. (1971)
Man	~270	Radiocardiographic and dye dilution	Lewis et al. (1970)
Man	~458 cc	Morphometric modeling	Review, Milnor (1989)
Dog	~200 cc	Morphometric modeling	Review, Milnor (1989)
Dog	~260 cc	Double indicator dilution	Maseri et al. (1972)
Dog	~232 cc	Double indicator dilution	Wanner et al. (1978)
Mouse	~141 μl	Evans blue dye	Zarbock et al. (2009)
Rat	~0.87 ml/g LW	Radiolabeled red blood cells	Ou et al. (1993)
Pig	~225 ml	Magnetic resonance imaging	Ugander et al. (2010)

* Values corrected for m^2 of body surface area; values are presented as approximations. For simplicity, standard errors around each mean has been ignored. In rats, LW is lung weight.

years, dual-energy contrast tomography has been used to quantify pulmonary blood volume. Using this technique, pulmonary blood volume in humans has been determined to be around 28.7 Hounsfield Unit (HU) (Sueyoshi et al.,). In addition, a novel method of determining pulmonary blood volume by injecting a contrast dye in peripheral vein and determining its mean transit time (MTT) to passage between the right ventricle and left atrium is detected noninvasively by an ultrasound transducer. Pulmonary blood volume is quantified by multiplying MTT by the cardiac output (Mischi and Korsten, 2006). Using the MTT technique, pulmonary blood volume in humans was determined to be around 322 ml/m^2 body surface area (Dock et al., 1961). The other major species in which rigorous determinations of pulmonary blood volume have been made are the dog, mouse, and pig. There have been a number of techniques used in describing this volume in these species (Table 4) (Milnor, 1989; Giuntini et al., 1963; Dock et al., 1961; Harris and Heath, 1986; Perasolo and Heisanken, 1971; Lewis et al., 1962, 1970; Backmann and Hartung, 1969; Milnor et al., 1960; Varnauskas et al., 1971; Maseri et al., 1972; Wanner et al., 1978; Zarbock et al., 2009; Ugander et al., 2010; Ou et al., 1993). Although these values are approximations, they represent a fair sampling of the values in the literature and have provided a starting point for a number of studies that have described the distribution of blood volumes between the pulmonary arteries, capillaries, and veins, and the factors that alter both the overall pulmonary blood volume and the partitioning of that volume as well.

9.2 Distribution of Pulmonary Blood Volume

The distribution of pulmonary blood volume has been examined in a number of ways, and estimates of these volumes have been described in both dogs and humans (Table 5) (Brody et al., 1968; Milnor et al., 1960; Wanner et al., 1978; Sackner and DuBois, 1964; Piiper, 1970; Dawson et al., 1988). The majority of these studies indicate that the pulmonary capillaries are the vessels that contain the largest proportion of the blood in the lung (Brudin et al., 1987; Thorvaldson et al., 1984; Smiseth et al., 1988), although data exist that suggest either an even distribution of this volume between arteries, capillaries, and veins (Piiper, 1970) or a volumetric distribution that favors the veins (Wanner et al., 1978; Sackner and DuBois, 1964). As is apparent, it is difficult to find unified opinions as to the distribution of pulmonary blood volume. This is understandable based on the differences in the assumptions upon which these studies were founded, and based on the different methodologies employed to examine this issue as

TABLE 5 Distribution of Pulmonary Blood Volume (cc, or % of total)

Species	Arteries	Capillaries	Veins	Test Segment	Zone	References
Man	120 cc	250 cc	150 cc	Lung	–	Milnor et al. (1960)
Dog	6.2 cc	12.8 cc	7.4 cc	Lobe	III	Brody et al. (1968)
	23%	48%	28%			
Dog	27%	15%	58%	Lung	–	Sackner and DuBois (1964)
Dog	27%	38%	35%	Lobe	–	Piiper (1970)
Dog	~64 cc	~48 cc	~120 cc	Lung	–	Wanner et al. (1978)
	26%	22%	52%			
Dog	19%	59%	22%	Lobe	III	Dawson et al. (1988)

well. Furthermore, Table 5 only presents data from studies that have tried to define the partitioning of pulmonary blood volume, despite the fact that information on the volume of the individual compartments abounds. For example, the pulmonary arterial blood volume in the dog has been estimated to range between 2.3 and 3.3 cc/kg (Julien et al., 1985; Smiseth et al., 1988; Stewart, 1921; Hamilton et al., 1928; Donato et al., 1962; Giuntini et al., 1963), and two thirds of this volume is thought to be in the extraparenchymal vessels (Friedman and Wanner, 1977). The pulmonary arterial blood volume in humans has been estimated to be slightly less, or between 120 and 150 cc (Milnor et al., 1960; Singhal et al., 1973) or 1.7–2.1 cc/kg in a 70-kg man.

A total of 15 orders of arteries are found between the main pulmonary artery and the capillaries in the left lung and a total of 15 orders of veins between the capillaries and the left atrium in the right lung. The total blood volume in all orders of the arteries was found to be 150.06 ml in the left lung of a 44-year-old man versus 86.59 ml in the veins in the right lung of a 24-year-old man (Huang et al., 1996). The arterial volumes have also been demonstrated to exceed the venous volumes by a substantial amount in the dog (Dawson et al., 1988), cat (Zhuang et al., 1983), and human (Singhal et al., 1973; Horsfield, 1978; Horsfield and Gordon, 1981), thus volumetrically relegating the pulmonary venous compartment to the least important of the three. Interestingly, this is not a wholly unanimous finding, for studies in the rabbit (Engelberg and DuBois, 1959), and in some studies in the dog, the venous compartment is thought to be where a major portion of pulmonary blood volume is stored (Wanner et al., 1978; Sackner and DuBois, 1964).

The vascular compartment that has unquestionably received the most attention has been the pulmonary capillaries, due to their relevance in the exchange of gases across the alveolar capillary membrane. Indeed, estimates of pulmonary capillary blood volume have repeatedly been made by the method that quantifies the overall diffusing capacity of the alveolar capillary membrane by the carbon monoxide/oxygen method (Roughton, 1945; Roughton and Forster, 1957). The use of this method in humans has indicated the capillary blood volume to be about 100 cc (Lewis et al., 1958; Bates et al., 1960), about 48 cc in the dog (Wanner et al., 1978), and about 0.77 ml in rats (Yilmaz et al., 2010). Importantly, there have been suggestions that the capillary blood volume determined by this method may be significantly underestimated and that the true volume in humans may be as high as 150–213 cc (Gehr et al., 1978; Cumming et al., 1969) and in excess of 100 cc in the dog (Smiseth et al., 1988; Wanner et al., 1978; Zarbock et al., 2009). Additionally, studies in the cat have shown that the capillary blood volume measured by the carbon monoxide method (3.7 cc) is also an underestimate when compared to morphometric measurements in the same preparation (4.5 cc) (Vreim and Staub, 1973). The reason for these differences seems to rest, in part, on the different hematocrit values in the pulmonary capillaries and in the larger pulmonary vessels due to plasma skimming. As a result, this method of measurement can result in a halving of the true value of pulmonary capillary blood volume (Brudin et al., 1986; Dawson et al., 1988).

9.3 Determinants of Pulmonary Blood Volume

Pulmonary blood volumes shift during the respiratory cycle and the cardiac cycle. The alveolar vessel blood volume decreases and extra-alveolar vessel volumes increase during inspiration, with the reverse occurring during expiration (Tournoux et al., 2011; Glazier et al., 1969; Mead and Whittenberger, 1964; Milic-Emili et al., 1966; Glenny et al., 1991; Sackner and DuBois, 1964). During inspiration, the expansion of the lungs increases pulmonary blood volume, which transiently decreases the flow of blood from the lungs to the left side of the heart. In contrast, during expiration, lung deflation causes flow to increase from the lungs to the left atrium. In addition, during cardiac systole, ejection of blood from right ventricle is more pulsatile compared to the return of blood to the left atria (Carlsson et al., 2004). Therefore, there is a transitory increase in the pulmonary blood volume during systole (Nitzan et al., 1994).

There are a number of other factors that also influence the overall blood volume in the lung. For example, simple mechanical changes in the pulmonary circulation can have dramatic effects on pulmonary blood volume. Such changes include elevating pulmonary arterial pressure, which increases pulmonary blood volume (Permutt et al., 1962; Julien et al., 1985; Perasolo and Heisanken, 1971; Piiper, 1970; Dawson et al., 1988; Friedman and Wanner, 1977; Singhal et al., 1973; Huang et al., 1996); elevating lung outflow pressures by either raising left atrial or pulmonary venous pressures also increases pulmonary blood volume (Zhuang et al., 1983; Horsfield, 1978; Horsfield and Gordon, 1981; Engelberg and DuBois, 1959), and simple increments in pulmonary blood flow have also been reported to increase pulmonary blood volume (Maseri et al., 1972; Tsang et al., 1986).

The most critical element in these hemodynamically induced changes is the zonal conditions at the time these observations are made. It would appear that an increase in vascular outflow pressure will increase blood volume in all of the vascular segments if the lung is in zone III conditions. If, however, the lung is in zone II conditions, then the increase in volume seems to be confined to the vascular segment that is downstream from the Starling resistor (Brower et al., 1986). Importantly, raising venous or outflow pressures in zone II conditions may raise pulmonary blood volume without the coincident increase in pulmonary arterial pressure, as is observed if this same maneuver is performed when the lung is in zone III conditions (Singhal et al., 1973; Horsfield, 1978; Engelberg and DuBois, 1959).

Studies on the effects of inflation on pulmonary blood volume have demonstrated that inflation of the lung over a wide range of alveolar pressures increases overall pulmonary blood volume (Wang et al., 1985; Raj et al., 1989; Horsfield, 1978; Roughton, 1945). The segment, or compartment, of the vasculature that accommodates this increased blood volume, however, is somewhat controversial. Some have suggested that the volume increase occurs in the pulmonary arteries (Quebbeman and Dawson, 1977), while other studies indicate that this occurs in the alveolar vessel compartment (Zhuang et al., 1983; Horsfield, 1978; Engelberg and DuBois, 1959).

These results, like the results with outflow pressures, seem to depend on the zonal conditions of the lung during the study. If the lung is in zone III conditions, inflation works to decrease, rather than increase, pulmonary blood volume This latter point is of consequence, for when one reviews the other conditions that have an impact on pulmonary blood volume, it is clear that any condition that changes the distensibility or compliance of the pulmonary vasculature will also have an appropriate effect on the vascular blood volume. Thus, pulmonary blood volume changes can be related to the ability of these conditions and/or stimuli to change the stiffness or compliance of part or all of the pulmonary vasculature. This alteration in vascular elasticity will, in turn, change the capacity of the pulmonary vasculature to accommodate changes in blood volume. This situation is collectively referred to as the capacitance or compliance of the pulmonary vasculature and can be defined as the change in volume per change in the pressure in the system.

10. EFFECT OF VASCULAR COMPLIANCE ON PULMONARY BLOOD FLOW

The most important vascular feature of the pulmonary circulation is the ability to sustain low pressure levels, even in the presence of high flow. Therefore, as mentioned earlier, pulmonary vessels constitute a low-resistance/high-compliance system. Compliance of pulmonary vasculature is the change in the blood volume for a give change in the pressure and can be denoted by the equation:

$$C = \Delta V / \Delta P \quad \text{or the equation can be rewritten as} \quad \Delta V = C \times \Delta P \tag{5}$$

where C is the compliance, ΔV is the change in pulmonary blood volume, and ΔP is the change in pressure. Equation (5) indicates that the change in the volume contained within a vascular segment will be great in a vessel with high compliance and a large change in intravascular pressure. Clinically, compliance can be estimated by determining the ratio of stroke volume and pulse pressure. A decrease in vascular compliance is as important a hemodynamic feature as an increase in PVR during the early phase of pulmonary dysfunction seen in pulmonary hypertension (Lankhaar et al., 2006; Barman et al., 1989). In recent years, however, the use of magnetic resonance has gained favor in assessing pulmonary vascular stiffness and predict outcome of disease (Ben-Yehuda and Barnett, 2009). A vast number of studies have been concerned with the location of the vascular segment that is responsible for the compliance of the pulmonary vasculature. The vast majority of these studies have attributed this compliance to the same vascular segment, whether it is described as the pulmonary capillaries (Smiseth et al., 1988; Horsfield, 1978; Roughton and Forster, 1957), the alveolar vessels

(Permutt et al., 1962; Engelberg and DuBois, 1959; Roughton, 1945), or the pulmonary microvasculature (Morgan et al., 1966b; Bhattacharya et al., 1989; Zhuang et al., 1983; Lewis et al., 1958; Bates et al., 1960). A significant change in compliance is observed when the pulmonary circulation is exposed to a number of vasoactive agents, such as histamine, 5HT, acetylcholine, and norepinephrine (Linehan et al., 1982; Rippe et al., 1987; Barman et al., 1989; Barman and Taylor, 1989). However, it is still unclear whether these agents change compliance directly by constricting the compliance vessels, or indirectly by constricting vessel segments that are upstream or downstream of the compliant segment.

The understanding of how the compliance of the pulmonary microvasculature relates to the anatomy of and the blood flow through this segment has evolved from a number of studies on the elasticity of this microvasculature. Initial notions of the alveolar capillary segment were portrayed as a network of tubes surrounding the alveoli. Unfortunately, this correlate does not entirely account for the observations of pressures, flows, and compliances in the lung. A conceptual break in this common representation of the pulmonary microvasculature occurred in 1969 with the development of the theory of sheet flow in the alveolar microvasculature (Fung and Sobin, 1969). In its essence, this theory suggested (and the morphological studies supported) that the pulmonary microvasculature consisted of two thin sheets of endothelium through which the blood flowed. The only interruptions in this continuous sheet were posts of endothelial tissue that connected the membranes together (Sobin et al., 1970). This configuration would then give the vasculature a distensibility that was only limited by the large number of posts holding the two layers together and the prevailing alveolar pressures on either side of the membrane (Sobin et al., 1972). Within these limitations, the net transmural pressure, or the difference between the blood pressure and the alveolar pressure, will determine the deflection of the alveolar-capillary membrane (Fung, 1972). For example, when the blood pressure exceeds the alveolar pressure, the membranes deflect outward and impinge on the alveolar space. On the other hand, when the alveolar pressure increases as the lung inflates, these membranes stiffen or become taut and can limit blood flow. Thus, the elasticity is determined by the arrangement of posts, membranes, and the prevailing transmural pressure. Furthermore, this arrangement provides for the response to changes in transmural pressure to be a highly localized event (Fung, 1972).

This concept has been extensively mathematically modeled, and the resulting equations have been solved using a number of studies in the literature, which have been referred to in the section on the effects of inflation on the vasculature. The results of these analyses indicate that there is reasonable agreement between this sheet-flow theory, morphological correlates, and the experimental data describing the relation between pressures and flows in the lung (Zhuang et al., 1983; Fung and Sobin, 1972a,b).

11. VESSEL RECRUITMENT VERSUS DISTENSION

Pulmonary vasculature maintains low PVR under high flow and pressure conditions. There are two major mechanisms responsible for this phenomenon. Initially, as PAP increases from low levels, some capillaries which were earlier closed, or open with no blood flow, start to conduct blood, thus lowering the overall resistance. This is termed *recruitment* and is the major mechanism responsible for the drop in PVR that occurs with an initial rise in PAP. However, upon significant increase in PAP, widening of individual capillary segments occur. This increase in caliber is defined as distension. The distension of capillaries is possible due to the presence of a very thin membrane separating the capillary from the alveolar space (Glazier et al., 1969; Fung and Sobin, 1969; Fung, 1972). Even though distension is the predominant mechanism responsible for the fall in PVR at relatively high vascular pressures, recruitment may also participate in the process. Many of the ideas as to how the pulmonary vasculature adapts to changes in blood flows, pressures, and volumes have their origins in studies in which measurements of the carbon monoxide diffusing capacity were made during rest and during exercise (Quebbeman and Dawson, 1977; Lankhaar et al., 2006; Barman et al., 1989; Ben-Yehuda and Barnett, 2009; Barman and Taylor, 1989; Fung and Sobin, 1969). Essentially, these studies demonstrated that as the flow through the pulmonary circulation increased during exercise, so did the diffusing capacity for carbon monoxide. This finding suggested that a larger circulatory bed was being perfused and that the recruitment of previously closed vessels had occurred (Permutt et al., 1969; Wagner and Latham, 1975; West et al., 1975; Wagner et al., 1979; Capen et al., 1981, 1987; Capen and Wagner, 1982; Hanson et al., 1989).

The recruitment and the distension of pulmonary vessels upon increased PAP (Permutt et al., 1969) suggest that, firstly, the pulmonary microvasculature is highly distensible under normal circumstances (Fung and Sobin, 1972a; Giuntini et al., 1966; Seibold et al., 1988). Second, as the result of this distensibility, the pulmonary microvasculature, or capillary segment, is the locus of such changes in flow and/or volume (Glazier et al., 1969; Wagner and Latham, 1975; West et al., 1975; Capen et al., 1981, 1987; Capen and Wagner, 1982; Hanson et al., 1989; Warrell et al., 1972). This finding correlates well with studies that indicate that the compliance or capacitance of the pulmonary vasculature is located in this same vascular segment. Third, the zone in which the lung is examined will determine whether the flow or volume increases will be accommodated by recruitment or by distension (Glazier et al., 1969; Warrell et al., 1972). For example, if the lung is in zone III conditions, then the open capillaries will distend as flow or volume increases. When the lung is in zone II or zone I conditions, then the increased flow or volume will fully open partially perfused vessels and begin to open closed vessels. This zonal dependence of recruitment/distension mirrors the zonal dependence of the blood volume changes already described.

An important criterion for the recruitment of microvessels is the maturity of the lungs. In species with large-enough lungs to develop hydrostatic perfusion zones, increased PAP redistributes the blood flow from the lower to the upper lung regions. This increase in blood flow in the upper zones recruits capillaries and increases gas exchange surface area. However, Means et al. demonstrated that, as compared to older lambs, capillary recruitment does not occur in newborn lambs with immature lungs when PAP is increased (Means et al., 1993). To determine how rapidly pulmonary capillaries are recruited after sudden changes in blood flow, Jaryszak et al. used an isolated canine lung lobe perfused by two pumps running in parallel. When one pump was turned off, flow was rapidly halved; when it was turned on again, flow immediately doubled. They found that by doubling the flow, the time for capillary recruitment decreased by half, suggesting that the pulmonary microcirculation responds rapidly to changes in flow (Jaryszak et al., 2000). Apart from keeping the PVR low, the recruitment of capillaries also increases the transit time for the blood in these vessels, allowing for a larger fraction of gas exchange. In the absence of recruitment or distension, the capillary transit time will decrease rapidly as pulmonary blood flow or pressure increases, resulting in ventilation perfusion mismatch. Indeed, Presson et al. found that in small animal species with limited reserves of recruitable capillaries, like rats, the transit time decreased with an increase in blood flow (Presson et al., 1997). This would inhibit the ability of these animals with high resting metabolic rates to raise oxygen consumption.

To summarize, it is difficult to ascribe increases in pulmonary vascular volume solely to the recruitment or distension of segments of the pulmonary vasculature. These volume changes most likely depend on both mechanisms for the final result, and thus to attribute such changes to either recruitment or distensibility may not be correct. Because such clear-cut assignments are difficult, the term *capacitance* can be used to encompass both mechanisms without attention to the proportion to which they occur (Shoukas, 1975). This, then, is a more functional description of the ability of the pulmonary vasculature to increase circulatory size, thus availing more gas-exchanging surfaces in an attempt to maintain a relatively constant state of respiratory exchange.

12. PRESSURE–FLOW RELATIONSHIPS

The understanding of the relation between pulmonary blood pressure and blood flow, as described by pressure/flow curves, is an important aspect of pulmonary hemodynamics. The pressure–flow relationship in human pulmonary circulation has previously been explored by various techniques to alter pulmonary blood flow such as exercise, balloon occlusion, or crystalloid infusion (Fowler, 1969). The simple passive relationship between a given flow rate and the resulting pressure has been presented graphically by numerous investigators; diagrammatic examples of this relationship are presented in Figure 4(A) and (B). Although these figures are arbitrarily plotted with the flow (Q) on the abscissa and the pressure (P) on the ordinate, other studies have chosen to represent these variables on the opposite axes. Despite this variable preference, most studies have indicated that there are two components to the passive pressure/flow relation in the lung (Figure 4(A)): a curvilinear portion, which is observed at low-flow conditions, and a linear portion, which is observed over much of the range of relatively higher flows (West et al., 1964; West and Dollery, 1965; Younes et al., 1987; Hyman, 1969; Mitzner and Sylvester, 1981; Graham et al., 1982; Guntheroth et al., 1982; Moffatt et al., 1982; Sylvester

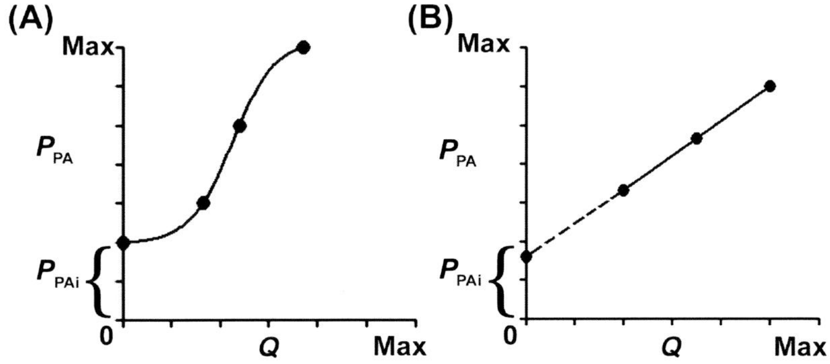

FIGURE 4 Representative pressure/flow curves in the lung. (A) Complete pressure/flow curve in an isolated-perfused lung in which zero-flow conditions can be studied. (B) In vivo model in which zero- and low-flow conditions cannot be achieved. These conditions are determined by extrapolating the linear portion of the curve. In each case, P_{PA} = pulmonary arterial pressure, Q = flow, and P_{Ai} = pulmonary arterial pressure at zero flow or the critical closing pressure. P_{Ai} is used to prevent confusion with interstitial pressure, which is at times, represented as P_i.

et al., 1983; Hakim et al., 1985; SooHoo et al., 1987). It should be pointed out that this type of curve, which presents conditions over a wide range of flows, including zero flow, is generally derived from highly controlled isolated perfused lung preparations in which such flow conditions can be achieved. In intact preparations, where zero flow and other low-flow conditions are not obtainable, the pressure/flow relation is generally presented as a discontinuous curve from which low- and zero-flow conditions are extrapolated from the linear portion the curve (Mitzner and Sylvester, 1981; Sylvester et al., 1983) (Figure 4(B)).

12.1 Curve Shape

The overall shape of a complete pressure/flow curve is dependent upon a number of interrelated conditions, which have already been discussed in a different context. These include alveolar pressures, pulmonary venous pressures, interstitial pressures, the structural integrity of the vasculature, and the vascular tone. As a result, a detailed and comprehensive definition of pressure/flow relations in gravity-dependent, independent, and interzonal regions of the lung would be very complicated and would be dependent upon a myriad of conditions too numerous to measure and to relate to one another. Therefore, a whole-lung pressure/flow curve, as represented in Figure 4(A), can be viewed as being a simplistic composite description of all of the zonal conditions as the flow increases from the zero-flow condition.

The most notable observation at zero flow is that the pressure in the lung is not zero. Such a pressure, termed the *critical closing pressure* (represented in Figure 4 as P_{Ai}), has been defined in peripheral vascular beds as the positive transmural pressure at which the vessel will close and flow will cease (Burton, 1951; Burton and Yamada, 1951; Nichol et al., 1951; Mitzner, 1974). As already noted, critical closing pressures have been described in the lung (Permutt et al., 1962; West and Dollery, 1965; Permutt and Riley, 1963); however, the vessel segment(s) that are critical in this regard, and the contribution of the forces that determine this pressure in the lung, are still a matter of some debate (Graham et al., 1982; Sylvester et al., 1983; Hakim et al., 1985; SooHoo et al., 1987). However, studies in rats have shown that under hypoxic conditions, the increase in critical closing pressure in isolated perfused lungs is attributed to remodeling of arteries less than 50 μm (Jeffery and Wanstall, 1999), suggesting an important role of microvasculature in maintaining this pressure. Despite the controversy, the value of the critical closing pressure in the pulmonary vasculature probably depends on a number of conditions, including alveolar pressure, vascular tone, and to some extent, vascular geometry.

The evidence for this suggestion, although not complete, can be found in previous studies that have specifically examined this issue. For example, studies indicate that alveolar pressure is an important element of the critical closing pressure (Graham et al., 1982; Sylvester et al., 1983) and is, therefore, dependent upon the zonal conditions at the time of the observation (Hakim et al., 1985; SooHoo et al., 1987; Ducas et al., 1988). With regard to vascular tone, the pressor response of acute hypoxia increased the critical closing pressure in the perfused lung (Burton, 1951; Mitzner and Sylvester, 1981), whereas hyperoxia did not affect the closing pressure (Emery et al., 1981). The effects of chronic hypoxia parallels those observed with acute hypoxia in that the critical closing pressures were elevated in this model as well (Emery et al., 1981; Bee and Wach, 1984). These changes in the chronically hypoxic lung have been attributed to both the anatomical narrowing of the vessels, as well as a decrease in vessel compliance (Emery et al., 1981; Bee and Wach, 1984), both of which are hallmarks of this model (Hislop and Reid, 1976; Barer et al., 1982). There have been similar findings with other structural changes in the vasculature as well. Emphysema, for example, tends to increase critical closing pressures (Mink et al., 1985). Additionally, a decrease in carotid sinus pressures augments pulmonary critical closing pressure, suggesting that the level of innervation in the vasculature may play a role as well (Shoukas et al., 1984). Lastly, vasodilators such as isoproterenol and hydralazine both decrease the critical closure of vessels in the lung (Ducas et al., 1987, 1986).

Once critical pressures are exceeded and flow commences, the initial segment of the pressure/flow curve becomes slightly curvilinear, which is then followed by the rectilinear portion of this relationship (Mitzner, 1983). The curvilinear "toe" of this curve is most likely reflective of a combination of events, such as the recruitment of underperfused vessels, the distention of vessels that already have blood flow, and possibly the pattern of filling or recruitment of the capillary bed itself. However, despite the fact that events occurring in the pulmonary capillary bed are also likely contributors to the overall shaping of the pressure/flow relation in the lung, the details of these contributions are not clear cut. For example, throughout the microcirculation, a number of closely adjacent vessel segments may have a wide variety of structural and functional characteristics that make their response to increased pressures and/or flows unpredictable (West et al., 1975). Unlike the larger vessels that are either upstream or downstream to this vascular segment, filling of these vessels may depend, in large measure, on conditions within this vessel segment, rather than on the vessels supplying and/or draining them (Warrell et al., 1972). It stands to reason that as the pressure rises in this segment, more vessels will have blood flowing through them. However, whether this results from recruitment and distention in the standard sense, or from the evening out of flow characteristics through a complex arrangement of alveolar capillary sheets (Fung and Yen, 1986), remains to be determined.

12.2 Determinants of Curve Placement

Once the vasculature is filled to capacity, any additional increments in flow produce step increases in pressure, thus producing the linear portion of this relationship. A number of conditions influence the placement of this portion of the curve relative to the pressure and flow axes. In general, any stimulus that contracts a given segment of the pulmonary vasculature, or that stiffens the vessel wall, will predictably shift the pressure/flow curve up and to the left in a parallel fashion; that is, for a given flow rate, the resulting pressure under such conditions will be higher. Furthermore, any agent that dilates the vasculature or reduces the stiffness of the vascular wall will similarly shift the curve downward.

The most notable of these examples is alveolar pressure, for increases in this pressure in excess of the prevailing venous pressure (zone II conditions) will increase the pressure/flow relation in an amount that is proportional to the change in alveolar pressure (Permutt et al., 1962; Graham et al., 1982; Sylvester et al., 1983; Hakim et al., 1985; Mitzner, 1983; DeBono and Caro, 1963; Lopez-Muniz et al., 1968; Janicki et al., 1985). Another passive influence on the pressure/flow relation is left atrial pressure. Studies have shown that fixed elevations in left atrial or pulmonary venous pressures will shift the pressure/flow relation upward in a parallel fashion (Moffatt et al., 1982; Ducas et al., 1988; Sylvester et al., 1980). A number of acute changes in pulmonary vascular tone will also produce shifts in the pressure/flow relation in the lung. The acute hypoxic pressor response has been shown to shift the relation upwards (Burton, 1951; Mitzner and Sylvester, 1981; Sylvester et al., 1983; Furman et al., 1982; Kennedy and Summer, 1982; Graham et al., 1983; Kennedy et al., 1984; Gibson et al., 1988; Lejeune et al., 1988), as does chronic hypoxia (Emery et al., 1981). Other vasoconstrictors, such as dopamine and dobutamine, demonstrated the same pattern, and this effect was blocked with phentolamine (Murray et al., 1986). In contrast, the vasodilators such as isoproterenol and hydralazine shift the pressure/flow relation in the opposite direction, downwards and to the right (Ducas et al., 1987, 1986).

Another influential factor, which most likely plays an underlying role in shaping the pulmonary vascular pressure/flow curve, is the level of neurogenic tone. This is suggested by studies that have indicated that stimulation of the carotid baroreceptor reflex shifts the pressure/flow curve up and to the left (Shoukas et al., 1984). Additionally, total autonomic ganglionic blockade with hexamethonium chloride and complete alpha and beta sympathetic blockade both increase this relation in a parallel fashion (Selkurt, 1985). This latter observation with sympathetic blockade was further examined, and it was found that beta-adrenergic blockade alone with propranolol shifted this relation upwards, whereas alpha-adrenergic blockade with phentolamine had the opposite effect (Selkurt, 1985). Lastly, cholinergic blockade with atropine was without dramatic effects on the pressure/flow relation in this study (Selkurt, 1985). Although other studies do not confirm the role of the autonomic nervous system in the pressure/flow relation (Wang et al., 1985), the differences in these studies suggest that such a role may depend on not only the presence or absence of a particular type of innervation but also the level of activity of that system as well.

Although it is difficult to discriminate between the direct effects of neurogenic tone and the role of circulating agonists on the pulmonary vascular pressure/flow relationship, data such as these are provocative in that they indicate that the normal hemodynamics in the lung may be a delicate balance between a number of the aforementioned determinants of pressures and flows. When any one or a number of these factors are altered, abnormal pulmonary hemodynamics can result. Indeed, a variety of pulmonary disease entities have that precise effect on the pressure/flow relationships in the lung. For example, patients with pulmonary vascular disease (Sylvester et al., 1980), emphysema (Mink et al., 1985), pulmonary edema (Michael et al., 1984; Boiteau et al., 1986), cystic fibrosis (Seibold et al., 1988), chronic obstructive pulmonary disease (Kloss et al., 1986), and severe hemorrhagic shock with pulmonary histologic lesions (Sarnoff and Berglund, 1952) all demonstrate pressure/flow curves that are shifted upwards and to the left. The mechanisms that underlie these observations are probably similar in nature to the experimental models already described. These would include mechanical limitation of the vascular bed, changes in the geometry of the blood vessels, and the probable superimposition of the effects of acute hypoxia as well.

12.3 Hysteresis

One last point in reference to the dynamics of the pressure/flow and pressure/volume relations in the pulmonary vasculature is that these relations seem to demonstrate hysteresis. Hysteresis is defined as the failure of one of two related phenomena to keep pace with the other. For instance, the relationship between the lung pressure—volume curves (p—v) is nonlinear, such that the ascending part of this curve (inspiration) does not correspond to the descending part of the curve (expiration) (Yuan et al., 2000). Additionally, the lung volume at the end of the expiration does not coincide with lung volume determined during inspiration. This anomaly in the p—v curve is attributed to the dissipation of energy form the lung: part of the energy generated during inspiration is not fully recovered during expiration. Similar hysteresis between the pressure/flow curves was demonstrated in dogs and cats early on; at the time, it was attributed to the "stress relaxation" or stress history of the vessel walls during the decreasing flow limb of the pulmonary blood pressure/flow cycle (Fung and Sobin, 1972a; Frank et al., 1959; Beck and Lai-Fook, 1983). This finding was not a universal one because other studies did not demonstrate such hysteresis (Maseri et al., 1972; Rosenzweig et al., 1970). The differences between these studies seem to be as much a result of technique as they are physiology. For example, hysteresis is abolished if, as one raises the flow, the pressure is allowed to stabilize for up to 5 min (Maseri et al., 1972) or alternately, if one studies lower ranges of pressures and flows (Rosenzweig et al., 1970). Another technique that abolishes hysteresis is to "cycle" or repeat the maneuver a number of times (Barer and Nusser, 1957b; Smith and Mitzner, 1980). In contrast, one can amplify pulmonary vascular hysteresis by increasing the time spent at either low- or zero-flow conditions (Smith and Mitzner, 1980).

The mechanisms proposed to underlie these observations are many and varied. There are indications that the initial zero- or low-flow conditions at the beginning of the pressure/flow cycle are important. This is a likely element, since it has long been suggested that increases in vascular resistance occur at low- or zero-flow conditions and that this change continues once normal flows are achieved (Mead et al., 1970). Studies have suggested a mechanism for this in that as the vasculature collapses at zero flow, the endothelial cells lining the vessel lumen seem to adhere to each other (Fung and Yen, 1986). This adds to the critical, vascular opening pressure as the flow commences and probably contributes to hysteresis. It would then be this aspect of hysteresis that the cycling maneuver would be most effective in preventing. In an important study, Schrijen et al. demonstrated the pressure/flow relationship in the peripheral pulmonary vascular bed of anaesthetized supine dogs using a specially designed double distal lumen Swan-Ganz catheter. They used one lumen for the pressure measurement and the other as a saline pump with variable flow from 0.1 to 1.0 ml/s. During infusion of saline through the catheter, the pulmonary arterial and balloon occlusion wedge pressures did not change. The pressure/flow curve of the occluded vascular bed was similar in shape as that of collapsible tubes with a pressure plateau at high flow. However, interestingly, the pressure/flow curve exhibited hysteresis when flow was decreased (Schrijen et al., 1992).

Another likely contributor to vascular hysteresis is the interdependence between the vessel walls and the parenchymal tissues of the rest of the lung (Barer and Nusser, 1957b; Lambert and Wilson, 1973; Benjamin et al., 1974; Lai-Fook, 1979). Simply stated, as the lung inflates not only will the intravascular pressures change but the vascular pressure/flow relationships can change as well. Thus, the observed hysteresis may result from the parenchymal stresses and strains as they are differentially transmitted to the vessel walls. These forces would apply during both inflation and deflation, as well as during either the increasing or decreasing limb of the pressure/flow maneuver. Although studies have shown that vascular geometry changes as the lung inflates (Macklin, 1946; Howell et al., 1961; Permutt et al., 1961), whether these result from the homogeneous (Barer and Nusser, 1957b) or nonhomogeneous (Hales, 1733; Hoppin et al., 1975; Lai-Fook et al., 1976) application of parenchymal forces to the wall of the vasculature, and/or whether these forces are equally applied during inflation and deflation in a closed-chested model, remains to be defined. As such, whether these forces have an impact on the hysteresis of the pressure/flow relationships also remains to be determined.

13. MODELING OF THE PULMONARY CIRCULATION

13.1 Electrical Analogues

The pulmonary vasculature has been historically modeled from both anatomical and mathematical perspectives, and many of these models have provided substantial insight into the fundamental biology of this vasculature (Wiebel, 1973; Collins and Maccarro, 1979; Rippe et al., 1987; Singhal et al., 1973; Zhuang et al., 1983; Horsfield, 1978; Horsfield and Gordon, 1981; Cumming et al., 1969; Fung and Sobin, 1969, 1972; Sobin et al., 1970, 1972; Fung, 1972; West et al., 1975; Shoukas, 1975; Guntheroth et al., 1982; Grimm et al., 1977). However, another useful modeling system is increasingly being used to describe pulmonary hemodynamics. This genre groups together the electrical analogue models of the pulmonary circulation, and they have the advantage of allowing for the ready study and prediction of hemodynamic outcomes as steady-state conditions are changed (Milnor, 1989). Furthermore, such modeling provides a simple framework into which experimental results can be placed and then compared.

The most elementary mathematical model describes the relationship between pulmonary blood pressure (P), pulmonary blood flow (Q), and PVR (R):

$$R = P/Q \tag{6}$$

Thus, the calculation of vascular resistance can, in its simplest form, be determined by dividing the pressure in the system by the prevailing flow. One can rewrite the Eqn (6) in terms of the Poiseuille-Hagan formula, as presented earlier:

$$\begin{aligned} R &= P/Q = (P_{PA} - P_{PV})/Q \\ &= 8 \times \text{vessel length } (L) \times \text{blood viscosity } (\eta)/\pi \\ &\quad \times \text{radius } (r)^4 \end{aligned} \tag{7}$$

This formulation is analogous to electrical resistance (R), which is calculated by dividing voltage (E) by current (I). This electrical counterpart of Eqn (7) is known as Ohm's law and is represented as:

$$R = E/I \tag{8}$$

where the voltage (E) is analogous to the pressure (P), and the current (I) is analogous to the blood flow (Q).

13.2 Vascular Capacitance

The analogies do not end here, for the concept of vascular compliance, or capacitance, has an electrical analogue as well. Capacitance is the volume of blood contained in a vessel for a given transmural pressure and is calculated as

$$\text{Capacitance} = \text{volume}/\text{transmural pressure mmHg} \tag{9}$$

Indeed, the capacitance of the vessel wall can be considered to be the circulatory equivalent of the electrical capacitor, which itself can be defined as a device that stores electrical energy that can be released at a later time. This concept of capacitance in a hemodynamic model was first described by Stephen Hales in his treatise on hemodynamics, entitled "Haemastatics" (Hales, 1733) and is known as the Windkessel effect. Hales attributed the constancy of flow through a circulatory bed to the distension of vessels during systole and to the recoil of these same vessels during diastole. The leveling of flow over the entire cardiac cycle was similar to the maintenance of water pressure in the fire engines by a dome-shaped, air-filled compression chamber (*Windkessel* in German translation). In fact, the pressure and volume that is ejected into the circulation during systole is "stored" by the stretching (elasticity) of the vessel walls, and during diastole, pressure and flow are maintained by the elastic recoil of these same vessel walls. This is in contrast to what is observed with rigid tubes in which flow ceases and pressure drops to nearly zero during similarly pulsatile events. Capacitance is therefore not factored into the Poiseuille-Hagen equation because this equation considers the vessel through which the fluid is flowing to be rigid.

Since Hales' time, this concept has been rethought, remodeled, and applied to the pulmonary circulation (Engelberg and DuBois, 1959; Wasserman et al., 1966). In its purest form, this concept has been presented as an electrical circuit with a resistor and a capacitor arranged in parallel (Figure 5(A)). This model of the pulmonary circulation has been the departure point for a number of similar constructs that have included the more recent findings concerning pulmonary vascular resistances and compliances. These models have been built on the same basic circuit and differ only in the placement and the numbers of components, which themselves reflect the individual experimental results (Figure 5(B) and (C)). The most notable rearrangement of this model has been its representation in a linear format (Figure 5(B)) (Dawson et al., 1982), the separation of vascular resistances into arterial and venous components, and the attribution of compliance to the middle, or capillary, segment of the vasculature (Parker et al., 1983; Dawson et al., 1982). This representation is still simplistic in that each vascular compartment most probably has its own resistance and capacitance characteristics (Linehan et al., 1982). Indeed, as one anatomically and physiologically partitions the vasculature, the elements of this scheme continue to grow (Figure 5(C)), with the larger arteries and veins differing from the smaller vessels, and they, in turn, differing from the capillary segment (Linehan et al., 1982).

13.3 Vascular Impedance

Another term that is used in the description of pulmonary hemodynamics is impedance (Milnor, 1989; Taylor, 1969; McDonald, 1974). Impedance is the impediment to changing flow as opposed to resistance, which is the obstruction to constant flow. Impedance (Z) is denoted by

$$Z = \rho \times \text{wave velocity}/\pi \times (r)^2 \tag{10}$$

where ρ (rho) is blood density, wave velocity is related to arterial elastance, and r is the radius of the vessel. Therefore, impedance is directly proportional to blood density and the elastic properties of the arterial tree and is inversely proportional to the second power of the vessel radius. In electrical parlance, impedance collectively refers to all the factors that oppose the flow of electrical current. In terms of hemodynamics, impedance refers to all the factors that oppose the flow of blood through a given vascular bed and is usually expressed as some ratio of blood pressures to blood flow. Input impedance, Z, is generally measured at the input end of a vascular bed, and specifically relates the amplitude and relative timing of the sinusoidal wave forms of pressure to flow (McDonald, 1974). This measurement is in contrast to the simple calculations of pulmonary vascular resistance, which relate the absolute values of pressure and flow without regard to the characteristics of the two wave forms as seen in Eqn (2). Thus, impedance is a dynamic measure of vascular function and is dependent upon a collage of factors. In the pulmonary circulation, these factors would include all of the structural elements that contribute to the overall vascular resistance, such as vessel length and radius, as well as the physical characteristics of the blood itself, such as the viscosity or hematocrit. Other important contributors to pulmonary vascular impedance include the branching pattern(s) of the vasculature itself, the compliance or distensibility of the vessel walls, and the actual magnitude and frequencies of the pressure pulses themselves (Milnor, 1989; Taylor, 1969; McDonald, 1974; Attinger, 1963; Hopkins et al., 1980). All of these elements

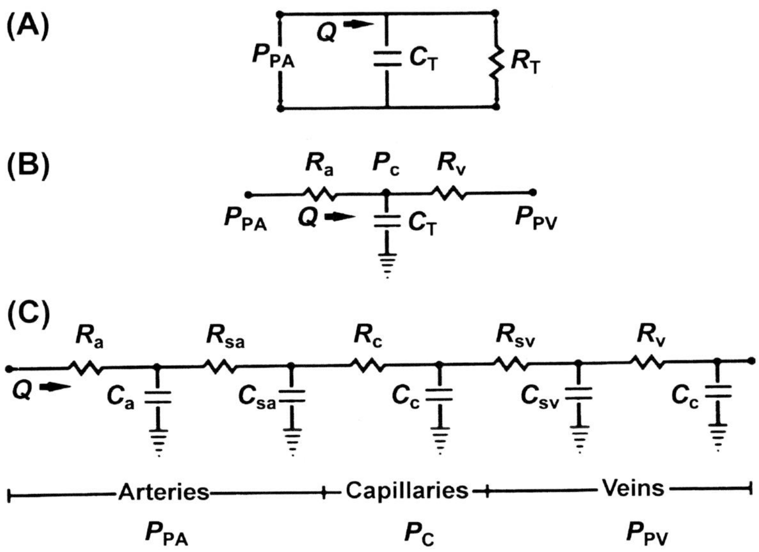

FIGURE 5 Selected electrical analogues of the pulmonary circulation. Respectfully redrawn from (A) Engelberg and DuBois (1959), (B) Dawson et al. (1982), and (C) Linehan et al. (1982). In each case, Q = flow; P = pressure in the pulmonary arteries (P_{PA}), capillaries (P_C), and veins (P_{PV}); R = resistance in the pulmonary arteries (R_a), small arteries (R_{sa}), capillaries (R_c), small veins (R_{sv}), and veins (R_v), and; C = capacitance in the pulmonary arteries (C_a), small arteries (C_{sa}), capillaries (C_c), small veins (C_{sv}), and veins (C_v).

work in concert to dampen the pressure pulses, reflect pressure pulses in a retrograde fashion, and alter the orthograde transmission of these same pulses; in so doing, blood flow is impeded. Another frequently used term, characteristic impedance (Z0), is used to define the same impedance parameters without the effects of the reflected pressure waves on the pressure and flow wave forms. This impedance term is then only a reflection of the geometry of the system and the viscoelastic properties of the vessel wall (McDonald, 1974; Attinger, 1963; Hopkins et al., 1980).

These definitions of impedance are, indeed, fundamental ones and are not meant to overly simplify either the concept or the calculation of these measures. However, impedance determinations have been useful in understanding certain aspects of pulmonary vascular function where pulse pressures and frequencies are important. A number of studies have quantified pulmonary vascular impedance and how that value changes as the vasculature is exposed to a variety of experimental conditions (Table 6) (Murray et al., 1986; Hopkins et al., 1980, 1979; Caro and McDonald, 1961; Caro and Harrison, 1962; Patel et al., 1963; Bergel and Milnor, 1965; Milnor et al., 1969; Elkins and Milnor, 1971; Reuben et al., 1971; Pace et al., 1972; Elkins et al., 1974; Dujardin et al., 1982; Lucas et al., 1983, 1986; Murgo and Westerhof, 1984; Greenwald et al., 1985; Crouch et al., 1987; Tuchscherer et al., 2006; Wauthy et al., 2004). The studies presented in this table represent a chronological sampling of the available data on characteristic impedances in the pulmonary circulation. What is apparent is that there are substantial species differences, in that impedance values in humans average about 22 dyn s/cm^5 (range: 20–29) (Milnor et al., 1969; Murgo and Westerhof, 1984), in dogs average about 165 dyn s/cm^5 (range: 60–279) (Murray et al., 1986; Hopkins et al., 1980, 1979; Patel et al., 1963; Bergel and Milnor, 1965; Elkins and Milnor, 1971; Reuben et al., 1971; Pace et al., 1972; Elkins et al., 1974; Dujardin et al., 1982; Lucas et al., 1983; Crouch et al., 1987), and are dramatically higher in the rabbit and pig (1100 and 1900 dyn s/cm^5, respectively) (Caro and McDonald, 1961; Greenwald et al., 1985). These values seem to indicate that the observed species differences may result from body size being inversely proportional to impedance. However, this conclusion is difficult to make in the absence of substantial pulmonary circulatory data that are normalized for body weight.

These studies also suggest that there are a number of other factors that influence the measured vascular impedance in the lung. Age seems to be important in that impedance values are dramatically low in the neonates (Lipsett and Gannon, 1991; Gilbert et al., 1972; Morin et al., 1988), and this has been variably attributed to either the increased vasculature dimensions (Greenwald et al., 1985) and/or the increased elasticity of the vasculature that occur postnatally and with age (Lucas et al., 1986). It is also clear that the remarkable anatomic rearrangement that is occurring in the neonate is not without a coincident change in impedance, despite the mechanism(s) by which it occurs. Other important conditions that dramatically increase pulmonary vascular impedance are pulmonary hypertension (Hopkins et al., 1980; Caro and McDonald, 1961; Milnor et al., 1969; Elkins et al., 1974), exercise (Elkins and Milnor, 1971), and pneumonectomy (Sobin et al., 1970; Sobin et al., 1972; Fung, 1972). Each of these situations probably induces their effect on impedance by either changing the inherent elasticity of the vasculature or by volume filling of the circulation to the point where the vessels reach their elastic limit. This action is subtly different from the effect of simple increments in flow,

TABLE 6 Pulmonary Vascular Impedence and Vasoactivity

Species	Stimulus or condition	Impedence (dyn s/cm^5)	Impedence Change	References
Rabbit	Norepinephrine, acetylcholine	1600–400	62%, 0	Caro and McDonald (1961)
Man	Pulmonary hypertension		2x	Caro and Harrison (1962)
Dog	Norepinephrine, vagal stimulation, isoproterenol		0	Patel et al. (1963)
Dog	Lung inflation, heart rate, 5-hydroxytryptamine	3094 ± 270	0	Bergel and Milnor (1965)
Man	Mitral stenosis, pulmonary hypertension	23 (range = 20–29)	2x	Milnor et al. (1969)
Dog	Exercise	147	35%	Elkins and Milnor (1971)
Dog	Acute hypoxia and 5HT	150	0, 17%	Reuben et al. (1971)
Dog	Stellate ganglion stimulation	60–70	~80%	Pace et al. (1972)
Dog	Pulmonary hypertension (glass bad embolism)	132	~30%	Elkins et al. (1974)
Dog	Increased flow	200	↓ 50%	Hopkins et al. (1979)
Dog	Acute pulmonary venous hypertension	178 ± 14	0	Hopkins et al. (1980)
	Chronic pulmonary venous hypertension	178 ± 14	361 ± 11	
Dog	Hemorrhage	121 ± 15	36 ± 11%	Dujardin et al. (1982)
	Volume expansion	127 ± 17	~ −18% ± 5%	
Dog	Pneumonectomy	204 ± 76	543 ± 273	Lucas et al. (1983)
Man	Quiet respiration	20 ± 1	0	Murgo and Westerhof (1984)
Pig	4 weeks postpartum (neonate)	1900	360	Greenwald et al. (1985)
Man	<1–7 years	~490–58	~−50%	Lucas et al. (1986)
Dog	Acute hypoxia	204 ± 76	179 ± 77	Murray et al. (1986)
Dog	Pneumonectomy	279	296	Crouch et al. (1987)
Dog	Hypoxia	343 ± 35	446 ± 51	Wauthy et al. (2004)
Goat	Hypoxia	515 ± 63	717 ± 82	Wauthy et al. (2004)
Pig	Hypoxia	558 ± 47	936 ± 81	Wauthy et al. (2004)
Mouse	Embolism	2.1 ± 0.4 mmHg min/ml	~7.5 mmHg min/ml	Tuchscherer et al. (2006)

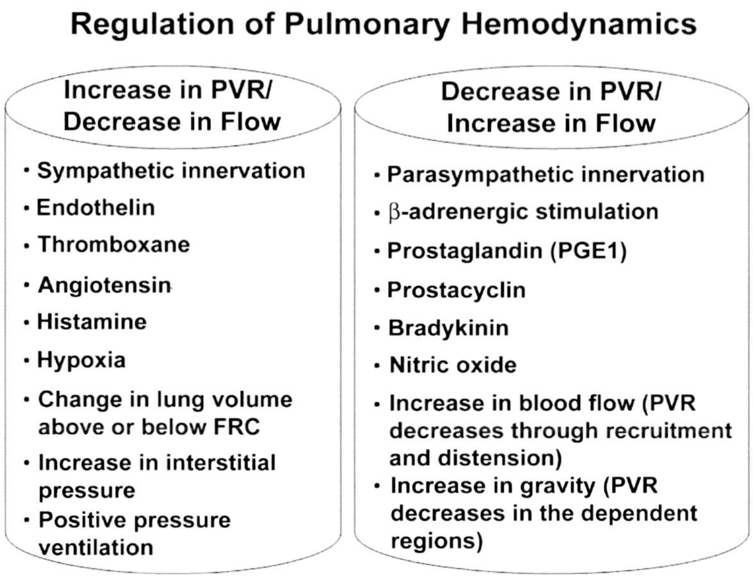

FIGURE 6 The diagrammatic representation of the effects of various biological, chemical, and physical factors influencing pulmonary vascular resistance and pulmonary blood flow. In each case, PVR = peripheral vascular resistance and FRC = functional residual volume.

which tend to decrease measured impedance (Hopkins et al., 1979; Dujardin et al., 1982) by opening previously closed vessels or by fully perfusing underperfused vessels. One would expect, however, that substantial increases in flow would also increase impedance once the vasculature was fully perfused and distended to capacity.

Much less clear are the results with a number of vasoactive agents. One would expect vascular impedance to increase as the pulmonary vasculature is constricted and to decrease as it is dilated. However, the results summarized in Table 6 are somewhat conflicting. These observations may result from the different sites at which these agents have their vasoactive effects or from the fact that the calculation of characteristic impedance, which eliminates the effects of reflective pressure pulses, is not a sensitive measure in this regard (Milnor, 1989). Indeed, many of these same studies report that both overall input impedance and vascular resistance increase with vasoconstriction and decrease upon vasodilation, as would be expected. However, these events are not necessarily reflected in the calculated characteristic impedance values in these studies. Despite these differences in outcome, impedance measures of pulmonary vascular function are useful and may be critical to the understanding of other pulmonary vascular phenomenon. For example, the hypoxic pressure responses in the isolated dog lung are quantitatively smaller with pulsatile flow than with steady flow (Gregory et al., 1982), whereas in the ferret lung no such flow-related differences have been observed (Gregory et al., 1984). Although these data were not analyzed from the perspective of input or characteristic impedances, these are examples of the types of data in which such an analysis may be focal to the interpretation of the findings.

In as much as these models attempt to describe the linear distribution of resistances, capacitances, and impedances of the pulmonary circulation, one has to consider a number of other conditions that will superimpose additional constraints on these models. Specifically, one cannot presume that the actual pulmonary circulation can be represented as a large array of similar linear analogues stacked in parallel. This vision neglects the possible contribution of gravity and the resulting development of Starling, rather than Ohmic, resistance within the different zones of the lung (Rock et al., 1985; Brower et al., 1986). Second, these analogue elements are not equal or equally distributed, and can change as a result of disease-induced anatomical changes, or with changes in baseline tone and reactivity of the vasculature itself (Horsfield and Gordon, 1981; Yilmaz et al., 2010; West et al., 1975; Wagner et al., 1979). As a result, future analogue models should include, in some fashion, variable resistance, capacitance, and impedance elements to be more realistic. Lastly, the strict ascription of these model elements to a given circulatory segment has to be viewed with some caution because both the anatomical and the physiological boundaries of these segments are not completely understood and will undoubtedly be redefined as time goes on.

14. CONCLUDING REMARKS

In conclusion, it is safe to state that across the species, the principles of pulmonary circulation hold up and many functional analogues are conserved. However, it is also evident that the concepts of pulmonary vascular circulation are complex and depend upon the morphological, biophysical, and physiological aspects of the pulmonary vasculature. Equally overwhelming is the description of how these biological constraints relate to each other and to the ability of this vasculature to maintain its normal function. Figure 6 summarizes these concepts and rather provides a brief overview of the fundamentals on which one can base a more detailed understanding of the normal pulmonary circulation, as well as provide some insights into how many pathophysiological conditions alter this normal steady state. Our knowledge is most assuredly incomplete, and much work still needs to be done.

REFERENCES

Anthonisen, N.R., Milic-Emili, J., 1966. Distribution of pulmonary perfusion in erect man. J. Appl. Physiol. 21, 760–766.

Agostoni, E., D'Angelo, E., 1970. Comparative features of the transpulmonary pressure. Respir. Physiol. 11, 76.

Albert, R.K., Lamm, W.J.E., Henderson, W.R., Bolin, R.W., 1989. Effect of leukotrienes B4, C4, and D4 on segmental pulmonary vascular pressures. J. Appl. Physiol. 66, 458.

Alexopoulos, D., Horowitz, S.F., Ambrose, J.A., Machac, J., 1988. Serial pulmonary blood volume changes with supine exercise in normal subjects and in coronary artery disease. Eur. Heart J. 9, 1223.

Alexopoulos, D., Machac, J., Arora, R.R., Horowitz, S.F., 1989. Exercise-induced pulmonary blood volume changes and diastolic dysfunction of the aged heart. Clin. Cardiol. 12, 209.

Attinger, E.O., 1963. Pressure transmission in pulmonary arteries related to frequency and geometry. Circ. Res. 12, 623.

Barer, G.R., Nusser, E., 1957a. Pulmonary blood flow in the cat; the effect of positive pressure respiration. J. Physiol. 138, 103–118.

Baxter, I.G., Cunningham, D.J., Pearce, J.W., 1952. Comparison of cardiac output determinations in the cat by direct Fick and flowmeter methods. J. Physiol. 118, 299–309.

Brudin, L.H., Rhodes, C.G., Valind, S.O., Jones, T., Hughes, J.M., 1994. Interrelationships between regional blood flow, blood volume, and ventilation in supine humans. J. Appl. Physiol. 76, 1205–1210.

Banister, J., Torrance, R.W., 1960. The effects of the tracheal pressure upon flow: pressure relations in the vascular bed of isolated lungs. Q. J. Exp. Physiol. 45, 352.

Burton, A.C., 1951. On the physical equilibrium of small blood vessels. Am. J. Physiol. 164, 319.

Bhattacharya, S., Glucksberg, M.R., Bhattacharya, J., 1989. Measurement of lung microvascular pressure in the intact anesthetized rabbit by the micropuncture technique. Circ. Res. 64, 167.

Bhattacharya, I., Nanjo, S., Staub, N.C., 1982. Micropuncture measurement of lung microvascular pressure during 5HT infusion. J. Appl. Physiol. Respir. Environ. Exerc. Physiol. 52, 634.

Brody, J.S., Stemmler, E.J., DuBois, A.B., 1968. Longitudinal distribution of vascular resistance in the pulmonary arteries, capillaries and veins. J. Clin. Invest. 47, 783.

Bhattacharya, J., Staub, N.C., 1980. Direct measurement of microvascular pressures in the isolated perfused dog lung. Science 210, 327.

Barnard, J.W., Barman, S.A., Adkins, W.K., Longenecker, G.L., Taylor, A.E., 1991. Sustained effects of endothelin-1 on rabbit, dog, and rat pulmonary circulations. Am. J. Physiol. 261, H479–H486.

Bhattacharya, J., Gropper, M.A., Staub, N.C., 1984. Interstitial fluid pressure gradient measured by micropuncture in excised dog lung. J. Appl. Physiol. Respir. Environ. Exerc. Physiol. 56, 271.

Burri, P.H., 1985. Development and growth of the human lung. In: Handbook of Physiology: The Respiratory System. American Physiological Society, Bethesda, MD, p. 1.

Brannon, T.S., North, A.J., Wells, L.B., Shaul, P.W., 1994. Prostacyclin synthesis in ovine pulmonary artery is developmentally regulated by changes in cyclooxygenase-1 gene expression. J. Clin. Invest. 93, 2230–2235.

Bayliss, L.E., 1959. The axial drift of the red cells when blood flows in a narrow tube. J. Physiol. 149, 593.

Bayliss, L.E., 1965. Flow of suspensions of red blood cells in capillary tubes. Changes in the "cell-free" marginal sheath with changes in the shearing stress. J. Physiol. 179, 1.

Bugliarello, G., Hayden, J.W., 1962. High-speed microcinematographic studies of blood in vitro. Science 138, 981.

Barbee, J.H., Cokelet, G.R., 1971. The fahraeus effect. Microvasc. Res. 3, 6.

Brudin, L.H., Valind, S.O., Rhodes, C.G., Turton, D.R., Hughes, J.M.B., 1986. Regional lung hematocrit in humans using positron emission tomography. J. Appl. Physiol. 60, 1155.

Brudin, L.H., Rhodes, C.G., Valind, S.O., Wollmer, P., Hughes, J.M., 1987. Regional lung density and blood volume in non-smoking and smoking subjects measured by PET. J. Appl. Physiol. 63, 1324.

Backmann, R., Hartung, W., 1969. Differentiating measurements of blood volumes in isolated human lungs. Progr. Resp. Res. 5, 327.

Bates, D.V., Varvis, C.J., Donevan, R.E., Cristie, R.V., 1960. Variations in the pulmonary capillary blood volume and membrane diffusion component in health and disease. J. Clin. Invest. 39, 1401.

Brower, R.G., Wise, R.A., Hassapoyannes, C., Hausknecht, M., Permutt, S., 1986. Longitudinal distribution of compliance in the canine lung. J. Appl. Physiol. 61, 240.

Barman, S.A., Senteno, E., Smith, S., Taylor, A.E., 1989. Acetylcholine's effect on vascular resistance and compliance in the pulmonary circulation. J. Appl. Physiol. 67, 1495–1503.

Ben-Yehuda, O., Barnett, C., 2009. Magnetic resonance assessment of pulmonary artery compliance: a promising diagnostic and prognostic tool in pulmonary hypertension? JACC Cardiovasc Imaging 2, 296–298.

Barman, S.A., Taylor, A.E., 1989. Histamine's effect on pulmonary vascular resistance and compliance at elevated tone. Am. J. Physiol. 257, H618.

Burton, A.C., Yamada, S., 1951. Relation between blood pressure and flow in the human forearm. J. Appl. Physiol. 4, 329.

Bee, D., Wach, R.A., 1984. Hypoxic pulmonary vasoconstriction in chronically hypoxic rats. Respir. Physiol. 56, 91.

Barer, G.R., Finlay, M., Bee, D., Wach, R.A., 1982. Pulmonary hypertension of chronic hypoxia in a rat model. Anatomic and rheological factors. Bull. Eur. Physiopathol. Respir. 18, 69.

Boiteau, P., Ducas, J., Schick, U., Girling, L., Prewitt, R.M., 1986. Pulmonary vascular pressure-flow relationship in canine oleic acid pulmonary edema. Am. J. Physiol. 251, H1163.

Beck, K.C., Lai-Fook, S.J., 1983. Effect of height on alveolar liquid pressure in isolated edematous dog lung. J. Appl. Physiol. Respir. Environ. Exerc. Physiol. 54, 619.

Barer, G.R., Nusser, E., 1957b. Pulmonary blood flow in the cat. The effect of positive pressure respiration. J. Physiol. (London) 138, 103.

Benjamin, J.J., Murtagh, P.S., Proctor, D.F., Menkes, H.A., Permutt, S., 1974. Pulmonary vascular interdependence in excised dog lobes. J. Appl. Physiol. 37, 887.

Bergel, D.H., Milnor, W.R., 1965. Pulmonary vascular impedance in the dog. Circ. Res. 16, 401.

Chen, C.W., Wang, C.M., Lee, C.H., Hsiue, T.R., Kuo, Y.L., Chang, H.Y., 2005. Macro- and micropulmonary hemodynamic changes to nitric oxide blockade in monocrotaline-induced pulmonary hypertension in rats—an in vivo approach. J. Formos. Med. Assoc. 104, 301–307.

Chick, T.W., Scotto, P., Icenogle, M.V., Sikes, C.W., Doyle, M.P., Riedel, C.E., Wood, S.C., Loeppky, J.A., 1988. Effects of pentoxifylline on pulmonary hemodynamics during acute hypoxia in anesthetized dogs. Am. Rev. Respir. Dis. 137, 1099–1103.

Chaliki, H.P., Hurrell, D.G., Nishimura, R.A., Reinke, R.A., Appleton, C.P., 2002. Pulmonary venous pressure: relationship to pulmonary artery, pulmonary wedge, and left atrial pressure in normal, lightly sedated dogs. Catheter. Cardiovasc. Interv. 56, 432–438.

Culver, B.H., Butler, J., 1980. Mechanical influences on the pulmonary microcirculation. Annu. Rev. Physiol. 42, 187.

Collins, R., Maccarro, J.A., 1979. Blood flow in the lung. J. Biomech. 12, 373.

Cornfield, D.N., Chatfield, B.A., McQueston, J.A., McMurtry, I.F., Abman, S.H., 1992. Effects of birth-related stimuli on L-arginine-dependent pulmonary vasodilation in ovine fetus. Am. J. Physiol. 262, H1474–H1481.

Crapo, R.O., Morris, A.H., Gardner, R.M., 1982. Reference values for pulmonary tissue volume, membrane diffusing capacity, and pulmonary capillary blood volume. Bull. Eur. Physiopathol. Respir. 18, 893.

Cumming, G., Henderson, R., Horsfield, K., Singhal, S.S., 1969. The functional morphology of the pulmonary circulation. In: The Pulmonary Circulation and Interstitial Space. University of Chicago Press, Chicago, p. 327.

Carlsson, M., Cain, P., Holmqvist, C., Stahlberg, F., Lundback, S., Arheden, H., 2004. Total heart volume variation throughout the cardiac cycle in humans. Am. J. Physiol. Heart Circ. Physiol. 287, H243–H250.

Capen, R.L., Latham, L.P., Wagner Jr, W.W., 1981. Diffusing capacity of the lung during hypoxia: role of capillary recruitment. J. Appl. Physiol. 50, 165.

Capen, R.L., Latham, L.P., Wagner Jr, W.W., 1987. Comparison of direct and indirect measurements of pulmonary capillary transit times. J. Appl. Physiol. 62, 1150.

Capen, R.L., Wagner Jr, W.W., 1982. Intrapulmonary blood flow redistribution during hypoxia increases gas exchange surface area. J. Appl. Physiol. 52, 1575.

Caro, C.G., McDonald, D.A., 1961. The relation of pulsatile pressure and flow in the pulmonary vascular bed. J. Physiol. (London) 157, 426.

Caro, C.G., Harrison, G.K., 1962. Observations on pulse wave velocity and pulsatile blood pressure in the human pulmonary circulation. Clin. Sci. 23, 317.

Crouch, J.D., Lucas, C.L., Keagy, B.A., Wilcox, B.R., Ha, B., 1987. The acute effects of pneumonectomy on pulmonary vascular impedance in the dog. Ann. Thorac. Surg. 43, 613.

Drougas, J.G., Barnard, S.E., Wright, J.K., Sika, M., Lopez, R.R., Stokes, K.A., Williams, P.E., Pinson, C.W., 1996. A model for the extended studies of hepatic hemodynamics and metabolism in swine. Lab. Anim. Sci. 46, 648–655.

Duke, H.N., Stedeford, R.D., 1960. Pulmonary vasomotor responses to epinephrine and norepinephrine in the cat. Influence of the sympathetic nervous system. Circ. Res. 8, 640–648.

Daly, I.D.B., Hebb, C., 1966. Pulmonary and Bronchial Vascular Systems. William and Wilkins, Baltimore, MD.

Dahm, P.L., Blomquist, S., De Robertis, E., Jonson, B., Myhre, E., Svantesson, C., Thorne, J., 2000. Effects of NO inhalation on pulmonary leukocyte sequestration and blood volume in porcine endotoxaemia. Intensive Care Med. 26, 336–343.

Donato, L., Giuntini, C., Lewis, M.L., Durand, J., Rochester, D.F., Harvey, R.M., Cournand, A., 1962. Quantitative radiocardiography. I. Theoretical considerations. Circulation 26, 174.

Dock, D.S., Kraus, W.L., McGuire, L.B., Hyland, J.W., Haynes, F.W., Dexter, L., 1961. The pulmonary blood volume in man. J. Clin. Invest. 40, 317.

Dawson, C.A., Rickaby, D.A., Linehan, J.H., Bronikowski, T.A., 1988. Distributions of vascular volume and compliance in the lung. J. Appl. Physiol. 64, 266.

Ducas, J., Schick, U., Girling, L., Prewitt, R.M., 1988. Effects of altered left atrial pressure on pulmonary vascular pressure-flow relationships. Am. J. Physiol. 255, H19.

Ducas, J., Duval, D., Dasilva, H., Boiteau, P., Prewitt, R.M., 1987. Treatment of canine pulmonary hypertension: effects of norepinephrine and isoproterenol on pulmonary vascular pressure-flow characteristics. Circulation 75, 235.

Ducas, J., Girling, L., Schick, U., Prewitt, R.M., 1986. Pulmonary vascular effects of hydralazine in a canine preparation of pulmonary thromboembolism. Circulation 73, 1050.

DeBono, E.F., Caro, C.G., 1963. Effect of lung-inflating pressure on pulmonary blood pressure and flow. Am. J. Physiol. 208, 1178.

Dawson, C.A., Linehan, J.H., Rickaby, D.A., 1982. Pulmonary microcirculatory hemodynamics. Ann. N. Y. Acad. Sci. 384, 90.

Dujardin, J.P.L., Stone, D.N., Fortino, C.D., Paul, L.T., Pieper, H.P., 1982. Effects of blood volume changes on characteristic impedance of the pulmonary artery. Am. J. Physiol. 242, H197.

Emery, C.J., Fang, L., Laude, E.A., Morcos, S.K., 2001. Effects of radiographic contrast media on pulmonary vascular resistance of normoxic and chronically hypoxic pulmonary hypertensive rats. Br. J. Radiol. 74, 1109–1117.

Engelberg, J., DuBois, A.B., 1959. Mechanics of pulmonary circulation in isolated rabbit lungs. Am. J. Physiol. 196, 401.

Emery, C.J., Bee, D., Barer, G.R., 1981. Mechanical properties and reactivity of vessels in isolated perfused lungs of chronically bypoxic rats. Clin. Sci. 62, 569.

Elkins, R.C., Milnor, W.R., 1971. Pulmonary vascular response to exercise in the dog. Circ. Res. 29, 591.

Elkins, R.C., Peyton, M.D., Greenfield, L.J., 1974. Pulmonary vascular impedance in chronic pulmonary hypertension. Surgery 76, 57.

Fesler, P., Pagnamenta, A., Rondelet, B., Kerbaul, F., Naeije, R., 2006. Effects of sildenafil on hypoxic pulmonary vascular function in dogs. J. Appl. Physiol. 101, 1085–1090.

Franceschi, R.C., Malbouisson, L., Yoshinaga, E., Auler Jr, J.O., Figueiredo, L.F., Carmona, M.J., 2013. Evaluation of hemodynamic effects of xenon in dogs undergoing hemorrhagic shock. Clinics (Sao Paulo) 68, 231–238.

Fishman, A.P., 1985. Pulmonary circulation. In: Handbook of Physiology: The Respiratory System, I. American Physiological Society, Bethesda, MD, p. 93.

Fahraeus, R., 1921. The suspension-stability of the blood. Acta Med. Scand. 55, 158.

Fahraeus, R., Lindqvist, R., 1931. The viscosity of the blood in narrow capillary tubes. Am. J. Physiol. 96, 562.

Friedman, M., Wanner, A., 1977. Volume characteristics of extra- and intraparenchymal segments of the canine pulmonary artery. J. Appl. Physiol. 42, 519.

Fung, Y.C., Sobin, S.S., 1969. Theory of sheet flow in the lung alveoli. J. Appl. Physiol. 26, 472.

Fung, Y.C., Sobin, S.S., 1972a. Pulmonary alveolar blood flow. Circ. Res. 30, 470.

Fung, Y.C., Sobin, S.S., 1972b. Elasticity of the pulmonary alveolar sheet. Circ. Res. 30, 451.

Fung, Y.C.B., 1972. Stress-strain-history relations of soft tissues in simple elongation. In: Biomechanics, its Foundations and Objectives. Prentice-Hall, Englewood Cliffs, NJ, p. 181.

Fowler, N.O., 1969. The normal pulmonary arterial pressure-flow relationships during exercise. Am. J. Med. 47, 1–6.

Fung, Y.C., Yen, R.T., 1986. A new theory of pulmonary blood flow in zone 2 condition. J. Appl. Physiol. 60, 1638.

Furman, W.R., Summer, W.R., Kennedy, T.P., Sylvester, J.T., 1982. Comparison of the effects of dobutamine, dopamine, and isoproterenol on hypoxic pulmonary vasoconstriction in the pig. Crit. Care Med. 10, 371.

Frank, N.R., Radford, E.D., Whittenberg, J.L., 1959. Static volume-pressure interrelationships of the lungs and pulmonary blood vessels in excised cats' lungs. J. Appl. Physiol. 14, 167.

Grant, B.J., Canty Jr, J.M., 1989. Effect of cardiac output on pulmonary hemodynamics. Respir. Physiol. 76, 303–317.

Goetz, T.E., Manohar, M., Magid, J.H., 1999. Repeated administration of frusemide does not offer an advantage over single dosing in attenuating exercise-induced pulmonary hypertension in thoroughbred horses. Equine Vet. J. Suppl. 30, 539–545.

Glenny, R.W., Lamm, W.J., Bernard, S.L., An, D., Chornuk, M., Pool, S.L., Wagner Jr, W.W., Hlastala, M.P., Robertson, H.T., 2000. Selected contribution: redistribution of pulmonary perfusion during weightlessness and increased gravity. J. Appl. Physiol. 89, 1239–1248.

Glenny, R.W., Bernard, S., Robertson, H.T., Hlastala, M.P., 1999. Gravity is an important but secondary determinant of regional pulmonary blood flow in upright primates. J. Appl. Physiol. 86, 623–632.

Glazier, J.B., Hughes, J.M.B., Maloney, J.E., West, J.B., 1969. Measurements of capillary dimensions of blood volume in rapidly frozen lungs. J. Appl. Physiol. 26, 65.

Glenny, R.W., Lamm, W.J., Albert, R.K., Robertson, H.T., 1991. Gravity is a minor determinant of pulmonary blood flow distribution. J. Appl. Physiol. 71, 620–629.

Greenleaf, J.F., Ritman, E.L., Sass, D.J., Wood, E.H., 1974. Spatial distribution of pulmonary blood flow in dog in left decubitus position. Am. J. Physiol. 227, 230.

Gil, J., Weibel, E.R., 1972. Morphological study of pressure-volume hysteresis in rat lungs fixed by vascular perfusion. Respir. Physiol. 15, 190.

Garr, K.A.J., Taylor, A.E., Owens, L.T., Guyton, A.C., 1967. Pulmonary capillary pressure and filtration coefficient in the isolated perfused lung. Am. J. Physiol. 213, 910.

Grimm, D.J., Dawson, C.A., Hakim, T.S., Linehan, J.H., 1978. Pulmonary vasomotion and the distribution of vascular resistance in a dog lung lobe. J. Appl. Physiol. Respir. Environ. Exerc. Physiol. 45, 545.

Gilbert, R.D., Hessler, J.R., Eitzman, D.V., Cassin, S., 1972. Site of pulmonary vascular resistance in fetal goats. J. Appl. Physiol. 32, 47.

Gehr, P., Bachofen, M., Weibel, E.R., 1978. The normal human lung: ultrastructural and morphometric estimation of diffusion capacity. Respir. Physiol. 32, 121.

Giuntini, C., Lewis, M.L., Luis, A.S., Harvey, R.M., 1963. A study of the pulmonary blood volume in man by quantitative radiocardiography. J. Clin. Invest. 42, 1589.

Giuntini, C., Maseri, A., Bianchi, R., 1966. Pulmonary vascular distensibility and lung compliance as modified by dextran infusion and subsequent atropine injection in normal subjects. J. Clin. Invest. 45, 1770.

Graham, R., Skoog, C., Oppenheimer, L., Rabson, J., Goldberg, H.S., 1982. Critical closure in the canine pulmonary vasculature. Circ. Res. 50, 566.

Guntheroth, W.G., Luchtel, D.L., Kawabori, I., 1982. Pulmonary microcirculation: tubules rather than sheet and post. J. Appl. Physiol. Respir. Environ. Exerc. Physiol. 53, 510.

Graham, R., Skoog, C., Macedo, W., Carter, J., Oppenheimer, L., Rabson, J., Goldberg, H.S., 1983. Dopamine, dobutamine, and phentolamine effects on pulmonary vascular mechanics. J. Appl. Physiol. Respir. Environ. Exerc. Physiol. 54, 1277.

Gibson, R.L., Truog, W.E., Redding, G.J., 1988. Hypoxic pulmonary vasoconstriction during and after infusion of group B. Streptococcus in neonatal piglets. Vascular pressure-flow analysis. Am. Rev. Respir. Dis. 137, 774.

Grimm, D.J., Linehan, J.H., Dawson, C.A., 1977. Longitudinal distribution of vascular resistance in the lung. J. Appl. Physiol. Respir. Environ. Exerc. Physiol. 43, 1093.

Greenwald, S.E., Johnson, R.J., Haworth, S.G., 1985. Pulmonary vascular input impedance in newborn and infant pig. Cardiovasc. Res. 18, 44.

Gregory, T.J., Newell, J.C., Hakim, T.S., Levitzky, M.G., Sedransk, N., 1982. Attenuation of hypoxic pulmonary vasoconstriction by pulsatile flow in dog lungs. J. Appl. Physiol. Respir. Environ. Exerc. Physiol. 53, 1583.

Gregory, T.J., Ellsworth, M.L., Newell, J.C., 1984. Hypoxic pulmonary vasoconstriction during steady and pulsatile flow in ferrets. J. Appl. Physiol. Respir. Environ. Exerc. Physiol. 57, 205.

Holm, P., Liska, J., Franco-Cereceda, A., 1998. The ETA receptor antagonist, BMS-182874, reduces acute hypoxic pulmonary hypertension in pigs in vivo. Cardiovasc Res. 37, 765–771.

Hinder, F., Meyer, J., Booke, M., Ehardt, J.S., Salsbury, J.R., Traber, L.D., Traber, D.L., 1998. Endogenous nitric oxide and the pulmonary microvasculature in healthy sheep and during systemic inflammation. Am. J. Respir. Crit. Care Med. 157, 1542–1549.

Hlastala, M.P., Bernard, S.L., Erickson, H.H., Fedde, M.R., Gaughan, E.M., McMurphy, R., Emery, M.J., Polissar, N., Glenny, R.W., 1996. Pulmonary blood flow distribution in standing horses is not dominated by gravity. J. Appl. Physiol. 81, 1051–1061.

Hughes, J.M., 1996. Pulmonary blood flow distribution in exercising and in resting horses. J. Appl. Physiol. 81, 1049–1050.

Hughes, J.M., Glazier, J.B., Maloney, J.E., West, J.B., 1968a. Effect of lung volume on the distribution of pulmonary blood flow in man. Respir. Physiol. 4, 58–72.

Hughes, J.M.B., Glazier, J.B., Maloney, J.E., West, J.B., 1968b. Effect of lung volume on the distribution of pulmonary blood flow in man. Respir. Physiol. 4, 58.

Hakim, T.S., Lisbona, R., Dean, G.W., 1987. Gravity-independent inequality in pulmonary blood flow in humans. J. Appl. Physiol. 63, 1114–1121.

Herenstierna, G., White, F.C., Wagner, P.D., 1979. Spatial distribution of pulmonary blood flow in the dog with PEEP ventilation. J. Appl. Physiol. Respir. Environ. Exerc. Physiol. 47, 938.

Hakim, T.S., Lisbona, R., Dean, G.W., 1989. Effect of cardiac output on gravity-dependent and nondependent inequality in pulmonary blood flow. J. Appl. Physiol. 66, 1570.

Howell, J.B.L., Permutt, S., Proctor, D.F., Riley, R.L., 1961. Effect of inflation on the lung on different parts of pulmonary vascular bed. J. Appl. Physiol. 16, 71.

Hedenstierna, G., Baehrendtz, S., Klingstedt, C., Santesson, J., Soderborg, B., Dahlborn, M., Bindslev, L., 1984. Ventilation and perfusion of each lung during differential ventilation with selective PEEP. Anesthesiology 61, 369–376.

Hughes, J.M.B., Rosenzweig, D.Y., 1970. Factors affecting trapped gas volume in perfused dog lungs. J. Appl. Physiol. 29, 332.

Hakim, T.S., Michel, R.P., Chang, H.K., 1982. Partitioning of pulmonary vascular resistance in dogs by arterial and venous occlusion. J. Appl. Physiol. Respir. Environ. Exerc. Physiol. 52, 710.

Hakim, T.S., Dawson, C.A., Linehan, J.H., 1979. Hemodynamic responses of dog lung lobe to lobar venous occlusion. J. Appl. Physiol. Respir. Environ. Exerc. Physiol. 47, 145.

Hakim, T.S., Kelly, S., 1989. Occlusion pressures vs. micropipette pressures in the pulmonary circulation. J. Appl. Physiol. 67, 1277.

Hughes, J.M., Glazier, J.B., Maloney, J.E., West, J.B., 1967. Effect of interstitial pressure on pulmonary blood-flow. Lancet 1, 192–193.

Hughes, J.M.B., 1971. Pulmonary interstitial pressure. Bull. Eur. Physio Path. Respir. 7, 1095.

Haworth, S.G., Hislop, A.A., 1981. Adaptation of the pulmonary circulation to extra-uterine life in the pig and its relevance to the human infant. Cardiovasc. Res. 15, 108.

Hislop, A., Reid, L., 1973. Pulmonary arterial development during childhood: branching pattern and structure. Thorax 28, 129.

Hislop, A., Reid, L., 1981. Growth and development of the respiratory system-anatomical development. In: Dobbing, J. A. D. a. J. (Ed.), Scientific Foundations of Paediatric. Heinemann, London, p. 390.

Hamilton, W.F., Moore, J.W., Kinsman, J.W., Spurling, R.G., 1928. Simultaneous determination of the greater and lesser circulation times, of the mean velocity of blood flow through the heart and lungs, of the cardiac output and an approximation of the amount of blood actively circulating in the heart and lungs. Am. J. Physiol. 85, 337.

Harris, P., Heath, D., 1986. The Measurement of Blood Volume in the Lungs. Churchill Livingstone, Edinburgh.

Huang, W., Yen, R.T., McLaurine, M., Bledsoe, G., 1996. Morphometry of the human pulmonary vasculature. J. Appl. Physiol. 81, 2123–2133.

Horsfield, K., 1978. Morphometry of small pulmonary arteries in man. Circ. Res. 42, 593.

Horsfield, K., Gordon, W.I., 1981. Morphometry of pulmonary veins in man. Lung 159, 211.

Hanson, W.L., Emhardt, J.D., Bartek, J.P., Latham, L.P., Check-ley, L.L., Capen, R.L., Wagner Jr, W.W., 1989. Site of recruitment in the pulmonary microcirculation. J. Appl. Physiol. 66, 2079.

Hyman, A.L., 1969. Effect of large increases in pulmonary blood flow on pulmonary venous pressure. J. Appl. Physiol. 27, 179.

Hakim, T.S., Chang, H.K., Michel, R.P., 1985. The rectilinear pressure-flow relationship in the pulmonary vasculature: zones 2 and 3. Respir. Physiol. 61, 115.

Hislop, A., Reid, L., 1976. New findings in pulmonary arteries of rats with hypoxia-induced pulmonary hypertension. Br. J. Exp. Pathol. 57, 542.

Hales, S., 1733. Haemastaticks, experiment III, note 26. In: Statistical Essays: Containing Haemastaticks, Library of the New York Academy of Medicine. Hafner Publishing, New York, p. 22.

Hoppin, F.G.J., Lee, G.C., Dawson, S.V., 1975. Properties of lung parenchyma in distortion. J. Appl. Physiol. 39, 742.

Hopkins, R.A., Hammon Jr, J.W., McHale, P.A., Smith, P.K., Anderson, R.W., 1980. An analysis of the pulsatile hemodynamic responses of the pulmonary circulation to acute and chronic pulmonary venous hypertension in the awake dog. Circ. Res. 47, 902.

Hopkins, R.A., Hammon Jr, J.W., McHale, P.A., Smith, P.K., Anderson, R.W., 1979. Pulmonary vascular impedance analysis of adaptation to chronically elevated blood flow in the awake dog. Circ. Res. 45, 267.

Inselman, L.S., Mellins, R.B., 1981. Growth and development of the lung. J. Pediatr. 98, 1.

Johnson, D.H., Chang, P.C., Hurst, T.S., Reynolds, F.B., Lang, S.A., Mayers, I., 1992. Changes in PETCO2 and pulmonary blood flow after bronchial occlusion in dogs. Can. J. Anaesth. 39, 184–191.

Jarvis, K.A., Steffey, E.P., Tyler, W.S., Willits, N., Woliner, M., 1992. Pulmonary blood flow distribution in anesthetized ponies. J. Appl. Physiol. 72, 1173–1178.

Julien, M., Hakim, T.S., Vahi, R., Chang, H.K., 1985. Effect of hematocrit on vascular pressure profile in dogs lungs. J. Appl. Physiol. 58, 743.

Jaryszak, E.M., Baumgartner Jr, W.A., Peterson, A.J., Presson Jr, R.G., Glenny, R.W., Wagner Jr, W.W., 2000. Selected contribution: measuring the response time of pulmonary capillary recruitment to sudden flow changes. J. Appl. Physiol. 89, 1233–1238.

Jeffery, T.K., Wanstall, J.C., 1999. Perindopril, an angiotensin converting enzyme inhibitor, in pulmonary hypertensive rats: comparative effects on pulmonary vascular structure and function. Br. J. Pharmacol. 128, 1407–1418.

Janicki, J.S., Weber, K.T., Likoff, M.J., Fishman, A.P., 1985. The pressure-flow response of the pulmonary circulation in patients with heart failure and pulmonary vascular disease. Circulation 72, 1270.

Krebs, J., Aebli, N., Goss, B.G., Wilson, K., Williams, R., Ferguson, S.J., 2007. Cardiovascular changes after pulmonary cement embolism: an experimental study in sheep. AJNR Am. J. Neuroradiol. 28, 1046–1050.

Keilholz, S.D., Knight-Scott, J., Christopher, J.M., Mai, V.M., Berr, S.S., 2001. Gravity-dependent perfusion of the lung demonstrated with the FAIRER arterial spin tagging method. Magn. Reson Imaging 19, 929–935.

Kosuda, S., Kobayashi, H., Kusano, S., 2000. Change in regional pulmonary perfusion as a result of posture and lung volume assessed using technetium-99m macroaggregated albumin SPET. Eur. J. Nucl. Med. 27, 529–535.

Koyama, S., Lamm, W.J., Hildebrandt, J., Albert, R.K., 1989. Flow characteristics of open vessels in zone 1 rabbit lungs. J. Appl. Physiol. 66, 1817–1823.

Krumpe, P., Gorin, A.B., 1981. Continuous negative external chest pressure decreases transvascular lung water transport in sheep after pseudomonas bacteremia. J. Clin. Invest 67, 264–273.

Kadowitz, P.J., Hyman, A.L., 1984. Analysis of responses to leukotriene D4 in the pulmonary vascular bed. Circ. Res. 55, 707.

Kennedy, T.P., Summer, W., 1982. Inhibition of hypoxic pulmonary vasoconstriction by nifedipine. Am. J. Cardiol. 50, 864.

Kennedy, T.P., Michael, J.R., Huang, C.K., Kaltman, C.H., Zahka, K., Schlott, W., Summer, W., 1984. Nifedipine inhibits hypoxic pulmonary vasoconstriction during rest and exercise in patients with chronic obstructive disease. Am. Rev. Respir. Dis. 129, 544.

Kloss, T., Bruckner, U.B., Leinberger, H., 1986. Pulmonary pressure-flow relation after trauma and hemorrhagic shock. Res. Exp. Med. 186, 325.

Lee, G.D., Dubois, A.B., 1955. Pulmonary capillary blood flow in man. J. Clin. Invest 34, 1380–1390.

Liu, C.P., Zhang, Y.J., Lu, W.X., Ji, Y.Q., Zhang, W.H., Wang, C., 2005. [The change of pulmonary surfactant associated protein A in acute pulmonary embolism]. Zhonghua Jie He He Hu Xi Za Zhi 28, 600–603.

Li, T.Z., Yuan, M., Chen, Z.H., Guo, Y.H., Kang, C.Y., Wang, J.Y., Li, Z.L., Wang de, S., Wang, H.J., Liu, C.T., 2013. Effect of simulated microgravity and its associated mechanism on pulmonary circulation in rats. Biomed. Environ. Sci. 26, 118–127.

Lamm, W.J., Obermiller, T., Hlastala, M.P., Albert, R.K., 1995. Perfusion through vessels open in zone 1 contributes to gas exchange in rabbit lungs in situ. J. Appl. Physiol. 79, 1895–1899.

Lloyd, T.C., Wright, G.W., 1960. Pulmonary vascular resistance and vascular transmural gradient. J. Appl. Physiol. 15, 241.

Linehan, J.H., Dawson, C.A., Rickaby, D.A., 1982. Distribution of vascular resistance and compliance in a dog lung lobe. J. Appl. Physiol. Respir. Environ. Exerc. Physiol. 53, 158.

Lai-Fook, S.J., 1982. Perivascular interstitial fluid pressure measured by micropipettes in isolated dog lung. J. Appl. Physiol. Respir. Environ. Exerc. Physiol. 52, 9.

Lewis, A.B., Heymann, M.A., Rudolph, A.M., 1976. Gestational changes in pulmonary vascular responses in fetal lambs in utero. Circ. Res. 39, 536.

Levin, D.L., Rudolph, A.M., Heymann, M.A., Phibbs, R.H., 1976. Morphological development of the pulmonary vascular bed in fetal lambs. Circulation 53, 144.

Lipsett, J., Hunt, K., Carati, C., Gannon, B., 1989. Changes in the spatial distribution of pulmonary blood flow during the fetal/neonatal transition: an in vivo study in the rabbit. Pediatr. Pulmonol. 6, 212.

Lipsett, J., Gannon, B., 1991. Regional cycles of perfusion and non-perfusion in the lung of the term fetal rabbit. Pediatr. Pulmonol. 11, 153–160.

Lewis, M.L., Giuntini, C., Donato, L., Harvey, R.M., Cournand, A., 1962. Quantitative radiocardiography. III. Results and validation of theory and method. Circulation 26, 189.

Lewis, M.L., Gnoj, J., Fisher, V.J., Christianson, L.C., 1970. Determinants of pulmonary blood volume. J. Clin. Invest. 49, 170.

Lewis, B.M., Tai-Hon, L., Noe, F.E., Komisaruk, R., 1958. The measurement of pulmonary capillary blood volume and pulmonary membrane diffusing capacity in normal subjects; the effects of exercise and position. J. Clin. Invest. 37, 1061.

Lankhaar, J.W., Westerhof, N., Faes, T.J., Marques, K.M., Marcus, J.T., Postmus, P.E., Vonk-Noordegraaf, A., 2006. Quantification of right ventricular afterload in patients with and without pulmonary hypertension. Am. J. Physiol. Heart Circ. Physiol. 291, H1731–H1737.

Lopez-Muniz, R., Stephens, N.L., Bromberger-Barnea, B., Per-mutt, S., Riley, R.L., 1968. Critical closure of pulmonary vessels analyzed in terms of Starling resistor model. J. Appl. Physiol. 24, 625.

Lejeune, P., Deloof, T., Leeman, M., Melot, C., Naeije, R., 1988. Multipoint pulmonary pressure/flow relationships in hypoxic and in normoxic dogs: effects of nitrous oxide with and without cyclooxygenase inhibition. Anesthesiology 68, 92.

Lambert, R.K., Wilson, T.A., 1973. A model for the elastic properties of the lung and their effect on expiratory flow. J. Appl. Physiol. 34, 34.

Lai-Fook, S.J., 1979. A continuum mechanics analysis of pulmonary vas¬cular interdependence in isolated dog lobes. J. Appl. Physiol. Respir. Environ. Exerc. Physiol. 46, 419.

Lai-Fook, S.J., Wilson, T.A., Hyatt, R.E., Rodarte, J.R., 1976. Elastic constants of inflated lobes of dog lung. J. Appl. Physiol. 40, 508.

Lucas, C.L., Murray, G.F., Wilcox, B.R., Shallal, J.A., 1983. Effects of pneumonectomy on pulmonary input impedance. Surgery 94, 807.

Lucas, C.L., Radke, N.F., Wilcox, B.R., Henry, G.W., Keagy, B.A., 1986. Maturation of pulmonary input impedance spectrum in infants and children with ventricular septal defect. Am. J. Cardiol. 57, 821.

Morris, K.P., Cox, P.N., Mazer, C.D., Frndova, H., McKerlie, C., Wolfe, R., 2000. Distribution of pulmonary blood flow in the perfluorocarbon-filled lung. Intensive Care Med. 26, 756–763.

Michels, D.B., West, J.B., 1978. Distribution of pulmonary ventilation and perfusion during short periods of weightlessness. J. Appl. Physiol. 45, 987–998.

Melsom, M.N., Flatebo, T., Kramer-Johansen, J., Aulie, A., Sjaastad, O.V., Iversen, P.O., Nicolaysen, G., 1995. Both gravity and non-gravity dependent factors determine regional blood flow within the goat lung. Acta Physiol. Scand. 153, 343–353.

Mead, J., Whittenberger, J.L., 1964. Lung inflation and hemodynamics. In: Handbook of Physiology Respiration. American Physiological Society, Washington, DC, p. 477.

Milic-Emili, J., Henderson, J.A., Dolovich, M.B., Trop, D., Kaneko, K., 1966. Regional distribution of inspired gas in the lung. J. Appl. Physiol. 21, 749.

Macklin, C.C., 1946. Evidences of increase in the capacity of the pulmonary arteries and veins of dogs, cats, and rabbits during inflation of the freshly excised lung. Rev. Can. Biol. 5, 199.

Mure, M., Domino, K.B., Robertson, T., Hlastala, M.P., Glenny, R.W., 1998. Pulmonary blood flow does not redistribute in dogs with reposition from supine to left lateral position. Anesthesiology 89, 483–492.

Murray, J.F., 1978. Effects of lung inflation on pulmonary arterial pressure in dogs with pulmonary edema. J. Appl. Physiol. Respir. Environ. Exerc. Physiol. 45, 442.

Muir, A.L., Hall, D.L., Despas, P., Hogg, J.C., 1972. Distribution of blood flow in the lungs in acute pulmonary edema in dogs. J. Appl. Physiol. 33, 736.

Muir, A.L., Hogg, J.C., Naimark, A., Hall, D.L., Chemecki, W., 1975. Effect of alveolar liquid on distribution of blood flow in dog lungs. J. Appl. Physiol. 39, 885.

Milnor, W.R., 1989. Hemodynamics. William and Wilkins, Baltimore, MD.

Morgan, B.C., Abel, F.L., Mullins, G.L., Guntheroth, W.G., 1966a. Flow patterns in cavae, pulmonary artery, pulmonary vein, and aorta in intact dogs. Am. J. Physiol. 210, 903.

Morgan, B.C., Dillard, D.H., Guntheroth, W.G., 1966b. Effect of cardiac and respiratory cycle on pulmonary vein flow, pressure and diameter. J. Appl. Physiol. 21, 1276.

Michel, R.P., Hakim, T.S., Chang, H.K., 1984. Pulmonary arterial and venous pressures measured with small catheters in dogs. J. Appl. Physiol. Respir. Environ. Exerc. Physiol. 57, 309.

Morin 3rd, F.C., Egan, E.A., Ferguson, W., Lundgren, C.E., 1988. Development of pulmonary vascular response to oxygen. Am. J. Physiol. 254, H542–H546.

Mischi, M., Korsten, H.H., 2006. Transpulmonary dilution system identification for pulmonary blood volume measurements by contrast echocardiography. Conf. Proc. IEEE Eng. Med. Biol. Soc. 1, 1818–1821.

Milnor, W.R., Jose, A.D., McGaff, C.J., 1960. Pulmonary vascular volume, resistance and compliance in man. Circulation 22, 130.

Maseri, A., Caldini, P., Harward, P., Joshi, R.C., Permutt, S., Zierler, K., 1972. Determinants of pulmonary vascular volume. Recruitment versus distensibility. Circ. Res. 31, 218.

Means, L.J., Hanson, W.L., Mounts, K.O., Wagner Jr, W.W., 1993. Pulmonary capillary recruitment in neonatal lambs. Pediatr. Res. 34, 596–599.

Mitzner, W., Sylvester, J.T., 1981. Hypoxic vasoconstriction and fluid filtration in pig lungs. J. Appl. Physiol. Respir. Environ. Exerc. Physiol. 51, 1065.

Moffatt, D.S., Guyton, A.C., Adair, T.H., 1982. Functional diagrams of flow and volume for the dog lung. J. Appl. Physiol. Respir. Environ. Exerc. Physiol. 52, 1035.

Mitzner, W., 1974. Hepatic outflow resistance, sinusoid pressure, and the vascular waterfall. Am. J. Physiol. 227, 513.

Mink, S.N., Unruh, H.W., Oppenheimer, L., 1985. Vascular and interstitial mechanics in canine pulmonary emphysema. J. Appl. Physiol. 59, 1704.

Mitzner, W., 1983. Resistance in the pulmonary circulation. Clin. Chest Med. 4, 127.

Murray, G.F., Lucas, C.L., Wilcox, B.R., Shallal, J.A., 1986. Cardiopulmonary hypoxic response 5 years post-pneumonectomy in beagles. J. Surg. Res. 41, 236.

Michael, J.R., Kennedy, T.P., Fitzpatrick, S., Rosenstein, B.J., 1984. Nifedipine inhibits hypoxic pulmonary vasoconstriction during rest and exercise in patients with cystic fibrosis and cor pulmonale. Am. Rev. Respir. Dis. 130, 516.

Mead, J., Takishima, T., Leith, D., 1970. Stress distribution in the lungs: a model of pulmonary elasticity. J. Appl. Physiol. 28, 596.

McDonald, D.A., 1974. Blood Flow in Arteries. Williams and Wilkins, Baltimore, MD.

Milnor, W.R., Conti, C.R., Lewis, K.B., O'Rourke, M.F., 1969. Pulmonary arterial pulse wave velocity and impedance in man. Circ. Res. 25, 637.

Murgo, J.P., Westerhof, N., 1984. Input impedance of the pulmonary arterial system in man. Circ. Res. 54, 666.

Newman, J.H., Cochran, C.P., Roselli, R.J., Parker, R.E., King, L.S., 1993. Pressure and flow changes in the pulmonary circulation in exercising sheep: evidence for elevated microvascular pressure. Am. Rev. Respir. Dis. 147, 921–926.

Nachar, R.A., Pastene, C.M., Herrera, E.A., Riquelme, R.A., Sanhueza, E.M., Troncoso, S., Llanos, A.J., 2001. Low-dose inhaled carbon monoxide reduces pulmonary vascular resistance during acute hypoxemia in adult sheep. High Alt. Med. Biol. 2, 377–385.

Nieman, G.F., Hakim, T.S., Bredenberg, C.E., 1988. Effect of increased alveolar surface tension on segmental pulmonary vascular resistance. J. Appl. Physiol. 64, 154.

Nagasaka, Y., Bhattacharya, J., Nanjo, S., Gropper, M.A., Staub, N.C., 1984. Micropuncture measurement of lung microvascular pressure profile during hypoxia. Circ. Res. 54, 90.

Noonan, T.C., Malik, A.B., 1986. Pulmonary vascular response to leukotriene D4 in unanesthetized sheep: role of thromboxane. J. Appl. Physiol. 60, 765.

Noonan, T.C., Selig, W.M., Kern, D.F., Malik, A.B., 1986. Mechanism of peptidoleukotriene-induced increases in pulmonary transvascular fluid filtration. J. Appl. Physiol. 61, 1928.

Nitzan, M., Mahler, Y., Schechter, D., Yaffe, S., Bocher, M., Chisin, R., 1994. A measurement of pulmonary blood volume increase during systole in humans. Physiol. Meas. 15, 489–498.

Nichol, J., Girling, F., Jerrard, W., Claxton, E.B., Burton, A.C., 1951. Fundamental instability of the small blood vessels and critical closing pressures in vascular beds. Am. J. Physiol. 164, 330.

Ortiz, L.A., Champion, H.C., Lasky, J.A., Gambelli, F., Gozal, E., Hoyle, G.W., Beasley, M.B., Hyman, A.L., Friedman, M., Kadowitz, P.J., 2002. Enalapril protects mice from pulmonary hypertension by inhibiting TNF-mediated activation of NF-kappaB and AP-1. Am. J. Physiol. Lung Cell. Mol. Physiol. 282, L1209–L1221.

Ou, L.C., Sardella, G.L., Hill, N.S., Thron, C.D., 1993. Possible role of pulmonary blood volume in chronic hypoxic pulmonary hypertension. J. Appl. Physiol. (1985) 74, 3020–3026.

Polglase, G.R., Wallace, M.J., Morgan, D.L., Hooper, S.B., 2006. Increases in lung expansion alter pulmonary hemodynamics in fetal sheep. J. Appl. Physiol. 101, 273–282.

Pennington, L.R., Weiskopf, R.B., Hannon, J.F., 1992. Swine as models in biomedical research. In: Swindle, M.M. (Ed.), Swine as Models in Biomedical Research.

Patschova, M., Kabes, R., Krisova, S., 2010. The effects of inhalation salbutamol administration on systemic and pulmonary hemodynamic, pulmonary mechanics and oxygen balance during general anaesthesia in the horse. Vet. Med. 55, 445–456.

Permutt, S., Bromberger-Barnea, B., Bane, H.N., 1962. Alveolar pressure, pulmonary venous pressure, and the vascular waterfall. Med. Thorac. 19, 239–260.

Parker, J.C., Ardell, J.L., Hamm, C.R., Barman, S.A., Coker, P.J., 1995. Regional pulmonary blood flow during rest, tilt, and exercise in unanesthetized dogs. J. Appl. Physiol. 78, 838–846.

Permutt, S., Riley, R.L., 1963. Hemodynamics collapsible vessels with tone: the vascular waterfall. J. Appl. Physiol 18, 924.

Pain, M.C., West, J.B., 1966. Effect of the volume history of the isolated lung on distribution of blood flow. J. Appl. Physiol. 21, 1545–1550.

Pinkerson, A.L., 1967. Pulse-wave propagation through the pulmonary vascular bed of dogs. Am. J. Physiol. 213, 450.

Piiper, J., 1958. Eine methode zur lokalisierung des stromungswiderstandes. Pflügers Arch. Ges. Physiol. 266, 199.

Parker, J.C., Kvietys, P.R., Ryan, K.P., Taylor, A.E., 1983. Comparison of isogravimetric and venous occlusion capillary pressures in isolated dog lungs. J. Appl. Physiol. 55, 964.

Perasolo, J., Heisanken, T., 1971. Right heart, pulmonary, and left heart blood volumes determined by analogue computer analysis of radiocardiograms in normal subjects and patients with mitral stenosis. Cardiovasc. Res. 5, 260.

Piiper, D., 1970. Attempts to determine volume, compliance and resistance to flow of pulmonary vascular compartments. Progr. Resp. Res. 5, 40.

Permutt, S., Caldini, P., Maseri, A., Palmer, W.H., Sasamori, T., Zierler, K., 1969. Recruitment versus distensibility in the pulmonary vascular bed. In: Pulmonary Circulation and Interstitial Space. University of Chicago Press, Chicago, p. 375.

Presson Jr, R.G., Todoran, T.M., De Witt, B.J., McMurtry, I.F., Wagner Jr, W.W., 1997. Capillary recruitment and transit time in the rat lung. J. Appl. Physiol. 83, 543–549.

Permutt, S., Howell, J.B.L., Proctor, D.F., Riley, R.L., 1961. Effect of lung inflation on static pressure-volume characteristics of pulmonary vessels. J. Appl. Physiol. 16, 64.

Patel, D.J., Defreitas, F.M., Fry, D.L., 1963. Hydraulic input impedance to aorta and pulmonary artery in dogs. J. Appl. Physiol. 18, 134–140.

Pace, J.B., Cox, R.H., Alvarez-Vara, F., Karreman, G., 1972. Influence of sympathetic nerve stimulation on pulmonary hydraulic input power. Am. J. Physiol. 222, 196.

Quebbeman, E.J., Dawson, C.A., 1977. Effect of lung inflation and hypoxia on pulmonary arterial blood volume. J. Appl. Physiol. Respir. Environ. Exerc. Physiol. 43, 8.

Romand, J.A., Pinsky, M.R., Firestone, L., Zar, H.A., Lancaster Jr, J.R., 1994. Effect of inhaled nitric oxide on pulmonary hemodynamics after acute lung injury in dogs. J. Appl. Physiol. 76, 1356–1362.

Ravikumar, P., Dane, D.M., McDonough, P., Yilmaz, C., Estrera, A.S., Hsia, C.C., 2011. Long-term post-pneumonectomy pulmonary adaptation following all-trans-retinoic acid supplementation. J. Appl. Physiol. 110, 764–773.

Reddy, V.M., Meyrick, B., Wong, J., Khoor, A., Liddicoat, J.R., Hanley, F.L., Fineman, J.R., 1995. In utero placement of aortopulmonary shunts. A model of postnatal pulmonary hypertension with increased pulmonary blood flow in lambs. Circulation 92, 606–613.

Roos, A., Thomas, L.J., Nagel, E.L., Prommas, D.C., 1961. Pulmonary vascular resistance as determined by lung inflation and vascular pressures. J. Appl. Physiol. 16, 77.

Ritchie, B.C., Schauberger, G., Staub, N.C., 1969. Inadequacy of perivascular edema hypothesis to account for distribution of pulmonary blood flow in lung edema. Circ. Res. 24, 807.

Raj, J.U., Bland, R.D., Lai-Fook, S.J., 1986a. Microvascular pressures measured by micropipettes in isolated edematous rabbit lungs. J. Appl. Physiol. 60, 539.

Raj, J.U., Chen, P., Navazo, L., 1986b. Micropuncture measurement of lung microvascular pressure profile in 3- to 4-week-old rabbits. Pediatr. Res. 20, 1107.

Rock, P., Patterson, G.A., Permutt, S., Sylvester, J.T., 1985. The nature and distribution of vascular resistance in hypoxic pig lungs. J. Appl. Physiol. 59, 1891.

Rippe, B., Parker, J.C., Townsley, M.I., Mortillaro, N.A., Taylor, A.E., 1987. Segmental vascular resistances and compliances in dog lung. J. Appl. Physiol. 62, 1206.

Ressmeyer, A.R., Larsson, A.K., Vollmer, E., Dahlen, S.E., Uhlig, S., Martin, C., 2006. Characterisation of guinea pig precision-cut lung slices: comparison with human tissues. Eur. Respir. J. 28, 603–611.

Rudolph, A.M., 1979. Fetal and neonatal pulmonary circulation. Ann. Rev. Physiol. 41, 383.

Rudolph, A.M., 1977. Fetal and neonatal pulmonary circulation. Am. Rev. Resp. Dis. 115, 11.

Rasanen, J., Wood, D.C., Debbs, R.H., Cohen, J., Weiner, S., Huhta, J.C., 1998. Reactivity of the human fetal pulmonary circulation to maternal hyperoxygenation increases during the second half of pregnancy: a randomized study. Circulation 97, 257–262.

Rendas, A., Branthwaite, M., Reid, L., 1978. Growth of pulmonary circulation in normal pig-structural analysis and cardiopulmonary function. J. Appl. Physiol. 45, 806.

Rosenthal, A., Fyler, D.C., 1974. Effect of red cell volume reduction on pulmonary blood flow in polycythemia of cyanotic congenital heart disease. Am. J. Cardiol. 33, 410.

Raj, J.U., Ramanathan, R., Chen, P., Anderson, J., 1989. Effect of hematocrit on microvascular pressure in 3- to 5-week-old rabbit lungs. Am. J. Physiol. 256, 766.

Raj, J.U., Anderson, J., 1991. Comparison of dextran and hematocrit effects in the pulmonary microcirculation. Circ. Res. 68, 1108–1116.

Roughton, F.J.W., 1945. The average time spent by the blood in the human lung capillary and its relation to the rates of CO uptake and elimination in man. Am. J. Physiol. 143, 621.

Roughton, F.J.W., Forster, R.E., 1957. Relative importance of diffusion and chemical reaction rates in determining rate of exchange of gases in the human lung, with special reference to true diffusing capacity of pulmonary membrane and volume of blood in lung capillaries. J. Appl. Physiol. 11, 290.

Rosenzweig, D.Y., Hughes, J.M.B., Glazier, J.B., 1970. Effects of transpulmonary and vascular pressures on pulmonary blood volume in isolated lung. J. Appl. Physiol. 28, 553.

Reuben, S.R., Swadling, J.P., Gersh, B.J., Lee, G.D., 1971. Impedance and transmission properties of the pulmonary arterial system. Cardiovasc. Res. 5, 1.

Swan, H.J., Ganz, W., Forrester, J., Marcus, H., Diamond, G., Chonette, D., 1970. Catheterization of the heart in man with use of a flow-directed balloon-tipped catheter. N Engl. J. Med. 283, 447–451.

Stickland, M.K., Lovering, A.T., Eldridge, M.W., 2007. Exercise-induced arteriovenous intrapulmonary shunting in dogs. Am. J. Respir. Crit. Care Med. 176, 300–305.

Sun, R.Y., Nieman, G.F., Hakim, T.S., Chang, H.K., 1987. Effects of lung volume and alveolar surface tension on pulmonary vascular resistance. J. Appl. Physiol. 62, 1622.

Szidon, J.P., Ingram, R.H., Fishman, A.P., 1968. Origin of the pulmonary venous flow pulse. Am. J. Physiol. 214, 10.

Shepard, J.M., Gropper, M.A., Nicolaysen, G., Staub, N.C., Bhattacharya, J., 1988. Lung microvascular pressure profile measured by micropuncture in anesthetized dogs. J. Appl. Physiol. 64, 874.

Siegel, L.C., Pearl, R.G., Shafer, S.L., Ream, A.K., Prielipp, R.C., 1989. The longitudinal distribution of pulmonary vascular resistance during unilateral hypoxia. Anesthesiology 70, 527.

Shaul, P.W., Pace, M.C., Chen, Z., Brannon, T.S., 1999. Developmental changes in prostacyclin synthesis are conserved in cultured pulmonary endothelium and vascular smooth muscle. Am. J. Respir. Cell. Mol. Biol. 20, 113–121.

Scheeren, T.W., Loer, S.A., Ding, Z.P., Arndt, J.O., 1998. Pulmonary blood volume and its effects on pressure/flow relations and flow resistance in isolated lungs of rabbits. Pflugers Arch. 435, 247–253.

Smiseth, A., Scott-Douglas, N.W., Manyari, D., Kingma, I., Smith, E.R., Tyberg, J.V., 1988. Increased pulmonary vascular capacitance with beta-adrenergic receptor stimulation: an experimental study of the effect of isoproterenol on the pulmonary vascular volume-pressure relationship. Can. J. Physiol. Pharmacol. 66, 85.

Stewart, G.N., 1921. The pulmonary circulation time, the quantity of blood in the lungs and the output of the heart. Am. J. Physiol. 58, 20–44.

Sueyoshi, E., Tsutsui, S., Hayashida, T., Ashizawa, K., Sakamoto, I., Uetani, M., 2011. Quantification of lung perfusion blood volume (lung PBV) by dual-energy CT in patients with and without pulmonary embolism: preliminary results. Eur. J. Radiol. 80, e505–e509.

Sackner, M.A., DuBois, A.B., 1964. Functional volumes of the pulmonary circulation. Med. Thorac. 22, 146.

Singhal, S., Henderson, R., Horsfield, K., Harding, K., Cumming, G., 1973. Morphometry of the human pulmonary arterial tree. Circ. Res. 33, 190.

Sobin, S., Tremer, H.M., Fung, Y.C., 1970. Morphometric basis of the sheet flow concept of the pulmonary alveolar microcirculation in the cat. Circ. Res. 26, 394.

Sobin, S.S., Fung, Y.C., Tremer, H.M., Rosenquist, T.H., 1972. Elasticity of pulmonary alveolar microvascular sheet in the cat. Circ. Res. 30, 440.

Seibold, H., Wieshammer, S., Kress, P., 1988. Relation of pressure and flow of pulmonary circulation in patients with chronic obstructive pulmonary disease. Clin. Physiol. Biochem. 6, 29.

Shoukas, A.A., 1975. Pressure-flow and pressure-volume relations in the entire pulmonary vascular bed of the dog determined by two-port analysis. Circ. Res. 37, 809.

Sylvester, J.T., Mitzner, W., Ngeow, Y., Permutt, S., 1983. Hypoxic constriction in extra-alveolar vessels in rat. J. Appl. Physiol. Respir. Environ. Exerc. Physiol. 54, 1660.

SooHoo, S.L., Goldberg, H.S., Graham, R., Jasper, A.C., 1987. Zone 2 and zone 3 pulmonary blood flow. J. Appl. Physiol. 62, 1982.

Shoukas, A.A., Brunner, M.J., Frankle, A.E., Greene, A.S., Kallman, C.H., 1984. Carotid sinus baroreceptor reflex control and the role of autoregulation in the systemic and pulmonary arterial pressure-flow relationships of the dog. Circ. Res. 54, 674.

Sylvester, J.T., Harabin, A.L., Peake, M.D., Frank, R.S., 1980. Vasodilator and constrictor responses to hypoxia in isolated pig lungs. J. Appl. Physiol. Respir. Environ. Exerc. Physiol. 49, 820.

Selkurt, E.E., 1985. Effect of acute pulmonary hypertension on pressure/flow in the canine pulmonary vascular bed. Respir. Physiol. 60, 169.

Sarnoff, S.J., Berglund, E., 1952. Pressure-volume characteristics and stress relaxation in the vascular bed of the dog. Am. J. Physiol. 171, 238.

Smith, J.C., Mitzner, W., 1980. Analysis of pulmonary vascular interdependence in excised dog lungs. J. Appl. Physiol. Respir. Environ. Exerc. Physiol. 48, 450.

Schrijen, F., Saunier, C., Henriquez, A., 1992. Pressure/flow relation in the peripheral pulmonary vascular bed in dogs: a preliminary study. Acta Physiol. Scand. 145, 395–399.

Takeda, K., Knapp, M.J., Wolfe, W.G., Crapo, J.D., 1987. Hypoxia enhances unilateral lung injury by increasing blood flow to the injured lung. J. Appl. Physiol. 63, 2516–2523.

Tournoux, F., Petersen, B., Thibault, H., Zou, L., Raher, M.J., Kurtz, B., Halpern, E.F., Chaput, M., Chao, W., Picard, M.H., Scherrer-Crosbie, M., 2011. Validation of noninvasive measurements of cardiac output in mice using echocardiography. J. Am. Soc. Echocardiogr. 24, 465–470.

Thomas, L.J., Griffo, Z.J., Roos, A., 1961. Effect of negative-pressure inflation of the lung on pulmonary vascular resistance. J. Appl. Physiol. 16, 451.

Tod, M.L., Sylvester, J.T., 1989. Distribution of pulmonary vascular pressure as a function of perinatal age in lambs. J. Appl. Physiol. 66, 79.

Tod, M.L., McGeady, M.L., Rock, P., Sylvester, J.T., 1987. Effects of arterial ligation and embolism on pulmonary vascular pressure distribution. J. Appl. Physiol. 63, 1387.

Townsley, M.I., Korthuis, R.J., Taylor, A.E., 1985a. Specific hemodynamic effects of PG E2 and PG Fl alpha in canine lung. Prostaglan. Leukotr. Med. 20, 275.

Townsley, M.I., Korthuis, R.J., Taylor, A.E., 1985b. Effects of arachidonate on permeability and resistance distribution in canine lungs. J. Appl. Physiol. 58, 206.

Thorvaldson, J., Stokland, Molaug, M., Ilebekk, A., 1984. Cardiopulmonary blood volume during acute blood pressure elevations in dogs. Acta Physiol. Scand. 122, 137.

Tsang, J.Y., Montaner, J.S., Hogg, J.C., 1986. Effects of pulmonary edema on regional blood volume and red blood cell transit time. Comparison of high pressure and oleic acid-induced edema. J. Clin. Invest. 77, 1780.

Taylor, M.G., 1969. Arterial impedance and distensibility. In: Pulmonary Circulation and Interstitial Space. University of Chicago Press, Chicago, p. 341.

Tuchscherer, H.A., Webster, E.B., Chesler, N.C., 2006. Pulmonary vascular resistance and impedance in isolated mouse lungs: effects of pulmonary emboli. Ann. Biomed. Eng. 34, 660–668.

Ullrich, R., Bloch, K.D., Ichinose, F., Steudel, W., Zapol, W.M., 1999. Hypoxic pulmonary blood flow redistribution and arterial oxygenation in endotoxin-challenged NOS2-deficient mice. J. Clin. Invest. 104, 1421–1429.

Ugander, M., Kanski, M., Engblom, H., Gotberg, M., Olivecrona, G.K., Erlinge, D., Heiberg, E., Arheden, H., 2010. Pulmonary blood volume variation decreases after myocardial infarction in pigs: a quantitative and noninvasive MR imaging measure of heart failure. Radiology 256, 415–423.

Vreim, C.E., Staub, N.C., 1973. Indirect and direct pulmonary capillary blood volume in anesthetized open-thorax cats. J. Appl. Physiol. 34, 452.

Varnauskas, E., Luepker, R., Liander, B., 1971. Pulmonary extravascular water and pulmonary blood volume at rest and in exercise in cardiac patients. In: Giuntini, C. (Ed.), Central Hemodynamics and Gas Exchange. Minerva Medka, Torino, Italy, p. 309.

Wasserman, K., Comroe Jr, J.H., 1962. A method for estimating instantaneous pulmonary capillary blood flow in man. J. Clin. Invest 41, 401–410.

West, J.B., Dollery, C.T., 1960. Distribution of blood flow and ventilation-perfusion ratio in the lung, measured with radioactive carbon dioxide. J. Appl. Physiol. 15, 405–410.

West, J.B., Dollery, C.T., Naimark, A., 1964. Distribution of blood flow in isolated lung: relation to vascular and alveolar pressures. J. Appl. Physiol. 19, 713.

West, J.B., Dollery, C.T., 1965. Distribution of blood flow and the pressure-flow relations of the whole lung. J. Appl. Physiol. 20, 175.

West, J.B., 1977. Blood flow. In: Regional Differences in the Lung. Academic Press, New York, p. 85.

Walther, S.M., Domino, K.B., Glenny, R.W., Polissar, N.L., Hlastala, M.P., 1997. Pulmonary blood flow distribution has a hilar-to-peripheral gradient in awake, prone sheep. J. Appl. Physiol. 82, 678–685.

Wiebel, E.R., 1973. Morphological basis of alveolar-capillary gas exchange. Physiol. Rev. 53, 419.

Whittenberger, J.L., Mc, G.M., Berglund, E., Borst, H.G., 1960. Influence of state of inflation of the lung on pulmonary vascular resistance. J. Appl. Physiol. 15, 878–882.

Wedgwood, S., Bekker, J.M., Black, S.M., 2001. Shear stress regulation of endothelial NOS in fetal pulmonary arterial endothelial cells involves PKC. Am. J. Physiol. Lung Cell. Mol. Physiol. 281, L490–L498.

Wang, C.G., Hakim, T.S., Michel, R.P., Chang, H.K., 1985. Segmental pulmonary vascular resistance in progressive hydrostatic and permeability edema. J. Appl. Physiol. 59, 242.

Wollmer, P., Rozkovec, A., Rhodes, C.G., Allan, R.M., Maseri, A., 1984. Regional pulmonary blood volume in patients with abnormal blood pressure or flow in the pulmonary circulation. Eur. Heart J. 5, 924.

Wanner, A., Begin, R., Cohn, M., Sackner, M.A., 1978. Vascular volumes of the pulmonary circulation in intact dogs. J. Appl. Physiol. Respir. Environ. Exerc. Physiol. 44, 956.

Wagner Jr, W.W., Latham, L.P., 1975. Pulmonary capillary recruitment during airway hypoxia in the dog. J. Appl. Physiol. 39, 900.

West, J.B., Schneider, A.M., Mitchell, M.M., 1975. Recruitment in networks of pulmonary capillaries. J. Appl. Physiol. 39, 976.

Wagner Jr, W.W., Latham, L.P., Capen, R.L., 1979. Capillary recruitment during airway hypoxia: role of pulmonary arterial pressure. J. Appl. Physiol. 47, 383.

Warrell, D.A., Evans, J.W., Clarke, R.O., Kingaby, G.P., West, J.B., 1972. Pattern of filling in the pulmonary capillary bed. J. Appl. Physiol. 32, 346.

Wasserman, K., Butler, J., Van Kessel, A., 1966. Factors affecting the pulmonary capillary flow pulse in man. J. Appl. Physiol. 21, 890.

Wauthy, P., Pagnamenta, A., Vassalli, F., Naeije, R., Brimioulle, S., 2004. Right ventricular adaptation to pulmonary hypertension: an interspecies comparison. Am. J. Physiol. Heart Circ. Physiol. 286, H1441–H1447.

Yuan, P., Wu, W.H., Liu, D., Zhang, R., Jing, Z.C., 2011. [Determination of pulmonary vascular resistance by improved right heart catheter in rat]. Zhonghua Xin Xue Guan Bing Za Zhi 39, 901–904.

Younes, M., Bshouty, Z., Ali, J., 1987. Longitudinal distribution of pulmonary vascular resistance with very high pulmonary blood flow. J. Appl. Physiol. 62, 344.

Yilmaz, C., Ravikumar, P., Bellotto, D.J., Unger, R.H., Hsia, C.C., 2010. Fatty diabetic lung: functional impairment in a model of metabolic syndrome. J. Appl. Physiol. 109, 1913–1919.

Yuan, H., Kononov, S., Cavalcante, F.S., Lutchen, K.R., Ingenito, E.P., Suki, B., 2000. Effects of collagenase and elastase on the mechanical properties of lung tissue strips. J. Appl. Physiol. 89, 3–14.

Zarbock, A., Distasi, M.R., Smith, E., Sanders, J.M., Kronke, G., Harry, B.L., von Vietinghoff, S., Buscher, K., Nadler, J.L., Ley, K., 2009. Improved survival and reduced vascular permeability by eliminating or blocking 12/15-lipoxygenase in mouse models of acute lung injury (ALI). J. Immunol. 183, 4715–4722.

Zhuang, F.Y., Fung, Y.C., Yen, R.T., 1983. Analysis of blood flow in cat's lung with detailed anatomical and elasticity data. J. Appl. Physiol. 55, 1341.

Chapter 15

Comparative Control of Respiration

Mehdi Saeed Hazari and Aimen K. Farraj

Chapter Outline

1. INTRODUCTION

Breathing is an automatic, rhythmic, and highly integrated process in mammals. Although it originates with impulses in the brainstem, the mechanism of breathing involves a complex interplay between the spinal cord, cranial and spinal nerves, diaphragm, intercostal muscles, structures in the larynx and pharynx, the lungs, and the vasculature. The purpose of these activities in virtually all mammals is the uptake of oxygen and release of carbon dioxide. Even though the mechanisms are assumed to be essentially the same, occasionally it is worthwhile to compare results from different species to see if this assumption is indeed valid and if species differences that may be found could actually be helpful in refining our theories about the physiology of the control of breathing. Thus, the goal of this chapter is to provide a comparative review of the control of breathing using data from specific mammalian species, examining not only the global mechanisms but also touching upon the species' differences and the utility of studying them in the context of respiration control. Following an overview of mammalian respiratory control, which will include a discussion of breathing pattern generation and regulation of inspiratory and expiratory processes, this chapter will examine the factors (e.g., arterial carbon dioxide [CO_2] tension [Pa_{CO2}]) that stimulate changes in ventilation and the appropriate functional measures of such alterations. Because the objective here is to provide a broad cross-species examination of the subject, data from both conscious unrestrained animals and those under anesthesia will be provided as proof of concept; however, to gain an appreciation of the evolution of this system, particularly with respect to nonmammalian vertebrate air-breathers, the reader should consult other relevant reviews (e.g., Milsom and Boutilier, 1989).

2. OVERVIEW OF RESPIRATORY CONTROL

The rate and depth of breathing, as learned from various mammalian species and known about humans, are regulated so that Pa_{CO2} is maintained within a relatively tight range. This mechanism, which operates on a breath-by-breath basis, is the main controller of breathing. There are two distinct patterns involved in breathing: metabolic (automatic) control pattern, which is concerned with oxygen delivery to the mitochondria and acid-base balance, and behavioral (voluntary) control, which coordinates breathing with other complex functions, such as eating or talking. Although metabolic breathing can be overridden, it quickly reasserts its control of the system.

2.1 Components of the System

The production of an optimal breathing pattern for any given circumstance is the result of the integration within the brainstem of many inputs from both central and peripheral sites (Figure 1). As described by Cunningham et al. (1986), these inputs, or afferents, can be classified as *feed-forward*

Comparative Biology of the Normal Lung. http://dx.doi.org/10.1016/B978-0-12-404577-4.00015-1

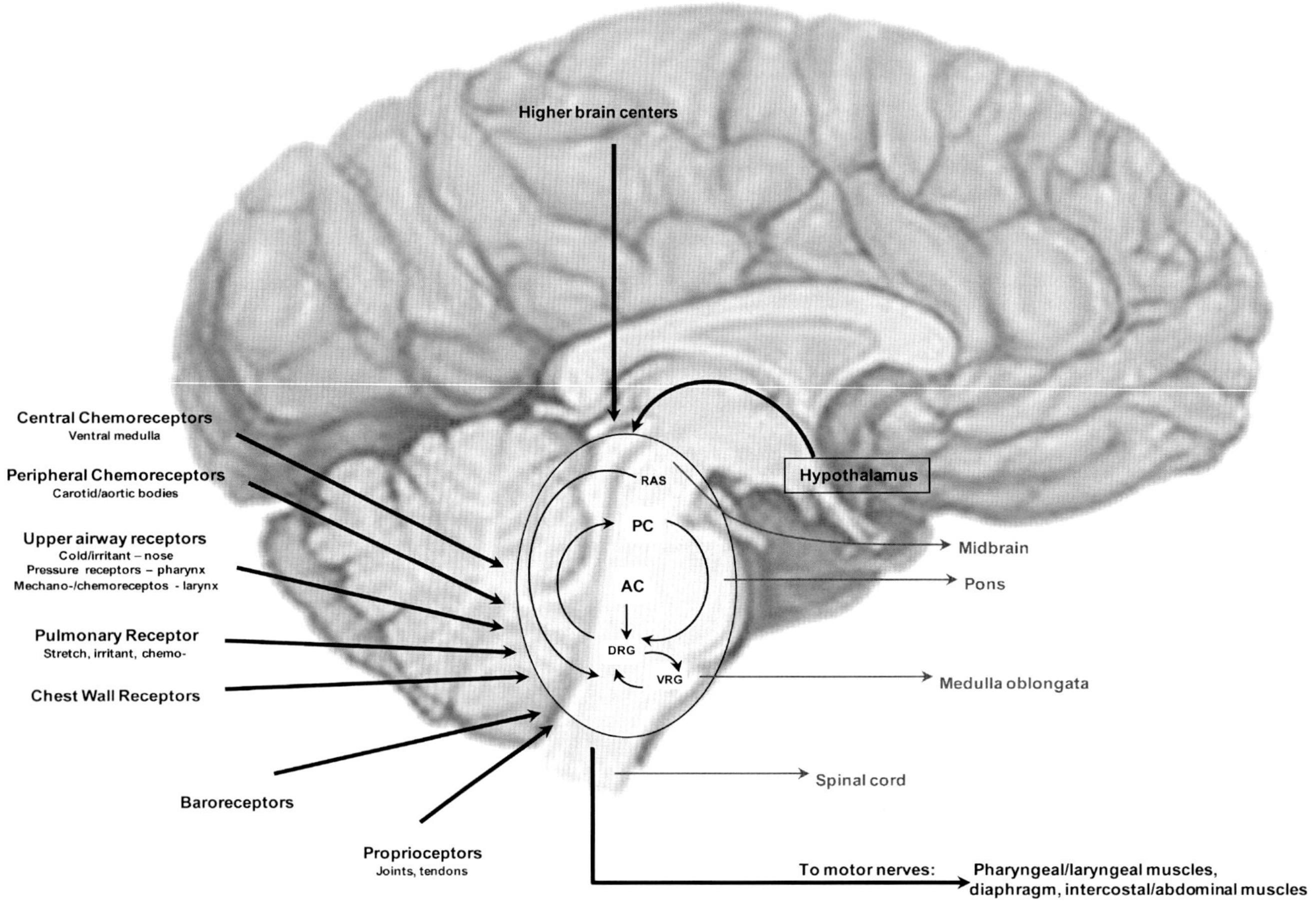

FIGURE 1 Schematic diagram of the various inputs that modulate the activity of the brainstem respiratory neurons. AC = apneustic center; DRG = Dorsal respiratory group; PC = Pneumotaxic center; RAS = Reticular activating system; VRG = Ventral respiratory group.

or *feedback,* depending on whether they are setting the needed level of ventilation in an anticipatory or predictive way (e.g., according to metabolic rate, skin temperature, or the onset of locomotion) or are giving modulating information back on how the system is doing (e.g., from chemoreceptors responding to blood gases and pH or from lung-volume monitoring mechanisms). The central pattern generator (CPG) for breathing, which is located in the medulla, receives many of the afferents carrying this sensory information and also discharges efferent signals to the respiratory muscles.

Various experimental techniques (e.g., ablation, stimulation, and recording) have been used to study the brain and to locate the areas that control breathing. Within the brainstem, two main regions function as intrinsic breathing controllers. The medullary respiratory area, which comprises the neuronal groups in the nucleus tractus solitarius (dorsal respiratory group, or DRG) and the nucleus retroambiguus, nucleus ambiguus, nucleus paraambiguus, and the Bötzinger complex (ventral respiratory group, or VRG; von Euler, 1986), provides metabolic control of breathing and even subserves some cardiovascular control functions. Within this area, the dorsal group of neurons discharges mainly during inspiration, whereas the ventral group contains neurons that become excited during both inspiration and expiration. Efferent fibers from the dorsal and ventral groups travel contralaterally in the dorsolateral columns of the spinal cord to the motor neurons innervating the diaphragm and intercostals, and the accessory muscles of breathing found in the larynx and abdominal muscles. Afferent fibers emanating from peripheral chemoreceptors, baroreceptors, and pulmonary mechanoreceptors synapse with neurons in the dorsal motor group, which also sends fibers to and receives input from the VRG.

Rhythmic breathing depends on a tonic inspiratory drive originating in the dorsal motor group, and on phasic expiratory inputs from the "higher" cerebral regions, thalamus, cranial nerves, and ascending spinal sensory tracts. In this manner, interconnected neuronal networks exert reciprocal inhibitory signals on each other to generate normal breathing. Voluntary control over breathing can be exerted by the cerebral cortex for such acts as speech, sniffing, coughing, or even expulsive efforts during the performance of ventilatory function tests. There is some evidence now that suggests the neurons controlling this mechanism may completely bypass the respiratory center and act directly on respiratory muscles (Corfield et al., 1998). In fact, there are three groups of upper motor neurons that converge on the anterior horn spinal neurons that supply the respiratory muscles: one from the DRG and VRG, which is concerned with inspiration and expiration output from the CPG; a second group concerned with voluntary control of breathing (e.g. speech); and a third group that deals with involuntary nonrhythmic control (e.g., swallowing, hiccup). In addition, spinal integration with intersegmental and intrasegmental activity leads to modulation and thereby rhythmic increases and decreases in the excitability of motor cells. This produces excitation and inhibition of respiratory muscles, which is coordinated by a system of interneurons in the spinal cord. Spinal reflexes are also important because they augment muscle activity within a given breath when resistance to airflow is increased or compliance is decreased.

Finally, the pontine respiratory group or the pneumotaxic center (PC), which is located in the pons, is the other main breathing control region; there appear to be numerous multisynaptic connections between the PC and both the DRG and VRG, which influence the medullary respiratory activity (e.g., setting the lung volume at which inspiration is terminated). Thus, the neurons of the PC influence the switching between inspiration and expiration; when the PC is inactivated, inspiration becomes prolonged (apneusis). Thus, the PC's primary role seems to be to contribute to an inspiratory off-switch mechanism, similar to that coming from pulmonary stretch receptors. The periodic inhibition of inspiration by the PC acts either directly on the DRG or by way of neurons in the caudal pons that make up an ill-defined region known as the apneustic center (AC). Without these neurons (in the AC), intact apneusis is not observed.

Other components include the reticular activating system, which plays an important role in modulating the activity of respiratory neurons during the different states of sleep or arousal. During sleep, the effect upon respiratory neurons is largely a depressive one. Also residing in the medulla, on the ventral surface, are the all-important central chemoreceptors responsive to changes in CO_2 and pH, which are essential in the matching of ventilation to metabolism. The carotid bodies and the aortic bodies are the most well known. Both of these structures monitor the partial pressure of oxygen in arterial blood, as well as the pH and partial pressure of carbon dioxide. Carotid bodies are paramount in driving the ventilatory response to hypoxia and are also important in the response to metabolic acidosis. There is anatomical evidence for other chemoreceptive structures, but their functional characteristics are far from understood. For example, small masses of such chemoreceptive highly vascularized and innervated tissue can also be found in the thorax and abdomen (Howe, 1984), and neuroepithelial bodies have long been described in the lungs (Lauweryns and Cokelaere, 1973; Michiels, 1984), but their functional significance has not been completely clarified.

Peripheral Inputs to the Respiratory Centers

The upper airways possess numerous types of receptive structures. There are cold and irritant receptors in the nose that inhibit respiration in response to cold inspired air (Burgess and Whitelaw, 1988) (see Section 5.3 on temperature effects) and inhalation of irritants (Widdicombe, 1999), respectively, and pressure receptors in the pharynx that contribute to the temporary inhibition of breathing during the swallowing reflex. The larynx contains chemoreceptive structures (Bradley, 2000), which have a potent inhibitory effect upon the respiratory neurons, particularly in newborn mammals. Among these are the irritant receptors, which respond to substances such ascigarette smoke or other air toxicants, much like what happens during mechanical stimulation of the larynx (i.e., cough, closure of the larynx, bronchoconstriction). Cold receptors are found more superficially on the vocal folds and when activated lead to depression of breathing (Sant'Ambrogio et al., 1995).

Within the lungs, there are three discrete populations of receptors. The pulmonary stretch receptors are slowly adapting receptors in the airways that increase their discharge frequency with increasing stretch (increasing lung volume) and have an inhibitory influence upon the DRG inspiratory neurons, thus contributing to an inspiratory *off-switch* mechanism. Thus, they are responsible for the Hering-Breuer reflex, which produces apnea in response to large lung inflations and augmented expiratory muscle contraction. This reflex can be easily demonstrated in laboratory animals; however, it is relatively weak in adult humans (Widdicombe, 1961). The activity of these receptors is enhanced by hypocapnia and diminished by hypercapnia (Coleridge and Coleridge, 1986). The irritant or rapidly adapting stretch receptors respond to various irritants in the airways, such as ammonia, dust, and smoke, as well as to lung inflation and deflation. Those in the primary bronchi are involved in the cough reflexes. A third category of pulmonary receptors have been called *J* (for juxtacapillary) receptors in the past, but now are collectively referred to as *C* fibers. They are found in pulmonary tissue and are often associated with capillaries. They respond to lung inflation, pulmonary congestion, and various chemicals (depending on receptor location), including prostaglandins (Es), histamine, and bradykinin, as well as foreign substances such as capsaicin, acrolein, and formaldehyde. Their stimulation tends to induce a transient apnea followed by a rapid, shallow breathing pattern.

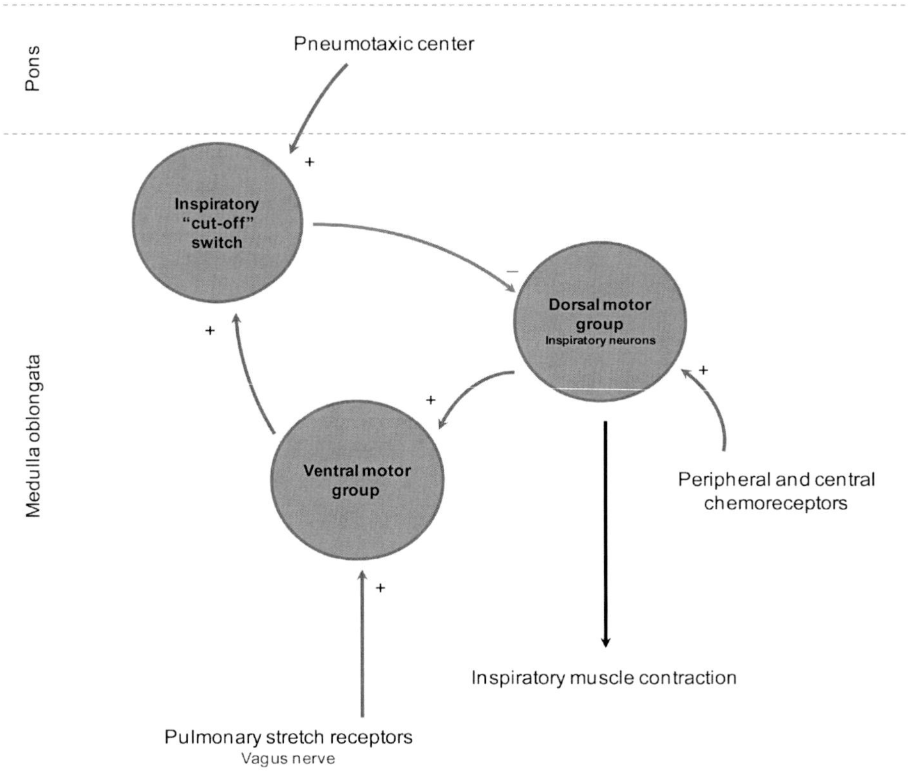

FIGURE 2 Neural circuitry of the brainstem respiratory controller. Arrows with a (+) sign denote excitatory output whereas arrows with a (−) sign denote inhibitory output. The dorsal motor group provides tonic inspiratory stimuli to the muscles of breathing and to the ventral motor group, which further stimulate breathing. The ventral motor group stimulates the inspiratory cut-off switch, which also receives inputs from the PC; once activated, the inspiratory cut-off switch sends inhibitory signals to the dorsal motor group. *Adapted from The Respiratory System in Physiology, Staub, NC Basic Respiratory Physiology, Ed.1. Staub, NC. Elsevier Health Sciences, April 1991.*

Several reflexes arise from outside the airways and lungs. These include phrenic afferent signals that inhibit respiratory efferent activity; however, the physiological role of this mechanism is largely unknown and it is not believed to contribute to normal breathing. Instead, it is assumed to provide information about inspiratory load. Other external reflexes include the baroreceptor reflex, which originates in the carotid sinus and aortic arch and produces hyperventilation during large decreases in blood pressure and vice versa. Lastly, mechanical stimulation of musculoskeletal afferents in dogs produces a reflex increase in ventilation (Kalia et al., 1972), which also likely does not play a role in normal breathing but rather contributes to exercise-induced hyperventilation.

Overall Model of Brainstem Mechanisms that Generate Respiratory Rhythm

Figure 2 shows the brainstem mechanisms and circuits that generate breathing rhythm:

1. Inputs from peripheral and central chemoreceptors travel to the inspiratory motor DRG and cause an increase in activity, which thereby induces an efferent signal that results in inspiratory muscle contraction. These changes are usually marked by changes in tidal volume.
2. The inspiratory activity in the DRG neurons stimulates neurons in the ventral motor group, which receives additional input signals from pulmonary stretch receptors via the vagus nerves.
3. The ventral motor group activates a group of neurons that represent the inspiratory cutoff switch because their activation inhibits inspiration by "turning off" the inspiratory motor DRG; this initiates expiration. It is also here that the PC feeds in to enhance the turning-off activity.

This represents a model of breathing in which tonic inspiratory activity is inhibited only when sufficient sensory signals are received; this is overlaid with chemoreceptor and reflex modulation, which then cause changes in pattern.

2.2 Central Pattern Generator

Anatomical approaches in which specific areas of the brainstem were either stimulated or removed were initially used to determine the source of respiratory rhythm and the areas of respiratory control. In humans, modern approaches have used imaging techniques such as magnetic resonance imaging (MRI) (Gozal et al., 1994) and positron emission tomography (PET) (Corfield et al., 1995) to localize the respiratory regions, and the results indicate that historical animal work applies to humans.

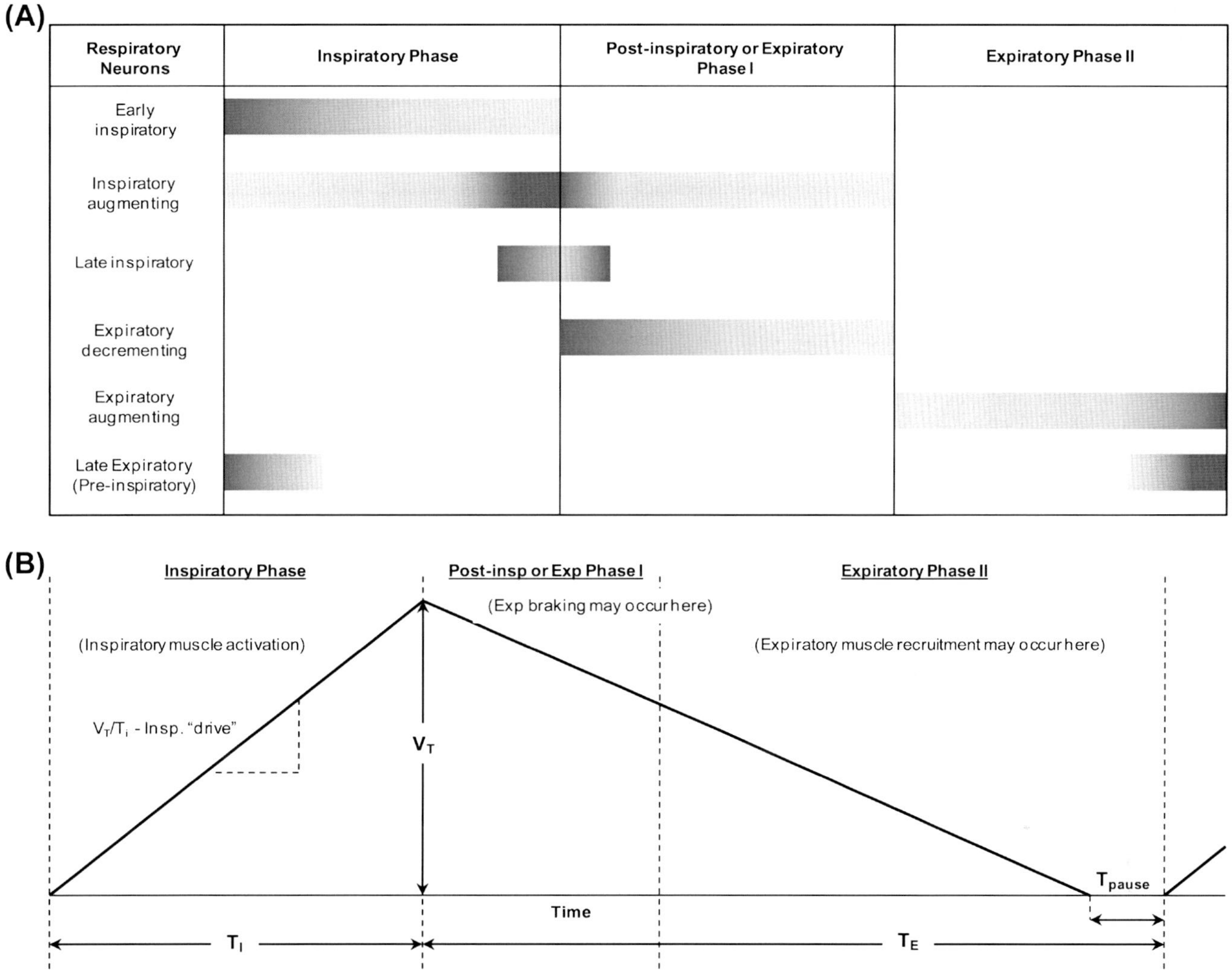

FIGURE 3 (A) Respiratory neuron firing patterns involved in the generation of central respiratory rhythm. Adapted from Lumb, 2005. (B) A simplified schematic representation of the volume–time profile of normal breathing.

In mammals, the respiratory rhythm has three phases (Richter et al., 1992): the inspiratory phase, postinspiratory or expiratory phase I, and expiratory phase II (Figure 3). The initial work that demonstrated this phenomenon was done in anesthetized, paralyzed cats, and the mechanisms hold true for humans. The phases of the cycle are generated by a network of at least six groups of neurons with distinct firing patterns, including early inspiratory neurons, inspiratory augmenting neurons, late inspiratory interneurons ("off-switch"), early expiratory decrementing neurons, expiratory augmenting neurons, and late expiratory (pre-inspiratory) neurons (Lumb, 2005). During the inspiratory phase, rapid firing of early inspiratory neurons initiates the cycle followed by a ramp increase in inspiratory augmenting neurons, which results in activation of the inspiratory muscles (note that pharyngeal dilator muscles are already activated during this period, possibly by the firing of late expiratory neurons). Declining activity of the inspiratory augmenting neurons marks the start of expiratory phase I; during this phase, motor discharge to the inspiratory muscles decreases, which aids in the gradual decrease in inspiratory tone. At this point, early expiratory decrementing neurons also reduce activity in the laryngeal adductor muscles as passive expiratory proceeds. Expiratory phase II begins as the inspiratory muscles become inactive; expiratory augmenting neurons become active if necessary to produce a gradual rise in expiratory muscle activity.

There are an infinite number of breathing patterns that can be achieved due to the fact that there are varying rates of spontaneous neuronal discharge, whether increasing or decreasing, and because the point at which the next set of neurons discharges can change as well. For example, during exercise, expiratory activation at the end of inspiration is immediate and almost completely active. During quiet breathing, early expiratory neurons decrease activity slowly and expiratory augmenting neurons activate only briefly to allow for a passive exhalation (Lumb, 2005).

The neural activity generating the respiratory rhythm is highly dependent on the intrinsic membrane properties of the neurons and both excitatory and inhibitory feedback mechanisms that require specific neurotransmitters. This is particularly true of the spontaneously firing neurons, which recruit other cells after their own activation by generating excitatory postsynaptic potentials, producing a ramped-up response. Calcium-dependent potassium channels then activate to repolarize the cell. This process can be either excitatory or inhibitory, as in the case of the expiratory augmenting neurons (Bianchi et al., 1995). There are also several neurotransmitters required for the excitatory and inhibitory activities of central pattern generation. Glutamate activates excitatory *N*-methyl-D-aspartate (NMDA) and non-NMDA receptors, which are fast-acting ion channels and slower-acting G-protein coupled receptors, respectively. Inhibitory neurotransmitters include glycine and γ-aminobutyric acid (GABA), which hyperpolarize the neuron but act during different phases of the cycle. There are also numerous neuromodulators, which influence the discharge of the CPG but are not directly involved in the generation of the rhythm. An example of this is the opioids, which have a tendency to depress respiratory activity (Bonham, 1995); others include acetylcholine, substance P, and serotonin.

2.3 Eupneic Breathing Pattern

Regulation of Expiratory Time (or Inspiratory Onset) and Expiratory Flow

Expiratory time (T_E) is generally defined as the period from the end of one inspiration to the beginning of the next (Figure 3). Although in interspecific comparisons it would be more useful to analyze expiration in terms of both the flow and pause phases, T_E will be used here in the usual sense to designate the whole period, unless otherwise specified. In the section on interspecific constants and systematic variations with body size (see Section 3), it is noted that the ratios of inspiratory and expiratory times to total breath time tend to be approximately the same interspecifically ($T_I/T_{TOT} = 0.35$, $T_E/T_{TOT} = 0.65$ (Boggs and Tenney, 1984)) (Table 1). These ratios were predictable on the basis of the scaling of respiratory frequency and the deflationary time constant. However, although expiration during eupnea (normal quiet breathing) may be passively driven by the elastic recoil properties of the lung, its duration may be prolonged beyond the passive relaxation time by two braking mechanisms (Figure 3): postinspiratory diaphragmatic activity and vocal cord adduction. Postinspiratory diaphragmatic activity has been demonstrated in cats (Gautier et al., 1973), humans (Petit et al., 1960), horses (Koterba et al., 1988), and dogs (Smith et al., 1989). A variation in laryngeal resistance due to vocal-cord abduction during inspiration and adduction during expiration was demonstrated in cats (Bartlett et al., 1973; Gautier et al., 1973) and subsequently in humans (England et al., 1982).

There appears to be interspecific variation, however, in the relative importance of expiratory braking mechanisms, as well as differences between adult and newborn animals. Remmers and Bartlett (1977) demonstrated the importance of this mechanisms in adult cats and showed that the brain monitors the rate of collapse of the respiratory system via a vagal pathway, such that if the upper airway resistance is suddenly bypassed, the diaphragm and upper-airway adductor muscles will increase their activity as they try to compensate for the lost braking of expiratory flow by increasing upper-airway resistance and diaphragmatic braking. That reflex response is lost in the vagotomized animal. This mechanism for the control of expiratory duration is probably more important in those species or neonates that lack an end-expiratory pause and those that have relatively high compliance chest walls. In newborn animals such as puppies (England et al., 1985), lambs (Harding, 1980), and humans (Mortola et al., 1982; Fisher et al., 1982) and small adult mammals such as mice (Vinegar et al., 1979), and even cats that have chest walls with high compliance, this regulation of expiratory airflow, particularly by regulating laryngeal resistance, may represent a dynamic mechanism for defending end-expiratory lung volume to keep it above the airway closing volume. This mechanism, enabling newborns of some species to keep their end-expiratory lung volume above the relaxation volume (V_{rx}), helps them to maintain lung O_2 stores, as well as to avoid airway closures. However, the absence of an effect of vagotomy on functional reserve capacity (FRC), or the volume of air present in the lungs, in neonatal calves suggests that vagal reflexes regulating upper-airway resistance are not important in determining FRC in neonates of this larger species (Slocombe et al., 1982). In adults of some moderately large species, such as the dog, who have relatively stiff chest walls, these mechanisms are also not required to defend end-expiratory lung volume; therefore, experiments in bypassing the upper airway do not induce either increased laryngeal adductor activity or postinspiratory diaphragmatic activity, as they do in cats or some neonates.

Beyond these variations in the importance of controlling the rate of passive deflation on expiration, there are differences between species in the extent to which there is an active phase of eupneic expiration and a corresponding passive phase to inspiration. Although most simple models of the breathing mechanism describe inspiration as active

TABLE 1 Average Ventilatory Parameters of Some Unanesthetized Mammals

Species	V_T(ml)	T_I(sec)	T_E	T_{TOT}	F (min^{-1})	V (l/min)	V_T/T_I (ml/sec)	VO_2 (ml/min)	V/VO_2	T_I/T_{TOT}	Reference
Horse (520 kg, 2)	5780	2.19	3.46	5.65	11	65.14	2692			0.39	Boggs and Tenney (1984)
Horse (586 kg, 13)	7400	2.8	4.3		10	74.0	2689			0.40	Gillespie et al. (1966)
Human (77.8 kg, 4)	426	1.55	2.52	4.08	14.7	8.7	407			0.38	Bechbache et al. (1979)
Goat (31–51 kg)	43				20	8.8		206	42.7		Boggs and Tenney (1984)
Dog (26 kg, 3)	416	1.4	2.53	3.93	15.3	6.4	297			0.36	Bowes et al. (1982)
Coatimundi (5.5 kg, 6)	55	1.18	1.44	2.68	24	1.29	49	47	30	0.44	Boggs et al. (1992)
Porcupine (5.9 kg, 4)	34				30	0.965	38	29.4	32	0.39	Boggs and Birchard (1989)
Woodchuck (4.5 kg, 6)	35	0.7	2.4	3.1	23	0.743	53	38	21	0.26	Boggs et al. (1992)
Cat (3.8 kg, 4)	30	0.98	1.77	2.75	22	0.664	32	23	29	0.36	Fordyce and Tenney (1984)
Rat (SD) (0.305 kg, 8)	2.05	0.25	0.34	0.6	103	0.209	8			0.41	Lai et al. (1978)
Rat (Wistar) (0.300 kg, 6)	1.32	0.19	0.27	0.46	130	0.172	7			0.41	Walker et al. (1985)
Syrian hamster (0.142 kg, 7)	0.87	0.29	0.74	1.03	58.3	0.051	3			0.28	Walker et al. (1985)

and expiration as passive in mammals (as it is in humans, cats, rabbits, woodchucks, and no doubt many others species), there are some who have active and passive components to each phase. This has been described for horses (Koterba et al., 1988) and dogs (Smith et al., 1989). After passive collapse to V_{rx}, these species actively exhale further, below V_{rx}, with the subsequent inspiration then commencing with passive recoil to V_{rx} before active inspiratory mechanisms commence. Inspiratory airflow can be observed to commence before diaphragmatic activity is initiated. In some species, such as the horse, midexpiratory pauses in airflow may be observed between the passive and active phases of expiration (Derksen et al., 1982).

The regulation of the duration of expiration, and therefore of the onset of inspiration, is the primary determinant of respiratory frequency in many species. In species that have an end-expiratory pause, this phase of breathing is the one that varies the most both within and between individuals, as noted for dogs by deVries and Bates (1988). They observed in dogs that once an inspiration was initiated, the course of inspiratory and expiratory flows was fairly fixed, while subtle adjustments in ventilation were achieved by varying the time of the pause before the next inspiration was initiated. This is also apparent in woodchucks (Boggs and Birchard, 1989). Interspecific comparisons of eupneic breathing patterns also revealed that the primary variation is often in the relative duration of the expiratory time. In comparing the fossorial Syrian hamster to the laboratory Wistar rat, Walker et al. (1985) found that the Syrian hamster is less sensitive to hypercapnia or hypoxia. They also note that the primary difference between the eupneic breathing patterns of these two species is in the longer T_E/T_{TOT} in the hamster (0.7) compared to the rat (0.59). This is also pointed out in the comparison of two larger rodents, the arboreal porcupine and the semi-fossorial woodchuck (Boggs and Birchard, 1989). Again, the semi-fossorial species, with a relatively reduced overall respiratory "drive" represented by higher CO_2 set point (Table 2) and a lower resting ventilation (Table 1), as well as reduced ventilatory responsiveness to inspired CO_2, has a larger T_E/T_{TOT} without a significant difference in the mean inspiratory flow (V_T/T_I).

Carotid Body Input

Arterial chemoreceptors contribute a tonic drive to ventilation under normoxic, normocapnic conditions, but this is quite variable between species, as indicated by the degree of depression of ventilation observed after carotid-body denervation (Table 3(A)). Another means of assessing the role of these receptors in determining normoxic drive is by the transient depression of ventilation associated with brief exposure to pure oxygen (i.e., the O_2 chemoreflex drive; Table 3(B)). The difference between the effects of these two tests is essentially due to differences in the contribution of aortic bodies or other chemoreceptive tissue to the tonic drive related to arterial O_2, since carotid-body denervation removes only carotid-body input, while the O_2 test should suppress all hypoxia-sensitive afferents.

In comparing the carotid-body-denervated breathing pattern to that of intact animals, there seems to be a consistent pattern of prolongation of T_E and reduction in V_T in the denervated animals. This has been observed in the awake dog (Bowes et al., 1982) and the awake cat (Gautier and Bonora, 1979), where the denervated animal shows both a reduction in V_T/T_I (due to a decrease in V_T) and an increase in T_E. Phillipson et al. (1981) also supported the notion that the onset of inspiration is dependent on the level of afferent respiratory stimulation. In awake sheep on an external membrane lung, which removes CO_2 at the rate it is metabolically produced, there is a reduction in frequency (eventually to apnea), which is due to a prolongation of the end-expiratory pause or an increase in the delay in the onset of inspiration. At the same time that this delay in inspiratory onset is observed, inspiratory flow rates are also diminishing.

This correlation between inspiratory drive (V_T/T_I) and T_E was also reported in cats (Gautier et al., 1973). Between species, however, such an inverse relationship between V_T/T_I and T_E (or T_E/T_{TOT}) is not always apparent, as observed, for example, in comparisons of the coati and woodchuck in Table 1 or the rat and Syrian hamster in Section 5.1. The prolongation of end-expiratory pauses with reduced chemical or metabolic drive is also apparent in hibernating mammals that develop long periods of apnea between breaths or bouts of breathing (Milsom, 1988).

Nasal and Pulmonary Afferents

The regulation of the duration of expiration appears to be dependent on intact vagi (Table 4); loss of these afferents has variable effects quantitatively between species, but in most cases it tends to prolong T_E. This is especially true of the neonatal rat compared to the adult rat (Davies, 1978) and again reflects the importance of vagal lung-volume tracking mechanisms in the control of breathing and lung volume in newborns. When stretch receptors alone are blocked by SO_2 in the rabbit (Davies, 1978), T_E is shortened, as compared to the lengthening of T_E observed with vagotomy. It has been demonstrated in dogs (Fishman et al., 1973) and rabbits (Davies, 1978) that the expiratory discharge of slowly adapting stretch receptors regulates T_E and thus influences resting eupneic frequency. The shortening of T_E with stretch-receptor block is consistent with its role in prolonging T_E when stimulated by lung inflation. The increase in T_E with vagotomy (removing all pulmonary afferents) suggests that rapidly adapting irritant receptors may play an important role in shortening T_E (see Coleridge and Coleridge, 1986 for a review).

TABLE 2 Normal Values of Blood Gases and Blood Buffering

Species	Body Weight (kg)	T_B (°C)	Pa_{CO_2} (torr)	pHa	Pa_{O_2} (torr)	HCO_3-(lmeq/l)	Δlog Pa_{CO_2} ΔpH	Reference
Mice	0.028	37.5	20.1			12.3	−1.75	Lahiri (1975)
Gerbil	0.078	36.4	31.2			19.0	−1.45	Lahiri (1975)
Hamster	0.135	35.8	45.1			27.5	−1.53	Lahiri (1975)
Rat	0.555	37.6	40.7	7.426	83.5	24.9	−1.48	Lahiri (1975)
Rat	0.531		33.1	7.425	87.1			Birchard and Tenney (1986)
Rat	0.297		39.8	7.467	91	28.7		Lai et al. (1978)
Guinea pig	0.669	37.8	40.2	7.395	90	23.5	−1.43	Lahiri (1975)
Cat	2.64	38.8	29.4	7.384	102	18	−1.39	Lahiri (1975)
Cat		38.9	32.5	7.426	108	21		Herbert and Mitchell (1971)
Rabbit	3.1	38.8	32.8	7.388	86	21		Lahiri (1975)
Porcupine	5.87	37.8	34.6	7.383	95	23	−1.35	Boggs and Birchard (1989)
Coatimundi	6	37.6	23.4	7.411	99	15		Boggs et al. (1992)
Baboon	8–11		37.6	7.388	106			Lewis (1975)
Mini-pig	11–19		47	7.432	94.3	30.5		Verbrugge et al. (1982)
Dog	20	38.3	41.6	7.386	93	23.2	−1.53	Lahiri (1975)
Goat	25	39	41.0	7.40	94.6	25.7	−1.30	Lahiri (1975)
Sheep	33	39	40.9	7.44	96	27.6	−1.35	Lahiri (1975)
Sheep	32–37	40	40.2	7.46	107			Hales and Finley, 1968
Calf	45–73		38.7	7.340	85.7	20.6		Bisgard and Vogel, 1971
Man	67	37	40.5	7.38	93	23.3	−1.55	Hales and Finley (1968)
Horse	387–543			7.408		25.8		Evans and Rose (1988)
Burrowers and Hibernators								
Syrian hamster	0.142		52.3	7.419	70.9			Walker et al. (1985)
Woodchuck	4.5	37.3	48	7.357	72	25.7	−1.29	Boggs and Birchard (1989)

TABLE 3A Effects of Carotid Body Denervation on Ventilation

Species	% Depression of Ventilation	Reference
Goat	10	Bisgard et al. (1980)
Dog	15	Bouverot and Bureau (1975)
Rat	25	Favier and LaCaisse (1978)
Rabbit	0	Bouverot (1976)
Cat	32	Gautier and Bonora (1979)

TABLE 3B Oxygen Chemoreflex Ventilatory Drive

Species	% Depression of Ventilation by 100% O_2	Reference
Human	10	Dempsey et al. (1978)
Rat	50	Favier and LaCaisse (1978)
Rabbit	5	Bouverot (1976)
Dog	35	Bouverot and Bureau (1975)
Mini-pig	40	Verbrugge et al. (1982)

TABLE 4 Effects of Removing Vagal Afferents on Breathing Pattern

Species	% ΔVT	% Δf	% ΔV	% ΔT_I	% ΔT_E	Reference
Rat (adult)	+94		+9	+91	+75	England et al. (1988)
Rat (newbom)	+107		−35	+83	+313	England et al. (1988)
Pig (adult)	+30	−24	−7(NS)	+98	+28(NS)	Coleridge et al. (1986)
Pig (newborn)	+22	−25	−14(NS)	+101	+52(NS)	Coleridge et al. (1986)
Rabbit (anesth)	+28	−10	+10			Kashani and Haigh (1975)
Rabbit (awake)	+26	−41	−26			Widdicombe and Winning (1974)
Rabbit (anesth)	+48	−29	−1.2			Widdicombe and Winning (1974)
Rabbit (anesth)[a]	+68			+177	+33	Davies (1978)
Dogs (anesth)	+101	−47	−0.5			Kashani and Haigh (1975)
Dogs (awake)	+81	−35		+74	+50	Fishman et al. (1973)
Sheep (anesth)	+56	−16.6	+21.3			Kashani and Haigh (1975)
Coatimundi (anesth)	+91.5	−44.5	−8	+111	+266	Boggs et al. (1992)
Woodchuck (anesth)	+3	−1	+0.9	+15	−3.3	Boggs et al. (1992)

Similar responses are observed with irritant reflex activation in rats (Hazari et al., 2008). When peripheral sensory irritation of the trigeminal nerve occurs, respiratory rate decreases secondary to an increased expiratory (T_E) phase (Alarie, 1973), which increases with the concentration of the irritant. This pattern of inhibition has been observed with numerous chemicals in several mammalian species.

Regulation of Inspiratory Time and Flow

Among the determinants of inspiratory time and tidal volume are the feedback signals from pulmonary stretch receptors tracking changes in lung volume and central control from the PC. The relative importance of the vagal afferents to eupneic breathing pattern seems to vary among species (as noted long ago by Widdicombe, 1964) and also between adult and neonatal animals (Tables 4 and 5). There have been three methods used to assess the role of vagal afferent information in the regulation of breathing pattern. One has been to look at the nature of the change in patterns when that afferent traffic is interrupted (e.g., by vagotomy or vagal cooling), and the other two approaches attempt to assess the strength of the Hering-Breuer inflation-inhibiting reflex. This is sometimes expressed in terms of the *inhibition ratio,* which is the ratio of the duration of the apnea induced by inflation to the average duration of the previous breath cycles, as represented in Table 5 (except for the data on pigs and baboons (Huszczuk et al., 1977), which are uniquely expressed as the ratio of the duration of the apnea following occlusion at peak inspiration to the duration of the control expiratory pause). Younes et al. (1974) pointed out that the duration of apnea following lung inflation is not a good quantifier of strength of the Hering-Breuer reflex, because it will be affected not only by the inspiration-inhibiting activity of pulmonary stretch receptors, but also by the interaction of that inhibition with rising inspiratory excitation influences, such as the rising Pa_{CO2} during the breath-hold period. The time course of these interactions may vary between individuals or species. Hence, a better approach may be to occlude the airway at end expiration so that the subsequent inspiration will lack afferent information from lung expansion. The extent to which the inspiratory effort following occlusion is prolonged compared to control inspirations that are terminated by inhibitory volume feedback becomes a measure of the strength of those inhibitory mechanisms (expressed as $1-T_I$ control/T_I occluded $\times$ 100, or simply as T_Io/T_Ic). This approach was used, for example, in comparing the strength of the Hering-Breuer reflex between premature and full-term human infants by Olinsky et al. (1974), who found that this index was greater in premature (58%) than in full-term infants (25%).

The data in Table 4 comparing neonatal and adult rats suggest that the vagal afferents are very important in the newborn mammal, not only for the regulation of breathing pattern, but also for the maintainance of eupneic levels of ventilation. The responses of newborn humans

TABLE 5 Inflation Reflex Inhibition Ratios

Species	Inhibition Ratio	Pressure or Volume	Reference
Rabbit	35	5 cm H_2O	Widdicombe (1964)
Rat	15	5 cm H_2O	Widdicombe (1964)
Guinea pig	7	5 cm H_2O	Widdicombe (1964)
Mouse	6	5 cm H_2O	Widdicombe (1964)
Monkey	7.5	5 cm H_2O	Widdicombe (1964)
Dog	5	5 cm H_2O	Widdicombe (1964)
Cat	3	5 cm H_2O	Widdicombe (1964)
Human	1.2	5 cm H_2O	Widdicombe (1964)
Human	1.3	800 ml	Widdicombe (1964)
Pig	6.23	1 V_T	Huszczuk et al. (1977)
Baboon	2.7	1 V_T	Huszczuk et al. (1977)
Coatimundi	7.85	10 cm H_2O	Boggs et al. (1992)
Woodchuck	3.4	10 cm H_2O	Boggs et al. (1992)
Mini-pig	5.4	5 cm H_2O	Adams et al. (1987)
Adult pig	2.35		Clement et al. (1986)
Newborn pig	1.47		Gaultier and Mortola (1981)

to lung inflation have also been found to be more vigorous than those of adults (Olinsky et al., 1974). However, Gaultier and Mortola (1981) pointed to some problems with the use of inflation reflexes for comparisons between neonates and adults, or between species of different size because of the differences in mechanical characteristics of their respiratory systems, and the need, therefore, for some normalization of the stimulus. Changes in lung volume only parallel changes in transpulmonary pressure if lung compliance is constant; but, of course, lung compliance is quite different between neonates and adults, or between small and large mammals. If the inhibition ratio was expressed in terms of transpulmonary pressure as a percent of the value at FRC, then they found that the inhibition ratio is smaller for any given relative change in transpulmonary pressure in newborn rats and rabbits than adults, while no difference was observed in puppies compared to dogs. Using the end-expiratory occlusion technique, they still found that the ratio of $T_{I}o/T_{I}c$ was lower in newborn rats (1.43) than adult rats (1.68) and in newborn pigs compared to adult pigs.

Bearing in mind these potential shortcomings of comparative inflation reflex data, Tables 4 and 5 suggest that, compared to some other species, humans, baboons, seals, and woodchucks seem to have a weaker Hering-Breuer reflex, or one that has comparatively little influence on the resting breathing pattern. These data must be viewed with some skepticism, however, because there is a great deal of variation among individuals within a species, and these mean values are taken from very few individuals. There are not sufficient data to indicate that a mechanism for such differences may exist, although the possibilities seem to be either that there are qualitative or quantitative differences in the signals coming from the pulmonary stretch receptors or there are differences in central thresholds for incoming inhibitory impulses. Angell-James et al. (1981) suggested that there seem to be relatively few fibers showing respiratory rhythm in the seal vagi, compared to cat or rabbit, which would suggest that there could be quantitatively less inhibitory input from pulmonary receptors to the CPG in this species. Bartlett and St John (1979) studied pulmonary stretch receptors in hamster, rat, guinea pig, cat, rabbit, and dog, and found both their static and dynamic response characteristics to be very similar among species. Contrary to the expectation that adaptation time (i.e., slowing of firing rate with constant stimulus) would be inversely related to breathing frequency, no differences were found in the adaptation rates of stretch receptors in these six species.

Lastly, it seems apparent from current research on the CPG that the most useful information on responses to various stimuli is that which is expressed in terms of the entire volume–time profile, so that it can be analyzed in terms of the three phases occurring centrally during each breath. If we can collect all our respiratory response data—not just in terms of tidal volume and frequency but also in terms of the timing of the inspiratory flow phase, the expiratory flow phase, and the pause between the end of expiratory flow and the beginning of the next breath—it will be more useful in the effort to understand the mechanisms that interact centrally to switch inspiratory activity on and off, and to control inspiratory and expiratory flows.

3. INTERSPECIFIC CONSTANTS AND SYSTEMATIC VARIATIONS WITH BODY SIZE

Allometric analyses have been useful in helping us to understand the principles of physiological engineering by revealing those parameters that are held constant over the full range of body size and those parameters that vary with body size in systematic ways. Although there is a lot of variation in the data upon which these relationships are based that may be considered methodological "noise," some species may exhibit real differences from the general mammalian pattern that represent adaptations worthy of further study.

In general, physiological frequencies have been found to be higher in smaller mammals, scaling to the −0.25 exponent of body mass, while volumes scale directly with body mass and rate functions vary with the 0.75 exponent of body mass (Lindstedt and Calder, 1981). These apply specifically for respiratory parameters as well. The volumes and capacities—total lung capacity, vital capacity, functional residual capacity, tidal volume (V_T), and dead-space volume—tend to increase in direct proportion to body size, with allometric exponents close to 1.0. Respiratory frequency, on the other hand, varies inversely with body mass by an allometric exponent of −0.26 or −0.28. There continues to be some debate about the proper exponent for metabolic rate, whether it is closer to a 0.66 power of weight or to the empirically derived 0.76 value; however, the latter seems to be more agreed upon (White and Seymour, 2003).

Consistent with its task of delivering oxygen to the gas exchanger at a rate appropriate to its utilization rate, total ventilation scales with body mass in approximately the same way as metabolism, with an exponent of 0.8. Less data have been available on alveolar ventilation, but it has been inferred to maintain a similar proportion to metabolic rate as total ventilation does (Travis and Bowers, 1991). This is based on the fact that tracheal volume, and therefore presumably dead-space volume, increases in direct proportion to body mass, with an exponent of 1.05 (Tenney and Bartlett, 1967), similar to tidal volume and total lung volume. It follows that dead-space volume is an approximately constant fraction of tidal volume and lung volume, and therefore that alveolar ventilation may be a nearly constant fraction of total ventilation interspecifically. If that is the case, we can expect alveolar oxygen and CO_2 tensions to be interspecific constants. Assuming arterial CO_2 reflects alveolar CO_2, in general, the partial pressure of CO_2 is maintained within a fairly narrow range around 40 ± 5 torr in most mammals. There are some obvious outliers at both extremes, however, because the range of reported values extends from 20 to 56 torr. Lahiri (1975) took exception to this notion of a fairly constant alveolar CO_2 in mammals on the basis of his multispecies study that showed fairly low Pa_{CO2} (20 and 31 torr) values for the smallest species, the mouse and gerbil. Given the difficulties of obtaining blood samples from such small mammals, these values could reflect some hyperventilation induced by handling for cardiac puncture. Nonetheless, there are also some larger species that have relatively low resting (awake, unrestrained, unstressed) Pa_{CO2} values, such as the cat (29 torr) and the coatimundi (24 torr). Depledge (1985) has suggested that mice have higher convection requirements ($\dot{V}/\dot{V}_{O2}$, 45–49) than other mammals, and this could account for the low Pa_{CO2} found in mice (Lahiri, 1975). The values for respiratory frequency (308–395) and $\dot{V}/\dot{V}_{O2}$ (45–49) in mice in the Depledge study are, however, over twice as great as those of several other investigators (97–225 breaths/min; Crosfill and Widdicombe, 1961; Crossland et al., 1977; Fairchild, 1972; Schlenker, 1985), which may again suggest a stress hyperventilation.

Related to this difficult problem of estimating alveolar gases of small mammals is the apparent dilemma that there does not appear to be the expected match between the scaling of the lung's O_2 diffusing capacity and either resting $\dot{V}_{O2}$ or $\dot{V}_{O2\,max}$ (Weibel et al., 1981). (This is contrary to the data from Tenney and Remmers (1963), which may have been biased by the inclusion of marine mammals whose pulmonary morphology and physiology seems to be quite different from terrestrial mammals.) This implies that the driving force for diffusion of O_2 across the gas-exchange membrane in the lung is not a constant but falls as body mass increases (Weibel et al., 1981; Weibel, 1987). This could be due to differences in PA_{O2} or Pc_{O2} (mean capillary) or both. The possibility that PA_{O2} may be size dependent is raised by Depledge's contention that $\dot{V}/\dot{V}_{O2}$ is greater in smaller mammals rather than being an interspecific constant of approximately 32.7 (Stahl, 1967), and by the fact that P_{O2} in the lung influences PA_{O2}. Rodriguez et al. (1987) found some structural basis for the latter concept in the longer acinar path length, along which O_2 must diffuse in the rabbit compared to the smaller rat. The other possibility—that mean capillary O_2 pressure is lower in smaller mammals—cannot be readily answered because it is dependent on a number of variables for which good data are not available. Data on mixed venous blood gases are not available for small mammals, but subcutaneous gas-pocket P_{O2} values of 42.1 torr for mice (Birchard et al., 1984); 32.9 torr for rats; and 37, 44, and 40 torr for shrew, pocket mouse, and bat, respectively (Tenney and Morrison, 1967), suggest that mixed venous P_{O2} values are similar in small and large mammals. This again suggests that body mass plays a role in this process.

Given that aerobic metabolism scales to the three-quarter power of body mass, a small mammal, such as a 10-g shrew, will have a $\dot{V}_{O2}$ per unit mass that is about 50 times greater than a 100-ton whale, which implies that ventilation will have to be 50 times greater in relation to body size in the smaller mammal as well. The only way the small mammal can achieve these relatively large ventilations is with very high respiratory frequencies. These must be determined by both the mechanical properties of the lung and chest wall, and the characteristics of the central rhythm generator, which tend to be harmonized to achieve the most mechanically efficient frequencies (Mead, 1960). The duration of inspiration (T_I) must be set centrally, but the duration of expiration (T_E) is determined in part by the elastic properties of the lung and chest wall, airway resistance, and the extent of postinspiratory diaphragmatic activity (Gautier et al., 1973). Bennett and Tenney (1982) studied the comparative respiratory mechanics of mammals; unlike previous investigations, this one included upper airway resistance. The deflationary time constants for the system were calculated with and without upper airway resistance included. They found that the reciprocal of the deflationary time constant scales with body mass in much the same way as respiratory frequency, suggesting that passive respiratory mechanics play a substantial role in determining T_E, and therefore T_{TOT}. They also predicted that if expiration is largely passive during normal quiet breathing, it should be approximated by four times the deflationary time constant for the whole system, including the upper airways. Combining that with estimates of frequency, the ratio of expiratory time to total breath time should be an interspecific constant of approximately 67%. Subsequently, Boggs and Tenney (1984) demonstrated that the fraction of time spent in each phase of the cycle is indeed an interspecific constant, with values of 0.65 (±0.004) for T_E/T_{TOT} and 0.35 for T_I/T_{TOT} among 11 species of mammals spanning a weight range from 30-g mice to 520-kg horses. Because this was predicted only on the basis of mechanical properties, it would seem to imply that active expiratory braking mechanisms (i.e., postinspiratory diaphragmatic activity and regulating laryngeal resistance) may be proportionately similar among mammals of various sizes.

Algebraic manipulations of allometric equations predict that V_T/T_I should scale with body mass as metabolism does, with an exponent of 0.72. Considerations of optimal physiological design would predict this as well if V_T/T_I is an index of respiratory drive. It turns out that it does scale with an exponent of 0.74 (Boggs and Tenney, 1984); therefore, on a weight-specific basis, this measure of "drive" increases with decreasing body size in the same fashion that oxygen consumption does. This may imply that the central respiratory controller is set at a higher gain in small mammals than in large mammals. However, another study (Leiter et al., 1986) pointed out again the difficulties of interpreting the mechanistic significance of allometric relationships and the problems with the use of V_T/T_I as an index of "drive" interspecifically. The authors removed the central component of the system by stimulating the phrenic nerve with a supramaximal, and therefore, equivalent stimulus in 13 species over the same range of body size. V_T/T_I still bore virtually the same relationship to body weight (exponent 0.82) without the central drive component of the system in operation. Hence, equivalent neural signals to the diaphragm do not induce equivalent flow rates. Furthermore, this study demonstrated that the interspecific changes in V_T/T_I could be accounted for by interspecific differences in the mechanical characteristics of the respiratory system and the velocity of diaphragmatic contraction. The time constant for maximal diaphragmatic contraction is shorter in smaller species. This is probably related to the higher proportion of fast-twitch fibers in the diaphragms of small mammals reported by Gautier and Padykula (1966). It was studied more systematically by Blank et al. (1988), who found 88.6% fast myosin isoforms in mouse diaphragm compared to 7.5% in cattle diaphragms. Also, the activities of enzymes involved in adenosine triphosphate (ATP) hydrolysis are higher in smaller mammals (Syrovy and Gutman, 1975). Citrate synthetase and phosphofructokinase activities and substrate oxidation rates are all significantly higher in the diaphragms of small mammals than in large mammals (Blank et al., 1988).

The possibility that the central drive is set at a higher gain in small mammals than in large ones was not fully excluded by the Leiter study, but it needs to be demonstrated directly. The fact that small mammals have a higher ventilatory output for a given level of a hypoxic stimulus (Boggs and Tenney, 1984) (see Section 5.2) lends some support to the proposition of a higher gain in smaller mammals being proportionate to their higher weight-specific metabolic rates. Above all, perhaps the lesson from these and many other studies is that the link between metabolic rate and the "gain" setting of respiratory control is not localized in any one set of cells, but rather is necessarily ubiquitous among both the sensory cells and the effector motor cells throughout the system, as reflected in these interspecific comparisons.

4. MEASUREMENTS OF RESPIRATORY DRIVE AND VENTILATORY RESPONSES

Measurement of respiratory drive is usually considered to be an assessment of the output of the system and how that output is changed in response to various stimuli or "inputs." Two methodological concerns arise in such studies:

(1) Is the stimulus really what we think it is (i.e., is it the direct cause of breathing change), or are multiple interrelated variables (secondary or indirect factors) actually responsible for altering the system's output in the observed fashion?

(2) What is the best measure of the output?

4.1 Considerations of Stimulus Definition

The two primary stimuli of the respiratory system are CO_2 and hypoxia. Thus, the primary concern is whether the measured stimulus is a reasonable and consistent index of the true stimulus and whether or not this stimulus is inducing secondary changes that also act upon the control mechanism. Looking first at CO_2, the stimulus is often defined as fractional concentration of inspired CO_2 (Fi_{CO2}), or end-tidal (PET_{CO2}), alveolar (PA_{CO2}), or arterial (Pa_{CO2}) partial pressures of CO_2. However, we know that the primary stimulus to respiration from changing partial pressures of CO_2 comes from changes in brain extracellular fluid H+ concentration (Loeschcke, 1974). Hence, one question that arises is how closely Pa_{CO2} represents changes in brain extracellular fluid or the intracelluar pH of chemosensitive cells. These will be affected by the change in cerebral blood flow during CO_2 breathing (Santiago and Edelman, 1986). Cerebrovascular resistance is quite sensitive to variations in Pa_{CO2}; hypercapnia increases brain blood flow and hypocapnia decreases it. In humans, the cerebral blood flow increases approximately 3% per mmHg increase in Pa_{CO2}, but there may be substantial interindividual variation in this factor (Dempsey, 1976), and certainly between species this aspect of the response could vary considerably. Early on, Dejours et al. (1965) pointed out as well that the time course of changes in pH and CO_2 and partial pressures of other compartments (e.g., cerebral venous blood, cerebrospinal fluid [CSF], intracellular) will be different from that of arterial CO_2 and pH. Ideally, of course, one would want to know what is happening to CO_2, [HCO_3^-], and [H+] in the chemoreceptive areas of the brain. However, because this is rarely possible, these should at least be measured in the blood. Peripheral chemoreceptors respond to both CO_2 and [H+], and in comparative studies there may be differences between species in the blood-buffering capacities and therefore in the effect of Pa_{CO2} on pH (Table 2).

Other changes that may take place during CO_2 inhalation may affect the actual stimulus. It is often assumed that oxygen consumption and CO_2 production are unchanged during a CO_2 test, but this is a rather important assumption because the alveolar gas tension is determined by the ratio of CO_2 production to alveolar ventilation, as well as by the concentration of inspired gas. The depressive effects of CO_2 on metabolism in mammals may be quite variable between species. The range of CO_2 concentrations used and the brief time of application may minimize the complicating metabolic effects of CO_2 (Schaefer, 1974), but any tests that exceed approximately 20 min may necessitate their consideration. The ventilatory response is usually thought to reach steady state within 12–15 min, at least in humans (Dejours et al., 1965). Another consideration is what is happening to arterial oxygen levels as the CO_2 is manipulated. The increased ventilation will elevate the O_2 in arterial blood, and this will have a minor depressing effect on arterial chemoreceptor stimulation of ventilation. This effect will be quite small, but can be slightly greater if pure oxygen is used in combination with the CO_2 mixtures (Dejours et al., 1965). The converse, where the ventilatory response to hypoxia is altering arterial pH and CO_2, is much more significant, as will be discussed later. Finally, the CO_2 may be activating the sympatho-adrenal system, which in turn exerts its own effects on the ventilation system, inducing an overestimate of the CO_2 stimulus (Nahas, 1974). This may not be very important until or unless high levels of CO_2 are reached (10% or higher), but there may be conditions that would lower the threshold for this response (e.g., hyperthyroidism), and the threshold could certainly be different among species. Another example of a confounding parameter in CO_2 responses is blood pressure. Grunstein et al. (1973) reported that increased arterial blood pressure results in a decreased slope of the ventilatory response to CO_2. The well-documented interdependence of circulatory and respiratory control should encourage the simultaneous measurement of parameters of both systems whenever possible (De Burgh Daly, 1986). Of course, prolonged and/or substantial increases in ventilation can alter the animal's temperature, which, in turn, can alter the apparent ventilatory response to the stimulus of interest (Cooper et al., 1986). Ambient temperatures that are different from those to which the animal is normally exposed may have effects on ventilation (Section 5.3).

Similar admonitions apply to hypoxic ventilatory response measurements. Like hypercapnia, hypoxia increases brain blood flow, but the amount of this increase will be reduced by any concomitant hypocapnia. Hypoxia may also alter the metabolic rate to different degrees among species. Particularly among small species and in the newborn, it may depress metabolism (Rosenmann and Morrison, 1975). In stimulating increased ventilation, the hypoxic stimulus will reduce the alveolar, arterial, and ultimately the cerebral CO_2 and elevate the pH, which will, in turn, inhibit ventilation. The extent of this secondary effect of hypoxia on ventilation may be quite variable among individuals and species. If the interest is only in the hypoxic stimulus, then CO_2 should be held constant (isocapnic) while the level of O_2 is being changed.

The use of the fractional concentration of inspired gas, whether CO_2 or O_2, as an index of the stimulus is more easily accomplished; however, it is generally the least desirable because it is a poor representation of the actual stimulus. The partial pressures of the gases in the blood are more representative of the actual stimulus, and the poor correlations between inspired levels of gases and blood levels of those gases reflect the substantial differences that may exist between total ventilation and alveolar or "effective" ventilation (total ventilation minus dead-space ventilation). Differences between individuals or species in physiological dead space and breathing patterns may account for considerable variation in the relationship between inspired and arterial gases, as will differences in the ventilatory responses themselves.

4.2 Measures of System Output

4.2.1 Ventilation

When the focus of interest is the integrated activity of the whole system (i.e., central controller, effector organs, and conducting airways), total minute ventilation ($\dot{V}$, ml/min) is often considered to be the appropriate measure. However, body weight and or surface area need to be considered in the case of interspecies comparisons, such as $\dot{V}/\text{kg}$ or $\dot{V}/\text{m}^2$. The body mass alone may not offer a consistent index of metabolically active tissue, however, and using the percent change in ventilation (which normalizes the data to a resting ventilation that is already a function of size and metabolism) may provide a better mechanism for comparison among individuals or species of varying body size. Its relative simplicity makes this a fairly widely used index of response, and it is expressed either as a ventilation ratio (stimulated ventilation, $\dot{V}'$, divided by resting ventilation, $\dot{V}^{\circ}$) or as percent change in ventilation ($\dot{V}' - \dot{V}^{\circ}/\dot{V}^{\circ} \times 100$). Alternatively, because the system tends to function to match ventilation to metabolism and because the metabolism may be variably affected by the respiratory stimulant, the use of the air convection requirement or specific ventilation, $\dot{V}/\dot{V}_{O2}$, may be preferable and is advocated by some comparative physiologists.

Because alveolar ventilation (or the more general term, *effective ventilation*) is more pertinent to the respiratory homeostatic adjustments and more directly reflects the ventilation of the gas exchanger, its use is recommended by some comparative physiologists as the most appropriate measure of system output (Bouverot, 1985; Milsom and Boutilier, 1989). Indeed, effective specific ventilation

($\dot{V}_{eff}/\dot{V}_{O2}$) may be the most useful indicator of efficacy of an animal's response to a given stimulus. The debate about the best index of output and the growing complexity of analysis emphasizes the value of reporting as many components of the response as possible, including tidal volume (V_T), inspiratory and expiratory times (T_I and T_E), frequency ($f = 0.60/T_{TOT}$, where T_{TOT} is total breath time), $\dot{V}$, $\dot{V}_{O2}$, $\dot{V}/\dot{V}_{O2}$, $\dot{V}_A$, V_T/T_I, blood pressure, and body temperature.

Integrated Phrenic Activity

If the output of the central controller is the point of interest, then the closest one can get to the integrated output of that motor neuron pool may be the activity of the phrenic nerve. Hence, integrated phrenic activity is often taken as the most proximate indicator of central drive. The advantage of using this approach is that it allows for the study of separate effects of specific inputs given that it bypasses some of the feedback circuits, which tightly regulate the function of the whole system. One such method is to use phrenic nerve activity in a paralyzed animal, ventilated to keep blood gases and pH constant, and perhaps vagotomized to remove volume feedback. Then single inputs can be altered without changing others. Eldridge (1971) established that there is an adequate correlation between phrenic activity and tidal volume, as long as it is the peak 0.1 s of integrated activity that is employed. Of course, phrenic motor impulses driving the diaphragm are not the only neuromotor output from the central controllers. There is evidence that activation of intercostal and other extradiaphragmatic muscles produces about 12% of the volume change during eupnea (Magnoni et al., 1969). Although the relative contribution of extradiaphragmatic muscles increases with larger tidal volumes, it does so in a linear fashion. Thus, Eldridge (1971) found that removal of the thoracic muscle contribution to tidal volume did not alter the linear relationship between phrenic activity and V_T.

Diaphragmatic Electromyography

Direct measurement of phrenic activity may not always be possible, but integrated diaphragmatic diaphragmatic electromyography (EMG) is the next step in the transformation from central inspiratory activity to tidal volume and may be accomplished either through direct electrode implant or with the less invasive esophageal tube electrodes. Maarsingh et al. (2000) showed that the EMG signals derived from the diaphragm are, in different age groups both with and without asthma, reproducible during tidal breathing. However, early on, Grassino et al. (1976) drew attention to potential problems with this procedure, showing that the diaphragmatic EMG may be altered by changes in lung volume, despite constant phrenic stimulation. They suggested that the wires must be inserted in the paratendinous region, an area that maintains a constant relationship to the lung during changes in lung volume. Moreover, if esophageal electrodes are used, they must be stabilized, which can be difficult (and unreliable) for only one electrode pair. An esophageal electrode catheter has been introduced for use in human studies that has a multipair electrode; the diaphragm EMG recorded from a multipair esophageal electrode has been found to be useful to assess neural respiratory drive in various patient groups (Luo et al., 2008).

Airway Occlusion Pressure ($P_{0.1}$)

The pressure generated in an occluded airway can be used as a measure of the force generated by the contracting diaphragm in response to the neural drive from the central respiratory controllers. This relatively noninvasive index of inspiratory drive has been used in humans and cats (Grunstein et al., 1973; Cherniack et al., 1976; Milic-Emili et al., 1981; Whitelaw et al., 1975). The airway is occluded at end expiration, with the system at its relaxation volume, and the pressure generated by the isometric contraction of the diaphragm against the closed airway during the first 0.1 s of the succeeding inspiratory effort provides a good index of the phrenic nerve activity (Evanich et al., 1976). This is referred to as $P_{0.1}$. Because there are only negligible changes in lung volume and there is no flow, this measure of inspiratory drive should be insensitive to the effects of airway resistance or of respiratory system compliance. Instead, it gives a weighted sum of the contribution of all respiratory muscles active at any given time and when applied to anesthetized subjects gives a tracing of the time course of respiratory neuromuscular output through the cycle. Several factors need to be considered when making the measurements and interpreting the data, the volume at which occlusion takes effect, distortions of the chest wall during the maneuver, tonic vagal input, and the strength of the muscles doing the work (Whitelaw and Derenne, 1993). The measurement is taken within the first 100–200 ms of the occluded inspiration before the subject becomes aware of, and reacts to, the airway occlusion. This method also has the advantage over diaphragmatic EMG in that it reflects the activity of all the inspiratory muscles. However, the validity of $P_{0.1}$ as an index of inspiratory drive depends on the shape of the occlusion pressure wave, which can be affected by such things as anesthetic, age, and body temperature, and there may be differences between species (Milic-Emili et al., 1981). This fairly noninvasive technique was developed to measure drive in ventilated patients for clinical use and it remains easily applied and useful due to its simplicity; however, it needs to be used with care in comparative interspecific studies.

Mean Inspiratory Flow (V_T/T_I)

Another index of respiratory drive, or at least of inspiratory drive, that has been suggested is the mean inspiratory airflow rate (V_T/T_I) (Milic-Emili and Grunstein, 1976). It results from the central inspiratory neural activity and the phrenic and diaphragmatic activity, which have been described as indices of central drive. This measurement, however, incorporates mechanical aspects of the respiratory system that may not be fixed and can therefore change the total output of the system and alter one's impression of the extent of the drive component. This potential pitfall can be used to an advantage if V_T/T_I is used in combination with occlusion pressure. This was done in a study of acclimatization to high altitude in humans (Gautier et al., 1973), which revealed that while V_T/T_I continued to increase beyond day 3 at altitude, occlusion pressure only rose on day 1 and remained constant thereafter. This suggested that the continuous increase in ventilation during altitude acclimatization was not due to changing central neural inspiratory drive but to changing respiratory mechanics such that the effective inspiratory impedance, $P_{0.1}/(V_T/T_I)$ (the ratio of airway occlusion pressure to mean inspiratory flow), declined from the first to the fourth or sixth day.

Mean inspiratory flow, as well as occlusion pressure or diaphragmatic EMG or phrenic activity, all only reflect inspiratory drive. Without simultaneous indications of what is happening to expiratory timing, they cannot be good indices of total respiratory drive, particularly in interspecific comparisons. The relationships between any of these and total ventilation depends on the relative durations of the inspiratory and expiratory phases; hence, Milic-Emili and Grunstein (1976) noted the need to use T_I/T_{TOT} as an index of the timing component of the output. This may be seen clearly in the interspecific comparisons from data in Tables 1 and 2. The coatimundi and woodchuck, two mammals of similar body size, have similar V_T/T_I values, and yet they have different total ventilations and maintain quite different resting Pa_{CO2} levels. Indeed, the woodchuck, which has the higher mean inspiratory flow value, actually must be taken to have the lower total respiratory drive because its Pa_{CO2} is high and total ventilation is relatively low. The key difference between these species can be seen in the patterns of respiratory timing. The coati spends a relatively longer time in inspiration, thus achieving larger tidal volumes with similar mean inspiratory flow, while the woodchuck has a relatively longer time in expiration, which includes the often prolonged end-expiratory pause characteristic of this species.

Phillipson et al. (1981) also noted that, contrary to the proposition of Milic-Emili and Grunstein (1976), the drive and timing components of the control of breathing cannot be conceptually separated, for changes in timing are often the first manifestation of changes in drive. As drive was reduced in their sheep by the removal of metabolically produced CO_2 by an external membrane exchanger, there was a prolongation of the pause between breaths, as well as a decrease in V_T/T_I. Again, the point to be made here is that in order to understand the regulation of breathing, we must try to understand the control of both the durations of its phases (inspiratory, expiratory, and pauses) and the control of the rates of air movement in and out of the respiratory system. All of the components of the output of the system need to be examined (see Figure 3 and Section 2.3).

5. FACTORS THAT ALTER BREATHING PATTERNS

5.1 Responses to Carbon Dioxide

Although the central chemoreceptors are located relatively close to the respiratory neurons of the medulla, contrary to previous understanding, they are a separate and distinct unit. Approximately 75–85% of the total respiratory response to inhaled CO_2 originates in the chemoreceptors located in the medulla (Cunningham, 1974); they also respond to changes in hydrogen ion concentration ($[H^+]$) of the surrounding brainstem interstitial fluid. The peripheral chemoreceptors, the carotid bodies, account for the remaining CO_2-induced increases in ventilation. Both central and peripheral chemoreceptors respond to an elevation of Pa_{CO2}, which leads to a rise in the P_{CO2} of CSF, cerebral tissue and jugular venous blood; these compartments all have a P_{CO2} that is about 10 mmHg higher than Pa_{CO2}. The rise in CSF P_{CO2} causes a fall in CSF pH, which then directly affects ventilation; this is valid for both respiratory and metabolic acidosis (Fencl et al., 1966).

Total Ventilation

The ventilatory response to changes in Pa_{CO2} can be characterized over most of its physiological range as a linear one, the slope of which varies considerably between species. This slope is generally reported in terms of the increase in ventilation in liters per minute per mmHg increase in Pa_{CO2}. Although CO_2 quickly reaches equilibrium with arterial blood and the brain, time is still required for ventilation to reach its final steady state; this is mainly due to the fact that adjustments of various hydrogen ion buffer systems occur very slowly due to the blood–brain barrier. However, the slope (S) of the response expressed in those units is of little utility for comparative purposes among mammals of a wide range of body size. Hence, these responses have been expressed here in terms of the percent change in ventilation per mmHg rise in $Paco_2$ (Tables 6 and 7), and they are taken from reports on responses of unanesthetized animals to inspired CO_2 (as opposed to venous CO_2 loading). It is apparent from Table 7 and

TABLE 6 Allometric Equations for Respiratory Variables in Mammals

Variable	Coefficient (a)	Exponent (b)	Reference
Tidal volume (ml)	7.69	1.04	Stahl (1967)
Dead space volume (ml)	2.8	0.96	Stahl (1967)
		1.05	Tenney and Bartlett (1967)
Lung volume		1.02	Tenney and Remmers (1963)
Total lung capacity (ml)	53.5	1.06	Stahl (1967)
Vital capacity (ml)	56.7	1.03	Stahl (1967)
Functional residual capacity (ml)	24.1	1.13	Stahl (1967)
Wet lung weight (g)	11.3	0.99	Stahl (1967)
	7.72	1.03	Bennett and Tenney (1982)
Lung compliance (ml/cm H_2O)	2.10	1.08	Stahl (1967)
Inflationary	1.94	1.08	Bennett and Tenney (1982)
Deflationary Po_{Max}20*	3.09	1.15	Bennett and Tenney (1982)
Deflationary Po_{Max}8*	2.2	1.09	Bennett and Tenney (1982)
Chest wall compliance	4.52	0.86	Stahl (1967)
Inflationary	5.59	1.2	Bennett and Tenney (1982)
Deflationary Po_{Max} 20	4.61	1.07	Bennett and Tenney (1982)
Deflationary Po_{Max} 8	4.9	1.16	Bennett and Tenney (1982)
Respiratory system comp.	1.56	1.04	Stahl (1967)
Inflationary	1.34	1.1	Bennett and Tenney (1982)
Deflationary Po_{Max} 20	1.8	1.11	Bennett and Tenney (1982)
Deflationary Po_{Max} 8	1.56	1.06	Bennett and Tenney (1982)
Ventilation (ml/min)	379	0.8	Stahl (1967)
Oxygen consumption (VO_2) (ml/min)	11.6	0.76	Stahl (1967)
Oxygen consumption (VO_2) (ml/s)	0.188	0.75	Kleiber, 1961
VO_2 Max (ml/s)	1.92	0.809	Taylor et al. (1981)
V_T/T_I (inspiratory flow)(ml/s)	13.5	0.74	Boggs and Tenney (1984)
Resistance (raw) (cm H_2O/l/s)	24.4	−0.70	Stahl (1967)
Resistance (raw) (cm H_2O/ml/s)	0.078	−0.819	Bennett and Tenney (1982)
Reciprocal time const. (s^{-1})	8.97	−0.298	Bennett and Tenney (1982)
Upper airway R (cm H_2O/ml/s)	0.056	−0.702	Bennett and Tenney (1982)
Reciprocal time constant lower airway + upper airway**	5.272	−0.326	Bennett and Tenney (1982)
Frequency (min^{-1})	53.5	−0.26	Stahl (1967)

**POMax = maximum airway pressure*
***law = lower airways; uaw = upper airways*

Figure 4 that there is considerable interspecific variation in the slope of the response. It is also apparent that the Pa_{CO2} set points, or Pa_{CO2} intercepts, are quite variable among species as well.

Some of this variation may be accounted for by differences in habitat. For example, fossorial and semi-fossorial species are chronically exposed to CO_2 in their environments and have adapted to it with a reduced

TABLE 7 Slopes of Ventilatory Responses to Carbon Dioxide

Species	Slope	Reference
Rat (Sprague Dawley)	6.6	Lai et al. 1978
Rat (Wistar)	6.4	Walker et al. (1985)
Rabbit	23.0	Honda 1968
Cat	12.8	Gautier and Bonora (1979)
Porcupine	33.0	Boggs and Birchard (1989)
Woodchuck	12.0	Boggs and Birchard (1989)
Coatimundi	23.0	Boggs et al. (1992)
Baboon	34.0	Lewis (1975)
Dog	38.3	Greco et al. (1978)
Dog	13.4	Reischl et al. (1978)
Goat	32.7	Mines and Sorensen (1970)
Human	28.0	Schaefer (1958)
Pony	38.0	Forster et al. (1976)
Weddell seal	16.7	Parkos and Wahrenbrock (1987)
Harbor seal	7.7	Robin et al. (1963)
Harbor seal	23, 12	Craig and Pasche (1980)
Hooded seal	6.52	Pasche (1976)
Harp seal	11.1	Pasche (1976)

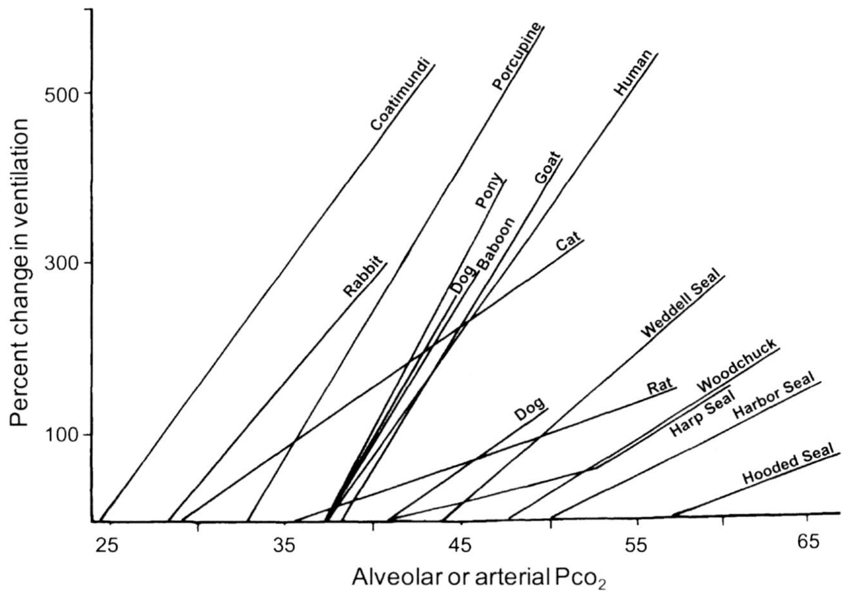

FIGURE 4 Ventilatory responses to changes in alveolar or arterial carbon dioxide in unanesthetized animals. *The source for each species is indicated in Table 7.*

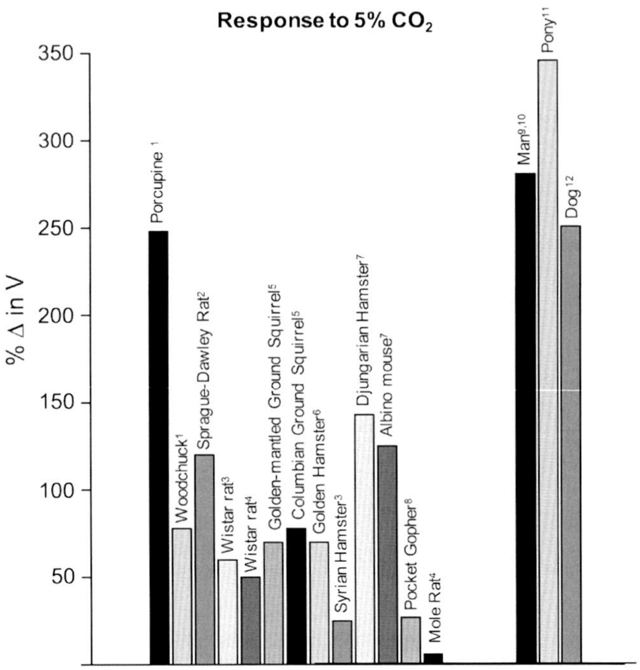

FIGURE 5 Ventilatory responses to a 5% CO_2 in the inspired air of 11 rodents and three other species. *Redone from Figure 5 in original edition.*

sensitivity to CO_2. This is illustrated in Figure 5, which represents the responses to a single level of inspired CO_2 (5%) of several species. All of the rodents except the porcupine show relatively smaller ventilatory responses to 5% inspired CO_2. Most of these are species that live in subterranean burrows part or all of the time, where the ambient CO_2 levels may often exceed 5% (Boggs and Tenney, 1984; Tenney, 1954). The porcupine, on the other hand, is more like the human, pony, or dog, sharing a life in the normal low CO_2 atmosphere. A semifossorial mammal, such as the woodchuck, has both a higher CO_2 set point and a lower slope in the CO_2 ventilatory response than a nonfossorial rodent, such as the porcupine (in the same study) or many of the other mammals. The laboratory mouse and rat (Figures 4 and 5) are interesting exceptions, although it seems likely that their ancestral wild types were semifossorial creatures as well.

This variation in CO_2 sensitivity appears likely to be genetically determined—a product of natural selection over many generations. However, ontogenetic acclimatization may be involved. Birchard et al. (1984) were unable to demonstrate any change in adult ventilatory response to CO_2 by prolonged fetal or neonatal exposure to elevated CO_2 (5%) environments in mice or rats. However, Rezzonico and Mortola (1988) did find persistent reductions in the ventilatory response to CO_2 of rats exposed as newborns to 7% CO_2 for 7 days. Findings support the former (Steggerda et al., 2010), but these authors also show a significant effect of hypercapnia on the subsequent ventilatory response to hypoxia. The different results in these studies may be due to the higher levels of CO_2 in both the initial exposure period (7% vs 5%) and the subsequent testing (10% vs 5%) in the Rezzonico and Mortola study versus the other two.

As yet, there is no good mechanistic explanation for the reduction in CO_2 sensitivity of burrowers compared to other mammals, although differences both in blood buffering and in chemoreceptor characteristics suggest themselves. There is not, however, a consistent difference in buffering capacity (Table 2). Some burrowing mammals do show an elevated buffering capacity, such as the completely fossorial pocket gopher (Chapman and Bennet, 1954), but others such as the woodchuck (Boggs and Birchard, 1989) or the totally fossorial mole rat (Ar et al., 1977) do not. The ground squirrels and echidna show distinct elevations in blood bicarbonate levels, while in other fossorial species this is not much above that found in other mammals. Human CO_2 retainers, such as patients with emphysema (Tenney, 1954), exhibit the same pattern of elevated Pa_{CO2} and $[HCO_3^-]$ and reduced sensitivity to inspired CO_2 observed in some burrowers. This pattern is also typical of normal humans acclimatized to elevated atmospheric CO_2 (Schaefer, 1974). Hence, some fraction of the reduced sensitivity to CO_2 in these animals must be attributable to elevated buffer base or noncarbonic buffering capacity. However, other adaptations at the cellular level of the central or peripheral chemoreceptors must have also taken place, particularly in those species that show little change in buffering and are operating at low blood and tissue pH (Boggs and Tenney, 1984; Ar et al., 1977). The eupneic breathing pattern and elevated Pa_{CO2} of some burrowers (e.g., woodchuck) is reminiscent of the pattern observed in other mammals of similar size that have had their carotid bodies denervated.

The reported ventilatory responses to CO_2 in diving mammals are quite variable, although they all have consistently high CO_2 set points. Unlike some of the other seals, the Weddell seal (Parkos and Wahrenbrock, 1987) seems to have a slope, or sensitivity, similar to some of the terrestrial mammals, such as the cat, rabbit, or dog. This is the case for the Harbor seals as well, at least in one study. The difference in results between two studies on Harbor seals may be related to the fact that the seals in the Robin study were restrained, whereas those of the Craig and Pasche (Craig and Pasche, 1980) study were not.

The slope of the ventilatory response to CO_2 in full-term human infants is apparently similar to that observed in adults, although they start from about a 4 torr lower Pa_{CO2} (Krauss et al., 1975). In studies of preterm and term monkeys, there is a progressive increase in the slope of the CO_2 response over the first 3 weeks of life. Preterm human infants respond less to CO_2 than term infants, which Rigatto et al. (1975) suggested is not due to a less effective pump but to a centrally mediated change.

Breathing Pattern

CO_2 tends to increase ventilation by both an increase in inspiratory drive and a decrease in expiratory time. The increase in inspiratory drive is generally reflected in increased tidal volume, with little change in inspiratory time until fairly high tidal volumes are reached. In most mammalian species, there is little change in T_I, but a significant decrease in T_E as V_T increases in response to CO_2. There are exceptions, however, represented by the woodchuck and Syrian hamster, both of which consistently increase T_E with increasing V_T in response to CO_2. Associated with this difference in the pattern of response is the smaller ventilatory response to CO_2 in these fossorial species (Figures 4 and 5), and a resting pattern that is also marked by longer T_E/T_{TOT}.

The increased tidal volume in response to CO_2 is a consistent response interspecifically and is due largely to the input from central chemoreceptors (Loeschcke and Widdicombe, 1974). The timing effects of CO_2 seem to have more to do with peripheral inputs to the CPG. Therefore, it would be interesting to see if differences in peripheral inputs can be associated with the interspecific differences in breathing pattern response to CO_2. There appear to be roles played by both chemoreceptor input and vagal afferents in controlling the pattern of response to CO_2. Davies (1978) demonstrated in anesthetized rabbits that blocking pulmonary stretch receptors with sulfur dioxide changed the pattern of response to CO_2 such that T_I was shortened as well as T_E, and that removing the other vagal afferents by vagotomy eliminated the shortening effect of CO_2 on expiratory time, suggesting that irritant receptors are responsible for the T_E shortening response to CO_2. Perhaps such receptors play a diminished role in the control of breathing in burrowers, such as the woodchuck and Syrian hamster. The literature on the role of vagal afferents in breathing-pattern regulation is replete with contradictions, possibly due in part to the use of transient versus steady-state responses, different methods of differential elimination of vagal fibers from different types of receptors, and different species and anesthetics. For example, in spite of the numerous studies indicating a role for vagal afferents in controlling the breathing pattern response to CO_2, a qualitatively similar pattern of response to CO_2 is reported to be seen in chlorolose anesthetized dogs with vagal influences removed (Ledlie et al., 1981).

Other pulmonary receptors that may function independent of phasic volume-related feedback have been postulated repeatedly. Suggestions of the existence of a mixed venous chemoreceptor have been made from investigations in the rat (Yamamoto and Edwards, 1960), rabbit (Linton et al., 1976), cat (Ponte and Purves, 1978), and dog (Stremel et al., 1978), but others have found no evidence of a pulmonary vascular chemoreceptor in the cat (Lambertson, 1980), baboon (Fordyce and Grodins, 1980), or dog (Fordyce and Grodins, 1980). Both Phillipson (1974) and Sheldon and Green (1982) suggested a role in the ventilatory response to CO_2 in dogs for some CO_2-sensitive pulmonary receptors with afferent fibers in the vagus nerve. Many reports had shown that increased airway CO_2 could decrease the response of pulmonary stretch receptors to a given transpulmonary pressure in opossums (Bystrzycka and Nail, 1980), dogs (Coleridge et al., 1978), and cats (Kunz et al., 1976), for example, but this effect was always greatest over the lower range of CO_2, and therefore seemed unlikely to play a significant role in the ventilatory response to inspired CO_2 or CO_2 loading. However, the latter had not been tested until Green et al. (1986) studied the effect of increasing the partial pressure of CO_2 in the pulmonary arterial blood. They found that even while ventilating the lung with room air, increased CO_2 in the pulmonary arterial blood depressed pulmonary stretch-receptor firing, and it did so both above and below the normal levels of CO_2. This depression of stretch-receptor firing was proportionately greater during expiration and was not dependent on changes in lung mechanics. Depression of stretch-receptor activity by CO_2 would have the dual effect of allowing lung volume to reach a higher volume before the inhibitory influence of stretch receptors would be sufficient to shut off inspiration, and it would reduce their T_E-prolonging effects and thus contribute to both the increased V_T and the reduction in expiratory time in response to CO_2. This may resolve the question of a venous CO_2 receptor in the lungs. Most investigators continue to believe that changes in ventilation specifically due to changes in pulmonary CO_2 can be accounted for by systemic arterial or central chemoreceptors and inhibition of pulmonary stretch receptors by CO_2 (Coleridge and Coleridge, 1986; Green et al., 1986; Fitzgerald and Lahiri, 1986).

Bowes et al. (1982) demonstrated a role for carotid-body chemoreceptors in shortening T_E in response to a single breath of CO_2 in awake dogs with vagal blockade. Other experiments with humans (Gardner, 1977; Ward et al., 1979) have also shown that carotid-body stimulation by CO_2 contributes to shortening T_E and can do so independent of changes in T_I or V_T, contrary to the earlier Clark and von Euler model (Clark and von Euler, 1972) that considered T_E to be dependent on the preceding T_I. Ledlie et al. (1981) demonstrated that even in the absence of either vagal or chest-wall afferents or carotid-sinus afferents, the response to CO_2 is still characterized by increased peak phrenic activity (and thus "intended" V_T) and decreased expiratory time, with little change in inspiratory time. Hence, although the increased chemical drive from CO_2 inhalation is quantitatively less, it seems to produce a similar qualitative pattern of response without peripheral inputs.

The decreased expiratory time may be due to one or all of three mechanisms: decreased expiratory braking, shortening end-expiratory pauses, and recruitment of abdominal and intercostal expiratory muscles (Saupe et al., 1992). Bartlett et al. (1973) demonstrated that the posterior cricoarytenoid muscle, the vocal cord abductor, shows prolonged activity into the expiratory phase in cats exposed to CO_2, thus reducing laryngeal resistance to expiratory airflow. Postinspiratory diaphragmatic activity also tended to be reduced in cats. This effect of CO_2 has been demonstrated in dogs by Smith et al. (1989), who also demonstrated the increased activity of expiratory muscles in dogs in response to 6.5% inspired CO_2. The increased tidal volume is thus recruited from both the inspiratory and expiratory reserve volumes. In a study of awake dogs (Bowes et al., 1982), the shortening of T_E in response to single-breath CO_2 tests was due entirely to the reduction of the end-expiratory pause.

5.2 Responses to Hypoxia

Even though CO_2 is the main variable that needs to be controlled in breathing regulation, oxygen also needs to be considered, particularly when arterial Pa_{O2} decreases significantly. When this happens, oxygen sensors in the carotid bodies are stimulated and ventilation increases. It should be noted that stimulation is not decreased by oxygen content but rather by the sensing of P_{O2}.

Total Ventilation

Although the variations in the response to CO_2 seemed to be more related to variations in habitat than to body size, there does appear to be a relationship between body size and hypoxic ventilatory response such that smaller mammals exhibit a greater response to a given level of hypoxia than large mammals (Boggs and Tenney, 1984; Figure 6). The slope of this relationship is essentially the same as that relating weight-specific metabolic rate to body size, which seems an entirely reasonable and predictable design feature of the respiratory control system. It is perhaps worth noting, however, that if this response were expressed in terms of the change in the convection ($\dot{V}/\dot{V}_{O2}$) requirement, or the process by which environmental air is brought in contact with the body's cells directly, it would probably have a greater negative slope. The design of the convection system is intimately related to both the metabolic needs of the organism and other nonrespiratory functions of the respiratory structures. Thus, as indicated by Figure 7, the $\dot{V}/\dot{V}_{O2}$ response in small mammals such as the mouse is greater than the $\dot{V}$ response, because at the same time that hypoxia is stimulating an increase in $\dot{V}$, it is inducing a decline in the metabolic rate. These are poikilocapnic or hypocapnic hypoxic responses (i.e., Pa_{CO2} is uncontrolled) (Figure 8). It would be of great interest to know what the interspecific differences in hypocapnic inhibition of breathing may be; however, due to the difficulty in obtaining such data in smaller species, little information is available. It might be

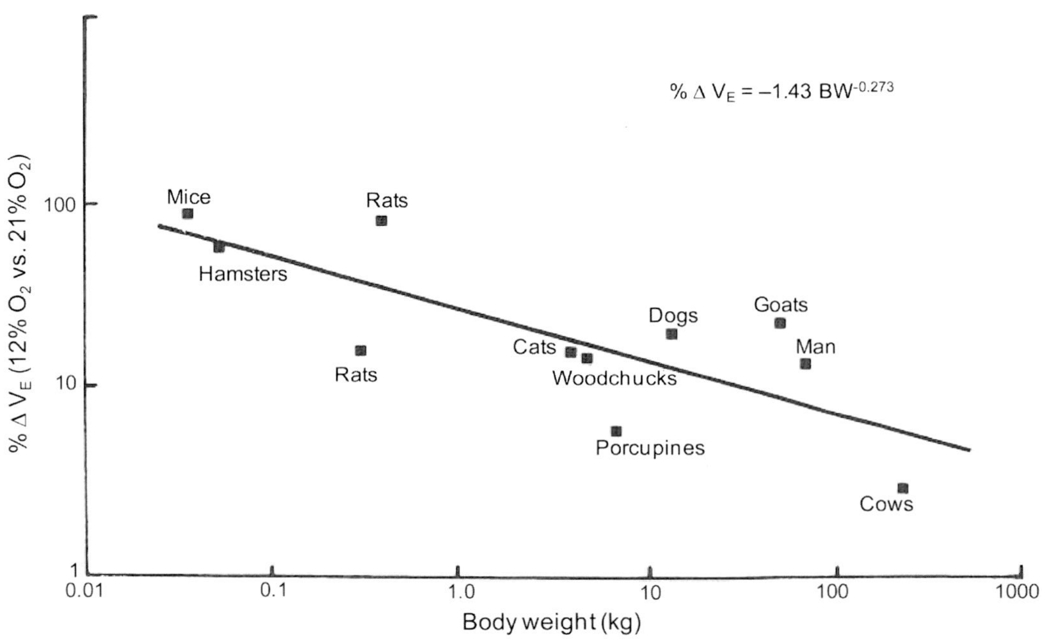

FIGURE 6 Ventilatory response to 12% inspired O_2 (or PIO_2 of 90 torr) in 10 species over a wide range of body size. *Redone from Figure 7 in original edition.*

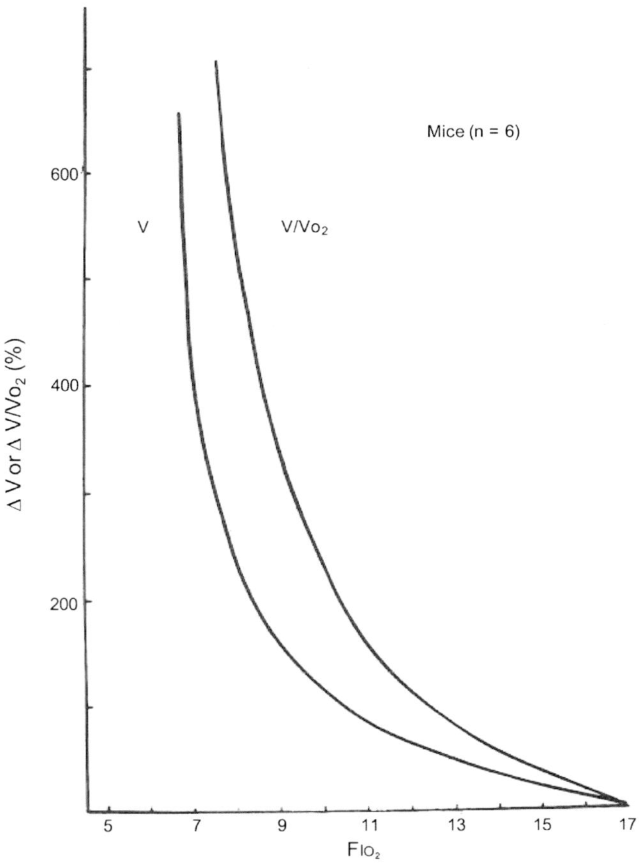

FIGURE 7 Comparison of the percent change in ventilation with hypoxia to the percent change in convection requirement with hypoxia in 30-g mice. Convection requirement changes more at any given level of hypoxia because oxygen consumption is depressed by hypoxia. *Redone from Figure 8 in original edition.*

less in smaller mammals that have a lower apparent slope to their responses to CO_2 (e.g., mouse, hamster, and rat in Figure 5) and greater responses to hypoxia than larger mammals.

There also tends to be a relationship between the O_2 affinity of hemoglobin (Hb) and body size, such that smaller mammals have a higher P_{50}, the P_{O2} at which Hb is 50% saturated, or lower affinity (Schmidt–Nielsen and Larimer, 1958), although this can also be affected by habitat as well; that is, higher affinity hemoglobins appearing in species adapted to the chronic hypoxia of high-altitude or subterranean habitats (Boggs and Tenney, 1984). It has been well known in humans for some time that there is a similarity in the shape of the hypoxic ventilatory response curve and the oxyhemoglobin dissociation curve, such that the rising phase of the response curve occurs over the same range of Pa_{O2} as the descending phase of the dissociation curve (Weil et al., 1970). This is also apparent in interspecific comparisons among mammals, as indicated in Figure 9 (Van Nice et al., 1980). The threshold of the hypoxic ventilatory response curve (i.e., the highest Pa_{O2} at which ventilation significantly increases over normoxic values) tends to occur at the level of Pa_{O2} at which the inflection of the dissociation curve is found. Hence, the ventilatory response curves (Figure 9(B)) of the cat and llama are shifted to lower Pa_{O2} values, as are their dissociation curves when compared to the rat (Figure 9(A)). On the other hand, the llama dissociation curve, being shifted so far to the left, is inevitably more hyperbolic than sigmoid, and hence descends quite steeply. Likewise, the slope of the llama's hypoxic response plotted against Sa_{O2} (percent saturation of arterial blood) is about twice as steep

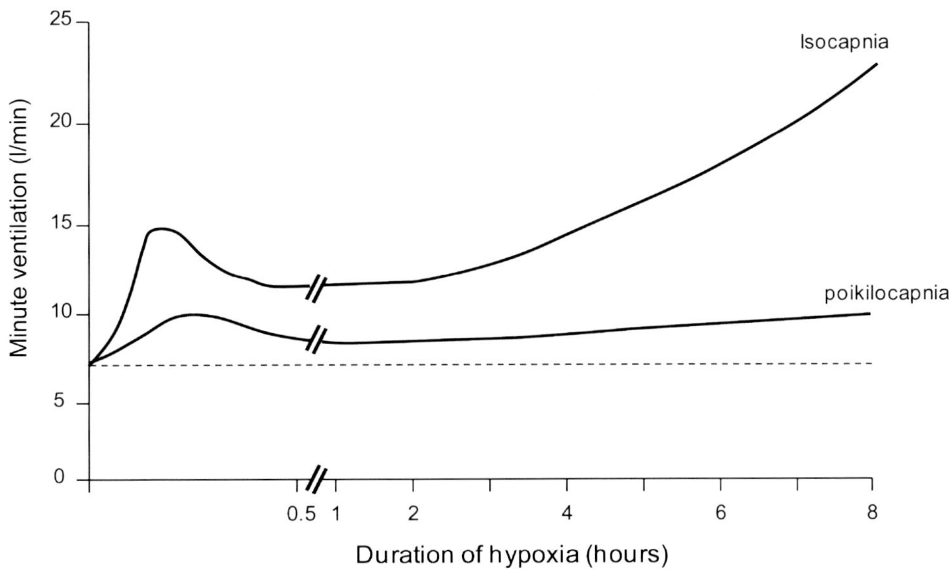

FIGURE 8 Example of the roll-off of the hypoxic ventilatory response under poikilocapnic and isocapnic conditions.

FIGURE 9 (A) Oxy-hemoglobin dissociation curves of llama, cat, and rat. (B) Ventilatory response to declining partial pressure of oxygen in arterial blood. (C) Ventilatory response to declining arterial O_2 saturation. *Redone from Figure 9 in original edition.*

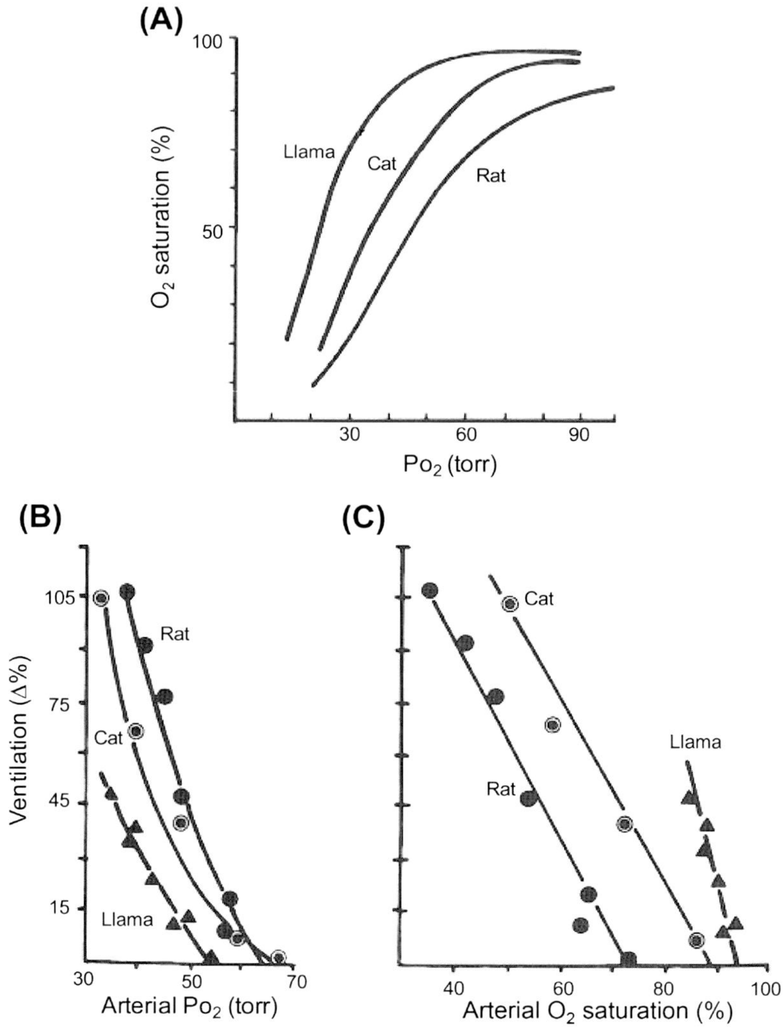

as that of either the rat or cat (Figure 9(C)). Again, these relationships between O_2 affinity and hypoxic ventilatory response seem quite parsimonious; it would be a waste of energy to increase ventilation at Pao_2 above the level of any significant desaturation and quite hazardous not to increase ventilation at Pa_{O2} that corresponded to levels of significant desaturation.

Numerous studies indicate that the primary sensors driving the hypoxic ventilatory response are the carotid bodies, which are responsive to changes in the partial pressure of oxygen in arterial blood, not its content (Lahiri et al., 1980; Torrance, 1974). Hence, these interspecific differences in hypoxic ventilatory threshold corresponding to differences in hemoglobin O_2 affinity are most likely a manifestation of natural selection, rather than being indicative of another physiological mechanism regulating hypoxic ventilatory responsiveness. However, the aortic bodies do appear to be sensitive to arterial O_2 content (Lahiri et al., 1980) and therefore could represent a mechanistic link in this observed interspecific relationship. In fact, studies demonstrate that in the absence of carotid bodies, compensatory upregulation of aortic bodies and other peripheral chemoreceptors contributes to the hypoxic ventilatory response (Prabhakar and Peng, 2004). More specifically, data from human subjects with bilateral carotid body resection indicated that carotid bodies are responsible for ventilatory stimulation during progressive hypoxia, whereas during transient hypoxia, the aortic bodies contribute 5–10% of the response (Honda, 1992). Yet few studies have been able to document a similar intraspecific relationship between Pa_{O2} and hypoxic threshold. Dawson and Evans (1966) demonstrated that sheep with a higher affinity HbA had a reduced hypoxic ventilatory responsiveness compared to sheep with the lower affinity HbB. However, Birchard and Tenney (1986) found no difference in the hypoxic ventilatory response of rats treated with sodium cyanate to lower their Pa_{O2} compared to control rats. The dissociation curve shifts to the left with low

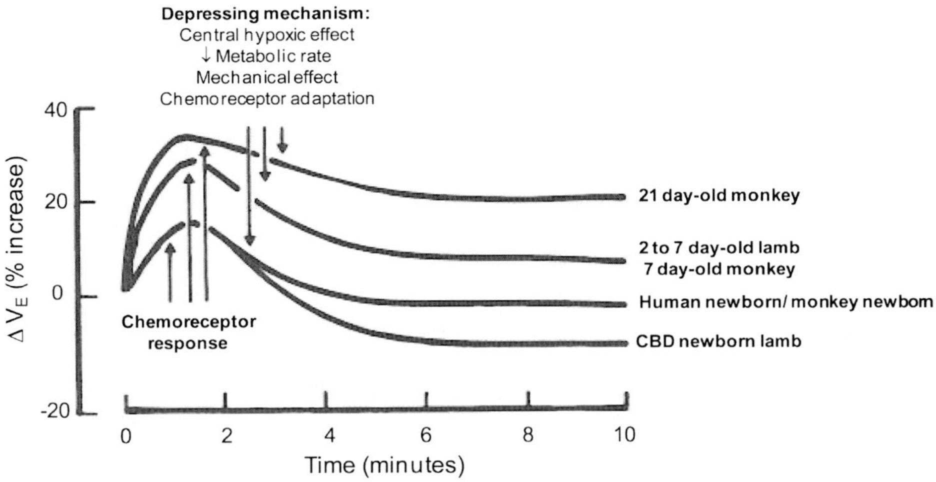

FIGURE 10 Schematic representation of the time course of the ventilatory response to hypoxia in newborns of different species and ages, with possible mechanisms of ventilatory depression. *Redone from Figure 10 in original edition.*

temperature, and if one compares the hypoxic ventilatory response of hibernating golden-mantled ground squirrels (7°C) to normothermic (37°C) ground squirrels, the threshold shifts to the lower Pa_{O2} of the descending portion of the dissociation curve at 7°C (Milsom et al., 1986).

Although burrow-dwelling mammals have a reduced ventilatory response to CO_2 (as pointed out in Section 5.1), there is less difference in their response to hypoxia. The Syrian hamster increases its ventilation in response to 10% O_2 to about the same extent as the white rat (Walker et al., 1985), and the woodchuck showed no difference in its hypoxic ventilatory response curve compared to the porcupine (Boggs and Birchard, 1989). The semifossorial golden-mantled and Columbian ground squirrels also show no difference in the threshold or sensitivity of their hypoxic response curves when compared to the white rat (Milsom et al., 1986). The totally fossorial mole rat, however, does have a lower threshold to its hypoxic ventilatory response when compared to the white rat (Arieli and Ar, 1979), which corresponds to the higher affinity hemoglobin in the mole rat compared to the white rat (Ar et al., 1977).

Studies of the ventilatory response to hypoxia of diving mammals have revealed variable and inconsistent results, though the majority of them seem to indicate a response similar to that of humans. A yearling harbor seal increased its ventilation almost threefold when PA_{O_2} fell to 55 torr (Bainton et al., 1973), and the hooded seal increased its ventilation 33% when FI_{O_2} was reduced to 0.10, which is similar to the human response (Pasche, 1976). The porpoise was observed to double its respiratory rate with an inspired O_2 between 9% and 10% (McCormick, 1969). In the studies of harbor and hooded seals (Pasche, 1976), the increases in ventilation as inspired O_2 declines between 15% and 10% are due mainly to a shortening of dive time, rather than any change in tidal volume or breath duration. On the other hand, Parkos and Wahrenbrock (1987) could not find any consistent relationship between minute ventilation and end-tidal O_2 in Weddell seals, although two of the five did show a twofold increase in ventilation when alveolar oxygen pressure dropped to the very low level of 28 torr. Hypoxia did not seem to effect the duration of dives either. They suggest that this species is quite insensitive to hypoxia.

Some newborn mammals respond to hypoxia in a distinctly biphasic fashion. The response is characterized by an initial increase in ventilation that is quite transitory and is followed by a sustained depression of ventilation back to or below control levels (Figure 10). This has been described in newborn humans (Cross and Oppe, 1952), monkeys (Woodrum et al., 1981), kittens (Blanco et al., 1984; Bonora et al., 1984a), rabbits (Grunstein et al., 1981), lambs (Bureau et al., 1984), and puppies (Haddad et al., 1982). In kittens, the response matures to the adult form at 2 months of age. Other reports on lambs indicate an ability to maintain a sustained ventilatory response to hypoxia (Cote et al., 1988). Davis et al. (1988) reported that a biphasic response is not apparent in awake newborn piglets either, and Mortola and Lanthier (1989) found that 2- to 3-day-old grey seals have a substantial hypoxic ventilatory response (Figure 11). This may reflect a difference in maturity at birth between species, where the more precocial are better able to sustain a response to hypoxia. However, it is interesting to note that Farber et al. (1972) found a greater response to hypoxia in newborn opossums than adults, despite their relative immaturity at birth, and a depression of metabolism by hypoxia.

The mechanism of this biphasic response to hypoxia in newborns remains the subject of some debate and considerable investigation. In adults, there is also a phenomenon of "roll-off" in the hypoxic ventilatory response described in humans (Easton et al., 1983), cats (Vizek et al., 1987), and dogs (Schaeffer and Haddad, 1988), such that after the

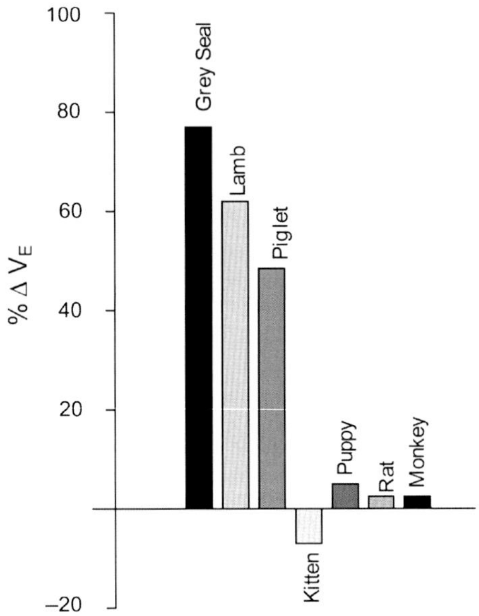

FIGURE 11 Comparative responses of newborn mammals to 5−10 min of 10−12% O_2 in inspired air. *Redone from Figure 11 in original edition.*

first few minutes of increased ventilation, there is a diminution of ventilation below the initial peak response, but still above the normoxic level. Vizek et al. (1987 posed the question as to whether this phenomenon in adults may represent a diminished manifestation of the same mechanism responsible for hypoxic ventilatory depression in newborns, but there appear to be some distinct differences. For instance, data suggest that neonatal hypoxia causes decreased levels of glutamate and aspartate and increased levels of GABA, which has respiratory depressing effects (Lumb, 2005; Teppema and Dahan, 2010).

Vizek et al. (1987) demonstrated in adult cats that the roll-off cannot be attributed to either changes in carotid body discharge or mechanical efficiency of the respiratory pump or changes in metabolic rate, and therefore postulated a "central mechanism." One suggested cause of "roll-off" is alkalinization of the central nervous system. Even though the phenomenon is still apparent in isocapnic hypoxia, as well as in hypocapnic hypoxia, there may be a sufficient increase in cerebral blood flow stimulated by hypoxia to induce a CO_2 washout from the area of the central chemoreceptors, resulting in respiratory depression.

Blanco et al. (1984) also found unchanged chemoreceptor activity in the kitten over time with hypoxia and thus attributed the depression in newborns to central mechanisms, but in this case mechanisms are retained from fetal life when hypoxia terminates fetal breathing movements. Some doubt is cast on direct generalized depression of central nervous system function by hypoxia in either newborns or adults by the fact that the phenomenon is apparent in cats and kittens, even with relatively mild hypoxia, and because in both cases there are periodic large breaths (sighs), which require a more intense discharge of inspiratory neurons (Blanco et al., 1984; Vizek et al., 1987). Also in newborn rats, although ventilation is depressed by hypoxia alone, the response to hypoxia, combined with elevated CO_2, is substantially greater than the response to CO_2 alone (Saetta and Mortola, 1987). Thus, the respiratory control system can respond with excitation to further chemical drive during the drop in ventilation of the biphasic response to hypoxia.

In contrast to Blanco et al. (1984), Bureau et al. (1984) found that there is little hypoxic depression of ventilation in carotid-body-denervated lambs and therefore attributed the phenomenon not to central mechanisms, but to immaturity of O_2-sensitive chemoreceptors, perhaps combined with only small depressive effects centrally of hypoxia and/or alkalosis, and mechanical changes. LaFramboise et al. (1983) suggested that the ventilatory depression is due to a change in respiratory mechanics rather than a reduction in central respiratory drive, because occlusion pressure (Po_2) was above the control level in both the hyperventilation and the hypoventilation phases. Grunstein et al. (1981) attributed hypoxic ventilatory depression in newborn rabbits to the release of endorphins with hypoxia, which depresses both metabolism and ventilation. Naloxone (an endorphin antagonist) was effective in reversing hypoxic respiratory depression, and the degree of depression by endorphins diminished with age in their study. Schaeffer and Haddad (1988) also suggested that endorphins are released in adult dogs and act to restrain ventilation, but only in response to severe hypoxia. Martin-Body and Johnston (1988) have demonstrated in newborn rabbits that decerebration at the midbrain/pontine junction abolishes the depressive component of the hypoxic response, which corresponds to studies in fetal lambs (Gluckman and Johnston, 1987) that identified an area of the lateral pons as essential for the depressive response to hypoxia at that stage.

An important factor that is sometimes overlooked is the fact that hypoxia also depresses metabolism in newborns (Haddad et al., 1982; Rezzonico and Mortola, 1988), and this must make some contribution to the decreased ventilation in newborns with hypoxia. Rezzonico and Mortola (1988) have demonstrated in unanesthetized newborn kittens that, despite the drop in total ventilation and alveolar ventilation to normoxic levels or below during the biphasic response to hypoxia, the kitten is actually hyperventilating with respect to CO_2 production, because metabolic rate falls more than ventilation in response to hypoxia. This drop in metabolic rate in response to hypoxia is likely to represent a protective response and may be similar to the response of small adult mammals, divers, and hibernators. The authors found a consistent drop in rectal temperature in kittens exposed to 10% O_2 for 10 min, and a larger drop was observed with more prolonged hypoxia in the study of kittens by Bonora et al. (1984b). Mortola suggested that this fall in body temperature may not be the result of the reduced metabolism but the cause of it. The loss of the hypoxic depression in decerebrate newborn rats observed by Martin-Body and Johnston (1988) could thus reflect the loss of pontine connections to the hypothalamus that play a role in the protective hypothermic hypometabolic response to hypoxia. Rezzonico and Mortola (1988) also noted that because the CO_2 production rate represents an important element in maintaining ventilation (Phillipson et al., 1981), this fall in metabolism with hypoxia may explain the periodic breathing observed in hypoxic infants as metabolic CO_2 load falls below the normal ventilatory threshold.

Hence, although hypoxia stimulates carotid bodies and increases ventilation, it can also have depressive effects on respiration in both adults and newborns. There are more postulated mechanisms for that depression in newborns than adults, and its consequences may be potentially more hazardous for the neonate. Regardless, these mechanisms probably include elements of all of the following: changes in pulmonary mechanics, a time-dependent decrease in carotid body function, impaired integration of afferent carotid sinus nerve input, decreased metabolism, and lastly, active inhibition by pontine and mesencephalic structures involving inhibitory neurotransmitters (Teppema and Dahan, 2010).

Breathing Pattern Responses

Although some significant differences were noted in the pattern of response to CO_2 among species, the pattern of response to hypoxia is more stereotyped (Figure 12), with the primary effect being an increase in frequency due to shortening of both inspiratory and expiratory times. Tidal volume is variously affected, showing small increases or decreases or no change, depending in part on the level of hypoxia. This pattern is observed even in those species that have unusual patterns of response to CO_2 and/or unusual eupneic patterns. Even in some of the newborns with a biphasic response to hypoxia, the decline in ventilation to normoxic levels is due to a fall in tidal volume, but frequency remains elevated, such as in rats (Saetta and Mortola, 1987) and kittens (Rezzonico and Mortola, 1988).

There are some important differences in the pattern of response to hypoxia alone (as distinct from isocapnic hypoxia), which have the effect of increasing the end-expiratory lung volume. This was elegantly demonstrated by Smith et al. (1989) in the awake dog; they found that with hypoxia inspiratory muscle activity increased, but expiratory muscle activity either decreased from control or did not change (the dog has an active phase to expiration in normoxia). This finding corresponds well with another report by Fregosi et al. (1987) that, in the decerebrate vagotomized cat, hypoxia inhibits abdominal expiratory nerve activity (as distinct from hypercapnia, which

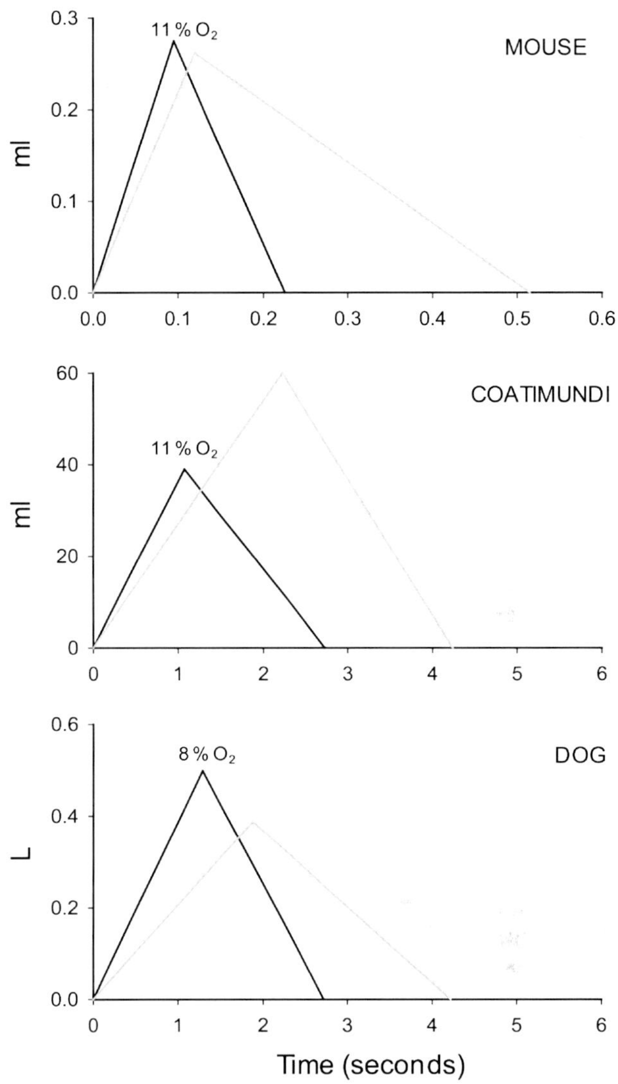

FIGURE 12 Comparative breathing pattern responses to hypoxia. All three species show a decrease in both inspiratory and expiratory times with variable effects on V_T. *Redone from Figure 13 in original edition.*

augments it). Smith et al. (1989) also found that post-inspiratory diaphragmatic activity that is reduced by hypercapnia is unchanged or lengthened by hypoxia, even to the extent of overlapping with expiratory muscle activity. Bartlett (1989) has demonstrated in the cat that while laryngeal resistance is decreased by hypercapnia and isocapnic hypoxia, it is increased by hypoxia alone. However, in humans, even isocapnic hypoxia induces glottic narrowing (England et al., 1982). The effect of the increased laryngeal resistance and airflow, along with the decreased expiratory time, is an elevation of end-expiratory lung volume. This elevation of FRC with hypoxia has been observed in humans (Garfinkel and Fitzgerald, 1978) and rats (Barer et al., 1978), in addition to dogs (Smith et al., 1989). Just as we speculated in Section 2.3 on eupneic breathing patterns, vis-à-vis newborns that use expiratory braking and high respiratory frequencies to elevate FRC above the relaxed volume of the respiratory system, here too in hypoxic adults, this response has the advantage of preserving lung oxygen stores and preventing small-airway narrowing or closure.

More extreme manifestations of these same response mechanisms result not just in retardation of expiratory flow, but also in cessation of it by glottic closure. Such end-inspiratory breath-holding patterns are seen in infant opossums (Farber et al., 1972), newborn humans (Mortola et al., 1982), newborn lambs (Harding, 1980), and certain diving mammals (Mortola and Lanthier, 1989). The response to hypoxia in reptiles such as the garter snake (Bartlett and Birchard, 1983) is also characterized by increased lung volume and prolonged end-inspiratory breath holding with glottic closure. This important post-inspiratory phase of the respiratory control system, then, is one that mammals have in common with other air-breathing vertebrates and that they all may use in elevating lung volume and oxygen stores in response to hypoxia.

High-Altitude Acclimatization

As a person ascends from sea level to high altitude, PAO_2, and therefore Pa_{O2}, falls, which results in hypoxemia and several compensatory responses thereafter. Within the first minutes of exposure to the hypobaric (decreased barometric pressure) hypoxia of high altitude, there is an acute increase in ventilation (Figure 13, point 1) that is followed, over the succeeding hours or days, by a further increase in ventilation (Figure 13, point 2) (and fall in Pa_{CO2}) that reflects the process of *ventilatory acclimatization*. This increased ventilation is not completely reversed by the administration of oxygen or a return to sea level (Figure 13, points 3–4). The ventilatory response to CO_2 also tends to be initiated at

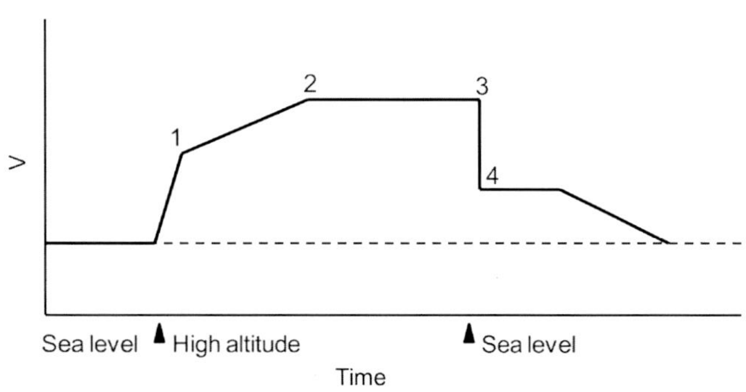

FIGURE 13 Schematic representation of changes in ventilation (V) with time at altitude. The acute response is represented by the rise to point 1. The further increase in ventilation with time at altitude (1–2) reflects the process of acclimatization. Some component of that same process is reflected as well in the failure of V to return to the pre-altitude acclimatization value with normoxia or upon return to sea level (3–4). *Redone from Figure 15 in original edition.*

lower values of Pa_{CO2}. As with so many physiological processes, the rate of ventilatory acclimatization is more rapid in small than in large mammals. This is suggested by data presented by Bouverot et al. (1973), wherein rats acclimatized to an altitude of 3550 m in 60 min, dogs in 3 h, and humans in 2–3 days. Bouverot suggests that these different rates of acclimatization to hypoxia are dependent upon the interspecific differences in the strength of the arterial O_2 chemoreceptor drive. However, it could relate to the more generalized phenomenon of higher physiological rates in small mammals as well, such as the faster interstitial fluid turnover in smaller mammals, which could affect cerebral acid-base changes associated with altitude acclimatization.

The initial hyperventilatory response to altitude is caused by stimulation of the peripheral chemoreceptors (i.e., the carotid bodies; Powell et al., 1998), which send afferent projections onto the brainstem respiratory nuclei. This initial response reduces arterial P_{CO2} and $[H^+]$, which in turn decreases the stimulation of the carotid bodies and thus limits the hyperventilation. Over time, ventilation increases steadily as part of the acclimatization process; the suggested mechanisms and structures involved include a time-dependent increase in carotid-body afferent activity, acidification of brain interstitial fluids surrounding the central chemoreceptors, or at least resolution of respiratory alkalosis in the blood, CSF, or cerebral interstitium; changes in monoamine metabolism in the brain; a suprapontine facilitatory influence upon respiratory control centers; and changes in respiratory mechanics that facilitate the expression of the drive to breathe over time.

Several studies on dogs, rabbits, cats, sheep, goats, and ponies have suggested that ventilatory acclimatization to high altitude is either prevented or reduced by carotid-body denervation (Bouverot et al., 1973; Bouverot and Bureau, 1975; Fordyce and Tenney, 1984; Forster et al., 1976, 1981; Lahiri et al., 1981). A few others, however, have indicated that ventilatory acclimatization may proceed in the absence of carotid-body input (Sorensen, 1970; Sorensen and Mines, 1970; Steinbrook et al., 1983). It is difficult to resolve these differences, although one possibility is that with carotid-body denervation there may be sufficient cerebral hypoxia to produce a cerebral acidosis that stimulates ventilation and reduces Pa_{CO2}. Differences in technique may contribute to some of the discrepancies in results, particularly, for example, in terms of attention to changes in blood acid-base state and ion concentrations in ruminants postoperatively (Steinbrook et al., 1983). Nevertheless, the majority opinion seems to be that the carotid bodies have a major role to play in the process, even though they are almost certainly not the sole actors (Joseph and Pequignot, 2009). A study by Bisgard et al. (1989) lends new support for this concept by demonstrating a time-dependent increase in carotid-body activity during hypoxia.

The next question, of course, is what is responsible for this change in carotid-body response to continuous hypoxia? Their data suggested that it is probably not the result of changing efferent input to the carotid bodies, because those were not intact, but there could be effects from circulating substances, such as catecholamines, or intrinsic changes in metabolism or neurotransmitters within the carotid bodies (Joseph and Pequignot, 2009). As such, studies have shown changes in peripheral neurotransmission linked to this effect at the gene and protein levels, not to mention by demonstrating neurotransmitter turnover and receptor expression (Joseph and Pequignot, 2009). Specifically, the action of acetylcholine (ACh), which mediates transmission between the glomus cell and the sensory nerve, has been found to be enhanced in rats during chronic hypoxia (He et al., 2005). Consequently, other neurotransmitters and neuromodulators such as dopamine and ATP, which is a co-transmitter with ACh, have also been found to be involved in the response (Huey et al., 2000; Joseph et al., 2002; He et al., 2006).

Attendant to the immediately increased ventilation in response to hypoxia is a hypocapnic alkalosis that inhibits breathing and hence constrains the full expression of the hypoxic ventilatory response. Therefore, the phenomenon of acclimatization could reflect an acid−base adjustment that removes that alkalotic constraint. Renal compensation of arterial hypocapnic alkalosis through bicarbonate excretion does occur, but the time course of that process does not correspond well to the time course of ventilatory acclimatization. Adjustments in the pH of arterial blood lag behind the pace of ventilatory acclimatization (Dempsey and Forster, 1982), and even when plasma bicarbonate levels are maintained constant, ventilatory acclimatization is unperturbed (Dempsey et al., 1978). It appears that hypoxia and hypocapnic alkalosis together may induce a lactic acidosis in central nervous system interstitial fluid (Fend et al., 1979) that would stimulate an increase in ventilation via the central chemoreceptors (Powell et al., 2000). Weil et al. (1986) summarized that this effect of hypoxia with hypocapnic alkalosis could be due to the reduction in cerebral blood flow attendant to hypocapnic alkalosis, the left shift of the oxyhemoglobin dissociation curve that impedes O_2 unloading, or the facilitation of glycolytic enzyme activities by hypocapnic alkalosis. It was challenged to some extent by a study by Bisgard et al. (1986) in which goats experienced hypoxia only in the blood perfusing the carotid bodies, while the rest of the body, including the brain, received normoxic normocapnic blood. These goats still exhibited the progressive increase in ventilation over the 4-h period, which is typical of acclimatization. They did not, however, exhibit the persistent hyperventilation after a return to normoxia, which is generally included as a component of the phenomenon of acclimatization and which may reflect inputs from some other component of the system. More recent data suggest that the involvement of central chemoreceptors is based on greater responsiveness to peripheral chemosensory inputs (Dwinell and Powell, 1999). Thus, just as the mediators of peripheral chemosensory neurotransmission are modified under hypoxic conditions, so too are those involved in central neurotransmission; these include not only the obvious catecholamines and dopamine, but also signaling molecules such as nitric oxide (NO) (Haxhiu et al), which may not have been previously considered.

To assess the role of suprapontine structures, Ou et al. (1983) studied the effects of decerebration on the process of high-altitude ventilatory acclimatization in cats. Decerebration of cats who were already acclimatized to high altitude had the effect of reducing their ventilation to the level of the "pre-acclimatized" acute hypoxic response. This suggested some facilitatory role for suprapontine structures in the process of acclimatization. However, cats that were decerebrated prior to exposure to chronic hypoxia were still able to exhibit an increased ventilation of acclimatization comparable to that observed in intact cats. Hence, while suprapontine structures may normally play some role in the process of acclimatization, it is not an obligatory one.

5.3 Responses to High and Low Temperatures

Both heat and cold can alter the breathing pattern in mammals. Central thermoreceptors monitor the temperature of the blood by using a servo-mechanism with a setpoint at normal body temperature. Hence, when the body temperature goes out of normal range (i.e., error signal), the body tends to respond with a negative feedback mechanism that restores temperature back to normal. The hypothalamus serves as the deep-body temperature control sensor. Thermoreceptors that respond to surface temperature are also present in the skin; these include cold and warm receptors, which are active over a broad range of temperatures. Lastly, there are receptors in the airways that respond to both warm and cold temperatures, as well as nociceptive stimuli; these have recently been discovered to be the transient receptor potential (TRP) ion channels (Banner et al., 2011). Temperature changes isolated to any of these sites can alter the breathing pattern independently.

Increased drive to breathe due to hyperthermia from the hypothalamus appears to function independently of central chemoreceptor activity, as it survives the elimination of central chemosensitivity; however, studies have demonstrated that respiratory pacemakers in the medulla also contribute to hyperthermic frequency modulation (Tryba and Marino-Ramirez, 2004). Temperature can also affect the response characteristics of other receptors normally involved in the control of breathing, such as the increase in neuronal activity in the carotid bodies induced by an increase in the temperature of the blood perfusing them and a decreased activity with decreased temperature (Bingman, 1977; Alcayaga et al., 1993); or the decreased response of vagal afferents with decreased temperature (Schoener and Frankel, 1972; Jonzon et al., 1988). Grunstein et al. (1973) found that hyperthermia affected both the bulbopontine and pulmonary vagal mechanisms controlling respiratory frequency (in the nonpanting anesthetized cat). Although there are qualitative similarities among mammals in the respiratory response to changes in temperature, there are also some notable differences, which will be discussed in the next sections.

Heat

In general, increased ambient temperature or body temperature have the effect of increasing breathing frequency and decreasing tidal volume, but to a much greater extent among those mammals that pant. Species that do not pant, but rely more on sweating than respiratory evaporative heat loss (such as man or horse), still exhibit an increased ventilation in excess of that attributable to the increased metabolic rate that may be associated with an increased body temperature (Robinson, 1949; Kaminski et al., 1985). In humans the increase in frequency is attributable to equivalent decrements in inspiratory and expiratory times (Petersen and Vejby-Christensen, 1977). Some desert species, such as the camel and eland, rely on heat storage, letting their body temperature fall at night and gradually rise during the day. To reduce evaporative heat loss in its dry environment, the eland reduces ventilation and increases O_2 extraction (Taylor, 1969). The camel was formerly thought not to pant, but apparently can be induced to do so with heat stress (if increasing from approximately 12 to 60 breaths/min can be considered panting). However, a dehydrated camel will not increase respiratory frequency nearly as much (Schroter et al., 1987). Dehydration reduces panting frequencies in dogs and rabbits as well. Elevations in osmolality of both carotid blood (Baker and Dawson, 1985) and cerebrospinal fluid (Turlejska and Baker, 1986) have been shown to reduce panting frequencies in dogs and rabbits, suggesting that brain osmoreceptors are involved in regulating panting. Another study (Baker and Dawson, 1985) also suggested a role for some other receptors in regulating panting frequency because it could be restored immediately after drinking, prior to changes in either blood volume or body fluid osmolality. Regardless, some degree of "true" hyperventilation may exist in mammals due to the development of hypocapnia during heat exposure (Calder and Schmidt–Nielsen, 1966; Jennings, 1984).

Kaminski et al. (1985) noted that, even with no change in deep body temperature, there is still a close relationship between changes in ambient temperature and changes in skin temperature in ponies, and changes in respiratory frequency are closely related to changes in skin temperature. They pointed out that it is not the absolute value of the ambient temperature that is important in this response, but the change from the acclimatization temperature. Hence, there is little difference in total ventilation or frequency between summer and winter (T, 5°C or 21°C), because the thermoneutral zone has changed. However, in either season, if T_A is increased by 9–18°C, skin temperature changes about 2°C and the frequency increases by approximately 15 breaths/min. Hence, studying the respiration of an animal at a temperature substantially different from the one it has been living in may present confounding stimuli.

At the same time that frequency increased in response to elevated ambient temperature, V_T decreased so that the ratio of dead-space volume to tidal volume increased from 0.43 to 0.7 in those ponies. This allows for an increase in ventilation without much change in alveolar ventilation and hence the maintenance of blood-gas homeostasis. Camels and other species also seem to maintain a constant V_T while increasing *f*. The absence of carotid-body input does not alter this protection of the blood-gas level through increased dead-space ventilation in ponies or in sheep (Hales et al., 1975). In contrast, Saxton (1975) found in humans a tendency for the rate to fall and V_T to increase, creating a hyperventilation hypocapnia when body temperature was elevated more than 1.5°C above normal.

In awake ponies, if inspired CO_2 is combined with elevated temperature, the slope of the relationship between Pa_{CO2} and ventilation remains the same, but is displaced to lower Pa_{CO2} and higher V_T values at the higher temperatures. A similar sort of displacement of the CO_2 response line without a change in slope is reported for awake humans in hyperthermia (Vejby-Christensen and Petersen, 1973). On the other hand, Cunningham and O'Riordan (1957) reported a near doubling of the slope of the CO_2 response curve, with a 1.1–2.9°C increase in body temperature in humans. Olievier et al. (1982) found no effect of temperature on the CO_2 response of anesthetized cats at body temperatures between 34 and 40°C, which is below the panting threshold, whereas above 41°C the slope was decreased. This is contrary to the increased slope found in hyperthermic cats (Widdicombe and Winning, 1974). These contradictory results point to the problems attendant to the methods used to raise body temperature, and therefore which receptors are stimulated, because some investigators use warm air, for example, whereas others use radiant heat. There are also problems introduced by the use of anesthesia because it depresses ventilatory responses to chemosensory stimuli (Paulin and Hornbien, 1986). There appeared to be a potentiation of the hypoxic ventilatory response with hyperthermia in humans, perhaps mediated through the effects of temperature on the discharge of the carotid chemoreceptors alluded to above (Vejby-Christensen and Petersen, 1973).

Many mammalian species respond to heat stress by panting, which refers to a tachypnea or very rapid shallow breathing. Panting may be initiated with increased ambient temperature before there is any change in body temperature, as in dogs, sheep, or cattle, or only after body temperature has reached some threshold temperature, as in cats (40 °C). In dogs, the panting frequency of highest incidence seemed to be 300 breaths/min (Iscoe et al., 1983), compared to a normal frequency of 16 breaths/min. Crawford (1962) also found that dogs often pant at approximately 320 breaths/min, which is roughly equivalent to the resonant frequency of the dog's respiratory system. At this rate, therefore, the respiratory impedance would be minimal, which should minimize the work of breathing. Iscoe et al. (1983) found that during panting T_I exceeds T_E. This may relate to the airflow patterns during panting in dogs. Schmidt–Nielsen and Larimer (1958) described the airflow in panting dogs as entering through the nose and leaving through the mouth. Because the nose has a countercurrent heat exchanger designed to conserve heat in the cold, this pattern of airflow bypasses that exchanger as it exits the respiratory system, and thus can dissipate more heat. Because the nose is a higher resistance pathway, it might explain the observation by Iscoe et al. that inspiratory time exceeds expiratory time.

Although the rapid, shallow pattern of panting is intended to ventilate the upper airways for evaporative heat loss without affecting gas exchange, it also increases alveolar ventilation and induces a respiratory alkalosis to varying degrees among species. Dogs (Albers et al., 1975), cats (Morgan and Adams, 1972), and wildebeest (Taylor et al., 1969) experience little or no alkalosis during moderate heat stress, and oxen (Hales and Findlay, 1968), sheep (Hales and Webster, 1967), and Tammar Wallaby (Dawson and Rose, 1970) experience only modest alkalosis. However, the goat and dikdik (Maskrey et al., 1978; Maskrey, 1984) tolerate considerable alkalosis while panting. With severe heat stress, however, all these species experience levels of hypocapnia and alkalosis beyond those tolerated by humans (e.g., Pa_{CO2} as low as 10 and a pH of 7.76 in the dog; Albers et al., 1975).

With severe heat stress, a second phase of respiratory response occurs in most mammals that are characterized by an increase in V_T and a decrease in frequency. The reason for this remains uncertain, but elevated tissue lactate seems a likely mechanism. The mechanisms elevating tissue lactate can be summarized as follows. It has been demonstrated in a number of studies that acidosis increases oxygen consumption. The increasing body temperature also increases oxygen demand, while the increased work of breathing raises metabolic rate as well. The redistribution of blood flow in response to heat stress leaves some tissues poorly perfused. Hypocapnia shifts the dissociation curve to the left and thus hinders off-loading of oxygen to the tissues. This, combined with the elevated O_2 demand, may cause hypoxia at the level of the tissues, even though Pa_{O2} is normal or above. In experiments with carotid-body-denervated cross-perfused dogs, Levine (1975) presented evidence for receptors that might respond to tissue hypoxia, later findings showed that it is the presence of a heme group in the receptors of several tissue types that allows oxygen sensation (Acker, 2008). These would play a physiological role during conditions when arterial oxygen tensions may be normal, but tissue hypoxia could nonetheless coexist, as in shock, severe exercise, or severe heat stress. In addition to the lactate-producing effect of alkalosis in tissues, in general there is the reduction of cerebral blood flow caused by hypocapnia, which may induce an acidosis in the vicinity of the central chemoreceptors. Hales et al. (1970) observed a reduced CSF [HCO_3^-] in sheep in the second phase of panting. Whether this was a cause or effect of the changed breathing pattern was uncertain, although the former seems more likely.

Cold

Although there tends to be a linear increase in ventilation with increasing metabolic rate in normothermia, this is not the case in the cold. In general, deep hypothermia causes the respiratory rate, V_T, and V_E to decrease; however, in some mammals the response is linked to oxygen consumption. In sheep, when oxygen consumption is doubled to increase heat production, ventilation is only increased by 40% and there is a fall in the percent O_2 in the exhaled air (Joyce and Baxter, 1964). This same phenomenon was reported in pigs by Ingram and Legge (1970), who also calculated that this reduced convection requirement and increased O_2 extraction from the air resulted in a significant savings in evaporative heat loss (approximately 15%). This is not the case for humans who proportionally accommodate ventilation to oxygen consumption; the Tasmanian devil is another mammal that responds in this fashion (Nicol and Maskrey, 1980). The reason some species respond to cold-induced metabolic demands by increasing pulmonary extraction while others respond by increasing pulmonary convection is not entirely clear. Increasing V_E proportional to oxygen consumption, especially if an increase in V_T contributes, would serve to maintain the proportional ratio between metabolic rate and ventilation, and therefore alveolar blood gases. Conversely, hyperpnea (i.e., increased V_E) would favor heat loss because ambient temperature is almost always lower than body temperature; this would greatly impact heat production.

As indicated earlier, there are cold receptors in the nasal passages that affect the breathing pattern. In their studies of these receptors in humans, Burgess and Whitelaw (1988) confirmed previous observations that cold air through the nose inhibits ventilation, and they showed that this was not related to an increase in flow resistance. The decrease in total ventilation was due to a decline in both V_T and T_E (i.e., pause) without a significant change in frequency. Furthermore, cold air through the nose reduced the average slope of the ventilatory response to CO_2 by 27%. They speculated that the afferents from these receptors run in the trigeminal nerve that terminates in the solitary tract nucleus, which is intimately related with the dorsal group of medullary respiratory neurons. In fact, studies have shown that sensory nerves sensitive to cold through TRP channels travel in the trigeminal nerve and alter breathing (Eccles, 2000, 2003). The utility of such a response to cold air is in the reduction of respiratory evaporative heat loss. Similar trigeminal nerve receptors in the face and nose have long been implicated in the dive reflex that can inhibit breathing in response to submergence of the face in cold water, which is clearly an important protective reflex.

A reduction of body temperature depresses respiration, not only because of the associated reduction in metabolism but also because of the impact of cold on neural function, particularly the central chemoreceptors. Hypothermia of 25°C or less eliminates the sensitivity of central chemoreceptors to CO_2 in dogs (Natsui, 1969; Ruiz, 1975). Peripheral chemoreceptors continue to function at a reduced level at that temperature but cease to function at all at 22°C in the dog (Oruc and Terzioglu, 1979). Some species, however, clearly retain neural function at much lower temperatures. Hibernators allow their body temperatures to fall as low as 7°C, but still respond to inspired CO_2 with an increase in ventilation. The maintenance of a ventilatory response to CO_2 is reported for hibernating hamsters (Lyman and Hastings, 1951) and hedgehogs (Tahti, 1975). However, hibernating ground squirrels do not respond to hypoxia (1−3% O_2; Milsom et al., 1986), and denervation of carotid bodies does not alter their response to hypoxia. These structures appear to play no role in the control of breathing in hibernating ground squirrels (Milsom et al., 1986). At these low temperatures, their oxyhemoglobin dissociation curves are shifted so far to the left that the P_{50} for a Columbian ground squirrel, for example, is only 4−8 torr. In this situation, when exposed to very low levels of oxygen, respiratory depression and death occur before any significant respiratory stimulation or arousal takes place.

The breathing pattern during hibernation is characterized by its periodic nature. This takes two basic forms. Either single breaths are separated by apneic periods of 1−6 min, as in the marmot (Malan et al., 1973) or Columbian ground squirrel (Milsom et al., 1986); or there are bursts of continuous breathing separated by longer periods of apnea up to 45 or 150 min, as observed in the hedgehog (Tahti, 1975) or golden-mantled ground squirrel (Milsom et al., 1986). In the latter species, however, its pattern can be switched to the more continuous single-breath pattern simply by decreasing the ambient temperature from above 5°C to 4°C. When these animals are stimulated to increase their ventilation with 5−7% CO_2, tidal volume increases but the breath duration remains unchanged. Frequency is increased by reducing the apneas or *nonventilatory periods*. These periods between breaths, then, during normal unstimulated breathing are probably regulated by changes in Pa_{CO_2}. This sort of periodic breathing that is regulated by changing the duration of the nonventilatory period is reminiscent of reptiles, which also do not have high enough metabolic rates to require continuous breathing (Milsom, 1988).

The control of respiration in hibernators is fascinating because it appears to play a key role in the processes of entrance into and arousal from hibernation (Malan, 1982, 1986). During the early stages of entrance into hibernation, respiration is depressed more than metabolism, resulting in hypoxemia, acidosis, and an increase in CO_2 stores. The hypoxia may depress shivering thermogenesis, as it does in cats. The respiratory acidosis, in turn, inhibits metabolism and nonshivering thermogenesis, and perhaps facilitates the lowering of the temperature setpoint by depressing the hypothalamic neuronal activity involved in thermoregulation (Wunnenberg and Baltruschat, 1982). Hence, this hypoventilation appears to facilitate entry into hibernation.

While in hibernation, the animal remains in a state of relative respiratory acidosis, with Pa_{CO_2} on the order of 150 torr (corrected to 37°C). Because temperature affects the dissociation constant of water, the pH of neutrality is 6.9 at 37°C and 7.4 at 5°C. If a constant $[HCO_3^-]/[H^+]$, or a constant relative alkalinity were to be maintained, then at a lower temperature the pH should be higher (e.g., 7.8−8.0), and this is observed in some poikilothermic vertebrates. Hence, a theory about the regulation of ventilation in poikilotherms was developed suggesting that as their body temperature varies, their respiration is regulated to maintain a constant relative alkalinity, which allows for the maintenance of a constant fractional dissociation of the most important protein buffer group, the imidazole of histidine residues. This maintains the conformation, and therefore the activity, of such regulatory proteins as enzymes, receptors, and ion channels. This so-called imidazole alpha-stat hypothesis for the regulation of ventilation in poikilothermic vertebrates was first developed by Reeves (1972), but with further work it appeared that it was too simple, as it was not apparent in all species or under all conditions (Jackson, 1989; Truchot, 1987). Nonetheless, the basic chemistry explains why hibernating mammals are said to be in a state of relative respiratory acidosis when they maintain a pH near 7.4 when their body temperature is 7°C. Malan, (1982, 1986) and Hand and Somero (1983) suggested that this lack of alpha-stat (or constant relative alkalinity) regulation of respiration does indeed cause some important metabolic enzymes to be inactivated, which is important in maintaining the depressed metabolism of hibernation. For example, phosphofructokinase is changed from its active tetrameric form to the inactive dimeric form as a result of the changed ionic state of histidine residues with the relative acidosis of hibernation (Hand and Somero, 1983). A quick way to reverse that inactivation, then, when the animal needs to arouse, is to hyperventilate and reverse the respiratory acidosis. It is quite a remarkable hypothetical system, but it is currently unknown what initiates this sudden hyperventilation for arousal or what depresses ventilation ahead of metabolism during the process of entrance into hibernation.

5.4 Entrainment of Respiration to Locomotion

Entrainment refers to the matching of breathing and exercise rhythm. With exercise, the frequency of the breathing cycle may become coordinated with the frequency of limb movements. Hence, entrainment can occur at various ratios, which depend on factors such as mode of exercise, intensity, and experience (Bernasconi and Kohl, 1993; Seebauer et al., 2003). This occurs to a variable degree among mammals with different forms of locomotion, as well as over the range of exercise intensities within a species. Various forms of entrainment have been documented in humans during walking, running (Paterson et al., 1987), and bicycling (Bechbache and Duffin, 1977; Jasinskas et al., 1981; Bonsignore et al., 1998). Some investigators find that entrainment is more pronounced at higher exercise intensities (Bechbache and Duffin, 1977; Loring et al., 1976), whereas others do not (Jasinskas et al., 1981), and still others find no evidence of entrainment in humans (Kay et al., 1975). It appears that bicycling may not be a good test of this phenomenon in humans, perhaps because there is so little mechanical impact of limb oscillations on the thorax. Because of its low incidence and the observation of a significant enhancement by a metronome set at pedal frequency, entrainment during cycling may be due to voluntary setting of the breathing pattern, rather than to any mechanical effect upon the thorax or a direct influence of neural feedback from the limbs to the respiratory controller. However, in running humans, Bramble and Carrier (1983) found considerable evidence of entrainment, particularly among experienced runners, and noted that most deny any awareness of breathing pattern coordination or changes with running speed.

The patterns of locomotor to respiratory cycle coordination vary considerably among human subjects (Table 8), but the most experienced runners tend to favor a 2:1 ratio (Bramble, 1983). Among those quadrupedal mammals studied thus far (hare, dog, horse, and cat), a 1:1 ratio is more commonly observed, particularly at faster gaits, such as galloping (Table 8). Bramble (1989) noted that it is the faster asymmetrical gaits that generally are associated with the more rigid 1:1 synchronization, whereas the slower symmetrical gaits seem to show greater flexibility in the patterns of coordination between stepping and breathing rates. (Asymmetrical gaits are defined as those in which the footfalls of the two feet of a pair are unevenly spaced in time, and these gaits increase stride length by introducing a period of suspension when all feet are off the ground.) Iscoe (1981) found that only two out of six cats demonstrated entrainment in treadmill studies, but this may be

TABLE 8 Entrainment Patterns in Mammals

Species	Locomotor to Respiratory Cycle Ratio	Reference
Human	3:2, 2:1, 5:2, 3:1, 7:2, 4:1	Bramble and Carrier (1983); Bramble (1983); Paterson et al. (1987)
Horse	Trot—1:1, 3:2, 2:1, 5:1 Canter—1:1 Gallop—1:1	Attenburrow (1982); Bramble and Carrier (1983) Hornicke et al. (1983)
Cat (decerebrate)	Gallop—1:1	Kawahara et al. (1989)
Dog	Trot—1:1 Gallop—1:1	Bramble and Carrier (1983) Bramble (1983)
Hare	Half-bound—1:2, 1:1 Gallop—1:1	Bramble and Carrier (1983)
Wallaby	Hop—1:1	Baudinette et al. (1987)
Bats	Flight—1:1	Suthers et al. (1972); Thomas (1981); Carpenter (1986)

attributable to the fact that his cats only walked and would not run. Kawahara et al. (1989) found a 1:1 entrainment of breathing rhythm to stepping rhythm in decerebrate cats induced to gallop on a treadmill by a tonic stimulus to the mesencephalic locomotor region. The strength of the locomotor–respiratory coupling increased as the mean stepping interval was shortened. They suggested that gait may be more important than stepping interval, however, because there was a decline in coupling when the cat switched from a gallop to a trot, even though the stepping interval or gait cycle time did not change.

The bipedal hop of the wallaby is in 1:1 linkage with respiratory frequency, regardless of speed (Baudinette et al., 1987). The phase relationship of the constant 1:1 linkage between hop frequency and breathing frequency is fixed, with inspiration beginning as the animal leaves the ground. A "visceral piston" mechanism has been postulated to explain this synchrony. It is suggested that as the hind limbs leave the ground, the rib cage is accelerated with respect to the abdominal contents, which pull the diaphragm down by the large central tendon, and later as the viscera catch up with the thorax they will tend to push the diaphragm forward, contributing to expiration. A similar visceral piston mechanism has been postulated to be operating in running dogs and horses (Bramble, 1989), in addition to other mechanical forces acting on the thorax during a stride.

Bramble (1989) described a four-component biomechanical model to explain the locomotor–respiratory cycle coupling in galloping horses. Inspiration occurs as the thorax is expanded when the forelimbs are stretched forward, and the flexion of the neck in lowering the head may also pull the ribs forward, tending to expand the thorax. At the same time, the thrust from the lumbo-pelvic unit results in the rearward displacement of the visceral piston, favoring the descent of the diaphragm. Exhalation is associated with the time when the forelimbs hit the ground, and the impact is transmitted to the thorax with a compressive effect. Simultaneously, the deceleration forces on the trunk cause anterior displacement of the visceral piston, contributing to the reduction of the thoracic volume. Bramble also suggested that locomotor–respiratory coupling could provide a basis for matching lung ventilation to running speed if respiratory frequency increases in proportion to stride frequency and tidal volume increases in proportion to stride length. Data from five horses support the concept of a proportionate increase in tidal volume with increasing stride length in the gait transitions from walk to trot to cantor to fast gallop.

Flying mammals also appear to have a strict 1:1 synchrony of wing beat and respiratory rate, regardless of flight speed (Thomas, 1981). This synchrony has been observed in two species of bats of widely differing body size, *Pteropus gouldii,* or flying foxes (800 g), and *Phyllostomus hastatus,* a fruit bat of about 80 g. The phase relationship is such that inspiration occurs during the downstroke (Thomas, 1981; Suthers et al., 1972). Here, too, it is postulated that the mechanical impact of locomotor muscles and appendages on the thorax plays an important role in phase linkage (Vaughan, 1959). It is interesting to note, however, that in birds, with their very different respiratory apparatus, this strict 1:1 synchrony is not observed, but rather some 12 different patterns of coordination between wing beat and respiratory cycle have been reported, and some birds readily switch from one to another (Berger et al., 1970).

The control mechanisms responsible for the synchrony of respiratory and limb movements in mammals are still not entirely clear, but several possibilities have been suggested involving feedback and feed-forward pathways, and the activity of spinal locomotor pattern generators (Dubayle and Viala). The idea that the biomechanical impact of locomotion upon the thorax may induce entrainment through feedback mechanisms derives some support from the observation of phase locking of phrenic output and/or diaphragmatic EMG to mechanical ventilator frequencies in cats (Petrillo et al., 1983) and humans (Graves et al., 1986). Bilateral vagotomy, in the animal experiments, abolishes the entrainment of respiratory motor neuron activity to mechanical ventilator frequencies (Petrillo et al., 1983). Although the vagal afferent information necessary for effective synchronization of the respiratory rhythm to an exogenous oscillator has generally been attributed to the slowly adapting receptors considered responsible for the Hering-Breuer reflexes, Graves et al. (1986) suggested that in their human studies the rapidly adapting receptors may be responsible. They would be stimulated by the rapid onset of mechanical lung inflation, and they are implicated because of the observation in humans with 1:1 phase locking that T_E is shortened with inconsistent changes in T_I. Studies in cats and rabbits (Davies et al., 1982) suggested that rapidly adapting reflexes may be responsible for shortening T_E, and augmenting inspiration. It is also possible that intercostal afferents (Remmers and Martilla, 1975) contribute to the synchronization of respiration to exogenous mechanical oscillations of the thorax, either by mechanical ventilators, by impact loading when the forelimbs hit the ground, or by a visceral piston.

Another source of afferent information and possible feedback mechanisms for entrainment is, of course, the proprioceptors of exercising limb muscles, which are well known to alter both the rate and depth of breathing, whether the limb movement is active or passive (Agostoni and D'Angelo, 1976). Iscoe and Polosa (1976) demonstrated the existence of reflex mechanisms that are capable of locking respiratory frequency to the frequency of a periodic afferent input from cutaneous (radial) and muscle (hamstring) afferents in cats. The entrainment of phrenic activity to the frequency of repetitive stimulation of such somatic afferents occurred at ratios of 4:3, 4:5, 1:2, 1:3, 1:4, and 1:7.

Morin and Viala (2002) showed that sensory input from the lumbar spinal cord entrained respiratory rhythm and evoked inhibitory postsynaptic potentials in phrenic motor neurons, which served to modulate their activity to coincide with descending command signals from the higher respiratory centers. Similarly, Potts et al. (2005) demonstrated that entrainment pathways from somatic afferents to the ventral respiratory group via the lateral parabrachial nucleus cause resetting of respiratory rhythm through excitation of expiratory-augmenting neurons and consequent inhibition of postinspiratory neurons.

In addition to these feedback mechanisms for entrainment, there is evidence for feed-forward mechanisms or direct interactions of CPGs for locomotion and respiration. Eldridge et al. (1981, 1985) presented compelling data indicating that locomotor and respiratory motor outputs are driven in a parallel and proportionate fashion by a hypothalamic locomotor command center in cats deprived of any peripheral afferent feedback. Viala et al. (1979) also described a locomotor pattern generator in the lumbar spinal cord of rabbits that appears capable of driving coordinated phrenic activity following high cervical transection.

5.5 Anesthesia

Although most of the studies of the control of breathing examined here have been in unanesthetized mammals, it seems appropriate to point out at least briefly some of the effects of anesthetics upon breathing. It is well known that anesthetics depress ventilation, although there is considerable variation in the extent and pattern of this effect. One of two approaches is usually applied to measure the effects of drugs on breathing: effect on resting ventilation or effect on end-tidal arterial P_{CO2}.

Anesthetics have multiple membrane targets including not only the ionotropic $GABA_A$, glycine, NMDA, and nicotinic acetylcholine receptors, which are crucial in the respiratory neural network, but also the metabotropic receptors that mediate cell signaling in these neurons (Teppema and Baby, 2011). Among the inhalation anesthetics, for example, fluroexene, cyclopropane, ether, and methoxyflurane depress ventilation less than halothane, enflurane, and isoflurane. The inhalation anesthetics reduce tidal volume but may actually increase the frequency of breathing. Although it would seem likely that they could be having this effect via pulmonary receptors, the response appears to be mediated centrally because vagal blockade does not change the pattern induced by halothane (Mazzarelli et al., 1979). On the other hand, there is some evidence that halothane reduces the ventilatory response to hypoxia by reducing the discharge frequency of carotid bodies (Davies et al., 1982). For brief overviews on the effects of anesthetics on the respiratory system using ex vivo and in vivo systems, please refer to studies by Stuth et al. (2008) and Takita and Morimoto (2010).

A different breathing pattern is induced by narcotics, such as morphine, that tend to depress breathing by decreasing frequency more than tidal volume. They also reduce ventilation, in part through their inhibitory effect on metabolic rate (Bellville et al., 1960). The sedatives tend to fall between the narcotics and inhalation anesthetics in the nature of their effects upon tidal volume and frequency, but they too depress breathing and the ventilatory responsiveness to respiratory stimuli. Diazepam, for example, decreases the ventilatory response to CO_2 by 20–30% (Gasser and Bellville, 1976). In comparing two barbiturates (thiopental and pentobarbital) at relatively light anesthetic levels to ketamine, Hirshman et al. (1975) found that although all three drugs reduced the hypoxic ventilatory response of dogs, ketamine had a far smaller effect. On the other hand, the ventilatory response to CO_2 was greatly reduced by all three anesthetics.

In addition to their central depression of breathing, barbiturates, sedatives, and alcohol depress the phasic respiratory activity of the hypoglossal nerve (to the tongue and pharynx), and this can contribute to the development of obstructive apneas (Bonora et al., 1984a,b). Finally, studies looking into the effects of opioids have demonstrated that this class of drug depresses respiration by not only changing rhythm generation, but also chemoreception; although the focus has been on brainstem-derived effects, some data on cortical mechanisms are now also available.

6. CONCLUSION

We have seen that the output of the central pattern generator controlling the rhythmic act of breathing can be modified by a great many influences, both within and between species. Interspecifically, variation in body size, which includes both apparent and subtle differences in metabolism and respiratory system mechanics, creates somewhat predictable effects upon the breathing pattern. However, perhaps not all variations associated with differences in body size may be explained directly in terms of mechanics and metabolism. The greater ventilatory response to hypoxia in small versus large mammals, for example, may also be related to a relatively lower sensitivity to CO_2, and therefore to the hypocapnic inhibition of breathing that is a consequence of hypoxic hyperventilation. Similarly, although there are some general consistencies in volume–time profiles across species, some differences in volume–time profiles have been documented both during eupnea and in response to hypercapnia. Further study of the mechanisms underlying such differences in breathing patterns between species would contribute to a better understanding of the essential underlying mechanisms of the control of breathing in mammals in general. This will be particularly relevant as new data on the role of proteins, such hypoxia-inducible factor 1 (HIF-1 alpha) and heme oxygenase 1 (HO-1), becomes available.

We have also seen many apparently universal features of the response of the respiratory control system to hypoxic, hypercapnic, and thermal stimuli, as well as to exercise. In response to hypocapnic hypoxia, for example, both adults and newborns appear to employ a pattern of increasing respiratory frequency more than tidal volume, and use both this pattern and upper-airway resistance to raise end-expiratory lung volume and to defend lung O_2 stores. This response is, in fact, shared with other air-breathing vertebrates. The response to CO_2 shows quantitative variations among species, related in some cases to adaptations to habitat, but the physiological mechanisms for such differences remain uncertain. The qualitative response to CO_2 is quite consistent interspecifically in terms of the increase in tidal volume. The effect upon timing is generally to decrease expiratory time, although a few exceptions to that pattern were seen. The distribution of responsibility for the effects of CO_2 upon breathing pattern among all the CO_2-receptive structures (central chemoreceptors, peripheral chemoreceptors, and pulmonary stretch receptors) is still being clarified, however.

The response of the respiratory system to changes in temperature is also consistent in the qualitative pattern among species but varies quantitatively. The response to heat is generally biphasic, with a decrease in volume and an increase in frequency representing the pattern of the first phase, while during a second phase, particularly with severe heat stress, there is ultimately a transition to a greater volume and lower frequency pattern. The first phase is highly variable quantitatively between species that pant, which try to effectively avoid hypocapnic alkalosis, and those that do not pant. The mechanisms driving the second phase are still not entirely well understood, this is especially true when making interspecies comparisons. For that matter, responses to cold are no different. For instance, the response of some hibernating species to cold has raised many intriguing questions about the nature of chemoreceptor function, such as whether it is an alpha-imidazole or pH mechanism, and how it may or may not vary between hibernating and nonhibernating species.

Lastly, there emerges from numerous studies a consistent pattern of some form of entrainment of the respiratory pattern generator to the locomotor pattern generator. The relative contributions of feed-forward and feedback neural circuits to this phenomenon remain to be elucidated further, but the complexity of this coordinated entrainment process, which includes not only direct input from somatic afferents but highly integrated spinal inputs as well, is becoming increasing clear. As we marvel at the intricate mechanisms that control breathing in mammals, it would be impossible to ignore the interactive nature of all the body systems, which work together by both direct impact and indirect and subtle influence, to insure the proper responsiveness to changing conditions and maintenance of physiological function. Although great progress has been made in unraveling it, the control of breathing pattern remains a marvelously complex puzzle.

DISCLAIMER

The research described in this article has been reviewed by the National Health and Environmental Effects Research Laboratory, U.S. Environmental Protection Agency, and approved for publication. Approval does not signify that the contents necessarily reflect the views and the policies of the agency nor does mention of trade names or commercial products constitute endorsement or recommendation for use.

REFERENCES

Adams, L., Schneider, D.A., Schertel, E.R., Strong, E.B., Green, J.F., 1987. Respiratory reflexes in the anesthetized miniature swine. Respir. Physiol. 70, 343–357.

Alarie, Y., 1973. Sensory irritation by airborne chemicals. CRC Crit. Rev. Toxicol. 2 (3), 299–363.

Angell-James, J.E., Elsner, R., deBurgh Daly, M., 1981. Lung inflation: effects on heart rate, respiration and vagal afferent activity in seals. Am. J. Physiol. 240, H190–H195.

Ar, A., Arieli, R., Shkolnik, A., 1977. Blood-gas properties and function in the fossorial mole rat under normal and hypoxic-hypercapnic atmospheric conditions. Respir. Physiol. 30, 201–218.

Arieli, R., Ar, A., 1979. Ventilation of a fossorial mammal *(Spalax ehrenbergi)* in hypoxic and hypercapnic conditions. J. Appl. Physiol. 47, 1011–1017.

Alcayaga, J., Sanhueza, Y., Zapata, P., 1993. Thermal dependence of chemosensory activity in the carotid body superfused in vitro. Brain Res. 600, 103–111.

Albers, C., Usinger, W., Scholand, C., 1975. Intracellular pH in unanesthetized dogs during panting. Respir. Physiol. 23, 59–70.

Acker, H., 2008. Cellular oxygen sensors. Ann. N. Y. Acad. Sci. 718, 3–12.

Agostoni, E., D'Angelo, E., 1976. The effect of limb movements on the regulation of depth and rate of breathing. Respir. Physiol. 27, 33–52.

Attenburrow, D.P., 1982. Time relationships between the respiratory cycle and the limb cycle in the horse. Equine Vet. J. 14, 69–72.

Bechbache, R.R., Chow, H.H.K., Duffin, J., Orsini, E.C., 1979. The effects of hypercapnia, hypoxia, exercise and anxiety on the pattern of breathing in man. J. Physiol. Lond. 293, 285–300.

Burgess, K.R., Whitelaw, W.W., 1988. Effects of nasal cold receptors on pattern of breathing. J. Appl. Physiol. 64, 371–376.

Bradley, R.M., 2000. Sensory receptors of the larynx. Am. J. Med. 108 (Suppl 4a), 47S–50S.

Bianchi, A.L., Denavit-Saubie, M., Champagnat, J., 1995. Central control of breathing in mammals: neuronal circuitry, membrane properties, and neurotransmitters. Physiol. Rev. 75, 1–31.

Bonham, A.C., 1995. Neurotransmitters in the CNS control of breathing. Respir. Physiol. 101, 219–230.

Boggs, D.F., Tenney, S.M., 1984. Scaling respiratory pattern and respiratory 'drive'. Respir. Physiol. 58, 245–251.

Bartlett Jr., D., Remmers, J.E., Gautier, H., 1973. Laryngeal regulation of respiratory airflow. Respir. Physiol. 18, 194–204.

Boggs, D.F., Birchard, G.F., 1989. Cardio-respiratory responses of the woodchuck and porcupine to CO, and hypoxia. J. Comp. Physiol. 159B, 641–648.

Bowes, G., Andrey, S.M., Kozar, L.F., Phillipson, E.A., 1982. Role of carotid chemoreceptors in regulation of inspiratory onset. J. Appl. Physiol. 52, 863–868.

Bartlett Jr., D., St John, W.M., 1979. Adaptation of pulmonary stretch receptors in different mammalian species. Respir. Physiol. 37, 303–312.

Birchard, G.F., Boggs, D.F., Tenney, S.M., 1984. Effects of perinatal CO, exposure on adult ventilatory responses. Respir. Physiol. 57, 234–341.

Bennett, F.M., Tenney, S.M., 1982. Comparative mechanics of mammalian respiratory system. Respir. Physiol. 49, 131–140.

Blank, S., Chen, V., Ianuzzo, C.D., 1988. Biochemical characteristics of mammalian diaphragms. Respir. Physiol. 74, 115–126.

De Burgh Daly, M., 1986. Interactions between respiration and circulation. In: Chemiack, N.S., Widdicombe, J.G. (Eds.), Handbook of Physiology. Section 3, Respiration. Vol. II, Control of Breathing. American Physiological Society, Bethesda, MD, pp. 529–593.

Bouverot, P., 1985. Adaptation to Altitude Hypoxia in Vertebrates. Springer Verlag, New York.

Bystrzycka, E.W., Nail, B.S., 1980. CO_2-sensitivity of stretch receptors in the marsupial lung. Respir. Physiol. 39, 111–119.

Birchard, G., Tenney, S.M., 1986. The hypoxic ventilatory response of rats with increased HB-0_2 affinity. Respir. Physiol. 66, 225–234.

Bainton, C.R., Elsner, R.E., Matthews, R.C., 1973. Inhaled CO, and progressive hypoxia: ventilatory response in a yearling and newborn harbor seal. Life Sci. 12, 527–533.

Blanco, C.E., Hanson, M.A., Johnson, P., Rigatto, H., 1984. Breathing pattern of kittens during hypoxia. J. Appl. Physiol. 56, 12–17.

Bonora, M., Shields, G.I., Knuth, S.L., Bartlett Jr., D., St John, W., 1984a. Selective depression by ethanol of upper airway respiratory motor activity in cats. Am. Rev. Respir. Dis. 130, 156–161.

Bureau, M.A., Foulon, P., Zinman, R., Begin, R., 1984. Diphasic ventilatory response to hypoxia in the newborn lamb. J. Appl. Physiol. 56, 84–90.

Bonora, M., Marlot, D., Gautier, H., Duron, B., 1984b. Effects of hypoxia on ventilation during postnatal development in conscious kittens. J. Appl. Physiol. 56, 1464–1471.

Bartlett Jr., D., 1989. Respiratory functions of the larynx. Physiol. Rev. 69, 33–57.

Barer, G.R., Herget, J., Sloan, P.J.M., Suggett, A.J., 1978. The effect of acute and chronic hypoxia on thoracic gas volumes in anesthetized rats. J. Physiol. Lond. 277, 177–192.

Bartlett, D., Birchard, G.F., 1983. Effects of hypoxia on lung volume in the garter snake. Respir. Physiol. 53 (1), 63–70.

Bouverot, P., Candas, V., Libert, J.P., 1973. Role of arterial chemoreceptors in ventilatory adaptation to hypoxia of awake dogs and rabbits. Respir. Physiol. 17, 209–219.

Bouverot, P., Bureau, M., 1975. Ventilatory acclimatization and CSF acid base balance in carotid chemodenervated dogs at 3550m. Pflugers Arch. 361, 17–23.

Bisgard, G.E., Nielsen, A., Vidruk, E., Daristotle, L., Engwall, M., Forster, H.V., 1989. Mechanisms of ventilatory acclimatization to hypoxia in goats. In: Lahiri, S., Forster II, R.E., Davies, R.O., Pack, A.I. (Eds.), Chenwreceptors and Reflexes in Breathing. Oxford University Press, Oxford, pp. 208–214.

Bisgard, G.E., Vogel, J.H.K., 1971. Hypoventilation and pulmonary hypertension in calves after carotid body excision. J. Appl. Physiol. 31, 431–437.

Bisgard, G.E., Busch, M.A., Forster, H.V., 1986. Ventilatory acclimatization to hypoxia is not dependent on cerebral hypocapnic alkalosis. J. Appl. Physiol. 60, 1011–1015.

Bisgard, G.E., Forster, H.V., Klein, J.P., Manohar, M., Bullard, V.A., 1980. Depression of ventilation by dopamine in goats-effects of carotid body excision. Respir. Physiol. 41, 379–392.

Banner, K.H., Igney, F., Poll, C., 2011. TRP channels: emerging targets for respiratory disease. Pharmacol. Ther. 130 (3), 371–384.

Bingman, D., 1977. Effects of temperature on steady-state activity and dynamic responses of carotid baro- and chemoreceptors. In: Acker, H., Fidone, S., Pallot, D., Eyzaquirre, C., Luebbers, D.W., Torrance, R.W. (Eds.), Chemoreception in the Carotid Body. Springer-Verlag, Berlin, pp. 277–285.

Baker, M.A., Dawson, D.D., 1985. Inhibition of thermal panting by intracarotid infusion of hypertonic saline in dogs. Am. J. Physiol. 249, R787–R791.

Bernasconi, P., Kohl, J., 1993. Analysis of coordination between breathing and exercise rhythms in man. J. Physiol. 471, 693–706.

Bechbache, R.R., Duffin, J., 1977. The entrainment of breathing frequency to exercise rhythm. J. Physiol. Lond. 272, 553–561.

Bouverot, P., 1976. Role of ventilatory acclimatization to altitude and strength of the O2-chemoreflex drive. In: Duron, B. (Ed.), Les Colloques de INSERM. INSERM, Paris, pp. 213–219.

Boggs, D.F., Irvin, C.G., 1992. Respiratory mechanics of the coatimundi and woodchuck. Respir. Physiol. 89, 147–155.

Bonsignore, M.R., Morici, G., Abate, P., Romano, S., Bonsignore, G., 1998. Ventilation and entrainment of breathing during cycling and running in triathletes. Med. Sci. Sports Exerc. 30, 239–245.

Bramble, D.M., Carrier, D.R., 1983. Running and breathing in mammals. Science 219, 251–256.

Bramble, D.M., 1983. Respiratory patterns and control during unrestrained human running. In: Whipp, B.J., Wiberg, D.M. (Eds.), Modelling and Control of Breathing. Elsevier, New York, pp. 213–220.

Bramble, D.M., 1989. Axial-appendicular dynamics and the integration of breathing and gait in mammals. Am. Zool. 29, 171–186.

Baudinette, R.V., Gannon, J.B., Runciman, W.B., Wells, S., 1987. Do cardiorespiratory frequencies show entrainment with hopping in the tammar wallaby? J. Exp. Biol. 129, 251–263.

Berger, M., Roy, O.Z., Hart, J.S., 1970. The coordination between respiration and wing beats in birds. Z. Vgl. Physiol. 66, 190200.

Bellville, J.W., Escarraga, L.A., Wallenstein, S.L., Houde, R.W., Howland, W.S., 1960. Relative respiratory depressant effect of oxymorphone (Numorphan) and morphine. Anesthiology 21, 397–400.

Carpenter, R.E., 1986. Flight physiology of intermediate-sized fruit bats (Pteropododae). J. Exp. Biol. 120, 79–103.

Clement, M.G., Mortola, J.P., Albertini, M., Aguggini, G., 1986. Effects of vagotomy on respiratory movements in newborn and adult pigs. J. Appl. Physiol. 60, 1992–1999.

Cunningham, D.J.C., Robbins, P.A., Wolff, C.B., 1986. Integration of respiratory responses to changes in alveolar partial pressures of CO_2 and O_2 and pH. In: Cherniack, N.S., Widdicombe, J.G. (Eds.), Handbook of Physiology. Section 3, Respiration. Volume II, Control of Breathing. American Physiological Society, Bethesda, MD, pp. 475–527.

Corfield, D.R., Murphy, K., Guz, A., 1998. Does the motor cortical control of the diaphragm 'bypass' the brain stem respiratory centres in man? Respir. Physiol. 114, 109–117.

Coleridge, H.M., Coleridge, J.C.G., 1986. Reflexes evoked from tracheobronchial tree and lungs. In: Chemiack, N.S., Widdicombe, J.G. (Eds.), Handbook of Physiology. Section 3, Respiration, Vol. II, Control of Breathing. American Physiological Society, Bethesda, MD, pp. 395–429.

Corfield, D.R., Fink, G.R., Ramsay, S.C., et al., 1995. Evidence for limbic system activation during CO_2-stimulated breathing in man. J. Physiol. 488, 77–84.

Crosfill, M.L., Widdicombe, J.G., 1961. Physical characteristics of the chest and lungs and the work of breathing in different mammalian species. J. Physiol. Lond. 158, 1–14.

Crossland, N.J., Horsfall, G.B., Oxenham, S.T., Shaw, J.S., Turnbull, M.S., 1977. A simple device for measurement of respiratory rate in the mouse. Br. J. Pharmacol. 61, 490P–491P.

Cooper, K.E., Veale, W.L., 1986. Effects of temperature on breathing. In: Chemiack, N.S., Widdicombe, J.G. (Eds.), Handbook of Physiology. Section 3, Respiration. Volume 2, Control of Breathing. American Physiological Society, Bethesda, MD, pp. 691–701.

Cherniack, N.S., Lederer, D.H., Altose, M.D., Kelsen, S.G., 1976. Occlusion pressure as a technique in evaluating respiratory control. Chest 70, S137–S141.

Cunningham, D.J.C., 1974. The control system regulating breathing in man. Q. Rev. Biophys. 6, 433–483.

Chapman, R.C., Bennet, A.F., 1954. Physiological correlates of burrowing in rodents. Comp. Biochem.. Physiol. 51, 599–603.

Craig, A.B., Pasche, A., 1980. Respiratory physiology of freely diving harbor seals (Phoca vitulina). Physiol. Zool. 53, 419–432.

Coleridge, H.M., Coleridge, J.C.G., Banzett, R.B., 1978. Effect of CO2 on afferent vagal endings in canine lung. Respir. Physiol. 34, 135–151.

Clark, F.J., von Euler, C., 1972. On the regulation of depth and rate of breathing. J. Physiol. Lond. 222, 267–295.

Cross, K.W., Oppe, T.E., 1952. The effect of inhalation of high and low concentration of oxygen on the respiration of the premature infant. J. Physiol. Lond. 117, 217–237.

Cote, A., Yunis, K., Blanchard, P.W., Mortola, J.P., Bureau, M.A., 1988. Dynamics of breathing in the hypoxic awake lamb. J. Appl. Physiol. 64, 354–359.

Calder, W.A., Schmidt-Nielsen, K., 1966. Evaporative cooling and respiratory alkalosis in the pigeon. Proc. Natl. Acad. Sci. U.S.A. 55, 750–756.

Cunningham, D.J.C., O'Riordan, J.L.H., 1957. The effect of a rise in the temperature of the body on the respiratory response to carbon dioxide at rest. J. Exp. Physiol. 42, 329–337.

Crawford Jr., E.C., 1962. Mechanical aspects of panting in dogs. J. Appl. Physiol. 17, 249–251.

Derksen, F.J., Robinson, N.E., Slocombe, R.F., Riebold, T.W., Brunson, D.B., 1982. Pulmonary function tests in standing ponies: reproducibility and effect of vagal blockade. Am. J. Vet. Res. 43, 598–602.

deVries, W., Bates, J.H.T., 1988. Breath-to-breath variability in different phases of the breathing cycle in anesthetized dogs at rest. Physiologist 31, A174.

Davies, D.G., 1978. Evidence for cerebral extracellular fluid [H+] as a stimulus during acclimatization to hypoxia. Respir. Physiol. 32, 167–182.

Depledge, M.H., 1985. Respiration and lung function in the mouse, *Mus musculus*, (with a note on mass exponents and respiratory variable). Respir. Physiol. 60, 83–94.

Dempsey, J.A., 1976. CO, response: stimulus definition and limitations. Chest 70, S114–S118.

Dejours, P., Puccineli, R., Armand, J., Dickarry, M., 1965. Concept and measurement of ventilatory sensitivity to carbon dioxide. J. Appl. Physiol. 20, 890–897.

Dawson, T.J., Evans, J.V., 1966. Effect of hypoxia on oxygen transport in sheep with different hemoglobin types. Am. J. Physiol. 210, 1021–1025.

Davis, G.M., Bureau, M.A., Gaultier, Cl, 1988. The sustained ventilatory response to hypoxic challenge in the awake newborn piglet with an intact upper airway. Respir. Physiol. 71, 307–314.

Dempsey, J.A., Forster, H.V., 1982. Mediation of ventilatory adaptations. Physiol. Rev. 62, 262–346.

Dempsey, J.A., Forster, H.V., Chosy, L.W., Hanson, P.G., Reddan, W.G., 1978. Regulation of CSF [HCO_3] during long term hypoxic hypocapnia in man. J. Appl. Physiol. 44, 175–182.

Dwinell, M.R., Powell, F.L., 1999. Chronic hypoxia enhances the phrenic nerve response to arterial chemoreceptor stimulation in anesthetized rats. J. Appl. Physiol. 87, 817–823.

Dawson, T.J., Rose, R.W., 1970. Influence of the respiratory response to moderate and severe heat on the blood gas values of a macropodid marsupial. Comp. Biochem. Physiol. 37, 59–66.

Dubayle, D., Viala, D. Entrainment of the medullary respiratory generators by electrical stimulation in the cervical grey matter on in vitro preparations of newborn rats. Neurosci. Lett. 248, 204–208.

Davies, R.O., Edwards, M.W., Lahiri, S., 1982. Halothane depresses the response of carotid body chemoreceptors to hypoxia and hypercapnia in the cat. Anesthesiology 57, 153–159.

von Euler, C., 1986. Brain stem mechanisms for generation and control of breathing pattern. In: Cherniack, N.S., Widdicombe, J.G. (Eds.), Handbook of Physiology. Section 3. The Respiratory System, Volume II. Control of Breathing. American Physiological Society, Bethesda, MD, pp. 1–67.

England, S.J., Bartlett Jr., D., Daubenspeck, J.A.S., 1982. Influence of human vocal cord movements on airflow and resistance during eupnia. J. Appl. Physiol. 52, 773–779.

England, S.J., Kent, G., Stogryn, H., 1985. Laryngeal muscle and diaphragmatic activities in conscious dog pups. Respir. Physiol. 60, 95–108.

England, S.J., Fedorko, L., Kelly, E., 1988. Vagotomy results in a substantial decrease in ventilation in newborn rats. Am. Rev. Resp. Dis. 137, 262.

Eldridge, F.L., 1971. Relationship between phrenic nerve activity and ventilation. Am. J. Physiol. 221, 535–543.

Evanich, M.J., Lopata, M., Lourenco, R.V., 1976. Phrenic nerve activity and occlusion pressure changes during CO, rebreathing in cats. J. Appl. Physiol. 41, 536–543.

Evans, D.L., Rose, R.J., 1988. Dynamics of cardiorespiratory function in standard bred horses during different intensities of constant-load exercise. J. Comp. Physiol. B157, 791–799.

Easton, P.A., Slykerman, L.J., Anthonisen, N.R., 1983. Ventilatory response to sustained hypoxia. Fed. Proc. 42, 739.

Eccles, R., 2000. Role of cold receptors and menthol in thirst, the drive to breathe and arousal. Appetite 34, 29–35.

Eccles, R., 2003. Menthol: effects on nasal sensation of airflow and the drive to breathe. Curr. Allergy Asthma Rep. 3, 210–214.

Eldridge, F.L., Millhorn, D.E., Waldrop, T.G., 1981. Exercise hypemea and locomotion: parallel activation from hypothalamus. Science 211, 844–846.

Eldridge, F.L., Millhorn, D.E., Kiley, J.P., Waldrop, T.G., 1985. Stimulation by central command of locomotion, respiration and circulation during exercise. Respir. Physiol. 59, 313–337.

Fisher, J.T., Mortola, J.P., Smith, J.B., Fox, G.S., 1982. Respiration in newborns: development of control of breathing. Am. Rev. Respir. Dis. 125, 650–657.

Fishman, N.H., Phillipson, E.A., Nadel, J.A., 1973. Effect of differential vagal cold blackade on breathing pattern in conscious dogs. J. Appl. Physiol. 34, 754–758.

Fairchild, G.A., 1972. Measurement of respiratory volume for virus retention studies in mice. Appl. Microbiol. 24, 812–818.

Favier, R., LaCaisse, A., 1978. Stimulus oxygene de la ventilation chez le rat eveille. J. Physiol. (Paris) 74, 411–417.

Fencl, V., Miller, T.B., Pappenheimer, J.R., 1966. Studies on the respiratory response to disturbances of acid-base balance, with deductions concerning the ionic composition of cerebral spinal fluid. Am. J. Physiol. 210, 459–472.

Fordyce, W.E., Grodins, G.S., 1980. Ventilatory responses to intravenous and airway CO2 administration in anesthetized dogs. J. Appl. Physiol. 48, 337–346.

Fitzgerald, R.S., Lahiri, S., 1986. Reflex responses to chemoreceptor stimulation. In: Cherniack, N.S., Widdicombe, J.G. (Eds.), Handbook of Physiology. Section 3, Respiration. Volume II, Control of Breathing. American Physiological Society, Bethesda, MD, pp. 313–361.

Farber, J.P., Hultgren, H.N., Tenney, S.M., 1972. Development of the chemical control of breathing in the Virginia opossum. Respir. Physiol. 14, 267–277.

Fregosi, R.F., Knuth, S.L., Ward, D.K., Bartlett Jr., D., 1987. Hypoxia inhibits abdominal expiratory nerve activity. J. Appl. Physiol. 63, 211–220.

Fordyce, W.E., Tenney, S.M., 1984. Role of carotid bodies in ventilatory acclimation to chronic hypoxia by the awake cat. Respir. Physiol. 58, 207–221.

Forster, H.V., Bisgard, G.E., Rasmussen, B., Orr, J.A., Buss, D.D., Manohar, M., 1976. Ventilatory control in peripheral chemoreceptordenervated ponies during chronic hypoxemia. J. Appl. Physiol. 41, 878–885.

Forster, H.V., Bisgard, G.E., Klein, J.P., 1981. Effect of peripheral chemoreceptor denervation on acclimatization of goats during hypoxia. J. Appl. Physiol. 50, 392–398.

Fend, V., Gabel, R.A., Wolfe, D., 1979. Composition of cerebral fluids in goats adapted to high altitude. J. Appl. Physiol. 47, 508–513.

Gozal, D., Hathout, G.M., Konrad, A.T., et al., 1994. Localization of putative neural respiratory regions in the human by functional magnetic resonance imaging. J. Appl. Physiol. 76, 2076–2083.

Gautier, H., Remmers, J.E., Bartlett Jr., D., 1973. Control of the duration of expiration. Respir. Physiol. 18, 205–221.

Gautier, H., Bonora, M., 1979. Effects of carotid body denervation on respiratory pattern of awake cats. J. Appl. Physiol. 46, 11271131.

Gaultier, C., Mortola, J.P., 1981. Hering-Breuer inflation reflex in young and adult mammals. Can. J. Physiol. Pharmacol. 59, 1017–1021.

Gautier, H., Padykula, H.A., 1966. Cytological studies of fiber types in skeletal muscle: a comparative study of the mammalian diaphragm. J. Cell. Biol. 28, 333–354.

Gillespie, J.R., Tyler, W.S., Eberly, V.E., 1966. Pulmonary ventilation and resistance in emphysematous and control horses. J. Appl. Physiol. 21, 416–422.

Grunstein, M.M., Younes, M., Milic-Emili, J., 1973. Control of tidal volume and respiratory frequency in anesthetized cats. J. Appl. Physiol. 35, 463–476.

Grassino, A.E., Whitelaw, W., Milic-Emili, J., 1976. Influence of lung volume and electrode position on electromyography of the diaphragm. J. Appl. Physiol. 40, 971–975.

Green, J.F., Schertel, E.R., Coleridge, H.M., Coleridge, J.C., 1986. Effect of pulmonary arterial PCO_2 on slowly adapting pulmonary stretch receptors. J. Appl. Physiol. 60, 2048–2055.

Greco, E.C., Fordyce, W.E., Gonzalez, F., Reischl, P., Grodins, F.S., 1978. Respiratory response to intravenous and intrapulmonary CO2 in awake dogs. J. Appl. Physiol. 45, 109–114.

Gardner, W.N., 1977. The relation between tidal volume and inspiratory and expiratory times during steady state carbon dioxide inhalation in man. J. Physiol. Lond. 272, 591–611.

Grunstein, M.M., Hazinski, T.A., Schlueter, M.A., 1981. Respiratory control during hypoxia in newborn rabbits: implied action of endorphins. J. Appl. Physiol. 51, 122–130.

Gluckman, P.D., Johnston, B.M., 1987. Lesions in the upper lateral pons abolish the hypoxic depression of breathing in unanesthetized fetal lambs in utero. J. Physiol. Lond. 382, 373–383.

Garfinkel, F., Fitzgerald, R.S., 1978. The effect of hyperoxia, hypoxia and hypercapnia on FRC and occlusion pressure in human subjects. Respir. Physiol. 33, 241–250.

Graves, C., Glass, L., Laporta, D., Meloche, R., Grassino, A., 1986. Respiratory phase locking during mechanical ventilation in anesthetized human subjects. Am. J. Physiol. 250, R902–R909.

Gasser, J.C., Bellville, J.W., 1976. The respiratory effects of hydroxyzine, diazepam and pentazocine in man. Anesthesia 31, 718–723.

Herbert, D.A., Mitchell, R.A., 1971. Blood gas tensions and acid-base balance in awake cats. J. Appl. Physiol. 30, 434–436.

Honda, Y., 1968. Ventilatory response to CO2 during hypoxia and hyperoxia in awake anesthetized rabbits. Respir. Physiol. 5, 279–287.

Howe, A., 1984. The distribution of glomus tissue: a reappraisal of the peripheral arteial chemoreceptor system and the question of abdominal chemoreceptors. In: Pallot, D.J. (Ed.), The Peripheral Arterial Chemoreceptors. Oxford University Press, New York, pp. 153–160.

Harding, R., 1980. State-related and developmental changes in laryngeal function. Sleep 3, 307.

Hazari, M.S., Rowan, W.H., Winsett, D.W., Ledbetter, A.D., Haykal-Coates, N., Watkinson, W.P., Costa, D.L., February 1, 2008. Potentiation of pulmonary reflex response to capsaicin 24h following whole-body acrolein exposure is mediated by TRPV1. Respir. Physiol. Neurobiol. 160 (2), 160–171.

Huszczuk, A., Jankowska, L., Kulesza, J., Ryba, M., 1977. Studies on reflex control of breathing in pigs and baboons. Acta Neurobiol. Exp. 37, 275–298.

Honda, Y., 1992. Respiratory and circulatory activities in carotid body-resected humans. J. Appl. Physiol. 73, 1–8.

Hornicke, H., Meixner, R., Pollman, V., 1983. Respiration in exercising horses. In: Snow, D.H., Person, S.G.B., Rose, R.J. (Eds.), Equine Exercise Physiology. Burlington Press, Cambridge, pp. 7–16.

Haddad, G.G., Gandhi, M.R., Mellins, R.B., 1982. Maturation of ventilatory response to hypoxia in puppies during sleep. J. Appl. Physiol. 52, 309–314.

He, L., Dinger, B., Fidone, S., 2005. Effect of chronic hypoxia on cholinergic chemotransmission in rat carotid body. J. Appl. Physiol. 98, 614–619.

Huey, K.A., Brown, I.P., Jordan, M.C., Powell, F.L., 2000. Changes in dopamine D(2)-receptor modulation of the hypoxic ventilatory response with chronic hypoxia. Respir. Physiol. 123, 177–187.

He, L., Chen, J., Dinger, B., Stensaas, L., Fidone, S., 2006. Effect of chronic hypoxia on purinergic synaptic transmission in rat carotid body. J. Appl. Physiol. 100, 157–162.

Haxhiu, M.A., Chang, C.H., Dreshaj, I.A., Erokwu, B., Prabhakar, N.R., Cherniack, N.S. Nitric oxide and entilatory response to hypoxia. Respir. Physiol. 101, 257–266.

Hales, J.R.S., Dampney, R.A.L., Bennett, J.W., 1975. Influence of chronic denervation of the carotid bifurcation regions on panting in sheep. Pflugers Arch. 360, 243–253.

Hales, J.R.S., Findlay, J.D., 1968. The oxygen cost of thermally induced and CO2-induced hyperventilation in the ox. Respir. Physiol. 4, 353–362.

Hales, J.R.S., Webster, M.E.D., 1967. Respiratory function during thermal tachypnoea in sheep. J. Physiol. Lond. 190, 241–260.

Hales, J.R.S., Bligh, J., Maskrey, M., 1970. Cerebrospinal fluid acid-base balance during respiratory alkalosis in the panting animal. Am. J. Physiol. 219, 469–473.

Hand, S.C., Somero, G.N., 1983. Phosphofructokinase of the hibernator *Citellus beeecheyi*: temperature and pH regulation of activity influences on the tetramer-dimer equilibrium. Physiol. Zool. 56, 380–388.

Hirshman, C., McCullough, R.E., Cohen, P.J., Weil, J.V., 1975. Hypoxic ventilatory drive in dogs during thiopental, ketamine or pentobarbital anesthesia. Anesthesiology 43, 628–634.

Iscoe, S., Young, R.B., Jennings, D.B., 1983. Control of respiratory pattern in conscious dogs: effects of heat and carbon dioxide. J. Appl. Physiol. 54, 623–631.

Ingram, D.L., Legge, K.F., 1970. The effect of environmental temperature on respiratory ventilation in the pig. Respir. Physiol. 8, 112.

Iscoe, S., 1981. Respiratory and stepping frequencies in conscious exercising cats. J. Appl. Physiol. 51, 835–839.

Iscoe, S., Polosa, C., 1976. Synchronization of respiratory frequency by somatic afferent stimulation. J. Appl. Physiol. 40, 138–148.

Joseph, V., Pequignot, J.M., 2009. Breathing at high altitude. Cell. Mol. Life Sci. 66, 3565–3573.

Joseph, V., Soliz, J., Soria, R., Pequignot, J., Favier, R., Spielvogel, H., Pequignot, J.M., 2002. Dopaminergic metabolism of carotid bodies and high altitude acclimatization in female rats. Am. J. Physiol. Regul. Integr. Comp. Physiol. 282, R773–R795.

Jonzon, A., Pisarri, T.E., Roberts, A.M., Coleridge, J.C.G., Coleridge, H.M., 1988. Attentuation of pulmonary afferent input by vagal cooling in dogs. Respir. Physiol. 72, 19–34.

Jennings, D.B., 1984. Breathing patterns and drives during thermal and non-thermal panting. In: Hales, J.R.S. (Ed.), Thermal Physiology. Raven, New York, pp. 335–340.

Joyce, J.P., Baxter, K.L., 1964. Respiration in sheep in cold environments. Res. Vet. Sci. 5, 506–516.

Jackson, D.C., 1989. Control of breathing: effects of temperature. In: Wood, S.C. (Ed.), Comparative Pulmonary Physiology. Marcel Dekker, New York, pp. 621–641.

Jasinskas, C.L., Wilson, B.A., Hoare, J., 1981. Entrainment of breathing rate to movement frequency during work at two intensities. Respir. Physiol. 42, 199–209.

Kalia, M., Senapati, J.M., Parida, B., Panda, A., 1972. Reflex increase in ventilation by muscle receptors with nonmedullated fibers (C-fibers). J. Appl. Physiol. 32, 189–193.

Kashani, M., Haigh, A.L., 1975. The effects of vagotomy on ventilation and blood gas composition in dog, sheep and rabbit. Q. J. Exp. Physiol. 60, 285–298.

Koterba, A.M., Kosch, P.C., Beech, J., Whitlock, T., 1988. Breathing strategy of the adult horse *(Equus caballus)* at rest. J. Appl. Physiol. 64, 337–346.

Kleiber, M., 1961. The fire of life: An introduction to animal energetics. Wiley, New York.

Krauss, A.N., Klain, D.B., Waldman, S., Auld, P.A.M., 1975. Ventilatory response to carbon dioxide in newborn infants. Pediatr. Res. 9, 46–50.

Kunz, A.L., Kawashiro, T., Scheid, P., 1976. Study of CO, sensitive vagal afferents in the cat lung. Respir. Physiol. 27, 347–355.

Kaminski, R.P., Forster, H.V., Bisgard, G.E., Pan, L.G., Dorsey, S.M., 1985. Effect of altered ambient temperature on breathing in ponies. J. Appl. Physiol. 58, 1585–1591.

Kay, J.D.S., Petersen, E.S., Vejby-Christensen, H., 1975. Breathing in man during steady state exercise on the bicycle at two pedalling frequencies and during treadmill walking. J. Physiol. Lond. 251, 645–656.

Kawahara, K., Kumagi, S., Nakazono, Y., Miyamoto, Y., 1989. Coupling between respiratory and stepping rhythms during locomotion in decerebrate cats. J. Appl. Physiol. 67, 110–115.

Lauweryns, J.M., Cokelaere, M., 1973. Hypoxia-sensitive neuro-epithelial bodies: intrapulmonary secretory neuroreceptors modulated by CNS. Z. Zellforsch. Mikrosk. Anat. 145, 521–540.

Lumb, A.B., 2005. Nunn's Applied Respiratory Physiology, sixth ed. Elsevier Ltd, Philadelphia, PA.

Lindstedt, S.L., Calder Jr., W.A., 1981. Body size, physiological time and longevity of homeothermic animals. Q. Rev. Biol. 56, 1–16.

Lahiri, S., 1975. Blood oxygen affinity and alveolar ventilation in relation to body weight in mammals. Am. J. Physiol. 229, 529–536.

Lai, Y.L., Tsuya, Y., Hidebrandt, J., 1978. Ventilatory response to acute CO2 exposure in the rat. J. Appl. Physiol. 45, 611–618.

Leiter, J.C., Mortola, J.P., Tenney, S.M., 1986. A comparative analysis of contractile characteristics of the diaphragm and of respiratory system mechanics. Respir. Physiol. 64, 267–276.

Loeschcke, H.H., 1974. Central nervous chemoreceptors. In: Widdicombe, J.G. (Ed.), Respiratory Physiology. Butterworths, London, p. 167196.

Luo, Y.M., Moxham, J., Polkey, M.I., 2008. Diaphragm electromyography using an esophageal catheter: current concepts. Clin. Sci. (Lond.) 115 (8), 233–244.

Ledlie, J.F., Kelsen, S.G., Cherniack, N.S., Fishman, A.P., 1981. Effects of hypercapnia and hypoxia on phrenic nerve activity and respiratory timing. J. Appl. Physiol. 51, 732–738.

Linton, R.A., Miller, F.R., Cameron, I.R., 1976. Ventilatory response to CO2 inhalation and intravenous infusion of hypercapnic blood. Respir. Physiol. 26, 383–394.

Lambertson, C.J., 1980. Physical, chemical and nervous interactions in respiratory control. In: Mountcastle, V.B. (Ed.), Medical Physiology. CV Mosby, St. Louis, pp. 1873–1875.

Lewis, S.M., 1975. Awake baboon's ventilatory response to venous and inhaled CO, loading. J. Appl. Physiol. 39, 417–422.

Lahiri, S., Mulligan, E., Nishino, R., Mokashi, A., Davies, R.O., 1980. Relative responses of aortic body and carotid body chemoreceptors to carboxyhemoglobinemia. J. Appl. Physiol. 50, 580–586.

LaFramboise, W.A., Guthrie, R.D., Standaert, T.A., Woodrum, D.E., 1983. Pulmonary mechanics during the ventilatory response to hypoxemia in the newborn monkey. J. Appl. Physiol. 55, 10081014.

Lahiri, S., Edelman, N.H., Cherniack, N.S., Fishman, A.P., 1981. Role of carotid chemoreflex in respiratory acclimatization to hypoxemia in goat and sheep. Respir. Physiol. 46, 367–382.

Levine, S., 1975. Non-peripheral chemoreceptor stimulation of ventilation by cyanide. J. Appl. Physiol. 39, 199–204.

Lyman, C.P., Hastings, A.B., 1951. Total CO_2, plasma pH and pCO_3 of hamsters and ground squirrels during hibernation. Am. J. Physiol. 167, 633–637.

Loring, S., Gulesian, P., Goldman, M., Mead, J., 1976. Entrainment of respiratory frequency to stepping rate during walking. Physiologist 19, 276.

Milsom, W.K., 1989. Control and co-ordination of gas exchange in air breathers. In: Boutilier, R.G. (Ed.), Vertebrate Gas Exchange from Environment to Cell. Springer-Verlag, New York.

Michiels, C., 1984. Physiological and pathological responses to hypoxia. Am. J. Pathol. 164 (6), 1875–1882.

Mortola, J.P., Fisher, J.T., Smith, J.B., Fox, G.S., Weeks, S., Willis, D., 1982. Onset of respiration in infants delivered by caesarean section. J. Appl. Physiol. 52, 716.

Milsom, W.K., 1988. Control of arrhythmic breathing in aerial breathers. Can. J. Zool. 66, 99–108.

Mead, J., 1960. Control of breathing frequency. J. Appl. Physiol. 15, 325336.

Magnoni, P., Saibene, F., Sant'Ambrogio, G., 1969. Contribution of the diaphragm and the other inspiratory muscles to different levels of tidal volume and static inspiratory effort in the rabbit. J. Physiol. Lond. 202, 517–534.

Maarsingh, E.J.W., van Eykern, L.A., Sprikkelman, A.B., Hoekstra, M.O., van Aalderen, W.M.C., 2000. Respiratory muscle activity measured with a noninvasive EMG technique: technical aspects and reproducibility. J. Appl. Physiol. 88, 1955–1961.

Milic-Emili, J., Whitelaw, W.A., Grassino, A.E., 1981. Measurement and testing of respiratory drive. In: Hornbein, T.F. (Ed.), Regulation of Breathing. Marcel Dekker, New York, pp. 675–743.

Milic-Emili, J., Grunstein, M.M., 1976. Drive and timing components of ventilation. Chest 70, S131–S133.

Milsom, W.K., McArthur, M.D., Webb, C.L., 1986. Control of breathing in hibernating ground squirrels. In: Heller, H.C., Musacchia, X.J., Wang, L.C.H. (Eds.), Living in the Cold: Physiological and Biochemical Adaptations. Elsevier, New York, pp. 469–475.

Mines, A.H., Sorensen, S.C., 1970. Ventilatory response of awake normal goats during acute and chronic hypoxia. J. Appl. Physiol. 28, 826–831.

McCormick, J.G., 1969. Relationship of sleep, respiration and anesthesia in the porpoise: a preliminary report. Proc. Natl. Acad. Sci. U.S.A. 62, 697–703.

Mortola, J.P., Lanthier, C., 1989. Normoxia and hypoxic breathing pattern in newborn grey seals. Can. J. Zool. 67, 483–487.

Martin-Body, R.L., Johnston, B.M., 1988. Central origin of the hypoxic depression of breathing in the newborn. Respir. Physiol. 71, 2532.

Morgan, M.L., Adams, T., 1972. Respiratory acid-base balance in the unanesthetized cat during acute heat stress. Physiologist 15, 221.

Maskrey, M., Hoppe, P.P., Bamford, O.S., 1978. Blood gas changes during panting in a small East African antelope, the dik-dik. J. Appl. Physiol. 44, 534–537.

Maskrey, M., 1984. Metabolic and acid-base implications of thermal panting. In: Hales, J.R.S. (Ed.), Thermal Physiology. Raven Press, New York, pp. 347–352.

Malan, A., Arens, H., Waechter, A., 1973. Pulmonary respiration and acid-base state in hibernating marmots and hamsters. Respir. Physiol. 17, 45–61.

Malan, A., 1982. Respiration and acid-base state in hibernation. In: Lyman, C.P., Wills, J.S., Malan, A., Wang, L.C.H. (Eds.), Hibernation and Torpor in Mammals and Birds. Academic Press, New York, pp. 237–282.

Malan, A., 1986. pH as a control factor in hibernation. In: Heller, H.C., Musacchia, X.J., Wang, L.C.H. (Eds.), Living in the Cold: Physiological and Biochemical Adaptations. Elsevier, New York, pp. 61–70.

Morin, D., Viala, D., 2002. Coordination of locomotor and respiratory rhythms in vitro are critically dependent upon hindlimb sensory inputs. J. Neurosci. 22, 4756–4765.

Mazzarelli, M., Haberer, J.P., Jaspar, N., Miserocchi, G., 1979. Mechanism of halothane-induced tachypnea in cats. Anesthesiology 51, 522–527.

Nahas, G.G., 1974. Mechanisms of carbon dioxide and pH effects on metabolism. In: Nahas, G., Schaefer, K.E. (Eds.), Carbon Dioxide and Metabolic Regulations. Springer Verlag, New York, pp. 107–117.

Van Nice, P., Black, C.P., Tenney, S.M., 1980. A comparative study of ventilatory responses to hypoxia with reference to hemoglobin 0, affinity in llama, cat, rat, duck and goose. Comp. Biochem.. Physiol. 66A, 347–350.

Nicol, S.C., Maskrey, M., 1980. Thermoregulation, respiration and sleep in the Tasmanian devil, *Sarcophilus harrisii* (Masupialia: Dasyuridae). J. Comp. Physiol. 140, 241–248.

Natsui, T., 1969. Respiratory response to hypoxia with hypocapnia or normocapnia and to CO, in hypothermic dogs. Respir. Physiol. 7, 188–202.

Olinsky, A., Bryan, M.H., Bryan, A.C., 1974. Influence of lung inflation on respiratory control in neonates. J. Appl. Physiol. 36, 426429.

Ou, L.C., St John, W.M., Tenney, S.M., 1983. The contribution of central mechanisms rostra to the pons in high altitude acclimatization. Respir. Physiol. 54, 343–351.

Olievier, C.N., Berkenbosch, A., DeGoede, J., 1982. Effect of temperature on the ventilatory response curve to CO, in anesthetized cats. Respir. Physiol. 47, 365–377.

Oruc, T., Terzioglu, M., 1979. The sensitivity to hypoxia of peripheral and central respiratory control mechanisms under normo- and hypothermic conditions. Bull. Eur. Physiopath. Respir. 15, 1117–1128.

Petit, J.M., Milic-Emili, J., Delhez, L., 1960. Role of diaphragm in breathing in conscious normal man: an electromyographic study. J. Appl. Physiol. 15, 1101–1106.

Phillipson, E.A., Duffin, J., Cooper, J.D., 1981. Critical dependence of respiratory rhythmicity on metabolic CO, load. J. Appl. Physiol. 50, 45–54.

Parkos, C.A., Wahrenbrock, E.A., 1987. Acute effects of hypercapnia and hypoxia on minute ventilation in unrestrained Weddell seals. Respir. Physiol. 67, 197.

Ponte, J., Purves, M.J., 1978. Carbon dioxide and venous return and their interaction as stimuli to ventilation in the cat. J. Physiol. Lond. 274, 455–475.

Phillipson, E.A., 1974. Vagal control of breathing pattern independent of lung inflation in conscious dogs. J. Appl. Physiol. 37, 183–189.

Prabhakar, N.R., Peng, Y.J., 2004. Peripheral chemoreceptors in health and disease. J. Appl. Physiol. 96, 359–366.

Pasche, A., 1976. The effect of hypercapnia on respiratory characteristics and diving behavior of freely diving seals. Respir. Physiol. 26, 183–194.

Powell, F.L., Milsom, W.K., Mitchell, G.S., 1998. Time domains of the hypoxic ventilatory response. Respir. Physiol. 112 (2), 123–134.

Powell, F.L., Huey, K.A., Dwinell, M.R., 2000. Central nervous system mechanisms of ventilatory acclimatization to hypoxia. Respir. Physiol. 121, 223–236.

Petersen, E.S., Vejby-Christensen, H., 1977. Effects of body temperature on ventilatory response to hypoxia and breathing pattern in man. Respir. Physiol. 19, 322–332.

Paulin, E.G., Hornbien, T.F., 1986. Anesthesia and the control of ventilation. In: Cherniack, N.S., Widdicombe, J.G. (Eds.), Handbook of Physiology. Section 3, Respiration. Volume II, Control of Breathing. American Physiological Society, Bethesda, MD, pp. 793–813.

Paterson, D.J., Wood, G.A., Marshall, R.N., Morton, A.R., Harrison, A.B.C., 1987. Entrainment of respiratory frequency to exercise rhythm during hypoxia. J. Appl. Physiol. 62, 1767–1771.

Petrillo, G.A., Glass, L., Trippenbach, T., 1983. Phase locking of the respiratory rhythm in cats to a mechanical ventilator. Can. J. Physiol. Pharmacol. 61, 599–607.

Potts, J.T., Rybak, I.A., Paton, J.F.R., 2005. Respiratory rhythm entrainment by somatic afferent stimulation. J. Neurosci. 25, 1965–1978.

Richter, D.W., Ballanyi, K., Schwarzacher, S., 1992. Mechanisms of respiratory rhythm generation. Curr. Opin. Neurobiol. 2 (6), 788–793.

Reischl, P., Staver, D.M., Lewis, S.M., Mrdock, L.C., O'Loughlin, B.J., 1978. End-tidal CO2 response to low levels of inspired CO2 in awake beagle dogs. J. Appl. Physiol. 48, 1077–1082.

Remmers, J.E., Bartlett Jr., D., 1977. Reflex control of expiratory airflow and duration. J. Appl. Physiol. 42, 80–87.

Robin, E.D., Murdaugh, H.V., Pyron, W., Weiss, W., Stores, J.P., 1963. Adaptations to diving in the harbor seal – gas exchange and ventilatory response to CO2. Am. J. Physiol. 205, 1175–1177.

Rodriguez, M., Bur, S., Favre, A., Weibel, E.R., 1987. Pulmonary acinus: geometry and morphometry of the peripheral airway system in rat and rabbit. Am. J. Anat. 180, 143–155.

Rosenmann, M., Morrison, P.R., 1975. Metabolic response of highland and lowland rodents to stimulated high altitudes and cold. Comp. Biochem.. Physiol. 51A, 523–530.

Rezzonico, R., Mortola, J.P., 1988. Respiratory adaptation to hypercapnia in newborn rats and its long term effects. Physiologist 31, A171.

Rigatto, H., Brady, J.P., De La Tone Verduzco, R., 1975. Chemoreceptor reflexes in preteen infants. I. The effect of gestational and postnatal age on the ventilatory response to inhalation of 100% and 15% oxygen. Pediatrics 55, 604–613.

Robinson, S., 1949. Physiological adjustments to heat. In: Newburgh, L.H. (Ed.), The Physiology of Heat Regulation and the Science of Clothing. WB Saunders, Philadelphia, pp. 193–231.

Ruiz, A.V., 1975. Carbon dioxide response curves during hypothermia. Pflugers Arch. 358, 125–133.

Reeves, R.B., 1972. An imidazole alphastat hypothesis for vertebrate acid-base regulation: tissue carbon dioxide content and body temperature in bullfrogs. Respir. Physiol. 14, 219–236.

Remmers, J.E., Martilla, J., 1975. Action of intercostal muscle afferents on the respiratory rhythm of anesthetized cats. Respir. Physiol. 24, 31–41.

Sant'Ambrogio, G., Tsubone, H., Sant'Ambrogio, F.B., 1995. Sensory information from the upper airway: role in the control of breathing. Respir. Physiol. 102, 1–16.

Smith, C.A., Ainsworth, D.M., Henderson, K.S., Dempsey, J.A., 1989. Differential timing of respiratory muscles in response to chemical stimuli in awake dogs. J. Appl. Physiol. 66, 392–399.

Slocombe, R.F., Robinson, N.E., Derksen, F.J., 1982. Effect of vagotomy on respiratory mechanisms and gas exchange in neonatal calf. Am. J. Vet. Res. 43, 1168–1173.

Schlenker, E.H., 1985. Ventilation and metabolism of Djungarian Hamster (*Phodopus* sungorus) and the albino mouse. Comp. Biochem. Physiol. 82A, 293–295.

Schaefer, K.E., 1958. Respiratory pattern and response to CO2. J. Appl. Physiol. 13, 1–14.

Stahl, W.R., 1967. Scaling of respiratory variables in mammals. J. Appl. Physiol. 22, 453–460.

Syrovy, I., Gutman, E., 1975. Myosin from fast and slow skeletal muscles of mammals of different size. Physiol. Bohemoslov. 24, 325–334.

Santiago, T.D., Edelman, N.H., 1986. Brain blood flow and control of breathing. In: Cherniack, N.S., Widdicome, J.G. (Eds.), Handbook of Physiology. Section 3. Volume II, Control of Breathing. American Physiological Society, Bethesda, MD, pp. 163–180.

Schaefer, K.E., 1974. Metabolic aspects of adaptation to carbon dioxide. In: Nahas, G., Schaefer, K.E. (Eds.), Carbon Dioxide and Metabolic Regulations. Springer Verlag, New York, pp. 253–264.

Steggerda, J.A., Mayer, C.A., Martin, R.J., Wilson, C.G., 2010. Effect of intermittent hypercapnia on respiratory control in rat pups. Neonatology 97, 117–123.

Stremel, R.E., Huntsman, D.J., Casaburi, R., Whipp, B.J., Wasserman, K., 1978. Control of ventilation during intravenous CO, loading in the awake dog. J. Appl. Physiol. 44, 311–316.

Sheldon, M.I., Green, J.F., 1982. Evidence for pulmonary CO_2 chemosensitivity: effects on ventilation. J. Appl. Physiol. 52, 1192–1197.

Saupe, K.W., Smith, C.A., Henderson, K.S., Dempsey, J.A., 1992. Respiratory muscle recruitment during selective central and peripheral chemoreceptor stimulation in awake dogs. J. Physiol. 448, 613–631.

Schmidt-Nielsen, K., Larimer, J.L., 1958. Oxygen dissociation curves of mammalian blood in relation to body size. Am. J. Physiol. 195, 424–428.

Schaeffer, J.I., Haddad, G.G., 1988. Ventilatory response to moderate and severe hypoxia in adult dogs: role of endorphins. J. Appl. Physiol. 65, 1383–1388.

Saetta, M., Mortola, J.P., 1987. Interaction of hypoxic and hypercapnic stimuli on breathing pattern in the newborn rat. J. Appl. Physiol. 62, 506–512.

Sorensen, S.C., 1970. Ventilatory acclimatization to hypoxia in rabbits after denervation of peripheral chemoreceptors. J. Appl. Physiol. 29, 836–839.

Sorensen, S.C., Mines, A.H., 1970. Ventilatory responses to acute and chronic hypoxia in goats after sinus nerve section. J. Appl. Physiol. 28, 832–835.

Steinbrook, R.A., Donovan, J.C., Gabel, R.A., Leith, D.E., Fencl, V., 1983. Acclimatization to high altitude in goats with ablated carotid bodies. J. Appl. Physiol. 55, 16–21.

Schoener, E.P., Frankel, H.M., 1972. Effect of hyperthermia and Pa_{CO2} on the slowly adapting pulmonary stretch receptor. Am. J. Physiol. 222, 68–72.

Schroter, R.C., Robertshaw, D., Baker, M.A., Shoemaker, V.H., Holmes, R., Schmidt-Nielsen, K., 1987. Respiration in heat stressed camels. Respir. Physiol. 70, 97–102.

Saxton, C., 1975. Respiration during heat stress. Aviat. Space Environ. Med. 46, 41–46.

Seebauer, M., Sidler, M., Kohl, J., 2003. Gender differences in workload effect on coordination between breathing and cycling. Med. Sci. Sports Exerc. 35, 495–499.

Suthers, G.K., Thomas, S.P., Suthers, B.S., 1972. Respiration, wing-beat and ultrasonic pulse emission in an echo locating bat. J. Exp. Biol. 56, 37–48.

Stuth, E.A., Stucke, A.G., Brandes, I.F., Zuperku, E.J., 2008. Anesthetic effects on synaptic transmission and gain control in respiratory control. Respir. Physiol. Neurobiol. 164, 151–159.

Taylor, C.R., Maloiy, G.M.O., Weibel, E.R., Langman, V.A., Kamau, J.M.Z., Seeherman, H.J., Heglund, N.C., 1981. Design of the mammalian respiratory system III: scaling maximum aerobic capacity to body mass: wild and domestic mammals. Respir. Physiol. 44, 25–38.

Travis, C.C., Bowers, J.C., 1991. Interspecies scaling of anesthetic potency. Toxicol. Ind. Health 7 (4), 249–260.

Tenney, S.M., Bartlett Jr., D., 1967. Comparative quantitative morphology of the mammalian lung: trachea. Respir. Physiol. 3, 130–135.

Tenney, S.M., Remmers, J.E., 1963. Comparative quantitative morphology of the mammalian lung: diffusing area. Nature 197, 54–56.

Tenney, S.M., Morrison, D.H., 1967. Tissue gas tensions in small wild mammals. Respir. Physiol. 3, 160–165.

Tenney, S.M., 1954. Ventilatory response to carbon dioxide in pulmonary emphysema. J. Appl. Physiol. 6, 477–484.

Torrance, R.W., 1974. Arterial chemoreceptors. In: Widdicombe, J.G. (Ed.), Respiratory Physiology, pp. 247–271. MTP Int. Rev. Sci.

Teppema, L.J., Dahan, A., 2010. The ventilatory response to hypoxia in mammals: mechanisms, measurements and analysis. Phys. Rev. 90, 675–754.

Tryba, A.K., Marino-Ramirez, J., 2004. Hyperthermia modulates respiratory pacemaker bursting properties. J. Neurophysiol. 92, 2844–2852.

Taylor, C.R., 1969. Metabolism, respiratory changes and water balance of an antelope, the eland. Am. J. Physiol. 217, 317–320.

Turlejska, E., Baker, M.A., 1986. Elevated CSF osmolarity inhibits thermoregulatory heat loss responses. Am. J. Physiol. 251, R749754.

Taylor, C.R., Robertshaw, D., Hofmann, R., 1969. Thermal panting: a comparison of wildebeest and zebu cattle. Am. J. Physiol. 217, 907–910.

Tahti, H., 1975. Effects of changes in CO, and 0, concentrations in the inspired gas on respiration in the hibernating hedgehog, *Erinaceus europaeus* L. Ann. Zool. Fenn. 12, 183–187.

Truchot, J.P., 1987. Comparative Aspects of Extracellular Acid-base Balance. Springer-Verlag, Berlin.

Thomas, S.P., 1981. Ventilation and oxygen extraction in the bat *Pteropus gouldii* during rest and steady flight. J. Exp. Biol. 63, 1–9.

Teppema, L.J., Baby, S., 2011. Anesthetics and control of breathing. Respir. Physiol. Neurobiol. 177, 80–92.

Takita, K., Morimoto, Y., 2010. Effects of sevoflurane on respiratory rhythm oscillators in the medulla oblongata. Respir. Physiol. Neurobiol. 173, 86–94.

Verbrugge, C., Laurent, P., Bouverot, P., 1982. Chemoreflex drive of ventilation in the awake miniature pig. Respir. Physiol. 47, 379–391.

Vinegar, A., Sinnett, E.E., Leith, D.E., 1979. Dynamic mechanisms determine functional residual capacity in mice, Mus musculus. J. Appl. Physiol. 46, 867–871.

Vizek, M., Pickett, C.K., Weil, J.V., 1987. Biphasic ventilatory response of adult cats to sustained hypoxia has central origin. J. Appl. Physiol. 63, 1658–1664.

Vejby-Christensen, H., Petersen, E.S., 1973. Effect of body temperature and hypoxia on the ventilatory CO, response in man. Respir. Physiol. 19, 322–332.

Vaughan, T.A., 1959. Functional Morphology of Three Bats: eumops, myotis, Macrotus. University of Kansas, Lawrence.

Viala, D., Vidal, C., Freton, E., 1979. Coordinated rhythmic bursting in respiratory and locomotor muscle nerves in the spinal rabbit. Neurosci. Lett. 11, 155–159.

Widdicombe, J., 1999. Respiratory reflexes. In: Holgate, S.T., Samet, J.M., Koren, H.S., Maynard, R.L. (Eds.), Air Pollution and Health. Academic Press, San Diego, CA.

Widdicombe, J., 1961. Respiratory reflexes in man and other mammalian species. Clin. Sci. 21, 163–170.

Walker, B.R., Adams, E.M., Voelkel, N.F., 1985. Ventilatory responses of hamsters and rats to hypoxia and hypercapnia. J. Appl. Physiol. 59, 1955–1960.

Widdicombe, J.G., 1964. Respiratory reflexes. In: Fenn, W.O., Rahn, H. (Eds.), Handbook of Physiology. Respiration. American Physiological Society, Washington D.C., p. 585.

White, C.R., Seymour, R.S., 2003. Mammalian basal metabolic rate is proportional to body mass$^{2/3}$. PNAS 100, 4046–4049.

Weibel, E.R., Taylor, C.R., Gehr, P., Hoppeler, H., Mathieu, O., Maloiy, G.M.O., 1981. Design of the mammalian respiratory system. IX. Functional and structural limits for oxygen flow. Respir. Physiol. 44, 151–164.

Weibel, E.R., 1987. Scaling of structural and functional variables in the respiratory system. Ann. Rev. Physiol. 49, 147–159.

Whitelaw, W.A., Derenne, J.-P., Milic-Emili, J., 1975. Occlusion pressure as a measure of respiratory center output in conscious man. Respir. Physiol. 23, 181–199.

Whitelaw, W.A., Derenne, J.P., 1993. Airway occlusion pressure. J. Appl. Physiol. 74 (4), 1475–1483.

Ward, S.A., Drysdale, D.B., Cunningham, D.J.C., Petersen, E.S., 1979. Inspiratory-expiratory responses to alternate-breath oscillation of $PA_{,02}$ and $P,4_{02}$. Respir. Physiol. 36, 311–325.

Weil, J.V., Byrne-Quinn, E., Sodal, I.E., Friesen, W.O., Lander-hill, B., Filley, G.F., Grover, R.F., 1970. Hypoxic ventilatory drive in normal man. J. Clin. Invest. 49, 1061–1072.

Woodrum, D.E., Standaert, T.A., Mayock, D.E., Guthrie, R.D., 1981. Hypoxic ventilatory response in the newborn monkey. Pediatr. Res. 15, 367–370.

Weil, J.V., 1986. Ventilatory control at high altitude. In: Cherniack, N.S., Widdicombe, J.G. (Eds.), Handbook of Physiology: Section 3. The Respiratory System, Vol. II, Control of Breathing. American Physiological Society, Bethesda, MD, pp. 703–728.

Widdicombe, J.G., Winning, A., 1974. Effects of hypoxia, hypercapnia and changes in body temperature on the pattern of breathing in cats. Respir. Physiol. 21, 203–221.

Wunnenberg, W., Baltruschat, D., 1982. Temperature regulation of golden hamsters during acute hypercapnia. J. Therm. Biol. 7, 83–86.

Younes, M., Vaillencourt, P., Milic-Emili, J., 1974. Interaction between chemical factors and duration of apnea following lung inflation. J. Appl. Physiol. 36, 190–201.

Yamamoto, W.S., Edwards, M.W., 1960. Homeostatis of carbon dioxide during intravenous infusion of carbon dioxide. J. Appl. Physiol. 15, 807–818.

Chapter 16

Mechanical Properties of the Lung

Jason H.T. Bates

Chapter Outline

SYMBOLS AND ABBREVIATIONS

A*, *B*, *K Parameters of the exponential pressure–volume relationship (Eqn (1))

a*, *b*, *c*, *d Parameters of the sigmoidal pressure–volume relationship (Eqn (2))

α Exponent of frequency in the constant-phase model (Eqn (8))

COPD Chronic obstructive pulmonary disease

E_1, E_2 Parameters of nonlinear pressure–volume relationship (Eqn (7))

E_L Lung elastance

f Frequency in Hertz (Hz)

FEV_1 Maximal volume of gas expelled in the first second of a forced expiration

FVC Forced vital capacity

G Tissue damping—a parameter of the constant-phase model (Eqn (8))

H Tissue stiffness—a parameter of the constant-phase model (Eqn (8))

Iaw Airway inertance—a parameter of the constant-phase model (Eqn (8))

i Positive square root of −1

P Pressure

Pao Pressure at the airway opening

Pes Pressure in the esophagus

Ptp Transpulmonary pressure, equal to *Pao* − *Pes*

P_0 Baseline pressure in the single-compartment model (Eqn (4))

ΔP Pressure drop across a flow resistor

R Resistance

R_1, R_2 The constants in a nonlinear equation for resistance (Eqn (6))

Raw Airway resistance

R_L Lung resistance

R_N Newtonian resistance—a parameter of the constant-phase model (Eqn (8))

Rt Tissue resistance

V Gas volume

$\dot{V}$ Flow of gas

t Time

Z_L Impedance of the lung

Zrs Impedance of the respiratory system

Zw Impedance of the chest wall

1. INTRODUCTION

The principal purpose of the lungs is to supply oxygen (O_2) to the tissues of the body in order to fuel metabolic processes and to remove the waste product carbon dioxide (CO_2). This two-way exchange is achieved through a chain of events, one of which is the cyclic inflation and deflation of the lungs known as breathing. This brings O_2-rich gas into the lungs during inspiration and expels CO_2-rich gas during expiration. Breathing is essentially a mechanical event that relies on the respiratory muscles (principally the diaphragm) being able to drive gas through the numerous conduits of the airway tree and to stretch the tissues of the lungs and chest wall against their natural tendency to recoil. Consequently, the mechanical properties of the lungs have a major influence on the ease and comfort of breathing.

Breathing for a healthy individual is virtually effortless at rest and is generally not a limiting factor even during severe exercise. In disease, however, this may no longer be the case, perhaps the classic example being the breathing difficulties caused by an attack of asthma. A variety of other important pulmonary diseases also lead to substantial derangements in lung mechanical function. Emphysema, for example, causes the lung tissue to lose its normal level of elastic recoil, while pulmonary fibrosis makes the lung

Comparative Biology of the Normal Lung. http://dx.doi.org/10.1016/B978-0-12-404577-4.00016-3

tissues much stiffer than normal. However, the mechanical properties of the lungs are not only affected by disease, they are also diagnostic of them. Thus, by measuring quantities such as the stiffness of the lung tissues, or the flow resistance of the airway tree, one can potentially monitor the progression of disease or the results of therapy. This applies not only to patients in the clinic but also to animal models of pulmonary disease in the research laboratory.

2. EMPIRICAL ASSESSMENT OF LUNG MECHANICAL FUNCTION

2.1 Relationships Between Pressure, Flow, and Volume

At its most basic level, the mechanical properties of the lung can be inferred empirically by observing how certain physical variables depend upon each other. For example, if an excessive pressure must be applied to the mouth or tracheal opening in order to inflate the lungs by a standard volume, then one can conclude that the tissues of the lung are abnormally stiff. Similarly, the requirement of an excessive pressure to inflate the lungs at a standard rate can be taken as an indication that the pulmonary airways are abnormally narrow. These notions can be formalized by considering the mathematical relationships between pressure, flow, and volume.

The stiffness of the lungs is embodied in the quasi-static relationship between the pressure applied across the lung tissues, known as transpulmonary pressure (*Ptp*), and the volume (*V*) of the lung relative to its elastic equilibrium value, known as functional residual capacity. *V* is relatively easy to measure, either by measuring inspired volume directly with a spirometer connected to the mouth or by measuring mouth flow with a pneumotachograph and then integrating it with respect to time (Bates, 2009). *Ptp* is somewhat more problematic to obtain in that it consists of the difference in pressure between the airway opening (*Pao*) and the pleural surface of the lungs. *Pao* is readily obtained with a pressure transducer placed at the entrance to the airways. However, unless the chest is open so that the pleural surface is exposed to atmospheric pressure, it is necessary to measure pleural pressure with an esophageal balloon (Baydur et al., 1982). This provides the pressure inside the esophagus (*Pes*), which is usually a good reflection of the pressure in the pleural space (Baydur et al., 1982; Dechman et al., 1992), allowing the calculation of *Ptp* as the difference between *Pao* and *Pes*.

Figure 1 shows the normalized relationship between *Ptp* and *V* for a normal lung (Venegas et al., 1998). The relationship is relatively steep and straight throughout its midportion, which corresponds to the normal range of breathing, but curves toward the horizontal at its extremes. The upward concavity of the plot at low values of *V*

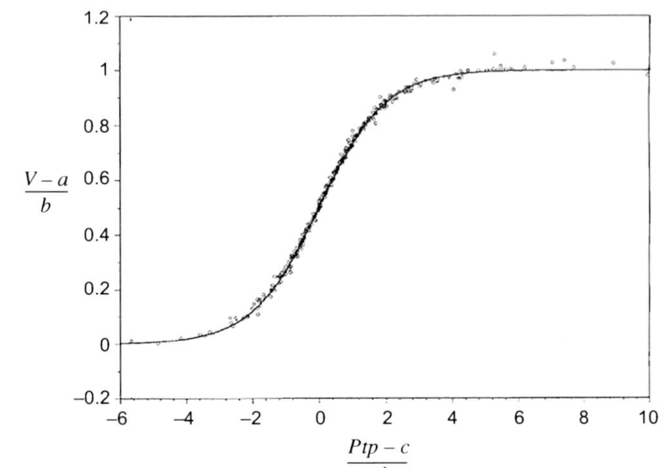

FIGURE 1 Normalized *Ptp*–*V* relationship for the lungs, including data from both humans and dogs, and from both normal and injured lungs. *(Reproduced with permission from Venegas et al. (1998).)*

indicates a stiffening of the lung due to progressive closure (known as derecruitment) of distal airspaces that occurs as *V* decreases. The downward concavity of the plot at high values of *V* indicates progressive strain stiffening of the lung tissues. Most analyses of the *Ptp*–*V* relationship have focused on the portion above functional residual capacity, which excludes the lower upward concave portion. This upper section of the relationship is well described by an exponential expression of the form (Salazar and Knowles, 1964)

$$V = A - Be^{-KP} \tag{1}$$

where *A*, *B*, and *K* are constants. Although this is a purely empirical expression, the parameter *K* has been used to characterize *Ptp*–*V* relationships in various diseases (Osborne et al., 1988; Colebatch et al., 1985; Pare et al., 1982). The value of *K* is independent of gender but increases with age (Colebatch et al., 1979) and usually has a value around 0.16 in normal humans (Osborne et al., 1988). *K* can become elevated in emphysema, where the *Ptp*–*V* relationship becomes more sharply curved, and has been postulated to reflect mean airspace size (Osborne et al., 1988). By contrast, the *Ptp*–*V* relationship becomes flatter in pulmonary fibrosis (Sansores et al., 1996).

More recently, the sigmoidal expression (Venegas et al., 1998)

$$V = a + \left(\frac{b}{1 + e^{-(Ptp-c)/d}}\right) \tag{2}$$

for the entire *Ptp*–*V* relationship has gained some traction. Again, this is an empirical description that does not purport to reflect anything mechanistic, but it does lead to an intriguing unification of different data sets. Specifically, when plotted as $(V - a)/b$ versus $(Ptp - c)/d$, data from humans and dogs with both normal and injured lungs collapse on to a single relationship (Figure 1), even though the individual values of *a*, *b*, *c*, and *d* may vary.

The dynamic behavior of the lung is reflected in relationships that involve rates of change of V, or flow ($\dot{V}$), the latter being readily measured with a flow meter at the airway opening (Bates, 2009). Several different types of flow meters have been used for respiratory applications, but the simplest and most common is the pneumotachograph, which consists of a conduit containing a resistive element through which the subject breathes. A differential pressure transducer measures the pressure drop (ΔP) across the resistive element, which is then converted into $\dot{V}$ using the known resistance (R) of the element as

$$\dot{V} = \frac{\Delta P}{R} \tag{3}$$

2.2 Spirometry

Most clinical testing of lung function in patients only involves measuring either V or $\dot{V}$ at the airway opening using a spirometer or pneumotachograph (Irvin, 2005). (Note that V and $\dot{V}$ can be calculated one from the other by numerical differentiation and integration, respectively, with respect to time.) Measuring these quantities in the clinic is referred to as spirometry and is noninvasive, easy, and requires only simple equipment. However, spirometry alone does not provide any information about the magnitude of the effort (i.e., pressure) generated by the subject. Fortunately, there are some key characteristics of the lung that are essentially independent of effort while still being diagnostic of its mechanical properties. One of these is the maximum volume of gas that can be forcibly exhaled following a maximal inspiration, termed the *forced vital capacity* (FVC). The magnitude of the FVC is normally closely related to height and gender (and, to a certain extent, ethnicity), so the value of FVC measured in a particular subject can be compared to its expected value. FVC is typically abnormally low in restrictive lung diseases such as pulmonary fibrosis (Allen et al., 2009).

The other effort-independent aspect of the lung concerns the maximum $\dot{V}$ that can be achieved during a forced expiration (Allen et al., 2009). Normal individuals above the age of about 5 years can voluntarily generate sufficient expiratory muscle pressure to reach their maximal flows over about the lower 80% of the vital capacity range. Expiratory flow reaches its maximum value as a result of dynamic collapse of the airways, and so airway compliance is a critical factor in determining how rapidly one can expire. The collapsibility of the airways is greatly influenced by the outward tethering forces applied to the airway wall by the parenchyma in which it is embedded, and these forces increase with the volume of air in the lungs (Bates, 2009). Consequently, maximum flow decreases with decreasing V. It also decreases when the caliber of the airways is reduced. Thus, the relationship between maximum expiratory flow and V is diagnostic of obstructive lung diseases such as asthma and chronic obstructive lung disease (COPD). Clinically, the nature of flow limitation is usually captured in the volume of air that leaves the lungs during the first second of a forced expiration starting from total lung capacity, known as FEV_1. Again, FEV_1 has a normal range that depends on subject demographics and so is typically expressed in terms of its percent predicted value. Abnormally low values of FEV_1 and the ratio FEV_1/FVC are commonly found in asthma and COPD (Irvin, 2005).

3. PARAMETERS OF LUNG MECHANICAL FUNCTION

3.1 The Single-Compartment Linear Model

The approaches to assessing lung mechanics described above are all empirical in that they consider relationships between pressures, flow, and volumes purely from the perspective of whether or not they appear normal. This is certainly useful for clinical diagnosis of pulmonary disease, but it does not allow a mechanistic basis to be ascribed to any abnormalities that might be observed (Mead, 1979). Making the link between structure and mechanical function in the lung is, however, an essential goal in the research laboratory and indeed would be beneficial in the clinic. Linking structure to function requires that a mathematical model of the lung be invoked so that the functional consequences of putative structural mechanisms can be calculated (Bates, 2009). However, not all situations call for the same mathematical model. Generally speaking, the more varied the experimental data the more complicated is the model needed to account for them. In fact, the field of lung mechanics progresses through a never-ending series of refinements and extensions to existing models based on new data from experiments designed to test the existing models. Nevertheless, much of lung mechanics can be understood based on rather simple mathematical models, which can be arranged in a hierarchy of complexity (Bates, 2009).

The simplest and most widely invoked model of lung mechanics is the single-compartment linear model consisting of an elastic compartment served by a flow-resistive airway. This model is commonly conceptualized as a balloon sealed over the end of a pipe, but for reasons that shall become apparent shortly, we represent the model here as a pair of telescoping canisters that are coupled mechanically via a spring and that communicate with the environment via a single conduit (Figure 2(A)). This

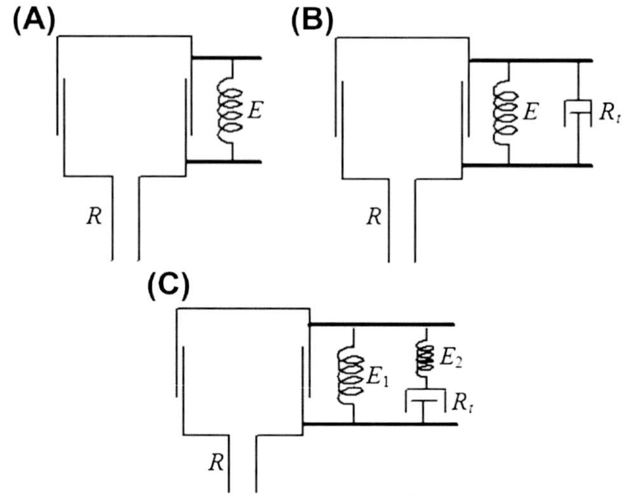

FIGURE 2 (A) Single-compartment linear model with resistance R and elastance E. (B) Model with addition of tissue resistance Rt. (C) Model extended to account for tissue viscoelasticity by the inclusion of a Kelvin body comprised of resistance Rt and elastances E_1 and E_2.

construct has a single mechanical degree of freedom because its state at any point in time is specified by a single variable, namely the volume V contained inside the canisters. If the spring is Hookean (has a linear force–tension relationship with slope E_L) and the conduit has a constant resistance (linear pressure–flow relationship with slope R_L), then Ptp across the model is given by the equation

$$Ptp(t) = R_L\dot{V}(t) + E_LV(t) + P_0 \tag{4}$$

where P_0 is a constant that allows for Ptp to be nonzero when V and $\dot{V}$ are both zero (as is the case when the lung is in situ within the chest cavity). Note that Ptp, V, and $\dot{V}$ are all functions of time (t) and thus are written explicitly as such in Eqn (4). Also, $\dot{V}$ is the derivative of V with respect to t.

3.2 Fitting the Single-Compartment Model to Data

Equation (4) is a first-order linear differential equation in which a single dependent variable (Ptp) is related to three parameters (R_L, E_L, and P_0) and their corresponding independent variables. Inspection of Eqn (4) reveals that the independent variables corresponding to R_L, E_L, are $\dot{V}$ and V, respectively. The parameter P_0 might at first glance not seem to be linked to an independent variable, but in fact it is multiplied by 1. Accordingly, the independent variable corresponding to P_0 is the trivial variable that has a value of 1 at every time point. The function of Eqn (4) is to predict the dependent variable from time-varying measurements of the independent variables based on particular values for the parameters. The usefulness of Eqn (4) thus lies in the accuracy with which it can predict $Ptp(t)$ measured in a particular lung from corresponding measurements of $\dot{V}(t)$ and $V(t)$. The accuracy of the prediction is maximized through choice of appropriate values for R_L, E_L, and P_0, something that is achieved using the numerical technique of multiple linear regression (Bates, 2009). The resulting prediction of the measured $Ptp(t)$ is the one that minimizes the sum of the squared differences between measurement and prediction over the duration of the data. This prediction is never perfect, of course, because the model in Figure 2(A) is not a perfect representation of a real lung. Indeed, it is a gross oversimplification of what is actually a highly complex multicomponent structure. Nevertheless, Eqn (4) manages to account for the vast majority of the data when measurements are made during normal breathing or mechanical ventilation.

3.3 Physiological Interpretation of the Single-Compartment Model

The parameters R_L and E_L in Eqn (4) are generally referred to as the resistance and elastance, respectively, of the lung. It is important to bear in mind, however, that they are actually just the parameters of a model that is thought to be a useful representation of the lung. The physiological interpretation of R_L and E_L requires experimental verification despite the apparent obviousness of their identities. In the case of R_L this is particularly important. Figure 2(A) compels one to interpret R_L as the flow resistance of the airway tree (Raw). However, invasive experimentation in animal models using the alveolar capsule technique (which allows the direct measurement of Raw) has shown that R_L evaluated during normal breathing contains a major component due to the dissipation of energy in the parenchymal tissues of the lung as they expand and contract (Bates et al., 1988; Ludwig et al., 1990). In other words,

$$R_L = Raw + Rt \tag{5}$$

where Rt represents tissue resistance. The parameter E_L, by contrast, reflects the temporary storage of energy in the tissues due to their elastic nature. We can thus see that the simple model in Figure 2(A) must be extended to allow for a more accurate interpretation of the structure–function link through the addition of a tissue resistance. This is represented as a dashpot (sliding friction element having resistance Rt) in Figure 2(B) connected between the two canisters, demonstrating why the sliding canister representation of the model has advantages over the conventional balloon-on-a-pipe representation. However, the value of Rt as depicted in Figure 2(A) depends strongly and inversely on the frequency of breathing, as discussed below.

It must also be pointed out that the parameters in Eqn (4) pertain to the lungs alone only when the independent variable is Ptp, the measurement of which requires either that the chest be open or that pleural pressure (or its surrogate Pes) be

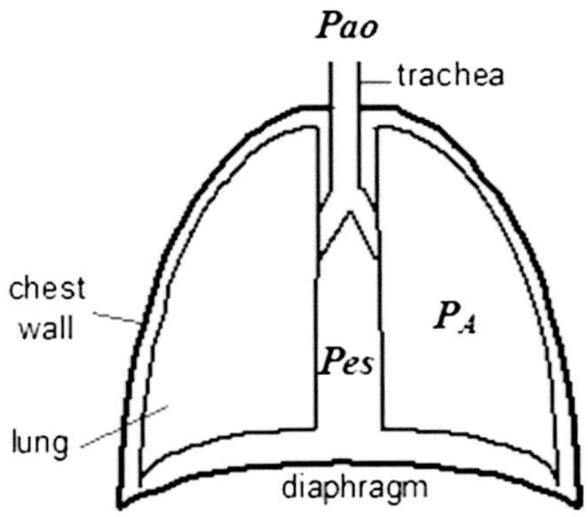

FIGURE 3 Schematic representation of the key respiratory system pressures. Pressure at the airway opening (*Pao*) is measured with a pressure transducer located at the entrance to the mouth or trachea. Pressure in the esophagus (*Pes*) serves as a surrogate for pressure in the pleural space surrounding the lungs inside the thoracic cavity that is bordered by the chest wall and the diaphragm. Alveolar pressure (P_A) can be measured with an alveolar capsule in open-chest animals. Transpulmonary pressure (*Ptp*) is the difference between *Pao* and *Pes*. *Pes* itself is the pressure drop across the chest wall. The resistive pressure drop across the conducting airways is the difference between *Pao* and P_A. All pressures are expressed relative to the atmospheric pressure that impinges on the outside of the chest wall.

measured. Practically, it is easier to record only *Pao*(t), in which case the resistive and elastic parameters in Eqn (4) pertain to the entire respiratory system. Alternatively, if the pressure in Eqn (4) is the difference between *Pao* and alveolar pressure measured by alveolar capsule, then the resistance is *Raw* and the elastance is zero. As a final variation on this theme, the pressure in Eqn (4) can be pleural pressure, in which case the resistance and elastance parameters refer to the chest wall tissues alone. Figure 3 gives a schematic illustration of the various pressure measurement scenarios.

3.4 Nonlinear Extensions of the Single-Compartment Model

The single-compartment model shown in Figure 2(A) can be extended by allowing its properties to become nonlinear. For example, direct measurement of the flow resistance of the airways shows that it exhibits a marked dependence on $\dot{V}$ due to the development of turbulence and other aspects of gas flow through complex geometries. The flow-dependence of resistance can be rather accurately, albeit empirically, described by the quadratic Rohrer relationship (Bates, 2009), which can be incorporated into Eqn (4) to yield the extended equation

$$Ptp(t) = R_1\dot{V}(t) + R_2\dot{V}(t)|\dot{V}(t)| + E_LV(t) + P_0 \quad (6)$$

where the vertical bars denote the absolute value of the enclosed quantity. Now the resistive behavior of the lungs is not captured in a single value but rather requires two independent parameters, R_1 and R_2, for its specification.

Similarly, the elastic recoil pressure of the lung tissue is not actually a perfectly linear function of inflation pressure but instead exhibits curvilinearities (see Figure 1) that can often be captured to a useful degree of accuracy by a quadratic function. Incorporating this into Eqn (4) yields

$$Ptp(t) = R_L\dot{V}(t) + E_1V(t) + E_2V^2(t) + P_0 \quad (7)$$

where again the two parameters E_1 and E_2 are required to account for behavior that previously was ascribed to the single parameter E_L (Bersten, 1998). The four independent parameters in either Eqn (6) or Eqn (7) can all be evaluated by multiple linear regression, and indeed the two equations can combined to yield a five-parameter equation. Which of Eqns (4), (6), or (7) should be invoked in any particular application depends on the nature of the data to which the equation is to be fit. Generally speaking, data that involve modest excursions in $\dot{V}$ and V, such as during typical tidal breathing, will tend to be well accounted for by the linear model (Eqn (4)). By contrast, when the excursions in $\dot{V}$ and V are large, such as during exercise or when the subject hyperventilates due to stress, the effects of nonlinearities are more likely to become apparent (Eqn (6) for large amplitudes in $\dot{V}$, and Eqn (7) for large amplitudes in V).

3.5 Multi-Compartment Extensions of the Single-Compartment Model

Making the constitutive properties of the single-compartment model nonlinear, as discussed above, is not the only way to make the model more representative of a real lung. The other general direction to go in is to add additional compartments (Bates, 2009). This makes perfect intuitive sense because the lung itself is clearly compartmentalized anatomically at every level of scale, from the right and left lungs, to the individual lobes, all the way down to its myriad individual acini. It is therefore not surprising that even the normal lung behaves much more like two independent compartments than a single uniformly ventilated alveolar compartment. Interestingly, this two-compartment behavior does not primarily reflect the existence of two distinct alveolar regions arranged either in parallel (Otis et al., 1956) or series (Mead, 1969) but rather is due to the viscoelastic properties of the lung tissues (Bates et al., 1988). This can be represented mathematically by further extending the model shown in Figure 2(B) to that shown in Figure 2(C). Now the lung tissues are represented mechanically as a collection of three elements that together constitute what is known as a Kelvin body, or linear viscoelastic solid (Fung, 1993).

The Kelvin body in Figure 2(C) imbues the model with the ability to exhibit stress adaptation to a change in lung volume (Bates, 2009), something that is characteristic of the lung. That is, if V is suddenly increased by a fixed amount, *Pao* does not stay at a fixed elevated level but rather falls gradually as if asymptoting to a level intermediate between its initial value and the level reached immediately after inflation. Conversely, if V is suddenly reduced, *Pao* falls immediately but then recovers asymptotically. These effects are exhibited by the model in Figure 2(C) due to the series spring (E_2) and dashpot (Rt) in the Kelvin body. A sudden extension of this second spring upon inflation of the alveolar compartment adds to the pressure inside the compartment generated by spring E_1, but then E_2 relaxes exponentially by moving the dashpot so that eventually the compartmental pressure is due only to the fixed extension in E_1. The converse occurs upon suddenly decreasing the volume of the alveolar compartment.

3.6 Frequency Dependence of Resistance and Elastance

The viscoelastic nature of the lung tissues also gives rise to another extremely important aspect of lung mechanics known as the frequency dependence of resistance and elastance (Bersten, 1998; Otis et al., 1956; Mead, 1969; Hantos et al., 1990, 1987a, 1992a). If the single-compartment model is invoked to identify values for R_L and E_L from measurements of $Pao(t)$, $V(t)$, and $\dot{V}(t)$ made during normal breathing or mechanical ventilation, as described above, one finds that the values obtained depend markedly on the frequency at which lung volume changes during the measurements. R_L decreases by several fold as frequency increases from below 1 Hz up to about 5 Hz, while E_L increases with increasing frequency over the same range although by a somewhat smaller relative amount (Hantos et al., 1995; Sato et al., 1993). Alveolar capsule studies have demonstrated that the dependencies of R_L and E_L on frequency below 5 Hz in the normal lung are entirely attributable to the viscoelastic behavior of the lung tissues. For example, the component of R_L due to Rt (Eqn (5)) in rabbits at a frequency of 0.16 Hz can be greater than that due to *Raw* by a factor of two or more, while at 5 Hz *Raw* is essentially unchanged but Rt has become negligible (Tepper et al., 1992). The frequency dependence of these parameters in dogs is qualitatively similar, but due to the large airways in this species, the value of Rt at the lower end of the frequency spectrum can be as much as 10 times the value of *Raw* (Sato et al., 1993).

The frequency dependence of resistance and elastance can also be recapitulated by the Kelvin body in Figure 2(C). For example, when the volume of the alveolar compartment is oscillated at a low frequency by driving flow in and out of the airway, spring E_1 will necessarily follow the volume changes and generate a recoil pressure inside the compartment accordingly. However, as soon as spring E_2 starts to change length, the force that results will move the dashpot piston so that the piston position rather than the spring length follows the volume change. Thus, in the limit of very low frequencies the Kelvin body appears to have an elastance only of E_1 and a resistance of Rt. Conversely, in the limit of high oscillation frequencies the dashpot piston has no time to move in either direction before the spring starts pulling it in the other direction, so it is the extension of the spring E_2 that now follows the changes in volume, producing an effective elastance of $E_1 + E_2$ and an effective resistance of zero.

This is not to say, however, that the normal lung behaves exactly like a single alveolar compartment surrounded by viscoelastic tissue. Regional heterogeneities do exist and their effects on R_L and E_L can become quite apparent at oscillation frequencies above those associated with normal breathing (Fredberg et al., 1984). These effects also become more pronounced in disease such as during simulated attacks of asthma that can be induced even in a normal lung by the application of a smooth muscle agonist (Gillis and Lutchen, 1999; Kaczka et al., 1997; Lutchen et al., 1996a, 1994). The effects of pathologies are thus to increase the negative dependence of R_L and the positive dependence of E_L on the oscillation frequency of the lungs.

4. LUNG IMPEDANCE

4.1 Forced Oscillation Technique

So long as the lung behaves mechanically like a linear dynamic system (i.e., nonlinear effects like the dependence of R_L on $\dot{V}$, and E_L on V, can be neglected) then its relationships between pressure, flow, and volume can be completely characterized by the way in which R_L and E_L vary with oscillation frequency (f). In other words, R_L and E_L are functions of f and together constitute what is known as the mechanical impedance of the lungs (Z_L) (Bates, 2009; Bates et al., 2011). The mathematical convention is to define Z_L as having two independent components. One component, the so-called real part, is termed resistance and is simply equal to $R_L(f)$. The other component, the so-called imaginary part, is termed reactance and is equal to $-E_L(f)/2\pi f$. These naming conventions arise from the theory of complex numbers, which is central to the mathematical theory of impedance and may be of interest to those pursuing basic research in the field of lung mechanics (Bates, 2009). However, if one merely wishes to use Z_L as a tool to investigate biological questions, then it suffices to understand how the real and imaginary parts of Z_L relate to R_L and E_L as just explained.

The measurement of Z_L requires that the lungs be perturbed with an appropriate pressure or flow while the resultant flow or pressure is recorded. In principle, any oscillatory signal will do. For example, a conventional mechanical ventilator, or even a subject's own respiratory muscles, cause lung volume to oscillate. Thus, if *Ptp* and $\dot{V}$

are measured continuously during either mechanical ventilation or spontaneous breathing while this is happening, one can determine Z_L at the oscillation frequency in question. Varying the frequency of oscillation over a certain range is also easily achieved through either mechanical ventilation or spontaneous breathing, allowing Z_L to be determined as function of *f*. However, most formal investigations of Z_L are now undertaken using what is known as the forced oscillation technique. Here, a flow signal containing a range of different frequencies is applied to the lungs while the resulting pressure is measured. Most applications for the forced oscillation technique only measure *Pao* rather than *Ptp* because of the practical advantages of avoiding the use of an esophageal balloon. In this case, however, one obtains the impedance of the entire respiratory system (*Zrs*) consisting of the sum of Z_L and the impedance of the chest wall (*Zw*). In larger species, the contribution of *Zw* to *Zrs* can be considerable. For example, the chest wall in dogs contributes at least as much to the mechanical properties of the entire respiratory system as does the lung (Bates et al., 1989). In addition, the chest wall in this species contributes a Newtonian resistance (that is, a conventional resistance directly proportional to flow) that is comparable to that of the airways (Bates et al., 1989), which means that only about half the real part of *Zrs* at frequencies above about 5 Hz corresponds to *Raw*. In rats, the real part of *Zrs* over this higher frequency range is virtually entirely due to *Raw*, but the viscoelastic properties of the chest wall tissues are still at least as dominant as those of the lung tissues (Hirai et al., 1999). In mice, the chest wall makes a small to negligible contribution to both the real and imaginary parts of *Zrs* (Sly et al., 2003). Humans behave more like larger animals, with the elastance of the chest wall and lungs being comparable, while the resistance of the lung is roughly double that of the chest wall (Dechman et al., 1995).

Once pressure and flow have been recorded, digital signal processing techniques based around the use of the fast Fourier transform are used to calculate impedance simultaneously at all the different frequencies involved (Bates, 2009). There are a variety of different devices that have been used to apply multifrequency (broadband) oscillations in $\dot{V}$ to the lungs, the most common being loudspeakers (Landser et al., 1976) and piston pumps (Kaczka and Lutchen, 2004; Schuessler and Bates, 1995), but in all cases the basic principles of impedance calculation remain the same.

4.2 Physiological Interpretation of Impedance

The physiological information embodied in a measurement of impedance depends on the frequency range over which it is measured, and this has varied substantially among different studies from a few Hz spanning the range of normal breathing (Lutchen et al., 1996a, 1996b; Kaczka et al., 1999) all the way up to hundreds of Hz in the acoustic range (Jackson et al., 1989). Early applications of the forced oscillation technique in humans tended to focus on frequencies from about 4 to 40 Hz (Landser et al., 1976; Clement et al., 1983; van de Woestijne et al., 1981) largely because available pressure and flow transducers had adequate frequency responses over this range and because most of the power in the spontaneous breathing flow waveform is below 4 Hz. When measuring *Zrs*, it is crucial to ensure that the measured *Pao* signal is due entirely to the imposed $\dot{V}$ waveform and contains no contribution from other unknown sources. A spontaneously breathing subject contributes in a variable and unknown manner to the generation of *Pao* and $\dot{V}$ within the breathing frequency range, making the accurate calculation of *Zrs* over this range impossible. By limiting the determination of *Zrs* to frequencies outside the breathing range, one avoids this problem. Of course, determining *Zrs* at frequencies below 4 Hz is possible if the subject is apneic during the measurements (Lutchen et al., 1990), but this generally requires some training and may be difficult for subjects with lung disease. A clever compromise is to deliver oscillations in flow that serve both to identify *Zrs* and simultaneously ventilate the lungs while the subject's respiratory muscles remains passive (Lutchen et al., 1993).

Most studies of *Zrs* in experimental animals have been done under anesthesia because this allows the forced oscillation technique to be applied directly to the lungs via a tracheostomy, thereby avoiding the complications created when some unknown fraction of the applied oscillations in $\dot{V}$ becomes shunted into the compliance of the upper airways (Farre et al., 1989; Peslin et al., 1985). Also, with the breathing pattern of the animal completely under the control of the experimenter, sometimes ensured by the use of a paralytic agent, one knows that the observed relationships between *Pao* and $\dot{V}$ are entirely reflective of *Zrs* (Allen et al., 2009; Lundblad et al., 2007; Wagers et al., 2004). Alternatively, anesthesia allows Z_L to be measured by opening the chest to eliminate the contribution that *Zw* makes to *Zrs*. This is readily achieved in animals and has also been performed in humans during surgery (Dechman et al., 1995; Babik et al., 2003).

Many recent investigations of impedance in animal models of lung disease have been performed over a relatively low frequency range that includes that of normal breathing because this provides useful information about both the viscoelastic properties of the tissues of the respiratory system and the degree of regional ventilation heterogeneity within the lungs. The mathematical model that has become the standard for interpreting low frequency impedance is the so-called constant-phase model. This model considers the lungs to be a single uniformly ventilated unit served by a single airway, but the construct

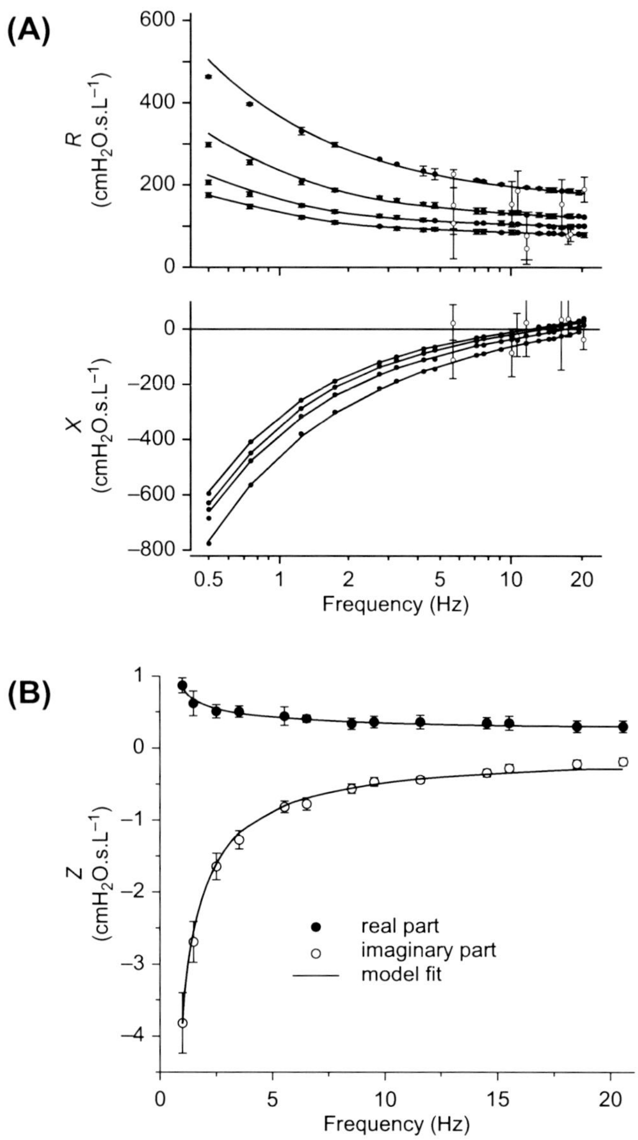

FIGURE 4 Fit of constant-phase model to impedance data from (A) rats *(reproduced with permission from Petak et al. (1997)* and (B) mice. R (resistance) is the real part of impedance, while X (reactance) is the imaginary part of impedance. *(Reproduced with permission from Wagers et al. (2004).)*

giving rise to the mechanical properties of the tissues is not expressible in terms of springs and dashpots as in the case of the Kelvin body (Figure 2(C)). Like the Kelvin body, the impedance of the tissues in the constant-phase model has both real and imaginary parts that vary with f, but unlike the Kelvin body, their ratio does not vary with f (in fact, it is this property that gives rise to the term *constant-phase*). The equation for the impedance of the constant-phase model is (Hantos et al., 1992a, 1987b)

$$Zrs(f) = R_N + i2\pi f \cdot Iaw + \frac{G - iH}{(2\pi f)^{\alpha}} \tag{8}$$

where

$$\alpha = \frac{2}{\pi}\tan^{-1}\left(\frac{H}{G}\right) \tag{9}$$

Fitting this equation to measurements of Zrs from about 1 to 20 Hz yields values for the four parameters R_N, Iaw, G, and H. R_N is a Newtonian resistance (i.e., a conventional resistance that dissipates energy in proportion to flow) that has been shown to provide a useful approximation to the overall flow resistance of the airway tree (Tomioka et al., 2002). Iaw is an inertance related to the mass of the gas in the central airways. In mice, Iaw is so small that its effects can be safely ignored below 20 Hz (Gomes et al., 2000). G and H characterize, respectively, the resistive and elastic properties of the respiratory tissues. G and H do not have exactly the same units as conventional tissue resistance and elastance, but their values can nevertheless be taken to embody the same physical phenomena. The constant-phase model has been used extensively to characterize Zrs and Z_L in a variety of species (Hantos et al., 1987a, 1992a, 1992b; Hirai et al., 1999; Gomes et al., 2000) and has been found to fit measurements of impedance below about 20 Hz extremely well (Figure 4) despite having only four free parameters (three in the case of mice because Iaw can be neglected). Of course, this does not mean the model is perfect; obviously, no real lung behaves exactly like a single uniformly ventilated compartment. However, its degree of parsimony and the convenient physiological interpretation of its parameters make the constant-phase model extremely attractive as a tool for characterizing lung mechanical function.

4.3 Applications of Impedance Measurement

The measurement of Z_L or Zrs has become the most widely used approach to studying animal models of lung disease because it is possible to use such measurements to link structure to function within the lung on the basis of well-accepted mathematical models, aided by the availability of specially designed hardware and software for performing the measurements (Schuessler and Bates, 1995). This contrasts with the recently popular technique of unrestrained plethysmography (Hamelmann et al., 1997) with which it is impossible to link structure to function in the lung because only a single variable (pressure within the plethysmograph) is measured (Lundblad et al., 2002; Bates et al., 2008); the determination of lung mechanics parameters such as resistance or elastance requires the measurement of two independent variables, that is, pressure and either flow or volume, respectively (Bates et al., 2004).

The widest application of impedance measurement in animal models of lung disease is in the study of airways

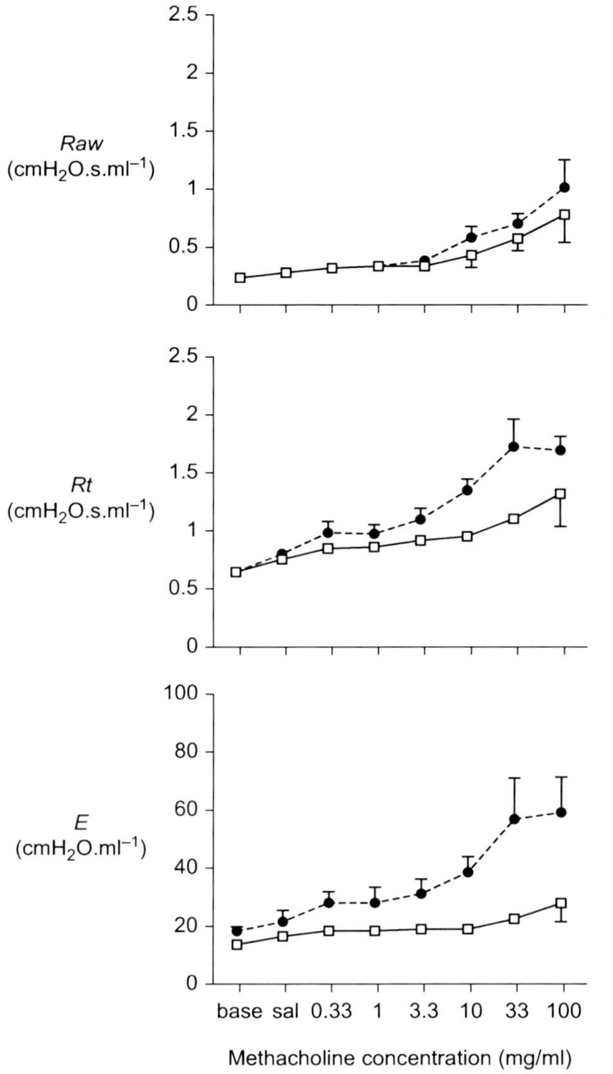

FIGURE 5 Methacholine dose–response curve in open-chest control mice (open symbols) and mice that were allergically inflamed (closed symbols). *Raw* is airway resistance determined from the pressure difference between *Pao* and alveolar pressure measured with an alveolar capsule. *Rt* is lung tissue resistance determined from alveolar pressure alone. *E* is lung tissue elastance. *(Reproduced with permission from Tomioka et al. (2002).)*

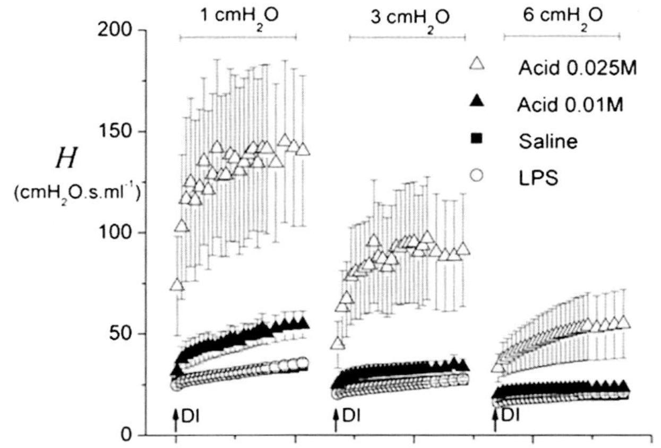

FIGURE 6 Evolution of lung tissue stiffness (*H*) in mice under control conditions during mechanical ventilation following a deep lung inflation, as well as in mice treated with lipopolysaccharide and hydrochloric acid to induce graded levels of acute lung injury. Note how the rates of rise of *H* depend on PEEP. *(Reproduced with permission from Allen and Bates (2004).)*

responsiveness, driven by questions related to the pathophysiology of asthma. Typically, progressively increasing doses of the smooth muscle agonist methacholine are either inhaled as an aerosol (Wagers et al., 2004, 2007; Tomioka et al., 2002) or injected intravenously (Bates et al., 2007; Cojocaru et al., 2008), and the resulting alterations in impedance are measured using the forced oscillation technique. Most studies use this approach to construct a methacholine dose–response relationship (Figure 5), the steepness of which reflects the responsiveness of the airways to challenge, analogous to methods used in the clinic with human patients. There are numerous possible variations on this theme, however, including examining the time course of the response to a given dose of agonist (Bates and Lauzon, 2007; Lauzon and Bates, 2000) or the relative responses when the agonist is given as an aerosol versus intravenously (Homma et al., 2005), all illuminating different physiological aspects of airways responsiveness.

Impedance, either Z_L or Zrs, has also been used to assess alterations in lung mechanics in numerous other animal models of lung disease. For example, the degree and rapidity of lung derecruitment following a deep inflation in models of acute lung injury can be assessed by following the changes in the stiffness parameter H of the constant-phase model as illustrated in Figure 6 (Allen and Bates, 2004; Allen et al., 2002; Massa et al., 2008). The relative changes in both G and H have also been used to characterize the parenchymal changes taking place in interstitial lung disease (Lundblad et al., 2005) and in emphysema (Takubo et al., 2002). Impedance is now gaining ground as a clinical method for assessing lung function in patients on the heels of a great deal of preclinical work establishing how it can be used to relate human lung structure to function (Bates et al., 2011; Farre et al., 1989, 1998, 2001).

4.4 Comparative Values of Impedance Parameters Across Species

Comparative studies of respiratory mechanics in different species are often confounded by the use of different measurement techniques and protocols between different studies. Parameters of lung mechanical function depend not only on the usual experimental variables such as the equipment and approach used to make the measurements but also on factors such as body position, depth, and type of anesthesia, surgical procedures, etc. Most important, however, are the effects of frequency, which as illustrated above have major effects on the apparent resistances and elastances of the tissues in both the lung and the chest wall. It is thus not useful to compare resistance or elastance between species unless the conditions of the measurement, most especially the frequency, are standardized.

On the other hand, the manner in which impedance varies with frequency is well preserved across species. This is demonstrated in Figure 7, which shows that both the real and imaginary parts of impedance below 10 Hz collapse onto single relationships of frequency when they are normalized by body weight (BW) in mice, rats, guinea pigs, and rabbits (Gomes et al., 2000). Therefore, lung mechanics between different species, or between conditions within a given species, could be meaningfully compared if some means of capturing the frequency dependence of impedance were available. Fortunately, the constant-phase model provides precisely the tool that is needed for this purpose for impedance measurements below about 20 Hz. Impedance over this range of frequency achieves a good balance between mechanical effects of flow of gas through airways and effects of movement and distortion of the tissues. The parameters R_N, *Iaw*, G, and H, being derived from the fit of the model over this entire frequency range, are thus themselves inherently independent of frequency. These parameters do, however, exhibit strong inverse dependencies on the size of the animal. Figure 8 shows the values of R_N, *Iaw*, G, and H versus BW from published reports on mice (Bozanich et al., 2005; Hantos et al., 2008; Johnston et al., 1812–1819; Petak et al., 2001; Shalaby, Gold, Schuessler, Martin, Robichaud, 2010), rats (Hirai et al., 1999; Habre et al., 2001; Petak et al., 1998, 1997; Sakai et al., 2001), guinea pigs (Adamicza et al., 1999, 2001), rabbits (Habre et al., 2002, 2006; Hantos et al., 1994), cats (Hantos et al., 1992b), minipigs (Petak et al., 2006), dogs (Hantos et al., 1995; Lutchen et al., 1994; Suki et al., 1997a, 1997b), adult humans (Kaczka et al., 1997; Babik et al., 2003; Hantos et al., 1986; Bachofen, 1968), and children (von Ungern-Sternberg et al., 2009). The results in mice were actually derived from *Zrs*, but this can be equated to Z_L in this species because of the negligible contribution made by the chest wall (Sly et al., 2003). The contribution of the lung parenchyma to R_N has been experimentally demonstrated to be negligible in mice (Tomioka et al., 2002) and dogs (Hantos et al., 1992a; Petak et al., 1993), which motivates the notion that R_N is, in fact, an accurate reflection of airway resistance. In three of the studies in human adults (Kaczka et al., 1997; Hantos et al., 1986; Fredberg and Stamenovic, 1989) R_N includes the flow resistance of the endotracheal or tracheostomy tube, and in the remaining pediatric (von Ungern-Sternberg et al., 2009) and adult (Babik et al., 2003) studies, the subjects were undergoing cardiac surgery and so may have had abnormal pulmonary function.

Figure 8(A) demonstrates that R_N (i.e., airway resistance) decreases more slowly with BW than would be expected if airway size scaled with body dimensions (i.e., the magnitude of the slope of the relationship in Figure 8(A) is less than unity). In other words, the airway resistance of a mouse is substantially less, relative to the size of its lung, than is the airway resistance of a human. Indeed, the relative large diameters of mouse airways are apparent even from visual inspection of micro-CT images (Lundblad et al., 2007). Rodents as an isolated group exhibit a similar kind of inverse dependence of R_N on BW (Gomes et al., 2000). Nevertheless, the relationship is not perfect across all species, with dogs providing anomalously low values of R_N (Figure 8) that may reflect the fact that it is a panting animal that often needs to move air rapidly in and out of the lungs for extended periods in order to thermoregulate. It must also be remembered that R_N obtained from *Zrs* in intact animals may contain a contribution from the chest wall. These contributions are minor in many species, but in the human the contribution to R_N from the chest wall is comparable to that from the lung itself and so cannot be neglected (Hantos et al., 1986; Nagels et al., 1980).

The tissue parameters G (Figure 8(B)) and H (Figure 8(C)) exhibit close to inverse dependencies on BW, meaning that the slopes of the relationships for G and H shown in Figure 8 are close to -1. This is what would be expected if the intrinsic mechanical properties of the lung tissues are identical across species so that bulk tissue resistance and elastance simply reflect lung size. The similarity of the slopes of the relationships for G and H also means that their ratio, termed *hysteresivity*, is species independent, also supporting the uniformity of intrinsic tissue properties across species. The average value of G/H is about 0.2 for the data shown in Figure 8, with a range of roughly 0.1–0.3. Part of the reason for this variation may be differences in the transpulmonary pressures at which Z_L was measured in these various studies (Hirai et al., 1999; Sly et al., 2003; Hantos et al., 2003) and in the amplitudes of the forced oscillations in flow and volume that were used to make the measurements (Hantos et al., 1992b; Barnas et al., 1989). Also, G tends to be increased by the presence of peripheral airway inhomogeneity (Hantos et al., 1992a; Kaczka et al., 1997; Lutchen et al., 1996a, 1996b), something that is affected by airway tone and thus may have varied between studies. In the intact animal, G and H include contributions from the chest wall and may also be affected by the altered configuration of the lung tissue imposed by the boundaries of the thoracic cavity (Sly et al., 2003). These effects are relatively minor in mice and can often be neglected, but they may be substantial in larger species. In the rat, for example, chest wall G and H have been found to be roughly comparable to their pulmonary counterparts (Hantos et al., 1987a; Tomioka et al., 2002; Petak et al., 1998), with similar proportions having been

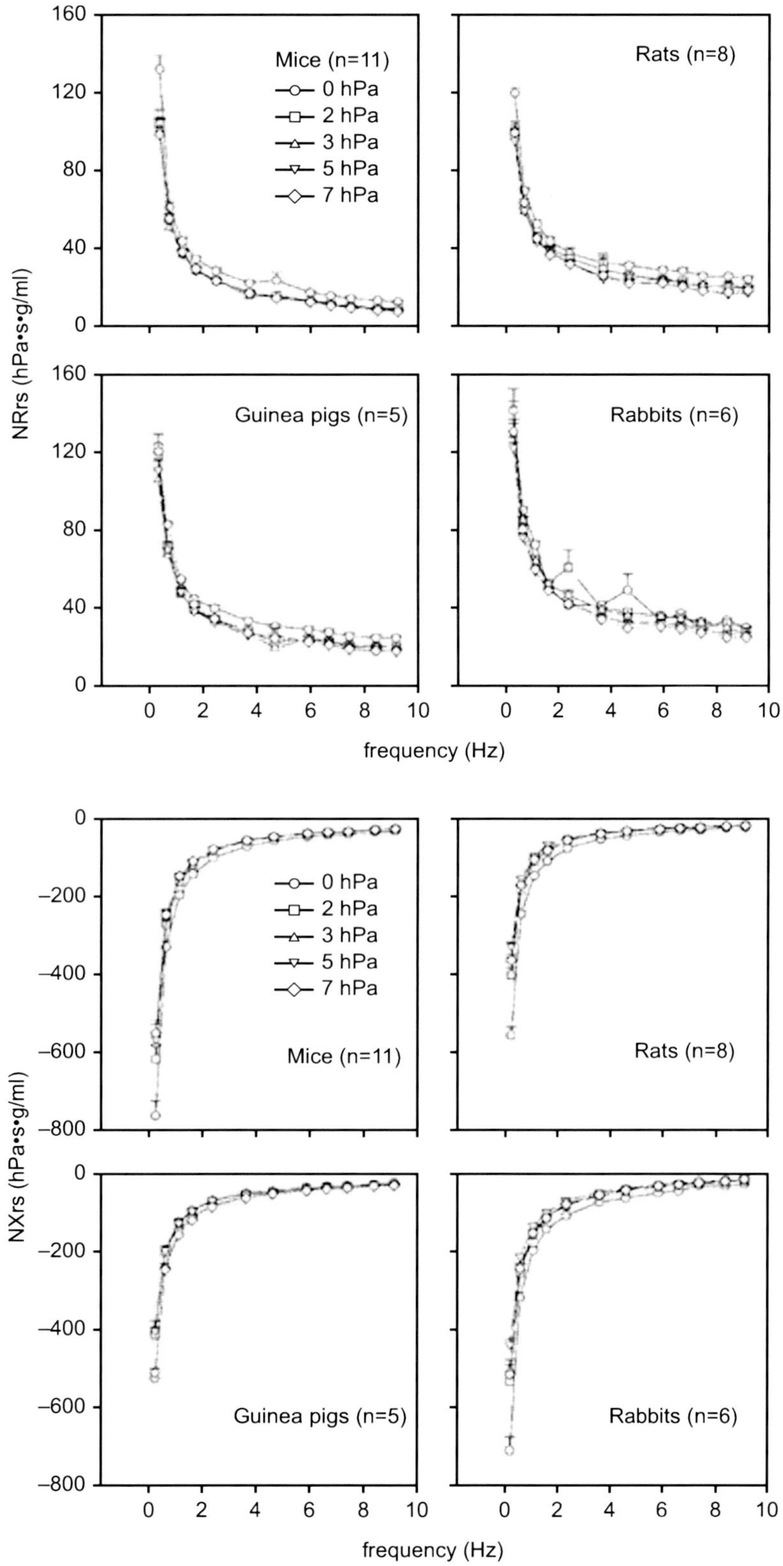

FIGURE 7 Impedance normalized to body weight in four different rodent species. NRrs is the normalized real part of respiratory system impedance, and NXrs is the normalized imaginary part. *(Reproduced with permission from Gomes et al. (2000).)*

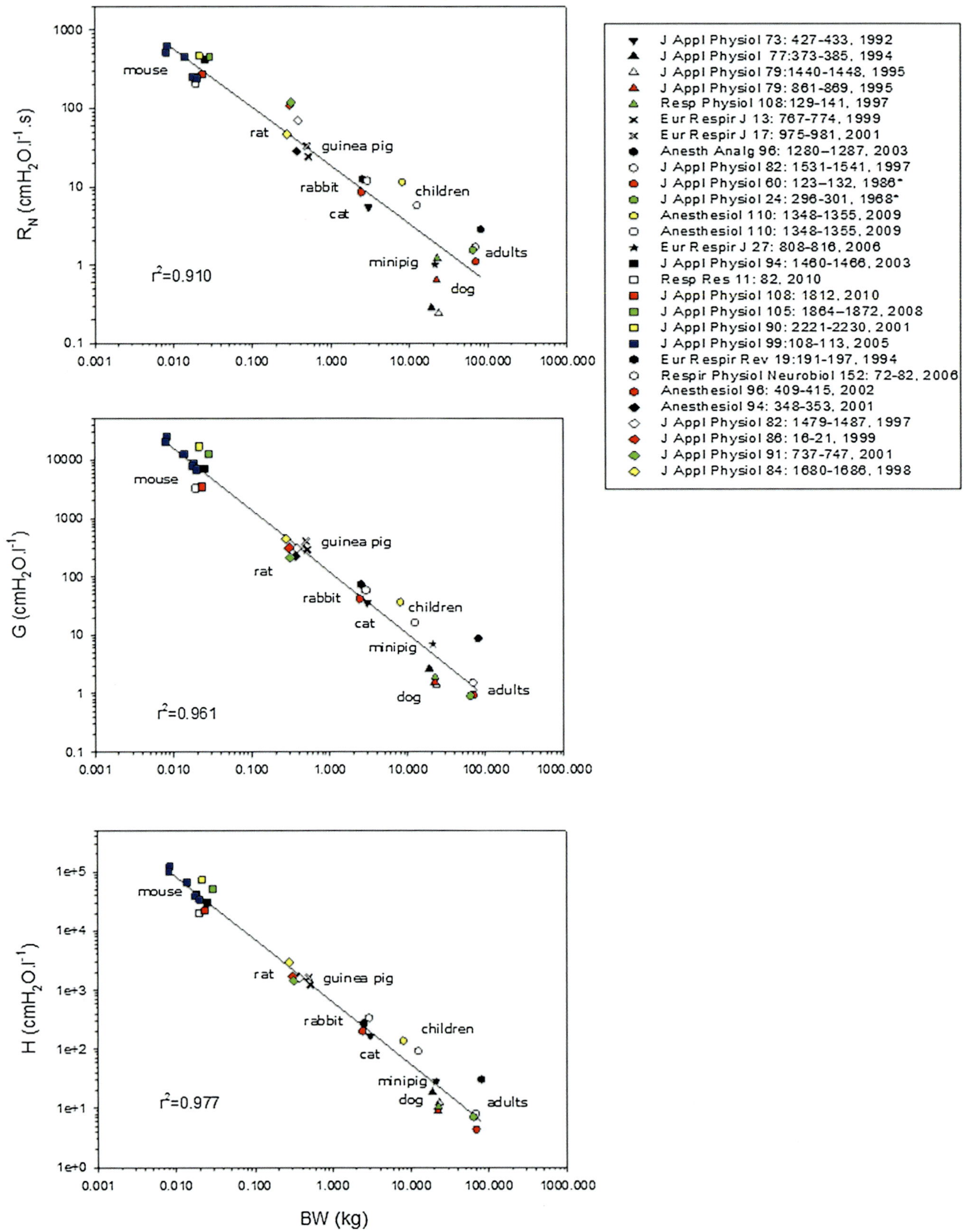

FIGURE 8 Newtonian resistance (R_N), tissue damping (G), and tissue stiffness (H) as functions of body weight (BW) in different mammalian species. *Original impedance data fitted by Kaczka et al. (1997). *(Reproduced with permission from Bates et al. (2011).)*

observed in the cat (Hantos et al., 1992b) and dog (Suki et al., 1995; Barnas and Sprung, 1993).

R_N, *Iaw*, *G*, and *H* also vary with the inflation pressure of the lungs, as illustrated in Figure 9 for four species of rodent. R_N decreases slightly with increasing pressure as the airways become distended and their resistance decreases. *G* and *H*, by contrast, exhibit a nadir around 3–5 cmH_2O. Below this range some regions of the lung start to close, effectively reducing the size of the available lung and so increasing its overall stiffness commensurately. Above the nadir, the lung tissues start to encroach on the nonlinear portion of their pressure–volume relationship, and so both stiffness and tissue resistance start to increase. The ratio *G/H* remains relatively constant throughout.

It should also be pointed out that as well as variations of R_N, *Iaw*, *G*, and *H* between species, within-species differences have also been reported both in terms of conventional measures of respiratory mechanics (Schulz, 2005; Tankersley et al., 1999) and in some cases in terms of measurements of impedance (Shore et al., 2003, 2008). Impedance parameters at baseline tend to be rather similar between different strains of mouse (Wagers et al., 2007; Shore et al., 2003, 2008), but substantial differences can arise in how these parameters change with various interventions. The A/J strain of mouse, for example, exhibits significantly greater increases in R_N following methacholine challenge compared to the BALB/c strain (Wagers et al., 2007). Substantial differences in lung volumes between strains have also been reported. For example, the total lung capacity of the C3H/HeJ strain is about 50% larger than either the A/J or C57BL/6J strains (Tankersley et al., 1999).

5. SUMMARY AND CONCLUSIONS

In contrast to the assessment of lung function in human patients, which is typically performed using spirometry, the study of lung mechanics in animal models is usually based on the observed relationships between pressures applied across the lungs and the corresponding flows and volumes that enter the airways. Under static conditions, this gives rise to the pressure–volume relationship that is roughly straight over its intermediate range corresponding to normal breathing where the lung is reasonably compliant, but curves toward lower compliance at low lung volumes due to derecruitment and at high volumes where the lung tissues become progressively stiffer. Under dynamic conditions, the mechanical properties of the lungs are embodied in measurements of resistance and elastance and their generalization as the complex function of frequency known as impedance. In recent years, the constant-phase model of lung impedance has become standard for the interpretation of impedance below about 20 Hz. Reported values of the parameters of the constant-phase model define allometric relationships that span more than five orders of magnitude of animal size. The slope of the relationship of airway

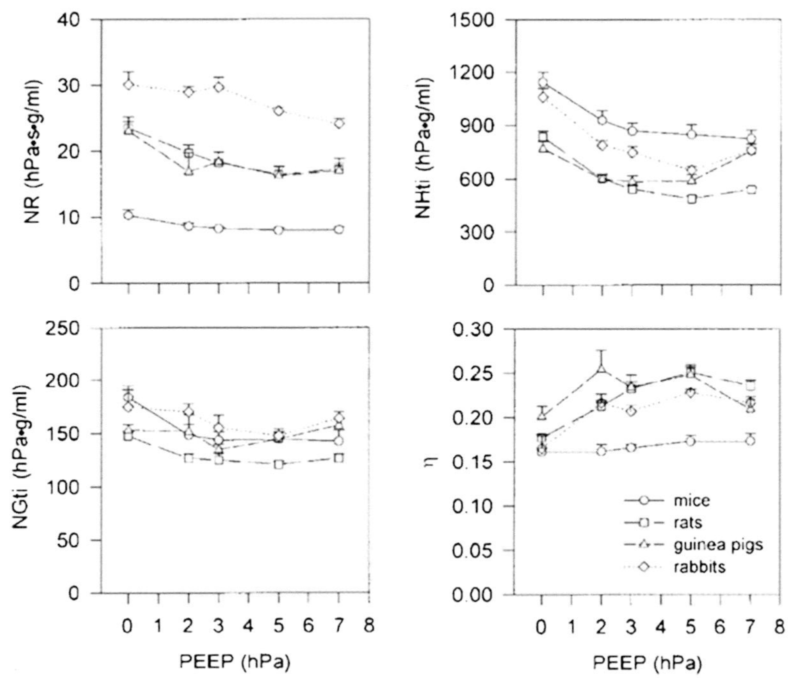

FIGURE 9 Relationships of constant-phase model parameters to body weight in different species. NR is the parameter R_N normalized to body weight. NGti and NHti are the normalized parameters *G* and *H*. η is hysteresivity, defined as the ratio *G/H*. *(Reproduced with permission from Gomes et al. (2000).)*

resistance to body weight in a log–log plot is between −0.5 and −1.0, showing that smaller animals have relatively larger airways than larger animals. By contrast, lung tissue resistance and elastance both describe relationships with slopes close to −1.0, showing that the intrinsic mechanical properties of lung tissue properties are relatively preserved across species.

REFERENCES

Allen, G.B., Leclair, T.R., von Reyn, J., Larrabee, Y.C., Cloutier, M.E., Irvin, C.G., Bates, J.H., 2009. Acid aspiration-induced airways hyperresponsiveness in mice. J. Appl. Physiol. 107, 1763–1770.

Allen, G., Bates, J.H., 2004. Dynamic mechanical consequences of deep inflation in mice depend on type and degree of lung injury. J. Appl. Physiol. 96, 293–300.

Allen, G., Lundblad, L.K., Parsons, P., Bates, J.H., 2002. Transient mechanical benefits of a deep inflation in the injured mouse lung. J. Appl. Physiol. 93, 1709–1715.

Adamicza, A., Petak, F., Asztalos, T., Hantos, Z., 1999. Effects of endothelin-1 on airway and parenchymal mechanics in guinea-pigs. Eur. Respir. J. 13, 767–774.

Adamicza, A., Petak, F., Asztalos, T., Tiszlavicz, L., Boros, M., Hantos, Z., 2001. Endothelin-1-induced airway and parenchymal mechanical responses in guinea-pigs: the roles of ETA and ETB receptors. Eur. Respir. J. 17, 975–981.

Bates, J.H.T., 2009. Lung Mechanics. An Inverse Modeling Approach. Cambridge University Press, Cambridge, UK.

Baydur, A., Behrakis, P.K., Zin, W.A., Jaeger, M., Milic-Emili, J., 1982. A simple method for assessing the validity of the esophageal balloon technique. Am. Rev. Respir. Dis. 126, 788–791.

Bates, J.H., Ludwig, M.S., Sly, P.D., Brown, K., Martin, J.G., Fredberg, J.J., 1988. Interrupter resistance elucidated by alveolar pressure measurement in open-chest normal dogs. J. Appl. Physiol. 65, 408–414.

Bersten, A.D., 1998. Measurement of overinflation by multiple linear regression analysis in patients with acute lung injury. Eur. Respir. J. : Off. J. Eur. Soc. Clin. Respir. Physiol. 12, 526–532.

Bates, J.H.T., Irvin, C.G., Farre, R., Hantos, Z., 2011. Oscillation mechanics of the respiratory system. In: Fredberg, J.J. (Ed.), Comprehensive Physiology. American Physiological Society, Bethesda, pp. 1233–1272.

Bates, J.H., Brown, K.A., Kochi, T., 1989. Respiratory mechanics in the normal dog determined by expiratory flow interruption. J. Appl. Physiol. 67, 2276–2285.

Babik, B., Asztalos, T., Petak, F., Deak, Z.I., Hantos, Z., 2003. Changes in respiratory mechanics during cardiac surgery. Anesth. Analgesia 96, 1280–1287 (table of contents).

Bates, J.H., Thompson-Figueroa, J., Lundblad, L.K., Irvin, C.G., 2008. Unrestrained video-assisted plethysmography: a noninvasive method for assessment of lung mechanical function in small animals. J. Appl. Physiol. 104, 253–261.

Bates, J., Irvin, C., Brusasco, V., Drazen, J., Fredberg, J., Loring, S., Eidelman, D., Ludwig, M., Macklem, P., Martin, J., Milic-Emili, J., Hantos, Z., Hyatt, R., Lai-Fook, S., Leff, A., Solway, J., Lutchen, K., Suki, B., Mitzner, W., Pare, P., Pride, N., Sly, P., 2004. The use and misuse of Penh in animal models of lung disease. Am. J. Respir. Cell Mol. Biol. 31, 373–374.

Bates, J.H., Cojocaru, A., Lundblad, L.K., 2007. Bronchodilatory effect of deep inspiration on the dynamics of bronchoconstriction in mice. J. Appl. Physiol. 103, 1696–1705.

Bates, J.H., Lauzon, A.M., 2007. Parenchymal tethering, airway wall stiffness, and the dynamics of bronchoconstriction. J. Appl. Physiol. 102, 1912–1920.

Bozanich, E.M., Collins, R.A., Thamrin, C., Hantos, Z., Sly, P.D., Turner, D.J., 2005. Developmental changes in airway and tissue mechanics in mice. J. Appl. Physiol. 99, 108–113.

Bachofen, H., 1968. Lung tissue resistance and pulmonary hysteresis. J. Appl. Physiol. 24, 296–301.

Barnas, G.M., Mackenzie, C.F., Skacel, M., Hempleman, S.C., Wicke, K.M., Skacel, C.M., Loring, S.H., 1989. Amplitude dependency of regional chest wall resistance and elastance at normal breathing frequencies. Am. Rev. Respir. Dis. 140, 25–30.

Barnas, G.M., Sprung, J., 1993. Effects of mean airway pressure and tidal volume on lung and chest wall mechanics in the dog. J. Appl. Physiol. 74, 2286–2293.

Colebatch, H.J., Greaves, I.A., Ng, C.K., 1985. Pulmonary distensibility and ventilatory function in smokers. Bull. Eur. Physiopathol. Respir. 21, 439–447.

Colebatch, H.J., Greaves, I.A., Ng, C.K., 1979. Exponential analysis of elastic recoil and aging in healthy males and females. J. Appl. Physiol.: Respir. Environ. Exerc. Physiol. 47, 683–691.

Clement, J., Landser, F.J., Van de Woestijne, K.P., 1983. Total resistance and reactance in patients with respiratory complaints with and without airways obstruction. Chest 83, 215–220.

Cojocaru, A., Irvin, C.G., Haverkamp, H.C., Bates, J.H., 2008. Computational assessment of airway wall stiffness in vivo in allergically inflamed mouse models of asthma. J. Appl. Physiol. 104, 1601–1610.

Dechman, G., Sato, J., Bates, J.H., 1992. Factors affecting the accuracy of esophageal balloon measurement of pleural pressure in dogs. J. Appl. Physiol. 72, 383–388.

Dechman, G.S., Chartrand, D.A., Ruiz-Neto, P.P., Bates, J.H., 1995. The effect of changing end-expiratory pressure on respiratory system mechanics in open- and closed-chest anesthetized, paralyzed patients. Anesth. Analgesia 81, 279–286.

Fung, Y.C., 1993. Biomechanics. Mechanical Properties of Living Tissues. Springer, New York.

Fredberg, J.J., Keefe, D.H., Glass, G.M., Castile, R.G., Frantz 3rd, I.D., 1984. Alveolar pressure nonhomogeneity during small-amplitude high-frequency oscillation. J. Appl. Physiol. 57, 788–800.

Farre, R., Peslin, R., Oostveen, E., Suki, B., Duvivier, C., Navajas, D., 1989. Human respiratory impedance from 8 to 256 Hz corrected for upper airway shunt. J. Appl. Physiol. 67, 1973–1981.

Farre, R., Ferrer, M., Rotger, M., Torres, A., Navajas, D., 1998. Respiratory mechanics in ventilated COPD patients: forced oscillation versus occlusion techniques. Eur. Respir. J. 12, 170–176.

Farre, R., Rigau, J., Montserrat, J.M., Ballester, E., Navajas, D., 2001. Evaluation of a simplified oscillation technique for assessing airway obstruction in sleep apnoea. Eur. Respir. J. 17, 456–461.

Fredberg, J.J., Stamenovic, D., 1989. On the imperfect elasticity of lung tissue. J. Appl. Physiol. 67, 2408–2419.

Gillis, H.L., Lutchen, K.R., 1999. How heterogeneous bronchoconstriction affects ventilation distribution in human lungs: a morphometric model. Ann. Biomed. Eng. 27, 14–22.

Gomes, R.F., Shen, X., Ramchandani, R., Tepper, R.S., Bates, J.H., 2000. Comparative respiratory system mechanics in rodents. J. Appl. Physiol. 89, 908–916.

Hantos, Z., Daroczy, B., Csendes, T., Suki, B., Nagy, S., 1990. Modeling of low-frequency pulmonary impedance in dogs. J. Appl. Physiol. 68, 849–860.

Hantos, Z., Daroczy, B., Suki, B., Nagy, S., 1987. Low-frequency respiratory mechanical impedance in the rat. J. Appl. Physiol. 63, 36–43.

Hantos, Z., Daroczy, B., Suki, B., Nagy, S., Fredberg, J.J., 1992. Input impedance and peripheral inhomogeneity of dog lungs. J. Appl. Physiol. 72, 168–178.

Hantos, Z., Petak, F., Adamicza, A., Daroczy, B., Fredberg, J.J., 1995. Differential responses of global airway, terminal airway, and tissue impedances to histamine. J. Appl. Physiol. 79, 1440–1448.

Hirai, T., McKeown, K.A., Gomes, R.F., Bates, J.H., 1999. Effects of lung volume on lung and chest wall mechanics in rats. J. Appl. Physiol. 86, 16–21.

Hantos, Z., Suki, B., Csendes, T., Daroczy, B., 1987. Constant-phase modelling of pulmonary tissue impedance. Bull. Eur. Physiopathol. Respir. 12, 326s.

Hantos, Z., Adamicza, A., Govaerts, E., Daroczy, B., 1992. Mechanical impedances of lungs and chest wall in the cat. J. Appl. Physiol. 73, 427–433.

Hamelmann, E., Schwarze, J., Takeda, K., Oshiba, A., Larsen, G.L., Irvin, C.G., Gelfand, E.W., 1997. Noninvasive measurement of airway responsiveness in allergic mice using barometric plethysmography. Am. J. Respir. Crit. Care Med. 156, 766–775.

Homma, T., Bates, J.H., Irvin, C.G., 2005. Airway hyperresponsiveness induced by cationic proteins in vivo: site of action. Am. J. Physiol. 289, L413–418.

Hantos, Z., Adamicza, A., Janosi, T.Z., Szabari, M.V., Tolnai, J., Suki, B., 2008. Lung volumes and respiratory mechanics in elastase-induced emphysema in mice. J. Appl. Physiol. 105, 1864–1872.

Habre, W., Petak, F., Sly, P.D., Hantos, Z., Morel, D.R., 2001. Protective effects of volatile agents against methacholine-induced bronchoconstriction in rats. Anesthesiology 94, 348–353.

Habre, W., Babik, B., Chalier, M., Petak, F., 2002. Role of endogenous histamine in altered lung mechanics in rabbits. Anesthesiology 96, 409–415.

Habre, W., Scalfaro, P., Schutz, N., Stucki, P., Petak, F., 2006. Measuring end-expiratory lung volume and pulmonary mechanics to detect early lung function impairment in rabbits. Respir. Physiol. Neurobiol. 152, 72–82.

Hantos, Z., Petak, F., Adamicza, A., Daroczy, B., Suki, B., KR, L., 1994. Optimum ventilator waveform for the estimation of respiratory impedance: an animal study. Eur. Respir. Rev. 4, 191–197.

Hantos, Z., Daroczy, B., Suki, B., Galgoczy, G., Csendes, T., 1986. Forced oscillatory impedance of the respiratory system at low frequencies. J. Appl. Physiol. 60, 123–132.

Hantos, Z., Collins, R.A., Turner, D.J., Janosi, T.Z., Sly, P.D., 2003. Tracking of airway and tissue mechanics during TLC maneuvers in mice. J. Appl. Physiol. 95, 1695–1705.

Irvin, C., 2005. Guide to the evaluation of pulmonary function. In: Hamid, Q., Shannon, J., Martin, J. (Eds.), Physiologic Basis of Respiratory Disease. BC Decker, Hamilton, Ontario.

Jackson, A.C., Giurdanella, C.A., Dorkin, H.L., 1989. Density dependence of respiratory system impedances between 5 and 320 Hz in humans. J. Appl. Physiol. 67, 2323–2330.

Johnston, R.A., Zhu, M., Hernandez, C.B., Williams, E.S., Shore, S.A., 1812-1819. Onset of obesity in carboxypeptidase E-deficient mice and effect on airway responsiveness and pulmonary responses to ozone. J. Appl. Physiol. 108.

Kaczka, D.W., Ingenito, E.P., Suki, B., Lutchen, K.R., 1997. Partitioning airway and lung tissue resistances in humans: effects of bronchoconstriction. J. Appl. Physiol. 82, 1531–1541.

Kaczka, D.W., Lutchen, K.R., 2004. Servo-controlled pneumatic pressure oscillator for respiratory impedance measurements and high-frequency ventilation. Ann. Biomed. Eng. 32, 596–608.

Kaczka, D.W., Ingenito, E.P., Lutchen, K.R., 1999. Technique to determine inspiratory impedance during mechanical ventilation: implications for flow limited patients. Ann. Biomed. Eng. 27, 340–355.

Ludwig, M.S., Romero, P.V., Sly, P.D., Fredberg, J.J., Bates, J.H., 1990. Interpretation of interrupter resistance after histamine-induced constriction in the dog. J. Appl. Physiol. 68, 1651–1656.

Lutchen, K.R., Greenstein, J.L., Suki, B., 1996. How inhomogeneities and airway walls affect frequency dependence and separation of airway and tissue properties. J. Appl. Physiol. 80, 1696–1707.

Lutchen, K.R., Suki, B., Zhang, Q., Petak, F., Daroczy, B., Hantos, Z., 1994. Airway and tissue mechanics during physiological breathing and bronchoconstriction in dogs. J. Appl. Physiol. 77, 373–385.

Landser, F.J., Nagles, J., Demedts, M., Billiet, L., van de Woestijne, K.P., 1976. A new method to determine frequency characteristics of the respiratory system. J. Appl. Physiol. 41, 101–106.

Lutchen, K.R., Hantos, Z., Petak, F., Adamicza, A., Suki, B., 1996. Airway inhomogeneities contribute to apparent lung tissue mechanics during constriction. J. Appl. Physiol. 80, 1841–1849.

Lutchen, K.R., Giurdanella, C.A., Jackson, A.C., 1990. Inability to separate airway from tissue properties by use of human respiratory input impedance. J. Appl. Physiol. 68, 2403–2412.

Lutchen, K.R., Yang, K., Kaczka, D.W., Suki, B., 1993. Optimal ventilation waveforms for estimating low-frequency respiratory impedance. J. Appl. Physiol. 75, 478–488.

Lundblad, L.K., Thompson-Figueroa, J., Allen, G.B., Rinaldi, L., Norton, R.J., Irvin, C.G., Bates, J.H., 2007. Airway hyperresponsiveness in allergically inflamed mice: the role of airway closure. Am. J. Respir. Crit. Care Med. 175, 768–774.

Lundblad, L.K., Irvin, C.G., Adler, A., Bates, J.H., 2002. A reevaluation of the validity of unrestrained plethysmography in mice. J. Appl. Physiol. 93, 1198–1207.

Lauzon, A.M., Bates, J.H., 2000. Kinetics of respiratory system elastance after airway challenge in dogs. J. Appl. Physiol. 89, 2023–2029.

Lundblad, L.K., Thompson-Figueroa, J., Leclair, T., Sullivan, M.J., Poynter, M.E., Irvin, C.G., Bates, J.H., 2005. Tumor necrosis factor-alpha overexpression in lung disease: a single cause behind a complex phenotype. Am. J. Respir. Crit. Care Med. 171, 1363–1370.

Mead, J., 1979. Problems in interpreting common tests of pulmonary mechanical function. In: Macklem, P., Permutt, S. (Eds.), The Lung in Transition Between Health and Disease. Marcel Dekker, New York.

Mead, J., 1969. Contribution of compliance of airways to frequency-dependent behavior of lungs. J. Appl. Physiol. 26, 670–673.

Massa, C.B., Allen, G.B., Bates, J.H., 2008. Modeling the dynamics of recruitment and derecruitment in mice with acute lung injury. J. Appl. Physiol. 105, 1813–1821.

Nagels, J., Landser, F.J., van der Linden, L., Clement, J., Van de Woestijne, K.P., 1980. Mechanical properties of lungs and chest wall during spontaneous breathing. J. Appl. Physiol. 49, 408–416.

Osborne, S., Hogg, J.C., Wright, J.L., Coppin, C., Pare, P.D., 1988. Exponential analysis of the pressure-volume curve. Correlation with mean linear intercept and emphysema in human lungs. Am. Rev. Respir. Dis. 137, 1083–1088.

Otis, A.B., McKerrow, C.B., Bartlett, R.A., Mead, J., McIlroy, M.B., Selver-Stone, N.J., Radford Jr., E.P., 1956. Mechanical factors in distribution of pulmonary ventilation. J. Appl. Physiol. 8, 427–443.

Pare, P.D., Brooks, L.A., Bates, J., Lawson, L.M., Nelems, J.M., Wright, J.L., Hogg, J.C., 1982. Exponential analysis of the lung pressure-volume curve as a predictor of pulmonary emphysema. Am. Rev. Respir. Dis. 126, 54–61.

Peslin, R., Duvivier, C., Didelon, J., Gallina, C., 1985. Respiratory impedance measured with head generator to minimize upper airway shunt. J. Appl. Physiol. 59, 1790–1795.

Petak, F., Habre, W., Donati, Y.R., Hantos, Z., Barazzone-Argiroffo, C., 2001. Hyperoxia-induced changes in mouse lung mechanics: forced oscillations vs. barometric plethysmography. J. Appl. Physiol. 90, 2221–2230.

Petak, F., Hall, G.L., Sly, P.D., 1998. Repeated measurements of airway and parenchymal mechanics in rats by using low-frequency oscillations. J. Appl. Physiol. 84, 1680–1686.

Petak, F., Hantos, Z., Adamicza, A., Asztalos, T., Sly, P.D., 1997. Methacholine-induced bronchoconstriction in rats: effects of intravenous vs. aerosol delivery. J. Appl. Physiol. 82, 1479–1487.

Petak, F., Habre, W., Babik, B., Tolnai, J., Hantos, Z., 2006. Crackle-sound recording to monitor airway closure and recruitment in ventilated pigs. Eur. Respir. J. 27, 808–816.

Petak, F., Hantos, Z., Adamicza, A., Daroczy, B., 1993. Partitioning of pulmonary impedance: modeling vs. alveolar capsule approach. J. Appl. Physiol. 75, 513–521.

Salazar, E., Knowles, J.H., 1964. An analysis of pressure-volume characteristics of the lungs. J. Appl. Physiol. 19, 97–104.

Sansores, R.H., Ramirez-Venegas, A., Perez-Padilla, R., Montano, M., Ramos, C., Becerril, C., Gaxiola, M., Pare, P., Selman, M., 1996. Correlation between pulmonary fibrosis and the lung pressure-volume curve. Lung 174, 315–323.

Sato, J., Suki, B., Davey, B.L., Bates, J.H., 1993. Effect of methacholine on low-frequency mechanics of canine airways and lung tissue. J. Appl. Physiol. 75, 55–62.

Sly, P.D., Collins, R.A., Thamrin, C., Turner, D.J., Hantos, Z., 2003. Volume dependence of airway and tissue impedances in mice. J. Appl. Physiol. 94, 1460–1466.

Schuessler, T.F., Bates, J.H., 1995. A computer-controlled research ventilator for small animals: design and evaluation. IEEE Trans. Biomed. Eng. 42, 860–866.

Shalaby, K.H., Gold, L.G., Schuessler, T.F., Martin, J.G., Robichaud, A., 2010. Combined forced oscillation and forced expiration measurements in mice for the assessment of airway hyperresponsiveness. Respir. Res. 11, 82.

Sakai, H., Ingenito, E.P., Mora, R., Abbay, S., Cavalcante, F.S., Lutchen, K.R., Suki, B., 2001. Hysteresivity of the lung and tissue strip in the normal rat: effects of heterogeneities. J. Appl. Physiol. 91, 737–747.

Suki, B., Petak, F., Adamicza, A., Daroczy, B., Lutchen, K.R., Hantos, Z., 1997. Airway and tissue constrictions are greater in closed than in open-chest conditions. Respir. Physiol. 108, 129–141.

Suki, B., Yuan, H., Zhang, Q., Lutchen, K.R., 1997. Partitioning of lung tissue response and inhomogeneous airway constriction at the airway opening. J. Appl. Physiol. 82, 1349–1359.

Suki, B., Petak, F., Adamicza, A., Hantos, Z., Lutchen, K.R., 1995. Partitioning of airway and lung tissue properties: comparison of in situ and open-chest conditions. J. Appl. Physiol. 79, 861–869.

Schulz, H., 2005. Mouse Phenome Database. The Jackson Laboratory.

Shore, S.A., Rivera-Sanchez, Y.M., Schwartzman, I.N., Johnston, R.A., 2003. Responses to ozone are increased in obese mice. J. Appl. Physiol. 95, 938–945.

Shore, S.A., Williams, E.S., Zhu, M., 2008. No effect of metformin on the innate airway hyperresponsiveness and increased responses to ozone observed in obese mice. J. Appl. Physiol. 105, 1127–1133.

Tepper, R., Sato, J., Suki, B., Martin, J.G., Bates, J.H., 1992. Low-frequency pulmonary impedance in rabbits and its response to inhaled methacholine. J. Appl. Physiol. 73, 290–295.

Tomioka, S., Bates, J.H., Irvin, C.G., 2002. Airway and tissue mechanics in a murine model of asthma: alveolar capsule versus forced oscillations. J. Appl. Physiol. 93, 263–270.

Takubo, Y., Guerassimov, A., Ghezzo, H., Triantafillopoulos, A., Bates, J.H., Hoidal, J.R., Cosio, M.G., 2002. Alpha1-antitrypsin determines the pattern of emphysema and function in tobacco smoke-exposed mice: parallels with human disease. Am. J. Respir. Crit. Care Med. 166, 1596–1603.

Tankersley, C.G., Rabold, R., Mitzner, W., 1999. Differential lung mechanics are genetically determined in inbred murine strains. J. Appl. Physiol. 86, 1764–1769.

von Ungern-Sternberg, B.S., Petak, F., Hantos, Z., Habre, W., 2009. Changes in functional residual capacity and lung mechanics during surgical repair of congenital heart diseases: effects of preoperative pulmonary hemodynamics. Anesthesiology 110, 1348–1355.

Venegas, J.G., Harris, R.S., Simon, B.A., 1998. A comprehensive equation for the pulmonary pressure-volume curve. J. Appl. Physiol. 84, 389–395.

van de Woestijne, K.P., Franken, H., Cauberghs, M., Landser, F.J., Clement, J., 1981. A modification of the forced oscillation technique. In: In 28th Int Cong of Physiol Sci Ed. Akademiai Kiado, Budapest, Hungary, pp. 655–660.

Wagers, S., Lundblad, L.K., Ekman, M., Irvin, C.G., Bates, J.H., 2004. The allergic mouse model of asthma: normal smooth muscle in an abnormal lung? J. Appl. Physiol. 96, 2019–2027.

Wagers, S.S., Haverkamp, H.C., Bates, J.H., Norton, R.J., Thompson-Figueroa, J.A., Sullivan, M.J., Irvin, C.G., 2007. Intrinsic and antigen-induced airway hyperresponsiveness are the result of diverse physiological mechanisms. J. Appl. Physiol. 102, 221–230.

Chapter 17

Methods, Measurements, and Interpretation of Animal Lung Function in Health and Disease

Jeffrey S. Tepper and Daniel L. Costa

Chapter Outline

LIST OF ABBREVIATIONS

alv Alveolar space
C_{cw} Chest wall compliance
C_{dyyn} Dynamic compliance
cm Centimeter
CO Carbon monoxide
CO_2 Carbon dioxide
COHb Carboxyhemoglobin
DL_{CO} Diffusing capacity for carbon monoxide
EEV End-expiratory volume
EPA U.S. Environmental Protection Agency
EPP Equal pressure point
f Breathing frequency
FDA U.S. Food and Drug Administration
FEV_1 Forced expiratory volume in 1 s
FRC Functional residual capacity
FVC Forced vital capacity
$[H^+]$ Hydrogen ion
H_2O Water
HCHO Formaldehyde
He Helium
Hg Mercury
Hz Hertz
ICH International Conference on Harmonization
Kg Kilogram(s)
MEFV Maximum expiratory flow volume
MetHb Methemoglobin
MIC Methyl isocyanate
M Mass (body weight)
ml Milliliter(s)
mm Millimeter(s)
N_2 Nitrogen
O_2 Oxygen
O_3 Ozone
P Pressure(s)
Pa_{O_2} Arterial partial pressure of oxygen
Pa_{CO_2} Arterial partial pressure of carbon dioxide
P_{ao} Pressure at the air opening or mouth
P_{bs} Pressure at the body surface
P_{cw} Pressure at the chest wall
P_{es} Esophageal pressure
Penh Enhanced pause
P_L Lung tissue pressure
P_{pl} Pleural pressure
P_{rs} Respiratory system pressure
P_{tp} Transpulmonary pressure
R_{aw} Airway resistance
RD_{50} Respiratory depression of 50%
RfC Inhalation reference concentration
R_L Lung resistance
R_n Newtonian resistance
RV Residual volume
Sa_{O_2} Arterial oxygen saturation

Comparative Biology of the Normal Lung. http://dx.doi.org/10.1016/B978-0-12-404577-4.00017-5

Sp_{O_2} Oxygen saturation by pulse oximetry
sR_{aw} Specific airway resistance
TDI Toluene diisocyanate
TLC Total lung capacity
TLV Threshold limit value
VC Vital capacity
V_D Dead space volume
$\dot{V}_E$ Expiratory minute volume
$\dot{V}_{O_2}$ Aerobic metabolism
$\dot{V}_{O_{2max}}$ Maximal O_2 consumption
VP Volume pressure
$\dot{V}/\dot{Q}$ Ventilation/perfusion
V_T Tidal volume
y Biological variable

1. INTRODUCTION

Previous chapters in this section have discussed several aspects of lung function, including factors or conditions that control, influence, or otherwise alter function, as well as some tests that may be utilized to assess function under these varied factors. This chapter examines the rational interpretation and practical application of lung function tests that are or can be used in laboratory animals that provide analogues of clinical pulmonary function tests conducted on humans. In this context, lung function tests have provided basic physiological information on the normal lungs of many species of mammals and have also served as tools in the assessment of injury or disease in experimental pulmonary biology (Snider and Sherter, 1977; Harkema et al., 1982; Raub et al., 1982; Mauderly, 1984; Costa et al., 1985; Hoymann, 2012). Thus, in much the same manner as function tests are used diagnostically in the clinic, the animal-adapted tests attempt to distinguish debilities or responses that predominate in the central or peripheral airways from those that exist in the gas-exchanging units of the lung. The careful examination of linkages between functional attributes and underlying macro and microstructure can reveal considerable information on the status of lung health and the pathogenesis of disease should it exist. As such, the interpretation of functional changes detected in animals after experimental intervention remains predicated on our knowledge and experience in human pulmonary medicine yet also serves to illuminate and refine our understanding of normal and altered lung function in all mammals, including humans.

2. MEASUREMENT TECHNIQUES

2.1 Basic Principles and Methodologies

The essential function of the lung is to deliver oxygen (O_2) to the blood stream in exchange for carbon dioxide (CO_2). Other ancillary functions include acid–base balance, host defense, and select endocrine functions due to its direct interface with the outside environment and intimate communication with other organ systems. To achieve this end, air must move into and out of the lung, and as such, the physics of moving air and tissues lies at the center of our discussion. Because the lung is a mechanical system, forces must act on it to overcome its inherent resistance to motion or deformation as well as to achieve and maintain a variety of equilibrium states with minimized energy consumption, e.g., inflation. Lung intransigence to change can be described in terms of (1) static elastic forces that oppose changes in volume, (2) resistive forces that affect the rate of volume change (flow), and (3) inertial forces that influence acceleration of air and tissue. Under most (physiological) circumstances, the contribution of acceleration to the mechanical properties of the lung of laboratory animals is negligible and thus can be ignored. Hence, the mechanical properties of the lungs can be essentially defined by its elastic and flow resistive properties. These properties are defined further by the dynamic interactions of tissue and airway infrastructure, viscous air–fluid interface, neurally mediated smooth muscle activity, skeletal muscle strength, cartilage, and bone. The most common way to measure these changes in static and flow resistive properties is to measure changes in pressure (calibrated as pressure, volume, or flow) using differential pressure transducers. The methods for making these measurements, which serve as the basis for all pulmonary function testing, are briefly described below.

2.2 Transpulmonary Pressure

Any assessment of lung function is predicated upon an understanding of the forces that drive changes in lung volume. Measurement of these static and dynamic changes in pressure can be used to calculate most of the commonly used pulmonary function tests. These driving forces are the cumulative effect of various linked and serial pressures that are imposed upon the respiratory system to move air and tissue, which result in a change in lung volume. The contributing pressures (P) emanate from the mouth or airway opening (ao), the alveolar space (alv), the intrapleural space (pl, space between the lung and chest wall), and the pressure at the body surface (bs). The relationships among these various pressures (P) can be described by the following expression:

$$P_{rs} = P_C + P_R + P_{CW} \quad (1)$$

Equation (1) reveals that the total respiratory system pressure ($P_{rs} = P_{ao} - P_{bs}$) is the summation of the lung compliant ($P_C = P_{pl} - P_{alv}$) and resistive ($P_R = P_{ao} - P_{alv}$) pressures of the lung tissue and the pressures associated with movement of the chest wall ($P_{CW} = P_{bs} - P_{pl}$). For example, at end expiration, air is not moving, P_{ao} equals P_{alv}, and thus P_R drops out of the equation. The remaining forces counteract each other with the chest wall ($P_{bs} - P_{pl}$), holding the lung open while equal but opposing pressures exerted by lung tissue ($P_{pl} - P_{alv}$) contract the lung. Because the remaining forces (P_C and P_{CW}) cancel each other out, there is no airflow and $P_{rs} = 0$.

For inflation to occur, the retractive forces of the lung tissue (resistive and compliant) must be overcome by the pressure exerted by the diaphragm and the tendency of the chest wall to keep the lung open at low lung volumes. As lung volume increases, eventually the chest wall adds, rather than subtracts, to the retractive forces of the lung tissue, requiring larger increases in pressure to incrementally increase volume (see Volume–Pressure Relationship). From understanding the component pressures in Eqn (1), it would appear that a direct measure of intrapleural pressure would be needed to measure static changes in pulmonary function. Intrapleural pressure (P_{pl}) can be measured directly in the pleural space after surgical insertion of a catheter or less invasively by measuring the pressure in the distal third of the highly compliant esophagus (P_{es}) as a surrogate P_{pl}. Thankfully, these two approaches yield similar estimates of P_{pl}, and thus surgical intervention is not required to make these measurements. Previously, pleural pressure measurements were made with a fluid-filled catheter attached to a pressure transducer; however, miniaturization of pressure sensors now allows the catheter tip to house an embedded pressure sensor. It is also possible to get around the invasive measures P_{pl} or P_{es} by momentarily stopping airflow with a flutter valve (in larger animals) or during ventilation in the rodent using a transoral or tracheal catheter. This maneuver allows measurement of pressure at the airway opening to be used to estimate total respiratory system pressure at a specific volume (as measured in a plethysmograph, see below) enabling the measurement of static lung volumes and static lung tissue elasticity.

To measure dynamic airflow mechanics (resistance and compliance), pressure measurements must be made while air flows in and out of the lung. The driving force controlling the ventilatory movements of the lung itself is typically measured as the transpulmonary pressure (P_{tp}). The P_{tp} is defined as the pressure drop from the intrapleural space (P_{pl}) to the airway opening (P_{ao}), a pressure that encompasses the combined forces inherent in lung tissue elasticity and resistance of the airway tubes. As nature always tries to minimize the loss of energy and extraneous work, it should not be a surprise that the tidal P_{ao} is about atmospheric, and thus, in many instances, can be neglected. Sly et al. (2003) demonstrated this lack of contribution of P_{ao} to the respiratory mechanics in mice by showing that there was no volume dependency due to the recoil forces associated with the chest wall that would have been observed if P_{ao} influenced the measurement. However, when evaluating larger animals with stiff chest walls or if restrictions or positive pressures are present at the airway, such as might be experienced with the insertion of a tracheal cannula, measurement of P_{ao} via a side port in the cannula is advisable. If not included in the measurement, then the impact of any downstream resistance from the cannula will be included in the value measured. As with the measurement of static compliance and lung volumes, airway or tracheal pressure can, in some instances, be used to measure airway mechanics without having to resort to the invasive measurement of P_{pl} or P_{es}. Thus, P_{ao} can be used for breathing mechanics measurements in animals when a known volume or flow is imposed upon the airways (see Maximum Expiratory Flow–Volume Curves or Forced Oscillation). However, in this circumstance, the measurement of P_{ao} can be compromised by gas compression due to dead space volume (Schuessler and Bates, 1995; Ewart et al., 1995); thus dead space volume should be minimized whenever possible.

2.3 Techniques for Measurement of Volume or Flow

As driving pressure changes (P_{rs}), resultant changes in lung volume and flow occur. Body plethysmography and pneumotachography are two techniques that are often employed to measure these changes in volume or flow rate of gas (usually air) in and out of the lung. These methods are adequate for measuring static lung volumes, ventilation, and other dynamic changes in flow and/or volume (e.g., forced expiratory spirometry). If volume or flow rate are measured while simultaneously measuring driving pressure, additional respiratory mechanical variables, such as compliance and resistance, can be derived (Chapter 16).

Plethysmography

Conceptually, the simplest plethysmograph is the pressure plethysmograph (sometimes referred to as a constant volume plethysmograph), which seals the animal's body either at the neck, snout, or at an extension of the trachea (via trans-oral or intratracheal cannula). With the body sealed inside the plethysmograph, inhalation of gas (from outside the plethysmograph) causes an increase in lung volume (chest expansion) and thus results in an increase in pressure inside the fixed dimension plethysmograph. On the other hand, exhalation to the atmosphere (outside the plethysmograph) results in a decrease in the plethysmographic pressure. The magnitude of lung volume change can be obtained via a calibrated measurement of the change in plethysmographic pressure as detected using a sensitive pressure transducer. The plethysmograph is calibrated by injecting or withdrawing a known volume of gas from the box and measuring the change in box pressure. To avoid an adiabatic artifact, the rate of air injection or withdrawal attempts to mimic the same frequency as that of chest expansion in the species of interest, indicated by the same change in pressure with time. These changes in box pressure, when calibrated as volume, permit the determination of ventilation during tidal breathing or can be used to estimate volume changes during an imposed maneuver allowing the determination of static or dynamic changes in lung volume (discussed below).

A variation on the pressure plethysmograph is the constant-pressure or flow plethysmograph. This plethysmograph is essentially a pressure plethysmograph with a calibrated leak (i.e., opening of known resistance). When flow across the leak is laminar, flow rate is proportional to the pressure difference across the fixed resistance, according to the Poiseuille equation. Flow rate is obtained by a differential pressure transducer sensing the pressure difference between the inside and outside of the box, and volume changes are computed via integration of the flow signal. There is an outward flow when the chest expands during inspiration that causes air to be pushed out of the plethysmograph and an inward flow that occurs during expiration. Unlike the pressure plethysmograph, flow plethysmographs are more tolerant of small leaks and present fewer problems with maintaining isothermality.

For both types of plethysmographs, dynamic accuracy requires an adequate frequency response. Both the transducer and plethysmograph, as a system, must be capable of responding to volume changes in a linear fashion within the volume range studied. A fast integrated flow plethysmograph, with a flat amplitude response for sinusoidal inputs up to 240 Hz, has been designed for small mammals (Sinnett et al., 1981). Similar plethysmographs for use with various sizes of laboratory mammals are commercially available.

A variation of the flow plethysmograph concept is commonly used in larger animals. When a fixed resistance is attached directly to the airway, either sealed at the mouth/snout or connected to a tracheal tube, the flow in and out of the lungs can be measured. This device, known as a pneumotachograph, measures the differential pressure (atmospheric vs airway) as flow passes through it. Pneumotachographs, placed at the airway opening, are generally not used for measuring flow in small animals because the dead space is too large. To get around this problem for small animals, a constant bias flow can be flushed through the pneumotachograph to eliminate high CO_2 accumulation (Lucey et al., 1978); however, the setup is cumbersome. Pneumotachographs may also need to be heated to prevent breath condensation from altering the fixed resistance.

Barometric Plethysmography

Barometric plethysmography, although simple in design, must be approached with caution to obtain accurate measurements of volume and frequency of breathing. This method requires only a sealed box, pressure transducer, and temperature and humidity sensors. Unlike the constant volume plethysmograph described above, the animal is placed entirely inside the sealed box, and thus this technique allows the measurement of ventilation in minimally restrained, unanesthetized animals. When the subject breathes inside a sealed chamber, the total number of air molecules in the chamber does not change during the ventilatory cycle. The pressure inside the chamber, however, varies during ventilation—an increase during inspiration and a decrease during expiration. This variation in chamber pressure is related to two factors: (1) gas compression and rarefaction as described by Boyle's law, and (2) warming and humidification of inhaled air during inspiration, as well as cooling and condensation of exhaled air during expiration. If the water vapor saturation is 100% within the chamber, and temperatures of the chamber and the animal are known, the tidal volume can be calculated according to the equation of Drorbaugh and Fenn (1955). From the fluctuations in chamber pressure, the breathing frequency can be obtained. Subsequently, minute ventilation can be obtained via multiplying tidal volume by frequency.

However, Drorbaugh and Fenn's initial formula (Drorbaugh and Fenn, 1955) for the calculation of tidal volume has been challenged. On theoretical grounds, Epstein and Epstein proposed that expired gas must be considered at nasal conditions instead of at chamber conditions (Epstein and Epstein, 1978). They reasoned that after the expired gas leaves the nose, further cooling and condensation of the expirate occurs over an interval that is long compared to the duration of expiration and thus cannot contribute to the phasic pressure signal from which tidal volume is calculated. As a result, they concluded that the observed pressure fluctuations are of smaller magnitude than would be seen if temperature and humidity equilibration of the expirate were rapid. Epstein and Epstein predicted that the classic formula may underestimate the true tidal volume by 10–15%. Their predictions have been confirmed by Jacky (1980). In addition, Jacky has suggested a method that corrects the tidal volume calculated from the formula of Drorbaugh and Fenn. The correction requires knowledge of the expired gas temperature at the nares and the ratio of inspiratory duration to total breath duration. Note also that for measurement of animals over time, the CO_2 must be scrubbed and the oxygen content needs to remain stable, which results in further technical issues in using this type of plethysmograph.

On the other hand, the advantages of using the barometric method are that ventilation can be determined repeatedly and noninvasively in unanesthetized, spontaneously breathing animals, with minimal physical restraint and without the need for tracheal cannulation or having an airtight seal at the neck. The barometric method, however, can only be used to measure limited respiratory parameters from the box pressure signal, lacking further invasive modifications (e.g., implantation of a pleural pressure catheter). Without some measurement of driving pressure (e.g., P_{tp}), barometric plethysmography cannot be used to determine respiratory system mechanics during ventilation, including such measurements as dynamic compliance or resistance in unanesthetized, minimally restrained animals. In addition, thermal drift due to small changes in room temperature, animal body-heat variation, room pressure fluctuation, and animal motion can make the respiratory signal noisy. Thus, application of the barometric plethysmograph has been limited.

Whole Body Plethysmography (WBP)

Recent adaptations of the barometric method have been found useful in the semiquantitative measurement of tidal volume (V_T) and respiratory frequency (f), as well as other derived measures, to assess the effects of aging, drugs, environmental agents, and chemicals both acutely and, perhaps more importantly, repeatedly over time. This technique, generally known as WBP, utilizes the same pressure changes as described above for barometric plethysmography except that the chamber has a fixed resistance so that it functions more like a flow plethysmograph. Since the chamber is open to the atmosphere and temperature and humidity are free to vary, the resulting differential pressure signal has been dubbed pseudo-flow. Therefore, the data generated by this plethysmograph are more appropriately used in a qualitative manner (i.e., comparison to baseline measurements and/or a similarly treated control group). In part, this technique is feasible due to availability of economical, sensitive, linear pressure transducers and electronics capable of compensating for the inherent drift in signal due to pressure and temperature change in the environment where the plethysmograph is placed. These characteristics have made this technique useful in measuring relative changes in ventilation acutely and repeatedly over time in unanesthetized, minimally restrained animals. Because of these attributes, various contract research organizations use WBP to evaluate changes in ventilatory function of rodents to fulfill the recommendation for respiratory safety pharmacology testing of new drugs (Guidance for Industry, 2001). In addition to measurements of ventilation using WBP, attempts have been made to mine the waveform for information pertaining to the flow resistive and compliant changes occurring in the lung (breathing mechanics). The controversy over the use of the derived measurement enhanced pause (Penh) to measure airway mechanics is discussed below (Section Enhanced Pause).

Double Chamber Plethysmography

As described above, there is only a single primary signal (volume or flow) obtained from plethysmographic methods previously mentioned, and, thus, respiratory mechanics cannot be evaluated because there is no measurement of driving pressure (P_{tp}). Although parameters from WBP attempt to provide such information, the method lacks a theoretical basis (Section Enhanced Pause). However, other methods are available to measure physically but not chemically restrained animals to obtain measurements of airway resistance with less method-associated artifact. One such method, double chamber plethysmography, can accurately measure ventilation and breathing patterns and can be used to determine specific airway resistance (sR_{aw}) corrected for differences in the volume at the end of expiration. This technique was originally described by Johanson and Pierce (1971), but technical requirements made this technique particularly cumbersome. The technique introduced by Pennock et al. (1979) and later modified by Agrawal (1981) to eliminate the temperature-humidity artifact that is usually observed in the Pennock method has been the more useful procedure. In this method, the flow generated by chest excursions and nasal airflow are measured separately by two independent plethysmographs. A latex dam(s) is used to seal the animal at the neck or nose, isolating the thoracic flow signal from the nasal signal. The delay in time between the airflow in the thoracic and nasal signals alters the slope (ratio of the flow signal over the change in volume signal) at end expiration and with correction for barometric pressure and water vapor pressure at body temperature yields sR_{aw} (Agrawal, 1981) There are several limitations to this technique that have greatly reduced its practical application in pharmacology, physiology, and toxicology studies, however. Discomfort and stress due to restraint in the plethysmograph and the latex collar may cause the animal to struggle and reposition its neck, affecting calibration and hence reproducibility of the measurement over time. Additionally, the thoracic gas volume remaining at end expiration (FRC) is estimated and is known to be influenced by various factors including temperature, pressure, changes in ventilation, and gas trapping (Castile et al., 2004).

Telemetry: Inductance and Impedance

Besides plethysmography or the use of a pneumotachograph for measurement of ventilation in animals, new telemetric methods have been developed to obtain such measurements. These methods have been refined sufficiently such that they are now practical, cost-effective, and appropriate for some types of animal studies, especially in larger animals. Inductive and impedance types of telemetry systems are available. Inductive plethysmography was first described by Konno and Mead (1967) and has been used for many years, primarily in sedated, mechanically ventilated humans. The technique relies on two inductive coils placed externally around the thorax and abdomen. The sensing coils are typically fixed into place within an elastic jacket to which the subject must be acclimated. A low voltage current is passed through the inductive coil, and expansion and contraction of the chest and abdomen are sensed as a change in current. The change in current is fed to a telemetry transmitter and broadcast to a receiver connected to a computer. Integration of the chest and abdominal signal, when properly calibrated, provides acceptable data relative to pneumotachography in restrained and unrestrained lambs (Warren et al., 1989), dogs (Murphy et al., 2010), and cynomolgus monkeys (Wang et al., 2010). The obvious advantage of this technique is that long-term continuous measurements can be made, and when coupled with external cardiovascular telemetry measures, simultaneous cardiopulmonary functional assessments are possible.

Impedance plethysmography is similar but requires the implantation of biometric leads, similar or identical to standard ECG electrodes, between the sternum and spine at the mid-thorax level. The impedance leads sense changes by discharging a small, undetectable current across the chest. Expansion and contraction of the chest result in modulation of the voltage, which is converted to an impedance waveform that is calibrated against a primary volume standard such as that obtained from a plethysmograph or attached pneumotachograph. Biosensor manufacturers have integrated this biopotential sensor with other implantable sensors, allowing simultaneous cardiopulmonary assessment. Reports demonstrating the accuracy and validation of these methods using known positive control drugs in restrained and freely moving monkeys (Authier et al., 2010) and dogs (Kearney et al., 2010) have been published. A telemetry method for measuring P_{pl} and f in smaller mammals, using an implanted pressure transducer, has been developed (Murphy et al., 1998), but difficulties with the technique have not led to widespread use (Ewart et al., 2010). The primary problem associated with the use of these two telemetry techniques (inductance and impedance) in laboratory mammals is artifacts related to movement of the test subject. However, 24 h of continuous collection of ventilation data are possible.

3. INDICES OF PULMONARY FUNCTION

3.1 Structure/Function Relationships

In the clinical setting, respiratory disorders are typically portrayed by signs, symptoms, and/or physical limitations. These disorders, in reality, are often problems resulting from impaired lung function. Thus, one could argue that inadequate performance, and not underlying pathology, is the ultimate criterion of the seriousness of lung disease (Macklem and Permutt, 1979). Nevertheless, virtually all lung diseases are classically defined and distinguished in terms of morphology. This would emphasize the need to understand how the design of the pulmonary infrastructure influences its mechanical properties and diffusionary functions.

Impairment of ventilatory function is often not apparent until substantial and largely irreversible damage has occurred to the lung tissues or airways. For this reason, emphasis has been placed on the development of sensitive pulmonary function tests that can differentiate impairments to specific lung compartments—large versus small airways versus distal alveolar regions, and perhaps even apical versus basal lung. As one might suspect, it is in the development and interpretation of these sensitive function tests for which the correlation with structure is critical. For end-stage diseases with their consummate damage of airway and/or alveolar regions, the correlations between structure and function can be clear-cut (Macklem and Permutt, 1979). By comparison, subtle or focal lesions are difficult to detect because the lung has substantial functional reserve. The lung is designed as an interdependent system whereby deficiencies can be compensated across and between lung compartments to ensure appropriate gas exchange while minimizing work or energy loss. Thus, while a lung function test may target a specific lung compartment, the structural and functional interdependence of the lung diminishes the test's sensitivity and, thereby, correlation with a specific structural change. On the other hand, there are circumstances where function may be altered with little or no apparent morphologic correlate due to the diffusiveness of the lesion (e.g., interstitial edema) or transience of the effect (e.g., localized reflex airway constriction). Nevertheless, the development of general axioms of structure/function relationships must be pursued to improve and better target the diagnosis and interpretation of lung function decrements.

Most pulmonary function tests were originally devised as noninvasive, clinical tools to ascertain the presence, nature, and severity of lung impairment. Though few would consider lung function tests as pathognomonic for any specific structural lesion, the information gleaned from these tests can be used to construct a pattern of impairment or disability that "infers" a structural correlate. This inference, though imperfect, draws more upon an understanding of the structure of the organ and basic engineering concepts than any other branch of physiology (Sackner, 1980). These concepts, discussed in greater detail in earlier chapters of this section, are mechanical analogues of the electrical principles that relate capacitance (compliance of the lung tissue), resistance (airflow resistance of conducting pathways), voltage (pressure), and amperage (flow). Besides these concepts, the mechanics of interdependent structural units and the hydraulic principles that govern fluid flow through tubes of variable size and rigidity are also readily applied in the analysis of lung function data.

Studies of healthy lung function generally rely on designed patterns of breathing, so-called academic breathing.

Breathing patterns are controlled or timed with the use of visual aids that can be programmed to instruct the subject to breath at varying depths or expansions of lung volume. These superimposed patterns allow a virtual separation of upper, central, and peripheral regions of the lung. However, more has been learned about structure/function relationships of the lung from studies of the diseased organ, particularly when specific test data are compared to those of the healthy lung (Macklem and Permutt, 1979). Despite the maxim of defining lung disease anatomically, the basic descriptors of lung disease, restriction or obstruction, are functional rather than structural descriptors perhaps because clinical function tests are among the primary diagnostic tools. Nevertheless, while lung function tests are not definitive as to the precise nature of the lesion or its etiology, hypotheses developed from engineering principles and simple models predicting changes in function with structural alterations have been substantiated by specific disease states that reflect distortions of intrinsic functional properties of the healthy lung.

Although an extensive overview of the physics and physiology of lung design and composition is covered in more detail in other sections of the book (Chapters 12,14 and 16), an attempt will be made to compare and illustrate some major correlates of lung structure/function among laboratory animal species as compared to those of the human. The goal is to provide some foundation for the interpretation of the data emanating from those lung function tests that have been adapted to laboratory animals. The reader is encouraged to consult any number of extensive reviews of basic lung physiology and morphology, and the application of lung function tests in humans, for additional background (Murray and Nadel, 2010). With regard to the specific methodologies of determining the various parameters of lung function in laboratory mammals, the reader is referred to several reviews that introduce and describe systems that have been used for this purpose (O'Neil and Raub, 1984; Costa, 1985; Costa and Tepper, 1988; Mauderly, 1989).

Many animal models of classic end-stage lung disease have been described, although most have been developed in rodent species. Virtually all of these models have been induced by the imposition or treatment of the animal with an exogenous substance or stress (Costa et al., 1985; Karol, 1983; Snider et al., 1986; Gross et al., 1984). Most exhibit surprisingly close functional and morphological resemblance to their human counterpart diseases (Antunes and Rocco, 2011). More recently, a number of genetic murine models resulting in specific morphological and functional lesions have been described and continue to be developed, not for the sake of structure—function comparisons but to further knowledge in basic cell biology and clinical medicine. However, in comparison to humans that typically have lung diseases bearing mixed functional attributes because of heterogeneous underlying disease, having animal models with uncomplicated and well-defined lesions simplifies function—structure comparisons (Mauderly, 1984, 1989; Costa, 1985; Snider et al., 1986; Costa et al., 1984). The interest in structure—function relationships has evolved in recent years to focus on ever-smaller lung regions and subcompartments. These studies attempt to understand better the biomechanics of specific tissues or areas as may relate to biochemical changes within that tissue during disease pathogenesis (Hamakawa et al., 2011; Nichols et al., 2013).

3.2 Static Lung Volume Measurements

Lung Volumes

In humans, total lung capacity (TLC) is defined as the amount of air residing in the lungs and airways after a maximal inspiration, whereas residual volume (RV) is the amount of gas in the respiratory tract after maximal expiration (Figure 1). These lung volumes can be measured by dilution of an inert gas like helium or neon or by plethysmography (Andersson et al., 1988). The intrapleural pressures required to achieve these extremes of volume can vary between subjects, depending on thoracic and diaphragmatic strength, particularly in the presence of disease. Often the maximal inspiratory pressure a subject can generate at TLC is used as a measure of compliance (McCarthy et al., 1980).

Given that the human must self-generate a P_{tp} force of about -25 cm H_2O P_{tp} to achieve maximum or full lung inflation (TLC), the analogue TLC in an anesthetized laboratory animal is conventionally an imposed positive inflationary pressure of $+30$ cm H_2O (P_{rs}) to nominally overcome the elastic recoil ($+5$ cm H_2O) contribution of the chest wall. The critical point here is that lung inflation is defined by the P_{tp} differential across the lung tissue itself. Presuming at rest the P_{bs} is small (near atmospheric) in a properly designed plethysmograph, P_{ao} can be used as a

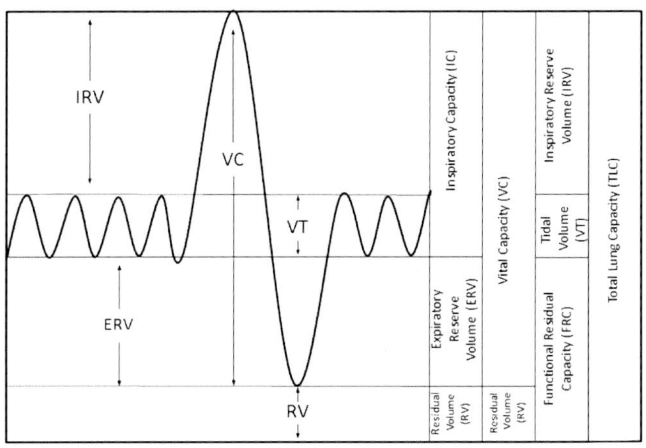

FIGURE 1 Subdivisions of lung volume.

surrogate for P_{rs}, when inflating the lung with a positive P. The RV is analogously defined, except that P_{ao} is lowered to −10 to −20 cm H_2O below barometric (atmospheric) pressure—the lower P minimizes artifact complications of the compliant chest wall at low lung volumes (Leith, 1983b). That differential volume between TLC and RV is the vital capacity (VC)—functional equivalent in all mammals. Measurement of TLC in animals is typically done by dilution of a VC volume of 20%/80% mixture of oxygen/inert gas such as neon or helium. From RV, this VC equivalent is injected slowly to the maximal lung volume (presumably +30 cm H_2O) and then withdrawn repeatedly several times to dilute the volume of inert gas concentration for use in the proportional calculation of TLC. This same dilution procedure can be modified for use in larger species such as the dog (Rozanski et al., 2010). With pressure as the independent variable defining lung volume, morphometric and physiological correlation can be made in the same animal when the lungs are fixed at an inflation pressure equivalent to that used in vivo to achieve TLC. (Costa et al., 1983a, 1986; Mauderly et al., 1985).

Another frequently measured subcompartment of TLC is the functional residual capacity (FRC). This subdivision of TLC is defined as that volume of air remaining in the lung after a normal tidal expiration. It is most often measured directly using a technique based on Boyle's law (Palecek, 1969). Unlike TLC and RV, which have fairly well-defined structural constraints, FRC in rodents is a complex function of breathing dynamics, neural control, and, to some extent, chest wall elasticity. Thus, in rodents, which have compliant chest walls, anesthesia can markedly reduce FRC by decreasing tidal rhythm and suppressing "respiratory braking." These effects underscore the importance of standardizing depth of anesthesia during a study and have prompted some investigators to limit use of the term FRC for animals with stiff chest walls and end-expiratory volume (EEV) for animals, such as rodents, with compliant chest walls (Raub et al., 1982; Vinegar et al., 1979; Lamm et al., 1982). In contrast, in humans FRC is the stabilized volume of the contracting lung and the stiff, noncompliant chest wall, which prevents lung collapse and hence a stable volume. While EEV may be the most appropriate term in the laboratory rodents, for the purposes of consistency and familiarity of terms, FRC will be used rather than EEV, with the understanding that they are not technically equivalent.

When estimating FRC in animals using an application of Boyle's law ($P_1V_1 = P_2V_2$), care must be taken to capture the respiratory cycle immediately at end expiration. This can be accomplished manually by rapidly occluding the tracheal cannula at the nadir point of the breathing cycle using a low dead space valve through which the animal is breathing. A more accurate and precise method uses a computer to track tidal breathing and to occlude the breath through a similar low volume automated valve exactly at end expiration. One potential complication of the Boyle's law method is the inclusion of abdominal gas that may be present in an artifactually enlarged volume in an end-stage diseased or sick animal. This problem, however, does not seem to be common in relatively healthy animals, as has been shown by Lamm et al. (1982) but should be considered in studies of animals with possible digestive impairments secondary to pulmonary debility. Dilution estimates of FRC, another commonly used technique (Takezawa et al., 1980; Raub et al., 1983) generally compare closely with the Boyle's law estimates in healthy animals and underestimate FRC (the difference being a measure of gas trapped in the lung) in animals with significant airway disease (Costa et al., 1983b; Stevens et al., 1988).

Volume–Pressure Relationship

Volume-pressure (VP) curves, referred to as lung compliance, express the integrated static volume-elastic properties of the lung. When plotted, the relationship between volume and pressure is sigmoidal rather than linear, as shown by Leith (1976) and Schroter (1980), demonstrating the interdependence of the lung matrix and its complex structure. The lung opens easily at lower lung volumes, exhibits a period during which there is a linear increase in recoil with increasing volume, and then volume change is limited at higher pressures as structural elements prevent further lung expansion. This shape appears to have been conserved across most mammalian species, as illustrated in Figure 2, where pressure is plotted conventionally, with a positive sign denoting a transmural pressure difference from the inside to the outside of the lung. This positive pressure

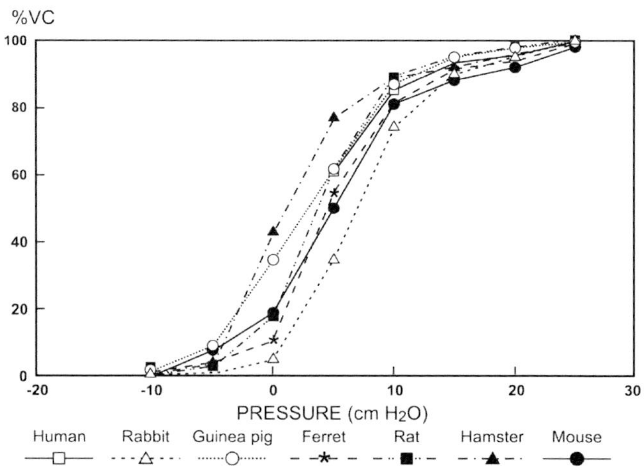

FIGURE 2 Quasi-static compliance curves (deflation limb only) for seven species of mammal, including humans. Volume data are represented as percent vital capacity (VC) to adjust for lung size. Data for humans (Pengelly, 1977), rabbit (Berend et al., 1985), guinea pig (Lai, 1988), ferret (Vinegar et al., 1982), rat (Costa et al., 1986), hamster (Koo et al., 1976), and mouse (Sinnett et al., 1981) are presented.

would thus promote inflation. In animals, when volume is plotted versus pressure during a slow inflation of the lung to a predetermined airway or P_{tp} (+25 to +30 cm H_2O) and then slowly deflated to a point where no further volume change occurs (−10 to −20 cm H_2O), two nonoverlapping sigmoidal curves are created, one for inflation and one for deflation. The displacement of the expiratory curve to the left of the inspiratory curve is termed *hysteresis*, which reflects the structural interdependence of the lung and the surface-tension lowering properties of surfactant in lung-lining fluid (Smith et al., 2013), which distributes the pressure across the entire lung. More often, VP relationships are described using the deflation wing of the compliance curve because this curve is not as subject to experimenter technique and is more reproducible and better understood (representing essential elastic recoil) than with the inflation wing (West, 1979; Harris, 2005).

Actual measures or indices of lung compliance have been reported in several forms. Often reported is a chord slope drawn between two discrete tidal breathing pressures (e.g., 0−10 cm H_2O) or a tangent slope at the steepest portion of the curve (Costa and Tepper, 1988; Mauderly, 1989). The indices depict essentially the same functional attribute. Because compliance increases with the size of the lung, sometimes these values are normalized for lung volume (TLC or FRC), even within a given species; the resultant value is expressed as "specific" lung compliance. Computer-derived exponential models capable of describing the entire VP curve have also been used (Pengelly, 1977). The advantage of these mathematical models is that they intrinsically normalize for lung volume and provide a coefficient of shape that better defines the entire curve. The merits of one approach over another are arbitrary (Snider and Sherter, 1977; Mauderly, 1989; Costa et al., 1983a, 1983b; Raub et al., 1985) and may be obscured where disease has led to compensatory lung growth or, in contrast, where lung growth has been retarded. Unremarkable differences in sensitivity were found in the only study that compared the above methods (Lai and Diamond, 1986).

Compliance can also be measured in excised rodent lungs. One advantage of the excised lung technique is that VP curves can be easily compared between air-filled and saline-filled lungs allowing differentiation between tissue recoil forces and the role of surfactant as an alveolar stabilizer (Clements and Tierney, 1964). Hayatdavoudi et al. (1981) reported air and saline VP curves on lungs excised from rats that were exposed to 60% O_2 for 7 days. They demonstrated changes in the elastic properties of the lung tissue as opposed to changes in surface forces due to changes in surfactant function. Of note, they confirmed changes in lung volumes that were measured using the in vivo gas dilution techniques of Takezawa et al. (1980) for which there was no apparent pathologic correlate suggesting lung growth retardation. Early on, Freeman et al. (1972) reported changes in the VP relationships of lungs excised from rats exposed to NO_2 that suggested mild emphysema, as did Bartlett et al. (1974) in a study of chronic O_3 exposure. Clearly, this technique has had a long life and broad application in lung toxicology, although currently the method sees more limited use for the preferential assessment of the in vivo functioning lung. The choice of in vivo versus excised lung assessment of VP curves is discretionary based on the objectives of the study, with the latter method offering some advantage in its simplicity if the pulmonary VP relationships, especially tissue versus lung lining, constitute a singular aim of a study.

3.3 Dynamic Lung Measurements

Dynamic measurements of lung function occur when gas is flowing in the lung. For gas to flow, forces are required to overcome the flow-resistive, inertial, and elastic properties of the lungs and chest wall. In this section, the characteristics of gas flow during normal breathing at rest and during maximal expiration will be discussed.

Ventilation

During normal breathing, the magnitude of minute ventilation ($\dot{V}_E$) depends on the metabolic needs of the body. The matching of ventilation with metabolism can thus maintain the blood gas tensions within the normal physiological range. Since small animals have higher O_2 consumption per unit body weight than large animals (Taylor et al., 1981), they are expected to have higher $\dot{V}_E$/body weight ratios. $\dot{V}_E$ has two components: tidal volume (V_T) and breathing frequency (f), or $\dot{V}_E = V_T \times f$. Tidal volume increases in an almost linear manner with increases in body weight both inter- and intraspecies, while f decreases exponentially as body weight increases between species (see allometry section, see Table 3 and Chapter 15). On the other hand, within a species, there is much less variation in f as the animal grows, with f remaining largely constant over the range of adult body weights and little difference in f between genders within the same species (Alexander et al., 2008; Tepper et al., 1988; Asgharian et al., 2012).

The mechanism for the optimal frequency of breathing has been hypothesized to be the frequency that causes minimal respiratory work (Crossfill and Widdicombe, 1961; Otis et al., 1950) or minimal average force of the respiratory muscle (Mead, 1961). In terms of minimal respiratory work, the main determining factor is the time constant (resistance × compliance); the lower the time constant, the higher the frequency. Crossfill and Widdicombe (1961) found a close correlation between the observed frequency and the predicted frequency derived from the work equation of Otis et al. (1950) in various mammalian species. In the guinea pig and in humans at rest and during exercise, Mead (1963) found, on the other hand, that natural breathing takes place at a frequency approximating the minimal average muscle force. Bates and Milic-Emili (1993) further explored this question applying a more sophisticated two-compartment model rather than the one-compartment model assumptions of Otis et al. (1950) and Mead (1963). Bates and Milic-Emili set out to determine if the viscoelastic properties of the lung, not accounted for previously, could better refine the determinants of optimal breathing frequency. To their surprise, they concluded that the viscoelastic properties of the respiratory system tissues do not significantly alter the prediction of optimal breathing frequency in humans or dogs and that inspiratory work was a better predictor of optimal breathing frequency than was the inspiratory pressure-time integral.

Blood Gas/Oximetry

Blood gases are, in fact, the only direct measure of the adequacy of ventilation. Table 1 reveals that values for various blood gas parameters remain conserved across species (also Table 1 Chapter 14, Table 2 Chapter 15. Unfortunately, to obtain these parameters, surgical placement of an indwelling, intra-arterial catheter is usually required. Furthermore, such measurements are typically insensitive due to the large dynamic reserve of the lung and the ability of the pulmonary circulatory system to make up for ventilation perfusion inequalities. To be useful, measurements should be made in awake but calm animals (Lucey et al., 1982; Silbaugh and Horvath, 1982; Burt et al., 1980; Steffens, 1969). New systems have been developed that allow automated blood sample collection via a catheter (Litwin et al., 2011), which could allow long-term measurement in undisturbed animals, although their use for measuring arterial blood gas has not been reported. Other than direct blood gas measurement via intra-arterial catheter, less invasive measurements are now available to assess adequacy of ventilation. Pulse oximetry is a technique in which a sensor is placed on a thin part of the body (e.g., fingertip, tongue, tail, ear) and two wavelengths of light are transmitted on one side and detected across the body part by a photosensor. In this way arterial oxygen saturation (Sa_{O_2}) can be approximated without invasive surgery in laboratory animals. The measurement indicates the amount of oxygen bound to hemoglobin; however, because of the shape of the O_2-hemoglobin dissociation curve, large changes in Pa_{O_2} (pressure drops below 70 mmHg) are required before small changes in saturation occur. In addition to the technique of transmissive oximetry, better results may be obtained with reflective pulse oximetry because this technique can be used on larger anatomical structures and may work better in veterinary applications. In all cases, some restraint of the animal, chemical or physical, is usually required to obtain reliable readings, although newer pulse oximeters with signal extraction technology have made the instruments less prone to movement artifacts. Further improvements, by adding additional wavelengths, now allow detection of carboxyhemoglobin and methemoglobin (co-oximetry). Carboxyhemoglobin (COHb) is formed when carbon monoxide binds hemoglobin. Similarly, methemoglobin (MetHb) is formed when the iron in hemoglobin is in the oxidized Fe^{3+} (ferric) state, not the more normal Fe^{2+} (ferrous) state. In either case, large quantities of abnormal hemoglobin can inhibit delivery of oxygen to the body. Both COHb and MetHb are altered by environmental pollutants (e.g., tobacco smoke) as well as pharmaceuticals (e.g., nitrites and nitrates). No significant differences in the absorbance spectra of hemoglobin in humans and other mammals suggest that the human algorithms used in co-oximetry and pulse oximetry are appropriate for use in animals (Grosenbaugh et al., 1997). However, note that hypotension, low hemoglobin, and anemia as well as high levels of MetHb can artifactually reduce oxygen saturation measurements.

Methods for Enhancing the Sensitivity of Ventilatory Tests

Woolcock et al. described the small airways (<2 mm in humans) as the "silent zone" of the lungs, since overall ventilatory function would likely remain within normal limits until airway obstruction of these airways is far advanced (Woolcock et al., 1969). Similarly, the composite nature of the distal ventilatory structure suggests that isolated lesions, within a limited group of ventilatory units, would likely have minimal, if any, impact on breathing patterns or total airway resistance, despite some misalignment of ventilation and perfusion. Even dynamic compliance, which is influenced by both small airway function and parenchymal elasticity, may be insensitive to such lesions, due to the large ventilatory reserve of the lung (Woolcock et al., 1969). For this reason, manipulations that involve exercise or other hyperventilatory stimulation have been utilized in an attempt to overcome this reserve and to demonstrate more subtle ventilatory deficits or disturbances (Tepper et al., 1988; Mautz et al., 1985a, 1985b; Wong et al., 1985).

Exercise techniques (e.g., treadmill) have been used in humans to stimulate ventilation and increase the time-constant heterogeneity of ventilatory units and have been useful in the evaluation of disease. This approach affirms the notion that performance, or the lack thereof, is at the essence of respiratory disease assessments. Similarly, exercise could also be used to limit an experimental animal's ability to use functional reserves to evaluate disease

TABLE 1 Range of Normal Blood Gas Values in Various Species

Species	Strain	Pa_{O_2} (mm Hg)	Pa_{CO_2} (mm Hg)	pH	HCO_3 (mEq/l)	Reference
Cynomolgus	*Macaca fascicularis*	135 ± 8	38.8 ± 0.6	7.38 ± 0.01	23 ± 0.5	Masuda et al. (1989)
Dog	Beagle	95 ± 9	33.6 ± 1.91	7.4 ± 0.022	20 ± 1	Ozaki et al. (2000)
Guinea pig	Hartley	93.6 ± 5.9	32 ± 2.4	7.45 ± 0.012	22.0 ± 1.8	Bar-Ilan and Marder (1980)
Human		80–100	35–45	7.35–7.45	22–26	www.globalrph.com/abg_analysis.htm
Mice	C57/BL6	88 ± 3	39 ± 3	7.41 ± 0.01		Lee et al. (2009)
Rabbit	NZW	76–117	28.8–42.4	7.37 ± 0.03	20.4 ± 1.8	Schroeder and Smith (2011)
Rat	Sprague Dawley	80.7–109.6	37.9–56.38	7.27–7.4	21.5–28.1	Polianski et al. (1984)
Rhesus	*Macaca mulatta*	81–96	34–42	7.36–7.42		Hom et al. (1999)

status as well as to directly enhance deposition of inhaled chemicals. Currently, however, there are few suitable methods for evaluating pulmonary function in small animals during exercise (Mautz et al., 1985b, 1985a; Harbison and Brain, 1983). While elegant systems have been described for larger animals, only a limited number of publications have reported such exercise systems to aid in the enhancement of inhaled exposure and/or the detection of lung damage (Kleinman and Mautz, 1991).

Fortunately, CO_2-induced hyperventilation, as an analogue of exercise, relies on similar mechanisms, and perhaps other factors related to cardiovascular and airway receptor responsiveness, to aid in the assessment of ventilatory capacity and improve the detection of functional lesions (Tepper et al., 1988). In terms of the assessment of altered ventilatory function during exposure to drugs, pollutants, or chemicals, CO_2 challenge has been used most frequently to improve detection of respiratory deficits. Carbon dioxide monotonically increases ventilation until the CO_2 exposure reaches approximately 8–10%; thereafter, further increases in CO_2 inhibit ventilation (Tepper et al., 1988). Such challenge concentrations maximally increase expiratory minute volume (VE) three to five times baseline levels in the rat. Inability to achieve these ventilatory increases suggests underlying pathology secondary to such factors as fibrosis, obstruction, central or peripheral nervous system damage (either afferent or efferent), muscle weakness, or fatigue. For example, 4 months after a single 3-h exposure to 10 ppm methyl isocyanate, a potent respiratory irritant, alterations in ventilatory patterns, reflecting both obstructive and restrictive lung lesions, were observed in rats challenged with CO_2 (Tepper et al., 1987, 1991). Similar results have been reported in guinea pigs up to 1 year after a single 19-ppm exposure (Ferguson and Alarie, 1991). Carbon dioxide challenge has also been used to enhance the detection of lung injury produced by sulfuric acid mist (Wong and Alarie, 1982), quartz dust (Chvalova et al., 1974), cotton dust (Ellakkani et al., 1984), and wood smoke (Wong et al., 1984). Furthermore, decreases in CO_2 responsiveness in the rat also have been detected after lung surgical manipulation (insertion of a pleural catheter), as well as with increasing animal age (Tepper et al., 1988).

Carbon dioxide-induced hyperventilation can also be used to enhance the deposition of inhaled pollutants. Concurrent CO_2 exposure increases ventilation analogous to the use of exercise in human study protocols (Folinsbee and Raven, 1984). As with exercise in humans, the detection of O_3-induced pulmonary irritation in rats was enhanced by CO_2-induced hyperventilation (Tepper et al., 1988, 1990). Interestingly, the alterations in breathing pattern appeared to be greater than would have been predicted based on the increase in dose associated with the CO_2-induced hyperventilation. The data suggest that enhanced deposition at critical lung sites may have occurred (Tepper et al., 1989). This observation is in apparent agreement with the findings of Mautz et al. (1985a) who demonstrated increased histopathology in the bronchoalveolar duct region of O_3 exposed rats that were exercised. Increased aerosol deposition has also been reported in animals rebreathing CO_2 (Kimmel et al., 2002). In contrast, however, studies in exercised hamsters exposed to a sulfur-colloid aerosol showed no difference in the particle deposition pattern compared to hamsters that were not active during exposure (Harbison and Brain, 1983). Thus, periods of altered ventilation may modify exposure, an additional factor that needs to be considered when assessing risk in evaluations of inhalation toxicity studies.

Measurement of Dynamic Breathing Mechanics

The ease of movement of air into and out of the lung with each tidal breath imparts a particularly critical role for the organization of the air conduits that lie between the ambient environment and the alveoli. The gradual decrease in conduit diameter from the trachea to the terminal bronchioles implies that some systematic design of the conduit structure that ensures the conductance of air to critical alveolar sites for the exchange of O_2/CO_2 with minimal wasted energy. Because the cross-sectional area of each individual airway of the lung decreases as one proceeds peripherally from the trachea, it was once believed that resistance also increases as one moves distally through the lung. However, the total cross-sectional diameter of the peripheral airways as one moves peripherally increases, and with the total volume of air being moved remaining constant, flow decreases in each airway resulting in an overall decrease in resistance. The trachea and large bronchi have mural support from cartilaginous rings or smooth muscle, while the frail distal airways have no intrinsic supporting structure. Rather they are structured interdependently, relying on their interconnectedness and on the negative intrathoracic pressure to maintain the airways in a state of partial expansion. The morphometric assessment of the human lung by Weibel (1963) and the measurement of regional airways resistance by Macklem and Mead (1967) provide structural and functional evidence, respectively, that a visualization of the cross-sectional area of the pulmonary airways from the trachea to the alveoli of the inflated lung (transverse sections) is that of a trumpet. In this model, the total airway diameter increases peripherally with a reduction in flow through each airway and a corresponding fall in airflow resistance. By partitioning total airflow resistance into a central and peripheral component, Macklem and Mead demonstrated that the distal airways account for less than 10% of the total resistance (Macklem and Mead, 1967). The morphometric trumpeting of the rodent lung (Phalen and Prasad, 1989) is analogous to that of higher species, leading to the assumption that the fall in resistance peripherally is similar across all mammals.

A technique pioneered by Amdur and Mead (1958) has been used extensively to evaluate breathing mechanics (dynamic compliance (C_{dyn}) and lung resistance (R_L)) in laboratory animals. The technique can be applied to either anesthetized or unanesthetized, restrained animals. This method requires simultaneous measurement of flow or volume and pressure. Flow or volume is usually measured by plethysmography, while pressure in the pleural space is measured directly by an indwelling catheter or indirectly via an esophageal catheter as described above. The advantage of using breathing mechanics over simple measures of ventilation is that these parameters can help localize central airway (R_L) versus peripheral airway (C_{dyn}) alterations. Newer, more sophisticated methods, such as forced oscillation (discussed below), can further partition lung mechanical changes to help better localize the site of lung damage.

Forced Oscillation

Another technique that has gained considerable traction in recent years is the forced oscillation technique, although the technique was first described in the 1950s (Dubois et al., 1956). There are two variations of this technique: input impedance and transfer impedance. In either case a small oscillating pressure signal is imposed upon the subject either at the opening of the airways (input impedance) or across the body surface (transfer impedance). The sinusoidal pressure oscillations may consist of a single frequency or spectrum of frequencies, but must exceed normal tidal breathing frequency. The perturbations in the resulting volume and flow signals can be used to calculate impedance. Lung impedance is composed of two parts, one made up of airway and tissue resistance and the second part, termed reactance, which indicates the compliance of the parenchyma. Using the impedance frequency spectrum data as input to the constant phase model (Hantos et al., 1995; Bates et al., 1971; Chapter 16) allows airway and tissue resistive, elastic, and inertial components of the lung to be differentiated. Many of the previous technical and instrumentation hurdles have been resolved for measuring input impedance in anesthetized tracheostomized laboratory animals from about 10 g to 40 kg (Glaab et al., 2007; LaPrad and Lutchen, 2008) using commercially available equipment. A commercially available device for the measurement of transfer impedance in conscious animals is still under development (Liah et al., 2013).

Enhanced Pause

The desire to have a less invasive measurement of airway mechanics that allows repeated measurement in the same animal over time has led to an increase in the use of WBP. Enhanced pause (Penh), a unitless measurement derived from the whole body plethysmograph signal, was described as a surrogate parameter for evaluating the mechanical properties of the lung (Hamelmann et al., 1997). The technique has found particular favor with scientists evaluating airway responsiveness to bronchoconstrictors in models of allergic inflammation. However, these surrogate measurements of breathing mechanics made with this technique, albeit even relative measurements, have been criticized to the point where at least one publication suggests that the use of Penh "be avoided completely" (Bates et al., 2009). Despite the repeated call for Penh to not be accepted solely as a measure of airway caliber, its use continues unabated (370 publications in PubMed, August, 2013). The reason for the continued use, despite the unknown theoretical basis for this parameter, is the published correlation of this parameter with measurements of resistance and airway pathological assessments.

Table 2 indicates a number of the well-conducted studies where breathing mechanics and Penh were compared in the same study. Overall, there has been a good correlation between Penh and resistance, especially in allergen-sensitized mice challenged with methacholine. Importantly, there have been notable exceptions where the correlation was poor or nonexistent. For example, when changes were made in the caliber of the tracheal tube in anesthetized ventilated rats, resistance increased, as expected, but Penh did not (Kirschvink et al., 2005). Both on theoretical grounds (Mitzner and Tankersley, 1998; Lundblad et al., 2002) and experimental observations (Table 2), it appears that Penh is not a measure directly related to changes in airway size.

However, that is not to say that this measure is of no value. In the authors' experience, increased Penh has been a reliable correlate of lung allergic inflammation in mice, rats, and guinea pigs after challenge with either inhaled and intravenous bronchoconstrictors. It has also been useful in screening for, and repeated measurements of, allergic inflammation in mice over days and weeks. In a pilot study conducted by Tepper et al. (1997) attempting to validate Penh, simultaneous measurements of breathing mechanics (R_{rs} and C_{dyn}) and Penh were made in anesthetized, spontaneously breathing tracheostomized mice placed in a double plethysmograph. The double plethysmograph method allowed independent measurements of thoracic expansion and nasal flow. P_{tp} was obtained from an esophageal cannula. Equivalent values of R_{rs} and C_{dyn} were obtained from either the thoracic or nasal flow signals. Physically combining the thoracic and nasal flows, as occurs in a whole body plethysmograph, resulted in a signal, approximating both the magnitude and shape of the flow signal obtained in the whole body plethysmograph. In this contrived configuration, changes in Penh correlated significantly ($r = 0.7$) with resistance and demonstrated a

TABLE 2 Concordance of Penh and Airway Mechanics Measurements

Comparison Variable	Manipulation	Species	Concordance	Comment	Reference
R_L, P_{es}	Methacholine	Mouse	Yes	Penh correlated with anesthetized R_L and P_{es} but not with breathing pattern or rate in same animals on following day	Hamelmann et al. (1997)
sR_{aw}	Histamine	Guinea pig	Yes	sR_{aw} and Penh produced statistically equivalent PC100 values with IH Hist	Chong et al. (1998)
sR_{aw}	Methacholine	Guinea pig	Yes	Penh and sR_{aw} were correlative measurements with IH Mch. $r = 0.72$	Bergren (2001)
R_T, C_{dyn}	Methacholine	Aqp5−/− mice	Yes	Compared R_T, C_{dyn} & APTI with intravenous acetylcholine to Penh after IH Mch. Both measurements indicated Aqp5−/− mice were hyperresponsive	Krane et al. (2001)
R_L, C_{dyn}	Methacholine	ANXA1−/− mice	Yes	Measured R_L and C_{dyn} with IH Mch. Found similar results with Penh in ANXA−/− naïve and sensitized mice compared to control mice	Ng et al. (2011)
R_{aw}	Hyperoxia	C57BL6 mice	No	Compared forced oscillation to Penh in a hyperoxia model. Found R_{aw} decreased as Penh increased at early time points.	Petak et al. (2001)
R_L, C_{dyn}	Methacholine	C57BL6 mice	Yes	Good correlation between R_L and Penh on days 14 and 15	Albertine et al. (2002)
R_L, C_{dyn}	Methacholine	C57BL6 mice	No	Correlation lost on days 18 and 22 (note R_L measured 24 h post)	Albertine et al. (2002)
R_L	Methacholine	A/J mice	Yes	R_L and Penh correlated	Lundblad et al. (2002)
R_L	Tracheal restriction	BALB/c mice	No	Correlation between R_L and Penh lost when plethysmograph environment adjusted to body temperature and humidity	Lundblad et al. (2002)
sRaw	Somatic growth	BALB/c C57BL6 mice	No	Lack of correlation found comparing somatic growth in double plethysmograph vs Penh in two strains of mice.	Flandre et al. (2003)
R_L, C_L	Methacholine	BALB/c C57BL6 mice	No	Compared R_L and C_L in BALB/c and BL6 mice and showed lack of correlation with Penh between two strains	Adler et al. (2004)
R_L, P_{pl}, C_{dyn}	Methacholine endotoxin	Piglets	Yes	Good correlation with R_L, P_{pl}, and Penh in endotoxin airway narrowing and inverse correlation w/C_{dyn} in methacholine- induced airway constriction	Halloy et al. (2004)

R_L	Carbachol	BALB/c mice	Yes	Penh and R_L increased with carbachol in anesthetized mice	Lai et al. (2004)
R_L	Histamine	BALB/c mice	No	Penh did not correlated with R_L with Hist or Hist + bradykinin	Lai et al. (2004)
R_L	Carbachol	Sprague −Dawley rat	Yes	Penh and R_L detected carbachol induced constriction	Kirschvink et al. (2005)
R_L	Tracheal restriction	Sprague −Dawley rat	No	Tracheal caliber reductions were only detected by R_L	Kirschvink et al. (2005)
R_L, C_{dyn}, P_{es}	Carbachol	Beagle dog	Yes	Penh correlated with R_L in sequential carbachol evaluation	Hirt et al. (2008)
R_L, C_{dyn}, P_{es}	Tracheal restriction	Beagle dog	Yes	Tracheal cannula reduction only increased R_L not Penh	Hirt et al. (2008)
R_L	Methacholine	BALB/c mice	No	Penh indicated airway responsiveness on day 1 of antigen challenge but RL did not	Zhang et al. (2009)
R_L	Methacholine	BALB/c mice	Yes	Penh and R_L indicated airway responsiveness after 3 days of antigen challenge	Zhang et al. (2009)
R_L	Methacholine	BALB/c mice	Yes	The patterns of Penh and lung resistance responses to methacholine were quite similar.	Midoro-Horiuti et al. (2010)
R_L	Acetylcholine histamine	Guinea pig	Yes	Good correlation of R_L (ventilated) and Penh (anesthetized) with iv Ach. Similar dose response with iv Hist or Ach in anesthetized animals with or without tracheal tube	Vargas et al. (2010)
R_n, H	Methacholine	29 mice strains	No	Forced oscillation R_n and H had different IH Mch slopes in different strains of mice compared to Penh	Berndt et al. (2011)
R_L	Methacholine	Guinea pig	Yes	Forced oscillation R_L with IH Mch show similar inhibition as Penh with a PDE4 inhibitor HDME	Yang et al. (2011)
R_L	Methacholine	gp91 phox −/− mice	Yes	Good correlation between results in KO mice using Penh and R_L	Banerjee and Henderson (2012)

Ach = acetylcholine, C_{dyn} = dynamic compliance, H = elastic tissue properties, Hist = Histamine, IH = inhaled, iv = intravenous, Mch = methacholine, PC_{100} = provocative concentration 100% increase, Penh = Enhanced Pause, P_{es} = esophageal pressure, P_{pl} = pleural pressure, R_{aw} = airway resistance, R_L = pulmonary resistance, R_n = Newtonian resistance, sR_{aw} = specific airway resistance.

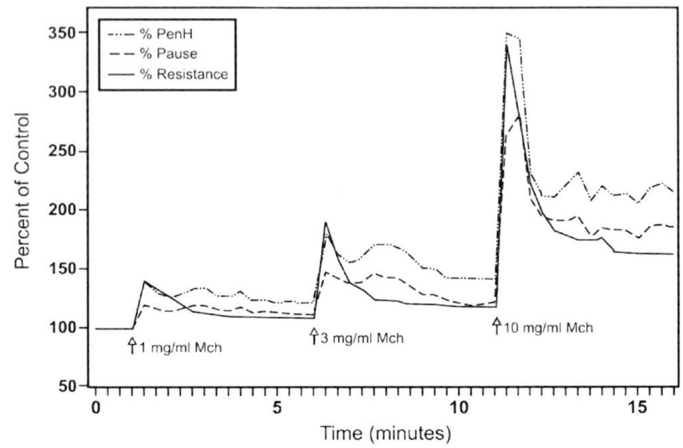

FIGURE 3 Simultaneous recording of Penh, pause, and resistance from the same anesthetized, tracheotomized rat. Data are expressed as a percent of the control value. After control measurements were obtained, the rat was challenged for 1 min with each of three doses of nebulized methacholine (Mch), and measurements were recorded for 5 min before receiving the next dose.

similar monotonic dose–response relationship (equivalent dose where R_{rs} or Penh doubled) when challenged with increasing doses of inhaled methacholine. However, Penh (and a parameter in the calculation of Penh termed Pause) did not correctly track (magnitude and duration) breath-by-breath changes in resistance at various doses of inhaled methacholine. After challenge with inhaled methacholine, Penh largely overestimated the time to recovery of R_{rs}, while the peak change in Pause was small and recovery slower compared to the bronchoconstrictor response measured by R_{rs} (Figure 3).

In conclusion, WBP can be useful for the measurement of ventilation in laboratory animals. However, the measurement of Pause and Penh are not direct measures of airway mechanics but can be useful as tools to investigate airway challenge repeatedly in the same animal over prolonged (days or months) periods. Indeed, its use among respiratory scientists is regaining interest as recent studies have shown coherence in Penh with other measures of airway mechanics in rodents (Frazer et al., 2011; Raemdonck et al., 2012; Fedulov and Kobzik, 2011). However, it would be prudent, and likely required for publication, to validate Penh measurements by measuring airway mechanics in a parallel group using standard techniques as described above.

Maximum Expiratory Flow–Volume Curves

The most commonly used test to evaluate small airway disease in humans is the forced expirogram (maximum forced expiratory maneuver). The measurement in laboratory animals, an analogue of the voluntary forced expirogram in humans, is typically plotted as maximal expiratory flow versus volume or time. However, in laboratory animals the maximum expiratory flow–volume curve (MEFV) is an imposed rather than a voluntary maneuver. Generally, the anesthetized animal is inflated to TLC and then is rapidly deflated to RV by exposing the airway opening to a negative pressure (about −40 cm H_2O). Although the methodology is different from that used in human subjects, the resultant curve qualitatively mimics the human curve, and flow limitation characteristics are achieved (Harkema et al., 1982; Costa et al., 1983a, 1983b; Diamond and O'Donnell, 1977; Lucey, 1983). In theory, the relative pressures from the inside to the outside (transmural) of the airways that ultimately limit flow in this maneuver should be similar in humans and in the rodent (Lucey, 1983; Mead, 1967). This flow limitation is thought to occur due to collapse or compression of intrathoracic airways distal (upstream) to the equal-pressure point (EPP) across the small flaccid airway wall lacking intrinsic supporting structure (as described above). Once this event occurs, no additional expiratory effort can enhance the expiratory flow since more force simply collapses more of the airway. Hence, flow during the latter portion of the maneuver is considered "effort independent" and is a direct function of intrinsic elastic recoil pressure of the distal lung moving air through the resistance encountered at the EPP. By definition, then, regardless of the source of expiratory pressure (voluntary thoracic drive in humans or involuntary evacuation in anesthetized, compliant-chested rodents), the prime determinant of effort-independent flow in the normal lung is the elastic recoil of the pulmonary tissues. Figure 4 reveals that at the end of forced expiration, the period where effort independence exists, humans and various mammalian species are qualitative similar as indicated by

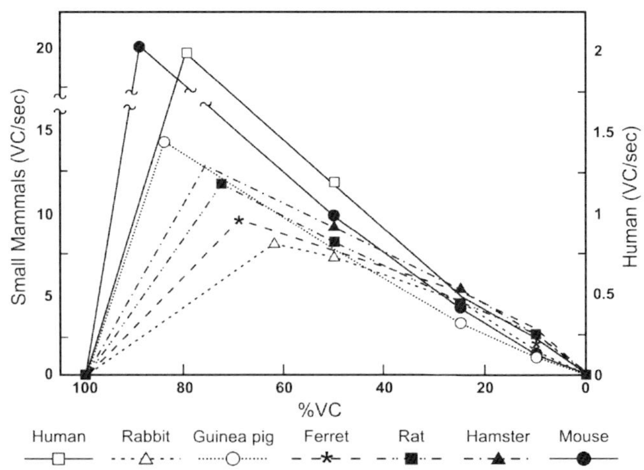

FIGURE 4 MEFV curves represented for seven animal species, including humans. Flows and volumes are expressed in terms of vital capacity (VC) to adjust for differences in lung size. Data for humans (Bouhuys et al., 1969), rabbit (Berend et al., 1985), guinea pig (Lai, 1988), ferret (Vinegar et al., 1982), rat (Costa et al., 1986), hamster (Lucey et al., 1978), and mouse (Sinnett et al., 1981) are presented.

the overlap of the species curves between approximately 40 and 0 percent of VC.

Generally, the reductions in flow in the effort-independent region of the curve in emphysema agree with subsequent histopathological examinations by light microscopy. On occasion, however, decrements in this measure of airway function have preceded overt pathology, suggesting that subtle changes in connective tissues may play a significant role in the maintenance of airway patency during this maneuver (Costa et al., 1983b). This degree of sensitivity, when combined with its good reproducibility for a given animal, makes the MEFV maneuver a powerful tool for the study of the progression of toxicant-induced disease. Several parameters obtained from the acquisition of the MEFV curve have been found to be useful for assessment of toxicant-induced lung injury, including forced vital capacity (FVC) as well as peak and various flow rates as a percentage of FVC (Stevens et al., 1988; Tepper et al., 1987; Costa et al., 1995). A variant of this technique, which has not been widely applied, is the partial flow—curve where expiration is "forced" with a spectrum of smaller driving pressures than typically used (Mauderly, 1989, 1988). Further examination of this technique may help address problems related to regional airway impairments.

Analogous forced expirograms after breathing an 80% helium (He), 20% O_2 mixture may help localize the site and degree of small airway obstruction (Dosman et al., 1975). Because He is less dense than air, the frictional resistance to airflow due to turbulence is reduced. Hence, the transition between turbulent and laminar flow can be determined by comparing the He-MEFV curve to the air-MEFV curve. In the presence of small airways disease, the volume at which these curves overlap, termed the *volume of isoflow*, would then be expected to increase when using He. The application of this technique to date (Costa and Tepper, 1988; Tepper et al., 1991; Diamond and O'Donnell, 1977; Lai, 1988) indicates considerable interanimal variability; thus, this technique needs refinement if it is to be used in studies that do not serially follow individual animal function.

Lung Diffusion

The cardinal function of the lung is to provide adequate oxygenation of blood for use in tissue metabolism while eliminating the metabolic waste product, CO_2. However, determining blood gas composition is generally not especially sensitive to underlying lung function impairments until function is severely compromised as previously discussed. Compensating mechanisms, both cardiovascular and respiratory, can mask diffusional deficiencies as the body systematically seeks homeostasis. That is not to say, however, that blood gas balance does not fluctuate in response to challenges where local circulation or rapid shifts in ventilation distribution in the lung might have immediate effects on blood gas levels (Belenkiy et al., 2013; Krishnamoorthy et al., 2012). Steady state diffusion assessment in chronic disease or in health is more frequently of interest.

In the absence of severe abnormalities in the distribution of inhaled air into the lung, the ability of oxygen to diffuse across the air—blood membrane of the alveoli can be approximated by the diffusion of carbon monoxide (CO). Carbon monoxide has about the same diffusion coefficient as O_2, and because it binds so strongly to hemoglobin, its disappearance from the lung air can be used to compute an index of O_2 diffusion. The strong binding of CO to hemoglobin minimizes problems associated with back diffusion as would occur with O_2. Although several methods have been published for measuring the diffusion of CO (DL_{CO}) into the blood in laboratory mammals (Costa et al., 1983a; Johanson and Pierce, 1973; Turek et al., 1978), the most common procedure is the single-breath DL_{CO} described by Takezawa et al. where a defined test gas containing trace amounts of CO and neon are inspired from RV to TLC, withdrawn to 50% volume after 5 s and then with a separate syringe, withdrawn to RV after 10 s (Takezawa et al., 1980). Measuring neon concentration provides a dilution factor and the loss of CO over the 10 s provides an uncorrected kinetic diffusion measure. These gases are readily measured using a gas chromatograph. A new method appears to give comparable results when tested in disease models (Fallica et al., 2011). Other approaches to assess diffusion integrity have been attempted, but most are much more involved technically (Mansson et al., 2003). (For a more complete discussion of the methods, use, and interpretation of lung diffusion, see Chapter 12.)

Distribution of Ventilation

As noted earlier, ventilation distribution throughout the lung is not perfectly homogeneous. Rather, a combination of static and dynamic factors, which include the distribution of P_{pl} around the lungs (influenced by gravity and forces associated with the viscera or chest wall), the distribution of regional resistances and compliances within the lungs, and possible airway closure during the respiratory cycle, imposes a pattern of intrapulmonary ventilation distribution. These regional differences in distribution of inhaled air are most marked when inhalation begins at lung volumes near RV (Bouhuys, 1965). The pattern of ventilation distribution approaches some uniformity when inhalation begins at mid-lung volumes (e.g., FRC), a distribution that has been termed *homogeneous*; in other words, the various pulmonary compartments fill or ventilate at "nearly" the same rate. However, in the human lung the alveoli of upper (apical) lung regions tend to be more open due to the pull of gravity, while the bottom alveoli are somewhat less open. Naturally, this difference in ventilation would be accentuated at RV as noted while breathing at mid volumes, it would be minimized. Nevertheless, under normal breathing at FRC, there is a slight preference for air to distribute to the basal region of the lungs where most of the pulmonary circulation is passing—hence the region is sometimes termed the dependent region of the lung. In the apical region of the lung, ventilation there is less dependent on lung volume and depth of breathing. Many diseases may alter the relative contributions of static and dynamic factors within the lung that can lead to a maldistribution of inspired air, even in the presence of apparent normal resting minute ventilation. When such maldistributions occur, the ventilation is said to become *inhomogeneous*. It is most noticeable in advanced obstructive disease and can sometimes be seen as a subtle change in early restrictive disease. Two techniques have been used in laboratory animals to assess inhomogeneity of distribution, the multibreath and single-breath nitrogen washout tests.

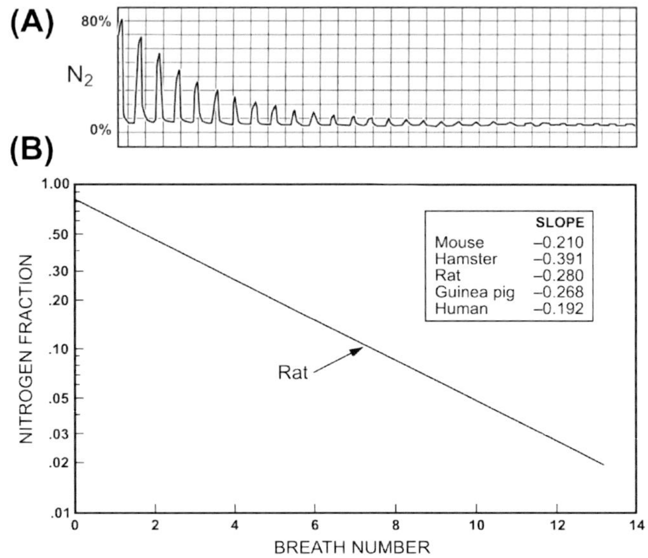

FIGURE 5 (A) Tracing of multibreath N_2 washout of the rat demonstrating the exponential pattern of the curve. (B) Plot of Ln (f/[N_2] vs breath number). Data for humans (Lewis, 1978), guinea pig (unpublished data), rat (Stevens et al., 1987), hamster (Raub et al., 1983), and mouse (Miller et al., 1987).

Multibreath Nitrogen Washout

During tidal breathing, the distribution of ventilation reflects an integration of regional resistances and compliances. As such, the multibreath nitrogen (N_2) washout can be used to assess deviations in regional mechanics. This test involves the breath-by-breath measurement of end-expiratory N_2 concentration during tidal breathing from a continuous source of 100% O_2. In theory, resident N_2 (80%) is exponentially diluted out of the lung with each tidal breath and replaced with O_2. The typical exponential washout pattern obtained in normal animals and humans can be characterized or modeled by at least three compartments, which, when expressed as a semilog plot (Figure 5) of end-expiratory N_2 fraction versus breath number, approximates a straight line. Various analyses based on the number of breaths, washout time, or volume of pure O_2 required to displace the residual N_2 from the lungs are used in the assessment of distal lung impairment (West, 1979; Mauderly, 1977; Lewis, 1978).

Typically, small airway obstruction results in delayed or otherwise impaired N_2 turnover due to poor airway communication of incoming O_2 and resident pulmonary N_2 laden air (Raub et al., 1982; Lewis, 1978; Likens and Mauderly, 1982). Because these lesions often have associated increases in FRC, evaluation of the N_2 turnover rate should be corrected to offset the effect of the enlarged volume of N_2 to be displaced. This adjustment is simply a turnover rate of FRC given as a fractional cumulative turnover, where the cumulative tidal volume at a given breath number is divided by FRC. A reasonable approximation of this cumulative volume can be determined by multiplying breath number by the average V_T:

$$\text{Breath number} \times V_T/\text{FRC} \tag{2}$$

This turnover ratio is used in place of breath number on the abscissa of the washout plot. This adjustment for volume will aid in the assessment of the distal lung impairment. If the lesion is predominantly airway or focal in the parenchyma, then volume-adjusted washouts should remain abnormal (i.e., the slope of the curve will be less negative). When the lesion is diffuse through the lung or is associated with significant volume changes, the adjustment may return the curve to within normal limits. A normal N_2 washout slope upon FRC adjustment, as indicated, may merely suggest that there has been a ventilatory compensation, not that there is no underlying disease or change. The application of a moment-analysis computation weighs the distal tail of the curve where early disease is most expected, to reflect small airway or regional compliance abnormalities (Saidel et al., 1975). However, the benefit of this approach has not been unequivocally demonstrated (Costa et al., 1986; Fleming et al., 1980). Additional analytical approaches have been suggested (Mauderly, 1977), but often the mere demonstration of the inhomogeneity is sufficient evidence of distal airway/lung disease as a toxicological end point (Raub et al., 1982; Costa et al., 1983a, 1983b, 1986).

Single-breath N_2 Washout

Although the multibreath N_2 washout may give a more representative view of gas distribution and mixing characteristics during tidal breathing, the variability of V_T/FRC ratio and potential shifting of FRC during the washout sometimes complicate interpretations (Mauderly, 1989). The single-breath washout relies on the same gas-mixing principles as the multibreath test, except that it is applied to a single-breath washout maneuver after a full VC inspiration of 100% O_2 from RV. Sometimes called the *closing-volume test*, this test is represented by plotting the slowly exhaled N_2 concentration against the VC volume from TLC to RV. Four phases are usually identifiable (Figure 6). The first gas washed out of the lungs is dead space gas that contains little or no N_2 (phase I). Nitrogen concentration then rises sharply as dead space volume (V_D) is washed out and alveolar gas begins to appear at the airway opening (phase II). This phase is followed by a slow increase in the concentration of N_2 (the alveolar plateau) that continues over most of the expired volume (phase III) and is composed of the expired gas from many alveolar units. It is the slope of this phase III plateau that is considered representative of the homogeneity of ventilation distribution. An upward shift in this slope is thought to represent greater variability in intrapulmonary compartmentalization—inhomogeneity—with some regions being relatively poorly ventilated (i.e., containing functionally trapped air). After this region of the curve, there is usually a final increase in N_2 (phase IV), most notable in larger mammals, such as the dog (Mink et al., 1984) or human (Gold, 1982) where the size of the lung imposes a great hydrostatic gradient that more noticeably exhibits the impaired distal alveoli. The onset of phase IV, though often poorly defined on the rodent curve (Mauderly, 1989), is believed to signal progressive closure of more flaccid small airways, though this interpretation of added value as a measure of small airways dysfunction is debated, even in humans (Gold, 1982; Timmins et al., 2012). This volume difference from RV is the so-called closing volume, which should increase with obstructive airways diseases. This single-breath method has been shown to be of value as an approach to assessing ventilation distribution in large animals; it has also been used in smaller laboratory animals (Mauderly, 1989). The test was shown effective in detecting emphysema in rats (Likens and Mauderly, 1982) and guinea pigs (Loscutoff et al., 1985), but reference to its use in rodents today cannot be found in the literature. Its use in humans is also not routine in part because of the demand on subject performance and questions as to its added value to other tests.

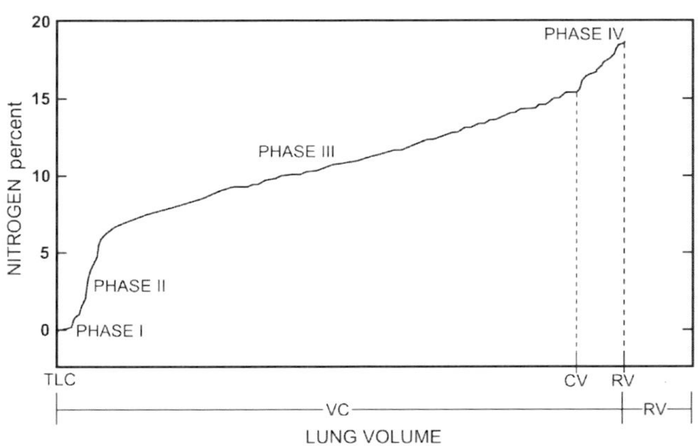

FIGURE 6 Single-breath N_2 washout from a rat showing phases I to IV and closing volume. The slope of phase III can be used to represent the overall homogeneity of distribution. The CV is determined from the inflection point of phase IV (O'Neil and Raub, 1984).

4. CAVEATS

Clearly, the ideal system for measuring pulmonary function in animals would approximate the human situation, in which tests are performed with cooperation and without physical or chemical restraint. In fact, this ideal environment for testing is recommended by ICH-S7a for safety pharmacology testing of all new drugs, including respiratory assessments. Unfortunately, this ideal is often difficult to achieve during all but the simplest measurements of pulmonary function in laboratory animals. For some measurements that are not influenced by sensory reflex or muscular contraction (e.g., certain lung volumes, diffusion capacity), acceptable measurements can be made with little regard to such factors. For other measurements, however, physical restraint and anesthesia (chemical restraint) have been shown to have pronounced effects that can alter the conduct or outcome of experiments as well as the interpretation of results. Therefore, to help put the understanding of these measurements in perspective, a brief discussion follows on the potential influence of these two caveats—anesthesia and restraint—on lung function studies.

4.1 Anesthesia (Chemical Restraint)

Anesthesia has been shown to alter measurements of pulmonary function in humans (Rehder et al., 1975) and laboratory animals (Lamm et al., 1982; Skornik and Brain, 1990; Raub and Gillespie, 1979; Rich et al., 1979). In general, minute ventilation is reduced and tidal ventilatory patterns are affected, but the specific type of alteration is generally a function of the anesthetic class. For example, inhaled anesthetics reduce tidal volume (V_T) but may increase frequency of breathing (f) (Pavlin and Hornbein, 1986); narcotics reduce f more than V_T (Bellville et al., 1960) and sedatives tend to depress V_T and f equally (Hunter et al., 1968; Siafakas et al., 1983). Urethane anesthesia generally reduced breathing frequency in neonatal rats and mice compared to unanesthetized estimates of this parameter in these species (Zehendner et al., 2013). Pentobarbital induces a mild hypothermia and decreases ventilation in rats without affecting arterial oxygenation until core temperature is reduced to 31 °C (Torbati et al., 2000). At comparable planes of pentobarbital-induced anesthesia, male and female rats had divergent responses to hyperoxia and hypercapnia (Torbati et al., 1999). A recent report indicated that the late asthmatic bronchoconstrictor response was completely blocked by intraperitoneal ketamine and xylazine, while the early response, after antigen challenge, was unaltered (Raemdonck et al., 2012).

Anesthesia decreases compliance and FRC, and increases lung recoil and airways resistance in humans (Rehder et al., 1975, 1974). In laboratory animals, the fall in FRC is more dramatic than that in the human, approaching a 50% decrease in volume (Vinegar et al., 1979; Lamm et al., 1982). This fall appears to be related to the loss of chest recoil in rodents and a combination of reduced expiratory "braking," diaphragmatic tone, and increased expiratory time available for deflation to lower lung volumes (Lamm et al., 1982). Atelectasis (closure of dependent lung regions) has been demonstrated in anesthetized humans (Rehder et al., 1975) with resulting ventilatory redistribution and shunts. A similar phenomenon may occur in anesthetized rats (Lamm et al., 1982), especially when in the dorsally recumbent position. Hence, studies of particle or gas deposition within the respiratory system may be altered substantially due to these volume and dead space shifts (Raab et al., 1975; Sweeney et al., 1983). Clearly, lung responses associated with aerosolized particulates, gases (e.g., CO_2), or irritant chemicals may be significantly affected by these changes in ventilation and distribution (Mauderly, 1989; Tepper et al., 1991; Wiester et al., 1995). The implication of these findings should be carefully considered prior to their incorporation as inputs to pharmacokinetic models.

Other confounding anesthesia-related effects not necessarily related to ventilation or volume changes include blockade of irritant reflexes (Weissberg et al., 1976), depression (Dennis and Douglas, 1970), and augmentation (Advenier et al., 1978) of the response to bronchoconstrictors, alteration of ventilatory patterns in response to inhaled CO_2 (Martin-Body and Sinclair, 1985; Hirshman et al., 1975; Gasser and Bellville, 1976; Vizek and Palecek, 1977), and reduction of stimulatory input to efferent responses (Barret, 1971). Airway responsiveness has also been shown to be depressed by pentobarbital, urethane, atropine, hexamethonium, and mepyramine (Advenier et al., 1978). In studies where animals are both anesthetized and paralyzed, and thus must be ventilated, the lung volume at which the animals are ventilated can alter the responsiveness to acetylcholine challenge (Benichou et al., 1990). Thus, studies of lung function per se, or studies dependent on normal lung function (e.g., aerosol studies) that involve anesthesia, should be carefully considered because of the influence of anesthesia on the outcome. That is not to say that these studies are thereby invalidated, but rather that anesthesia can profoundly influence the response.

4.2 Physical Restraint/Activity

If anesthesia is not used, some type of restraint is usually required for the measurement of pulmonary function in laboratory animals. Restraint can be minimal (i.e., confinement inside a cage), such as used with a whole body plethysmograph (see Section 6.1). One study in the literature shows that with appropriate training (4–8 weeks), measurements of V_T, O_2 uptake, and CO_2 output can be obtained in unrestrained small laboratory mammals using a mask outfitted with a nonrebreathing valve (Mauderly, 1979). For other measurements, a device that forcibly holds the animal in place may be required. The obvious advantages of restraint are that movement artifacts are reduced and delicate probes or sensors can be attached and monitored without disruption. However, physical restraint methods may also induce physiological disturbances, such as gastric lesions (Toraason et al., 1980), changes in core temperature (Nagasaka et al., 1979) and metabolism (Nagasaka et al., 1980), and stress-related increased expiratory CO_2 (Jaeger and Gearhart, 1982). Furthermore, changes in ventilation and ventilatory pattern have been observed (Jaeger and Gearhart, 1982; Lai and Hildebrandt, 1978; Mauderly, 1986; Berendt, 1968). Several approaches to restraint for small laboratory mammals that avoid anesthesia have been described in the literature (Amdur and Mead, 1958; Matijak-Schaper et al., 1983; Wiester et al., 1987).

Recently, the advent of very small, implantable telemetric sensors has opened a new approach to physiologic measurements in unrestrained, free-roaming laboratory animals (Murphy et al., 1998; Sandler and Meindl, 1983; Watkinson et al., 1990). Although there are few reports of respiratory measurements being obtained in small laboratory animals using telemetry (Murphy et al., 1998; Ewart et al., 2010), other related measures, such as heart rate, electrocardiogram, blood pressure, and core temperature, can now be readily obtained without disturbing animals in their cages, or even during inhalation exposure challenge (Watkinson et al., 1990). Furthermore, respiratory function measurements using minimal restraint (WBP) can be evaluated simultaneously with telemetric measurements to allow a more comprehensive assessment of cardiopulmonary function during drug and chemical exposure (Murphy et al., 1998; Harris, 2005; Perez et al., 2013). Similarly, for larger animals telemetric methods (inductance and impedance methods, see section on Telemetry) can be used to measure ventilation as well as in combination with telemetered cardiovascular measurements (Authier et al., 2010).

Stress responses are so poorly understood that their influence on toxicologic responses may well pass unnoticed until specifically examined. One such example is that of rat cardiac and temperature responses to inhaled ozone (O_3). With restraint (head-out plethysmography), rats exhibited no changes in body core temperature (Tepper et al., 1989), while similarly exposed unrestrained rats experienced substantial decreases in heart rate and body temperature (Watkinson et al., 1990).

5. ALLOMETRY

5.1 Interspecific Allometry

As the current and previous chapters from this section have demonstrated, many tests of lung function, originally developed for human subjects, have been adapted to laboratory animals. The cumulative evidence presented in these chapters suggests that the theories describing and accounting for the ventilatory, mechanical, and diffusional behavior of the human lung hold reasonably well in most laboratory mammals (Leith, 1984). In fact, lung function variables, such as aerobic metabolism ($\dot{V}_{O_2}$), scales systematically to body mass over nearly seven orders of magnitude (see Appendix A for additional references on scaling) (Leith, 1976; Stahl, 1967).

Relationships between body mass and functional variables are typically not linear, but the data can be transformed logarithmically to be expressed by linear equations of the form:

$$y = aM^b \tag{3}$$

where "y" is the biological variable, "M" is body mass (in kilograms), "a" is the mass coefficient, and "b" is the mass exponent. Such equations are not just empirical curves fitted to experimental data; rather, the parameters from these power functions have specific meanings for analyzing structure and function relationships in animals of different sizes (Heusner, 1982a, 1982b, 1983a, 1983b; Feldman and McMahon, 1983). For example, respiratory properties of the lung generally vary with body size in a similar manner, as shown in Table 3 (additional allometric scaling variable can be found in Table 6, Chapter 15, and Table 3, 9, 10 and 12, Chapter 12). Comparison of the parameters in this table reveals a degree of variability from study to study, particularly in the mass coefficient. To avoid misuse of these power functions as reference standards, Leith (1983a, 1983b, 1976, 1984) recommended that approximate mass exponents be emphasized in assessing relationships of lung function variables with body size across mammalian species. When used in this manner, generalized scaling of lung variables indicates that (1) lung inflation pressures are invariant with body size; (2) lung weight, volume, and compliance vary with body mass to the first power; (3) ventilation and maximal O_2 consumption ($\dot{V}_{O_{2max}}$) vary with body mass to the 3/4 power; and (4) breathing frequency (f) varies with body mass to the −1/4 power. The existence of such relationships suggests that the governing principles of respiration hold for most mammals and that the evolution of structure to meet physiologic demands has been intrinsically conserved. In spite of the intrinsic biological differences, such as the high chest wall compliance and low FRC of small mammals relative to that of humans, the architectural constraints of the parenchyma and chest wall limit lung volume in all species. Similarly, conducting airway size and infrastructure control the dynamics of airflow. Furthermore, these relationships give credence to the interpretation of alterations in pulmonary function measurements after toxic insult, based on the accumulated evidence acquired primarily from human pathophysiology.

In recent years, the original intention of comparative physiologists to demonstrate the relative scalability of respiratory functional parameters across mammalian species has been repurposed in an attempt to help accurately determine respiratory exposure to inhaled drugs and chemicals in laboratory animals and extrapolate findings to human exposure. With inhalation exposure, the dose received by a subject is directly related to the amount of air inhaled. In this context, the most common equation for dose to the respiratory system has been defined as:

$$DD = \frac{C \times \dot{V}_E \times D \times F}{BWT} \tag{4}$$

where DD = delivered dose (mg/kg); C = concentration of substance in exposure atmosphere (mg/l); $\dot{V}_E$ = expired volume in 1 min (l/min); D = duration of exposure (min); BWT = body weight (kg); and (F) = the inhalable fraction or the estimated deposited fraction. This latter parameter, F, is sometimes included when deposited dose is to be estimated rather than delivered dose. In general, C, D, and BWT are known or measured, while $\dot{V}_E$ and F are estimated. A discussion of F is beyond the scope of this review and is discussed in more detail in Chapters 27 and 28 (deposition discussion in this book).

However, $\dot{V}_E$ is rarely directly measured and usually estimated by application of an allometric equation. Two references are widely used in this regard by regulatory authorities for pharmaceutical application. In the first paper by Bide et al. (2000), data were collected from 18 adult, mammalian terrestrial species ranging from mice at 0.012 kg to horses and a giraffe at 480−550 kg. The data were separated to estimate $\dot{V}_E$ in sedated versus unanesthetized animals in an attempt to extrapolate to human exposure under different ventilatory conditions (resting versus light exercise). Very young animals and gender were not included in the analysis. Log−log normalized data were linear and followed a power law distribution. Although the extreme upper and lower tails of the body weight versus $\dot{V}_E$ regression fell somewhat below and above the line, respectively, the overall fit of the equation

TABLE 3 Power Function Predictions

Variable	Units	*a*	*b*	Reference
BSA	m^2	0.11	0.65	Stahl (1967)
S_A	m^2	1.87	0.88	Lechner (1978)[a]
LWt	g	7.72	1.03	Bennett and Tenney (1982)
W/breath	g-cm	17.1	1.08	Stahl (1967)
$\dot{V}_{O_{2max}}$	ml/min	53.96	0.678	Lechner (1978)[a]
$\dot{V}_{O_{2std}}$	ml/min	9.13	0.727	Lechner (1978)[a]
$\dot{V}_E$	ml/min	0.499	0.809	Bide et al. (2000)
f	Min-1	47.7	−0.28	Gunther (1975)[b]
V_T	ml	7.69	1.04	Stahl (1967)
V_D	ml	2.76	0.96	Stahl (1967)
VC	ml	56.7	1.03	Stahl (1967)
FRC	ml	24.1	1.13	Stahl (1967)
TLC	ml	40	1.021	Lechner (1978)[a]
P_{pl}	cm H_2O	4.38	0.004	Gunther (1975)[b]
C_L	ml/cm H_2O	1.469	1.2	Spells (1969)
C_{cw}	ml/cm H_2O	4.693	0.898	Spells (1969)
C_{rs}	ml/cm H_2O	1.471	1.005	Spells (1969)
R_{aw}	cm H_2O/1/s	21.418	−0.862	Spells (1969)
R_L	cm H_2O/1/s	41.163	−0.903	Spells (1969)
R_{rs}	cm H_2O/1/s	21.41	−0.393	Spells (1969)
DL_{CO}	ml/min/mm/Hg	0.614	0.985	O'Neil and Leith (1980)
DL_{CO_2}	ml/min/mm/Hg	3.92	0.991	Gehr et al. (1981)

Note: Based on Eqn (2). $y = aM^b$, where y = biological variable; M = body mass in kg; a = mass coefficient and b = mass exponent.
In some cases, the mass coefficient was recalculated to be consistent with body mass in kg.
Other abbreviations: BSA = body surface area; S_A = alveolar surface area; LWt = lung weight; W = work of breathing; $\dot{V}_{O_{2max}}$ = maximal oxygen consumption; $\dot{V}_{O_{2std}}$ = whole body oxygen consumption; $\dot{V}_E$ = minute ventilation; f = breathing frequency; V_T = tidal volume; V_D = dead space volume; VC = vital capacity; FRC = functional residual capacity; TLC = total lung capacity; P_{pl} = pleural pressure; C_L = lung compliance; C_{CW} = chest wall compliance; C_{rs} = total respiratory system compliance; R_{aw} = airway resistance: R_L = pulmonary resistance; R_{rs} = total respiratory resistance; DL_{CO} = carbon monoxide diffusing capacity. DL_{O_2} = oxygen diffusing capacity.
[a]*For the combined group of laboratory and wild animals.*
[b]*Calculated from Gunther and Leon de la Barra (1966).*

was good ($r^2 = 97.5\%$) but did not fit the idealized relationship:

$$\dot{V}_E = k \times BWT^{0.75} \quad (5)$$

based on an application of dimensional analysis as discussed above. The mouse showed the poorest agreement for the common laboratory species. For unanesthetized animals, the allometric equation ($r^2 = 97.5\%$) was:

$$\dot{V}_E = 0.499 \times BWT^{0.809} \quad (6)$$

While for sedated animals, the equation was similar, although the fit was not as good ($r^2 = 92.7\%$):

$$\dot{V}_E = 0.325 \times BWT^{0.789} \quad (7)$$

Practical experience with rodents and monkeys suggested that Eqn (6) underestimated $\dot{V}_E$ and thus underestimated exposure in inhalation toxicology studies. This implied that the NOAEL and the effective dose found in regulatory toxicology studies may have been underestimated by as much as 40% (oral presentation, T.

Sweeney, ACT 2007). This underestimate could make a significant difference in the starting and maximum dose allowed in a clinical trial. Because of these concerns, the Association of Inhalation Toxicologists (Alexander et al., 2008) reexamined the basis and methodology used to develop four of the most commonly used allometric scaling equations for the estimation of $\dot{V}_E$ in laboratory animals. They concluded that the published algorithms were inadequate for estimating animal exposure in FDA-regulatory toxicology studies. The equation of Guyton (1947) was faulted as modern analytical methods were not used. McMahon (1977) used anesthetized animals and Bide et al. (2000) and Stahl (1967) included many species not commonly used in inhalation toxicology studies. Based on these deficiencies, Alexander et al. (2008) collected data from the four most commonly used species (dog, monkey, mouse, and rat) to develop a better validated allometric relationship for ventilation in these species. All of the animals were unanesthetized, physically restrained control animals (962 males, 1021 females). Respiratory measurements were made as part of either inhalation toxicology or safety pharmacology studies using good laboratory practice (GLP) validated methods (rodents in head-out or whole body plethysmographs and nonrodents outfitted with pneumotachographs). Animals were acclimatized to the apparatus prior to measurement, and all measurements occurred during control exposure (in most cases filtered air). Body weight for all species ranged from 0.02 to 16.7 kg. A best fit was obtained by log−log normalization and then fitting the data using nonlinear regression modeling. Based on these well-characterized data, the following allometric scaling equation was proposed for estimating ventilation in common laboratory mammals used for inhalation studies:

$$\dot{V}_E = 0.608 \times BWT^{0.852} \quad (8)$$

Note the exponent further deviated from the "ideal" 0.75 and Eqn (6) ($b = 0.809$) for unanesthetized animals. Minute ventilation, when calculated with both Eqns (6) and (8), predicts the same ventilation in a 10 g mouse, but a 35% larger $\dot{V}_E$ is predicted in a 10 kg dog by Eqn (8).

Using Eqn (8) to extrapolate a ventilation rate in a 70 kg adult human would be equivalent to the ventilation rate of a human performing light to moderate activity (20 l/min) (U.S. EPA, 1988; U.S. EPA, 2011). Whereas Eqn (6) seemed to be a better fit for young adult males (70 kg) doing mild activity and predicted ventilation was slightly lower than values obtained from soldiers carrying light loads while walking on a treadmill. Refinement of such equations to take into account different activity levels (sedation, rest, light exercise, heavy exercise) would be useful because inhaled dose is directly proportional to the volume of air inhaled.

5.2 Intraspecific Allometry

Intraspecific allometry: EPA also uses allometric equations as part of the determination of a reference concentration (RfC), which is, in part, a methodology used to predict inhalation exposure to humans from animal studies (U.S. EPA, 1994). The estimate of RfC for various types of gases, vapors, and soluble and insoluble particles is beyond the scope of this review but can be found in the following references (U.S. EPA, 2011; U.S. EPA, 1994). Unlike FDA, EPA has taken the approach to use intraspecific estimates of minute ventilation when appropriate (same species, sex, and weight range). Interspecies equations are used when intraspecific equations are inappropriate. The RfC methodology lists another report (U.S. EPA, 1988) that catalogs various biological values to be used in risk assessment when extrapolating exposure in animals to humans. The report suggests that when possible intraspecific allometric equations should be used because it best reflects the data for that species and because growing animals and older animals gain weight for different reasons, which results in different metabolic requirements (e.g., ventilation) within a species (e.g., 1 kg young monkey versus a 1 kg old rat). On the other hand, Bide et al. (2000) concluded that these intraspecific allometric equations appeared to result in greater errors of minute volume estimation as the correlations and fit of the individual allometric equations were not as good.

Table 4 uses EPA intraspecific equations (U.S. EPA, 1988) that were transformed to calculate $\dot{V}_E$ in l/min in the form of Eqn (3). Approximate body weights for various species were included in the table to allow comparison between Eqns (6) and (8) and the EPA allometric equations for determining $\dot{V}_E$. A comparison of the $\dot{V}_E$ between different methods is provided. Note that Eqn (6) (Bide) and Eqn (8) (Alexander) are in fairly good agreement with Eqn (8) predicting a higher $\dot{V}_E$ than Eqn (6), a discrepancy that becomes more apparent as body weight increases. Compared to the EPA intraspecific equations, Eqns (6) and (8) underpredict the mouse, but in most other cases, these equations overpredict EPA's estimates of $\dot{V}_E$. However, this may be appropriate since Eqn (8) was derived to represent the ventilation of a physically restrained animal during an inhalation exposure, while the EPA data are meant to capture average ventilation during a 24 h period.

6. PRACTICAL USE AND INTERPRETATION OF PULMONARY FUNCTION TESTS

In the evaluation of the effects of disease, chemical exposure, toxins, pollutants, and drugs, measurement of pulmonary function in animals can help discover, localize, and assess the magnitude of lesions in the lung. The use of a particular lung

TABLE 4 Minute Ventilation Comparing Intraspecies Versus Two Commonly Used Interspecies Allometric Equations

				Ventilation (l/min)			% Discrepancy From EPA	
Species	**Intercept a^a**	**Exponent b^a**	**M^a BWT (Kg)**	**EPA(Barret, 1971) 1988**	**Alexander et al., 2008**	**Bide(Hirshman et al., 1975) 2000**	**Alexander/ EPA**	**Bide/ EPA**
Mice	1.382	1.050	0.03	0.035	0.031	0.029	−12%	−16%
Hamster	0.347	0.902	0.1	0.044	0.085	0.077	96%	78%
Rats	0.556	0.821	0.25	0.178	0.187	0.163	5%	−9%
GP	0.306	0.516	1	0.306	0.608	0.499	99%	63%
Rabbits	0.319	0.831	2	0.568	1.097	0.874	93%	54%
Monkey	0.563	0.486	3	0.960	1.550	1.214	62%	26%
Cats	0.222	0.595	4	0.507	1.981	1.532	291%	202%
Dogs	0.465	0.709	8	2.033	3.575	2.683	76%	32%
All species	0.458	0.758	70	11.470	22.695	15.516	98%	35%

[a]*Based on Eqn (2).* $y = aM^b$, *where* y = *biological variable;* a = *mass coefficient;* M = *body mass in kg;* b = *mass exponent.*

function test or tests may be best for evaluating different types of lesions. The following section delineates which test(s) is most appropriate for assessment of various common lesions of the lung. Normal values of selected pulmonary function tests for various species are listed in Table 5.

6.1 Ventilation

One of the most common effects from many centrally acting drugs including opiates, benzodiazepines, sedatives, and barbiturates is respiratory depression or hypoventilation. As ventilation is critical to the maintenance of normal acid–base balance in the blood and tissues, disturbances in ventilation can result in systemic problems. Hypoventilation is defined by an increase in carbon dioxide (hypercapnia) beyond normal levels ($Pa_{CO_2} > 45$ mm Hg) and can result in respiratory acidosis (pH < 7.35). Impairment of respiratory muscle function or control can also cause hypoventilation. Respiratory diseases of conditions associated with decreased breathing rate and ensuant hypoventilation include emphysema, chronic bronchitis, asthma, severe pneumonia, aspiration, irritant chemical exposure, and pneumothorax.

In contrast, hyperventilation is defined by a decrease in blood Pa_{CO_2} below normal levels (<35 mm Hg) and can result in respiratory alkalosis (pH > 7.45). Symptoms of hyperventilation include dizziness, paresthesias, headache, fainting, and seizures. Hypo- and hyperventilation should be distinguished from the terms dyspnea (subjective experience of breathing discomfort), labored breathing (increased effort to breath), tachypnea (rapid shallow breathing), and hyperpnea (rapid and deep ventilation), all of which can occur without changes in the partial pressure of CO_2 in blood. Common medical causes of hyperventilation and increased rate of breathing include asthma and allergies, excess blood loss, emphysema and chronic bronchitis, congestive heart failure, myocardial infarction, pneumonia, and pulmonary embolism. Note that many of the same diseases may cause increased or decreased ventilation, thus making diagnostic judgments, based solely on changes in ventilation, suspect. Stimulants and deep lung irritants are drug and toxicant examples that may result in increased rate of breathing (e.g., theophylline and O_3).

The measurement of minute ventilation in conjunction with blood gases can provide important additional diagnostic information on the respiratory status of exposed animals (see Chapter 14). However, respiratory measurements of V_T and f as well as their product, minute ventilation ($\dot{V}_E$), are frequently more sensitive to disease-, drug-, or toxicant-induced changes in ventilation. The most severe lung lesions are characterized by both decreases in Pa_{O_2} and increases in Pa_{CO_2}, while in more mild disease, alterations in ventilation and buffering ability can compensate for most diffusion impairments, resulting in apparently normal blood gas values. Since changes in blood Pa_{O_2} and Pa_{CO_2} do not always indicate impairment of the respiratory system, the use of pH and bicarbonate values can assist in the interpretation of acid–base status by partitioning the origin of blood gas disturbance into primary respiratory, metabolic, or combined disorders.

Currently there is a published guideline, ICH-S7a (Guidance for industry, 2001), indicating the need for the assessment of respiratory function as part of the safety pharmacology evaluation of new drugs prior to entry into humans. The primary goal of these safety pharmacology studies is "to identify undesirable pharmaco-dynamic properties of a substance that may have relevance to human safety" on "functions which are acutely critical for life" prior to the first human exposure (Guidance for industry, 2001). For respiratory safety pharmacology, the guidelines indicate that "Respiratory rate and other measures of respiratory function (e.g., tidal volume or hemoglobin oxygen saturation) should be evaluated." It is our recommendation that respiratory rate (f) and tidal volume (V_T) should be measured, thus allowing the calculation of minute ventilation ($\dot{V}_E = V_T \times f$). ICH-S7a (Guidance for industry, 2001) recommends that these tests be performed in unanesthetized and unrestrained animals; however, if restrained animals are used, the animals should be acclimated to the restraining apparatus. Although measurement of ventilation (including the use of newer less invasive techniques such as telemetry and WBP) are relatively insensitive to subtle changes in pulmonary function, they are sufficiently sensitive for detecting potential life-threatening responses and, thus, meet the goal of protecting clinical trial subjects prior to the first human exposures. Typically, these animal studies are performed as acute single-dose studies using a Latin square design but parallel, escalating, or even de-escalating dose designs may be appropriate (Sarazan et al., 2011). More recently, these safety pharmacology studies have been incorporated into repeat-dose GLP toxicology studies allowing the assessment of drug effects that may accumulate with time. Either rodent or non-rodent species are allowed as long as group size is sufficient for detection of moderate effects (N = 6–8 rodent, ≥4 non-rodent). Typically, the most relevant route of administration is the intended route for clinical administration, and it is best practice to use the formulated drug product, not just the active pharmaceutical ingredient, intended for use in humans. Selection of evaluation time points and interdose intervals needs to reflect the pharmacodynamics and pharmacokinetics (including tissue persistence) of the drug candidate. The ideal window of time for capturing safety pharmacology endpoints would encompass the period during which plasma concentration, target tissue concentration, and/or pharmacological activity are maximum; although in practice capturing all three is not always feasible. The use of a pharmacologically responsive

TABLE 5 Selected Functional Variables in Normal Small Laboratory Mammals

Anesthesia	Strain[a]	Weight (g)	No.	Age (weeks)	Sex	TLC^b	FRC^b	DL_{co}^b	Cst^b	FEF_{25}^b	Reference
Rat											
Nembutal	F-344	330 ± 3	22	23	M	11.0 ± 0.2	2.9 ± 0.1	0.1999 ± 0.005	0.66 ± 0.02	38.5 ± 3.0	Costa and Tepper (1991)
Nembutal	F-344	360 ± 13	23	46	M	12.1 ± 0.2	2.5 ± 0.1	0.233 ± 0.007	0.81 ± 0.02	55.9 ± 3.3	Costa and Tepper (1991)
Nembutal	F-344	480 ± 9	22	92	M	12.5 ± 0.1	2.2 ± 0.1	0.242 ± 0.005	0.94 ± 0.01	55.2 ± 2.3	Costa and Tepper (1991)
Nembutal	F-344	33 ± 1	28	3	M	1.98 ± 0.02	0.66 ± 0.01	0.023 ± 0.001	0.118 ± 0.002	–	Stevens et al. (1988)
Nembutal	F-344	133 ± 3	26	6	M	5.59 ± 0.05	1.12 ± 0.04	0.113 ± 0.002	0.397 ± 0.006		Stevens et al. (1988)
Nembutal	F-344	234 ± 2	20	10	M	8.63 ± 0.12	1.91 ± 0.04	0.167 ± 0.004	0.583 ± 0.013		Stevens et al. (1988)
Nembutal	F-344	297 ± 3	24	16	M	11.19 ± 0.1	2.11 ± 0.03	0.172 ± 0.005	0.675 ± 0.013	–	Stevens et al. (1988)
Halothane	F-344	222 ± 61	10/10	14		11.9 ± 1.7	2.1 ± 0.3	0.119 ± 0.036	0.59 ± 0.27	–	Mauderly and Hahn (1982)
Halothane	F-344	334 ± 10	10/10	77		13.9 ± 2.2	1.7 ± 0.3	0.278 ± 0.042	0.94 ± 0.34	–	Mauderly and Hahn (1982)
Halothane	F-344	332 ± 71	10/10	116		14.4 ± 1.9	2.7 ± 0.4	0.114 ± 0.028	0.86 ± 0.19	–	Mauderly and Hahn (1982)
Halothane	F-344	258 ± 22	10	12–14	M	14.2 ± 1.2	3.2 ± 0.3	0.27 ± 0.04	0.78 ± 0.18	50 ± 3.0	Harkema et al. (1984)
Urethane	F-344	233 ± 32	32	–	F	8.2 ± 1.0	4.6 ± 0.6	–	0.25 ± 0.08	50	Diamond and O'Donnell (1977)
Urethane	SD	450 ± 11	27	15–17	M	17.6 ± 0.35	3.9 ± 0.12	0.27 ± 0.016	1.8 ± 0.72	–	Loscutoff et al. (1985)

Nembutal	SD	307 ± 10	15			12.23 ± 0.55	2.51 ± 0.20	–	0.57 ± 0.03	–	Lai and Hildebrandt (1978)
Nembutal	SD	128	21	–	M	9.2	–	0.146	–	–	Takezawa et al. (1980)
Nembutal	SD	186	23	–	F	10.4	–	0.151		–	Takezawa et al. (1980)
Nembutal	SD	313 ± 50	14		M	–	1.69 ± 0.12	–		–	Lamm et al. (1982)
Awake	SD	313 ± 50	14		M	–	3.19 ± 0.22	–		–	Lamm et al. (1982)
Methoxyflurane	-	337 ± 30	25	–	M	–	3.0 ± 1.2	0.083 ± 0.19	0.74 ± 0.19	15.8 ± 2.7	Holub and Frank (1979)
Chloralose	Wistar	135 ± 31	18	4.5–5.5	M	5.4 ± 1.0	2.2 ± 0.5	–	0.25 ± 0.06	–	Yokoyama (1983)
Chloralose	Wistar	248 ± 19	21	7.5–8	M	8.8 ± 1.1	3.2 ± 0.8	–	0.50 ± 0.12		Yokoyama (1983)
Chloralose	Wistar	338 ± 27	27	8.5–12	M	11.3 ± 1.4	3.9 ± 0.8	–	0.61 ± 0.16		Yokoyama (1983)
Chloralose	Wistar	457 ± 37	30	12 –19.5	M	14.9 ± 1.4	4.6 ± 0.9	–	1.00 ± 0.23		Yokoyama (1983)
Mice											
Nembutal	Blotchy	23 ± 1	6	18		1.8 ± 0.03	0.44 ± 0.01	0.025 ± 0.002	0.103 ± 0.006	–	Stevens et al. (1987)
Nembutal	–	25 ± 30	–	18			–		–	4	Sinnett et al. (1981)
Nembutal	Swiss CD	–	13	37	M	1.8 ± 0.02	0.032 ± 0.02	0.032 ± 0.002	0.188 ± 0.006		Miller et al. (1987)
Nembutal	Swiss CD	–	15	57	M	1.8 ± 0.02	0.46 ± 0.02	0.035 ± 0.002	0.111 ± 0.006		Miller et al. (1987)
Hamsters											
Nembutal	SGH	123 ± 2	16	–	M	5.83 ± 0.12	1.75 ± 0.04	0.123 ± 0.009	0.41 ± 0.14	–	Raub et al. (1982)
Halothane	SGH	126 ± 13	5/5	15	M/F	9.6 ± 1.3	3.5 ± 0.5				Mauderly (1979)
Halothane	SGH	125 ± 7	5/5	65	M/F	11.1 ± 1.0	4.3 ± 0.4				Mauderly (1979)
Nembutal	SGH	114–145	–	–	M	–	–			35	Lucey et al. (1978)
Nembutal	SGH	122 ± 3	23	6–18	M	7.2 ± 0.14	2.4 ± 0.06		0.63 ± 0.03		Koo et al. (1976)

Continued

TABLE 5 Selected Functional Variables in Normal Small Laboratory Mammals—cont'd

Anesthesia	Strain[a]	Weight (g)	No.	Age (weeks)	Sex	TLC[b]	FRC[b]	DL^{b}_{co}	Cst[b]	FEF^{b}_{25}	Reference
Guinea pig											
Urethane	Hartley GP	564 ± 11	27	11–13	M	21.2 ± 0.42	7.7 ± 0.21	0.36 ± 0.085	2.13 ± 0.085	–	Loscutoff et al. (1985)
Nembutal/ gallium trieth.	Albino GP	310–380	17		M	13.5–18.0	–	–	–	39	Lai (1988)
Nembutal	GP	301–855	–	–		–	24.2% TLC		–	–	Raub and Gillespie (1979)
Awake	GP	301–855	–		–	–	38.8% TLC	–	–	–	Raub and Gillespie (1979)
Rabbit											
Nembutal		4100 ± 480	15	26.52	M/F	–	52.5 ± 7.5	–	6.4 ± 0.14	–	Davidson et al. (1966)
Nembutal	N. Zealand	3400 ± 100	8	24	M	123 ± 4	48 ± 1	1.7 ± 0.1	0.19 ± 0.01c		Berend et al. (1985)
Ferret											
Nembutal		624 ± 99	6	12	M	89 ± 5	20.7 ± 0.43	5.39 ± 0.43	5.4 ± 4.3(rs)	461 ± 42	Vinegar et al. (1982)
									10.8 ± 1.35 (lung)		Vinegar et al. (1982)

species is advantageous but not required. If a nonresponsive species is selected, it should be noted that the study can only evaluate off-target effects.

6.2 Irritation

Anger estimated that 60% of all threshold limit values (TLVs) have been established based on their potency to cause irritation in humans (Anger, 1984). For many airborne exposures, changes in ventilatory pattern are the first and most sensitive method of detection of an irritant response in animals. Detection of irritating chemicals has typically been examined by evaluating changes in f during head-only exposure to toxicants using a mouse model described by Alarie (1966, 1973). This test method has been endorsed by the American Industrial Hygiene Association (Ferguson et al., 1986; ASTM., 1984) and is based on the high correlation between the concentration at which the respiratory rate of the mouse is reduced by 50% (RD_{50}) and the current TLVs. An assessment by the California Environmental Protection Agency further validated this relationship finding a high correlation between the RD_{50} in mice and the lowest observed adverse effect levels (LOAELs) or the reference exposure levels (RELs) for many chemicals (Figure 7) (Kuwabara et al., 2007). They concluded that the test was sufficient to protect the public from experiencing chemicals that cause a burning sensation of the eyes, nose, or throat. Although any discomfort or acute disability that occurs during an exposure may be deemed adverse (Buckley et al., 1984), the relationship between such acute irritant responses, as reflected in altered breathing patterns, and subsequent health effects, is not well established (Tepper and Costa, 1992). Buckley et al. provided some evidence to establish a relationship between sensory irritation and cellular responses (Buckley et al., 1984). Histopathological evaluation of 10 different irritant chemicals at the RD_{50} concentration indicated that all irritants produced lesions in the nasal cavity that ranged from slight hypertrophy to severe epithelial erosion and necrosis. Additionally, sensory irritants also can produce damage to the lower respiratory tract. Thus, although the RD_{50} did not predict the magnitude or location of injury, it did generally indicate a concentration that resulted in cellular damage. Furthermore, the link between the acute response and chronic disease is even less well established.

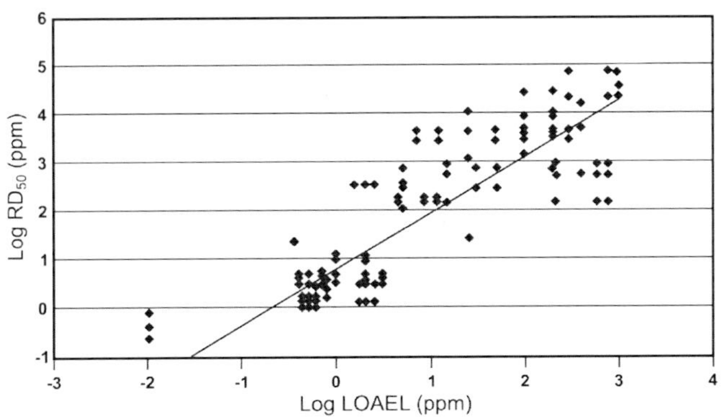

FIGURE 7 Linear least-squares regression analysis for log RD_{50} (for all mouse strains) versus log LOAEL (human irritation endpoint) for 25 compounds using 195 data points. Log RD_{50} = 1.16 (Log LOAEL) + 0.77; R2 = 0.80 (Kuwabara et al., 2007). *Redrawn with permission from Environmental Health Perspectives.*

Two types of irritant responses have been identified (Alarie, 1973). The first, usually produced by highly water-soluble chemicals, such as ammonia, chlorine, and formaldehyde (HCHO), has been classified as "sensory" or upper airway irritation. These effects are believed to be caused by stimulation of the trigeminal nerve that innervates the eyes, nose, and mouth areas. Although it should be noted that these irritant reflexes occur despite olfactory stimulation, many sensory irritants do, in fact, stimulate olfactory receptors and, thus, may be considered noxious. Exposures to sensory irritants typically produce a burning sensation, often precipitating lacrimation. The larynx also can be stimulated by these chemicals, producing cough. In addition to the symptoms of irritation and lacrimation reported for humans, the sensory irritants are readily identified in small laboratory mammals by a distinct prolongation of the expiratory phase of the tidal breath due to phrenic nerve inhibition that interrupts expiratory relaxation (Alarie, 1973). Other physiological consequences of sensory irritation include sympathetically induced vasoconstriction, which results in increased blood pressure, and vagally mediated reflex bradycardia (Alarie, 1973).

"Pulmonary" irritation, a second type of irritant response, usually occurs with less water-soluble chemicals. Prototypic examples are O_3, phosgene, and nitrogen dioxide. Subjectively, these chemicals produce cough and substernal soreness without the burning sensation of the eyes and upper airways produced by the sensory irritants. Exposure to sufficient concentrations of these chemicals has been shown to induce reflex reactions in several mammalian species (e.g., guinea pigs, cats, dogs, monkeys, and humans). Most notably, the response to this type of irritant is observed as an increase in f and as a decrease in V_T (tachypnea), in contrast to the sensory irritant pattern of increased V_T and decreased f. Although evidence exists to indicate these reflexes are of vagal origin (Alarie, 1973; Lee et al., 1979), other investigators could not block the reflex after vagotomy (Gordon et al., 1984). Species difference in response also may occur. For example, rather than tachypnea, mice exhibit a decrease in respiration that is qualitatively different from the pattern of response seen with sensory irritation. In mice, pulmonary irritation is manifested by an increase in apnea time between breaths rather than a prolongation of expiration, as seen with sensory irritants.

Knowledge of irritant responses can be important in toxicology studies, not only because of the potential for these chemicals to cause discomfort but because irritant responses to chemicals can also alter the deposition, and therefore the toxicity, of inhaled chemicals. For example, Chang et al. (1983) could explain the difference between rat and mouse susceptibility to formaldehyde (HCHO)-induced squamous cell carcinoma by examining the sensory irritant response of these species to the exposure. In their experiments, mice, upon repeated exposure, could reduce their minute ventilation more than rats and thus reduced the inhaled dose, which resulted in protection from histological damage and tumors.

Acute responses to irritants may be different from those observed with repeated exposures. The literature indicates that repeated exposure to irritants may produce an increase, a decrease, or little change in the reflex response. Kane and Alarie observed a typical sensory irritant response in mice upon initial exposure to various concentrations of acrolein and HCHO (Kane and Alarie, 1977). However, during the 3-h exposure, HCHO exposed rats developed tachyphylaxis, whereas the response of acrolein-exposed rats was sustained. With repeated exposure (4 days), both HCHO- and acrolein-exposed rats showed a progressively exaggerated immediate response, which was maintained as exposure continued. Thus, exposure to some inhaled irritants seems to sensitize the airway receptors such that the reflex response appears stronger and more sustained. This could result in more significant acute hypoxia secondary to bronchoconstriction, but might result in fewer chronic health problems, as suggested by the previously discussed experiment by Chang et al. (1983).

On the other hand, attenuated responses to repeated exposures to irritants have also been observed for both sensory and pulmonary irritants. Tepper et al. have shown that with exposure to O_3, rats displayed an initial pulmonary irritant response (tachypnea) that attenuated during five consecutive days of exposure (Tepper et al., 1989). The study also revealed that several indices of lung damage increased despite the loss of the ventilatory response over the 5-day (2 h/day) exposure. The tachypneic response in guinea pigs was also diminished with repeated exposure to the pulmonary irritant paraquat when inhaled as an aerosol (Burleigh-Flayer and Alarie, 1988). Tolerance has also been observed to the sensory irritant chlorine (Barrow and Steinhagen, 1982) and the mixed sensory and pulmonary irritants cotton dust (Ellakkani et al., 1984) and hexamethylene diisocyanate trimer (Ferguson et al., 1987). These studies suggest that important host defense mechanisms, such as the ability to detect a noxious environment and to appropriately alter breathing patterns to reduce pollutant deposition, may be lost with continued exposure, while cellular damage may progress unabated.

6.3 Bronchoconstriction

In many instances the assessment of overall ventilation and pattern of breathing is inadequate to detect potentially serious debilitating effects of compounds (chemicals, pollutants, drugs) on the function of the lung. Studies reported in the literature indicate that in rat, dog, monkey (Murphy, 2013), guinea pig (Wiester et al., 2005), and humans (Savoy et al., 1984), a two- to three-fold increase in resistance did not result in a change in ventilation. Thus tests to evaluate alterations in airflow obstruction are particularly important for the protection of human health. In humans, airflow obstruction is primarily assessed as a decrease in forced expiration (e.g., FEV_1, FVC). These measurements are more technically challenging and less useful for measuring rapid changes in airflow in animals but are possible even in animals as small as mice (Sinnett et al., 1981; Lin et al., 2002). For animal research, increased resistance (R) is the parameter most used to assess decreases in airflow and is typically considered a direct measure of bronchoconstriction. However, note as discussed further below, R can increase for reasons other than airway smooth muscle contraction. Ideally the evaluation of bronchoconstriction is made in unanesthetized animals, where anesthesia does not influence the response (Amdur and Mead, 1958). Among the various other changes produced by bronchoconstrictors, an increase in driving pressure (P_{TP}) and a decrease in C_{dyn} are functional changes typically observed during bronchoconstriction. In fact, several investigators have found changes in C_{dyn} to be more sensitive than R in demonstrating bronchoconstriction to parenteral and inhaled histamine in unanesthetized guinea pigs. The greater drop in compliance in this species may suggest more significant air trapping and, thus, involvement of smaller, peripheral airways in this response (Dennis and Douglas, 1970; Silbaugh et al., 1981; Wiester et al., 1990). In concordance with this finding, Wiester et al. has shown that Sa_{O_2}/Sp_{O_2} changes correlate well with the changes in breathing mechanics and air trapping after the administration of graded concentrations of known bronchoconstrictors (Wiester et al., 2005, 1990).

The exact site of the bronchoconstriction in the guinea pig has not been determined, but may extend through much of the animal's airway due to the preponderance of airway smooth muscle. The assessment of these variables in the awake, restrained guinea pig after inhaled chemical exposure was widely applied during the 1970s and 1980s because this rodent exhibits particular pulmonary sensitivity to bronchoconstrictors and still finds use as an animal model of allergic inflammation subsequent to antigen exposure and challenge. Additionally, guinea pigs are easy to handle and remain quiescent during testing. However, recently the use of guinea pigs has declined, in part due to the concern that they overpredict bronchoconstrictor responses but primarily because of the availability of reagents for other rodent species and the use of genetically modified mice and rats.

Rats and mice, on the other hand, are generally less responsive to similar doses of bronchoconstrictors, in part due to less airway smooth muscle and differences in receptor distribution and function (Eidelman et al., 1993). However, data obtained in the unanesthetized rat utilizing modifications of the Amdur and Mead (1958) procedure assist in the animal-to-human extrapolation of pulmonary toxicity data because of the better understanding and extensive use of the rat in toxicological research (Tepper et al., 1987; Jaeger et al., 1983; Dorato et al., 1983). Unfortunately, difficulties in using unanesthetized rats have limited publications with this model. The continued and growing importance of the mouse as a species for pharmacology and toxicology evaluation requires pulmonary function testing methods suitable to its size and associated flows and volumes. Few techniques for measuring pulmonary mechanics in unanesthetized mice have been published (Flandre et al., 2003; Hessel et al., 1995; Bates et al., 2008) and none is in widespread use.

These limitations with the Amdur and Mead (1958) technique in unanesthetized animals to measure bronchoconstriction have prompted investigators to use two other techniques in unanesthetized animals: whole body and double-chamber plethysmography. The use and caveats associated with the use of whole body and double-chamber plethysmography have been previously discussed (Section 2.3 respectively). Because of the limitations of these methods in unanesthetized animals, many investigators have chosen to gain improved experimental control by using anesthetized and sometimes paralyzed animals for these types of studies. Respiratory mechanics can be measured using plethysmography for volume or flow and a measure of driving pressure (e.g., P_{tp}, P_{pl}, or P_{ao}) or using an imposed airway pressure signal as in the forced oscillation technique (see Chapter 16) (Bates et al., 1971).

Airway Hyperresponsiveness

The worldwide incidence of asthma is increasing. As a result, the motivation for greater understanding of this disease, which is characterized by reversible bronchoconstriction, airway hyperresponsiveness (AHR), and inflammation, has grown considerably. The development of animal models of asthma has encouraged the use of pulmonary function tests as tools for evaluating drugs, as well as for examining the mechanisms of bronchoconstriction. To date, evidence of AHR has been reported in all 14 species of animals tested (www.Quertle.info, August, 2013) (Kirschvink and Reinhold, 2008) The growing appreciation of AHR in some patients with chronic obstructive pulmonary disease has similarly spurred interest in this research area (O'Connor et al., 1989).

AHR is characterized by an abnormal increase in airway narrowing and airflow limitation to various direct and indirect bronchoconstrictors. The pulmonary measurements used to assess AHR are the same as those used for evaluating bronchoconstriction. However, since the assessment of AHR may be the most common use for pulmonary function tests in animals, a more thorough description of the technique is warranted. Exposures causing AHR may not result in any detectable lung damage or physiological changes on their own. An exposure resulting in AHR manifests as an increased sensitivity to bronchoconstrictors. Although the mechanism(s) causing AHR is not fully known, inflammation and altered neurohumoral mechanisms are believed to be primarily responsible (Baroffio et al., 2009). In the clinic and in many animal studies, heightened airway responsiveness is determined during challenge with progressively increasing amounts of known pharmacologic bronchoconstrictors, such as methacholine or histamine. Many other direct bronchoconstrictors are known, and, in some cases, these agents have also been useful in revealing hyperresponsive airways not detected by cholinergic or histamine agonists. For example, Dusser et al. have shown that guinea pigs are hyperresponsive to substance P and capsaicin, but not to acetylcholine after viral infection (Dusser et al., 1989). Similarly, Szarek (1989) showed that rats are hyperresponsive to serotonin, but not to bethanechol or potassium chloride after a 7-day hyperoxic (85% O_2) exposure. Additionally, many indirect stimuli such as inhaled cold air (Jammes et al., 1986), dry air (Freed, 1989), hyperventilation (Chapman and Danko, 1985), numerous antigens (Chan-Yeung and Lam, 1986; Zosky and Sly, 2007), and viral infections (Rosenthal et al., 2009) have been used to produce airway responsiveness (bronchoconstriction) in laboratory animals. In general, these indirect stimuli release inflammatory mediators and/or stimulate neural pathways that ultimately result in AHR.

The general term *airway hyperresponsiveness* reflects the three attributes of the typically sigmoid-shaped dose–response curve to direct and indirect bronchoconstrictors. Airway hypersensitivity refers to the leftward shift in the dose–response curve characteristic of asthmatics (i.e., less dose to cause the same response), while hyperreactivity refers to the increase in slope of the dose–response curve. The third attribute is the maximal response that typically plateaus in normal individuals and may not be ethically achieved in asthmatics. Of note is that independent changes in each of these three parameters can occur in asthmatics (Baroffio et al., 2009).

Unlike the evaluation of airway responsiveness in humans, where there has been some agreement in testing methodology, there currently exists no such standardization for animal studies. This makes comparisons among animal studies difficult. Measurements are performed in awake or anesthetized animals, even though anesthesia is known to affect the bronchoconstrictor response (see Section 4). Furthermore, anesthetized animals are sometimes evaluated when breathing spontaneously or when paralyzed and artificially ventilated. Typically, breathing mechanics (C_{dyn} and R) are assessed, although some investigators have used other indices, such as flow-derived parameters obtained from a forced expiratory volume maneuver (Holroyde et al., 1980), trapped gas in excised lungs (Wiester et al., 2005; Silbaugh et al., 1987), blood gas measurements (Wiester et al., 2005, 1990), or cough (Clay and Thompson, 1985). Forced oscillatory techniques used to measure frequency-dependent changes in lung mechanics have becoming increasingly popular (Schuessler and Bates, 1995). For anesthetized and ventilated animals, the tidal volume, frequency, and level of positive end-expiratory pressure may alter the results (Sly et al., 2003; Bates and Lauzon, 2007). Complicating these models further is that both intravenous and inhalation challenge approaches have been used, which may produce different results (Roum and Murlas, 1984; Bethel and McClure, 1990). Additionally, agonist dose is increased in some studies, whereas dose rate or exposure duration is altered in other experiments. Timing between doses is also not standardized, so that in some studies cumulative effects of bronchoconstrictors are reported, while in others, challenge responses are essentially independent. Similarly, a comparison of interlaboratory data must be carefully scrutinized, since mean, median, and maximal/minimal responses after challenge may be reported as well as relative data (percent or baseline subtraction). Group data are often reported as an effective dose producing a variable amount of responsiveness (e.g., the dose that causes between 50% and 300% increase in resistance). Clearly, the broad application of varying lung function methodologies can be confusing and, sometimes, misleading, further complicating the reliable determination of AHR. As methodological approaches in this area evolve with mechanistic understanding, the hope is that some standardization and associated interpretation will come forth. Meanwhile, caution is warranted in the use of data from any one study purporting to demonstrate AHR.

Late-phase Responses

It is now recognized that a delayed bronchoconstrictor response, sometimes called the *late-asthmatic* or *late-phase* response, also can occur in experimental animals as is observed in ~50% of asthmatics. After repeated antigen challenge, late-phase responses have been observed in 11 of 12 species examined (Kirschvink and Reinhold, 2008), although there are also studies in many of these same species in which a response was not attained. Additionally, as with humans, delayed-onset responses can occur in animals when reexposed to industrial chemicals (Karol et al., 1981). In either case, the delayed response typically occurs 2–12 h post challenge and can last from 1 to 24 h. The response is observed only in a portion of subjects in all laboratory animal species as well as in humans with demonstrated AHR. The mechanism responsible and the variable penetrance of the response are not known. Evidence is accumulating that inflammatory cells are important sources of mediators associated with antigen-mediated late-phase responses (Abraham et al., 1988, 1989; Okayama et al., 1989). A recent article pointed to a role for sensory nerves and the TRPA1 receptor in modulating the LAR (Raemdonck et al., 2012). The study showed that in both unanesthetized rats and mice, LAR occurred but was completely blocked when the animals were anesthetized with intraperitoneal ketamine and xylazine. Further investigations demonstrated that WBP measurements of increased Penh, corroborated by wheeze and gasping, were diminished by drugs that blocked LAR, but the early asthmatic response was not altered. The TRPA1 blocker, HC-030031, also abrogated LAR, implicating sensory nerves in the mechanism of LAR in both species. Although whole body or double-chamber plethysmograph methods allow repeated assessment of animals or prolonged monitoring, they are fraught with either theoretical and/or methodological issues, respectively, as previously described. Such studies could now be performed with miniature pressure-sensing telemetry probes to capture driving pressure while volume or flow could be fairly accurately measured by WBP, which could allow repeated long-term study in animals (Murphy et al., 1998), but to date, the technique has seen limited use (Murphy et al., 1998; Ewart et al., 2010).

6.4 Obstructive and Restrictive Lesions

The small airways of the lung (<2 mm in humans, but anatomically undefined in rodents) are held open by an interdependent framework of adjacent airways and alveolar parenchyma. Destruction of the interdependent network of tissues would result in an increase in flaccidity of the airway tissues, while damage that led to scarring of these tissues would result in stiffening. Thus, measurements of peripheral airway function should be revealing. The challenge is to devise and interpret testing procedures that target this so-called "silent" zone of the lung where air moves at a relatively low flow rate due to the cross-sectional area of the lung.

The distinction of lesions as obstructive versus restrictive arises from studies of diseased lungs that yield rather distinctive patterns of dysfunction in tests that theoretically isolate the distal lung (Table 6). Restrictive dysfunctions would include diseases in which changes in lung parenchyma or pleura interfere or "restrict" the normal expansion of the lung (i.e., resulting in reduced compliance or enhanced "stiffness" of the tissue). In contrast, the obstructive dysfunctions encompass diseased lungs with lesions that impede or "obstruct" normal airflow through the airways. Figure 8 represents a composite summary of selected published data derived from conventional pulmonary function tests performed in two rodent species and in humans with clinically defined or, in the case of the rodents, induced obstructive or restrictive lung disease. Although these are of more qualitative than quantitative value, the figure illustrates that rodents can serve as reasonable models of chronic lung disease, bearing quite analogous functional manifestations of the respective disease entities in all three species.

The following sections attempt to characterize the functional changes that occur when relatively "pure" lesions are present; however, in real disease, animal models of disease, or in toxicant-induced lung injury, this often is

TABLE 6 Typical Functional Deficits in Obstructive and Restrictive Diseases

Parameter	Obstructive Type 1[a]	Obstructive Type 2[b]	Restrictive
$\dot{V}_E$	↓	↓	N/↓
VC	↓/N	N[c]/↓	↓
FRC	↑	↑	↓
RV	↑	↑	↓/N
TLC	↑/N	↑	↓
Compliance	↓	↑	↓
R	↑	↑	N
FEV_1	↓	↓	N/↑
FVC	N/↓	↓	↓
FEV_1/FVC	↓	↓	↑
MEFR	↓	↓	N/↓
DL_{CO}	↓	↓	↓
N_2 washout	↑	↑	N/↑

$\dot{V}_E$ = minute ventilation; VC = vital capacity; FRC = functional residual capacity; TLC = total lung capacity; C_{rs} = total respiratory system compliance; R_{aw} = airway resistance: FEV_1 = forced expiratory volume in 1 s; FVC = forced vital capacity; MEFR = maximum expiratory flow rate; DL_{CO} = carbon monoxide diffusing capacity; N_2 = nitrogen washout.
[a]*asthma-like lesion involving bronchoconstriction, airway inflammation, and excess mucus production.*
[b]*emphysemic-like lesion with loss of alveolar septum.*
[c]*N = may not change with disease.*

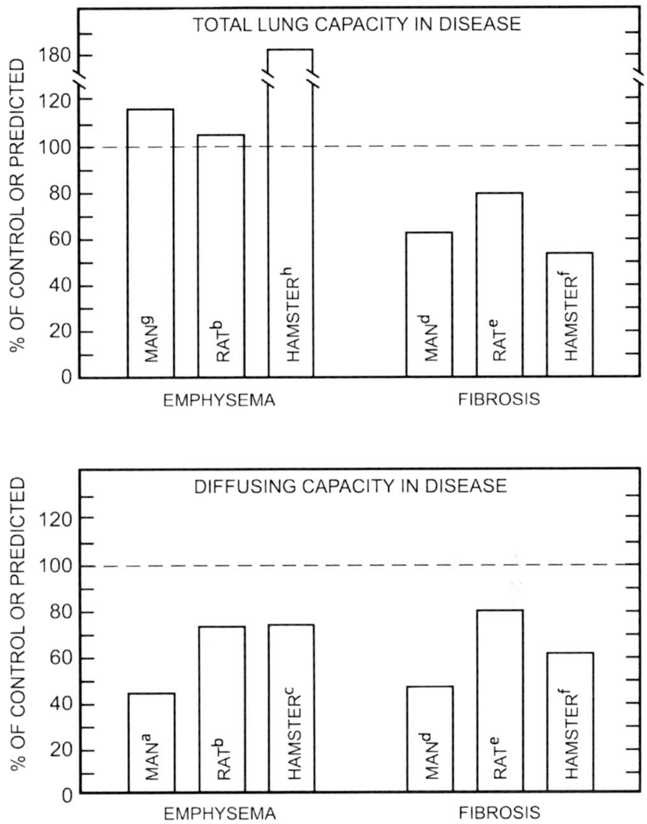

FIGURE 8 Comparison of TLC and DL_{CO} in emphysema and fibrosis models in rats and hamsters with the corresponding disease in humans. *(a) Flick and Block (1977); (b) Damon et al. (1982); (c) Raub et al. (1982); (d) Crystal et al. (1976); (e) Costa, unpublished data on CdCl2; (f) Snider et al. (1978); (g) Bates et al. (1971); (h) Lucey (1983).*

not the case, and thus, the prototypic changes are not absolute. For example, ventilatory pattern is normally centrally controlled in an effort to maintain normal blood-gas values with the least expenditure of energy (Mead, 1961). In the toxicant-injured lung, resultant lesions may be distributed in both the airways and deep lung, and thus changes in ventilation may reflect functional elements of both obstructive and restrictive disease. Because of this complexity, ventilatory patterns alone may provide little discriminatory information. Nevertheless, techniques have been reported useful to distinguish obstructive and restrictive lung dysfunction by assessing the pattern or phase angle of tidal-flow volume or flow—flow plots (Johanson and Pierce, 1971; Pennock et al., 1979; Agrawal, 1981; Wong and Alarie, 1982; Hamelmann et al., 1997).

Animal models of lung disease can often be designed to have relatively pure lesions, but like humans, when induced by some exogenous agent they may show analogous paradoxical manifestations of impairment. In these complex lung disease models, multiple lesions in central and/or peripheral regions coexist (Costa et al., 1983a, 1986). Hence, there may appear a "predominant" dysfunction, but interpretation of the underlying structural lesion(s) may not be clearly distinguishable. For example, intratracheal instillation of materials, such as silica dust (Gross et al., 1984; Raub et al., 1985), fly ash, or volcanic ash (Raub et al., 1985), may result in considerable airway occlusion due to the focal nature of the dust distribution. In each case, FRC increased in these rats. However, the DL_{CO} values decreased with intratracheal instillation of silica where the parenchyma was severely diseased but increased with mechanical instillation of a bolus of fly ash where the alveolar regions were not substantially damaged but were hyperinflated due to the airway obstruction. This response contrasts with tissue damage due to inhaled toxicants such as silica where damage is more likely to be to peripheral tissue rather than airways, which in the latter case obstructs airflow. Such damage in the rat appears to stimulate compensatory lung hypertrophy to maintain diffusion capacity. As such, FRC is not increased following inhalation damage, and the fibrotic changes in the parenchyma scar restrict lung tissue, which may over time decrease FRC and DL_{CO} (Kutzman, 1984). Such a paradoxical response could well mislead the interpretation if postmortem or other visualization of lung pathology is not carefully compared with the functional lesion. With this caveat in mind, the following is a discussion of the assessment of more pure type lung lesions and the functional alterations associated with that lesion.

Restrictive Lesions

Chronic restrictive diseases, where the lung is functionally stiffer due to fibrosis/scarring, are generally characterized by inadequate ventilation as a result of the loss of functional alveolar-capillary membrane, reduced static, and forced lung volume and capacity, and a stiffer less compliant lung. In severe fibrotic lesions, V_T and Pa_{O_2} may be decreased while f and Pa_{CO_2} increase. This tachypneic pattern disproportionately ventilates the conducting airways and proximal alveoli at the expense of dependent regional alveoli, thereby decreasing effective gas exchange (apart from any diffusion abnormality that also is likely present due to the thickening alveolar—blood barrier). Furthermore, since air is being diverted to the less-damaged independent (upper) region of the lung while gravity is harboring blood flow in the lung bases, there may arise mismatches in ventilation perfusion so blood does not get adequately oxygenated. Due to this ventilation-perfusion mismatch, chemoreceptors in the brain and aorta can trigger the above-noted changes in breathing pattern. To maintain adequate alveolar ventilation and normal blood-gas values, minute ventilation may increase resulting in increased energy expenditure just to achieve normal values of gas exchange. Examples of this restrictive tachypnea in laboratory animals have been observed in radiation pneumonitis (Travis et al., 1979, 1983; Mauderly et al., 1980), cadmium-induced fibrosis (Oberdorster and Hochrainer, 1980), and with paraquat exposure (Lam et al., 1980; Vizek et al., 1975; Burleigh-Flayer and Alarie, 1987). In general, alterations in ventilation are not sufficient, and more extensive characterization of lung function is needed to differentiate regional structural changes within the lung.

A number of conditions may involve restrictive parenchymal injury sufficient to induce alterations in static and forced lung volumes, diffusion, homogeneity, and the VP curve (Table 6). Among these are radiation poisoning, acute respiratory distress syndrome (ARDS), hypersensitivity pneumonitis, drug reactions (amiodarone, bleomycin, and methotrexate), as well as toxicant exposures (mineral dusts, high level O_3, or hyperoxia). All these challenges lead to fluid infiltration or over the longer term result in the accumulation of lung collagen, both of which restrict lung expansion. This scarring of lung tissue usually results in a loss of TLC, and in severe cases, RV will also be reduced. With fibrosis and the reduction in lung volumes, compliance is reduced, as can be seen from a shift downward and to the right from the normal VP curve (Gross et al., 1984; Raub et al., 1985; Hatch et al., 1984). Lung restriction can occur also as a result of interstitial edema or inflammation induced by single or repeated exposures to irritant gases, but such defects may resolve with time, depending on the extent of tissue injury (Stevens et al., 1988; Gross and White, 1985). Changes in lung volume and its apportionment typically coexist with shifts in the VP curve (Table 6). Analysis of the hysteresis of the VP curve during inflation and deflation has been used to recognize underlying changes in tissue elasticity in fibrotic hamsters (Fedullo et al., 1980) as well as damage to the functional integrity of lung lining fluid during acute lung injury (Proquitte et al., 2013), although as a rule this technique has not been widely applied in assessing the functional implications of experimental lung injuries. Lung fibrosis clinically is defined as moderate when VC or FVC fall below 80%; as disease progresses, most lung volumes decrease due to the retraction of scar tissue. Surprisingly, the smaller lung can sometimes have a forced peak expiratory flow that exceeds normal since the lung is so stiff its retraction actually pushes air out more rapidly at higher lung volumes (effort dependent). This apparent increase in function is misleading in that the lung overall is quite diseased and dysfunctional.

Obstructive Lesions

In contrast to restrictive type lesions, obstructive lesions alter breathing by virtue of airway narrowing. In general, there are two causes of lung obstruction. In the first type, intrinsic factors such as airway narrowing due to disease (loss of diameter due to physical pathology) or airway smooth muscle contraction (bronchoconstriction) cause the reduction in airflow. In the second type, extrinsic factors that overcome lung interdependency result in dynamic airway compression (e.g., emphysema). Intrinsic obstructions (type 1), such as those produced by an inflammatory response, cellular debris, mucus hypersecretion, and/or edema, all reduce the radius of the airway, thereby reducing flow. In contrast, bronchoconstriction narrows the airways due to muscle contraction and increased resistance to flow. Reduction of flow in the large airways during tidal breathing generally results in a compensatory increase in f as PaO_2 decreases. In severe disease, no amount of compensated breathing can overcome the hypoxemia due to the narrowing of the airways. However, such ventilatory changes are not diagnostic, as f can also be increased in restrictive-type lesions since less work is required to increase f than to stretch noncompliant lung tissue. With more peripheral airway obstruction, ventilation may increase as a means to get more air into the distal lung. Patients with peripheral obstruction also might breathe at a higher lung volume so that airways are "held" open by residual interdependency, which stretches the airways and thereby reduces resistance. Similarly, pulmonary irritants may produce a similar ventilatory pattern as type 1 obstruction due to diffuse peripheral congestion that stimulates deep lung C fibers. It has been suggested that this pulmonary congestion may be the common stimulus that initiates the pulmonary irritant and the obstructive tachypneic response, rather than direct stimulation of the irritant on nerve endings (Lee et al., 1979; Paintal, 1969; Schelegle and Walby, 2012).

Reduction in airway diameter resulting from smooth muscle contraction is typically seen in acute, reversible asthma and with the inhaled exposure to compounds that cause acute bronchoconstriction. As previously discussed, severe bronchoconstriction can cause a decrease in f and V_T that may progress to apnea. If the stimulus is removed and bronchoconstriction resolves, f and V_T may increase to repay the oxygen debt. Thus, there is considerable variability in the ventilatory pattern with obstructive-type lesions, and therefore the identification of obstruction based solely on changes in ventilatory pattern alone is controvertible (Section Enhanced Pause).

Both type 1 and type 2 obstructive diseases of the lung are reflected in enlarged volumes, primarily due to trapped air behind narrowed/closed small airways or in collapsed alveolar pockets within the lung parenchyma. Thus, TLC may be enlarged because RV is occupying a disproportionately larger share of the lung volume. Other lung volumes may change but not to the extent of RV. Shifts in the VP curve (Table 6) will result from these changes in lung volume and its apportionment. Unlike the downward and right shifts in the VP curve characteristic of fibrosis, the direction of shift in compliance is dependent on the underlying etiology of the lesion (type 1 vs 2). For type 1 lesions for the reasons discussed above, compliance is decreased because the lung is functionally stiffer (trapped air/edema).

In contrast, when degenerative alterations in the lung matrix occur, such as in the lesions responsible for emphysema, the lung becomes more compliant and the VP curve shifts upward and to the left. This increased expandability of the lung is thought to be the result of degenerative tissue changes, most notably in its elastic components, with loss of interdependency. Intratracheally instilled elastase, which enzymatically degrades these fibers, has been shown to produce this functional lesion in several species (Snider et al., 1986). Again, hysteresis of the VP curve during inflation and deflation has been shown to be useful in evaluating changes in tissue elasticity in emphysematous hamster lungs (Fedullo et al., 1980). Subchronic exposure to O_3 (Bartlett et al., 1974) and NO_2 (Stevens et al., 1988) in young rats, as well as instilled cadmium chloride in lathyrogen-fed hamsters (Niewoehner and Hoidal, 1982), also result in an increase in lung compliance consistent with emphysema-like disease.

As with type 1 obstructive lesions, type 2 lesions yield increases in lung volumes. However, as opposed to the narrowing of airways preventing air in the peripheral lung from moving out, lung volume is increased as a result of the disintegrating, highly compliant alveolar acini with a resultant loss of structural interdependency and untethering of airways. These collapsed airways trap air behind them. Homogeneity of ventilation is similarly disrupted due to the heterogeneity of damage and untethering, while diffusion capacity is reduced due mainly to loss of lung surface area.

Obstruction affects the MEFV maneuver and is the primary technique used to diagnose obstruction in humans. In the MEFV maneuver, once peak flow has been reached and the lung continues to empty under maximal effort, a condition of "effort independence" is achieved. This means that regardless of increased effort, the airways collapse as a result of elastic recoil and loss of tissue interdependence, which ultimately limits flow. In emphysema-like degeneration, obstruction to airflow due to airway flaccidity results in premature compression. Figure 9 displays MEFV curves for human subjects, rats, and hamsters with or without emphysema and indicates the analogous involvement of the small airways in all three species. Often flow is normalized for FVC or TLC to minimize the effect of lung volume. In emphysema-like diseases, which are characterized by enlarged lung volumes due to loss of elastic recoil, this normalization process will enhance the detection of minimal dysfunction.

Airflow obstruction can also be caused by intraluminal blockage or intrinsic narrowing of airways, such as might occur in uncomplicated bronchiolitis or asthma (type 1). Intrinsic obstruction or narrowing of airways should

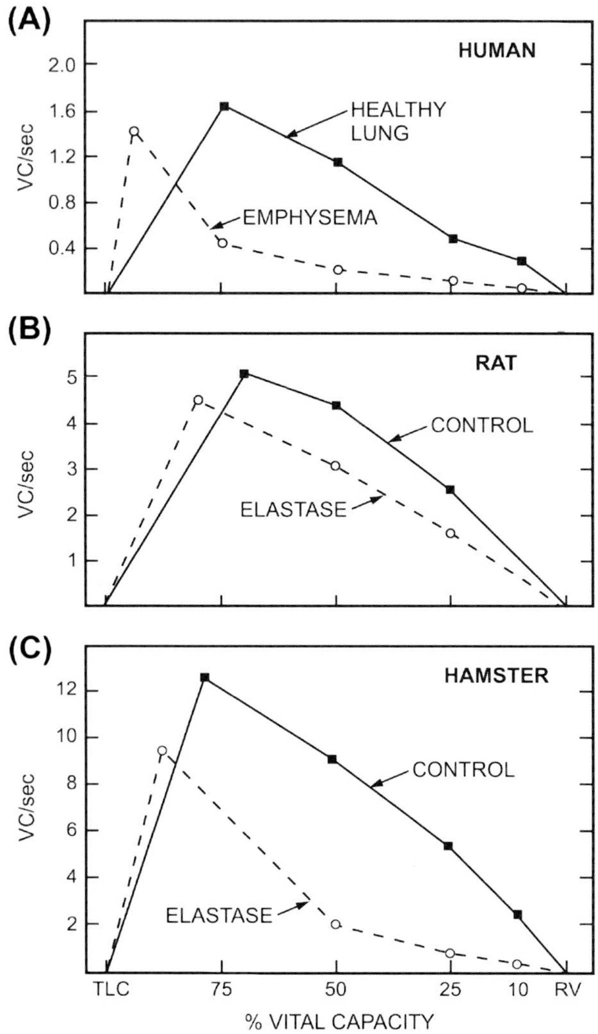

FIGURE 9 Flow–volume curves in airway obstruction. Comparison of flow–volume curves from normal and emphysematous (A) humans (Bates et al., 1971); (B) rats (Damon et al., 1982); and (C) hamsters (Lucey, 1983). Note the qualitative similarity for the normal and diseased lungs for all species. Effort independence of flow is achieved in each case (see text).

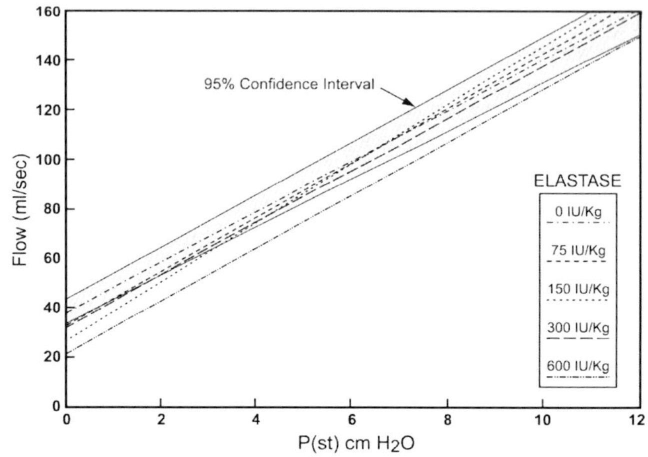

FIGURE 10 Maximum-flow static–recoil curves for graded emphysema states in the rat (unpublished data). Despite the severity of the disease, the functional lesion is loss of elastic recoil and not intraluminal obstruction. Note the lack of slope difference (representing upstream airway resistance) among the treatment doses.

theoretically be distinguishable from airway compression (type 2 obstruction) by creating a plot of resistance using effort-independent airflow versus elastic recoil pressure from isovolume comparisons of the VP and MEFV curves (Diamond and O'Donnell, 1977). In one example of a "pure" lesion, such as elastase-induced emphysema in the rat, which has no apparent airway component, plots of these maximum-flow static–recoil curves (Figure 10) over a wide range of elastase doses and functional impairment show no evidence of intrinsically altered upstream (peripheral) airway resistance (i.e., the slopes are similar). The forced expiratory decrement is thus a simple function of diminished elastic recoil due to the loss of alveolar tissue. However, in real world disease or subsequent to toxicant exposure, the functional changes observed are rarely the result of a single pathological process, and, therefore, such comparisons do not always clearly distinguish lesion types.

6.5 Lesions That Alter Diffusion

Measurements of DL_{CO} would appear to be a direct measurement of the prime function of the lung. In general, this interpretation is correct. Anything that impairs this process stands in the way of ensuring adequate exchange of respiratory gases. A decreased DL_{CO} can indicate a thickening or disturbance of the alveolar epithelial–endothelial barrier. Loss of alveolar surface area (common in emphysema and late fibrosis) and mismatching of ventilation and perfusion (common in early fibrotic disease) are frequent interpretations of a reduced DL_{CO} especially if there is other corroborative functional evidence (e.g., impairment of the flow–volume curve). However, diminished capillary blood volume or red blood cell mass (anemia, chronic CO poisoning) can also reduce this diffusion index. Certain circumstances, such as hyperinflation or hypertrophy of the lung (Costa et al., 1984, 1983a; Raub et al., 1983; Stevens et al., 1988; Raub et al., 1985), polycythemia (Weinberger et al., 1980), or reduced pleural pressures with asthma (Keens et al., 1979) may actually increase the DL_{CO}. For this reason, DL_{CO} without complementary data on lung volume and hematology can be misleading (see Chapter 12).

Even with the complementary data noted above, the DL_{CO} can at times be one of the most difficult functional end points to interpret clinically. The difficulties arise because both restrictive disease, in which the blood–air barrier is thickened, and obstructive disease, in which there is an effective loss of alveolar surface area, can reduce the DL_{CO}. Conventionally, this problem is addressed by normalizing the DL_{CO} for lung volume, which typically increases the index in restrictive disease (decreased lung volume) and decreases the index in obstructive disease (increased lung volume). The result is that the diffusion impairment is accentuated in the obstructive disease, where loss of surface area is thought to be the essential cause of the deficiency. In restrictive disease, the decrease in DL_{CO} is somewhat compensated and relates better to the loss of lung volume. Because diffusion is thought to be less impaired by barrier thickness than loss of alveolar surface (West, 1978), obstructive disease shows somewhat greater impairment. In addition, early in the pathogenesis of restrictive lung diseases, there is often a shift in the distribution of ventilation due to the affected areas being less compliant. If the disease (as in silicosis) is in the independent basal regions of the lung where blood circulation is most prominent due to gravity, there may be ventilation-perfusion imbalance, which would yield a lower DL_{CO}.

In rodents, the difficulty in interpretation of the DL_{CO} may be compounded by an apparent ability of the rodent lung to compensate for diffusion inadequacies after injury that irreversibly damage lung parenchyma (Costa et al., 1985, 1983a; Raub et al., 1983, 1985). The maintenance of near-normal organ diffusion levels when decreased diffusion efficiency is expected appears to correlate with hypertrophic or hyperplastic lung changes that result in expanded lung surface area (Costa et al., 1985, 1983a). The driving force for this apparent "compensatory" response is unclear, but may relate to O_2 consumption demands on the lungs (Hugonnaud et al., 1977). Highland Indian tribes of the Peruvian Andes who are reared and sustained in a hypobaric environment (−0.5 atm) for life clearly show adaptations of lung volume (barrel chests) that maximize O_2 diffusion. Diffusion measurements obtained from these highlanders are significantly higher than normal DL_{CO} values, even when adjusted for increased hemoglobin, which itself may affect the DL_{CO} (Clegg et al., 1970). Rats maintained under similar hypobaric conditions exhibit the same increases in lung volumes and DL_{CO} (Costa et al., 1984; Cunningham et al., 1974) with clear evidence of compensatory lung growth and expanded lung surface areas. However, the analogy of hypoxia-induced lung growth in humans and rats is not clearly established for toxicant-induced responses where injury may not be so dramatic. The rat, which is quite capable of regenerating lung tissue after pneumonectomy, may respond uniquely to severe toxicant injury. There is some evidence that catecholamines may play a role in lung hypertrophy in the rat (Rassler et al., 2012), a response not observed in humans. For example, in spite of the qualitative similarity of the airway lesions in humans and rats surviving an acute exposure to methyl isocyanate (MIC), only rats clearly exhibited enlarged DL_{CO} values after an extended "recovery" period (Stevens et al., 1988). In contrast, humans showed a progressive loss of diffusion, with resultant disability or death after MIC poisoning (Kamat et al., 1985). Humans who undergo pneumonectomy, especially at young ages, on the other hand, have remarkable recovery with near normal lung volume and related function values 30 years after the procedure (Laros and Westermann, 1987). These ostensibly conflicting phenomena are not well understood but warn of a potential pitfall in the interpretation of DL_{CO} in rodents after some forms of severe lung injury with a standard human clinical perspective.

7. CONCLUSION

Pulmonary function tests are tools available to the researcher and clinician to evaluate the ability of the lung to perform its essential function of gas exchange. To meet this principal function, the lung needs to operate efficiently with minimal mechanical work, as well as to provide an optimally minimal barrier to the diffusion of oxygen. Tests of lung function permit the nondestructive assessment of a continuum of lung performance, a continuum that may be influenced temporarily or permanently by structural and nonstructural alterations. In theory, any function test conducted in human subjects can be applied to animals and largely interpreted in the same manner. Fortunately, those physiologic principles that form the basis for clinical assessments of function seem to apply to animals in a quantitative as well as qualitative manner. Nevertheless, being aware of species differences and technical pitfalls is critical to the ultimate interpretation of the tests, especially when comparing them to their human analogues in health and disease. Although, in general, no single test is indicative of a specific lesion, an appropriate grouping of tests can be skillfully combined such that most lung physiological dysfunctions can be delineated. Thus, care and expertise in test and species selection, measurement technique, and interpretation are important considerations. It is our hope that the discussion of these factors in this chapter provides guidance to lung biology researchers investigating normal and diseased lungs as well as those investigators providing safety assessments of chemical, environmental, and pharmaceutical exposure.

REFERENCES

Abraham, W.M., Sielczak, M.W., Wanner, A., Perruchoud, A.P., Blinder, L., Stevenson, J.S., Ahmed, A., Yerger, L.D., 1988. Cellular markers of inflammation in the airways of allergic sheep with and without allergen-induced late responses. Am. Rev. Respir. Dis. 138, 1565−1571.

Abraham, W.M., Stevenson, J.S., Garrido, R., 1989. A possible role for PAF in allergen-induced late responses: modification by a selective antagonist. J. Appl. Physiol. 66, 2351−2357.

Adler, A., Cieslewicz, G., Irvin, C.G., 2004. Unrestrained plethysmography is an unreliable measure of airway responsiveness in BALB/c and C57BL/6 mice. J. Appl. Physiol. (1985) 97, 286−292.

Advenier, C., Boissier, J.R., Ho, S., Mallard, B., Ruff, F., 1978. The effects of pentobarbitone and urethane on pulmonary airway resistance in guinea-pigs and their interactions with drugs. Br. J. Pharmacol. 64, 519–525.

Agrawal, K.P., 1981. Specific airways conductance in guinea pigs: normal values and histamine induced fall. Respir. Physiol. 43, 23–30.

Alarie, Y., 1966. Irritating properties of airborne materials to the upper respiratory tract. Arch. Environ. Health 13, 433–449.

Alarie, Y., 1973. Sensory irritation by airborne chemicals. CRC Crit. Rev. Toxicol. 2, 299–363.

Albertine, K.H., Wang, L., Watanabe, S., Marathe, G.K., Zimmerman, G.A., McIntyre, T.M., 2002. Temporal correlation of measurements of airway hyperresponsiveness in ovalbumin-sensitized mice. Am. J. Physiol. Lung Cell. Mol. Physiol. 283, L219–L233.

Alexander, D.J., Collins, C.J., Coombs, D.W., Gilkison, I.S., Hardy, C.J., Healey, G., Karantabias, G., Johnson, N., Karlsson, A., Kilgour, J.D., McDonald, P., 2008. Association of inhalation toxicologists (AIT) working party recommendation for standard delivered dose calculation and expression in non-clinical aerosol inhalation toxicology studies with pharmaceuticals. Inhal. Toxicol. 20, 1179–1189.

Amdur, M.O., Mead, J., 1958. Mechanics of respiration in unanesthetized guinea pigs. Am. J. Physiol. 192, 364–368.

Andersson, L.G., Ringqvist, I., Walker, A., 1988. Total lung capacity measured by body plethysmography and by the helium dilution method. A comparative study in different patient groups. Clin. Physiol. 8, 113–119.

Anger, W.K., 1984. Neurobehavioral testing of chemicals: impact on recommended standards. Neurobehav. Toxicol. Teratol. 6, 147–153.

Antunes, M.A., Rocco, P.R., 2011. Elastase-induced pulmonary emphysema: insights from experimental models. An. Acad. Bras. Cienc. 83, 1385–1396.

Asgharian, B., Price, O., McClellan, G., Corley, R., Einstein, D.R., Jacob, R.E., Harkema, J., Carey, S.A., Schelegle, E., Hyde, D., Kimbell, J.S., Miller, F.J., 2012. Development of a rhesus monkey lung geometry model and application to particle deposition in comparison to humans. Inhal. Toxicol. 24, 869–899.

ASTM, 1984. Standard Test Method for Estimating Sensory Irritancy of Airborne Chemicals. Designation: E981-84. American Society for Testing and Materials, Philadelphia.

Authier, S., Haefner, P., Fournier, S., Troncy, E., Moon, L.B., 2010. Combined cardiopulmonary assessments with implantable telemetry device in conscious freely moving cynomolgus monkeys. J. Pharmacol. Toxicol. Methods 62, 6–11.

Banerjee, E.R., Henderson Jr., W.R., 2012. Defining the molecular role of gp91phox in the immune manifestation of acute allergic asthma using a preclinical murine model. Clin. Mol. Allergy 10, 2.

Bar-Ilan, A., Marder, J., 1980. Acid base status in unanesthetized, unrestrained guinea pigs. Pflugers Arch. 384, 93–97.

Baroffio, M., Barisione, G., Crimi, E., Brusasco, V., 2009. Noninflammatory mechanisms of airway hyper-responsiveness in bronchial asthma: an overview. Ther. Adv. Respir. Dis. 3, 163–174.

Barret, A.M., 1971. The effects of some autonomic blocking agents on the heart rate of the anesthetized and pithed rats. Eur. J. Pharmacol. 15, 267–273.

Barrow, C.S., Steinhagen, W.H., 1982. Sensory irritation tolerance development to chlorine in F-344 rats following repeated inhalation. Toxicol. Appl. Pharmacol. 65, 383–389.

Bartlett Jr., D., Faulkner 2nd, C.S., Cook, K., 1974. Effect of chronic ozone exposure on lung elasticity in young rats. J. Appl. Physiol. 37, 92–96.

Bates, J.H., Lauzon, A.M., 2007. Parenchymal tethering, airway wall stiffness, and the dynamics of bronchoconstriction. J. Appl. Physiol. (1985) 102, 1912–1920.

Bates, J.H., Milic-Emili, J., 1993. Influence of the viscoelastic properties of the respiratory system on the energetically optimum breathing frequency. Ann. Biomed. Eng. 21, 489–499.

Bates, D.V., Macklem, P.T., Christie, R.V., 1971. Respiratory Function in Disease. W.B. Saunders, Toronto.

Bates, J.H., Thompson-Figueroa, J., Lundblad, L.K., Irvin, C.G., 2008. Unrestrained video-assisted plethysmography: a noninvasive method for assessment of lung mechanical function in small animals. J. Appl. Physiol. (1985) 104, 253–261.

Bates, J.H., Rincon, M., Irvin, C.G., 2009. Animal models of asthma. Am. J. Physiol. Lung Cell. Mol. Physiol. 297, L401–L410.

Belenkiy, S., Ivey, K.M., Batchinsky, A.I., Langer, T., Necsoiu, C., Baker, W., Salinas, J., Cancio, L.C., 2013. Noninvasive carbon dioxide monitoring in a porcine model of acute lung injury due to smoke inhalation and burns. Shock 39, 495–500.

Bellville, J.W., Escarraga, L.A., Wallenstein, S.L., Houde, R.W., Howland, W.S., 1960. Relative respiratory depresant effects of oxymorphone (numorphan) and morphine. Anesthesiology 21, 397–400.

Benichou, M., Lorino, A.M., Lorino, H., Macquin-Mavier, I., Istin, N., Harf, A., 1990. Influence of tidal volume on histamine-induced bronchoconstriction in guinea pigs. J. Appl. Physiol. 68, 1634–1639.

Berend, N., Feldsien, D., Cederbaums, D., Cherniack, R.M., 1985. Structure–function correlation of early stages of lung injury induced by intratracheal bleomycin in the rabbit. Am. Rev. Respir. Dis. 132, 582–589.

Berendt, R.F., 1968. The effect of physical and chemical restraint on selected respiratory parameters of *Macaca mulatta*. Lab. Anim. Care 18, 391–394.

Bergren, D.R., 2001. Chronic tobacco smoke exposure increases airway sensitivity to capsaicin in awake guinea pigs. J. Appl. Physiol. (1985) 90, 695–704.

Berndt, A., Leme, A.S., Williams, L.K., Von Smith, R., Savage, H.S., Stearns, T.M., Tsaih, S.W., Shapiro, S.D., Peters, L.L., Paigen, B., Svenson, K.L., 2011. Comparison of unrestrained plethysmography and forced oscillation for identifying genetic variability of airway responsiveness in inbred mice. Physiol. Genomics 43, 1–11.

Bethel, R.A., McClure, C.L., 1990. Differential effect of ozone on the response of canine airways to nebulized and parenteral acetylcholine. Am. Rev. Respir. Dis. 141, A658.

Bide, R.W., Armour, S.J., Yee, E., 2000. Allometric respiration/body mass data for animals to be used for estimates of inhalation toxicity to young adult humans. J. Appl. Toxicol. 20, 273–290.

Bouhuys, A., Hunt, V.R., Kim, B.M., Zapletal, A., 1969. Maximum expiratory flow rates in induced bronchoconstriction in man. J. Clin. Invest. 48, 1159–1168.

Bouhuys, A., 1965. Distribution of inspired gas in the lungs. In: Handbook of Physiology-Respiration. American Physiological Society, Washington, D.C, pp. 715–733.

Buckley, L.A., Jiang, X.Z., James, R.A., Morgan, K.T., Barrow, C.S., 1984. Respiratory tract lesions induced by sensory irritants at the RD_{50} concentration. Toxicol. Appl. Pharmacol. 74, 417–429.

Burleigh-Flayer, H., Alarie, Y., 1987. Concentration-dependent respiratory response of guinea pigs to paraquat aerosol. Arch. Toxicol. 59, 391–396.

Burleigh-Flayer, H., Alarie, Y., 1988. Pulmonary effects of repeated exposures to paraquat aerosol in guinea pigs. Fundam. Appl. Toxicol. 10, 717–729.

Burt, M.E., Arbeit, J., Brennan, M.F., 1980. Chronic arterial and venous access in the unrestrained rat. Am. J. Physiol. 238, H599–H603.

Castile, R.G., Iram, D., McCoy, K.S., 2004. Gas trapping in normal infants and in infants with cystic fibrosis. Pediatr. Pulmonol. 37, 461–469.

Chang, J.C., Gross, E.A., Swenberg, J.A., Barrow, C.S., 1983. Nasal cavity deposition, histopathology, and cell proliferation after single or repeated formaldehyde exposures in B6C3F1 mice and F-344 rats. Toxicol. Appl. Pharmacol. 68, 161–176.

Chan-Yeung, M., Lam, S., 1986. Occupational asthma. Am. Rev. Respir. Dis. 133, 686–703.

Chapman, R.W., Danko, G., 1985. Hyperventilation-induced bronchoconstriction in guinea pigs. Int. Arch. Allergy Appl. Immunol. 78, 190–196.

Chong, B.T., Agrawal, D.K., Romero, F.A., Townley, R.G., 1998. Measurement of bronchoconstriction using whole-body plethysmograph: comparison of freely moving versus restrained guinea pigs. J. Pharmacol. Toxicol. Methods 39, 163–168.

Chvalova, M., Kuncova, M., Havrankova, J., Palecek, F., 1974. Regulation of respiration in experimental silicosis. Physiol Bohemoslov. 23, 539–547.

Clay, T.P., Thompson, M.A., 1985. Irritant induced cough as a model of intrapulmonary airway reactivity. Lung 163, 183–191.

Clegg, E.J., Harrison, G.A., Baker, P.T., 1970. The impact of high altitudes on human populations. Hum. Biol. 42, 486–518.

Clements, J.A., Tierney, D.F., 1964. Alveolar instability associated with altered surface tension. In: Rahn, W.O.F.a. H. (Ed.), Handbook of Physiology, Sect. 3, Respiration, vol. II. American Physiological Society, Washington, D.C, pp. 1565–1583.

Costa, D.L., Tepper, J.S., 1988. Approaches to lung function assessment in small animals. In: Gardner, D.E., Crapo, J.D., Massaro, E.J. (Eds.), Toxicology of the Lung. Raven Press, New York.

Costa, D.L., Lehmann, J.R., Slatkin, D.N., Popenoe, E.A., Drew, R.T., 1983a. Chronic airway obstruction and bronchiectasis in the rat after intratracheal bleomycin. Lung 161, 287–300.

Costa, D.L., Kutzman, R.S., Lehmann, J.R., Popenoe, E.A., Drew, R.T., 1983b. A subchronic multidose ozone study in rats. Adv. Mod. Environ. Toxicol. 28, 369–393.

Costa, D.L., Lehmann, J.R., Osheroff, M.R., Bergofsky, E.H., 1984. Lung enlargement in rats after chronic hypoxia. Am. Rev. Respir. Dis. 129, 211.

Costa, D.L., Kutzman, R.S., Drew, R.T., 1985. Relationship between lung function and composition in rats subchronically exposed to chlorine. Toxicologist 5, 280.

Costa, D.L., Kutzman, R.S., Lehmann, J.R., Drew, R.T., 1986. Altered lung function and structure in the rat after subchronic exposure to acrolein. Am. Rev. Respir. Dis. 133, 286–291.

Costa, D.L., Tepper, J.S., Oct. 1991. Methods for lung function assessment in small mammals. 10th Annual Meeting of the Comparative Respiratory Society, Lansing, MI.

Costa, D.L., Tepper, J.S., Stevens, M.A., Watkinson, W.P., Doerfler, D.L., Gelzleichter, T.R., Last, J.A., 1995. Restrictive lung disease in rats exposed chronically to an urban profile of ozone. Am. J. Respir. Crit. Care Med. 151, 1512–1518.

Costa, D.L., 1985. Interpretation of new techniques used in the determination of pulmonary function in rodents. Fundam. Appl. Toxicol. 5, 423–434.

Crossfill, M.L., Widdicombe, J.G., 1961. Physiological characteristics of the chest and lungs and the work of breathing in different mammalian species. J. Physiol. 158, 1–14.

Crystal, R.G., Fulmer, J.D., Roberts, W.C., Moss, M.L., Line, B.R., Reynolds, H.Y., 1976. Idiopathic pulmonary fibrosis. Clinical, histologic, radiographic, physiologic, scintigraphic, cytologic, and biochemical aspects. Ann. Intern. Med. 85, 769–788.

Cunningham, E.L., Brody, J.S., Jain, B.P., 1974. Lung growth induced by hypoxia. J. Appl. Physiol. 37, 362–366.

Damon, E.G., Mauderly, J.L., Jones, R.K., 1982. Early effects of intratracheal instillation of elastase on mortality, respiratory function, and pulmonary morphometry of F-344 rats. Toxicol. Appl. Pharmacol. 64, 465–475.

Davidson, J.T., Wasserman, K., Lillington, G.A., Schmidt, R.W., 1966. Effect of aging on respiratory mechanics and gas exchange in rabbits. J. Appl. Physiol. 21 (3), 837–842.

Dennis, M.W., Douglas, J.S., 1970. Control of bronchomotor tone in spontaneous breathing, aunanesthetized guinea pigs. In: Bouhuys, A. (Ed.), Airway Dynamics, pp. 253–262.

Diamond, L., O'Donnell, M., 1977. Pulmonary mechanics in normal rats. J. Appl. Physiol. 43, 942–948.

Dorato, M.A., Carlson, K.H., Copple, D.L., 1983. Pulmonary mechanics in conscious Fischer 344 rats: multiple evaluations using nonsurgical techniques. Toxicol. Appl. Pharmacol. 68, 344–353.

Dosman, J., Bode, F., Urbanetti, J., Martin, R., Macklem, P.T., 1975. The use of a helium–oxygen mixture during maximum expiratory flow to demonstrate obstruction in small airways in smokers. J. Clin. Invest. 55, 1090–1099.

Drorbaugh, J.E., Fenn, W.O., 1955. A barometric method for measuring ventilation in newborn infants. Pediatrics 16, 81–87.

Dubois, A.B., Brody, A.W., Lewis, D.H., Burgess Jr., B.F., 1956. Oscillation mechanics of lungs and chest in man. J. Appl. Physiol. 8, 587–594.

Dusser, D.J., Jacoby, D.B., Djokic, T.D., Rubinstein, I., Borson, D.B., Nadel, J.A., 1989. Virus induces airway hyperresponsiveness to tachykinins: role of neutral endopeptidase. J. Appl. Physiol. 67, 1504–1511.

Eidelman, D.H., Lei, M., Ghezzo, R.H., 1993. Morphometry of methacholine-induced bronchoconstriction in the rat. J. Appl. Physiol. (1985) 75, 1702–1710.

Ellakkani, M.A., Alarie, Y.C., Weyel, D.A., Mazumdar, S., Karol, M.H., 1984. Pulmonary reactions to inhaled cotton dust: an animal model for byssinosis. Toxicol. Appl. Pharmacol. 74, 267–284.

Epstein, M.A., Epstein, R.A., 1978. A theoretical analysis of the barometric method for measurement of tidal volume. Respir. Physiol. 32, 105–120.

Ewart, S., Levitt, R., Mitzner, W., 1995. Respiratory system mechanics in mice measured by end-inflation occlusion. J. Appl. Physiol. (1985) 79, 560–566.

Ewart, L.C., Haley, M., Bickerton, S., Bright, J., Elliott, K., McCarthy, A., Williams, L., Ricketts, S.A., Holland, T., Valentin, J.P., 2010. Pharmacological validation of a telemetric model for the measurement of bronchoconstriction in conscious rats. J. Pharmacol. Toxicol. Methods 61, 219–229.

Fallica, J., Das, S., Horton, M., Mitzner, W., 2011. Application of carbon monoxide diffusing capacity in the mouse lung. J. Appl. Physiol. (1985) 110, 1455–1459.

Fedullo, A.J., Jung-Legg, Y., Snider, G.L., Karlinsky, J.B., 1980. Hysteresis ratio: a measure of the mechanical efficiency of fibrotic and emphysematous hamster lung tissue. Am. Rev. Respir. Dis. 122, 47–52.

Fedulov, A., Kobzik, L., 2011. The Penh police are not infallible. Am. J. Respir. Cell Mol. Biol. 45, 1272–1273.

Feldman, H.A., McMahon, T.A., 1983. The 3/4 mass exponent for energy metabolism is not a statistical artifact. Respir. Physiol. 52, 149–163.

Ferguson, J.S., Alarie, Y., 1991. Long term pulmonary impairment following a single exposure to methyl isocyanate. Toxicol. Appl. Pharmacol. 107, 253–268.

Ferguson, J.S., Schaper, M., Stock, M.F., Weyel, D.A., Alarie, Y., 1986. Sensory and pulmonary irritation with exposure to methyl isocyanate. Toxicol. Appl. Pharmacol. 82, 329–335.

Ferguson, J.S., Schaper, M., Alarie, Y., 1987. Pulmonary effects of a polyisocyanate aerosol: hexamethylene diisocyanate trimer (HDIt) or Desmodur-N (DES-N). Toxicol. Appl. Pharmacol. 89, 332–346.

Flandre, T.D., Leroy, P.L., Desmecht, D.J., 2003. Effect of somatic growth, strain, and sex on double-chamber plethysmographic respiratory function values in healthy mice. J. Appl. Physiol. (1985) 94, 1129–1136.

Fleming, G.M., Chester, E.H., Saniie, J., Saidel, G.M., 1980. Ventilation inhomogeneity using multibreath nitrogen washout: comparison of moment ratios and other indexes. Am. Rev. Respir. Dis. 121, 789–794.

Flick, M.R., Block, A.J., 1977. Chronic oxygen therapy. Med. Clin. North Am. 61, 1397–1406.

Folinsbee, L.J., Raven, P.B., 1984. Exercise and air pollution. J. Sports Sci. 2, 57–75.

Frazer, D.G., Reynolds, J.S., Jackson, M.C., 2011. Determining when enhanced pause (Penh) is sensitive to changes in specific airway resistance. J. Toxicol. Environ. Health A 74, 287–295.

Freed, A.N., 1989. Regional and temporal variation in canine peripheral lung responses to dry air. J. Appl. Physiol. 67, 1727–1733.

Freeman, G., Crane, S.C., Furiosi, N.J., Stephens, R.J., Evans, M.J., Moore, W.D., 1972. Covert reduction in ventilatory surface in rats during prolonged exposure to subacute nitrogen dioxide. Am. Rev. Respir. Dis. 106, 563–579.

Gasser, J.C., Bellville, J.W., 1976. The respiratory effects of hydroxyzine, diazepam and pentazocine in man. Anaesthesia 31, 718–723.

Glaab, T., Taube, C., Braun, A., Mitzner, W., 2007. Invasive and noninvasive methods for studying pulmonary function in mice. Respir. Res. 8, 63.

Gold, P.M., 1982. Single breath nitrogen test: closing volume and distribution of ventilation. In: Clausen, J.L. (Ed.), Pulmonary Function Testing: Guidelines and Controversies. Academic Press, New York, pp. 105–114.

Gordon, T., Venugopalan, C.S., Amdur, M.O., Drazen, J.M., 1984. Ozone-induced airway hyperreactivity in the guinea pig. J. Appl. Physiol. 57, 1034–1038.

Grosenbaugh, D.A., Alben, J.O., Muir, W.W., 1997. Absorbance spectra of inter-species hemoglobins in the visible and near infrared regions. J. Vet. Emerg. Crit. Care 7, 36–42.

Gross, K.G., White, H.J., 1985. Functional and pathological consequences of a 52-week exposure to 0.5 ppm ozone followed by a clean air recovery period. Am. Rev. Respir. Dis. 131, 183.

Gross, K.B., White, H.J., Smiler, K.L., 1984. Functional and morphologic changes in the lungs after a single intratracheal instillation of silica. Am. Res. Respir. Dis. 129, 833–839.

Guidance for industry, 2001. S7A Safety Pharmacology Studies for Human Pharmaceuticals. Corp Author(s): Center for Drug Evaluation and Research (U.S.) ; Center for Biologics Evaluation and Research (U.S.) Conf Author(s): International Conference on Harmonisation, U.S. Dept. of Health and Human Services Food and Drug Administration Center for Drug Evaluation and Research : Center for Biologics Evaluation and Research, Rockville, MD.

Guyton, A.C., 1947. Measurement of the respiratory volumes of laboratory animals. Am. J. Physiol. 150, 70–77.

Halloy, D.J., Kirschvink, N.A., Vincke, G.L., Hamoir, J.N., Delvaux, F.H., Gustin, P.G., 2004. Whole body barometric plethysmography: a screening method to investigate airway reactivity and acute lung injuries in freely moving pigs. Vet. J. 168, 276–284.

Hamakawa, H., Bartolak-Suki, E., Parameswaran, H., Majumdar, A., Lutchen, K.R., Suki, B., 2011. Structure–function relations in an elastase-induced mouse model of emphysema. Am. J. Respir. Cell. Mol. Biol. 45, 517–524.

Hamelmann, E., Schwarze, J., Takeda, K., Oshiba, A., Larsen, G.L., Irvin, C.G., Gelfand, E.W., 1997. Noninvasive measurement of airway responsiveness in allergic mice using barometric plethysmography. Am. J. Respir. Crit. Care Med. 156, 766–775.

Hantos, Z., Petak, F., Adamicza, A., Daroczy, B., Fredberg, J.J., 1995. Differential responses of global airway, terminal airway, and tissue impedances to histamine. J. Appl. Physiol. (1985) 79, 1440–1448.

Harbison, M.L., Brain, J.D., 1983. Effects of exercise on particle deposition in Syrian golden hamsters. Am. Rev. Respir. Dis. 128, 904–908.

Harkema, J.R., Mauderly, J.L., Hahn, F.F., 1982. The effects of emphysema on oxygen toxicity in rats. Am. Rev. Respir. Dis. 126, 1058–1065.

Harkema, J.R., Mauderly, J.L., Gregory, R.E., Pickrell, J.A., 1984. A comparison of starvation and elastase models of emphysema in the rat. Am. Rev. Respir. Dis. 129 (4), 584–591.

Harris, R.S., 2005. Pressure-volume curves of the respiratory system. Respir. Care 50, 78–98; discussion 98–99.

Hatch, G.E., Raub, J.A., Graham, J.A., 1984. Functional and biochemical indicators of pneumoconiosis in mice: comparison with rats. J. Toxicol. Environ. Health 13, 487–497.

Hayatdavoudi, G., O'Neil, J.J., Barry, B.E., Freeman, B.A., Crapo, J.D., 1981. Pulmonary injury in rats following continuous exposure to 60% O_2 for 7 days. J. Appl. Physiol. 51, 1220–1231.

Hessel, E.M., Zwart, A., Oostveen, E., Van Oosterhout, A.J., Blyth, D.I., Nijkamp, F.P., 1995. Repeated measurement of respiratory function and bronchoconstriction in unanesthetized mice. J. Appl. Physiol. (1985) 79, 1711–1716.

Heusner, A.A., 1982a. Energy metabolism and body size. I. Is the 0.75 mass exponent of Kleiber's equation a statistical artifact? Respir. Physiol. 48, 1–12.

Heusner, A.A., 1982b. Energy metabolism and body size. II. Dimensional analysis and energetic non-similarity. Respir. Physiol. 48, 13–25.

Heusner, A.A., 1983a. Body size, energy metabolism, and the lungs. J. Appl. Physiol. Respir. Environ. Exerc. Physiol. 54, 867–873.

Heusner, A.A., 1983b. Mathematical expression of the effects of changes in body size on pulmonary function and structure. Am. Rev. Respir.

Hirshman, C.A., McCullough, R.E., Cohen, P.J., Weil, J.V., 1975. Hypoxic ventilatory drive in dogs during thiopental, ketamine, or pentobarbital anesthesia. Anesthesiology 43, 628–634.

Hirt, R.A., Leinker, S., Mosing, M., Wiederstein, I., 2008. Comparison of barometric whole body plethysmography and its derived parameter enhanced pause (PENH) with conventional respiratory mechanics in healthy beagle dogs. Vet. J. 176, 232–239.

Holroyde, M.C., Smith, S.Y., Holme, G., 1980. Evaluation of pulmonary mechanics in guinea pigs during respiratory anaphylaxis. J. Pharmacol. Exp. Ther. 212, 162–166.

Holub, D., Frank, R., 1979. A system for rapid measurement of lung function in small animals. J. Appl. Physiol. Respir. Environ. Exerc. Physiol. 46 (2), 394–398.

Hom, G.J., Bach, T.J., Carroll, D., Forrest, M.J., Mariano, M.A., Trainor, C.E., Wang, P.R., MacIntyre, D.E., March 1999. Comparison of cardiovascular parameters and/or serum chemistry and hematology profiles in conscious and anesthetized rhesus monkeys (*Macaca mulatta*). Contemp. Top. Lab. Anim. Sci. 38 (2), 60–64. PubMed PMID: pmid:12086436.

Hoymann, H.G., 2012. Lung function measurements in rodents in safety pharmacology studies. Front. Pharmacol. 3, 156.

Hugonnaud, C., Gehr, P., Weibel, R., Burri, P.H., 1977. Adaptation of the growing lung to increased oxygen consumption. II. Morphometric analysis. Respir. Physiol. 29, 1–10.

Hunter, A.R., Pleuvry, B.J., Rees, J.M., 1968. The respiratory depressant effects of barbiturates and narcotic analgesics in the unanaesthetized rabbit. Br. J. Anaesth. 40, 927–935.

Jacky, J.P., 1980. Barometric measurement of tidal volume: effects of pattern and nasal temperature. J. Appl. Physiol. Respir. Environ. Exercise Physiol. 49, 319–325.

Jaeger, R.J., Gearhart, J.M., 1982. Respiratory and metabolic response of rats and mice to formalin vapor. Toxicology 25, 299–309.

Jaeger, R.J., Sussman, R.G., Gearhart, J.M., Bowers, A.B., Shore, R.E., Sellakumar, A.R., 1983. Repeated pulmonary function evaluation following bleomycin treatment. J. Toxicol. Environ. Health 11, 415–423.

Jammes, Y., Barthelemy, P., Fornaris, M., Grimaud, C., 1986. Cold-induced bronchospasm in normal and sensitized rabbits. Respir. Physiol. 63, 347–360.

Johanson Jr., W.G., Pierce, A.K., 1971. A noninvasive technique for measurement of airway conductance in small animals. J. Appl. Physiol. 30, 146–150.

Johanson Jr., W.G., Pierce, A.K., 1973. Lung structure and function with age in normal rats and rats with papain emphysema. J. Clin. Invest. 52, 2921–2927.

Kamat, S.R., Mahashur, A.A., Tiwari, A.K., Potdar, P.V., Gaur, M., Kolhatkar, V.P., Vaidya, P., Parmar, D., Rupwate, R., Chatterjee, T.S., et al., 1985. Early observations on pulmonary changes and clinical morbidity due to the isocyanate gas leak at Bhopal. J. Postgrad. Med. 31, 63–72.

Kane, L.E., Alarie, Y., 1977. Sensory irritation to formaldehyde and acrolein during single and repeated exposures in mice. Am. Ind. Hyg. Assoc. J. 38, 509–522.

Karol, M.H., Stadler, J., Underhill, D., Alarie, Y., 1981. Monitoring delayed-onset pulmonary hypersensitivity in guinea pigs. Toxicol. Appl. Pharmacol. 61, 277–285.

Karol, M.H., 1983. Concentration-dependent immunologic response to toluene diisocyanate (TDI) following inhalation exposure. Toxicol. Appl. Pharmacol. 68, 229–241.

Kearney, K., Metea, M., Gleason, T., Edwards, T., Atterson, P., 2010. Evaluation of respiratory function in freely moving beagle dogs using implanted impedance technology. J. Pharmacol. Toxicol. Methods 62, 119–126.

Keens, T.G., Mansell, A., Krastins, I.R., Levison, H., Bryan, A.C., Hyland, R.H., Zamel, N., 1979. Evaluation of the single-breath diffusing capacity in asthma and cystic fibrosis. Chest 76, 41–44.

Kimmel, E.C., Whitehead, G.S., Reboulet, J.E., Carpenter, R.L., 2002. Carbon dioxide accumulation during small animal, whole body plethysmography: effects on ventilation, indices of airway function, and aerosol deposition. J. Aerosol Med. 15, 37–49.

Kirschvink, N., Reinhold, P., 2008. Use of alternative animals as asthma models. Curr. Drug Targets 9, 470–484.

Kirschvink, N., Vincke, G., Onclinx, C., Peck, M.J., Gustin, P., 2005. Comparison between pulmonary resistance and Penh in anaesthetised rats with tracheal diameter reduction and after carbachol inhalation. J. Pharmacol. Toxicol. Methods 51, 123–128.

Kleinman, M.T., Mautz, W.J., 1991. The effects of exercise on dose and dose distribution of inhaled automotive pollutants. Res. Rep. Health Eff. Inst. 41–50; discussion 1–40.

Konno, K., Mead, J., 1967. Measurement of the separate volume changes of rib cage and abdomen during breathing. J. Appl. Physiol. 22, 407–422.

Koo, K.W., Leith, D.E., Sherter, C.B., Snider, G.L., 1976. Respiratory mechanics in normal hamsters. J. Appl. Physiol. 40, 936–942.

Krane, C.M., Fortner, C.N., Hand, A.R., McGraw, D.W., Lorenz, J.N., Wert, S.E., Towne, J.E., Paul, R.J., Whitsett, J.A., Menon, A.G., 2001. Aquaporin 5-deficient mouse lungs are hyperresponsive to cholinergic stimulation. Proc. Natl. Acad. Sci. U.S.A. 98, 14114–14119.

Krishnamoorthy, V., Hiller, D.B., Ripper, R., Lin, B., Vogel, S.M., Feinstein, D.L., Oswald, S., Rothschild, L., Hensel, P., Rubinstein, I., Minshall, R., Weinberg, G.L., 2012. Epinephrine induces rapid deterioration in pulmonary oxygen exchange in intact, anesthetized rats: a flow and pulmonary capillary pressure-dependent phenomenon. Anesthesiology 117, 745–754.

Kutzman, R.S., 1984. A Study of Fischer 344 Rats Exposed to Silica Dust at Concentrations of 0, 2, 10 OR 20 MG/M3, then Maintained for Six Months Prior to Assessment. BNL–35735.

Kuwabara, Y., Alexeeff, G.V., Broadwin, R., Salmon, A.G., 2007. Evaluation and application of the RD_{50} for determining acceptable exposure levels of airborne sensory irritants for the general public. Environ. Health Perspect. 115, 1609–1616.

Lai, Y.L., Diamond, L., 1986. Comparison of five methods of analyzing respiratory pressure–volume curves. Respir. Physiol. 66, 147–155.

Lai, Y.L., Hildebrandt, J., 1978. Respiratory mechanics in the anesthetized rat. J. Appl. Physiol. 45, 255–260.

Lai, Y.L., Lee, K.C., Kuo, P.H., 2004. Detection of mediator-induced airway constriction by barometric plethysmography in mice. Chin. J. Physiol. 47, 161–167.

Lai, Y.L., 1988. Maximal expiratory flow in the guinea pig. Lung 166, 303–313.

Lam, H.F., Takezawa, J., Gupta, B.N., van Stee, E.W., 1980. A comparison of the effects of paraquat and diquat on lung compliance, lung volumes and single breath diffusing capacity in the rat. Toxicology 18, 111–123.

Lamm, W.J., Hildebrandt, J.R., Hildebrandt, J., Lai, Y.L., 1982. End-expiratory volume in the rat: effects of consciousness and body position. J. Appl. Physiol. 53, 1071–1079.

LaPrad, A.S., Lutchen, K.R., 2008. Respiratory impedance measurements for assessment of lung mechanics: focus on asthma. Respir. Physiol. Neurobiol. 163, 64–73.

Laros, C.D., Westermann, C.J., 1987. Dilatation, compensatory growth, or both after pneumonectomy during childhood and adolescence. A thirty-year follow-up study. J. Thorac. Cardiovasc. Surg. 93, 570–576.

Lee, L.Y., Dumont, C., Djokic, T.D., Menzel, T.E., Nadel, J.A., 1979. Mechanism of rapid, shallow breathing after ozone exposure in conscious dogs. J. Appl. Physiol. 46, 1108–1114.

Lee, E.J., Woodske, M.E., Zou, B., O'Donnell, C.P., 2009. Dynamic arterial blood gas analysis in conscious, unrestrained C57BL/6J mice during exposure to intermittent hypoxia. J. Appl. Physiol. (1985) 107, 290–294.

Leith, D.E., 1976. Comparative mammalian respiratory mechanics. Physiologist 19, 485–510.

Leith, D.E., 1983a. Comparative mammalian respiratory mechanics. Am. Rev. Respir. Dis. 128, S77–S82.

Leith, D.E., 1983b. Mammalian tracheal dimensions: scaling and physiology. J. Appl. Physiol. 55, 196–200.

Leith, D.E., 1984. Mass transport in mammalian lungs: comparative physiology. J. Toxicol. Environ. Health 13, 251–271.

Lewis, S.M., 1978. Emptying patterns of the lung studied by multiple-breath N_2 washout. J. Appl. Physiol. 44, 424–430.

Liah, F., Annette, R., Thomas, F.S., 2013. Raised-volume low frequency forced oscillation measurements of respiratory mechanics in conscious and anaesthetized mice. In: A60. Assessing Pulmonary Function: Airways, Mechanics, and Gas Exchange, vol. 276. American Thoracic Society, p. A1957.

Likens, S.A., Mauderly, J.L., 1982. Effect of elastase or histamine on single-breath N_2 washouts in the rat. J. Appl. Physiol. 52, 141–146.

Lin, C.C., Chang, C.F., Liaw, S.F., Lin, C.Y., 2002. Maximal forced expiratory maneuver to measure airway obstruction in allergen challenged mice. Respir. Physiol. Neurobiol. 130, 79–87.

Litwin, D.C., Lengel, D.J., Kamendi, H.W., Bialecki, R.A., 2011. An integrative pharmacological approach to radio telemetry and blood sampling in pharmaceutical drug discovery and safety assessment. Biomed. Eng. Online 10, 5.

Loscutoff, S.M., Cannon, W.C., Buschbom, R.L., Busch, R.H., Killand, B.W., 1985. Pulmonary function in elastase-treated guinea pigs and rats exposed to ammonium sulfate or ammonium nitrate aerosols. Environ. Res. 36, 170–180.

Lucey, E.C., Celli, B.R., Snider, G.L., 1978. Maximum expiratory flow and transpulmonary pressure in the hamster. J. Appl. Physiol. 45, 840–845.

Lucey, E.C., Snider, G.L., Javaheri, S., 1982. Pulmonary ventilation and blood gas values in emphysematous hamsters. Am. Rev. Respir. Dis. 125, 299–303.

Lucey, E.C., 1983. Experimental emphysema. Clin. Chest Med. 4, 389–403.

Lundblad, L.K., Irvin, C.G., Adler, A., Bates, J.H., 2002. A reevaluation of the validity of unrestrained plethysmography in mice. J. Appl. Physiol. (1985) 93, 1198–1207.

Macklem, P.T., Mead, J., 1967. Resistance of central and peripheral airways measured by a retrograde catheter. J. Appl. Physiol. 22, 395–401.

Macklem, P.T., Permutt, S., 1979. The Lung in the Transition between Health and Disease. Marcel Dekker, New York.

Mansson, S., Wolber, J., Driehuys, B., Wollmer, P., Golman, K., 2003. Characterization of diffusing capacity and perfusion of the rat lung in a lipopolysaccaride disease model using hyperpolarized 129Xe. Magn. Reson. Med. 50, 1170–1179.

Martin-Body, R.L., Sinclair, J.D., 1985. Analysis of respiratory patterns in the awake and in the halothane anaesthetised rat. Respir. Physiol. 61, 105–113.

Masuda, H., Ogata, T., Kikuchi, K., 1989. Physiological changes during temporary occlusion of the superior vena cava in cynomolgus monkeys. Ann. Thorac. Surg. 47, 890–896.

Matijak-Schaper, M., Wong, K.L., Alarie, Y., 1983. A method to rapidly evaluate the acute pulmonary effects of aerosols in unanesthetized guinea pigs. Toxicol. Appl. Pharmacol. 69, 451–460.

Mauderly, J.L., Muggenburg, B.A., Hahn, F.F., Boecker, B.B., 1980. The effects of inhaled 144Ce on cardiopulmonary function and histopathology of the dog. Radiat. Res. 84, 307–324.

Mauderly, J.L., Gillet, N.A., Henderson, R.F., Jones, R.K., McClellan, R.O., 1985. Relationships of lung structure and functional changes to accumulation of diesel exhaust particles. In: Proceedings of Sixth International Symposium on Inhaled Particles, Cambridge, U.K, pp. 659–669.

Mauderly, J.L., 1977. A new technique for evaluating nitrogen washout efficiency. Am. J. Vet. Res. 38, 69–74.

Mauderly, J.L., 1979. Ventilation, lung volumes and lung mechanics of young adult and old Syrian hamsters. Exp. Aging Res. 5, 497–508.

Mauderly, J.L., 1984. Respiratory function responses of animals and man to oxidant gases and to pulmonary emphysema. J. Toxicol. Environ. Health 13, 345–361.

Mauderly, J.L., 1986. Respiration of F344 rats in nose-only inhalation exposure tubes. J. Appl. Toxicol. 6, 25–30.

Mauderly, J.L., 1988. Comparison of respiratory function responses of laboratory animals and humans. In: Mohr, U. (Ed.), Inhalation Toxicology: The Design and Interpretation of Inhalation Studies and Their Use in Risk Assessment. Springer Verlag, New York, pp. 243–262.

Mauderly, J.L., 1989. Effect of inhaled toxicants on pulmonary function. In: McClellan, R.O., Henderson, R.F. (Eds.), Concepts in Inhalation Toxicology. Hemisphere Publishing Co., New York, pp. 347–462.

Mauderly, J.L., Hahn, F.F., 1982. The effects of age on lung function and structure of adult animals. Adv. Vet. Sci. Comp. Med. 26, 35–77.

Mautz, W.J., McClure, T.R., Reischl, P., Phalen, R.F., Crocker, T.T., 1985a. Enhancement of ozone-induced lung injury by exercise. J. Toxicol. Environ. Health 16, 841–854.

Mautz, W.J., Phalen, R.F., McClure, T.R., Bufalino, C., 1985b. A rodent treadmill for inhalation toxicological studies and respirometry. J. Appl. Physiol. 58, 673–679.

McCarthy, D.S., Ostrow, D.N., Hershfield, E.S., 1980. Chronic obstructive pulmonary disease following idiopathic pulmonary fibrosis. Chest 77, 473–477.

McMahon, T.A., 1977. Allometry. In: Yearbook of Science and Technology. McGraw-Hill, New York.

Mead, J., 1961. Mechanical properties of lungs. Physiol. Rev. 41, 281–330.

Mead, J., 1963. The control of respiratory frequency. Ann. N. Y. Acad. Sci. 109, 724–729.

Mead, J., 1967. Mechanical properties of the lung. In: The Lung. Williams & Wilkins, Baltimore, MD, pp. 48–53.

Midoro-Horiuti, T., Tiwari, R., Watson, C.S., Goldblum, R.M., 2010. Maternal bisphenol a exposure promotes the development of experimental asthma in mouse pups. Environ. Health Perspect. 118, 273–277.

Miller, F.J., Graham, J.A., Raub, J.A., Illing, J.W., Menache, M.G., House, D.E., Gardner, D.E., 1987. Evaluating the toxicity of urban patterns of oxidant gases. II. Effects in mice from chronic exposure to nitrogen dioxide. J. Toxicol. Environ. Health 21, 99–112.

Mink, S.N., Coalson, J.J., Whitley, L., Greville, H., Jadue, C., 1984. Pulmonary function tests in the detection of small airway obstruction in a canine model of bronchiolitis obliterans. Am. Rev. Respir. Dis. 130, 1125–1133.

Mitzner, W., Tankersley, C., 1998. Noninvasive measurement of airway responsiveness in allergic mice using barometric plethysmography. Am. J. Respir. Crit. Care Med. 158, 340–341.

Murphy, D.J., Renninger, J.P., Gossett, K.A., 1998. A novel method for chronic measurement of pleural pressure in conscious rats. J. Pharmacol. Toxicol. Methods 39, 137–141.

Murphy, D.J., Renninger, J.P., Schramek, D., 2010. Respiratory inductive plethysmography as a method for measuring ventilatory parameters in conscious, non-restrained dogs. J. Pharmacol. Toxicol. Methods 62, 47–53.

Murphy, D.J., 2013. Respiratory safety pharmacology - current practice and future directions. Regul. Toxicol. Pharmacol, pp. 135–140.

Murray, Nadel, 2010. Textbook of Respiratory Medicine, fifth ed. Saunders/Elsevier, USA.

Nagasaka, T., Hirata, K., Sugano, Y., Shibata, H., 1979. Heat balance during physical restraint in rats. Jpn. J. Physiol. 29, 383–392.

Nagasaka, T., Hirata, K., Shibata, H., Sugano, Y., 1980. Metabolic and cardiovascular changes during physical restrains in rats. Jpn. J. Physiol. 30, 799–803.

Ng, F.S., Wong, K.Y., Guan, S.P., Mustafa, F.B., Kajiji, T.S., Bist, P., Biswas, S.K., Wong, W.S., Lim, L.H., 2011. Annexin-1-deficient mice exhibit spontaneous airway hyperresponsiveness and exacerbated allergen-specific antibody responses in a mouse model of asthma. Clin. Exp. Allergy 41, 1793–1803.

Nichols, J.E., Niles, J., Riddle, M., Vargas, G., Schilagard, T., Ma, L., Edward, K., La Francesca, S., Sakamoto, J., Vega, S., Ogadegbe, M., Mlcak, R., Deyo, D., Woodson, L., McQuitty, C., Lick, S., Beckles, D., Melo, E., Cortiella, J., 2013. Production and assessment of decellularized pig and human lung scaffolds. Tissue Eng. Part A 19, 2045–2062.

Niewoehner, D.E., Hoidal, J.R., 1982. Lung fibrosis and emphysema: divergent responses to a common injury? Science 217, 359–360.

Oberdorster, G., Hochrainer, D., 1980. Generation of Cd aerosols and their effects after inhalation. GAF 6, 230–235.

O'Connor, G.T., Sparrow, D., Weiss, S.T., 1989. The role of allergy and nonspecific airway hyperresponsiveness in the pathogenesis of chronic obstructive pulmonary disease. Am. Rev. Respir. Dis. 140, 225–252.

Okayama, H., Aikawa, T., Ohtsu, H., Sasaki, H., Takishima, T., 1989. Leukotriene C_4 and B_4 in bronchoalveolar lavage fluid during biphasic allergic bronchoconstriction in sheep. Am. Rev. Respir. Dis. 139, 725–731.

O'Neil, J.J., Raub, J.A., 1984. Pulmonary function testing in small laboratory mammals. Environ. Health Perspect. 56, 11–22.

Otis, A.B., Fenn, W.O., Rahn, H., 1950. Mechanics of breathing in man. J. Appl. Physiol. 2, 592–607.

Ozaki, J., Tanimoto, N., Kuse, H., Hori, M., 2000. Comparison of arterial blood gases and acid–base balance in young and aged beagle dogs, with regard to postprandial alkaline tide. J. Toxicol. Sci. 25, 205–211.

Paintal, A.S., 1969. Mechanism of stimulation of type J pulmonary receptors. J. Physiol. 203, 511–532.

Palecek, F., 1969. Measurement of ventilatory mechanics in the rat. J. Appl. Physiol. 27, 149–156.

Pavlin, E.G., Hornbein, T.F., 1986. Anesthesia and the control of ventilation. In: Cherniack, N.S., Widdicombe, J.G. (Eds.), Handbook of Physiology. Sect. 3, Respiration. Vol. II, Control of Breathing. American Physiological Society, Bethesda, MD, pp. 793–813.

Pengelly, L.D., 1977. Curve-fitting analysis of pressure-volume characteristics of the lungs. J. Appl. Physiol. 42, 111–116.

Pennock, B.E., Cox, C.P., Rogers, R.M., Cain, W.A., Wells, J.H., 1979. A noninvasive technique for measurement of changes in specific airway resistance. J. Appl. Physiol. 46, 399–406.

Perez, C.M., Ledbetter, A.D., Hazari, M.S., Haykal-Coates, N., Carll, A.P., Winsett, D.W., Costa, D.L., Farraj, A.K., 2013. Hypoxia stress test reveals exaggerated cardiovascular effects in hypertensive rats after exposure to the air pollutant acrolein. Toxicol. Sci. 132, 467–477.

Petak, F., Habre, W., Donati, Y.R., Hantos, Z., Barazzone-Argiroffo, C., 2001. Hyperoxia-induced changes in mouse lung mechanics: forced oscillations versus barometric plethysmography. J. Appl. Physiol. (1985) 90, 2221–2230.

Phalen, R.F., Prasad, S.B., 1989. Morphology of the respiratory tract. In: McClellan, R.O., Henderson, R.F. (Eds.), Concepts in Inhalation Toxicology. Hemisphere Publishing, New York, pp. 123–140.

Polianski, J.M., Brun-Pascaud, M.C., Jelazko, P.R., Pocidalo, J.J., 1984. Ventilation in awake rats with permanent arterial catheters. Comp. Biochem. Physiol. A Comp. Physiol. 77, 319–324.

Proquitte, H., Hartenstein, S., Koelsch, U., Wauer, R.R., Rudiger, M., Schmalisch, G., 2013. A comparison of conventional surfactant treatment and partial liquid ventilation on the lung volume of injured ventilated small lungs. Physiol. Meas. 34, 915–924.

Raab, O.G., Yeh, H.C., Newton, G.J., Phalen, R.F., Velasquez, D.J., 1975. Deposition of inhaled monodisperse aerosols in small rodents. Inhaled Part 4 (Pt 1), 3–21.

Raemdonck, K., de Alba, J., Birrell, M.A., Grace, M., Maher, S.A., Irvin, C.G., Fozard, J.R., O'Byrne, P.M., Belvisi, M.G., 2012. A role for sensory nerves in the late asthmatic response. Thorax 67, 19–25.

Rassler, B., Marx, G., Schierle, K., Zimmer, H.G., 2012. Catecholamines can induce pulmonary remodeling in rats. Cell. Physiol. Biochem. 30, 1134–1147.

Raub, J.A., Gillespie, J.R., 1979. Effects of sodium pentobarbital anesthesia on respiratory function in normal guinea pigs. Fed. Proc. 38, 9096.

Raub, J.A., Mercer, R.R., Miller, F.J., Graham, J.A., O'Neil, J.J., 1982. Dose response of elastase-induced emphysema in hamsters. Am. Rev. Respir. Dis. 125, 432–435.

Raub, J.A., Miller, F.J., Graham, J.A., Gardner, D.E., O'Neil, J.J., 1983. Pulmonary function in normal and elastase-treated hamsters exposed to a complex mixture of olefin-ozone-sulfur dioxide reaction products. Environ. Res. 31, 302–310.

Raub, J.A., Hatch, G.E., Mercer, R.R., Grady, M., Hu, P.C., 1985. Inhalation studies of Mt. St. Helens volcanic ash in animals. II. Lung function, biochemistry, and histology. Environ. Res. 37, 72–83.

Rehder, K., Mallow, J.E., Fibuch, E.E., Krabill, D.R., Sessler, A.D., 1974. Effects of isoflurane anesthesia and muscle paralysis on respiratory mechanics in normal man. Anesthesiology 41, 477–485.

Rehder, K., Sessler, A.D., Marsh, H.M., 1975. General anesthesia and the lung. Am. Rev. Respir. Dis. 112, 541–563.

Rich, C.R., Rehder, K., Knopp, T.J., Hyatt, R.E., 1979. Halothane and enflurane anesthesia and respiratory mechanics in prone dogs. J. Appl. Physiol. 46, 646–653.

Rosenthal, L.A., Amineva, S.P., Szakaly, R.J., Lemanske Jr., R.F., Gern, J.E., Sorkness, R.L., 2009. A rat model of picornavirus-induced airway infection and inflammation. Virol. J. 6, 122.

Roum, J.H., Murlas, C., 1984. Ozone-induced changes in muscarinic bronchial reactivity by different testing methods. J. Appl. Physiol. 57, 1783–1789.

Rozanski, E.A., Bedenice, D., Lofgren, J., Abrams, J., Bach, J., Hoffman, A.M., 2010. The effect of body position, sedation, and thoracic bandaging on functional residual capacity in healthy deep-chested dogs. Can. J. Vet. Res. 74, 34–39.

Sackner, M.A., 1980. Diagnostic Techniques in Pulmonary Disease: Lung Biology in Health and Disease, vol. 16. Marcel Dekker, New York. Part I.

Saidel, G.M., Salmon, R.B., Chester, E.H., 1975. Moment analysis of multibreath lung washout. J. Appl. Physiol. 38, 328–334.

Sandler, H., Meindl, J.D., 1983. Telemetry methods: animal and man. Physiologist 26, 47–50.

Sarazan, R.D., Mittelstadt, S., Guth, B., Koerner, J., Zhang, J., Pettit, S., 2011. Cardiovascular function in nonclinical drug safety assessment: current issues and opportunities. Int. J. Toxicol. 30, 272–286.

Savoy, J., Allgower, E., Courteheuse, C., Junod, A.F., 1984. Ventilatory response to bronchospasm induced by methylcholine and histamine in man. Respir. Physiol. 56, 195–203.

Schelegle, E.S., Walby, W.F., 2012. Vagal afferents contribute to exacerbated airway responses following ozone and allergen challenge. Respir. Physiol. Neurobiol. 181, 277–285.

Schroeder, C.A., Smith, L.J., 2011. Respiratory rates and arterial blood-gas tensions in healthy rabbits given buprenorphine, butorphanol, midazolam, or their combinations. J. Am. Assoc. Lab. Anim. Sci. 50, 205–211.

Schroter, R.C., 1980. Quantitative comparisons of mammalian lung pressure volume curves. Respir. Physiol. 42, 101–107.

Schuessler, T.F., Bates, J.H., 1995. A computer-controlled research ventilator for small animals: design and evaluation. IEEE Trans. Biomed. Eng. 42, 860–866.

Siafakas, N.M., Bonora, M., Duron, B., Gautier, H., Milic-Emili, J., 1983. Dose effect of pentobarbital sodium on control of breathing in cats. J. Appl. Physiol. 55, 1582–1592.

Silbaugh, S.A., Horvath, S.M., 1982. Effect of acute carbon monoxide exposure on cardiopulmonary function of the awake rat. Toxicol. Appl. Pharmacol. 66, 376–382.

Silbaugh, S.A., Mauderly, J.L., Macken, C.A., 1981. Effects of sulfuric acid and nitrogen dioxide on airway responsiveness of the guinea pig. J. Toxicol. Environ. Health 8, 31–45.

Silbaugh, S.A., Stengel, P.W., Rinkema, L.E., 1987. Pulmonary aerosol actions of LY188695 (KB2413), a new potent H1-receptor antagonist. Agents Actions 21, 83–88.

Sinnett, E.E., Jackson, A.C., Leith, D.E., Butler, J.P., 1981. Fast integrated flow plethysmograph for small mammals. J. Appl. Physiol. 50, 1104–1110.

Skornik, W.A., Brain, J.D., 1990. Breathing and lung mechanics in hamsters: effect of pentobarbital anesthesia. J. Appl. Physiol. 68, 2536–2541.

Sly, P.D., Collins, R.A., Thamrin, C., Turner, D.J., Hantos, Z., 2003. Volume dependence of airway and tissue impedances in mice. J. Appl. Physiol. (1985) 94, 1460–1466.

Smith, L.S., Zimmerman, J.J., Martin, T.R., 2013. Mechanisms of acute respiratory distress syndrome in children and adults: a review and suggestions for future research. Pediatr. Crit. Care Med. 14, 631–643.

Snider, G.L., Sherter, C.B., 1977. A one-year study of the evolution of elastase-induced emphysema in hamsters. J. Appl. Physiol. 43, 721–729.

Snider, G.L., Hayes, J.A., Korthy, A.L., 1978. Chronic interstitial pulmonary fibrosis produced in hamsters by endotracheal bleomycin: pathology and stereology. Am. Rev. Respir. Dis. 117, 1099–1108.

Snider, G.L., Lucey, E.C., Stone, P.J., 1986. Animal models of emphysema. Am. Rev. Respir. Dis. 133, 149–169.

Stahl, W.R., 1967. Scaling of respiratory variables in mammals. J. Appl. Physiol. 22, 453–460.

Steffens, A.B., 1969. A method for frequent sampling of blood and continuous infusion of fluids in the rat without disturbing the animal. Physiol. Behav. 4, 833–836.

Stevens, M.A., Fitzgerald, S., Menache, M.G., Costa, D.L., Bucher, J.R., 1987. Functional evidence of persistent airway obstruction in rats following a two-hour inhalation exposure to methyl isocyanate. Environ. Health Perspect. 72, 89–94.

Stevens, M.A., Menache, M.G., Crapo, J.D., Miller, F.J., Graham, J.A., 1988. Pulmonary function in juvenile and young adult rats exposed to low-level NO_2 with diurnal spikes. J. Toxicol. Environ. Health 23, 229–240.

Sweeney, T.D., Brain, J.D., LeMott, S., 1983. Anesthesia alters the pattern of aerosol retention in hamsters. J. Appl. Physiol. 54, 37–44.

Szarek, J.L., 1989. In vivo exposure to hyperoxia increases airway responsiveness in rats. Demonstration in vivo and in vitro. Am. Rev. Respir. Dis. 140, 942–947.

Takezawa, J., Miller, F.J., O'Neil, J.J., 1980. Single-breath diffusing capacity and lung volumes in small laboratory mammals. J. Appl. Physiol. 48, 1052–1059.

Taylor, C.R., Maloiy, G.M., Weibel, E.R., Langman, V.A., Kamau, J.M., Seeherman, H.J., Heglund, N.C., 1981. Design of the mammalian respiratory system. III scaling maximum aerobic capacity to body mass: wild and domestic mammals. Respir. Physiol. 44, 25–37.

Tepper, J.S., Costa, D.L., 1992. Will the mouse bioassay for estimating sensory irritancy of airborne chemicals (ASTM E981-84) be useful for evaluation of indoor air contaminants. Indoor Built Environ. 1, 367–372.

Tepper, J.S., Wiester, M.J., Costa, D.L., Watkinson, W.P., Weber, M.F., 1987. Cardiopulmonary effects in awake rats four and six months after exposure to methyl isocyanate. Environ. Health Perspect. 72, 95–103.

Tepper, J.S., Wiester, M.J., King, M.E., Weber, M.F., Costa, D.L., 1988. Use of CO_2 challenge to detect toxicant induced changes in cardiopulmonary function of awake rats. Inhal. Toxicol. 1, 79–95.

Tepper, J.S., Costa, D.L., Lehmann, J.R., Weber, M.F., Hatch, G.E., 1989. Unattenuated structural and biochemical alterations in the rat lung during functional adaptation to ozone. Am. Rev. Respir. Dis. 140, 493–501.

Tepper, J.S., Wiester, M.J., Weber, M.F., Menache, M.G., 1990. Measurements of cardiopulmonary response in awake rats during acute exposure to near-ambient concentrations of ozone. J. Appl. Toxicol. 10, 7–15.

Tepper, J.S., Wiester, M.J., Weber, M.F., Fitzgerald, S., Costa, D.L., 1991. Chronic exposure to a simulated urban profile of ozone alters ventilatory responses to carbon dioxide challenge in rats. Fundam. Appl. Toxicol. 17, 52–60.

Tepper, J., Blanchard, J., Pfeiffer, J., 1997. Validation of noninvasive measures of bronchoconstriction in mice. Am. J. Respir. Crit. Care Med. 155, A160.

Timmins, S.C., Diba, C., Farrow, C.E., Schoeffel, R.E., Berend, N., Salome, C.M., King, G.G., 2012. The relationship between airflow obstruction, emphysema extent, and small airways function in COPD. Chest 142, 312–319.

Toraason, M.A., Badger, D.W., Wright, G.L., 1980. Gastrointestinal response in rats to vibration and restraint. Environ. Res. 23, 341–347.

Torbati, D., Ramirez, J., Hon, E., Camacho, M.T., Sussmane, J.B., Raszynski, A., Wolfsdorf, J., 1999. Experimental critical care in rats: gender differences in anesthesia, ventilation, and gas exchange. Crit. Care Med. 27, 1878–1884.

Torbati, D., Camacho, M.T., Raszynski, A., Sussmane, J.B., Totapally, B.R., Hultquist, K., Wolfsdorf, J., 2000. Effect of hypothermia on ventilation in anesthetized, spontaneously breathing rats: theoretical implications for mechanical ventilation. Intensive Care Med. 26, 585–591.

Travis, E.L., Vojnovic, B., Davies, E.E., Hirst, D.G., 1979. A plethysmographic method for measuring function in locally irradiated mouse lung. Br. J. Radiol. 52, 67–74.

Travis, E.L., Parkins, C.S., Down, J.D., Fowler, J.F., Thames Jr., H.D., 1983. Repair in mouse lung between multiple small doses of X rays. Radiat. Res. 94, 326–339.

Turek, Z., Kreuzer, F., Ringnalda, B.E., 1978. Blood gases at several levels of oxygenation in rats with a left-shifted blood oxygen dissociation curve. Pflugers Arch. 376, 7–13.

U.S. EPA, 1988. Recommendations for and Documentation of Biological Values for Use in Risk Assessment. U.S. Environmental Protection Agency, Office of Research and Development, Washington, D.C. EPA/600/6–87/008 (NTIS PB88179874).

U.S. EPA, 1994. Methods for Derivation of Inhalation Reference Concentrations and Application of Inhalation Dosimetry. U.S. Environmental Protection Agency, Office of Research and Development, Office of Health and Environmental Assessment, Washington, DC. EPA/600/8–90/066F.

U.S. EPA, 2011. Exposure Factors Handbook. National Center for Environmental Assessment, U.S. Environmental Protection Agency, Office of Research and Development, Washington, D.C. EPA/600/R-09/052F.

Vargas, M.H., Sommer, B., Bazan-Perkins, B., Montano, L.M., 2010. Airway responsiveness measured by barometric plethysmography in guinea pigs. Vet. Res. Commun. 34, 589–596.

Vinegar, A., Sinnett, E.E., Leith, D.E., 1979. Dynamic mechanisms determine functional residual capacity in mice, *Mus musculus*. J. Appl. Physiol. 46, 867–871.

Vinegar, A., Sinnett, E.E., Kosch, P.C., 1982. Respiratory mechanics of a small carnivore: the ferret. J. Appl. Physiol. 52, 832–837.

Vizek, M., Palecek, F., 1977. Effect of anaesthesia on the pattern of breathing. Physiol. Bohemoslov. 26, 417–423.

Vizek, M., Holusa, R., Palecek, F., 1975. Lung function in acute paraquat intoxication. Physiol. Bohemoslov. 24, 559–563.

Wang, X., Reece, S., Olmstead, S., Wardle, R.L., Van Scott, M.R., 2010. Nocturnal thoracoabdominal asynchrony in house dust mite-sensitive nonhuman primates. J. Asthma Allergy 3, 75–86.

Warren, R.H., Fewell, J.E., Alderson, S.H., 1989. Calibration of respiratory inductive plethysmography during quiet and active sleep in lambs. J. Dev. Physiol. 12, 347–352.

Watkinson, W.P., Aileru, A.A., Dowd, S.M., Tepper, J.T., Gordon, C.J., 1990. Effect of ozone on body temperature and heart rate in the unanesthetized, unrestrained rat. FASEB 4, A752.

Weibel, E.R., 1963. Morphometry of the Human Lung. Springer, Berlin.

Weinberger, S.E., Johnson, T.S., Weiss, S.T., 1980. Use and interpretation of the single-breath diffusing capacity. Chest 78, 483–488.

Weissberg, R.M., Marian, J., Bradshaw, J., 1976. Respiratory and cardiovascular effects of prostaglandins in the conscious guinea pig. J. Pharmacol. Exp. Ther. 198, 197–208.

West, J.B., 1978. Pulmonary Pathophysiology-the Essentials. Williams & Wilkins, Baltimore, MD.

West, J.B., 1979. Respiratory Physiology: The Essentials. Williams & Wilkins, Baltimore, MD.

Wiester, M.J., Tepper, J.S., Weber, M.F., Setzer, C.J., Schutt Jr., W.A., 1987. A restraining system for plethysmography in small animals. Lab. Anim. Sci. 37, 235–238.

Wiester, M.J., Tepper, J.S., Weber, M.F., Menache, M.G., 1990. Histamine and methacholine aerosol bronchial challenge in awake guinea pigs. J. Pharmacol. Exp. Ther. 253, 27–33.

Wiester, M.J., Tepper, J.S., Doerfler, D.L., Costa, D.L., 1995. Ozone adaptation in rats after chronic exposure to a simulated urban profile of ozone. Fundam. Appl. Toxicol. 24, 42–51.

Wiester, M.J., Costa, D.L., Tepper, J.S., Winsett, D.W., Slade, R., 2005. Agonist-mediated airway challenge: cardiopulmonary interactions modulate gas exchange and recovery. Respir. Physiol. Neurobiol. 145, 183–199.

Wong, K.L., Alarie, Y., 1982. A method for repeated evaluation of pulmonary performance in unanesthetized, unrestrained guinea pigs and its application to detect effects of sulfuric acid mist inhalation. Toxicol. Appl. Pharmacol. 63, 72–90.

Wong, K.L., Stock, M.F., Malek, D.E., Alarie, Y., 1984. Evaluation of the pulmonary effects of wood smoke in guinea pigs by repeated CO_2 challenges. Toxicol. Appl. Pharmacol. 75, 69–80.

Wong, K.L., Karol, M.H., Alarie, Y., 1985. Use of repeated CO_2 challenges to evaluate the pulmonary performance of guinea pigs exposed to toluene diisocyanate. J. Toxicol. Environ. Health 15, 137–148.

Woolcock, A.J., Vincent, N.J., Macklem, P.T., 1969. Frequency dependence of compliance as a test for obstruction in the small airways. J. Clin. Invest. 48, 1097–1106.

Yang, Y.L., Hsu, H.T., Wang, K.H., Han, C.Y., Chen, C.M., Chen, C.M., Ko, W.C., 2011. Hesperetin-7,3'-O-dimethylether selectively inhibits phosphodiesterase 4 and effectively suppresses ovalbumin-induced airway hyperresponsiveness with a high therapeutic ratio. J. Biomed. Sci. 18, 84.

Yokoyama, E., 1983. Determination of ventilatory capacity in rats. Nihon. Kyobu. Shikkan. Gakkai. Zasshi. 21 (4), 357–363.

Zehendner, C.M., Luhmann, H.J., Yang, J.W., 2013. A simple and novel method to monitor breathing and heart rate in awake and urethane-anesthetized newborn rodents. PLoS One 8, e62628.

Zhang, Q., Lai, K., Xie, J., Chen, G., Zhong, N., 2009. Does unrestrained single-chamber plethysmography provide a valid assessment of airway responsiveness in allergic BALB/c mice? Respir. Res. 10, 61.

Zosky, G.R., Sly, P.D., 2007. Animal models of asthma. Clin. Exp. Allergy 37, 973–988.

APPENDIX A. SELECTED REFERENCES ON COMPARATIVE MAMMALIAN RESPIRATORY PHYSIOLOGY

Adolph, E.F., 1949. Quantitative relations in the physiological constitutions of mammals. Science 109, 579–585.

Agostoni, E., Thimm, F.F., Fenn, W.O., 1959. Comparative features of the mechanics of breathing. J. Appl. Physiol. 14, 679–683.

Bennett, F.M., Tenney, S.M., 1982. Comparative mechanics of mammalian respiratory system. Respir. Physiol. 49, 131–140.

Bertalanffy, L.von, 1957. Quantitative laws in metabolism and growth. Q. Rev. Biol. 32, 217–231.

Boggs, D.F., Tenney, S.M., 1984. Scaling respiratory pattern and respiratory "drive." Respir. Physiol. 58, 245–251.

Brody, S., 1945. Bioenergetics and Growth, with Special Reference to the Efficiency Complex in Domestic Animals, Reinhold, New York.

Crossfill, M.L., Widdicombe, J.G., 1961. Physiological characteristics of the chest and lungs and the work of breathing in different mammalian species. J. Physiol. 158, 1–14.

Dejours, P., 1975. Principles of Comparative Respiratory Physiology, vols. 1–253. American Elsevier, New York.

Drorbaugh, J.E., 1960. Pulmonary function in different animals. J. Appl. Physiol. 15, 1069–1072.

Feldman, H.A., McMahon, T.A., 1983. The 3/4 mass exponent for energy metabolism is not a statistical artifact. Respir. Physiol. 52, 149–163.

Fisher, J.T., Mortola, J.P., 1981. Statics of the respiratory system and growth: an experimental and allometric approach. Am. J. Physiol. 241, R336–R341.

Fishman, A., 1983. Comparative biology of the lung. Am. Rev. Respir. Dis. 128, S90–S91.

Gehr, P., Mwangi, D.K., Ammann, A., Maloiy, G.M.O., Taylor, C.R., Weibel, E.R., 1981. Design of the mammalian respiratory system. V. Scaling morphometric pulmonary diffusing capacity to body mass: wild and domestic mammals. Respir. Physiol. 44, 61–86.

Gillespie, J.R., 1983. Mechanisms that determine functional residual capacity in different species. Am. Rev. Respir. Dis. 128, S74–S77.

Gunther, B., 1975. Dimensional analysis and theory of biological similarity. Physiol. Rev. 55, 659–699.

Gunther, B., Guerra, E., 1955. Biological similarities. Aeta Physiol. Lat. Amer. 5, 169–186.

Gunther, B., Leon De La Barra, B., 1966. A unified theory of biological similarities. J. Theor. Biol. 13, 48–59.

Gunther, B., Leon De La Barra, B., 1966. On the space-time continuum in biology. Aeta Physiol. Lat. Amer. 16, 221–231.

Gunther, B., Leon De La Barra, B., 1966. Physometry of the mammalian circulatory system. Aeta Physiol. Lat. Amer. 16, 32–42.

Gunther, B., Leon De La Barra, B., 1966c. Theory of biological similarities nondimensional parameters and invariant numbers. Bull. Math. Biophys. 28, 91–102.

Guyton, A.C., 1947. Measurement of the respiratory volumes of laboratory animals. Am. J. Physiol. 150, 70–77.

Hill, A.V., 1950. The dimensions of animals and their muscular dynamics. Sci. Prog. 38, 209–230.

Huxley, J.S., 1932. Problems of Relative Growth. Dial Press, New York.

Jones, J.D., 1972. Comparative Physiology of Respiration. Edward Arnold, London, pp. 1–202.

Kleiber, M., 1932. Body size and metabolism. Hilgardia 6, 315–358.

Kleiber, M., 1961. The Fire of Life: An Introduction to Animal Energetics. John Wiley and Sons, New York.

Krogh, A., 1941. The Comparative Physiology of Respiratory Mechanisms, vols. 1–172. University of Pennsylvania Press, Philadelphia.

Lambert, R., Teissier, G., 1927. Theory de la similitude biologique. Ann. Physiol. 3, 212–246.

Lechner, A.J., 1978. The scaling of maximal oxygen consumption and pulmonary dimensions in small mammals. Resp. Physiol. 34, 2944.

Leith, D.E., 1984. Mammalian tracheal dimensions: scaling and physiology. J. Appl. Physiol. 55, 196–200.

Lindstedt, L.S., Calder Jr., W.A., 1981. Body size, physiological time and longevity of homeothermic animals. Q. Rev. Biol. 56, 1–16.

McMahon, T.A., 1973. Size and shape in biology. Science 179, 1201–1204.

McMahon, T.A., 1975. Using body size to understand the structural design of animals: quadrupedal locomotion. J. Appl. Physiol. 39, 619–627.

McMahon, T.A., 1977. Allometry. In: Yearbook of Science and Technology. McGraw-Hill, New York.

Mortola, J.P., 1983. Some functional mechanical implications of the structural design of the respiratory system in newborn mammals. Am. Rev. Respir. Dis. 128, S69–S72.

Mortola, J.P., 1983. Dysanaptic lung growth: an experimental and allometric approach. J. Appl. Physiol. Respir. Environ. Exerc. Physiol. 54, 1236–1241.

O'Neil, J.J., Leith, D.E., 1980. Lung diffusing capacity scaled in mammals from 25 g to 500 kg. Fed. Proc. 39, 972.

Schmidt-Nielsen, K., 1970. Energy metabolism, body size, and problems of scaling. Fed. Proc. 29, 1524–1532.

Schmidt-Nielsen, K., 1975. Scaling in biology: the consequences of size. J. Exp. Zool. 194, 287–308.

Schmidt-Nielsen, K., 1984. Scaling: Why is Animal Size So Important? Cambridge University Press, Cambridge.

Seeherman, H.J., Taylor, C.R., Maloiy, G.M.O., Armstrong, R.B., 1981. Design of the mammalian respiratory system. II. Measuring maximum aerobic capacity. Respir. Physiol. 44, 11–23.

Spells, K.E., 1969. Comparative studies in lung mechanics based on a survey of literature data. Respir. Physiol. 8, 37–57.

Stahl, W.R., 1961. Dimensional analysis in mathematical biology. I. Bull. Math. Biophys. 23, 355–376.

Stahl, W.R., 1962. Dimensional analysis in mathematical biology. II. Bull. Math. Biophys. 24, 81.

Stahl, W.R., 1962. Similarity and dimensional methods in biology. Science 137, 205–211.

Stahl, W.R., 1963. Similarity analysis of physiological systems. Perspect. Biol. Med. 6, 291.

Stahl, W.R., 1963. The analysis of biological similarity. Adv. Biol. Med. Phys. 9, 355–364.

Stahl, W.R., 1967. Scaling of respiratory variables in mammals. J. Appl. Physiol. 22, 453–460.

Stahl, W.R., Gummerson, J.Y., 1967. Systematic allometry in five species of adult primates. Growth 31, 21–34.

Steen, J.B., 1971. Comparative Physiology of Respiratory Mechanisms, vols. 1–182. Academic Press, New York.

Taylor, C.R., Weibel, E.R., 1981. Design of the mammalian respiratory system. I. Problem and strategy. Respir. Physiol. 44, 1–10.

Taylor, C.R., Maloiy, G.M.O., Weibel, E.R., Langman, V.A., Kamau, J.M.Z., Seeherman, H.J., Heglund, N.C., 1981. Design of the mammalian respiratory system. III. Scaling maximum aerobic capacity to body mass: wild and domestic mammals. Respir. Physiol. 44, 25–37.

Tenney, S.M., 1967. Some aspects of the comparative physiology of muscular exercise in mammals. Circ. Res. 20-21, 1–7.

Tenney, S.M., Remmers, J.E., 1963. Comparative quantitative morphology of the mammalian lung: diffusing area. Nature 4862, 54–56.

Tenney, S.M., Bartlett Jr., D., 1967. Comparative quantitative morphology of the mammalian lung: trachea. Respir. Physiol. 3, 130–135.

Thompson, D'Arcy W., 1959. On Growth and Form, vols. I and II. Cambridge University Press, London.

Weibel, E.R., 1987. Scaling of structural and functional variables in the respiratory system. Annu. Rev. Physiol. 49, 147–159.

Section III

Comparative Biology of the Normal Lung

Debra L. Laskin and Andrew J. Gow

Chapter 18

Oxygen Metabolism in the Lung

Andrew J. Gow and Henry J. Forman

Chapter Outline

1. OXYGEN CHEMISTRY

The chemical reactions of elemental atoms are governed by energetics that drive atoms to either fill or lose electrons until they reach the electronic configuration of their nearest inert element (e.g., the noble gases, helium, neon, etc.). This can be accomplished by either transferring electrons from one atom to another or by combining the atoms to form molecules in which sharing of electrons allows each atom to attain its ideal electronic configuration. Reactions of molecules almost always (there are exceptions to every rule) yield molecules having all of their atoms in an inert gas configuration. However, the path between two stable molecules can involve unstable intermediate molecules that do not have all of their atoms in the inert gas configuration.

Electrons around the nucleus of each atom are arranged in orbitals. The core electrons, i.e., those that are more tightly bound, are arranged in orbitals around the nuclei that comprise a molecule. When filled, these orbitals contain two electrons that must be paired, i.e., spin in opposite directions. The electrons that bind the atoms together are found in the valence orbitals. The electrons in these orbitals are shared so that each atom can achieve the noble gas configuration. However, atoms and molecules do not need to have noble gas configurations to have a stable existence in solutions or in crystals. For example, most electropositive elements, like the metals sodium and potassium, exist as positive ions, while the more electronegative elements, such as chlorine and fluorine, are often found as their anions, e.g., chloride and fluoride. Ions have a net charge due to an unequal number of electrons and protons. In addition, molecules that are normally most stable in a neutral, spin-paired state may react by mechanisms that give ions or radicals as transient intermediates.

Radicals are often defined as molecules that contain an unpaired electron. In chemical nomenclature, a superscript dot is often used to designate a free radical; e.g., $CH_3^\bullet$ (methyl radical), as distinct from the negative or positive designations that characterize ions. Nevertheless, a molecule can be both a free radical and an ion. The hydroxyl anion (OH^-), hydroxyl radical ($\bullet OH$), and superoxide anion radical ($O_2^{\bullet -}$) are examples of an ion, an uncharged free radical, and an ionic free radical, respectively. The spin state of a molecule is calculated as $2 \times (n/2) + 1$, where n equals the number of unpaired electrons (electrons whose spin is not matched by one of the opposite spin, regardless of orbital location). Most molecules are therefore singlet $(2 \times (0/2) + 1)$, while O_2 is a triplet $(2 \times (2/2) + 1)$ and $O_2^{\bullet -}$ is a doublet $(2 \times (1/2) + 1)$. Molecular oxygen's two unpaired electrons spin in the same direction. The reason for this oddity (most molecules have all of their electrons paired) is that oxygen has two degenerate valence orbitals (i.e., they are of equal energy) and only two electrons to share between these orbitals. Hund's rules, describing how electrons fill orbitals, imply that lower energy states are obtained when electrons are placed in separate orbitals of equal energy. In those reactions of molecules that result from a transfer of a pair of electrons, these electrons must either have opposite spins or the reaction must proceed by a series of one-electron steps. If the redox potential difference between oxygen and an organic molecule would give a large, negative free energy change ($-\Delta G$) for a one-electron transfer, stepwise electron transfer will occur to yield oxidized products. Most bioorganic molecules are

Comparative Biology of the Normal Lung. http://dx.doi.org/10.1016/B978-0-12-404577-4.00018-7

sufficiently stable and do not undergo rapid electron transfer with oxygen. For oxygen to undergo a reaction where two electrons move simultaneously, one of the electron spins must be inverted. The energy required to cause spin pairing is relatively high and, therefore, oxygen does not normally undergo simultaneous two-electron reactions. Thus, it is important to remember that within most biological reactions, including those of the electron transport chain, oxygen only participates in single electron oxidation and reduction reactions.

Light can supply sufficient energy to cause the spin inversion of one of the spins in ground-stage O_2. A photosensitizing molecule, such as methylene blue, on absorbing a photon of light, is converted from its ground singlet state into an excited singlet state (no spin inversion). A slower process, called intersystem crossing, allows the spin-forbidden singlet-to-triplet state interconversion. This triplet can transfer its energy to triplet O_2 in a spin-allowed process that gives spin-inverted O_2, singlet oxygen, and ground-state singlet photosensitizer. There are two ways to form singlet oxygen by pairing these electrons. The first involves the pairing of now opposite-spinning electrons into a single orbital, while the second, lower energy state has them in different orbitals. Both of these forms of singlet oxygen are much more reactive than the triplet ground state. Although there have been suggestions of a role for singlet O_2 in biology, such as in microbicidal action of phagocytes, the evidence is not totally convincing.

Oxygen can be readily reduced in water. One-electron reduction yields O_2^- or its conjugate acid (HO_2). Two-electron reduction in the presence of metal gives H_2O_2. When O_2^- reacts, it can either accept a second electron to fill all of its orbitals or give up its electronic charge to become uncharged O_2. The reduced product, O_2^{-2}, has such high pKs that it is immediately protonated to form hydrogen peroxide (H_2O_2). While, as stated above, the major driving force in chemistry is the tendency to obtain an inert gas electronic configuration, the chemistry of O_2 is unique in that O_2^- more easily reduces than oxidizes other molecules and, thus, more easily returns to the diradical O_2 state than to H_2O_2. The evidence indicates that O_2^- appears to oxidize only a limited number of compounds, while it easily reduces numerous others. As a result, its production is tightly regulated by superoxide dismutases within biological systems (Fridovich, 1981).

Addition of another electron to H_2O_2 results in the splitting of the hydrogen peroxide molecule. Such a reaction occurs, for example, when H_2O_2 reacts with ferrous iron in the Fenton reaction:

$$H_2O_2 + Fe^{2+} \rightarrow HO\bullet + OH^- + Fe^{3+} \quad (1)$$

The $HO\bullet$ radical is an extremely potent oxidant, which will rapidly abstract electrons, to become OH^-, in reaction with almost any substance. Production of free hydroxyl radical is usually considered a toxicological rather than a normal metabolic process, although it may be involved in killing of microorganisms by phagocytic cells. Nevertheless, reactions in which oxygen is activated to a state resembling hydroxyl radical must be involved in some of the oxidases that will be discussed in this chapter. However, H_2O_2 is generated in numerous biological processes and is considered to be a major transmitter of redox signaling within cells (Forman et al., 2010). Critically important in these processes is the fact that, although H_2O_2 is a potent oxidant, there are considerable activation energy barriers to its reaction, meaning that it possesses a specific set of biological targets (Winterbourn, 2013). Although reactive with thiols, such reactions generally require transition metal catalysis. In contrast, selenoproteins such as most glutathione peroxidases and the nonselenium glutathione peroxidases and peroxiredoxins, which have unusually reactive cysteine residues, react with H_2O_2 at rates of reaction orders of magnitude higher than other thiol-containing compounds and proteins (Ferrer-Sueta et al., 2011). This strongly suggests that nonenzymatic oxidation of other thiol-containing signal transduction proteins is unlikely (Winterbourn, 2013).

Finally, oxygen can accept a fourth electron. In biology, the major use of oxygen is in a reaction in which molecular oxygen accepts four electrons in a single reaction catalyzed by cytochrome *c* oxidase. The single reaction, however, does involve multiple steps, as will be discussed later in the chapter. The result of molecular oxygen accepting four electrons is the production of two molecules of hydroxyl anion. Of course, the protonated form of hydroxyl anion is water.

2. BIOENERGETICS

The lung, like other biological systems, requires an efficient mechanism in which energy contained in chemical bonds of food substances can be used for reactions ranging from the biochemical, such as the synthesis of surfactant phospholipids, to the mechanical, such as the contraction of respiratory muscles. Oxidation of glucose, fatty acids, and other nutrients produces CO_2 and H_2O, and releases energy; however, the energy from oxidation of these nutrients cannot be used directly for most physiological processes. Catabolic processes transfer energy in a series of steps, which allows efficient transfer of energy from the breakdown of nutrients to the production of relatively stable compounds containing "high-energy" bonds. These bonds, such as the acyl phosphate bonds in adenosine triphosphate (ATP), yield relatively large quantities of free energy upon hydrolysis. Thus, the misnomer "high-energy bond" actually indicates that breaking such a bond generally results in an exothermic reaction. Hydrolysis of these bonds can be directly coupled, in an efficient manner, to reactions that produce chemical or physical work. The catabolic steps in the metabolism of nutrients are covered in other chapters. Here, we will focus on the major pathway for ATP generation in eukaryotic cells in which electrons are released from the breakdown of nutrients and where the reduction of oxygen to water is the major mechanism of oxygen metabolism. These processes are known as *electron transport* and *oxidative phosphorylation*, and occur in mitochondria.

2.1 Reducing Equivalents

To transfer energy efficiently from biological oxidations to mitochondrial production of the "high-energy" compound ATP, the energy from the bonds in nutrients must be transferred to compounds that can directly drive ATP synthesis. The reduction of nucleotide cofactors—NAD^+ (nicotine adenine dinucleotide), FAD (flavin adenine dinucleotide), and FMN (flavin mononucleotide)—is by far the most common mechanism for transferring energy. Electrons are directly transferred from the cofactors into the process that provides energy for ATP generation in mitochondria. NAD^+ can be a cosubstrate for a vast number of biological reactions. In some of these, the cosubstrate is tightly bound so that the same molecule essentially functions as a part of the enzyme. FAD and FMN are cofactors of dehydrogenases and are almost always tightly bound to these enzymes. FAD is tightly bound to fatty acyl coenzyme A (acyl-CoA) dehydrogenase, and a covalently bound FAD is in the active site of succinate dehydrogenase (Singer and Johnson, 1985). Thus, the same molecule of FAD can be involved in repetitive redox cycles of dehydrogenases. Such FAD and FMN binding proteins function in electron transfer from dehydrogenases to the respiratory chain and, as will be discussed, are involved in a number of biological oxidations. NAD^+ and FAD are reduced to NADH and FADH in the catabolic reactions of intermediary metabolism. These reactions occur in both the cytoplasmic and mitochondrial compartments of the cell, but much of the NADH is generated in the mitochondrial matrix.

The inner mitochondrial membrane is not permeable to NADH so that NADH produced in the cytosol cannot be oxidized directly. So-called shuttle mechanisms are present to transfer reducing equivalents through the inner mitochondrial membrane to the mitochondrial matrix. In actuality, these shuttles involve four steps: reduction in the cytosol of a mitochondrial-permeable metabolite by NADH; passage of the reduced metabolite into the mitochondrion; dehydrogenation of the metabolite in the mitochondrion to generate NADH or $FADH_2$ plus the original metabolite; and passage of the oxidized metabolite to the cytosol. Examples of the substrate pairs used in these shuttle mechanisms are aspartate conversion to malate and glycerol-3-phosphate oxidation to dihydroxyacetone phosphate.

Nicotinamide adenine dinucleotide phosphate (NADPH), the other major carrier of reducing equivalents, is not used directly by the electron-transport chain. NADPH provides reducing potential for a number of biosynthetic reactions and is involved in biological oxidations, such as cytochrome-P450-catalyzed reactions and the respiratory burst of phagocytic cells. While the midpoint potentials for $NADP^+$/NADPH and NAD^+/NADH are identical, the redox pairs do not equilibrate because of enzyme specificity for either NADH or NADPH. In most cells the $NADP^+$/NADPH ratio is low, while the steady-state NAD^+/NADH ratio favors the oxidized state. While NADPH is not used to generate NADH, a mitochondrial transhydrogenase can catalyze the essentially irreversible energy-dependent reduction of $NADP^+$ by NADH, in which one ATP is hydrolyzed to adenosine diphosphate (ADP) plus inorganic phosphate per molecule of NADPH produced. This reaction presumably allows an intramitochondrial pool of NADPH to be maintained.

2.2 Electron Transport

The inner membrane of mitochondria contains a number of proteins with cofactors that can undergo cyclic reduction and oxidation. The transfer of electrons from NADH to molecular oxygen occurs by a series of electron transfers through these proteins. The net reaction of mitochondrial electron transport, starting from NADH, is as follows:

$$2NADH + O_2 + 2H^+ \rightarrow 2NAD^+ + 2H_2O \quad (2)$$

The electron transport chain can be divided into groups of carriers that are physically, as well as chemically, linked. Such complexes are NADH dehydrogenase, which contains FMN and six nonheme iron-sulfur proteins (Albracht et al., 1972; Salerno et al., 1979) and succinate dehydrogenase, which contains a covalently bound FAD and three nonheme iron-sulfur proteins (Singer and Johnson, 1985; Albracht et al., 1972). The exact nature of electron passage through these complexes is a tightly controlled process that can be interrogated through spectroscopic studies (Ohnishi, 1998). A protein called the *electron transferring flavoprotein* is an intermediate carrier of electrons from auxiliary dehydrogenases, such as fatty acyl-CoA dehydrogenase, which are not physical components of the respiratory chain. In the ubiquinone-cytochrome b-cytochrome c_1 complex, two different cytochrome b's are involved; cytochrome b_K and cytochrome b_T (Erecinska and Wilson, 1976). The subscript K stands for kinetic, since it is more rapidly reduced. This form has also been referred to as cytochrome b_{561}, as the reduced form has an absorption maximum at 561 nm. In current textbooks, it is referred to as cytochrome b_L. The subscript T stands for thermodynamic, since this form is thermodynamically closer to the midpoint potentials of the complexes from which the ubiquinone-cytochrome b-cytochrome c complex receives electrons. This form is also referred to as cytochrome b_{566} or cytochrome b_H.

Although the initial entry of electrons to the respiratory chain occurs in pairs, most of the carriers in the chain can only function as one-electron carriers because the active components are iron atoms in the center of porphyrin rings or clusters of iron and sulfur. These iron atoms can only cycle between the Fe^{3+} and Fe^{2+} forms. Both flavin and ubiquinone can form radicals easily, and thereby could function as points at which two-to-one electron transport occurs. Although both one- and two-electron carriers are involved in the chain, the terminal cytochrome c oxidase must receive four electrons for its concerted reduction of O_2 to water (Capaldi et al., 1983). How electrons move through the chain and the importance of mobility within the mitochondrial membrane is critical to the regulation of electron transport. Recently, it has been demonstrated that the individual complexes of the chain can directly interact, forming super complexes that may be critical in regulation (Lapuente-Brun et al., 2013).

2.3 Reduction of Oxygen in Mitochondria

The entry of O_2 into the electron transport chain occurs at the terminal end. This is composed of the complex known as cytochrome c oxidase (Capaldi et al., 1983). It contains cytochromes a and a_3, as well as two copper atoms. Cytochrome c oxidase catalyzes the reduction of molecular oxygen to two molecules of water, a process requiring the transfer of four electrons to O_2. While intermediates may be formed within the active site of cytochrome c oxidase, none of these escape under normal conditions. This transfer of electrons probably involves sequential transfers of two electrons from cytochrome a_3 and one of the copper atoms first, followed by a rapid transfer of electrons from cytochrome a and the other copper atom (Erecinska and Wilson, 1978). The oxygen molecule is probably bound as a bridged complex between the cytochrome a_3 and one of the copper atoms (Chance, 1972). Electrons going to O_2 in cytochrome c oxidase initially come from cytochrome c_1, another carrier that is capable of carrying only one electron. Because the reduction of oxygen appears to occur by four electrons nearly simultaneously, the enzyme must act as a sort of biological capacitor, accumulating the four electrons from cytochrome c before transferring them to O_2. The net reaction for cytochrome c oxidase can thus be written:

$$4\text{Cyt.c-Fe}^{2+} + O_2 + 4H^+ \rightarrow 4\text{Cyt.c-Fe}^{3+} + 2H_2O \quad (3)$$

Cytochrome c oxidase catalyzes the major use of O_2 in eukaryotic organisms. To a much lesser extent, O_2 can be reduced at earlier points in the electron transport chain. These "leaks" occur within the NADH dehydrogenase (complex I) and from the ubiquinone-cytochrome b-cytochrome c (complex III) (Forman and Kennedy, 1974; Jastroch et al., 2010). At both of these sites, the major product is one electron reduction of oxygen, consistent with its thermodynamic property. Because of the presence of superoxide dismutase (Forman and Fridovich, 1973) in mitochondria, which catalyzes the reaction (4), these leaks were initially reported as producing hydorgen peroxide (Lochsen et al., 1971).

$$2O_2^- + 2H^+ \rightarrow H_2O_2 + O_2 \quad (4)$$

These leaks represent an inefficiency of mitochondrial bioenergetics, since they will not result in the production of as much ATP as would normally occur should electrons go through the complete chain. Nevertheless, these leaks do appear to occur physiologically and are not the result of damage to isolated mitochondria. Indeed, when isolated mitochondria are damaged and thereby become uncoupled (see below), O_2^- and H_2O_2 production decline markedly. It has become increasingly accepted that the "leak" of partially reduced O_2 may be a significant signaling mechanism from the organelle to the rest of the cell (Forman et al., 2010; Droge, 2002; Murphy, 2009).

It has proved difficult to directly estimate the rate of O_2^-/H_2O_2 production from mitochondria in vivo. The reductive potential that drives O_2^- formation is dependent upon the relative concentrations of O_2 and O_2^- (Murphy, 2009), and within the matrix of the mitochondria the presence of MnSOD (Manganese Superoxide Dismutase) keeps the free O_2^- at essentially zero as the rate constant for dismutation is near the limit of diffusion. Therefore, the driving redox potential is almost exclusively determined by the concentration of molecular O_2 and, therefore, the "leak" rate is a direct function of tissue oxygenation. The major source of superoxide production in mitochondria, especially in unstressed cells, is dependent upon the energetic balance within that cell. In situations where the $NADH/NAD^+$ ratio is high but the ATP production is low, it allows for the buildup of reducing equivalents in complex I, providing a driving force for one electron reduction of O_2 via the bound FMN. Alternatively if there is a buildup of mitochondrial membrane potential and ATP production is low, then there will be a buildup of bound reduced coenzyme Q within complex I, providing an alternate mechanism of O_2^- generation. Finally, when electron throughput in the electron transport chain is high, in times of high metabolic demand, then there is the opportunity of electron "leak" from a number of sites, including complex I, Co Q, and α-ketoglutarate dehydrogenase. While it has long been known that certain pathological conditions such as hyperoxia can increase O_2^- production (Boveris, 1977; Boveris and Chance, 1973; Freeman and Crapo, 1981), the release of oxidative molecules is more than just a toxic response. It has been proposed that these different pathways of formation of O_2^- lead to different rates of H_2O_2 efflux from the mitochondria and that, in this way, the cell is capable of sensing what kind of oxidative stress it may be under. Leading to a more finely tuned response to such oxidative signaling (Murphy, 2009).

2.4 Oxidative Phosphorylation

The electron transport scheme provides the path for electrons to pass from the reducing equivalents generated during metabolism via the dehydrogenases to O_2. However, the transport of electrons, while releasing energy, does not directly generate ATP. It does not indicate a mechanism for the production of ATP. Nevertheless, the scheme contains three places where the transfer of electrons involves sufficient energy to drive phosphorylation of ADP. The change in free energy ($\Delta G°$) required to drive ATP synthesis is 7 kcal/mol. A transfer of electrons from carriers with midpoint potential differences of 156 mV would provide the required 7 kcal/mol. The three reactions that exceed this requirement are the NADH dehydrogenase to the ubiquinone-cytochrome b-cytochrome c complex ($\Delta G°$ 12.2 kcal/mol), the potential drop across the ubiquinone-cytochrome b-cytochrome c complex itself ($\Delta G°$ 10 kcal/mol), and the potential drop from cytochrome c to O_2 mediated by cytochrome c oxidase ($\Delta G°$ 24 kcal/mol). Thus, pairs of electrons entering the respiratory chain through NADH can drive the synthesis of three ATP, whereas entrance through the succinate dehydrogenase complex or electron transfer flavoprotein only produces two ATP.

The reactions involved within the electron transport chain do not actually catalyze the phosphorylation of ADP. For many years, the general concept that phosphorylation must involve a "high-energy" intermediate drove investigators to search for such a compound. This concept is consistent with both thermodynamic considerations and the known mechanisms involved with substrate-level phosphorylation in the glycolytic pathway. Nevertheless, failure to find this intermediate led Mitchell (Mitchell, 1966) to propose that the chemical energy of electron transport could produce both an osmotic gradient of protons and electrical potential across the inner mitochondrial membrane, which in turn could drive ATP synthesis. ATP formation does appear to have a strict requirement for an intact inner mitochondrial membrane, which is consistent with this hypothesis. Mechanisms for proton movement via the electron transport chain were not originally obvious (Villalobo and Lehninger, 1979); however, it was known that the mitochondrial matrix is basic compared to the cytosol and that a large potential could be measured across its membrane. Since that time, a number of studies have established that complex I pumps four protons across the inner membrane; although the mechanism is not entirely clear, it appears that a protein conformational change is involved rather than direct coupling to electron transport. Within complex III, four further protons are pumped via the intermediacy of coenzyme Q. Finally, two protons are pumped by cytochrome c oxidase in conjunction with the terminal transport of electrons to molecular oxygen. This gives a total of 10 potential protons pumped per NADH fed into the electron transport chain. The chemiosmotic mechanism provides a mechanism for connecting the energy derived from the flow of electrons down the respiratory electron transport chain to ATP synthase (or as it is sometimes referred to, complex V) (Mazat et al., 2013).

The information we now have regarding the coupling of electron transport and ATP generation appear to follow the principles outlined by Mitchell (Mitchell, 1961). However, despite detailed crystallographic information, the precise mechanisms involved in this coupling remain unclear (Mazat et al., 2013). The unusual arrangement in terms of the biochemical synthesis being indirectly linked to the generation of potential and proton movement makes the system appear inexact and thereby to violate chemical principles (Wilson and Forman, 1982). Yet, the very obvious inefficiency of electron transport in driving ATP synthesis (46 kcal energy generated to drive 27 kcal worth of ATP synthesis in reaction two) may in part be explained by leaks of protons and other ions across the mitochondrial membrane. Such leaks would produce what would appear as a nonstoichiometric relationship. Once the proton and electrical potential across the inner mitochondrial membrane is produced, the energy from these gradients is used to provide the energy for the enzymatic generation of ATP. In the conformational coupling mechanism proposed by Boyer, the proton gradient is used to produce an energy-dependent conformational change in the mitochondrial ATPase, which in turn releases that energy in catalyzing ATP production (Boyer et al., 1977). The energetics of this process are an important area of investigation and have been the subject of a number of computer modeling approaches that have sought to encompass both electron transport and ATP generation (Beard, 2005; Cortassa et al., 2003).

2.5 Regulation

Regulation of ATP synthesis is complex process, and much of what we know has come from the study of cardiac mitochondria (Yaniv et al., 2010). Originally, it was proposed that ATP production was regulated directly by the utilization of ATP (Chance and Williams, 1955a), implying that one of the principal regulators of ATP production is the supply of ADP and inorganic phosphate. The synthesis of ATP by mitochondria has been shown to vary with the availability of substrate (those that generate either NADH or succinate), ADP, inorganic phosphate, and O_2, and the extent of coupling between the respiratory chain and ADP phosphorylation (Brown, 1992). The extent of coupling is defined as the efficiency of ATP production per mole of O_2 consumed (Jastroch et al., 2010). The concept of ATP utilization as a major controlling factor was developed from the concept that ADP (and inorganic phosphate) were the limiting factors in oxidative phosphorylation (Chance and Williams, 1955a, 1956). Theoretically, any of the substrate components of the respiratory chain can be rate limiting under particular conditions. The control of ATP production by these factors was described as several mitochondrial states (Chance and Williams, 1955a). Normally, neither substrates for mitochondria nor O_2 would be limiting, particularly in the lung. Instead, the primary regulation would be the amount of ADP and, therefore, the energy state of a cell could be accurately described by the ratio of the participants in ATP hydrolysis. This results in the formulation of the phosphate potential (F):

$$F = (ATP)/(ADP) + (P) \tag{5}$$

Since most of the ATP is made in the mitochondria but used in the cytosol, transport of ATP and ADP in and out of the mitochondria is another potential point at which regulation of ATP synthesis and utilization could occur (Chance and Williams, 1955b; Kunz et al., 1981). Nevertheless, under normal physiologic conditions, the translocase responsible for these movements of ATP and ADP is saturated and would not, therefore, contribute to regulation.

This single equilibrium concept was extended to the *near-equilibrium hypothesis*, in which one considers that all the reactions of the respiratory chain from NADH to ATP operate at close to equilibrium, except for cytochrome c oxidase (Erecinska and Wilson, 1982). In this context, ATP production is regulated by the $NADH/NAD^+$ ratio, the F potential, or any effector of Cytochrome c oxidase, including O_2. Therefore, the rate of ATP production can be controlled by O_2 concentration. However, a decrease in ATP production is not observed until below 1 mmHg with isolated mitochondria (Chance and Williams, 1955b). Even if the critical PO_2 were 10 mmHg, as has been suggested (Wilson et al., 1979), the lung would not be a likely place for the regulation of ATP by oxygen tension because, while normal respiration is maintained, the lowest PO_2 observed is 40 mmHg within the pulmonary arteriole. Although, in some tissues a gradient might be produced in which oxygen tension falls below 10 mmHg, this would be rather unlikely in the cells composing the alveolar spaces of the lung. The only situations where hypoxia may be a critical regulator of cellular function in the lung would be during embryological development and in disease such as within a tumor or in bronchiectasis.

It has become clear that cytochrome c oxidase and the matrix dehydrogenases, as well as many of the enzymes of the glycolytic pathway, are allosterically regulated (Kadenbach, 1986; Glancy and Balaban, 2012). Critically important among the regulators of ATP production appear to be calcium and the whole range of adenine nucleotides (Brown, 1992; Glancy and Balaban, 2012; Wilson and Vinogradov, 2014). In these situations, it is worth considering the energy charge equation, which is a more complicated form of the F potential that takes into account the formation of AMP that occurs under metabolic stress.

$$\text{Energy charge} = (ATP) + 0.5(ADP)/(ATP) + (ADP) + (AMP) \tag{6}$$

The formation of AMP is critical in the regulation of the AMP-dependent kinase (AMPK). AMPK is found throughout the lung and is present as different isoforms within small and larger artery endothelia. AMPK is an important cellular regulator, and its activity can directly alter cellular metabolism as well as alter downstream expression targets such as mTOR. It is regulated by both the SIRT1 and LKB1 kinases in an AMP-dependent manner. It has been proposed that down-regulation of AMPK function via SIRT1 can lead to reduced mitochondrial activity and sensitivity to chemotherapeutic drugs in non-small cell lung cancers (Shin et al., 2014).

2.6 Lung Bioenergetics

The lung is the passageway for the entire amount of O_2 consumed by the body. As such, the utilization of O_2 by the lung itself is a very small fraction of that total and could easily be overlooked. Nevertheless, it is now well established that the lung has metabolic requirements that must be satisfied to maintain its structural integrity. Thus, these fairly small metabolic requirements determine the metabolic fate of the entire organism.

Techniques developed for metabolic studies of the lung must account for the unique structure–function relationships of the organ. The lung has a large volume and yet constitutes only 0.5% of body weight, and approximately 33% of that is blood. While most of the pulmonary circulation is involved with the major function of the lung—gas exchange between the blood and atmosphere—a second circulation, the bronchial circulation, must also be considered. Finally, since the entire cardiac output flows through the lung, the small differences in metabolite concentration due to the lungs' own small requirements are difficult to measure. Nevertheless, as demonstrated by the very existence of this book, lung metabolism is not an impossible area of investigation. Indeed, in terms of oxidative metabolism, the lack of diffusion limitation for O_2 presents an opportunity for metabolic study of an intact organ.

The consumption of oxygen by the lung itself has been measured in tissue slices (103 ± 37 μL/min/g) and compared with that of other organs. The lung is similar to a group of organs that are lower in oxygen consumption, namely the intestine (116 μL/min/g), pancreas (60 μL/min/g), and spleen (166 μL/min/g); and is considerably lower than the brain (327/μL/min/g), kidney cortex (405 μL/min/g), and heart (204 μL/min/g) (Fisher and Forman, 2011). However, the lung consumes about 1% of the total oxygen uptake, while it is only about 0.5% of total body weight (Fisher, 1976). If these statements appear to be somewhat inconsistent, the most likely explanation is the difficulty in measuring oxygen consumption in the intact lung and the imperfections of the tissue-slice measurement for metabolism. Nevertheless, these measurements indicate that the lung is a respectable consumer of oxygen.

Just as the rate of total body consumption is inversely related to total body weight, the lungs show a similar relationship. Energy demands of the rest of the body would not necessarily require increased oxygen consumption by the lung, since the work of respiration, in terms of lung inflation, is supplied by respiratory muscles rather than the lung itself. It is more likely that relative lung metabolism reflects the general inverse relationship between metabolic rate and size. Oxygen consumption by the lung should be related to the metabolic activities of the lung itself, and its specialized energy-dependent function. Table 1 gives a list of these energy-dependent functions of lung cells. It is possible that some of these processes may be indirectly stimulated by lung inflation and thus are somewhat dependent on the metabolic rate of the whole animal (Faridy and Naimark, 1971). Histochemical approaches show that staining of both the horse and rat lung for glycolytic and pentose shunt pathway enzymes was similar to that for the rat kidney and liver, while the mitochondrial enzymes were weak in comparison. Within the lungs, alveolar and free cells (presumably type II epithelial cells and macrophages, respectively) were found to have comparatively greater activities for glycolytic, pentose shunt, and mitochondrial enzymes than either the

TABLE 1 Special Energy-Dependent Functions of Lung Cells

Function	Cell	Reaction	References[a]
Secretion			
Mucus	Bronchial gland epithelium	Cytoskeletal elements	
Chloride	Tracheobronchial epithelium	Linked to Na^+-K^+-ATPase	Widdicombe and Welsh (1980)
Surfactant	Granular pneumocyte	Cytoskeletal elements	Marino and Rooney (1980)
Superoxide release	Alveolar macrophage	Activation of NADPH oxidase	
Synthesis			
Surfactant	Granular pneumocyte	Generation of fatty acyl-CoA and CDP choline	
Collagen	Fibroblast	Activation of amino acyl-tRNA	Alper et al. (1982b)
Transport			
Amine uptake	Capillary endothelium	Linked to Na^+-K^+-ATPase	Junod (1973) and Steinberg et al. (1975c)
Na^+ absorption	Alveolar epithelium	Linked to Na^+-K^+-ATPase	Goodman and Crandall (1982) and Mason et al. (1982)
Motility			
Phagocytosis	Alveolar macrophages	Cytoskeletal elements	Oren et al. (1983)
Ciliary clearance	Ciliated bronchial cell	Actin–tubulin interaction	Dahlman and Rosengren (1968)
Bronchoconstriction	Bronchial smooth muscle	Actin–myosin cross-bridge cycling	Stephens et al. (1968)
Vasoconstriction	Vascular smooth muscle	Actin–myosin cross-bridge cycling	Lloyd (1964)

[a]*Studies in which a presumed energy-dependent reaction has been demonstrated in the lung.*
Reproduced from Fisher, A.B., Forman, H.J. (1985). Oxygen utilization and toxicity. In: Handbook of Physiology. Section 3: The Respiratory System, vol. I. American Physiology Society, Baltimore, 231–254. With permission.

pulmonary epithelium (type I cells) or endothelium (Tyler et al., 1985). For the bronchial epithelium, a similar study was made to differentiate the pattern in human, cat, and mouse bronchial glands, goblet cells, and bronchial or tracheal epithelium. The patterns were approximately the same in these areas (Azzopardi and Thurlbeck, 1968, 1969).

2.7 Lung Mitochondria

As noted in the previous section, most oxygen consumption is dependent upon mitochondria, and one would expect that the volume density of mitochondria would reflect the relative metabolic rate. On that basis, the granular type II pneumocyte, having approximately three times greater mitochondrial volume per cell than other cells, such as the endothelium and type I pneumocyte (Massaro et al., 1975), would be expected to have a much greater oxygen consumption. Alveolar macrophages also have a high mitochondrial volume. While surfactant metabolism and secretion of granular pneumocytes and the phagocytic activity of alveolar macrophages could consume considerable amounts of energy compared to the activities of other lung cells, the energy requirement for transport across the large surfaces of the endothelium and type I pneumocyte cannot be disregarded a priori. However, in agreement with the mitochondrial volume measurements, electron microscopic analysis indicates that type I cells may be deficient in cytochrome c oxidase (Hirai and Ogawa, 1986), implying that the energy demands of this cell are relatively low and/or are supplied primarily from glycolysis rather than mitochondrial respiration.

Isolation of lung mitochondria is difficult because of the high elastic fiber content of the lung, which hinders homogenization. Nevertheless, intact mitochondria showing respiratory control have been isolated from lungs by several investigators (Fisher et al., 1973; Fisher, 1975; Mustafa and Cross, 1974; Spear and Lumeng, 1978; Sayeed and Baue, 1971). As with other mitochondria, they accumulate divalent cations by an energy-dependent process (Fisher et al., 1973) and have a high affinity for oxygen. Lung mitochondria can oxidize pyruvate, fatty acids, Krebs cycle intermediates, glycerol-3-phosphate, and glutamate. Oxidation of acetoacetate and 13-hydroxybutyrate appears to vary with species (Fisher et al., 1973, 1975; Mustafa and Cross, 1974).

2.8 ATP Generation

As with other organs, O_2 consumption is related to energy-dependent functions through the generation of ATP. The ATP content of the lung is approximately half of that of the heart, but is very close to that of the brain, liver, and kidney (Bassett and Fisher, 1976a; Williamson and Corkey, 1979). The ATP/ADP ratio of the lung (8.5) is similar to the heart and brain, but is approximately double that of liver. The energy charge (see above; 0.92) is also similar to the value obtained from other organs. Thus, the lung maintains a similar balance between energy utilization and generation as other tissues.

Evidence that the majority of the ATP is generated in mitochondria rather than through glycolysis has been provided through the use of mitochondrial respiration inhibitors, such as CO and the uncoupler of oxidative phosphorylation, dinitrophenol (Bassett and Fisher, 1976a,b). As expected for a tissue that depends on mitochondrial respiration for the major fraction of its ATP, CO decreased ATP content and the ATP/ADP ratio, while inclusion of dinitrophenol resulted in a large increase in glucose oxidation while decreasing cellular ATP and the ATP/ADP ratio. These experiments showed that mitochondrial oxidative phosphorylation accounted for 86% of the total ATP produced from glucose metabolism. Reducing equivalents from the glycolytic pathway are transported to the mitochondrial matrix via the shuttle mechanisms. Interference with the malate-aspartate shuttle caused increased lactate/pyruvate ratios in the lung (Fisher et al., 1976). Glycerol-3-phosphate dehydrogenase (essential for the glycerol-3-phosphate/dihydroxyacetone phosphate shuttle) is present in lung cells in both the mitochondrial and cytoplasmic compartments (Fisher et al., 1973, 1975; Mustafa and Cross, 1974; Fisher and Chander, 1982).

3. OXYGEN-DEPENDENT METABOLIC REACTIONS

As in most tissues, the cytochrome c oxidase reaction in lung mitochondria accounts for most O_2 utilization. Nevertheless, a great number of other metabolic reactions utilize O_2 as a substrate. These reactions can be classified into three general groups: dioxygenases, oxidoreductases, and mixed-function oxidases that were originally proposed by Keevil and Mason (Keevil and Mason, 1978). Dioxygenases catalyze transfer both oxygen atoms within molecular oxygen into the substrate as exemplified by indoleamine 2,3 dioxygenase (Capece et al., 2012) and the lipoxygenases (Smyrniotis et al., 2014). Oxidoreductases catalyze electron transfer to molecular oxygen to generate O_2^-, H_2O_2, or H_2O. Examples of this class include cytochrome c oxidase (see Electron Transport) and the NADPH oxidase of phagocytes, which catalyzes the formation of two superoxide molecules from NADPH. The oxidases also include a special group in which oxygen is incorporated into the product, but from H_2O rather than molecular O_2. An example of this subclass is amine oxidase (flavin containing), which catalyzes in transfer of electrons to FAD with an oxidative deamination. The third class is the mixed-function oxidases, which share the properties of the other two groups (one atom of the O_2 molecule is incorporated into the substrate while the other atom is reduced). The mixed-function oxidase reactions involve cosubstrates, in addition to O_2 and the organic substrate. These reactions result in the incorporation of O_2 into an organic product, with electrons transferred to O_2, resulting in H_2O and the simultaneous oxidation of the cosubstrate (most commonly NADPH).

$$SH_2 + NADPH + O_2 \rightarrow POH + NADP^+ + OH^- \quad (7)$$

Other members of the mixed-function oxidase class are the cytochrome-P450-dependent steroid (or drug) hydroxylases (Gillette et al., 1972) and phenylalanine 4-monooxygenase, which generates tyrosine. Another group of mixed-function oxidases uses ascorbate instead of NADPH as the cosubstrate. An example is the neural tissue dopamine (3-monooxygenase (hydroxylase)), which converts dopamine to norepinephrine (Kaufman, 1966).

3.1 Mechanisms

All enzymes that use O_2 as a substrate require the participation of prosthetic cofactors that can cycle between reduced and oxidized forms. The cofactors bind and then transfer electrons to O_2. They have been classified into four groups: copper containing, iron-porphyrin, nonheme-iron sites, and flavin sites, including the similar tetrahydrobiopterin, a pteridine derivative (Keevil and Mason, 1978). Some O_2 utilizing enzymes have more than one cofactor, such as the cytochrome-P450-containing mixed-function oxidases that involve FAD, FMN, and an iron-porphyrin (Poulos, 1988). Xanthine oxidase (which is an altered form of xanthine dehydrogenase) and aldehyde oxidase are flavoproteins that pass electrons through molybdenum and an iron-sulfur center during catalysis (Coughlan et al., 1969; Fridovich and Handler, 1962; Gardlik and Rajagopalan, 1990; Rajagopalan and Handler, 1967). These enzymes convert their substrate to a product at one site, while reducing O_2 at another. In this regard, these enzymes are similar to the electron transport system in mitochondria.

When metal prosthetic groups are involved, reduction of the metal is often required for the binding of O_2. This means that the chemical mechanism involving the metal cofactor can be described as sequential with some reductant initially reducing the metal center, and itself becoming oxidized, followed by the formation of a reduced metal–oxygen complex. If this complex decomposes at this stage, O_2^- would be the product. Addition of a second electron to the metal–oxygen complex before release would result in H_2O_2 formation. If two cofactors simultaneously bind O_2, it is possible to reduce the O_2 by more than two electrons. Such a bridge-type complex is involved in the action of cytochrome c oxidase. The action of flavins, and biopterin, can be similar to that of metals in that flavins can exist in both one-electron and two-electron reduced forms. Thus, flavoproteins can catalyze the direct formation of either O_2^- or H_2O_2.

Dioxygenases incorporate O_2 into an organic product without production of additional partially reduced oxygen species. These require the formation of an enzyme substrate complex into which O_2 is incorporated to form the final product. The sequence of binding of substrate and O_2 depends to a large extent upon the requirement for O_2 binding. In those enzymes in which the O_2 binding site is a metal, which must be in the reduced state to bind O_2, the substrate must bind first and provide electrons for reduction of the metallic prosthetic group. For the heme dioxygenase enzymes, such as indoleamine 2,3-dioxygenase, oxygen binding is the initial step in the reaction mechanism. However, the mechanism of catalysis has been a topic of some considerable investigation, with proposals varying from base-catalyzed abstraction (Hamilton, 1969; Sugimoto et al., 2006) to electrophilic and radical addition (Chung et al., 2008), and recently an alternative mechanism involving a ferryl heme species (Efimov et al., 2011).

For the oxidoreductases, there is a requirement for both an electron-transference and a substrate oxidation, and so the general reaction scheme will involve the formation of a substrate/enzyme/O_2 complex that is transitioned to an intermediate oxygen complexed product. Transfer of a single electron to this complex would allow for the formation of product and O_2^-, while two-electron transfer allows for H_2O_2 production. There is a wide range of proteins that can assemble into oxidoreductases (Sumimoto, 2008). This biological complexity is best exemplified by the Nox family of enzymes, which subserve functions from bactericidal to cell regulatory (Lambeth, 2004; Altenhofer et al., 2012).

For the mixed-function oxidases, the kinetic scheme is complex, since two enzymes are actually involved. The complexity of these reaction mechanisms is exemplified by the nitric oxide synthases (NOSs), which convert arginine and oxygen to citrulline and NO with the concomitant consumption of NADPH. These enzymes involve the full range of cofactors as they incorporate FAD, FMN, tetrahydrobiopterin, and heme iron (Hemmens and Mayer, 1998). They catalyze a two-stage enzymatic process; in the first step, the ferric heme iron binds O_2, which is reduced to form a $Fe(III)O_2^-$. Upon binding of arginine, the O–O bond is cleaved to allow for the release of H_2O and the formation of ferrylepoxide, which immediately transfers the O to form N-hydroxyarginine (NOHA). In the second step, a further round of O_2 binding and reduction, is followed by release of the NO, citrulline, and H_2O (Stuehr, 1999). Thus, in the complete cycle, these enzymes consume to generate two H_2O molecules and two oxidized products, NO and citrulline.

Although their contribution to total oxygen uptake is relatively small, the dioxygenases, oxidoreductases, and mixed-function oxidase, other than cytochrome c oxidase, catalyze important cellular reactions. One manner in which their contribution has been estimated is by inhibiting cytochrome c oxidase with cyanide. Nevertheless, as stated above, cyanide is not specific, so that measurement of cyanide-insensitive respiration does not include a number of enzymes. Inhibition of ATP synthesis with cyanide may also affect metabolism, such that cyanide-insensitive enzymes are affected through allosteric effects or competing reactions. Considering these limitations in the method, the estimation of oxygen utilization that is independent of cytochrome c oxidase is 14% of the total O_2 utilization in the lung (Rossouw and Engelbrecht, 1978).

Oxygen-utilizing enzymes, as most enzymes, can be competitively inhibited by analogs of their organic substrates. These enzymes are also subject to the same type of regulation, product inhibition, allosterism, etc. A unique factor affecting heme-containing dioxygenases, however, is that they can be strongly inhibited by small molecules or ions, such as NO, CO, cyanide, and H_2S. These compounds can bind tightly to iron in the nonheme complex. Inhibition by such molecules results from blocking the necessary interaction of O_2 with the transition metal center and may represent a critical form of cellular regulation of these processes (Kolluru et al., 2013; Foresti and Motterlini, 2010; Pinakoulaki et al., 2004). Since there is a great variety in the reactions catalyzed by oxygenases that have the same or similar cofactors, how and if these small molecular weight inhibitors can act with specificity is not clear, a factor that is critical to their role as signal transduction elements.

4. OXYGEN-DEPENDENT REACTIONS IN THE LUNG

Having considered the various types of O_2 consumption reactions that occur in the lung, in this section, we will consider some of the important uses of O_2-utilizing enzymes. Many of the important enzymes that require O_2 as a substrate are heme-dependent enzymes, and the structure and function of these enzymes has been extensively reviewed elsewhere (Poulos, 2014). The flavin-containing amine oxidase is important for the metabolism of vasoactive amines such as serotonin. Cytochrome-P450 monooxygenases are a group of enzymes that are important in the metabolism of steroids and drugs, and can be involved in both detoxification or conversion of nontoxic agents into potent toxins, including carcinogens (Boyd, 1984; Boyd et al., 1980; Forkert et al., 2001; Rowland and Yost, 2010; Warren et al., 1982). Proline hydroxylase is responsible for the conversion of proline to hydroxyproline, which is important in collagen synthesis and is a significant regulator of the hypoxia-inducible transcription factor HIF-1α. Lysine hydroxylase is responsible for the conversion of lysine into an aldehyde, which can cross-link protein chains and is involved in connective tissue modification. The arachidonic acid metabolizing enzymes, cyclooxygenases and lipoxygenases, are involved in the synthesis of prostaglandins and leukotrienes, respectively. These are enormously important reactions both as utilizers of oxygen and as inflammatory regulators. As such, they are discussed in their own chapter of this text. Finally, we will consider the NADPH oxidase of the macrophage and the neutrophil, which is involved in the antibacterial activity of these cells.

4.1 Amine Oxidation

One of the most important roles of the lung, in terms of whole-body metabolism, is removal from circulation of amines such as 5-hydroxytryptamine (serotonin) and norepinephrine. These are removed by the pulmonary endothelial cells through active transport (Hughes et al., 1969; Steinberg et al., 1975a,b). In the lung, serotonin is converted, through the action of the flavin-containing amine oxidase, to 5-hydroxy-indole-acetaldehyde. This enzyme is located on the outer membrane of the mitochondria. Electrons from the serotonin are transferred to FAD. In this unusual mixed-function oxidase reaction, oxygen is incorporated into the product from H_2O, rather than O_2. However, O_2 is involved in the reoxidation of the flavin cofactor, and in that process produces H_2O_2. Norepinephrine is converted in a similar process to vanillyl-mandelic acid. Another isoenzyme of amine oxidase (flavin-containing) shows specificity for phenylethylamine (Bakhle and Youdim, 1976) but has a similar mechanism.

The amine oxidases are members of the FAD monooxygenase (FMO) class, which are important sources of H_2O_2 within the lung and play a critical role in xenobiotic metabolism (Francois et al., 2009; Siddens et al., 2014; Zhang et al., 2006). The FMOs of lung are distinctly different from liver, with FMO2 being the predominant form in adult lung; and there appears to be a distinct developmental aspect to their expression (Henderson et al., 2014; Hines, 2006). In addition, there is a considerable variation in FMO2 allelic frequencies between ethnic groups that may have implications for xenobiotic metabolism (Krueger et al., 2002). The overlap between the FMOs and the cytochrome-P450 monooxygenases in specificity is considerable. These H_2O_2-producing enzymes oxidize thiourea, sulfides, thioamides, thioacetamide, thiols, cysteamine, disulfides, secondary and tertiary amines, hydrazine, and phosphine substituents, and have an endogenous NADPH oxidase activity (Sabourin et al., 1986; Tynes et al., 1985).

4.2 Cytochrome-P450 Monooxygenases

The largest class of mixed-function oxidases utilizes the cofactor cytochrome-P450. (The 450 refers to an absorption maximum that is observed upon binding of carbon monoxide to this cytochrome.) The information regarding these enzymes is being continually updated as new isoforms are being identified (at this point in time that are 57 known human P450s, for constant updates consult this Website: http://drnelson.uthsc.edu/CytochromeP450.html). A systemized nomenclature has been developed for the P450 enzymes based on sequence homology. Enzymes with at least 40% homology are considered to part of the same superfamily, denoted by an Arabic numeral; while greater than 55% homology denotes presence in a particular subfamily, denoted by a letter (Nebert et al., 2013). The final numeral in the naming of these enzymes indicates the individual isotype concerned. These enzymes use NADPH as a cofactor, and early studies indicated that all mammalian enzymes utilized a classical FAD/FMN reductase system to supply electrons for the reaction. However, as the discovered complexity of these enzymes has increased, it has become apparent that there are a number of different reductase systems that are capable of transferring electrons to these monooxygenases (Hannemann et al., 2007). Electrons from both NADPH and the substrate are used to reduce O_2 in these reactions (Poulos, 2014). One of the atoms from O_2 is incorporated into the product, while the other is reduced to H_2O. Binding of O_2 to the heme of cytochrome-P450 is dependent on both the oxidation and spin state of the heme iron. Reduction of the iron is dependent on the supply of electrons, while the spin state is determined by the binding of substrate. Therefore, both the substrate and reducing equivalents must be present before O_2 can bind. The apparent affinity for O_2 varies markedly with the different cytochrome-P450 monooxygenases, primarily because of the differences in substrate effect upon the spin state (Bernhardt et al., 1973). The cytochrome P450-dependent enzymes have a very broad specificity toward organic substrates. Among the reactions catalyzed by these enzymes are dealkylation of oxygen- and nitrogen-bound alkyl groups, sulfoxidation, N-oxidation, and hydroxylation (Gillette et al., 1972; Korzekwa et al., 1995). Paranitroanisole demethylase is an example of a dealkylation reaction that results in the generation of formaldehyde from a methyl group. Substrates for the hydroxylases include aryl hydrocarbons (Gillette et al., 1972), steroids (Cooper et al., 1973), and unsaturated fatty acids (Capdevila et al., 1981).

Cytochrome-P450 monooxygenase activity can be studied in the isolated perfused lung (Itakura et al., 1977). The lung can metabolize a number of drugs, including phenobarbital, methadone, lidocaine, and several steroids (Devereux et al., 1989; Philpot and Smith, 1984; Blase and Loomis, 1976; Hartiala, 1976; Post et al., 1978; Nicholas and Kim, 1975), and much of this metabolism occurs in both Clara and type II epithelial cells (Devereux et al., 1981). The lung is one of the major sites for cytochrome-P450 activity outside of the liver (Gasser et al., 1988; Lawton et al., 1990). Glucose is required in this system to provide NADPH (Fisher et al., 1981). However, mitochondrial substrates can substitute for glucose, suggesting a role for the transhydrogenation reaction in supplying NADPH. It appears that, regardless of substrate specificity, all cytochrome-P450 monooxygenases are inhibited by compounds that will bind or react with the active site heme, such as CO and metyrapone (Devereux et al., 1981). The affinity for CO relative to oxygen is much greater (approximately 30-fold) for cytochrome-P450 monooxygenases than it is for cytochrome oxidase (Fisher et al., 1981), emphasizing the importance of heme pocket structure in determining ligand binding (Poulos, 1988, 2014).

These enzymes have been far better characterized in the liver and adrenal glands. The enzymes are usually associated with the endoplasmic reticulum, although some, particularly the steroid hydroxylases in the adrenal glands, are associated with the mitochondria. Lung cytochrome-P450 monooxygenase activity is approximately one-twentieth to one-tenth (per unit weight) that of the liver. Also, compared with the liver, the lung shows a relatively small inducibility for cytochrome-P450 monooxygenase activity with drug substrates such as phenobarbital (Philpot and Smith, 1984). Nevertheless, aryl hydrocarbon hydroxylase activity can be induced up to 15-fold in rat lung by treatment with 3-methylcholanthrene (Matsubara et al., 1974) and its induction is associated with exposure-related mitochondrial dysfunction (Bansal et al., 2014). The inducible aryl hydrocarbon hydroxylase activity of the lung may be particularly important in the formation of carcinogens from cigarette smoke components, and polymorphisms in these enzymes may be a risk factor for the development of lung cancer (Hussain et al., 2014; Hussein et al., 2014; Kim et al., 2004). The lung-specific activity for metabolism of some inhaled agents, particularly benzene and *p*-xylene, exceeds the activity in the liver (Philpot and Smith, 1984) and is critical to the bioactivation of a range of pulmonary toxicants (Kartha and Yost, 2008).

Different isoforms of P450 monooxygenases have been identified by western blotting, functional, and mRNA expression studies within the lung including CYP1A1/2, 2A6, 1B1, 2B6/7, 2D6, 2E1, 3A4/5 (Bernauer et al., 2006;

Castell et al., 2005). There is considerable variability in the expression of individual isoforms within human tissue samples (Raunio et al., 1999), and it is often confusing to ascribe individual expression patterns. For this reason, publications often refer to enzyme activities rather than to specific isoforms. Examples are benzpyrene hydroxylase activity, which can be induced in rat lung (Wattenberg et al., 1968), the increase of O-deethylase activity, primarily in alveolar type II cells, and the increase in coumarin hydroxylase in Clara cells (Kitagawa et al., 1986). In attempting to define the cellular sites of cytochrome-P450 activity in the lung, investigators using rabbit lungs first identified the Clara cell, a bronchiolar epithelial cell, and the cuboidal alveolar cell (also called the granular pneumocyte or type II cell) as the principal sites of cytochrome-P450 activity (Devereux et al., 1981; Serabjit-Singh et al., 1980). The Clara cell has higher specific activity for some of the cytochrome-P450 activities; however, the type II cell may contribute more to the overall lung activity because of its relatively greater abundance. Alveolar macrophages appear to have only a small level of cytochrome-P450 activity (Castell et al., 2005; Hook et al., 1972). The evidence regarding expression has shown that certain enzymes are constitutive, such as 3A5, while others are inducible, such as 3A6, and it is these forms where the most variety in expression is observed (Raunio et al., 2005).

The P450s form part of a complex network of drug-metabolizing enzymes whose expression is controlled by an array of nuclear receptors, such as those that respond to polycyclic aryl hydrocarbons. These receptors and their target enzymes are an area of intense of investigation with regard to pharmacogenomics and disease susceptibility (Rushmore and Kong, 2002). The role that these enzymes play in controlling oxygen tension within the lung is probably minimal, but it has been suggested that they could play a role in hypoxic pulmonary vasoconstriction (Miller and Hales, 1979).

4.3 Prolyl and Lysyl Oxidases

Fibroblasts synthesize and secrete collagen and elastin. These two products, along with the other proteins that compose the core of the lung connective tissue, provide most of the lung's elastic properties (Hance et al., 1976). The epithelial and endothelial cells of the lung also synthesize type IV collagen as a component of their basement membranes (Sage et al., 1983). One of the essential steps in the synthesis of structurally functional collagen is hydroxylation of prolyl and lysyl residues (Prockop et al., 1979a,b). Lysyl oxidase, a copper-containing enzyme (Shieh and Yasunobu, 1976) of the flavin-containing oxidase class, works in a manner similar to other amine oxidases (Prockop et al., 1979a,b).

Prolyl residues are oxidized to either 4-hydroxyprolyl or 3-hydroxyprolyl residues by prolyl-4-hydroxylase and prolyl-3-hydroxylase, respectively. These reactions are catalyzed by mixed-function oxidases that utilize α-ketoglutarate plus coenzyme A. Since the iron in the active site of these enzymes must be in the Fe^{2+} state to begin the catalytic cycle, a reducing agent is also required prior to initiation of the reaction. This requirement is served by ascorbate. Ascorbate is not used in each catalytic cycle and is thus not a substrate. The apparent K_M for O_2, in terms of alveolar PO_2, for prolyl hydroxylation is approximately 5 mmHg (Alper et al., 1982a).

The hydroxylation of prolyl and oxidative coupling of the lysyl residues of collagen are necessary for formation of a stable triple helix, which provides the structural stability of collagen (Prockop et al., 1979a,b; Last and Reiser, 1984). Oligosaccharides are attached to the hydroxyl moieties of hydroxylysine in collagen. Collagen without sufficient hydroxylation is gelatinous in structure. In elastin, three or four hydroxylysine residues are converted to the extraordinary amino acids isodesmosine and desmosine, which form the cross-links responsible for much of the elastic properties of this protein. The collagen content within the lung is controlled by constant cycles of deposition and degradation, which are critical both to normal lung function and to development (McKleroy et al., 2013; Spooner et al., 1986). Regulation of these processes is controlled by a complex set of intracellular signals as maintenance of this balancing act is critical to normal lung function (Bienkowski and Gotkin, 1995).

Another critical function of prolyl hydroxylases is the regulation of the transcription factor, HIF-1. HIF-1 is heterodimer of α and β subunits that regulates a wide range of genes, with a particular emphasis on oxidative stress response genes (Olson et al., 2011). The β subunit is constitutively expressed, but the expression of the α subunit is dependent on O_2 tension in a manner regulated by prolyl-2-hydroxylase (Wang et al., 1995). When O_2 tension is high, the HIF-1α subunit is hydroxylated at two potential target prolines, which mediates association with von Hippel–Lindau protein and degradation by the proteasome (Semenza, 2009). Thus, when the O_2 tension is high, there is little functional HIF-1 protein, but in hypoxia this transcription factor becomes a major cellular regulator. While this process may not represent a major target of oxygen consumption within the lung, it is a major mechanism of cellular regulation in hypoxia.

4.4 The NOX Enzymes

As has been mentioned throughout this chapter, the production of partially reduced oxygen species can lead to cellular signaling through redox reactions. These processes are critical to normal cellular homeostasis, and the concept of redox signaling has now been well developed (Forman et al., 2014). The importance of redox signaling is further demonstrated by the NADPH oxidases (Nox), a group of enzymes whose main function appears to be the generation of partially reduced oxygen species using NADPH as an electron donor (Bedard and Krause, 2007). These enzymes not only participate in redox signaling, but are also used in hormone production, matrix regulation, and, critically, host defense. There are seven homologues termed Nox 1 to 5 and Duox 1 and 2, and these proteins all differ in their expression, tissue distribution, and induction/activation (Brandes et al., 2014a,b).

Of particular interest within the lung is inducible Nox activation within phagocytes, such as macrophages and neutrophils, in response to noxious stimuli. Indeed, in activation of Nox 2 via the Fc receptor, phagocytes can increase O_2^- to approximately 10 nmol/min/10^6 cells (Brandes et al., 2014b). This sudden elevation in O_2 consumption by stimulated phagocytes is called the *respiratory burst* and is generated by the prototypic leukocyte NADPH oxidase, Nox 2. The activation of Nox 2 is regulated by assembly of its constitutive components, including p47phox, p67phox, and p22phox, with the unique membrane-bound cytochrome b_{558} (Dang et al., 2002). Activation also requires association with the GTP-binding protein p21rac1 (Abo et al., 1991) and is regulated by phosphorylation at series of critical serine residues (Ago et al., 1999; el Benna et al., 1994). The central cytochrome b_{558} is at the center of each of the Nox isoforms, and has a unique reductive potential (−245 mV) (Segal and Jones, 1978), making it ideally suited for O_2^- generation.

The Nox enzymes utilize heme centers in their catalytic cycle and, as such, are at least theoretically regulated by gaseous second messengers. CO generated from HO-1 has been proposed to regulate Nox 1 in smooth muscle cells (Rodriguez et al., 2010), while H_2S regulates Nox 1 expression and p21rac function (Muzaffar et al., 2008). NO, which may regulate Nox function directly, or by induction of HO-1, may be an important regulator in endothelial cells (Jiang et al., 2006). NO also participates in the modification of key cysteines within the Nox proteins by S-nitrosation (Selemidis et al., 2007); however, the mechanism, which involves an oxidation of uncertain chemistry, may be of more importance within inflammation where NO production is maximized and nitrosothiol degradation may be reduced (Qian et al., 2012).

Once formed, O_2^- can react to produce H_2O_2. This can lead to the generation of other potent oxidants, as described above (see Oxygen Chemistry). Species of partially reduced oxygen that have been proposed as responsible for the microbicidal action associated with the respiratory burst are O_2^-, H_2O_2, •OH, HOCl, and HO_2, the protonated form of superoxide (pK = 4.69). Singlet O_2 has also been proposed as a participant in microbial action (Forman and Thomas, 1986). While the process of Nox function is exclusively reliant upon NADPH, the process of phagocytosis requires ATP. Many phagocytes rely on glycolysis for this energetic requirement, but the alveolar macrophage relies on mitochondrial ATP generation (Cohen and Chovaniec, 1978a). Just about every element of signal transduction that has been investigated in other systems appears to be involved in the regulation of O_2^- production by phagocytes. Although it can be biochemically dissociated from other functions of phagocytes, the respiratory burst is intimately associated with phagocytosis and represents most of the O_2-dependent microbial activity of the immune system.

4.5 Nitric Oxide Synthases

Nitric oxide is a critical regulator within the lung. It controls a wide variety of physiological processes, from airway and blood vessel tone to innate immune defense. Furthermore, dysregulation of NO production has been implicated in a multitude of pulmonary diseases. There are three isoforms of the enzyme that synthesize NO, nitric oxide synthase (NOS). NOS1 and 3 are both constitutively expressed and were originally identified within endothelial and neurological tissue, respectively (Griffith and Stuehr, 1995); while NOS2 is inducible in its expression and was originally found in stimulated macrophages (Lancaster and Hibbs, 1990). Importantly, all three isoforms are expressed within the lung and are critical to the maintenance of pulmonary homeostasis (Kobzik et al., 1993). NOS catalyzes the conversion of arginine and O_2 into citrulline, NO, and H_2O. All three isoforms operate as specialist P450-type enzymes in that they utilize NADPH as a source of reducing equivalents and have a FAD/FMN-containing reductase domain, and that they act as mixed-function oxidases incorporating one atom of oxygen into the guanidino group of arginine to generate NO and the other into water. The NOSs are all homodimers with each subunit possessing a reductase and an oxidase domain. However, the passage of electrons goes from the reductase of one subunit to the oxidase domain of the other, hence NOSs are only functional as dimers (Griffith and Stuehr, 1995). A critical aspect of this connection is the biopterin cofactor, which is critical to linking reductase function and substrate binding. In the absence of a biopterin cofactor, the enzyme becomes "uncoupled" and will generate O_2^- from the direct reduction of O_2 (Crabtree et al., 2013); such uncoupling can also occur when there is a lack of arginine substrate.

4.6 Xanthine Oxidase

There has been much speculation about a role for xanthine oxidase in the injury to tissues that occurs during reperfusion after ischemia (Parks and Granger, 1983). Xanthine oxidase activity is the result of conversion of xanthine dehydrogenase, which normally uses NAD^+ as an acceptor of electrons. In contrast, xanthine oxidase transfers electrons to O_2, producing either O_2^-, or H_2O_2. The enzyme contains FAD, nonheme iron (an iron–sulfur complex), and a molybdenum–pterin complex. Conversion to the oxidase can be due to either oxidation or proteolysis (McCord, 1985). Xanthine oxidase activity in the lung of some species has been localized in capillary endothelium and to a small extent in alveolar macrophages (Bruder et al., 1983). Recently, nitrite reduction has been proposed as a secondary function for xanthine oxidase. In this mechanism, NO_2^- can substitute for the O_2 at the molybdenum–pterin complex and be reduced to NO. This provides for an O_2 tension regulated NO generation mechanism (Cantu-Medellin and Kelley, 2013). This may have important implications for the control of blood flow and ventilation/perfusion matching within the lung (Bueno et al., 2013).

4.7 Antioxidant Defenses

Although this volume concerns the normal physiology of the lung, the antioxidant enzymes constitute an important part of the normal utilization of oxygen in the lung. As described in previous sections of this chapter, normal oxygen metabolism can result in the partial reduction of oxygen to species that are often referred to as *reactive oxygen species*. It is, however, far more preferable and accurate to refer to the actual species, as they are markedly different in chemistry and biological relevance. O_2^-, H_2O_2, and •OH, which is generated in the Fenton reaction, are implicated in oxidant injury. The normal physiologic role of the antioxidant enzymes, superoxide dismutase, catalase, and glutathione peroxidase and peroxiredoxins is to ensure that steady-state concentrations of these molecules remain low. Alterations in the concentrations of these enzymes in a positive or negative direction have been correlated with increases in resistance or susceptibility to oxidant injury, respectively (Forman and Fisher, 1981). However, the role of these reactive molecules is not merely to act as injurious agents. In particular, H_2O_2 also is a significant cellular signaling molecule. Within the lung, redox signaling provides a mechanism for sensing and adapting to the environment (Forman et al., 2010, 2014).

5. CONCLUSIONS

The lung, as most other aerobic tissues, requires oxygen for the generation of ATP and for much of its intermediary metabolism. The lung is unique in some respects of oxygen metabolism and in having the highest oxygen concentration available to it of any internal organ. Nevertheless, the lung appears to be much like other tissues in both its utilization of oxygen and its defenses against the potential deleterious consequences of oxygen metabolism.

ACKNOWLEDGMENT

This chapter is an update of the same chapter that appeared in the previous edition of this book authored by Dr Henry Forman and Dr Michael Thomas.

REFERENCES

Albracht, S.P., van Heerikhuizen, H., Slater, E.C., 1972. Iron-sulphur proteins in the succinate oxidase system. Biochim. Biophys. Acta 256 (1), 1–13. Epub 1972/01/21. PubMed PMID: 4333297.

Azzopardi, A., Thurlbeck, W.M., 1968. Oxidative enzyme pattern of the bronchial mucous glands. Am. Rev. Respir. Dis. 97 (6), 1038–1045. PubMed PMID: 4172077.

Azzopardi, A., Thurlbeck, W.M., 1969. The histochemistry of the nonciliated bronchiolar epithelial cell. Am. Rev. Respir. Dis. 99 (4), 516–525. PubMed PMID: 5767585.

Altenhofer, S., Kleikers, P.W., Radermacher, K.A., Scheurer, P., Rob Hermans, J.J., Schiffers, P., et al., 2012. The NOX toolbox: validating the role of NADPH oxidases in physiology and disease. Cell. Mol. Life Sci. 69 (14), 2327–2343. Epub 2012/06/01. http://dx.doi.org/10.1007/s00018-012-1010-9. PubMed PMID: 22648375; PubMed Central PMCID: PMCPMC3383958.

Alper, R., Kerr, J.S., Kefalides, N.A., Fisher, A.B., 1982a. Relation between reduced alveolar PO_2 and collagen biosynthesis in the perfused rat lung. J. Lab. Clin. Med. 99 (3), 442–450. PubMed PMID: 7057067.

Alper, R., Kerr, J.S., Kefalides, N.A., Fisher, A.B., 1982b. Relation between alveolar PO, and collagen biosynthesis in the perfused rat lung. J. Lab. Clin. Med. 99, 442.

Abo, A., Pick, E., Hall, A., Totty, N., Teahan, C.G., Segal, A.W., 1991. Activation of the NADPH oxidase involves the small GTP-binding protein p21rac1. Nature 353 (6345), 668–670. http://dx.doi.org/10.1038/353668a0. PubMed PMID: 1922386.

Ago, T., Nunoi, H., Ito, T., Sumimoto, H., 1999. Mechanism for phosphorylation-induced activation of the phagocyte NADPH oxidase protein p47(phox). Triple replacement of serines 303, 304, and 328 with aspartates disrupts the SH3 domain-mediated intramolecular interaction in p47(phox), thereby activating the oxidase. J. Biol. Chem. 274 (47), 33644–33653. PubMed PMID: 10559253.

Boveris, A., 1977. Mitochondrial production of superoxide radical and hydrogen peroxide. Adv. Exp. Med. Biol. 78, 67–82. Epub 1977/01/01. PubMed PMID: 197811.

Boveris, A., Chance, B., 1973. The mitochondrial generation of hydrogen peroxide. General properties and effect of hyperbaric oxygen. Biochem. J. 134 (3), 707–716. Epub 1973/07/01. PubMed PMID: 4749271; PubMed Central PMCID: PMCPMC1177867.

Boyer, P.D., Chance, B., Ernster, L., Mitchell, P., Racker, E., Slater, E.C., 1977. Oxidative phosphorylation and photophosphorylation. Annu. Rev. Biochem. 46, 955–966. http://dx.doi.org/10.1146/annurev.bi.46.070177.004515. PubMed PMID: 18361775.

Beard, D.A., 2005. A biophysical model of the mitochondrial respiratory system and oxidative phosphorylation. PLoS Comput. Biol. 1 (4), e36. http://dx.doi.org/10.1371/journal.pcbi.0010036.

Brown, G.C., 1992. Control of respiration and ATP synthesis in mammalian mitochondria and cells. Biochem. J. 284 (Pt 1), 1–13. Epub 1992/05/15. PubMed PMID: 1599389; PubMed Central PMCID: PMCPMC1132689.

Bassett, D.J., Fisher, A.B., 1976a. Metabolic response to carbon monoxide by isolated rat lungs. Am. J. Physiol. 230 (3), 658–663. PubMed PMID: 1266968.

Bassett, D.J., Fisher, A.B., 1976b. Stimulation of rat lung metabolism with 2,4-dinitrophenol and phenazine methosulfate. Am. J. Physiol. 231 (3), 898–902. PubMed PMID: 970473.

Boyd, M.R., April 1984. Metabolic-activation and lung toxicity – a basis for cell-selective pulmonary damage by foreign chemicals. Environ. Health Perspect. 55, 47–51. http://dx.doi.org/10.2307/3429691. PubMed PMID: WOS: A1984SU06800005.

Boyd, M.R., Statham, C.N., Longo, N.S., 1980. The pulmonary clara cell as a target for toxic chemicals requiring metabolic activation; studies with carbon tetrachloride. J. Pharmacol. Exp. Ther. 212 (1), 109–114. PubMed PMID: 7351618.

Bakhle, Y.S., Youdim, M.B., 1976. Metabolism of phenylethylamine in rat isolated perfused lung: evidence for monoamine oxidase 'type B' in lung. Br. J. Pharmacol. 56 (1), 125–127. PubMed PMID: 1252659; PubMed Central PMCID: PMC1666762.

Bernhardt, F.H., Erdin, N., Staudinger, H., Ullrich, V., 1973. Interactions of substrates with a purified 4-methoxybenzoate monooxygenase system (*O*-demethylating) from *Pseudomonas putida*. Eur. J. Biochem. 35 (1), 126–134. Epub 1973/05/01. PubMed PMID: 4351526.

Blase, B.W., Loomis, T.A., 1976. The uptake and metabolism of carbaryl by isolated perfused rabbit lung. Toxicol. Appl. Pharmacol. 37 (3), 481–490. PubMed PMID: 823669.

Bansal, S., Leu, A.N., Gonzalez, F.J., Guengerich, F.P., Chowdhury, A.R., Anandatheerthavarada, H.K., et al., 2014. Mitochondrial targeting of cytochrome P450 (CYP) 1B1 and its role in polycyclic aromatic hydrocarbon-induced mitochondrial dysfunction. J. Biol. Chem. 289 (14), 9936–9951. Epub 2014/02/06. http://dx.doi.org/10.1074/jbc.M113.525659. PubMed PMID: 24497629; PubMed Central PMCID: PMCPMC3975038.

Bernauer, U., Heinrich-Hirsch, B., Tönnies, M., Peter-Matthias, W., Gundert-Remy, U., 2006. Characterisation of the xenobiotic-metabolizing cytochrome P450 expression pattern in human lung tissue by immunochemical and activity determination. Toxicol. Lett. 164 (3), 278–288. http://dx.doi.org/10.1016/j.toxlet.2006.01.007.

Bienkowski, R.S., Gotkin, M.G., 1995. Control of collagen deposition in mammalian lung. Proc. Soc. Exp. Biol. Med. Soc. Exp. Biol. Med. 209 (2), 118–140. Epub 1995/06/01. PubMed PMID: 7770462.

Bedard, K., Krause, K.H., 2007. The NOX family of ROS-generating NADPH oxidases: physiology and pathophysiology. Physiol. Rev. 87 (1), 245–313. http://dx.doi.org/10.1152/physrev.00044.2005. PubMed PMID: 17237347.

Brandes, R.P., Weissmann, N., Schröder, K., 2014. Redox-mediated signal transduction by cardiovascular Nox NADPH oxidases. J. Mol. Cell. Cardiol. 73, 70–79. http://dx.doi.org/10.1016/j.yjmcc.2014.02.006.

Brandes, R.P., Weissmann, N., Schroder, K., 2014. Nox family NADPH oxidases: molecular mechanisms of activation. Free Radic. Biol. Med. 76C, 208–226. Epub 2014/08/27. http://dx.doi.org/10.1016/j.freeradbiomed.2014.07.046. PubMed PMID: 25157786.

el Benna, J., Faust, L.P., Babior, B.M., 1994. The phosphorylation of the respiratory burst oxidase component p47phox during neutrophil activation. Phosphorylation of sites recognized by protein kinase C and by proline-directed kinases. J. Biol. Chem. 269 (38), 23431–23436. PubMed PMID: 8089108.

Bruder, G., Heid, H.W., Jarasch, E.D., Mather, I.H., 1983. Immunological identification and determination of xanthine oxidase in cells and tissues. Differ. Res. Biol. Divers. 23 (3), 218–225. PubMed PMID: 6687872.

Bueno, M., Wang, J., Mora, A.L., Gladwin, M.T., 2013. Nitrite signaling in pulmonary hypertension: mechanisms of bioactivation, signaling, and therapeutics. Antioxid. Redox Signal. 18 (14), 1797–1809. Epub 2012/08/09. http://dx.doi.org/10.1089/ars.2012.4833. PubMed PMID: 22871207; PubMed Central PMCID: PMCPMC3619206.

Capaldi, R.A., Malatesta, F., Darley-Usmar, V.M., 1983. Structure of cytochrome c oxidase. Biochim. Biophys. Acta 726 (2), 135–148. Epub 1983/07/15. PubMed PMID: 6307356.

Chance, B., 1972. The nature of electron transfer and energy coupling reactions. FEBS Lett. 23 (1), 3–20. Epub 1972/06/01. PubMed PMID: 4343618.

Cortassa, S., Aon, M.A., Marban, E., Winslow, R.L., O'Rourke, B., 2003. An integrated model of cardiac mitochondrial energy metabolism and calcium dynamics. Biophys. J. 84 (4), 2734–2755. Epub 2003/04/02. http://dx.doi.org/10.1016/s0006-3495(03)75079-6. PubMed PMID: 12668482; PubMed Central PMCID: PMCPMC1201507.

Chance, B., Williams, G.R., 1955a. Respiratory enzymes in oxidative phosphorylation. III. The steady state. J. Biol. Chem. 217 (1), 409–427. PubMed PMID: 13271404.

Chance, B., Williams, G.R., 1956. The respiratory chain and oxidative phosphorylation. Adv. Enzym. Relat. Subj. Biochem. 17, 65–134. PubMed PMID: 13313307.

Chance, B., Williams, G.R., 1955b. Respiratory enzymes in oxidative phosphorylation. I. Kinetics of oxygen utilization. J. Biol. Chem. 217 (1), 383–393. PubMed PMID: 13271402.

Capece, L., Lewis-Ballester, A., Yeh, S.R., Estrin, D.A., Marti, M.A., 2012. Complete reaction mechanism of indoleamine 2,3-dioxygenase as revealed by QM/MM simulations. J. Phys. Chem. B 116 (4), 1401–1413. http://dx.doi.org/10.1021/jp2082825. PubMed PMID: 22196056; PubMed Central PMCID: PMC3304497.

Coughlan, M.P., Rajagopalan, K.V., Handler, P., 1969. The role of molybdenum in xanthine oxidase and related enzymes. Reactivity with cyanide, arsenite, and methanol. J. Biol. Chem. 244 (10), 2658–2663. PubMed PMID: 4306033.

Chung, L.W., Li, X., Sugimoto, H., Shiro, Y., Morokuma, K., 2008. Density functional theory study on a missing piece in understanding of heme chemistry: the reaction mechanism for indoleamine 2,3-dioxygenase and tryptophan 2,3-dioxygenase. J. Am. Chem. Soc. 130 (37), 12299–12309. Epub 2008/08/21. http://dx.doi.org/10.1021/ja803107w. PubMed PMID: 18712870.

Cooper, D.Y., Schleyer, H., Levin, S.S., Rosenthal, O., 1973. Studies on the partially purified heme protein P-450 from the adrenal cortex. Ann. N. Y. Acad. Sci. 212, 227–242. PubMed PMID: 4155923.

Capdevila, J., Chacos, N., Werringloer, J., Prough, R.A., Estabrook, R.W., 1981. Liver microsomal cytochrome P-450 and the oxidative metabolism of arachidonic acid. Proc. Natl. Acad. Sci. U.S.A. 78 (9), 5362–5366. PubMed PMID: 6795631; PubMed Central PMCID: PMC348745.

Castell, J.V., Teresa Donato, M., Gómez-Lechón, M.J., 2005. Metabolism and bioactivation of toxicants in the lung. The in vitro cellular approach. Exp. Toxicol. Pathol. 57 (Suppl. 1), 189–204. http://dx.doi.org/10.1016/j.etp.2005.05.008.

Cohen, H.J., Chovaniec, M.E., 1978a. Superoxide production by digitonin-stimulated guinea pig granulocytes. The effects of *N*-ethyl maleimide, divalent cations; and glycolytic and mitochondrial inhibitors on the activation of the superoxide generating system. J. Clin. Invest. 61 (4), 1088–1096. http://dx.doi.org/10.1172/JCI109008. PubMed PMID: 207722; PubMed Central PMCID: PMC372626.

Crabtree, M.J., Brixey, R., Batchelor, H., Hale, A.B., Channon, K.M., 2013. Integrated redox sensor and effector functions for tetrahydrobiopterin- and glutathionylation-dependent endothelial nitric-oxide synthase uncoupling. J. Biol. Chem. 288 (1), 561–569. http://dx.doi.org/10.1074/jbc.M112.415992. PubMed PMID: 23139420; PubMed Central PMCID: PMC3537053.

Cantu-Medellin, N., Kelley, E.E., 2013. Xanthine oxidoreductase-catalyzed reduction of nitrite to nitric oxide: insights regarding where, when and how. Nitric Oxide Biol. Chem. 34, 19–26. Epub 2013/03/05. http://dx.doi.org/10.1016/j.niox.2013.02.081. PubMed PMID: 23454592; PubMed Central PMCID: PMCPMC3706534.

Droge, W., 2002. Free radicals in the physiological control of cell function. Physiol. Rev. 82 (1), 47–95. PubMed PMID: ISI:000173028700003.

Devereux, T.R., Domin, B.A., Philpot, R.M., 1989. Xenobiotic metabolism by isolated pulmonary cells. Pharmacol. Ther. 41 (1–2), 243–256. Epub 1989/01/01. PubMed PMID: 2652151.

Devereux, T.R., Serabjit-Singh, C.J., Slaughter, S.R., Wolf, C.R., Philpot, R.M., Fouts, J.R., 1981. Identification of cytochrome P-450 isozymes in nonciliated bronchiolar epithelial (Clara) and alveolar type II cells isolated from rabbit lung. Exp. Lung Res. 2 (3), 221–230. PubMed PMID: 7028473.

Dahlman, T., Rosengren, A., 1968. The effect of oxygen lack on the tracheal ciliary activity. Arch. Environ. Health 16, 371.

Dang, P.M., Cross, A.R., Quinn, M.T., Babior, B.M., 2002. Assembly of the neutrophil respiratory burst oxidase: a direct interaction between p67PHOX and cytochrome b558 II. Proc. Natl. Acad. Sci. U. S. A. 99 (7), 4262–4265. http://dx.doi.org/10.1073/pnas.072345299. PubMed PMID: 11917128; PubMed Central PMCID: PMC123636.

Erecinska, M., Wilson, D.F., 1976. The effect of antimycin A on cytochromes b561, b566, and their relationship to ubiquinone and the iron-sulfer centers S-1 (+N-2) and S-3. Arch. Biochem. Biophys. 174 (1), 143–157. Epub 1976/05/01. PubMed PMID: 180891.

Erecinska, M., Wilson, D.F., 1978. Cytochrome c oxidase: a synopsis. Arch. Biochem. Biophys. 188 (1), 1–14. Epub 1978/05/01. PubMed PMID: 209738.

Erecinska, M., Wilson, D.F., 1982. Regulation of cellular energy metabolism. J. Membr. Biol. 70 (1), 1–14. PubMed PMID: 6226798.

Efimov, I., Basran, J., Thackray, S.J., Handa, S., Mowat, C.G., Raven, E.L., 2011. Structure and reaction mechanism in the heme dioxygenases. Biochemistry 50 (14), 2717–2724. Epub 2011/03/03. http://dx.doi.org/10.1021/bi101732n. PubMed PMID: 21361337; PubMed Central PMCID: PMCPMC3092302.

Fridovich, I., 1981. The biology of superoxide and of superoxide dismutases – in brief. Prog. Clin. Biol. Res. 51, 153–172. Epub 1981/01/01. PubMed PMID: 6267615.

Forman, H.J., Maiorino, M., Ursini, F., 2010. Signaling functions of reactive oxygen species. Biochemistry 49 (5), 835–842. http://dx.doi.org/10.1021/bi9020378. PubMed PMID: 20050630.

Ferrer-Sueta, G., Manta, B., Botti, H., Radi, R., Trujillo, M., Denicola, A., 2011. Factors affecting protein thiol reactivity and specificity in peroxide reduction. Chem. Res. Toxicol. 24 (4), 434–450. http://dx.doi.org/10.1021/tx100413v. PubMed PMID: 21391663.

Forman, H.J., Kennedy, J.A., 1974. Role of superoxide radical in mitochondrial dehydrogenase reactions. Biochem. Biophys. Res. Commun. 60 (3), 1044–1050. Epub 1974/10/08. PubMed PMID: 4372996.

Forman, H.J., Fridovich, I., 1973. Superoxide dismutase: a comparison of rate constants. Arch. Biochem. Biophys. 158 (1), 396–400. Epub 1973/09/01. PubMed PMID: 4354035.

Freeman, B.A., Crapo, J.D., 1981. Hyperoxia increases oxygen radical production in rat lungs and lung mitochondria. J. Biol. Chem. 256 (21), 10986–10992. Epub 1981/11/10. PubMed PMID: 7287745.

Fisher, A.B., Forman, H.J., 2011. Oxygen Utilization and Toxicity in the Lungs. Comprehensive Physiology. Wiley-Blackwell, 231–254.

Fisher, A.B., 1976. Normal and pathologic biochemistry of the lung. Environ. Health Perspect. 16, 3–9. PubMed PMID: 797566; PubMed Central PMCID: PMC1475232.

Faridy, E.E., Naimark, A., 1971. Effect of distension on metabolism of excised dog lung. J. Appl. Physiol. 31 (1), 31–37. PubMed PMID: 5556960.

Fisher, A.B., Scarpa, A., LaNoue, K.F., Bassett, D., Williamson, J.R., 1973. Respiration of rat lung mitochondria and the influence of Ca^{2+} on substrate utilization. Biochemistry 12 (7), 1438–1445. PubMed PMID: 4348834.

Fisher, A.B., 1975. Functional evaluation of lung mitochondria. Chest 67 (2 Suppl.), 24S–26S. PubMed PMID: 163171.

Fisher, A.B., Huber, G.A., Bassett, D.J., 1975. Oxidation of alpha-glycerophosphate by mitochondria from lungs of rabbits, sheep and pigeons. Comp. Biochem. Physiol. B 50 (1), 5–8. PubMed PMID: 804374.

Fisher, A.B., Furia, L., Chance, B., 1976. Evaluation of redox state of isolated perfused rat lung. Am. J. Physiol. 230 (5), 1198–1204. PubMed PMID: 5896.

Fisher, A.B., Chander, A., 1982. Glycerol kinase activity and glycerol metabolism of rat granular pneumocytes in primary culture. Biochim. Biophys. Acta 711 (1), 128–133. PubMed PMID: 6279165.

Fridovich, I., Handler, P., 1962. Xanthine oxidase. V. Differential inhibition of the reduction of various electron acceptors. J. Biol. Chem. 237, 916–921. PubMed PMID: 13895169.

Foresti, R., Motterlini, R., 2010. Interaction of carbon monoxide with transition metals: evolutionary insights into drug target discovery. Curr. Drug Targets 11 (12), 1595–1604. Epub 2010/08/14. PubMed PMID: 20704543.

Forkert, P.G., Boyd, S.M., Ulreich, J.B., 2001. Pulmonary bioactivation of 1,1-dichloroethylene is associated with CYP2E1 levels in A/J, CD-1, and C57BL/6 mice. J. Pharmacol. Exp. Ther. 297 (3), 1193–1200. PubMed PMID: WOS:000169005200045.

Francois, A.A., Nishida, C.R., de Montellano, P.R., Phillips, I.R., Shephard, E.A., 2009. Human flavin-containing monooxygenase 2.1 catalyzes oxygenation of the antitubercular drugs thiacetazone and ethionamide. Drug Metab. Dispos. 37 (1), 178–186. Epub 2008/10/25. http://dx.doi.org/10.1124/dmd.108.024158. PubMed PMID: 18948378; PubMed Central PMCID: PMCPMC2683658.

Fisher, A.B., Itakura, N., Dodia, C., Thurman, R.G., 1981. Pulmonary mixed-function oxidation: stimulation by glucose and the effects of metabolic inhibitors. Biochem. Pharmacol. 30 (4), 379–383. PubMed PMID: 7213425.

Forman, H.J., Ursini, F., Maiorino, M., 2014. An overview of mechanisms of redox signaling. J. Mol. Cell. Cardiol. 73, 2–9. Epub 2014/02/12. http://dx.doi.org/10.1016/j.yjmcc.2014.01.018. PubMed PMID: 24512843; PubMed Central PMCID: PMCPMC4048798.

Forman, H.J., Thomas, M.J., 1986. Oxidant production and bactericidal activity of phagocytes. Annu. Rev. Physiol. 48, 669–680. http://dx.doi.org/10.1146/annurev.ph.48.030186.003321. PubMed PMID: 3010830.

Forman, H.J., Fisher, A.B., 1981. Antioxidant enzymes of rat granular pneumocytes. Constitutive levels and effect of hyperoxia. Lab. Invest. 45 (1), 1–6. PubMed PMID: 6894775.

Glancy, B., Balaban, R.S., 2012. Role of mitochondrial Ca^{2+} in the regulation of cellular energetics. Biochemistry 51 (14), 2959–2973. Epub 2012/03/27. http://dx.doi.org/10.1021/bi2018909. PubMed PMID: 22443365; PubMed Central PMCID: PMCPMC3332087.

Gillette, J.R., Davis, D.C., Sasame, H.A., 1972. Cytochrome P-450 and its role in drug metabolism. Annu. Rev. Pharmacol. 12, 57–84. Epub 1972/01/01. http://dx.doi.org/10.1146/annurev.pa.12.040172.000421. PubMed PMID: 4114185.

Gardlik, S., Rajagopalan, K.V., 1990. The state of reduction of molybdopterin in xanthine oxidase and sulfite oxidase. J. Biol. Chem. 265 (22), 13047–13054. PubMed PMID: 2376587.

Gasser, R., Negishi, M., Philpot, R.M., 1988. Primary structures of multiple forms of cytochrome P-450 isozyme 2 derived from rabbit pulmonary and hepatic cDNAs. Mol. Pharmacol. 33 (1), 22–30. Epub 1988/01/01. PubMed PMID: 2826996.

Goodman, B.E., Crandall, E.D., 1982. Dome formation in primary cultured monolayers of alveolar epithelial cells. Am. J. Physiol. 243, C96 (Cell Physiol. 12).

Griffith, O.W., Stuehr, D.J., 1995. Nitric oxide synthases: properties and catalytic mechanism. Annu. Rev. Physiol. 57, 707–736. PubMed PMID: 7539994.

Hirai, K., Ogawa, K., 1986. Cytochemical quantitation of cytochrome oxidase activity in rat pulmonary alveolar epithelial cells and possible defect in type I cells. J. Electron Microsc. 35 (1), 19–28. PubMed PMID: 3027222.

Hamilton, G.A., 1969. Mechanisms of two- and four-electron oxidations catalyzed by some metalloenzymes. Adv. Enzym. Relat. Areas Mol. Biol. 32, 55–96. PubMed PMID: 4978050.

Hemmens, B., Mayer, B., 1998. Enzymology of nitric oxide synthases. Methods Mol. Biol. 100, 1–32. PubMed PMID: 10906988.

Hughes, J., Gillis, C.N., Bloom, F.E., 1969. The uptake and disposition of dl-norepinephrine in perfused rat lung. J. Pharmacol. Exp. Ther. 169 (2), 237–248. PubMed PMID: 5824606.

Henderson, M.C., Siddens, L.K., Krueger, S.K., Stevens, J.F., Kedzie, K., Fang, W.K., et al., 2014. Flavin-containing monooxygenase *S*-oxygenation of a series of thioureas and thiones. Toxicol. Appl. Pharmacol. 278 (2), 91–99. Epub 2014/04/15. http://dx.doi.org/10.1016/j.taap.2014.04.002. PubMed PMID: 24727368; PubMed Central PMCID: PMCPMC4110205.

Hines, R.N., 2006. Developmental and tissue-specific expression of human flavin-containing monooxygenases 1 and 3. Expert Opin. Drug Metab. Toxicol. 2 (1), 41–49. Epub 2006/07/26. http://dx.doi.org/10.1517/17425255.2.1.41. PubMed PMID: 16863467.

Hannemann, F., Bichet, A., Ewen, K.M., Bernhardt, R., 2007. Cytochrome P450 systems–biological variations of electron transport chains. Biochim. Biophys. Acta 1770 (3), 330–344. http://dx.doi.org/10.1016/j.bbagen.2006.07.017. PubMed PMID: 16978787.

Hartiala, J., 1976. Steroid metabolism in adult lung. Agents Actions 6 (4), 522–526. PubMed PMID: 785982.

Hussain, T., Al-Attas, O.S., Al-Daghri, N.M., Mohammed, A.A., De Rosas, E., Ibrahim, S., et al., 2014. Induction of CYP1A1, CYP1A2, CYP1B1, increased oxidative stress and inflammation in the lung and liver tissues of rats exposed to incense smoke. Mol. Cell. Biochem. 391 (1–2), 127–136. Epub 2014/02/22. http://dx.doi.org/10.1007/s11010-014-1995-5. PubMed PMID: 24557852.

Hussein, A.G., Pasha, H.F., El-Shahat, H.M., Gad, D.M., Toam, M.M., 2014. CYP1A1 gene polymorphisms and smoking status as modifier factors for lung cancer risk. Gene 541 (1), 26–30. Epub 2014/03/13. http://dx.doi.org/10.1016/j.gene.2014.03.003. PubMed PMID: 24613751.

Hook, G.E., Bend, J.R., Fouts, J.R., 1972. Mixed-function oxidases and the alveolar macrophage. Biochem. Pharmacol. 21 (24), 3267–3277. PubMed PMID: 4405368.

Hance, A.J., Bradley, K., Crystal, R.G., 1976. Lung collagen heterogeneity. Synthesis of type I and type III collagen by rabbit and human lung cells in culture. J. Clin. Invest. 57 (1), 102–111. http://dx.doi.org/10.1172/JCI108250. PubMed PMID: 1245595; PubMed Central PMCID: PMC436630.

Itakura, N., Fisher, A.B., Thurman, R.G., 1977. Cytochrome P450-linked *p*-nitroanisole *O*-demethylation in the perfused lung. J. Appl. Physiol. Respir. Environ. Exerc. Physiol. 43 (2), 238–245. PubMed PMID: 561048.

Jastroch, M., Divakaruni, A.S., Mookerjee, S., Treberg, J.R., Brand, M.D., 2010. Mitochondrial proton and electron leaks. Essays Biochem. 47, 53–67. http://dx.doi.org/10.1042/bse0470053. PubMed PMID: 20533900; PubMed Central PMCID: PMC3122475.

Jiang, F., Roberts, S.J., Datla, S.R., Dusting, G.J., 2006. NO modulates NADPH oxidase function via heme oxygenase-1 in human endothelial cells. Hypertension 48 (5), 950–957.

Junod, A.F., 1973. Uptake, metabolism, and efflux of ^{14}C-5-hydroxytryptamine in isolated perfused rat lungs. J. Pharmacol. Exp. Ther. 183, 341.

Kunz, W., Bohnensack, R., Bohme, G., Kuster, U., Letko, G., Schonfeld, P., 1981. Relations between extramitochondrial and intramitochondrial adenine nucleotide systems. Arch. Biochem. Biophys. 209 (1), 219–229. PubMed PMID: 6456693.

Kadenbach, B., 1986. Regulation of respiration and ATP synthesis in higher organisms: hypothesis. J. Bioenerg. Biomembr. 18 (1), 39–54. PubMed PMID: 3009427.

Keevil, T., Mason, H.S., 1978. Molecular oxygen in biological oxidations–an overview. Methods Enzym. 52, 3–40. PubMed PMID: 353444.

Kaufman, S., 1966. Coenzymes and hydroxylases: ascorbate and dopamine-beta-hydroxylase; tetrahydropteridines and phenylalanine and tyrosine hydroxylases. Pharmacol. Rev. 18 (1), 61–69. PubMed PMID: 5904174.

Kolluru, G.K., Shen, X., Bir, S.C., Kevil, C.G., 2013. Hydrogen sulfide chemical biology: pathophysiological roles and detection. Nitric Oxide Biol. Chem. 35, 5–20. Epub 2013/07/16. http://dx.doi.org/10.1016/j.niox.2013.07.002. PubMed PMID: 23850632; PubMed Central PMCID: PMCPMC4077051.

Krueger, S.K., Williams, D.E., Yueh, M.F., Martin, S.R., Hines, R.N., Raucy, J.L., et al., 2002. Genetic polymorphisms of flavin-containing monooxygenase (FMO). Drug Metab. Rev. 34 (3), 523–532. Epub 2002/09/07. http://dx.doi.org/10.1081/dmr-120005653. PubMed PMID: 12214664.

Korzekwa, K.R., Gillette, J.R., Trager, W.F., 1995. Isotope effect studies on the cytochrome P450 enzymes. Drug Metab. Rev. 27 (1–2), 45–59. Epub 1995/01/01. http://dx.doi.org/10.3109/03602539509029814. PubMed PMID: 7641584.

Kim, J.H., Sherman, M.E., Curriero, F.C., Guengerich, F.P., Strickland, P.T., Sutter, T.R., 2004. Expression of cytochromes P450 1A1 and 1B1 in human lung from smokers, non-smokers, and ex-smokers. Toxicol. Appl. Pharmacol. 199 (3), 210–219. http://dx.doi.org/10.1016/j.taap.2003.11.015.

Kartha, J.S., Yost, G.S., 2008. Mechanism-based inactivation of lung-selective cytochrome P450 CYP2F enzymes. Drug Metab. Dispos. 36 (1), 155–162. Epub 2007/10/27. http://dx.doi.org/10.1124/dmd.107.017897. PubMed PMID: 17962375.

Kitagawa, Y., Okamoto, M., Sugiyama, T., Kitamura, H., Nakahara, K., Kawashima, Y., et al., 1986. 3-Methylcholanthrene-induced ethoxyresorufin *O*-deethylase activity in alveolar type II cells of rabbit lung. Arch. Toxicol. 59 (1), 1–3. PubMed PMID: 3741136.

Kobzik, L., Bredt, D.S., Lowenstein, C.J., Drazen, J., Gaston, B., Sugarbaker, D., et al., 1993. Nitric oxide synthase in human and rat lung: immunocytochemical and histochemical localization. Am. J. Respir. Cell Mol. Biol. 9 (4), 371–377. PubMed PMID: 419.

Lapuente-Brun, E., Moreno-Loshuertos, R., Acín-Pérez, R., Latorre-Pellicer, A., Colás, C., Balsa, E., et al., 2013. Supercomplex assembly determines electron flux in the mitochondrial electron transport chain. Science 340 (6140), 1567–1570.

Loschen, G., Flohe, L., Chance, B., 1971. Respiratory chain linked H(2) O(2) production in pigeon heart mitochondria. FEBS Lett. 18 (2), 261–264. Epub 1971/11/01. PubMed PMID: 11946135.

Lambeth, J.D., 2004. NOX enzymes and the biology of reactive oxygen. Nat. Rev. Immunol. 4 (3), 181–189. Epub 2004/03/25. http://dx.doi.org/10.1038/nri1312. PubMed PMID: 15039755.

Lawton, M.P., Gasser, R., Tynes, R.E., Hodgson, E., Philpot, R.M., 1990. The flavin-containing monooxygenase enzymes expressed in rabbit liver and lung are products of related but distinctly different genes. J. Biol. Chem. 265 (10), 5855–5861. Epub 1990/04/05. PubMed PMID: 2318837.

Last, J.A., Reiser, K.M., 1984. Collagen biosynthesis. Environ. Health Perspect. 55, 169–177. Epub 1984/04/01. PubMed PMID: 6428877; PubMed Central PMCID: PMCPMC1568368.

Lancaster Jr., J.R., Hibbs Jr., J.B., 1990. EPR demonstration of iron-nitrosyl complex formation by cytotoxic activated macrophages. Proc. Natl. Acad. Sci. U. S. A. 87 (3), 1223–1227. PubMed PMID: 327.

Lloyd Jr., T.C., 1964. Effect of alveolar hypoxia on pulmonary vascular resistance. J. Appl. Physiol. 19, 1086.

Marino, P.A., Rooney, S.A., 1980. Surfactant secretion in a newborn rabbit lung slice model. Biochim. Biophys. Acta 620, 509.

Mason, R.J., Williams, M.C., Widdicombe, J.H., Sanders, M.J., Misfeldt, D.S., Berry Jr., L.C., 1982. Transepithelial transport by pulmonary alveolar type II cells in primary culture. Proc. Natl. Acad. Sci. U. S. A. 79, 6033.

Murphy, M.P., 2009. How mitochondria produce reactive oxygen species. Biochem. J. 417 (1), 1–13. http://dx.doi.org/10.1042/BJ20081386. PubMed PMID: 19061483; PubMed Central PMCID: PMC2605959.

Mitchell, P., 1966. Chemiosmotic coupling in oxidative and photosynthetic phosphorylation. Biol. Rev. Camb. Philos. Soc. 41 (3), 445–502. PubMed PMID: 5329743.

Mazat, J.P., Ransac, S., Heiske, M., Devin, A., Rigoulet, M., 2013. Mitochondrial energetic metabolism-some general principles. IUBMB Life 65 (3), 171–179. http://dx.doi.org/10.1002/iub. 1138. PubMed PMID: 23441039.

Mitchell, P., 1961. Coupling of phosphorylation to electron and hydrogen transfer by a chemi-osmotic type of mechanism. Nature 191, 144–148. PubMed PMID: 13771349.

Massaro, G.D., Gail, D.B., Massaro, D., 1975. Lung oxygen consumption and mitochondria of alveolar epithelial and endothelial cells. J. Appl. Physiol. 38 (4), 588–592. PubMed PMID: 1141087.

Mustafa, M.G., Cross, C.E., 1974. Effects of short-term ozone exposure on lung mitochondrial oxidative and energy metabolism. Arch. Biochem. Biophys. 162 (2), 585–594. PubMed PMID: 4366422.

Matsubara, T., Prough, R.A., Burke, M.D., Estabrook, R.W., 1974. The preparation of microsomal fractions of rodent respiratory tract and their characterization. Cancer Res. 34 (9), 2196–2203. PubMed PMID: 4367284.

Miller, M.A., Hales, C.A., 1979. Role of cytochrome P-450 in alveolar hypoxic pulmonary vasoconstriction in dogs. J. Clin. Invest. 64 (2), 666–673. http://dx.doi.org/10.1172/JCI109507. PubMed PMID: 457876; PubMed Central PMCID: PMC372164.

McKleroy, W., Lee, T.H., Atabai, K., 2013. Always cleave up your mess: targeting collagen degradation to treat tissue fibrosis. Am. J. Physiol. Lung Cell. Mol. Physiol. 304 (11), L709–L721. Epub 2013/04/09. http://dx.doi.org/10.1152/ajplung.00418.2012. PubMed PMID: 23564511; PubMed Central PMCID: PMCPMC3680761.

Muzaffar, S., Shukla, N., Bond, M., Newby, A.C., Angelini, G.D., Sparatore, A., et al., 2008. Exogenous hydrogen sulfide inhibits superoxide formation, NOX-1 expression and Rac1 activity in human vascular smooth muscle cells. J. Vasc. Res. 45 (6), 521–528.

McCord, J.M., 1985. Oxygen-derived free radicals in postischemic tissue injury. N. Engl. J. Med. 312 (3), 159–163. Epub 1985/01/17. http://dx.doi.org/10.1056/nejm198501173120305. PubMed PMID: 2981404.

Nebert, D.W., Wikvall, K., Miller, W.L., 2013. Human cytochromes P450 in health and disease. Philos. Trans. R. Soc. B Biol. Sci. 368 (1612) http://dx.doi.org/10.1098/rstb.2012.0431.

Nicholas, T.E., Kim, P.A., 1975. The metabolism of 3-H-cortisone and 3-H-cortisol by the isolated perfused rat and guinea pig lungs. Steroids 25 (3), 387–402. PubMed PMID: 1145675.

Ohnishi, T., 1998. Iron-sulfur clusters/semiquinones in complex I. Biochim. Biophys. Acta 1364 (2), 186–206. Epub 1998/06/19. PubMed PMID: 9593887.

Olson, N., Hristova, M., Heintz, N.H., Lounsbury, K.M., van der Vliet, A., 2011. Activation of hypoxia-inducible factor-1 protects airway epithelium against oxidant-induced barrier dysfunction. Am. J. Physiol. Lung Cell. Mol. Physiol. 301 (6), L993–L1002. Epub 2011/09/20. http://dx.doi.org/10.1152/ajplung.00250.2011. PubMed PMID: 21926263; PubMed Central PMCID: PMCPMC3233830.

Oren, R., Farnham, A.E., Saito, K., Milofsky, E., Karnovsky, M.L., 1983. Metabolic patterns in three types of phagocytizing cells. J. Cell. Biol. 17, 487.

Poulos, T.L., 1988. Cytochrome P450: molecular architecture, mechanism, and prospects for rational inhibitor design. Pharm. Res. 5 (2), 67–75. PubMed PMID: 3073382.

Pinakoulaki, E., Vamvouka, M., Varotsis, C., 2004. Resonance Raman detection of the Fe^{2+}-C-N modes in heme-copper oxidases: a probe of the active site. Inorg. Chem. 43 (16), 4907–4910. Epub 2004/08/03. http://dx.doi.org/10.1021/ic035216r. PubMed PMID: 15285666.

Poulos, T.L., 2014. Heme enzyme structure and function. Chem. Rev. 114 (7), 3919–3962. http://dx.doi.org/10.1021/cr400415k.

Philpot, R.M., Smith, B.R., 1984. Role of cytochrome P-450 and related enzymes in the pulmonary metabolism of xenobiotics. Environ. Health Perspect. 55, 359–367. Epub 1984/04/01. PubMed PMID: 6376107; PubMed Central PMCID: PMCPMC1568372.

Post, C., Andersson, R.G., Ryrfeldt, A., Nilsson, E., 1978. Transport and binding of lidocaine by lung slices and perfused lung of rats. Acta Pharmacol. Toxicol. 43 (2), 156–163. PubMed PMID: 696345.

Prockop, D.J., Kivirikko, K.I., Tuderman, L., Guzman, N.A., 1979. The biosynthesis of collagen and its disorders (second of two parts). N. Engl. J. Med. 301 (2), 77–85. http://dx.doi.org/10.1056/NEJM197907123010204. PubMed PMID: 36559.

Prockop, D.J., Kivirikko, K.I., Tuderman, L., Guzman, N.A., 1979. The biosynthesis of collagen and its disorders (first of two parts). N. Engl. J. Med. 301 (1), 13–23. http://dx.doi.org/10.1056/NEJM197907053010104. PubMed PMID: 449904.

Parks, D.A., Granger, D.N., 1983. Ischemia-induced vascular changes: role of xanthine oxidase and hydroxyl radicals. Am. J. Physiol. 245 (2), G285–G289. PubMed PMID: 6309018.

Qian, J., Chen, F., Kovalenkov, Y., Pandey, D., Moseley, M.A., Foster, M.W., et al., 2012. Nitric oxide reduces NADPH oxidase 5 (Nox5) activity by reversible *S*-nitrosylation. Free Radic. Biol. Med. 52 (9), 1806–1819.

Rajagopalan, K.V., Handler, P., 1967. Purification and properties of chicken liver xanthine dehydrogenase. J. Biol. Chem. 242 (18), 4097–4107. PubMed PMID: 4294045.

Rossouw, D.J., Engelbrecht, F.M., 1978. The effect of paraquat on the respiration of lung cell fractions. South Afr. Med. J. 54 (26), 1101–1104. Epub 1978/12/23. PubMed PMID: 746468.

Rowland, A.M., Yost, G.S., 2010. Selected pneumotoxic agents. In: Comprehensive Toxicology Respiratory Toxicology, second ed., vol. 8, pp. 511–547.

Raunio, H., Hakkola, J., Hukkanen, J., Lassila, A., Paivarinta, K., Pelkonen, O., et al., 1999. Expression of xenobiotic-metabolizing CYPs in human pulmonary tissue. Exp. Toxicol. Pathol. 51 (4–5), 412–417. http://dx.doi.org/10.1016/S0940-2993(99)80031-1. PubMed PMID: 10445407.

Raunio, H., Hakkola, J., Pelkonen, O., 2005. Regulation of CYP3A genes in the human respiratory tract. Chem. Biol. Interact. 151 (2), 53–62. Epub 2005/02/09. http://dx.doi.org/10.1016/j.cbi.2003.12.007. PubMed PMID: 15698577.

Rushmore, T.H., Kong, A.N., 2002. Pharmacogenomics, regulation and signaling pathways of phase I and II drug metabolizing enzymes. Curr. Drug Metab. 3 (5), 481–490. Epub 2002/10/09. PubMed PMID: 12369894.

Rodriguez, A.I., Gangopadhyay, A., Kelley, E.E., Pagano, P.J., Zuckerbraun, B.S., Bauer, P.M., 2010. HO-1 and CO decrease platelet-derived growth factor-induced vascular smooth muscle cell migration via inhibition of Nox1. Arterioscler. Thromb. Vasc. Biol. 30 (1), 98–104.

Singer, T.P., Johnson, M.K., 1985. The prosthetic groups of succinate dehydrogenase: 30 years from discovery to identification. FEBS Lett. 190 (2), 189–198. Epub 1985/10/14. PubMed PMID: 2995133.

Salerno, J.C., Blum, H., Ohnishi, T., 1979. The orientation of iron-sulfur clusters and a spin-coupled ubiquinone pair in the mitochondrial membrane. Biochim. Biophys. Acta 547 (2), 270–281. Epub 1979/08/14. PubMed PMID: 223637.

Shin, D.H., Choi, Y.J., Park, J.W., 2014. SIRT1 and AMPK mediate hypoxia-induced resistance of non-small cell lung cancers to cisplatin and doxorubicin. Cancer Res. 74 (1), 298–308. http://dx.doi.org/10.1158/0008-5472.CAN-13-2620. PubMed PMID: 24240701.

Spear, R.K., Lumeng, L., 1978. A method for isolating lung mitochondria from rabbits, rats, and mice with improved respiratory characteristics. Anal. Biochem. 90 (1), 211–219. PubMed PMID: 31819.

Sayeed, M.M., Baue, A.E., 1971. Mitochondrial metabolism of succinate, beta-hydroxybutyrate, and alpha-ketoglutarate in hemorrhagic shock. Am. J. Physiol. 220 (5), 1275–1281. PubMed PMID: 4324939.

Smyrniotis, C.J., Barbour, S.R., Xia, Z., Hixon, M.S., Holman, T.R., 2014. ATP allosterically activates the human 5-lipoxygenase molecular mechanism of arachidonic acid and 5(S)-hydroperoxy-6(E),8(Z),11(Z),14(Z)-eicosatetraenoic acid. Biochemistry 53 (27), 4407–4419. http://dx.doi.org/10.1021/bi401621d. PubMed PMID: 24893149.

Stephens, N.L., Meyers, J.L., Cherniack, R.M., 1968. Oxygen, carbon dioxide, H± ion, and bronchial length-tension relationships. J. Appl. Physiol. 25, 276.

Sugimoto, H., Oda, S., Otsuki, T., Hino, T., Yoshida, T., Shiro, Y., 2006. Crystal structure of human indoleamine 2,3-dioxygenase: catalytic mechanism of O_2 incorporation by a heme-containing dioxygenase. Proc. Natl. Acad. Sci. U. S. A. 103 (8), 2611–2616. http://dx.doi.org/10.1073/pnas.0508996103. PubMed PMID: 16477023; PubMed Central PMCID: PMC1413787.

Sumimoto, H., 2008. Structure, regulation and evolution of Nox-family NADPH oxidases that produce reactive oxygen species. FEBS J. 275 (13), 3249–3277. http://dx.doi.org/10.1111/j.1742-4658.2008.06488.x.

Stuehr, D.J., 1999. Mammalian nitric oxide synthases. BBA-Bioenergetics 1411 (2–3), 217–230. http://dx.doi.org/10.1016/S0005-2728(99)00016-X.

Steinberg, H., Bassett, D.J., Fisher, A.B., 1975a. Depression of pulmonary 5-hydroxytryptamine uptake by metabolic inhibitors. Am. J. Physiol. 228 (5), 1298–1303. PubMed PMID: 1130533.

Steinberg, H., Bassett, D.J., Fisher, A.B., 1975b. Metabolic requirements for 5-hydroxytryptamine uptake by the isolated lung. Chest 67 (2 Suppl.), 59S–60S. PubMed PMID: 1112116.

Steinberg, H., Bassett, D.J.P., Fisher, A.B., 1975c. Depression of pulmonary 5-hydroxytryptamine uptake by metabolic inhibitors. Am. J. Physiol. 228, 1298.

Siddens, L.K., Krueger, S.K., Henderson, M.C., Williams, D.E., 2014. Mammalian flavin-containing monooxygenase (FMO) as a source of hydrogen peroxide. Biochem. Pharmacol. 89 (1), 141–147. Epub 2014/02/25. http://dx.doi.org/10.1016/j.bcp.2014.02.006. PubMed PMID: 24561181; PubMed Central PMCID: PMCPMC4116332.

Sabourin, P.J., Tynes, R.E., Smyser, B.P., Hodgson, E., 1986. The FAD--containing monooxygenase of lung and liver tissue from rabbit, mouse and pig: species and tissue differences. Adv. Exp. Med. Biol. 197, 263–272. Epub 1986/01/01. PubMed PMID: 3766262.

Serabjit-Singh, C.J., Wolf, C.R., Philpot, R.M., Plopper, C.G., 1980. Cytochrome p-450: localization in rabbit lung. Science 207 (4438), 1469–1470. PubMed PMID: 6767272.

Sage, H., Farin, F.M., Striker, G.E., Fisher, A.B., 1983. Granular pneumocytes in primary culture secrete several major components of the extracellular matrix. Biochemistry 22 (9), 2148–2155. PubMed PMID: 6860655.

Shieh, J.J., Yasunobu, K.T., 1976. Purification and properties of lung lysyl oxidase, a copper-enzyme. Adv. Exp. Med. Biol. 74, 447–463. PubMed PMID: 8967.

Spooner, B.S., Thompson-Pletscher, H.A., Stokes, B., Bassett, K.E., 1986. Extracellular matrix involvement in epithelial branching morphogenesis. Dev. Biol. (N. Y. 1985) 3, 225–260. Epub 1986/01/01. PubMed PMID: 3077966.

Semenza, G.L., 2009. Regulation of oxygen homeostasis by hypoxia-inducible factor 1. Physiology (Bethesda), 24, 97–106. htttp://dx.doi.org/10.1152/physiol.00045.2008.

Segal, A.W., Jones, O.T., 1978. Novel cytochrome b system in phagocytic vacuoles of human granulocytes. Nature 276 (5687), 515–517. PubMed PMID: 723935.

Selemidis, S., Dusting, G.J., Peshavariya, H., Kemp-Harper, B.K., Drummond, G.R., 2007. Nitric oxide suppresses NADPH oxidase-dependent superoxide production by *S*-nitrosylation in human endothelial cells. Cardiovasc. Res. 75 (2), 349–358.

Tyler, W.S., Dungworth, D.L., Plopper, C.G., Hyde, D.M., Tyler, N.K., 1985. Structural evaluation of the respiratory system. Fund. Appl. Toxicol. 5 (3), 405–422. PubMed PMID: 2408953.

Tynes, R.E., Sabourin, P.J., Hodgson, E., 1985. Identification of distinct hepatic and pulmonary forms of microsomal flavin-containing monooxygenase in the mouse and rabbit. Biochem. Biophys. Res. Commun. 126 (3), 1069–1075. Epub 1985/02/15. PubMed PMID: 3919719.

Villalobo, A., Lehninger, A.L., 1979. The proton stoichiometry of electron transport in Ehrlich ascites tumor mitochondria. J. Biol. Chem. 254 (11), 4352–4358. PubMed PMID: 35536.

Winterbourn, C.C., 2013. The biological chemistry of hydrogen peroxide. Methods Enzym. 528, 3–25. Epub 2013/07/16. http://dx.doi.org/10.1016/b978-0-12-405881-1.00001-x. PubMed PMID: 23849856.

Wilson, D.F., Forman, N.G., 1982. Mitochondrial transmembrane pH and electrical gradients: evaluation of their energy relationships with respiratory rate and adenosine 5'-triphosphate synthesis. Biochemistry 21 (6), 1438–1444. PubMed PMID: 7074098.

Wilson, D.F., Vinogradov, S.A., 2014. Mitochondrial cytochrome c oxidase: mechanism of action and role in regulating oxidative phosphorylation. J. Appl. Physiol. (Bethesda, Md: 1985). Epub 2014/10/18. http://dx.doi.org/10.1152/japplphysiol.00737.2014. PubMed PMID: 25324518.

Williamson, J.R., Corkey, B.E., 1979. Assay of citric acid cycle intermediates and related compounds—update with tissue metabolite levels and intracellular distribution. Methods Enzym. 55, 200–222. PubMed PMID: 459841.

Warren, D.L., Brown, D.L., Buckpitt, A.R., 1982. Evidence for cytochrome-p-450 mediated metabolism in the bronchiolar damage by naphthalene. Chem. Biol. Interact. 40 (3), 287–303. http://dx.doi.org/10.1016/0009-2797(82)90152-1. PubMed PMID: WOS: A1982NW99400004.

Wattenberg, L.W., Leong, J.L., Galbraith, A.R., 1968. Induction of increased benzpyrene hydroxylase activity in pulmonary tissue in vitro. Proc. Soc. Exp. Biol. Med. Soc. Exp. Biol. Med. 127 (2), 467–469. PubMed PMID: 5645036.

Wang, G.L., Jiang, B.H., Rue, E.A., Semenza, G.L., 1995. Hypoxia-inducible factor 1 is a basic-helix-loop-helix-PAS heterodimer regulated by cellular O_2 tension. Proc. Natl. Acad. Sci. U. S. A. 92 (12), 5510–5514. PubMed PMID: 7539918; PubMed Central PMCID: PMC41725.

Widdicombe, J.H., Welsh, M.J., 1980. Anion selectivity of the chloride-transport process in dog tracheal epithelium. Am. J. Physiol. 239, C112 (Cell Physiol. 8).

Wilson, D.F, Erecinska, M., Drown, C., Silver, I.A., 1979. The oxygen dependence of cellular energy metabolism. Arch. Biochem. Biophys. 195 (2), 485–493.

Yaniv, Y., Juhaszova, M., Nuss, H.B., Wang, S., Zorov, D.B., Lakatta, E.G., et al., 2010. Matching ATP supply and demand in mammalian heart: in vivo, in vitro, and in silico perspectives. Ann. N. Y. Acad. Sci. 1188, 133–142. Epub 2010/03/06. http://dx.doi.org/10.1111/j.1749-6632.2009.05093.x. PubMed PMID: 20201896; PubMed Central PMCID: PMCPMC2943203.

Zhang, J.Y., Wang, Y., Prakash, C., 2006. Xenobiotic-metabolizing enzymes in human lung. Curr. Drug Metab. 7 (8), 939–948. Epub 2006/12/16. PubMed PMID: 17168693.

Chapter 19

The Biochemistry of Endothelial Cells

Rashmin C. Savani

Chapter Outline

1. INTRODUCTION

The endothelium, comprised of a single cell layer of cells lining all blood vessels, is the interface between blood and the vascular wall, providing integrity to the vessel to prevent permeability, but also inhibiting platelet aggregation and leukocyte adhesion. It also acts as a mechanosensor, transducing intracellular signals in response to changes in stretch, shear stress, and pressure. In general, endothelial cells (ECs) maintain quiescence, responding to a number of exogenous molecules such as reactive oxygen species, high density lipoprotein (HDL), and growth factors, and secreting a wide variety of molecules, such as nitric oxide (NO), prostacyclin (PGI_2), and endothelin 1 (ET1) to regulate EC function, in particular vascular tone. Loss of this homeostasis is a key component of a variety of pathological conditions where vasomotor tone is lost, permeability is increased, and leukocyte adhesion and platelet aggregation are prominent, thereby leading to hypertension, inflammation, and thrombosis. EC also respond to a balance of pro- and antiangiogenic stimuli to regulate the formation of vasculature. These responses occur in both normal and abnormal circumstances such as with microvascular changes required for normal alveolarization during lung development, or in abnormal circumstances such as dysregulated angiogenesis as seen in retinopathies and tumor vascularization. A key regulator of both EC homeostatic and pathologic responses is NO, produced by three mammalian forms of nitric oxide synthases (NOS).

2. NO AND NOS

The principal vasoregulator produced by EC is undoubtedly NO, which is synthesized by the conversion of L-arginine to L-citrulline, a reaction catalyzed by three mammalian isoforms of NOS, with molecular oxygen and reduced nicotinamide-adenine-dinucleotide phosphate (NADPH) as cosubstrates (Allen et al., 2009; De Cruz et al., 2009; Hollenberg and Cinel, 2009). The three forms, neuronal NOS (nNOS, NOS1), inducible NOS (iNOS, NOS2), and endothelial NOS (eNOS, NOS3), are all homodimers and transfer electrons from NADPH to the enzyme, via the cofactors flavin adenine dinucleotide (FAD), flavin mononucleotide (FMN), and tetrahydro-L-biopterin (BH_4), to reduce and activate oxygen and oxidize L-arginine to L-citrulline and NO (Figure 1) (Forstermann and Sessa, 2012; Feng et al., 2013). All forms of NOS bind calmodulin. For nNOS and eNOS, this binding is promoted with an increase in intracellular calcium, whereas for iNOS, the calmodulin binds at very low concentrations of intracellular calcium, resulting in full activation of the enzyme upon stimulation. Many actions of NO are mediated via the activation of soluble guanylate cyclase (sGC) to produce cyclic $3'-5'$-guanosine monophosphate (cGMP), which is, in turn, degraded by phosphodiesterase V, thereby providing mechanisms for the fine control of NO action (Figure 1) (Martinez-Ruiz et al., 2011; Potter, 2011). Further, NO can react with a number of molecules with significant beneficial and detrimental downstream effects. Thus, its reaction with superoxide occurs at near diffusion limited rates and

Comparative Biology of the Normal Lung. http://dx.doi.org/10.1016/B978-0-12-404577-4.00019-9

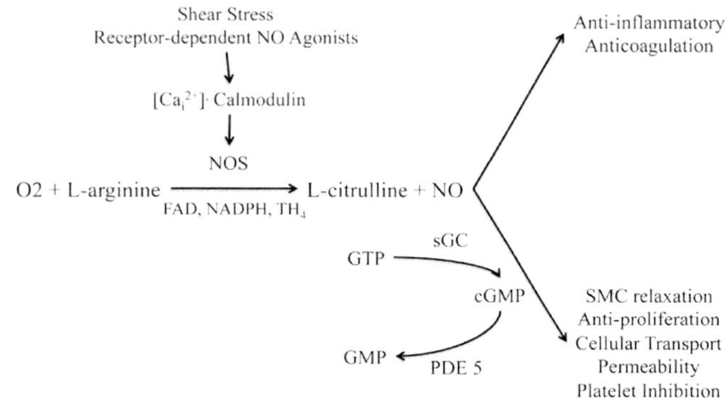

FIGURE 1 The synthesis and regulation of nitric oxide (NO) and its diverse functions. NO is produced by the conversion of molecular oxygen and L-arginine to L-citrulline and NO, a reaction that is catalyzed by three nitric oxide synthases (NOS). This reaction is stimulated by vascular shear stress and a variety of receptor-dependent NO agonists that alter intracellular calcium concentrations that are critical for nNOS and eNOS activity. Critical functions of NO are mediated by the activation of soluble guanylate cyclase that produces cyclic 3′−5′-guanosine monophosphate (cGMP), which, in turn, mediates smooth muscle cell relaxation, inhibits proliferation, increases permeability, and inhibits platelet adherence. Levels of cGMP are regulated by the activity of phosphodiesterase-5 that converts cGMP to GMP. Further, NO also has anti-inflammatory and anticoagulation functions.

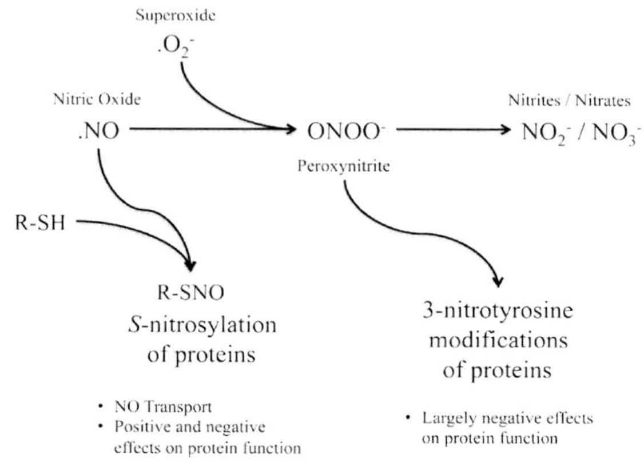

FIGURE 2 The reactions of nitric oxide (NO). NO reactivity results in posttranslational modifications of proteins that significantly impact both homeostatic and pathologic function. Thus, direct nitrosylation of cysteine residues in proteins, by transfer of NO from *S*-nitrosoglutathione, is involved in the regulation of a wide array of cellular signaling events as normal physiologic activity, as well as inhibition of protein function in pathologic situations. The reaction of NO with superoxide produces peroxynitrite ($OONO^-$), a highly damaging species that results in 3-nitrotyrosine (3NT) modifications of proteins with largely negative effects on protein function. More stable compounds to measure NO presence and reactivity include nitrates and nitrites.

produces peroxynitrite ($OONO^-$), a highly damaging species that causes the nitration of tyrosine residues at the 3-position to produce 3-nitrotyrosine (3NT). Nitration has been shown to result in functional alterations of modified proteins and has been demonstrated in a wide variety of human diseases and in animal models of tissue injury (Beckman and Koppenol, 1996; Pacher et al., 2007; Calcerrada et al., 2011; Sugiura and Ichinose, 2011). Alternatively, direct nitrosylation of cysteine residues in proteins, by transfer of NO from *S*-nitrosoglutathione, is involved in the regulation of a wide array of cellular signaling events, for example, the activity of NFκB and AP1 transcription factors (Figure 2) (Hess et al., 2005; Atochina-Vasserman et al., 2010).

Each of the NOS enzymes has distinct localization and function as has been elucidated by in vitro knockdown and in vivo knockout mouse experiments (Tsutsui et al., 2006). As the name would suggest, nNOS was initially found in specific neurons in the brain and spinal cord. However, it is also expressed in epithelial cells, pancreatic islet cells, and in vascular and skeletal smooth muscle cells (Zhou and Zhu, 2009). Among its diverse functions, nNOS modulates learning, memory, and neurogenesis in the central nervous system, but also regulates smooth muscle tone peripherally, most distinctly in the corpus cavernosum to allow for penile erection (Zhou and Zhu, 2009; Forstermann and Sessa, 2012; Zhang and Casadei, 2012). Conversely, with cerebrovascular injury and intracellular influx of calcium, NO production by nNOS is substantially increased, leading to peroxynitrite-induced neuronal damage as seen in stroke (Guix et al., 2005; Willmot et al., 2005). Inducible NOS expression is increased by specific stimuli, such as lipopolysaccharide (LPS) in macrophages where it is a marker of the M1 or classically activated macrophage phenotype (Bogdan et al., 2000; Coleman, 2001). The physiologic function of iNOS appears to be on the cytotoxicity to microbial invasion and tumor cells, as mediated by its damaging effects on mitochondrial function and cellular DNA (Bian et al., 2012). However, excess NO production by iNOS can also lead to the formation of $ONOO^-$ and affect surrounding healthy cells, as well as cause the extreme vasodilatation and severe hypotension that is seen in septic shock (Wolkow, 1998; Andrades et al., 2011). The most abundant NOS isoform in EC is eNOS.

3. ENDOTHELIAL NOS

The localization and regulation of eNOS activity has been studied extensively. The enzyme is localized to caveolae via specific posttranslational modifications such as myristoylation, and this localization is critical for its function (Shaul, 2002). Caveolae are specialized plasma membrane invaginations found in abundance in EC. The key molecule that determines the formation of these organelles is caveolin-1, a protein that spans the inner plasma membrane and has a unique intracellular sequence that mediates binding to a large number of molecules, including eNOS. This binding prevents eNOS activation, and upon appropriate stimulus, eNOS dissociates from caveolin-1 and becomes activated (Mineo and Shaul, 2012). Further, in

TABLE 1 Positive and Negative Regulators of eNOS

Positive	Negative
Acetylcholine	C-Reactive protein (CRP)
Bradykinin	27-Hydroxycholesterol
Estradiol	Antiphospholipid antibodies
HDL	
Insulin	
VEGF	

models of pulmonary hypertension where there is a deficiency of eNOS-mediated NO production, nitrosylation of *N*-ethylmaleimide-sensitive factor (NSF), an ATPase critical for all intracellular vesicular trafficking, results in a block of intracellular delivery of eNOS at the Golgi level, and decreased caveolar eNOS content and activity (Sehgal and Lee, 2011).

A large number of positive and negative regulators of eNOS activation have been studied (Table 1) (Mineo and Shaul, 2006). Agonists include acetylcholine, insulin, estradiol, vascular endothelial growth factor (VEGF), and HDL. While each agonist interacts with its specific receptor to initiate signaling, the pathways converge to the activation of Src, which, in turn, results in the sequential tyrosine phosphorylation of caveolin-1, phosphoinositide-3-kinase (PI3K), and Akt (also known as protein kinase B, PKB). Activation of Akt, in turn, results in the phosphorylation of serine 1117 and dephosphorylation of serine 495 by the phosphatase PP1, which are both required for the activation of eNOS (Figure 3) (Michel and Vanhoutte, 2010). Inhibitors of this pathway include C-reactive protein (Schwartz et al., 2007; Tanigaki et al., 2009; Sundgren et al., 2011), antiphospholipid antibodies (Mineo and Shaul, 2011; Ramesh et al., 2011), and 27-hydroxycholesterol (Umetani et al., 2007; Umetani and Shaul, 2011), all of which also have specific receptors that mediate the signals that result in the inhibition of the eNOS signaling pathway, and worsen atherosclerosis, thrombosis, and pregnancy loss in elegant mouse models.

Another potential mechanism regulating eNOS inactivation, particularly in situations of excess oxidative stress, is the process of eNOS uncoupling (Kietadisorn et al., 2012; Li and Forstermann, 2013). Essentially, the enzyme changes from its normal function to produce NO to the generation of superoxide (O_2^-). Several mechanisms have been proposed for this and include the loss of NO due to its reaction with O_2^-, the oxidation of BH_4 by the formation of $ONOO^-$, and the elaboration of arginases that deplete arginine as a substrate for the reaction (Alkaitis and Crabtree, 2012; Li and Forstermann, 2013; Yang and Ming, 2013). Since BH_4 stabilizes the formation of NOS homodimers, uncoupling of eNOS is associated with a decrease in homodimerization of the enzyme (Alkaitis and Crabtree, 2012).

NO produced by EC has a number of key functions that result in EC quiescence. In addition to vasodilatation by activating sGC and the production of cGMP to relax vascular smooth muscle and control blood pressure (Arnal et al., 1999; Li and Forstermann, 2000), NO is a potent inhibitor of platelet aggregation and adhesion, thereby preventing thrombosis and degranulation (Achneck et al., 2008; Dixon et al., 2012). Further, NO decreases the expression of leukocyte adhesion molecules (CD11/CD18) and the activation of NFkB, and is therefore

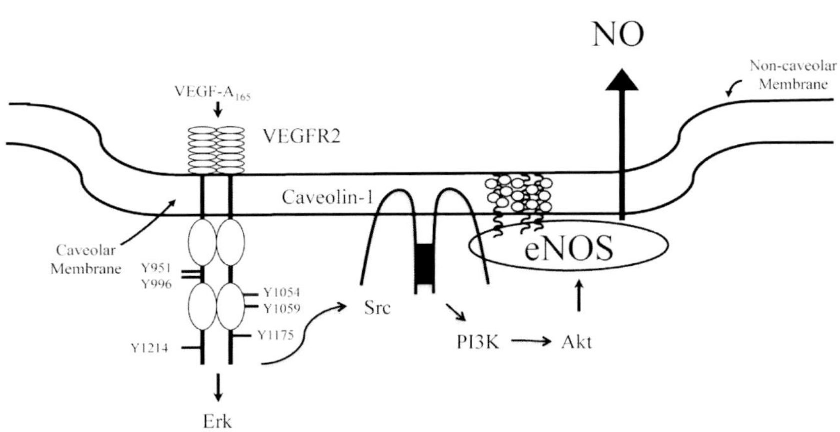

FIGURE 3 The signaling pathway to activate eNOS is localized to caveolae. Receptor-mediated agonists such as vascular endothelial growth factor (VEGF) stimulate a signaling cascade that includes sequential activation of Src, Caveolin-1, PI3 Kinase, Akt, and finally endothelial NOS (eNOS). Specific phosphorylation and dephosphorylation events are necessary that ultimately lead to dissociation of eNOS from Caveolin-1 to activate eNOS and produce NO. These proteins are localized in characteristic small invaginations of the cell membrane called caveolae, abundant in endothelial cells that are both signaling hubs and also mediate endocytosis.

anti-inflammatory (Li and Forstermann, 2000; Ramesh et al., 2011).

More recently, another key role for NO has been described in the regulation of lung angiogenesis, a process that is important for lung vascular development and appropriate secondary septation required for alveolarization (Shaul, 1999; Lin et al., 2005; Bachiller et al., 2013). Indeed, eNOS knockout mice have abnormal lung development consistent with alveolar capillary dysplasia and delayed alveolarization, which is improved with postnatal inhaled NO (iNO) treatment (Balasubramaniam et al., 2003, 2006; Han and Stewart, 2006). Figure 4 shows the effects of iNO in mice exposed to either 21% or 85% oxygen and examined on postnatal day (PN) 15. As quantified by radial alveolar counts, hyperoxia exposure from PN3 to PN15 is associated with a simplified distal lung structure and a lack of secondary septation indicative of significantly delayed alveolarization. Treatment with iNO rescues the alveolarization defect.

Preterm infants are at high risk of developing bronchopulmonary dysplasia (BPD), a chronic debilitating lung condition thought to involve oxidative and nitrative stresses on an immature lung sustaining ventilator-dependent damage that results in an inflammatory response and ends in decreased alveolarization (Stenmark and Abman, 2005; Chess et al., 2006; Coalson, 2006; Speer, 2006; Thebaud, 2007; Ryan et al., 2008; Jobe, 2011; Mosca et al., 2011). Interestingly, agents that activate eNOS, such as estradiol and vitamin D, also promote alveolarization (McCurnin et al., 2009; Nadeau et al., 2010). Further, in the preterm baboon model of prematurity, compared to controls, administration of iNO significantly decreased pulmonary artery pressures, improved lung compliance, decreased airway resistance, and showed some improvement in secondary septation after 2 weeks of ventilation and appropriate oxygen therapy (McCurnin et al., 2005). Based on such positive preclinical results, trials of iNO therapy have been performed in human preterm infants to prevent BPD (Ballard et al., 2006; Kinsella et al., 2006). These have yielded mixed results, and current consensus suggests that iNO has potential in the treatment of premature infants but that considerably more research is needed (Barrington and Finer, 2010; Walsh et al., 2010; Donohue et al., 2011).

4. THE GLYCOCALYX, MECHANOSENSATION AND MECHANOTRANSDUCTION

While traditional consideration of the endothelium is that it represents the interface between circulating blood and the vascular smooth muscle, the surface of EC is decorated with a layer called the glycocalyx that contributes to EC integrity, mechanotransduction, leukocyte adherence, and coordination of EC signaling (van den Berg et al., 2006; Reitsma et al., 2007; Fu and Tarbell, 2013). This cell-free layer, ranging from 0.5 to 4.5 μm in thickness depending on the vessel, is composed of a multitude of proteoglycans and glycoproteins that make it a relatively viscous layer that protects the endothelium from changes in shear stress and pressure, but also from undue leukocyte and platelet adherence and sequestration of NO by hemoglobin (Brouland et al., 1999; Gao and Lipowsky, 2010; Hightower et al., 2011).

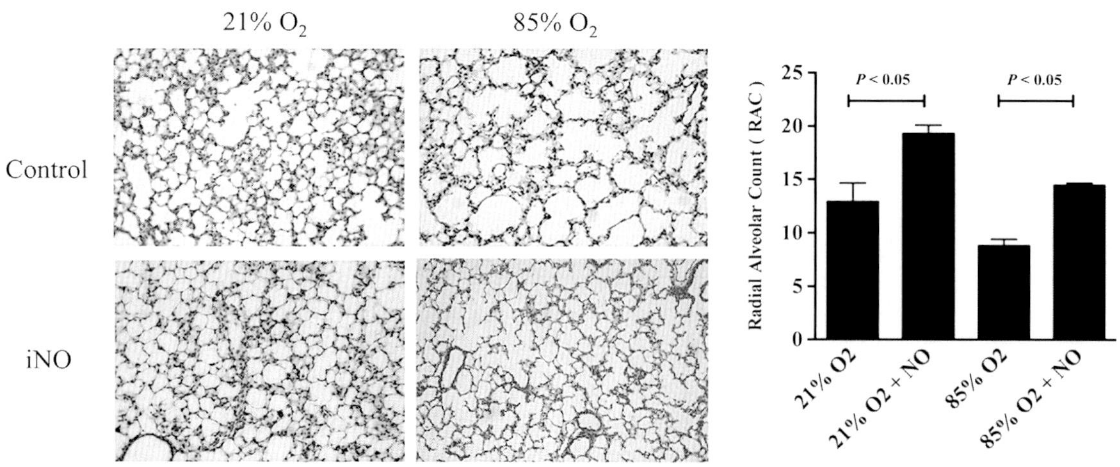

FIGURE 4 Nitric oxide promotes alveolarization both in normal and hyperoxia-exposed neonatal mice. Angiogenesis plays a critical role in lung vascular development and contributes significantly to alveolarization, the final stage of lung development. In rodents, the process of alveolarization occurs exclusively in the postnatal period (PN3 to PN21). Disruption of normal vascular development results in a delay or failure of alveolarization. The effects of inhaled NO (iNO) in mice exposed to either 21% or 85% oxygen were examined on postnatal day PN15. As quantified by radial alveolar counts, hyperoxia exposure from PN3 to PN15 was associated with a simplified distal lung structure and a lack of secondary septation indicative of significantly delayed alveolarization. Treatment with iNO rescues the defect in alveolarization due to hyperoxia.

The composition of the glycocalyx has been studied in vitro both by specific enzymatic degradation experiments and biochemical analysis of composition. The glycocalyx is enriched with glycosaminoglycans (GAGs), including heparan sulfate (HS), chondroitin sulfate (CS), sialic acids, and hyaluronan (hyaluronic acid, HA), giving this structure a negative charge and large hydration capacity (Wheeler-Jones et al., 2010; Ebong et al., 2011; Zeng et al., 2012). Membrane proteins providing structural context to the glycocalyx include syndecans, glypicans, and hyaluronan receptors such as CD44, all of which are covalently attached to GAG side chains (Bourguignon et al., 2004; Lee et al., 2008; Tarbell and Ebong, 2008; Midgley et al., 2013). The transmembrane proteins have cytoplasmic tails that interact with the cytoskeleton and signaling elements to transmit cues from the extracellular environment to orchestrate the endothelial response. Under normal circumstances, the fluid shear stress sensed at the surface of the glycocalyx is dampened such that the endothelium itself does not sense these changes. However, in situations where there is abnormally increased stress or the glycocalyx is diminished, the mechanical signal is now sensed by the endothelium with signaling changes that alter EC behavior (Tarbell et al., 2005; Tarbell and Pahakis, 2006). In elegant studies by Tarbell and Ebong (2008) and others (Mochizuki et al., 2003; Pahakis et al., 2007; Kumagai et al., 2009), compared to static conditions, the application of steady shear stress on EC in culture results in the phosphorylation of eNOS and rapid production of NO. However, individual treatments of cultures with heparinase, hyaluronidase, or neuraminidase (to remove sialic acid) inhibit this response, but not the bradykinin-induced activation of eNOS (Tarbell and Ebong, 2008). These findings have been confirmed in intact vessels (Yen et al., 2012). Interestingly, these GAGs did not contribute to the production of PGI_2 in response to shear stress. The principal organelles contributing to the regulation of glycocalyx-mediated signal transduction are lipid rafts in general and caveolae in particular. Syndecans (with HS and CS side chains), glypicans (with HS side chains), and CD44 (with HS side chains) are concentrated in these organelles, which is also where the relevant signaling complexes that mediate pathways to activate eNOS also reside (Reitsma et al., 2007; Weinbaum et al., 2007; Zeng et al., 2013). Considerable work from our laboratory and those of others has implicated HA in the process of angiogenesis and, by inference, the activation of eNOS.

5. HYALURONAN, HA RECEPTORS, AND ANGIOGENESIS

Hyaluronan (HA), an important constituent of the extracellular matrix, is a polymer of repeating disaccharides of D-glucuronic acid and *N*-acetyl glucosamine linked together by alternating beta-1,4 and beta-1,3 glycosidic bonds (Figure 5(A)) (Fraser et al., 1997; Hascall and Laurent, 1997; Toole, 2000, 2004). Most tissues in the body contain high molecular weight (HMW) HA with

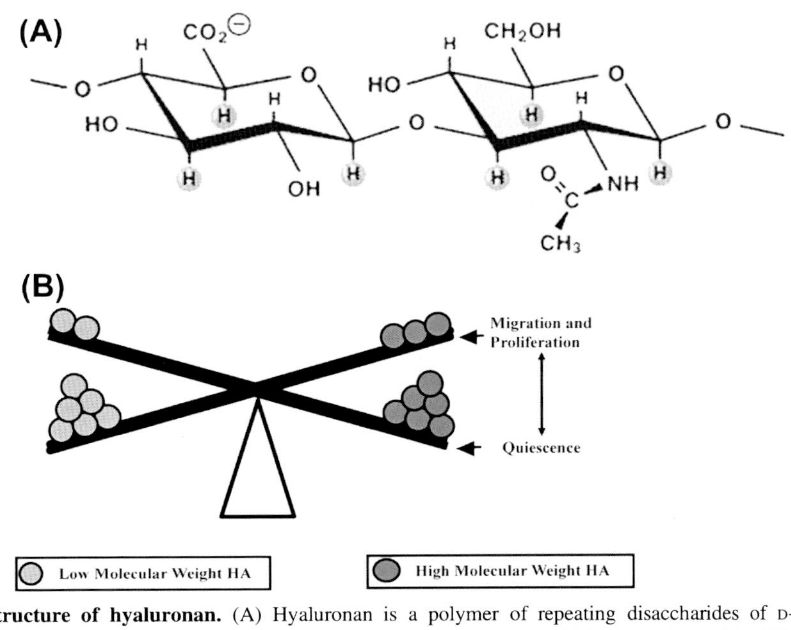

FIGURE 5 Biochemical structure of hyaluronan. (A) Hyaluronan is a polymer of repeating disaccharides of D-glucuronic acid and *N*-acetyl glucosamine linked together by alternating beta-1,4 and beta-1,3 glycosidic bonds. It exists in high molecular weight (HMW) form consisting of 10,000 disaccharide units or more (>1 million Dalton), in low molecular weight (LMW) form (<500 kDa), and in oligomeric form consisting of 4–25 disaccharide units. HMW HA is inert, providing structure to tissues, and is also anti-inflammatory and antiangiogenic. LMW, in particular oligomeric HA, interacts with cell-associated receptors and promotes migration, proliferation, and intracellular signaling. (B) The relative contents of HMW versus LMW or oligomeric HA within tissues determines whether the environment is quiescent or pro-migratory and proliferative.

10,000 or more disaccharides reaching molecular masses of about 4 million Daltons. In addition, low molecular weight (LMW) HA regulates cellular events such as cell proliferation and locomotion that are required for a variety of biological processes including tumorigenesis, morphogenesis, inflammation, and host response to injury (Savani and DeLisser, 2003). HA has been implicated in both vasculogenesis and angiogenesis. Its actions, however, are molecular size- and concentration-dependent. Thus, in chick embryo limb buds, HMW HA localizes to avascular areas, transplantation of HA-rich mesoderm from normally avascular to vascular areas causes avascularity, and exogenous administration of HMW HA also results in a lack of vascular development (Feinberg and Beebe, 1983). HMW HA also either has no effect or inhibits EC migration, proliferation, and tube formation (West and Kumar, 1989).

Conversely, HA fragments (3−25 oligomers) are potent and cell-specific stimulators of EC proliferation, tube formation, and angiogenesis in chick chorioallantoic membranes (Montesano et al., 1996; Rahmanian et al., 1997). In vivo studies also support the angiogenic nature of LMW HA. Application of HA oligosaccharides to rat skin results in an increased number of blood vessels per mm skin length and accelerates wound healing in a delayed revascularization model (Sattar et al., 1994; Lees et al., 1995). Interestingly, quiescent EC do not synthesize HA whereas activated cells do, thereby providing for more complex regulation of the HA content in any particular microenvironment (Amanuma and Mitsui, 1991). In addition, the generation of superoxide and/or peroxynitrite, or increased expression of hyaluronidases by tumor or stromal cells, may be responsible for an accumulation of LMW HA at sites of inflammation or angiogenesis (Li et al., 1997; Csoka et al., 2001; Hrabarova et al., 2011; Østerholt et al., 2012). Thus, angiogenesis can be regulated by the relative quantity of LMW and HMW in a given tissue (Figure 5(B)).

In order to determine whether embryonic lung vascular development was dependent on LMW HA, we cultured E12.5 lung explants obtained from Flk.LacZ mice that are heterozygous for targeted insertion of bacterial beta-galactosidase (LacZ) into the *Flk* locus (Jackson Labs, Bar Harbor, Maine). The other *Flk* locus is normal and homozygous insertion has an embryonic lethal phenotype. This results in a single copy of the marker gene LacZ regulated by the flk-1 promoter, per EC, thus allowing for simple histologic identification of EC. Lung buds (E12.5) were cultured in serum-free medium and exposed to 8−30 mer HA oligosaccharides (5−10 μg/ml) for 3 days with medium refreshed each day. Explants were then stained with X-gal to demonstrate EC. A dose-dependent increase in staining was noted in explants, suggesting that LMW HA stimulates vascularization of embryonic lung explants in vitro (Figure 6(A)). We then sought to determine whether the effects of VEGF on embryonic lung vasculature were dependent on HA. Mouse lung explants were exposed to VEGF with either 200 μg/ml of HA-binding peptide

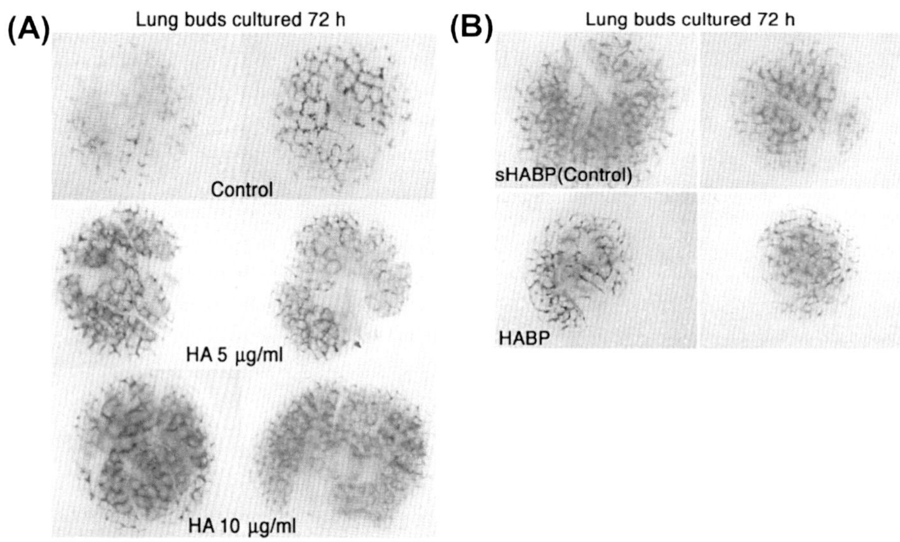

FIGURE 6 Hyaluronan (HA) stimulates lung vascular development. (A) In order to determine the effects of HA on embryonic lung vascular development, we used Flk.LacZ mice that express bacterial beta-galactosidase in an endothelial-specific manner. Lung buds (E12.5) were exposed to 8-30-mer HA oligosaccharides for 3 days. Explants were then stained with X-gal to demonstrate EC. A dose-dependent increase in staining was noted in explants, suggesting that oligomeric HA stimulates vascularization of embryonic lung explants in vitro. (B) Exogenous vascular endothelial growth factor (VEGF) stimulates vascularization of lung buds. Mouse lung explants were exposed to VEGF with either 200 μg/ml of HA-binding peptide (RGGGRGRRR, HABP) or the same concentration of the scrambled peptide control consisting of the same amino acids arranged in a non-HA-binding sequence (RGRRRGRGRG, sHABP). HABP treatment decreased vascular integrity of the explants, suggesting that growth factor-stimulated vascular development is dependent on HA.

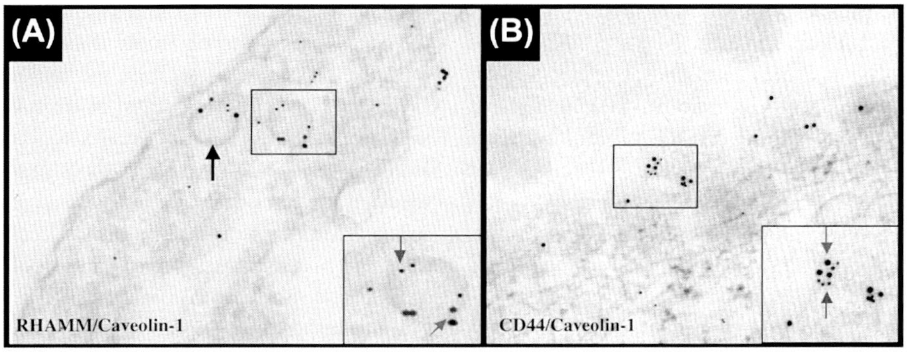

FIGURE 7 Localization of CD44 and RHAMM to caveolae in endothelial cells. Double immunogold-labeled electron microscopy was used to determine the localization of HA receptors in cultured human umbilical vein endothelial cells (HUVEC). Both CD44 and RHAMM were colocalized with caveolin-1.

(RGGGRGRRR, HABP) or the same concentration of the scrambled peptide control consisting of the same amino acids arranged in a non-HA-binding sequence (RGRRRGRGRG, sHABP). Treatments were repeated daily for the 72-h period. We have previously demonstrated the efficacy of this peptide in blocking cell migration of inflammatory cells and fibroblasts (Savani et al., 1995, 2000b). Figure 6(B) shows that HABP treatment decreased the integrity of the fine vascular plexus of the explants, suggesting that growth factor-stimulated vascular development is dependent on HA.

The activities of HA on cell behavior are regulated, at least in part, by receptor and binding protein interactions (Toole, 1990; Turley et al., 2002). A number of proteins that interact with HA have been described and consist of extracellular and cell-associated proteins that bind to HA with high affinity (Savani et al., 2000a; Savani and DeLisser, 2003). Interaction of HA with cell-associated receptors regulates cell adhesion, motility, and proliferation, whereas extracellular interactions are thought to contribute to structure and maintenance of tissue architecture (Lennon and Singleton, 2011). Cellular HA-binding proteins consist of two classes of molecules. CD44, a prototypic type I transmembrane receptor, is predominantly present on the cell surface and binds to HA via a complex tertiary structure defined as the link module (Sneath and Mangham, 1998). Nontransmembrane proteins, of which RHAMM (Receptor for HA-Mediated Motility) is a prototype, are found in multiple locations including the cell surface, cytoplasm, and nucleus (Cheung et al., 1999; Slevin et al., 2007). In previous studies, we demonstrated differential functions of CD44 and RHAMM in EC, such that blockade of CD44 specifically inhibited EC proliferation and adhesion to HA, and blockade of RHAMM specifically inhibited EC migration. Both receptors contributed to in vitro tube formation with additive effects suggesting that their intracellular signaling occurs via distinct pathways (Savani et al., 2001). Interestingly, in vivo blockade of RHAMM resulted in an almost complete inhibition of angiogenesis (Savani et al., 2001). Further, Matrigel-induced wound and tumor angiogenesis was inhibited in CD44 knockout mice, suggesting that CD44 is also important in this process (Cao et al., 2006). Interestingly, electron micrographs of blood vessels from CD44 null mice showed decreased integrity of the endothelium and a propensity for hemorrhage (Cao et al., 2006). The precise mechanisms by which HA receptors regulate EC function and angiogenesis remain to be elucidated. As a first step, we examined the localization of HA receptors to caveolae. Using double immunogold labeled electron microscopy, we colocalized caveolin-1 with both CD44 and RHAMM in cultured human umbilical vein EC (HUVEC) (Figure 7). These data suggest that HA receptors are integral to the signaling organelle that regulates eNOS function. How these receptors influence the activation of the NO pathway is currently being investigated.

6. OTHER SIGNALING PATHWAYS REGULATING ENDOTHELIAL FUNCTIONS

6.1 Permeability

The endothelium serves a key function in the transport of solutes, large molecules, and cells across the normally impervious monolayer (Levick, 1997; Pries and Kuebler, 2006). This transport can occur by two mechanisms, namely by paracellular and transcellular routes, and is also influenced by surrounding cells such as pericytes and inflammatory cells (Vandenbroucke et al., 2008; Komarova and Malik, 2010). EC form intercellular junctions between cells to maintain barrier function. These consist of tight and adherens junctions that consist of specific proteins that can regulate the controlled permeability of the intercellular space (Vandenbroucke et al., 2008; Le Guelte and Gavard, 2011). Adherens junctions consist of VE-cadherin–protein interactions between cells that are regulated by phosphorylation (Vestweber, 2008; Dejana and Giampietro, 2012).

Stimulation with permeability promoting agents such as VEGF results in the phosphorylation of specific tyrosine residues, internalization of the protein, and disruption of the barrier integrity (Dejana and Giampietro, 2012). This is counterbalanced with the activity of VE-protein tyrosine phosphatase (VE-PTP), a phosphatase specific for VE-cadherin that prevents or reverses the phosphorylation of VE-cadherin (Liu et al., 2009). A number of proteins, including p190RhoGAP and β-catenin, connect the VE-cadherin complex with downstream signaling molecules including RhoGTPases that help organize the cytoskeleton (Zebda et al., 2013). Tight junctions consist of another set of protein complexes that include caludin-5, occluding and junctional adhesion molecule C (JAM-C), that are also regulated by phosphorylation after agonist stimulation (Goddard and Iruela-Arispe, 2013). A number of upstream agonists increase cell permeability of EC, including growth factors such as VEGF (also known as vascular permeability factor) and inflammatory mediators such as LPS and TNFα, as well as negative regulators of permeability such as Angiopoetin-1/Tie-2 (Senger et al., 1993; Bates, 2010; Parikh, 2013).

Transcellular transport occurs by transfer of macromolecules through the cell, an active process that involves a number of mechanisms including internalization of caveolin-1, transcellular channels, and vesicular trafficking (Minshall et al., 2003). Interestingly, caveolin-1 negatively regulates paracellular permeability by preventing the uncoupling of eNOS to produce peroxynitrite, which results in nitration of p190RhoGAP to loosen adherens junctions (Siddiqui et al., 2011). Thus, crosstalk between paracellular and transcellular pathways exists and appears also to involve the NO pathway.

6.2 Leukocyte Recruitment

Innate immunity involves several mechanisms that protect the host. These include mechanical barriers such as skin integrity and ciliary movement, biologic factors such as gut commensal organisms, chemical factors such as lysozyme in tears and defensins, and humoral factors like complement and the coagulation pathway. Microbes and other environmental challenges display unique structures called pathogen-associated molecular patterns (PAMPs), while tissue injury results in the production of endogenous ligands called danger-associated molecular patterns (DAMPs) that are recognized by specific pattern recognition receptors (PRRs) (Xiang and Fan, 2010). These include Toll-like receptors, named after Toll, a protein identified in *Drosophila* that interacted with fungi and activated NFκB (Becker and O'Neill, 2007). There are at least 12 known mammalian TLRs with varying ligand specificities. Activation of the innate immune system results in a robust, yet nonspecific, response with the elaboration of cytokines and growth factors.

The recruitment of inflammatory cells to areas of injury requires activation of bone marrow-derived cells, circulating leukocytes, as well as the endothelium (Ley, 1996). Initiation of injury results in initial rolling, then firm adhesion of inflammatory cells to the activated endothelium. This is followed by transmigration of leukocytes through the endothelium and their accumulation in areas of injury. Initially recruited neutrophils undergo apoptosis and alveolar macrophages engulf these cells, a process that results in the production of active TGFβ (Fadok et al., 1998). A number of key molecules have been identified in the process of inflammatory cell recruitment and include selectins, integrins, cytokines, and growth factors, the innate immune system including TLRs, surfactant protein A (SPA) and SPD, as well as components of the extracellular matrix (Reidy and Wright, 2003; Reutershan and Ley, 2004; Craig et al., 2009; Murugan and Peck, 2009).

Leukocyte-endothelial interactions are generally prevented by the glycocalyx, but with the appropriate signals, these interactions are promoted by specific ligands and receptors (Ley et al., 2007; Chappell et al., 2010). Thus, P- and E-selectin, as well as CD44 on endothelium and PSGL-1 and ESL-1 on leukocytes, together with degradation of the glycocalyx by matrix metalloproteinases, heparinases, and hyaluronidases as well as oxidative/nitrative stress, promote initial adhesion and rolling of leukocytes on the endothelium (DeGrendele et al., 1996; Williams et al., 2011). Firmer adhesion is mediated by a more long-term elaboration of molecules such as ICAM-1 (for neutrophils) and VCAM-1 (for monocytes) (Albelda, 1991). Finally, the endothelium plays crucial roles in the trans-endothelial migration of adherent cells, which occurs both via paracellular as well as transcellular mechanisms as described above.

6.3 Anticoagulation and Antiplatelet Function

Another key role for the endothelium is the prevention of both thrombosis and platelet activation (van Hinsbergh, 2012). Coagulation is a highly regulated and coordinated process involving both intrinsic and extrinsic pathways and consisting of multiple cascades of proteases. Key in the regulation of the extrinsic pathway is the activation of tissue factor (TF). Its interaction with Factor VII enhances the proteolysis of Factor X to activate the conversion of prothrombin to thrombin, which subsequently converts fibrinogen to fibrin for clot formation (Okajima, 2004; Finigan, 2009; Mackman, 2009). The endothelium expresses tissue factor pathway inhibitor (TFPI) that binds to Factors X and VII to inhibit them (Osterud et al., 1995). In counterbalance, Antithrombin III (ATIII) is a potent inhibitor of thrombin (Nieuwdorp et al., 2006). In addition, the endothelium also generates the powerful anticoagulant protein activated Protein C, which cleaves Factors Va and VIIIa, which are cofactors required to activate Factors X and IX (Okajima, 2004). Thrombomodulin, synthesized by EC, is required for thrombin-mediated conversion of Protein C to aPC, a process that also requires the endothelial Protein C receptor. This is a key step to control of the coagulation pathway since it limits continued coagulation (Okajima, 2004). Thus, patients with AT III, Protein C, or Factor V Leiden (a variant that is poorly degraded by aPC) have the clinical presentation of thrombosis. Finally, NO and PGI_2 have activities that prevent the aggregation and degranulation of platelets (Desjardins and Balligand, 2006; Dixon et al., 2012). The importance of NO as an antithrombotic mediator is exemplified in studies elucidating the mechanisms of antiphospholipid syndrome, a condition in which circulating antibodies are associated with increased thrombosis and pregnancy loss. The circulating antibodies in antiphospholipid syndrome are directed against a specific dimer of β2-glycoprotein 1 that, via the ApoER2 lipoprotein receptor, activates PP2A, a phosphatase that inactivates eNOS phosphorylation (Ramesh et al., 2011).

7. SUMMARY

The endothelium, though largely quiescent in steady state, is a highly metabolically active tissue consisting of cells that influence a wide variety of homeostatic and pathologic responses. Understanding of the physiology and biochemistry of EC has resulted in dramatic translational and therapeutic advances that have revolutionized the management of patients, as well as provided tangible approaches to the potential treatment of seemingly disparate diseases such as systemic and pulmonary hypertension, atherosclerosis, diabetes, retinopathies, and immune-mediated pregnancy loss. It is clear that further advances in our knowledge of EC biochemistry will yield important new targets and therapies in the future.

ACKNOWLEDGMENTS

I want to thank the following for giving permission to use schematics or photomicrographs that were either generated or modified from their work: Dr Philip Shaul for the schematic of localization of eNOS in caveolae (Figure 3); Dr Vincent Hascall (Cleveland Clinic) for the schematic of HA structure (Figure 5(A)); Dr Horace M. DeLisser for the electron micrograph images (Figure 7). I also gratefully acknowledge funding from the National Institutes of Health (HL62868, HL62472, HL075930, HL075900, HL093535), Children's Medical Center Foundation (Dallas), and The William Buchanan Chair in Pediatrics of the University of Texas Southwestern Medical Center that shaped much of the work from my laboratory that is cited in this chapter. I also wish to acknowledge the donation of NO to my laboratory by Ikaria Therapeutics through an independent research agreement.

REFERENCES

Achneck, H.E., et al., 2008. Review of the biology of bleeding and clotting in the surgical patient. Vascular 16 (Suppl. 1), S6–S13.

Albelda, S.M., 1991. Endothelial and epithelial cell adhesion molecules. Am. J. Respir. Cell Mol. Biol. 4, 195–203.

Alkaitis, M.S., Crabtree, M.J., 2012. Recoupling the cardiac nitric oxide synthases: tetrahydrobiopterin synthesis and recycling. Curr. Heart Fail Rep. 9, 200–210.

Allen, B.W., et al., 2009. Two faces of nitric oxide: implications for cellular mechanisms of oxygen toxicity. J. Appl. Physiol. 106, 662–667.

Amanuma, K., Mitsui, Y., 1991. Hyaluronic acid synthesis is absent in normal human endothelial cells irrespective of hyaluronic acid synthetase inhibitor activity, but is significantly high in transformed cells. Biochem. Biophys. Acta 1092, 336–340.

Andrades, M.E., et al., 2011. Bench-to-bedside review: sepsis—from the redox point of view. Crit. Care 15, 230.

Arnal, J.F., et al., 1999. Endothelium-derived nitric oxide and vascular physiology and pathology. Cell. Mol. Life Sci. 55, 1078–1087.

Atochina-Vasserman, E.N., et al., 2010. Review: chemical and structural modifications of pulmonary collectins and their functional consequences. Innate Immun. 16, 175–182.

Bachiller, P.R., et al., 2013. Soluble guanylate cyclase modulates alveolarization in the newborn lung. Am. J. Physiol. Lung Cell. Mol. Physiol. 305 (8), L569–L581.

Balasubramaniam, V., et al., 2003. Mild hypoxia impairs alveolarization in the endothelial nitric oxide synthase-deficient mouse. Am. J. Physiol. Lung Cell. Mol. Physiol. 284, L964–L971.

Balasubramaniam, V., et al., 2006. Inhaled NO restores lung structure in eNOS-deficient mice recovering from neonatal hypoxia. Am. J. Physiol. Lung Cell. Mol. Physiol. 291, L119–L127.

Ballard, R.A., et al., 2006. Inhaled nitric oxide in preterm infants undergoing mechanical ventilation. N. Engl. J. Med. 355, 343–353.

Barrington, K.J., Finer, N., 2010. Inhaled nitric oxide for respiratory failure in preterm infants. Cochrane Database Syst. Rev. CD000509.

Bates, D.O., 2010. Vascular endothelial growth factors and vascular permeability. Cardiovasc. Res. 87, 262–271.

Becker, C.E., O'Neill, L.A., 2007. Inflammasomes in inflammatory disorders: the role of TLRs and their interactions with NLRs. Semin. Immunopathol. 29, 239–248.

Beckman, J.S., Koppenol, W.H., 1996. Nitric oxide, superoxide, and peroxynitrite: the good, the bad, and ugly. Am. J. Physiol. 271, C1424–C1437.

Bian, K., et al., 2012. NOS-2 signaling and cancer therapy. IUBMB Life 64, 676–683.

Bogdan, C., et al., 2000. The role of nitric oxide in innate immunity. Immunol. Rev. 173, 17–26.

Bourguignon, L.Y., et al., 2004. CD44 interaction with Na+–H+ exchanger (NHE1) creates acidic microenvironments leading to hyaluronidase-2 and cathepsin B activation and breast tumor cell invasion. J. Biol. Chem. 279, 26991–27007.

Brouland, J.P., et al., 1999. Macro and microheterogeneity in normal endothelial cells: differential composition of luminal glycocalyx and functional implications. Endothelium 6, 251–262.

van den Berg, B.M., et al., 2006. Glycocalyx and endothelial (dys) function: from mice to men. Pharmacol. Rep. 58 (Suppl.), 75–80.

Calcerrada, P., et al., 2011. Nitric oxide-derived oxidants with a focus on peroxynitrite: molecular targets, cellular responses and therapeutic implications. Curr. Pharm. Des. 17, 3905–3932.

Cao, G., et al., 2006. Involvement of endothelial CD44 during in vivo angiogenesis. Am. J. Pathol. 169, 325–336.

Chappell, D., et al., 2010. Glycocalyx protection reduces leukocyte adhesion after ischemia/reperfusion. Shock 34, 133–139.

Chess, P.R., et al., 2006. Pathogenesis of bronchopulmonary dysplasia. Semin. Perinatol. 30, 171–178.

Cheung, W.F., et al., 1999. Receptor for hyaluronan-mediated motility (RHAMM), a hyaladherin that regulates cell responses to growth factors. Biochem. Soc. Trans. 27, 135–142.

Coalson, J.J., 2006. Pathology of bronchopulmonary dysplasia. Semin. Perinatol. 30, 179–184.

Coleman, J.W., 2001. Nitric oxide in immunity and inflammation. Int. Immunopharmacol. 1, 1397–1406.

Craig, A., et al., 2009. Neutrophil recruitment to the lungs during bacterial pneumonia. Infect. Immun. 77, 568–575.

Csoka, A.B., et al., 2001. The six hyaluronidase-like genes in the human and mouse genomes. Matrix Biol. 20, 499–508.

De Cruz, S.J., et al., 2009. Bench-to-bedside review: the role of nitric oxide in sepsis. Expert Rev. Respir. Med. 3, 511–521.

DeGrendele, H.C., et al., 1996. CD44 and its ligand hyaluronate mediate rolling under physiologic flow: a novel lymphocyte-endothelial cell primary adhesive pathway. J. Exp. Med. 183, 1119–1130.

Dejana, E., Giampietro, C., 2012. Vascular endothelial-cadherin and vascular stability. Curr. Opin. Hematol. 19, 218–223.

Desjardins, F., Balligand, J.L., 2006. Nitric oxide-dependent endothelial function and cardiovascular disease. Acta Clin. Belg. 61, 326–334.

Dixon, J.T., et al., 2012. Platelet-mediated vascular dysfunction during acute lung injury. Arch. Physiol. Biochem. 118, 72–82.

Donohue, P.K., et al., 2011. Inhaled nitric oxide in preterm infants: a systematic review. Pediatrics 127, e414–422.

Ebong, E.E., et al., 2011. Imaging the endothelial glycocalyx in vitro by rapid freezing/freeze substitution transmission electron microscopy. Arterioscler. Thromb. Vasc. Biol. 31, 1908–1915.

Fadok, V.A., et al., 1998. Macrophages that have ingested apoptotic cells in vitro inhibit proinflammatory cytokine production through autocrine/paracrine mechanisms involving TGF-beta, PGE2, and PAF. J. Clin. Invest. 101, 890–898.

Feinberg, R.N., Beebe, D.C., 1983. Hyaluronate in vasculogenesis. Science 220, 1177–1179.

Feng, C., et al., 2014. Dissecting regulation mechanism of the FMN to heme interdomain electron transfer in nitric oxide synthases. J. Inorg. Biochem. 130, 130–140.

Finigan, J.H., 2009. The coagulation system and pulmonary endothelial function in acute lung injury. Microvasc. Res. 77, 35–38.

Forstermann, U., Sessa, W.C., 2012. Nitric oxide synthases: regulation and function. Eur. Heart J. 33, 829–837, 837a–837d.

Fraser, J.R., et al., 1997. Hyaluronan: its nature, distribution, functions and turnover. J. Intern Med. 242, 27–33.

Fu, B.M., Tarbell, J.M., 2013. Mechano-sensing and transduction by endothelial surface glycocalyx: composition, structure, and function. Wiley Interdiscip. Rev. Syst. Biol. Med. 5, 381–390.

Gao, L., Lipowsky, H.H., 2010. Composition of the endothelial glycocalyx and its relation to its thickness and diffusion of small solutes. Microvasc. Res. 80, 394–401.

Goddard, L.M., Iruela-Arispe, M.L., 2013. Cellular and molecular regulation of vascular permeability. Thromb. Haemost. 109, 407–415.

Guix, F.X., et al., 2005. The physiology and pathophysiology of nitric oxide in the brain. Prog. Neurobiol. 76, 126–152.

Han, R.N., Stewart, D.J., 2006. Defective lung vascular development in endothelial nitric oxide synthase-deficient mice. Trends Cardiovasc. Med. 16, 29–34.

Hascall, V.C., Laurent, T.C., 1997. Hyaluronan: sturcture and physical properties, first ed. Glycoforum Group.

Hess, D.T., et al., 2005. Protein S-nitrosylation: purview and parameters. Nat. Rev. Mol. Cell. Biol. 6, 150–166.

Hightower, C.M., et al., 2011. Integration of cardiovascular regulation by the blood/endothelium cell-free layer. Wiley Interdiscip. Rev. Syst. Biol. Med. 3, 458–470.

Hollenberg, S.M., Cinel, I., 2009. Bench-to-bedside review: nitric oxide in critical illness—update 2008. Crit. Care 13, 218.

Hrabarova, E., et al., 2011. Pro-oxidative effect of peroxynitrite regarding biological systems: a special focus on high-molar-mass hyaluronan degradation. Gen. Physiol. Biophys. 30, 223–238.

van Hinsbergh, V.W., 2012. Endothelium—role in regulation of coagulation and inflammation. Semin. Immunopathol. 34, 93–106.

Jobe, A.H., 2011. The new bronchopulmonary dysplasia. Curr. Opin. Pediatr. 23, 167–172.

Kietadisorn, R., et al., 2012. Tackling endothelial dysfunction by modulating NOS uncoupling: new insights into its pathogenesis and therapeutic possibilities. Am. J. Physiol. Endocrinol. Metab. 302, E481–E495.

Kinsella, J.P., et al., 2006. Early inhaled nitric oxide therapy in premature newborns with respiratory failure. N. Engl. J. Med. 355, 354–364.

Komarova, Y., Malik, A.B., 2010. Regulation of endothelial permeability via paracellular and transcellular transport pathways. Annu. Rev. Physiol. 72, 463–493.

Kumagai, R., et al., 2009. Role of glycocalyx in flow-induced production of nitric oxide and reactive oxygen species. Free Radic. Biol. Med. 47, 600–607.

Le Guelte, A., Gavard, J., 2011. Role of endothelial cell-cell junctions in endothelial permeability. Methods Mol. Biol. 763, 265–279.

Lee, J.L., et al., 2008. CD44 engagement promotes matrix-derived survival through the CD44-SRC-integrin axis in lipid rafts. Mol. Cell. Biol. 28, 5710–5723.

Lees, V.C., et al., 1995. Angiogenesis in a delayed revascularization model is accelerated by angiogenic oligosaccharides of hyaluronan. Lab. Invest. 73, 259–266.

Lennon, F.E., Singleton, P.A., 2011. Role of hyaluronan and hyaluronan-binding proteins in lung pathobiology. Am. J. Physiol. Lung Cell. Mol. Physiol. 301, L137–L147.

Levick, J.R., 1997. Fluid exchange across endothelium. Int. J. Microcirc. Clin. Exp. 17, 241–247.

Ley, K., 1996. Molecular mechanisms of leukocyte recruitment in the inflammatory process. Cardiovasc. Res. 32, 733–742.

Ley, K., et al., 2007. Getting to the site of inflammation: the leukocyte adhesion cascade updated. Nat. Rev. Immunol. 7, 678–689.

Li, H., Forstermann, U., 2000. Nitric oxide in the pathogenesis of vascular disease. J. Pathol. 190, 244–254.

Li, H., Forstermann, U., 2013. Uncoupling of endothelial NO synthase in atherosclerosis and vascular disease. Curr. Opin. Pharmacol. 13, 161–167.

Li, M., et al., 1997. Degradation of hyaluronan by peroxynitrite. Arch. Biochem. Biophys. 341, 245–250.

Lin, Y.J., et al., 2005. Inhaled nitric oxide enhances distal lung growth after exposure to hyperoxia in neonatal rats. Pediatr. Res. 58, 22–29.

Liu, A., et al., 2009. The counteradhesive proteins, thrombospondin 1 and SPARC/osteonectin, open the tyrosine phosphorylation-responsive paracellular pathway in pulmonary vascular endothelia. Microvasc. Res. 77, 13–20.

Mackman, N., 2009. The role of tissue factor and factor VIIa in hemostasis. Anesth. Analg. 108, 1447–1452.

Martinez-Ruiz, A., et al., 2011. Nitric oxide signaling: classical, less classical, and nonclassical mechanisms. Free Radic. Biol. Med. 51, 17–29.

McCurnin, D.C., et al., 2005. Inhaled NO improves early pulmonary function and modifies lung growth and elastin deposition in a baboon model of neonatal chronic lung disease. Am. J. Physiol. Lung Cell. Mol. Physiol. 288, L450–L459.

McCurnin, D.C., et al., 2009. Postnatal estradiol up-regulates lung nitric oxide synthases and improves lung function in bronchopulmonary dysplasia. Am. J. Respir. Crit. Care Med. 179, 492–500.

Michel, T., Vanhoutte, P.M., 2010. Cellular signaling and NO production. Pflugers Arch. 459, 807–816.

Midgley, A.C., et al., 2013. Transforming growth factor-beta1 (TGF-beta1)-stimulated fibroblast to myofibroblast differentiation is mediated by hyaluronan (HA)-facilitated epidermal growth factor receptor (EGFR) and CD44 co-localization in lipid rafts. J. Biol. Chem. 288, 14824–14838.

Mineo, C., Shaul, P.W., 2006. Circulating cardiovascular disease risk factors and signaling in endothelial cell caveolae. Cardiovasc. Res. 70, 31–41.

Mineo, C., Shaul, P.W., 2011. New insights into the molecular basis of the antiphospholipid syndrome. Drug Discov. Today Dis. Mech. 8, e47–e52.

Mineo, C., Shaul, P.W., 2012. Regulation of eNOS in caveolae. Adv. Exp. Med. Biol. 729, 51–62.

Minshall, R.D., et al., 2003. Caveolin regulation of endothelial function. Am. J. Physiol. Lung Cell. Mol. Physiol. 285, L1179–L1183.

Mochizuki, S., et al., 2003. Role of hyaluronic acid glycosaminoglycans in shear-induced endothelium-derived nitric oxide release. Am. J. Physiol. Heart Circ. Physiol. 285, H722–H726.

Montesano, R., et al., 1996. Synergistic effect of hyaluronan oligosaccharides and vascular endothelial growth factor on angiogenesis in vitro. Lab. Invest. 75, 249–262.

Mosca, F., et al., 2011. BPD: old and new problems. J. Matern. Fetal Neonatal. Med. 24 (Suppl. 1), 80–82.

Murugan, V., Peck, M.J., 2009. Signal transduction pathways linking the activation of alveolar macrophages with the recruitment of neutrophils to lungs in chronic obstructive pulmonary disease. Exp. Lung Res. 35, 439–485.

Nadeau, K., et al., 2010. Modulation of Lgl1 by steroid, retinoic acid, and vitamin D models complex transcriptional regulation during alveolarization. Pediatr. Res. 67, 375–381.

Nieuwdorp, M., et al., 2006. Loss of endothelial glycocalyx during acute hyperglycemia coincides with endothelial dysfunction and coagulation activation in vivo. Diabetes 55, 480–486.

Okajima, K., 2004. Prevention of endothelial cell injury by activated protein C: the molecular mechanism(s) and therapeutic implications. Curr. Vasc. Pharmacol. 2, 125–133.

Østerholt, H.C., et al., 2012. Antioxidant protects against increases in low molecular weight hyaluronan and inflammation in asphyxiated newborn pigs resuscitated with 100% oxygen. PLoS One 7, e38839.

Osterud, B., et al., 1995. Sites of tissue factor pathway inhibitor (TFPI) and tissue factor expression under physiologic and pathologic conditions. On behalf of the subcommittee on tissue factor pathway inhibitor (TFPI) of the scientific and standardization committee of the ISTH. Thromb. Haemost. 73, 873–875.

Pacher, P., et al., 2007. Nitric oxide and peroxynitrite in health and disease. Physiol. Rev. 87, 315–424.

Pahakis, M.Y., et al., 2007. The role of endothelial glycocalyx components in mechanotransduction of fluid shear stress. Biochem. Biophys. Res. Commun. 355, 228–233.

Parikh, S.M., 2013. Dysregulation of the angiopoietin-tie-2 axis in sepsis and ARDS. Virulence 4, 517–524.

Potter, L.R., 2011. Guanylyl cyclase structure, function and regulation. Cell. Signal. 23, 1921–1926.

Pries, A.R., Kuebler, W.M., 2006. Normal endothelium. Handb. Exp. Pharmacol. 1–40.

Rahmanian, M., et al., 1997. Hyaluronan oligosaccharides induce tube formation of a brain endothelial cell line in vitro. Exp. Cell. Res. 237, 223–230.

Ramesh, S., et al., 2011. Antiphospholipid antibodies promote leukocyte-endothelial cell adhesion and thrombosis in mice by antagonizing eNOS via beta2GPI and apoER2. J. Clin. Invest. 121, 120–131.

Reidy, M.F., Wright, J.R., 2003. Surfactant protein A enhances apoptotic cell uptake and TGF-beta1 release by inflammatory alveolar macrophages. Am. J. Physiol. Lung Cell. Mol. Physiol. 285, L854–L861.

Reitsma, S., et al., 2007. The endothelial glycocalyx: composition, functions, and visualization. Pflugers Arch. 454, 345–359.

Reutershan, J., Ley, K., 2004. Bench-to-bedside review: acute respiratory distress syndrome—how neutrophils migrate into the lung. Crit. Care 8, 453–461.

Ryan, R.M., et al., 2008. Inflammatory mediators in the immunobiology of bronchopulmonary dysplasia. Clin. Rev. Allergy Immunol. 34, 174–190.

Sattar, A., et al., 1994. Application of angiogenic oligosaccharides of hyaluronan increases blood vessel numbers in rat skin. J. Invest. Dermatol. 103, 576–579.

Savani, R.C., DeLisser, H.M., 2003. Hyaluronan and its receptors in lung health and disease. In: Garg, H.G., Roughley, P.J., Hales, C.A. (Eds.), Proteoglycans and Lung Disease. Marcel Dekker, New York, pp. 73–106.

Savani, R.C., et al., 1995. Hyaluronan receptor antagonists alter skin inflammation and fibrosis following injury. Proc. West Pharmacol. Soc. 38, 131–136.

Savani, R.C., et al., 2000a. The role of hyaluronan-receptor interactions in wound repair. In: Garg, H.G., Longaker, M.T. (Eds.), Scarless Wound Healing, 19th ed. Marcel Dekker, New York-Basel, pp. 115–142.

Savani, R.C., et al., 2001. Differential involvement of the hyaluronan (HA) receptors CD44 and receptor for HA-mediated motility in endothelial cell function and angiogenesis. J. Biol. Chem. 276, 36770–36778.

Savani, R.C., et al., 2000b. A role for hyaluronan (HA) in macrophage accumulation and collagen deposition after bleomycin-induced lung injury. Am. J. Respir. Cell. Mol. Biol. 23, 475–484.

Schwartz, R., et al., 2007. C-reactive protein downregulates endothelial NO synthase and attenuates reendothelialization in vivo in mice. Circ. Res. 100, 1452–1459.

Sehgal, P.B., Lee, J.E., 2011. Protein trafficking dysfunctions: role in the pathogenesis of pulmonary arterial hypertension. Pulm. Circ. 1, 17–32.

Senger, D.R., et al., 1993. Vascular permeability factor (VPF, VEGF) in tumor biology. Cancer Metastasis Rev. 12, 303–324.

Shaul, P.W., 1999. Regulation of vasodilator synthesis during lung development. Early Hum. Dev. 54, 271–294.

Shaul, P.W., 2002. Regulation of endothelial nitric oxide synthase: location, location, location. Annu. Rev. Physiol. 64, 749–774.

Siddiqui, M.R., et al., 2011. Caveolin-1-eNOS signaling promotes p190RhoGAP-A nitration and endothelial permeability. J. Cell Biol. 193, 841–850.

Slevin, M., et al., 2007. Hyaluronan-mediated angiogenesis in vascular disease: uncovering RHAMM and CD44 receptor signaling pathways. Matrix Biol. 26, 58–68.

Sneath, R.J., Mangham, D.C., 1998. The normal structure and function of CD44 and its role in neoplasia. Mol. Pathol. 51, 191–200.

Speer, C.P., 2006. Inflammation and bronchopulmonary dysplasia: a continuing story. Semin. Fetal Neonatal Med. 11, 354–362.

Stenmark, K.R., Abman, S.H., 2005. Lung vascular development: implications for the pathogenesis of bronchopulmonary dysplasia. Annu Rev. Physiol. 67, 623–661.

Sugiura, H., Ichinose, M., 2011. Nitrative stress in inflammatory lung diseases. Nitric Oxide: Biology and Chemistry/Official journal of the Nitric Oxide Society 25, 138–144.

Sundgren, N.C., et al., 2011. Coupling of fcgamma receptor I to fcgamma receptor IIb by SRC kinase mediates C-reactive protein impairment of endothelial function. Circ. Res. 109, 1132–1140.

Tanigaki, K., et al., 2009. C-reactive protein inhibits insulin activation of endothelial nitric oxide synthase via the immunoreceptor tyrosine-based inhibition motif of FcgammaRIIB and SHIP-1. Circ. Res. 104, 1275–1282.

Tarbell, J.M., Pahakis, M.Y., 2006. Mechanotransduction and the glycocalyx. J. Intern. Med. 259, 339–350.

Tarbell, J.M., Ebong, E.E., 2008. The endothelial glycocalyx: a mechano-sensor and -transducer. Sci. Signal. 1 (Pt 8), 1–5.

Tarbell, J.M., et al., 2005. Cellular fluid mechanics and mechanotransduction. Ann. Biomed. Eng. 33, 1719–1723.

Thebaud, B., 2007. Angiogenesis in lung development, injury and repair: implications for chronic lung disease of prematurity. Neonatology 91, 291–297.

Toole, B.P., 1990. Hyaluronan and its binding proteins, the hyaladherins. Curr. Opin. Cell. Biol. 2, 839–844.

Toole, B.P., 2000. Hyaluronan. In: Izozzo, R.V. (Ed.), Proteoglycans: Structure, Biology and Molecular Interactions. Marcel Dekker, New York, pp. 61–92.

Toole, B.P., 2004. Hyaluronan: from extracellular glue to pericellular cue. Nat. Rev. Cancer 4, 528–539.

Tsutsui, M., et al., 2006. Development of genetically engineered mice lacking all three nitric oxide synthases. J. Pharmacol. Sci. 102, 147–154.

Turley, E.A., et al., 2002. Signaling properties of hyaluronan receptors. J. Biol. Chem. 277, 4589–4592.

Umetani, M., Shaul, P.W., 2011. 27-Hydroxycholesterol: the first identified endogenous SERM. Trends Endocrinol. Metab. 22, 130–135.

Umetani, M., et al., 2007. 27-Hydroxycholesterol is an endogenous SERM that inhibits the cardiovascular effects of estrogen. Nat. Med. 13, 1185–1192.

Vandenbroucke, E., et al., 2008. Regulation of endothelial junctional permeability. Ann. N. Y. Acad. Sci. 1123, 134–145.

Vestweber, D., 2008. VE-cadherin: the major endothelial adhesion molecule controlling cellular junctions and blood vessel formation. Arterioscler. Thromb. Vasc. Biol. 28, 223–232.

Walsh, M.C., et al., 2010. Two-year neurodevelopmental outcomes of ventilated preterm infants treated with inhaled nitric oxide. J. Pediatr. 156, 556–561 e551.

Weinbaum, S., et al., 2007. The structure and function of the endothelial glycocalyx layer. Annu. Rev. Biomed. Eng. 9, 121–167.

West, D.C., Kumar, S., 1989. The effect of hyaluronate and its oligosaccharides on endothelial cell proliferation and monolayer integrity. Exp. Cell. Res. 183, 179–196.

Wheeler-Jones, C.P., et al., 2010. Targeting hyaluronan of the endothelial glycocalyx for therapeutic intervention. Curr. Opin. Investig. Drugs 11, 997–1006.

Williams, M.R., et al., 2011. Emerging mechanisms of neutrophil recruitment across endothelium. Trends Immunol. 32, 461–469.

Willmot, M., et al., 2005. Nitric oxide synthase inhibitors in experimental ischemic stroke and their effects on infarct size and cerebral blood flow: a systematic review. Free Radic. Biol. Med. 39, 412–425.

Wolkow, P.P., 1998. Involvement and dual effects of nitric oxide in septic shock. Inflamm. Res.: Official Journal of the European Histamine Research Society [et al] 47, 152–166.

Xiang, M., Fan, J., 2010. Pattern recognition receptor-dependent mechanisms of acute lung injury. Mol. Med. 16, 69–82.

Yang, Z., Ming, X.F., 2013. Arginase: the emerging therapeutic target for vascular oxidative stress and inflammation. Front. Immunol. 4, 149.

Yen, W.Y., et al., 2012. Quantification of the endothelial surface glycocalyx on rat and mouse blood vessels. Microvasc. Res. 83, 337–346.

Zebda, N., et al., 2013. Interaction of p190RhoGAP with C-terminal domain of p120-catenin modulates endothelial cytoskeleton and permeability. J. Biol. Chem. 288, 18290–18299.

Zeng, Y., et al., 2012. The structural stability of the endothelial glycocalyx after enzymatic removal of glycosaminoglycans. PLoS One 7, e43168.

Zeng, Y., et al., 2013. Fluid shear stress induces the clustering of heparan sulfate via mobility of glypican-1 in lipid rafts. Am. J. Physiol. Heart Circ. Physiol. 305, H811–H820.

Zhang, Y.H., Casadei, B., 2012. Sub-cellular targeting of constitutive NOS in health and disease. J. Mol. Cell. Cardiol. 52, 341–350.

Zhou, L., Zhu, D.Y., 2009. Neuronal nitric oxide synthase: structure, subcellular localization, regulation, and clinical implications. Nitric Oxide: Biology and Chemistry/Official Journal of the Nitric Oxide Society 20, 223–230.

Chapter 20

Comparative Biology of the Normal Lung Extracellular Matrix

Stephanie A. Matthes, Ryan Hadley, Jesse Roman and Eric S. White

Chapter Outline

1. EXTRACELLULAR MATRIX BASIC ARCHITECTURE

Extracellular matrix (ECM) is a highly dynamic complex of fibrous proteins, glycoproteins, and proteoglycans that comprises the noncellular aspect of tissues and varies in composition according to its tissue localization and physiological circumstances. In addition to providing structural integrity, ECM delivers important spatial and contextual cues to drive cellular phenotypes. The ECM of organs is typically restricted to two basic compartments: basement membranes and the interstitial spaces. Basement membranes are thin, specialized layers of ECM found under all epithelial and endothelial cell layers, whereas interstitial spaces form the parenchyma of organs. Despite our ever-increasing knowledge of ECM constituents and functions, relatively little is known about the human lung ECM.

Conceptually, the lung ECM is arranged in a "sandwich" configuration, in which the epithelial layer (sitting upon a basement membrane) and the endothelial layer (also sitting upon a basement membrane) are separated by the interstitial space (Figure 1(A)). However, this layout varies based on location within the lung. For example, large airways often have a prominent epithelial basement membrane and an interstitial space that harbors smooth muscle, elastic fibers, and vasculature, whereas alveolar structures have very thin epithelial and endothelial basement membranes that are usually in apposition to each other to facilitate gas transfer (West and Mathieu-Costello, 1999), with relatively little intervening tissue. Transmission electron microscopy of a normal human lung alveolus following removal of cells illustrates the spatial relationships of basement membranes and interstitium (Figure 1(B)).

2. ECM MOLECULES

Broadly, the molecules that compose the ECM can be divided into two groups: *fibrous proteins* (such as collagen, elastin, and fibronectin) and *glycosaminoglycans* (GAGs), either attached to a protein core (proteoglycans) or free floating (hyaluronan) (Frantz et al., 2010; Schaefer and Schaefer, 2010). Together, these molecules form an interwoven network essential for organ structure and function. We will first briefly review the basic constituents of the ECM.

2.1 Fibrous Proteins

Collagens

Collagens are the major fibrous ECM proteins in the lung. The structure of collagen, first proposed in 1940 (Astbury, 1940) and then refined by 1955 (Cowan et al., 1955; Pauling and Corey, 1951; Ramachandran and Kartha, 1954; Ramachandran and Kartha, 1955; Rich and Crick, 1955), resembles a triple helix of three polypeptide strands composed of various α chains, each in a left-handed helical

Comparative Biology of the Normal Lung. http://dx.doi.org/10.1016/B978-0-12-404577-4.00020-5

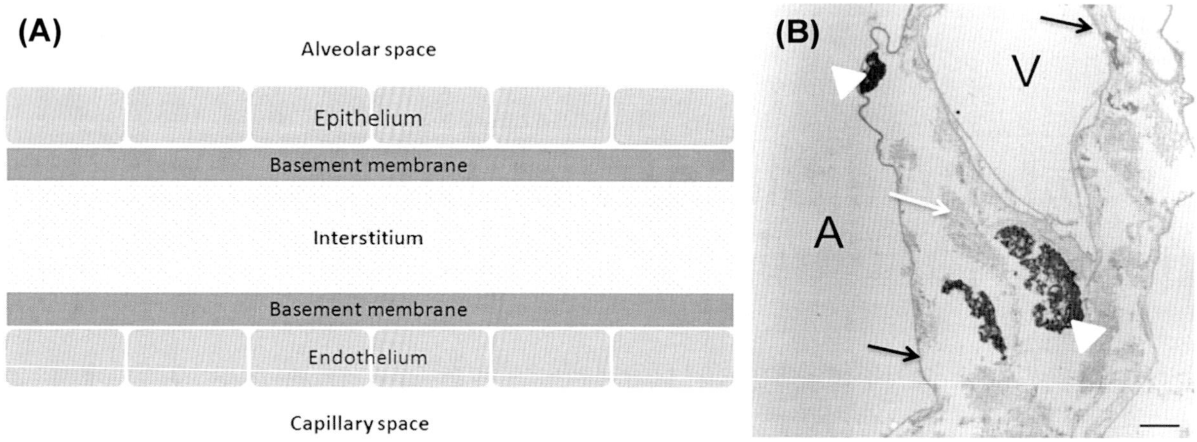

FIGURE 1 (A) Schematic of ECM compartmentalization in the lung. Both epithelial and endothelial cell layers reside on basement membranes composed of various ECM molecules (see text for detail). The interstitial space also contains ECM molecules that may be specific for various regions within the lung. (B) Transmission electron microscopy of normal human lung matrix following decellularization. Black arrows denote basement membranes. White arrow shows collagen bundle in the interstitium. White arrowheads depict elastin bundles. A – alveolar space; V – pulmonary capillary. Scale bar = 2 μm. 8520 × original magnification. *Reprinted with permission of the American Thoracic Society. Copyright © 2012 American Thoracic Society.*

coil. The three individual α chains are combined to form a right-handed supercoiled triple helix that is supported by two hydrogen bonds (Orgel et al., 2006; Orgel et al., 2001). Each macromolecular structure is composed of identical (homotrimer) or different (heterotrimer) α chains (Shoulders and Raines, 2009).

The variety in α chain composition and arrangement results in the identification of 26 collagen types, which can be classified as either fibrillar or nonfibrillar. However, the fibril-forming interstitial collagens I and III (occurring at approximately a 2:1 ratio in the human lung) are the most abundant types of collagen, distributed within the airways, blood vessels, and the interstitium (Gay et al., 1976; Hance et al., 1976; Seyer et al., 1976); fibrils have a characteristic striated appearance on electron microscopy (Figure 2), and individual fibrils most often aggregate in bundles. Even though collagen provides structural support to the surrounding tissues in human lung, type I collagen is unstable at body temperature. However, stability of the triple helix domains in type I collagen is established during the process of fibrillogenesis (Leikina et al., 2002; Shoulders and Raines, 2009). Fibrillogenesis occurs by self-assembly, which occurs in two stages: nucleation and fiber growth (Shoulders and Raines, 2009; Silver et al., 2003). In fibril-forming collagens, helical domains predominantly contain a (Gly-X-Y) repeating motif where X and Y can be any amino acid, although X is most often proline and Y is hydroxyproline (Gelse et al., 2003; Hofmann et al., 1978). The assembly of the triple helix requires the N- and C-terminal polypeptides of procollagen but must be cleaved by procollagen peptidases, such as pronase, in order for fibrillogenesis to occur (Kadler et al., 1987; Last and Reiser, 1984; Silver et al., 2003). Vast quantities of procollagens can be degraded within minutes prior to being secreted from the cell. Even though cross-linking leaves mature collagens less susceptible to degradation, a variety of collagenases are still capable of degrading mature collagen (Laurent, 1986).

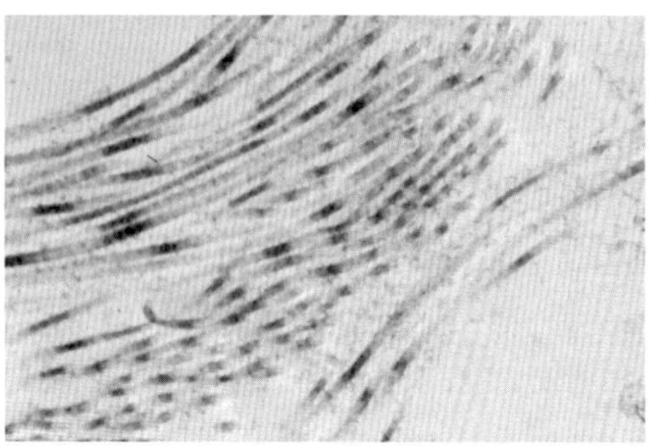

FIGURE 2 **Collagen fibers within the human lung interstitial space.** The characteristic alternating dark and white regions reflect the normal periodic structure of collagen. Individual fibers are incorporated into a bundle. 42,200 × original magnification.

Biomechanically, tensile strength of the ECM is dictated by the amount of cross-linking between the self-assembled collagen fibers (Danielsen, 1981). The degree of cross-linking, length, and diameter of collagen fibril assembly all affect the viscoelastic properties of the lung. This is important in organs such as lung, where reversible stretching can occur (Silver et al., 2000; Silver et al., 2003). Evidence suggests that the rod-like structure of type I collagen is quite rigid whereas type III collagen is less so. The difference in stiffness accommodates both storing and transmitting (collagen I) or storing and reflecting (collagen III) the elastic energy within the lung (Silver et al., 2002).

Elastins

During respiration, regions of the lung expand and then passively recoil. This dynamic capability is primarily provided by elastic fibers that exist within the vasculature as well as the airway and alveolar interstitium. Elastic fibers and collagen comprise a majority of the lung ECM, although the functionality of the matrix is dictated by the interaction of all of its components (Starcher, 2000). Elastic fibers are primarily composed of tropoelastin molecules, as well as microfibrils, which include both fibrillin and fibulin (Cleary and Gibson, 1983; Reinhardt et al., 1996).

Elastin is a durable and highly insoluble protein that has little turnover in the healthy individual, with an estimated half-life of 74 years (Shapiro et al., 1991). Elastin within the lung is primarily secreted by fibroblasts, although chondroblasts, mesothelial cells, and smooth muscle cells are all capable of synthesizing tropoelastin (Cantor et al., 1986; Rennard et al., 1984; Starcher, 2000; Starcher and Galione, 1976). Tropoelastin is secreted from the rough endoplasmic reticulum (Sandberg et al., 1969) and then chaperoned and secreted from the cell, where coacervation occurs (Toonkool et al., 2001). Coacervation is a process that results in the self-annealing of the hydrophobic residues at an increase in temperature of 37 °C (Urry, 1978). This process supports proper alignment and cross-linking between tightly associated lysine residues in tropoelastin molecules and, along with microfibril incorporation, results in mature elastin fiber formation (Hinek and Rabinovitch, 1994; Partridge et al., 1963). It is the cross-linking of the tropoelastin polymers that establishes a functional elastin protein (Mecham, 2008) (Figure 3).

Laminins

Basement membranes are thin sheets of specialized ECM that underlie epithelial and endothelial cell monolayers. This organized scaffold is composed of a variety of molecules including laminins, nonfibrillar type IV collagen, proteoglycans, and nidogen/entactin (Colognato and Yurchenco, 2000; Nguyen and Senior, 2006). Basement membrane stability relies on the self-assembly of both type IV collagen and laminin, which polymerize in a honeycomb-like structure. The proteoglycans and nidogen link the collagen and laminin structures in the basement membrane, which further stabilizes and strengthens the matrix (LeBleu et al., 2007). Laminins were first isolated in 1979 and are large heterotrimeric glycoproteins (Chung et al., 1979; Timpl et al., 1979). Laminins are synthesized by alveolar type II cells (Rannels et al., 1992).

Structurally, laminins resemble an asymmetrical crucifix and contain one α, one β, and one γ chain (Burgeson et al., 1994) (Figure 4). To date, five α chains and three of each the β and γ chain isoforms have been identified, and at least three alternately spliced forms of the α2, α3, and γ3 chains are known (Colognato and Yurchenco, 2000). The variety in chain identification and partnering results theoretically in the assembly of over 45 different heterotrimers, although only 18 have been described thus far (Durbeej, 2010); this may be due in part to chain assembly restrictions. Specificity of laminin isoforms is largely directed by the α chain (Sanes et al., 1990). In the normal adult lung, α3, α4, and α5 chains are the most prevalent (Miner et al., 1997; Yurchenco et al., 1997).

The laminin cross-like structure consists of one long arm in addition to three short arms and forms via self-assembly of the α, β, and γ chains (Figure 4). The long arm is derived from the α chain and varies considerably in length. It contains five C-terminal globular domains that act as docking sites for other molecules. The short arms are highly homologous and are primarily composed of the β and γ chains that contain both central and N-terminal globular domains (Beck et al., 1990; Tzu and Marinkovich, 2008). Each of the N-termini of the three short arms can bind to a neighboring N-terminal globular domain to create a lattice-like structural network characteristic of basement membranes (Cheng et al., 1997; Tzu and Marinkovich, 2008).

Laminins serve dual roles by acting as an architectural support and by participating in cell-binding activities. They provide a docking site to several other molecules, which

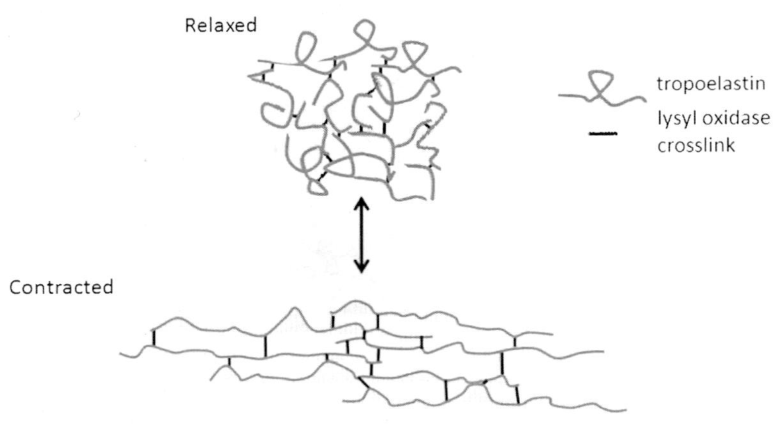

FIGURE 3 Schematic of elastin fiber. Individual tropoelastin molecules (gray curved lines) are linked together via lysyl oxidase cross-linkers (black lines). Tropoelastin molecules are retained in a globular form (top) until stretched (bottom). Release of tension results in molecules readopting a globular configuration.

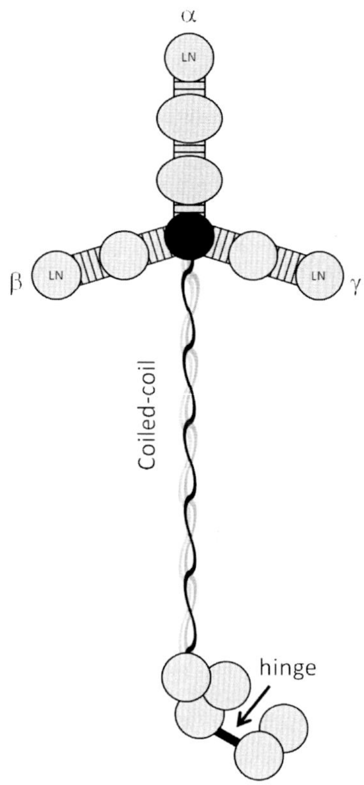

FIGURE 4 Schematic of generic laminin molecule. Laminin is composed of one α, one β, and one γ chain configured in a cross-like pattern. The C-termini of each chain join in a coiled coil to form the long arm, which ends in five homologous globular domains (gray circled) joined by a hinge region. LN: Laminin N-terminal.

include a variety of integrins, heparan sulfate proteoglycans, and dystroglycan (Nguyen and Senior, 2006). Laminins help create an environment within the basement membrane with the purpose of transmitting morphologic and differentiation cues, particularly for type I or type II alveolar epithelial cellular switching. Laminins also provide adhesion and migratory cues (Kleinman et al., 1985; Sannes et al., 1998; Sannes and Wang, 1997; Sugrue and Hay, 1981).

Fibronectin

Fibronectin (FN) is an ubiquitous glycoprotein that exists as either an insoluble multimeric fibrillar component of the ECM (cellular FN or cFN) or as a soluble monomeric form that is produced by hepatocytes and secreted directly into plasma (plasma FN or pFN) (Hynes, 1973; Morrison et al., 1948; Mosesson and Umfleet, 1970; Tamkun et al., 1984; Xu and Mosher, 2011). Rotary shadowing electron microscopy of pFN reveals two identically flexible strands (Engel et al., 1981; Erickson et al., 1981) each of 230–250 kDa, giving the disulfide-linked dimer a relatively high molecular weight (Kornblihtt and Gutman, 1988). A single gene encodes for FN, although alternative splicing accounts for more than 20 different isoforms possible in humans (Pankov and Yamada, 2002; Singh et al., 2010; White et al., 2008; White and Muro, 2011) (Figure 5). FN is structurally composed of repeating, highly homologous amino acid domains termed Type I, Type II, and Type III (Odermatt

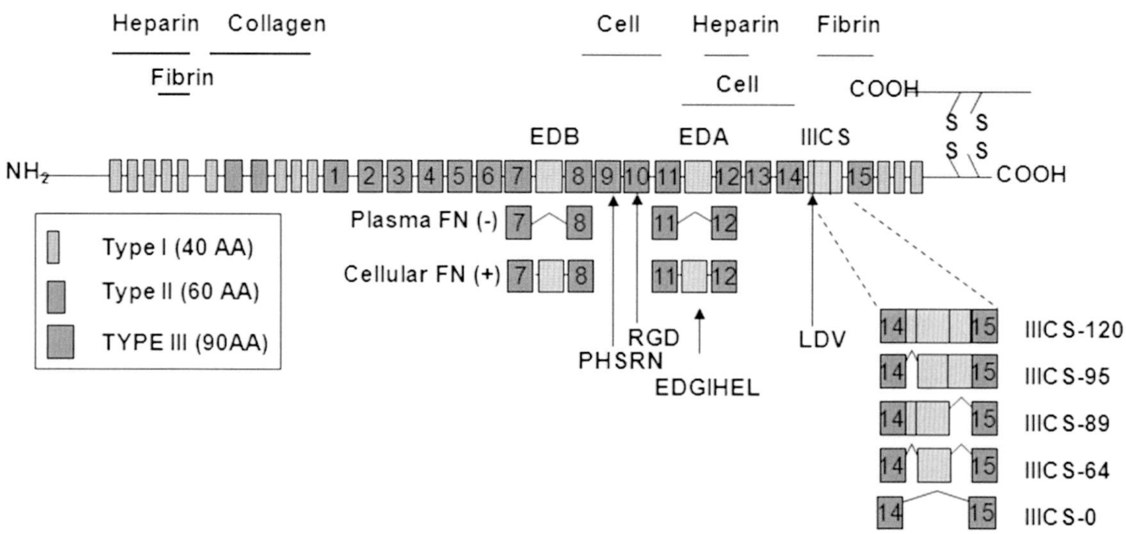

FIGURE 5 Schematic of a fibronectin dimer. The different domains (12 Type I, 2 Type II, and 15 Type III) are represented. Numbering of Type III domains excludes extra domain A (EDA) and EDB. Type I, II, and III domains consist of 40, 60, and 90 amino acids, respectively. Constitutive (RGD) and alternatively spliced (LDV), synergy (PHSRN), and EDA (EDGIHEL) cell-binding sites are indicated, together with their integrin receptor partners. EDA and EDB splicing is similar in all species (either total inclusion or exclusion), while that of the IIICS region is species specific (five variants in humans, three in rodents, and two in chickens). *Modified from White et al. (2008) with permission.*

et al., 1985; Patel et al., 1987; White et al., 2008; Xu and Mosher, 2011). Alternative splicing of the polypeptide results in two extra domains found within the Type III repeats. The first splice variant, Extra Domain A (EDA) is located between the 11th and 12th Type III repeat, and the second, Extra Domain B (EDB) is inserted between the 7th and 8th Type III repeats (Kornblihtt and Gutman, 1988). A third variable region (termed IIICS for Type III connecting segment or V for variable region) is located between the last two Type III repeats. The V region in humans is further divided into five separate variants that are labeled based on the number of amino acids in each variant (V0, V64, V89, V95, and V120) (Pankov and Yamada, 2002; Umezawa et al., 1985; Xu and Mosher, 2011). pFN lacks the EDA and EDB segments, whereas cFN has variable expression of EDA and EDB. The complexity of the V region may be responsible for enhancing the binding activity of cFN (Kornblihtt and Gutman, 1988; Ruoslahti, 1988).

2.2 Proteoglycans and Glycosaminoglycans

GAGs are composed of repeating disaccharide units that form long, filamentous structures (Scott, 1992) and are classified as sulfated (heparan sulfate, chondroitin sulfate, dermatan sulfate, keratan sulfate) or nonsulfated (hyaluronan) (Schaefer and Schaefer, 2010). Both sulfated and nonsulfated GAGs are highly anionic, which attracts hydration and provides a filler between the structural fibrous proteins, but some have also been shown to provide potent signaling for cellular function (Iozzo and Murdoch, 1996).

Hyaluronan

Hyaluronan (HA) is the predominant nonsulfated GAG in the lung (Schmid et al., 1982), although all human tissues contain HA distributed in the extravascular space. Hyaluronan was first isolated from the vitreous humor of the eye (Meyer and Palmer, 1934) and forms an extremely viscous fluid in solution that serves important functions in absorbing compressive forces (Laurent and Fraser, 1992; Lennon and Singleton, 2011). At the molecular level, hyaluronan is a series of repeating disaccharide units of D-glucuronic acid and *N*-acetyl-D-glucosamine arranged in linear, nonbranching arrays to form a polysaccharide of massive molecular weight when compared to other GAG molecules (Laurent and Fraser, 1992; Weissman and Meyer, 1954). Given that the molecule is defined by the repeating sugar units and not length, the molecular weight is variable and is of different length in different tissues, with the HA in synovial fluid and cartilage being 10 times larger than in lung (Holmes et al., 1988; Nettelbladt et al., 1989). The substantial ionic charge of hyaluronan imparted from the repeating sugar units generates an enormous avidity for water making it an immensely viscous filler between structural ECM proteins (Lennon and Singleton, 2011).

Synthesis of hyaluronan is unique among the GAG molecules as it is constructed at the plasma membrane (Toole, 1990). The synthetic proteins, known collectively as hyaluronan synthases, are membrane-spanning proteins that assemble the polysaccharide polymer on the inner face of the plasma membrane (Weigel and DeAngelis, 2007; Weigel et al., 1997). Notably, hyaluronan is synthesized and expelled into the extracellular environment simultaneously, allowing the polymer to attain its enormous length (Weigel and DeAngelis, 2007; Weigel et al., 1997). Turnover of hyaluronan is achieved in part by hyaluronidases, which are specific for hyaluronan, and other proteases (Csoka et al., 2001; Girish and Kemparaju, 2007; Stern et al., 2007).

Although hyaluronan is an efficient space-filling molecule, it is not biologically inert. Collagen, FN, laminins, and elastin all may have close associations with hyaluronan, possibly influencing their mechanical properties (Turino and Cantor, 2003). For example, elastin molecules bound to hyaluronan may be stabilized and protected from degradation; experimental evidence suggests that intratracheal administration of hyaluronan abrogates elastase-induced animal models of emphysema, whereas hyaluronidase-induced depletion of endogenous hyaluronan exacerbates the emphysematous changes (Cantor et al., 1993; Cantor et al., 1997; Cantor et al., 1995; Cantor et al., 2005; Cantor et al., 2000).

In addition to interactions with ECM proteins, hyaluronan also directly influences cell behavior through membrane binding proteins, such as CD44 and RHAMM (receptor for hyaluronic acid-mediated mobility) (Toole, 1990; Turley et al., 2002). Surfactant activity may be augmented directly by hyaluronan secreted into the alveolus by alveolar epithelial cells (Bray, 2001; Lu et al., 2005a; Lu et al., 2005b), and it also binds growth factors in the interstitial spaces to direct cellular responses to tissue injury (Smits et al., 2004).

Surprisingly, hyaluronan of different molecular weights can generate vastly different signals; for example, alveolar macrophages exposed to low-molecular-weight hyaluronan secreted increased amounts of inflammatory cytokines, whereas high-molecular-weight hyaluronan did not have similar effects (McKee et al., 1996). Hyaluronan involvement (and its binding proteins) are implicated in a number of human pulmonary pathologic conditions (Lennon and Singleton, 2011).

Proteoglycans and GAGs

GAGs, similar to hyaluronan, are composed of a backbone of repeating disaccharide units, which are linear, negatively charged, and unbranching (Schaefer and Schaefer, 2010; Scott, 1992). All GAGs other than hyaluronan are modified by addition of sulfate moieties, giving them additional anionic charge (Schaefer and Schaefer, 2010). Whilst hyaluronan is not covalently bound to protein during synthesis, all other GAGs are fused to protein cores to form proteoglycans, which may have one or more GAG molecules attached.

Proteoglycans are named based on the sequence of their protein core, but the specific type and number of GAG attachments may be variable even among molecules that bear the same name (Iozzo and Murdoch, 1996). A useful classification of proteoglycans divides them into three groups: (1) small leucine-rich proteoglycans (SLRPs); (2) modular proteoglycans; and (3) cell surface proteoglycans (Iozzo and Murdoch, 1996; Schaefer and Schaefer, 2010). SLRPs, as the name implies, are of low molecular weight (36–42 kDa) and are defined by a series of conserved leucine repeats (Iozzo and Murdoch, 1996). Modular proteins are comparatively larger (44–467 kDa) with "modular" motifs that are homologous to other unrelated proteins and are further subdivided into those that bind hyaluronan and those that do not (Iozzo and Murdoch, 1996).

In human lung, few studies exist pertaining to the proteoglycans expressed. One such study of GAG distribution in human lung by two-dimensional gel electrophoresis demonstrated compositional differences in GAG signatures among various compartments of the same organ; airway GAG composition is markedly different from alveolar, pleural, or vascular GAGs (Dolhnikoff et al., 1998; Schmid et al., 1982). Other studies using immunohistochemistry have identified various proteoglycans in airways and vessels, or in diseased tissues, but were insensitive to alveolar expression (Bensadoun et al., 1996). In normal human decellularized lung, unbiased liquid chromatography with tandem mass spectrometry revealed expression of perlecan, biglycan, decorin, agrin, mimecan, and lumican (Booth et al., 2012).

3. ECM IN LUNG DEVELOPMENT

In addition to providing structural support for cells, intrinsic elasticity, and tensile strength, the lung ECM is a critical force that provides temporal and spatial contextual cues to cells and guides behavior. Perhaps nowhere is this more evident than during lung development. Human lung development is often separated into five stages: (1) the *embryonic stage* (spanning from conception to ~5 weeks); (2) the *pseudoglandular stage* (spanning from ~5 to 17 weeks postconception), during which branching morphogenesis occurs; (3) the *canalicular stage* (from ~17 to 25 weeks postconception), during which respiratory bronchioles appear and peripheral mesenchyme is vascularized; (4) the *alveolarization stage* (from ~25 weeks to birth) during which peripheral airspaces elongate and become saccular and secondary septae appear; and (5) the *postnatal stage* (beginning at birth) when alveoli continue to mature and grow in size and number. However, much of what is known about lung ECM in development occurs from observational studies spanning these stages, making inferences about direct roles in development much more difficult. For this reason, animal modeling (primarily in the form of transgenic mice) is often employed to better understand the role of ECM in development.

Human lung development begins when the embryo is approximately 4 weeks old (Sadler, 2012). During the fourth week, the primitive endoderm-derived lung bud emerges from the ventral foregut tube wall between the 4th and 6th pharyngeal arches (Sadler, 2012). Although lung bud appearance is critically dependent on expression of numerous growth factors, including retinoic acid (Desai et al., 2004), sonic hedgehog (Shh) (Miller et al., 2004; Pepicelli et al., 1998), and fibroblast growth factor (FGF)-10 (Gibby et al., 2009; Sekine et al., 1999), it is clear that basement membrane formation during this time period is also crucial for continued lung development; indeed, the absence of basement membranes results in embryonic lethality (Miner and Yurchenco, 2004). Mechanistically, laminin isoforms are essential for basement membrane assembly, cell proliferation, and migration. Organotypic cocultures of murine embryonic epithelial and mesenchymal cells result in cyst-like formation by epithelium and subsequent development of the laminin-rich basement membrane (Schuger et al., 1998). The assembly of laminin into the basement membrane is a vital step; in the presence of function-blocking antibodies to globular domains of laminin β1 (clone AL-5) chain or a proteolytic fragment of laminin (termed E4, which also comprises the outer globular domain of laminin β1), both of which inhibit laminin assembly, basement membrane formation does not occur appropriately, epithelial polarity is lost, and lung development is arrested (Schuger et al., 1995; Schuger et al., 1998).

Immunohistochemical studies of human lung evaluating specific laminin isoforms at the pseudoglandular and canalicular stages as well as in adult lung found that basement membranes underlying epithelial buds and forming bronchi were reactive for α1, α3, and β3 chains at all stages. Conversely, α2 chains were expressed only underlying epithelial buds and β2 chains underlying epithelial buds and bronchi in the pseudoglandular stage; these disappeared from the bronchial basement membrane before the canalicular stage. Interestingly, β1 chains first appeared in the bronchial basement membrane in mature lung (Lallemand et al., 1995; Virtanen et al., 1996), suggesting β1 chains may be dispensable for normal human lung development. In total, these findings support a significant role for laminin in lung development. Throughout postnatal life, laminin remains stably expressed within the basement membranes of the lung, where it serves to maintain epithelial cell polarity, maintain tissue barriers, and promote cell viability.

Collagen IV, as a second major constituent of basement membranes, would also be predicted to be critically important in human lung development. Again, using immunohistochemical techniques it is readily apparent that collagen IV localizes to epithelial basement membranes and diffusely stains throughout the lung (Haidar et al., 1990). In hypoplastic human lungs, however, collagen IV sparsely populates such basement membranes (Haidar et al., 1990), raising a possibility that collagen IV is also necessary for normal lung development. Interestingly, mutations in various murine Col4 α chain genes do not cause appreciable lung abnormalities (Khoshnoodi et al., 2008). However, simultaneous deletion of two Col4 α chains ($\alpha1$ and $\alpha2$) in mice results in embryonic lethality due to structural deficiencies in basement membranes (Poschl et al., 2004). Thus, coordinated expression of collagen IV α chains is important for basement membrane integrity and strength, and this holds true throughout the postnatal period as well.

Nidogen-1 and -2 (also known as entactin-1 and -2; Carlin et al., 1981; Kimura et al., 1998) are important basement membrane glycoproteins that are widely distributed in embryonic tissues (Bader et al., 2005). In adult tissues, nidogen-2 becomes more narrowly expressed (Miosge et al., 1999), whereas nidogen-1 remains ubiquitous. Developmentally, deletion of nidogen-1 or nidogen-2 in mice has minimal effects on basement membrane formation (Miosge et al., 1999). However, simultaneous deletion of both nidogen-1 and nidogen-2 shows a phenotype of normal lung size and lobe number but with delayed development characterized by pseudoglandular and canalicular stage-like morphology and extensive retention of mesenchyme even in late development (Bader et al., 2005). At birth, nidogen-1/nidogen-2 double knockout mice also had improperly expanded saccules, smaller alveoli, and thicker alveolar septae resulting in perinatal lethality associated with hypoxia (Bader et al., 2005). Although localization studies for nidogen-1 or -2 using immunohistochemistry in human lung have not been reported, recent mass spectrometric evaluation of acellular adult human lung matrix reveals the continued presence of both nidogen isoforms in the distal lung (Booth et al., 2012).

Branching morphogenesis, the dichotomous bifurcation of proximal airways to generate distal airways down to the level of the terminal bronchioles, primarily occurs during the pseudoglandular stage and is characterized by proliferation and migration of epithelial cells distally toward newly developing buds (Kim and Nelson, 2012). Through elegant embryonic studies in rodents, certain constituents of the basement membrane and other ECM molecules have emerged as key players in branching morphogenesis. Further investigation in humans refines these observations and sheds light on the complexities of lung development.

Early studies demonstrated a role for GAGs in mouse lung branching morphogenesis. As mentioned previously, GAGs are long, unbranched polysaccharides consisting of a repeating disaccharide unit containing an amino sugar (*N*-acetylglucose amine or *N*-acetylgalactose amine) along with a uronic sugar (glucuronic acid or iduronic acid) or galactose (Esko et al., 2009). Sulfated GAGs, principally chondroitin sulfate and heparan sulfate proteoglycans, localize to the interstitium and basement membrane during branching morphogenesis (Spooner et al., 1993). Moreover, when chondroitin sulfate proteoglycan synthesis is pharmacologically disrupted in animals, branching morphogenesis is attenuated, suggesting a key role for this GAG in lung development (Spooner et al., 1993).

In addition to GAGs, ECM fibrous proteins (such as FN, tenascin-C, collagen IV, laminin) and proteoglycans (nidogen) have all been shown to play functional roles in branching morphogenesis (Kaarteenaho-Wiik et al., 2001; Roman, 1997; Schuger et al., 1995; Thomas and Dziadek, 1994; Wu and Santoro, 1996). In addition, rat studies suggest that fibrillin-2, an ECM protein necessary for elastin fiber formation, is also critical in branching morphogenesis. Mechanistic studies using antisense oligonucleotide technology demonstrated that fibrillin-2 inhibition results in rudimentary lung bud formation and collapsed airways (Yang et al., 1999), suggesting a critical role for this protein in lung formation.

Indirect evidence for the importance of ECM proteins in human lung development comes from immunohistochemical studies of integrins, heterodimeric transmembrane receptors that bind and transduce signals from the ECM. Such examination of integrin subunits in human lungs during the early pseudoglandular stage shows uniform expression of $\alpha2$, $\alpha5$, $\alpha6$, αv, and $\beta1$ subunits in epithelial cell membranes of the distal airways (Coraux et al., 1998). In contrast, tracheal epithelial cells expressed $\alpha3$, $\alpha6$, and $\beta1$ integrins on cell surfaces facing the basement membrane (Coraux et al., 1998). Basal cells expressed $\alpha6$ and $\beta4$ integrins (known laminin receptors) on the basement membrane-facing side, whereas $\alpha4$ integrins were ubiquitously expressed throughout the lung (Coraux et al., 1998). These data suggest that numerous ECM targets are likely of importance during human lung development. More recently, a study characterized expression of collagens (I, III, and VI) and the proteoglycans decorin, biglycan, and lumican using immunohistochemistry in developing human lung (Godoy-Guzman et al., 2012). This study showed that all proteoglycans (save biglycan) were expressed within basement membranes of proximal airways but also at the epithelial—mesenchymal interface along branch points (Godoy-Guzman et al., 2012). Biglycan was localized primarily to subendothelial regions of vessels. Together, these data suggest that human lung branching morphogenesis is critically influenced by numerous cell—ECM interactions.

Alveolarization (or septation) is a key aspect of lung development, during which primary and then secondary septae form to generate and then increase the gas-exchanging surface area of the lung and promote alveolar capillary development by angiogenesis. Whether occurring via sprouting or by intussusception (Risau, 1997), vascular development in the lung is dependent upon many factors, including ECM. For example, sprouting angiogenesis occurs in part due to myofibroblast precursors that begin to produce elastin and tenascin-C at sites of branching (Roth-Kleiner and Post, 2005). Elastic fibers are thought to provide the driving force of secondary septum outpouching into the airspace, taking capillaries with them, in order to vascularize the newly forming alveoli. Although data are scant in humans, elegant rat studies suggest that the elastin precursor tropoelastin and elastin itself both increase in expression through postnatal days 9—11, which corresponds to the peak of septation (Bruce, 1991; Bruce and Lo, 1991; Goncalves et al., 2001; Noguchi and Samaha, 1991). Similarly, tenascin-C expression also peaks in rats during septation, and colocalizes to elastin fibers at tips of newly formed septae (Young et al., 1994; Zhao and Young, 1995). Additional data suggest that proteoglycans (primarily chondroitin sulfate and decorin) are concentrated at sites of secondary septum formation in rat lung (Wang et al., 2002). Together, these data suggest that elastin, tenascin-C, GAGs, and proteoglycans are essential for normal alveologenesis.

4. COMPARTMENTALIZATION OF ECM IN THE POSTNATAL HUMAN LUNG

One important role of the lung ECM is to assist with cellular organization within compartmentalized structures. It is this compartmentalization, among other things, that allows for appropriate development (as discussed previously) and specialization of organs during embryogenesis and for sustaining postnatal organ function. The exact mechanisms responsible for these events remain incompletely elucidated, but cell—matrix interactions must influence these processes since cells are capable of recognizing the ECM through integrins and other cell-surface receptors capable of signal transduction. Thus, the ECM provides a "roadmap" for cellular organization, while influencing a number of key cellular functions including proliferation, migration, differentiation, and the secretion of soluble mediators, among others (Daley et al., 2008). Once the overall architecture of the lung is established, the lung ECM continues to provide a support "skeleton" upon which cells rest, while separating epithelium and endothelium from the interstitium.

Because of their tight spatial relationship, understanding how cell—matrix interactions influence cellular behavior is key to gaining insights into the role of the ECM in lung development, injury, and repair. By extension, it is important to define the relative composition of the ECM in each of the pulmonary compartments to better understand how the ECM influences organ structure and function in lung health and disease. Despite this perceived importance, however, studies designed to identify the individual components of ECM in pulmonary structures are relatively scarce and outdated, and many focus on nonhuman species. Much of the earlier information was generated through descriptive studies using electron microscopy. This work was later advanced by the use of immunohistochemistry and molecular biology and by studies evaluating the ECM components secreted by cultured lung cells (Dunsmore and Rannels, 1996).

Similar to what was observed in embryologic studies, postnatal basement membranes consist of type IV collagen, laminin, sulfate and heparan proteoglycans, entactin (also known as nidogen), fibulins, and collagen XVIII (Halfter et al., 1998; Pan et al., 1993; Yurchenco and Schittny, 1990), while the postnatal pulmonary interstitium consists mainly of fibrillar collagens, elastic fibers, and proteoglycans. Since basement membranes and interstitial matrices are present in all pulmonary structures, the abovementioned components are shared within airways, pulmonary vessels, lymphatics, and the alveoli. However, variations in the relative content of these matrices have been noted in different pulmonary structures. For example, the content of proteoglycans differs in the various compartments of the lung such that chondroitin sulfate/keratin sulfate cartilaginous proteoglycans predominate in the tracheobronchial tree, dextran sulfate proteoglycans predominate in the pleural lining, and heparan sulfate proteoglycans like perlecan and agrin predominate in alveolar basement membranes (Smits et al., 2004). The functional implications of these differences are not always clear.

4.1 The ECM in Airways

The primary function of airways is to conduct air into the gas-exchanging units of the lung and back, and the ECM is assembled throughout the length of the airway to support that function. In the proximal airways, cartilage provides structural support to prevent airway collapse, while the connective tissue in the interstitium allows for the fluctuations in lung volume that occur during inspiration and expiration. Human lung cartilage is composed of collagen fibers (collagens type I, III) and hydrophilic proteoglycans like aggrecan; this composition is responsible for its biomechanical properties (Hardingham et al., 1994).

Hypertrophic chondrocytes are also known to produce type X collagen (Kielty et al., 1985) and tenascin C (Pacifici et al., 1993). More distally, cartilage is absent in small bronchi, bronchioles, and alveoli of the human lung.

The airway basement membrane contains chondroitin sulfate and heparan sulfate proteoglycans, entactin, laminin (Sannes et al., 1993), FN (Stenman and Vaheri, 1978), and type IV collagen (Madri and Furthmayr, 1979). Cultured bronchial epithelial cells produce many of these components including FN (Shoji et al., 1989).

4.2 The ECM in the Lung Interstitium

The pulmonary interstitium contains interstitial cells including fibroblasts, myofibroblasts, and pericytes, as well as resident and incoming immune cells like macrophages and mast cells (Weibel and Crystal, 1991). Thus, the composition of the ECM in the pulmonary interstitium is dependent on the products secreted by these cells. Human lung interstitium is composed of fibroblast-derived fibrillar collagens (type I and III) (Hance et al., 1976); elastin (Campagnone et al., 1987); chondroitin, dermatan sulfate, and heparan sulfate proteoglycans (Castor et al., 1983; Heremans et al., 1988; Vogel and Peterson, 1981); hyaluronic acid (Skinner et al., 1989); entactin (Aratani and Kitagawa, 1988); and FN (Roberts et al., 1988; Roman et al., 2004).

The most prominent structural ECM components in the pulmonary interstitium are collagen and elastin, which provide structural strength and integrity as well as elasticity, respectively (Nakamura et al., 2002).

4.3 The ECM of the Pulmonary Vessels and Lymphatics

The mucosal vascular plexus of the lung is arranged in parallel to the airway tree and lies just beneath the epithelial basement membrane, extending into the smooth muscle layer (Bai et al., 1994). Similarly, the adventitial vascular plexus extends from the smooth muscle layer to the alveolar border (Blosser et al., 1994). The vascular endothelium is separated from the underlying interstitium by a basement membrane that contains essentially the same ECM components listed previously. These are produced by smooth muscle cells as well as endothelial cells known to secrete collagen type VIII (Sage et al., 1980), entactin (Katz et al., 1994), versican (Kenagy et al., 2006), and decorin (Williams, 2001).

The pulmonary lymphatics are immersed in the connective tissue that underlies the pleural lining cells and that surrounds lung lobules, within the interlobular septa, and within the peribronchial and perivascular sheaths (Lauweryns, 1970). The connective tissue surface of the lymphatic capillary endothelium lacks a well-defined basal lamina and contains hyaluronate, chondroitin sulfate, and FN (Leak, 1987).

4.4 The ECM of the Alveolus

The pulmonary alveolus is the most specialized structure in the lung and is responsible for its most important function, gas exchange. Consequently, it is expected that the ECM organization in the alveolus differs from those in other pulmonary structures. For example, in contrast to the airways, the basement membrane of the epithelium and capillary endothelium in the alveolus oftentimes become fused (with no intervening interstitium) to facilitate gas exchange (Vaccaro and Brody, 1981).

The structure of the alveolar wall was characterized by electron microscopy in the 1950s, but that work focused on the major components and the organization of cells (Karrer, 1956; Low, 1953). It is now known that alveolar basement membranes consist of laminin (Gil and Martinez-Hernandez, 1984; Sannes et al., 1993), entactin (Sannes et al., 1993), type IV collagen (Raghu et al., 1985), type V collagen (Madri and Furthmayr, 1980), chondroitin sulfate and heparan sulfate proteoglycans (Sannes et al., 1993). FN is also found in alveolar basement membrane especially in the developing lung (Rosenkrans et al., 1983; Sannes et al., 1993). This is likely due to the ability of alveolar epithelial cells to secrete and organize these matrices. The ECM molecules secreted by type I pneumocytes are largely unknown; however, cultured type II pneumocytes secrete FN (Torikata et al., 1985), laminin (Maniscalco et al., 1994), type IV collagen (Simon et al., 1993), heparan, chondroitin sulfate, and dermatan sulfate proteoglycans (Rannels et al., 1987), and hyaluronic acid (Maniscalco and Campbell, 1994).

Early studies using tissue staining with cationic dyes (which react with negatively charged regions of carbohydrates and GAGs) unveiled variations in the composition of basement membranes regarding charged molecules. Anionic sites, for example, were found to be more prevalent in the basement membrane of the alveolar epithelium when compared to the capillary basement membrane (Skinner et al., 1987). Anionic sites are denser under type I pneumocytes cells in the rat (Grant et al., 1983; Sannes, 1984), rabbit (Van Kuppevelt et al., 1984b), mouse, and human (van Kuppevelt et al., 1984a).

The alveolar basement membrane is dense and amorphous, has filaments in the lamina rara externa, and has discontinuities beneath type II pneumocytes, but not beneath type I pneumocytes, that allow penetration of cytoplasmic processes into the interstitium. In contrast, the capillary basement membrane is fibrillar and less compact than the alveolar basement membrane, has no lamina rara filaments, and has less anionic sites overlying the lamina densa. Alveolar basement membrane anionic sites represent heparan sulfate proteoglycans, whereas capillary basement membrane anionic sites contain these as well as sulfated proteoglycans (Vaccaro and Brody, 1981). Regional variations in the subepithelial basement membrane are also suggested by immunohistochemical studies showing changes in staining intensity for laminin, entactin, and chondroitin sulfate in the alveoli as opposed to airways (Sannes et al., 1993).

FN is also found in basement membranes. Although epithelial cells, endothelial cells, fibroblasts, and smooth muscle cells produce FN (Roberts et al., 1988; Shoji et al., 1989; Stenman and Vaheri, 1978), the origin of FN demonstrated in tissues is not always clear. This is because hepatocyte-derived FN is present in great concentrations in plasma and is deposited in extracellular tissues (Moretti et al., 2007). Additionally, plasma FN could become trapped within the basement membranes during the processing of tissues. Studies using more specific reagents capable of detecting EDA-containing cellular FN show increased production in the developing lung, whereas plasma FN is found in large vascular structures and alveolar macrophages in the adult uninjured lung (Torikata et al., 1985).

5. THE ECM OF THE AGING LUNG

The composition of the lung ECM remains relatively stable in health. We do know, however, that there is continuous synthesis, deposition, and degradation (so-called "turnover") of ECM components in the adult lung; the latter is due to the concomitant expression of matrix metalloproteinases and other proteases. However, differences exist between the rate of turnover depending on the matrix component considered. As mentioned previously, elastin is very stable and some elastic fibers may last throughout the animal's lifespan (Mecham, 1991).

Physiological lung aging is associated with the enlargement of alveolar spaces, reduced surface area for gas exchange, and the loss of alveolar attachments supporting peripheral airways. These changes are mild but may lead to measurable functional changes characterized by reduced static elastic recoil, and decreased compliance of the chest wall and the strength of the respiratory muscles (Faner et al., 2012; Janssens et al., 1999). Although the factors that lead to these physiological changes are unclear, they are likely driven, at least in part, by alterations in lung ECM deposition and relative composition.

Immunohistochemical studies reveal decreased elastic fibers along the alveolar walls, while type III collagen increased in elderly subjects compared to controls (D'Errico et al., 1989). These findings, however, are not consistent across studies. For example, when examining the tissues of lungs harvested postmortem after sudden death from subjects ages 15 to 83, others found decreased collagen content but few changes in elastin content (Andreotti et al., 1983). Similar inconsistencies have been found in animal studies. As a case in point, researchers found increased lung volume and respiratory system compliance in 20-month-old C57BL/6 mice compared to 2-month-old mice. These changes were associated with decreased elastic fiber content, while collagen content increased (Huang et al., 2007). In contrast, other investigators found no changes in elastin content in 24-month-old Balb/c mice compared to 3-month-old animals. Collagen content increased, but there were no significant changes in hydroxyproline content per dry lung weight or the proportion of type III to type I collagen. The authors concluded that, in terms of ECM, the lungs of aged mice are not very different from the lungs of young mice and that the changes observed were probably the simple consequence of growth (Takubo et al., 1999). These seemingly paradoxical observations are likely related to the different methods of analysis used and/or genetic differences between strains of mice.

More recently, when examining 2-month-old versus 24-month-old C57BL/6 mouse lungs, investigators found increased expression of EDA-containing cellular FN as well as in the expression of fibrillar collagens type I and III (Sueblinvong et al., 2012). These changes were associated with increased gelatinolytic activity and expression of the matrix metalloproteinases (MMP)-2 and -9. This is interesting in light of studies showing that impaired lung function in the elderly was associated with alterations in the serum levels of MMP-9, TIMP-1 (for tissue inhibitor of metalloproteinase-1), and their ratio in a recent population-based study of 70-year-old subjects. The authors concluded that impaired lung function might be related to ECM remodeling; this association was stronger in men than in women (Olafsdottir et al., 2010). It should be noted, however, that a significant percentage of subjects with smoking history was included in this study, which may have influenced the results. Changes in proteoglycans in cartilage have also been demonstrated with age (Roberts and Pare, 1991) with progressive increase in tensile stiffness of the tracheal cartilage concomitant to decreases in hydroxyproline and water content (Rains et al., 1992).

When considered together, the studies described above suggest that aging is indeed associated with alterations in the expression and deposition of ECM. This may be due to changes in production or in degradation, which also appears to be affected by changes in expression and activity of MMPs and their inhibitors (TIMPs). Recently, another group of proteases capable of matrix degradation (granzymes) has been proposed to be instrumental in aging ECM turnover (Hendel et al., 2010).

Aging lungs not only present with alterations in the ECM content but, indeed, the proteins themselves may be altered. Aging is associated with protein modifications induced by the reaction of proteins with reducing sugars resulting in the production of advanced glycation end products (AGEs; Bellmunt et al., 1995). AGEs increase in aged tissues including the lung. Because of their longevity, collagens are characterized by AGE modifications through glycation, glycoxidation, and cross-links, and these changes may influence immune and tumor cell migration, among other processes (Bartling et al., 2009; Sell et al., 2001).

As mentioned previously, the relative composition of the ECM in any given compartment is driven by the cells present in that particular compartment. Thus, alterations in the ECM of the aging lung are likely related to alterations in the composition or phenotype of cells at a particular age. It has long been established that the proliferative and migratory capabilities of human lung and skin fibroblasts decrease with aging (Kondo and Yonezawa, 1992). Further, aging is associated with a decline in the capacity for regeneration after pneumonectomy in most mammals. This has been linked to a loss of fibroblast clonogenicity and progressive myofibroblastic differentiation in murine lungs (Paxson et al., 2011). Further work designed to evaluate the phenotypic changes in fibroblasts and other lung cells that take place during aging, and how they impact connective tissue composition, lung mechanics, and cellular functions after injury, should be considered in order to determine the true role of these cells in lung aging.

6. STUDY OF NORMAL LUNG ECM

Although great strides have been made in our understanding of normal lung ECM through animal modeling, it is likely that lung ECM composition varies across species with varying degrees of homology. Thus, the only way to better understand human lung ECM is to examine human lungs. To be sure, many studies outlined above have done just that, at various stages of embryonic or fetal development, as well as in postnatal human lungs, using various techniques such as in situ hybridization, immunohistochemistry, immunogold electron microscopy, and the like. However, in each of these cases one must first identify the protein(s) of interest a priori.

To gain a more thorough understanding of lung ECM, one must take an unbiased approach, which affords the opportunity to explore previously unidentified matrix proteins and matrix-related molecules. This has recently been undertaken in the mouse lung (Naba et al., 2012), in which lungs were first decellularized to enrich for ECM, and then they were subjected to mass spectrometry evaluation. In so doing, a core "matrisome" emerged, consisting of the ECM proteins and related molecules that comprise the noncellular portion of lung (Naba et al., 2012). More recently, the matrisome of acellular human distal lung as identified by mass spectrometric evaluation has been reported (Booth et al., 2012) (Table 1). These two studies demonstrate the role that newer technologies will play in enhancing our understanding of lung ECM and the role it plays in development, homeostasis, injury, and repair.

TABLE 1 Extracellular Matrix (ECM) and ECM Associated/Modifying Proteins Identified in Normal Human Distal lung

Fibrous proteins and glycoproteins
Abi3bp
Agrin
Collagen alpha-1(I) chain
Collagen alpha-2(I) chain
Collagen alpha-1(IV) chain
Collagen alpha-2(IV) chain
Collagen alpha-3(IV) chain
Collagen alpha-1(V) chain
Collagen alpha-1(VI) chain
Collagen alpha-2(VI) chain
Collagen alpha-3(VI) chain
Collagen alpha-5(VI) chain
Collagen alpha-6(VI) chain
Collagen alpha-1(VIII) chain

Continued

TABLE 1 Extracellular Matrix (ECM) and ECM Associated/Modifying Proteins Identified in Normal Human Distal lung—cont'd

Fibrous proteins and glycoproteins (cont'd)
Collagen alpha-1(XII) chain
Collagen alpha-1(XIV) chain
Collagen alpha-1(XV) chain
Collagen alpha-1(XVIII) chain
Deleted in malignant brain tumors 1 protein
Dermatopontin
EGF-containing fibulin-like extracellular matrix protein 1
EMILIN-1
Fibrillin-1
Fibrillin-2
Fibrinogen alpha chain
Fibrinogen beta chain
Fibrinogen gamma chain
Fibronectin
Fibulin-1
Fibulin-2
Fibulin-5
Laminin subunit alpha-2
Laminin subunit alpha-3
Laminin subunit alpha-4
Laminin subunit alpha-5
Laminin subunit beta-1
Laminin subunit beta-2
Laminin subunit beta-3
Laminin subunit gamma-1
Laminin subunit gamma-2
Microfibril-associated glycoprotein 4
Microfibrillar-associated protein 2
Microfibrillar-associated protein 5
Multimerin-2
Nephronectin
Nidogen-1
Nidogen-2
Periostin
Tenascin
Tenascin-X

Continued

TABLE 1 Extracellular Matrix (ECM) and ECM Associated/Modifying Proteins Identified in Normal Human Distal lung—cont'd

Fibrous proteins and glycoproteins (cont'd)
Tubulointerstitial nephritis antigen-like
Vitronectin
von Willebrand factor
von Willebrand factor A domain-containing protein 5A
Proteoglycans
Aggrecan core protein
Asporin
Biglycan
Decorin
Heparan sulfate proteoglycan core protein
Hyaluronan and proteoglycan link protein 1
Lumican
Mimecan
Versican core protein
Matrisome-associated/regulating proteins
Annexin A1
Annexin A2
Annexin A3
Annexin A4
Annexin A5
Annexin A6
Annexin A7
Annexin A11
ADAM-10
Galectin-1
Galectin-3
Galectin-7
Inter-alpha-trypsin inhibitor heavy chain H1
Inter-alpha-trypsin inhibitor heavy chain H4
Latent-transforming growth factor β-binding protein 2
Latent-transforming growth factor β-binding protein 4
Matrix metalloproteinase-9
Mucin-1
Neutrophil elastase
Plasminogen
Prolargin

Continued

TABLE 1 Extracellular Matrix (ECM) and ECM Associated/Modifying Proteins Identified in Normal Human Distal lung—cont'd

Matrisome-associated/regulating proteins (cont'd)
Protein AMBP
Protein-glutamine gamma-glutamyltransferase 2
Protein-glutamine gamma-glutamyltransferase E
Protein-glutamine gamma-glutamyltransferase K
Pulmonary surfactant-associated protein A1
Pulmonary surfactant-associated protein B
Pulmonary surfactant-associated protein D
Serpin B3
Serpin B6
Serpin B12
TIMP-3
Transforming growth factor-β-induced protein ig-h3

Modified from Booth et al. (2012) with permission of the American Thoracic Society. Copyright © 2013 American Thoracic Society. This modified table is based on the original table available from www.atsjournals.org.

7. CONCLUSIONS

Our understanding of the normal lung ECM has been enhanced in recent years, although we still have much to learn. Animal studies have revealed the ECM to be an important and active contributor to normal lung development and postnatal function, with localization of specific ECM proteins to various compartments. It is this localization that likely contributes to organized cell function, although it is not yet clear what other role(s) the lung ECM may play in normal homeostasis, aging, and in disease states. For example, while it is clear that basement membranes play a direct role in maintaining epithelial cell polarity, it is as yet unknown whether the ECM directly contributes to the pluripotency of lung-resident progenitor cells. Newer techniques have allowed us to begin to identify the constituents of the lung ECM in an unbiased fashion, and ongoing and future studies will allow for direct comparisons among various species. Through these types of investigations, we have already identified proteins never-before described in the lung. Future study is now necessary to elucidate the function(s) of these new proteins and enrich our comprehension of the true contributions of ECM to lung function.

REFERENCES

Andreotti, L., et al., 1983. Gerontology 29 (6), 377–387.

Aratani, Y., Kitagawa, Y., 1988. J. Biol. Chem. 263 (31), 16163–16169.

Astbury, W.T., 1940. The Molecular Structure of the Fibres of the Collagen Group. International Society of Leather Trades' Chemists.

Bader, B.L., et al., 2005. Mol. Cell. Biol. 25 (15), 6846–6856.

Bai, A., et al., 1994. J. Appl. Physiol. 77 (2), 1011–1014.

Bartling, B., et al., 2009. FASEB J. 23 (5), 1510–1520.

Beck, K., et al., 1990. FASEB J. 4 (2), 148–160.

Bellmunt, M.J., et al., 1995. Lung 173 (3), 177–185.

Bensadoun, E.S., et al., 1996. Am. J. Respir. Crit. Care Med. 154 (6 Pt 1), 1819–1828.

Blosser, S., et al., 1994. J. Appl. Physiol. 76 (4), 1624–1629.

Booth, A.J., et al., 2012. Am. J. Respir. Crit. Care Med. 186 (9), 866–876.

Bray, B.A., 2001. J. Theor. Biol. 210 (1), 121–130.

Bruce, M.C., 1991. Am. J. Respir. Cell. Mol. Biol. 5 (4), 344–350.

Bruce, M.C., Lo, P.Y., 1991. J. Lab. Clin. Med. 117 (3), 226–233.

Burgeson, R.E., et al., 1994. Matrix Biol. 14 (3), 209–211.

Campagnone, R., et al., 1987. Lab. Invest 56 (2), 224–230.

Cantor, J.O., et al., 1986. Proc. Soc. Exp. Biol. Med. 181 (3), 387–391.

Cantor, J.O., et al., 1993. Exp. Lung Res. 19 (2), 177–192.

Cantor, J.O., et al., 1995. Exp. Lung Res. 21 (3), 423–436.

Cantor, J.O., et al., 1997. Exp. Lung Res. 23 (3), 229–244.

Cantor, J.O., et al., 2000. Proc. Soc. Exp. Biol. Med. 225 (1), 65–71.

Cantor, J.O., et al., 2005. Exp. Lung Res. 31 (4), 417–430.

Carlin, B., et al., 1981. J. Biol. Chem. 256 (10), 5209–5214.
Castor, C.W., et al., 1983. Vitro 19 (6), 462–470.
Cheng, Y.S., et al., 1997. J. Biol. Chem. 272 (50), 31525–31532.
Chung, A.E., et al., 1979. Cell 16 (2), 277–287.
Cleary, E.G., Gibson, M.A., 1983. Int. Rev. Connect. Tissue Res. 10, 97–209.
Colognato, H., Yurchenco, P.D., 2000. Dev. Dyn. 218 (2), 213–234.
Coraux, C., et al., 1998. J. Histochem. Cytochem. 46 (7), 803–810.
Cowan, P.M., et al., 1955. Nature 176 (4492), 1062–1064.
Csoka, A.B., et al., 2001. Matrix Biol. 20 (8), 499–508.
D'Errico, A., et al., 1989. Virchows Arch. A Pathol. Anat. Histopathol. 415 (2), 137–144.
Daley, W.P., et al., 2008. J. Cell. Sci. 121 (Pt 3), 255–264.
Danielsen, C.C., 1981. Connect. Tissue Res. 9 (1), 51–57.
Desai, T.J., et al., 2004. Dev. Biol. 273 (2), 402–415.
Dolhnikoff, M., et al., 1998. Am. J. Respir. Cell. Mol. Biol. 19 (4), 582–587.
Dunsmore, S.E., Rannels, D.E., 1996. Am. J. Physiol. 270 (1 Pt 1), L3–L27.
Durbeej, M., 2010. Cell. Tissue Res. 339 (1), 259–268.
Engel, J., et al., 1981. J. Mol. Biol. 150 (1), 97–120.
Erickson, H.P., et al., 1981. J. Cell. Biol. 91 (3 Pt 1), 673–678.
Esko, J.D., et al., 2009. Proteoglycans and sulfated glycosaminoglycans. In: Varki, A. (Ed.), Essentials of Glycobiology. Cold Spring Harbor Laboratory Press, p. 784.
Faner, R., et al., 2012. Am. J. Respir. Crit. Care Med. 186 (4), 306–313.
Frantz, C., et al., 2010. J. Cell. Sci. 123 (Pt 24), 4195–4200.
Gay, S., et al., 1976. Proc. Natl. Acad. Sci. U. S. A. 73 (11), 4037–4040.
Gelse, K., et al., 2003. Adv. Drug Deliv. Rev. 55 (12), 1531–1546.
Gibby, K.A., et al., 2009. Proc. Natl. Acad. Sci. U. S. A. 106 (21), 8585–8590.
Gil, J., Martinez-Hernandez, A., 1984. J. Histochem. Cytochem. 32 (2), 230–238.
Girish, K.S., Kemparaju, K., 2007. Life Sci. 80 (21), 1921–1943.
Godoy-Guzman, C., et al., 2012. Eur. J. Histochem. 56 (3), e29.
Goncalves, C., et al., 2001. Exp. Lung Res. 27 (7), 533–545.
Grant, M.M., et al., 1983. Dev. Biol. 97 (1), 173–183.
Haidar, A., et al., 1990. Early Hum. Dev. 21 (3), 175–180.
Halfter, W., et al., 1998. J. Biol. Chem. 273 (39), 25404–25412.
Hance, A.J., et al., 1976. J. Clin. Invest. 57 (1), 102–111.
Hardingham, T.E., et al., 1994. Eur. J. Clin. Chem. Clin. Biochem. 32 (4), 249–257.
Hendel, A., et al., 2010. Cell. Death Differ. 17 (4), 596–606.
Heremans, A., et al., 1988. J. Biol. Chem. 263 (10), 4731–4739.
Hinek, A., Rabinovitch, M., 1994. J. Cell. Biol. 126 (2), 563–574.
Hofmann, H., et al., 1978. J. Mol. Biol. 125 (2), 137–165.
Holmes, M.W., et al., 1988. Biochem. J. 250 (2), 435–441.
Huang, K., et al., 2007. J. Appl. Physiol. 102 (1), 200–206.
Hynes, R.O., 1973. Proc. Natl. Acad. Sci. U. S. A. 70 (11), 3170–3174.
Iozzo, R.V., Murdoch, A.D., 1996. FASEB J. 10 (5), 598–614.
Janssens, J.P., et al., 1999. Eur. Respir. J. 13 (1), 197–205.
Kaarteenaho-Wiik, R., et al., 2001. Am. J. Respir. Cell. Mol. Biol. 25 (3), 341–346.
Kadler, K.E., et al., 1987. J. Biol. Chem. 262 (32), 15696–15701.
Karrer, H.E., 1956. J. Biophys. Biochem. Cytol. 2 (3), 241–252.
Katz, A., et al., 1994. Invest Ophthalmol. Vis. Sci. 35 (2), 495–502.
Kenagy, R.D., et al., 2006. Trends Cardiovasc Med. 16 (6), 209–215.
Khoshnoodi, J., et al., 2008. Microsc. Res. Tech. 71 (5), 357–370.
Kielty, C.M., et al., 1985. Biochem. J. 227 (2), 545–554.
Kim, H.Y., Nelson, C.M., 2012. Organogenesis 8 (2), 56–64.
Kimura, N., et al., 1998. Exp. Cell. Res. 241 (1), 36–45.
Kleinman, H.K., et al., 1985. J. Cell. Biochem. 27 (4), 317–325.
Kondo, H., Yonezawa, Y., 1992. Mech. Ageing Dev. 63 (3), 223–233.
Kornblihtt, A.R., Gutman, A., 1988. Biol. Rev. Camb Philos. Soc. 63 (4), 465–507.
Lallemand, A.V., et al., 1995. Anat. Rec. 242 (2), 233–241.
Last, J.A., Reiser, K.M., 1984. Environ. Health Perspect. 55, 169–177.
Laurent, G.J., 1986. Thorax 41 (6), 418–428.
Laurent, T.C., Fraser, J.R., 1992. FASEB J. 6 (7), 2397–2404.
Lauweryns, J.M., 1970. Am. Rev. Respir. Dis. 102 (6), 877–885.
Leak, L.V., 1987. Lymphology 20 (4), 196–204.
LeBleu, V.S., et al., 2007. Exp. Biol. Med. (Maywood) 232 (9), 1121–1129.
Leikina, E., et al., 2002. Proc. Natl. Acad. Sci. U. S. A. 99 (3), 1314–1318.
Lennon, F.E., Singleton, P.A., 2011. Am. J. Physiol. Lung Cell. Mol. Physiol. 301 (2), L137–L147.
Low, F.N., 1953. Anat. Rec. 117 (2), 241–263.
Lu, K.W., et al., 2005a. Pediatr. Res. 57 (2), 237–241.
Lu, K.W., et al., 2005b. Pediatr. Res. 58 (2), 206–210.
Madri, J.A., Furthmayr, H., 1979. Am. J. Pathol. 94 (2), 323–331.
Madri, J.A., Furthmayr, H., 1980. Hum. Pathol. 11 (4), 353–366.
Maniscalco, W.M., Campbell, M.H., 1994. Am. J. Physiol. 266 (6 Pt 1), L672–L680.
Maniscalco, W.M., et al., 1994. Am. J. Physiol. 267 (5 Pt 1), L569–L577.
McKee, C.M., et al., 1996. J. Clin. Invest 98 (10), 2403–2413.
Mecham, R.P., 1991. Ann. N. Y. Acad. Sci. 624, 137–146.
Mecham, R.P., 2008. Methods 45 (1), 32–41.
Meyer, K., Palmer, J.W., 1934. J. Biol. Chem. 107, 629–634.
Miller, L.A., et al., 2004. Dev. Dyn. 231 (1), 57–71.
Miner, J.H., et al., 1997. J. Cell. Biol. 137 (3), 685–701.
Miner, J.H., Yurchenco, P.D., 2004. Annu. Rev. Cell. Dev. Biol. 20, 255–284.
Miosge, N., et al., 1999. Lab. Invest. 79 (12), 1591–1599.
Moretti, F.A., et al., 2007. J. Biol. Chem. 282 (38), 28057–28062.
Morrison, P.R., et al., 1948. J. Am. Chem. Soc. 70 (9), 3103–3108.
Mosesson, M.W., Umfleet, R.A., 1970. J. Biol. Chem. 245 (21), 5728–5736.
Naba, A., et al., 2012. Mol. Cell. Proteomics 11 (4). M111 014647.
Nakamura, T., et al., 2002. Nature 415 (6868), 171–175.
Nettelbladt, O., et al., 1989. Am. J. Physiol. 257 (6 Pt 1), L379–L384.
Nguyen, N.M., Senior, R.M., 2006. Dev. Biol. 294 (2), 271–279.
Noguchi, A., Samaha, H., 1991. Am. J. Respir. Cell. Mol. Biol. 5 (6), 571–578.
Odermatt, E., et al., 1985. Proc. Natl. Acad. Sci. U. S. A. 82 (19), 6571–6575.
Olafsdottir, I.S., et al., 2010. Respirology 15 (3), 530–535.
Orgel, J.P., et al., 2006. Proc. Natl. Acad. Sci. U. S. A. 103 (24), 9001–9005.
Orgel, J.P., et al., 2001. Structure 9 (11), 1061–1069.
Pacifici, M., et al., 1993. Dev. Dyn. 198 (2), 123–134.
Pan, T.C., et al., 1993. Eur. J. Biochem. 215 (3), 733–740.
Pankov, R., Yamada, K.M., 2002. J. Cell. Sci. 115 (Pt 20), 3861–3863.
Partridge, S.M., et al., 1963. Nature 197, 1297–1298.
Patel, R.S., et al., 1987. EMBO J. 6 (9), 2565–2572.
Pauling, L., Corey, R.B., 1951. Proc. Natl. Acad. Sci. U. S. A. 37 (5), 272–281.

Paxson, J.A., et al., 2011. PLoS One 6 (8), e23232.
Pepicelli, C.V., et al., 1998. Curr. Biol. 8 (19), 1083–1086.
Poschl, E., et al., 2004. Development 131 (7), 1619–1628.
Raghu, G., et al., 1985. Am. Rev. Respir. Dis. 131 (2), 281–289.
Rains, J.K., et al., 1992. J. Appl. Physiol. 72 (1), 219–225.
Ramachandran, G.N., Kartha, G., 1954. Nature 174 (4423), 269–270.
Ramachandran, G.N., Kartha, G., 1955. Nature 176 (4482), 593–595.
Rannels, D.E., et al., 1992. Am. J. Physiol. 262 (5 Pt 1), L582–L589.
Rannels, S.R., et al., 1987. Am. J. Physiol. 253 (6 Pt 1), C759–C765.
Reinhardt, D.P., et al., 1996. J. Biol. Chem. 271 (32), 19489–19496.
Rennard, S.I., et al., 1984. Am. Rev. Respir. Dis. 130 (2), 267–274.
Rich, A., Crick, F.H., 1955. Nature 176 (4489), 915–916.
Risau, W., 1997. Nature 386 (6626), 671–674.
Roberts, C.J., et al., 1988. J. Biol. Chem. 263 (10), 4586–4592.
Roberts, C.R., Pare, P.D., 1991. Am. J. Physiol. 261 (2 Pt 1), L92–L101.
Roman, J., 1997. Exp. Lung Res. 23 (2), 147–159.
Roman, J., et al., 2004. FASEB J. 18 (12), 1436–1438.
Rosenkrans Jr., W.A., et al., 1983. Cell. Tissue Res. 234 (1), 165–177.
Roth-Kleiner, M., Post, M., 2005. Pediatr. Pulmonol. 40 (2), 113–134.
Ruoslahti, E., 1988. Annu Rev. Biochem. 57, 375–413.
Sadler, T.W., 2012. Langman's Medical Embryology. Lippincott Williams & Wilkins, Philadelphia.
Sage, H., et al., 1980. Biochemistry 19 (25), 5747–5755.
Sandberg, L.B., et al., 1969. Biochemistry 8 (7), 2940–2945.
Sanes, J.R., et al., 1990. J. Cell. Biol. 111 (4), 1685–1699.
Sannes, P.L., 1984. J. Histochem. Cytochem. 32 (8), 827–833.
Sannes, P.L., et al., 1993. Am. J. Respir. Cell. Mol. Biol. 8 (3), 245–251.
Sannes, P.L., et al., 1998. Am. J. Physiol. 275 (4 Pt 1), L701–L708.
Sannes, P.L., Wang, J., 1997. Exp. Lung Res. 23 (2), 101–108.
Schaefer, L., Schaefer, R.M., 2010. Cell. Tissue Res. 339 (1), 237–246.
Schmid, K., et al., 1982. Biochim. Biophys. Acta 716 (2), 178–187.
Schuger, L., et al., 1995. Dev. Biol. 169 (2), 520–532.
Schuger, L., et al., 1998. Int. J. Dev. Biol. 42 (2), 217–220.
Scott, J.E., 1992. FASEB J. 6 (9), 2639–2645.
Sekine, K., et al., 1999. Nat. Genet. 21 (1), 138–141.
Sell, D.R., et al., 2001. J. Gerontol. A Biol. Sci. Med. Sci. 56 (9), B405–B411.
Seyer, J.M., et al., 1976. J. Clin. Invest. 57 (6), 1498–1507.
Shapiro, S.D., et al., 1991. J. Clin. Invest. 87 (5), 1828–1834.
Shoji, S., et al., 1989. Am. J. Respir. Cell. Mol. Biol. 1 (1), 13–20.
Shoulders, M.D., Raines, R.T., 2009. Annu Rev. Biochem. 78, 929–958.
Silver, F.H., et al., 2000. Connect. Tissue Res. 41 (2), 155–164.
Silver, F.H., et al., 2002. J. Theor. Biol. 216 (2), 243–254.
Silver, F.H., et al., 2003. J. Biomech. 36 (10), 1529–1553.
Simon, R.H., et al., 1993. Am. J. Respir. Cell. Mol. Biol. 8 (6), 640–646.
Singh, P., et al., 2010. Annu Rev. Cell. Dev. Biol. 26, 397–419.
Skinner, S.J., et al., 1987. Exp. Lung Res. 12 (3), 253–264.
Skinner, S.J., et al., 1989. Exp. Lung Res. 15 (2), 269–283.
Smits, N.C., et al., 2004. Am. J. Respir. Cell. Mol. Biol. 30 (2), 166–173.
Spooner, B.S., et al., 1993. Trans. Kans Acad. Sci. 96 (1–2), 46–55.
Starcher, B.C., 2000. CHEST J. 117 (5 Suppl. 1), 229S–234S.
Starcher, B.C., Galione, M.J., 1976. Anal. Biochem. 74 (2), 441–447.
Stenman, S., Vaheri, A., 1978. J. Exp. Med. 147 (4), 1054–1064.
Stern, R., et al., 2007. Biotechnol. Adv. 25 (6), 537–557.
Sueblinvong, V., et al., 2012. Am. J. Med. Sci. 344 (1), 41–51.
Sugrue, S.P., Hay, E.D., 1981. J. Cell. Biol. 91 (1), 45–54.
Takubo, Y., et al., 1999. Exp. Gerontol. 34 (3), 353–364.
Tamkun, J.W., et al., 1984. Proc. Natl. Acad. Sci. U. S. A. 81 (16), 5140–5144.
Thomas, T., Dziadek, M., 1994. Mech. Dev. 45 (3), 193–201.
Timpl, R., et al., 1979. J. Biol. Chem. 254 (19), 9933–9937.
Toole, B.P., 1990. Curr. Opin. Cell. Biol. 2 (5), 839–844.
Toonkool, P., et al., 2001. J. Biol. Chem. 276 (30), 28042–28050.
Torikata, C., et al., 1985. Lab. Invest. 52 (4), 399–408.
Turino, G.M., Cantor, J.O., 2003. Am. J. Respir. Crit. Care Med. 167 (9), 1169–1175.
Turley, E.A., et al., 2002. J. Biol. Chem. 277 (7), 4589–4592.
Tzu, J., Marinkovich, M.P., 2008. Int. J. Biochem. Cell. Biol. 40 (2), 199–214.
Umezawa, K., et al., 1985. FEBS Lett. 186 (1), 31–34.
Urry, D.W., 1978. Perspect. Biol. Med. 21 (2), 265–295.
Vaccaro, C.A., Brody, J.S., 1981. J. Cell. Biol. 91 (2 Pt 1), 427–437.
van Kuppevelt, T.H., et al., 1984a. Histochem. J. 16 (6), 671–686.
Van Kuppevelt, T.H., et al., 1984b. Histochem. J. 16 (6), 657–669.
Virtanen, I., et al., 1996. Am. J. Respir. Cell. Mol. Biol. 15 (2), 184–196.
Vogel, K.G., Peterson, D.W., 1981. J. Biol. Chem. 256 (24), 13235–13242.
Wang, Y., et al., 2002. Am. J. Physiol. Lung Cell. Mol. Physiol. 282 (3), L484–L490.
Weibel, E.R., Crystal, R.G., 1991. Structural organization of the pulmonary interstitium. In: Crystal, R.G., West, J.B. (Eds.), The Lung. Raven, New York, pp. 369–380.
Weigel, P.H., DeAngelis, P.L., 2007. J. Biol. Chem. 282 (51), 36777–36781.
Weigel, P.H., et al., 1997. J. Biol. Chem. 272 (22), 13997–14000.
Weissman, B., Meyer, K.J., 1954. J. Am. Chem. Soc. 76 (7), 1753–1757.
West, J.B., Mathieu-Costello, O., 1999. Annu Rev. Physiol. 61, 543–572.
White, E.S., et al., 2008. J. Pathol. 216 (1), 1–14.
White, E.S., Muro, A.F., 2011. IUBMB Life 63 (7), 538–546.
Williams, K.J., 2001. Curr. Opin. Lipidol. 12 (5), 477–487.
Wu, J.E., Santoro, S.A., 1996. Dev. Dyn. 206 (2), 169–181.
Xu, J., Mosher, D., 2011. Fibronectin and other adhesive glycoproteins. In: Mecham, R.P. (Ed.), The Extracellular Matrix: An Overview. Springer Berlin Heidelberg, pp. 41–75.
Yang, Q., et al., 1999. Dev. Biol. 212 (1), 229–242.
Young, S.L., et al., 1994. Dev. Biol. 161 (2), 615–625.
Yurchenco, P.D., et al., 1997. Proc. Natl. Acad. Sci. U. S. A. 94 (19), 10189–10194.
Yurchenco, P.D., Schittny, J.C., 1990. FASEB J. 4 (6), 1577–1590.
Zhao, Y., Young, S.L., 1995. Am. J. Physiol. 269 (4 Pt 1), L482–L491.

Chapter 21

Lipid Mediators and Lung Function

Barry Weinberger, Daniel Hirsch, Kingsley Yin and Bernd W. Spur

Chapter Outline

Arachidonic acid is a ubiquitous polyunsaturated fatty acid that is found almost exclusively esterified to phospholipids at the secondary alcohol of glycerol. Phospholipids containing arachidonic acid are present in all mammalian membranes. It is an essential fatty acid that is not synthesized de novo within cells but must be incorporated through diet or by chain elongation and desaturation of linoleic acid, another essential fatty acid (Jeffcoat, 1979). Cells in the lung most likely extract arachidonic acid from the blood, where it is present in relatively high concentrations (1−5 μg/ml) (Manku et al., 1983). However, "free" concentrations are much lower since it is tightly bound to albumin and other serum proteins (Irvine, 1982). Arachidonic acid is the initial substrate for lipid-derived mediators that are important in the maintenance of lung homeostasis and inflammatory tone, including prostaglandins (PG), leukotrienes (LT), lipoxins (LX), and epoxyeicosatrienoic acids (EET) (Figure 1). These mediators are formed subsequent to the hydrolytic action of phospholipase A2 (PLA2) on cell membrane phospholipids to release arachidonic acid. In addition, 8-isoprostanes are produced when arachidonic acid undergoes nonenzymatic peroxidation by reactive oxygen intermediates (Clarke et al., 2009). PLA2 also catalyzes the hydrolysis of cellular phospholipids to produce the lysophospholipids, another class of lipid mediators that incorporate lysophosphatidic acid and sphingosine 1-phosphate. Interestingly, several isoenzymes of PLA2 with different substrate affinities are detected in lung cells and fluid under pathologic conditions such as acute respiratory distress syndrome. These have distinct roles in generating inflammatory eicosanoids, lung surfactant recycling, and/or antioxidant activity, and their succession may be important in disease progression and restoration of lung function (Kitsiouli et al., 2009).

1. PROSTANOIDS

Prostanoids include PGD_2, PGE_2, $PGF_{2\alpha}$, and PGI_2 (prostacyclin), as well as thromboxane A_2 (TXA_2). These are key mediators and modulators of responses in both inflammatory and structural cells in the lung, with PGE2 being the major prostanoid produced by airway smooth muscle cells (Belvisi et al., 1997; Delamere et al., 1994; Rolin et al., 2006). Prostanoids are generated from arachidonic acid via the cyclooxygenase (COX)/prostaglandin G/H synthase pathway; many intermediate forms are chemically unstable, with half-lives of less than 1 min in the blood (Clarke et al., 2009; Vane, 1971) (Figure 2). Synthesized prostanoids exit the cell via a carrier-mediated process and activate specific G protein-coupled receptors (GPCR) in target cells (Carey et al., 2003). Specific receptors have been identified—DP for PGD_2, EP for PGE_2, FP for $PGF_{2\alpha}$, IP for PGI_2, and TP for TXA_2, each with subtypes (Coleman et al., 1994)—which are all present to varying extents in lung tissues and are known to play a role in regulating airway smooth muscle contractility, leukocyte activity, and platelet function. Despite the classification of prostanoid receptors based on their selectivity, agonists variably cross-react with each of the receptors.

Prostanoids mediate both contraction and relaxation of airway smooth muscle, with PGE_2 and PGI_2 generally exerting bronchodilatory effects and TXA_2, PGD_2, and

Comparative Biology of the Normal Lung. http://dx.doi.org/10.1016/B978-0-12-404577-4.00021-7

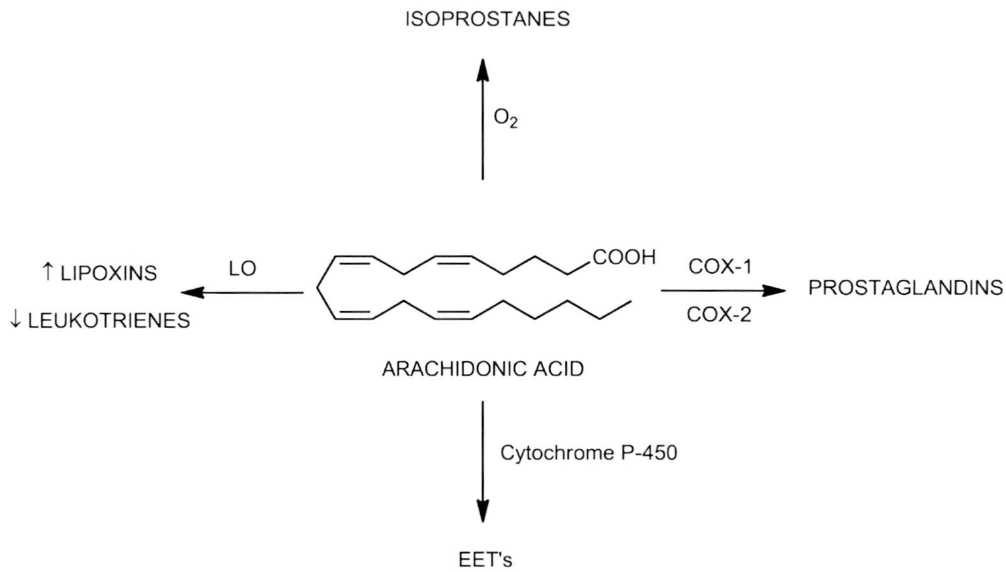

FIGURE 1 Arachidonic acid is the initial substrate for lipid-derived mediators that are important in the maintenance of lung homeostasis and inflammatory tone, including prostaglandins, leukotrienes, lipoxins, isoprostanes, and epoxyeicosatrienoic acids (EETs).

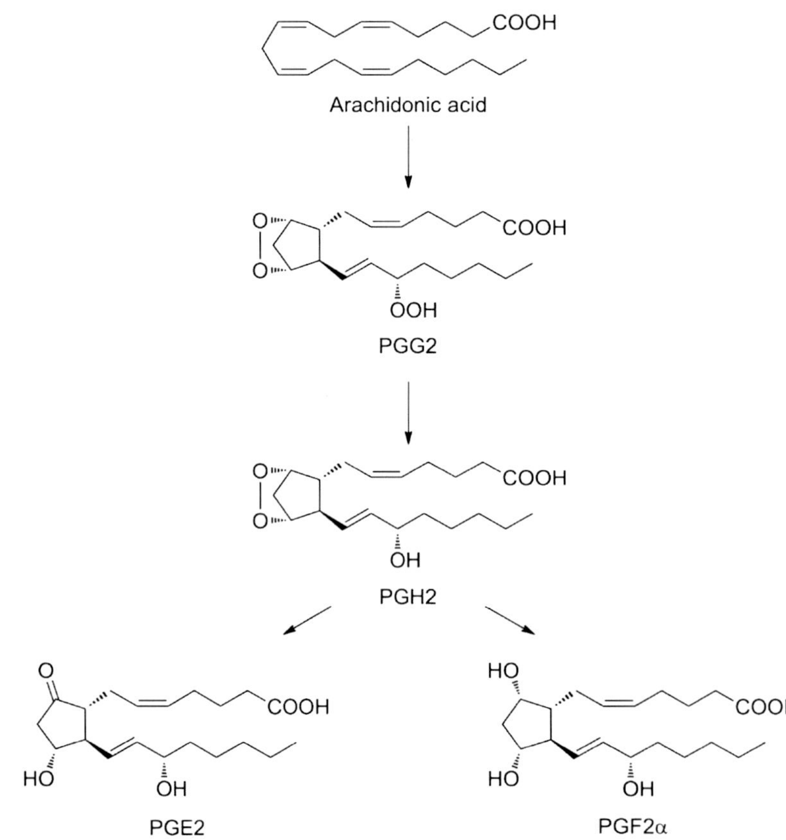

FIGURE 2 **Generation of prostaglandins.** Prostanoids, such as prostaglandin E2 (PGE2) and prostaglandin F2α (PGF2α), are generated from arachidonic acid via the cyclooxygenase-prostaglandin G/H synthase pathway. Many intermediate forms, such as PGG2 and PGH2, are chemically unstable.

$PGF_{2\alpha}$ inducing bronchoconstriction (Clarke et al., 2009). PGE_2 acts via EP_2 and EP_4 receptors expressed on airway smooth muscle cells (Hartney et al., 2006). Activation of these receptors leads to protein kinase A-mediated phosphorylation of myosin light-chain kinase, which inhibits actin–myosin interactions and decreases muscle tone. PGE_2 also acts via EP_3 receptors to reduce the release of contractile neurotransmitters from parasympathetic neurons in the lung in animal models (Spicuzza et al., 1998). Similarly, prostacyclin relaxes airway smooth muscle. The stable prostacyclin analog, Beraprost, has been shown to oppose acetylcholine-mediated contraction through cAMP production and stimulation of $Na^+–K^+$ ATPase and to inhibit acetylcholine release from cholinergic nerve terminals (Tamaoki et al., 1993). However, both PGE_2 and prostacyclin can promote bronchoconstriction under certain conditions by activating alternative receptors and pathways. Activation of EP1 and EP3 receptors can be coupled to increased Ca^{2+} mobilization and inhibition of adenylate cyclase, leading to increased tone. Both PGE_2 and prostacyclin may also cause reflex bronchoconstriction by stimulating afferent vagal C-fibers in the lower airway in animals (Roberts et al., 1981).

TXA_2, PGD_2, and $PGF_{2\alpha}$ increase airway tone by activating TP receptors, which are abundantly expressed in lung vasculature (Armour et al., 1989; Johnston et al., 1995). TXA_2 is produced mainly by platelets, monocytes, macrophages, neutrophils, and lung parenchymal cells (Widdicombe et al., 1989). Physiologically, TXA_2 stimulates platelet aggregation and shape change. It acts on endothelial cells and induces expression of adhesion molecules, polymerization of actin, and contraction (Zarbock and Ley, 2009). Activation of TP receptors in bronchial smooth muscle cells by TXA_2 induces intracellular calcium mobilization leading to bronchoconstriction. Under pathologic conditions, such as exposure to infectious, allergic, or noxious stimuli, this can contribute to airway hyperresponsiveness (Capra et al., 2003). TXA_2 can also induce mucosal barrier disruption, and ultimately airway smooth muscle cell proliferation (Maniatis et al., 2008). As a result, it plays a role in the pathogenesis of asthma. TXA_2 and its metabolites are elevated in the airways, urine, and plasma of asthma patients (Devillier and Bessard, 1997; Oosterhoff et al., 1995). TP receptor activation also contributes to bronchial smooth muscle hyperplasia and airway remodeling in chronic asthma (Vignola et al., 2003). TXA_2 released by platelet-neutrophil aggregates has also been implicated in acute lung injury (Zarbock and Ley, 2009), and elevated levels have been reported in primary pulmonary hypertension (Christman, 1998). Therapeutic strategies for inhibiting TXA_2 are under investigation. These include inhibitors COX/prostaglandin G/H synthase and thromboxane synthase, the enzyme that catalyzes the synthesis of TXA_2 from PGH_2. TP receptor antagonists are particularly promising since they may block the activity of several bronchoconstrictor PGs in addition to TXA_2 (Rolin et al., 2006; Devillier and Bessard, 1997; O'Brien et al., 2007).

2. ISOPROSTANES

Isoprostanes are PG-like compounds that are derived from peroxidation of polyunsaturated fatty acids (usually arachidonic acid) by reactive oxygen intermediates in tissues (Voynow and Kummarapurugu, 2011). These peroxidation reactions lead to formation of structures named for the prostane ring structure that they closely resemble, with different regioisomers ("series") differing by the placement of a hydroxyl group on the alkyl chain (most commonly, on carbons 5, 8, 12, or 15) (Figure 3). Chemically, there are a few key differences between isoprostanes and PG. Isoprostane side chains are mainly *cis* to the cyclopentane ring, while PG side chains are *trans*. The F_2-isoprostanes (F_2-isoP) are stable molecules that are formed in vivo under conditions of oxidative stress, and other classes are similarly named for the PG with a common prostane ring structure, including (D_2-isoP), (E_2-isoP), (J_2-isoP), and (A_2-isoP). Isoprostanes (including "neuroprostanes") can also be generated from the ω-3 fatty acids DHA and EPA (Musiek et al., 2008; Song et al., 2009). The production and metabolism of isoprostanes has been reported in most detail for 15-F_{2t}-isoP. It is formed in situ in lipid membranes and released into the circulation as the free molecule by phospholipase A2 or platelet activating factor acetylhydrolase. 15-F_{2t}-isoP and its degradation products are excreted in urine, where they have been used as markers of oxidative stress (Morrow and Roberts, 2002; Weinberger et al., 2004).

Isoprostanes exert important biologic effects in both health and disease (reviewed in Janssen et al., 2005). They are present at nanomolar concentrations in blood under normal physiologic conditions (Janssen, 2008), and have excitatory effects in vascular tissue, including the aorta and pulmonary arteries. These effects are mediated primarily via TP receptors, but likely also in part via EP and FP prostaglandin receptors (Janssen and Tazzeo, 2002; Kang et al., 1993). Isoprostanes regulate human airway smooth muscle (ASM) tone and hyperresponsiveness to stimuli (Janssen, 2008). Both 8-iso-PGE_2 and 8-iso-$PGF_{2\alpha}$ induce airway smooth muscle contraction via TP receptor-mediated calcium influx. In contrast, 8-iso-$PGF_{3\alpha}$ relaxes human ASM via a non-TP receptor (Janssen et al., 2000). Isoprostanes also regulate airway smooth muscle tone by modulating prejunctional neuronal acetylcholine secretion in animal models, but this mechanism has not been demonstrated in human airways (Spicuzza et al., 2001).

Isoprostanes also regulate inflammatory tone in the lung. Lung macrophages play an important role in

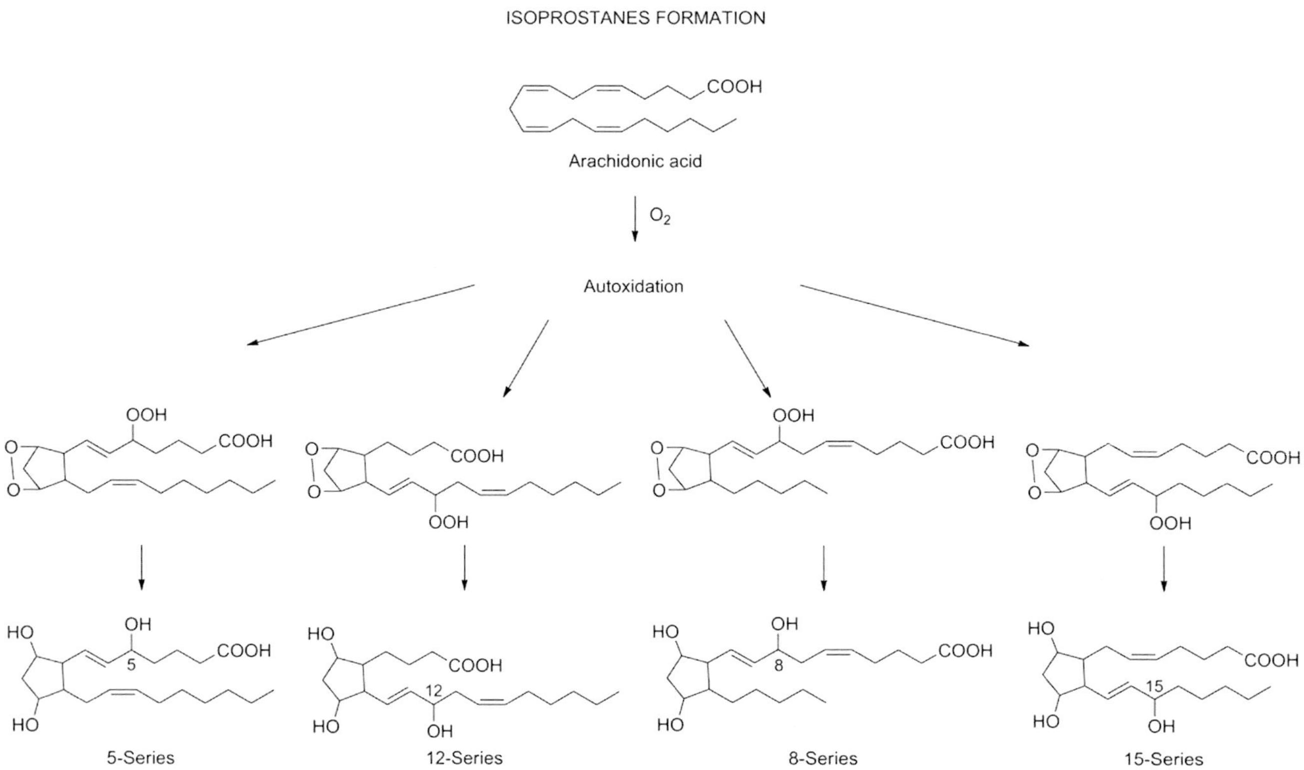

FIGURE 3 Generation of isoprostanes. Isoprostanes are PG-like compounds that are derived from peroxidation of arachidonic acid or other polyunsaturated fatty acids. These compounds are named for the prostane ring structure that they closely resemble, with different regioisomers ("series") differing by the placement of a hydroxyl group on the alkyl chain (most commonly, on carbons 5, 8, 12, or 15).

modulating innate immune responses to environmental pollutants and pathogens in the lungs (Hogg and van Eeden, 2009). Exposure of human alveolar macrophages to elemental carbon particles induces 8-iso-$PGF_{2\alpha}$ production, which stimulates macrophage binding to endothelial cells and production of interleukin (IL)-8 (Leitinger et al., 2001; Scholz et al., 2003). Isoprostanes may further contribute to inflammation by increasing neutrophil adhesion to human venous endothelial cells and by increasing endothelial cell permeability in response to oxidative stress (Hart et al., 1998; Zahler and Becker, 1999). In contrast to 8-iso-$PGF_{2\alpha}$, cyclopentenone isoprostanes block NF-κB-mediated inflammatory cascades, most likely by attacking exposed Cys residues and inactivating IκK, thereby impairing IκBα degradation. The cyclopentenone isoprostanes similarly modify Cys residues and inactivate Keap1, a protein that inactivates the transcription factor Nrf2. Inactivation of Keap1 results in the release of Nrf2, which increases expression of genes for antioxidant proteins. Therefore, while some isoprostanes induce inflammatory cytokines and neutrophil chemotaxis, the cyclopentenone isoprostanes enhance antioxidant capacity and inhibit inflammation in the lung (Rossi et al., 2000; Sawyer et al., 2010).

3. LEUKOTRIENES

5-lipoxygenase (5-LO) catalyzes the oxidation of arachidonic acid at the 5-position to yield 5-hydroperoxyeicosatetraenoic acid (5-HPETE), and then the conversion of 5-HPETE to leukotriene (LT) A_4. LTA4 is then metabolized to LTB_4, LTC_4, LTD_4, and LTE_4, which exert generally proinflammatory effects (Figure 4). 5-LO also acts on eicosapentaenoic acid (EPA) to generate the 5-series LT (LTB_5, LTC_5, LTD_5, LTE_5), which are anti-inflammatory. LTC_4, LTD_4, and LTE_4 are referred to as the cysteinyl leukotrienes (CysLTs) because of the presence of cystgeine and three conjugated double bonds, in contrast to the non-cysteine-containing LTB_4 (Singh et al., 2010). Inflammatory cells, including alveolar macrophages, express the integral membrane protein LTC_4 synthase and synthesize CysLTs in response to biological and nonbiological stimuli (Thien and Walters, 1995). LTC_4 synthase conjugates reduced glutathione at position C6 of LTA_4 to produce LTC_4, which is transferred to the extracellular space, where a γ-glutamyl transpeptidase cleaves the glutamic acid moiety to form LTD_4. This can then be transformed to LTE_4 by cleavage of the glycine moiety by dipeptidase enzyme (Orning and Hammarstrom, 1980).

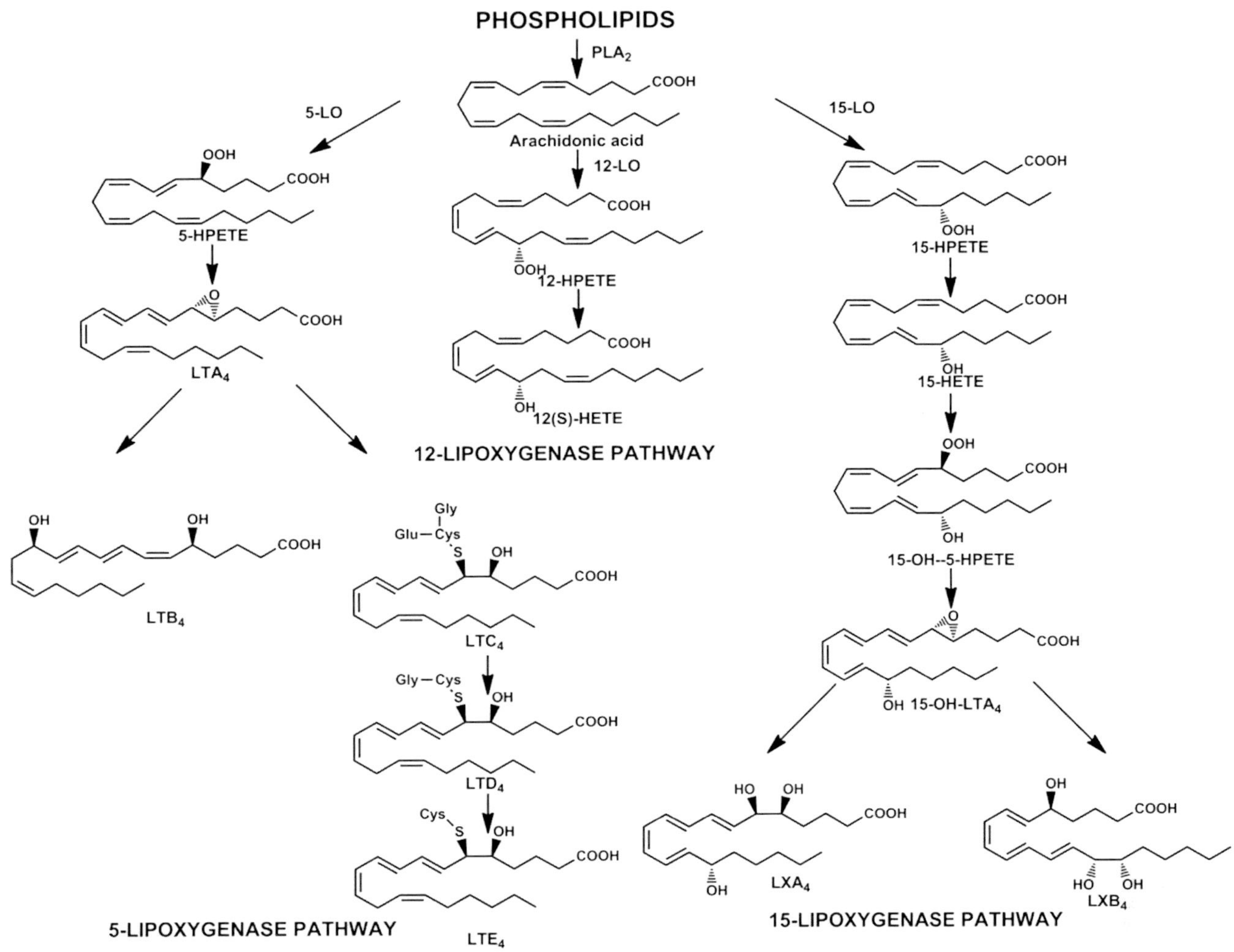

FIGURE 4 Actions of lipoxygenases on arachidonic acid. 5-lipoxygenase (5-LO) catalyzes the oxidation of arachidonic acid to yield leukotriene (LTA_4). LTA4 is then metabolized to LTB_4, LTC_4, LTD_4, and LTE_4, which exert generally proinflammatory effects. Alternatively, 15-lipoxygenase (15-LO) is the first step in a series of reactions leading to the generation of lipoxins (LXA_4, LXB_4), which exert anti-inflammatory effects.

CysLTs exert their effects through cell surface receptors. Two distinct receptor proteins have been characterized, CysLT1 and CysLT2; both are G-protein-coupled receptors and are coded on the X chromosome and chromosome 13, respectively (Heise et al., 2000; Lynch et al., 1999). CysLT1 receptor was first identified in normal lung smooth muscle cells and interstitial macrophages, with little or no expression in normal airway epithelial cells (Lynch et al., 1999). CysLT1 receptor is expressed by airway mucosal inflammatory cells—eosinophils, neutrophils, mast cells, macrophages, B-lymphocytes, and plasma cells—and their numbers are increased in pathologic conditions such as asthma. In contrast, CysLT2 is only weakly expressed in the lung, although elevated expression has been detected in macrophages near smooth muscle cells (Heise et al., 2000; Evans, 2003). The existence of additional classes of CysLT receptors is suggested by the fact that the pan-CysLTR inhibitor BAYu9773 does not abrogate all CysLT activity, and by the observation that LTE_4 is the most stable CysLT but only a weak agonist for the classical CysLT1/CysLT2 receptors.

CysLTs are the most potent bronchoconstrictors identified in humans. The basal tone of human airways is maintained by a balance between the effects of contractile mediators including CysLTs, LTB_4, and histamine, with vasodilating agents such as PGE_2 (Ellis and Undem, 1994). The contractile activity of LTC_4 and LTD_4 has been demonstrated in vivo in human subjects (Ramsay et al., 2009). CysLTs also mediate inflammatory effects in asthma, causing increased microvascular permeability leading to pulmonary edema, increased mucus secretion, impaired ciliary function, airway smooth muscle hyperplasia, and increased eosinophil survival in response to paracrine signals from mast cells and lymphocytes (Arm, 2004; Hallstrand and Henderson, 2010; Mehrotra and Henderson, 2009). In allergic asthma, CysLTs are produced primarily by eosinophils and mast cells in the lung, and markedly increased levels are found in bronchoalveolar lavage fluid in asthmatics (Lam et al., 1988). Elevated levels of CysLTs are also found in exhaled breath condensate in patients with pneumoconiosis and expression of CysLT1 receptors is increased in COPD exacerbations, suggesting a role in these pathologies (Pelclova et al., 2012; Zhu et al., 2012). CysLTs also promote leukocyte maturation and migration from the bone marrow into the circulatory system, and they are chemoattractant for eosinophils, increasing their cellular adhesion and transendothelial migration across the vessel wall into the airways (Hui and Funk, 2002). Consequently, blocking LT pathways may be beneficial in asthma, and this has led to the development of 5-LO enzyme inhibitors (e.g., zileutin) and CysLT1 receptor antagonists (e.g., montelukast, zafirlukast, and pranlukast) (Barnes, 2011; Peters-Golden and Henderson, 2007). Montelukast has been shown to reduce lung inflammation in animal models (Al-Amran et al., 2012). However, recent meta-analyses have shown that corticosteroids are superior for the treatment of chronic and acute asthma, and the role of these agents in the treatment of asthma remains poorly defined (Chauhan and Ducharme, 2012; Watts and Chavasse, 2012).

4. ENDOCANNABINOIDS

Endogenous cannabinoid receptor agonists have recently been recognized as an additional class of lipid-derived mediators with effects on pulmonary physiology and disease. Two cannabinoid receptors have been extensively described: CB1 and CB2. While CB1 is expressed primarily in neural tissues, CB2 is highly expressed in immune cells, including those in the lung. The first endogenous cannabinoid receptor agonist identified was anandamide (AEA), a derivative of arachidonic acid (Croxford and Yamamura, 2005). Another identified endocannabinoid ligand is 2-arachidonoylethanolamide (2-AG) (Mechoulam et al., 1995). While the neurologic effects of these agents have been studied most, they also appear to play a robust role in immunoregulation by activating CB2 receptors on leukocytes (Pacher and Mechoulam, 2011). Endocannabinoids modulate T and B lymphocyte proliferation and apoptosis, cytokine production, macrophage cytotoxicity, chemotaxis, and inflammatory cell migration (Cencioni et al., 2010; Klein et al., 2003). However, effects are somewhat inconsistent and dependent on specific experimental conditions (Miller and Stella, 2008). Endocannabinoids may also modulate inducible nitric oxide synthase expression and production of reactive nitrogen and oxygen species by inflammatory cells, which are important in immune responses, cytotoxicity, and pulmonary vascular tone. In animal models, AEA increases expression of heat shock proteins, Hsp70 and Hsp25, suggesting a role in dampening inflammation (Kopczynska et al., 2011). Synthetic CB2 receptor agonists have also been shown to inhibit airway sensory nerve function and lipopolysaccharide (LPS)-induced pulmonary inflammation (Belvisi et al., 2008; Berdyshev et al., 1998).

5. PLATELET-ACTIVATING FACTOR

Platelet-activating factor (PAF) is a membrane-derived glycerophospholipid that has a wide variety of physiologic and pathophysiologic actions in the lung. At the molecular level, its unique biologic properties are attributed to the presence of an ether linkage at the *sn*-1 position and an acetate ester linkage at the *sn*-2 position (Toyoshima et al., 1995). PAF is recognized by a G-protein coupled, seven-membrane-spanning surface receptor that is constitutively expressed on platelets and myeloid leukocytes and mediates the activation of intracellular kinases and calcium mobilization (Giri, 2003). Animal studies have shown that PAF receptors are also present on type II pneumocytes and stimulate the production of dipalmitoylphosphatidylcholine, the major active component of pulmonary surfactant (Eguchi et al., 1994), suggesting that PAF plays a role in fetal lung maturation. Postnatally, PAF is synthesized by endothelial and innate immune effector cells in response to inflammatory and thrombotic agonists. PAF signaling probably evolved as a component of innate host—defense responses. Surface expression of PAF on human endothelial cells leads to binding and activation of neutrophils in response to infection or injury (Zimmerman and McIntyre, 2004). Endogenously generated PAF causes fever and the cardinal signs of inflammation: pain, erythema, and edema (Ivanov et al., 2003). The concurrent generation of inflammatory cytokines by locally activated neutrophils and monocytes amplifies these responses.

The most prominent effects of PAF in the lung are altered vascular tone and the promotion of lung edema. PAF contracts airway and vascular smooth muscle and causes vascular gaps in capillaries, venules, and veins (Uhlig et al., 2005; Yang et al., 2010). Decreased PAF receptor binding and downregulation of PAF receptor gene expression in the immediate perinatal period may facilitate the fall in pulmonary vascular resistance that is essential for successful cardiorespiratory transition (Ibe et al., 2000). Several mechanisms have been proposed for these effects. PAF stimulates lipoxygenase, which triggers smooth muscle contraction by upregulating TXA_2, LTC_4, and LTD_4; it also increases cyclooxygenase activity in the lung, which results in increases vascular permeability via PGE_2 (Uhlig et al., 2005, 1994; Goggel et al., 2002). Lung edema is mediated directly by alterations in hydrostatic pressure and blood flow, increased vascular permeability, and alveolar epithelial desquamation. In addition, PAF stimulates the extracellular release of acid sphingomyelinase, possibly from platelets (Goggel et al., 2004; Zimmerman et al., 2002). Acid sphingomyelinase hydrolyzes sphingomyelin, generating ceramide in pulmonary epithelial cells. Intracellular ceramide directly promotes the formation of plasma membrane lipid rafts and caveolae, affecting cell function and membrane permeability by altering vesicular trafficking and surface receptor distribution (van Blitterswijk et al., 2003). Endogenous ceramide also alters cytoskeletal and integrin function in leukocytes and may contribute to PAF-induced alterations in nitric oxide synthesis (Feldhaus et al., 2002; Igarashi et al., 1999; Kuebler et al., 2010).

The actions of PAF in lung are best understood in the context of disease. Increased levels of PAF are found in patients with asthma, acute respiratory distress syndrome, hydrostatic pulmonary edema, trauma, and sepsis. Chronically, PAF induces vascular remodeling, decreased pulmonary vascular compliance, loss of hydroxyproline and vascular matrix, and fibrosis (Uhlig et al., 2005; Ohar et al., 1994a). PAF receptor antagonists and agents that interfere with intracellular calcium mobilization are under investigation for treatment of both acute respiratory distress syndrome and idiopathic pulmonary fibrosis (Giri, 2003).

6. PEROXISOME PROLIFERATOR-ACTIVATED RECEPTOR-γ

Peroxisome proliferator-activated receptors (PPARs) are transcription factors that belong to the nuclear hormone receptor superfamily and exist as three subtypes: PPARα, PPARβ/δ, and PPARγ. PPARs were first identified for their role in the control of lipid and glucose metabolism, but PPARα and PPARγ have been implicated in the regulation of inflammation (Becker et al., 2006). PPARs are constitutively associated with the retinoic X receptor (RXR) as a heterodimer, bound to corepressor proteins in the inactive state. When endogenous (or synthetic) agonists bind to the ligand binding domain (LBD), coactivator proteins displace corepressors at their binding sites, allowing binding of activated PPAR complexes to the peroxisome proliferator response element (PPRE) in the promoter regions of target genes. Transcriptional responses are determined by ligand-specific recruitment of coactivators to the transcription factor complex (A IJ et al., 1997; Tudor et al., 2007).

PPARs are mechanistically important in the biologic activity of fatty acids and their lipid derivatives in the lung (Standiford et al., 2005). Known endogenous ligands for PPARα include 8S-hydroxyeicosatetraenoic acid (8S-HETE), LTB4, and dihydroepiandrosterone. PPARγ agonists include the cyclopentenone prostaglandin 15-deoxy-Δ12,14-prostaglandin J2 (15d-PGJ2) and the 12/15 lipoxygenase products 15-hydroxyeicosatetraenoic acid (15-HETE) and 13-hydroxyoctadecadienoic acid (13-HODE). Among these, 15d-PGJ2 was widely considered to be a selective agonist, but has also been shown to exhibit anti-inflammatory effects that are PPARγ independent, including in the lung (Nosjean and Boutin, 2002). Dietary ω-3 polyunsaturated fatty acids EPA and docosahexaenoic acid (DHA), and nitrated fatty acids are also PPAR-γ agonists, exerting anti-inflammatory effects at physiologic concentrations (Villacorta et al., 2009).

PPARγ expression has been detected in the airway epithelium (Sundvold et al., 1997; Wang et al., 2001), bronchial submucosa (Ueki et al., 2004), airway smooth muscle cells (Patel et al., 2003), and alveolar macrophages (Asada et al., 2004; Huang et al., 2005). Deletion of PPARγ in animals leads to abnormally enlarged airspaces and other phenotypic changes, suggesting that it plays a role in lung maturation (Simon et al., 2006). The most prominent biologic activity of PPARγ is immunomodulation, which may be particularly relevant in the lung because it contains a vast epithelial-capillary interface that is constantly exposed to inhaled pathogens. Under some conditions, alveolar macrophages play a role in dampening inflammatory responses to these stimuli. Signaling via PPAR-γ is involved in reducing production of inflammatory cytokines and nitric oxide, as well as promoting phagocytosis of apoptotic neutrophils during the resolution phase of inflammation (Asada et al., 2004; Aronoff et al., 2007; Chinetti et al., 2000; Majai et al., 2007). Treatment of murine alveolar macrophages with PPARγ agonists inhibits oxidative burst activity after treatment with phorbol myristate acetate, expression of inducible nitric oxide synthase after treatment with LPS and interferon (IFN)-γ, and production of IL-12 after LPS (Reddy, 2008; Reddy et al., 2004). These activities are thought to prevent chronic lung injury in the face of ongoing exposure of the lung epithelium to environmental antigens. In animals, PPARγ and PPARα have been shown to downregulate pathologic activity in models of human inflammatory disease, including asthma (Delayre-Orthez et al., 2005; Heikkinen et al., 2007; Woerly et al., 2003). Consequently, investigations are ongoing to define roles for synthetic agonists of PPARα (fibrates) and PPARγ (glitazones, indomethacin, ibuprofen), or endogenous fatty acids and their derivatives, for the prevention and/or treatment of pulmonary diseases such as asthma, sarcoidosis, chronic obstructive pulmonary disease, acute lung injury, and alveolar proteinosis (Becker et al., 2006; Reddy, 2008).

7. LIPOXINS

Lipoxins (lipoxygenase interaction products), which are generated by the conjugate activity of both 5- and 15 lipoxygenase, were discovered by Serhan and Samuelsson in 1984 after they incubated 15L-hydroperoxy-5,8,11,13-eicosatetraenoic acid with human leukocytes (Serhan, Hamberg, and Samuelsson, 1984). Platelet-derived

12-lipoxygenase can also serve as a lipoxin synthetase. Aspirin, a cyclooxygenase inhibitor that acetylates cyclooxygenase-2 (COX-2), induces the synthesis of "aspirin-triggered" proresolving mediators in humans, including 15-epi-lipoxin A4 from lipoxin A4 (Mitchell et al., 2002) (Figure 5). All of these anti-inflammatory eicosanoids have been identified in humans using mass spectroscopy (Lee et al., 1990). Lipoxins have been identified in lung lavage fluid in humans, with concentrations inversely associated with the amount of leukotrienes recovered.

Lipoxins exert effects through a G-protein coupled receptor, formyl peptide receptor-2 (FPR2) (Serhan et al., 2011). The receptor is shared by several lipid and peptide/protein ligands, including bacterial peptides and serum amyloid. Interestingly, responses are both ligand specific and cell type specific. For example, lipoxins decrease neutrophil migration, but increase macrophage motility. In general, lipoxins are potent suppressors of inflammatory activity, inducing the resolution of leukocyte activity (Psychogios et al., 2011; Serhan and Savill, 2005). They block responses to peptidoleukotrienes and chemotactic agents such as leukotriene B4, platelet-activating factor, and the chemotactic peptide fMLP (Grandordy et al., 1990; Lee et al., 1989, 1991; Papayianni et al., 1996). These effects inhibit the infiltration of proinflammatory cells. The resolution of inflammation involves reduction in neutrophil migration, increased macrophage migration, and augmented phagocytosis of apoptotic cells. The full range of immune-modulatory activity of lipoxins has been reviewed elsewhere (Chiang et al., 2005; Drazen et al., 1999; McMahon et al., 2001). Consistent with a role of lipoxins in preventing or ameliorating inflammatory disease, they have been shown to be reduced or absent in the lung in asthma, cystic fibrosis, and other types of severe respiratory illness (Mitchell et al., 2002; Planaguma et al., 2008). In a clinical trial in asthmatic patients, inhaled lipoxin decreased leukotriene-induced pulmonary bronchoconstriction (Drazen et al., 1999; Christie et al., 1992).

The anti-inflammatory potency of lipoxin A4 is comparable with corticosteroids but without the typical side effect of immunosuppression. This suggests that lipoxins may have therapeutic value in treating or preventing inflammation-mediated diseases. However, the effects of lipoxin in models of infection are not as clear as the beneficial effects seen in sterile models of inflammation. In the *Toxoplasma gondii* model of parasitic infection, lipoxin decreases overall inflammation but increases infection (Aliberti et al., 2002). Similarly, lipoxin reduces inflammatory responses but increases pulmonary *Mycobacterium tuberculosis* infection (Bafica et al., 2005). A possible explanation for this is that lipoxin increases macrophage necrosis, allowing mycobacteria to evade killing (Romano, 2010). Alternatively, it is possible that the lack of protection from infection is an artifact related to the use of a stable lipoxin analog in these studies, rather than the natural

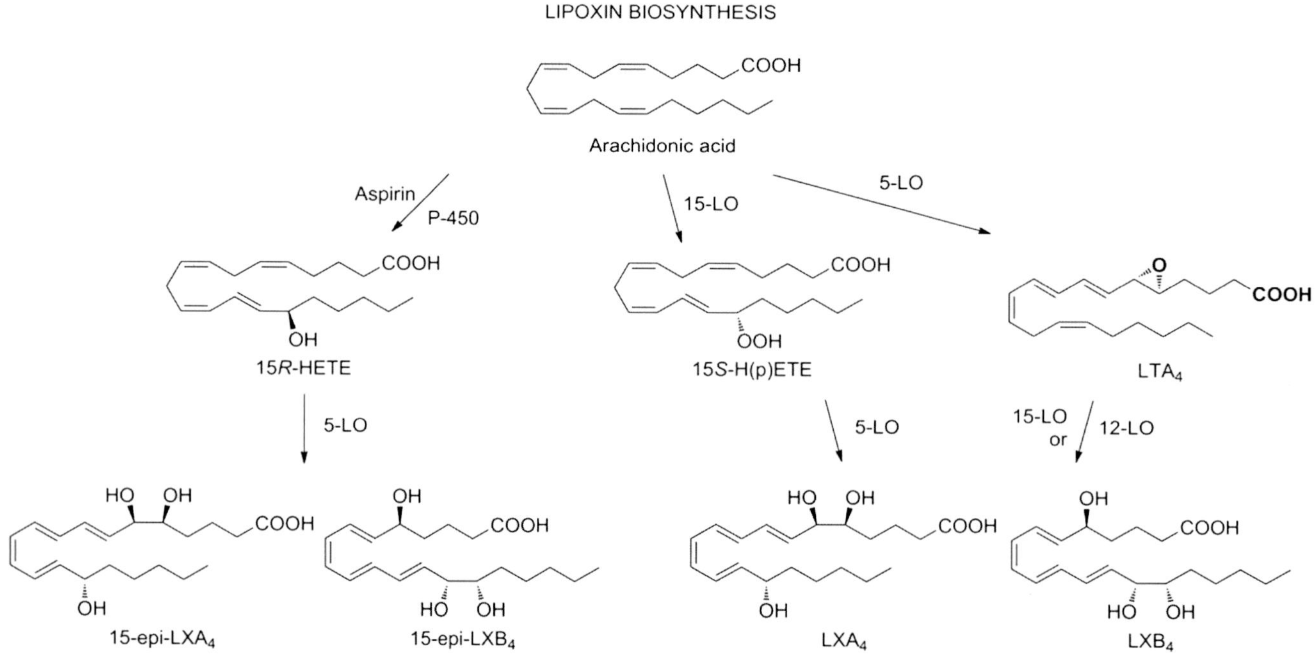

FIGURE 5 Generation of lipoxins. Sequential actions by lipoxygenases (15-LO, 5-LO, 12-LO) lead to the formation of lipoxins (LXA_4, LXB_4) from arachidonic acid. Alternatively, in the presence of aspirin, the "aspirin-triggered" proresolving mediators 15-epi-lipoxin A4 and 15-epi-lipoxin B4 (15-epi-LXA_4, 15-epi-LXB_4) are generated.

compound. Consistent with this idea, we recently found that natural lipoxin A4 increases survival, decreases systemic inflammation, and reduces blood bacterial load in polymicrobial sepsis in rats (Walker et al., 2011).

8. RESOLVINS E_1, E_2, AND E_3

In addition to lipoxins, Serhan and collaborators also identified bioactive derivatives of two major ω-3 polyunsaturated fatty acids from fish oil, EPA and DHA: the resolvins, protectin D1 and maresin 1. EPA is converted to 18-(R)-hydroxy eicosapentaenoic acid, the precursor of the E-type resolvins. The 5-lipoxygenase pathways in neutrophils generates resolvin (Rv)E_1 and resolvin E_2 (Figure 6). Eosinophils convert 18-(R)-hydroxy eicosapentaenoic acid into 8,18-dihydroxyeicosapentaenoic acid as well as into 11,18- and 12,18-dihydroxyeicosapentaenoic acids in addition to resolvin E_3 (Serhan et al., 2000). The biological activities of the E-series resolvins are similar to the lipoxins (Oh et al., 2011; Schwab et al., 2007; Seki et al., 2010; Tjonahen et al., 2006; Xu et al., 2010). Resolvin E1 and analogs are more potent than lipoxin A_4 on a molar basis. Resolvin E_1 binds to the leukotriene receptor BLT1 (but not to BLT2) and blocks TNF-α-stimulated NF-kB activation at the ChemR23 receptor. In the lung, it reduces airway hyperresponsiveness and bronchoconstriction in response to allergen challenges (Aoki et al., 2010; Arita et al., 2005, 2007; Haas-Stapleton et al., 2007; Hasturk et al., 2007; Haworth et al., 2008). Resolvin E_1 also increases the synthesis of lipoxin A_4 (Haworth et al., 2008), and administration of resolvin E_1 significantly improves survival in bacteria-induced acute lung injury (Seki et al., 2010). The potential therapeutic uses of exogenous resolvin E_1 are currently under investigation.

9. RESOLVINS D_1, D_2, D_3, D_4, D_5, D_6, MARESIN 1, PROTECTIN D_1

DHA is converted via 15-lipoxygenase into a 17(S)-hydroperoxy-docosahexaenoic acid intermediate that then undergoes reduction of the hydroperoxide to produce 17(S) hydroxydocosahexaenoic acid (17S-HDHA). Further action

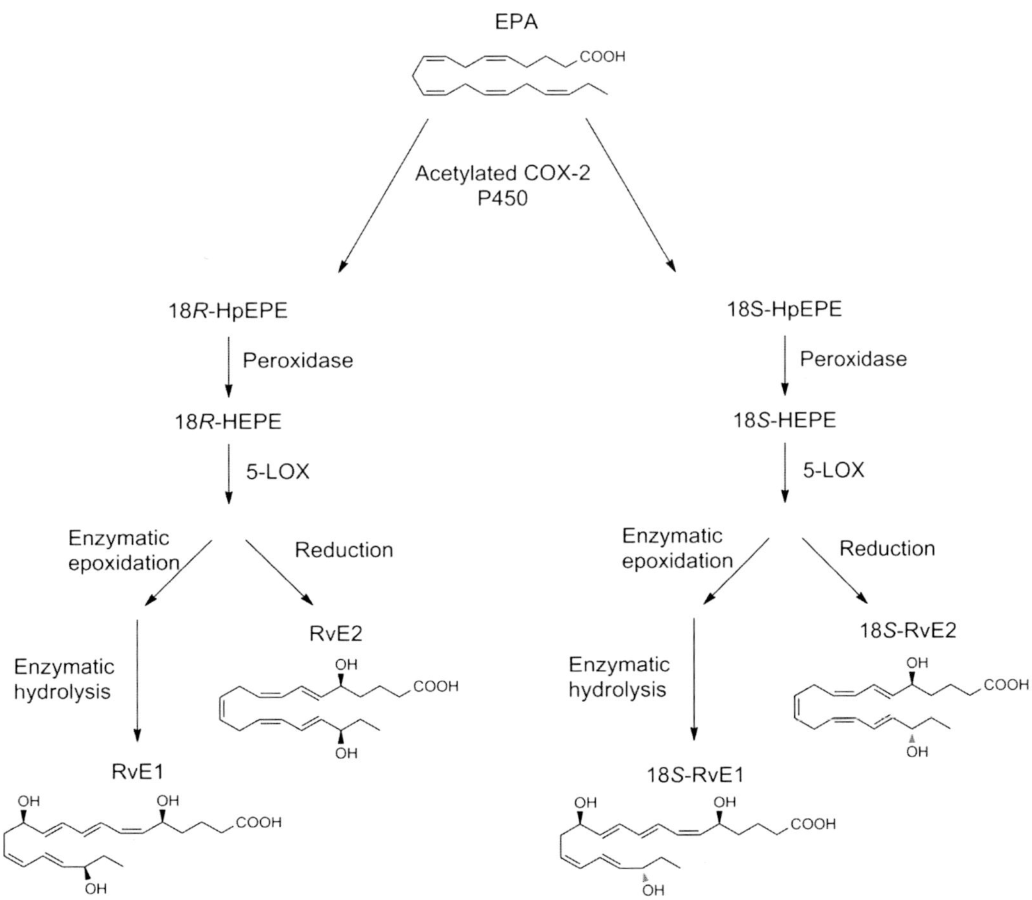

FIGURE 6 Generation of E-series resolvins. E-series resolvins are generated by conversion of the ω-3 polyunsaturated fatty acid, eicosapentaenoic acid (EPA) to 18-(R)-hydroxy eicosapentaenoic acid (18R-HpEPE) or 18-(S)-hydroxy eicosapentaenoic acid (18S-HpEPE) by aspirin-acetylated COX-2. Resolvin E1 (RvE1), RvE2, 18S-RvE1, and 18S-RvE2 are then generated by the action of 5-lipoxygenase (5-LO) in neutrophils.

via 5-lipoxygenase gives rise to the D-series resolvins. Alternatively, 7(S),8(S)-epoxide formation followed by hydrolysis of 17S-HDHA gives rise to protectin D1 (Serhan et al., 2000, 2007, 2006; Sun et al., 2007). In a similar fashion, DHA is converted to maresin 1 (macrophage mediator in resolving inflammation) via the platelet 12/15 lipoxygenase (Serhan et al., 2006, 2009) (Figure 7). In general, these novel eicosanoids of the D series show potent anti-inflammatory and immunoregulatory actions, and the full spectrum of their biological activities has been summarized in several reviews (Levy et al., 2012; Mas et al., 2012; Serhan et al., 2008). In a recent report, it was shown that resolvin D_1 promotes phagocytosis of *Escherichia coli* by macrophages, accelerates the resolution of gram-positive and gram-negative bacteria infections, and enhances survival when compared to antibiotics alone. Similarly, resolvin D_2 improves outcomes in sepsis (Spite et al., 2009), and resolvin D5 has been shown to lower the antibiotic levels needed to treat *E. coli* infections (Chiang et al., 2012). In summary, the proresolving lipid mediators (lipoxins, resolvins, protectin D_1, maresin 1) may shorten the times required for cessation of inflammation, possibly leading to decreased tissue injury while maintaining the efficacy of bacterial clearance. Further clinical trials are required. However, it is likely that several existing classes of drugs, including the statins, act in part via these mediators.

10. SUMMARY AND CONCLUSIONS

Bioactive derivatives of polyunsaturated fatty acids—most importantly, arachidonic acid, EPA, and DHA—regulate essential aspects of lung function (Table 1). These mediators exert synergistic and opposing effects on bronchial constriction and relaxation, pulmonary vascular tone, and the initiation and cessation of immune and inflammatory

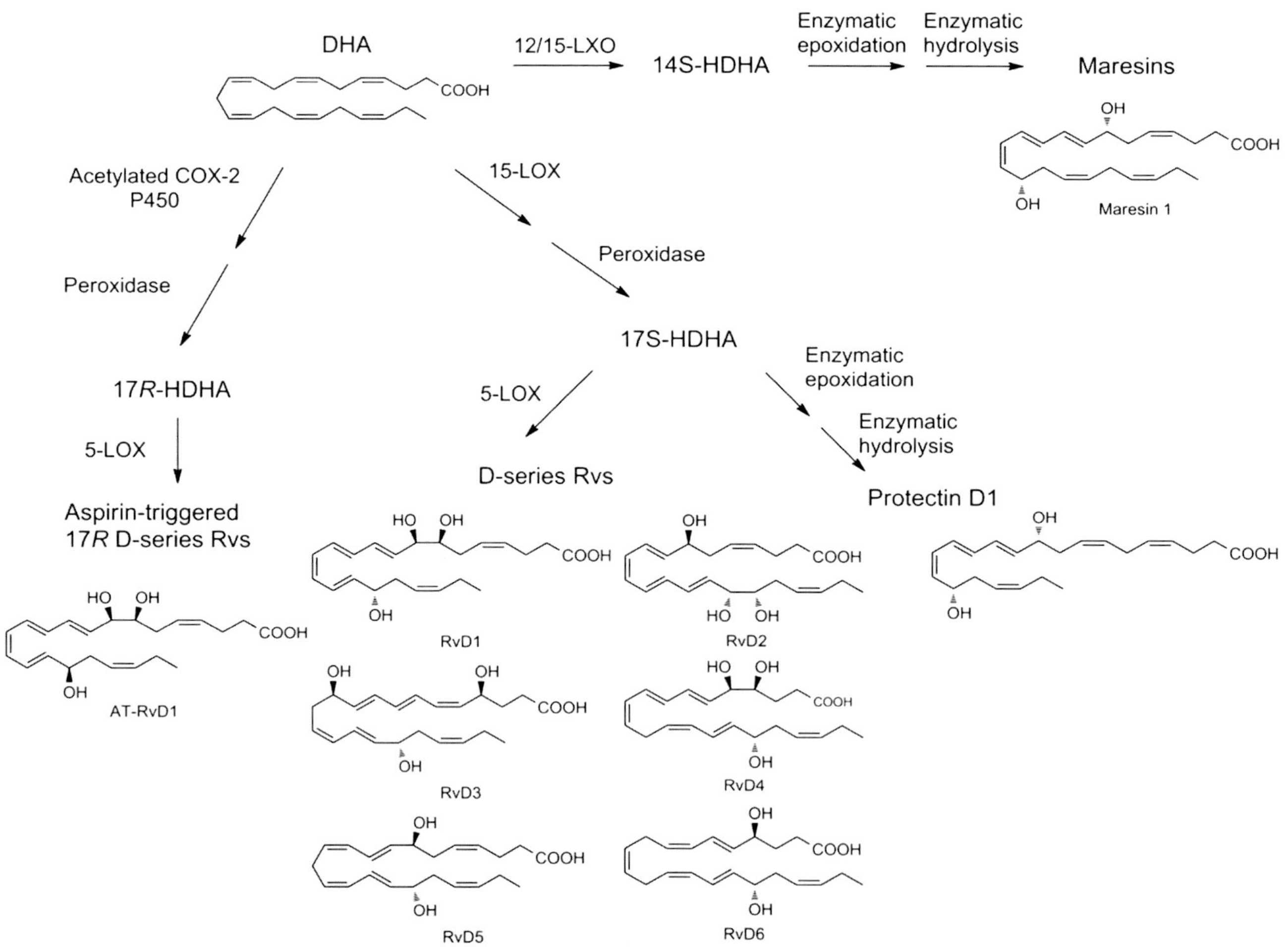

FIGURE 7 Generation of aspirin-triggered resolvins, D-series resolvins, maresins, and protectins. The ω-3 polyunsaturated fatty acid docosahexaenoic acid (DHA) is converted via 15-lipoxygenase into 17(S) hydroxydocosahexaenoic acid (17S-HDHA). Further action via 5-lipoxygenase gives rise to the D-series resolvins (RvD1, RvD2, RvD3, RvD4, RvD5, RvD6). Alternatively, 7(S),8(S)-epoxide formation followed by hydrolysis of 17S-HDHA gives rise to protectin D1. In a similar fashion DHA is converted to maresin 1 (macrophage mediator in resolving inflammation) via the platelet 12/15 lipoxygenase and to aspirin-triggered D-series resolvins via aspirin-acetylated COX-2.

TABLE 1 Vascular, Airway, and Inflammatory Responses to Lipid Mediators

Mediator	Effects	References
Prostaglandin (PG)D2	Bronchoconstriction, vasodilation, induces eosinophil migration	(Braun and Schror, 1992; Sampson et al., 1997; Schratl et al., 2007; Selg et al., 2009)
PGE2	Bronchodilation, bronchoconstriction, induces vascular leakage	(Hartney et al., 2006; Roberts et al., 1981; Chan and Ford-Hutchinson, 1985; Sheller et al., 2000; Tilley et al., 2003)
PGF2α	Bronchoconstriction, vasoconstriction	(Fish et al., 1984; Yu et al., 2009)
PGI2 (prostacyclin)	Bronchodilation, bronchoconstriction, pulmonary vasodilation	(Tamaoki et al., 1993; Kadowitz et al., 1978; Pepke-Zaba et al., 1991; Wasserman et al., 1980)
Thromboxane A2 (TXA_2)	Bronchoconstriction, airway hyperresponsiveness, airway smooth muscle proliferation, disruption of airway mucosa, endothelial cell expression of adhesion molecules and actin polymerization	(Zarbock and Ley, 2009; Capra et al., 2003; Maniatis et al., 2008; Jones et al., 1992)
Thromboxane B2 (TXB_2)	Induces neutrophil chemotaxis	(Kitchen et al., 1978)
Lipoxin A4 (LXA_4)	Deficiency of LXA_4 associated with severe asthma	(Planaguma et al., 2008; Levy et al., 2005; Wu et al., 2010)
Lipoxin B4 (LXB_4)	Reduces neutrophil infiltration	(Takano et al., 1998)
Leukotriene B4	Induces vascular leakage and leukocyte adhesion	(Dahlen et al., 1981; Rosengren et al., 1991; Thureson-Klein et al., 1986)
Leukotriene C4	Bronchoconstriction, increased vascular permeability	(Ramsay et al., 2009; Adelroth et al., 1986; Hallstrand et al., 2005; Smith et al., 1985)
Leukotriene D4	Bronchoconstriction, increased vascular permeability	(Ramsay et al., 2009; Adelroth et al., 1986; Hallstrand et al., 2005; Smith et al., 1985; Gauvreau et al., 2001)
Leukotriene E4	Bronchoconstriction, increased vascular permeability	(Adelroth et al., 1986; Hallstrand et al., 2005; Smith et al., 1985; Gauvreau et al., 2001)
Resolvins E1-3	Decreased airway responsiveness, decreased bronchoconstriction, reduced neutrophil infiltration, decreased leukocyte recruitment, decreased cytokine production	(Aoki et al., 2010; Arita et al., 2005, 2007; Haas-Stapleton et al., 2007; Hasturk et al., 2007; Haworth et al., 2008; Oh et al., 2011)
Resolvins D1-5	Enhanced bacterial killing by antibiotics, enhanced neutrophil phagocytosis, decreased neutrophil chemotaxis	(Schwab et al., 2007; Spite et al., 2009; Chiang et al., 2012)
Protectin D1	Decreased airway eosinophils, T-lymphocytes and inflammatory mediators	(Schwab et al., 2007; Levy et al., 2007)
Maresins	Reduced neutrophil infiltration, enhanced macrophage phagocytosis of neutrophils	(Serhan et al., 2009)
Platelet-activating factor	Increased pulmonary vascular permeability, pulmonary vasoconstriction, increased LOX activity, bronchoconstriction	(Yang et al., 2010; Uhlig et al., 1994; Ohar et al., 1994b; Samapati et al., 2012)
Isoprostanes	Bronchoconstriction, bronchodilation, pulmonary vasoconstriction	(Janssen and Tazzeo, 2002; Janssen et al., 2000)
8-iso-PGF2α	Bronchoconstriction, macrophage binding to endothelial cells and production of IL-8	(Leitinger et al., 2001; Scholz et al., 2003; Kawikova et al., 1996)
Cyclopentenone isoprostanes	Increased gene expression of antioxidant proteins	(Rossi et al., 2000; Sawyer et al., 2010)

Continued

TABLE 1 Vascular, Airway, and Inflammatory Responses to Lipid Mediators—cont'd

Mediator	Effects	References
Endocannabinoids	Primarily via CB2R receptors: decreased responsiveness of alveolar macrophages, Th1 suppression, and Th2 activation	(Baldwin et al., 1997; Shay et al., 2003; Yuan et al., 2002)
Peroxisome proliferator-activated receptor-γ (PPAR-γ)	Enhanced lung maturation, enhanced phagocytosis of neutrophils, decreased proinflammatory cytokine secretion, decreased respiratory burst activity	(Simon et al., 2006; Aronoff et al., 2007; Majai et al., 2007; Reddy et al., 2004; Christie et al., 1992; Jiang et al., 1998)

responses in the lung. The identification in lung tissues of specific receptors for these mediators, as well as the lipid-responsive PPAR-α and PPAR-γ transcription factors, suggests that they are important in maintaining lung homeostasis. Several fatty acid-derived mediators (or their synthetic agonists or antagonists) have demonstrated therapeutic value in correcting derangements of lung function in asthma, pulmonary hypertension, inflammatory lung disease, fibrosis, or acute lung injury. However, the triggers and mechanisms determining the relative rates of synthesis of these compounds in health are not known. Decoding this complex regulatory network may suggest a coherent and targeted approach to the prevention and treatment of diverse pulmonary diseases.

REFERENCES

IJpenberg, A., Jeannin, E., Wahli, W., Desvergne, B., August 8, 1997. Polarity and specific sequence requirements of peroxisome proliferator-activated receptor (PPAR)/retinoid X receptor heterodimer binding to DNA. A functional analysis of the malic enzyme gene PPAR response element. J. Biol. Chem. 272 (32), 20108–20117. PubMed PMID: 9242684. Epub 1997/08/08. Eng.

Adelroth, E., Morris, M.M., Hargreave, F.E., O'Byrne, P.M., August 21, 1986. Airway responsiveness to leukotrienes C4 and D4 and to methacholine in patients with asthma and normal controls. N. Engl. J. Med. 315 (8), 480–484. PubMed PMID: 3526153.

Al-Amran, F.G., Hadi, N.R., Hashim, A.M., July 31, 2012. Cysteinyl leukotriene receptor antagonist montelukast ameliorates acute lung injury following haemorrhagic shock in rats. Eur. J. Cardiothorac. Surg. PubMed PMID: 22851661. Epub 2012/08/02. Eng.

Aliberti, J., Serhan, C., Sher, A., Nov 4, 2002. Parasite-induced lipoxin A4 is an endogenous regulator of IL-12 production and immunopathology in *Toxoplasma gondii* infection. J. Exp. Med. 196 (9), 1253–1262. PubMed PMID: 12417634. PubMed Central PMCID: 2194099. Epub 2002/11/06. Eng.

Aoki, H., Hisada, T., Ishizuka, T., Utsugi, M., Ono, A., Koga, Y., et al., September 10, 2010. Protective effect of resolvin E1 on the development of asthmatic airway inflammation. Biochem. Biophys. Res. Commun. 400 (1), 128–133. PubMed PMID: 20708601. Epub 2010/08/17. Eng.

Arita, M., Bianchini, F., Aliberti, J., Sher, A., Chiang, N., Hong, S., et al., March 7, 2005. Stereochemical assignment, antiinflammatory properties, and receptor for the omega-3 lipid mediator resolvin E1. J. Exp. Med. 201 (5), 713–722. PubMed PMID: 15753205. Pubmed Central PMCID: 2212834. Epub 2005/03/09. Eng.

Arita, M., Ohira, T., Sun, Y.P., Elangovan, S., Chiang, N., Serhan, C.N., Mar 15, 2007. Resolvin E1 selectively interacts with leukotriene B4 receptor BLT1 and ChemR23 to regulate inflammation. J. Immunol. 178 (6), 3912–3917. PubMed PMID: 17339491. Epub 2007/03/07. Eng.

Arm, J.P., January–February, 2004. Leukotriene generation and clinical implications. Allergy Asthma Proc. 25 (1), 37–42. PubMed PMID: 15055561. Epub 2004/04/02. Eng.

Armour, C.L., Johnson, P.R., Alfredson, M.L., Black, J.L., June 20, 1989. Characterization of contractile prostanoid receptors on human airway smooth muscle. Eur. J. Pharmacol. 165 (2–3), 215–222. PubMed PMID: 2776830. Epub 1989/06/20. Eng.

Aronoff, D.M., Serezani, C.H., Carstens, J.K., Marshall, T., Gangireddy, S.R., Peters-Golden, M., et al., 2007. Stimulatory effects of peroxisome proliferator-activated receptor-gamma on Fcgamma receptor-mediated phagocytosis by alveolar macrophages. PPAR Res. 2007, 52546. PubMed PMID: 18253476. Pubmed Central PMCID: 2220046.

Asada, K., Sasaki, S., Suda, T., Chida, K., Nakamura, H., January 15, 2004. Antiinflammatory roles of peroxisome proliferator-activated receptor gamma in human alveolar macrophages. Am. J. Respir. Crit. Care Med. 169 (2), 195–200. PubMed PMID: 14563653. Epub 2003/10/18. Eng.

Bafica, A., Scanga, C.A., Serhan, C., Machado, F., White, S., Sher, A., et al., June 2005. Host control of *Mycobacterium tuberculosis* is regulated by 5-lipoxygenase-dependent lipoxin production. J. Clin. Invest. 115 (6), 1601–1606. PubMed PMID: 15931391. Pubmed Central PMCID: 1136995. Epub 2005/06/03. Eng.

Baldwin, G.C., Tashkin, D.P., Buckley, D.M., Park, A.N., Dubinett, S.M., Roth, M.D., November 1997. Marijuana and cocaine impair alveolar macrophage function and cytokine production. Am. J. Respir. Crit. Care Med. 156 (5), 1606–1613. PubMed PMID: 9372683.

Barnes, P.J., September 23, 2011. Biochemical basis of asthma therapy. J. Biol. Chem. 286 (38), 32899–32905. PubMed PMID: 21799015. Pubmed Central PMCID: 3190895. Epub 2011/07/30. Eng.

Becker, J., Delayre-Orthez, C., Frossard, N., Pons, F., October 2006. Regulation of inflammation by PPARs: a future approach to treat lung inflammatory diseases? Fundam. Clin. Pharmacol. 20 (5), 429–447. PubMed PMID: 16968414. Epub 2006/09/14. Eng.

Belvisi, M.G., Saunders, M.A., Haddadel, B., Hirst, S.J., Yacoub, M.H., Barnes, P.J., et al., March 1997. Induction of cyclo-oxygenase-2 by cytokines in human cultured airway smooth muscle cells: novel inflammatory role of this cell type. Br. J. Pharmacol. 120 (5),

910–916. PubMed PMID: 9138698. Pubmed Central PMCID: 1564529. Epub 1997/03/01. Eng.

Belvisi, M.G., Patel, H.J., Freund-Michel, V., Hele, D.J., Crispino, N., Birrell, M.A., October 2008. Inhibitory activity of the novel CB(2) receptor agonist, GW833972A, on guinea-pig and human sensory nerve function in the airways. Br. J. Pharmacol. 155 (4), 547–557. PubMed PMID: 18695648.

Berdyshev, E., Boichot, E., Corbel, M., Germain, N., Lagente, V., July 17, 1998. Effects of cannabinoid receptor ligands on LPS-induced pulmonary inflammation in mice. Life Sci. 63 (8), Pl125–Pl129. PubMed PMID: 9718090.

Braun, M., Schror, K., December 1992. Prostaglandin D2 relaxes bovine coronary arteries by endothelium-dependent nitric oxide-mediated cGMP formation. Circulation Res. 71 (6), 1305–1313. PubMed PMID: 1385004.

Capra, V., Habib, A., Accomazzo, M.R., Ravasi, S., Citro, S., Levy-Toledano, S., et al., August 8, 2003. Thromboxane prostanoid receptor in human airway smooth muscle cells: a relevant role in proliferation. Eur. J. Pharmacol. 474 (2–3), 149–159. PubMed PMID: 12921856. Epub 2003/08/19. Eng.

Carey, M.A., Germolec, D.R., Langenbach, R., Zeldin, D.C., August–September, 2003. Cyclooxygenase enzymes in allergic inflammation and asthma. Prostagl. Leukot. Essent. Fat. Acids 69 (2–3), 157–162. PubMed PMID: 12895598. Epub 2003/08/05. Eng.

Cencioni, M.T., Chiurchiu, V., Catanzaro, G., Borsellino, G., Bernardi, G., Battistini, L., et al., January 14, 2010. Anandamide suppresses proliferation and cytokine release from primary human t-lymphocytes mainly via CB2 receptors. PLoS One 5 (1), e8688. PubMed PMID: 20098669.

Chan, C.C., Ford-Hutchinson, A., February 1985. Effects of synthetic leukotrienes on local blood flow and vascular permeability in porcine skin. J. Invest. Dermatol. 84 (2), 154–157. PubMed PMID: 2981933.

Chauhan, B.F., Ducharme, F.M., 2012. Anti-leukotriene agents compared to inhaled corticosteroids in the management of recurrent and/or chronic asthma in adults and children. Cochrane Database Syst. Rev. 5, CD002314. PubMed PMID: 22592685. Epub 2012/05/18. Eng.

Chiang, N., Arita, M., Serhan, C.N., September–October 2005. Anti-inflammatory circuitry: lipoxin, aspirin-triggered lipoxins and their receptor ALX. Prostagl. Leukot. Essent. Fat. Acids 73 (3–4), 163–177. PubMed PMID: 16125378. Epub 2005/08/30. Eng.

Chiang, N., Fredman, G., Backhed, F., Oh, S.F., Vickery, T., Schmidt, B.A., et al., April 26, 2012. Infection regulates pro-resolving mediators that lower antibiotic requirements. Nature 484 (7395), 524–528. PubMed PMID: 22538616. Pubmed Central PMCID: 3340015. Epub 2012/04/28. Eng.

Chinetti, G., Fruchart, J.C., Staels, B., October 2000. Peroxisome proliferator-activated receptors (PPARs): nuclear receptors at the crossroads between lipid metabolism and inflammation. Inflamm. Res. Offi. J. Eur. Histamine Res. Soc. 49 (10), 497–505. PubMed PMID: 11089900.

Christie, P.E., Spur, B.W., Lee, T.H., June 1992. The effects of lipoxin A4 on airway responses in asthmatic subjects. Am. Rev. Respir. Dis. 145 (6), 1281–1284. PubMed PMID: 1595991. Epub 1992/06/01. Eng.

Christman, B.W., September 1998. Lipid mediator dysregulation in primary pulmonary hypertension. Chest 114 (Suppl. 3), 205S–207S. PubMed PMID: 9741570. Epub 1998/09/19. Eng.

Clarke, D.L., Dakshinamurti, S., Larsson, A.K., Ward, J.E., Yamasaki, A., October 2009. Lipid metabolites as regulators of airway smooth muscle function. Pulm. Pharmacol. Ther. 22 (5), 426–435. PubMed PMID: 19114116. Epub 2008/12/31. Eng.

Coleman, R.A., Smith, W.L., Narumiya, S., June 1994. International Union of pharmacology classification of prostanoid receptors: properties, distribution, and structure of the receptors and their subtypes. Pharmacol. Rev. 46 (2), 205–229. PubMed PMID: 7938166. Epub 1994/06/01. Eng.

Croxford, J.L., Yamamura, T., September 2005. Cannabinoids and the immune system: potential for the treatment of inflammatory diseases? J. Neuroimmunol. 166 (1–2), 3–18. PubMed PMID: 16023222. Epub 2005/07/19. Eng.

Dahlen, S.E., Bjork, J., Hedqvist, P., Arfors, K.E., Hammarstrom, S., Lindgren, J.A., et al., June 1981. Leukotrienes promote plasma leakage and leukocyte adhesion in postcapillary venules: in vivo effects with relevance to the acute inflammatory response. Proc. Natl. Acad. Sci. USA 78 (6), 3887–3891. PubMed PMID: 6267608. Pubmed Central PMCID: 319678.

Delamere, F., Holland, E., Patel, S., Bennett, J., Pavord, I., Knox, A., April 1994. Production of PGE2 by bovine cultured airway smooth muscle cells and its inhibition by cyclo-oxygenase inhibitors. Br. J. Pharmacol. 111 (4), 983–988. PubMed PMID: 8032626. Pubmed Central PMCID: 1910129. Epub 1994/04/01. Eng.

Delayre-Orthez, C., Becker, J., Guenon, I., Lagente, V., Auwerx, J., Frossard, N., et al., 2005. PPARalpha downregulates airway inflammation induced by lipopolysaccharide in the mouse. Respir. Res. 6, 91. PubMed PMID: 16091136. Pubmed Central PMCID: 1199625. Epub 2005/08/11. Eng.

Devillier, P., Bessard, G., 1997. Thromboxane A2 and related prostaglandins in airways. Fundam. Clin. Pharmacol. 11 (1), 2–18. PubMed PMID: 9182072. Epub 1997/01/01. Eng.

Drazen, J.M., Israel, E., O'Byrne, P.M., January 21, 1999. Treatment of asthma with drugs modifying the leukotriene pathway. N. Engl. J. Med. 340 (3), 197–206. PubMed PMID: 9895400. Epub 1999/01/23. Eng.

Eguchi, H., Frenkel, R.A., Johnston, J.M., February 1, 1994. Binding and metabolism of platelet-activating-factor (PAF) by isolated rat type-II pneumonocytes. Arch. Biochem. Biophys. 308 (2), 426–431. PubMed PMID: 8109971.

Ellis, J.L., Undem, B.J., January 1994. Role of cysteinyl-leukotrienes and histamine in mediating intrinsic tone in isolated human bronchi. Am. J. Respir. Crit. Care Med. 149 (1), 118–122. PubMed PMID: 8111568. Epub 1994/01/01. Eng.

Evans, J.F., August–September 2003. The cysteinyl leukotriene receptors. Prostagl. Leukot. Essent. Fat. Acids 69 (2–3), 117–122. PubMed PMID: 12895594. Epub 2003/08/05. Eng.

Feldhaus, M.J., Weyrich, A.S., Zimmerman, G.A., McIntyre, T.M., February 8, 2002. Ceramide generation in situ alters leukocyte cytoskeletal organization and beta(2)-integrin function and causes complete degranulation. J. Biological Chem. 277 (6), 4285–4293. PubMed PMID: 11706024. Eng.

Fish, J.E., Jameson, L.S., Albright, A., Norman, P.S., October 1984. Modulation of the bronchomotor effects of chemical mediators by prostaglandin F2 alpha in asthmatic subjects. Am. Rev. Respir. Dis. 130 (4), 571–574. PubMed PMID: 6592997.

Gauvreau, G.M., Parameswaran, K.N., Watson, R.M., O'Byrne, P.M., October 15, 2001. Inhaled leukotriene E(4), but not leukotriene D(4), increased airway inflammatory cells in subjects with atopic asthma. Am. J. Respir. Crit. Care Med. 164 (8 Pt 1), 1495–1500. PubMed PMID: 11704602.

Giri, S.N., 2003. Novel pharmacological approaches to manage interstitial lung fibrosis in the twenty-first century. Annu. Rev. Pharmacol. 43, 73–95. PubMed PMID: 12540741.

Goggel, R., Hoffman, S., Nusing, R., Narumiya, S., Uhlig, S., September 1, 2002. Platelet-activating factor-induced pulmonary edema is partly mediated by prostaglandin E-2, E-prostanoid 3-receptors, and potassium channels. Am. J. Resp. Crit. Care 166 (5), 657–662. PubMed PMID: 12204861.

Goggel, R., Winoto-Morbach, S., Vielhaber, G., Imai, Y., Lindner, K., Brade, L., et al., February 2004. PAF-mediated pulmonary edema: a new role for acid sphingomyelinase and ceramide. Nat. Med. 10 (2), 155–160. PubMed PMID: 14704790.

Grandordy, B.M., Lacroix, H., Mavoungou, E., Krilis, S., Crea, A.E., Spur, B.W., et al., March 30, 1990. Lipoxin A4 inhibits phosphoinositide hydrolysis in human neutrophils. Biochem. Biophys. Res. Commun. 167 (3), 1022–1029. PubMed PMID: 2157419. Epub 1990/03/30. Eng.

Haas-Stapleton, E.J., Lu, Y., Hong, S., Arita, M., Favoreto, S., Nigam, S., et al., 2007. Candida albicans modulates host defense by biosynthesizing the pro-resolving mediator resolvin E1. PLoS One 2 (12), e1316. PubMed PMID: 18091990. Pubmed Central PMCID: 2134765. Epub 2007/12/20. Eng.

Hallstrand, T.S., Henderson Jr., W.R., February 2010. The evolving role of intravenous leukotriene modifiers in acute asthma. J. Allergy Clin. Immunol. 125 (2), 381–382. PubMed PMID: 20159248. Epub 2010/02/18. Eng.

Hallstrand, T.S., Moody, M.W., Wurfel, M.M., Schwartz, L.B., Henderson Jr., W.R., Aitken, M.L., September 15, 2005. Inflammatory basis of exercise-induced bronchoconstriction. Am. J. Respir. Crit. Care Med. 172 (6), 679–686. PubMed PMID: 15947280. Pubmed Central PMCID: 2041799.

Hart, C.M., Karman, R.J., Blackburn, T.L., Gupta, M.P., Garcia, J.G., Mohler 3rd., E.R., January 1998. Role of 8-epi PGF2alpha, 8-isoprostane, in H_2O_2-induced derangements of pulmonary artery endothelial cell barrier function. Prostagl. Leukot. Essent. Fat. Acids 58 (1), 9–16. PubMed PMID: 9482161. Epub 1998/03/03. Eng.

Hartney, J.M., Coggins, K.G., Tilley, S.L., Jania, L.A., Lovgren, A.K., Audoly, L.P., et al., January 2006. Prostaglandin E2 protects lower airways against bronchoconstriction. Am. J. Physiol. Lung Cell. Mol. Physiol. 290 (1), L105–L113. PubMed PMID: 16113047. Epub 2005/08/23. Eng.

Hasturk, H., Kantarci, A., Goguet-Surmenian, E., Blackwood, A., Andry, C., Serhan, C.N., et al., November 15, 2007. Resolvin E1 regulates inflammation at the cellular and tissue level and restores tissue homeostasis in vivo. J. Immunol. 179 (10), 7021–7029. PubMed PMID: 17982093. Epub 2007/11/06. Eng.

Haworth, O., Cernadas, M., Yang, R., Serhan, C.N., Levy, B.D., August 2008. Resolvin E1 regulates interleukin 23, interferon-gamma and lipoxin A4 to promote the resolution of allergic airway inflammation. Nat. Immunol. 9 (8), 873–879. PubMed PMID: 18568027. Pubmed Central PMCID: 2784998. Epub 2008/06/24. Eng.

Heikkinen, S., Auwerx, J., Argmann, C.A., August 2007. PPARgamma in human and mouse physiology. Biochim. Biophys. Acta 1771 (8), 999–1013. PubMed PMID: 17475546. Pubmed Central PMCID: 2020525. Epub 2007/05/04. Eng.

Heise, C.E., O'Dowd, B.F., Figueroa, D.J., Sawyer, N., Nguyen, T., Im, D.S., et al., September 29, 2000. Characterization of the human cysteinyl leukotriene 2 receptor. J. Biol. Chem. 275 (39), 30531–30536. PubMed PMID: 10851239. Epub 2000/06/14. Eng.

Hogg, J.C., van Eeden, S., April 2009. Pulmonary and systemic response to atmospheric pollution. Respirology 14 (3), 336–346. PubMed PMID: 19353772. Epub 2009/04/09. Eng.

Huang, T.H., Razmovski-Naumovski, V., Kota, B.P., Lin, D.S., Roufogalis, B.D., 2005. The pathophysiological function of peroxisome proliferator-activated receptor-gamma in lung-related diseases. Respir. Res. 6, 102. PubMed PMID: 16153299. Pubmed Central PMCID: 1242255. Epub 2005/09/13. Eng.

Hui, Y., Funk, C.D., December 1, 2002. Cysteinyl leukotriene receptors. Biochem. Pharmacol. 64 (11), 1549–1557. PubMed PMID: 12429344. Epub 2002/11/14. Eng.

Ibe, B.O., Sander, F.C., Raj, J.U., April 2000. Platelet-activating factor receptors in lamb lungs are downregulated immediately after birth. Am. J. Physiol. Heart Circ. Physiol. 278 (4), H1168–H1176. PubMed PMID: 10749711.

Igarashi, J., Thatte, H.S., Prabhakar, P., Golan, D.E., Michel, T., October 26, 1999. Calcium-independent activation of endothelial nitric oxide synthase by ceramide. Proc. Natl. Acad. Sci. USA 96 (22), 12583–12588. PubMed PMID: 10535965.

Irvine, R.F., April 15, 1982. How is the level of free arachidonic acid controlled in mammalian cells? Biochem. J. 204 (1), 3–16. PubMed PMID: 6810878. Pubmed Central PMCID: 1158309. Epub 1982/04/15. Eng.

Ivanov, A.I., Patel, S., Kulchitsky, V.A., Romanovsky, A.A., November 15, 2003. Platelet-activating factor: a previously unrecognized mediator of fever. J. Physiol-London 553 (1), 221–228. PubMed PMID: 14565987.

Janssen, L.J., October 2008. Isoprostanes and lung vascular pathology. Am. J. Respir. Cell. Mol. Biol. 39 (4), 383–389. PubMed PMID: 18441279. Epub 2008/04/29. Eng.

Janssen, L.J., Tazzeo, T., June 2002. Involvement of TP and EP3 receptors in vasoconstrictor responses to isoprostanes in pulmonary vasculature. J. Pharmacol. Exp. Ther. 301 (3), 1060–1066. PubMed PMID: 12023538. Epub 2002/05/23. Eng.

Janssen, L.J., Premji, M., Netherton, S., Catalli, A., Cox, G., Keshavjee, S., et al., November 2000. Excitatory and inhibitory actions of isoprostanes in human and canine airway smooth muscle. J. Pharmacol. Exp. Ther. 295 (2), 506–511. PubMed PMID: 11046082. Epub 2000/10/25. Eng.

Janssen, L.J., Catalli, A., Helli, P., January–February 2005. The pulmonary biology of isoprostanes. Antioxid. Redox Signal. 7 (1–2), 244–255. PubMed PMID: 15650412. Epub 2005/01/15. Eng.

Jeffcoat, R., 1979. The biosynthesis of unsaturated fatty acids and its control in mammalian liver. Essays Biochem. 15, 1–36. PubMed PMID: 42533. Epub 1979/01/01. Eng.

Jiang, C., Ting, A.T., Seed, B., January 1, 1998. PPAR-gamma agonists inhibit production of monocyte inflammatory cytokines. Nature 391 (6662), 82–86. PubMed PMID: 9422509.

Johnston, S.L., Freezer, N.J., Ritter, W., O'Toole, S., Howarth, P.H., March 1995. Prostaglandin D2-induced bronchoconstriction is mediated only in part by the thromboxane prostanoid receptor. Eur. Respir. J. 8 (3), 411–415. PubMed PMID: 7789486. Epub 1995/03/01. Eng.

Jones, G.L., Saroea, H.G., Watson, R.M., O'Byrne, P.M., June 1992. Effect of an inhaled thromboxane mimetic (U46619) on airway function in human subjects. Am. Rev. Respir. Dis. 145 (6), 1270–1274. PubMed PMID: 1595990.

Kadowitz, P.J., Chapnick, B.M., Feigen, L.P., Hyman, A.L., Nelson, P.K., Spannhake, E.W., September 1978. Pulmonary and systemic vasodilator effects of the newly discovered prostaglandin, PGI2. J. Appl. Physiol. 45 (3), 408–413. PubMed PMID: 359514.

Kang, K.H., Morrow, J.D., Roberts 2nd., L.J., Newman, J.H., Banerjee, M., January 1993. Airway and vascular effects of 8-epi-prostaglandin F2 alpha in isolated perfused rat lung. J. Appl. Physiol. 74 (1), 460–465. PubMed PMID: 8444728. Epub 1993/01/01. Eng.

Kawikova, I., Barnes, P.J., Takahashi, T., Tadjkarimi, S., Yacoub, M.H., Belvisi, M.G., February 1996. 8-Epi-PGF2 alpha, a novel noncyclooxygenase-derived prostaglandin, constricts airways in vitro. Am. J. Respir. Crit. Care Med. 153 (2), 590–596. PubMed PMID: 8564103.

Kitchen, E.A., Boot, J.R., Dawson, W., August 1978. Chemotactic activity of thromboxane B2, prostaglandins and their metabolites for polymorphonuclear leucocytes. Prostaglandins 16 (2), 239–244. PubMed PMID: 704931.

Kitsiouli, E., Nakos, G., Lekka, M.E., October 2009. Phospholipase A(2) subclasses in acute respiratory distress syndrome. BBA-Mol. Basis Dis. 1792 (10), 941–953. PubMed PMID: 19577642.

Klein, T.W., Newton, C., Larsen, K., Lu, L., Perkins, I., Nong, L., et al., October 2003. The cannabinoid system and immune modulation. J. Leukoc. Biol. 74 (4), 486–496. PubMed PMID: 12960289. Epub 2003/09/10. Eng.

Kopczynska, B., Sulejczak, D., Welniak-Kaminska, M., Gietka, A., Grieb, P., October 1, 2011. Anandamide enhances expression of heat shock proteins Hsp70 and Hsp25 in rat lungs. Eur. J. Pharmacol. 668 (1–2), 257–263. PubMed PMID: 21756900.

Kuebler, W.M., Yang, Y., Samapati, R., Uhlig, S., 2010. Vascular barrier regulation by PAF, ceramide, caveolae, and no – an intricate signaling network with discrepant effects in the pulmonary and systemic vasculature. Cell. Physiol. Biochem. 26 (1), 29–40. PubMed PMID: 20502002.

Lam, S., Chan, H., LeRiche, J.C., Chan-Yeung, M., Salari, H., April 1988. Release of leukotrienes in patients with bronchial asthma. J. Allergy Clin. Immunol. 81 (4), 711–717. PubMed PMID: 3356850. Epub 1988/04/01. Eng.

Lee, T.H., Horton, C.E., Kyan-Aung, U., Haskard, D., Crea, A.E., Spur, B.W., August 1989. Lipoxin A4 and lipoxin B4 inhibit chemotactic responses of human neutrophils stimulated by leukotriene B4 and N-formyl-L-methionyl-L-leucyl-L-phenylalanine. Clin. Sci. (Lond). 77 (2), 195–203. PubMed PMID: 2548801. Epub 1989/08/01. Eng.

Lee, T.H., Crea, A.E., Gant, V., Spur, B.W., Marron, B.E., Nicolaou, K.C., et al., June 1990. Identification of lipoxin A4 and its relationship to the sulfidopeptide leukotrienes C4, D4, and E4 in the bronchoalveolar lavage fluids obtained from patients with selected pulmonary diseases. Am. Rev. Respir. Dis. 141 (6), 1453–1458. PubMed PMID: 2161628. Epub 1990/06/01. Eng.

Lee, T.H., Lympany, P., Crea, A.E., Spur, B.W., November 14, 1991. Inhibition of leukotriene B4-induced neutrophil migration by lipoxin A4: structure-function relationships. Biochem. Biophys. Res. Commun. 180 (3), 1416–1421. PubMed PMID: 1659416. Epub 1991/11/14. Eng.

Leitinger, N., Huber, J., Rizza, C., Mechtcheriakova, D., Bochkov, V., Koshelnick, Y., et al., May 2001. The isoprostane 8-iso-PGF(2alpha) stimulates endothelial cells to bind monocytes: differences from thromboxane-mediated endothelial activation. FASEB J. 15 (7), 1254–1256. PubMed PMID: 11344105. Epub 2001/05/10. Eng.

Levy, B.D., Bonnans, C., Silverman, E.S., Palmer, L.J., Marigowda, G., Israel, E., et al., October 1, 2005. Diminished lipoxin biosynthesis in severe asthma. Am. J. Respir. Crit. Care Med. 172 (7), 824–830. PubMed PMID: 15961693. Pubmed Central PMCID: 2718403.

Levy, B.D., Kohli, P., Gotlinger, K., Haworth, O., Hong, S., Kazani, S., et al., January 1, 2007. Protectin D1 is generated in asthma and dampens airway inflammation and hyperresponsiveness. J. Immunol. 178 (1), 496–502. PubMed PMID: 17182589. Pubmed Central PMCID: 3005704.

Levy, B.D., Vachier, I., Serhan, C.N., September 2012. Resolution of inflammation in asthma. Clin. Chest Med. 33 (3), 559–570. PubMed PMID: 22929102. Pubmed Central PMCID: 3431599. Epub 2012/08/30. Eng.

Lynch, K.R., O'Neill, G.P., Liu, Q., Im, D.S., Sawyer, N., Metters, K.M., et al., June 24, 1999. Characterization of the human cysteinyl leukotriene CysLT1 receptor. Nature 399 (6738), 789–793. PubMed PMID: 10391245. Epub 1999/07/03. Eng.

Majai, G., Sarang, Z., Csomos, K., Zahuczky, G., Fesus, L., May 2007. PPARgamma-dependent regulation of human macrophages in phagocytosis of apoptotic cells. Eur. J. Immunol. 37 (5), 1343–1354. PubMed PMID: 17407194.

Maniatis, N.A., Kotanidou, A., Catravas, J.D., Orfanos, S.E., October–December 2008. Endothelial pathomechanisms in acute lung injury. Vasc. Pharmacol. 49 (4–6), 119–133. PubMed PMID: 18722553. Epub 2008/08/30. Eng.

Manku, M.S., Horrobin, D.F., Huang, Y.S., Morse, N., December 1983. Fatty acids in plasma and red cell membranes in normal humans. Lipids 18 (12), 906–908. PubMed PMID: 6664259. Epub 1983/12/01. Eng.

Mas, E., Croft, K.D., Zahra, P., Barden, A., Mori, T.A., October 2012. Resolvins D1, D2, and other mediators of self-limited resolution of inflammation in human blood following n-3 fatty acid supplementation. Clin. Chem. 58 (10), 1476–1484. PubMed PMID: 22912397. Epub 2012/08/23. Eng.

McMahon, B., Mitchell, S., Brady, H.R., Godson, C., August 2001. Lipoxins: revelations on resolution. Trends Pharmacol. Sci. 22 (8), 391–395. PubMed PMID: 11478982. Epub 2001/08/02. Eng.

Mechoulam, R., Ben-Shabat, S., Hanus, L., Ligumsky, M., Kaminski, N.E., Schatz, A.R., et al., June 29, 1995. Identification of an endogenous 2-monoglyceride, present in canine gut, that binds to cannabinoid receptors. Biochem. Pharmacol. 50 (1), 83–90. PubMed PMID: 7605349. Epub 1995/06/29. Eng.

Mehrotra, A.K., Henderson Jr., W.R., April 2009. The role of leukotrienes in airway remodeling. Curr. Mol. Med. 9 (3), 383–391. PubMed PMID: 19355919. Epub 2009/04/10. Eng.

Miller, A.M., Stella, N., January 2008. CB2 receptor-mediated migration of immune cells: it can go either way. Br. J. Pharmacol. 153 (2), 299–308. PubMed PMID: 17982478.

Mitchell, S., Thomas, G., Harvey, K., Cottell, D., Reville, K., Berlasconi, G., et al., Oct 2002. Lipoxins, aspirin-triggered epi-lipoxins, lipoxin stable analogues, and the resolution of inflammation: stimulation of macrophage phagocytosis of apoptotic neutrophils in vivo. J. Am. Soc. Nephrol. 13 (10), 2497–2507. PubMed PMID: 12239238. Epub 2002/09/20. Eng.

Morrow, J.D., Roberts, L.J., Dec 15, 2002. The isoprostanes: their role as an index of oxidant stress status in human pulmonary disease. Am. J.

Respir. Crit. Care Med. 166 (12 Pt 2), S25–S30. PubMed PMID: 12471085. Epub 2002/12/10. Eng.

Musiek, E.S., Brooks, J.D., Joo, M., Brunoldi, E., Porta, A., Zanoni, G., et al., July 18, 2008. Electrophilic cyclopentenone neuroprostanes are anti-inflammatory mediators formed from the peroxidation of the omega-3 polyunsaturated fatty acid docosahexaenoic acid. J. Biol. Chem. 283 (29), 19927–19935. PubMed PMID: 18490445. Pubmed Central PMCID: 2459280. Epub 2008/05/21. Eng.

Nosjean, O., Boutin, J.A., July 2002. Natural ligands of PPARgamma: are prostaglandin J(2) derivatives really playing the part? Cell. Signal 14 (7), 573–583. PubMed PMID: 11955950. Epub 2002/04/17. Eng.

O'Brien, J.J., Ray, D.M., Spinelli, S.L., Blumberg, N., Taubman, M.B., Francis, C.W., et al., January 2007. The platelet as a therapeutic target for treating vascular diseases and the role of eicosanoid and synthetic PPARgamma ligands. Prostagl. Other Lipid Mediat. 82 (1–4), 68–76. PubMed PMID: 17164134. Epub 2006/12/14. Eng.

Oh, S.F., Pillai, P.S., Recchiuti, A., Yang, R., Serhan, C.N., February 2011. Pro-resolving actions and stereoselective biosynthesis of 18S E-series resolvins in human leukocytes and murine inflammation. J. Clin. Invest. 121 (2), 569–581. PubMed PMID: 21206090. Pubmed Central PMCID: 3026718. Epub 2011/01/06. Eng.

Ohar, J.A., Waller, K.S., Pantano, J., Demello, D.E., Dahms, T.E., June 1994. Chronic platelet-activating-factor induces a decrease in pulmonary vascular compliance, hydroxyproline, and loss of vascular matrix. Am. J. Resp. Crit. Care Med. 149 (6), 1628–1634. PubMed PMID: 8004322.

Ohar, J.A., Waller, K.S., Pantano, J., Demello, D.E., Dahms, T.E., June 1994. Chronic platelet-activating factor induces a decrease in pulmonary vascular compliance, hydroxyproline, and loss of vascular matrix. Am. J. Respir. Crit. Care Med. 149 (6), 1628–1634. PubMed PMID: 8004322.

Oosterhoff, Y., Kauffman, H.F., Rutgers, B., Zijlstra, F.J., Koeter, G.H., Postma, D.S., August 1995. Inflammatory cell number and mediators in bronchoalveolar lavage fluid and peripheral blood in subjects with asthma with increased nocturnal airways narrowing. J. Allergy Clin. Immunol. 96 (2), 219–229. PubMed PMID: 7636059. Epub 1995/08/01. Eng.

Orning, L., Hammarstrom, S., September 10, 1980. Inhibition of leukotriene C and leukotriene D biosynthesis. J. Biol. Chem. 255 (17), 8023–8026. PubMed PMID: 6106018. Epub 1980/09/10. Eng.

Pacher, P., Mechoulam, R., April 2011. Is lipid signaling through cannabinoid 2 receptors part of a protective system? Prog. Lipid Res. 50 (2), 193–211. PubMed PMID: 21295074.

Papayianni, A., Serhan, C.N., Brady, H.R., March 15, 1996. Lipoxin A4 and B4 inhibit leukotriene-stimulated interactions of human neutrophils and endothelial cells. J. Immunol. 156 (6), 2264–2272. PubMed PMID: 8690917. Epub 1996/03/15. Eng.

Patel, H.J., Belvisi, M.G., Bishop-Bailey, D., Yacoub, M.H., Mitchell, J.A., March 1, 2003. Activation of peroxisome proliferator-activated receptors in human airway smooth muscle cells has a superior anti-inflammatory profile to corticosteroids: relevance for chronic obstructive pulmonary disease therapy. J. Immunol. 170 (5), 2663–2669. PubMed PMID: 12594295. Epub 2003/02/21. Eng.

Pelclova, D., Fenclova, Z., Vlckova, S., Lebedova, J., Syslova, K., Pecha, O., et al., 2012. Leukotrienes B4, C4, D4 and E4 in the exhaled breath condensate (EBC), blood and urine in patients with pneumoconiosis. Ind. Health 50 (4), 299–306. PubMed PMID: 22785421. Epub 2012/07/13. Eng.

Pepke-Zaba, J., Higenbottam, T.W., Dinh-Xuan, A.T., Stone, D., Wallwork, J., November 9, 1991. Inhaled nitric oxide as a cause of selective pulmonary vasodilatation in pulmonary hypertension. Lancet 338 (8776), 1173–1174. PubMed PMID: 1682593.

Peters-Golden, M., Henderson Jr., W.R., November 1, 2007. Leukot. N. Engl. J. Med. 357 (18), 1841–1854. PubMed PMID: 17978293. Epub 2007/11/06. Eng.

Planaguma, A., Kazani, S., Marigowda, G., Haworth, O., Mariani, T.J., Israel, E., et al., September 15, 2008. Airway lipoxin A4 generation and lipoxin A4 receptor expression are decreased in severe asthma. Am. J. Respir. Crit. Care Med. 178 (6), 574–582. PubMed PMID: 18583575. Pubmed Central PMCID: 2542432. Epub 2008/06/28. Eng.

Psychogios, N., Hau, D.D., Peng, J., Guo, A.C., Mandal, R., Bouatra, S., et al., 2011. The human serum metabolome. PLoS One 6 (2), e16957. PubMed PMID: 21359215. Pubmed Central PMCID: 3040193. Epub 2011/03/02. Eng.

Ramsay, C.F., Sullivan, P., Gizycki, M., Wang, D., Swern, A.S., Barnes, N.C., et al., July 2009. Montelukast and bronchial inflammation in asthma: a randomised, double-blind placebo-controlled trial. Respir. Med. 103 (7), 995–1003. PubMed PMID: 19249198. Epub 2009/03/03. Eng.

Reddy, R.C., Feb 2008. Immunomodulatory role of PPAR-gamma in alveolar macrophages. J. Investig. Med. 56 (2), 522–527. PubMed PMID: 18317435. Epub 2008/03/05. Eng.

Reddy, R.C., Keshamouni, V.G., Jaigirdar, S.H., Zeng, X., Leff, T., Thannickal, V.J., et al., March 2004. Deactivation of murine alveolar macrophages by peroxisome proliferator-activated receptor-gamma ligands. Am. J. Physiol. Lung Cell. Mol. Physiol. 286 (3), L613–L619. PubMed PMID: 14617516. Epub 2003/11/18. Eng.

Roberts, A.M., Kaufman, M.P., Baker, D.G., Brown, J.K., Coleridge, H.M., Coleridge, J.C., Aug 1981. Reflex tracheal contraction induced by stimulation of bronchial C-fibers in dogs. J. Appl. Physiol. 51 (2), 485–493. PubMed PMID: 7263455. Epub 1981/08/01. Eng.

Rolin, S., Masereel, B., Dogne, J.M., Mar 8, 2006. Prostanoids as pharmacological targets in COPD and asthma. Eur. J. Pharmacol. 533 (1–3), 89–100. PubMed PMID: 16458293. Epub 2006/02/07. Eng.

Romano, M., 2010. Lipoxin and aspirin-triggered lipoxins. Scientific World Journal 10, 1048–1064. PubMed PMID: 20526535. Epub 2010/06/09. Eng.

Rosengren, S., Olofsson, A.M., von Andrian, U.H., Lundgren-Akerlund, E., Arfors, K.E., October 1991. Leukotriene B4-induced neutrophil-mediated endothelial leakage in vitro and in vivo. J. Appl. Physiol. 71 (4), 1322–1330. PubMed PMID: 1661722.

Rossi, A., Kapahi, P., Natoli, G., Takahashi, T., Chen, Y., Karin, M., et al., January 6, 2000. Anti-inflammatory cyclopentenone prostaglandins are direct inhibitors of IkappaB kinase. Nature 403 (6765), 103–108. PubMed PMID: 10638762. Epub 2000/01/19. Eng.

Samapati, R., Yang, Y., Yin, J., Stoerger, C., Arenz, C., Dietrich, A., et al., January 15, 2012. Lung endothelial Ca^{2+} and permeability response to platelet-activating factor is mediated by acid sphingomyelinase and transient receptor potential classical 6. Am. J. Respir. Crit. Care Med. 185 (2), 160–170. PubMed PMID: 22246702.

Sampson, S.E., Sampson, A.P., Costello, J.F., June 1997. Effect of inhaled prostaglandin D2 in normal and atopic subjects, and of pretreatment with leukotriene D4. Thorax 52 (6), 513–518. PubMed PMID: 9227716. Pubmed Central PMCID: 1758586.

Sawyer, K., Mundandhara, S., Ghio, A.J., Madden, M.C., 2010. The effects of ambient particulate matter on human alveolar macrophage oxidative and inflammatory responses. J. Toxicol. Environ. Health A 73 (1), 41–57. PubMed PMID: 19953419. Epub 2009/12/03. Eng.

Scholz, H., Yndestad, A., Damas, J.K., Waehre, T., Tonstad, S., Aukrust, P., et al., October 1, 2003. 8-isoprostane increases expression of interleukin-8 in human macrophages through activation of mitogen-activated protein kinases. Cardiovasc Res. 59 (4), 945–954. PubMed PMID: 14553834. Epub 2003/10/14. Eng.

Schratl, P., Royer, J.F., Kostenis, E., Ulven, T., Sturm, E.M., Waldhoer, M., et al., October 1, 2007. The role of the prostaglandin D2 receptor, DP, in eosinophil trafficking. J. Immunol. 179 (7), 4792–4799. PubMed PMID: 17878378.

Schwab, J.M., Chiang, N., Arita, M., Serhan, C.N., June 14, 2007. Resolvin E1 and protectin D1 activate inflammation-resolution programmes. Nature 447 (7146), 869–874. PubMed PMID: 17568749. Pubmed Central PMCID: 2757086. Epub 2007/06/15. Eng.

Seki, H., Fukunaga, K., Arita, M., Arai, H., Nakanishi, H., Taguchi, R., et al., January 15, 2010. The anti-inflammatory and proresolving mediator resolvin E1 protects mice from bacterial pneumonia and acute lung injury. J. Immunol. 184 (2), 836–843. PubMed PMID: 20007539. Pubmed Central PMCID: 3000830. Epub 2009/12/17. Eng.

Selg, E., Andersson, M., Lastbom, L., Ryrfeldt, A., Dahlen, S.E., April 2009. Two different mechanisms for modulation of bronchoconstriction in guinea-pigs by cyclooxygenase metabolites. Prostagl. Other Lipid Mediat 88 (3–4), 101–110. PubMed PMID: 19103301.

Serhan, C.N., Savill, J., December 2005. Resolution of inflammation: the beginning programs the end. Nat. Immunol. 6 (12), 1191–1197. PubMed PMID: 16369558. Epub 2005/12/22. Eng.

Serhan, C.N., Hamberg, M., Samuelsson, B., September 1984. Lipoxins: novel series of biologically active compounds formed from arachidonic acid in human leukocytes. Proc. Natl. Acad. Sci. USA 81 (17), 5335–5339. PubMed PMID: 6089195. Pubmed Central PMCID: 391698. Epub 1984/09/01. Eng.

Serhan, C.N., Clish, C.B., Brannon, J., Colgan, S.P., Chiang, N., Gronert, K., October 16, 2000. Novel functional sets of lipid-derived mediators with antiinflammatory actions generated from omega-3 fatty acids via cyclooxygenase 2-nonsteroidal antiinflammatory drugs and transcellular processing. J. Exp. Med. 192 (8), 1197–1204. PubMed PMID: 11034610. Pubmed Central PMCID: 2195872. Epub 2000/10/18. Eng.

Serhan, C.N., Gotlinger, K., Hong, S., Lu, Y., Siegelman, J., Baer, T., et al., Feb 1, 2006. Anti-inflammatory actions of neuroprotectin D1/protectin D1 and its natural stereoisomers: assignments of dihydroxy-containing docosatrienes. J. Immunol. 176 (3), 1848–1859. PubMed PMID: 16424216. Epub 2006/01/21. Eng.

Serhan, C.N., Brain, S.D., Buckley, C.D., Gilroy, D.W., Haslett, C., O'Neill, L.A., et al., February 2007. Resolution of inflammation: state of the art, definitions and terms. FASEB J. 21 (2), 325–332. PubMed PMID: 17267386. Pubmed Central PMCID: 3119634. Epub 2007/02/03. Eng.

Serhan, C.N., Chiang, N., Van Dyke, T.E., May 2008. Resolving inflammation: dual anti-inflammatory and pro-resolution lipid mediators. Nat. Rev. Immunol. 8 (5), 349–361. PubMed PMID: 18437155. Pubmed Central PMCID: 2744593. Epub 2008/04/26. Eng.

Serhan, C.N., Yang, R., Martinod, K., Kasuga, K., Pillai, P.S., Porter, T.F., et al., January 16, 2009. Maresins: novel macrophage mediators with potent antiinflammatory and proresolving actions. J. Exp. Med. 206 (1), 15–23. PubMed PMID: 19103881. Pubmed Central PMCID: 2626672. Epub 2008/12/24. Eng.

Serhan, C.N., Krishnamoorthy, S., Recchiuti, A., Chiang, N., 2011. Novel anti-inflammatory–pro-resolving mediators and their receptors. Curr. Top. Med. Chem. 11 (6), 629–647. PubMed PMID: 21261595. Pubmed Central PMCID: 3094721. Epub 2011/01/26. Eng.

Shay, A.H., Choi, R., Whittaker, K., Salehi, K., Kitchen, C.M., Tashkin, D.P., et al., February 15, 2003. Impairment of antimicrobial activity and nitric oxide production in alveolar macrophages from smokers of marijuana and cocaine. J. Infect. Dis. 187 (4), 700–704. PubMed PMID: 12599091.

Sheller, J.R., Mitchell, D., Meyrick, B., Oates, J., Breyer, R., June 2000. EP(2) receptor mediates bronchodilation by PGE(2) in mice. J. Appl. Physiol. 88 (6), 2214–2218. PubMed PMID: 10846038.

Simon, D.M., Arikan, M.C., Srisuma, S., Bhattacharya, S., Tsai, L.W., Ingenito, E.P., et al., July 2006. Epithelial cell PPAR[gamma] contributes to normal lung maturation. FASEB J. 20 (9), 1507–1509. PubMed PMID: 16720732. Epub 2006/05/25. Eng.

Singh, R.K., Gupta, S., Dastidar, S., Ray, A., 2010. Cysteinyl leukotrienes and their receptors: molecular and functional characteristics. Pharmacology 85 (6), 336–349. PubMed PMID: 20516735. Epub 2010/06/03. Eng.

Smith, L.J., Greenberger, P.A., Patterson, R., Krell, R.D., Bernstein, P.R., March 1985. The effect of inhaled leukotriene D4 in humans. Am. Rev. Respir. Dis. 131 (3), 368–372. PubMed PMID: 3883862.

Song, W.L., Paschos, G., Fries, S., Reilly, M.P., Yu, Y., Rokach, J., et al., August 28, 2009. Novel eicosapentaenoic acid-derived F3-isoprostanes as biomarkers of lipid peroxidation. J. Biol. Chem. 284 (35), 23636–23643. PubMed PMID: 19520854. Pubmed Central PMCID: 2749138. Epub 2009/06/13. Eng.

Spicuzza, L., Giembycz, M.A., Barnes, P.J., Belvisi, M.G., March 1998. Prostaglandin E2 suppression of acetylcholine release from parasympathetic nerves innervating guinea-pig trachea by interacting with prostanoid receptors of the EP3-subtype. Br. J. Pharmacol. 123 (6), 1246–1252. PubMed PMID: 9559911. Pubmed Central PMCID: 1565272. Epub 1998/04/29. Eng.

Spicuzza, L., Barnes, P.J., Di Maria, G.U., Belvisi, M.G., March 30, 2001. Effect of 8-iso-prostaglandin F(2 alpha) on acetylcholine release from parasympathetic nerves in guinea pig airways. Eur. J. Pharmacol. 416 (3), 231–234. PubMed PMID: 11290373. Epub 2001/04/06. Eng.

Spite, M., Norling, L.V., Summers, L., Yang, R., Cooper, D., Petasis, N.A., et al., October 29, 2009. Resolvin D2 is a potent regulator of leukocytes and controls microbial sepsis. Nature 461 (7268), 1287–1291. PubMed PMID: 19865173. Pubmed Central PMCID: 2779525. Epub 2009/10/30. Eng.

Standiford, T.J., Keshamouni, V.G., Reddy, R.C., 2005. Peroxisome proliferator-activated receptor-{gamma} as a regulator of lung inflammation and repair. Proc. Am. Thorac. Soc. 2 (3), 226–231. PubMed PMID: 16222042.

Sun, Y.P., Oh, S.F., Uddin, J., Yang, R., Gotlinger, K., Campbell, E., et al., March 30, 2007. Resolvin D1 and its aspirin-triggered 17R epimer. Stereochemical assignments, anti-inflammatory properties, and enzymatic inactivation. J. Biol. Chem. 282 (13), 9323–9334. PubMed PMID: 17244615. Epub 2007/01/25. Eng.

Sundvold, H., Brzozowska, A., Lien, S., October 29, 1997. Characterisation of bovine peroxisome proliferator-activated receptors gamma 1 and gamma 2: genetic mapping and differential expression of the two isoforms. Biochem. Biophys. Res. Commun. 239 (3), 857–861. PubMed PMID: 9367859. Epub 1997/11/21. Eng.

Takano, T., Clish, C.B., Gronert, K., Petasis, N., Serhan, C.N., February 15, 1998. Neutrophil-mediated changes in vascular permeability are inhibited by topical application of aspirin-triggered 15-epi-lipoxin A4 and novel lipoxin B4 stable analogues. J. Clin. Invest. 101 (4), 819–826. PubMed PMID: 9466977. Pubmed Central PMCID: 508630.

Tamaoki, J., Chiyotani, A., Takeyama, K., Yamauchi, F., Tagaya, E., Konno, K., April 1993. Relaxation and inhibition of contractile response to electrical field stimulation by Beraprost sodium in canine airway smooth muscle. Prostaglandins 45 (4), 363–373. PubMed PMID: 8388118. Epub 1993/04/01. Eng.

Thien, F.C., Walters, E.H., May 1995. Eicosanoids and asthma: an update. Prostagl. Leukot. Essent. Fat. Acids 52 (5), 271–288. PubMed PMID: 7630914. Epub 1995/05/01. Eng.

Thureson-Klein, A., Hedqvist, P., Lindbom, L., 1986. Leukocyte diapedesis and plasma extravasation after leukotriene B4: lack of structural injury to the endothelium. Tissue Cell 18 (1), 1–12. PubMed PMID: 2421456.

Tilley, S.L., Hartney, J.M., Erikson, C.J., Jania, C., Nguyen, M., Stock, J., et al., April 2003. Receptors and pathways mediating the effects of prostaglandin E2 on airway tone. Am. J. Physiol. Lung Cell. Mol. Physiol. 284 (4), L599–L606. PubMed PMID: 12618422.

Tjonahen, E., Oh, S.F., Siegelman, J., Elangovan, S., Percarpio, K.B., Hong, S., et al., November 2006. Resolvin E2: identification and anti-inflammatory actions: pivotal role of human 5-lipoxygenase in resolvin E series biosynthesis. Chem. Biol. 13 (11), 1193–1202. PubMed PMID: 17114001. Epub 2006/11/23. Eng.

Toyoshima, K., Narahara, H., Furukawa, M., Frenkel, R.A., Johnston, J.M., June 1995. Platelet-activating-factor – role in fetal lung development and relationship to normal and premature labor. Clin. Perinatol. 22 (2), 263–280. PubMed PMID: 7671539.

Tudor, C., Feige, J.N., Pingali, H., Lohray, V.B., Wahli, W., Desvergne, B., et al., February 16, 2007. Association with coregulators is the major determinant governing peroxisome proliferator-activated receptor mobility in living cells. J. Biol. Chem. 282 (7), 4417–4426. PubMed PMID: 17164241. Epub 2006/12/14. Eng.

Ueki, S., Matsuwaki, Y., Kayaba, H., Oyamada, H., Kanda, A., Usami, A., et al., June 2004. Peroxisome proliferator-activated receptor gamma regulates eosinophil functions: a new therapeutic target for allergic airway inflammation. Int. Arch. Allergy Immunol. 134 (Suppl. 1), 30–36. PubMed PMID: 15166481. Epub 2004/05/29. Eng.

Uhlig, S., Wollin, L., Wendel, A., July 1994. Contributions of thromboxane and leukotrienes to PAF-induced impairment of lung-function in the rat. J. Appl. Physiol. 77 (1), 262–269. PubMed PMID: 7961243.

Uhlig, S., Goggel, R., Engel, S., 2005. Mechanisms of platelet-activating factor (PAF)-mediated responses in the lung. Pharmacol. Rep. 57, 206–221. PubMed PMID: 16415501.

van Blitterswijk, W.J., van der Luit, A.H., Veldman, R.J., Verheij, M., Borst, J., January 15, 2003. Ceramide: second messenger or modulator of membrane structure and dynamics? Biochem. J. 369, 199–211. PubMed PMID: 12408751.

Vane, J.R., June 23, 1971. Inhibition of prostaglandin synthesis as a mechanism of action for aspirin-like drugs. Nat. New. Biol. 231 (25), 232–235. PubMed PMID: 5284360. Epub 1971/06/23. Eng.

Vignola, A.M., Mirabella, F., Costanzo, G., Di Giorgi, R., Gjomarkaj, M., Bellia, V., et al., March 2003. Airway remodeling in asthma. Chest 123 (Suppl. 3), 417S–422S. PubMed PMID: 12629009. Epub 2003/03/12. Eng.

Villacorta, L., Schopfer, F.J., Zhang, J., Freeman, B.A., Chen, Y.E., February 2009. PPARgamma and its ligands: therapeutic implications in cardiovascular disease. Clin. Sci. (Lond) 116 (3), 205–218. PubMed PMID: 19118492. Pubmed Central PMCID: 2677102. Epub 2009/01/03. Eng.

Voynow, J.A., Kummarapurugu, A., Nov 2011. Isoprostanes and asthma. Biochim. Biophys. Acta 1810 (11), 1091–1095. PubMed PMID: 21596100. Pubmed Central PMCID: 3192308. Epub 2011/05/21. Eng.

Walker, J., Dichter, E., Lacorte, G., Kerner, D., Spur, B., Rodriguez, A., et al., October 2011. Lipoxin a4 increases survival by decreasing systemic inflammation and bacterial load in sepsis. Shock 36 (4), 410–416. PubMed PMID: 21701419. Epub 2011/06/28. Eng.

Wang, A.C., Dai, X., Luu, B., Conrad, D.J., June 2001. Peroxisome proliferator-activated receptor-gamma regulates airway epithelial cell activation. Am. J. Respir. Cell. Mol. Biol. 24 (6), 688–693. PubMed PMID: 11415933. Epub 2001/06/21. Eng.

Wasserman, M.A., Ducharme, D.W., Wendling, M.G., Griffin, R.L., Degraaf, G.L., August 22, 1980. Bronchodilator effects of prostacyclin (PGI2) in dogs and guinea pigs. Eur. J. Pharmacol. 66 (1), 53–63. PubMed PMID: 6997057.

Watts, K., Chavasse, R.J., 2012. Leukotriene receptor antagonists in addition to usual care for acute asthma in adults and children. Cochrane Database Syst. Rev. 5, CD006100. PubMed PMID: 22592708. Epub 2012/05/18. Eng.

Weinberger, B., Anwar, M., Henien, S., Sosnovsky, A., Hiatt, M., Jochnowitz, N., et al., 2004. Association of lipid peroxidation with antenatal betamethasone and oxygen radial disorders in preterm infants. Biol. Neonate. 85 (2), 121–127. PubMed PMID: 1463115. Epub 2003/11/25. Eng.

Widdicombe, J.H., Ueki, I.F., Emery, D., Margolskee, D., Yergey, J., Nadel, J.A., December 1989. Release of cyclooxygenase products from primary cultures of tracheal epithelia of dog and human. Am. J. Physiol. 257 (6), L361–L365. PubMed PMID: 2514604.

Woerly, G., Honda, K., Loyens, M., Papin, J.P., Auwerx, J., Staels, B., et al., August 4, 2003. Peroxisome proliferator-activated receptors alpha and gamma down-regulate allergic inflammation and eosinophil activation. J. Exp. Med. 198 (3), 411–421. PubMed PMID: 12900517. Pubmed Central PMCID: 2194090. Epub 2003/08/06. Eng.

Wu, S.H., Yin, P.L., Zhang, Y.M., Tao, H.X., April 2010. Reversed changes of lipoxin A4 and leukotrienes in children with asthma in different severity degree. Pediatr. Pulmonol. 45 (4), 333–340. PubMed PMID: 20232472.

Xu, Z.Z., Zhang, L., Liu, T., Park, J.Y., Berta, T., Yang, R., et al., May 2010. Resolvins RvE1 and RvD1 attenuate inflammatory pain via central and peripheral actions. Nat. Med. 16 (5), 592–597, 1p following 7. PubMed PMID: 20383154. Pubmed Central PMCID: 2866054. Epub 2010/04/13. Eng.

Yang, Y., Yin, J., Baumgartner, W., Samapati, R., Solymosi, E.A., Reppien, E., et al., August 2010. Platelet-activating factor reduces endothelial nitric oxide production: role of acid sphingomyelinase. Eur. Respir. J. 36 (2), 417–427. PubMed PMID: 19926744.

Yu, Y., Lucitt, M.B., Stubbe, J., Cheng, Y., Friis, U.G., Hansen, P.B., et al., May 12, 2009. Prostaglandin F2alpha elevates blood pressure and promotes atherosclerosis. Proc. Natl. Acad. Sci. USA 106 (19),

7985–7990. PubMed PMID: 19416858. Pubmed Central PMCID: 2673134.

Yuan, M., Kiertscher, S.M., Cheng, Q., Zoumalan, R., Tashkin, D.P., Roth, M.D., December 2002. Delta 9-Tetrahydrocannabinol regulates Th1/Th2 cytokine balance in activated human T cells. J. Neuroimmunol. 133 (1–2), 124–131. PubMed PMID: 12446015.

Zahler, S., Becker, B.F., July 1999. Indirect enhancement of neutrophil activity and adhesion to cultured human umbilical vein endothelial cells by isoprostanes (iPF2alpha-III and iPE2-III). Prostagl. Other Lipid Mediat. 57 (5–6), 319–331. PubMed PMID: 10480486. Epub 1999/09/10. Eng.

Zarbock, A., Ley, K., 2009. The role of platelets in acute lung injury (ALI). Front. Biosci. 14, 150–158. PubMed PMID: 19273059. Pubmed Central PMCID: 2745111. Epub 2009/03/11. Eng.

Zhu, J., Bandi, V., Qiu, S., Figueroa, D., Evans, J.F., Barnes, N., et al., February 23, 2012. CysLT1 receptor expression associated with bronchial inflammation in severe exacerbations of COPD. Chest 142 (2), 347–357. PubMed PMID: 22362876. Epub 2012/03/01. Eng.

Zimmerman, G.A., McIntyre, T.M., June 2004. PAF, ceramide and pulmonary edema: alveolar flooding and a flood of questions. Trends Mol. Med. 10 (6), 245–248. PubMed PMID: 15177186.

Zimmerman, G.A., McIntyre, T.M., Prescott, S.M., Stafforini, D.M., May 2002. The platelet-activating factor signaling system and its regulators in syndromes of inflammation and thrombosis. Crit. Care Med. 30 (5), S294–S301. PubMed PMID: 12004251.

Chapter 22

Lung Lipid Composition and Surfactant Biology

Aron B. Fisher

Chapter Outline

1. INTRODUCTION

This chapter was originally written by Dr. Seamus A. Rooney, now retired but then in the Department of Pediatrics, Yale University School of Medicine; it has been revised and updated by the present author.

The lung is composed of a variety of cell types, all of which (like other cells) synthesize their cellular membranes that contain a high percentage of phospholipids. It is generally assumed that the pathways for synthesis of lung cell membranes is similar to that occurring in other organs. On the other hand, the lung surfactant is a specialized product that is synthesized and secreted by lung epithelium. Thus, the phospholipid composition, biosynthesis, and secretion of surfactant by the normal lung is the topic of this chapter. Related reviews that supplement the material herein include regulation of surfactant secretion (Chander and Fisher, 1990; Dietl and Haller, 2005), the influence of hormones on fetal lung maturation and pulmonary surfactant production (Possmayer, 1988b; Post and van Golde, 1988; Rooney, 1983, 1985; Van Golde et al., 1988), the physical chemistry of pulmonary surfactant (Bangham, 1987; Notter and Finkelstein, 1984; Possmayer, 1988b), the clearance and reutilization of surfactant (Fisher and Chander, 1985a; Fisher, 1998a; Fisher et al., 1990; Wright and Clements, 1987), the use of surfactant in the treatment of respiratory distress syndrome (Jobe and Ikegami, 1987; Robertson and Lachmann, 1988), the influence of toxic agents on surfactant and lung phospholipid biochemistry (Akino and Ohno, 1981; Haagsman and van Golde, 1985; Rooney, 1985), and host defense properties of surfactant (Claypool and Fisher, 1983; Wright, 2004).

Lung surfactant is more easily described functionally than chemically. It is material that lines the alveolar surface of the lung and promotes alveolar stability by lowering surface tension. Chemically, surfactant consists largely of phospholipids and proteins. The major lipid is phosphatidylcholine, a large portion of which is disaturated (mainly dipalmitoyl). Surfactant contains three specific proteins that are important for surfactant function (SP-A, SP-B, SP-C), although many other proteins, including one called SP-D, co-isolate with it. More than 500 proteins have been identified in the lung surfactant storage organelles (Possmayer, 1988a; Ridsdale, et al., 2011; Weaver, 1988; Wright and Clements, 1987). Surfactant exists in the alveolar lumen in a number of different physical forms. These include newly secreted material; a lattice-like structure called tubular myelin; a monolayer at the air—liquid interface, which is the physiologically active form of surfactant; and micellar lipid forms possibly representing "spent" material (Williams, 1977; Wright and Clements, 1987). Different physical forms of surfactant have been isolated from lung lavage by differential centrifugation (Magoon et al., 1983; Stevens et al., 1987). The phospholipid compositions of these different

Comparative Biology of the Normal Lung. http://dx.doi.org/10.1016/B978-0-12-404577-4.00022-9

physical forms were generally similar, although protein content varied (Magoon et al., 1983).

Surfactant is synthesized in the endoplasmic reticulum of type II pneumocyte (one of the major types of epithelial cells lining the alveolar space) and stored in morphologically distinct organelles, called the lamellar inclusion bodies, prior to secretion into the alveolar lumen by the process of exocytosis. Lamellar bodies have been isolated usually from whole lung (Gil and Reiss, 1973; Chander et al., 1986), but also from isolated type II cells (Chander et al., 1983; Crecelius and Longmore, 1983). Purity has been assessed by absence of other subcellular organelles as determined by electron microscopy and marker enzymes. Identification of lamellar bodies has been greatly facilitated by the demonstration that the ATP-binding cassette transporter protein A3 (ABCA3) and lysosomal membrane-associated protein 3 (LAMP3) are greatly enriched in these organelles (Mulugeta et al., 2002; Wade et al., 2006; Zen et al., 1998). Methods for the isolation of type II cells were first developed by Kikkawa and Yoneda (1974). These methods and subsequent modifications have been utilized to isolate type II cells from the adult rat (Dobbs et al., 1980; Dobbs et al., 1986a; Fisher and Furia, 1977; Fisher et al., 1980), mouse (Bortnick et al., 2003), rabbit (Finkelstein and Shapiro, 1982; Kikkawa et al., 1975), human (Robinson et al., 1984), guinea pig (Sikpi et al., 1986), hamster (Pfleger, 1977; Smith et al., 1987), and bovine (Augustin-Voss et al., 1989) lung, as well as from fetal rat (Batenburg et al., 1988; Kresch et al., 1987), rabbit (Aeberhard et al., 1985), and human (Ballard et al., 1986) lung.

This chapter will consider the biochemistry and cell biology of lung surfactant from the standpoint of surfactant composition, synthesis of lipid components, packing in the storage/secreting organelle (lamellar bodies), phospholipid secretion, and phospholipid reuptake and intracellular processing. These studies have relied on the preparations described above (surfactant isolation from lung lavage, isolated lamellar bodies, type II cells in primary culture), as well as the isolated perfused lung preparation (Bassett et al., 1974; Brown and Longmore, 1981; Fisher, 1985; Fisher et al., 1985a; Nicholas and Barr, 1981).

The values for surfactant composition reported in the literature have been obtained from a variety of preparations. Surfactant is easily obtained from the alveolar lumen by lavaging the lungs with saline or from the amniotic fluid because the fetus secretes lung fluid into the amniotic cavity. Cellular material is removed by a brief low-speed centrifugation resulting in a preparation that is enriched in surfactant. Such material has been taken to be representative of surfactant in many studies. In others, lung lavage surfactant has been further purified and fractionated by density gradient centrifugation. Surfactant has also been purified from whole lung homogenate by density gradient methods. In this article, no attempt is made to distinguish between these various preparations in presenting the properties of lung surfactant.

2. LIPID COMPOSITION

2.1 Lipid Classification and Nomenclature

Lipids can be defined as biological chemicals that are insoluble in water but extractable into organic solvents. They form a heterogeneous group of compounds ranging from simple fatty acids and cholesterol to esters of glycerol and other alcohols, as well as more complex compounds containing an alcohol, fatty acids, phosphate, sugars, amino alcohols or acids, and other moieties. Lipids can be classified in a number of ways, depending on the purpose of the exercise. For the purpose of this discussion, lung lipids are divided into three groups: neutral lipids, phospholipids, and glycolipids. The major neutral lipids are free fatty acids, glycerides (monoacylglycerols, diacylglycerols, and triacylglycerols), and cholesterol, including cholesterol esters; the phospholipids are glycerophosphatides and sphingosylphosphatides; and glycolipids are carbohydrate-containing lipids devoid of phosphorus.

Fatty acids commonly encountered in lung lipids are listed in Table 1. Phospholipids are listed in Table 2. In most of these phospholipids, the fatty acid moiety is linked to the alcohol (glycerol or sphingosine) via an ester (as in glycerophosphatides) or an amide (as in sphingomyelin) bond. Phospholipids containing alkyl or alk-1-enyl groups attached to the glycerol moiety via an ether linkage have also been reported as minor components of lung glycerophospholipids (Body, 1971; Kumar et al., 1985; Okano et al., 1978).

TABLE 1 Some Common Fatty Acids of Lung Lipids

Shorthand Notations[a]	Common Name	Systemic Name
12:0	Lauric acid	Dodecanoic acid
14:0	Myristic acid	Tetradecanoic acid
16:0	Palmitic acid	Hexadecanoic acid
16:1	Pamitoleic acid	Hexadecanoic acid
18:0	Stearic acid	Octadecanoic acid
18:1	Oleic acid	Octadecanoic acid
18:2	Linoleic acid	Octadecadienoic acid
18:3	Linolenic acid	Octadecadienoic acid
20:4	Arachidonic acid	Eicosatetraenoic acid

[a] *The number preceding the colon is the number of carbon atoms in the molecule, while that after the colon designates the number of double bonds. Although it should be indicated to fully describe the molecule, the position of the double bonds are not indicated in this simplified system.*

TABLE 2 Phospholipids Commonly Encountered in the Lung

Abbreviation	Name[a]	Chemical Structure
Glycerophosphatides		1,2-diacyl-*sn*-glycerol-3-phospho-X
PA	Phosphatide acid	X = H
PC	Phosphatidylcholine (lecithin)	X = choline
PE	Phosphatidylethanolamine	X = ethanolamine
PG	Phosphatidylglycerol	X = glycerol
PI	Phosphatidylinositol	X = inositol
PS	Phosphatidylserine	X = serine
CL	Cardiolipin (diphosphatidylglycerol)	X = phosphatidylglycerol
LBPA	Lysobisphosphatidic acid	X = 1-acylglycerol
Lysoglycerophosphatides		1-acyl-sn-glycerol-3-phospho-X or 2-acyl-sn-glycerol-3-phospho-X
LPC	Lysophosphatidylcholine (lysolecithin)	X = choline
Sphingosyl phosphatides		
Sph	Sphingomyelin	*N*-acylsphingosine-l-phosphorylcholine

[a]*See Strickland (1973) for a detailed discussion of phospholipid nomenclature.*

2.2 Methods of Lipid Extraction and Separation

Procedures for the extraction, separation, and analysis of lipids are detailed in a number of excellent methodology books (Kates, 1986; Lowenstein, 1969), so only a brief overview is given here. Lipids are usually extracted from biological tissues or liquids with a mixture of chloroform and methanol. The two most widely used extraction methods are those of Folch et al. (1957) and Bligh and Dyer (1959). Because chloroform is highly toxic and must be used in a fume hood, methods of lipid extraction employing less toxic solvents have been devised (Radin, 1981) but have not been widely used so far. The method of Folch et al. (1957) is more suitable when relatively large amounts of lung tissue are to be extracted, whereas the method of Bligh and Dyer (1959) is more suitable when lipids are to be extracted from homogenates of small amounts of tissue or from lung liquid or amniotic fluid. In both methods, the lipids are finally partitioned in a biphasic mixture of chloroform, methanol, and water so that the lipids remain in the lower organic solvent phase and water-soluble contaminants remain in the upper aqueous phase. It should be noted, however, that highly polar lipids, such as gangliosides and lysophosphatidic acid, remain in the upper aqueous phase. When it is necessary to remove all water-soluble contaminants, such as in experiments involving radiolabeled precursors, this can be largely achieved by repeated washing of the lower organic phase with additional aliquots of upper-phase solution (Folch et al., 1957). The purified crude lipid mixture is recovered after evaporation of the solvent either on a rotary evaporator or under a flow of nitrogen.

Total lipids can be separated into neutral lipids, glycolipids, and phospholipids by chromatography on columns of silicic acid (Rooney et al., 1974). Each lipid group can then be separated into individual classes, such as triacylglycerols or phosphatidylcholines. This is usually achieved by thin-layer chromatography, although high-pressure liquid chromatography methods are also available (Shukla, 1988). Each such lipid class consists of a mixture of individual molecular species, differing only in fatty acid composition. Separation of each individual molecular species, especially of complex lipids, is usually not necessary, although this can be accomplished readily by high-pressure liquid chromatography (Caesar et al., 1988; Schlame et al., 1988). Lipids are sometimes separated into groups of molecular species with a common degree of fatty acid saturation. In the case of the lung, this is particularly valuable for evaluating disaturated phosphatidylcholines. A number of methods for separation of the disaturated from unsaturated species of lung phosphotidylcholine have been used. The most widely used method is to oxidize the unsaturated phosphatidylcholines with osmium tetroxide and recover the unchanged disaturated species by column or thin-layer chromatography (Mason et al., 1976). This method has been adapted so that disaturated and unsaturated species of phosphatidylcholine, as well as all other lung phospholipids, can be resolved on a single thin-layer plate (Gilfillan et al., 1983). Other methods used in studies of lung phospholipid fatty acid saturation have included cryochromatography (Henderson and Clayton, 1976) or separation of disaturated phosphatidylcholine after mercuric acetate adduction (Mason et al., 1972) or permanganate-periodate oxidation (Shimojo et al., 1974) of the unsaturated species.

Fatty acid composition is determined by gas liquid chromatographic analysis of fatty acid methyl esters after their release from the lipids by methanolysis or by saponification followed by methylation. Considerable progress has been made in analyzing lipid composition by mass spectroscopy and the advent of electrospray and laser desorption ionization (ESI) and matrix-assisted laser desorption ionization (MALDI) has resulted in the ability to analyze essentially all biological lipids (Bernhard et al., 2007; Murphy and Gaskell, 2011; Postle and Hunt, 2009). The use of stable isotype-labeled derivative tagging and ozone reactivity has enabled the determination of specific phospholipids and double-bond positions in fatty acyl groups in complex lipid mixtures (Berry et al., 2011; Murphy and Gaskell, 2011).

2.3 Lung Lipid Composition

The lipid compositions of lung tissue and lung surfactant from a number of animal species are shown in Table 3. Neutral lipids account for as much as 60% of total lipids in lung tissue, whereas phospholipids account for at least 80% of the total in surfactant. Glycolipids account for ~5% of the total in lung tissue and ~10% in surfactant. A number of glycolipids have been identified in lung tissue and lung lavage from animals (Banerjee, 1982; Sahu et al., 1976; Slomiany et al., 1979a,b) and humans (Hanqing et al., 1986; Narasimhan and Murray, 1979; Rauvala and Hallman, 1984; Yoda et al., 1982), including patients with adult respiratory distress syndrome (Rauvala and Hallman, 1984) and alveolar proteinosis (Sahu et al., 1976). These include neutral glycosphingolipids and gangliosides (Banerjee, 1982; Hanqing et al., 1986; Narasimhan and Murray, 1979; Rauvala and Hallman, 1984; Sahu et al., 1976; Yoda et al., 1982) as well as neutral (Slomiany et al., 1979b) and sulfated (Slomiany et al., 1979a) glyceroglucolipids.

The neutral lipid compositions of lung tissue and surfactant-enriched fractions are shown in Tables 4 and 5, respectively. Although there is considerable variation in the reported values, cholesterol and triacylglycerols are generally the major neutral lipids in both surfactant and whole lung. The phospholipid compositions of lung tissue and lung surfactant from several species are shown in Tables 6 and 7, respectively. Surfactant contains no unique phospholipids. All of the phospholipids found in surfactant are also found in whole lung and other organs. There are distinct compositional differences, however, between the phospholipids in lung tissue and those in surfactant. Phosphatidylcholine accounts for approximately 50% of the total phospholipids in lung tissue, but for as much as 70–90% in surfactant. Phosphatidylglycerol is a minor phospholipid in lung tissue, but it accounts for up to 13% of the total in surfactant, where it is the second most abundant phospholipid. Phosphatidylethanolamine accounts for approximately 20% of the total in the tissue, but for only 2–6% in surfactant. Sphingomyelin and phosphatidylserine are also relatively much more abundant in lung tissue than in surfactant. Disaturated phosphatidylcholine accounts for less that 40% of total phosphatidylcholine in lung tissue but for at least 60% in surfactant. Thus, surfactant lipid composition clearly differs from that of lung tissue in containing relatively large percentages of disaturated phosphatidylcholine and phosphatidylglycerol, and relatively small percentages of phosphatidylethanolamine, sphingomyelin, and phosphatidylserine.

As lamellar bodies are the organelle for surfactant storage, their phospholipid composition (Table 8) is generally similar to that of surfactant and different from that of lung tissue. The phospholipid composition of isolated type II cells (Table 9) is consistent with this cell being the site of surfactant production; it contains relatively large percentages of phosphatidylcholine and phosphatidylglycerol and relatively small percentages of sphingomyelin and phosphatidylserine. The percentages of phosphatidyl ethanolamine and disaturated phosphatidylcholine in type II cells are intermediate between those in whole lung and those in surfactant. The composition of the phospholipids secreted by the type II cell (Table 10) is clearly characteristic of surfactant.

The fatty acid compositions of phosphatidylcholine from lung tissue and surfactant are shown in Tables 11 and 12, respectively. Compared to that of lung tissue, surfactant phosphatidylcholine has more palmitic acid and less of both 18-carbon fatty acids and arachidonic acid. As shown in Table 13, lung tissue phosphatidylethanolamine contains much less palmitic acid but much more 18-carbon and arachidonic acids than tissue phosphatidylcholine (Table 11). Tissue phosphatidylglycerol (Table 14) is also rich in 18-carbon fatty acids, especially stearic and oleic acids, but it contains more palmitic and less arachidonic acid as compared to phosphatidylethanolamine. The fatty acid composition of phosphatidylglycerols from lung tissue (Table 14) and surfactant (Table 15) are not strikingly different.

Phospholipid molecular species in lung tissue and surfactant are shown in Tables 16 and 17, respectively. The disaturated is the most abundant species of phosphatidylcholine from both tissue and surfactant. As also shown in Tables 6 and 7, it accounts for 53–85% of total phosphatidylcholine in surfactant and 30–40% in the tissue. In contrast, the most abundant species of phosphatidylethanolamine from lung

TABLE 3 Lipid Composition of Lung Tissue and Lung Surfactant (% of Total)

	Bovine (Yu et al., 1983)	Dog (Pfleger and Thomas, 1971)	Guinea Pig (Khan et al., 1985)	Hamster (Henderson and Escobedo, 1976)	Human (Nakamura et al., 1980)	Human (Hallman et al., 1983)	Pig (Body, 1971)	Rabbit (Rooney et al., 1974)	Rat (Toshima and Akino, 1972)
Lung tissue									
Neutral lipids	–	28	–	35	35	–	60	50	38
Glycolipids	–	–	–	–	–	–	–	4	–
Phospholipids	–	72	–	65	65	–	40	46	62
Surfactant									
Neutral lipids	3	17	20	–	–	10	15	5	22
Glycolipids	–		–	–	–	–		10	–
Phospholipids	97	83	80	–	–	90	85	85	78

TABLE 4 Neutral Lipid Composition of Lung Tissue (% of Total)

	Bovine (Hardwood et al., 1975)	Bovine (Clements, 1971)	Dog (Pfleger and Thomas, 1971)	Dog (Clements, 1971)	Guinea Pig (Goadby, 1974)	Guinea Pig (Clements, 1971)	Human (Nakamura et al., 1980)	Human (Clements, 1971)	Hamster (Henderson and Escobedo, 1976)	Mouse (Clements, 1971)	Rabbit	Rat (Clements, 1971)
Cholesterol	7.6[a]	62.6	52.0	73.0	50.0	70.7	63.0	46.1	29.0	60.8	75.3	48.5
Cholesterol esters	–	3.6	8.0	7.9	2.8	4.1	13.4	28.8	3.7	3.4	3.8	4.7
Free fatty acids	15.8	3.9	11.0	6.5	3.8	0.7	12.0	33.3	4.9	3.6	4.8	
Triacylglycerol	76.6	29.8	29.0	12.7	30.3	21.4	22.9	13.1	35.0	30.8	17.3	41.9
Unidentified	–	–	–	–	16.9	–	–	–	–	–	–	–

[a]*Includes cholesterol esters.*

TABLE 5 Neutral Lipid Composition of Lung Surfactant (% of Total)

	Lung Lavage Surfactant							Amniotic Fluid Surfactant	Lamellar Bodies		
	Bovine (Yu et al., 1983)	Bovine (Hardwood et al., 1975)	Dog (King and Clements, 1972)	Rabbit (Hardwood et al., 1975)	Rat (Hardwood et al., 1975)	Rat (Hass and Longmore, 1979)	Sheep (Hardwood et al., 1975)	Human (Hallman et al., 1983)	Human (Post et al., 1982b)	Rat (Gil and Reiss, 1973)	Rat (Hass and Longmore, 1979)
Cholesterol	89.0	6.4[a]	57.9	36.7[a]	30.8	52.2	15.0	59.9	52.1	19.5	34.6
Cholesterol esters	<1	–	3.4	–	–	6.2	–	2.6	17.7	10.4	22.7
Free fatty acids	0.0	15.9	6.9	17.8	25.0	1.7	51.3	16.8	11.6	36.4	5.8
Triacylglycerol	0.0	77.8	22.1	45.6	44.2	24.7	33.7	6.1	6.9	23.4	22.6
Diacylglycerols	9.8	–	4.1	–	–	12.9	–	12.4	5.1	10.4	11.0
Monoacylglycerols	<1	–	1.4	–	–	2.2	–	2.2	6.6	–	3.3
Unidentified	–	–	4.1	–	–		–	–	–	–	–

[a]*Includes cholesterol esters.*

TABLE 6 Phospholipid Composition of Lung Tissue (% of Total)

	Bovine (Baxter et al., 1969)	Dog (Pfleger and Thomas, 1971)	Guinea Pig (Trzeciak et al., 1977)	Hamster (Henderson and Escobedo, 1976)	Human (Baxter et al., 1969)	Rhesus Monkey (Rooney et al., 1974)	Mouse (Prueitt et al., 1978)	Pig (Body, 1971)	Rabbit (Gilfillan et al., 1983a)	Rat (Gilfillan et al., 1983a)	Squirrel (Aloia, 1981)
PC[a]	41.6	48.0	38.9	51.6	50.3	42.4	55.1	49.0	50.5	57.5	50.8
PE	22.3	24.0	20.9	14.9	18.5	22.0	18.9	36.0	21.2	16.6	18.5
PG	2.1 3.5	3.0 4.0	1.1 4.8	5.0 5.2	2.6 3.4	3.4 –	4.6 –	0.8 –	1.7 4.6	3.3 2.6	2.4 3.0
PS	9.9	8.0	9.1	5.8	7.4	–	11.4[c]	–	7.3	9.3	8.0
Sph	17.0	7.0[d]	13.3	12.2	11.8	26.6[b]	7.2	12.0	10.9	7.5	13.0
LPC	0.3	–	–	5.4	2.1	–	–	–	0.6	0.5	–
LBPA	0.5	–	–	–	1.6	–	–	0.8	1.0	1.0	–
CL	1.1		–	–	1.1	–	–	1.5	0.8	1.4	1.0
Others	1.7	5.0	12.0	–	1.2	5.6	2.5	–	1.8	0.3	3.3
DSPC (% of PC)	–	–	–	–	–	–	38.3	–	36.6	–	–

[a]*DSPC, disaturated PC; other abbreviations as in Table 2.*
[b]*Includes PA and PS.*
[c]*PI and PS combined.*
[d]*Includes LPC.*

TABLE 7 Phospholipid Composition of Lung Surfactant (% of Total)

	Bovine (Yu et al., 1983)	Cat (Shelley et al., 1984)	Dog (King, 1982)	Guinea Pig (Khan et al., 1985)	Human (Gilfillan et al., 1983a)	Rhesus Monkey (Egberts et al., 1987)	Mouse (Prueitt et al., 1978)	Pig (Body, 1971)	Rabbit (Gilfillan et al., 1983a)	Rat (Hallman and Gluck, 1975)	Sheep (Benson et al., 1983)	Squirrel (Melling and Keough, 1981)
PC[a]	79.2	86.3	87.6	67.1	67.5	83.2	77.0	78.9	80.9	75.5	79.1	81.6
PE	3.5	3.2	3.8	5.5	5.3	2.1	2.5	8.4[b]	3.2	4.6	3.5	–
PG	11.1	2.1	6.3	4.3	10.0	1.4	10.8	2.4	7.1	11.0	12.8	9.5
PI	1.8	–	–	6.7	3.6	10.3	5.5	3.9	3.2	3.0	1.2	–
PS	0.0	–		6.3	1.6	–	–	–	0.5	1.9	1.2	–
Sph	2.6	–	–	6.3	4.0	1.3	3.1	3.2	1.4	1.5	2.3	–
LPC	–		–	3.9	1.7	0.7	–	0.6	1.0	0.2	–	–
LBPA	1.5	–	–	–	3.6	–	–	1.8	1.1	1.7	–	–
CL	–	–	–	–	0.9	–	–	–	0.2	0.1	–	–
Others	–	8.4[c]	2.5	–	1.8	1.2	1.1	0.9	1.5	0.5	–	8.9[d]
DSPC (% of PC)	52.9	73.0	64.3	79.5	73.1	–	–	–	61.6	–	85.3	–

[a]*Abbreviations as in Tables 2 and 6.*
[b]*Includes PS.*
[c]*Includes PI, PS, and Sph and LPBA.*
[d]*Includes PI, PS, and Sph and others.*

TABLE 8 Phospholipid Composition of Isolated Lamellar Bodies (% of Total)

	Human (Post et al., 1982a)	Rhesus Monkey (Egberts et al., 1987)	Rabbit (Dethloff et al., 1986)	Rat (Hallman and Gluck, 1975)	Rat (Xu and Rooney, 1989)
PC[a]	71.2	73.4	89.0	77.0	73.2
PE	7.7	8.0	0.6	4.2	6.4
PG	9.9	1.2	3.9	11.2	9.5
PI	3.8	8.7	0.3	3.2	2.8
PS	2.3	–	1.4	1.2	2.4
Sph	2.2	6.7	2.8	0.7	2.4
LPC	2.9	0.5	2.0	0.4	1.4
LBPA	–	–	–	1.5	1.5
CL	0.0	–	–	0.3	0.6
Others	–	0.7	–	0.3	–
DSPC (% of PC)	67.9	–	66.4	–	–

[a]*Abbreviations as in Tables 2 and 6.*

tissue is polyenoic, and no disaturated species are measureable. The monoenoic species is the most common species of phosphatidylglycerol in both tissue and surfactant, but appreciable amounts of disaturated phosphatidylglycerol are also present. The fatty acid composition of lung disaturated phosphatidylcholine is shown in Table 18. Because palmitic acid is by far the most abundant fatty acid, most of the disaturated phosphatidylcholine is dipalmitoylphosphatidylcholine.

3. LIPID BIOSYNTHESIS

3.1 Biochemical Pathways

For the purpose of discussion, pathways of phospholipid biosynthesis can be conveniently divided into the following discrete reaction sequences, as illustrated in Figures 1–5: biosynthesis of phosphatidic acid from glycolytic intermediates and fatty acids (Figure 1); de novo fatty acid biosynthesis (Figure 2); biosynthesis of phosphatidylcholine, phosphatidylethanolamine, phosphatidyl serine, and sphingomyelin (Figure 3); disaturated phosphatidylcholine biosynthesis (Figure 4); and acidic phospholipid biosynthesis (Figure 5). Enzymes catalyzing individual reactions are listed in Table 19.

Phosphatidic acid occupies a central position in glycerolipid biosynthesis, being the common lipid precursor of all other glycerophosphatides, as well as acylglycerols. The biosynthesis of phosphatidic acid from nonlipid glycolytic intermediates and fatty acids is illustrated in Figure 1. The glycerophosphate backbone originates from the triose phosphate, dihydroxyacetone phosphate. Carbons derived from dietary glucose and delivered via the blood or from glycogen degradation can be incorporated into phospholipids at this step. Similarly, glycerol can be incorporated after its phosphorylation to glycerol 3-phosphate (Fisher and Chander, 1982). Dihydroxyacetone phosphate is metabolized to 1-acylglycerol 3-phosphate by either acylation followed by reduction or reduction followed by acylation. There is evidence that both of these pathways exist in the lung (Fisher et al., 1976) and are involved in the synthesis of phosphatidylcholine and phosphatidylglycerol in type II cells (Mason, 1978).

Fatty acids may be synthesized in the lung de novo from acetyl-CoA or supplied from extrapulmonary sources via the blood (Figure 1). In either case, the free fatty acids are activated by conversion to acyl-CoAs before their incorporation into lipids. The steps in the biosynthesis of fatty acids starting from Kreb's cycle intermediates in the mitochondria are illustrated in Figure 2.

The pathways in the biosynthesis of phosphatidylcholine, phosphatidylethanolamine, and phosphatidylserine are illustrated in Figure 3. Phosphatidic acid is first dephosphorylated. The resulting diacylglyerol can be further acylated to form triacylglycerol or can react with CDPcholine or CDPethanolamine to form phosphatidylcholine or phosphatidylethanolamine, respectively. CDPcholine and CDPethanolamine arise from choline and ethanolamine, respectively, after

TABLE 9 Phospholipid Composition of Isolated Type II Pneumocytes

Species	Rat (Mason and Williams, 1980)	Rat (Gilfillan et al., 1986)	Rat (Dobbs et al., 1980)	Rat (Kikkawa et al., 1975)	Rabbit (Finkelstein and Shapiro, 1982)	Rabbit (Smith and Kikkawa, 1979)	Human (Robinson et al., 1984)
Method of isolation[a]	I	II	III	IV	V	IV	III
Purity (%)	90	>95	94	>90	>80	80–90	84
Method of phospholipid measurements[b]	VII	VIII, IX	VII	VII	X	VII	XI
	Percent of Total Phospholipid						
PC[c]	71.8	71.7	68.0	76.1	66.6	66.7	63.4
PE	12.1	11.8	13.6	4.8	17.8	13.5	11.8
PG	10.4	6.0	9.6	6.4	4.4	3.5	8.9
PI	0.4	2.8	1.6	2.9	–	3.1	7.1
PS	0.1	1.5	2.0	1.1	–	1.9	3.6
Sph	1.3	3.2	2.3	4.2	10.3	6.1	5.2
LPC	0.3	0.3	–	1.5	–	0.4	–
LBPA	0.8	0.5		1.8	0.8	–	–
CL	1.3	1.1	–	1.1	–	3.7	–
Others	0.5	1.7	2.1	–	–	1.1	–
DSPC (% of PC)	44.4	43.8	44.7	52.5	–	–	–

[a] *Type II cell isolation method: I, trypsin digestion, density gradient centrifugation, elutriation (Mason et al., 1976); II, trypsin digestion, density gradient configuration, differential adherence (Brown and Longmore, 1981; Gilfillan et al., 1986); III, elastase digestion, density gradient centrifugation, differential adherence (Dobbs et al., 1980); IV, trypsin digestion, density gradient centrifugation (Kikkawa et al., 1975); V, elastase/trypsin digestions, density gradient centrifugation (Finkelstein and Shapiro, 1982).*

[b] *Phospholipids were quantitated by phosphorus assay (VII) or by measurement of phospholipid radioactivity after incubation of the cells with [^{14}C]acetate (VIII), [^{3}H]glycerol (IX), [^{3}H]acetate (X) or ^{32}P, (XI).*

[c] *Abbreviations as in Tables 2 and 6.*

TABLE 10 Phospholipid Composition of Material Secreted by Type II Pneumocytes

Species	Rat (Gilfillan et al., 1986)	Rat (Dobbs et al., 1982)	Rat (Dobbs et al., 1982)	Human (Robinson et al., 1984)
Isolation method[a]	II	III	III	III
Analysis method[a]	VIII	VIII	IX	XI
	Percent of Total Phospholipid			
PC[k]	80.1	87.0	73.8	68.5
PE	2.8	1.6	1.9	4.7
PG	6.5	8.7	21.2	11.8
PI	1.7	1.1	1.3	6.5
PS	0.6	0.3	0.7	3.2
Sph	3.5	0.8	0.2	5.4
LPC	1.0	0.3	0.3	—
LBPA	1.5	—	—	—
CL	0.9	—	—	—
Others	1.4	0.2	0.6	—
DSPC (% of PC)	65.6	69.0	64.0	—

[a]*Methods of type II cell isolation and phospholipid analysis as in Table 9.*
[b]*Abbreviations as in Tables 2 and 6.*

TABLE 11 Fatty Acid Composition of Lung Tissue Phosphatidylcholine (% of Total)

	Hamster (Tombropoulos and Hadley, 1976)	Human (Nikolasev, 1974)	Rhesus Monkey (Kerr and Helmuth, 1974)	Mouse (Oldenborg and Van Golde, 1976)	Rabbit (Abe et al., 1973)	Rat (Okano et al., 1978)	Squirrel (Melling and Keough, 1981)
14:0[a]	1.5	4.0	—	3.8	—	3.0	2.3
16:0	53.2	51.8	46.9	55.4	65.8	54.2	49.5
16:1	4.7	5.5	—	9.6	1.4	8.3	13.2
18:0	10.0	14.6	12.8	10.0	2.4	6.8	11.0
18:1	14.1	16.9	28.1	8.8	6.6	9.9	10.7
18:2	6.7	4.0	12.2	7.1	19.7	9.2	12.3
20:4	4.8	3.0	—	5.4	4.0	5.2	—
Others	5.2	—	—	—		3.2	—

[a]*Fatty acid shorthand notation as in Table 1.*

TABLE 12 Fatty Acid Composition of Surfactant Phosphatidylcholine (% of Total)

	Bovine (Yu et al., 1983)	Cat (Shelley et al., 1984)	Dog (Shelley et al., 1984)	Guinea Pig (Khan et al., 1985)	Human (Hallman et al., 1982)	Pig (Body, 1971)	Rabbit (Rooney et al., 1975)	Rat (Harwood et al., 1975)	Sheep (Harwood et al., 1975)	Squirrel (Melling and Keough, 1981)
14.0[a]	3.1	3.8	2.6	5.4	3.6	5.4	2.2	3.3	0.4	2.7
16:0	64.0	65.7	65.8	68.0	68.0	73.4	62.8	72.8	81.9	62.8
16:1	7.5	2.8	7.4	4.2	4.6	6.3	5.9	7.5	4.3	21.4
18:0	2.4	3.9	6.7	5.1	4.5	3.0	2.5	6.5	3.7	3.2
18:1	17.3	13.9	12.8	9.2	10.9	10.4	14.8	6.5	6.9	4.7
18:2	1.6	3.0	1.8	6.1	3.6	0.8	10.6	2.4	0.5	5.2
18:3	0.3	–	–	0.4	3.2	–	1.3	–	0.1	–
20:4	0.5	–	–	0.6	1.6	–	–	–	–	–
Others	3.1	6.9	2.9	1.0	–	0.7	–	1.0	2.2	–

[a]*Fatty acid shorthand notation as in Table 1.*

TABLE 13 Fatty Acid Composition of Lung Tissue Phosphatidylethanolamine (% of Total)

	Hamster (Tombropoulos and Hadley, 1976)	Human (Nikolasev, 1974)	Rhesus Monkey (Kerr and Helmuth, 1975)	Rat (Okano et al., 1978)	Squirrel (Melling and Keough, 1981)
14:0[a]	1.8	3.4	–	–	1.3
16:0	24.1	18.4	14.3	9.2	11.7
16:1	3.9	2.8	–	2.3	2.4
18:0	24.7	20.1	25.8	18.5	17.6
18:1	26.9	18.1	47.8	12.4	21.3
18:2	11.9	3.4	12.1	4.8	12.0
20:4	–	9.7	–	30.3	15.5
Others	6.7	20.4	–	22.5	18.1

[a]*Fatty acid shorthand notation as in Table 1.*

TABLE 14 Fatty Acid Composition on Lung Tissue Phosphatidylglycerol (% of Total)

	Bovine (Okano and Akino, 1979)	Guinea Pig (Okano and Akino, 1979)	Pig (Okano and Akino, 1979)	Pig (Body, 1971)	Rabbit (Okano and Akino, 1979)	Rat (Okano et al., 1978)
14:0[a]	0.6	1.6	0.6	–	0.4	–
16:0	34.2	36.8	27.1	28.0	29.0	46.9
16:1	1.0	6.4	2.3	2.0	2.5	2.5
18:0	15.2	17.7	21.1	17.0	19.3	7.3
18:1	37.2	23.7	34.2	45.0	27.0	18.5
18:2	2.5	4.9	6.5	6.0	7.5	16.0
20:4	2.7	3.2	3.4	–	3.9	5.1
Others	6.6	5.6	4.7	2.0	10.4	3.6

[a]*Fatty acid shorthand notation as in Table 1.*

phosphorylation and transfer to cytidine triphosphate (CTP). These pathways represent de novo synthesis of the respective phospholipids.

In some organs, phosphatidylcholine is also synthesized from phosphatidylethanolamine by triple *N*-methylation. In this pathway, phosphatidylethanolamine is first converted to phosphadityl-*N*-methylethanol amine by phosophatidylethanolamine methyltransferase (EC 2.1.1.17). This is then converted to phosphatidyl-*N*-dimethylethanolamine and finally phosphatidylcholine and phosphatidyl-N-methylethanolamine methyltransferase (EC 2.1.1.71). There are also reports that the entire methylation sequence is accomplished by a single enzyme (Mato et al., 1984). The methylation pathway of phosphatidylcholine biosynthesis is of quantitative importance in the liver (Pelech and Vance, 1984), and it was at one time considered an important mechanism of surfactant choline biosynthesis in the lung (Morgan, 1969; Morgan et al., 1965), especially during fetal development (Gluck and Kulovich, 1973; Gluck et al., 1970). However, the experimental evidence in support of this pathway in the lung was largely (Van Golde, 1976), although not entirely (Morgan, 1969), based on the misidentification of phosphatidylglycerol as phosphatidyl-N-dimethylethanolamine (Hallman and Gluck, 1974; Rooney et al., 1974). Subsequent studies found no evidence in support of the methylation pathway of phosphatidylcholine synthesis in the normal lung (Epstein and Farrell, 1975; Rooney and Motoyama, 1976), but it may be active in conditions of choline deficiency (Yost et al.,

TABLE 15 Fatty Acid Composition of Surfactant Phosphatidylglycerol (% of Total)

	Bovine (Harwood et al., 1975)	Bovine (Yu et al., 1983)	Cat (Shelley et al., 1984)	Dog (Pfleger and Thomas, 1971	Dog (Shelley et al., 1984)	Guinea Pig (Okano and Akino, 1979)	Human (Shelley et al., 1984)	Pig (Body, 1971)	Rabbit (Rooney et al., 1975)	Rabbit (Oulton et al., 1986)	Rat (Shelley et al., 1984)	Sheep (Harwood et al., 1975)	Squirrel (Melling and Keough, 1981)
14:0[a]	8.6	0.8	–	3.0	–	1.6	–	2.5	–	0.7	–	1.0	0.9
16:0	42.8	51.2	36.0	58.0	54.5	52.8	22.2	39.5	26.5	55.3	58.9	30.6	53.3
16:1	5.7	4.3	0.8	5.0	3.3	6.6	3.0	4.8	3.7	5.6	4.0	13.4	12.4
18:0	19.4	4.7	6.3	11.0	12.6	8.3	16.7	8.5	9.4	6.2	5.8	14.4	6.4
18:1	18.9	33.0	31.0	20.0	23.4	26.0	51.3	37.8	36.7	26.4	19.2	31.5	17.0
18:2	–	2.9	11.2	3.0	1.6	3.6	4.0	4.2	12.1	3.8	8.7	1.6	8.3
18:3	–	0.5	–	–	–	0.3	–	–	2.0	–	–	1.5	–
20:4	–	1.0	–	–	–	–	–	0.9	1.4	–	–	–	1.7
Others	4.6	1.6	14.7	-	4.6	0.8	2.8	1.8	8.2	2.0	3.4	6.0	–

[a]*Fatty acid shorthand notation as in Table 1.*

TABLE 16 Molecular Species of Phosphatidylcholine, Phosphatidylethanolamine, and Phosphatidylglycerol in Lung Tissue

	PC[a]		PE[a]	PG[a]				
	Human (Hanamura et al., 1980)	Rat (Okano et al., 1978)	Rat (Okano et al., 1978)	Bovine (Okano and Akino, 1979)	Guinea Pig (Okano and Akino, 1979)	Pig (Okano and Akino, 1979)	Rabbit (Okano and Akino, 1979)	Rat (Okano et al., 1978)
	Percent of the Total Phospholipid							
Disaturated	31.4	33.5	0	15.8	19.0	15.5	13.2	22.3
Monoenoic	29.4	23.8	13.3	59.1	53.4	57.5	54.9	28.5
Dienoic	21.2	16.2	13.1	14.6	16.5	12.6	15.4	21.2
Polyenoic	18.0	26.5	73.5	10.4	10.1	14.9	16.5	28.1

[a]*Abbreviations as in Table 2.*

TABLE 17 Molecular Species of Surfactant Phosphatidylcholine and Phosphadityglycerol

	Phosphatidylcholine			Phosphaditylglycerol			
	Bovine (Yu et al., 1983)	Human[a]	Rat (Toshima and Akino, 1972)	Bovine (Yu et al., 1983)	Guinea Pig (Okano and Akino, 1979)	Human[a]	Rabbit (Okano and Akino, 1979)
Disaturated	52.9	67.1	70.5	17.2	32.7	24.2	28.0
Monoenoic	45.0	14.1	14.7	76.9	58.4	36.2	58.1
Dienoic	1.5	11.1	9.0	4.2	6.2	–	8.6
Polyenoic	0.5	7.7	5.7	1.6	2.7	39.1[b] (Toshima and Akino, 1972)	5.3

[a]*Lamellar bodies isolated from human lung: Post et al., 1982a.*
[b]*Includes dienoic species.*

TABLE 18 Fatty Acid Composition of Lung Disaturated Phosphatidylcholine

Species	Guinea Pig (Khan et al., 1985)	Hamster (Henderson and Clayton, 1976)	Rat (Mason et al., 1976)	Rat (Shimojo et al., 1974)	Rabbit (Gilfillan et al., 1983a)
Method[a]	I	II	III	IV	V
	Percent of Total Fatty Acids				
14:0[b]	6.8	2.0	6.3	–	3.3
15:0	1.3	1.0	1.9	–	–
16:0	88.1	90.5	87.9	96.1	93.1
18:0	3.8	4.8	3.9	3.9	2.4
Others	–	1.7	–	–	1.2

[a]*I, osmium tetroxide oxidation of lung lavage phosphatidylcholine; II, cry-ochromatography of lung tissue total lipid; III, osmium tetroxide oxidation of lung tissue total lipid; IV, permanganate-periodate oxidation of lung tissue phosphatidylcholine; V, osmium tetroxide oxidation of phosphatidylcholine after thin layer chromatography of lung tissue total lipid.*
[b]*Fatty acid shorthand notation as in Table 1.*

1986). Because this pathway is not operative in the normal lung and choline is not synthesized by mammals, dietary choline is essential for de novo synthesis of phosphatidylcholine.

Lung surfactant characteristically contains primarily disaturated phosphatidylcholine in contrast to cell membranes, where the phosphatidylcholines are characteristically monounsaturated. Lung disaturated phosphatidylcholine may be synthesized de novo by incorporation of saturated fatty acids into diacylglycerol (Figure 3) or by the remodeling mechanisms shown in Figure 4. The de novo pathway accounts for a major portion of disaturated phosphatidylcholine both in whole lung (Ide and Weinhold, 1982; Rustow and Kunze, 1985; Van Heusden et al., 1981a; Van Heusden and Van den Bosch, 1982) as well as in type II cells (Post et al., 1983). In the remodeling mechanism, 1-acyl(saturated)-2-acyl(unsaturated) phosphatidylcholine is first deacylated to 1-acyl-2-lysophosphatidylcholine. This can then be reacylated by reaction with a saturated acyl-CoA. An alternate pathway is transacylation, in which two molecules of lysophosphatidylcholine react to form one molecule each of phosphatidylcholine and glycerophosphocholine. Of the two remodeling mechanisms, only the reacylation pathway is of any quantitative importance in the lung (Rooney, 1985).

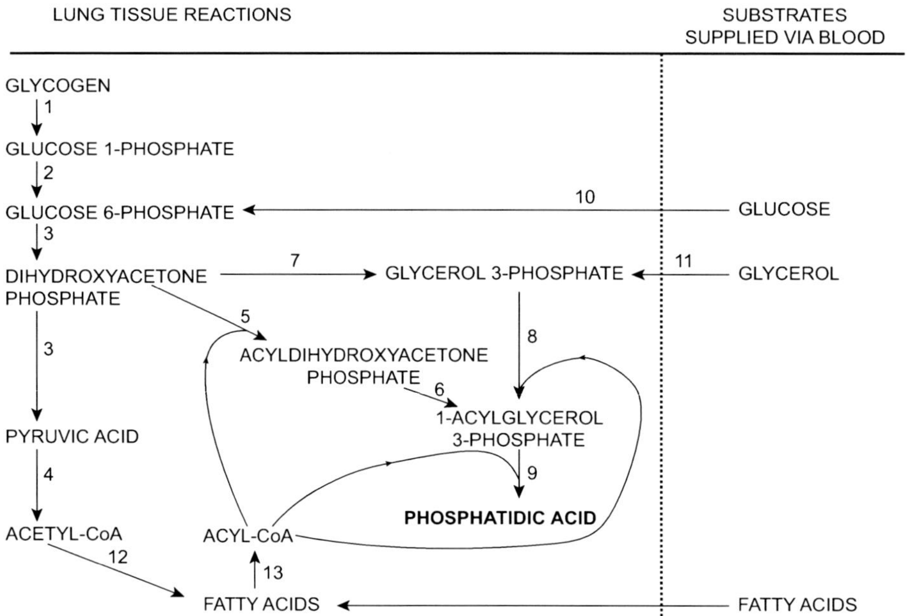

FIGURE 1 Pathways in the biosynthesis of phosphatidic acid from nonlipid precursors and fatty acids. The numbers refer to the reactions and enzymes listed in Table 19.

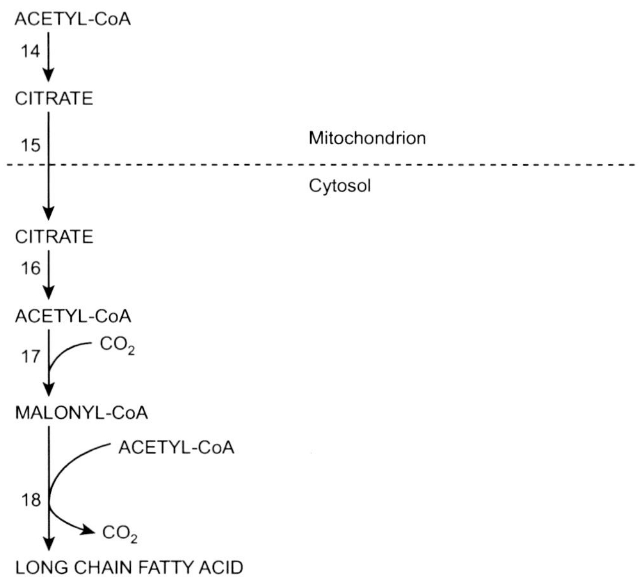

FIGURE 2 De novo fatty acid biosynthesis.

Reacylation requires the sequential activity of a phospholipase A_2 to generate a lysophospholipid (lysophosphatidylcholine) and an acyltransferase to reacylate the lysophospholipid. Enzymes of relative specificity for these functions in type II cells have recently been identified, although it is possible that related enzymes may fulfill a similar function. The phospholipase A_2 activity that is involved in the lung reacylation pathway is associated with the protein peroxiredoxin 6, which is present in lung lamellar bodies and is maximally active at an acidic pH (Fisher and Dodia, 2001; Fisher et al., 2005; Wu et al., 2009b). The pH of the lamellar body contents has been estimated as ~5 (Chander et al., 1986). Of note, the lung lamellar bodies contain a broad spectrum of lysosomal-like proteins, and lamellar bodies have been characterized as lysosomal-related organelles, analogous to melanosomes and platelet-dense granules (Ridsdale et al., 2011; Wang et al., 2008). An acyltransferase with localization to the lung and some specificity for saturated fatty acids to generate disaturated phosphatidylcholine has been described (Bridges et al., 2010; Chen et al., 2006; Nakanishi et al., 2006), and called lung lysophosphatidylcholine acyltransferase (LPCAT1). This enzyme is restricted to the endoplasmic reticulum. These findings suggest that there is cooperativity for disaturated phosphatidylcholine synthesis by the reacylation pathway between the endoplasmic reticulum containing LPCAT1 and the lysosomal-related organelles (lamellar bodies) containing phospholipase A_2. Studies using agents that neutralize the acidic pH of lamellar bodies significantly diminish dipalmitoylphosphatidylcholine synthesis, providing evidence that the lamellar bodies do play an important role in synthesis of surfactant phospholipid (Chander et al., 1982; Post et al., 1983).

The major pathway of phosphatidylserine biosynthesis in most mammalian organs appears to be a base exchange reaction (Figure 3) in which serine replaces the choline or ethanolamine moiety of phosphatidylcholine or phosphatidylethanolamine, respectively (Ansell and Spanner, 1982; Kuge et al., 1985). There is evidence that phosphatidylethanolamine can be synthesized from phosphatidylserine by the action of phosphatidylserine decarboxylase

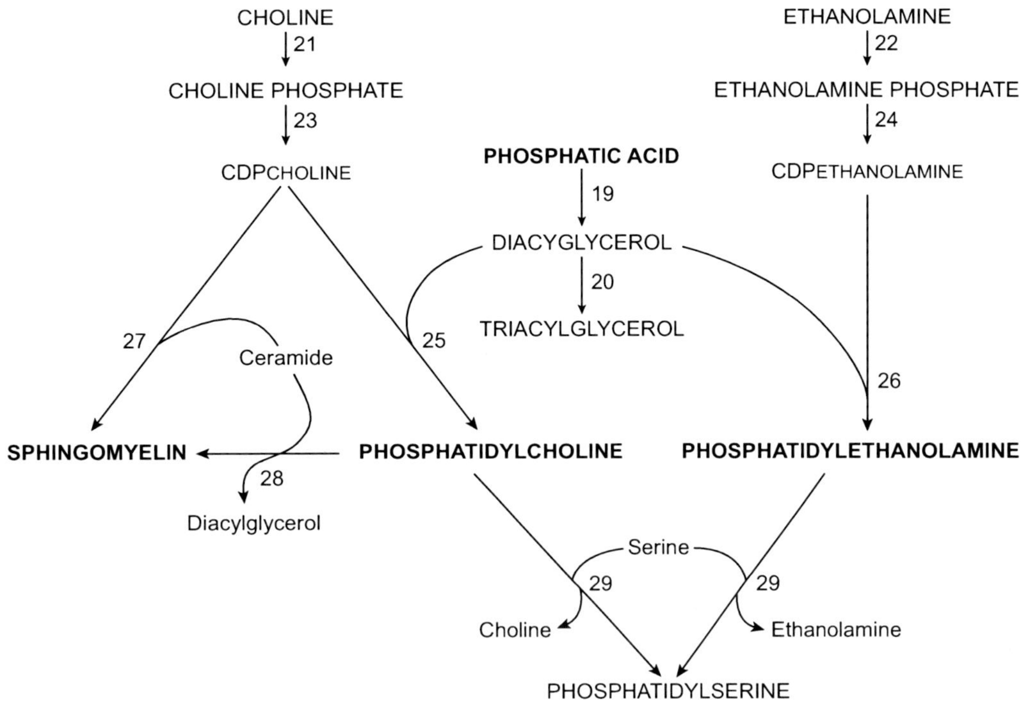

FIGURE 3 **Pathways in the biosynthesis of phosphatidylcholine, phosphatidylethanolamine, phosphatidiylserine, and sphingomyelin.**

(EC 4.1.1.65), and that this mechanism is quantitatively more important than synthesis of phosphatidylethanolamine via the CDPethanolamine pathway (Miller and Kent, 1986; Voelker, 1984). The extent to which these pathways exist in the lung has not been investigated.

Sphingomyelin can be synthesized from CDPcholine in a reaction analogous to that involved in the biosynthesis of phosphatidylcholine (Figure 3). In addition, sphingomyelin can be synthesized from phosphatidylcholine in an exchange reaction in which the diacylglycerol moiety of phosphatidylcholine is replaced with a ceramide (acylsphingosine). Although the latter reaction has been reported to predominate in rat lung lamellar bodies (Lecerf et al., 1987), the relative importance of these as well as other pathways of sphingomyelin biosynthesis (Barenholz and Gatt, 1982) has not been adequately evaluated in the whole lung.

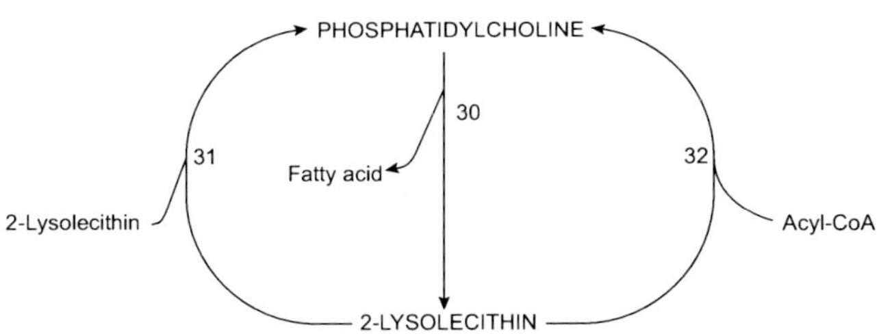

FIGURE 4 **Biosynthesis of disaturated phosphatidylcholine by remodeling unsaturated phosphatidylcholine.**

Pathways in the biosynthesis of phosphatidylglycerol, phosphatidylinositol, lysobisphosphatidic acid, and cardiolipin are shown in Figure 5. The first step in this series of reactions is the synthesis of CDPdiacylglycerol from phosphatidic acid and CTP. Phosphatidylinositol and phosphatidylglycerophosphate are then formed in exchange reactions with inositol and glycerol 3-phosphate, respectively. Phosphatidylglycerophosphate does not accumulate to any appreciable extent, but it is immediately

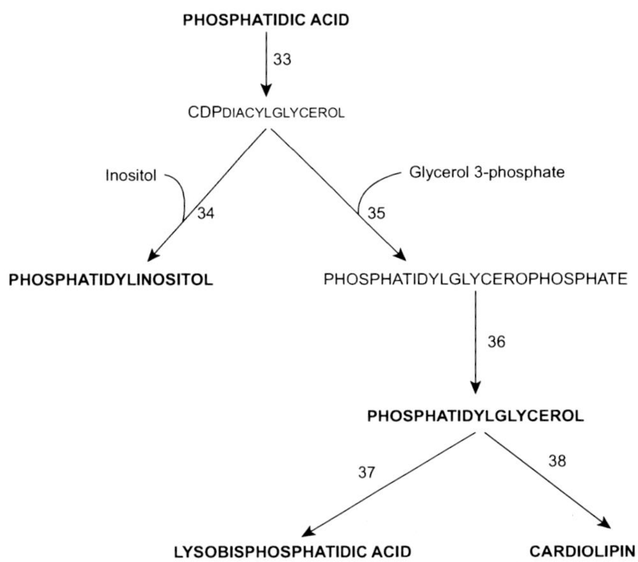

FIGURE 5 **Pathways in the biosynthesis of acidic phospholipids.**

TABLE 19 Enzymes Involved in Lipid Biosynthesis

Number[a]	Enzyme Name[b] or Metabolic Reaction	EC Number[b]	Whole Lung	Type II Cells	References[c]
1	Glycogen Phosphorylase	2.4.1.1	Rat		Bourbon and Jost (1982), Maniscalco et al. (1978), Pope and Rooney (1987)
			Rabbit		Bhavani (1983), Brandstrup and Kretchmer (1965), Khosla et al. (1983)
			Sheep		Warburton et al. (1987b)
2	Phosphoglucomutase	5.4.2.2			
3	Glycolysis				
4	Pyruvate dehydrogenase complex				
5	Glycer one-phosphate acyltransferase (dihydroxyacetone-phosphate acyltransferase)	2.3.1.42	Rat		Schlossman and Bell (1977)
				Mouse[d]	Snyder and Malone (1975)
6	Acylglycerone-phosphate reductase (acyldihydroxy acetone-phosphate reductase)	1.1.1.101			
7	Glycerol-3-phoshphate dehydrogenase	1.1.1.8	Rabbit		Fisher et al. (1976)
				Rat	Uhal and Longmore (1988)
8	Glycerol-3-phosphate acyltransferase	2.3.1.15	Rat		Schlossman and Bell (1977), Haq et al. (1987b), Yamada and Okuyama (1979)
			Rabbit		Hendry and Possmayer (1974)
			Guinea pig		Das et al. (1984)
				Rat	Batenburg et al. (1986), Uhal and Longmore (1988)
				Rabbit	Finkelstein et al. (1983)
9	1-Acylglycerol-3-phosphate actytransferase (lysophosphatidic acid acyltransferase_	2.3.1.51	Rat		Gross et al., 1979, Yamada and Okuyama (1979)
			Rabbit		Khosla et al. (1980), Rooney et al. (1976)
10	Hexokinase	2.7.1.30	Rat		Bargoni and Tazartes (1983), Rijksen et al. (1985), Rinaudo et al. (1981)
11	Glycerol kinase	2.7.1.30		Rat	Fisher and Chander (1982), Uhal and Longmore (1988)
				Mouse[d]	Wykle and Kraemer (1977)

12	De novo fatty acid biosynthesis		Rat		Buechler and Rhoades (1980), Patterson et al. (1988), Rooney et al. (1986)
			Rabbit		Gross and Warshaw (1974), Maniscalco et al. (1985), Schiller and Bensch (1971)
			Mouse	Mouse[d]	Voelker et al. (1976)
				Rat	Geppert and Elstein (1983)
				Rabbit	Maniscalco et al. (1983)
13	Long-chain-fatty-acid-CoA ligase	6.2.1.3	Rat		Das (1982), Haq (1987b)
			Rabbit		Ayromlooi et al. (1981)
				Rat	Haq et al. (1987a)
14	Citric acid cycle				
15	Tricarboxylate anion carrier				
16	ATP citrate (pro-3S)-lyase	4.1.3.8	Rat		Batenburg et al. (1992), Rami et al. (1991)
				Rat	Rami et al. (1991)
17	Acetyl-CoA carboxylase	6.4.1.2	Rat		Das and Ganguly (1982), Maniscalco et al. (1982), Pope and Rooney (1987)
			Rabbit		Gross and Warshaw (1974), Gross et al. (1975)
			Mouse	Mouse[d]	Voelker et al. (1976)
				Rat	Rami et al. (1991)
				Rabbit	Maniscalco et al. (1983)
18	Fatty-acid synthase	2.3.1.85	Rat		Das and Ganguly (1982), Kumar (1977), Pope and Rooney (1987)
			Rabbit		Das (1980), Gross and Warshaw (1974), Gross et al. (1975)
			Mouse	Mouse[d]	Voelker et al. (1976)
				Human	Gonzales et al. (1990)
				Rat	Kresch et al. (1988), Rami et al. (1991)

Continued

TABLE 19 Enzymes Involved in Lipid Biosynthesis—cont'd

Number[a]	Enzyme Name[b] or Metabolic Reaction	EC Number[b]	Whole Lung	Type II Cells	References[c]
19	Phosphatidate phosphatase	3.1.3.4	Rat		Casola and Possmayer (1981a), Maniscalco et al. (1978), Ravinuthala et al. (1978)
			Rabbit		Casola and Possmaye (1981b), Khosla et al. (1980), Schultz et al. (1974)
			Mouse		Brehier and Rooney (1981), Oosterhuis et al. (1984)
			Pig		Delahunty et al. (1979), Spitzer and Johnston (1978)
			Dog		Benson (1980)
			Sheep		Rosenfeld et al. (1980)
			Human		Bleasdale et al. (1978), Herbert et al. (1978)
				Rat	Crecelius and Longmore (1983), Finkelstein and Mavis (1979), Gilfillan et al. (1986)
				Human[e]	Walton and Possmayer (1986)
20	Diacylglycerol acyltransferase	2.3.1.20			
21	Choline kinase	2.7.1.32	Rat		Gross et al. (1984), Ishidate et al. (1985), Maniscalco et al. (1978)
			Rabbit		Khosla et al. (1980), Possmayer et al. (1981), Tsao and Zachman (1977)
			Mouse		Oldenborg and van Gold (1977)
			Monkey[f]		Ulane et al. (1978)
			Human		Zachman (1971)
				Rat	Gilfillan et al. (1986), Hoehn et al. (1983), Tesan et al. (1985)
22	Ethanolamine kinase	2.7.7.15	Rat		Ishidate et al. (1985)
			Monkey[f]		Ulane et al. (1978)
23	Choline-phosphate cytidylyltransferase	2.7.7.15	Rat		Hoehn et al. (1983), Maniscalco et al. (1978), Weinhold et al. (1984)
			Rabbit		Chu and Rooney (1985b), Freese and Hallman (1983), Possmayer et al. (1981)
			Mouse		Brehier and Rooney (1981), Oldenborg and van Golde, (1977), Oosterhuis et al. (1984)
			Human		Thom and Zachman (1975)
				Rat	Chander and Fisher (1988), Gilfillan et al. (1986), Tesan et al. (1985)

24	Ethanolamine-phosphate cytidylyltransferase	2.7.7.14	Rabbit		Khosla et al. (1980)
25	Cholmephosphotransferase	2.7.8.2	Rat		Harding et al. (1983), Ide and Weinhold (1982), Maniscalco et al. (1978)
			Rabbit		Possmayer et al. (1977), Rooney and Wai-Lee (1977), Tsao (1986)
			Mouse		Brehier and Rooney (1981), Oldenborg and van Golde (1976)
			Guinea pig		Sikpi and Das (1987), Stith and Das (1982)
			Human		Sass et al. (1978), Zachman (1973)
				Rat	Crecelius and Longmore (1984b), Gilfillan et al. (1986), Tesan et al. (1985)
				Rabbit	Finkelstein et al. (1983)
26	Ethanolaminephosphotransferase	2.7.8.1	Rat		Ide et al. (1988)
			Rabbit		Possmayer et al. (1977)
				Rat	Crecelius and Longmore (1984b)
27	Ceramide cholinephosphotransferase	2.7.8.3	Rat		Lecerf and Bodin (1984), Lecerf et al. (1987)
28	Phospholipase C?	3.1.4.3			Kanfer and Spielvogel (1975)
29	Base exchange pathway				Ansell and Spanner (1982)
30	Phospholipase A_2	3.1.1.4	Rat		Garcia et al. (1975), Longmore et al. (1979), Nijssen et al. (1986), Fisher et al. (1992a)
			Rabbit		Filguerias and Possmayer (1987), Franson and Weir (1982), Scott et al. (1987)
			Mouse		Scott et al. (1988), Fisher et al. (2005)
				Rat	Fisher and Dodia (1997)
			Human		Sahu and Lynn (1977)
31	Lysolecithin: lysolecithin acyltransferase		Rat		Abe et al. (1974), Brumley and van den Bosch (1977), Van Heudsen et al. (1981b)
			Rabbit		Cassals et al. (1982), Rooney et al. (1979), Tsao and Zachman (1977)
			Mouse		Oldenborg and van Golde (1976)
			Several[g]		van Heusden and Van den Bosch (1981)
				Rat	Batenburg et al. (1979), Chen et al. (2006), Crecelius and Longmore, (1984a)
				Mouse[d]	Wykle et al. (1980)

Continued

TABLE 19 Enzymes Involved in Lipid Biosynthesis—cont'd

Number[a]	Enzyme Name[b] or Metabolic Reaction	EC Number[b]	Whole Lung	Type II Cells	References[c]
32	1-Acylglycerophosphocholine acyltransferase (lysolecithin acyltransferase)	2.3.1.23	Rat		Gross et al. (1979), Van Heudsen and van den Bosch (1981)
			Rabbit		Yamada and Okuyama (1979)
			Mouse		Arche et al. (1987), Khosla et al. (1980), Tsao and Zachman (1977)
			Several[g]		Oldenborg and van Golde (1976)
				Rat	Van Heudsen and van den Bosch (1981)
					Batenburg et al. (1979), Chen et al. (2006)
				Rabbit	Crecelius and Longmore (1984a), Gilfillan et al. (1986)
				Mouse[d]	Finkelstein et al. (1983)
					Renooij and Snyder (1981)
33	Phosphatidate cytidylyltransferase	2.7.7.41	Rat		Gross et al. (1984), Hallman and Gluck (1975)
			Rabbit		Harding et al. (1983)
				Rat	Hallman and Gluck (1980). Khosla et al. (1983), Longmuir and Johnston (1980)
					Gilfillan et al. (1986)
34	CDPdiacylglycerol-inositol 3-phosphatidyltransferase	2.7.8.11	Rat		Gross et al. (1984), Hallman and Gluck (1975)
			Rabbit		Harding et al. (1983)
				Rat	Bleasdale et al. (1979), Hallman and Gluck (1975), Khosla et al. (1983), Batenburg et al. (1985), Bleasdale et al. (1985), Gilfillan et al. (1986)

35	CDPdiacylglycerol-glycerol-3-phosphate 3-phosphatidytransferase (glyerophosphate phosphatidy transferase)	2.7.8.5	Rat		Gross et al. (1984), Harding et al. (1983), Mavis and Vang (1981)
			Rabbit		Brehier et al. (1977), Hallman and Gluck (1980), Rooney et al. (1975)
				Rat	Batenburg et al. (1985), Bleasdale et al. (1985)
					Gilfillan et al. (1986)
36	Phosphatidylglycerophosphatase	3.1.3.27	Rat		Casola et al. (1980)
			Pig		Johnston et al. (1978)
			Dog		Benson (1980), Hostetler (1982)
37	Lysobisphosphatidic acid synthesis		Rabbit[h]		Waite et al. (1987)
38	Cardiolipin biosynthesis				Hostetler (1982)

[a]The reaction steps are shown in Figures 1–5.
[b]Enzymes names and EC number are those recommended by the Nomenclature Committee of the International Union of Biochemistry (1984). Other names are given in parentheses.
[c]References are restricted to three per item. Studies on both adult and fetal animals are included.
[d]Type II cell adenoma induced by urethane injection (Snyder et al., 1973).
[e]A549 tumor cells (Lieber et al., 1976).
[f]African green monkey.
[g]Rat, rabbit, mouse, guinea pig, bovine, pig, horse, and sheep.
[h]Alveolar macrophages.

dephosphorylated to phosphatidylglycerol. Reaction of a further molecule of CDPdiaylclycerol with phosphatidylglycerol leads to the synthesis of cardiolipin (Hostetler, 1982). Lysobisphosphatidic is also formed from phosphatidylglycerol in alveolar macrophages and other systems through the combined activity of a phospholipase A and a transacylase (Amidon et al., 1996; Hostetler, 1982; Waite et al., 1987).

4. CHANGES IN LUNG PHOSPHOLIPIDS AND RELATED PARAMETERS DURING FETAL DEVELOPMENT

Towards the end of gestation, as the fetus prepares for extrauterine life, several morphological (Burd, 1984; Perelman et al., 1981) and biochemical (Post and van Golde, 1988; Rooney, 1983, 1985) changes occur in the lung. A number of these relate to increased production of pulmonary surfactant, which is essential for normal ventilation. Therefore, marked changes occur in lung phospholipid content, composition, and biosynthesis during this period.

A number of hormones, including glucocorticoids, estrogen, thyroid hormone, and thyrotropin-releasing hormone, have been reported to accelerate lung maturation and to stimulate surfactant synthesis in the late gestation fetus; others, such as insulin and androgens, may have the opposite effect. Glucocorticoids are widely used clinically to accelerate fetal lung maturation and to prevent respiratory distress syndrome in premature newborns (Avery et al., 1986; Ballard, 1986), whereas a combination of glucocorticoids and thryrotropin-releasing hormone was reported to be even better in this respect (Liggins et al., 1988; Morales et al., 1989).

As shown in Table 20, developmental changes in phospholipid content and composition have been documented in several species. Although there is a developmental increase in the amount of total and disaturated phosphatidylcholine in lung tissue, a more dramatic change occurs in the lipids obtained by lung lavage. As shown in Figure 6, the amount of phosphatidylcholine in lung lavage from the fetal rabbit increases from less than 30% of the total phospholipid at 27 day gestation to almost 70% at term (31 day) and to 80% or more after birth. At the same time, sphingomyelin decreases from almost 40% of the total at 27 day gestation to less than 10% at term and to 1−2% after birth. This results in a marked increase in the ratio of phosphatidylcholine (lecithin, L) to sphingomyelin (S). Because fetal lung fluid contributes to amniotic fluid, measurement of the L-to-S ratio in human amniotic fluid obtained by amniocentesis has been used to predict the degree of fetal lung maturity, and hence the optimum time for elective delivery with respect to adequacy of lung surfactant (Tsao and Zachman, 1982). There is a reciprocal relationship between phosphatidylgylcerol and phosphatidylinositol in fetal lung lavage and amniotic fluid at late gestation (Hallman and Gluck, 1980; Kulovich et al., 1979). As phosphatidylglycerol increases, phosphatidylinositol decreases (Table 20). Measurement of these acidic phospholipids in human amniotic fluid, especially the increase in the level of phosphatidylglycerol, leads to greater accuracy as compared to the L-to-S ratio in the prediction of fetal lung maturity (Kulovich et al., 1979).

The rate of radiolabeled choline incorporation into phosophatidylcholine or disaturated phosphatidylcholine has been widely used to assess the rate of biosynthesis of these phospholipids in the developing fetal lung. Despite possible developmental differences in the pool sizes of choline and related intermediates, such measurements have correlated well with other measures of phospholipid synthesis, such as increased phospholipid content and increased enzyme activities (Rooney, 1983, 1985). As shown in Figures 7−9, there are developmental increases in the rates of choline incorporation into phosphatidylcholine in the fetal rat, rabbit, and mouse lung. Similar increases in choline incorporation, as well as in the incorporation of other precursors, have been reported for other species (Table 21). As shown in Figure 7, there is also a developmental increase in the rate of de novo fatty acid biosynthesis during late fetal life, followed by a decrease after birth. Concomitantly with the increase in rates of substrate incorporation, developmental increases in the activities of choline-phosphate cytidylyltransferase, phosphatidate phosphatase, and fatty-acid synthase have been reported (Table 21).

As shown in Figures 7−9, there is a developmental increase, followed by a subsequent decrease, in the glycogen content of the fetal rat, rabbit, and mouse lung. Similar changes were reported in the guinea pig and rhesus monkey (Table 22). A corresponding increase, followed by a decrease, in the activity of glycogen synthase—the enzyme that catalyzes the final step in glycogen biosynthesis (Huijing, 1975)—was reported in the rat and rabbit lung, whereas the activities of glycogen phosphorylase and 6-phosphofructokinase—an important regulatory enzyme in glycolysis (Hers and Hue, 1983)—were reported to increase concomitantly with the decrease in glycogen content (Table 22). Because the decrease in glycogen occurs at the time of increased choline incorporation into phosphatidylcholine (Figures 7−9), this has led to speculation that glycogen provides substrate or energy for phospholipid biosynthesis in the fetal lung (Farrell and Bourbon, 1986; Maniscalco et al., 1978). A direct relationship between glycogen depletion and increased phospholipid synthesis has not been proven, however. It is noteworthy that the temporal relationship that exists between glycogen depletion and increased phosphatidylcholine biosynthesis also exists between glycogen depletion and increased fatty acid biosynthesis (Figure 7). Therefore, it is just as likely that

TABLE 20 Parameters of Fetal Lung Maturation and Surfactant Production Which Change During Late Fetal and/or Early Postnatal Life: Surfactant Phospholipid Content

Lung Tissue			Lung Lavage or Amniotic Fluid		
Increased PC/DSPC[a] Content		Increased % PC/DSPC and PC/Sph Ratio	Decreased % Sph	Increased % PG	Increased and Later Decreased % PI
Rat[b]	Funkhouser and Hughes (1978), Maniscalco et al. (1978)				
Rabbit	Oulton and Dolphin, (1988), Rooney et al. (1976)	Oulton et al. (1986), Rooney et al. (1976)	Oulton et al. (1986), Rooney et al. (1976)	Oulton et al. (1986), Freese and Hallman (1983)	Freese and Hallman (1983)
Guinea pig	Sosenko and Frank (1987)	Benson et al. (1983)	Benson et al. (1983)	Ikegami and Jobe (1981)	
Sheep	Brumley et al. (1967)	Ikegami and Jobe (1981)	Ikegami and Jobe (1981)		
Horse	Arvidson et al. (1975)	Arvidson et al. (1975)			
Macaque Monkey Rhesus	Epstein et al. (1976), Perelman et al. (1982)	Epstein and Farrell (1975), Epstein et al. (1976)	Gluck et al. (1974)		
Pigtail	Palmer et al. (1977)	Palmer et al. (1977)			
Stumptail		Feijen et al. (1979)			
Baboon		Gluck et al. (1974), Kling and Kotas (1975)			
Human	Adams et al. (1970)	Kulovich et al. (1979), Oulton et al. (1982)	Oulton et al. (1982)	Kulovich et al. (1979), Muneschige et al. (1983)	Kulovich et al. (1979), Oulton et al. (1982)

[a] *Phospholipid abbreviations as in Tables 2 and 6.*
[b] *References are restricted to two per species/parameter.*

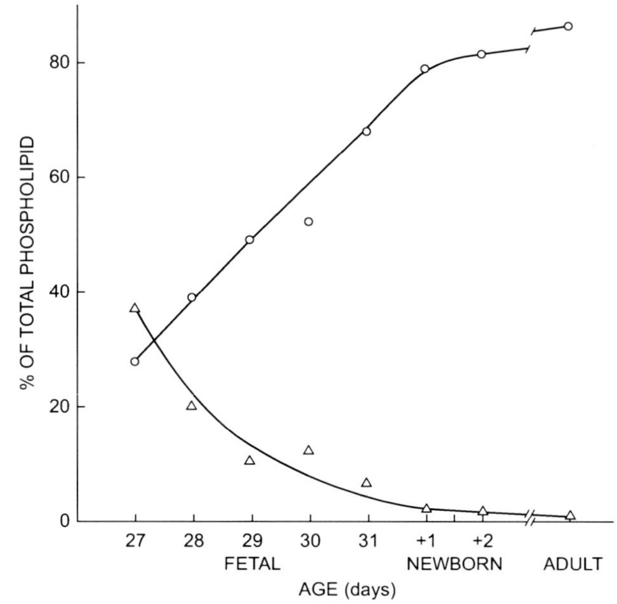

FIGURE 6 Developmental changes in the amount of phosphatidylcholine (O) and sphingomyelin (Δ) in lung lavage from fetal, newborn, and adult rabbits. *From Rooney, S.A. 1979. The Surfactant System and the Neonatal Lung, Mead Johnson Co., Evansville, IN, 17–24, Used by permission.*

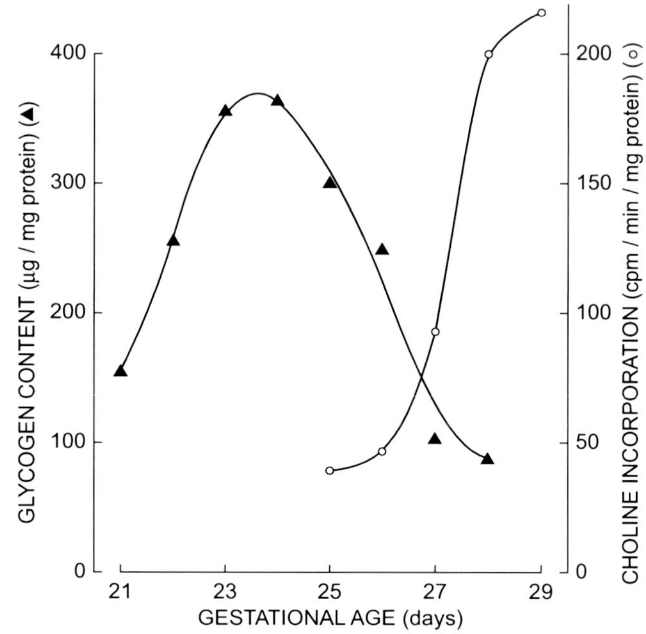

FIGURE 8 Developmental changes in glycogen content and in the rate of choline incorporation into phosphatidylcholine in fetal rabbit lung. *From Rooney, S.A., 1983. The Biological Basis of Reproductive and Developmental Medicine. In: Warshaw, J.B., (Ed.), 239–287, Copyright, 1983, Elsevier Science Publishing Co., Inc. Used by permission.*

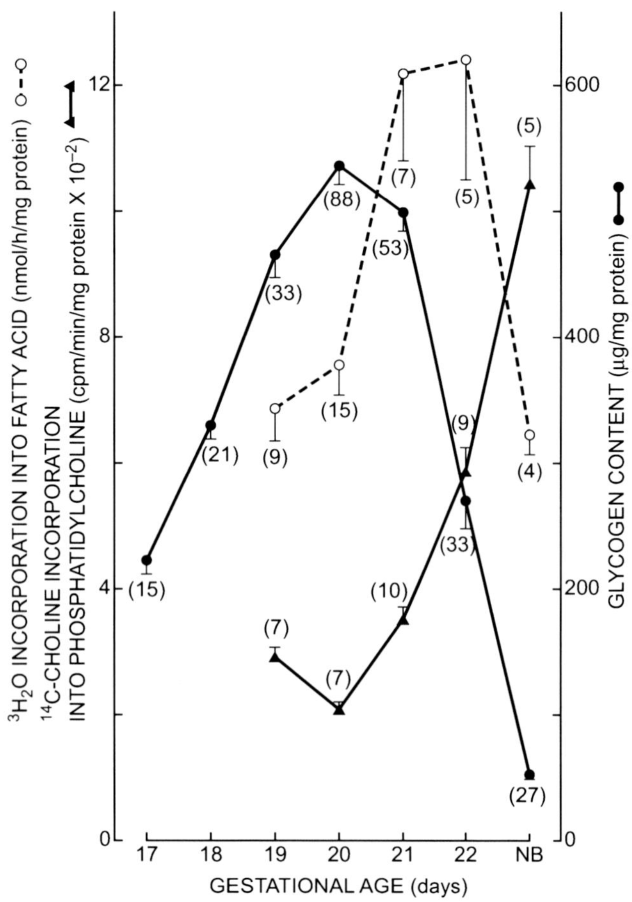

FIGURE 7 Developmental changes in glycogen content and in rates of choline incorporation into phosphatidylcholine and of ^3H_2O incorporation into fatty acids in fetal rat lung. *From Rooney, S.A., et al. 1986. Pediatr. Res. 20, 545, Used by permission.*

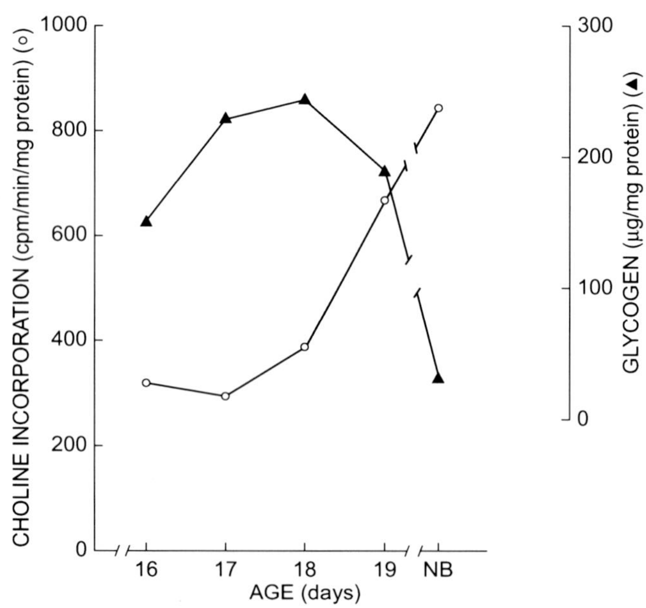

FIGURE 9 Developmental changes in glycogen content and in the rate of choline incorporation into phosphatidylcholine in fetal mouse lung. *From Brehier, A., Rooney S.A., 1981. Exp. Lung Res. 2, 883, Used by permission.*

TABLE 21 Parameters of Fetal Lung Maturation and Surfactant Production Which Change During Late Fetal and/or Early Postnatal Life: Phosphatidylcholine and Fatty Acid Biosynthesis[a]

	Increased Rate of Substrate Incorporation			Increased Enzyme Activities		
	Choline into PC/DSPC	Others Substrates into PC/DSPC	^3H_2O into Fatty Acid	Choline-Phosphate Cytidylyltransferase (23)[b]	Phosphatidate Phosphatase (19)	Fatty-Acid Synthase (18)
Rat	Maniscalco et al. (1978), Rooney et al. (1986)	Harding et al. (1983)	Maniscalco et al. (1982), Rooney et al. (1986)	Casola et al. (1980), Maniscalco et al. (1978)	Casola and Possmayer (1981a), Oosterhuis et al. (1984)	Pope and Rooney (1987)
Rabbit	Gross et al. (1983), Rooney et al. (1979)			Heath and Jacobson (1984), Rooney et al. (1976)	Casola and Possmayer (1981b), Schultz et al. (1974)	Das (1980)
Mouse	Brehier and Rooney (1981)			Brehier and Rooney (1981), Olderborg and van Golde (1977)	Oosterhuis et al. (1984)	
Sheep	Chida and Adams (1967)	Chida and Adams (1967)				
Rhesus monkey	Epstein and Farrell (1975)					
Human					Jimenez et al. (1975)	

[a]*See footnotes to Table 20.*
[b]*Number in parenthesis is the reaction described in Figures 1–5.*

TABLE 22 Parameters of Fetal Lung Maturation and Surfactant Production Which Change During Late Fetal and/or Early Postnatal Life: Glycogen Content and Metabolism[a]

		Enzyme Activities		
	Lung Glycogen Content (Increase; Later Decrease)	**Glycogen Synthase (EC 2.4.1.11)[b] (Increase; Later Decrease)**	**Glycogen Phosporylase (1) (Increase)**	**6-Phophofructokinase (EC 2.7.1.11) (Increase)**
Rat	Maniscalco et al. (1978)	Maniscalco et al. (1978)	Maniscalco et al. (1978)	Pope and Rooney (1987)
	Rooney et al. (1986)	Pope and Rooney (1987)	Pope and Rooney (1987)	Rijksen et al. (1985)
Rabbit	Brandstrup and Kretchmer (1965)	Bhavani (1983)	Bhavani (1983)	
	Khosla et al. (1983)		Brandstrup and Kretchmer (1965)	
Mouse	Brehier and Rooney (1981)			
Guinea pig	Sosenko and Frank (1987)			
Rhesus monkey	Perelman et al. (1982)			

[a] *See footnotes to Table 20.*
[b] *EC number are given for reactions not described in Figures 1–5.*

glycogen provides substrate or energy for the biosynthesis of fatty acids as for that of phospholipids.

5. REGULATION OF LIPID BIOSYNTHESIS

Discussion of all of the possible regulatory reactions in the metabolic pathways illustrated in Figures 1–5 is beyond the scope of this chapter. The focus will be on steps that are of particular importance and that have been well studied in the lung. Because of the developmental and hormone-induced increases in surfactant synthesis that occur during late gestation, the fetal lung has been extensively used to study the regulation of lung phospholipid biosynthesis.

5.1 Fatty Acid Biosynthesis

The relationship between fatty acid biosynthesis and surfactant production is of interest because fatty acids are obligatory components of phospholipids. As shown in Table 19, the lung is clearly capable of de novo fatty synthesis. The type II cell has been reported to have a relatively high rate of de novo fatty acid biosynthesis (Maniscalco et al., 1983; Voelker, et al., 1976), as well as of acetyl-CoA carboxylase (Voelker et al., 1976) and fatty-acid synthase (Gonzales et al., 1990; Voelker et al., 1976) activities. Lung de novo fatty acid synthesis has been shown to increase developmentally toward the end of gestation in the fetal rat (Maniscalco et al., 1982; Rooney et al., 1986) and to be increased by glycocorticoids in the fetal rat (Patterson et al., 1986; Rooney et al., 1986) and rabbit (Maniscalco et al., 1985). Compared to exogenous palmitate, fatty acids synthesized de novo from acetate were reported to be preferentially incorporated into surfactant total and disaturated phosphatidylcholine in newborn and adult rabbit lung (Jobe, 1979; Jobe et al., 1980). However, fatty acids can be supplied to the lung in the form of glycerides (Gal et al., 1982) as well as free acids (Darrah et al., 1973; Thomas et al., 1970), so the quantitative importance for phospholipid synthesis of de novo fatty acid biosynthesis in the lung versus the exogenous supply might vary with metabolic conditions.

De novo fatty acid biosynthesis (Figure 2) can be inhibited by agaric acid, which inhibits citrate transfer from the mitochondrion to the cytosol (Freedland and Newton, 1981); by hydroxycitrate, which inhibits ATP citrate (pro-3S)-lyase (Lowenstein and Brunengraber, 1981); and by 5-(tetradecyloxy)-2-furoic acid, which inhibits acetyl-CoA carboxylase (Harris and McCune, 1981). Inhibition of de novo fatty acid synthesis with such agents in adult (Patterson and Rhoades, 1988) and fetal (Patterson et al., 1988) rat and in fetal rabbit (Maniscalco et al., 1989) lung resulted in decreased disaturated phosphatidylcholine synthesis. This suggests an important role for fatty acid synthesis in surfactant production. More than 85% of newly synthesized fatty acids in fetal rat lung were recovered in the phospholipid fraction, and 80% of these were in phosphatidylcholine (Xu and Rooney, 1989). Dexamethasone increased the distribution of newly synthesized fatty acids into disaturated phosphatidylcholine and phosphatidylglycerol. However, there was no preferential incorporation into lamellar bodies compared to a residual fraction not enriched in surfactant. This shows that newly synthesized fatty acids are not preferentially incorporated into surfactant. On the other hand, inhibition of de novo fatty acid biosynthesis by the addition of agaric acid or hydroxycitrate to the culture medium abolished the stimulatory effect of dexamethasone on choline phosphate cytidylyltransferase activity in fetal rat lung explants (Xu et al., 1990). As discussed further below, this suggests that fatty acids mediate the stimulatory effect of dexamethasone on this important regulatory enzyme in phosphatidylcholine biosynthesis. Taken together, these data indicate that fatty acids directly regulate surfactant phospholipid synthesis as well as provide substrate. In addition to stimulating choline-phosphate cytidylyltransferase, it is of interest that fatty acids have also been reported to stimulate cholinephosphotransferase (Radominska-Pyrek et al., 1976; Sribney and Lyman, 1973; Whitlon et al., 1985) and phosphatidate phosphatase (Brindley, 1984) activities in some systems.

Acetyl-CoA carboxylase is considered to be the rate-regulatory enzyme in fatty acid biosynthesis in liver and adipose tissue (Goodridge, 1985; Rosenthal, 1987; Wakil et al., 1983). Its catalytic activity is rapidly altered by covalent modification (phosphorylation) and allosteric factors (citrate, acyl-CoA). In addition to such short-term regulation, the amount of the enzyme is also altered in some conditions. The activity of acetyl-CoA carboxylase was reported to decrease developmentally in the fetal rat lung (Pope and Rooney, 1987), although a 1-day peak in the activity of this enzyme was reported at 21 days gestation in the same species (Maniscalco et al., 1982). Fetal lung acetyl-CoA carboxylase activity did not increase developmentally in the rabbit (Gross and Warshaw, 1974) and was not increased by glucocorticoids in the rat (Pope and Rooney, 1987).

In contrast to acetyl-CoA carboxylase, there is no evidence that fatty acid synthase plays a role in the short-term regulation of fatty acid biosynthesis (Goodridge, 1985), although it clearly does have a role in long-term regulation. The amount of fatty acid synthase and its mRNA are increased in association with increased fatty acid biosynthesis in liver (Back et al., 1986; Goodridge, 1986; Nepokroeff et al., 1984; Paulauskis and Sul, 1988, 1989) and mammary gland (Braddock and Hardie, 1988). Lung fatty acid synthase activity has been reported to increase developmentally in the fetal rat (Pope and Rooney, 1987) and rabbit (Das, 1980) and to be increased by glucocorticoids in the fetal rat (Pope and Rooney, 1987; Pope et al., 1988) and human (Gonzales et al., 1990). Glucocorticoid treatment induced both fatty acid synthase protein (Pope et al., 1988) and mRNA (Palayoor et al., 1988) in the fetal lung. The activity and amount of fatty acid synthase also increased in the lungs of hypophysectomized, diabetic, or starved adult rats after refeeding or treatment with hormones (Das and Ganguly, 1982; Kumar, 1977), establishing a role for fatty acid synthase in the regulation of pulmonary fatty acid biosynthesis.

5.2 Phosphatidylcholine Biosynthesis

Because phosphatidylcholine is the major component of pulmonary surfactant and is also an important membrane phospholipid, regulation of its biosynthesis has been the subject of investigation in many laboratories. Such studies have established that the de novo pathway for the biosynthesis of phosphatidylcholine is regulated both by the availability of substrate (choline) and by the rate-limiting enzyme, CTP:choline-phosphate cytidyl transferase (CCT, #23 in Table 19) in the lung (Possmayer, 1988b; Rooney, 1985) and in type II cell (Post et al., 1982a; Post et al., 1984a,b), as it is in other animal (Pelech and Vance, 1984; Sleight and Kent, 1983) and plant (Price-Jones and Harwood, 1986) systems. Free choline is accumulated in lung cells by both Na^+-dependent and Na^+-independent transport processes (Dodia et al., 1992; Fisher et al., 1989a, 1992b). Choline that is generated intracellularly by enzymatic degradation of phosphatidylcholine is avidly retained and reutilized for resynthesis of phosphatidylcholine (Chander et al., 1987; Fisher et al., 1987b; Nakamura, et al., 2010). The evidence for the important role of CTT has come from measurements of intracellular levels of choline and choline metabolites, as well as from pulse-chase studies that have established that the choline phosphate pool is larger than that of choline or CDPcholine. In addition, the activity of choline-phosphate cytidylyltransferase is almost always increased in association with increased phosphatidylcholine biosynthesis. In the fetal lung, there is a developmental increase in choline-phosphate cytidylyltransferase activity in association with the increased phosphatidylcholine biosynthesis at the end of gestation, and glucocorticoids, estrogen, and thyroid hormone, which increase fetal lung phosphatidylcholine biosynthesis, have been shown to also increase choline-phosphate cytidylyltransferase activity (Rooney, 1983, 1985).

A number of mechanisms for the activation of choline-phosphate cytidylyltransferase have been proposed (Goss et al., 2013). Cytidylyltransferase activity can be increased by translocation from the cytosol to the endoplasmic reticulum (Pelech and Vance, 1984; Sleight and Kent, 1983;

Terce et al., 1988; Weinhold et al., 1984), by dephosphorylation (Pelech and Vance, 1982; Radika and Possmayer, 1985), as well as by phospholipids (Choy and Vance, 1978; Chu and Rooney, 1985b; Fiscus and Schneider, 1966; Stern et al., 1976) and free fatty acids (Chander and Fisher, 1988; Feldman et al., 1981; Pelech et al., 1983; Weinhold et al., 1984). Clearly, these mechanisms are not mutually exclusive. Activation of choline-phosphate cytidylyltransferase by free fatty acids, for instance, has been reported to be mediated by translocation of the enzyme from cytosol to microsomes (Pelech et al., 1983; Weinhold et al., 1984). Dephosphorylation has also been suggested to promote translocation (Pelech and Vance, 1984). There is evidence that each of these mechanisms can operate in the lung. Choline-phosphate cytidylyltransferase activity has been reported to be increased by fatty acids in adult rat lung (Feldman et al., 1981), as well as in fetal rabbit (Aeberhard et al., 1985) and adult rat (Burkhardt et al., 1988; Chander and Fisher, 1988) type II cells. Fatty acids were reported to promote translocation of the enzyme from cytosol to microsomes in fetal and newborn rat lung (Weinhold et al., 1984). Phospholipid activation of choline-phosphate cytidylyltransferase has been reported in adult and fetal rabbit (Chu and Rooney, 1985a,b) and rat (Stern et al., 1976) lung and in adult rat type II cells (Gilfillan et al., 1985; Rosenberg et al., 1987). Dephosphorylation has been reported to increase lung choline-phosphate cytidylyltransferase activity in the fetal rabbit (Radika and Possmayer, 1985). It is not clear, however, if all of these activation mechanisms have a physiological role.

In addition to choline-phosphate cytidylyltransferase, it is possible that other enzymes have a limited regulatory role in de novo phospholipid biosynthesis. The activities of phosphatidate phosphatase (Table 21) and other enzymes of phospholipid biosynthesis (Rooney, 1983) have been reported to increase developmentally in fetal lung. There are scattered reports that the activities of phosphatidate phosphatase, cholinephosphotransferase, lysolecithin acyltransferase, lysolecithin:lysolecithin acyltransferase, and CDPdiacylglycerol-glycerol-3-phosphate 2-phosphatidyltransferase are increased by hormones (Rooney, 1983). However, there is considerable disagreement with respect to these effects in that there are as many, and often more, reports of no effects of stimulation with hormones (Rooney, 1983, 1985; Van Golde et al., 1988).

The remodeling pathway for phosphatidylcholine biosynthesis has not been as well studied and the rate-limiting steps are not as well understood. Lysophosphatidylcholine, the substrate for reacylation, is a potentially toxic agent in high concentration (membrane lysis), and its concentration is therefore maintained at relatively low levels in cells. Logically then, the availability of lysophosphatidylcholine would represent an important step in the regulation of the reacylation pathway. Lysophosphatidylcholine is generated by the activity of phospholipase A_2, a relatively large family of enzymes with both extracellular and intracellular forms. As indicated in Table 19, several members of this family of enzymes have been implicated in lung phosphatidylcholine biosynthesis. However, it is now clear that the major phospholipase A_2 activity is that from peroxiredoxin 6 (Fisher and Dodia, 2001; Fisher et al., 2005). This enzyme is present in the lung lamellar bodies and generates lysophosphatidylcholine from its major substrate, phosphatidylcholine, but shows no specificity for the fatty acyl composition of the phosphatidylcholine substrate (Akiba et al., 1998; Kim et al., 1997). Thus, activity of this enzyme as part of the reacylation cycle can result in degradation and resynthesis of dipalmitoylphosphatidylcholine or degradation of monounsaturated phosphatidylcholine (an unsaturated fatty acid in the sn-2 position) with resynthesis of the disaturated species. Because the major acyltransferase (LPCAT1) in lung type II cells shows preference for palmitate (Chen et al., 2006), synthesis of disaturated phosphatidylcholine would be the major product of this deacylation−reacylation cycle. Thus, the physiological role of this pathway appears to be geared toward the generation of dipalmitoylphosphatidylcholine, and activity of this pathway in part represents degradation of "old" dipalmitoylphosphatidylcholine with resynthesis of "new" phosphatidylcholine. The physiological role of this perhaps futile cycle is unclear, but some possibilities are discussed below in the section on surfactant recycling. As distinguishing features, peroxiredoxin 6 PLA_2 activity is pH-dependent, with its activity maximal at approximately pH 4, consistent with the acidic pH of lamellar bodies; also, unlike many other members of the PLA_2 family, it is calcium independent (Fisher, 2011). PLA_2 activity is markedly increased by phosphorylation of the protein, but that reaction occurs in the cytosol (Rahaman et al., 2012; Wu et al., 2009a) and there is no evidence as yet that phosphorylation of Prdx6 has a role in lung phospholipid synthesis.

6. LAMELLAR BODY GENESIS

As indicated previously, lamellar bodies are considered lysosomal-related organelles (Weaver et al., 2002) and are characterized by a high content of lysosomal type enzymes and an acidic pH (Chander et al., 1986; Ridsdale et al., 2011). These organelles represent both the storage site of lung surfactant phospholipids (and proteins) and also the secretory organelle responsible for exocytosis of lung surfactant. Also as indicated above, the lung surfactant phospholipids are synthesized in the endoplasmic reticulum and then transferred through a process that is not well defined as of yet to the storage organelle. Lamellar bodies are also the storage site for lipids that are endocytosed from the alveolar space through a recycling process. Lamellar bodies possibly represent a heterogenous population with respect to their content of endogenously synthesized phospholipids versus those obtained through endocytosis, although that is still speculative. In any event, nascent lamellar bodies appear to acquire lipids and mature as they move toward the periphery of the cell prior to secretion. These more mature lamellar bodies demonstrate an increase in their calcium content that possibly plays a role in their subsequent exocytosis (Eckenhoff, 1989). Data suggest that the Rab38 protein may play a role in regulating the size (and maturity) of lamellar bodies (Zhang et al., 2011).

Although the intracellular transport of various proteins has been well studied, the mechanism for transport of lipids between organelles is less well understood. The recent demonstration of the presence of ABCA3 (ATP-binding cassette subfamily A, member 3), a member of a highly conserved multispan transmembrane ABC superfamily of transporters that use the energy of ATP hydrolysis to translocate a variety of substrates across cell membranes, has provided important insights into lung lamellar body formation. The human ABCA3 gene has been mapped to chromosome 16p13.3 and encodes a 1704-amino acid protein that contains 12 membrane spanning helices with two cytosolic ATP-binding domains (Connors et al., 1997; Klugbauer and Hofmann, 1996). While the ABCA3 transporter is expressed in many tissues, it is expressed at especially high levels in alveolar type II (AT2) cells, localizing at the limiting membrane of lamellar bodies (Mulugeta et al., 2002; Yamano et al., 2001). ABCA3 expression in the lung is developmentally regulated in that its production increases with advancing gestational age and peaks just before delivery (Mulugeta et al., 2002; Stahlman et al., 2007).

ABCA3 functions as a lipid and phospholipid transporter; it also has been recognized as a critical regulator of lung surfactant metabolism. Located at the limiting membrane of lamellar bodies of AT2 cells, ABCA3 plays an essential role in lipid transport including phosphatidylcholine (PC), phosphatidylglycerol (PG), sphingomyelin, and cholesterol (Ban et al., 2007; Cheong et al., 2006). The importance of ABCA3 in lamellar body biogenesis and surfactant regulation is underscored by an abundance of reports linking mutations in the ABCA3 gene with neonatal lung disorders. ABCA3 deficiency is inherited as an autosomal recessive mutation, and more than 150 distinct mutations have been identified; the vast majority cause severe neonatal lung disease, including respiratory distress syndrome and fatal surfactant deficiency (Doan et al., 2008; Hamvas, 2010; Shulenin et al., 2004; Wert et al., 2009; Whitsett et al., 2010). Surfactant from patients with ABCA3-related lung disease is deficient in both PG and PC and has little surface activity (Garmany et al., 2006). Mouse models of ABCA3 deficiency recapitulate the human phenotype. While homozygous ABCA3 null mice died within a few hours following birth, primarily due to acute respiratory failure, heterozygotic mice had diminished levels of surfactant PC and PG (Ban et al., 2007; Cheong et al., 2007). In both mouse models and patients with ABCA3 deficiency, lamellar bodies in AT2 cells at the ultrastructural level are abnormal, showing small intracellular organelles containing electron-dense material (Ban et al., 2007; Cheong et al., 2007; Shulenin et al., 2004), consistent with the abnormal lipid transport mechanism.

Phospholipid exchange proteins are another group of proteins that have been proposed to have a role in the transfer of phospholipids between organelles. In vitro, these proteins transfer phospholipid from a donor to an acceptor organelle or artificial vesicle. The presence of phospholipid exchange proteins in supernatant from lung homogenates (Engle et al., 1978; Robinson et al., 1978) and in isolated granular pneumocytes (Pool et al., 1983) has been demonstrated. The lung phospholipid exchange protein had the ability to transfer phosphatidylcholine to acceptor organelles, such as microsomes and mitochondria, but less efficiently to lamellar bodies. Furthermore, a mouse with knockout of the phosphatidylcholine transfer protein had no abnormalities of surfactant metabolism (Kanno et al., 2007). Thus, at this time, there does not appear to be an important role for the phosphatidylcholine exchange proteins in the transfer of newly synthesized lipids to lamellar bodies.

Although phospholipids are the predominant lipids in surfactant, cholesterol comprises most of the neutral lipid at 10–15% of the total lipids (Yu et al., 1983). Cholesterol is a critical component in the determination of surfactant physical properties (Veldhuizen et al., 1998). Low-density lipoproteins (LDL), taken up by type II cells via receptor-mediated endocytosis, can serve as the source of cholesterol for the pneumocytes with subsequent transfer into the lamellar bodies (Hass and Longmore, 1979; Voyno-Yasenetskaya et al., 1993). Experimentally, labeled cholesterol added to the perfusate of an isolated lung preparation was recovered in surfactant obtained by lung lavage (Hass and Longmore, 1979). Homeostasis of exogenous cholesterol in cells is complex. For most cells to use the cholesterol derived from exogenous lipoproteins, the endocytosed LDL passes through lysosomes, where the cholesteryl ester is hydrolyzed by lysosomal acid lipase, releasing unesterfied cholesterol and fatty acid (Kwon et al., 2009). Lysosomes then regulate the trafficking of unesterfied cholesterol out of the organelles into the cytoplasm through the Niemann-Pick C (NPC) pathway that consists of two cholesterol-binding proteins: Niemann-Pick C1, a large 170–190 kDa transmembrane protein, and NPC2, a smaller 16–22 kDa soluble lysosomal protein (Carstea et al., 1997; Liou et al., 2006; Loftus et al., 1997; Naureckiene et al., 2000). The accepted current model predicts that unesterfied cholesterol is bound by NPC2 protein that then docks to the second luminal domain of NPC1 (Deffieu et al., 2011; Infante et al., 2008b; Xu, et al., 2007), where cholesterol is exchanged in a "hydrophobic handoff" between NPC2 and the N-terminal domain of NPC1 (Infante et al., 2008a; Wang, et al., 2010). Thus, cholesterol is exported from the lysosome (lamellar body) by NPC1 to either the plasma membrane or the endoplasmic reticulum, possibly through oxysterol-binding protein-related protein 5 (Du et al., 2011) or by Rab9-dependent vesicular transport back to the Golgi (Kaptzan et al., 2009; Walter et al., 2003).

The possibility that the NPC pathway plays a major role in the regulation of surfactant cholesterol composition was suggested by a demonstration that lamellar bodies in isolated type II cells contain all of the constituents of the NPC pathway normally present in lysosomes (Roszell et al., 2012). NPC1 protein in isolated lamellar bodies colocalized with the lamellar body limiting membrane protein ABCA3, while lysosomal acid lipase and NPC2 were found in the lumen of the lamellar bodies (Roszell et al., 2012). Thus, in the specialized case of type II cell lamellar bodies, the lipoprotein cholesterol delivered to the organelle can remain for use in surfactant, solubilized by the excess phospholipid, or be removed by the NPC pathway. Evidence that the NPC pathway is functional in lamellar bodies was obtained with U18666A, an amphiphile that blocks cellular cholesterol trafficking by the NPC pathway (Cenedella, 2009). This inhibitor resulted in cholesterol accumulation in lamellar bodies (Roszell et al., 2012). Genetic disruption of the NPC pathway due to dysfunctional NPC2 in a patient with Niemann-Pick type C2 disease resulted in a marked increase in surfactant cholesterol (50% of total lipid), supporting a link between the NPC pathway and the regulation of surfactant cholesterol levels (Griese et al., 2010).

7. SURFACTANT SECRETION

Secretion of surfactant contained in the intracellular lamellar bodies to the alveolar lumen occurs by exocytosis (Ryan et al., 1975). This is an energy-dependent process that requires intact microtubular and microfilament systems (Hollingsworth and Gilfillan., 1984; Rooney, 1985; Tsilibary and Williams, 1983a,b). The exocytotic process is Ca^{2+}-dependent and requires the mediation of several proteins (SNARE, SNAP, annexins) to complete the membrane fusion events that are required for this process (Chander and Fisher, 1990; Chander et al., 2013; Dietl and Haller, 2005; Frick et al., 2004; Gerelsaikhan et al., 2012). Lamellar body exocytosis is a relatively slow event that is characteristic of epithelial secretion as contrasted with the rapid events in neuronal tissue. Details of the secretory process are beyond the scope of this chapter, but physiological regulation of secretion will be discussed.

The regulation of surfactant secretion has been studied in a variety of experimental models, including the intact animal in vivo (Corbet et al., 1976; Lawson et al., 1979; Oyarzun and Clements, 1978), the isolated perfused lung (Brown and Longmore, 1981; Nicholas and Barr, 1981), lung slices (Gilfillan et al., 1983b; Marino and Rooney, 1980), and isolated type II cells (Table 23). Intact lung and whole animal studies are clearly essential in elucidating the influence of physiological factors. Using such models, it has been shown that secretion of surfactant is stimulated by ventilation in adult (Massaro and Massaro, 1983; Oyarzun and Clements, 1978), and newborn (Lawson et al., 1979; Rooney and Gobran, 1988) animals and by labor in the newborn (Rooney et al., 1977). It has been proposed that a periodic deep breath or sigh provides the stimulus for the continuous secretion of lung surfactant. In the intact lung, however, it is often difficult to distinguish between direct effects on secretion and those that are secondary to increased synthesis or other effects. In addition, especially when intact animals are used, effects may not be directly on the lung but may be mediated via other organs or endocrine systems. The isolated type II cell in tissue culture overcomes many of the problems associated with the intact lung, and it is used to study the mechanism of surfactant secretion.

In this model, freshly isolated cells are usually cultured in the presence of [^{3}H]choline for 18–24 h to label intracellular phosphatidylcholine. After transfer to fresh medium, the agent of interest is added. After a further 1–3 h incubation, the amount of [^{3}H]phosphatidylcholine in the medium is measured. Secretion is usually expressed as the amount of disaturated [^{3}H]phosphatidylcholine in the medium as a percentage of the total in the cells and medium combined. Basal rates of secretion in type II cells from adult animals are usually 0.5–1% per hour, although this underestimates the true rate of surfactant phospholipid secretion because not all of the intracellular phosphatidylcholine is associated with surfactant. Studies with isolated cells have confirmed that a mechanical stimulus, such as stretch, leads to increased exocytosis (Frick et al., 2004; Wirtz and Dobbs, 1990), providing support for the role of ventilation in directly modulating surfactant secretion.

Phosphatidylcholine secretion in type II cells is stimulated by a variety of agents, including beta-adrenergic agonists and other agents that elevate intracellular cAMP levels, vasopressin, endothelin-1, prostaglandins and leukotrienes, purinoceptor agonists, agents that activate protein kinase C, agents that increase the level of intracellular calcium, and antihistamines (Table 23). There is evidence that some of these agents have a physiological (as contrasted with pharmacological) role in the regulation of surfactant secretion. A role for beta-agonists was suggested by the findings that the stimulatory effects of labor in newborn rabbit lung slices (Marino and Rooney, 1981) and

TABLE 23 Agents That Have Been Reported to Stimulate Phosphatidylcholine Secretion in Cultured Type II Cells[a]

	Increase (Control = 1.0)	Molar Concentration Used	EC_{50}[b] (Molar)	References[c]
β-Adrenergic Agonists				
Epinephrine	2.4	1×10^{-5}		Dobbs and Mason (1979)
Terbutaline	3.1	1×10^{-5}	8×10^{-7}	Dobbs and Mason (1979)
	2.1	1×10^{-5}	1×10^{-7}	Gilfillan et al. (1987)
	3.5	1×10^{-5}		Rice et al. (1985)
Terbutaline[d]	2.0	5×10^{-6}		Robinson et al. (1984)
Terbutaline[e]	1.8	1×10^{-5}		Kresch et al. (1987)
Terbutaline[f]	6.8[g]	1×10^{-5}		Ballard et al. (1986)
Isoproterenol	3.0	1×10^{-5}	4×10^{-9}	Dobbs and Mason (1979)
	2.7	1×10^{-6}	1×10^{-9}	Brown and Longmore (1981)
	2.1	1×10^{-7}	3×10^{-9}	Gilfillan et al. (1987)
Isoproterenol[h]	1.7	1×10^{-5}		Saito et al. (1985)
Other Agents Elevating cAMP Levels				
Forskolin	4.5	5×10^{-6}	3×10^{7}	Rice et al. (1985)
	3.2	2×10^{-5}		Sano et al. (1987)
	3.6	5×10^{-6}		Gobran and Rooney (1990)
Cholera toxin	2.0	1×10^{-9}	$\sim 10^{-12}$	Mescher et al. (1983)
P_1 Purinoceptor Agonists				
Adenosine	1.7	1×10^{-5}	5×10^{-6}	Gilfillan and Rooney (1987a)
NECA[1]	2.0	1×10^{-5}	9×10^{-8}	Gilfillan and Rooney (1987b)
L-PIA[1]	2.2	1×10^{-5}		Gilfillan and Rooney (1987b)
Arachidonic Acid Metabolites				
Arachidonic acid	1.8	8×10^{-6}		Gilfillan and Rooney (1985)
Leukotriene C_4	1.6	1×10^{-6}		Gilfillan and Rooney (1986)
Leukotriene D_4	1.8	1×10^{-6}	5×10^{-11}	Gilfillan and Rooney (1986)
Leukotriene E_4	2.1	1×10^{-6}	4×10^{-13}	Gilfillan and Rooney (1986)
Prostaglandin $F_{2\alpha}$[j]	2-5	1×10^{-6}		Anderson et al. (1978)
Agents Activating Protein Kinase C				
TPA[i]	8.4	1×10^{-8}		Dobbs and Mason (1978)
	8.1	1×10^{-8}		Dobbs et al. (1986b)
TPA[d]	4.5	6×10^{-8}		Robinson et al. (1984)
TPA[f]	10.9[g]	1×10^{-8}		Ballard et al. (1986)

Continued

TABLE 23 Agents That Have Been Reported to Stimulate Phosphatidylcholine Secretion in Cultured Type II Cells[a]—cont'd

	Increase (Control = 1.0)	Molar Concentration Used	EC_{50}[b] (Molar)	References[c]
Diacylycerols				
1-Oleoyl-2-acetyl glycerol	5.5	2×10^{-4}		Sano et al. (1985)
Dioctanoylglycerol	6.0	1×10^{-4}		Sano et al. (1987)
Calcium Ionophores				
A23187	1.8	5×10^{-6}		Mason et al. (1977)
	2.5	1×10^{-7}		Dobbs et al. (1986b)
A23187[k]	8.9	1×10^{-5}		Shapiro et al. (1978)
A23187[f]	2.9	1×10^{-7}		Ballard et al. (1986)
Ionomycin	2.5	1×10^{-4}		Sano et al. (1987)
P_2 Purinoceptor Agonists				
ATP	5.1	1×10^{-3}	4×10^{-6}	Gilfillan and Rooney (1987a)
	5.9	1×10^{-5}	10^{-6}–10^{-7}	Rice and Singleton (1986b)
				Rice and Singleton (1987)
ADP	6.7	1×10^{-3}	2×10^{-5}	Gilfillan and Rooney (1987a)
Antihistamines				
Thiazinamium chloride	1.5	1×10^{-6}	3×10^{-8}	Gilfillan et al. (1987)
Pyrilamine	1.5	1×10^{-5}	6×10^{-9}	Gilfillan et al. (1987)
Promethazine	1.5	1×10^{-6}		Gilfillan et al. (1987)

[a]*Unless indicated otherwise all cells were isolated from the adult rat.*
[b]*The concentration at which 50% of the maximum effect is achieved.*
[c]*References are restricted to three per item.*
[d]*Adult human lung cells.*
[e]*Fetal rat lung cells.*
[f]*Fetal human lung cells.*
[g]*Basal secretion was very low in this study.*
[h]*Cells isolated from 6-day-old rats.*
[i]*Abbreviations: NECA, 5′-N-ethylcarboxyamidoadenosine; PIA, N^6-phenylisopropyldenosine; TPA, 12-O-tetradecanoyl-13-phorbol acetate.*
[j]*L-21-cloned adult rat type II cells (Douglas and Kaighn, 1974).*
[k]*A549 Human tumor cells (Lieber et al., 1976).*

of ventilation in intact adult rabbits (Oyarzun and Clements, 1978) were blocked by beta-adrenergic antagonists. Beta-adrenergic receptors have been demonstrated in type II cells (Das et al., 1987; Jones et al., 1987; Sommers Smith and Giannopoulos, 1983). There is a developmental increase in lung beta receptors toward the end of gestation in the fetus (Cheng et al., 1980; Warburton et al., 1987a; Whitsett et al., 1981), and this is accelerated by glucocorticoids (Cheng et al., 1980; Maniscalco and Shapiro, 1983), estrogen (Moawad et al., 1985), and thyroid hormone (Whitsett et al., 1980). Administration of beta-agonists in intact fetal and newborn animals stimulated surfactant secretion (Abdellatif and Hollingsworth, 1980; Corbet et al., 1977; Enhorning et al., 1977; Lawson et al., 1978), and these agents have also been reported to reduce the incidence of respiratory distress syndrome when given to pregnant women to delay labor (Bergman and Hedner, 1978; Boog et al., 1975; Kero et al., 1973).

A physiological role for arachidonic acid metabolites and their conversion to prostaglandins in surfactant secretion was suggested by the findings that cyclooxygenase inhibitors blocked the increase in secretion produced by labor (Marino and Rooney, 1981) and ventilation (Oyarzun and Clements, 1978; Rooney and Gobran, 1988). A role for leukotrienes in the regulation of surfactant secretion was suggested by the finding that the ventilation-induced increase in surfactant in lung lavage in the newborn rabbit was decreased by the leukotriene antagonists FPL-55712 and FPL-57231, and by the lipoxygenase inhibitor nordihydroguaiaretic acid (Rooney and Gobran, 1988). A physiological role for cholinergic agonists in surfactant secretion was suggested by the finding that the ventilation induced increase in secretion in adult and newborn rabbits was blocked by atropine (Lawson et al., 1979; Oyarzun and Clements, 1977). Cholinergic stimulation of surfactant secretion in intact newborn rabbits was also blocked by a beta-agonist and by adrenalectomy (Abdellatif and Hollingsworth, 1980; Corbet et al., 1977), suggesting that the mechanism for the effect of cholinergic agents was to stimulate the adrenal medulla to produce catecholamines, which in turn act directly on the type II cell. Support for this indirect effect was given by the findings that cholinergic agonists did not stimulate phosphatidylcholine secretion in the isolated perfused adult rat lung (Nicholas and Barr, 1981), in fetal rabbit lung slices (Marino and Rooney, 1980), or in isolated type II cells from the adult rat (Dobbs and Mason, 1979) or human (Robinson et al., 1984).

Purinoceptor agonists also affect surfactant secretion. Purinoceptors have been subdivided into two major classes, P_1 (adenosine) and P_2 (ATP) receptors (Burnstock and Buckley, 1985), and there is functional evidence that both exist on the type II cell. Phosphatidylcholine secretion in type II cells was stimulated approximately twofold by adenosine and nonmetabolizable adenosine analogs (Gilfillan and Rooney, 1987a) and by as much as eightfold by ATP and its analogs (Gilfillan and Rooney, 1987b, 1988; Rice and Singleton, 1986a, 1987). Each of the two major purinoceptors have been further subdivided. The P_1 receptor is divided into A, and A_2 subclasses (Burnstock and Buckley, 1985), while at least five subgroups of the P_2 receptor have been proposed (Burnstock and Kennedy, 1985; Gordon, 1986; Wiklund and Gustafsson, 1988). There is evidence that the stimulatory effect of adenosine in type II cells is mediated by the A_1 subtype of the P_1, purinoceptor and involves an increase in the level of intracellular cAMP (Gilfillan and Rooney, 1987b). A physiological role for this receptor in the regulation of surfactant secretion was suggested by the finding that an adenosine receptor antagonist, 8-phenyltheophylline, diminished the stimulatory effect of ventilation on surfactant secretion in the newborn rabbit (Rooney and Gobran, 1988). Secretion of phosphatidylcholine in type II cells is inhibited by substance P (Rice and Singleton, 1986b), possibly acting through an A_1 adenosine receptor (Gobran and Rooney, 1990).

Another physiologically important regulator of surfactant secretion is surfactant protein A (SP-A) (Bates et al., 2003; Dobbs et al., 1987; Kuroki et al., 1988; Rice and Singleton, 1987), although the precise mechanism is not understood. Inhibition of secretion by SP-A requires binding of the protein to its cell membrane receptor (Bates, 2010) and presumably involves an as-yet-unidentified cell signaling pathway or possibly direct regulation of the fusion pore that is necessary for secretion (Bates et al., 2003). The SP-A receptor on the type II cell membrane has been identified as a protein called P63 (Bates et al., 2008b; Gupta et al., 2006).

There are various pharmacologic agents that inhibit surfactant secretion when studied in vitro. Colchicine and vinblastine, agents that disrupt microtubules (Wilson et al., 1974), have been reported to inhibit secretion in intact fetal rabbits (Corbet et al., 1988), adult hamster (Delahunty and Johnston, 1976), and newborn rabbit (Marino and Rooney, 1980) lung slices and in adult rat type II cells (Brown et al., 1985; Dobbs and Mason, 1978). Disruption of microfilaments with cytochalasins has been reported to inhibit surfactant secretion in some studies (Marino and Rooney, 1980; Tsilibary and Williams, 1983a) but to stimulate it in others (Brown et al., 1985; Rice et al., 1984). The reason for these discrepant effects is not clear, but it may be related to the concentration of cytochalasin used (Rice et al., 1984). Cytochalasin has been reported to have both stimulatory and inhibitory effects on secretion in other systems (Davies and Allison, 1978). Inhibition of secretion by some plant lectins (Rice and Singleton, 1988) also has been reported. However, none of these agents can be considered physiologic.

8. SURFACTANT LIPID RECYCLING

What is the fate of the lung surfactant phospholipids after secretion? Clearly, the rate of removal of phospholipids from the alveolar space has to equal the rate of secretion in order to maintain a constant amount of extracellular lipids. There are several possible pathways for removal, including the transport up the bronchial tree where surfactant could be swallowed, in situ degradation in the alveolar space, or pericellular transport through the epithelium for removal by lymphatics. Each of these potential pathways for surfactant removal has been shown to be negligible in terms of surfactant turnover (Fisher, 1998b). It is now known that the primary pathway for removal is reuptake (endocytosis) by the lung epithelial type II cells (Chander et al., 1983; Chinoy et al., 1993, 1994; Fisher et al., 1987a, 1985b, 1989b, 1991; Muller et al., 1995; Poelma et al., 2004; Rider et al., 1992; Tsuzuki et al., 1993; Wissel et al., 2001). Alveolar macrophages can also phagocytize and degrade surfactants, but this process appears to be important only where there has been excessive accumulation of lipids, usually associated with disease (Poelma et al., 2004; Rider et al., 1992). Under normal circumstances, the secretion of surfactant and the reuptake process are linked in order to maintain constant extracellular surfactant. This "linkage" has been demonstrated with hyperventilation (Jain et al., 2003) and with secretagogues (Fisher et al., 1985b, 1989b, 1991) where increased secretion is associated with increased uptake.

Type II cells endocytose surfactant phospholipids by both clathrin-dependent and non-clathrin-dependent pathways (Jain et al., 2005; Muller et al., 1995). Under normal circumstances, the clathrin-dependent pathway predominates (Bates et al., 2008a). This pathway involves surfactant protein A (SP-A) (Bates et al., 2008a). SP-A binds to phosphatidylcholines in the alveolar space with especially high affinity for dipalmitoylphosphatidylcholine, the major phospholipid component of surfactant, and in turn is bound to its receptor on the type II cell membrane. The receptor-SP-A-phosphatidylcholine complex is internalized by the clathrin-dependent pathway of receptor-mediated endocytosis. As described above, the SP-A membrane receptor is P63 protein (Bates et al., 2008b; Gupta, et al., 2006). The effect of secretagogues on phospholipid uptake is mediated by recruitment of SP-A receptors to the cell surface (Chen et al., 1996, 1998), a process that is regulated by protein kinase C activity (Kazi et al., 2010). Although precise details are not known, it is assumed that the internalized P63/SP-A/phospholipid complex dissociates and the P63 protein is returned to the cell surface and the SP-A is resecreted. The processing of the internalized phospholipid is more complex. There is good evidence that some of the phospholipid is directed toward lamellar bodies and is resecreted intact without further processing (Jacobs et al., 1983, 1982). This may be the preferred route when the demand for surfactant is great as in the newborn. Under normal circumstances, the major fraction of the internalized lipid is directed to lamellar bodies, where it is degraded and the components reutilized for lipid resynthesis (Chander et al., 1987; Fisher et al., 1987b, 1991, 1992a). The pathways for degradation of phosphatidylcholine, the role of phospholipase A_2, and the resynthesis of phosphatidylcholine by reacylation have been described above.

Whether there are specific lamellar bodies for direct recycling of surfactant lipids versus those for degradation and resynthesis is not known. As pointed out above, degradation of phospholipids can occur in lamellar bodies, but the reacylation pathway requires enzymes that are present in the endoplasmic reticulum. Thus, it is possible that lamellar bodies close to the endoplasmic reticulum and the cell nucleus are involved in the degradation/resynthesis of surfactant lipids, while the more mature lamellar bodies closer to the cell surface are the site for recycling.

Evidence for reutilization of surfactant components has been obtained by analysis of the turnover time (biological half-time) in surfactant obtained by lung lavage following the intravenous injection of radiolabeled precursors. The biological half-time of radiolabeled dipalmitoylphosphatidylcholine in multiple studies of rat and rabbit surfactant following intravenous injection of radiolabeled palmitate was about 16 h (Gross et al., 1988; Hallman et al., 1981; Jacobs et al., 1982; Jobe, 1977; Jobe et al., 1978; Spitzer and Norman, 1971; Thomas and Rhoades, 1970; Tierney et al., 1967; Toshima et al., 1972). The biological half-time was about the same for labeled surfactant phosphatidylcholine following injection of glycerol or glucose. On the other hand, the calculated PC turnover time with labeled choline as substrate was considerably longer at 42 h. The longer turnover time for choline is compatible with degradation of phosphatidylcholine with liberation of choline (or choline phosphate) and its retention in the type 2 cell for phosphatidylcholine resynthesis.

Both the synthesis as well as the reuptake of surfactant is energy (ATP) dependent (Fisher et al., 1985b). It is reasonable to question why the lung spends significant resources in the process of surfactant recycling. The answer is partly philosophical and must be speculative in the absence of data. One possible explanation is that this process represents quality control with the opportunity to reprocess damaged surfactant components. Another, perhaps more reasonable, explanation might be that the turnover process represents a mechanism for assaying and controlling the amount of surfactant present in the extracellular space.

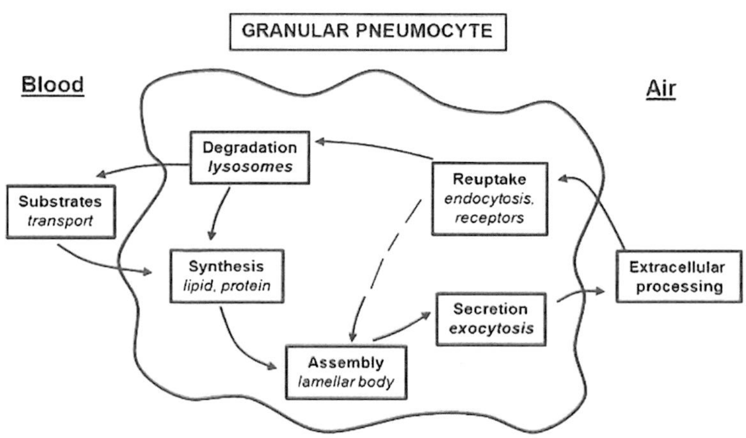

FIGURE 10 Schematic of a type II alveolar epithelial cell and its role in the surfactant cycle.

To summarize, surfactant phospholipids have a complex lifecycle with the major steps of intracellular synthesis, storage in the lamellar bodies, secretion by exocytosis, reuptake by receptor-mediated endocytosis, and intracellular degradation with reutilization of components for phospholipid resynthesis. This cycle is illustrated in Figure 10.

ACKNOWLEDGMENT

I thank Drs. Sandra Bates and Surufel Mulugeta for assistance with the sections on cholesterol metabolism and ABCA3 function, respectively; Susan Turbitt and Molly McGoey for preparation of the manuscript; and Dr. Seamus Rooney for his encouragement.

REFERENCES

Abdellatif, M.M., Hollingsworth, M., 1980. Pediatr. Res. 14, 916–920.
Abe, M., et al., 1973. Tohoku J. Exp. Med. 109, 163–172.
Abe, M., et al., 1974. J. Bacteriol. 119, 543–546.
Adams, F.H., et al., 1970. J. Pediatr. 77, 833–841.
Aeberhard, E.E., et al., 1985. Biochim. Biophys. Acta 833, 473–483.
Akiba, S., et al., 1998. Comp. Biochem. Physiol. B Biochem. Mol. Biol. 120, 393–404.
Akino, T., Ohno, K., 1981. Crit. Rev. Toxicol. 9, 201–274.
Aloia, R.C., 1981. Comp. Biochem. Physiol. 68B, 203–208.
Amidon, B., Brown, A., Waite, M., 1996. Biochemistry 35, 13995–14002.
Anderson, G.G., et al., 1978. Prostaglandins 16, 923–929.
Ansell, G.B., Spanner, S., 1982. In: Hawthorne, J.N., Ansell, G.B. (Eds.), Phosphatidylserine, Phosphatidylethanolamine and Phosphatidylcholine. Elsevier, Amsterdam, pp. 1–49.
Arche, R., et al., 1987. Arch. Biochem. Biophys. 257, 131–139.
Arvidson, G., et al., 1975. J. Reprod. Fertil. 23 (Suppl.), 663–665.
Augustin-Voss, H.G., et al., 1989. Lung 167, 1–10.
Avery, M.E., et al., 1986. Am. J. Obstet. Gynecol. 155, 2–5.
Ayromlooi, J., et al., 1981. Pediatr. Pharmacol. 1, 209–214.
Back, D.W., et al., 1986. J. Biol. Chem. 261, 4190–4197.
Ballard, P.L., 1986. Hormones and Lung Maturation. Springer-Verlag, Berlin.
Ballard, P.L., et al., 1986. Biochim. Biophys. Acta 883, 335–344.
Ban, N., et al., 2007. J. Biol. Chem. 282, 9628–9634.
Banerjee, D.K., 1982. Science 218, 569–571.
Bangham, A.D., 1987. Lung 165, 17–25.
Barenholz, Y., Gatt, S., 1982. In: Hawthorne, J.N., Ansell, G.B. (Eds.), Sphingomyelin: Metabolism, Chemical Synthesis, Chemical and Physical Properties. Elsevier Biomedical, Amsterdam, pp. 129–177.
Bargoni, N., Tazartes, O., 1983. Ital. J. Biochem. 32, 84–91.
Bassett, D.J., Fisher, A.B., Rabinowitz, J.L., 1974. Am. J. Physiol. 227, 1103–1108.
Batenburg, J.J., Elfring, R.H., 1992. FEBS Lett. 307, 164–168.
Batenburg, J.J., et al., 1979. Biochim. Biophys. Acta 573, 136–144.
Batenburg, J.J., et al., 1985. Biochim. Biophys. Acta 833, 17–24.
Batenburg, J.J., et al., 1986. Biochim. Biophys. Acta 878, 201–209.
Batenburg, J.J., et al., 1988. Biochim. Biophys. Acta 960, 441–453.
Bates, S.R., 2010. Cell. Physiol. Biochem. 25, 41–54.
Bates, S.R., et al., 2003. Am. J. Physiol. Lung Cell. Mol. Physiol. 285, L1055–L1065.
Bates, S.R., et al., 2008a. Am. J. Physiol. Lung Cell. Mol. Physiol. 294, L325–L333.
Bates, S.R., et al., 2008b. Am. J. Physiol. Lung Cell. Mol. Physiol. 295, L658–L669.
Baxter, C.F., et al., 1969. Lipid 3, 243–244.
Benson, B.J., 1980. Proc. Natl. Acad. U. S. A. 77, 808–811.
Benson, B.J., et al., 1983. Biochim. Biophys. Acta 753, 83–88.
Bergman, B., Hedner, T., 1978. Acta Obstet. Gynecol. Scand. 57, 217–221.
Bernhard, W., et al., 2007. Pediatr. Pulmonol. 42, 794–804.
Berry, K.A., et al., 2011. J. Lipid Res. 52, 1551–1560.
Bhavani, B.R., 1983. Can. J. Biochem. Cell. Biol. 61, 191–197.
Bleasdale, J.E., et al., 1978. Biochim. Biophys. Acta 528, 331–343.
Bleasdale, J.E., et al., 1979. Biochim. Biophys. Acta 575, 135–147.
Bleasdale, J.E., et al., 1985. Biochem. J. 232, 539–545.
Bligh, E.G., Dyer, W.J., 1959. Can. J. Biochem. Physiol. 37, 911–917.
Body, D.R., 1971. Lipids 6, 625–629.
Boog, G., Brahim, M.B., Gandar, R., 1975. Br. J. Obstet. Gynaecol. 82, 285–288.
Bortnick, A.E., et al., 2003. Am. J. Physiol. Lung Cell. Mol. Physiol. 285, L869–L878.
Bourbon, J., Jost, A., 1982. Pediatr. Res. 16, 50–56.
Braddock, M., Hardie, D.G., 1988. Biochem. J. 249, 603–607.
Brandstrup, N., Kretchmer, N., 1965. Dev. Biol. 11, 202–216.
Brehier, A., et al., 1977. Biochem. Biophys. Res. Commun. 77, 883–890.
Brehier, A., Rooney, S.A., 1981. Exp. Lung Res. 2, 273–287.
Bridges, J.P., et al., 2010. J. Clin. Invest. 120, 1736–1748.
Brindley, D.N., 1984. Prog. Lipid Res. 23, 115–133.
Brown, L.A., Longmore, W.J., 1981. J. Biol. Chem. 256, 66–72.
Brown, L.A., Pasquale, S.M., Longmore, W.J., 1985. J. Appl. Physiol. 58, 1866–1873.
Brumley, G., van Den Bosch, H., 1977. J. Lipid Res. 18, 523–532.
Brumley, G.W., et al., 1967. J. Clin. Invest. 46, 863–873.
Buechler, K.F., Rhoades, R.A., 1980. Biochim. Biophys. Acta 619, 186–195.
Burd, P.H., 1984. Annu. Rev. Physiol. 46, 617.
Burkhardt, R., et al., 1988. Biochem. J. 254, 495–500.
Burnstock, G., Buckley, N.J., 1985. Methods Pharmacol. 6, 193.
Burnstock, G., Kennedy, C., 1985. Gen. Pharmacol. 16, 433–440.
Caesar, P.A., et al., 1988. Biochem. J. 253, 451–457.
Carstea, E.D., et al., 1997. Science 277, 228–231.
Casola, P.G., et al., 1980. Biochem. Biophys. Res. Commun. 96, 1209–1215.
Casola, P.G., Possmayer, F., 1981a. Biochim. Biophys. Acta 665, 177–185.
Casola, P.G., Possmayer, F., 1981b. Biochim. Biophys. Acta 664, 298–315.
Cassals, C., et al., 1982. Arch. Biochem. Biophys. 217, 422–433.
Cenedella, R.J., 2009. Lipids 44, 477–487.
Chander, A., et al., 1983. Biochim. Biophys. Acta 753, 119–129.
Chander, A., et al., 1986. J. Biol. Chem. 261, 6126–6131.
Chander, A., et al., 2013. Biochim. Biophys. Acta.
Chander, A., Fisher, A.B., 1988. Biochim. Biophys. Acta 958, 343–351.
Chander, A., Fisher, A.B., 1990. Am. J. Physiol 258, L241–L253.
Chander, A., Fisher, A.B., Strauss 3rd, J.F., 1982. Biochem. J. 208, 651–658.
Chander, A., Reicherter, J., Fisher, A.B., 1987. J. Clin. Invest. 79, 1133–1138.
Chen, Q., Bates, S.R., Fisher, A.B., 1996. J. Biol. Chem. 271, 25277–25283.
Chen, Q., et al., 1998. Am. J. Physiol. 275, L38–L46.
Chen, X., et al., 2006. Proc. Natl. Acad. Sci. U. S. A. 103, 11724–11729.

Cheng, J.B., et al., 1980. Endocrinology 107, 1646–1648.

Cheong, N., et al., 2006. J. Biol. Chem. 281, 9791–9800.

Cheong, N., et al., 2007. J. Biol. Chem. 282, 23811–23817.

Chida, N., Adams, F.H., 1967. Pediatr. Res. 1, 364–371.

Chinoy, M.R., Dodia, C., Fisher, A.B., 1993. Am. J. Physiol. 264, L300–L307.

Chinoy, M.R., Fisher, A.B., Shuman, H., 1994. Am. J. Physiol. 266, L713–L721.

Choy, P.C., Vance, D.E., 1978. J. Biol. Chem. 253, 5163–5167.

Chu, A.J., Rooney, S.A., 1985a. Biochim. Biophys. Acta 835, 132–140.

Chu, A.J., Rooney, S.A., 1985b. Biochim. Biophys. Acta 834, 346–356.

Claypool, W.D., Fisher, A.B., 1983. Surv. Synth. Path. Res. 2, 34–40.

Clements, J.A., 1971. Arch. Intern. Med. 127, 387–389.

Connors, T.D., et al., 1997. Genomics 39, 231–234.

Corbet, A.J., et al., 1988. J. Appl. Physiol. 65, 1710–1715.

Corbet, A.J., Flax, P., Rudolph, A.J., 1976. J. Appl. Physiol. 41, 7–14.

Corbet, A.J., Flax, P., Rudolph, A.J., 1977. J. Appl. Physiol.: Respir. Environ. Exerc. Physiol. 43, 1039–1045.

Crecelius, C.A., Longmore, W.J., 1983. Biochim. Biophys. Acta 750, 447–456.

Crecelius, C.A., Longmore, W.J., 1984a. Biochim. Biophys. Acta 795, 247–256.

Crecelius, C.A., Longmore, W.J., 1984b. Biochim. Biophys. Acta 795, 238–246.

Darrah, H.K., Hedley-Whyte, J., 1973. J. Appl. Physiol. 34, 205–213.

Das, D.K., 1980. Biochem. Biophys. Res. Commun. 92, 867–875.

Das, D.K., 1982. Lung 160, 207–218.

Das, D.K., Ganguly, M., 1982. Arch. Biochem. Biophys. 218, 142–155.

Das, S.K., et al., 1984. Biochim. Biophys. Acta 802, 423–427.

Das, S.K., Sikpi, M.O., Skolnick, P., 1987. Biochem. Biophys. Res. Commun. 142, 898–903.

Davies, P., Allison, A.C., 1978. Front. Biol. 46, 143–160.

Deffieu, M.S., Pfeffer, S.R., 2011. Proc. Natl. Acad. Sci. U. S. A. 108, 18932–18936.

Delahunty, R.J., et al., 1979. Am. Rev. Respir. Dis. 119, 75–80.

Delahunty, T.J., Johnston, J.M., 1976. J. Lipid Res. 17, 112–116.

Dethloff, L.A., et al., 1986. Biochem. J. 233, 111–118.

Dietl, P., Haller, T., 2005. Annu. Rev. Physiol. 67, 595–621.

Doan, M.L., et al., 2008. Thorax 63, 366–373.

Dobbs, L.G., et al., 1980. Biochim. Biophys. Acta 618, 510–523.

Dobbs, L.G., et al., 1982. Biochim. Biophys. Acta 713, 118–127.

Dobbs, L.G., et al., 1986b. Biochim. Biophys. Acta 877, 305–313.

Dobbs, L.G., et al., 1987. Proc. Natl. Acad. Sci. U. S. A. 84, 1010–1014.

Dobbs, L.G., Gonzalez, R., Williams, M.C., 1986a. Am. Rev. Respir. Dis. 134, 141–145.

Dobbs, L.G., Gonzalez, R.F., Marinari, L.A., Mescher, E.J., Hawgood, S., 1986. Biochim. Biophys. Acta 877, 305–313.

Dobbs, L.G., Mason, R.J., 1978. Am. Rev. Respir. Dis. 118, 705–733.

Dobbs, L.G., Mason, R.J., 1979. J. Clin. Invest. 63, 378–387.

Dodia, C., et al., 1992. Am. J. Respir. Cell Mol. Biol. 6, 426–429.

Douglas, W.H., Kaighn, M.E., 1974. Vitro 10, 230–237.

Du, X., et al., 2011. J. Cell Biol. 192, 121–135.

Eckenhoff, R.G., 1989. J. Clin. Invest. 84, 1295–1301.

Egberts, J., et al., 1987. Biochim. Biophys. Acta 919, 90–92.

Engle, M.J., van Golde, L.M., Wirtz, K.W., 1978. FEBS Lett. 86, 277–281.

Enhorning, G., et al., 1977. Am. J. Obstet. Gynecol. 129, 197–202.

Epstein, M.F., et al., 1976. Am. J. Obstet. Gynecol. 125, 545–549.

Epstein, M.F., Farrell, P.M., 1975. Pediatr. Res. 9, 658–665.

Farrell, P.M., Bourbon, J.R., 1986. Biochim. Biophys. Acta 878, 159–167.

Feijen, H.W., et al., 1979. Gynecol. Obstet. Invest. 10, 186–192.

Feldman, D.A., Brubaker, P.G., Weinhold, P.A., 1981. Biochim. Biophys. Acta 665, 53–59.

Filguerias, O.M., Possmayer, F., 1987. Lipids 22, 731–735.

Finkelstein, J.D., et al., 1983. J. Nutr. 113, 519–521.

Finkelstein, J.N., Mavis, R.D., 1979. Lung 156, 243–254.

Finkelstein, J.N., Shapiro, D.L., 1982. Lung 160, 85–98.

Fiscus, W.G., Schneider, W.C., 1966. J. Biol. Chem. 241, 3324–3330.

Fisher, A., Chander, A., 1985. Ann. Rev. Physiol. 47, 789–802.

Fisher, A.B., 1985. In: Witschi, H., Brain, J. (Eds.), The Isolated Perfused Lung. Springer-Verlag, Heidelberg, pp. 149–179.

Fisher, A.B., 1998. Surfactant Catabolism and Recycling. Landes Bioscience, Austin, TX.

Fisher, A.B., 2011. Antioxid. Redox Signal. 15, 831–844.

Fisher, A.B., Chander, A., 1982. Biochim. Biophys. Acta 711, 128–133.

Fisher, A.B., Chander, A., Dodia, C., 1990. Prog. Resp. Res. 25, 149–157.

Fisher, A.B., Chander, A., Reicherter, J., 1987a. Am. J. Physiol. 253, C792–C796.

Fisher, A.B., Dodia, C., 1997. Am. J. Physiol. 272, L238–L243.

Fisher, A.B., Dodia, C., 2001. Am. J. Physiol. Lung Cell. Mol. Physiol. 280, L748–L754.

Fisher, A.B., Dodia, C., Chander, A., 1985a. J. Appl. Physiol. 59, 743–748.

Fisher, A.B., Dodia, C., Chander, A., 1985b. Am. J. Physiol. 249, E89–E93.

Fisher, A.B., Dodia, C., Chander, A., 1987b. J. Appl. Physiol. 62, 2295–2299.

Fisher, A.B., Dodia, C., Chander, A., 1989b. Am. J. Physiol. 257, L248–L252.

Fisher, A.B., Dodia, C., Chander, A., 1991. Am. J. Physiol. 261, L334–L340.

Fisher, A.B., et al., 1976. J. Laboratory Clin. Med. 87, 1033–1040.

Fisher, A.B., et al., 1980. J. Appl. Physiol. Respir. Environ. Exerc. Physiol. 49, 743–750.

Fisher, A.B., et al., 1989a. Am. J. Respir. Cell. Mol. Biol. 1, 455–462.

Fisher, A.B., et al., 1992a. Biochem. J. 288 (Pt 2), 407–411.

Fisher, A.B., et al., 1992b. Am. J. Physiol. 263, C1250–C1257.

Fisher, A.B., et al., 2005. J. Lipid Res. 46, 1248–1256.

Fisher, A.B., Furia, L., 1977. Lung 154, 155–165.

Folch, J., Lees, M., Sloane Stanley, G.H., 1957. J. Biol. Chem. 226, 497–509.

Franson, R.C., Weir, D.L., 1982. Lung 160, 275–284.

Freedland, R.A., Newton, R.S., 1981. Methods Enzymol. 72, 497–506.

Freese, W.B., Hallman, M., 1983. Biochim. Biophys. Acta 750, 47–59.

Frick, M., et al., 2004. Am. J. Physiol. Lung Cell. Mol. Physiol. 286, L210–L220.

Funkhouser, J.D., Hughes, E.R., 1978. Biochim. Biophys. Acta 530, 9–16.

Gal, S., et al., 1982. Biochim. Biophys. Acta 713, 222–229.

Garcia, A., et al., 1975. Biochem. Biophys. Res. Commun. 64, 128–135.

Garmany, T.H., et al., 2006. Pediatr. Res. 59, 801–805.

Geppert, E.F., Elstein, K.H., 1983. Exp. Lung Res. 4, 281–291.

Gerelsaikhan, T., Vasa, P.K., Chander, A., 2012. Biochim. Biophys. Acta 1823, 1796–1806.

Gil, J., Reiss, O.K., 1973. J. Cell Biol. 58, 152–171.

Gilfillan, A.M., et al., 1983. J. Lipid Res. 24, 1651–1656.
Gilfillan, A.M., et al., 1983a. J. Lipid Res. 24, 1651–1656.
Gilfillan, A.M., et al., 1986. Biochim. Biophys. Acta 877, 151–157.
Gilfillan, A.M., et al., 1987. Biochem. Pharmacol. 36, 277–281.
Gilfillan, A.M., Hollingsworth, M., Jones, A.W., 1983b. Br. J. Pharmacol. 79, 363–371.
Gilfillan, A.M., Rooney, S.A., 1985. Biochim. Biophys. Acta 833, 336–341.
Gilfillan, A.M., Rooney, S.A., 1987a. J. Pharmacol. Exp. Ther. 241, 907–914.
Gilfillan, A.M., Rooney, S.A., 1987b. Biochim. Biophys. Acta 917, 18–23.
Gilfillan, A.M., Rooney, S.A., 1988. Biochim. Biophys. Acta 959, 31–37.
Gilfillan, A.M., Smart, D.A., Rooney, S.A., 1985. Biochim. Biophys. Acta 835, 141–146.
Gilfillian, A.M., Rooney, S.A., 1986. Biochim. Biophys. Acta 876, 22–27.
Gluck, L., et al., 1974. Am. J. Obstet. Gynecol. 120, 524–530.
Gluck, L., Kulovich, M.V., 1973. Pediatr. Clin. North Am. 20, 367–379.
Gluck, L., Landowne, R.A., Kulovich, M.V., 1970. Pediatr. Res. 4, 352–364.
Goadby, P., 1974. Br. J. Pharmacol. 52, 144P.
Gobran, L.I., Rooney, S.A., 1990. Am. J. Physiol. 258, L45–L51.
Gonzales, L.W., et al., 1990. Biochim. Biophys. Acta 1042, 1–12.
Goodridge, A.G., 1985. In: Vance, D.E., Vance, J.E. (Eds.), Fatty Acid Synthesis in Eucaryotes. Benjamin/Cummings, Menlo Park, pp. 143–180.
Goodridge, A.G., 1986. Fed. Proc. 45, 2399–2405.
Gordon, J.L., 1986. Biochem. J. 233, 309–319.
Goss, V., Hunt, A.N., Postle, A.D., 2013. Biochim. Biophys. Acta 1831, 448–458.
Griese, M., et al., 2010. Clin. Genet. 77, 119–130.
Gross, I., et al., 1979. Biochim. Biophys. Acta 575, 375–383.
Gross, I., et al., 1983. Endocrinology 112, 829–837.
Gross, I., et al., 1984. Pediatr. Res. 18, 191–196.
Gross, I., Rooney, S.A., Warshaw, J.B., 1975. Pediatr. Res. 9, 752–755.
Gross, I., Warshaw, J.B., 1974. Pediatr. Res. 8, 193–199.
Gross, N.J., Barnes, E., Narine, K.R., 1988. J. Appl. Physiol. 64, 2017–2025.
Gupta, N., et al., 2006. Am. J. Physiol. Lung Cell. Mol. Physiol. 291, L436–L446.
Haagsman, H.P., van Golde, L.M., 1985. Lung 163, 275–303.
Hallman, M., Epstein, B.L., Gluck, L., 1981. J. Clin. Invest. 68, 742–751.
Hallman, M., et al., 1982. J. Clin. Invest. 70, 673–683.
Hallman, M., et al., 1983. Pediatrics 71, 473–482.
Hallman, M., Gluck, L., 1974. Biochem. Biophys. Res. Commun. 60, 1–7.
Hallman, M., Gluck, L., 1975. Biochim. Biophys. Acta 409, 172–191.
Hallman, M., Gluck, L., 1980. Pediatr. Res. 14, 1250–1259.
Hallman, M., Merritt, T.A., Schneider, H., Epstein, B.L., Mannino, F., Edwards, D.K., Gluck, L., 1983. Pediatrics 71, 473–482.
Hamvas, A., 2010. Chin. Med. J. Engl. 123, 2943–2947.
Hanamura, H., et al., 1980. Clin. Orthop. Relat. Res. 148, 281–290.
Hanqing, M., et al., 1986. Biochim. Biophys. Acta 878, 360–370.
Haq, R.U., Tsao, F., Shrago, E., 1987a. Biochim. Biophys. Acta 918, 36–39.
Haq, R.U., Tsao, F., Shrago, E., 1987b. J. Lipid Res. 28, 216–220.
Harding, P.G., et al., 1983. Biochim. Biophys. Acta 750, 373–382.
Harris, R.A., McCune, S.A., 1981. Methods Enzymol. 72, 552–559.
Harwood, J.L., Desai, R., Hext, P., Tetly, T., Richards, R., 1975. Biochem. J. 151, 707–714.
Harwood, J.L., et al., 1975. Biochem. J. 151, 707–714.
Hass, M.A., Longmore, W.J., 1979. Biochim. Biophys. Acta 573, 166–174.
Heath, M.F., Jacobson, W., 1984. Pediatr. Res. 18, 395–401.
Henderson, R.F., Clayton, M.H., 1976. Anal. Biochem. 70, 440–446.
Henderson, R.F., Escobedo, L.G., 1976. Lab. Anim. Sci. 26, 899–901.
Hendry, A.T., Possmayer, F., 1974. Biochim. Biophys. Acta 369, 156–172.
Herbert, W.N., et al., 1978. Am. J. Obstet. Gynecol. 132, 373–379.
Hers, H.G., Hue, L., 1983. Annu. Rev. Biochem. 52, 617–653.
Hoehn, S.K., et al., 1983. Lung 161, 229–234.
Hollingsworth, M., Gilfillan, A.M., 1984. Pharmacol. Rev. 36, 69–90.
Hostetler, K.Y., 1982. In: Hawthorne, J.N., Ansell, G.B. (Eds.), Polyglycerophospholipids: phosphatidylglycerol, diphosphatidylglycerol and bis(monacylglycero)phosphate. Elsevier, Amsterdam, pp. 215–261.
Huijing, F., 1975. Physiol. Rev. 55, 609–658.
Ide, H., et al., 1988. Biochim. Biophys. Acta 960, 119–124.
Ide, H., Weinhold, P.A., 1982. J. Biol. Chem. 257, 14926–14931.
Ikegami, M., Jobe, A., 1981. Am. J. Obstet. Gynecol. 141, 227–229.
Infante, R.E., et al., 2008a. J. Biol. Chem. 283, 1064–1075.
Infante, R.E., et al., 2008b. Proc. Natl. Acad. Sci. U. S. A. 105, 15287–15292.
Ishidate, K., et al., 1985. Biochim. Biophys. Acta 833, 1–8.
Jacobs, H., et al., 1982. J. Biol. Chem. 257, 1805–1810.
Jacobs, H., et al., 1983. J. Biol. Chem. 258, 4159–4165.
Jain, D., et al., 2003. Am. J. Physiol. Lung Cell. Mol. Physiol. 284, L759–L765.
Jain, D., et al., 2005. Am. J. Physiol. Lung Cell. Mol. Physiol. 289, L1011–L1018.
Jimenez, J.M., et al., 1975. Obstet. Gynecol. 46, 588–590.
Jobe, A., 1977. Biochim. Biophys. Acta 489, 440–453.
Jobe, A., 1979. Biochim. Biophys. Acta 572, 404–412.
Jobe, A., Ikegami, M., 1987. Am. Rev. Respir. Dis. 136, 1256–1275.
Jobe, A., Ikegami, M., Sarton-Miller, I., 1980. Biochim. Biophys. Acta 617, 65–75.
Jobe, A., Kirkpatrick, E., Gluck, L., 1978. J. Biol. Chem. 253, 3810–3816.
Johnston, J.M., et al., 1978. Biochim. Biophys. Acta 531, 65–71.
Jones, L.G., et al., 1987. Lung 165, 201–211.
Kanfer, J.N., Spielvogel, C.H., 1975. Lipids 10, 391–394.
Kanno, K., et al., 2007. Biochim. Biophys. Acta 1771, 654–662.
Kaptzan, T., et al., 2009. Am. J. Pathol. 174, 14–20.
Kates, M. (Ed.), 1986. Techniques of Lipidology. Elsevier.
Kazi, A.S., et al., 2010. Am. J. Physiol. Lung Cell. Mol. Physiol. 299, L794–L807.
Kero, P., Hirvonen, T., Valimaki, I., 1973. Lancet 2, 198 (passim).
Kerr, G.R., Helmuth, A.C., 1974. Biol. Neonate 251, 10–22.
Khan, A.Q., et al., 1985. Lipids 20, 7–10.
Khosla, S.S., et al., 1980. Biochim. Biophys. Acta 617, 282–290.
Khosla, S.S., et al., 1983. Biochim. Biophys. Acta 750, 112–126.
Kikkawa, M.J., Kikkawa, Y., Yoneda, K., Smith, F., Packard, B., Suzuki, K., 1975. Lab. Invest. 32, 295–302.
Kikkawa, Y., et al., 1975. Lab. Invest. 32, 295–302.
Kikkawa, Y., Yoneda, K., 1974. Lab. Invest. 30, 76–84.
Kikkawa, Y., Yoneda, K., Smith, F., Packard, B., Suzuki, K., 1975. Lab. Invest. 32, 295–302.

Kim, T.S., et al., 1997. J. Biol. Chem. 272, 2542–2550.
King, R.J., 1982. J. Appl. Physiol. Respir. Environ. Exerc. Physiol. 53, 1–8.
King, R.J., Clements, J.A., 1972. Am. J. Physiol. 223, 715–726.
Kling, O.R., Kotas, R.V., 1975. Am. J. Obstet. Gynecol. 121, 664–668.
Klugbauer, N., Hofmann, F., 1996. FEBS Lett. 391, 61–65.
Kresch, M.J., Dynia, D.W., Gross, I., 1987. Biochim. Biophys. Acta 930, 19–32.
Kresch, M.J., et al., 1988. Biochim. Biophys. Acta 962, 173–177.
Kuge, O., Nishijima, M., Akamatsu, Y., 1985. Proc. Natl. Acad. Sci. U. S. A. 82, 1926–1930.
Kulovich, M.V., Hallman, M.B., Gluck, L., 1979. Am. J. Obstet. Gynecol. 135, 57–63.
Kumar, R., King, R.J., Hanahan, D.J., 1985. Biochim. Biophys. Acta 836, 19–26.
Kumar, S., 1977. Arch. Biochem. Biophys. 183, 625–637.
Kuroki, Y., Mason, R.J., Voelker, D.R., 1988. Proc. Natl. Acad. Sci. U. S. A. 85, 5566–5570.
Kwon, H.J., et al., 2009. Cell 137, 1213–1224.
Lawson, E.E., et al., 1978. Am. Rev. Respir. Dis. 118, 1023–1026.
Lawson, E.E., et al., 1979. Pediatr. Res. 13, 611–614.
Lecerf, J., Bodin, J.L., 1984. J. Physiol. Paris 79, 345–351.
Lecerf, J., Fouilland, L., Gagniarre, J., 1987. Biochim. Biophys. Acta 918, 48–59.
Lieber, M., et al., 1976. Int. J. Cancer 17, 62–70.
Liggins, G.C., et al., 1988. In: A Randomized Double Blind Trial of Antipartum TRH and Steroids in the Prevention of Newborn RDS (Aukland).
Liou, H.L., et al., 2006. J. Biol. Chem. 281, 36710–36723.
Loftus, S.K., et al., 1997. Science 277, 232–235.
Longmore, W.J., et al., 1979. Biochim. Biophys. Acta 572, 452–460.
Longmuir, K.J., Johnston, J.M., 1980. Biochim. Biophys. Acta 620, 500–508.
Lowenstein, J.M. (Ed.), 1969. Lipids. Academic Press.
Lowenstein, J.M., Brunengraber, H., 1981. Methods Enzymol. 72, 486–497.
Magoon, M.W., et al., 1983. Biochim. Biophys. Acta 750, 18–31.
Maniscalco, W.M., et al., 1978. Biochim. Biophys. Acta 530, 333–346.
Maniscalco, W.M., Finkelstein, J.N., Parkhurst, A.B., 1982. Biochim. Biophys. Acta 711, 49–58.
Maniscalco, W.M., Finkelstein, J.N., Parkhurst, A.B., 1983. Biochim. Biophys. Acta 751, 462–469.
Maniscalco, W.M., Finkelstein, J.N., Parkhurst, A.B., 1985. Pediatr. Res. 19, 1272–1277.
Maniscalco, W.M., Finkelstein, J.N., Parkhurst, A.B., 1989. Exp. Lung Res. 15, 473–489.
Maniscalco, W.M., Shapiro, D.L., 1983. Pediatr. Res. 17, 274–277.
Marino, P.A., Rooney, S.A., 1980. Biochim. Biophys. Acta 620, 509–519.
Marino, P.A., Rooney, S.A., 1981. Biochim. Biophys. Acta 664, 389–396.
Mason, R.J., 1978. J. Biol. Chem. 253, 3367–3370.
Mason, R.J., et al., 1977. Fed. Proc. 36, 2697–2702.
Mason, R.J., Huber, G., Vaughan, M., 1972. J. Clin. Invest. 51, 68–73.
Mason, R.J., Nellenbogen, J., Clements, J.A., 1976. J. Lipid Res. 17, 281–284.
Mason, R.J., Williams, M.C., 1980. Biochim. Biophys. Acta 617, 36–50.
Massaro, G.D., Massaro, D., 1983. Am. Rev. Respir. Dis 127, 235–236.
Mato, J.M., Pajares, M.A., Varela, I., 1984. Trends Biochem. Sci. 9, 471.
Mavis, R.D., Vang, M.J., 1981. Toxicol. Appl. Pharmacol. 58, 262–268.
Melling, J., Keough, K.M.W., 1981. Comp. Biochem. Physiol. 69B, 797–802.
Mescher, E.J., et al., 1983. Exp. Lung Res. 5, 173–182.
Miller, M.A., Kent, C., 1986. J. Biol. Chem. 261, 9753–9761.
Moawad, A.H., River, L.P., Lin, C.C., 1985. Am. J. Obstet. Gynecol. 151, 514–519.
Morales, W.J., et al., 1989. Obstet. Gynecol. 73, 111–116.
Morgan, T.E., 1969. Biochim. Biophys. Acta 178, 21–34.
Morgan, T.E., Finley, T.N., Fialkow, H., 1965. Biochim. Biophys. Acta 106, 403–413.
Muller, W.J., et al., 1995. Am. J. Physiol. 269, L11–L19.
Mulugeta, S., et al., 2002. J. Biol. Chem. 277, 22147–22155.
Muneshige, A., et al., 1983. Am. J. Obstet. Gynecol 145, 474–480.
Murphy, R.C., Gaskell, S.J., 2011. J. Biol. Chem. 286, 25427–25433.
Nakamura, M., et al., 1980. Biochim. Biophys. Acta 6, 24–36.
Nakamura, T., et al., 2010. Biol. Pharm. Bull. 33, 691–696.
Nakanishi, H., et al., 2006. J. Biol. Chem. 281, 20140–20147.
Narasimhan, R., Murray, R.K., 1979. Biochem. J. 179, 199–211.
Naureckiene, S., et al., 2000. Science 290, 2298–2301.
Nepokroeff, C.M., et al., 1984. Eur. J. Biochem. 140, 441–445.
Nicholas, T.E., Barr, H.A., 1981. J. Appl. Physiol. Respir. Environ. Exerc. Physiol. 51, 90–98.
Nijssen, J.G., et al., 1986. Biochim. Biophys. Acta 876, 611–618.
Nikolasev, V., 1974. Lipids 9, 827–830.
Notter, R.H., Finkelstein, J.N., 1984. J. Appl. Physiol. Respir. Environ. Exerc. Physiol. 57, 1613–1624.
Okano, G., Akino, T., 1979. Lipids 14, 541–546.
Okano, G., Kawamoto, T., Akino, T., 1978. Biochim. Biophys. Acta 528, 385–393.
Oldenborg, V., van Golde, L.M., 1976. Biochim. Biophys. Acta 441, 433–442.
Oldenborg, V., Van Golde, L.M., 1977. Biochim. Biophys. Acta 489, 454–465.
Oosterhuis, W.P., et al., 1984. Biol. Neonate 45, 236–243.
Oulton, M., Dolphin, M., 1988. Lipids 23, 55–61.
Oulton, M., et al., 1982. Am. J. Obstet. Gynecol. 142, 684–691.
Oulton, M., et al., 1986. J. Lipid Res. 27, 602–612.
Oyarzun, M.J., Clements, J.A., 1977. J. Appl. Physiol. Respir. Environ. Exerc. Physiol. 43, 39–45.
Oyarzun, M.J., Clements, J.A., 1978. Am. Rev. Respir. Dis. 117, 879–891.
Palayoor, T., Smart, D.A., Rooney, S.A., 1988. FASEB J. 2, A1540.
Palmer, S., et al., 1977. Pediatr. Res. 11, 1057–1063.
Patterson, C.E., Davis, K.S., Rhoades, R.A., 1988. Biochim. Biophys. Acta 958, 60–69.
Patterson, C.E., et al., 1986. Biochim. Biophys. Acta 878, 110–126.
Patterson, C.E., Rhoades, R.A., 1988. FASEB J. 2.
Paulauskis, J.D., Sul, H.S., 1988. J. Biol. Chem. 263, 7049–7054.
Paulauskis, J.D., Sul, H.S., 1989. J. Biol. Chem. 264, 574–577.
Pelech, S.L., et al., 1983. J. Biol. Chem. 258, 6782–6788.
Pelech, S.L., Vance, D.E., 1982. J. Biol. Chem. 257, 14198–14202.
Pelech, S.L., Vance, D.E., 1984. Biochim. Biophys. Acta 779, 217–251.
Perelman, R.H., Engle, M.J., Farrell, P.M., 1981. Perspectives on Fetal Lung Development. Springer-Verlag, Berlin, p. 53.
Perelman, R.H., et al., 1982. J. Appl. Physiol. Respir. Environ. Exerc. Physiol. 53, 230–235.
Pfleger, R.C., 1977. Exp. Mol. Pathol. 27, 152–166.

Pfleger, R.C., Thomas, H.G., 1971. Arch. Intern. Med. 127, 863–872.

Poelma, D.L., et al., 2004. Am. J. Respir. Cell. Mol. Biol. 30, 751–758.

Pool, G.L., et al., 1983. Biochem. J. 215, 637–642.

Pope, T.S., Rooney, S.A., 1987. Biochim. Biophys. Acta 918, 141–148.

Pope, T.S., Smart, D.A., Rooney, S.A., 1988. Biochim. Biophys. Acta 959, 169–177.

Possmayer, F. (Ed.), 1988b. Pulmonary Phosphatidate Phosphohydrolase and its Relation to the Surfactant System of the Lung. CRC Press.

Possmayer, F., 1988a. Am. Rev. Respir. Dis. 138, 990–998.

Possmayer, F., Duwe, G., Metcalfe, R., Stewart-DeHaan, P.J., Wong, C., Heras, J.L., Harding, P.G., 1977. Biochem. J. 166, 485–494.

Possmayer, F., et al., 1977. Biochem. J. 166, 485–494.

Possmayer, F., et al., 1981. Biochim. Biophys. Acta 664, 10–21.

Post, M., et al., 1982a. Biochim. Biophys. Acta 712, 390–394.

Post, M., et al., 1982b. Exp. Lung Res. 3, 17–28.

Post, M., et al., 1983. Biochim. Biophys. Acta 750, 68–77.

Post, M., et al., 1984a. Biochim. Biophys. Acta 795, 552–557.

Post, M., et al., 1984b. Biochim. Biophys. Acta 795, 558–563.

Post, M., van Golde, L.M., 1988. Biochim. Biophys. Acta 947, 249–286.

Postle, A.D., Hunt, A.N., 2009. J. Chromatogr. B Anal. Technol. Biomed. Life Sci. 877, 2716–2721.

Price-Jones, M.J., Harwood, J.L., 1986. Biochem. J. 240, 837–842.

Prueitt, J.L., et al., 1978. J. Lipid Res 19, 410–415.

Radika, K., Possmayer, F., 1985. Biochem. J. 232, 833–840.

Radin, N.S., 1981. Methods Enzymol. 72, 5–7.

Radominska-Pyrek, A., et al., 1976. J. Lipid Res. 17, 657–662.

Rahaman, H., et al., 2012. Biochemistry 51, 5521–5530.

Rami, J., Sasic, et al., 1991. Am. J. Physiol. 260, L577–L585.

Rauvala, H., Hallman, M., 1984. J. Lipid Res. 25, 1257–1262.

Ravinuthala, H.R., et al., 1978. Biochim. Biophys. Acta 530, 347–356.

Renooij, W., Snyder, F., 1981. Biochim. Biophys. Acta 666, 468–474.

Rice, W.R., et al., 1985. Exp. Lung Res. 9, 135–149.

Rice, W.R., Osterhoudt, K.C., Whitsett, J.A., 1984. Biochim. Biophys. Acta 805, 12–18.

Rice, W.R., Singleton, F.M., 1986a. Br. J. Pharmacol. 89, 485–491.

Rice, W.R., Singleton, F.M., 1986b. Biochim. Biophys. Acta 889, 123–127.

Rice, W.R., Singleton, F.M., 1987. Br. J. Pharmacol. 91, 833–838.

Rice, W.R., Singleton, F.M., 1988. Biochim. Biophys. Acta 958, 205–210.

Rider, E.D., Ikegami, M., Jobe, A.H., 1992. Am. J. Physiol. 263, L201–L209.

Ridsdale, R., et al., 2011. PLoS One 6, e16482.

Rijksen, G., et al., 1985. Biochim. Biophys. Acta 838, 114–121.

Rinaudo, M.T., et al., 1981. Int. J. Biochem 13, 571–575.

Robertson, B., Lachmann, B., 1988. Exp. Lung Res. 14, 279–310.

Robinson, M.E., et al., 1978. FEBS Lett. 87, 41–44.

Robinson, P.C., Voelker, D.R., Mason, R.J., 1984. Am. Rev. Respir. Dis 130, 1156–1160.

Rooney, S.A. (Ed.), 1983. Biochemical Development of the Lung. Elsevier.

Rooney, S.A., 1985. Am. Rev. Respir. Dis. 131, 439–460.

Rooney, S.A., Canavan, P.M., Motoyama, E.K., 1974. Biochim. Biophys. Acta 360, 56–67.

Rooney, S.A., et al., 1975. Biochim. Biophys. Acta 398, 433–441.

Rooney, S.A., et al., 1976. Biochim. Biophys. Acta 431, 447–458.

Rooney, S.A., et al., 1976b. Biochim. Biophys. Acta 43, 447–458.

Rooney, S.A., et al., 1979. Biochim. Biophys. Acta 572, 64–76.

Rooney, S.A., Gobran, L.I., 1988. Biochim. Biophys. Acta 960, 98–106.

Rooney, S.A., Gobran, L.I., Chu, A.J., 1986. Pediatr. Res. 20, 545–550.

Rooney, S.A., Gobran, L.I., Wai-Lee, T.S., 1977. J. Clin. Invest. 60, 754–759.

Rooney, S.A., Gross, I., Gassenheimer, L.N., Motoyama, E.K., 1975. Biochim. Biophys. Acta 398, 433–441.

Rooney, S.A., Motoyama, E.K., 1976. Clin. Chim. Acta 69, 525–531.

Rooney, S.A., Wai-Lee, T.S., 1977. Lung 154, 201–211.

Rosenberg, I.L., et al., 1987. Biochim. Biophys. Acta 921, 473–480.

Rosenfeld, C.R., et al., 1980. Pediatr. Res. 14, 891–893.

Rosenthal, M.D., 1987. Prog. Lipid Res. 26, 87.

Roszell, B.R., et al., 2012. Am. J. Physiol. Lung Cell. Mol. Physiol. 302, L919–L932.

Rustow, B., Kunze, D., 1985. Biochim. Biophys. Acta 835, 273–278.

Ryan, U.S., Ryan, J.W., Smith, D.S., 1975. Tissue Cell. 7, 587–599.

Sahu, S., DiAugustine, R.P., Lynn, W.S., 1976. Am. Rev. Respir. Dis. 114, 177–185.

Sahu, S., Lynn, W.S., 1977. Biochim. Biophys. Acta 489, 307–317.

Saito, K., et al., 1985. Exp. Lung Res. 8, 213–225.

Sano, K., et al., 1985. J. Biol. Chem. 260, 12725–12729.

Sano, K., et al., 1987. Am. J. Physiol. 253, C679–C686.

Sass, N.L., et al., 1978. Pediatr. Res. 12, 729–730.

Schiller, H., Bensch, K., 1971. J. Lipid Res. 12, 248–255.

Schlame, M., et al., 1988. Biochem. J. 253, 209–215.

Schlossman, D.M., Bell, R.M., 1977. Arch. Biochem. Biophys. 182, 732–742.

Schultz, F.M., et al., 1974. Gynecol. Invest. 5, 222–229.

Schultz, F.M., Jimenez, J.M., McDonald, P.C., Johnston, J.M., 1974. Gynecol. Invest. 5, 222–229.

Scott, J.E., et al., 1987. Biochim. Biophys. Acta 921, 75–81.

Scott, J.E., et al., 1988. Exp. Mol. Pathol. 49, 141–150.

Shapiro, D.L., et al., 1978. Biochim. Biophys. Acta 530, 197–207.

Shelley, S.A., et al., 1984. Lipids 19, 857–862.

Shimojo, T., Abe, M., Ota, M., 1974. J. Lipid Res. 15, 525–527.

Shukla, V.K., 1988. Prog. Lipid Res. 27, 5–38.

Shulenin, S., et al., 2004. N. Engl. J. Med. 350, 1296–1303.

Sikpi, M.O., Das, S.K., 1987. Biochim. Biophys. Acta 899, 35–43.

Sikpi, M.O., et al., 1986. Biochim. Biophys. Acta 877, 20–30.

Sleight, R., Kent, C., 1983. J. Biol. Chem. 258, 836–839.

Slomiany, A., Smith, F.B., Slomiany, B.L., 1979a. Eur. J. Biochem 98, 47–51.

Slomiany, B.L., Smith, F.B., Slomiany, A., 1979b. Biochim. Biophys. Acta 574, 480–486.

Smith, F.B., Kikkawa, Y., 1979. Lab. Invest. 40, 172–177.

Smith, S.K., et al., 1987. Life Sci. 41, 1097–1107.

Snyder, C., et al., 1973. Cancer Res. 33, 2437–2443.

Snyder, F., Malone, B., 1975. Biochem. Biophys. Res. Commun. 66, 914–919.

Sommers Smith, S.K., Giannopoulos, G., 1983. Life Sci. 33, 2071–2078.

Sosenko, I.R., Frank, L., 1987. Am. J. Physiol. 252, R693–R698.

Spitzer, H.L., Johnston, J.M., 1978. Biochim. Biophys. Acta 53, 275–285.

Spitzer, H.L., Norman, J.R., 1971. Arch. Intern. Med. 127, 429–435.

Sribney, M., Lyman, E.M., 1973. Can. J. Biochem. 51, 1479–1486.

Stahlman, M.T., et al., 2007. J. Histochem. Cytochem. 55, 71–83.

Stern, W., Kovac, C., Weinhold, P.A., 1976. Biochim. Biophys. Acta 441, 280–293.

Stevens, P.A., Wright, J.R., Clements, J.A., 1987. J. Appl. Physiol. 63, 1049–1057.

Stith, I.E., Das, S.K., 1982. Biochim. Biophys. Acta 714, 250–256.
Strickland, K.P., 1973. The chemistry of phospholipids. In: Ansell, G.B., Hawthorne, J.N. (Eds.), Form and Function of Phospholipids. Elsevier Scientific, Amsterdam, pp. 9–42.
Terce, F., et al., 1988. J. Biol. Chem. 263, 3142–3149.
Tesan, M., et al., 1985. Biochem. J. 232, 705–713.
Thom, M.L., Zachman, R.D., 1975. Pediatr. Res. 9, 201–205.
Thomas Jr, T., Rhoades, R.A., 1970. Am. J. Physiol. 219, 1535–1538.
Tierney, D.F., Clements, J.A., Trahan, H.J., 1967. Am. J. Physiol. 213, 671–676.
Tombropoulos, E.G., Hadley, J.G., 1976. Lipids 11, 491–497.
Toshima, N., Akino, T., 1972. Tohoku J. Exp. Med 108, 253–263.
Toshima, N., Akino, T., Ono, K., 1972. Tohoku J. Exp. Med. 108, 265–277.
Trzeciak, H.I., et al., 1977. Environ. Res. 14, 87–91.
Tsao, F.H., 1986. Lipids 21, 498–502.
Tsao, F.H., Zachman, R.D., 1977. Pediatr. Res. 11, 858–861.
Tsao, F.H., Zachman, R.D., 1982. Clin. Chim. Acta 118, 109–120.
Tsilibary, E.C., Williams, M.C., 1983a. J. Histochem. Cytochem. 31, 1298–1304.
Tsilibary, E.C., Williams, M.C., 1983b. J. Histochem. Cytochem. 31, 1289–1297.
Tsuzuki, A., Kuroki, Y., Akino, T., 1993. Am. J. Physiol. 265, L193–L199.
Uhal, B.D., Longmore, W.J., 1988. Biochim. Biophys. Acta 958, 279–288.
Ulane, R.E., et al., 1978. Biochim. Biophys. Acta 53, 295–300.
Van Golde, L.M., 1976. Am. Rev. Respir. Dis. 114, 977–1000.
Van Golde, L.M., Batenburg, J.J., Robertson, B., 1988. Physiol. Rev. 68, 374–455.
Van Heusden, G.P., et al., 1981a. Biochim. Biophys. Acta 666, 313–321.
Van Heusden, G.P., et al., 1981b. Biochim. Biophys. Acta 664, 49–60.
Van Heusden, G.P., Van den Bosch, H., 1981. Biochim. Biophys. Acta 666, 508–511.
Van Heusden, G.P., Van den Bosch, H., 1981. Biochim. Biophys. Acta 666, 508–511.
Van Heusden, G.P., Van den Bosch, H., 1982. Biochim. Biophys. Acta 711, 361–368.
Veldhuizen, R., et al., 1998. Biochim. Biophys. Acta 1408, 90–108.
Voelker, D.R., 1984. Proc. Natl. Acad. Sci. U. S. A. 81, 2669–2673.
Voelker, D.R., Lee, T.C., Snyder, F., 1976. Arch. Biochem. Biophys. 176, 753–756.
Voyno-Yasenetskaya, T.A., et al., 1993. Proc. Natl. Acad. Sci. U. S. A. 90, 4256–4260.
Wade, K.C., et al., 2006. Am. J. Respir. Cell Mol. Biol. 34, 727–737.
Waite, M., et al., 1987. Faseb J. 1, 318–325.
Waite, M., Roddick, V., Thornburg, T., King, L., Cochran, F., 1987. FASEB J. 1, 318–325.
Wakil, S.J., Stoops, J.K., Joshi, V.C., 1983. Annu. Rev. Biochem. 52, 537–579.
Walter, M., Davies, J.P., Ioannou, Y.A., 2003. J. Lipid Res. 44, 243–253.
Walton, P.A., Possmayer, F., 1986. Biochem. Cell. Biol. 64, 1135–1140.
Wang, M.L., et al., 2010. Cell. Metab. 12, 166–173.
Wang, P., et al., 2008. BMC Cell. Biol. 9, 34.
Warburton, D., et al., 1987a. J. Appl. Physiol. 63, 828–833.
Warbuton, D., et al., 1987b. J. Appl. Physiol. 63, 1750–1756.
Weaver, T.E., 1988. Gen. Pharmacol. 19, 361–368.
Weaver, T.E., Na, C.L., Stahlman, M., 2002. Semin. Cell. Dev. Biol. 13, 263–270.
Weinhold, P.A., et al., 1984. J. Biol. Chem. 259, 10315–10321.
Wert, S.E., Whitsett, J.A., Nogee, L.M., 2009. Pediatr. Dev. Pathol. 12, 253–274.
Whitlon, D.S., Anderson, K.E., Mueller, G.C., 1985. Biochim. Biophys. Acta 835, 369–377.
Whitsett, J.A., et al., 1980. Biochem. Biophys. Res. Commun. 97, 913–917.
Whitsett, J.A., et al., 1981. Am. J. Physiol. 240, E351–E357.
Whitsett, J.A., Wert, S.E., Weaver, T.E., 2010. Annu Rev. Med. 61, 105–119.
Wiklund, N.P., Gustafsson, L.E., 1988. Eur. J. Pharmacol. 148, 361–370.
Williams, M.C., 1977. J. Cell. Biol. 72, 260–277.
Wilson, L., et al., 1974. Fed. Proc. 33, 158–166.
Wirtz, H.R., Dobbs, L.G., 1990. Science 250, 1266–1269.
Wissel, H., et al., 2001. Am. J. Physiol. Lung Cell. Mol. Physiol. 281, L345–L360.
Wright, J.R., 2004. Biol. Neonate 85, 326–332.
Wright, J.R., Clements, J.A., 1987. Am. Rev. Respir. Dis. 136, 426–444.
Wu, M., et al., 2009a. Int. J. Cancer 124, 1907–1913.
Wu, Y., et al., 2009b. Biochem. J. 419, 669–679.
Wykle, R.L., et al., 1980. Arch. Biochem. Biophys 199, 526–537.
Wykle, R.L., Kraemer, W.F., 1977. FEBS Lett. 78, 83–85.
Xu, S., et al., 2007. J. Biol. Chem. 282, 23525–23531.
Xu, Z.X., Rooney, S.A., 1989. Biochim. Biophys. Acta 1005, 209–216.
Xu, Z.X., Smart, D.A., Rooney, S.A., 1990. Biochim. Biophys. Acta 1044, 70–76.
Yamada, K., Okuyama, H., 1979. Arch. Biochem. Biophys. 196, 209–219.
Yamano, G., et al., 2001. FEBS Lett. 508, 221–225.
Yoda, Y., Makita, A., Gasa, S., 1982. J. Biochem. 91, 627–632.
Yost, R.W., et al., 1986. Biochim. Biophys. Acta 875, 122–125.
Yu, S., et al., 1983. Lipids 18, 522–529.
Zachman, R.D., 1971. Biol. Neonate 19, 211–219.
Zachman, R.D., 1973. Pediatr. Res. 7, 632–637.
Zen, K., et al., 1998. Am. J. Physiol. 275, L172–L183.
Zhang, L., et al., 2011. Am. J. Physiol. Lung Cell. Mol. Physiol. 301, L461–L477.

ADDITIONAL REFERENCES

Bates, S.R., et al., 2002. Am. J. Physiol. Lung Cell. Mol. Physiol. 282, L267–L276.
Finkelstein, J.N., et al., 1983. Biochim. Biophys. Acta 762, 398–404.
Fisher, A.B., Dodia, C., 1996. J. Lipid Res. 37, 1057–1064.
Kim, H.S., et al., 2002. Am. J. Respir. Cell. Mol. Biol. 27, 227–233.
Kim, T.S., 1998. Am. J. Physiol. 274, 750–761.
Nanjundan, M., Possmayer, F., 2003. Am. J. Physiol. Lung Cell. Mol. Physiol. 284, L1–L23.

Chapter 23

Regulation of Airway Lining Fluid in Health and Disease

James D. Londino, James F. Collawn and Sadis Matalon

Chapter Outline

1. OVERVIEW

The conducting airways comprise the nose, trachea, and bronchi. The human lung consists of a series of highly branched tubes ending in the alveoli (Figure 1). The nasal cavity, trachea, and bronchi are lined with a pseudostratified epithelium consisting of basal, ciliated, and mucous secretory (goblet) cells. The smaller bronchioles are also lined by a columnar pseudostratified epithelium, which transitions into a shorter columnar and cuboidal epithelium. The alveolar epithelium itself is composed of two types of cuboidal cells that cover the vast majority ($\sim$99%) of the surface area of the lung. Gas exchange occurs in alveoli located in the respiratory bronchioles, alveolar ducts, and sacs. The close proximity of the alveolar epithelium to the vascular endothelium ($\sim$0.1–0.2 μm) facilitates efficient gas exchange. Although this barrier is very thin, alveolar cells form a tight barrier that prevents movement of liquid and proteins though the epithelium (Matthay et al., 2002).

2. MUCOCILIARY CLEARANCE

Mucociliary clearance in the lung epithelium is a critical and physiologically regulated protective function of the airways and lung that is essential for the clearance of respiratory pathogens. The airways are covered by a layer of fluid and mucus that is constantly being propelled by cilia from the distal to proximal lung airways. The number of ciliated cells is higher in the upper airways (there are an estimated 109 cilia/cm^2 in the ciliated cells of the nasal and tracheal/bronchial epithelium (Livraghi and Randell, 2007)), while nonciliated cells and basal cells are more frequently found in the distal lung. Highly glycosylated mucins, lipids, and secreted proteins bind to inhaled respiratory irritants and pathogens. In addition to mucus secretion by the epithelium, the proximal airways also contain submucosal glands that secrete fluid and mucus. If mucociliary clearance is impaired, the excess mucus forms plaques and plugs that allow invasive bacteria to adhere and colonize the airways (Livraghi and Randell, 2007).

3. ION TRANSPORT ACROSS LUNG EPITHELIAL CELLS

Although originally fluid regulation in the lung was believed to be controlled by Starling forces (Mason et al., 1982), it became apparent that differences in hydrostatic and protein osmotic pressure accumulation could not adequately explain the rates of fluid transport (Berthiaume et al., 1987). Furthermore, when a single lung of a sheep

Comparative Biology of the Normal Lung. http://dx.doi.org/10.1016/B978-0-12-404577-4.00023-0

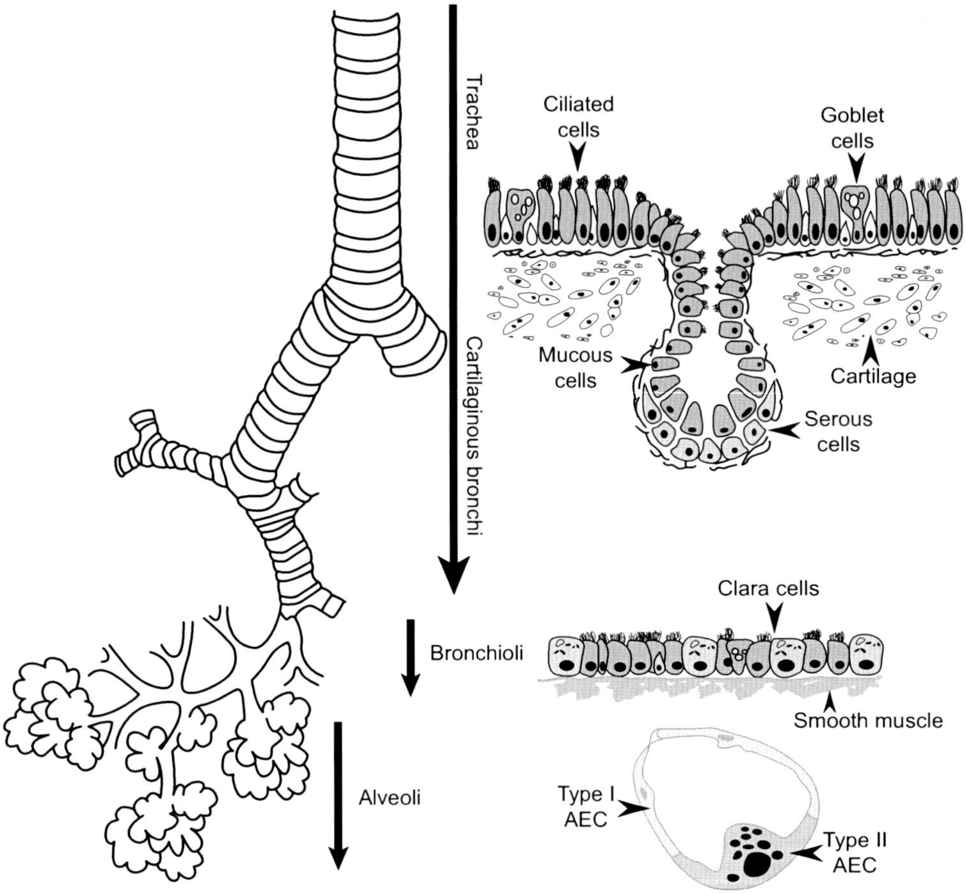

FIGURE 1 Schematic illustration of human lower airways. The trachea and bronchi are lined with a pseudostratified epithelium consisting of basal, ciliated, and mucous secretory (goblet) cells. Submucosal glands are present throughout the cartilaginous airways (trachea, bronchi). The smaller bronchioles are also lined by a columnar pseudostratified epithelium, which transitions into a shorter, columnar to cuboidal epithelium. Clara cells only appear in bronchioles. The alveolar epithelium itself is composed of two types of cuboidal cells that cover the vast majority (~99%) of the surface area of the lung. *Adapted from Collawn, J.F., Lazrak, A., Bebok, Z., Matalon, S., 1 Jun 2012. The CFTR and ENaC debate: how important is ENaC in CF lung disease? Am. J. Physiol. Lung Cell Mol. Physiol. 302 (11).*

was ventilated, the rate of fluid clearance in the opposite lung remained the same (Sakuma et al., 1993). Meanwhile, evidence was quickly accumulating that fluid transport must be controlled by an active process. First, the temperature dependence of fluid clearance suggested that protein activity was necessary for fluid clearance (Serikov et al., 1993). Second, ouabain, an inhibitor of the Na^+/K^+-ATPase (which provides the proper ionic gradients for Na^+ sodium entry into cells), inhibited fluid clearance across the alveolar and airway spaces of several different species including the human lung (Basset et al., 1987). Finally, isolated cells from various epithelia of the lung were shown to actively transport Na^+, Cl^-, and other ions (Cheek et al., 1989; Nielsen et al., 1998).

Together with cilia and mucus, an adequate periciliary layer (PCL) is required for effective mucociliary clearance. This PCL fluid provides a layer in which cilia can freely beat to propel overlying mucus. Maintenance of the PCL requires the careful coordination of Na^+ absorption mainly through the amiloride-sensitive epithelial Na^+ channel (ENaC) and Cl^- secretion through the cystic fibrosis transmembrane conductance regulator (CFTR) as well as other anion transporters (Ca^{2+}-activated Cl^- channels, etc.) (Figure 2). Early experiments showed that active Na^+ transport across the distal lung epithelium was driven by the Na^+/K^+-ATPase at the basolateral membrane (reviewed by Matthay et al., 2002). The Na^+/K^+-ATPase provides an electrochemical gradient for Na^+ transport into the cell by transporting three Na^+ ions out of the basolateral epithelium in exchange for potassium ions. The importance of this transporter is demonstrated by the Na^+/K^+-ATPase inhibitor ouabain, which completely inhibits all active ion channel activity in the epithelium.

4. ENaC AND CFTR

ENaC was cloned in 1994 and characterized as a highly selective unidirectional Na^+ transporter composed of α, β,

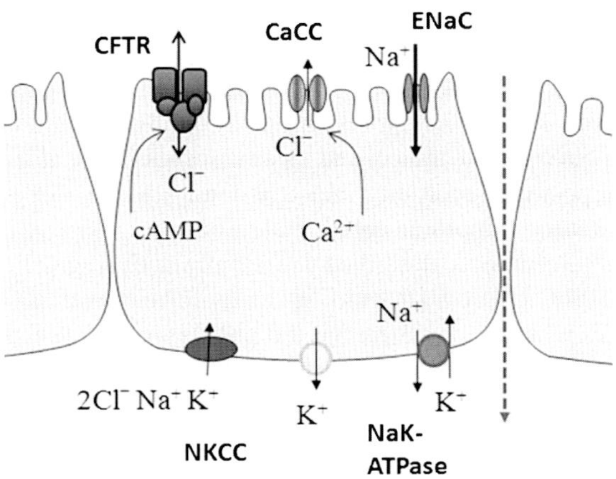

FIGURE 2 Ion transport in epithelial cells. The basolateral Na^+/K^+-ATPase provides an electrochemical gradient for Na^+ transport into the cell through ENaC at the apical membrane. The $Na^+/K^+/2Cl^-$ (NKCC) symporter transports Cl^- through the basolateral membrane. Cl^- exits through the cAMP-activated channel CFTR or through calcium-activated chloride channels (CaCC).

and γ subunits ($\sim$4–5 pS) (Collier and Snyder, 2011; Snyder et al., 1994). Other combinations of subunits are reputed to form channels that have unique conductances, decreased amiloride sensitivity, and less Na^+ specificity (Johnson et al., 2006). Recently, other ENaC associated subunits (such as δ) have also been described (Ji et al., 2012). A number of studies have demonstrated the relevance of ENaC-mediated Na^+ transport across distal lung epithelia in vivo. Patients hospitalized for acute respiratory distress syndrome (ARDS) with inhibited amiloride-sensitive fluid clearance have a higher incidence of mortality (Ware and Matthay, 2001). Furthermore, Na^+ transport through ENaC is also essential for the resolution of high altitude edema (Scherrer et al., 1999). In mouse models, inhibition of α-ENaC by siRNA in the adult lung decreased basal fluid clearance as well as terbutaline-stimulated clearance (Li and Folkesson, 2006). Knockout of α-ENaC entirely is lethal because the mice are unable to clear lung fluid at birth (Hummler et al., 1996).

The alveolar epithelium makes up 99% of the surface area of the lung. Alveolar type I (ATI) cells make up $\sim$95% of the alveolar space despite only representing one-third of the total cell number, while the remaining alveolar type II (ATII) cells constitute the remaining 5% of the surface area (Matthay et al., 2002). Studies by Johnson et al. show that freshly isolated ATI and ATII cells contain similar channel density of highly selective cation channels most likely composed of the $\alpha\beta\gamma$ ENaC subunits ($\sim$4–5 pS) and nonselective ENaC cation channels probably composed of α subunits ($\sim$21 pS). Nonselective cyclic nucleotide-gated cation channels were also observed, but to a much lesser extent (Johnson et al., 2006). Although it was originally believed that ATI cells lacked expression of the Na^+/K^+-ATPase and therefore active ion transport, the discovery of the expression of both the α_1 and α_2 subunits as well as confirmation of ouabain-sensitive transport suggests that both cells contribute to fluid clearance. Interestingly, differing expression of Na^+/K^+-ATPase subunits in ATI and ATII may mean that these cells respond differently to stressors (Matalon and Davis, 2003).

Chloride secretion has been demonstrated in vivo, in air-filled (Lindert et al., 2007) and fluid-filled lungs, in the presence of a Cl^- secretory gradient, and after cAMP activation (Nielsen et al., 1998). Cl^- is believed to enter the basolateral membranes through $Na^+/K^+/2Cl^-$ (NKCC) transporters (Barker et al., 1995), down an electrochemical gradient generated by the basolateral Na^+/K^+/ATPase, and exit through apical anion channels, including CFTR. The CFTR is a cAMP-activated, nonrectifying Cl^- channel located at the apical surface of epithelial cells that plays an essential role in fluid homeostasis and pH regulation (Riordan, 2008). This channel transports Cl^- in parallel with Na^+ in order to maintain electroneutrality. However, CFTR is also responsible for Cl^- secretion in submucosal and other upper airway epithelium (Ballard et al., 1999). It is possible that CFTR may switch between absorptive and secretory roles depending on changes in the electrochemical gradients (Solymosi et al., 2013).

5. ROLE OF OTHER ION CHANNELS

Several other channels also play a role in the maintenance of ion transport. As previously mentioned, a nonselective nucleotide-gated cation channel contributes to apical Na^+ absorption in the rat airway epithelium (Johnson et al., 2006). In addition to Cl^- transport through CFTR, there are a number of Ca^{2+} activated Cl^- channels (CaCC). TMEM16A, a member of the anoctamin family, is thought to compensate for CFTR in the mouse lung, potentially providing a rationale for the lack of a lung phenotype in CFTR knockout mice (Rock et al., 2009). However, other CaCC must play a role since up to 40% of the UTP-stimulated CaCC activity persists in TMEM16A knockout mice (Rock et al., 2009). Interestingly, recent experiments have shown CFTR to play a direct role in calcium-activated Cl^- transport. Although Bove et al. found no expression of known CaCC channels in human ATII cells, these cells show that Cl^- induced currents in response to UTP and a calcium ionophore. These currents were also sensitive to specific CFTR inhibitors (Bove et al., 2010).

6. CYSTIC FIBROSIS DISEASE

In 1983, Paul Quinton demonstrated that cystic fibrosis (CF) is caused by a Cl^- transport defect (Quinton, 1983). The CF gene was cloned and named in 1989 (Riordan et al., 1989). After translation, CFTR is transported through the endoplasmic reticulum where core glycosylation occurs.

Normally, properly folded CFTR is maturely glycosylated in the Golgi apparatus and transported to the plasma membrane, however, the most common CF mutation, ΔF508, leads to premature degradation of the channel and inhibited Cl^- transport (Riordan, 2008). Defective CFTR expression and activity leads to a variety of symptoms such as elevated levels of sweat Cl^- as well as chronic bacterial respiratory infections that are the result of compromised mucociliary clearance. These symptoms can cause life-threatening complications including airway obstruction, bronchiectasis, and respiratory failure. Although there is no clear consensus on how altered CFTR expression and activity causes disease, it is hypothesized that inhibited Cl^- transport through CFTR may alter the PCL layer hydration, pH (bicarbonate transport), or viscosity (Riordan, 2008). CFTR expression may also indirectly alter Na^+ transport through ENaC by inhibiting ENaC activity (Collawn et al., 2012).

7. BETA ENaC MOUSE

Mall et al. sought to address the role of fluid regulation in CF disease by creating an ENaC overexpressing mouse. If CF disease is due to PCL dehydration caused by loss of ENaC regulation by CFTR, then overexpression of ENaC should recapitulate the effects of CF disease. All three subunits of ENaC (*Scnn1a*, *Scnn1b*, and *Scnn1c*) were introduced in mice under the control of the Clara cell secretory protein (CCSP) promoter (Mall et al., 2004). Mouse lower airways (trachea and bronchi) contain greater than 50% Clara cells, whereas these regions in the human airways contain mostly ciliated cells. Interestingly, only the β-ENaC-overexpressing animals had elevated ENaC activity. Airway surface liquid (ASL) volume in cultured epithelial monolayers was reduced in β-ENaC animals compared to the controls (Figure 3). All cAMP-activated and CaCC activities were normal in the β-ENaC mice. Consistent with a fluid regulation hypothesis, β-ENaC mice displayed a number of CF-like symptoms including dehydrated mucus, reduced mucociliary transport, neutrophilic inflammation, poor bacterial clearance, and airway obstruction that led to a 50% postnatal mortality rate at around 4 weeks due to the respiratory symptoms (Mall et al., 2004).

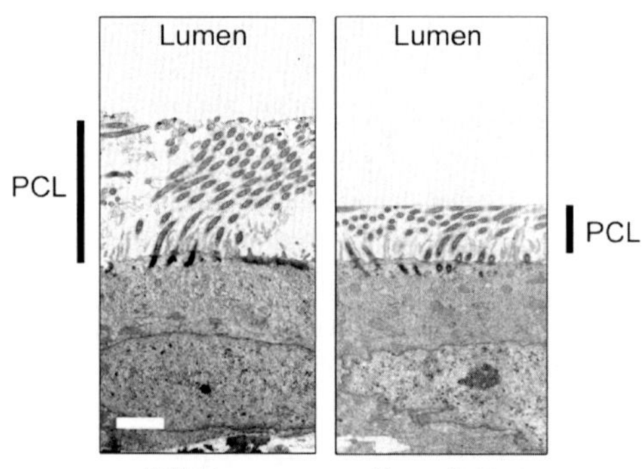

FIGURE 3 Periciliary layer (PCL) in bronchi from wild-type and *Scnn1b*-transgenic mice, as visualized by electron microscopy. *Zhou, Z., Duerr, J., Johannesson, B., Schubert, S.C., Treis, D., Harm, M., Graeber, S.Y., Dalpke, A., Schultz, C., Mall, M.A., Jun 2011. The ENaC-overexpressing mouse as a model of cystic fibrosis lung disease. J. Cyst. Fibros. 10 (Suppl. 2), S172–S182.*

If the defects described are simply a dysregulation of the ENaC and CFTR ratio, it stands to reason that CFTR overexpression could rescue the phenotype. To test this, a transgenic mouse strain expressing human CFTR (hCFTR) using the same CCSP promoter as for the β-ENaC was developed (Grubb et al., 2012). In these mice, hCFTR mRNA was expressed at higher levels (~5-fold) than the endogenous murine CFTR mRNA expressed in these animals. Interestingly, despite higher CFTR mRNA expression, the forskolin (CFTR) and UTP (CaCC)-activated currents were much smaller in the hCFTR transgenic mice. Because basal activity of CFTR was dramatically increased in the tracheal epithelia of CFTR overexpressing mice, it seems plausible that there is a maximum threshold of CFTR activation despite elevated CFTR expression. Even with the observed changes in basal activity, no respiratory phenotype for the CFTR overexpressers was observed. When this hCFTR expressing strain was crossed with the β-ENaC mouse, the double transgenic mice (hCFTR/β-ENaC) demonstrated many of the same phenotypic traits found in β-ENaC mice including significant mucus plugging, decreased ASL height, and reduced survival rate (Grubb et al., 2012). One plausible explanation is that the ratio of ENaC to CFTR is too great to overcome the enhanced Na^+ transport. Since levels of β-ENaC were elevated 25- to 100-fold over endogenous ENaC, a 5-fold elevation of CFTR expression may not be sufficient to inhibit ENaC (Collawn et al., 2012). Further evidence for this conclusion is provided by Liddle's syndrome patients. Liddle's syndrome is a mutation in the β-subunit of ENaC gene that results in a hereditable form of salt-sensitive hypertension caused by a disrupted ubiquitination and degradation ENaC (Abriel et al., 1999). Although these patients suffer from hypertension, they do not suffer from lung disease. A possible explanation for this discrepancy is that CFTR expression in humans is higher in both the upper and lower airways than in the mouse. Mice have approximately 20-fold higher CFTR expression in nasal versus tracheal tissues, and while ENaC is hyperactive in the tracheal epithelium of Liddle mice, ENaC activity is normal in the nasal epithelium (Mall et al., 2010).

8. CF PIG

In order to generate a model of CF that more closely resembles the disease found in humans, Welsh and colleagues generated a porcine model of *CFTR-null* and

CFTR-ΔF508 homozygote pigs that more successfully mimics many of the CF disease phenotypes (Rogers et al., 2008). As expected, CF pig airways show almost no cAMP-stimulated Cl^- and $HCO3^-$ transport. In the porcine lung (Li et al., 2012), and unlike in the mouse lung (Rock et al., 2009), the CaCC inhibitor DIDS had no effect on the short-circuit currents, suggesting the absence of compensatory Cl^- channels in the pig. Interestingly, stimulated Na^+ absorption in $CFTR^{+/-}$ ATII was identical to $CFTR^{+/+}$ cells, suggesting that expression of one normal allele of CFTR is sufficient (Li et al., 2012). In addition, although an adenoviral vector expressing CFTR rescues fluid transport in $CFTR^{-/-}$ epithelium, overexpression of CFTR in wild-type cells did not increase Cl^- secretion (Potash et al., 2013).

Similar to what is seen in human ATII cells, the basal depth of periciliary liquid observed in vitro and in vivo is the same in porcine CF alveolar epithelium and in normal epithelium. In addition, both human and porcine CF ATII cells have impaired cAMP-stimulated fluid secretion (Bove et al., 2010). Although there were no observed differences in basal PCL depth in the airway epithelium between the CF pigs and control littermates, basal transepithelial voltage was enhanced in the nasal epithelia of CF pigs over the non-CF control in both excised and cultured epithelia. In the tracheal/bronchial epithelium, these differences were not observed (Chen et al., 2010). When human nasal and tracheal CF epithelial cells were also studied for this defect, similar results were found (Itani et al., 2011). Altered transepithelial voltage suggests that enhanced amiloride sensitive current in the nasal cells is due to an alteration of the electrical potential of the epithelium and not an increase in the conduction of the ENaC channel. However, a recent study by O'Donoghue et al. modeling epithelial ion transport suggests that although hyperpolarization does occur, loss of apical Cl^- permeability alone is insufficient to explain the increased Na^+ transport observed and that a parallel increase in ENaC permeability must also occur (O'Donoghue et al., 2013).

If decreased PCL fluid by loss of Cl^- transport through CFTR is not the primary cause of CF what potential mechanisms could be responsible? Recent evidence from the porcine model points to loss of bicarbonate transport, rather than Cl^- transport, as a primary cause of disease. Pezzulo et al. showed that the epithelial lining fluid (ELF) of newborn pigs is able to kill bacteria within seconds of contact. However, in the CF lung, basal ELF pH was decreased in comparison to controls. Alteration of the lung ELF pH decreased bacterial clearance in newborn CF pigs. Even more remarkably, this defect occurs prior to the onset of inflammation in the CF lung and can be corrected by restoring ASL pH to normal levels (Pezzulo et al., 2012). This deregulation of immune function could be responsible for bacterial colonization of the CF lung and therefore much of the pathology caused by the inflammatory responses.

9. ALTERATION OF ION TRANSPORT BY INFLUENZA

Altered fluid regulation in the airway epithelium is consistently observed in many respiratory infections. During influenza infection, rhinorrhea, the accumulation of fluid in the eustachian tubes and middle ear, as well as fluid accumulation in the respiratory tract are commonly described symptoms in humans and animals (Buchman et al., 1995; Doyle et al., 1994; Kudlacz et al., 1993; Suarez et al., 1998). Severe infection by influenza can lead to pneumonia, pulmonary edema, and ARDS. Since inhibition of fluid clearance occurs prior to the onset of morphological changes in the respiratory epithelium, it is likely that alteration of active ion active transport is responsible (Traylor et al., 2012).

The earliest evidence that influenza virus directly alters ion transport comes from experiments performed by Kunzelmann et al. in which the apical membrane of mouse tracheal epithelium, mouse colonic epithelium, and mouse cortical collecting ducts were exposed to WSN and PR8 influenza A virus, and ion transport was measured in an Ussing chamber system. One hour after infection, influenza-infected tissue had decreased transepithelial potential difference, total short-circuit current, and amiloride-sensitive current. Significant inhibition of ENaC activity occurred as early as 15 min postinfection and continued for at least 2 h. This effect was concentration dependent and washout of virus from the epithelium did not inhibit the effect. No alterations of cAMP-activated or CaCC currents were observed (Kunzelmann et al., 2000).

Influenza is a negative sense RNA virus that contains eight gene segments that code for approximately 13 proteins. The influenza A virus contains three transmembrane proteins: hemagglutinin (HA), neuraminidase, and matrix protein 2 (M2). During infection, influenza HA binds to the sialic acid residues on the surface of the lung epithelium (Matlin et al., 1981). The virus is then internalized through a cytochalasin D-dependent mechanism (Gottlieb et al., 1993). Postinternalization, the virus uncoats by a pH-dependent, M2 proton channel—mediated mechanism (Takeda et al., 2002). Because of the early time point of observed inhibition, Kunzelmann et al. tested a number of inhibitors of these initial entry pathways. Blocking viral binding by cleaving sialic acid residues or by targeting HA with a blocking antibody prevented the influenza-mediated alteration of ENaC activity, while inhibition of the subsequent steps had no effect. In addition, split virus preparation that retains HA activity but lacks the replicative capacity was able to alter amiloride sensitive currents. Paradoxically, the authors also claim that UV-inactivated PR8, which should also retain HA binding, had no effect on ENaC. To test the hypothesis that the binding of HA alone can initiate the signals necessary for ENaC inhibition, the authors incubated lung slices with a hemagglutinating lectin concanavalin A. This lectin caused significant ENaC inhibition with a similar time course as influenza virus infection (Kunzelmann et al., 2000). Based on this information, the authors proceeded to inhibit signaling pathways known to be upregulated during influenza binding. These experiments were the first to show that activation of protein kinase C (PKC) inhibits ENaC activity, a now widely recognized pathway of sodium transport regulation (Kunzelmann et al., 2000).

10. ALTERATION OF FLUID CLEARANCE BY INFLUENZA IN VIVO

Although the earlier experiments suggested influenza could alter Na^+ transport, studies by Chen et al. confirmed these findings in vivo (Chen et al., 2004). As seen in the tracheal epithelium, in ATII cells, virus binding, but not internalization, decreased ENaC activity. In addition, since Chen et al. had previously characterized the single channels expressed in ATII cells as Na^+ selective, inwardly rectifying cation (HSC) channels with high amiloride sensitivity, the authors were able to confirm that the channels altered by influenza were likely $\alpha\beta\gamma$ encoded canonical ENaC (Jain et al., 2001). The researchers also further examined the influenza binding/internalization step and discovered that in addition to PLC and PKC, activation of the cytosolic tyrosine kinase Src, an early step in the clathrin-mediated internalization process, was required for ENaC inhibition (Chen et al., 2004).

If in vitro inhibition of ENaC activity is consistent with what occurs in the in vivo infected lung, alveolar flooding due to inhibited Na^+ and water transport across the alveolar epithelium would be expected. When Chen et al. instilled influenza virus intratracheally, lung wet–dry ratios, a marker of alveolar flooding, were increased at 2 h. Importantly, the combined effect of influenza and amiloride instillation was not additive, suggesting that alveolar flooding caused by influenza in vivo is due to the inhibition of amiloride-sensitive channels (Chen et al., 2004).

11. ALTERATION OF FLUID CLEARANCE THROUGHOUT INFECTION

In the course of normal infection, the number of viruses initially binding to the epithelium is extremely low. However, since viruses are constantly budding and reinfecting during the length of an infection, consistent alteration of ion transport by virus binding is certainly plausible. However, the analysis of ion transport in the infected lung over time is complicated by influenza- and inflammation-mediated damage to the epithelium. Although fluid clearance in the intact lung is known to be controlled almost exclusively by active ion transport, fluid deposition in the influenza-infected lung is also enhanced by increased passive fluid transport through compromised epithelium and tight junctions (Mori et al., 1995; Pittet et al., 2010).

Wolk et al. sought to separate the effects of virally induced cell death and passive fluid influx from active Na^+ and Cl^- transport induced through ion channels. Infected mice began to lose weight and showed decreased peripheral arterial O_2 saturation approximately 2 days postinfection onward. This corresponded to an increase in the wet–dry ratio of the lung that persisted throughout infection and was maximal by day 8. By correcting for murine albumin that leaked through the alveolar epithelium into the instilled BSA, the authors were able to produce a measure of the rate of active transport-mediated alveolar fluid clearance (AFC) throughout infection. AFC is maximally inhibited in infected lungs ~2 days postinfection but remains compromised until around day 18. Although increased epithelial permeability and changes in gross lung morphology did occur later in infection, at day 2 differences in AFC were almost exclusively due to changes in amiloride-sensitive components (Wolk et al., 2008). Consistent with prior influenza infection experiments, the authors observed a PKC-mediated inhibition in AFC early after infection. In contrast to the observations of Kunzelmann et al., UV-inactivated virus alters amiloride-sensitive fluid clearance to the same extent as wild-type virus (Kunzelmann et al., 2000). Given that the binding ability of UV-inactivated virus should remain intact, these results are not surprising. More interestingly, decreased AFC by UV-inactivated virus persisted ~4 days postinfection, albeit at a lower level than observed with the wild-type virus (Wolk et al., 2008).

Seeking to determine the mechanism responsible for long-term ENaC inhibition, Wolk et al. measured ATP and UTP levels after influenza infection. Both ATP and UTP were significantly increased 1 and 2 days postinfection, and decreased by day 4. Previous studies have shown that elevated bronchial–alveolar UTP concentrations are responsible for the inhibition of ENaC during RSV infection (Davis et al., 2004). Inhibition of de novo pyrimidine synthesis as well as inhibition of channels known to transport ATP into the alveolar space prevented the influenza-mediated decrease in fluid transport. Degradation of circulating UTP and ATP to the monophosphate form also increased the rate of fluid clearance to near control levels, while conversion of UTP to its diphosphate form was ineffective, suggesting that alteration of AFC is mediated by UDP. These treatments had no effect in the presence of amiloride, suggesting that UDP acts on amiloride-sensitive Na^+ transport rather than through an alternative pathway.

Although ENaC inhibition is responsible for the majority of decreased fluid clearance, a small but significant portion appears to be mediated through CFTR. Under control conditions, in the presence of CFTR inhibitors, AFC was significantly reduced. These results are consistent with other reports (Fang et al., 2002; Solymosi et al., 2013). However, 2 days after the instillation of WSN influenza virus into the airways, CFTR inhibitors, and not inhibitors of CaCC, appeared to significantly enhance fluid clearance (i.e., decrease in the amount of fluid in the lung, suggesting that CFTR Cl^- transport had a detrimental effect on the influenza-infected lung). Since the ATP was elevated at this time point, and since the degradation product adenosine is known to increase CFTR activity through adrenergic receptor stimulation of fluid clearance, the authors sought to determine the role of adenosine receptors in the flu-mediated enhancement of CFTR activity. Both blocking adenosine production and inhibition of the A1-AR receptor pathway increased fluid clearance, suggesting that CFTR activity was enhanced by adrenergic stimulation (Wolk et al., 2008). Prior studies have shown that physiologic concentrations (micromolar) of adenosine reduce the lung's ability to clear alveolar fluid in rats and mice via an A_1R-mediated pathway (Factor et al., 2007).

While adrenergic receptor stimulation is sufficient to explain increased CFTR activity, additional explanation is needed to determine how CFTR switches from absorption in the normal lung to secretion in the infected lung. It is possible that separate populations of CFTR-expressing cells, some secretory and others absorptive, are preferentially activated in response to influenza infection. However, it is difficult to reconcile a small subset of epithelial cells drastically altering fluid transport in the vast surface area of the alveolar epithelium. A second intriguing explanation is mirrored by alteration of CFTR transport observed during the onset of hydrostatic pulmonary edema (Londino and Matalon, 2013). Hydrostatic pulmonary edema is caused by chronic congestive heart failure, acute myocardial infarction, or a number of other causes. Increased hydrostatic pulmonary pressure decreased active Na^+ transport acutely and reversibly with little change in lung permeability (Saldias et al., 2001). Although, ~70% of the pressure-induced net fluid shift into the alveolar space was attributable to impaired fluid reabsorption, the authors noticed that at high pulmonary pressure, alveolar transport was no longer absorptive, but mostly secretory, despite little change in the integrity of the epithelial monolayer. Cl^- secretion by CFTR was responsible for the bulk of this fluid accumulation. They proposed that inhibition of apical Na^+ transport through ENaC may change the electrochemical gradient, driving CFTR to switch from an absorptive to secretory phenotype. In an independent experiment, when amiloride was added to the lung instillate to block ENaC activity at normal arterial pressure, CFTR Cl^- transport shifted from absorptive to secretory. Consistent with these observations, studies of small airways using a miniature Ussing chamber system also showed a switch from Cl^- absorption to secretion through CFTR when Na^+ transport was inhibited by amiloride (Shamsuddin and Quinton, 2012).

12. ALTERATION OF ENaC ACTIVITY BY THE INFLUENZA ION CHANNEL M2 PROTEIN

Influenza M2 protein is one of ~13 known proteins expressed by eight influenza gene segments. M2 is a 97−amino acid integral membrane protein that forms a homotetrameric proton (H^+) channel. During viral entry, the M2 protein is activated by low pH in the endosome, allowing acidification of the virion, which facilitates M1 uncoating, fusion, and release of ribonucleoprotein into the cytoplasm for nuclear transport. This step is essential for viral replication and can be blocked by treatment with the M2 ion transport inhibitor amantadine (Lamb et al., 1985; Takeda et al., 2002). Transport of M2 through the secretory pathway is also known to alter the pH of the secretory pathway (Henkel et al., 1999; Sakaguchi et al., 1996; Takeuchi and Lamb, 1994). In addition, influenza M2 has been shown to alter autophagy (Gannage et al., 2009), activate the inflammasome immune signaling complex (Ichinohe et al., 2010), and play a role in "endosomal sorting complex required for transport" independent budding (Rossman et al., 2010).

Lazrak et al. determined that influenza M2 protein inhibits amiloride-sensitive currents both in *Xenopus* oocytes injected with cRNA encoding the α, β, and γ subunits of ENaC and in the Clara cell-like lung epithelial cell line H441 transfected with GFP tagged M2. M2 also decreased single channel conductance, confirming that ENaC is responsible for the decreased Na^+ transport. Expression of another influenza protein, M1, had no effect on ENaC current. To determine which regions of M2 were responsible for the inhibition of ENaC, the authors created a series of C-terminal truncations. All M2 mutant proteins larger than 62AA of the original 97AA construct reached the plasma membrane and exhibited ion channel activity. Although the last 30AA appeared to play a minor role in the inhibition of ENaC, the first 62AA were essential (Lazrak et al., 2009).

In addition to decreased ENaC activity, expression of all three ENaC subunits was decreased in M2 co-injected oocytes compared to controls. Lazrak et al. were able to prevent the degradation of ENaC by inhibiting the proteosomal pathway. To determine how ENaC is targeted for degradation, the authors coexpressed ENaC channels composed of subunits containing Liddle mutations with M2. Liddle ENaC lacks the ability to interact with ubiquitin and was therefore insensitive to M2-mediated degradation. Since PKC had been previously shown to play a role in viral inhibition of ENaC by influenza and both SARS E and

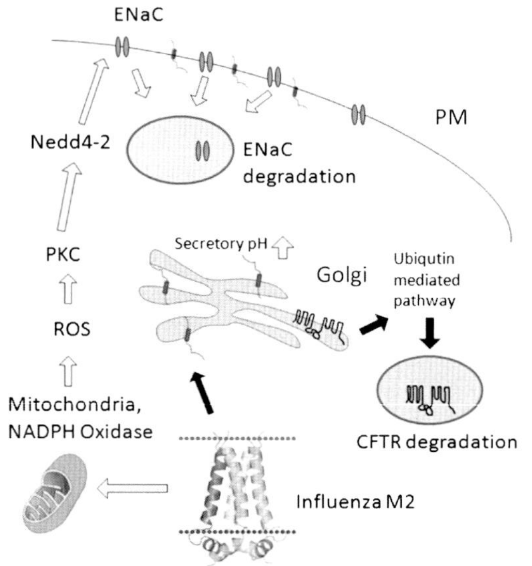

FIGURE 4 Inhibition of epithelial Na^+ channel (ENaC) and cystic fibrosis transmembrane conductance regulator (CFTR) by influenza M2. ENaC; Influenza M2 expression increases Reactive oxygen species (ROS) production either through NADPH oxidases or mitochondria signaling. ROS activates through protein kinase C (PKC), which phosphorylates E3 ligase Nedd4-2, enhancing its interaction with ENaC subunits. Ubiquitination of ENaC targets the protein for proteosomal degradation instead of recycling. CFTR; Influenza M2 protein is trafficked through the secretory pathway where it is activated by low pH. M2 transports protons out of acidified organelles, increasing their pH. CFTR trafficked through acidified organelles fails to mature properly and is improperly trafficked or degraded by a ubiquitin-mediated pathway.

CoV protein (Ji et al., 2009; Kunzelmann et al., 2000), the authors next examined the role of PKC on M2-mediated inhibition. While inhibition of α- and β-PKC rescued ENaC activity in H441 cells, PKC inhibition also greatly enhanced basal ENaC currents. In contrast, inhibition of the ζ-PKC isoform also rescues ENaC activity, but had no effect on basal currents.

Having shown that M2 is able to enhance the degradation of ENaC, Lazrak et al. present evidence that M2 expression enhanced the production reactive oxygen species (ROS). When cytosolic and mitochondrial ROS in A549 cells were measured using DCF-DA and Mitosox, respectively, transfection with M2 greatly enhanced the ROS levels in both compartments (Lazrak et al., 2009). Reactive oxidants have been shown to alter ENaC activity by either post-translational modifications (such as oxidation, nitration, nitrosylation, and glutathionylation of critical residues) or activation of signal transduction pathways (Song et al., 2010). To determine whether ROS were involved in the inhibition of ENaC, cells were incubated with a cell-permeable glutathione ethyl ester. Supplementation of GSH increases the concentration of reduced thiols in the cell in order to return the cell to a less oxidized state. ENaC inhibition was significantly attenuated in GSH ester-treated oocytes and cells (Figure 4) (Lazrak et al., 2009).

13. ALTERATION OF CFTR ACTIVITY BY INFLUENZA M2 PROTEIN

Since both ENaC and CFTR play major roles in the maintenance of the PCL, it was also important to determine whether M2 had any effect on the activity of CFTR. Surprisingly, CFTR activity was inhibited by M2 in *Xenopus* oocytes, cells overexpressing CFTR, and a human bronchial epithelial cell line expressing endogenous CFTR. Interestingly, although ER core glycosylated CFTR protein expression was enhanced in M2 injected oocytes, the maturely glycosylated form responsible for apical ion transport was significantly reduced, suggesting a posttranslational defect (Londino et al., 2013).

M2 is known to raise the pH of the secretory pathway by transporting protons out of acidified organelles (Henkel et al., 1999; Sakaguchi et al., 1996; Takeuchi and Lamb, 1994). Since the maturation of CFTR is dependent upon proper trafficking, it appeared that M2 proton channel activity could be altering CFTR trafficking and processing. Consistent with this hypothesis, inhibition of the M2 ion transport inhibitor amantadine was able to rescue CFTR expression and activity. In addition, M2 mutant proteins lacking ion transport activity and proteins that were not trafficked through the acidified portions of the secretory pathway did not alter CFTR activity. Since M2 ion channel activity was necessary for defective processing of CFTR, Londino et al. also showed that when *Xenopus* oocytes were incubated with agents that raised the pH of organelles in the secretory pathway, CFTR activity was dose-dependently inhibited, suggesting that alteration of the pH of acidified organelles was sufficient. Finally, because improperly folded/trafficked CFTR is known to be sensitive to ubiquitination and degradation, Londino et al. blocked ubiquitination using an E1 ubiquitin-activating enzyme inhibitor and were able to rescue CFTR activity in M2 co-injected oocytes (Figure 4). Whether CFTR is directly targeted by ubiquitin due to M2 or if its activity is altered through a secondary ubiquitinated mediator remains to be determined (Londino et al., 2013).

14. NET EFFECT OF ENaC AND CFTR INHIBITION?

It appears that both ENaC and CFTR inhibit influenza M2 protein albeit by two different mechanisms. Although ion channel activity was essential for the inhibition of CFTR by M2, ENaC inhibition was ion channel independent. Further, ROS scavengers were able to rescue ENaC but had no effect on CFTR activity. This is especially surprising since CFTR is known to be inhibited by high concentrations of ROS (Bebok et al., 2002).

The alteration of both CFTR and ENaC begs the question of the net effect of influenza transport on PCL fluid. Based on the data obtained in CF pigs, it appears that the Na^+ transport defect would predominate and lead to decreased fluid depth. However, the inhibition of CFTR could have profound effects on opportunistic viral infection through alteration in ASL pH. Secondary infections are commonly observed after initial influenza infection (Pittet et al., 2010). It is also plausible that while individual virus-infected cells may have inhibited CFTR activity, the net effect of the inflammatory response during infection might be enhanced CFTR activity and decreased fluid clearance as suggested by Wolk et al. Further research is needed to resolve these questions and should lead to a better understanding of the role of fluid regulation during influenza infection.

ACKNOWLEDGMENTS

This work was supported by 2R01HL031197, 5R21ES024027, and 1R21ES02542.

REFERENCES

Abriel, H., Loffing, J., Rebhun, J.F., Pratt, J.H., Schild, L., Horisberger, J.D., Rotin, D., Staub, O., 1999. Defective regulation of the epithelial Na^+ channel by Nedd4 in Liddle's syndrome. J. Clin. Invest. 103, 667–673.

Ballard, S.T., Trout, L., Bebok, Z., Sorscher, E.J., Crews, A., 1999. CFTR involvement in chloride, bicarbonate, and liquid secretion by airway submucosal glands. Am. J. Physiol. 277, L694–L699.

Barker, P.M., Brigman, K.K., Paradiso, A.M., Boucher, R.C., Gatzy, J.T., 1995. Cl- secretion by trachea of CFTR (+/−) and (−/−) fetal mouse. Am. J. Respir. Cell. Mol. Biol. 13, 307–313.

Basset, G., Crone, C., Saumon, G., 1987. Significance of active ion transport in transalveolar water absorption: a study on isolated rat lung. J. Physiol. (Lond) 384, 311–324.

Bebok, Z., Varga, K., Hicks, J.K., Venglarik, C.J., Kovacs, T., Chen, L., Hardiman, K.M., Collawn, J.F., Sorscher, E.J., Matalon, S., 2002. Reactive oxygen nitrogen species decrease cystic fibrosis transmembrane conductance regulator expression and cAMP-mediated Cl^- secretion in airway epithelia. J. Biol. Chem. 277, 43041–43049.

Berthiaume, Y., Staub, N.C., Matthay, M.A., 1987. Beta-adrenergic agonists increase lung liquid clearance in anesthetized sheep. J. Clin. Invest. 79, 335–343.

Bove, P.F., Grubb, B.R., Okada, S.F., Ribeiro, C.M., Rogers, T.D., Randell, S.H., O'Neal, W.K., Boucher, R.C., 2010. Human alveolar type II cells secrete and absorb liquid in response to local nucleotide signaling. J. Biol. Chem. 285, 34939–34949.

Buchman, C.A., Doyle, W.J., Skoner, D.P., Post, J.C., Alper, C.M., Seroky, J.T., Anderson, K., Preston, R.A., Hayden, F.G., Fireman, P., 1995. Influenza A virus-induced acute otitis media. J. Infect. Dis. 172, 1348–1351.

Cheek, J.M., Kim, K.J., Crandall, E.D., 1989. Tight monolayers of rat alveolar epithelial cells: bioelectric properties and active sodium transport. Am. J. Physiol. 256, C688–C693.

Chen, J.H., Stoltz, D.A., Karp, P.H., Ernst, S.E., Pezzulo, A.A., Moninger, T.O., Rector, M.V., Reznikov, L.R., Launspach, J.L., Chaloner, K., Zabner, J., Welsh, M.J., 2010. Loss of anion transport without increased sodium absorption characterizes newborn porcine cystic fibrosis airway epithelia. Cell 143, 911–923.

Chen, X.J., Seth, S., Yue, G., Kamat, P., Compans, R.W., Guidot, D., Brown, L.A., Eaton, D.C., Jain, L., 2004. Influenza virus inhibits ENaC and lung fluid clearance. Am. J. Physiol. Lung Cell. Mol. Physiol. 287, L366–L373.

Collawn, J.F., Lazrak, A., Bebok, Z., Matalon, S., 2012. The CFTR and ENaC debate—how important is ENaC in CF lung disease? Am. J. Physiol. Lung Cell. Mol. Physiol. 302, L1141–L1146.

Collier, D.M., Snyder, P.M., 2011. Identification of epithelial Na^+ channel (ENaC) intersubunit Cl^- inhibitory residues suggests a trimeric alpha gamma beta channel architecture. J. Biol. Chem. 286, 6027–6032.

Davis, I.C., Sullender, W.M., Hickman-Davis, J.M., Lindsey, J.R., Matalon, S., 2004. Nucleotide-mediated inhibition of alveolar fluid clearance in BALB/c mice after respiratory syncytial virus infection. Am. J. Physiol. Lung Cell. Mol. Physiol. 286, L112–L120.

Doyle, W.J., Skoner, D.P., Hayden, F., Buchman, C.A., Seroky, J.T., Fireman, P., 1994. Nasal and otologic effects of experimental influenza A virus infection. Ann. Otol. Rhinol. Laryngol. 103, 59–69.

Factor, P., Mutlu, G.M., Chen, L., Mohameed, J., Akhmedov, A.T., Meng, F.J., Jilling, T., Lewis, E.R., Johnson, M.D., Xu, A., Kass, D., Martino, J.M., Bellmeyer, A., Albazi, J.S., Emala, C., Lee, H.T., Dobbs, L.G., Matalon, S., 2007. Adenosine regulation of alveolar fluid clearance. Proc. Natl. Acad. Sci. U. S. A. 104, 4083–4088.

Fang, X., Fukuda, N., Barbry, P., Sartori, C., Verkman, A.S., Matthay, M.A., 2002. Novel role for CFTR in fluid absorption from the distal airspaces of the lung. J. Gen. Physiol. 119, 199–208.

Gannage, M., Dormann, D., Albrecht, R., Dengjel, J., Torossi, T., Ramer, P.C., Lee, M., Strowig, T., Arrey, F., Conenello, G., Pypaert, M., Andersen, J., Garcia-Sastre, A., Munz, C., 2009. Matrix protein 2 of influenza A virus blocks autophagosome fusion with lysosomes. Cell Host Microbe 6, 367–380.

Gottlieb, T.A., Ivanov, I.E., Adesnik, M., Sabatini, D.D., 1993. Actin microfilaments play a critical role in endocytosis at the apical but not the basolateral surface of polarized epithelial cells. J. Cell. Biol. 120, 695–710.

Grubb, B.R., O'Neal, W.K., Ostrowski, L.E., Kreda, S.M., Button, B., Boucher, R.C., 2012. Transgenic hCFTR expression fails to correct beta-ENaC mouse lung disease. Am. J. Physiol. Lung Cell. Mol. Physiol. 302, L238–L247.

Henkel, J.R., Popovich, J.L., Gibson, G.A., Watkins, S.C., Weisz, O.A., 1999. Selective perturbation of early endosome and/or trans-Golgi network pH but not lysosome pH by dose-dependent expression of influenza M2 protein. J. Biol. Chem. 274, 9854–9860.

Hummler, E., Barker, P., Gatzy, J., Beermann, F., Verdumo, C., Schmidt, A., Boucher, R., Rossier, B.C., 1996. Early death due to defective neonatal lung liquid clearance in *aENaC*-deficient mice. Nat. Genet. 12, 325–328.

Ichinohe, T., Pang, I.K., Iwasaki, A., 2010. Influenza virus activates inflammasomes via its intracellular M2 ion channel. Nat. Immunol. 11, 404–410.

Itani, O.A., Chen, J.H., Karp, P.H., Ernst, S., Keshavjee, S., Parekh, K., Klesney-Tait, J., Zabner, J., Welsh, M.J., 2011. Human cystic fibrosis airway epithelia have reduced Cl^- conductance but not increased Na^+ conductance. Proc. Natl. Acad. Sci. U. S. A. 108, 10260–10265.

Jain, L., Chen, X.J., Ramosevac, S., Brown, L.A., Eaton, D.C., 2001. Expression of highly selective sodium channels in alveolar type II cells is determined by culture conditions. Am. J. Physiol. Lung Cell. Mol. Physiol. 280, L646–L658.

Ji, H.L., Song, W., Gao, Z., Su, X.F., Nie, H.G., Jiang, Y., Peng, J.B., He, Y.X., Liao, Y., Zhou, Y.J., Tousson, A., Matalon, S., 2009. SARS-CoV proteins decrease levels and activity of human ENaC via activation of distinct PKC isoforms. Am. J. Physiol. Lung Cell. Mol. Physiol. 296, L372–L383.

Ji, H.L., Zhao, R.Z., Chen, Z.X., Shetty, S., Idell, S., Matalon, S., 2012. Delta ENaC: a novel divergent amiloride-inhibitable sodium channel. Am. J. Physiol. Lung Cell. Mol. Physiol. 303, L1013–L1026.

Johnson, M.D., Bao, H.F., Helms, M.N., Chen, X.J., Tigue, Z., Jain, L., Dobbs, L.G., Eaton, D.C., 2006. Functional ion channels in pulmonary alveolar type I cells support a role for type I cells in lung ion transport. Proc. Natl. Acad. Sci. U. S. A. 103, 4964–4969.

Kudlacz, E.M., Baugh, L.E., Porter, W.P., Kenny, M.T., Farrell, A.M., 1993. A time-course study of airway hyperresponsiveness in conscious parainfluenza virus type 3-infected guinea pigs. Lab. Anim. Sci. 43, 445–453.

Kunzelmann, K., Beesley, A.H., King, N.J., Karupiah, G., Young, J.A., Cook, D.I., 2000. Influenza virus inhibits amiloride-sensitive Na^+ channels in respiratory epithelia. Proc. Natl. Acad. Sci. U. S. A. 97, 10282–10287.

Lamb, R.A., Zebedee, S.L., Richardson, C.D., 1985. Influenza virus M2 protein is an integral membrane protein expressed on the infected-cell surface. Cell 40, 627–633.

Lazrak, A., Iles, K.E., Liu, G., Noah, D.L., Noah, J.W., Matalon, S., 2009. Influenza virus M2 protein inhibits epithelial sodium channels by increasing reactive oxygen species. FASEB J. 23, 3829–3842.

Li, T., Folkesson, H.G., 2006. RNA interference for alpha-ENaC inhibits rat lung fluid absorption in vivo. Am. J. Physiol. Lung Cell. Mol. Physiol. 290, L649–L660.

Li, X., Comellas, A.P., Karp, P.H., Ernst, S.E., Moninger, T.O., Gansemer, N.D., Taft, P.J., Pezzulo, A.A., Rector, M.V., Rossen, N., Stoltz, D.A., McCray Jr., P.B., Welsh, M.J., Zabner, J., 2012. CFTR is required for maximal transepithelial liquid transport in pig alveolar epithelia. Am. J. Physiol. Lung Cell. Mol. Physiol. 303, L152–L160.

Lindert, J., Perlman, C.E., Parthasarathi, K., Bhattacharya, J., 2007. Chloride-dependent secretion of alveolar wall liquid determined by optical-sectioning microscopy. Am. J. Respir. Cell. Mol. Biol. 36, 688–696.

Livraghi, A., Randell, S.H., 2007. Cystic fibrosis and other respiratory diseases of impaired mucus clearance. Toxicol. Pathol. 35, 116–129.

Londino, J.D., Lazrak, A., Jurkuvenaite, A., Collawn, J.F., Noah, J.W., Matalon, S., 2013. Influenza matrix protein 2 alters CFTR expression and function through its ion channel activity. Am. J. Physiol. Lung Cell. Mol. Physiol. 304, L582–L592.

Londino, J.D., Matalon, S., 2013. Chloride secretion across adult alveolar epithelial cells contributes to cardiogenic edema. Proc. Natl. Acad. Sci. U. S. A. 110, 10055–10056.

Mall, M., Grubb, B.R., Harkema, J.R., O'Neal, W.K., Boucher, R.C., 2004. Increased airway epithelial Na^+ absorption produces cystic fibrosis-like lung disease in mice. Nat. Med. 10, 487–493.

Mall, M.A., Button, B., Johannesson, B., Zhou, Z., Livraghi, A., Caldwell, R.A., Schubert, S.C., Schultz, C., O'Neal, W.K., Pradervand, S., Hummler, E., Rossier, B.C., Grubb, B.R., Boucher, R.C., 2010. Airway surface liquid volume regulation determines different airway phenotypes in liddle compared with betaENaC-overexpressing mice. J. Biol. Chem. 285, 26945–26955.

Mason, R.J., Williams, M.C., Widdicombe, J.H., Sanders, M.J., Misfeldt, D.S., Berry Jr., L.C., 1982. Transepithelial transport by pulmonary alveolar type II cells in primary culture. Proc. Natl. Acad. Sci. U. S. A. 79, 6033–6037.

Matalon, S., Davis, I.C., 2003. Vectorial sodium transport across the mammalian alveolar epithelium: it occurs but through which cells? Circ. Res. 92, 348–349.

Matlin, K.S., Reggio, H., Helenius, A., Simons, K., 1981. Infectious entry pathway of influenza virus in a canine kidney cell line. J. Cell. Biol. 91, 601–613.

Matthay, M.A., Folkesson, H.G., Clerici, C., 2002. Lung epithelial fluid transport and the resolution of pulmonary edema. Physiol. Rev. 82, 569–600.

Mori, I., Komatsu, T., Takeuchi, K., Nakakuki, K., Sudo, M., Kimura, Y., 1995. In vivo induction of apoptosis by influenza virus. J. Gen. Virol. 76 (Pt 11), 2869–2873.

Nielsen, V.G., Duvall, M.D., Baird, M.S., Matalon, S., 1998. cAMP activation of chloride and fluid secretion across the rabbit alveolar epithelium. Am. J. Physiol. 275, L1127–L1133.

O'Donoghue, D.L., Dua, V., Moss, G.W., Vergani, P., 2013. Increased apical Na^+ permeability in cystic fibrosis is supported by a quantitative model of epithelial ion transport. J. Physiol. 591, 3681–3692.

Pezzulo, A.A., Tang, X.X., Hoegger, M.J., Alaiwa, M.H., Ramachandran, S., Moninger, T.O., Karp, P.H., Wohlford-Lenane, C.L., Haagsman, H.P., van, E.M., Banfi, B., Horswill, A.R., Stoltz, D.A., McCray Jr., P.B., Welsh, M.J., Zabner, J., 2012. Reduced airway surface pH impairs bacterial killing in the porcine cystic fibrosis lung. Nature 487, 109–113.

Pittet, L.A., Hall-Stoodley, L., Rutkowski, M.R., Harmsen, A.G., 2010. Influenza virus infection decreases tracheal mucociliary velocity and clearance of Streptococcus pneumoniae. Am. J. Respir. Cell. Mol. Biol. 42, 450–460.

Potash, A.E., Wallen, T.J., Karp, P.H., Ernst, S., Moninger, T.O., Gansemer, N.D., Stoltz, D.A., Zabner, J., Chang, E.H., 2013. Adenoviral gene transfer corrects the ion transport defect in the sinus epithelia of a porcine CF model. Mol. Ther. 21, 947–953.

Quinton, P.M., 1983. Chloride impermeability in cystic fibrosis. Nature 301, 421–422.

Riordan, J.R., 2008. CFTR function and prospects for therapy. Annu. Rev. Biochem. 77, 701–726.

Riordan, J.R., Rommens, J.M., Kerem, B., Alon, N., Rozmahel, R., Grzelczak, Z., Zielenski, J., Lok, S., Plavsic, N., Chou, J.L., 1989. Identification of the cystic fibrosis gene: cloning and characterization of complementary DNA. Science 245, 1066–1073.

Rock, J.R., O'Neal, W.K., Gabriel, S.E., Randell, S.H., Harfe, B.D., Boucher, R.C., Grubb, B.R., 2009. Transmembrane protein 16A (TMEM16A) is a Ca_2^+-regulated Cl^- secretory channel in mouse airways. J. Biol. Chem. 284, 14875–14880.

Rogers, C.S., Abraham, W.M., Brogden, K.A., Engelhardt, J.F., Fisher, J.T., McCray Jr., P.B., McLennan, G., Meyerholz, D.K., Namati, E., Ostedgaard, L.S., Prather, R.S., Sabater, J.R., Stoltz, D.A., Zabner, J., Welsh, M.J., 2008. The porcine lung as a potential model for cystic fibrosis. Am. J. Physiol. Lung Cell. Mol. Physiol. 295, L240–L263.

Rossman, J.S., Jing, X., Leser, G.P., Lamb, R.A., 2010. Influenza virus M2 protein mediates ESCRT-independent membrane scission. Cell 142, 902–913.

Sakaguchi, T., Leser, G.P., Lamb, R.A., 1996. The ion channel activity of the influenza virus M2 protein affects transport through the golgi apparatus. J. Cell. Biol. 133, 733–747.

Sakuma, T., Pittet, J.F., Jayr, C., Matthay, M.A., 1993. Alveolar liquid and protein clearance in the absence of blood flow or ventilation in sheep. J. Appl. Physiol. 74, 176–185.

Saldias, F.J., Azzam, Z.S., Ridge, K.M., Yeldandi, A., Rutschman, D.H., Schraufnagel, D., Sznajder, J.I., 2001. Alveolar fluid reabsorption is impaired by increased left atrial pressures in rats. Am. J. Physiol. Lung Cell. Mol. Physiol. 281, L591–L597.

Scherrer, U., Sartori, C., Lepori, M., Allemann, Y., Duplain, H., Trueb, L., Nicod, P., 1999. High-altitude pulmonary edema: from exaggerated pulmonary hypertension to a defect in transepithelial sodium transport. Adv. Exp. Med. Biol. 474, 93–107.

Serikov, V.B., Grady, M., Matthay, M.A., 1993. Effect of temperature on alveolar liquid and protein clearance in an in situ perfused goat lung. J. Appl. Physiol. 75, 940–947.

Shamsuddin, A.K., Quinton, P.M., 2012. Surface fluid absorption and secretion in small airways. J. Physiol. 590, 3561–3574.

Snyder, P.M., McDonald, F.J., Stokes, J.B., Welsh, M.J., 1994. Membrane topology of the amiloride-sensitive epithelial sodium channel. J. Biol. Chem. 269, 24379–24383.

Solymosi, E.A., Gembrardt, S.M., Vadasz, I., Wang, L., Ng, C.F., Chupin, C., Rozowsky, S., Ruehl, R., Tabuchi, A., Schulz, H., Kapus, A., Morty, R.E., Kuebler, W.M., 2013. Chloride-transport driven alveolar fluid secretion as a novel mechanism in cardiogenic lung edema. Proc. Natl. Acad. Sci. U. S. A. 110, E2308–E2316.

Song, W., Wei, S., Zhou, Y., Lazrak, A., Liu, G., Londino, J.D., Squadrito, G.L., Matalon, S., 2010. Inhibition of lung fluid clearance and epithelial Na^+ channels by chlorine, hypochlorous acid, and chloramines. J. Biol. Chem. 285, 9716–9728.

Suarez, D.L., Perdue, M.L., Cox, N., Rowe, T., Bender, C., Huang, J., Swayne, D.E., 1998. Comparisons of highly virulent H5N1 influenza A viruses isolated from humans and chickens from Hong Kong. J. Virol. 72, 6678–6688.

Takeda, M., Pekosz, A., Shuck, K., Pinto, L.H., Lamb, R.A., 2002. Influenza a virus M2 ion channel activity is essential for efficient replication in tissue culture. J. Virol. 76, 1391–1399.

Takeuchi, K., Lamb, R.A., 1994. Influenza virus M2 protein ion channel activity stabilizes the native form of fowl plague virus hemagglutinin during intracellular transport. J. Virol. 68, 911–919.

Traylor, Z.P., Aeffner, F., Davis, I.C., 2012. Influenza A H1N1 induces declines in alveolar gas exchange in mice consistent with rapid post-infection progression from acute lung injury to ARDS. Influenza. Other Respi. Viruses. 7, 472–479.

Ware, L.B., Matthay, M.A., 2001. Alveolar fluid clearance is impaired in the majority of patients with acute lung injury and the acute respiratory distress syndrome. Am. J. Respir. Crit. Care Med. 163, 1376–1383.

Wolk, K.E., Lazarowski, E.R., Traylor, Z.P., Yu, E.N., Jewell, N.A., Durbin, R.K., Durbin, J.E., Davis, I.C., 2008. Influenza a virus inhibits alveolar fluid clearance in BALB/c mice. Am. J. Respir. Crit. Care Med. 178, 969–976.

Chapter 24

Zinc Homeostasis in Lung

Hui-Hua Li, Kalidasan Thambiayya, Murat Kaynar, Karla A. Wasserloos, Claudette M. St Croix and Bruce R. Pitt

Chapter Outline

1. INTRODUCTION

Although the role of zinc in biology was first reported by Raulin in 1869 (Raulin, 1869), it was not recognized to be important for human life until about a hundred years later when zinc deficiency was revealed by Ananda Prasad and colleagues as a major contributing factor in nutritional dwarfism syndrome and hypogonadism (Prasad et al., 1961) in a cohort of young male Egyptians that was reminiscent of previous experience in Iran. The profound zinc deficiency was likely due to consumption of diet restricted to certain groups of cereal proteins and geophagia (clay eating). Ultimately, in 1974, recommended dietary allowances for humans for zinc were established in United States (Prasad, 2009). Outten and O'Halloran (2001) coined the term *zinc quota* to indicate that total intracellular zinc $[Zn]_i$ is maintained in a tight range (100−500 μM) across numerous cell types. Zinc is considered a trace metal, however, because more than 99% of intracellular zinc is protein bound. It is generally assumed that a vanishingly small (10^{-12}−10^{-9} M; Bozym et al., 2006; Colvin et al., 2008; Li and Maret, 2009) component of intracellular zinc ($[Zn]_i$), referred to as free or labile zinc, and operationally defined as the pool sensitive to chelation (by agents such as *N,N,N′,N′*-tetrakis(2-pyridylmethyl) ethylenediamine (TPEN)) and capable of detection by a variety of chemical and genetic sensors, participates in signal transduction pathways (Williams, 1984; Yamasaki et al., 2007) in a fashion well supported for other divalent cations such as calcium. Recent approaches using bioinformatics methods to mine existing protein databases indicate that approximately 10% of the human proteome is zinc dependent (Andreini et al., 2006). As such, zinc is: (1) a structural component of at least 3000 proteins (Maret and Li, 2009), including transcription factors, cytokines and receptors; (2) regulator of enzyme activity (Mocchegiani et al., 2000) including acting as catalytic component of more than 300 enzymes (Gamsjaeger et al., 2007; Barceloux, 1999; Mocchegiani et al., 2000; Carter et al., 2002); and (3) regulator of enzyme activity by acting as an activator or inhibitor ion. It is now well established that zinc is important for numerous cellular functions including cell differentiation and division, DNA synthesis, RNA transcription, and maintaining plasma membrane integrity (Vallee and Falchuk, 1993; Prasad, 1995; Truong-Tran et al., 2001; Rink and Haase, 2007; Prasad, 2008).

In this chapter, we discuss molecular fundamentals of zinc homeostasis and the role of zinc in physiological signaling in lung (primarily epithelium, endothelium, and macrophages). We will review the impact of zinc deficiency or excess on lung pathophysiology and then consider the role of zinc as a therapeutic agent. More general and complete reviews of extrapulmonary aspects of these topics are available, including molecular biology of zinc transporters (Lichten and Cousins, 2009; Lichten et al., 2009; Huang and Tepaamorndech, 2013; Jeong and Eide, 2013), zinc signaling (Yamasaki et al., 2007; Murakami and Hirano, 2008; Fukada et al., 2011), zinc and metallochemistry (Maret and Li, 2009), and zinc in health and disease including innate, adaptive (Hirano et al., 2008; Haase and Rink, 2009) and tumor immunology (John et al., 2010), neurobiology (Sensi et al., 2009) and endocrinology

Comparative Biology of the Normal Lung. http://dx.doi.org/10.1016/B978-0-12-404577-4.00024-2

(Rutter, 2010; Li, 2013), childhood diarrhea and pneumonia in low-to-middle income countries (Yakoob et al., 2011), and human health in general (Saper and Rash, 2009; Prasad, 2013). We have previously considered relevant aspects in the context of acute lung injury (Li et al., 2010) and pulmonary endothelial cell biology (Thambiayya et al., 2012a), and parts of these efforts helped formulate the current review.

2. MOLECULAR FUNDAMENTALS OF ZINC HOMEOSTASIS

At an organismal level in mammals, zinc is an essential metal absorbed in small intestines from food sources including red meat, poultry, seafood (highly enriched in oysters), fortified breakfast cereals, and a variety of beans, nuts, whole grains, and dairy products (U.S. Department of Agriculture, Agricultural Research Service, 2011). As such, dietary zinc deficiencies in North America are less of a concern than in lower-to-middle income countries with a more restricted diet of cereal substances (often times high in phytic acid that makes zinc unavailable for absorption). At a cellular level, zinc homeostasis is maintained by combined and coordinated activities of two large families of transporters (Lichten et al., 2009) and numerous binding proteins (Andreini et al., 2006) (see Figure 1). Solute-linked carrier 30A (SLC30A) or ZnT contains 10 family members that function in zinc efflux and compartmentalization (Palmiter and Huang, 2004; Huang and Tepaamorndech, 2013). ZnT1 is expressed at the plasma membrane and serves to transport zinc to extracellular space, and its expression appears to be affected by zinc levels and dietary zinc (Dufner-Beattie et al., 2003). The other ZnTs are located on intracellular organelles and account for discrete intracellular trafficking of zinc. The other family, SLC39A or ZIP (Zrt-, Irt-like protein), consists of 14 members of zinc importers that are primarily involved in zinc uptake and sequestration (Jeong and Eide, 2013). These zinc transporters and importers are expressed in a tissue- and cell- and organelle-specific fashion. Although efforts have not been exhaustive, the lungs of mice (Lang et al., 2007) and humans (Besecker et al., 2008) appear to express almost all SLC30A and 39A genes.

The prototypic autosomal recessive life-threatening disorder, acrodermatitis enteropathica (AE), has provided our greatest insight into the vital role of zinc transporters in human health, and abundant experimental genetic information (primarily in mice) has greatly expanded our sense of the importance of these molecules in health and disease. Zinc deficiency in human neonates secondary to AE results in skin lesions, alopecia, and gastrointestinal dysfunction,

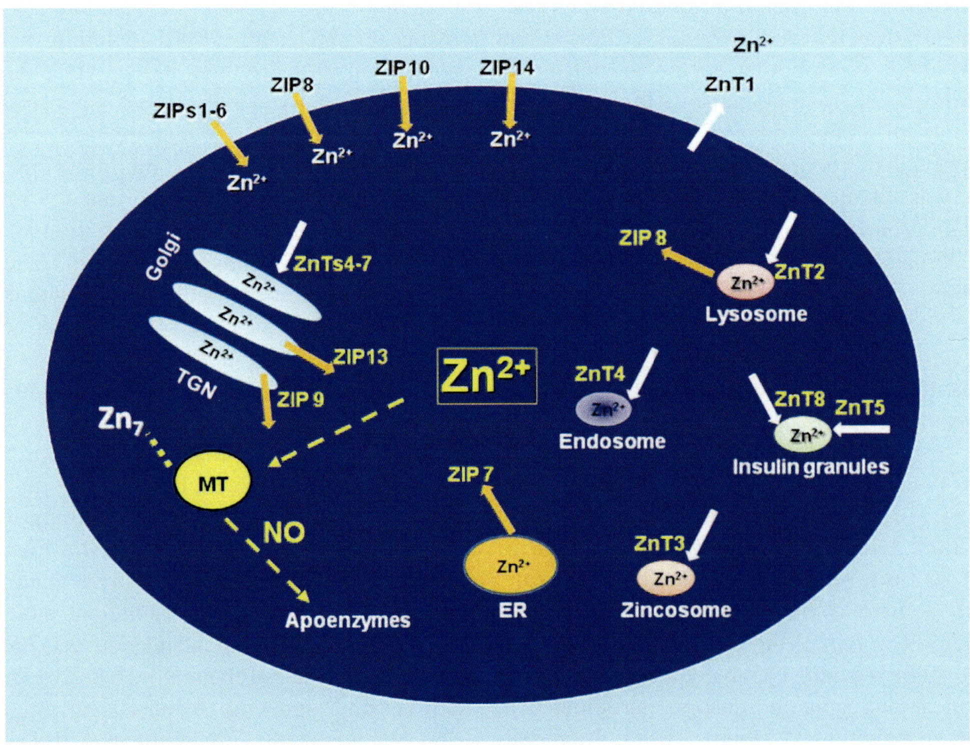

FIGURE 1 **Regulators of intracellular zinc homeostasis.** The labile pool of intracellular zinc is tightly controlled by zinc importers (ZIPs), zinc exporters (ZnTs), zinc storing vesicles and zinc binding proteins such as metallothionein (MT). MT plays a critical role in zinc homeostasis acting as a buffer in the steady state while controlling the cellular distribution of transiently elevated zinc in response to perturbations and/or agonists such as nitric oxide (NO) (Maret, 2011). *This figure was reproduced from Thambiayya et al. (2012a) with permission from Pulmonary Circulation, which was originally modified from Fukada et al. (2011) and Maret (2011).*

and if untreated (with zinc supplementation) can include advanced disease with growth delay, neurodevelopmental delay, poor wound healing, anemia, photophobia, hypogeusia, and hypogonadism in boys and men (Maverakis et al., 2007). ZIP4 Knockout (KO) mice embryos die during early development, whereas heterozygous mice exhibit a phenotype similar to AE secondary to impairment of intestinal absorption of zinc (Kury et al., 2002; Wang et al., 2002; Dufner-Beattie et al., 2007). Oral clioquinol and zinc supplementation can successfully rescue an AE phenotype in conditional ZIP4 KO mice (Geiser et al., 2013) by stimulating stem cell proliferation and differentiation of intestinal epithelium. Knockout (KO) mice lacking ZIP1, ZIP2, ZIP3 are reported to have abnormal embryogenesis under zinc-limiting conditions (Dufner-Beattie et al., 2005, 2006; Peters et al., 2007). ZIP13 KO mice recapitulate the disorganized hard connective tissue phenotype (Fukada et al., 2008) that typifies spondylocheiro dysplasia, a form of Ehlers–Danlos syndrome (Fukada et al., 2008; Giunta et al., 2008) in humans with mutations in ZIP13. Mice lacking ZIP14 (Table 1) exhibit retarded growth and impaired gluconeogenesis and have impaired G-protein coupled receptor (GPCR) signaling (Hojyo et al., 2011).

As with *ZIP* genes, disruption of the *ZnT* genes yields diverse phenotypes providing insight into the biological function and specificity of the various family members (Table 1). ZnT1 knockout mice are embryonic lethal (Andrews et al., 2004). Mutations in zinc transporters (ZnT2 in mice; (Chowanadisai et al., 2006); ZnT4 in humans (Huang and Gitschier, 1997)) affect critical enrichment of breast milk with zinc. Phenotypic changes in ZnT3 and ZnT5 knockout mice included seizures (Chasapis et al., 2012) and growth retardation and osteogenic problems (Inoue et al., 2002) and impaired cytokine production in mast cells (Nishida et al., 2009), respectively. Deletion of the ZnT8 gene results in impaired insulin secretion in mice (Pound et al., 2009) and genome-wide association studies in human populations (Table 2) show associations with single-nucleotide polymorphisms in ZnT8 and type 2 diabetes (Sladek et al., 2007).

Decreases in intracellular zinc lead to homeostatic increased expression of members of ZIP family and decreased expression of ZnT family members (Cao et al., 2001; Truong-Tran et al., 2002; Devergnas et al., 2004; Lang et al., 2007; Overbeck et al., 2008b). Our recent data in cultured sheep pulmonary artery endothelial cells (SPAECs) showed that ZIP14 is sensitive to changes in labile zinc, and further, that zinc-mediated protection against LPS-induced apoptosis is dependent upon the homeostatic increase in ZIP14 subsequent to LPS-induced decrease in zinc (Thambiayya et al., 2012a). Forceful overexpression of ZIP6 inhibits LPS-induced decrease in zinc and consequent maturation in mouse dendritic cells (Kitamura et al., 2006). Relatively little is known about the uptake, efflux, and changes in the intracellular compartmentalization of zinc in lung disease, though reported increases in ZIP1 and ZIP14 mRNA in a mouse model of acute lung inflammation (Lang et al., 2007) were shown to be normalized by dietary zinc supplementation, and zinc-mediated cytoprotection against TNF-α-induced damage in human lung epithelial cells was shown to be dependent upon expression of ZIP8 (Besecker et al., 2008).

Metallothioneins (MT) are small (<7 kd), cysteine-rich (approximately mol 30%) metal binding proteins capable of coordinating up to 7 mol Zn^{2+}/mol MT via dynamic thiolate clusters (Lazo and Pitt, 1995; Bell and Vallee, 2009). Although the function of MT is unclear, it is involved in detoxification of nonessential heavy metals (mercury, cadmium, etc.) and contributes to antioxidant cellular defense mechanisms (Lazo and Pitt, 1995; Vallee, 1995; Davis and Cousins, 2000; Higashimoto et al., 2009). Although its importance was long ascribed to its intracellular location (including compartmentalization to mitochondria and nucleus (Ye et al.,2001)), it recently has been noted to be distributed to extracellular space as well (Hao et al., 2007). Its low reduction potential (<−366 mV) (Bell and Vallee, 2009), is the chemical basis for its role in oxidant stress, and the uniqueness of its thiolate clusters underscores its lability in nitrosative stress (Liu et al., 2000; Zangger et al., 2001). We (Pearce et al., 2000) and Maret and colleagues (Maret and Vallee, 1998; Maret et al., 1999; Maret, 2003; Krezel et al., 2007; Krezel and Maret, 2007) have shown that MT is sensitive to changes in cellular redox state and demonstrated that increases in reactive oxygen or nitrogen intermediates can oxidize or transnitrosate cysteines in its zinc sulfur clusters leading to liberation of zinc (see below). MT is thus an important buffer for $[Zn]_i$ (Li and Maret, 2009) and also a sensor and switch connecting changes in cellular redox status with alterations in labile zinc (Figure 1).

3. ZINC AS A SIGNALING MOLECULE IN LUNG

As proposed by Hirano and colleagues (Fukada et al., 2011), changes in the free or labile pool of intracellular zinc have the capacity to transduce extracellular stimuli to a myriad of physiological pathways. Such changes can be acute and transient and are highly analogous to the role of calcium. These effects are transcription independent and were considered as early Zn signaling. The prototype of these is the zinc wave and the allergic response in mast cells (Kabu et al., 2006). Zinc signaling that is dependent on modifying expression of zinc transporters is induced with a more considerable time delay. An example of such late zinc signaling is the coupling of LPS-stimulated dendritic cells, a decrease in ZIP6, and expression of MHC (and hence maturation) of these cells (Kitamura et al., 2006).

TABLE 1 Phenotypical Profile (in Mouse, *Human,Drosophila,***Sheep) of Mutants of Zinc Transporters**

Gene Name	Protein Name	Type	Phenotypes
SLC39A1	ZIP1	−/−	Abnormal embryogenesis (Dufner-Beattie, Huang et al., 2005, Dufner-Beattie, Huang et al., 2006)
SLC39A2	ZIP2	−/−	Abnormal embryogenesis (Peters, Dufner-Beattie et al., 2007)
SLC39A3	ZIP3	−/−	Abnormal embryogenesis (Dufner-Beattie, Huang et al., 2005, Dufner-Beattie, Huang et al., 2006)
SLC39A4	ZIP4	−/−	Death of embryos during development (Dufner-Beattie, Weaver et al., 2007, Geiser et al., 2013)
SLC39A4	ZIP4*	+/−	Inherited disorder acrodermatitis enteropathica (AE), in which the intestines ability to absorb zinc is impaired (Kury, Dreno et al., 2002, Wang, Zhou et al., 2002)
SLC39A6	ZIP6	+/−	**Abnormal gonad formation (Mathews, Ong et al., 2006) and E-cadherin expression (Van Doren, Mathews et al., 2003)
SLC39A7	ZIP7	+/−	Impaired melanin synthesis (Stathakis, Burton et al., 1999)
SLC39A8	ZIP8	+/−	Resistance to cadmium-induced testicular damage (Dalton, He et al., 2005)
SLC39A13	ZIP13	−/−	(1) Disorganization in hard connective tissue, including bone, teeth, skin, and eyes (Fukada, Civic et al., 2008) (2)*Spondylocheiro dysplastic, a form of Ehlers-Danlos syndrome (Dalton, He et al., 2005, Bin, Fukada et al., 2011)
SLC39A14	ZIP14	−/−	Impaired GPCR signaling, growth retardation, and impaired gluconeogenesis (Hojyo, Fukada et al., 2011)
SLC39A14	ZIP14	KD	(1)***Blocked extracellular zinc-mediated protection against apoptosis in SPAECs (Thambiayya, Wasserloos et al., 2012a) (2)***Lowered base level zinc in SPAECs (Thambiayya, Wasserloos et al., 2012b)
SLC30A1	ZnT1	−/−	Embryonic lethal (Andrews, Wang et al., 2004)
SLC30A2	ZnT2	+/−	*Zinc deficiency in milk (Chowanadisai, Lonnerdal et al., 2006)
SLC30A3	ZnT3	−/−	Seizures (Cole, Robbins et al., 2000, Chasapis, Loutsidou et al., 2012)
SLC30A4	ZnT4	−/−	Zinc deficiency in milk (Huang and Gitschier, 1997)
SLC30A5	ZnT5	−/−	(1) Growth retardation and osteogenic problem (Inoue, Matsuda et al., 2002) (2) Impaired cytokine production in mast cells (Nishida, Hasegawa et al., 2009)
SLC30A7	ZnT7	−/−	Growth retardation, low body zinc status, and low fat accumulation (Huang, Yu et al., 2007)
SLC30A8	ZnT8	−/−	Impaired insulin secretion (Pound, Sarkar et al., 2009)
SLC30A8	ZnT8	SNPs	*Type 2 diabetes (Sladek, Rocheleau et al., 2007)

TABLE 2 Phenotypical Profile (in Humans) of SNPs of Zinc Transporters

Gene Name	Protein Name	Phenotypes
SLC39A2	ZIP2	(1) Associated with carotid artery disease in elderly population from central Italy (Giacconi, Muti et al., 2008) (2) Contributes to arsenic-associated bladder cancer (Karagas, Andrew et al., 2012)
SLC39A8	ZIP8	(1) Associated with schizophrenia in European populations (Carrera, Arrojo et al., 2012) (2) Elevated lipid levels in the circulation (dyslipidemias) and risk of coronary artery disease (Waterworth, Ricketts et al., 2010)
SLC30A2	ZnT2	Affects zinc homeostasis in mammary epithelial cells and contributes in the development and progression of breast disease (Seo and Kelleher, 2010)
SLC30A8	ZnT8	Associated with Type 2 diabetes mellitus (Sladek, Rocheleau et al., 2007, Kifagi, Makni et al., 2011, Xu, Wang et al., 2012)

Airway epithelium: Zalewski and colleagues developed useful fluorophores for zinc and were the first to image labile zinc in the lung (Truong-Tran et al., 2000; Carter et al., 2002). As reviewed by Troung-Tran et al. (2001), intracellular zinc has also been shown to be important for ciliary function, wound healing (via reepithelialization), and suppression of oxidative stress and apoptosis (Truong-Tran et al., 2000), and in general, acts as an anti-inflammatory molecule in the airway epithelium (Truong-Tran et al., 2002; Lang et al., 2007). Initial insight into zinc suppression of caspase-3 and apoptosis led to a series of subsequent studies on altered zinc homeostasis in murine allergic airway inflammation including an important role for ZnT4 in human airway epithelium and its loss in a murine model of allergic airway inflammation (Murgia et al., 2011), as well as zinc affecting inhibitor of apoptosis proteins (IAPs; Roscioli et al., 2013). Knoell and colleagues have provided additional insight into the role of zinc in airway inflammation, epithelial cell injury, and apoptosis, and the role of zinc transporters in the lung and innate immunity. They initially showed that Zip8 was uniquely critical in promoting zinc-mediated survival of cultured human airway epithelium to TNFα-induced apoptosis (Besecker et al., 2008). Following their original observations that zinc deficiency increased inflammation and morbidity from polymicrobial sepsis (e.g., cecal ligation and puncture) through regulation of NF-κB (Bao et al., 2010), they identified ZIP8 as a transcriptional target of NF-κB that negatively regulates proinflammatory responses through zinc-mediated down-modulation of IκB kinase in cultured primary human airway epithelium (Liu et al., 2013). These studies in cultured human airways, monocytes, and intact mice provided critical links to how zinc affects innate immunity and infection (Rink and Haase, 2007). Samet and colleagues have considered the response of airway epithelium to exogenous zinc (Wu et al., 2013) in the context of mechanistic studies on toxicity of zinc-laden particle in the airway and have demonstrated how zinc affects cytokine production via modification of phosphoprotein metabolism (Wu et al., 2003), EGF receptor-mediated pathways (Wu et al., 2005), and MAPK (Kim et al., 2006) and NF-κB (Wu et al., 2002) pathways. They have outlined a model to show interactive pathways leading to prooxidative stress and toxic effects of large concentrations of extracellular zinc (Wu et al., 2013). Other aspects of airway epithelial cell function and zinc have been considered, including barrier function (Bao and Knoell, 2006) and ion channel effects, including sodium channel (Chen et al., 2012) and chloride transport (Zsembery et al., 2004).

Pulmonary endothelium: Depletion of zinc causes spontaneous apoptosis in pulmonary endothelium (Tang et al., 2001). Elevations in labile zinc via ZIP14 (Thambiayya et al., 2012b) or iNOS-induced nitric oxide (NO) or NO donors (Tzeng et al., 1997; Ceneviva et al., 1998; Spahl et al., 2003; St Croix et al., 2005; Cortese-Krott et al., 2009; Thambiayya et al., 2012b) inhibits apoptosis in cultured pulmonary endothelial cells. A role for zinc inhibition of caspase-3 in such antiapoptotic effects was noted (Thambiayya et al., 2011, 2012b). Moreover, we have shown that zinc chelation (via TPEN) exacerbates LPS-induced apoptosis in cultured sheep pulmonary artery endothelial cells (SPAECs; Tang et al., 2001). TPEN also reversed the protective effect of NO donors on LPS-induced apoptosis (Tang et al., 2002). In separate studies in SPAEC, siRNA knockdown of ZIP14 decreased basal levels of intracellular labile zinc and blocked zinc uptake, and abrogated zinc-mediated protection against LPS-induced apoptosis (Thambiayya et al., 2012b). Collectively, these data suggest that endogenous levels of $[Zn]_i$ can modulate the sensitivity of pulmonary endothelium to LPS, and ZIP14 appears critical in mediating the influx of extracellular zinc to inhibit LPS-induced apoptosis in SPAEC (Thambiayya et al., 2012a).

Our laboratories have a long-standing interest in NO-(MT)-Zn^{2+} signaling in pulmonary endothelium (Li et al., 2010; Thambiayya et al., 2012a). We, and others, have shown that the zinc-thiolate moieties of MT are critical targets for NO (Kroncke et al., 1994; Pearce et al., 2000; Liu et al., 2001; St Croix et al., 2004) with resultant effects on intracellular zinc homeostasis (Berendji et al., 1997; St Croix et al., 2002; Spahl et al., 2003). Subsequent investigations supported the potential for MT to participate in intracellular signal transduction pathways affecting gene expression in pulmonary endothelium in that exposure of mouse lung endothelial cells (MLEC) to the NO donor, S-nitroso-N-acetylpenicillamine, caused nuclear translocation of the zinc-dependent transcription factor, MTF-1, and corresponding increases in MT gene expression (Stitt et al., 2006). These effects were not apparent in MT null cells. More recently, we reported that hypoxia-induced increases in eNOS-derived NO disrupts intracellular zinc homeostasis that in turn modulate vascular reactivity of small (>50 μm) microvessels and intra-acinar arteries of lung (Bernal et al., 2008). This anatomic site was composed primarily of endothelial cells, confirming the potential (Hillier et al., 1997; Clough et al., 2000) for endothelium to actively regulate pulmonary blood flow distribution and supporting a role for zinc thiolate signaling in pulmonary vasoregulation. These findings were reminiscent of our original observations regarding a role for NO-MT-Zn^{2+} signaling in the myogenic reflex in systemic circulation (Pearce et al., 2000) in that the NO formed in response to increased intraluminal pressure caused greater relaxation in superior mesenteric arteries isolated from MT null mice than in vessels from wild-type animals. This enhanced relaxing effect of NO during myogenic reflex in MT null vessels may be direct via increased availability of NO to affect presumptive targets such as soluble guanylate cyclase or calcium-activated potassium channels (Pearce et al., 2000). Alternatively, the enhanced relaxing effect of NO during myogenic reflex in MT null vessels may be indirect due to the absence of releasable metals (i.e., Zn) from MT that would normally produce contraction. Endothelial cell contraction is initiated by phosphorylation of the 20 kDa regulatory myosin light chain (MLC) at S19/T18 (Goeckeler and Wysolmerski, 1995). The phosphorylation of MLC is dependent upon a balance between the activities of calcium/calmodulin-dependent MLC kinase (MLCK) and myosin light chain phosphatase (MLCP). Remodeling of the endothelial actin cytoskeletal in response to either hypoxia, or the procoagulant protein, thrombin, involves MLC and actin-related proteins (Gavard and Gutkind, 2008). One potential connection between these signaling pathways and Zn is the relationship between divalent metal ions and sulfhydryl residues in the activation of type 1 and type 2A serine/threonine phosphoprotein phosphatases (PPases), with Zn having been shown to be a potent inhibitor of lambda-PPase (Zhuo and Dixon, 1997). We recently established a link between acute hypoxia, elevations in labile Zn, and activation of protein kinase C (PKC), which in turn acted through CPI-17 to inhibit MLCP activity and promote MLC phosphorylation, ultimately inducing stress fiber formation and endothelial cell contraction (Bernal et al., 2011). Considered in light of earlier data from our group (Bernal et al., 2008) and others (Hillier et al., 1997; Yamaguchi et al., 1998; Tabuchi et al., 2008) demonstrating hypoxia-induced contraction of nonmuscularized (30–50 μm) pulmonary arterioles and venules, these data suggest that endothelial cytoskeletal rearrangement may regulate capillary diameter in vivo (Sylvester et al., 2012).

Monocytes: In addition to airway epithelium, the immunomodulatory and antimicrobial properties of zinc have been considered in the context of monocytes including alveolar macrophages (Stafford et al., 2013). Indeed a parallel set of information on the importance of ZIP8 and zinc and NF-κB was reported in human monocytes in addition to airway epithelium (Liu et al., 2013). In a series of studies by Guidot and colleagues (Mehta and Guidot, 2012), chronic alcohol ingestion decreases zinc levels and alveolar macrophage function in rats (Joshi et al., 2009) and humans (Mehta et al., 2013), including decreases in phagocytic activity, GM-CSF receptors signaling via PU.1, glutathione, and activation of Nrf2 and other antioxidants (Mehta et al., 2011).

4. ZINC TOXICITY IN LUNG

At supraphysiological levels, zinc is often toxic, and this is clearly the case in lung where both chronic and acute exposure to elevated levels of zinc by inhalation are of concern. In this regard, exposure to extraordinarily high concentrations of zinc (e.g., accidental inhalation of zinc chloride (Homma et al., 1992)) can produce fatal acute lung injury in humans. Alternatively, acute and/or chronic exposure to moderate levels of zinc oxide (especially in welders) produces a form of metal fume fever associated with decrements in lung function, infiltration of neutrophils in the airways, and a cytokine-mediated proinflammatory state within the lung (Fine et al., 1997). In addition, zinc appears to account for a significant aspect of particulate matter cardiopulmonary toxicity (Adamson et al., 2000; Kodavanti et al., 2002; Adamson et al., 2003; Gavett et al., 2003). Okeson et al. (2004) reported that zinc is the most toxic among three transitional metals present in particulate matter (zinc, iron, and vanadium) for cytotoxicity on confluent monolayers of rat alveolar epithelial cells, with half-maximal effect concentration of 0.1 mM in serum-free media and 0.6 mM in media containing 10% serum. The molecular determinants of zinc toxicity in airways have recently been reviewed by Wu et al. (2013).

4.1 Zinc Deficiency and Acute Lung Injury: Complexities of Integrated Response

Several studies have demonstrated that zinc deficiency sensitizes the lung to acute injury. In particular, nutritionally induced zinc deficiency led to enhanced sensitivity to polymicrobial sepsis (Knoell et al., 2009; Bao et al., 2010) and hyperoxic lung injury in mice (Taylor et al., 1990, 1997). Alcohol-fed rats (with similar caloric intake) became zinc deficient and had impaired alveolar macrophage and epithelial cell function (Joshi et al., 2009). Zinc repletion reversed phenotype in all of the above conditions. Nonetheless, hypozincemia in septic or aseptic (Cousins et al., 2006; Lichten et al., 2009) conditions is a somewhat underappreciated phenomenon. Transmigration of zinc from tissues, including lung, to liver has been noted in hyperoxia (Levy et al., 2001; Lee et al., 2013), bacterial sepsis (Philcox et al., 1995), and turpentine injury (Liuzzi et al., 2005), and has been presumed to subserve: (1) gluconeogenesis in liver; (2) new protein synthesis in acute phase response; or (3) host defense in an analogous fashion to hypoferremia in bacterial pneumonia (Liuzzi et al., 2005; Thambiayya et al., 2012a). IL-6-mediated hepatic expression of ZIP14 appears critical in this phenomenon (Lichten et al., 2009) and IL-8-mediated changes of ZIP8 in airways is important in the integrated response to sepsis and the lung (Liu et al., 2013). We recently (Lee et al., 2013) noted that hepatic expression of metallothionein was important in transmigration of zinc from lung to liver during hyperoxic lung injury, apparently contributing to the unexpected observation that MT null mice were resistant to hyperoxia. Joshi and Guidot (2011) noted that HIV-1-related protein expression in rats caused liver and zinc sequestration. This model limited zinc bioavailability in the lung and impaired alveolar macrophage phagocytosis—a phenotype reversible with exogenous zinc therapy. Understanding the mechanisms underlying zinc transmigration may ultimately support the use of exogenous zinc (and other approaches to affect hypozincemia) as a rational therapeutic agent in acute lung injury (Thambiayya et al., 2012a).

5. ZINC AS A THERAPEUTIC AGENT

Zalewski and colleagues (Zalewski et al., 2005; Zalewski, 2006) have suggested that in the respiratory tract, zinc may be an antioxidant, membrane and cytoskeletal stabilizer, essential component of DNA synthesis and cellular growth, participate in wound healing, and in general act as an anti-inflammatory molecule (St Croix et al., 2005). Perhaps because of its pleiotropic nature and/or simplified approaches for supplementation, zinc has become an interesting candidate for pharmacotherapy of a number of pulmonary disorders including upper airway viral infection (Hulisz, 2004) and asthma (McKeever and Britton, 2004). The efficacy of extracellular zinc in restoring chloride secretion across cystic fibrosis (CF) airway epithelia (Zsembery et al., 2004) underscores its potential as adjunct therapy in CF (Cantin, 2004). As upregulated epithelial sodium channel (ENaC) is an important component of CF pathophysiology, our recent observations (Chen et al., 2012) that at low-to-moderate concentrations extracellular zinc can open ENaC (via inhibition of Na^+-self inhibition) add complexities to this direct approach for this complex disorder.

Zinc supplementation has been shown to be therapeutic and preventive and appears to be a uniquely beneficial monotherapy for bacterial, parasitic, and viral infections (Overbeck et al., 2008a), including H1N1 influenza (Sandstead and Prasad, 2010). When used on the background of presumed profound zinc deficiency secondary to malnutrition in children, zinc reduces mortality from diarrheal and pneumonia disorders by 13% and 15%, respectively (Yakoob et al., 2011). In an elderly population, zinc supplementation sufficient to replete zinc stores was effective in reducing incidence of infections and lowering circulating TNFα and markers of oxidative stress (Prasad et al., 2007). A multiple dosage regimen of zinc by gavage was effective against H1N1 (and H5N1) virus infection in mice (Barnard et al., 2007).

In its most familiar form, zinc is the presumptive active component in over-the-counter homeopathic formulations for curtailing the course of human rhinovirus infection (Singh and Das, 2013). Although the epidemiological data are compelling (Science et al., 2012), the mechanisms remain unclear and have been ascribed to nonspecific effects of zinc on the immune system, antiviral properties of zinc itself, or zinc interfering with ICAM receptors critical for viral uptake. More critical is potential loss of smell (anosmia) first noted after intranasal administration of zinc gluconate (Jafek et al., 2004). Damage to the olfactory system can be long term (Lim et al., 2009), leading the FDA to warn consumers of certain nasal and gel swab formulations of zinc.

6. INTERACTIONS OF ZINC AND OTHER METALS

Both the physiological, therapeutic, and adverse effects of zinc can be greatly influenced by interactions at multiple steps with other metals. Indeed from work in sheep (Schouwink, 1961) with copper toxicosis and human sickle cell patients (Brewer and Hill, 1983), it became apparent that zinc was able to induce a state of hypocupremia (perhaps secondary to induction of MT in intestines) that could subserve therapeutic purposes in Wilson disease, a genetic disorder of excess copper. As such, Brewer (Brewer et al., 1990) and others (Hoogenraad, 1998) have put forth zinc acetate (and other formulations) as primary therapy for this disorder. Alternatively, we now know that abuse of dental adhesives (previously containing significant amounts of zinc) produced profound neurological symptoms secondary to hypocupremia (Barton et al., 2011). An increased risk of cancer mortality associated with cadmium exposures in older Americans with low zinc intake was recently found (Lin et al., 2013). In such individuals, zinc can be used as a potential detoxifying agent to reduce cadmium-induced toxicity. Zip8 plays an important role in import of both zinc and cadmium, a nonessential highly toxic heavy metal, in respiratory epithelium. As such, Knoell and colleagues (Napolitano et al., 2012) recently reported that ZIP8 plays a critical role at the interface between zinc and cadmium in cigarette smoke exposure. ZIP8 is also an iron transporter whose cell-surface expression is upregulated by cellular iron loading (Wang et al., 2012). High heme iron intake may increase the risk of lung cancer and thus interactions of iron, and zinc transport may be important for smokers for whom high dietary zinc supplementation (and vitamin C) is recommended (Lee and Jacobs, 2005). This finding may be of particular importance to smokers, for whom vitamin C and zinc supplementation is a common recommendation (Lee and Jacobs, 2005). In the human pathogen *Streptococcus pneumoniae*, acquisition of the essential trace metal, manganese, is facilitated by PsaA, a high affinity manganese-specific cluster A-1 substrate binding protein and direct inhibition of PsaA--mediated manganese transport by zinc has recently been shown to be a component of host defense (McDevitt et al., 2011). The structural details of such competition have recently been reported by Counago et al. (2013).

7. SUMMARY

As essential information emerges from simple organisms to complex animal models on the determinants of zinc homeostasis and its role as an effector molecule in physiological signaling, we are now able to better define the role of zinc in the lung in health and disease. Challenges remain regarding the need for better zinc sensors including those capable of helping quantify minute concentrations of the presumptive labile pool of zinc. In the same regard, physiological studies on the role of free zinc will be advanced with development of more selective zinc chelators. The pleiotropic role of zinc in lung biology underscores its ubiquitous importance and impacts on cell death and differentiation, wound repair, immunology and inflammation, cytoskeletal function and contractility, and permeability. As an essential nutrient, zinc couples nutrition with lung function and underscores the public health importance of zinc deficiency in low-to-moderate income countries and defines the current uncertainties in the United States where zinc deficiency for the vast majority is unlikely. Nonetheless, zinc is a common part of almost all multivitamin and mineral supplements (MedlinePlus, 2013) that are used by approximately half the U.S. adult population (Bailey et al., 2013) on a routine basis. As a common feature of sepsis and shock, hypozincemia is an underappreciated component of acute lung injury. Although it is not without its toxicities (especially to lung after direct exposure to inhalation), a relatively wide margin of therapeutic index and low cost make zinc an attractive pharmacological agent in the context of lung infections and airway disease and may be used rationally in susceptible groups that are relatively zinc deficient secondary to age, alcohol, or comorbidities.

REFERENCES

Adamson, I.Y., et al., 2000. Toxicol. Appl. Pharmacol. 166 (2), 111–119.

Adamson, I.Y., et al., 2003. Exp. Lung Res. 29 (6), 375–388.

Andreini, C., et al., 2006. J. Proteome Res. 5 (1), 196–201.

Andrews, G.K., et al., 2004. Genesis 40 (2), 74–81.

Bailey, R.L., et al., 2013. JAMA Intern. Med. 173 (5), 355–361.

Bao, S., Knoell, D.L., 2006. Am. J. Physiol. Lung Cell. Mol. Physiol. 291 (6), L1132–L1141.

Bao, S., et al., 2010. Am. J. Physiol. Lung Cell. Mol. Physiol. 298 (6), L744–L754.

Barceloux, D.G., 1999. J. Toxicol. Clin. Toxicol. 37 (2), 279–292.

Barnard, D.L., et al., 2007. Antivir. Chem. Chemother. 18 (3), 125–132.

Barton, A.L., et al., 2011. Ann. Clin. Biochem. 48 (Pt 4), 383–385.

Bell, S.G., Vallee, B.L., 2009. Chembiochem 10 (1), 55–62.

Berendji, D., et al., 1997. FEBS Lett. 405 (1), 37–41.

Bernal, P.J., et al., 2011. Am. J. Physiol. Lung Cell. Mol. Physiol. 300 (6), L874–L886.

Bernal, P.J., et al., 2008. Circ. Res. 102 (12), 1575–1583.

Besecker, B., et al., 2008. Am. J. Physiol. Lung Cell. Mol. Physiol. 294 (6), L1127–L1136.

Bin, B.H., Fukada, T., Hosaka, T., Yamasaki, S., Ohashi, W., Hojyo, S., Miyai, T., Nishida, K., Yokoyama, S., Hirano, T., 2011 Nov 18. J Biol Chem. 286 (46), 40255–40265.

Bozym, R.A., et al., 2006. ACS Chem. Biol. 1 (2), 103–111.

Brewer, G.J., Hill, G.M., 1983. Prog. Clin. Biol. Res. 127, 97–113.

Brewer, G.J., et al., 1990. J. Am. Coll. Nutr. 9 (5), 487–491.

Cantin, A.M., 2004. Curr. Opin. Pulm. Med. 10 (6), 531–536.

Cao, J., et al., 2001. J. Leukoc. Biol. 70 (4), 559–566.

Carrera, Arrojo, et al., 2012 Jan 15. Biol Psychiatry 71 (2), 169–177.

Carter, J.E., et al., 2002. Biochem. Biophys. Res. Commun. 297 (4), 1062–1070.

Ceneviva, G.D., et al., 1998. Am. J. Physiol. 275 (4 Pt 1), L717–L728.

Chasapis, C.T., Loutsidou, A.C., Spiliopoulou, C.A., Stefanidou, M.E., 2012 Apr. Arch Toxicol 86 (4), 521–534. http://dx.doi.org/10.1007/s00204-011-0775-1. Epub 2011 Nov 10. Review.

Chasapis, C.T., et al., 2012. Arch. Toxicol. 86 (4), 521–534.

Chen, J., et al., 2012. J. Biol. Chem. 287 (42), 35589–35598.

Chowanadisai, W., et al., 2006. J. Biol. Chem. 281 (51), 39699–39707.

Clough, A.V., et al., 2000. Am. J. Physiol. Heart Circ. Physiol. 279 (3), H1274–H1282.

Cole, T.B., Robbins, C.A., Wenzel, H.J., Schwartzkroin, P.A., Palmiter, R.D., 2000 Apr. Epilepsy Res. 39 (2), 153–169.

Colvin, R.A., et al., 2008. Am. J. Physiol. Cell. Physiol. 294 (3), C726–C742.

Cortese-Krott, M.M., et al., 2009. Am. J. Physiol. Cell. Physiol. 296 (4), C811–C820.

Counago, R.M., et al., 2013. Nat. Chem. Biol. 10, 35–41. http://dx.doi.org/10.1038/nchembio.1382.

Cousins, R.J., et al., 2006. J. Biol. Chem. 281 (34), 24085–24089.

Davis, S.R., Cousins, R.J., 2000. J. Nutr. 130 (5), 1085–1088.

Dalton, T.P., He, L., Wang, B., Miller, M.L., Jin, L., Stringer, K.F., Chang, X., Baxter, C.S., Nebert, D.W., 2005 Mar 1. Proc Natl Acad Sci U S A 102 (9), 3401–3406.

Devergnas, S., et al., 2004. Biochem. Pharmacol. 68 (4), 699–709.

Dufner-Beattie, J., et al., 2005. Mol. Cell. Biol. 25 (13), 5607–5615.

Dufner-Beattie, J., et al., 2006. Genesis 44 (5), 239–251.

Dufner-Beattie, J., et al., 2003. J. Biol. Chem. 278 (50), 50142–50150.

Dufner-Beattie, J., et al., 2007. Hum. Mol. Genet. 16 (12), 1391–1399.

Fine, J.M., et al., 1997. J. Occup. Environ. Med. 39 (8), 722–726.

Fukada, T., et al., 2008. PLoS One 3 (11), e3642.

Fukada, T., et al., 2011. J. Biol. Inorg. Chem. 16 (7), 1123–1134.

Gamsjaeger, R., et al., 2007. Trends Biochem. Sci. 32 (2), 63–70.

Gavard, J., Gutkind, J.S., 2008. J. Biol. Chem. 283 (44), 29888–29896.

Gavett, S.H., et al., 2003. Environ. Health Perspect. 111 (12), 1471–1477.

Geiser, J., et al., 2013. PLoS One 8 (8), e72543.

Giacconi, R., Muti, E., Malavolta, M., Cardelli, M., Pierpaoli, S., Cipriano, C., Costarelli, L., Tesei, S., Saba, V., Mocchegiani, E., 2008 Apr. Rejuvenation Res. 11 (2), 297–300.

Giunta, C., et al., 2008. Am. J. Hum. Genet. 82 (6), 1290–1305.

Goeckeler, Z.M., Wysolmerski, R.B., 1995. J. Cell. Biol. 130 (3), 613–627.

Haase, H., Rink, L., 2009. Annu. Rev. Nutr. 29, 133–152.

Hao, Q., et al., 2007. J. Cell. Physiol. 210 (2), 428–435.

Higashimoto, M., et al., 2009. Life Sci. 84 (17–18), 569–575.

Hillier, S.C., et al., 1997. J. Appl. Physiol. (1985) 82 (4), 1084–1090.

Hirano, T., et al., 2008. Adv. Immunol. 97, 149–176.

Hojyo, S., et al., 2011. PLoS One 6 (3), e18059.

Homma, S., et al., 1992. Hum. Pathol. 23 (1), 45–50.

Hoogenraad, T.U., 1998. J. Lab. Clin. Med. 132 (4), 240–241.
Huang, L., Gitschier, J., 1997. Nat. Genet. 17 (3), 292–297.
Huang, L., Tepaamorndech, S., 2013. Mol. Asp. Med. 34 (2–3), 548–560.
Huang, L., Yu, Y.Y., Kirschke, C.P., Gertz, E.R., Lloyd, K.K., 2007 Dec 21. J Biol Chem 282 (51), 37053–37063.
Hulisz, D., 2004. J. Am. Pharm. Assoc. (2003) 44 (5), 594–603.
Inoue, K., et al., 2002. Hum. Mol. Genet. 11 (15), 1775–1784.
Jafek, B.W., et al., 2004. Am. J. Rhinol. 18 (3), 137–141.
Jeong, J., Eide, D.J., 2013. Mol. Asp. Med. 34 (2–3), 612–619.
John, E., et al., 2010. J. Transl. Med. 8 (118). http://dx.doi.org/10.1186/1479-5876-8-118.
Joshi, P.C., Guidot, D.M., 2011. AIDS Res. Ther. 8, 36.
Joshi, P.C., et al., 2009. Am. J. Respir. Cell. Mol. Biol. 41 (2), 207–216.
Kabu, K., et al., 2006. J. Immunol. 177 (2), 1296–1305.
Karagas, M.R., Andrew, A.S., Nelson, H.H., Li, Z., Punshon, T., Schned, A., Marsit, C.J., Morris, J.S., Moore, J.H., Tyler, A.L., Gilbert-Diamond, D., Guerinot, M.L., Kelsey, K.T., 2012 Mar. Hum Genet 131 (3), 453–461.
Kim, Y.M., et al., 2006. Am. J. Physiol. Lung Cell. Mol. Physiol. 290 (5), L1028–L1035.
Kifagi, C., Makni, K., Boudawara, M., Mnif, F., Hamza, N., Abid, M., Granier, C., 2011 Jun. Ayadi H. Genet Test Mol Biomarkers 15 (6), 399–405.
Kitamura, H., et al., 2006. Nat. Immunol. 7 (9), 971–977.
Knoell, D.L., et al., 2009. Crit. Care Med. 37 (4), 1380–1388.
Kodavanti, U.P., et al., 2002. Toxicol. Sci. 70 (1), 73–85.
Krezel, A., et al., 2007. Arch. Biochem. Biophys. 463 (2), 188–200.
Krezel, A., Maret, W., 2007. Biochem. J. 402 (3), 551–558.
Kroncke, K.D., et al., 1994. Biochem. Biophys. Res. Commun. 200 (2), 1105–1110.
Kury, S., et al., 2002. Nat. Genet. 31 (3), 239–240.
Lang, C., et al., 2007. Am. J. Physiol. Lung Cell. Mol. Physiol. 292 (2), L577–L584.
Lazo, J.S., Pitt, B.R., 1995. Annu. Rev. Pharmacol. Toxicol. 35, 635–653.
Lee, D.H., Jacobs Jr., D.R., 2005. Nutr. Cancer 52 (2), 130–137.
Lee, S.M., et al., 2013. Am. J. Physiol. Lung Cell. Mol. Physiol. 304 (5), L350–L360.
Levy, M.A., et al., 2001. Am. J. Physiol. Lung Cell. Mol. Physiol. 281 (1), L172–L182.
Li, H., et al., 2010. Ann. N. Y. Acad. Sci. 1203, 73–78.
Li, Y., Maret, W., 2009. Exp. Cell. Res. 315 (14), 2463–2470.
Li, Y.V., 2013. Endocrine 45 (2), 178–189.
Lichten, L.A., Cousins, R.J., 2009. Annu. Rev. Nutr. 29, 153–176.
Lichten, L.A., et al., 2009. Am. J. Physiol. Gastrointest. Liver Physiol. 296 (4), G860–G867.
Lim, J.H., et al., 2009. PLoS One 4 (10), e7647.
Lin, Y.S., et al., 2013. J. Toxicol. Environ. Health A 76 (1), 1–15.
Liu, M.J., et al., 2013. Cell. Rep. 3 (2), 386–400.
Liu, S., et al., 2001. Biochem. J. 354 (Pt 2), 397–406.
Liu, S.X., et al., 2000. Chem. Res. Toxicol. 13 (9), 922–931.
Liuzzi, J.P., et al., 2005. Proc. Natl. Acad. Sci. U. S. A. 102 (19), 6843–6848.
Maret, W., 2003. J. Nutr. 133 (5 Suppl. 1), 1460S–1462S.
Maret, W., 2011. J. Biol. Inorg. Chem. 16 (7), 1079–1086.
Maret, W., et al., 1999. Proc. Natl. Acad. Sci. U. S. A. 96 (5), 1936–1940.
Maret, W., Li, Y., 2009. Chem. Rev. 109 (10), 4682–4707.
Maret, W., Vallee, B.L., 1998. Proc. Natl. Acad. Sci. U. S. A. 95 (7), 3478–3482.
Mathews, W.R., Ong, D., Milutinovich, A.B., Van, Doren M., 2006 Mar. Development 133 (6), 1143–1153.
Maverakis, E., et al., 2007. J. Am. Acad. Dermatol. 56 (1), 116–124.
McDevitt, C.A., et al., 2011. PLoS Pathog. 7 (11), e1002357.
McKeever, T.M., Britton, J., 2004. Am. J. Respir. Crit. Care Med. 170 (7), 725–729.
MedlinePlus. US National Library of Medicine, 2013. NIH. Zinc Diet. Available at: http://www.nlm.nih.gov/medlineplus/ency/article/002416.htm.
Mehta, A.J., Guidot, D.M., 2012. Am. J. Med. Sci. 343 (3), 244–247.
Mehta, A.J., et al., 2011. Alcohol Clin. Exp. Res. 35 (8), 1519–1528.
Mehta, A.J., et al., 2013. Am. J. Respir. Crit. Care Med. 188 (6), 716–723.
Mocchegiani, E., et al., 2000. Trends Pharmacol. Sci. 21 (6), 205–208.
Murakami, M., Hirano, T., 2008. Cancer Sci. 99 (8), 1515–1522.
Murgia, C., et al., 2011. Nutrients 3 (11), 910–928.
Napolitano, J.R., et al., 2012. Am. J. Physiol. Lung Cell. Mol. Physiol. 302 (9), L909–L918.
Nishida, K., et al., 2009. J. Exp. Med. 206 (6), 1351–1364.
Okeson, C.D., et al., 2004. Toxicol. Vitro. 18 (5), 673–680.
Outten, C.E., O'Halloran, T.V., 2001. Science 292 (5526), 2488–2492.
Overbeck, S., et al., 2008a. Arch. Immunol. Ther. Exp. Warsz. 56 (1), 15–30.
Overbeck, S., et al., 2008b. J. Leukoc. Biol. 83 (2), 368–380.
Palmiter, R.D., Huang, L., 2004. Pflugers Arch. 447 (5), 744–751.
Pearce, L.L., et al., 2000. Proc. Natl. Acad. Sci. U. S. A. 97 (1), 477–482.
Peters, J.L., et al., 2007. Genesis 45 (6), 339–352.
Philcox, J.C., et al., 1995. Biochem. J. 308 (Pt 2), 543–546.
Pound, L.D., et al., 2009. Biochem. J. 421 (3), 371–376.
Prasad, A.S., 1995. Nutrition 11 (1 Suppl.), 93–99.
Prasad, A.S., 2008. Mol. Med. 14 (5–6), 353–357.
Prasad, A.S., 2009. J. Am. Coll. Nutr. 28 (3), 257–265.
Prasad, A.S., 2013. Adv. Nutr. 4 (2), 176–190.
Prasad, A.S., et al., 2007. Am. J. Clin. Nutr. 85 (3), 837–844.
Prasad, A.S., et al., 1961. Am. J. Med. 31, 532–546.
Raulin, J., 1869. Ann. Sci. Nat. 11, 93–99 (in French).
Rink, L., Haase, H., 2007. Trends Immunol. 28 (1), 1–4.
Roscioli, E., et al., 2013. Biometals 26 (2), 205–227.
Rutter, G.A., 2010. Islets 2 (1), 49–50.
Sandstead, H.H., Prasad, A.S., 2010. Am. J. Public Health 100 (6), 970–971.
Saper, R.B., Rash, R., 2009. Am. Fam. Physician 79 (9), 768–772.
Schouwink, G., 1961. De hepatocerebrale degeneratie met eenonderzoek naar de zinkstofwisseling proefschrift (Ph.D. thesis). University of Amsterdam.
Science, M., et al., 2012. CMAJ 184 (10), E551–E561.
Sensi, S.L., et al., 2009. Nat. Rev. Neurosci. 10 (11), 780–791.
Seo, Y.A., Kelleher, S.L., 2010 Nov 29. Physiol Genomics 42A (4), 219–227.
Singh, M., Das, R.R., 2013. Cochrane Database Syst. Rev. 6. http://dx.doi.org/10.1002/14651858.CD001364.pub4.
Sladek, R., et al., 2007. Nature 445 (7130), 881–885.
Spahl, D.U., et al., 2003. Proc. Natl. Acad. Sci. U. S. A. 100 (24), 13952–13957.
St Croix, C.M., et al., 2005. Proc. Am. Thorac. Soc. 2 (3), 236–242.
St Croix, C.M., et al., 2004. Free Radic. Biol. Med. 37 (6), 785–792.
St Croix, C.M., et al., 2002. Am. J. Physiol. Lung Cell. Mol. Physiol. 282 (2), L185–L192.
Stathakis, D.G., Burton, D.Y., McIvor, W.E., Krishnakumar, S., Wright, T.R., O'Donnell, J.M., 1999 Sep. Genetics 153 (1), 361–382.
Stafford, S.L., et al., 2013. Biosci. Rep. 33 (4).
Stitt, M.S., et al., 2006. Vasc. Pharmacol. 44 (3), 149–155.
Sylvester, J.T., et al., 2012. Physiol. Rev. 92 (1), 367–520.

Tabuchi, A., et al., 2008. J. Appl. Physiol. (1985) 104 (2), 338–346.

Tang, Z.L., et al., 2001. Am. J. Physiol. Lung Cell. Mol. Physiol. 281 (1), L243–L249.

Tang, Z.L., et al., 2002. Mol. Cell. Biochem. 234–235 (1–2), 211–217.

Taylor, C.G., et al., 1997. Free Radic. Biol. Med. 22 (3), 543–550.

Taylor, C.G., et al., 1990. Free Radic. Biol. Med. 9 (3), 229–233.

Thambiayya, K., et al., 2012a. Pulm. Circ. 2 (4), 443–451.

Thambiayya, K., et al., 2012b. Am. J. Physiol. Lung Cell. Mol. Physiol. 302 (12), L1287–L1295.

Thambiayya, K., et al., 2011. Am. J. Physiol. Lung Cell. Mol. Physiol. 300 (4), L624–L632.

Truong-Tran, A.Q., et al., 2001. Immunol. Cell. Biol. 79 (2), 170–177.

Truong-Tran, A.Q., et al., 2002. Am. J. Respir. Cell. Mol. Biol. 27 (3), 286–296.

Truong-Tran, A.Q., et al., 2000. Am. J. Physiol. Lung Cell. Mol. Physiol. 279 (6), L1172–L1183.

Tzeng, E., et al., 1997. Surgery 122 (2), 255–263.

U.S. Department of Agriculture, Agricultural Research Service, 2011. USDA National Nutrient Database for Standard Reference. Release 24. Nutrient Data Laboratory Home Page. http://www.ars.usda.gov/ba/bhnrc/ndl.

Vallee, B.L., 1995. Neurochem. Int. 27 (1), 23–33.

Van Doren, M., Mathews, W.R., Samuels, M., Moore, L.A., Broihier, H.T., Lehmann, R., 2003 Jun. Development 130 (11), 2355–2364.

Vallee, B.L., Falchuk, K.H., 1993. Physiol. Rev. 73 (1), 79–118.

Wang, C.Y., et al., 2012. J. Biol. Chem. 287 (41), 34032–34043.

Wang, K., et al., 2002. Am. J. Hum. Genet. 71 (1), 66–73.

Waterworth, Ricketts, et al., 2010 Nov. Arterioscler Thromb Vasc Biol 30 (11), 2264–2276.

Williams, R.J., 1984. Endeavour 8 (2), 65–70.

Wu, W., et al., 2013. Free Radic. Biol. Med. 65C, 57–69.

Wu, W., et al., 2002. Am. J. Physiol. Lung Cell. Mol. Physiol. 282 (5), L1040–L1048.

Wu, W., et al., 2005. Am. J. Physiol. Lung Cell. Mol. Physiol. 289 (5), L883–L889.

Wu, W., et al., 2003. J. Biol. Chem. 278 (30), 28258–28263.

Xu, J., Wang, J., Chen, B., 2012 May 24. Genet Mol Res 11 (2), 1592–1598.

Yakoob, M.Y., et al., 2011. BMC Public Health 11 (Suppl. 3), S23. http://dx.doi.org/10.1186/1471-2458-11-S3-S23.

Yamaguchi, K., et al., 1998. Circ. Res. 82 (6), 722–728.

Yamasaki, S., et al., 2007. J. Cell. Biol. 177 (4), 637–645.

Ye, B., et al., 2001. Proc. Natl. Acad. Sci. U. S. A. 98 (5), 2317–2322.

Zalewski, P.D., 2006. Curr. Opin. Pharmacol. 6 (3), 237–243.

Zalewski, P.D., et al., 2005. Pharmacol. Ther. 105 (2), 127–149.

Zangger, K., et al., 2001. FASEB J. 15 (7), 1303–1305.

Zhuo, S., Dixon, J.E., 1997. Protein Eng. 10 (12), 1445–1452.

Zsembery, A., et al., 2004. J. Biol. Chem 279 (11), 10720–10729.

Chapter 25

Antioxidant Defenses in the Lung

Albert van der Vliet

Chapter Outline

1. INTRODUCTION

The lung represents the body's largest interface with the external environment, with an estimated surface area of approximately 150 m^2. While this large exposure area is important for effective oxygen delivery and gas exchange, it also renders the lung vulnerable to airborne microbes, allergens, and common air pollutants such as ozone (O_3), nitrogen oxides including nitrogen dioxide (NO_2), automobile emissions, tobacco smoke, and particulate matter (PM) from geological or industrial sources. Therefore, the respiratory tract must be equipped with adequate antimicrobial and antioxidant defense systems to prevent infection or injury and maintain appropriate lung function. Central to such defense is the respiratory epithelium, which coats the entire lung surface and acts as a physical barrier against airborne pollutants and microorganisms, forming a molecular sieve that excludes inhaled antigens and pathogens. Throughout the respiratory tract, the respiratory epithelium also maintains a thin fluid layer that covers the epithelial surface, termed respiratory tract lining fluids (RTLF), and contains high-molecular-weight glycoproteins (mucins), iron-binding proteins (e.g., lactoferrin and transferrin), and other antimicrobial factors, as a first-line innate defense shield. In addition, RTLF also contain various secreted antioxidant enzymes and low-molecular-weight antioxidant molecules (ascorbic acid, uric acid, tocopherols, glutathione), that collectively provide a complex antioxidant defense network to protect the underlying epithelium against exogenous oxidative stress and to control local redox status at the respiratory surface.

The respiratory epithelium is also instrumental in mediating acute responses to a wide variety of luminal triggers, such as allergens, infectious agents, and other pollutants, and plays a central role in regulating innate or adaptive inflammatory responses to these commonly inhaled environmental triggers (Hammad and Lambrecht, 2008; Holgate et al., 2004). A common critical component of such acute responses is the regulated production of reactive oxygen species (ROS) by activation of NADPH oxidases in epithelial and endothelial cells, alveolar macrophages, and extravasated granulocytes. While such controlled ROS production is critical in appropriate regulation of inflammatory responses (Segal et al., 2012; van der Vliet, 2011), inappropriate production of ROS during acute or chronic lung inflammation or excessive exposure to exogenous pollutants may lead to oxidative injury in critical cell constituents and impaired lung cell function (Fahy, 2009; Jacobsen et al., 2007; Shaw et al., 2007). Indeed, abundant evidence for such deleterious properties of endogenously produced ROS during chronic airway inflammation has been obtained by detection of irreversible oxidation products in proteins, nucleic acids, or lipids within chronically inflamed airways (Andreadis et al., 2003; Wedes et al., 2009).

To cope with these various oxidant challenges, the lung is endowed with various antioxidant systems to protect against cellular dysfunction by environmental and endogenous oxidants and to coordinate the important biological functions of endogenous ROS in, for example, inflammatory signaling and tissue regeneration. Because of its anatomical location and exposure to the environment, the respiratory

Comparative Biology of the Normal Lung. http://dx.doi.org/10.1016/B978-0-12-404577-4.00025-4

epithelium is uniquely challenged by both environmental and endogenous ROS and equipped with a range of antioxidant enzyme systems, but other lung cell types similarly contain a spectrum of antioxidant systems to cope with endogenous ROS production related to metabolism, activation of specific NADPH oxidase isoforms, or other sources of ROS. This review is a brief overview of the general sources of exogenous and endogenous ROS that are encountered by lung, with some focus on NADPH oxidase systems, and a summary of the available antioxidant systems within the lung and their importance in preventing lung disease or dysfunction.

2. EXOGENOUS AND ENDOGENOUS ROS PRODUCTION IN THE LUNG

Although genome-wide association studies have provided insight into genetic determinants of lung diseases such as asthma or chronic obstructive pulmonary disease (COPD), relatively limited contribution of genetic variance to pulmonary disease has been identified, and environmental factors are recognized to play an important role (Budinger and Mutlu, 2011; Ober et al., 2010). Many of these environmental factors include air pollutants that often exert their effects by causing oxidative stress (Ciencewicki et al., 2008). These include gaseous pollutants that compromise photochemical smog such as ozone (O_3) and nitrogen dioxide (NO_2), which are highly oxidizing molecules are reactive with several molecules at the respiratory surface to generate secondary ROS, including superoxide anion ($O_2^{\bullet-}$), hydrogen peroxide (H_2O_2), and hydroxyl radical ($OH^{\bullet}$). Other air pollutants that induce oxidative stress include PM with variable size and dimensions originating from, for example, industrial sources, diesel exhaust, or asbestos mining, which generate ROS largely due to their contained redox-active transition metal ions (iron, copper, etc.) (Shannahan et al., 2012). In addition to their common exposure to an oxidative environment, lung cells are also continuously subjected to ROS generation as by-product of various enzymatic reactions, including production of $O_2^{\bullet-}$ and H_2O_2 at complex I and III in the mitochondrial respiratory chain, the major site of cellular O_2 consumption, and by other enzymatic sources within mitochondria, such as p66shc, amine oxidases, α-ketoglutarate dehydrogenase, and pyruvate dehydrogenase (Drose and Brandt, 2012; Lenaz, 2012). Peroxisomes are another major cellular source of endogenous ROS; they are organelles that are specialized in peroxide metabolism, lipid biosynthesis, and β-oxidation of amino acids, via the action of acyl-CoA-oxidases, amine oxidases, xanthine oxidase, etc. (Antonenkov et al., 2010; Schrader and Fahimi, 2006). Furthermore, ROS are also produced as by-products of cytochrome P450 enzymes, cyclooxygenases, and lipoxygenases, which are expressed in most lung cell types (Bae et al., 2011), and the spectrum of endogenous ROS production is further expanded by interactions with nitric oxide (NO•) production from widely expressed nitric oxidase synthases (NOS) to produce reactive nitrogen species such as NO_2 and peroxynitrite ($ONOO^-$) (van der Vliet et al., 1999a).

While most of these endogenous sources are thought to generate ROS as by-products of metabolism, ROS such as $O_2^{\bullet-}$ and H_2O_2 are also produced by many lung cell types in a regulated manner as a primary mediator to support various biological functions such as antimicrobial host defense as well as oxidant-dependent signaling pathways that regulate gene expression, and cell proliferation or differentiation. The biological importance of ROS is most well appreciated with respect to inflammatory immune cells including phagocytic cells, which produce high levels of ROS by assembly and activation of the NADPH oxidase complex in the so-called respiratory burst (Babior et al., 2002; Bedard and Krause, 2007). This NADPH oxidase complex comprises two membrane-associated proteins, the flavocytochrome b_{558} (gp91^{phox}) and p22^{phox}, as well as various cytosolic cofactors (p47^{phox}, p67^{phox}, and p40^{phox}, and the GTPase, Rac1) that are recruited to the membrane-bound component upon appropriate cell activation, primarily during bacterial phagocytosis, and are activated to produce intraphagosomal ROS to kill ingested microorganisms by oxidative mechanisms. Initial ROS production involves the following reactions:

$$\text{NADPH} + O_2 \rightarrow O_2^{\bullet-} + H^+ + \text{NADP}$$

$$O_2^{\bullet-} + O_2^{\bullet-} \rightarrow O_2 + H_2O_2$$

Thus produced $O_2^{\bullet-}$ and H_2O_2 can subsequently give rise to more potent antimicrobial oxidants by reaction with locally secreted heme peroxidases such as myeloperoxidase or eosinophil peroxidase (Winterbourn et al., 2006).

In addition to promoting bacterial killing, activation of NADPH oxidase in phagocytic cells also serves to regulate production of inflammatory mediators and phagocyte apoptosis as a key mechanism in terminating inflammation (Segal et al., 2012). These actions are mediated by oxidative regulation of cell signaling pathways involved in activation of transcription factors such as NF-κB and AP-1 (Fialkow et al., 2007) or by ROS-dependent oxidation of phospholipids that promotes externalization of phosphatidylserine as a critical event neutrophil apoptosis and its clearance to terminate inflammation (Fadeel et al., 1998, 2010). Indeed, functional defects in NADPH oxidase in patients with chronic granulomatous disease (CGD) lead to increased susceptibility to various infections but also in dysregulated phagocyte clearance mechanisms, thereby resulting in hyperinflammatory lesions that are often unrelated to infection (Lee et al., 2006; Staple et al., 2005).

Over the past 15 years, several homologs of the main NADPH oxidase component, gp91^{phox}, have been discovered and found to be expressed in a wide range of cell types. To date, seven mammalian NADPH oxidase (NOX) homologs have been described, comprising NOX1-5 (NOX2 being the new name for gp91^{phox}) and two larger Dual Oxidases, DUOX1 and DUOX2, and their presence in many diverse cell types suggests their involvement in a wide variety of biological functions. Analogous to NOX2 (gp91^{phox}), other NOX members also require association with p22^{phox} and assembly with Rac and cytosolic cofactors (p47^{phox} and p67^{phox} or their homologs, NOX organizer 1 (NOXO1) and NOX activator 1 (NOXA1)) (Bedard and Krause, 2007; Geiszt and Leto, 2004; Lambeth, 2004). Some NOXs (NOX5 and DUOX1/2) contain additional intracellular Ca^{2+}-binding EF-hand domain regions and are activated by Ca^{2+} signaling, without the need for association with p22^{phox} or other cytosolic cofactors (Ameziane-El-Hassani et al., 2005; Banfi et al., 2004; Dupuy et al., 1999; Morand et al., 2003). In addition, the two DUOX proteins also contain a unique extracellular domain that is homologous to heme peroxidases but appears to lack functional peroxidase activity, and its functional importance is unknown (Geiszt and Leto, 2004; Lambeth, 2004; Meitzler and Ortiz de Montellano, 2009).

Activation of NOX/DUOX generates superoxide anion ($O_2^{\cdot-}$) and/or hydrogen peroxide (H_2O_2) as primary products (Ameziane-El-Hassani et al., 2005; Geiszt and Leto, 2004; Martyn et al., 2006). In analogy to cooperative actions between NOX2 in granulocytes and locally secreted peroxidases to generate potent antimicrobial oxidants, similar cooperative actions exist between DUOX proteins within the respiratory epithelium and locally secreted lactoperoxidase (LPO) within the airway mucosa, as an innate local host defense mechanism (El Hassani et al., 2005; Geiszt and Leto, 2004; Wijkstrom-Frei et al., 2003). However, most biological actions of NOX-derived $O_2^{\cdot-}/H_2O_2$ appear to be mediated by direct reaction with susceptible cysteine residues in target proteins or with other redox-sensitive metal centers, through which NOX enzymes control a variety of biological processes ranging from cell proliferation and differentiation to inflammatory signaling and immune regulation. These signaling mechanisms of NOX enzymes are largely related to regulation of common signaling processes involving protein tyrosine phosphorylation and Ca^{2+} signaling (Finkel, 2003; Forman et al., 2004; Janssen-Heininger et al., 2008; Terada, 2006; van der Vliet, 2008), and specificity in such oxidative signaling is achieved by strict spatial localization of NOX activation and its target proteins, in association with the cytoskeleton or in membrane lipid rafts (Terada, 2006; van der Vliet, 2008). Such spatially confined oxidative signaling is also ensured by the global presence of cytosolic antioxidant systems (such as Cu/Zn SOD, GSH peroxidases, peroxiredoxins), which restrict ROS diffusion and thereby avoid more distant oxidative effects. Indeed, ROS can have opposing effects on certain biological processes depending on the location or extent of ROS production, as illustrated by the ability of ROS to both activate and inhibit transcription factors such as nuclear factor (NF)-κB (Brar et al., 2003; Janssen-Heininger et al., 2008; Li et al., 2006).

The biological properties of ROS from, for example, NOX enzymes indicate that antioxidant defenses must be carefully regulated to allow for effective NOX-dependent ROS-mediated signaling while affording appropriate defense against other endogenous ROS sources associated with cellular stress or environmental sources ROS that challenge the lung. Before discussing the available antioxidant systems that are present within the lung, and their importance in preventing pulmonary disease, the next section will first briefly overview the current state of knowledge regarding the distribution of NOX enzymes in the lung and their roles in physiology or pathology.

2.1 NOX Enzymes in the Lung

Respiratory Epithelium

Although the respiratory epithelium is the primary target for inhaled oxidant pollutants, respiratory tract epithelial cells themselves generate extracellular ROS in response to infectious or inflammatory stimuli (Kinnula et al., 1992), which originate primarily from locally expressed DUOX1 and DUOX2 (Gattas et al., 2009; Geiszt et al., 2003; Harper et al., 2006; Moskwa et al., 2007; Wesley et al., 2007). DUOX1 is mostly localized at the apical surface of tracheobronchial and alveolar epithelial cells, and DUOX2 is primarily found in salivary and submucosal glands (Fischer et al., 2007; Forteza et al., 2005; Geiszt et al., 2003; Harper et al., 2005; Schwarzer et al., 2004). While airway epithelial or alveolar DUOX1 is constitutively expressed in differentiated epithelia (Fischer et al., 2007; Gattas et al., 2009; Harper et al., 2005; Linderholm et al., 2010), epithelial DUOX2 expression is highly inducible by various bacterial and viral stimuli (Gattas et al., 2009; Harper et al., 2005), indicating its importance in host defense in association with locally secreted LPO from submucosal glands (Allaoui et al., 2009; Conner et al., 2007; Moskwa et al., 2007; Wijkstrom-Frei et al., 2003). Both DUOX1 and DUOX2 also contribute to innate responses to a variety of microbial and environmental stimuli, by participating in H_2O_2-dependent signaling pathways that regulate expression of inflammatory mediators, such as interleukin-8 (CXCL8) (Boots et al., 2009; Koff et al., 2008; Nakanaga et al., 2007; Yu et al., 2011), mucin genes such as *MUC1* and *MUC5AC* (Kuwahara et al., 2007; Shao and Nadel, 2005), or matrix remodeling enzymes (Sham et al., 2013; Wesley et al., 2007), and DUOX1 is also critical in epithelial wound responses and maintenance of epithelial integrity and barrier function (Gorissen et al., 2013; Koff et al., 2006; Wesley et al., 2007).

Airway and alveolar epithelial cells also express other NOX isoforms such as NOX1, NOX2, or NOX4 (Lavigne and Eppihimer, 2005; Schwarzer et al., 2004), which are important in epithelial responses to viral infection (Comstock et al., 2011; Soucy-Faulkner et al., 2010; Yoboua et al., 2010) or environmental stress (Amara et al., 2007; Kim et al., 2008), but can also mediate hyperoxia-induced apoptosis and acute lung injury (Carnesecchi et al., 2009) and bleomycin-induced pulmonary fibrosis (Carnesecchi et al., 2011).

Hemapoietic Cell Types

In addition to granulocytes, lung macrophages and dendritic cells also express NOX2 to mediate similar direct antimicrobial properties (Huang et al., 2009) but also to regulate antigen processing and cross-presentation (Jancic et al., 2007; Rybicka et al., 2012, Rybicka et al., 2010) and proinflammatory signaling (Cruz et al., 2007; Forman and Torres, 2002; Rinna et al., 2006). Activation of NOX2 has also been implicated in inflammasome activation by environmental stimuli (Dostert et al., 2008; Moore and MacKenzie,2009), although this is still controversial (Meissner et al., 2010; van de Veerdonk et al., 2010). Similarly, NOX2 also contributes to antigen-dependent signaling in lymphocytes in response to T cell receptor activation (Jackson et al., 2004; Kwon et al., 2005; Reth, 2002). In spite of the common association of NOX2 with proinflammatory signaling, genetic deficiency of NOX2 or its cofactors in fact often results in enhanced inflammation and injury in response to infection or other stimuli (Gao et al., 2002; Kassim et al., 2005; Morgenstern et al., 1997; Segal et al., 2007), indicating its importance in regulating inflammation and its resolution (Segal et al., 2010).

Pulmonary Endothelium

Several NOX isoforms (NOX1, NOX2, and NOX4) are also expressed in pulmonary endothelial cells and are involved in ROS production in response to pulmonary ischemia (Al-Mehdi et al., 1998; Fisher et al., 2003) or hyperoxia (Carnesecchi et al., 2009; Parinandi et al., 2003), resulting in decreased alveolar-capillary barrier function as a mechanism of hyperoxia-induced lung injury (Chowdhury et al., 2005; Usatyuk et al., 2007). Activation of NOX2 by pulmonary endothelial cell depolarization during ischemia/hypoxia has also been linked to ROS-mediated endothelial proliferation (Milovanova et al., 2006).

Smooth Muscle Cells/Fibroblasts

Pulmonary fibroblasts and smooth muscle cells express several NOX cofactors ($p47^{phox}$, $p67^{phox}$, $p22^{phox}$) and NOX4 in response to inflammatory mediators such as tumor necrosis factor α (TNF-α) or transforming growth factor β (TGF-β) (Dhaunsi et al., 2004; Sturrock et al., 2006; Thabut et al., 2002; Waghray et al., 2005). Expression of NOX4 in smooth muscle cells is closely associated with smooth muscle differentiation markers such as smooth muscle α-actin (SMA) or myosin heavy chain (Clempus et al., 2007; Sturrock et al., 2006), and induction of NOX4-mediated ROS production in human pulmonary artery smooth muscle cells by TGF-β mediates smooth muscle cell proliferation and hypertrophy (Sturrock et al., 2006, Sturrock et al., 2007). NOX4 expression is enhanced in lung tissues from patients with pulmonary arterial hypertension (Mittal et al., 2007) and pulmonary fibrosis (Amara et al., 2010; Hecker et al., 2009), and may contribute to development of pulmonary fibrosis by promoting differentiation of fibroblasts to pathological myofibroblasts (Amara et al., 2010; Hecker et al., 2009).

The wide spectrum of exogenous and endogenous sources of ROS in different lung cell types, and their variable chemical and biological properties in both lung physiology and pathology, support the important presence of a variety of cellular or extracellular antioxidant systems to control or counter their actions. Expression of various antioxidant enzymes is ubiquitous throughout the lung, to cope with endogenous ROS production by NOX isozymes as well as other ROS sources, but is especially critical in airway or alveolar epithelial cells and other luminal cell types that are directly exposed to the external environment. The remainder of this overview will summarize the major antioxidant systems that are present in the lung, with specific focus on antioxidant mechanisms that exist at the epithelial surface, and discuss their importance in preventing environmental and inflammatory lung diseases.

3. ANTIOXIDANT DEFENSES IN THE LUNG

To curtail unwanted effects of cellular or extracellular ROS production, all cell types contain a spectrum of antioxidant enzyme systems that specifically detoxify superoxide ($O_2^{\bullet-}$) or hydrogen peroxide (H_2O_2) or contribute to defense against ROS by more indirect mechanisms.

Detoxification of $O_2^{\bullet-}$ is catalyzed by superoxide dismutases (SOD), which exist in three isoforms with specific subcellular distribution, and detoxification of H_2O_2 depends on cellular localization and is mediated by catalase or several isoforms of glutathione peroxidase (GPX) or peroxiredoxin (Prx). The combined actions of SOD and H_2O_2-metabolizing enzymes maintain low overall steady-state levels of $O_2^{\bullet-}$ and H_2O_2 and prevent unwanted actions of these ROS while allowing for appropriate localized ROS-mediated redox signaling by, for example, NOX enzymes. Furthermore, since these primary ROS can also react with available redox-active transition metals to produce the highly reactive hydroxyl radical ($OH^{\bullet}$) (Cross et al., 1994), homeostasis of these transition metals must be carefully regulated, and alterations in iron homeostasis genes in various lung pathologies (Ghio, 2009; Lagan et al., 2008) may contribute to ROS-dependent injury in these cases. Additional enzymatic antioxidant systems include thioredoxins (Trx) and glutaredoxins (GRX), which act primarily by reversing protein cysteine oxidation, enzymes involved in synthesis or maintenance of the low-molecular-weight thiol glutathione (GSH) as a critical cofactor for GPX or GRX, and other enzymes such as heme oxygenase and metallothionein, which are thought to contribute to antioxidant defense by more indirect mechanisms. While some of these basic antioxidant enzyme systems are ubiquitously expressed at high levels to cope with continued ROS production during normal aerobic metabolism, expression of some of these antioxidant systems is also cell- or tissue-specific and is inducible as part of adaptive responses to inflammation, hyperoxia, or other oxidative challenges. In addition to this spectrum of antioxidant enzymes, nonenzymatic low-molecular-weight antioxidants such as the vitamins C and E (ascorbic acid and α-tocopherol), carotenoids, the purine metabolite uric acid, and other dietary antioxidants contribute to detoxification of ROS or other environmental oxidants for which no enzymatic antioxidant enzymes exist. The combined antioxidant systems at the lung surface are schematically illustrated in Figure 1, and the following sections will summarize the specific features of these various antioxidant systems and their known functions in the lung and potential roles in lung pathology.

3.1 Enzymatic Antioxidants: Superoxide Dismutases

Three isoforms of SOD exist: cytoplasmic CnZnSOD (SOD1), mitochondrial MnSOD (SOD2), and extracellular EC-SOD (SOD3). All SODs are expressed by almost all lung cell types, but with substantial variability in cell-specific expression and localization (Kinnula and Crapo, 2003), and their primary function is to catalyze the dismutation of superoxide to hydrogen peroxide and oxygen:

$$2O_2^{\bullet-} + 2H^+ + SOD \rightarrow H_2O_2 + O_2$$

This accelerates the spontaneous dismutation of $O_2^{\bullet-}$ by about 10,000-fold, from a second-order rate constant of 10^5/M/s to 10^9/M/s. Thus, generated H_2O_2 is subsequently detoxified by various enzymatic systems.

SOD1

The most abundant Cu,Zn-SOD isoform, SOD1, is a heterodimeric protein of 32 kDa that contains a catalytic Cu ion and a Zn ion that is important for structural stability. SOD1 is primarily localized in the cytosol of most lung cell types but is also localized in the nucleus and in some organelles such as lysosomes (Kinnula and Crapo, 2003). Overall SOD1 expression is lower in the lung compared to several other organ systems, but SOD1 is abundantly expressed in bronchial and alveolar epithelial cells, as well as mesenchymal cells, fibroblasts, and capillary endothelial cells (Kinnula and Crapo, 2003). SOD1 is constitutively expressed within the lung, is not significantly induced by oxidative stress or inflammation, and is consistently expressed during lung development but appears to increase toward adulthood (Kinnula and Crapo, 2003; Zelko et al., 2002). The functional importance of SOD1 in the lung is not fully established but is most likely related to regulating cytosolic $O_2^{\bullet-}$ levels (e.g., to minimize production of $ONOO^-$) NOX-dependent proinflammatory signaling (Oakley et al., 2009). Genetic SOD1 deficiency is not associated with developmental defects or compensation by other SODs (Rahman et al., 2006), but fetal fibroblasts from SOD1-deficient mice show increased susceptibility to ROS-mediated injury induced by the herbicide paraquat (St Clair et al., 1991). Although SOD1 overexpression is protective against allergen-induced changes in airway

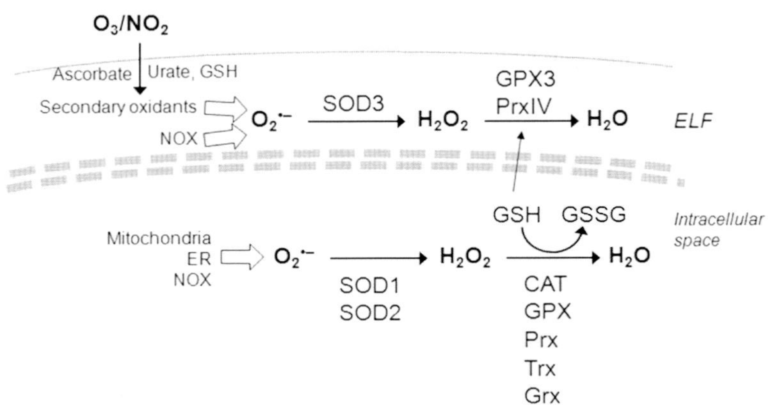

FIGURE 1 Cellular and extracellular antioxidant systems in the lung. Schematic overview of both cellular and extracellular mechanisms involved in ROS breakdown. Nonenzymatic antioxidants such as ascorbate, urate, and GSH are the first reactants with reactive oxidant gases such as ozone (O_3) or nitrogen dioxide (NO_2), which results in the formation of secondary oxidants that are detoxified by other enzymatic antioxidant systems in the lung.

control (Larsen et al., 2000) and hyperoxia in young animals (White et al., 1991), it does not significantly affect susceptibility to lung inflammation or injury by hyperoxia or endotoxemia (Kinnula and Crapo, 2003).

Lung levels of SOD1 are not significantly altered during pulmonary disease (Kinnula and Crapo, 2003), although overall lung tissue SOD activity (which is primarily accounted for by SOD1) is markedly reduced in asthmatic airways (Comhair et al., 2005; Smith et al., 1997) and corresponds with increased airway remodeling and reactivity (Comhair et al., 2005). This inactivation may be related to increased susceptibility to inactivation due to enhanced thiol availability in SOD1 in asthmatic subjects (Ghosh et al., 2013). To date there are no published studies indicating changes in SOD1 in COPD, and no associations have been reported between functional SOD1 variants and COPD (Bentley et al., 2008; Young et al., 2006).

SOD2

The mitochondrial MnSOD isoform SOD2 is a Mn-containing enzyme and exists as a tetramer (86–88 kDa), and its expression in the lung is proportional to the mitochondrial content of lung cells, and is highest in bronchial epithelium, alveolar type II epithelial cells, and alveolar macrophages (Kinnula and Crapo, 2003). While overall SOD2 expression is lower in lung tissue compared to other major organs, its expression is high in lung cell types with high metabolic activity, such as alveolar type II cells and macrophages. Similar to SOD1, lung SOD2 expression does not appear to differ much between neonatal and adult lung, but SOD2 mRNA increases toward adulthood. In contrast to SOD1, SOD2 is readily inducible by oxidative stress, hyperoxia, environmental pollutants such as cigarette smoke or ozone, and by inflammatory cytokines such as TNFα. SOD2 is essential for aerobic life, and mice deficient in SOD2 die within 2–3 weeks after birth from cardiomyopathy, metabolic acidosis, and neurodegeneration (Lebovitz et al., 1996). Heterozygous SOD2 knockout mice (with 50% SOD2 activity) are not sensitized to hyperoxia-induced lung injury (Tsan et al., 1998), but overexpression of SOD2 has been found to increase resistance to hyperoxia, although findings are somewhat inconsistent (Ho et al., 1998; Wispe et al., 1992).

Lung SOD2 levels are increased in response to hyperoxia and inflammatory cytokines, and lung SOD2 expression is elevated in bronchial and alveolar epithelium of smokers with or without COPD (Harju et al., 2004). However, genetic variants of SOD2 have not been associated with COPD (Bentley et al., 2008; Young et al., 2006), although some association may exist between SOD2 haplotypes and smoking-related lung cancer (Gallagher et al., 2009). Oxidative and nitrative modifications have been detected in SOD2 from asthmatic airways, which may contribute to the loss of overall SOD activity in these patients (Comhair et al., 2005).

SOD3

The third isoform, extracellular (EC)-SOD, is a tetrameric glycoprotein (MW 135 kDa) that is 60% homologous to SOD1 and similarly contains one Cu and one Zn atom per subunit that are required for activity (Gao et al., 2008). As a secreted protein, SOD3 is localized primarily within the extracellular matrix and at cell surfaces, through interactions with a heparin/matrix-binding domain within the enzyme. In contrast to SOD1 and SOD2, expression of SOD3 is highly cell- and tissue-specific and highest in lung, heart, blood vessels, placenta, and kidney. Within the lung, SOD3 is primarily expressed in bronchial and alveolar type II epithelial cells, alveolar macrophages, and pulmonary endothelial cells (Folz et al., 1997; Su et al., 1997), and local SOD3 mRNA expression correlates with locally secreted enzyme activity. Lung expression of SOD3 mRNA and protein varies during lung development, although the developmental profile appears to differ between humans and animals. While SOD3 can be detected in the arterial intima, media, and bronchial and alveolar epithelium of neonatal lungs (Kinnula and Crapo, 2003), SOD3 mRNA expression is low in rodent lungs until gestation and rises sharply after birth (Zelko et al., 2002). In rabbit lungs, SOD3 activity increases markedly from preterm to adult, which is largely related to relocalization from intracellular SOD3 to the extracellular space, without significant changes in overall SOD3 protein expression (Zelko et al., 2002). The increased expression or activity of SOD3 post birth is important for adaptation to the high-oxygen environment after birth.

Since SOD3 represents the only $O_2^{\bullet -}$-scavenging enzyme within the extracellular environment, it is likely of major importance in protection against exogenous and environmental stress. Indeed, animal studies using either SOD3 transgenic mice or SOD3-deficient mice have demonstrated a protective role of SOD3 in models of acute lung injury (Bowler et al., 2004), and in models of interstitial lung disease, including bleomycin-induced pulmonary fibrosis (Bowler et al., 2002; Fattman et al., 2003) or asbestos-induced lung inflammation and injury (Fattman et al., 2006; Tan et al., 2004). Common in these models of fibrosis is the marked loss of interstitial SOD3 due to proteolytic loss of its heparin-binding domain and resulting accumulation of SOD3 in the alveolar lining fluid (Fattman et al., 2003; Tan et al., 2004). The protective effects of interstitial SOD3 are mostly related to preventing oxidative fragmentation of collagen, which generates neutrophil chemoattractants, cell surface syndecan, which regulates matrix metalloproteinase function, or hyaluronan (Gao et al., 2008). The importance of SOD3 in human lung diseases is illustrated by reports showing induction of lung SOD3 in, for example, COPD (Folz et al., 1997), and findings that SOD3 haplotypes are associated with development of COPD (Bentley et al., 2008; Young et al., 2006) and with acute lung injury and mortality

(Arcaroli et al., 2009). Recent studies also suggest a relationship between loss of SOD3 and lung cancer (Svensk et al., 2004), which involves its epigenetic silencing and promotes tumor invasiveness by altered ECM homeostasis (Teoh-Fitzgerald et al., 2012).

3.2 H_2O_2-Metabolizing Enzymes

Catalase

One common enzyme involved in H_2O_2 decomposition is catalase, a ubiquitous homotetrameric heme protein (MW 240 kDa) that catalyzes H_2O_2 decomposition as follows:

$$2H_2O_2 \rightarrow 2H_2O + O_2$$

Because the decomposition is second order with respect to H_2O_2 concentration, catalase is most effective at removing high concentrations of H_2O_2 that are produced in specific organelles such as peroxisomes, which therefore contain high levels of catalase (Subramani, 1993), although catalase is also found in the cytoplasm and other cell components. In the lung, catalase is primarily present within alveolar macrophages and type II epithelial cells (Rahman et al., 2006). Catalase expression increases during fetal development and through adulthood (Asikainen et al., 1998), and appears to be inducible by environmental challenges such as hyperoxia or inflammation, although results are inconsistent (Rahman et al., 2006). Studies of asthmatic airways indicate suppressed catalase activity in asthmatics (Comhair and Erzurum, 2010), which may be associated with oxidative inactivation of the enzyme (Ghosh et al., 2006), but the overall importance of catalase in asthma or other pulmonary diseases remains unclear. Studies with acatalasemic mice or catalase inhibitors also revealed anti-inflammatory effects of H_2O_2 on phagocyte signaling and suggest that catalase inhibition may be protective in acute lung injury (Zmijewski et al., 2009).

GSH Peroxidase

The GPX family consists of eight isoforms (GPX1-8) of which four are expressed in the lung. All known GPX enzymes catalyze the reduction of H_2O_2 or other peroxides via initial reaction with a selenocysteine in GPX isoforms 1-4 and GPX6, which is subsequently reduced by cellular glutathione (GSH) according to the overall reaction

$$H_2O_2 + 2GSH \rightarrow GSSG + H_2O$$

Glutathione disulfide (GSSG) is subsequently reduced back to GSH by glutathione reductase (GR) using NADPH as reducing equivalent (Figure 2). GPX isoforms 1-4 is known to be expressed in the lung with variable location.

GPX1 (cellular GPx) is a cytosolic enzyme that is expressed in most cell types and is thought to be responsible for 95% of overall lung tissue GPX activity (Ho et al., 1997). GPX1-deficient mice develop normally but are more susceptible to lung inflammation by influenza infection or cigarette smoke exposure (Duong et al., 2010; Yatmaz et al., 2013). However, GPX1 can also promote proinflammatory cytokine production and neutrophilia in response to endotoxin (Bozinovski et al., 2012), and can contribute to development of ovalbumin-induced allergic asthma, due to GPX1-dependent regulation of T helper cell proliferation and differentiation toward Th2 and Th17 phenotypes (Won et al., 2010).

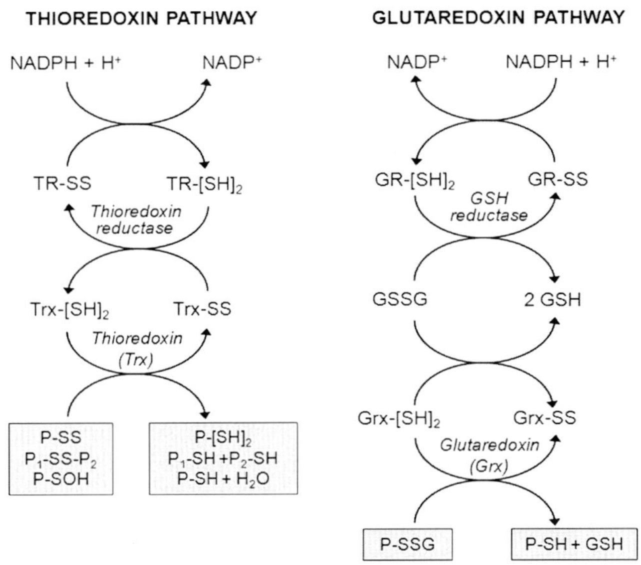

FIGURE 2 Thioredoxin and glutaredoxin systems. Thioredoxins (Trx) and glutaredoxins (Grx) primarily serve to reduce oxidized protein cysteine residues (P-SS-P, P-SOH, P-SSG) to regulate or restore their function, and Trx and Grx are in turn maintained by thioredoxin reductase (TR) and glutathione reductase (GR), using reducing equivalents from NADPH.

GPX2 is also known as gastrointestinal GPx (giGPx) based on its initial discovery in the colon and liver (Chu et al., 1993); it is only marginally expressed in the normal lung, although several studies have indicated marked increases in GPX2 mRNA and protein with the airway epithelium in response to cigarette smoke exposure (Singh et al., 2006) or allergen-induced inflammation (Dittrich et al., 2010), suggesting a protective function in these conditions. Indeed, GPX2 deficiency enhances allergen-induced airway inflammation (Dittrich et al., 2010), which appears to contrast with the contribution of GPX1 to Th2/Th17 development and allergic inflammation (Meyer et al., 2011). The protective effects of GPX2 may be related to its substantial induction during airway inflammation, as opposed to the more constitutive presence of GPX1 and its role in T cell responses or inflammatory signaling (Bozinovski et al., 2012; Won et al., 2010), but may also involve enzymatic effects beyond antioxidant function, such as inhibition of prostaglandin synthesis as an anti-inflammatory mechanism (Meyer et al., 2011).

GPX3, the extracellular isoform (eGPx), was first identified as a distinct enzyme in plasma and is also expressed by bronchial epithelial cells and alveolar macrophages that contribute to the presence of GPX3 as a major extracellular GPX within the airway lumen (Avissar et al., 1996; Comhair and Erzurum, 2005). Airway GPX3 mRNA and extracellular GPX3 protein levels are increased in response to hyperoxia (Kim et al., 1999) and GPX3 is elevated in asthmatic airways, primarily by oxidant-induced activation of AP-1 (Comhair and Erzurum, 2005). However, GPX3 protein levels and activity in lung lining fluids are decreased upon exposure to ozone, although similar exposure to nitrogen dioxide did not affect GPX3 status (Avissar et al., 2000). Collectively, these findings support a major role of GPX3 in extracellular antioxidant defense within the airway lumen.

GPX4 is also known as phospholipid hydroperoxide GPx (PHGPx), is ubiquitously expressed, and exists as cytosolic, mitochondrial, and nuclear isoforms, all encoded by the same gene (Schneider et al., 2009). GPX4 is primarily involved in reduction of phospholipid hydroperoxides in membranes and is primarily involved in regulating oxidative lipid metabolism in the plasma membrane and in mitochondria and nuclear compartments. GPX4 is critical for development and GPX4-deficient embryos die at mid-gestation, and cells of GPX4-deficient mice are increasingly sensitive to oxidative insults (Yant et al., 2003). The specific role of GPX4 in lung biology is, however, not exactly known.

Peroxiredoxins

The Prxs were initially described as thiol-specific antioxidant based on their capacity to protect proteins from oxidative damage in the presence of reducing agents such as DTT, and comprise a family of six mammalian Prx enzymes (MW 21–28 kDa) that act as homodimers and utilize redox-active cysteine residues to reduce H_2O_2 or other peroxides. Prxs are classified as 2-Cys (Prx I-V) or 1-Cys Prx (Prx VI), and 2-Cys Prx reduces H_2O_2 to H_2O by rapid reaction with a peroxidatic Cys on one monomer to form a sulfenic acid (-S-OH), which subsequently reacts with a "resolving" Cys on the other monomer to form a mixed disulfide, which in turn is reduced by thioredoxin (Trx) mechanisms to recycle active Prx (Wood et al., 2003); peroxiredoxins have therefore also been referred to as thioredoxin peroxidases. Mammalian 2-Cys Prx is also susceptible to overoxidation of the peroxidatic Cys to a sulfinic acid (-S-O_2H), which inactivates the enzyme and is thought to regulate H_2O_2-dependent redox signaling (Rhee et al., 2005a, Rhee et al., 2012). This oxidized inactive form of Prx can be regenerated by sulfiredoxins (Rhee et al., 2012). Prxs can undergo structural alterations between dimeric and decameric complexes, depending on their oxidation state or other posttranslational modifications, which further regulates function. Prx VI is a 1-Cys peroxiredoxin and its oxidation by H_2O_2 is reversed by GSH or by other cellular reductants and involves heterodimerization with pi-glutathione S-transferase (Fisher, 2011). The kinetics of Prx reaction with H_2O_2 is extremely fast due to the low pK_a of the peroxidatic Cys (Winterbourn and Hampton, 2008) and indicates that Prx is important in cellular metabolism of H_2O_2. The relative importance of Prx enzymes in H_2O_2 or peroxide metabolism compared to GPX or catalase depends not only on subcellular localization but also on additional regulation of their enzymatic activities by, for example, phosphorylation (Rhee et al., 2005b), further illustrating the multiple levels of regulation of H_2O_2-dependent redox events in cell signaling pathways, not only by tight control of cellular H_2O_2 production by, for example, NOX enzymes, but also by localized regulation of antioxidant enzyme function. Moreover, rather than simply detoxifying H_2O_2, Prxs may also act as proximal sensors of localized H_2O_2 production and transmit redox signals to other target proteins by thiol–disulfide exchange (Fomenko et al., 2011; Gutscher et al., 2009; Winterbourn and Hampton, 2008).

All six mammalian Prxs are expressed in the lung, with moderate to high expression of Prx I, III, V, and VI in the bronchial epithelium, and primarily Prx V and VI in the alveolar epithelium and Prx I and III in alveolar macrophages (Kinnula et al., 2002; Park et al., 2006). Several lines of evidence indicate the importance of Prx I, a major cytosolic isoform, in environmentally induced lung inflammation or injury, although findings seem paradoxical. For example, mice deficient in Prx I show increased sensitivity toward bleomycin-induced lung injury and fibrosis (Kikuchi et al., 2011), and appear to be more susceptible to epithelial-to-mesenchymal transition as a putative feature of fibrosis (Ha et al., 2012). Prx I is also important in suppressing Th2 immune responses and allergic inflammation (Inoue et al., 2009). However, inflammatory cytokine production and neutrophil influx in response to ozone exposure were found to be suppressed in Prx I null mice (Yanagisawa et al., 2012), suggesting that Prx I does not simply act as an antioxidant but rather participates in regulating inflammatory signaling. Much less is known about the specific importance of Prx II, III and IV in lung biology, although Prx III (a mitochondrial enzyme) was found to be important in preventing oxidative stress induced by bacterial endotoxin (Li et al., 2007). Prx V is important in protecting alveolar epithelial cells from oxidative stress (Avila et al., 2008), and a genetic Prx V variant has been associated with reduced lung function in some smoking populations (Bentley et al., 2012). Prx VI is the highest expressed isoform in the lung and is more abundant in the lung compared to other organs. It is primary localized in the cytosol and in acidic organelles of Clara cells of the conducting airways, as well as in type II alveolar epithelial cells and alveolar macrophages (Fisher, 2011). Prx VI has peroxidase activity against phospholipid hydroperoxides and other organic hydroperoxides as well as H_2O_2 and also expresses Ca^{2+}-independent phospholipase A_2 activity, which may be important in lung surfactant homeostasis. Models of altered Prx VI expression at the cellular, organ, and whole animal levels have demonstrated that Prx VI functions as an important antioxidant enzyme (Fisher, 2011; Liu et al., 2010) and also protects against lung inflammation by cigarette smoke or LPS (Sundar et al., 2010; Yang et al., 2011). Most Prx enzymes, especially Prx I, V, and VI, are inducible by oxidative stress or during inflammation, largely mediated by the redox-sensitive transcription factor Nrf2 (see below), and Prx I and VI also show developmental induction of transcription at birth. Expression of Prxs is also elevated in several lung diseases including lung cancer, mesothelioma, and sarcoidosis (Aracena-Parks et al., 2006; Kinnula et al., 2002; Rostila et al., 2012), most likely related to increased constitutive activation of Nrf2 in many cancers with poor prognosis.

3.3 Other Redox Regulating Enzymes

Many of the antioxidant systems described above depend on the presence of GSH and are maintained by additional redox enzymes such as glutathione reductase, which keeps GSH in its primarily reduced state to support the actions of GPX and some Prx enzymes; they also depend on thioredoxins, which support the actions of Prxs and also directly regulate thiol status in other target proteins, and are maintained by thioredoxin reductases (Figure 2).

Thioredoxins are small (MW 10−12 kDa) dithiol-disulfide oxidoreductases that largely act by reducing protein disulfides to the corresponding sulfhydryls. Trx may contribute to direct ROS detoxification, but this is likely of minor importance because of its relatively low abundance compared to, for example, GSH, and the main antioxidant role of Trx is related to its ability to regenerate oxidatized forms of Prx. Mammalian thioredoxin exists as three isoforms: a cytosolic Trx-1, a mitochondrial Trx-2, and a third SpTrx isoform that is highly expressed in spermatozoa (Xu et al., 2012). The reduced status of Trx is maintained by thioredoxin reductase-1 or -2, utilizing reducing equivalents from NADPH. Trx can also be released from cells, particularly during oxidative stress, and has extracellular chemokine/cytokine properties, and may also exert protective actions by cell reentry (Nakamura et al., 2005). Trx expression is upregulated in various pulmonary diseases, primarily in lung cancer (Fernandes et al., 2009; Park et al., 2006; Soini et al., 2001), pulmonary fibrosis and interstitial lung disease (Iwata et al., 2010; Tiitto et al., 2003), and COPD (Lehtonen et al., 2008), and increased extracellular levels of Trx have been reported in subjects with acute lung injury and asthma exacerbations (Callister et al., 2006; Yamada et al., 2003), indicating increased oxidative stress in these cases (Xu et al., 2012). Transgenic overexpression of Trx protects alveolar epithelial cells from hyperoxia-induced injury (Chen et al., 2010), is protective against bleomycin-induced injury (Hoshino et al., 2003), and suppresses allergic inflammation by a mechanism independent of Th1/Th2 immune modulation (Torii et al., 2010), and administration of recombinant Trx proteins has shown protective effects against various lung pathologies associated with oxidative stress (Nakamura et al., 2005).

The *glutaredoxins* (Grxs) are members of the Trx superfamily that also include protein disulfide isomerases and GPx enzymes, and are generally involved in the reduction of oxidative modifications involving GSH (Lillig et al., 2008; Stroher and Millar, 2012). As such Grxs do not act as direct antioxidant enzymes but as thioltransferases to support overall antioxidant systems by reversing protein *S*-glutathionylation or other disulfides, using GSH as a cofactor. In this manner, Grx may also be involved in recycling Prx VI (Fisher, 2011), thereby more directly regulating antioxidant function. Mammalian Grxs largely comprise cytosolic Grx-1 and mitochondrial Grx-2, and lung Grx-1 is mostly expressed in airway epithelial cells and alveolar macrophages (Fernandes et al., 2009; Peltoniemi et al., 2006). Lung tissue expression of Grx-1 is increased in models of acute lung injury and allergic asthma (Aesif et al., 2011; Kuipers et al., 2013), although it appears to be decreased in patients with COPD (Peltoniemi et al., 2006). Increased levels of extracellular Grx-1 have also been detected in sputum samples from subjects with asthma and COPD (Kuipers et al., 2013; Peltoniemi et al., 2006). The functional importance of Grx-1 in preventing oxidative lung injury is not clear (Ho et al., 1997), and several studies indicate that Grx-1 mediates and prolongs proinflammatory signaling in response to LPS or during allergic inflammation, potentially by regulating *S*-glutathionylation of NF-κB or other proinflammatory signaling pathways (Aesif et al., 2011; Hoffman et al., 2012). The importance of Grx-1 in regulating smoking-induced lung inflammation and injury is less clear (Chung et al., 2010; Kuipers et al., 2012).

The large diversity in these various antioxidant systems suggests that these enzymes do not merely act as general antioxidants against oxidative stress but play more selective roles in regulating localized actions of biologically generated ROS in diverse cell compartments, and are thereby critically important in controlling localized redox signaling pathways induced by, for example, activation of NOX enzymes or by other biological sources of ROS. In addition to the enzyme systems listed above, other putative antioxidant systems exist such as heme oxygenase and proteins involved in transport and regulation of redox-active transition metals to minimize unwanted reactions with $O_2^{\cdot-}$ or H_2O_2 to generate damaging hydroxyl radicals (OH•). Furthermore, the various antioxidant enzyme systems rely on the presence of sufficient reducing equivalents from NADPH, which is supported by additional enzymatic pathways that sustain overall cellular antioxidant capacity.

3.4 Adaptive Responses to Oxidative Stress: Nrf2

The sections above indicate that a number of antioxidant systems are highly inducible under conditions of oxidative stress or by inflammatory cytokines or other environmental stimuli. Likewise, the biosynthesis of GSH is subject to induction by oxidative stress, as a critical adaptive mechanism to cope with continued oxidative stress. A key mediator of these adaptive responses is the transcription factor NF-E2-related factor 2 (Nrf2), which regulates the transcription of many of these antioxidant genes as well as genes involved in phase II detoxification pathways (Muller and Hengstermann, 2012; Osburn and Kensler, 2008). Activation of Nrf2-mediated transcription relies on initial oxidation or modification of its cytosolic regulator, Kelch-like ECH-associated protein 1 (Keap1), which sequesters Nrf2 within the cytoplasm and targets its rapid proteolytic degradation. The Keap1 protein contains a number of conserved reactive cysteines that function as a sensor for various oxidants and electrophilic compounds (Holland and Fishbein, 2010; McMahon et al., 2010), and its oxidation/modification disrupts its interaction with Nrf2, allowing for its accumulation and activation as a transcription factor. Indeed, genetic models of Keap1 or Nrf2 deficiency have indicated the overall importance of the Nrf2-mediated adaptive response against lung injury induced by environmental oxidants and in chronic inflammatory lung diseases (Cho and Kleeberger, 2010; Muller and Hengstermann, 2012). During chronic pulmonary diseases such as COPD or asthma, activation of Nrf2-mediated antioxidant gene expression appears to be defective, which is related to alterations in redox status or expression of regulatory molecules (Fitzpatrick et al., 2011; Malhotra et al., 2008). Based on these considerations, synthetic or food-derived electrophiles are currently being promoted as activators of Nrf2 to minimize environmental lung pathologies and correct these defects in chronic disease (Dinkova-Kostova and Kostov, 2012; Liby and Sporn, 2012; Sussan et al., 2009).

Contrasting the protective properties of Nrf2 against inflammatory and oxidative stress are observations in lung cancer tissues, which frequently harbor mutations within the *Keap1* or *Nrf2* genes, which lead to constitutive Nrf2 activation to offer cancer cells enhanced resistance against oxidative stress and ability to maintain the highly metabolic demands of tumor growth (Kensler and Wakabayashi, 2010; Muller and Hengstermann, 2012). This constitutive Nrf2 activation is largely responsible for the common overexpression of Trx, Prx, and other redox systems in lung cancers and may minimize the efficacy of exogenous Nrf2 activation in treatment of lung pathologies.

3.5 Antioxidant Systems at the Respiratory Surface

Nonenzymatic Antioxidants in Respiratory Tract Lining Fluids

The major branching of the conducting airways results in large expansion of the surface area from the proximal airways to the distal airways and poses interesting challenges for innate airway defenses against inhaled pathogens, particulates, or other pollutants. Instrumental in these defenses is the presence of a thin fluid layer, commonly termed airway surface liquid (ASL) or epithelial lining fluid (ELF), that covers the epithelial surface throughout the respiratory tract and is critically important for normal function. In addition to providing appropriate surface hydration to allow for efficient mucociliary clearance of inhaled pathogens or particles (Knowles and Boucher, 2002), ASL also contains a range of secreted antibacterial factors, such as lactoferrin, lysozyme, defensins, and LPO that are critical in innate immunity (Gattas et al., 2009), as well as a range of enzymatic and nonenzymatic antioxidant systems that are critical as a first-line defense against environmental oxidant challenges and regulation of local oxidative events at the respiratory surface. The volume and thickness of ELF varies widely throughout the respiratory tract, ranging from approximately 10 μm in the proximal airways to less than 1 μm in the distal airways and 0.2 μm in alveoli (Cross et al., 1994). Likewise, ELF composition also varies between different lung regions, although such information is still very incomplete due to technical limitations with collecting and analyzing ELF from different areas of the lung (Cross et al., 1994; van der Vliet et al., 1999b).

As discussed above, the ELF contains a number of extracellular antioxidant enzymes that are secreted from epithelial cells or other lung cell types, such as SOD3 (EC-SOD), eGpx (GPX3), as well as secreted forms of Trx and Prx, but the main proximal antioxidant defenses against environmental oxidants such as O_3 or NO_2 and other common pollutants are small nonenzymatic low-molecular-weight antioxidant molecules that act as first-line sacrificial antioxidant defenses against these toxicants. The major nonenzymatic antioxidants in ELF are ascorbic acid (vitamin C), uric acid, and GSH, whereas alveolar ELF also contain significant amounts of the lipid-soluble antioxidant α-tocopherol (vitamin E), which is actively cosecreted from alveolar type II cells with surfactant lipids (Kolleck et al., 2002). Levels of ASL ascorbate and urate are relatively uniform throughout the respiratory tract and are comparable to concentrations present in other extracellular compartments such as blood plasma (van der Vliet et al., 1999b), suggesting that these antioxidants may be largely maintained by passive diffusion from interstitial fluids or from the circulation. Indeed, supplementation experiments with ascorbate appear to support this notion, although its local concentrations in saliva or in upper airway secretions may also be maintained by local secretion mechanisms and/or reducing mechanisms that maintain ascorbate in its reduced state within ASL (Schock et al., 2004). Similarly, ASL uric acid is also largely maintained by active secretion from submucosal nasal glands, which take up extracellular urate to concentrate and secrete it in response to, for example, cholinergic stimulation (Peden et al., 1993). ELF concentrations of GSH vary widely between different lung compartments and are relatively high within alveoli but much lower in upper airway or nasal secretions (van der Vliet et al., 1999b). Indeed, bronchoalveolar ELF contains GSH at concentrations ranging from 100 to 500 μM, about 100-fold higher than in circulating plasma (Cantin et al., 1987; van der Vliet et al., 1999b), which is sustained primarily by local secretion from alveolar epithelial cells. Moreover, ELF GSH reaches even higher concentrations under certain acute stress conditions (Fitzpatrick et al., 2012; Gould and Day, 2011).

Role of Antioxidants in the Response to Environmental Pollutants

Studies with surrogate extracellular fluids indicate that ascorbate and urate are the most prominent antioxidants that react with reactive oxidant gases such as O_3 and NO_2 (Cross et al., 1994). Accordingly, in vivo O_3 exposure of human subjects results in transient decreases in extracellular levels of ascorbate and urate at the airway surface, which are rapidly restored over time (Behndig et al., 2009; Mudway et al., 1999). Exposure to O_3 or to cigarette smoke also appears to affect factors that regulate lung α-tocopherol homeostasis (Valacchi et al., 2007). The prominent role of ELF ascorbate and urate as proximal reactants for inhaled O_3 or NO_2 is related to their relative low aqueous solubility and the fact that they are largely taken up by "reactive absorption." These initial reactions result in the formation of secondary oxidants and can actually augment oxidative injury to other ELF constituents or underlying epithelium (Ballinger et al., 2005). Therefore, effective defense against inhaled O_3 and other reactive gases relies on the combined presence of sacrificial antioxidants at the airway surface (ascorbate, urate) and other (enzymatic) antioxidant systems within the ELF or within targeted lung cells. Consistent with these findings, some studies with antioxidant supplementation (either alone or in combination) have indicated some benefit in suppressing O_3-induced inflammation or lung function changes (Samet et al., 2001; Sienra-Monge et al., 2004) and support the overall benefit of antioxidant supplementation against lung function changes associated with air pollution (Tashakkor et al., 2011), although available data are still inconsistent. Similarly, observational studies have suggested beneficial associations between dietary antioxidants (such as vitamins C and E) and chronic lung diseases (such as asthma and COPD), but supplementation strategies have shown minimal overall benefit (Allan et al., 2010; Tsiligianni and van der Molen, 2010).

Extracellular GSH at the Respiratory Surface

Bronchoalveolar ELF maintains extracellular GSH at high μM levels, largely as a direct antioxidant defense component against inhaled environmental pollutants and also to support the function of extracellular GPX3 or Prx. Extracellular GSH in the airway exists largely (<95%) in the reduced form, maintaining a redox potential of −200 mV (Fitzpatrick et al., 2012). ELF GSH levels are maintained largely by secretion from alveolar type II cells, which synthesize GSH from glutamate, glycine, and cysteine in two enzymatic steps, by γ-glutamylcysteine ligase (GCL) and glutathione synthetase (GS), and may also take up circulating GSH that originates from the liver (Gould and Day, 2011). Transport of cellular GSH into the apical lumen occurs in part via the cystic fibrosis transmembrane conductance regulator, a member of the multidrug resistance family of transporters, although the precise mechanisms involved in such GSH transport are not fully understood and other transport systems may contribute as well (Gould and Day, 2011). ELF GSH functions as a direct local antioxidant but also serves to regulate local extracellular redox status in the lung, which is critical in proper maintenance of epithelial function (Fitzpatrick et al., 2012), and can regulate inflammatory signaling by alveolar macrophages (Gould et al., 2011). Lung ELF GSH concentrations decrease acutely after oxidant or cigarette smoke exposure but then rapidly increase due to increased local GSH synthesis by upregulation of GCL and GS due to Nrf2 activation. In chronic pulmonary diseases, such as COPD, asthma, or pulmonary fibrosis, ELF GSH levels are often suppressed and exist in a more oxidized state, possibly related to defects in GSH synthesis and Nrf2 activation and the presence of a more oxidizing environment. This has encouraged a number of strategies to correct local GSH status using Nrf2 activation, GSH prodrugs, or other thiol-based antioxidants, with some success (Fitzpatrick et al., 2012; Gould and Day, 2011).

4. SUMMARY AND FINAL CONCLUSIONS

Like other aerobic organs, the lung requires O_2 for the generation of ATP and for much of its intermediary metabolism. Since the lung is exposed to high O_2 tensions compared to most other organs, which increases the potential of local ROS production during normal respiration, and is also continuously challenged by inhaled environmental oxidant pollutants, it is well endowed with an intricate network of antioxidant systems to effectively cope with these various oxidant challenges and to simultaneously allow for appropriate pathophysiological functioning of endogenous ROS generation from, for example, NOX enzymes in antimicrobial host defense and immune regulation. Because of the variable roles of endogenous or environmental ROS in both lung physiology as well as pathology, global antioxidant-based strategies have had limited success in treatment of lung disease, and more effective antioxidant-based therapeutic strategies should consider more specifically targeted approaches to affect localized actions of ROS or to selectively augment antioxidant systems that regulate specific functional outcomes in the lung. Recent efforts to develop immunotargeting approaches to deliver antioxidant enzymes (Han et al., 2012) or synthetic mimetics of specific antioxidant enzymes (Day, 2008) are likely to be more advantageous in using antioxidant-based approaches in treatment of pulmonary disease.

ACKNOWLEDGMENT

The author wishes to thank NIH (grants HL085646 and ES021476) and the Flight Attendant Medical Research Institute (FAMRI) for research support.

REFERENCES

Aesif, S.W., Anathy, V., Kuipers, I., Guala, A.S., Reiss, J.N., Ho, Y.S., Janssen-Heininger, Y.M., 2011. Ablation of glutaredoxin-1 attenuates lipopolysaccharide-induced lung inflammation and alveolar macrophage activation. Am. J. Respir. Cell. Mol. Biol. 44 (4), 491–499.

Al-Mehdi, A.B., Zhao, G., Dodia, C., Tozawa, K., Costa, K., Muzykantov, V., Ross, C., Blecha, F., Dinauer, M., Fisher, A.B., 1998. Endothelial NADPH oxidase as the source of oxidants in lungs exposed to ischemia or high K^+. Circ. Res. 83 (7), 730–737.

Allan, K., Kelly, F.J., Devereux, G., 2010. Antioxidants and allergic disease: a case of too little or too much? Clin. Exp. Allergy 40 (3), 370–380.

Allaoui, A., Botteaux, A., Dumont, J.E., Hoste, C., De Deken, X., 2009. Dual oxidases and hydrogen peroxide in a complex dialogue between host mucosae and bacteria. Trends Mol. Med. 15 (12), 571–579.

Amara, N., Bachoual, R., Desmard, M., Golda, S., Guichard, C., Lanone, S., Aubier, M., Ogier-Denis, E., Boczkowski, J., 2007. Diesel exhaust particles induce matrix metalloprotease-1 in human lung epithelial cells via a NADP(H) oxidase/NOX4 redox-dependent mechanism. Am. J. Physiol. Lung Cell. Mol. Physiol. 293 (1), L170–L181.

Amara, N., Goven, D., Prost, F., Muloway, R., Crestani, B., Boczkowski, J., 2010. NOX4/NADPH oxidase expression is increased in pulmonary fibroblasts from patients with idiopathic pulmonary fibrosis and mediates TGFbeta1-induced fibroblast differentiation into myofibroblasts. Thorax 65 (8), 733–738.

Ameziane-El-Hassani, R., Morand, S., Boucher, J.L., Frapart, Y.M., Apostolou, D., Agnandji, D., Gnidehou, S., Ohayon, R., Noel-Hudson, M.S., Francon, J., et al., 2005. Dual Oxidase-2 has an intrinsic Ca^{2+}-dependent H_2O_2-generating activity. J. Biol. Chem. 280 (34), 30046–30054.

Andreadis, A.A., Hazen, S.L., Comhair, S.A., Erzurum, S.C., 2003. Oxidative and nitrosative events in asthma. Free Radic. Biol. Med. 35 (3), 213–225.

Antonenkov, V.D., Grunau, S., Ohlmeier, S., Hiltunen, J.K., 2010. Peroxisomes are oxidative organelles. Antioxid. Redox Signal. 13 (4), 525–537.

Aracena-Parks, P., Goonasekera, S.A., Gilman, C.P., Dirksen, R.T., Hidalgo, C., Hamilton, S.L., 2006. Identification of cysteines involved in *S*-nitrosylation, *S*-glutathionylation, and oxidation to disulfides in ryanodine receptor type 1. J. Biol. Chem. 281 (52), 40354–40368.

Arcaroli, J.J., Hokanson, J.E., Abraham, E., Geraci, M., Murphy, J.R., Bowler, R.P., Dinarello, C.A., Silveira, L., Sankoff, J., Heyland, D., et al., 2009. Extracellular superoxide dismutase haplotypes are associated with acute lung injury and mortality. Am. J. Respir. Crit. Care Med. 179 (2), 105–112.

Asikainen, T.M., Raivio, K.O., Saksela, M., Kinnula, V.L., 1998. Expression and developmental profile of antioxidant enzymes in human lung and liver. Am. J. Respir. Cell. Mol. Biol. 19 (6), 942–949.

Avila, P.C., Kropotov, A.V., Krutilina, R., Krasnodembskay, A., Tomilin, N.V., Serikov, V.B., 2008. Peroxiredoxin V contributes to antioxidant defense of lung epithelial cells. Lung 186 (2), 103–114.

Avissar, N., Finkelstein, J.N., Horowitz, S., Willey, J.C., Coy, E., Frampton, M.W., Watkins, R.H., Khullar, P., Xu, Y.L., Cohen, H.J., 1996. Extracellular glutathione peroxidase in human lung epithelial lining fluid and in lung cells. Am. J. Physiol. 270 (2 Pt 1), L173–L182.

Avissar, N.E., Reed, C.K., Cox, C., Frampton, M.W., Finkelstein, J.N., 2000. Ozone, but not nitrogen dioxide, exposure decreases glutathione peroxidases in epithelial lining fluid of human lung. Am. J. Respir. Crit. Care Med. 162 (4 Pt 1), 1342–1347.

Babior, B.M., Lambeth, J.D., Nauseef, W., 2002. The neutrophil NADPH oxidase. Arch. Biochem. Biophys. 397 (2), 342–344.

Bae, Y.S., Oh, H., Rhee, S.G., Yoo, Y.D., 2011. Regulation of reactive oxygen species generation in cell signaling. Mol. Cells 32 (6), 491–509.

Ballinger, C.A., Cueto, R., Squadrito, G., Coffin, J.F., Velsor, L.W., Pryor, W.A., Postlethwait, E.M., 2005. Antioxidant-mediated augmentation of ozone-induced membrane oxidation. Free Radic. Biol. Med. 38 (4), 515–526.

Banfi, B., Tirone, F., Durussel, I., Knisz, J., Moskwa, P., Molnar, G.Z., Krause, K.H., Cox, J.A., 2004. Mechanism of Ca2+ activation of the NADPH oxidase 5 (NOX5). J. Biol. Chem. 279 (18), 18583–18591.

Bedard, K., Krause, K.H., 2007. The NOX family of ROS-generating NADPH oxidases: physiology and pathophysiology. Physiol. Rev. 87 (1), 245–313.

Behndig, A.F., Blomberg, A., Helleday, R., Duggan, S.T., Kelly, F.J., Mudway, I.S., 2009. Antioxidant responses to acute ozone challenge in the healthy human airway. Inhal. Toxicol. 21 (11), 933–942.

Bentley, A.R., Emrani, P., Cassano, P.A., 2008. Genetic variation and gene expression in antioxidant related enzymes and risk of COPD: a systematic review. Thorax 63 (11), 956–961.

Bentley, A.R., Kritchevsky, S.B., Harris, T.B., Newman, A.B., Bauer, D.C., Meibohm, B., Clark, A.G., Cassano, P.A., 2012. Genetic variation in antioxidant enzymes and lung function. Free Radic. Biol. Med. 52 (9), 1577–1583.

Boots, A.W., Hristova, M., Kasahara, D.I., Haenen, G.R., Bast, A., van der Vliet, A., 2009. ATP-mediated activation of the NADPH oxidase DUOX1 mediates airway epithelial responses to bacterial stimuli. J. Biol. Chem. 284 (26), 17858–17867.

Bowler, R.P., Nicks, M., Tran, K., Tanner, G., Chang, L.Y., Young, S.K., Worthen, G.S., 2004. Extracellular superoxide dismutase attenuates lipopolysaccharide-induced neutrophilic inflammation. Am. J. Respir. Cell. Mol. Biol. 31 (4), 432–439.

Bowler, R.P., Nicks, M., Warnick, K., Crapo, J.D., 2002. Role of extracellular superoxide dismutase in bleomycin-induced pulmonary fibrosis. Am. J. Physiol. Lung Cell. Mol. Physiol. 282 (4), L719–L726.

Bozinovski, S., Seow, H.J., Crack, P.J., Anderson, G.P., Vlahos, R., 2012. Glutathione peroxidase-1 primes pro-inflammatory cytokine production after LPS challenge in vivo. PloS one 7 (3), e33172.

Brar, S.S., Kennedy, T.P., Quinn, M., Hoidal, J.R., 2003. Redox signaling of NF-kappaB by membrane NAD(P)H oxidases in normal and malignant cells. Protoplasma 221 (1–2), 117–127.

Budinger, G.R., Mutlu, G.M., 2011. Update in environmental and occupational medicine 2010. Am. J. Respir. Crit. Care Med. 183 (12), 1614–1619.

Callister, M.E., Burke-Gaffney, A., Quinlan, G.J., Nicholson, A.G., Florio, R., Nakamura, H., Yodoi, J., Evans, T.W., 2006. Extracellular thioredoxin levels are increased in patients with acute lung injury. Thorax 61 (6), 521–527.

Cantin, A.M., North, S.L., Hubbard, R.C., Crystal, R.G., 1987. Normal alveolar epithelial lining fluid contains high levels of glutathione. J. Appl. Physiol. 63 (1), 152–157.

Carnesecchi, S., Deffert, C., Donati, Y., Basset, O., Hinz, B., Preynat-Seauve, O., Guichard, C., Arbiser, J.L., Banfi, B., Pache, J.C., et al., 2011. A key role for NOX4 in epithelial cell death during development of lung fibrosis. Antioxid. Redox Signal. 15 (3), 607–619.

Carnesecchi, S., Deffert, C., Pagano, A., Garrido-Urbani, S., Metrailler-Ruchonnet, I., Schappi, M., Donati, Y., Matthay, M.A., Krause, K.H., Barazzone Argiroffo, C., 2009. NADPH oxidase-1 plays a crucial role in hyperoxia-induced acute lung injury in mice. Am. J. Respir. Crit. Care Med. 180 (10), 972–981.

Chen, Y., Chang, L., Li, W., Rong, Z., Liu, W., Shan, R., Pan, R., 2010. Thioredoxin protects fetal type II epithelial cells from hyperoxia-induced injury. Pediatr. Pulmonol. 45 (12), 1192–1200.

Cho, H.Y., Kleeberger, S.R., 2010. Nrf2 protects against airway disorders. Toxicol. Appl. Pharmacol. 244 (1), 43–56.

Chowdhury, A.K., Watkins, T., Parinandi, N.L., Saatian, B., Kleinberg, M.E., Usatyuk, P.V., Natarajan, V., 2005. Src-mediated tyrosine phosphorylation of p47phox in hyperoxia-induced activation of NADPH oxidase and generation of reactive oxygen species in lung endothelial cells. J. Biol. Chem. 280 (21), 20700–20711.

Chu, F.F., Doroshow, J.H., Esworthy, R.S., 1993. Expression, characterization, and tissue distribution of a new cellular selenium-dependent glutathione peroxidase, GSHPx-GI. J. Biol. Chem. 268 (4), 2571–2576.

Chung, S., Sundar, I.K., Yao, H., Ho, Y.S., Rahman, I., 2010. Glutaredoxin 1 regulates cigarette smoke-mediated lung inflammation through differential modulation of I{kappa}B kinases in mice: impact on histone acetylation. Am. J. Physiol. Lung Cell. Mol. Physiol. 299 (2), L192–L203.

Ciencewicki, J., Trivedi, S., Kleeberger, S.R., 2008. Oxidants and the pathogenesis of lung diseases. J. Allergy Clin. Immunol. 122 (3), 456–468 quiz 469–470.

Clempus, R.E., Sorescu, D., Dikalova, A.E., Pounkova, L., Jo, P., Sorescu, G.P., Schmidt, H.H., Lassegue, B., Griendling, K.K., 2007. Nox4 is required for maintenance of the differentiated vascular smooth muscle cell phenotype. Arterioscler. Thromb. Vasc. Biol. 27 (1), 42–48.

Comhair, S.A., Erzurum, S.C., 2005. The regulation and role of extracellular glutathione peroxidase. Antioxid. Redox Signal. 7 (1–2), 72–79.

Comhair, S.A., Erzurum, S.C., 2010. Redox control of asthma: molecular mechanisms and therapeutic opportunities. Antioxid. Redox Signal. 12 (1), 93–124.

Comhair, S.A., Xu, W., Ghosh, S., Thunnissen, F.B., Almasan, A., Calhoun, W.J., Janocha, A.J., Zheng, L., Hazen, S.L., Erzurum, S.C., 2005. Superoxide dismutase inactivation in pathophysiology of asthmatic airway remodeling and reactivity. Am. J. Pathol. 166 (3), 663–674.

Comstock, A.T., Ganesan, S., Chattoraj, A., Faris, A.N., Margolis, B.L., Hershenson, M.B., Sajjan, U.S., 2011. Rhinovirus-induced barrier dysfunction in polarized airway epithelial cells is mediated by NADPH oxidase 1. J. Virol. 85 (13), 6795–6808.

Conner, G.E., Wijkstrom-Frei, C., Randell, S.H., Fernandez, V.E., Salathe, M., 2007. The lactoperoxidase system links anion transport to host defense in cystic fibrosis. FEBS Lett. 581 (2), 271–278.

Cross, C.E., van der Vliet, A., O'Neill, C.A., Louie, S., Halliwell, B., 1994. Oxidants, antioxidants, and respiratory tract lining fluids. Environ. Health Perspect. 102 (Suppl 10), 185–191.

Cruz, C.M., Rinna, A., Forman, H.J., Ventura, A.L., Persechini, P.M., Ojcius, D.M., 2007. ATP activates a reactive oxygen species-dependent oxidative stress response and secretion of proinflammatory cytokines in macrophages. J. Biol. Chem. 282 (5), 2871–2879.

Day, B.J., 2008. Antioxidants as potential therapeutics for lung fibrosis. Antioxid. Redox Signal. 10 (2), 355–370.

Dhaunsi, G.S., Paintlia, M.K., Kaur, J., Turner, R.B., 2004. NADPH oxidase in human lung fibroblasts. J. Biomed. Sci. 11 (5), 617–622.

Dinkova-Kostova, A.T., Kostov, R.V., 2012. Glucosinolates and isothiocyanates in health and disease. Trends Mol. Med. 18 (6), 337–347.

Dittrich, A.M., Meyer, H.A., Krokowski, M., Quarcoo, D., Ahrens, B., Kube, S.M., Witzenrath, M., Esworthy, R.S., Chu, F.F., Hamelmann, E., 2010. Glutathione peroxidase-2 protects from allergen-induced airway inflammation in mice. Eur. Respir. J. 35 (5), 1148–1154.

Dostert, C., Petrilli, V., Van Bruggen, R., Steele, C., Mossman, B.T., Tschopp, J., 2008. Innate immune activation through Nalp3 inflammasome sensing of asbestos and silica. Science 320 (5876), 674–677.

Drose, S., Brandt, U., 2012. Molecular mechanisms of superoxide production by the mitochondrial respiratory chain. Adv. Exp. Med. Biol. 748, 145–169.

Duong, C., Seow, H.J., Bozinovski, S., Crack, P.J., Anderson, G.P., Vlahos, R., 2010. Glutathione peroxidase-1 protects against cigarette smoke-induced lung inflammation in mice. Am. J. Physiol. Lung Cell. Mol. Physiol. 299 (3), L425–L433.

Dupuy, C., Ohayon, R., Valent, A., Noel-Hudson, M.S., Deme, D., Virion, A., 1999. Purification of a novel flavoprotein involved in the thyroid NADPH oxidase. Cloning of the porcine and human cdnas. J. Biol. Chem. 274 (52), 37265–37269.

El Hassani, R.A., Benfares, N., Caillou, B., Talbot, M., Sabourin, J.C., Belotte, V., Morand, S., Gnidehou, S., Agnandji, D., Ohayon, R., et al., 2005. Dual oxidase2 is expressed all along the digestive tract. Am. J. Physiol. Gastrointest. Liver Physiol. 288 (5), G933–G942.

Fadeel, B., Ahlin, A., Henter, J.I., Orrenius, S., Hampton, M.B., 1998. Involvement of caspases in neutrophil apoptosis: regulation by reactive oxygen species. Blood 92 (12), 4808–4818.

Fadeel, B., Xue, D., Kagan, V., 2010. Programmed cell clearance: molecular regulation of the elimination of apoptotic cell corpses and its role in the resolution of inflammation. Biochem. Biophys. Res. Commun. 396 (1), 7–10.

Fahy, J.V., 2009. Eosinophilic and neutrophilic inflammation in asthma: insights from clinical studies. Proc. Am. Thorac. Soc. 6 (3), 256–259.

Fattman, C.L., Chang, L.Y., Termin, T.A., Petersen, L., Enghild, J.J., Oury, T.D., 2003. Enhanced bleomycin-induced pulmonary damage in mice lacking extracellular superoxide dismutase. Free Radic. Biol. Med. 35 (7), 763–771.

Fattman, C.L., Tan, R.J., Tobolewski, J.M., Oury, T.D., 2006. Increased sensitivity to asbestos-induced lung injury in mice lacking extracellular superoxide dismutase. Free Radic. Biol. Med. 40 (4), 601–607.

Fernandes, A.P., Capitanio, A., Selenius, M., Brodin, O., Rundlof, A.K., Bjornstedt, M., 2009. Expression profiles of thioredoxin family proteins in human lung cancer tissue: correlation with proliferation and differentiation. Histopathology 55 (3), 313–320.

Fialkow, L., Wang, Y., Downey, G.P., 2007. Reactive oxygen and nitrogen species as signaling molecules regulating neutrophil function. Free Radic. Biol. Med. 42 (2), 153–164.

Finkel, T., 2003. Oxidant signals and oxidative stress. Curr. Opin. Cell. Biol. 15 (2), 247–254.

Fischer, H., Gonzales, L.K., Kolla, V., Schwarzer, C., Miot, F., Illek, B., Ballard, P.L., 2007. Developmental regulation of DUOX1 expression and function in human fetal lung epithelial cells. Am. J. Physiol. Lung Cell. Mol. Physiol. 292 (6), L1506–L1514.

Fisher, A.B., 2011. Peroxiredoxin 6: a bifunctional enzyme with glutathione peroxidase and phospholipase A(2) activities. Antioxid. Redox Signal. 15 (3), 831–844.

Fisher, A.B., Al-Mehdi, A.B., Wei, Z., Song, C., Manevich, Y., 2003. Lung ischemia: endothelial cell signaling by reactive oxygen species. A progress report. Adv. Exp. Med. Biol. 510, 343–347.

Fitzpatrick, A.M., Jones, D.P., Brown, L.A., 2012. Glutathione redox control of asthma: from molecular mechanisms to therapeutic opportunities. Antioxid. Redox Signal. 17 (2), 375–408.

Fitzpatrick, A.M., Stephenson, S.T., Hadley, G.R., Burwell, L., Penugonda, M., Simon, D.M., Hansen, J., Jones, D.P., Brown, L.A., 2011. Thiol redox disturbances in children with severe asthma are associated with posttranslational modification of the transcription factor nuclear factor (erythroid-derived 2)-like 2. J. Allergy Clin. Immunol. 127 (6), 1604–1611.

Folz, R.J., Guan, J., Seldin, M.F., Oury, T.D., Enghild, J.J., Crapo, J.D., 1997. Mouse extracellular superoxide dismutase: primary structure, tissue-specific gene expression, chromosomal localization, and lung in situ hybridization. Am. J. Respir. Cell. Mol. Biol. 17 (4), 393–403.

Fomenko, D.E., Koc, A., Agisheva, N., Jacobsen, M., Kaya, A., Malinouski, M., Rutherford, J.C., Siu, K.L., Jin, D.Y., Winge, D.R., et al., 2011. Thiol peroxidases mediate specific genome-wide regulation of gene expression in response to hydrogen peroxide. Proc. Natl. Acad. Sci. U. S. A. 108 (7), 2729–2734.

Forman, H.J., Fukuto, J.M., Torres, M., 2004. Redox signaling: thiol chemistry defines which reactive oxygen and nitrogen species can act as second messengers. Am. J. Physiol. Cell Physiol. 287 (2), C246–C256.

Forman, H.J., Torres, M., 2002. Reactive oxygen species and cell signaling: respiratory burst in macrophage signaling. Am. J. Respir. Crit. Care Med. 166 (12 Pt 2), S4–S8.

Forteza, R., Salathe, M., Miot, F., Conner, G.E., 2005. Regulated hydrogen peroxide production by Duox in human airway epithelial cells. Am. J. Respir. Cell. Mol. Biol. 32 (5), 462–469.

Gallagher, C.J., Ahn, K., Knipe, A.L., Dyer, A.M., Richie Jr., J.P., Lazarus, P., Muscat, J.E., 2009. Association between haplotypes of manganese superoxide dismutase (SOD2), smoking, and lung cancer risk. Free Radic. Biol. Med. 46 (1), 20–24.

Gao, F., Kinnula, V.L., Myllarniemi, M., Oury, T.D., 2008. Extracellular superoxide dismutase in pulmonary fibrosis. Antioxid. Redox Signal. 10 (2), 343–354.

Gao, X.P., Standiford, T.J., Rahman, A., Newstead, M., Holland, S.M., Dinauer, M.C., Liu, Q.H., Malik, A.B., 2002. Role of NADPH oxidase in the mechanism of lung neutrophil sequestration and microvessel injury induced by gram-negative sepsis: studies in p47phox-/- and gp91phox-/- mice. J. Immunol. 168 (8), 3974–3982.

Gattas, M.V., Forteza, R., Fragoso, M.A., Fregien, N., Salas, P., Salathe, M., Conner, G.E., 2009. Oxidative epithelial host defense is regulated by infectious and inflammatory stimuli. Free Radic. Biol. Med. 47 (10), 1450–1458.

Geiszt, M., Leto, T.L., 2004. The Nox family of NAD(P)H oxidases: host defense and beyond. J. Biol. Chem. 279 (50), 51715–51718.

Geiszt, M., Witta, J., Baffi, J., Lekstrom, K., Leto, T.L., 2003. Dual oxidases represent novel hydrogen peroxide sources supporting mucosal surface host defense. FASEB J. 17, 1502–1504.

Ghio, A.J., 2009. Disruption of iron homeostasis and lung disease. Biochim. Biophys. Acta 1790 (7), 731–739.

Ghosh, S., Janocha, A.J., Aronica, M.A., Swaidani, S., Comhair, S.A., Xu, W., Zheng, L., Kaveti, S., Kinter, M., Hazen, S.L., et al., 2006. Nitrotyrosine proteome survey in asthma identifies oxidative mechanism of catalase inactivation. J. Immunol. 176 (9), 5587–5597.

Ghosh, S., Willard, B., Comhair, S.A., Dibello, P., Xu, W., Shiva, S., Aulak, K.S., Kinter, M., Erzurum, S.C., 2013. Disulfide bond as a switch for copper-zinc superoxide dismutase activity in asthma. Antioxid. Redox Signal. 18 (4), 412–423.

Gorissen, S.H., Hristova, M., Habibovic, A., Sipsey, L.M., Spiess, P.C., Janssen-Heininger, Y.M., van der Vliet, A., 2013. DUOX1 is required for airway epithelial cell migration and bronchiolar reepithelialization following injury. Am. J. Respir. Cell Mol. Biol. 48 (3), 337–345.

Gould, N.S., Day, B.J., 2011. Targeting maladaptive glutathione responses in lung disease. Biochem. Pharmacol. 81 (2), 187–193.

Gould, N.S., Min, E., Day, B.J., 2011. Macropinocytosis of extracellular glutathione ameliorates tumor necrosis factor alpha release in activated macrophages. PloS one 6 (10), e25704.

Gutscher, M., Sobotta, M.C., Wabnitz, G.H., Ballikaya, S., Meyer, A.J., Samstag, Y., Dick, T.P., 2009. Proximity-based protein thiol oxidation by H_2O_2-scavenging peroxidases. J. Biol. Chem. 284 (46), 31532–31540.

Ha, B., Kim, E.K., Kim, J.H., Lee, H.N., Lee, K.O., Lee, S.Y., Jang, H.H., 2012. Human peroxiredoxin 1 modulates TGF-beta1-induced epithelial-mesenchymal transition through its peroxidase activity. Biochem. Biophys. Res. Commun. 421 (1), 33–37.

Hammad, H., Lambrecht, B.N., 2008. Dendritic cells and epithelial cells: linking innate and adaptive immunity in asthma. Nat. Rev. Immunol. 8 (3), 193–204.

Han, J., Shuvaev, V.V., Muzykantov, V.R., 2012. Targeted interception of signaling reactive oxygen species in the vascular endothelium. Ther. Deliv. 3 (2), 263–276.

Harju, T., Kaarteenaho-Wiik, R., Sirvio, R., Paakko, P., Crapo, J.D., Oury, T.D., Soini, Y., Kinnula, V.L., 2004. Manganese superoxide dismutase is increased in the airways of smokers' lungs. Eur. Respir. J. 24 (5), 765–771.

Harper, R.W., Xu, C., Eiserich, J.P., Chen, Y., Kao, C.Y., Thai, P., Setiadi, H., Wu, R., 2005. Differential regulation of dual NADPH oxidases/peroxidases, Duox1 and Duox2, by Th1 and Th2 cytokines in respiratory tract epithelium. FEBS Lett. 579 (21), 4911–4917.

Harper, R.W., Xu, C., McManus, M., Heidersbach, A., Eiserich, J.P., 2006. Duox2 exhibits potent heme peroxidase activity in human respiratory tract epithelium. FEBS Lett. 580 (22), 5150–5154.

Hecker, L., Vittal, R., Jones, T., Jagirdar, R., Luckhardt, T.R., Horowitz, J.C., Pennathur, S., Martinez, F.J., Thannickal, V.J., 2009. NADPH oxidase-4 mediates myofibroblast activation and fibrogenic responses to lung injury. Nat. Med. 15 (9), 1077–1081.

Ho, Y.S., Magnenat, J.L., Bronson, R.T., Cao, J., Gargano, M., Sugawara, M., Funk, C.D., 1997. Mice deficient in cellular glutathione peroxidase develop normally and show no increased sensitivity to hyperoxia. J. Biol. Chem. 272 (26), 16644–16651.

Ho, Y.S., Vincent, R., Dey, M.S., Slot, J.W., Crapo, J.D., 1998. Transgenic models for the study of lung antioxidant defense: enhanced manganese-containing superoxide dismutase activity gives partial protection to B6C3 hybrid mice exposed to hyperoxia. Am. J. Respir. Cell Mol. Biol. 18 (4), 538–547.

Hoffman, S.M., Tully, J.E., Lahue, K.G., Anathy, V., Nolin, J.D., Guala, A.S., van der Velden, J.L., Ho, Y.S., Aliyeva, M., Daphtary, N., et al., 2012. Genetic ablation of glutaredoxin-1 causes enhanced resolution of airways hyperresponsiveness and mucus metaplasia in mice with allergic airways disease. Am. J. Physiol. Lung Cell. Mol. Physiol. 303 (6), L528–L538.

Holgate, S.T., Holloway, J., Wilson, S., Bucchieri, F., Puddicombe, S., Davies, D.E., 2004. Epithelial-mesenchymal communication in the pathogenesis of chronic asthma. Proc. Am. Thorac. Soc. 1 (2), 93–98.

Holland, R., Fishbein, J.C., 2010. Chemistry of the cysteine sensors in Kelch-like ECH-associated protein 1. Antioxid. Redox Signal. 13 (11), 1749–1761.

Hoshino, T., Nakamura, H., Okamoto, M., Kato, S., Araya, S., Nomiyama, K., Oizumi, K., Young, H.A., Aizawa, H., Yodoi, J., 2003. Redox-active protein thioredoxin prevents proinflammatory cytokine- or bleomycin-induced lung injury. Am. J. Respir. Crit. Care Med. 168 (9), 1075–1083.

Huang, J., Canadien, V., Lam, G.Y., Steinberg, B.E., Dinauer, M.C., Magalhaes, M.A., Glogauer, M., Grinstein, S., Brumell, J.H., 2009. Activation of antibacterial autophagy by NADPH oxidases. Proc. Natl. Acad. Sci. U. S. A. 106 (15), 6226–6231.

Inoue, K., Takano, H., Koike, E., Warabi, E., Yanagawa, T., Yanagisawa, R., Ishii, T., 2009. Peroxiredoxin I is a negative regulator of Th2-dominant allergic asthma. Int. Immunopharmacol. 9 (11), 1281–1288.

Iwata, Y., Okamoto, M., Hoshino, T., Kitasato, Y., Sakazaki, Y., Tajiri, M., Matsunaga, K., Azuma, K., Kawayama, T., Kinoshita, T., et al., 2010. Elevated levels of thioredoxin 1 in the lungs and sera of idiopathic pulmonary fibrosis, non-specific interstitial pneumonia and cryptogenic organizing pneumonia. Intern. Med. 49 (22), 2393–2400.

Jackson, S.H., Devadas, S., Kwon, J., Pinto, L.A., Williams, M.S., 2004. T cells express a phagocyte-type NADPH oxidase that is activated after T cell receptor stimulation. Nat. Immunol. 5 (8), 818–827.

Jacobsen, E.A., Taranova, A.G., Lee, N.A., Lee, J.J., 2007. Eosinophils: singularly destructive effector cells or purveyors of immunoregulation? J. Allergy Clin. Immunol. 119 (6), 1313–1320.

Jancic, C., Savina, A., Wasmeier, C., Tolmachova, T., El-Benna, J., Dang, P.M., Pascolo, S., Gougerot-Pocidalo, M.A., Raposo, G., Seabra, M.C., et al., 2007. Rab27a regulates phagosomal pH and NADPH oxidase recruitment to dendritic cell phagosomes. Nat. Cell. Biol. 9 (4), 367–378.

Janssen-Heininger, Y.M., Mossman, B.T., Heintz, N.H., Forman, H.J., Kalyanaraman, B., Finkel, T., Stamler, J.S., Rhee, S.G., van der Vliet, A., 2008. Redox-based regulation of signal transduction: principles, pitfalls, and promises. Free Radic. Biol. Med. 45 (1), 1–17.

Kassim, S.Y., Fu, X., Liles, W.C., Shapiro, S.D., Parks, W.C., Heinecke, J.W., 2005. NADPH oxidase restrains the matrix metalloproteinase activity of macrophages. J. Biol. Chem. 280 (34), 30201–30205.

Kensler, T.W., Wakabayashi, N., 2010. Nrf2: friend or foe for chemoprevention? Carcinogenesis 31 (1), 90–99.

Kikuchi, N., Ishii, Y., Morishima, Y., Yageta, Y., Haraguchi, N., Yamadori, T., Masuko, H., Sakamoto, T., Yanagawa, T., Warabi, E., et al., 2011. Aggravation of bleomycin-induced pulmonary inflammation and fibrosis in mice lacking peroxiredoxin I. Am. J. Respir. Cell Mol. Biol. 45 (3), 600–609.

Kim, H.J., Park, Y.D., Moon, U.Y., Kim, J.H., Jeon, J.H., Lee, J.G., Bae, Y.S., Yoon, J.H., 2008. The role of Nox4 in oxidative stress-induced MUC5AC overexpression in human airway epithelial cells. Am. J. Respir. Cell Mol. Biol. 39 (5), 598–609.

Kim, K.K., Whitin, J.C., Sukhova, N.M., Cohen, H.J., 1999. Increase in extracellular glutatione peroxidase in plasma and lungs of mice exposed to hyperoxia. Pediatr. Res. 46 (6), 715–721.

Kinnula, V.L., Adler, K.B., Ackley, N.J., Crapo, J.D., 1992. Release of reactive oxygen species by guinea pig tracheal epithelial cells in vitro. Am. J. Physiol. 262 (6 Pt 1), L708–L712.

Kinnula, V.L., Crapo, J.D., 2003. Superoxide dismutases in the lung and human lung diseases. Am. J. Respir. Crit. Care Med. 167 (12), 1600–1619.

Kinnula, V.L., Lehtonen, S., Kaarteenaho-Wiik, R., Lakari, E., Paakko, P., Kang, S.W., Rhee, S.G., Soini, Y., 2002. Cell specific expression of peroxiredoxins in human lung and pulmonary sarcoidosis. Thorax 57 (2), 157–164.

Knowles, M.R., Boucher, R.C., 2002. Mucus clearance as a primary innate defense mechanism for mammalian airways. J. Clin. Invest. 109 (5), 571–577.

Koff, J.L., Shao, M.X., Kim, S., Ueki, I.F., Nadel, J.A., 2006. Pseudomonas lipopolysaccharide accelerates wound repair via activation of a novel epithelial cell signaling cascade. J. Immunol. 177 (12), 8693–8700.

Koff, J.L., Shao, M.X., Ueki, I.F., Nadel, J.A., 2008. Multiple TLRs activate EGFR via a signaling cascade to produce innate immune responses in airway epithelium. Am. J. Physiol. Lung Cell. Mol. Physiol. 294 (6), L1068–L1075.

Kolleck, I., Sinha, P., Rustow, B., 2002. Vitamin E as an antioxidant of the lung: mechanisms of vitamin E delivery to alveolar type II cells. Am. J. Respir. Crit. Care Med. 166 (12 Pt 2), S62–S66.

Kuipers, I., Bracke, K.R., Brusselle, G.G., Aesif, S.W., Krijgsman, R., Arts, I.C., Wouters, E.F., Reynaert, N.L., 2012. Altered cigarette smoke-induced lung inflammation due to ablation of Grx1. PloS one 7 (6), e38984.

Kuipers, I., Louis, R., Manise, M., Dentener, M.A., Irvin, C.G., Janssen-Heininger, Y.M., Brightling, C.E., Wouters, E.F., Reynaert, N.L., 2013. Increased glutaredoxin-1 and decreased protein *S*-glutathionylation in sputum of asthmatics. Eur. Respir. J. 41 (2), 469–472.

Kuwahara, I., Lillehoj, E.P., Koga, T., Isohama, Y., Miyata, T., Kim, K.C., 2007. The signaling pathway involved in neutrophil elastase-stimulated MUC1 transcription. Am. J. Respir. Cell. Mol. Biol. 37 (6), 691–698.

Kwon, J., Qu, C.K., Maeng, J.S., Falahati, R., Lee, C., Williams, M.S., 2005. Receptor-stimulated oxidation of SHP-2 promotes T-cell adhesion through SLP-76-ADAP. EMBO J. 24 (13), 2331–2341.

Lagan, A.L., Quinlan, G.J., Mumby, S., Melley, D.D., Goldstraw, P., Bellingan, G.J., Hill, M.R., Briggs, D., Pantelidis, P., du Bois, R.M., et al., 2008. Variation in iron homeostasis genes between patients with ARDS and healthy control subjects. Chest 133 (6), 1302–1311.

Lambeth, J.D., 2004. NOX enzymes and the biology of reactive oxygen. Nat. Rev. Immunol. 4 (3), 181–189.

Larsen, G.L., White, C.W., Takeda, K., Loader, J.E., Nguyen, D.D., Joetham, A., Groner, Y., Gelfand, E.W., 2000. Mice that overexpress Cu/Zn superoxide dismutase are resistant to allergen-induced changes in airway control. Am. J. Physiol. Lung Cell. Mol. Physiol. 279 (2), L350–L359.

Lavigne, M.C., Eppihimer, M.J., 2005. Cigarette smoke condensate induces MMP-12 gene expression in airway-like epithelia. Biochem. Biophys. Res. Commun. 330 (1), 194–203.

Lebovitz, R.M., Zhang, H., Vogel, H., Cartwright Jr., J., Dionne, L., Lu, N., Huang, S., Matzuk, M.M., 1996. Neurodegeneration, myocardial injury, and perinatal death in mitochondrial superoxide dismutase-deficient mice. Proc. Natl. Acad. Sci. U. S. A. 93 (18), 9782–9787.

Lee, P.L., West, C., Crain, K., Wang, L., 2006. Genetic polymorphisms and susceptibility to lung disease. J. Negat. Results Biomed. 5 (5).

Lehtonen, S.T., Ohlmeier, S., Kaarteenaho-Wiik, R., Harju, T., Paakko, P., Soini, Y., Kinnula, V.L., 2008. Does the oxidative stress in chronic obstructive pulmonary disease cause thioredoxin/peroxiredoxin oxidation? Antioxid. Redox Signal. 10 (4), 813–819.

Lenaz, G., 2012. Mitochondria and reactive oxygen species. Which role in physiology and pathology? Adv. Exp. Med. Biol. 942, 93–136.

Li, L., Shoji, W., Takano, H., Nishimura, N., Aoki, Y., Takahashi, R., Goto, S., Kaifu, T., Takai, T., Obinata, M., 2007. Increased susceptibility of MER5 (peroxiredoxin III) knockout mice to LPS-induced oxidative stress. Biochem. Biophys. Res. Commun. 355 (3), 715–721.

Li, Q., Harraz, M.M., Zhou, W., Zhang, L.N., Ding, W., Zhang, Y., Eggleston, T., Yeaman, C., Banfi, B., Engelhardt, J.F., 2006. Nox2 and Rac1 regulate H_2O_2-dependent recruitment of TRAF6 to endosomal interleukin-1 receptor complexes. Mol. Cell. Biol. 26 (1), 140–154.

Liby, K.T., Sporn, M.B., 2012. Synthetic oleanane triterpenoids: multifunctional drugs with a broad range of applications for prevention and treatment of chronic disease. Pharmacol. Rev. 64 (4), 972–1003.

Lillig, C.H., Berndt, C., Holmgren, A., 2008. Glutaredoxin systems. Biochim. Biophys. Acta 1780 (11), 1304–1317.

Linderholm, A.L., Onitsuka, J., Xu, C., Chiu, M., Lee, W.M., Harper, R.W., 2010. All-trans retinoic acid mediates DUOX2 expression and function in respiratory tract epithelium. Am. J. Physiol. Lung Cell. Mol. Physiol. 299 (2), L215–L221.

Liu, G., Feinstein, S.I., Wang, Y., Dodia, C., Fisher, D., Yu, K., Ho, Y.S., Fisher, A.B., 2010. Comparison of glutathione peroxidase 1 and peroxiredoxin 6 in protection against oxidative stress in the mouse lung. Free Radic. Biol. Med. 49 (7), 1172–1181.

Malhotra, D., Thimmulappa, R., Navas-Acien, A., Sandford, A., Elliott, M., Singh, A., Chen, L., Zhuang, X., Hogg, J., Pare, P., et al., 2008. Decline in NRF2-regulated antioxidants in chronic obstructive pulmonary disease lungs due to loss of its positive regulator, DJ-1. Am. J. Respir. Crit. Care Med. 178 (6), 592–604.

Martyn, K.D., Frederick, L.M., von Loehneysen, K., Dinauer, M.C., Knaus, U.G., 2006. Functional analysis of Nox4 reveals unique characteristics compared to other NADPH oxidases. Cell. Signal 18 (1), 69–82.

McMahon, M., Lamont, D.J., Beattie, K.A., Hayes, J.D., 2010. Keap1 perceives stress via three sensors for the endogenous signaling molecules nitric oxide, zinc, and alkenals. Proc. Natl. Acad. Sci. U. S. A. 107 (44), 18838–18843.

Meissner, F., Seger, R.A., Moshous, D., Fischer, A., Reichenbach, J., Zychlinsky, A., 2010. Inflammasome activation in NADPH oxidase defective mononuclear phagocytes from patients with chronic granulomatous disease. Blood 116 (9), 1570–1573.

Meitzler, J.L., Ortiz de Montellano, P.R., 2009. *Caenorhabditis elegans* and human dual oxidase 1 (DUOX1) "peroxidase" domains: insights into heme binding and catalytic activity. J. Biol. Chem. 284 (28), 18634–18643.

Meyer, H.A., Dittrich, A.M., Hamelmann, E., 2011. Different isoforms of glutathione peroxidase cause opposing effects during the development of allergic asthma in mice. Antioxid. Redox Signal 14 (1), 169–170 author reply 170–161.

Milovanova, T., Chatterjee, S., Manevich, Y., Kotelnikova, I., Debolt, K., Madesh, M., Moore, J.S., Fisher, A.B., 2006. Lung endothelial cell proliferation with decreased shear stress is mediated by reactive oxygen species. Am. J. Physiol. Cell Physiol. 290 (1), C66–C76.

Mittal, M., Roth, M., Konig, P., Hofmann, S., Dony, E., Goyal, P., Selbitz, A.C., Schermuly, R.T., Ghofrani, H.A., Kwapiszewska, G., et al., 2007. Hypoxia-dependent regulation of nonphagocytic NADPH oxidase subunit NOX4 in the pulmonary vasculature. Circ. Res. 101, 258–267.

Moore, S.F., MacKenzie, A.B., 2009. NADPH oxidase NOX2 mediates rapid cellular oxidation following ATP stimulation of endotoxin-primed macrophages. J. Immunol. 183 (5), 3302–3308.

Morand, S., Chaaraoui, M., Kaniewski, J., Deme, D., Ohayon, R., Noel-Hudson, M.S., Virion, A., Dupuy, C., 2003. Effect of iodide on nicotinamide adenine dinucleotide phosphate oxidase activity and Duox2 protein expression in isolated porcine thyroid follicles. Endocrinology 144 (4), 1241–1248.

Morgenstern, D.E., Gifford, M.A., Li, L.L., Doerschuk, C.M., Dinauer, M.C., 1997. Absence of respiratory burst in X-linked chronic granulomatous disease mice leads to abnormalities in both host defense and inflammatory response to *Aspergillus fumigatus*. J. Exp. Med. 185 (2), 207–218.

Moskwa, P., Lorentzen, D., Excoffon, K.J., Zabner, J., McCray Jr., P.B., Nauseef, W.M., Dupuy, C., Banfi, B., 2007. A novel host defense system of airways is defective in cystic fibrosis. Am. J. Respir. Crit. Care Med. 175 (2), 174–183.

Mudway, I.S., Blomberg, A., Frew, A.J., Holgate, S.T., Sandstrom, T., Kelly, F.J., 1999. Antioxidant consumption and repletion kinetics in nasal lavage fluid following exposure of healthy human volunteers to ozone. Eur. Respir. J. 13 (6), 1429–1438.

Muller, T., Hengstermann, A., 2012. Nrf2: friend and foe in preventing cigarette smoking-dependent lung disease. Chem. Res. Toxicol. 25 (9), 1805–1824.

Nakamura, T., Nakamura, H., Hoshino, T., Ueda, S., Wada, H., Yodoi, J., 2005. Redox regulation of lung inflammation by thioredoxin. Antioxid. Redox Signal. 7 (1–2), 60–71.

Nakanaga, T., Nadel, J.A., Ueki, I.F., Koff, J.L., Shao, M.X., 2007. Regulation of interleukin-8 via an airway epithelial signaling cascade. Am. J. Physiol. Lung Cell. Mol. Physiol. 292 (5), L1289–L1296.

Oakley, F.D., Abbott, D., Li, Q., Engelhardt, J.F., 2009. Signaling components of redox active endosomes: the redoxosomes. Antioxid. Redox Signal. 11 (6), 1313–1333.

Ober, C., Butte, A.J., Elias, J.A., Lusis, A.J., Gan, W., Banks-Schlegel, S., Schwartz, D., 2010. Getting from genes to function in lung disease: a National Heart, Lung, and Blood Institute workshop report. Am. J. Respir. Crit. Care Med. 182 (6), 732–737.

Osburn, W.O., Kensler, T.W., 2008. Nrf2 signaling: an adaptive response pathway for protection against environmental toxic insults. Mutat. Res. 659 (1–2), 31–39.

Parinandi, N.L., Kleinberg, M.A., Usatyuk, P.V., Cummings, R.J., Pennathur, A., Cardounel, A.J., Zweier, J.L., Garcia, J.G., Natarajan, V., 2003. Hyperoxia-induced NAD(P)H oxidase activation and regulation by MAP kinases in human lung endothelial cells. Am. J. Physiol. Lung Cell. Mol. Physiol. 284 (1), L26–L38.

Park, J.H., Kim, Y.S., Lee, H.L., Shim, J.Y., Lee, K.S., Oh, Y.J., Shin, S.S., Choi, Y.H., Park, K.J., Park, R.W., et al., 2006. Expression of peroxiredoxin and thioredoxin in human lung cancer and paired normal lung. Respirology 11 (3), 269–275.

Peden, D.B., Swiersz, M., Ohkubo, K., Hahn, B., Emery, B., Kaliner, M.A., 1993. Nasal secretion of the ozone scavenger uric acid. Am. Rev. Respir. Dis. 148 (2), 455–461.

Peltoniemi, M.J., Rytila, P.H., Harju, T.H., Soini, Y.M., Salmenkivi, K.M., Ruddock, L.W., Kinnula, V.L., 2006. Modulation of glutaredoxin in the lung and sputum of cigarette smokers and chronic obstructive pulmonary disease. Respir. Res. 7, 133.

Rahman, I., Biswas, S.K., Kode, A., 2006. Oxidant and antioxidant balance in the airways and airway diseases. Eur. J. Pharmacol. 533 (1–3), 222–239.

Reth, M., 2002. Hydrogen peroxide as second messenger in lymphocyte activation. Nat. Immunol. 3 (12), 1129–1134.

Rhee, S.G., Kang, S.W., Jeong, W., Chang, T.S., Yang, K.S., Woo, H.A., 2005a. Intracellular messenger function of hydrogen peroxide and its regulation by peroxiredoxins. Curr. Opin. Cell. Biol. 17 (2), 183–189.

Rhee, S.G., Woo, H.A., Kil, I.S., Bae, S.H., 2012. Peroxiredoxin functions as a peroxidase and a regulator and sensor of local peroxides. J. Biol. Chem. 287 (7), 4403–4410.

Rhee, S.G., Yang, K.S., Kang, S.W., Woo, H.A., Chang, T.S., 2005b. Controlled elimination of intracellular H(2)O(2): regulation of peroxiredoxin, catalase, and glutathione peroxidase via post-translational modification. Antioxid. Redox Signal. 7 (5–6), 619–626.

Rinna, A., Torres, M., Forman, H.J., 2006. Stimulation of the alveolar macrophage respiratory burst by ADP causes selective glutathionylation of protein tyrosine phosphatase 1B. Free Radic. Biol. Med. 41 (1), 86–91.

Rostila, A., Puustinen, A., Toljamo, T., Vuopala, K., Lindstrom, I., Nyman, T.A., Oksa, P., Vehmas, T., Anttila, S.L., 2012. Peroxiredoxins and tropomyosins as plasma biomarkers for lung cancer and asbestos exposure. Lung Cancer 77 (2), 450–459.

Rybicka, J.M., Balce, D.R., Chaudhuri, S., Allan, E.R., Yates, R.M., 2012. Phagosomal proteolysis in dendritic cells is modulated by NADPH oxidase in a pH-independent manner. EMBO J. 31 (4), 932–944.

Rybicka, J.M., Balce, D.R., Khan, M.F., Krohn, R.M., Yates, R.M., 2010. NADPH oxidase activity controls phagosomal proteolysis in macrophages through modulation of the lumenal redox environment of phagosomes. Proc. Natl. Acad. Sci. U. S. A. 107 (23), 10496–10501.

Samet, J.M., Hatch, G.E., Horstman, D., Steck-Scott, S., Arab, L., Bromberg, P.A., Levine, M., McDonnell, W.F., Devlin, R.B., 2001. Effect of antioxidant supplementation on ozone-induced lung injury in human subjects. Am. J. Respir. Crit. Care Med. 164 (5), 819–825.

Schneider, M., Forster, H., Boersma, A., Seiler, A., Wehnes, H., Sinowatz, F., Neumuller, C., Deutsch, M.J., Walch, A., Hrabe de Angelis, M., et al., 2009. Mitochondrial glutathione peroxidase 4 disruption causes male infertility. FASEB J. 23 (9), 3233–3242.

Schock, B.C., Koostra, J., Kwack, S., Hackman, R.M., Van Der Vliet, A., Cross, C.E., 2004. Ascorbic acid in nasal and tracheobronchial airway lining fluids. Free Radic. Biol. Med. 37 (9), 1393–1401.

Schrader, M., Fahimi, H.D., 2006. Peroxisomes and oxidative stress. Biochim. Biophys. Acta 1763 (12), 1755–1766.

Schwarzer, C., Machen, T.E., Illek, B., Fischer, H., 2004. NADPH oxidase-dependent acid production in airway epithelial cells. J. Biol. Chem. 279 (35), 36454–36461.

Segal, B.H., Davidson, B.A., Hutson, A.D., Russo, T.A., Holm, B.A., Mullan, B., Habitzruther, M., Holland, S.M., Knight 3rd, P.R., 2007. Acid aspiration-induced lung inflammation and injury are exacerbated in NADPH oxidase-deficient mice. Am. J. Physiol. Lung Cell. Mol. Physiol. 292 (3), L760–L768.

Segal, B.H., Grimm, M.J., Khan, A.N., Han, W., Blackwell, T.S., 2012. Regulation of innate immunity by NADPH oxidase. Free Radic. Biol. Med. 53 (1), 72–80.

Segal, B.H., Han, W., Bushey, J.J., Joo, M., Bhatti, Z., Feminella, J., Dennis, C.G., Vethanayagam, R.R., Yull, F.E., Capitano, M., et al., 2010. NADPH oxidase limits innate immune responses in the lungs in mice. PloS one 5 (3), e9631.

Sham, D., Wesley, U.V., Hristova, M., van der Vliet, A., 2013. ATP-mediated Transactivation of the epidermal growth factor receptor in airway epithelial cells involves DUOX1-dependent oxidation of Src and ADAM17. PloS one 8 (1), e54391.

Shannahan, J.H., Nyska, A., Cesta, M., Schladweiler, M.C., Vallant, B.D., Ward, W.O., Ghio, A.J., Gavett, S.H., Kodavanti, U.P., 2012. Subchronic pulmonary pathology, iron overload, and transcriptional activity after Libby amphibole exposure in rat models of cardiovascular disease. Environ. Health Perspect. 120 (1), 85–91.

Shao, M.X., Nadel, J.A., 2005. Dual oxidase 1-dependent MUC5AC mucin expression in cultured human airway epithelial cells. Proc. Natl. Acad. Sci. U. S. A. 102 (3), 767–772.

Shaw, D.E., Berry, M.A., Hargadon, B., McKenna, S., Shelley, M.J., Green, R.H., Brightling, C.E., Wardlaw, A.J., Pavord, I.D., 2007.

Association between neutrophilic airway inflammation and airflow limitation in adults with asthma. Chest 132 (6), 1871–1875.

Sienra-Monge, J.J., Ramirez-Aguilar, M., Moreno-Macias, H., Reyes-Ruiz, N.I., Del Rio-Navarro, B.E., Ruiz-Navarro, M.X., Hatch, G., Crissman, K., Slade, R., Devlin, R.B., et al., 2004. Antioxidant supplementation and nasal inflammatory responses among young asthmatics exposed to high levels of ozone. Clin. Exp. Immunol. 138 (2), 317–322.

Singh, A., Rangasamy, T., Thimmulappa, R.K., Lee, H., Osburn, W.O., Brigelius-Flohe, R., Kensler, T.W., Yamamoto, M., Biswal, S., 2006. Glutathione peroxidase 2, the major cigarette smoke-inducible isoform of GPX in lungs, is regulated by Nrf2. Am. J. Respir. Cell. Mol. Biol. 35 (6), 639–650.

Smith, L.J., Shamsuddin, M., Sporn, P.H., Denenberg, M., Anderson, J., 1997. Reduced superoxide dismutase in lung cells of patients with asthma. Free Radic. Biol. Med. 22 (7), 1301–1307.

Soini, Y., Kahlos, K., Napankangas, U., Kaarteenaho-Wiik, R., Saily, M., Koistinen, P., Paaakko, P., Holmgren, A., Kinnula, V.L., 2001. Widespread expression of thioredoxin and thioredoxin reductase in non-small cell lung carcinoma. Clin. Cancer Res. 7 (6), 1750–1757.

Soucy-Faulkner, A., Mukawera, E., Fink, K., Martel, A., Jouan, L., Nzengue, Y., Lamarre, D., Vande Velde, C., Grandvaux, N., 2010. Requirement of NOX2 and reactive oxygen species for efficient RIG-I-mediated antiviral response through regulation of MAVS expression. PLoS Pathog. 6 (6), e1000930.

St Clair, D.K., Oberley, T.D., Ho, Y.S., 1991. Overproduction of human Mn-superoxide dismutase modulates paraquat-mediated toxicity in mammalian cells. FEBS Lett. 293 (1–2), 199–203.

Staple, L., Andrews, T., McDonald-McGinn, D., Zackai, E., Sullivan, K.E., 2005. Allergies in patients with chromosome 22q11.2 deletion syndrome (DiGeorge syndrome/velocardiofacial syndrome) and patients with chronic granulomatous disease. Pediatr. Allergy Immunol. 16 (3), 226–230.

Stroher, E., Millar, A.H., 2012. The biological roles of glutaredoxins. Biochem. J. 446 (3), 333–348.

Sturrock, A., Cahill, B., Norman, K., Huecksteadt, T.P., Hill, K., Sanders, K., Karwande, S.V., Stringham, J.C., Bull, D.A., Gleich, M., et al., 2006. Transforming growth factor-beta1 induces Nox4 NAD(P)H oxidase and reactive oxygen species-dependent proliferation in human pulmonary artery smooth muscle cells. Am. J. Physiol. Lung Cell. Mol. Physiol. 290 (4), L661–L673.

Sturrock, A., Huecksteadt, T.P., Norman, K., Sanders, K., Murphy, T.M., Chitano, P., Wilson, K., Hoidal, J.R., Kennedy, T.P., 2007. Nox4 mediates TGF-beta1-induced retinoblastoma protein phosphorylation, proliferation, and hypertrophy in human airway smooth muscle cells. Am. J. Physiol. Lung Cell. Mol. Physiol. 292 (6), L1543–L1555.

Su, W.Y., Folz, R., Chen, J.S., Crapo, J.D., Chang, L.Y., 1997. Extracellular superoxide dismutase mRNA expressions in the human lung by in situ hybridization. Am. J. Respir. Cell Mol. Biol. 16 (2), 162–170.

Subramani, S., 1993. Protein import into peroxisomes and biogenesis of the organelle. Annu. Rev. Cell Biol. 9, 445–478.

Sundar, I.K., Chung, S., Hwang, J.W., Arunachalam, G., Cook, S., Yao, H., Mazur, W., Kinnula, V.L., Fisher, A.B., Rahman, I., 2010. Peroxiredoxin 6 differentially regulates acute and chronic cigarette smoke-mediated lung inflammatory response and injury. Exp. Lung Res. 36 (8), 451–462.

Sussan, T.E., Rangasamy, T., Blake, D.J., Malhotra, D., El-Haddad, H., Bedja, D., Yates, M.S., Kombairaju, P., Yamamoto, M., Liby, K.T., et al., 2009. Targeting Nrf2 with the triterpenoid CDDO-imidazolide attenuates cigarette smoke-induced emphysema and cardiac dysfunction in mice. Proc. Natl. Acad. Sci. U. S. A. 106 (1), 250–255.

Svensk, A.M., Soini, Y., Paakko, P., Hiravikoski, P., Kinnula, V.L., 2004. Differential expression of superoxide dismutases in lung cancer. Am. J. Clin. Pathol. 122 (3), 395–404.

Tan, R.J., Fattman, C.L., Watkins, S.C., Oury, T.D., 2004. Redistribution of pulmonary EC-SOD after exposure to asbestos. J. Appl. Physiol. 97 (5), 2006–2013.

Tashakkor, A.Y., Chow, K.S., Carlsten, C., 2011. Modification by antioxidant supplementation of changes in human lung function associated with air pollutant exposure: a systematic review. BMC Public Health 11, 532.

Teoh-Fitzgerald, M.L., Fitzgerald, M.P., Jensen, T.J., Futscher, B.W., Domann, F.E., 2012. Genetic and epigenetic inactivation of extracellular superoxide dismutase promotes an invasive phenotype in human lung cancer by disrupting ECM homeostasis. Mol. Cancer Res. 10 (1), 40–51.

Terada, L.S., 2006. Specificity in reactive oxidant signaling: think globally, act locally. J. Cell Biol. 174 (5), 615–623.

Thabut, G., El-Benna, J., Samb, A., Corda, S., Megret, J., Leseche, G., Vicaut, E., Aubier, M., Boczkowski, J., 2002. Tumor necrosis factor-alpha increases airway smooth muscle oxidants production through a NADPH oxidase-like system to enhance myosin light chain phosphorylation and contractility. J. Biol. Chem. 277 (25), 22814–22821.

Tiitto, L., Kaarteenaho-Wiik, R., Sormunen, R., Holmgren, A., Paakko, P., Soini, Y., Kinnula, V.L., 2003. Expression of the thioredoxin system in interstitial lung disease. J. Pathol. 201 (3), 363–370.

Torii, M., Wang, L., Ma, N., Saito, K., Hori, T., Sato-Ueshima, M., Koyama, Y., Nishikawa, H., Katayama, N., Mizoguchi, A., et al., 2010. Thioredoxin suppresses airway inflammation independently of systemic Th1/Th2 immune modulation. Eur. J. Immunol. 40 (3), 787–796.

Tsan, M.F., White, J.E., Caska, B., Epstein, C.J., Lee, C.Y., 1998. Susceptibility of heterozygous MnSOD gene-knockout mice to oxygen toxicity. Am. J. Respir. Cell Mol. Biol. 19 (1), 114–120.

Tsiligianni, I.G., van der Molen, T., 2010. A systematic review of the role of vitamin insufficiencies and supplementation in COPD. Respir. Res. 11, 171.

Usatyuk, P.V., Romer, L.H., He, D., Parinandi, N.L., Kleinberg, M.E., Zhan, S., Jacaobson, J.R., Dudek, S., Pendyala, S., Garcia, J.G., et al., 2007. Regulation of hyperoxia-induced NADPH oxidase activation in human lung endothelial cells by the actin cytoskeleton and cortactin. J. Biol. Chem. 282 (32), 23284–23295.

Valacchi, G., Vasu, V.T., Yokohama, W., Corbacho, A.M., Phung, A., Lim, Y., Aung, H.H., Cross, C.E., Davis, P.A., 2007. Lung vitamin E transport processes are affected by both age and environmental oxidants in mice. Toxicol. Appl. Pharmacol. 222 (2), 227–234.

van de Veerdonk, F.L., Smeekens, S.P., Joosten, L.A., Kullberg, B.J., Dinarello, C.A., van der Meer, J.W., Netea, M.G., 2010. Reactive oxygen species-independent activation of the IL-1beta inflammasome in cells from patients with chronic granulomatous disease. Proc. Natl. Acad. Sci. U. S. A. 107 (7), 3030–3033.

van der Vliet, A., 2008. NADPH oxidases in lung biology and pathology: host defense enzymes, and more. Free Radic. Biol. Med. 44 (6), 938–955.

van der Vliet, A., 2011. Nox enzymes in allergic airway inflammation. Biochim. Biophys. Acta 1810 (11), 1035–1044.

van der Vliet, A., Eiserich, J.P., Shigenaga, M.K., Cross, C.E., 1999a. Reactive nitrogen species and tyrosine nitration in the respiratory tract: epiphenomena or a pathobiologic mechanism of disease? Am. J. Respir. Crit. Care Med. 160 (1), 1–9.

van der Vliet, A., O'Neill, C.A., Cross, C.E., Koostra, J.M., Volz, W.G., Halliwell, B., Louie, S., 1999b. Determination of low-molecular-mass antioxidant concentrations in human respiratory tract lining fluids. Am. J. Physiol. 276 (2 Pt 1), L289–L296.

Waghray, M., Cui, Z., Horowitz, J.C., Subramanian, I.M., Martinez, F.J., Toews, G.B., Thannickal, V.J., 2005. Hydrogen peroxide is a diffusible paracrine signal for the induction of epithelial cell death by activated myofibroblasts. FASEB J. 19 (7), 854–856.

Wedes, S.H., Khatri, S.B., Zhang, R., Wu, W., Comhair, S.A., Wenzel, S., Teague, W.G., Israel, E., Erzurum, S.C., Hazen, S.L., 2009. Noninvasive markers of airway inflammation in asthma. Clin. Transl. Sci. 2 (2), 112–117.

Wesley, U.V., Bove, P.F., Hristova, M., McCarthy, S., van der Vliet, A., 2007. Airway epithelial cell migration and wound repair by ATP-mediated activation of dual oxidase 1. J. Biol. Chem. 282 (5), 3213–3220.

White, C.W., Avraham, K.B., Shanley, P.F., Groner, Y., 1991. Transgenic mice with expression of elevated levels of copper-zinc superoxide dismutase in the lungs are resistant to pulmonary oxygen toxicity. J. Clin. Invest. 87 (6), 2162–2168.

Wijkstrom-Frei, C., El-Chemaly, S., Ali-Rachedi, R., Gerson, C., Cobas, M.A., Forteza, R., Salathe, M., Conner, G.E., 2003. Lactoperoxidase and human airway host defense. Am. J. Respir. Cell Mol. Biol. 29 (2), 206–212.

Winterbourn, C.C., Hampton, M.B., 2008. Thiol chemistry and specificity in redox signaling. Free Radic. Biol. Med. 45 (5), 549–561.

Winterbourn, C.C., Hampton, M.B., Livesey, J.H., Kettle, A.J., 2006. Modeling the reactions of superoxide and myeloperoxidase in the neutrophil phagosome: implications for microbial killing. J. Biol. Chem. 281 (52), 39860–39869.

Wispe, J.R., Warner, B.B., Clark, J.C., Dey, C.R., Neuman, J., Glasser, S.W., Crapo, J.D., Chang, L.Y., Whitsett, J.A., 1992. Human Mn-superoxide dismutase in pulmonary epithelial cells of transgenic mice confers protection from oxygen injury. J. Biol. Chem. 267 (33), 23937–23941.

Won, H.Y., Sohn, J.H., Min, H.J., Lee, K., Woo, H.A., Ho, Y.S., Park, J.W., Rhee, S.G., Hwang, E.S., 2010. Glutathione peroxidase 1 deficiency attenuates allergen-induced airway inflammation by suppressing Th2 and Th17 cell development. Antioxid. Redox Signal. 13 (5), 575–587.

Wood, Z.A., Schroder, E., Robin Harris, J., Poole, L.B., 2003. Structure, mechanism and regulation of peroxiredoxins. Trends Biochem. Sci. 28 (1), 32–40.

Xu, J., Li, T., Wu, H., Xu, T., 2012. Role of thioredoxin in lung disease. Pulm. Pharmacol. Ther. 25 (2), 154–162.

Yamada, Y., Nakamura, H., Adachi, T., Sannohe, S., Oyamada, H., Kayaba, H., Yodoi, J., Chihara, J., 2003. Elevated serum levels of thioredoxin in patients with acute exacerbation of asthma. Immunol. Lett. 86 (2), 199–205.

Yanagisawa, R., Warabi, E., Inoue, K., Yanagawa, T., Koike, E., Ichinose, T., Takano, H., Ishii, T., 2012. Peroxiredoxin I null mice exhibits reduced acute lung inflammation following ozone exposure. J. Biochem. 152 (6), 595–601.

Yang, D., Song, Y., Wang, X., Sun, J., Ben, Y., An, X., Tong, L., Bi, J., Bai, C., 2011. Deletion of peroxiredoxin 6 potentiates lipopolysaccharide-induced acute lung injury in mice. Crit. Care Med. 39 (4), 756–764.

Yant, L.J., Ran, Q., Rao, L., Van Remmen, H., Shibatani, T., Belter, J.G., Motta, L., Richardson, A., Prolla, T.A., 2003. The selenoprotein GPX4 is essential for mouse development and protects from radiation and oxidative damage insults. Free Radic. Biol. Med. 34 (4), 496–502.

Yatmaz, S., Seow, H.J., Gualano, R.C., Wong, Z.X., Stambas, J., Selemidis, S., Crack, P.J., Bozinovski, S., Anderson, G.P., Vlahos, R., 2013. Glutathione peroxidase-1 reduces influenza a virus-induced lung inflammation. Am. J. Respir. Cell Mol. Biol. 48 (1), 17–26.

Yoboua, F., Martel, A., Duval, A., Mukawera, E., Grandvaux, N., 2010. Respiratory syncytial virus-mediated NF-kappa B p65 phosphorylation at serine 536 is dependent on RIG-I, TRAF6, and IKK beta. J. Virol. 84 (14), 7267–7277.

Young, R.P., Hopkins, R., Black, P.N., Eddy, C., Wu, L., Gamble, G.D., Mills, G.D., Garrett, J.E., Eaton, T.E., Rees, M.I., 2006. Functional variants of antioxidant genes in smokers with COPD and in those with normal lung function. Thorax 61 (5), 394–399.

Yu, M., Lam, J., Rada, B., Leto, T.L., Levine, S.J., 2011. Double-stranded RNA induces shedding of the 34-kDa soluble TNFR1 from human airway epithelial cells via TLR3-TRIF-RIP1-dependent signaling: roles for dual oxidase 2- and caspase-dependent pathways. J. Immunol. 186 (2), 1180–1188.

Zelko, I.N., Mariani, T.J., Folz, R.J., 2002. Superoxide dismutase multigene family: a comparison of the CuZn-SOD (SOD1), Mn-SOD (SOD2), and EC-SOD (SOD3) gene structures, evolution, and expression. Free Radic. Biol. Med. 33 (3), 337–349.

Zmijewski, J.W., Lorne, E., Zhao, X., Tsuruta, Y., Sha, Y., Liu, G., Abraham, E., 2009. Antiinflammatory effects of hydrogen peroxide in neutrophil activation and acute lung injury. Am. J. Respir. Crit. Care Med. 179 (8), 694–704.

Section IV

Physical and Immunological Defenses of the Respiratory Tract

Gary R. Burleson and Richard B. Schlesinger

Chapter 26

Comparative Defense Capability of the Respiratory System: Introduction and Section Overview

Gary R. Burleson and Richard B. Schlesinger

Chapter Outline

"Immunological and Physical Defenses of the Respiratory Tract" examines the various inherent defense systems of the respiratory tract. Cells and tissues will be described both individually and as integrated, orchestrated networks that protect the pulmonary system from attack by allergens, bacteria, viruses, fungi, and parasites. Comparative biology of these defenses will be discussed as well as differences observed from young to aged individuals.

The respiratory system consists of the upper and lower respiratory tracts. The former consists of the nasal cavity, the pharynx, and the larynx, while the latter consists of the trachea, the bronchial tree, and the lungs. The components of the upper and lower respiratory systems together act to perform two major and complementary functions: (1) gas exchange and (2) physical and immunological defenses.

Discussions regarding the physical defenses of the respiratory tract are found in Chapters 27–29. Chapter 27, "Deposition of Particles" by J. Brown discusses mechanisms affecting particle deposition and the resulting effects on the respiratory tract. Chapter 28, "Deposition of Gases and Vapors" by J. Morris, details the means by which the deposition of gases and vapors affects gas exchange within the lungs, from which transport to the circulatory system may then easily occur. Chapter 29, "Mucociliary Clearance" by M. Foster, details the importance of mucociliary clearance in the removal of deposited particulates from the respiratory tract.

Discussions regarding the immunological defenses of the respiratory tract are found in Chapters 30–36. Infection stimulates a cascade of both nonspecific innate and antigen-specific adaptive immunological functions. The major cell subsets within the pulmonary immune system include alveolar macrophages, dendritic cells, T helper lymphocytes, cytotoxic T lymphocytes, memory lymphocytes, natural killer cells, natural killer T lymphocytes, eosinophils, mast cells, B cells, and neutrophils. The immunological defenses mounted by these cell types and their mediators occur in a kinetic fashion and are initiated following infection. While the overall response to infection is discussed in Chapter 30, there are in-depth descriptions of individual components of the immune response in Chapters 31–36. Chapter 30, "Pulmonary Immunology of Infectious Disease," by G.R. Burleson, S.C.M. Burleson, and F.G. Burleson discusses pulmonary immune responses following insult with infectious agents and during defense against neoplastic disease. Chapter 31, "The Effect of Particle Deposition on Immunological Response as Measured by Cytokine Production" by M. Singal, discusses pulmonary defense systems with a focus on inflammatory and oxidant injury responses in the lung following exposure to airborne particulate irritants and/or allergenic materials. Chapter 32, "Pulmonary Macrophages" by D.L. Laskin, R. Malaviya, and J.D. Laskin, discusses the different pulmonary macrophage subsets; location in the alveoli, interstitium, airways, pleura, and vasculature; unique phenotypes, functions, and characteristics; as well as their dual role with regard to inflammatory responses during tissue injury or infection, and anti-inflammatory responses during wound repair. Chapter 33, "Pulmonary Dendritic Cells" by D.N. Cook and H. Nakano, discusses the characteristics and important functions of pulmonary dendritic cells, including

Comparative Biology of the Normal Lung. http://dx.doi.org/10.1016/B978-0-12-404577-4.00026-6

their potent antigen presenting ability, ability to migrate from peripheral tissue to draining lymph nodes containing large numbers of naïve T cells, and protective role against responses to self-antigens and autoimmunity through interactions with regulatory T cells (T_{Regs}). In Chapter 34, P. Katwa and J.M. Brown discuss the immunological contributions of pulmonary mast cells, including their roles as sentinels of innate immunity, regulators of adaptive immune responses, and classic mediators of allergic disease. Katwa and Brown also outline and discuss new research that introduces mast cells as important mediators of immunity in addition to their well-described function in allergic diseases. Chapter 35, "The Immune Basis of Allergic Lung Disease" by S.C.M. Burleson, R. Fick, M.D. Mannie, S.G. Olmstead, and M.R. Van Scott, describes the cytokines, cell types, and mechanisms underlying immunopathology related to allergic asthma, as well as the effects of aging on allergic responses. Chapter 36, "Role of Genetic Factors in Pulmonary Disease Susceptibility" by B. Yucesoy, V.J. Johnson, and M.I. Luster, details the genetic contribution to immune system response, with a focus on immunogenetic predisposition to pulmonary disorders.

Comparative pulmonary immune functions between humans and various species is presented in these chapters when available. However, as it is unethical to deliberately expose healthy human volunteers to agents that may result in mortality or impairment of health, the Food and Drug Administration enacted the Animal Rule to allow studies in animals for efficacy determinations for new drugs or biologic products, resulting in animal research accounting for the vast majority of new insights into immune system function (FDA, Department of Health and Human Services, 2002). Furthermore, the vast majority of knowledge regarding pulmonary immunology has been obtained specifically in rodents due to ethical reasons associated with using higher-order species such as dogs or nonhuman primates, as well as the relative lack of transgenics and immunological reagents available for some of the higher-order species. Additionally, the cost and lack of containment facilities for larger animals, particularly nonhuman primates, impedes research in these species. For these reasons, much of the knowledge gained relating to pulmonary immunology has derived from studies in rodents, particularly mice, though species comparisons are discussed wherever possible.

Recognition of the immunological differences in different species is important, however, as ignorance of differences in expression or function can have disastrous consequences. A well-known example is that of the poor outcome of the TeGenero clinical trial due to an unrecognized difference in Siglec (sialic acid-binding immunoglobulin-type lectins) expression between monkey and man. While such difficulties have occasionally been observed in translating animal research to the clinic, when one considers the vast number of single nucleotide polymorphisms (SNPs) in different human individuals, one might conclude that even humans themselves may not be good models for humans. An exciting new and recent alternative for studies of immunological evaluations, however, involves humanized mice. Humanized mice offer the advantage of a human immune system within a mouse, the use of human reagents and biomarkers for research, a uniform population lacking differences due to SNPs, and low cost of housing. This will allow studies of human immune functions as well as the interaction of infectious agents with the human immune system. Indeed, humanized mice have been used in studies with HIV-1, HTLV-1, and EBV, as reviewed by Ito et al. (2012) and Shultz et al. (2012). While still expensive due to the relative newness of the technology, as the field of immunology moves forward, exciting progress can be expected.

REFERENCES

FDA, Department of Health and Human Services, 2002. Fed. Regist.: Rules Regul. 37988–37998.

Ito, R., Takahashi, T., Katano, I., Ito, M., 2012. Current advances in humanized mouse models. Cell. Mol. Immunol. 9 (3), 208–214.

Shultz, L.D., Brehm, M.A., Garcia-Martinez, J.V., Greiner, D.L., 2012. Humanized mice for immune system investigation: progress, promise and challenges. Nat. Rev. Immunol. 12 (11), 786–798.

Chapter 27

Deposition of Particles

James S. Brown

Chapter Outline

1. INTRODUCTION

Knowledge of sites where particles of different sizes deposit in the respiratory tract and the amount of deposition therein is necessary for understanding and interpreting the health effects associated with exposure to particles. The site of particle deposition within the respiratory tract affects particle retention and the surface dose of particles as well as potential systemic distribution of particles or their constituents. This chapter focuses on the mechanisms of particle deposition and on issues that may lead to differences in total and regional deposition of particles among individuals and species.

A particle is deposited when it moves from the airway lumen to the wall of an airway. The deposition of spherical particles occurs mainly by three mechanisms: diffusion, sedimentation, and impaction. The efficiency of particle deposition in the respiratory tract depends, in large part, on inhaled particle size and shape, route of breathing (nasal or oronasal), tidal volume (V_T), breathing frequency (f), and respiratory tract morphology. The route of breathing—the distinction between air passing through the nose versus the mouth—is important because the nasal passages more effectively remove inhaled particles than the oral passage. Respiratory tract morphology, which affects particle transport and deposition, varies between species, the size of an animal or human, and health status.

Before discussing particle deposition further, the remainder of this introduction will cover the fundamentals related to particle size characterization, respiratory tract structure and function, particle inhalability, and dose metrics. Because the mechanisms of deposition and major factors affecting deposition do not differ between species, most of the discussion in this chapter focuses on deposition in humans. However, some important differences between various species are described. Particle clearance from the respiratory tract is discussed within this chapter only as it relates to the interpretation of studies examining regional particle deposition.

1.1 Size Characterization of Inhaled Particles

Particle size is a major determinant of the fraction of inhaled particles depositing in and cleared from various regions of the respiratory tract. The distribution of particle sizes in an aerosol is typically described by the lognormal distribution (i.e., the situation in which the logarithms of particle diameter are distributed normally). The geometric mean is the median of the distribution and the variability around the median is the geometric standard deviation (GSD or σ_g), which is commonly calculated as

$$\text{GSD} = \sigma_g = \frac{d_{84.13\%}}{d_{50\%}} = \frac{d_{50\%}}{d_{15.87\%}} \tag{1}$$

where $d_{15.87\%}$, $d_{50\%}$, and $d_{84.13\%}$ are the particle diameters approximately associated with the 16th, 50th (i.e., the median), and the 84th percentiles from the cumulative frequency distribution of particle sizes. By definition, the GSD is always greater than 1. The particle size associated with any percentile of the distribution, d_i, is given by

$$d_i = d_{50\%}\sigma_g^{z(P)} \tag{2}$$

where $z(P)$ is the normal standard deviate for a given cumulative probability provided by a standard normal table or Z table. In most cases, the aerosols to which people are

Comparative Biology of the Normal Lung. http://dx.doi.org/10.1016/B978-0-12-404577-4.00027-8

naturally exposed are polydisperse (GSD > 1.15). By contrast, most experimental studies of particle deposition and clearance in the lung use monodisperse aerosols (GSD ≤ 1.15). Ambient aerosols may also be composed of multiple size modes; each mode should be described individually by its specific median diameter and GSD.

Aerosol size distributions may be measured and described in various ways. When a distribution is described by counting particles, the median is called the count median diameter (CMD). On the other hand, the median of a distribution based on particle mass in an aerosol is the mass median diameter. Impaction and sedimentation of particles in the respiratory tract depend on a particle's aerodynamic diameter (d_{ae}), which is the size of a sphere of unit density that has the same terminal settling velocity as the particle of interest. The size distribution is frequently described in terms of d_{ae} as the mass median aerodynamic diameter (MMAD), which is the median of the distribution of mass with respect to aerodynamic equivalent diameter. Airborne fibers (length/diameter ratio ≥3) can exceed 150 μm in length and appear to be relatively stable in air. This is because their aerodynamic size is determined predominantly by their diameter, not their length. Fibers longer than 10 μm can deposit by interception, and when aligned with the direction of airflow, may penetrate deep into the respiratory tract.

Alternative descriptions should be used for particles with actual physical sizes below ≈0.5 μm because, for these sized particles, aerodynamic properties become less important and diffusion becomes more important. For these smaller particles, diffusive deposition depends on their physical diameter or CMD because diffusivity is not a function of particle density. For small, irregularly shaped particles and aggregates, diffusive deposition depends on the thermodynamic equivalent diameter (i.e., the diameter of a spherical particle that has the same diffusion coefficient in air as the particle in question).

The term "ultrafine particle" has traditionally been used by the aerosol research and occupational and environmental health communities to describe airborne particles or other laboratory-generated aerosols used in toxicological studies that are <100 nm (0.1 μm) in size (based on physical size, diffusivity, or electrical mobility). Generally consistent with the definition of an ultrafine particle, the International Organization for Standardization (ISO) defined a nanoparticle as an object with all three external dimensions in the nanoscale—that is, from approximately 1 to 100 nm (ISO, 2008). The ISO also defined a nano-object as a material with one or more external dimensions in the nanoscale. The terms nanoparticle and ultrafine particle have been used rather synonymously in the recent literature. However, the terms nanoparticle and nano-object are more commonly associated with engineered materials that are created for consumer products and industrial applications. With the current interest in nanotechnologies, many nano-objects have been created by manipulating materials at the atomic or molecular scale for the purpose of forming new materials, structures, and devices that exploit the unique physical and chemical properties associated with their nanoscale. In general, the deposition of unique nano-objects (e.g., dots, hollow spheres) depends on their thermodynamic equivalent diameter. For nano-rods, fibers, and tubes, the ratio of length to diameter may also influence the efficiency of deposition.

1.2 Structure of the Respiratory Tract Related to Particle Deposition

Various aspects of the respiratory tract relevant to particle deposition are discussed in other chapters of this volume. Some basic terminology and concepts are provided herein for completeness. Briefly, the basic structure of the human respiratory tract is illustrated in Figure 1. In the literature, the terms extrathoracic (ET) region and upper airways are generally used synonymously. The term lower airways is used to refer to the intrathoracic airways—that is, the combination of the tracheobronchial (TB) region which is the conducting airways and the alveolar region, which is the functional part or parenchyma of the lung. A review of interspecies similarities and differences in the structure and function of the respiratory tract was provided by Phalen et al. (2008). Although the structure varies, the illustrated anatomic regions are common to all mammalian species, with the exception of the respiratory bronchioles. Respiratory bronchioles—the transition region between ciliated and fully alveolated airways—are found in humans, dogs, ferrets, cats, and monkeys. They are absent in rats and mice and abbreviated in hamsters, guinea pigs, oxen, sheep, and pigs. The branching structure of the ciliated bronchi and bronchioles also differs between species, from being a rather symmetric and dichotomous branching network of airways in humans to a more monopodial branching network in other mammals.

Numerous models of TB airway morphology have been described for humans and several animal species for use in predicting particle deposition. For instance, Weibel (1963) provided a simplified description of the intrathoracic airway structures in an adult male having the lungs inflated to 4.8 l. In this model, the lung consists of 23 symmetric and dichotomously branching airway generations. Given that the dimensions of unsupported airways (i.e., those not having cartilaginous rings) in the lung vary with lung inflation, airway dimensions are adjusted depending on the degree of lung inflation. In general, these changes in airway dimensions are thought to be invariant to direction; hence, changes in length and radius are proportional to the cube root of changes in lung volume (Ultman, 1985). When modeling particle deposition, the airways are typically

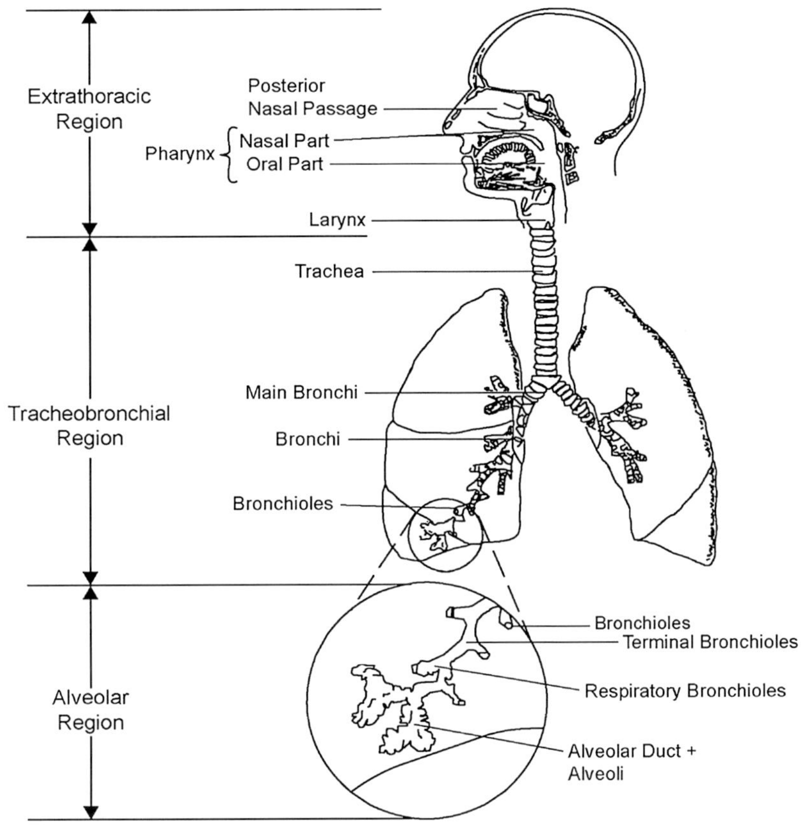

FIGURE 1 **Diagrammatic representation of respiratory tract regions in humans.**

scaled to a lung volume of functional residual capacity plus half of the tidal volume.

Morphologic models vary in complexity from that of an idealized whole lung (typical or single path), such as the Weibel (1963) model, to lobar specific morphologies (multipath) or stochastic models. Yeh (1980) provided whole lung and lobar morphologies and predicted particle deposition for rats, Syrian hamsters, beagle dogs, and humans. Human and rat portions of the Yeh (1980) report are provided in Yeh et al. (1979) and Yeh and Schum (1980), respectively. Several mouse models have also become available. Oldham and Robinson (2007) provided a typical path lung morphology and predicted particle deposition in the Balb/c mouse. However, for the monopodial airway branching structure of the mouse and rat, some have recommended characterizing the properties of a given airway by its diameter rather than by a theoretically assigned generation number, as done in most morphologic models. This is due to greatly differing airway dimensions and branching angles between major and minor daughter airways following a given bifurcation in the monopodial structure of the rat and mouse lung. Additionally, the length of the path that is traversed from the trachea to a terminal bronchiole also varies widely across the lung. Madl et al. (2010) provided a stochastic lung model developed from the same lung casts as Oldham and Robinson (2007). This model allows for independent probabilities for reaching an acinar unit, branching angles, and ratios of daughter-to-parent diameters. These stochastic models may overcome the limitation observed by Mercer et al. (1994), whereby not accounting for heterogeneity in airway number and branching patterns causes morphology models to overestimate by twofold the luminal surface area of the small bronchi and bronchioles.

1.3 Route of Breathing

Another species difference relevant to particle dosimetry is the route of breathing. Rodents are obligate nose breathers, whereas most humans are oronasal breathers who breathe through the nose when at rest and increasingly through the mouth with increasing activity level. There is interindividual variability in the route by which people breathe. For example, in one study (Niinimaa et al., 1981), most people, 87% (26 of 30) breathed through their nose until an activity level was reached where they switched to oronasal breathing. Thirteen percent (4 of 30) of the subjects, however, were oronasal breathers even at rest. These two subject groups are commonly referred to in the literature (e.g., ICRP, 1994) as "normal augmenters" and "mouth breathers," respectively. Others have reported a more gradual increase in oronasal breathing, with trends for effects of age and sex. Bennett et al. (2003) reported that males ($n = 11$; 22 ± 4 years) had a greater oral contribution than females ($n = 11$; 22 ± 2 years) at rest (87 vs 100% nasal, respectively) and during exercise (45 vs 63% nasal at 60% maximum workload, respectively). Bennett et al. (2008) reported that children ($n = 12$; 6–10 years) have a greater oral contribution than adults ($n = 11$; 18–27 years) at rest (68 vs 88% nasal, respectively) and during exercise (47 vs 59% nasal at 40% maximum workload, respectively). Becquemin et al. (1999) have also reported children to have a greater fraction of oral breathing than adults at rest and during exercise. In a study of 37 subjects from 7 to 72 years of age, James et al. (1997) reported that two subjects (5.4%) breathed purely through the mouth, but provided no other characteristics of these individuals. Disease can affect preferred breathing route. For example, Chadha et al. (1987) found that the majority (11 of 12) of patients with asthma or allergic rhinitis breathe oronasally, even at rest. In humans, the penetration of particles greater than 1 μm d_{ae} into the lower respiratory tract is more affected by route of breathing than age, sex, activity level, or breathing pattern (Brown et al., 2013).

1.4 Particle Inhalability

To potentially become deposited in the respiratory tract, particles must first be inhaled. The inspirable or inhalable particulate mass fraction of an aerosol is the fraction of ambient airborne particles that can enter the uppermost respiratory tract compartment, the head (Soderholm, 1985). Inhalability is a generally a function of d_{ae}. Thus, inhalability is negligible and has not been characterized for small particles (those <1 μm) where particle inertia minimally affects the ability of particles to follow changes in the direction of airflow.

The American Conference of Governmental Industrial Hygienists (ACGIH) and the International Commission on Radiological Protection (ICRP) have established inhalability criteria for humans (ACGIH, 2005; ICRP, 1994). For the ACGIH criterion, inhalability is 97% for $d_{ae} = 1$ μm, 87% for $d_{ae} = 5$ μm, and 77% for $d_{ae} = 10$ μm; this plateaus at 50% for d_{ae} greater than ~40 μm. The ICRP criterion, which also plateaus at 50% for very large d_{ae}, does not become of real importance until $d_{ae} = 5$ μm, where inhalability is 97%. The ACGIH and ICRP criteria are based on experimental inhalability data for $d_{ae} \leq 100$ μm at wind speeds between 1 and 8 m/s. These criteria are indifferent to the route of breathing and assume random orientation with respect to wind direction.

The route of breathing affects inhalability in relatively calm air (velocity ≤ 0.3 m/s). Dai et al. (2006) reported

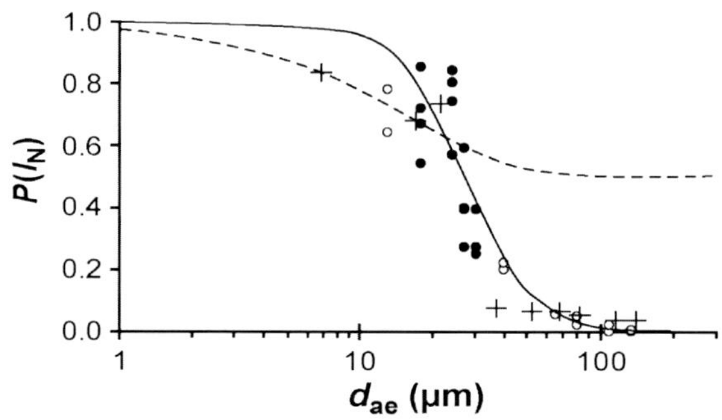

FIGURE 2 Nasal inhalability, $P(I_N)$, as a function of aerodynamic particle diameter, d_{ae}. For $d_{ae} > 10$ μm, the experimental data and the Ménache et al. (1995) inhalability function show a rapid decrease in $P(I_N)$ to near zero for $d_{ae} \geq 100$ μm ● Breysse and Swift (1990), + Hinds et al. (1998), ○ Hsu and Swift (1999), —— Ménache et al. (1995), - - - ACGIH (2005).

slightly lower nasal particle inhalability in humans during moderate exercise than rest (e.g., 89.2 vs 98.1% for 13 μm particles, respectively). Nasal particle inhalability was similar between an adult and 7-year-old child (Hsu and Swift, 1999). Inhalability into the mouth from calm air in humans also becomes important for $d_{ae} > 10$ μm (Anthony and Flynn, 2006; Brown, 2005). Unlike the inhalability from high wind speeds, which plateaus at 50% for d_{ae} greater than ~40 μm, particle inhalability from calm air continues to decrease toward zero with increasing d_{ae}. Human nasal inhalability in calm air versus higher wind speed criterion is illustrated in Figure 2. Human oral inhalability shows a similar decline, but at larger particle sizes than observed for nasal inhalability (Brown, 2005). Regardless of the particle size being considered, the accurate assessment of delivered doses clearly necessitates the use of inhalability criteria appropriate for ambient wind speeds and direction relative to the exposed individual. A review of particle inhalability data for humans is available from Millage et al. (2010).

Inhalability data in laboratory animals, such as rats, are only available for breathing from relatively calm air. For nasal breathing, inhalability becomes an important consideration for d_{ae} of above 1 μm in rodents and 10 μm in humans (Ménache et al., 1995). The inhalability of particles having d_{ae} of 2.5, 5, and 10 μm is 80, 65, and 44% in rats, respectively, whereas it only decreases to 96% for d_{ae} of 10 μm in humans during nasal breathing (Ménache et al., 1995). Figure 3 compares estimates of nasal inhalability in humans versus rodents. Asgharian et al. (2003) suggested that inhalability by rats decreased more rapidly with increasing d_{ae} than reported by Ménache et al. (1995) and that inhalability decreased with increasing breathing rate. For example, Asgharian et al. (2003) reported the inhalability of 10 μm particles as 23% and 10% at minute ventilations of 3.33 ml/s and 14 ml/s, respectively, whereas Ménache et al. (1995) reported 44% inhalability. Thus, inhalability is a

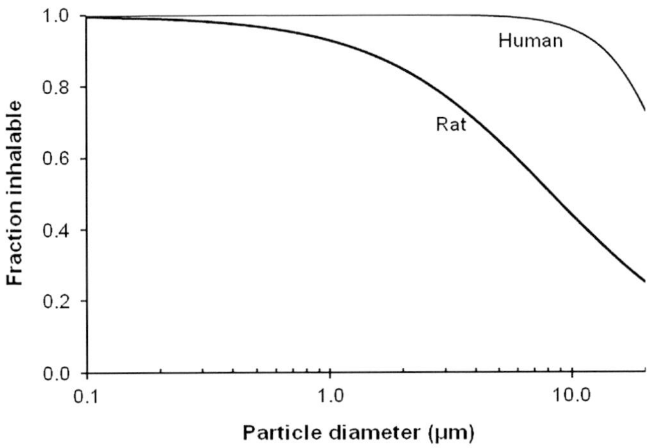

FIGURE 3 **Nasal inhalability curves for human and rat showing the fraction particles that enter the nose.** *Based on Table 1 from Ménache et al. (1995).*

particularly important consideration for rodent exposures and, therefore, for interspecies comparisons.

As discussed in Section 1.1, particle size distributions are typically log-normal. On a mass basis, the total probability of inhalability, $I(d_{50\%}, \sigma_g)$ for a polydisperse aerosol having a median particle diameter ($d_{50\%}$, typically the MMAD) and geometric standard deviation (σ_g), is the integral of the inhalability across all particle sizes in the aerosol:

$$I(d_{50\%}, \sigma_g) = \int_0^\infty I(d_i)\rho(d_i)\delta d_i \qquad (3)$$

where $I(d_i)$ is the inhalability of particles having an aerodynamic diameter of d_i and $\rho(d_i)$ is the fraction aerosol distribution associated with the interval δd_i.

Equation (3) can be approximated by summing the inhalability probabilities for particles at known intervals or percentiles of the aerosol size distribution. Using a 1% interval, the approximation is

$$I(d_{50\%}, \sigma_g) \approx \frac{1}{100} \sum_{P=0.01}^{0.99} I(d_i) \qquad (4)$$

where $I(d_i)$ is the inhalability of the particle size associated with a given percentile P of the size distribution, as calculated by Eqn (2).

1.5 Dose and Dose Metrics

Assuming a constant exposure concentration, breathing rate, and aerosol particle size distribution, the total particle exposure or intake dose (ID) is given by

$$\text{ID} = C \times f \times V_T \times I(d_{50\%}, \sigma_g) \times t \qquad (5)$$

where C is the mass concentration of the aerosol, f is breathing frequency, V_T is tidal volume, $I(d_{50\%}, \sigma_g)$ is aerosol inhalability, and t is the duration of exposure. As illustrated in Figure 3, $I(d_{50\%}, \sigma_g)$ should be considered for comparisons across species (e.g., human vs rat), although this parameter should be negligible for particles under 1 μm d_{ae}. Intake doses characterized by Eqn (5) are commonly normalized to body mass (Alexander et al., 2008). This may be particularly appropriate for soluble particles or materials expected to have systemic effects. Although C was specified as having units of particle mass per unit volume, other metrics, such as particle surface area or number of particles per unit volume, may be desired, especially for smaller particle sizes (e.g., <0.1 μm). Equation (5) is limited in that it does not recognize that there are within-species differences as a function of particle size in total deposition (whole lung) and regional deposition (e.g., between TB and alveolar region) of particles.

The particle mass dose in a specific region (D_r) of the respiratory tract resulting from the inhalation of aerosol may be given as follows:

$$D_r = \text{ID} \times \text{DF}_r \qquad (6)$$

where ID is the intake dose from Eqn (5) and DF_r is the fraction of inhaled particles depositing in region r of the respiratory tract. Analogous to Eqn (4) for inhalability, the DF_r in Eqn (6) can be calculated for a polydisperse aerosol by estimating the deposition fractions for a series of monodisperse aerosols as

$$\text{DF}_r(d_{50\%}, \sigma_g) \approx \frac{1}{100} \sum_{P=0.01}^{0.99} \text{DF}_r(d_i) \qquad (7)$$

where $\text{DF}_r(d_i)$ in the summation is the deposition fraction in a region of the particle size associated with a given percentile P of the size distribution, as calculated by Eqn (2). Depending on health endpoints and particle size, the most appropriate dose metric choice for D_r may be mass, particle surface area, or number of particles deposited. D_r may also be normalized to factors such as lung weight or surface area of specific regions of the respiratory tract. Because all of the variables potentially change over time, Eqns (5) and (6) are most appropriate for short duration exposures. Within an individual, the variability in DF_r over time is largely attributable to variations in inhaled particle size, f, V_T, and route of breathing (ICRP, 1994). Intersubject and interspecies variability in DF_r is additionally affected by morphologic differences in the size and structure of the respiratory tract.

Some general species information relevant to particle dosimetry (e.g., breathing parameters and respiratory surface areas) is provided in Table 1. The data in this table are for gross comparison among resting adults because specific strains are not individually characterized, nor are changes

TABLE 1 Typical Respiratory Parameters and Body Weights Among Animals and Humans

Species	Breathing Frequency (1/min)	Tidal Volume (ml)	Minute Ventilation (ml/min)	Functional Residual Capacity (ml)	Alveolar Surface Area (m^2)	Body Weight (kg)
Mouse (restrained)	290[a]	0.22[a]	64[a]	0.5[e]	0.05[f]	0.02[f]
Mouse (unrestrained)	145[a]	0.19[a]	27[a]	0.5[e]	0.05[f]	0.02[f]
Rat	102[b]	2.1[b]	214[b]	3.5[e]	0.4[f]	0.4[f]
Dog	22[c]	175[c]	3600[c]	500[c]	52[f]	16[f]
Human (male)	12[d]	625[d]	7500[d]	3300[d]	140[d]	73[d]
Human (female)	12[d]	444[d]	5330[d]	2700[d]	100[g]	60[d]

[a]*Méndez et al. (2010).*
[b]*de Winter-Sorkina and Cassee (2002).*
[c]*Mauderly (1979).*
[d]*ICRP (1994).*
[e]*Tankezawa et al. (1980), anesthetized animals.*
[f]*Stone et al. (1992).*
[g]*Alveolar surface area of male scaled by ratio of total lung capacity, i.e., 4.97 ÷ 6.98.*

with animal age characterized. Across species, ventilation rates increase with increases in activity. Within species, there are also differences among strains in breathing patterns and rates. Furthermore, stress due to experimental protocols may alter breathing patterns differentially among species. In rats, Mauderly and Kritchevsky (1979) reported restraint to cause increased f and decreased V_T, while minimally affecting overall minute ventilation. In mice, Méndez et al. (2010) reported restrained animals to have approximately 2.4 times the minute ventilation of unstrained animals (27 and 64 ml/min, respectively). Most of this increase in minute ventilation came from a doubling of f from 145/min to 290/min. In unrestrained but apparently stressed mice, Alessandrini et al. (2008) reported a minute ventilation of 106 ml/min and f of 500/min. Thus, even within one species, there can be large differences in breathing conditions between studies. Breathing patterns and minute ventilation must both be considered to accurately assess particle deposition fractions and dose rates.

For chronic exposures, it is necessary to consider the retained dose. The particle dose retained in a region of the lung is determined by the balance between rate of input and the rate of removal. The particle burden (B_r) in a region of lung may be expressed as

$$B_r(t) = \dot{D}_r(t - \Delta t)\Delta t + B_r(t - \Delta t)\left[\exp(-\lambda_r \Delta t)\right] \quad (8)$$

where $\dot{D}_r$ is the rate of deposition per unit time in region r, t is time, and λ_r is the clearance rate constant for region r, Δt is the time increment for the calculations [~1% (or less) of the clearance halftime (i.e., $0.693/\lambda_r$) of the region]. $\dot{D}_r$ is calculated as D_r in Eqn (6), except it is calculated for discrete Δt where parameters (namely, f, V_T, route of breathing, and DF_r) are relatively constant.

It should be noted that transfer into region r from another region may also occur. Such situations in which a region receives a portion of its burden from another region are common in the lung, for example, mucus clearance of the segmental bronchi into the lobar bronchi, which clear into the main bronchi, which in turn clear into the trachea. In addition, the clearance from one region can transfer burden into more than one other compartment; for example, soluble particles in the airways may be cleared into the blood as well as via the mucus. Here, the focus is on particle deposition; however, multiple pathways for clearance of insoluble particles exist, such as from the alveoli into the lymph and into the terminal bronchioles via macrophages. There are also considerable species differences in rates of clearance that should be considered for interspecies extrapolations evaluating chronic exposure scenarios. Some limited discussion of particle clearance will follow the discussion of particle deposition, mainly for the interpretation of regional deposition data.

2. PARTICLE DEPOSITION

Inhaled particles may be either exhaled or deposited in the ET, TB, or alveolar region. The fraction of inhaled aerosol becoming deposited in the respiratory tract of humans and other species has been measured experimentally. By using light scattering or particle counting techniques to quantify the amount of aerosol in inspired and expired breaths, studies have characterized total particle deposition for varied breathing conditions and particle sizes. The vast majority of in vivo data on regional particle deposition has been obtained by scintigraphic methods, where external monitors are used to measure gamma emissions from radiolabeled particles. These scintigraphic data have shown highly variable regional deposition, with sites of highly localized deposition or "hot spots" in the obstructed lung relative to the healthy lung. However, even in the healthy lung, "hot spots" occur in the region of airway bifurcations. Mathematical models aid in predicting the mixed effects of particle size, breathing conditions, and lung volume on total and regional deposition. Experimentally, however, there is considerable interindividual variability in total and regional deposition, even when inhaled particle size and breathing conditions are strictly controlled. Section 2.4 provides more detailed information on factors affecting deposition among individuals.

2.1 Mechanisms of Deposition

Particle deposition in the lung is predominantly governed by diffusion, impaction, and sedimentation. Most discussion herein focuses on these three dominant mechanisms of deposition. Simple interception is an important mechanism of macro- and nanosized fiber deposition. Electrostatic and thermophoretic forces as mechanisms of deposition have not been thoroughly evaluated. Both experimental studies and mathematical models have demonstrated that breathing patterns can dramatically alter regional and total deposition for all sized particles. The combined processes of aerodynamic and diffusive (or thermodynamic) deposition are important for particles in the range of 0.1−1 μm. Aerodynamic processes predominate above and thermodynamic processes predominate below this range. Some generalizations with regard to deposition by these mechanisms follows, but these should not be viewed as absolute rules.

Diffusive deposition, by the process of Brownian diffusion, is the primary mechanism of deposition for particles having physical diameters of less than 0.1 μm. Diffusion—or more specifically, a particle's diffusion coefficient—is a function of the physical diameter of a particle. Particle density does not affect the diffusion coefficient. For irregularly shaped particles, deposition is dependent on the thermodynamic equivalent diameter (i.e., the diameter of a spherical particle that has the same diffusion coefficient in air as the particle in question). Diffusive deposition efficiency in a region of the respiratory tract is a function of $(Dt)^{0.5}$, where D is the diffusion coefficient and t is time in the region. For particles having physical diameters of roughly between 0.05 and 0.1 μm, diffusive deposition occurs mainly in the small distal bronchioles and the pulmonary region of the lung. However, with further decreases in particle diameter below ~0.05 μm, increases in particle diffusivity shift more deposition proximally to the bronchi and ET regions.

Governed by inertial or aerodynamic properties, impaction and sedimentation increase with d_{ae}^2 times the mean linear velocity and residence time within an airway, respectively. Recall that d_{ae} is the size of a unit density sphere that has the same terminal settling velocity as the particle of interest. More generally, because cross-sectional areas of the airways and ventilation patterns are not necessarily known or homogeneous, the probability of impaction is also considered to be a function of d_{ae}^2 times inspiratory flow. When a particle has sufficient inertia, it is unable to follow changes in flow direction and strikes a surface, thus depositing by the process of impaction. Impaction occurs predominantly at bifurcations in the proximal airways, where linear velocities and secondary eddies are at their highest. Sedimentation, caused by the gravitational settling of a particle, is most important in the distal airways and pulmonary region of the lung. In these regions, residence time is the greatest, and the distances that a particle must travel to reach the wall of an airway are minimal.

A few studies have investigated the deposition of particles in the "transition range" of 0.1−1 μm (e.g., Gebhart, 1992; Gebhart and Heyder, 1985; Heyder et al., 1985). Across this range, both diffusive and aerodynamic processes are rather minimal, with diffusivity decreasing and inertial forces increasing with increasing particle size. Heyder et al. (1985) noted that the particle size of minimal deposition is the diameter of a sphere that is equally deposited by diffusion and sedimentation. They observed that this size of minimal deposition is independent of time in a system but dependent on particle density. The efficiency of deposition (η) of a given particle size in this range is given by

$$\eta = \eta_D + \eta_G - \frac{\eta_D \eta_G}{\eta_D + \eta_G} \tag{9}$$

where η_D and η_G are the efficiency of deposition by diffusion and gravitational settling, respectively. Diffusive and sedimentary deposition mechanisms compete in the distal airways and pulmonary region where residence time is sufficient for particle losses from inhaled air to occur.

Figure 4 shows predicted deposition of 0.1-, 1-, and 5-μm particles in the TB airways and alveolar region as a function of mechanism of deposition. Predictions in Figure 4 used the deposition equations and modeling approach of Hofmann et al. (1989) and Weibel's (1963) standard lung morphology scaled to a lung volume of 3375 ml. Deposition in the proximal airways occurs mainly by impaction, which increases with particle size and linear flow velocity. Highly localized sites of deposition within the bronchi are also predicted (see Section 2.2). For the size particles and breathing pattern illustrated in Figure 4, sedimentation and diffusion predominate in the distal airways, where linear velocities are low and particle residence times are high. The total intrathoracic deposition (area under the curve) for 5-μm particles is nearly six times greater than 1-μm particles; deposition of 0.1-μm particles is 2.4 times greater than for 1-μm particles.

Interception is an important mechanism of deposition for large fibers. Airborne fibers (length/diameter ratio ≥3)

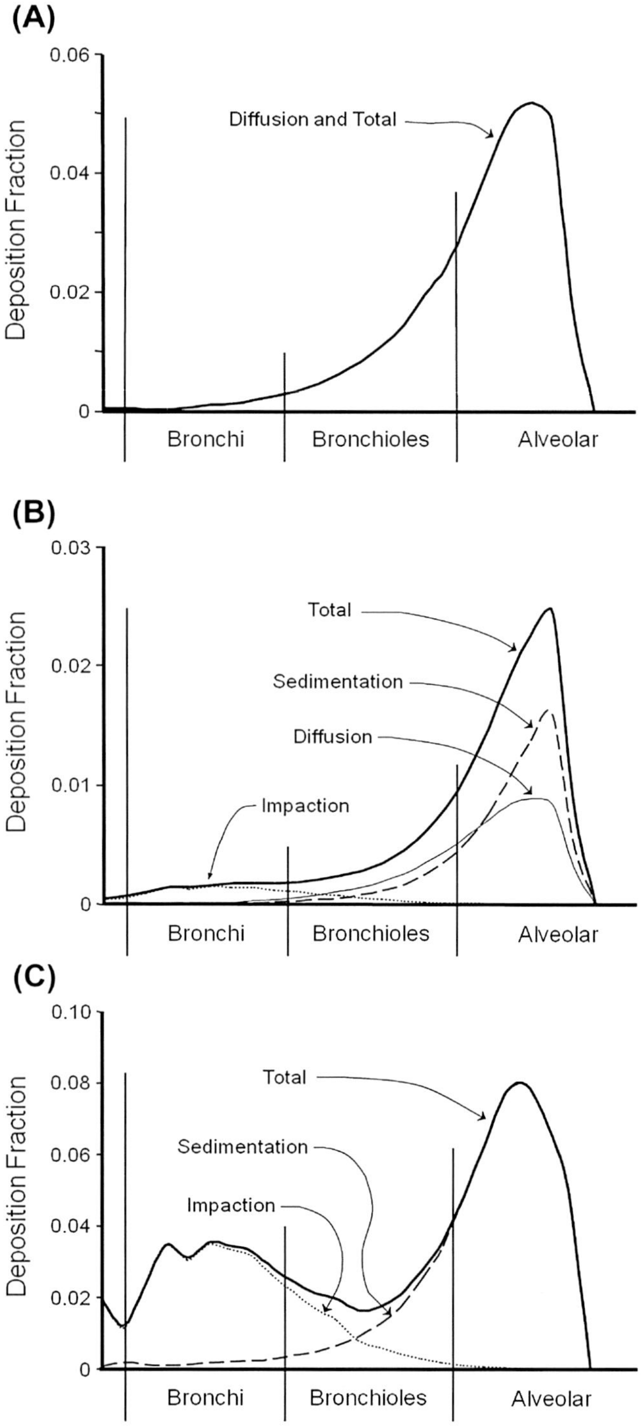

FIGURE 4 Predicted intrathoracic deposition of 0.1 μm (A), 1 μm (B), and 5 μm (C) particles (V_T = 750 ml; f = 15/min). The respiratory tract regions are designated as bronchi, bronchioles, and alveolar and refer to generations 1–8, 9–15, and 16–23, respectively.

can exceed 150 μm in length and appear to be relatively stable in air. This is because their aerodynamic size is determined predominantly by their diameter, not their length. On inhalation, fibers become aligned with the direction of airflow and may penetrate deep into the respiratory tract. Asgharian and Yu (1989a, 1989b) theoretically evaluated fiber (length of 0.5–100 μm) deposition in humans and rats. In the lower airways of humans, fiber deposition efficiency increased with distal progression and with fiber length. In the TB airways, neglecting the effect of interception resulted in an underestimation of deposition, particularly for long (20–100 μm) and narrow (<1 μm) fibers. In the pulmonary region, mainly due to removal in the TB airways by interception, neglecting the effect of interception resulted in overestimation in the deposition of long (20–100 μm) fibers. The aspect ratio (length/diameter) of nanofibers may also affect deposition. Relative to spherical particles (e.g., 20 nm), perhaps due to Brownian rotation decreasing distance to respiratory surfaces, small aspect ratios (e.g., 4; 80-nm length, 20 nm diameter) will likely increase nanofiber deposition in the head and TB airways. For longer aspect ratios (10–100; 200- to 2000-nm length, 20-nm diameter), deposition efficiency will likely be decreased relative to spherical particles (e.g., 20 nm) as nanofibers become aligned with the direction of airflow.

The electrical charge on some particles may result in an enhanced deposition over what would be expected based on size alone. With an estimated charge of 10–50 negative ions per 0.5-μm particle, Scheuch et al. (1990) found total deposition in humans (V_T = 500 ml, f = 15/min) to increase from 13.4% (no charge) to 17.8% (charged). This increase in deposition is thought to result from image charges induced on the surface of the airway by charged particles. Yu (1985) estimated a charge threshold level of 12, 30, and 54 negative ions for 0.3, 0.6, and 1.0 μm particles, respectively, above which deposition would be enhanced relative to no charge. Employing hollow airway casts of the human TB tree to assess deposition of ultrafine (0.02 μm) and fine (0.125 μm) particles, Cohen et al. (1998) found the deposition of singly charged particles to be five to six times that of particles having no charge and two to three times that of particles at Boltzmann equilibrium. Electrostatic deposition is generally considered negligible for particles below 0.01 μm because so few of these particles carry a charge at Boltzmann equilibrium. This mechanism is also thought to be a minor contributor to overall particle deposition, but it may be important in some laboratory studies due to specific aerosol generation techniques, such as nebulization. Laboratory methods, such as the passage of aerosols through a Kr-85 charge neutralizer prior to inhalation, are commonly used to mitigate this effect.

Thermophoretic forces on particles occur due to temperature differences between respired air and respiratory tract surfaces. Temperature gradients of around 20°C are thought to produce sufficient thermophoretic force to oppose diffusive and electrostatic deposition during inspiration, and to perhaps augment deposition by these mechanisms during expiration (Jeffers, 2005). Thermophoresis is only relevant in the extrathoracic and large bronchi airways and reduces to zero as the temperature gradient decreases deeper in the lung. Theoretical analysis of thermophoresis has been done for smooth-walled tubes and is important over distances that are several orders of magnitude smaller than the diameter of the trachea. The alteration of the flow patterns by airway surface features, such as cartilaginous rings, may affect particle transport and deposition over far greater distances than does thermophoretic force.

2.2 Deposition Patterns

Predicted total and regional particle deposition in several mammalian species are illustrated in Figure 5. For all the species illustrated in Figure 5, ET deposition was based on experimental data at specific particle sizes or empirical fits to experimental data, while TB and pulmonary deposition were based on theoretical losses by diffusion, sedimentation, and impaction in species specific models of lower airways morphology. The predicted deposition for the human (male), mouse (unrestrained), and rat are for respiratory parameters in Table 1 using the Multi-Path Particle Dosimetry model (MPPD; Version 3.0, ©2013).[a] The effects of physiologic parameters on deposition in humans and rats free of respiratory disease were also described by de Winter-Sorkina and Cassee (2002). The predicted deposition for the dog ($V_T = 170$ ml, $f = 11.7$/min) and hamster ($V_T = 0.72$ ml, $f = 59$/min) are based on Yeh (1980). The trends and magnitude of particle deposition are quite similar between the illustrated species. In the mouse and rat, due to particle inhalability, there is a gradual decrease in total and ET deposition for particles greater than approximately 2.5−3 μm. In the human, a similar decline in total deposition due to particle inhalability starts becoming apparent for particles greater than 7−8 μm.

Total Respiratory Tract Deposition

Across mammalian species, the efficiency of deposition in the respiratory tract may generally be described as a "U-shaped" curve on a plot of total particle deposition efficiency versus the of log particle diameter (see total deposition in Figure 5). Total and regional deposition in the healthy lung are determined by a complex interaction between the mechanisms of aerosol deposition, lung volume, and breathing conditions. Total deposition shows a minimum for particle diameters in the range of 0.1−1.0 μm,

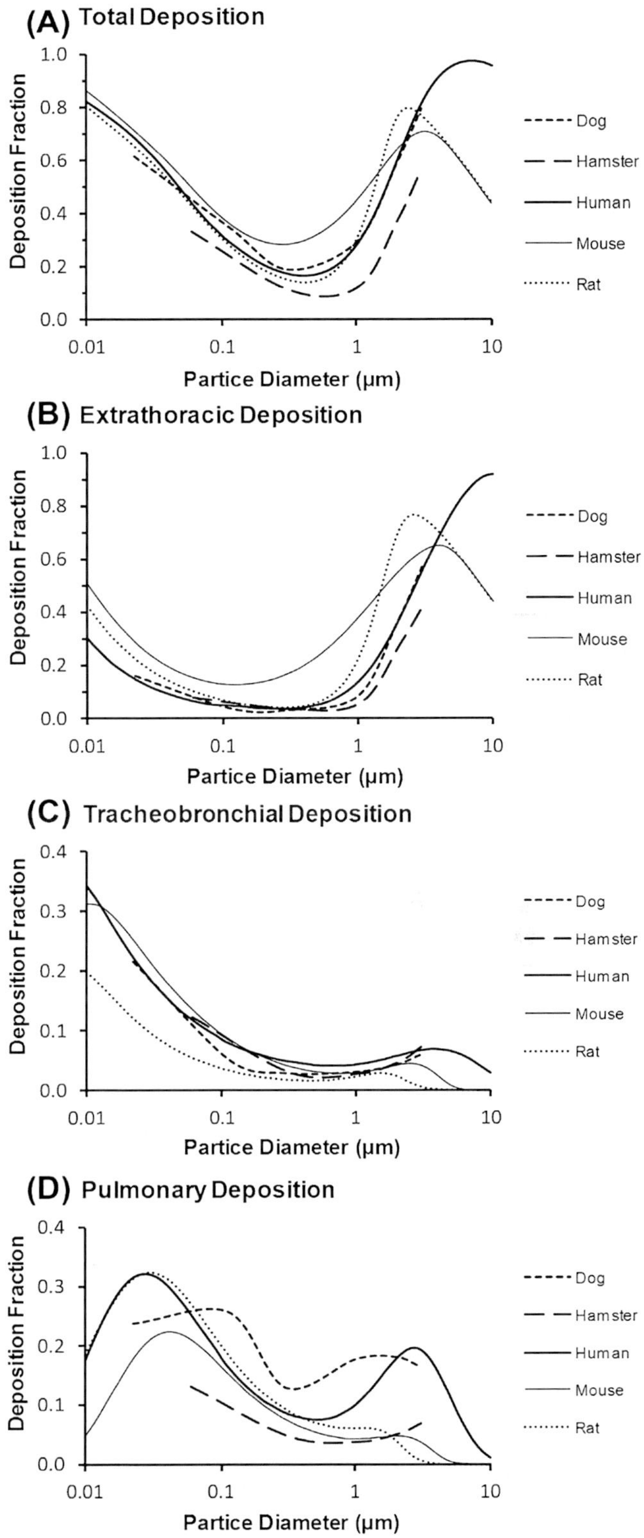

FIGURE 5 Predicted total and regional particle deposition in select mammalian species. (A) total deposition; (B) extrathoracic deposition (nasal breathing); (C) tracheobronchial deposition; and (D) pulmonary deposition.

[a] The MPPD model was developed by the CIIT Centers for Health Research with support from the Dutch National Institute of Public Health and the Environment. For more information about this model, the reader is referred to: http://www.ara.com/products/mppd.htm.

where particles are small enough to have minimal sedimentation or impaction and sufficiently large so as to have minimal diffusive deposition. Total deposition does not decrease to zero for any sized particle, in part because of mixing between particle-laden tidal air and residual lung air. The particles mixed into residual air remain in the lung following a breath and are removed on subsequent breaths or gradually deposited. Total deposition approaches 100% for particles of roughly 0.01 μm (physical diameter) due to diffusive deposition and for particles of around 10 μm (d_{ae}) due to sedimentation and impaction.

Total human lung deposition, as a function of particle size, is depicted in Figure 6. These experimental data were obtained by using monodisperse spherical test particles in healthy adults during controlled breathing on a mouthpiece. Despite the control of inhaled particle size and breathing conditions, this figure illustrates variability in deposition efficiencies due to interindividual differences in lung size and anatomical variability in airway dimensions and branching patterns. For the same breathing conditions, deposition can be expected to increase with decreasing airway dimensions, which decrease with decreasing lung inflation and lung size.

Extrathoracic Region

The first line of defense for protecting the lower respiratory tract from inhaled particles is the nose and mouth. Particle deposition in the ET region, especially the nasal passages, reduces the amount available for deposition in the TB and alveolar regions. Predicted ET particle deposition in several mammalian species was provided in Figure 5. The mouse and rat are predicted to have greater nasal deposition of particles (particularly those >1 μm) than other species. Recent data were largely derived from computational fluid dynamics (CFD) modeling and experimental measurements in casts. Because many of these studies do not substantially improve our understanding of deposition in the ET region, they are not discussed here.

For particles >1 μm d_{ae}, deposition efficiency in the oral and nasal passages has been generally described as a function of an impaction parameter (Stokes number) with the addition of a flow regime parameter (Reynolds number) for the oral passages (Finlay and Martin, 2008; Grgic et al., 2004; Kelly et al., 2005; Schroeter et al., 2006). For an adult male, the CFD simulations of Schroeter et al. (2006) predicted that nasal deposition of 10 μm d_{ae} particles was 90% and 100% for a minute ventilation of 7.5 l/min (rest) and 15 l/min (light activity), respectively. Based on experimental human data, the ICRP (1994) model predicts 94% (rest, 7.5 l/min) and 97% (light activity, 15 l/min) nasal deposition during inhalation of 10 μm d_{ae} particles by an adult male. Thus, relatively few large coarse particles will pass through the nasal passages into the lungs. Because the nasal passages are more efficient at removing inhaled particles than the oral passage, an individual's mode of breathing (i.e., oral vs nasal) influences the quantity of particles penetrating to the lung. Deposition efficiency of the nasal passages is less in children than adults (see Section 2.2).

TB and Alveolar Region

Inhaled particles passing the ET region enter and may become deposited in the lungs. Much of the available regional deposition data for the TB and alveolar regions have been obtained from experiments with radioactively labeled, poorly soluble particles or by use of aerosol bolus techniques. In general, the ability of these experimental data to define specific sites of particle deposition is limited to anatomic regions of the respiratory tract such as the head, larynx, and the TB and alveolar regions. Mathematical modeling can provide more refined predictions of deposition sites (see Figure 5). Both experimental and modeling techniques are based on many assumptions related to anatomy and ventilation distribution that may be relatively good, but not perfect, for the healthy lung (see Anatomical Variability and Ventilation Distribution Sections).

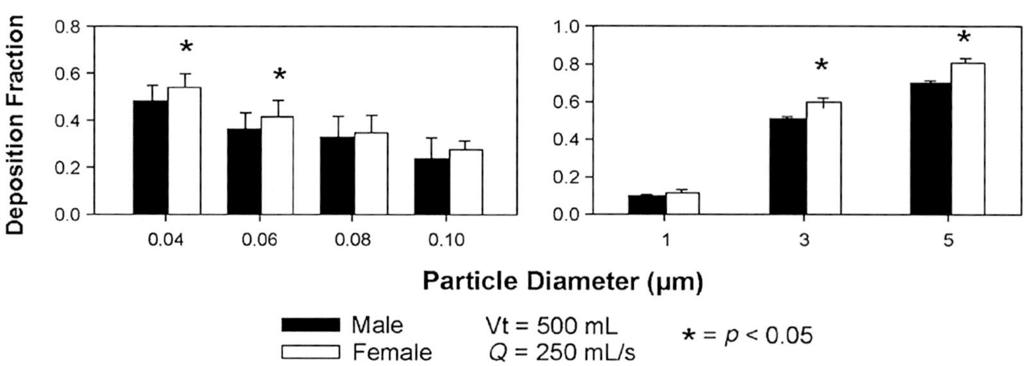

FIGURE 6 Total lung deposition measured in healthy adults (≤0.1 μm particles, 11 M, 11 F, 31 ± 4 years; ≥1 μm particles, 11 M, 11 F, 25 ± 4 years) during controlled breathing on a mouthpiece. Deposition calculated from aerosol bolus measurements between 50 and 500 ml into a breath with 50 ml increments. Illustrated data are means and standard errors from and Kim and Hu (1998) and Kim and Jaques (2000). Asterisk indicates significantly greater total deposition in females versus males.

In general, during oral breathing, the human TB deposition of particles (>2–3 μm) increases with increasing flow due to impaction. This phenomenon is less readily apparent in humans and smaller mammals during nasal breathing due to removal of particles >1 μm in the ET region. Across species, the deposition of particles (>1 μm) in the alveolar region increases mainly with increasing residence time. Lower inspiratory flow rates and increased V_T also increase deposition in the alveolar region by promoting penetration of particles (>1 μm) through the ET and TB airways. Below 0.1 μm, deposition in the TB region increases with decreasing particle size. Alveolar deposition also increases with decreasing particle size below 0.1 μm until around 0.02–0.03 μm, when removal of particles in the ET and TB regions causes a decline in alveolar deposition.

Experimental estimates of regional deposition in the TB and alveolar regions from scintigraphic data assume, perhaps problematically, the rapid clearance of particles from the TB airways. Clearance kinetics may be generally characterized as having two phases, the rapid clearance of materials deposited in the ciliated TB airways (hours to days) and the much slower process of alveolar clearance (months to years). It is this distinction in kinetics between rapid TB clearance and the far slower alveolar clearance that has been used experimentally to differentiate between particles deposited in these regions of the lung. TB clearance is generally assumed complete with 4–6 h in rats and 24–48 h in humans (Brown et al., 2005). However, it has been suggested that part of the TB airways is slowly cleared. Estimates of TB deposition based on short periods (i.e., a day in humans) may underestimate the actual TB deposition of particles due to the potential existence of a slow-cleared TB fraction.

The slow-cleared TB fraction was thought to increase with decreasing particle size. For instance, Roth et al. (1993) showed approximately 93% retention of ultrafine particles (30-nm median diameter) thought to be deposited in the TB region at 24 h postinhalation. The slow phase clearance of these ultrafine particles continued with an estimated half-time ($t_{1/2}$) of around 40 days. Using a technique to target inhaled particles (monodisperse 4.2 μm MMAD) to the TB airways, Möller et al. (2004) observed that 49% ± 9% of particles cleared rapidly ($t_{1/2}$ of 3.0 ± 1.6 h), whereas the remaining fraction cleared considerably slower ($t_{1/2}$ of 109 ± 78 days). The ICRP (1994) human respiratory tract model assumes particles ≤2.5 μm (physical diameter) to have a slow-cleared TB fraction of 50%. The slow-cleared fraction assumed by the ICRP (1994) decreases with increasing particle size to <1% for 9-μm particles. Considering the ultrafine particle data of Roth et al. (1993) in addition to data considered by the ICRP (1994), Bailey et al. (1995) estimated a slow-cleared TB fraction of 75% for ultrafine particles. Bailey et al. (1995) also estimated the slow-cleared fraction to decrease with increasing particle size to 0% for particles ≥6 μm. Additional experimental evidence from the same group (Smith et al., 2008) showed no difference in TB clearance among humans for particles with geometric sizes of 1.2 versus 5 μm, but having the same d_{ae} (5 μm) so as to deposit similarly in the TB airways. For at least micron-sized particles, these findings do not support the particle size dependence of a slow-cleared TB fraction.

The apparent slow-cleared TB fraction may be accounted for by differences in deposition patterns—that is, greater deposition in the alveolar region than expected. A portion of the slow-cleared fraction from the TB region appears to be associated with small bronchioles. For large particles (d_{ae} = 6.2 μm) inhaled at a very slow rate to theoretically deposit mainly in small ciliated airways, 50% had cleared by 24 h postinhalation. Of the remaining particles, 20% cleared with a $t_{1/2}$ of 2.0 days and 80% with a $t_{1/2}$ of 50 days (Falk et al., 1997). Using the same techniques, Svartengren et al. (2005) also reported the existence of long-term clearance from the small airways of humans. It should be noted that the clearance rates for the slow-cleared TB fraction still exceeds the clearance rate of the alveolar region in humans. Kreyling et al. (1999) targeted inhaled particle (d_{ae} = 2.2 and 2.5 μm) deposition to the TB airways of adult beagle dogs and subsequently quantified particle retention using scintigraphic and morphometric analyses. Despite the use of shallow aerosol bolus inhalation to a volumetric lung depth of less than the anatomic dead space, 2.5–25% of inhaled particles deposited in alveoli. At 24 and 96 h postinhalation, more than 50% of the retained particles were in alveoli. However, 40% of particles present at 24 and 96 h were localized to small TB airways of between 0.3 and 1 mm in diameter. In humans, Bennett et al. (1998) found 44% retention (corrected for isotope leaching from particles) at 24 h after inhalation of monodisperse aerosol (3.5-μm mean diameter) boluses inhaled to a maximum volumetric depth of 70 ml beyond the lips. Of the material retained at 24 h, an additional 10% was cleared by 48 h. Collectively, these studies suggest that although mucociliary clearance is fast and effective in healthy large airways, it is less effective and sites of longer retention exist in the smaller TB airways; also, some particles penetrate into the alveolar region even with very shallow inhalation.

Estimates of TB deposition from scintigraphic data are also potentially subject to effects of age. Ho et al. (2001) demonstrated that nasal mucociliary clearance rates were approximately 40% lower in old (age >40–90 year) versus young (age 11–40 year) men and women. Tracheal mucus velocities in elderly (or aged) humans and beagle dogs are approximately 50% that of young adults (Goodman et al., 1978; Whaley et al., 1987). Several human studies have demonstrated decreasing rates of mucociliary particle clearance from the large and small bronchial airways with increasing age (Puchelle et al., 1979; Svartengren et al., 2005; Vastag et al., 1985). Linear fits to the data show that rapid clearance (within 1 h) from large bronchi and prolonged clearance (between 1 and 21 days) from the small bronchioles in an 80-year-old is only about 50% of that in 20-year-old (Svartengren et al., 2005; Vastag et al., 1985). Although data are somewhat limited, they consistently show a depression of clearance throughout the respiratory tract, with increasing age from young adulthood in humans and laboratory animals.

Localized Deposition Sites

From a toxicological perspective, it is important to realize that not all epithelial cells in an airway will receive the same dose of deposited particles. Localized deposition in the vicinity of airway bifurcations has been analyzed using experimental and mathematical modeling techniques. Experimental data in physical models (lung generations 3–5) show the peak deposition of coarse particles (3, 5, and 7 μm d_{ae}) in daughter airways during inspiration and the parent airway during expiration, but always near the carinal ridge—that is, the ridge separating daughter airways (Kim and Iglesias, 1989; Kim et al., 1989). Mathematical models of the segmental bronchi (lung generations 4–5) also predict distinct "hot spots" of deposition in the vicinity of the carinal ridge for both coarse (10 μm) and ultrafine (0.01 μm) particles (Heistracher and Hofmann, 1997; Hofmann et al., 1996). Hot spots occurred during inspiration at the carinal ridge for 10 μm d_{ae} particles due to inertial impaction and for 0.01 μm particles due to secondary flow patterns formed at the bifurcation. During expiration, preferential sites of deposition for both particle sizes occurred approaching the juncture of daughter airways on the walls forming and across the lumen from the carinal ridge, and at the top and bottom (visualizing the Y-shaped geometry laying horizontal) of the parent airway downstream of the bifurcation.

Other studies further support these findings (Balashazy et al., 2003; Farkas et al., 2006; Farkas and Balásházy, 2008; Isaacs et al., 2006). Most of these studies quantified localized deposition in terms of an enhancement factor. Typically, the enhancement factor is the ratio of the deposition in a prespecified surface area (e.g., 100 × 100 μm, which corresponds to ~10 × 10 epithelial cells) to the average deposition density for the whole airway geometry. These enhancement factors are very sensitive to the size of the surface considered (Balashazy et al., 1999). The studies by Farkas et al. (2006) and Farkas and Balásházy (2008) investigated the phenomena of localized deposition down to 0.001-μm particles. The deposition of 0.001 μm was rather uniform; however, the deposition pattern became increasingly less uniform with increasing particle size. These studies indicate that, for particles greater than ~0.01 μm, some cells located near the carinal ridge of bronchial bifurcations may receive hundreds to thousands times the average dose (particles per unit surface area) of the parent and daughter airways. Furthermore, the inertial impaction of particles ≥1 μm d_{ae} at the carinal ridge of large bronchi will increase with increasing inspiratory flow. In a comparison of constricted versus healthy airways, Farkas et al. (2006) also reported that the overall deposition efficiency of 10 μm d_{ae} particles at bifurcations downstream of a constriction may be increased by 18 times. Given these considerations, Phalen and Oldham (2006) noted that substantial doses of particles (≥1 μm d_{ae}) may be justified for in vitro studies using TB epithelial cell cultures.

2.3 Interspecies Patterns of Deposition

As illustrated in Figure 5, the general pattern of total particle deposition efficiency is similar between laboratory animals and humans: deposition increases on both sides (increasing and decreasing particle size) of a minimum that occurs at 0.2–1 μm. Similar deposition efficiencies have been reported in nasal casts of humans and rhesus monkeys for 1–10 μm d_{ae} for inspiratory flows mimicking resting breathing patterns (Kelly et al., 2005). There are, however, marked interspecies differences in uptake into the respiratory tract and regional deposition. For instance, the nasal inhalability of 10 μm d_{ae} particles is predicted to be 96% in humans, whereas it is only 44% in rats (Ménache et al., 1995). Additionally, deposition in the ET region is near 100% for particles >5 μm d_{ae} in most laboratory animals (rat, mouse, hamster, guinea pig, and dogs), indicating greater efficiency than that seen in humans (Raabe et al., 1988). There may also be localized differences in the pattern of deposition within the thoracic airways of rodents versus humans. Hofmann et al. (1996) reported that deposition was more uniformly distributed within the asymmetrically branching airways (generations 4–5) of rodents as compared to distinct hot spots in the more symmetrically branching airways (generations 4–5) of humans.

Figure 7 illustrates experimental regional particle deposition in mice, rats, dogs, and humans normalized to total deposition. Within a given species, considerable interstudy variability is apparent. Some of the within-species variability may be attributable to breathing pattern. Of the illustrated studies, only Kuehl et al. (2012) reported breathing patterns for mice ($V_T = 0.20$ ml, $f = 275$/min) and rats ($V_T = 1.71$ ml, $f = 181$/min). The f reported by Kuehl et al. (2012) for mice are similar to those of restrained mice in Table 1. On average, Cuddihy et al. (1969) reported a V_T of 164 ml and f of 12/min in dogs. However, there was considerable within dog variability among the aerosol

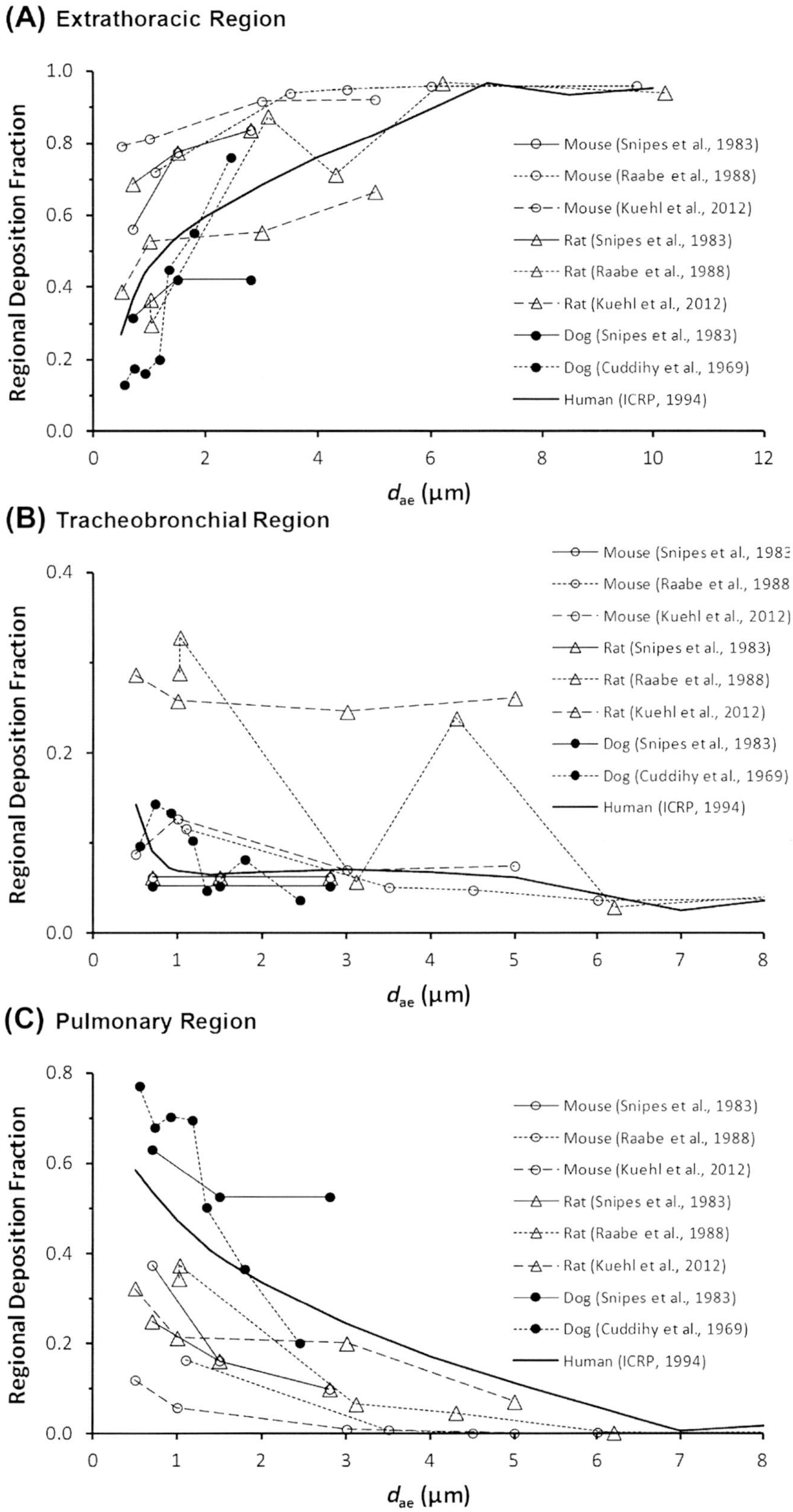

FIGURE 7 Experimental regional particle deposition (normalized to total deposition) in select mammalian species. (A) extrathoracic deposition (nasal breathing); (B) tracheobronchial deposition; and (C) pulmonary deposition.

exposures in the Cuddihy et al. (1969) study, with V_T ranging from 130 to 200 ml and f ranging from 8 to 20/min. The human data are for a resting breathing pattern ($V_T = 625$ ml, $f = 12$/min). Despite the within- and between-species differences, some trends become apparent from this figure. First, the ET fraction generally increases with decreasing species size and increasing particle size. Second, the pulmonary fraction generally decreases with decreasing species size and increasing particle size. Third, the TB fraction is a small component of the overall deposition. With respect to this third observation, however, it should be noted that due to relatively small surface area of the TB region, delivered surface doses can be quite high.

Detailed presentations of the dosimetric differences between rats and humans are available elsewhere (Brown et al., 2005; Jarabek et al., 2005). Brown et al. (2005) conducted an evaluation of extrapolations between rats and humans in relation to particle exposures. One of many factors they considered was the choice of a dose metric appropriate for comparison between species. For example, deposited mass may be an appropriate particle indicator for health effects associated with soluble particle constituents. For health effects associated with insoluble particles, the particle number, surface area, or mass may be appropriate indicators. Given the interspecies differences in deposition patterns and clearance rates, the question of retained versus deposited dose was also discussed. It was concluded that for acute effects, the incremental dose may be the appropriate type of dose metric. For chronic effects, long-term burden may be more appropriate. For various dose metrics, estimates of particle concentration and exposure duration required for a rat to receive the same dose as received by a human were obtained with consideration of activity levels, mode of breathing, and particle size distributions. The dosimetric calculations indicated that to achieve nominally similar acute doses per surface area in rats, relative to humans undergoing moderate to high exertion, particle exposure concentrations for rats would need to be somewhat higher than for humans. Because particle clearance from the lungs of rats is much faster than humans (e.g., Hofmann and Asgharian, 2003), much higher exposure concentrations are required for the rat to simulate retained burdens of humans. Illustrating the complexity of such analyses, in some cases, rats were found to require lower exposures than humans to have comparable doses (generally when considering a scenario of humans at rest).

2.4 Factors Modulating Deposition

Evaluation of factors affecting particle deposition is important to help understand potentially susceptible subpopulations. Differences in biological response following pollutant exposure may be caused by dosimetry differences as well as by differences in innate sensitivity. The effects of different factors on deposition are summarized briefly here.

Physical Activity

The activity level of an individual is well recognized to affect their minute ventilation and route of breathing. Changes in minute ventilation during exercise are accomplished by increasing both V_T and f (Table 2). Humans are oronasal breathers tending to breathe through the nose when at rest and increasingly through the mouth with increasing activity level. There is considerable interindividual variability in both the route by which people breathe and the way breathing pattern changes occur.

Individuals typically breathe through their nose while at rest, switching to oronasal breathing as ventilation increases (Bennett et al., 2003; Niinimaa et al., 1981). The role of the nose in filtering particles is diminished as airflow is diverted from the nose to the mouth during exercise, bringing more particles to the lower respiratory tract. Particle ($d_{ae} > 1$ μm) penetration into the lower respiratory tract is more affected by route of breathing than age, sex, activity level, or breathing pattern (Brown et al., 2013). In adults, Bennett et al. (2003) found that nasal ventilation during exercise varied as a function of both race and sex. African Americans possessed a greater nasal contribution to breathing during exercise than Caucasians. At similar exercise efforts (i.e., normalized to a percent of maximum work capacity), the females also had a greater nasal contribution to breathing during exercise than males.

Predicted total and regional deposition for an adult human male during rest and light exercise are illustrated in

TABLE 2 Breathing Patterns with Activity Level in Adult Human Male

Activity	Awake Rest[a]	Slow Walk[a]	Light Exertion[a]	Moderate Exertion[a]	Heavy Exertion[b]
Breaths/min	12	16	19	28	26
Tidal volume, ml	625	813	1000	1429	1923
Minute ventilation, l/min	7.5	13	19	40	50

[a]*de Winter-Sorkina and Cassee (2002).*
[b]*ICRP (1994).*

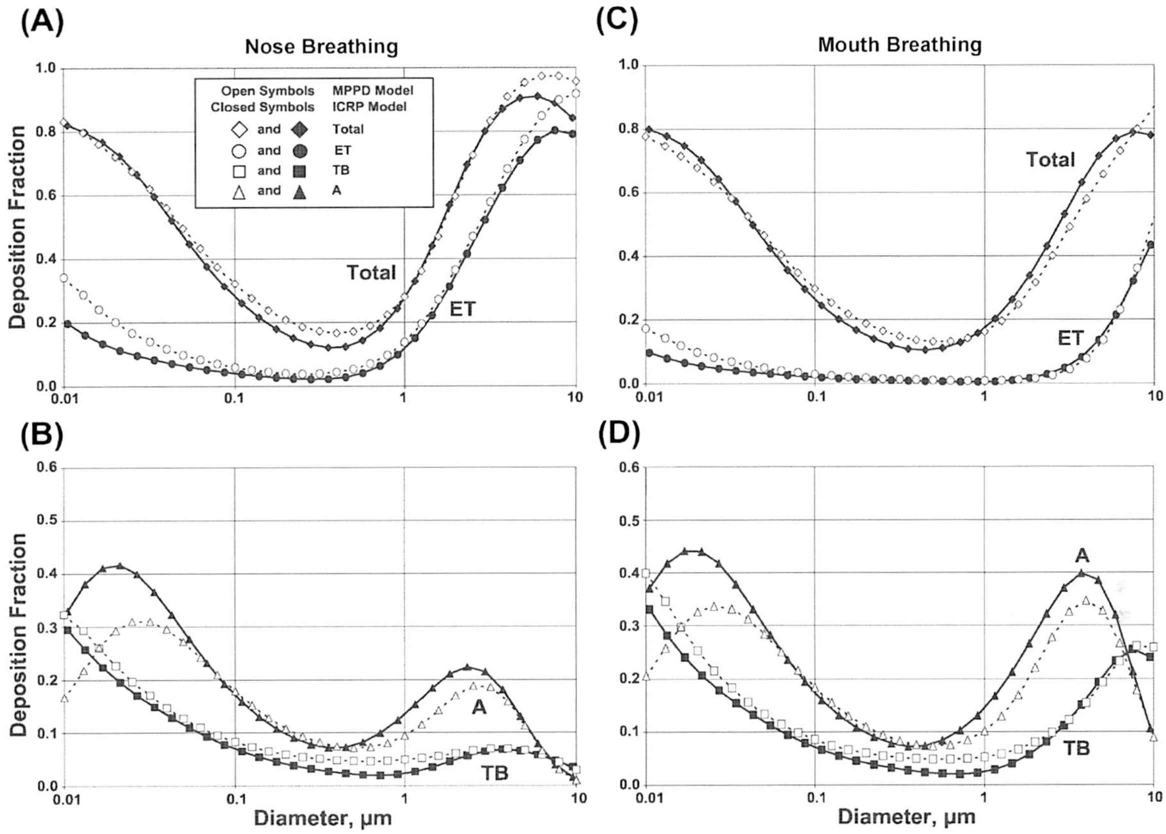

FIGURE 8 **Comparison of total and regional deposition results from the ICRP and MPPD models for a resting breathing pattern (V_T = 625 ml, f = 12/min) and corrected for particle inhalability.** Regions are extrathoracic, ET; tracheobronchial, TB; and alveolar, (A). Panels (A)–(B) are for nose breathing; panels (C)–(D) are for mouth breathing.

Figures 8 and 9, respectively. These figures illustrate changes in deposition as a function of particle size and breathing conditions. There is dramatically more deposition of particles larger than 1 μm in the lower airways when breathing through the mouth relative to nasal breathing. Particularly during mouth breathing, a large proportion of inhaled coarse particles in the 3–6 μm (d_{ae}) range can reach and deposit in the lower respiratory tract. The predictions were based on two widely available particle deposition models, the ICRP Model (ICRP, 1994) and the Multi-Path Particle Dosimetry model (MPPD; Version 1.0, ©2002). The ICRP model was implemented by Lung Dose Evaluation Program (LUDEP; Version 2.07, June 2000).

In addition, when individuals increase their ventilation with activity, not only does the total number of particles inhaled per unit time (i.e., exposure rate) increase, but the fractional deposition of particles in each breath also changes with the breathing pattern. During exercise, both V_T and f increase. Fractional deposition for all particles increases with increased V_T. Increasing the f, however, decreases the fractional deposition of fine and ultrafine particles due to decreased time for gravitational and diffusive deposition. For particles of larger than roughly 3 μm d_{ae}, increasing f can increase the deposition fraction due to increased impaction in the ET and TB airways. Thus, it should be expected that the change in deposition fraction with activity will vary among individuals depending on the relative influences of these two variables (i.e., V_T and f) in a given subject and the particle size to which they are exposed. Experimentally, the lung deposition fractions of fine particles during moderate exercise and mouth breathing are unchanged between rest and exercise (Bennett et al., 1985; Morgan et al., 1984). Löndahl et al. (2007) also found no difference in deposition fractions of particles (hygroscopic and hydrophobic; 0.013–0.290 μm mobility diameter of dry particles) between rest ($V_T = 0.72 \pm 0.15$ l; $f = 12 \pm 2$/min) and exercise ($V_T = 2.1 \pm 0.5$ l; $f = 17 \pm 4$/min). Kim (2000) evaluated differences in deposition of 1, 3, and 5 μm (MMAD) particles under varying breathing patterns (simulating breathing conditions of sleep, resting, and mild exercise). Total lung deposition increased with increasing V_T at a given flow rate and with increasing flow rate at a given breathing period (which also leads in increases in V_T). These experimental studies suggest that the total deposited dose rate (i.e., deposition per unit time), but not necessarily the breath-by-breath deposition, of

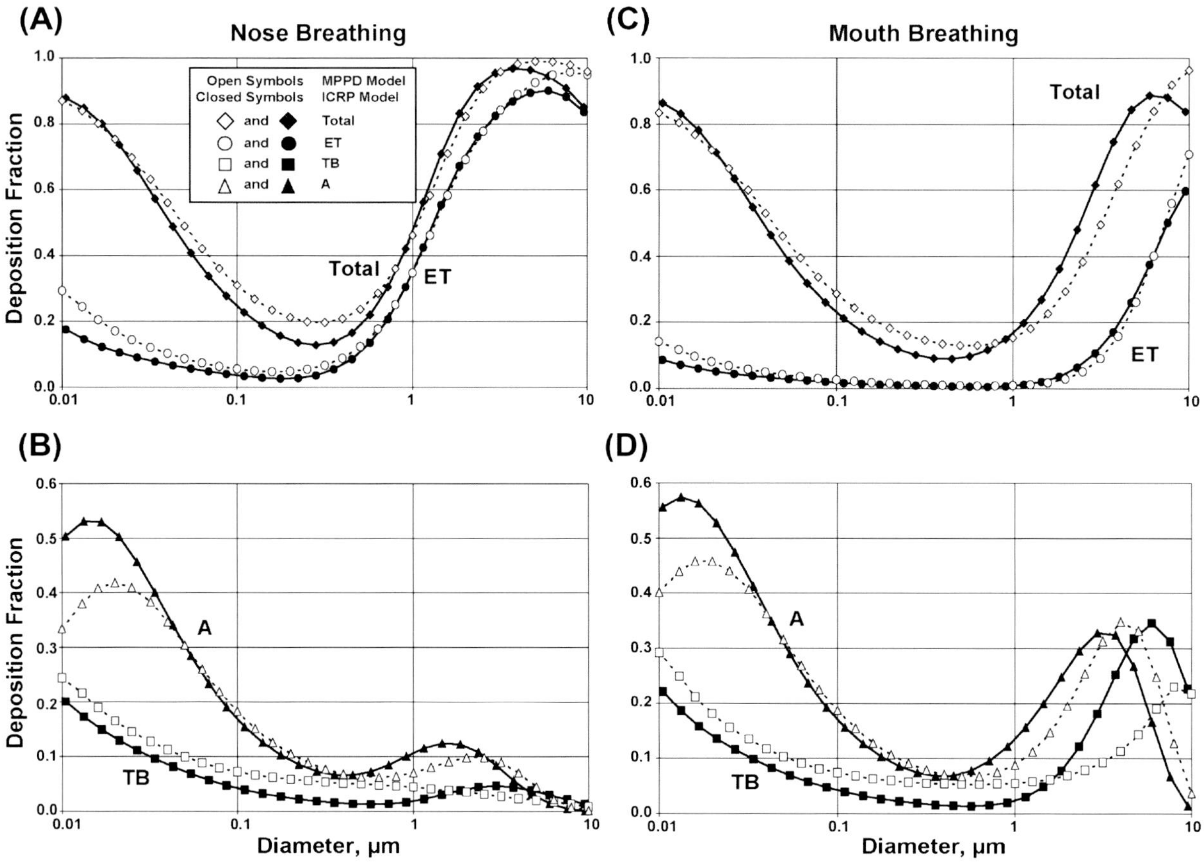

FIGURE 9 **Comparison of total and regional deposition results from the ICRP and MPPD models for a light exercise breathing pattern (V_T = 1250 ml, f = 20/min) and corrected for particle inhalability.** Regions are extrathoracic, ET; tracheobronchial, TB; and alveolar, (A). Panels (A) and (B) are for nose breathing; panels (C) and (D) are for mouth breathing.

particles will increase in direct proportion to the increase in minute ventilation associated with exercise.

The changes in ventilation (i.e., breathing pattern and flow rate) may also alter the regional deposition of particles. Coarse particle deposition increases in the TB and ET regions during exercise due to increased flow rates and associated increased impaction. A rapid-shallow breathing pattern during exercise may result in more bronchial airway versus alveolar deposition, while a slow-deep pattern will shift deposition to deeper lung regions (Valberg et al., 1982). Bennett et al. (1985) showed for 2.6-μm particles that moderate exercise shifted deposition from the lung periphery toward ET and larger, bronchial airways. Similarly, Morgan et al. (1984) showed that even for fine particles (0.7 μm), TB deposition was enhanced with exercise. This shift in deposition toward the bronchial airways results in a much greater dose per unit surface area of tissue in those regions. Morgan et al. (1984) also found that the apical-to-basal distribution of fine particles increased with exercise (i.e., a shift toward increased deposition in the lung apices). This shift may be less likely for larger particles, however, whose deposition at large airway bifurcations may preclude their transport to these more apical regions.

Age

Airway structure and respiratory conditions vary with age, and these variations may alter the amount and site of particle deposition in the respiratory tract. A few studies have attempted to measure oronasal breathing in children as compared to adults (Becquemin et al., 1999; Bennett et al., 2008; James et al., 1997). This is important because particles deposit with greater efficiency in the nose relative to the mouth, thereby affecting exposure of the lower respiratory tract. James et al. (1997) found that children (age 7–16 year, $n = 10$) displayed more variability than adults with respect to their oronasal pattern of breathing with exercise. However, it was not possible to predict the pattern of the partitioning of ventilation during exercise based on age, sex, or nasal airway resistance. Furthermore, in a limited number of children (age 8–16 year, $n = 10$), Becquemin et al. (1999) found that the children tended to display more oral breathing both at rest and during exercise than the adults. The highest oral fractions were also found in the youngest children. None of these studies, however, was able to show a relationship between nasal resistance and the relative contribution of nasal breathing in children. Bennett et al. (2008) made preliminary measurements of the relative contributions of oral versus nasal breathing at rest and during incrementally graded submaximal exercise on the cycle ergometer for children (age 6–10 year, $n = 12$) and adults (age 18–27 year, $n = 11$). Children tended to have a greater oral contribution than adults at rest (68 vs 88% nasal, respectively) and during exercise (47 vs 59% nasal at 40% maximum workload, respectively), although these differences were not statistically significant.

There are limited data on the deposition of particles for oral or nasal breathing in children relative to that available for adults. Theoretical calculations by Xu and Yu (1986) predicted enhanced deposition of particles (>2 μm) in the head region for children when compared to adults. Studies of fine particle deposition in physical models of the nose, scaled to adult versus child sizes, predicted that deposition efficiency in the nose is a function of pressure drop across the nose (Phalen et al., 1989). Consequently, these model analyses suggest that, when properly scaled physiological flows are used in the calculation of nasal deposition, children (who have higher nasal resistance than adults) should have higher nasal deposition compared to adults (ICRP, 1994). Surprisingly, the few experimental studies reporting measures of nasal deposition in children found lower nasal deposition (particularly during light-to-moderate exercise) for fine particles (1–3 μm d_{ae}) as compared to adults, despite their higher nasal resistance (Becquemin et al., 1991; Bennett et al., 2008). For 1-, 2-, and 3-μm particles inhaled during exercise, children (<12 years) had 18% nasal deposition, on average, whereas adults had 34% nasal deposition (Becquemin et al., 1991). These findings of lesser nasal versus oral breathing and less efficient nasal deposition suggest that children's lower respiratory tract (i.e., the TB and alveolar regions) may receive a higher dose of particles compared to adults. Normalized to lung surface area, the dose rate to the lower airways of children versus adults is increased further because children breathe at higher minute ventilations relative to their lung volumes than do adults.

Breathing patterns are well recognized to change with increasing age from childhood into adulthood—that is, V_T increases and respiratory rates decrease (Tabachnik et al., 1981; Tobin et al., 1983). Bennett and Zeman (1998) measured the deposition fraction of inhaled, fine particles (2 μm d_{ae}) in children (age 7–14 year, $n = 16$) and adults (age 19–35 year, $n = 12$) as they breathed the aerosol with their natural, resting breathing pattern. Among the children, variation in deposition fractions measured by photometry at the mouth was highly dependent on intersubject variation in V_T. On the other hand, they found no difference in deposition fractions between children versus adults for these fine particles. This finding and the modeling predictions (Hofmann et al., 1989) are explained in part by the smaller V_T and faster breathing rate of children relative to adults for natural breathing conditions. Bennett et al. (2008) also reported measures of fine particle (1 and 2 μm d_{ae}) deposition at ventilation rates typical of light exercise in children (age 6–10 year, $n = 12$) and adults (age 18–27 year, $n = 11$); the authors showed that, like with resting breathing, deposition fractions were predicted by breathing pattern and did not differ or tended to be less in children compared to adults. On the other hand, because children breathe at higher minute ventilations relative to their lung volumes, the rate of deposition of fine particles normalized to lung surface area may be greater in children versus adults (Bennett and Zeman, 1998).

Bennett and Zeman (2004) expanded their measures of fine particle deposition during resting breathing to a larger group of healthy children (6–13 year; 20 boys, 16 girls) and found again that the variation in total deposition was best predicted by V_T ($r = 0.79$, $p < 0.001$). But both V_T and resting minute ventilation increased with both height and body mass index (BMI) of the children. Interestingly, these data suggest that for a given height and age, children with higher BMI have larger minute ventilations and V_T at rest than those with lower BMI. These differences in breathing patterns as a function of BMI translated into increased deposition of fine particles in the heaviest children. The rate of deposition (i.e., particles depositing per unit time) in the overweight children was 2.8 times that of the leanest children ($p < 0.02$). Among all children, the rate of deposition was significantly correlated with BMI ($r = 0.46$, $p < 0.004$). Some of the increase in deposition fractions of heavier children may be due to their elevated V_T, which was well correlated with BMI ($r = 0.72$, $p < 0.001$).

In 62 healthy adults with normal lung function aged 18–80 year, Bennett et al. (1996) showed there was no effect of age on the whole lung deposition fractions of 2-μm particles under natural breathing conditions. Across all subjects, the deposition fractions were found to be independent of age, depending on breathing period ($r = 0.58$, $p < 0.001$) and airway resistance ($r = 0.46$, $p < 0.001$). In the same adults breathing with a fixed pattern (360 ml V_T, 3.4 s breathing period), there was a mild decrease in deposition with increasing age, which could be attributed to increased peripheral airspace dimensions in the elderly.

Sex

Males and females differ in body size, conductive airway size, and ventilatory parameters; therefore, sex differences in deposition might be expected. In some of the controlled studies, however, the men and women were constrained to breathe at the same V_T and f. Because women are generally smaller than men, the increased minute ventilation of women compared to their normal ventilation could affect deposition patterns. This may help explain why sex-related effects on deposition have been observed in some studies.

Kim and Hu (1998) assessed the regional deposition patterns of 1-, 3-, and 5-μm MMAD particles in healthy adult males and females using controlled breathing. The total fractional deposition in the lungs was similar for both sexes with the 1-μm particle size, but was greater in women for the 3- and 5-μm particles. Deposition also appeared to be more localized in the lungs of females compared to those of males. Kim and Jaques (2000) measured deposition in healthy adults using sizes in the ultrafine mode (0.04–0.1 μm). Total fractional lung deposition was greater in females than in males for 0.04- and 0.06-μm particles. The region of peak fractional deposition was shifted closer to the mouth and peak height was slightly greater for women than for men for all exposure conditions. Total lung deposition data from these studies are illustrated in Figure 6.

In individuals aged 18–80 year breathing at their normal resting breathing pattern, Bennett et al. (1996) found a tendency for a greater total deposition fraction of 2-μm particles in females compared to males. However, because males had greater minute ventilation, the deposition rate (i.e., deposition per unit time) was greater in males than in females. More recently, Bennett and Zeman (2004) found no difference in the deposition of 2-μm particles in boys versus girls aged 6–13 year ($n = 36$). In individuals between 20 and 40 year (median 25 year), Löndahl et al. (2007) also found no difference between males and females in the deposition fractions of particles (hygroscopic and hydrophobic) between 0.013 and 0.290 μm (mobility diameter of dry particles) inhaled by mouth during rest and exercise with spontaneous breathing.

Sex differences in deposition fractions are generally observed in studies using experimental matched breathing patterns, i.e., constrained to breathe at the same V_T and f. This finding is not unexpected given that females have smaller-sized upper and lower airways, particularly the laryngeal structure, relative to males. During spontaneous and natural breathing, minimal differences in deposition fractions are observed between the sexes.

Anatomical Variability

Anatomical variability, even in the absence of respiratory disease, can affect deposition throughout the respiratory tract. The ET region is the first to be exposed to inhaled particles; therefore, deposition within this region would reduce the amount of particles available for deposition in the lungs. Variations in relative deposition within the ET region will therefore propagate through the rest of the respiratory tract, creating differences in calculated doses among individuals.

The influence of variations in nasal airway geometry on particle deposition has been investigated. Cheng et al. (1996) examined nasal airway deposition in healthy adults using particles ranging in size from 0.004 to 0.15 μm and at two constant inspiratory flow rates, 167 and 333 ml/s. Interindividual variability in deposition was correlated with the wide variation of nasal dimensions, such that greater surface area, smaller cross-sectional area, and increasing complexity of airway shape were all associated with enhanced deposition. Bennett and Zeman (2005) have also shown that nasal anatomy influences the efficiency of particle deposition in the noses of adults. For light exercise breathing conditions in adults, their study demonstrated that nasal deposition efficiencies for both 1- and 2-μm monodisperse particles were significantly less in African Americans versus Caucasians. The lesser nasal efficiencies in African Americans were associated with both lower nasal resistance and less elliptical nostrils compared to Caucasians.

Within the lungs, the branching structure of the airways may also differ between individuals. Zhao et al. (2009) examined the bronchial anatomy of the left lung in patients (132 men, 84 women; mean age 47 year) who underwent conventional thoracic computed tomography scans for various reasons. At the level of the segmental bronchus in the upper and lower lobes, a bifurcation occurred in the majority of patients. A trifurcation, however, was observed in 23% of the upper and 18% of the lower lobes. Other more unusual findings were also reported, such as four bronchi arising from the left upper lobe bronchus. As described in Section 2.4, deposition can be highly localized near the carinal ridge of bifurcations. The effect of a bifurcation versus other branching patterns on airflow patterns and particle deposition has not been described in the literature. Martonen et al. (1994) showed that a wide blunt carinal ridge shape dramatically affected the flow stream lines relative to a narrower and more rounded ridge shape. Specifically, there were high flow velocities across the entire area of the blunt carinal ridge versus a smoother division of the airstream in the case of the narrow rounded ridge shape. The implication may be that localized particle deposition on the carinal ridge would increase with ridge width. A similar situation might be expected for a trifurcation versus a bifurcation. These differences in branching patterns provide a clear example of anatomical variability among individuals that might affect both air flow patterns and, therefore, sites of particle deposition.

Ventilation Distribution

Ventilation distribution, which can affect the partitioning of inhaled aerosols between lung regions and the residence time within these regions, may be an important determinant of particle deposition in the lung. In large mammals such as humans, there is a gravity-induced gradient that causes the volume of alveoli in dependent lung regions (i.e., the lowest areas in the lungs) to be smaller than those in nondependent lung regions. During normal tidal breathing, dependent regions may have somewhat increased ventilation relative to nondependent regions. As a breath is distributed, associated airborne particles may also be distributed. Some experimental data are available on the association between regional deposition of ultrafine, fine, and coarse particles and regional ventilation in the healthy and diseased lung. Ventilatory inhomogeneity due to obstructive disease generally exceeds normal gravity-induced gradients.

The distribution of ventilation has been studied in a number of animal species. There is a pronounced gravitation gradient in the ventilation distribution of standing horses with the dependent (ventral) regions receiving more of each breath than the nondependent (dorsal) regions (Amis et al., 1984). In standing Shetland ponies, late-term pregnancy has been reported to increase ventilation to the nondependent regions, possibly due to intra-abdominal pressure on the dependent (ventral) regions (Schramel et al., 2012). In contrast to horses, data out to 20 days postpartum showed equal ventral–dorsal ventilation in these ponies. In the supine position, dogs and sloths show increased ventilation of the dependent (dorsal) regions relative to the nondependent (ventral) regions (Hoffman and Ritman, 1985). However, in the prone position, there is essentially uniform ventral–dorsal ventilation in both the dogs and sloths. The position in which rats are exposed may also influence the regional delivery and deposition of inhaled aerosols. In rats, the nondependent region of the lung has been reported to be better ventilated, whether positioned supine, prone, or on either side (Rooney et al., 2009; Dunster et al., 2012). In humans, ventilation patterns are affected by both body position and lung inflation.

Milic-Emili et al. (1966) showed that apical (nondependent) to basal (dependent) differences in pleural pressure can affect ventilation distribution in healthy individuals. In upright humans, the apical lung receives the majority of an inhaled air at low lung volumes (less than 20% vital capacity). Above this volume, the vertical proportioning of ventilation is relatively constant across a breath with basal regions (dependent part) having somewhat increased ventilation relative to apical regions (Milic-Emili et al., 1966). The effect of gravity is shifted by changes in body position. For instance, while lying on the left side, aerosols inhaled at low lung volumes will be preferentially transported into and deposited in the right lung (Bennett et al., 2002). In upright individuals, at high lung volumes (70% or more of total lung capacity), particles are transported preferentially into and deposit in the left lung (Bennett et al., 2002). A more uniform left–right distribution of particle deposition is observed for inhalations closer to functional residual capacity. Left–right asymmetry in particle deposition at high lung volumes is primarily due to differences in ventilation between the lungs (Möller et al., 2009). The effect of gravity-induced gradients on ventilation and left–right asymmetry in upright individuals described here for healthy individuals, however, are small relative to the ventilatory heterogeneity caused by obstructive lung disease (Suga et al., 1995).

Regional deposition in excess of regional ventilation to poorly ventilated areas has been reported for aerosols in the 0.5- to 1.0-μm size range and attributed to increased residence time in obstructed areas (Susskind et al., 1986; Trajan et al., 1984). However, others show increasing deposition with increasing ventilation. For instance, a significant association of increased aerosol (1.2 μm MMAD) deposition in better ventilated regions has been observed in lung transplant patients with bronchiolitis obliterans (O'Riordan et al., 1995). The trend for increased aerosol (0.78 μm MMAD) deposition with increasing ventilation has also been reported in normal individuals and asymptomatic smokers (Chamberlain et al., 1983). Other studies using similar-sized aerosols have found no association between ventilation distribution and particle deposition (O'Riordan and Smaldone, 1994; Smaldone et al., 1991). All of these studies compared regional ventilation to the regional particle deposition using scintigraphic methods. The mixed results in these studies may be due to deposition not having a simple monotonic relationship with ventilation.

Brown et al. (2001) examined the relationship of 5-μm MMAD particles in healthy individuals and patients with cystic fibrosis using scintigraphic techniques. Deposition of particles in the TB airways followed the pattern of ventilation in the healthy individuals, whereas it was inversely related to ventilation in the patients. This is consistent with Kim et al. (1983), who found that the pattern of particle deposition (3.0 μm MMAD) followed ventilation distribution in a three-generation model, but it was enhanced in the vicinity of obstructions. In the alveolar region, Brown et al. (2001) found deposition to be very strongly associated with ventilation distribution in the patients; that is, the well-ventilated regions received increased alveolar deposition of particles relative to poorly ventilated regions. A similar trend was observed in healthy individuals, but a more uniform pattern of ventilation lead to smaller differences in ventilation and deposition between lung regions. A later modeling study of ventilation distribution and particle deposition in the patients and the healthy individuals confirmed these experimental observations (Brown and Bennett, 2004). The patient data strongly support the obvious fact that particles must be transported into a region in order to become deposited there. This phenomenon can be overtly obvious in patients with mild-to-moderate asthma, where large wedge-shaped regions of the lung were observed absent the deposition of 0.12 μm (equivalent diameter) particles (King et al., 1998).

Aerosol Size Distribution

Most aerosols present in ambient air and workplace are polydisperse and may consist of multiple modes. Ambient particles are recognized as being distributed into at least three modes: Aitken or nucleation, accumulation, and coarse (Whitby, 1978). When estimating deposition of multimode aerosols, it is preferable to assess each mode individually. This is necessitated, in part, by the choice of dose metrics, because the nucleation mode predominates the count distribution, whereas larger coarse particles predominate the mass distribution.

The deposition of a unimodal, log-normally distributed polydisperse aerosol in the lung is commonly estimated from the deposition of a series of monodisperse aerosols, as described by Eqn (7). In general, the deposition of a unimodal polydisperse aerosol in the lung is expected to be greater than that of a monodisperse aerosol for median particle diameters between 0.04 and 2 μm (Diu and Yu, 1983). For median particle diameters above 2 μm and below 0.04 μm, the polydisperse aerosol is expected to have less deposition than a monodisperse aerosol.

Hygroscopicity of Aerosols

Hydrophobic particles are typically used experimentally to investigate particle deposition. Ambient aerosols, such as sulfates, nitrates, some organics, and aerosols laden with sodium or potassium, are also hygroscopic. The high relative humidity in the lungs contributes to rapid growth of hygroscopic particles and dramatically alters their deposition relative to hydrophobic aerosols inhaled at the same initial size. In an individual during controlled breathing ($V_T = 0.75-1.0$ l; $f = 15$/min), Tu and Knutson (1984) found minimal deposition in the range of 0.06–0.09 μm for hygroscopic particles, whereas it was in the range 0.3–0.6 μm for hydrophobic particles. Hygroscopic 0.3 μm (dry diameter) salt particles will grow to nearly 2 μm in the respiratory tract and deposit to a far greater extent than hydrophobic 0.3 μm particles (Anselm et al., 1990). The deposition curves for hygroscopic and hydrophobic particles intersected at approximately 0.15 μm in the Tu and Knutson (1984) study. This implies that hygroscopic growth reduced diffusive deposition below 0.15 μm and increased aerodynamic deposition above this particle size.

Löndahl et al. (2007) measured particle deposition in 29 individuals (20 men, 9 women; median age, 25 years) who inhaled hygroscopic and hydrophobic particles between 0.013 and 0.290 μm (mobility diameter of dry particles) by mouth during spontaneous breathing while engaged in rest ($V_T = 0.72 \pm 0.15$ l; $f = 12 \pm 2$/min) or exercise ($V_T = 2.1 \pm 0.5$ l; $f = 17 \pm 4$/min). Deposition fractions for each particle type were minimally affected by sex or activity. The prior study by Tu and Knutson (1984) found that the deposition curves for hygroscopic and hydrophobic particles were also generally unaffected by route of breathing. Figure 10 illustrates deposition curves for hygroscopic and hydrophobic particles inhaled during rest in the Löndahl et al. (2007) study. From this figure, it is seen that the growth of 0.02-μm hygroscopic particles lowers their diffusive deposition to that of 0.07-μm hydrophobic particles. Deposition of the hygroscopic particles reached a minimum in the range of 0.09–0.14 μm. Hygroscopic growth reduced diffusive deposition below 0.21 μm and increased aerodynamic deposition above this particle size.

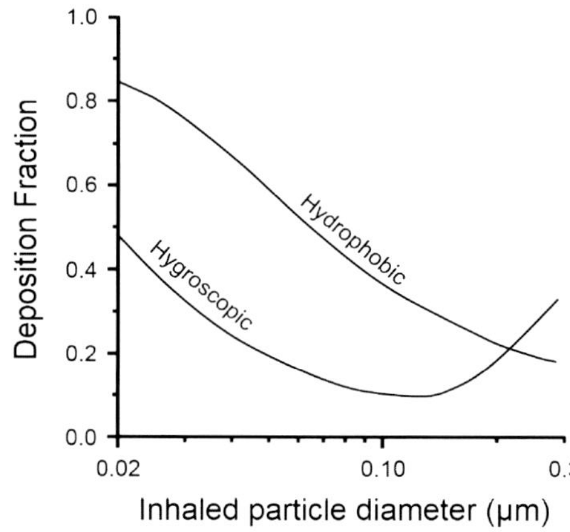

FIGURE 10 **Total deposition of hygroscopic sodium chloride and hydrophobic diethylhexylsebacate aerosols inhaled orally during rest ($V_T = 0.72 \pm 0.15$ l; $f = 12 \pm 2$/min).**

2.5 Summary

Particle deposition in the respiratory tract occurs predominantly by diffusion, impaction, and sedimentation. Deposition is minimal for particle diameters in the range of 0.1–1.0 μm, where particles are small enough to have minimal sedimentation or impaction and sufficiently large so as to have minimal diffusive deposition. Depending on breathing rates and species, total respiratory tract deposition approaches 100% for particles of roughly 0.01 μm (physical diameter) due to diffusive deposition and for particles of around 5–10 μm d_{ae} due to the efficiency of sedimentation and impaction.

To be deposited, particles must first be able to enter the respiratory tract. The inhalability of particles having d_{ae} of 2.5, 5, and 10 μm is 80, 65, and 44% in rats, respectively, whereas it remains near 100% for a d_{ae} of 10 μm in humans. Once inhaled, the first line of defense for protecting the lower respiratory tract from inhaled particles is the nose and mouth. In most laboratory animal species (rat, mouse, hamster, guinea pig, and dogs), deposition in the extrathoracic region is near 100% for particles greater than 5 μm d_{ae}. By contrast, in humans, nasal deposition approaches 100% for 10-μm d_{ae}. Oronasal breathing versus obligate nasal breathing further contributes to greater penetration of coarse particles into the lower respiratory tract of humans than rodents.

Experimental studies show lower nasal particle deposition in human children than adults. Relative to adults, children also tend to breathe more through their mouth, which is less efficient for removing inhaled particles than the nose. These findings suggest that the lower respiratory tract of children may receive a higher dose of ambient particles compared to adults. Because children breathe at higher minute ventilations relative to their lung volumes, the rate of particle deposition normalized to lung surface area may be further increased relative to adults.

Modeling studies indicate that, for particles greater than ~0.01 μm, some cells located near the carinal ridge of bronchial bifurcations may receive hundreds to thousands of times the average dose (particles per unit surface area) of the parent and daughter airways. The inertial impaction of particles ≥1 μm d_{ae} at the carinal ridge of large bronchi increases with increasing inspiratory flows. Airway constriction can further augment the overall deposition efficiency of coarse particles at downstream bifurcations. These findings suggest that substantial doses of particles (≥1 μm d_{ae}) may be justified for in vitro studies using TB epithelial cell cultures.

3. DISCLAIMER

This document has been reviewed in accordance with U.S. Environmental Protection Agency policy and approved for publication. Mention of trade names or commercial products does not constitute endorsement or recommendation for use. The views expressed in this article are those of the authors and do not necessarily reflect the views or policies of the U.S. Environmental Protection Agency.

REFERENCES

ACGIH, 2005. TLVs and BEIs: Based on the Documentation of the Threshold Limit Values for Chemical Substances and Physical Agents and Biological Exposure Indices. American Conference of Governmental Industrial Hygienists, Cincinnati, Ohio.

Alessandrini, F., Semmler-Behnke, M., Jakob, T., Schulz, H., Behrendt, H., Kreyling, W., 2008. Total and regional deposition of ultrafine particles in a mouse model of allergic inflammation of the lung. Inhal. Toxicol. 20, 585–593.

Alexander, D.J., Collins, C.J., Coombs, D.W., Gilkison, I.S., Hardy, C.J., Healey, G., Karantabias, G., Johnson, N., Karlsson, A., Kilgour, J.D., McDonald, P., 2008. Association of inhalation toxicologists (AIT) working party recommendation for standard delivered dose calculation and expression in non-clinical aerosol inhalation toxicology studies with pharmaceuticals. Inhal. Toxicol. 20 (13), 1179–1189.

Amis, T.C., Pascoe, J.R., Hornof, W., 1984. Topographic distribution of pulmonary ventilation and perfusion in the horse. Am. J. Vet. Res. 45 (8), 1597–1601.

Anselm, A., Heibel, T., Gebhart, J., Ferron, G.A., 1990. In vivo studies of growth factors of sodium chloride particles in the human respiratory tract. J. Aerosol Sci. 21, S427–S430.

Anthony, T.R., Flynn, M.R., 2006. Computational fluid dynamics investigation of particle inhalability. J. Aerosol Sci. 37, 750–765.

Asgharian, B., Yu, C.P., 1989a. A simplified model of interception deposition of fibers at airway bifurcations. J. Aerosol Sci. Technol. 11, 80–88.

Asgharian, B., Yu, C.P., 1989b. Deposition of fibers in the rat lung. J. Aerosol Sci. 20, 355–366.

Asgharain, B., Kelly, J.T., Tewksbury, E.W., 2003. Respiratory deposition and inhalability of monodisperse aerosols in Long-Evans rats. Toxicol. Sci. 71, 104–111.

Bailey, M., Dorrian, M., Birchall, A., 1995. Implications of airway retention for radiation doses from inhaled radionuclides. J. Aerosol Med. 8, 373–390.

Balashazy, I., Hofmann, W., Heistracher, T., 1999. Computation of local enhancement factors for the quantification of particle deposition patterns in airway bifurcations. J. Aerosol Sci. 30, 185–203.

Balashazy, I., Hofmann, W., Heistracher, T., 2003. Local particle deposition patterns may play a key role in the development of lung cancer. J. Appl. Physiol. 94, 1719–1725.

Becquemin, M.H., Swift, D.L., Bouchikhi, A., Roy, M., Teillac, A., 1991. Particle deposition and resistance in the noses of adults and children. Eur. Respir. J. 4, 694–702.

Becquemin, M.M., Bertholon, J.F., Bouchikhi, A., Malarbet, J.L., Roy, M., 1999. Oronasal ventilation partitioning in adults and children: effect on aerosol deposition in airways. Radiat. Prot. Dosim. 81, 221–228.

Bennett, W.D., Zeman, K.L., 1998. Deposition of fine particles in children spontaneously breathing at rest. Inhal. Toxicol. 10, 831–842.

Bennett, W.D., Zeman, K.L., 2004. Effect of body size on breathing pattern and fine-particle deposition in children. J. Appl. Physiol. 97, 821–826.

Bennett, W.D., Zeman, K.L., 2005. Effect of race on fine particle deposition for oral and nasal breathing. Inhal. Toxicol. 17, 641–648.

Bennett, W., Messina, M., Smaldone, G., 1985. Effect of exercise on deposition and subsequent retention of inhaled particles. J. Appl. Physiol. 59, 1046–1054.

Bennett, W.D., Zeman, K.L., Kim, C., 1996. Variability of fine particle deposition in healthy adults: effect of age and gender. Am. J. Respir. Crit. Care Med. 153, 1641–1647.

Bennett, W.D., Scheuch, G., Zeman, K.L., Brown, J.S., Kim, C., Heyder, J., Stahlhofen, W., 1998. Bronchial airway deposition and retention of particles in inhaled boluses: effect of anatomic dead space. J. Appl. Physiol. 85 (2), 685–694.

Bennett, W.D., Brown, J.S., Zeman, K.L., Hu, S.C., Scheuch, G., Sommerer, K., 2002. Targeting delivery of aerosols to different lung regions. J. Aerosol Med. 15 (2), 179–188.

Bennett, W., Zeman, K., Jarabek, A., 2003. Nasal contribution to breathing with exercise: effect of race and gender. J. Appl. Physiol. 95, 497–503.

Bennett, W.D., Zeman, K.L., Jarabek, A.M., 2008. Nasal contribution to breathing and fine particle deposition in children versus adults. J. Toxicol. Environ. Health A 71, 227–237.

Breysse, P.N., Swift, D.L., 1990. Inhalability of large particles into the human nasal passage: in vivo studies in still air. Aerosol Sci. Technol. 13, 459–464.

Brown, J.S., 2005. Particle inhalability at low wind speeds. Inhal. Toxicol. 17, 831–837.

Brown, J., Bennett, W., 2004. Deposition of coarse particles in cystic fibrosis: model predictions versus experimental results. J. Aerosol Med. 17, 239–248.

Brown, J.S., Zeman, K.L., Bennett, W.D., 2001. Regional deposition of coarse particles and ventilation distribution in healthy subjects and patients with cystic fibrosis. J. Aerosol Med. 14 (4), 443–454.

Brown, J.S., Wilson, W.E., Grant, L.D., 2005. Dosimetric comparisons of particle deposition and retention in rats and humans. Inhal. Toxicol. 17, 355–385.

Brown, J.S., Gordon, T., Price, O., Asgharian, B., 2013. Thoracic and respirable particle definitions for human health risk assessment. Part Fibre Toxicol. 10, 12.

Chadha, T.S., Birch, S., Sackner, M.A., 1987. Oronasal distribution of ventilation during exercise in normal subjects and patients with asthma and rhinitis. Chest 92, 1037–1041.

Chamberlain, M.J., Morgan, W.K., Vinitski, S., 1983. Factors influencing the regional deposition of inhaled particle in man. Clin. Sci. 64, 69–78.

Cheng, K.-H., Cheng, Y.-S., Yeh, H.-C., Guilmette, R.A., Simpson, S.Q., Yang, Y.-H., Swift, D.L., 1996. In vivo measurements of nasal airway dimensions and ultrafine aerosol deposition in the human nasal and oral airways. J. Aerosol Sci. 27, 785–801.

Cohen, B.S., Xiong, J.Q., Fang, C.-P., Li, W., 1998. Deposition of charged particles on lung airways. Health Phys. 74, 554–560.

Cuddihy, R.G., Kanapilly, G.M., Raabe, O.G., November 1, 1969. Relationship between particles size and respiratory deposition of ^{140}La-labeled aerosols in the beagle dog. In: Fission Product Inhalation Program Annual Report 1968–1969. Inhalation Toxicology Research Institute, Albuquerque, NM, pp. 82–87. Report LF-41. Available at: http://dspace.lrri.org:8080/xmlui/handle/123456789/58.

Dai, Y.T., Juang, Y.J., Wu, Y., Breysse, P.N., Hsu, D.J., 2006. In vivo measurements of inhalability of ultralarge aerosol particles in calm air by humans. J. Aerosol Sci. 37, 967–973.

Diu, C.K., Yu, C.P., 1983. Respiratory tract deposition of polydisperse aerosols in humans. Am. Ind. Hyg. Assoc. J. 44, 62–65.

Dunster, K.R., Friese, M., Fraser, J.F., Cowin, G.J., Schibler, A., 2012. Ventilation distribution in rats: part I—the effect of gas composition as measured with electrical impedance tomography. Biomed. Eng. Online 11, 64.

Falk, R., Philipson, K., Svartengren, M., Jarvis, N., Bailey, M., Camner, P., 1997. Clearance of particles from small ciliated airways. Exp. Lung Res. 23, 495–515.

Farkas, A., Balashazy, I., SzQcs, K., 2006. Characterization of regional and local deposition of inhaled aerosol drugs in the respiratory system by computational fluid and particle dynamics methods. J. Aerosol Med. 19, 329–343.

Farkas, A., Balásházy, I., 2008. Quantification of particle deposition in asymmetrical tracheobronchial model geometry. Comput. Biol. Med. 38, 508–518.

Finlay, W.H., Martin, A.R., 2008. Recent advances in predictive understanding of respiratory tract deposition. J. Aerosol Med. Pulm. Drug Deliv. 21, 189–206.

Gebhart, J., 1992. To the relevant diameter of aerosol particles in the 0.1 to 1 μm transition range. J. Aerosol Sci. 23 (Suppl.), S305–S308.

Gebhart, J., Heyder, J., 1985. Removal of aerosol particles from stationary air within porous media. J. Aerosol Sci. 16, 175–187.

Goodman, R.M., Yergin, B.M., Landa, J.F., Golinvaux, M.H., Sackner, M.A., 1978. Relationship of smoking history and pulmonary function tests to tracheal mucous velocity in nonsmokers, young smokers, ex-smokers, and patients with chronic bronchitis. Am. Rev. Respir. Dis. 117, 205–214.

Grgic, B., Finlay, W.H., Burnell, P.K.P., Heenan, A.F., 2004. In vitro intersubject and intrasubject deposition measurements in realistic mouth–throat geometries. J. Aerosol Sci. 35, 1025–1040.

Heistracher, T., Hofmann, W., 1997. Flow and deposition patterns in successive airway bifurcations. Presented at. In: Cherry, N., Ogden, T. (Eds.), Inhaled Particles VIII: Proceedings of an International Symposium on Inhaled Particles Organised by the British Occupational Hygiene Society; August 1996, vol. 41, pp. 537–542. Cambridge, UK. Ann. Occup. Hyg. Suppl.

Heyder, J., Gebhart, J., Scheuch, G., 1985. Interaction of diffusional and gravitational particle transport in aerosols. Aerosol Sci. Technol. 4, 315–326.

Hinds, W.C., Kennedy, N.J., Tatyan, K., 1998. Inhalability of large particles from mouth and nose breathing. J. Aerosol Sci. 29 (S1), S277–S278.

Ho, J.C., Chan, K.N., Hu, W.H., Lam, W.K., Zheng, L., Tipoe, G.L., Sun, J., Leung, R., Tsang, K.W., 2001. The effect of aging on nasal mucociliary clearance, beat frequency, and ultrastructure of respiratory cilia. Am. J. Respir. Crit. Care Med. 163, 983–988.

Hofmann, W., Asgharian, B., 2003. The effect of lung structure on mucociliary clearance and particle retention in human and rat lungs. Toxicol. Sci. 73, 448–456.

Hoffman, E.A., Ritman, E.L., 1985. Effect of body orientation on regional lung expansion in dog and sloth. J. Appl. Physiol. 59 (2), 481–491.

Hofmann, W., Martonen, T.B., Graham, R.C., 1989. Predicted deposition of nonhygroscopic aerosols in the human lung as a function of subject age. J. Aerosol Med. 2, 49–68.

Hofmann, W., Balashazy, I., Heistracher, T., Koblinger, L., 1996. The significance of particle deposition patterns in bronchial airway bifurcations for extrapolation modeling. Aerosol Sci. Technol. 25, 305–327.

Hsu, D.J., Swift, D.L., 1999. The measurements of human inhalability of ultralarge aerosols in calm air using mannikins. J. Aerosol Sci. 30, 1331–1343.

ICRP, 1994. Human respiratory tract model for radiological protection: a report of a task group of the International Commission on Radiological Protection. Ann. ICRP 24, 1–482.

ISO, 2008. Nanotechnologies – Terminology and Definitions for Nano-objects – Nanoparticle, Nanofibre and Nanoplate. International Organization for Standardization, Geneva.

Isaacs, K.K., Schlesinger, R.B., Martonen, T.B., 2006. Three-dimensional computational fluid dynamics simulations of particle deposition in the tracheobronchial tree. J. Aerosol Med. 19, 344–352.

James, D.S., Stidley, C.A., Lambert, W.E., Chick, T.W., Mermier, C.M., Samet, J.M., 1997. Oronasal distribution of ventilation at different ages. Arch. Environ. Occup. Health 52, 118–123.

Jarabek, A.M., Asgharian, B., Miller, F.J., 2005. Dosimetric adjustments for interspecies extrapolation of inhaled poorly soluble particles (PSP). Inhal. Toxicol. 17, 317–334.

Jeffers, D.E., 2005. Relative magnitudes of the effects of electrostatic image and thermophoretic forces on particles in the respiratory tract. Radiat. Prot. Dosim. 113, 189–194.

Kelly, J.T., Asgharian, B., Wong, B.A., 2005. Inertial particle deposition in a monkey nasal mold compared with that in human nasal replicas. Inhal. Toxicol. 17, 823–830.

Kim, C.S., 2000. Methods of calculating lung delivery and deposition of aerosol particles. Respir. Care 45, 695–711.

Kim, C.S., Hu, S.C., 1998. Regional deposition of inhaled particles in human lungs: comparison between men and women. J. Appl. Physiol. 84, 1834–1844.

Kim, C.S., Iglesias, A.J., 1989. Deposition of inhaled particles in bifurcating airway models: I inspiratory deposition. J. Aerosol Med. 2, 1–14.

Kim, C.S., Jaques, P.A., 2000. Respiratory dose of inhaled ultrafine particles in healthy adults. Philos. Trans. A Math. Phys. Eng. Sci. 358, 2693–2705.

Kim, C.S., Brown, L.K., Lewars, G.G., Sackner, M.A., 1983. Deposition of aerosol particles and flow resistance in mathematical and experimental airway models. J. Appl. Physiol. Respir. Environ. Exersice Physiol. 55 (1), 154–163.

Kim, C.S., Iglesias, A.J., Garcia, L., 1989. Deposition of inhaled particles in bifurcating airway models: II. Expiratory deposition. J. Aerosol Med. 2, 15–27.

King, G.G., Eberl, S., Salome, C.M., Young, I.H., Woolcock, A.J., 1998. Differences in airway closure between normal and asthmatic subjects measured with single-photon emission computed tomography and technegas. Am. J. Respir. Crit. Care Med. 158, 1900–1906.

Kreyling, W.G., Blanchard, J.D., Godleski, J.J., Haeussermann, S., Heyder, J., Hutzler, P., Schulz, H., Sweeney, T.D., Takenaka, S., Ziesenis, A., 1999. Anatomic localization of 24- and 96-h particle retention in canine airways. J. Appl. Physiol. 87, 269–284.

Kuehl, P.J., Anderson, T.L., Candelaria, G., Gershman, B., Harlin, K., Hesterman, J.Y., Holmes, T., Hoppin, J., Lackas, C., Norenberg, J.P., Yu, H., McDonald, J.D., 2012. Regional particle size dependent deposition of inhaled aerosols in rats and mice. Inhal. Toxicol. 24, 27–35.

Löndahl, J., Massling, A., Pagels, J., Swietlicki, E., Vaclavik, E., Loft, S., 2007. Size-resolved respiratory-tract deposition of fine and ultrafine hydrophobic and hygroscopic aerosol particles during rest and exercise. Inhal. Toxicol. 19 (2), 109–116.

Madl, P., Hofmann, W., Oldham, M.J., Asgharian, B., 2010. Stochastic morphometric model of the BALB/c mouse lung. Anat. Rec. (Hoboken) 293 (10), 1766–1775.

Martonen, T.B., Yang, Y., Xue, Z.Q., 1994. Effects of carinal ridge shapes on lung airstreams. Aerosol Sci. Technol. 21, 119–136.

Mauderly, J.L., 1979. Effect of age on pulmonary structure and function of immature and adult animals and man. Fed. Proc. 38, 173–177.

Mauderly, L.J., Kritchevsky, J., 1979. Respiration of Unsedated Fischer 344 Rats and the Effect of Confinement in Exposure Tubes. Inhalation Toxicology Research Institute, Albuquerque, NM. Annual Report, 1978–1979. Report LF-69.

Ménache, M.G., Miller, F.J., Raabe, O.G., 1995. Particle inhalability curves for humans and small laboratory animals. Ann. Occup. Hyg. 39, 317–328.

Méndez, L.B., Gookin, G., Phalen, R.F., 2010. Inhaled aerosol particle dosimetry in mice: a review. Inhal. Toxicol. 22 (Suppl. 2), 15–20.

Mercer, R.R., Russell, M., Crapo, J.D., 1994. Alveolar septal structure in different species. J. Appl. Physiol. 77, 1060–1066.

Milic-Emili, J., Henderson, J.A.M., Dolovich, B.M., Trop, D., Kaneko, K., 1966. Regional distribution of inspired gas in the lung. J. Appl. Physiol. 21, 749–759.

Millage, K.K., Bergman, J., Asgharian, B., McClellan, G., 2010. A review of inhalability fraction models: discussion and recommendations. Inhal. Toxicol. 22 (2), 151–159.

Möller, W., Haussinger, K., Winkler-Heil, R., Stahlhofen, W., Meyer, T., Hofmann, W., Heyder, J., 2004. Mucociliary and long-term particle clearance in the airways of healthy nonsmoker subjects. J. Appl. Physiol. 97, 2200–2206.

Möller, W., Meyer, G., Scheuch, G., Kreyling, W.G., Bennett, W.D., 2009. Left-to-right asymmetry of aerosol deposition after shallow bolus inhalation depends on lung ventilation. J. Aerosol Med. Pulm. Drug Deliv. 22 (4), 333–339.

Morgan, W., Ahmad, D., Chamberlain, M., Clague, H., Pearson, M., Vinitski, S., 1984. The effect of exercise on the deposition of an inhaled aerosol. Respir. Physiol. 56, 327–338.

Niinimaa, V., Cole, P., Mintz, S., Shephard, R.J., 1981. Oronasal distribution of respiratory airflow. Respir. Physiol. Neurobiol. 43, 69–75.

Oldham, M.J., Robinson, R.J., 2007. Predicted tracheobronchial and pulmonary deposition in a murine asthma model. Anat. Rec. (Hoboken) 290, 1309–1314.

O'Riordan, T.G., Smaldone, G.C., 1994. Regional deposition and regional ventilation during inhalation of pentamidine. Chest 105, 395–401.

O'Riordan, T.G., Iaconc, A., Keenan, R.J., Duncan, S.R., Burckart, G.J., Griffith, B.P., Smaldone, G.C., 1995. Delivery and distribution of aerosolized cyclosporine in lung allograft recipients. Am. J. Resp. Crit. Care Med. 151 (2), 516–521.

Phalen, R.F., Oldham, M.J., 2006. Aerosol dosimetry considerations. Clin. Occup. Environ. Med. 5, 773–784.

Phalen, R.F., Oldham, M.J., Mautz, W.J., 1989. Aerosol deposition in the nose as a function of body size. Health Phys. 57, 299–305.

Phalen, R.F., Oldham, M.J., Wolff, R.K., 2008. The relevance of animal models for aerosol studies. J. Aerosol Med. Pulm. Drug Deliv. 21, 113–124.

Puchelle, E., Zahm, J.-M., Bertrand, A., 1979. Influence of age on bronchial mucociliary transport. Scand. J. Respir. Dis. 60, 307–313.

Raabe, O.G., Al-Bayati, M.A., Teague, S.V., Rasolt, A., 1988. Regional deposition of inhaled monodisperse, coarse, and fine aerosol particles in small laboratory animals. In: Inhaled Particles VI: Proceedings of an International Symposium and Workshop on Lung Dosimetry. Pergamon Press, Cambridge, United Kingdom, pp. 53–63.

Rooney, D., Friese, M., Fraser, J.F., R Dunster, K., Schibler, A., 2009. Gravity-dependent ventilation distribution in rats measured with electrical impedance tomography. Physiol. Meas. 30 (10), 1075–1085.

Roth, C., Scheuch, G., Stahlhofen, W., 1993. Clearance of the human lungs for ultrafine particles. J. Aerosol Sci. 24, S95–S96.

Scheuch, G., Gebhart, J., Roth, C., 1990. Uptake of electrical charges in the human respiratory tract during exposure to air loaded with negative ions. J. Aerosol Sci. 21, S439–S442.

Schramel, J., Nagel, C., Auer, U., Palm, F., Aurich, C., Moens, Y., 2012. Distribution of ventilation in pregnant Shetland ponies measured by electrical impedance tomography. Respir. Physiol. Neurobiol. 180 (2–3), 258–262.

Schroeter, J.D., Kimbell, J.S., Asgharian, B., 2006. Analysis of particle deposition in the turbinate and olfactory regions using a human nasal computational fluid dynamics model. J. Aerosol Med. Pulm. Drug Deliv. 19, 301–313.

Smaldone, G.C., Fuhrer, J., Steigbigel, R.T., McPeck, M., 1991. Factors determining pulmonary deposition of aerosolized pentamidine in patients with human immunodeficiency virus infection. Am. Rev. Respir. Dis. 143, 727–737.

Smith, J.R., Bailey, M.R., Etherington, G., Shutt, A.L., Youngman, M.J., 2008. Effect of particle size on slow particle clearance from the bronchial tree. Exp. Lung Res. 34, 287–312.

Snipes, M.B., Boecker, B.B., McClellan, R.O., 1983. Retention of monodisperse or polydisperse aluminosilicate particles inhaled by dogs, rats, and mice. Toxicol. Appl. Pharmacol. 69, 345–362.

Soderholm, S.C., 1985. Size-selective sampling criterial for inspirable mass fraction. In: Particle Size-selective Sampling in the Workplace: Report of the ACGIH Technical Committee on Air Sampling Procedures. American Conference of Governmental Industrial Hygienists, Cincinnati, Ohio, pp. 27–32.

Stone, K.C., Mercer, R.R., Gehr, P., Stockstill, B., Crapo, J.D., 1992. Allometric relationships of cell numbers and size in the mammalian lung. Am. J. Respir. Cell. Mol. Biol. 6, 235–243.

Suga, K., Nishigauchi, K., Kume, N., Koike, S., Takano, K., Matsunaga, N., 1995. Regional ventilatory evaluation using dynamic SPET imaging of xenon-133 washout in obstructive lung disease: an initial study. Eur. J. Nucl. Med. 22, 220–226.

Susskind, H., Brill, A.B., Harold, W.H., 1986. Quantitative comparison of regional distributions of inhaled Tc-99m DTPA aerosol and Kr-81m gas in coal miners' lungs. Am. J. Physiol. Imaging 1, 67–76.

Svartengren, M., Falk, R., Philipson, K., 2005. Long-term clearance from small airways decreases with age. Eur. Respir. J. 26, 609–615.

Tabachnik, E., Muller, N., Toye, B., Levison, H., 1981. Measurement of ventilation in children using the respiratory inductive plethysmograph. J. Pediatr. 99, 895–899.

Takezawa, J., Miller, F.J., O'Neil, J.J., 1980. Single-breath diffusing capacity and lung volumes in small laboratory mammals. J. Appl. Physiol. Respir. Environ. Exerc. Physiol. 48, 1052–1059.

Tobin, M.J., Chadha, T.S., Jenouri, G., Birch, S.J., Gazeroglu, H.B., Sackner, M.A., 1983. Breathing patterns. 1. Normal subjects. Chest 84, 202–205.

Trajan, M., Logus, J.W., Enns, E.G., Man, S.F.P., 1984. Relationship between regional ventilation and aerosol deposition in tidal breathing. Am. Rev. Respir. Dis. 130, 64–70.

Tu, K.W., Knutson, E.O., 1984. Total deposition of ultrafine hydrophobic and hygroscopic aerosols in the human respiratory system. Aerosol Sci. Technol. 3, 453–465.

Ultman, J.S., 1985. Gas transport in the conduction airways. In: Engel, L.A., Paiva, M. (Eds.), Gas Mixing and Distribution in the Lung, vol. 25. Dekker, New York, pp. 63–136.

Valberg, P., Brain, J., Sneddon, S., LeMott, S., 1982. Breathing patterns influence aerosol deposition sites in excised dog lungs. J. Appl. Physiol. 53, 824–837.

Vastag, E., Matthys, H., Kohler, D., Gronbeck, L., Daikeler, G., 1985. Mucociliary clearance and airways obstruction in smokers, ex-smokers and normal subjects who never smoked. Eur. J. Respir. Dis. 139, 93–100.

de Winter-Sorkina, R., Cassee, F.R., 2002. From Concentration to Dose: Factors Influencing Airborne Particulate Matter Deposition in Humans and Rats. National Institute of Public Health and the Environment RIVM, Bilthoven, The Netherlands. Report 650010031/2002. Available at: http://www.rivm.nl/bibliotheek/rapporten/650010031.html.

Weibel, E.R., 1963. Morphology of the Lung. Springer Verlag and Academic Press, Berlin, New York.

Whaley, S.L., Muggenburg, B.A., Seiler, F.A., Wolff, R.K., 1987. Effect of aging on tracheal mucociliary clearance in beagle dogs. J. Appl. Physiol. 62, 1331–1334.

Whitby, K.T., 1978. The physical characteristics of sulfur aerosols. Atmos. Environ. 12, 135–159.

Xu, G.B., Yu, C.P., 1986. Effects of age on deposition of inhaled aerosols in the human lung. Aerosol Sci. Technol. 5, 349–357 (rere).

Yeh, H.-C., March 1, 1980. Respiratory Tract Deposition Models. Inhalation Toxicology Research Institute, Albuquerque, NM. Report LF-72. Available at: http://dspace.lrri.org:8080/xmlui/handle/123456789/103.

Yeh, H.C., Schum, G.M., 1980. Models of human lung airways and their application to inhaled particle deposition. Bull. Math. Biol. 42, 461–480.

Yeh, H.C., Schum, G.M., Duggan, M.T., 1979. Anatomic models of the tracheobronchial and pulmonary regions of the rat. Anat. Rec. 195, 483–492.

Yu, C.P., 1985. Theories of electrostatic lung deposition of inhaled aerosols. Ann. Occup. Hyg. 29, 219–227.

Zhao, X., Ju, Y., Liu, C., Li, J., Huang, M., Sun, J., Wang, T., 2009. Bronchial anatomy of left lung: a study of multi-detector row CT. Surg. Rad. Anat. 31, 85–91.

Chapter 28

Deposition of Inhaled Gases and Vapors

John B. Morris

Chapter Outline

1. INTRODUCTION

Inhalation dosimetry is an essential aspect of the inhalation toxicology of inspired materials as it describes the relationship between the inspired concentration and the dose delivered to critical targets within the respiratory tract (USEPA, 2009; Morris, 2012). Absorption of gases and vapors in the proximal airways limits penetration to the distal airways and, thus, serves as an important respiratory defense mechanism (Brain, 1979). Indeed, the absence of deleterious effects from an inspired gas or vapor in the bronchiolar airways may simply be due to the fact that the chemical is completely absorbed in the large airways and does not penetrate to the distal bronchioli (Gloede et al., 2011). That the upper respiratory tract (URT, all airways proximal to the larynx) provides an essential defense mechanism has long been appreciated (Jordan and Carlson, 1913). While absorption of inspired gases or vapors in the nose serves to protect the lower airways, it places the nose at risk of injury. In fact, the nasal passages are an extremely common target for inhaled gas or vapor-induced injury in the rodent. A large database exists on the ability of the nose to scrub (i.e., absorb) inspired gases and vapors from the inspired airstream. This is likely because of the importance of this process and because it is straightforward to isolate the URT (via a tracheotomy or other means) and directly study vapor absorption in that site. This chapter summarizes the rich database on gas and vapor absorption in the URT, with an emphasis on comparisons among species. In this chapter, the terms *gas* and *vapor* are considered interchangeable. Technically, a vapor is the gaseous state of a substance that is liquid or solid under ordinary conditions. There is, however, no significant physical or chemical difference between a vapor and a gas, and vapor can be considered to be a particular type of gas.

All reports of URT vapor absorption rely on a fundamentally identical approach (Morris, 1999). A sampling tube is inserted into the nasopharynx, which allows for measurement of vapor concentration in that site. By comparison of the nasopharyngeal concentration to the inspired concentration, a measurement of URT absorption efficiency is obtained. The nasopharyngeal sampling tube can be inserted via many approaches. In laboratory animals it can be done by performing a tracheotomy and inserting a tube anteriorly through the larynx into the nasopharynx. This requires the use of long-acting anesthesia. In larger species, including humans, a tube can be inserted through the mouth into the posterior pharynx. Alternatively, a small sampling tube can be inserted through the external nares and nasal passages until its tip lies in the nasopharynx. However, this latter approach can induce physiological effects within the nose, and often local anesthetics and/or decongestants are required that may alter absorption pathways.

Various air sampling protocols have been used to assess URT vapor absorption. The most straightforward approach is to draw air continuously from the external nares to the nasopharynx at a constant velocity. The term *one-way* is used to describe this flow regime in this chapter. Although not physiologic, this provides definitive data on URT absorption efficiency at specific flow rates. Data so obtained can be used to validate a model that can then be used to estimate absorption under normal cyclic flow conditions (Gloede et al., 2011; Morris, 2012). Alternatively, air can be bled from the nasopharynx at low flow rates in animals (or humans) respiring normally. Data so obtained can be

Comparative Biology of the Normal Lung. http://dx.doi.org/10.1016/B978-0-12-404577-4.00028-X

used to validate modeling approaches, but it is difficult to directly assess the effect of inspiratory flow rate on uptake by this approach. This approach has been used in persons respiring normally and also has been used during a prolonged breath hold during which time air is drawn through the nose. A somewhat similar approach in rodents relies on performing a tracheotomy and connecting the sampling tube to a small animal respirator, such that air is drawn through the nose and then pushed (exhaled) back through the nose with a small air sampling rate (e.g., bleed) from the nasopharynx (e.g., Morris, 1997). A novel approach has been used in humans in which the "penetration volume" of a bolus of inhaled gas or vapor is assessed via sophisticated monitoring of the test gas concentration profile during expiration (Nodelman and Ultman, 1999).

Proper interpretation of URT uptake/absorption data requires knowledge of the mechanisms of uptake. The uptake/absorption process is quite complicated, but the fundamental processes involved have long been appreciated (Brain, 1979; Haggard, 1924). Unlike aerosol deposition, vapor/gas absorption is reversible, with the potential for vapor molecules to absorb/desorb from the air tissue interface, e.g., the mucous lining layer. In normal breathing, this absorption/desorption cycling occurs with absorption occurring during inspiration and desorption occurring during expiration when air that has been scrubbed of vapor in the deep lung passes back through the nose. It is important to recognize that desorption during expiration lowers the vapor concentration in nasal mucus, and this deficit is then made up during the next inspiration. Thus, the absorption/desorption cycling serves to increase uptake efficiencies in the nose and limit penetration to the deep lung (Dahl et al., 1991). This process cannot be mimicked by unidirectional (constant inspiration) flow regimes. However, data collected under unidirectional flow can be used to validate models that incorporate absorption/desorption cycling (Gloede et al., 2011; Hinderliter et al., 2005; Plowchalk et al., 1997).

The physical chemical properties of gases and vapors control the uptake process (Brain, 1979; Morris, 2012; USEPA, 2009). As described by Henry's law, gases and vapors reach saturation in fluids, with the concentration in the fluid phase being directly proportional to the concentration in the gas phase over the fluid. The proportionality constant can be expressed in many forms; the most common form for dosimetry studies is as a partition coefficient. Typically, the blood:air partition coefficient is used. This is the ratio of the concentration of vapor in the blood to that in the gas phase at equilibrium. Experimental studies using constant inspiratory flow regimes have revealed that vapor uptake in the nose quickly achieves a plateau value that is maintained for long periods of time. During this quasi-steady state, the amount of vapor entering nasal tissue must exactly equal the amount leaving, or being cleared, from nasal tissues (Morris, 1999; Morris et al., 2010). Thus, tissue clearance pathways are essential determinants of nasal vapor absorption efficiency. For nonreactive gases and/or vapors that are not metabolized, the only tissue clearance pathway is via the circulation (under constant inspiratory velocity flow). Thus, for these vapors, absorption efficiencies depend on the nasal blood flow rate and the solubility of vapor (e.g., blood:air partition coefficient). Nasal tissues are rich in xenobiotic metabolizing enzymes (Morris et al., 2010; Morris and Shusterman, 2010). Local vapor metabolism serves to enhance nasal vapor absorption because it clears vapor from nasal tissues; this steepens the air-to-tissue concentration gradient, which favors more absorption. Chemical reactivity, by decreasing tissue vapor concentrations, enhances vapor uptake in a manner analogous to local metabolism.

Since solubility, metabolism, and reactivity all influence the potential for vapor absorption in the nose, it is important to consider all three factors when interpreting nasal vapor uptake data. The most efficient uptake occurs for highly soluble vapors that are reactive and/or quickly metabolized. Indeed, essentially total absorption in the nose can occur for such vapors. This has been observed in many species, including humans. Table 1 summarizes published data on reactive gases. Low solubility vapors demonstrate much lower uptake potential; essentially, no nasal absorption occurs for low solubility vapors that are neither reactive nor metabolized. For nonmetabolized, nonreactive vapors, vapor solubility and nasal perfusion control uptake. Such vapors provide an assessment of the intrinsic ability of the nose to absorb vapor. Acetone vapor is an example of this type of vapor. Nasal acetone uptake has been measured in a variety of species and provides a data set for interspecies comparisons. Table 2 summarizes the published data on nasal acetone uptake in multiple species.

Local metabolism can profoundly enhance nasal uptake efficiencies. This phenomenon can be easily demonstrated by showing that administration of an enzymatic inhibitor diminishes uptake efficiencies. This was first recognized for ester vapors, which are substrates for carboxylesterase (Stott and McKenna, 1984). Carboxylesterases are high capacity enzymes that are expressed in high levels in the nose (Bodganffy et al., 1997). Table 3 summarizes the published data on nasal absorption of ester vapors. Table 4 summarizes the published data on organic vapors that are not directly reactive but are metabolized within nasal tissues. Data in normal animals and animals pretreated with enzymatic inhibitors are provided. In the inhibited animals it would be anticipated that uptake efficiency would mimic that for nonmetabolized vapors and be simply reflective of solubility (partition coefficient) and nasal perfusion. This has been observed (Gloede et al., 2011; Morris et al., 1993). Finally, it should be noted that metabolism is saturable; thus, vapor uptake of metabolized vapors should

TABLE 1 Reactive Inorganic Gases

Species/ Strain	Sex	Body Wt (kg) or age (year)	Est. Min Vent[a] (ml/min)	Exposure Concentration	Flow Regime[b]	MV[c] (cyclic) (ml/min)	Inspiratory Flow Rate[d] (ml/min)	Uptake[e] %	Length[f]	Method[g]	Ref
Sulfur Dioxide											
Human	M	27–39 year		16 ppm	Normal breathing	8500		99%	min	Oral tube	Speizer and Frank (1966)
Dog	–	–		7–87 ppm	One-way		3500	>90%	min	Tracheotomized	Yokoyama (1968)
Dog	–	15–18 kg		1 ppm	One-way		3500	99%	min	Tracheotomized	Frank et al. (1969)
				10 ppm	One-way		3500	99%	min	Tracheotomized	
				1 ppm	One-way		35,000	97%	min	Tracheotomized	
Rabbit	–	–		7–87 ppm	One-way		750	>90%	min	Tracheotomized	Yokoyama (1968)
				100–300 ppm	One-way		1000	93%	min		Dalhamn and Sjoholm (1963)
Chlorine											
Human	M/F	18–40 year		3 ppm	One-way		9000	92%	sec	Bolus inhalation, controlled breathing	Nodelman and Ultman (1999)
							15,000	93%			
Mouse	F	0.022 kg		5 ppm	One-way		25	97.5%	min	Tracheotomized	Morris et al. (2005)

Continued

TABLE 1 Reactive Inorganic Gases—cont'd

Species/ Strain	Sex	Body Wt (kg) or age (year)	Est. Min Vent[a] (ml/min)	Exposure Concentration	Flow Regime[b]	MV[c] (cyclic) (ml/min)	Inspiratory Flow Rate[d] (ml/min)	Uptake[e] %	Length[f]	Method[g]	Ref
Ammonia											
Human	–	–		50 mg/m^3	One-way		18,000	92%	sec	Oral tube	Landahl and Herrmann (1950)
Human	–	–			One-way		30,000	100%	sec	Oral tube	Pattle (1961)
Rabbit	–	–		2000 ppm	One-way		1000	95%	min	Tracheotomized	Dalham and Sjoholm (1963)
Hydrogen Fluoride											
Rat (Long–Evans)	M	0.29		36–176 mg/m^3	One-way		140	99.9%	min	Tracheotomized	Morris and Smith (1982)
Ozone											
Human	M/F	75 kg		0.1–0.4 ppm	Normal breathing	10,000	20,000	41%	sec	Intranasal tube to nasopharynx	Gerrity et al. (1988)
					Fast breathing	20,000	40,000	38%			
	M/F	18–40 year		1 ppm	One-way		9000	90%	sec	Bolus inhalation, controlled breathing	
							15,000	85%			
							60,000	45%			
	M/F	29–40 year		0.4 ppm	One-way		3000	78%	sec	Flow in on nostril, out other during breath hold	Santiago et al. (2001)
							5000	65%			
							8000	50%			
							15,000	33%			

Dog (beagle)	–	10 kg		2–8 ppm	One-way		3000	70%	min	Tracheotomized	Vaughan (1969)
Dog (mongrel)	M/F	13–18 kg		0.03 ppm	One-way		5000	72%	min	Tracheotomized	Yokoyama and Frank (1972)
				0.08 ppm			5000	59%			
				0.03 ppm			40,000	37%			
				0.08 ppm			40,000	27%			
Rabbit (New Zealand)	M/F	3 kg		0.1–2 ppm	One-way		160	50%	min	Tracheotomized	Miller et al. (1979)
Guinea pig (Hartley)	M/F	0.4		0.1–2 ppm	One-way		1000	50%	min	Tracheotomized	Miller et al. (1979)
Nitrogen Dioxide											
Dog	–	–		4–41 ppm	One-way		3500	40%	Min	Tracheotomized	Yokoyama (1968)
Rabbit	–	–		4–41 ppm	One-way		3500	40%	min	Tracheotomized	Yokoyama (1968)
Rabbit	–	–		20–400 ppm	One-way		1000	50%	min	Tracheotomized	Dalhamn and Sjoholm (1963)

[a]*Minute ventilation (min. vent) was estimated by the equation of Guyton as described by Phalen (xx) for laboratory animals; the ICRP value of 9000 ml/min for the sitting man was used for the human (xx).*
[b]*Various flow regimes have been used including constant velocity inspiratory flow (one-way) or methods based on sampling from the pharynx during normal breathing.*
[c]*The minute ventilation is listed for those procedures that relied on cyclic flow regimes.*
[d]*The inspiratory flow rate was used for those procedures that relied on one-way flow regimes.*
[e]*The reported average value is given. Standard errors typically average ~5%. Not all data reported in specific references are necessarily provided in this table. The reviewer is referred to the original reference for additional details.*
[f]*Length refers to the durations over which measurements were made. Sec refers to seconds and is listed for experiments of 1 min or less duration. Min refers to minutes and refers to experiments that used durations of measurement in excess of 1 min.*
[g]*The method column provides a brief description of the method used to insert an air sampling tube into the nasopharynx. A listing of oral indicates that the air sampling tube was placed in the oral cavity and not the nasopharynx.*

TABLE 2 Acetone

Species/ Strain	Sex	Body Wt (kg) or Age (year)	Est. Min Vent[a] (ml/min)	Exposure Concentration	Flow Regime[b]	Inspiratory Flow Rate[c] (ml/min)	Uptake[d] %	Length[e]	Method[f]	Ref
Human	M/F	80,000	9000	1 ppm	Normal breathing		55%	min	Nasopharyngeal tube	Thrall et al. (2003)
	M	–	9000	300–3000 mg/m^3	One-way	18,000	24%	sec	Oral tube, oral sampling	Landahl and Herrmann (1950)
						54,000	8%			
Guinea Pig										
Hartley	M	430	206	1200 mg/m^3	One-way	100	30%	min	Tracheotomized	Morris et al. (1986)
						200	13%			
						575	7%			
Rat										
Sprague–Dawley	M	320	165	1200 mg/m^3	One-way	150	40%	min	Tracheotomized	Morris and Cavanagh (1986)
						300	19%			
						500	21%			
Sprague–Dawley	M	240	133	100 mg/m^3	One-way	50	63%	min	Tracheotomized	Morris et al. (1991)
						100	43%			
						300	19%			
	F	180	107	100 mg/m^3	One-way	50	60%	min	Tracheotomized	
						100	42%			
						300	21%			

F344	M	180	107	1200 mg/m^3	One-way	50	45%	min	Tracheotomized	Morris et al. (1986)
						100	34%			
						300	12%			
F344	M	200	116	400 mg/m^3	One-way	50	38%	min	Tracheotomized	Morris (1993)
						100	25%			
						300	9%			
Hamster										
Syrian	M	110	74	1000 mg/m^3	One-way	40	23%	min	Tracheotomized	Morris and Cavanagh (1987)
						70	19%			
						200	5%			
Syrian	M	130	84	400 mg/m^3	One-way	35	27%	min	Tracheotomized	Morris (1993)
						70	19%			
						200	12%			
Mouse										
B6C3F1	M	20	21	>1500 mg/m^3	One-way	20	25%	min	Tracheotomized	Morris (1991)
						30	20%			
						70	14%			

[a]*Minute ventilation (min. vent) was estimated by the equation of Guyton as described by Phalen (1984) for laboratory animals; the ICRP value of 9000 ml/min for the sitting man was used for the human.*
[b]*Various flow regimes have been used including constant velocity inspiratory flow (one-way) or methods based on sampling from the pharynx during normal breathing.*
[c]*The inspiratory flow rate that was used for those procedures relied on one-way flow regimes.*
[d]*The reported average value is given. Standard errors typically average ~5%. Not all data reported in a specific references are necessarily provided in this table. The reviewer is referred to the original reference for additional details.*
[e]*Length refers to the durations over which measurements were made. Sec refers to seconds and is listed for experiments of 1 min or less duration. Min refers to minutes and refers to experiments that used durations of measurement in excess of 1 min.*
[f]*The method column provides a brief description of the method used to insert an air sampling tube into the nasopharynx. A listing of oral indicates that the air sampling tube was placed in the oral cavity and not the nasopharynx.*

TABLE 3 Ester Vapors

Species/ Strain	Sex	Body Wt (g)	Est. Min Vent[a] (ml/min)	Vapor	Part. Coeff.[b]	Concentration	Regime[c]	Insp. Flow Rate[d,e] (ml/ min)	Uptake[f] % normal	Uptake[g,h] % Min-metab	Method[i]	Ref[j]
Rat												
Sprague–Dawley	M	240	133	Dimethyl-succinate	–	50–100 mg/m^3	One-way	100	>98%		Tracheotomized	Morris et al. (1991)
				Dimethyl-glutarate	–	50–100 mg/m^3		100	>98%			
				Dimethyl-adipate	–	50–100 mg/m^3		100	>98%			
				Ethyl acetate	90	33 mg/m^3			51%			
	F	180	107	Dimethyl-succinate	–	50–100 mg/m^3	One-way	100	>98%			
				Dimethyl-glutarate	–	50–100 mg/m^3		100	>98%			
				Dimethyl-adipate	–	50–100 mg/m^3		100	>98%			
				Ethyl acetate	90	33 mg/m^3			57%			
Sprague–Dawley	M	300	157	Vinyl acetate	29	100 ppm	One-way	100	95%		Tracheotomized	Plowchalk et al. (1997)
						300 ppm		100	50%			
						500 ppm		100	40%			
						1000 ppm		100	45%			
						2000 ppm		100	30%			
F344	M	200	116	Ethyl acrylate	30	6 ppm	One-way	200	16%	13%	Tracheotomized	Morris and Frederick (1995)
						24 ppm		200	22%	15%		
						88 ppm		200	21	10		
						6 ppm	Cyclic	207[j]	24	16		
						24 ppm		207[j]	25	16		
						88 ppm		207[j]	14%	11%		

				Methyl methacrylate	33	6 ppm	One-way	200	18%	13%		
						24 ppm		200	20%	13%		
						88 ppm		200	9%	7%		
						6 ppm	Cyclic	207[j]	18%	13%		
						24 ppm		207[j]	20%	13%		
						88 ppm		207[j]	15%	9%		
F344	M	200	116	Ethyl acetate	74	50–800 mg/m^3	One-way	50	35%	25%	Tracheotomized	Morris (1990)
								100	21%	16%		
								300	10%	8%		
F344	M	230	129	Propylene glycol mono-methyl ether acetate	141	1000 ppm	One-way	53	94%[j]	55%[j]	Tracheotomized	Stott and McKenna (1984)
								100	95%[j]			

[a]Minute ventilation (min. vent) was estimated by the equation of Guyton as described by Phalen (1984) for laboratory animals; the ICRP value of 9000 ml/min for the sitting man was used for the human.
[b]The partition coefficient is listed if known. The value is the blood:air coefficient for nonreactive vapor and the water:air for reactive vapors.
[c]Various flow regimes have been used including constant velocity inspiratory flow (one-way) or methods based on sampling from the pharynx during normal breathing.
[d]The minute ventilation is listed for those procedures that relied on cyclic flow regimes.
[e]The inspiratory flow rate was used for those procedures that relied on one-way flow regimes. For cyclic flow the value is the average inspiratory flow rate. The cyclic flow regime consisted of connecting a rodent respirator to the tracheal tube and operating the pump at a tidal volume of 1 ml and 100 breaths per minute with a bleed (air sampling) of 7 ml/min being drawn from the nasopharynx. The average inspiratory flow rate under these conditions was 207 ml/min.
[f]The reported average value is given under normal (metabolically competent) conditions. Standard errors typically average ~5%. Not all data reported in specific references are necessarily provided in this table. The reviewer is referred to the original reference for additional details.
[g]Average uptake under conditions in which the effect of metabolism is diminished. This can be either by the use of specific enzymatic inhibitors or by the use of inspired concentrations sufficiently high to overwhelm local metabolic pathways.
[h]Length refers to the durations over which measurements were made. Sec refers to seconds and is listed for experiments of 1 min or less duration. Min refers to minutes and refers to experiments that used durations of measurement in excess of 1 min.
[i]The method column provides a brief description of the method used to insert an air sampling tube into the nasopharynx. A listing of oral indicates the air sampling tube was placed in the oral cavity and not the nasopharynx.
[j]Nonheated, nonhumidified air was used and no correction was made for the effect of heating/humidification on the nasopharyngeal vapor concentration.

TABLE 4 Non-reactive Organic Vapors

Species/ Strain	Sex	Body Wt (g)	Est. Min Vent[a] (ml/min)	Vapor	Part. Coef[b]	Conc.[c]	Regime[d]	Insp. Flow Rate[e] (ml/min)	Uptake % normal[f]	Uptake% Min-metab.[g]	Method[h]	Ref
Dog												
Beagle	M/F	7–12 kg	2000	Ethanol	1940	–	Normal breathing			88%	Oral-pharyngeal tube	Snipes et al. (1991)
				Dioxolane	500	–				58%		
				Butanone	200	–				50%		
				Propyl ether	10					7%		
				Dimethyl-pentane	1					0%		
Guinea Pig												
Hartley	M	320	206	Ethanol	1800	1200 mg/m^3	One-way	100		62%	Tracheotomized	Morris et al. (1986)
								200		54%		
								575		28%		
Rat												
Sprague–Dawley	M	320	165	Ethanol	1700	1200 mg/m^3	One-way	70		89%	Tracheotomized	Morris and Cavanagh (1986)
								100		85%		
								150		79%		
								300		61%		
								500		50%		
Sprague–Dawley	M	240	133	Isoamyl alcohol	1300	30 mg/m^3	One-way	100	71%		Tracheotomized	Morris et al. (1991)
	F	240	133					100	73%			

Sprague–Dawley	M	220	125	Styrene	40	50 ppm	One-way	70	17%		Tracheotomized	Morris (2000)
								150	9%			
								450	5.7%			
						5 ppm		70	24%	14%		
						10 ppm		70	22%	11%		
						25 ppm		70	15%	10%		
						50 ppm		70	13%	8%		
						100 ppm		70	9%	8%		
						200 ppm		70	10%	9%		
F344	M	320	165	Ethanol	1700	1200 mg/m^3	One-way	70		89%	Tracheotomized	Morris and Cavanagh (1986)
								100		85%		
								150		79%		
								300		60%		
								500		50%		
F344	M	240	133	Isoamyl alcohol	1300	30 mg/m^3	One-way	100		71%	Tracheotomized	Morris et al. (1991)
	F	180	107					100		43%		
	M	230	129	Naphthalene	725	1 ppm	One-way	150	57%	32%	Tracheotomized	Morris and Buckpitt (2009)
						3 ppm		150	49%	38%		
						10 ppm		150	37%	29%		

Continued

TABLE 4 Non-reactive Organic Vapors—cont'd

Species/ Strain	Sex	Body Wt (g)	Est. Min Vent[a] (ml/min)	Vapor	Part. Coef[b]	Conc.[c]	Regime[d]	Insp. Flow Rate[e] (ml/ min)	Uptake % normal[f]	Uptake% Min-metab.[g]	Method[h]	Ref
						30 ppm		150	36%	32%		
						1 ppm		300	32%	17%		
						3 ppm		300	29%	18%		
						10 ppm		300	20%	14%		
						30 ppm		300	20%	15%		
	F	140	89	Naphthalene	725	1 ppm	One-way	150	56%	26%		
						3 ppm		150	40%	25%		
						10 ppm		150	35%	29%		
						30 ppm		150	27%	26%		
						1 ppm		300	32%	19%		
						3 ppm		300	24%	8%		
						10 ppm		300	21%	16%		
						30 ppm		300	16%	41%		
	M	230	129	Propylene glycol mono-methyl ether	403	1000 ppm	One-way	53	96%[i]		Tracheotomized	Stott and McKenna (1984)
								105	95%[i]			
	M	200	116	Bromobenzene	53	170 mg/m^3	One-way	50	12%	9%	Tracheotomized	Morris (1993)
								300	4%	1%		
	M	230	129	Styrene	39	200 ppm	One-way	53	57%[i]		Tracheotomized	Stott and McKenna (1984)
								105	35%[i]			

	M	200	116	*o*-Xylene	25	100–500 mg/m^3	One-way	50	10%	6%	Tracheotomized	Morris (1993)
								300	2%	0.4%		
	M	230	129	Nitroethane	23	1000 ppm	One-way	53	65%[i]		Tracheotomized	Stott and McKenna (1984)
								105	50%[i]			
				Methylene chloride	10	500 ppm	One-way	53	13%[i]			
								105	5%[i]			
				Ethylene dibromide	4	75 ppm	One-way	53	49%[i]			
								105	30%[i]			
Hamster												
Syrian	M	110	74	1-propanol	1350	500 mg/m^3	One-way	40	65%		Tracheotomized	Morris and Cavanagh (1987)
								70	55%			
								200	20%			
	M	130	84	Isoamyl alcohol	1100	16–400 mg/m^3	One-way	35	63%	55%	Tracheotomized	Morris (1993)
								200	24%	21%		
				Bromobenzene	53	170 mg/m^3		35	12%	9%		
								200	3%	3%		
				o-Xylene	24	100–500 mg/m3		35	12%	7%		
								200	6%	2%		

Continued

TABLE 4 Non-reactive Organic Vapors—cont'd

Species/ Strain	Sex	Body Wt (g)	Est. Min Vent[a] (ml/min)	Vapor	Part. Coef[b]	Conc.[c]	Regime[d]	Insp. Flow Rate[e] (ml/min)	Uptake % normal[f]	Uptake% Min-metab.[g]	Method[h]	Ref
Mouse												
B6C3F1	M	30	28	Styrene	40	50 ppm		12	17%			
								24	10%			
								70	4%			
						5 ppm		12	42%	19%		
						10 ppm		12	30%	19%		
						25 ppm		12	24%	17%		
						50 ppm		12	18%	16%		
						100 ppm		12	13%	10%		
						200 ppm		12	10%	13%		

[a]*Minute ventilation (min. vent) was estimated by the equation of Guyton as described by Phalen (1984) for laboratory animals.*
[b]*The blood:air partition coefficient is listed if known.*
[c]*Exposure concentration in either ppm (v/v) or mg/m^3.*
[d]*Various flow regimes have been used including constant velocity inspiratory flow (one-way) or methods based on sampling from the pharynx during normal breathing.*
[e]*The inspiratory flow rate was used for those procedures that relied on one-way flow regimes.*
[f]*The reported average value is given under normal (metabolically competent) conditions. Standard errors typically average ~5%. Not all data reported in specific references are necessarily provided in this table. The reviewer is referred to the original reference for additional details. All measurements were made for durations of several minutes or longer.*
[g]*Average uptake under conditions in which the effect of metabolism is diminished. This can be either by the use of specific enzymatic inhibitors or by the use of inspired concentrations sufficiently high to overwhelm local metabolic pathways.*
[h]*The method column provides a brief description of the method used to insert an air sampling tube into the nasopharynx. A listing of oral indicates the air sampling tube was placed in the oral cavity and not the nasopharynx.*
[i]*Nonheated, nonhumidified air was used and no correction was made for the effect of heating/humidification on the nasopharyngeal vapor concentration.*

demonstrate saturation characteristics, with high uptake efficiencies being observed at low (nonsaturating) concentrations and low uptake efficiencies occurring at high (saturating) concentrations. This effect has also been observed. Many organic vapors are both directly metabolized and directly reactive. Published data on these vapors are summarized in Table 5. Perhaps the best studied in this regard is acetaldehyde, the uptake of which has been measured over a 1000-fold concentration range in multiple species under constant flow inspiratory and cyclic flow conditions (Table 6).

Two additional factors warrant consideration when interpreting nasal uptake data. A primary function of the nose is to heat and humidify the inspired airstream. This process, particularly the addition of water vapor, serves to dilute the airborne concentration of inspired vapors even if no absorption occurs. Precise measurement of nasal vapor absorption requires consideration of this factor. This can be controlled either by having the test animal inspire heated and humidified air (which also prevents nasal dehydration) and/or by correcting for the effect of heating/humidification. The duration of exposure is also an important consideration relative to interpretation of nasal vapor uptake. At the onset of exposure there is no vapor in nasal tissue; thus there is very efficient vapor uptake while the nasal tissue concentration rises to the plateau value. This can take seconds to minutes. The plateau value for uptake is then maintained and is controlled by the clearance pathways from the nose (see above). For vapors that are nonreactive and slowly metabolized, nasal vapor uptake will begin to diminish when the entire body begins to accumulate vapor molecules. This may take several hours. Theoretically, vapor uptake will diminish to zero when the entire body is fully saturated with vapors. This could occur, for example, with 24 h/7 day exposure to environmental air pollutants (Morris, 2012; Morris et al., 2010; Henderson and Haggard, 1943). The data reported in this chapter were all obtained during the initial wash-in period or during the plateau. The tables list the duration of exposure during which measurements were made, with listings of either "seconds" or "minutes." Measurements taken over minutes reflect the plateau value; measurements over seconds may, to some (unknown) degree, reflect the initial wash-in behavior as well.

2. REACTIVE INORGANIC GASES

For aid of interpretation, the summarized data on reactive inorganic vapors are itemized on the basis of gas property, with data for highly soluble, reactive vapors (sulfur dioxide, chlorine, ammonia, and hydrogen fluoride) given first, followed by data for low solubility reactive vapors (ozone, nitrogen dioxide) given second and third, respectively. Based on theoretical considerations (see above), highly efficient uptake would be anticipated for highly soluble reactive vapors. This has been observed in every species examined to date. For example, greater than 90% nasal uptake efficiency has been observed in multiple species for sulfur dioxide, chlorine, ammonia, and hydrogen fluoride. In some cases, undetectable amounts of vapor penetrated to the nasopharynx. If so, the measured uptake efficiency is based on the limit of detection rather than the ability of the nose to absorb vapor; such data are indicated with a ">" sign. Uptake of less soluble reactive gases (ozone, nitrogen dioxide) is less efficient, averaging between 30% and 70%, depending on the flow regime used for measurement. This same general trend (i.e., greater uptake for soluble than nonsoluble gases) is seen in all species. Decreased uptake efficiencies are observed at increasing flow rates. This is probably best demonstrated by the data for ozone uptake.

Direct comparisons among species are very difficult due to the differing flow regimes and durations that were used. Moreover, uptake is dependent on the inspiratory flow rate and, therefore, will depend on the flow rates used in each species relative to the flow rates that occur during normal breathing. For this reason, Table 1 provides the body weights (if known) of the subjects/experimental animals and an estimate of the normal minute ventilation. For laboratory animals the minute ventilation is estimated from the equation of Guyton, as described by Phalen (1984), and the ICRP value is used for humans (Gloede et al., 2011). Marked species differences are not apparent from this database. Even the human nose is capable of scrubbing 99% or more of soluble reactive vapors (e.g., sulfur dioxide). Based on the chlorine and ammonia data, the human nose may be less efficient than that of the mouse or rabbit, but the difference is subtle. It should be noted that modeling efforts for the soluble reactive vapor formaldehyde suggest that the human nose is less efficient than the rodent (Kimbell et al., 1993). Many studies report the use of both males and females, yet none reports a sex difference. To the authors' knowledge, no comprehensive data exist on nasal uptake of reactive gases in juveniles.

The following sections provide a discussion of comparative dosimetry for specific gases for which there is a fairly robust interspecies database.

3. ACETONE

Acetone is a nonreactive, nonmetabolized vapor with relatively high solubility (blood air partition coefficient of ~260). Uptake of this vapor has been measured in many species, including humans (Table 2). Because uptake of acetone is dependent simply on Henry's law partitioning, the uptake efficiency is independent of inspired concentration; no study has reported a concentration dependence for acetone uptake efficiency.

TABLE 5 Reactive Organic Vapors

Species/ Strain	Sex	Body Wt (g)	Est. Min Vent[a] (ml/min)	Vapor	Part. Coef[b]	Conc.[c]	Regime[d]	MV[e] (cyclic) (ml/min	Insp. Flow Rate[f] (ml/min)	Uptake % normal[g]	Uptake % Min-metab.[h]	Method[i]	Ref
Monkey													
Rhesus	–	–		Phosgene		90–4000 mg/m^3	Cyclic	500		35%		Tracheotomized	Cameron et al. (1946)
				Nitrogen mustard		1000–2000 mg/m^3				2%			
				Mustard gas		40–350 mg/m^3				80%			
Rabbit													
–	–	–	–	Phosgene		50–800 mg/m^3	Cyclic	1000		20%		Tracheotomized	
				Mustard gas		15–200 mg/m^3				85%			
				Nitrogen mustard		20–400 mg/m^3				87%			
Rat													
F344	M	200	116	Nicotine		4–400 mg/m^3	One-way		50	>90%		Tracheotomized	Morris (1999)
									300	>90%			
	M	200	116	Acrylic acid		131 mg/m^3			200	97%		Tracheotomized	Morris and Frederick (1995)
				Methacrylic acid		133 mg/m^3			200	95%			
	M	200	116	Acetaldehyde	140	6 ppm	One-way		100	72%	38%	Tracheotomized	Stanek and Morris (1999)
						10 ppm			100	55%	30%		
						50 ppm			100	50%	37%		
						200 ppm			100	40%	39%		
						300 ppm			100	35%	28%		
						1500 ppm			100	33%	31%		

	M	200	116	Acrolein	88	0.9 ppm	One-way		100	88%		Tracheotomized	Morris (1996)
						4.5 ppm			100	58%			
						9.0 ppm			100	47%			
						0.9 ppm			200	62%			
						4.5 ppm			200	38%			
						9.0 ppm			200	30%			
						0.9 ppm		100	207	75%			
						4.5 ppm		100	207	58%			
						9.0 ppm		100	207	50%			
						0.9 ppm			300	49%			
						4.5 ppm			300	33%			
						9.0 ppm			300	28%			
	M	190	112	Acrolein	88	0.6 ppm	One-way		100	98%	*	Tracheotomized	Struve et al. (2008)
						1.8 ppm			100	62%	*		
						3.6 ppm			100	50%	*		
						0.6 ppm			300	8%	*		
						1.8 ppm			300	48%	*		
						3.6 ppm			300	37%	*		
	M	200	116	Propylene oxide		25–300 ppm	One-way		50	23%		Tracheotomized	Morris et al. (2004)
									200	12%			
Sprague–Dawley	M	210	16	Butyric acid		30 ppm	One-way		400	>90%		Tracheotomized	Morris and Hubbs (2009)
				Diacetyl	550	100–300 ppm			100	75%			

Continued

TABLE 5 Reactive Organic Vapors—cont'd

Species/Strain	Sex	Body Wt (g)	Est. Min Vent[a] (ml/min)	Vapor	Part. Coef[b]	Conc.[c]	Regime[d]	MV[e] (cyclic) (ml/min	Insp. Flow Rate[f] (ml/min)	Uptake % normal[g]	Uptake % Min-metab.[h]	Method[i]	Ref
									200	44%			
									300	33%			
Mouse													
C56Bl/6J	F	23	21	Acrolein		1.1 ppm	One-way		25	>92%		Tracheotomized	Morris et al. (2003)
				Acetic acid		330 ppm			25	97%			
B6C3F1	M	30	28	Propylene oxide		25–500 ppm	One-way		12	29%		Tracheotomized	Morris and Pottenger (2006)
									50	14%			

Nonreactive Organic Vapors.
[a]*Minute ventilation (min. vent) was estimated by the equation of Guyton as described by Phalen (1984) for laboratory animals; the ICRP value of 9000 ml/min for the sitting man was used for the human.*
[b]*The water:air partition coefficient is listed if known; because of reactivity the blood:air partition coefficient is often unknown.*
[c]*Exposure concentration in either ppm (v/v) or mg/m*^{3}
[d]*Various flow regimes have been used including constant velocity inspiratory flow (one-way) or methods based on sampling from the pharynx during normal breathing.*
[e]*The minute ventilation is listed for those procedures that relied on cyclic flow regimes.*
[f]*The inspiratory flow rate was used for those procedures that relied on one-way flow regimes. For cyclic flow the value is the average inspiratory flow rate. The cyclic flow regime consisted of connecting a rodent respirator to the tracheal tube and operating the pump at a tidal volume of 1 ml and 100 breaths/min with a bleed (air sampling) of 7 ml/min being drawn from the nasopharynx. The average inspiratory flow rate under these conditions was 207 ml/min.*
[g]*The reported average value is given under normal (metabolically competent) conditions. Standard errors typically average ~5%. Not all data reported in specific references are necessarily provided in this table. The reviewer is referred to the original reference for additional details. All measurements were made for durations of several minutes or longer.*
[h]*Average uptake under conditions in which the effect of metabolism is diminished. This can be either by the use of specific enzymatic inhibitors or by the use of inspired concentrations sufficiently high to overwhelm local metabolic pathways.*
[i]*The method column provides a brief description of the method used to insert an air sampling tube into the nasopharynx. A listing of oral indicates the air sampling tube was placed in the oral cavity and not the nasopharynx.*

TABLE 6 Acetaldehyde Vapor

Species/ Strain	Sex	Body Wt (g)	Est. Min Vent[a] (ml/min)	Conc.[b]	Regime[c]	MV[d] (cyclic) (ml/min	Insp.[d] Flow Rate (ml/min)	Uptake[e] % normal	Ref
Guinea pig									
Harley	M	440	210	10 ppm	One-way		400	45	Morris (1997)
				100 ppm			400	11	
				1000 ppm			400	8	
					Cyclic	200	407	40	
						200	407	60	
						200	407	15	
Rat									
F344	M	200	116	1 ppm	One-way		200	65	Morris and Blanchard (1992)
				10 ppm			200	40	
				100 ppm			200	25	
				1000 ppm			200	24	
				1 ppm	Cyclic	100	207	65	
				10 ppm		100	207	45	
				100 ppm		100	207	40	
				1000 ppm		100	207	25	
Sprague–Dawley	M	275	150	1 ppm	One-way		200	58	Morris (1997)
				10 ppm			200	38	
				100 ppm			200	25	
				1000 ppm			200	15	
				1 ppm	Cyclic	100	207	57	
				10 ppm		100	207	50	
				100 ppm		100	207	27	
				1000 ppm		100	207	32	

Continued

TABLE 6 Acetaldehyde Vapor—cont'd

Species/ Strain	Sex	Body Wt (g)	Est. Min Vent[a] (ml/min)	Conc.[b]	Regime[c]	MV[d] (cyclic) (ml/min	Insp.[d] Flow Rate (ml/min)	Uptake[e] % normal	Ref
Hamster									
Syrian	M	140	88	1 ppm	One-way		150	62	Morris (1997)
				10 ppm			150	50	
				100 ppm			150	33	
				1000 ppm			150	15	
				1 ppm	Cyclic	70	147	72	
				10 ppm		70	147	68	
				100 ppm		70	147	50	
				1000 ppm		70	147	28	
Mouse									
B6C3F1	M	24	23	1 ppm	One-way		35	65	Morris (1997)
				10 ppm			35	42	
				100 ppm			35	44	
				1000 ppm			35	25	

Nonreactive Organic Vapors.
[a]*Minute ventilation (min. vent) was estimated by the equation of Guyton as described by Phalen (1984) for laboratory animals; the ICRP value of 9000 ml/min for the sitting man was used for the human.*
[b]*Exposure concentration in either ppm (v/v).*
[c]*Various flow regimes have been used including constant velocity inspiratory flow (one-way) or connection of the tracheotomy tube to a rodent respirator. All measurements were made in tracheotomized animals.*
[d]*The minute ventilation is listed for those procedures that relied on cyclic flow regimes. In all studies a nasopharyngeal bleed rate of 7 ml/min was used and a pump stroke rate of 100 breaths/min. Thus at a tidal (stroke) volume of 1.0 ml and 100 breaths/min the average minute ventilation was 100 ml/min and the average inspiratory flow rate (inspiration is $\frac{1}{2}$ of the breathing cycle) was 200 ml/min plus the 7 ml/min bleed or 207 ml/min.*
[e]*The reported average value is given under normal (metabolically competent) conditions. Standard errors typically average ~5%. Not all data reported in specific references are necessarily provided in this table. The reviewer is referred to the original reference for additional details. All measurements were made for durations of several minutes or longer.*

In all species, including humans, acetone uptake efficiency is strongly dependent on inspiratory flow rate, with less efficient uptake occurring at higher flows. Most studies report interindividual variability in uptake efficiency with standard errors of only ~5%. This complicates comparisons at high flow rates where uptake efficiencies are in the 5−10% range. Nonetheless, general trends across species lines are apparent by comparing uptake efficiencies at constant proportions for each species minute ventilation. Uptake appears to be the most efficient in the rat and least efficient in the guinea pig and hamster, with the mouse being intermediate. The same generalized trends are seen for most other vapors.

Because of the differing time scales and techniques used, direct comparison of animal data to humans is difficult, making precise comparisons impossible. The human data obtained via the nasopharyngeal catheter technique (Thrall et al., 2003) represent a promising new approach, but the method requires local nasal anesthesia as well as administration of a decongestant. In addition, data obtained by this method indicate an acetone concentration of zero in the pharynx at the end of inspiration, whereas multiple studies indicate that end expiratory concentrations are considerably higher (due to vapor desorption during exhalation, see Kumagai and Matsunaga (2000). In addition, the manner in which uptake efficiency was corrected for the dilution due to heating/humidification of the airstream is not clear. Thus, this technique may slightly overestimate absorption of this nonreactive, nonmetabolized vapor. Probably the most appropriate human data for comparison to animals are that obtained by Landahl and Herrman (1950). These authors relied on an oral cannula and measured uptake at constant flow during a 15−60 s breath hold. These data likely reflect the plateau period in animals. Acetone uptake in the human nose, as reported by this technique, is certainly within the range observed in animal species and appears similar to that reported for the rat. Modeling efforts do suggest, however, that nasal uptake of acetone is likely to be less efficient in humans than in rodents (Gloede et al., 2011).

Acetone uptake has been measured in male and female rats, and no sex differences were observed (Morris et al., 1991). Thrall et al. (2003) measured acetone uptake in a male and female human subjects and did not report a sex difference. To the authors' knowledge, no comprehensive data exist on nasal uptake of acetone in juveniles.

4. ESTER VAPOR

Organic esters are substrates for carboxylesterase, a high capacity enzyme that is expressed in the nose (Bogdanffy et al., 1987). Uptake of a variety of ester vapors has been examined because they all are nasal cytotoxicants (Table 3). Indeed the first study on the potential contribution of metabolism to vapor uptake was performed on propylene glycol monomethyl ether acetate (Stott and McKenna, 1984). All ester vapor uptake studies have been done using one-way inspiratory flow conditions. Use of carboxylesterase inhibitors has revealed that in all animal models (i.e., Sprague−Dawley rat, F344 rat, Syrian hamster), uptake is diminished by inhibitor treatment. This effect is seen under both one-way and cyclic flow conditions. At the doses used, the inhibitors may not have produced total inhibition of the high capacity esterase pathways, so carboxylesterase metabolism may still contribute to some degree in uptake in inhibitor treated animals.

The dibasic esters are water soluble and highly metabolized by carboxylesterase; greater than 95% uptake efficiencies are observed. No sex difference is apparent in uptake of these vapors, nor in uptake of ethyl acetate (Morris et al., 1991). The affinity of carboxylesterase for many esters is low (i.e., high Km); thus it is difficult to achieve saturating concentrations of esters in nasal tissue and, consequently, saturation of uptake is not always observed.

5. NONREACTIVE ORGANIC VAPORS

Extensive data exist on the nasal uptake of organic vapors that are not directly reactive but are substrates for nasal xenobiotic metabolizing enzymes (Table 4). Data have been reported for vapors metabolized by alcohol dehydrogenase (e.g., isoamyl alcohol, propanol), cytochrome P450 monoxygenases (styrene, naphthalene, bromobenzene), and aldehyde dehydrogenase (see below). Due to its toxicological importance, many studies have focused on P450 substrate vapors. Ideally, two data sets are obtained for such vapors, one in normal conditions and the other in conditions in which the effect of metabolism is minimized. This latter condition can be obtained by administration of appropriate enzyme inhibitors and/or by use of an inspired concentration sufficiently large that any metabolic pathway is overwhelmed. Invariably, the uptake efficiency is greater in the normal condition than the minimal metabolism condition providing strong evidence that local nasal metabolism serves to increase nasal vapor uptake. This has been observed experimentally in multiple species including the rat, hamster, and mouse. Xenobiotic metabolizing enzymes are expressed in the human nose, and studies using vinyl acetate indicate that local nasal metabolism occurs and serves a similar role in the human nose as well (Hinderliter et al., 2005).

Since metabolism is a saturable process, it follows that saturation-like behavior should be observed in nasal uptake of metabolized vapors. This effect has been confirmed in the rat and mouse. Perhaps the most illustrative vapor in this regard is styrene, where decreased uptake efficiencies are observed at high exposure concentrations (e.g., saturation) in both the rat and mouse, with the saturation behavior being eliminated by administration of a metabolic inhibitor (Table 3, 24).

Comparison among species of uptake of metabolized vapors must be made with caution. Species differences may simply be the result of differences in xenobiotic metabolizing enzyme expression levels. In that case, any observed species difference will be specific to an individual vapor based on its metabolism and not generalizable to other vapors that are not metabolized via the same pathway. Comparison of uptake efficiencies in the minimized metabolism condition reveals that uptake efficiency is directly related to solubility, with increased uptake occurring with increasing partition coefficient. The same species difference patterns are observed as for acetone (Table 2). Uptake efficiency appears highest in the rat, equivalent or slightly lower in the mouse, and lowest in the hamster and guinea pig. Although precise comparisons are not possible due to the differing flow regimes, it appears that nasal uptake is more efficient in the dog than in other species.

Sex differences in uptake are not apparent. They were not reported for the dog and not observed for isoamyl alcohol vapor uptake in the rat. Naphthalene vapor uptake was lower in the female than male rat, but the difference was due to the body weight difference in the animals in the study and the fact that the inspiratory flow rate was a larger fraction of the minute ventilation in the females than males (Morris, 2000). To the authors' knowledge, no comprehensive data exist on nasal uptake of nonreactive, metabolized vapor uptake in juveniles.

6. REACTIVE ORGANIC VAPORS

Table 5 summarizes nasal uptake of reactive organic vapors. Directly reactive organic vapors cover a broad spectrum of reaction pathways. The term is used to describe vapors that react directly (e.g., nonenzymatically) with water or other tissue substrates to form reaction products (that may or may not be volatile). Since the partitioning of a vapor between air and tissue phases depends on the Henry's law behavior of the parent compound, any process that removes parent compound from the tissue will serve to enhance partitioning into the tissue. Reaction pathways may reflect simple reactions with water, such as ionization of weak acids, or covalent binding to low- or high-molecular-weight substrates, such as binding of acrolein to sulfhydryl residues. Many reactive vapors are directly cytotoxic and induce immediate nasal sensory nerve reflex responses that may alter uptake patterns; this is perhaps best studied for acrolein vapor (Morris et al., 1999).

As for inorganic gases, water-soluble reactive organic vapors are scrubbed with high efficiency in the nose. This is shown by weak organic acids (e.g., acetic acid, acrylic acid, methacrylic acid, butyric acid) for which greater than 90% uptake is observed regardless of the species. Reaction rate kinetics are critical in determining the uptake efficiency of reactive vapors, so generalized trends among vapors (with differing reaction kinetics) are difficult to discern. Nevertheless, it should be noted that an inspiratory flow rate dependence has been observed for all reactive vapors whose uptake has been examined at multiple flow rates. Saturation kinetics have been observed for reactive vapors (e.g., acrolein, acetaldehyde). This may be due to depletion of tissue substrates or, since these vapors are also metabolized, to saturation of metabolic pathways.

Uptake of phosgene and nitrogen mustard has been examined in the monkey and rabbit, and appears similar in both species. This provides another vapor for which uptake in the primate nose does not appear to differ greatly from that in a lower species. Propylene oxide vapor uptake appears to be fairly similar in the rat and mouse nose, providing another instance in which these two species do not differ greatly. Insufficient data exist to draw a conclusion relative to sex differences in nasal uptake of reactive organic vapors. To the authors' knowledge, no comprehensive data exist on nasal uptake of reactive, metabolized vapor uptake in juveniles.

7. ACETALDEHYDE

Acetaldehyde is a reactive aldehyde that is metabolized in nasal tissues (via aldehyde dehydrogenase). Its water:air partition coefficient is 88. A rich database exists on this vapor, which includes measurement of nasal uptake under one-way and cyclic flow conditions in multiple species over a 1000-fold concentration range (Table 6). The same generalized trends are seen under both one-way and cyclic flow, indicating that these trends are not dependent on the flow regime used for uptake measurement. Saturable uptake behavior was seen in all species, and in all species the effect was quite large. For example, in the rat, uptake efficiency ranged between 65% and 24% at 1 ppm compared to 1000 ppm. That this effect occurs in all species indicates it is a generalizable phenomenon. This has important implications relative to interpreting animal inhalation toxicity data obtained at high concentration to predict potential effects that may occur at low, environmentally relevant levels. Certainly, a simple linear extrapolation would appear unwarranted.

At low exposure concentrations, species differences in uptake of acetaldehyde are strongly dependent on the metabolism kinetics across species (Morris, 1997). In contrast, at high concentrations, the potential quantitative contribution of metabolism is minimal because metabolic pathways are overwhelmed. At 1000 ppm, the same general species differences are observed, with uptake being most efficient in the rat and mouse and less efficient in the hamster and guinea pig. Insufficient data exist to draw a conclusion relative to sex differences in nasal uptake of acetaldehyde, or on acetaldehyde uptake in juveniles.

8. CONCLUSIONS

A rich database exists on gas/vapor uptake in the upper respiratory tract of numerous species, including humans. Theoretical considerations indicate that vapor uptake will depend on three physicochemical properties of the vapor: solubility (as measured by partition coefficients), reactivity, and propensity for local metabolism. Extremely efficient nasal uptake is expected for soluble vapors that are highly reactive and/or efficiently metabolized, while low uptake efficiency is expected for nonreactive, slowly metabolized vapors of limited solubility. These theoretic predictions are fully confirmed by experiments in every species that has been tested to date (including humans). In addition, saturable uptake behavior should be seen for reactive/metabolized vapors due to depletion of tissue reactive substrates and/or saturation of metabolic pathways. This has also been widely observed in multiple species. Theoretical considerations indicate that vapor uptake efficiency will be dependent on the air flow regime and air flow rates at which it is measured. Again, this has been confirmed in every species that has been examined to date.

Since vapor uptake is dependent on the physicochemistry of the vapor, the flow conditions and inspiratory flow rate at which it is measured, and also upon the inspired concentration (for reactive/metabolized vapors), the elucidation of universal differences among species is problematic. Nonetheless, examination of the database as a whole provides some generalized trends. In all species, including humans, vapor solubility is an important determinant of uptake efficiency, with greater uptake being observed for vapors of greater solubility. This has been observed for all types of gases/vapors, ranging from reactive inorganic gases to nonreactive organic vapors.

The accumulated database by itself is not sufficient to make precise comparisons between laboratory animals and humans. The data suggest that uptake in the human nose is similar to that in the rodent but perhaps somewhat less efficient. Modeling efforts suggest that this is true, with less uptake in the human than rodent nose (Gloede et al., 2011; Frederick et al., 1998, 2002). Given the much greater complexity and proportionately higher surface area in the rodent nose, the lack of profound species differences may seem perplexing. However, the complexity of the rodent nose is within the ethmoturbinate region. Only ~15% of the inspired airstream passes over these turbinates (Kimbell et al., 1997). Approximately 85% of the inspired airstream passes over the nasomaxillary turbinates whose fundamental structure is similar in rodents and humans. The less efficient vapor absorption in the human nose is not a function of surface area but is likely due to the thicker epithelial layer, lower perfusion, and wider airways in the human than rodent nose (Gloede et al., 2011). Lower uptake in the human nose would suggest that the human nose is less susceptible to vapor-induced nasal damage because the delivered dosages are lower in the human than animal nose. In contrast, less uptake in the human nose indicates that there is greater penetration of inspired vapor through the nose to the lower airways of the human than would occur in laboratory animals. This would place the human at greater risk for vapor-induced lower airway damage. This difference is obviously increased by the fact that humans can bypass their nose by mouth breathing (Gloede et al., 2011).

Studies have been performed in both males and females in multiple species for several gases/vapors. Often the publications that include both sexes make no comment on sex-related differences implying that none exists. In those publications in which direct comparisons were made between sexes, no difference was observed. To the authors' knowledge, there have been no comprehensive studies on vapor uptake in the developing nose or in juveniles for any vapor. This represents a significant gap in the database.

REFERENCES

Bogdanffy, M.S., Randall, H.W., Morgan, K.T., 1987. Toxicol. Appl. Pharmacol. 88, 183–194.

Brain, J.D., 1979. Ann. Otolaryngol. 79, 529–539.

Cameron, G.R., Gaddum, J.H., Short, R.H.D., 1946. The absorption of war gases by the nose. J. Pathol. 58, 449–455.

Dahl, A.R., Snipes, M.B., Gerde, P., 1991. Toxicol. Appl. Pharmacol. 109, 263–275.

Dalham, T., Sjoholm, J., 1963. Studies on SO_2, NO_2 and NH_3, effect on cilicary activity in rabbit trachea of single in vitro exposure and resorption in rabbit nasal cavity. Acta Physiol. Scand. 58, 287–291.

Frederick, C.B., Bush, M.L., Lomax, L.G., et al., 1998. Toxicol. Appl. Pharmacol. 152, 211–231.

Frederick, C.B., Lomax, L.G., Black, K.A., et al., 2002. Toxicol. Appl. Pharmacol. 183, 23–40.

Frank, N.R., Yoder, R.E., Brain, J.D., Yokoyama, E., 1969. SO_2 (35S labeled) absorption by the nose and mouth under conditions of varying concentration and flow. Arch. Environ. Health 18 (3), 315–322.

Gloede, E., Cichocki, J.A., Baldino, J.B., et al., 2011. Toxicol. Sci. 123, 231–246.

Gerrity, T.R., Weaver, R.A., Berntsen, J., House, D.E., O'Neill, J.J., 1988. Extrathoracic and intrathoracic removal of O_3 in tidal-breathing humans. J. Appl. Physiol. 65, 393–400.

Haggard, H.W., 1924. J. Biol. Chem. 59, 795–802.

Henderson, Y., Haggard, H.W., 1943. Noxious Gases and the Principles of Respiration Influencing Their Action. Reinhold, New York pp. 71–89.

Hinderliter, P.M., Thrall, K.D., Corley, R.A., et al., 2005. Toxicol. Sci. 85, 460–467.

Jordan, E.O., Carlson, A.J., 1913. J. Am. Med. Assoc. 61, 1007–1012.

Kimbell, J.S., Godo, M.N., Gross, E.A., et al., 1997. Toxicol. Appl. Pharmacol. 145, 388–398.

Kimbell, J.S., Gross, E.A., Joyner, D.R., et al., 1993. Toxicol. Appl. Pharmacol. 121, 253–263.

Kumagai, S., Matsunaga, I., 2000. Inhal. Toxicol. 12, 491–510.

Landahl, H.D., Herrmann, R.G., 1950. Retention of vapors and gases in the human nose and lung. Arch. Indust. Hyg. Occup. Med. 1, 36–45.

Miller, F.J., McNeal, C.A., Kirtz, J.M., Gardner, D.E., Coffin, D.L., Menzel, D.B., 1979. Nasopharyngeal removal of ozone in rabbits and guinea pigs. Toxicol. 14 (3), 273–281.

Morris, J.B., 2012. Pharmacol. Ther. 136, 401–413.

Morris, J.B., Hassett, D.N., Blanchard, K.T., 1993. Toxicol. Appl. Pharmacol. 123, 120–129.

Morris, J.B., Kimbell, J.S., Asgharian, B., 2010. In: Morris, J.B., Shusterman, D.J. (Eds.), Toxicology of the Nose and Upper Airway. Informa Healthcare, New York, pp. 99–115.

Morris, J.B., Shusterman, D.J., 2010. In: Morris, J.B., Shusterman, D.J. (Eds.), Toxicology of the Nose and Upper Airways. Informa Healthcare, New York, pp. 82–98.

Morris, J.B., Stanek, J., Gianutsos, G., 1999. J. Appl. Physiol. 87, 1877–1886.

Morris, J.B., Smith, F.A., 1982. Regional deposition and absorption of inhaled hydrogen fluoride in the rat. Toxicol. Appl. Pharmacol. 62 (1), 81–89.

Morris, J.B., Wilkie, W.S., Shusterman, D.J., 2005. Acute respiratory responses of the mouse to chlorine. Toxicol. Sci. 83 (2), 380–387.

Morris, J., Cavanagh, D.G., 1986. Deposition of ethanol and acetone vapors in the upper respiratory tract of the rat. Fundam. Appl. Toxicol. 6 (1), 78–88.

Morris, J.B., 1991. Deposition of acetone vapor in the upper respiratory tract of the B6C3F1 mouse. Toxicol. Lett. 56 (1–2), 187–196.

Morris, J.B., 1993. Upper respiratory tract metabolism of inspired alcohol dehydeogrnase and mixed function oxidase substrate vapors under defined airflow conditions. Inhal. Toxicol. 5, 203–221.

Morris, J.B., Clay, R.J., Trela, B.A., Bogdanffy, M.S., 1991. Deposition of dibasic esters in the upper respiratory tract of the male and female Sprague-Dawley rat. Toxicol. Appl. Pharmacol. 108 (3), 538–546.

Morris, J.B., Clay, R.J., Cavanagh, D.G., 1986. Species differences in upper respiratory tract deposition of acetone and ethanol vapors. Fundam. Appl. Toxicol. 7, 671–680.

Morris, J.B., 1990. First-pass metabolism of inspired ethyl acetate in the upper respiratory tracts of the F344 rat and Syrian hamster. Toxicol. Appl. Pharmacol. 102 (2), 331–345.

Morris, J.B., Frederick, C.B., 1995. Upper respiratory tract uptake of acrylate ester and acid vapors. Inhal. Toxicol. 7, 557–574.

Morris, J.B., 2000. Uptake of styrene in the upper respiratory tract of the CD mouse and Sprague-Dawley rat. Toxicol. Sci. 54 (1), 222–228.

Morris, J.B., Cavanagh, D.G., 1987. Metabolism and deposition of propanol and acetone vapors in the upper respiratory tract of the hamster. Fundam. Appl. Toxicol. 9 (1), 34–40.

Morris, J.B., Buckpitt, A.R., 2009. Upper respiratory tract uptake of naphthalene. Toxicol. Sci. 111, 383–391.

Morris, J.B., 1999. A method for measuring upper respiratory tract vapor uptake and its applicability to quantitative inhalation risk assessment. Inhal. Toxicol. 11 (10), 943–965.

Morris, J.B., 1996. Uptake of acrolein in the upper respiratory tract of the F344 rat. Inhal. Toxicol. 8, 387–403.

Morris, J.B., Banton, M.I., Pottenger, L.H., 2004. Uptake of inspired propylene oxide in the upper respiratory tract of the f344 rat. Toxicol. Sci. 81 (1), 216–224.

Morris, J.B., Hubbs, A.F., 2009. Inhalation dosimetry of diacetyl and butyric acid, two components of butter flavoring vapors. Toxicol. Sci. 108 (1), 173–183.

Morris, J.B., Pottenger, L.H., 2006. Nasal NPSH depletion and propylene oxide uptake in the upper respiratory tract of the mouse. Toxicol. Sci. 92 (1), 228–234.

Morris, J.B., Symanowicz, P.T., Olsen, J.E., Thrall, R.S., Cloutier, M.M., Hubbard, A.K., 2003. Immediate sensory nerve-mediated respiratory responses to irritants in healthy and allergic airway-diseased mice. J. Appl. Physiol. 94 (4), 1563–1571.

Morris, J.B., 1997. Uptake of acetaldehyde vapor and aldehyde dehydrogenase levels in the upper respiratory tracts of the mouse, rat, hamster, and guinea pig. Fundam. Appl. Toxicol. 35 (1), 91–100.

Morris, J.B., Blanchard, K.T., 1992. Upper respiratory tract deposition of inspired acetaldehyde. Toxicol. Appl. Pharmacol. 114 (1), 140–146.

Nodelman, V., Ultman, J.S., 1999. Longitudinal distribution of chlorine absorption in human airways: a comparison to ozone absorption. J. Appl. Physiol. 87, 2073–2080.

Phalen, R.F., 1984. Inhalation Studies: Foundations and Techniques. CRC Press, Boca Raton, FL.

Pattle, R.E., 1961. The retention of gases and particles in the human nose. In: Davies, C.N. (Ed.), Inhaled Particles and Vapors. Pergammon Press, Oxford, UK, pp. 302–311.

Plowchalk, D.R., Andersen, M.E., Bogdanffy, M.S., 1997. Physiologically based modeling of vinyl acetate uptake, metabolism, and intracellular pH changes in the rat nasal cavity. Toxicol. Appl. Pharmacol. 142 (2), 386–400.

Santiago, L.Y., Hann, M.C., Ben-Jebria, A., Ultman, J.S., 2001. Ozone absorption in the human nose during unidirectional airflow. J. Appl. Physiol. 91, 725–732.

Speizer, F.E., Frank, N.R., 1966. The uptake and release of SO_2 by the human nose. Arch. Environ. Health 12 (6), 725–728.

Stott, W.T., McKenna, M.J., 1984. The comparative absorption an excretion of chemical vapors by the upper, lower, and intact respiratory tract of rats. Fundam. Appl. Toxicol. 4, 594–692.

Snipes, M.B., Spoo, J.W., Brookins, L.K., Jones, S.E., Mauderly, J.L., Orwat, T.B., et al., 1991. A method for measuring nasal and lung uptake of inhaled vapor. Fundam. Appl. Toxicol. 16 (1), 81–91.

Stanek, J.J., Morris, J.B., 1999. The effect of inhibition of aldehyde dehydrogenase on nasal uptake of inspired acetaldehyde. Toxicol. Sci. 49 (2), 225–231.

Struve, M.F., Wong, V.A., Marshall, M.W., Kimbell, J.S., Schroeter, J.D., Dorman, D.C., 2008. Nasal uptake of inhaled acrolein in rats. Inhal. Toxicol. 20 (3), 217–225.

Thrall, K.D., Schwartz, R.E., Weitz, K.K., Soelberg, J.J., Foureman, G.L., Prah, J.D., et al., 2003. A real-time method to evaluate the nasal deposition and clearance of acetone in the human volunteer. Inhal. Toxicol. 15 (6), 523–538.

USEPA, 2009. Advances in Inhalation Dosimetry of Gases and Vapors with Portal of Entry Effects in the Upper Respiratory Tract. edn, vol. EPA/600/R-09/072. EPA, Washington, DC.

Vaughan, T.R., Jennelle, L.F., Lewis, T.R., 1969. Long-term exposure to low levels of air pollutants. Effects on pulmonary function in the beagle. Arch. Environ. Health 19, 45–50.

Yokoyama, E., 1968. Uptake of SO_2 and NO_2 by the isolated upper airways. Bull. Inst. Publ. Health 17, 302–306.

Yokoyama, E., Frank, R., 1972. Respiratory uptake of ozone in dogs. Arch. Environ. Health 25, 132–138.

Chapter 29

Mucociliary Function

W. Michael Foster

Chapter Outline

1. INTRODUCTION

Surfaces of the upper and lower respiratory tract of mammals are coated by a fluid lining layer composed of mucus, which is a mixture of water, ions, glycoproteins, and lipid molecules. Mucus serves as a protective barrier against pathogens and toxins for epithelial surface membranes from the nares to the respiratory bronchioles (Ewert and Martensson, 1967; Barnett and Miller, 1966; Rose and Voynow, 2006). In the peripheral air spaces of the lung, a film layer also overlies the air-exchange units of the distal lung, providing a liquid covering for epithelial cells that line the surfaces of alveolar ducts and alveoli (Kilburn, 1968; Gross and Detreville, 1972; Macklin, 1954). The fluid lining of the alveolar epithelial surfaces is approximately 0.2 μm thick (Macklin, 1954) and is derived from alveolar cells and transudation of fluid from alveolar capillaries (Gross and Detreville, 1972). The alveolar film moves directionally from the alveoli into the respiratory bronchioles, and in transport with it, may be deposited xenobiotic matter, necrotic tissues, phagocytic-type cells, and dissolved gases and secretions from distal secretory epithelial, such as Clara cells. Mucin glycoproteins (airway mucins) synthesized and secreted by epithelial goblet cells and submucosal glands are large, highly glycosylated, viscoelastic macromolecules and thus arise from cellular sources located within the more proximal ciliated airways (Kaliner et al., 1986; Rose and Voynow, 2006). Secretory mucins (stored as granules) are released onto apical airway surfaces, whereas membrane-tethered mucins are integrated external to the cell membrane, and thus both become major constituents of the protective mucus layer lining the airway surfaces. The thickness of this protective liquid layer may increase focally 20-fold and approach mucus layer depths ranging between 5.0 and 10.0 μm (Hilding, 1963; Kilburn, 1968; Adler et al., 1973; Gross and Detreville, 1972). Fluid transudates from the capillaries of the bronchial mucosa are also believed to make a contribution to this mucus layer (Florey et al., 1932).

2. EPITHELIAL MEMBRANE LINING

The liquid mucus layer lining the tracheobronchial airways conceptually is believed to be stratified into two components: a gel layer above and an underlying sol layer in close proximity to the epithelium (Lucas and Douglas, 1934; Adler et al., 1973; Kaliner et al., 1986). Although it was initially suggested that the mucus layer lining of epithelial surfaces formed a continuous surface layer (Kilburn, 1968) from distal bronchiolar airways to proximal stem bronchi, evidence that such a fluid layer is a noncontinuous, nonuniform mucus layer has been shown for some regions of the bronchial airways (Iravani and van As, 1972; Adler et al., 1973; Mercer et al., 1992). However, accumulating airway mucus is transported dynamically from the peripheral airways to the larynx by kinetic activity of ciliated cells lining epithelial surfaces.

Comparative Biology of the Normal Lung. http://dx.doi.org/10.1016/B978-0-12-404577-4.00029-1

Based upon model assumptions of a 2-component mucus liquid layer system, cilia of the underlying surface epithelial cells are believed to beat and recover in the sol portion of the mucus layer, which is less viscous and has a high sheer rate (Lucas and Douglas, 1934; Blake, 1975), whereas the upper gel portion, which was found to be more viscoelastic (Denton et al., 1968), acts as the transport layer being propelled proximally by the tips of the beating cilia in a directional manner (Asgharian et al., 2001). However, challenges to this traditional two-layer sol–gel model, whereby the gel-like mucus layer is transported by cilia beating in a less viscous periciliary fluid layer, are alternative models that propose a system in which the periciliary layer is occupied by membrane-spanning mucins and large mucopolysaccharides that are tethered to the cilia and epithelial cell surfaces (Verdugo, 1991; Button et al., 2012). For this model, conceptually densely tethered macromolecules occupy the periciliary layer and provide structure and stability to the two-layer mucus clearance system, enabling a distinct layer of mucus to form that is prevented from penetrating the interciliary space layer.

3. CILIUM STRUCTURE

There are two general types of cilia: primary cilia serve as sensory organelle, whereas motile cilia can exert mechanical force and for example can generate flows of mucus (Satir and Christenson, 2007; Marshall, 2008; Broekhuis et al., 2013). The motile cilium are membrane-bounded, approximately 0.2–0.3 μm diameter, and emerge as projections from the cellular surface of luminal epithelial cells that line human airways and transport the protective gel-like mucus layer that is positioned above the surface of epithelial lining cells. Postmortem collections from human airways have established that the percent density of ciliated cells at the airway surface decreases as one moves distally along the airway from the trachea toward the smaller bronchial airways, averaging 53.0% of the surface epithelial cells in the trachea, and 23.5% and 15.1% in third- and fifth-generation bronchi, respectively. In addition, the length of the cilia have been found to decrease in size as one moves distally down the airways; for example, the mean cilium length decreased by approximately 20% from the tracheal airway to third-generation airways (mean cilium lengths of 6.03 and 4.70 μm, respectively) and by 46% from the trachea to fifth-generation bronchi (Serafini and Michaelson, 1977).

In addition to the mucus layer, xenobiotic materials entrapped within the mucus layer, which may have deposited onto airway surfaces during inspiratory and expiratory breathing, may also be directionally transported from distal to hilar lung regions, and to the larynx where such materials are either swallowed into the digestive tract or expectorated. These motile cilia, which supply the kinetic energy that is necessary for transporting the mucus layer, contain a microtubule cytoskeleton known as the ciliary axoneme; it has a characteristic internal organization consisting of an array of nine peripheral microtubule doublets and a pair of central, singlet microtubules. A diagram of a transverse section of a cilium, based upon ultrastructure of a human nasal cilium by electron microscopy (Rutland and de Iongh, 1990), is illustrated in Figure 1. The microtubules are enclosed within a membrane that is continuous with the plasma membrane of the epithelial cell (Sleigh, 1962; Satir and Christensen, 2007). Protein complexes (outer and inner dynein arms, radial spokes, and central pair projections) within the axoneme attach to the microtubules. Motile power of the cilium is provided by two sets of dynein arms: the inner and outer arms, and motility is regulated by radial spokes that interact with the central singlet microtubules (Salathe, 2007; Satir and Christensen, 2007; Marshall, 2008). Movement of a motile cilium involves an active effective stroke and a recovery stroke (Sleigh, 1968; Sleigh et al., 1988), which are initiated by bending of the base of the cilium's shaft to its tip, with the active vertical stroke extending the cilia into the overlying mucus layer (Brokow, 1966; Satir and Christensen, 2007). During the recovery stroke, the ciliary motion is directed horizontally beneath the mucus layer and within the periciliary layer (Wanner et al., 1996; Smith et al., 2008; Sears et al., 2013).

Except perhaps during transient in vivo conditions of rapid changes in luminal airflow (e.g., during coughing), it has been suggested that the mucus layer is likely only propelled during the maximum velocity of the effective ciliary stroke (Hilding, 1963). Mechanisms leading to

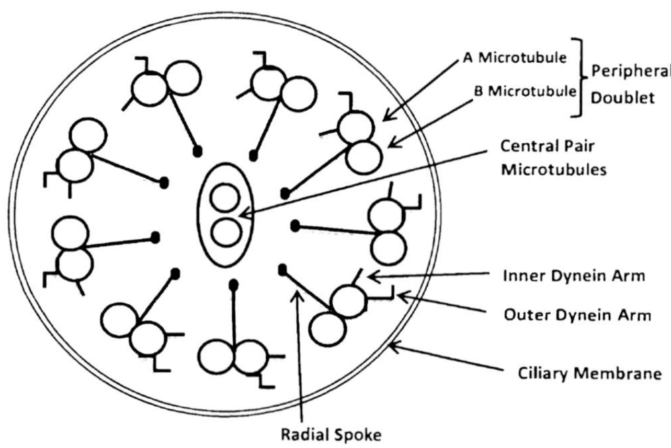

FIGURE 1 **Illustration of main structural components of a motile cilium axoneme.** On cross sections of electron microscopic images, ciliary microtubules are arranged in classic 9 + 2 configuration [9 outer microtubule doublets (A and B) and a middle pair of singlet microtubules]. Motile power is provided by two sets of dynein arms: the inner and outer arms and motility is regulated by radial spokes that interact with central pair of microtubules. *Adapted from Rutland and de Iongh, 1990.*

controlled bending of the axoneme shaft are not entirely clear; however, the peripheral microtubule doublets sliding past one another in a manner similar to the movement of the contractile proteins of striated muscle has been suggested (Sleigh, 1971; Rutland and de Iongh, 1990). Although the mechanism(s) is not fully understood, the ciliary beat frequency correlates with the number of outer dynein arms within the axoneme and beat frequency is increased by enhancement of the intracellular calcium concentration ($[Ca^{2+}]_i$) (Verdugo et al., 1983; Salathe et al., 1993).

The molecular characterization of Ca^{2+} entry pathways (ion channels) within epithelial ciliated cells, which are activated in response to different stimuli, have now been linked to calcium-permeable cationic channels of the transient receptor potential (TRP) vanilloid superfamily (Clapham, 2003). Specifically, a role for TRPV4 channels has been demonstrated for the generation of Ca^{2+} signaling required to maintain ciliary beat frequency in oviductal ciliated cells under conditions of mechanical (high viscous load) stress (Andrade et al., 2005). The energy for ciliary motility has been shown to be derived from adenosine triphosphate (ATP; Brokow, 1966); in vivo evidence in support of this concept was obtained from early studies in animal models whereby metabolic inhibition led to depression of mucus flow within the tracheal airway, followed by the demonstration that administration of ATP was able to directly reverse a slowing of mucus layer transport induced by hypoxic conditions (Laurenzi et al., 1968).

Increases in intracellular levels of cyclic adenosine monophosphate (cAMP) have been associated with stimulation of mammalian ciliary motility in the rat, rabbit, dog, ovine, bovine, and human; similarly, cyclic guanosine monophosphate (cGMP) has also been shown to elevate mammalian ciliary beat frequency (Wyatt et al., 2005). These second messengers, cAMP and cGMP, and others such as calcium and nitric oxide, have been shown to target protein kinases (protein kinase A, PKA) and induce an increase in ciliary beating (Salathe, 2007; Wyatt et al., 2012). An opposing observation is that decreases in ciliary beat frequency have been found with agents known to associate with signal transduction of protein kinase C (PKC). Maximum beat frequencies can approach 100 Hz, although most observations with mammalian beat frequency are lower, with a range of 10–20 Hz (Rutland et al., 1982; Satir and Christensen, 2007) and a kinetic ratio between effective to recovery stroke of 1:3 (Kilburn, 1968).

Cilia are resistant to many noxious and xenobiotic influences and continue to beat under many of the conditions encountered in normal environments. It has been demonstrated in vitro that differentiated human airway epithelial cells express sensory receptors, which were specifically localized to the cilia. Bitter compounds upon application to the cells were found to increase calcium ion concentration and stimulated ciliary beat frequency in a dose-related manner (Shah et al., 2009). This would suggest that classical motile cilia also have a sensory function and can provide a cell-autonomous defense system by which airway epithelial cilia are capable of both sensing a noxious substance and initiating a mechanical response. Additional observations in cilia within ovarian and oviduct tissues add further support for the possible presence of an active sensory/membrane receptor function on motile airway cilia (Teilmann et al., 2005).

Within the respiratory tract, the motile cilia tend to beat in the same direction and one after another (i.e., metachronously; Sleigh, 1966; Sleigh et al., 1988), setting up waves of motion of a uniform pattern and rate. This degree of organization has been studied in depth to determine the source and mechanism(s) of control, whether it be mechanical, neuronal, or humoral. It is well established that cilia are sensitive to tactile stimuli (Asmundsson and Kilburn, 1970), and that the activity of each cilium has a mechanically stimulating effect on the succeeding cilia (Brokow, 1966; Sleigh, 1966). Effector proteins and signal transduction pathways associated with primary and motile cilia in vertebrate cells have been reviewed by Christensen et al. (2007).

Near the ciliary base, there is a transmembrane complex referred to as the ciliary necklace (Figure 2), which consists of multiple strands of intramembrane particles that are clearly prominent and found on respiratory tract cilia of mammalian species (Wanner et al., 1996; Thai et al., 2002).

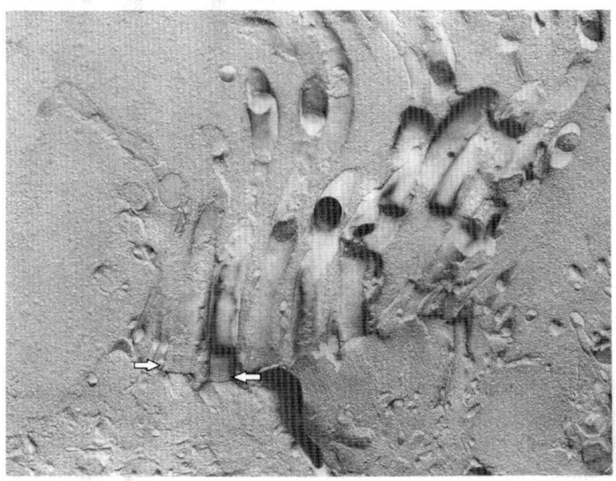

FIGURE 2 Transmission electron micrograph (TEM) of freeze fracture preparation of airway epithelium. Tracheal epithelial tissue from a rodent animal model with ciliary necklace (arrows) located on ciliary membrane near to the basal area of individual cilia. Ciliary necklace consists of multiple strands of intramembrane particles and found on mammalian 9 + 2 cilia. TEM prepared in collaboration with Peter R. Brink, Department of Physiology and Biophysics, Stony Brook University, NY.

A function associated with these necklace particles has not been clearly defined, although a guanine nucleotide exchange factor protein has been localized to the necklace region of motile cilia of the trachea (Satir and Christensen, 2007).

From the perspective of genes essential to cilia formation (or function), a comparison by Ross et al. (2007) between gene expression data acquired during a time course of human ciliated cell differentiation, which overlapped with genes encoding proteins from proteomic analysis of axonemes purified from human cilia, identified 37 different genes. These 37 genes and their corresponding proteins are listed in Table 1. The list includes several genes with acknowledged roles in the structure of cilia, such as genes that encode protein components of the axoneme such as axonemal dynein (*DNAH9, DNAI1, DNAI2, and DNALI1*) and genes associated with calcium homeostasis and transport that encode proteins essential to ciliary beating (*CENT2*).

4. CILIARY HOMEOSTASIS

Conditions that seemingly are favorable to ciliary beat are a moist fluid layer (Sade et al., 1970), a sodium chloride concentration between 0.9% and 2.0% (Scudi et al., 1951), a pH between 6.5 and 8.5 (Fabricant, 1955; Kensler and Battista, 1966), the presence of cations such as potassium and ammonium salts (Lillie, 1903–1904; Gosselin, 1966), and a temperature range of 28–35 °C (Delmotte and Sanderson, 2006). The depth of the sol periciliary layer is critical for the normal ciliary activity; if the sol layer is too thick, then the cilia will not be able to make contact with the transport layer, and if it is too shallow, the cilia will be impeded by the increased viscosity (Barnett and Miller, 1966; Sleigh et al., 1988). Periciliary fluid volume is dependent on the integrity of the underlying epithelium and by ion channels on epithelial cells that control the mass of salt and water at airway surfaces. At steady state, this fluid volume and hydration of the airway is regulated primarily by epithelial cell chloride (Cl^-) secretion and the cystic fibrosis transmembrane regulator (CFTR). Investigations suggest that mutations in the CFTR gene contribute to dehydration of the airway surfaces and dynamically uncouple mucociliary function, resulting in mucus stasis and epithelial mucus adhesion (Schmid et al., 2011). Osmolytes, such as hypertonic saline, when added to airway surfaces can re-establish periciliary fluid thickness and restore or enhance airway mucociliary transport (Pavia et al., 1978; Clunes et al., 2012). With respect to compression of the depth of the periciliary fluid layer, utilization of optical tomography to view the surface microanatomy of differentiated human bronchial epithelial cultures has shown that autoregulatory mechanism(s) related to periciliary layer depth are interdependent with ciliary beat frequency and mucus transport and are compromised when the depth of the periciliary layer is decreased below ~2 μm (Liu et al., 2014). In situ investigations in animal model tissues suggest that ciliary beat frequency (CBF) of intrapulmonary airways is at maximal rates and exceeds rates of CBF established for cultured ciliated tracheal cells (Delmotte and Sanderson, 2006). The efficiency of transport of the mucus layer depends not only upon the magnitude and synchronization of ciliary beating but also on the quantity and rheological properties of the mucus layer (Dulfano and Adler, 1975; Lai et al., 2009). The elasticity of mucus has been shown to be an essential property for transport; under conditions of physiologic pH and ionic strength, the mucin concentration appears to be a determining parameter for elasticity of mucus and hence transportability (Litt, 1981). Early studies based upon the pharyngeal ciliated epithelium of amphibian models demonstrated that, with respect to load, transport velocity was relatively constant for loads that ranged between 0.0008 and 20 mg/mm^2 (Stewart, 1948). Similarly, the thickness of the transport layer can reduce the efficiency of the ciliary beat; at a thickness of 25 μm, the movement of the mucus layer has been curtailed even though the ciliary motility appeared to be normal (Dalhamn and Rhodin, 1956).

5. AIRWAY SECRETIONS

Pure tracheobronchial secretions are difficult to collect from the normal human bronchial tree, thus making estimates of the volume of airway mucus produced difficult to assess. However, the quality of the secretions is believed to be critical for effective ciliary motility in the healthy state and in the diseased state when hypersecretion of the submucosal glands leads to the formation of excessive volumes of mucus (Reid, 1954). The total volume of secretions of the tracheobronchial airways represents the difference between the volume of fluid secreted into the respiratory airways and that which is reabsorbed before being transported to the tracheal airway and swallowed into the digestive tract at the level of the larynx. Evaporative water loss also occurs as secretions are transported through the tracheal airway. In the human airway, the cellular sources of the mucin component of the airway liquid layer are the serous and mucous cells of the submucosal glands, the epithelial goblet cells, and perhaps Clara cells within distal airways. Submucosal glands are present throughout the lower respiratory tract in the airways containing cartilage, whereas goblet cells are located within all airways and extend to the level of the alveolar ducts, at which site Clara cells are found. Due to their prominence in airway biological section, submucosal glands are considered be the major contributors

TABLE 1 Genes Encoding Proteins from the Cilia Proteome

	Gene	Protein
1	C10orf63	Mus musculus male testis cDNA
2	C6orf97	cDNA FLj23305 FIS
3	CAPS	CAYP human calcyphosine
4	CETN2	Human caltractin
5	CGI-38	Brain-specific protein
6	CLUAP1	KIAA0643 protein
7	DNAH9	Dynein heavy chain
8	DNAI1	Dynein intermediate chain
9	DNAI2	Dynein intermediate chain
10	DNALI1	Dynein light chain
11	DYDC2	Similar to Riken 4933428D01
12	EFHC1	FLj10466 FIS
13	EFHC2	FLj22601 FIS
14	FLj13946	FLj13946
15	FLj23577	KIA1770
16	GHR	Growth hormone receptor
17	HSPA1A	70-kD heat shock protein
18	IFT57	MHS4R2
19	IFT81	CDV-1 protein
20	IFT88	TG737
21	IQCE	KIAA1023
22	LRRC23	B7 protein
23	MLF1	Fusion gene
24	MNS1	Unnamed protein product
25	NME5	Human nucleoside diphosphate kinase homolog
26	RABL4	Putative GTP-binding protein
27	RABL5	Unknown
28	RIBC2	Unknown
29	ROPN1L	AKAP-associated sperm protein
30	RTDR1	Rhabdoid tumor deletion region protein
31	SPA17	HU sperm surface protein Sp17
32	SPAG6	Sperm flagellar protein, SPAG6
33	TEKT2	H-Tektin-T
34	TUBA3	Alpha-Tubulin
35	TUBA4	Alpha-Tubulin
36	TUBB2	Beta-Tubulin
37	ZMYND12	cDNA DKFZp434N2435

Table based on information from Ross et al. (2007).

to mucus in airway surface fluid. Reid (1960) had estimated that the volume of the glands in the airway mucosa was 40 times greater than the volume of goblet cells. Reid's calculation was based upon several assumptions; the current interpretation is that the relative contributions of goblet cell and glands to the mucin component of airway mucus are uncertain and likely to vary with airway level and state of health. The daily output of tracheobronchial secretions for humans has been estimated under normal physiological conditions to be $\leq$0.5 ml/kg body weight (Toremalm, 1960).

Because of the heterogeneity in the cell types that release secretory products onto the airway surfaces, control factors are not fully defined for the rate(s) of secretion or for the composition of products secreted by each cell type. Mechanisms that regulate the quality and volume of the respiratory secretions involve, for example, the transepithelial secretion of the chloride ion across the airway epithelium with passive diffusion of water, the stimulation of secretion by a number of mediators such as arachidonic acid metabolite, and the overall stability in the numbers of mucus-secreting epithelial cells present in the airways (Jacquot et al., 1992). The final product of respiratory mucus can be complex, with differing degrees of hydration and composition of ion and sugar content, as well as variation in amino acid, mucin glycoprotein, and lipid moieties.

Mucus is viscoelastic with properties of both a liquid and solid state (Denton et al., 1968), and consistency and adhesiveness are strongly influenced by the structure of the acid mucin glycoproteins (White and Elmes, 1960). Histochemical and autoradiographic investigations have provided evidence that the mucous glands can contain at least four types of acid mucin glycoproteins (two sulfoproteins and two sialoproteins) (Lamb and Reid, 1970); goblet cells have also been shown to produce acid mucin glycoproteins (Reid, 1960). These mucin glycoproteins are a high-molecular-weight mixture of gene products.

Sixteen mucin genes are currently recognized to be expressed in normal respiratory tract tissues: MUC1, MUC2, MUC4, MUC5AC, MUC5B, MUC7, MUC8, MUC11, MUC13, MUC15, MUC16, MUC18, MUC19, MUC20, MUC21, and MUC22 (Rose and Voynow, 2006; Lilliehoj et al., 2013). Other mucin genes, such as MUC3 and MUC6, are expressed in some adenocarcinomas (Copin et al., 2001). MUC5B mRNA (transcriptional messenger RNA) has been demonstrated to mainly be expressed and secreted by the mucosal cells of submucosal glands in healthy human airways, whereas expression and secretion of MUC5AC has been shown to generally be restricted to surface goblet cells in the upper and lower respiratory tracts (Reid et al., 1997). MUC5AC and MUC5B, as the predominant gel-forming airway mucins, constitute the main line of protection against inhalable and deposited irritants by providing the essential defenses of mucociliary clearance and a macromolecular barrier. The current understanding is that within respiratory epithelial goblet cells, mucins are synthesized and stored in cytoplasmic membrane-bound granules. Upon appropriate stimulation, the granules are released by an exocytotic process in which the granules translocate to the cell periphery and fuse with the plasma membrane of the cell, followed by mucin release onto the epithelial surface. Intracellular protein kinases (protein kinase C and gCMP-dependent protein kinase) appear to be key determinants in the exocytotic release of mucin (Li et al., 2001).

In representative animal models, the blockade of mucus release has been achieved with the inhibition of myristoylated alanine-rich C kinase substrate (MARCKS) protein, which has been shown to be an essential molecule regulating mucin secretion by airway epithelium in vitro (Li et al., 2001) and in vivo (Singer et al., 2004; Foster et al., 2010). There are also epithelial membrane-associated mucins that are non-gel-forming; the three main airway membrane-bound mucins are MUC1, MUC4, and MUC16. These mucins are less involved with barrier functions and rather integrated with the control of inflammation and immune responses (Hollingsworth and Swanson, 2004). Table 2 lists the currently identified transmembrane and secreted mucins that are located and secreted within the human respiratory tract (Thornton et al., 2000; Lilliehoj et al., 2013).

A variety of inhalable irritants and inflammatory/humoral mediators can provoke mucin secretion; therapeutic agonists to control the flow of secretions also have been evaluated (reviewed by Boyd, 1954; Wanner et al.,

TABLE 2 Airway Transmembrane and Secreted Mucin Proteins

Transmembrane Mucins	Secreted Mucins
MUC1	MUC2
MUC4	MUC5AC
MUC8	MUC5B
MUC11	MUC7
MUC13	MUC19
MUC15	
MUC16	
MUC18	
MUC20	
MUC21	

1996; Rogers, 2002). In healthy subjects, the peripheral airways contain few goblet cells, but goblet cell metaplasia may occur in respiratory disorders. It is also recognized that surface goblet cells can quickly discharge vast quantities of mucus in response to an acute insult; although integral to protection, this outcome may also precipitate airway diseases associated with hypersecretion of mucus, including bronchitis (Rogers, 1994). A uniform airway response following exposure to a respirable irritant, such as oxidants or acid aerosols, is a neuronally mediated increase in the rate or volume of secretions from epithelial cells and/or submucosal glands (Costa and Schelegle, 1999). In part, this reflex is viewed as a protective mechanism to enhance the depth of the airway mucus layer and modify the sensitivity of airway irritant receptors and ameliorate potential bronchoconstrictive stimuli at airway epithelial surfaces. Rapid secretion of mucus is largely under parasympathetic nervous system control, whereby efferent postganglionic parasympathetic fibers have been identified in association with submucosal glands (Laitinen, 1985). Airway surface epithelial cells and submucosal gland cells express muscarinic receptors (Barnes et al., 1983; Basbaum et al., 1983, 1984), and there is also a high density of β-adrenergic receptors (Barnes and Basbaum, 1983; Kelsen et al., 1992). However, submucosal glands also express α-adrenergic receptors, which have been found to localize mainly to serous cells (Barnes and Basbaum, 1983). Peptidergic receptors (for vasoactive intestinal peptide (VIP) and substance P) are also located on the airway epithelial cells (Carstairs and Barnes, 1986; Bausbaum, 1995). Mucus glands preserved from bronchial surgical specimens of smokers with chronic bronchitis demonstrate a significant increase in the density of nerve fibers immunoreactive for VIP (Lucchini et al., 1997) and support the concept that neuropeptides are involved in hypersecretion of airway mucus.

Three airway pathologies that lead to excess airway mucus are increased synthesis of mucus as a result of overexpression of mucin genes; enhanced production of mucus secondary to hyperplasia, hypertrophy, or even metaplasia of the secretory cells; and hypersecretion of stored mucin granules form surface goblet cells and submucosal glands. Treatments for modulating the level of mucus secretion, and therapies designed to improve airway mucus transport and clearance, have been re-evaluated by Fahy and Dickey (2010).

6. AIRWAY MUCUS TRANSPORT

With respect to mucociliary clearance within a defined airway, a wide margin of mucus layer transport rates has been reported for the tracheal airway of awake, healthy humans. Observations range between 4–5 and 20–22 mm/min (Ewert and Martensson, 1967; Yeates et al., 1975; Santa Cruz et al., 1974; Foster et al., 1980); however, mucus layer transport rates in more distal major bronchi were found to be slightly slower at <4 mm/min (Foster et al., 1980). However, in anesthetized patients (propofol and morphine), transport velocities observed directly and invasively within lobar bronchi were found to be enhanced (9–10 mm/min range; Ledowski et al., 2006). Due to the difficulty of acquiring accurate measurements within peripheral small airways of the intact lung, mucus layer transport rates have not been reported in humans for the smaller, more distal airways.

A number of investigators have modeled mucous velocity within more distal airways by making calculations based on the mucous velocity of the tracheal airway and properties of the mucous flow in the airways (Yu et al., 1986). Asgharian et al. (2001), using various symmetric and stochastically generated asymmetric models of the lower respiratory tract, calculated mucus velocities in all airways based upon mass balance for the mucus. Indeed, in investigations with large animal models (e.g., dog) the velocity of mucus flow within peripheral, smaller airways are found to exhibit lower values (slower) in the velocity of mucus flow, in comparison to velocities (more rapid) observed within the more central, bronchial airways (Asmundsson and Kilburn, 1970).

Table 3 lists mucus transport velocities observed in vivo on a comparative basis in popular laboratory animal species. A major factor for consideration is that tracheal and bronchial mucus transport velocities can be recorded noninvasively in awake humans, whereas for animal models mucus transport velocities are monitored during anesthesia. It has been generally assumed that transport rates of the mucus layer within individual airway segments of a healthy human lung will parallel the clearance of the mucus layer from all tracheobronchial airways when considered in aggregate (Foster et al., 1980). However, in the compromised or remodeled airway, this assumption has not been observed to hold true, such as in habituated cigarette smokers, in which the airways are repeatedly exposed to toxic cigarette smoke. Although within central dependent lung airways the velocity of mucus transport has been observed to be comparable with age-matched nonsmokers, mucus flow from peripheral lung regions was significantly reduced as compared to nonsmokers (Foster et al., 1985). Thus, although the transport rate of the mucus layer by the ciliated airways has been confirmed by in vitro and in vivo studies (Hilding, 1963; Morrow et al. 1968; Yeates et al., 1975; Goodman et al., 1978; Foster et al., 1980), the actual motion and directional flow in vivo of the more distal alveolar surface liquid film is somewhat inferential and the actual kinetics of transport for this film are based more on conjecture (Macklin, 1954; Kilburn,

TABLE 3 Species Comparison: Airway Mucus Transport Velocity[a]

Species	Tracheal (mm/min)	Bronchial (mm/min)
Human (Age Range: 24–26 years)		
Healthy	5.5 ± 0.4 (Foster et al., 1980)	2.4 ± 0.5 (Foster et al., 1980)
Smoker		4.1 ± 1.6 (Foster et al., 1985)
Rodent Models		
Rat	1.9 ± 0.2 (Felicetti et al., 1981)	0.33 ± 0.19 (Rodriquez et al., 2001)
Guinea pig	2.7 ± 1.4 (Felicetti et al., 1981)	
Ferret	10.7 ± 3.7 (Voter et al., 1992)	
Mouse	2.3 ± 0.5 (Brownstein, 1987)	
Rabbit	3.2 ± 1.1 (Felicetti et al., 1981)	
Dog	9.8 ± 2.1 (Felicetti et al., 1981)	
Sheep	9.6 ± 1.0 (Abraham et al., 2004)	

[a]Note: values are means ± SEM; for human and sheep, the velocities were acquired in an awake state; and for the rodent species, and rabbit and dog, the animals were anesthetized.

1968; Gross and Detreville, 1972; Morrow, 1973; Green, 1973) than direct observation and measurement (Moller et al., 2004).

7. BIOLOGIC CONTROL FACTORS

The kinetics of the mucociliary transport system do not appear to be under direct autonomic control. However, for animal model investigations, mucociliary clearance was observed to be impaired in most transplanted and denervated lungs for periods up to several months after surgery (Edmunds et al., 1969, 1971). Denervation, which essentially affects submucosal glands that atrophy during the period after bronchial transection (Brody et al., 1972), was not thought to be the major pathogenic cause due to a rapid rate of recovery during the period of persistent denervation. In healthy humans, atropine (a muscarinic antagonist that apparently has no direct effect on the rheological properties of mucus) can significantly reduce the rate of mucociliary clearance in humans (Groth et al., 1991; Foster et al., 1976). Both vagal and sympathetic nerve stimulation increase the rate of glandular secretion in several species. Direct stimulation of the airway surface by exogenous agonists (i.e., adrenergic and cholinergic) and mediators, such as histamine, alter transepithelial secretory processes (Kaliner et al., 1986; Rogers, 1994) and in humans increase airway mucociliary clearance (Camner et al., 1974; Foster et al., 1980; Mussato et al., 1988).

The effects of oxygen and metabolite delivery to the submucosal and epithelial tissues via the bronchial blood flow supply upon the mucociliary system have been evaluated in animal models (Wagner and Foster, 1996) but not directly in human subjects. Agents with low vapor pressures, such as alcohol when exhaled as vapor from the respiratory tract, are derived mainly in humans from the bronchial circulation (Sisson, 2007). In investigations of humans with mild to moderate ethanol consumption, where mucociliary clearance of the conducting airways was assessed directly, faster clearance was demonstrated (Venizelos et al., 1981). Consistent with these studies in humans, investigations that evaluated ciliary motility of isolated airway axonemes found low-dose alcohol to rapidly stimulate axoneme beating to rates 40% above baseline (Sisson et al., 2009). Although this ciliary response is likely alcohol dependent, the human studies demonstrate the capability of the bronchial circulation to affect mucociliary-airway homeostasis. Similarly, the lung transplant literature (experimental and clinical) provides additional insight on the importance of bronchial blood flow to mucociliary clearance. For example, animal models of allo- and autotransplantation of lung lobes have shown that during the immediate postoperative period (a few days to duration in weeks), mucociliary clearance was impaired (Brody et al., 1972; Paul et al., 1989). Initially, this may have reflected a combination of factors, such as local obstruction at the site of bronchial anastomosis, which are known to resolve within several days, and a longer-term dysregulation of clearance. Several studies in dogs have suggested that the bronchial vascular and lymphatic supplies of reimplanted lungs re-establish at ~1 month postimplant (Baile et al., 1992; Eraslan et al., 1964; Wright et al., 1971), although (Baile et al., 1985) have shown blood flow to exceed control levels by 1 week posttransplantation. In humans, however, reversal appears to be complicated by additional factors, including immunosuppression therapy. Also, at least with respect to long-term survivors (1.5-years posttransplantation) of heart and/or lung or double-lung transplantation, bronchial mucus clearance is suppressed as compared to normal control subjects (Herve et al., 1993).

8. COUGH CLEARANCE OF AIRWAY MUCUS

Cough can be considered as a respiratory reflex with protective capabilities. In general, its presence is a sentinel of an abnormal condition or an irritant exposure to the respiratory system. By example, an important role in epithelial membrane homeostasis and continuous effective clearance of the mucus layer is the interaction between mucus viscosity, periciliary fluid, and effective ciliary beating. Cough can signal dysfunction of these components, especially in chronic bronchitis. Although the pathophysiology of chronic cough is not certain, in bronchial biopsies from subjects (nonasthmatics) with chronic cough, a number of consistent observations have been found, including increased epithelial desquamation and the presence of inflammatory cells, submucosal fibrosis, squamous cell metaplasia, and loss of cilia. The high intrathoracic pressures generated during cough serve to compress intrathoracic airways, thereby giving rise to high gas velocities within the airway (Scherer, 1981). These high velocities of airflow supply the shearing forces necessary to dislodge material adherent to the airway epithelial surfaces. It thus appears likely that two-phase gas−liquid flow can occur in the human airway under certain conditions. During coughing and rapid breathing (e.g., with exercise) and maximal expiratory flow maneuvers, the Reynolds numbers in the large airways will be sufficiently high to result in gas−liquid interaction if the thickness of the mucus layer at the epithelial surface is in excess of approximately 300 μm (Clarke et al., 1970). These depths of mucus are believed to occur and manifest in airway diseases with excessive secretions of mucins. Although the pathophysiology of persistent cough remains difficult to unravel, cough can be a manifestation of mucin hypersecretory airway disease (Banner, 1986).

Several investigations have evaluated cough as an adjunct to the mucociliary clearance system, particularly in the more central airways (Foster, 2002). It has generally been presumed that voluntary cough maneuvers are ineffective in the normal airway, where moderate depth of the mucus layer ($<$20 μm) exists as compared to the excessive airway secretions and mucus layer thickness found in bronchitis ($>$200 μm) (Bennett et al., 1990). Camner et al. (1979) observed no abrupt changes in the clearance of airway mucus during periods of short vigorous coughing. In a similar fashion, Yeates et al. (1975) demonstrated that voluntary coughing had an effect on mucus flow rates through the tracheal airway of healthy human subjects. In contradiction to these experimental approaches, Scherer and Burtz (1978) used a physical model and non-Newtonian fluid, proposing that cough-related ciliary mucus transport can be enhanced within small airways (12th-generation airway), which should prevail even within normal airway free of abnormal tracheobronchial cellular topography. In this connection, Bennett et al. (1990) reported that frequent episodes of voluntary coughing in healthy subjects without airway disease significantly improves the kinetics of mucociliary clearance. This response was found to be related to airflow within large conducting airways, independent of the direction of the airflow. For example, voluntary maneuvers that were designed to control and attain equivalent levels of airflow demonstrated that inspiratory airflow rates were equally as effective in stimulating mucus layer transport within central airways as mucus transport induced by increasing expiratory airflow. Wolff et al. (1977) have demonstrated that both exercise and resting eucapnic hyperpnea can enhance the clearance of airway mucus layer in healthy subjects, although exercise induced a more robust effect. The response to hyperpnoea is consistent with the findings of Bennett et al. (1990)—that is, independent of the direction of the airflow, transport and clearance of the airway mucus layer was stimulated by high velocities of airflow. Tracheobronchial neuronal reflexes may be responsible, whereby the velocity of airflow through large central airway may act as a stimulus via rapidly adapting receptors whose terminals lie within and under the epithelium of the larger bronchi and which responds to changes in lung volume (inflation and deflation). Excitation of these receptors leads to responses including cough or deep inspirations, bronchoconstriction, mucus secretion, and airway vasodilation (Widdicombe, 1995). For example, a rapid lung inflation/deflation or exposure of the airway epithelial surfaces to a respirable irritant (dust, ozone, smoke, capsaicin) induces mucus release from submucosal glands onto the epithelial surface and would dilute the specific activity of a deposited irritant. Such response as well would serve to enhance/thicken the gel mucus layer and favor annular-type radial flow of the mucus layer during expiratory phase of breathing with increased velocities of airflow passing through bronchoconstricted airway lumens. Animal model studies support that tracheobronchial gland secretions are stimulated in response to rapid deflation—an action that is abolished by vagal cooling, vagal section, or anticholinergic treatment

(Yu et al., 1989). Thus, during expiratory directed airflow, the stimulatory action on airway mucus layer transport may be of a reflex basis and interdependent with cholinergic mechanisms.

9. MUCOCILIARY CLEARANCE AND AIRWAY TOXICOLOGY

From toxicity and injury standpoints, the removal of cellular debris, xenobiotic materials, and inhaled ambient particulate matter that have deposited upon and within the mucus layer is a dynamic process. In part, it represents a complex lung defense system that involves a series of interrelated mechanisms, including particle solubility, macrophage phagocytosis, mucociliary transport, cellular endocytosis and intercellular sieving, and lymph and capillary blood flow. Complete reviews are available on nasal physiology and transport and retention of particulate species deposited onto epithelial surfaces of the upper respiratory tract (Fry and Black, 1973; Proctor, 1977; Rusznak et al., 1994). Thus, the remaining information will pertain to the protective systems of the lower respiratory tract.

It has been generally accepted that relatively insoluble particulate matter interspersed within the mucus layer lining of the respiratory epithelium are removed during two phases (Langenback et al., 1990; Asgharian et al., 2001; Brown et al., 2002; Moller et al., 2008). The initial phase, identified as a "tracheobronchial" phase, is thought to have clearance half-times between 3 and 12 h and by convention usually concludes after 24 h. The second phase, slow in comparison to the tracheobronchial phase, requires days to months to be completed and has been identified as the "alveolar" phase. The early phase is believed to represent clearance of particles that have deposited onto the ciliated epithelial surfaces of the tracheobronchial airways and that, if insoluble, are removed by mucociliary clearance. Particle clearance during the delayed phase is thought to reflect phagocytic activity, lymphatic drainage, and dissolution of particles that deposited onto epithelial surfaces distal to the respiratory bronchioles. For the alveolar phase, individual mechanisms have not been precisely characterized; thus, physical and cellular clearance routes are frequently considered in aggregate. It is clear that neither mucociliary nor alveolar phases and removal are all encompassing as protection, and that some fraction of deposited particulate can become endocytosed and sequestered within the airway and alveolar epithelia and submucosal tissues (Churg, 1996).

With regard to the temporal aspects of particulate clearance, the following concepts evolved from early in vivo studies of particle deposition and clearance. Human investigations between 1955 and 1970 (Albert and Arnett, 1955; Booker et al., 1955; Lippmann and Albert, 1969; Morrow et al., 1968) established the following concepts

1. After deposition by inhalation upon epithelial surfaces of the lower respiratory tract, insoluble particles are rapidly removed from the lung via the tracheal airway during an initial 24-h clearance period. The fraction of particles cleared during this period, expressed as a percentage of the amount of particulate initially deposited, increases with particle diameter. That is, larger-sized particles inhaled with nasal and/or mouth breathing penetrate less effectively to the distal epithelial respiratory surfaces and are cleared by mucociliary activity more completely.
2. After 24 h, the removal of insoluble particulate from the lung is minimal. For example, following the initial fast phase of tracheobronchial clearance, the daily rate of insoluble particles being removed averaged 5% or less of the fraction of particles cleared during the initial 24-h period.

Other investigations of insoluble particle clearance have tended to support the earlier concepts. Studies designed to evaluate the initial fast phase of clearance have demonstrated in healthy subjects that particle clearance concludes after 24 h with little measureable removal of retained particulate during 2–3 days of additional measurement. Additional early evidence in support of the concept that the fast phase of particle clearance immediately apparent after deposition of aerosolized particulate concludes within 24 h was provided by performing replicate studies in healthy human subjects, clearly demonstrating repeatability of particle deposition within the lung and uniformity within the levels of particle retention during consecutive follow-up measurements accomplished 2–4 days after the initial deposition of the particles (Camner and Philipson, 1978). A subsequent investigation by Stahlhofen et al. (1980) in human subjects used several particle sizes (diameters ranged between 2 and 9 μm) and demonstrated that the initial fast phase of clearance ended within 24–34 h of particle deposition. Therefore, considered in context, these early investigations in healthy humans affirmed that the initial fast phase of particle removal from epithelial surfaces of the lower respiratory tract represents mucus layer transport and can be used to quantify mucociliary activity. This process of airway defense is complete within 2–3 days of deposition for insoluble particles $\geq$2 μm in diameter. Investigations in healthy human subjects have used very small-sized aerosols, with size distributions in the ultrafine range (diameter $\leq$100 nm), which deposit more efficiently onto epithelial surfaces of conducting airways than penetrating into the peripheral air spaces of the lung. These studies demonstrated postinhalation lung retention levels consistent with models and experimental data, showing that particles cleared during the initial 24 h had deposited initially within the conducting airways and were effectively cleared by mucocilary clearance (Brown et al., 2002; Moller et al., 2008).

The percentage of particles retained within the lower respiratory tract at 1 day postinhalation, which is the practical endpoint of the initial mucociliary clearance phase, correlates with penetration and deposition of inhaled particulate onto lung epithelial surfaces. For instance, low values of the 24-h retention (expressed as a percentage of the initial particle lung burden) are interpreted to indicate that a smaller fraction of particles inhaled penetrated the lower respiratory tract and deposited within smaller, more peripheral airways (Foster, 1988). Indirect support of this concept arises from investigations of subjects with obstructive airway disease (Dolovich et al., 1976) who were found to have lower values of 24-h retention (Greening et al., 1980; Pavia et al., 1977, Poe et al., 1977). As described above, the major physical adjunct to mucociliary transport process, and equally important for insoluble particle clearance from central dependent airways, is coughing. For example, in elderly subjects with obstructive airway disease, a group considered to be at high risk from particulate air pollution, those subjects with periods of spontaneous coughing had significant improvement in insoluble particle clearance from the lower respiratory tract over the short term (several hours), indicative of the physical positive influence of cough on reducing and limiting particle retention upon epithelial surfaces (Groth et al., 1997).

Several studies in humans have used long-lived radiolabeling of insoluble particles or magnetopneumogrpahic methods (employing ferromagnetic particles) to evaluate the kinetics of the slower, alveolar clearance phase of particle removal by the lung. The results generally support the paradigm that for healthy subjects following an initial mucociliary clearance phase, the retention of particle burdens in the lung can be described by a single time constant over the next 15–20 days postinhalation (Stahlhofen et al., 1980). Characteristic and replicable half-times of retention have been reported for a variety of particle species (Camner and Philipson, 1978; Stahlhofen et al., 1981); for at least three studies in humans, particle retentions within the pulmonary system were monitored noninvasively for durations of 9–12 months (Bailey et al., 1982; Bohning et al., 1982; Moller et al., 2004). This slow phase of lung clearance is proposed to represent particles that had initially deposited onto epithelial surfaces located distal to the terminal bronchioles and whereby particle removal from these regions is dependent upon endocytosis, phagocytosis, lymphatic drainage, and particle dissolution.

Additional air pollutants that are frequently present in urban environments, along with inhalable particulate, include gas-phase pollutants, such as ozone, and acid sulfur-based aerosols. Ozone is extremely reactive with unsaturated fatty acids at the airway surface liquid-epithelial cell membrane interface and does not penetrate the cell membrane. Ozone reacts with mucins and phospholipids that comprise the plasma membrane to generate lipid ozonation products, including aldehydes and hydrogen peroxide (Pryor et al., 1995). Interleukin 8 (IL-8) gene expression, a pro-inflammatory cytokine known to associate with cellular neutrophilic infiltration and airway mucus cell metaplasia, is upregulated acutely in human bronchial epithelial cells treated with ozone in vitro (Voynow et al., 2009). Ozone's influence in vivo upon the mucociliary clearance system has been studied in healthy adult human subjects and dose–response alterations of the mucociliary system have suggested that the smaller, more peripheral airways are more sensitive than larger airway units (stem bronchi) (Foster et al., 1987). An additional investigation in human subjects that combined ozone with a substantial exercise load found the response to ozone was transient (Gerrity et al., 1993); however, exercise alone merits consideration and numerous investigations in human subjects have demonstrated that exercise increases clearance of both soluble and insoluble particles from the tracheobronchial airways (Bennett et al., 1985; Lorino et al., 1989; Wolff et al., 1977). A number of epithelial-derived factors may be responsible for the ozone-induced increase in mucociliary clearance of particulate; a potential stimulus includes epithelial cell-derived prostanoid mediators. For example, prostaglandin $PGF_{2\alpha}$ is known to enhance the secretion of human respiratory secretions in vitro (Shelhamer et al., 1980) and is present at increased levels systemically in subjects exposed to ozone (Schelegle et al., 1989), and within bronchoalveolar lavage fluids sampled directly from the lower respiratory tract between 3 and 18 h postexposure (Devlin et al., 1991; Seltzer et al., 1986).

Investigations of insoluble particle clearance in vivo have clearly demonstrated that exposure to acid sulfate aerosol impairs tracheobronchial mucociliary function in the human airway in a dose-related fashion (Leikauf et al., 1984). Although directly opposite in direction of effect on the kinetics of mucociliary clearance of exposure to ozone, similar to ozone in regional influence, the distal ciliated airways appeared to be more sensitive to exposure to H_2SO_4 aerosol (0.5-μm diameter) than the larger, more central bronchi (Leikauf et al., 1981). Potential causative factors for response to H_2SO_4 were investigated in vitro, indicating that ciliary activity of cells sampled from healthy subjects was significantly impaired by exposure to H_2SO_4 at conditions of low pH (Hastie et al., 1997); however, impaired ciliary activity was reversible to normal levels upon re-establishment of conditions of physiologic pH. The in vitro responses suggest that H^+ may be the effector for H_2SO_4-induced decrease of mucociliary clearance rather than SO_4 leading to airway injury (Balmes et al., 1989; Schlesinger, 1989).

10. INTEGRITY OF THE EPITHELIAL BARRIER

Overall, investigations of the initial phase of tracehobronchial clearance and the delayed phase of alveolar clearance support the general concept that clearance of insoluble, compact particles from the respiratory tract of healthy subjects occurs in two phases: a fast, tracheobronchial phase that requires approximately 24 h to complete; and a slow, alveolar phase that may require several months for completion (Foster, 1988). The interplay between fluid transfer across the airway epithelium and the efficiency of particle clearance by the mucociliary apparatus along with respiratory secretions and cellular debris atop the ciliated epithelium have often been postulated as essential components of epithelial barrier defense and function (Gail and Lenfant, 1982; Ganesan et al., 2013). An association between epithelial permeability and the kinetics of particle clearance from the epithelial surfaces has not been firmly established in vivo; however, in this connection, it is known that individuals habituated to cigarette smoking have increased permeability within epithelia at parenchymal surfaces (O'Brodovich and Coates, 1987) and abnormalities in mucociliary clearance of particles from the peripheral airways (Foster et al., 1985).

Airborne, ambient particulate matter deposited onto airway and parenchymal epithelial surfaces may undergo dissolution and release soluble and chemical species. Even though protective fluid layers lining epithelial surfaces may be adequate, xenobiotic materials translocate to submucosal tissues. For example, solutes present at the airway and parenchymal surfaces are known to undergo temporal, size-dependent clearance by passage through paracellular pathways and cell to cell junctional complexes that are positioned at the apical-lateral borders between neighboring epithelial cells (Huchon et al., 1987). Once interstitial, solutes are eventually removed by endothelial transfer and entry into bronchial or pulmonary blood supplies (Jones et al., 1982). A direct in vivo approach for characterizing epithelial integrity of the lower respiratory tract is to deposit hydrophilic, low-molecular-weight, radio-labeled tracers by inhalation or instillation onto epithelial surfaces and monitor translocation dynamically using gamma-scintigraphy (Figure 3). Translocation kinetics of radioligands is repeatable (Huchon et al., 1987; Que et al., 2011) and easily quantitated noninvasively in humans; however, for animal models, anesthesia is usually required. Sterile soluble solute, such as technetium^{99m}-labeled diethyltriaminepentaacetate (Tc^{99m}-DTPA, 492 Da) has been utilized in humans and animal models to evaluate paracellular clearance of TC^{99m}-DTPA as an index of epithelial integrity (O'Brodovich and Coates, 1987; Foster and Stetkiewicz, 1996).

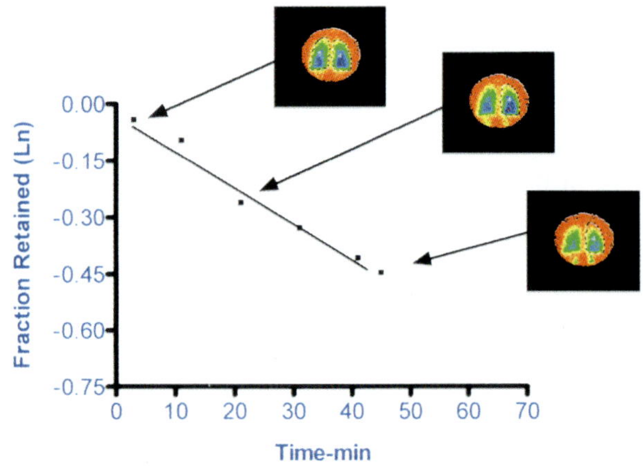

FIGURE 3 Lung retention of Tc^{99m}-DTPA in a human with superimposed gamma-scintigraphy imaging. Tc^{99m}-DTPA aerosol by inhalation deposits onto epithelial surfaces and clears airway surfaces via paracellular pathways. Dynamic images collected from the posterior aspect are superimposed and represent lung retention of Tc^{99m}-DTPA at indicated points (activity levels of Tc^{99m}-DTPA from high to low levels by color: dark blue (dark gray in print versions) > green (pale gray in print versions) > yellow (white in print versions) > orange (light gray in print versions)). Lung retention is expressed as natural logarithm of radioactivity deposited in the lung at time zero; as clearance progresses, retention levels decrease in the lung and increase bilaterally (below the diaphragm) after vascular filtration within the kidneys. Lung retention points can be fit to a linear regression to provide clearance metric.

In Table 4, a species comparison is listed for in vivo epithelial clearance of TC^{99m}-DTPA in humans and a variety of animal models, with consideration upon regional differences in surface area, by separately comparing solute permeation across bronchial (airway) and alveolar (parenchymal) surfaces. A review by Widdicombe (1997) on the permeability characteristics of the respiratory epithelium is available, along with an overview by Rezaee and Georas (2014) of the molecular regulation of apical junctional complexes of airway epithelia. Several of the investigations characterizing epithelial integrity have relied upon small-molecular-weight (<1000 Da) hydrophilic solutes that can be delivered and inhaled as an aerosol. Conceptually, clearance of these low-molecular-weight markers occurs through paracellular pathways, although a fraction of the clearance will also be dependent upon solubility within lipid cellular membranes at apical epithelial surfaces and intracellular penetrance of small molecules (Nathanson and Nadel, 1984). Acute periods of exercise and hyperventilation in healthy subjects alters epithelial barrier integrity and leads to an increase in soluble particle clearance from peripheral airway surfaces (Lorino et al., 1989). Soluble particle clearance is also known to be extremely rapid in acute lung injury syndromes and deficiencies in lung surfactant, injury to alveolar ducts and bronchioles, and presence of oxidizing agents—all of which may exert an influence (Groth, 1991; O'Brodovich and Coates, 1987).

TABLE 4 Species Comparison: Epithelial Paracellular Permeability for Hydrophilic, Low-Molecular-Weight Solute[a]

Species	Bronchial Region (T1/2 − min: Tc-DTPA)	Alveolar Region (T1/2 − min: Tc-DTPA)
Human		
Healthy	118.1 ± 3.5 (Que et al., 2011)	110.0 ± 22.2 (Kennedy et al., 1984)
Smoker		42.4 ± 5.3 (Kennedy et al., 1984)
Asthma	126.0 ± 20.5 (Ilowite et al., 1989)	78.2 ± 17.9 (O'Byrne et al., 1984)
Rodent Model		
Rat	50.4 ± 4.7 (McGowan et al., 2002)	
Guinea pig	58.6 ± 4.5 (Grieff et al., 1990)	24.8 ± 3.3 (Grieff et al., 1990)
Mouse		53.0 ± 2.0 (Bishai et al., 2006)
Rabbit		71.0 ± 9.0 (Nilsson et al., 1997)
Dog	23.6 ± 2.6 (Foster and Freed, 1999)	
Piglet		57.0 ± 4.4 (O'Brodovich et al., 1989)
Sheep	14.8 ± 2.7 (Foster and Wagner, 2001)	
Horse	39.2 ± 2.9 (Votion et al., 1998)	

[a]Note: values are means ± SEM. The $T^{1}/_{2}$ is time in minutes required for 50% of the amount of initially deposited or instilled radiotracer (Tc-DTPA: technetium^{99m} labeled diethylene-triamine pentaacetate, 492 Da) to be transferred from airway and parenchymal surfaces via paracellular pathways between neighboring epithelial cells; for the human, clearance was characterized non-invasively in awake subjects; for all of the animal models, the animals were anesthetized.

From an occupational injury viewpoint, soluble particle clearance through the respiratory epithelium was shown to be enhanced in both smoking and nonsmoking retired coal miners as compared to individuals unexposed to coal dust (Suskind et al., 1988). The biologic mechanisms related to persistence in epithelial barrier injury in the absence of continued exposure to dusts (duration since last exposure to coal dust in some cases exceeded >2 years) are unknown, although the continued presence of inflammatory processes in the lung were apparent, and different from the rapid improvement and re-establishment of epithelial barrier function observed in asymptomatic smokers with smoking cessation (Mason et al., 1983).

With respect to soluble particle clearance and the influence of co-pollutants commonly found in urban environments, for humans only ozone's effects have been described to date. In a controlled laboratory setting and concentrations relevant to ambient levels (200−250 ppb), ozone was not found to alter epithelial barrier function and permeability of a small hydrophilic molecule (<400 Da) during the acute period immediately following exposure of healthy young adults, or even asymptomatic asthmatic subjects (Laube et al., 1996). However, when epithelial integrity of peripheral airways were assessed in healthy subjects at a delayed time, 18−20 h postexposure, a dramatic increase in airway permeability and injury to barrier integrity was observed (Foster and Stetkiewicz, 1996). However, more recently and in a substantial cohort of healthy young adult subjects, exposure to ozone at 220 ppb demonstrated that sensitivity (i.e., susceptibility) to ozone was not uniform, wherein approximately 25% of the subjects were found to be highly responsive to ozone exposure and developed a significant reduction in airway epithelial integrity at 20 h postexposure (Que et al., 2011).

Impairment of epithelial barrier function is of concern with respect to risk of injury and submucosal airway tissue remodeling, as these data in humans suggest that airway epithelial injury as a temporary outcome from exposure permits xenobiotic materials present at airway surfaces (e.g., soluble components of inhaled ambient particulate) to gain access to endogenous basal and progenitor cells of airway epithelia (Teisanu et al., 2011). The alterations in epithelial integrity also coincide with inflammatory changes (cellular and biochemical) that have been reported for airway and bronchoalveolar lavage fluids (Devlin et al., 1991) that were collected at similar time points (~20 h postexposure) and support the concept for translocation of soluble and small molecules from airway surfaces into bronchial and pulmonary capillary blood supplies.

11. CONCLUSIONS

This chapter attempted to summarize the science of mucociliary clearance of the conducting airways of the lower respiratory tract. The traditional view is that the respiratory epithelial surfaces are largely protected by a liquid lining layer composed of a periciliary phase that is adjacent to the luminal surface of epithelial cells, and a gel phase superimposed atop the lower less-viscous sol phase. The thickness of the sol–gel layer is dependent upon transepithelial ion secretion (transport) across epithelial cells with passive diffusive flow of water, and steady state and stimulated mucin secretion. Several of the secreted mucins remain tethered or cell membrane bound. Effective and dynamic, directional clearance of the gel phase to the larynx removes gel-forming mucins from the epithelial surfaces reflects the coordinated ciliary activity of the underlying ciliated epithelial cells. Lung inflation/deflation reflexes and shearing forces at the airway surfaces during high velocity airflow (e.g., cough) are stimulatory factors for increasing the rate of mucociliary clearance.

Although poorly studied in humans at present, the importance of bronchial blood flow to nutritional support of the mucociliary system is largely unstudied; at present, it is only inferred from the lung transplant literature. As an innate defense to toxicology insults, this clearance system serves as a protective lining layer, and also provides continuous removal of xenobiotic materials and inhaled irritants deposited and retained at airway surfaces or sequestered within resident phagocytic cells. Clearance is dynamic and separated into a tracheobronchial phase, representing the mucociliary defense system with an approximate turnover time of 24 h, and a slower alveolar phase that is dependent upon macrophage scavenger biology and lymphatic drainage from respiratory bronchiolar alveolar surfaces and has a duration extending from weeks to months.

As tools and techniques of cell biology continue to develop (i.e., flow cytometry and proteomic approaches), a clearer understanding will evolve of progenitor cell origin and repair mechanisms (Teisanu et al., 2011). The molecular biology associated with assembly and disassembly of paracellular junctional complexes (Rezaee and Georas, 2014), as well as the interactions between phagocyte recruitment within distal alveolar surfaces and the mucocilary defense system of the proximal airways, will continue to be more clearly defined.

REFERENCES

Abraham, W.M., Bourdelais, A.J., Satater, J.R., Ahmed, A., Lee, T.A., Serebriakov, I., Baden, D.G., 2004. Airway responses to aerosolized brevetoxins in an animal model of asthma. Am. J. Respir. Crit. Care Med. 171, 26–34.

Andrade, Y.N., Fernandes, J., Vazquez, E., Fernandez-Fernandez, J.M., Arniges, M., Sanchez, T.M., Villalon, M., Valverde, M.A., 2005. TRPV4 channel is involved in the coupling of fluid viscosity to epithelial ciliary activity. J. Cell Biol. 168, 869–874.

Adler, K.B., Wooten, O., Dulfano, M.J., 1973. Mammalian respiratory mucociliary clearance. Arch. Environ. Health 27, 364–369.

Albert, R.E., Arnett, L.C., 1955. Clearance of radioactive dust from the human lung. Arch. Ind. Health 12, 99–106.

Asgharian, B., Hofmann, W., Miller, F.J., 2001. Mucociliary clearance of insoluble particles from the tracheobronchial airways of the human lung. J. Aerosol Sci. 32, 817–832.

Asmundsson, T., Kilburn, K.H., 1970. Mucociliary clearance rates at various levels in dog lungs. Am. Rev. Respir. Dis. 102, 388–397.

Baile, E.M., Dahlby, R.W., Wiggs, B.R., Pare, P.D., 1985. Role of tracheal and bronchial circulation in respiratory heat exchange. J. Appl. Physiol. 58, 217–222.

Baile, E.M., Jasso-Victoria, R., Sotres-Vega, A., Selman-Lamma, M., Villalba, J., Arreola, J.L., Olmos, R., Pare, P.D., Santillan-Doherty, P., 1992. Tracheobronchial blood flow after a modified canine lung autotransplantation of omental wrapping. Transplant Proc. 24, 2024–2029.

Bailey, M.R., Fry, F.A., James, A.C., 1982. The long term clearance kinetics of insoluble particles from the human lung. Ann. Occup. Hyg. 26, 273–290.

Balmes, J.R., Fine, J.M., Gordon, T., Sheppard, D., 1989. Potential bronchoconstrictor stimuli in acid fog. Environ. Health Perspect. 79, 163–166.

Banner, A.S., 1986. Cough: physiology, evaluation, and treatment. Lung 164, 79–92.

Barnes, P.J., Nadel, J.A., Roberts, J.M., Basbaum, C.B., 1983. Muscarinic receptors in lung and trachea: autoradiographic localization using quinuclidinyl benzilate. Eur. J. Pharmacol. 86, 103–106.

Barnes, P.J., Basbaum, C.B., 1983. Mapping of adrenergic receptors in mammalian trachea using an autoradiograpjhic method. Exp. Lung Res. 5, 183–192.

Barnett, B., Miller, C.E., 1966. Flow induced by biological mucociliary systems. Ann. N. Y. Sci. 130, 891–901.

Basbaum, C.B., Barnes, P.J., Grillo, M., 1983. Adrenergic and cholinergic receptors in submucosal glands in ferret trachea: autoradiographic localization. Eur. J. Respir. Dis. 64, 433–435.

Basbaum, C.B., Grillo, M., Widdicombe, J.H., 1984. Muscarinic receptors: evidence for a non-uniform distribution in tracheal smooth muscle and exocrine glands. J. Neurol. Sci. 4, 508–520.

Basbaum, C.B., 1995. Innervation of the airway mucosa and submucosa. Semin. Respir. Med. 308, 313.

Bennett, W.D., Foster, W.M., Chapman, W.F., 1990. Cough-enhanced mucus clearance in the normal lung. J. Appl. Physiol. 69, 1670–1675.

Bennett, W.D., Messina, M., Smaldone, G.C., 1985. Effect of exercise on deposition and subsequent retention of inhaled particles. J. Appl. Physiol. 59, 1046–1054.

Bishai, J.M., Mitzner, W., Tankersley, C.G., Wagner, E.M., 2006. PEEP-induced changes in epithelial permeability in inbred mouse strains. Respir. Physiol. Neurobiol. 156, 340–344.

Blake, J., 1975. On the movement of mucus in the lung. J. Biomech. 8, 179–190.

Bohning, D.E., Atkins, H.L., Cohen, S.H., 1982. Long-term particle clearance in man: normal and impaired. Ann. Occup. Hyg. 26, 259–271.

Booker, D.V., Chamberlain, A.C., Rundo, J., Muir, D.C., Thomson, M.L., 1955. Elimination of 5 micron particles from the human lung. Nature 215, 30–33.

Boyd, E.M., 1954. Expectorants and respiratory tract fluid. Pharmacol. Rev. 6, 521–542.

Brody, J.S., Klempfner, G., Staum, M.M., Vidyasagar, D., Kuhl, D.E., Waldhausen, J.A., 1972. Mucociliary clearance after lung denervation and bronchial transection. J. Appl. Physiol. 32, 160–164.

Brokow, C.J., 1966. Mechanics and energetics of cilia. Am. Rev. Resp. Dis. 93 (Suppl.), 32–40.

Broekhuis, J.R., Leong, W.Y., Jansen, G., 2013. Regulation of cilium length and intraflagellar transport. Int. Rev. Cell Mol. Biol. 303, 101–138.

Brown, J.S., Zeman, K.L., Bennett, W.D., 2002. Ultrafine particle deposition and clearance in the healthy and obstructed lung. Am. J. Respir. Crit. Care Med. 166, 1240–1247.

Brownstein, D.G., 1987. Tracheal mucociliary transport in laboratory mice: evidence for genetic polymorphism. Exp. Lung Res. 13, 185–191.

Button, B., Cai, L.-H., Ehre, C., Kesimer, M., Hill, D.B., Sheehan, J.K., Boucher, R.C., Rubinstein, M., 2012. A periciliary brush promotes the lung health by separating the mucus layer from airway epithelia. Science 337, 937–941.

Camner, P., Strandberg, K., Philipson, K., 1974. Increased mucociliary transport by cholinergic stimulation. Arch. Environ. Health 29, 220–224.

Camner, P., Philipson, K., 1978. Human alveolar deposition of 4 micron teflon particles. Arch. Environ. Health 33, 181–185.

Camner, P., Mossberg, M., Philipson, K., Strandberg, K., 1979. Elimination of test particles form the human tracheobronchial tract by voluntary coughing. Scand. J. Respir. Dis. 60, 56–62.

Carstairs, J.R., Barnes, P.J., 1986. Visualization of vasoactive intestinal peptide receptors in human and guinea-pig lung. J. Pharmacol. Exp. Ther. 239, 240–255.

Churg, A., 1996. The uptake of mineral particles by pulmonary epithelial cells. Am. J. Resp. Crit. Care Med. 154, 1124–1140.

Christensen, S.T., Pedersen, L.B., Schneider, L., Satir, P., 2007. Sensory cilia and integration of signal transduction in human health and disease. Traffic 8, 97–109.

Clapham, D.E., 2003. TRP channels as cellular sensors. Nature 426, 517–524.

Clarke, S.W., Jones, J.G., Oliver, D.R., 1970. Resistance to two-phase gas-liquid flow in airways. J. Appl. Physiol. 29, 464–471.

Clunes, L.A., Davies, C., Coakley, R.D., Aleksandrov, A.A., Henderson, A.G., Zeman, K.L., Worthington, E.N., Gentzsch, M., Kreda, S.M., Cholon, D., Bennet, W.D., Riordan, J.R., Boucher, R.C., Tarran, R., 2012. Cigarette smoke exposure induces CFTR internalization and insolubility, leading to airway surface liquid dehydration. FASEB J. 26, 533–545.

Copin, M.C., Buisine, M.-P., Leteurtre, E., Marquette, C.H., Porte, H., Aubert, J.P., Gosselin, B., Porchet, N., 2001. Mucinous bronchoalveolar carcinomas display a specific pattern of mucin gene expression among primary lung adenocarcinomas. Hum. Pathol. 32, 274–281.

Costa, D.L., Schelegle, E.S., 1999. Air pollutants and the respiratory tract. In: Swift, D.L., Foster, W.M. (Eds.), Inhaled Air Pollutants. Marcel Dekker, N.Y., pp. 119–145.

Dalhamn, T., Rhodin, J., 1956. Mucus flow and ciliary activity in the trachea of rats exposed to pulmonary irritant gases. Br. J. Ind. Med. 13, 110–113.

Delmotte, P., Sanderson, M.J., 2006. Ciliary beat frequency is maintained at a maximal rate in the small airways of mouse lung slices. Am. J. Respir. Cell Mol. Biol. 35, 110–117.

Denton, R., Forsman, W., Hwang, S.H., Litt, M., Miller, C., 1968. Viscoelasticity of mucus. Am. Rev. Resp. Dis. 98, 380–391.

Devlin, R.B., McDonnell, W.F., Mann, R., Becker, S., House, D.E., Schreinemachers, D., Koren, H.S., 1991. Exposure of humans to ambient levels of ozone for 6.6 hr causes cellular and biochemical changes in the lung. Am. J. Resp. Cell Mol. Biol. 4, 72–81.

Dolovich, M.B., Sanchis, J., Rossman, C., Newhouse, M.T., 1976. Aerosol penetrance: a sensitive index of peripheral airways obstruction. J. Appl. Physiol. 40, 468–471.

Dulfano, M.J., Adler, K.B., 1975. Physical properties of sputum. VII. Rheological properties and mucocilary transport. Am. Rev. Resp. Dis. 112, 341–347.

Edmunds, L.H., Stallone, R.J., Graf, P.D., Sagel, S.S., Greenspan, R.H., 1969. Mucus transport in transplanted lungs of dogs. Surgery 66, 15–22.

Edmunds, L.H., Nadel, J.A., Graf, P.D., 1971. Reinervation of the re-implanted canine lung. J. Appl. Physiol. 31, 722–727.

Eraslan, S., Turner, M.D., Hardy, J.D., 1964. Lymphatic regeneration following lung transplantation in dogs. Surg. St. Louis 56, 970–973.

Ewert, G., Martensson, B., 1967. Mucous flow in the subglottic region studied by transonioscopy. Acta. Otol. 224 (Suppl.), 515–517.

Fabricant, N.D., 1955. The physiological approach to nasal medication. Am. J. Med. Sci. 230, 436–440.

Fahy, J.V., Dickey, B.F., 2010. Airway mucus function and dysfunction. N. Engl. J. Med. 363, 2233–2247.

Felicetti, S.A., Wolff, R.K., Muggenburg, B.A., 1981. Comparison of tracheal mucous transport in rats, guniea pigs, rabbits, and dogs. J. Appl. Physiol. 51, 1612–1617.

Florey, H., Careleton, H.M., Wells, A.Q., 1932. Mucous secretion in the trachea. Br. J. Exp. Pathol. 13, 269–284.

Foster, W.M., 1988. Is the 24 hour lung retention an index of alveolar deposition? J. Aer. Med. 1, 1–8.

Foster, W.M., 2002. Mucociliary transport and cough in humans. Pulm. Pharmacol. Ther. 15, 277–282.

Foster, W.M., Adler, K.B., Crews, A.L., Potts, E.N., Fischer, B.M., Voynow, J.A., 2010. MARCKS-related peptide modulates in vivo the secretion of airway Muc5ac. Am. J. Physiol. Lung Cell. Mol. Physiol. 299, L345–L352.

Foster, W.M., Bergofsky, E.H., Bohning, D., Lippmann, M., Albert, R.E., 1976. Effect of adrenergic agents and their mode of action on mucociliary clearance in man. J. Appl. Physiol. 41, 146–152.

Foster, W.M., Costa, D.L., Langenback, E.G., 1987. Ozone exposure alters tracheobronchial mucocilary function. J. Appl. Physiol. 63, 996–1002.

Foster, W.M., Freed, A.N., 1999. Regional clearance of solute from peripheral airway epithelia: recovery after sublobar exposure to ozone. J. Appl. Physiol. 86, 641–646.

Foster, W.M., Langenback, E.G., Bergofsky, E.H., 1980. Measurement of tracheal and bronchial mucus velocities in man: relation to lung clearance. J. Appl. Physiol. 48, 965–971.

Foster, W.M., Langenback, E.G., Bergofsky, E.H., 1985. Disassociation in the mucociliary function of central and peripheral airways of asymptomatic smokers. Am. Rev. Respir. Dis. 132, 633–639.

Foster, W.M., Stetkiewicz, P., 1996. Regional clearance of solute from the respiratory epithelia: 18-20 hours postexposure to ozone. J. Appl. Physiol. 81, 1143–1149.

Foster, W.M., Wagner, E.M., 2001. Cilia and mucus from development to respiratory defense. In: Salathe, M. (Ed.), Mechanisms of Clearance of

Soluble Substances from the Intrathoracic Airways. Marcel Dekker, N.Y., pp. 385–398.

Fry, F.A., Black, A., 1973. Regional deposition and clearance of particles in the human nose. J. Aerosol Sci. 4, 113–124.

Gail, D.B., Lenfant, C.J.M., 1982. Cells of the lung: biology and clinical implications. Am. Rev. Respir. Dis. 127, 366–387.

Ganesan, S., Comstock, A.T., Sajjan, U., 2013. Barrier function of airway tract epithelium. Tissue Barriers 1, e24997.

Gerrity, T.R., Bennett, W.D., Kehrl, H., DeWitt, P., 1993. Mucociliary clearance of inhaled particles measured 2 h after ozone exposure in humans. J. Appl. Physiol. 74, 2984–2989.

Goodman, R.M., Yergin, B.M., Landa, J.F., Golivanux, M.H., Sackner, M.A., 1978. Relationship of smoking history and pulmonary function tests to tracheal mucous velocity in nonsmokers, young smokers, ex-smokers, and patients with chronic bronchitis. Am. Rev. Respir. Dis. 117, 205–214.

Gosselin, R.E., 1966. Physiologic regulators of ciliary motion. Am. Rev. Respir. Dis. 93 (Suppl.), 41–59.

Grieff, L., Wollmer, P., Erjefalt, I., Pipkorn, U., Persson, C.G.A., 1990. Clearance of ^{99m}Tc DTPA from guinea pig nasal, tracheobronchial, and bronchoalveolar airways. Thorax 45, 841–845.

Green, G.M., 1973. Alveolar-bronchiolar transport mechanisms. Arch. Intern. Med. 131, 109–114.

Greening, A.P., Miniati, M., Fazio, F., 1980. Regional deposition of aerosols in health and in airways obstruction: a comparison with krypton-81m ventilation scanning. Bull. Eur. Physiopathol. Respir. 16, 287–298.

Gross, P., Detreville, R.T.P., 1972. The lung as an embattled domain against inanimate pollutants. Am. Rev. Respir. Dis. 106, 684–691.

Groth, M.L., Macri, K., Foster, W.M., 1997. Cough and mucociliary transport of airway particulate in chronic obstructive disease. Ann. Occup. Hyg. 41 (S1), 515–521.

Groth, M.L., Langenback, E.G., Foster, W.M., 1991. Influence of inhaled atropine on lung mucociliary function in humans. Am. Rev. Respir. Dis. 144, 1042–1047.

Groth, S., 1991. Pulmonary clearance of ^{99m}Tc-DTPA. Dan. Med. J. 38, 101–113.

Hastie, A.T., Everts, K.B., Zangrilli, J., Shaver, J.R., Pollice, M.B., Fish, J.E., Peters, S.P., 1997. HSP27 elevated in mild allergic inflammation protects airway epithelium form H_2SO_4 effects. Am. J. Physiol. Lung Cell. Mol. Physiol. 17, L401–L409.

Herve, P., Silbert, D., Cerrina, J., Simonneau, G., Dartevelle, P., 1993. Impairment of bronchial mucociliary clearance in long-term survivors of heart/lung and double lung transplantation. Chest 103, 59–63.

Hilding, A.C., 1963. Phagocytosis, mucous flow, and ciliary action. Arch. Environ. Health 6, 67–69.

Hollingsworth, M.A., Swanson, B.J., 2004. Mucins in cancer: protection and control of the cell surface, Nature Reviews Cancer 4, 45–60.

Huchon, G.J., Montgomery, A.B., Lipavsky, A., Hoeffel, J.M., Murray, J.F., 1987. Respiratory clearance of aerosolized radioactive solutes of varying molecular weight. J. Nucl. Med. 28, 894–902.

Ilowite, J.S., Bennett, W.D., Sheetz, M.S., Groth, M.L., Nierman, D.M., 1989. Permeability of the bronchial mucosa to ^{99m}Tc-DTPA in asthma. Am. Rev. Respir. Dis. 139, 1139–1143.

Iravani, J., van As, A., 1972. Mucous transport in the tracheobronchial tree of normal and bronchitic rats. J. Pathol. 106, 81–93.

Jacquot, J., Hayem, A., Galabert, C., 1992. Functions of proteins and lipids in airway sections. Eur. Respir. J. 5, 343–358.

Jones, J.G., Minty, B.D., Royston, D., 1982. The physiology of leaky lungs. Br. J. Anaesth. 54, 705–721.

Kaliner, M., Shelhamer, J.H., Borson, B., Nadel, J., Patow, C., Marom, Z., 1986. Human respiratory mucus. Am. Rev. Respir. Dis. 134, 612–621.

Kelsen, S.G., Mardini, I.A., Zhou, S., 1992. A technique to harvest viable trachea-bronchial epithelial cells from living human donors. Am. J. Respir. Cell Mol. Biol. 67, 66–72.

Kennedy, S.M., Elwood, R.K., Wiggs, B.J.R., Pare, P.D., Hogg, J.C., 1984. Increased airway mucosal permeability of smokers. Am. Rev. Respir. Dis. 129, 143–148.

Kensler, C.J., Battista, S.P., 1966. Chemical and physical factors affecting mammalian ciliary activity. Am. Rev. Respir. Dis. 93 (Suppl.), 93–102.

Kilburn, K.H., 1968. A hypothesis for pulmonary clearance and it implications. Am. Rev. Respir. Dis. 98, 449–463.

Lai, S.K., Wang, Y.-Y., Wirtz, D., Hanes, J., 2009. Micro- and macrorheology of mucus. Adv. Drug Deliv. Rev. 61, 86–100.

Laitinen, A., 1985. Ultrastructural organization of intraepithelial nerves in human airway tract. Thorax 40, 488–492.

Lamb, D., Reid, L., 1970. Histochemical and autoradiographic investigation of the serous cells of the human bronchial glands. J. Pathol. 100, 127–138.

Langenback, E.G., Bergofsky, E.H., Halpern, J.G., Foster, W.M., 1990. Supramicorn-sized particle clearance from alveoli: route and kinetics. J. Appl. Physiol. 69, 1302–1308.

Laube, B., Kurian, V., Foster, W.M., 1996. Fine particle deposition and epithelial clearance after ozone exposure in normal and asthma subjects. Am. J. Respir. Crit. Care Med. 153, A700.

Laurenzi, G.A., Yin, S., Guarneri, J.J., 1968. Adverse effects of oxygen on tracheal mucous flow. N. Eng. J. Med. 279, 333–339.

Ledowski, T., Hilmi, S., Paech, M.J., 2006. Bronchial mucus transport velocity in patients receiving anaesthesia with propofol and morphine or propofol and remifentanil. Anaesthesia 61, 747–751.

Leikauf, G., Spektor, D.M., Albert, R.E., Lippmann, M., 1984. Dose-dependent effects of submicrometer sulfuric acid aerosol on particle clearance form ciliated human lung airways. Am. Ind. Hyg. Assoc. J. 45, 285–292.

Leikauf, G., Yeates, D.B., Wales, K.A., Spektor, D., Albert, R.E., Lippmann, M., 1981. Effects of sulfuric acid aerosol on respiratory mechanics and mucociliary particle clearance in healthy non-smoking adults. Am. Ind. Hyg. Assoc. J. 42, 273–282.

Li, Y., Martin, L.D., Spizz, G., Adler, K.B., 2001. MARKS protein is a key molecule regulating mucin secretion in human airway epithelia cells vitro. J. Biol. 276, 40982–40990.

Lillie, R.S., 1903–1904. The relation of ions to ciliary movement. J. Physiol. 10, 419–443.

Lilliehoj, E.P., Kato, K., Lu, W., Kim, K.C., 2013. Cellular and molecular biology of airway mucins. Int. Rev. Cell Mol. Biol. 303, 139–202.

Lippmann, M., Albert, R.E., 1969. The effects of particle size on the regional deposition of inhaled aerosols in the human respiratory tract. Am. Ind. Hyg. Assoc. J. 30, 257–275.

Litt, M., 1981. Physiochemical determinants of mucociliary flow. Chest 80 (6 Suppl), 846–849.

Liu, L., Shastry, S., Byan-Parker, S., Houser, G., Chu, K., Birket, S.E., Fernandez, C.M., Gardecki, J.A., Grizzle, W., Wilsterman, E.J., Sorscher, E.J., Rowe, S.M., Tearney, G.J., 2014. An autoregulatory mechanism governing mucociliary transport in sensitive to mucus load. Am. J. Respir. Cell Mol. Biol. June 17 [Epub ahead of print].

Lorino, A.M., Meignan, M., Bouissou, P., Atlan, G., 1989. Effects of sustained exercise on pulmonary clearance of aerosolized ^{99m}Tc-DTPA. J. Appl. Physiol. 67, 2055–2059.

Lucas, A.M., Douglas, L.C., 1934. Principles underlying ciliary activity in the respiratory tract. II. A comparison of nasal clearance in man, monkey, and other mammals. Arch. Otolaryngol. 20, 518–541.

Lucchini, R.E., Facchini, F., Turato, G., 1997. Increased VIP-positive nerve fibers in the mucous glands of subjects with chronic bronchitis. Am. J. Crit. Care Med. 156, 1963–1968.

Macklin, C.C., 1954. The pulmonary alveolar mucoid film and the pneumonocytes. Lancet 1, 1099–1104.

Mason, G.R., Uszler, J.M., Effros, R.M., Reid, E., 1983. Rapidly reversible alterations of pulmonary epithelial permeability induced by smoking. Chest 83, 6–11.

Marshall, W.F., 2008. The cell biological basis of ciliary disease. J. Cell Biol. 180, 17–20.

McGowan, S.E., Smith, J., Holmes, A.J., Smith, L.A., Businga, T.R., Madsen, M.T., Kopp, U.C., Kline, J.N., 2002. Vitamin A deficiency promotes bronchial hyperreactivity in rats by altering muscarinic M_2 receptor function. Am. J. Physiol. Lung Cell. Mol. Physiol. 282, L1031–L1039.

Mercer, R.R., Russell, M.L., Crapo, J.D., 1992. Mucous lining layers in human and rat airways. Ann. Rev. Respir. Dis. 145, 355.

Moller, W., Haussinger, K., Winkler-Heil, R., Stahlhofen, W., Meyer, T., Hofmann, W., Heyder, J., 2004. Mucociliary and long-term particle clearance in the airways of healthy non-smokers. J. Appl. Physiol. 97, 2200–2206.

Moller, W., Felten, K., Sommerer, K., Scheuch, G., Meyer, G., Meyer, P., Haussinger, K., Kreyling, W.G., 2008. Deposition, retention, and translocation of ultrafine particles from the central airways and lung periphery. Am. J. Respir. Crit. Care Med. 177, 426–432.

Morrow, P.E., 1973. Alveolar clearance of aerosols. Arch. Intern. Med. 131, 101–108.

Morrow, P.E., Gibb, F.R., Gazioglu, K.M., 1968. A study of particulate clearance from the human lungs. Am. Rev. Respir. Dis. 96, 1209–1221.

Mussato, D.J., Garrard, C.S., Lourenco, R.V., 1988. The effects of inhaled histamine on human tracheal mucus velocity and bronchial mucociliary clearance. Am. Rev. Respir. Dis. 138, 775–779.

Nathanson, I., Nadel, J.A., 1984. Movement of electrolytes and fluid across airways. Lung 162, 125–137.

Nilsson, K., Lachmann, J.B., Robertson, B., Wollmer, P., 1997. Pulmonary clearance of ^{99m}Tc-DTPA in experimental surfactant dysfunction treated with surfactant instillation. Acta Anaesthesiol. Scand. 41, 297–303.

O'Brodovich, H., Coates, G., 1987. Pulmonary clearance of Tc-DTPA: a non-invasive assessment of epithelial integrity. Lung 165, 1385–1389.

O'Brodovich, H., Coates, G., Kay, J., Muysson, D., 1989. Simultaneous measurement of lung clearance rates for Tc- and In-DTPA in normal and damaged lungs. J. Appl. Physiol. 66, 2293–2297.

O'Byrne, P.M., Dolovich, M., Dirks, R., Roberts, R.S., Newhouse, M.T., 1984. Lung epithelial permeability: relation to nonspecific airway responsiveness. J. Appl. Physiol. 57, 77–84.

Paul, A., Marclli, D., Shenni, H., King, M., Wang, N.S., Wilson, J.A., Mulder, D.S., Chiu, R.C., 1989. Mucociliary function in autotransplanted, allotransplanted, and sleeve resected lungs. J. Thorac. Cardiovasc. Surg. 98, 523–528.

Pavia, D., Thomson, M.L., Clarke, S.W., 1978. Enhanced clearance of secretions from the human lung after the administration of hypertonic saline aerosol. Am. Rev. Respir. Dis. 117, 199–203.

Pavia, D., Thomson, M.L., Clarke, S.W., Sahnnon, H.S., 1977. Effect of lung function and mode of inhalation on penetration of aerosol into the human lung. Thorax 32, 194–197.

Poe, N.D., Cohen, M.B., Yanda, R.L., 1977. Application of delayed lung imaging following radioaerosol inhalation. Radiology 122, 739–746.

Proctor, D.F., 1977. The upper airways. I. Nasal physiology and defense of the lungs. Am. Rev. Respir. Dis. 115, 97–129.

Pryor, W.A., Squadrito, G.L., Friedman, M., 1995. The cascade mechanism to explain ozone toxicity: the role of lipid ozonation products. Free Radic. Biol. Med. 19, 935–941.

Que, L.G., Stiles, J.V., Sundy, J.S., Foster, W.M., 2011. Pulmonary function, bronchial reactivity, and epithleial permeability are response phenotypes to ozone and develop differentially in healthy humans. J. Appl. Physiol. 111, 679–687.

Reid, C., Gould, S., Harris, A., 1997. Developmental expression of mucin genes in the human respiratory tract. Am. J. Respir. Cell Mol. Biol. 17, 592–598.

Reid, L., 1954. Pathology of chronic bronchitis. Lancet 1, 275–278.

Reid, L., 1960. Measurement of the bronchial mucous gland layer: a diagnostic yardstick in chronic bronchitis. Thorax 15, 132–141.

Rezaee, F., Georas, S.N., 2014. Breaking barriers: new insights into airway epithelial barrier function in health and disease. Am. J. Respir. Crit. Care Med. 50, 857–869.

Rodriquez, D.H., Rivero, F., Lorenzi-Filho, G., Pazetti, R., Saldiva, P.H., 2001. Effects of bronchial transection and re-anastomosis on mucociliary system. Chest 119, 1510–1515.

Rogers, D.F., 1994. Airway goblet cell: responsive and adaptable front line defenders. Eur. Respir. J. 7, 1690–1706.

Rogers, D.F., 2002. Pharmacological regulation of the neuronal control of airway mucus secretion. Curr. Opin. Pharmacol. 2, 249–255.

Rose, M.C., Voynow, J.A., 2006. Respiratory tract mucin genes and mucin glycoproteins in health and disease. Physiol. Rev. 86, 245–278.

Ross, A.J., Dailey, L.A., Brighton, L.E., Devlin, R.B., 2007. Transcriptional profiling of mucociliary differentiation in human airway epithelial cells. Am. J. Respir. Cell Mol. Biol. 37, 169–185.

Rusznak, C., Devalia, J.L., Lozewicz, S., Davies, R.J., 1994. The assessment of nasal mucociliary clearance and the effect of drugs. Respir. Med. 88, 89–101.

Rutland, J., de Iongh, R.U., 1990. Random ciliary orientation. N. Engl. J. Med. 323, 1681–1684.

Rutland, J., Griffin, W.M., Cole, P.J., 1982. Am. Rev. Respir. Dis. 125, 100–105.

Sade, J., Eliezer, N., Silberberg, A., Nevo, A.C., 1970. The role of mucus in transport by cilia. Am. Rev. Respir. Dis. 102, 48–52.

Salathe, M., 2007. Regulation of mammalian cilary beating. Annu. Rev. Physiol. 69, 401–422.

Salathe, M., Pratt, M.M., Wanner, A., 1993. Protein kinase C-dependent phosphorylation of a ciliary membrane protein and inhibition of ciliary beating. J. Cell Sci. 106, 1211–1220.

Santa Cruz, R., Landa, J.F., Hirsch, J., Sackner, M.A., 1974. Tracheal mucous velocity in normal man and patients with obstructive lung disease: effects of terbutaline. Am. Rev. Respir. Dis. 109, 458–463.

Satir, P., Christensen, S.T., 2007. Overview of structure and function of mammalian cilia. Annu. Rev. Physiol. 69, 377–400.

Schmid, A., Clunes, L.A., Salathe, M., Verdugo, P., Dietl, P., Davis, C.W., Tarran, R., 2011. Nucleotide-mediated airway clearance. Subcell. Biochem. 55, 95–138.

Scudi, J.V., Kimur, E.T., Reinhard, J.F., 1951. Study of drug action on mammalian ciliated epithelium. J. Pharmacol. Exp. Ther. 102, 132–137.

Sears, P.R., Thompson, K., Knowles, M.R., Davis, C.W., 2013. Am. J. Physiol. Lung Cell. Mol. Physiol. 304, L170–L183.

Serafini, S.M., Michaelson, E.D., 1977. Length and distribution of cilia in human and canine airways. Bull. Eur. Physiopathol. Respir. 13, 551–559.

Shah, A.S., Ben-Shahar, Y., Moninger, T.O., Kline, J.N., Welsh, M.J., 2009. Motile cilia of human airway epithelial are chemosensory. Science 325, 1131–1134.

Schelegle, E.S., Adams, W.C., Giri, S.N., Siefkin, A.D., 1989. Acute ozone exposure increases plasma PGF2α in ozone sensitive human subjects. Am. Rev. Respir. Dis. 140, 211–216.

Scherer, P., 1981. Mucus transport by cough. Chest 6S, 830–833.

Scherer, P., Burtz, L., 1978. Fluid mechanical experiments relevant to coughing. J. Biomech. 11, 183–187.

Schlesinger, R.B., 1989. Factors affecting the response of lung clearance systems to acid aerosols: role of exposure concentration, exposure time, and relative acidity. Environ. Health Perspect. 79, 121–126.

Seltzer, J., Bigby, B.G., Stulbarg, M., Holtzman, M.J., Nadel, J.A., 1986. Ozone induced change in bronchial reactivity to methacholine and airway inflammation in humans. J. Appl. Physiol. 60, 1321–1326.

Shelhamer, J.H., Marom, Z., Kaliner, M., 1980. Immunologic and neuropharmacologic stimulation of mucous glycoprotein release from human airways in vitro. J. Clin. Invest. 66, 1400–1408.

Singer, M., Martin, L.D., Vargaftig, B.B., Park, J., Gruber, A.D., Li, Y., Adler, K.B., 2004. A MARCKS-related peptide blocks mucus hypersecretion in a mouse model of asthma. Nat. Med. 10, 193–196.

Sisson, J.H., 2007. Alcohol and airways function in health and disease. Alcohol 41, 293–307.

Sisson, J.H., Pavlik, J.A., Wyatt, T.A., 2009. Alcohol stimulates ciliary motility of isolated airway axonemes through a nitric oxide, cyclase, and cyclic nucleotide-dependent kinase mechanism. Alcohol Clin. Exp. Res. 33, 610–616.

Sleigh, M.A., 1962. International series of monographs on pure and applied biology. In: Kerkut, G.A. (Ed.), The Biology of the Cilia and Flagella, vol. 12. Macmillan Co., N.Y., pp. 1–231.

Sleigh, M.A., 1966. Some aspects of the comparative physiology of cilia. Am. Rev. Respir. Dis. 93 (Suppl.), 16–31.

Sleigh, M.A., 1968. Patterns of ciliary beating. Symp. Soc. Exp. Biol. 22, 131–150.

Sleigh, M.A., 1971. Cilia. Endeavor 30, 11–17.

Sleigh, M.A., Blake, J.R., Liron, N., 1988. The propulsion of mucus by cilia. Am. Rev. Respir. Dis. 137, 726–741.

Smith, D.J., Faffney, E.A., Blake, J.R., 2008. Modelling mucociliary clearance. Respir. Physiol. Neurobiol. 163, 178–188.

Stahlhofen, W., Gebhart, J., Heyder, J., 1980. Experimental determination of the regional deposition of aerosol particles in the human respiratory tract. Am. Ind. Hyg. Assoc. J. 41, 385–398.

Stahlhofen, W., Gebhart, J., Heyder, J., Philipson, K., Camer, P., 1981. Intercomparison of regional deposition of aerosol particles in the human respiratory tract and their long-term elimination. Exp. Lung Res. 10, 223–233.

Stewart, W.C., 1948. Weight-carrying capacity and excitability of excised ciliated epithelium. Am. J. Physiol. 152, 1–5.

Suskind, H., Brill, A.B., Harold, W.H., 1988. Clearance of Tc-99m aerosol from coal miners' lungs. Ann. Occup. Hyg. 32 (S1), 157–169.

Teilmann, S.C., Byskov, A.G., Pedersen, P.A., Wheatley, D.N., Pajour, G.J., Christensen, S.Y., 2005. Localization of transient receptor potential ion channels in primary and motile cilia of the female murine reproductive organs. Mol. Reprod. Dev. 71, 444–452.

Teisanu, R.M., Chen, H., Matsumoto, K., McQualter, J.L., Potts, E., Foster, W.M., Bertoncello, I., Stripp, B.R., 2011. Functional analysis of two distinct bronchiolar progenitor during lung injury and repair. Am. J. Respir. Cell Mol. Biol. 44, 794–803.

Thai, C.H., Gambling, T.M., Carson, J.L., 2002. Freeze fracture study of epithelium from patients with primary ciliary dyskinesia. Thorax 57, 363–365.

Thornton, D.J., Gray, T., Nettesheim, P., Howard, J., Koo, J.S., Sheehan, J.K., 2000. Characterization of mucins from cultured normal human tracheobronchial epithelial cells. Am. J. Physiol. Lung Cell. Mol. Physiol. 278, L1118–L1128.

Toremalm, N.G., 1960. The daily amount of trachea-bronchial secretions in man. A method for continuous tracheal aspiration in laryngectomized and tracheotomized patients. Acta Otolaryngol. 158 (Suppl.), 48–53.

Venizelos, P.C., Gerrity, T.R., Yeates, D.B., 1981. Response of human mucociliary clearance to acute alcohol administration. Arch. Environ. Health 36, 194–201.

Verdugo, P., 1991. Mucin exocytosis. Am. Rev. Respir. Dis. 144, S33–S37.

Verdugo, P., Raess, B.V., Villalon, M., 1983. The role of calmodulin in the regulation of ciliary movement in mammalian epithelial cilia. J. Submicrosc. Cytol. 15, 95–96.

Voter, K.Z., Leigh, M.W., Boat, T.F., Carson, J.L., Wood, R.E., 1992. Development of mucociliary transport in the postnatal ferret trachea. J. Appl. Physiol. 73, 1500–1503.

Votion, D., Vandenput, S., Duvivier, D.H., Lambert, P., Art, T., Lekeux, P., 1998. Scintigraphical evaluation of alveolar clearance in horses. Vet. J. 156, 51–58.

Voynow, J.A., Fischer, B.M., Xheng, S., Potts, E.N., Grover, A.R., Jaiswal, A.K., Ghio, A.J., Foster, W.M., 2009. NAD(P)H quinone oxidoreductase 1 is essential for ozone-induced oxidative stress in mice and humans. Am. J. Respir. Cell Mol. Biol. 41, 107–113.

Wagner, E.M., Foster, W.M., 1996. Importance of airway blood flow on particle clearance from the lung. J. Appl. Physiol. 81, 1878–1883.

Wanner, A., Salathe, M., O'Riordan, T.G., 1996. Mucociliary clearance in the airways. Am. J. Respir. Crit. Care. Med. 154, 1868–1902.

White, J.C., Elmes, P.C., 1960. Flow properties of blood and other biological systems. In: Copley, A.L. (Ed.), Some Rheological Properties of Bronchial Mucus and Mucoprotein. Pergamon Press, N.Y., pp. 259–286.

Widdicombe, J., 1997. Airway and alveolar permeability and surface liquid thickness: theory. J. Appl. Physiol. 82, 3–12.

Widdicombe, J.G., 1995. Neurophysiology of the cough reflex. Eur. Respir. J. 8, 1193–1202.

Wolff, R.K., Dolovich, M.B., Obminski, G., Newhouse, M.T., 1977. Effects of exercise and eucapnic hyperventilation on bronchial clearance in man. J. Appl. Physiol. 43, 46–50.

Wright, W.B., Berg, P., Seki, Y., 1971. An evaluation of surgical interruption and spontaneous return of the systemic arterial circulation of the lung. J. Thorac. Cardiovasc. Surg. 62, 473–476.

Wyatt, T.A., Forget, M.A., Adams, J.M., Sisson, J.H., 2005. Both cAMP and cGMP are required for maximal ciliary beat stimulation in a cell-free model of bovine ciliary axonemes. Am. J. Physiol. Lung Cell. Mol. Physiol. 288, L546–L551.

Wyatt, T.A., Sisson, J.H., Allen-Gipson, D.S., McCaskill, M.L., Boten, J.A., DeVasure, J.M., Bailey, K.L., Poole, J.A., 2012. Co-exposure to cigarette smoke and alcohol decreases airway epithelial cell cilia beating in a protein kinase C-dependent manner. Am. J. Pathol. 181, 431–440.

Yeates, D.B., Aspin, N., Levinson, H., Jones, M.J., Bryan, A.C., 1975. Mucociliary tracheal transport rates in man. J. Appl. Physiol. 39, 487–495.

Yu, C.P., Asgharian, B., Yen, B.M., 1986. Impaction and sedimentation deposition of fibers in airways. Am. Ind. Hyg. Assoc. J. 4, 72–77.

Yu, J., Schultz, H.D., Goodman, J.C., Coleridge, J.C.G., Coleridge, H.M., Davis, B., 1989. Pulmonary rapidly adapting receptors reflexly increase airway secretion in dogs. J. Appl. Physiol. 67, 682–687.

Chapter 30

Pulmonary Immunology of Infectious Disease

Gary R. Burleson, Stefanie C.M. Burleson and Florence G. Burleson

Chapter Outline

1. INTRODUCTION

The immune system is comprised of an intricate network of lymphoid and nonlymphoid cells communicating via molecular messengers to provide surveillance against foreign agents and to protect against disease. This network, while connected via molecular mediators, contains compartmentalized units uniquely equipped to defend the host against insults to the integrity of the immune system. The pulmonary immune system, capable of engaging in localized immunological functions as well as communicating with the systemic immune system, is an example of such a specialized system. The protective mechanisms employed by the pulmonary immune system must be present and functional in order to prevent infectious disease despite exposure to an infinite variety of antigenic agents. Unless compromised, these mechanisms are extraordinarily efficient, and under normal conditions, the lower respiratory tract is devoid of infectious microorganisms and generally sterile.

The port of entry for the respiratory tract is the nasal cavity, and the epithelial cells lining the nasal luminal cavity and various regions of the respiratory tract are a target of inhaled toxicants and numerous infectious microorganisms (Harkema, 1992). The respiratory tract provides significant opportunity for exposure given that the surface area of the nasal airway is 160 cm^2 (International Commission, 1975), that the volume of the human nasal cavity is an estimated 16 cm^3 (Guilmette et al., 1989), and that the pulmonary air exchange alveolar surface area is an estimated 150 m^2, approximately the area of a tennis court (Reynolds, 1991a,b; Weibel and Taylor, 1988). Moreover, these large respiratory surfaces process an average human inspiratory exchange of 10,000 liters of air per day (Reynolds, 1991a,b). Therefore, exposure of the respiratory system to aerosolized infectious microorganisms in such large quantities of daily inspired air is not a rare event, and the large surface area provides an ample target for infection. These frequent challenges, however, must be overcome in such a way as to limit impediments to the lung function of gas exchange; thus, pathogens must be eradicated without eliciting superfluous inflammation. This unique balance underlies the necessity for the specialized and effective nature of pulmonary immune function.

Knowledge of pulmonary immunology is best understood through antigenic challenge with infectious agents. Infection with influenza virus and *Mycobacterium tuberculosis*, for example, has provided in-depth and comprehensive knowledge of the pulmonary immune system. Infection stimulates a cascade of both nonspecific innate and antigen-specific adaptive immunological functions and molecular mediators, responses designed to defend against infectious disease and to return the system to

Comparative Biology of the Normal Lung. http://dx.doi.org/10.1016/B978-0-12-404577-4.00030-8

homeostasis. The major cell subsets that confer protection to the pulmonary immune system include alveolar macrophages, dendritic cells (DCs), T helper (T_H) lymphocytes, cytotoxic T lymphocytes (CTLs, or T_C), memory lymphocytes, natural killer (NK) cells, natural killer T lymphocytes (NKT), eosinophils, mast cells, B cells, and neutrophils. Each of these immune cell subsets and their mediators, as discussed in this chapter, plays an important role in the prevention and/or control of disease in this immunological compartment.

2. PULMONARY IMMUNOLOGICAL MEDIATORS: SOLUBLE IMMUNE MESSENGERS

Cytokines and chemokines are immunological mediators that serve in cell–cell communication within the immune system. They are the language used by cells to orchestrate complex immune responses through the recruitment of particular cell subsets to perform specialized functions, and this language is influenced by the type of challenge encountered by the system. The complexity of cellular communication is compounded by the nature of cytokine–receptor interactions, which are responsible for subsequent intracellular signaling and downstream effects, including cell differentiation, growth, movement, activation, generation of various products, and death. These cytokine–receptor interactions are promiscuous, as cytokine ligands can interact with multiple receptors and receptor binding sites can bind multiple ligands; functionally diverse, as the hundreds of genes that may be activated within a cell via cytokine signaling may express differently based on cell type; and interactive, as the effects of particular cytokines may augment, inhibit, or otherwise influence the effects of other cytokines (Toews, 2001). While a detailed review of each mediator involved in infectious pulmonary pathogenesis is outside the scope of this chapter due to the enormous complexity of the cytokine signaling network, a brief overview of select mediators with regard to pulmonary infectious disease is presented below.

2.1 Cytokines

Interleukins (ILs) and interferons (IFNs) are often classified together as cytokines as both classes represent soluble immunomodulators, and many can be classified as proinflammatory, type-1 (T_H1/T_C1), type-2 (T_H2/T_C2), T_H17/T_C17, or regulatory. Proinflammatory cytokines such as tumor necrosis factor (TNF)-α, IL-1α, and IL-1β are typically involved in inflammatory responses; type-1 cytokines such as IL-12 and IFN-γ are primarily involved in microbicidal responses, can counter type-2 responses, and have been implicated in immune dysfunction with regard to autoimmune disease; type-2 cytokines such as IL-4, IL-5, IL-9, and IL-13 are generally involved in countering type-1 responses and contribute to allergic inflammation (see Chapter 35 for pulmonary allergic responses); T_H17/T_C17 cytokines such as IL-17A and IL-17F have been implicated in proinflammatory responses as well as autoimmunity; and regulatory cytokines such as IL-10 generally exert immunosuppressive, anti-inflammatory effects (Berger, 2000; Dinarello, 2010; Jin and Dong, 2013; Toews, 2001). Numerous additional cytokines have been discovered, described, and placed into these and other classifications, though distinct classifications are difficult due to individual cytokine functional diversity. Within the pulmonary compartment, a comprehensive range of cytokines and lipid mediators has been reported in bronchoalveolar lavage fluid (BALF) from mice following infection, including early increases in interleukin (IL)-lα, IL-1β, IL-6, IL-12 TNF-α, IFN-γ, GM-CSF, and leukotriene B4 (LTB4), and later increases in M-CSF and G-CSF (Hennet et al., 1992; Toews, 2001). Each of these immune mediators and their inhibitors aids in controlling the immune response and is important to pulmonary immunocompetence.

Macrophage activation, one of the earliest phases of the innate immune response, is dependent upon macrophage–NK cell cytokine signaling. TNF-α and IL-12, produced by macrophages following the recognition of microbial pathogens via toll-like receptors (TLRs), act synergistically to induce NK cells to produce IFN-γ. IFN-γ, in turn, augments IL-12 production in a positive feedback loop, and together with IL-12, induces the differentiation of IFN-γ-producing T_H1 cells (Toews, 2001). In addition to these synergistic effects, TNF-α, IL-12, and IFN-γ exert multiple accompanying effects as well. TNF-α contributes to the amplification of the immune response through the stimulation of chemokine expression and adhesion molecule expression on leukocyte and vascular endothelial cell surfaces, resulting in leukocyte recruitment and infiltration into the lung tissue; enhancement of phagocyte activation; and maintenance of granuloma formation, aiding in the containment of infectious pathogens (Cooper et al., 2011; Standiford et al., 1999). IL-12 augments the immune response to pathogens through increasing macrophage antimicrobial activity, NK cell activation, T_H1 adhesion molecule expression, CTL generation and activity, and the generation of T_H1 cells from naïve $CD4^+$ cells (Hamza et al., 2010). IFN-γ activates macrophage antimicrobial capabilities; activates NK cells with further enhancement of IFN-γ production; upregulates major histocompatibility class (MHC)-I, MHC-II, and costimulatory molecule expression to enhance antigen presentation; and upregulates the expression of T_H1 chemokines. IFN-γ has also been shown to play a key role in the control of acute, systemic infections caused by influenza virus, vaccinia virus, ectromelia virus, lymphocytic choriomeningitis virus,

M. tuberculosis, and *Listeria monocytogenes*, among others (Buchmeier and Schreiber, 1985; Huang et al., 1993; Karupiah et al., 1993; Kolls, 2013; Sarawar et al., 1994; Spellberg and Edwards, 2001; Toews, 2001; Wille et al., 1989).

In these ways, TNF-α and the positive feedback loop between IL-12 and IFN-γ help to shape an early T_H1 response to a pathogen. Negative feedback mechanisms are equally important, however, as uncontrolled T_H1 inflammation could damage the host tissues it was elicited to protect. IL-10, a regulatory cytokine produced by macrophages as well as other cell types, potently inhibits TNF-α, IL-12, and related inflammation (Toews, 2001). While studies have shown increased survival for mice overexpressing IL-12 in the lung during bacterial pneumonia infection, and decreased survival for mice with attenuated IL-12 activity, it has also been reported that IL-10-deficient mice produce excess inflammatory IL-12 and have higher mortality rates following infection, highlighting the importance of control within the system (Gazzinelli et al., 1996; Greenberger et al., 1996; Toews, 2001).

TLR signaling pathways, among others, also stimulate the production of cytokines including proinflammatory IL-1α and IL-1β. The IL-1α precursor exists naturally in the cytoplasm of resting mesenchymal and epithelial cells but must be synthesized de novo in monocytes and macrophages. The IL-1α precursor translocates to the nucleus following cellular activation by products such as lipopolysaccharide (LPS), a component of gram-negative bacteria, where it can then induce expression of IL-8, IL-6, IL-1β, and added IL-1α; as well as reduce the cell activation threshold to TNF-α and IFN-γ, further enhancing cellular cytokine production (Cohen et al., 2010; Dinarello et al., 1987; Kaplanski et al., 1995; Weber et al., 2010; Werman et al., 2004). IL-8 and IL-6 induced by IL-1α aid in the response to pathogens as IL-8 is known for neutrophil recruitment, and IL-6 is involved in monocyte and T cell recruitment, and T and B cell differentiation (Scheller et al., 2011; van Eeden and Terashima, 2000). Upon necrotic (though not natural apoptotic) cell death and release of IL-1α precursor into the extracellular space, the IL-1α precursor may then continue effects on neighboring cells via IL-1R1 or via pattern recognition receptors (PRRs) recognizing IL-1α precursor as a damage-associated molecular pattern. IL-1α has been shown to potently enhance neutrophil infiltration in the innate immune response, as well as to prime T cell antigen-specific responses, providing support in adaptive immunity (Barry et al., 2013; Werman et al., 2004). IL-1β is found secreted rather than primarily intracellular or membrane bound as with IL-1α, though it is induced by similar stimuli as IL-1α and by IL-1α itself, and functional similarity to IL-1α has been reported (Contassot et al., 2012; Dinarello, 2011; Dinarello et al., 1987). In addition to some of the effects elicited by IL-1α, IL-1β has been reported to drive T_H17 differentiation, which has been implicated in bacterial and viral clearance, as well as T cell-dependent antibody responses (Dinarello, 2011). IL-1α and β signaling is regulated by the IL-1 receptor antagonist (IL-1Ra), which binds the type I IL-1 receptor (IL-1RI) with higher affinity than IL-1α and IL-1β, downregulating proinflammatory effects (Schmitz et al., 2005; Strieter et al., 2003).

Colony-stimulating factors such as granulocyte macrophage colony stimulating factor (GM-CSF) and granulocyte colony stimulating factor (G-CSF) contribute to the inflammatory response as they are necessary for the maturation and proliferation of hematopoietic stem cells and also influence mature leukocyte activity. GM-CSF participates in promoting the maturation and proliferation of neutrophils and macrophages, and the activation of mature macrophages (Standiford et al., 1999). In a study with GM-CSF-deficient mice, alveolar macrophages had diminished phagocytic and cytokine production ability that could be restored following GM-CSF administration. In another study comparing influenza infection survival in wild-type versus transgenic mice expressing GM-CSF in the lung, only mice expressing GM-CSF survived infection with the chosen influenza strains (Huang et al., 2011). G-CSF, produced by alveolar macrophages among other cells, enhances neutrophil survival, phagocytic capability, and respiratory burst function. In rat studies with *Pseudomonas*-induced pneumonia, impaired G-CSF expression was correlated with decreased neutrophil infiltration and decreased bacterial clearance (Standiford et al., 1999).

2.2 Chemokines

Chemotactic cytokines, or chemokines, are largely responsible for recruiting specific cell types to target tissues during homeostatic conditions as well as inflammatory responses and are also involved in cell maturation, homing, and activation (Ono et al., 2003). These small proteins interact with cells bearing appropriate chemokine receptors, and receptor expression differs based on cell type, maturation status, and activation status. Chemokines are named and grouped based on the number and position of their cysteines. The C chemokines have one N-terminal cysteine, the CC chemokines have two adjacent N-terminal cysteines, the CXC chemokines have two N-terminal cysteines that are separated by one amino acid, and the CX_3C chemokine (CX_3CL_1, or fractalkine) has two N-terminal cysteines with three amino acids in between (Le et al., 2004). Chemokine receptors for each group are denoted by "R" following the CC, CXC, or CX_3C designation. It should be noted that many chemokines often have function- or origin-based names in addition to structural names. CXCL10, for example, is also known as IFNγ-induced protein 10 (IP-10), named for the cytokine that induces its expression.

Inflammatory chemokines, induced by inflammatory cytokines such as IL-1β or TNF-α, or microbial products such as LPS, play a large role in cell migration, allowing immune cells to reach and enter target tissues. IL-1β, generated in response to antigens such as LPS, induces the production of chemokines such as cytokine-induced neutrophil chemoattractant-1 (CINC-1) and macrophage inflammatory protein (MIP)-2, resulting in neutrophil infiltration (Calkins et al., 2002). TNF-α has been shown to induce chemokines such as monokine induced by gamma interferon (MIG), a chemoattractant for T cells; RANTES (regulated on activation normal T cell expressed and secreted), a chemoattractant for T cells and DCs; and IP-10, a potent T_H1 chemoattractant for T cells, NK cells, macrophages, and DCs (Algood et al., 2004; Scolletta et al., 2013). IFN-γ, generated in response to T_H1-inducing antigens, contributes to the production of chemokines such as IP-10, MIG, MIP-1β, and interferon-inducible T-cell α chemoattractant (I-TAC), enhancing the migration of T_H1 cells with corresponding receptors into inflammatory tissues (Olson and Ley, 2002; Widney et al., 2000). In addition, IFN-γ induces the expression of chemokine receptors such as CXC R3 on IgG^+ memory B cells, expression that is maintained following differentiation into plasma cells, allowing the homing of antibody secreting cells to sites of inflammation (Muehlinghaus et al., 2005). TNF-α also contributes to related adaptive immune responses through the stimulation of B lymphocyte chemoattractant (BLC) and secondary lymphoid tissue chemokine (SLC) production, involved in B cell homing to B cell follicles and lymphocyte and DC homing to secondary lymphoid tissues, respectively (Moser et al., 2004). Beyond their powerful role in cell migration, chemokines are also pivotal in cell development through cell homing, particularly with regard to B and T cells (D'Apuzzo et al., 1997; Forster et al., 1994; Germain, 2002; Olson and Ley, 2002; Seagal and Melamed, 2003; Stein and Nombela-Arrieta, 2005; Takahama, 2006).

As with cytokines, chemokine signaling is complex, and selective agonistic and antagonistic balances exist. CXCR3, an inflammation-associated chemokine receptor primarily found on T_H1 cells, interacts with chemokine ligands including IP-10, MIG, and I-TAC. CXCR3 ligands serve as antagonists for the CCR3 chemokine receptor, often found on T_H2 cells. The chemotaxis of T_H1 cells may then block the migration of T_H2 cells, contributing to a polarized immune response. Some chemokines that are elevated as a result of inflammation do not distinguish between T_H1 and T_H2-associated chemokine receptors, however, as observed with RANTES signaling through CCR5 (generally T_H1 cells) and CCR3 (generally T_H2 cells), illustrating signaling overlap and ligand promiscuity (Moser et al., 2004; Muehlinghaus et al., 2005). In addition, the particular pathogen encountered appears to selectively influence chemokine and chemokine receptor expression. CCR1 and CCR5, receptors for RANTES and/or MIP-1α, for example, facilitate DC recruitment following inhalation of heat-killed *Moraxella catarrhalis* in a rat model; while CCR2, the receptor for many monocyte chemotactic protein (MCP)-family chemokines, seems to orchestrate DC recruitment during murine *M. tuberculosis* infection (Vermaelen and Pauwels, 2005).

While cytokine and chemokine signaling is intended to aid in host defense and resolution of infection through various pathways, uncontrolled cellular activation and cytokine and chemokine release has the potential to lead to a lethal "cytokine storm." Severe cytokine storms are characterized by sudden, massive, systemic inflammation, tissue damage and multiple organ dysfunction, edema, and hypotension. The mechanisms underlying the initiation of a cytokine storm are poorly understood, though it is thought that highly virulent pathogens, such as the one responsible for the unusually high mortality rates among healthy young adults during the 1918 influenza A pandemic, and superantigens, which can cause nonspecific mass T cell activation, are capable of triggering its activation (Canna and Behrens, 2012; Iwasaki and Medzhitov, 2011; Morens and Fauci, 2007; Tisoncik et al., 2012). Investigational therapeutics such as TGN1412, a humanized anti-CD28 antibody manufactured by TeGenero, have also been implicated in initiating cytokine storms. TGN1412, a superagonist that circumvented T cell receptor (TCR)-mediated activation and activated T cells via coreceptor CD28, was intended to expand T cell and regulatory T cell populations to treat B cell chronic lymphocytic leukemia and autoimmune disease, respectively. While seemingly innocuous in preclinical animal studies including those in nonhuman primates, nonspecific T cell activation led to the hospitalization of six healthy volunteers through the induction of a cytokine storm and resulting multiorgan failure in the Phase I clinical trial (St. Clair, 2008). Due to the beneficial as well as potentially deleterious effects of cytokine and chemokine signaling, an understanding of the pathways involved in health, disease, and dysfunction is critical.

3. PULMONARY INNATE IMMUNITY

Innate immunity represents the first of three arms of the immune response and occurs early (hours to days) after stimulation by foreign substances or infectious agents. Innate immunity is nonspecific, includes physical barriers such as that provided by the airway mucosa (see Chapters 26–29 for detailed information on physical defenses of the respiratory tract), and is spearheaded by macrophages, neutrophils, mast cells, neutrophils, NK cells, and NKT cells. Innate immune cells utilize PRRs to recognize various pathogen-associated molecular patterns from infectious agents as well as damage-associated molecular patterns from cellular components released during cell injury. Such receptors include TLRs, receptors involved in the identification of bacterial, fungal, and viral products and structures. These innate recognition mechanisms initiate cell-signaling cascades, resulting in cytokine and chemokine production to direct initial host defenses and aid in the destruction of the pathogen. These early responses, while independent of the adaptive immune response, are critical for the priming of adaptive immunity and the subsequent generation of immunological memory to specific pathogens (Martin, 2005; Mogensen, 2009).

Early innate responses and subsequent adaptive responses are also tailored to the challenge presented, as observed in studies comparing the dominant cell types in the lungs and airways during infection with different pathogens. Three to four days following viral infection, NK cells are the most abundant responders, responsible, in part, for the recruitment of $CD4^+$ and $CD8^+$ T cells during the peak of influenza virus infection on day 7. During the peak of a bacterial infection with *Streptococcus pneumoniae* on day 4, macrophages and neutrophils are among the most abundant responders. Alternatively, fungal infection with *Cryptococcus neoformans* has been shown to generate a strong eosinophilic response during peak infection on day 12 (Wissinger et al., 2008). The cell types activated and recruited by the cytokine and chemokine milieu generated by the invading pathogen are determined by the nature of the pathogen, highlighting the carefully orchestrated and specialized nature of the immune system.

3.1 Macrophages

Pulmonary macrophages (see Chapter 32 for detailed information on pulmonary macrophages) are important in the phagocytosis and clearance of foreign material and function as antigen-presenting cells (APCs) but possess diminished APC activity compared with DCs. Pulmonary macrophages may develop from bone marrow-derived monocytes that differentiate following entry into the lung tissue and may be classified as alveolar macrophages (AM), interstitial macrophages (IM), or pulmonary intravascular macrophages (PIM), each with emphasis on different functions. Pulmonary macrophages may also be classified as M1 or M2, based on inflammatory involvement (Balhara and Gounni, 2012; Benoit et al., 2008; Haley, 2003; Wynn et al., 2013).

IMs, found in the interstitium, interact with interstitial lymphocytes and have more involvement in antigen presentation. AMs, which have been shown to differentiate from IMs and are dominant in number, inhabit the lung alveoli and have higher involvement in phagocytosis (Balhara and Gounni, 2012). PIMs are largely phagocytic and are located within the pulmonary vasculature of cats, calves, goats, sheep, and pigs (Haley, 2003). Pulmonary macrophages may be M1 or M2, with M1 macrophages exerting greater proinflammatory effects in response to encounters with intracellular antigens, and M2 macrophages having greater implication in anti-inflammatory and tissue repair roles. Studies have suggested that macrophages have great plasticity with regard to M1 and M2 classification, however, and that there exists a range of pro- and anti-inflammatory effects bridging these classifications (Benoit et al., 2008).

M1 macrophages, induced by type-1 cytokines such as IFN-γ and microbial components such as LPS, are characterized by the release of proinflammatory cytokines such as IL-12, TNF-α, and IL-6, and antimicrobial products such as nitric oxide (NO) (Lugo-Villarino et al., 2011). Upon infection with *M. tuberculosis*, a bacterial pathogen that replicates in pulmonary macrophages and causes tuberculosis (TB), macrophages typically adopt a proinflammatory M1 phenotype (Benoit et al., 2008; Chacon-Salinas et al., 2005; Raju et al., 2008). Increased IL-10 dampens antimicrobial and proinflammatory macrophage function, likely through inhibition of IL-12, and has been associated with reactivation of TB in human patient populations. Indeed, in transgenic mice with IL-10 overexpressing macrophages, macrophages are primarily M2 in nature, and mice were found to have increased susceptibility to *M. tuberculosis* infection, indicating an important role for correct macrophage polarization during infection (Schreiber et al., 2009). Interestingly, while the M1 phenotype has been shown to be important in initial encounter with this pathogen, approximately 3 weeks following infection macrophages are predominately M2, perhaps promoting resolution of inflammation and restoration of balance to the lung tissue (Lugo-Villarino et al., 2011). A similar pattern has been observed in certain viral infections such as HIV-1, with an early phase M1 response followed by a later- or chronic-phase M2 response, potentially as a means of mediating inflammation in the host (Lugo-Villarino et al., 2011; Sica and Mantovani, 2012). The targeting of AMs by HIV-1, due to their prominent role in phagocytosis and pathogenic clearance, may explain the increased risk of respiratory infection in HIV-infected individuals (Jambo et al., 2014).

M2 macrophages, the dominant phenotype under homeostatic and reparative conditions, can be induced by type-2 cytokines such as IL-4 and IL-13, immunosuppressive cytokines such as IL-10 and transforming growth factor-β (TGF-β), and agonists of toll-like and IL-1 receptors. M2 macrophages are characterized by the release of immunosuppressive IL-10, as well as agents involved in tissue repair such as TGF-β and ornithine (Lugo-Villarino et al., 2011; Mills, 2012).

Due to their APC and phagocytic function, pulmonary macrophages are one of the first lines of pulmonary defense and play an important role in the nonspecific immunological defense against infectious diseases. An impaired phagocytic response or the production of inappropriate cytokines by pulmonary macrophages may result in health problems manifested by more frequent and more severe pulmonary infectious disease, pneumoconiosis from inhaled dusts (such as silicosis from silica inhalation), or pulmonary fibrosis. Impairment of AM function in smoking populations, for example, has been shown to increase susceptibility to pulmonary infection due to reduced phagocytic and clearance mechanisms (Marti-Lliteras et al., 2009).

3.2 Neutrophils

Neutrophils are the most abundant of the circulating leukocytes, although among the most short-lived (Mocsai, 2013). They are the first of the immune cells recruited to sites of infection and tissue damage, and play a critical role in innate immunity and clearance of microbial pathogens. Clearance mechanisms include phagocytosis, degranulation, respiratory burst and release of reactive oxygen species, as well as the formation of neutrophil extracellular traps (NETs). NETs, composed of chromatin and antimicrobial products such as bactericidal permeability increasing protein, myeloperoxidase, and neutrophil elastase, allow the entrapment and killing of pathogens that would otherwise be too large to phagocytose (Drescher and Bai, 2013; Urban et al., 2006). Neutrophils also have ties to adaptive immunity, as they have the ability to contribute to DC maturation and B cell differentiation and maturation, and ties to regulatory aspects of the immune response, with the ability to affect T_H17 responses, IL-10 generation, and NK cell responses (Klesney-Tait et al., 2013).

Neutrophils are central to an effective early response against bacterial and fungal pathogens. In a murine model of necrotizing pneumonic plague due to *Yersinia pestis* infection, in vivo ablation of neutrophils led to higher numbers of bacterial colonies within 24 h of exposure. In addition, it was shown that human neutrophils, but not other peripheral blood mononuclear cells (PBMCs), added to in vitro cultures were capable of reducing *Y. pestis* numbers 5-fold (Laws et al., 2010). In another study examining TREM-1 (triggering receptor expressed on myeloid cells-1), a neutrophil cell surface receptor that contributes to neutrophil phagocytosis, degranulation, and respiratory burst; cytokine and chemokine release; and neutrophil cell adhesion and migration; TREM-1 deficiency in a murine pneumonia model led to decreased clearance of *Pseudomonas aeruginosa* and increased mortality, likely due to diminished neutrophil transepithelial migration (Arts et al., 2013; Klesney-Tait et al., 2013). In the event of a fungal infection with agents such as *Candida albicans*, neutrophils have been shown to mediate clearance through the generation of NETs to capture and kill the invading pathogen (Urban et al., 2006).

The role of the neutrophil in pulmonary viral pathogenesis is a current area of investigation and has not yet been fully elucidated. While one study in mice reported that a reduction in neutrophil numbers preceding influenza A infection led to higher rates of mortality, another study showed that excessive neutrophil migration into lung tissue during influenza infection caused acute inflammation, perhaps suggesting that a certain balance of activity is important in controlling infection without causing damage to host tissues (Urban et al., 2006). Contributing toward a role in adaptive immunity, Hufford et al. found that neutrophils were capable of presenting antigen to antiviral effector $CD8^+$ T cells during influenza infection (Hufford et al., 2012).

While the vast majority of cells in BALF obtained from healthy subjects consists of alveolar macrophages, neutrophils may also be observed, and neutrophil cell counts may provide an indicator of inflammatory response to infection or to inhaled xenobiotics. An early and large influx of neutrophils into the airways with recovery in the BALF is a sensitive biomarker and index of an inflammatory response. Driscoll and Schlesinger reported that in vitro exposure of alveolar macrophages to ozone stimulated the release of chemotactic activity for neutrophils (Driscoll and Schlesinger, 1988). Driscoll et al. reported that the neutrophil chemotactic protein, macrophage inflammatory protein 2 (MIP-2, a member of the family of cytokines containing IL-8, a neutrophil chemotactic factor in humans), and MCP-l (a monocyte chemotactic protein characterized in humans and rodents) was produced in a temporal and exposure-response manner with the influx of neutrophils to the lungs following exposure to the environmental air pollutant ozone (Driscoll et al., 1993).

3.3 Mast Cells

Mast cells (MCs) (see Chapter 34 for detailed information on mast cells), one of the airway cell subsets that is well known for its involvement in allergic responses, have emerged in recent years as contributors to the innate immune response. Classical allergic activation of MCs occurs via the antigenic cross-linking of IgE bound to the high-affinity IgE receptor, FcεRI, on the surface of mast cells. Cross-linking induces activation via a complex intracellular signaling cascade, leading to degranulation and the release of mediators such as histamine, a chemical responsible for increasing vascular permeability and edema (Metcalfe et al., 1997; Saluja et al., 2012). Activation may occur through other means during encounter with parasites, bacteria, viruses, and fungi, however, due to the PRRs located on MCs. In addition, MCs are phagocytic, have direct antimicrobial ability through the release of proteases and cathelicidins, and similar to NET formation by neutrophils, MCs form MCETs (mast cell extracellular traps) coated with cathelicidin to ensnare and kill pathogens (Saluja et al., 2012).

One of the PRRs expressed by MCs is TLR2, capable of recognizing diverse bacterial, fungal, and viral pathogens and related components. In a study of *M. tuberculosis* infection in $TLR2^{-/-}$ mice by Carlos et al., the adoptive transfer of $TLR2^{+/+}$ MCs decreased *M. tuberculosis* bacterial load and reversed the lack of $CD8^+$ T cell homing to the lungs and the lack of T_H1 cytokines observed in $TLR2^{-/-}$ mice, demonstrating a protective role for MCs in bacterial infection (Carlos et al., 2009). In a study of *Mycoplasma pulmonis* infection in MC-deficient mice by Xu et al., bacterial burden was increased 162-fold in MC-deficient mice compared with wild-type mice on day 4 following infection, accompanied by progressive pneumonia and increased mortality (Xu et al., 2006).

Due to the newly explored roles of MCs in various aspects of innate immunity, limited information exists with regard to the role of MCs in viral infection. It has been shown, however, that MCs can undergo virus-induced activation via toll-like receptor 3 (TLR3). TLR3 recognizes double-stranded RNA (dsRNA) viruses. As a result of TLR3 activation, MCs have been shown to produce type-I interferons (IFNs), which are implicated in antiviral responses, as well as to strongly enhance antiviral $CD8^+$ T cell recruitment, indicating potential roles for the mast cell in viral clearance (Orinska et al., 2005; Saluja et al., 2012). As with macrophages, a detrimental role in viral infection, however, has been proposed for progenitor mast cells as targets of HIV-1 infection. Progenitor mast cells are more susceptible to viral infection than mature mast cells, and upon maturation, may harbor infectious virus long-term despite antiretroviral therapy (Sundstrom et al., 2007).

As with the role of mast cells in viral responses, there is currently a similar lack of in vivo studies examining the role of MCs in pulmonary fungal infection. It could be hypothesized, however, that MCs would participate in fungal infection with pathogens such as *Aspergillus fumigatus*, the causative agent of the disease aspergillosis, due to their likelihood of being in close proximity with the pathogen in the lung tissue, and due to the ability of their TLR2 receptors to interact with fungal agents. Since immunocompromised individuals, such as those with HIV/AIDS, are more susceptible to fungal infection, the role of the mast cell in fungal clearance would be an area worthy of study (Saluja et al., 2012).

3.4 Natural Killer Cells

NK cells are critical players in the immune response due to rapid cytotoxic capabilities executed via the release of lytic granules or through antibody-dependent cell-mediated cytotoxicity (ADCC); the ability to produce numerous cytokines, chemokines, and mediators including IFN-γ, which exerts antiviral effects; the ability to recognize "self" cells with reduced MHC-I expression, caused by certain viruses in an attempt to escape recognition from $CD8^+$ T cells by reducing MHC-I antigen presentation; and the ability to express costimulatory and regulatory ligands (Orange, 2013). NK cells may be recruited to the lung within days of infection or within hours of activation (Culley, 2009). These cells express the CD56 cell surface marker as $CD56^{bright}$ or $CD56^{dim}$, and may be NK1 or NK2 depending on the cytokines produced (T_H1 cytokines or T_H2 cytokines, respectively). Approximately 10% of NK cells are $CD56^{bright}$ and $CD16^{-/dim}$ and contribute to cytokine production more than their $CD56^{dim}$ counterparts. The majority of NK cells (80–90%) are $CD56^{dim}$ and $CD16^+$, are more cytotoxic than the $CD56^{bright}$ subset at rest, and are found circulating until signaled to migrate toward inflamed tissues via chemotactic stimuli (Agaugue et al., 2008; Bellora et al., 2010; Timmons and Cieslak, 2008).

Challenge with antigen, such as *M. tuberculosis*, herpesvirus, or influenza virus, has shed light on the innate NK functional response. Active *M. tuberculosis* infection results in increased numbers of $CD56^{bright}$ NK cells, resulting in lowered cytotoxic capability but increased release of IFN-γ, which supports numbers of $CD8^+$ IFN-γ^+ T effector cells with the ability to lyse *M. tuberculosis*-infected cells (Culley, 2009; Vankayalapati et al., 2004). NK cells also play a specialized role during infection with certain viruses such as herpesviruses. Herpesvirus infection leads to a virus-induced downregulation of MHC-I in infected cells as a mechanism for avoiding $CD8^+$ T cell recognition. In such instances NK cells, which have self-MHC-I-specific inhibitory receptors, become activated in response to "missing self" and destroy the cell in question (Yawata et al., 2008). In the event of influenza virus infection, mouse studies have shown an enhanced pulmonary NK activity with intranasal infection, while intravenous infection resulted in enhanced splenic NK activity, each contributing to the innate response (Burleson and Burleson, 2008a; Burleson et al., 2008; Leung and Ada, 1981). In Fischer-344 rats, this enhanced activity is abrogated upon exposure to respiratory toxicants such as phosgene, resulting in a more severe and prolonged influenza virus infection, likely related to phosgene-induced suppression of pulmonary NK activity (Burleson and Keyes, 1989; Ehrlich and Burleson, 1991). Species-specific differences have been observed, however, as this alteration in influenza virus replication was not observed in mice (Ehrlich and Burleson, 1991; Selgrade et al., 1989). It should be noted, however, that differences in species, age, and the dose and regimen of phosgene exposure make direct comparisons between these two studies difficult.

In addition to its innate immune functions, NK cell activity also influences the adaptive immune response. NK cells are necessary for the generation of the CTL response to influenza, and the release of IL-2 following T cell activation intensifies the NK release of IFN-γ. Furthermore, during influenza infection, NK cells are critical to ADCC responses. Following the production and coating of infected cells with influenza-specific antibody, the NK cell surface FcγRIII receptor recognizes the Fc domain of the antibody on the surface of the infected cell, leading to the release of cytotoxic granules and killing of the infected cell (Culley, 2009).

The importance of pulmonary NK activity in immunological defenses is readily observed in human populations with various NK deficiency (NKD) disorders. Such patients have increased susceptibility to viral infection. Patients with classical NK deficiency (CNKD), for example, lack NK cells, and NK functions in peripheral blood lymphocytes, and of those afflicted, 42% experience premature mortality and 53% experience severe herpesvirus infections (Orange, 2013). Equally important is the regulation of heightened NK responses, as the delivery of IL-2 and IL-18, an activator of NK IFN-γ production and cytotoxicity, caused systemic NK activation and lethal lung inflammation in a murine model (Culley, 2009).

3.5 Natural Killer T Cells

Natural killer T cells express the NK cell-associated CD161 (NK1.1 in mice) surface marker, and share some qualities of NK cells such as natural cytotoxic ability and an involvement in both innate and adaptive immune responses. Unlike NK cells, however, NKT cells also express $\alpha\beta$ TCR (Curtis, 2005). NKT cells are a glycolipid-responsive T cell subset that recognize antigen in the context of CD1d and express an activated or memory phenotype. They are known for their ability to rapidly and heavily secrete various T_H1 and T_H2 cytokines, induce cytotoxicity, and activate or influence macrophages, NK cells, DCs, T cells, and B cells (Chiba et al., 2008; Hanna et al., 2004; Juno et al., 2012; Moretta et al., 2006). NKT cells may be type I (invariant NKT cells or iNKT cells), or type II, depending on differences in the TCR. Type I invariant NKT (iNKT) cells have an invariant TCR α-chain and are the more widely studied NKT cells, while type II NKT cells have wider TCR diversity.

iNKT cells can express CD4 or CD8α and may be $CD4^+$, $CD8^+$, or $CD4^-$, $CD8^-$. $CD4^+$ iNKTs generally express CD62L for homing to the lymph nodes and produce both T_H1 and T_H2 cytokines. iNKT cells, particularly the $CD4^+$ subset, show decreased number and function with HIV-1 infection. $CD4^-$ iNKT cells are $CD62L^{low}$ but express CD11a for migration into tissues, and typically produce only T_H1 cytokines.

During influenza virus infection, iNKT cells may counteract the immunosuppressive effects of myeloid-derived suppressor cells (MDSCs) and augment anti-influenza responses. iNKT cell activation in early infection has been shown to reduce viral titer and heighten the innate immune response, fitting with the ability of iNKT to directly lyse influenza-infected monocytes, produce relevant cytokines, and activate innate immune cells (Juno et al., 2012). iNKT cell activation with the glycolipid α-galactosylceramide (*α-GalCer*), a specific activator of iNKT cells, has also been linked to enhanced antibody responses, $CD8^+$ memory responses, and antigen recall responses, demonstrating a role for iNKT cells in adaptive immunity and perhaps supporting investigation into the use of αGalCer as a potential vaccine adjuvant (Juno et al., 2012). Similar to infection with influenza virus, activated NKT cells expedite clearance of bacterial infections with *S. pneumoniae* as well as *M. tuberculosis*, while NKT cell-deficient mice are more susceptible to infection with *S. pneumoniae* (Chiba et al., 2008).

In a study by Chiba et al., iNKT cells were shown to be activated prior to activation of MHC-restricted T cells. Interestingly, activated iNKT cells that did not undergo cell death following expansion became unresponsive to TCR engagement, despite ongoing infection, as the MHC-restricted T cell proliferative and IFN-γ response mounted (Chiba et al., 2008). This perhaps represents a mechanism of mediating the level of the cytotoxic and inflammatory response, as such responses can be detrimental to the host if left unchecked. It has been reported, for example, that iNKT cells exert protective effects during bacterial infection with lower doses of *Novosphingobium capsulata*, but during infection at higher doses, iNKT cell responses are overwhelming and capable of inducing a lethal cytokine storm. It should be noted that differences in responses have been observed in different species, however, as iNKT cells have been shown to be protective against respiratory syncytial virus (RSV) in BALB/c mouse studies but to contribute to pathogenesis in C57BL/6 mice (Wu and Van Kaer, 2011).

4. PULMONARY ANTIGEN PRESENTATION: A GATEWAY TO ADAPTIVE IMMUNITY

APCs such as macrophages and B cells, and to a much greater extent dendritic cells, are a main link between the innate and adaptive immune responses. APCs present antigen to $CD8^+$ or $CD4^+$ T cells in the context of MHC Class I and MHC Class II, respectively. Should recognition of the antigen occur, soluble mediators (interferons, cytokines, and chemokines) are released in order to recruit and activate the appropriate effector cells of the adaptive immune response, expediting clearance of the pathogen.

4.1 Dendritic Cells

Dendritic cells (DCs) (see Chapter 33 for detailed information on pulmonary dendritic cells) are considered one of the few "professional" antigen presenting cells due to high efficiency in the uptake and presentation of antigen. Typically, DCs found in peripheral tissues are in an immature state, characterized by an enhanced ability to sense, uptake, and process antigen, but with reduced ability to activate T cells due to the decreased costimulatory molecule expression prior to maturation (Dauer et al., 2003; Steinman and Hemmi, 2006). Monocyte-derived immature dendritic cells express inflammatory chemokine receptors such as CXCR1, CCR1, CCR2, and CCR5, which encourage attraction and migration toward sites of inflammation (Dauer et al., 2003; Olson and Ley, 2002). Interaction of immature DCs with peptides derived from endogenous or intracellular sources are presented by MHC-I molecules on the surface of DCs, and peptides derived from exogenous or extracellular sources are presented by MHC-II molecules on the DC surface. Cross-presentation (a process whereby exogenous antigen is loaded onto MHC-I) may also occur with certain DC subsets (Fonteneau et al., 2003). Following encounter with antigen and subsequent antigen processing, immature DCs undergo maturation influenced by exposure to microbial or T cell-derived stimuli, or proinflammatory signals such as TNF-α and IFN-γ. DC maturation results in decreased antigen uptake and processing, and increased MHC expression and antigen presentation. During maturation, DCs transiently express MIP-1α and MIP-1β, and express IP-10, MCP-2, and RANTES, potentiating the recruitment of immature DCs, monocytes, and T cells to the inflammatory tissue. CCR7 expression then allows interaction with SLC and MIP-3β, and subsequent DC migration to secondary lymphatic tissues for antigen presentation to T cells (Dauer et al., 2003; Olson and Ley, 2002; Vermaelen and Pauwels, 2005). The DC peptide-MHC-I or peptide-MHC-II complex, if recognized by CD8$^+$ or CD4$^+$ T cell receptors, respectively, induces T cell synthesis and expression of CD40 ligand. CD40 ligand interacts with CD40 on the surface of DCs, resulting in DC synthesis of B7 costimulatory molecules, ligands for T cell CD28 receptors. Sufficient costimulatory signaling induces the production of IL-2, and subsequent T cell proliferation and polarization of CD4$^+$ cells toward T_H1, T_H2, T_H17, or T_{Reg} phenotypes; or CD8$^+$ CTL toward T_C1, T_C2, T_C17, or regulatory phenotypes (Betts and Kemeny, 2009; Dauer et al., 2003; Steinman and Hemmi, 2006). T cells then secrete a number of cytokines to activate other immune cell types, resulting in an inflammatory cascade (Dauer et al., 2003; Steinman and Hemmi, 2006).

While DCs are clearly involved in the initial stages of an immune reaction, it should be noted that not all DC subsets are purely proinflammatory. Certain subsets are capable of inducing tolerance to antigen and decreasing inflammatory immune responses. Conventional DCs (cDC) generally play a role in proinflammatory responses while plasmacytoid (pDC) are capable of anti-inflammatory roles. cDCs in the conducting airways can project dendrites through bronchial epithelial cell tight junctions for the detection of inspired antigen, and upon activation and maturation, migrate to the lymph nodes for antigen presentation and initiation of the inflammatory response described above (Hammad and Lambrecht, 2011; Hammad et al., 2010). pDCs, found in alveolar septa and large conducting airways on the other hand, may mediate anti-inflammatory responses to innocuous inhaled antigen due to low levels of T cell costimulatory molecules and increased expression of inhibitory programmed death ligand 1 (PD-L1) (Hammad and Lambrecht, 2011; Kool et al., 2009; Vermaelen and Pauwels, 2005). PD-L1, when highly expressed on pDCs, has been correlated with elevated T_{Reg} numbers (Hammad and Lambrecht, 2011; Kool et al., 2009). A dual role has been proposed for pDCs, however, as studies have shown that following exposure to virus, pDCs aid in viral clearance through the production of large amounts of type I IFN (Vermaelen and Pauwels, 2005; Wissinger et al., 2008).

As a key antigen presenting cell, negative impacts on DCs can result in negative overall effects on immune system function. In one study investigating the effects of infection with *M. tuberculosis*, infection caused reduced DC migration to the lymph nodes for antigen presentation. In addition, lung DCs that were infected at high frequency had a reduced capacity to engage antigen-specific T cells due to decreased MHC-II presentation, contributing to disease prolongation (Wolf et al., 2007). In another study examining the effects of viral pathogenesis on lung DC, infection with paramyxoviruses such as human metapneumovirus or RSV resulted in diminished pDC IFN-α, IL-6, and TNF-α cytokine production, as well as reduced cDC antigen presentation, effects that lingered after the acute phase of infection (Guerrero-Plata et al., 2009). In addition, while DC numbers typically return to baseline following abatement of inflammation, RSV infection has been shown to cause increased pulmonary DC numbers for weeks following inflammatory resolution (Vermaelen and Pauwels, 2005).

5. PULMONARY ADAPTIVE IMMUNITY

As previously discussed, the pulmonary immune system is an example of a specialized and compartmentalized system. The compartmentalized pulmonary immune system, while capable of performing all immunological functions locally, is also capable of communicating with the systemic immune system. An example of this communication is the immunization of humans by intramuscular influenza vaccination in the arm, resulting in a systemic immune response measured by specific antibody production in the serum, leading to subsequent protection against infection in the lungs. Other examples can be observed in studies with mice and rats, in which pulmonary infection with influenza infection results in a systemic immune response including stimulation of NK activity (innate immune activation), CTL activity (cell-mediated immunity) in the spleen and lung, and influenza specific antibody (humoral-mediated immunity (HMI)) in the serum and lung. Adaptive immunity is primed by innate immune responses, involves cell-mediated (T-lymphocyte-mediated) and humoral-mediated (B-lymphocyte-mediated) immunity, and results in long-lived antigen-specific recognition and rapid protection against invading pathogens.

5.1 Cell-Mediated Immunity

Cell-mediated immunity (CMI) represents the second of three arms of the immune response, and along with the humoral immune response (third arm of the immune response), comprises the adaptive immune system. Adaptive immune responses are delayed compared to innate immune responses, with initial activity appearing 4–7 days following pulmonary infection (Janeway et al., 2001). CMI represents a major component of the adaptive host response that is specific to a stimulating event, such as tumor cells or infectious agents. T lymphocytes, drivers of the CMI response, are capable of differentiating into various $CD4^+$ T-helper (T_H) subsets, $CD8^+$ CTLs, or into memory T cells (Betts and Kemeny, 2009; Coffman, 2006). These cell types help control the immune response to particular antigens and play important roles in pulmonary immunological protection against infectious disease.

$CD4^+$ T Cells

$CD4^+$ T cells respond to antigen presented on MHC-II, located primarily on APCs. Exposure to antigen can induce naïve $CD4^+$ T-cell expansion and differentiation into effector cell T_H1, T_H2, T_H17, or T_{Reg} phenotypes, some undergoing further differentiation into memory cells that reactivate rapidly upon antigen reexposure (Coffman, 2006). The expansion of $CD4^+$ T cells is vital not only for effector and memory functions but also in serving a key role as follicular T helper cells, responsible for providing essential B cell help within germinal centers in secondary lymphoid organs (MacLeod et al., 2010). B cell help provided by CD4+ T cells is required for B cell germinal center formation and the generation of B cell memory (Tsai and Yu, 2014). Additionally, $CD4^+$ T cells are required for efficient formation of $CD8^+$ memory T cells (MacLeod et al., 2010).

While T_H1 and T_H2 cells both develop from naïve $CD4^+$ T cells, each subset reciprocally inhibits the expansion of the other due to the release of opposing cytokine factors, particularly IFN-γ and IL-4, respectively (Coffman, 2006). T_H1 differentiation, typically observed in response to infectious pathogens, is promoted by IL-12, and requires the expression of transcription factor T-bet. Cytokines classically released by T_H1 cells include IL-12 and IFN-γ. T_H2 differentiation, generally associated with allergy and some aspects of humoral immunity, is promoted by IL-4 and requires the expression of transcription factor GATA-3. Cytokines typically released by T_H2 cells include IL-4, IL-5, and IL-13 (Sallusto and Lanzavecchia, 2009). Generally, T_H1 responses play a greater role in the clearance of pulmonary pathogens while T_H2 responses have been associated with ineffective clearance. It has also been noted that certain pathogens, particularly fungal pathogens, have evolved mechanisms to elicit T_H2 responses as a means to evade host defenses (Chai et al., 2009).

T_H17 cells can develop from naïve $CD4^+$ T cells in the presence of IL-1β, IL-6, and IL-23 (Sallusto and Lanzavecchia, 2009). They may also develop in the presence of IL-6 and TGF-β, although in the absence of IL-6, T_{Regs} are generated from the precursor cells instead of T_H17 cells (Robinson, 2010; Thorburn and Hansbro, 2010). T_H17 cells secrete IL-17 and IL-22, though upon stimulation with IL-12, may be induced to produce IFN-γ as well as IL-17 (Sallusto and Lanzavecchia, 2009). IL-17A produced by T_H17 cells enhances neutrophil phagocytic ability, antimicrobial peptide production, as well as the expression of Ig receptors on airway epithelial cells (Wright et al., 2013). Additionally, IL-17 has been reported to act synergistically with B cell activating factor, contributing to B cell survival and proliferation. T_H17 cytokines have been shown to have a protective role during bacterial, fungal, and parasitic infections, as deficiencies have resulted in impaired neutrophil recruitment, perhaps due to T_H17-induced chemokine expression (Tsai et al., 2013).

$CD4^+$ $CD25^+$ T-regulatory lymphocytes, or T_{Regs}, are immunosuppressive and may develop naturally in the thymus following exposure to self-antigens (natural T_{Regs}), or may be induced from $CD4^+$ cells in the presence of IL-2 and TGF-β (induced T_{Regs}) (Thorburn and Hansbro, 2010). In addition to having the potential of being induced from $CD4^+$ cells, recent observations have shown that T_{Regs} have the ability to then differentiate into other effector cell subsets, suggesting that T_{Regs} have great plasticity (Thorburn and Hansbro, 2010). T_{Regs} exert immunosuppressive effects in various ways, including the release of anti-inflammatory cytokines such as IL-10 and TGF-β, the initiation of cell-contact mediated mechanisms resulting in T effector cell cycle arrest, apoptosis, or APC immaturity, as well as through engaging in competition for T effector cell growth factors such as IL-2 (Vock et al., 2010). As previously discussed, immunoregulatory cytokines and mechanisms are essential in preventing uncontrolled pathologic inflammation in response to infectious antigen.

$CD8^+$ Cytotoxic T Lymphocytes

$CD8^+$ CTLs respond to antigen presented on MHC-I, expressed on all nucleated cells. CTLs are classified in subsets resembling those for $CD4^+$ cells: T_C1, T_C2, T_C17, regulatory, and memory (Betts and Kemeny, 2009). Each subset has the capacity to produce different cytokines, some T_H1-oriented and others T_H2-oriented, and thus may be proinflammatory or anti-inflammatory depending on the dominant subset. T_C1 cells, for example, capable of efficient migration to sites of inflammation, have been implicated in viral clearance via general cytotoxic abilities and IFN-γ production. Contrary to T_C1 cell responses, T_C2 cells, characterized by IL-4, IL-5, IL-10, and TGF-β release, have often been associated with pathologic responses and disease severity (Iezzi et al., 2006). T_C17 cells, characterized by IL-17 and IL-22 production, have been shown to confer protection against viral and fungal pathogens and have been observed in the lung after challenge with influenza (Chen and Kolls, 2013).

Following activation, CTLs proliferate, acquire effector capacity, migrate to inflammatory tissues, and execute targeted killing of infected cells (Williams and Bevan, 2007). Effector CTL killing functions are mediated via direct contact with the target cell and initiation of the CTL granzyme/perforin pathway or activation of target cell death receptors such as FAS (CD95), and are crucial to clearance of viral and intracellular pathogens (Trapani and Smyth, 2002). The importance of the CTL defense mechanism has been extensively studied in influenza virus infection (Burleson et al., 2010). In general, a suppression of early nonspecific antiviral defense mechanisms, such as NK activity, places a greater emphasis on the need for a properly functioning CTL as an additional antiviral defense. If the CTL is also impaired, the host's ability to respond to viral infection will be further compromised (Burleson and Burleson, 2008a,b; Burleson et al., 2010). IFN-γ also plays an important role in the CTL response, as demonstrated by the fact that anti-IFNγ monoclonal antibody blocked not only the generation of virus-specific CTLs but also the elimination of lymphocytic choriomeningitis virus (Wille et al., 1989). CTL responses also provide protection during bacterial challenge. *S. pneumoniae* infection in $CD8^{-/-}$ mice caused an increase in bacterial dissemination, and in perforin$^{-/-}$ and IFN$\gamma^{-/-}$ mice, mortality was increased. Furthermore, reduction in the $CD8^+$ cell repertoire due to HIV infection has been correlated with a higher probability of developing bacterial pneumonia (Chen and Kolls, 2013).

A reduction in the antiviral CTL response in a host exposed to inhaled immunotoxicants may result in an enhanced and prolonged viral infection. In one study, exposure of Fischer-344 rats to phosgene resulted in an enhanced and prolonged influenza virus infection (Ehrlich and Burleson, 1991), as well as a small, but statistically significant, decrease in pulmonary CTL activity (Ehrlich et al., 1989). In addition to a reduction in antiviral CTL activity, phosgene-exposed animals also experience a suppression of pulmonary NK activity (Burleson and Burleson, 2008b, 2007; Burleson and Keyes, 1989; Kim et al., 2008).

Memory T Cells

Memory T cells, key components of the acquired immune response, are antigen-specific, are developed after exposure and recognition of a particular antigen, and are long-lived to provide rapid protection upon reexposure to the antigen due to enhanced function and lower activation threshold. Memory T cells may be $CD4^+$ or $CD8^+$ and are divided into central memory (T_{CM}) or effector memory (T_{EM}) subsets, depending on CD62L and CCR7 expression, receptors allowing migration into high endothelial venules and to secondary lymphoid tissues (Kaech et al., 2002; Sallusto et al., 2004; Zielinski et al., 2011). T_{CM} express high levels of CD62L and CCR7, tend to be found in the lymph nodes and tonsils, and have higher proliferative ability than the CD62L- and CCR7-low T_{EM}, which tend to be found in the lungs, gut, and liver and have more potent effector function (Sallusto et al., 2004). T_{CM} and T_{EM} phenotypes may be flexible, however, as each phenotype has been observed to convert to the other; some T_{CM} resulting from T_{CM} proliferation and expansion have been shown to convert to T_{EM} cell types, and T_{EM} have been shown to convert to T_{CM} following $CD4^+$ or $CD8^+$ T cell activation (Lees and Farber, 2010). Although T_{CM} and T_{EM} may be either $CD4^+$ or $CD8^+$, the T_{CM} phenotype is more common within the $CD4^+$ subset, while the T_{EM} phenotype is more common within the $CD8^+$ subset (Sallusto et al., 2004).

In studies of viral infection in mice, virus-specific memory T cells in the local pulmonary tissue increase viral clearance and host survival upon secondary encounter with a particular viral pathogen. Memory $CD8^+$ T cells specific for highly conserved viral proteins are capable of generating a quick and effective response against a range of viral pathogens. Such memory in the CMI compartment is an important tool for viral clearance as certain viruses may mutate as a means to escape antibody-mediated attack (Piet et al., 2011). Memory T cells are also important in the host response to bacterial pathogens, as the rapid expansion of *L. monocytogenes*-specific $CD8^+$ T cells are critical for clearance (Wong and Pamer, 2003). Fungal challenge with pathogens such as *Aspergillus* results in the generation of $CD8^+$ memory T cells as well (Chen and Kolls, 2013). In general, more is known about the $CD8^+$ memory T cell compartment than the $CD4^+$ memory compartment with regard to infectious disease, due in part to the fact that while $CD8^+$ memory T cells remain steady in number for years following the initial encounter, numbers of $CD4^+$ T cells specific for the same pathogen are not as long-lived. Remarkably, $CD8^+$ memory T cells have been shown to survive for up to 75 years following vaccination (Williams and Bevan, 2007). It is thought that this longevity may be related to the ability of $CD8^+$ memory T cells to respond to IL-15, which acts as a survival signal, while $CD4^+$ memory T cells do not respond and slowly die out. Despite the issue of long-term survival, however, $CD4^+$ memory T cells are known to contribute to the adaptive response, as studies with *L. monocytogenes* infection in mice show that $CD4^+$ memory T cells enhance CD8-mediated clearance of pathogen (MacLeod et al., 2010).

5.2 Humoral-Mediated Immunity

HMI, driven by B lymphocytes and plasma cells, is the counterpart to CMI that comprises the adaptive immune response. B lymphocytes and plasma cells contribute to HMI through the production of antigen-specific antibodies. As with CMI, HMI follows the innate immune response and may appear days after infection. Damage to the humoral-mediated arm of the immune response increases the susceptibility to, and decreases the recovery from, infectious agents.

B Lymphocytes and Plasma Cells

B cells and plasma cells are primarily involved in producing antibody to target antigen, though B cells may serve an additional role as an APC, and certain subsets participate in innate-like and regulatory responses. Antigen recognition by naive B cells may be thymus dependent (TD) or thymus independent (TI), and leads to activation of the B cell. TI antigens are classified as TI-type I, including LPS, CpG, or polyinosinic:polycytidylic acid (poly I:C), and TI-type II, including bacterial polysaccharides, and are capable of eliciting B cell activation in the absence of T cell help (Obukhanych and Nussenzweig, 2006). TI antigens presented by MZ DCs and macrophages generally induce responses by MZ B cells through triggering cell surface PRRs, including TLRs, and/or BCRs. MZ B cells then rapidly differentiate into plasmablasts, producing high levels of low-affinity, broad-specificity IgM as a first line of defense (Cerutti et al., 2013; Garraud et al., 2012; Pieper et al., 2013; Toellner et al., 2002). This innate-like response yielding low-affinity antibody production aids in defense during the temporal gap between initial antigen encounter and the generation of responses with higher antigen specificity. Few, if any, memory B cells develop as a result of TI antigen response, however (Stein, 1992).

TD antigens such as viral pathogens require T cell help to generate an antibody response. T cell—B cell signaling and B cell activation induces the formation of intrafollicular germinal centers, B cell proliferation, high levels of somatic hypermutation of Ig variable domains, and Ig class switching from IgM to IgG, IgA, or IgE. Progeny are selected to differentiate into plasma cells, dedicated to upregulated antibody production and secretion, or memory B cells, dedicated to long-lasting protection against the antigen. Progeny producing Ig with the highest affinity for the antigen are favored for survival (Seagal and Melamed, 2003). Primary encounter with pulmonary viral, bacterial, or fungal antigen generally results in an IgM-dominant response, important in the neutralization and complement-mediated eradication of pathogen. Secondary encounter with antigen commonly generates an IgG-dominant response, contributing to phagocytosis- and opsonization-mediated pathogenic clearance (Cerutti et al., 2013).

Memory B cells are capable of surviving for decades following initial antigen exposure, as observed with survivors of the 1918 influenza pandemic (Pieper et al., 2013). Memory B cells are responsible for long-term maintenance of anti-antigen antibody and prevention of recurrent pulmonary infections, and are recruited and/or produced locally after lung immunization (Bice and Muggenburg, 1988; Bice et al., 1991; Mason et al., 1985). The importance of specific antibody for resistance to influenza virus infection was previously suggested by Ada et al. in studies measuring virus-specific antibody-secreting cells in the mouse lung following primary infection. IgM antibody-secreting cells were detected 3 days after infection in the spleen and on day 5 in the lung. IgG-secreting cells appeared in the spleen on day 6 and in the lung on day 10 (Ada et al., 1983; Jones and Ada, 1986). In mice with deficiencies in the B cell compartment, challenge with influenza results in increased susceptibility to influenza infection and a reduction in IFN-γ-producing T effector cells (Kopf et al., 2002). The importance of antibody in resistance to bacterial infection is well documented, as antibody-related deficiencies result in increased susceptibility to infection, particularly with regard to encapsulated bacterial pathogens such as *S. pneumoniae* and *Haemophilus influenzae* type b (Ballow, 2002). Indeed, in patients with primary antibody deficiencies, pneumonia following infection with *S. pneumoniae*, *H. influenzae* type b, *H. parainfluenzae*, *Mycoplasma* spp., *Pseudomonas* spp., and *Staphylococcus* spp. is relatively common, as are resulting complications such as chronic lung disease (Fried and Bonilla, 2009).

6. ALTERATIONS IN PULMONARY IMMUNOLOGICAL RESPONSES

Exposure to a variety of stimuli, including environmental and occupational pollutants, various therapeutics, as well as certain pathogens, may enhance susceptibility or severity of viral, bacterial, and fungal diseases. Xenobiotics, chemical compounds or pollutants not naturally produced in an organism, may initiate or potentiate hypersensitivity diseases of the upper and lower respiratory tracts due to a direct or indirect consequence of species, genetics, epigenetics, or the microbiome or immunome. Inhaled particulates cleared to the lung-associated lymph nodes (LALN) may affect populations and numbers of lymphocytes in the LALN, particularly because insoluble particles may remain in the LALN for long periods, and these toxic materials may alter APC and immune function (Bice et al., 1987b; Snipes et al., 1983). Epithelial cell damage caused by inhaled xenobiotics may alter lymphatic clearance of antigens from the lung to the LALN, and alterations in the numbers and/or functions of alveolar macrophages may alter clearance of antigen as well (Harmsen et al., 1987, 1985; Hillam et al., 1983). Furthermore, changes in vascular permeability, important in the recruitment of lymphoid cells and APC to the lung, may be affected (Bice et al., 1982). Thus, inflammation and lung damage as a result of exposure to xenobiotics may alter the recruitment of immune cells and antibody to the lung from the blood. With regard to the memory lymphocyte response, plasma cells in the alveoli and interstitial tissues may be damaged by inhaled xenobiotics with a resultant decrease in antibody production and decreased resistance to infectious agents (Bice et al., 1987a). Most of the memory cells are located in the lung interstitium and are therefore susceptible to a variety of inhaled xenobiotics (Bice et al., 1993). These cells provide local immunological memory against infectious diseases. Damage to memory cells or to AFC will adversely affect localized immune memory, thereby increasing the number of infectious microorganisms to which an individual is susceptible.

In addition to alterations caused by inhaled xenobiotics and toxicants, alterations may occur due to exposure to particular pathogens. Severe RSV infection during infancy, for example, has been correlated with a 3.5-fold increase in the development of allergy and a more than 10-fold increase in the development of asthma within 3 years of infection when compared with uninfected controls. It is thought that this may be related to the propensity of RSV to induce an atypical T_H2 response to virus, that, if present during an early encounter with inhaled antigen, could result in a bystander activation of T_H2 cells by the inhaled antigen (Holt and Sly, 2002). Moreover, respiratory viral infections have been shown to exacerbate asthma in children and adults, and are an underlying cause of morbidity and mortality for asthmatic individuals (Tan, 2005). Another example of altered immune responses due to pathogenic exposure is the observed increased susceptibility to bacterial infection following initial infection with influenza virus. Together, influenza and bacterial pneumonia are a deadly combination, contributing to worldwide morbidity and mortality. Indeed, during the 1918 Spanish influenza, the majority of the deaths were related to secondary bacterial pneumonia infection. While the association between respiratory viral infection and increased incidence of secondary bacterial infection is well known, the mechanisms involved have not yet been fully elucidated (Small et al., 2010). It has been reported, however, that damage to the respiratory epithelium incurred by influenza infection may increase bacterial adherence to injured cells, resulting in increased colonization. NK cell function and NK TNF-α release is also decreased following influenza infection, contributing to decreased macrophage-mediated phagocytosis of bacteria (Hament et al., 1999; Small et al., 2010). Furthermore, influenza infection promotes apoptosis of alveolar macrophages, decreased macrophage function directly due to infection as well as factors produced due to infection, and decreased macrophage activation through induction of CD200 receptor expression (Braciale et al., 2012).

Therapeutics, particularly immunomodulatory and immunosuppressant drugs, cause alterations to the immune response. The use of infliximab, an anti-TNF antibody prescribed for patients diagnosed with autoimmune disease and rheumatoid arthritis, has been correlated with a 5- to 10-fold increased susceptibility to *M. tuberculosis* reactivation, likely due to the requirement of TNF for the maintenance of granulomas containing the bacterium (Ehlers, 2005; Hanauer, 1999). Not surprisingly, infliximab has also been correlated with an increase in upper respiratory tract infections (Hanauer, 1999). Cyclosporin, an immunosuppressant used to prevent transplant rejection as well as rheumatoid arthritis and psoriasis, exerts its effects through inhibition of T cell activation. The potent inhibition of T cell activation, however, results in increased susceptibility to viral infections (Graham, 1994). Natalizumab (Tysabri), a humanized monoclonal antibody against $\alpha 4$ subunit of $\alpha 4\beta 1$ and $\alpha 4\beta 7$ integrin, a cell adhesion molecule expressed on activated T cells and other cell types, is prescribed for relapse-remitting multiple sclerosis. Use of this therapeutic, however, is associated with an increased risk of progressive multifocal leukoencephalopathy and opportunistic infections (Goodin et al., 2008). While there are simply too many to list, immunomodulatory and immunosuppressant therapeutics should be prescribed with care. Modification of the immune response in the treatment of immune-related disorders may carry a risk—benefit ratio with regard to immune homeostasis as well as response to infectious disease.

7. COMPARATIVE PULMONARY IMMUNOLOGY

While we have gained a great deal of information regarding the pulmonary immunology of infectious diseases from animal research, care must be exercised in the extrapolation of data from one species to another. The species, strain, age, and sex of the animal are all important considerations. Differences in the kinetics of the immune response for different species may be incorrectly interpreted as an insensitivity of a particular animal species if assayed at an inappropriate time. Genetic susceptibility must be considered when comparing responses observed with different strains and species. Repair rates and reversibility of lesions should be identified in order to make interspecies comparisons. For these reasons, recognition of the differences observed between species is an important aspect of comparative immunology.

Structural differences have been observed with regard to immune-related tissues for different species. While thymic structure and thymic involution, the process of thymus tissue replacement by adipose tissue occurring with age, is generally consistent across species, other structures such as the lymph nodes and spleen have dissimilarities across species. Compared with larger species, mice have fewer lymph nodes that are organized into uncomplicated chains. The number of lymph nodes and the complexity of the lymphatic system increase with the size of the species, and the area of tissue that requires draining by the lymphatic system. Rats, for instance, have two lung-draining lymph nodes, dogs have three to five, and humans have 35 or more that are divided into five distinct classification groups. The general organization of the lymph node interior is similar among species, although in swine, the cortical and medullary compartments of the lymph node are reversed. Aging also seems to have different effects among the species, as lymph node atrophy and fibrosis are common in dogs (Haley, 2003).

The spleen differs between species in its structure and function. The three classes of splenic structure and function have been described as storage, defensive, and intermediate. The storage spleen is typically found in dogs and cats, is capable of storing up to one-third of the total volume of circulating blood, and contains smooth muscle and limited lymphoid tissue. The defensive spleen is typically found in mice, rats, rabbits, and humans, and has less smooth muscle and more developed lymphoid tissue. The intermediate spleen has characteristics in between the storage and defensive spleen with regard to lymphoid tissue, and is found in swine and ruminants. With regard to hematopoietic activity, the spleen is a significant contributor in mice and, to a lesser extent, rats, while adult rabbit and human spleens do not participate in hematopoietic activity outside of disease states (Haley, 2003).

Comparative differences in pulmonary immunology also exist within the nasal lavage fluid (NALF) and BALF, which serve as windows for viewing numerous inflammatory and immunological events occurring in the upper and lower respiratory tract in vivo. Each allows for the measurement of many inflammatory, biochemical, and immunological cells and mediators, and may also contain immunological indicators of lung damage, including antioxidants, proteins, enzymes, cytokines, growth factors, and arachidonic acid metabolites. Protein constituents of nasal secretions include albumin, lysozyme, lactoferrin, kallikrein, antiproteases, β-glucuronidase, α-galactosidase, succinic dehydrogenase, IgA, IgG, IgE, secretory component, glutamate dehydrogenase, and leucine aminopeptidase (Raphael et al., 1991). Constituents measured in BALF include albumin, lysozyme, sIgA, IgG, IgA, lactoferrin, transferrin, alpha I-antitrypsin, complement components C3 and C4, haptoglobin, A2-macroglobulin, hemopexin, A2-HS-glycoprotein, and a1-acid glycoprotein (Hatch, 1992). Pulmonary cell composition of the BALF as well as concentrations of extracellular antioxidant substances in NALF and BALF can vary widely among species and are presented in Table 1 and Table 2, respectively.

Variations in cell function have been observed across species, particularly with regard to NK cells, macrophages, and AFC. NK cell activity in mice peaks between 4 and 10 weeks of age, and although activity is generally high within mouse lungs, there exists great variability in activity between strains. Human NK cell activity, on the other hand, does not peak at a particular age as in mice, and activity tends to be lower within the lungs. With regard to alveolar macrophages, AMs are morphologically similar among mice, rats, and dogs, whereas human and nonhuman primate AMs are comparatively larger, with human AMs being largest of all. In addition, macrophage phagocytosis and killing capabilities were the lowest in the rat, intermediate in hamsters and rabbits, and highest in humans. With regard to AFC, the introduction of antigen directly into the lungs of mice, rats, hamsters, guinea pigs, and rabbits results in elevated numbers of AFC in the LALN, though AFC are not typically noted in lung lavage fluid samples. Following the introduction of antigen into dog and nonhuman primate lungs, however, elevated numbers of AFC have been noted in the LALN (dogs only; not evaluated in nonhuman primates) as well as lavage fluid. It has also been observed that the regulation of the IgG subclass differs in mice and humans, and that rabbits and mice do not produce all of the IgG subclasses that are observed in humans (Haley, 2003).

TABLE 1 Relative Proportions of Immunocompetent Cell Populations Obtained by Bronchoalveolar Lavage (Murray and Driscoll, 1992)

	Bronchoalveolar Lavage Cells (%)			Lymphocytes (%)	
Species	**Macrophages**	**Lymphocytes**	**Granulocytes**	**T**	**B**
Human	78–91	9–20	1–3	47	15–17
Monkey	90–91	3–6	3–6	62	4
Dog	59–75	22–39	0–9	–	–
Swine	60–70	30–33	0–6	–	–
Guinea pig	50–80	12–50	2–19	68–76	10–20
Rabbit	84–98	2–16	0	–	–
Hamster	89	3	10	–	–
Mouse	45–96	3–39	1–6	–	–
Rat	93	2	5	50	12

TABLE 2 Concentrations of Extracellular Antioxidant Substances (Hatch, 1992)

Nasal Lavage	**Human (Bice et al., 1991)**	**Guinea Pig (Cerutti et al., 2013)**	**Rat (Berger, 2000)**
Protein (μg/ml)	147 ± 23.8	24.9 ± 0.95	11.8 ± 2.4
Ascorbic acid (μg/ml)	0.22 ± 0.16	1.28 ± 0.09	1.39 ± 0.24
Uric acid (μg/ml)	2.85 ± 0.22	<0.04	<0.04
Glutathione (μg/ml)	<0.30	<0.30	<0.9
Bronchoalveolar Lavage	**Human (Coffman, 2006)**	**Guinea Pig (Balhara and Gounni, 2012)**	**Rat (Balhara and Gounni, 2012)**
Protein (μg/ml)	81.0 ± 6.6	202 ± 7.4	110 ± 5.6
Ascorbic acid (μg/ml)	0.153 ± 0.008	0.907 ± 0.123	3.815 ± 0.064
Uric acid (μg/ml)	0.147 ± 0.008	0.182 ± 0.050	0.144 ± 0.006
Glutathione (μg/ml)	2.067 ± 0.538	0.821 ± 0.133	0.108 ± 0.067

Note: Number of animals or human subjects is shown in parenthesis. Means ± SE are shown.

Intriguing differences between species have also been noted in studies investigating immune signaling. In studies surrounding the previously mentioned TeGenero incident, it was hypothesized that the lack of response observed in preclinical studies in primates did not correlate with the rapid, devastating response in man and may have been influenced by differential T cell Siglec expression (Horvath and Milton, 2009; Nguyen et al., 2006). Siglecs (Sialic acid-binding Ig-type lectins) are expressed on immune cells, and CD33-related Siglecs (Siglec-3 and Siglec 5–11) are thought to act as "brakes" on innate immune cell activation. Human T cells, unlike primate T cells (particularly $CD4^+$ T cells), have little-to-no Siglec expression, perhaps explaining the widespread T cell activation and ensuing cytokine storm that occurred in the TGN1412 clinical trial (Nguyen et al., 2006). Distinctions in signaling with regard to TLR activation have also been observed. Chimpanzees and humans, for example, are more sensitive to LPS than baboons, monkeys, and certain strains of mice, due to variations in TLR4 signaling (Horvath and Milton, 2009; Smirnova et al., 2000).

In addition to species-related differences, gender leads to variation in immune responses as well. Females tend to produce heightened cellular and humoral immune responses and tend to have a higher resistance to bacterial infection than males, though females are also more prone to the development of autoimmune disease. These tendencies toward gender-related response differences may be tied to basic differences in cell number and function. While human male and female total lymphocyte counts are similar, females have a higher percentage of T lymphocytes, perhaps due to the observation that elevated testosterone levels may increase T cell apoptosis. It has also been observed that females have higher serum levels of IgM and IgG, in agreement with in vitro data linking estrogen to B cell activation and production of IgM and IgG, and inhibition of IgM and IgG production following testosterone exposure (Bouman, 2005). Though differences exist, continued insight into immune function may yet be obtained from study with various animal models as long as due diligence is exercised to account for comparative variation in responses.

REFERENCES

Ada, G.L., Mak, N.K., Sweet, C., 1983. The regulation of influenza virus infection. In: Yamanura, T., Tada, T. (Eds.), Progress in Immunology, vol. 5. Academic Press, New York.

Agaugue, S., Marcenaro, E., Ferranti, B., Moretta, L., Moretta, A., 2008. Human natural killer cells exposed to IL-2, IL-12, IL-18, or IL-4 differently modulate priming of naive T cells by monocyte-derived dendritic cells. Blood 112 (5), 1776–1783.

Algood, H.M.S., Lin, P.L., Yankura, D., Jones, A., Chan, J., Flynn, J.L., 2004. TNF influences chemokine expression of macrophages in vitro and that of CD11b+ cells in vivo during Mycobacterium tuberculosis infection. J. Immunol. 172 (11), 6846–6857.

Arts, R.J., Joosten, L.A., van der Meer, J.W., Netea, M.G., 2013. TREM-1: intracellular signaling pathways and interaction with pattern recognition receptors. J. Leukoc. Biol. 93 (2), 209–215.

Balhara, J., Gounni, A.S., 2012. The alveolar macrophages in asthma: a double-edged sword. Mucosal Immunol. 5 (6), 605–609.

Ballow, M., 2002. Primary immunodeficiency disorders: antibody deficiency. J. Allergy Clin. Immunol. 109 (4), 581–591.

Barry, K.C., Fontana, M.F., Portman, J.L., Dugan, A.S., Vance, R.E., 2013. IL-1alpha signaling initiates the inflammatory response to virulent *Legionella pneumophila* in vivo. J. Immunol. 190 (12), 6329–6339.

Bellora, F., Castriconi, R., Dondero, A., Reggiardo, G., Moretta, L., Mantovani, A., Moretta, A., Bottino, C., 2010. The interaction of human natural killer cells with either unpolarized or polarized macrophages results in different functional outcomes. Proc. Natl. Acad. Sci. U. S. A. 107 (50), 21659–21664.

Benoit, M., Desnues, B., Mege, J., 2008. Macrophage polarization in bacterial infections. J. Immunol. 181 (6), 3733–3739.

Berger, A., 2000. Science commentary: Th1 and Th2 responses: what are they? BMJ 321 (7258), 424.

Betts, R.J., Kemeny, D.M., 2009. CD8+ T cells in asthma: friend or foe? Pharmacol. Ther. 121 (2), 123–131.

Bice, D.E., Degen, M.A., Harris, D.L., Muggenburg, B.A., 1982. Recruitment of antibody-forming cells in the lung after local immunization is nonspecific. Am. Rev. Respir. Dis. 126 (4), 635–639.

Bice, D.E., Gray, R.H., Evans, M.J., Muggenburg, B.A., 1987a. Identification of plasma cells in lung alveoli and interstitial tissues after localized lung immunization. J. Leukoc. Biol. 41 (1), 1–7.

Bice, D.E., Hahn, F.F., Benson, J., Carpenter, R.L., Hobbs, C.H., 1987b. Comparative lung immunotoxicity of inhaled quartz and coal combustion fly ash. Environ. Res. 43 (2), 374–389.

Bice, D.E., Jones, S.E., Muggenburg, B.A., 1993. Long-term antibody production after lung immunization and challenge: role of lung and lymphoid tissues. Am. J. Respir. Cell. Mol. Biol. 8 (6), 662–667.

Bice, D.E., Muggenburg, B.A., 1988. Localized immune memory in the lung. Am. Rev. Respir. Dis. 138, 565–571.

Bice, D.E., Weissman, D.N., Muggenburg, B.A., 1991. Long-term maintenance of localized antibody responses in the lung. Immunology 74 (2), 215–222.

Bouman, A., 2005. Sex hormones and the immune response in humans. Hum. Reprod. Update 11 (4), 411–423.

Braciale, T.J., Sun, J., Kim, T.S., 2012. Regulating the adaptive immune response to respiratory virus infection. Nat. Rev. Immunol. 12 (4), 295–305.

Buchmeier, N.A., Schreiber, R.D., 1985. Requirement of endogenous interferon-gamma production for resolution of *Listeria monocytogenes* infection. Proc. Natl. Acad. Sci. U. S. A. 82 (21), 7404–7408.

Burleson, G.R., Burleson, F.G., 2007. Influenza virus host resistance model. Methods 41 (1), 31–37.

Burleson, G.R., Burleson, F.G., 2008a. Animal models of host resistance. In: Herzyk, D.J., Bussiere, J.L. (Eds.), Immunotoxicology Strategies for Pharmaceutical Safety Assessment. John Wiley & Sons, Inc., Hoboken, NJ, pp. 163–178.

Burleson, G.R., Burleson, F.G., 2008b. Testing human biologicals in animal host resistance models. J. Immunotoxicol. 5 (1), 23–31.

Burleson, G.R., Burleson, F.G., Dietert, R.R., 2008. Juvenile immunotoxicity (DIT) testing in SD rats: effect of dexamethasone on influenza viral clearance, NK activity, CTL activity, and T-dependent viral-specific IgM and IgG antibody production (TDAR). Toxicol. Sci. 102 (1), 12.

Burleson, G.R., Burleson, F.G., Dietert, R.R., 2010. The cytotoxic T lymphocyte assay for evaluating cell-mediated immune function. Methods Mol. Biol. 598, 195–205.

Burleson, G.R., Keyes, L.L., 1989. Natural killer activity in Fischer-344 rat lungs as a method to assess pulmonary immunocompetence: immunosuppression by phosgene inhalation. Immunopharmacol. Immunotoxicol. 11 (2–3), 421–443.

Calkins, C.M., Bensard, D.D., Shames, B.D., Pulido, E.J., Abraham, E., Fernandez, N., Meng, X., Dinarello, C.A., McIntyre Jr, R.C., 2002. IL-1 regulates in vivo C-X-C chemokine induction and neutrophil sequestration following endotoxemia. J. Endotoxin Res. 8 (1), 59–67.

Canna, S.W., Behrens, E.M., 2012. Making sense of the cytokine storm: a conceptual framework for understanding, diagnosing, and treating hemophagocytic syndromes. Pediatr. Clin. North Am. 59 (2), 329–344.

Carlos, D., Frantz, F.G., Souza-Junior, D.A., Jamur, M.C., Oliver, C., Ramos, S.G., Quesniaux, V.F., Ryffel, B., Silva, C.L., Bozza, M.T., Faccioli, L.H., 2009. TLR2-dependent mast cell activation contributes to the control of mycobacterium tuberculosis infection. Microbes Infect. 11 (8–9), 770–778.

Cerutti, A., Cols, M., Puga, I., 2013. Marginal zone B cells: virtues of innate-like antibody-producing lymphocytes. Nat. Rev. Immunol. 13 (2), 118–132.

Chacon-Salinas, R., Serafin-Lopez, J., Ramos-Payan, R., Mendez-Aragon, P., Hernandez-Pando, R., Van Soolingen, D., Flores-Romo, L., Estrada-Parra, S., Estrada-Garcia, I., 2005. Differential pattern of cytokine expression by macrophages infected in vitro with different mycobacterium tuberculosis genotypes. Clin. Exp. Immunol. 140 (3), 443–449.

Chai, L.Y., Netea, M.G., Vonk, A.G., Kullberg, B.J., 2009. Fungal strategies for overcoming host innate immune response. Med. Mycol. 47 (3), 227–236.

Chen, K., Kolls, J.K., 2013. T cell-mediated host immune defenses in the lung. Annu. Rev. Immunol. 31, 605–633.

Chiba, A., Dascher, C.C., Besra, G.S., Brenner, M.B., 2008. Rapid NKT cell responses are self-terminating during the course of microbial infection. J. Immunol. 181 (4), 2292–2302.

Coffman, R.L., 2006. Origins of the T(H)1-T(H)2 model: a personal perspective. Nat. Immunol. 7 (6), 539–541.

Cohen, I., Rider, P., Carmi, Y., Braiman, A., Dotan, S., White, M.R., Voronov, E., Martin, M.U., Dinarello, C.A., Apte, R.N., 2010. Differential release of chromatin-bound IL-1alpha discriminates between necrotic and apoptotic cell death by the ability to induce sterile inflammation. Proc. Natl. Acad. Sci. U. S. A. 107 (6), 2574–2579.

Contassot, E., Beer, H.D., French, L.E., 2012. Interleukin-1, inflammasomes, autoinflammation and the skin. Swiss Med. Wkly. 142, w13590.

Cooper, A.M., Mayer-Barber, K.D., Sher, A., 2011. Role of innate cytokines in mycobacterial infection. Mucosal Immunol. 4 (3), 252–260.

Culley, F.J., 2009. Natural killer cells in infection and inflammation of the lung. Immunology 128 (2), 151–163.

Curtis, J.L., 2005. Cell-mediated adaptive immune defense of the lungs. Proc. Am. Thorac. Soc. 2 (5), 412–416.

D'Apuzzo, M., Rolink, A., Loetscher, M., Hoxie, J.A., Clark-Lewis, I., Melchers, F., Baggiolini, M., Moser, B., 1997. The chemokine SDF-1, stromal cell-derived factor 1, attracts early stage B cell precursors via the chemokine receptor CXCR4. Eur. J. Immunol. 27 (7), 1788–1793.

Dauer, M., Obermaier, B., Herten, J., Haerle, C., Pohl, K., Rothenfusser, S., Schnurr, M., Endres, S., Eigler, A., 2003. Mature dendritic cells derived from human monocytes within 48 hours: a novel strategy for dendritic cell differentiation from blood precursors. J. Immunol. 170 (8), 4069–4076.

Dinarello, C.A., 2010. Anti-inflammatory agents: present and future. Cell 140 (6), 935–950.

Dinarello, C.A., 2011. Interleukin-1 in the pathogenesis and treatment of inflammatory diseases. Blood 117 (14), 3720–3732.

Dinarello, C.A., Ikejima, T., Warner, S.J., Orencole, S.F., Lonnemann, G., Cannon, J.G., Libby, P., 1987. Interleukin 1 induces interleukin 1. I. Induction of circulating interleukin 1 in rabbits in vivo and in human mononuclear cells in vitro. J. Immunol. 139 (6), 1902–1910.

Drescher, B., Bai, F., 2013. Neutrophil in viral infections, friend or foe? Virus Res. 171 (1), 1–7.

Driscoll, K.E., Schlesinger, R.B., 1988. Alveolar macrophage-stimulated neutrophil and monocyte migration: effects of in vitro ozone exposure. Toxicol. Appl. Pharmacol. 93 (2), 312–318.

Driscoll, K.E., Simpson, L., Carter, J., Hassenbein, D., Leikauf, G.D., 1993. Ozone inhalation stimulates expression of a neutrophil chemotactic protein, macrophage inflammatory protein 2. Toxicol. Appl. Pharmacol. 119 (2), 306–309.

van Eeden, S.F., Terashima, T., 2000. Interleukin 8 (IL-8) and the release of leukocytes from the bone marrow. Leuk. Lymphoma 37 (3–4), 259–271.

Ehlers, S., 2005. Tumor necrosis factor and its blockade in granulomatous infections: differential modes of action of infliximab and etanercept? Clin. Infect. Dis. 41 (Suppl. 3), S199–S203.

Ehrlich, J.P., Burleson, G.R., 1991. Enhanced and prolonged pulmonary influenza virus infection following phosgene inhalation. J. Toxicol. Environ. Health 34 (2), 259–273.

Ehrlich, J.P., Gunnison, A.F., Burleson, G.R., 1989. Influenza virus-specific cytotoxic T lymphocyte activity in Fischer 344 rat lungs as a method to assess pulmonary immunocompetence: effect of phosgene inhalation. Inhalation Toxicol. 1 (2), 129–138.

Fonteneau, J.F., Kavanagh, D.G., Lirvall, M., Sanders, C., Cover, T.L., Bhardwaj, N., Larsson, M., 2003. Characterization of the MHC class I cross-presentation pathway for cell-associated antigens by human dendritic cells. Blood 102 (13), 4448–4455.

Forster, R., Emrich, T., Kremmer, E., Lipp, M., 1994. Expression of the G-protein–coupled receptor BLR1 defines mature, recirculating B cells and a subset of T-helper memory cells. Blood 84 (3), 830–840.

Fried, A.J., Bonilla, F.A., 2009. Pathogenesis, diagnosis, and management of primary antibody deficiencies and infections. Clin. Microbiol. Rev. 22 (3), 396–414.

Garraud, O., Borhis, G., Badr, G., Degrelle, S., Pozzetto, B., Cognasse, F., Richard, Y., 2012. Revisiting the B-cell compartment in mouse and humans: more than one B-cell subset exists in the marginal zone and beyond. BMC Immunol. 13 (1), 63.

Gazzinelli, R.T., Wysocka, M., Hieny, S., Scharton-Kersten, T., Cheever, A., Kuhn, R., Muller, W., Trinchieri, G., Sher, A., 1996. In the absence of endogenous IL-10, mice acutely infected with toxoplasma gondii succumb to a lethal immune response dependent on CD4+ T cells and accompanied by overproduction of IL-12, IFN-gamma and TNF-alpha. J. Immunol. 157 (2), 798–805.

Germain, R.N., 2002. T-cell development and the CD4-CD8 lineage decision. Nat. Rev. Immunol. 2 (5), 309–322.

Goodin, D.S., Cohen, B.A., O'Connor, P., Kappos, L., Stevens, J.C., 2008. Assessment: the use of natalizumab (Tysabri) for the treatment of multiple sclerosis (an evidence-based review): report of the therapeutics and technology assessment subcommittee of the American academy of neurology. Neurology 71 (10), 766–773.

Graham, R.M., 1994. Cyclosporine: mechanisms of action and toxicity. Cleve Clin. J. Med. 61 (4), 308–313.

Greenberger, M.J., Kunkel, S.L., Strieter, R.M., Lukacs, N.W., Bramson, J., Gauldie, J., Graham, F.L., Hitt, M., Danforth, J.M., Standiford, T.J., 1996. IL-12 gene therapy protects mice in lethal Klebsiella pneumonia. J. Immunol. 157 (7), 3006–3012.

Guerrero-Plata, A., Kolli, D., Hong, C., Casola, A., Garofalo, R.P., 2009. Subversion of pulmonary dendritic cell function by paramyxovirus infections. J. Immunol. 182 (5), 3072–3083.

Guilmette, R.A., Wicks, J.D., Wolff, R.K., 1989. Morphometry of human nasal airways in vivo using magnetic resonance imaging. J. Aerosol Med. 2, 365–377.

Haley, P.J., 2003. Species differences in the structure and function of the immune system. Toxicology 188 (1), 49–71.

Hament, J., Kimpen, J.L.L., Fleer, A., Wolfs, T.F.W., 1999. Respiratory viral infection predisposing for bacterial disease: a concise review. FEMS Immunol. Med. Microbiol. 26 (3–4), 189–195.

Hammad, H., Lambrecht, B.N., 2011. Dendritic cells and airway epithelial cells at the interface between innate and adaptive immune responses. Allergy 66 (5), 579–587.

Hammad, H., Plantinga, M., Deswarte, K., Pouliot, P., Willart, M.A., Kool, M., Muskens, F., Lambrecht, B.N., 2010. Inflammatory dendritic cells—not basophils—are necessary and sufficient for induction of Th2 immunity to inhaled house dust mite allergen. J. Exp. Med. 207 (10), 2097–2111.

Hamza, T., Barnett, J.B., Li, B., 2010. Interleukin 12 a key immunoregulatory cytokine in infection applications. Int. J. Mol. Sci. 11 (3), 789–806.

Hanauer, S.B., 1999. Review article: safety of infliximab in clinical trials. Aliment. Pharmacol. Ther. 13 (Suppl. 4), 16–22, 38.

Hanna, J., Gonen-Gross, T., Fitchett, J., Rowe, T., Daniels, M., Arnon, T.I., Gazit, R., Joseph, A., Schjetne, K.W., Steinle, A., Porgador, A., Mevorach, D., Goldman-Wohl, D., Yagel, S., LaBarre, M.J., Buckner, J.H., Mandelboim, O., 2004. Novel APC-like properties of human NK cells directly regulate T cell activation. J. Clin. Invest 114 (11), 1612–1623.

Harkema, J.R., 1992. Epithelial cells of the nasal passages. In: Parent, R.A. (Ed.), Comparative Biology of the Normal Lung. CRC Press, Boca Raton, FL, pp. 27–36.

Harmsen, A.G., Mason, M.J., Muggenburg, B.A., Gillett, N.A., Jarpe, M.A., Bice, D.E., 1987. Migration of neutrophils from lung to tracheobronchial lymph node. J. Leukoc. Biol. 41 (2), 95–103.

Harmsen, A.G., Muggenburg, B.A., Snipes, M.B., Bice, D.E., 1985. The role of macrophages in particle translocation from lungs to lymph nodes. Science 230 (4731), 1277–1280.

Hatch, G.E., 1992. Comparative biochemistry of airway lining fluid. In: Parent, R.A. (Ed.), Comparative Biology of the Normal Lung. CRC Press, Boca Raton, FL, pp. 617–632.

Hennet, T., Ziltener, H.J., Frei, K., Peterhans, E., 1992. A kinetic study of immune mediators in the lungs of mice infected with influenza A virus. J. Immunol. 149 (3), 932–939.

Hillam, R.P., Bice, D.E., Hahn, F.F., Schnizlein, C.T., 1983. Effects of acute nitrogen dioxide exposure on cellular immunity after lung immunization. Environ. Res. 31 (1), 201–211.

Holt, P.G., Sly, P.D., 2002. Interactions between RSV infection, asthma, and atopy: unraveling the complexities. J. Exp. Med. 196 (10), 1271–1275.

Horvath, C.J., Milton, M.N., 2009. The TeGenero incident and the Duff Report conclusions: a series of unfortunate events or an avoidable event? Toxicol. Pathol. 37 (3), 372–383.

Huang, F.F., Barnes, P.F., Feng, Y., Donis, R., Chroneos, Z.C., Idell, S., Allen, T., Perez, D.R., Whitsett, J.A., Dunussi-Joannopoulos, K., Shams, H., 2011. GM-CSF in the lung protects against lethal influenza infection. Am. J. Respir. Crit. Care Med. 184 (2), 259–268.

Huang, S., Hendriks, W., Althage, A., Hemmi, S., Bluethmann, H., Kamijo, R., Vilcek, J., Zinkernagel, R.M., Aguet, M., 1993. Immune response in mice that lack the interferon-gamma receptor. Science 259 (5102), 1742–1745.

Hufford, M.M., Richardson, G., Zhou, H., Manicassamy, B., García-Sastre, A., Enelow, R.I., Braciale, T.J., 2012. Influenza-infected neutrophils within the infected lungs act as antigen presenting cells for anti-viral CD8+ T cells. PLoS One 7 (10), e46581.

Iezzi, G., Boni, A., Degl'Innocenti, E., Grioni, M., Bertilaccio, M.T., Bellone, M., 2006. Type 2 cytotoxic T lymphocytes modulate the activity of dendritic cells toward type 2 immune responses. J. Immunol. 177 (4), 2131–2137.

International Commission on Radiological Protection (ICRP), 1975. Task Group, Snyder WS. Report of the Task Group on Reference Man, Volume 23 of ICRP Publication. Pergamon Press, Elmsford, NY.

Iwasaki, A., Medzhitov, R., 2011. A new shield for a cytokine storm. Cell 146 (6), 861–862.

Jambo, K.C., Banda, D.H., Kankwatira, A.M., Sukumar, N., Allain, T.J., Heyderman, R.S., Russell, D.G., Mwandumba, H.C., 2014. Small alveolar macrophages are infected preferentially by HIV and exhibit impaired phagocytic function. Mucosal Immunol. Advance online publication 29 January.

Janeway, C.A., Travers, P., Walport, M., Shlomchik, M.J., 2001. Immunobiology: The Immune System in Health and Disease. Garland, New York, p. 732.

Jin, W., Dong, C., 2013. IL-17 cytokines in immunity and inflammation. Emerging Microbes Infect. 2 (9), e60.

Jones, P.D., Ada, G.L., 1986. Influenza virus-specific antibody-secreting cells in the murine lung during primary influenza virus infection. J. Virol. 60 (2), 614–619.

Juno, J.A., Keynan, Y., Fowke, K.R., 2012. Invariant NKT cells: regulation and function during viral infection. PLoS Pathog. 8 (8), e1002838.

Kaech, S.M., Wherry, E.J., Ahmed, R., 2002. Effector and memory T-cell differentiation: implications for vaccine development. Nat. Rev. Immunol. 2 (4), 251–262.

Kaplanski, G., Teysseire, N., Farnarier, C., Kaplanski, S., Lissitzky, J.C., Durand, J.M., Soubeyrand, J., Dinarello, C.A., Bongrand, P., 1995. IL-6 and IL-8 production from cultured human endothelial cells stimulated by infection with Rickettsia conorii via a cell-associated IL-1 alpha-dependent pathway. J. Clin. Invest 96 (6), 2839–2844.

Karupiah, G., Fredrickson, T.N., Holmes, K.L., Khairallah, L.H., Buller, R.M., 1993. Importance of interferons in recovery from mousepox. J. Virol. 67 (7), 4214–4226.

Kim, T., Henry, S.P., Burleson, F.G., Burleson, G.R., 2008. Evaluation of the effects of antisense oligonucleotides on influenza clearance and macrophage function in the mouse. Toxicologist 102(1), 441.

Klesney-Tait, J., Keck, K., Li, X., Gilfillan, S., Otero, K., Baruah, S., Meyerholz, D.K., Varga, S.M., Knudson, C.J., Moninger, T.O., Moreland, J., Zabner, J., Colonna, M., 2013. Transepithelial migration of neutrophils into the lung requires TREM-1. J. Clin. Invest 123 (1), 138–149.

Kolls, J.K., 2013. CD4(+) T-cell subsets and host defense in the lung. Immunol. Rev. 252 (1), 156–163.

Kool, M., van Nimwegen, M., Willart, M.A., Muskens, F., Boon, L., Smit, J.J., Coyle, A., Clausen, B.E., Hoogsteden, H.C., Lambrecht, B.N., Hammad, H., 2009. An anti-inflammatory role for plasmacytoid dendritic cells in allergic airway inflammation. J. Immunol. 183 (2), 1074–1082.

Kopf, M., Brombacher, F., Bachmann, M.F., 2002. Role of IgM antibodies versus B cells in influenza virus-specific immunity. Eur. J. Immunol. 32 (8), 2229–2236.

Laws, T.R., Davey, M.S., Titball, R.W., Lukaszewski, R., 2010. Neutrophils are important in early control of lung infection by *Yersinia pestis*. Microbes Infect. 12 (4), 331–335.

Le, Y., Zhou, Y., Iribarren, P., Wang, J., 2004. Chemokines and chemokine receptors: their manifold roles in homeostasis and disease. Cell. Mol. Immunol. 1 (2), 95–104.

Lees, J.R., Farber, D.L., 2010. Generation, persistence and plasticity of CD4 T-cell memories. Immunology 130 (4), 463–470.

Leung, K.N., Ada, G.L., 1981. Induction of natural killer cells during murine influenza virus infection. Immunobiology 160 (3–4), 352–366.

Lugo-Villarino, G., Verollet, C., Maridonneau-Parini, I., Neyrolles, O., 2011. Macrophage polarization: convergence point targeted by mycobacterium tuberculosis and HIV. Front. Immunol. 2 (43).

MacLeod, M.K., Kappler, J.W., Marrack, P., 2010. Memory CD4 T cells: generation, reactivation and re-assignment. Immunology 130 (1), 10–15.

Marti-Lliteras, P., Regueiro, V., Morey, P., Hood, D.W., Saus, C., Sauleda, J., Agusti, A.G., Bengoechea, J.A., Garmendia, J., 2009. Nontypeable *Haemophilus influenzae* clearance by alveolar macrophages is impaired by exposure to cigarette smoke. Infect. Immun. 77 (10), 4232–4242.

Martin, T.R., Frevert, C.W., 2005. Innate immunity in the lungs. Proc. Am. Thorac. Soc. 2 (5), 403–411.

Mason, M.J., Bice, D.E., Muggenburg, B.A., 1985. Local pulmonary immune responsiveness after multiple antigenic exposures in the cynomolgus monkey. Am. Rev. Respir. Dis. 132 (3), 657–660.

Metcalfe, D.D., Baram, D., Mekori, Y.A., 1997. Mast cells. Physiol. Rev. 77 (4), 1033–1079.

Mills, C.D., 2012. M1 and M2 macrophages: oracles of health and disease. Crit. Rev. Immunol. 32 (6), 463–488.

Mocsai, A., 2013. Diverse novel functions of neutrophils in immunity, inflammation, and beyond. J. Exp. Med. 210 (7), 1283–1299.

Mogensen, T.H., 2009. Pathogen recognition and inflammatory signaling in innate immune defenses. Clin. Microbiol. Rev. 22 (2), 240–273.

Morens, D.M., Fauci, A.S., 2007. The 1918 influenza pandemic: insights for the 21st Century. J. Infect. Dis. 195 (7), 1018–1028.

Moretta, L., Ferlazzo, G., Bottino, C., Vitale, M., Pende, D., Mingari, M.C., Moretta, A., 2006. Effector and regulatory events during natural killer-dendritic cell interactions. Immunol. Rev. 214, 219–228.

Moser, B., Wolf, M., Walz, A., Loetscher, P., 2004. Chemokines: multiple levels of leukocyte migration control. Trends Immunol. 25 (2), 75–84.

Muehlinghaus, G., Cigliano, L., Huehn, S., Peddinghaus, A., Leyendeckers, H., Hauser, A.E., Hiepe, F., Radbruch, A., Arce, S., Manz, R.A., 2005. Regulation of CXCR3 and CXCR4 expression during terminal differentiation of memory B cells into plasma cells. Blood 105 (10), 3965–3971.

Murray, M.J., Driscoll, K.E., 1992. Immunology of the respiratory system. In: Parent, R.A. (Ed.), Comparative Biology of the Normal Lung. CRC Press, Boca Raton, FL.

Nguyen, D.H., Hurtado-Ziola, N., Gagneux, P., Varki, A., 2006. Loss of siglec expression on T lymphocytes during human evolution. Proc. Natl. Acad. Sci. U. S. A. 103 (20), 7765–7770.

Obukhanych, T.V., Nussenzweig, M.C., 2006. T-independent type II immune responses generate memory B cells. J. Exp. Med. 203 (2), 305–310.

Olson, T.S., Ley, K., 2002. Chemokines and chemokine receptors in leukocyte trafficking. Am. J. Physiol. Regul. Integr. Comp. Physiol. 283 (1), R7–R28.

Ono, S.J., Nakamura, T., Miyazaki, D., Ohbayashi, M., Dawson, M., Toda, M., 2003. Chemokines: roles in leukocyte development, trafficking, and effector function. J. Allergy Clin. Immunol. 111 (6), 1185–1199.

Orange, J.S., 2013. Natural killer cell deficiency. J. Allergy Clin. Immunol. 132 (3), 515–525.

Orinska, Z., Bulanova, E., Budagian, V., Metz, M., Maurer, M., Bulfone-Paus, S., 2005. TLR3-induced activation of mast cells modulates CD8+ T-cell recruitment. Blood 106 (3), 978–987.

Pieper, K., Grimbacher, B., Eibel, H., 2013. B-cell biology and development. J. Allergy Clin. Immunol. 131 (4), 959–971.

Piet, B., de Bree, G.J., Smids-Dierdorp, B.S., van der Loos, C.M., Remmerswaal, E.B., von der Thusen, J.H., van Haarst, J.M., Eerenberg, J.P., ten Brinke, A., van der Bij, W., Timens, W., van Lier, R.A., Jonkers, R.E., 2011. CD8(+) T cells with an intraepithelial phenotype upregulate cytotoxic function upon influenza infection in human lung. J. Clin. Invest 121 (6), 2254–2263.

Raju, B., Hoshino, Y., Belitskaya-Lévy, I., Dawson, R., Ress, S., Gold, J.A., Condos, R., Pine, R., Brown, S., Nolan, A., Rom, W.N., Weiden, M.D., 2008. Gene expression profiles of bronchoalveolar cells in pulmonary TB. Tuberculosis 88 (1), 39–51.

Raphael, G.D., Baraniuk, J.N., Kaliner, M.A., 1991. How and why the nose runs. J. Allergy Clin. Immunol. 87 (2), 457–467.

Reynolds, H.Y., 1991a. Immunologic system in the respiratory tract. Physiol. Rev. 71 (4), 1117–1133.

Reynolds, H.Y., 1991b. Integrated host defense against infection. In: Crystal, R.G., West, J.B., Barnes, P.J., Chemiack, N.S., Weibel, E.R. (Eds.), The Lung: Scientific Foundations, vol. 2. Raven Press, New York.

Robinson, D.S., 2010. The role of the T cell in asthma. J. Allergy Clin. Immunol. 126 (6), 1081–1091.

Sallusto, F., Geginat, J., Lanzavecchia, A., 2004. Central memory and effector memory T cell subsets: function, generation, and maintenance. Annu. Rev. Immunol. 22, 745–763.

Sallusto, F., Lanzavecchia, A., 2009. Heterogeneity of CD4+ memory T cells: functional modules for tailored immunity. Eur. J. Immunol. 39 (8), 2076–2082.

Saluja, R., Metz, M., Maurer, M., 2012. Role and relevance of mast cells in fungal infections. Front. Immunol. 3, 146.

Sarawar, S.R., Sangster, M., Coffman, R.L., Doherty, P.C., 1994. Administration of anti-IFN-gamma antibody to beta 2-microglobulin-deficient mice delays influenza virus clearance but does not switch the response to a T helper cell 2 phenotype. J. Immunol. 153 (3), 1246–1253.

Scheller, J., Chalaris, A., Schmidt-Arras, D., Rose-John, S., 2011. The pro- and anti-inflammatory properties of the cytokine interleukin-6. Biochim. Biophys. Acta 1813 (5), 878–888.

Schmitz, N., Kurrer, M., Bachmann, M.F., Kopf, M., 2005. Interleukin-1 is responsible for acute lung immunopathology but increases survival of respiratory influenza virus infection. J. Virol. 79 (10), 6441–6448.

Schreiber, T., Ehlers, S., Heitmann, L., Rausch, A., Mages, J., Murray, P.J., Lang, R., Holscher, C., 2009. Autocrine IL-10 induces hallmarks of alternative activation in macrophages and suppresses antituberculosis effector mechanisms without compromising T cell immunity. J. Immunol. 183 (2), 1301–1312.

Scolletta, S., Colletti, M., Di Luigi, L., Crescioli, C., 2013. Vitamin D receptor agonists target CXCL10: new therapeutic tools for resolution of inflammation. Mediators Inflamm. 2013, 876319.

Seagal, J., Melamed, D., 2003. Selection events in directing B cell development. Histol. Histopathol. 18 (2), 519–527.

Selgrade, M.K., Starnes, D.M., Illing, J.W., Daniels, M.J., Graham, J.A., 1989. Effects of phosgene exposure on bacterial, viral, and neoplastic lung disease susceptibility in mice. Inhal. Toxicol. 1 (3), 243–259.

Sica, A., Mantovani, A., 2012. Macrophage plasticity and polarization: in vivo veritas. J. Clin. Invest 122 (3), 787–795.

Small, C.L., Shaler, C.R., McCormick, S., Jeyanathan, M., Damjanovic, D., Brown, E.G., Arck, P., Jordana, M., Kaushic, C., Ashkar, A.A., Xing, Z., 2010. Influenza infection leads to increased susceptibility to subsequent bacterial superinfection by impairing NK cell responses in the lung. J. Immunol. 184 (4), 2048–2056.

Smirnova, I., Poltorak, A., Chan, E.K., McBride, C., Beutler, B., 2000. Phylogenetic variation and polymorphism at the toll-like receptor 4 locus (TLR4). Genome Biol. 1 (1), 2.1–2.10.

Snipes, M.B., Boecker, B.B., McClellan, R.O., 1983. Retention of monodisperse or polydisperse aluminosilicate particles inhaled by dogs, rats, and mice. Toxicol. Appl. Pharmacol. 69 (3), 345–362.

Spellberg, B., Edwards, J.E., 2001. Type 1/Type 2 immunity in infectious diseases. Clin. Infect. Dis. 32 (1), 76–102.

St Clair, E.W., 2008. The calm after the cytokine storm: lessons from the TGN1412 trial. J. Clin. Invest 118 (4), 1344–1347.

Standiford, T.J., Mehrad, B., Tsai, W.C., Moore, T.A., 1999. Pulmonary host defenses: the role of cytokines in mediating lung inflammation [Online]. Medscape Pulm. Med. eJournal.

Stein, J.V., Nombela-Arrieta, C., 2005. Chemokine control of lymphocyte trafficking: a general overview. Immunology 116 (1), 1–12.

Stein, K.E., 1992. Thymus-independent and thymus-dependent responses to polysaccharide antigens. J. Infect. Dis. 165 (Suppl. 1), S49–S52.

Steinman, R.M., Hemmi, H., 2006. Dendritic cells: translating innate to adaptive immunity. Curr. Top. Microbiol. Immunol. 311, 17–58.

Strieter, R.M., Belperio, J.A., Keane, M.P., 2003. Host innate defenses in the lung: the role of cytokines. Curr. Opin. Infect. Dis. 16 (3), 193–198.

Sundstrom, J.B., Ellis, J.E., Hair, G.A., Kirshenbaum, A.S., Metcalfe, D.D., Yi, H., Cardona, A.C., Lindsay, M.K., Ansari, A.A., 2007. Human tissue mast cells are an inducible reservoir of persistent HIV infection. Blood 109 (12), 5293–5300.

Takahama, Y., 2006. Journey through the thymus: stromal guides for T-cell development and selection. Nat. Rev. Immunol. 6 (2), 127–135.

Tan, W.C., 2005. Viruses in asthma exacerbations. Curr. Opin. Pulm. Med. 11 (1), 21–26.

Thorburn, A.N., Hansbro, P.M., 2010. Harnessing regulatory T cells to suppress asthma. Am. J. Respir. Cell. Mol. Biol. 43 (5), 511–519.

Timmons, B.W., Cieslak, T., 2008. Human natural killer cell subsets and acute exercise: a brief review. Exerc Immunol. Rev. 14, 8–23.

Tisoncik, J.R., Korth, M.J., Simmons, C.P., Farrar, J., Martin, T.R., Katze, M.G., 2012. Into the eye of the cytokine storm. Microbiol. Mol. Biol. Rev. 76 (1), 16–32.

Toellner, K.M., Jenkinson, W.E., Taylor, D.R., Khan, M., Sze, D.M., Sansom, D.M., Vinuesa, C.G., MacLennan, I.C., 2002. Low-level hypermutation in T cell-independent germinal centers compared with high mutation rates associated with T cell-dependent germinal centers. J. Exp. Med. 195 (3), 383–389.

Toews, G.B., 2001. Cytokines and the lung. Eur. Respir. J. Suppl. 34 (3s–17s).

Trapani, J.A., Smyth, M.J., 2002. Functional significance of the perforin/granzyme cell death pathway. Nat. Rev. Immunol. 2 (10), 735–747.

Tsai, H.C., Velichko, S., Hung, L.Y., Wu, R., 2013. IL-17A and Th17 cells in lung inflammation: an update on the role of Th17 cell differentiation and IL-17R signaling in host defense against infection. Clin. Dev. Immunol. 2013 (267971).

Tsai, L.M., Yu, D., 2014. Follicular helper T-cell memory: establishing new frontiers during antibody response. Immunol. Cell. Biol. 92 (1), 57–63.

Urban, C.F., Reichard, U., Brinkmann, V., Zychlinsky, A., 2006. Neutrophil extracellular traps capture and kill *Candida albicans* yeast and hyphal forms. Cell. Microbiol. 8 (4), 668–676.

Vankayalapati, R., Klucar, P., Wizel, B., Weis, S.E., Samten, B., Safi, H., Shams, H., Barnes, P.F., 2004. NK cells regulate CD8+ T cell effector function in response to an intracellular pathogen. J. Immunol. 172 (1), 130–137.

Vermaelen, K., Pauwels, R., 2005. Pulmonary dendritic cells. Am. J. Respir. Crit. Care Med. 172 (5), 530–551.

Vock, C., Hauber, H.P., Wegmann, M., 2010. The other T helper cells in asthma pathogenesis. J. Allergy (Cairo) 180 (8), 1–14.

Weber, A., Wasiliew, P., Kracht, M., 2010. Interleukin-1 (IL-1) pathway. Sci. Signal. 3 (105), cm1.

Weibel, E.R., Taylor, C.R., 1988. Design and structure of the human lung. In: Fishman, A.P. (Ed.), Pulmonary Diseases and Disorders. McGraw-Hill, New York.

Werman, A., Werman-Venkert, R., White, R., Lee, J.K., Werman, B., Krelin, Y., Voronov, E., Dinarello, C.A., Apte, R.N., 2004. The precursor form of IL-1alpha is an intracrine proinflammatory activator of transcription. Proc. Natl. Acad. Sci. U. S. A. 101 (8), 2434–2439.

Widney, D.P., Xia, Y.R., Lusis, A.J., Smith, J.B., 2000. The murine chemokine CXCL11 (IFN-inducible T cell alpha chemoattractant) is an IFN-gamma- and lipopolysaccharide-inducible glucocorticoid-attenuated response gene expressed in lung and other tissues during endotoxemia. J. Immunol. 164 (12), 6322–6331.

Wille, A., Gessner, A., Lother, H., Lehmann-Grube, F., 1989. Mechanism of recovery from acute virus infection. VIII. Treatment of lymphocytic choriomeningitis virus-infected mice with anti-interferon-γ monoclonal antibody blocks generation of virus-specific cytotoxic T lymphocytes and virus elimination. Eur. J. Immunol. 19, 1283–1288.

Williams, M.A., Bevan, M.J., 2007. Effector and memory C.T.L. differentiation. Annu. Rev. Immunol. 25 (1), 171–192.

Wissinger, E.L., Saldana, J., Didierlaurent, A., Hussell, T., 2008. Manipulation of acute inflammatory lung disease. Mucosal Immunol. 1 (4), 265–278.

Wolf, A.J., Linas, B., Trevejo-Nunez, G.J., Kincaid, E., Tamura, T., Takatsu, K., Ernst, J.D., 2007. Mycobacterium tuberculosis infects dendritic cells with high frequency and impairs their function in vivo. J. Immunol. 179 (4), 2509–2519.

Wong, P., Pamer, E.G., 2003. CD8 T cell responses to infectious pathogens. Annu. Rev. Immunol. 21, 29–70.

Wright, A.K., Bangert, M., Gritzfeld, J.F., Ferreira, D.M., Jambo, K.C., Wright, A.D., Collins, A.M., Gordon, S.B., 2013. Experimental human pneumococcal carriage augments IL-17A-dependent T-cell defence of the lung. PLoS Pathog. 9 (3), e1003274.

Wu, L., Van Kaer, L., 2011. Natural killer T cells in health and disease. Front. Biosci. Schol. Ed. 3, 236–251.

Wynn, T.A., Chawla, A., Pollard, J.W., 2013. Macrophage biology in development, homeostasis and disease. Nature 496 (7446), 445–455.

Xu, X., Zhang, D., Lyubynska, N., Wolters, P.J., Killeen, N.P., Baluk, P., McDonald, D.M., Hawgood, S., Caughey, G.H., 2006. Mast cells protect mice from mycoplasma pneumonia. Am. J. Respir. Crit. Care Med. 173 (2), 219–225.

Yawata, M., Yawata, N., Draghi, M., Partheniou, F., Little, A.M., Parham, P., 2008. MHC class I-specific inhibitory receptors and their ligands structure diverse human NK-cell repertoires toward a balance of missing self-response. Blood 112 (6), 2369–2380.

Zielinski, C.E., Corti, D., Mele, F., Pinto, D., Lanzavecchia, A., Sallusto, F., 2011. Dissecting the human immunologic memory for pathogens. Immunol. Rev. 240 (1), 40–51.

Chapter 31

The Effect of Particle Deposition on Immunological Response as Measured by Cytokine Production

Madhuri Singal

Chapter Outline

1. THE RESPIRATORY SYSTEM AND PARTICULATE EXPOSURE

The cardiorespiratory system is a dynamic organ system in mammalian systems responsible for many functions beyond gas exchange, including fluid and electrolyte and acid—base balance. It serves as an important barrier against airborne pathogens and foreign material (Matthews, 1996). For the purposes of this text, however, the focus will remain on pulmonary defense mechanisms as specifically applied to inflammatory and oxidant injury responses in the lung following exposure to particulate irritants and/or allergens (in the form of particulate, airborne materials). The goal is to provide a basic understanding regarding how foreign particulate irritants, and potentially allergenic materials, can induce inflammation pathways leading to a sequelae of events including the possible causation of asthma and/or allergy in individuals with no prior history of either condition and, more likely, exacerbation of a preexisting asthma/allergy condition.

Particulate matter is defined as a complex mixture of anthropogenic and naturally occurring airborne particles (Oberdorster et al., 2000). In urban environments, airborne particles are primarily derived from combustion of fuels in the exhaust of motor vehicles and power plants (Oberdorster et al., 2000). Ultrafine and fine fractions of particulates result mainly from combustion processes and are composed of carbonaceous materials in addition to metals, sulfate, nitrate, and ammonium ions; coarse particulates arise from a variety of noncombustion sources and are comprised of mechanically generated particles consisting of insoluble minerals (Agency, 1996).

Exposure to inhaled particulate matter has been correlated with the development of inflammatory lung diseases leading to fibrotic tissue changes in the lung (Pryhuber et al., 2003). Studies have shown that exposure to particulate matter, in both human and mouse alveolar epithelial cells, in vitro, can induce expression of a variety of inflammatory cytokines (Barrett et al., 1998; Bonvallot et al., 2001; Aust et al., 2002; Adamson et al., 2004; Singal and

Comparative Biology of the Normal Lung. http://dx.doi.org/10.1016/B978-0-12-404577-4.00031-X

Finkelstein, 2005a). There have been great strides in elucidating the mechanism by which particulate matter and allergenic materials promote the transcription of proinflammatory genes in alveolar epithelial cells, as well as other cells related to, or in proximity of, the respiratory tract (Glaab et al., 2005; Keinan et al., 2005; Laumbach et al., 2005; Singal and Finkelstein, 2005a; Barnes, 2008; Boverhof et al., 2008; Ciprandi and Passalacqua, 2008; Lewis et al., 2008; Matsuda et al., 2008; Pantano et al., 2008; Passalacqua and Ciprandi, 2008; Siddiqui et al., 2008; Verstraelen et al., 2008; Desai and Brightling, 2009; Ferguson et al., 2009; Hammad et al., 2009; Prefontaine et al., 2009; Qian et al., 2009; Volckens et al., 2009; Woodruff et al., 2009; Bhattacharyya et al., 2010; Burgess et al., 2010; Cao et al., 2010; Corps et al., 2010; Lambrecht and Hammad, 2010; Olin et al., 2010; Raja et al., 2010; Trivedi and Caughey, 2010; Chu and Chiang, 2011; Chustz et al., 2011; Fuchimoto et al., 2011; Jiang et al., 2011; Kelada et al., 2011; Maillet et al., 2011; Parker and Prince, 2011; Plant et al., 2011; Poole et al., 2011; Qiao et al., 2011; Rahman and Fazal, 2011; Scott et al., 2011; Vandebriel et al., 2011; Wegmann, 2011; Bobic et al., 2012; Dai et al., 2012; Fitz et al., 2012; Kilsgard et al., 2012; McDonald et al., 2012; Royce et al., 2012; Scarino et al., 2012; Yamabayashi et al., 2012). In vivo studies exposing rodents to high concentrations of aerosolized particulate matter have shown significant gross histological changes in lung morphology (Pryhuber et al., 2003). In particular, the airways and lungs were found to have noticeable granulomatous lesions throughout the parenchyma and increased vascular swelling—characteristic definitions of inflammatory injury. Furthermore, these inflammatory changes correlated with changes in other physiological parameters in the animals (Oberdorster et al., 2000).

Carbon black (CB) is used as its effects tend to be benign in small doses and are most representative of ambient air pollution, urban ultrafine particles (Oberdorster et al., 2000). CB is generated from the combustion of gasoline and/or diesel fuel and is emitted from motor vehicles through exhaust fumes. Because urban ultrafine particles also have many other constituents, such as various transition metals (Hughes et al., 1998), particularly iron and iron oxides (Aust et al., 2002), fine and ultrafine carbon/iron (C/Fe) particles have also been produced for study purposes (Singal and Finkelstein, 2005b). The percentage of iron composition in the particles generated in the lab can vary according to the needs of a particular study and to determine the extent to which the iron and/or iron oxides may promote inflammatory effects. Silica (silicon dioxide) is a naturally occurring substance that comes in many different forms (Table 1). There are two main forms of silica: crystalline and amorphous silica. Crystalline silica can be found in nature as quartz, granite, sandstones, or in sand and/or soil; cristobalite and tridymite are forms of crystalline silica that are found in volcanic rocks and are formed from quartz exposed to temperatures exceeding 1400°C and from some amorphous silicas (e.g., diatomaceous earth). Quartz is the most abundant and is widely used in a variety of industries and occupations; respirable quartz levels greater than 0.1 mg/m^3 are found in metal, nonmetal, and coal mines/mills, granite quarrying/processing, stone crushing, and related industries (IARC, 1998). Cristobalites are particles generated in the production of glass and electrical components in the ceramics and computer industries, respectively. Amorphous silica is found in nature as a biogenic silica (diatomaceous earth) and as silica glass of volcanic origins (opal, flint) (IARC, 1998). Diatomaceous earth originates from the deposition of diatom skeletons on the sea floors and contains small amounts of cristobalite and quartz. After calcination, to increase the cristobalite content, diatomaceous earth is used as a filtration agent, carrier for pesticides, and a filler for paper (IARC, 1998). Significant occupational exposure has been reported for both amorphous and crystalline silica during the production and use of diatomaceous earth (IARC, 1998; Hessel et al., 2000). It has also been reported that fibers of amorphous silica are produced by a variety of plants (sugar cane and rice), and may be inhaled when released into the air during farming (IARC, 1998). Large quantities of synthetic amorphous silica are produced as pyrogenic, or fumed, silica, and are used frequently as a binding agent in paints and inks, as an anticaking agent in cosmetics, and in the production and reinforcement of fiberoptic materials (IARC, 1998). The National Aeronautics and Space Administration (NASA) has taken advantage of the unique properties of amorphous silica and produced the lightest solid on earth, Aerogel, a substance that was used to collect comet debris on the Stardust spacecraft, and is currently in use on the Mars Pathfinder mission for thermal insulation (NASA). For the multiple uses and methods of amorphous silica production, high exposure rates in the human population are likely to occur.

While cristobalite and amorphous silica are both silicon dioxide particles with similar origins, each has unique characteristics. Cristobalite is comprised of particles that are larger and coarser than amorphous silica, and, by consequence, cristobalite particles are insoluble in media. Furthermore, research in animals and humans has shown that cristobalite is capable of inducing a chronic inflammatory condition that precedes the development of fibrotic lung disease; cristobalite has known carcinogenic potential and is categorized by the International Agency for Research on Cancer (IARC) as a Group 1 agent (see the Appendix for IARC Group classifications) (IARC, 1998). Amorphous silica is an ultrafine/nanoparticle (nanoparticles are defined as a particle having at least one dimension <100 nm), with a diameter of approximately 12 nm, that is classified by the IARC as a Group 3 agent. Although not regarded as a primary occupational health hazard, as it is less

TABLE 1 Silica Types, Characteristics, and Carcinogenic Potential[a]

Type	Source/Formation	Uses	Carcinogenic Potential
Crystalline Silica			Class I agent: Sufficient evidence exists to classify crystalline silica is carcinogenic to humans
Cristobalite	Formed from quartz at high temperatures; found in volcanic rocks and liberated by quarrying	Calcination of amorphous silicas	
Quartz	Most common in nature (granite, sandstones, soils); liberated by quarrying/mining		
Tridymite	Found in volcanic rocks		
Amorphous silica			Class III agent: Insufficient evidence exist to classify amorphous silica as carcinogenic to humans
Pyrogenic (fumed) amorphous silica	Synthetically formed during metallurgical processes and is released by plants (sugar cane, rice) in farming	Elastomer reinforcement, thickening resins and paints,and as an anticaking agent in foods and toothpastes	
Precipitated silica; silica gel	Synthetically formed during wet processes	Elastomer reinforcement, thickening resins and paints,and as an anticaking agent in foods and toothpastes	
Diatomaceous earth (uncalcined)	Formed from diatom skeletons on the sea floor		
Flux-calcined diatomaceous earth	Calcination increases cristobalite content	Filtration agent, carrier for pesticides, filler in paints and paper, and as an abrasive agent in industries	

[a]*Notter et al., 2005. Lung Injury: Mechanisms, Pathophysiology, and Therapy, vol. 196.*

biopersistent than cristobalite and has not been shown to cause silicosis, in vitro exposure to high doses of amorphous silica appear to induce significant cell death (Johnston et al., 2000); in vivo exposures in rats resulted in acute inflammatory responses that subsided immediately following the removal of the rats from the source of aerosolized amorphous silica (Johnston et al., 2000). Aerosil200, a commercially available, hydrophilic form of amorphous silica, is mass produced for the manufacture of tires (rubber amalgamation), binding agent for paints and printing inks, anticaking agent in cosmetics and antiperspirants, and the construction of fiberoptic materials and reinforcing elastomers. Aerosil200 particles have no defined crystal structure based on X-ray diffraction analysis and have a mean diameter of less than 12 nm and surface area of $200 \pm 25\ m^2/g$ particles. Because of its size and general molecular characteristics Aerosil200 is highly soluble in water, and it is this characteristic that makes Aerosil200 particles unique in comparison to crystalline types of silica, such as cristobalite, that have been previously studied (Raghavan and Khan, 1997; Raghavan et al., 2000).

It had been previously established that a murine alveolar Type II epithelial cell line, MLE15, when exposed to crystalline silica in vitro, could be induced to express proinflammatory cytokine proteins and mRNAs (Barrett et al., 1998; Barrett et al., 1999b). It has also been shown that cristobalite-induced MIP-2 protein expression is dependent on TNF-α-induced oxidant production (Barrett et al., 1998; Barrett et al., 1999a). It was next established by what mechanism silica initiated the inflammatory signaling pathway that resulted in MIP-2 protein production (Singal and Finkelstein, 2005a).

Last, but not least, asbestos has been a well-studied particulate that has been the cause of a pulmonary disease known as asbestosis. Asbestos is a fibrous-type silicate particle that was once commonly used to create fire retardant building and electrical insulation. Asbestos particles embed within the cellular architecture of the respiratory tract and cause a chronic inflammatory response that leads to a well-characterized fibrotic lung disease (Demers et al., 1990; Choe et al., 1998; Dai et al., 1998; Mossman and Churg, 1998; Geist et al., 2000; Lasky et al., 2001;Iwagaki et al., 2003). The widespread use/production of different types of particles in the environment has caused concern for both occupational and consumer exposure.

2. STRUCTURE AND FUNCTION OF NORMAL RESPIRATORY DEFENSES

Normal respiratory defenses are comprised of the nose, mucociliary clearance, high-velocity clearance and reflex mechanisms (sneezing, coughing, and sighing), nonciliated airway secretions, blood, and lymph clearance, and, most importantly, immunological defenses including protective mechanisms involving responses to oxidant injury (Matthews, 1996). In order to describe the role of each component of normal respiratory defense, it is important to consider them in the context of particulate matter exposure.

Under normal circumstances, inhaled particulate materials are first intercepted by the vestibular hairs of the nose, which function as a filter to trap large particles. This causes turbulence that allows for a directional change in the flow of air such that any particles that follow are prevented from entering the lung by inertial impaction (Matthews, 1996; Egan et al., 1999; Salem and Katz, 2006). Nasal breathing provides the best defense against airborne pathogens and particulate material, while oropharyngeal breathing, which allows for the inhalation of much larger, and potentially more hazardous, materials, does not provide the same degree of protection.

In particular, particle deposition in the airways can occur by different mechanisms throughout the upper airway (nose, mouth, pharynx) and is highly dependent upon the nature of the flow of air inspired and the physical characteristics of the particles inhaled (Matthews, 1996; Egan et al., 1999; Salem and Katz, 2006). Particles can settle in the airway under the influence of gravity, which is determined by the size and density of the particles inspired, known as gravitational settling. Particles greater than 30 μm in diameter encounter inertial impaction in the nasal passages; particles less than 30 μm in diameter, interface with the movement of gas molecules and travel to the lower airways (Matthews, 1996; Egan et al., 1999; Salem and Katz, 2006). When this occurs, particle size becomes the primary determinant of entry into the lower airways. Additionally, the mucosal lining of the respiratory tract, prior to entry into the lower airways, can also serve as a preventive barrier to particles of 30 μm in diameter or less (Matthews, 1996; Egan et al., 1999; Salem and Katz, 2006) as well as the mucociliary escalator that facilitates the fluid upward movement of trapped debris so that the material can be expectorated through high-velocity clearance mechanisms. Therefore, the hierarchy of particle deposition in the airways is as follows: ≥ 30 μm deposit in the mouth, larynx, and trachea; 3−10 μm deposit on the mucus layer of the bronchial tree; particles ≥ 3 μm can reach the alveoli (Matthews, 1996; Egan et al., 1999; Salem and Katz, 2006). The types of material that can reach this depth in the airways include viruses (i.e., respiratory syncytial virus, etc.), endotoxins, gas mixtures (i.e., ozone), and ultrafine particulate matter (Driscoll et al., 1996; Elder et al., 2000; Johnston et al., 2001; Zhang et al., 2001).

Unlike the ciliated airways, clearance of foreign material from the peripheral lung, the alveolar surface, is slow due to the depth of deposition as well as the absence of cilia. The alveolar unit is comprised of four different cell types: Type I epithelial cells, a squamous cell that forms the majority of the alveolus and allows for efficient exchange of gases across the alveolar–capillary membrane; Type II epithelial cells, the progenitor cell type for the development of Type I epithelial cells, the predominant cell type in the lung, and are responsible for the production of surfactant, a material that decreases alveolar surface tension and allows for diffusion of gases; alveolar macrophages, the resident phagocytic cell type of the lung; and endothelial cells that form the capillaries surrounding the alveolus, forming the functional alveolar–capillary unit (Matthews, 1996; Egan et al., 1999; Salem and Katz, 2006).

As important as the understanding of the behavior of the distinct cell types found throughout the respiratory tract, it is equally imperative to correlate dosimetrics, actual particulate deposition kinetics, with the effects observed. The size of the particles inhaled will dictate how far into the respiratory tract the particles will travel and where the interaction between particle and tissue surface may occur (Conolly et al., 2000, 2002, 2004; Hofmann et al., 2000; Asgharian, 2001; Kelly et al., 2001 a and b; Overton et al., 2001; Cassee et al., 2002; Frederick et al., 2002; Georgieva et al., 2003; Asgharian et al., 2004, 2006, 2011; Asgharian and Price, 2006; Schroeter et al., 2006; Carey et al., 2007; Kimbell et al., 2007; Garcia and Kimbell, 2009; Garcia et al., 2009; Asgharian et al., 2012 a and b; Schroeter et al., 2012).

3. INFLAMMATION IN THE LUNG

The primary mechanism of clearance, at the alveolar surface, is via the activity of the alveolar macrophages (Forman et al., 1980, 1982; Hartmann et al., 1984; Stringer et al., 1995; Bates and Fisher, 1996; Kaul and Forman, 1996; Yagisawa et al., 1996; Tang et al., 2001; Iles et al., 2002; Jimenez et al., 2002; Quintero and Wright, 2002; Tao and Kobzik, 2002; Chin et al., 2003; Hjort et al., 2003; Laskin et al., 2003; Fakhrzadeh et al., 2004a,b; Rao et al., 2004; Kanj et al., 2005, 2006; Rothen-Rutishauser et al., 2005; Hu et al., 2006, 2007; Sunil et al., 2007; Baker et al., 2008; Gaschler et al., 2008; Geiser et al., 2008; Manzer et al., 2008; Naito et al., 2008; Yamasaki et al., 2008; Jordan et al., 2009; Kuronuma et al., 2009; Pauluhn, 2009; Ryman-Rasmussen et al., 2009; Woodruff et al., 2009; Hoogerwerf et al., 2010; Snyder et al., 2010; Fuchimoto et al., 2011; Hristova et al., 2011; Poole et al., 2011; Groves et al., 2012; Johnston et al., 2012), a component of the mononuclear phagocyte family of the immune system (Matthews, 1996; Zhang et al., 2001; Poynter et al., 2003; Janeway, 2005). A significant number of resident tissue macrophages make up the alveolar walls and actively phagocytize entrapped particles, often digesting the particles and releasing the remaining resultant material, free or encapsulated, into the lymphatic system or, most often, moving out of the airways via mucociliary escalator (Matthews, 1996; Egan et al., 1999; Zhang et al., 2001; Salem and Katz, 2006). It is through absorption into the lymphatic and circulatory system that many large molecule materials, which cannot move across the epithelium to be inactivated, must be eliminated (Matthews, 1996; Egan et al., 1999; Salem and Katz, 2006). Immunological defenses begin the moment the lymphatic and circulatory system become involved in the destruction of foreign inhaled materials and are dependent on the release of chemical mediators, cytokines and chemokines, from alveolar macrophages and alveolar epithelial cells (Matthews, 1996; Egan et al., 1999; Salem and Katz, 2006). Aside from physical debris that can promote an inflammatory response in the lung, exposure to noxious agents, such as ozone, can also be responsible for inflammatory damage, particularly through the generation of reactive oxygen species (ROS) (Prokhorova et al., 1998; Johnston et al., 2001; Deshpande et al., 2002; Baulig et al., 2003). In recent years, the lymph nodes in close proximity to the respiratory tract have been suggested as an important source of information for assessing the potential allergenicity of materials by inhalation (Arts et al., 1998, 2006, 2008; Arts and Frieke Kuper, 2003; Boverhof et al., 2008; Adenuga et al., 2012). It is also suggested that materials inducing an allergenic inflammatory response through dermal exposure may result in elicitation of chemokine/cytokine patterns leading to a respiratory allergic response following inhalation exposure (Arts et al., 1998, 2006, 2008; Plitnick, 2002a,b; Arts and Frieke Kuper, 2003; Holsapple et al., 2006; Selgrade et al., 2006; Farraj et al., 2007; Boverhof et al., 2008; Adenuga et al., 2012). At this time, there is no consensus on the matter (Selgrade et al., 2006; Farraj et al., 2007).

Chemokines function as a signaling mechanism to recruit circulating leukocytes, mononuclear cells, and lymphocytes into the lung; chemokines mediate immune cell chemotaxis into the airways to eliminate pathogenic and/or particulate materials (Matthews, 1996; Zhang et al., 2001; Laskin et al., 2007). Chemokines exist as a family of proteins divided into four functionally distinct groups of basic (heparin-binding) proteins (Table 2): C, CC (i.e., MIP-1α and β, MCP-1 and RANTES), CXC (i.e., IL-8, IP-10 and MIP-2), and CX3C (i.e., Fractalkine) (Zhang et al., 2001). The nomenclature is based on the number and spacing of conserved NH_2-terminal cysteine moieties; CX3C has been so named as it is composed of large, membrane-bound glycoproteins connected through a COOH-mucin-like domain (Zhang et al., 2001). Receptors exist for those chemokines that function with high specificity in a cell-type restricted manner: C-chemokines activate lymphocyte chemotaxis; CC-chemokines induce monocytes, lymphocytes, and eosinophil chemotaxis; and CXC-chemokines initiate neutrophil chemotaxis (Zhang et al., 2001).

The stimulated production of various chemokines in inflamed tissues results in inflammatory cell recruitment and initiation of a cascade of signaling events that can

TABLE 2 Human C, CC, and CXC Chemokine Families of Chemotactic Cytokines

The C chemokines	
XCL1	Lymphotactin
XCL2	SCM-1β
The CC chemokines	
CCL1	I-309
CCL2	Monocyte chemotactic protein-1 (MCP-1)
CCL3	Macrophage inflammatory protein-1 alpha (MIP-1α)
CCL4	Macrophage inflammatory protein-1 beta (MIP-1β)
CCL5	Regulated on activation normal T-cell expressed and secreted (RANTES)
CCL7	Monocyte chemotactic protein-3 (MCP-3)
CCL8	Monocyte chemotactic protein-2 (MCP-2)
CCL9	Macrophage inflammatory protein-1 delta (MIP-1δ)
CCL11	Eotaxin
CCL13	Monocyte chemotactic protein-4
CCL14	HCC-1
CCL15	HCC-2
CCL16	HCC-4
CCL17	Thymus and activation-regulated chemokine (TARC)
CCL18	DC-CK-1
CCL19	Macrophage inflammatory protein-3 beta (MIP-3β)
CCL20	Macrophage inflammatory protein-3 alpha (MIP-3α)
CCL21	6Ckine
CCL22	MDC
CCL23	MPIF-1
CCL24	MPIF-2
CCL25	TECK
CCL26	Eotaxin-3
CCL27	CTACK
The CXC chemokines	
CXCL1	Growth-related oncogene alpha (GRO-α)
CXCL2	Growth-related oncogene beta (GRO-β)
CXCL3	Growth-related oncogene gamma (GRO-γ)
CXCL4	Platelet factor-4 (PF4)
CXCL5	Epithelial neutrophil activating protein-78 (ENA-78)
CXCL6	Granulocyte chemotactic protein-2 (GCP-2)
CXCL7	Neutrophil activating protein-2 (NAP-2)
CXCL8	Interleukin-8 (IL-8)
CXCL9	Monokine induced by interferon-γ (MIG)

TABLE 2 Human C, CC, and CXC Chemokine Families of Chemotactic Cytokines—cont'd

CXCL10	Interferon-γ inducible protein (IP-10)
CXCL11	Interferon inducible T-cell alpha chemoattractant (ITAC)
CXCL12	Stromal cell-derived factor-1 (SDF-1)
CXCL13	B-cell-attracting chemokine-1 (BCA-1)

[a] *Notter et al., 2005. Lung Injury: Mechanisms, Pathophysiology, and Therapy, vol. 196.*

culminate in dysregulated tissue repair leading to tissue fibrosis. Cellular edema and increased cellular permeability progressively decrease the efficient exchange of gases across the alveolar–capillary membrane. Fibrosis can occur following massive accumulation of fibroblasts that deposit collagen to repair damaged tissue at the site of inflammation. The combination of cellular swelling, permeability, and collagen deposition causes the alveolar surfaces of the lung to become thickened and stiff and restrict the ventilatory/respiratory capacity of the lung by decreasing lung compliance and decreasing gas exchange capabilities (Mossman and Churg, 1998; Faffe et al., 2001).

4. PARTICLE DEPOSITION KINETICS

As described earlier, deposition of particulate matter in the lung is dependent on not only the geometry of the airways but also the physicochemical properties of the material inhaled. Properties affecting deposition include surface area and electrostatic charge (Alarie et al., 1995; Kimber and Dearman, 2005; Long et al., 2005; Card et al., 2008; Hess, 2008; Pietruska et al., 2011). An excellent example of deposition kinetics and dosimetry patterns for particulates can be found through the efforts in the pharmaceutical industry and the development of nebulizer devices as well as metered dose inhalers for respiratory medication delivery (Hess, 2008). To that end, in silico models have been developed by which to assess deposition characteristics and correlate the predicted effects (based on cell types in location of significant deposition) to actual in vivo observed data.

The deposition of particulates has been studied for both the upper and lower respiratory tract of humans and rats (Morris, 1990; Frederick et al., 2002; Heck et al., 2002; Asgharian and Price, 2006; Kimbell et al., 2007; Morris and Hubbs, 2009; Gloede et al., 2011). Coupled human–rat deposition modeling allows for interspecies data extrapolation using laboratory measurements to assess human health risk from exposure to inhaled materials.

There are two main approaches for conducting this analysis. The first approach is site-specific modeling for which the precise geometry of the region of interest as well as information on airflow and compound transport at the boundaries (inlet and outlets) of the geometry are known. This approach is most useful at the entry port to the respiratory tract (i.e., nasal and oral airways) for which airway passages are reconstructed from scanned images of airway coronal sections and combined with breathing information to form anatomically and physiologically realistic geometric models.

The second approach is whole-lung modeling in which airways are described dimensionally but precise detail configuration (e.g., curvature and other artifacts) is neglected. By simple description of the geometry, the entire respiratory tract can be studied. The whole-lung modeling approach is ideal when considering deposition in the entire lung.

Deposition of inhaled materials in the nasal passages is needed not only to determine the amount entering the lung but also to assess nasal tissue dose since many inhaled compounds have been shown to elicit effects in rodent nasal passages (Kimbell et al., 2007). Anatomically accurate three-dimensional reconstructions of the rat and human nasal passages have been constructed for use in airflow and mass transfer studies (Kimbell et al., 2001).

Methods are currently in place to compute mass transfer of inhaled material in the nasal passages of rats and humans. Deposition of inhaled particles assumes a perfect particle absorption condition on nasal airway walls where deposition is a function of diffusion and inertial impaction for ultrafine and coarse particles, respectively.

Lung airway dimensions span over multiple scales ranging from centimeters to micrometers. Lung reconstruction from scanned images can be made only for the first few airway generations of the lung conducting airways. Outlet boundary conditions that incorporate lung compliance are constructed based on pressure information in the pleural cavity. Pressure in the pleural cavity is gravity dependent and varies vertically. In addition, pleural pressure varies with the time into the breathing cycle and breathing frequency, adding complications to detailed computations of lung ventilation. Hence, fluid dynamics analysis (site-specific modeling) in the deep lung is severely limited by the lack of detailed information on lung geometry and pressure information.

To facilitate particle deposition computations in the lung, lung geometry is assumed to be a collection of cylindrically shaped tubes arranged as a dichotomous branching network. Airway dimensions and orientations are calculated from available measurements. Additional volumes are added to respiratory bronchial airways in the acinar region to account for the alveolar space. Mathematical models of airflow and particle transport in the model geometry are developed and solved to calculate particle deposition.

The most popular lung geometry for humans is the typical-path lung structure obtained from limited airway measurements in the lung (Weibel, 1973). The lung geometry is assumed to consist of 23 airway generations, of which the first 16 generations comprise the tracheobronchial region (conducting airways), and the last 7 generations form the alveolar region. Using established measurements (Raabe et al., 1976), a similar typical path was constructed. In addition, a five-lobe lung model, with each lobe being symmetric but different in dimensions, was developed (Yeh and Schum, 1980). This human geometric model was further expanded to create stochastic lung geometries that were asymmetric and closely resembled the anatomical structure of the lung (Koblinger and Hofmann, 1985). However, airway shapes were still cylindrical and deviated from reality. In rats, an asymmetric lung geometry was created (Anjilvel and Asgharian, 1995) based on previously established measurements (Raabe et al., 1976).

A number of deposition models are available in the literature for the calculations of particle deposition. Early deposition models were compartmental and limited in predictive capability. More realistic deposition models assumed the lung to expand and contract uniformly and resembled a trumpet in shape when symmetric lung geometry was assumed (Yu, 1978). These models have been validated for regional deposition fractions. The human lung has an asymmetric branching structure that affects both airflow and particle deposition. A multiple-path analogy in 10 stochastically generated lungs of human adults to calculate site-specific and regional deposition of particles is now used (Asgharian, 2001).

Deposition models based on asymmetric lung geometries offer an advantage over previous models in that lung physiology and morphometric variability within each generation are accounted for. These two parameters are shown to be most critical in accurate microdosimetry predictions of particles in the lung (Asgharian, 2001; Subramaniam et al., 2003; Asgharian et al., 2004).

Modeling efforts have focused on two components—the upper and lower respiratory tracts—as deposition and potential absorption into airway tissues are likely to occur in both sites. In addition, particle deposition in the nasal passages is needed to determine the amount that enters the lung.

As has been noted, the physicochemical properties of a material will be instrumental in determining deposition and consequent effects in the respiratory tract. Different approaches are needed in the nose and lung depending on the vapor pressure of the material inhaled in a particulate form. For example, as opposed to a solid particle, medium vapor pressure liquids must be assessed based on the mixed particle/vapor phase in which it exists.

Deposition and dosimetry is important to defining the total lung burden of particles, or liquids delivered in a particulate manner via aerosol, as it is essential to assessing the inflammatory effects that occur in the lungs following exposure (Morris, 1990; Donaldson et al., 2008).

5. PARTICULATE-INDUCED INFLAMMATORY INJURY IN THE LUNG

Particulate matter, such as crystalline silica, amorphous silica, titanium dioxide, residual oil fly ash, laboratory-generated CB or carbon/iron particles, and those generated by cigarette smoking, have not only been shown to increase the production of cytokines/chemokines by alveolar macrophages but the data also suggest that an increase in ROS may contribute to the upregulation of cytokine/chemokine production (Oberdorster et al., 2000). The use of various types of particulate matter in the laboratory has served as a relatively reliable, and reproducible, model of inflammatory pathways in the lungs (Singal and Finkelstein, 2005 a and b; Laskin et al., 2007; Donaldson et al., 2008). Additionally, there was increasing evidence that Type II cells are a potential mediator of pulmonary inflammatory cell recruitment and activation via the release of specific chemokines (Holley et al., 1992; Johnston et al., 2000; van der Vliet and Cross, 2000; Li et al., 2002). The focus on Type II epithelial cells arises due to their unique location at the alveolar–capillary membrane. At this location, exposure to airborne materials and epithelial cell interaction with a variety of different cell types such as endothelial cells, alveolar macrophages, and fibroblasts, takes place (Crystal and West, 1991).

Various inflammatory mediators are expressed in the lung following particulate exposure including macrophage inflammatory protein (MIP)-2, monocyte chemoattractant protein (MCP)-1, RANTES, interferon-γ-inducible protein (IP)-10, and tumor necrosis factor (TNF)-α. The transcription of many of these mediators is regulated by the activity of specific proinflammatory transcription factors, nuclear factor (NF)-κB and Activator Protein (AP)-1, an alternative to the NF-κB pathway in the upregulation of proinflammatory cytokine transcription via the MAPK (mitogen-activated protein kinase) pathway (Ye et al., 2001; Jimenez et al., 2002; Li et al., 2002). A study done by Driscoll et al. (1996) demonstrated that the nuclear translocation of the proinflammatory transcription factor,

NF-κB, is increased in rat lung epithelial cells upon exposure to α-quartz silica in vitro (Driscoll et al., 1996). This activity is effectively abrogated by the addition of extracellular antioxidants, suggesting that silica exposure resulted in NF-κB activation in an oxidant-dependent manner (Singal and Finkelstein, 2005a). However, it has also been shown that many of the mediators expressed by alveolar epithelial cells in response to particulate exposure contain DNA binding sequences for both transcription factors, NF-κB and AP-1 (Kim et al., 2003; Zampetaki et al., 2004; Sohn et al., 2005).

6. REACTIVE OXYGEN AND NITROGEN-DEPENDENT MECHANISMS OF PULMONARY INJURY/INFLAMMATION

The alveolar epithelium is constantly in the presence of reactive oxygen and nitrogen species (ROS/RNS) from endogenous and exogenous sources (Table 3). It is exposure to these reactive intermediates that can potentially cause damage to the alveolar epithelium, specifically affecting the surfactant system and the cell membrane such that increased protein leakage can occur, resulting in pulmonary atelectasis (alveolar collapse) and hypoxemia/hypoxia (Halliwell and Gutteridge, 1999; Notter et al., 2005). Many particles may have reactive surfaces charges or induce an ROS/RNS-mediated inflammatory pathway.

ROS are generated as a result of the mitochondrial electron transport system and during the microsomal metabolism of endogenous cytoplasmic compounds and xenobiotics (drugs and environmental pollutants). During an innate inflammatory response to injury, neutrophils, alveolar macrophages, and other inflammatory cells can generate and release ROS via NADPH (nicotinamide adenine dinucleotide phosphate) oxidase (Cross et al., 1994; Halliwell and Gutteridge, 1999; Notter et al., 2005). Superoxide anion radical ($O_2^{\bullet -}$) and H_2O_2 are long-lived compounds and both can enter cells; ($O_2^{\bullet -}$) enters via anion channels, while H_2O_2 can pass through cell membranes via simple diffusion (Halliwell and Gutteridge, 1999; van der Vliet and Cross, 2000; Notter et al., 2005). However, the most reactive metabolite of $O_2^{\bullet -}$ that is generated is the hydroxyl radical (•OH); $O_2^{\bullet -}$ can directly reduce H_2O_2 to produce •OH (along with molecular O_2 and hydroxide ion (OH^-)), or, conversely, $O_2^{\bullet -}$ can reduce redox active metals (such as iron (Fe^{3+})) to form molecular O_2. The reduced form of the metal can now react with H_2O_2 to regenerate an oxidized metal ion with coproduction of OH^- and •OH (Haber–Weiss and Fenton reactions) (Halliwell and Gutteridge, 1999).

Several factors can induce the production of ROS in acute and chronic lung diseases. High oxygen therapy used to treat hypoxemia in patients with cardiopulmonary disease, can also result in the induction of proinflammatory cytokines and chemokines expressed by cells in response to

TABLE 3 Reactive Oxygen and Reactive Nitrogen Species[a]

Name	Formula	Comments/Process of generation
Hydrogen atom	$H^\bullet$	Simplest free radical
Trichloromethyl	$CCl_3^\bullet$	A carbon-centered radical formed during metabolism of CCl_4 in the liver and contributes to its toxic effects. Carbon radicals react rapidly with O_2 to make peroxyl radicals, e.g., $CCl_3^\bullet + O_2 \rightarrow CCl_3O_2^\bullet$.
Superoxide	$O_2^{\bullet -}$	An oxygen-centered radical that is less reactive than hydroxyl radicals
Hydroxyl	$OH^\bullet$	A highly reactive oxygen-centered radical; attacks all biomolecules
Thiyl/perthiyl	$RS^\bullet/RSS^\bullet$	A group of radicals that have unpaired electrons residing on sulfur; formed by the oxidation of sulfur compounds in the presence of transition metal ions
Peroxyl, alkoxyl	$RO_2^\bullet$, $RO^\bullet$	Oxygen-centered radicals formed during the breakdown of organic peroxides and reaction of carbon radicals with $O_2(RO_2^\bullet)$
Oxides of nitrogen	$NO^\bullet$, $NO_2^\bullet$	Nitric oxide is formed in vivo from the amino acid L-arginine; nitrogen dioxide is made when $NO^\bullet$ reacts with O_2; both are found in polluted air and smoke from burning organic materials, e.g., cigarette smoke.
Nitrogen-centered radicals	$C_6H_5N{=}N^\bullet$	Formed during oxidation of phenylhydrazine by erythrocytes, e.g., phenyldiazine radical
Transition-metal ions	Fe, Cu, etc.	Ability to change oxidation numbers by one allows transition metals to accept/donate single electrons; hence they are often powerful catalysts of free radical reactions

[a]*Halliwell, B. and Gutteridge, J.M.C., 1999. Free Radicals in Biology and Medicine, third ed.*

cellular injury, triggering the influx of neutrophils (Allen et al., 2000; Hjort et al., 2003; Notter et al., 2005).

This initiates a cascade of events that is propagated through a receptor-mediated activation signaling pathway through protein kinase (PK)-C and phospholipase C, leading to the translocation of the cytosolic components of the NADPH oxidase complex (gp^{40phox}, gp^{47phox}, gp^{67phox}, and the Rho-family GTPase, Rac2) to the membrane-bound complex (gp^{91phox}/gp^{22phox}/Rap1a) that carries cytochrome *c* (Notter et al., 2005). The membrane-bound NADPH oxidase complex is responsible for the production of $O_2^{\bullet -}$. Finally, ischemia, due to low perfusion, decreased arterial oxygen tension, or trauma, can induce xanthine dehydrogenase to secure oxidase activity by utilizing xanthine and molecular oxygen to create a partially reduced oxygen species (Notter et al., 2005).

RNS are a product of nitric oxide (•NO)-induced formation of other reactive oxygen–nitrogen intermediates (Cross, van der Vliet et al., 1994; Halliwell and Gutteridge, 1999; van der Vliet and Cross, 2000; Notter et al., 2005). Lung tissues can be exposed to •NO through the inhalation of increasing concentration of polluted air, the overproduction of •NO by alveolar macrophages, pulmonary epithelial, interstitial, and/or endothelial cells; •NO and other RNS have been blamed for many inflammatory lung diseases, including acute lung injury and the acute respiratory distress syndrome (van der Vliet and Cross, 2000; Notter et al., 2005). More specifically, •NO is generated by nitric oxide synthase (NOS), which involves a 5-electron oxidation of the guanidine nitrogen of L-arginine. Molecular O_2 and NADPH are cosubstrates where flavin nucleotides (FMN and FAD) and thiols act as enzyme cofactors (Halliwell and Gutteridge, 1999). •NO is an important mediator of many physiological processes in different cell types including neurons, smooth and skeletal muscle cells, hepatocytes, neutrophils, macrophages, and epithelial cells (Halliwell and Gutteridge, 1999; Notter et al., 2005). Pathologically, the toxic effects of •NO are the result of its reaction with $O_2^{\bullet -}$ producing peroxynitrite ($ONOO^-$), which functions as a strong oxidizing and nitrating agent.

In particular, with respect to nitric oxide synthases, there are three distinct classes: neuronal (nNOS, isoform 1), inducible (iNOS, isoform II), and endothelial (eNOS, isoform III) (Notter et al., 2005). However, as RNS and NOS are not the primary focus, it is important only to conclude this discussion by noting certain details. First, gene transcription of iNOS (Nos2) can be regulated by PKC, and the activity of iNOS is induced in a similar manner to the NADPH oxidase complex by proinflammatory stimuli; both pathways are believed to operate simultaneously to generate reactive intermediates during an inflammatory response (Notter et al., 2005). Second, the main mechanism of RNS cytotoxicity is the ability to inhibit eukaryotic gene expression. This can occur through various means: (1) •NO and $ONOO^-$ can induce nucleotide deamination, causing abasic sites and subsequent DNA strand breaks; (2) Alveolar macrophages can produce enough •NO to hinder the activity of ribonucleotide reductase, the enzyme required for ribonucleotide conversion to deoxyribonucleotides for DNA synthesis; (3) $ONOO^-$ and high concentrations of •NO can inactivate mitochondrial enzymes, such as cytochrome c oxidase, which results in the disruption of the electrochemical proton-motive force, reduced ATP production, and consequently, reduced protein synthesis; and (4) •NO can, positively or negatively, regulate the activity of redox-sensitive transcription factors such as NF-κB and AP-1 components (Notter et al., 2005).

The normal lung is well protected from toxic concentrations of ROS/RNS by various antioxidant systems. There are three forms of superoxide dismutase (SOD), an enzyme that catalyzes the dismutation of $O_2^{\bullet -}$ to molecular O_2 and H_2O_2: CuZnSOD (cytoplasm and peroxisomes), MnSOD (mitochondria), and ECSOD (extracellular form identified in the lung matrix). Additionally, peroxisomes carry catalase, which can degrade H_2O_2 (Cross et al., 1994; Halliwell and Gutteridge, 1999; van der Vliet and Cross, 2000; Notter et al., 2005). Lung tissue also contains high amounts of nonenzymatic antioxidants such as Vitamin E, reduced glutathione, and ascorbate (Cross et al., 1994; Halliwell and Gutteridge, 1999; Notter et al., 2005). Normally, in eukaryotic cells, intracellular $O_2^{\bullet -}$ concentrations are low (<10 pM), and formation of $ONOO^-$ is prevented by large concentrations of SOD (4–10 μM) (Halliwell and Gutteridge, 1999; Notter et al., 2005). In the case of silica and asbestos exposures, SOD and MnSOD are particularly important in mediating an inflammatory response in the lungs (Holley et al., 1992; Fubini and Hubbard, 2003; Xiao et al., 2005).

In addition to the enzymatic antioxidant defense mechanisms present in the lung, the roles of PKC and NADPH oxidase are pivotal for the involvement of the heme oxygenase system (Mossman and Churg, 1998; Halliwell and Gutteridge, 1999; Notter et al., 2005). Particularly, heme oxygenase-1 (HO-1), also known as heat shock protein-32 (hsp32), is an enzyme that catalyzes the conversion of heme proteins to bilirubin (Halliwell and Gutteridge, 1999). Byproducts of this process are carbon monoxide (CO) and iron. Increases in HO-1 protein abundance in the lungs of animals exposed to asbestos or crystalline silica have been correlated with the pathogenesis of asbestos- or silica-induced pulmonary fibrosis (Mossman and Churg, 1998).

HO-1 protein transcription is regulated by the transcription factor Nrf2, a member of the nuclear factor-2 derived transcription factor family. Nrf2 is sequestered in the cytosol by an inhibitory protein, Keap1 (Kelch-like associated protein-1), which releases Nrf2 under conditions of oxidative stress. Nrf2, once free of Keap1, translocates to the nucleus to promote the transcription of Nrf2 responsive genes. HO-1 has been suggested as both cytoprotective and proinflammatory (Mossman and Churg, 1998; Halliwell and Gutteridge, 1999).

7. PULMONARY TISSUE REPAIR MECHANISMS

In the previous sections, the data suggest that the proinflammatory, potentially deleterious, effects of particle exposure may also serve in a protective function by initiating a signaling cascade that may induce minor cellular damage that, when in the process of regulated tissue repair, prevents greater structural damage to the organ. Conversely, when repair mechanisms fail, dysregulated repair functions will cause greater likelihood of disease development (Finkelstein et al., 1997). An example of regulated repair would be an interruption of the cell cycle such that cellular replication is halted to promote repair (O'Reilly et al., 1998, 2001, 2003; Koike et al., 2002; Mroz et al., 2007). Dysregulated tissue repair occurs when normal repair mechanisms fail either due to change in cellular maintenance or cellular injury. Dysregulated tissue repair can result in either loss of tissue integrity or cellular hyperplasia leading to fibrosis (Finkelstein et al., 1997; Su et al., 2000; Koike et al., 2002; Zhang et al., 2004).

Disease states that are associated with inhalation exposure of particulate matter are emphysema (under the umbrella of chronic obstructive pulmonary diseases), asbestosis, "coal worker's lung," and other conditions related to fibrotic changes in the lung parenchyma (Dai et al., 1998; Mossman and Churg, 1998; Choe et al., 1999; Huang et al., 1999; Geist et al., 2000; Lasky et al., 2001; Huang et al., 2002a; Pei et al., 2002; Zhang and Huang, 2002, 2003; Costa and Kodavanti, 2003; Fubini and Hubbard, 2003; Huang and Zhang, 2003; Iwagaki et al., 2003; Panduri et al., 2004; Swain et al., 2004; Kilsgard et al., 2012). Additionally, these conditions may also impact the deposition and dose of particulates in the lung which, in turn, will affect the development of these disease states (Kim and Kang, 1997). These conditions are often found to be attributed to activities of the cytokine signaling networks described (Elder et al., 2000; Oberdorster et al., 2000; Selman et al., 2003; Rangasamy et al., 2004; Barnes, 2008).

In addition to those conditions that are initiated by the inhalation of particulate matter, there are a number of preexisting, or genetically driven, respiratory disorders that can impact the severity of the inflammatory effects following particulate inhalation. As previously noted, the deposition of particulate matter is affected by the underlying status of the airways (Kim and Kang, 1997). This includes particulate-induced conditions, already described, as well as conditions such as nonallergic asthma and other disorders that predispose an individual to bronchial hypersensitivity (Nel et al., 1998; Hamada et al., 2000; Nicolai, 2002; Peden, 2002; Weisel, 2002; Whitekus et al., 2002; Hao et al., 2003; Ryman-Rasmussen et al., 2009; Scarino et al., 2012; Thevenot et al., 2012). For example, beryllium sensitivity is believed to be the result of a genetic polymorphism (Santo Tomas, 2009).

The inflammatory signaling processes associated with allergy-induced asthma symptoms can also be triggered, or exacerbated, by the inhalation of various types of particulate matter (Nel et al., 1998; Lambert et al., 1999; Diaz-Sanchez et al., 2000; Hamada et al., 2000; Whitekus et al., 2002; Ryman-Rasmussen et al., 2009). Nonasthmatic respiratory allergy symptoms may manifest following induction of similar cytokine signaling profiles as with allergic asthma and present with an influx of eosinophils (antigen-specific or not) in the lower respiratory tract (Kimber and Dearman, 2002; Siddiqui et al., 2008). The main profiles on which research is focused are the distinction between the T-helper cell type 1 (Th_1) and T-helper cell type 2 (Th_2) cytokines (Dearman and Kimber, 1999; Herrick et al., 2000, 2003; Tang et al., 2001; Dearman et al., 2002; Kuperman et al., 2002; Plitnick et al., 2002b; Dearman et al., 2003a,b; Duramad et al., 2004; Eisenbarth et al., 2004; Bisset and Schmid-Grendelmeier, 2005; Piggott et al., 2005; Tsiligianni et al., 2005; Holsapple et al., 2006; Selgrade et al., 2006; Farraj et al., 2007; Barrera et al., 2008; Ciprandi and Passalacqua, 2008; Matsuda et al., 2008; Passalacqua and Ciprandi, 2008; Turato et al., 2008; Zindler et al., 2008; Harada et al., 2009; Holgate and Davies, 2009; Hollams et al., 2009; Prefontaine et al., 2009; Woodruff et al., 2009; Hardy et al., 2010; Lambrecht and Hammad, 2010; Chu and Chiang, 2011; Fuchimoto et al., 2011; Hacha et al., 2012; Krishnaswamy et al., 2012; Makino et al., 2012; Morita et al., 2012; Munitz et al., 2012; Schroeder et al., 2012; Thaikoottathil et al., 2012). Th_1 cytokines, in general, are associated with contact dermatitis and allergic rhinitis symptoms, while Th_2 cytokines are associated with the etiology of allergic asthma.

8. MODEL SYSTEMS FOR THE STUDY OF PARTICULATE-INDUCED PULMONARY INJURY AND INFLAMMATION

There has been a variety of models developed—in vivo, in vitro, and in silico—for the study of pulmonary inflammation (Table 4) (Notter et al., 2005). Many current models focus on the distinction between Th_1 and Th_2 phenotypes.

While each model is useful for different research purposes, an in vitro tracheobronchial, or alveolar, epithelial cell model system has been more actively pursued in recent years based on the desire to answer questions regarding the mechanism of inflammatory signaling pathways in cell types specific to the pulmonary system. For this reason, this

TABLE 4 Models of Pulmonary Injury[a]

I. Animal Models of Acute Pulmonary Injury
Antibody-induced lung injury
Aspiration lung injury (e.g., acid, gastric particulates, meconium)
Airway-instilled microorganisms (e.g., bacteria, fungi)
Bacterial sepsis or endotoxin-induced lung injury
Viral-induced lung injury (e.g., influenza A, respiratory syncytial virus)
Fatty acid lung injury (e.g., oleic acid lung injury)
Oxidant-induced lung injury (e.g., severe acute exposure to oxygen, ozone)
In vivo lung lavage (with or without mechanical ventilation)
Neurogenic edema (e.g., vagotomy-induced lung injury)
II. Animal models of chronic pulmonary injury
Oxidant-induced lung injury (e.g., prolonged lower level exposure to oxygen, ozone)
Radiation lung injury
Drug-induced fibroproliferative lung injury (e.g., BCNU, Cytoxan, bleomycin, and methotrexate)
Toxicant-induced lung injury (e.g., paraquat, phorbol myristate acetate, butylated hydroxytoluene)
Particulate/fiber-induced lung injury (e.g., silica, asbestos)
III. Cell types used for the study of pulmonary injury
Alveolar macrophage
Tracheobronchial epithelium
Alveolar epithelium (Type II cells)
Vascular endothelium
Pulmonary interstitium (e.g., fibroblasts)

[a]*Notter et al., 2005. Lung Injury: Mechanisms, Pathophysiology, and Therapy, vol. 196.*

discussion will focus on the various epithelial cell models and, more specifically, illustrate the benefit of utilizing selected cell lines in the study of cellular responses to injury (Table 5) (Notter et al., 2005; Boverhof et al., 2008; Huang et al., 2011; Adenuga et al., 2012).

9. STUDYING PARTICULATE-INDUCED INFLAMMATION IN IN VITRO SYSTEMS

Primary alveolar Type II epithelial cells are typically isolated for lung injury research using intratracheally instilled proteases (Wikenheiser et al., 1993; Notter et al., 2005). Characteristically, these cells are ideal for the study of surfactant metabolism and ion transport (presence of tight junctions in culture). However, primary cells are difficult to work with as they are unstable in culture and exhibit a tendency to lose many differentiated functions such as the expression of specific surfactant proteins and cell cycle regulation activities (Notter et al., 2005).

In an effort to maintain these characteristic functions through multiple generations, cell lines have been developed through overexpression of a viral oncogene, such as simian virus (SV)-40 large tumor antigen, as a means of immortalizing cells (Wikenheiser et al., 1993; Notter et al., 2005). These viral-transformed cell lines are often better characterized than many tumor-derived cell lines as far as cell-specific origin and growth control are concerned (Notter et al., 2005). Two of the most utilized cell lines today are the A549 and MLE15 cell lines. The A549 cell line is a pulmonary epithelial cell line derived from a human alveolar cell carcinoma (Notter et al., 2005). At either early or late passage levels, A549 cells contain multilamellar cytoplasmic inclusion bodies, characteristic of primary Type II alveolar epithelial cells, and are also capable of producing and secreting pulmonary surfactant components (Notter et al., 2005). However, A549 cells have a cellular lipid composition that mirrors fibroblasts and

TABLE 5 Cell Types Used to Study Injury-induced Cellular Responses[a,b]

Type of Cell Line	Derivation	Selected Qualities	Considerations
Epithelial			
A549	Human lung	Type II-like	Lack of SP genes
Mv1Lu	Mink lung	Responsive to TGF-β	p53 pathway is defective
MLE15	Mouse lung	Type II-like	Immortalized with SV40
C10	Mouse lung	Type II-like	Spontaneous transformed
SV40-T2	Rat lung	Derived from type II epithelial cells	Immortalized with SV40
RLE-6TN	Rat lung	Type II-like	Spontaneous transformation
Tracheobronchial			
BEAS-2B	Human lung	Squamous differentiation	Immortalized with SV40
16HBE14o-	Human lung	Express CFTR solute transport	Immortalized with SV40
MucilAir™	Human nose and bronchi	Fully differentiated, produces mucous and functional cilia	Does not represent the lower respiratory tract
Endothelial			
HMEC	Human lung	Venous	Immortalized with SV40
Monocyte/macrophage			
RAW 264.7	Mouse leukemia	Macrophage/monocyte	Derived from peritoneal cavity
MHS-C	Mouse	Macrophage	SV40 immortalized alveolar macrophage
NR8383	Rat macrophage	Alveolar macrophage	Spontaneous transformation
U937	Human	Monocyte/macrophage	Nonalveolar undifferentiated
THP-1	Human	Monocyte/macrophage	Nonalveolar undifferentiated

[a]*Notter et al., 2005. Lung Injury: Mechanisms, Pathophysiology, and Therapy, vol. 196.*
[b]*Huang et al., 2011. The Use of In vitro 3D Cell Models in Drug Development for Respiratory Diseases.*

not Type II epithelial cells (Notter et al., 2005). Furthermore, A549 cells do not express the surfactant proteins, under any culture condition, to a high extent (Notter et al., 2005). These details make A549 cells inappropriate for the study of surfactant metabolism though they have been extensively utilized, successfully, for the study of cytokine production, inflammatory cell-epithelial cell interactions, and oxidative stress.

MLE15 cells are among the recently developed cell model of alveolar Type II epithelial cells. MLE15 cells were developed from lung tumors produced in transgenic mice harboring the SV40 large tumor antigen operating under transcriptional control of the promoter for the human surfactant protein (SP)-C gene (Wikenheiser et al., 1993; Barrett et al., 1998; Barrett et al., 1999 a and b; Notter et al., 2005). Unlike A549 cells, MLE15 cells exhibit many characteristics, in culture, that resemble Type II alveolar epithelial cells. MLE15 cells have epithelial cell microvilli, multilamellar inclusion bodies, and, unlike A549 cells, the ability to express surfactant proteins and mRNA, as well as secrete phospholipids (Notter et al., 2005).

MLE15 cells have been successfully utilized for the purpose of studying cytokine expression and injury-related cellular responses (Wikenheiser et al., 1993; Barrett et al., 1998; Barrett et al., 1999 a and b; Notter et al., 2005; Singal and Finkelstein, 2005a). MLE15 cells appear to most resemble and behave as primary alveolar Type II epithelial cells in culture. Additionally, optimal culture conditions have been previously developed for the exposure of MLE15 cells to various types of particulate stimuli. Specifically, the process of serum starvation was determined necessary for particle–cell interactions. Furthermore, all particle types were sonicated in a solution of DMEM/F12 culture medium supplemented with 1% fetal bovine serum (FBS). This concentration of FBS sustains the cells in culture and allows for minimal particle-serum protein binding that has been shown to inhibit particle-induced inflammatory gene expression (Barrett et al., 1999a).

10. ROLE OF TYPE II EPITHELIAL CELLS IN PULMONARY INFLAMMATION

Particulate inhalation studies have used at least five distinct types of either commercially available or laboratory-generated particulate matter: CB, carbon/iron, crystalline silica, amorphous silica, and asbestos. This diverse array of particles allows for adequate comparison of the different inflammatory effects/pathways of each in alveolar Type II epithelial cells in culture; this is also useful as a way of isolating components of particle mixtures that are encountered in the environment.

Unlike particles possessing a defined crystal structure and known reactive surface, such as cristobalite or carbon/iron (C/Fe), Aerosil200 is amorphous and does not have free radicals or redox active metals, such as iron, on its surface (Jimenez et al., 2000; Johnston et al., 2000; Oberdorster et al., 2000; Aust et al., 2002). Because of its size and chemical composition, Aerosil200 has a large surface area (per mass [specific surface area]) capable of interacting with various cell types in the lower airways, and is highly soluble in aqueous and organic media (Degussa, Iler, 1979; Raghavan et al., 2000). This solubility feature makes Aerosil200 a truly unique particle type to study. Aerosil200 exposure in Type II alveolar epithelial cells results in an increase in the abundance of inflammatory mediators, particularly MIP-2, the murine homolog of human interleukin (IL)-8 (Singal and Finkelstein, 2005 a and b; Donaldson et al., 2008).

The promoter region of the MIP-2 gene contains DNA-binding sites for multiple proinflammatory transcription factors including NF-κB, NF-IL-6, AP-3, and AP-1 (Kim et al., 2003; Zampetaki et al., 2004; Sohn et al., 2005). It has been shown by other investigators that various isoforms of PKC are involved in proinflammatory signaling (Kaul et al., 1998; Gopalakrishna and Jaken, 2000; Catley et al., 2004; Storz et al., 2004), specifically, that PKC activity is a target of redox modulation and that PKC plays a significant role in the propagation of inflammatory signaling events through AP-1 and NF-κB. With respect to particulate matter exposure, it has been shown that cigarette smoke exposure can alter the distribution of PKC cytosol-to-membrane ratio in a certain type of lung carcinoma (Gopalakrishna and Jaken, 2000) and that asbestos, or crystalline silica, exposure can induce PKC-regulated ROS formation in alveolar macrophages (Kim et al., 2002). Of particular significance is that oxidative regulation of PKC is crucial to the cellular processes required for metastatic conversion in cancer (Gopalakrishna and Jaken, 2000).

11. AGE-RELATED IMMUNOLOGICAL RESPONSES TO PARTICULATE INHALATION EXPOSURE

It is important to note that while various aspects have been described regarding inhalation exposure of particulates, one must consider how the immunological response to particulate matter will be affected by the age of the individual. Starting with the anatomy and function of the airways, infants and children will experience a different deposition pattern and, therefore, a different exposure dose. Infants in particular preferentially breathe through their noses, and particle size will be a major determinant of deposition location following inhalation. With adults, oral breathing is more frequent, and hence the dose of particulates inhaled may be higher in the lower respiratory tract (Asgharian et al., 2004; Salem and Katz, 2006).

The immune status of the individual will change with age, and the immature nature of an infant's or child's immune system will respond to the inhalation of particulate matter differently from that of an adult. Compounding the complexity of such an interaction with the body is the presence of ROS/RNS in conjunction with the particulates or if the individual has a compromised cardiopulmonary system (Van Loveren et al., 1998; Bruckner, 2000; Hu et al., 2000; Gehring et al., 2002; Kheradmand et al., 2002; Gern et al., 2009; Hollams et al., 2009).

12. CONCLUSIONS

Ambient air pollution and occupational exposure to particulate matter is a growing concern in both the commercialized and developing parts of the world (Demers et al., 1990; Ball et al., 2000; Elder et al., 2000; Oberdorster et al., 2000; Aust et al., 2002). Inhalation of aerosolized particulate matter has been shown to cause increased incidences of cardiopulmonary disease (Elder et al., 2000; Oberdorster et al., 2000; Aust et al., 2002). The research presented in this chapter depicts an evolving picture of the mechanism by which specific types of particulate matter and/or allergenic materials can induce an inflammatory response in various cell types of the respiratory system.

However, although the data present information regarding possible signaling mechanisms and cytokine profiling, questions remain with regard to the particle−cell interaction that induces the initial signal that is consequently propagated. Is there a physical change in cell membrane conformation caused by particle−cell interactions? For example, because of the unique water-soluble properties of Aerosil200, and its ability to form a gel at high concentrations, is it possible that Aerosil200 induces an inflammatory cascade in Type II epithelial cells in response to hypoxia? Studies have shown that MAPK/AP-1-regulated inflammatory signaling can be initiated, in vitro, by exposure to hypoxic conditions (Kamp et al., 2002; Schroedl et al., 2002; Dada et al., 2003; Panduri et al., 2003, 2004; Emerling et al., 2005).

In conjunction with this idea, Aerosil200, similar to C/Fe particles, can promote heme oxygenase (HO)-1 protein expression in a time- and dose-dependent manner. It has also been shown that cigarette smoke exposure can also induce HO-1 protein expression in vitro and in vivo (Lee et al., 1996; Maestrelli et al., 2001; Morse, 2003). HO-1 protein abundance is regulated by many factors, in particular AP-1, Nrf2, and hypoxia inducible factor (HIF)-1α (Venugopal and Jaiswal, 1996, 1998; Aoki et al., 2001; Hur et al., 2001; Schroedl et al., 2002; Baulig et al., 2003; Yang et al., 2003; Goyal et al., 2004; Motohashi and Yamamoto, 2004; Papaiahgari et al., 2004; Zhang et al., 2004). Nrf2 has shown particular promise as it is an antioxidant response factor that binds to the antioxidant response element in the promoter of a variety of antioxidant genes, including HO-1. For example, studies have suggested that Nrf2 has an integral role in the development of pulmonary fibrosis and the ability to affect carcinogenesis and is strongly linked to the transcriptional regulation of cytokines/chemokines that are involved in chronic inflammation in the lung and other organ systems (Thimmulappa et al., 2002; Tirumalai et al., 2002; Rangasamy et al., 2004). Specifically, a variety of studies have been conducted using microarray technology to determine Nrf2-regulated genes in mice, either overexpressing Nrf2 (Nrf2$^{+/+}$), heterozygous for Nrf2 expression (Nrf2$^{+/-}$), or in which Nrf2 gene expression was conditionally removed (Nrf2$^{-/-}$), upon exposure to sulforaphane, a strong chemopreventive agent (Thimmulappa et al., 2002; Rangasamy et al., 2004). Nrf2 has been shown to be responsible for the expression, or silencing, of a great number of genes that encode antioxidant enzymes, cytokine/chemokine proteins, and other proteins necessary for cellular proliferation and extracellular matrix construction (Thimmulappa et al., 2002; Tirumalai et al., 2002; Rangasamy et al., 2004). The data suggest a mechanism by which some individuals, who may or may not be smokers, develop fibrotic lung disease or cancer, and others, who may or may not be smokers, do not. Chronic inflammation due to cigarette smoke exposure plus a genetic predisposition may be a more plausible explanation for this dichotomy and suggests that Nrf2 is an important regulator of fibroproliferative lung diseases.

The role of PKC, specifically PKCδ, deserves closer future examination in the relationship between MIP-2 and HO-1 protein expression. This may hold promise for translation to human exposures as well. Nrf2 has been studied in great detail in the context of cigarette smoke–induced emphysematous/fibrotic tissue changes in the lung and in the assessment of chemopreventive agents on antioxidant and inflammatory gene expression (Thimmulappa et al., 2002; Rangasamy et al., 2004). Nrf2 is a unique transcription factor that is stabilized in the presence of an oxidant stimulus or chemopreventive agents (Chanas et al., 2002; Huang, 2002 b; Balogun et al., 2003; Cho et al., 2004; Jaiswal, 2004; Li et al., 2004; Zhang and Gordon, 2004). In particular, PKC activity can enhance Nrf2 protein stability allowing for the transcription of Nrf2-dependent genes. When considering that PKC is necessary for metastatic conversion of tumor cells, the fact that chemopreventive agents can inactivate PKC becomes more substantial, and therefore the role of PKC-dependent Nrf2 protein stability in MLE15 cells deserves greater attention (Gopalakrishna and Jaken, 2000; Bloom and Jaiswal, 2003). While the role of Nrf2 in MLE15 cells has not been clearly defined, the data suggest that Nrf2 is responsible for transcriptional regulation of HO-1 protein expression and may also play a role in regulating particle-induced inflammatory signaling (Alcaraz et al., 2003, 2004; Li et al., 2004). The data indicate that HO-1 and MIP-2 may be related through the action of Nrf2.

Particulate matter, either independently, or acting as a vector by which other materials are carried in the airways, in addition to inducing oxidant and inflammatory injury, also presents a common route of mammalian exposure to pathogenic and hazardous materials into the more vulnerable lower airways (Elder et al., 2000; Laskin et al., 2003).

For example, diesel exhaust particles contain various transition, or redox-active, metals, such as iron, that can participate in Fenton reactions to produce hydroxyl radicals. In addition to the prooxidant and proinflammatory properties of "clean," laboratory-generated particles, often used in order to isolate the effect(s) of certain fractions of urban mixed particle compositions, it has been suggested that particulate exposure can act as a vector for the transmission of other environmental contaminants (Boland et al., 1999; Bonvallot et al., 2001; Ernst et al., 2002). Furthermore, these other contaminants may be the predisposing factor for particle-induced lung injury. Laskin et al. (2000) conducted a comprehensive study evaluating the effect of peroxide-coated particles in an animal model of aerosolized particulate inhalation injury (Laskin et al., 2003). That study found that chemicals, carried by particles, serve to either initiate or potentiate particle-induced inflammatory responses in alveolar macrophages, in the lower airways. It is possible that this effect is occurring in epithelial cells as well (Ernst et al., 2002; Skerrett et al., 2004).

In addition to the role particle exposure may play in vector transmission of pathogenic and/or hazardous materials, particles may serve as an adjuvant for chemicals (allergenic or irritant) and/or radiation exposures (Nel et al., 1998; Ernst et al., 2002; Whitekus et al., 2002). Particulate matter as an adjuvant has been shown to exert effects that cause, or exacerbate, allergic asthma through enhanced oxidative stress (Nel et al., 1998; Lambert et al.,

1999; Whitekus et al., 2002; Bornehag and Nanberg, 2010). While particle–adjuvant activity has often been studied for its negative effects, biotechnology companies and the Center for Cancer Research have attempted to use this property in a beneficial manner to produce efficient drug delivery systems and imaging technology (www.nanoparticles.org).

Nanoparticles, particles in which at least one dimension is <100 nm, have been incorporated into many personal consumer products and advances in medical technology (Murdock et al., 2008; Stern and McNeil, 2008). Specifically, studies in which nanoparticles are tagged with tumor-specific antibodies, for radiation therapy in cancer patients, or tissue-specific antibodies for use in magnetic resonance imaging or computed tomography scanning, are in progress (www.nanoparticles.org). While the use of nanotechnology shows much promise, there remains more to be understood regarding the potential toxicity, and proinflammatory effects, of nanoparticles in mammalian systems because few studies have been conducted to evaluate the hazards of nanoparticles. Aerosil200, although often described as an ultrafine-type particle, with a mean diameter ≤12 nm, is more appropriately described as a nanoparticle. In this respect, the Aerosil200-induced inflammatory and cytotoxicity studies serve as a basis for comparison to evaluate the potential toxicity of emerging nanoparticle-based technologies (Singal and Finkelstein, 2005a).

13. FUTURE DIRECTIONS FOR PARTICLE AND ALLERGEN INFLAMMATION RESEARCH

Studies have been performed to determine the mechanism of particle-induced cytotoxicity and inflammatory protein expression in various cell types (Hetland et al., 2001; Aust et al., 2002; Long et al., 2005). However, a fundamental question regarding particle–cell interactions remains to be understood. Aust et al. (2002) conducted studies to ascertain whether A549 cells were capable of endocytosing coal fly ash (CFA) particles. The study illustrated that CFA particle exposure resulted in an increase in IL-8 protein secretion by A549 cells and that the expression of IL-8 protein can be inhibited by cotreatment with an actin polymerization inhibitor, cytochalasin D (Aust et al., 2002). Particulate interactions with epithelial cells may involve a direct conformational change in the cell membrane, such that a particle may enter the cell and initiate signaling events preceding the transcription of proinflammatory mediators. An alternative to endocytosis, particle passage through the phospholipid bilayer of the cell membrane is possible through diffusion, provided that the particle type lends itself to solubility in aqueous or organic media (Iler, 1979; Raghavan et al., 2000). Aerosil200 is one such type of particle and may exert its proinflammatory effect in alveolar Type II epithelial cells, not through a direct, physical, particle cell membrane interaction but through diffusion into the cell or other paracellular pathway.

A topic of considerable interest resides with the concept of the prooxidant capacity of a particulate stimulus. While a variety of previous studies have shown that different types of genotoxic stresses (UV, IR, hyperoxia) operate through enhanced production of free radical species, nongenotoxic cellular insults may also promote free radical accumulation through hypoxia (Hessel et al., 2000; Kamp et al., 2002; Schroedl et al., 2002; Panduri et al., 2004). Hypoxic injury results from the lack of adequate oxygen transfer to the cell. Recent literature suggests that hypoxia and/or ischemia can induce the generation of reactive oxygen and/or nitrogen species within the cell (Hessel et al., 2000; Kamp et al., 2002; Schroedl et al., 2002; Panduri et al., 2004). The fact that ROS/RNS can initiate a proinflammatory and/or antioxidant response in various cell types is widely accepted, however, the concept that the lack of oxidative processes can also lead to free radical–dependent signaling is still under investigation.

One could speculate, for example, that the soluble characteristics of Aerosil200 allow for gel formation on the surface of MLE15 cells in culture when high concentrations of Aerosil200 are mixed with aqueous/organic media, and that this gel may create an environment that is not conducive to adequate cellular oxygenation; lack of proper cellular oxygenation can result in the activation of the MAPK pathway and thus commence a proinflammatory response that may be dependent on the transcriptional activity of AP-1 (Benjamin et al., 1990; Leeper-Woodford and Detmer, 1999; Craig and Dorscheid, 2002; Yang et al., 2002; Fantozzi et al., 2003; Goyal et al., 2004; Zampetaki et al., 2004; Singal and Finkelstein, 2005 a).

Both crystalline silica and amorphous silica promote MIP-2 protein expression in MLE15 cells, however, though they are both silica derivatives, the mechanism by which a proinflammatory/prooxidant response occurs appears to be very different. This difference may first be attributed to the variation in structure of each type of particle. Crystalline silica possesses a lattice formation that is not soluble in aqueous/organic media. This attribute is important as far as showcasing the impact of physical chemical structure. Amorphous silica has no defined crystal structure and has been shown, experimentally, to be highly soluble in aqueous/organic media (the phrase "highly soluble" refers to the fact that large concentrations of Aerosil200 are required for gel formation; it is not meant to imply the rate of Aerosil200 solubility in organic/aqueous media) (Iler, 1979; Raghavan et al., 2000). The ability of Aerosil200 to form hydrogen bonds allows for the formation of a gel-like substance that maybe impermeable to sufficient oxygenation. While this may not be the sole contributing factor to Aerosil200 induced cytotoxicity, it is an interesting hypothesis as to why LDH release, in Aerosil200-treated MLE15 cells, increases rapidly within the first 3 h post-exposure and continues to rise until, by 24 h, a majority of the cell population has died, either by necrosis or apoptosis (Singal and Finkelstein, 2005 a). A more expansive assessment of oxidant-induced damage (lipid peroxidation, etc.) needs to be conducted in addition to measurements of hypoxia-regulated changes.

In summary, particle exposure can result in a significant inflammatory response that operates through the activation of both NF-κB and AP-1 transcription factors through various pathways such as PKCδ. In addition to the proinflammatory signaling events, the data suggest that an alternative signaling pathway, leading to HO-1 protein expression, may operate simultaneously.

HO-1 has emerged as an important cellular redox endpoint and, with respect to particle research, has proven to be an excellent distinguishing factor when comparing the effects of different types of particles (Venugopal and Jaiswal, 1996; Aoki et al., 2001; Ma and Ma, 2002; Hirano, 2003; Laskin et al., 2003; Li et al., 2003; Risom et al., 2003; Papaiahgari et al., 2004). For example, high dose Aerosil200 exposure significantly promotes HO-1 protein abundance in MLE15 cells, and this increase in HO-1 is easily inhibited by NADPH oxidase inhibition, which also inhibits Aerosil200-induced MIP-2 protein expression. While this effect does not confirm a relationship between HO-1 and MIP-2 regulation by a common transcriptional pathway, it does suggest that a connection exists between the prooxidant and proinflammatory effects of Aerosil200 exposure in an alveolar epithelial cell line (Singal, 2005). This may also be important to the proinflammatory process of other types of particles.

This information, in addition to the nature of inflammatory signaling events, will be important to the development of measures to prevent and/or treat particle-induced lung injury more effectively in the future.

REFERENCES

Adamson, I.Y., Prieditis, H., Vincent, R., 2004. Soluble and insoluble air particle fractions induce differential production of tumor necrosis factor alpha in rat lung. Exp. Lung Res. 30 (5), 355–368.

Adenuga, D., Woolhiser, M.R., Gollapudi, B.B., Boverhof, D.R., 2012. Differential gene expression responses distinguish contact and respiratory sensitizers and nonsensitizing irritants in the local lymph node assay. Toxicol. Sci. 126 (2), 413–425.

Agency, E.P., 1996. Air Quality Criteria for Particulate Matter.

Alarie, Y., Nielsen, G.D., Andonian-Haftvan, J., Abraham, M.H., 1995. Physicochemical properties of nonreactive volatile organic chemicals to estimate RD50: alternatives to animal studies. Toxicol. Appl. Pharmacol. 134 (1), 92–99.

Alcaraz, M.J., Fernandez, P., Guillen, M.I., 2003. Anti-inflammatory actions of the heme oxygenase-1 pathway. Curr. Pharm. Des. 9 (30), 2541–2551.

Alcaraz, M.J., Vicente, A.M., Araico, A., Dominguez, J.N., Terencio, M.C., Ferrandiz, M.L., 2004. Role of nuclear factor-kappaB and heme oxygenase-1 in the mechanism of action of an anti-inflammatory chalcone derivative in RAW 264.7 cells. Br. J. Pharmacol. 142 (7), 1191–1199.

Allen, G.L., Menendez, I.Y., Ryan, M.A., Mazor, R.L., Wispe, J.R., Fiedler, M.A., Wong, H.R., 2000. Hyperoxia synergistically increases TNF-alpha-induced interleukin-8 gene expression in A549 cells. Am. J. Physiol. Lung Cell. Mol. Physiol. 278 (2), L253–L260.

Anjilvel, S., Asgharian, B., 1995. A multiple-path model of particle deposition in the rat lung. Fundam. Appl. Toxicol. 28 (1), 41–50.

Aoki, Y., Sato, H., Nishimura, N., Takahashi, S., Itoh, K., Yamamoto, M., 2001. Accelerated DNA adduct formation in the lung of the Nrf2 knockout mouse exposed to diesel exhaust. Toxicol. Appl. Pharmacol. 173 (3), 154–160.

Arts, J.H., de Jong, W.H., van Triel, J.J., Schijf, M.A., de Klerk, A., van Loveren, H., Kuper, C.F., 2008. The respiratory local lymph node assay as a tool to study respiratory sensitizers. Toxicol. Sci. 106 (2), 423–434.

Arts, J.H., Frieke Kuper, C., 2003. Approaches to induce and elicit respiratory allergy: impact of route and intensity of exposure. Toxicol. Lett. 140-141, 213–222.

Arts, J.H., Kuper, C.F., Spoor, S.M., Bloksma, N., 1998. Airway morphology and function of rats following dermal sensitization and respiratory challenge with low molecular weight chemicals. Toxicol. Appl. Pharmacol. 152 (1), 66–76.

Arts, J.H., Mommers, C., de Heer, C., 2006. Dose-response relationships and threshold levels in skin and respiratory allergy. Crit. Rev. Toxicol. 36 (3), 219–251.

Asgharian, B., Hofmann, B., Bergmann, R., 2001. Particle deposition in a multiple-path model of the human lung. Aerosol Sci. Technol. 34 (4), 332–339.

Asgharian, B., Menache, M.G., Miller, F.J., 2004. Modeling age-related particle deposition in humans. J. Aerosol Med. 17 (3), 213–224.

Asgharian, B., Price, O., Oberdorster, G., 2006. A modeling study of the effect of gravity on airflow distribution and particle deposition in the lung. Inhal. Toxicol. 18 (7), 473–481.

Asgharian, B., Price, O.T., 2006. Airflow distribution in the human lung and its influence on particle deposition. Inhal. Toxicol. 18 (10), 795–801.

Asgharian, B., Price, O.T., Schroeter, J.D., Kimbell, J.S., Jones, L., Singal, M., 2011. Derivation of mass transfer coefficients for transient uptake and tissue disposition of soluble and reactive vapors in lung airways. Ann. Biomed. Eng. 39 (6), 1788–1804.

Asgharian, B., Price, O., McClellan, G., Corley, R., Einstein, D.R., Jacob, R.E., Harkema, J., Carey, S.A., Schelegle, E., Hyde, D., Kimbell, J.S., Miller, F.J., 2012a. Development of a rhesus monkey lung geometry model and application to particle deposition in comparison to humans. Inhal. Toxicol. 24 (13), 869–899.

Asgharian, B., Price, O.T., Schroeter, J.D., Kimbell, J.S., Singal, M., 2012b. A lung dosimetry model of vapor uptake and tissue disposition. Inhal. Toxicol. 24 (3), 182–193.

Aust, A.E., Ball, J.C., Hu, A.A., Lighty, J.S., Smith, K.R., Straccia, A.M., Veranth, J.M., Young, W.C., 2002. Particle characteristics responsible for effects on human lung epithelial cells. Res. Rep. Health Eff. Inst. (110), 1–65; discussion 67–76.

Baker, G.L., Gupta, A., Clark, M.L., Valenzuela, B.R., Staska, L.M., Harbo, S.J., Pierce, J.T., Dill, J.A., 2008. Inhalation toxicity and lung toxicokinetics of C60 fullerene nanoparticles and microparticles. Toxicol. Sci. 101 (1), 122–131.

Ball, J.C., Straccia, A.M., Young, W.C., Aust, A.E., 2000. The formation of reactive oxygen species catalyzed by neutral, aqueous extracts of NIST ambient particulate matter and diesel engine particles. J. Air Waste Manag. Assoc. 50 (11), 1897–1903.

Balogun, E., Hoque, M., Gong, P., Killeen, E., Green, C.J., Foresti, R., Alam, J., Motterlini, R., 2003. Curcumin activates the haem oxygenase-1 gene via regulation of Nrf2 and the antioxidant-responsive element. Biochem. J. 371 (Pt 3), 887–895.

Barnes, P.J., 2008. The cytokine network in asthma and chronic obstructive pulmonary disease. J. Clin. Invest 118 (11), 3546–3556.

Barrera, L., Mendoza, F., Zuniga, J., Estrada, A., Zamora, A.C., Melendro, E.I., Ramirez, R., Pardo, A., Selman, M., 2008. Functional diversity of T-cell subpopulations in subacute and chronic hypersensitivity pneumonitis. Am. J. Respir. Crit. Care Med. 177 (1), 44–55.

Barrett, E.G., Johnston, C., Oberdorster, G., Finkelstein, J.N., 1998. Silica-induced chemokine expression in alveolar type II cells is mediated by TNF-alpha. Am. J. Physiol. 275 (6 Pt 1), L1110–L1119.

Barrett, E.G., Johnston, C., Oberdorster, G., Finkelstein, J.N., 1999a. Silica binds serum proteins resulting in a shift of the dose-response for silica-induced chemokine expression in an alveolar type II cell line. Toxicol. Appl. Pharmacol. 161 (2), 111–122.

Barrett, E.G., Johnston, C., Oberdorster, G., Finkelstein, J.N., 1999b. Silica-induced chemokine expression in alveolar type II cells is mediated by TNF-alpha-induced oxidant stress. Am. J. Physiol. 276 (6 Pt 1), L979–L988.

Bates, S.R., Fisher, A.B., 1996. Surfactant protein A is degraded by alveolar macrophages. Am. J. Physiol. 271 (2 Pt 1), L258–L266.

Baulig, A., Garlatti, M., Bonvallot, V., Marchand, A., Barouki, R., Marano, F., Baeza-Squiban, A., 2003. Involvement of reactive oxygen species in the metabolic pathways triggered by diesel exhaust particles in human airway epithelial cells. Am. J. Physiol. Lung Cell. Mol. Physiol. 285 (3), L671–L679.

Benjamin, I.J., Kroger, B., Williams, R.S., 1990. Activation of the heat shock transcription factor by hypoxia in mammalian cells. Proc. Natl. Acad. Sci. U. S. A. 87 (16), 6263–6267.

Bhattacharyya, S., Solakyildirim, K., Zhang, Z., Chen, M.L., Linhardt, R.J., Tobacman, J.K., 2010. Cell-bound IL-8 increases in bronchial epithelial cells after arylsulfatase B silencing due to sequestration with chondroitin-4-sulfate. Am. J. Respir. Cell. Mol. Biol. 42 (1), 51–61.

Bisset, L.R., Schmid-Grendelmeier, P., 2005. Chemokines and their receptors in the pathogenesis of allergic asthma: progress and perspective. Curr. Opin. Pulm. Med. 11 (1), 35–42.

Bloom, D.A., Jaiswal, A.K., 2003. Phosphorylation of Nrf2 at Ser40 by protein kinase C in response to antioxidants leads to the release of Nrf2 from INrf2, but is not required for Nrf2 stabilization/accumulation in the nucleus and transcriptional activation of antioxidant response element-mediated NAD(P)H:quinone oxidoreductase-1 gene expression. J. Biol. Chem. 278 (45), 44675–44682.

Bobic, S., Seys, S., De Vooght, V., Callebaut, I., Hox, V., Dooms, C., Vinckier, S., Jonckx, B., Saint-Remy, J.M., Stassen, J.M., Bullens, D.M., Ceuppens, J.L., Carmeliet, P., Hellings, P.W., 2012. Placental growth factor contributes to bronchial neutrophilic inflammation and edema in allergic asthma. Am. J. Respir. Cell. Mol. Biol.

Boland, S., Baeza-Squiban, A., Fournier, T., Houcine, O., Gendron, M.C., Chevrier, M., Jouvenot, G., Coste, A., Aubier, M., Marano, F., 1999. Diesel exhaust particles are taken up by human airway epithelial cells in vitro and alter cytokine production. Am. J. Physiol. 276 (4 Pt 1), L604–L613.

Bonvallot, V., Baeza-Squiban, A., Baulig, A., Brulant, S., Boland, S., Muzeau, F., Barouki, R., Marano, F., 2001. Organic compounds from diesel exhaust particles elicit a proinflammatory response in human airway epithelial cells and induce cytochrome p450 1A1 expression. Am. J. Respir. Cell. Mol. Biol. 25 (4), 515–521.

Bornehag, C.G., Nanberg, E., 2010. Phthalate exposure and asthma in children. Int. J. Androl. 33 (2), 333–345.

Boverhof, D.R., Billington, R., Gollapudi, B.B., Hotchkiss, J.A., Krieger, S.M., Poole, A., Wiescinski, C.M., Woolhiser, M.R., 2008. Respiratory sensitization and allergy: current research approaches and needs. Toxicol. Appl. Pharmacol. 226 (1), 1–13.

Bruckner, J.V., 2000. Differences in sensitivity of children and adults to chemical toxicity: the NAS panel report. Regul. Toxicol. Pharmacol. 31 (3), 280–285.

Burgess, J.K., Boustany, S., Moir, L.M., Weckmann, M., Lau, J.Y., Grafton, K., Baraket, M., Hansbro, P.M., Hansbro, N.G., Foster, P.S., Black, J.L., Oliver, B.G., 2010. Reduction of tumstatin in asthmatic airways contributes to angiogenesis, inflammation, and hyperresponsiveness. Am. J. Respir. Crit. Care Med. 181 (2), 106–115.

Cao, D., Bromberg, P.A., Samet, J.M., 2010. Diesel particle-induced transcriptional expression of p21 involves activation of EGFR, Src, and Stat3. Am. J. Respir. Cell. Mol. Biol. 42 (1), 88–95.

Card, J.W., Zeldin, D.C., Bonner, J.C., Nestmann, E.R., 2008. Pulmonary applications and toxicity of engineered nanoparticles. Am. J. Physiol. Lung Cell. Mol. Physiol. 295 (3), L400–L411.

Carey, S.A., Minard, K.R., Trease, L.L., Wagner, J.G., Garcia, G.J., Ballinger, C.A., Kimbell, J.S., Plopper, C.G., Corley, R.A., Postlethwait, E.M., Harkema, J.R., Einstein, D.R., 2007. Three-dimensional mapping of ozone-induced injury in the nasal airways of monkeys using magnetic resonance imaging and morphometric techniques. Toxicol. Pathol. 35 (1), 27–40.

Cassee, F.R., Muijser, H., Duistermaat, E., Freijer, J.J., Geerse, K.B., Marijnissen, J.C., Arts, J.H., 2002. Particle size-dependent total mass deposition in lungs determines inhalation toxicity of cadmium chloride aerosols in rats. Application of a multiple path dosimetry model. Arch. Toxicol. 76 (5–6), 277–286.

Catley, M.C., Cambridge, L.M., Nasuhara, Y., Ito, K., Chivers, J.E., Beaton, A., Holden, N.S., Bergmann, M.W., Barnes, P.J., Newton, R., 2004. Inhibitors of protein kinase C (PKC) prevent activated transcription: role of events downstream of NF-kappaB DNA binding. J. Biol. Chem. 279 (18), 18457–18466.

Chanas, S.A., Jiang, Q., McMahon, M., McWalter, G.K., McLellan, L.I., Elcombe, C.R., Henderson, C.J., Wolf, C.R., Moffat, G.J., Itoh, K., Yamamoto, M., Hayes, J.D., 2002. Loss of the Nrf2 transcription factor causes a marked reduction in constitutive and inducible expression of the glutathione S-transferase Gsta1, Gsta2, Gstm1, Gstm2, Gstm3 and Gstm4 genes in the livers of male and female mice. Biochem. J. 365 (Pt 2), 405–416.

Chin, B.Y., Trush, M.A., Choi, A.M., Risby, T.H., 2003. Transcriptional regulation of the HO-1 gene in cultured macrophages exposed to model airborne particulate matter. Am. J. Physiol. Lung Cell. Mol. Physiol. 284 (3), L473–L480.

Cho, H.Y., Reddy, S.P., Yamamoto, M., Kleeberger, S.R., 2004. The transcription factor NRF2 protects against pulmonary fibrosis. Faseb J. 18 (11), 1258–1260.

Choe, N., Tanaka, S., Kagan, E., 1998. Asbestos fibers and interleukin-1 upregulate the formation of reactive nitrogen species in rat pleural mesothelial cells. Am. J. Respir. Cell. Mol. Biol. 19 (2), 226–236.

Choe, N., Zhang, J., Iwagaki, A., Tanaka, S., Hemenway, D.R., Kagan, E., 1999. Asbestos exposure upregulates the adhesion of pleural leukocytes to pleural mesothelial cells via VCAM-1. Am. J. Physiol. 277 (2 Pt 1), L292–L300.

Chu, K.H., Chiang, B.L., 2011. Regulatory T cells induced by mucosal B cells alleviate allergic airway hypersensitivity. Am. J. Respir. Cell. Mol. Biol.

Chustz, R.T., Nagarkar, D.R., Poposki, J.A., Favoreto Jr., S., Avila, P.C., Schleimer, R.P., Kato, A., 2011. Regulation and function of the IL-1 family cytokine IL-1F9 in human bronchial epithelial cells. Am. J. Respir. Cell. Mol. Biol. 45 (1), 145–153.

Ciprandi, G., Passalacqua, G., 2008. Allergy and the nose. Clin. Exp. Immunol. 153 (Suppl. 1), 22–26.

Conolly, R.B., Kimbell, J.S., Janszen, D., Schlosser, P.M., Kalisak, D., Preston, J., Miller, F.J., 2004. Human respiratory tract cancer risks of inhaled formaldehyde: dose-response predictions derived from biologically-motivated computational modeling of a combined rodent and human dataset. Toxicol. Sci. 82 (1), 279–296.

Conolly, R.B., Kimbell, J.S., Janszen, D.B., Miller, F.J., 2002. Dose response for formaldehyde-induced cytotoxicity in the human respiratory tract. Regul. Toxicol. Pharmacol. 35 (1), 32–43.

Conolly, R.B., Lilly, P.D., Kimbell, J.S., 2000. Simulation modeling of the tissue disposition of formaldehyde to predict nasal DNA-protein cross-links in Fischer 344 rats, rhesus monkeys, and humans. Environ. Health Perspect. 108 (Suppl. 5), 919–924.

Corps, K.N., Islam, Z., Pestka, J.J., Harkema, J.R., 2010. Neurotoxic, inflammatory, and mucosecretory responses in the nasal airways of mice repeatedly exposed to the macrocyclic trichothecene mycotoxin roridin A: dose-response and persistence of injury. Toxicol. Pathol. 38 (3), 429–451.

Costa, D.L., Kodavanti, U.P., 2003. Toxic responses of the lung to inhaled pollutants: benefits and limitations of lung-disease models. Toxicol. Lett. 140–141, 195–203.

Craig, K., Dorscheid, D., 2002. Understanding the roles of the transcription factors nuclear factor-kappaB and hypoxia-inducible factor-1alpha in lung injury. Crit. Care 6 (6), 471–472.

Cross, C.E., van der Vliet, A., O'Neill, C.A., Louie, S., Halliwell, B., 1994. Oxidants, antioxidants, and respiratory tract lining fluids. Environ. Health Perspect. 102 (Suppl. 10), 185–191.

Crystal, R.G., West, J.B., 1991. The Lung: Scientific Foundations. Raven Press, New York.

Dada, L.A., Chandel, N.S., Ridge, K.M., Pedemonte, C., Bertorello, A.M., Sznajder, J.I., 2003. Hypoxia-induced endocytosis of Na,K-ATPase in alveolar epithelial cells is mediated by mitochondrial reactive oxygen species and PKC-zeta. J. Clin. Invest 111 (7), 1057–1064.

Dai, C., Yao, X., Keeran, K.J., Zywicke, G.J., Qu, X., Yu, Z.X., Dagur, P.K., McCoy, J.P., Remaley, A.T., Levine, S.J., 2012. Apolipoprotein a-i attenuates ovalbumin-induced neutrophilic airway inflammation via a g-CSF-dependent mechanism. Am. J. Respir. Cell. Mol. Biol.

Dai, J., Gilks, B., Price, K., Churg, A., 1998. Mineral dusts directly induce epithelial and interstitial fibrogenic mediators and matrix components in the airway wall. Am. J. Respir. Crit. Care Med. 158 (6), 1907–1913.

Dearman, R.J., Betts, C.J., Humphreys, N., Flanagan, B.F., Gilmour, N.J., Basketter, D.A., Kimber, I., 2003a. Chemical allergy: considerations for the practical application of cytokine profiling. Toxicol. Sci. 71 (2), 137–145.

Dearman, R.J., Kimber, I., 1999. Cytokine fingerprinting: characterization of chemical allergens. Methods 19 (1), 56–63.

Dearman, R.J., Skinner, R.A., Humphreys, N.E., Kimber, I., 2003b. Methods for the identification of chemical respiratory allergens in rodents: comparisons of cytokine profiling with induced changes in serum IgE. J. Appl. Toxicol. 23 (4), 199–207.

Dearman, R.J., Warbrick, E.V., Skinner, R., Kimber, I., 2002. Cytokine fingerprinting of chemical allergens: species comparisons and statistical analyses. Food Chem. Toxicol. 40 (12), 1881–1892.

Degussa. From http://www.degussa.com.

Demers, R.Y., Neale, A.V., Robins, T., Herman, S.C., 1990. Asbestos-related pulmonary disease in boilermakers. Am. J. Ind. Med. 17 (3), 327–339.

Desai, D., Brightling, C., 2009. Cytokine and anti-cytokine therapy in asthma: ready for the clinic? Clin. Exp. Immunol. 158 (1), 10–19.

Deshpande, A., Narayanan, P.K., Lehnert, B.E., 2002. Silica-induced generation of extracellular factor(s) increases reactive oxygen species in human bronchial epithelial cells. Toxicol. Sci. 67 (2), 275–283.

Diaz-Sanchez, D., Jyrala, M., Ng, D., Nel, A., Saxon, A., 2000. In vivo nasal challenge with diesel exhaust particles enhances expression of the CC chemokines rantes, MIP-1alpha, and MCP-3 in humans. Clin. Immunol. 97 (2), 140–145.

Donaldson, K., Borm, P.J., Oberdorster, G., Pinkerton, K.E., Stone, V., Tran, C.L., 2008. Concordance between in vitro and in vivo dosimetry in the proinflammatory effects of low-toxicity, low-solubility particles: the key role of the proximal alveolar region. Inhal. Toxicol. 20 (1), 53–62.

Driscoll, K.E., Howard, B.W., Carter, J.M., Asquith, T., Johnston, C., Detilleux, P., Kunkel, S.L., Isfort, R.J., 1996. Alpha-quartz-induced chemokine expression by rat lung epithelial cells: effects of in vivo and in vitro particle exposure. Am. J. Pathol. 149 (5), 1627–1637.

Duramad, P., McMahon, C.W., Hubbard, A., Eskenazi, B., Holland, N.T., 2004. Flow cytometric detection of intracellular TH1/TH2 cytokines using whole blood: validation of immunologic biomarker for use in epidemiologic studies. Cancer Epidemiol. Biomarkers Prev. 13 (9), 1452–1458.

Egan, D.F., Scanlan, C.L., Wilkins, R.L., Stoller, J.K., 1999. Egan's Fundamentals of Respiratory Care. Mosby, St. Louis.

Eisenbarth, S.C., Zhadkevich, A., Ranney, P., Herrick, C.A., Bottomly, K., 2004. IL-4-dependent Th2 collateral priming to inhaled antigens independent of Toll-like receptor 4 and myeloid differentiation factor 88. J. Immunol. 172 (7), 4527–4534.

Elder, A.C., Gelein, R., Finkelstein, J.N., Cox, C., Oberdorster, G., 2000. Pulmonary inflammatory response to inhaled ultrafine particles is modified by age, ozone exposure, and bacterial toxin. Inhal. Toxicol. 12 (Suppl. 4), 227–246.

Emerling, B.M., Platanias, L.C., Black, E., Nebreda, A.R., Davis, R.J., Chandel, N.S., 2005. Mitochondrial reactive oxygen species activation of p38 mitogen-activated protein kinase is required for hypoxia signaling. Mol. Cell. Biol. 25 (12), 4853–4862.

Ernst, H., Rittinghausen, S., Bartsch, W., Creutzenberg, O., Dasenbrock, C., Gorlitz, B.D., Hecht, M., Kairies, U., Muhle, H., Muller, M., Heinrich, U., Pott, F., 2002. Pulmonary inflammation in rats after intratracheal instillation of quartz, amorphous SiO2, carbon black, and coal dust and the influence of poly-2-vinylpyridine-N-oxide (PVNO). Exp. Toxicol. Pathol. 54 (2), 109–126.

Faffe, D.S., Silva, G.H., Kurtz, P.M., Negri, E.M., Capelozzi, V.L., Rocco, P.R., Zin, W.A., 2001. Lung tissue mechanics and extracellular matrix composition in a murine model of silicosis. J. Appl. Physiol. 90 (4), 1400–1406.

Fakhrzadeh, L., Laskin, J.D., Gardner, C.R., Laskin, D.L., 2004a. Superoxide dismutase-overexpressing mice are resistant to ozone-induced tissue injury and increases in nitric oxide and tumor necrosis factor-alpha. Am. J. Respir. Cell. Mol. Biol. 30 (3), 280–287.

Fakhrzadeh, L., Laskin, J.D., Laskin, D.L., 2004b. Ozone-induced production of nitric oxide and TNF-alpha and tissue injury are dependent on NF-kappaB p50. Am. J. Physiol. Lung Cell. Mol. Physiol. 287 (2), L279–L285.

Fantozzi, I., Zhang, S., Platoshyn, O., Remillard, C.V., Cowling, R.T., Yuan, J.X., 2003. Hypoxia increases AP-1 binding activity by enhancing capacitative Ca2+ entry in human pulmonary artery endothelial cells. Am. J. Physiol. Lung Cell. Mol. Physiol. 285 (6), L1233–L1245.

Farraj, A.K., Boykin, E., Haykal-Coates, N., Gavett, S.H., Doerfler, D., Selgrade, M., 2007. Th2 cytokines in skin draining lymph nodes and serum IgE do not predict airway hypersensitivity to intranasal isocyanate exposure in mice. Toxicol. Sci. 100 (1), 99–108.

Ferguson, H.E., Kulkarni, A., Lehmann, G.M., Garcia-Bates, T.M., Thatcher, T.H., Huxlin, K.R., Phipps, R.P., Sime, P.J., 2009. Electrophilic peroxisome proliferator-activated receptor-gamma ligands have potent antifibrotic effects in human lung fibroblasts. Am. J. Respir. Cell. Mol. Biol. 41 (6), 722–730.

Finkelstein, J.N., Johnston, C., Barrett, T., Oberdorster, G., 1997. Particulate-cell interactions and pulmonary cytokine expression. Environ. Health Perspect. 105 (Suppl. 5), 1179–1182.

Fitz, L.J., DeClercq, C., Brooks, J., Kuang, W., Bates, B., Demers, D., Winkler, A., Nocka, K., Jiao, A., Greco, R.M., Mason, L.E., Fleming, M., Quazi, A., Wright, J., Goldman, S., Hubeau, C., Williams, C.M., 2012. Acidic mammalian chitinase is not a critical target for allergic airway disease. Am. J. Respir. Cell. Mol. Biol. 46 (1), 71–79.

Forman, H.J., Nelson, J., Fisher, A.B., 1980. Rat alveolar macrophages require NADPH for superoxide production in the respiratory burst. Effect of NADPH depletion by paraquat. J. Biol. Chem. 255 (20), 9879–9883.

Forman, H.J., Williams, J.J., Nelson, J., Daniele, R.P., Fisher, A.B., 1982. Hyperoxia inhibits stimulated superoxide release by rat alveolar macrophages. J. Appl. Physiol. 53 (3), 685–689.

Frederick, C.B., Lomax, L.G., Black, K.A., Finch, L., Scribner, H.E., Kimbell, J.S., Morgan, K.T., Subramaniam, R.P., Morris, J.B., 2002. Use of a hybrid computational fluid dynamics and physiologically based inhalation model for interspecies dosimetry comparisons of ester vapors. Toxicol. Appl. Pharmacol. 183 (1), 23–40.

Fubini, B., Hubbard, A., 2003. Reactive oxygen species (ROS) and reactive nitrogen species (RNS) generation by silica in inflammation and fibrosis. Free Radic. Biol. Med. 34 (12), 1507–1516.

Fuchimoto, Y., Kanehiro, A., Miyahara, N., Koga, H., Ikeda, G., Waseda, K., Tanimoto, Y., Ueha, S., Kataoka, M., Gelfand, E.W., Tanimoto, M., 2011. Requirement for chemokine receptor 5 in the development of allergen-induced airway hyperresponsiveness and inflammation. Am. J. Respir. Cell. Mol. Biol. 45 (6), 1248–1255.

Garcia, G.J., Kimbell, J.S., 2009. Deposition of inhaled nanoparticles in the rat nasal passages: dose to the olfactory region. Inhal. Toxicol. 21 (14), 1165–1175.

Garcia, G.J., Schroeter, J.D., Segal, R.A., Stanek, J., Foureman, G.L., Kimbell, J.S., 2009. Dosimetry of nasal uptake of water-soluble and reactive gases: a first study of interhuman variability. Inhal. Toxicol. 21 (7), 607–618.

Gaschler, G.J., Zavitz, C.C., Bauer, C.M., Skrtic, M., Lindahl, M., Robbins, C.S., Chen, B., Stampfli, M.R., 2008. Cigarette smoke exposure attenuates cytokine production by mouse alveolar macrophages. Am. J. Respir. Cell. Mol. Biol. 38 (2), 218–226.

Gehring, U., Cyrys, J., Sedlmeir, G., Brunekreef, B., Bellander, T., Fischer, P., Bauer, C.P., Reinhardt, D., Wichmann, H.E., Heinrich, J., 2002. Traffic-related air pollution and respiratory health during the first 2 yrs of life. Eur. Respir. J. 19 (4), 690–698.

Geiser, M., Casaulta, M., Kupferschmid, B., Schulz, H., Semmler-Behnke, M., Kreyling, W., 2008. The role of macrophages in the clearance of inhaled ultrafine titanium dioxide particles. Am. J. Respir. Cell. Mol. Biol. 38 (3), 371–376.

Geist, L.J., Powers, L.S., Monick, M.M., Hunninghake, G.W., 2000. Asbestos stimulation triggers differential cytokine release from human monocytes and alveolar macrophages. Exp. Lung Res. 26 (1), 41–56.

Georgieva, A.V., Kimbell, J.S., Schlosser, P.M., 2003. A distributed-parameter model for formaldehyde uptake and disposition in the rat nasal lining. Inhal. Toxicol. 15 (14), 1435–1463.

Gern, J.E., Visness, C.M., Gergen, P.J., Wood, R.A., Bloomberg, G.R., O'Connor, G.T., Kattan, M., Sampson, H.A., Witter, F.R., Sandel, M.T., Shreffler, W.G., Wright, R.J., Arbes Jr., S.J., Busse, W.W., 2009. The Urban Environment and Childhood Asthma (URECA) birth cohort study: design, methods, and study population. BMC Pulm. Med. 9, 17.

Glaab, T., Ziegert, M., Baelder, R., Korolewitz, R., Braun, A., Hohlfeld, J.M., Mitzner, W., Krug, N., Hoymann, H.G., 2005. Invasive versus noninvasive measurement of allergic and cholinergic airway responsiveness in mice. Respir. Res. 6, 139.

Gloede, E., Cichocki, J.A., Baldino, J.B., Morris, J.B., 2011. A validated hybrid computational fluid dynamics-physiologically based pharmacokinetic model for respiratory tract vapor absorption in the human and rat and its application to inhalation dosimetry of diacetyl. Toxicol. Sci. 123 (1), 231–246.

Gopalakrishna, R., Jaken, S., 2000. Protein kinase C signaling and oxidative stress. Free Radic. Biol. Med. 28 (9), 1349–1361.

Goyal, P., Weissmann, N., Grimminger, F., Hegel, C., Bader, L., Rose, F., Fink, L., Ghofrani, H.A., Schermuly, R.T., Schmidt, H.H., Seeger, W., Hanze, J., 2004. Upregulation of NAD(P)H oxidase 1 in hypoxia activates hypoxia-inducible factor 1 via increase in reactive oxygen species. Free Radic. Biol. Med. 36 (10), 1279–1288.

Groves, A.M., Gow, A.J., Massa, C.B., Laskin, J.D., Laskin, D.L., 2012. Prolonged injury and altered lung function after ozone inhalation in mice with chronic lung inflammation. Am. J. Respir. Cell. Mol. Biol. 47 (6), 776–783.

Hacha, J., Tomlinson, K., Maertens, L., Paulissen, G., Rocks, N., Foidart, J.M., Noel, A., Palframan, R., Gueders, M., Cataldo, D.D., 2012. Nebulized anti-IL-13 monoclonal antibody Fab' fragment reduces allergen-induced asthma. Am. J. Respir. Cell. Mol. Biol. 47 (5), 709–717.

Halliwell, B., Gutteridge, J.M.C., 1999. Free Radicals in Biology and Medicine. Clarendon Press; Oxford University Press, Oxford, New York.

Hamada, K., Goldsmith, C.A., Goldman, A., Kobzik, L., 2000. Resistance of very young mice to inhaled allergen sensitization is overcome by coexposure to an air-pollutant aerosol. Am. J. Respir. Crit. Care Med. 161 (4 Pt 1), 1285–1293.

Hammad, H., Chieppa, M., Perros, F., Willart, M.A., Germain, R.N., Lambrecht, B.N., 2009. House dust mite allergen induces asthma via Toll-like receptor 4 triggering of airway structural cells. Nat. Med. 15 (4), 410–416.

Hao, M., Comier, S., Wang, M., Lee, J.J., Nel, A., 2003. Diesel exhaust particles exert acute effects on airway inflammation and function in murine allergen provocation models. J. Allergy Clin. Immunol. 112 (5), 905–914.

Harada, M., Obara, K., Hirota, T., Yoshimoto, T., Hitomi, Y., Sakashita, M., Doi, S., Miyatake, A., Fujita, K., Enomoto, T., Taniguchi, M., Higashi, N., Fukutomi, Y., Nakanishi, K., Nakamura, Y., Tamari, M., 2009. A functional polymorphism in IL-18 is associated with severity of bronchial asthma. Am. J. Respir. Crit. Care Med. 180 (11), 1048–1055.

Hardy, C.L., Lemasurier, J.S., Olsson, F., Dang, T., Yao, J., Yang, M., Plebanski, M., Phillips, D.J., Mollard, R., Rolland, J.M., O'Hehir, R.E., 2010. Interleukin-13 regulates secretion of the tumor growth factor-{beta} superfamily cytokine activin A in allergic airway inflammation. Am. J. Respir. Cell. Mol. Biol. 42 (6), 667–675.

Hartmann, D.P., Georgian, M.M., Kagan, E., 1984. Enhanced alveolar macrophage Ia antigen expression after asbestos inhalation. J. Immunol. 132 (6), 2693–2695.

Heck, J.D., Gaworski, C.L., Rajendran, N., Morrissey, R.L., 2002. Toxicologic evaluation of humectants added to cigarette tobacco: 13-week smoke inhalation study of glycerin and propylene glycol in Fischer 344 rats. Inhal. Toxicol. 14 (11), 1135–1152.

Herrick, C.A., MacLeod, H., Glusac, E., Tigelaar, R.E., Bottomly, K., 2000. Th2 responses induced by epicutaneous or inhalational protein exposure are differentially dependent on IL-4. J. Clin. Invest. 105 (6), 765–775.

Herrick, C.A., Xu, L., McKenzie, A.N., Tigelaar, R.E., Bottomly, K., 2003. IL-13 is necessary, not simply sufficient, for epicutaneously induced Th2 responses to soluble protein antigen. J. Immunol. 170 (5), 2488–2495.

Hess, D.R., 2008. Aerosol delivery devices in the treatment of asthma. Respir. Care 53 (6), 699–723 discussion 723–725.

Hessel, P.A., Gamble, J.F., Gee, J.B., Gibbs, G., Green, F.H., Morgan, W.K., Mossman, B.T., 2000. Silica, silicosis, and lung cancer: a response to a recent working group report. J. Occup. Environ. Med. 42 (7), 704–720.

Hetland, R.B., Schwarze, P.E., Johansen, B.V., Myran, T., Uthus, N., Refsnes, M., 2001. Silica-induced cytokine release from A549 cells: importance of surface area versus size. Hum. Exp. Toxicol. 20 (1), 46–55.

Hirano, S., Furuyama, A., Koike, E., Kobayashi, T., 2003. Oxidative-stress potency of organic extracts of diesel exhaust and urban fine particles in rat heart microvessel endothelial cells. Toxicology 187, 161–170.

Hjort, M.R., Brenyo, A.J., Finkelstein, J.N., Frampton, M.W., LoMonaco, M.B., Stewart, J.C., Johnston, C.J., D'Angio, C.T., 2003. Alveolar epithelial cell-macrophage interactions affect oxygen-stimulated interleukin-8 release. Inflammation 27 (3), 137–145.

Hofmann, W., Asgharian, B., Bergmann, R., Anjilvel, S., Miller, F.J., 2000. The effect of heterogeneity of lung structure on particle deposition in the rat lung. Toxicol. Sci. 53 (2), 430–437.

Holgate, S.T., Davies, D.E., 2009. Rethinking the pathogenesis of asthma. Immunity 31 (3), 362–367.

Hollams, E.M., Deverell, M., Serralha, M., Suriyaarachchi, D., Parsons, F., Zhang, G., de Klerk, N., Holt, B.J., Ladyman, C., Sadowska, A., Rowe, J., Loh, R., Sly, P.D., Holt, P.G., 2009. Elucidation of asthma phenotypes in atopic teenagers through parallel immunophenotypic and clinical profiling. J. Allergy Clin. Immunol. 124 (3), 463–470, 470 .e1–16.

Holley, J.A., Janssen, Y.M., Mossman, B.T., Taatjes, D.J., 1992. Increased manganese superoxide dismutase protein in type II epithelial cells of rat lungs after inhalation of crocidolite asbestos or cristobalite silica. Am. J. Pathol. 141 (2), 475–485.

Holsapple, M.P., Jones, D., Kawabata, T.T., Kimber, I., Sarlo, K., Selgrade, M.K., Shah, J., Woolhiser, M.R., 2006. Assessing the potential to induce respiratory hypersensitivity. Toxicol. Sci. 91 (1), 4–13.

Hoogerwerf, J.J., de Vos, A.F., van't Veer, C., Bresser, P., de Boer, A., Tanck, M.W., Draing, C., van der Zee, J.S., van der Poll, T., 2010. Priming of alveolar macrophages upon instillation of lipopolysaccharide in the human lung. Am. J. Respir. Cell. Mol. Biol. 42 (3), 349–356.

Hristova, M., Spiess, P.C., Kasahara, D.I., Randall, M.J., Deng, B., van der Vliet, A., 2011. The tobacco smoke component acrolein suppresses innate macrophage responses by direct alkylation of c-Jun-N-terminal kinase. Am. J. Respir. Cell. Mol. Biol. 46 (1), 23–33.

Hu, H.L., Forsey, R.J., Blades, T.J., Barratt, M.E., Parmar, P., Powell, J.R., 2000. Antioxidants may contribute in the fight against ageing: an in vitro model. Mech. Ageing Dev. 121 (1–3), 217–230.

Hu, S., Zhao, H., Al-Humadi, N.H., Yin, X.J., Ma, J.K., 2006. Silica-induced apoptosis in alveolar macrophages: evidence of in vivo thiol depletion and the activation of mitochondrial pathway. J. Toxicol. Environ. Health A 69 (13), 1261–1284.

Hu, S., Zhao, H., Yin, X.J., Ma, J.K., 2007. Role of mitochondria in silica-induced apoptosis of alveolar macrophages: inhibition of apoptosis by rhodamine 6G and N-acetyl-L-cysteine. J. Toxicol. Environ. Health A 70 (17), 1403–1415.

Huang, C., Li, J., Zhang, Q., Huang, X., 2002a. Role of bioavailable iron in coal dust-induced activation of activator protein-1 and nuclear factor of activated T cells: difference between Pennsylvania and Utah coal dusts. Am. J. Respir. Cell. Mol. Biol. 27 (5), 568–574.

Huang, H.C., Nguyen, T., Pickett, C.B., 2002b. Phosphorylation of Nrf2 at Ser-40 by protein kinase C regulates antioxidant response element-mediated transcription. J. Biol. Chem. 277 (45), 42769–42774.

Huang, S., Wiszniewski, L., Constant, S., 2011. The Use of In vitro 3D Cell Models in Drug Development for Respiratory Diseases. InTech, Rijeka, Croatia.

Huang, X., Zalma, R., Pezerat, H., 1999. Chemical reactivity of the carbon-centered free radicals and ferrous iron in coals: role of bioavailable $Fe2+$ in coal workers pneumoconiosis. Free Radic. Res. 30 (6), 439–451.

Huang, X., Zhang, Q., 2003. Coal-induced interleukin-6 gene expression is mediated through ERKs and p38 MAPK pathways. Toxicol. Appl. Pharmacol. 191 (1), 40–47.

Hughes, L.S., Cass, G.R., Gone, J., Ames, M., Olmez, I., 1998. Physical and chemical characterization of atmospheric ultrafine particles in the Los Angeles area. Environ. Sci. Technol. 32 (9), 1153–1161.

Hur, E., Chang, K.Y., Lee, E., Lee, S.K., Park, H., 2001. Mitogen-activated protein kinase kinase inhibitor PD98059 blocks the trans-activation but not the stabilization or DNA binding ability of hypoxia-inducible factor-1alpha. Mol. Pharmacol. 59 (5), 1216–1224.

IARC, 1998. Silica and Some Silicates. World Health Organization, 42.

Iler, R.K., 1979. The Chemistry of Silica : Solubility, Polymerization, Colloid and Surface Properties, and Biochemistry. Wiley, New York.

Iles, K.E., Dickinson, D.A., Watanabe, N., Iwamoto, T., Forman, H.J., 2002. AP-1 activation through endogenous H(2)O(2) generation by alveolar macrophages. Free Radic. Biol. Med. 32 (12), 1304–1313.

Iwagaki, A., Choe, N., Li, Y., Hemenway, D.R., Kagan, E., 2003. Asbestos inhalation induces tyrosine nitration associated with extracellular signal-regulated kinase 1/2 activation in the rat lung. Am. J. Respir. Cell. Mol. Biol. 28 (1), 51–60.

Jaiswal, A.K., 2004. Nrf2 signaling in coordinated activation of antioxidant gene expression. Free Radic. Biol. Med. 36 (10), 1199–1207.

Janeway, C., 2005. Immunobiology. The Immune System in Health and Disease. Garland Science, New York.

Jiang, D., Liang, J., Guo, R., Xie, T., Kelly, F.L., Martinu, T., Yang, T., Lovgren, A.K., Chia, J., Liu, N., Jung, Y., Palmer, S.M., Noble, P.W., 2011. Long-term exposure of chemokine CXCL10 causes bronchiolitis-like inflammation. Am. J. Respir. Cell. Mol. Biol. 46 (5), 592–598.

Jimenez, L.A., Drost, E.M., Gilmour, P.S., Rahman, I., Antonicelli, F., Ritchie, H., MacNee, W., Donaldson, K., 2002. PM(10)-exposed macrophages stimulate a proinflammatory response in lung epithelial cells via TNF-alpha. Am. J. Physiol. Lung Cell. Mol. Physiol. 282 (2), L237–L248.

Jimenez, L.A., Thompson, J., Brown, D.A., Rahman, I., Antonicelli, F., Duffin, R., Drost, E.M., Hay, R.T., Donaldson, K., MacNee, W., 2000. Activation of NF-kappaB by PM(10) occurs via an iron-mediated mechanism in the absence of IkappaB degradation. Toxicol. Appl. Pharmacol. 166 (2), 101–110.

Johnston, C.J., Driscoll, K.E., Finkelstein, J.N., Baggs, R., O'Reilly, M.A., Carter, J., Gelein, R., Oberdorster, G., 2000. Pulmonary chemokine and mutagenic responses in rats after subchronic inhalation of amorphous and crystalline silica. Toxicol. Sci. 56 (2), 405–413.

Johnston, C.J., Oberdorster, G., Finkelstein, J.N., 2001. Recovery from oxidant-mediated lung injury: response of metallothionein, MIP-2, and MCP-1 to nitrogen dioxide, oxygen, and ozone exposures. Inhal. Toxicol. 13 (8), 689–702.

Johnston, L.K., Rims, C.R., Gill, S.E., McGuire, J., Manicone, A.M., 2012. Pulmonary macrophage subpopulations in induction and resolution of acute lung injury. Am. J. Respir. Cell. Mol. Biol. 47 (4), 417–426.

Jordan, J.A., Verhoff, A.M., Morgan, J.E., Fischer, D.G., 2009. Assessing the in vitro toxicity of the lunar dust environment using respiratory cells exposed to Al(2)O(3) or SiO(2) fine dust particles. In vitro Cell. Dev. Biol. Anim. 45 (10), 602–613.

Kamp, D.W., Panduri, V., Weitzman, S.A., Chandel, N., 2002. Asbestos-induced alveolar epithelial cell apoptosis: role of mitochondrial dysfunction caused by iron-derived free radicals. Mol. Cell. Biochem. 234-235 (1–2), 153–160.

Kanj, R.S., Kang, J.L., Castranova, V., 2005. Measurement of the release of inflammatory mediators from rat alveolar macrophages and alveolar type II cells following lipopolysaccharide or silica exposure: a comparative study. J. Toxicol. Environ. Health A 68 (3), 185–207.

Kanj, R.S., Kang, J.L., Castranova, V., 2006. Interaction between primary alveolar macrophages and primary alveolar type II cells under basal conditions and after lipopolysaccharide or quartz exposure. J. Toxicol. Environ. Health A 69 (11), 1097–1116.

Kaul, N., Forman, H.J., 1996. Activation of NF kappa B by the respiratory burst of macrophages. Free Radic. Biol. Med. 21 (3), 401–405.

Kaul, N., Gopalakrishna, R., Gundimeda, U., Choi, J., Forman, H.J., 1998. Role of protein kinase C in basal and hydrogen peroxide-stimulated NF-kappa B activation in the murine macrophage J774A.1 cell line. Arch. Biochem. Biophys. 350 (1), 79–86.

Keinan, E., Alt, A., Amir, G., Bentur, L., Bibi, H., Shoseyov, D., 2005. Natural ozone scavenger prevents asthma in sensitized rats. Bioorg. Med. Chem. 13 (2), 557–562.

Kelada, S.N., Wilson, M.S., Tavarez, U., Kubalanza, K., Borate, B., Whitehead, G.S., Maruoka, S., Roy, M.G., Olive, M., Carpenter, D.E., Brass, D.M., Wynn, T.A., Cook, D.N., Evans, C.M., Schwartz, D.A., Collins, F.S., 2011. Strain-dependent genomic factors affect allergen-induced airway hyperresponsiveness in mice. Am. J. Respir. Cell. Mol. Biol. 45 (4), 817–824.

Kelly, J.T., Bobbitt, C.M., Asgharian, B., 2001a. In vivo measurement of fine and coarse aerosol deposition in the nasal airways of female Long-Evans rats. Toxicol. Sci. 64 (2), 253–258.

Kelly, J.T., Kimbell, J.S., Asgharian, B., 2001b. Deposition of fine and coarse aerosols in a rat nasal mold. Inhal. Toxicol. 13 (7), 577–588.

Kheradmand, F., Rishi, K., Corry, D.B., 2002. Environmental contributions to the allergic asthma epidemic. Environ. Health Perspect. 110 (Suppl. 4), 553–556.

Kilsgard, O., Andersson, P., Malmsten, M., Nordin, S.L., Linge, H.M., Eliasson, M., Sorenson, E., Erjefalt, J.S., Bylund, J., Olin, A.I., Sorensen, O.E., Egesten, A., 2012. Peptidylarginine deiminases present in the airways during tobacco smoking and inflammation can citrullinate the host defense peptide LL-37, resulting in altered activities. Am. J. Respir. Cell. Mol. Biol. 46 (2), 240–248.

Kim, C.S., Kang, T.C., 1997. Comparative measurement of lung deposition of inhaled fine particles in normal subjects and patients with obstructive airway disease. Am. J. Respir. Crit. Care Med. 155 (3), 899–905.

Kim, D.S., Han, J.H., Kwon, H.J., 2003. NF-kappaB and c-Jun-dependent regulation of macrophage inflammatory protein-2 gene expression in response to lipopolysaccharide in RAW 264.7 cells. Mol. Immunol. 40 (9), 633–643.

Kim, K.A., Kim, Y.H., Seok Seo, M., Kyu Lee, W., Won Kim, S., Kim, H., Lee, K.H., Shin, I.C., Han, J.S., Joong Kim, H., Lim, Y., 2002. Mechanism of silica-induced ROS generation in Rat2 fibroblast cells. Toxicol. Lett. 135 (3), 185–191.

Kimbell, J.S., Overton, J.H., Subramaniam, R.P., Schlosser, P.M., Morgan, K.T., Conolly, R.B., Miller, F.J., 2001. Dosimetry modeling of inhaled formaldehyde: binning nasal flux predictions for quantitative risk assessment. Toxicol. Sci. 64 (1), 111–121.

Kimbell, J.S., Segal, R.A., Asgharian, B., Wong, B.A., Schroeter, J.D., Southall, J.P., Dickens, C.J., Brace, G., Miller, F.J., 2007. Characterization of deposition from nasal spray devices using a computational fluid dynamics model of the human nasal passages. J. Aerosol Med. 20 (1), 59–74.

Kimbell, J.S., Subramaniam, R.P., Gross, E.A., Schlosser, P.M., Morgan, K.T., 2001b. Dosimetry modeling of inhaled formaldehyde: comparisons of local flux predictions in the rat, monkey, and human nasal passages. Toxicol. Sci. 64 (1), 100–110.

Kimber, I., Dearman, R.J., 2002. Chemical respiratory allergy: role of IgE antibody and relevance of route of exposure. Toxicology 181-182, 311–315.

Kimber, I., Dearman, R.J., 2005. What makes a chemical a respiratory sensitizer? Curr. Opin. Allergy Clin. Immunol. 5 (2), 119–124.

Koblinger, L., Hofmann, W., 1985. Analysis of human lung morphometric data for stochastic aerosol deposition calculations. Phys. Med. Biol. 30 (6), 541–556.

Koike, E., Hirano, S., Shimojo, N., Kobayashi, T., 2002. cDNA microarray analysis of gene expression in rat alveolar macrophages in response to organic extract of diesel exhaust particles. Toxicol. Sci. 67 (2), 241–246.

Krishnaswamy, J.K., Jirmo, A.C., Baru, A.M., Ebensen, T., Guzman, C.A., Sparwasser, T., Behrens, G.M., 2012. TLR2 Agonist-allergen Coupling Efficiently Redirects Th2 cell responses and inhibits allergic airway eosinophilia. Am. J. Respir. Cell. Mol. Biol.

Kuperman, D.A., Huang, X., Koth, L.L., Chang, G.H., Dolganov, G.M., Zhu, Z., Elias, J.A., Sheppard, D., Erle, D.J., 2002. Direct effects of interleukin-13 on epithelial cells cause airway hyperreactivity and mucus overproduction in asthma. Nat. Med. 8 (8), 885–889.

Kuronuma, K., Mitsuzawa, H., Takeda, K., Nishitani, C., Chan, E.D., Kuroki, Y., Nakamura, M., Voelker, D.R., 2009. Anionic pulmonary surfactant phospholipids inhibit inflammatory responses from alveolar macrophages and U937 cells by binding the lipopolysaccharide-interacting proteins CD14 and MD-2. J. Biol. Chem. 284 (38), 25488–25500.

Lambert, A.L., Dong, W., Winsett, D.W., Selgrade, M.K., Gilmour, M.I., 1999. Residual oil fly ash exposure enhances allergic sensitization to house dust mite. Toxicol. Appl. Pharmacol. 158 (3), 269–277.

Lambrecht, B.N., Hammad, H., 2010. The role of dendritic and epithelial cells as master regulators of allergic airway inflammation. Lancet 376 (9743), 835–843.

Laskin, D.L., Morio, L., Hooper, K., Li, T.H., Buckley, B., Turpin, B., 2003. Peroxides and macrophages in the toxicity of fine particulate matter in rats. Res. Rep. Health Eff. Inst. 117, 1–51; discussion 53–63.

Laskin, D.L., Sunil, V.R., Laumbach, R., Kipen, H., 2007. Inflammatory cytokines and lung toxicity. In: House, R.V., Descotes, J. (Eds.), Cytokines in Human Health: Immunotoxicology, Pathology, and Therapeutic Applications. Humana Press, Totowa, New Jersey, pp. 83–112.

Lasky, J.A., Friedman, M., Zhuo, Y., Liu, J.Y., Poovey, H., Brody, A., 2001. Platelet-derived growth factor receptors are essential in the development of asbestos-induced fibrosis. Chest 120, 61S–62S.

Laumbach, R.J., Fiedler, N., Gardner, C.R., Laskin, D.L., Fan, Z.H., Zhang, J., Weschler, C.J., Lioy, P.J., Devlin, R.B., Ohman-Strickland, P., Kelly-McNeil, K., Kipen, H.M., 2005. Nasal effects of a mixture of volatile organic compounds and their ozone oxidation products. J. Occup. Environ. Med. 47 (11), 1182–1189.

Lee, P.J., Alam, J., Sylvester, S.L., Inamdar, N., Otterbein, L., Choi, A.M., 1996. Regulation of heme oxygenase-1 expression in vivo and in vitro in hyperoxic lung injury. Am. J. Respir. Cell. Mol. Biol. 14 (6), 556–568.

Leeper-Woodford, S.K., Detmer, K., 1999. Acute hypoxia increases alveolar macrophage tumor necrosis factor activity and alters NF-kappaB expression. Am. J. Physiol. 276 (6 Pt 1), L909–L916.

Lewis, C.C., Yang, J.Y., Huang, X., Banerjee, S.K., Blackburn, M.R., Baluk, P., McDonald, D.M., Blackwell, T.S., Nagabhushanam, V., Peters, W., Voehringer, D., Erle, D.J., 2008. Disease-specific gene expression profiling in multiple models of lung disease. Am. J. Respir. Crit. Care Med. 177 (4), 376–387.

Li, J., Kartha, S., Iasvovskaia, S., Tan, A., Bhat, R.K., Manaligod, J.M., Page, K., Brasier, A.R., Hershenson, M.B., 2002. Regulation of human airway epithelial cell IL-8 expression by MAP kinases. Am. J. Physiol. Lung Cell. Mol. Physiol. 283 (4), L690–L699.

Li, N., Alam, J., Venkatesan, M.I., Eiguren-Fernandez, A., Schmitz, D., Di Stefano, E., Slaughter, N., Killeen, E., Wang, X., Huang, A., Wang, M., Miguel, A.H., Cho, A., Sioutas, C., Nel, A.E., 2004. Nrf2 is a key transcription factor that regulates antioxidant defense in macrophages and epithelial cells: protecting against the proinflammatory and oxidizing effects of diesel exhaust chemicals. J. Immunol. 173 (5), 3467–3481.

Li, N., Sioutas, C., Cho, A., Schmitz, D., Misra, C., Sempf, J., Wang, M., Oberley, T., Froines, J., Nel, A., 2003. Ultrafine particulate pollutants induce oxidative stress and mitochondrial damage. Environ. Health Perspect. 111 (4), 455–460.

Long, J.F., Waldman, W.J., Kristovich, R., Williams, M., Knight, D., Dutta, P.K., 2005. Comparison of ultrastructural cytotoxic effects of carbon and carbon/iron particulates on human monocyte-derived macrophages. Environ. Health Perspect. 113 (2), 170–174.

Ma, J.Y., Ma, J.K., 2002. The dual effect of the particulate and organic components of diesel exhaust particles on the alteration of pulmonary immune/inflammatory responses and metabolic enzymes. J. Environ. Sci. Health C. Environ. Carcinog. Ecotoxicol. Rev. 20 (2), 117–147.

Maestrelli, P., El Messlemani, A.H., De Fina, O., Nowicki, Y., Saetta, M., Mapp, C., Fabbri, L.M., 2001. Increased expression of heme oxygenase (HO)-1 in alveolar spaces and HO-2 in alveolar walls of smokers. Am. J. Respir. Crit. Care Med. 164 (8 Pt 1), 1508–1513.

Maillet, I., Schnyder-Candrian, S., Couillin, I., Quesniaux, V.F., Erard, F., Moser, R., Fleury, S., Kanda, A., Dombrowicz, D., Szymkowski, D.E., Ryffel, B., 2011. Allergic lung inflammation is mediated by soluble tumor necrosis factor (TNF) and attenuated by dominant-negative TNF biologics. Am. J. Respir. Cell. Mol. Biol. 45 (4), 731–739.

Makino, F., Ito, J., Abe, Y., Harada, N., Kamachi, F., Yagita, H., Takahashi, K., Okumura, K., Akiba, H., 2012. Blockade of CD70-CD27 interaction inhibits induction of allergic lung inflammation in mice. Am. J. Respir. Cell. Mol. Biol. 47 (3), 298–305.

Manzer, R., Dinarello, C.A., McConville, G., Mason, R.J., 2008. Ozone exposure of macrophages induces an alveolar epithelial chemokine response through IL-1alpha. Am. J. Respir. Cell. Mol. Biol. 38 (3), 318–323.

Matsuda, A., Fukuda, S., Matsumoto, K., Saito, H., 2008. Th1/Th2 cytokines reciprocally regulate in vitro pulmonary angiogenesis via CXC chemokine synthesis. Am. J. Respir. Cell. Mol. Biol. 38 (2), 168–175.

Matthews, L.R., 1996. Cardiopulmonary Anatomy & Physiology. Lippincott, Philadelphia.

McDonald, J.D., Doyle-Eisele, M., Kracko, D., Lund, A., Surratt, J.D., Hersey, S.P., Seinfeld, J.H., Rohr, A.C., Knipping, E.M., 2012. Cardiopulmonary response to inhalation of secondary organic aerosol derived from gas-phase oxidation of toluene. Inhal. Toxicol. 24 (11), 689–697.

Morita, H., Arae, K., Ohno, T., Kajiwara, N., Oboki, K., Matsuda, A., Suto, H., Okumura, K., Sudo, K., Takahashi, T., Matsumoto, K., Nakae, S., 2012. ST2 requires Th2-, but not Th17-, type airway inflammation in epicutaneously antigen- sensitized mice. Allergol. Int. 61 (2), 265–273.

Morris, J.B., 1990. First-pass metabolism of inspired ethyl acetate in the upper respiratory tracts of the F344 rat and Syrian hamster. Toxicol. Appl. Pharmacol. 102 (2), 331–345.

Morris, J.B., Hubbs, A.F., 2009. Inhalation dosimetry of diacetyl and butyric acid, two components of butter flavoring vapors. Toxicol. Sci. 108 (1), 173–183.

Morse, D., 2003. The role of heme oxygenase-1 in pulmonary fibrosis. Am. J. Respir. Cell. Mol. Biol. 29 (3 Suppl.), S82–S86.

Mossman, B.T., Churg, A., 1998. Mechanisms in the pathogenesis of asbestosis and silicosis. Am. J. Respir. Crit. Care Med. 157 (5 Pt 1), 1666–1680.

Motohashi, H., Yamamoto, M., 2004. Nrf2-Keap1 defines a physiologically important stress response mechanism. Trends Mol. Med. 10 (11), 549–557.

Mroz, R.M., Schins, R.P., Li, H., Drost, E.M., Macnee, W., Donaldson, K., 2007. Nanoparticle carbon black driven DNA damage induces growth arrest and AP-1 and NFkappaB DNA binding in lung epithelial A549 cell line. J. Physiol. Pharmacol. 58 (Suppl. 5(Pt 2)), 461–470.

Munitz, A., Cole, E.T., Atar-Karo, D., Finkelman, F.D., Rothenberg, M.E., 2012. Resistin-like molecule alpha regulates IL-13-Induced chemokine production but not allergen-induced airway responses. Am. J. Respir. Cell. Mol. Biol. 46 (5), 703–713.

Murdock, R.C., Braydich-Stolle, L., Schrand, A.M., Schlager, J.J., Hussain, S.M., 2008. Characterization of nanomaterial dispersion in solution prior to in vitro exposure using dynamic light scattering technique. Toxicol. Sci. 101 (2), 239–253.

Naito, T., Suda, T., Suzuki, K., Nakamura, Y., Inui, N., Sato, J., Chida, K., Nakamura, H., 2008. Lung dendritic cells have a potent capability to induce production of immunoglobulin A. Am. J. Respir. Cell. Mol. Biol. 38 (2), 161–167.

NASA. From http://stardust.jpl.nasa/tech/aerogel.html.

Nel, A.E., Diaz-Sanchez, D., Ng, D., Hiura, T., Saxon, A., 1998. Enhancement of allergic inflammation by the interaction between diesel exhaust particles and the immune system. J. Allergy Clin. Immunol. 102 (4 Pt 1), 539–554.

Nicolai, T., 2002. Pollution, environmental factors and childhood respiratory allergic disease. Toxicology 181-182, 317–321.

Notter, R.H., Finkelstein, J.N., Holm, B.A., 2005. Lung Injury: Mechanisms, Pathophysiology, and Therapy. Taylor & Francis, Boca Raton.

O'Reilly, M.A., Staversky, R.J., Finkelstein, J.N., Keng, P.C., 2003. Activation of the G2 cell cycle checkpoint enhances survival of epithelial cells exposed to hyperoxia. Am. J. Physiol. Lung Cell. Mol. Physiol. 284 (2), L368–L375.

O'Reilly, M.A., Staversky, R.J., Stripp, B.R., Finkelstein, J.N., 1998. Exposure to hyperoxia induces p53 expression in mouse lung epithelium. Am. J. Respir. Cell. Mol. Biol. 18 (1), 43–50.

O'Reilly, M.A., Staversky, R.J., Watkins, R.H., Reed, C.K., de Mesy Jensen, K.L., Finkelstein, J.N., Keng, P.C., 2001. The cyclin-dependent kinase inhibitor p21 protects the lung from oxidative stress. Am. J. Respir. Cell. Mol. Biol. 24 (6), 703–710.

Oberdorster, G., Finkelstein, J.N., Johnston, C., Gelein, R., Cox, C., Baggs, R., Elder, A.C., 2000. Acute pulmonary effects of ultrafine particles in rats and mice. Res. Rep. Health Eff. Inst. 96, 5–74 disc 75–86.

Olin, A.C., Rosengren, A., Thelle, D.S., Lissner, L., Toren, K., 2010. Increased fraction of exhaled nitric oxide predicts new-onset wheeze in a general population. Am. J. Respir. Crit. Care Med. 181 (4), 324–327.

Overton, J.H., Kimbell, J.S., Miller, F.J., 2001. Dosimetry modeling of inhaled formaldehyde: the human respiratory tract. Toxicol. Sci. 64 (1), 122–134.

Panduri, V., Weitzman, S.A., Chandel, N., Kamp, D.W., 2003. The mitochondria-regulated death pathway mediates asbestos-induced alveolar epithelial cell apoptosis. Am. J. Respir. Cell. Mol. Biol. 28 (2), 241–248.

Panduri, V., Weitzman, S.A., Chandel, N.S., Kamp, D.W., 2004. Mitochondrial-derived free radicals mediate asbestos-induced alveolar epithelial cell apoptosis. Am. J. Physiol. Lung Cell. Mol. Physiol. 286 (6), L1220–L1227.

Pantano, C., Ather, J.L., Alcorn, J.F., Poynter, M.E., Brown, A.L., Guala, A.S., Beuschel, S.L., Allen, G.B., Whittaker, L.A., Bevelander, M., Irvin, C.G., Janssen-Heininger, Y.M., 2008. Nuclear factor-kappaB activation in airway epithelium induces inflammation and hyperresponsiveness. Am. J. Respir. Crit. Care Med. 177 (9), 959–969.

Papaiahgari, S., Kleeberger, S.R., Cho, H.Y., Kalvakolanu, D.V., Reddy, S.P., 2004. NADPH oxidase and ERK signaling regulates hyperoxia-induced Nrf2-ARE transcriptional response in pulmonary epithelial cells. J. Biol. Chem. 279 (40), 42302–42312.

Parker, D., Prince, A., 2011. Innate immunity in the respiratory epithelium. Am. J. Respir. Cell. Mol. Biol. 45 (2), 189–201.

Passalacqua, G., Ciprandi, G., 2008. Allergy and the lung. Clin. Exp. Immunol. 153 (Suppl. 1), 12–16.

Pauluhn, J., 2009. Subchronic 13-week inhalation exposure of rats to multiwalled carbon nanotubes: toxic effects are determined by density of agglomerate structures, not fibrillar structures. Toxicol. Sci. 113 (1), 226–242.

Peden, D.B., 2002. Pollutants and asthma: role of air toxics. Environ. Health Perspect. 110 (Suppl. 4), 565–568.

Pei, X.H., Nakanishi, Y., Inoue, H., Takayama, K., Bai, F., Hara, N., 2002. Polycyclic aromatic hydrocarbons induce IL-8 expression through nuclear factor kappaB activation in A549 cell line. Cytokine 19 (5), 236–241.

Pietruska, J.R., Liu, X., Smith, A., McNeil, K., Weston, P., Zhitkovich, A., Hurt, R., Kane, A.B., 2011. Bioavailability, intracellular mobilization of nickel, and HIF-1alpha activation in human lung epithelial cells exposed to metallic nickel and nickel oxide nanoparticles. Toxicol. Sci. 124 (1), 138–148.

Piggott, D.A., Eisenbarth, S.C., Xu, L., Constant, S.L., Huleatt, J.W., Herrick, C.A., Bottomly, K., 2005. MyD88-dependent induction of allergic Th2 responses to intranasal antigen. J. Clin. Invest 115 (2), 459–467.

Plant, P.J., North, M.L., Ward, A., Ward, M., Khanna, N., Correa, J., Scott, J.A., Batt, J., 2011. Hypertrophic airway smooth muscle mass correlates with increased airways responsiveness in a mouse model of asthma. Am. J. Respir. Cell. Mol. Biol.

Plitnick, L.M., Loveless, S.E., Ladics, G.S., Holsapple, M.P., Selgrade, M.J., Sailstad, D.M., Smialowicz, R.J., 2002a. Cytokine profiling for chemical sensitizers: application of the ribonuclease protection assay and effect of dose. Toxicol. Appl. Pharmacol. 179 (3), 145–154.

Plitnick, L.M., Loveless, S.E., Ladics, G.S., Holsapple, M.P., Selgrade, M.J., Sailstad, D.M., Smialowicz, R.J., 2002b. Cytokine profiling and chemical allergy. Toxicol. Appl. Pharmacol. 185, 228–230.

Poole, J.A., Wyatt, T.A., Kielian, T., Oldenburg, P., Gleason, A.M., Bauer, A., Golden, G., West, W.W., Sisson, J.H., Romberger, D.J., 2011. Toll-like receptor 2 regulates organic dust-induced airway inflammation. Am. J. Respir. Cell. Mol. Biol. 45 (4), 711–719.

Poynter, M.E., Irvin, C.G., Janssen-Heininger, Y.M., 2003. A prominent role for airway epithelial NF-kappa B activation in lipopolysaccharide-induced airway inflammation. J. Immunol. 170 (12), 6257–6265.

Prefontaine, D., Lajoie-Kadoch, S., Foley, S., Audusseau, S., Olivenstein, R., Halayko, A.J., Lemiere, C., Martin, J.G., Hamid, Q., 2009. Increased expression of IL-33 in severe asthma: evidence of expression by airway smooth muscle cells. J. Immunol. 183 (8), 5094–5103.

Prokhorova, S., Patel, N., Laskin, D.L., 1998. Regulation of alveolar macrophage and type II cell DNA synthesis: effects of ozone inhalation. Am. J. Physiol. 275 (6 Pt 1), L1200–L1207.

Pryhuber, G.S., Huyck, H.L., Baggs, R., Oberdorster, G., Finkelstein, J.N., 2003. Induction of chemokines by low-dose intratracheal silica is reduced in TNFR I (p55) null mice. Toxicol. Sci. 72 (1), 150–157.

Qian, Z., Lin, H.M., Chinchilli, V.M., Lehman, E.B., Duan, Y., Craig, T.J., Wilson, W.E., Liao, D., Lazarus, S.C., Bascom, R., 2009. Interaction of ambient air pollution with asthma medication on exhaled nitric oxide among asthmatics. Arch. Environ. Occup. Health 64 (3), 168–176.

Qiao, J., Li, A., Jin, X., Wang, J., 2011. Mastic alleviates allergic inflammation in asthmatic model mice by inhibiting recruitment of eosinophils. Am. J. Respir. Cell. Mol. Biol. 45 (1), 95–100.

Quintero, O.A., Wright, J.R., 2002. Clearance of surfactant lipids by neutrophils and macrophages isolated from the acutely inflamed lung. Am. J. Physiol. Lung Cell. Mol. Physiol. 282 (2), L330–L339.

Raabe, O.G., Yeh, H.C., Schum, G.M., Phalen, R.F., 1976. Tracheobronchial Geometry: Human, Dog, Rat, Hamster LF-53. Lovelace Foundation, Albuquerque.

Raghavan, S.R., Khan, S.A., 1997. Shear-thickening response of fumed silica suspensions under steady and oscillatory shear. J. Colloid Interface Sci. 185 (1), 57–67.

Raghavan, S.R., Walls, H.J., Khan, S.A., 2000. Rheology of silica dispersions in organic liquids: new evidence for solvation forces dictated by hydrogen bonding. Langmuir 16 (21), 7920–7930.

Rahman, A., Fazal, F., 2011. Blocking NF-kappaB: an inflammatory issue. Proc. Am. Thorac. Soc. 8 (6), 497–503.

Raja, S., Xu, Y., Ferro, A.R., Jaques, P.A., Hopke, P.K., 2010. Resuspension of indoor aeroallergens and relationship to lung inflammation in asthmatic children. Environ. Int. 36 (1), 8–14.

Rangasamy, T., Cho, C.Y., Thimmulappa, R.K., Zhen, L., Srisuma, S.S., Kensler, T.W., Yamamoto, M., Petrache, I., Tuder, R.M., Biswal, S., 2004. Genetic ablation of Nrf2 enhances susceptibility to cigarette smoke-induced emphysema in mice. J. Clin. Invest. 114 (9), 1248–1259.

Rao, K.M., Porter, D.W., Meighan, T., Castranova, V., 2004. The sources of inflammatory mediators in the lung after silica exposure. Environ. Health Perspect. 112 (17), 1679–1686.

Risom, L., Dybdahl, M., Bornholdt, J., Vogel, U., Wallin, H., Moller, P., Loft, S., 2003. Oxidative DNA damage and defence gene expression in the mouse lung after short-term exposure to diesel exhaust particles by inhalation. Carcinogenesis 24 (11), 1847–1852.

Rothen-Rutishauser, B.M., Kiama, S.G., Gehr, P., 2005. A three-dimensional cellular model of the human respiratory tract to study the interaction with particles. Am. J. Respir. Cell. Mol. Biol. 32 (4), 281–289.

Royce, S.G., Lim, C., Muljadi, R.C., Samuel, C.S., Ververis, K., Karagiannis, T.C., Giraud, A.S., Tang, M.L., 2012. Trefoil factor 2 reverses airway remodeling changes in allergic airways disease. Am. J. Respir. Cell. Mol. Biol.

Ryman-Rasmussen, J.P., Tewksbury, E.W., Moss, O.R., Cesta, M.F., Wong, B.A., Bonner, J.C., 2009. Inhaled multiwalled carbon nanotubes potentiate airway fibrosis in murine allergic asthma. Am. J. Respir. Cell. Mol. Biol. 40 (3), 349–358.

Salem, H., Katz, S.A., 2006. Inhalation Toxicology. CRC/Taylor & Francis, Boca Raton, FL.

Santo Tomas, L.H., 2009. Beryllium hypersensitivity and chronic beryllium lung disease. Curr. Opin. Pulm. Med. 15 (2), 165–169.

Scarino, A., Noel, A., Renzi, P.M., Cloutier, Y., Vincent, R., Truchon, G., Tardif, R., Charbonneau, M., 2012. Impact of emerging pollutants on pulmonary inflammation in asthmatic rats: ethanol vapors and agglomerated TiO(2) nanoparticles. Inhal. Toxicol. 24 (8), 528–538.

Schroeder, B.W., Verhaeghe, C., Park, S.W., Nguyenvu, L.T., Huang, X., Zhen, G., Erle, D.J., 2012. AGR2 is induced in asthma and promotes allergen-induced mucin overproduction. Am. J. Respir. Cell. Mol. Biol. 47 (2), 178–185.

Schroedl, C., McClintock, D.S., Budinger, G.R., Chandel, N.S., 2002. Hypoxic but not anoxic stabilization of HIF-1alpha requires mitochondrial reactive oxygen species. Am. J. Physiol. Lung Cell. Mol. Physiol. 283 (5), L922–L931.

Schroeter, J.D., Kimbell, J.S., Andersen, M.E., Dorman, D.C., 2006. Use of a pharmacokinetic-driven computational fluid dynamics model to predict nasal extraction of hydrogen sulfide in rats and humans. Toxicol. Sci. 94 (2), 359–367.

Schroeter, J.D., Kimbell, J.S., Asgharian, B., Tewksbury, E.W., Singal, M., 2012. Computational fluid dynamics simulations of submicrometer and micrometer particle deposition in the nasal passages of a Sprague-Dawley rat. J. Aerosol Sci. 43 (1), 31–44.

Schroeter, J.D., Kimbell, J.S., Bonner, A.M., Roberts, K.C., Andersen, M.E., Dorman, D.C., 2006b. Incorporation of tissue reaction kinetics in a computational fluid dynamics model for nasal extraction of inhaled hydrogen sulfide in rats. Toxicol. Sci. 90 (1), 198–207.

Scott, J.A., North, M.L., Rafii, M., Huang, H., Pencharz, P., Subbarao, P., Belik, J., Grasemann, H., 2011. Asymmetric dimethylarginine is increased in asthma. Am. J. Respir. Crit. Care Med. 184 (7), 779–785.

Selgrade, M., Boykin, E.H., Haykal-Coates, N., Woolhiser, M.R., Wiescinski, C., Andrews, D.L., Farraj, A.K., Doerfler, D.L., Gavett, S.H., 2006. Inconsistencies between cytokine profiles, antibody responses, and respiratory hyperresponsiveness following dermal exposure to isocyanates. Toxicol. Sci. 94 (1), 108–117.

Selman, M., Cisneros-Lira, J., Gaxiola, M., Ramirez, R., Kudlacz, E.M., Mitchell, P.G., Pardo, A., 2003. Matrix metalloproteinases inhibition attenuates tobacco smoke-induced emphysema in Guinea pigs. Chest 123 (5), 1633–1641.

Siddiqui, S., Morris, J., Avery, N., Wyand, S., Rood, D., Silbart, L.K., 2008. Pulmonary eosinophilia correlates with allergen deposition to the lower respiratory tract in a mouse model of asthma. Clin. Exp. Allergy 38 (8), 1381–1390.

Singal, M., 2005. Doctoral Dissertation, Mechanism of Aerosil200 Particulate Induced Inflammatory Gene Expression in Alveolar Epithelial Cells.

Singal, M., Finkelstein, J.N., 2005a. Amorphous silica particles promote inflammatory gene expression through the redox sensitive transcription factor, AP-1, in alveolar epithelial cells. Exp. Lung Res. 31 (6), 581–597.

Singal, M., Finkelstein, J.N., 2005b. Use of indicator cell lines for determining inflammatory gene changes and screening the inflammatory potential of particulate and non-particulate stimuli. Inhal. Toxicol. 17 (9), 415–425.

Skerrett, S.J., Liggitt, H.D., Hajjar, A.M., Ernst, R.K., Miller, S.I., Wilson, C.B., 2004. Respiratory epithelial cells regulate lung inflammation in response to inhaled endotoxin. Am. J. Physiol. Lung Cell. Mol. Physiol. 287 (1), L143–L152.

Snyder, J.C., Reynolds, S.D., Hollingsworth, J.W., Li, Z., Kaminski, N., Stripp, B.R., 2010. Clara cells attenuate the inflammatory response through regulation of macrophage behavior. Am. J. Respir. Cell. Mol. Biol. 42 (2), 161–171.

Sohn, W.J., Lee, K.W., Lee, Y., Han, J.H., Choe, Y.K., Kim, D.S., Kwon, H.J., 2005. Pyrrolidine dithiocarbamate-induced macrophage inflammatory protein-2 gene expression is NF-kappaB-independent but c-Jun-dependent in macrophage cell line RAW 264.7. Mol. Immunol. 42 (10), 1165–1175.

Stern, S.T., McNeil, S.E., 2008. Nanotechnology safety concerns revisited. Toxicol. Sci. 101 (1), 4–21.

Storz, P., Doppler, H., Toker, A., 2004. Protein kinase Cdelta selectively regulates protein kinase D-dependent activation of NF-kappaB in oxidative stress signaling. Mol. Cell. Biol. 24 (7), 2614–2626.

Stringer, B., Imrich, A., Kobzik, L., 1995. Flow cytometric assay of lung macrophage uptake of environmental particulates. Cytometry 20 (1), 23–32.

Su, W.Y., Jaskot, R.H., Richards, J., Abramson, S.R., Woessner Jr., J.F., Yu, W.H., Dreher, K.L., 2000. Induction of pulmonary matrilysin expression by combustion and ambient air particles. Am. J. Physiol. Lung Cell. Mol. Physiol. 279 (1), L152–L160.

Subramaniam, R.P., Asgharian, B., Freijer, J.I., Miller, F.J., Anjilvel, S., 2003. Analysis of lobar differences in particle deposition in the human lung. Inhal. Toxicol. 15 (1), 1–21.

Sunil, V.R., Laumbach, R.J., Patel, K.J., Turpin, B.J., Lim, H.J., Kipen, H.M., Laskin, J.D., Laskin, D.L., 2007. Pulmonary effects of inhaled limonene ozone reaction products in elderly rats. Toxicol. Appl. Pharmacol. 222 (2), 211–220.

Swain, W.A., O'Byrne, K.J., Faux, S.P., 2004. Activation of p38 MAP kinase by asbestos in rat mesothelial cells is mediated by oxidative stress. Am. J. Physiol. Lung Cell. Mol. Physiol. 286 (4), L859–L865.

Tang, C., Inman, M.D., van Rooijen, N., Yang, P., Shen, H., Matsumoto, K., O'Byrne, P.M., 2001. Th type 1-stimulating activity of lung macrophages inhibits Th2-mediated allergic airway inflammation by an IFN-gamma-dependent mechanism. J. Immunol. 166 (3), 1471–1481.

Tao, F., Kobzik, L., 2002. Lung macrophage-epithelial cell interactions amplify particle-mediated cytokine release. Am. J. Respir. Cell. Mol. Biol. 26 (4), 499–505.

Thaikoottathil, J.V., Martin, R.J., Di, P.Y., Minor, M., Case, S., Zhang, B., Zhang, G., Huang, H., Chu, H.W., 2012. SPLUNC1 deficiency enhances airway eosinophilic inflammation in mice. Am. J. Respir. Cell. Mol. Biol. 47 (2), 253–260.

Thevenot, P.T., Saravia, J., Jin, N., Giaimo, J.D., Chustz, R.E., Mahne, S., Kelley, M.A., Hebert, V.Y., Dellinger, B., Dugas, T.R., Demayo, F.J., Cormier, S.A., 2012. Radical-Containing PM0.2 initiates epithelial-to-mesenchymal transitions in airway epithelial cells. Am. J. Respir. Cell. Mol. Biol.

Thimmulappa, R.K., Mai, K.H., Srisuma, S., Kensler, T.W., Yamamoto, M., Biswal, S., 2002. Identification of Nrf2-regulated genes induced by the chemopreventive agent sulforaphane by oligonucleotide microarray. Cancer Res. 62 (18), 5196–5203.

Tirumalai, R., Rajesh Kumar, T., Mai, K.H., Biswal, S., 2002. Acrolein causes transcriptional induction of phase II genes by activation of Nrf2 in human lung type II epithelial (A549) cells. Toxicol. Lett. 132 (1), 27–36.

Trivedi, N.N., Caughey, G.H., 2010. Mast cell peptidases: chameleons of innate immunity and host defense. Am. J. Respir. Cell. Mol. Biol. 42 (3), 257–267.

Tsiligianni, I., Antoniou, K.M., Kyriakou, D., Tzanakis, N., Chrysofakis, G., Siafakas, N.M., Bouros, D., 2005. Th1/Th2 cytokine pattern in bronchoalveolar lavage fluid and induced sputum in pulmonary sarcoidosis. BMC Pulm. Med. 5, 8.

Turato, G., Barbato, A., Baraldo, S., Zanin, M.E., Bazzan, E., Lokar-Oliani, K., Calabrese, F., Panizzolo, C., Snijders, D., Maestrelli, P., Zuin, R., Fabbri, L.M., Saetta, M., 2008. Nonatopic children with multitrigger wheezing have airway pathology comparable to atopic asthma. Am. J. Respir. Crit. Care Med. 178 (5), 476–482.

van der Vliet, A., Cross, C.E., 2000. Oxidants, nitrosants, and the lung. Am. J. Med. 109 (5), 398–421.

Van Loveren, H., De Jong, W.H., Vandebriel, R.J., Vos, J.G., Garssen, J., 1998. Risk assessment and immunotoxicology. Toxicol. Lett. 102–103, 261–265.

Vandebriel, R., Callant Cransveld, C., Crommelin, D., Diamant, Z., Glazenburg, B., Joos, G., Kuper, F., Natsch, A., Nijkamp, F., Noteborn, H., Pieters, R., Roberts, D., Roggen, E., Rorije, E., Seed, M., Sewald, K., van den Heuvel, R., van Engelen, J., Verstraelen, S., van Loveren, H., 2011. Respiratory sensitization: advances in assessing the risk of respiratory inflammation and irritation. Toxicol. vitro 25 (7), 1251–1258.

Venugopal, R., Jaiswal, A.K., 1996. Nrf1 and Nrf2 positively and c-Fos and Fra1 negatively regulate the human antioxidant response element-mediated expression of NAD(P)H:quinone oxidoreductase1 gene. Proc. Natl. Acad. Sci. U. S. A. 93 (25), 14960–14965.

Venugopal, R., Jaiswal, A.K., 1998. Nrf2 and Nrf1 in association with Jun proteins regulate antioxidant response element-mediated expression and coordinated induction of genes encoding detoxifying enzymes. Oncogene 17 (24), 3145–3156.

Verstraelen, S., Bloemen, K., Nelissen, I., Witters, H., Schoeters, G., Van Den Heuvel, R., 2008. Cell types involved in allergic asthma and their use in in vitro models to assess respiratory sensitization. Toxicol. vitro 22 (6), 1419–1431.

Volckens, J., Dailey, L., Walters, G., Devlin, R.B., 2009. Direct particle-to-cell deposition of coarse ambient particulate matter increases the production of inflammatory mediators from cultured human airway epithelial cells. Environ. Sci. Technol. 43 (12), 4595–4599.

Wegmann, M., 2011. Targeting eosinophil biology in asthma therapy. Am. J. Respir. Cell. Mol. Biol. 45 (4), 667–674.

Weibel, E.R., 1973. Morphometry of the Human Lung. Springer Verlag, Berlin, Germany.

Weisel, C.P., 2002. Assessing exposure to air toxics relative to asthma. Environ. Health Perspect. 110 (Suppl. 4), 527–537.

Whitekus, M.J., Li, N., Zhang, M., Wang, M., Horwitz, M.A., Nelson, S.K., Horwitz, L.D., Brechun, N., Diaz-Sanchez, D., Nel, A.E., 2002. Thiol antioxidants inhibit the adjuvant effects of aerosolized diesel exhaust particles in a murine model for ovalbumin sensitization. J. Immunol. 168 (5), 2560–2567.

Wikenheiser, K.A., Vorbroker, D.K., Rice, W.R., Clark, J.C., Bachurski, C.J., Oie, H.K., Whitsett, J.A., 1993. Production of immortalized distal respiratory epithelial cell lines from surfactant protein C/simian virus 40 large tumor antigen transgenic mice. Proc. Natl. Acad. Sci. U. S. A. 90 (23), 11029–11033.

Woodruff, P.G., Modrek, B., Choy, D.F., Jia, G., Abbas, A.R., Ellwanger, A., Koth, L.L., Arron, J.R., Fahy, J.V., 2009. T-helper type 2-driven inflammation defines major subphenotypes of asthma. Am. J. Respir. Crit. Care Med. 180 (5), 388–395.

Xiao, G.G., Nel, A.E., Loo, J.A., 2005. Nitrotyrosine-modified proteins and oxidative stress induced by diesel exhaust particles. Electrophoresis 26 (1), 280–292.

Yagisawa, M., Yuo, A., Yonemaru, M., Imajoh-Ohmi, S., Kanegasaki, S., Yazaki, Y., Takaku, F., 1996. Superoxide release and NADPH oxidase components in mature human phagocytes: correlation between functional capacity and amount of functional proteins. Biochem. Biophys. Res. Commun. 228 (2), 510–516.

Yamabayashi, C., Koya, T., Kagamu, H., Kawakami, H., Kimura, Y., Furukawa, T., Sakagami, T., Hasegawa, T., Sakai, Y., Matsumoto, K., Nakayama, M., Gelfand, E.W., Suzuki, E., Narita, I., 2012. A novel prostacyclin agonist protects to airway hyperresponsiveness and remodeling in mice. Am. J. Respir. Cell. Mol. Biol. 47 (2), 170–177.

Yamasaki, M., Kang, H.R., Homer, R.J., Chapoval, S.P., Cho, S.J., Lee, B.J., Elias, J.A., Lee, C.G., 2008. P21 regulates TGF-beta1-induced pulmonary responses via a TNF-alpha-signaling pathway. Am. J. Respir. Cell. Mol. Biol. 38 (3), 346–353.

Yang, X., Sheares, K.K., Davie, N., Upton, P.D., Taylor, G.W., Horsley, J., Wharton, J., Morrell, N.W., 2002. Hypoxic induction of cox-2 regulates proliferation of human pulmonary artery smooth muscle cells. Am. J. Respir. Cell. Mol. Biol. 27 (6), 688–696.

Yang, Z.Z., Zhang, A.Y., Yi, F.X., Li, P.L., Zou, A.P., 2003. Redox regulation of HIF-1alpha levels and HO-1 expression in renal medullary interstitial cells. Am. J. Physiol. Renal. Physiol. 284 (6), F1207–F1215.

Ye, J., Zeidler, P., Young, S.H., Martinez, A., Robinson, V.A., Jones, W., Baron, P., Shi, X., Castranova, V., 2001. Activation of mitogen-activated protein kinase p38 and extracellular signal-regulated kinase is involved in glass fiber-induced tumor necrosis factor-alpha production in macrophages. J. Biol. Chem. 276 (7), 5360–5367.

Yeh, H.C., Schum, G.M., 1980. Models of human lung airways and their application to inhaled particle deposition. Bull. Math. Biol. 42 (3), 461–480.

Yu, C.P., 1978. Exact analysis of aerosol deposition during steady state breathing. Powder Technol. 21, 55–62.

Zampetaki, A., Mitsialis, S.A., Pfeilschifter, J., Kourembanas, S., 2004. Hypoxia induces macrophage inflammatory protein-2 (MIP-2) gene expression in murine macrophages via NF-kappaB: the prominent role of p42/p44 and PI3 kinase pathways. FASEB J. 18 (10), 1090–1092.

Zhang, Q., Huang, X., 2002. Induction of ferritin and lipid peroxidation by coal samples with different prevalence of coal workers' pneumoconiosis: role of iron in the coals. Am. J. Ind. Med. 42 (3), 171–179.

Zhang, Q., Huang, X., 2003. Induction of interleukin-6 by coal containing bioavailable iron is through both hydroxyl radical and ferryl species. J. Biosci. 28 (1), 95–100.

Zhang, Q., Kleeberger, S.R., Reddy, S.P., 2004. DEP-induced fra-1 expression correlates with a distinct activation of AP-1-dependent gene transcription in the lung. Am. J. Physiol. Lung Cell. Mol. Physiol. 286 (2), L427–L436.

Zhang, Y., Gordon, G.B., 2004. A strategy for cancer prevention: stimulation of the Nrf2-ARE signaling pathway. Mol. Cancer Ther. 3 (7), 885–893.

Zhang, Y., Luxon, B.A., Casola, A., Garofalo, R.P., Jamaluddin, M., Brasier, A.R., 2001. Expression of respiratory syncytial virus-induced chemokine gene networks in lower airway epithelial cells revealed by cDNA microarrays. J. Virol. 75 (19), 9044–9058.

Zindler, E., Gehrke, N., Luft, C., Reuter, S., Taube, C., Finotto, S., Reske-Kunz, A.B., Sudowe, S., 2008. Divergent effects of biolistic gene transfer in a mouse model of allergic airway inflammation. Am. J. Respir. Cell. Mol. Biol. 38 (1), 38–46.

Chapter 32

Pulmonary Macrophages

Debra L. Laskin, Rama Malaviya and Jeffrey D. Laskin

Chapter Outline

1. MACROPHAGES AND INNATE IMMUNITY

Macrophages are cellular effectors of the innate immune system. They are resident in nearly all tissues of the body, where they acquire specialized function depending on the needs of the tissue. Macrophages play an essential role in ridding the body of pathogens, as well as worn-out cells, debris, apoptotic cells, and some tumor cells (Stefater et al., 2011; Davies et al., 2013; Wynn et al., 2013). They are also one of the most active secretory cells in the body, releasing a multitude of mediators that regulate all aspects of host defense, inflammation, and homeostasis. In addition, they function as professional antigen presenting cells, initiating specific immune responses of T lymphocytes. Accumulating evidence suggests that the diverse activities of macrophages are mediated by phenotypically distinct subpopulations that develop in response to specific inflammatory mediators they encounter in the tissue microenvironment. Based on their phenotype and function, they have been broadly classified into two major subpopulations: classically activated proinflammatory M1 macrophages and alternatively activated anti-inflammatory/ wound repair M2 macrophages (Martinez et al., 2008; Laskin et al., 2011; Sica and Mantovani, 2012; Mantovani et al., 2013). M1 macrophages develop in response to infectious or injurious signals including proinflammatory cytokines (e.g., interferon [IFN]γ, tumor necrosis factor [TNF]α) or after recognition of pathogen-associated molecular patterns (e.g., lipopolysaccharide [LPS], lipoproteins, dsRNA, lipoteichoic acid) or endogenous "danger" signals (e.g., heat shock proteins, HMGB1). M1 macrophages exhibit potent microbicidal activity and release cytokines such as interleukin (IL)-12 and IL-23, which promote strong proinflammatory Th1 immune responses. They also generate cytotoxic reactive oxygen species (ROS) and reactive nitrogen species (RNS), proteolytic enzymes, and proinflammatory cytokines (e.g., TNFα, IL-1, IL-6), which are thought to be key in macrophage-mediated tissue injury (Laskin et al., 2011; Murray and Wynn, 2011; Wynn et al., 2013). M2 macrophages are primarily involved in downregulating inflammation and initiating wound repair. This is accomplished through the release of anti-inflammatory cytokines such as IL-4, IL-10, and IL-13. M2 macrophages also contribute to the resolution of inflammation by phagocytizing apoptotic neutrophils and synthesizing mediators important in tissue remodeling, angiogenesis, and wound repair including transforming growth factor (TGF)β, vascular endothelial growth factor (VEGF), and epidermal growth factor (EGF). M2 macrophages support Th2-associated effector functions and play a key role in regulating T cell functioning. Because of their diverse activities, alternatively activated M2 macrophages have been further subdivided into M2a (activated by IL-4 and IL-13), M2b (activated by immune complexes in combination with IL-1β or LPS), and M2c (activated by IL-10, TGFβ, or glucocorticoids) subpopulations. However, it has been suggested that classification of macrophages into two polarized states (M1 vs M2) oversimplifies the complex functional activity of these cells. Macrophage activation is known to be a dynamic process; thus the same cells may initially take part in proinflammatory and cytotoxic reactions, and later participate

Comparative Biology of the Normal Lung. http://dx.doi.org/10.1016/B978-0-12-404577-4.00032-1

in the resolution of inflammation and wound healing (Stout and Suttles, 2004; Porcheray et al., 2005; Martinez et al., 2008; Mosser and Edwards, 2008). This illustrates the plasticity of macrophages and their ability to modulate their responses due to a changing microenvironment. To exemplify this more clearly, an alternative grouping of macrophages has been proposed based on three different homeostatic functions of the cells: host defense macrophages, wound healing macrophages, and regulatory macrophages, which "blend" together into various states of activation (Mosser and Edwards, 2008).

2. REGULATION OF PULMONARY MACROPHAGE ACTIVITY IN THE NORMAL LUNG

Pulmonary macrophages are among the most prominent macrophage populations in the body. They are major sentinels of the lung, serving as the first line of defense against inhaled xenobiotics. However, they are faced with a unique challenge because of the distinct environment of the lung, with high oxygen tension and constant exposure to inhaled pathogens and other foreign substances, most of which are innocuous (Wissinger et al., 2009; Snelgrove et al., 2011; Gwyer Findlay and Hussell, 2012). In order to prevent overreaction and collateral tissue damage, the response of pulmonary macrophages must be carefully controlled. Under homeostatic conditions, this is accomplished by mediators derived mainly from lung epithelial cells including surfactant protein (SP)-A and SP-D, TGFβ, granulocyte macrophage-colony stimulating factor (GM-CSF), IL-10, and the mucin MUC-1 (Snelgrove et al., 2011). SP-A and SP-D are pulmonary collectins that suppress macrophage functioning by binding to transmembrane receptor signal inhibitory regulatory protein α (SIRP α) (Janssen et al., 2008). This inhibits p38 mitogen-activated protein kinase activation and blocks nuclear factor (NF)-κB-mediated transcription of proinflammatory genes. SP-A and SP-D, along with mucin MUC-1, also block Toll-like receptor (TLR) signaling, which is important in macrophage recognition of microbial pathogens (Yamada et al., 2006; Liu et al., 2010; Snelgrove et al., 2011). Another key negative regulator of pulmonary macrophages is the cell surface receptor CD200R; activation of CD200R on macrophages transmits a suppressive signal preventing production of proinflammatory mediators (Snelgrove et al., 2008; Jiang-Shieh et al., 2010). Pulmonary macrophages express high basal levels of CD200R, allowing them to maintain a strong threshold for response (Gwyer Findlay and Hussell, 2012). CD200R levels are maintained in the normal lung by the epithelial cell membrane-associated ligand CD200, as well as by epithelial cell-derived IL-10 and TGFβ. These findings suggest that epithelial-macrophage cross-talk is an important mechanism to balance pulmonary host defense and excessive inflammation (Wissinger et al., 2009; Herold et al., 2011).

Following exposure to noxious, infectious, or injurious agents, suppressive signals in the lung must be rapidly overcome, allowing pulmonary macrophages to become activated and release inflammatory mediators aimed at protecting the respiratory tract and restoring homeostasis (Munger et al., 1999; Bingisser and Holt, 2001; Snelgrove et al., 2011). Epithelial integrity is thought to be key to this process (Wissinger et al., 2009). Disruption of the epithelium causes the loss of negative regulatory pathways including TGFβ, IL-10, and CD200 (Wissinger et al., 2009). A number of other immune potentiators have been identified in the respiratory tract, which have the capacity to overcome the blunted activity of pulmonary macrophages. Most prominent are TLRs, which are abundant in the airways and lung. These pattern recognition receptors recognize danger signals associated with invading pathogens, foreign substances, and dead or dying cells, promoting strong inflammatory responses. Of note, during the resolution phase of inflammatory responses in the lung, levels of CD200R and IL-10 increase far above those observed during homeostasis (Moore et al., 2001; Snelgrove et al., 2008). This overshoot mechanism is thought to represent an attempt to regain control of innate immunity in the respiratory tract (Wissinger et al., 2009). Evidence suggests that there is also long-term direct or indirect desensitization of lung macrophages to TLR ligands following infection or injury, which may contribute to restoration of homeostasis.

3. PULMONARY MACROPHAGE RECEPTORS

Pulmonary macrophage functioning is controlled in large part by cell surface and intracellular receptors. A multiplicity of receptors have been identified on these cells including cytokine/chemokine receptors, scavenger receptors, and apoptotic receptors, as well as receptors for Fcγ, complement, extracellular matrix components, and adhesion molecules (Holt et al., 1982; du Bois, 1985; Chandler et al., 1986c; Bilyk et al., 1988; Miller et al., 1992; Lohmann-Matthes et al., 1994; Prieto et al., 1994; Johansson et al., 1997; Grundy and Sentman, 2005; Garn et al., 2006; Geiser, 2010; Misharin et al., 2013) (Table 1). Lung macrophages also express a variety of pattern recognition receptors including TLRs, C-type lectin receptors, helicase RIG-like receptors, and biosensor Nod-like receptors (Taylor et al., 2002; Martin and Frevert, 2005; Liu et al., 2010; Opitz et al., 2010; Herold et al., 2011; Kovach and Standiford, 2011; Schneberger et al., 2011a). These receptors play important roles in activation of innate immunity by instructing lung macrophages to engulf and destroy foreign particles and bacteria, migrate to sites of injury and infection, and release pro- and anti-inflammatory mediators, as well as mitogens and angiogenic factors. They also activate signaling by the

TABLE 1 Phenotypic Characteristics of Pulmonary Macrophage Subpopulations

		AM	IM	PLM	IVM	Airway	References
General Markers							
CD45	Pan-leukocyte	+++	+++			+++	Johansson et al., 1997; Taylor et al., 2000; Lay et al., 2009; Johnston et al., 2012; Misharin et al., 2013; Zaynagetdinov et al., 2013
F4/80	Mature mouse MP	+	++		++	+	Nibbering et al., 1987; Franke-Ullmann et al., 1996; Bedoret et al., 2009; Schneberger et al., 2012; Misharin et al., 2013; Zaynagetdinov et al., 2013
Ly6C	Mouse inflammatory MP	–	–				Misharin et al., 2013
CD33	Sialic acid binding lectin (Siglec)	++	+/–	+			Frankenberger et al., 2000; Feng and Mao, 2012; Misharin et al., 2013
CD40	T cell costimulatory molecule	++	+++				Pons et al., 2005; Johnston et al., 2012
CD80/86	T cell costimulatory molecule	++	+++			++	Kradin et al., 2004; Pons et al., 2005; Lay et al., 2009; Misharin et al., 2013; Zaynagetdinov et al., 2013
MHC class II	Antigen presenting cells	+	++	+++		+	Miller et al., 1992; Lohmann-Matthes et al., 1994; Franke-Ullmann et al., 1996; Zetterberg et al., 1998; Frankenberger et al., 2000; Steinmuller et al., 2000; Garn et al., 2006; Bedoret et al., 2009; Lay et al., 2009; Johnston et al., 2012; Misharin et al., 2013; Zaynagetdinov et al., 2013
MRP-14	Calcium binding inflammatory protein (S100A)	++			++		Takahashi et al., 1984; Morbini et al., 2006
Mac-2 (Galectin-3)	Inflammatory MP	++	+				Nibbering et al., 1987; Kasper and Hughes, 1996
CD107b (Mac-3)	Lysosome-associated membrane protein-2 (LAMP-2) Mature MP	+	+				Nibbering et al., 1987
CD172α	Signal regulatory protein α (SIRPα)	+++		+			Gjomarkaj et al., 1999; Garn et al., 2006; Janssen et al., 2008
CCL18	Chemokine ligand	++					Kollert et al., 2009
CX3CL1	Chemokine ligand	++	–				McComb et al., 2008
α-naphthyl acetate	Nonspecific esterase	++	+/–	+		+	Rouzer et al., 1982; Gant and Hamblin, 1985; Lehnert et al., 1990; Lehnert, 1992

Continued

TABLE 1 Phenotypic Characteristics of Pulmonary Macrophage Subpopulations—cont'd

		AM	IM	PLM	IVM	Airway	References
Adhesion Molecules/Integrins							
CD9	Motility-related protein (MRP)-1 Tetraspanin family	+++					Taylor et al., 2000
CD54	ICAM-1	++	+++	+++			Prieto et al., 1994; Johansson et al., 1997; Frankenberger et al., 2000; Taylor et al., 2000; Garn et al., 2006; Johnston et al., 2012
CD61	β3 integrin subunit	+/−			++		Pons et al., 2005; Jimenez-Marin et al., 2008; Schneberger et al., 2012
CD11b/CD18	CR3 (Mac-1)	++	+	+++		+	Miller et al., 1992; Lohmann-Matthes et al., 1994; Prieto et al., 1994; Franke-Ullmann et al., 1996; Johansson et al., 1997; Gjomarkaj et al., 1999; Frankenberger et al., 2000; Zetterberg et al., 2000; Lagranderie et al., 2003; Pons et al., 2005; Lay et al., 2009; Misharin et al., 2013; Zaynagetdinov et al., 2013
CD11c	CR4 α chain	+++	+				Taylor et al., 2000; Lagranderie et al., 2003; Grundy and Sentman, 2005; Bedoret et al., 2009; Misharin et al., 2013; Zaynagetdinov et al., 2013
Receptors							
CD14	LPS receptor	+	+	+		+	Frankenberger et al., 2000; Lay et al., 2009; Misharin et al., 2013; Zayngetdinov et al., 2013
CD44	Hyaluronan receptor	+++	+				Kasper et al., 1995; Taylor et al., 2000; Pons et al., 2005; Garn et al., 2006
CD200R	CD200 receptor	+++					Snelgrove et al., 2008; Jiang-Shieh et al., 2010; Herold et al., 2011; Gwyer Findlay and Hussell, 2012
CD115	CSFR1	+++	+++				Trapnell and Whitsett, 2002
CD71	Transferrin receptor	++	++				Lohmann-Matthes et al., 1994; Johansson et al., 1997; Taylor et al., 2000; Zetterberg et al., 2000; Johnston et al., 2012; Misharin et al., 2013

Dectin-1/2	β-glucan receptor	++	+				Taylor et al., 2002
CD205/206	Mannose receptor	+++	+/−				Misharin et al., 2013; Zaynagetdinov et al., 2013
SR-A	Scavenger receptor A	+++					Gough et al., 1999
CD36	Scavenger receptor B	+/−	+				Prieto et al., 1994; Taylor et al., 2000; Misharin et al., 2013
MARCO	Scavenger receptor	+++					Gough et al., 1999; Palecanda and Kobzik, 2001; Arredouani et al., 2004; Thakur et al., 2008
CD163	Scavenger receptor ED2 (rat)	+	+++	+++		−	Lehnert et al., 1990; Lehnert, 1992; Miller et al., 1992; Van den Heuvel et al., 1999
CD204	Scavenger receptor (MSR1)	+++	++				Zaynagetdinov et al., 2013
CD68	Scavenger receptor Macrosialin ED1 (rat)	+++	++	++	+	+++	Lehnert et al., 1990; Lehnert, 1992; Miller et al., 1992; Gjomarkaj et al., 1999; Parbhakar et al., 2004; Bedoret et al., 2009; Schneberger et al., 2012; Zaynagetdinov et al., 2013
TLR4	Pattern recognition receptor	+++	+++		++		Singh Suri et al., 2006; Bedoret et al., 2009; Kovach and Standiford, 2011; Schneberger et al., 2011a; Schneberger et al., 2012
TLR9	Pattern recognition receptor	++			++		Schneberger et al., 2011a; Schneberger et al., 2012
CD64	FcγRI	+++	+++	++		+++	Bilyk et al., 1988; Lohmann-Matthes et al., 1994; Frankenberger et al., 2000; Lay et al., 2009; Misharin et al., 2013
CD32	FcγRII	+		++			Lohmann-Matthes et al., 1994; Frankenberger et al., 2000
CD16	FcγRIII	++		+			Lohmann-Matthes et al., 1994; Frankenberger et al., 2000
CX3CR1	Chemokine receptor	−	++				McComb et al., 2008; Lee, 2012
CCR2	Chemokine receptor	++	−				Maus et al., 2002
CD47	SIRPα receptor	+++					Han et al., 2000

Note: AM, alveolar macrophages; IM, interstitial macrophages; PLM, pleural macrophages; IVM, intravascular macrophages; Airway, airway macrophages; MP, macrophages. +++, High expression; ++, Intermediate expression; +, Low expression; +/−, Low/no expression; −, No expression.

adaptor MyD88 and regulate inflammasome formation (Martinon et al., 2002; Opitz et al., 2010), which can enhance antimicrobial activity by stimulating production of cytokines such as TNFα and IL-1β.

4. PULMONARY MACROPHAGE SUBPOPULATIONS

Macrophages are among the most numerous cells in the respiratory tract. The most prominent populations are found in alveoli (alveolar macrophages) and in the space between the alveolar epithelium and the vascular epithelium (interstitial macrophages). Other subpopulations of macrophages have been identified in lung pleural spaces, in the vasculature, and in the airways, although less is known about the functioning of these cells due to inherent difficulties in their isolation. A summary of differences in the phenotype of lung macrophage subpopulations and their relative functioning is provided in Tables 1 and 2.

4.1 Alveolar Macrophages

Alveolar macrophages represent ~3–5% of all cells in the lungs of healthy humans, rats, dogs, and baboons (Crapo et al., 1983; Lehnert et al., 1985; Haley et al., 1991; Crowell et al., 1992; Sebring and Lehnert, 1992; Krombach et al., 1997). They are the most numerous cell type in the alveolar space and can readily be recovered in high purity (>95%) by bronchoalveolar lavage. Alveolar macrophages "float" in the lipid layer covering the inside of the alveolus and are often observed protruding from the alveolar epithelial walls into the lumen of the lung (Perez-Arellano et al., 1990). They are strategically located to function as the first line of defense of the lung against inhaled microorganisms, particulate matter, and environmental toxins, and damage to these cells is an important factor in increased host susceptibility to airborne bacterial infection and toxicants (Martin and Frevert, 2005; Wissinger et al., 2009; Laskin et al., 2011; Gwyer Findlay and Hussell, 2012; Davies et al., 2013). This is due to constitutive expression of numerous cell surface receptors and their strong phagocytic and microbicidal capacity (Zwilling et al., 1982; Nibbering et al., 1987; Lavnikova et al., 1993; Prokhorova et al., 1994; Wu et al., 2005; Guilliams et al., 2013b). Alveolar macrophages are readily activated by infectious and injurious agents to release large quantities of ROS and RNS, as well as TNFα, which aid in pathogen destruction; chemokines like CXCL8 (IL-8), CCL2 (MCP-1), and CX3CL1, which promote neutrophil and macrophage recruitment to the lung; and growth factors such as macrophage-colony stimulating factor (CSF-1), which augment monocyte survival (Becker et al., 1989; Lohmann-Matthes et al., 1994; Prokhorova et al., 1994; Wizemann and Laskin, 1994a; Franke-Ullmann et al., 1996; Marsh et al., 1996, 1997; Capelli et al., 1999; Kelley et al., 1999; Lane et al., 1999; Steinmuller et al., 2000; Maus et al., 2002; Trapnell and Whitsett, 2002; Fujimori et al., 2003; Morgan et al., 2005; Gaschler et al., 2008; McComb et al., 2008; Lee, 2012). For humans, who have a slow mechanical clearance from the alveolar region (Bailey et al., 1982; Philipson et al., 1985; Phillipson et al., 1996), alveolar macrophages are especially important because they can efficiently dissolve many poorly water-soluble inorganic particles due to the low pH in their phagolysosome (Lundborg et al., 1984, 1985, 1995; Andre et al., 1987; Marafante et al., 1987; Kreyling et al., 1990; Nyberg et al., 1992). Alveolar macrophages also function to help maintain the dampened immune characteristics of the airways by producing IL-10 and CCL18 (Salez et al., 2001; Oltmanns et al., 2003; Prasse et al., 2006; Kollert et al., 2009; Herold et al., 2011; Gwyer Findlay and Hussell, 2012), and directly suppressing dendritic cells and T cells via secretion of prostaglandins and TGFβ (Rouzer et al., 1982; Thepen et al., 1989; Holt et al., 1993; Lohmann-Matthes et al., 1994; Blumenthal et al., 2001; Bedoret et al., 2009; Herold et al., 2011). In the resting state, alveolar macrophages, like other resident tissue macrophages, possess an alternative M2-like phenotype, which contributes to their role in maintaining tissue homeostasis (Day et al., 2009; Alber et al., 2012; Chen et al., 2012). The major immunologic role of alveolar macrophages is to protect pulmonary lymphoid cells from allergenic overload (Lay et al., 2009; Liu et al., 2010). They also express MHC class II antigens and exhibit antigen presenting activity, although this is reduced relative to pulmonary dendritic cells and interstitial macrophages (Lohmann-Matthes et al., 1994; Franke-Ullmann et al., 1996; Steinmuller et al., 2000; Bedoret et al., 2009; Misharin et al., 2013; Zaynagetdinov et al., 2013).

Alveolar macrophages are characterized by considerable cytoplasm, irregular and acentrically located indented nuclei, numerous spherical mitochondria, and an abundance of vacuoles, lipid droplets, phagocytic vesicles, and lysosomes containing hydrolytic enzymes, which are transferred to phagocytic vesicles during phagocytosis (Pratt et al., 1971; Nakstad et al., 1989; Sebring and Lehnert, 1992; Herold et al., 2011) (Figures 1 and 2). Elastases, collagenases, and various other proteolytic enzymes have also been identified in lysosomal vesicles, many of which are secreted externally (Harris et al., 1975; Rouzer et al., 1982; Bowden, 1984; du Bois, 1985; Bilyk et al., 1988; Lehnert et al., 1990). Alveolar macrophages are also strongly positive for nonspecific esterase and acid phosphatase (Rouzer et al., 1982; Bowden, 1984; Gant and Hamblin, 1985; Lavnikova et al., 1993). Flow cytometric analysis has shown that murine alveolar macrophages are unique, relative to macrophages in other tissues, in that they express high levels of CD11c or integrin alpha X (Taylor et al., 2000; Lagranderie et al., 2003;

TABLE 2 Functional Comparison of Pulmonary Macrophage Subpopulations

	AM	IM	PLM	IVM	Airway	References
Function						
Phagocytosis	+++	++	++	+++	++	Meuret et al., 1980; Lehnert et al., 1988; Chitko-McKown et al., 1991; Lehnert, 1992; Peao et al., 1992; Staub, 1994; Longworth, 1997; Gjomarkaj et al., 1999; Frankenberger et al., 2000; Zetterberg et al., 2000; Fathi et al., 2001; Lagranderie et al., 2003; Grundy and Sentman, 2005
Chemotaxis	+	++				Holt et al., 1982; Prokhorova et al., 1994; Wizemann and Laskin, 1994b; Steinmuller et al., 2000
Cytotoxicity	+++	+		++++		Chitko-McKown et al., 1991; Staub, 1994; Longworth, 1997; Gjomarkaj et al., 1999; Steinmuller et al., 2000
Dendritic cell suppression	+	+++				Bedoret et al., 2009
Antigen presentation	+	+++	++		+	Miller et al., 1992; Gjomarkaj et al., 1999; Zetterberg et al., 2000; Lay et al., 2009
Modulate T cell proliferation	+	+++	++	+++		Holt, 1979a; Holt et al., 1982; Sestini et al., 1984; Weissler et al., 1986; Chitko-McKown et al., 1992; Bilyk and Holt, 1993; Franke-Ullmann et al., 1996; Gjomarkaj et al., 1999
Inflammatory Mediators						
ROS	+++	+	++			Lohmann-Matthes et al., 1994; Franke-Ullmann et al., 1996; Steinmuller et al., 2000; Dorger et al., 2001
RNS	+++	+	++	+		Chitko-McKown et al., 1992; Lohmann-Matthes et al., 1994; Wizemann et al., 1994; Franke-Ullmann et al., 1996; Choe et al., 1997; Arias-Diaz et al., 2000; Steinmuller et al., 2000; Werneck-Barroso et al., 2000; Dorger et al., 2001
TNFα	+++	+	++	++		Chitko-McKown et al., 1991; Chitko-McKown et al., 1992; Lohmann-Matthes et al., 1994; Choe et al., 1997; Longworth, 1997; Arias-Diaz et al., 2000; Frankenberger et al., 2000; Steinmuller et al., 2000; Dorger et al., 2001; Salez et al., 2001; Chen et al., 2003; Oltmanns et al., 2003; Parbhakar et al., 2005; Cailhier et al., 2006; Schneberger et al., 2012; Duke-Novakovski et al., 2013
IFNα/β	+++	+				Lohmann-Matthes et al., 1994; Franke-Ullmann et al., 1996
IL-1	++	+++	++	+++		Weissler et al., 1986; Bilyk et al., 1988; Chitko-McKown et al., 1991; Lohmann-Matthes et al., 1994; Franke-Ullmann et al., 1996; Longworth, 1997; Arias-Diaz et al., 2000; Steinmuller et al., 2000; Chen et al., 2003; Oltmanns et al., 2003; Parbhakar et al., 2005
IL-6	+	++	+	++		Lohmann-Matthes et al., 1994; Franke-Ullmann et al., 1996; Arias-Diaz et al., 2000; Steinmuller et al., 2000; Chen et al., 2003; Cailhier et al., 2006; Shimotakahara et al., 2007
IL-10	+	+++	+++	++		Frankenberger et al., 2000; Salez et al., 2001; Oltmanns et al., 2003; Cailhier et al., 2006; Garn et al., 2006; Shimotakahara et al., 2007; Mosser and Edwards, 2008; Bedoret et al., 2009; Herold et al., 2011; Gwyer Findlay and Hussell, 2012; Johnston et al., 2012
Fibronectin	++	+++				Miller et al., 1992; Johnston et al., 2012

Continued

TABLE 2 Functional Comparison of Pulmonary Macrophage Subpopulations—cont'd

	AM	IM	PLM	IVM	Airway	References
Eicosanoids						
PGE_2	++	++	+	++		Rouzer et al., 1982; Sestini et al., 1984; Bertram et al., 1988; Miller et al., 1992; Lohmann-Matthes et al., 1994; Staub, 1994; Longworth, 1997
PGI_2	–	+				Rouzer et al., 1982; Chandler et al., 1988
TxA_2	+	+				Rouzer et al., 1982; Chandler et al., 1988; Lohmann-Matthes et al., 1994
TxB_2	+			++		Bertram et al., 1988; Miyamoto et al., 1988; Lohmann-Matthes et al., 1994; Staub, 1994; Longworth, 1997
LTB_4	+			+++		Bertram et al., 1988; Lohmann-Matthes et al., 1994; Longworth, 1997
LTC	+	++		++		Rouzer et al., 1982; Bertram et al., 1988
HETEs	++	++		+++		Rouzer et al., 1982; Bertram et al., 1988; Staub, 1994; Longworth, 1997
Growth Factors						
PDGF	++	++				Brody et al., 1992; Herold et al., 2011
TGFβ	+	+				Lohmann-Matthes et al., 1994; Bedoret et al., 2009; Herold et al., 2011
CSF-1	++	++				Becker et al., 1989; Lohmann-Matthes et al., 1994
Chemokines						
CCL2 (MCP-1)	++		++			Marsh et al., 1997; van den Heuvel et al., 1998; Capelli et al., 1999; Oltmanns et al., 2003
CCL3/4 (MIP-1α/β)	++					Lohmann-Matthes et al., 1994; Capelli et al., 1999
CCL5 (RANTES)	++					Lane et al., 1999; Gaschler et al., 2008
CCL18	++					Prasse et al., 2006; Kollert et al., 2009
CXCL2 (MIP-2)	++		++			Lohmann-Matthes et al., 1994; Cailhier et al., 2006
CXCL8 (IL-8)	++	++		++		Lohmann-Matthes et al., 1994; Marsh et al., 1996; Chen et al., 2003; Fujimori et al., 2003; Lee, 2012
CXCL16	++					Morgan et al., 2005
CX3CL1	++					McComb et al., 2008
Proteolytic enzymes	++	++			+	Harris et al., 1975; Rouzer et al., 1982; Bowden, 1984; du Bois, 1985; Gant and Hamblin, 1985; Lehnert et al., 1990

Note: AM, alveolar macrophages; IM, interstitial macrophages; PLM, pleural macrophages; IVM, intravascular macrophages; Airway, airway macrophages. PG, prostaglandin; Tx, thromboxane; LT, leukotriene; HETE, hydroxyeicosatetraenoic acid; PDGF, platelet-derived growth factor. +++, High expression; ++, Intermediate expression; +, Low expression; +/–, Low/no expression; –, No expression.

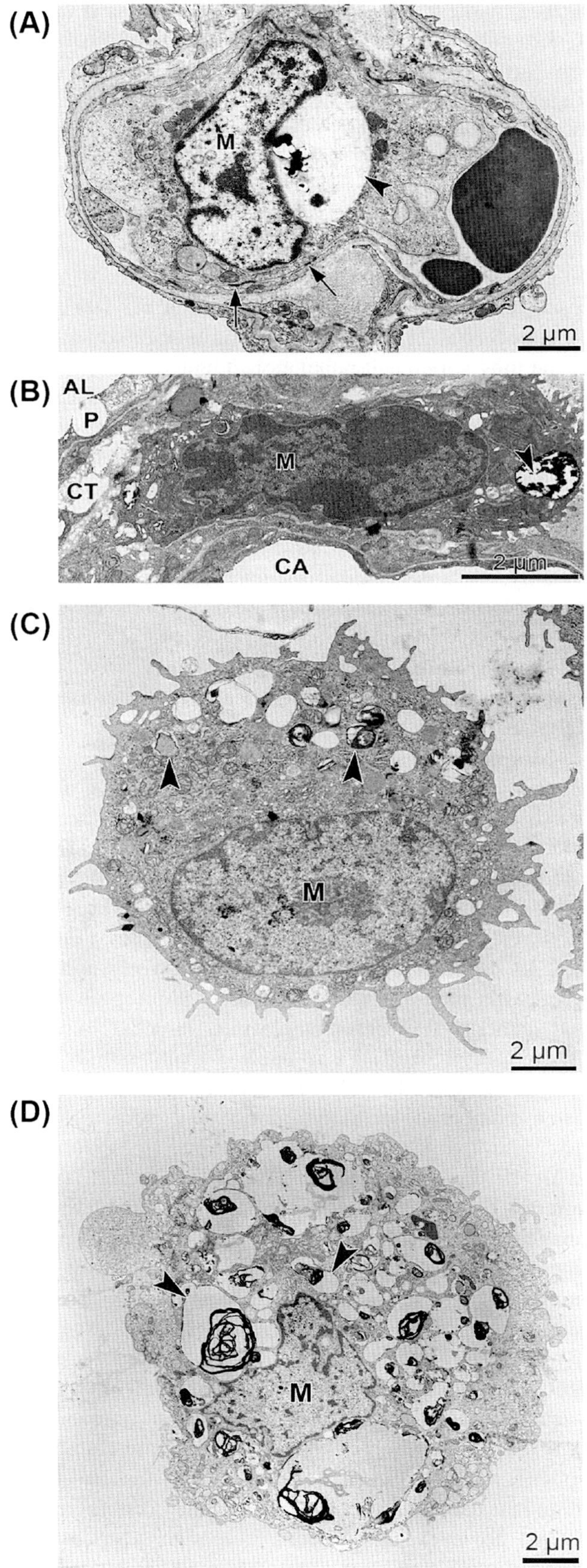

Bedoret et al., 2009; Misharin et al., 2013; Zaynagetdinov et al., 2013) (Table 2). This is thought to be due to high GM-CSF and SP-D levels in the alveoli (Guth et al., 2009). They also express high levels of scavenger receptors (e.g., MARCO, CD68, CD204, CD205/206, SR-A) and pattern recognition receptors (e.g., TLR4), as well as receptors for Fcγ (CD64), hyaluronan (CD44), and SIRP-α (CD47), along with receptors for various cytokines, chemokines, and signaling molecules (Bilyk et al., 1988; Lehnert et al., 1990; Miller et al., 1992; Franke et al., 1994; Lohmann-Matthes et al., 1994; Kasper et al., 1995; Franke-Ullmann et al., 1996; Gjomarkaj et al., 1999; Gough et al., 1999; Van den Heuvel et al., 1999; Han et al., 2000; Taylor et al., 2000; Palecanda and Kobzik, 2001; Maus et al., 2002; Taylor et al., 2002; Trapnell and Whitsett, 2002; Arredouani et al., 2004; Grundy and Sentman, 2005; Pons et al., 2005; Garn et al., 2006; Singh Suri et al., 2006; Janssen et al., 2008; McComb et al., 2008; Thakur et al., 2008; Kovach and Standiford, 2011; Feng and Mao, 2012; Misharin et al., 2013; Zaynagetdinov et al., 2013) (Table 1). Conversely, they are negative for CD11b (Mac-1) and express low levels of the proinflammatory M1 macrophage activation marker Ly6C (Misharin et al., 2013; Zaynagetdinov et al., 2013).

Alveolar macrophages are long-lived cells with a relatively slow turnover rate of ~40% per year (Maus et al., 2006). They populate the lung during early embryogenesis and are maintained in the tissue for months with minimal replenishment from hematopoietic precursors (Murphy et al., 2008; Janssen et al., 2011; Yona et al., 2013). Moreover, resident alveolar macrophages have been shown to persist after acute lung injury, participating in the resolution of inflammation by clearing cellular debris and apoptotic neutrophils, recruiting anti-inflammatory/wound repair monocyte-derived macrophages, and aiding in remodeling of the lung parenchyma (Knapp et al., 2003; Herold et al., 2011; Janssen et al., 2011; Xiong et al., 2011). However, if alveolar macrophages are depleted experimentally using clodronate-loaded liposomes or by administration of diphtheria toxin (DT) to CD11c-DT receptor mice, they are restored in large part by blood monocytes and interstitial macrophages (Landsman and Jung, 2007).

FIGURE 1 TEM micrographs of lung macrophages. (A) Typical intravascular macrophage within a pulmonary capillary; the cell is attached by adhesion complexes (arrows) to the underlying capillary endothelial cell. Arrowheads point to phagosome containing particles, which were injected intravenously. Micrograph by A. Warner, from Geiser, 2002. (B) Macrophage in the connective tissue (CT), that is, in the interstitium between the alveoli (AL)—with a polystyrene particle (P) retained in the surface lining layer—and the pulmonary capillary (CA). (C) and (D) show surface macrophages recovered by BAL from rats (C) and humans (D). M, macrophage nucleus; arrowheads, phago(lyso)somes. *Reprinted from Geiser (2010); the publisher for this copyrighted material is Mary Ann Liebert, Inc.*

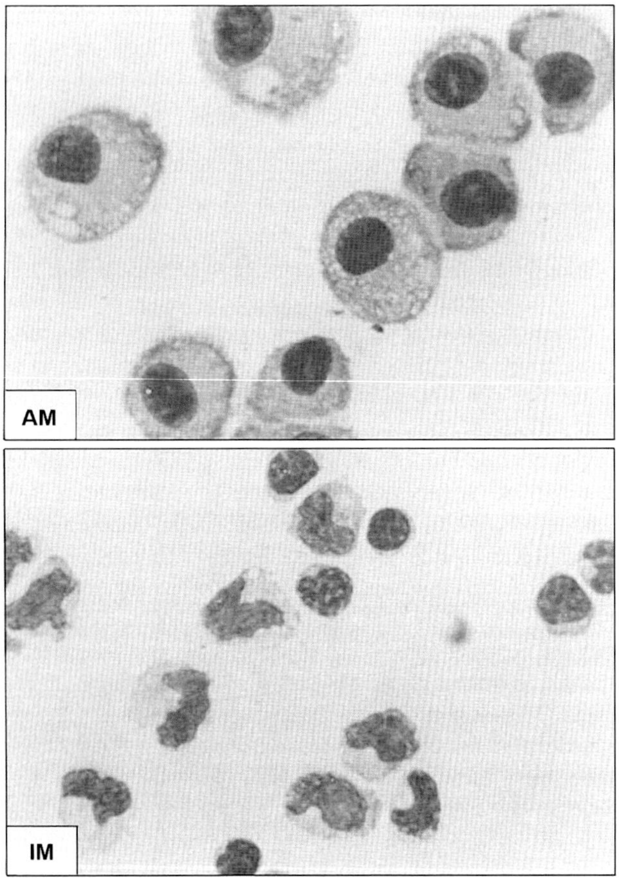

FIGURE 2 Photomicrographs of rat lung macrophages. Alveolar macrophages (AM, Upper panel); interstitial macrophages (IM, Lower panel). (Magnification ×1000). *Reprinted from Lavnikova et al. (1993) with permission of the American Thoracic Society. Copyright © 2014 American Thoracic Society.*

Evidence suggests that there is considerable functional heterogeneity within the alveolar macrophage subpopulation (Johnston et al., 2012). Thus, alveolar macrophage subpopulations have been identified based on different procedures used for their isolation or on density gradient centrifugation. These subpopulations differ morphologically and have distinct capacities for phagocytosis, cytotoxicity, chemotactic responsiveness, and production of ROS and proteolytic enzymes, as well as FcγR1 expression, suppression of lymphocyte proliferation, and production of IL-1 (Holt, 1979a, 1979b; Holt et al., 1982; Zwilling et al., 1982; Dauber et al., 1983; Holian et al., 1983; Bowden, 1984; O'Neill et al., 1984; Shellito and Kaltreider, 1984; Gant and Hamblin, 1985; Chandler et al., 1986a, 1986b; Sandron et al., 1986; Chandler and Fulmer, 1987; Oghiso, 1987; Brannen and Chandler, 1988; Nakstad et al., 1989; DiGregorio et al., 1991; Spiteri and Poulter, 1991; Lavnikova et al., 1993; Lohmann-Matthes et al., 1994; Prokhorova et al., 1994). It has also been reported that alveolar macrophages isolated by lavage differ in size, surface markers, and expression of MHC class II antigens (Cohen and Cline, 1971; Rhodes, 1975; van oud Alblas and van Furth, 1979). Whether these differences reflect distinct stages of differentiation or maturation, or specific recruitment of individual subpopulations into the lung, remains to be determined.

4.2 Interstitial Macrophages

Interstitial macrophages are found within the interstitial spaces of the lung parenchyma (Crapo et al., 1982, 1983; Pinkerton et al., 1982). They constitute approximately 2% of the lung's total cell population (Lehnert et al., 1985). In rat lung, ultrastructure studies have suggested that 15–40% of pulmonary macrophages are in the interstitial compartment, and morphologic studies have shown that the number of macrophages within the interstitium of the normal lung approximates or exceeds the number of alveolar macrophages (Crowell et al., 1992; Sebring and Lehnert, 1992). Notably, because interstitial macrophages are in direct contact with matrix and other pulmonary connective tissue components, activation of these cells and the release of cytotoxic/proinflammatory mediators may have greater pathological effects than those released by macrophages in the alveolar compartment.

Morphologic studies indicate that interstitial macrophages are smaller in surface area and volume and less dense than alveolar macrophages but have a greater nuclear-to-cytoplasm ratio (Warren et al., 1987; Dethloff and Lehnert, 1988; Sebring and Lehnert, 1992; Lavnikova et al., 1993; Johansson et al., 1997; Zetterberg et al., 1998; Fathi et al., 2001; Geiser, 2002) (Table 3; Figures 1 and 2). Interstitial macrophages are relatively uniform in size, with little membrane ruffling, blunt pseudopodia, and contain few intracytoplasmic lamellar inclusions (Warren et al., 1987). The nucleus of interstitial macrophages is typically indented, kidney shaped, and large, when compared to blood monocytes, with prominent heterochromatin (Sebring and Lehnert, 1992; Johansson et al., 1997; Zetterberg et al., 1998; Fathi et al., 2001). In contrast, heterochromatin is sparse in alveolar macrophage nuclei, and it generally appears as thin, marginal lines at the periphery. Lysosomes are relatively few in number in interstitial macrophages and mainly consist of large spherical primary lysosomes and/or large phagolysosomes (Sebring and Lehnert, 1992; Johansson et al., 1997). Interstitial macrophages also generally contain few vacuoles, negligible granular endoplasmic reticulum, and only occasional Golgi apparatus. Their mitochondria are spherical to ovoid in shape, have few cristae, and are less abundant than in alveolar macrophages (Zetterberg et al., 1998); however, they contain numerous polyribosomes. In addition, although interstitial macrophages are weakly positive for nonspecific esterase, they are strongly

TABLE 3 Numbers of Cells, Recovery and Size of Alveolar and Interstitial Macrophages

	Mouse	Hamster	Rat	Human	References
Alveolar Macrophages					
Avg. Recovery (10^6)	1.4 ± 0.4	4.2 ± 1.9	7.4 ± 0.2	8.8 ± 1.9	van oud Alblas and van Furth, 1979; Kobzik et al., 1988; Spiteri and Poulter, 1991; Wizemann and Laskin, 1994b
Diameter (μm)	19.0 ± 0.08	13.6 ± 0.4	13.1 ± 0.2	21.2 ± 0.3	Haley et al., 1991; Krombach et al., 1997
Size (μm^3)		1328.0 ± 123.0	1166.0 ± 42.0	4990.0 ± 174.0	Krombach et al., 1997
Interstitial Macrophages					
Avg. Recovery (10^6)	0.4 ± 0.4	3.0 ± 1.4	2.1 ± 0.3		Rouzer et al., 1982; Kobzik et al., 1988; Wizemann and Laskin, 1994b
Diameter (μM)			6.6 ± 0.2	7.6 ± 1.8	Brody et al., 1992; Fathi et al., 2001
Size (μm^3)			303.8 ± 15.4		Lavnikova et al., 1993

Note: Values are mean ± SE.

positive for peroxidase (Rouzer et al., 1982; Dethloff and Lehnert, 1988; Lavnikova et al., 1993). Phenotypically, interstitial macrophages express higher levels of CD163, CD54, and CX3CR1 than alveolar macrophages, but lower levels of CD11c, CD44, and CD206 (mannose receptor) (Miller et al., 1992; Kasper et al., 1995; Franke-Ullmann et al., 1996; Johansson et al., 1997; Lagranderie et al., 2003; Dioszeghy et al., 2008; Bedoret et al., 2009; Johnston et al., 2012; Lee, 2012; Misharin et al., 2013; Zaynagetdinov et al., 2013). Other cell surface antigens including CD64 (FcγR1), CD68, CD11b, CD115 (CSFR1), and CD71 (transferrin receptor) have also been identified on interstitial macrophages at levels approximately equivalent to or less than alveolar macrophages (Table 1).

In response to inflammatory stimuli, interstitial macrophages exhibit phagocytic, cytotoxic, tumoricidal, and antimicrobial activity, and release RNS and ROS, as well as TNFα and IFNα/β, however, these activities are generally reduced when compared to alveolar macrophages (Corrin, 1970; Sorokin and Brain, 1975; Stein-Streilein et al., 1983; Chandler et al., 1986c; Dethloff and Lehnert, 1988; Kobzik et al., 1988; Chandler and Brannen, 1990; Lohmann-Matthes et al., 1994; Prokhorova et al., 1994; Wizemann et al., 1994; Wizemann and Laskin, 1994b; Franke-Ullmann et al., 1996; Johansson et al., 1997; Zetterberg et al., 1998; Arias-Diaz et al., 2000; Steinmuller et al., 2000; Zetterberg et al., 2000; Fathi et al., 2001; Lagranderie et al., 2003) (Table 2). Interstitial macrophages also produce varying amounts of eicosanoids, including PGE_2, PGI_2, leukotriene C and TxA_2, and growth factors (e.g., PDGF, TGFβ) depending on the stimulus at levels comparable to alveolar macrophages (Rouzer et al., 1982; Chandler et al., 1988; Brody et al., 1992; Lohmann-Matthes et al., 1994; Bedoret et al., 2009). In contrast, interstitial macrophages exhibit markedly greater immunoregulatory and accessory functions, releasing more IL-1 and IL-6 (Lohmann-Matthes et al., 1994; Arias-Diaz et al., 2000; Lagranderie et al., 2003; Bedoret et al., 2009), and express higher levels of MHC class II antigens (Lohmann-Matthes et al., 1994; Franke-Ullmann et al., 1996; Johansson et al., 1997; Zetterberg et al., 1998; Steinmuller et al., 2000; Misharin et al., 2013). They are also more effective in functioning as accessory cells for mitogen-stimulated lymphocyte proliferation and maintaining specific immune reactions (Holt et al., 1982; Weissler et al., 1986; Bilyk and Holt, 1993; Lohmann-Matthes et al., 1994; Franke-Ullmann et al., 1996; Steinmuller et al., 2000). Additionally, interstitial macrophages inhibit dendritic cell migration and maturation and produce high levels of IL-10, negatively regulating Th2 responses in the lung and contributing to pulmonary immune homeostasis (Kradin et al., 2004; Bedoret et al., 2009; Johnston et al., 2012). Proliferative capacity is also significantly increased in interstitial macrophages when compared to alveolar macrophages (Bilyk et al., 1988; Kobzik et al., 1988; Steinmuller et al., 2000). Taken together, these observations suggest that alveolar macrophages play a more prominent role in nonspecific first-line host defense against infectious and injurious agents, whereas interstitial macrophages are equipped to cooperate with interstitial lymphocytes in inducing an adaptive immune reaction (Lohmann-Matthes et al., 1994; Franke-Ullmann et al., 1996; Geiser, 2010).

4.3 Pleural Macrophages

The pleural space is found between the visceral pleura that covers the surface of the lung and the parietal pleura that covers the mediastinum, diaphragm, and inner surface of the thoracic cage. A variety of cell types are found in the pleural space, most notably macrophages, which depending on the species comprise up to 80% of the cells in the pleural fluid (Peao et al., 1992; Shimotakahara et al., 2007). It has been suggested that pleural macrophages serve as tiny "roller bearings" that facilitate movement of the parietal and visceral pleura (Agostoni, 1972); they are also thought to play a role in host defense in the visceral and parietal pleura, and may be involved in some lung disease processes (Lehnert, 1992).

Pleural macrophages exhibit typical morphology of mononuclear phagocytes including enlarged horseshoe-shaped nucleus, vacuolated cytoplasm, numerous microvilli, and prevalent intermediate filaments (Baumann et al., 1993). They express higher levels of CD11b/CD18 and CD163 than alveolar macrophages but lower levels of CD68 and CD64 (Table 1). In general, pleural macrophages are smaller than alveolar macrophages and have less nonspecific esterase activity, as well as reduced bactericidal and fungicidal activity (Lehnert, 1992; Gjomarkaj et al., 1999; Frankenberger et al., 2000). However, like interstitial macrophages, pleural macrophages are highly enriched for markers associated with accessory cell functioning including MHC class II antigens, and they exhibit greater antigen-presenting cell activity than alveolar macrophages (Miller et al., 1992; Gjomarkaj et al., 1999). These observations suggest that pleural macrophages are key in cell-mediated immune reactions in the pleural spaces.

Pleural macrophages are phagocytic, bactericidal, and tumoricidal, and suppress mitogen-stimulated lymphocyte proliferation (Basic et al., 1979; Meuret et al., 1980; Zlotnik and Crowle, 1982; Zlotnik et al., 1982; Nakahashi et al., 1984; Sestini et al., 1984; Nagashima et al., 1987; Peao et al., 1992). They also respond to inflammatory mediators, particles, and pathogens, and release eicosanoids, RNS, macrophage inflammatory protein (MIP)-2, MCP-1, IL-1, IL-6, IL10, TNFα, and GM-CSF (Sestini et al., 1984; Nagashima et al., 1987; van den Heuvel et al., 1998; Frankenberger et al., 2000; Dorger et al., 2001; Cailhier et al., 2006; Shimotakahara et al., 2007). However, these activities are generally reduced when compared to alveolar macrophages (Table 2). They appear to function to clear the pleural space of degenerated, exfoliated mesothelial cells and other foreign materials. Findings that this is associated with a heightened propensity to secrete TNFα and generate RNS suggest that pleural macrophages are capable of orchestrating local inflammatory responses (Bozza et al., 1994; Choe et al., 1997; Werneck-Barroso et al., 2000; Dorger et al., 2001; Cailhier et al., 2006).

4.4 Pulmonary Intravascular Macrophages

Pulmonary intravascular macrophages are a resident population of mature mononuclear phagocytes localized in the lung capillaries. They have been identified in various mammalian species including cows, horses, pigs, sheep, goats, cats, and whales (Warner and Brain, 1984; Warner et al., 1986; Warner and Brain, 1986; Warner et al., 1987; Staub, 1994; Longworth, 1997). It remains uncertain, however, if pulmonary intravascular macrophages are present in healthy humans or rodents; it has been suggested that if they are resident in normal lung, they likely represent a relatively small population of cells (Dehring and Wismar, 1989; Lehnert, 1992). Conversely, significant numbers of macrophages have been noted in pulmonary capillaries of patients with pulmonary arterial hypertension, as well as in rodents with experimentally induced pulmonary hypertension, and they are thought to contribute to the development of the disease (Tuder et al., 1994; Longworth, 1997; Sone et al., 1999; El Chami and Hassoun, 2012; Gerasimovskaya et al., 2012; Schneberger et al., 2012). Pulmonary intravascular macrophages have also been reported to increase in numbers in humans suffering from liver disease/abnormalities, and in experimental rodent models of sepsis (Garty et al., 1984; Schneberger et al., 2012). Species differences in numbers of constitutive pulmonary intravascular macrophages likely reflect the prominence of hepatic Kupffer cell clearance of blood-derived materials in animals lacking these cells (Staub, 1994; Brain et al., 1999).

Pulmonary intravascular macrophages are large cells ranging in diameter from 20 to 80 μm. They have characteristic morphological features of differentiated macrophages including an irregular shape, an indented nucleus, lysosomal granules, and phagolysosomes (Warner and Brain, 1986; Winkler and Cheville, 1987) (Figure 1). A unique feature of pulmonary intravascular macrophages is their electron-dense surface coat (Brain et al., 1999). This "fuzzy" glycocalyx consists mainly of lipid or lipoprotein globules of 50−200 nm in size, which decorate the cell surface (Atwal et al., 1992; Singh and Atwal, 1997). It is the globular surface coat of pulmonary intravascular macrophages that complexes with blood-borne materials such as endotoxin, facilitating their uptake (Schneberger et al., 2012). Pulmonary intravascular macrophages also have multiple pseudopodia and membrane invaginations and are highly phagocytic, removing foreign particles, cellular debris, fibrin, and pathogens, as well as erythrocytes and immune cells from the pulmonary circulation (Warner and Brain, 1984, 1986; Atwal and Saldanha, 1985; Warner et al., 1986, 1987; Whiteley et al., 1990, 1991; Longworth, 1997). Findings that pulmonary intravascular macrophages contain large phagosomes filled with cellular debris indicate that a normal function of these cells is clearance of effete cells from the circulation (Winkler and Cheville,

1984, 1987; Atwal and Saldanha, 1985; Warner et al., 1986; Wheeldon and Hansen-Flaschen, 1986). Pulmonary intravascular macrophages are firmly attached to the interior of the capillary endothelium, forming membrane–adhesion complexes that are thought to be key for anchoring them to the microvasculature (Warner et al., 1986; Brain et al., 1999; Geiser, 2002). They attach preferentially to the thick portion of the air–blood barrier, minimizing interference with gas exchange at the air–blood barrier (Winkler and Cheville, 1987; Winkler, 1988).

Like other macrophage subpopulations in the respiratory tract, pulmonary intravascular macrophages have the capacity to generate IL-1β, IL-6, IL-8, IL-10, TNFα, and RNS in response to inflammatory stimuli as well as lipoxygenase and cyclooxygenase metabolites including PGE_2, PGF_2, PGD_2, TXB_2, and LTB_4 (Bertram et al., 1988; Miyamoto et al., 1988; Chitko-McKown et al., 1991, 1992; Longworth, 1997; Chen et al., 2003; Parbhakar et al., 2004, 2005; Schneberger et al., 2012). Whereas intravascular macrophages release less RNS than alveolar macrophages, they generate more eicosanoids (Table 2). In contrast, they exhibit similar phagocytic, bactericidal, tumoricidal, and antibody-dependent cytotoxic activity, and release comparable levels of IL-1 and TNFα (Chitko-McKown et al., 1991; Staub, 1994). Studies using gadolinium chloride and clodronate liposomes to deplete pulmonary intravascular macrophages have demonstrated a proinflammatory role of these cells in lung inflammation induced by bacteria, viruses, and endotoxins that is likely facilitated by high constitutive expression of β3 integrin subunit (CD61), as well as TLR4, TLR9, and MRP-14, an endogenous activator of TLR (Parbhakar et al., 2004; Singh Suri et al., 2006; Jimenez-Marin et al., 2008; Schneberger et al., 2011b, 2012; Duke-Novakovski et al., 2013). Because of the communication between the alveolar and capillary environments across the alveolar septum, activated pulmonary intravascular macrophages may amplify these signals resulting in higher levels of lung vascular inflammation (Schneberger et al., 2012). Additionally, in species with large numbers of reactive pulmonary intravascular macrophages, there are notable hemodynamic and acute lung injury responses to intravenous infusions of endotoxin or gram negative bacteria (Brigham et al., 1974, 1979; Bottoms et al., 1981; Crocker et al., 1981; Staub, 1989).

4.5 Airway Macrophages

Macrophages are present on the intraluminal surfaces of both the large and small conducting airways (Lehnert, 1992). They are found suspended in the mucus lining or beneath the mucous and serous lining adjacent to underlying airway epithelial cells (Sorokin and Brain, 1975). It has been estimated that the relative abundance of airway macrophages is similar to alveolar macrophages (Lehnert, 1992). Airway macrophages are generally larger than alveolar macrophages and have a smoother cell surface and larger and more regularly shaped nuclei (Lehnert et al., 1990). However, they are less phagocytic than alveolar macrophages and have reduced lysosomal enzymes, acid phosphatase, nonspecific esterase, and β-glucuronidase. It has been suggested that airway macrophages are essentially alveolar macrophages that have translocated out of alveoli and transitioned into a functionally less active or effete state (Brain et al., 1984; Lehnert, 1984, 1990, 1992; Powdrill et al., 1989). This is supported by findings that airway macrophages are phenotypically similar to alveolar macrophages expressing comparable levels of F4/80, CD64, MHC class II antigens, and CD68, and lacking CD163 (Lehnert, 1992; Maus et al., 2002; Lay et al., 2009) (Table 1).

Airway macrophages are thought to serve a similar protective function as alveolar macrophages, isolating and removing potentially noxious materials, including allergens, from the airways (Lehnert, 1992; Lay et al., 2009). They also release mediators that attract lymphocytes, neutrophils, and mast cells into the airways, where they regulate their activity (Brain, 1988). Airway macrophages can ingest and degrade antigens deposited in the airways, thereby suppressing their antigenicity or process and present them to cells of the adaptive immune system. They have also been implicated in the pathogenesis and progression of allergic airways disease and may contribute to increased risk of disease exacerbation associated with particulate exposure (Lay et al., 2009).

5. ORIGIN OF PULMONARY MACROPHAGES

A question arises regarding the origin of the different pulmonary macrophage subpopulations. Early literature proposed that alveolar macrophages are part of the mononuclear phagocyte system and are routinely replenished by blood monocyte and bone marrow precursors (van oud Alblas and van Furth, 1979; Lohmann-Matthes et al., 1994). Moreover, it was suggested that blood monocytes do not migrate directly into the air sacs but rather pass into an interstitial compartment where they undergo division and maturation into alveolar macrophages (Bowden et al., 1969; Bowden and Adamson, 1972; Holt et al., 1982; Bilyk et al., 1988; Dethloff and Lehnert, 1988; Sebring and Lehnert, 1992). One difficulty with this idea is that it is based on animal studies of macrophage replenishment after experimental whole body irradiation or in humans undergoing bone marrow transplantation, which do not reflect the response during homeostasis (Pinkett et al., 1966; Virolainen, 1968; Brunstetter et al., 1971; Godleski and Brain, 1972; Thomas et al., 1976). In this regard, recent studies using chimeric mice and parabiosis models have demonstrated that while adult circulating monocytes contribute to pools of alveolar macrophages in irradiated mice, this does not occur in the normal lung (Tarling et al., 1987; Guilliams et al., 2013a). Rather, it appears that most pulmonary macrophages are descendants of hematopoietic precursor cells that migrate into the lung during embryogenesis and develop into long-lived macrophages shortly after birth in response to instructive signals provided by cytokines such as GM-CSF (Tarling et al., 1987; Guilliams et al., 2013a; Hashimoto et al., 2013). Additionally, while local proliferation contributes to macrophage replenishment following injury or infection, under steady-state conditions, lung macrophages are relatively quiescent, exhibiting minimal growth in vivo (Evans et al., 1973; Golde et al., 1974; Bowden and Adamson, 1980; Bitterman et al., 1983; Sorokin et al., 1984, 1989; Shellito et al., 1987; Sorokin and Hoyt, 1987; Kobzik et al., 1988; Sebring and Lehnert, 1992; Murphy et al., 2008).

6. CONCLUSIONS

The respiratory tract is the largest surface area of the body in contact with the outside environment. Consequently, like the skin and gastrointestinal mucosa, it is continuously exposed to a diverse array of microbes and particles. Thus, it is essential that effective host defense mechanisms be in place in the respiratory tract. The fact that distinct subpopulations of macrophages are present throughout the airways and lung suggests that their contribution to host defense and homeostasis is essential. It is now well established that macrophages are also key initiators of adaptive immune responses in the lung. Further evaluation of the unique functions of the different macrophage subpopulations and their role in disease pathogenesis may lead to more efficacious approaches for treating respiratory diseases and disorders.

ACKNOWLEDGMENTS

This work was supported by NIH Grants U54AR055073, K08HL096426, R01ES004738, R01CA132624, and P30ES005022.

REFERENCES

Agostoni, E., 1972. Mechanics of the pleural space. Physiol. Rev. 52, 57–128.

Alber, A., Howie, S.E., Wallace, W.A., Hirani, N., 2012. The role of macrophages in healing the wounded lung. Int. J. Exp. Pathol. 93, 243–251.

van oud Alblas, A.B., van Furth, R., 1979. Origin, kinetics, and characteristics of pulmonary macrophages in the normal steady state. J. Exp. Med. 149, 1504–1518.

Andre, S., Metivier, H., Lantenois, G., Boyer, M., Nolibe, D., Masse, R., 1987. Beryllium metal solubility in the lung, comparison of metal and hot-pressed forms by in vivo and in vitro dissolution bioassays. Hum. Toxicol. 6, 233–240.

Arias-Diaz, J., Garcia-Verdugo, I., Casals, C., Sanchez-Rico, N., Vara, E., Balibrea, J.L., 2000. Effect of surfactant protein A (SP-A) on the production of cytokines by human pulmonary macrophages. Shock 14, 300–306.

Arredouani, M., Yang, Z., Ning, Y., Qin, G., Soininen, R., Tryggvason, K., Kobzik, L., 2004. The scavenger receptor MARCO is required for lung defense against pneumococcal pneumonia and inhaled particles. J. Exp. Med. 200, 267–272.

Atwal, O.S., Saldanha, K.A., 1985. Erythrophagocytosis in alveolar capillaries of goat lung: ultrastructural properties of blood monocytes. Acta. Anat. (Basel). 124, 245–254.

Atwal, O.S., Singh, B., Staempfli, H., Minhas, K., 1992. Presence of pulmonary intravascular macrophages in the equine lung: some structuro-functional properties. Anat. Rec. 234, 530–540.

Bailey, M.R., Fry, F.A., James, A.C., 1982. The long-term clearance kinetics of insoluble particles from the human lung. Ann. Occup. Hyg. 26, 273–290.

Basic, I., Rode, B., Kastelan, A., Milas, L., 1979. Activation of pleural macrophages by intrapleural application of *Corynebacterium parvum*. Adv. Exp. Med. Biol. 121, 333–341.

Baumann, M.H., Heinrich, K., Sahn, S.A., Green, C., Harley, R., Strange, C., 1993. Electron microscopic analysis of the normal and the activated pleural macrophage. Exp. Lung Res. 19, 731–742.

Becker, S., Devlin, R.B., Haskill, J.S., 1989. Differential production of tumor necrosis factor, macrophage colony stimulating factor, and interleukin 1 by human alveolar macrophages. J. Leukoc. Biol. 45, 353–361.

Bedoret, D., Wallemacq, H., Marichal, T., Desmet, C., Quesada Calvo, F., Henry, E., Closset, R., Dewals, B., Thielen, C., Gustin, P., de Leval, L., Van Rooijen, N., Le Moine, A., Vanderplasschen, A., Cataldo, D., Drion, P.V., Moser, M., Lekeux, P., Bureau, F., 2009. Lung interstitial macrophages alter dendritic cell functions to prevent airway allergy in mice. J. Clin. Invest. 119, 3723–3738.

Bertram, T.A., Overby, L.H., Danilowicz, R., Eling, T.E., Brody, A.R., 1988. Pulmonary intravascular macrophages metabolize arachidonic acid in vitro. Comparison with alveolar macrophages. Am. Rev. Respir. Dis. 138, 936–944.

Bilyk, N., Holt, P.G., 1993. Inhibition of the immunosuppressive activity of resident pulmonary alveolar macrophages by granulocyte/macrophage colony-stimulating factor. J. Exp. Med. 177, 1773–1777.

Bilyk, N., Mackenzie, J.S., Papadimitriou, J.M., Holt, P.G., 1988. Functional studies on macrophage populations in the airways and the lung wall of SPF mice in the steady-state and during respiratory virus infection. Immunology 65, 417–425.

Bingisser, R.M., Holt, P.G., 2001. Immunomodulating mechanisms in the lower respiratory tract: nitric oxide mediated interactions between alveolar macrophages, epithelial cells, and T-cells. Swiss Med. Wkly. 131, 171–179.

Bitterman, P., Adelberg, S., Goodman, S., Saltzman, L., Crystal, R.G., 1983. Role of macrophage replication in modulating the increased number of alveolar macrophages in chronic inflammatory lung disorders. Am. Rev. Respir. Dis. 127 (Suppl.), A60.

Blumenthal, R.L., Campbell, D.E., Hwang, P., DeKruyff, R.H., Frankel, L.R., Umetsu, D.T., 2001. Human alveolar macrophages induce functional inactivation in antigen-specific CD4 T cells. J. Allergy Clin. Immunol. 107, 258–264.

du Bois, R.M., 1985. The alveolar macrophage. Thorax 40, 321–327.

Bottoms, G.D., Fessler, J.F., Roesel, O.F., Moore, A.B., Frauenfelder, H.C., 1981. Endotoxin-induced hemodynamic changes in ponies: effects of flunixin meglumine. Am. J. Vet. Res. 42, 1514–1518.

Bowden, D.H., 1984. The alveolar macrophage. Environ. Health Perspect. 55, 327–341.

Bowden, D.H., Adamson, I.Y., 1972. The pulmonary interstitial cell as immediate precursor of the alveolar macrophage. Am. J. Pathol. 68, 521–537.

Bowden, D.H., Adamson, I.Y., 1980. Role of monocytes and interstitial cells in the generation of alveolar macrophages I. Kinetic studies of normal mice. Lab. Invest. 42, 511–517.

Bowden, D.H., Adamson, I.Y., Grantham, W.G., Wyatt, J.P., 1969. Origin of the lung macrophage. Evidence derived from radiation injury. Arch. Pathol. 88, 540–546.

Bozza, P.T., Castro-Faria-Neto, H.C., Penido, C., Larangeira, A.P., das Gracas, M., Henriques, M.O., Silva, P.M., Martins, M.A., dos Santos, R.R., Cordeiro, R.S., 1994. Requirement for lymphocytes and resident macrophages in LPS-induced pleural eosinophil accumulation. J. Leukoc. Biol. 56, 151–158.

Brain, J.D., 1988. Lung macrophages: how many kinds are there? What do they do? Am. Rev. Respir. Dis. 137, 507–509.

Brain, J.D., Gehr, P., Kavet, R.I., 1984. Airway macrophages. The importance of the fixation method. Am. Rev. Respir. Dis. 129, 823–826.

Brain, J.D., Molina, R.M., DeCamp, M.M., Warner, A.E., 1999. Pulmonary intravascular macrophages: their contribution to the mononuclear phagocyte system in 13 species. Am. J. Physiol. 276, L146–L154.

Brannen, A.L., Chandler, D.B., 1988. Alveolar macrophage subpopulations' responsiveness to chemotactic stimuli. Am. J. Pathol. 132, 161–166.

Brigham, K.L., Bowers, R., Haynes, J., 1979. Increased sheep lung vascular permeability caused by *Escherichia coli* endotoxin. Circ. Res. 45, 292–297.

Brigham, K.L., Woolverton, W.C., Blake, L.H., Staub, N.C., 1974. Increased sheep lung vascular permeability caused by Pseudomonas bacteremia. J. Clin. Invest. 54, 792–804.

Brody, A.R., Bonner, J.C., Overby, L.H., Badgett, A., Kalter, V., Kumar, R.K., Bennett, R.A., 1992. Interstitial pulmonary macrophages produce platelet-derived growth factor that stimulates rat lung fibroblast proliferation in vitro. J. Leukoc. Biol. 51, 640–648.

Brunstetter, M.A., Hardie, J.A., Schiff, R., Lewis, J.P., Cross, C.E., 1971. The origin of pulmonary alveolar macrophages. Studies of stem cells using the Es-2 marker of mice. Arch. Intern. Med. 127, 1064–1068.

Cailhier, J.F., Sawatzky, D.A., Kipari, T., Houlberg, K., Walbaum, D., Watson, S., Lang, R.A., Clay, S., Kluth, D., Savill, J., Hughes, J., 2006. Resident pleural macrophages are key orchestrators of neutrophil recruitment in pleural inflammation. Am. J. Respir. Crit. Care Med. 173, 540–547.

Capelli, A., Di Stefano, A., Gnemmi, I., Balbo, P., Cerutti, C.G., Balbi, B., Lusuardi, M., Donner, C.F., 1999. Increased MCP-1 and MIP-1β in bronchoalveolar lavage fluid of chronic bronchitics. Eur. Respir. J. 14, 160–165.

Chandler, D.B., Bayles, G., Fuller, W.C., 1988. Prostaglandin synthesis and release by subpopulations of rat interstitial macrophages. Am. Rev. Respir. Dis. 138, 901–907.

Chandler, D.B., Brannen, A.L., 1990. Interstitial macrophage subpopulations: responsiveness to chemotactic stimuli. Tissue Cell 22, 427–434.

Chandler, D.B., Fuller, W.C., Jackson, R.M., Fulmer, J.D., 1986a. Fractionation of rat alveolar macrophages by isopycnic centrifugation: morphological, cytochemical, biochemical, and functional properties. J. Leukoc. Biol. 39, 371–383.

Chandler, D.B., Fuller, W.C., Jackson, R.M., Fulmer, J.D., 1986b. Studies of membrane receptors and phagocytosis in subpopulations of rat alveolar macrophages. Am. Rev. Respir. Dis. 133, 461–467.

Chandler, D.B., Fulmer, J.D., 1987. Prostaglandin synthesis and release by subpopulations of rat alveolar macrophages. J. Immunol. 139, 893–898.

Chandler, D.B., Kennedy, J.I., Fulmer, J.D., 1986c. Studies of membrane receptors, phagocytosis, and morphology of subpopulations of rat lung interstitial macrophages. Am. Rev. Respir. Dis. 134, 542–547.

Chen, W.H., Toapanta, F.R., Shirey, K.A., Zhang, L., Giannelou, A., Page, C., Frieman, M.B., Vogel, S.N., Cross, A.S., 2012. Potential role for alternatively activated macrophages in the secondary bacterial infection during recovery from influenza. Immunol. Lett. 141, 227–234.

Chen, Z.T., Li, S.L., Cai, E.Q., Wu, W.L., Jin, J.S., Zhu, B., 2003. LPS induces pulmonary intravascular macrophages producing inflammatory mediators via activating NF-κB. J. Cell. Biochem. 89, 1206–1214.

Chitko-McKown, C.G., Chapes, S.K., Brown, R.E., Phillips, R.M., McKown, R.D., Blecha, F., 1991. Porcine alveolar and pulmonary intravascular macrophages: comparison of immune functions. J. Leukoc. Biol. 50, 364–372.

Chitko-McKown, C.G., Reddy, D.N., Chapes, S.K., McKown, R.D., Blecha, F., 1992. Immunological characterization of pulmonary intravascular macrophages. Reg. Immunol. 4, 236–244.

Choe, N., Tanaka, S., Xia, W., Hemenway, D.R., Roggli, V.L., Kagan, E., 1997. Pleural macrophage recruitment and activation in asbestos-induced pleural injury. Environ. Health Perspect. 105 (Suppl. 5), 1257–1260.

Cohen, A.B., Cline, M.J., 1971. The human alveolar macrophage: isolation, cultivation in vitro, and studies of morphologic and functional characteristics. J. Clin. Invest. 50, 1390–1398.

Corrin, B., 1970. Phagocytic potential of pulmonary alveolar epithelium with particular reference to surfactant metabolism. Thorax 25, 110–115.

Crapo, J.D., Barry, B.E., Gehr, P., Bachofen, M., Weibel, E.R., 1982. Cell number and cell characteristics of the normal human lung. Am. Rev. Respir. Dis. 126, 332–337.

Crapo, J.D., Young, S.L., Fram, E.K., Pinkerton, K.E., Barry, B.E., Crapo, R.O., 1983. Morphometric characteristics of cells in the alveolar region of mammalian lungs. Am. Rev. Respir. Dis. 128, S42–S46.

Crocker, S.H., Eddy, D.O., Obenauf, R.N., Wismar, B.L., Lowery, B.D., 1981. Bacteremia: host-specific lung clearance and pulmonary failure. J. Trauma 21, 215–220.

Crowell, R.E., Heaphy, E., Valdez, Y.E., Mold, C., Lehnert, B.E., 1992. Alveolar and interstitial macrophage populations in the murine lung. Exp. Lung Res. 18, 435–446.

Dauber, J.H., Holian, A., Rosemiller, M.E., Daniele, R.P., 1983. Separation of bronchoalveolar cells from the guinea pig on continuous density gradients of Percoll: morphology and cytochemical properties of fractionated lung macrophages. J. Reticuloendothel. Soc. 33, 119–126.

Davies, L.C., Jenkins, S.J., Allen, J.E., Taylor, P.R., 2013. Tissue-resident macrophages. Nat. Immunol. 14, 986–995.

Day, J., Friedman, A., Schlesinger, L.S., 2009. Modeling the immune rheostat of macrophages in the lung in response to infection. Proc. Natl. Acad. Sci. U. S. A. 106, 11246–11251.

Dehring, D.J., Wismar, B.L., 1989. Intravascular macrophages in pulmonary capillaries of humans. Am. Rev. Respir. Dis. 139, 1027–1029.

Dethloff, L.A., Lehnert, B.E., 1988. Pulmonary interstitial macrophages: isolation and flow cytometric comparisons with alveolar macrophages and blood monocytes. J. Leukoc. Biol. 43, 80–90.

DiGregorio, K.A., Cilento, E.V., Lantz, R.C., 1991. Heterogeneity in superoxide production as measured by nitro blue tetrazolium reduction from individual PAM. Am. J. Physiol. 260, L464–L470.

Dioszeghy, V., Rosas, M., Maskrey, B.H., Colmont, C., Topley, N., Chaitidis, P., Kuhn, H., Jones, S.A., Taylor, P.R., O'Donnell, V.B., 2008. 12/15-Lipoxygenase regulates the inflammatory response to bacterial products in vivo. J. Immunol. 181, 6514–6524.

Dorger, M., Munzing, S., Allmeling, A.M., Messmer, K., Krombach, F., 2001. Phenotypic and functional differences between rat alveolar, pleural, and peritoneal macrophages. Exp. Lung Res. 27, 65–76.

Duke-Novakovski, T., Singh-Suri, S., Kajikawa, O., Caldwell, S., Charavaryamath, C., Singh, B., 2013. Immuno-phenotypic and functional characterization of rabbit pulmonary intravascular macrophages. Cell Tissue Res. 351, 149–160.

El Chami, H., Hassoun, P.M., 2012. Immune and inflammatory mechanisms in pulmonary arterial hypertension. Prog. Cardiovasc. Dis. 55, 218–228.

Evans, M.J., Cabral, L.J., Stephens, R.J., Freeman, G., 1973. Cell division of alveolar macrophages in rat lung following exposure to NO_2. Am. J. Pathol. 70, 199–208.

Fathi, M., Johansson, A., Lundborg, M., Orre, L., Skold, C.M., Camner, P., 2001. Functional and morphological differences between human alveolar and interstitial macrophages. Exp. Mol. Pathol. 70, 77–82.

Feng, Y.H., Mao, H., 2012. Expression and preliminary functional analysis of Siglec-F on mouse macrophages. J. Zhejiang Univ. Sci. B. 13, 386–394.

Franke-Ullmann, G., Pfortner, C., Walter, P., Steinmuller, C., Lohmann-Matthes, M.L., Kobzik, L., 1996. Characterization of murine lung interstitial macrophages in comparison with alveolar macrophages in vitro. J. Immunol. 157, 3097–3104.

Franke, G., Freihorst, J., Steinmuller, C., Verhagen, W., Hockertz, S., Lohmann-Matthes, M.L., 1994. Interaction of alveolar macrophages and respiratory syncytial virus. J. Immunol. Methods 174, 173–184.

Frankenberger, M., Passlick, B., Hofer, T., Siebeck, M., Maier, K.L., Ziegler-Heitbrock, L.H., 2000. Immunologic characterization of normal human pleural macrophages. Am. J. Respir. Cell. Mol. Biol. 23, 419–426.

Fujimori, Y., Kataoka, M., Tada, S., Takehara, H., Matsuo, K., Miyake, T., Okahara, M., Yamadori, I., Tanimoto, M., 2003. The role of interleukin-8 in interstitial pneumonia. Respirology 8, 33–40.

Gant, V.A., Hamblin, A.S., 1985. Human bronchoalveolar macrophage heterogeneity demonstrated by histochemistry, surface markers and phagocytosis. Clin. Exp. Immunol. 60, 539–545.

Garn, H., Siese, A., Stumpf, S., Wensing, A., Renz, H., Gemsa, D., 2006. Phenotypical and functional characterization of alveolar macrophage subpopulations in the lungs of NO_2-exposed rats. Respir. Res. 7, 4.

Garty, I., Tal, I., Kaynan, A., 1984. Tc-99m colloid lung uptake in a rare case of toxoplasmosis with liver involvement. Clin. Nucl. Med. 9, 310–313.

Gaschler, G.J., Zavitz, C.C., Bauer, C.M., Skrtic, M., Lindahl, M., Robbins, C.S., Chen, B., Stampfli, M.R., 2008. Cigarette smoke exposure attenuates cytokine production by mouse alveolar macrophages. Am. J. Respir. Cell. Mol. Biol. 38, 218–226.

Geiser, M., 2002. Morphological aspects of particle uptake by lung phagocytes. Microsc. Res. Tech. 57, 512–522.

Geiser, M., 2010. Update on macrophage clearance of inhaled micro- and nanoparticles. J. Aerosol Med. Pulm. Drug Delivery 23, 207–217.

Gerasimovskaya, E., Kratzer, A., Sidiakova, A., Salys, J., Zamora, M., Taraseviciene-Stewart, L., 2012. Interplay of macrophages and T cells in the lung vasculature. Am. J. Physiol. Lung Cell. Mol. Physiol. 302, L1014–L1022.

Gjomarkaj, M., Pace, E., Melis, M., Spatafora, M., Profita, M., Vignola, A.M., Bonsignore, G., Toews, G.B., 1999. Phenotypic and functional characterization of normal rat pleural macrophages in comparison with autologous peritoneal and alveolar macrophages. Am. J. Respir. Cell Mol. Biol. 20, 135–142.

Godleski, J.J., Brain, J.D., 1972. The origin of alveolar macrophages in mouse radiation chimeras. J. Exp. Med. 136, 630–643.

Golde, D.W., Byers, L.A., Finley, T.N., 1974. Proliferative capacity of human alveolar macrophage. Nature 247, 373–375.

Gough, P.J., Greaves, D.R., Suzuki, H., Hakkinen, T., Hiltunen, M.O., Turunen, M., Herttuala, S.Y., Kodama, T., Gordon, S., 1999. Analysis of macrophage scavenger receptor (SR-A) expression in human aortic atherosclerotic lesions. Arterioscler. Thromb. Vasc. Biol. 19, 461–471.

Grundy, M., Sentman, C.L., 2005. GFP transgenic mice show dynamics of lung macrophages. Exp. Cell Res. 310, 409–416.

Guilliams, M., De Kleer, I., Henri, S., Post, S., Vanhoutte, L., De Prijck, S., Deswarte, K., Malissen, B., Hammad, H., Lambrecht, B.N., 2013a. Alveolar macrophages develop from fetal monocytes that differentiate into long-lived cells in the first week of life via GM-CSF. J. Exp. Med. 210, 1977–1992.

Guilliams, M., Lambrecht, B.N., Hammad, H., 2013b. Division of labor between lung dendritic cells and macrophages in the defense against pulmonary infections. Mucosal Immunol. 6, 464–473.

Guth, A.M., Janssen, W.J., Bosio, C.M., Crouch, E.C., Henson, P.M., Dow, S.W., 2009. Lung environment determines unique phenotype of alveolar macrophages. Am. J. Physiol. Lung Cell. Mol. Physiol. 296, L936–L946.

Gwyer Findlay, E., Hussell, T., 2012. Macrophage-mediated inflammation and disease: a focus on the lung. Mediators Inflammation 2012, 140937.

Haley, P.J., Muggenburg, B.A., Weissman, D.N., Bice, D.E., 1991. Comparative morphology and morphometry of alveolar macrophages from six species. Am. J. Anat. 191, 401–407.

Han, X., Sterling, H., Chen, Y., Saginario, C., Brown, E.J., Frazier, W.A., Lindberg, F.P., Vignery, A., 2000. CD47, a ligand for the macrophage fusion receptor, participates in macrophage multinucleation. J. Biol. Chem. 275, 37984–37992.

Harris, J.O., Olsen, G.N., Castle, J.R., Maloney, A.S., 1975. Comparison of proteolytic enzyme activity in pulmonary alveolar macrophages and blood leukocytes in smokers and nonsmokers. Am. Rev. Respir. Dis. 111, 579–586.

Hashimoto, D., Chow, A., Noizat, C., Teo, P., Beasley, M.B., Leboeuf, M., Becker, C.D., See, P., Price, J., Lucas, D., Greter, M., Mortha, A., Boyer, S.W., Forsberg, E.C., Tanaka, M., van Rooijen, N., Garcia-Sastre, A., Stanley, E.R., Ginhoux, F., Frenette, P.S., Merad, M., 2013. Tissue-resident macrophages self-maintain locally throughout adult life with minimal contribution from circulating monocytes. Immunity 38, 792–804.

Herold, S., Mayer, K., Lohmeyer, J., 2011. Acute lung injury: how macrophages orchestrate resolution of inflammation and tissue repair. Front. Immunol. 2, 65.

van den Heuvel, M.M., Smit, H.J., Barbierato, S.B., Havenith, C.E., Beelen, R.H., Postmus, P.E., 1998. Talc-induced inflammation in the pleural cavity. Eur. Respir. J. 12, 1419–1423.

Holian, A., Dauber, J.H., Diamond, M.S., Daniele, R.P., 1983. Separation of bronchoalveolar cells from the guinea pig on continuous gradients of Percoll: functional properties of fractionated lung macrophages. J. Reticuloendothel. Soc. 33, 157–164.

Holt, P.G., 1979a. Alveolar macrophages. II. Inhibition of lymphocyte proliferation by purified macrophages from rat lung. Immunology 37, 429–436.

Holt, P.G., 1979b. Alveolar macrophages. III. Studies on the mechanisms of inhibition of T-cell proliferation. Immunology 37, 437–445.

Holt, P.G., Oliver, J., Bilyk, N., McMenamin, C., McMenamin, P.G., Kraal, G., Thepen, T., 1993. Downregulation of the antigen presenting cell function(s) of pulmonary dendritic cells in vivo by resident alveolar macrophages. J. Exp. Med. 177, 397–407.

Holt, P.G., Warner, L.A., Papadimitriou, J.M., 1982. Alveolar macrophages: functional heterogeneity within macrophage populations from rat lung. Aust. J. Exp. Biol. Med. Sci. 60, 607–618.

Janssen, W.J., Barthel, L., Muldrow, A., Oberley-Deegan, R.E., Kearns, M.T., Jakubzick, C., Henson, P.M., 2011. Fas determines differential fates of resident and recruited macrophages during resolution of acute lung injury. Am. J. Respir. Crit. Care Med. 184, 547–560.

Janssen, W.J., McPhillips, K.A., Dickinson, M.G., Linderman, D.J., Morimoto, K., Xiao, Y.Q., Oldham, K.M., Vandivier, R.W., Henson, P.M., Gardai, S.J., 2008. Surfactant proteins A and D suppress alveolar macrophage phagocytosis via interaction with SIRP α. Am. J. Respir. Crit. Care Med. 178, 158–167.

Jiang-Shieh, Y.F., Chien, H.F., Chang, C.Y., Wei, T.S., Chiu, M.M., Chen, H.M., Wu, C.H., 2010. Distribution and expression of CD200 in the rat respiratory system under normal and endotoxin-induced pathological conditions. J. Anat. 216, 407–416.

Jimenez-Marin, A., Yubero, N., Esteso, G., Moreno, A., de las Mulas, J.M., Morera, L., Llanes, D., Barbancho, M., Garrido, J.J., 2008. Molecular characterization and expression analysis of the gene coding for the porcine β_3 integrin subunit (CD61). Gene 408, 9–17.

Johansson, A., Lundborg, M., Skold, C.M., Lundahl, J., Tornling, G., Eklund, A., Camner, P., 1997. Functional, morphological, and phenotypical differences between rat alveolar and interstitial macrophages. Am. J. Respir. Cell Mol. Biol. 16, 582–588.

Johnston, L.K., Rims, C.R., Gill, S.E., McGuire, J.K., Manicone, A.M., 2012. Pulmonary macrophage subpopulations in the induction and resolution of acute lung injury. Am. J. Respir. Cell Mol. Biol. 47, 417–426.

Kasper, M., Gunthert, U., Dall, P., Kayser, K., Schuh, D., Haroske, G., Muller, M., 1995. Distinct expression patterns of CD44 isoforms during human lung development and in pulmonary fibrosis. Am. J. Respir. Cell Mol. Biol. 13, 648–656.

Kasper, M., Hughes, R.C., 1996. Immunocytochemical evidence for a modulation of galectin 3 (Mac-2), a carbohydrate binding protein, in pulmonary fibrosis. J. Pathol. 179, 309–316.

Kelley, T.W., Graham, M.M., Doseff, A.I., Pomerantz, R.W., Lau, S.M., Ostrowski, M.C., Franke, T.F., Marsh, C.B., 1999. Macrophage colony-stimulating factor promotes cell survival through Akt/protein kinase B. J. Biol. Chem. 274, 26393–26398.

Knapp, S., Leemans, J.C., Florquin, S., Branger, J., Maris, N.A., Pater, J., van Rooijen, N., van der Poll, T., 2003. Alveolar macrophages have a protective antiinflammatory role during murine pneumococcal pneumonia. Am. J. Respir. Crit. Care Med. 167, 171–179.

Kobzik, L., Godleski, J.J., Barry, B.E., Brain, J.D., 1988. Isolation and antigenic identification of hamster lung interstitial macrophages. Am. Rev. Respir. Dis. 138, 908–914.

Kollert, F., Probst, C., Muller-Quernheim, J., Zissel, G., Prasse, A., 2009. CCL18 production is decreased in alveolar macrophages from cigarette smokers. Inflammation 32, 163–168.

Kovach, M.A., Standiford, T.J., 2011. Toll like receptors in diseases of the lung. Int. Immunopharmacol. 11, 1399–1406.

Kradin, R.L., Sakamoto, H., Jain, F., Zhao, L.H., Hymowitz, G., Preffer, F., 2004. IL-10 inhibits inflammation but does not affect fibrosis in the pulmonary response to bleomycin. Exp. Mol. Pathol. 76, 205–211.

Kreyling, W.G., Godleski, J.J., Kariya, S.T., Rose, R.M., Brain, J.D., 1990. In vitro dissolution of uniform cobalt oxide particles by human and canine alveolar macrophages. Am. J. Respir. Cell Mol. Biol. 2, 413–422.

Krombach, F., Munzing, S., Allmeling, A.M., Gerlach, J.T., Behr, J., Dorger, M., 1997. Cell size of alveolar macrophages: an interspecies comparison. Environ. Health Perspect. 105 (Suppl. 5), 1261–1263.

Lagranderie, M., Nahori, M.A., Balazuc, A.M., Kiefer-Biasizzo, H., Lapa e Silva, J.R., Milon, G., Marchal, G., Vargaftig, B.B., 2003. Dendritic cells recruited to the lung shortly after intranasal delivery of *Mycobacterium bovis* BCG drive the primary immune response towards a type 1 cytokine production. Immunology 108, 352–364.

Landsman, L., Jung, S., 2007. Lung macrophages serve as obligatory intermediate between blood monocytes and alveolar macrophages. J. Immunol. 179, 3488–3494.

Lane, B.R., Markovitz, D.M., Woodford, N.L., Rochford, R., Strieter, R.M., Coffey, M.J., 1999. TNF-α inhibits HIV-1 replication in peripheral blood monocytes and alveolar macrophages by inducing the production of RANTES and decreasing C-C chemokine receptor 5 (CCR5) expression. J. Immunol. 163, 3653–3661.

Laskin, D.L., Sunil, V.R., Gardner, C.R., Laskin, J.D., 2011. Macrophages and tissue injury: agents of defense or destruction? Annu. Rev. Pharmacol. Toxicol. 51, 267–288.

Lavnikova, N., Prokhorova, S., Helyar, L., Laskin, D.L., 1993. Isolation and partial characterization of subpopulations of alveolar macrophages, granulocytes, and highly enriched interstitial macrophages from rat lung. Am. J. Respir. Cell Mol. Biol. 8, 384–392.

Lay, J.C., Alexis, N.E., Zeman, K.L., Peden, D.B., Bennett, W.D., 2009. In vivo uptake of inhaled particles by airway phagocytes is enhanced in patients with mild asthma compared with normal volunteers. Thorax 64, 313–320.

Lee, J.S., 2012. Heterogeneity of lung mononuclear phagocytes in chronic obstructive pulmonary disease. J. Innate Immun. 4, 489–497.

Lehnert, B.E., 1984. Lung clearance and retention of particles: mechanisms involving pulmonary macrophages. J. Aerosol Med. 1, A214–A215.

Lehnert, B.E., 1992. Pulmonary and thoracic macrophage subpopulations and clearance of particles from the lung. Environ. Health Perspect. 97, 17–46.

Lehnert, B.E., Dethloff, L.A., Valdez, Y.E., 1988. Leukocytic responses to the intrapleural deposition of particles, particle-cell associations, and the clearance of particles from the pleural space compartment. J. Toxicol. Environ. Health 24, 41–66.

Lehnert, B.E., Valdez, Y.E., Holland, L.M., 1985. Pulmonary macrophages: alveolar and interstitial populations. Exp. Lung Res. 9, 177–190.

Lehnert, B.E., Valdez, Y.E., Sebring, R.J., Lehnert, N.M., Saunders, G.C., Steinkamp, J.A., 1990. Airway intra-luminal macrophages: evidence of origin and comparisons to alveolar macrophages. Am. J. Respir. Cell Mol. Biol. 3, 377–391.

Liu, C.F., Rivere, M., Huang, H.J., Puzo, G., Wang, J.Y., 2010. Surfactant protein D inhibits mite-induced alveolar macrophage and dendritic cell activations through TLR signalling and DC-SIGN expression. Clin. Exp. Allergy 40, 111–122.

Lohmann-Matthes, M.L., Steinmuller, C., Franke-Ullmann, G., 1994. Pulmonary macrophages. Eur. Respir. J. 7, 1678–1689.

Longworth, K.E., 1997. The comparative biology of pulmonary intravascular macrophages. Front. Biosci. 2, d232–241.

Lundborg, M., Eklund, A., Lind, B., Camner, P., 1985. Dissolution of metals by human and rabbit alveolar macrophages. Br. J. Ind. Med. 42, 642–645.

Lundborg, M., Johard, U., Johansson, A., Eklund, A., Falk, R., Kreyling, W., Camner, P., 1995. Phagolysosomal morphology and dissolution of cobalt oxide particles by human and rabbit alveolar macrophages. Exp. Lung Res. 21, 51–66.

Lundborg, M., Lind, B., Camner, P., 1984. Ability of rabbit alveolar macrophages to dissolve metals. Exp. Lung Res. 7, 11–22.

Mantovani, A., Biswas, S.K., Galdiero, M.R., Sica, A., Locati, M., 2013. Macrophage plasticity and polarization in tissue repair and remodelling. J. Pathol. 229, 176–185.

Marafante, E., Lundborg, M., Vahter, M., Camner, P., 1987. Dissolution of two arsenic compounds by rabbit alveolar macrophages in vitro. Fundam. Appl. Toxicol. 8, 382–388.

Marsh, C.B., Anderson, C.L., Lowe, M.P., Wewers, M.D., 1996. Monocyte IL-8 release is induced by two independent Fc γ R- mediated pathways. J. Immunol. 157, 2632–2637.

Marsh, C.B., Wewers, M.D., Tan, L.C., Rovin, B.H., 1997. Fc(gamma) receptor cross-linking induces peripheral blood mononuclear cell monocyte chemoattractant protein-1 expression: role of lymphocyte Fc(γ)RIII. J. Immunol. 158, 1078–1084.

Martin, T.R., Frevert, C.W., 2005. Innate immunity in the lungs. Proc. Am. Thorac. Soc. 2, 403–411.

Martinez, F.O., Sica, A., Mantovani, A., Locati, M., 2008. Macrophage activation and polarization. Front. Biosci. 13, 453–461.

Martinon, F., Burns, K., Tschopp, J., 2002. The inflammasome: a molecular platform triggering activation of inflammatory caspases and processing of proIL-β. Mol. Cell 10, 417–426.

Maus, U.A., Janzen, S., Wall, G., Srivastava, M., Blackwell, T.S., Christman, J.W., Seeger, W., Welte, T., Lohmeyer, J., 2006. Resident alveolar macrophages are replaced by recruited monocytes in response to endotoxin-induced lung inflammation. Am. J. Respir. Cell Mol. Biol. 35, 227–235.

Maus, U.A., Koay, M.A., Delbeck, T., Mack, M., Ermert, M., Ermert, L., Blackwell, T.S., Christman, J.W., Schlondorff, D., Seeger, W., Lohmeyer, J., 2002. Role of resident alveolar macrophages in leukocyte traffic into the alveolar air space of intact mice. Am. J. Physiol. Lung Cell. Mol. Physiol. 282, L1245–L1252.

McComb, J.G., Ranganathan, M., Liu, X.H., Pilewski, J.M., Ray, P., Watkins, S.C., Choi, A.M., Lee, J.S., 2008. CX3CL1 up-regulation is associated with recruitment of $CX3CR1^{+}$ mononuclear phagocytes and T lymphocytes in the lungs during cigarette smoke-induced emphysema. Am. J. Pathol. 173, 949–961.

Meuret, G., Schildknecht, O., Joder, P., Senn, H., 1980. Proliferation activity and bacteriostatic potential of human blood monocytes, macrophages in pleural effusions, ascites, and of alveolar macrophages. Blut 40, 17–25.

Miller, K., Hudspith, B.N., Meredith, C., 1992. Secretory and accessory cell functions of the alveolar macrophage. Environ. Health Perspect. 97, 85–89.

Misharin, A.V., Morales-Nebreda, L., Mutlu, G.M., Budinger, G.R., Perlman, H., 2013. Flow cytometric analysis of macrophages and dendritic cell subsets in the mouse lung. Am. J. Respir. Cell Mol. Biol. 49, 503–510.

Miyamoto, K., Schultz, E., Heath, T., Mitchell, M.D., Albertine, K.H., Staub, N.C., 1988. Pulmonary intravascular macrophages and hemodynamic effects of liposomes in sheep. J. Appl. Physiol. (1985) 64, 1143–1152.

Moore, K.W., de Waal Malefyt, R., Coffman, R.L., O'Garra, A., 2001. Interleukin-10 and the interleukin-10 receptor. Annu. Rev. Immunol. 19, 683–765.

Morbini, P., Villa, C., Campo, I., Zorzetto, M., Inghilleri, S., Luisetti, M., 2006. The receptor for advanced glycation end products and its ligands: a new inflammatory pathway in lung disease? Mod. Pathol. 19, 1437–1445.

Morgan, A.J., Guillen, C., Symon, F.A., Huynh, T.T., Berry, M.A., Entwisle, J.J., Briskin, M., Pavord, I.D., Wardlaw, A.J., 2005. Expression of CXCR6 and its ligand CXCL16 in the lung in health and disease. Clin. Exp. Allergy 35, 1572–1580.

Mosser, D.M., Edwards, J.P., 2008. Exploring the full spectrum of macrophage activation. Nat. Rev. Immunol. 8, 958–969.

Munger, J.S., Huang, X., Kawakatsu, H., Griffiths, M.J., Dalton, S.L., Wu, J., Pittet, J.F., Kaminski, N., Garat, C., Matthay, M.A., Rifkin, D.B., Sheppard, D., 1999. The integrin α v β 6 binds and activates latent TGF β 1: a mechanism for regulating pulmonary inflammation and fibrosis. Cell 96, 319–328.

Murphy, J., Summer, R., Wilson, A.A., Kotton, D.N., Fine, A., 2008. The prolonged life-span of alveolar macrophages. Am. J. Respir. Cell Mol. Biol. 38, 380–385.

Murray, P.J., Wynn, T.A., 2011. Protective and pathogenic functions of macrophage subsets. Nat. Rev. Immunol. 11, 723–737.

Nagashima, A., Yasumoto, K., Nakahashi, H., Takeo, S., Yano, T., Nomoto, K., 1987. Antitumor activity of pleural cavity macrophages and its regulation by pleural cavity lymphocytes in patients with lung cancer. Cancer Res. 47, 5497–5500.

Nakahashi, H., Yasumoto, K., Nagashima, A., Yaita, H., Takeo, S., Motohiro, A., Furukawa, T., Inokuchi, K., Nomoto, K., 1984. Antitumor activity of macrophages in lung cancer patients with special reference to location of macrophages. Cancer Res. 44, 5906–5909.

Nakstad, B., Lyberg, T., Skjorten, F., Boye, N.P., 1989. Subpopulations of human lung alveolar macrophages: ultrastructural features. Ultrastruct. Pathol. 13, 1–13.

Nibbering, P.H., Leijh, P.C., van Furth, R., 1987. Quantitative immunocytochemical characterization of mononuclear phagocytes. II. Monocytes and tissue macrophages. Immunology 62, 171–176.

Nyberg, K., Johansson, U., Johansson, A., Camner, P., 1992. Phagolysosomal pH in alveolar macrophages. Environ. Health Perspect. 97, 149–152.

O'Neill, S.J., Hoehn, S.K., Lesperance, E., Klass, D.J., 1984. Functional heterogeneity of isopycnic fractions of rat alveolar macrophages. Infect. Immun. 46, 282–284.

Oghiso, Y., 1987. Heterogeneity in immunologic functions of rat alveolar macrophages—their accessory cell function and IL-1 production. Microbiol. Immunol. 31, 247–260.

Oltmanns, U., Schmidt, B., Hoernig, S., Witt, C., John, M., 2003. Increased spontaneous interleukin-10 release from alveolar macrophages in active pulmonary sarcoidosis. Exp. Lung Res. 29, 315–328.

Opitz, B., van Laak, V., Eitel, J., Suttorp, N., 2010. Innate immune recognition in infectious and noninfectious diseases of the lung. Am. J. Respir. Crit. Care Med. 181, 1294–1309.

Palecanda, A., Kobzik, L., 2001. Receptors for unopsonized particles: the role of alveolar macrophage scavenger receptors. Curr. Mol. Med. 1, 589–595.

Parbhakar, O.P., Duke, T., Townsend, H.G., Singh, B., 2004. Immunophenotypic characterization and depletion of pulmonary intravascular macrophages of horses. Vet. Res. 35, 39–51.

Parbhakar, O.P., Duke, T., Townsend, H.G., Singh, B., 2005. Depletion of pulmonary intravascular macrophages partially inhibits lipopolysaccharide-induced lung inflammation in horses. Vet. Res. 36, 557–569.

Peao, M.N., Aguas, A.P., Grande, N.R., 1992. Cellular kinetics of inflammation in the pleural space of mice in response to the injection of exogenous particles. Exp. Lung Res. 18, 863–876.

Perez-Arellano, J.L., Alcazar-Montero, M.C., Jimenez-Lopez, A., 1990. Alveolar macrophage: origin, kinetics and relationship with cells of the alveolo-interstitial region. Allergol. Immunopathol. Madr. 18, 175–183.

Philipson, K., Falk, R., Camner, P., 1985. Long-term lung clearance in humans studied with teflon particles labeled with chromium-51. Exp. Lung Res. 9, 31–42.

Philipson, K., Falk, R., Gustafsson, J., Camner, P., 1996. Long-term lung clearance of 1^{95}Au-labeled teflon particles in humans. Exp. Lung Res. 22, 65–83.

Pinkerton, K.E., Barry, B.E., O'Neil, J.J., Raub, J.A., Pratt, P.C., Crapo, J.D., 1982. Morphologic changes in the lung during the lifespan of Fischer 344 rats. Am. J. Anat. 164, 155–174.

Pinkett, M.O., Cowdrey, C.R., Nowell, P.C., 1966. Mixed hematopoietic and pulmonary origin of 'alveolar macrophages' as demonstrated by chromosome markers. Am. J. Pathol. 48, 859–867.

Pons, A.R., Noguera, A., Blanquer, D., Sauleda, J., Pons, J., Agusti, A.G., 2005. Phenotypic characterisation of alveolar macrophages and peripheral blood monocytes in COPD. Eur. Respir. J. 25, 647–652.

Porcheray, F., Viaud, S., Rimaniol, A.C., Leone, C., Samah, B., Dereuddre-Bosquet, N., Dormont, D., Gras, G., 2005. Macrophage activation switching: an asset for the resolution of inflammation. Clin. Exp. Immunol. 142, 481–489.

Powdrill, J., Buckley, C., Valdez, Y.E., Lehnert, B.E., 1989. Airway intraluminal macrophages: origin and role in lung clearance. Toxicology 9, A306.

Prasse, A., Pechkovsky, D.V., Toews, G.B., Jungraithmayr, W., Kollert, F., Goldmann, T., Vollmer, E., Muller-Quernheim, J., Zissel, G., 2006. A vicious circle of alveolar macrophages and fibroblasts perpetuates pulmonary fibrosis via CCL18. Am. J. Respir. Crit. Care Med. 173, 781–792.

Pratt, S.A., Smith, M.H., Ladman, A.J., Finley, T.N., 1971. The ultrastructure of alveolar macrophages from human cigarette smokers and nonsmokers. Lab. Invest. 24, 331–338.

Prieto, J., Eklund, A., Patarroyo, M., 1994. Regulated expression of integrins and other adhesion molecules during differentiation of monocytes into macrophages. Cell. Immunol. 156, 191–211.

Prokhorova, S., Lavnikova, N., Laskin, D.L., 1994. Functional characterization of interstitial macrophages and subpopulations of alveolar macrophages from rat lung. J. Leukoc. Biol. 55, 141–146.

Rhodes, J., 1975. Macrophage heterogeneity in receptor activity: the activation of macrophage Fc receptor function in vivo and in vitro. J. Immunol. 114, 976–981.

Rouzer, C.A., Scott, W.A., Hamill, A.L., Cohn, Z.A., 1982. Synthesis of leukotriene C and other arachidonic acid metabolites by mouse pulmonary macrophages. J. Exp. Med. 155, 720–733.

Salez, L., Balloy, V., van Rooijen, N., Lebastard, M., Touqui, L., McCormack, F.X., Chignard, M., 2001. Surfactant protein A suppresses lipopolysaccharide-induced IL-10 production by murine macrophages. J. Immunol. 166, 6376–6382.

Sandron, D., Reynolds, H.Y., Laval, A.M., Venet, A., Israel-Biet, D., Chretien, J., 1986. Human alveolar macrophage subpopulations isolated on discontinuous albumin gradients. Cytological data in normals and sarcoid patients. Eur. J. Respir. Dis. 68, 177–185.

Schneberger, D., Aharonson-Raz, K., Singh, B., 2011a. Monocyte and macrophage heterogeneity and toll-like receptors in the lung. Cell Tissue Res. 343, 97–106.

Schneberger, D., Aharonson-Raz, K., Singh, B., 2012. Pulmonary intravascular macrophages and lung health: what are we missing? Am. J. Physiol. Lung Cell. Mol. Physiol. 302, L498–L503.

Schneberger, D., Lewis, D., Caldwell, S., Singh, B., 2011b. Expression of toll-like receptor 9 in lungs of pigs, dogs and cattle. Int. J. Exp. Pathol. 92, 1–7.

Sebring, R.J., Lehnert, B.E., 1992. Morphometric comparisons of rat alveolar macrophages, pulmonary interstitial macrophages, and blood monocytes. Exp. Lung Res. 18, 479–496.

Sestini, P., Tagliabue, A., Boraschi, D., 1984. Modulation of macrophage suppressive activity and prostaglandin release by lymphokines and interferon: comparison of alveolar, pleural and peritoneal macrophages. Clin. Exp. Immunol. 58, 573–580.

Shellito, J., Esparza, C., Armstrong, C., 1987. Maintenance of the normal rat alveolar macrophage cell population. The roles of monocyte influx and alveolar macrophage proliferation in situ. Am. Rev. Respir. Dis. 135, 78–82.

Shellito, J., Kaltreider, H.B., 1984. Heterogeneity of immunologic function among subfractions of normal rat alveolar macrophages. Am. Rev. Respir. Dis. 129, 747–753.

Shimotakahara, A., Kuebler, J.F., Vieten, G., Metzelder, M.L., Petersen, C., Ure, B.M., 2007. Pleural macrophages are the dominant cell population in the thoracic cavity with an inflammatory cytokine profile similar to peritoneal macrophages. Pediatr. Surg. Int. 23, 447–451.

Sica, A., Mantovani, A., 2012. Macrophage plasticity and polarization: in vivo veritas. J. Clin. Invest. 122, 787–795.

Singh, B., Atwal, O.S., 1997. Ultrastructural and immunocytochemical study of the pulmonary intravascular macrophages of *Escherichia coli* lipopolysaccharide-treated sheep. Anat. Rec. 247, 214–224.

Singh Suri, S., Janardhan, K.S., Parbhakar, O., Caldwell, S., Appleyard, G., Singh, B., 2006. Expression of toll-like receptor 4 and 2 in horse lungs. Vet. Res. 37, 541–551.

Snelgrove, R.J., Godlee, A., Hussell, T., 2011. Airway immune homeostasis and implications for influenza-induced inflammation. Trends Immunol. 32, 328–334.

Snelgrove, R.J., Goulding, J., Didierlaurent, A.M., Lyonga, D., Vekaria, S., Edwards, L., Gwyer, E., Sedgwick, J.D., Barclay, A.N., Hussell, T., 2008. A critical function for CD200 in lung immune homeostasis and the severity of influenza infection. Nat. Immunol. 9, 1074–1083.

Sone, Y., Serikov, V.B., Staub Sr., N.C., 1999. Intravascular macrophage depletion attenuates endotoxin lung injury in anesthetized sheep. J. Appl. Physiol. (1985) 87, 1354–1359.

Sorokin, S.P., Brain, J.D., 1975. Pathways of clearance in mouse lungs exposed to iron oxide aerosols. Anat. Rec. 181, 581–625.

Sorokin, S.P., Hoyt, R.F., 1987. Pure population of nonmonocyte derived macrophages arising in organ cultures of embryonic rat lungs. Anat. Rec. 217, 35–52.

Sorokin, S.P., Hoyt Jr., R.F., Grant, M.M., 1984. Development of macrophages in the lungs of fetal rabbits, rats, and hamsters. Anat. Rec. 208, 103–121.

Sorokin, S.P., Kobzik, L., Hoyt Jr., R.F., Godleski, J.J., 1989. Development of surface membrane characteristics of "premedullary" macrophages in organ cultures of embryonic rat and hamster lungs. J. Histochem. Cytochem. 37, 365–376.

Spiteri, M.A., Poulter, L.W., 1991. Characterization of immune inducer and suppressor macrophages from the normal human lung. Clin. Exp. Immunol. 83, 157–162.

Staub, N.C., 1989. Pulmonary Vascular Reactivity: A Status Report. Futura Pub. Co., Mount Kisco, NY.

Staub, N.C., 1994. Pulmonary intravascular macrophages. Annu. Rev. Physiol. 56, 47–67.

Stefater, J.A., Ren, S., Lang, R.A., Duffield, J.S., 2011. Metchnikoff's policemen: macrophages in development, homeostasis and regeneration. Trends Mol. Med. 17, 743–752.

Stein-Streilein, J., Bennett, M., Mann, D., Kumar, V., 1983. Natural killer cells in mouse lung: surface phenotype, target preference, and response to local influenza virus infection. J. Immunol. 131, 2699–2704.

Steinmuller, C., Franke-Ullmann, G., Lohmann-Matthes, M.L., Emmendorffer, A., 2000. Local activation of nonspecific defense against a respiratory model infection by application of interferon-γ: comparison between rat alveolar and interstitial lung macrophages. Am. J. Respir. Cell Mol. Biol. 22, 481–490.

Stout, R.D., Suttles, J., 2004. Functional plasticity of macrophages: reversible adaptation to changing microenvironments. J. Leukoc. Biol. 76, 509–513.

Takahashi, K., Isobe, T., Ohtsuki, Y., Sonobe, H., Takeda, I., Akagi, T., 1984. Immunohistochemical localization and distribution of S-100 proteins in the human lymphoreticular system. Am. J. Pathol. 116, 497–503.

Tarling, J.D., Lin, H.S., Hsu, S., 1987. Self-renewal of pulmonary alveolar macrophages: evidence from radiation chimera studies. J. Leukoc. Biol. 42, 443–446.

Taylor, M.L., Noble, P.W., White, B., Wise, R., Liu, M.C., Bochner, B.S., 2000. Extensive surface phenotyping of alveolar macrophages in interstitial lung disease. Clin. Immunol. 94, 33–41.

Taylor, P.R., Brown, G.D., Reid, D.M., Willment, J.A., Martinez-Pomares, L., Gordon, S., Wong, S.Y., 2002. The β-glucan receptor, dectin-1, is predominantly expressed on the surface of cells of the monocyte/macrophage and neutrophil lineages. J. Immunol. 169, 3876–3882.

Thakur, S.A., Hamilton Jr., R.F., Holian, A., 2008. Role of scavenger receptor A family in lung inflammation from exposure to environmental particles. J. Immunotoxicol. 5, 151–157.

Thepen, T., Van Rooijen, N., Kraal, G., 1989. Alveolar macrophage elimination in vivo is associated with an increase in pulmonary immune response in mice. J. Exp. Med. 170, 499–509.

Thomas, E.D., Ramberg, R.E., Sale, G.E., Sparkes, R.S., Golde, D.W., 1976. Direct evidence for a bone marrow origin of the alveolar macrophage in man. Science 192, 1016–1018.

Trapnell, B.C., Whitsett, J.A., 2002. GM-CSF regulates pulmonary surfactant homeostasis and alveolar macrophage-mediated innate host defense. Annu. Rev. Physiol. 64, 775–802.

Tuder, R.M., Groves, B., Badesch, D.B., Voelkel, N.F., 1994. Exuberant endothelial cell growth and elements of inflammation are present in plexiform lesions of pulmonary hypertension. Am. J. Pathol. 144, 275–285.

Van den Heuvel, M.M., Tensen, C.P., van As, J.H., Van den Berg, T.K., Fluitsma, D.M., Dijkstra, C.D., Dopp, E.A., Droste, A., Van Gaalen, F.A., Sorg, C., Hogger, P., Beelen, R.H., 1999. Regulation of CD163 on human macrophages: cross-linking of CD163 induces signaling and activation. J. Leukoc. Biol. 66, 858–866.

Virolainen, M., 1968. Hematopoietic origin of macrophages as studied by chromosome markers in mice. J. Exp. Med. 127, 943–952.

Warner, A.E., Barry, B.E., Brain, J.D., 1986. Pulmonary intravascular macrophages in sheep. Morphology and function of a novel constituent of the mononuclear phagocyte system. Lab. Invest. 55, 276–288.

Warner, A.E., Brain, J.D., 1984. Intravascular pulmonary macrophages in ruminants participate in reticuloendothelial clearance of particles. Fed. Proc. 43, 1001.

Warner, A.E., Brain, J.D., 1986. Intravascular pulmonary macrophages: a novel cell removes particles from blood. Am. J. Physiol. 250, R728–R732.

Warner, A.E., Molina, R.M., Brain, J.D., 1987. Uptake of bloodborne bacteria by pulmonary intravascular macrophages and consequent inflammatory responses in sheep. Am. Rev. Respir. Dis. 136, 683–690.

Warren, J.S., Kunkel, R.G., Johnson, K.J., Ward, P.A., 1987. Comparative O_2-. responses of lung macrophages and blood phagocytic cells in the rat. Possible relevance to IgA immune complex induced lung injury. Lab. Invest. 57, 311–320.

Weissler, J.C., Lyons, C.R., Lipscomb, M.F., Toews, G.B., 1986. Human pulmonary macrophages. Functional comparison of cells obtained from whole lung and by bronchoalveolar lavage. Am. Rev. Respir. Dis. 133, 473–477.

Werneck-Barroso, E., Novaes Moura, A.C., Dias Fernandes, P., Cruz Rosas, E., Assreuy, J., das Gracas, M.O.H.M., 2000. Nitric oxide production by BCG-infected pleural macrophages from C57Bl/6 and DBA-2 mice. Int. J. Mol. Med. 5, 95–99.

Wheeldon, E.B., Hansen-Flaschen, J.H., 1986. Intravascular macrophages in the sheep lung. J. Leukoc. Biol. 40, 657–661.

Whiteley, L.O., Maheswaran, S.K., Weiss, D.J., Ames, T.R., 1990. Immunohistochemical localization of *Pasteurella haemolytica* A1--derived endotoxin, leukotoxin, and capsular polysaccharide in experimental bovine *Pasteurella* pneumonia. Vet. Pathol. 27, 150–161.

Whiteley, L.O., Maheswaran, S.K., Weiss, D.J., Ames, T.R., 1991. Alterations in pulmonary morphology and peripheral coagulation profiles caused by intratracheal inoculation of live and ultraviolet light-killed *Pasteurella haemolytica* A1 in calves. Vet. Pathol. 28, 275–285.

Winkler, G.C., 1988. Pulmonary intravascular macrophages in domestic animal species: review of structural and functional properties. Am. J. Anat. 181, 217–234.

Winkler, G.C., Cheville, N.F., 1984. The neonatal porcine lung: ultrastructural morphology and postnatal development of the terminal airways and alveolar region. Anat. Rec. 210, 303–313.

Winkler, G.C., Cheville, N.F., 1987. Postnatal colonization of porcine lung capillaries by intravascular macrophages: an ultrastructural, morphometric analysis. Microvasc. Res. 33, 224–232.

Wissinger, E., Goulding, J., Hussell, T., 2009. Immune homeostasis in the respiratory tract and its impact on heterologous infection. Semin. Immunol. 21, 147–155.

Wizemann, T.M., Gardner, C.R., Laskin, J.D., Quinones, S., Durham, S.K., Goller, N.L., Ohnishi, S.T., Laskin, D.L., 1994. Production of nitric oxide and peroxynitrite in the lung during acute endotoxemia. J. Leukoc. Biol. 56, 759–768.

Wizemann, T.M., Laskin, D.L., 1994a. Effects of acute endotoxemia on production of cytokines and nitric oxide by pulmonary alveolar and interstitial macrophages. Ann. N. Y. Acad. Sci. 730, 336–337.

Wizemann, T.M., Laskin, D.L., 1994b. Enhanced phagocytosis, chemotaxis, and production of reactive oxygen intermediates by interstitial lung macrophages following acute endotoxemia. Am. J. Respir. Cell Mol. Biol. 11, 358–365.

Wu, H.M., Jin, M., Marsh, C.B., 2005. Toward functional proteomics of alveolar macrophages. Am. J. Physiol. Lung Cell. Mol. Physiol. 288, L585–L595.

Wynn, T.A., Chawla, A., Pollard, J.W., 2013. Macrophage biology in development, homeostasis and disease. Nature 496, 445–455.

Xiong, Z., Leme, A.S., Ray, P., Shapiro, S.D., Lee, J.S., 2011. $CX3CR1^+$ lung mononuclear phagocytes spatially confined to the interstitium produce TNF-α and IL-6 and promote cigarette smoke-induced emphysema. J. Immunol. 186, 3206–3214.

Yamada, C., Sano, H., Shimizu, T., Mitsuzawa, H., Nishitani, C., Himi, T., Kuroki, Y., 2006. Surfactant protein A directly interacts with TLR4 and MD-2 and regulates inflammatory cellular response. Importance of supratrimeric oligomerization. J. Biol. Chem. 281, 21771–21780.

Yona, S., Kim, K.W., Wolf, Y., Mildner, A., Varol, D., Breker, M., Strauss-Ayali, D., Viukov, S., Guilliams, M., Misharin, A., Hume, D.A., Perlman, H., Malissen, B., Zelzer, E., Jung, S., 2013. Fate mapping reveals origins and dynamics of monocytes and tissue macrophages under homeostasis. Immunity 38, 79–91.

Zaynagetdinov, R., Sherrill, T.P., Kendall, P.L., Segal, B.H., Weller, K.P., Tighe, R.M., Blackwell, T.S., 2013. Identification of myeloid cell subsets in murine lungs using flow cytometry. Am. J. Respir. Cell Mol. Biol. 49, 180–189.

Zetterberg, G., Elmberger, G., Johansson, A., Lundahl, J., Lundborg, M., Skold, C.M., Tornling, G., Camner, P., Eklund, A., 2000. Rat alveolar and interstitial macrophages in the fibrosing stage following quartz exposure. Hum. Exp. Toxicol. 19, 402–411.

Zetterberg, G., Johansson, A., Lundahl, J., Lundborg, M., Skold, C.M., Tornling, G., Camner, P., Eklund, A., 1998. Differences between rat alveolar and interstitial macrophages 5 wk after quartz exposure. Am. J. Physiol. 274, L226–L234.

Zlotnik, A., Crowle, A.J., 1982. Lymphokine-induced mycobacteriostatic activity in mouse pleural macrophages. Infect. Immun. 37, 786–793.

Zlotnik, A., Vatter, A., Hayes, R.L., Blumenthal, E., Crowle, A.J., 1982. Mouse pleural macrophages: characterization and comparison with mouse alveolar and peritoneal macrophages. J. Reticuloendothel. Soc. 31, 207–220.

Zwilling, B.S., Campolito, L.B., Reiches, N.A., 1982. Alveolar macrophage subpopulations identified by differential centrifugation on a discontinuous albumin density gradient. Am. Rev. Respir. Dis. 125, 448–452.

Chapter 33

Pulmonary Dendritic Cells

Donald N. Cook and Hideki Nakano

Chapter Outline

1. OVERVIEW OF DENDRITIC CELLS

The lung is particularly vulnerable to infection, and it is essential that vertebrates are able to rapidly recognize and clear inhaled pathogens before they establish overwhelming infections. The ability to generate robust immune and inflammatory responses to infection is therefore critical to health, but these responses can also damage delicate lung tissue and thereby prevent efficient gas exchange. Accordingly, vertebrates must generate effective immune responses to pathogens while also remaining tolerant to self-antigens and exogenous antigens that do not pose a threat to health. The cellular and molecular mechanisms that control immunologic responses to inhaled antigens are therefore of upmost importance in maintaining homeostasis. A wealth of evidence over the last several decades has shown that a bone marrow-derived cell known as the dendritic cell (DC) is largely responsible for the initiation of immune responses to both exogenous and self-antigens.

DCs were identified by Ralph Steinman as cells whose morphology and ability to stimulate naïve T cells distinguished them from monocytes and macrophages (Steinman and Cohn, 1973, 1974; Steinman et al., 1974; Steinman and Witmer, 1978). In subsequent years, the importance of DCs for initiating adaptive immune responses gained increasing appreciation, culminating in the 2011 Noble Prize in Physiology or Medicine for Steinman (shared with Bruce Beutler and Jules Hoffmann). We now know that DCs are present in virtually all tissues but are particularly abundant in organs that are in intimate contact with the environment, such as the skin, gut, and lung (Sertl et al., 1986). In keeping with their role of sensing the antigenic content of inhaled air, DCs reside close to or within the respiratory epithelium of the entire airway, including the nose and nasopharynx, trachea, bronchi, bronchioles, and alveoli. Like DCs in other tissues, pulmonary DCs take up antigens and process them into peptides that are displayed on the cell surface in association with molecules of the major histocompatibility (MHC) locus. There are two types of MHC, class I and class II, which stimulate $CD8^+$ cytotoxic T cells and $CD4^+$ T helper cells, respectively. Class I is displayed on most cells in the body, whereas display of class II is restricted to antigen presenting cells, including DCs, macrophages, and B cells.

In addition to their potent antigen presenting ability, DCs also possess the ability to migrate from peripheral tissue to draining lymph nodes (LNs), which contain large numbers of naïve T cells. Thus, pulmonary DCs function as sentinels of the pulmonary adaptive immune system by taking up inhaled antigens and transporting them to thoracic LNs. In this way, pulmonary DCs present to naïve T cells a dynamic array of antigens that are present within the lung (Cook and Bottomly, 2007; Lambrecht and Hammad, 2009; Randolph et al., 2005). It is likely that the migratory ability of DCs evolved to increase the frequency of their encounters with naïve T cells. The number of naïve T cells specific for any given antigen is very small ($\sim$1/100,000), and chance encounters between such a cell and a DC displaying that antigen are likely to be very rare. The ability of DCs to migrate from peripheral

Comparative Biology of the Normal Lung. http://dx.doi.org/10.1016/B978-0-12-404577-4.00033-3

tissue to naïve T cell-rich draining LNs undoubtedly increases the frequency of encounters between those two cell types.

Efficient T cell activation by DCs requires more than antigen presentation. Otherwise, self-antigens and harmless exogenous antigens would continually trigger immune responses that would ultimately lead to organ failure and death. The ability of DCs to initiate productive immune responses depends on the innate immune system that includes cells such as macrophages, neutrophils, epithelial cells, and DCs themselves (Iwasaki and Medzhitov, 2010). In the absence of signals from these cells, DCs do not become sufficiently activated to provide strong costimulation to T cells. Under these circumstances, antigen presentation by DCs leads to either T cell anergy (the inability to respond to subsequent stimuli) or to the proliferation of regulatory T cells that negatively regulate effector immune responses. Thus, the activation state of DCs determines whether or not antigen inhalation leads to the induction of effector T cell responses.

2. DC SUBSETS IN THE LUNG

2.1 Anatomic Distribution of DCs

DCs are sometimes classified according to whether they are located in the large airways or in the lung parenchyma, which is comprised of small airways and alveoli. Confocal microscopic analysis of rat trachea has revealed that approximately 20% of resident are in close contact with the resting epithelium, whereas 80% of DCs reside within the subepithelial region (Jahnsen et al., 2006). DCs in the airways of the rat and mouse turn over rapidly with an estimated half-life of 12 h, whereas parenchymal DCs have a longer half-life, about several days (Holt et al., 1988; von Garnier et al., 2005). The replenishment of DCs during both steady-state and inflammatory conditions is mediated in part by the chemokine receptors, CCR1 and CCR5 (McWilliam et al., 1994; Stumbles et al., 2001). DCs prepared from the airway are more endocytic and present peptides more efficiently to $CD4^+$ T cells than DCs prepared from the lung parenchyma (von Garnier et al., 2005). These findings suggest that DCs residing in the airway and parenchyma of the lung might fulfill different functions. In addition to the mucosa and parenchyma, the vasculature of the lung also contains DCs that are able to take up antigens in the blood and present them to T cells in mediastinal LNs (Willart et al., 2009). Interestingly, T cells that have been activated in this way have a propensity to return to the lungs. Thus, DCs in the vasculature of the lung can also be considered as bona fide pulmonary DCs, although relatively little is known about these cells compared to mucosal or parenchymal DCs.

2.2 Classification by Cell Surface Markers

Although classification of DCs according to their anatomical position is reasonable and has been instructive, most investigators classify lung DCs based on their display of cell surface proteins. In the mouse, all DCs express the integrin CD11c, and for most tissues, expression of this marker is sufficient to identify DCs by immunohistochemistry or flow cytometry. However, use of this marker alone is insufficient to identify DCs in the lung because alveolar macrophages also display CD11c (Figure 1). However, macrophages are highly autofluorescent, whereas DCs are not (Vermaelen and Pauwels, 2004), and at least for flow cytometry-based experiments, autofluorescence is a useful parameter for distinguishing between these two cell types. There are two major conventional (c)DC subsets in the lung, which are defined according to their display of the CD11b and α_E(CD103)β_7 integrins: $CD11b^{hi}$ $CD103^-$ ($CD11b^{hi}$

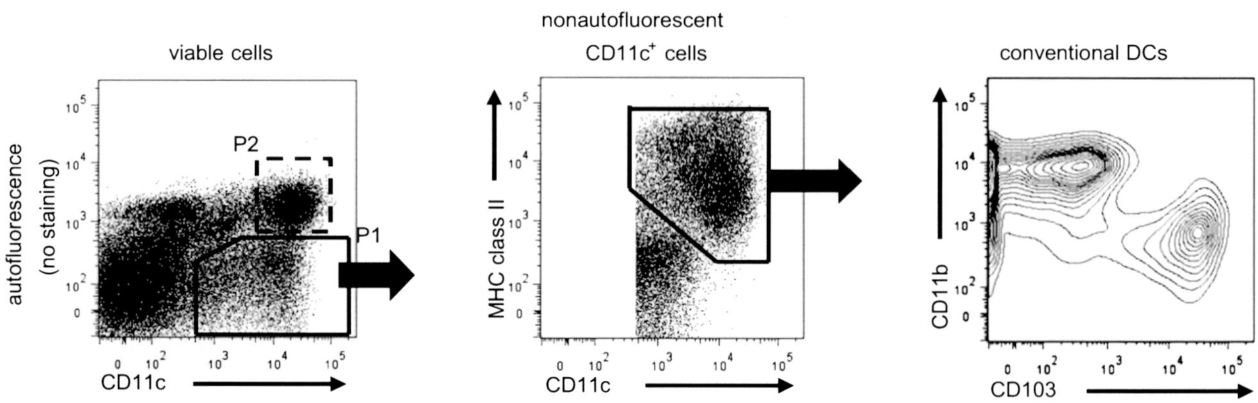

FIGURE 1 Resolution of the two major pulmonary DC populations by flow cytometry. (left panel) Nonautofluorescent $CD11c^+$ cells are shown within the black rectangle (P1). (middle panel) Staining and gating for MHC class II^+ cells. (right panel) Resolution of nonautofluorescent, $CD11c^+MHC^+$ cells into the two major pulmonary DC subsets, expressing CD11b and CD103, respectively.

DCs) and CD11b$^{low\ (lo)}$ CD103^{+} DCs (CD103^{+} DCs) (Sung et al., 2006). The lung also contains smaller populations of CD11c$^{intermediate\ (int)}$ CD11b^{lo} Ly-6C^{+} B220^{+} plasmacytoid DCs (pDC), and CD11c^{int} CD11b^{hi} Ly-6C^{+} monocyte-derived inflammatory DCs. The assignment of specialized functions to these various DC subsets is a rapidly evolving field of pulmonary DC biology, as discussed below.

2.3 CD103^{+} DCs

CD103^{+} DCs reside in the lung mucosa, in close proximity to the basal surface of bronchial epithelial cells (Sung et al., 2006). Because CD103 binds E-cadherin, which is displayed on the basolateral side of epithelial cells, the positioning of CD103^{+} DCs in close contact with the airway epithelia is likely mediated by interactions between these two molecules (Cepek et al., 1994; Taraszka et al., 2000). In addition, CD103^{+} DCs also express several tight junction proteins, including Claudin-1, Claudin-2, and ZO-2. These proteins are thought to allow CD103^{+} DCs to extend processes between epithelial cells and gain access to the lumen of the airway (Jahnsen et al., 2006; Sung et al., 2006). In this way, CD103^{+} DCs can sample luminal antigens in airways with an intact epithelial barrier. CD103^{+} DCs also express high levels of the chemokine receptor XCR-1, although the functional significance of this observation is not currently known. CD103^{+} DCs are not present in large numbers in alveolar septa, although pDCs can be found there.

Recent reports have revealed that in addition to their differences in display of cell surface markers, CD103^{+} DCs and CD11b^{hi} DCs also have different functions. For example, CD103^{+} DCs can efficiently cross-prime CD8^{+} T cells, a process whereby exogenous antigens are displayed on MHC class I. By contrast, CD11b^{hi} DCs are more proficient at priming CD4^{+} T cells (del Rio et al., 2007). In agreement with this report, CD103^{+} DCs can also acquire and transport apoptotic cells to the draining LNs and cross present apoptotic cell-associated antigens to CD8^{+} T cells (Desch et al., 2011).

CD103^{+} DCs also express the protein langerin, and mice that express the diphtheria toxin receptor (DTR) under control of the langerin promoter have been generated. This allows CD103^{+} DCs to be depleted by the administration of diphtheria toxin (DT) (Bennett et al., 2005). Using this system, it has been shown that the development of influenza virus-specific T cells is delayed, with consequent increased clinical severity and delayed viral clearance following influenza infection (GeurtsvanKessel et al., 2008). Depleting CD103^{+} DCs in langerin-DTR mice also reduces eosinophil recruitment to the lung following LPS-treatment of animals that had been previously sensitized and challenged with the LACK (*Leishmania* analog of the receptors of activated C kinase) antigen (Ortiz-Stern et al., 2010). Although selective depletion of DC subsets in this way has great potential for improving our understanding of their function in vivo, we and others (Zammit et al., 2005) have found that DT can cause massive cell death and fatalities in mice. Thus, it remains to be determined whether some of the effects of DT attributed to depletion of specific DC subsets might actually be due to nonspecific effects of DT that cause cell death and inflammation.

Although CD103^{+} DCs do not generally produce large amounts of chemokines, they do produce CCL17 and CCL22, which bind the chemokine receptor CCR4 (Beaty et al., 2007). Interestingly, this receptor is on T helper (Th) 2 cells, which produce the cytokines IL-4, IL-5, and IL-13 and are associated with allergic responses. We have also found that eosinophil chemoattractant, CCL11 (eotaxin), is also more highly expressed by CD103^{+} DCs than by CD11b^{hi} DCs (Nakano et al., 2012). Thus, one of the functions of CD103^{+} DCs might be to recruit Th2 cells to the lung during allergen challenge.

CD103^{+} DCs are also present in the intestinal mucosa, but the extent to which these DCs differ from their counterparts in the lung remains to be determined. Gut-associated CD103^{+} DCs promote T regulatory (Treg) differentiation by producing aldehyde dehydrogenase 1a2 (ALDH1a2), which converts retinol to retinoic acid (RA) (Annacker et al., 2005; Coombes et al., 2007). However, to date, a similar role for lung CD103^{+} DCs in the induction of tolerance to inhaled antigens has not been shown.

2.4 CD11b^{hi} DCs

Unlike CD103^{+} DCs, CD11b^{hi} DCs in the lung do not directly contact the airway epithelium but lie just beneath it. Consequently, the latter cells probably have limited access to antigens in the airway. However, CD11b^{hi} DCs produce very high levels of chemokines ex vivo and can sustain allergic pulmonary inflammation during the challenge phase of allergic responses (Beaty et al., 2007; Raymond et al., 2009). Together, these findings suggest that CD11b^{hi} DCs can respond to antigen rechallenge and produce many chemokines that serve to amplify inflammatory responses. The chemokines produced by CD11b^{hi} DCs include the C-X-C and C-C categories, which can collectively recruit leukocytes of all types to the lung.

In addition to their differences in chemokine production, CD103^{+} DCs and CD11b^{hi} DCs also have distinct radiosensitivities and turnover rates. Repopulation of the CD11b^{hi} DC subsets is complete after 10 days, whereas CD103^{+} DCs require 28 days (Hahn et al., 2011). It is unclear if this difference relates to differences in anatomic location, intrinsic resistance to radiation, or the ability of monocytes to replenish CD11b^{hi} DCs but not CD103^{+} DCs.

Although the vast majority of DCs in the lung are either CD11b^{hi} or CD103^{+}, a small percentage of DCs are positive for both markers (Figure 1). In part because of their relatively small numbers, it is not currently known whether these double positive DCs are similar in function to one of the major DC subsets or if they possess unique functions.

2.5 Inflammatory DCs

Blood-borne monocytes can crawl along the vascular endothelium of healthy tissue in a manner dependent on the integrin LFA-1 and the chemokine receptor CX$_3$CR1. During infection, these cells can rapidly invade peripheral tissue and LNs and develop into inflammatory DCs. These DCs are particularly adept at priming Th1 responses to proinflammatory stimuli (Cheong et al., 2010; Leon et al., 2007; Nakano et al., 2009), although a subset of inflammatory DCs that display the high-affinity IgE receptor (FcεR1) is reported to have Th2 stimulating activity (Hammad et al., 2010). Several lines of evidence suggest that inflammatory DCs might contribute to the exacerbation of asthma symptoms that are often seen following viral infection. First, large numbers of monocyte-derived inflammatory DCs are recruited to the lungs of virus-infected mice (Lin et al., 2008). Second, FcεR1 is displayed at high levels on circulating monocytes during asthma exacerbations. Third, expression of this receptor is increased by type I interferons (Grayson et al., 2007), which are produced at high levels during virus infection. Thus, viruses might trigger asthma exacerbations by increasing FcεR1 on circulating monocytes and by promoting the recruitment of FcεR1-expressing monocytes and inflammatory DCs to the lung. Under this scenario, allergen/IgE complexes cause DCs to release cytokines and chemokines that can amplify existing pulmonary inflammation.

Upon their maturity, monocyte-derived DCs display increased levels of DC SIGN and CD14, which distinguishes them from CD11b^{hi} cDCs (Cheong et al., 2010). This differential display of CD14 might provide a basis to separate monocyte-derived CD11b^{hi} cDCs from CD11b^{hi} cDCs that arise from pre-DCs. Using this strategy, it should be possible to determine if these two types of CD11b^{hi} DCs are functionally different.

2.6 pDCs

The lung contains a minor population of pDCs, which display either moderate levels of CD11c (murine pDCs) or are negative for CD11c (human pDCs). In addition to their relatively low levels of CD11c, murine pDCs can also be distinguished phenotypically from other DC subsets by their display of Siglec H, Ly6C, bone marrow stromal cell antigen 2 (BST-2, PDCA-1, 120G8), and B220 (Asselin-Paturel et al., 2001; Bjorck, 2001; Nakano et al., 2001). Human pDCs can be identified by their expression of CD4, IL-3Rα (CD123), BDCA-2, and BDCA-4.

One of the most important features of pDCs is their ability to produce very high levels of type I interferons in response to viruses or their products (Siegal et al., 1999). Recent work has shown that this response results from the activation of TLR7 or TLR9, which are highly expressed in pDCs but not in cDCs. TLR7 and TLR9 are intracellular receptors for single-stranded RNA and unmethylated CpG DNA, respectively, nucleic acids that are not normally seen in the cytoplasm of healthy cells. The transcription factor IRF7 is critically important for the induction of type I interferon production following stimulation of either of these TLRs (Honda et al., 2005).

Despite their ability to produce large amounts of type I interferons, pDCs are relatively poor inducers of effector T cell responses. In fact, pDCs have been shown to have suppressive effects on immune responses to inhaled antigens (de Heer et al., 2004). Depletion of pDCs with an anti-Ly-6C/G (Gr-1) antibody resulted in allergic sensitization to inhaled ovalbumin under conditions that normally induce tolerization. Because Gr-1 antibodies also bind other cell types, including monocytes and inflammatory DCs recognized by Ly-6C, and neutrophils recognized by Ly-6G, this experiment alone does not prove that pDCs suppress allergic responses. However, adoptive transfer of pDCs before sensitization resulted in reduced levels of allergic inflammation (de Heer et al., 2004). Moreover, recent findings with mice in which the DTR gene was inserted into the 3′ untranslated region of the *Siglech* gene have supported the concept that pDCs suppress the proliferation of CD4^{+} T cells (Takagi et al., 2011). Interestingly, however, the latter study also showed that pDCs could support the proliferation of CD8^{+} T cells. This finding suggests that pDCs can promote viral clearance in two ways: by producing type I interferons and by inducing virus-specific, cytotoxic CD8^{+} T cells.

3. DC DEVELOPMENT

All DCs are derived from the bone marrow-derived, monocyte and DC precursor (MDP), which can also give rise to monocytes (Figure 2). Phenotypically, MDPs are identified as lineage (lin)$^{-}$CD117^{hi}CX$_3$CR1^{+}CD115^{+} CD135^{+} (Fogg et al., 2006). MDPs can differentiate into the common DC precursor (CDP), whose progeny is restricted to pDCs and the immediate precursor of cDCs, known as the pre-DC (Naik et al., 2007; Onai et al., 2007). Recent studies have suggested that CD103^{+} DCs and the highly related CD8α^{+} lymphoid DCs are derived exclusively from pre-DCs via a developmental pathway dependent on interactions between fms-like tyrosine kinase 3 (Flt3) and its ligand (Flt3L) (Bogunovic et al.,

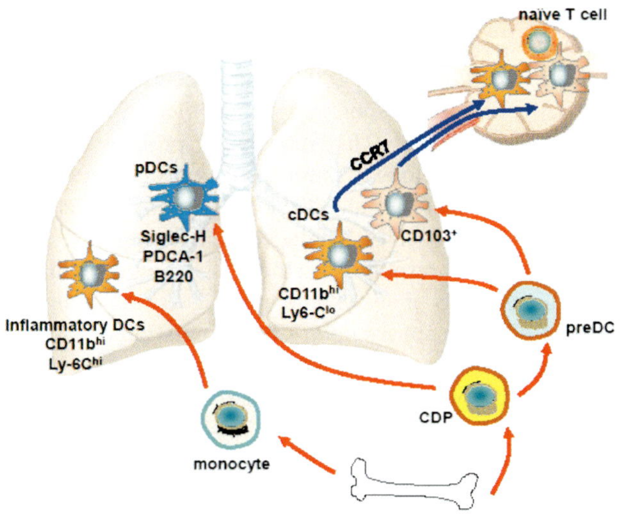

FIGURE 2 DC development and induction of adaptive immune responses. Pre-DCs and monocytes from the bone marrow are circulating precursors of mature DCs. Pre-DCs give rise to both CD11b^{hi} DCs and CD103^{+} DCs, whereas monocytes can differentiate into only CD11b^{hi} DCs. Inhaled antigens are taken up by lung DCs that migrate to draining LNs in a CCR7-dependent manner. Depending on the signals they receive from DCs, antigen-specific naïve T cells can differentiate into CD8^{+} cytotoxic T cells or CD4^{+} helper T cells.

2009; Ginhoux et al., 2009). Transcription factors that are required for the development of these DC subsets include IRF8, which is mutated in BXH2 recombinant inbred mice (Tailor et al., 2008), BATF3 (Hildner et al., 2008), and ID2 (Hacker et al., 2003). The development of CD11b^{hi} DCs in the lung is also at least partly dependent on FLT3L, probably because it also contributes to the development of MDPs, which give rise to all DCs. Intraperitoneal delivery of FLT3L to mice blunts allergic responses and inhibits their ability to take up inhaled antigens and migrate to draining LNs (Shao et al., 2009a,b), suggesting that this cytokine might be of therapeutic benefit. However, because FLT3L is a growth factor, its effects might result from the expansion of immature DCs, which might not yet be capable of eliciting effector immune responses. It is conceivable, however, that these DCs could eventually acquire immunostimulatory activity, and that these cells might exacerbate inflammation in the longer term.

Although CD11b^{hi} DCs have been reported to be absent in *Flt3L*-deficient mice (Ginhoux et al., 2009), others have reported that CD11b^{hi} DCs can develop from monocytes (moDCs), which do not require FLT3 or FLT3L for their differentiation into DCs (Bogunovic et al., 2009; Jakubzick et al., 2008a,b; Varol et al., 2007). This might explain the finding that although *Flt3L*-deficient mice have drastic reductions in lymphoid tissue DCs and parenchymal DCs, they display no defect in the total number of MHC class II^{+} airway DCs or in their ability to reactivate CD4^{+} T cells that had been previously primed with antigen (Walzer et al., 2005).

It is possible that the relative abundance of monocyte-derived DCs in the lung, and therefore the total number of DCs in lungs of *Flt3L*-deficient mice, depends on the conditions used to house the mice. Very clean conditions might not promote the recruitment and development of monocyte-derived DCs, and *Flt3L*-deficient mice might therefore have a larger deficit in total DCs than when mice are housed under dirtier conditions that might promote monocyte recruitment to the lung.

Peripheral blood monocytes are comprised of two phenotypically distinct subsets. One subset is CCR2^{hi} Gr1^{hi}C$_{X}$3CR1^{int} inflammatory monocytes and the other CCR2^{lo}Gr1^{lo}CX3CR1^{hi} noninflammatory monocytes, respectively. Landsman and colleagues showed that when adoptively transferred into recipient mice, both subsets of monocytes can give rise to pulmonary DCs, even under steady-state conditions (Landsman et al., 2007). Under the conditions studied, only Gr1^{lo}C$_{X}$3CR1^{hi} monocytes had the potential to differentiate into lung macrophages. However, Gr1^{hi}C$_{X}$3CR1^{int} monocytes could acquire this potential upon conversion into Gr1^{lo}C$_{X}$3CR1^{hi} cells. Thus, the lung can contain CDP-derived, *Flt3L*-dependent classical CD103^{+} DCs and CD11b^{hi} DCs, as well as monocyte-derived CD11b^{hi} DCs. Whether CD11b^{hi} DCs that arise from these alternative developmental pathways possess different capacities and functions is currently unknown.

4. DC ACTIVATION

Efficient pathogen recognition and elimination requires the coordinated interactions of many cells of the innate and adaptive immune systems. It has been more than 20 years since Charles Janeway first suggested that the innate immune system might recognize infectious organisms through germline-encoded genes and that this recognition would activate the adaptive immune system. We now know that this hypothesis was correct and that there are several families of such pattern recognition receptors (PRRs), including the TLR family, C-type lectins, nucleotide-binding oligomerization domain (NOD)-like receptors (NLRs), and retinoic acid-inducible gene (RIG)-I-like receptors that can fulfill this function (Joffre et al., 2009). These receptors have evolved to directly recognize pathogen-associated molecular patterns (PAMPs) and to trigger innate immune responses. Many of these receptors are expressed in DCs, which also initiate adaptive immune responses. Thus, DCs are strategically positioned at the crossroads of the innate and adaptive immune systems. Normally residing in a relatively quiescent state, DCs are able to sense danger signals from pathogen products and respond by stimulating adaptive immune responses.

4.1 DC Activation by TLRs

The TLR family is probably the best-studied family of PRRs (Iwasaki and Medzhitov, 2010). Following binding of PAMPs, individual receptors form homo- or heterodimers that transduce signals to their downstream partners. Except for TLR3, all TLRs signal through the adaptor molecule MyD88. Many TLRs are present on the cell surface, including TLR1, TLR2, TLR4, TLR5, TLR6, and TLR11. One of the best studied of these receptors is TLR4, which recognizes lipopolysaccharide (LPS), a membrane-bound lipid found in both Gram^{-ve} and Gram^{+ve} bacteria. Other cell-surface TLRs recognize different conserved molecular patterns, including bacterial peptidoglycan (TLR2) and flagellin (TLR5). Some TLRs, including TLR3, TLR7, TLR8, and TLR9, are found primarily within intracellular vesicles. As noted in the discussion on pDCs, these receptors recognize nucleic acids, such as double-stranded RNA (TLR3) and unmethylated CpG motifs (TLR9). The importance of TLR7 in antiviral responses is underscored by the finding that mice lacking this receptor fail to generate protective immunity following vaccination with inactivated virus (Koyama et al., 2007). Using a microarray-based comparison of gene expression in CD103$^+$ and CD11b^{hi} DC subsets prepared from lungs of untreated mice, we have found that that *Tlr2* is highly expressed in CD11b^{hi} DCs, whereas *Tlr3* is highly expressed in CD103$^+$ DCs. The latter finding is consistent with the observation that CD103$^+$ DCs are important for clearance of influenza virus (GeurtsvanKessel et al., 2008). However, *Tlr7* and *Tlr8* are both higher in CD11b^{hi} DCs, although a requirement for this cell type in clearing viral infections has not yet been reported.

Signaling responses of individual TLRs can be sufficient to activate both innate and adaptive immune response, but activation of more than one TLR can result in complementary, synergistic, or antagonistic effects (Trinchieri and Sher, 2007). It is likely, therefore, that multiple TLRs can also cooperate in the lung, but the consequences of that cooperation remain to be determined. In addition to interactions within the TLR family, these signaling pathways can also combine with non-TLR pathways to elicit stronger signals than from either one alone. For example, TLR signaling combined with CD40 signaling leads to particularly strong DC activation (Ahonen et al., 2004). It is likely that the signals that pulmonary DCs receive are from multiple sources and the consequences are still poorly understood.

4.2 DC Activation by C-type Lectins

Dectin-1

Dectin-1 is a member of the C-type lectin family, which recognizes β-glucans from fungal pathogens such as *Candida albicans* or *Aspergillus fumigatus* (Taylor et al., 2007). *C. albicans* preferentially induces Th17 responses (Acosta-Rodriguez et al., 2007), and mice lacking Dectin-1 have reduced Th17 responses to *A. fumigatus*. These mice also have increased Th1 responses, probably because Dectin-1 on wild-type DCs reduces the production of IL-12 and IFN-γ (Rivera et al., 2011). Transient depletion of monocyte-derived DCs reduced Th1 responses and increased Th17 responses, which is in agreement with previous findings that inflammatory DCs are potent inducers of Th1 responses upon LPS stimulation or influenza infection (Cheong et al., 2010; Nakano et al., 2009). However, Ly-6C$^+$CD11b^{hi} DCs are also reported to promote IL-17A production by CD4$^+$ T cells through a mechanism that requires TNF-α production, and this pathway is abolished in Dectin-1-deficient mice (Fei et al., 2011). Together, these findings suggest that inflammatory DCs can promote either Th1 or Th17 responses, depending on experimental conditions. The ability of Dectin-1 to promote Th17 responses to fungi exemplifies the ability of PPRs to induce immune responses specialized at clearing specific types of pathogens.

DC-SIGN

The DC-specific intercellular adhesion molecule-3 grabbing nonintegrin (DC-SIGN) is another C-type lectin that is expressed by DCs in the lung. DC-SIGN was originally described as a molecule on DCs that binds ICAM-3 on resting T cells, thereby promoting T cell: DC interactions (Geijtenbeek et al., 2000). Subsequently, DC-SIGN was shown to bind also both endogenous and exogenous antigens. For example, DC-SIGN mediates the binding and internalization of *Mycobacterium tuberculosis* (Tailleux et al., 2003) and *A. fumigatus* (Serrano-Gomez et al., 2004). Interestingly, binding to DC-SIGN by a cell membrane component, ManLAM, of *M. bovis* suppressed DC function by interfering with TLR signaling (Geijtenbeek et al., 2003). This suppression is mediated through Raf-1 and NF-kB, and leads to increased IL-10 production and anti-inflammatory responses (Gringhuis et al., 2007). Other IL-10-inducing pathways, such as those including ERK and PI3K, can also be triggered by DC-SIGN (Caparros et al., 2006). Whether this molecule is relevant to the induction of diseases such as asthma remains to be determined.

4.3 DC Activation by RIG-I-Like Receptors

In addition to TLR3 and TLR7, another family of proteins known as retinoic acid-inducible gene-I (RIG-I)-like receptors are capable of recognizing viral RNA by a TLR-independent mechanism (Yoneyama et al., 2005). These proteins are RNA helicases that are expressed ubiquitously at low levels but are highly induced by type I interferons or viral infection. In addition to RIG-I itself, this family contains the melanoma differentiation-associated 5 (MDA5) and the laboratory of genetics and physiology 2 (LGP2) receptors that can respond to viral infection. Interestingly, RIG-I is important for the induction of type I interferons in cDCs, but not in pDCs, even though the latter produce very large quantities of these proteins (Kato et al., 2005).

5. DC TRAFFICKING

The ability of pulmonary DCs to take up antigen and migrate to draining LNs provides naïve T cells with a dynamic representation of antigens in the lung. DC migration proceeds at low levels during steady-state conditions, but the rate and volume of that migration is markedly increased following exposure to proinflammatory stimuli (Vermaelen et al., 2001). This migration of DCs from the lung to draining LNs is thought to be an important event in the induction of effector responses to inhaled antigens. However, DC migration might also be important to maintain stimulation of memory T cells because after influenza infection, some DCs continue to transport viral antigen to LNs long after virus has been cleared from the lung (Kim et al., 2010).

The transition from resident to migratory DCs is a multistep process that involves the MHC-II invariant chain CD74 (Faure-Andre et al., 2008). When DCs are stationary, CD74 sequesters the actin-based motor protein myosin II in the endosomal compartment and promotes loading of peptide onto the MHC. Upon DC activation, antigen uptake increases transiently and is then markedly reduced, in concert with the acquisition of mobility. Once DCs are mobile, their migration to draining LNs is guided by interactions between the chemokine receptor CCR7 and its cognate chemokine ligands CCL19 and CCL21, which are produced in the lymphatic vessels and in T cell areas of LNs (Forster et al., 1999; Gunn et al., 1998). Many $CD103^{+}$ DCs display CCR7 and are dependent on this receptor for their migration to lung-draining mediastinal LNs (Hintzen et al., 2006). Consequently, $CD103^{+}$ DCs are virtually absent in LNs of $Ccr7^{-/-}$ mice. However, $CD11b^{hi}$ DCs can nonetheless be found in LNs of these mice, albeit at reduced levels compared to those of wild-type mice (Hintzen et al., 2006; Nakano et al., 2012). The latter observation suggests that $CD11b^{hi}$ DCs can gain access to mediastinal LNs by both CCR7-dependent and -independent pathways. Other studies have confirmed that under some circumstances, $CD11b^{hi}$ DCs can migrate to LNs (Jakubzick et al., 2006; Raymond et al., 2009), but it is unclear whether all $CD11b^{hi}$ DCs in the lung have this potential. The factors that induce migration of DCs to draining LNs are poorly characterized, and it is not known if all DCs have the potential for migration. Some DCs might reside in an anatomic position that effectively shields them from stimuli that would otherwise promote their mobilization. Alternatively, DCs might gain the ability to migrate on a stochastic basis that ensures an appropriate distribution of DCs in the lung and draining LNs. A third possibility is that some DCs are programmed to remain in the lung, even when exposed to stimuli that promote migration of other DC subsets. Future studies will be required to resolve these questions.

It is well documented that the aging process includes a decline in immune function, and the elderly are more susceptible to infectious pathogens, including respiratory viruses. This segment of the population also displays diminished responses to vaccines, suggesting that the initiation of immune responses declines with age. Some insight into the underlying causes of this phenomenon was provided by a recent report showing that migration of lung DCs to draining LNs in response to viral infection diminishes with increasing age (Zhao et al., 2011). The impaired migration of lung DCs in aged mice is due at least in part to increased levels of the arachidonic acid metabolite, prostaglandin D2 (PDG_2), because blockade of PDG_2 synthesis with small molecule antagonists increased levels of *Ccr7* expression and consequent DC migration to draining LNs. The constitutively expressed cyclooxygenase (COX)-1 and the inducible COX-2 are responsible for the synthesis of PDG_2, but it remains to be seen whether nonsteroidal anti-inflammatory drugs such as aspirin can increase DC migration in the elderly and thereby confer more robust responses to pathogens.

6. DC-MEDIATED INDUCTION OF T HELPER CELL DIFFERENTIATION

DC stimulation of naïve $CD4^{+}$ T cells directs their differentiation into various Th cell lineages having specialized functions for pathogen clearance (Pulendran et al., 2010). Thus, Th1 cells provide protective immunity against viruses and intracellular bacteria, whereas Th17 cells promote clearance of extracellular bacteria (Zhou et al., 2009). Th2 cells help to clear helminths by promoting IgE production, eosinophil proliferation, and mucus production, but these same responses to Th2 cytokines are pathogenic in the setting on allergic diseases such as asthma. An improved understanding of the APCs and molecules that drive Th2 cell differentiation might provide new opportunities for therapeutic intervention in multiple disease settings.

Naïve $CD4^{+}$ T cell differentiation into specific T cell lineages is determined largely by local cytokine levels. Th1 differentiation is driven by IL-12, which is produced in very high amounts by monocyte-derived, $Ly6C^{+}CD11b^{hi}$ inflammatory DCs (Leon et al., 2007; Nakano et al., 2009). Th17 cell differentiation is directed by TGF-β together with either IL-6 or IL-1 (Dong, 2008), whereas TGF-β on its own, or together with retinoic acid, promotes the differentiation of regulatory T cells (Coombes et al., 2007). Unlike these relatively well-characterized mechanisms, the cells and molecules that promote Th2 differentiation remain poorly understood. Although IL-4 can promote Th2 differentiation in vitro by activating STAT6 and promoting GATA3 expression, DCs do not produce IL-4 (Lambrecht and Hammad, 2009; Paul and Zhu, 2010). Accordingly, identifying the APC and cellular source of IL-4 that directs Th2 differentiation in vivo has been the focus of considerable effort and controversy during the past several years. Basophils have been shown to act as APCs and as a source of IL-4, and injection of anti-FcεRIα-specific antibodies suppress Th2 responses in mice (Perrigoue et al., 2009; Sokol et al., 2007, 2009; Yoshimoto et al., 2009). Based largely on these findings, it has been suggested that basophils are both necessary and sufficient for inducing Th2 responses in vivo. However, genetically engineered mice lacking basophils are still able to initiate Th2 cell differentiation, including allergic pulmonary inflammation (Ohnmacht et al., 2010). Moreover, several recent studies have shown a requirement for DCs in Th2 induction (Hammad et al., 2010; Lambrecht and Hammad, 2009; Ohnmacht et al., 2010; Phythian-Adams et al., 2010), and lung DCs are still widely considered to be required for the induction of allergic responses to inhaled allergens.

$CD103^+$ DCs and $CD11b^{hi}$ DCs were originally thought to possess similar T helper inducing activities (Beaty et al., 2007). However, this conclusion was based on experiments in which exogenous peptides were added to DC subsets ex vivo, a procedure that bypasses events related to in vivo uptake of inhaled antigens. Our own experiments in which allergen was instilled directly into the airways to allow lung resident DCs to take up the allergen in vivo revealed that $CD103^+$ DCs from the lung or lung-draining LNs efficiently primed Th2 differentiation ex vivo, whereas $CD11b^{hi}$ DCs primed Th1 differentiation (Nakano et al., 2012). Moreover, mice lacking $CD103^+$ DCs failed to undergo Th2 priming to various inhaled allergens and did not develop asthma-like responses following subsequent allergen challenge. Paradoxically, however, when stimulated with TLR ligands or PMA and ionomycin, $CD103^+$ DCs produce the Th1-inducing cytokine IL-12, suggesting that these DCs have the potential to prime Th1 responses. Indeed, when cockroach antigen or house dust mite preparations (that include both PAMPs and allergens) are instilled into the airway, $CD103^+$ DCs are able to prime both Th1 and Th2 responses, whereas $CD11b^{hi}$ DCs exclusively prime Th1 responses (Nakano et al., 2012). Thus, depending on the experimental setting, $CD103^+$ DCs can prime either Th1 or Th2 responses. These findings might be related to the observation that low levels of TCR stimulation lead to Th2 responses by upregulating GATA3 expression, whereas strong TCR stimulation suppresses GATA3 expression through activation of extracellular signal-regulated kinase (ERK) and thereby primes Th1 responses (Iezzi et al., 2009; Yamane et al., 2005). In agreement with these observations, we found that when the two major DC subsets were purified from the lung and cultured with low levels of peptide, $CD103^+$ DCs primed Th2 responses, but at high levels of peptide, they primed Th1 responses. Together, these findings show that $CD103^+$ DCs have a significant role in priming both Th1 and Th2 responses to inhaled allergens.

Differential expression of Notch ligands has been proposed to contribute to T-helper cell differentiation, with Jagged-1 and Jagged-2 promoting Th2 differentiation and Delta-like 4 promoting Th1 differentiation (Amsen et al., 2004). However, a requirement of Jagged-2 for Th2 responses in vivo has been questioned (Worsley et al., 2008), and it remains to be determined whether Jagged is required for allergic sensitization through the airway. Nonetheless, Jagged-1 is upregulated on bone marrow-derived DCs (BMDCs) pulsed with allergen, and inhibition of Notch signaling results in decreased cytokine production when antigen-pulsed DCs and naïve $CD4^+$ T cells are transferred into IL-4-deficient mice. Thus, interactions between Notch on $CD4^+$ T cells and Jagged-1 on APCs might contribute to IL-4 production by T cells and promote Th2 differentiation for the development of airway hyperresponsiveness and allergic airway inflammation. In agreement with this concept, it has been reported that Jagged-2 is blunted in DCs from mice lacking c-Kit and that they are unable to induce robust Th2 and Th17 responses when adoptively transferred into mice (Krishnamoorthy et al., 2008). We have observed that c-Kit is much more highly expressed in $CD103^+$ DCs than in $CD11b^{hi}$ DCs, but we have been unable to confirm that c-Kit-deficient $CD103^+$ DCs are impaired in their ability to prime Th2 responses (Nakano et al., 2012).

7. ACTIVATION OF MEMORY T CELLS DURING ANTIGEN CHALLENGE

In addition to their primary role in driving the differentiation of naïve T cells to $CD8^+$ cytotoxic cells or $CD4^+$ T helper cells, DCs are also important in the effector phase of immune responses. Following allergen challenge, DCs in the lung increase rapidly (McWilliam et al., 1994), with most of this increase being due to $CD11b^{hi}$ DCs (Beaty et al., 2007). The recruitment of these cells is mediated largely by the chemokine receptor CCR2 (Robays et al., 2007), which is highly expressed on inflammatory monocytes. DCs are necessary during the time of allergen challenge because their depletion by delivery of DTX to CD11c-DTR mice prevents development of asthma-like features, including eosinophilic inflammation, goblet cell hyperplasia, and airway hyperresponsiveness. Moreover, intratracheal delivery of bone marrow-derived CD11b^{hi} DCs is sufficient to restore these asthma-like features (Lambrecht et al., 1998; van Rijt et al., 2005). In a similar experiment, selective depletion of CD11b^{+} myeloid cells in the lung by administration of DTX to CD11b-DTR mice prevented production of the chemokines CCL17 and CCL22, which likely recruit Th2 cells by binding to their receptor, CCR4 (Medoff et al., 2009). Taken together, these data suggest that CD11b^{hi} DCs might be important to activate memory or effector T cells arriving in the lung in response to allergen challenge. Interestingly, production of these chemokines was also found to be dependent on the presence of STAT6, which also is required for the development of Th2 cells. The requirement of DCs for activating effector cells is not unique to challenges with allergen because lung resident DCs are also required to trigger release of inflammatory mediators by influenza-specific CD8^{+} T cells during the late phase of influenza infection, after most of the virus has been cleared (Hufford et al., 2011). It remains to be shown which specific subset of lung DCs in most important in this regard, although they might be monocyte-derived because CCR2-deficient mice have less pathology and increased survival following influenza infection (Lin et al., 2008).

8. INTERACTION OF DCs WITH OTHER CELL TYPES

8.1 Interaction of DCs with Epithelial Cells

The epithelium is widely considered to provide an effective barrier to most inhaled molecules, but it has long been known that whole proteins can penetrate the epithelium because they can be found in the blood within 24 h of instillation (Bensch et al., 1967). It is unclear whether this occurs by active transport of proteins through epithelial cells, or if some gaps between epithelial cells allow molecules to gain access to the subepithelial space. Thus, although the ability of CD103^{+} DCs to extend processes into the lumen of the airway might suggest that they have greater access to it, relatively small molecules, including ovalbumin, can be readily taken up by DCs that do not extend processes into the lumen. In fact, CD11b^{hi} DCs take up higher levels of ovalbumin than do CD103^{+} DCs (Nakano et al., 2012). Nonetheless, recent findings have revealed that the TLR3 ligand, polyinosinic:polycytidylic acid (poly(I:C)), can markedly increase the paracellular permeability of the airway epithelium in a manner dependent on protein kinase D (Rezaee et al., 2011). This finding suggests that poly(I:C) might act as an adjuvant by exposing increased numbers of subepithelial DCs to antigens that might otherwise be shielded from them.

In addition to its barrier function, the airway epithelium might also sense the environment and respond by providing signals to DCs. For example, human epithelial cells can activate DCs through the production of cytokines, such as thymic stromal lymphopoietin (TSLP) (Soumelis et al., 2002). In response to TSLP, DCs upregulate MHC class II, CD40, CD80, CD86, and OX40-L (Wang et al., 2006). However, TSLP does not upregulate IL-12, the Th1-promoting cytokine, which likely favors the induction of Th2 responses by DCs. Airway epithelial cell lines also increase levels of TLR3 and TLR4 on human monocyte-derived DCs, which consequently develop increased sensitivity to LPS (Rate et al., 2009). This axis might be highly relevant to the effects of air pollution on allergic responses. Although isolated monocyte-derived DCs fail to undergo phenotypic (CD80, CD83, CD86) or functional (T cell activation) maturation in response to exposure to diesel exhaust particles (DEP), these DCs can be activated if they are cultured with primary human bronchial epithelial cells treated with DEP (Bleck et al., 2006). It is likely that soluble molecules convey some of these signals because conditioned medium from DEP-treated epithelial cells can also promote DC maturation, and this activity is inhibited by blockade of GM-CSF. Together, these findings suggest that airway epithelial cells might promote DC maturation through production of multiple soluble cytokines, including TSLP and GM-CSF. In this way, epithelial cells can exert some control over the initial events that ultimately give rise to immune responses to inhaled antigens.

8.2 Interaction of DCs with T Cells

Another cell type that can promote the ability of DCs to stimulate T cells is the T cell itself. In the rat, "immature" DCs can acquire antigen, but they lack the ability to effectively present antigen to T cells, probably because levels of costimulatory molecules such as CD86 are too low. However, levels of CD86 on mucosal DCs (not parenchymal DCs) are increased by their interaction with antigen-specific T cells (Huh et al., 2003). Interaction of allergen-specific, CD4^{+} T cells with DCs is also linked to enhanced allergen uptake and CD40 expression by CD11b^{lo} (CD103^{+}) DCs (von Garnier et al., 2007). Thus, peripheral memory T cells might indirectly promote the activation of naïve T cell or central memory T cell activation by activating airway DCS.

8.3 Interaction of DCs with Neutrophils

Like T cells, neutrophils can also promote DC activation. In the setting of *Aspergillus* infection, neutropenia results in the increased accumulation of DCs in the lung, but these DCs have a reduced ability to migrate to draining LNs (Park et al., 2012). Neutrophils are also reported to increase

FcεRI expression on lung DCs (Cheung et al., 2010) and to stimulate IL-12 production by DCs (Kreisel et al., 2011). These findings reveal yet another link between innate and adaptive immune responses and show that the rapid neutrophilic response to infection contributes to DC activation, and by extension, the development of adaptive immune responses.

8.4 Interaction of DCs with Macrophages

The lung contains two major types of macrophages, alveolar macrophages that express high levels of CD11c but relatively low levels of F4/80 and CD11b, and interstitial macrophages that have low levels of CD11c but high levels of F4/80 and CD11b. It has long been known that macrophages can regulate the function of DCs (Holt et al., 1993, 1988), suggesting that these cells might be responsible for preventing the induction of allergic responses in healthy individuals. Recently, the inhibitory function of lung macrophages was shown to be conferred interstitial macrophages, not alveolar macrophages. The former cell type can suppress the function of bone marrow-derived DCs by an IL-10-dependent mechanism (Bedoret et al., 2009). Although this report did not directly study interactions between macrophages and lung DCs, interstitial macrophages were preferentially depleted with an anti-F4/80 antibody, and these mice developed increased allergic responses following sensitization with allergen.

9. SUMMARY AND FUTURE DIRECTIONS

The past 20 years have witnessed a dramatic increase in our understanding of pulmonary DCs. This improved understanding includes knowledge of how DCs develop, are activated, traffic to draining LNs, and stimulate antigen-specific T cells to become effector and memory T cells. Specialized functions have begun to be assigned to specific DC populations in the lung, which is important for the eventual design of therapies designed to alter DC function. Because DCs play such a pivotal role in the development and control of immune responses to inhaled antigens and pathogens, it is likely that they might eventually be used in therapeutic applications to either increase responses to pathogens or decrease immune responses to harmless antigens. Currently, the relatively low number of pulmonary DCs limits their use in this regard. Consequently, bone marrow DCs are often used in experiments because they can be obtained in much higher numbers. However, it is not clear that bone marrow-derived DCs have the capacity to interact with other cell types in the lung and in the way that bone fide lung DCs are able to do. The development of procedures to expand specific populations of lung DCs in a way that maintains their functions would be an important advance. Similarly, the development of small molecules that specifically target select subsets of DCs in the lung would also provide a major advance in modulating immune responses to inhaled antigens by targeting the actions of pulmonary DCs.

REFERENCES

Acosta-Rodriguez, E.V., Rivino, L., Geginat, J., Jarrossay, D., Gattorno, M., Lanzavecchia, A., Sallusto, F., Napolitani, G., 2007. Surface phenotype and antigenic specificity of human interleukin 17-producing T helper memory cells. Nat. Immunol. 8, 639–646.

Ahonen, C.L., Doxsee, C.L., McGurran, S.M., Riter, T.R., Wade, W.F., Barth, R.J., Vasilakos, J.P., Noelle, R.J., Kedl, R.M., 2004. Combined TLR and CD40 triggering induces potent CD8$^+$ T cell expansion with variable dependence on type I IFN. J. Exp. Med. 199, 775–784.

Amsen, D., Blander, J.M., Lee, G.R., Tanigaki, K., Honjo, T., Flavell, R.A., 2004. Instruction of distinct CD4 T helper cell fates by different notch ligands on antigen-presenting cells. Cell 117, 515–526.

Annacker, O., Coombes, J.L., Malmstrom, V., Uhlig, H.H., Bourne, T., Johansson-Lindbom, B., Agace, W.W., Parker, C.M., Powrie, F., 2005. Essential role for CD103 in the T cell-mediated regulation of experimental colitis. J. Exp. Med. 202, 1051–1061.

Asselin-Paturel, C., Boonstra, A., Dalod, M., Durand, I., Yessaad, N., Dezutter-Dambuyant, C., Vicari, A., O'Garra, A., Biron, C., Briere, F., Trinchieri, G., 2001. Mouse type I IFN-producing cells are immature APCs with plasmacytoid morphology. Nat. Immunol. 2, 1144–1150.

Beaty, S.R., Rose Jr., C.E., Sung, S.S., 2007. Diverse and potent chemokine production by lung CD11bhigh DCs in homeostasis and in allergic lung inflammation. J. Immunol. 178, 1882–1895.

Bedoret, D., Wallemacq, H., Marichal, T., Desmet, C., Quesada Calvo, F., Henry, E., Closset, R., Dewals, B., Thielen, C., Gustin, P., de Leval, L., Van Rooijen, N., Le Moine, A., Vanderplasschen, A., Cataldo, D., Drion, P.V., Moser, M., Lekeux, P., Bureau, F., 2009. Lung interstitial macrophages alter dendritic cell functions to prevent airway allergy in mice. J. Clin. Invest. 119, 3723–3738.

Bennett, C.L., van Rijn, E., Jung, S., Inaba, K., Steinman, R.M., Kapsenberg, M.L., Clausen, B.E., 2005. Inducible ablation of mouse langerhans cells diminishes but fails to abrogate contact hypersensitivity. J. Cell Biol. 169, 569–576.

Bensch, K.G., Dominguez, E., Liebow, A.A., 1967. Absorption of intact protein molecules across the pulmonary air-tissue barrier. Science 157, 1204–1206.

Bjorck, P., 2001. Isolation and characterization of plasmacytoid DCs from Flt3 ligand and granulocyte-macrophage colony-stimulating factor-treated mice. Blood 98, 3520–3526.

Bleck, B., Tse, D.B., Jaspers, I., Curotto de Lafaille, M.A., Reibman, J., 2006. Diesel exhaust particle-exposed human bronchial epithelial cells induce dendritic cell maturation. J. Immunol. 176, 7431–7437.

Bogunovic, M., Ginhoux, F., Helft, J., Shang, L., Hashimoto, D., Greter, M., Liu, K., Jakubzick, C., Ingersoll, M.A., Leboeuf, M., Stanley, E.R., Nussenzweig, M., Lira, S.A., Randolph, G.J., Merad, M., 2009. Origin of the lamina propria dendritic cell network. Immunity 31, 513–525.

Caparros, E., Munoz, P., Sierra-Filardi, E., Serrano-Gomez, D., Puig-Kroger, A., Rodriguez-Fernandez, J.L., Mellado, M., Sancho, J., Zubiaur, M., Corbi, A.L., 2006. DC-SIGN ligation on DCs results in ERK and PI3K activation and modulates cytokine production. Blood 107, 3950–3958.

Cepek, K.L., Shaw, S.K., Parker, C.M., Russell, G.J., Morrow, J.S., Rimm, D.L., Brenner, M.B., 1994. Adhesion between epithelial cells and T lymphocytes mediated by E-cadherin and the alpha E beta 7 integrin. Nature 372, 190–193.

Cheong, C., Matos, I., Choi, J.H., Dandamudi, D.B., Shrestha, E., Longhi, M.P., Jeffrey, K.L., Anthony, R.M., Kluger, C., Nchinda, G., Koh, H., Rodriguez, A., Idoyaga, J., Pack, M., Velinzon, K., Park, C.G., Steinman, R.M., 2010. Microbial stimulation fully differentiates monocytes to DC-SIGN/CD209(+) DCs for immune T cell areas. Cell 143, 416–429.

Cheung, D.S., Ehlenbach, S.J., Kitchens, R.T., Riley, D.A., Thomas, L.L., Holtzman, M.J., Grayson, M.H., 2010. Cutting edge: $CD49d^+$ neutrophils induce FcepsilonRI expression on lung DCs in a mouse model of postviral asthma. J. Immunol. 185, 4983–4987.

Cook, D.N., Bottomly, K., 2007. Innate immune control of pulmonary dendritic cell trafficking. Proc. Am. Thorac. Soc. 4, 234–239.

Coombes, J.L., Siddiqui, K.R., Arancibia-Carcamo, C.V., Hall, J., Sun, C.M., Belkaid, Y., Powrie, F., 2007. A functionally specialized population of mucosal $CD103^+$ DCs induces $Foxp3^+$ regulatory T cells via a TGF-beta and retinoic acid-dependent mechanism. J. Exp. Med. 204, 1757–1764.

de Heer, H.J., Hammad, H., Soullie, T., Hijdra, D., Vos, N., Willart, M.A., Hoogsteden, H.C., Lambrecht, B.N., 2004. Essential role of lung plasmacytoid DCs in preventing asthmatic reactions to harmless inhaled antigen. J. Exp. Med. 200, 89–98.

del Rio, M.L., Rodriguez-Barbosa, J.I., Kremmer, E., Forster, R., 2007. $CD103^-$ and $CD103^+$ bronchial lymph node DCs are specialized in presenting and cross-presenting innocuous antigen to $CD4^+$ and $CD8^+$ T cells. J. Immunol. 178, 6861–6866.

Desch, A.N., Randolph, G.J., Murphy, K., Gautier, E.L., Kedl, R.M., Lahoud, M.H., Caminschi, I., Shortman, K., Henson, P.M., Jakubzick, C.V., 2011. $CD103^+$ pulmonary DCs preferentially acquire and present apoptotic cell-associated antigen. J. Exp. Med. 208, 1789–1797.

Dong, C., 2008. TH17 cells in development: an updated view of their molecular identity and genetic programming. Nat. Rev. Immunol. 8, 337–348.

Faure-Andre, G., Vargas, P., Yuseff, M.I., Heuze, M., Diaz, J., Lankar, D., Steri, V., Manry, J., Hugues, S., Vascotto, F., Boulanger, J., Raposo, G., Bono, M.R., Rosemblatt, M., Piel, M., Lennon-Dumenil, A.M., 2008. Regulation of dendritic cell migration by CD74, the MHC class II-associated invariant chain. Science 322, 1705–1710.

Fei, M., Bhatia, S., Oriss, T.B., Yarlagadda, M., Khare, A., Akira, S., Saijo, S., Iwakura, Y., Fallert Junecko, B.A., Reinhart, T.A., Foreman, O., Ray, P., Kolls, J., Ray, A., 2011. TNF-alpha from inflammatory DCs (DCs) regulates lung IL-17A/IL-5 levels and neutrophilia versus eosinophilia during persistent fungal infection. Proc. Natl. Acad. Sci. U.S.A. 108, 5360–5365.

Fogg, D.K., Sibon, C., Miled, C., Jung, S., Aucouturier, P., Littman, D.R., Cumano, A., Geissmann, F., 2006. A clonogenic bone marrow progenitor specific for macrophages and DCs. Science 311, 83–87.

Forster, R., Schubel, A., Breitfeld, D., Kremmer, E., Renner-Muller, I., Wolf, E., Lipp, M., 1999. CCR7 coordinates the primary immune response by establishing functional microenvironments in secondary lymphoid organs. Cell 99, 23–33.

Geijtenbeek, T.B., Torensma, R., van Vliet, S.J., van Duijnhoven, G.C., Adema, G.J., van Kooyk, Y., Figdor, C.G., 2000. Identification of DC-SIGN, a novel dendritic cell-specific ICAM-3 receptor that supports primary immune responses. Cell 100, 575–585.

Geijtenbeek, T.B., Van Vliet, S.J., Koppel, E.A., Sanchez-Hernandez, M., Vandenbroucke-Grauls, C.M., Appelmelk, B., Van Kooyk, Y., 2003. Mycobacteria target DC-SIGN to suppress dendritic cell function. J. Exp. Med. 197, 7–17.

GeurtsvanKessel, C.H., Willart, M.A., van Rijt, L.S., Muskens, F., Kool, M., Baas, C., Thielemans, K., Bennett, C., Clausen, B.E., Hoogsteden, H.C., Osterhaus, A.D., Rimmelzwaan, G.F., Lambrecht, B.N., 2008. Clearance of influenza virus from the lung depends on migratory langerin+CD11b- but not plasmacytoid DCs. J. Exp. Med. 205, 1621–1634.

Ginhoux, F., Liu, K., Helft, J., Bogunovic, M., Greter, M., Hashimoto, D., Price, J., Yin, N., Bromberg, J., Lira, S.A., Stanley, E.R., Nussenzweig, M., Merad, M., 2009. The origin and development of nonlymphoid tissue $CD103^+$ DCs. J. Exp. Med. 206, 3115–3130.

Grayson, M.H., Cheung, D., Rohlfing, M.M., Kitchens, R., Spiegel, D.E., Tucker, J., Battaile, J.T., Alevy, Y., Yan, L., Agapov, E., Kim, E.Y., Holtzman, M.J., 2007. Induction of high-affinity IgE receptor on lung DCs during viral infection leads to mucous cell metaplasia. J. Exp. Med. 204, 2759–2769.

Gringhuis, S.I., den Dunnen, J., Litjens, M., van Het Hof, B., van Kooyk, Y., Geijtenbeek, T.B., 2007. C-type lectin DC-SIGN modulates Toll-like receptor signaling via Raf-1 kinase-dependent acetylation of transcription factor NF-kappaB. Immunity 26, 605–616.

Gunn, M.D., Tangemann, K., Tam, C., Cyster, J.G., Rosen, S.D., Williams, L.T., 1998. A chemokine expressed in lymphoid high endothelial venules promotes the adhesion and chemotaxis of naive T lymphocytes. Proc. Natl. Acad. Sci. U.S.A. 95, 258–263.

Hacker, C., Kirsch, R.D., Ju, X.S., Hieronymus, T., Gust, T.C., Kuhl, C., Jorgas, T., Kurz, S.M., Rose-John, S., Yokota, Y., Zenke, M., 2003. Transcriptional profiling identifies Id2 function in dendritic cell development. Nat. Immunol. 4, 380–386.

Hahn, I., Klaus, A., Maus, R., Christman, J.W., Welte, T., Maus, U.A., 2011. Dendritic cell depletion and repopulation in the lung after irradiation and bone marrow transplantation in mice. Am. J. Respir. Cell Mol. Biol. 45, 534–541.

Hammad, H., Plantinga, M., Deswarte, K., Pouliot, P., Willart, M.A., Kool, M., Muskens, F., Lambrecht, B.N., 2010. Inflammatory DCs—not basophils—are necessary and sufficient for induction of Th2 immunity to inhaled house dust mite allergen. J. Exp. Med. 207, 2097–2111.

Hildner, K., Edelson, B.T., Purtha, W.E., Diamond, M., Matsushita, H., Kohyama, M., Calderon, B., Schraml, B.U., Unanue, E.R., Diamond, M.S., Schreiber, R.D., Murphy, T.L., Murphy, K.M., 2008. Batf3 deficiency reveals a critical role for $CD8alpha^+$ DCs in cytotoxic T cell immunity. Science 322, 1097–1100.

Hintzen, G., Ohl, L., del Rio, M.L., Rodriguez-Barbosa, J.I., Pabst, O., Kocks, J.R., Krege, J., Hardtke, S., Forster, R., 2006. Induction of tolerance to innocuous inhaled antigen relies on a CCR7-dependent dendritic cell-mediated antigen transport to the bronchial lymph node. J. Immunol. 177, 7346–7354.

Holt, P.G., Oliver, J., Bilyk, N., McMenamin, C., McMenamin, P.G., Kraal, G., Thepen, T., 1993. Downregulation of the antigen presenting cell function(s) of pulmonary DCs in vivo by resident alveolar macrophages. J. Exp. Med. 177, 397–407.

Holt, P.G., Schon-Hegrad, M.A., Oliver, J., 1988. MHC class II antigen-bearing DCs in pulmonary tissues of the rat. Regulation of antigen presentation activity by endogenous macrophage populations. J. Exp. Med. 167, 262–274.

Honda, K., Yanai, H., Negishi, H., Asagiri, M., Sato, M., Mizutani, T., Shimada, N., Ohba, Y., Takaoka, A., Yoshida, N., Taniguchi, T., 2005. IRF-7 is the master regulator of type-I interferon-dependent immune responses. Nature 434, 772–777.

Hufford, M.M., Kim, T.S., Sun, J., Braciale, T.J., 2011. Antiviral $CD8^+$ T cell effector activities in situ are regulated by target cell type. J. Exp. Med. 208, 167–180.

Huh, J.C., Strickland, D.H., Jahnsen, F.L., Turner, D.J., Thomas, J.A., Napoli, S., Tobagus, I., Stumbles, P.A., Sly, P.D., Holt, P.G., 2003. Bidirectional interactions between antigen-bearing respiratory tract DCs (DCs) and T cells precede the late phase reaction in experimental asthma: DC activation occurs in the airway mucosa but not in the lung parenchyma. J. Exp. Med. 198, 19–30.

Iezzi, G., Sonderegger, I., Ampenberger, F., Schmitz, N., Marsland, B.J., Kopf, M., 2009. CD40-CD40L cross-talk integrates strong antigenic signals and microbial stimuli to induce development of IL-17-producing $CD4^+$ T cells. Proc. Natl. Acad. Sci. U.S.A. 106, 876–881.

Iwasaki, A., Medzhitov, R., 2010. Regulation of adaptive immunity by the innate immune system. Science 327, 291–295.

Jahnsen, F.L., Strickland, D.H., Thomas, J.A., Tobagus, I.T., Napoli, S., Zosky, G.R., Turner, D.J., Sly, P.D., Stumbles, P.A., Holt, P.G., 2006. Accelerated antigen sampling and transport by airway mucosal DCs following inhalation of a bacterial stimulus. J. Immunol. 177, 5861–5867.

Jakubzick, C., Bogunovic, M., Bonito, A.J., Kuan, E.L., Merad, M., Randolph, G.J., 2008a. Lymph-migrating, tissue-derived DCs are minor constituents within steady-state lymph nodes. J. Exp. Med. 205, 2839–2850.

Jakubzick, C., Tacke, F., Ginhoux, F., Wagers, A.J., van Rooijen, N., Mack, M., Merad, M., Randolph, G.J., 2008b. Blood monocyte subsets differentially give rise to $CD103^+$ and $CD103^-$ pulmonary dendritic cell populations. J. Immunol. 180, 3019–3027.

Jakubzick, C., Tacke, F., Llodra, J., van Rooijen, N., Randolph, G.J., 2006. Modulation of dendritic cell trafficking to and from the airways. J. Immunol. 176, 3578–3584.

Joffre, O., Nolte, M.A., Sporri, R., Reis e Sousa, C., 2009. Inflammatory signals in dendritic cell activation and the induction of adaptive immunity. Immunol. Rev. 227, 234–247.

Kato, H., Sato, S., Yoneyama, M., Yamamoto, M., Uematsu, S., Matsui, K., Tsujimura, T., Takeda, K., Fujita, T., Takeuchi, O., Akira, S., 2005. Cell type-specific involvement of RIG-I in antiviral response. Immunity 23, 19–28.

Kim, T.S., Hufford, M.M., Sun, J., Fu, Y.X., Braciale, T.J., 2010. Antigen persistence and the control of local T cell memory by migrant respiratory dendritic cells after acute virus infection. J. Exp. Med. 207 (6), 116–172. http://dx.doi.org/10.1084/jem.20092017. Epub 2010 May 31.

Koyama, S., Ishii, K.J., Kumar, H., Tanimoto, T., Coban, C., Uematsu, S., Kawai, T., Akira, S., 2007. Differential role of TLR- and RLR-signaling in the immune responses to influenza A virus infection and vaccination. J. Immunol. 179, 4711–4720.

Kreisel, D., Sugimoto, S., Zhu, J., Nava, R., Li, W., Okazaki, M., Yamamoto, S., Ibrahim, M., Huang, H.J., Toth, K.A., Ritter, J.H., Krupnick, A.S., Miller, M.J., Gelman, A.E., 2011. Emergency granulopoiesis promotes neutrophil-dendritic cell encounters that prevent mouse lung allograft acceptance. Blood 118, 6172–6182.

Krishnamoorthy, N., Oriss, T.B., Paglia, M., Fei, M., Yarlagadda, M., Vanhaesebroeck, B., Ray, A., Ray, P., 2008. Activation of c-Kit in DCs regulates T helper cell differentiation and allergic asthma. Nat. Med. 14, 565–573.

Lambrecht, B.N., Hammad, H., 2009. Biology of lung DCs at the origin of asthma. Immunity 31, 412–424.

Lambrecht, B.N., Salomon, B., Klatzmann, D., Pauwels, R.A., 1998. DCs are required for the development of chronic eosinophilic airway inflammation in response to inhaled antigen in sensitized mice. J. Immunol. 160, 4090–4097.

Landsman, L., Varol, C., Jung, S., 2007. Distinct differentiation potential of blood monocyte subsets in the lung. J. Immunol. 178, 2000–2007.

Leon, B., Lopez-Bravo, M., Ardavin, C., 2007. Monocyte-derived DCs formed at the infection site control the induction of protective T helper 1 responses against Leishmania. Immunity 26, 519–531.

Lin, K.L., Suzuki, Y., Nakano, H., Ramsburg, E., Gunn, M.D., 2008. $CCR2^+$ monocyte-derived DCs and exudate macrophages produce influenza-induced pulmonary immune pathology and mortality. J. Immunol. 180, 2562–2572.

McWilliam, A.S., Nelson, D., Thomas, J.A., Holt, P.G., 1994. Rapid dendritic cell recruitment is a hallmark of the acute inflammatory response at mucosal surfaces. J. Exp. Med. 179, 1331–1336.

Medoff, B.D., Seung, E., Hong, S., Thomas, S.Y., Sandall, B.P., Duffield, J.S., Kuperman, D.A., Erle, D.J., Luster, A.D., 2009. $CD11b^+$ myeloid cells are the key mediators of Th2 cell homing into the airway in allergic inflammation. J. Immunol. 182, 623–635.

Naik, S.H., Sathe, P., Park, H.Y., Metcalf, D., Proietto, A.I., Dakic, A., Carotta, S., O'Keeffe, M., Bahlo, M., Papenfuss, A., Kwak, J.Y., Wu, L., Shortman, K., 2007. Development of plasmacytoid and conventional dendritic cell subtypes from single precursor cells derived in vitro and in vivo. Nat. Immunol. 8, 1217–1226.

Nakano, H., Free, M.E., Whitehead, G.S., Maruoka, S., Wilson, R.H., Nakano, K., Cook, D.N., 2012. Pulmonary CD103(+) DCs prime Th2 responses to inhaled allergens. Mucosal Immunol. 5, 53–65.

Nakano, H., Lin, K.L., Yanagita, M., Charbonneau, C., Cook, D.N., Kakiuchi, T., Gunn, M.D., 2009. Blood-derived inflammatory DCs in lymph nodes stimulate acute T helper type 1 immune responses. Nat. Immunol. 10, 394–402.

Nakano, H., Yanagita, M., Gunn, M.D., 2001. CD11c(+)B220(+)Gr-1(+) cells in mouse lymph nodes and spleen display characteristics of plasmacytoid DCs. J. Exp. Med. 194, 1171–1178.

Ohnmacht, C., Schwartz, C., Panzer, M., Schiedewitz, I., Naumann, R., Voehringer, D., 2010. Basophils orchestrate chronic allergic dermatitis and protective immunity against helminths. Immunity 33, 364–374.

Onai, N., Obata-Onai, A., Schmid, M.A., Ohteki, T., Jarrossay, D., Manz, M.G., 2007. Identification of clonogenic common Flt3+M-CSFR+ plasmacytoid and conventional dendritic cell progenitors in mouse bone marrow. Nat. Immunol. 8, 1207–1216.

Ortiz-Stern, A., Kanda, A., Mionnet, C., Cazareth, J., Lazzari, A., Fleury, S., Dombrowicz, D., Glaichenhaus, N., Julia, V., 2010. Langerin(+) DCs are responsible for LPS-induced reactivation of allergen-specific Th2 responses in postasthmatic mice. Mucosal Immunol. 4 (3), 343–353.

Park, S.J., Burdick, M.D., Mehrad, B., 2012. Neutrophils mediate maturation and efflux of lung DCs in response to *Aspergillus* germ tubes. Infect. Immun. 80 (5), 1759–1765.

Paul, W.E., Zhu, J., 2010. How are T(H)2-type immune responses initiated and amplified? Nat. Rev. Immunol. 10, 225–235.

Perrigoue, J.G., Saenz, S.A., Siracusa, M.C., Allenspach, E.J., Taylor, B.C., Giacomin, P.R., Nair, M.G., Du, Y., Zaph, C., van Rooijen, N., Comeau, M.R., Pearce, E.J., Laufer, T.M., Artis, D.,

2009. MHC class II-dependent basophil-CD4$^+$ T cell interactions promote T(H)2 cytokine-dependent immunity. Nat. Immunol. 10, 697–705.

Phythian-Adams, A.T., Cook, P.C., Lundie, R.J., Jones, L.H., Smith, K.A., Barr, T.A., Hochweller, K., Anderton, S.M., Hammerling, G.J., Maizels, R.M., MacDonald, A.S., 2010. CD11c depletion severely disrupts Th2 induction and development in vivo. J. Exp. Med. 207, 2089–2096.

Pulendran, B., Tang, H., Manicassamy, S., 2010. Programming DCs to induce T(H)2 and tolerogenic responses. Nat. Immunol. 11, 647–655.

Randolph, G.J., Angeli, V., Swartz, M.A., 2005. Dendritic-cell trafficking to lymph nodes through lymphatic vessels. Nat. Rev. Immunol. 5, 617–628.

Rate, A., Upham, J.W., Bosco, A., McKenna, K.L., Holt, P.G., 2009. Airway epithelial cells regulate the functional phenotype of locally differentiating DCs: implications for the pathogenesis of infectious and allergic airway disease. J. Immunol. 182, 72–83.

Raymond, M., Rubio, M., Fortin, G., Shalaby, K.H., Hammad, H., Lambrecht, B.N., Sarfati, M., 2009. Selective control of SIRP-alpha-positive airway dendritic cell trafficking through CD47 is critical for the development of T(H)2-mediated allergic inflammation. J. Allergy Clin. Immunol. 124, 1333–1342 e1331.

Rezaee, F., Meednu, N., Emo, J.A., Saatian, B., Chapman, T.J., Naydenov, N.G., De Benedetto, A., Beck, L.A., Ivanov, A.I., Georas, S.N., 2011. Polyinosinic:polycytidylic acid induces protein kinase D-dependent disassembly of apical junctions and barrier dysfunction in airway epithelial cells. J. Allergy Clin. Immunol. 128, 1216–1224 e1211.

Rivera, A., Hohl, T.M., Collins, N., Leiner, I., Gallegos, A., Saijo, S., Coward, J.W., Iwakura, Y., Pamer, E.G., 2011. Dectin-1 diversifies *Aspergillus* fumigatus-specific T cell responses by inhibiting T helper type 1 CD4 T cell differentiation. J. Exp. Med. 208, 369–381.

Robays, L.J., Maes, T., Lebecque, S., Lira, S.A., Kuziel, W.A., Brusselle, G.G., Joos, G.F., Vermaelen, K.V., 2007. Chemokine receptor CCR2 but not CCR5 or CCR6 mediates the increase in pulmonary DCs during allergic airway inflammation. J. Immunol. 178, 5305–5311.

Serrano-Gomez, D., Dominguez-Soto, A., Ancochea, J., Jimenez-Heffernan, J.A., Leal, J.A., Corbi, A.L., 2004. Dendritic cell-specific intercellular adhesion molecule 3-grabbing nonintegrin mediates binding and internalization of *Aspergillus* fumigatus conidia by DCs and macrophages. J. Immunol. 173, 5635–5643.

Sertl, K., Takemura, T., Tschachler, E., Ferrans, V.J., Kaliner, M.A., Shevach, E.M., 1986. DCs with antigen-presenting capability reside in airway epithelium, lung parenchyma, and visceral pleura. J. Exp. Med. 163, 436–451.

Shao, Z., Bharadwaj, A.S., McGee, H.S., Makinde, T.O., Agrawal, D.K., 2009a. Fms-like tyrosine kinase 3 ligand increases a lung DC subset with regulatory properties in allergic airway inflammation. J. Allergy Clin. Immunol. 123, 917–924 e912.

Shao, Z., Makinde, T.O., McGee, H.S., Wang, X., Agrawal, D.K., 2009b. Fms-like tyrosine kinase 3 ligand regulates migratory pattern and antigen uptake of lung dendritic cell subsets in a murine model of allergic airway inflammation. J. Immunol. 183, 7531–7538.

Siegal, F.P., Kadowaki, N., Shodell, M., Fitzgerald-Bocarsly, P.A., Shah, K., Ho, S., Antonenko, S., Liu, Y.J., 1999. The nature of the principal type 1 interferon-producing cells in human blood. Science 284, 1835–1837.

Sokol, C.L., Barton, G.M., Farr, A.G., Medzhitov, R., 2007. A mechanism for the initiation of allergen-induced T helper type 2 responses. Nat. Immunol. 9 (3), 310–318.

Sokol, C.L., Chu, N.Q., Yu, S., Nish, S.A., Laufer, T.M., Medzhitov, R., 2009. Basophils function as antigen-presenting cells for an allergen-induced T helper type 2 response. Nat. Immunol. 10, 713–720.

Soumelis, V., Reche, P.A., Kanzler, H., Yuan, W., Edward, G., Homey, B., Gilliet, M., Ho, S., Antonenko, S., Lauerma, A., Smith, K., Gorman, D., Zurawski, S., Abrams, J., Menon, S., McClanahan, T., de Waal-Malefyt Rd, R., Bazan, F., Kastelein, R.A., Liu, Y.J., 2002. Human epithelial cells trigger dendritic cell mediated allergic inflammation by producing TSLP. Nat. Immunol. 3, 673–680.

Steinman, R.M., Cohn, Z.A., 1973. Identification of a novel cell type in peripheral lymphoid organs of mice. I. Morphology, quantitation, tissue distribution. J. Exp. Med. 137, 1142–1162.

Steinman, R.M., Cohn, Z.A., 1974. Identification of a novel cell type in peripheral lymphoid organs of mice. II. Functional properties in vitro. J. Exp. Med. 139, 380–397.

Steinman, R.M., Lustig, D.S., Cohn, Z.A., 1974. Identification of a novel cell type in peripheral lymphoid organs of mice. 3. Functional properties in vivo. J. Exp. Med. 139, 1431–1445.

Steinman, R.M., Witmer, M.D., 1978. Lymphoid DCs are potent stimulators of the primary mixed leukocyte reaction in mice. Proc. Natl. Acad. Sci. U.S.A. 75, 5132–5136.

Stumbles, P.A., Strickland, D.H., Pimm, C.L., Proksch, S.F., Marsh, A.M., McWilliam, A.S., Bosco, A., Tobagus, I., Thomas, J.A., Napoli, S., Proudfoot, A.E., Wells, T.N., Holt, P.G., 2001. Regulation of dendritic cell recruitment into resting and inflamed airway epithelium: use of alternative chemokine receptors as a function of inducing stimulus. J. Immunol. 167, 228–234.

Sung, S.S., Fu, S.M., Rose Jr., C.E., Gaskin, F., Ju, S.T., Beaty, S.R., 2006. A major lung CD103 (alphaE)-beta7 integrin-positive epithelial dendritic cell population expressing Langerin and tight junction proteins. J. Immunol. 176, 2161–2172.

Tailleux, L., Schwartz, O., Herrmann, J.L., Pivert, E., Jackson, M., Amara, A., Legres, L., Dreher, D., Nicod, L.P., Gluckman, J.C., Lagrange, P.H., Gicquel, B., Neyrolles, O., 2003. DC-SIGN is the major Mycobacterium tuberculosis receptor on human DCs. J. Exp. Med. 197, 121–127.

Tailor, P., Tamura, T., Morse 3rd, H.C., Ozato, K., 2008. The BXH2 mutation in IRF8 differentially impairs dendritic cell subset development in the mouse. Blood 111, 1942–1945.

Takagi, H., Fukaya, T., Eizumi, K., Sato, Y., Sato, K., Shibazaki, A., Otsuka, H., Hijikata, A., Watanabe, T., Ohara, O., Kaisho, T., Malissen, B., 2011. Plasmacytoid DCs are crucial for the initiation of inflammation and T cell immunity in vivo. Immunity 35, 958–971.

Taraszka, K.S., Higgins, J.M., Tan, K., Mandelbrot, D.A., Wang, J.H., Brenner, M.B., 2000. Molecular basis for leukocyte integrin alpha(E) beta(7) adhesion to epithelial (E)-cadherin. J. Exp. Med. 191, 1555–1567.

Taylor, P.R., Tsoni, S.V., Willment, J.A., Dennehy, K.M., Rosas, M., Findon, H., Haynes, K., Steele, C., Botto, M., Gordon, S., Brown, G.D., 2007. Dectin-1 is required for beta-glucan recognition and control of fungal infection. Nat. Immunol. 8, 31–38.

Trinchieri, G., Sher, A., 2007. Cooperation of Toll-like receptor signals in innate immune defence. Nat. Rev. Immunol. 7, 179–190.

van Rijt, L.S., Jung, S., Kleinjan, A., Vos, N., Willart, M., Duez, C., Hoogsteden, H.C., Lambrecht, B.N., 2005. In vivo depletion of lung

CD11c$^+$ DCs during allergen challenge abrogates the characteristic features of asthma. J. Exp. Med. 201, 981–991.

Varol, C., Landsman, L., Fogg, D.K., Greenshtein, L., Gildor, B., Margalit, R., Kalchenko, V., Geissmann, F., Jung, S., 2007. Monocytes give rise to mucosal, but not splenic, conventional DCs. J. Exp. Med. 204, 171–180.

Vermaelen, K., Pauwels, R., 2004. Accurate and simple discrimination of mouse pulmonary dendritic cell and macrophage populations by flow cytometry: methodology and new insights. Cytometry A 61, 170–177.

Vermaelen, K.Y., Carro-Muino, I., Lambrecht, B.N., Pauwels, R.A., 2001. Specific migratory DCs rapidly transport antigen from the airways to the thoracic lymph nodes. J. Exp. Med. 193, 51–60.

von Garnier, C., Filgueira, L., Wikstrom, M., Smith, M., Thomas, J.A., Strickland, D.H., Holt, P.G., Stumbles, P.A., 2005. Anatomical location determines the distribution and function of DCs and other APCs in the respiratory tract. J. Immunol. 175, 1609–1618.

von Garnier, C., Wikstrom, M.E., Zosky, G., Turner, D.J., Sly, P.D., Smith, M., Thomas, J.A., Judd, S.R., Strickland, D.H., Holt, P.G., Stumbles, P.A., 2007. Allergic airways disease develops after an increase in allergen capture and processing in the airway mucosa. J. Immunol. 179, 5748–5759.

Walzer, T., Brawand, P., Swart, D., Tocker, J., De Smedt, T., 2005. No defect in T-cell priming, secondary response, or tolerance induction in response to inhaled antigens in Fms-like tyrosine kinase 3 ligand-deficient mice. J. Allergy Clin. Immunol. 115, 192–199.

Wang, Y.H., Ito, T., Homey, B., Watanabe, N., Martin, R., Barnes, C.J., McIntyre, B.W., Gilliet, M., Kumar, R., Yao, Z., Liu, Y.J., 2006. Maintenance and polarization of human TH2 central memory T cells by thymic stromal lymphopoietin-activated DCs. Immunity 24, 827–838.

Willart, M.A., Jan de Heer, H., Hammad, H., Soullie, T., Deswarte, K., Clausen, B.E., Boon, L., Hoogsteden, H.C., Lambrecht, B.N., 2009. The lung vascular filter as a site of immune induction for T cell responses to large embolic antigen. J. Exp. Med. 206, 2823–2835.

Worsley, A.G., LeibundGut-Landmann, S., Slack, E., Phng, L.K., Gerhardt, H., Reis e Sousa, C., MacDonald, A.S., 2008. Dendritic cell expression of the Notch ligand jagged2 is not essential for Th2 response induction in vivo. Eur. J. Immunol. 38, 1043–1049.

Yamane, H., Zhu, J., Paul, W.E., 2005. Independent roles for IL-2 and GATA-3 in stimulating naive CD4$^+$ T cells to generate a Th2-inducing cytokine environment. J. Exp. Med. 202, 793–804.

Yoneyama, M., Kikuchi, M., Matsumoto, K., Imaizumi, T., Miyagishi, M., Taira, K., Foy, E., Loo, Y.M., Gale Jr., M., Akira, S., Yonehara, S., Kato, A., Fujita, T., 2005. Shared and unique functions of the DExD/H-box helicases RIG-I, MDA5, and LGP2 in antiviral innate immunity. J. Immunol. 175, 2851–2858.

Yoshimoto, T., Yasuda, K., Tanaka, H., Nakahira, M., Imai, Y., Fujimori, Y., Nakanishi, K., 2009. Basophils contribute to T(H)2-IgE responses in vivo via IL-4 production and presentation of peptide-MHC class II complexes to CD4$^+$ T cells. Nat. Immunol. 10, 706–712.

Zammit, D.J., Cauley, L.S., Pham, Q.M., Lefrancois, L., 2005. DCs maximize the memory CD8 T cell response to infection. Immunity 22, 561–570.

Zhao, J., Legge, K., Perlman, S., 2011. Age-related increases in PGD(2) expression impair respiratory DC migration, resulting in diminished T cell responses upon respiratory virus infection in mice. J. Clin. Invest. 121, 4921–4930.

Zhou, L., Chong, M.M., Littman, D.R., 2009. Plasticity of CD4$^+$ T cell lineage differentiation. Immunity 30, 646–655.

Chapter 34

Pulmonary Mast Cells

Pranita Katwa and Jared M. Brown

Chapter Outline

1. INTRODUCTION

The expansion of mast cell biology in recent years has transformed the perception of mast cells as elementary players in immunoglobulin (Ig)-driven anaphylaxis and allergic disease to an integral cell type in a multitude of distinct immune responses and pathophysiologies. Derived from bone marrow hematopoietic stem cells, mast cells mature and reside broadly in tissue types throughout the body. They are mostly located around connective tissue and mucosal membranes, often optimally primed to interface with external environments. The plasticity of mast cell function and phenotype is attestable to its microenvironment, enabling them to perform a vast number of functions.

This chapter discusses the role of mast cells as sentinels of innate immunity, regulators of adaptive immune responses, and classic mediators of allergic disease. Mast cells are distinguished by their granular contents and can be activated through IgE-dependent and IgE-independent pathways, eliciting the release of preformed mediators or de novo synthesized mediators tailored to the microenvironment. As active players in numerous pulmonary disease pathologies including asthma, fibrosis, and chronic obstructive pulmonary disease (COPD), mast cells produce a multitude of various cytokines, chemokines, eicosanoids, and other mediators, which facilitate activation of pulmonary tissue cells or immune effector cells. The involvement of mast cells in pulmonary disorders is not limited to intrinsic disease etiology, but it also extends to extrinsic and xenobiotic etiological factors. The adverse pulmonary responses to many toxicant exposures are modulated by mast cell activation and function. It is increasingly important to understand and recognize the versatility of mast cell function in both the normal and diseased lung in order to adequately assess potential therapeutics. Mast cell therapeutics may provide a means of counteracting many of the mast cell-mediated immunotoxicologic effects observed in the lung.

2. MAST CELL CHARACTERIZATION

2.1 Growth and Differentiation

Mast cells develop from $CD34^+/CD117^+$ (c-Kit^+) hematopoietic stem cells that originate in the bone marrow (Kirshenbaum et al., 1991). These pluripotent hematopoetic stem cells may also give rise to other lineages, but unlike many involved in myeloid and lymphoid differentiation, mast cells do not downregulate c-Kit receptor expression. Therefore, in the presence of stem cell factor (SCF), a ligand for the c-Kit receptor, progenitor cells differentiate and mature within various tissues, such as mast cells (Brown et al., 2008). Another cytokine known to promote growth of mast cells is interleukin (IL)-9. IL-9, along with SCF, has been shown not only to increase mast cell growth and promote the release of mast cell progenitors from bone marrow, but also to facilitate the production of mast cell proteases and other cytokines (Matsuzawa et al., 2003; Wiener et al., 2004). Mast cell maturation and migration from the peripheral vasculature into tissues require the influence of several mediators, including IL-3, α4 integrins, and vascular cell adhesion molecule-1 (VCAM-1). Mast cell recruitment to the lung, for example, is dependent upon expression of α4β1 and α4β7 integrins, which are able to interact with pulmonary endothelial VCAM-1 to enable migration into lung tissue

Comparative Biology of the Normal Lung. http://dx.doi.org/10.1016/B978-0-12-404577-4.00034-5

(Abonia et al., 2006). In addition, the chemokine receptor CXCR2 is known to modulate the expression of VCAM-1 and is required for proper migration of mast cell progenitors to host tissues (Hallgren et al., 2007).

2.2 Mast Cells in Host Tissues

Mast cells are found in a variety of tissue types and are easily distinguished by the metachromatic staining of their granules, as first discovered by the scientist Paul Erlich (Beaven, 2009). In humans, they are most commonly known to reside around mucosal membranes and connective tissues and thus are found in higher density in tissues of the gastrointestinal tract and skin (Yong, 1997). However, mast cells do exist in most tissue types, including the stomach, colon, lymph nodes, brain, heart, and lungs (Ghanem et al., 1988; Weidner et al., 1991). They are found in the thalamus, hypothalamus, and various nuclei of the brain, as mast cell mediators have been known to impact neuronal responses as well as vascular function, altering permeability of the blood−brain barrier (Wilhelm et al., 2005; Zhuang et al., 1996, 1997). Mast cells have long been identified in cardiac and arterial tissues, but the potential role of mast cells in the cardiovascular system has been elusive until recent years, when cardiac mast cells have been implicated in several pathophysiological conditions (Bot et al., 2007; Dvorak, 1986; Ghanem et al., 1988; Gilles et al., 2003). Similarly, mast cells have been known to be involved in several pulmonary pathophysiological conditions and are able to modulate immune responses due to their anatomical locations. In the lung, mast cells reside throughout the lung parenchyma, in the surrounding connective tissue, as well as in the mucosal regions of the respiratory tract (Patterson and Suszko, 1971). The close proximity of mast cells to the smooth muscle layer within the respiratory tract ideally positions them to be able to influence airway hyperreactivity in the pathophysiology of asthma, as will be discussed later in greater detail (Brightling et al., 2003b). As is evident with asthma, mast cells are able to contribute to several disease pathologies due to their unique locations in certain tissues.

2.3 Mast Cell Phenotypes

There are two commonly recognized phenotypes of mast cells, which are delineated by their intracellular protease content. Mast cells that only contain tryptase are referred to as the MC_T phenotype, whereas mast cells that contain tryptase, chymase, carboxypeptidase, and cathepsin G are denoted as the MC_{TC} phenotype (Pejler et al., 2009). A third phenotype, MC_C, defined by mast cells that only contain chymase, has also been reported, although it is not often identified in tissues (Bradding et al., 1995). In the human lung, MC_{TC}, the phenotype generally associated with connective tissue mast cells, have been predominantly seen in pulmonary vessels and the pleura (Andersson et al., 2009). In contrast, the MC_T mucosal phenotype mast cells were discovered in higher proportion near the epithelium of the conducting airways, as well as throughout the lung parenchyma (Andersson et al., 2009; Schwartz et al., 1987). The heterogeneity of mast cells throughout the human body, currently characterized by protease content into the MC_T and MC_{TC} phenotypes, can be further categorized in numerous ways due to varying attributes within phenotypes. Although the identification of specific proteases within mast cells is a widely accepted means of phenotype classification, it is also understood that mast cell phenotype is driven by the microenvironment.

Mast cells are able to respond to mediators in their microenvironments that contribute to a specific phenotype. Cord blood-derived mast cells cultured with IL-4 and Th2 cytokines in addition to the required mast cell growth factor, SCF, demonstrated a MC_{TC} connective tissue phenotype in vitro (Hsieh et al., 2005). However, when co-cultured with human airway epithelial cells, the mast cells, originating with a MC_{TC} phenotype, eventually developed a MC_T phenotype (Hsieh et al., 2005). Classically, mast cells with an MC_T phenotype have been known to develop an eicosanoid phenotype resulting in increased cys-leukotriene (cys-LT) production following IgE-dependent activation. However, in culture, mast cells were only able to display an eicosanoid phenotype following IgE-dependent high-affinity IgE receptor (FcεRI) stimulation and priming with IL-4 and other Th2 cytokines (Hsieh et al., 2005). In addition to epithelial cell-derived Th2 cytokines, SCF expression by epithelial cells has been shown to directly impact mast cell survival in co-cultures in vitro. Thus epithelial cells may play a significant role in directing mast cell-mediated responses, potentially through the expression of SCF (Hsieh et al., 2005). Similarly, it has been proposed that endothelial cells may regulate mast cell function because both cell types are in close proximity to connective tissues of blood vessels. In the presence of pro-inflammatory factors such as tumor necrosis factor (TNF)-α, endothelial cells also upregulate SCF, promoting mast cell survival and proliferation (Konig et al., 1997). Furthermore, increases in mast cell numbers were observed in co-culture systems with endothelial cells, as well as increases in VCAM-1 and E-selectin, enabling mast cell adhesion (Mierke et al., 2000). Mast cells display great phenotypic plasticity, and the microenvironments driving these mast cell phenotypes contribute greatly to mast cell function in specific biological compartments.

3. MAST CELLS IN LUNG IMMUNOLOGY

3.1 Mast Cells in Innate Immunity

Mast cells play an important role as key effector cells in innate immunity. Once activated, mast cells serve to regulate functions of other effector cells, including dendritic cells and neutrophils, which act to propagate an innate immune response or to bridge the gap between innate and adaptive immunity, as is the case with dendritic cells. The release of TNF-α and IL-1 by mast cells has been shown to induce chemotaxis and differentiation of dendritic cells and neutrophils to a targeted location of infection or inflammation (Cumberbatch et al., 2001; Zhang et al., 1992). Other mediators released from mast cells, such as histamine, stimulate antigen presentation by dendritic cells through increasing expression of the major histocompatibility complex (MHC) class II molecules and facilitate further release of cytokines like macrophage chemoattractant protein (MCP)-1 to recruit additional immune cells and potentially activate the adaptive immune system (Caron et al., 2001).

During an innate immune response, the activation of mast cells and the release of their mediators can occur through several pathways. It has long been established that mast cells have the ability to directly bind and phagocytose or opsonize pathogens. In doing so, mast cells have been shown to process pathogenic proteins and act as antigen presenting cells to T lymphocytes to further facilitate the immune response (Frandji et al., 1996; Malaviya et al., 1996). Another important facet of mast cell function in innate immunity and host defense is the ability of mast cells to recognize numerous pathogens as well as endogenous danger signals through the direct activation of pathogen-associated molecular pattern (PAMP) receptors, such as toll-like receptors (TLRs), scavenger receptors, and several others. The TLRs belong to the IL-1 superfamily of receptors, many of which are expressed on mast cells (McCurdy et al., 2003; Varadaradjalou et al., 2003). Activation of the various TLRs enables the release of specific cytokine or chemokine profiles, formulating differential immune responses to the particular pathogen or to the endogenous insults at hand. Additionally, mast cells can also be activated through the complement system, which is traditionally initiated when C1 complement factors bind to IgG-antigen or IgM-antigen complexes to propagate opsonization, chemotaxis, and other means of fighting infection. Mast cells not only play a role in bacterial infections, but similarly in parasitic and viral infections, acting as sentinels of the innate immune system (Marshall, 2004).

3.2 Mast Cells in Adaptive Immunity

Although historically viewed as cells of the innate immune system, mast cells are often associated with Th2 responses and adaptive immunity due to their classical activation and degranulation through antigen-dependent cross-linking of the high-affinity IgE receptor, FcεRI. This allergic response results from the release of preformed mediators, such as proteases, cytokines, and chemokines, from the granular stores within the mast cell (Metcalfe et al., 2009). This early response is then followed by a late-phase response 4–6 h later, wherein mast cells synthesize additional mediators for further release (Tkaczyk et al., 2006). The mast cell can also be activated through other immunoglobulin receptors, such as the high-affinity IgG receptor (FcγRI) known to bind monomeric IgG_1 and IgG_3 isoforms with high affinity, in the presence of interferon-γ (IFN-γ). With IFN-γ stimulation, it has been demonstrated that human mast cells express higher levels of FcγRI and that aggregates of FcγRI are capable of inducing mast cell degranulation and enhancing cytokine release including TNF-α, IL-3, IL-4, IL-13, and granulocyte-macrophage colony-stimulating factor (GM-CSF) (Okayama et al., 2000). Interestingly, it has also been shown that aggregation of IgG_1, unlike IgG_3 aggregates, can still activate FcγRI on mast cells to elicit degranulation, and that subsequent IgG complexes can initiate C3a formation to synergistically facilitate mast cell activation through the C3a G-protein coupled receptor (GPCR) expressed on mast cells (Ali et al., 2000; Woolhiser et al., 2004).

Along with immunoglobulin receptor activation, mast cells are able to modulate another key function of adaptive immunity through antigen presentation. Like most cell types, mast cells express MHC class I and are able to process bacterial antigens and present to T cells for further clonal expansion (Malaviya et al., 1996). Although not constitutively expressed, MHC class II expression can be induced in mast cells via TNF-α or IFN-γ stimulation, enabling antigen presentation to T cells (Frandji et al., 1996). These mast cell and T-cell interactions are supported by mast cell expression of the co-stimulatory molecule OX40 ligand (OX40L), known to be involved with T-cell proliferation and polarization (Kashiwakura et al., 2004; Ohshima et al., 1998). The ability of mast cells to express OX40L also enables interactions with B cells and has been shown to induce B-cell production of IgE in the presence of IL-4, thus driving many allergic and Th2 responses (Gauchat et al., 1993; Yoshikawa et al., 2001). The production of IgG_1 by IL-4 stimulated B cells can also be facilitated by mast cell protease I (Yoshikawa et al., 2001). With the release of cytokines such as IL-4 and IL-13, mast cells influence the polarization of naïve T-helper cells.

Another cell type crucial to an adaptive immune response is the dendritic cell. The dendritic cell is foremost an antigen presenting cell, expressing high levels of MHC class II at full maturity following exposure to an antigen. Investigation of mast cell and dendritic cell interactions has provided evidence suggesting that mast cell components such as TNF, IL-1, and prostaglandin E_2 (PGE_2) propagate dendritic cell migration and maturation (Cumberbatch et al., 2001; Kabashima et al., 2003). While inducing the maturation of dendritic cells, PGE_2 has also been shown to promote the transformation of these cells into the DC2 phenotype, which can facilitate a Th2 shift in T-cell phenotype (Kalinski et al., 1997). Thus, mast cells can either directly or indirectly influence the polarization of naïve T cells to favor an adaptive immune response.

3.3 Activation Pathways

Mast cells have diverse functions in both innate and adaptive immunity and can be activated through a variety of means. Mostly commonly, mast cells are recognized for their classical activation of the IgE-dependent FcεRI receptor in allergic disease. However, they can be alternately activated through many IgE-independent pathways, which are critical to the innate and adaptive immune functions of mast cells.

IgE-Dependent Pathways

One of the most common and best-studied mechanisms of mast cell activation is the cross-linking of the high-affinity IgE receptor, FcεRI, by IgE and antigen, leading to an adaptive immune-driven Th2 response. The binding of antigen-IgE complexes to FcεRI initiates a signal transduction pathway beginning with the phosphorylation of the FcεRI cytosolic domains, immunoreceptor tyrosine-based activation motifs, and eventually leading to downstream activation of phospholipase C isoforms, resulting in calcium release through store operated calcium channels in the endoplasmic reticulum, and eventually degranulation of the mast cell (Metcalfe et al., 2009). This activation pathway is a feature of type I hypersensitivity or allergic responses, largely driven by mast cell activation and the release of preformed mediators, such as proteases and cytokines, from cytoplasmic granules during degranulation (Metcalfe et al., 2009).

IgE-Independent Pathways

Mast cells are also involved in innate immune responses, several of which are initiated by alternative activation pathways including the TLRs, scavenger receptors, and other PAMP receptors. The TLRs are a family of receptors that are specific to the binding of different pathogen components. In the case of TLR-4, binding of specific pathogen components, such as lipopolysaccharide (LPS) from gram-negative bacteria, activates the mast cell and promotes release of a cytokine profile specific for the receptor (Varadaradjalou et al., 2003).

An additional mast cell-induced innate immune response to pathogen is the activation of mast cells through the complement system. Although the complement system itself plays an important role in host defense of bacterial infections through opsonization and chemotaxis of other effector cells, it is also known for its ability to activate mast cells. Complement factors, particularly C3 and C4 proteins of the classical pathway, have been shown to be crucial in mast cell activation, as mice deficient in C3 and C4 proteins displayed significantly reduced degranulation. When reconstituted with human C3 proteins, $C3^{-/-}$ mice in a model of bacterial infection exhibited levels of mast cell degranulation and cytokine release comparable to wild-type animals (Prodeus et al., 1997). Activation of mast cells by C3a and C5a complement proteins occurs through the corresponding G-protein coupled receptor (GPCR) expressed by the mast cell, initiating the phospholipase C or phosphoinositide 3 kinase (PI3K) signaling cascades to mobilize calcium for degranulation and stimulate cytokine and chemokine production (Ali et al., 2000). Activation of the complement pathway can also be simultaneously stimulated by the binding of IgG to a mast cell IgG receptor FcγRIII, resulting in direct opsonization of pathogens by bacteria and triggering of the classical complement pathway with mast cell activation and subsequent initiation of the inflammatory cascade, both facilitated through the C3 complement protein (Baumann et al., 2001).

The activation of mast cells by IL-33 and the ST2 receptor may occur as an innate or adaptive immune response, as the cytokine itself has been shown to mediate pathogenic conditions involving both (Haraldsen et al., 2009; Liew et al., 2010). IL-33 initiates activation of the mast cell through its receptor, ST2. Upon binding of its ligand, the ST2 receptor dimerizes with IL-1RAcP to begin recruitment of adaptor proteins and propagate signal transduction downstream to promote synthesis of pro-inflammatory cytokines and other mediators imperative to mast cell activation during an immune response (Chackerian et al., 2007; Schmitz et al., 2005; Suzuki et al., 2002).

Activation of the mast cell can facilitate the aforementioned effector cell functions, such as opsonization or clearance of pathogens or release of mediators to induce vascular permeability, smooth muscle cell contraction, or even collagen deposition by fibroblasts. However, it is prudent to recognize that activation of mast cells can also facilitate immunomodulatory functions of the mast cell described previously through the release of certain mediators, which influence functions of other immune cells.

3.4 Mast Cells and Aging

The aging process is often associated with detrimental effects on immunity and complex components of the immune system, leading to immunosenescence. Due to the varied location, distribution, and function, any alteration in the normal immunogenic response of mast cells can significantly affect both innate and adaptive immunity. In humans, the number of dermal mast cells is significantly increased in individuals with progression of age (Gunin et al., 2011). Similarly, C3H/HeN and BALB/c mice exhibiting photoaging induced by ultraviolet B exposure also resulted in an increased number of dermal mast cells (Hart et al., 1999; Kochevar et al., 1994). Dermal mast cells have also been shown to be more reactive to inflammatory mediators such as PGE_2, causing inadvertent degranulation of cells in aging animals (Nguyen et al., 2005). Although several studies have documented altered numbers and distribution of mast cells during aging in both humans and animals, few have investigated pulmonary mast cells in regard to aging. It is well known that with age, the respiratory system faces a decline in function associated with structural, physiological, and immunological changes, including the dilation of airspaces, decreased pulmonary β-adrenoceptor affinity, and increased susceptibility to pulmonary inflammation (Sharma and Goodwin, 2006). Increases and altered function of several cell types, including neutrophils and lymphocytes, have been implicated in innate and adaptive immune defects in aged individuals (Gomez et al., 2008; Sharma and Goodwin, 2006), while little remains known about mast cells in aging.

3.5 Mast Cell and Effector Cell Interactions

Mast cells are recognized as quintessential and indispensible elements of the immune response. With the release of mediators (Table 1), mast cells influence complex interactions with several types of effector cells, including epithelial cells, polymorphonuclear leukocyte (PMNs), neutrophils, smooth muscle cells, T lymphocytes, and macrophages, modulating critical functions in pulmonary immunology.

Epithelial Cells

Pulmonary epithelial cells can have significant impacts on the function of many immune cells, including mast cells. As the primary cell type in responding to external insult, airway epithelial cells are important regulators of innate and adaptive immunity. In response to pathogens, pulmonary epithelial cells act to counter the infection and amplify the immune response through the production cytokines and chemokines (Kato and Schleimer, 2007). Airway epithelial cells are also important in adaptive immunity and may be able to manipulate Th2 cellular environments. When stimulated with a virus, bronchial epithelial cells were shown to release IL-1 and thymic stromal lymphopoietin, inducing Th2 cytokine production in mast cells (Nagarkar et al., 2012). The products of mast cells, such as tryptase, can subsequently modulate airway epithelial cells. Mast cell-derived tryptase, a known bronchoconstrictor, has been shown to induce PGE_2 release from epithelial cells through activation of PLA_2, resulting in the inhibition of bronchoconstriction, induction of vasodilation, and enhancement of leukocyte infiltration (Rastogi et al., 2008). In addition to viral infection or allergen exposure, airway epithelial cells also release mediators due to cell damage. IL-33, a novel cytokine termed an "alarmin" for its ability to alert the immune system during injury, was found to be highly expressed in pulmonary epithelial cells (Pichery et al., 2012; Schmitz et al., 2005; Willart et al., 2012). It has been described as a mediator in several disease pathologies, including fibrosis, asthma, allergic responses, and anaphylaxis (Haraldsen et al., 2009). As the receptor for IL-33, ST2 is highly expressed on mast cells; stimulated release of IL-33 by epithelial cells may directly affect mast cells in modulating these disease pathologies.

PMNs and Eosinophils

The recruitment of effector cells plays a major role in mast cell initiation or facilitation of allergic inflammatory responses. The recruitment of PMNs, particularly through the vascular endothelium, is dependent on mast cell activation and was inhibited in the presence of mast cell stabilizing chromone drugs (Yazid et al., 2010). Mast cell mediators such as histamine can also elicit migration of neutrophils and eosinophils. Histamine is known to increase vascular permeability and expression of adhesion molecules such as P-selectin, which mediates neutrophil recruitment to the site of inflammation (Burns et al., 1999). During bacterial infections, neutrophil recruitment by mast cells occurs with the release of IL-1α, IL-1β, and TNF-α (Lin et al., 2002). Eosinophilia is often associated with the pathophysiology of asthma and other allergic diseases (Bousquet et al., 1990). Thus, the recruitment of eosinophils often occurs in the presence of a variety of mediators, including CCL11 or eotaxin, RANTES, and cysteinyl leukotrienes, several of which are products released by mast cells (Laitinen et al., 1993; Lamkhioued et al., 1997). Patients with asthma who were treated with Montelukast, which inhibits leukotrienes, a major mast cell product, demonstrated a decrease in eosinophil counts in sputum, indicating the involvement of mast cells in eosinophilia (Pizzichini et al., 1999). Furthermore, IgE-mediated degranulation of mast cells, following multiple cycles of sensitization and challenge in rats, has been shown to increase numbers of eosinophils associated with elevated eotaxin and leukotrienes, suggesting a mast cell-mediated mechanism of eosinophilic recruitment (Calheiros et al., 2001).

Smooth Muscle Cells

Several mast cell mediators are known to impact the smooth muscle cells of the vasculature and airways. The location of mast cells along the connective tissue of the vasculature and airways is ideal for modulating vascular permeability bronchoconstriction and other vasoactive effects, which may significantly aid in facilitating allergic

TABLE 1 Examples of Key Mast Cell Mediators in the Lung

Class	Mast Cell Mediators	Pulmonary Effects
Preformed	Histamine	Bronchoconstriction Vasoconstriction Dendritic cell activation Immunosuppression of Tregs ↑Expression of MHC II
	Tryptase	Fibroblast proliferation Collagen synthesis Vasoconstriction ↑Airway hyperresponsiveness
	Chymase	Activation of TGF-β Synthesis of Ang II
	Secretory PLA_2	Macrophage activation Leukocyte extravastation
Lipid	PGE_2	Dendritic cell maturation Bronchodilation Vasodilation Leukocyte extravastation ↑Th2 phenotype in T cells and dendritic cells
Cytokine	IL-1	Dendritic cell differentiation and chemotaxis Neutrophil chemoattractant Inflammation
	IL-4	Airway inflammation B-cell antibody production Polarization of naïve T cells Inflammation
	IL-13	B-cell antibody production Polarization of naïve T cells Inflammation
	TNF-α	Leukocyte chemotaxis T-cell proliferation Dendritic cell differentiation and chemotaxis Inflammation
	TGF-β	Fibroblast growth factor Epithelial cell proliferation
Chemokines	CX3CL1 (Fractalkine)	Eosinophil chemotaxis
	CXCL10	Leukocyte chemotaxis
	CCL2 (MCP-1)	Monocyte and leukocyte chemotaxis
	CCL5 (RANTES)	Leukocyte chemotaxis
	CCL11	Eosinophil chemotaxis
Growth factors	bFGF	Fibroblast proliferation Collagen synthesis

PLA_2, phospholipase A_2; TNF-α, tumor necrosis factor-α; TGF-β, transforming growth factor-β; MCP-1, monocyte chemoattractant protein-1; bFGF, basic fibroblast growth factor.

and inflammatory responses. Histamine has been shown to directly affect constriction of small and large airway smooth muscle (ASM), leading to increased airway resistance and decreased pulmonary compliance (Wasserman, 1980). In addition, histamine induces vasoconstriction and leakage of bronchial venules likely to enable extravastation of leukocytes, contributing to edema (Wasserman, 1984). Mast cell tryptase promotes similar bronchoconstrictive responses, but conversely also stimulates the release of cytokines by airway smooth muscle cells (ASMC) (Sekizawa et al., 1989). Lipid metabolites of mast cell-derived arachadonic acid can have a multitude of effects on pulmonary smooth muscle, many of which have contrasting functions capable of both vasoconstriction and vasodilation (Wasserman, 1984). Dysfunction of the ASM and the pulmonary vasculature is recognized as a significant contributor to the pathophysiology of asthma, COPD, and other pulmonary diseases. Mast cells have a great versatility in modulating function of smooth muscle cells and, subsequently, physiologic function and respiratory health.

T Lymphocytes

Mast cell and T-cell interactions are varied and quite complex. Studies have demonstrated that mast cells can directly and indirectly influence T-cell function and activity. As we have previously discussed in this chapter, mast cells can act as antigen-presenting cells to directly stimulate T-effector cell activity through OX40L expression and presentation of both MHC I and II, resulting in proliferation of $CD4^+$ and $CD8^+$ T cell proliferation (Frandji et al., 1996; Kashiwakura et al., 2004; Ohshima et al., 1998). Antigen presentation and cross-talk between mast cells and T cells can also promote immunoglobulin production by B cells, particularly IgE (Gauchat et al., 1993). IgE can indirectly affect T-helper cell proliferation by regulating antigen-presenting cells through FcγRIII receptor activation, particularly on dendritic cells, to promote a Th2 skewed environment (Blink and Fu, 2010). Mediators of mast cells can also contribute to the indirect enhancement of T-cell proliferation and cytokine production. TNF release following FcεR1 activation of mast cells results in proliferation of both $CD4^+$ Th1 and Th2 subtypes as well as $CD8^+$ Tc1 and Tc2 subtypes, emphasizing the importance of mast cell-derived TNF in contributing to a variety of immune responses (Nakae et al., 2005).

These functions are positive modulators of immune function; however, mast cells also possess negative immunomodulatory functions, most commonly by means of IL-10 production leading to immunosuppression (Frossi et al., 2010). In addition to influencing T-effector cells, mast cells exhibit the potential to suppress T-regulatory cells (Tregs). Mast cell-derived histamine has been shown to bind to H1 histamine receptors on Tregs, inhibiting their suppressor functions. This response was associated with decreased expression in CD25, an IL-2 receptor component, and Foxp3, a transcription factor that influences gene expression of IL2, a cytokine crucial to Treg function (Forward et al., 2009). Mast cells play a large immunomodulatory role through interactions with Tregs and other T-cell subtypes.

Macrophages

As one of the predominant immune cell types within the lungs, alveolar macrophages are vital due to their phagocytic and secretory functions. Like neutrophils and eosinophils, macrophages respond to a number of mast cell-derived chemokines. Several chemokines are produced by lung epithelial cells under normal conditions; however, cell types like mast cells are known to facilitate the infiltration of immune cells during lung injury through chemokine production. The MCP family of chemokines predominantly targets chemotaxis of macrophages to sites of inflammation or cellular injury. A study examined the interaction between mast cells and macrophages mediated by secretory phospholipases A_2 ($sPLA_2$), found to be elevated in inflammatory lung diseases and involved in the induction of downstream pro-inflammatory factors (Kim et al., 1995). While $sPLA_2$s facilitate signaling and metabolism of arachadonic acid, a precursor to mast cell-derived eicosanoids, they can also directly impact immune cell function through the binding and activation of M-type (mannose) receptors, expressed by several immune effector cells (Triggiani et al., 2005). As with many other mediators, $sPLA_2$s can be secreted by and target both mast cells and macrophages. However, it has been suggested that $sPLA_2$s act in a paracrine fashion, preferentially released by mast cells and targeted to macrophages, inducing release of macrophage-derived cytokines, such as IL-6 and TNF-α, among many more (Granata et al., 2009; Triggiani et al., 2000, 2002). Macrophages display a complex interaction with the cellular environment in which mast cells induce not only chemotaxis but also macrophage involvement in innate and adaptive responses in the lung with the release of mediators.

4. MAST CELLS AND PULMONARY DISEASE

4.1 Mast Cells in Asthma

Mast cells have been known to play a role in several pulmonary diseases, including asthma and fibrosis. Asthma is defined by thickening of the basement membrane, airway hyperresponsiveness, and eosinophilic airway inflammation, resulting in the release of Th2 cytokines—all responses that can be notably affected by mast cells (Alkhouri et al., 2011). The symptoms and pathology of asthma, with the exception of airflow obstruction and airway hyperresponsiveness, are almost identical to those of eosinophilic bronchitis (Brightling et al., 2002). Additionally, significant infiltration of mast cells in the ASMC layers is found in asthmatic patients and markedly absent in patients with eosinophilic bronchitis or normal healthy individuals (Brightling et al., 2003a). Mast cell implication in the pathophysiology of asthma is mainly centered on the presence and location of mast cells in the lung tissue as well as the mediators they release. Mast cells are known to release Th2 cytokines such as IL-4 and IL-13, which have been shown to contribute to airway inflammation during asthma through activation of receptors on ASMCs specific for these cytokines (Brightling et al., 2003b). The ability of mast cells to modulate ASM is further enhanced by co-localization in the ASM during asthma (Brightling et al., 2003b). The ASM can also secrete mediators to stimulate migration of immune cells, particularly mast cells, with the release of chemotactic factors such as CXCL10 and CX3CL1 (also known as Fractalkine or FKN) (Brightling et al., 2005; El-Shazly et al., 2006). Both of these mediators have been known to induce chemotaxis through the CXCR3, a chemokine receptor on mast cells, which demonstrates a significantly increased expression in patients with asthma compared to nonasthmatics (Brightling et al., 2005; El-Shazly et al., 2006).

Several studies have investigated different mast cell phenotypes and the potential phenotype-specific release of mediators in contributing to the pathogenesis of asthma. Increased numbers of mast cells found associated with ASMCs, previously identified as the MC_T phenotype, have been associated with increased levels of tryptase mRNA and protein in bronchoalveolar lavage fluid (BALF) of patients with asthma (Brightling et al., 2005, 2002). Mast cell tryptase is known to potentiate ASM hyperresponsiveness, an important and distinguishing feature in asthma pathophysiology (Sekizawa et al., 1989). It is important to note that the MC_T mast cell phenotype is typically expressed within the lungs, particularly the submucosa and epithelium of normal, nonasthmatic subjects (Balzar et al., 2011). For delineating mild and severe asthma, it was found that mild asthmatics displayed this high infiltration of the normal MC_T mast cell phenotype, whereas severe asthmatics displayed lower levels of total mast cells with a switch to the MC_{TC} phenotype, supported by a characteristic increase in carboxypeptidase 3A (CP3A) and prostaglandin D_2 (PGD_2) levels (Balzar et al., 2011; Lewis et al., 1982). Despite this decrease in total mast cells in patients with severe asthma, levels of tryptase remained high in these individuals as mast cells of the MC_{TC} phenotype are known to produce two to three times more tryptase compared to MC_T phenotypes (Balzar et al., 2011; Schwartz et al., 1987). Unlike many tissue types, which exhibited a reduction in the number of mast cells in patients with severe asthma, the pulmonary epithelium continued to display increased mast cell numbers, suggesting that the epithelium may contribute to the differentiation and survival of the MC_{TC} phenotype.

It has been suggested that epithelial cells may release mediators known to promote the development of the MC_{TC} phenotype, such as transforming growth factor beta (TGF-β) and other Th2 cytokines (Balzar et al., 2011; Veldhoen et al., 2008). Thus, the location of mast cells, their phenotypes, and the mediators they release are all aspects that contribute to the pathogenesis of asthma.

4.2 Mast Cells in Pulmonary Fibrosis

Pulmonary fibrosis is often a characteristic of interstitial diseases and can manifest in several forms. One of the most common and severe forms of pulmonary fibrosis is idiopathic pulmonary fibrosis, which is attributed to an unknown cause. In pulmonary fibrotic conditions, an increase in mast cells has been identified and correlated with the degree of severity in fibrosis (Inoue et al., 1996; Pesci et al., 1993). Furthermore, tissue taken from pulmonary fibrosis patients displayed mast cells in young interstitial and alveolar connective tissues containing fibroblasts and were determined to have fewer granules present (Pesci et al., 1993). The absence of normal numbers of granules within mast cells suggests a correlation with increased levels of histamine, tryptase, and other mast cell mediators found in BALF of patients with pulmonary fibrosis, often released during the process of chronic degranulation (Pesci et al., 1993).

Several mast cell mediators, including tryptase, basic fibroblast growth factor (bFGF), and chymase, have been implicated in advancing profibrotic pulmonary pathologies. The release of tryptase by mast cells has been shown to stimulate fibroblast proliferation and collagen synthesis through protease-activated receptor 2 (PAR-2) (Akers et al., 2000). Like tryptase, bFGF is a mediator released by pulmonary mast cells and largely associated with lung fibrosis (Inoue et al., 1996). Interestingly, bFGF release from the HMC-1 human mast cell line was determined to be an IgE-independent mechanism because HMC-1 cells lack complete and functional IgE receptors (Inoue et al., 1996). Another important factor in mast cell-mediated fibrosis is chymase. Chymase is involved in modulating the pathogenesis of fibrosis, partially through the activation of TGF-β, a profibrotic mediator involved in the induction of fibrosis (Tomimori et al., 2003). Bleomycin-induced fibrosis in rats was significantly increased, as indicated by elevated collagen accumulation, but significantly diminished in the presence of a chymase inhibitor and in a mast cell-deficient (i.e., Ws/Ws) rat strain (Tomimori et al., 2003). In addition, it has also been suggested that the pulmonary fibrotic toxicities induced by paraquat dichloride, a widely used herbicide, is also mediated through mast cell chymase activation of TGF-β and angiotensin II (Ang II) (Lang et al., 2010; Orito et al., 2004). The role of mast cell chymase has been established in the conversion of Ang I to Ang II in several tissue types including pulmonary tissue (Lindberg et al., 1997). Furthermore, Ang II induces a mitogenic response in human lung fibroblasts modulated by TGF-β, facilitating profibrotic conditions (Marshall et al., 2000).

These studies provide potential roles for mast cells and their mediators in modulating the pathogenesis of pulmonary fibrosis.

4.3 Mast Cells in Chronic Obstructive Pulmonary Disease

Little is known about the involvement of mast cells in COPD. However, some studies investigating the pathology of COPD have examined mast cell recruitment, alterations in mast cell populations, and the implication of mast cells on functional changes in the lung. In patients with late-stage COPD, the density of total mast cells is significantly reduced in small airways and pulmonary vessel walls, surmised to be a result of the severity of the disease (Andersson et al., 2010). Interestingly, there is an observed increase and shift in mast cell phenotype from MC_T to MC_{TC}, particularly in the small airways and alveolar parenchyma, with the progression of severe COPD (Andersson et al., 2010). In addition, patients with COPD demonstrate increased tryptase levels and activity in BALF and sputum compared to normal subjects, a factor further implicating mast cells in COPD (Zhang et al., 2011). The increase in mast cells with the MC_{TC} phenotype in the small airways and alveolar parenchyma was shown to be positively correlated to altered lung function in patients with COPD (Andersson et al., 2010). In fact, a weak negative correlation was reported with COPD between mast cell density in ASM and airway obstruction reversibility with treatment using salbutamol, a short-acting β2-adrenergic receptor agonist often used to alleviate difficulty in breathing (Liesker et al., 2007). Although these studies provide support for the potential involvement of mast cells in COPD pathogenesis, the underlying mechanisms remain unknown, prompting the need for further studies in this area.

5. MAST CELLS AND IMMUNOTOXICOLOGY

5.1 Exacerbation of Pre-Existing Pulmonary Conditions

Exacerbation of pre-existing pulmonary conditions, such as asthma, can result from pathogen encounters from infection or allergen encounters from occupational and environmental exposures. In fact, pathogen- and allergen-induced activation of mast cells can lead to a synergistic increase in the production of mast cell mediators. The activation of TLRs by bacterial pathogens in conjunction with FcεR1 receptor activation following allergen cross-linking of IgE has been shown to enhance production of TNF-α, IL-6, IL-13, MCP-1, and other cytokines (Qiao et al., 2006). This synergistic activation of mast cells can have great implications in the progression of allergic disease.

Mast cells have been identified as contributors in several toxicant-induced pulmonary conditions, both occupational and environmental in nature. In this chapter, we have discussed the role of mast cells in both asthma and interstitial lung disease. Exposure to multiple toxicants can exacerbate these conditions. A study examining exposure to dog allergens and environmental smoke reported significantly increased the risk of asthma (Carlsten et al., 2011). Similarly, allergen-sensitized children with viral infections were found to be at increased risk for hospital admissions due to the exacerbations of asthma (Murray et al., 2006).

Pulmonary disorders induced by environmental and occupational exposure to toxicants, such as grain dust, asbestos, silica, and other particulates, have identified mast cells as contributing factors in the pathogenesis of the condition. Environmental exposures to diesel exhaust particulates have been shown to aggravate asthma-like inflammation in a mouse model of asthma, demonstrating the potential for environmental factors in exacerbating pre-existing conditions (Kim et al., 2011). Furthermore, increased airway hyperresponsiveness and exacerbation of allergic asthma in sensitized mice was observed with multiwalled carbon nanotubes (CNTs) inhalation, a potential occupational exposure in nanotechnology manufacturing (Ryman-Rasmussen et al., 2009). The involvement of mast cells in the adverse pulmonary effects of diesel exhaust particles (DEP), CNTs, and other toxicant exposures will be discussed in later in this chapter.

5.2 Exposure to Toxicants

The location of mast cells within various tissues has been proven to be pivotal, not only in immune responses to pathogens but also in xenobiotic insults. As discussed previously, mast cells play a distinct role in innate immunity, often acting as the first line of defense in pathogen recognition and subsequent initiation of host defense mechanisms. Given the importance of mast cells in innate immunity, mast cells may also be critical immune cells in response to environmental exposures. Respiratory exposures to toxins, pesticides, and particulates, such as ambient particles, DEP, asbestos, silica, and engineered particles, have all implicated the involvement of mast cells in the resulting immune responses.

Toxins

Brevetoxin

Brevetoxins (PbTx) are produced by the dinoflagellate *Karenia brevis* during red tides. They are sodium channel neurotoxins that can induce adverse pulmonary responses when aerosolized and subsequently inhaled by marine animals or humans. Humans, who are generally exposed in areas with red tide events off the coasts of Florida and North Carolina or in the Gulf of Mexico, exhibit a variety of upper and lower respiratory symptoms, including coughing, sneezing, rhinorrhea, and bronchoconstriction (Backer et al., 2003, 2005). Exposure to PbTxs has been shown to induce more severe respiratory effects in individuals with asthma (Fleming et al., 2007). Similarly, studies using animal models of nonasthmatic sheep and sheep with allergic asthma demonstrated a nonspecific increase in airway hyperresponsiveness with PbTx exposure. This effect was blocked by the use of several pharmacological agents, including the mast cell stabilizer cromolyn sodium and the histamine H_1-antagonist diphenhydramine, indicating the role of mast cells in PbTx pulmonary toxicity (Abraham et al., 2005a,b).

In vitro, PbTx has been shown to activate mast cells, resulting in degranulation and production of proinflammatory cytokines such as IL-6 (Hilderbrand et al., 2011). The PbTx activation of mast cells occurred in conjunction with an increase in calcium, an integral factor in mast cell degranulation. However, activation of mast cells following exposure to PbTx was shown to be independent of altered intracellular sodium levels, demonstrating that calcium signaling may be the means by which PbTx can contribute to the mast cell involvement in the allergic and pulmonary responses induced following PbTx exposure.

Pesticides

Malathion

The organophosphorous insecticide malathion is largely used due to its low toxicity in mammalian species and its conversion to nontoxic derivatives by carboxylesterases. However, malathion also contains toxic impurities, such as O,S,S-trimethyl phosphorodithioate (OSS). Toxicity following oral exposure to OSS in rats presents with increased levels of lactate dehydrogenase (LDH) levels, a cytotoxic marker, and altered pulmonary morphologies resulting in decreased numbers of Clara cells in the bronchial epithelium and destruction of type I pneumocytes (Imamura et al., 1983; Konno et al., 1984). Malathion itself has been shown to produce oxidative damage in the lungs of exposed rats (Possamai et al., 2007). Although there are limited reports on human pulmonary toxicity to malathion, agricultural workers reported an increase in wheezing, a symptom of adverse respiratory outcomes, associated with malathion use on farms (Hoppin et al., 2006).

In addition to the respiratory toxicities already discussed, malathion and its metabolites can also affect immune responses through mast cell activation and subsequent actions of mast cell mediators, such as arachadonic acid metabolites and TNF, in altering macrophage function (Rodgers and Xiong, 1996). Studies with mice exposed to malathion demonstrate an increase in mast cell degranulation and in macrophage respiratory burst functions, which are significantly decreased following treatment with the mast cell stabilizer, disodium cromoglycate (Rodgers et al., 1996; 1996b).

Paraquat

Paraquat, a fast-acting, nonselective herbicide, is widely used in agriculture worldwide. The hazards of paraquat through various routes of exposure are well recognized and exposure by ingestion is known to result in pulmonary fibrosis due to selective accumulation in the lung (Dinis-Oliveira et al., 2008). Paraquat-induced pulmonary injury and fibrosis occur in two stages: (1) the destruction of type I and type II alveolar epithelial cells, and (2) collagen deposition following proliferation of fibroblasts and myofibroblasts (Hampson and Pond, 1988; Hemmati and Hicks, 1999). In addition, it has been shown that paraquat can significantly augment allergic reactions as demonstrated by increased mast cell histamine release with antigen sensitization (Sato et al., 1998).

Mast cell involvement in paraquat-induced pulmonary toxicity was also implicated in several animal studies. In rats, treatment with the mast cell stabilizer cromolyn sodium attenuated the fibrogenic effects of paraquat on the lung (Hemmati et al., 2002). Furthermore, chymase, a mediator released by mast cells known to promote fibrogenic responses through the cleavage of angiotensin I, also modulates profibrotic effects following paraquat administration (Lang et al., 2010; Orito et al., 2004). Increased chymase expression was found in human and animal lung fibroblasts and corroborated with angiotension II production, with subsequent increases in collagen expression (Lang et al., 2010).

Particulates

Ambient Particles

The adverse effects of ambient particulate matter (PM; Mossman et al., 2011) are well known and have been reported to affect several physiological aspects including the respiratory system. Studies have shown a positive correlation between PM exposures and hospital visits for respiratory diseases, including asthma and COPD (Atkinson et al., 2001; Peel et al., 2005). Several pulmonary and immune cell types, such as alveolar macrophages and bronchial epithelial cells, have been implicated in the production of IL-6, GM-CSF, macrophage inflammatory protein-1β (MIP-1β), monocyte chemotactic protein-1 (MCP-1), and other pro-inflammatory mediators in response to PM exposure (Ishii et al., 2005).

In addition to these cell types, mast cells have been shown to play an important role in amplifying the PM-induced inflammatory response (Alfaro-Moreno et al.,

2008). In vitro, co-cultures of macrophages, alveolar epithelial cells, and endothelial cells demonstrate an increased granulocyte colony-stimulating factor (G-CSF), MIP-1β, IL-8, and IL-6 production in the presence of mast cells (Alfaro-Moreno et al., 2008). Of the mediators that are notably released, GM-CSF, G-CSF, MIP-1β, and MCP-1 are well known to induce recruitment or proliferation of monocytes, macrophages, and other granulocytes. Therefore, mast cells may be an important component in the initiation and sustained recruitment of inflammatory cells to PM exposure. As many of the adverse respiratory effects of PM exposure include asthma and other allergic responses, several studies have investigated mast cells in PM-induced allergic responses. Ambient PM in conjunction with airborne allergen exposure has been reported to exacerbate allergic asthmatic responses in mice potentiated by sustained mast cell activation (Jin et al., 2011). Furthermore, organic dust-exposed horses with equine heaves (similar to human occupationally or environmentally induced asthma, characterized by neutrophilic airway inflammation and obstruction) have demonstrated increased mast cell numbers in bronchial epithelium and tryptase production in BALF following an allergen challenge (Dacre et al., 2007).

Diesel Exhaust

Urbanization in many countries throughout the world has resulted in the rise of motor vehicle-generated air pollutants. DEPs contribute significantly to the particulate matter emissions in urban areas, with particles that range from 20 to 200 nm in size, composed of hydrocarbons, metals, gases, and other materials (Riedl and Diaz-Sanchez, 2005). Several studies have examined the health effects of DEP exposures. In healthy human volunteers from low-particulate environments, BALF and bronchial tissue samples display neutrophil and mast cell infiltration, associated with increased pro-inflammatory mediators following short-term DEP exposures (Behndig et al., 2006; Salvi et al., 1999). Moreover, an increase in mast cell activation was observed with DEP exposure, as determined by a rise in BALF methyl histamine, a mast cell-specific mediator known to mediate endothelial cell permeability that most likely contributes to the observed inflammatory cell infiltrates in peripheral blood (Salvi et al., 1999). In sensitized humans, DEP exposure augments allergic responses resulting in mast cell production of IL-4 and subsequently driving a Th2 response (Wang et al., 1999). The exacerbated Th2 response with DEP and antigenic challenge has also been observed in mice, particularly resulting in allergic eosinophilic airway inflammation. However, this response was notably absent in mast cell-deficient mice, indicating the involvement of mast cells in mediating DEP inflammation and exacerbation of allergic responses (Ichinose et al., 2002).

Asbestos

The inhalation of asbestos fiber has been well characterized to induce pulmonary inflammation, fibrosis, and mesothelioma (Mossman et al., 2011; Padilla-Carlin et al., 2011). In rats, intratracheal exposure to chrysotile asbestos fibers resulted in peribronchial pulmonary fibrosis and an increase in mast cells within or surrounding the fibrotic tissue projections, associated with a subsequent increase of serotonin and histamine release, a response that is ablated in rats not exposed to chrysotile (Keith et al., 1987). Similar data have been reported with crocidolite, amosite, and anthophyllite asbestos fibers (Wagner et al., 1984). Although the exact role of mast cells in asbestos-induced pulmonary toxicity is undefined, mast cells have been implicated in several pulmonary fibrotic diseases and thus may be involved in mediating the fibrotic toxicities observed following exposure to asbestos fibers.

Silica

The inhalation of crystalline silica, most commonly through occupational exposures, results in the development of silicosis, an interstitial lung disease characterized by fibrotic nodules, inflammatory cell infiltrates, thickening of the alveolar interstitium, and production of pro-inflammatory mediators such as TNF-α, TGF-β, and IL-1 (Vanhee et al., 1995). Although the production of these cytokines has often been attributed to alveolar macrophages, mast cells have been shown to produce a similar array of mediators in the presence of silica, including MCP-1, IL-13, and TNF-α (Brown et al., 2007). Furthermore, it has been established that mast cells are able to modulate the pulmonary effects of silica through the activation of scavenger receptors (Brown et al., 2007). In a clinical study, the lung tissue of subjects occupationally exposed to silica not only demonstrated a significant increase in total pulmonary mast cells found to be near fibrotic nodules, but it also found these mast cells to be positive for bFGF, correlating with the severity of fibrosis (Hamada et al., 2000). Additional studies have reported similar results with increased mast cell numbers in acute and intermediary lesions representative of various fibrotic remodeling stages (Delgado et al., 2006).

Mast cells may mediate the development of fibrosis and the remodeling of lung tissue through increased release and activity of mast cell proteases such as chymase, known to be involved in tissue remodeling and shown to decrease the severity of silica-induced pulmonary toxicity when inhibited (Brown et al., 2007; Takato et al., 2011). The role of mast cells in mediating the pulmonary toxicities of silica exposure includes the induction of inflammation through the production of pro-inflammatory cytokines, as well as the modulation of fibrotic tissue through the increase in protease activity.

Engineered Carbon Nanoparticles

Fullerenes (C60), CNTs, and other engineered carbon nanoparticles can be functionalized and subsequently used in a plethora of applications, such as drug delivery. The potential for the use of C60 in the medical field is due primarily to the capability of adding modified side-chain functional groups to permit greater biocompatibility and thus allow for use in the treatment of a variety of disease pathologies (Satoh and Takayanagi, 2006). Interestingly, C60 has displayed uniquely favorable outcomes in modulating allergic responses for potential therapeutic purposes (Norton et al., 2010; Ryan et al., 2007). Water-soluble derivatives of C60, such as those with tetraglycolic acid or N-ethyl moieties, were able to inhibit mast cell cytokine production and degranulation, due to the trafficking of fullerenes in mast cell endoplasmic reticulum and subsequent inhibition of calcium release, required for FcεRI receptor-mediated degranulation (Dellinger et al., 2010; Norton et al., 2010; Ryan et al., 2007). In vivo, these fullerene derivatives were able to attenuate mast cell-driven anaphylaxis reactions and possibly provide a means for targeting the activation of mast cells and subsequent release of mediators during allergic responses (Norton et al., 2010; Ryan et al., 2007).

In contrast, exposure to CNTs has demonstrated a significant number of adverse pulmonary, cardiovascular, and immune effects. Like C60, the increased interest in CNTs is due to their unique physical and chemical properties, including high mechanical strength, high electrical conductivity, and lightweight structure (Ajayan et al., 1999; Endo et al., 2008; Lin et al., 2004). These characteristics have made CNTs ideal for use in conductive and high-strength composites, sensors, energy conversion devices, drug delivery systems, as well as other biomedical applications (Baughman et al., 2002; Martin and Kohli, 2003). However, exposure to CNTs, particularly pristine CNTs, induces pulmonary inflammation, pulmonary fibrosis, and exacerbation of cardiac ischemia reperfusion injury (Katwa et al., 2012; Wang et al., 2011). Furthermore, it was determined that activation of mast cells by IL-33 and the ST2 receptor plays an integral role in orchestrating the observed toxicities induced by exposure to CNTs (Katwa et al., 2012).

6. MAST CELL THERAPEUTICS

6.1 Anti-IgE Agents

Several therapeutics have been developed and used to target mast cell activation and downstream effects. The activation of mast cells, specifically through the classical IgE cross-linking and FcεR1 receptor signaling, has been studied in great detail in the interest of understanding the pathway and identifying potential mast cell therapeutics. Anti-IgE drugs are often monoclonal antibodies, such as omalizumab, which act to inhibit FcεR1 receptor activation to ensure insensitivity of the mast cell to allergen stimulation (Chang and Shiung, 2006). Omalizumab not only neutralizes IgE, but also downregulates expression of the FcεR1 receptor because unbound receptors free of IgE are gradually degraded by the mast cell (Borkowski et al., 2001). It has been approved in the United States for the treatment of severe asthma. In clinical studies, the use of anti-IgE drugs for asthma resulted in decreased eosinophilia and attenuation of both the early- and late-phase allergic responses, demonstrating a potential for IgE levels in mediating asthmatic airway inflammation (Fahy, 2000).

6.2 Cromoglycate-Based Drugs

Cromoglycate-based drugs were first discovered by Roger E. C. Altounyan in the early 1960s and were derived from the khellin, a well-known Middle Eastern remedy. Originally developed for asthma as protectant against antigen challenge, cromolyn sodium, also known as sodium cromoglycate, proved to be effective for allergic asthma treatment (Howell, 2005). Approved by the U.S. Food and Drug Administration in 1973 as Intal, cromolyn sodium was indicated for use in asthma, but has since been studied and approved in various formulations for the treatment of several allergic conditions, including allergic rhinitis and allergic conjunctivitis (Berman et al., 1975). Although the exact mechanism of action remains elusive, cromolyn sodium acts to prevent cytokine and mediator release by stabilization of mast cells and subsequent prevention of degranulation through the inhibition of changes in intracellular calcium levels (Foreman et al., 1977; Orr and Cox, 1969).

Although structurally distinct from cromolyn sodium, nedocromil is another drug that similarly acts to stabilize mast cells. The ability of nedocromil to stabilize mast cells may be due to its inhibition of chloride channels, suggested to be involved in the release of mast cell mediators such as histamine (Alton and Norris, 1996). Effective in the treatment of asthma, cromolyn sodium and nedocromil are also used for the treatment of rhinoconjunctivitis, mastocytosis, and other conditions. While the inhaler formulations of both cromolyn sodium and nedocromil have been phased out and are currently off the market, nebulizer and other formulations for these drugs are still commercially available. Cromoglycate drugs are reported to have very minor adverse effects and few drug interactions, making them safe and effective therapeutics to supplement prophylactic treatment and management of allergic disease (Storms and Kaliner, 2005).

6.3 Leukotriene Antagonists

Leukotriene antagonists, such as zafirlukast and montelukast, have been proven to be effective as treatments for allergic indications. Because leukotrienes are products of arachadonic acid metabolism and released by mast cells as potent mediators capable of inducing a plethora of physiological actions, inhibition of leukotrienes can directly affect symptoms associated with allergic disease. In particular, treatment with 4-(5-cyclopentyloxycarbonylamino-1-methylindol-3-ylmethy l)-3-methoxy-N-o-tolylsulfonylbenzamide (ICI), a nonselective cysteinyl-leukotriene receptor antagonist, has demonstrated decreased bronchoconstriction in patients with asthma (Hui and Barnes, 1991).

Montelukast, a selective leukotriene (LT) D4 inhibitor, and zafirlukast, a selective LTD4 and LTDE inhibitor, target the specific leukotrienes associated with allergic responses following antigen challenges. They have been shown to be most effective in improving pulmonary function in patients with asthma, as well as nasal congestion and rhinoconjunctivitis in patients with allergic rhinitis (Creticos et al., 1984; Garcia-Marcos and Schuster, 1999; Nayak, 2004). As previously discussed in this chapter, mast cells are critical effector cells not only in allergic disease, but also in several other pulmonary pathophysiologies. In a rat model of chronic obstructive lung disease induced by exposure to cigarette smoke, treatment with montelukast demonstrated protective effects against morphological changes in the lung, such as enlargement of the alveolar airspaces, which implicates the involvement of mast cells in pulmonary disease as mast cells are a contributing source of leukotrienes in the lung parenchyma (Yuksel et al., 2003).

6.4 Histamine Receptor Antagonists

The H1 histamine receptor plays a significant role in modulation of allergic lung disease because the actions of histamine have long been recognized in hypersensitivity responses. Allergic mice models that lack H1 receptor expression exhibit attenuated inflammatory responses, cytokine production, goblet cell metaplasia, and airway hyperresponsiveness (Bryce et al., 2006). Similar effects were observed with the use of the H1 histamine receptor antagonist desloratadine in a mouse model of asthma (Bryce et al., 2003). Ketotifen, one of many antihistamines developed following their discovery in 1937, is a second-generation, nonsedating histamine H1 receptor antagonist. Widely used in the prophylactic treatment of allergic asthma, ketotifen acts by blocking the effects of the H1 histamine receptor to alleviate pulmonary inflammation, but it is an ideal as an antiasthmatic therapeutic drug due to its unique secondary mechanism of action. Ketotifen can also promote bronchodilation by increasing concentrations of β_2-adrenergic receptors and preventing desensitization of β-adrenergic receptors with the use of beta-adrenergic therapeutics in asthma (Brodde et al., 1992; Poller et al., 1998). Although ketotifen is well recognized for its use in asthma and other allergic diseases, several other H1 antihistamines are effective at neutralizing the downstream effects of mast cell mediators in further facilitating an allergic response (Okayama et al., 1994).

REFERENCES

Abonia, J.P., Hallgren, J., Jones, T., et al., 2006. Alpha-4 integrins and VCAM-1, but not MAdCAM-1, are essential for recruitment of mast cell progenitors to the inflamed lung. Blood 108, 1588–1594.

Abraham, W.M., Bourdelais, A.J., Ahmed, A., et al., 2005a. Effects of inhaled brevetoxins in allergic airways: toxin-allergen interactions and pharmacologic intervention. Environ. Health Perspect. 113, 632–637.

Abraham, W.M., Bourdelais, A.J., Sabater, J.R., et al., 2005b. Airway responses to aerosolized brevetoxins in an animal model of asthma. Am. J. Respir. Crit. Care Med. 171, 26–34.

Ajayan, P.M., Charlier, J., Rinzler, A.G., 1999. Carbon nanotubes: from macromolecules to nanotechnology. Proc. Natl. Acad. Sci. U. S. A. 96, 14199–14200.

Akers, I.A., Parsons, M., Hill, M.R., et al., 2000. Mast cell tryptase stimulates human lung fibroblast proliferation via protease-activated receptor-2. Am. J. Physiol. Lung Cell. Mol. Physiol. 278, L193–L201.

Alfaro-Moreno, E., Nawrot, T.S., Vanaudenaerde, B.M., et al., 2008. Co-cultures of multiple cell types mimic pulmonary cell communication in response to urban PM10. Eur. Respir. J. 32, 1184–1194.

Ali, H., Ahamed, J., Hernandez-Munain, C., et al., 2000. Chemokine production by G protein-coupled receptor activation in a human mast cell line: roles of extracellular signal-regulated kinase and NFAT. J. Immunol. 165, 7215–7223.

Alkhouri, H., Hollins, F., Moir, L.M., et al., 2011. Human lung mast cells modulate the functions of airway smooth muscle cells in asthma. Allergy 66, 1231–1241.

Alton, E.W., Norris, A.A., 1996. Chloride transport and the actions of nedocromil sodium and cromolyn sodium in asthma. J. Allergy Clin. Immunol. 98, S102–S105; discussion S105–S106.

Andersson, C.K., Mori, M., Bjermer, L., et al., 2009. Novel site-specific mast cell subpopulations in the human lung. Thorax 64, 297–305.

Andersson, C.K., Mori, M., Bjermer, L., et al., 2010. Alterations in lung mast cell populations in patients with chronic obstructive pulmonary disease. Am. J. Respir. Crit. Care Med. 181, 206–217.

Atkinson, R.W., Anderson, H.R., Sunyer, J., et al., 2001. Acute effects of particulate air pollution on respiratory admissions: results from APHEA 2 project. Air pollution and health: a European approach. Am. J. Respir. Crit. Care Med. 164, 1860–1866.

Backer, L.C., Fleming, L.E., Rowan, A., et al., 2003. Recreational exposure to aerosolized brevetoxins during Florida red tide events. Harmful Algae 2, 19–28.

Backer, L.C., Kirkpatrick, B., Fleming, L.E., et al., 2005. Occupational exposure to aerosolized brevetoxins during Florida red tide events: effects on a healthy worker population. Environ. Health Perspect. 113, 644–649.

Balzar, S., Fajt, M.L., Comhair, S.A., et al., 2011. Mast cell phenotype, location, and activation in severe asthma. Data from the Severe Asthma Research Program. Am. J. Respir. Crit. Care Med. 183, 299–309.

Baughman, R.H., Zakhidov, A.A., de Heer, W.A., 2002. Carbon nanotubes—the route toward applications. Science 297, 787–792.

Baumann, U., Chouchakova, N., Gewecke, B., et al., 2001. Distinct tissue site-specific requirements of mast cells and complement components C3/C5a receptor in IgG immune complex-induced injury of skin and lung. J. Immunol. 167, 1022–1027.

Beaven, M.A., 2009. Our perception of the mast cell from Paul Ehrlich to now. Eur. J. Immunol. 39, 11–25.

Behndig, A.F., Mudway, I.S., Brown, J.L., et al., 2006. Airway antioxidant and inflammatory responses to diesel exhaust exposure in healthy humans. Eur. Respir. J. 27, 359–365.

Berman, B.A., Fenton, M.M., Girsch, L.S., et al., 1975. Cromolyn sodium in the treatment of children with severe, perennial asthma. Pediatrics 55, 621–629.

Blink, S.E., Fu, Y.X., 2010. IgE regulates T helper cell differentiation through FcgammaRIII mediated dendritic cell cytokine modulation. Cell. Immunol. 264, 54–60.

Borkowski, T.A., Jouvin, M.H., Lin, S.Y., et al., 2001. Minimal requirements for IgE-mediated regulation of surface Fc epsilon RI. J. Immunol. 167, 1290–1296.

Bot, I., de Jager, S.C., Zernecke, A., et al., 2007. Perivascular mast cells promote atherogenesis and induce plaque destabilization in apolipoprotein E-deficient mice. Circulation 115, 2516–2525.

Bousquet, J., Chanez, P., Lacoste, J.Y., et al., 1990. Eosinophilic inflammation in asthma. N. Engl. J. Med. 323, 1033–1039.

Bradding, P., Okayama, Y., Howarth, P.H., et al., 1995. Heterogeneity of human mast cells based on cytokine content. J. Immunol. 155, 297–307.

Brightling, C.E., Ammit, A.J., Kaur, D., et al., 2005. The CXCL10/CXCR3 axis mediates human lung mast cell migration to asthmatic airway smooth muscle. Am. J. Respir. Crit. Care Med. 171, 1103–1108.

Brightling, C.E., Bradding, P., Symon, F.A., et al., 2002. Mast-cell infiltration of airway smooth muscle in asthma. N. Engl. J. Med. 346, 1699–1705.

Brightling, C.E., Symon, F.A., Birring, S.S., et al., 2003a. Comparison of airway immunopathology of eosinophilic bronchitis and asthma. Thorax 58, 528–532.

Brightling, C.E., Symon, F.A., Holgate, S.T., et al., 2003b. Interleukin-4 and -13 expression is co-localized to mast cells within the airway smooth muscle in asthma. Clin. Exp. Allergy 33, 1711–1716.

Brodde, O.E., Petrasch, S., Bauch, H.J., et al., 1992. Terbutaline-induced desensitization of beta 2-adrenoceptor in vivo function in humans: attenuation by ketotifen. J. Cardiovasc Pharmacol. 20, 434–439.

Brown, J.M., Swindle, E.J., Kushnir-Sukhov, N.M., et al., 2007. Silica-directed mast cell activation is enhanced by scavenger receptors. Am. J. Respir. Cell. Mol. Biol. 36, 43–52.

Brown, J.M., Wilson, T.M., Metcalfe, D.D., 2008. The mast cell and allergic diseases: role in pathogenesis and implications for therapy. Clin. Exp. Allergy 38, 4–18.

Bryce, P.J., Geha, R., Oettgen, H.C., 2003. Desloratadine inhibits allergen-induced airway inflammation and bronchial hyperresponsiveness and alters T-cell responses in murine models of asthma. J. Allergy Clin. Immunol. 112, 149–158.

Bryce, P.J., Mathias, C.B., Harrison, K.L., et al., 2006. The H1 histamine receptor regulates allergic lung responses. J. Clin. Invest. 116, 1624–1632.

Burns, A.R., Bowden, R.A., Abe, Y., et al., 1999. P-selectin mediates neutrophil adhesion to endothelial cell borders. J. Leukoc. Biol. 65, 299–306.

Calheiros, A.S., Aguiar Pires, A.L., Pereira da Silva, J., et al., 2001. Role of the IgE-mediated system in eosinophil recruitment triggered by two consecutive cycles of sensitisation and challenge in rats. Int. Arch. Allergy Immunol. 126, 325–334.

Carlsten, C., Brauer, M., Dimich-Ward, H., et al., 2011. Combined exposure to dog and indoor pollution: incident asthma in a high-risk birth cohort. Eur. Respir. J. 37, 324–330.

Caron, G., Delneste, Y., Roelandts, E., et al., 2001. Histamine induces CD86 expression and chemokine production by human immature dendritic cells. J. Immunol. 166, 6000–6006.

Chackerian, A.A., Oldham, E.R., Murphy, E.E., et al., 2007. IL-1 receptor accessory protein and ST2 comprise the IL-33 receptor complex. J. Immunol. 179, 2551–2555.

Chang, T.W., Shiung, Y.Y., 2006. Anti-IgE as a mast cell-stabilizing therapeutic agent. J. Allergy Clin. Immunol. 117, 1203–1212; quiz 1213.

Creticos, P.S., Peters, S.P., Adkinson Jr., N.F., et al., 1984. Peptide leukotriene release after antigen challenge in patients sensitive to ragweed. N. Engl. J. Med. 310, 1626–1630.

Cumberbatch, M., Dearman, R.J., Antonopoulos, C., et al., 2001. Interleukin (IL)-18 induces Langerhans cell migration by a tumour necrosis factor-alpha- and IL-1beta-dependent mechanism. Immunology 102, 323–330.

Dacre, K.J., McGorum, B.C., Marlin, D.J., et al., 2007. Organic dust exposure increases mast cell tryptase in bronchoalveolar lavage fluid and airway epithelium of heaves horses. Clin. Exp. Allergy 37, 1809–1818.

Delgado, L., Parra, E.R., Capelozzi, V.L., 2006. Apoptosis and extracellular matrix remodelling in human silicosis. Histopathology 49, 283–289.

Dellinger, A., Zhou, Z., Norton, S.K., et al., 2010. Uptake and distribution of fullerenes in human mast cells. Nanomedicine 6, 575–582.

Dinis-Oliveira, R.J., Duarte, J.A., Sanchez-Navarro, A., et al., 2008. Paraquat poisonings: mechanisms of lung toxicity, clinical features, and treatment. Crit. Rev. Toxicol. 38, 13–71.

Dvorak, A.M., 1986. Mast-cell degranulation in human hearts. N. Engl. J. Med. 315, 969–970.

El-Shazly, A., Berger, P., Girodet, P.O., et al., 2006. Fraktalkine produced by airway smooth muscle cells contributes to mast cell recruitment in asthma. J. Immunol. 176, 1860–1868.

Endo, M., Tsuruoka, S., Ichihara, G., 2008. Carbon Nanotubes in historical and future perspective Summary of an Extended Session at Carbon 2008 in Nagano (JP). Part Fibre Toxicol. 5, 21.

Fahy, J.V., 2000. Reducing IgE levels as a strategy for the treatment of asthma. Clin. Exp. Allergy 30 (Suppl. 1), 16–21.

Fleming, L.E., Kirkpatrick, B., Backer, L.C., et al., 2007. Aerosolized red-tide toxins (brevetoxins) and asthma. Chest 131, 187–194.

Foreman, J.C., Hallett, M.B., Mongar, J.L., 1977. Site of action of the antiallergic drugs cromoglycate and doxantrazole [proceedings]. Br. J. Pharmacol. 59, 473P–474P.

Forward, N.A., Furlong, S.J., Yang, Y., et al., 2009. Mast cells down-regulate $CD4^{+}CD25^{+}$ T regulatory cell suppressor function via histamine H1 receptor interaction. J. Immunol. 183, 3014–3022.

Frandji, P., Tkaczyk, C., Oskeritzian, C., et al., 1996. Exogenous and endogenous antigens are differentially presented by mast cells to $CD4^{+}$ T lymphocytes. Eur. J. Immunol. 26, 2517–2528.

Frossi, B., Gri, G., Tripodo, C., et al., 2010. Exploring a regulatory role for mast cells: 'MCregs'? Trends Immunol. 31, 97–102.

Garcia-Marcos, L., Schuster, A., 1999. New perspectives for asthma treatment: anti-leukotriene drugs. Pediatr. Allergy Immunol. 10, 77–88.

Gauchat, J.F., Henchoz, S., Mazzei, G., et al., 1993. Induction of human IgE synthesis in B cells by mast cells and basophils. Nature 365, 340–343.

Ghanem, N.S., Assem, E.S., Leung, K.B., et al., 1988. Cardiac and renal mast cells: morphology, distribution, fixation and staining properties in the guinea pig and preliminary comparison with human. Agents Actions 23, 223–226.

Gilles, S., Zahler, S., Welsch, U., et al., 2003. Release of TNF-alpha during myocardial reperfusion depends on oxidative stress and is prevented by mast cell stabilizers. Cardiovasc. Res. 60, 608–616.

Gomez, C.R., Nomellini, V., Faunce, D.E., et al., 2008. Innate immunity and aging. Exp. Gerontol. 43, 718–728.

Granata, F., Nardicchi, V., Loffredo, S., et al., 2009. Secreted phospholipases A(2): a proinflammatory connection between macrophages and mast cells in the human lung. Immunobiology 214, 811–821.

Gunin, A.G., Kornilova, N.K., Vasilieva, O.V., et al., 2011. Age-related changes in proliferation, the numbers of mast cells, eosinophils, and cd45-positive cells in human dermis. J. Gerontol. A Biol. Sci. Med. Sci. 66, 385–392.

Hallgren, J., Jones, T.G., Abonia, J.P., et al., 2007. Pulmonary CXCR2 regulates VCAM-1 and antigen-induced recruitment of mast cell progenitors. Proc. Natl. Acad. Sci. U. S. A. 104, 20478–20483.

Hamada, H., Vallyathan, V., Cool, C.D., et al., 2000. Mast cell basic fibroblast growth factor in silicosis. Am. J. Respir. Crit. Care Med. 161, 2026–2034.

Hampson, E.C., Pond, S.M., 1988. Ultrastructure of canine lung during the proliferative phase of paraquat toxicity. Br. J. Exp. Pathol. 69, 57–68.

Haraldsen, G., Balogh, J., Pollheimer, J., et al., 2009. Interleukin-33-cytokine of dual function or novel alarmin? Trends Immunol. 30, 227–233.

Hart, P.H., Grimbaldeston, M.A., Hosszu, E.K., et al., 1999. Age-related changes in dermal mast cell prevalence in BALB/c mice: functional importance and correlation with dermal mast cell expression of Kit. Immunology 98, 352–356.

Hemmati, A.A., Hicks, R., 1999. Increased myofibroblast contractile sensitivity in paraquat pretreated rat lung tissue. Life Sci. 65, 2325–2332.

Hemmati, A.A., Nazari, Z., Motlagh, M.E., et al., 2002. The role of sodium cromolyn in treatment of paraquat-induced pulmonary fibrosis in rat. Pharmacol. Res. 46, 229–234.

Hilderbrand, S.C., Murrell, R.N., Gibson, J.E., et al., 2011. Marine brevetoxin induces IgE-independent mast cell activation. Arch. Toxicol. 85, 135–141.

Hoppin, J.A., Umbach, D.M., London, S.J., et al., 2006. Pesticides and adult respiratory outcomes in the agricultural health study. Ann. N. Y. Acad. Sci. 1076, 343–354.

Howell, J., 2005. Roger Altounyan and the discovery of cromolyn (sodium cromoglycate). J. Allergy Clin. Immunol. 115, 882–885.

Hsieh, F.H., Sharma, P., Gibbons, A., et al., 2005. Human airway epithelial cell determinants of survival and functional phenotype for primary human mast cells. Proc. Natl. Acad. Sci. U. S. A. 102, 14380–14385.

Hui, K.P., Barnes, N.C., 1991. Lung function improvement in asthma with a cysteinyl-leukotriene receptor antagonist. Lancet 337, 1062–1063.

Ichinose, T., Takano, H., Miyabara, Y., et al., 2002. Enhancement of antigen-induced eosinophilic inflammation in the airways of mast-cell deficient mice by diesel exhaust particles. Toxicology 180, 293–301.

Imamura, T., Gandy, J., Fukuto, T.R., et al., 1983. An impurity of malathion alters the morphology of rat lung bronchiolar epithelium. Toxicology 26, 73–79.

Inoue, Y., King Jr., T.E., Tinkle, S.S., et al., 1996. Human mast cell basic fibroblast growth factor in pulmonary fibrotic disorders. Am. J. Pathol. 149, 2037–2054.

Ishii, H., Hayashi, S., Hogg, J.C., et al., 2005. Alveolar macrophage-epithelial cell interaction following exposure to atmospheric particles induces the release of mediators involved in monocyte mobilization and recruitment. Respir. Res. 6, 87.

Jin, C., Shelburne, C.P., Li, G., et al., 2011. Particulate allergens potentiate allergic asthma in mice through sustained IgE-mediated mast cell activation. J. Clin. Invest 121, 941–955.

Kabashima, K., Sakata, D., Nagamachi, M., et al., 2003. Prostaglandin E2-EP4 signaling initiates skin immune responses by promoting migration and maturation of Langerhans cells. Nat. Med. 9, 744–749.

Kalinski, P., Hilkens, C.M., Snijders, A., et al., 1997. Dendritic cells, obtained from peripheral blood precursors in the presence of PGE2, promote Th2 responses. Adv. Exp. Med. Biol. 417, 363–367.

Kashiwakura, J., Yokoi, H., Saito, H., et al., 2004. T cell proliferation by direct cross-talk between OX40 ligand on human mast cells and OX40 on human T cells: comparison of gene expression profiles between human tonsillar and lung-cultured mast cells. J. Immunol. 173, 5247–5257.

Kato, A., Schleimer, R.P., 2007. Beyond inflammation: airway epithelial cells are at the interface of innate and adaptive immunity. Curr. Opin. Immunol. 19, 711–720.

Katwa, P., Wang, X., Urankar, R.N., et al., 2012. A carbon nanotube toxicity paradigm driven by mast cells and the IL-33/ST2 Axis. Small 8, 2904–2912.

Keith, I., Day, R., Lemaire, S., et al., 1987. Asbestos-induced fibrosis in rats: increase in lung mast cells and autacoid contents. Exp. Lung Res. 13, 311–327.

Kim, D.K., Fukuda, T., Thompson, B.T., et al., 1995. Bronchoalveolar lavage fluid phospholipase A2 activities are increased in human adult respiratory distress syndrome. Am. J. Physiol. 269, L109–L118.

Kim, J., Natarajan, S., Vaickus, L.J., et al., 2011. Diesel exhaust particulates exacerbate asthma-like inflammation by increasing CXC chemokines. Am. J. Pathol. 179, 2730–2739.

Kirshenbaum, A.S., Kessler, S.W., Goff, J.P., et al., 1991. Demonstration of the origin of human mast cells from $CD34^+$ bone marrow progenitor cells. J. Immunol. 146, 1410–1415.

Kochevar, I.E., Moran, M., Granstein, R.D., 1994. Experimental photoaging in C3H/HeN, C3H/HeJ, and Balb/c mice: comparison of changes in extracellular matrix components and mast cell numbers. J. Invest. Dermatol. 103, 797–800.

Konig, A., Corbacioglu, S., Ballmaier, M., et al., 1997. Downregulation of c-kit expression in human endothelial cells by inflammatory stimuli. Blood 90, 148–155.

Konno, N., Fukuto, T.R., Imamura, T., 1984. Lung injury and delayed toxicity produced by O, S,S-trimethyl phosphorodithioate, an impurity of malathion. Toxicol. Appl. Pharmacol. 75, 219–228.

Laitinen, L.A., Laitinen, A., Haahtela, T., et al., 1993. Leukotriene E4 and granulocytic infiltration into asthmatic airways. Lancet 341, 989–990.

Lamkhioued, B., Renzi, P.M., Abi-Younes, S., et al., 1997. Increased expression of eotaxin in bronchoalveolar lavage and airways of asthmatics contributes to the chemotaxis of eosinophils to the site of inflammation. J. Immunol. 159, 4593–4601.

Lang, Y.D., Chang, S.F., Wang, L.F., et al., 2010. Chymase mediates paraquat-induced collagen production in human lung fibroblasts. Toxicol. Lett. 193, 19–25.

Lewis, R.A., Soter, N.A., Diamond, P.T., et al., 1982. Prostaglandin D2 generation after activation of rat and human mast cells with anti-IgE. J. Immunol. 129, 1627–1631.

Liesker, J.J., Ten Hacken, N.H., Rutgers, S.R., et al., 2007. Mast cell numbers in airway smooth muscle and PC20AMP in asthma and COPD. Respir. Med. 101, 882–887.

Liew, F.Y., Pitman, N.I., McInnes, I.B., 2010. Disease-associated functions of IL-33: the new kid in the IL-1 family. Nat. Rev. Immunol. 10, 103–110.

Lin, T.J., Garduno, R., Boudreau, R.T., et al., 2002. Pseudomonas aeruginosa activates human mast cells to induce neutrophil transendothelial migration via mast cell-derived IL-1 alpha and beta. J. Immunol. 169, 4522–4530.

Lin, Y., Taylor, S., Li, H., et al., 2004. Advances toward bioapplications of carbon nanotubes. J. Mater. Chem. 14, 527–541.

Lindberg, B.F., Gyllstedt, E., Andersson, K.E., 1997. Conversion of angiotensin I to angiotensin II by chymase activity in human pulmonary membranes. Peptides 18, 847–853.

Malaviya, R., Twesten, N.J., Ross, E.A., et al., 1996. Mast cells process bacterial Ags through a phagocytic route for class I MHC presentation to T cells. J. Immunol. 156, 1490–1496.

Marshall, J.S., 2004. Mast-cell responses to pathogens. Nat. Rev. Immunol. 4, 787–799.

Marshall, R.P., McAnulty, R.J., Laurent, G.J., 2000. Angiotensin II is mitogenic for human lung fibroblasts via activation of the type 1 receptor. Am. J. Respir. Crit. Care Med. 161, 1999–2004.

Martin, C.R., Kohli, P., 2003. The emerging field of nanotube biotechnology. Nat. Rev. Drug Discov. 2, 29–37.

Matsuzawa, S., Sakashita, K., Kinoshita, T., et al., 2003. IL-9 enhances the growth of human mast cell progenitors under stimulation with stem cell factor. J. Immunol. 170, 3461–3467.

McCurdy, J.D., Olynych, T.J., Maher, L.H., et al., 2003. Cutting edge: distinct Toll-like receptor 2 activators selectively induce different classes of mediator production from human mast cells. J. Immunol. 170, 1625–1629.

Metcalfe, D.D., Peavy, R.D., Gilfillan, A.M., 2009. Mechanisms of mast cell signaling in anaphylaxis. J. Allergy Clin. Immunol. 124, 639–646; quiz 647–648.

Mierke, C.T., Ballmaier, M., Werner, U., et al., 2000. Human endothelial cells regulate survival and proliferation of human mast cells. J. Exp. Med. 192, 801–811.

Mossman, B.T., Lippmann, M., Hesterberg, T.W., et al., 2011. Pulmonary endpoints (lung carcinomas and asbestosis) following inhalation exposure to asbestos. J. Toxicol. Environ. Health B Crit. Rev. 14, 76–121.

Murray, C.S., Poletti, G., Kebadze, T., et al., 2006. Study of modifiable risk factors for asthma exacerbations: virus infection and allergen exposure increase the risk of asthma hospital admissions in children. Thorax 61, 376–382.

Nagarkar, D.R., Poposki, J.A., Comeau, M.R., et al., 2012. Airway epithelial cells activate TH2 cytokine production in mast cells through IL-1 and thymic stromal lymphopoietin. J. Allergy Clin. Immunol. 130, 225–232 e4.

Nakae, S., Suto, H., Kakurai, M., et al., 2005. Mast cells enhance T cell activation: Importance of mast cell-derived TNF. Proc. Natl. Acad. Sci. U. S. A. 102, 6467–6472.

Nayak, A., 2004. A review of montelukast in the treatment of asthma and allergic rhinitis. Expert Opin. Pharmacother. 5, 679–686.

Nguyen, M., Pace, A.J., Koller, B.H., 2005. Age-induced reprogramming of mast cell degranulation. J. Immunol. 175, 5701–5707.

Norton, S.K., Dellinger, A., Zhou, Z., et al., 2010. A new class of human mast cell and peripheral blood basophil stabilizers that differentially control allergic mediator release. Clin. Transl. Sci. 3, 158–169.

Ohshima, Y., Yang, L.P., Uchiyama, T., et al., 1998. OX40 costimulation enhances interleukin-4 (IL-4) expression at priming and promotes the differentiation of naive human CD4(+) T cells into high IL-4-producing effectors. Blood 92, 3338–3345.

Okayama, Y., Benyon, R.C., Lowman, M.A., et al., 1994. In vitro effects of H1-antihistamines on histamine and PGD2 release from mast cells of human lung, tonsil, and skin. Allergy 49, 246–253.

Okayama, Y., Kirshenbaum, A.S., Metcalfe, D.D., 2000. Expression of a functional high-affinity IgG receptor, Fc gamma RI, on human mast cells: up-regulation by IFN-gamma. J. Immunol. 164, 4332–4339.

Orito, K., Suzuki, Y., Matsuda, H., et al., 2004. Chymase is activated in the pulmonary inflammation and fibrosis induced by paraquat in hamsters. Tohoku J. Exp. Med. 203, 287–294.

Orr, T.S., Cox, J.S., 1969. Disodium cromoglycate, an inhibitor of mas cell degranulation and histamine release induced by phospholipase A. Nature 223, 197–198.

Padilla-Carlin, D.J., Schladweiler, M.C., Shannahan, J.H., et al., 2011. Pulmonary inflammatory and fibrotic responses in Fischer 344 rats after intratracheal instillation exposure to Libby amphibole. J. Toxicol. Environ. Health A 74, 1111–1132.

Patterson, R., Suszko, I.M., 1971. Primate respiratory mast cells. Reactions with Ascaris antigen and anti-heavy chain sera. J. Immunol. 106, 1274–1283.

Peel, J.L., Tolbert, P.E., Klein, M., et al., 2005. Ambient air pollution and respiratory emergency department visits. Epidemiology 16, 164–174.

Pejler, G., Knight, S.D., Henningsson, F., et al., 2009. Novel insights into the biological function of mast cell carboxypeptidase A. Trends Immunol. 30, 401–408.

Pesci, A., Bertorelli, G., Gabrielli, M., et al., 1993. Mast cells in fibrotic lung disorders. Chest 103, 989–996.

Pichery, M., Mirey, E., Mercier, P., et al., 2012. Endogenous IL-33 is highly expressed in mouse epithelial barrier tissues, lymphoid organs, brain, embryos, and inflamed tissues: in situ analysis using a novel Il-33-LacZ gene trap reporter strain. J. Immunol. 188, 3488–3495.

Pizzichini, E., Leff, J.A., Reiss, T.F., et al., 1999. Montelukast reduces airway eosinophilic inflammation in asthma: a randomized, controlled trial. Eur. Respir. J. 14, 12–18.

Poller, U., Fuchs, B., Gorf, A., et al., 1998. Terbutaline-induced desensitization of human cardiac beta 2-adrenoceptor-mediated positive inotropic effects: attenuation by ketotifen. Cardiovasc Res. 40, 211–222.

Possamai, F.P., Fortunato, J.J., Feier, G., et al., 2007. Oxidative stress after acute and sub-chronic malathion intoxication in Wistar rats. Environ. Toxicol. Pharmacol. 23, 198–204.

Prodeus, A.P., Zhou, X., Maurer, M., et al., 1997. Impaired mast cell-dependent natural immunity in complement C3-deficient mice. Nature 390, 172–175.

Qiao, H., Andrade, M.V., Lisboa, F.A., et al., 2006. FcepsilonR1 and toll-like receptors mediate synergistic signals to markedly augment production of inflammatory cytokines in murine mast cells. Blood 107, 610–618.

Rastogi, P., Young, D.M., McHowat, J., 2008. Tryptase activates calcium-independent phospholipase A2 and releases PGE2 in airway epithelial cells. Am. J. Physiol. Lung Cell. Mol. Physiol. 295, L925–L932.

Riedl, M., Diaz-Sanchez, D., 2005. Biology of diesel exhaust effects on respiratory function. J. Allergy Clin. Immunol. 115, 221–228; quiz 229.

Rodgers, K., St Amand, K., Xiong, S., 1996. Effects of malathion on humoral immunity and macrophage function in mast cell-deficient mice. Fundam. Appl. Toxicol. 31, 252–258.

Rodgers, K., Xiong, S., 1996. Contribution of mast cell mediators to alterations in macrophage function after malathion administration. Fundam. Appl. Toxicol. 33, 100–108.

Ryan, J.J., Bateman, H.R., Stover, A., et al., 2007. Fullerene nanomaterials inhibit the allergic response. J. Immunol. 179, 665–672.

Ryman-Rasmussen, J.P., Tewksbury, E.W., Moss, O.R., et al., 2009. Inhaled multiwalled carbon nanotubes potentiate airway fibrosis in murine allergic asthma. Am. J. Respir. Cell. Mol. Biol. 40, 349–358.

Salvi, S., Blomberg, A., Rudell, B., et al., 1999. Acute inflammatory responses in the airways and peripheral blood after short-term exposure to diesel exhaust in healthy human volunteers. Am. J. Respir. Crit. Care Med. 159, 702–709.

Sato, T., Taguchi, M., Nagase, H., et al., 1998. Augmentation of allergic reactions by several pesticides. Toxicology 126, 41–53.

Satoh, M., Takayanagi, I., 2006. Pharmacological studies on fullerene (C60), a novel carbon allotrope, and its derivatives. J. Pharmacol. Sci. 100, 513–518.

Schmitz, J., Owyang, A., Oldham, E., et al., 2005. IL-33, an interleukin-1-like cytokine that signals via the IL-1 receptor-related protein ST2 and induces T helper type 2-associated cytokines. Immunity 23, 479–490.

Schwartz, L.B., Irani, A.M., Roller, K., et al., 1987. Quantitation of histamine, tryptase, and chymase in dispersed human T and TC mast cells. J. Immunol. 138, 2611–2615.

Sekizawa, K., Caughey, G.H., Lazarus, S.C., et al., 1989. Mast cell tryptase causes airway smooth muscle hyperresponsiveness in dogs. J. Clin. Invest 83, 175–179.

Sharma, G., Goodwin, J., 2006. Effect of aging on respiratory system physiology and immunology. Clin. Interv. Aging 1, 253–260.

Storms, W., Kaliner, M.A., 2005. Cromolyn sodium: fitting an old friend into current asthma treatment. J. Asthma 42, 79–89.

Suzuki, N., Suzuki, S., Duncan, G.S., et al., 2002. Severe impairment of interleukin-1 and Toll-like receptor signalling in mice lacking IRAK-4. Nature 416, 750–756.

Takato, H., Yasui, M., Ichikawa, Y., et al., 2011. The specific chymase inhibitor TY-51469 suppresses the accumulation of neutrophils in the lung and reduces silica-induced pulmonary fibrosis in mice. Exp. Lung Res. 37, 101–108.

Tkaczyk, C., Jensen, B.M., Iwaki, S., et al., 2006. Adaptive and innate immune reactions regulating mast cell activation: from receptor-mediated signaling to responses. Immunol. Allergy Clin. North Am. 26, 427–450.

Tomimori, Y., Muto, T., Saito, K., et al., 2003. Involvement of mast cell chymase in bleomycin-induced pulmonary fibrosis in mice. Eur. J. Pharmacol. 478, 179–185.

Triggiani, M., Granata, F., Giannattasio, G., et al., 2005. Secretory phospholipases A2 in inflammatory and allergic diseases: not just enzymes. J. Allergy Clin. Immunol. 116, 1000–1006.

Triggiani, M., Granata, F., Oriente, A., et al., 2000. Secretory phospholipases A2 induce beta-glucuronidase release and IL-6 production from human lung macrophages. J. Immunol. 164, 4908–4915.

Triggiani, M., Granata, F., Oriente, A., et al., 2002. Secretory phospholipases A2 induce cytokine release from blood and synovial fluid monocytes. Eur. J. Immunol. 32, 67–76.

Vanhee, D., Gosset, P., Boitelle, A., et al., 1995. Cytokines and cytokine network in silicosis and coal workers' pneumoconiosis. Eur. Respir. J. 8, 834–842.

Varadaradjalou, S., Feger, F., Thieblemont, N., et al., 2003. Toll-like receptor 2 (TLR2) and TLR4 differentially activate human mast cells. Eur. J. Immunol. 33, 899–906.

Veldhoen, M., Uyttenhove, C., van Snick, J., et al., 2008. Transforming growth factor-beta 'reprograms' the differentiation of T helper 2 cells and promotes an interleukin 9-producing subset. Nat. Immunol. 9, 1341–1346.

Wagner, M.M., Edwards, R.E., Moncrieff, C.B., et al., 1984. Mast cells and inhalation of asbestos in rats. Thorax 39, 539–544.

Wang, M., Saxon, A., Diaz-Sanchez, D., 1999. Early IL-4 production driving Th2 differentiation in a human in vivo allergic model is mast cell derived. Clin. Immunol. 90, 47–54.

Wang, X., Katwa, P., Podila, R., et al., 2011. Multi-walled carbon nanotube instillation impairs pulmonary function in C57BL/6 mice. Part Fibre Toxicol. 8, 24.

Wasserman, S.I., 1980. The lung mast cell: its physiology and potential relevance to defense of the lung. Environ. Health Perspect. 35, 153–164.

Wasserman, S.I., 1984. The human lung mast cell. Environ. Health Perspect. 55, 259–269.

Weidner, N., Austen, K.F., 1991. Ultrastructural and immunohistochemical characterization of normal mast cells at multiple body sites. J. Invest. Dermatol. 96, 26S–30S; discussion 30S–31S.

Wiener, Z., Falus, A., Toth, S., 2004. IL-9 increases the expression of several cytokines in activated mast cells, while the IL-9-induced IL-9 production is inhibited in mast cells of histamine-free transgenic mice. Cytokine 26, 122–130.

Wilhelm, M., Silver, R., Silverman, A.J., 2005. Central nervous system neurons acquire mast cell products via transgranulation. Eur. J. Neurosci. 22, 2238–2248.

Willart, M.A., Deswarte, K., Pouliot, P., et al., 2012. Interleukin-1alpha controls allergic sensitization to inhaled house dust mite via the epithelial release of GM-CSF and IL-33. J. Exp. Med. 209, 1505–1517.

Woolhiser, M.R., Brockow, K., Metcalfe, D.D., 2004. Activation of human mast cells by aggregated IgG through FcgammaRI: additive effects of C3a. Clin. Immunol. 110, 172–180.

Yazid, S., Leoni, G., Getting, S.J., et al., 2010. Antiallergic cromones inhibit neutrophil recruitment onto vascular endothelium via annexin-A1 mobilization. Arterioscler. Thromb. Vasc. Biol. 30, 1718–1724.

Yong, L.C., 1997. The mast cell: origin, morphology, distribution, and function. Exp. Toxicol. Pathol. 49, 409–424.

Yoshikawa, T., Imada, T., Nakakubo, H., et al., 2001. Rat mast cell protease-I enhances immunoglobulin E production by mouse B cells stimulated with interleukin-4. Immunology 104, 333–340.

Yuksel, H., Ozbilgin, K., Coskun, S., et al., 2003. Protective effect of leukotriene receptor antagonist montelukast on smoking-induced lung injury in Wistar rats. Acta Med. Okayama 57, 13–19.

Zhang, X., Zheng, H., Ma, W., et al., 2011. Tryptase enzyme activity is correlated with severity of chronic obstructive pulmonary disease. Tohoku J. Exp. Med. 224, 179–187.

Zhang, Y., Ramos, B.F., Jakschik, B.A., 1992. Neutrophil recruitment by tumor necrosis factor from mast cells in immune complex peritonitis. Science 258, 1957–1959.

Zhuang, X., Silverman, A.J., Silver, R., 1996. Brain mast cell degranulation regulates blood–brain barrier. J. Neurobiol. 31, 393–403.

Zhuang, X., Silverman, A.J., Silver, R., 1997. Mast cell number and maturation in the central nervous system: influence of tissue type, location and exposure to steroid hormones. Neuroscience 80, 1237–1245 (Scopus Exact).

Chapter 35

The Immune Basis of Allergic Lung Disease

Stefanie C.M. Burleson, Robert B. Fick, Mark D. Mannie, Stephen G. Olmstead and Michael R. Van Scott

Chapter Outline

1. INTRODUCTION

Exposure to specific aeroallergens in sensitized human subjects most often manifests in the upper airways as allergic rhinoconjunctivitis and in the lower airways as allergic asthma. The two conditions often present as comorbidities, with approximately 80% of allergic asthmatics exhibiting rhinitis (de Groot et al., 2012). Both conditions exhibit epithelial, mesenchymal, smooth muscle, and endothelial pathologies with functional ramifications that are site-specific. However, rhinitis is characterized by sneezing, nasal itching, and nasal congestion leading to airflow obstruction, rhinorrhea, and postnasal drainage (Long et al., 2002), whereas allergic asthma is associated with the pathophysiology of wheezing, excessive bronchial secretions, reduced forced expiratory volume in 1 s (FEV_1) normalized to vital capacity, and irritability or nonspecific hyperresponsiveness of the bronchial smooth muscle. Both atopic conditions are initiated when immunoglobulin (Ig) E that is bound to its high-affinity receptor (FcεRI) on the surface of mast cells is cross-linked by specific inhaled allergen. Degranulation of the cells and release of preformed mediators, including histamine, results in an acute, early-phase allergic response (EAR). The EAR subsides over 30–120 min and is followed by a late allergic response (LAR) that can last for several days. Chronic disease can manifest as intermittent episodes or as more sustained, ongoing airway dysfunction. The majority of research on allergic responses of the respiratory tract has focused on asthma. In this chapter, we provide a brief look at the history of asthma, the canonical cells, and mediators with well-established roles in pathogenesis, and recently identified cytokines, cell types, and regulatory pathways that contribute to complexity of the disease and represent both exciting opportunity and additional challenge to the development of effective new therapies.

In contrast to assessing functional aspects of the pulmonary system, delineation of immunologic processes is dependent on reagents (e.g., antibodies and recombinant proteins) and tools (e.g., transgenic animals and gene arrays) that are species-specific. Over the last two decades, far more resources have been developed for mice than any other species, and a repetitive pattern has emerged in which basic discoveries from murine studies are examined for relevance in human tissues and cells; the validated findings have been tested for relevance in clinical trials. This pattern is changing as tools for analyzing small tissue samples are developed, allowing for more of the discovery research to be conducted with human samples. Still, the relative ease of safely manipulating genes and controlling environmental conditions has led to landmark discoveries using mice. Consequently, in this chapter, we discuss what is known about asthma in humans and then present selected observations from mice that provide mechanistic insights into the immune basis of the human disease.

1.1 Historical Perspective

References to asthma-like respiratory dysfunction can be found in written records dating back thousands of years, but significant insights into the pathogenic mechanisms, as indicated by effective targeted therapy, are limited to the last century (Figure 1) (Cockcroft, 1999). The term "asthma" comes from the Greek word *aazein*, which means

Comparative Biology of the Normal Lung. http://dx.doi.org/10.1016/B978-0-12-404577-4.00035-7

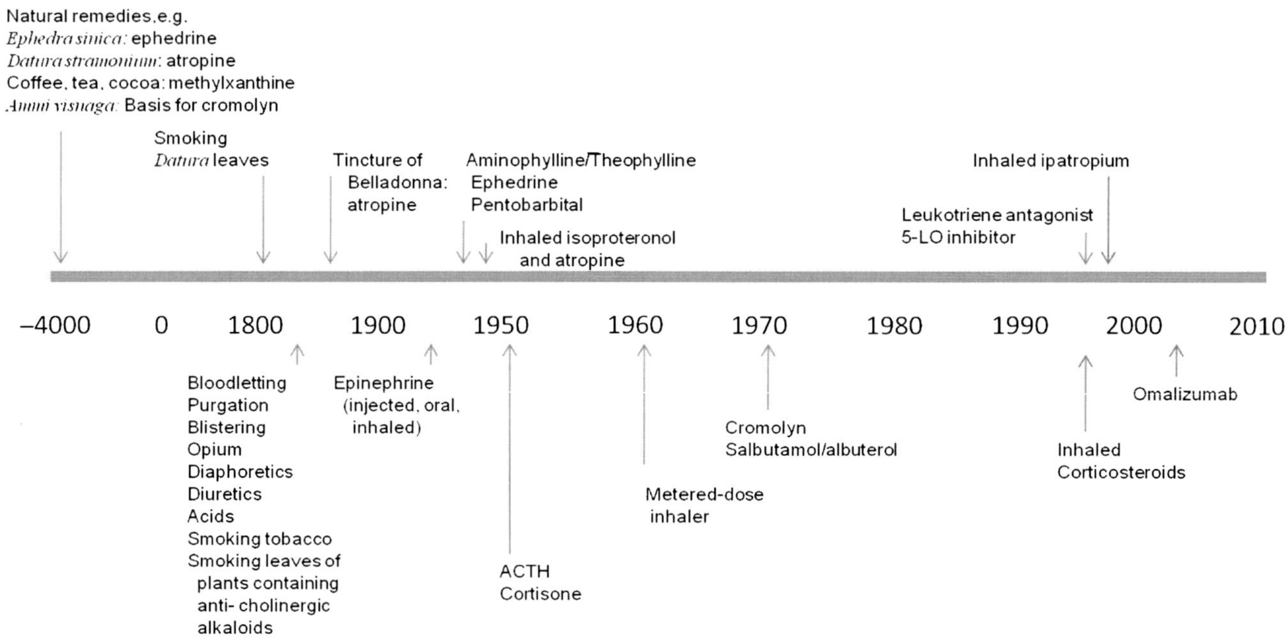

FIGURE 1 Timeline of asthma therapeutics.

to exhale with open mouth, or to pant. Reference to asthma is found in ancient documents, including hieroglyphic scrolls; the Iliad, which was written around the eighth century BC; and the Hippocratic Corpus, written in the four and fifth centuries BC. In the twelfth century, Moses Maimonides wrote a treatise that linked environmental conditions to lung function, and in the sixteenth century, Jean Baptiste Van Helmont described asthma as a disease of the airways. Bronchodilators became common in the early 1900's and remain a mainstay of acute asthma therapy today.

In the 1960s, a link to inflammation was made, corticosteroids became a primary intervention, and mast cell stabilizers (chromones) came into common use. Emergence of these therapies indicated recognition of the role of atopy, cross-linking IgE, triggering release of preformed mediators, inducing EAR and adaptive immunity in asthma pathogenesis. Development of recombinant DNA technology in the 1970s led to the creation of recombinant proteins and transgenic and knockout mice, which became primary research tools for dissecting immune mechanisms. The technology led to production of highly specific reagents, such as antibodies and oligonucleotides, which could be used in humans for proof-of-concept studies and as novel therapeutics.

In 1986, the T_H2 hypothesis was formulated (Mosmann et al., 1986), and studies defining the adaptive immune response in the lungs dominated asthma research for the next two decades. Leukotriene inhibitors that antagonize the slow-reacting substance of anaphylaxis (now known to be LTC4, which is released during exposure to inhaled allergens) were introduced in 1997. In 2003, omalizumab entered the market, specifically targeting IgE. Early in the last decade, clinical trials targeting relevant T_H2 cytokines did not yield robust results, but more recent efforts targeting distinct asthma phenotypes have shown promise. An example is a trial using an interleukin (IL)-13 blocking antibody (lebrikizumab, Genentech). Lebrikizumab was most effective in a subset of patients who exhibited high levels of serum periostin and were poorly controlled with glucocorticoids (Corren et al., 2011). Periostin is an integrin ligand involved in cell adhesion and regulated by transforming growth factor (TGF)-β. In another study, antagonizing IL-4Rα with AMG-317 (Amgen) exhibited some efficacy in patients with severe asthma (Corren et al., 2010). Consistent with the function of IL-5, reslizumab/ mepolizumab (anti-IL-5, Cephalon) reduced eosinophilia and FEV_1 in patients with eosinophilic asthma (Castro et al., 2011). These findings emphasize the diversity underlying asthma and the potential importance of asthma phenotypes in seeking subsets of patients responsive to drug effects. Recent findings have added to the diversity in pathogenic mechanisms, ascribing roles for unique lymphocyte subpopulations, epithelial, dendritic, smooth muscle, and natural killer cells.

Today, asthma is recognized as a heterogenous disease or syndrome, characterized by airflow obstruction that varies both spontaneously and with treatment. Resistance to airflow increases periodically or episodically, causing symptomatic wheezing and dyspnea, which can be alleviated with therapy. In addition to this therapeutically reversible component, some patients manifest an element of irreversible airflow obstruction. A component of reversible obstructive airways disease with wheezing is seen in a wide variety of asthma-like conditions, such as cardiac asthma accompanying congestive heart failure; acute and chronic bronchitis associated with tobacco use, environmental pollutant exposure, or a range of respiratory infectious agents; drug exposure (e.g., aspirin sensitive asthma, treatment with β blockers); and sensitivity to occupational agents (e.g., toluene diisocyanate and trimellitic anhydride). Consequently, a precise, full diagnosis of "asthma" requires an understanding of the etiologic mechanism.

Allergic asthma, or atopic asthma, is the most common form of the disease and is often associated with a family history of allergic diseases. In fact, genetically determined production of specific IgE antibodies accompanied by expression of high-affinity IgE receptors (FcεRI) cells in the airway mucosa is fundamental to the onset of asthma in an atopic individual (Humbert et al., 1996). The distinguishing clinical features of allergic asthma are a history of younger age of onset co-existent with allergic rhinitis and summer exacerbations (Romanet-Manent et al., 2002; Appleton et al., 2006).

2. PATHOPHYSIOLOGY OF ALLERGIC ASTHMA

2.1 Pathogenesis

The process by which an individual develops asthma is not understood. The observation that allergen exposure is a common trigger and the fact that corticosteroid and anti-IgE therapies are effective in many patients provide evidence that humoral immunity plays a central role in the disease; but the incidence of asthma in developed countries has increased at a rate faster than can be explained by simple inheritance of T-cell repertoires and single nucleotide polymorphisms (SNPs) in genes regulating immune responses (see National Center for Health Statistics for data on the incidence of asthma; Michael E. March, 2011). Furthermore, the major SNPs associated with asthma in genome-wide screens do not appear to be affected by environmental conditions that reduce the incidence of asthma (Ege et al., 2011b). Specifically, SNPs identified in the GABRIELA (Advanced Multidisciplinary Study to Identify the Genetic and Environmental Causes of Asthma in the European Community) asthma meta-analysis do not demonstrate significant interaction with the farm environment, which is associated with reduced incidence of asthma. In addition, factors not traditionally associated with allergy, such as viral and bacterial infections and epithelial activation, can promote the development of asthma and increase exacerbations. Therefore, although genetically determined predisposition to humoral immune responses is a major factor in asthma, there are other environmental or developmental factors contributing to the high, and increasing, incidence of the disease.

A recurring theme in asthma pathogenesis is the importance of the prenatal and childhood periods in the development of T_H2-dominant immune responses. Neonates are thought to be predisposed to T_H2 immune responses; however, in normal development, the phenotype changes to T_H1 dominant and antigen exposure promotes tolerization to allergen (Strickland et al., 2012; Strickland and Holt, 2011). Why asthmatics exhibit sustained sensitivity to allergens and T_H2 dominant responses is unclear, but studies in animals have revealed the importance of the prenatal (Hollingsworth et al., 2008; Brand et al., 2011; Whitrow et al., 2009) and neonatal (Arnold et al., 2011; Bisgaard et al., 2007; Minshall et al., 1996; Van Scott et al., 2004) environment in both immunological memory and physiological responses to stress (Coe and Laudenslager, 2007; Coe and Ershler, 2001; Coe and Erickson, 1997; Coe, 1993) later in life. The integration of innate and adaptive immune responses is also becoming better appreciated. As discussed in more detail below, immune functions of dendritic cells (DCs; Strickland et al., 2012), regulatory T lymphocytes (Strickland et al., 2012), and T-helper lymphocytes (Eriksson et al., 2005) are influenced by the airway mucosa, which is inherently responsive to the environment and has the capacity to transduce environmental stimuli into immune responses (Raemdonck et al., 2012; Banner et al., 2011; Caceres et al., 2009; Bartemes and Kita, 2012; Hackett, 2012; Hammad and Lambrecht, 2011; Schleimer et al., 2007).

2.2 Acute/Early-Phase Asthmatic Response

In approximately 60% of asthmatics, exposure to a relevant allergen results in a two phase response: acute and late (Figure 2). In the acute/early response to allergen (EAR), IgE bound to FcεRI on the surface of mast cells is cross-linked by allergen, resulting in the release of histamine, tryptase, leukotrienes, and prostaglandins that directly contracts bronchial smooth muscle (Busse and Lemanske, 2007). As the disease progresses, the endothelium and epithelium become involved, leading to edema of the airway wall, mucus hypersecretion, and changes in mucus rheology. In severe cases, acute aeroallergen exposure can lead to anaphylaxis—a severe, generalized type I allergic reaction. Manifestation and severity of EAR is predicable on the basis of allergen-specific IgE titers and skin testing.

2.3 Late-Phase Asthmatic Response

Approximately 2 h after an acute allergen challenge, inflammatory cells begin to accumulate in the airways. Neutrophil and lymphocyte migration into human airways peaks within 24 h of allergen exposure, and then diminishes

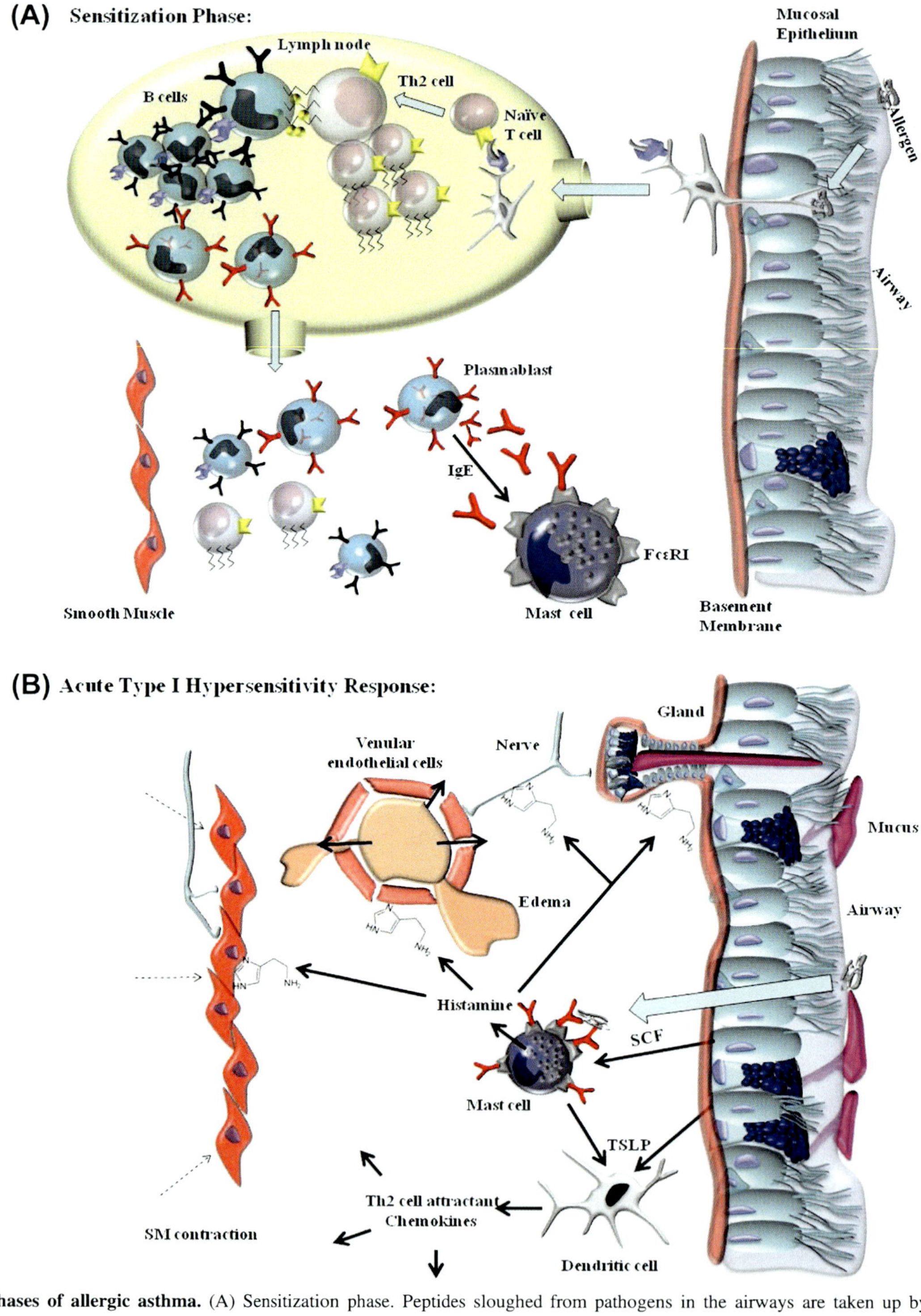

FIGURE 2 **Phases of allergic asthma.** (A) Sensitization phase. Peptides sloughed from pathogens in the airways are taken up by dendritic cells, processed, and fragments are expressed on MHCII. The immature dendritic cells migrate to the regional lymph nodes, where they present the antigen to naïve T-helper lymphocytes. Activation of the T-cell receptor (TCR) by the antigen, in combination with co-stimulation via B7 and CD28, induces IL-2 secretion, which acts as an autocoid to stimulate T-cell proliferation. Interaction of the dendritic cell and T-cell upregulates expression of CD40 on the antigen-presenting cell (APC) and CD40 ligand on the T cell, which further stimulates proliferation of the T cell. In the environment of the draining lymph node, IL-4 levels are elevated, which promotes T_H2 differentiation. Naïve B cells encountering activated T_H2 cells and soluble or membrane-bound antigen are induced to proliferate and differentiate into plasmablasts that circulate to the mucosal tissues and release antigen-specific IgE. (B) Acute type I hypersensitivity response (early asthmatic response). The IgE binds to the high-affinity IgE receptor (FcεRI) on mast cells. Contact of the mast cell with antigen causes cross-linking of the FcεRI receptors and activation of the mast cell, a key component of subsequent acute responses to allergen exposure. Activation of mast cells causes release of histamine, which (1) permeabilizes blood vessels causing plasma extravasation and mucosal edema; (2) activates epithelial cells and glands to secrete mucus and electrolytes onto the surface of the airway; and (3) activates smooth muscle cells leading to bronchoconstriction. The epithelia can be activated via pattern recognition receptors and/or protease activated receptors to secrete stem cell factor (SCF), a

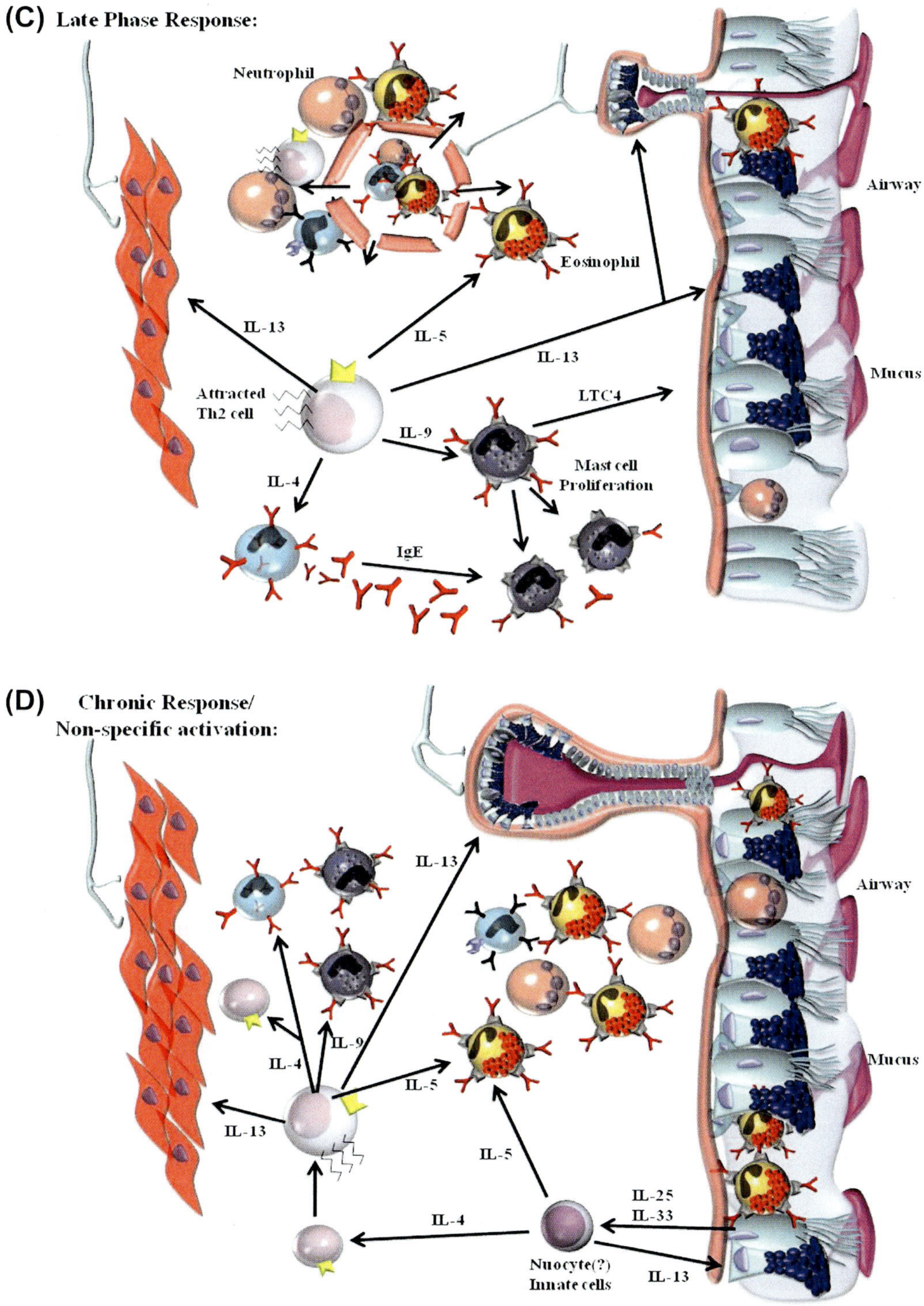

FIGURE 2 cont'd

growth factor for mast cells. Both mast cells and epithelial cells are stimulated by antigen to release thymic stromal lymphopoietin (TSLP), which stimulates dendritic cell maturation. (C) Late-phase asthmatic responses. Neutrophils, eosinophils, and lymphocytes are recruited to the mucosa. Secretion of IL-5 by T_H2 cells, and other cell types including mast cells, supports eosinophil growth. IL-9 promotes mast cell recruitment and growth, and IL-13 promotes remodeling of the airway wall. (D) Chronic responses. Repeated antigen stimulation leads to chronic inflammation and the growth of smooth muscle cells, submucosal glands, and surface goblet cells. In addition, innate cells become sensitized, and epithelial cells can respond to allergens with secretion of IL-25 and IL-33, promoting T_H2 responses, including activation of nuocytes, a recently described lineage negative cell population capable of producing T_H2 cytokines.

(Lommatzsch et al., 2006). Eosinophil migration lags behind the neutrophils and lymphocytes, peaking approximately 2 days after allergen exposure. Release of inflammatory mediators induces a second phase of bronchoconstriction that is associated with hyperresponsiveness of the airway smooth muscle to nonspecific agents, such as adenosine, methacholine, and histamine. Airway hyperresponsiveness (AHR) to nonspecific stimuli is often a surrogate measure for LAR in experimental protocols, but it cannot be considered solely a property of LAR or chronic asthma because nonasthmatic individuals can exhibit AHR (Berend et al., 2008). For example, AHR is observed in other respiratory diseases, including chronic obstructive pulmonary disease (COPD; Hassan et al., 2010), where differential expression of β-adrenergic and muscarinic M3 receptor mRNA can be observed (Selivanova et al., 2012). However, AHR in asthma is observed in response to a broader array of provocative agents than in COPD patients, such as histamine, the β blocker propranolol, environmental SO_2, and cold air.

2.4 Chronic Disease

In addition to ongoing bronchoconstriction, AHR, and inflammation, chronic asthma involves remodeling of the airway wall. Repetitive injury, epithelial damage, and mechanical stress leads to epithelial fragility, epithelial–mesenchymal transformation, subepithelial fibrosis, smooth muscle hypertrophy, goblet cell hyperplasia, angiogenesis, and nerve growth and redifferentiation (Johnson et al., 2011; Cutz et al., 1978). Remodeling—particularly smooth muscle hyperplasia, goblet cell growth, mucous gland expansion, and subepithelial fibrosis—and its functional ramifications have been recognized for decades and are targets of therapeutic interventions, such as bronchial thermoplasty.

Manifestations of Remodeling

Airway remodeling has been implicated as playing a role in persistent AHR (McParland et al., 2003), excessive airway narrowing (Pare et al., 2007), and ultimately fixed airflow obstruction, especially in subjects with severe asthma (Royce et al., 2009). Smooth muscle growth, collagen deposition, and edema reduce the lumen diameter, increasing resistance to airflow, leading to imbalance in ventilation and perfusion across the lungs. Stiffening of the airway wall limits parenchymal-airway interdependence and the increase in airway diameter normally observed at high lung volumes. Reduced dilation of airways contributes to dynamic airway closure observed during forced expiratory maneuvers, which can be observed in asthmatic subjects that are otherwise asymptomatic. Scalloping of the expiratory limb of the flow-volume loop is often observed, indicative of an altered resistance profile along the airways, leading to premature dynamic airway closure.

Remodeling of bronchioles and alveolar ducts is common in human asthma and animal models of allergic lung disease and contributes to the clinical profile (Van Scott et al., 2004; Xisto et al., 2005; Tulic and Hamid, 2003; Tulic et al., 2001; Wagner et al., 1998; Kuwano et al., 1993; Carroll et al., 1993; Akhter et al., 1989; Dolhnikoff et al., 2009; Martin, 2008). Hypertrophy of smooth muscle in the bronchioles and alveolar ducts and narrowing of the airway lumen increases resistance in the distal lung, which can increase the time required for alveolar filling. Consequently, at high respiratory rates, alveolar filling is compromised and dynamic compliance is reduced. This effect is particularly evident and easily measured by pressure–flow analyses of spontaneously breathing allergic animals and humans, as well as with forced oscillatory techniques (Minshall et al., 1996; Van Scott et al., 2004; Van Scott et al., 2005; Metzger, 1990; Oppenheimer et al., 2009).

In severe asthma, mucus accumulates in the airways and contributes to obstruction (Turner and Jones, 2009). The mucus has a rubbery quality due to a high concentration of MUC5AC and MUC5B polymers and plasma proteins. Metaplasia of goblet cells in the superficial epithelium, and hyperplasia to a lesser extent, are typical (Hays and Fahy, 2003). Numerous cytokines upregulate MUC5 expression including IL-4, IL-9, IL-13, IL-17, IL-23, and IL-25, with IL-13 playing a major role (Turner and Jones, 2009; Zuhdi Alimam et al., 2000; Dabbagh et al., 1999). In mild asthma, mucin glycoproteins appear to originate mainly in goblet cells in the surface epithelium (Ordonez et al., 2001). However, in severe asthma, submucosal serous and mucus glands enlarge and make an increased contribution to mucus secreted onto the airway surface (Dunnill et al., 1969). Corticosteroids are more effective than anticholinergic agents in inhibiting mucus hypersecretion in asthma, indicating that inflammatory mediators are primarily responsible for mucus hypersecretion (Barnes, 2002).

Reflex control of respiration is altered in asthma, as evidenced by AHR, chronic cough, altered breathing patterns in the absence of bronchoconstriction, and the sensation of chest tightness (Saisch et al., 1996). Anatomical evidence in humans is lacking, but observations in rodents indicate that neurotrophins such as NGF are upregulated by allergic sensitization and promote neural remodeling within the airway wall (Joos et al., 2000). In monkeys, exposure to aeroallergens has been shown to reduce neural projections to the epithelium but increase lateral projections to the basal lamina (Larsen et al., 2012; Larson et al., 2004). Ozone augments the nerve remodeling in monkeys. Allergen challenge and environmental exposures can also alter expression of receptor and signal transduction molecules in nerves, changing their responsiveness to environmental stimuli. For example, Undem and colleagues have demonstrated that aeroallergen challenge induces expression of transient receptor potential channel TRPV1 on Aδ fibers, increasing the range of stimuli that can activate these fibers (Lieu et al., 2012). Neural remodeling and altered reflexes do not appear to drive bronchoconstriction in asthma, but they may account for other manifestations, such as cough (Barnes, 2002, 2001).

The extracellular matrix (ECM) is dynamic, and it is continually being synthesized and degraded. Persistent activation of fibroblasts in the airway wall leads to increased deposition of collagens I, III, and V, fibronectin, tenascin, lumican, and biglycan (Bergeron et al., 2009). Studies of matrix metalloproteinase (MMP)-9 and its inhibitor, tissue inhibitor of metalloproteinase (TIMP)-1, reveal an altered balance in protease and antiprotease activity in asthma (Suzuki et al., 2001). Alterations in ECM turnover result in thickening of the lamina reticularis and altered mechanical properties of the airway wall (Al-Muhsen et al., 2011). The overall effect of airway wall fibrosis on pulmonary function is unclear. As mentioned previously, increasing stiffness of the airway wall opposes airway dilation during lung expansion, although it would also oppose collapse of the intrapulmonary airway during exhalation.

Both the number and diameter of blood vessels in the airways are increased in asthma, contributing to fluid and plasma protein accumulation in the airway wall (Dunnill et al., 1969). Angiogenesis was described in severe asthma by Dunhill in the 1960s and has been documented to occur in mild asthma as well. Many of the cells linked to asthma pathogenesis—epithelial cells, eosinophils, lymphocytes, mast cells, basophils, macrophages, endothelial cells, and smooth muscle cells—secrete angiogenic factors, including vascular endothelial growth factor (VEGF), fibroblast growth factor-2 (FGF-2), tryptase, chymase, IL-8, TGF-β, tumor necrosis factor alpha (TNF-α), nerve growth factor (NGF), and angiogenin. Thus, strong angiogenic signals are expressed in the airways throughout the disease processes, emanating from infiltrating cells in the acute inflammation stage and continuing with structural cells in chronic stages.

Findings over the last decade have led to the realization that the epithelium and smooth muscle play important roles in immune processes underlying asthma pathogenesis. Epithelial cells can activate DCs, initiating and prolonging immune responses. Airway epithelial cells express pattern recognition receptors including TLRs and dectin, as well as protease activated receptors that directly respond to activators in the airway surface liquid. Upon activation, epithelial cells can produce granulocyte-macrophage colony-stimulating factor (GM-CSF), TSLP, IL-25, IL-33, and osteopontin. GM-CSF is a DC differentiation factor. TSLP upregulates CD80, CD86, MHC class II, and OX40 ligand, inducing maturation of DCs within the lung, promoting antigen presentation, and augmenting T_H2 responses. IL-25 is thought to enhance TSLP activity. IL-33 stimulates DC migration to lymph nodes, and osteopontin promotes B-cell production of immunoglobulin and augments IgE-mediated events. Genome-wide association studies have pointed to a role for the epithelium in asthma, identifying associations between asthma and variations in genes for TSLP and IL-33 ligand and receptor (Ober and Yao, 2011).

McKenzie and colleagues described a cell population that responds to IL-25 and IL-33 with increased T_H2 cytokine production. These cells, which McKenzie termed "nuocytes," are described as innate lymphoid cells that are negative for hematopoietic lineage markers, but express c-Kit, IL-17 receptor, and Sca-1 (Neill et al., 2010; Wong et al., 2012; Barlow and McKenzie, 2011; Kim et al., 2012). These innate helper cells could mediate T_H2 lymphocyte-independent responses to allergen. Similar cells have been reported in other tissues, including natural helper cells (Moro et al., 2010), innate type 2 helper cells (Price et al., 2010), and multipotent progenitor cells that promote T_H2 responses (Saenz et al., 2010).

There is increasing interest in the potential role that the epithelium plays in remodeling, specifically in whether epithelial–mesenchymal transition (EMT) occurs in asthma (Hackett, 2012). EMT is a form of epithelial plasticity in which epithelial cells redifferentiate into migratory fibroblastoid cells that control tissue fibrosis. The process involves loss of epithelial cell polarity and cell–cell adhesions, followed by expression of mesenchymal proteins, including vimentin and fibroblast-specific protein (FSP-1). Epithelial damage upregulates profibrotic factors, such as TGF, epidermal growth factor (EGF), and fibroblast growth factor (FGF), which facilitate wound healing. Exposure of bronchial epithelial cells in vitro to TGF-β downregulates E-cadherin and upregulates mesenchymal markers; in mice, chronic exposure to house dust mite (HDM) allergen increases TGF-β expression in airway epithelial cells, loss of tight junction proteins, and upregulation of mesenchymal markers (Johnson et al., 2011). Definitive evidence of these changes in human asthmatic airways has yet to be published, but the mechanism could be very important in asthma pathogenesis and therapy. In clinical trials with IL-13 antagonism, patients who responded best to therapy exhibited a high level of serum periostin (Corren et al., 2011). Periostin is secreted from epithelial cells in response to IL-13, and it can modulate fibroblast and epithelial functions.

The synthetic capacity of smooth muscle cells is well documented. Following activation via pattern recognition receptors such as TLRs; direct contact with mast cells, lymphocytes, and eosinophils; and T_H2 cytokines including IL-4 and IL-5, smooth muscle cells are stimulated to secrete chemokines and cytokines that can amplify the immune and inflammatory response. Chemokines secreted by smooth muscle cells include eotaxin, RANTES, monocyte chemoattractant proteins, IL-8, thymus- and activation-regulated chemokine (TARC). Secreted cytokines include stem cell factor, GM-CSF, IL-5, IL-6, IL-17 (Howarth et al., 2004; Tliba and Panettieri, 2009). In this way, the smooth muscle can be a significant factor in promoting an inflammatory response and maintaining an asthma phenotype.

3. THE IMMUNE BASIS OF ASTHMA

The incidence of allergy and asthma in developed countries has been increasing over the last 40 years (Bousquet et al., 2004). This alarming trend has stimulated efforts to understand the basic processes underlying sensitization of the airways, establishment of memory, and asthma exacerbations, as well as how these processes are affected by the environment. With a few notable exceptions, it is widely accepted that asthma results from imbalance in the immune system that leads to augmented T_H2 responses. The role of T_H2 cells in allergic responses has its foundation in the T_H2 hypothesis proposed by Coffman and Mosman in 1986 (Coffman, 2006). At the core of this hypothesis is a balance in the differentiation of CD3+ CD4+ lymphocytes into T_H1 or T_H2 cells, reciprocal inhibition of T_H1 cell differentiation by T_H2 cells via secretion of IL-4, and inhibition of T_H2 cell differentiation via secretion of interferon (IFN)-γ by T_H1 cells. Today, it is well accepted that antigen presentation by DCs leads to differentiation of CD3+ CD4+ T-helper lymphocytes into IL-4, IL-5, and IL-13 secretory cells; subsequent upregulation of IgE synthesis; activation of mast cells by cross-linking of FcεRI on mast cells; release of histamine and leukotrienes; and induction of bronchoconstriction and eosinophilic inflammation in the airways. The hypothesis integrates well with the hygiene hypothesis, which proposes that urbanization and increased sanitation standards in developed countries limit exposure to bacteria, decreasing environmental pressure to maintain strong T_H1 immune responses, tipping immune balance in favor of and T_H2 immunity and allergic responses.

This paradigm is supported by differential expression of GATA-3, a critical T_H2 transcription factor, over T-bet, a key T_H1 transcription factor in asthmatic airways; and corticosteroid-induced upregulation of FOXP3, a key transcription factor in regulatory T lymphocytes (T_{reg}). However, accumulated evidence also reveals a high level of cross-talk and redundancy across the immune system, and it is now recognized that numerous cell types other than T_H2 lymphocytes contribute to the production of IL-4, IL-5, and IL-13, including T_H17, CD8+ cytotoxic T-, natural killer (NK), invariant NKT (iNKT), γδT, innate helper, epithelial, smooth muscle, fibroblasts, and mast cells. In addition, as central as IL-4, IL-5, and IL-13 appear to be in asthma, targeting these cytokines in clinical trials has yielded limited therapeutic benefit. This finding has fueled efforts to elucidate redundant regulatory pathways that either converge on a common phenotype or lead to different asthma phenotypes, to identify major key regulatory cell and cytokine nodes that can be targeted therapeutically, and to optimize therapeutic approaches via pharmacogenetics.

3.1 Cell Network

Dendritic Cells

DCs express diverse receptors for taking up molecules from the extracellular space and synthesizing T-cell receptor (TCR) ligands. The ligands can be complexes of peptide antigens and MHC molecules, or lipid antigens complexed with CD1 MHC-like molecules (Steinman and Hemmi, 2006; Spinozzi and Porcelli, 2007). Presentation of extracellular/environmental peptides on the surface of DCs in the context of MHCII and in combination with a B7 co-stimulatory signal induces T_H2 cell growth, production of IL-4, immunoglobulin class switching in B lymphocytes, production of high-affinity IgE, and priming of mast cells. T_H2 lymphocyte secretion of IL-5 and IL-13 mediates eosinophilic inflammation, hyperresponsiveness to nonspecific stimuli, and airway remodeling. In similar fashion, presentation of amphipathic lipids from pollens and microbes in the context of CD1d can activate iNKT cells to secrete IL-4, IL-5, and IL-13 (Spinozzi and Porcelli, 2007). The proinflammatory role of some DC subsets is offset by other subsets that are capable of downregulating immune responses and inducing tolerance to allergens, thereby minimizing tissue damage induced by chronic inflammation.

DCs in the lungs can be divided into subtypes—conventional (cDC), inflammatory, plasmacytoid (pDC), alveolar, and regulatory DCs—based on their lineage, the receptors they express, and their location. Conventional and inflammatory DCs are thought to play an important role in airway allergic responses (Hammad and Lambrecht, 2011; Hammad et al., 2010). cDC express cadherins and CD103, which mediate attachment to the basolateral membrane of epithelial cells. Appendages extend through the epithelial cell tight junctions into the airway lumen to monitor the airway surface liquid for inhaled antigen. During inflammation, monocyte-derived inflammatory DCs are recruited to the lungs. These cells have similar characteristics as resident cDCs, but retain the monocyte marker, Ly6C (Hammad and Lambrecht, 2011). In contrast to the proinflammatory cDC and monocyte-derived DCs, plasmacytoid DCs oppose inflammation induced by aeroallergen via expression of programmed death ligand 1 (PD-L1) (Kool et al., 2009). The counterregulatory role of PD-L1 is discussed later in this chapter.

Under normal conditions, DCs in peripheral tissues exist in an immature state, expressing low levels of both co-stimulatory molecules necessary for efficient antigen presentation, and counterregulatory molecules that oppose antigen presentation. Activation by diverse stimuli leads to stimulus-specific maturation. Thus, exposure to soluble protein antigens or thymic stromal lymphopoietin (TSLP) stimulates myeloid DCs to promote growth of T_H2 cells, whereas maturation in response to lipid-associated antigens or CD40L promotes growth of T_H1 cells (Steinman and Hemmi, 2006).

During the stimulus-induced maturation process, antigen uptake is downregulated and expression of co-stimulatory and accessory molecules is upregulated. Chemotaxis is stimulated, resulting in migration of the DCs to lymph nodes where T and B cells reside. Conventional DCs in the airways express signal regulatory protein α (SIRPα) and CCR-like 2 (CCRL2), which have been implicated in migration from the airway mucosa to the lymph nodes (Raymond et al., 2010; Lambrecht and Hammad, 2009, 2012). The specific environmental signals determine the DC responses and the specific cytokines they secrete: IL-12 and INF-γ that support T_H1 immunity and cytotoxicity responses, IL-4 that supports T_H2 differentiation and allergic responses, and/or IL10 and TGF-β that are involved in differentiation of regulatory T cells (Steinman and Hemmi, 2006). These regulatory cytokines can also be produced by other cell types including lymphocytes, basophils, mast cells, eosinophils, NK cells, and NKT cells (Barlow and McKenzie, 2011; Lambrecht and Hammad, 2009).

T-Helper Lymphocytes

Following exposure to antigen, antigen-specific T-lymphocytes proliferate and differentiate into effector cells, some of which further differentiate into memory cells that can be reactivated by subsequent antigen exposure (Lees and Farber, 2010). Clearly, T_H2 cells are major effector and memory cells in allergic asthma, although other cell types can play these roles, leading to redundant pro-inflammatory pathways and complexity in pathogenesis. Depending on growth conditions, naïve CD4+ T-cells can differentiate into T_H1, T_H2, T_H9, T_H17, and T_{reg} phenotypes, all of which have been implicated in allergic responses and contribute to different phenotypes in asthma. It should be noted that much of the evidence for existence of these CD4+ subtypes and their involvement in allergic responses comes from in vitro studies and in vivo experiments using mice. Actual roles in human asthma have yet to be confirmed.

T_H1 immunity defends against bacteria and fungi, and it is promoted by IFN-γ mediated activation of macrophages (Vock et al., 2010). T_H1 and T_H2 cells develop from a common pool of naïve CD4+ cells, and growth of one subtype reduces growth of the other subtype. Thus, upregulation of T_H1 differentiation is expected to oppose differentiation of CD4+ cells into T_H2 cells. Although there is extensive evidence to support this paradigm, there is also evidence that the interplay between T_H1 and T_H2 cells involves additional complexity. For example, adoptive transfer of T_H1 cells into mice can induce neutrophilia in bronchoalveolar lavage fluid (BALF), and the presence of T_H1 cells in the airways may be a prerequisite for T_H2 cell recruitment (Huang et al., 2001). Furthermore, there is evidence that T_H1 and T_H2 cells are both upregulated in the airways of asthmatic patients (Vock et al., 2010). Thus, while the predominant effect of IFN-γ and T_H1 cells is inhibition of T_H2 responses, exceptions to this paradigm exist and the ultimate role of T_H1 cells in asthma remains unclear.

Naïve CD4+ cells exposed to IL-4 and TGF-β can be driven to secrete IL-9, a potent T- and mast-cell growth factor (Vock et al., 2010; Barnes, 2008). Genetic studies have linked IL-9 to human asthma, and clinical trials with anti-IL-9 antibody have produced modest therapeutic effects (Antoniu, 2010a). In the presence of IL-23, memory CD4+ cells can be induced to secrete IL-17, and are therefore classified as T_H17 cells. As with other lymphocytes, the effector functions of T_H17 cells are highly dependent on environmental factors. Thus, using different models, T_H17 cells in mice have been shown to induce steroid-resistant, neutrophilic airway inflammation and AHR; augment T_H2 responses; but also reduce eosinophilic inflammation (Robinson, 2010). Interestingly, in the presence of high levels of IL-4, T_H17 cells appear to be malleable and can be induced to express IL-4, IL-5, and IL-13 (Cosmi et al., 2010).

T_H17 cells express IL-22, a cytokine involved in promoting mucosal immunity (Vock et al., 2010). Another member of the IL-17 family, IL-25, promotes Th2 type responses, inducing expression of IL-4, IL-5, and IL-13 by invariant NKT cells.

T-Regulatory Lymphocytes

In addition to the pro-inflammatory T-cell subsets mentioned above, CD4+ cells in presence of TGF-β differentiate into a regulatory phenotype. These CD4+ CD25+ T_{reg} cells produce the anti-inflammatory cytokines IL-10 and TGF-β. Studies in humans have demonstrated reduced numbers of T_{reg} cells in blood and bronchoalveolar lavage (BAL) fluid of asthmatic subjects, a positive correlation between the number of T_{reg} cells in BAL fluid and FEV_1, and reduced responses of T_{reg} cells to chemokines that normally mediate recruitment to the airways (Vock et al., 2010). Reduced numbers of T_{reg} cells in the airways may underlie excessive upregulation of APCs and effector cells. Recent observations indicate that T_{reg} cells exhibit a high level of plasticity, being capable of differentiating into effector cells of various subsets.

Cytotoxic T-Lymphocytes

CD8+ cytotoxic T-lymphocytes can be divided into subsets that are functionally similar to those described for CD4+ cells, including Tc1, Tc2, Tc17, FOXP3+, γδ+, effector, and memory cells (Betts and Kemeny, 2009). Clinically, CD8+ T cell number correlates with asthma severity, but whether the association is causative or a result of pathogenesis remains to be proven. Extensive data from rodents reveal subset-specific anti-inflammatory and pro-inflammatory capacities of CD8+ cells consistent with ability to express T_H1- and/or T_H2-type cytokines. These studies indicate that different subsets of CD8+ cells localize to peripheral tissues and lymph nodes, resulting in different regulatory roles in systemic immune responses and peripheral inflammation. Studies in mice also indicate that virus-stimulated CD8+ cells in a T_H2 environment can be induced to produce IL-5, thereby augmenting the allergic asthmatic response and providing a potential mechanism for virus-induced asthma exacerbations (Coyle et al., 1995a).

Memory Lymphocytes

During sensitization to allergen, a portion of antigen-specific T and B cells develop longevity and proliferative capacity that maintain immunologic memory for long periods, allowing the immune system to respond rapidly upon re-exposure to pathogens. Memory T cells are divided into central (T_{CM}) and effector (T_{EM}) populations based on expression of CD62L and CCR7 (Kaech et al., 2002; Sallusto et al., 2004; Zielinski et al., 2011). T_{CM} express CD45RO, CD62L, and CCR7. CD62L and CCR7 mediate cell adhesion to endothelium in lymph nodes, resulting in homing of T_{CM} to nodes. T_{CM} exhibit a greater proliferative capacity than T_{EM}, which leads to expansion in the lymph nodes. They produce IL-2, a growth factor for T and B cells, which supports expansion of these cells in the nodes. Prior to proliferation, they produce low levels of effector cytokines, but after proliferation, production of effector cytokines such as IFN-γ and IL-4 increases (Sallusto et al., 2004). In contrast, T_{EM} are $CD62L^{lo}$ $CCR7^{-}$ cells. The majority of T_{EM} express the activation marker CD45RO, with a small percentage expressing CD45RA (Sallusto et al., 2004). T_{EM} express chemokine receptors and adhesion molecules for homing to peripheral tissues, including the lungs, where they secrete effector cytokines to promote T cell, B cell, and DC growth and differentiation (Sallusto et al., 2004; Zielinski et al., 2011; Robinson et al., 1993).

Both CD4+ and CD8+ memory cells exist, though there are distinct differences in their development. Commitment of CD8+ cells to a long-lived memory phenotype occurs early in the sensitization, after which the subset is maintained by homeostatic cytokines such as IL-15 and IL-7 (Lees and Farber, 2010). Maintenance is independent of TCR signaling and MHCI engagement. In contrast, the development of CD4+ memory cells occurs along less-well-defined pathways, through processes that are inherently plastic and affected by intensity, duration, and frequency of antigen exposure (Lees and Farber, 2010). Maintenance of CD4+ memory cells is more dependent on recurring TCR and MHCII engagement, and abolishing signaling through this pathway diminishes recall responses to antigen (Lees and Farber, 2010).

Type 1 Natural Killer T-Lymphocytes

Approximately 2−10% of the T-lymphocytes (i.e., 0.2−1% of the total white cells) in the airway of asthma patients are reported to be type 1 natural killer T-lymphocytes (invariant, iNKT cells). These cells express a repertoire of αβ TCRs that incorporate an invariant α chain (Vα14-Jα18 in mice and Vα24-Jα18 in humans), which is restricted to binding antigen-loaded CD1d (Meyer et al., 2008; Iwamura and Nakayama, 2010). CD1d presents glycolipids, such as α-galactosylceramide (α-GalCer), and stimulates iNKT cells to produce IL-4, IL-5, IL-13, and IFN-γ. In some, but not all, strains of mice, depletion of iNKT cells eliminates development of AHR and adoptive transfer of the cells restores hyperresponsiveness.

Although their involvement in human asthma has not been proven, iNKT have the capacity to secrete IL-4, IL-5, IL-9, IL-13, and IFN-γ and therefore can modulate DC, B cell, and T_H2 cell activity (Meyer et al., 2008; Iwamura and Nakayama, 2010). iNKT cells are activated by alternative pathways that do not require peptide antigens, and thereby increase the diversity of pathways leading to asthmatic responses. Activation by glycolipids expressed on CD1d and presented to the TCR is discussed above. iNKT cells can also be activated by cytokines, particularly IL-25 and IL-33, which are produced by epithelial cells following activation of protease-activated receptors (PRRs) and cell injury (Bartemes and Kita, 2012). The link to the epithelium could explain asthma exacerbation by numerous environmental factors such as chemical irritants, ozone, and cold air. Activation of iNKT cells may also induce secretion of IL-13 from macrophages (Watarai et al., 2012). In this scenario, IL-13 released by iNKT cells induces IL-13 from macrophages, leading to an autocrine positive feedback secretion of IL-13. Finally, iNKT cells are resistant to downregulation by corticosteroids and could thereby account for, or at least contribute to, steroid-resistant forms of asthma (Holtzman et al., 2009).

γδ T-Lymphocytes

CD1d-restricted γδ T cells that are responsive to pollen-derived phospholipids have been isolated from human airways and could play a role in asthma (Meyer et al., 2008). Two populations of γδ T cells are found in the airways, Vγ1+ and Vγ4+ γδ T cells. These cells respond to lipid antigens presented in the context of CD1d. $\gamma\delta$ T cells that express Vγ1+ TCR secrete IL-5 and IL-13 and promote AHR (Spinozzi and Porcelli, 2007; Lloyd and Hessel, 2010). In contrast, $\gamma\delta$ T cells that express Vγ4+ TCR secrete IL-17A and suppress inflammation (Afshar et al., 2008). Airways and nasal passages in patients with asthma exhibit increased numbers of $\gamma\delta$ T cells that express T_H2 cytokines, indicating that they may potentiate pathogenesis (Murdoch and Lloyd, 2010).

Natural Killer Cells

NK cells can produce T_H1 and T_H2 cytokines that affect DC maturation in lymphoid and peripheral tissue (Afshar et al., 2008). In addition, they can process and present target cell antigens on their surface in the context of MHC II, and in the presence of costimulatory molecules can present antigens to T_H2 cells (Moretta et al., 2006). Korsgren and colleagues demonstrated that NK cells contribute to ovalbumin (OVA)-induced eosinophilic inflammation in mice (Hanna et al., 2004). However, depletion of NK cells does not impair CD4+ T cell development and antibody responses to antigen (Korsgren et al., 1999). In other studies, Gelfand and colleagues demonstrated that a combination of IL-2 and IL-18 attenuated OVA-induced eosinophilic inflammation, goblet cell growth, and AHR in mice (Wang et al., 1998), and that the inhibitory effects were dependent on NK cells. An anti-inflammatory role for NK cells was supported by Haworth and colleagues who reported that NK cells are required for timely resolution of OVA-induced airway inflammation, possibly via Resolvin E1 binding to its receptor, CMKLR1, on NK cells (Matsubara et al., 2007). Observations in humans indicate that NK cell number and cytolytic activity of NK cells were reduced in peripheral blood from patients with respiratory allergies (Haworth et al., 2011).

NK cells comprise approximately 10% of the circulating lymphocytes and can be separated into subsets based on expression levels of CD56 adhesion molecule, CD16 low-affinity Fc receptor, and IFN-γ (Scordamaglia et al., 2008). $CD56^{dim}$ $CD16^+$ cells comprise 80–90% of the circulating NK cells, and are thought to be recruited readily to sites of inflammation (Moretta et al., 2006; Timmons and Cieslak, 2008; Deniz et al., 2002). Approximately 10% of the NK cells in peripheral blood are $CD56^{bright}$ $CD16^{+/-}$ cells. This subset expresses high levels of CD62L and tends to localize to spleen and lymphoid tissues. CD16 dim cells exhibit a higher level of spontaneous IFN-γ production than $CD16^+$ cells, whereas the CD16+ cells tended to produce less IFN-γ and greater amounts of IL-4 and IL-13 (Moretta et al., 2006; Bellora et al., 2010; Agaugue et al., 2008). The two cell populations are labeled NK1 and NK2 based on the similarities in cytokine production with T_H1 and T_H2 cells. NK2 cells are reported to be increased in asthma subjects (Deniz et al., 2002).

3.2 Cytokine Network

Numerous cytokines regulate the growth and activity of cells implicated in asthma and allergic diseases. The interactions between the cytokines and the cells that produce them, and interactions with components of the complement and innate immune systems are likewise numerous and complex. As demonstrated by the nodal analysis illustrated in Figure 3, asthma literature to date places the T_H2 cytokines in central roles in disease pathogenesis, with IL-13, IL-4, IL-5, IL-6, and TNF-α representing points of convergence in pathogenic pathways involving innate and adaptive immunity, the complement system, and regulatory cytokines.

3.3 General Growth and Acute Response Cytokines

IL-2

IL-2 is a general growth factor for lymphocytes. The primary source of IL-2 is activated T cells, and activation of the TCR induces both secretion of IL-2 and expression of IL-2R. Activation of the IL-2R induces differentiation and clonal expansion of antigen-specific CD4+ and CD8+ T cells, as well as growth and differentiation of B cells, NK cells, macrophages, and T_{reg} cells (Scordamaglia et al., 2008; Wei et al., 2005; Wingett and Nielson, 2003).

Consistent with its function as a general growth factor, IL-2 has been implicated in allergic responses, cytotoxicity, and peripheral tolerance (Stern and Smith, 1986; Beadling and Smith, 2002; Beadling et al., 1993; Letourneau et al., 2009; Chung and Barnes, 1999). Lack of IL-2 is associated with lymphoproliferative disorders and autoimmunity (Letourneau et al., 2009; Klebb et al., 1996; Fehervari et al., 2006; Thornton et al., 2004; Renauld, 2001; Smith, 1988). Thus, in the context of allergic disorders, IL-2 appears to play a supportive role opposed to a direct pathogenic role.

Granulocyte-Macrophage Colony Stimulating Factor

GM-CSF is a general growth factor for neutrophils, eosinophils, and basophils, in addition to myeloid APCs such as macrophages and DCs (Thornton et al., 2004; Smith, 1988; Sakaguchi et al., 1995; Thornton and Shevach, 1998; Shin et al., 2009). GM-CSF is primarily produced by T cells, as well as by diverse cell types found in the airway mucosa including epithelial cells, smooth muscle cells, macrophages, fibroblasts, endothelial cells,

FIGURE 3 A priori analysis of factors associated with canonical cytokines in allergy and asthma using the Ingenuity Pathway software. Solid lines represent paths of direct control. Dashed lines represent pathways involving intermediates that are not shown in the illustration. The analysis did not discriminate between positive and negative control pathways. IL, interleukin; IFN, interferon, C, complement; PAR, protease activated receptor; TLR, toll like receptor; GMCSF, granulocyte-macrophage colony-stimulating factor; TNF, tumor necrosis factor; TGF, transforming growth factor; T_H, T helper.

and eosinophils (Shin et al., 2009). Consistent with these mucosal sources, GM-CSF mRNA is increased in the airways of human asthmatics (Chung and Barnes, 1999) where it promotes maturation, activation, and growth of monocytes and DCs, resulting in increased antigen presentation and activation of antigen-specific B and T cells (Chung and Barnes, 1999). GM-CSF also contributes to T_H17-induced inflammatory responses (Ganesh et al., 2009; Blanchfield and Mannie, 2010; Hamilton and Anderson, 2004; Ritz et al., 2002).

In mouse lungs, adenoviral-mediated overexpression of GM-CSF upregulates macrophage proliferation and inflammatory cytokine production leading to eosinophil infiltration, fibrosis, and increased in macrophage numbers (McGeachy, 2011). Conversely, treatment with anti-GM-CSF antibody attenuates sensitization to aeroallergens, allergen-induced inflammation, and AHR (Shin et al., 2009). The importance of GM-CSF in macrophage function is illustrated in GM-CSF knockout mice, which develop alveolar proteinosis secondary to impaired processing of alveolar surfactant.

TNF-α

TNF-α is an initiator of the acute-phase response that is produced primarily by macrophages, but it is also expressed in T lymphocytes, mast cells, neutrophils, eosinophils, natural killer cells, and epithelial cells (Hamilton and Anderson, 2004). TNF-α secretion by monocytes and macrophages is amplified by other cytokines such as IL-1, GM-CSF, and IFN-α; and IgE has been shown to induce TNF-α expression in lung epithelial cells via FcεRI (Costa et al., 1993; Devalia et al., 1993). TNF-α receptors, TNF-R55 and TNF-R75, are found on most cells with the exception of red blood cells and unactivated T cells. Expression of TNF-R75 is particularly high on hematopoietic cells (Ohno et al., 1990; Hirshman et al., 1980; Ohkawara et al., 1992).

TNF-α augments allergic reactions by upregulating MHC class II expression and IL-1 secretion by APCs (Chung and Barnes, 1999) and by acting as a costimulatory molecule for activated T cells, allowing for increased T-cell proliferation and IL-2 receptor expression (Chung and Barnes, 1999; Kips et al., 1993; Shah et al., 1995). TNF-α induces chemotaxis, degranulation, and respiratory burst activity in neutrophils (Chung and Barnes, 1999) and stimulates production of intercellular adhesion molecule (ICAM)-1, vascular cell adhesion molecule (VCAM)-1, and E-selectin that mediate migration of granulocytes and mononuclear cells at sites of inflammation (Chung and Barnes, 1999).

TNF-α expression is elevated in BAL fluid (Commins et al., 2010) and bronchial biopsies from human asthmatics (Broide et al., 1992); airway responsiveness correlates with TNF-α levels in humans (Ying et al., 1991) and Brown-Norway rats (Yates et al., 1993). The effects of TNF-α in the airways may be secondary to increased production of GM-CSF, IL-8, and RANTES that promote growth and recruitment of inflammatory cells (Kips et al., 1992). Inhalation of TNF-α by normal human subjects causes neutrophil-induced AHR (Chung and Barnes, 1999). Blockade of TNF-α with etanercept (Embrel) and infliximab (Remicade) reduces AHR in patients with refractory and moderate asthma, respectively (Barnes, 2008).

3.4 Canonical T_H2 Cytokines

IL-4

IL-4 was originally described as a B-cell growth factor, but it is now known to also play a critical role in T-cell differentiation. IL-4 is produced by T_H0 and T_H2 T lymphocytes, NK T cells, eosinophils, basophils, and mast cells (Barnes, 2008). IL-4 synthesis is enhanced via costimulatory signals generated by cross-linking of CD40 ligand on human CD4+ T cells, activation of TCR, and cross-linking of FcεRI on basophil and mast cells (Commins et al., 2010).

IL-4 promotes differentiation of naive T_H0 into T_H2 cells and inhibits apoptotic pathways within T_H2 cells (Chung and Barnes, 1999). IL-4 is required for immunoglobulin isotype switching to IgE in B cells and stimulates expression of IgE receptors on mast cells (Commins et al., 2010). IL-4 also promotes antigen presentation by increasing MHC class II expression on APCs and increases expression of B7 (CD80/CD86), CD40, low-affinity IgE receptor (CD23), and surface IgM on B cells. Transgenic mice overexpressing IL-4 develop symptoms of allergic disease, including lymphocytic and eosinophilic inflammation and elevated IgE concentrations (Renauld, 2001). In humans, the severity of asthma has been associated with a gain of function mutation in the IL-4 receptor protein (IL-4Rα) (Renauld, 2001). Conversely, knockout of the IL-4 gene in mice attenuates ovalbumin-induced eosinophilia and production of allergen-specific IgE, but has little effect on AHR (Hershey et al., 1997). Yet, studies using anti-IL-4 receptor (IL-4R) antibodies or mice deficient in the IL-4R subunit IL-4Rα revealed reductions in AHR. A potential explanation for this discrepancy is that IL-4Rα can also bind IL-13, providing a parallel pathway for activation of IL-4Rα and induction of AHR (Chung and Barnes, 1999; Renauld, 2001). The overlap in IL-4 and IL-13 signaling through IL-4Rα has clinical significance, resulting in reduced effectiveness of therapies targeting IL-4 or IL-13 alone compared to therapies targeting IL-4Ra (Renauld, 2001).

IL-5

IL-5 is a primary growth factor for eosinophils and is produced by T_H2 T lymphocytes, eosinophils, and mast cells. As the major cytokine in eosinophil development, IL-5 induces eosinophil migration, activation, and survival (Corren et al., 2010; Slager et al., 2012). The primary source of IL-5 is T_H2 lymphocytes, but mast cells are also a source in the airways. IL-5 is increased in bronchial biopsies and BAL fluid and serum from symptomatic asthmatic patients, knockout of IL-5 in mice eliminates AHR and eosinophilia, and overexpression or treatment with rIL-5 results in asthma-like lung histopathology (Chung and Barnes, 1999). Blockade of IL-5 in humans with mepolizumab, an anti-IL-5 antibody, decreases eosinophils in the sputum and general circulation, and in asthma patients exhibiting hypereosinophilia and resistance to steroid therapy mepolizumab reduces exacerbations (Barnes, 2008; Chung and Barnes, 1999; Commins et al., 2010).

IL-9

IL-9 is secreted by CD4+ cells, typically T_H2 cells. Originally described as a T-cell growth factor, IL-9 induces not only T-cell proliferation but also proliferation and differentiation of mast cells and hematopoietic progenitor cells, as well as B-cell immunoglobulin synthesis. IL-9 has been shown to increase mast cell IgE receptor expression, upregulate IgE production, and promote eosinophil growth (Busse et al., 2010). In mice, mast cell proliferation and stimulation of B-cell immunoglobulin synthesis appears to be mediated by IL-9 directly, whereas IL-9 effects on eosinophilia and mucus production are through IL-13 (Commins et al., 2010).

Both IL-9 and the IL-9R are increased in the airways of asthmatic patients, and the expression IL-9 mRNA is correlated with increased airway responses to methacholine (Barnes, 2008). Transgenic mice with increased lung expression of IL-9 exhibit increased numbers of eosinophils and mast cells in the airways, excess mucus, airway remodeling, AHR, and elevated IgE. Conversely, antagonism of IL-9 suppresses allergen-induced airway eosinophilia, mucus production, and AHR (Renauld, 2001). The ability of IL-9 overexpression to elicit symptoms of airway disease may be due to IL-9-induced upregulation of IL-5 and IL-13 secretion (Barnes, 2008; Chung and Barnes, 1999). Clinical trials with a therapeutic antibody against IL-9, MEDI-528, have shown modest efficacy in asthma (Finkelman et al., 2010).

IL-13

IL-13 is produced by activated CD8+ T cells as well as CD4+ T_H1, T_H2, and T_H0 cells in response to antigenic stimuli (Antoniu, 2010a; Parker et al., 2011). The IL-13 receptor shares a common subunit with the IL-4 receptor, IL-4Rα, and the two cytokines exhibit some functional overlap. The other component of the heterodimeric IL-13 receptor complex, IL-13Rα, is prominently expressed on smooth muscle cells. Unlike IL-4, IL-13 has no effect on T-cell activation due to a lack of IL-13 receptors on T cells, although both IL-13 and IL-4 have been shown to increase IgE secretion and promote remodeling of the airway wall. IL-13 increases mucus secretion and AHR through STAT6 activation in the airway epithelium, and affects eosinophil activation and survival through CD69 (Chung and Barnes, 1999).

IL-13 is elevated in the airways of human asthmatics and correlates with the severity of allergen-induced eosinophilia. In mice, IL-13 causes airway eosinophilia and AHR, and neutralization attenuates mucus production, airway eosinophilia, and AHR (Chung and Barnes, 1999; Barnes, 2008). Pitrakinra, an IL-4 mutein that blocks binding of both IL-13 and IL-4 to IL-4Rα, decreases the late response to inhaled allergen in humans (Antoniu, 2010b), and efficacy is increased in asthmatics with specific polymorphisms IL-4Rα (Barnes, 2008; Chung and Barnes, 1999; Slager et al., 2012).

3.5 T_H1 Cytokines

IFN-γ

IFN-γ inhibits T_H2 cell responses, allergic inflammation, and IgE production, and it promotes cell-mediated cytotoxicity. IFN-γ production is limited to CD4+ and CD8+ T lymphocytes and NK cells, but its receptors are found on diverse cell types including T and B cells, DCs, granulocytes, monocytes and macrophages, epithelial and endothelial cells, and platelets (Antoniu, 2010b). IFN-γ is a potent inhibitor of IL-4-induced IgE production by B cells and IL-10 synthesis by monocytes, which results in increased production of TNF-α IL-1, platelet activating factor (PAF), and production of Reactive Oxygen Intermediates (ROI). IFN-γ also augments TNF-α effects on airway smooth muscle cells, upregulating RANTES synthesis.

The role of IFN-γ in asthma is not clearly understood. Consistent with its role as an inhibitor of T_H2 responses, production of IFN-γ is reported to be low in mild asthmatics (Chung and Barnes, 1999). Similarly, inhaled recombinant IFN-γ decreases the number of eosinophils in BAL fluid from asthmatic patients. In mice exposed to allergen, treatment with IFN-γ attenuates eosinophil infiltration in the airways and AHR, and knockout of the IFN-γ receptor increases airway eosinophil infiltration (Truyen et al., 2006). However, IFN-γ is not entirely protective and has been shown to stimulate the expression of cysteinyl leukotriene receptor CysLTR-2 on eosinophils, which can contribute to cysteinyl leukotriene-mediated bronchoconstrictive and inflammation (Chung and Barnes, 1999). In addition, sputum IFN-γ levels are reported to be higher in severe asthma than in mild asthma (Fujii et al., 2005).

IL-12

IL-12 is primarily produced by monocytes and macrophages, and to a lesser degree by B cells and DCs. IL-12 receptors are located mainly on T cells and NK cells, and stimulate T_H1 and NK cell growth, while inhibiting T_H2 cell responses. IL-12 augments cytotoxicity by T lymphocytes and NK cells, at least in part through the enhancement of IFN-γ and TNF-α secretion, and inhibition of IL-4 production. Consequently, IL-12 favors T_H1-type responses over T_H2-type and IgE inflammation, and it may serve as a regulator of IgE-related inflammation in pulmonary allergic disease (Truyen et al., 2006).

In patients with allergic asthma, levels of IL-12 in whole blood are reduced, and IL-12 stimulated IFN-γ production is less than in healthy controls. IL-12 mRNA is reduced in airway biopsies from allergic asthma patients, and the levels are increased by administration of corticosteroids. In mice, IL-12 treatment reduces eosinophilia, IgE release, and AHR, but only if administered during active sensitization. IL-12 administered postsensitization shows little or no difference from controls (Chung and Barnes, 1999). Thus, the effects of IL-12 are dependent upon the state of the allergic response.

3.6 Noncanonical Cytokines

IL-6

Originally described as a B-cell growth factor, IL-6 is now recognized as a pleiotropic cytokine synthesized by diverse cell types, including T and B cells, monocytes and macrophages, bone marrow stromal cells, keratinocytes, fibroblasts, eosinophils, endothelial cells, and epithelial cells. Production of IL-6 is augmented by IL-1, TNF-α, platelet-derived growth factor, and Lipopolysaccharides (LPS). IL-6 signals through a receptor complex comprised of gp130 and IL-6R, which is found primarily on leukocytes and hepatocytes (Chung and Barnes, 1999).

IL-6 plays a role in the differentiation, growth, and activation of T and B lymphocytes, and stimulates production of IgG, IgA, and IgM (Dienz and Rincon, 2009). IL-6 is crucial to IL-4-induced IgE production, enhances IL-2 production and CD4+ T cell survival, and promotes T_H2 cell differentiation (Akdis et al., 2011). In addition, IL-6 can potentially exert anti-inflammatory activity through the inhibition of IL-1 and TNF synthesis by macrophages (Chung and Barnes, 1999; Akdis et al., 2011).

In asthmatic patients, IL-6 secretion is elevated both at rest and after allergen challenge. In vitro, monocytes and macrophages release IL-6 in response to IgE. In transgenic mice expressing IL-6, mice show decreased AHR and increased numbers of airway lymphocytes (Chung and Barnes, 1999). A therapeutic antibody against IL-6, siltuximab, is in clinical trials for cancer, but its application to asthma has not yet been reported (Chung and Barnes, 1999).

Thymic Stromal Lymphopoietin

TSLP is a key initiator of allergic inflammation and considered to be one of the two major cytokines involved in mucosal T_H2 immune deviation along with IL-33. The cytokine is increased in asthmatic airways and mast cells (Hudes et al., 2012), and in the lungs is produced mainly by airway epithelial cells. Inflammation in allergic lung disease may be exacerbated by the ability of TSLP to react synergistically with IL-10 and TNF-α to stimulate mast cell growth and release of T_H2 cytokines. TSLP may also positively influence T_H2 cytokine release from CD4+ T cells. In addition, TSLP stimulates the production of T_H2 chemoattractants CCL17 and CCL22 from airway DCs, leading to T_H2 airway infiltration (Barnes, 2008; Bird, 2005).

In TSLP receptor-deficient mice, T_H2 cells, CCL11, and CCL17 in BAL were decreased and airway inflammation was reduced compared to wild-type mice. Adoptive transfer of wild-type CD4+ cells with normal TSLP receptors into the receptor-deficient mice restored airway inflammation, indicating TSLP involvement in T cell activation (Barnes, 2008). Overexpression of the Tslp transgene in lung epithelial cells results in airway infiltration by eosinophils and T_H2 cells, and development of AHR. Blocking OX40 ligand, an important mediator of TSLP signaling, inhibits allergic inflammation in mice and nonhuman primates (Bird, 2005). The findings with TSLP and OX40 ligand have not yet been translated into human therapy for asthma.

IL-22

In humans, IL-22 is primarily produced by T_H1 and T_H22 helper T cells, but it can also be produced by T_H17 cells, mast cells, and NK cells. In mice, however, IL-22 is predominately produced by T_H17 cells. IL-22 receptors are heterodimers composed of IL-22R1/IL-10R2 chains. IL22R1 is not expressed on immune cells, and the major role appears to be stimulation of hepatocyte production of acute phase proteins and stimulation of nonhematopoietic cell genes implicated in innate immunity, inflammation, chemotaxis, and cell proliferation (Barnes, 2008).

IL-22 can exhibit proinflammatory or anti-inflammatory activity depending on the tissue and the cytokine milieu (Sonnenberg et al., 2011). According to a study by Besnard et al. (2011), IL-22 can either promote or inhibit pulmonary allergic responses depending on the state of the allergic response (Barnes, 2008; Commins et al., 2010; Sonnenberg et al., 2011). Study investigators noted increased IL-22 protein concentrations in serum from asthmatic patients and increased IL-22 transcript and IL-22 protein in the lungs in an ovalbumin-induced murine model of asthma. Within the murine asthma model, Besnard et al. found that IL-22-KO mice exhibited significantly less airway inflammation accompanied by decreased IL-4, IL-13, IL-33, CCL11, and CCL17, as well as increased IL-17A compared to normal WT controls. Similarly, normal WT controls exhibited less inflammation if treated with an IL-22-neutralizing antibody during the sensitization phase, suggesting that IL-22 plays a pro-inflammatory role in the induction of airway inflammation. Once inflammation was established, however, administration of IL-22-neutralizing antibody amplified IL-17A and T_H2 cytokine production and cellular infiltration, whereas administration of exogenous IL-22 during this phase reduced IL-17A and intracellular infiltration. This suggests a protective, anti-inflammatory activity for IL-22 in established inflammation. In addition, there were indications of a reciprocal regulation between IL-22 and IL-17A, with IL-22 being required for IL-17A-induced inflammation.

IL-25

IL-25, or IL-17E, is produced by T_H2 cells, epithelial cells, activated mast cells, alveolar macrophages, activated eosinophils and basophils (Besnard et al., 2011). The receptor for IL-25 is IL-17BR (also known as EVI27), and activation results in airway inflammation mediated via CD4+ T cells and the STAT6 pathway. In vitro, IL-25 enhances the release of T_H2 cytokines from T_H2 cells both during and after differentiation. In mice, IL-25 increases expression of IL-4, IL-5, and IL-13; this leads to eosinophilic inflammation, elevation of IgE, mucus and epithelial cell hyperplasia, and AHR. Overexpression in the lung induces allergic T_H2 responses characterized by increased numbers of macrophages and eosinophils and mucus secretion. In contrast, IL-25 blockade attenuates allergen-induced AHR, lung eosinophilia, and antigen-specific T_H2 cells (Barnes, 2008; Pappu et al., 2008).

In patients with asthma, eosinophils and basophils are described as the major source of IL-25. IL-25 RNA is elevated in the lungs of chronic asthmatics, and IL-25 positive cells in the bronchial submucosa of these patients are reported to have eosinophil-like characteristics (Pappu et al., 2008). IL-25 and TSLP stimulate human T_H2 memory cells in the presence of DCs to produce IL-4, IL-5, and IL-13 (Monteleone et al., 2010). In addition to enhancing T_H2-related responses, IL-25 limits T_H1- and T_H17-related inflammatory responses associated with diseases such as multiple sclerosis, diabetes, and inflammatory bowel disease. Thus, while anti-IL-25 therapies have the potential to alleviate asthma and allergic inflammation, the approach might increase the risk for autoimmune-related activity (Pappu et al., 2008).

IL-33

IL-33 is considered to be one of major cytokines involved in T_H2 immune deviation. It is produced primarily by bronchial epithelial cells, but can also be produced by smooth muscle cells, macrophages, DCs, and fibroblasts (Monteleone et al., 2010). ST2, the IL-33 receptor, is located on T_H2 cells, mast cells, basophils, eosinophils, NK cells, and NKT cells. ST2 acts through an IL-1 receptor-related signaling cascade. The soluble form of the receptor, sST2, is produced in response to anti-inflammatory signals and acts to bind and reduce IL-33 activity. IL-33 is released during inflammation and tissue damage, and is therefore classified as an "alarmin" (Barnes, 2008; Commins et al., 2010).

IL-33 exerts both nuclear and extracellular effects. IL-33 aids in T_H2 cell differentiation by translocating to the nucleus, and altering chromatin structure and transcription. Extracellular effects of IL-33 include its role as a T_H2 activator and chemoattractant, and as a promoter of T_H2 cell production of IL-5 and IL-13. IL-33 is not limited to enhancement of T_H2 cytokine production from T_H2 cells; it can also induce IFN-γ production from T_H2 cells. The ability of IL-33 to enhance production of both IL-13 and IFN-γ provides a potential explanation for why some chronic asthma patients exhibit mixed T_H1/T_H2 cytokine profiles.

In addition to exerting effects on T_H2 T cell function, IL-33 aids in promoting eosinophil survival and adhesion to proteins in the ECM. IL-33 has also been shown to enhance mast cell survival and activation, as well as mast cell production of IL-6 and IL-13. These effects are further enhanced when IL-33 is accompanied by TSLP or IgE cross-linking (Smith, 2010).

In humans with asthma, there is a significant elevation in the concentration of sST2, possibly as a compensatory response to elevated IL-33 levels (Smith, 2010). In mice, treatment with IL-33 receptor antagonists downregulates T_H2 cytokines and allergen-induced inflammation, while treatment with IL-33 produces a T_H2 deviation, an escalation in T_H2 cytokines and IgE, and lung eosinophilia (Smith, 2010). Elucidation of IL-33's role in pulmonary allergic responses is an active area of research at this time.

3.7 Counterregulatory Mechanisms

Both adaptive and pathogenic immune responses are limited by dominant negative feedback pathways that regulate the duration and magnitude of inflammatory responses. The pivotal counterregulatory molecules that mediate these negative feedback pathways have been extensively studied as sources of potential therapeutics for chronic inflammatory disease, including asthma and other inflammatory diseases of the lung (Commins et al., 2010). Although counterregulatory molecules are numerous, several pathways have global impact in controlling the balance between normal immune homeostasis and pathogenic inflammatory disease and therefore have been the focus of intensive investigation. One centrally important set of counterregulatory molecules include the co-inhibitory CTLA-4 and PD-1 molecules of the CD28 family (Bosnjak et al., 2011; Saxon et al., 2008; Cook and Bochner, 2010; Chen and Shi, 2006). Another pivotal family of counterregulatory molecules includes cell surface proteins such as FcγRIIB, CD22, CD72, and Fc-receptor-like molecules (FCRLs), which possess immunoreceptor tyrosine-based inhibition motifs (ITIM) in their intracellular cytoplasmic domains (Singh et al., 2011; Bour-Jordan et al., 2011; Okazaki and Honjo, 2006; Kasagi et al., 2011). These receptors signal via recruitment and docking of tyrosine phosphatases such as SHP-1 (Src homology 2-containing protein tyrosine phosphatase 1) onto the ITIM. The CD200R-CD200 is another important counterregulatory receptor—ligand pair of cell surface proteins (Karra and Levi-Schaffer, 2011; Smith and Clatworthy, 2010; Zhu et al., 2010). CD200 specifically binds to the inhibitory CD200R receptor, which in turn trigger tyrosine phosphorylation and recruitment Dok2 and RasGAP to the intracellular domain of CD200R to limit cellular activation of lymphoid and myeloid subsets. Other counterregulatory systems include inflammation-associated induction of enzymes, such as inducible nitric oxide synthetase (NOS2), which elaborates the free radical nitric oxide; indoleamine 2,3-dioxygenase (IDO), which produces immunomodulatory catabolites of the tryptophan-kynurenine pathway; and prostaglandin-endoperoxide synthase 2 (PTGS2; also referred to as cyclooxygenase-2 or COX-2), which controls prostanoid biosynthesis (Minas and Liversidge, 2006; Gorczynski, 2005; Barclay et al., 2002).

Prostaglandins and other mediators that interact with seven-transmembrane domain (7TM) cell-surface receptors are coupled through heterotrimeric G proteins to adenylate cyclase and thereby regulate concentrations of intracellular cAMP. Elevation of intracellular cAMP inhibits cell-mediated immunity (Park and Christman, 2006; Tan and Bharath, 2009; Le and Broide, 2006; Prendergast et al., 2011; Sugiura and Ichinose, 2011; Coleman, 2002). These counterregulatory pathways are known to be important in modulating immune reactions in the lung and serve to balance adaptive clearance of infectious pathogens within the homeostatic bounds needed to preclude immunopathogenesis and hypersensitivity disease.

CTLA-4 and PD-1 Molecules of the CD28 Family

The CD28 family of cell surface proteins is comprised of both co-stimulatory molecules (e.g., CD28, ICOS) and co-inhibitory molecules (e.g., CTLA-4, PD-1) that are critical for induction and regulation of immune responses. CD28 is the prototypic co-stimulatory molecule that fulfills the requisite "secondary signal" to enable productive responses to the primary antigenic signal delivered through MHC-antigen interactions with the T-cell antigen receptor. CD28 is constitutively expressed on most $CD4^+$ T-helper cells as well as major subsets of $CD8^+$ T cells and interacts with B7.1 and B7.2 co-stimulatory molecules on professional antigen-presenting cells. CD28 is critical for stimulating activation and differentiation of naïve T cells and guides the differentiation of memory/effector T-cell subsets. During prolonged activation, CD28 is downregulated from the cell surface, and intracellular stores of CTLA4 (CD152) are mobilized to the cell surface. Because CTLA4 has higher affinity for B7.1 and B7.2, the activation-dependent expression the co-inhibitory CTLA4 preempts positive signaling through CD28 and limits the extent of T cell-dependent responses by blocking clonal expansion and cytokine production. The essential counterregulatory role of CTLA4 was revealed by the dramatic phenotype of CTLA4-deficient mice, which exhibited systemic lymphoproliferative disease, multiorgan autoimmunity, and death by 3–4 weeks of age. CTLA4-deficient mice, however, did not succumb to lethal lymphoproliferative disease when these mice also lacked both B7.1 and B7.2. The primary physiological ligand for CD28 and CTLA4 appears to be B7.2 and B7.1, respectively. This pairing is consistent with the observation that both CD28 and B7.2 have a high degree of constitutive expression, whereas both CTLA4 and B7.1 are largely inducible upon activation. Overall, the data support the concept that the balance of CD28/CTLA4 signaling is a central controller of immunity and tolerance.

PD-1 (PDCD-1, CD279) is another activation-dependent, co-inhibitory molecule that shares homology with CD28. PD-1 is expressed on activated T cells and B cells and interacts with two ligands, PD-L1 and PD-L2. The PD-L1 molecule is constitutively expressed on leukocytes and many nonhematopoietic cell types, whereas PD-L2 is primarily expressed as an activation-dependent molecule on myeloid-derived APC, such as monocytes and DCs. The role of PD-1 as a counterregulatory molecule was in part based on the phenotype of the PD-1 knockout mouse strain, which was characterized by heightened susceptibility to different organ-specific autoimmune diseases that varied based upon the genetic background of the strain. In the lung, PD-L2 is expressed abundantly on pulmonary macrophages and DCs in allergen-sensitized mice, and lack of PD-L2 expression or blockade of PD-L2 augments airway hyperreactivity. These and other studies have implicated the PD-1/PD-L2 pathway as a negative regulatory mechanism preventing pulmonary inflammatory disease and heightened airway hyperreactivity. The current view is that CTLA4 and PD-1 play complementary roles in limiting autoaggressive immune responses in that CTLA4 is hypothesized to halt the generation of autoreactive T cells, and PD-1 is thought to inhibit the responses of established effector cells. Overall, the CD28, CTLA4, and PD-1 molecules are being actively targeted for the development of new therapeutics to control chronic inflammatory disease (e.g., CTLA4-Ig fusion proteins abatacept and belatacept, humanized PD-1 monoclonal antibody MDX-1106).

FcγRIIB (CD32B)

FcγRIIB is a transmembrane Fc receptor that has preferential affinity for the Fc region of the IgG1 and IgG3 isotypes. FcγRIIB is expressed on many cell types including B cells, macrophages, activated neutrophils, mast cells, and basophils. Two isoforms of FcγRIIB exist in humans and mice (FcγRIIB1 and FcγRIIB2) that bear inhibitory ITIM motifs within the cytoplasmic domain. FcγRIIB1 is prominently expressed in B cells and serves to mediate feedback inhibition of antibody production so that the boosting of a memory B-cell response does not also stage an additional (and unnecessary) primary response. The FcγRIIB1 isoform may thereby contribute to the inhibitory efficacy of anti-RhD gamma globulin (prophylaxis for hemolytic disease of the newborn) and the mechanisms by which passive vaccination may inhibit active immunization. FcγRIIB may also contribute to the suppressive activity of intravenous immunoglobulin (IVIG).

The FcγRIIB2 isoform is prominently expressed on myeloid-derived phagocytes and may be pivotal in the noninflammatory clearance of immune complexes. Overall, the primary function of the inhibitory FcγRIIB is to block the stimulatory action of surface BCR, stimulatory Fcγ receptors, or other receptors that bear ITAM motifs in their cytoplasmic tail or that associate with stimulatory ITAM-containing signaling molecules such as the FcR common gamma chain. Cross-linking of the FcγRIIB with stimulatory cell surface BCR or Fcγ receptors via cell-bound immune complexes recruits SH2-domain-containing inositol phosphatases (SHIPs), particularly SHIP1, to dephosphorylate and inactivate signaling through these stimulatory surface receptors. By this mechanism, FcγRIIB increases the BCR activation threshold, suppresses antigen presentation by B cells and DC, decreases production of pro-inflammatory cytokines by macrophages, and mediates the noninflammatory clearance of antigen. In regard to mast cell and basophil physiology, IgG-antigen immune complexes that cross-link cell surface FcγRIIB with high-affinity, cell surface FcεR1/IgE complexes are known to inhibit IgE-mediated degranulation and release of T_H2 cytokines to profoundly mitigate allergen-driven type I hypersensitivity responses. Indeed, novel biologics are being designed to deliberately cross-link FcγRIIB with FcεR1--IgE-antigen complexes as potential therapeutic strategies for allergic asthma and other type I hypersensitivity disease.

CD200R–CD200

The CD200R–CD200 receptor–ligand system is emerging as a pivotal regulatory locus that broadly restrains inflammatory immune responses. CD200 has a relatively broad tissue and cell-type distribution; it is primarily expressed on B cells, DCs, and activated T cells as well as vascular endothelial cells and many nonhematopoietic cell types including cells (neurons) within the central nervous system and retina. CD200R has a more limited expression profile and is considered primarily a receptor expressed on leukocytes in both the myeloid and lymphoid lineages. CD200 binds exclusively to the inhibitory forms of the CD200R receptor, which lacks classical (ITIM) motifs within the intracellular domain but instead has alternative tyrosine-based motifs, which initiate inhibitory signal transduction circuits. Signaling through CD200R inhibits mast cell degranulation, cytokine release, and cellular infiltration together with increased attraction or expansion of T_{reg} subsets and diminution of T effector responses.

3.8 T_{reg} Cytokines

IL-10

IL-10, originally described as "cytokine synthesis inhibitor factor," is produced by T_H0, T_H1, T_H2, T_{reg}, cytotoxic T cells, mast cells, and activated monocytes. IL-10 receptors are located on lymphoid, myeloid, and NK cells. IL-10 decreases T_H2 signaling by downregulating MHC class II, B7.1/B7.2 and CD23 expression (Bjorgo et al., 2011; Bopp et al., 2009; Baumer et al., 2007; Heijink and Van Oosterhout, 2006). IL-10 suppresses monocyte and macrophage activity, including release of ROI, and downregulates secretion of pro-inflammatory cytokines such as IL-1β, TNF-α, IL-6, IL-8, and MIP-1α. In T_H1 cells, IFN-γ and IL-2 synthesis are inhibited, as is synthesis of IL-4 and IL-5 by T_H2 cells. In contrast, IL-10 can promote growth of mast cells, T cells, and B cells. However, even these responses can be bimodal, with inhibition observed at high concentrations of IL-10.

IL-10 mRNA and protein are reduced in alveolar macrophages from patients with asthma, although the number of IL-10 expressing T cells and macrophages may be increased. In Brown Norway rats, IL-10 is reported to attenuate late-phase responses and eosinophilia post-challenge. In mice, treatment with IL-10 at the time of sensitization decreases eosinophilia (Chung and Barnes, 1999). Studies involving antibodies against IL-10 and IL-10R, as well as IL-10 knockout, have revealed attenuation of AHR and eosinophilia. This apparent inconsistency in the typical T_H1 functioning of IL-10 might be explained by the finding that IL-10 can inhibit IL-13Rα2, a protein responsible for binding and inhibiting IL-13, thereby potentially increasing IL-13 effects. It has been noted that mice lacking both IL-10 and IL-13Rα2 exhibit a more severe manifestation of pulmonary allergic disease than mice deficient in IL-10 or IL-13Rα2 alone (Barnes, 2008).

Osteopontin

Osteopontin (OPN) is a T_H1 cytokine implicated as a potential regulator of allergic responses. OPN is produced by monocytes and monocyte-derived DCs and acts as a chemoattractant for monocytes, neutrophils, and lymphocytes (Finkelman et al., 2010). In vivo studies with patients with asthma have shown a significant correlation between OPN expression and eosinophilia in BAL fluid. OPN may also contribute to airway remodeling through the regulation of cellular interactions with ECM as well as the expression of MMP. In mice, OPN is positively correlated with airway remodeling and stimulates collagen deposition in cell cultures. In addition, ovalbumin sensitization and challenge of osteopontin-deficient mice results in an accentuated allergen-specific-IgE response. It has also been observed, however, that treatment with an anti-OPN antibody downregulates Ag-specific IgE and upregulates pDCs (Konno et al., 2011). This discrepancy in observations may have been due to the use of antibodies that recognized different domains on OPN, as it has been shown that the function of OPN can be dependent upon the ligated domain (Xanthou et al., 2007). In studies by Koguchi et al., the effects of OPN on IL-12 and IFN-γ production by PBMCs could be attenuated with a C-terminal-specific anti-OPN antibody, although not with an integrin-binding-domain anti-OPN Ab (Konno et al., 2011). It would seem, then, that effects exerted by OPN may change depending on the functional domain involved in the response. Although OPN may play a role in the allergic response pathway, more studies are needed to elucidate the diverse nature of these roles.

TGF-β

TGF-β is a pro-fibrotic growth factor produced by activated monocytes, pulmonary macrophages, eosinophils, neutrophils, airway epithelial cells, smooth muscle cells, mast cells, lung fibroblasts, T helper, and T_{reg} lymphocytes (Koguchi et al., 2003). In chronic allergic inflammation, eosinophils are a primary source of TGF-β (Chung and Barnes, 1999; Commins et al., 2010). TGF-β receptors include high-affinity (type I and type II) and low-affinity (type III) isoforms (Commins et al., 2010; Akdis and Akdis, 2009). TGF-β is a strong stimulating factor for epithelial cell differentiation and fibroblast secretion of ECM proteins, contributing to the airway remodeling observed in chronic pulmonary allergic disease pathology. In addition, TGF-β is also a powerful chemoattractant for monocytes, mast cells, and fibroblasts.

TGF-β has contrasting roles in allergic disease pathology. Depending on the cytokine milieu, it can either stimulate or inhibit proliferation of airway smooth muscle cells and fibroblasts (Chung and Barnes, 1999). Likewise, TGF-β can be pro-inflammatory by augmenting differentiation of T_H17 and T_H9/IL-9-producing-T_H2 cells, or anti-inflammatory by inhibiting IgE production, mast cell proliferation, T cell, B cell, NK cell, and cytotoxic lymphocyte functions (Chung and Barnes, 1999; Commins et al., 2010). One of the factors affecting the role played by TGF-β is peptidyl-prolyl isomerase Pin1 (protein NIMA-interacting one). Pin1, shown to regulate eosinophil GM-CSF production, has also been shown to regulate eosinophil TGF-β production. Compared with normal control animals, inhibition of Pin1 in rat models of allergic disease caused decreased expression of TGF-β1 and type I collagen, decreased eosinophilia in BAL fluid, and a decline in fibrotic deposits in the lungs (Akdis and Akdis, 2009). In human asthma patients, increases in TGF-β have been observed in BAL fluid following allergen challenge, and TGF-β expression seems to have a positive correlation with increased basement membrane thickness, subepithelial fibrosis, and severity of asthma symptoms (Shen et al., 2008).

4. THE ENVIRONMENT AND EPIGENETIC MODULATION OF IMMUNITY

The environment can have protective or deleterious effects on pulmonary immunity. As discussed previously, the prenatal and neonatal environments have ramifications on immune process through adulthood, with viral and antigen exposure setting the stage for allergic and asthma later in life. In contrast, microbial exposure in a farm setting can have protective effects (Renauld, 2001). Interestingly, SNPs commonly associated with asthma do not appear to be related to the protective effects of microbial exposure (Ege et al., 2011a). In recent years, epigenetic mechanisms have been invoked to explain long-lasting effects of the environment on immune function.

Epigenetic regulation includes DNA methylation, chromatin stabilization/destabilization (histone acetylation, methylation, phosphorylation, sumoylation, and ubiquitylation), and expression of microRNA (miRNA). In general, T_H cell differentiation is associated with a high degree of epigenetic control, and studies in humans and animals are beginning to provide direct evidence that environmental factors at critical times in utero, including diet and exposure to pathogens, increase the risk of asthma through epigenetic mechanisms (Ege et al., 2011b). Supplementation of the maternal diet with folic acid, a methyl donor, increases the risk of wheeze, lower respiratory tract infections, and diagnosis of asthma early in life (Hollingsworth et al., 2008; Whitrow et al., 2009). In a prospective study, Hollingsworth supplemented the diet of pregnant mice with methyl donors and found that methylation of Runx3, a gene that negatively impacts development of allergic asthma symptoms, was increased in the pups. Commensurate with the DNA modifications, the severity of allergen-induced respiratory symptoms was increased in the pups (Whitrow et al., 2009).

Once thought to be the result of mRNA degradation, miRNA is now recognized as a major regulatory mechanism that can lead to degradation of its target RNA via interaction with DICER, or sequestration of the target RNA in P-bodies. A preeminent role for miRNA in allergic airways disease is yet to be recognized, but diverse miRNA species have been associated with functional aspects of the disease, particularly immune responses and tissue remodeling (Table 1).

The microbiomes of the lung and gut may impact manifestation of asthma, with effects that either attenuate or augment symptoms depending on the flora. *Helicobacter pylori* in the gut has been shown to be protective against allergic airway inflammation in both mice and humans, with the resistance to allergic airway inflammation being linked to accumulation of T-regulatory cells in the lungs (Hollingsworth et al., 2008; Arnold et al., 2011). These and other similar findings have led to the speculation that

TABLE 1 miRNA Implicated in Asthma Allergic Airways Disease

miRNA	Observation
miR-1	Overexpressed in IL-13 transgenic mice
miR-16	Overexpressed in HDM stimulated mice
	Upregulated in stretch-induced hypertrophic human airway smooth muscle cells
miR-21	Overexpressed in IL-13 transgenic and HDM stimulated mice
miR-25	Downregulated in human airway smooth muscle cells stimulated with IL-1β, IFN-g, and TNF-α
	Antagomer inhibited RANTES production, increase TNF-α secretion in vitro
miR-26a	Upregulated in stretch-induced hypertrophic human airway smooth muscle cells
	Antagomer attenuated hypertrophy
	Overexpression induced hypertrophy
miR-29c	Decreased in OVA-stimulated mice
miR-126	Overexpressed in HDM stimulated mice
miR-133	Downregulates RhoA in human bronchial smooth muscle cells in vitro
	Downregulated in bronchial smooth muscle cells of OVA-stimulated mice
mi-R140	Downregulated in human airway smooth muscle cells stimulated with IL-1β, IFN-γ, and TNF-α
	Upregulated in stretch-induced hypertrophic human airway smooth muscle cells
miR-143	Decreased when MMP-12 is increased in OVA-stimulated mice
miR-146a	Upregulated in human bronchial epithelial cells stimulated with TGF-β, IL-1β, IFN-γ, and TNF-α epithelial cells in vitro
	Upregulated in human airway smooth muscle cells stimulated with IL-1β
	Expression promotes proliferation of human bronchial epithelial cells in vitro
miR-148a	Expression inhibits IL-1β-stimulated RNATES and IL-8 production in lung epithelial cells
miR-148b	Decreases HLA-G expression in vitro
miR-152	Decreases HLA-G expression in vitro
miR-188	Decreases HLA-G expression in vitro
miR-320	Downregulated in human airway smooth muscle cells stimulated with IL-1β, IFN-γ, and TNF-α
miR-672	Downregulated in human airway smooth muscle cells stimulated with IL-1β, IFN-γ, and TNF-α
	Decreased when MMP-12 is increased in OVA-stimulated mice

increased use of antibiotics and changes in diet have altered the gut micro flora and increased the prevalence of asthma in developed countries (Arnold et al., 2011).

Analysis of the lung microbiota has revealed that colonization by *Proteobacteria* spp, particularly *Haemophilus* species, is positively correlated with asthma (Beck et al., 2012). However, Brand and colleagues exposed pregnant mice to Gram-negative bacteria and demonstrated reduced prevalence of ovalbumin-induced asthma symptoms in the offspring that was associated with maintenance of histone 4 (H4) acetylation at the IFN-γ promoter, and decreased H4 acetylation at the IL-4 promoter (Bisgaard et al., 2007; Hilty et al., 2010; Han et al., 2012). As with other environmental influences, timing of the exposure and, in the case of bacteria, the location of the colonization may impact its effect on asthma pathogenesis.

5. AGING AND PULMONARY IMMUNE RESPONSES

Both lung function and immune responses decline in older adults, resulting in age-related changes in the manifestation of pulmonary diseases (Brand et al., 2011). During the first two decades of the life, the lungs physically grow and, as lung capacity increases, maximum forced expiratory flow (FEV_1) and volume (FVC) increase. From birth to adolescence, the lung tissue becomes stiffer, and elastic recoil increases. Thereafter, lung elastic recoil decreases, and the lungs become more flaccid as part of the normal aging process. In addition, with age, alveolar ventilation becomes more heterogeneous, physicochemical transfer of inspired gas to the blood (diffusing capacity) declines, and FEV_1 and FVC gradually decline. A variety of exogenous factors affect lung function throughout life, causing gradual and cumulative effects that are often not manifest until later life. These factors include exposure to environmental and industrial agents and residual effects of disease.

The normal decline in lung function is less well tolerated in asthmatic individuals who have a lower level of basal function due to chronic disease (Sharma and Goodwin, 2006; Colebatch et al., 1979; Wahba, 1983). In addition, responsiveness to pharmacologic agents changes, altering efficacy of therapeutic agents in the elderly. For example, bronchodilator efficacy decreases with age in some individuals as reversible bronchoconstriction transitions toward fixed obstruction. Also, zafirlukast, an oral leukotriene receptor antagonist, is less effective in improving lung function with age (Hanania et al., 2011).

Asthma-related inflammation becomes more severe with age, and the inflammatory cell profile is altered compared to the younger population (Busse and Mathur, 2010). Several factors contribute to this trend, including reduced capacity for mucociliary clearance of inhaled pathogens. Ciliary beating and particle clearance decreases from adolescence to old age, while production of mucins increases. The change in balance between mucus production and clearance increases the propensity for airway obstruction. Decreased mucus clearance also increases risk of infection (Busse and Mathur, 2010; Lee et al., 2012).

Immunosenescence that accompanies aging contributes to changes in the characteristics of pulmonary inflammation (Vignola et al., 2003). Immunosenescence results from somatic mutations and other forms of molecular damage coupled with decreased efficiency of DNA repair. Molecular damage affects cytokine expression and growth and function of diverse cell types, including epithelial cells, lymphocytes, eosinophils, neutrophils, natural killer cells, macrophages, and monocytes (Hanania et al., 2011; Busse and Mathur, 2010; Lee et al., 2012; Jones et al., 2011). Neutrophils are found in greater abundance in the airways of aging adults; comparisons of sputum reveals elevated neutrophil numbers in older asthmatic subjects, as well as increased mediators associated with neutrophilic inflammation including IL-8, neutrophil elastase, and MMP-9. Increased neutrophil elastase in the elderly may contribute to diminished recoil observed in the aging lung. Neutrophils from older adults also exhibit reduced chemotaxis, phagocytosis, and respiratory burst activity, which can affect killing and clearance of pathogens from the airways.

The contribution of eosinophils to asthma in the elderly is unclear. Eosinophils from older patients with asthma have been reported to exhibit decreased superoxide anion production. One study with aged mice exposed to antigenic sensitization and challenge reported a greater number of eosinophils in BAL fluid than in young mice, yet the aged mice experienced reduced airway responsiveness. In another study, the number of eosinophils in peripheral blood was elevated in aged animals, and the elevation correlated with AHR to methacholine. These discrepancies have not been fully explained at this time.

T-cell function also is affected by age. The thymus—the site of early T-cell development—undergoes involution and atrophies with age. It is currently unknown whether thymic involution causes age-dependent losses in cell-mediated immunity (Busse and Mathur, 2010; Vignola et al., 2003). The ratio of memory T cells compared to naïve T cells also increases as individuals grow older, most likely due to cumulative antigenic exposures. Age-dependent losses in T-cell function concur with reduced clonotypic diversity of the naïve T-cell repertoire and, in some cases, with oligoclonal expansions of memory T cells in the elderly. Potential reductions in clonotypic diversity may be a reason why certain infectious diseases, such as measles and chickenpox, are much more severe in adulthood.

The numbers of B-cell precursors in bone marrow decrease with age in humans (Lee et al., 2012; Vignola et al., 2003). Studies in mice indicate that antibody production is unaffected by age, but affinity and avidity of antibodies produced during a primary humoral immune response are decreased. It has been proposed that this may be due to limited somatic hypermutation, which plays a role in increasing antibody-antigen binding. Some of the age-related effects on B cells may be related to age-related changes in T-cell function because adoptive transfer of aged T-cells into younger recipient mice resulted in reduced B-cell activation, expansion, and IgG production.

Reports have indicated that the numbers of NK cells are increased in older adults, whereas the proliferative response of NKT cells is decreased with age, but both NK and NKT cells are reported to exhibit lower of cytotoxic capacity with age (Busse and Mathur, 2010). NK and NKT cell types are implicated in viral clearance, and impaired cytotoxicity and proliferative responses in the case of NKT cells may contribute to episodes of asthma induced by viral infection in elderly patients.

Monocyte and macrophage phagocytosis, chemotaxis, adherence, and antigen presentation show little or no difference with age (Busse and Mathur, 2010). However, toll-like receptor (TLR) expression and signaling, and ROI production are reduced. Monocytes from aged adults, when stimulated through TLR 1/2 heterodimer, exhibit decreased production of IL-6 and TNF-α (Vignola et al., 2003). Monocytes from older adults, when differentiated into macrophages in vitro, exhibit decreased TLR3 expression and response to virus, as well as less ROI and reactive nitrogen intermediates (RNI) production upon stimulation with LPS.

6. ANIMAL MODELS OF HUMAN ASTHMA

Respiratory allergies and asthma are considered human diseases and, with few exceptions, are not naturally occurring in laboratory animals. Yet, animal models have proved useful in investigating the pathways underlying pulmonary allergic responses, developing mechanistic hypotheses to explain pathogenesis, and testing novel therapeutics. Still, allergic reactions to aeroallergens in humans develop over the lifetime and change with environmental factors, and no animal model fully recapitulates the chronic human disease. Models of allergic asthma have been implemented in mice, rats, guinea pigs, rabbits, dogs, sheep, horses, and nonhuman primates, each with advantages and disadvantages (Table 2).

6.1 Mice

The murine model of asthma has been the most widely used in recent years due to the amount known about murine immunology, high availability of commercial species-specific reagents, ease of genetic manipulation leading to wide-availability of transgenic and knockout strains, low cost of husbandry, and relative low level of social sensitivity and federal regulatory oversight. Mice can be sensitized to a wide array of allergens such as OVA, and HDM, Aspergillus fumigatus, ragweed, and cockroach antigens (Baron et al., 2012). Mouse models incorporating AHR, bronchopulmonary inflammation, and increases in IgE and mucus production have been described, and some chronic models have shown airway remodeling—all key features of human asthma. Inflammation and AHR vary across murine models of asthma, due in some part to strain-dependent differences in airway smooth muscle and predisposition to T_H1 and T_H2 mediated inflammation (Zosky and Sly, 2007).

Disadvantages of the mouse model include distinct anatomical and immunological differences between mice and humans (Safholm et al., 2011). Mouse lungs have fewer airway generations, and most airways lack the smooth muscle bundles found in human airways, which may contribute to differences in bronchoconstriction and AHR. Second, mice can be manipulated to accentuate T_H1 or T_H2 phenotypes, and murine models of asthma are predominately T_H2 in nature. But there is a greater amount of overlap of these pathways in humans, and some human asthmatics have been observed to have increased levels of IL-12 and IFN-γ suggesting increased T_H1 involvement in these subjects (Wenzel and Holgate, 2006). Thirdly, while AHR has been observed in murine models of asthma, it is not clear whether or not mouse models reliably demonstrate late-phase responses as would be observed in humans (Wenzel and Holgate, 2006). In addition, many murine models of asthma lack chronicity in post-sensitization allergen responses. Following repeated allergen challenge, tolerance to the allergen and diminished immune responses have been observed in many of these models. Careful titration of allergen exposure is required to induce long-term hyperresponsiveness, airway remodeling, and eosinophilia in the mouse (Safholm et al., 2011).

Significant differences in FcεRI have been reported between humans and mice. In humans, IgE interacts with two "high-affinity" FcεRI receptors (Kumar and Foster, 2002). One FcεRI is tetrameric and is comprised of alpha, beta, and homodimeric disulfide-linked FcR-γ chains (αβγ2). This FcεRI is found primarily on mast cells and basophils, where IgE-mediated cross-linking of adjacent receptors leads to degranulation, eicosanoid production, cytokine secretion, and early and late-phase type I hypersensitivity reactions. The FcεRI alpha chain mediates high affinity binding of IgE, whereas the beta and gamma chains contain ITAM motifs in the cytoplasmic domains to mediate signal transduction and cellular activation.

A second "high-affinity" IgE receptor is comprised of the alpha chain and the FcR-γ homodimer but lacks the beta chain. This trimeric receptor (αγ2) is also expressed on mast cells and basophils and has broad expression on subsets of DCs, macrophages, eosinophils, neutrophils, and other non-leukocytic cell types including subsets of mucosal epithelial cells (Kraft and Kinet, 2007; Gould and Sutton, 2008; Galli et al., 2008). Engagement of FcεRI αγ2 on DCs and macrophages results in enhanced antigen presentation and cytokine production to bias T_H2 responses and allergic sensitization (Kraft and Kinet, 2007; Untersmayr et al., 2010).

The high-affinity FcεRI IgE receptors exhibit important species-specific differences (Kraft and Kinet, 2007; Sallmann et al., 2011; Maurer et al., 1996). In humans, surface expression of FcεRI does not require the beta chain and in many cases FcεRI is expressed in the absence of the beta chain. Nonetheless, the beta chain facilitates cell-surface FcεRI expression in human mast cells and basophils and amplifies signaling from the gamma-chain ITAM (Kraft and Kinet, 2007; Wu, 2011). In mice however, the beta chain appears to be an absolute requirement for expression of FcεRI. The mouse FcεRI αβγ2 receptor is restricted to mast cells and basophils, and FcεRI αγ2 trimer is not known to exist in mice.

TABLE 2 Advantages and Disadvantages of Animal Models of Asthma

Model		Advantages		Disadvantages
Small animal models	Mouse	Small and inexpensive species (Busse and Mathur, 2010)		Lack of chronicity in model (Karol, 1994)
		Many inbred strains available (Kumar and Foster, 2002)		Vasculature is the anaphylactic target (Karol, 1994)
		Wide variety of species-specific reagents available (Karol, 1994)		Poorly developed airway smooth muscle (Karol, 1994)
		IgE is the major class of anaphylactic antibody (Karol, 1994)		Poor response to histamine (Karol, 1994)
		Demonstrates AHR to carbachol (Martin et al., 1988)		
		Many transgenic and knockout strains available: (Garssen et al., 1991)		
		Cytokine TG/KO:	**Effects/findings:**	
		IL-13 TG Dox-inducible, driven by Clara cell secretory protein (CCSP)	Mucus metaplasia, accumulation of eosinophils in the airway, subepithelial fibrosis (Baron et al., 2012)	
		IL-9 TG driven by CCSP	Airway inflammation (eosinophils, lymphocytes), mast cell hyperplasia, bronchial hyperresponsiveness, subepithelial collagen deposition (Zhu et al., 1999)	
		IL-4 TG driven by CCSP	Mixed mononuclear and eosinophilic response, mucus metaplasia, mild fibrotic airway disease (Temann et al., 1998; McLane et al., 1998)	
		IL-4 KO	Lack of allergen-induced T_H2 cytokines, eosinophilia, Ag-specific IgE, and AHR following sensitization and challenge. (Rankin et al., 1996)	
		IL-5 TG driven by CCSP	Tissue eosinophilia, mucus metaplasia, subepithelial fibrosis, AHR (Brusselle et al., 1995, 1994; Coyle et al., 1995b)	
		IL-5 KO	Lack of antigen-induced AHR and eosinophilia (Lee et al., 1997)	
		IL-6 TG	Lymphocytic infiltration in the airways and decline in AHR (Foster et al., 1996)	
		IL-10 KO	Decreased AHR, IL-5, eosinophilia, and mucus production. Increased IL-12 and IFN-γ (DiCosmo et al., 1994)	
		IL-13Rα2 KO	Increased Th2 responses, and diminished AHR and mucus production (Yang et al., 2000; Justice et al., 2001)	

Continued

TABLE 2 Advantages and Disadvantages of Animal Models of Asthma—cont'd

Model	Advantages		Disadvantages
	IL-22 KO	Decreased airway inflammation, IL-4, IL-13, IL-33, CCL11, and CCL17; increased IL-17A (Ramalingam et al., 2008)	
	CD4-specific T_H2 locus control Region-KO	Decreased T_H2 cytokines, OVA-induced BAL eosinophils and lymphocytes, IgE, mucus production, and AHR. (Besnard et al., 2011)	
	IFN-γ receptor KO	Prolonged allergen-induced eosinophilia; increased production of IgE, IgG1, IL-4, IL-5; reduced IgG2a. (Koh et al., 2010)	
	TSLP TG overexpression in lung	Airway inflammation and spontaneous AHR following administration of methacholine, and lung infiltration of T_H2 lymphocytes and eosinophils (Coyle et al., 1996)	
	TSLP-receptor KO	Decreased T_H2 cells in BAL, allergic inflammation, CCL11 and CCL17. Reversed by adoptive transfer of CD4+ T cells (Zhou et al., 2005)	
	OPN KO	Decreased AHR, inflammation, collagen deposition, and mucus production (Al-Shami et al., 2005)	
	Transcription factor TG/KO:	**Effects/findings:**	
	T-bet KO	AHR, airway inflammation, BAL T_H2 cytokines, collagen deposition, and myofibroblast infiltration of airway wall (Simoes et al., 2009)	
	T-cell overexpression of dominant-negative (DN)-GATA-3	Decreased IL-4, IL-5, IL-13, airway eosinophilia, mucus secretion, and IgE production (Finotto et al., 2005)	
	Growth factor TG/KO:	**Effects/findings:**	
	Vascular endothelial growth factor (VEGF) TG Dox-inducible, driven by CCSP	Airway inflammation, remodeling, edema, mucus cell metaplasia, myocyte hyperplasia, AHR, enhanced Ag sensitization via IL-13-dependent and -independent pathways (Zhang et al., 1999)	
	GM-CSF TG	Lung macrophage proliferation, eosinophil infiltration, fibrosis, (Lee et al., 2004)	
	Allergen mediation/enzyme TG/KO:	**Effects/findings:**	
	BRP-39/Chitinase 3 KO	Decreased Ag-induced T_H2 inflammation, IL-13-induced inflammation and fibrosis, IgE production, DC accumulation and activation, and alternative macrophage activation (Shi et al., 2006)	

	Proteinase TG/KO:	**Effects/findings:**	
	Matrix metalloproteinase (MMP)-9 KO	Increased airway eosinophilia IL-4, IL-13, CCL11, CCL22 cytokine levels (Lee et al., 2009)	
	MMP-8 KO	Increased LPS-induced inflammation (McMillan et al., 2004)	
Rat	Small and inexpensive species (Quintero et al., 2010)		Lack of chronicity in model (Karol, 1994)
	IgE is the major class of anaphylactic antibody (Zosky and Sly, 2007)		Injection of antigen and adjuvants required for sensitization (Karol, 1994)
	Demonstrates EPR and LPR (Bellofiore and Martin, 1988; Watanabe and Hayashi, 1990)		Poor response to histamine (Watanabe and Hayashi, 1990)
	Demonstrates AHR lasting for days (Watanabe and Hayashi, 1990)		Poor response of tracheal smooth muscle to leukotrienes (Bellofiore and Martin, 1988)
	Responds to methacholine and cromolyn (Karol, 1994)		Serotonin-induced bronchoconstriction (Bellofiore and Martin, 1988; Watanabe and Hayashi, 1990)
Guinea pig	Small, relatively inexpensive (Watanabe and Hayashi, 1990)		Lack of chronicity in model (Karol, 1994)
	May be sensitized via inhalation (Andrew et al., 1984)		Fewer inbred strains available (Ratner et al., 1927)
	The lung is the major anaphylactic target (Shin et al., 2009)		Fewer species-specific reagents available (Karol, 1994)
	Airways respond to histamine (Shin et al., 2009)		IgG1 is the major class of anaphylactic antibody (Griffiths-Johnson and Karol, 1991)
	Both EPR and LPR are present (Griffiths-Johnson et al., 1993)		Poor response to cromolyn (Karol, 1983)
	Eosinophilic inflammation during LPR (Karol, 1994)		Serotonin-induced bronchoconstriction (Van Oosterhout et al., 1993)
	Neutrophilia in lung following LPR (Ressmeyer et al., 2006)		Acute/fatal responses to chemical irritants (Van Oosterhout et al., 1993)
	AHR following airway hypersensitivity response (Zosky and Sly, 2007)		Lack of species-specific reagents compared to mice (Van Oosterhout et al., 1993)
	Tracheal smooth muscle responds to histamine (Karol, 1994)		
Rabbit	The lung is the major anaphylactic target (Griffiths-Johnson and Karol, 1991)		

Continued

TABLE 2 Advantages and Disadvantages of Animal Models of Asthma—cont'd

Model		Advantages	Disadvantages
			Neonatal immunization is necessary for development of LPR (Karol, 1994)
		IgE is the major class of anaphylactic antibody (Larsen, 1985)	Lack of species-specific reagents compared to mice
		Development of EPR and LPR (Larsen, 1985)	
Large animal models	Dog	IgE is major class of anaphylactic antibody (Metzger, 1990)	Individual variability in response to *Ascaris suum* (Karol, 1994)
		Natural sensitivity to *A. suum* (Booth et al., 1970)	Lack of LPR (Booth et al., 1970)
		Development of AHR (Karol, 1994)	Lack of species-specific reagents compared to mice
	Horse	Natural sensitivity to "barn" environment (Chung et al., 1985)	Large animal, expensive (Lowell, 1964)
		Observable AHR (Karol, 1994)	Lack of species-specific reagents compared to mice
		Recurrent episodes of clinical disease (Karol, 1994)	
	Sheep	Natural sensitivity to *A. suum* (Karol, 1994)	Large animal, expensive (Abraham et al., 1983)
		Development of EPR and LPR (Karol, 1994)	Lack of species-specific reagents compared to mice (Abraham et al., 1983)
		Demonstrates AHR to carbachol (Zosky and Sly, 2007)	
	Non-human primate	Genetic and physiological similarity to humans (Lanes et al., 1986)	Expensive species (Coffman and Hessel, 2005)
		Consistent response to antigen (Karol, 1994)	
		Natural sensitivity to *A. suum* (Karol, 1994)	
		Allergic responsiveness may persist for years (Weiszer et al., 1968; Patterson and Harris, 1992)	
		IgE is the major class of anaphylactic antibody (Coffman and Hessel, 2005)	
		Wide variety of species-specific reagents available (Patterson and Harris, 1992)	

AHR, airway hyperresponsiveness; EPR, early-phase response; LPR, late phase response; IgE, immunoglobulin E; Ag, antigen.
Adapted from Karol (Coffman and Hessel, 2005) and Baron (Karol, 1994).

Mice appear to have other mechanisms, not present in humans, to enable binding of IgE immune complexes by a broad array of cell types (Donnadieu et al., 2000; On et al., 2004). For example, the "low-affinity" CD16 FcγRIII and CD32 FcγRII receptors together with another receptor not present in humans (FcγRIV) bind IgG but also serve as low-affinity IgE receptors. These low-affinity receptors cannot bind monomeric IgE but readily bind multivalent IgE in immune complexes. The mouse-specific FcγRIV receptor is of particular interest. FcγRIV is an activating receptor that is expressed with the FcR-γ homodimer, binds monomeric IgG2a or IgG2b with intermediate-affinity, and is primarily expressed on macrophages and neutrophils. Monomeric IgG bound to FcγRIV is readily displaced by multimeric IgE immune complexes to initiate antigen presentation and cellular responses. Given the convergence of function, the mouse FcγRIV and human FcεRI αγ2 receptors may have evolved to fulfill a similar physiological niche (Jonsson et al., 2012; Mancardi et al., 2008). Both species of receptor are present on macrophages and neutrophils, and when bound by IgE immune complexes, causes TNF-α production, production of inflammatory mediators, phagocytosis, antigen presentation, and IgE-dependent inflammatory responses. In the mouse, recognition of IgE-antigen immune complexes is facilitated by the FcγRIIB and FcγRIIIA receptors that also act as low-affinity receptors for multimeric IgE.

Immunological pathways shown to be important in mouse models have not easily translated to new treatments for human disease. Early clinical trials with anti-IL-4 and anti-IL-5 antibodies yielded disappointing results, and lack of efficacy in preventing asthma symptoms is often cited as lack of fidelity between mouse models and human asthma (Mancardi et al., 2008). However, more recent studies with altered experimental design, modified exclusion criteria, use of biomarkers to identify responsive asthmatic subjects, refinement of pharmacodynamic measures, and optimization of dosing has yielded encouraging results. For example, pitrakinra exhibits greater efficacy in asthmatics with specific SNPs in IL-4Rα (Wenzel and Holgate, 2006), (Slager et al., 2012), and Mepolizumab is somewhat more effective in steroid-refractory patients with hypereosinophilia than in the general population of mild and moderate asthmatics (Slager et al., 2012). This controversy aside, the mouse remains an important tool for elucidating pathways involved in immune responses and identifying potential novel targets.

6.2 Rats

Rats are small and inexpensive, though large enough to obtain ample serum and BAL fluid for analysis. In addition, they are large enough to be stable under anesthesia, which facilitates measurements of pulmonary resistance, compliance, and AHR (Wenzel, 2009; Nair et al., 2009; Haldar et al., 2009). While rats have less species-specific reagents than mice and offer fewer transgenic and knockout strains, available rat transgenic technologies and rat-specific reagents have increased more recently. Rats can be sensitized to a wide array of allergens including OVA, HDM, and Ascaris antigens resulting in a T_H2 eosinophilic inflammation and IgE responses, although the strength of the responses can be strain-specific. Brown Norway rats, for example, exhibit acute early and late-phase responses, and AHR, and have shown more robust inflammatory and IgE responses compared with other strains. As with mice, however, a shortcoming of the rat model is the relative ease with which the animals can develop tolerance to allergen. From another perspective, induction of tolerance in rat models has facilitated delineation of the role T_{reg} cells play in modulating allergic responses. Lack of reagents specific for rats ultimately limits their usefulness in asthma research.

6.3 Guinea Pigs

Guinea pigs, utilized in some of the earliest models of allergic airway responses, present early and late phase responses with concurrent AHR and eosinophilia, much like human manifestations of the disease, but few inbred strains and species-specific reagents are available (Zosky and Sly, 2007). Guinea pigs can be easily sensitized to OVA and exhibit AHR and eosinophilia (Shin et al., 2009). Studies comparing lung slices of guinea pigs to those of human and rodents found better correlation between guinea pig and human airway responses to pharmacological agents than in other rodents. Following sensitization, the response of guinea pigs to methacholine, histamine, and allergen are similar to responses observed in human airways, but serotonin also induces marked bronchoconstriction in guinea pigs that is not seen in humans. In addition, differences are observed in response to leukotrienes, indicating an underlying mechanistic distinction between humans and guinea pigs. As with mice and rats, guinea pigs develop tolerance to allergen after repeated exposure. Guinea pigs also have elevated numbers of eosinophils prior to sensitization and challenge (Zosky and Sly, 2007). Also noteworthy is that in chemical irritant studies, the immunological response of guinea pigs is often acute and sometimes fatal, and pretreatment with antihistamines is often required to dampen the severity of the immune response.

6.4 Rabbits

Rabbit asthma models exhibit early and late phase responses; and as with humans, the rabbit lung is the primary site of IgE-mediated anaphylactic responses (Zosky and Sly, 2007). An early phase response to allergen characterized by increased lung resistance and decreased dynamic compliance is observed within 15 min of challenge in this model, followed by a late phase response starting within 2 h. In this model, neonatal immunization is facilitates development of the late phase response.

6.5 Dogs

Dogs, unlike rodents and rabbits, may become naturally sensitized to antigens that also affect the human population (Shin et al., 2009; Karol, 1994; Zosky and Sly, 2007). Such naturally occurring allergic reactions in dogs do not typically induce asthma-like symptoms despite an elevation in airway eosinophilia. Instead, the responses present as dermatitis or conjunctivitis (Zosky and Sly, 2007). One factor that may influence the absence of airway responses in these cases is the anatomy of the dog airway. The dog airway is proportionally larger than that of many other mammals, and perhaps less-likely to constrict due to allergen exposure.

In contrast to the lack of an airway response in naturally occurring allergic reactions, there are canine models of asthma that show airway constriction and AHR in addition to increased IgE and eosinophilia. The Basenji Greyhound model, for example, shows AHR as well as responsiveness to methacholine, though baseline AHR is elevated in this model and AHR does not always mark clinical disease (Zosky and Sly, 2007). Allergic responses with elevated IgE levels can be optimized in dogs via selective breeding. Such models show AHR post-challenge and increased responsiveness for up to 5 months, a significant distinction from the aforementioned models (Karol, 1994). Although canine models have the potential for evaluating chronic responses to allergen, they are relatively expensive, require additional time for breeding and sensitization, and, as with other larger-animal models, there are fewer species-specific reagents.

6.6 Sheep

Sheep present early and late phase responses that involve AHR and eosinophilia (Zosky and Sly, 2007). As seen with some dogs, sheep can become naturally sensitized to *Ascaris suum,* though these responses vary from animal to animal (Karol, 1994). Some sheep respond with an early phase response only, while 30–50% respond with both early and late phase responses, mimicking the variability found in the human population. AHR is observed in sheep exhibiting a late phase response, within 24 h post–challenge. The allergic response in sheep and the timing of mediators observed during allergic pathogenesis, particularly LTE_4, shares many similarities with human asthma patients (Zosky and Sly, 2007). As in humans, cromolyn and corticosteroids are effective in sheep. Dissimilar to humans, however, is that allergic responses in sheep lungs may be ameliorated through the use of drugs targeting platelet activating factors (Zosky and Sly, 2007). Despite the similarities to aspects of human asthma pathogenesis, sheep, like dogs, are larger and much more costly to maintain than smaller species and fewer species-specific reagents are available.

6.7 Horses

Horses, naturally sensitized to a "barn environment," develop AHR and recurrent episodes of disease (Zosky and Sly, 2007). Following exposure to hay or barn dust, horses show elevated breathing rates and decreased tidal volume during an acute response (Shin et al., 2009). Increased lung resistance and decreased dynamic compliance have also been observed. In addition, sensitized horses show serum antibodies to allergen, goblet cell metaplasia, increases in airway smooth muscle, and inflammatory cell infiltration. Despite the similarities to aspects of human asthma pathogenesis especially the spontaneous recurrence of disease, horses, as with other larger animal species, are more costly and fewer species-specific reagents are available.

6.8 Non-Human Primates

Non-human primates (NHPs) arguably share the most similarity with humans and have been used in asthma studies for over 40 years. While some NHPs are naturally sensitive to Ascaris antigen, most protocols involve active sensitization with antigens implicated in human asthma (Karol, 1994). Humans and NHPs share many genetic, immunological, and physiological similarities, and human reagents can often be used in NHPs (Karol, 1994; Coffman and Hessel, 2005). NHP and human lungs are similar in size, anatomy, and ultrastructure; and pulmonary function measurements that can be made, the instruments that are used, and the techniques are similar as well. Many of the characteristics central to human asthma have been reported in monkeys, including allergen-specific IgE cutaneous and airway responses, elevated T_H2 cells in the airways, eosinophilia, goblet cell hyperplasia, airway remodeling, and histamine and methacholine sensitivity (Coffman and Hessel, 2005). In the NHP Ascaris and HDM models, immunological changes that occur due to an inhaled allergen are nearly identical to those that occur in similarly challenged human asthma patients (Hyde et al., 2006). In addition, in studies examining bronchoconstriction in precision cut lung slices, different NHP species (common marmosets, cynomolgus macaques, rhesus macaques, and anubis baboons) responded to agents pertinent to human airway diseases including methacholine, histamine, U46619, and endothelin-1 (Van Scott et al., 2004, 2005; Coffman and Hessel, 2005; Ayanoglu et al., 2011). Only cynomolgus macaques and baboons could generate responses to leukotrienes, perhaps indicating a closer connection between airway response mechanisms for human subjects and these two species than other NHP species. While NHPs offer many advantages in modeling human pathophysiology, in addition to cost and social sensitivities, their social environment is complex, and responses to social stress can be very different from humans, which can affect their responses to corticosteroids.

7. CONCLUSION

Extensive progress has been made in defining the immunological processes involved in allergic lung diseases, though complexities in development and regulation of immune processes, as well as complexities in the interaction of the immune system with cellular and molecular pathways controlling development and function of the pulmonary system, have thwarted development of novel efficacious therapeutics. Two factors that could enhance drug development in this area are: (1) emerging appreciation for distinct genotypes and phenotypes in asthma, and (2) elucidation of innate immune pathways that parallel, duplicate, enhance, and oppose adaptive immune pathways. Definition of phenotype-specific pathogenic factors has shown some promise in establishing effectiveness of agents targeting the IL-4 and IL-13 pathways. Elucidation of the role of innate immunity in asthma is leading to better understanding of how lipids and viruses can induce exacerbations in allergic asthmatics, even when their allergies are otherwise well controlled. However, other than some encouraging results from animal studies with agents that interrupt inflammation induced via the OX40 pathway (Seehase et al., 2011), and modest effects of Pitrakinra (Kaur and Brightling, 2012), MEDI-528 (Antoniu, 2010b), and Mepolizumab (Antoniu, 2010a; Parker et al., 2011), little has emerged in terms of translating the basic observations into effective and affordable therapeutics.

REFERENCES

Abraham, W.M., Delehunt, J.C., Yerger, L., Marchette, B., 1983. Characterization of a late phase pulmonary response after antigen challenge in allergic sheep. Am. Rev. Respir. Dis. 128 (5), 839–844.

Afshar, R., Medoff, B.D., Luster, A.D., 2008. Allergic asthma: a tale of many T cells. Clin. Exp. Allergy 38 (12), 1847–1857.

Agaugue, S., Marcenaro, E., Ferranti, B., Moretta, L., Moretta, A., 2008. Human natural killer cells exposed to IL-2, IL-12, IL-18, or IL-4 differently modulate priming of naive T cells by monocyte-derived dendritic cells. Blood 112 (5), 1776–1783.

Akdis, C.A., Akdis, M., 2009. Mechanisms and treatment of allergic disease in the big picture of regulatory T cells. J. Allergy Clin. Immunol. 123 (4), 735–746; quiz 747–748.

Akdis, M., Burgler, S., Crameri, R., Eiwegger, T., Fujita, H., Gomez, E., Klunker, S., Meyer, N., O'Mahony, L., Palomares, O., Rhyner, C., Ouaked, N., Schaffartzik, A., Van De Veen, W., Zeller, S., Zimmermann, M., Akdis, C.A., 2011. Interleukins, from 1 to 37, and interferon-gamma: receptors, functions, and roles in diseases. J. Allergy Clin. Immunol. 127 (3), 701–721; e1–e70.

Akhter, J., Gaspar, M.M., Newcomb, R.W., 1989. Persistent peripheral airway obstruction in children with severe asthma. Ann. Allergy 63 (1), 53–58.

Al-Muhsen, S., Johnson, J.R., Hamid, Q., 2011. Remodeling in asthma. J. Allergy Clin. Immunol. 128 (3), 451–462; quiz 463–464.

Al-Shami, A., Spolski, R., Kelly, J., Keane-Myers, A., Leonard, W.J., 2005. A role for TSLP in the development of inflammation in an asthma model. J. Exp. Med. 202 (6), 829–839.

Andrew, D.K., Schellenberg, R.R., Hogg, J.C., Hanna, C.J., Pare, P.D., 1984. Physiological and immunological effects of chronic antigen exposure in immunized guinea pigs. Int. Arch. Allergy Appl. Immunol. 75 (3), 208–213.

Antoniu, S.A., 2010a. MEDI-528, an anti-IL-9 humanized antibody for the treatment of asthma. Curr. Opin. Mol. Ther. 12 (2), 233–239.

Antoniu, S.A., 2010b. Pitrakinra, a dual IL-4/IL-13 antagonist for the potential treatment of asthma and eczema. Curr. Opin. Investig. Drugs 11 (11), 1286–1294.

Appleton, S.L., Adams, R.J., Wilson, D.H., Taylor, A.W., Ruffin, R.E., North West Adelaide Health Study Team, 2006. Central obesity is associated with nonatopic but not atopic asthma in a representative population sample. J. Allergy Clin. Immunol. 118 (6), 1284–1291.

Arnold, I.C., Dehzad, N., Reuter, S., Martin, H., Becher, B., Taube, C., Muller, A., 2011. *Helicobacter pylori* infection prevents allergic asthma in mouse models through the induction of regulatory T cells. J. Clin. Invest. 121 (8), 3088–3093.

Ayanoglu, G., Desai, B., Fick Jr., R.B., Grein, J., de Waal Malefyt, R., Mattson, J., McClanahan, T., Olmstead, S., Reece, S.P., Van Scott, M.R., Wardle, R.L., 2011. Modelling asthma in macaques: longitudinal changes in cellular and molecular markers. Eur. Respir. J. 37 (3), 541–552.

Banner, K.H., Igney, F., Poll, C., 2011. TRP channels: emerging targets for respiratory disease. Pharmacol. Ther. 130 (3), 371–384.

Barclay, A.N., Wright, G.J., Brooke, G., Brown, M.H., 2002. CD200 and membrane protein interactions in the control of myeloid cells. Trends Immunol. 23 (6), 285–290.

Barlow, J.L., McKenzie, A.N., 2011. Nuocytes: expanding the innate cell repertoire in type-2 immunity. J. Leukoc. Biol. 90 (5), 867–874.

Barnes, P.J., 2001. Neurogenic inflammation in the airways. Respir. Physiol. 125 (1–2), 145–154.

Barnes, P.J., 2002. Current and future therapies for airway mucus hypersecretion. Novartis Found. Symp. 248, 237–249; discussion 249–253, 277–282.

Barnes, P.J., 2008. The cytokine network in asthma and chronic obstructive pulmonary disease. J. Clin. Invest. 118 (11), 3546–3556.

Baron, R.M., Choi, A.J., Owen, C.A., Choi, A.M., 2012. Genetically manipulated mouse models of lung disease: potential and pitfalls. Am. J. Physiol. Lung Cell. Mol. Physiol. 302 (6), L485–L497.

Bartemes, K.R., Kita, H., 2012. Dynamic role of epithelium-derived cytokines in asthma. Clin. Immunol. 143 (3), 222–235.

Baumer, W., Hoppmann, J., Rundfeldt, C., Kietzmann, M., 2007. Highly selective phosphodiesterase 4 inhibitors for the treatment of allergic skin diseases and psoriasis. Inflamm. Allergy Drug Targets 6 (1), 17–26.

Beadling, C., Johnson, K.W., Smith, K.A., 1993. Isolation of interleukin 2-induced immediate-early genes. Proc. Natl. Acad. Sci. U.S.A 90 (7), 2719–2723.

Beadling, C., Smith, K.A., 2002. DNA array analysis of interleukin-2-regulated immediate/early genes. Med. Immunol. 1 (1), 2.

Beck, J.M., Young, V.B., Huffnagle, G.B., 2012. The microbiome of the lung. Transl. Res. 160 (4), 258–266.

Bellofiore, S., Martin, J.G., 1988. Antigen challenge of sensitized rats increases airway responsiveness to methacholine. J. Appl. Physiol. 65 (4), 1642–1646.

Bellora, F., Castriconi, R., Dondero, A., Reggiardo, G., Moretta, L., Mantovani, A., Moretta, A., Bottino, C., 2010. The interaction of human natural killer cells with either unpolarized or polarized macrophages results in different functional outcomes. Proc. Natl. Acad. Sci. U.S.A 107 (50), 21659–21664.

Berend, N., Salome, C.M., King, G.G., 2008. Mechanisms of airway hyperresponsiveness in asthma. Respirology 13 (5), 624–631.

Bergeron, C., Al-Ramli, W., Hamid, Q., 2009. Remodeling in asthma. Proc. Am. Thorac. Soc. 6 (3), 301–305.

Besnard, A.G., Sabat, R., Dumoutier, L., Renauld, J.C., Willart, M., Lambrecht, B., Teixeira, M.M., Charron, S., Fick, L., Erard, F., Warszawska, K., Wolk, K., Quesniaux, V., Ryffel, B., Togbe, D., 2011. Dual role of IL-22 in allergic airway inflammation and its cross-talk with IL-17A. Am. J. Respir. Crit. Care Med. 183 (9), 1153–1163.

Betts, R.J., Kemeny, D.M., 2009. CD8+ T cells in asthma: friend or foe? Pharmacol. Ther. 121 (2), 123–131.

Bird, L., 2005. Asthma and allergy: TSLP: key role in allergic responses confirmed. Nat. Rev. Immunol. 5, 747.

Bisgaard, H., Hermansen, M.N., Buchvald, F., Loland, L., Halkjaer, L.B., Bonnelykke, K., Brasholt, M., Heltberg, A., Vissing, N.H., Thorsen, S.V., Stage, M., Pipper, C.B., 2007. Childhood asthma after bacterial colonization of the airway in neonates. N. Engl. J. Med. 357 (15), 1487–1495.

Bjorgo, E., Moltu, K., Tasken, K., 2011. Phosphodiesterases as targets for modulating T-cell responses. Handb. Exp. Pharmacol. 204, 345–363.

Blanchfield, J.L., Mannie, M.D., 2010. A GMCSF-neuroantigen fusion protein is a potent tolerogen in experimental autoimmune encephalomyelitis (EAE) that is associated with efficient targeting of neuroantigen to APC. J. Leukoc. Biol. 87 (3), 509–521.

Booth, B.H., Patterson, R., Talbot, C.H., 1970. Immediate-type hypersensitivity in dogs: cutaneous, anaphylactic, and respiratory responses to ascaris. J. Lab. Clin. Med. 76 (2), 181–189.

Bopp, T., Dehzad, N., Reuter, S., Klein, M., Ullrich, N., Stassen, M., Schild, H., Buhl, R., Schmitt, E., Taube, C., 2009. Inhibition of cAMP degradation improves regulatory T cell-mediated suppression. J. Immunol. 182 (7), 4017–4024.

Bosnjak, B., Stelzmueller, B., Erb, K.J., Epstein, M.M., 2011. Treatment of allergic asthma: modulation of Th2 cells and their responses. Respir. Res. 12, 114.

Bour-Jordan, H., Esensten, J.H., Martinez-Llordella, M., Penaranda, C., Stumpf, M., Bluestone, J.A., 2011. Intrinsic and extrinsic control of peripheral T-cell tolerance by costimulatory molecules of the CD28/B7 family. Immunol. Rev. 241 (1), 180–205.

Bousquet, J., Jacot, W., Yssel, H., Vignola, A.M., Humbert, M., 2004. Epigenetic inheritance of fetal genes in allergic asthma. Allergy 59 (2), 138–147.

Brand, S., Teich, R., Dicke, T., Harb, H., Yildirim, A.O., Tost, J., Schneider-Stock, R., Waterland, R.A., Bauer, U.M., von Mutius, E., Garn, H., Pfefferle, P.I., Renz, H., 2011. Epigenetic regulation in murine offspring as a novel mechanism for transmaternal asthma protection induced by microbes. J. Allergy Clin. Immunol. 128 (3), 618–625 e1–e7.

Broide, D.H., Lotz, M., Cuomo, A.J., Coburn, D.A., Federman, E.C., Wasserman, S.I., 1992. Cytokines in symptomatic asthma airways. J. Allergy Clin. Immunol. 89 (5), 958–967.

Brusselle, G., Kips, J., Joos, G., Bluethmann, H., Pauwels, R., 1995. Allergen-induced airway inflammation and bronchial responsiveness in wild-type and interleukin-4-deficient mice. Am. J. Respir. Cell. Mol. Biol. 12 (3), 254–259.

Brusselle, G.G., Kips, J.C., Tavernier, J.H., van der Heyden, J.G., Cuvelier, C.A., Pauwels, R.A., Bluethmann, H., 1994. Attenuation of allergic airway inflammation in IL-4 deficient mice. Clin. Exp. Allergy 24 (1), 73–80.

Busse, W.W., Lemanske, R.F., 2007. Expert Panel Report 3: Guidelines for the Diagnosis and Management of Asthma. NIH Publication No. 07–4051. Bethesda, MD: U.S. Department of Health and Human Services; National Institutes of Health; National Heart, Lung, and Blood Institute; National Asthma Education and Prevention Program.

Busse, P.J., Mathur, S.K., 2010. Age-related changes in immune function: effect on airway inflammation. J. Allergy Clin. Immunol. 126 (4), 690–699; quiz 700–701.

Busse, W.W., Ring, J., Huss-Marp, J., Kahn, J.E., 2010. A review of treatment with Mepolizumab, an anti-IL-5 mAb, in hypereosinophilic syndromes and asthma. J. Allergy Clin. Immunol. 125 (4), 803–813.

Caceres, A.I., Brackmann, M., Elia, M.D., Bessac, B.F., del Camino, D., D'Amours, M., Witek, J.S., Fanger, C.M., Chong, J.A., Hayward, N.J., Homer, R.J., Cohn, L., Huang, X., Moran, M.M., Jordt, S.E., 2009. A sensory neuronal ion channel essential for airway inflammation and hyperreactivity in asthma. Proc. Natl. Acad. Sci. U.S.A 106 (22), 9099–9104.

Carroll, N., Elliot, J., Morton, A., James, A., 1993. The structure of large and small airways in nonfatal and fatal asthma. Am. Rev. Respir. Dis. 147 (2), 405–410.

Castro, M., Mathur, S., Hargreave, F., Boulet, L.P., Xie, F., Young, J., Wilkins, H.J., Henkel, T., Nair, P., Res-5-0010 Study Group, 2011. Reslizumab for poorly controlled, eosinophilic asthma: a randomized, placebo-controlled study. Am. J. Respir. Crit. Care Med. 184 (10), 1125–1132.

Chen, Y.Q., Shi, H.Z., 2006. CD28/CTLA-4–CD80/CD86 and ICOS–B7RP-1 costimulatory pathway in bronchial asthma. Allergy 61 (1), 15–26.

Chung, K.F., Barnes, P.J., 1999. Cytokines in asthma. Thorax 54 (9), 825–857.

Chung, K.F., Becker, A.B., Lazarus, S.C., Frick, O.L., Nadel, J.A., Gold, W.M., 1985. Antigen-induced airway hyperresponsiveness and pulmonary inflammation in allergic dogs. J. Appl. Physiol. 58 (4), 1347–1353.

Cockcroft, D.W., 1999. Pharmacologic therapy for asthma: overview and historical perspective. J. Clin. Pharmacol. 39 (3), 216–222.

Coe, C.L., 1993. Psychosocial factors and immunity in nonhuman primates: a review. Psychosom. Med. 55 (3), 298–308.

Coe, C.L., Erickson, C.M., 1997. Stress decreases lymphocyte cytolytic activity in the young monkey even after blockade of steroid and opiate hormone receptors. Dev. Psychobiol. 30 (1), 1–10.

Coe, C.L., Ershler, W.B., 2001. Intrinsic and environmental influences on immune senescence in the aged monkey. Physiol. Behav. 73 (3), 379–384.

Coe, C.L., Laudenslager, M.L., 2007. Psychosocial influences on immunity, including effects on immune maturation and senescence. Brain Behav. Immun. 21 (8), 1000–1008.

Coffman, R.L., 2006. Origins of the T(H)1-T(H)2 model: a personal perspective. Nat. Immunol. 7 (6), 539–541.

Coffman, R.L., Hessel, E.M., 2005. Nonhuman primate models of asthma. J. Exp. Med. 201 (12), 1875–1879.

Colebatch, H.J., Greaves, I.A., Ng, C.K., 1979. Exponential analysis of elastic recoil and aging in healthy males and females. J. Appl. Physiol. 47 (4), 683–691.

Coleman, J.W., 2002. Nitric oxide: a regulator of mast cell activation and mast cell-mediated inflammation. Clin. Exp. Immunol. 129 (1), 4–10.

Commins, S.P., Borish, L., Steinke, J.W., 2010. Immunologic messenger molecules: cytokines, interferons, and chemokines. J. Allergy Clin. Immunol. 125 (2 Suppl. 2), S53–S72.

Cook, M.L., Bochner, B.S., 2010. Update on biological therapeutics for asthma. World Allergy Organiz. J. 3 (6), 188–194.

Corren, J., Busse, W., Meltzer, E.O., Mansfield, L., Bensch, G., Fahrenholz, J., Wenzel, S.E., Chon, Y., Dunn, M., Weng, H.H., Lin, S.L., 2010. A randomized, controlled, phase 2 study of AMG 317, an IL-4Ralpha antagonist, in patients with asthma. Am. J. Respir. Crit. Care Med. 181 (8), 788–796.

Corren, J., Lemanske, R.F., Hanania, N.A., Korenblat, P.E., Parsey, M.V., Arron, J.R., Harris, J.M., Scheerens, H., Wu, L.C., Su, Z., Mosesova, S., Eisner, M.D., Bohen, S.P., Matthews, J.G., 2011. Lebrikizumab treatment in adults with asthma. N. Engl. J. Med. 365 (12), 1088–1098.

Cosmi, L., Maggi, L., Santarlasci, V., Capone, M., Cardilicchia, E., Frosali, F., Querci, V., Angeli, R., Matucci, A., Fambrini, M., Liotta, F., Parronchi, P., Maggi, E., Romagnani, S., Annunziato, F., 2010. Identification of a novel subset of human circulating memory CD4(+) T cells that produce both IL-17A and IL-4. J. Allergy Clin. Immunol. 125 (1), 222–230 e1–e4.

Costa, J.J., Matossian, K., Resnick, M.B., Beil, W.J., Wong, D.T., Gordon, J.R., Dvorak, A.M., Weller, P.F., Galli, S.J., 1993. Human eosinophils can express the cytokines tumor necrosis factor-alpha and macrophage inflammatory protein-1 alpha. J. Clin. Invest. 91 (6), 2673–2684.

Coyle, A.J., Erard, F., Bertrand, C., Walti, S., Pircher, H., Le Gros, G., 1995a. Virus-specific CD8+ cells can switch to interleukin 5 production and induce airway eosinophilia. J. Exp. Med. 181 (3), 1229–1233.

Coyle, A.J., Le Gros, G., Bertrand, C., Tsuyuki, S., Heusser, C.H., Kopf, M., Anderson, G.P., 1995b. Interleukin-4 is required for the induction of lung Th2 mucosal immunity. Am. J. Respir. Cell. Mol. Biol. 13 (1), 54–59.

Coyle, A.J., Tsuyuki, S., Bertrand, C., Huang, S., Aguet, M., Alkan, S.S., Anderson, G.P., 1996. Mice lacking the IFN-gamma receptor have impaired ability to resolve a lung eosinophilic inflammatory response associated with a prolonged capacity of T cells to exhibit a Th2 cytokine profile. J. Immunol. 156 (8), 2680–2685.

Cutz, E., Levison, H., Cooper, D.M., 1978. Ultrastructure of airways in children with asthma. Histopathology 2 (6), 407–421.

Dabbagh, K., Takeyama, K., Lee, H.M., Ueki, I.F., Lausier, J.A., Nadel, J.A., 1999. IL-4 induces mucin gene expression and goblet cell metaplasia in vitro and in vivo. J. Immunol. 162 (10), 6233–6237.

de Groot, E.P., Nijkamp, A., Duiverman, E.J., Brand, P.L., 2012. Allergic rhinitis is associated with poor asthma control in children with asthma. Thorax. 67 (7), 582–587.

Deniz, G., Akdis, M., Aktas, E., Blaser, K., Akdis, C.A., 2002. Human NK1 and NK2 subsets determined by purification of IFN-gamma-secreting and IFN-gamma-nonsecreting NK cells. Eur. J. Immunol. 32 (3), 879–884.

Devalia, J.L., Campbell, A.M., Sapsford, R.J., Rusznak, C., Quint, D., Godard, P., Bousquet, J., Davies, R.J., 1993. Effect of nitrogen dioxide on synthesis of inflammatory cytokines expressed by human bronchial epithelial cells in vitro. Am. J. Respir. Cell. Mol. Biol. 9 (3), 271–278.

DiCosmo, B.F., Geba, G.P., Picarella, D., Elias, J.A., Rankin, J.A., Stripp, B.R., Whitsett, J.A., Flavell, R.A., 1994. Airway epithelial cell expression of interleukin-6 in transgenic mice. Uncoupling of airway inflammation and bronchial hyperreactivity. J. Clin. Invest. 94 (5), 2028–2035.

Dienz, O., Rincon, M., 2009. The effects of IL-6 on CD4 T cell responses. Clin. Immunol. 130 (1), 27–33.

Dolhnikoff, M., da Silva, L.F., de Araujo, B.B., Gomes, H.A., Fernezlian, S., Mulder, A., Lindeman, J.H., Mauad, T., 2009. The outer wall of small airways is a major site of remodeling in fatal asthma. J. Allergy Clin. Immunol. 123 (5), 1090–1097, 1097.e1.

Donnadieu, E., Jouvin, M.H., Kinet, J.P., 2000. A second amplifier function for the allergy-associated Fc(epsilon)RI-beta subunit. Immunity 12 (5), 515–523.

Dunnill, M.S., Massarella, G.R., Anderson, J.A., 1969. A comparison of the quantitative anatomy of the bronchi in normal subjects, in status asthmaticus, in chronic bronchitis, and in emphysema. Thorax 24 (2), 176–179.

Ege, M.J., Mayer, M., Normand, A.C., Genuneit, J., Cookson, W.O., Braun-Fahrlander, C., Heederik, D., Piarroux, R., von Mutius, E., GABRIELA Transregio 22 Study Group, 2011a. Exposure to environmental microorganisms and childhood asthma. N. Engl. J. Med. 364 (8), 701–709.

Ege, M.J., Strachan, D.P., Cookson, W.O., Moffatt, M.F., Gut, I., Lathrop, M., Kabesch, M., Genuneit, J., Buchele, G., Sozanska, B., Boznanski, A., Cullinan, P., Horak, E., Bieli, C., Braun-Fahrlander, C., Heederik, D., von Mutius, E., GABRIELA Study Group, 2011b. Gene-environment interaction for childhood asthma and exposure to farming in Central Europe. J. Allergy Clin. Immunol. 127 (1), 138–144, 144.e1–144.e4.

Eriksson, U., Egermann, U., Bihl, M.P., Gambazzi, F., Tamm, M., Holt, P.G., Bingisser, R.M., 2005. Human bronchial epithelium controls TH2 responses by TH1-induced, nitric oxide-mediated STAT5 dephosphorylation: implications for the pathogenesis of asthma. J. Immunol. 175 (4), 2715–2720.

Fehervari, Z., Yamaguchi, T., Sakaguchi, S., 2006. The dichotomous role of IL-2: tolerance versus immunity. Trends Immunol. 27 (3), 109–111.

Finkelman, F.D., Hogan, S.P., Hershey, G.K., Rothenberg, M.E., Wills-Karp, M., 2010. Importance of cytokines in murine allergic airway disease and human asthma. J. Immunol. 184 (4), 1663–1674.

Finotto, S., Hausding, M., Doganci, A., Maxeiner, J.H., Lehr, H.A., Luft, C., Galle, P.R., Glimcher, L.H., 2005. Asthmatic changes in mice lacking T-bet are mediated by IL-13. Int. Immunol. 17 (8), 993–1007.

Foster, P.S., Hogan, S.P., Ramsay, A.J., Matthaei, K.I., Young, I.G., 1996. Interleukin 5 deficiency abolishes eosinophilia, airways hyperreactivity, and lung damage in a mouse asthma model. J. Exp. Med. 183 (1), 195–201.

Fujii, M., Tanaka, H., Abe, S., 2005. Interferon-gamma up-regulates expression of cysteinyl leukotriene type 2 receptors on eosinophils in asthmatic patients. Chest 128 (5), 3148–3155.

Galli, S.J., Tsai, M., Piliponsky, A.M., 2008. The development of allergic inflammation. Nature 454 (7203), 445–454.

Ganesh, B.B., Cheatem, D.M., Sheng, J.R., Vasu, C., Prabhakar, B.S., 2009. GM-CSF-induced CD11c+CD8a–dendritic cells facilitate Foxp3+ and IL-10+ regulatory T cell expansion resulting in suppression of autoimmune thyroiditis. Int. Immunol. 21 (3), 269–282.

Garssen, J., Nijkamp, F.P., Van Der Vliet, H., Van Loveren, H., 1991. T-cell-mediated induction of airway hyperreactivity in mice. Am. Rev. Respir. Dis. 144 (4), 931–938.

Gorczynski, R.M., 2005. CD200 and its receptors as targets for immunoregulation. Curr. Opin. Investig. Drugs 6 (5), 483–488.

Gould, H.J., Sutton, B.J., 2008. IgE in allergy and asthma today. Nat. Rev. Immunol. 8 (3), 205–217.

Griffiths-Johnson, D.A., Karol, M.H., 1991. Validation of a non-invasive technique to assess development of airway hyperreactivity in an animal model of immunologic pulmonary hypersensitivity. Toxicology 65 (3), 283–294.

Griffiths-Johnson, D., Jin, R., Karol, M.H., 1993. Role of purified IgG1 in pulmonary hypersensitivity responses of the guinea pig. J. Toxicol. Environ. Health 40 (1), 117–127.

Hackett, T.L., 2012. Epithelial-mesenchymal transition in the pathophysiology of airway remodelling in asthma. Curr. Opin. Allergy Clin. Immunol. 12 (1), 53–59.

Haldar, P., Brightling, C.E., Hargadon, B., Gupta, S., Monteiro, W., Sousa, A., Marshall, R.P., Bradding, P., Green, R.H., Wardlaw, A.J., Pavord, I.D., 2009. Mepolizumab and exacerbations of refractory eosinophilic asthma. N. Engl. J. Med. 360 (10), 973–984.

Hamilton, J.A., Anderson, G.P., 2004. GM-csf biology. Growth Factors 22 (4), 225–231.

Hammad, H., Lambrecht, B.N., 2011. Dendritic cells and airway epithelial cells at the interface between innate and adaptive immune responses. Allergy 66 (5), 579–587.

Hammad, H., Plantinga, M., Deswarte, K., Pouliot, P., Willart, M.A., Kool, M., Muskens, F., Lambrecht, B.N., 2010. Inflammatory dendritic cells—not basophils—are necessary and sufficient for induction of Th2 immunity to inhaled house dust mite allergen. J. Exp. Med. 207 (10), 2097–2111.

Han, M.K., Huang, Y.J., Lipuma, J.J., Boushey, H.A., Boucher, R.C., Cookson, W.O., Curtis, J.L., Erb-Downward, J., Lynch, S.V., Sethi, S., Toews, G.B., Young, V.B., Wolfgang, M.C., Huffnagle, G.B., Martinez, F.J., 2012. Significance of the microbiome in obstructive lung disease. Thorax. 67 (5), 456–463.

Hanania, N.A., King, M.J., Braman, S.S., Saltoun, C., Wise, R.A., Enright, P., Falsey, A.R., Mathur, S.K., Ramsdell, J.W., Rogers, L., Stempel, D.A., Lima, J.J., Fish, J.E., Wilson, S.R., Boyd, C., Patel, K.V., Irvin, C.G., Yawn, B.P., Halm, E.A., Wasserman, S.I., Sands, M.F., Ershler, W.B., Ledford, D.K., Asthma in elderly workshop participants, 2011. Asthma in the elderly: current understanding and future research needs—a report of a National Institute on Aging (NIA) workshop. J. Allergy Clin. Immunol. 128 (3 Suppl.), S4–S24.

Hanna, J., Gonen-Gross, T., Fitchett, J., Rowe, T., Daniels, M., Arnon, T.I., Gazit, R., Joseph, A., Schjetne, K.W., Steinle, A., Porgador, A., Mevorach, D., Goldman-Wohl, D., Yagel, S., LaBarre, M.J., Buckner, J.H., Mandelboim, O., 2004. Novel APC-like properties of human NK cells directly regulate T cell activation. J. Clin. Invest. 114 (11), 1612–1623.

Hassan, N.M., Hargreave, F.E., Nair, P., 2010. Airway responsiveness to indirect challenges in COPD. COPD 7 (2), 133–140.

Haworth, O., Cernadas, M., Levy, B.D., 2011. NK cells are effectors for resolvin E1 in the timely resolution of allergic airway inflammation. J. Immunol. 186 (11), 6129–6135.

Hays, S.R., Fahy, J.V., 2003. The role of mucus in fatal asthma. Am. J. Med. 115 (1), 68–69.

Heijink, I.H., Van Oosterhout, A.J., 2006. Strategies for targeting T-cells in allergic diseases and asthma. Pharmacol. Ther. 112 (2), 489–500.

Hershey, G.K., Friedrich, M.F., Esswein, L.A., Thomas, M.L., Chatila, T.A., 1997. The association of atopy with a gain-of-function mutation in the alpha subunit of the interleukin-4 receptor. N. Engl. J. Med. 337 (24), 1720–1725.

Hilty, M., Burke, C., Pedro, H., Cardenas, P., Bush, A., Bossley, C., Davies, J., Ervine, A., Poulter, L., Pachter, L., Moffatt, M.F., Cookson, W.O., 2010. Disordered microbial communities in asthmatic airways. PLoS One 5 (1), e8578.

Hirshman, C.A., Malley, A., Downes, H., 1980. Basenji-Greyhound dog model of asthma: reactivity to *Ascaris suum*, citric acid, and methacholine. J. Appl. Physiol. 49 (6), 953–957.

Hollingsworth, J.W., Maruoka, S., Boon, K., Garantziotis, S., Li, Z., Tomfohr, J., Bailey, N., Potts, E.N., Whitehead, G., Brass, D.M., Schwartz, D.A., 2008. In utero supplementation with methyl donors enhances allergic airway disease in mice. J. Clin. Invest. 118 (10), 3462–3469.

Holtzman, M.J., Byers, D.E., Benoit, L.A., Battaile, J.T., You, Y., Agapov, E., Park, C., Grayson, M.H., Kim, E.Y., Patel, A.C., 2009. Immune pathways for translating viral infection into chronic airway disease. Adv. Immunol. 102, 245–276.

Howarth, P.H., Knox, A.J., Amrani, Y., Tliba, O., Panettieri, Jr., R.A., Johnson, M., 2004. Synthetic responses in airway smooth muscle. J. Allergy Clin. Immunol. 114 (2 Suppl.), S32–S50.

Huang, T.J., MacAry, P.A., Eynott, P., Moussavi, A., Daniel, K.C., Askenase, P.W., Kemeny, D.M., Chung, K.F., 2001. Allergen-specific Th1 cells counteract efferent Th2 cell-dependent bronchial hyperresponsiveness and eosinophilic inflammation partly via IFN-gamma. J. Immunol. 166 (1), 207–217.

Hudes, G., Tagawa, S.T., Whang, Y.E., Qi, M., Qin, X., Puchalski, T.A., Reddy, M., Cornfeld, M., Eisenberger, M., 2012. A phase 1 study of a chimeric monoclonal antibody against interleukin-6, siltuximab, combined with docetaxel in patients with metastatic castration-resistant prostate cancer. Invest. New Drugs. 31 (3), 669–676.

Humbert, M., Grant, J.A., Taborda-Barata, L., Durham, S.R., Pfister, R., Menz, G., Barkans, J., Ying, S., Kay, A.B., 1996. High-affinity IgE receptor (FcepsilonRI)-bearing cells in bronchial biopsies from atopic and nonatopic asthma. Am. J. Respir. Crit. Care Med. 153 (6 Pt 1), 1931–1937.

Hyde, D.M., Miller, L.A., Schelegle, E.S., Fanucchi, M.V., Van Winkle, L.S., Tyler, N.K., Avdalovis, M.V., Evans, M.J., Kajekar, R., Buckpitt, A.R., Pinkerton, K.E., Joad, J.P., Gershwin, L.J., Wu, R., Plopper, C.G., 2006. Asthma: a comparison of animal models using stereological methods. Eur. Respir. Rev. 15 (1), 122–135.

Iwamura, C., Nakayama, T., 2010. Role of NKT cells in allergic asthma. Curr. Opin. Immunol. 22 (6), 807–813.

Johnson, J.R., Roos, A., Berg, T., Nord, M., Fuxe, J., 2011. Chronic respiratory aeroallergen exposure in mice induces epithelial-mesenchymal transition in the large airways. PLoS One 6 (1), e16175.
Jones, S.C., Iverson, D., Burns, P., Evers, U., Caputi, P., Morgan, S., 2011. Asthma and ageing: an end user's perspective–the perception and problems with the management of asthma in the elderly. Clin. Exp. Allergy 41 (4), 471–481.
Jonsson, F., Mancardi, D.A., Zhao, W., Kita, Y., Iannascoli, B., Khun, H., van Rooijen, N., Shimizu, T., Schwartz, L.B., Daeron, M., Bruhns, P., 2012. Human FcgammaRIIA induces anaphylactic and allergic reactions. Blood 119 (11), 2533–2544.
Joos, G.F., Germonpre, P.R., Pauwels, R.A., 2000. Role of tachykinins in asthma. Allergy 55 (4), 321–337.
Justice, J.P., Shibata, Y., Sur, S., Mustafa, J., Fan, M., Van Scott, M.R., 2001. IL-10 gene knockout attenuates allergen-induced airway hyperresponsiveness in C57BL/6 mice. Am. J. Physiol. Lung Cell. Mol. Physiol. 280 (2), L363–L368.
Kaech, S.M., Wherry, E.J., Ahmed, R., 2002. Effector and memory T-cell differentiation: implications for vaccine development. Nat. Rev. Immunol. 2 (4), 251–262.
Karol, M.H., 1983. Concentration-dependent immunologic response to toluene diisocyanate (TDI) following inhalation exposure. Toxicol. Appl. Pharmacol. 68 (2), 229–241.
Karol, M.H., 1994. Animal models of occupational asthma. Eur. Respir. J. 7 (3), 555–568.
Karra, L., Levi-Schaffer, F., 2011. Down-regulation of mast cell responses through ITIM containing inhibitory receptors. Adv. Exp. Med. Biol. 716, 143–159.
Kasagi, S., Kawano, S., Kumagai, S., 2011. PD-1 and autoimmunity. Crit. Rev. Immunol. 31 (4), 265–295.
Kaur, D., Brightling, C., 2012. OX40/OX40 ligand interactions in T-cell regulation and asthma. Chest 141 (2), 494–499.
Kim, H.Y., Chang, Y.J., Subramanian, S., Lee, H.H., Albacker, L.A., Matangkasombut, P., Savage, P.B., McKenzie, A.N., Smith, D.E., Rottman, J.B., DeKruyff, R.H., Umetsu, D.T., 2012. Innate lymphoid cells responding to IL-33 mediate airway hyperreactivity independently of adaptive immunity. J. Allergy Clin. Immunol. 129 (1), 216–227 e1–e6.
Kips, J.C., Tavernier, J., Pauwels, R.A., 1992. Tumor necrosis factor causes bronchial hyperresponsiveness in rats. Am. Rev. Respir. Dis. 145 (2 Pt 1), 332–336.
Kips, J.C., Tavernier, J.H., Joos, G.F., Peleman, R.A., Pauwels, R.A., 1993. The potential role of tumour necrosis factor alpha in asthma. Clin. Exp. Allergy 23 (4), 247–250.
Klebb, G., Autenrieth, I.B., Haber, H., Gillert, E., Sadlack, B., Smith, K.A., Horak, I., 1996. Interleukin-2 is indispensable for development of immunological self-tolerance. Clin. Immunol. Immunopathol. 81 (3), 282–286.
Koguchi, Y., Kawakami, K., Uezu, K., Fukushima, K., Kon, S., Maeda, M., Nakamoto, A., Owan, I., Kuba, M., Kudeken, N., Azuma, M., Yara, S., Shinzato, T., Higa, F., Tateyama, M., Kadota, J., Mukae, H., Kohno, S., Uede, T., Saito, A., 2003. High plasma osteopontin level and its relationship with interleukin-12-mediated type 1 T helper cell response in tuberculosis. Am. J. Respir. Crit. Care Med. 167 (10), 1355–1359.
Koh, B.H., Hwang, S.S., Kim, J.Y., Lee, W., Kang, M.J., Lee, C.G., Park, J.W., Flavell, R.A., Lee, G.R., 2010. Th2 LCR is essential for regulation of Th2 cytokine genes and for pathogenesis of allergic asthma. Proc. Natl. Acad. Sci. U.S.A 107 (23), 10614–10619.
Konno, S., Kurokawa, M., Uede, T., Nishimura, M., Huang, S.K., 2011. Role of osteopontin, a multifunctional protein, in allergy and asthma. Clin. Exp. Allergy 41 (10), 1360–1366.
Kool, M., van Nimwegen, M., Willart, M.A., Muskens, F., Boon, L., Smit, J.J., Coyle, A., Clausen, B.E., Hoogsteden, H.C., Lambrecht, B.N., Hammad, H., 2009. An anti-inflammatory role for plasmacytoid dendritic cells in allergic airway inflammation. J. Immunol. 183 (2), 1074–1082.
Korsgren, M., Persson, C.G., Sundler, F., Bjerke, T., Hansson, T., Chambers, B.J., Hong, S., Van Kaer, L., Ljunggren, H.G., Korsgren, O., 1999. Natural killer cells determine development of allergen-induced eosinophilic airway inflammation in mice. J. Exp. Med. 189 (3), 553–562.
Kraft, S., Kinet, J.P., 2007. New developments in FcepsilonRI regulation, function and inhibition. Nat. Rev. Immunol. 7 (5), 365–378.
Kumar, R.K., Foster, P.S., 2002. Modeling allergic asthma in mice: pitfalls and opportunities. Am. J. Respir. Cell. Mol. Biol. 27 (3), 267–272.
Kuwano, K., Bosken, C.H., Pare, P.D., Bai, T.R., Wiggs, B.R., Hogg, J.C., 1993. Small airways dimensions in asthma and in chronic obstructive pulmonary disease. Am. Rev. Respir. Dis. 148 (5), 1220–1225.
Lambrecht, B.N., Hammad, H., 2009. Biology of lung dendritic cells at the origin of asthma. Immunity 31 (3), 412–424.
Lambrecht, B.N., Hammad, H., 2012. Lung dendritic cells in respiratory viral infection and asthma: from protection to immunopathology. Annu. Rev. Immunol. 30, 243–270.
Lanes, S., Stevenson, J.S., Codias, E., Hernandez, A., Sielczak, M.W., Wanner, A., Abraham, W.M., 1986. Indomethacin and FPL-57231 inhibit antigen-induced airway hyperresponsiveness in sheep. J. Appl. Physiol. 61 (3), 864–872.
Larsen, G.L., 1985. The rabbit model of the late asthmatic response. Chest 87 (5), S184–S188.
Larsen, J.M., Steen-Jensen, D.B., Laursen, J.M., Sondergaard, J.N., Musavian, H.S., Butt, T.M., Brix, S., 2012. Divergent pro-inflammatory profile of human dendritic cells in response to commensal and pathogenic bacteria associated with the airway microbiota. PLoS One 7 (2), e31976.
Larson, S.D., Schelegle, E.S., Walby, W.F., Gershwin, L.J., Fanuccihi, M.V., Evans, M.J., Joad, J.P., Tarkington, B.K., Hyde, D.M., Plopper, C.G., 2004. Postnatal remodeling of the neural components of the epithelial-mesenchymal trophic unit in the proximal airways of infant rhesus monkeys exposed to ozone and allergen. Toxicol. Appl. Pharmacol. 194 (3), 211–220.
Le, A.V., Broide, D.H., 2006. Indoleamine-2,3-dioxygenase modulation of allergic immune responses. Curr. Allergy Asthma Rep. 6 (1), 27–31.
Lee, C.G., Hartl, D., Lee, G.R., Koller, B., Matsuura, H., Da Silva, C.A., Sohn, M.H., Cohn, L., Homer, R.J., Kozhich, A.A., Humbles, A., Kearley, J., Coyle, A., Chupp, G., Reed, J., Flavell, R.A., Elias, J.A., 2009. Role of breast regression protein 39 (BRP-39)/chitinase 3-like-1 in Th2 and IL-13-induced tissue responses and apoptosis. J. Exp. Med. 206 (5), 1149–1166.
Lee, C.G., Link, H., Baluk, P., Homer, R.J., Chapoval, S., Bhandari, V., Kang, M.J., Cohn, L., Kim, Y.K., McDonald, D.M., Elias, J.A., 2004. Vascular endothelial growth factor (VEGF) induces remodeling and enhances TH2-mediated sensitization and inflammation in the lung. Nat. Med. 10 (10), 1095–1103.

Lee, J.J., McGarry, M.P., Farmer, S.C., Denzler, K.L., Larson, K.A., Carrigan, P.E., Brenneise, I.E., Horton, M.A., Haczku, A., Gelfand, E.W., Leikauf, G.D., Lee, N.A., 1997. Interleukin-5 expression in the lung epithelium of transgenic mice leads to pulmonary changes pathognomonic of asthma. J. Exp. Med. 185 (12), 2143–2156.

Lee, N., Shin, M.S., Kang, I., 2012. T-cell biology in aging, with a focus on lung disease. J. Gerontol. A Biol. Sci. Med. Sci. 67 (3), 254–263.

Lees, J.R., Farber, D.L., 2010. Generation, persistence and plasticity of CD4 T-cell memories. Immunology 130 (4), 463–470.

Letourneau, S., Krieg, C., Pantaleo, G., Boyman, O., 2009. IL-2- and CD25-dependent immunoregulatory mechanisms in the homeostasis of T-cell subsets. J. Allergy Clin. Immunol. 123 (4), 758–762.

Lieu, T.M., Myers, A.C., Meeker, S., Undem, B.J., 2012. TRPV1 induction in airway vagal low-threshold mechanosensory neurons by allergen challenge and neurotrophic factors. Am. J. Physiol. Lung Cell. Mol. Physiol. 302 (9), L941–L948.

Lloyd, C.M., Hessel, E.M., 2010. Functions of T cells in asthma: more than just T(H)2 cells. Nat. Rev. Immunol. 10 (12), 838–848.

Lommatzsch, M., Julius, P., Kuepper, M., Garn, H., Bratke, K., Irmscher, S., Luttmann, W., Renz, H., Braun, A., Virchow, J.C., 2006. The course of allergen-induced leukocyte infiltration in human and experimental asthma. J. Allergy Clin. Immunol. 118 (1), 91–97.

Long, A., McFaddden, C., DeVine, D., 2002. Management of Allergic and Nonallergic Rhinitis. Evidence Reports/Technology Assessments US Department of Health and Human Services; Agnecy for Healthcare Research and Quality (US).

Lowell, F.C., 1964. Observations on heaves. An asthma-like syndrome in the horse. J. Allergy 35, 322–330.

Mancardi, D.A., Iannascoli, B., Hoos, S., England, P., Daeron, M., Bruhns, P., 2008. FcgammaRIV is a mouse IgE receptor that resembles macrophage FcepsilonRI in humans and promotes IgE-induced lung inflammation. J. Clin. Invest. 118 (11), 3738–3750.

Martin, R.J., 2008. Exploring the distal lung: new direction in asthma. Isr. Med. Assoc. J. 10 (12), 846–849.

Martin, T.R., Gerard, N.P., Galli, S.J., Drazen, J.M., 1988. Pulmonary responses to bronchoconstrictor agonists in the mouse. J. Appl. Physiol. 64 (6), 2318–2323.

Matsubara, S., Takeda, K., Kodama, T., Joetham, A., Miyahara, N., Koya, T., Swasey, C.H., Okamoto, M., Dakhama, A., Gelfand, E.W., 2007. IL-2 and IL-18 attenuation of airway hyperresponsiveness requires STAT4, IFN-gamma, and natural killer cells. Am. J. Respir. Cell. Mol. Biol. 36 (3), 324–332.

Maurer, D., Fiebiger, S., Ebner, C., Reininger, B., Fischer, G.F., Wichlas, S., Jouvin, M.H., Schmitt-Egenolf, M., Kraft, D., Kinet, J.P., Stingl, G., 1996. Peripheral blood dendritic cells express Fc epsilon RI as a complex composed of Fc epsilon RI alpha- and Fc epsilon RI gamma-chains and can use this receptor for IgE-mediated allergen presentation. J. Immunol. 157 (2), 607–616.

McGeachy, M.J., 2011. GM-CSF: the secret weapon in the T(H)17 arsenal. Nat. Immunol. 12 (6), 521–522.

McLane, M.P., Haczku, A., van de Rijn, M., Weiss, C., Ferrante, V., MacDonald, D., Renauld, J.C., Nicolaides, N.C., Holroyd, K.J., Levitt, R.C., 1998. Interleukin-9 promotes allergen-induced eosinophilic inflammation and airway hyperresponsiveness in transgenic mice. Am. J. Respir. Cell. Mol. Biol. 19 (5), 713–720.

McMillan, S.J., Kearley, J., Campbell, J.D., Zhu, X.W., Larbi, K.Y., Shipley, J.M., Senior, R.M., Nourshargh, S., Lloyd, C.M., 2004. Matrix metalloproteinase-9 deficiency results in enhanced allergen-induced airway inflammation. J. Immunol. 172 (4), 2586–2594.

McParland, B.E., Macklem, P.T., Pare, P.D., 2003. Airway wall remodeling: friend or foe? J. Appl. Physiol. 95 (1), 426–434.

Metzger, W., 1990. Late phase asthma in an allergic rabbit model. In: Dorsch, W. (Ed.), Late Phase Allergic Reactions. CRC Press, Boston, pp. 347–371 (Chapter 35).

Meyer, E.H., DeKruyff, R.H., Umetsu, D.T., 2008. T cells and NKT cells in the pathogenesis of asthma. Annu. Rev. Med. 59, 281–292.

Michael, E., March, 2011. The Genetics of Asthma and Allergic Disorders, Discovery Medicine 11 (56), 35–45.

Minas, K., Liversidge, J., 2006. Is the CD200/CD200 receptor interaction more than just a myeloid cell inhibitory signal? Crit. Rev. Immunol. 26 (3), 213–230.

Minshall, E., Spina, D., Page, C.P., 1996. Effects of neonatal immunization and repeated allergen exposure on airway responsiveness in the rabbit. J. Appl. Physiol. 80 (6), 2108–2119.

Monteleone, G., Pallone, F., Macdonald, T.T., 2010. Interleukin-25: a two-edged sword in the control of immune-inflammatory responses. Cytokine Growth Factor Rev. 21 (6), 471–475.

Moretta, L., Ferlazzo, G., Bottino, C., Vitale, M., Pende, D., Mingari, M.C., Moretta, A., 2006. Effector and regulatory events during natural killer-dendritic cell interactions. Immunol. Rev. 214, 219–228.

Moro, K., Yamada, T., Tanabe, M., Takeuchi, T., Ikawa, T., Kawamoto, H., Furusawa, J., Ohtani, M., Fujii, H., Koyasu, S., 2010. Innate production of T(H)2 cytokines by adipose tissue-associated c-Kit(+)Sca-1(+) lymphoid cells. Nature 463 (7280), 540–544.

Mosmann, T.R., Cherwinski, H., Bond, M.W., Giedlin, M.A., Coffman, R.L., 1986. Two types of murine helper T cell clone. I. Definition according to profiles of lymphokine activities and secreted proteins. J. Immunol. 136 (7), 2348–2357.

Murdoch, J.R., Lloyd, C.M., 2010. Resolution of allergic airway inflammation and airway hyperreactivity is mediated by IL-17-producing {gamma}{delta}T cells. Am. J. Respir. Crit. Care Med. 182 (4), 464–476.

Nair, P., Pizzichini, M.M., Kjarsgaard, M., Inman, M.D., Efthimiadis, A., Pizzichini, E., Hargreave, F.E., O'Byrne, P.M., 2009. Mepolizumab for prednisone-dependent asthma with sputum eosinophilia. N. Engl. J. Med. 360 (10), 985–993.

Neill, D.R., Wong, S.H., Bellosi, A., Flynn, R.J., Daly, M., Langford, T.K., Bucks, C., Kane, C.M., Fallon, P.G., Pannell, R., Jolin, H.E., McKenzie, A.N., 2010. Nuocytes represent a new innate effector leukocyte that mediates type-2 immunity. Nature 464 (7293), 1367–1370.

Ober, C., Yao, T.C., 2011. The genetics of asthma and allergic disease: a 21st century perspective. Immunol. Rev. 242 (1), 10–30.

Ohkawara, Y., Yamauchi, K., Tanno, Y., Tamura, G., Ohtani, H., Nagura, H., Ohkuda, K., Takishima, T., 1992. Human lung mast cells and pulmonary macrophages produce tumor necrosis factor-alpha in sensitized lung tissue after IgE receptor triggering. Am. J. Respir. Cell. Mol. Biol. 7 (4), 385–392.

Ohno, I., Ohkawara, Y., Yamauchi, K., Tanno, Y., Takishima, T., 1990. Production of tumor necrosis factor with IgE receptor triggering from sensitized lung tissue. Am. J. Respir. Cell. Mol. Biol. 3 (4), 285–289.

Okazaki, T., Honjo, T., 2006. The PD-1-PD-L pathway in immunological tolerance. Trends Immunol. 27 (4), 195–201.

On, M., Billingsley, J.M., Jouvin, M.H., Kinet, J.P., 2004. Molecular dissection of the FcRbeta signaling amplifier. J. Biol. Chem. 279 (44), 45782–45790.

Oppenheimer, B.W., Goldring, R.M., Berger, K.I., 2009. Distal airway function assessed by oscillometry at varying respiratory rate: comparison with dynamic compliance. COPD 6 (3), 162–170.

Ordonez, C.L., Khashayar, R., Wong, H.H., Ferrando, R., Wu, R., Hyde, D.M., Hotchkiss, J.A., Zhang, Y., Novikov, A., Dolganov, G., Fahy, J.V., 2001. Mild and moderate asthma is associated with airway goblet cell hyperplasia and abnormalities in mucin gene expression. Am. J. Respir. Crit. Care Med. 163 (2), 517–523.

Pappu, B.P., Angkasekwinai, P., Dong, C., 2008. Regulatory mechanisms of helper T cell differentiation: new lessons learned from interleukin 17 family cytokines. Pharmacol. Ther. 117 (3), 374–384.

Pare, P.D., McParland, B.E., Seow, C.Y., 2007. Structural basis for exaggerated airway narrowing. Can. J. Physiol. Pharmacol. 85 (7), 653–658.

Park, G.Y., Christman, J.W., 2006. Involvement of cyclooxygenase-2 and prostaglandins in the molecular pathogenesis of inflammatory lung diseases. Am. J. Physiol. Lung Cell. Mol. Physiol. 290 (5), L797–L805.

Parker, J.M., Oh, C.K., LaForce, C., Miller, S.D., Pearlman, D.S., Le, C., Robbie, G.J., White, W.I., White, B., 2011. Molfino NA and MEDI-528 Clinical trials group. Safety profile and clinical activity of multiple subcutaneous doses of MEDI-528, a humanized anti-interleukin-9 monoclonal antibody, in two randomized phase 2a studies in subjects with asthma. BMC Pulm. Med. 11 (14).

Patterson, R., Harris, K.E., 1992. IgE-mediated rhesus monkey asthma: natural history and individual variation. Int. Arch. Allergy Immunol. 97, 154–159.

Prendergast, G.C., Chang, M.Y., Mandik-Nayak, L., Metz, R., Muller, A.J., 2011. Indoleamine 2,3-dioxygenase as a modifier of pathogenic inflammation in cancer and other inflammation-associated diseases. Curr. Med. Chem. 18 (15), 2257–2262.

Price, A.E., Liang, H.E., Sullivan, B.M., Reinhardt, R.L., Eisley, C.J., Erle, D.J., Locksley, R.M., 2010. Systemically dispersed innate IL-13-expressing cells in type 2 immunity. Proc. Natl. Acad. Sci. U.S.A 107 (25), 11489–11494.

Quintero, P.A., Knolle, M.D., Cala, L.F., Zhuang, Y., Owen, C.A., 2010. Matrix metalloproteinase-8 inactivates macrophage inflammatory protein-1 alpha to reduce acute lung inflammation and injury in mice. J. Immunol. 184 (3), 1575–1588.

Raemdonck, K., de Alba, J., Birrell, M.A., Grace, M., Maher, S.A., Irvin, C.G., Fozard, J.R., O'Byrne, P.M., Belvisi, M.G., 2012. A role for sensory nerves in the late asthmatic response. Thorax 67 (1), 19–25.

Ramalingam, T.R., Pesce, J.T., Sheikh, F., Cheever, A.W., Mentink-Kane, M.M., Wilson, M.S., Stevens, S., Valenzuela, D.M., Murphy, A.J., Yancopoulos, G.D., Urban Jr., J.F., Donnelly, R.P., Wynn, T.A., 2008. Unique functions of the type II interleukin 4 receptor identified in mice lacking the interleukin 13 receptor alpha1 chain. Nat. Immunol. 9 (1), 25–33.

Rankin, J.A., Picarella, D.E., Geba, G.P., Temann, U.A., Prasad, B., DiCosmo, B., Tarallo, A., Stripp, B., Whitsett, J., Flavell, R.A., 1996. Phenotypic and physiologic characterization of transgenic mice expressing interleukin 4 in the lung: lymphocytic and eosinophilic inflammation without airway hyperreactivity. Proc. Natl. Acad. Sci. U.S.A. 93 (15), 7821–7825.

Ratner, B., Jackson, H.C., Gruehl, H.L., 1927. Respiratory anaphylaxis. Sensitization, shock, bronchial asthma and death induced in the guinea-pig by the nasal inhalation of dry horse dander. Am. J. Dis. Child. 34, 23–47.

Raymond, M., Van, V.Q., Rubio, M., Welzenbach, K., Sarfati, M., 2010. Targeting SIRP-alpha protects from type 2-driven allergic airway inflammation. Eur. J. Immunol. 40 (12), 3510–3518.

Renauld, J.C., 2001. New insights into the role of cytokines in asthma. J. Clin. Pathol. 54 (8), 577–589.

Ressmeyer, A.R., Larsson, A.K., Vollmer, E., Dahlen, S.E., Uhlig, S., Martin, C., 2006. Characterisation of guinea pig precision-cut lung slices: comparison with human tissues. Eur. Respir. J. 28 (3), 603–611.

Ritz, S.A., Stampfli, M.R., Davies, D.E., Holgate, S.T., Jordana, M., 2002. On the generation of allergic airway diseases: from GM-CSF to Kyoto. Trends Immunol. 23 (8), 396–402.

Robinson, D.S., 2010. The role of the T cell in asthma. J. Allergy Clin. Immunol. 126 (6), 1081–1091 quiz 1092-3.

Robinson, D.S., Bentley, A.M., Hartnell, A., Kay, A.B., Durham, S.R., 1993. Activated memory T helper cells in bronchoalveolar lavage fluid from patients with atopic asthma: relation to asthma symptoms, lung function, and bronchial responsiveness. Thorax 48 (1), 26–32.

Romanet-Manent, S., Charpin, D., Magnan, A., Lanteaume, A., Vervloet, D., EGEA Cooperative Group, 2002. Allergic versus nonallergic asthma: what makes the difference? Allergy 57 (7), 607–613.

Royce, S.G., Tan, L., Koek, A.A., Tang, M.L., 2009. Effect of extracellular matrix composition on airway epithelial cell and fibroblast structure: implications for airway remodeling in asthma. Ann. Allergy Asthma Immunol. 102 (3), 238–246.

Saenz, S.A., Siracusa, M.C., Perrigoue, J.G., Spencer, S.P., Urban Jr., J.F., Tocker, J.E., Budelsky, A.L., Kleinschek, M.A., Kastelein, R.A., Kambayashi, T., Bhandoola, A., Artis, D., 2010. IL25 elicits a multipotent progenitor cell population that promotes T(H)2 cytokine responses. Nature 464 (7293), 1362–1366.

Safholm, J., Lovdahl, C., Swedin, L., Boels, P.J., Dahlen, S.E., Arner, A., Adner, M., 2011. Inflammation-induced airway smooth muscle responsiveness is strain dependent in mice. Pulm. Pharmacol. Ther. 24 (4), 361–366.

Saisch, S.G., Wessely, S., Gardner, W.N., 1996. Patients with acute hyperventilation presenting to an inner-city emergency department. Chest 110 (4), 952–957.

Sakaguchi, S., Sakaguchi, N., Asano, M., Itoh, M., Toda, M., 1995. Immunologic self-tolerance maintained by activated T cells expressing IL-2 receptor alpha-chains (CD25). Breakdown of a single mechanism of self-tolerance causes various autoimmune diseases. J. Immunol. 155 (3), 1151–1164.

Sallmann, E., Reininger, B., Brandt, S., Duschek, N., Hoflehner, E., Garner-Spitzer, E., Platzer, B., Dehlink, E., Hammer, M., Holcmann, M., Oettgen, H.C., Wiedermann, U., Sibilia, M., Fiebiger, E., Rot, A., Maurer, D., 2011. High-affinity IgE receptors on dendritic cells exacerbate Th2-dependent inflammation. J. Immunol. 187 (1), 164–171.

Sallusto, F., Geginat, J., Lanzavecchia, A., 2004. Central memory and effector memory T cell subsets: function, generation, and maintenance. Annu. Rev. Immunol. 22, 745–763.

Saxon, A., Kepley, C., Zhang, K., 2008. "Accentuate the negative, eliminate the positive": engineering allergy therapeutics to block allergic reactivity through negative signaling. J. Allergy Clin. Immunol. 121 (2), 320–325.

Schleimer, R.P., Kato, A., Kern, R., Kuperman, D., Avila, P.C., 2007. Epithelium: at the interface of innate and adaptive immune responses. J. Allergy Clin. Immunol. 120 (6), 1279–1284.

Scordamaglia, F., Balsamo, M., Scordamaglia, A., Moretta, A., Mingari, M.C., Canonica, G.W., Moretta, L., Vitale, M., 2008. Perturbations of natural killer cell regulatory functions in respiratory allergic diseases. J. Allergy Clin. Immunol. 121 (2), 479–485.

Seehase, S., Schleputz, M., Switalla, S., Matz-Rensing, K., Kaup, F.J., Zoller, M., Schlumbohm, C., Fuchs, E., Lauenstein, H.D., Winkler, C., Kuehl, A.R., Uhlig, S., Braun, A., Sewald, K., Martin, C., 2011. Bronchoconstriction in nonhuman primates: a species comparison. J. Appl. Physiol. 111 (3), 791–798.

Selivanova, P.A., Kulikov, E.S., Kozina, O.V., Trofimenko, I.N., Freidin, M.B., Chernyak, B.A., Ogorodova, L.M., 2012. Differential expression of the beta2-adrenoreceptor and M3-cholinoreceptor genes in bronchial mucosa of patients with asthma and chronic obstructive pulmonary disease. Ann. Allergy Asthma Immunol. 108 (1), 39–43.

Shah, A., Church, M.K., Holgate, S.T., 1995. Tumour necrosis factor alpha: a potential mediator of asthma. Clin. Exp. Allergy 25 (11), 1038–1044.

Sharma, G., Goodwin, J., 2006. Effect of aging on respiratory system physiology and immunology. Clin. Interv. Aging 1 (3), 253–260.

Shen, Z.J., Esnault, S., Rosenthal, L.A., Szakaly, R.J., Sorkness, R.L., Westmark, P.R., Sandor, M., Malter, J.S., 2008. Pin1 regulates TGF-beta1 production by activated human and murine eosinophils and contributes to allergic lung fibrosis. J. Clin. Invest. 118 (2), 479–490.

Shi, Y., Liu, C.H., Roberts, A.I., Das, J., Xu, G., Ren, G., Zhang, Y., Zhang, L., Yuan, Z.R., Tan, H.S., Das, G., Devadas, S., 2006. Granulocyte-macrophage colony-stimulating factor (GM-CSF) and T-cell responses: what we do and don't know. Cell. Res. 16 (2), 126–133.

Shin, Y.S., Takeda, K., Gelfand, E.W., 2009. Understanding asthma using animal models. Allergy Asthma Immunol. Res. 1 (1), 10–18.

Simoes, D.C., Xanthou, G., Petrochilou, K., Panoutsakopoulou, V., Roussos, C., Gratziou, C., 2009. Osteopontin deficiency protects against airway remodeling and hyperresponsiveness in chronic asthma. Am. J. Respir. Crit. Care Med. 179 (10), 894–902.

Singh, A.K., Stock, P., Akbari, O., 2011. Role of PD-L1 and PD-L2 in allergic diseases and asthma. Allergy 66 (2), 155–162.

Slager, R.E., Otulana, B.A., Hawkins, G.A., Yen, Y.P., Peters, S.P., Wenzel, S.E., Meyers, D.A., Bleecker, E.R., 2012. IL-4 receptor polymorphisms predict reduction in asthma exacerbations during response to an anti-IL-4 receptor alpha antagonist. J. Allergy Clin. Immunol. 130 (2), 516–522 e4.

Smith, D.E., 2010. IL-33: a tissue derived cytokine pathway involved in allergic inflammation and asthma. Clin. Exp. Allergy 40 (2), 200–208.

Smith, K.A., 1988. Interleukin-2: inception, impact, and implications. Science 240 (4856), 1169–1176.

Smith, K.G., Clatworthy, M.R., 2010. FcgammaRIIB in autoimmunity and infection: evolutionary and therapeutic implications. Nat. Rev. Immunol. 10 (5), 328–343.

Sonnenberg, G.F., Fouser, L.A., Artis, D., 2011. Border patrol: regulation of immunity, inflammation and tissue homeostasis at barrier surfaces by IL-22. Nat. Immunol. 12 (5), 383–390.

Spinozzi, F., Porcelli, S.A., 2007. Recognition of lipids from pollens by CD1-restricted T cells. Immunol. Allergy Clin. North. Am. 27 (1), 79–92.

Steinman, R.M., Hemmi, H., 2006. Dendritic cells: translating innate to adaptive immunity. Curr. Top. Microbiol. Immunol. 311, 17–58.

Stern, J.B., Smith, K.A., 1986. Interleukin-2 induction of T-cell G1 progression and c-myb expression. Science 233 (4760), 203–206.

Strickland, D.H., Holt, P.G., 2011. T regulatory cells in childhood asthma. Trends Immunol. 32 (9), 420–427.

Strickland, D.H., Thomas, J.A., Mok, D., Blank, F., McKenna, K.L., Larcombe, A.N., Sly, P.D., Holt, P.G., 2012. Defective aeroallergen surveillance by airway mucosal dendritic cells as a determinant of risk for persistent airways hyper-responsiveness in experimental asthma. Mucosal Immunol. 5 (3), 332–341.

Sugiura, H., Ichinose, M., 2011. Nitrative stress in inflammatory lung diseases. Nitric Oxide 25 (2), 138–144.

Suzuki, R., Kato, T., Miyazaki, Y., Iwata, M., Noda, Y., Takagi, K., Nakashima, N., Torii, K., 2001. Matrix metalloproteinases and tissue inhibitors of matrix metalloproteinases in sputum from patients with bronchial asthma. J. Asthma 38 (6), 477–484.

Tan, P.H., Bharath, A.K., 2009. Manipulation of indoleamine 2,3 dioxygenase; a novel therapeutic target for treatment of diseases. Expert Opin. Ther. Targets 13 (8), 987–1012.

Temann, U.A., Geba, G.P., Rankin, J.A., Flavell, R.A., 1998. Expression of interleukin 9 in the lungs of transgenic mice causes airway inflammation, mast cell hyperplasia, and bronchial hyperresponsiveness. J. Exp. Med. 188 (7), 1307–1320.

Thornton, A.M., Shevach, E.M., 1998. CD4+CD25+ immunoregulatory T cells suppress polyclonal T cell activation in vitro by inhibiting interleukin 2 production. J. Exp. Med. 188 (2), 287–296.

Thornton, A.M., Donovan, E.E., Piccirillo, C.A., Shevach, E.M., 2004. Cutting edge: IL-2 is critically required for the in vitro activation of CD4+CD25+ T cell suppressor function. J. Immunol. 172 (11), 6519–6523.

Timmons, B.W., Cieslak, T., 2008. Human natural killer cell subsets and acute exercise: a brief review. Exerc. Immunol. Rev. 14, 8–23.

Tliba, O., Panettieri Jr., R.A., 2009. Noncontractile functions of airway smooth muscle cells in asthma. Annu. Rev. Physiol. 71, 509–535.

Truyen, E., Coteur, L., Dilissen, E., Overbergh, L., Dupont, L.J., Ceuppens, J.L., Bullens, D.M., 2006. Evaluation of airway inflammation by quantitative Th1/Th2 cytokine mRNA measurement in sputum of asthma patients. Thorax 61 (3), 202–208.

Tulic, M.K., Hamid, Q., 2003. Contribution of the distal lung to the pathologic and physiologic changes in asthma: potential therapeutic target Roger S. Mitchell lecture. Chest 123 (3 Suppl.), 348S–355S.

Tulic, M.K., Christodoulopoulos, P., Hamid, Q., 2001. Small airway inflammation in asthma. Respir. Res. 2 (6), 333–339.

Turner, J., Jones, C.E., 2009. Regulation of mucin expression in respiratory diseases. Biochem. Soc. Trans. 37 (Pt 4), 877–881.

Untersmayr, E., Bises, G., Starkl, P., Bevins, C.L., Scheiner, O., Boltz-Nitulescu, G., Wrba, F., Jensen-Jarolim, E., 2010. The high affinity IgE receptor Fc epsilonRI is expressed by human intestinal epithelial cells. PLoS One 5 (2), e9023.

Van Oosterhout, A.J., Ladenius, A.R., Savelkoul, H.F., Van Ark, I., Delsman, K.C., Nijkamp, F.P., 1993. Effect of anti-IL-5 and IL-5 on airway hyperreactivity and eosinophils in guinea pigs. Am. Rev. Respir. Dis. 147 (3), 548–552.

Van Scott, M.R., Aycock, D., Cozzi, E., Salleng, K., Stallings 3rd, H.W., 2005. Separation of bronchoconstriction from increased ventilatory drive in a nonhuman primate model of chronic allergic asthma. J. Appl. Physiol. 99 (6), 2080–2086.

Van Scott, M.R., Hooker, J.L., Ehrmann, D., Shibata, Y., Kukoly, C., Salleng, K., Westergaard, G., Sandrasagra, A., Nyce, J., 2004. Dust mite-induced asthma in cynomolgus monkeys. J. Appl. Physiol. 96 (4), 1433–1444.

Vignola, A.M., Scichilone, N., Bousquet, J., Bonsignore, G., Bellia, V., 2003. Aging and asthma: pathophysiological mechanisms. Allergy 58 (3), 165–175.

Vock, C., Hauber, H.P., Wegmann, M., 2010. The other T helper cells in asthma pathogenesis. J. Allergy (Cairo) 2010, (519298).

Wagner, E.M., Bleecker, E.R., Permutt, S., Liu, M.C., 1998. Direct assessment of small airways reactivity in human subjects. Am. J. Respir. Crit. Care Med. 157 (2), 447–452.

Wahba, W.M., 1983. Influence of aging on lung function—clinical significance of changes from age twenty. Anesth. Analg. 62 (8), 764–776.

Wang, M., Ellison, C.A., Gartner, J.G., HayGlass, K.T., 1998. Natural killer cell depletion fails to influence initial CD4 T cell commitment in vivo in exogenous antigen-stimulated cytokine and antibody responses. J. Immunol. 160 (3), 1098–1105.

Watanabe, A., Hayashi, H., 1990. Allergen-induced biphasic bronchoconstriction in rats. Int. Arch. Allergy Appl. Immunol. 93 (1), 26–34.

Watarai, H., Sekine-Kondo, E., Shigeura, T., Motomura, Y., Yasuda, T., Satoh, R., Yoshida, H., Kubo, M., Kawamoto, H., Koseki, H., Taniguchi, M., 2012. Development and function of invariant natural killer T cells producing T(h)2- and T(h)17-cytokines. PLoS Biol. 10 (2), e1001255.

Wei, H., Zhang, J., Xiao, W., Feng, J., Sun, R., Tian, Z., 2005. Involvement of human natural killer cells in asthma pathogenesis: natural killer 2 cells in type 2 cytokine predominance. J. Allergy Clin. Immunol. 115 (4), 841–847.

Weiszer, I., Patterson, R., Pruzansky, J.J., 1968. Ascaris hypersensitivity in the rhesus monkey. I. A model for the study of immediate type thypersensitity in the primate. J. Allergy 41 (1), 14–22.

Wenzel, S.E., 2009. Eosinophils in asthma—closing the loop or opening the door? N. Engl. J. Med. 360 (10), 1026–1028.

Wenzel, S., Holgate, S.T., 2006. The mouse trap: it still yields few answers in asthma. Am. J. Respir. Crit. Care Med. 174 (11), 1173–1176; discussion 1176–8.

Whitrow, M.J., Moore, V.M., Rumbold, A.R., Davies, M.J., 2009. Effect of supplemental folic acid in pregnancy on childhood asthma: a prospective birth cohort study. Am. J. Epidemiol. 170 (12), 1486–1493.

Wingett, D., Nielson, C.P., 2003. Divergence in NK cell and cyclic AMP regulation of T cell CD40L expression in asthmatic subjects. J. Leukoc. Biol. 74 (4), 531–541.

Wong, S.H., Walker, J.A., Jolin, H.E., Drynan, L.F., Hams, E., Camelo, A., Barlow, J.L., Neill, D.R., Panova, V., Koch, U., Radtke, F., Hardman, C.S., Hwang, Y.Y., Fallon, P.G., McKenzie, A.N., 2012. Transcription factor RORalpha is critical for nuocyte development. Nat. Immunol. 13 (3), 229–236.

Wu, L.C., 2011. Immunoglobulin E receptor signaling and asthma. J. Biol. Chem. 286 (38), 32891–32897.

Xanthou, G., Alissafi, T., Semitekolou, M., Simoes, D.C., Economidou, E., Gaga, M., Lambrecht, B.N., Lloyd, C.M., Panoutsakopoulou, V., 2007. Osteopontin has a crucial role in allergic airway disease through regulation of dendritic cell subsets. Nat. Med. 13 (5), 570–578.

Xisto, D.G., Farias, L.L., Ferreira, H.C., Picanco, M.R., Amitrano, D., Lapa, E., Silva, J.R., Negri, E.M., Mauad, T., Carnielli, D., Silva, L.F., Capelozzi, V.L., Faffe, D.S., Zin, W.A., Rocco, P.R., 2005. Lung parenchyma remodeling in a murine model of chronic allergic inflammation. Am. J. Respir. Crit. Care Med. 171 (8), 829–837.

Yang, X., Wang, S., Fan, Y., Han, X., 2000. IL-10 deficiency prevents IL-5 overproduction and eosinophilic inflammation in a murine model of asthma-like reaction. Eur. J. Immunol. 30 (2), 382–391.

Yates, D.H., Barnes, P.J., Thomas, P.S., 1993. Tumor necrosis factor a alters human bronchial reactivity and induces inflammatory cell influx. Am. Rev. Respir. Dis. 147, A1011.

Ying, S., Robinson, D.S., Varney, V., Meng, Q., Tsicopoulos, A., Moqbel, R., Durham, S.R., Kay, A.B., Hamid, Q., 1991. TNF alpha mRNA expression in allergic inflammation. Clin. Exp. Allergy 21 (6), 745–750.

Zhang, D.H., Yang, L., Cohn, L., Parkyn, L., Homer, R., Ray, P., Ray, A., 1999. Inhibition of allergic inflammation in a murine model of asthma by expression of a dominant-negative mutant of GATA-3. Immunity 11 (4), 473–482.

Zhou, B., Comeau, M.R., De Smedt, T., Liggitt, H.D., Dahl, M.E., Lewis, D.B., Gyarmati, D., Aye, T., Campbell, D.J., Ziegler, S.F., 2005. Thymic stromal lymphopoietin as a key initiator of allergic airway inflammation in mice. Nat. Immunol. 6 (10), 1047–1053.

Zhu, Z., Homer, R.J., Wang, Z., Chen, Q., Geba, G.P., Wang, J., Zhang, Y., Elias, J.A., 1999. Pulmonary expression of interleukin-13 causes inflammation, mucus hypersecretion, subepithelial fibrosis, physiologic abnormalities, and eotaxin production. J. Clin. Invest. 103 (6), 779–788.

Zhu, Z., Oh, S.Y., Cho, Y.S., Zhang, L., Kim, Y.K., Zheng, T., 2010. Tyrosine phosphatase SHP-1 in allergic and anaphylactic inflammation. Immunol. Res. 47 (1-3), 3–13.

Zielinski, C.E., Corti, D., Mele, F., Pinto, D., Lanzavecchia, A., Sallusto, F., 2011. Dissecting the human immunologic memory for pathogens. Immunol. Rev. 240 (1), 40–51.

Zosky, G.R., Sly, P.D., 2007. Animal models of asthma. Clin. Exp. Allergy 37 (7), 973–988.

Zuhdi Alimam, M., Piazza, F.M., Selby, D.M., Letwin, N., Huang, L., Rose, M.C., 2000. Muc-5/5ac mucin messenger RNA and protein expression is a marker of goblet cell metaplasia in murine airways. Am. J. Respir. Cell. Mol. Biol. 22 (3), 253–260.

Chapter 36

Role of Genetic Factors in Pulmonary Disease Susceptibility

Berran Yucesoy, Victor J. Johnson and Michael I. Luster

Chapter Outline

1. INTRODUCTION

Common human diseases are multifactorial in nature, being influenced by complex interactions between environmental, lifestyle, and genetic factors. Genetic variations and their interaction with these factors represent disease modifiers. In addition to influencing disease occurrence in the general population, they can affect disease severity or the response to specific treatment regimens. The human genome is intrinsically variable, including single nucleotide polymorphisms (SNPs), variable number tandem repeats, copy number variations, and structural changes such as deletions, duplications, and inversions. The majority of this variation is in the form of SNPs, which result from single base changes that substitute one nucleotide for another. They occur at a frequency of approximately 1 in 1000 bp throughout the genome. Approximately 50% of SNPs are located in the noncoding regions, whereas the remaining lead to either missense or synonymous (silent) mutations (Brookes, 1999; Halushka et al., 1999). SNPs that affect a phenotype are referred to as *functional* variants. Most functional SNPs do not directly affect gene expression but rather change the rate of mRNA synthesis and degradation (Collins et al., 1999; Shastry, 2009). Common SNPs with minor allele frequency greater than 1–5% in the human population generally possess low or incomplete penetrance and normally must interact with other genes or environmental factors to influence disease. Susceptibility profiles, which reflect a unique combination of multiple common risk variants, can help to define interindividual variability and more susceptible groups in the general population. Molecular epidemiological studies have identified a number of genetic variations within immune/inflammatory genes as susceptibility factors for common immune-mediated/chronic inflammatory diseases, including many pulmonary disorders. Data are now accumulating on their role in diseases of occupational and environmental origin. Some susceptibility factors are shared between different pulmonary disorders, indicating a common immunological and genetic background. This chapter focuses on immunogenetic predisposition to pulmonary disorders, particularly those that can be related to occupational exposures such as occupational asthma (OA), chronic beryllium disease (CBD), coal workers' pneumoconiosis (CWP), and silicosis.

2. SNP DISEASE ASSOCIATIONS

Genetic association studies identify genes that include inherited variants that influence disease risk and, in certain instances, responses to therapeutic treatments or environmental exposures. Linkage disequilibrium, which is a nonrandom association of genes between loci, gene–gene, and gene–environment interactions and the existence of multiple disease loci, are the major factors to be considered for the interpretation of results. The candidate gene

Comparative Biology of the Normal Lung. http://dx.doi.org/10.1016/B978-0-12-404577-4.00036-9

approach has been commonly used in searching for susceptibility variants. These studies use biological information with regard to disease pathogenesis and select candidate genes on the basis of prior information from animal models, in vitro/in vivo experiments, or data-mining for disease-associated genes (Hoh and Ott, 2004). On the other hand, genome-wide association studies (GWAS) provide an opportunity for studying dense sets of SNPs across the genome simultaneously for associations. Because there is no a priori hypothesis, this approach has the potential to identify and test new candidate SNPs and regions. Although GWAS platforms do not cover the entire genome and large sample sizes are needed to compensate for the multiple comparisons, using this approach, a number of candidate regions/genes for complex phenotypes have been identified (Michel et al., 2010). For example, chromosomal regions 4q, 5q31, 6p21, 9q21, 10p, 12q, 13q, 16p12, and 17q21 were found to be associated with asthma/allergy and related phenotypes (Blumenthal, 2005; Akhabir and Sandford, 2011; Hancock et al., 2009; Potaczek and Kabesch, 2011; Hirota et al., 2011), and genetic markers on chromosomes 4, 8, 12, and 19 were found to contribute to susceptibility to chronic obstructive pulmonary disease (COPD) (DeMeo et al., 2004; Hersh et al., 2006; Cho et al., 2011a; Pillai et al., 2009).

3. GENE—ENVIRONMENT INTERACTIONS

Complex diseases result from interactions between multiple genes, lifestyle, and environmental/occupational factors. Gene—environment studies investigate the interaction between common susceptibility variants and commonly encountered exposures. Exposure characteristics (e.g., dose, duration) may overwhelm differences associated with genetic variability; therefore, they are of primary consideration in genetic association studies. Occupational diseases may differ from other environmentally induced diseases because there is a quantifiable environmental component behind gene—environment interactions.

Environmental/occupational molecular epidemiology studies have primarily focused on examining hypothesis-driven associations between exposures and specific genetic variations. In this respect, a number of susceptibility markers have been identified in several occupational lung diseases, such as silicosis, asthma, CBD, and CWP (Yucesoy et al., 2001a; McCanlies et al., 2004; Ates et al., 2008; Bernstein et al., 2006). These studies have provided strong evidence that the pathogenesis of occupational lung diseases involves gene—environment interactions. The common link among these diseases are chronic inflammation and alterations in host immunity, which suggests that exposed individuals who have susceptibility variants that influence immune and inflammatory responses are more likely to develop disease than those without the susceptibility variants. The remainder of this chapter discusses lung diseases prompted by occupational exposures and gene—environment interactions using several examples. A simplified, theoretical model of how genetic and environmental, occupational, and lifestyle factors may interact to induce pulmonary disease is shown in Figure 1.

4. GENETIC SUSCEPTIBILITY AND GENE—ENVIRONMENT INTERACTIONS IN PULMONARY DISEASES

4.1 Occupational Asthma

Asthma is an inflammatory airway disease characterized by airway hyperreactivity, chronic eosinophilic inflammation, and mucus hypersecretion and is one of the most significant examples of gene—environment interaction causing disease. In addition to exposures to an allergen (e.g., diisocyanates), a number of triggers such as environmental pollutants (e.g., O_3, tobacco smoke, diesel exhaust), domestic exposures (e.g., bioaerosols containing endotoxin), and/or viral respiratory infections interact with host susceptibility factors (Kleeberger and Peden, 2005). Linkage studies have suggested a number of candidate genes that are related to asthma phenotypes. In particular, variants of interleukin (IL)-4, IL-4RA, IL-13, β-2 adrenergic receptor (ADRB2), human leukocyte antigen (HLA)-DR, the β-subunit of the high-affinity IgE receptor (FceRI), and CD14 genes have been consistently associated with asthma-related phenotypes in independent association studies (Ober and Hoffjan, 2006; Postma and Koppelman, 2009; Vercelli, 2008; Ober and Yao, 2011). The genetic variants identified to date had only small effects and unknown inheritance patterns indicating complex polygenic nature of disease.

Among occupational exposures, the diisocyanates are the most frequently reported cause of OA in the workplace. These agents are widely used in polymerization reactions for manufacturing surface coatings, varnishes, paints, urethane foams, insulation, and adhesives. Genetic studies investigating the immunopathogenesis of OA have focused largely on the HLA alleles. HLA class II molecules are involved in the presentation of intracellularly processed peptides to CD4+ T-helper cells. These molecules are highly polymorphic and the variations in their protein structure help to determine the specific epitopes presented to T cells. A number of significant associations have been found between OA and HLA class II molecules. The HLA DQB1*0503 and the allelic combination DQB1*0201/0301 are associated with susceptibility to diisocyanate asthma (DA), whereas the DQB1*0501 allele and the DQA1*0101-DQB1*0501-DR1 haplotype are believed to be protective (Bignon et al., 1994). Others

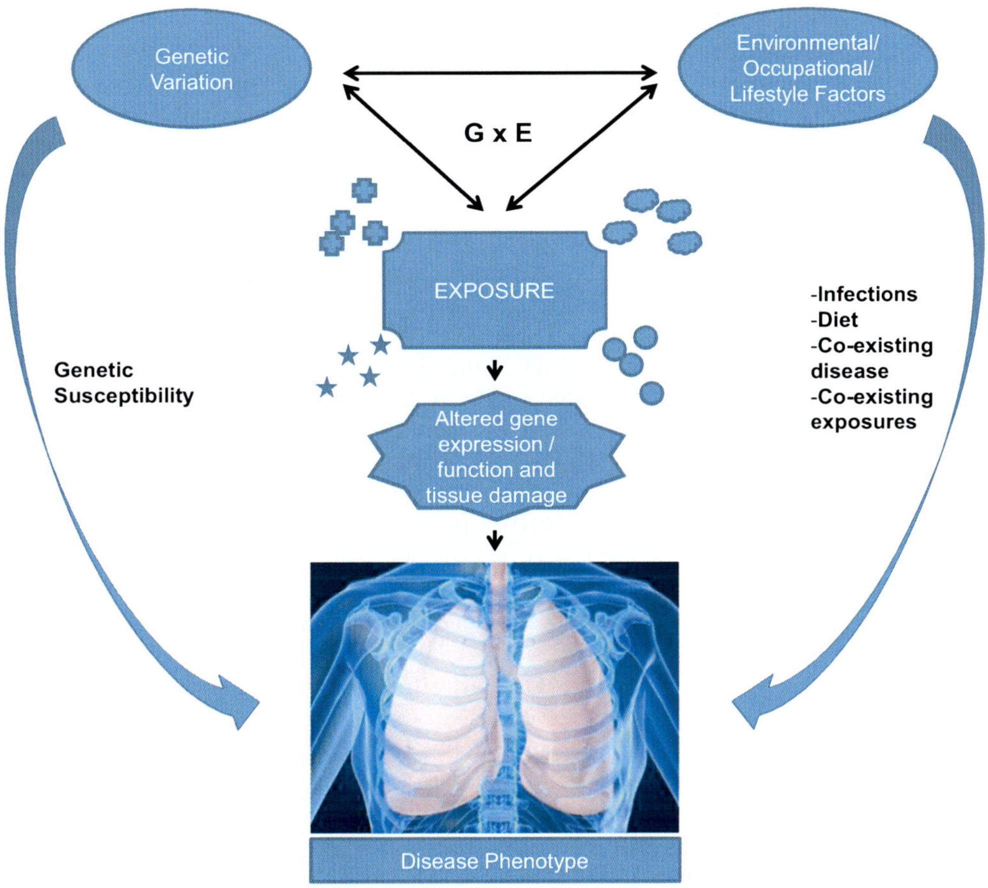

FIGURE 1 Interaction between genetic and environmental/occupational/lifestyle factors in the development of pulmonary disease.

reported increased frequency of the DQA1*0104 and DQA1*0101-HLA-DQB1*0501 in DA and asymptomatic exposed workers, respectively (Mapp et al., 2000). In another study, a significantly higher proportion of subjects with DA were found to express the HLA-DQB1*0503-associated aspartic acid at residue 57 (Balboni et al., 1996). On the other hand, the HLA haplotypes DRB1*15-DPB1*05, and HLA DRB1*1501-DQB1*0602-DPB1*0501 were reported as a susceptibility marker for the development of toluene diisocyanate-asthma in Koreans (Kim et al., 2006; Hur et al., 2009). In addition, an association was found between DRB1*0901-DQB1*0303-DPB1*0501 haplotype and the presence of serum-specific IgG in methylene diisocyanate-exposed workers (Hur et al., 2009).

Similarly, HLA class 2 alleles were found to be associated with OA caused by other low-molecular-weight agents. Workers with red cedar asthma had a higher frequency of HLA DQB1*0603 and DQB1*0302 alleles compared to exposed control subjects and a reduced frequency of DQB1*0501 allele (Horne et al., 2000). They also reported an increased frequency of the DRB1*0401-DQB1*0302 haplotype and a decreased frequency of the DRB1*0101-DQB1*0501 haplotype. In another study, the DQB1*0501 allele was found to be more prevalent in workers exposed to organic acid anhydrides than in controls (Jones et al., 2004). The protective effect of DQB1*0501 for other low-molecular-weight sensitizers (isocyanates and plicatic acid) suggests differential affinities of these chemicals for specific class II molecules. In addition, a significant excess of HLA-DR3 was reported in subjects with specific IgE antibody to trimellitic anhydride (Young et al., 1995) and positive skin prick tests to ammonium hexachloroplatinate (Newman Taylor et al., 1999).

Several studies reported associations between HLA alleles and OA caused by high-molecular-weight allergens. While a higher prevalence of HLA-B15 and DR4 was reported in workers with laboratory animal allergy (Low et al., 1988), the prevalence rate of HLA-B16 was higher in symptom-free subjects (Sjostedt et al., 1996). In another study, the HLA-DR7 was found to be associated with work-related respiratory symptoms and specific IgE to rat urinary proteins. HLA-DR3 was reported to be protective against sensitization (Jeal et al., 2003). Pacheco et al., reported a significant gene–environment interaction affecting airway function in laboratory animal workers (Pacheco et al., 2010). More highly endotoxin-exposed workers with CD14 -1619G allele have significantly lower forced expiratory volume in 1 s (FEV_1) and forced expiratory flow (FEF 25–75) percent predicted than those with CD14/-1619AA alleles. The gene–environment effect was marked in atopic workers.

Based on their role in allergen-induced airway inflammation, Bernstein et al. investigated SNPs in immune response genes (IL-4Rα, IL-13, and CD14) in a group of workers with DA as determined by specific inhalation challenge—the criterion standard for establishing allergic asthma. The results demonstrated increased frequencies of IL-4RA I50V allele and combinatorial genotypes of IL4RA (I50V), IL-13 (R110Q), and CD14 (C159T) in workers with DA, supporting the notion that immune mechanisms play an important role in the pathogenesis of DA (Bernstein et al., 2006, 2011). Many of the examples of OA gene–environment interactions have been collected and tabulated in Table 1.

4.2 Chronic Beryllium Disease

CBD is a granulomatous lung disease caused by beryllium (Be) exposure in the workplace. CBD continues to occur in industries where Be is manufactured and processed, such as aerospace, nuclear, automotive, and electronics. Be exposure leads to a cell-mediated hypersensitivity (delayed, type IV) reaction, in which Be forms haptens with native proteins, leading to the production of the specific allergen (Maier et al., 2003). Accumulation of Be-specific CD4 (+) T cells and persistent lung inflammation play a key role in the immunopathogenesis of CBD. A number of molecular epidemiology studies showed that the presence of glutamic acid in position 69 of the B1 chain of the HLA-DPB1 molecule (coded for by *HLA-DPB1E69* also known as HLA-DPB1Glu69 or simply Glu69) confers an increased risk for both BeS and CBD (McCanlies et al., 2004; Maier et al., 2003; Richeldi et al., 1993; Wang et al., 2001; Fontenot et al., 2000; Lombardi et al., 2001). Thus, the allele frequency for *HLA-DPB1E69* is highest in workers with CBD (F = 0.38–74), next highest in BeS (F = 0.20–0.60), and least in nondiseased/nonsensitized workers (F = 0.18–0.35) (McCanlies et al., 2004; Maier et al., 2003; Wang et al., 2001; Rossman et al., 2002; Saltini et al., 2001; Wang et al., 1999). Importantly, a dose-dependent effect of *HLA-DPB1E69* alleles has been demonstrated, suggesting that Glu69 is a potential marker of disease severity in addition to overall disease risk (Maier et al., 2003). Although *HLA-DPB169* is more frequent in individuals with BeS and CBD, 30–40% of exposed workers carrying *HLA-DPB1E69* do not develop CBD or BeS (McCanlies et al., 2004; Maier et al., 2003; Samuel and Maier, 2008). This suggests that other host and environmental factors likely play key roles in the etiology of CBD. Studies investigating the interaction between the *HLA-DPB1E69* and Be exposure have shown independent and additive effects of Glu69 carriage and Be exposure in the development of BeS and CBD (Richeldi et al., 1997; Van Dyke et al., 2011). The *HLA-DQB1G86* and *HLA-DRB1S11* alleles were reported to be more frequent in individuals with CBD (Rossman et al., 2002). Maier et al., found a reduced frequency of the *HLA-DRB1*01* and *DQB1*05* alleles in workers with CBD. They also reported that *DRB1*13* and *DQB1*06* were associated with CBD in the absence of Glu69 (Maier et al., 2003). One study showed that the *DRBE71* allele is a risk factor for both CBD and BeS in the absence of Glu69 and highlighted the importance of interactions between peptides and T cells in the development of CBD (Rosenman et al., 2010). Chemically specific Be–protein interactions were also investigated using a computational approach. Glu69 alleloforms with the greatest negative surface charge were found to confer the highest risk for CBD and, irrespective of allele, equal risk for BeS (Snyder et al., 2008). Current HLA research includes investigating whether risk is associated with any or only certain Glu69 alleles or allelic combinations. In a recent study, *HLA-DPB1* and *HLA-DRB1* loci were typed at the allele level in subjects with BeS, CBD, and nonaffected, BE-exposed controls. The *DRBE71* was reported to be a risk factor for CBD and BeS in the absence of *DPBE69* (Rosenman et al., 2011).

Non-HLA genetic studies have also identified some significant associations. Maier et al. investigated eight SNPs within the *CCR5* gene which is implicated in the chemotaxis and activation of leukocyte subsets. The results showed that *CCR5-5663* and *-3458* variants were associated with worsening pulmonary function over time in CBD (Sato et al., 2010). The *-509C* and codon *10T* variants of the transforming growth factor–beta1 (*TGFβ*) gene, a cytokine which helps control the fibrotic/Th1 response, were found to be associated with more severe granulomatous disease in CBD (Jonth et al., 2007). While Saltini et al. reported an association between the *TNFα-308*02* variant and BeS and CBD, this result was not confirmed in a large population-based study (Saltini et al., 2001; McCanlies et al., 2007). On the other hand, Dotti et al., reported that the *TNFα -308A* and *-857T* variants are significantly associated with BeS (Dotti et al., 2004). McCanlies et al., showed that *IL-1α-1142, -3769*, and *-4697* variants were significantly associated with CBD compared to individuals with BeS or nonsensitized workers after adjusting for Glu69 status (McCanlies et al., 2010). These results suggested that the formation of granulomas in CBD may require an independent inflammatory response controlled by genes unrelated to Be recognition.

TABLE 1 Genetic Variants of Inflammatory/Immune Response Genes Involved in Occupational Asthma

Disease	Gene/Variant	Population Case/Control	Effect Size (OR, RR, or *P*)	Reference
Occupational Asthma				
Diisocyanates (TDI)	HLADQB1/*503	56/32	RR 9.8	Bignon et al. (1994)
	HLA-DQB1/*0201/0301		RR 9.5	Bignon et al. (1994)
	HLA-DQB1/*501		RR 0.1	Bignon et al. (1994)
	HLA-DQA1/*0101-*0102		RR 0.1	Bignon et al. (1994)
Diisocyanates (TDI)	HLA-DQB1/*503	30/12/126[a]	RR 2.9	Balboni et al. (1996)
	HLA-DQB1/*501		RR 0.04	Balboni et al. (1996)
Diisocyanates (TDI)	HLA-DQB1/*0503	67/27	P = 0.009	Mapp et al. (2000)
	HLA-DQB1/*0501		P = 0.01	Mapp et al. (2000)
	HLA-DQA1/*0101		P = 0.004	Mapp et al. (2000)
	HLA-DQA1/*0104		P = 0.005	Mapp et al. (2000)
Diisocyanates (TDI)	HLA-DRB1/*1501- DQB1/*0602-DPB1/*0501	84/47/127[a]	OR 4.4 (1.5–13.1)	Choi et al. (2009)
Diisocyanates (HDI)	IL-4RA(I50V) II + IL-13(R110Q) RR + CD14 (C159T) CT	103/115/150[a]	OR 3.86 (1.26–12.0)	Bernstein et al. (2011)
Acid anhydrides	HLA-DR3	30/30	OR 6.0	Young et al. (1995)
	HLA-DQ5	52/73	OR 4.3 (1.7–11.0)	Jones et al. (2004)
	HLA-DQB1/*0501		OR 3.0 (1.2–7.4)	Jones et al. (2004)
	HLA-DQ-DR1		OR 3.0 (1.2–11.0)	Jones et al. (2004)
Laboratory animal	HLA-DR1/*07	109/397	OR 1.8 (1.1–2.9)	Jeal et al. (2003)
	HLA-DR1/*03		OR 0.5 (0.3–1.0)	Jeal et al. (2003)
Laboratory animal	TLR4/8551G	335	OR 2.5 (1.5–5.5)	Pacheco et al. (2008)
Platinum salts	HLA-DR3	44/57	OR 2.3 (1.0–5.6)	Newman Taylor et al. (1999)
	HLA-DR6		OR 0.4 (0.2–0.8)	Newman Taylor et al. (1999)
Red cedar	HLA-DQB1/*0302	56/63	OR 4.9 (1.3–18.6)	Horne et al. (2000)
	HLA-DQB1/*0603		OR 2.9 (1.0–8.2)	Horne et al. (2000)
	HLA-DQB1/*0501		OR 0.3 (0.1–0.8)	Horne et al. (2000)
	HLA-DRB1/*0401-DQB1*0302		OR 10.3	Horne et al. (2000)
	HLA-DRB1/*0101-DQB1*0501		OR 0.3	Horne et al. (2000)
Trimellitic anhydride	HLA-DR3	30/30	OR 16.0	Young et al. (1995)
Wheat flour	TLR4/-2027 GG	381	OR 0.16 (0.04–0.73)	Cho et al. (2011b)

[a]*Case/asymptomatic exposed/unexposed normal control.*

4.3 Silicosis

Silicosis is an interstitial lung disease resulting from inhalation of crystalline silica or quartz. The disease is characterized by chronic inflammation leading to severe pulmonary fibrotic changes and is prevalent in miners exposed chronically to dusts and other irritants. In a study conducted in underground coal miners participating in the National Coal Workers Autopsy Study (NCWAS), a strong association was found between severity and presence of silicosis and variants of TNFα (−308) and IL-1RA (+2018) genes (Yucesoy et al., 2001a). These results supported previous experimental and clinical data demonstrating the importance of TNFα and IL-1 in regulating the development of silicotic lesions. Another study was conducted in the NCWAS population to investigate associations between genetic variability within genes involved in inflammatory and fibrotic processes and susceptibility to progressive massive fibrosis (PMF). Although, there was no significant single−locus association, the polygenotype of VEGF +405/ICAM-1 +241/IL-6 -174 (C-A-G) conferred an increased risk for PMF (Yucesoy et al., 2008). These results suggested that particular SNPs in the VEGF, ICAM-1, and IL-6 genes may influence the interaction and amplification process between these genes, and play an important role in the pathogenesis of pulmonary fibrosis. In studies of South African miners, TNFα polymorphisms at positions -238, -376, -308 of the promoter region were found to be associated with severe silicosis (Corbett et al., 2002). IL-12p40 SNPs were also studied in silicosis for their potential influence on inflammatory response by affecting Th1/Th2 cytokine balance. Stanilova et al. found an association between IL-12BA/C 3′UTR SNP and silicosis severity in Bulgarian miners (Stanilova et al., 2007). The minor allele carriers of the IL-12Bpro promoter polymorphism were associated with a protection against susceptibility to silicosis. This was attributed to the higher production of IL-12p40 in homozygous wild-type patients (Stanilova et al., 2008). As demonstrated for other occupational lung diseases, HLA alleles were also found to be related to silicosis pathogenesis. Honda et al., reported an association between increased frequency of the HLA-Bw54 allele and silicosis (Honda et al., 1993). HLA-DR gene locus (HLA-DRB1 and DQB1 alleles) was also found to be related to silicosis susceptibility in Asian miners (Ueki et al., 2001; Yuan et al., 2002).

4.4 Coal Workers' Pneumoconiosis

CWP is characterized by chronic inflammation that usually leads to fibrosis due to irritation caused by dust exposure. Although, the mechanisms of susceptibility to disease development and progression are not clear, several studies have proposed a genetic component (Borm and Schins, 2001). In a study investigating associations between TNFα gene polymorphisms and development of CWP, the frequency of the TNFα -308 variant was significantly increased in miners with CWP (50%), as compared with miners without lung disease (25%) or non-miners (29%) (Zhai et al., 1998). Similarly, the frequency of TNFα -308 variant was greater in Korean CWP patients (20.6%) and in patients with a large opacity (28.2%) in comparison with patients with simple CWP (13.4%) (Kim et al., 2002). The presence of the minor allele was significantly associated with increased relative risk of complicated CWP. In another study, the -308 A allele frequency was found to be higher in patients with CWP and nodular CWP compared to controls and subjects with PMF (Wang et al., 2005). The TNFα -308 and lymphotoxin-α (LTA) NcoI polymorphisms were also investigated in a prospective epidemiological study in miners differentially exposed to coal dust and cigarette smoke. While the LTA NcoI polymorphism found associated with CWP prevalence in miners with low blood catalase activity, the TNFα -308 SNP showed an interaction with erythrocyte GSH-Px activity in individuals with high occupational exposure (Nadif et al., 2003). The TNFα -238 and -308 polymorphisms were associated with CWP development and severity in Turkish coal miners, respectively (Ates et al., 2008). They also reported a protective effect of the IL-6 -174 variant in the development and severity of CWP. Genetic variants in IL-18, chemokines and chemokine receptors have also been studied in relation to CWP. The IL18 -137C allele was found to be associated with lower and slower progression of the computed tomography (CT) score and lower pneumoconiosis prevalence in a group of miners (Nadif et al., 2006a). The same group also reported an association between the CX3CR1 V249I and lower CWP prevalence, especially in miners with high dust exposure. In addition, the CX3CR1 I249 allele was significantly associated with lower and slower progression of the CT score in CCR5 Delta32 carriers (Nadif et al., 2006b). Recently, the IL-4 -590 CT/CC genotypes were reported to be associated with a significantly decreased risk of CWP in a Chinese population, particularly among subgroups of age <65 years and dust exposure years ≥ 26 years (Wang et al., 2011).

4.5 Lung Function Decline

Endotoxin exposure confers an increased risk for non-atopic asthma and lung function decline in exposed workers. In a study investigating associations between TNFα -308 and LTA (252A/G rs909253) polymorphisms, endotoxin exposure and lung function in cotton workers, endotoxin exposure was associated with faster FEV1 decline among subjects carrying high producer TNFα -308 A allele. The annual changes in FEV1 were higher in G allele carriers of the LTA polymorphism. The effect modification of these variants was prominent in never

TABLE 2 Genetic Variants of Inflammatory/Immune Response Genes Involved in Metal, Mineral, and Organic Dust-Related Pulmonary Diseases

Disease	Gene/Variant	Population Case/Control	Effect Size (OR or P)	Reference
CBD				
	HLA-DPB1/E69	104/64/727[a]	9.4 (5.4–16.6)	McCanlies et al. (2004)
	HLA-DQB1/G86	22/23/82	P < 0.04	Rossman et al. (2002)
	HLA-DRB1/S11		P < 0.03	Rossman et al. (2002)
	CCR5/-3458	88/86/173[a]	P < 0.0001	Sato et al. (2010)
	TGFβ1/-509	130/135	P = 0.01	Jonth et al. (2007)
	IL-1A/-1142	85/61/730[a]	3.02 (1.36–6.70)	McCanlies et al. (2010)
	IL-1A/-3769		2.51 (1.21–5.19)	McCanlies et al. (2010)
	IL-1A/-4697		2.56 (1.24–5.29)	McCanlies et al. (2010)
Silicosis				
	TNFα/-308	325/164	2.25 (1.4–3.6)	Yucesoy et al. (2001a)
	TNFα/-238		4.00 (2.4–6.8)	Yucesoy et al. (2001a)
	IL-1Ra/+2018	318/163	2.15 (1.4–3.3)	Yucesoy et al. (2001b)
	TNFα/-308, -238	121/120	P = 0.022, 0.034	Corbett et al. (2002)
	IL-12/A/C-3′UTR	62/138	6.56 (1.7–25.6)	Stanilova et al. (2007)
	HLA/Bw54	43/315	P < 0.05	Honda et al. (1993)
CWP				
	TNFα/-308	78/56	3.0 (1.0–9.0)	Zhai et al. (1998)
	TNFα/-308	80/54	3.18 (1.25–8.10)	Kim et al. (2002)
	TNFα/-308	84/44/122[b]	P < 0.01	Wang et al. (2005)
	TNFα/-238	67/92	3.79 (1.37–10.46)	Ates et al. (2008)
	TNFα/-308		2.84 (1.08–7.39)	Ates et al. (2008)
	IL-6/-174		0.48 (0.21–0.93)	Ates et al. (2008)
	IL-18/-137C	66/134	P = 0.02	Nadif et al. (2006a)
	CCR5/Δ32 CX3CR1/V249I		P = 0.03	Nadif et al. (2006b)
	IL-4/-590	556/541	0.74 (0.58-0.05)	Wang et al. (2011)
Organic Dust Exposure/Lung Function				
Endotoxin	TNFα/-308	263/230		Zhang et al. (2007)
Grain dust	TNFα/-308	243	P < 0.05	Pahwa et al. (2009)
Endotoxin	CD14/-260	408	P ≤ 0.01	Smit et al. (2011)

[a]*CBD/Be-sensitized/control.*
[b]*CWP/PMF/control.*

smokers when joint effects of endotoxin exposure and smoking were considered (Zhang et al., 2007). Another study conducted in male grain handlers demonstrated that years in the grain industry is an effect modifier between TNFα -308 genotype and longitudinal decline in FEV1. The annual FEV1 decline for grain workers carrying the TNFα -308 G allele and worked in the grain industry for <10 years were lower by comparison to those of grain workers with AA genotype and had been in the industry for <10 years (Pahwa et al., 2009).

The association between occupational endotoxin exposure and wheeze in agricultural workers was found to be significantly modified by genetic variants in the CD14 (-260 C/T, rs2569190) and MD2 (lymphocyte antigen 96, rs10808798 T/C) genes. The prevalence of wheeze was increased with increasing endotoxin concentration in individuals with the CD14 and MD2 major alleles. The carriers of the functional CD14 -260C allele were more responsive to endotoxin exposure than T allele homozygotes (Smit et al., 2011). Some of the associations related to metal, mineral, and organic dust-induced pulmonary diseases are summarized in Table 2.

CONCLUSIONS

Molecular epidemiology offers a powerful approach to the identification of genetic variants that influence susceptibility to many common diseases. The guidance from human genetic studies may lead to better animal models and in vitro systems to investigate critical pathways and molecular regulators of disease processes. Despite the rapid growth of publications, some of the genetic associations lack consistency across different studies. The inconsistency can be explained by a number of factors including the differences in study populations, phenotype characterization, exposure assessment, characterization of other environmental exposure (e.g., air pollution, smoking), intermediate phenotypes (e.g., airway hyperresponsiveness), statistical inconsistencies, and other potentially modifiable risk factors such as lifestyle. In spite of sometimes contradictory findings in association studies, a growing number of susceptibility markers have been identified for complex diseases using large, well-characterized DNA collections, and high throughput genotyping technologies. Understanding how the genetic variability between individuals at the sequence level translates into variation in disease susceptibility and the interaction between genetic and environmental/occupational factors will provide new insights into disease pathogenesis as well as opportunities into developing novel therapeutic, preventive, and educational strategies.

DISCLAIMER

The findings and conclusions in this report are those of the authors and do not necessarily represent the views of the National Institute for Occupational Safety and Health.

REFERENCES

Akhabir, L., Sandford, A.J., 2011. Genome-wide association studies for discovery of genes involved in asthma. Respirology 16, 396–406.

Ates, I., Suzen, H.S., Yucesoy, B., Tekin, I.O., Karakaya, A., 2008. Association of cytokine gene polymorphisms in CWP and its severity in Turkish coal workers. Am. J. Ind. Med. 51, 741–747.

Balboni, A., Baricordi, O.R., Fabbri, L.M., Gandini, E., Ciaccia, A., et al., 1996. Association between toluene diisocyanate-induced asthma and DQB1 markers: a possible role for aspartic acid at position 57. Eur. Respir. J. 9, 207–210.

Bernstein, D.I., Wang, N., Campo, P., Chakraborty, R., Smith, A., et al., 2006. Diisocyanate asthma and gene-environment interactions with IL4RA, CD-14, and IL-13 genes. Ann. Allergy Asthma Immunol. 97, 800–806.

Bernstein, D.I., Kissling, G.E., Khurana Hershey, G., Yucesoy, B., Johnson, V.J., et al., 2011. Hexamethylene diisocyanate asthma is associated with genetic polymorphisms of CD14, IL-13, and IL-4 receptor alpha. J. Allergy Clin. Immunol. 128, 418–420.

Bignon, J.S., Aron, Y., Ju, L.Y., Kopferschmitt, M.C., Garnier, R., et al., 1994. HLA class II alleles in isocyanate-induced asthma. Am. J. Respir. Crit. Care Med. 149, 71–75.

Blumenthal, M.N., 2005. The role of genetics in the development of asthma and atopy. Curr. Opin. Allergy Clin. Immunol. 5, 141–145.

Borm, P.J., Schins, R.P., 2001. Genotype and phenotype in susceptibility to coal workers' pneumoconiosis. The use of cytokines in perspective. Eur. Respir. J. Suppl. 32, 127s–133s.

Brookes, A.J., 1999. The essence of SNPs. Gene 234, 177–186.

Cho, H.J., Kim, S.H., Kim, J.H., Choi, H., Son, J.K., et al., 2011. Effect of toll-like receptor 4 gene polymorphisms on work-related respiratory symptoms and sensitization to wheat flour in bakery workers. Ann. Allergy Asthma Immunol. 107, 57–64.

Cho, M.H., Castaldi, P.J., Wan, E.S., Siedlinski, M., Hersh, C.P., et al., 2011. A genome-wide association study of COPD identifies a susceptibility locus on chromosome 19q13. Hum. Mol. Genet. 21, 947–957.

Choi, J.H., Lee, K.W., Kim, C.W., Park, C.S., Lee, H.Y., et al., 2009. The HLA DRB1*1501-DQB1*0602-DPB1*0501 haplotype is a risk factor for toluene diisocyanate-induced occupational asthma. Int. Arch. Allergy Immunol. 150, 156–163.

Collins, A., Lonjou, C., Morton, N.E., 1999. Genetic epidemiology of single-nucleotide polymorphisms. Proc. Natl. Acad. Sci. U. S. A. 96, 15173–15177.

Corbett, E.L., Mozzato-Chamay, N., Butterworth, A.E., De Cock, K.M., Williams, B.G., et al., 2002. Polymorphisms in the tumor necrosis factor-alpha gene promoter may predispose to severe silicosis in black South African miners. Am. J. Respir. Crit. Care Med. 165, 690–693.

DeMeo, D.L., Celedon, J.C., Lange, C., Reilly, J.J., Chapman, H.A., et al., 2004. Genome-wide linkage of forced mid-expiratory flow in chronic obstructive pulmonary disease. Am. J. Respir. Crit. Care Med. 170, 1294–1301.

Dotti, C., D'Apice, M.R., Rogliani, P., Novelli, G., Saltini, C., et al., 2004. Analysis of TNF-alpha promoter polymorphisms in the susceptibility to beryllium hypersensitivity. Sarcoidosis Vasc. Diffuse Lung Dis. 21, 29–34.

Fontenot, A.P., Torres, M., Marshall, W.H., Newman, L.S., Kotzin, B.L., 2000. Beryllium presentation to CD4+ T cells underlies disease-susceptibility HLA-DP alleles in chronic beryllium disease. Proc. Natl. Acad. Sci. U. S. A. 97, 12717–12722.

Halushka, M.K., Fan, J.B., Bentley, K., Hsie, L., Shen, N., et al., 1999. Patterns of single-nucleotide polymorphisms in candidate genes for blood-pressure homeostasis. Nat. Genet. 22, 239–247.

Hancock, D.B., Romieu, I., Shi, M., Sienra-Monge, J.J., Wu, H., et al., 2009. Genome-wide association study implicates chromosome 9q21.31 as a susceptibility locus for asthma in mexican children. PLoS Genet. 5, e1000623.

Hersh, C.P., DeMeo, D.L., Raby, B.A., Litonjua, A.A., Sylvia, J.S., et al., 2006. Genetic linkage and association analysis of COPD-related traits on chromosome 8p. COPD 3, 189–194.

Hirota, T., Takahashi, A., Kubo, M., Tsunoda, T., Tomita, K., et al., 2011. Genome-wide association study identifies three new susceptibility loci for adult asthma in the Japanese population. Nat. Genet. 43, 893–896.

Hoh, J., Ott, J., 2004. Genetic dissection of diseases: design and methods. Curr. Opin. Genet. Dev. 14, 229–232.

Honda, K., Kimura, A., Dong, R.P., Tamai, H., Nagato, H., et al., 1993. Immunogenetic analysis of silicosis in Japan. Am. J. Respir. Cell. Mol. Biol. 8, 106–111.

Horne, C., Quintana, P.J., Keown, P.A., Dimich-Ward, H., Chan-Yeung, M., 2000. Distribution of DRB1 and DQB1 HLA class II alleles in occupational asthma due to western red cedar. Eur. Respir. J. 15, 911–914.

Hur, G.Y., Lee, K.W., Lee, H.Y., Choi, G.S., Park, H.J., et al., 2009. HLA class II allele and IgG sensitization to methylene diisocyanate in exposed workers. Ann. Allergy Asthma Immunol. 103, 174–175.

Jeal, H., Draper, A., Jones, M., Harris, J., Welsh, K., et al., 2003. HLA associations with occupational sensitization to rat lipocalin allergens: a model for other animal allergies? J. Allergy Clin. Immunol. 111, 795–799.

Jones, M.G., Nielsen, J., Welch, J., Harris, J., Welinder, H., et al., 2004. Association of HLA-DQ5 and HLA-DR1 with sensitization to organic acid anhydrides. Clin. Exp. Allergy 34, 812–816.

Jonth, A.C., Silveira, L., Fingerlin, T.E., Sato, H., Luby, J.C., et al., 2007. TGF-beta 1 variants in chronic beryllium disease and sarcoidosis. J. Immunol. 179, 4255–4262.

Kim, K.A., Cho, Y.Y., Cho, J.S., Yang, K.H., Lee, W.K., et al., 2002. Tumor necrosis factor-alpha gene promoter polymorphism in coal workers' pneumoconiosis. Mol. Cell. Biochem. 234–235, 205–209.

Kim, S.H., Oh, H.B., Lee, K.W., Shin, E.S., Kim, C.W., et al., 2006. HLA DRB1*15-DPB1*05 haplotype: a susceptible gene marker for isocyanate-induced occupational asthma? Allergy 61, 891–894.

Kleeberger, S.R., Peden, D., 2005. Gene-environment interactions in asthma and other respiratory diseases. Annu. Rev. Med. 56, 383–400.

Lombardi, G., Germain, C., Uren, J., Fiorillo, M.T., du Bois, R.M., et al., 2001. HLA-DP allele-specific T cell responses to beryllium account for DP-associated susceptibility to chronic beryllium disease. J. Immunol. 166, 3549–3555.

Low, B., Sjostedt, L., Willers, S., 1988. Laboratory animal allergy—possible association with HLA B15 and DR4. Tissue Antigens 31, 224–226.

Maier, L.A., McGrath, D.S., Sato, H., Lympany, P., Welsh, K., et al., 2003. Influence of MHC class II in susceptibility to beryllium sensitization and chronic beryllium disease. J. Immunol. 171, 6910–6918.

Mapp, C.E., Beghe, B., Balboni, A., Zamorani, G., Padoan, M., et al., 2000. Association between HLA genes and susceptibility to toluene diisocyanate-induced asthma. Clin. Exp. Allergy 30, 651–656.

McCanlies, E.C., Ensey, J.S., Schuler, C.R., Kreiss, K., Weston, A., 2004. The association between HLA-DPB1(Glu69) and chronic beryllium disease and beryllium sensitization. Am. J. Ind. Med. 46, 95–103.

McCanlies, E.C., Schuler, C.R., Kreiss, K., Frye, B.L., Ensey, J.S., et al., 2007. TNF-alpha polymorphisms in chronic beryllium disease and beryllium sensitization. J. Occup. Environ. Med. 49, 446–452.

McCanlies, E.C., Yucesoy, B., Mnatsakanova, A., Slaven, J.E., Andrew, M., et al., 2010. Association between IL-1A single nucleotide polymorphisms and chronic beryllium disease and beryllium sensitization. J. Occup. Environ. Med. 52, 680–684.

Michel, S., Liang, L., Depner, M., Klopp, N., Ruether, A., et al., 2010. Unifying candidate gene and GWAS approaches in asthma. PLoS One 5, e13894.

Nadif, R., Jedlicka, A., Mintz, M., Bertrand, J.P., Kleeberger, S., et al., 2003. Effect of TNF and LTA polymorphisms on biological markers of response to oxidative stimuli in coal miners: a model of gene-environment interaction. Tumour necrosis factor and lymphotoxin alpha. J. Med. Genet. 40, 96–103.

Nadif, R., Mintz, M., Marzec, J., Jedlicka, A., Kauffmann, F., et al., 2006a. IL18 and IL18R1 polymorphisms, lung CT and fibrosis: a longitudinal study in coal miners. Eur. Respir. J. 28, 1100–1105.

Nadif, R., Mintz, M., Rivas-Fuentes, S., Jedlicka, A., Lavergne, E., et al., 2006b. Polymorphisms in chemokine and chemokine receptor genes and the development of coal workers' pneumoconiosis. Cytokine 33, 171–178.

Newman Taylor, A.J., Cullinan, P., Lympany, P.A., Harris, J.M., Dowdeswell, R.J., et al., 1999. Interaction of HLA phenotype and exposure intensity in sensitization to complex platinum salts. Am. J. Respir. Crit. Care Med. 160, 435–438.

Ober, C., Hoffjan, S., 2006. Asthma genetics 2006: the long and winding road to gene discovery. Genes Immun. 7, 95–100.

Ober, C., Yao, T.C., 2011. The genetics of asthma and allergic disease: a 21st century perspective. Immunol. Rev. 242, 10–30.

Pacheco, K., Maier, L., Silveira, L., Goelz, K., Noteware, K., et al., 2008. Association of toll-like receptor 4 alleles with symptoms and sensitization to laboratory animals. J. Allergy Clin. Immunol. 122, 896–902 e894.

Pacheco, K.A., Rose, C.S., Silveira, L.J., Van Dyke, M.V., Goelz, K., et al., 2010. Gene–environment interactions influence airways function in laboratory animal workers. J. Allergy Clin. Immunol. 126, 232–240.

Pahwa, P., Nakagawa, K., Koehncke, N., McDuffie, H.H., 2009. Effects of polymorphic variations in tumor necrosis factor alpha and occupational exposure to grain dust on longitudinal decline in pulmonary function. J. Agromedicine 14, 215–221.

Pillai, S.G., Ge, D., Zhu, G., Kong, X., Shianna, K.V., et al., 2009. A genome-wide association study in chronic obstructive pulmonary disease (COPD): identification of two major susceptibility loci. PLoS Genet. 5, e1000421.

Postma, D.S., Koppelman, G.H., 2009. Genetics of asthma: where are we and where do we go? Proc. Am. Thorac. Soc. 6, 283–287.

Potaczek, D.P., Kabesch, M., 2011. Current concepts of IgE regulation and impact of genetic determinants. Clin. Exp. Allergy 42, 852–871.

Richeldi, L., Sorrentino, R., Saltini, C., 1993. HLA-DPB1 glutamate 69: a genetic marker of beryllium disease. Science 262, 242–244.

Richeldi, L., Kreiss, K., Mroz, M.M., Zhen, B., Tartoni, P., et al., 1997. Interaction of genetic and exposure factors in the prevalence of berylliosis. Am. J. Ind. Med. 32, 337–340.

Rosenman, K.D., Rossman, M., Hertzberg, V., Reilly, M.J., Rice, C., et al., 2010. HLA class II DPB1 and DRB1 polymorphisms associated with genetic susceptibility to beryllium toxicity. Occup. Environ. Med. 68, 487–493.

Rosenman, K.D., Rossman, M., Hertzberg, V., Reilly, M.J., Rice, C., et al., 2011. HLA class II DPB1 and DRB1 polymorphisms associated with genetic susceptibility to beryllium toxicity. Occup. Environ. Med. 68, 487–493.

Rossman, M.D., Stubbs, J., Lee, C.W., Argyris, E., Magira, E., et al., 2002. Human leukocyte antigen Class II amino acid epitopes: susceptibility and progression markers for beryllium hypersensitivity. Am. J. Respir. Crit. Care Med. 165, 788–794.

Saltini, C., Richeldi, L., Losi, M., Amicosante, M., Voorter, C., et al., 2001. Major histocompatibility locus genetic markers of beryllium sensitization and disease. Eur. Respir. J. 18, 677–684.

Samuel, G., Maier, L.A., 2008. Immunology of chronic beryllium disease. Curr. Opin. Allergy Clin. Immunol. 8, 126–134.

Sato, H., Silveira, L., Spagnolo, P., Gillespie, M., Gottschall, E.B., et al., 2010. CC chemokine receptor 5 gene polymorphisms in beryllium disease. Eur. Respir. J. 36, 331–338.

Shastry, B.S., 2009. SNPs: impact on gene function and phenotype. Methods Mol. Biol. 578, 3–22.

Sjostedt, L., Willers, S., Orbaek, P., 1996. Human leukocyte antigens in occupational allergy: a possible protective effect of HLA-B16 in laboratory animal allergy. Am. J. Ind. Med. 30, 415–420.

Smit, L.A., Heederik, D., Doekes, G., Koppelman, G.H., Bottema, R.W., et al., 2011. Endotoxin exposure, CD14 and wheeze among farmers: a gene–environment interaction. Occup. Environ. Med. 68, 826–831.

Snyder, J.A., Demchuk, E., McCanlies, E.C., Schuler, C.R., Kreiss, K., et al., 2008. Impact of negatively charged patches on the surface of MHC class II antigen-presenting proteins on risk of chronic beryllium disease. J. R. Soc. Interface 5, 749–758.

Stanilova, S., Miteva, L., Prakova, G., 2007. Interleukin-12B-3'UTR polymorphism in association with IL-12p40 and IL-12p70 serum levels and silicosis severity. Int. J. Immunogenet. 34, 193–199.

Stanilova, S., Miteva, L., Prakova, G., 2008. IL-12Bpro and GSTP1 polymorphisms in association with silicosis. Tissue Antigens 71, 169–174.

Ueki, A., Isozaki, Y., Tomokuni, A., Ueki, H., Kusaka, M., et al., 2001. Different distribution of HLA class II alleles in anti-topoisomerase I autoantibody responders between silicosis and systemic sclerosis patients, with a common distinct amino acid sequence in the HLA-DQB1 domain. Immunobiology 204, 458–465.

Van Dyke, M.V., Martyny, J.W., Mroz, M.M., Silveira, L.J., Strand, M., et al., 2011. Exposure and genetics increase risk of beryllium sensitisation and chronic beryllium disease in the nuclear weapons industry. Occup. Environ. Med. 68, 842–848.

Vercelli, D., 2008. Discovering susceptibility genes for asthma and allergy. Nat. Rev. Immunol. 8, 169–182.

Wang, Z., White, P.S., Petrovic, M., Tatum, O.L., Newman, L.S., et al., 1999. Differential susceptibilities to chronic beryllium disease contributed by different Glu69 HLA-DPB1 and -DPA1 alleles. J. Immunol. 163, 1647–1653.

Wang, Z., Farris, G.M., Newman, L.S., Shou, Y., Maier, L.A., et al., 2001. Beryllium sensitivity is linked to HLA-DP genotype. Toxicology 165, 27–38.

Wang, X.T., Ohtsuka, Y., Kimura, K., Muroi, M., Ishida, T., et al., 2005. Antithetical effect of tumor necrosis factor-alphagene polymorphism on coal workers' pneumoconiosis (CWP). Am. J. Ind. Med. 48, 24–29.

Wang, M., Wang, S., Song, Z., Ji, X., Zhang, Z., et al., 2011. Associations of IL-4, IL-4R, and IL-13 gene polymorphisms in coal workers' pneumoconiosis in China: a case-control study. PLoS One 6, e22624.

Young, R.P., Barker, R.D., Pile, K.D., Cookson, W.O., Taylor, A.J., 1995. The association of HLA-DR3 with specific IgE to inhaled acid anhydrides. Am. J. Respir. Crit. Care Med. 151, 219–221.

Yuan, B., Zhang, Z., Li, H., Chang, Y., Liu, Z., et al., 2002. [The relationship between silicosis and the polymorphism of HLA-DRB1 *, DQB1 * genes]. Zhonghua Laodong Weisheng Zhiyebing Zazhi 20, 93–96.

Yucesoy, B., Vallyathan, V., Landsittel, D.P., Sharp, D.S., Weston, A., et al., 2001a. Association of tumor necrosis factor-alpha and interleukin-1 gene polymorphisms with silicosis. Toxicol. Appl. Pharmacol. 172, 75–82.

Yucesoy, B., Vallyathan, V., Landsittel, D.P., Sharp, D.S., Matheson, J., et al., 2001b. Polymorphisms of the IL-1 gene complex in coal miners with silicosis. Am. J. Ind. Med. 39, 286–291.

Yucesoy, B., Johnson, V.J., Kissling, G.E., Fluharty, K., Kashon, M.L., et al., 2008. Genetic susceptibility to progressive massive fibrosis in coal miners. Eur. Respir. J. 31, 1177–1182.

Zhai, R., Jetten, M., Schins, R.P., Franssen, H., Borm, P.J., 1998. Polymorphisms in the promoter of the tumor necrosis factor-alpha gene in coal miners. Am. J. Ind. Med. 34, 318–324.

Zhang, H., Hang, J., Wang, X., Zhou, W., Sun, B., et al., 2007. TNF polymorphisms modify endotoxin exposure-associated longitudinal lung function decline. Occup. Environ. Med. 64, 409–413.

Appendix 1

Architecture of Nasal Passages and Larynx

TABLE 1 Summary of Available Morphometry Data on Nasal Passages, Paranasal Sinuses, and Larynx

	Nasal Cavity						Paranasal Sinuses			Larynx		
	Length	Maximum Vertical Height	Perimeter	Cross-Sectional Surface Area	Surface Area of Wall	Volume	Wall Width and Height	Surface Area of Wall	Volume	Linear Measurements	Diameter	Surface Area
Human[a]	+	+		+	+	+	+	+	+	+		
Rhesus monkey	+	+	+	+	+	+	+	+	+			
Beagle dog	+	+	+	+	+	+						
Rat: Sprague-Dawley	+	+	+	+	+	+						
F 344	+		+	+	+	+						
B6C3F1 mouse	+				+	+						
Guinea pig	+		+	+	+	+						
Other species					+[b]	+[c]				+[d]	+[e]	

[a]*Many additional measurements.*
[b]*Antelope, vole, dik dik, camel, elephant seal.*
[c]*Camel, antelope.*
[d]*Large cats, rabbit, sheep.*
[e]*Rabbit.*

TABLE 2 Interspecies Comparison of Nasal Cavity Characteristics

	Sprague-Dawley Rat	Guinea Pig	Beagle Dog	Rhesus Monkey	Man
Body weight	250 g	600 g	10 kg	7 kg	~70 kg
Naris cross-section	0.7 mm^2	2.5 mm^2	16.7 mm^2	22.9 mm^2	140 mm^2
Bend in naris	40°	40°	30°	30°	
Length	23 cm	3.4 cm	10 cm	5.3 cm	7–8 cm
Greatest vertical diameter	9.6 mm	12.8 mm	23 mm	27 mm	40–45 mm
Surface area (both sides of nasal cavity)	10.4 cm^2	27.4 cm^2	220.7 cm^2	61.6 cm^2	181 cm^2
Volume (both sides)	0.4 cm^3	0.9 cm^3	20 cm^3	8 cm^3	16–19 cm^3(does not include sinuses)
Bend in nasopharynx	15°	30°	30°	80°	~90°
Turbinate complexity	Complex scroll	Complex scroll	Very complex membranous	Simple scroll	Simple scroll

Reprinted with permission from: Schreider, 1983. Nasal airway anatomy and inhalation deposition in experimental animals and people. In: Reznik, G., Stinson, S.F. (Eds.), Nasal Tumors in Animals and Man. Vol. III. Experimental Nasal Carcinogenesis. Copyright CRC Press, Inc., Boca Raton, FL, pp. 1–26.

TABLE 3 Morphometric Analysis of Three Rhesus Monkey Noses

	#1	#2	#3
Weight	7.27 kg	6.99 kg	7.26 kg
Volume			
Nasal cavity	4.44 cm^2	5.94 cm^2	8.17 cm^2
Maxillary sinus	1.00 cm^2	3.75 cm^2	2.00 cm^2
Surface area			
Nasal cavity	48.70 cm^2	68.65 cm^2	75.55 cm^2
Maxillary sinus	7.48 cm^2	15.30 cm^2	10.4 cm^2

Source: Gross et al. (1987)

TABLE 4 Dimensions of Cross Sections of a Nasal Cast of a Rhesus Monkey

Distance from Anterior End of Nose (mm)	Perimeter (mm)	Area (mm^2)
0	45.8	55.6
3	43.3	51.0
8	62.3	61.0
13	73.8	86.2
18	95.6	105.6
23	119.3	161.1
25	121.0	155.0
28	169.1	180.8

Continued

TABLE 4 Dimensions of Cross Sections of a Nasal Cast of a Rhesus Monkey—cont'd

Distance from Anterior End of Nose (mm)	Perimeter (mm)	Area (mm^2)
33	192.9	165.6
38	211.8	165.3
40	171.8	164.1
43	143.4	161.7
48	116.1	152.6
53	58.5	91.5
58	45.0	53.2
63	37.9	38.0
68	24.0	18.8
73	29.0	27.3
78	37.9	66.0
80	86.5	224.9
83	73.6	253.1
88	90.5	263.0
88[a]	75.4	256.4
90	118.8	257.3
93	34.1	43.8
95	27.6	15.8
98	29.5	18.2
103	35.0	72.0
105	33.3	75.4
110	27.9	56.1

[a]*Main airway alone. Does not include separate pockets.*
Reprinted with permission from: Schreider, 1983. Nasal airway anatomy and inhalation deposition in experimental animals and people. In: Reznik, G., Stinson, S.F. (Eds.), Nasal Tumors in Animals and Man. Vol. III. Experimental Nasal Carcinogenesis. Copyright CRC Press, Inc., Boca Raton, FL, pp. 1–26.

TABLE 5 Nasal Fossa of Four Nonhuman Primates

	Tupaia glis	*Nycticebus coucang*	*Macaca fascicularis*	*Hylobates agilis*
Nasal fossa length expressed as nose-tip to occiput length, %	62	52	35	35
Ratio of height to width of nasal fossa at three-fifth distance	1:1.9	1:1.5	1:0.9	1:0.9
Nasoturbinal	1	1	nil	1 (reduced)
Ethmoturbinals	4	4	1	2 (fused)
Maxilloturbinal	1	1	1	1

Source: Loo (1973). (Courtesy S. Karger AG, Basel.)

TABLE 6 Dimensions (mm) of the Maxillary Sinuses in *Macaca* Measured from Casts

	Rhesus	Cynomolgus
Sagittal	7.4–21.5 mm	8.5–24.8 mm
Vertical	5.2–16.0 mm	5.0–22.2 mm
Transverse	4.8–11.8 mm	3.3–12.0 mm
Volume	0.56 cm^3	1.07 cm^3

Source: Kocon and Stepien (1967).

TABLE 7 Mean Volume (cm^3) of the Maxillary Sinuses in *Macaca* in Relation to Age

	Macaca Mulatta (Rhesus)		*Macaca fascicularis* (Cynomolgus)	
Age (years)	Right	Left	Right	Left
1–2	0.18	0.17	0.19	0.13
3–4	0.24	0.27	0.23	0.27
5–6	0.58	0.62	1.45	1.41
6–8	0.94	0.92	1.49	1.49

Kocon and Stepien (1967).

TABLE 8 Dimensions of a Cross Section of a Nasal Cast of a Beagle Dog[a]

Distance from Anterior End of Nose (mm)	Perimeter (mm)	Area (mm^2)
0	40.0	33.3
5	50.3	42.3
10	72.5	44.8
15	66.5	32.0
20	63.8	37.6
25	70.1	49.6
30	106.3	75.9
35	143.4	95.5
40	294.8	155.0
43	503.0	205.8
50	674.3	228.4
50[b]	665.6	227.3
55	470.6	267.3
60	333.9	312.5

Continued

TABLE 8 Dimensions of a Cross Section of a Nasal Cast of a Beagle Dog[a]—cont'd

Distance from Anterior End of Nose (mm)	Perimeter (mm)	Area (mm^2)
65	318.8	240.9
65[b]	64.3	91.2
70	424.1	277.5
70[b]	55.1	68.0
75	266.8	289.5
75[b]	109.1	143.8
80	200.9	272.5
80[a]	54.5	113.9
90	121.5	138.3
90[a]	52.6	105.3
100	44.4	95.2
100[a]	35.6	92.6
150	39.3	88.8
150	64.5	139.1
160	184.5	1022.5
170	131.0	871.3
180	210.0	1055.4
185	250.9	933.6
185[a]	53.1	89.1
190	50.8	112.6
195	50.6	175.4
205	57.5	251.3
275	57.0	237.1

[a]*7 kg body weight.*
[b]*Main airway alone. Does not include separate pockets.*
Reprinted with permission from: Schreider, 1983. Nasal airway anatomy and inhalation deposition in experimental animals and people. In: Reznik, G., Stinson, S.F. (Eds.), Nasal Tumors in Animals and Man. Vol. III. Experimental Nasal Carcinogenesis. Copyright CRC Press, Inc., Boca Raton, FL, pp. 1–26.

TABLE 9 Morphometric Data from the Nasal cast of Seven Male F 344 Rats

	N[d] (7)[e]			R[d] (14)			B[d] (22)			E[d] (28)			J[d] (35)			A[d] (50)			G[d] (441)		
Section Number	D[a]	P[b]	A[c]	D	P	A	D	P	A	D	P	A	D	P	A	D	P	A	D	P	A
1	0.9	12.85	2.28	1.4	13.6	2.77	1.8	23.66	6.85	1.6	13.16	2.88	1.70	27.27	6.6	1.7	13.04	2.92	1.70	29.17	8.86
2	2.0	11.78	2.18	3.2	19.18	4.07	4.0	26.8	6.70	3.7	19.32	3.41	4.00	20.79	4.21	4.1	22.99	5.02	3.9	26.01	8.31
3	3.0	20.06	2.98	4.5	27.52	5.25	6.2	35.17	7.02	6.2	31.67	5.84	7.00	34.69	6.81	6.9	31.67	9.51	6.1	31.66	11.00
4	4.3	25.66	6.49	6.4	25.21	7.26	8.4	33.61	8.54	8.95	37.2	6.72	9.40	39.53	6.80	8.6	49.36	10.15	8.35	47.91	13.68
5	5.5	23.3	3.55	8.3	35.21	9.66	10.6	47.1	10.24	11.15	59.34	11.79	12.10	61.42	13.93	10.2	51.51	12.43	10.15	55.97	14.46
6	7.6	5.09	1.04	9.8	23.28	4.30	12.6	22.68	10.78	12.75	57.96	10.01	14.75	59.22	12.46	12.2	38.11	14.84	12.35	57.59	16.78
7	9.0	3.99	0.92	11.3	5.17	1.43	15.3	6.10	2.10	15.55	42.7	7.06	17.75	9.06	3.87	14.0	78.16	18.74	14.65	73.08	33.88
8	10.7	4.07	1.05	13.7	10.33	3.41	17.2	5.04	1.68	17.85	5.04	1.38	20.85	6.31	2.72	16.1	65.33	17.24	17.15	52.7	21.74
9	12.2	4.39	1.16	15.2	13.29	5.14	19.5	4.70	1.40	20.15	5.66	2.12	24.15	6.86	3.20	18.1	36.09	6.41	19.15	52.0	13.54
10	14.2	4.83	1.54	16.9	4.18	1.45	21.8	5.50	2.21	22.15	6.66	3.33	27.05	22.17	9.23	20.5	21.75	5.21	22.25	7.79	4.3
11	15.7	7.82	2.05	19.5	5.07	1.56	23.7	5.90	2.42	23.35	17.50	8.4	28.65	20.15	7.85	22.5	6.33	2.61	24.55	7.13	3.81
12							25.6	8.19	3.02	25.95	17.30	5.51	30.85	15.93	8.13	24.9	6.31	2.81	28.15	8.70	5.66
13							28.4	14.52	4.25	26.95	12.35	4.53				26.9	7.54	3.64	29.85	9.08	5.76
14							31.2	16.05	5.03							29.2	7.74	3.83	32.05	11.04	5.93
15							33.6	16.05	5.03							31.2	19.33	8.73	34.45	22.12	13.20
16																34.1	21.17	7.34	36.65	13.3	12.25
17																36.5	17.27	6.11	39.85	18.11	10.40
18																39.5	13.4.6	6.11	44.25	19.33	13.19

[a] *Distance along centerline of nose model from the nose tip (in millimeters).*
[b] *Perimeter (in millimeters).*
[c] *Cross-sectional area (in square millimeters).*
[d] *individual animal identification.*
[e] *Age (days).*
Source: Patra (1987).

TABLE 10 Estimated Surface Area as a Function of Age and Location of Nasal Casts from Male F 344 Rats

	Age (days)						
Estimated Surface Area (mm^2)	**7**	**14**	**22**	**28**	**35**	**50**	**441**
Turbinate area	106	239	402	586	616	811	920
Nasopharynx area	51	73	210	116	203	242	337
Total	157	312	612	702	818	1053	1257
Body weight (g)	12.1	20.8	38.7	65.9	96.8	167	366

Source: Patra et al. (1987).

TABLE 11 Dimensions of a Cross Section of a Nasal Cast of a Sprague-Dawley Rat

Distance from Anterior End of Nose (mm)	Perimeter (mm)	Area (mm^2)
0	8.3	1.4
1	5.6	1.3
2	11.9	2.0
3	21.7	3.5
4	23.4	3.8
5	27.3	4.5
5.5	35.0	7.6
7	38.3	9.9
8	44.0	9.3
10	73.9	13.2
12	63.5	10.3
14	54.4	10.3
16	48.3	8.1
17	67.6	16.7
17[a]	57.8	13.0
18	58.4	21.7
19	104.1	25.8
19[a]	47.5	14.0
20	97.8	17.0
20[a]	9.0	5.0
21	44.5	12.5
21[a]	9.0	5.0
22	30.9	7.7
23	17.4	5.9
23[a]	8.9	4.9

Continued

TABLE 11 Dimensions of a Cross Section of a Nasal Cast of a Sprague-Dawley Rat—cont'd

Distance from Anterior End of Nose (mm)	Perimeter (mm)	Area (mm^2)
25	7.5	3.6
32	7.9	3.7
35	8.6	2.8
37	7.7	2.2
40	8.4	3.4
41	8.7	3.4
42	23.7	15.8
43	24.2	14.9
44	9.2	3.2
44.5	12.8	3.2
45	7.6	1.8
45.5	7.4	1.9
46	7.8	3.1
47	6.5	3.0
52	7.1	3.8

[a]*Main airway alone. Does not include separate pockets.*

Reprinted with permission from: Schreider, 1983. Nasal airway anatomy and inhalation deposition in experimental animals and people. In: Reznik, G., Stinson, S.F. (Eds.), Nasal Tumors in Animals and Man. Vol. III. Experimental Nasal Carcinogenesis. Copyright CRC Press, Inc., Boca Raton, FL, pp. 1–26.

TABLE 12 Morphometric Data on Male Rats and Mice

	Length (mm)	Surface Area (mm^2)	Volume (mm^3)
Rats			
F 344 rat (115 g)[a]	7.3	798.6	155.5
F 344 rat (288 g)[a]	9.1	1343.5	256.7
Sprague-Dawley (250 g)[b]	2.3	10.4 cm^2	
Mouse			
B6C3F1 (30 g)[a]	4.9	277.7	32.5
B6C3F1 (33 g)[a]	5.1	289.0	31.5
Deer mice (Peromyscus)[c] (16.7 g)	8.9–11.8	196.6	16.2

[a]*Gross et al. (1982).*
[b]*Schreider (1983).*
[c]*Adams (1972).*

TABLE 13 Dimensions of a Cross Section of a Nasal Cast of a Guinea Pig[a]

Distance from Anterior End of Nose (mm)	Perimeter (mm)	Area (mm^2)
0	18.0	5.0
1	14.0	4.4
2	15.3	4.9
2.5	12.4	3.6
3	13.9	4.2
3.5	18.5	5.4
4	32.9	13.0
5.0	19.8	13.4
8	66.4	20.4
11	88.0	21.5
16	91.5	38.6
18	91.2	31.1
20	82.0	36.2
22	62.7	30.1
24	75.1	38.1
26	109.4	44.4
26[a]	94.3	39.2
28	153.5	58.1
28[a]	14.9	11.2
31	184.6	73.4
31[a]	14.5	10.8
34	14.2	13.3
34[a]	8.2	11.7
39	13.9	13.7
45	13.2	13.2
50	12.0	8.1
55	8.0	4.1
56	18.9	8.8
57	20.7	12.7
59	8.9	2.8
62	9.1	5.1
70	8.6	5.7

[a]*Main airway alone. Does not include separate pockets.*

Reprinted with permission from: Schreider, 1983. Nasal airway anatomy and inhalation deposition in experimental animals and people. In: Reznik, G., Stinson, S.F. (Eds.), Nasal Tumors in Animals and Man. Vol. III. Experimental Nasal Carcinogenesis. Copyright CRC Press, Inc., Boca Raton, FL, pp. 1–26.

TABLE 14 Nasal Passages: Other Species

Species	Surface Area of Wall	Volume	References
Camel		>1000 cm^3	
Rhynchotragus kirkii	147 cm^2	—	Schmidt-Nielsen (1981)
Dik-dik antelope	95–200 cm^2	4.7–8.1 cm^3	Kamau et al. (1984)
Vole	317 mm^2	—	
Vole maxilloturbinate	22.8 mm^2	—	Kamau et al. (1984)
Bank vole *Clethrinomys*	3.177 cm^2	—	Kamau et al. (1984)
Narrow headed vole *Microtus gregalis*	1.938 cm^2	—	Gurtovoi (1966)
			Gurtovoi (1966)
Bats (4 species)	38.3–208	—	Bhatnagar and Kalien (1975)
Red toothed shrew	107–274	—	Larochelle and Baron (1989)

TABLE 15 Comparison of Various Nasal Parameters in the Giraffe and the Ox

	Giraffe	Ox
Nasal cross-sectional epithelial perimeter (at the level of the third premolar)	116 cm	58 cm
Distance from premaxilla to cribiform plate	40 cm	35 cm
Distance from cribiform plate to foramen occipitate magnum	18[a]	14[a]
Weighted of the animal	~600 kg	~600 kg

[a]*Estimated from the ruler given with the figure.*
Source: Langman et al. (1979).

TABLE 16 Human Larynx

	Mean of the Dist. in cm (±SE)		Significant Male-Female Differences[c]
	Male[a]	Female[b]	
1	2.15 ± 0.06	1.58 ± 0.04	+
2	4.90 ± 0.12	3.84 ± 0.14	+
3	5.01 ± 0.09	3.98 ± 0.11	+
4	4.65 ± 0.08	3.88 ± 0.13	+
5	3.75 ± 0.06	2.88 ± 0.09	+
6	2.53 ± 0.07	1.93 ± 0.06	+
7	4.19 ± 0.09	3.02 ± 0.08	+
8	3.91 ± 0.07	2.90 ± 0.07	+
9	4.06 ± 0.06	3.27 ± 0.07	+

Continued

TABLE 16 Human Larynx—cont'd

	Mean of the Dist. in cm (±SE)		Significant Male-Female Differences[c]
	Male[a]	Female[b]	
10	2.53 ± 0.04	2.07 ± 0.05	+
11	0.81 ± 0.03	0.68 ± 0.02	+
12	2.07 ± 0.06	1.60 ± 0.04	+
13	2.14 ± 0.06	1.59 ± 0.05	+
14	1.42 ± 0.04	1.08 ± 0.03	+
15	0.90 ± 0.03	0.81 ± 0.02	+
16	0.81 ± 0.03	0.56 ± 0.02	+
17	1.32 ± 0.04[d]	1.51 ±0.15[d]	+

The mean values of 17 sample sites identified in Figure 1(A) and standard error of the mean (SE). Sex differences are also presented.
[a]*11 male specimens.*
[b]*10 female specimens.*
[c]*Sex difference were examined with the student's* t *test. Significance,* $p < 0.05$ *are noted (+).*
[d]*Angle in radians.*
Source: Zrunek et al. (1988).

TABLE 17 Human Larynx - Length of the Vocal Cord

	Length of the Vocal Cord (mm): Mean (SEM)		References
Men	14.15	(0.37)	Stell et al. (1978)
Women	10.11	(0.11)	
Men	15.5	—	Grant (1947)
Women	11.5	—	

TABLE 18 Human Larynx - Heights of the Vocal Cord

		(Along the Vocal Cord)			
	Anterior Comissure	Quarter	Half	Three-Quarters	Posterior End of the Cord
Men	4.78 (0.40)	4.95 (0.32)	5.5 (0.26)	5.29 (0.27)	5.0 (0.22)
Women	2.78 (0.30)	3.35 (0.21)	4.23 (0.15)	4.32 (0.17)	4.18 (0.15)

Heights of the vocal cord (mm): mean (SEM)
Source: Stell et al. (1978).

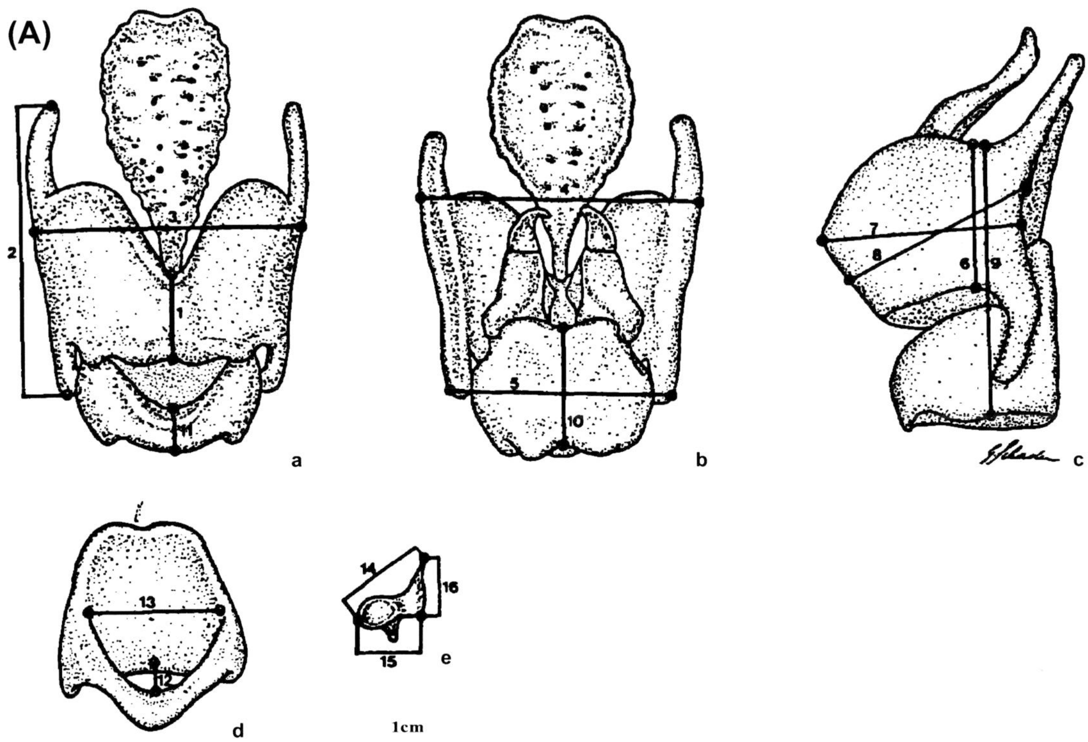

FIGURE 1 Drawing of several different views of laryngeal cartilages to show sites selected for measurement by Zrunek et al. (1988) in the male human (A) and female sheep (B). Numbers below correspond to the diagrams. Measurements are shown in Tables 16 and 21.

Thyroid cartilage (a)

1. Anterior thyroid height (distance from upper to lower thyroid incisure)
2. Dorsal thyroid height (distance between tips of upper and lower horns)
3. Maximum thyroid breadth (length of thyroid tubercula)
4. Upper thyroid breadth (outer distance between bases of upper horns)
5. Lower thyroid breadth (outer distance of inferior horns)
6. Height of thyroid lamina (distance of notches prior to horns)
7. Anteroposterior dimension (distance between laryngeal prominence and plane through posterior margins of laminae)
8. Length of thyroid lamina (distance from posterior margin to anterior edge through middle of distances 1 and 6)

Thyroid and cricoid cartilage (b)

9. Main height of larynx (extension of distance 6 to lower margin of cricoid cartilate)

Cricoid cartilage (c)

10. Height of cricoid lamina
11. Anterior height of cricoid arch
12. Internal anteroposterior diameter of cricoid cartilage
13. Internal transverse diameter of cricoid cartilage

Arytenoid cartilage (d)

14. Distance between tips of vocal and muscular process
15. Distance between tip of muscular process and dorsal end of vocal process
16. Length of vocal process

Thyroid angle (e)

17. Angle between thyroid laminae

Except for distance number 8 in sheep, all 17 values could be determined in both human and sheep larynges. This distance could not be evaluated due to the unique shape of the sheep thyroid cartilage at its laryngeal prominence. For measuring distance number 10 in sheep, it was necessary to cut off the inconstant process at the inferior margin of the cricoid cartilage.

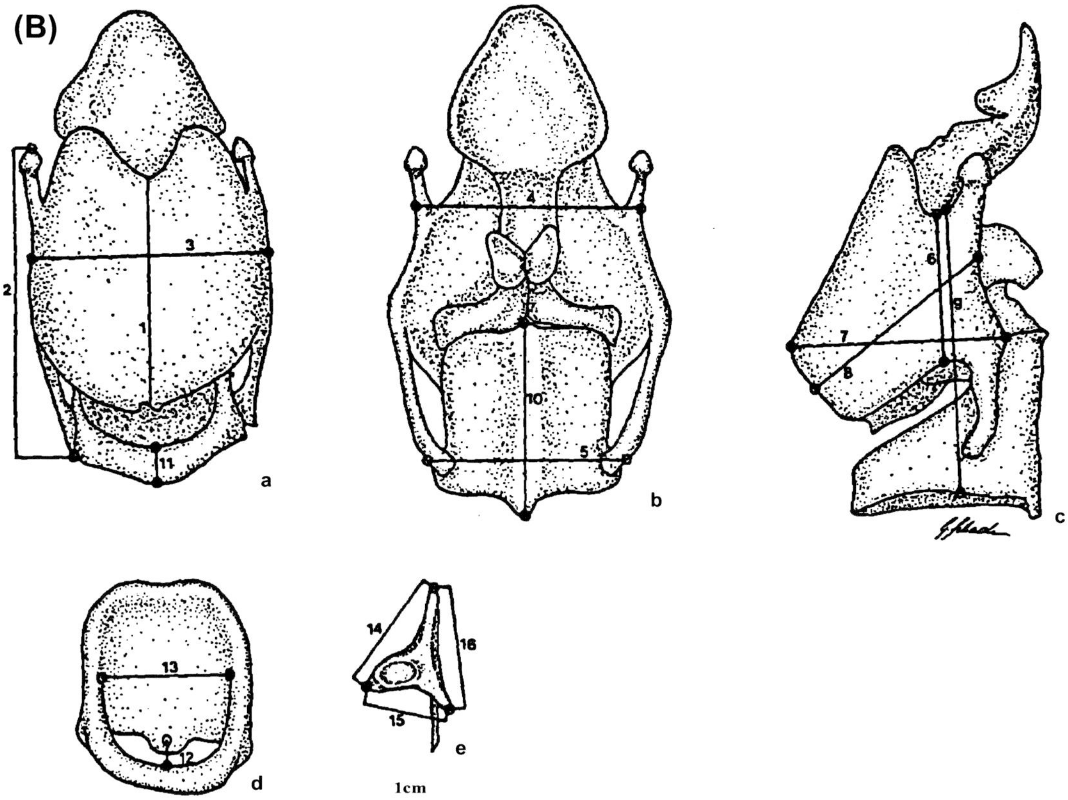

FIGURE 1 Cont'd

TABLE 19 Human Larynx - Area of the Vocal Cords

Area (mm^2) of the Vocal Cords: Mean (SEM)		
Men	70.5	(3.45)
Women	35.2	(1.32)

Source: Stell et al. (1978).

TABLE 20 Size of Human Intrinsic Laryngeal Muscles

	Length (mm)	Breadth (mm)	Thickness (mm)	Cross-Sectional Area (mm^2)	Volume (mm^3)	Calculated Weight (mg)	Weight (mg)
Cricothyroid	15	10	6	60	900	954	940
Vocal	15	4	5	20	300	318	357
Transverse arytenoid	15	7	5	35	525	556	558
Posterior cricoarytenoid	12	10	5	50	600	636	625

Source: Fink (1975) original data; Faaborg-Andersen (1957).

TABLE 21 Sheep Larynx

	Mean of the Dist. in cm (±SE)		Significant Male-Female Differences[b]
	Male[a]	Female[a]	
1	3.15 ± 0.09	2.96 ± 0.06	+
2	4.60 ± 0.10	4.42 ± 0.12	
3	3.95 ± 0.09	3.74 ± 0.09	+
4	3.73 ± 0.07	3.45 ± 0.09	+
5	3.17 ± 0.08	3.01 ± 0.11	
6	2.24 ± 0.05	2.13 ± 0.05	
7	3.63 ± 0.10	3.41 ± 0.11	
8[c]			
9	4.00 ± 0.10	3.82 ± 0.10	
10	2.38 ± 0.06	2.39 ± 0.05	
11	0.51 ± 0.02	0.51 ± 0.02	
12	2.55 ± 0.09	2.41 ± 0.07	
13	2.15 ± 0.07	1.97 ± 0.04	+
14	1.81 ± 0.06	1.71 ± 0.04	
15	1.41 ± 0.03	1.32 ± 0.03	
16	1.97 ± 0.06	1.91 ± 0.05	
17	1.11 ± 0.02[d]	1.03 ± 0.03[d]	+

Note: The mean values of sample sites identified in Figure 1(B) and standard errors of the means (SE). Sex differences are also presented.
[a]*11 specimens.*
[b]*Sex differences were examined with Student's* t *test. Significance,* $p < 0.05$ *are noted (+).*
[c]*Not measurable in sheep.*
[d]*Angle in radians.*
Zrunek et al. (1988).

TABLE 22 Dimensions of the Rabbit Larynx

Age	4 Weeks (n)	24 Weeks (n)
Subglottic Lumen Area (mm^2)	8.6 ± 1.4 (13)[a]	21.1 ± 2.9 (28)
Inner Diameter (mm)		
A_2P_2[b]	4.43 ± 0.21 (10)	7.24 ± 0.46 (25)
B_2C_2	3.54 ± 0.39 (13)	5.19 ± 0.43 (27)
D_2E_2	4.02 ± 0.30 (13)	6.07 ± 0.39 (26)
Thickness of Cricoid Cartilage (mm)		
A_1A_2	0.55 ± 0.06 (10)	0.58 ± 0.09 (24)
$B_1B_2 + C_1C_2$	0.76 ± 0.08 (13)	0.75 ± 0.08 (27)
$D_1D_2 + E_1E_2$	0.75 ± 0.08 (13)	0.87 ± 0.12 (26)
P_1P_2	0.69 ± 0.15 (13)	0.71 ± 0.13 (24)

[a]*Data are Means ± SEM (number of measurements).*
[b]*See Figure 9 for landmarks used in inner diameter and cricoid cartilage measurements.*
Source: Adriaansen et al. (1986).

TABLE 23 Laryngeal Morphometry in Big Cats[a]

	Longitudinal Length of Vocal Fold (at Rima Glottis) (mm)	Distance from Base of Cricoid Cartilage to Superior Border of Vocal Fold (mm)	PLVF[b] (%)
Lion *(P. leo)*	25	62	40
Tiger *(P. tigris)*	22	55	40
Leopard *(P. pardus)*	20	50	40
Jaguar *(P. onca)*	19	46	41
Snow leopard *(P. uncia)*	9	36	25
Cheetah *(A. jubatus)*	6	31	19
Clouded leopard *(N. nebulosa)*	5	25	20
Puma (cougar) *(F. concolor)*	6	34	18

[a]*Females.*
[b]*PLVF is the proportion of the longitudinal length of the vocal folds to the distance from the base of the cricoid cartilage to the superior border of the vocal folds (infraglottic-glottic vocal tract).*
Source: Hast (1989).

TABLE 24 Mean Values (±F SEM) and Relative Size Differences (M/F) of Male and Female *Rana pipiens* Laryngeal Characteristics

Characteristics	Male ($N = 6$)	Female ($N = 5$)	M/F
Body weight (g)	34.5 (±3.4)	42.7 (±6.2)	0.81
Larynx weight (g)	0.22 (±0.02)	0.09 (±0.01)	2.4 (++)[a]
Arytenoid cartilage (mm^3)	1.68 (±0.3)	0.64 (±0.09)	2.6 (++)
Vocal cord (mm^3)	0.8 (±0.2)	0.4 (±0.2)	2.0
Muscle mass (mm^3)	4.9 (±1.2)	1.5 (±0.2)	3.3 (+)
Bronchial process (mm^3)	0.76 (±0.11)	0.44 (±0.09)	1.7 (+)

[a]*Male-female differences were assessed using two-tailed Student's* t *test. Significance,* $p < 0.05$ *are noted (+) and* $p < 0.01$ *are noted (++).*
Source: McClelland and Wilczynski (1989).

REFERENCES

Adams, D.R., 1972. Olfactory and non-olfactory epithelia in the nasal cavity of the mouse. Peromyscus. Am. J. Anat. 133, 37.

Adriaansen, F.C.P.M., Verwoerd-Verhoef, H.L., Van der Heul, R.O., Verwoerd, C.D.A., 1986. A morphometric study of the growth of the subglottis after endolaryngeal trauma. Int. J. Pediatr. Otorhinolaryngol. 12, 217.

Bhatnagar, K.P., Kalien, F.C., 1975. Quantitative observations on the nasal epithelia and olfactory innervation in bats. Acta Anat. 91, 272.

Faaborg-Andersen, K., 1957. Electromyographic investigation of intrinsic laryngeal muscles in humans. Acta Physiol. Scand. 41 (Suppl. 41), 11.

Fink, B.R., 1975. The Human Larynx-A Functional Study. Raven Press Books, New York, pp. 31–52.

Grant, J.B.C., 1947. Cunningham's Textbook of Anatomy, eighth ed. Cumberlodge, Oxford. pp. 1–1604.

Gross, E.A., Swenberg, J.A., Fields, S., Popp, J.A., 1982. Comparative morphometry of the nasal cavity in rats and mice. J. Anat. 135, 83.

Gross, E.A., Starr, T.B., Randall, H.W., Morgan, K.T., 1987. Morphometric analysis of the primate nasal cavity. Toxicologist 7, 193.

Gurtovoi, N.N., 1966. Ecological-morphological differences in the structure of the nasal cavity in the representatives of the orders insectivore, chiroptera and rodentia. Zool. Zhurnal 45, 1536.

Hast, M.H., 1989. The larynx of roaring and non-roaring cats. J. Anat. 163, 117.

Kamau, J.M.Z., Maina, J.N., Maloiy, G.M.O., 1984. The design and the role of the nasal passages in temperature regulation in the Dik-Dik antelope (*Rhynchotragus kirkii*) with observations on the carotid rete. Respir. Physiol. 56, 183.

Kocon, T., Stepien, M., 1967. The paranasal sinuses in macaca. Folia Morphol. 24, 408.

Langman, V.A., Maloiy, G.M.O., Schmidt-Nielson, K., Schroter, R.C., 1979. Nasal heat exchange in the giraffe and other large mammals. Respir. Physiol. 37, 325.

Larochelle, R., Baron, G., 1989. Morphology and morphometry of the nasal fossae of four species of North American shrews (Soricinae). Am. J. Anat. 186, 306.

Loo, S.K., 1973. A comparative study of the nasal fossa of four nonhuman primates. Folia Primat. 20, 410.

McClelland, B.E., Wilczynski, W., 1989. Sexually dimorphic laryngeal morphology in *Rana pipiens*. J. Morph. 201, 293.

Patra, A.L., Menache, M.G., Shaka, N.B., Gooya, A., 1987. A morphometric study of nasal-pharyngeal growth for particle deposition in the rat. Am. Ind. Hyg. Assoc. J. 48, 556.

Schmidt-Nielsen, K., 1981. Countercurrent systems in animals. Sci. Am. 244 (5), 118.

Schreider, J.P., 1983. Nasal airway anatomy and inhalation deposition in experimental animals and people. In: Reznik, G., Stinson, S.F. (Eds.), Nasal Tumors in Animals and Man, Experimental Nasal Carcinogenesis, vol. III. CRC Press, Boca Raton, FL, pp. 1−26.

Stell, P.M., Gregory, I., Watt, J., 1978. Morphometry of the epithelial lining of the human larynx, I, the glottis. Clin. Otolaryngol. 3, 13.

Zrunek, M., Happak, W., Hermann, M., Strienzer, W., 1988. Comparative anatomy of human and sheep laryngeal skeleton. Acta Otolaryngol. (Stockholm) 105, 155.

Appendix 2

Kinetics of Pulmonary Cells

Susan G. Shami and Michael J. Evans

Chapter Outline

1. INTRODUCTION

Lung cell kinetics, or the study of cell proliferation in the lung, has led to an understanding of the functions of various cell types in the normal lung, as well as the differentiation patterns of these lung cells throughout growth and adulthood. The study of lung-cell kinetics is a useful tool for toxicologists in assessing the pulmonary toxicity of inhaled materials and the capacity of the lung to repair itself. Cell-kinetic studies have also shed light on the process of carcinogenesis and the development of other lung diseases, most of which involve abnormal or uncontrolled cellular growth. For previous reviews, the reader is referred to Bertalonby (1964), Boren and Paradise (1978), Kauffman (1980), Evans (1982), and Evans and Shami (1989).

A major problem for the scientist who studies this field for the first time is that the investigators use a great variety of techniques and methods to measure proliferation rates or indices (PI). The most commonly used technique for the study of cell kinetics is that of tritiated thymidine (^{3}H-TdR) incorporation by proliferating cells. This technique can yield information concerning all the parameters of the cell cycle. However, it is most frequently used to calculate the labeling index (LI), which is the proportion of cells of a chosen population that are synthesizing DNA (S). An example of difficulty in comparing data obtained from the use of this technique occurs when the LI represents a single pulse or injection of tritiated thymidine, as opposed to several pulses or a continuous infusion over a period of time as long as several weeks. Another method for the study of cell proliferation is the enumeration of cells in mitosis (M). This yields the mitotic index (MI) or the proportion of cells in a chosen population that are in the M phase of the cell cycle. Again, variations of this technique may include the use of stathmokinetic agents, such as colchicine, to arrest cells in metaphase and to allow an accumulation of these cells over time. This causes difficulty in comparing the information and, therefore, the conclusions presented by different investigators. In this chapter we will present the PIs for the various lung cell types observed in different animal species obtained from a variety of studies. We will also briefly describe commonly used techniques, the types of data they yield, and experimental variables that must be taken into account when interpreting these data.

2. THE CELL CYCLE

Proliferating cells go through four phases of growth, called the cell cycle (Baserga, 1976; Aherne et al., 1977;

Wright and Alison, 1984). The cell cycle begins after mitosis (M) with the first gap (G_1). This is the period of cell growth and preparation for DNA synthesis. The period of S or DNA synthesis follows G_1. S phase is followed by the second "gap" (G_2), during which the cell prepares for mitosis or M phase. Finally mitosis occurs. Cells in any proliferating compartment go through this cycle in an asynchronous manner, independent of each other.

Following mitosis, the cell may stay in the proliferative cycle, or it may leave and participate in tissue functions. These cells could become terminally differentiated and unable to divide or could enter a resting state (G_0). At this point, they can be recalled to the G_1 phase of the cell cycle after receiving the appropriate biochemical signals. Type I epithelial cells and ciliated cells are terminally differentiated in the lung and are unable to divide (Table 1). The rest of the cell types studied to date are capable of cell division following the appropriate biochemical signals and are most likely to be in a state of G_0. These are basal, nonciliated columnar, type **II** and endothelial cells, fibroblasts, and alveolar macrophages (Evans and Shami, 1989).

3. TECHNIQUES IN CELL KINETICS

3.1 Mitotic Index

The MI is the ratio of the number of cells in mitosis to the total number of cells in that cell population (CMI) or of cells in the tissue (TMI). This method is useful for calculating proliferation rates of rapidly dividing cell populations, such as prenatal and postnatal tissues, and tissues repairing after injury. However, for most cell types in the normal lung, there are too few mitoses for this method to yield statistically significant results. By using stathmokinetic agents, such as vinblastine, vincristine, colchicine, and colcecimed, which arrest cells in mitosis, the number of mitoses counted can be increased. This method has certain drawbacks, particularly potential toxicity to the treated animals or resistance of the animals to the agents.

3.2 Labeling Index

The labeling index (LI) is the most common measure of cell proliferation used in the lung. It is expressed as the ratio of the number of cells labeled with a DNA tracer, such as ^{3}H-TdR, to the total number of cells in the population, either as a tissue LI (TLI) or cell LI (CLI). The nucleotide thymidine (TdR) is incorporated into DNA during the S phase of the cell cycle. Because the S phase is usually longer than the M phase of the cell cycle, the LI may be up to 10 times greater than the MI (Cleaver, 1967).

The technique most commonly used to quantitate ^{3}H-TdR incorporation is autoradiography (AR). The cells or tissues are placed on a slide and coated with a photographic emulsion. The tritium isotope in the thymidine in the tissue emits beta particles, which expose latent silver grains in the overlying photographic emulsion. The slide is then developed as photographic film would be. ^{14}Carbon is also used as a label for thymidine and can be distinguished from ^{3}H-TdR in an AR by the longer path length of the beta

TABLE 1 System for Cell Proliferation in the Lung

Region	Cell Population: Progenitor		Differentiated		Terminal
Airway epithelium	Basal	↔	Basal		
	Mucous	↔	Goblet Serous Presecretory Clara	→	Ciliated
Alveolar epithelium	Type 2	↔	Type 2	→	Type 1
Vascular	Endothelial	↔	Endothelial		
	Intravascular macrophage	↔	Intravascular macrophage		
Interstitial	Fibroblast	↔	Fibroblast		
	Interstitial macrophage	↔	Interstitial macrophage		
Alveolar	Alveolar macrophage Monocyte	↔ →	Alveolar macrophage		

particles from ^{14}C. The relative amounts of ^{3}H-TdR incorporated into the DNA of a cell are indicated by the number of grains in the emulsion over the nucleus. Details of the techniques of AR may be found elsewhere (Rogers, 1979). A single injection of ^{3}H-TdR is equivalent to a pulse label, because it is available to cells for about 30 to 60 min before it is metabolized and excreted.

Scintillation counting of ^{3}H-TdR labeled cells has been used to obtain a rapid estimate of the proliferation rate of lung cells. This method is most appropriately applied to a homogenous population of cells, such as in tissue culture or cells that have been separated and isolated from intact tissue. When used for a whole lung that has been digested, it must be recognized that the population of cells being sampled is het-erogenous, and even if perfused previously, contains a significant number of circulating blood cells.

4. CLASSIFICATION OF LUNG CELL POPULATIONS

Three general types of cell populations are (1) renewing cell populations that exist in a dynamic equilibrium, proliferating continuously in order to replace cells lost from the tissue; (2) expanding cell populations that contain cells that proliferate during growth and may either proliferate or remain in a static phase of the cell cycle during adult life. A subdivision of the expanding cell population is the conditional renewal population, which contains cells that may divide under certain conditions such as injury; and (3) stable cell populations that contain cells that do not proliferate (Baserga, 1976; Aherne et al., 1977; Wright and Alison, 1984).

The lung cells were originally thought to be renewing cell populations (Bertalanffy, 1964). Based on more recent studies, Kauffman (1980) concluded that lung cells represent an expanding cell population. Under this interpretation, the majority of the cell proliferation observed in the lungs of normal animals is related to growth and not to the replacement of normally lost or damaged cells. Within this division, the lung was further classified as a conditional renewal system, because most lung cells respond to injury and other stimuli with increased cell proliferation (Wright and Alison, 1984; Evans and Shami, 1989).

5. LUNG CELL KINETICS

5.1 Airway Epithelium

The upper airway is lined by a pseudostratified columnar epithelium containing columnar cells and basal cells (Breeze and Wheeldon, 1977). Basal cells are attached to the basement membrane, and their apical surface does not reach the surface of the columnar epithelium. Basal cells are most numerous in the trachea and decrease in number from the trachea to the small bronchi. Thus, there are usually no basal cells in the terminal and respiratory bronchioles. The columnar cell population is composed of ciliated and nonciliated cells lying on a common basement membrane. The nonciliated cell population contains cells with and without secretory granules (goblet, small mucous granule, serous, Clara, indeterminate), both of which appear to act as one cell population with respect to cell proliferation. The nonciliated cell population can also differentiate into ciliated cells (Evans and Shami, 1989).

From the earliest studies of cell kinetics of airway epithelium, most investigators concluded that basal cells were the stem cell responsible for renewal of airway epithelium (Boren and Paradise, 1978; Kauffman, 1980; Evans, 1982; Bindreiter et al., 1968; Blenkinsopp, 1967; Donnelly et al., 1982; Jeffery and Reid, 1975; Jeffery et al., 1982; Reid and Jones, 1979; Wells, 1970). However, recent studies suggest this is not the case, and a different scheme for the renewal of the airway 147 epithelium has been proposed (McDowell et al., 1984; Evans and Shami, 1989). In this scheme, the nonciliated columnar cell is the primary progenitor cell, and the basal cell is not related to renewal of the columnar epithelium. Thus proliferation of basal cells is associated with the formation of basal cells and the proliferation of nonciliated cells with the formation of ciliated and nonciliated cells (Table 1).

5.1.1 *Basal Cell*

The frequency of basal cells may vary in different airways. This occurs because their frequency is related to the height of the epithelium and not to airway levels or animal species (Evans and Plopper, 1988; Evans et al., 1989). Because of this, the basal cell PI may also vary in different regions of the airway and between animal species. Recently it was shown that basal cells are involved with attachment of the epithelium to the basal lamina (Evans et al., 1989). Under this interpretation, most of the basal cell proliferation reported may be related to growth of the airway and attachment to the basal lamina. However, this has not been demonstrated experimentally.

Proliferation indices have been reported for only a few animal species, and most of these have been carried out on young rodents. The PI appear to be higher in the trachea than in the distal airways (Tables 2 and 3). This is probably due to there being fewer basal cells in the distal airway. Only one study has reported the effects of age (Bolduc and Reid, 1976). They found that the PI declined with age.

TABLE 2 Extrapulmonary Basal Cells Proliferative Indices (%) (Trachea and Bronchi)

		Cell		Tissue		
Animal	Age[a]	Mitotic	Labeling	Mitotic	Labeling	Authors[b]
Hamster	1 d	0.2	—	—	—	Blenkinsopp (1967)
	3 m	0.04–0.05	0.3–1.1	—	—	Bolduc and Ried (1976)
						Boren and Paradise (1978)
Monkey	4–9 y	—	—	—	0.7	Donnelly et al. (1982)
						Gordon and Lane (1977, 1980)
Rat	1 m	2.3	7.0	2.2	1.4	Keenon et al. (1982a,b,c)
	3 m	—	—	0.18	—	Lane and Gordon (1974)
						McDowell et al. (1984, 1985)
						Wells (1970)
						Wilson et al. (1984)

[a]*d = day; m = month; y = years.*
[b]*In alphabetical order.*

TABLE 3 Intrapulmonary Basal Cells Proliferative Indices (%) (Bronchi)

		Cell		Tissue		
Animal	Age[a]	Mitotic	Labeling	Mitotic	Labeling	Authors[b]
Hamster	3 m	—	1.5	—	—	Bolduc and Reid (1976)
						Breuer et al. (1990)
Rat	1 m	—	—	0.003–0.15	0.44	Evans et al. (1986a)
	3 m	—	—	0.25	—	

[a]*m = month.*
[b]*In alphabetical order.*

In response to injury, the PI responds less than the nonciliated columnar cells or not at all (Evans and Shami, 1989). This observation plus the decline in PI with age suggests that most basal cell proliferation is related to growth. This may be a very important concept in understanding basal cell proliferation. During growth the number of columnar cells/nun of basement membrane remains approximately constant. In contrast, the number of basal cells/mm of basement membrane increases. This fact could easily account for the increased proliferation rate of basal cells compared to nonciliated columnar cells that is often reported. The durations of phases of the basal cell cycle are S, 10.5 to 12.8 h; G_2, 3.2 h; M, 0.6 h (Boren and Paradise, 1978).

Questions that need further study concern the effects of (1) age, (2) sex, (3) basal cell cytodifferentiation, and (4) airway level on the CPI. In addition, the question of whether basal cell proliferation is for growth of the airway or another purpose should be studied.

5.1.2 Nonciliated Columnar Cells

This grouping of cells includes serous, goblet, indeterminate, small mucous granule cells, and Clara cells. All of these cell populations are capable of proliferation and appear to respond to the same stimuli (Evans and Shami, 1989). In this respect, they act as one population of dividing cells. They respond to injury with an increase in proliferation. Proliferation of these cells is associated with the formation of nonciliated columnar cells and ciliated cells. Ciliated cells do not divide and are formed through the differentiation of nonciliated cells (Evans and Shami, 1989).

Proliferation indices for nonciliated columnar cells have been reported for several animal species (Tables 4–6). The frequency of nonciliated columnar cells, ciliated cells, and basal cells may vary in different airway levels and animal species. For this reason, the PI may also vary between airway levels. In one study (Bolduc and Reid, 1976), the PI was found to decline with age. However, in another study

TABLE 4 Extrapulmonary Nonciliated Cells Proliferative Indices (%) (Trachea and Bronchi)

		Cell		Tissue		
Animal	**Age**[a]	**Mitotic**	**Labeling**	**Mitotic**	**Labeling**	**Authors**[b]
Hamster	1 d	0.2	—	—	—	Blenkensopp (1967)
	3 m	0.02–0.1	0.4	—	—	Bolduc and Reid (1976)
						Boron and Paradise (1978)
Rat	1 m	—	0.9	0.9	—	Donnelly et al. (1982)
	3 m	—	—	0.1	—	Keenan et al. (1982a,b,c)

[a]*d = day; m = month.*
[b]*In alphabetical order.*

TABLE 5 Intrapulmonary Nonciliated Cells Proliferative Indices (%) (Bronchi)

		Cell		Tissue		
Animal	**Age**[a]	**Mitotic**	**Labeling**	**Mitotic**	**Labeling**	**Authors**[b]
Hamster	3 m	—	1.0	—	—	Bolduc and Reid (1976)
						Breuer et al. (1990)
Rat	1 m	—	—	0.05–0.7	0.72	Evans et al. (1986a)
	3 m			0.05	—	

[a]*m = month.*
[b]*In alphabetical order.*

TABLE 6 Intrapulmonary Bronchiolar Cell Proliferative Indices (%) (Clara Cells)

		Cell		Tissue		
Animal	**Age**[a]	**Mitotic**	**Labeling**	**Mitotic**	**Labeling**	**Authors**[b]
Hamster	12 m	—	—	—	0.16	Castleman et al. (1980)
						Eustis et al. (1981)
						Breuer et al. (1990)
Monkey	2–7 y	—	—	—	0.10	Evans et al. (1976, 1977, 1978a Evans (1982)
						Hacket (1979, 1983)
Mouse	3 m	—	—	—	0.3	Lum et al. (1978)
						Shami et al. (1982a,b, 1986a,b)
Rat	1 m	—	—	—	0.2–1.1	
	3 m	—	—	—	0.1–1.1	
	11 m	—	—	—	0.1	
	19 m	—	—	—	0.2	
	25 m	—	—	—	0.8	

[a]*m = month; y = year.*
[b]*In alphabetical order.*

(Evans et al., 1986a), the PI remained the same or increased. Since the PI can increase to a very high rate in response to injury, these differences may reflect infection. Questions that need further study concern the effects of (1) age, (2) sex, (3) non-ciliated cell cytodifferentiation, and airway level on this process. To date, reported durations of phases of the cell cycle are S, 7.0 to 10.9 h; G_2, 2.5 h; M, 0.6 h (Boren and Paradise, 1978). As with basal cells, all future studies on nonciliated columnar cells must specify accurately the airway locations and diameters investigated.

5.2 Alveolar Epithelium

The alveolar epithelium is composed of type I and type II cells. Type I cells cover 93% to 97% of the alveolar surface. They are large squamous cells that are not capable of cell division. The cuboidal type II cells are dispersed throughout the alveoli, lying between type I cells, and cover the remaining 3% to 5% of the surface area. Type II cells are capable of cell division. They act as the progenitor cells for the alveolar epithelium, proliferating to replenish themselves and to form the terminally differentiated type I cells (Evans and Shami, 1989).

The kinetics of type II cell proliferation have been described for several animal species (Table 7). As in the airways, the actual number of cells in the process of division in the normal lung is very small. However, in response to injury or inflammation, there is a large increase in PI. In the aging animal, the PI declines with time, suggesting that most type III cell proliferation is associated with growth. Although much data have been produced on type II cell proliferation, very little is known about the effects of sex and age on this process. To date, the reported durations of phases of the cell cycle are S, 7.3 to 9.0 h; G_2, 1.0 to 2.0 h; M, 0.8 h (Kauffman, 1980). It should be noted that, until recently, the main stimuli for division of type II cells were thought to be growth and injury. However, it has recently been demonstrated that a mild inflammatory response in the lung can cause type II cell proliferation (Shami et al., 1986a,b). This finding emphasizes the importance of the health status of the animals studied for kinetic data.

TABLE 7 Type 2 Cell Proliferative Indices (%)

		Cell		Tissue		
Animal	**Age**[a]	**Mitotic**	**Labeling**	**Mitotic**	**Labeling**	**Authors**[b]
Hamster	3 m	—	0.6	—	—	Adamson et al. (1977)
	5 m	—	—	—	2.9	Evans and Bils (1969)
	12 m	—	—	—	1.1	Evans et al. (1972, 1973a, 1977, 1978b)
						Hacket (1979, 1983)
Monkey	Adult	—	—	—	0.3–0.8	Hackney et al. (1975a,b)
						Haschek et al. (1983)
Mouse	1 m	—	0.3	—	—	Kauffman (1972), Kauffman et al. (1974)
	3 m	—	0.3	—	—	Shami et al. (1982a,b, 1984, 1985, 1986a,b)
						Simnett and Heppleston (1966)
Rat	1 d	—	0.5	—	—	Tryka et al. (1986)
	4 d	—	3.0	—	—	White and Garg (1981)
	7 d	—	6.5	—	—	Witschi and Morse (1985)
	10 d	—	5.5	—	—	Witschi (1986)
	13 d	—	3.0	—	—	Wright et al. (1987)
	21 d	—	1.0	—	—	
	1 m	—	—	—	1.8–2.6	
	2 m	—	0.3	—	—	
	3 m	—	0.6–1.2	—	0.1–0.3	
	11 m	—	—	—	0.2	
	19 m	—	—	—	0.5	
	25 m	—	—	—	0.8	

[a]*d = day; m = month.*
[b]*In alphabetical order.*

5.3 Endothelium

Endothelial cells line the blood vessels in the lung. They are the most numerous resident cell type in the lung, averaging 40% of the total number of lung cells. Endothelial cells are capable of division and have the highest LI of any cell type in the normal lung (Table 8). In young mice examined with electron microscopic (EM) AR, about 30% of the labeled cells of the alveolar wall are endothelial cells (Evans and Bils, 1969). Proliferation is especially high during growth of the lung and during the repair phase following injury to the capillaries (Kauffman, 1980; Evans, 1982; Bowden and Adamson, 1974; Adamson et al., 1977; Haschek et al., 1983; Tryka et al., 1986). In the aging lung, the rate of proliferation declines. A major problem associated with the endothelial cell PI concerns the precise identification of endothelial cells with the light microscope (LM) (Crapo et al., 1980). With the LM, endothelial cells are often incorrectly classified as interstitial cells, unless they are associated with open capillary lumina. Since this problem of identification is inherent in lung tissue, the LI of endothelial cells, as determined by LM, underestimates the actual LI. Conversely, the size of the interstitial compartment is overestimated. Thus, precise studies of endothelial-cell kinetics are difficult to carry out in the capillary bed of the lung. In addition, very little is known concerning the effects of sex and age on endothelial cell kinetics or phases of the cell cycle.

5.4 Cells in the Interstitium

The resident cells of the interstitium are fibroblasts, smooth muscle cells, and pericytes. Fibroblasts are the most numerous of these, making up more than 95% of the resident interstitial cells in the alveoli. They are capable of active cell division. The fibroblast population in the lung is separated morphologically into two cell types, the contractile interstitial cell and the collagen-synthesizing fibroblasts (Kapanci et al., 1974; Kaplan et al., 1985). The proliferation characteristics of these two cell populations are different. Using EM, Brody and Kaplan (1983) found that the lipid-containing fibroblast (contractile interstitial cell) of the postnatal rat lung at 4 days of age has an LI of about 5%, and at 11 days, an LI of about 0.4%. In contrast, the nonlipid-containing fibroblast (fibroblast) has an LI of about 9.5% at 4 days and 2.5% at 11 days. The difference in the response of these two morphological forms of the fibroblast was interpreted as evidence of two distinct cell populations.

In the interstitial regions around pulmonary arteries, there are smooth muscle cells and fibroblasts. Meyrick and Reid (1979) showed that these two cell populations, which can be identified with the LM, have different proliferative characteristics. In response to hypoxia, the proliferative response of the fibroblast was much greater than that of the smooth muscle cell. In the alveoli, kinetic studies of smooth muscle cells and pericytes have not been conducted.

TABLE 8 Endothelial Cell Proliferation Indices (%)

		Cell		Tissue		
Animal	Age[a]	Mitotic	Labeling	Mitotic	Labeling	Authors[b]
Hamster	3 m	–	0.6	–	–	Adamson et al. (1977) Adamson and Bowden (1983)
Mouse	1 m	–	0.3	–	–	Bowden and Adamson (1974)
	3 m	–	0.3	–	–	Evans and Bils (1969) Haschek et al. (1983)
Rat	1 d	–	8.0	–	–	Kauffman et al. (1974)
	4 d	–	10.0	–	–	Meyrick and Reid (1979, 1982)
	7 d	–	8.0	–	–	Tryka et al. (1986)
	10 d	–	8.5	–	–	
	13 d	–	2.5	–	–	
	21 d	–	0	–	–	
	4 m	–	0.4	–	–	

[a]*d = day; m = month.*
[b]*In alphabetical order.*

In addition to fibroblasts, smooth muscle cells, and pericytes, the interstitial cell population of the normal lung contains small numbers of migratory cells, such as monocytes, plasma cells, polymorphonuclear leukocytes (PMN), and macrophages. During an inflammatory response, the numbers of interstitial migratory cells can increase dramatically. The problem of cell identification with the LM is quite serious when considering the interstitium. Using EM, Crapo et al. (1980) found that 27% of the cells in the interstitium of normal rats could not be identified. During an inflammatory response associated with exposure to 100% 0, 49% of the cells could not be positively identified with the EM (Crapo et al., 1980). Considering the lack of resolution of the LM compared with the EM, the number of cells that cannot be identified will be much higher with the LM.

The PI of interstitial cell populations reflect these same problems of cell identification. As pointed out previously, the PI also contain endothelial cells not associated with a capillary lumen. Because of the difficulty in identification of interstitial cells in the alveoli with the LM, the specific kinetic parameters of fibroblasts have not yet been calculated. Fibroblasts have, however, been studied extensively with in vitro systems. The PI for the cells in the interstitium of various animal species, as determined with the LM, are presented in Table 9. These data should be interpreted with respect to the problems just mentioned. Additional studies that allow clear identification of the interstitial cell population are needed before additional kinetic information can be obtained.

5.5 Macrophages

5.5.1 Alveolar

Pulmonary macrophages are found in three anatomically different compartments in the lung. Pulmonary alveolar macrophages (PAM) are present on the surface of the alveolar epithelium. Pulmonary interstitial macrophages (PIM) are present in the interstitium, primarily around large vessels and airways. Pulmonary intravascular macrophages (PVM) are found in the capillary lumen. All of these cell populations are capable of cell proliferation, but only the kinetics of PAM have been studied in detail. Currently there are two clearly defined mechanisms by which the PAM population is maintained: (1) migration of monocytes into the alveolar space with subsequent differentiation into PAM and (2) the division of existing PAM. The migration of monocytes into the lung is known to be a source of PAM, particularly during an inflammatory response (Blusse Van Oud Albus et al., 1981, 1983). The division of PAM was only recently demonstrated to be an important mechanism for maintaining the PAM population in the lung (Evans and Shami, 1989).

An important question arising from these studies concerns the significance of PAM proliferation compared with monocyte migration, as a source of PAM (Masse et al., 1976). Although there is a large influx of monocytes into the lung during an inflammatory response, the importance of this response in the normal lung is unknown. In the adult lung, investigators have used models of chronic leukopenia induced by several techniques, including cancer chemotherapeutic agents, internal and external bone marrow irradiation, and glucocorticoids to study this problem. In general, these studies demonstrate that the PAM population can be maintained through cell division in the absence of migratory monocytes (Coggle and Tarling, 1982, 1984; Evans et al., 1986b; Sawyer, 1986a,b; Sawyer et al., 1982). Under similar conditions of leukopenia, the capacity for PAM to respond to a carbon load was measured (Evans et al., 1986b). The proportion of proliferating PAM increased dramatically in response to the carbon, and the total PAM population gradually increased in number. Thus, there is ample evidence that PAM can

TABLE 9 Interstitial Cell Proliferation Indices (%)

Animal	Age[a]	Cell		Tissue		Authors[b]
		Mitotic	Labeling	Mitotic	Labeling	
Mouse	1 m	—	1.0	—	—	Adamson et al. (1977)
	3 m	—	0.2	—	—	Bowden and Adamson (1972, 1978, 1980)
						Brody and Kaplan (1983)
Rat	1 d	—	6.8	—	—	Evans et al. (1986a)
	4 d	—	14.2	—	—	Haschek et al. (1983)
	7 d	—	9.0	—	—	Kauffman et al. (1974)
	10 d	—	7.0	—	—	Shami et al. (1986a,b)
	13 d	—	1.0	—	—	Tryka et al. (1986)
	21 d	—	0	—	—	

[a]*d = days; m = month.*
[b]*In alphabetical order.*

divide to maintain or increase their numbers in the absence of migratory monocytes.

From their studies, Coggle and Tarling (1984) propose a model for PAM population renewal in normal mice in which 70% of the population is derived through PAM proliferation and 30% from the migration of monocytes. More recently, Sawyer (1986b) suggested that the entire PAM population of the normal lung is maintained through division of PAM. From existing data, it is clear that both division of PAM and migration of monocytes are important in maintaining the PAM population. What is not clear are the conditions and factors that stimulate division of existing PAM in the lung or migration of monocytes to the lung.

In the newborn rabbits, it was recently shown that the PAM population is established during the first weeks of life entirely through PAM divisions (Evans et al., 1987; Sherman et al., 1988). In this study, there was no evidence that interstitial macrophages or migratory monocytes were involved in the process. Although the source of the dividing PAM was not determined, this study stresses the importance of PAM division in the lung of the newborn. The kinetics of PAM cell division have been described for several animal species, in- eluding humans (Table 10). During early postnatal development, the PI are very high. However, following establishment of the PAM population, the PI decline. Since PAM respond to irritation and particle inhalation with proliferation, variability in PI is a common

TABLE 10 Alveolar Macrophage Proliferation Indices (%)

		Cell		Tissue		
Animal	Age[a]	Mitotic	Labeling	Mitotic	Labeling	Authors[b]
Hamster	3 m	–	–	–	–	Bitterman et al. (1984)
	6 m	–	2.1	–	–	Blusse Van Alblas et al. (1981, 1983)
	12 m	–	0.5	–	–	Bowden (1984)
						Coggle and Tarling (1982, 1984)
Human	Adult	–	0.4–1.25	–	–	Evans and Bils (1969)
						Evans et al. (1972, 1973b, 1977, 1986b, 1987)
Monkey	Adult	–	–	–	0.3–0.5	Golde et al. (1974)
						Hackett (1983)
Mouse	1 m		1.9–2.8	0.02	1.0	Hackney et al. (1975a,b)
	3 m		0.9–3.1	–	–	Harbison et al. (1984)
						Masse et al. (1976)
Rabbit	0.5 d	0.6	–	–	–	Sawyer (1986a,b)
	1 d	1.6	17.0	–	–	Sawyer et al. (1982)
	2 d	1.2	11.0	–	–	Shami et al. (1982a,b, 1984, 1985)
	3 d	1.3	17.0	–	–	Shellito et al. (1987)
	4 d	1.4	12.0	–	–	Sherman et al. (1988)
	5 d	1.6	–	–	–	Weiss et al. (1986)
	7 d	0.7	–	–	–	Wright et al. (1987)
	14 d	0.2	–	–	–	
	4 m	0.2	–	–	–	
Rat	1 m	–	0.2	–	0.1–1.0	
	3 m	0.1	0.6–0.7	–	0.1–0.2	
	11 m	–	–	–	0.1	
	19 m	–	–	–	0.2	
	25 m	–	–	–	1.2	
Sheep	1 d	–	17.5	–	–	
	4 d	–	23.9	–	–	
	8 d	–	14.1	–	–	
	21 d	–	6.3	–	–	
	42 d	–	5.0	–	–	
	90 d	–	3.7	–	–	
	180 d	–	2.8	–	–	

[a]*d = day; m = month.*
[b]*In alphabetical order.*

occurrence. Also, since there is a regular loss, of PAM from the lung and replacement through cell division, PAM should most likely be classified as a renewing cell pouplation (Coggle and Tarling, 1984; Sawyer, 1986a). Duration of phases of the cell cycle are S, 11.0 h; G_2, 3.0 h; M, 0.4 h (Sawyer, 1986a,b).

5.5.2 Interstitial

Proliferation of PIM was previously thought to be a precursor of PAM (Bowden and Adamson, 1972; Bowden, 1984). This hypothesis states that monocytes from the vascular system enter the interstitium and become macrophages. When PAM are required in the alveoli, the interstitial macrophage divides and one or both daughter cells migrate into the alveoli, where they are classified as PAM. A basic premise of this hypothesis is the assumption that PAM cannot divide. It is now clear that PAM can divide and that such division accounts for the establishment of PAM in the newborn and normal maintenance in the adult (Evans and Shami, 1989). Thus, PIM do not act as unique precursors of PAM. There are, however, macrophages in the interstitium (PIM), and they appear to divide at the same rate or greater tharn PAM (Blusse Van Oud Albus et al., 1981, 1983). Since PAM may migrate into the interstitium, it is possible PIM are the same cell as PAM. However, more research is necessary to determine whether or not this is true. It should also be noted that the original kinetic parameters of PIM were calculated from data obtained from the interstitium with the LM (Bowden and Adamson, 1972). As discussed previously, it is clear that this technique is inadequate for cell identification in the interstitium. The interstitial cell population in Table 8 includes the PIM cell population.

5.5.3 Intravascular

The recently discovered PVM exist in the vascular system and are capable of cell division. However, at this point there is little information concerning their kinetics (Winkler and Cheville, 1985).

6. CONCLUSIONS

Up to this time, most lung-cell kinetic studies have involved young growing rodents. There is very little information concerning other animal groups. These studies indicate that the proliferation of pulmonary cells declines with age. Very little is known concerning the effects of sex, circadian rhythms, or nutrition on cell proliferation in vivo (Simnett and Heppleston, 1966; Kauffman, 1980; Evans and Shami, 1989). The cell types most studied in vivo are nonciliated airway cells, type II cells, and alveolar macrophages. Very little is known of basal-cell, endothelial-cell, fibroblast, and interstitial and intravascular macrophage proliferation in vivo. It would require a large research effort to answer all of these questions. However, until these basic questions that remain concerning cell proliferation in the lung are answered, application of molecular biology to the problems of cell proliferation in the lung will be poorly understood.

REFERENCES

Adamson, I.Y.R., Bowden, D.H., Cote, M.G., Witschi, H., 1977. Lung injury by butylated hydroxytoluene-cytodynamic and biochemical studies in mice. Lab. Invest. 36, 26.

Adamson, I.Y.R., Bowden, D.H., 1983. Endothelial injury and repair in radiation-induced pulmonary fibrosis. Am. J. Pathol. 112, 224.

Aherne, W.A., Camplejohn, R.S., Wright, N.A., 1977. An Introduction to Cell Population Kinetics, vol. 203. Edward Arnold Ltd, London.

Baserga, R., 1976. Multiplication and Division in Mammalian Cells, vol. 382. Marcel Dekker, New York.

Bertalanffy, F.D., 1964. Respiratory tissue: structure, histophysiology, cytodynamics. Part I. Review and basic cytomorphology. Int. Rev. Cytol. 16, 233.

Bindreiter, M., Schuppler, J., Stokinger, L., 1968. Zellproliferation and Differenzierung im Trachealepithel der Ratte. Exp. Cell Res. 50, 377.

Bitterman, P.B., Saltzman, L.E., Adelberg, S., Ferrans, V.J., Crystal, R.G., 1984. Alveolar macrophage replication. One mechanism for the expansion of the mononuclear phagocyte population in the chronically inflamed lung. J. Clin. Invest. 74, 460.

Blenkinsopp, W.K., 1967. Proliferation of respiratory tract epithelium in the rat. Exp. Cell Res. 46, 144.

Blusse Van Oud Albus, A., Van Der Linden-Schrever, B., Van Furth, R., 1981. Origin and kinetics of pulmonary macrophages during an inflammatory reaction induced by intravenous administration of heat-killed bacillus Calmette-Guerin. J. Exp. Med. 154, 235.

Blusse Van Oud Albus, A., Van Der Linden-Schrever, B., Van Furth, R., 1983. Origin and kinetics of pulmonary macrophages during an inflammatory reaction induced by intra-alveolar administration of aerosolized heat-killed BCG. Am. Rev. Respir. Dis. 128, 276.

Bolduc, P., Reid, L., 1976. Mitotic index of the bronchial and alveolar lining of the normal rat lung. Am. Rev. Respir. Dis. 114 (6), 1121–1128.

Boren, H.G., Paradise, L.J., 1978. Cytokinetics of the lung. In: Harris, C.C. (Ed.), Pathogenesis and Therapy of Lung Cancer. Marcel Dekker, New York, pp. 369–395.

Bowden, D.H., 1984. The alveolar macrophages. Environ. Health Perspect. 55, 327.

Bowden, D.H., Adamson, I.Y.R., 1972. The pulmonary interstitial cell as immediate precursor of the alveolar macrophage. Am. J. Pathol. 68, 521.

Bowden, D.H., Adamson, I.Y.R., 1974. Endothelial regeneration as a marker of the differential vascular responses in oxygen-induced pulmonary edema. Lab. Invest. 30, 350.

Bowden, D.H., Adamson, I.Y.R., 1978. Adaptive responses of the pulmonary macrophagic system to carbon. I. Kinetic studies. Lab. Invest. 38, 422.

Bowden, D.H., Adamson, I.Y.R., 1980. Role of monocytes and interstitial cells in the generation of alveolar macrophages. I. Kinetic studies of normal mice. Lab. Invest. 42, 511.

Breeze, R.G., Wheeldon, E.B., 1977. The cells of the pulmonary airways. Am. Rev. Respir. Dis. 166, 705.

Breuer, R., Zajicek, G., Christensen, T.G., Lucey, E.C., Snider, G.L., 1990. Cell kinetics of normal adult hamster bronchial epithelium in the steady state. Am. J. Respir. Cell. Mol. Biol. 2, 51.

Brody, J.S., Kaplan, N.B., 1983. Proliferation of alveolar interstitial cells during postnatal lung growth—evidence for two distinct populations of pulmonary fibroblasts. Am. Rev. Respir. Dis. 127, 763.

Castleman, W.L., Dungworth, D.L., Schwartz, L.W., Tyler, W.S., 1980. Acute respiratory bronchiolitis. An ultrastructural and autoradiographic study of epithelial cell injury and renewal in Rhesus monkeys exposed to ozone. Am. J. Pathol. 98, 811.

Cleaver, J.E., 1967. Thymidine Metabolism and Cell Kinetics, vol. 292. John Wiley and Sons, New York.

Coggle, J.E., Tarling, J.D., 1982. Cell kinetics of pulmonary alveolar macrophages in the mouse. Cell Tissue Kinet. 15, 139.

Coggle, J.E., Tarling, J.D., 1984. The proliferation kinetics of pulmonary alveolar macrophages. J. Leuk. Biol. 35, 317.

Crapo, J.D., Barry, B.E., Foscue, H.A., Shelburne, J., 1980. Structural and biochemical changes in rat lungs occurring during exposures to lethal and adaptive doses of oxygen. Am. Rev. Respir. Dis. 122, 123.

Donnelly, G.M., Haach, D.G., Heird, C.S., 1982. Tracheal epithelium: cell kinetics and differentiation in normal rat tissue. Cell Tissue Kinet. 15, 119–130.

Eustis, S.L., Schwartz, L.W., Kosch, P.C., Dungworth, D.L., 1981. Chronic bronchiolitis in nonhuman primates after prolonged ozone exposure. Am. J. Pathol. 105, 121.

Evans, M.J., Bils, R.F., 1969. Identification of cells labeled with tritiated thymidine in the pulmonary alveolar walls of the mouse. Am. Rev. Respir. Dis. 100, 372.

Evans, M.J., Stephens, R.J., Cabral, L.J., Freeman, G., 1972. Cell renewal in the lungs of rats exposed to low levels of NO_2. Arch. Environ. Health 24, 180.

Evans, M.J., Cabral, L.J., Stephens, R.J., Freeman, G., 1973a. Renewal of alveolar epithelium in the rat following exposure to NO_2. Am. J. Pathol. 70, 175.

Evans, M.J., Cabral, L.J., Stephens, R.J., Freeman, G., 1973b. Cell division of alveolar macrophages in the rat lung following exposure to NO_2. Am. J. Pathol. 70, 199.

Evans, M.J., Johnson, L.V., Stephens, R.J., Freeman, G., 1976. Renewal of the terminal bronchiolar epithelium in the rat following exposure to NO_2, or O_3. Lab. Invest. 35, 246.

Evans, M.J., Cabral-Anderson, L.J., Freeman, G., 1977. Effects of NO_2, on the lungs of aging rats. II. Cell proliferation. Exp. Mol. Pathol. 27, 366.

Evans, M.J., Cabral-Anderson, L.J., Freeman, G., 1978a. Role of the Clara cell in renewal of the bronchiolar epithelium. Lab. Invest. 38, 648.

Evans, M.J., Dekker, N.P., Cabral-Anderson, L.J., Freeman, G., 1978b. Quantitation of damage to the alveolar epithelium by means of type 2 cell proliferation. Am. Rev. Respir. Dis. 168, 787.

Evans, M.J., 1982. Cell death and cell renewal in small airways and alveoli. In: Witchi, H., Nettesheim, P. (Eds.), Mechanisms in Respiratory Toxicology, vol. 189. CRC Press, Boca Raton, FL.

Evans, M.J., Shami, S.G., Cabral-Anderson, L.J., Dekker, N.P., 1986a. Role of nonciliated cells in renewal of the bronchial epithelium of rats exposed to NO_2. Am. J. Pathol. 123, 126.

Evans, M.J., Shami, S.G., Martinez, L.A., 1986b. Enhanced proliferation of pulmonary alveolar macrophages after carbon instillation in mice depleted of blood monocytes by strontium-89. Lab. Invest. 54, 154.

Evans, M.J., Sherman, M.P., Campbell, L.A., Shami, S.G., 1987. Proliferation of pulmonary alveolar macrophages during postnatal development of rabbit lungs. Am. Rev. Respir. Dis. 136, 384.

Evans, M.J., Plopper, C.G., 1988. The role of basal cells in adhesion of columnar epithelium to airway basement membrane. Am. Rev. Respir. Dis. 138, 481.

Evans, M.J., Cox, R.A., Shami, S.G., Wilson, B., Plopper, C.G., 1989. The role of basal cells in attachment of columnar cells to the basal lamina of the trachea. Am. J. Respir. Cell. Mol. Biol. 1, 463.

Evans, M.J., Shami, S.G., 1989. Lung cell kinetics. In: Massaro, D. (Ed.), Lung Cell Biology. Marcel Dekker, New York, pp. 1–36.

Golde, D.W., Byers, L.A., Finley, T.N., 1974. Proliferative capacity of human alveolar macrophages. Nature 247, 373.

Gordon, R.E., Lane, B.P., 1977. Cytolcinetics of rat tracheal epithelium stimulated by mechanical trauma. Cell Tissue Kinet. 10, 171.

Gordon, R.E., Lane, B.P., 1980. Wound repair in rat tracheal epithelium. Division of G_1 and G_2 arrested cells following injury. Lab. Invest. 42, 616.

Hackett, N.A., 1979. Proliferation of lung and airway cells induced by nitrogen dioxide. J. Toxicol. Environ. Health 5, 917.

Hackett, N.A., 1983. Cell proliferation in lung following acute fly ash exposure. Toxicology 27, 273.

Hackney, J.D., Evans, M.J., Christie, B.R., 1975a. Effects of 60% and 80% oxygen on cell division in lung alveoli of squirrel monkeys. Aviat. Space Environ. Med. 46, 791.

Hackney, J.D., Evans, M.J., Spier, C.E., 1975b. Effects of 100% oxygen on cell division in lung alveoli of squirrel monkeys. Aviat. Space Environ. Med. 46, 1340.

Harbison, M.L., Godleski, J.J., Mortara, M., Brain, J.D., 1984. Correlation of lung macrophage age and surface antigen in the hamster. Lab. Invest. 50, 653.

Haschek, W.M., Reiser, K.M., Klein-Szanto, A.J.P., Kehrer, J.P., Smith, L.H., Last, J.A., Witschi, H.P., 1983. Potentiation of butylated hydroxy-toluene-induced acute lung damage by oxygen. Am. Rev. Respir. Dis. 127, 28.

Jeffery, P.K., Reid, L., 1975. New observations of rat airway epithelium: a quantitative and electron microscopic study. J. Anat. 120, 295.

Jeffery, P.K., Ayers, M., Rogers, D., 1982. The mechanisms and control of bronchial mucus cell hyperplasia. Adv. Exp. Med. Biol. 144, 399.

Kapanci, Y., Assimacopoulos, A., Irle, C., Zwahlen, A., Gabbiani, G., 1974. "Contractile interstitial cells" in pulmonary alveolar septae. A possible regulator of ventilation-perfusion ratio? Ultra-structural, inununofluorescence, and in vitro studies. J. Cell Biol. 60, 375.

Kaplan, N.P., Grant, M.M., Brody, J.S., 1985. The lipid interstitial cell of the pulmonary alveolus-age and species differences. Am. Rev. Respir. Dis. 132, 1307.

Kauffman, S.L., 1972. Alterations in cell proliferation in mouse lung following urethane exposure. Am. J. Pathol. 68, 317.

Kauffman, S.L., Burri, P.H., Weibel, E.R., 1974. The postnatal growth of the rat lung. II. Autoradiography. Anat. Rec. 180, 63.

Kauffman, S.K., 1980. Cell proliferation in the mammalian lung. Int. Rev. Exp. Pathol. 22, 131.

Keenan, K.P., Combs, J.W., McDowell, E.M., 1982a. Regeneration of hamster tracheal epithelium after mechanical injury. I. Focal lesions: quantitative morphologic study of cell proliferation. Virchows Arch. 41, 193.

Keenan, K.P., Combs, J.W., McDowell, E.M., 1982b. Regeneration of hamster tracheal epithelium after mechanical injury. II. Multifocal lesions: stathmokinetic and autoradiographic studies of cell proliferation. Virchows Arch. 41, 215.

Keenan, K.P., Combs, J.W., McDowell, E.M., 1982c. Regeneration of hamster tracheal epithelium after mechanical injury. III. Large and small lesions: comparative stathmokinetic and single pulse and continuous thymidine labeling autoradiographic studies. Virchows Arch. 41, 231.

Lane, B.D., Gordon, R.G., 1974. Regeneration of rat tracheal epithelium after mechanical injury. I. The relationship between mitotic activity and cellular differentiation. Proc. Soc. Exp. Biol. Med. 145, 1139.

Lum, H., Schwartz, L.W., Dungworth, D.L., Tyler, W.S., 1978. A comparative study of cell renewal after exposure to ozone or oxygen. Response of terminal bronchiolar epithelium in the rat. Am. Rev. Respir. Dis. 118, 335.

Masse, R., Fritsch, P., Nolibe, D., Lafuma, J., Chretien, J., 1976. Cytokinetic study of alveolar macrophage renewal in rats. In: Pulmonary Macrophages and Epithelial Cells, proceedings of the 16th Annual Hanford Biology Symposium, Richland, Washington, pp. 106–114.

McDowell, E.M., Keenan, K.P., Huang, M., 1984. Effects of vitamin A deprivation in hamster tracheal epithelium. A quantitative morphologic study. Virchows Arch. 45, 197.

McDowell, E.M., Newkirk, C., Coleman, B., 1985. Development of hamster tracheal epithelium. II. Cell proliferation in the fetus. Anat. Rec. 213, 448.

Meyrick, B., Reid, L., 1979. Hypoxia and incorporation of ^{3}H-thymidine by cells of the rat pulmonary arteries and alveolar wall. Am. J. Pathol. 96, 51.

Meyrick, B., Reid, L., 1982. Pulmonary arterial and alveolar development in normal postnatal rat lung. Am. Rev. Respir. Dis. 125 (4), 468–473.

Reid, L., Jones, R., 1979. Bronchial mucosal cells. Fed. Proc. 38, 191.

Rogers, A.W., 1979. Techniques of Autoradiography, vol. 276. Elsevier/North-Holland Biomedical Press, New York.

Sawyer, R.T., 1986a. The significance of local resident pulmonary alveolar macrophages in monocytopenic mice. J. Leuk. Biol. 39, 77.

Sawyer, R.T., 1986b. The cytokinetic behavior of pulmonary alveolar macrophages in monocytopenic mice. J. Leuk. Biol. 39, 89.

Sawyer, R.T., Strausbach, P.H., Volkman, A., 1982. Resident macrophage proliferation in mice depleted of blood monocytes by strontium-89. Lab. Invest. 46, 165.

Shami, S.G., Thibodeau, L.A., Kennedy, A.R., Little, J.B., 1982a. Proliferative and morphological changes in the pulmonary epithelium of the Syrian golden hamster during carcinogenesis initiated by ^{210}Po alpha-radiation. Cancer Res. 42, 1405.

Shami, S.G., Thibodeau, L.A., Kennedy, A.R., Little, J.B., 1982b. Recovery from ozone-induced injury in the lungs of the Syrian golden hamster. Exp. Mol. Pathol. 36, 57.

Shami, S.G., Silbaugh, S.A., Hahn, F.F., Griffith, W.C., Hobbs, C.H., 1984. Cytokinetic and morphological changes in the lungs and lung-associated lymph nodes of rats after inhalation of fly ash. Environ. Res. 35, 373.

Shami, S.G., Wolff, R.K., Hahn, F.F., Brooks, A.L., Griffith, W.C., 1985. Early cytokinetic and morphological response of rat lungs to inhaled benzo(a)pyrene, gallium oxide, and SO_2. Environ. Res. 37, 12.

Shami, S.G., Evans, M.J., Martinez, L.A., 1986a. Type II cell proliferation related to migration of inflammatory cells into the lung. Exp. Mol. Pathol. 44, 344.

Shami, S.G., Martinez, L.A., Evans, M.J., 1986b. The role of migrating inflammatory cells in proliferation of lung interstitium and epithelium. Chest 89, 170S.

Shellito, J., Esparza, C., Armstrong, C., 1987. Maintenance of the normal rat alveolar macrophage cell population. Am. Rev. Respir. Dis. 135, 78.

Sherman, M.P., Evans, M.J., Campbell, L.A., 1988. Prevention of pulmonary alveolar macrophage proliferation in newborn rabbits by hyperoxia. J. Pediat. 112, 782.

Simnett, J.D., Heppleston, A.G., 1966. Cell renewal in the mouse lung-the influence of sex, strain, and age. Lab. Invest. 15, 1793.

Tryka, A.F., Witschi, H., Gosslee, D.G., McArthur, A.H., Clapp, N.K., 1986. Patterns of cell proliferation during recovery from oxygen injury-species differences. Am. Rev. Respir. Dis. 133, 1055.

Wells, A.B., 1970. The kinetics of cell proliferation in the tracheobronchial epithelia of rats with and without chronic respiratory disease. Cell Tissue Kinet. 3, 186.

Weiss, R.A., Chanana, A.D., Joel, O.D., 1986. Postnatal maturation of pulmonary antimicrobial defense mechanisms in conventional and germ-free lambs. Pediat. Res. 20, 496.

White, H.J., Garg, B.D., 1981. Early pulmonary response of the rat lung to inhalation of high concentration of diesel particles. J. Appl. Toxicol. 1, 104.

Wilson, D.W., Plopper, C.G., Dungworth, D.L., 1984. The response of macaque tracheobronchial epithelium to acute ozone injury. Am. J. Pathol. 116, 193.

Winkler, G.C., Cheville, N.F., 1985. Monocytic origin and postnatal mitosis of intravascular macrophages in the porcine lung. J. Leuk. Biol. 38, 471.

Witschi, H.P., Morse, C.C., 1985. Cell kinetics in mouse lung following administration of carcinogens and butylated hydroxytoluene. Toxicol. Appl. Pharmacol. 78, 464.

Witschi, H.P., 1986. Separation of early diffuse alveolar cell proliferation from enhanced tumor development in mouse lung. Cancer Res. 46, 2674.

Wright, N., Alison, M., 1984. The Biology of Epithelial Cell Populations, vol. 438. Clarendon Press, Oxford.

Wright, E.S., White, D.M., Brady, A.N., Li, L.C., D'Arcy, J.B., Smiler, K.L., 1987. DNA synthesis in pulmonary alveolar macrophages and type II cell: effects of ozone exposure and treatment with alpha-difluoro methylornithine. J. Toxicol. Environ. Health 21, 15.

Appendix 3

Blood Vessels of the Lung

J. Michael Kay

Chapter Outline

1. INTRODUCTION

The lungs are unique with respect to their blood supply, since they possess a double circulation. The bronchial arteries are responsible for the nutrition of parts of the lung, including the walls of the bronchi and larger blood vessels, the interstitial connective tissue, hilar lymph nodes, and visceral pleura. The bronchial circulation is part of the systemic circulation, so the bronchial arteries and veins are similar in structure to other systemic arteries and veins. The bronchial circulation is modest in terms of its contribution to the total blood supply of the lungs. Their major source of blood is the pulmonary circulation, which carries the entire output of the right ventricle. The pulmonary circulation is morphologically and functionally unique. Not only does the lung have a more extensive and elaborate circulation than any other organ, but the pulmonary blood vessels are particularly suited to maintaining a delicate balance of pressure and distribution of flow that optimizes gas exchange. Morphologically, there are great differences between virtually every part of the pulmonary vasculature, on one hand, and their counterparts in the systemic vasculature on the other hand. The structure of blood vessels is related to the pressure and flow of blood within them. In normal mature mammals, the pulmonary circulation is a low-pressure, high-flow, low-resistance system. The pulmonary arterial pressure is normally about one-sixth that of the systemic arterial pressure. Accordingly, the pulmonary arteries, which are faced with pressures that are much lower and blood flows much higher than in the systemic circulation, tend to have thinner walls and wider lumina than their systemic counterparts. The structure of the pulmonary blood vessels also depends on the age of the individual. Striking differences are observed when the lungs of newborn infants are compared with those of young adults. However, there are also characteristic changes in the pulmonary blood vessels of older individuals, which may be considered normal for that age group.

There are species differences in the course and structure of the pulmonary and bronchial blood vessels that need to be considered when planning studies of the lungs using animals. These differences in structure and their physiological implications must also be considered when interpreting morphological data derived from animals and extrapolating the results to humans. There are morphological studies of the postnatal growth and development of the pulmonary circulation (Reid, 1979), and of the pulmonary and bronchial blood vessels in amphibians, reptiles, and birds (Wagenvoort, 1981). However, this account is primarily concerned with the morphological characteristics of the pulmonary and bronchial circulations in mature mammals, since these are most frequently chosen for experiments. Tables 1 and 2 contain morphological data concerning the pulmonary and bronchial circulations, respectively. The following paragraphs amplify these data with regard to the pulmonary trunk, elastic pulmonary arteries, muscular pulmonary arteries, pulmonary arterioles, pulmonary veins, and bronchial arteries.

2. PULMONARY TRUNK

After infancy, the human pulmonary trunk is about half the thickness of the aorta and the ratio of the medial thickness of the pulmonary trunk to that of the aorta (PT/A ratio) ranges from 0.4 to 0.7. This ratio is similar in the other mammals listed in Table 1. However, there are qualitative species differences in the structure of the pulmonary trunk. In human subjects after infancy, the tunica media of the pulmonary trunk is composed of short, irregular, branching elastic fibrils interspersed with collagen and smooth muscle. This pattern is also seen in the dog, ferret, guinea pig, and monkey. In the rat the pulmonary trunk resembles the aorta in that it is composed of long, parallel concentric elastic lamellae. In the mountain vizcacha, the outer elastic laminae of the pulmonary trunk are concentric, whereas the inner are fragmented. In the llama and yak, which live at high altitude, the outer part of the wall of the pulmonary trunk contains abundant smooth muscle, whereas the inner part of the wall consists of thick elastic fibrils.

3. ELASTIC PULMONARY ARTERIES

The human pulmonary blood vessels are conveniently classified according to their external diameter and structure using the criteria of Brenner (1935). Elastic pulmonary arteries measure more than 1000 μm in external diameter and have a medial coat composed of concentric elastic laminae interspersed with circular smooth muscle. In other mammals, the basic structure of this type of vessel is similar to that in human subjects, but there is considerable variation in the medial thickness, expressed as a percentage of the external diameter (Best and Heath, 1961). In humans, deer, monkey, dog, cat, and mouse, the pulmonary arterial tree is smoothly tapering so that elastic pulmonary arteries blend imperceptibly and without irregularity of wall thickness into muscular pulmonary arteries (Ferencz, 1969). This contrasts with the rabbit, rat, and guinea pig, where the elastic segment is thin walled and branching is associated with abrupt changes in wall thickness and the size of the lumen. In all species, the length of the arterial segment distal to a point where the external diameter is 1000 μm is similar. The difference in the total length of the pulmonary arterial tree is accounted for by a variation in the length of the elastic pulmonary arterial segment. The increase in the length of the pulmonary arterial tree that occurs during postnatal growth is largely due to an increase in the length of the elastic pulmonary arteries. The elastic pulmonary arteries are the capacitance vessels of the lung, in contrast to the muscular pulmonary arteries and arterioles, which are the resistance vessels.

4. MUSCULAR PULMONARY ARTERIES

The basic structure of this type of vessel is an intima of endothelium, an internal elastic lamina, a media of circular smooth muscle and external elastic lamina, and an adventitia of fibrous tissue. In human subjects, the muscular pulmonary arteries range in external diameter from 100 to 1000 μm and the thickness of the tunica media ranges from 3% to 7% of the external diameter, with a mean value of 5% (Figure 1). In the baboon, ferret, goat, llama, monkey, and yak, the medial thickness of muscular pulmonary arteries is similar to that in human subjects. However, in most other animals the muscular pulmonary arteries are more muscular than those in humans. Accordingly, it is important that when histologic material from animals is examined, normal thick-walled muscular pulmonary arteries, such as those that occur in the cow (Figure 2), are not mistakenly diagnosed as being affected by medial hypertrophy associated with pulmonary hypertension.

Cats may suffer from a spontaneously arising condition known as pulmonary artery medial hyperplasia (PAMH). Hamilton (1970) concluded that the condition was a result of infestation with the lung worm *Aelurostrongylus abstrusus*. However, Rogers et al. (1971) observed PAMH with the same frequency in both specific-pathogen-free and conventional cats. Furthermore, none of their animals showed evidence of infection by lung worm. Pulmonary hypertension and right ventricular hypertrophy are not associated with PAMH (Rogers et al., 1971) or infestation with *A. abstrusus* (Rawlings et al., 1980). In the guinea pig, the tunica media of muscular pulmonary arteries is irregular. It is composed of concentric masses of circular smooth muscle disposed in discontinuous segments resembling sphincters (Figure 3). The functional significance of this unusual arrangement of pulmonary artery smooth muscle is unknown. Along the course of many of the muscular pulmonary arteries of the rat is a short segment where the tunica media is thickened and composed largely of obliquely oriented, closely packed smooth muscle cells with no external elastic lamina. In this so-called thick-walled oblique muscle segment, the medial thickness may be as much as 35% of the external diameter (Meyrick et al., 1978). The muscular pulmonary arteries of rats are also unusual in that the endothelial cells may contain large elliptical intracytoplasmic inclusions, called discoid bodies (Weibel and Palade, 1964). These are composed of dilated segments of rough endoplasmic reticulum, which contain granular material.

TABLE 1 Structure of Pulmonary Arteries and Veins in Mammals

	Pulmonary Trunk		Muscular Pulmonary Arteries				
Animal Species	Structure of Media	PT/A	Diameter Range (μm)	Medial Thickness (%)	Distinctive Features	Pulmonary Veins	References
Humans	Short irregular branching elastic fibrils	0.4–0.7	100–1000	2.0–6.9	—	Thin, mainly fibrous wall	Heath et al. (1959) and Best and Heath (1961)
Baboon (*Papio papio*)	—	0.43	—	4.1–6.8	—	—	Smith et al. (1973)
Cat	—	—	20–500	8.0–25.0	"Pulmonary artery medial hyperplasia" occurs spontaneously	Thin, mainly fibrous wall	Best and Heath (1961) and Rhodin (1978)
Civet	—	—	20–40	9.1–18.8	—	Thin, mainly fibrous wall	Best and Heath (1961)
Cow	Muscular outer wall with long, thick, wavy elastic fibrils	0.48–0.55	20–150	6.2–16.4	—	Follow bronchial tree. Thick spiral bundles of smooth muscle in intima	Heath et al. (1968), Alexander and Jensen (1963) and Wagenvoort and Wagenvoort (1969)
Beagle dog	Short irregular branching elastic fibrils	0.47–0.56	30–120	3.9–11.7	—	Thin, mainly fibrous wall	Heath et al. (1959), Kay et al. (1971) and Best and Heath (1961)
Ferret	Fragmented discon-tinuous elastic lamellae	0.47	20–225	2.3–9.0	—	Thin, mainly fibrous wall	Kay (1983)
Fox	—	—	30–100	7.3–10.0	—	Thin, mainly fibrous wall	Best and Heath (1961)
Goat	—	—	100–300	2.2–4.0	—	Thin, mainly fibrous wall	Best and Heath (1961)
Guinea pig	Fragmented discon-tinuous elastic lamellae	0.47	—	4.3 ± 0.2 SEM	Discontinuous seg-ments of medial smooth muscle	Thin, fibromuscular wall	Best and Heath (1961) and Tucker et al. (1975)
Horse	—	—	40–300	5.1–10.3	—	Follow bronchial tree. Thin, mainly fibrous wall	Kay (1983)

Continued

TABLE 1 Structure of Pulmonary Arteries and Veins in Mammals—cont'd

Animal Species	Pulmonary Trunk		Muscular Pulmonary Arteries			Pulmonary Veins	References
	Structure of Media	PT/A	Diameter Range (μm)	Medial Thickness (%)	Distinctive Features		
Llama (*Llama glama*), at high altitude	Muscular outer wall with long, thick, wavy elastic fibrils	0.62	28–230	4.8	—	Thin, muscular wall	Harris et al. (1982) and Heath et al. (1974)
Monkey (*Erythrocebus patas patas*)	Branched irregular elastic fibrils	0.38	50–160	2.1–5.2	—	Thin, mainly fibrous wall	Heath et al. (1970)
Mountain vizcacha (*Lagidium peruanum*)	Outer elastic lamellae concentric, inner fragmented	0.39–0.50	42–189	2.4–10.0	—	Cardiac muscle in media	Heath et al. (1981)
Mouse	—	—	20–300	4.4–11.7	—	Follow bronchial tree. Cardiac muscle in media	Best and Heath (1961)
Pig	—	—	25–70	7.4–17.6	—	Follow bronchial tree. Thick intimal pads of smooth muscle	Best and Heath (1961)
Rabbit	—	—	40–150	3.8–10.6	—	Thin, mainly fibrous wall	Best and Heath (1961)
Rat	Long parallel elastic lamellae	0.24–0.47	20–350	2.6–26.0	Thick walled oblique muscle segment. Discoid bodies in endothelium	Intimal pads of smooth muscle. Cardiac muscle in media	Kay and Heath (1969), Meyrick et al. (1978) and Weibel and Palade (1964)
Sheep	Muscular outer wall with fragmented discontinuous elastic lamellae	0.46	30–200	5.2–11.8	—	Follow bronchial tree. Thin, muscular wall	Best and Heath (1961)
Squirrel	—	—	—	10.3–22.1	—	Cardiac muscle in media	Best and Heath (1961)
Yak (*Bos grunniens*)	Nodular muscular outer wall with parallel thick elastic fibrils	0.59	76–228	2.8–5.5	—	—	Heath et al. (1984)

Definition of abbreviations: PT = medial thickness of pulmonary trunk; A = medial thickness of aorta.

TABLE 2 Distribution of Bronchial Arteries in Mammals

	Lung Type I (Thick Pleura)			Lung Type II (Thin Pleura)			Variant Type II (Thin Pleura)			Lung Type III (Thick Pleura)	
	Cow	Lamb	Pig	Dog	Cat	Rhesus Monkey	Rabbit	Rat	Guinea Pig	Horse	Humans
Pleura	+	+	+	−	−	−	−	−	−	+	+
Septa	+	+	+	−	−	−	−	−	−	+	+
Bronchi	+	+	+	+	+	+	+	+	+	+	+
Terminal bronchioles	+	+	+	+	+	+	−	+	+	+	+
Alveoli	−	−	−	−	−	−	−	−	−	+	+
Hilar nodes	+	+	+	+	+	+	+	+	+	+	+
Vasa vasorum	+	+	+	+	+	+	+	+	+	+	+
BPA	+	+	−	−	−	−	+	+	+	+	+

Definition of abbreviations: + = distribution to structure; − = not distributed to structure; BPA = bronchial artery-pulmonary artery anastomoses.
From McLaughlin, R.F., 1983. Am. Rev. Resp. Dis. 128, 557. With permission.

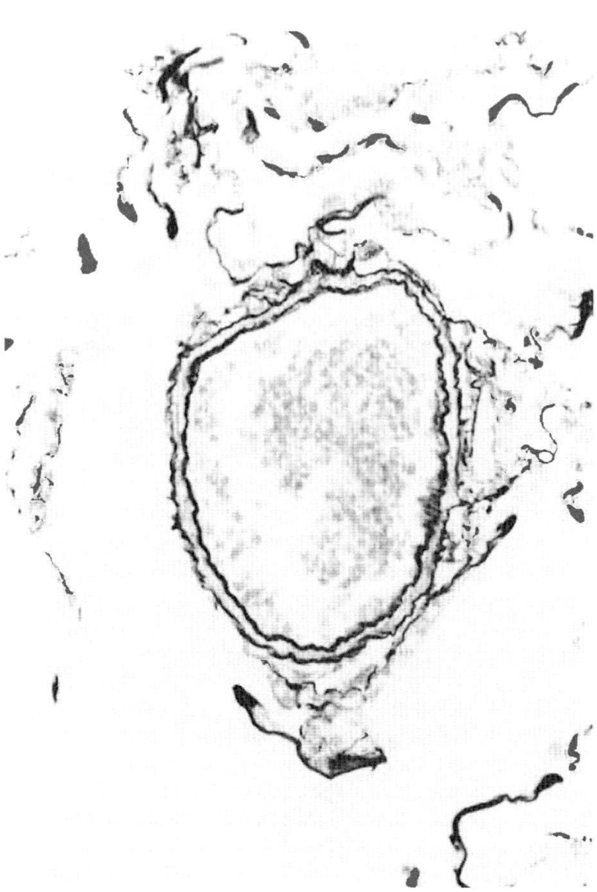

FIGURE 1 Muscular pulmonary artery from woman aged 37 years. The thin tunica media of the circular smooth muscle is bounded by internal and external elastic laminae. There is virtually no adventitia. Elastic-van Gieson stain; magnification × 500.

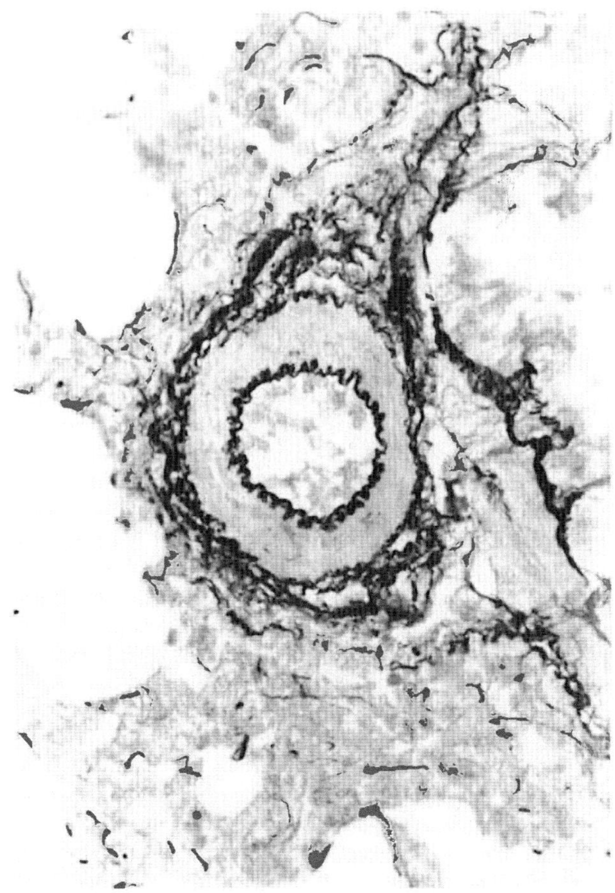

FIGURE 2 Cow. Muscular pulmonary artery. The tunics media of circular smooth muscle is thick and is located between internal and external elastic laminae. The wall of a bronchiole is on the right side of the picture. Elastic-van Gieson stain; magnification × 500.

There is evidence that the amount of pulmonary artery smooth muscle inherent within each species is a major determinant of the pulmonary hypertensive response to chronic alveolar hypoxia and contributes to the interspecies variability in this response. Tucker et al. (1975) exposed seven animal species to hypobaric hypoxia from 19 to 48 days. The pulmonary hypertension that developed in the hypoxic animals varied in the following order of decreasing severity: calf and pig (severe); rat and rabbit (moderate); sheep, guinea pig, and dog (mild). These interspecies variations in response did not correlate with the degree of systemic arterial hypoxemia, polycythemia, tachycardia, or postnatal age. However, there was a significant relation between the medial thickness of small pulmonary arteries in control animals, on the one hand, and the severity of pulmonary hypertension and the degree of right ventricular hypertrophy in hypoxic animals on the other. The susceptibility or resistance to high-altitude pulmonary hypertension in cattle may be determined genetically (Weir et al., 1974). Calves that are susceptible to high-altitude pulmonary hypertension have thicker muscular pulmonary arteries than calves that are resistant to high-altitude pulmonary hypertension (Weir et al., 1979). It is interesting that the llama, mountain vizcacha, and yak, which are genetically adapted to life at high altitude, have thin muscular pulmonary arteries. Although the amount of pulmonary artery smooth muscle may be a major determinant of chronic hypoxic pulmonary hypertension, it does not seem to be the determinant of the pulmonary vasoconstrictor response to acute alveolar hypoxia. The steady-state stimulus response relationships of the pulmonary circulation to graded hypoxia in the isolated lungs of pigs, dogs, rabbits, cats, and ferrets were studied by Peake et al. (1981). Pronounced species differences were apparent in both the severity of hypoxia required to elicit responses and the amplitude of the responses. The ferret and pig had the largest vasoconstrictor responses to hypoxia, the cat and rabbit were intermediate responders, and no significant response was obtained in the dog. The pig has relatively thick muscular pulmonary arteries, with the medial thickness ranging from 7.4% to 17.6%. On the other hand, the ferret has thin pulmonary arteries, with a medial thickness ranging from 2.3% to 9% (Figure 4).

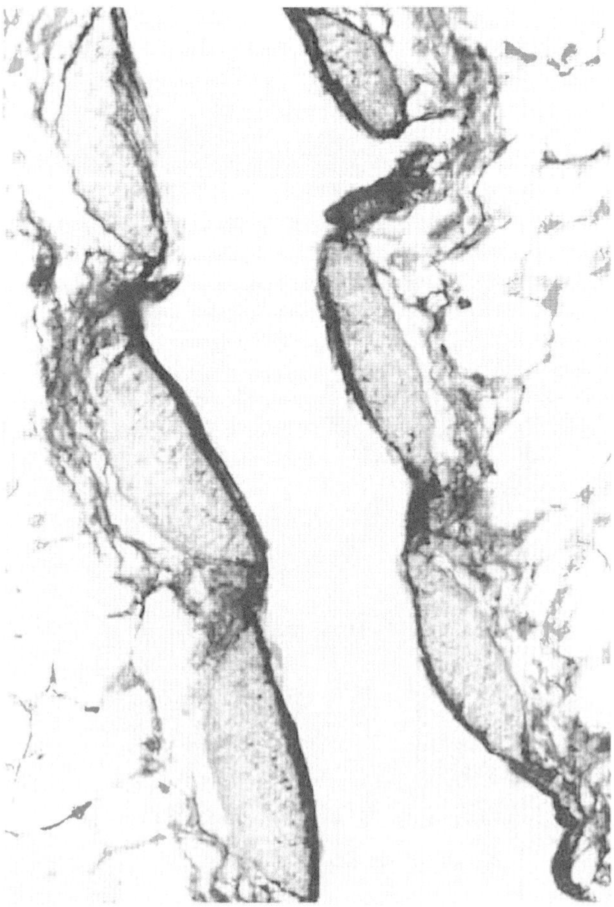

FIGURE 3 Guinea pig. Muscular pulmonary artery in longitudinal section. The tunica media is composed of crescentic masses of circular smooth muscle arranged in discontinuous segments. Elastic-van Gieson stain; magnification × 500.

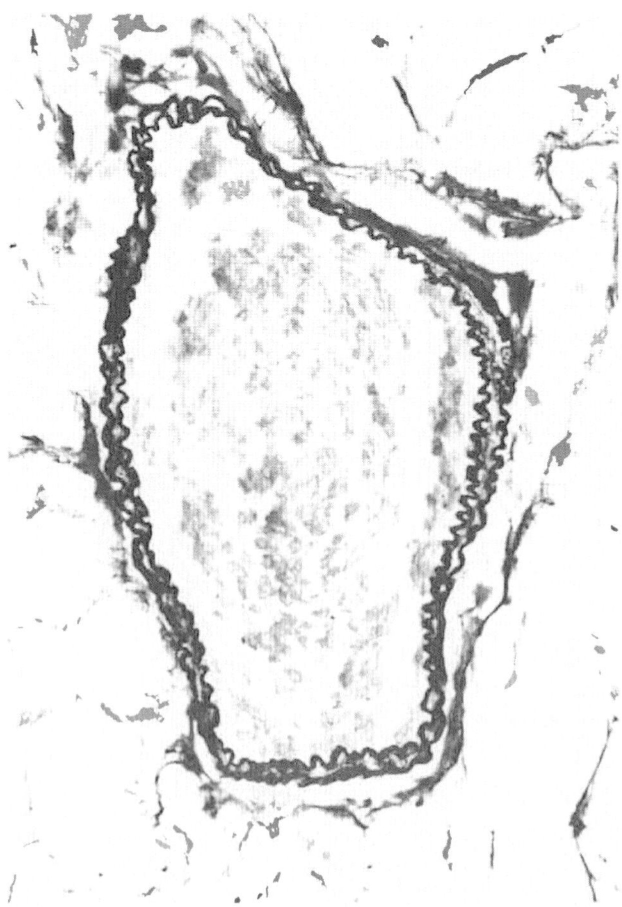

FIGURE 4 Ferret. Muscular pulmonary artery. The tunica media of circular smooth muscle is thin and is located between internal and external elastic laminae. Elastic-van Gieson stain; magnification × 500.

5. PULMONARY ARTERIOLES

In human subjects pulmonary arterioles are arbitrarily defined as pulmonary arterial vessels measuring less than 100 μm in diameter. The transition from the muscular to the nonmuscular part of the pulmonary arterial tree takes place over a diameter range of 100 to 70 μm. This transition is not sharp, because the media changes gradually from a complete layer of smooth muscle to a discontinuous one. The circularly arranged smooth muscle cells in this intermediate segment follow an ever-widening spiral course. The distal portions of the pulmonary arterioles that are nonmuscular have a wall consisting of a single elastic lamina, which is a continuation of the external elastic lamina, lined by endothelial cells. There is virtually no adventitia. The goat is similar to human subjects in that the muscular pulmonary arteries terminate at an external diameter of about 100 μm. However, in most other mammals smooth muscle is present in pulmonary arterial vessels of much smaller external diameter. Specifically, the pulmonary arterial vessels in the cat, cow, ferret, mouse, and rat are muscularized down to an external diameter of 20 μm. The earliest histologic evidence of pulmonary arterial hypertension is muscularization of pulmonary arterioles. Accordingly, in the human lung the finding of large numbers of small muscular arterial vessels measuring less than 70 μm in diameter is pathognomonic of hypertensive pulmonary vascular disease. Clearly, this criterion cannot be applied to the lungs of most other mammals, where such small muscular vessels are merely the smallest muscular pulmonary arteries and have no pathologic significance.

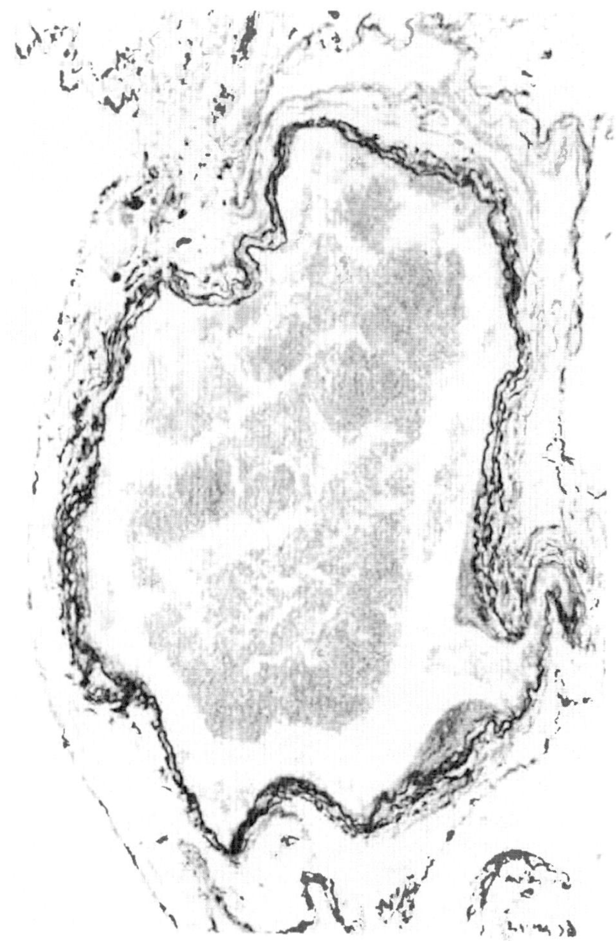

FIGURE 5 Pulmonary vein from woman aged 37 years. The wall is thin and is composed of a haphazard arrangement of elastic tissue, collagen, and smooth muscle. There is no distinct tunica media. A pulmonary venule joins the bottom right quadrant of the vessel. Elastic-van Gieson stain; magnification × 200.

6. PULMONARY VEINS

In human subjects the pulmonary veins are located in the fibrous interlobular septa, away from the pulmonary arteries and bronchioles, and they pursue an independent course towards the hilum of the lung. This anatomic relationship is not present in the cow, horse, mouse, pig, or sheep, where the pulmonary veins accompany the bronchial tree. In human subjects, the pulmonary veins measuring more than 100 μm in external diameter have walls that are thinner than those of an artery of comparable size (Figure 5). The intima is composed of an internal elastic lamina lined by endothelium. The tunica media is composed of a haphazard arrangement of elastic fibrils, smooth muscle, and collagen. There is no clear demarcation between the media and adventitia. Pulmonary veins with thin fibrous walls are also present in the cat, civet, dog, ferret, fox, goat, horse, monkey, and rabbit. The pulmonary veins are muscular in the cow, guinea pig, llama, pig, and rat. In the cow the pulmonary veins present a beaded appearance in longitudinal section, because of the presence of thick spiral bundles of smooth muscle in the intima (Figure 6). The functional significance of these prominent muscle bundles is not clear. They are absent in newborn calves, begin to develop at the age of 2 weeks, and are present in all animals aged 3 months or older (Wagenvoort and Wagenvoort, 1969). It is important not to misinterpret these thick muscular pulmonary veins adjacent to bronchi as pulmonary arteries. Most larger animals are similar to humans in that the intrapulmonary veins are devoid of cardiac muscle. However, the intrapulmonary veins of the mouse, rat (Figure 7), squirrel, and mountain vizcacha are characterized by a thick medial coat of striated muscle that is continuous with the left atrium. In the larger intrapulmonary veins, the cardiac muscle coat is arranged into an external longitudinal and an internal circular layer.

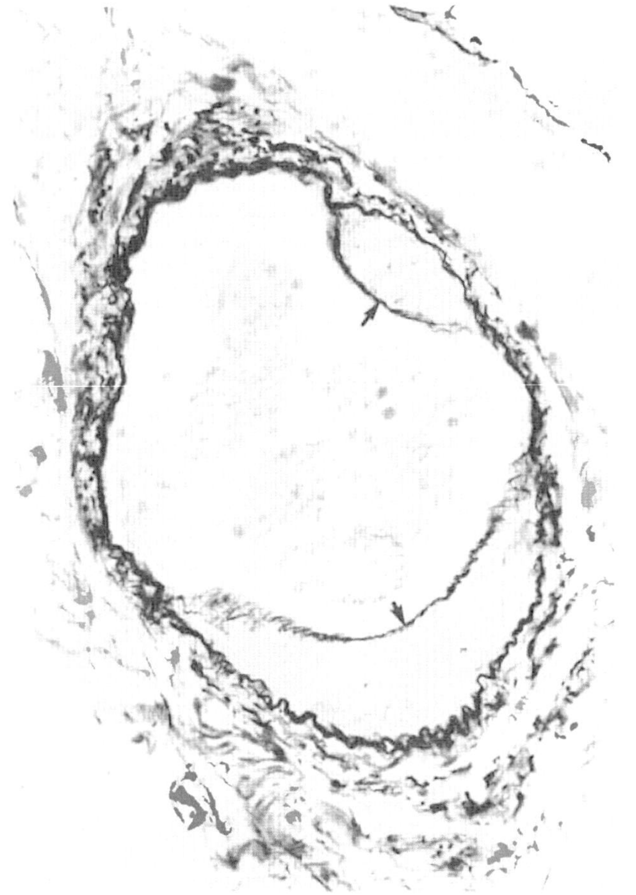

FIGURE 6 Cow. Pulmonary vein in transverse section. The intima contains a spiral bundle of smooth muscle that has been sectioned in two places (arrows). Elastic-van Gieson stain; magnification × 500.

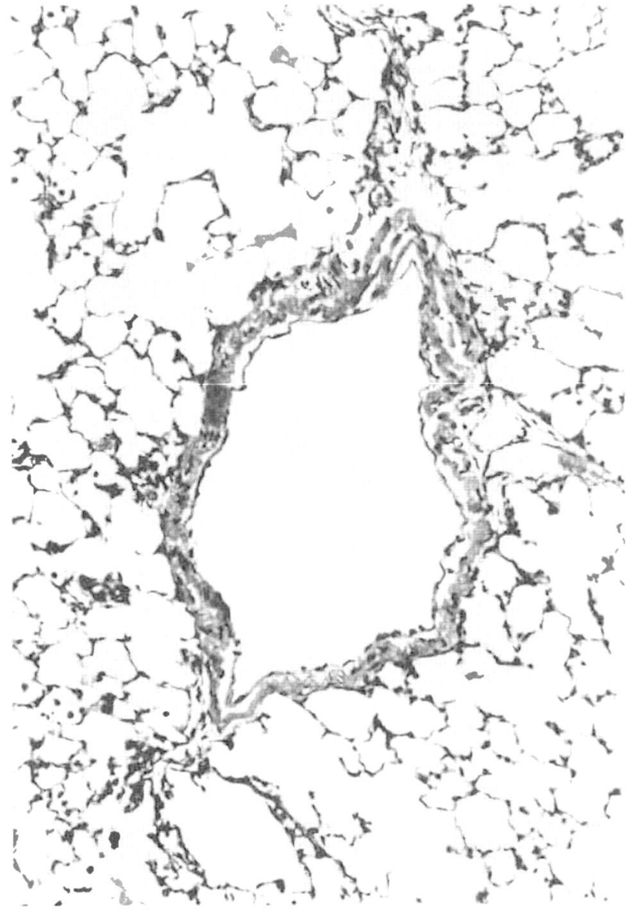

FIGURE 7 Rat. Pulmonary vein. The tunica media is composed of a thick coat of striated cardiac muscle. Hematoxylin and eosin stain; magnification × 200.

7. BRONCHIAL ARTERIES

Immediately after arising from the aorta, internal mammary, or intercostal arteries, the bronchial arteries have the characteristic structure of systemic arteries. They have a thick tunica media, composed of circular smooth muscle and a well-defined internal elastic lamina. The external elastic lamina is either absent or thin and fragmented (Figure 8). Both pulmonary and bronchial arteries accompany bronchi, but the latter arteries are always small compared to the large bronchi that they serve as nutrient vessels. The bronchial arteries usually lie adjacent to bronchial veins and nerve bundles. They may also be found in the adventitia of large elastic pulmonary arteries. When a bronchial artery enters the bronchial wall, it often shows crescent-shaped bundles or circular layers of longitudinal smooth muscle cells in the intima. These smooth muscle cells are surrounded by a reduplication of the internal elastic lamina. This intimal layer of longitudinal muscle may be so thick as to virtually occlude the lumen of the vessel (Figure 9). In such arteries, there may be virtual absence of the circular smooth muscle coat (Harris and Heath, 1986; Wagenvoort, 1981).

In human subjects there is considerable variation in the number and origin of the bronchial arteries (Cauldwell et al., 1948; Pump, 1972). The left bronchial artery usually arises from the aorta at about the level of the fifth thoracic vertebra, and the right bronchial artery also frequently originates from this source. However, one or more right bronchial arteries may arise from the intercostal and internal mammary arteries. The extreme variability of the origin and numbers of the bronchial arteries is demonstrated by the fact that Cauldwell et al. (1948) were able to distinguish no fewer than nine arrangements in 150 cadavers. There has been no systemic investigation of the origin and number of the bronchial arteries in animals (McLaughlin, 1983).

McLaughlin (1983) studied the distribution of the bronchial arteries in 11 mammalian species and related it to the type of subgross anatomy present in a particular species (Table 2). In previous studies McLaughlin et al. (1966) had distinguished three types of subgross lung anatomy. In type I, represented by the cow, sheep, and pig, the pleura

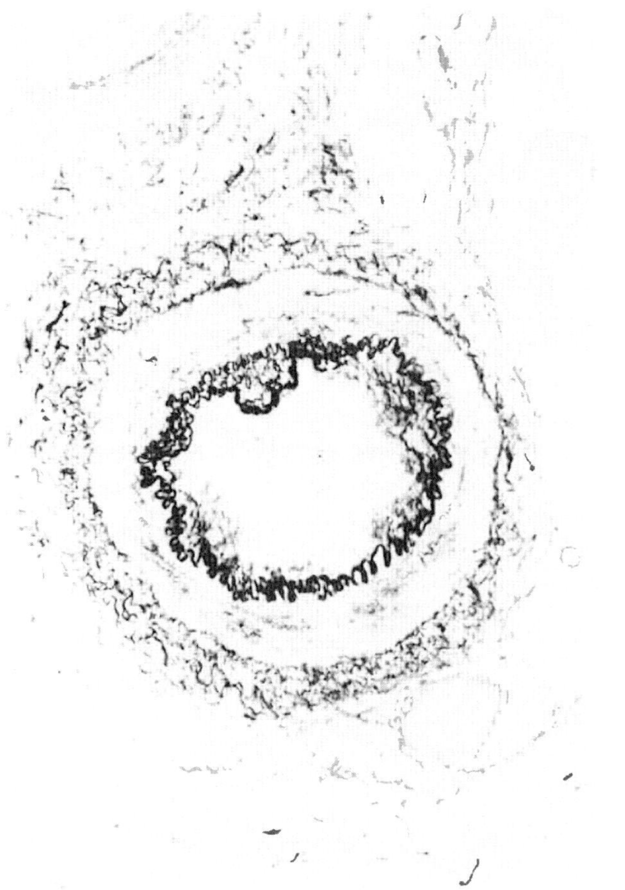

FIGURE 8 Bronchial artery from man aged 68 years. There is a prominent internal elastic lamina and a thick tunica media composed of circular smooth muscle. The external elastic lamina is indistinct. Elastic-van Gieson stain; magnification × 200.

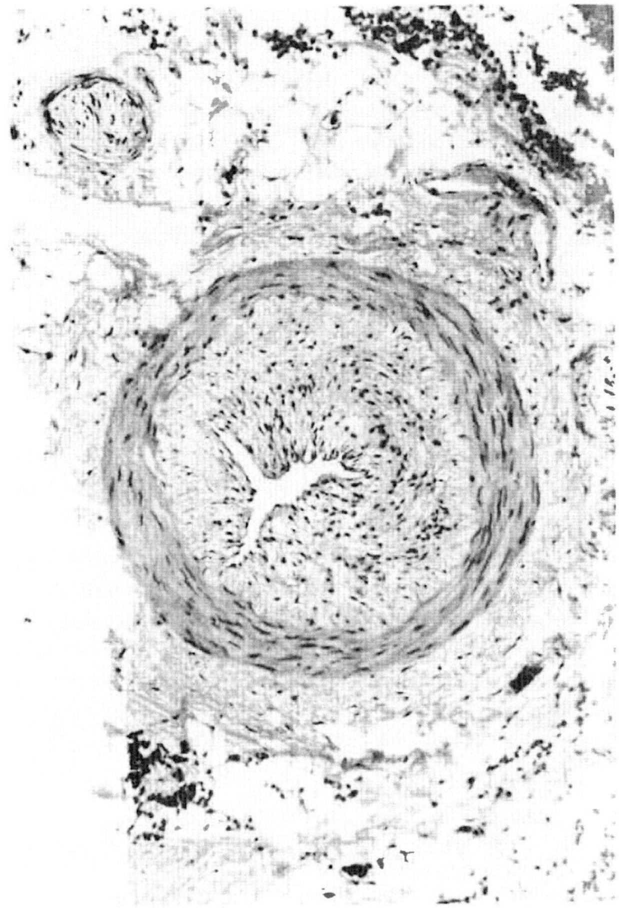

FIGURE 9 Bronchial artery from a man aged 68 years. The tunica media is composed of circular smooth muscle. The lumen is narrow and has an irregular outline. The intima is thick and is composed of longitudinal muscle. A nerve is situated at the top left corner of the picture. Hematoloxylin and eosin stain; magnification × 200.

and interlobular septa are thick and the secondary lung lobules are very well defined. In lungs of type II (dog, cat, and rhesus monkey), the pleura is thin and interlobular septa and secondary lung lobules are absent. In a variant of type II represented by the rabbit, rat, and guinea pig, the distal bronchioles do not possess as many alveoli as those in the dog, cat, and rhesus monkey. In type III lungs (horse and human), the pleura and interlobular septa are thick, but the secondary lung lobules are imperfectly defined.

In all three anatomical types, the bronchial arteries supply the bronchi, hilar lymph nodes, and vasa vasorum to the pulmonary arteries and pulmonary veins. In type III lungs the bronchial artery also provides arteriolar branches that connect directly to some alveolar capillaries. In contrast, it has been impossible to trace the bronchial artery beyond the second or third division of the main stem bronchus in the rabbit whose lungs are a variant of type II. In lungs of types I and III, the pleura and interlobular septa, which are thick, derive their arterial blood supply from the bronchial arteries. In lungs of type II, the pleura, which is thin, derives its arterial supply from the pulmonary artery. McLaughlin (1983) found anastomoses between the bronchial and pulmonary arteries in all species studied, except the pig, dog, cat, and rhesus monkey. In the variant type II lungs (rabbit, rat, and guinea pig), they measured from 400 to 600 μm in diameter and were found only near the hilum. In the horse and in human subjects, they were found at various levels along the bronchi and bronchioles, and also in the pleura. The largest measured 400 μm in diameter. A similar distribution was found in the cow and sheep.

8. SUMMARY

There are interspecies differences in the quantity and arrangement of collagen, smooth muscle, and elastic tissue in the pulmonary trunk. The medial thickness of muscular pulmonary arteries is similar in the human, baboon, ferret, goat, llama, monkey, and yak. In most other mammals these vessels are more muscular than those in humans. The goat is similar to the human because the muscular pulmonary arteries terminate at an external diameter of 100 μm. In most other mammals, smooth muscle is present in much smaller pulmonary arterial vessels. There is pronounced interspecies variation in the course and structure of the pulmonary veins. In human subjects the pulmonary veins are located in the fibrous interlobular septa, away from the pulmonary arteries and bronchioles. This anatomic relationship is not present in the cow, horse, mouse, pig, or sheep, where the pulmonary veins accompany the bronchial tree. The cat, civet, dog, ferret, fox, goat, horse, monkey, and rabbit have pulmonary veins with thin fibrous walls, such as occur in human subjects. However, the pulmonary veins are muscular in the cow, guinea pig, llama, pig, and rat. Most larger animals are similar to humans in that the intrapulmonary veins are devoid of cardiac muscle. However, the intrapulmonary veins in the mouse, rat, squirrel, and mountain vizcacha possess a thick medial coat of striated muscle that is continuous with the left atrium. The bronchial artery is considered to be the nutrient artery to the lung. Its distribution varies greatly from species to species and seems to depend upon the amount of supportive or non-gas-exchanging tissue present within the lung.

REFERENCES

Alexander, A.F., Jensen, R., 1963. Normal structure of bovine pulmonary vasculature. Am. J. Vet. Res. 24, 1083.

Best, P.V., Heath, D., 1961. Interpretation of the appearances of small pulmonary blood vessels in animals. Circ. Res. 9, 288.

Brenner, O., 1935. Pathology of the vessels of the pulmonary circulation. Arch. Intern. Med. 56, 211.

Cauldwell, E.W., Siekert, R.G., Lininger, R.E., Anson, B.J., 1948. The bronchial arteries. An anatomic study of 150 human cadavers. Surg. Gynecol. Obstet. 86, 395.

Ferencz, C., 1969. Pulmonary arterial design in animals. Johns Hopkins Med. J. 125, 207.

Hamilton, J.M., 1970. The influence of infestation by *Aelurostrongylus abstrusus* on the pulmonary vasculature of the cat. Br. Vet. J. 126, 202.

Harris, P., Heath, D., 1986. The Human Pulmonary Circulation, third ed. Churchill Livingstone, London, pp. 590–595.

Harris, P., Heath, D., Smith, P., Williams, D.R., Ramirez, A., Kriiger, H., Jones, D.M., 1982. Pulmonary circulation of the llama at high and low altitudes. Thorax 37, 38.

Heath, D., Harris, P., Castillo, Y., Arias-Stella, J., 1968. Histology, extensibility and chemical composition of the pulmonary trunk of dogs, sheep, cattle, and llamas living at high altitude. J. Pathol. Bacteriol. 96, 161.

Heath, D., Jones, E.L., Housley, E., 1970. Pulmonary arterial pressure and structure in the patas monkey (*Erythrocebus patas patas*). Cardiovasc. Res. 4, 515.

Heath, D., Smith, P., Williams, D., Harris, P., Arias-Stella, J., Krüger, H., 1974. The heart and pulmonary vasculature of the llama (*llama glama*). Thorax 29, 462.

Heath, D., Williams, D., Dickinson, J., 1984. The pulmonary arteries of the yak. Cardiovasc. Res. 18, 133.

Heath, D., Williams, D., Harris, P., Smith, P., Krüger, H., Ramirez, A., 1981. The pulmonary vasculature of the mountain-viscacha (*Lagidium peruanum*). The concept of adapted and acclimatized vascular smooth muscle. J. Comp. Pathol. 91, 293.

Heath, D., Wood, E.H., DuShane, J.W., Edwards, J.E., 1959. The structure of the pulmonary trunk at different ages and in cases of pulmonary hypertension and pulmonary stenosis. J. Pathol. Bacteriol. 77, 443.

Kay, J.M., 1983. Comparative morphologic features of the pulmonary vasculature in mammals. Am. Rev. Respir. Dis. 128, S53.

Kay, J.M., Heath, D., 1969. *Crotalaria spectabilis*: The Pulmonary Hypertension Plant. Charles C. Thomas, Springfield, IL, pp. 23–40.

Kay, J.M., Smith, P., Heath, D., 1971. Aminorex and the pulmonary circulation. Thorax 26, 262.

McLaughlin, R.F., 1983. Bronchial artery distribution in various mammals in humans. Am. Rev. Respir. Dis. 128, S57.

McLaughlin, R.F., Tyler, W.S., Canada, R.O., 1966. Subgross pulmonary anatomy of the rabbit, rat, and guinea pig, with additional notes on the human lung. Am. Rev. Respir. Dis. 94, 380.

Meyrick, B., Hislop, A., Reid, L., 1978. Pulmonary arteries of the normal rat: the thick walled oblique muscle segment. J. Anat. 125, 209.

Peake, M.D., Harabin, A.L., Brennan, N.J., Sylvester, J.T., 1981. Steady-state vascular responses to graded hypoxia in isolated lungs of five species. J. Appl. Physiol. 51, 1214.

Pump, K.K., 1972. Distribution of bronchial arteries in the human lung. Chest 62, 447.

Rawlings, C.A., Losonsky, J.M., Lewis, R.E., Hubble, J.J., Prestwood, A.K., 1980. Response of the feline heart to *Aelurostrongylus abstrusus*. J. Am. Anim. Hosp. Assoc. 16, 573.

Reid, L.M., 1979. The pulmonary circulation: remodeling in growth and disease. Am. Rev. Respir. Dis. 119, 531.

Rhodin, J.A.G., 1978. Microscopic anatomy of the pulmonary vascular bed of the cat lung. Microvasc. Res. 15, 169.

Rogers, W.A., Bishop, S.P., Rohovsky, M.W., 1971. Pulmonary artery medial hypertrophy and hyperplasia in conventional and specificpathogen-free cats. Am. J. Vet. Res. 32, 767.

Smith, P., Heath, D., Wright, J.S., McKendrick, C.S., 1973. Pulmonary arterial pressure and structure in the baboon *Papio papio*. J. Pathol. 111, 43.

Tucker, A., McMurtry, I.F., Reeves, J.T., Alexander, A.F., Will, D.H., Grover, R.F., 1975. Lung vascular smooth muscle as a determinant of pulmonary hypertension at high altitude. Am. J. Physiol. 228, 762.

Wagenvoort, C.A., 1981. The blood vessels of the lung. In: Schwartz, C.J., Werthessen, N.T., Wolf, S. (Eds.), Structure and Function of the Circulation. Plenum Press, New York, pp. 223–362.

Wagenvoort, C.A., Wagenvoort, N., 1969. The pulmonary vasculature in normal cattle at sea level at different ages. Pathol. Eur. 4, 265.

Weibel, E.R., Palade, G.E., 1964. New cytoplasmic components in arterial endothelia. J. Cell Biol. 23, 101.

Weir, E.K., Tucker, A., Reeves, J.T., Will, D.H., Grover, R.F., 1974. The genetic factor influencing pulmonary hypertension in cattle at high altitude. Cardiovasc. Res. 8, 745.

Weir, E.K., Will, D.H., Alexander, A.F., McMurtry, I.F., Looga, R., Reeves, J.T., Grover, R.F., 1979. Vascular hypertrophy in cattle susceptible to hypoxic pulmonary hypertension. J. Appl. Physiol. 46, 517.

Appendix 4

Comparative Ventilation of the Normal Lung

TABLE 1 Lung Volumes of Various Mammalian Species

Species	Body Weight (kg)	Conditions	TLC (ml)	TLC (ml/kg)	FRC (ml)	FRC (ml/kg)	RV (ml)	RV (ml/kg)	References
Horses	485.00	Anesthetized, prone	44800	92.4	17600	36.3	9200	19.0	Sorenson and Robinson (1980)
	485.00	Anesthetized, supine	34000	70.1	11700	24.1	5700	11.8	Sorenson and Robinson (1980)
Humans	78.40	Awake, supine	7210	92.0	3080	39.3	2230	28.4	Westbrook et al. (1973)
	71.00	Awake, sitting	7260	102.3	3952	55.7	1460	20.6	Rehder et al. (1975)
Dogs	18.60	Awake, prone	2090	112.4	997	53.6	310	16.7	Gillespie et al. (1973)
	16.50	Anesthetized, prone	1684	102.1	665	40.3	253	15.3	Lai et al. (1977)
Monkeys	5.56	Anesthetized, upright	503	90.5	275	49.5	92	16.6	Paré et al. (1978)
	7.77	Anesthetized, sitting	619	79.7	333	42.8	76	9.8	Kosch et al. (1979a)
Rabbits	3.55	Anesthetized, prone	123	34.7	37.7	10.6	5.2	1.5	Caldwell (1971)
	3.14	Anesthetized, prone	111	35.4	36.4	11.6	20.2	6.4	Caldwell and Fry (1969)
Guinea pigs	0.27	Anesthetized, prone	8.4	31.0	4.5	16.6	2.5	9.2	Lamm et al. (1982)
	0.52	Anesthetized, supine	20.8	40.9	8.6	17.0	4.8	9.5	Paré et al. (1979)

Continued

TABLE 1 Lung Volumes of Various Mammalian Species—cont'd

Species	Body Weight (kg)	Conditions	TLC (ml)	TLC (ml/kg)	FRC (ml)	FRC (ml/kg)	RV (ml)	RV (ml/kg)	References
Rats	0.23	Anesthetized, Lf. lat.	8.2	35.7	4.6	20.0	2.7	11.7	Diamond and O'Donnell (1977)
	0.31	Anesthetized, prone	12.2	39.4	2.1	6.8	1.3	4.2	Lai and Hildebrandt (1978)
Hamsters	0.122	Anesthetized, supine	7.2	59.0	2.4	19.7	2.0	16.4	Koo et al. (1976)
	0.118	Anesthetized, supine	5.6	47.5	1.1	9.3	0.5	4.2	Lai (1979)
Mice	0.020	Anesthetized	1.57	78.5	0.50	25.0	0.39	19.5	Hatch et al. (1984)
	—	Anesthetized	1.84	—	0.55	—	0.31	—	Miller et al. (1987)

TABLE 2 Recoil Pressures at 80% TLC (P_{80}) and at 30% TLC (P_{30}) of the Saline-Filled Lungs in Mammals

Species	P_{80} (cm H_2O)	P_{30} (cm H_2O)	References
Dogs	5.5	0.6	Faridy et al. (1966)
Cats	3.7	0.6	Bachofen et al. (1970)
Rabbits	4.5	1.3	Inoue et al. (1982)
Rats	5.0	1.4	Faridy (1981)
Hamsters	4.2	1.0	Snider et al. (1977)
Mice	4.0	2.0	Gross (1981)

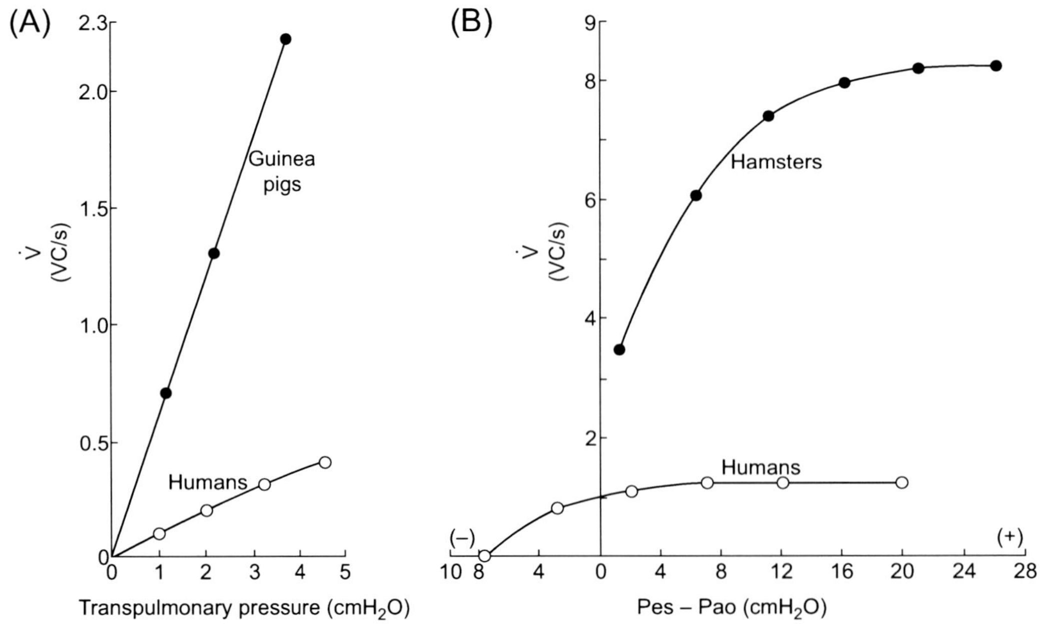

FIGURE 1 Pressure-flow curves in mammals. Flow is expressed as vital capacity/second (VC/s). (A): During spontaneous breathing. Data of humans are from Mead and Whittenberger (1953) and guinea pigs from Lamm et al. (1984). (B): During more forceful expiratory effort at 49% (humans) or at 50% (hamsters) VC. Values of humans are from Hyatt and Wilcox (1963) and hamsters from Lucey et al. (1978).

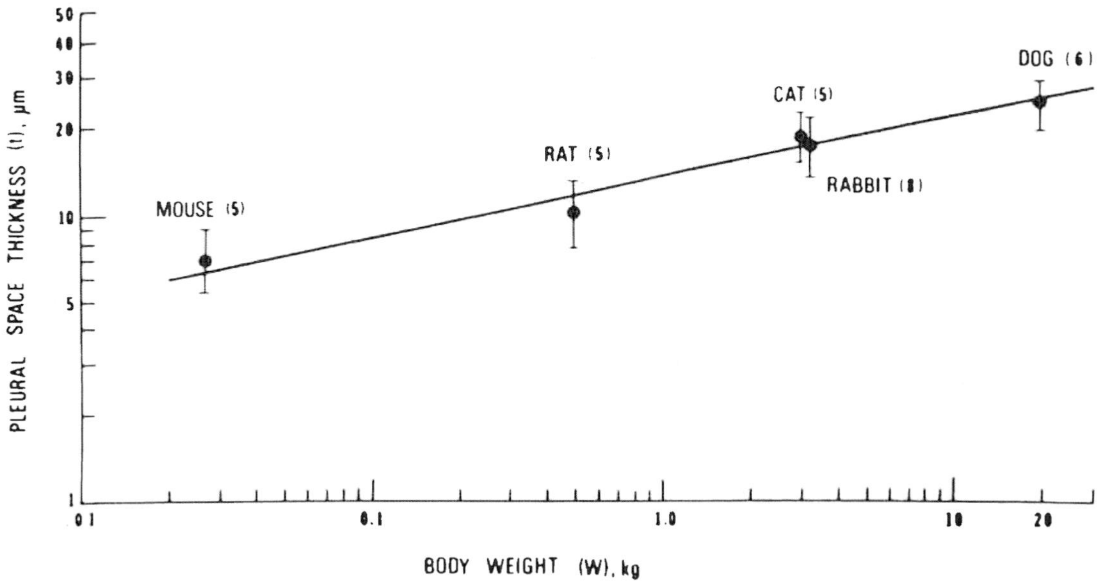

FIGURE 2 Pleural space thickness (t) versus body weight (W) measured using light microscopy in five mammalian species. The power curve fit to the data was $t = 13.1\ W^{0.2}$. *From Li-Fook, S.J., Kaplowitz, M.R., 1985. J. Appl. Physiol. 59, 603–610. Used by permission.*

REFERENCES

Bachofen, H., Hildebrandt, J., Bachofen, M., 1970. Pressure-volume curves of air- and liquid-filled excised lungs-surface tension in situ. J. Appl. Physiol. 29, 422–431.

Caldwell, E.J., 1971. Physiologic and anatomic effects of papain on the rabbit lung. J. Appl. Physiol. 31, 458–465.

Caldwell, E.J., Fry, D.L., 1969. Pulmonary mechanics in the rabbit. J. Appl. Physiol. 27, 280–285.

Diamond, L., O'Donnell, M., 1977. Pulmonary mechanics in normal rats. J. Appl. Physiol. 43, 942–948.

Faridy, E.E., 1981. Pressure-volume characteristics of lungs of rats during pregnancy and postpartum. Respir. Physiol. 43, 43–58.

Faridy, E.E., Permutt, S., Riley, R.L., 1966. Effect of ventilation of surface forces in excised dogs' lungs. J. Appl. Physiol. 21, 1453–1462.

Gillespie, D.J., Lai, Y.-L., Hyatt, R.E., 1973. Comparison of esophageal and pleural pressures in the anesthetized dog. J. Appl. Physiol. 35, 709–713.

Gross, N.J., 1981. Mechanical properties of mouse lungs: effects of degassing on normal, hyperoxic, and irradiated lungs. J. Appl. Physiol. 51, 391–398.

Hatch, G.E., Raub, J.A., Graham, J.A., 1984. Functional and biochemical indicators of pneumoconiosis in mice: comparison with rats. J. Toxicol. Environ. Health 13, 487–497.

Hyatt, R.E., Wilcox, R.E., 1963. The pressure-flow relationships of the intrathoracic airway in man. J. Clin. Invest. 42, 29–39.

Inoue, H., Inoue, C., Hildebrandt, J., 1982. Temperature effects on lung mechanics in air- and liquid-filled rabbit lungs. J. Appl. Physiol. 53, 567–575.

Koo, K.W., Leith, D.E., Sherter, C.B., Snider, G.L., 1976. Respiratory mechanics in normal hamsters. J. Appl. Physiol. 40, 936–942.

Kosch, P.C., Gillespie, J.R., Berry, J.D., 1979a. Respiratory mechanics in normal bonnet and rhesus monkeys. J. Appl. Physiol. 4, 166–175.

Lai, Y.-L., 1979. Lung volume and pleural pressure in the anesthetized hamster. J. Appl. Physiol. 46, 927–931.

Lai, Y.-L., Hildebrandt, J., 1978. Respiratory mechanics in the anesthetized rat. J. Appl. Physiol. 45, 255–260.

Lai, Y.-L., Rodarte, J.R., Hyatt, R.E., 1977. Effect of body position on lung emptying in recumbant anesthetized dogs. J. Appl. Physiol. 43, 983–987.

Lam, H.F., Peisch, R., Amdur, M.O., 1982. Changes in lung volumes and diffusing capacity in guinea pigs exposed to a combination of sulfur dioxide and submicron zinc oxide mixed in a humidified furnace. Toxicol. Appl. Pharmacol. 66, 427–433.

Lamm, W.J.E., Lai, Y.-L., Hildebrandt, J., 1984. Histamine and leukotrienes mediate pulmonary hypersensitivity to antigen in guinea pigs. J. Appl. Physiol. 56, 1032–1038.

Lamm, W.J.E., Hildebrandt, J.R., Hildebrandt, J., Lai, Y.-L., 1982. End-expiratory volume in the rat: effects of consciousness and body position. J. Appl. Physiol. 53, 1071–1079.

Lucey, E.C., Celli, B.R., Snider, G.L., 1978. Maximum expiratory flow and transpulmonary pressure in the hamster. J. Appl. Physiol. 45, 840–845.

Mead, J., Whittenberger, J.L., 1953. Physical properties of human lungs measured during spontaneous respiration. J. Appl. Physiol. 5, 779–796.

Miller, F.J., Graham, J.A., Raub, J.A., Ming, J.W., Menache, M.G., House, D.E., Gardner, D.E., 1987. Evaluating the toxicity of urban patterns of oxidant gases. II. Effects in mice from chronic exposure to nitrogen dioxide. J. Toxicol. Environ. Health 21, 99–112.

Paré, P.D., Boucher, R., Michoud, M.-C., Hogg, J., 1978. Static lung mechanics of intact and excised rhesus monkey lungs and lobes. J. Appl. Physiol. 44, 547–552.

Paré, P.D., Michoud, M.-C., Boucher, R.C., Hogg, J., 1979. Pulmonary effects of acute and chronic antigen exposure of immunized guinea pigs. J. Appl. Physiol. 46, 346–353.

Rehder, K., Abboud, N., Rodarte, J.R., Hyatt, R.E., 1975. Positive airway pressure and vertical transpulmonary pressure gradient in man. J. Appl. Physiol. 38, 896–899.

Snider, G.L., Sherter, C.B., Koo, K.W., Karlinsky, J.B., Hayes, J.A., Franzblau, C., 1977. Respiratory mechanics in hamsters following treatment with endotracheal elastase or collagenase. J. Appl. Physiol. 42, 206–215.

Sorenson, P.R., Robinson, N.E., 1980. Postural effects on lung volumes and asynchronous ventilation in anesthetized horses. J. Appl. Physiol. 48, 97–103.

Westbrook, P.R., Stubbs, S.E., Sessler, A.D., Rehder, K., Hyatt, R.E., 1973. Effects of anesthesia and muscle paralysis on respiratory mechanics in normal man. J. Appl. Physiol. 34, 81–86.

Appendix 5

Comparative Control of Respiration

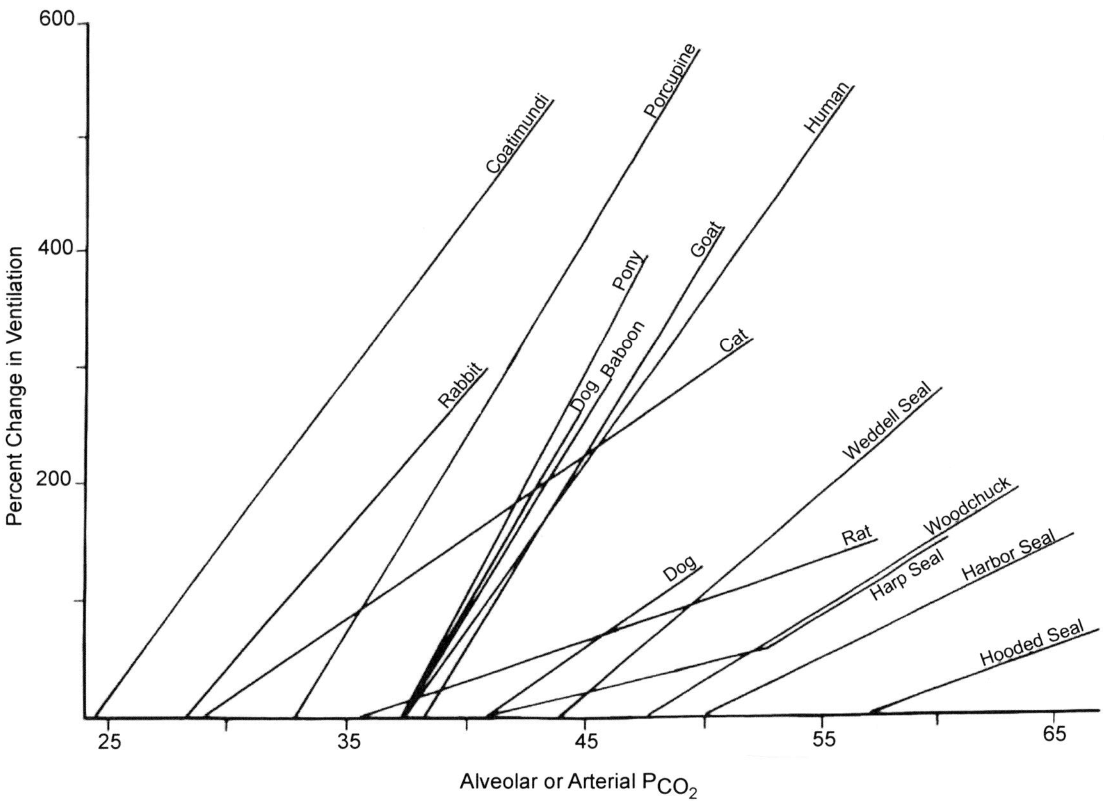

FIGURE 1 Ventilatory responses to changes in alveolar or arterial carbon dioxide in unanesthetized animals. *The source for each species is indicated in Table 7.*

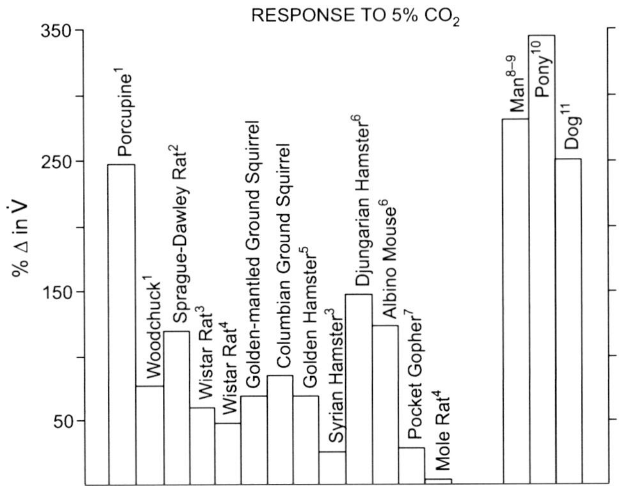

FIGURE 2 Ventilatory responses to a 5% CO_2 in the inspired air of 11 rodents and 3 other species. Data taken from 1 (Boggs and Birchard, 1989), 2 (Birchard et al., 1984), 3 (Walker et al., 1985), 4 (Arieli and Ar, 1979), 5 (Chapin, 1954), 6 (Schlenker, 1985), 7 (Darden, 1972), 8 (Haywood and Bloete, 1969), 9 (Schaefer, 1958), 10 (Klein et al., 1982), and 11 (Albers, 1961). *From Boggs, D.F., Birchard, G.F., 1989. J. Comp. Physiol. 159B, 641–648. Used by permission.*

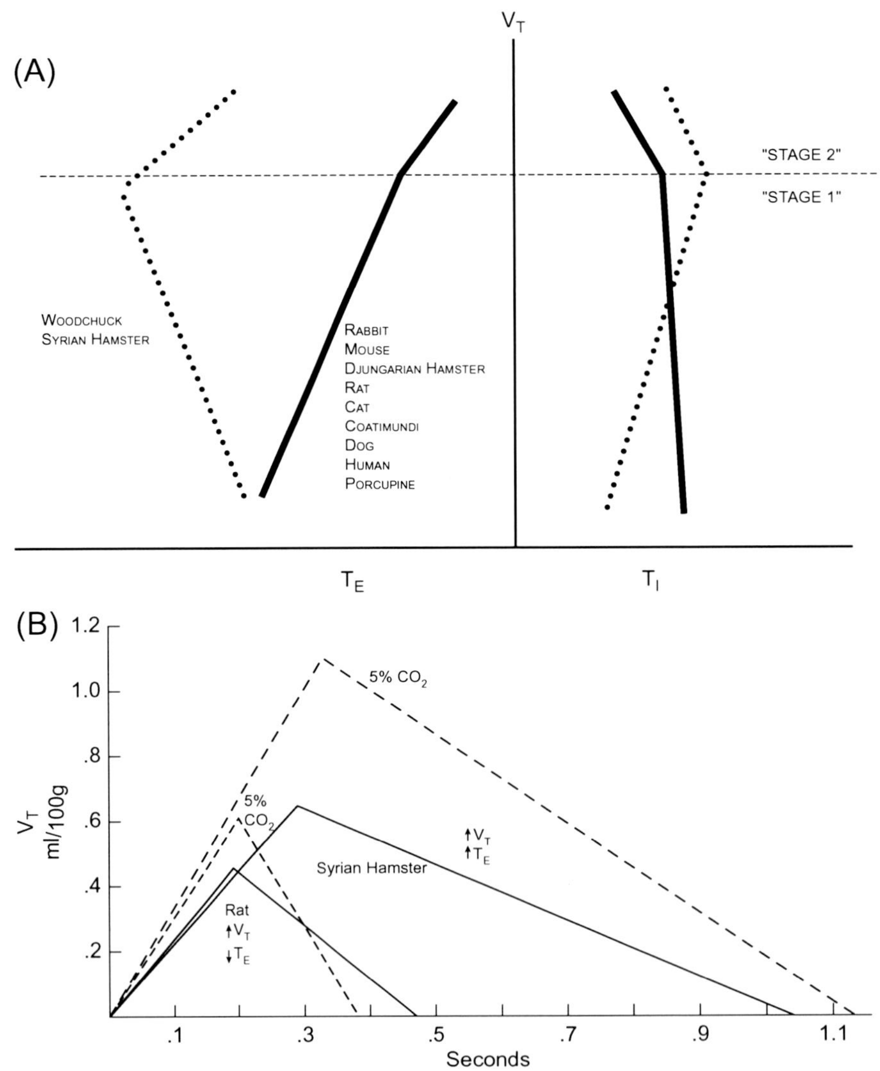

FIGURE 3 (A) Comparative schematic representation of changes in breathing pattern of several species in response to CO_2. T_I increases to left; T_E increases to right. Stages refer to Clark and von Euler's (1972) analysis in humans, where stage 1 refers to the range of increasing V_T over which there is little change in T_i. In stage 2 both T_i, and T_E decrease more steeply. Primary differences between species are in the nature of the effect of CO_2 on T_E. Trends in rabbits taken from Davies et al. (1978); mouse and Djungarian hamster, Schlenker (1988); rat and Syrian hamster, Walker et al. (1985); woodchuck and porcupine, Boggs and Birchard (1989); coati mundi, Boggs et al., submitted; cat, Gautier and Bonora (1979); dog, Hammouda and Wilson (1932) and Smith et al. (1989); human, Clark and von Euler (1972). (B) Different patterns of response to 5% CO_2. The fossorial Syrian hamster prolongs T_E while increasing V_T, whereas the lab rat reduces T_E while increasing V_T. *Data from Walker et al. (1985).*

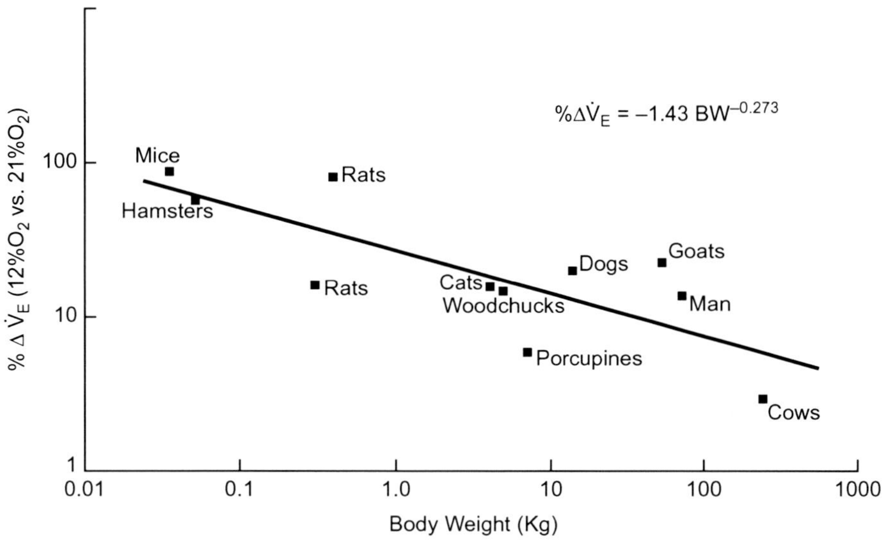

FIGURE 4 Ventilatory response to 12% inspired O_2 (or Pt_{O2} of 90 torr) in 10 species over a wide range of body size. Data from Olson and Dempsey et al. (1978), rats; Laciga and Koller (1976), man; Bouverot et al. (1981), dogs; Grover et al. (1963), cattle; Forster et al. (1981), goats. Remainder from Boggs and Tenney (1984). *From Boggs, D.F., Tenney, S.M., 1984. Respir. Physiol. 58, 245–251. Used by permission.*

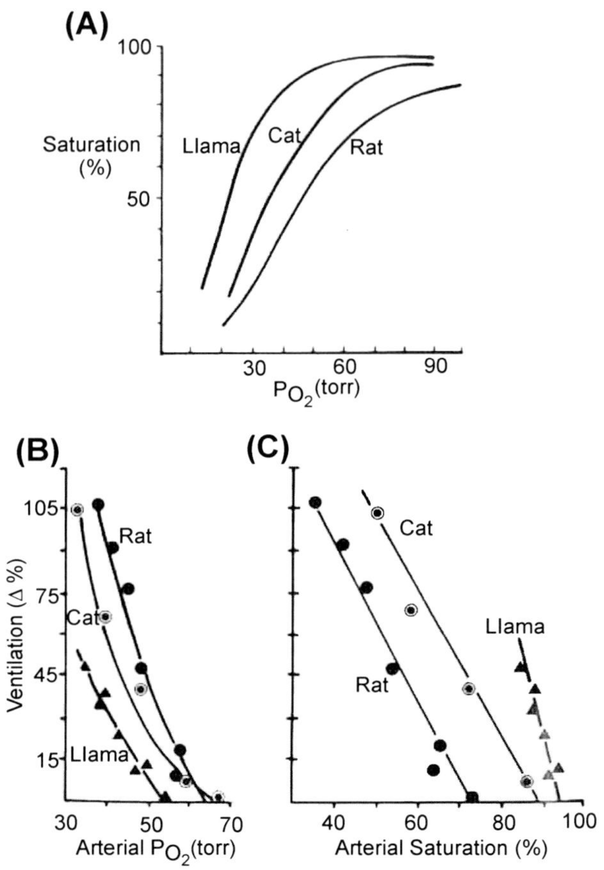

FIGURE 5 (A) Oxy-hemoglobin dissociation curves of llama, cat, and rat. (B) Ventilatory response to declining partial pressure of oxygen in arterial blood. (C) Ventilatory response to declining arterial O_2 saturation. *From Van Nice, P., et al. 1980. Comp. Biochem. Physiol. 60A, 347–350. Used by permission.*

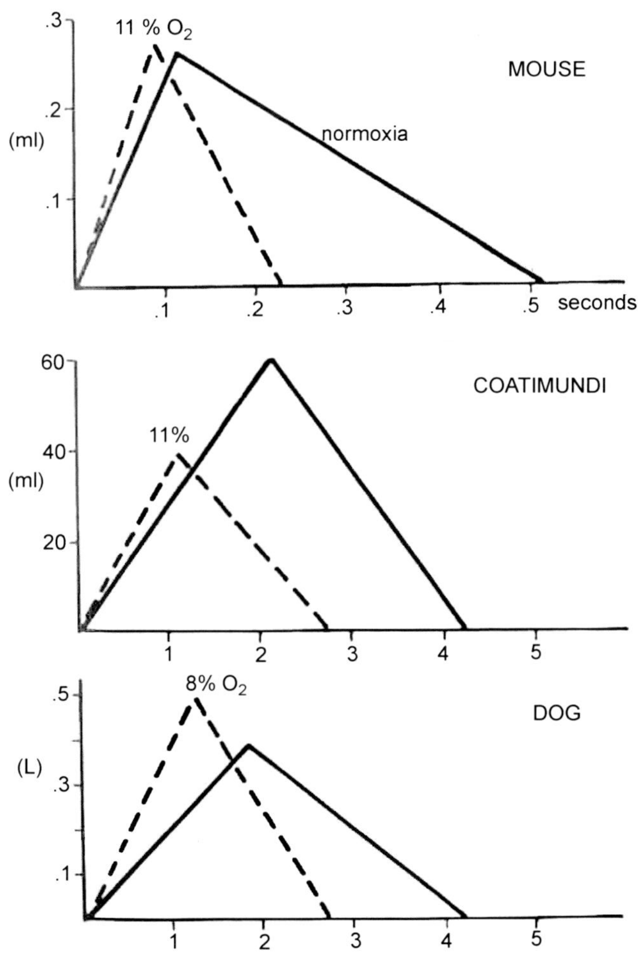

FIGURE 6 Comparative breathing pattern responses to hypoxia. All three species show a decrease in both inspiratory and expiratory times with variable effects on V_T. *Data on mice and coati from Boggs (unpublished) and dog data from Smith et al. (1989).*

REFERENCES

Albers, C., 1961. The mechanism of thermal stress in the dog. I. Ventilation and arterial blood gases during thermal stress. Pflügers Archiv für die gesamte Physiologie des Menschen und der Tiere 274, 125–147.

Arieli, R., Ar, A., 1979. Ventilation of a fossorial mammal (*Spalax ehrenbergi*) in hypoxic and hypercapnic conditions. J. Appl. Physiol. 47, 1011–1017.

Birchard, G.F., Boggs, D.F., Tenney, S.M., 1984. Effects of perinatal CO_2 exposure on adult ventilatory responses. Respir. Physiol. 57, 341–347.

Boggs, D.F., Colby C.C., Williams, B.R., Kilgore Jr., D.L. Comparative chemosensitivity and breathing pattern regulation of the coatimundi and woodchuck. Respir. Physiol., submitted for publication.

Boggs, D.F., Birchard, G.F., 1989. Cardio-respiratory responses of the woodchuck and porcupine to CO_2 and hypoxia. J. Comp. Physiol. 159B, 641–648.

Boggs, D.F., Tenney, S.M., 1984. Scaling respiratory pattern and respiratory 'drive'. Respir. Physiol. 58, 245–251.

Bouverot, P., Collins, R., Favier, R., Flandrois, R., Sebert, P., 1981. Carotid chemoreceptor function in ventilatory and circulatory O_2 convection of exercising dogs at low and high altitude. Respir. Physiol. 43, 147.

Chapin, J.L., 1954. Ventilatory response of the unrestrained and unanesthetized hamster to CO_2. Am. J. Physiol. 179, 146–148.

Clark, F.J., von Euler, C., 1972. On the regulation of depth and rate of breathing. J. Physiol. (Lond.) 222, 267–295.

Darden, T.R., 1972. Respiratory adaptations of a fossorial mammal, the pocket gopher (*Thomomys bottae*). J. Comp. Physiol. 78, 121–137.

Davies, A., Dixon, M., Callahan, D., Huszczuk, A., Widdicombe, J.G., Wise, J.L.M., 1978. Lung reflexes in rabbits during pulmonary stretch receptor block by sulphur dioxide. Respir. Physiol. 34, 83–101.

Dempsey, J.A., Forster, H.V., Chosy, L.W., Hanson, P.G., Reddan, W.G., 1978. Regulation of CSF [HCO3 during long term hypoxic hypocapnia in man. J. Appl. Physiol. 44, 175–182.

Forster, H.V., Bisgard, G.E., Klein, J.P., 1981. Effect of peripheral chemoreceptor denervation on acclimatization of goats during hypoxia. J. Appl. Physiol. 50, 392–398.

Gautier, H., Bonora, M., 1979. Effects of carotid body denervation on respiratory pattern of awake cats. J. Appl. Physiol. 46, 1127–1131.

Grover, R.F., Reeves, J.T., Will, D.H., Blount Jr., S.G., 1963. Pulmonary vasoconstriction in steers at high altitude. J. Appl. Physiol. 18, 567–574.

Hammouda, M., Wilson, W.H., 1932. The vagus influences giving rise to the phenomena accompanying expansion and collapse of the lungs. J. Physiol. (Lond.) 74, 81–114.

Haywood, D., Bloete, M., 1969. Respiratory responses of healthy young women to CO_2 inhalation. J. Appl. Physiol. 27, 32–35.

Klein, J.P., Forster, H.V., Bisgard, G.E., Kaminski, R.P., Pan, L.G., Hamilton, L.H., 1982. Ventilatory response to inspired CO_2 in normal and carotid body denervated ponies. J. Appl. Physiol. 52, 1614–1622.

Laciga, P., Koller, E.A., 1976. Respiratory, circulatory and EEG changes during acute exposure to high altitude. J. Appl. Physiol. 41, 159–167.

Olson Jr., E.B., Dempsey, J.A., 1978. Rat as a model for human like ventilatory adaptation to chronic hypoxia. J. Appl. Physiol. 44, 763.

Schaefer, K.E., 1958. Respiratory pattern and response to CO. J. Appl. Physiol. 13, 1–14.

Schlenker, E.H., 1985. Ventilation and metabolism of Djungarian hamster (*Phodopus sungorus*) and the albino mouse. Comp. Biochem. Physiol. 82A, 293–295.

Smith, C.A., Ainsworth, D.M., Henderson, K.S., Dempsey, J.A., 1989. Differential timing of respiratory muscles in response to chemical stimuli in awake dogs. J. Appl. Physiol. 66, 392–399.

Walker, B.R., Adams, E.M., Voelkel, N.F., 1985. Ventilatory responses of hamsters and rats to hypoxia and hypercapnia. J. Appl. Physiol. 59, 1955–1960.

Appendix 6

Aging of the Normal Lung

TABLE 1 **Mean Life-Span and Age at Puberty of Various Species**

	Mean Life-Span	Age at Puberty
House rat	30 months	2 months
Domestic dog	15 years	7 months
Golden hamster	24 months	2 months
European rabbit	5.5 years	1 year
Mouse	18 months	1.5 months

Adapted from Rockstein et al. (1977).

TABLE 2 **Body Weight and Lung Volumes in Fischer 344 Rats at Various Ages**

Parameter	3 Months	18 Months	27 Months
Body weight (g)	222 ± 61	334 ± 106	332 ± 71
Total lung capacity (TLC) (ml)	11.9 ± 1.7	13.9 ± 2.2	14.4 ± 1.9
TLC/body weight (ml/kg)	56 ± 8	42 ± 7	43 ± 6
Vital capacity (ml)	11.0 ± 1.8	13.4 ± 2.3	13.4 ± 1.7
Functional residual capacity (ml)	2.1 ± 0.3	1.7 ± 0.3	2.7 ± 0.4
Residual volume (RV) (ml)	1.0 ± 0.3	0.6 ± 0.2	1.1 ± 0.5
RV/TLC (ml/ml)	0.08 ± 0.03	0.04 ± 0.01	0.07 ± 0.03

Values are means ± SD.
Adapted from Mauderly (1982).

TABLE 3 Body Weight and Lung Volumes in Adult and Older Hamsters

Parameter	15 Weeks	65 Weeks	P Value
Body weight (g)	126 ± 12	125 ± 7	>0.20
Total lung capacity (ml)	9.6 ± 1.3	11.1 ± 1.0	<0.02
Vital capacity (ml)	6.9 ± 1.0	7.8 ± 0.9	<0.10
Functional residual capacity (ml)	3.5 ± 0.5	4.3 ± 0.3	<0.05
Residual volume (RV) (ml)	2.7 ± 0.6	3.3 ± 0.3	<0.05
RV/TLC (%)	28 ± 5	30 ± 5	>0.20

Values are means ± SD.
Adapted from Mauderly (1979a).

TABLE 4 Ventilatory Parameters in Fischer 344 Rats of Various Ages

Parameter	3 Months	18 Months	27 Months
Respiratory frequency (breath/min)	48 ± 6[a]	54 ± 7	54 ± 6
Tidal volume (ml)	1.1 ± 0.3[a]	1.5 ± 0.3	1.5 ± 0.3
Minute ventilation ($\dot{V}_E$) (ml/min)	54 ± 14[a]	82 ± 23	82 ± 18
V_E/body weight (ml/min/kg)	254 ± 48	251 ± 45	252 ± 52

Note: Values are means ± SD.
[a]*Significantly different from other 2 groups.*
Adapted from Mauderly (1982).

TABLE 5 Ventilatory Parameters in Hamsters at Various Ages

Parameter	15 Weeks	65 Weeks
Respiratory frequency (breath/min)	24 ± 2.7	25 ± 3.9
Tidal volume (ml)	1.2 ± 0.2	1.1 ± 0.2
Minute volume (ml/min)	27.8 ± 3.3	28.1 ± 4.0

Values are means ± SD.
Adapted from Mauderly (1979).

TABLE 6 Ventilatory Parameters in Young and Old Rabbits

Parameter	0.5–1 Years	3–5 Years	P Value
Body weight (kg)	4.1 ± 0.4	4.1 ± 0.5	NS
Respiratory frequency (breath/min)	36 ± 10	26 ± 6	<0.01
Tidal volume (ml)	21 ± 3	26 ± 1	<0.01
Minute ventilation/body weight (1/min/kg)	0.188 ± 0.04	0.170 ± 0.03	NS

Note: Values are means ± SD; NS, not significant.
Adapted from Davidson et al. (1966).

TABLE 7 Ventilatory Parameters of Female Beagle Dogs at Various Ages

Parameter	12–14 Months	3–4 Years	8–10.5 Years
Breathing frequency (breath/min)	20 ± 5[a]	25 ± 8	28 ± 11
Tidal volume (ml)	181 ± 52[a]	155 ± 48	150 ± 46
Minute volume (ml/min)	3.60 ± 0.8	3.61 ± 1.0	3.81 ± 1.0
% Alveolar ventilation (% min vol.)	71 ± 6[a]	69 ± 6	56 ± 5
Nitrogen clearance equivalent (l expired/1 FRC)	24 ± 5[b]	22 ± 4	34 ± 5

Note: Values are means ± SD.
[a]*Significant differences at all 3 age levels.*
[b]*Significant difference between youngest and oldest groups.*
Adapted from Mauderly (1974).

TABLE 8 Gas Exchange Indexes in Three Age Groups of Beagle Dogs

Parameter	12–14 Months	3.5–8.4 Years	8.5–11.3 Years
Alveolar P_{O_2} (mm Hg)	81 ± 6	85 ± 5	82 ± 6
Arterial P_{O_2} (mm Hg)	75 ± 6	77 ± 6	75 ± 6
Alveolar-arterial P_{O_2} (mm Hg)	7 ± 4	8 ± 5	7 ± 4
Arterial O_2 saturation (%)	91 ± 2	92 ± 3	91 ± 3
Alveolar P_{CO_2} (mm Hg)	37 ± 4	34 ± 4	34 ± 5
Arterial P_{CO_2} (mm Hg)	40 ± 4	37 ± 3	38 ± 5
Alveolar-arterial P_{CO_2} (mm Hg)	3 ± 2	3 ± 3	4 ± 3
Arterial pH	7.40 ± 0.03	7.40 ± 0.04	7.40 ± 0.04
O_2 uptake (ml/min)	98 ± 25	100 ± 32	89 ± 25[a]
Ventilatory equivalent for O_2 (V_EO_2/V_{O_2})	36 ± 7	40 ± 9[a]	41 ± 9[a]

Note: Values are means ± SD.
[a]*Significantly different 12–14 months group.*
Adapter from Mauderly, 1979, 1982.

TABLE 9 Morphometric Values in Sprague-Dawley Rats of Various Ages

Parameter	4 Months	8 Months	18 Months
V_L/body weight (ml/kg)[a]	21.7 ± 1.0	30.9 ± 1.5	38.4 ± 2.8
Lm (μm)[a]	54 ± 2	71 ± 2	87 ± 7
ISA (cm^2)	5571 ± 445[b]	7979 ± 318	8733 ± 721

Note: Values are means ± SEM. V_L, postfixation lung volume; Lm, mean chord length; ISA, internal surface area.
[a]*Significantly different among groups.*
[b]*Significantly different compared with other groups.*
Adapted from Johanson and Pierce (1973).

REFERENCES

Davidson, J.T., Wasserman, K., Lillington, G.A., Schmidt, R.W., 1966. Effect of aging on respiratory mechanics and gas exchange in rabbits. J. Appl. Physiol. 21, 837.

Johanson Jr., W.G., Pierce, A.K., 1973. Lung structure and function with age in normal rats and rats with papain emphysema. J. Clin. Invest 52, 2921.

Mauderly, J.L., 1974. Influence of sex and age on the pulmonary function of the unanesthetized Beagle dogs. J. Gerontal 29, 282.

Mauderly, J.L., 1979a. Ventilation, lung volumes and lung mechanics of young adult and old Syrian hamsters. Exp. Aging Res. 5, 497–508.

Mauderly, J.L., 1979b. Effect of age on pulmonary structure and function of immature and adult animals and man. Fed. Proc. 38, 173.

Mauderly, J.L., 1982. The effect of age on respiratory function of Fischer-344 rats. Exp. Aging Res. 8, 31.

Rockstein, M., Chesky, J., Sussman, R., 1977. In: Finch, C.E., Hayflick, L. (Eds.), Handbook of the Biology of Aging. Van Nostrand Reinhold, Princeton, NJ, pp. 3–34.

Appendix 7

Carbohydrate/Intermediary Metabolism

TABLE 1 Glucose Utilization and Lactate Production by Perfused Lung

Species	Perfusate Glucose (m*M*)	Glucose Utilized (μmol/g dry wt/h)	Lactate Produced	References
Rabbit	5.5	44–78	32–37	Koga (1956)
Rat	8.3	45	83	Levey and Gast (1966)
Dog	4.8	41	25	Weber and Vissher (1969)
Rat	4.2	70	75	O'Neil and Tierney (1974)
Rat	5.0	50	86	Shaw and Rhoades (1977)
Rat	4.0	57	76	Datta et al. (1980)
Rat	5.0	101	115	Datta et al. (1980)

REFERENCES

Datta, H., Stubbs, W.A., Alberti, K.G.M.M., 1980. Substrate utilization by the lung. In: Metabolic Activities of the Lung, CIBA Foundation 78. Excerpta Medica, pp. 85–104.

Koga, H., 1956. Studies on the function of isolated perfused mammalian lung. II. Carbohydrate metabolism of rabbit lung. Kumamoto Med. J. 9, 182–187.

Levey, S., Gast, R., 1966. isolated perfused rat lung preparation. J. Appl. Physiol. 21, 313–316.

O'Neil, J.J., Tierney, D.F., 1974. Rat lung metabolism: glucose utilization by tissue slices and the isolated perfused lung. Am. J. Physiol. 226, 867–873.

Shaw, M.F., Rhoades, R.A., 1977. Substrate metabolism in the perfused lung: response to changes in circulation glucose and palmitate levels. Lipids 12, 930–935.

Weber, K.C., Vissher, M.B., 1969. Metabolism of the isolated canine lung. Am. J. Physiol. 217, 1044–1052.

Appendix 8

Biochemistry of Lipid Mediators in the Lung

TABLE 1 Production of Lipid Mediators by Cells in the Lung

Cells	Prostanoid	Leukotriene	PAF	Other	Animal Species	References
Epithelial trachea	PGE_2				Dog	Milsap and Jusko (1983)
	PGE_2				Rat	Sivarajah et al. (1986)
	TxA_2				Rabbit	Sivarajah et al. (1986)
				15-HETE	Human	Kulkami et al. (1985)
Clara cell	PGI_2				Rat(l)	Sivarajah et al. (1986)
Fibroblast	TxA_2				Human(l)	Hopkins et al. (1978)
						Hunter et al. (1984)
						Fitzpatrick and Stringfellow (1983)
						Feinmark and Bailey (1982)
	PGE_2				Mouse(l)	Hassid and Levine (1977)
					Human	Burstein et al. (1982)
	PGI_2				Human	Murota et al. (1979)
Pneumocyte type II	PGI_2				Rat(l)	Sivarajah et al. (1986)
	PGE_2, $PGF_{2\alpha}$				Rat(l)	Taylor et al. (1979)
Endothelial artery	PGE_2, PGI_2, $PGF_{2\alpha}$				Human(l)	Johnson et al. (1981)
			PAF			Prescott et al. (1984)
Mast cell	PGD_2	LTC_4			Human	MacGlashan et al. (1982)
						Peters et al. (1984)
			PAF		Mouse(2)	Mencia-Huerta et al. (1983)

Continued

TABLE 1 Production of Lipid Mediators by Cells in the Lung—cont'd

Cells	Prostanoid	Leukotriene	PAF	Other	Animal Species	References
Macrophage alveolar	PGD_2, PGE_2, $PGF_{2\alpha}$, TxA_2				Human	Brown et al. (1988)
	TxA_2				Guinea pig	Morley et al. (1979)
		LTB_4			Human	McDermot et al. (1984)
						Martin et al. (1984)
						Fels et al. (1982)
			PAF		Rat	Albert and Snyder (1983)
Neutrophils		LTB_4			Human	Borgeat and Samuelsson (1979)
			PAF		Human	Clay et al. (1984)
			PAF		Rabbit	Lynch et al. (1979)
Eosinophil			PAF		Human	Lee et al. (1984)
		LTC_4			Horse	Jorg et al. (1982)
		LTC_4			Human	Weller et al. (1983)
				15-HETE	Human	Turk et al. (1982)
Platelet	TxA_2				Human	Hamberg and Samuelsson (1974b)
			PAF		Human	Benveniste et al. (1982)
		LTC_4(2)			Human	Maclouf and Murphy (1988)
				12-HETE	Human	Hamberg and Samuelsson (1974b)

REFERENCES

Albert, D.H., Snyder, F., 1983. Biosynthesis of 1-alkyl-2-acetyl-sn-glycero-3-phosphocholine (platelet activating factor) by rat alveolar macrophages. J. Biol. Chem. 258, 97.

Benveniste, J., Chignard, M., Le Couedic, J.P., Vargaftig, B.B., 1982. Biosynthesis of platelet activating factor (PAF-acether). II. Involvement of phospholipase A_2 in the formation of PAF-acether and lyso-PAF-acether from rabbit platelets. Thromb. Res. 25, 375.

Borgeat, P., Samuelsson, B., 1979. Arachidonic acid metabolism and polymorphonuclear leukocytes: unstable intermediate in formation of dihydroxy acids. Proc. Natl. Acad. Sci. U.S.A. 76, 3213.

Brown, G.P., Monick, M.M., Hunninghake, G.W., 1988. Human alveolar macrophage arachidonic acid metabolism. Am. J. Physiol. 254, C809.

Burstein, S., Hunter, S.A., Sedor, C., Chulman, S., 1982. Prostaglandins and cannabis. IX. Stimulation of prostaglandin E_2 synthesis in human lung fibroblasts by delta 1-tetrahydrocannabinol. Biochem. Pharmacol. 31, 2361.

Clay, K.L., Murphy, R.C., Andres, J.L., Lynch, J., Henson, P.M., 1984. Structure elucidation of platelet activating factor derived from human neutrophils. Biochem. Biophys. Res. Commun. 121, 815.

Feinmark, S.J., Bailey, J.M., 1982. Lipid metabolism in cultured cells. Activators of endogenous thromboxane A_2 synthesis in cultured lung fibroblasts. J. Biol. Chem. 257, 2816.

Fels, A.O.S., Pawlowski, N.A., Cramer, E.B., King, T.K.C., Cohn, Z.A., Scott, W.A., 1982. Human alveolar macrophage produce leukotriene B_4. Proc. Natl. Acad. Sci. U.S.A. 79, 7866.

Fitzpatrick, F.A., Stringfellow, D., 1983. Influence of thromboxane synthetase inhibitors on virus replication in human lung fibroblasts. in vitro. Biochem. Biophys. Res. Commun. 116, 264.

Hamberg, M., Samuelsson, B., 1974b. Prostaglandin endoperoxides. Novel transformations of arachidonic acid in human platelets. Proc. Natl. Acad. Sci. U.S.A. 71, 3400.

Hassid, A., Levine, L., 1977. Stimulation of phospholipase activity and prostaglandin biosynthesis by melittin in cell culture and *in vivo*. Res. Commun. Chem. Pathol. Pharmacol. 18, 507.

Hopkins, N.K., Sun, F.F., Gorman, R.R., 1978. Thromboxane A_2 Biosynthesis in human lung fibroblasts WI-38. Biochem. Biophys. Res. Commun. 85, 827.

Hunter, S.A., Burstein, S., Sedor, C., 1984. Stimulation of prostaglandin synthesis in WI-38 human lung fibroblasts following inhibition of phospholipid acylation by p-hydroxymercuribenzoate. Biochim. Biophys. Acta 793, 202.

Johnson, A.R., Callahan, K.S., Tasi, S.C., Campbell, W.B., 1981. Prostacyclin and prostaglandin biosynthesis in human pulmonary endothelial cells. Clin. Respir. Physiol. 17, 531.

Jorg, A., Henderson, W.R., Murphy, R.C., Klebanoff, S.J., 1982. Leukotriene generation by eosinophils. J. Exp. Med. 155, 390.

Kulkami, A.B., Reinke, R., Feigelson, P., 1985. Acute phase mediators and glucocorticoids elevate alpha-l-acid glycoprotein gene transcription. J. Biol. Chem. 260, 15386.

Lee, T.C., Lenihan, D.J., Malone, B., Roddy, L.L., Wasserman, S.I., 1984. Increased biosynthesis of platelet activating factor in activated human eosinophils. J. Biol. Chem. 259, 5526.

Lynch, J.M., Lotner, G.Z., Betz, S.J., Henson, P.M., 1979. The release of platelet activating factor by stimulated rabbit neutrophils. J. Immunol. 123, 1219.

MacGlashan, D.W., Schleimer, R.P., Peters, S.P., Schulman, E.S., Adams, G.K., Newball, H.H., Lichtenstein, L.M., 1982. Generation of leukotrienes by purified human lung mast cells. J. Clin. Invest. 70, 747.

Maclouf, J., Murphy, R.C., 1988. Transcellular metabolism of neutrophil-derived leukotriene A_4 by human platelets. J. Biol. Chem. 263, 174.

Martin, T.R., Altman, L.C., Albert, R.K., Henderson, W.R., 1984. Leukotriene B_4 production by the human alveolar macrophage: a potential mechanism for amplifying inflammation in the lung. Am. Rev. Respir. Dis. 129, 106.

McDermot, J., Kelsey, C.R., Waddell, K.A., Richmond, R., Blair, I.A., 1984. Synthesis of leukotriene B_4 and prostanoids by human alveolar macrophages: analysis by gas chromatography/mass spectrometry. Prostaglandins 27, 163.

Mencia-Huerta, J.M., Lewis, R.A, Razin, E., Austen, K.F., 1983. Antigen-initiated release of platelet-activating factor (PAF-acether) from mouse bone marrow-derived mast cells sensitized with monoclonal IgE. J. Immunol. 131 (6), 2964.

Milsap, R.L., Jusko, W.J., 1983. Binding of prednisolone to alpha 1-acid glycoprotein. J. Steroid Biochem. 18, 191.

Morley, J., Bray, M.A., Jones, R.W., Nugteren, D.H., Van Doys, D.A., 1979. Prostaglandin and thromboxane production by human and guinea pig macrophages and leukocytes. Prostaglandins 17, 730.

Murota, S.I., Mitsui, Y., Kawamura, M., 1979. Effect of in vitro aging on 6-ketoprostaglandin F1 alpha-producing activity in cultured human diploid lung fibroblasts. Biochim. Biophys. Acta. 30, 574 (2), 355.

Peters, S.P., MacGlashan, D.W., Schulamn, E.S., Schleimer, R.P., Hayes, E.C., Rokach, J., Adkinson, N.F., Lichtenstein, L.M., 1984. Arachidonic acid metabolism in purified human lung mast cells. J. Immunol. 132, 1972.

Prescott, S.M., Zimmerman, G.A., McIntyre, T.M., 1984. Human endothelial cells in culture produce platelet activating factor (1-alkyl2-acetyl-sn-glycero-3-phosphocholine) when stimulated with thrombin. Med. Sci. 81, 3534.

Sivarajah, K., Wu, R., Nettesheim, P., Eling, T., 1986. Biosynthesis of prostaglandins by isolated and cultured airway epithelial cells. Exp. Lung Res. 10, 101.

Taylor, L., Polgar, P., McAteer, J.A., Douglas, W.H., 1979. Prostaglandin production by type II alveolar epithelial cells. Biochim. Biophys. Acta 572, 502.

Turk, J., Maas, R.L., Brash, A.R., Roberts, L.J., Oates, J.A., 1982. Arachidonic acid 15-lipoxygenase products from human eosinophils. J. Biol. Chem. 257, 7068.

Weller, P.F., Lee, C.W., Foster, D.W., Corey, E.J., Austen, K.F., Lewis, R.A., 1983. Generation and metabolism of 5-lipoxygenase pathway leukotrienes by human eosinophils: predominant production of leukotriene C_4. Proc. Natl. Acad. Sci. U.S.A. 80, 7626.

Appendix 9

The Biology and Biochemistry of Pulmonary Alveolar Epithelial Cells

TABLE 1 Dependence of Phenotype on Substrata

Substrata	Shape	Disaturated Phosphatidylcholine Synthesis	Phosphatidylglycerol Synthesis
Tissue culture plastic	Spread	Decreased	Decreased
Bovine corneal extracellular matrix	Cuboidal	Decreased	Decreased
Human lung matrix	Spread	Decreased	Decreased
Amniotic membrane			
Interstitial side	Spread	Decreased	Decreased
Basement membrane side	Cuboidal	Maintained	Decreased
Engelbreth-Holm-Swarm matrix	Cuboidal	Maintained	Decreased
Floating collagen membranes	Cuboidal	Decreased	Decreased
Feeder cell layers	Cuboidal	Maintained	Decreased

TABLE 2 Secretory Products of Alveolar Epithelial Cells
Surfactant lipids and proteins
Complement components
Prostaglandins
Lysozyme
Glutathione (?)

GSH is a diffusable product of the airway lining cells. However, it is not known if GSH is actively secreted from epithelial cells. GSH is highly unlikely to be taken back up into cells.

TABLE 3 Epithelial Cell Enzymes
Alkaline phosphatase
Acid phosphatase
Aryl sulfatase
β-*N*-acetylglucosaminidase
Vitamin K-dependent carboxylase
Protein kinase C
Cyclic AMP-dependent protein kinase
Cytochrome P450 dependent monooxygenase
NADPH dependent cytochrome P450 reductase

Appendix 10

The Biochemical Pharmacology of Pulmonary Amine Disposition

TABLE 1 Experimental Acute Lung Injury and Pulmonary Amine Metabolism

Type of Injury or Agent	Condition	Observed Change	References
Hyperoxia	IPL, rat	5-HT ↓	Block and Fisher (1977a,b)
	IPL, rat	NE ↓	Block and Cannon (1978)
	IPL, rat	Imipramine→	Fisher, Block and Pietra (1980)
	Conscious rabbit	5-HT ↓	Dobuler et al. (1982)
	Conscious sheep	5-HT→propranolol ↓	Howell et al. (1989)
	BPAEC	5-HT ↓	Block and Stalcup (1981)
	BPAEC, vesicles	5-HT ↓	Sheridan and Block (1988a,b)
Phorbol myristate acetate	IPL, rabbit	5-HT ↓	Myers and Pitt (1988)
	IPL, rabbit	5-HT ↓ propranolol→	Merker and Gillis (1988)
	IPL, rat	5-HT ↓ (blocked with staurosporine)	
	Anesthetized dog	5-HT ↓→propranolol ↓	Gardaz et al. (1988)
	Anesthetized rabbit, acute	5-HT ↓	Riggs et al. (1988)
	Anesthetized rabbit, chronic	5-HT ↓	Havill et al. (1989)
	BPAEC	5-HT ↓ (blocked with staurosporine)	Myers et al. (1989)
Bleomycin	IPL, rabbit	5-HT ↓	Jeffrey et al. (1986)
	IPL, rat	Metaiodobenzyl ↓ guanidine	Slosman et al. (1989)
	Anesthetized rabbit, s.c.	5-HT ↓ NE ↓	Catravas et al. (1981)
	Anesthetized rabbit, i.t.	5-HT ↓	Catravas et al. (1983)
	Anesthetized rabbit	Reversible 5-HT	Lazo et al. (1983)
Monocrotaline	IPL, rat	5-HT ↓	Huxtable et al. (1978)
	IPL, rat	5-HT ↓	Gillis et al. (1972)
	IPL, rat	5-HT ↓	Hilliker et al. (1982)

Continued

TABLE 1 Experimental Acute Lung Injury and Pulmonary Amine Metabolism—cont'd

Type of Injury or Agent	Condition	Observed Change	References
Paraquat	IPL, rat	5-HT ↓	Roth et al. (1979)
	IPL, rat	5-HT ↓	Block and Schoen (1982)
Alphanaphthylthiourea	IPL, rat	5-HT ↓	Block and Schoen (1981)
	Anesthetized dog	5-HT→	Gardaz et al. (1988)
Oleic acid	Anesthetized dog	5-HT→	Dawson et al. (1985)
	Anesthetized dog	5-HT→propranolol ↓	Gardaz et al. (1988)
Oxygen free-radical generation	IPL, rabbit (xanthine)	5-HT ↓	Cook et al. (1982)
	IPL, rat (hypoxanthine)	5-HT→↓	Steinberg et al. (1982)
	Anesthetized dog (glucose oxidase)	5-HT ↓	Vigneswaran et al. (1989)
Other forms	IPL, dog (complement)	5-HT ↓	Bult et al. (1988)
	IPL, dog	5-HT ↓	White et al. (1975)

Note: IPL = isolated perfused lung; BPAEC = bovine pulmonary artery endothelial cell in culture; ↓ = decrease; ↓→ = little change; → = no change.

REFERENCES

Block, E.R., Cannon, J.K., 1978. Effect of oxygen exposure on lung clearance of amines. Lung 155, 287–295.

Block, E.R., Fisher, A.B., 1977a. Effect of hyperbaric oxygen exposure on pulmonary clearance of 5-hydroxytryptamine. J. Appl. Physiol. Respir. Environ. Exerc. Physiol. 43, 254–257.

Block, E.R., Fisher, A.B., 1977b. Prevention of hyperoxic-induced depression of pulmonary serotonin clearance by pretreatment with superoxide dismutase. Am. Rev. Resp. Dis. 116, 441–447.

Block, E.R., Schoen, F.J., 1981. Effect of alpha naphthylthiourea on uptake of 5-hydroxytryptamine from the pulmonary circulation. Am. Rev. Resp. Dis. 123, 69–73.

Block, E.R., Schoen, F.J., 1982. Depression of serotonin uptake by rat lungs exposed to paraquat. J. Pharmacol. Exp. Ther. 221, 254.

Block, E.R., Stalcup, S.A., 1981. Depression of serotonin uptake by cultured endothelial cells exposed to high 0, tension. J. Appl. Physiol. 50, 1212.

Bult, H., Heiremans, J.J., Herman, A.G., Malcorps, C.M.A., Peeters, A.M., 1988. Prostacyclin biosynthesis and reduced 5-HT uptake in complement-induced endothelial cell injury in the dog isolated lung. Pharmacology 93, 791.

Catravas, J.D., Lazo, J.S., Gillis, C.N., 1981. Biochemical markers of bleomycin toxicity: clearance of 5-hydroxytryptamine and norepinephrine by rabbit lung in vivo. J. Pharmacol. Exp. Ther. 217, 524.

Catravas, J.D., Lazo, J.S., Dobuler, K.J., et al., 1983. Pulmonary endothelial dysfunction in the presence or absence of interstitial injury induced by intratracheally injected bleomycin in rabbits. Am. Rev. Resp. Dis. 128, 740.

Cook, D.R., Howell, R.E., Gillis, C.N., Augest 1982. Xanthine oxidase-induced lung injury inhibits removal of 5-hydroxytryptamine from the pulmonary circulation. Anesth. Analg. 61 (8), 666–670.

Dawson, C.A., Christensen, C.W., Rickaby, D.A., et al., 1985. Lung damage and pulmonary uptake of serotonin in intact dogs. J. Appl. Physiol. 58, 1761.

Dobuler, K.J., Catravas, J.D., Gillis, C.N., 1982. Early detection of oxygen-induced lung injury in conscious rabbits: reduced activity of angiotensin converting enzyme and removal of 5-hydroxytryptamine in vitro. Am. Rev. Resp. Dis. 126, 534.

Fisher, A.B., Block, E.R., Pietra, G., 1980. Environmental influences on uptake of serotonin and other amines. Environ. Health Perspect. 35, 191–198.

Gardaz, J.P., Py, P., Suter, P.M., Junod, A.F., 1988. Effects of oleic acid, alpha naphthylthiourea and phorbol myristate acetate-induced microvascular damage on indexes of pulmonary endothelial function in anesthetized dogs. Am. Rev. Resp. Dis. 137, 1350.

Gardaz, J.P., Py, P., Suter, P.M., Junod, A.F., 1988. Effects of oleic acid, alpha naphthylthiourea and phorbol myristate acetate-induced microvascular damage on indexes of pulmonary endothelial function in anesthetized dogs. Am. Rev. Resp. Dis. 137, 1350.

Gillis, C.N., Greene, N.M., Cronau, L.H., Hammond, G.L., 1972. Pulmonary extraction of 5-hydroxytryptamine and norepinephrine before and after cardiopulmonary bypass in man. Circ. Res. 30, 666–674.

Havill, P.M., Riggs, D., Pitt, B.R., Gillis, C.N., 1989. Resolution of impaired pulmonary function and pulmonary hypertension after phorbol ester administration in rabbits. Am. Rev. Resp. Dis. 140, 782.

Hilliker, K.S., Bell, T.G., Roth, R.A., 1982. Pneumotoxicity and thrombocytopenia after single injection of monocrotaline. Am. J. Physiol. 242 (4), H573-H5739.

Howell, R.E., Lanken, P.N., Hansen-Flaschen, J.N., Haselton, F.R., Albelda, S.M., Fishman, A.P., 1989. Pulmonary extraction of propranolol in normal and oxygen-toxic sheep. J. Pharmacol. Exp. Ther. 250, 93.

Huxtable, R., Ciaramitaro, D., Eisenstein, D., 1978. The effect of a pyrolizidine alkaloid, monocrotaline, and a pyrrole, dehydroretronecine, on the biochemical functions of the pulmonary endothelium. Mol. Pharmacol. 14, 1189–1203.

Jeffrey, R., Doctor, R., Johnston, M., Stanbrook, H., 1986. Bleomycin induces changes in pulmonary endothelial cell function in rabbits. Fed. Proc. 45, 278.

Lazo, J.S., Catravas, J.D., Dobuler, J.K., 1983. Prolonged reduction in serum angiotensin-converting enzyme activity after treatment of rabbits with bleomycin. Toxicol. Appl. Pharmacol. 69, 276.

Merker, M., Gillis, C.N., 1988. Propranolol and serotonin removal in lung injury. J. Appl. Physiol. 65, 2579.

Myers, C.L., Pitt, B.R., 1988. Selective effect of phorbol ester on kinetics of serotonin removal of ACE activity of in situ perfused rabbit lungs. J. Appl. Physiol. 65, 377.

Myers, C.L., Lazo, J.S., Pitt, B.R., 1989. Translocation of protein kinase C is associated with inhibition of 5-HT uptake by cultured endothelial cells. Am. J. Physiol. 257, L253. Lung Cell. Mol. Physiol. 1.

Riggs, D., Havill, A.M., Pitt, D.R., et al., 1988. Pulmonary angiotensin converting enzyme kinetics following acute lung injury in the anesthetized rabbit. J. Appl. Physiol. 64, 2508.

Roth, R.A., Wallace, K.B., Alper, R.H., Bailie, M.D., 1979. Effect of paraquat treatment of rats on disposition of 5-hydroxytryptamine and angiotensin I by perfused lung. Biochem. Pharmacol. 28, 2349–2355.

Sheridan, N.P., Block, E.R., 1988a. Serotonin transport and fluidity in plasma membrane vesicles: effect of hyperoxia. Am. J. Physiol. 254, C781. Cell Physiol. 23.

Sheridan, N.P., Block, E.R., 1988b. Plasma membrane fluidity measurements in intact endothelial cells: effect of hyperoxia on fluorescence anisotropies of 1[4-(trimethylamino)pheny11-6-phenyl hexa1,3,5-triene. J. Cell. Physiol. 134, 117.

Slosman, D.O., Polla, B.S., Donath, A., 1989. Metaiodobenzylguanidine (MIBG): a potential non-invasive scintigraphic marker of the pulmonary endothelial cell function. Am. Rev. Resp. Dis. 139, A57.

Steinberg, H., Greenwald, R.A., Sciubba, J., 1982. The effect of oxygen free radicals on pulmonary endothelial cell function in the isolated perfused rat lung. Exp. Lung Res. 3, 163.

Vigneswaran, W.T., Stanbrook, H.S., Doctor, R., McCormick, S., Schwappach, J., Johnston, M.R., 1989. 5-hydroxytryptamine uptake in oxygen radical-mediated acute injury. Am. Rev. Resp. Dis. 139, 382.

White, M.K., Hechtman, H.B., Shepro, D., 1975. Canine lung uptake of plasma and platelet serotonin. Microvasc. Res. 9, 230–241.

Appendix 11

Comparative Biochemistry of Airway Lining Fluid

TABLE 1 General Composition of Directly Sampled Mucus

Source	% Dry	% Carbohydrate	% Protein	% Lipid	References
Human sputum	0.5–6.0	2.0–3.0	0.1–0.5	0.3–0.5	Masson and Heremans (1973)
Human, laryngectomized direct sample[a]	3.2	0.95	1.0	0.84	Boat and Matthews (1973)
Dog tracheal mucus directly absorbed			4.1		Reasor et al. (1978)
Human bronchial secretion		0.5–1.0	0.5–1.0		Creeth (1978)
Human resp. tract mucus		2.0–3.0	0.1–0.5	0.3–0.5	Lopez Vidriero et al. (1980)
Human sputum	5.0	1.0–2.0	0.25–1.25	0.05–0.15	Lopez Vidriero (1981)
Human nasal mucus (unstimulated)	1.8–2.2	0.28–0.29	0.16–0.25		Brofeldt et al. (1987)
Human nasal mucus (methacholine stim.)	2.8	0.19	0.65		Brofeldt et al. (1986)
Comparison with Other Body Fluids					
Plasma (human)	8.0	0.27[e]	6.5–8.0	0.35–0.85	Diem and Lentner (1974)
Saliva (human)	3.0–8.0	0.08–0.6	0.14–0.64	0.002–0.05[b]	Diem and Lentner (1974)
Synov. fluid (human)	3.4	0.25–0.40[c]	0.45–3.2	0.018–0.029[d]	Diem and Lentner (1974)

[a] *Also reported 1.13% ash and 1% dialyzable content.*
[b] *Cholestrol content.*
[c] *Reported as hyaluronic acid content.*
[d] *Cholesterol + phospholipid content.*
[e] *Protein-bound carbohydate.*

TABLE 2 Ionic Composition of Airway Secretions

Source	Humana[a] Plasma (meq/l)	Whole Human Secretions[b] (meq/l)	Cat Submucosal Gland Secretion[c] (meq/l)	Human Nasal Mucus[d]
Osmolality	289	359 ± 56	—	278–314
Sodium	132	165 ± 42	134 ± 2	
Potassium	4	13.2 ± 5.4	6.3 ± 0.2	
Chloride	106	162 ± 60	124 ± 2	
Calcium	5[e]	3.1 ± 1.0	0.7 ± 0.1	
Phosphorus	33.6[e]	27 ± 16	1.3 ± 0.1	

Note: Means ± SE are shown.
[a]*Diem and Lentner (1974).*
[b]*Matthews et al. (1963).*
[c]*Quinton (1979).*
[d]*Knowles et al. (1987).*
[e]*Not protein or lipid bound.*

REFERENCES

Boat, T.F., Matthews, L.W., 1973. Chemical composition of human tracheobronchial secretions. In: Dulfano, M.D. (Ed.), Sputum: Fundamentals and Clinical Pathology. Charles C Thomas, Springfield, IL, pp. 243–274.

Brofeldt, S., Mygind, N., Sorensen, C.H., Readman, A.S., Marriott, C., 1986. Biochemical analysis of nasal secretions induced by methacholine, histamine, and allergen provocations. Am. Rev. Respir. Dis. 133, 1138–1142.

Brofeldt, S., Ingstrup, M.H., Niebuhr, M., Mygind, N., Mar-riot, C., 1987. A study of nasal secretions sampled from the anterior and posterior parts of the human nose. Acta Otolcyngol. (Stockholm).

Creeth, J.M., 1978. Constituents of mucus and their separation. Br. Med. Bull. 34, 17–24.

Diem, K., Lentner, C. (Eds.), 1974. Scientific Tables. Geigy Pharmaceuticals, Ardsley, New York, p. 10502.

Knowles, M.R., Clark, C.E., Fischer, N.D., Fisher, S.R., Kenan, P.D., Pillsbury, H.C., Boucher, R.C., Brofeldt, S., 1987. Nasal secretions: role of epithelial ion transport. In: Mygind, N., Pipkom, U. (Eds.), Allergic and Vasomotor Rhinitis: Pathophysiological Aspects. Munksgaard, Copenhagen, pp. 77–90.

Lopez-Vidriero, M.T., Reid, L., 1980. Respiratory tract fluid-chemical and physical properties of airway mucus. Eur. J. Respir. Dis. 110, 21–26.

Lopez-Vidriero, M.T., 1981. Airway mucus production and composition. Chest 80 (6), 799–804.

Masson, P.L., Heremans, J.F., 1973. Sputum proteins. In: Dulfano, M.D. (Ed.), Sputum: Fundamentals and Clinical Pathology. Charles C Thomas, Springfield, IL, pp. 243–274.

Masson, P.L., Heremans, J.F., 1973. Sputum proteins. In: Dulfano, M.D. (Ed.), Sputum: Fundamentals and Clinical Pathology. Charles C Thomas, Springfield, IL, pp. 404–475.

Matthews, L.W., Spector, S., Lemm, J., Potter, J., 1963. Studies on pulmonary secretions. 1. The overall chemical composition of pulmonary secretions from patients with cystic fibrosis, bronchiectasis and laryngectomy. Am. Respir. Dis. 88, 199–204.

Quinton, P.M., 1979. Composition and control of secretions from tracheobronchial submucosal glands: implications in cystic fibrosis. Nature (London) 279, 551–552.

Reasor, M.J., Adams III, G.K., Proctor, D.F., Rubin, R.J., 1978. Tracheobronchial secretions collected from intact dogs. I. Protein and mucous glycoprotein composition. J. Appl. Physiol. Respir. Environ. Exerc. Physiol. 45 (2), 182–189.

Appendix 12

Mucociliary Function

TABLE 1 Tracheal Mucociliary Clearance

Species	Mucous Velocity[a] (mm/min)	References
Mouse	±	Snipes et al. (1983)
Rat	1.9 ± 0.7	Felicetti et al. (1981)
	5.1 ± 3.0	Patrick and Stirling (1977)
	5.9 ± 2.5	Giordano and Morrow (1972)
Ferret	+	Kyle and Widdicombe (1987)
	18.2 ± 5.1	Marin et al. (1989)
	10.7 ± 3.7	Voter et al. (1987)
Guinea pig	2.7 ± 1.4	Felicetti et al. (1981)
Rabbit	3.2 ± 1.1	Felicetti et al. (1981)
	+	Schlesinger et al. (1982)
Chicken	*	Battista et al. (1972)
Cat	2.5 ± 0.8	Carson et al. (1966)
Dog	21.6 ± 5.0	Laurenzi et al. (1968)
	9.8 ± 2.1	Felicetti et al. (1981)
	19.2 ± 1.6	Chopra et al. (1977)
	7.5 ± 3.7	Wolff et al. (1979)
	14.5 ± 6.3	Chopra (1978)
Baboon	+	Griffin et al. (1986)
Sheep	17.3 ± 6.2	Landa et al. (1975)
	10.5 ± 2.9	Ahmed et al. (1979)
Pig	*	Alexopoulos et al. (1984)
Cow	*	Hilding (1956)
Donkey	14.7 ± 3.8	Schlesinger et al. (1978)
Horse	16.6 ± 2.4	Nelson and Hampe (1983)
	17.8 ± 5.1	Coombs and Weddon (1987)

Continued

TABLE 1 Tracheal Mucociliary Clearance—cont'd

Species	Mucous Velocity[a] (mm/min)	References
Human	3.6 ± 1.5	Yeates et al. (1975)
	5.5 ± 0.4	Foster et al. (1980)
	5.1 ± 2.9	Yeates et al. (1981)
	11.5 ± 4.7	Friedman et al. (1977)
	10.1 ± 3.5	Goodman et al. (1978)
	21.5 ± 5.5	Santa Cruz et al. (1974)
	15.5 ± 1.7	Chopra et al. (1979)

Note: *, Transport studied but no velocity given; +, inhalation study, clearance measured but no tracheal velocities given.
[a]*Mean ± SD.*

TABLE 2 Nasal Mucociliary Clearance

Species	Velocity[a] (mm/min)	References
Rat	2.3 ± 0.8	Dahl et al. (1986)
Dog	3.7 ± 0.9	Whaley et al. (1987)
Man	5.2 ± 2.3	Van Ree (1962)
	5.5 ± 3.2	Ewert (1965)
	5.3 (0.5 − 23.6)	Proctor et al. (1977)
	8.4 ± 4.8	Andersen et al. (1971)
	6.8 ± 5.1	Black et al. (1974)
	7 ± 4	

[a]*Mean ± SD.*

REFERENCES

Ahmed, T., Januszkiewicz, A.J., Landa, J.F., Brown, A., Chapman, G.A., Kenny, P.J., Finn, R.D., Bondick, J., Sackner, M.A., 1979. Effect of local radioactivity on tracheal mucous velocity of sheep. Am. Rev. Respir. Dis. 120, 567.

Alexopoulos, C., Jansson, B., Lindholm, C.E., February 1984. Mucus transport and surface damage after endotracheal intubation and tracheostomy. An experimental study in pigs. Acta. Anaesthesiol. Scand. 28 (1), 68−76.

Andersen, I., Lundqvist, G.R., Proctor, D.F., 1971. Human nasal mucosal function in a controlled climate. Arch. Environ. Health 23, 408.

Battista, S.P., Denine, E.P., Kenstler, C.J., 1972. Restoration of tracheal mucosa and ciliary particle transport activity after mechanical denudation in the chicken. Toxicol. Pharmacol. 22, 59.

Black, A., Evans, J.C., Hadfield, E.H., Macbeth, R.G., Morgan, A., Walsh, M., 1974. Impairment of nasal mucociliary clearance in woodworkers in the furniture industry. Br. J. Ind. Med. 31, 10.

Carson, S., Goldhamer, R., Carpenter, R., 1966. Mucus transport in the respiratory tract. Am. Rev. Respir. Dis. 93 (Supplement), 86.

Chopra, S.K., Taplin, G.V., Simmons, D.H., Robinson Jr., G.D., Elam, D., Coulson, A., June 1977. Effects of hydration and physical therapy on tracheal transport velocity. Am. Rev. Respir. Dis. 115 (6), 1009−1014. PMID:262095.

Chopra, S.K., 1978. Effect of atropine on mucociliary transport velocity in anesthetized dogs. Am. Rev. Respir. Dis. 118, 367.

Chopra, S.K., Taplin, G.V., Elam, D., Carson, S.A., Golde, D., 1979. Measurement of tracheal mucociliary transport velocity in humans-smokers versus non-smokers. Am. Rev. Respir. Dis. 119, 205.

Coombs, S.L., Weddon, P.M., 1987. Observations on tracheal mucociliary clearance in horses. Tierarztl. Prax. 5-9 (2 (Suppl).

Dahl, A.R., Bice, D.E., Hahn, F.F., Henderson, R.F., Mauderly, J.L., Muggenburg, B.A., Pickrell, J.A., Wolff, R.K., Hobbs, C.H., 1986. Acute and Subcronic Toxicity of Methylphosphoric Difluoride (DF) by Inhalation in Rats. CRDC Report, Aberdeen Proving Grounds.

Ewert, G., 1965. On the mucus flow rate in the human nose. Acta Otolaryngol. Stockh. 200 (Suppl.), 1.

Felicetti, S.A., Wolff, R.K., Muggenburg, B.A., 1981. Comparison of tracheal mucous transport in rats, guinea pigs, rabbits, and dogs. J. Appl. Physiol. 51, 1612.

Foster, W.M., Langenback, E., Bergofsky, E.H., 1980. Measurement of tracheal and bronchial mucus velocities in man: relation to lung clearance. J. Appl. Physiol. 48, 965.

Friedman, M., Stott, F.D., Poole, D.O., Dougherty, R., Chapman, G.A., Watson, H., Sackner, M.A., 1977. A new roentgenographic method for estimating mucous velocity in airways. Am. Rev. Respir. Dis. 115, 67.

Giordano, A.M., Morrow, P.E., 1972. Chronic low-level nitrogen dioxide exposure and mucociliary clearance. Arch. Environ. Health 25, 443.

Goodman, R.M., Yergin, B.M., Sackner, M.A., December 1978. Effects of S-carboxymethylcysteine on tracheal mucus velocity. Chest 74 (6), 615–618. PMID:367726.

Griffin, C.L., Musselman, R.P., Yeates, D.B., Raju, T.N., Harshbarger, R.D., Lourenco, R.V., December 1986. Hand-rearing baboons for laboratory investigations. Lab. Anim. Sci. 36 (6), 686–690.

Hilding, A.C., 1956. On cigarette smoking, bronchial carcinoma and ciliary action. II. Experimental study on the filtering action of cows' lungs, the deposition of tar in the bronchial tree and removal by ciliary action. N. Engl. J. Med. 254, 1155.

Kyle, H., Widdicombe, J.G., 1987. The effects of peptides and mediators on mucus secretion rate and smooth muscle tone in the ferret trachea. Agents Actions 22, 86.

Landa, J.F., Hirsch, J.A., Lebeaux, M.I., 1975. Effects of topical and general anesthetic agents on tracheal mucous velocity of sheep. J. Appl. Physiol. 38, 946.

Laurenzi, G.A., Yen, S., Guarneri, J.M., 1968. Adverse effect of oxygen on tracheal mucus flow. N. Engl. J. Med. 279, 333.

Marin, M.G., Zwierzynski, D., McBride, R.K., Culp, D.J., 1989. Effects of soman and its antidotes on tracheal mucociliary transport of ferrets. J. Appl. Physiol. 66, 1390.

Nelson, R., Hampe, D.W., 1983. Measurement of tracheal mucous transport rate in the horse. Am. J. Vet. Res. 44, 1165.

Patrick, G., Stirling, C., 1977. Measurement of mucociliary clearance from the trachea of conscious and anesthetized rats. J. Appl. Physiol. Respir. Environ. Exerc. Physiol. 42, 451.

Proctor, D.F., Andersen, I., Lundqvist, G., 1977. Nasal mucociliary function in humans. In: Brain, J.D., Proctor, D.F., Reid, L.M. (Eds.), Respiratory Defense Mechanism, Part 1, Vol. 5. Marcel Dekker, New York, p. 427.

Santa Cruz, R., Landa, J., Hirsch, J., Sackner, M.A., 1974. Tracheal mucous velocity in normal man and patients with obstructive lung disease; effects of terbutaline. Am. Rev. Respir. Dis. 109, 458.

Schlesinger, R.B., Lippmann, M., Albert, R.E., April 1978. Effects of short-term exposures to sulfuric acid and ammonium sulfate aerosols upon bronchial airway function in the donkey. Am. Ind. Hyg. Assoc. J. 39 (4), 275–286.

Schlesinger, R.B., Naumann, B., Chen, L.C., 1982. A system for measurement of radioactive aerosol retention in the lungs of rabbits. Health Phys. 42, 73.

Snipes, M.B., Boecker, B.B., McClellan, R.O., 1983. Retention of monodisperse or polydisperse aluminosilicate particles inhaled by dogs, rats, and mice. Toxicol. Appl. Pharmacol. 69, 345.

van Ree, J.H.L., van Dishoeck, H.A.E., 1962. Some investigations on nasal ciliary activity. Pract. Ororhinolaryngol. 24, 383.

Voter, K.Z., Carson, J.L., Leigh, M.W., Wood, R.E., 1987. Postnatal development of mucociliary transport in ferrets. Am. Rev. Resp. Dis. 135, A25.

Whaley, S.L., Muggenburg, B.A., Seiler, F.A., Wolff, R.K., March 1987. Effect of aging on tracheal mucociliary clearance in beagle dogs. J. Appl. Physiol. (1985). 62 (3), 1331–1334.

Wolff, R.K., Muggenburg, B.A., 1979. Comparison of two methods of measuring tracheal mucous clearance in anesthetized beagle dogs. Am. Rev. Respir. Dis. 120, 137.

Yeates, D.B., Aspin, N., Levison, H., Jones, M.T., Bryan, A.C., 1975. Mucociliary tracheal transport rates in man. J. Appl. Physiol. 39, 487.

Yeates, D.B., Pitt, B.R., Spektor, D.M., Karron, G.A., Albert, R.E., 1981. Coordination of mucociliary transport in human trachea and intrapulmonary airways. J. Appl. Physiol. Respir. Environ. Exerc. Physiol. 51, 1057.

Yeates, D.B., Gerrity, T.R., Garrard, C.S., 1981. Particle deposition and clearance in the bronchial tree. Ann. Biomed. Eng. 9, 577.

Index

Note: Page numbers followed by "f" and "t" indicate figures and tables respectively.

B

C

D

E

F

G

H

N

O

P

Q

R

U

V

W

CPI Antony Rowe
Eastbourne, UK
August 18, 2015